PERIODIC TABLE OF THE ELEMENTS

	Atomic number
1	
H	Symbol
Hydrogen	Name
1.0079	Average atomic mass

Metals
Metalloids
Nonmetals

1 1A																	18 8A
1 **H** Hydrogen 1.0079	2 2A											13 3A	14 4A	15 5A	16 6A	17 7A	2 **He** Helium 4.0026
3 **Li** Lithium 6.941	4 **Be** Beryllium 9.0122											5 **B** Boron 10.811	6 **C** Carbon 12.011	7 **N** Nitrogen 14.007	8 **O** Oxygen 15.999	9 **F** Fluorine 18.998	10 **Ne** Neon 20.180
11 **Na** Sodium 22.990	12 **Mg** Magnesium 24.305	3 3B	4 4B	5 5B	6 6B	7 7B	8	9 8B	10	11 1B	12 2B	13 **Al** Aluminum 26.982	14 **Si** Silicon 28.086	15 **P** Phosphorus 30.974	16 **S** Sulfur 32.065	17 **Cl** Chlorine 35.453	18 **Ar** Argon 39.948
19 **K** Potassium 39.098	20 **Ca** Calcium 40.078	21 **Sc** Scandium 44.956	22 **Ti** Titanium 47.867	23 **V** Vanadium 50.942	24 **Cr** Chromium 51.996	25 **Mn** Manganese 54.938	26 **Fe** Iron 55.845	27 **Co** Cobalt 58.933	28 **Ni** Nickel 58.693	29 **Cu** Copper 63.546	30 **Zn** Zinc 65.38	31 **Ga** Gallium 69.723	32 **Ge** Germanium 72.63	33 **As** Arsenic 74.922	34 **Se** Selenium 78.96	35 **Br** Bromine 79.904	36 **Kr** Krypton 83.798
37 **Rb** Rubidium 85.468	38 **Sr** Strontium 87.62	39 **Y** Yttrium 88.906	40 **Zr** Zirconium 91.224	41 **Nb** Niobium 92.906	42 **Mo** Molybdenum 95.96	43 **Tc** Technetium [98]	44 **Ru** Ruthenium 101.07	45 **Rh** Rhodium 102.91	46 **Pd** Palladium 106.42	47 **Ag** Silver 107.87	48 **Cd** Cadmium 112.41	49 **In** Indium 114.82	50 **Sn** Tin 118.71	51 **Sb** Antimony 121.76	52 **Te** Tellurium 127.60	53 **I** Iodine 126.90	54 **Xe** Xenon 131.29
55 **Cs** Cesium 132.91	56 **Ba** Barium 137.33	57 **La** Lanthanum 138.91	72 **Hf** Hafnium 178.49	73 **Ta** Tantalum 180.95	74 **W** Tungsten 183.84	75 **Re** Rhenium 186.21	76 **Os** Osmium 190.23	77 **Ir** Iridium 192.22	78 **Pt** Platinum 195.08	79 **Au** Gold 196.97	80 **Hg** Mercury 200.59	81 **Tl** Thallium 204.38	82 **Pb** Lead 207.2	83 **Bi** Bismuth 208.98	84 **Po** Polonium [209]	85 **At** Astatine [210]	86 **Rn** Radon [222]
87 **Fr** Francium [223]	88 **Ra** Radium [226]	89 **Ac** Actinium [227]	104 **Rf** Rutherfordium [265]	105 **Db** Dubnium [268]	106 **Sg** Seaborgium [271]	107 **Bh** Bohrium [270]	108 **Hs** Hassium [277]	109 **Mt** Meitnerium [276]	110 **Ds** Darmstadtium [281]	111 **Rg** Roentgenium [280]	112 **Cn** Copernicium [285]	113 **Nh** Nihonium [284]	114 **Fl** Flerovium [289]	115 **Mc** Moscovium [288]	116 **Lv** Livermorium [293]	117 **Ts** Tennessine [294]	118 **Og** Oganesson [294]

6 Lanthanides

| 58 **Ce** Cerium 140.12 | 59 **Pr** Praseodymium 140.91 | 60 **Nd** Neodymium 144.24 | 61 **Pm** Promethium [145] | 62 **Sm** Samarium 150.36 | 63 **Eu** Europium 151.96 | 64 **Gd** Gadolinium 157.25 | 65 **Tb** Terbium 158.93 | 66 **Dy** Dysprosium 162.50 | 67 **Ho** Holmium 164.93 | 68 **Er** Erbium 167.26 | 69 **Tm** Thulium 168.93 | 70 **Yb** Ytterbium 173.05 | 71 **Lu** Lutetium 174.97 |

7 Actinides

| 90 **Th** Thorium 232.04 | 91 **Pa** Protactinium 231.04 | 92 **U** Uranium 238.03 | 93 **Np** Neptunium [237] | 94 **Pu** Plutonium [244] | 95 **Am** Americium [243] | 96 **Cm** Curium [247] | 97 **Bk** Berkelium [247] | 98 **Cf** Californium [251] | 99 **Es** Einsteinium [252] | 100 **Fm** Fermium [257] | 101 **Md** Mendelevium [258] | 102 **No** Nobelium [259] | 103 **Lr** Lawrencium [262] |

We have used the U.S. system as well as the system recommended by the International Union of Pure and Applied Chemistry (IUPAC) to label the groups in this periodic table. The system used in the United States includes a letter and a number (1A, 2A, 3B, 4B, etc.), which is close to the system developed by Mendeleev. The IUPAC system uses numbers 1–18 and has been recommended by the American Chemical Society (ACS). Although we show both numbering systems here, we use the IUPAC system exclusively in the book.

Element	Symbol	Atomic Number	Average Atomic Mass[a]
Actinium	Ac	89	[227]
Aluminum	Al	13	26.982
Americium	Am	95	[243]
Antimony	Sb	51	121.76
Argon	Ar	18	39.948
Arsenic	As	33	74.922
Astatine	At	85	[210]
Barium	Ba	56	137.33
Berkelium	Bk	97	[247]
Beryllium	Be	4	9.0122
Bismuth	Bi	83	208.98
Bohrium	Bh	107	[270]
Boron	B	5	10.811
Bromine	Br	35	79.904
Cadmium	Cd	48	112.41
Calcium	Ca	20	40.078
Californium	Cf	98	[251]
Carbon	C	6	12.011
Cerium	Ce	58	140.12
Cesium	Cs	55	132.91
Chlorine	Cl	17	35.453
Chromium	Cr	24	51.996
Cobalt	Co	27	58.933
Copernicium	Cn	112	[285]
Copper	Cu	29	63.546
Curium	Cm	96	[247]
Darmstadtium	Ds	110	[281]
Dubnium	Db	105	[268]
Dysprosium	Dy	66	162.50
Einsteinium	Es	99	[252]
Erbium	Er	68	167.26
Europium	Eu	63	151.96
Fermium	Fm	100	[257]
Flerovium	Fl	114	[289]
Fluorine	F	9	18.998
Francium	Fr	87	[223]
Gadolinium	Gd	64	157.25
Gallium	Ga	31	69.723
Germanium	Ge	32	72.63
Gold	Au	79	196.97
Hafnium	Hf	72	178.49
Hassium	Hs	108	[277]
Helium	He	2	4.0026
Holmium	Ho	67	164.93
Hydrogen	H	1	1.0079
Indium	In	49	114.82
Iodine	I	53	126.90
Iridium	Ir	77	192.22
Iron	Fe	26	55.845
Krypton	Kr	36	83.798
Lanthanum	La	57	138.91
Lawrencium	Lr	103	[262]
Lead	Pb	82	207.2
Lithium	Li	3	6.941
Livermorium	Lv	116	[293]
Lutetium	Lu	71	174.97
Magnesium	Mg	12	24.305
Manganese	Mn	25	54.938
Meitnerium	Mt	109	[276]
Mendelevium	Md	101	[258]
Mercury	Hg	80	200.59
Molybdenum	Mo	42	95.96
Moscovium[b]	Mc	115	[288]
Neodymium	Nd	60	144.24
Neon	Ne	10	20.180
Neptunium	Np	93	[237]
Nickel	Ni	28	58.693
Nihonium[b]	Nh	113	[284]
Niobium	Nb	41	92.906
Nitrogen	N	7	14.007
Nobelium	No	102	[259]
Oganesson[b]	Og	118	[294]
Osmium	Os	76	190.23
Oxygen	O	8	15.999
Palladium	Pd	46	106.42
Phosphorus	P	15	30.974
Platinum	Pt	78	195.08
Plutonium	Pu	94	[244]
Polonium	Po	84	[209]
Potassium	K	19	39.098
Praseodymium	Pr	59	140.91
Promethium	Pm	61	[145]
Protactinium	Pa	91	231.04
Radium	Ra	88	[226]
Radon	Rn	86	[222]
Rhenium	Re	75	186.21
Rhodium	Rh	45	102.91
Roentgenium	Rg	111	[280]
Rubidium	Rb	37	85.468
Ruthenium	Ru	44	101.07
Rutherfordium	Rf	104	[265]
Samarium	Sm	62	150.36
Scandium	Sc	21	44.956
Seaborgium	Sg	106	[271]
Selenium	Se	34	78.96
Silicon	Si	14	28.086
Silver	Ag	47	107.87
Sodium	Na	11	22.990
Strontium	Sr	38	87.62
Sulfur	S	16	32.065
Tantalum	Ta	73	180.95
Technetium	Tc	43	[98]
Tellurium	Te	52	127.60
Tennessine[b]	Ts	117	[294]
Terbium	Tb	65	158.93
Thallium	Tl	81	204.38
Thorium	Th	90	232.04
Thulium	Tm	69	168.93
Tin	Sn	50	118.71
Titanium	Ti	22	47.867
Tungsten	W	74	183.84
Uranium	U	92	238.03
Vanadium	V	23	50.942
Xenon	Xe	54	131.29
Ytterbium	Yb	70	173.05
Yttrium	Y	39	88.906
Zinc	Zn	30	65.38
Zirconium	Zr	40	91.224

[a]Average atomic mass values for most elements are from *Pure Appl. Chem.* (2011) 83, 359. Those for B, C, Cl, H, Li, N, O, Si, S, and Tl are from *Pure Appl. Chem.* (2009) 81, 2131, and are within the ranges cited in the first reference. Atomic masses in brackets are the mass numbers of the longest-lived isotopes of elements with no stable isotopes.
[b]Names and symbols for these elements were recommended by the International Union for Pure and Applied Chemistry (IUPAC) in June 2016.

FIFTH EDITION

Chemistry
The Science in Context

Thomas R. Gilbert

NORTHEASTERN UNIVERSITY

Rein V. Kirss

NORTHEASTERN UNIVERSITY

Natalie Foster

LEHIGH UNIVERSITY

Stacey Lowery Bretz

MIAMI UNIVERSITY

Geoffrey Davies

NORTHEASTERN UNIVERSITY

W. W. NORTON & COMPANY
NEW YORK • LONDON

W. W. Norton & Company has been independent since its founding in 1923, when William Warder Norton and Mary D. Herter Norton first published lectures delivered at the People's Institute, the adult education division of New York City's Cooper Union. The firm soon expanded its program beyond the Institute, publishing books by celebrated academics from America and abroad. By midcentury, the two major pillars of Norton's publishing program—trade books and college texts—were firmly established. In the 1950s, the Norton family transferred control of the company to its employees, and today—with a staff of four hundred and a comparable number of trade, college, and professional titles published each year—W. W. Norton & Company stands as the largest and oldest publishing house owned wholly by its employees.

Editor: Erik Fahlgren
Developmental Editor: Andrew Sobel
Associate Managing Editor, College: Carla L. Talmadge
Assistant Editor: Arielle Holstein
Production Manager: Eric Pier-Hocking
Managing Editor, College: Marian Johnson
Managing Editor, College Digital Media: Kim Yi
Media Editor: Christopher Rapp
Associate Media Editor: Julia Sammaritano
Media Project Editor: Marcus Van Harpen
Media Editorial Assistants: Victoria Reuter, Doris Chiu
Digital Production: Lizz Thabet
Marketing Manager, Chemistry: Stacy Loyal
Associate Design Director: Hope Miller Goodell
Photo Editor: Aga Millhouse
Permissions Manager: Megan Schindel
Composition: Graphic World
Illustrations: Imagineering—Toronto, ON
Manufacturing: Transcontinental

Permission to use copyrighted material is included at the back of the book.

Library of Congress Cataloging-in-Publication Data

Names: Gilbert, Thomas R. | Kirss, Rein V. | Foster, Natalie. | Bretz, Stacey
  Lowery, 1967- | Davies, Geoffrey, 1942-
Title: Chemistry. The science in context.
Description: Fifth edition / Thomas R. Gilbert, Northeastern University, Rein
  V. Kirss, Northeastern University, Natalie Foster, Lehigh University,
  Stacey Lowery Bretz, Miami University, Geoffrey Davies, Northeastern
  University. | New York : W.W. Norton & Company, Inc., [2018] | Includes
  index.
Identifiers: LCCN 2016048998 | ISBN 9780393264845 (hardcover)
Subjects: LCSH: Chemistry--Textbooks.
Classification: LCC QD33.2 .G55 2018 | DDC 540--dc23 LC record available at
https://lccn.loc.gov/2016048998

W. W. Norton & Company, Inc., 500 Fifth Avenue, New York, NY 10110

wwnorton.com

W. W. Norton & Company Ltd., 15 Carlisle Street, London W1D 3BS

2 3 4 5 6 7 8 9 0

Brief Contents

1 Particles of Matter: Measurement and the Tools of Science 2

2 Atoms, Ions, and Molecules: Matter Starts Here 44

3 Stoichiometry: Mass, Formulas, and Reactions 82

4 Reactions in Solution: Aqueous Chemistry in Nature 142

5 Thermochemistry: Energy Changes in Reactions 208

6 Properties of Gases: The Air We Breathe 272

7 A Quantum Model of Atoms: Waves, Particles, and Periodic Properties 330

8 Chemical Bonds: What Makes a Gas a Greenhouse Gas? 386

9 Molecular Geometry: Shape Determines Function 436

10 Intermolecular Forces: The Uniqueness of Water 496

11 Solutions: Properties and Behavior 536

12 Solids: Crystals, Alloys, and Polymers 588

13 Chemical Kinetics: Reactions in the Atmosphere 634

14 Chemical Equilibrium: How Much Product Does a Reaction Really Make? 694

15 Acid–Base Equilibria: Proton Transfer in Biological Systems 738

16 Additional Aqueous Equilibria: Chemistry and the Oceans 784

17 Thermodynamics: Spontaneous and Nonspontaneous Reactions and Processes 832

18 Electrochemistry: The Quest for Clean Energy 878

19 Nuclear Chemistry: Applications to Energy and Medicine 922

20 Organic and Biological Molecules: The Compounds of Life 960

21 The Main Group Elements: Life and the Periodic Table 1016

22 Transition Metals: Biological and Medical Applications 1052

Contents

List of Applications   xv

List of ChemTours   xvii

About the Authors   xviii

Preface   xix

1   Particles of Matter: Measurement and the Tools of Science   2

1.1   How and Why   4

1.2   Macroscopic and Particulate Views of Matter   5
Classes of Matter   5 • A Particulate View   7

1.3   Mixtures and How to Separate Them   9

1.4   A Framework for Solving Problems   11

1.5   Properties of Matter   12

1.6   States of Matter   14

1.7   The Scientific Method: Starting Off with a Bang   16

1.8   SI Units   18

1.9   Unit Conversions and Dimensional Analysis   20

1.10   Evaluating and Expressing Experimental Results   22
Significant Figures   23 • Significant Figures in Calculations   23 •
Precision and Accuracy   27

1.11   Testing a Theory: The Big Bang Revisited   32
Temperature Scales   32 • An Echo of the Big Bang   34

Summary   37 • Particulate Preview Wrap-Up   37 • Problem-Solving Summary   38 •
Visual Problems   38 • Questions and Problems   40

Just how small are these atoms?
(Chapter 1)

2   Atoms, Ions, and Molecules: Matter Starts Here   44

2.1   Atoms in Baby Teeth   46

2.2   The Rutherford Model   47
Electrons   47 • Radioactivity   49 • Protons and Neutrons   50

2.3   Isotopes   52

2.4   Average Atomic Mass   54

2.5   The Periodic Table of the Elements   55
Navigating the Modern Periodic Table   56

2.6   Trends in Compound Formation   59
Molecular Compounds   60 • Ionic Compounds   60

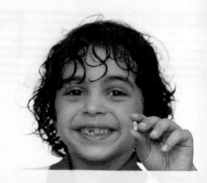

What can baby teeth tell us
about nuclear fallout? *(Chapter 2)*

2.7   Naming Compounds and Writing Formulas   62
Molecular Compounds   62 • Ionic Compounds   63 • Compounds of Transition
Metals   64 • Polyatomic Ions   65 • Acids   66

2.8   Organic Compounds: A First Look   67
Hydrocarbons   67 • Heteroatoms and Functional Groups   68

2.9   Nucleosynthesis: The Origin of the Elements   70
Primordial Nucleosynthesis   70 • Stellar Nucleosynthesis   72

Summary   74 • Particulate Preview Wrap-Up   74 • Problem-Solving Summary   75 •
Visual Problems   75 • Questions and Problems   77

How much medicine can be
isolated from the bark of a
yew tree? *(Chapter 3)*

3 Stoichiometry: Mass, Formulas, and Reactions   82

3.1   Air, Life, and Molecules   84
Chemical Reactions and Earth's Early Atmosphere   85

3.2   The Mole   87
Molar Mass   89 • Molecular Masses and Formula Masses   91 •
Moles and Chemical Equations   95

3.3   Writing Balanced Chemical Equations   96

3.4   Combustion Reactions   101

3.5   Stoichiometric Calculations and the Carbon Cycle   104

3.6   Determining Empirical Formulas from Percent Composition   108

3.7   Comparing Empirical and Molecular Formulas   113
Molecular Mass and Mass Spectrometry   116

3.8   Combustion Analysis   117

3.9   Limiting Reactants and Percent Yield   122
Calculations Involving Limiting Reactants   122 •
Actual Yields versus Theoretical Yields   126

Summary   129 • Particulate Preview Wrap-Up   130 • Problem-Solving Summary   130 •
Visual Problems   131 • Questions and Problems   134

How do antacid tablets relieve
indigestion? *(Chapter 4)*

4 Reactions in Solution: Aqueous Chemistry in Nature   142

4.1   Ions and Molecules in Oceans and Cells   144

4.2   Quantifying Particles in Solution   146
Concentration Units   147

4.3   Dilutions   154
Determining Concentration   156

4.4   Electrolytes and Nonelectrolytes   158

4.5   Acid–Base Reactions: Proton Transfer   159

4.6   Titrations   166

4.7   Precipitation Reactions   169
Making Insoluble Salts   170 • Using Precipitation in Analysis   174 •
Saturated Solutions and Supersaturation   177

4.8   Ion Exchange   178

4.9   Oxidation–Reduction Reactions: Electron Transfer   180
Oxidation Numbers   181 • Considering Changes in Oxidation Number in Redox
Reactions   183 • Considering Electron Transfer in Redox Reactions   184 •
Balancing Redox Reactions by Using Half-Reactions   185 •
The Activity Series for Metals   188 • Redox in Nature   190

Summary   194 • Particulate Preview Wrap-Up   195 • Problem-Solving Summary   195 •
Visual Problems   197 • Questions and Problems   198

5 Thermochemistry: Energy Changes in Reactions 208

5.1 Sunlight Unwinding 210

5.2 Forms of Energy 211
Work, Potential Energy, and Kinetic Energy 211 • Kinetic Energy and Potential Energy at the Molecular Level 214

5.3 Systems, Surroundings, and Energy Transfer 217
Isolated, Closed, and Open Systems 218 • Exothermic and Endothermic Processes 219 • P–V Work and Energy Units 222

5.4 Enthalpy and Enthalpy Changes 225

5.5 Heating Curves, Molar Heat Capacity, and Specific Heat 227
Hot Soup on a Cold Day 227 • Cold Drinks on a Hot Day 232

5.6 Calorimetry: Measuring Heat Capacity and Enthalpies of Reaction 235
Determining Molar Heat Capacity and Specific Heat 235 • Enthalpies of Reaction 238 • Determining Calorimeter Constants 241

5.7 Hess's Law 243

5.8 Standard Enthalpies of Formation and Reaction 246

5.9 Fuels, Fuel Values, and Food Values 252
Alkanes 252 • Fuel Value 255 • Food Value 257

Summary 260 • Particulate Preview Wrap-Up 261 • Problem-Solving Summary 261 • Visual Problems 262 • Questions and Problems 264

What reaction powers hydrogen-fueled vehicles?
(Chapter 5)

6 Properties of Gases: The Air We Breathe 272

6.1 Air: An Invisible Necessity 274

6.2 Atmospheric Pressure and Collisions 275

6.3 The Gas Laws 280
Boyle's Law: Relating Pressure and Volume 280 • Charles's Law: Relating Volume and Temperature 283 • Avogadro's Law: Relating Volume and Quantity of Gas 285 • Amontons's Law: Relating Pressure and Temperature 287

6.4 The Ideal Gas Law 288

6.5 Gases in Chemical Reactions 293

6.6 Gas Density 295

6.7 Dalton's Law and Mixtures of Gases 299

6.8 The Kinetic Molecular Theory of Gases 304
Explaining Boyle's, Dalton's, and Avogadro's Laws 304 • Explaining Amontons's and Charles's Laws 305 • Molecular Speeds and Kinetic Energy 306 • Graham's Law: Effusion and Diffusion 309

6.9 Real Gases 311
Deviations from Ideality 311 • The van der Waals Equation for Real Gases 313

Summary 315 • Particulate Preview Wrap-Up 316 • Problem-Solving Summary 317 • Visual Problems 318 • Questions and Problems 321

How is emergency oxygen generated on airplanes?
(Chapter 6)

7 A Quantum Model of Atoms: Waves, Particles, and Periodic Properties 330

7.1 Rainbows of Light 332

7.2 Waves of Energy 335

7.3 Particles of Energy and Quantum Theory 337
Quantum Theory 337 • The Photoelectric Effect 339 • Wave–Particle Duality 340

Why does a metal rod first glow red when being heated? *(Chapter 7)*

7.4 The Hydrogen Spectrum and the Bohr Model   341
The Hydrogen Emission Spectrum   341 • The Bohr Model of Hydrogen   343

7.5 Electron Waves   345
De Broglie Wavelengths   346 • The Heisenberg Uncertainty Principle   348

7.6 Quantum Numbers and Electron Spin   350

7.7 The Sizes and Shapes of Atomic Orbitals   355
s Orbitals   355 • *p* and *d* Orbitals   357

7.8 The Periodic Table and Filling the Orbitals of Multielectron Atoms   358

7.9 Electron Configurations of Ions   366
Ions of the Main Group Elements   366 • Transition Metal Cations   368

7.10 The Sizes of Atoms and Ions   369
Trends in Atom and Ion Sizes   369

7.11 Ionization Energies   372

7.12 Electron Affinities   375

Summary   377 • Particulate Preview Wrap-Up   377 • Problem-Solving Summary   377 •
Visual Problems   378 • Questions and Problems   380

Why is CO_2 considered a greenhouse gas? *(Chapter 8)*

8 Chemical Bonds: What Makes a Gas a Greenhouse Gas?   386

8.1 Types of Chemical Bonds and the Greenhouse Effect   388
Forming Bonds from Atoms   389

8.2 Lewis Structures   391
Lewis Symbols   391 • Lewis Structures   392 • Steps to Follow When Drawing
Lewis Structures   392 • Lewis Structures of Molecules with Double and Triple
Bonds   394 • Lewis Structures of Ionic Compounds   397

8.3 Polar Covalent Bonds   398
Polarity and Type of Bond   400
Vibrating Bonds and Greenhouse Gases   401

8.4 Resonance   403

8.5 Formal Charge: Choosing among Lewis Structures   407
Calculating Formal Charge of an Atom in a Resonance Structure   408

8.6 Exceptions to the Octet Rule   411
Odd-Electron Molecules   411 • Atoms with More than an Octet   413 •
Atoms with Less than an Octet   416 • The Limits of Bonding Models   418

8.7 The Lengths and Strengths of Covalent Bonds   419
Bond Length   419 • Bond Energies   420

Summary   424 • Particulate Preview Wrap-Up   424 • Problem-Solving Summary   424 •
Visual Problems   425 • Questions and Problems   427

How do some insects communicate chemically? *(Chapter 9)*

9 Molecular Geometry: Shape Determines Function   436

9.1 Biological Activity and Molecular Shape   438

9.2 Valence-Shell Electron-Pair Repulsion (VSEPR) Theory   439
Central Atoms with No Lone Pairs   440 • Central Atoms with Lone Pairs   444

9.3 Polar Bonds and Polar Molecules   450

9.4 Valence Bond Theory   453
Bonds from Orbital Overlap   453 • Hybridization   454 • Tetrahedral Geometry: sp^3
Hybrid Orbitals   455 • Trigonal Planar Geometry: sp^2 Hybrid Orbitals   456 • Linear
Geometry: sp Hybrid Orbitals   458 • Octahedral and Trigonal Bipyramidal Geometries:
sp^3d^2 and sp^3d Hybrid Orbitals   461

9.5   Shape and Interactions with Large Molecules   463
Drawing Larger Molecules   465 • Molecules with More than One Functional Group   467

9.6   Chirality and Molecular Recognition   468

9.7   Molecular Orbital Theory   470
Molecular Orbitals of Hydrogen and Helium   472 • Molecular Orbitals of Homonuclear Diatomic Molecules   474 • Molecular Orbitals of Heteronuclear Diatomic Molecules   478 • Molecular Orbitals of N_2^+ and Spectra of Auroras   480 • Metallic Bonds and Conduction Bands   480 • Semiconductors   482

Summary   485 • Particulate Preview Wrap-Up   486 • Problem-Solving Summary   486 • Visual Problems   487 • Questions and Problems   488

10 Intermolecular Forces: The Uniqueness of Water   496

10.1   Intramolecular Forces versus Intermolecular Forces   498

10.2   Dispersion Forces   499
The Importance of Shape   501

10.3   Interactions among Polar Molecules   502
Ion–Dipole Interactions   502 • Dipole–Dipole Interactions   503 • Hydrogen Bonds   504

10.4   Polarity and Solubility   510
Combinations of Intermolecular Forces   513

10.5   Solubility of Gases in Water   514

10.6   Vapor Pressure of Pure Liquids   517
Vapor Pressure and Temperature   518 • Volatility and the Clausius–Clapeyron Equation   519

10.7   Phase Diagrams: Intermolecular Forces at Work   520
Phases and Phase Transformations   520

10.8   Some Remarkable Properties of Water   523
Surface Tension, Capillary Action, and Viscosity   524 • Water and Aquatic Life   526

Summary   528 • Particulate Preview Wrap-Up   528 • Problem-Solving Summary   528 • Visual Problems   529 • Questions and Problems   530

Why does ice float on top of liquid water? *(Chapter 10)*

11 Solutions: Properties and Behavior   536

11.1   Interactions between Ions   538

11.2   Energy Changes during Formation and Dissolution of Ionic Compounds   542
Calculating Lattice Energies by Using the Born–Haber Cycle   545 • Enthalpies of Hydration   548

11.3   Vapor Pressure of Solutions   550
Raoult's Law   551

11.4   Mixtures of Volatile Solutes   553
Vapor Pressures of Mixtures of Volatile Solutes   553

11.5   Colligative Properties of Solutions   558
Molality   558 • Boiling Point Elevation   561 • Freezing Point Depression   562 • The van 't Hoff Factor   564 • Osmosis and Osmotic Pressure   568 • Reverse Osmosis   573

11.6   Measuring the Molar Mass of a Solute by Using Colligative Properties   575

Summary   580 • Particulate Preview Wrap-Up   580 • Problem-Solving Summary   580 • Visual Problems   582 • Questions and Problems   584

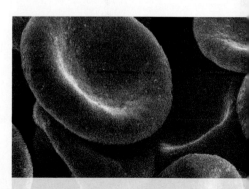

How is blood different from a pure liquid? *(Chapter 11)*

12 Solids: Crystals, Alloys, and Polymers 588

12.1 The Solid State 590

12.2 Structures of Metals 592
Stacking Patterns and Unit Cells 593 • Unit Cell Dimensions 596

12.3 Alloys and Medicine 599
Substitutional Alloys 600 • Interstitial Alloys 601

12.4 Ionic Solids and Salt Crystals 603

12.5 Allotropes of Carbon 606

12.6 Polymers 607
Small Molecules versus Polymers: Physical Properties 608 • Polymers of Alkenes 609 • Polymers Containing Aromatic Rings 612 • Polymers of Alcohols and Ethers 612 • Polyesters and Polyamides 615

Summary 622 • Particulate Preview Wrap-Up 622 • Problem-Solving Summary 622 • Visual Problems 623 • Questions and Problems 626

What materials are used in artificial joints? *(Chapter 12)*

13 Chemical Kinetics: Reactions in the Atmosphere 634

13.1 Cars, Trucks, and Air Quality 636

13.2 Reaction Rates 638
Experimentally Determined Reaction Rates 640 • Average Reaction Rates 642 • Instantaneous Reaction Rates 642

13.3 Effect of Concentration on Reaction Rate 645
Reaction Order and Rate Constants 645 • Integrated Rate Laws: First-Order Reactions 650 • Reaction Half-Lives 653 • Integrated Rate Laws: Second-Order Reactions 655 • Zero-Order Reactions 658

13.4 Reaction Rates, Temperature, and the Arrhenius Equation 659

13.5 Reaction Mechanisms 665
Elementary Steps 666 • Rate Laws and Reaction Mechanisms 667 • Mechanisms and Zero-Order Reactions 672

13.6 Catalysts 672
Catalysts and the Ozone Layer 672 • Catalysts and Catalytic Converters 676 • Enzymes: Biological Catalysts 677

Summary 680 • Particulate Preview Wrap-Up 680 • Problem-Solving Summary 680 • Visual Problems 682 • Questions and Problems 684

What causes smog? *(Chapter 13)*

14 Chemical Equilibrium: How Much Product Does a Reaction Really Make? 694

14.1 The Dynamics of Chemical Equilibrium 696

14.2 The Equilibrium Constant 698

14.3 Relationships between K_c and K_p Values 703

14.4 Manipulating Equilibrium Constant Expressions 706
K for Reverse Reactions 706 • K for an Equation Multiplied or Divided by a Number 707 • Combining K Values 708

What reactions produce nitrogen-based fertilizers? *(Chapter 14)*

14.5 Equilibrium Constants and Reaction Quotients   710

14.6 Heterogeneous Equilibria   713

14.7 Le Châtelier's Principle   714
Effects of Adding or Removing Reactants or Products   715 • Effects of Pressure and Volume Changes   717 • Effect of Temperature Changes   719 • Catalysts and Equilibrium   721

14.8 Calculations Based on K 721

Summary   729 • Particulate Preview Wrap-Up   729 • Problem-Solving Summary   729 • Visual Problems   730 • Questions and Problems   731

15 Acid–Base Equilibria: Proton Transfer in Biological Systems   738

15.1 Acids and Bases: A Balancing Act   740

15.2 Strong and Weak Acids and Bases   741
Strong and Weak Acids   742 • Strong and Weak Bases   745 • Conjugate Pairs   746 • Relative Strengths of Conjugate Acids and Bases   747

15.3 pH and the Autoionization of Water   748
The pH Scale   749 • pOH, pK_a, and pK_b Values   752

15.4 K_a, K_b, and the Ionization of Weak Acids and Bases   753
Weak Acids   753 • Weak Bases   756

15.5 Calculating the pH of Acidic and Basic Solutions   759
Strong Acids and Strong Bases   759 • Weak Acids and Weak Bases   760 • pH of Very Dilute Solutions of Strong Acids   762

15.6 Polyprotic Acids   764
Acid Rain   764 • Normal Rain   765

15.7 Acid Strength and Molecular Structure   768

15.8 Acidic and Basic Salts   770

Summary   775 • Particulate Preview Wrap-Up   775 • Problem-Solving Summary   776 • Visual Problems   778 • Questions and Problems   779

What's responsible for the color of hydrangeas? *(Chapter 15)*

16 Additional Aqueous Equilibria: Chemistry and the Oceans   784

16.1 Ocean Acidification: Equilibrium under Stress   786

16.2 The Common-Ion Effect   788

16.3 pH Buffers   791
Buffer Capacity   794

16.4 Indicators and Acid–Base Titrations   798
Acid–Base Titrations   799 • Titrations with Multiple Equivalence Points   805

16.5 Lewis Acids and Bases   809

16.6 Formation of Complex Ions   812

16.7 Hydrated Metal Ions as Acids   814

16.8 Solubility Equilibria   816
K_{sp} and Q 820

Summary   824 • Particulate Preview Wrap-Up   825 • Problem-Solving Summary   825 • Visual Problems   825 • Questions and Problems   827

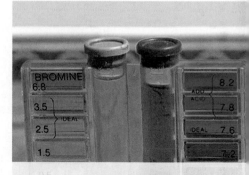

How can we test the water in swimming pools? *(Chapter 16)*

How do cold packs work?
(Chapter 17)

17 **Thermodynamics: Spontaneous and Nonspontaneous Reactions and Processes**   832

17.1   Spontaneous Processes   834
17.2   Thermodynamic Entropy   837
17.3   Absolute Entropy and the Third Law of Thermodynamics   841
        Entropy and Structure   844
17.4   Calculating Entropy Changes   845
17.5   Free Energy   846
17.6   Temperature and Spontaneity   852
17.7   Free Energy and Chemical Equilibrium   854
17.8   Influence of Temperature on Equilibrium Constants   859
17.9   Driving the Human Engine: Coupled Reactions   861
17.10   Microstates: A Quantized View of Entropy   865

        Summary   869 • Particulate Preview Wrap-Up   869 • Problem-Solving Summary   870 • Visual Problems   870 • Questions and Problems   872

What's inside an electric car?
(Chapter 18)

18 **Electrochemistry: The Quest for Clean Energy**   878

18.1   Running on Electrons: Redox Chemistry Revisited   880
18.2   Voltaic and Electrolytic Cells   883
        Cell Diagrams   884
18.3   Standard Potentials   887
18.4   Chemical Energy and Electrical Work   890
18.5   A Reference Point: The Standard Hydrogen Electrode   893
18.6   The Effect of Concentration on E_{cell} 895
        The Nernst Equation   895 • $E°$ and K 898
18.7   Relating Battery Capacity to Quantities of Reactants   899
        Nickel–Metal Hydride Batteries   900 • Lithium-Ion Batteries   901
18.8   Corrosion: Unwanted Electrochemical Reactions   903
18.9   Electrolytic Cells and Rechargeable Batteries   906
18.10   Fuel Cells   909

        Summary   913 • Particulate Preview Wrap-Up   913 • Problem-Solving Summary   914 • Visual Problems   914 • Questions and Problems   916

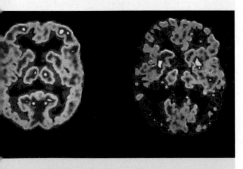

How are radioactive nuclei used in diagnostic medicine?
(Chapter 19)

19 **Nuclear Chemistry: Applications to Energy and Medicine**   922

19.1   Energy and Nuclear Stability   924
19.2   Unstable Nuclei and Radioactive Decay   926
19.3   Measuring Radioactivity   932
19.4   Rates of Radioactive Decay   934
19.5   Radiometric Dating   935
19.6   Biological Effects of Radioactivity   938
        Radiation Dosage   938 • Evaluating the Risks of Radiation   940

19.7 Medical Applications of Radionuclides   942
Therapeutic Radiology   943 • Diagnostic Radiology   943

19.8 Nuclear Fission   944

19.9 Nuclear Fusion and the Quest for Clean Energy   946

Summary   952 • Particulate Preview Wrap-Up   952 • Problem-Solving Summary   952 •
Visual Problems   953 • Questions and Problems   954

20 Organic and Biological Molecules: The Compounds of Life   960

20.1 Molecular Structure and Functional Groups   962
Families Based on Functional Groups   963

20.2 Organic Molecules, Isomers, and Chirality   965
Chirality and Optical Activity   969 • Chirality in Nature   972

20.3 The Composition of Proteins   974
Amino Acids   974 • Zwitterions   976 • Peptides   979

20.4 Protein Structure and Function   981
Primary Structure   982 • Secondary Structure   983 • Tertiary and Quaternary
Structure   985 • Enzymes: Proteins as Catalysts   986

20.5 Carbohydrates   988
Molecular Structures of Glucose and Fructose   988 • Disaccharides and
Polysaccharides   989 • Energy from Glucose   992

20.6 Lipids   992
Function and Metabolism of Lipids   994 • Other Types of Lipids   996

20.7 Nucleotides and Nucleic Acids   997
From DNA to New Proteins   1000

20.8 From Biomolecules to Living Cells   1001

Summary   1004 • Particulate Preview Wrap-Up   1004 •
Problem-Solving Summary   1004 • Visual Problems   1005 •
Questions and Problems   1007

Why is spider silk so strong?
(Chapter 20)

21 The Main Group Elements: Life and the Periodic Table   1016

21.1 Main Group Elements and Human Health   1018

21.2 Periodic and Chemical Properties of Main Group Elements   1021

21.3 Major Essential Elements   1022
Sodium and Potassium   1023 • Magnesium and Calcium   1026 •
Chlorine   1028 • Nitrogen   1029 • Phosphorus and Sulfur   1032

21.4 Trace and Ultratrace Essential Elements   1037
Selenium   1037 • Fluorine and Iodine   1038 • Silicon   1039

21.5 Nonessential Elements   1039
Rubidium and Cesium   1039 • Strontium and Barium   1039 •
Germanium   1039 • Antimony   1040 • Bromine   1040

21.6 Elements for Diagnosis and Therapy   1040
Diagnostic Applications   1041 • Therapeutic Applications   1043

Summary   1044 • Particulate Preview Wrap-Up   1045 •
Problem-Solving Summary   1045 • Visual Problems   1046 •
Questions and Problems   1048

What causes the smell of skunk spray? *(Chapter 21)*

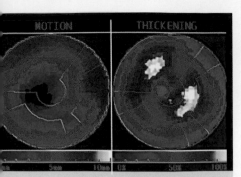

What drugs help doctors image the human heart? *(Chapter 22)*

22 Transition Metals: Biological and Medical Applications   1052

22.1 Transition Metals in Biology: Complex Ions   1054

22.2 Naming Complex Ions and Coordination Compounds   1058
Complex Ions with a Positive Charge   1058 • Complex Ions with a Negative Charge   1060 • Coordination Compounds   1060

22.3 Polydentate Ligands and Chelation   1062

22.4 Crystal Field Theory   1066

22.5 Magnetism and Spin States   1071

22.6 Isomerism in Coordination Compounds   1073
Enantiomers and Linkage Isomers   1075

22.7 Coordination Compounds in Biochemistry   1076
Manganese and Photosynthesis   1077 • Transition Metals in Enzymes   1078

22.8 Coordination Compounds in Medicine   1081
Transition Metals in Diagnosis   1082 • Transition Metals in Therapy   1084

Summary   1088 • Particulate Preview Wrap-Up   1088 • Problem-Solving Summary   1088 • Visual Problems   1089 • Questions and Problems   1091

Appendices   APP-1

Glossary   G-1

Answers to Particulate Review, Concept Tests, and Practice Exercises   ANS-1

Answers to Selected End-of-Chapter Questions and Problems   ANS-13

Credits   C-1

Index   I-1

Applications

Blood centrifugation 9

Electrophoresis of blood proteins 10

Air filtration 10

Seawater distillation 10

Driving the Mars rover *Curiosity* 36

Radioactivity and the Baby Tooth Survey 46

Big Bang and primordial nucleosynthesis 70

Star formation and stellar nucleosynthesis 72

Radioactivity and medical imaging 73

Miller–Urey experiment 84

Volcanic eruptions 86

Natural gas stoves 102

Photosynthesis, respiration, and the carbon cycle 104

Atmospheric carbon dioxide 104

Power plant emissions 106

Anticancer drugs (Taxol) 128

Evidence for water on Mars 144

Polyvinyl chloride (PVC) pipes 151

Great Salt Lake 151

Saline intravenous infusion 156

Stalactites and stalagmites 162

Chemical weathering 162

Drainage from abandoned coal mines 166

Antacids 168

Rock candy 178

Water softening and zeolites 178

Iron oxides in rocks and soils 190

Drug stability calculations 193

Rockets 217

Diesel engines and hot-air balloons 222

Resurfacing an ice rink 226

Instant cold packs 226

Car radiators 232

Chilled beverages 232

Comparing fuels 255

Calories in food 257

Recycling aluminum 259

Barometers and manometers 275

Hurricane Sandy 277

Bicycle tire pressure 286

Aerosol cans 287

Breathing 290

Weather balloons 291

Compressed oxygen for mountaineering 292

Dieng Plateau gas poisoning disaster 295

Nitrogen narcosis 302

Gas mixtures for scuba diving 314

Rainbows 332

Remote control devices 340

Lasers 355

Road flares 365

Fireworks 376

Greenhouse effect 388

Oxyacetylene torches 396

Atmospheric ozone 403

Moth balls 423

Ripening tomatoes 463

Polycyclic aromatic hydrocarbon (PAH) intercalation in DNA 468

Spearmint and caraway aromas 468

Auroras 470

Semiconductors in bar-code readers and DVD players 483

Insect pheromones 484

Hydrogen bonds in DNA 506

Anesthetics 511

Petroleum-based cleaning solvents 512

High-altitude endurance training 516

Supercritical carbon dioxide and dry ice 522

Water striders 524

Aquatic life in frozen lakes 526

Drug efficacy 527

Antifreeze 538

Fractional distillation of crude oil 553

Radiator fluid 562

Osmosis in red blood cells 568

Saline and dextrose intravenous solutions 573

Desalination of seawater via reverse osmosis 573

Eggs 578

Brass and bronze 600

Shape-memory alloys in stents 600

Stainless steel and surgical steel 601

Diamond and graphite 606

Graphene, carbon nanotubes, and fullerenes 607

Polyethylene: LDPE and HDPE plastics 609

Teflon in cookware and surgical tubing 610

Polypropylene products 611

Polystyrene and Styrofoam 612

Plastic soda bottles 613

Artificial skin and dissolving sutures 615

Synthetic fabrics: Dacron, nylon, and Kevlar 617

Thorite lantern mantles 621

Photochemical smog 636

Chlorofluorocarbons (CFCs) and ozone in the stratosphere 673

Catalytic converters 676

Smog simulations 679

Fertilizers 696

Hindenburg airship disaster 703

Manufacturing sulfuric acid 708

Limestone kilns 713

Manufacturing nitric acid 728

Colors of hydrangea blossoms 740

Lung disease and respiratory acidosis 741

Liquid drain cleaners 759

Carabid beetles 760

Acid rain and normal rain 764

Bleach 772

pH of human blood 774

Ocean acidification 786

Swimming pool test kits for pH 798

Sapphire Pool in Yellowstone National Park   808

Milk of magnesia   816

Climate change and seawater acidity   823

Instant cold packs   835

Engine efficiency   851

Energy from glucose; glycolysis   863

Prehistoric axes and copper refining   868

Hybrid and electric vehicles   880

Alkaline, nicad, and zinc–air batteries   889

Lead–acid car batteries   896

Nickel–metal hydride and lithium-ion batteries   900

Rusted iron via oxidation   904

Rechargeable batteries   906

Electroplating   908

Proton-exchange membrane (PEM) fuel cells   909

Alloys and corrosion at sea   912

Scintillation counters and Geiger counters   932

Radiometric dating   935

Biological effects of radioactivity: Chernobyl and Fukushima   940

Radon gas exposure   941

Therapeutic and diagnostic radiology   943

Nuclear weapons and nuclear power   944

Nuclear fusion in the Sun   946

Tokamak reactors and ITER   948

Radium paint and the Radium Girls   950

Perfect foods and complete proteins   974

Aspartame   979

Sickle-cell anemia and malaria   982

Silk   984

Alzheimer's disease   984

Lactose intolerance   986

Blood type and glycoproteins   988

Ethanol production from cellulose   991

Unsaturated fats, saturated fats, and trans fats   993

Olestra, a modified fat substitute   995

Cholesterol and coronary disease   997

DNA and RNA   997

Origin of life on Earth   1001

Hydrogenated oils   1003

Dietary reference intake (DRI) for essential elements   1020

Ion transport across cell membranes   1023

Osteoporosis and kidney stones   1026

Chlorophyll   1026

Teeth, bones, and shells   1027

Acid reflux and antacid drugs   1028

Bad breath, skunk odor, and smelly shoes   1035

Toothpaste and fluoridated water   1038

Goiter and Graves' disease   1038

Prussian blue pigment   1057

Food preservatives   1065

Anticancer drugs (cisplatin)   1073

Cytochromes   1079

Thalassemia and chelation therapy   1085

Organometallic compounds as drugs   1086

ChemTours

Dimensional Analysis 21
Significant Figures 23
Scientific Notation 23
Temperature Conversion 33
Cathode-Ray Tube 47
Millikan Oil-Drop Experiment 49
Rutherford Experiment 50
NaCl Reaction 60
Synthesis of Elements 72
Avogadro's Number 87
Balancing Equations 97
Carbon Cycle 104
Percent Composition 108
Limiting Reactant 122
Molarity 148
Dilution 154
Ions in Solution 158
State Functions and Path Functions 212
Internal Energy 221
Pressure–Volume Work 222
Heating Curves 228
Calorimetry 235
Hess's Law 243
The Ideal Gas Law 289
Dalton's Law 299
Molecular Speed 307
Molecular Motion 307
Electromagnetic Radiation 335
Emission Spectra and the Bohr Model of the Atom 343
De Broglie Wavelength 346
Quantum Numbers 351

Electron Configuration 358
Periodic Trends 371
Bonding 389
Lewis Structures 392
Bond Polarity and Polar Molecules 398
Vibrational Modes 401
Greenhouse Effect 402
Resonance 404
Lewis Structures: Expanded Valence Shells 414
Estimating Enthalpy Changes 420
Hybridization 462
Structure of Benzene 467
Molecular Orbitals 471
Intermolecular Forces 499
Henry's Law 516
Phase Diagrams 520
Capillary Action 523
Dissolution of Ammonium Nitrate 542
Raoult's Law 551
Fractional Distillation 554
Boiling and Freezing Points 563
Osmotic Pressure 568
Unit Cell 597
Allotropes of Carbon 607
Polymers 608
Reaction Rate 639
Reaction Order 645
Arrhenius Equation 661
Collision Theory 661
Reaction Mechanisms 665
Equilibrium 698

Equilibrium in the Gas Phase 700
Le Châtelier's Principle 715
Solving Equilibrium Problems 721
Acid–Base Ionization 743
Autoionization of Water 748
pH Scale 749
Acid Rain 764
Acid Strength and Molecular Structure 769
Buffers 791
Acid–Base Titrations 800
Titrations of Weak Acids 802
Entropy 837
Gibbs Free Energy 847
Equilibrium and Thermodynamics 854
Zinc–Copper Cell 881
Cell Potential 888
Alkaline Battery 889
Cell Potential, Equilibrium, and Free Energy 895
Fuel Cell 909
Balancing Nuclear Equations 928
Radioactive Decay Modes 929
Half-Life 934
Fusion of Hydrogen 946
Chiral Centers 969
Condensation of Biological Polymers 979
Fiber Strength and Elasticity 984
Formation of Sucrose 990
Crystal Field Splitting 1067

About the **Authors**

Thomas R. Gilbert has a BS in chemistry from Clarkson and a PhD in analytical chemistry from MIT. After 10 years with the Research Department of the New England Aquarium in Boston, he joined the faculty of Northeastern University, where he is currently an associate professor of chemistry and chemical biology. His research interests are in chemical and science education. He teaches general chemistry and science education courses, and he conducts professional development workshops for K–12 teachers. He has won Northeastern's Excellence in Teaching Award and Outstanding Teacher of First-Year Engineering Students Award. He is a fellow of the American Chemical Society and in 2012 was elected to the American Chemical Society Board of Directors.

Rein V. Kirss received both a BS in chemistry and a BA in history as well as an MA in chemistry from SUNY Buffalo. He received his PhD in inorganic chemistry from the University of Wisconsin, Madison, where the seeds for this textbook were undoubtedly planted. After two years of postdoctoral study at the University of Rochester, he spent a year at Advanced Technology Materials, Inc., before returning to academics at Northeastern University in 1989. He is an associate professor of chemistry with an active research interest in organometallic chemistry.

Natalie Foster is an emeritus professor of chemistry at Lehigh University in Bethlehem, Pennsylvania. She received a BS in chemistry from Muhlenberg College and MS, DA, and PhD degrees from Lehigh University. Her research interests included studying poly(vinyl alcohol) gels by NMR as part of a larger interest in porphyrins and phthalocyanines as candidate contrast enhancement agents for MRI. She taught both semesters of the introductory chemistry class to engineering, biology, and other nonchemistry majors and a spectral analysis course at the graduate level. She is a fellow of the American Chemical Society and the recipient of the Christian R. and Mary F. Lindback Foundation Award for distinguished teaching.

Stacey Lowery Bretz is a University Distinguished Professor in the Department of Chemistry and Biochemistry at Miami University in Oxford, Ohio. She earned her BA in chemistry from Cornell University, MS from Pennsylvania State University, and a PhD in chemistry education research from Cornell University. Stacey then spent one year at the University of California, Berkeley, as a postdoc in the Department of Chemistry. Her research expertise includes the development of assessments to characterize chemistry misconceptions and measure learning in the chemistry laboratory. Of particular interest is method development with regard to the use of multiple representations (particulate, symbolic, and macroscopic) to generate cognitive dissonance, including protocols for establishing the reliability and validity of these measures. She is a fellow of both the American Chemical Society and the American Association for the Advancement of Science. She has been honored with both of Miami University's highest teaching awards: the E. Phillips Knox Award for Undergraduate Teaching in 2009 and the Distinguished Teaching Award for Excellence in Graduate Instruction and Mentoring in 2013.

Geoffrey Davies holds BSc, PhD, and DSc degrees in chemistry from Birmingham University, England. He joined the faculty at Northeastern University in 1971 after doing postdoctoral research on the kinetics of very rapid reactions at Brandeis University, Brookhaven National Laboratory, and the University of Kent at Canterbury. He is now a Matthews Distinguished University Professor at Northeastern University. His research group has explored experimental and theoretical redox chemistry, alternative fuels, transmetalation reactions, tunable metal–zeolite catalysts and, most recently, the chemistry of humic substances, the essential brown animal and plant metabolites in sediments, soils, and water. He edits a column on experiential and study-abroad education in the *Journal of Chemical Education* and a book series on humic substances. He is a fellow of the Royal Society of Chemistry and was awarded Northeastern's Excellence in Teaching Award in 1981, 1993, and 1999, and its first Lifetime Achievement in Teaching Award in 2004.

Preface

Dear Student,

We wrote this book with three overarching goals in mind: to make chemistry interesting, relevant, and memorable; to enable you to see the world from a molecular point of view; and to help you become an expert problem-solver. You have a number of resources available to assist you to succeed in your general chemistry course. This textbook will be a valuable resource, and we have written it with you, and the different ways you may use the book, in mind.

If you are someone who reads a chapter from the first page to the last, you will see that *Chemistry: The Science in Context*, Fifth Edition, introduces the chemical principles within a chapter by using contexts drawn from daily life as well as from other disciplines, including biology, environmental science, materials science, astronomy, geology, and medicine. We believe that these contexts make chemistry more interesting, relevant, understandable, and memorable.

Chemists' unique perspective of natural processes and insights into the properties of substances, from high-performance alloys to the products of biotechnology, are based on understanding these processes and substances at the particulate level (the atomic and molecular level). A major goal of this book is to help you develop this microscale perspective and link it to macroscopic properties.

With that in mind, we begin each chapter with a **Particulate Review** and **Particulate Preview** on the first page. The goal of these tools is to prepare you for the material in the chapter. The Review assesses important prior knowledge you need to interpret particulate images in the chapter. The Particulate Preview asks you to speculate about new concepts you will see in the chapter and is meant to focus your reading.

PARTICULATE **REVIEW**

Acid and Base

In Chapter 5 we consider the energy changes that occur during reactions such as the combustion reactions from Chapter 3 and neutralization reactions from Chapter 4.

- Here we see the key molecules and ions involved in the titration of hydrochloric acid with sodium hydroxide. Name each molecule or ion and write its formula.

- The colorless solution in the flask on the left is hydrochloric acid. The colorless solution in the buret is sodium hydroxide. On the right is a picture of the titration after all the acid has been neutralized. Which of the illustrated particles are present in the buret, the flask on the left, and the flask on the right?

(Review Sections 4.5–4.6 if you need help.)

(Answers to Particulate Review questions are in the back of the book.)

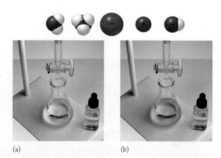

(a)   (b)

PARTICULATE **PREVIEW**

Breaking Bonds and Energy

When ozone molecules absorb ultraviolet rays (UV rays) from the Sun, the ozone falls apart into oxygen molecules and oxygen atoms according to the chemical reaction depicted here. As you read Chapter 5, look for ideas that will help you answer these questions:

- What role does energy play in breaking the bonds?
- Does bond breaking occur when energy is absorbed? Or does breaking a bond release energy?

$$O_3(g) \xrightarrow{\text{UV rays}} O_2(g) + O(g)$$

If you want a quick summary of what is most important in a chapter to direct your studying on selected topics, check the **Learning Outcomes** at the beginning of each chapter. Whether you are reading the chapter from first page to last, moving from topic to topic in an order you select, or reviewing material for an exam, the Learning Outcomes can help you focus on the key information you need to know and the skills you should acquire.

Learning Outcomes

LO1 Explain kinetic and potential energies at the molecular level
Sample Exercise 5.1

LO2 Identify familiar endothermic and exothermic processes
Sample Exercise 5.2

LO3 Calculate changes in the internal energy of a system
Sample Exercises 5.3, 5.4

LO4 Calculate the amount of heat transferred in physical or chemical processes
Sample Exercises 5.5, 5.6, 5.7, 5.8, 5.9

LO5 Calculate thermochemical values by using data from calorimetry experiments
Sample Exercises 5.10, 5.11

LO6 Calculate enthalpies of reaction
Sample Exercises 5.12, 5.13, 5.15

LO7 Recognize and write equations for formation reactions
Sample Exercise 5.14

LO8 Calculate and compare fuel and food values and fuel densities
Sample Exercises 5.16, 5.17

In every section, you will find **key terms** in boldface in the text and in a **running glossary** in the margin. We have inserted the definitions throughout the text, so you can continue reading without interruption but quickly find key terms when doing homework or reviewing for a test. All key terms are also defined in the Glossary in the back of the book.

Approximately once per section, you will find a **Concept Test**. These short, conceptual questions provide a self-check opportunity by asking you to stop and answer a question relating to what you just read. We designed them to help you see for yourself whether you have grasped a key concept and can apply it. You will find answers to Concept Tests in the back of the book.

CONCEPT TEST

Identify the following systems as isolated, closed, or open: (a) the water in a pond; (b) a carbonated beverage in a sealed bottle; (c) a sandwich wrapped in thermally conducting plastic wrap; (d) a live chicken.

(Answers to Concept Tests are in the back of the book.)

New concepts naturally build on previous information, and you will find that many concepts are related to others described earlier in the book. We point out these relationships with **Connection** icons in the margins. These reminders will help you see the big picture and draw your own connections between the major themes covered in the book.

At the end of each chapter is a group of **Visual Problems** that ask you to interpret atomic and molecular views of elements and compounds, along with graphs of experimental data. The last Visual Problem in each chapter contains a **visual problem matrix**. This grid consists of nine images followed by a series of questions that will test your ability to identify the similarities and differences among the macroscopic and particulate images.

If you're looking for additional help visualizing a concept, we have almost 100 **ChemTours**, denoted by the ChemTour icon. The ChemTours, available at digital .wwnorton.com/chem5, provide animations of physical changes and chemical reactions to help you envision events at the molecular level. Many ChemTours are interactive, allowing you to manipulate variables and observe resulting changes in

CONNECTION In Chapter 1 we discussed the arrangement of molecules in ice, water, and water vapor.

CHEMTOUR
Heating Curves

a graph or a process. Questions at the end of the ChemTour tutorials offer step-by-step assistance in solving problems and provide useful feedback.

Another goal of the book is to help you improve your problem-solving skills. Sometimes the hardest parts of solving a problem are knowing where to start and distinguishing between information that is relevant and information that is not. Once you are clear on where you are starting and where you are going, planning for and arriving at a solution become much easier.

To help you hone your problem-solving skills, we have developed a framework that is introduced in Chapter 1 and used consistently throughout the book. It is a four-step approach we call **COAST**, which is our acronym for (1) **C**ollect and **O**rganize, (2) **A**nalyze, (3) **S**olve, and (4) **T**hink About It. We use these four steps in *every* Sample Exercise and in the solutions to *odd-numbered* problems in the Student's Solutions Manual. They are also used in the hints and feedback embedded in the Smartwork5 online homework program. To summarize the four steps:

Collect and Organize helps you understand where to begin. In this step we often point out what you must find and what is given, including the relevant information that is provided in the problem statement or available elsewhere in the book.

Analyze is where we map out a strategy for solving the problem. As part of that strategy we often estimate what a reasonable answer might be.

Solve applies our strategy from the second step to the information and relationships identified in the first step to actually solve the problem. We walk you through each step in the solution so that you can follow the logic as well as the math.

Think About It reminds us that calculating or determining an answer is not the last step when solving a problem. Checking whether the solution is reasonable in light of an estimate is imperative. Is the answer realistic? Are the units correct? Is the number of significant figures appropriate? Does it make sense with our estimate from the Analyze step?

Many students use the **Sample Exercises** more than any other part of the book. Sample Exercises take the concept being discussed and illustrate how to apply it to solve a problem. We hope that repeated application of COAST will help you refine your problem-solving skills and become an expert problem-solver. When you finish a Sample Exercise, you'll find a **Practice Exercise** to try on your own. Notice that the Sample Exercises and the Learning Objectives are connected. We think this will help you focus efficiently on the main ideas in the chapter.

Students sometimes comment that the questions on an exam are more challenging than the Sample Exercises in a book. To address this, we have an **Integrating Concepts Sample Exercise** near the end of each chapter. These exercises require you to use more than one concept from the chapter and may expect you to use concepts from earlier chapters to solve a problem. Please invest your time working through these problems because we think they will further enhance your problem-solving skills and give you an increased appreciation of how chemistry is used in the world.

SAMPLE EXERCISE 5.2 Identifying Exothermic and Endothermic Processes   **LO2**

Describe the flow of energy during the purification of water by distillation (Figure 5.14), identify the steps in the process as either endothermic or exothermic, and give the sign of q associated with each step. Consider the water being purified to be the system.

Collect and Organize We are given that the water is the system. We must evaluate how the water gains or loses energy during distillation.

Analyze In distillation, energy flows in three steps: (1) liquid water is heated to the boiling point and (2) vaporizes. (3) The vapors are cooled and condense as they pass through the condenser.

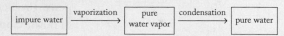

impure water → (vaporization) → pure water vapor → (condensation) → pure water

Solve Energy flows from the surroundings (hot plate) to heat the impure water (the system) to its boiling point and then to vaporize it. Therefore processes 1 and 2 are endothermic. The sign of q is positive for both. Because energy flows from the system (water vapor) into the surroundings (condenser walls), process 3 is exothermic. Therefore, the sign of q is negative.

Think About It *Endothermic* means that energy is transferred from the surroundings into the system—the water in the distillation flask. When the water vapor is cooled in the condenser, energy flows from the vapor as it is converted from a gas to a liquid; the process is exothermic.

Practice Exercise What is the sign of q as (a) a match burns, (b) drops of molten candle wax solidify, and (c) perspiration evaporates from skin? In each case, define the system and indicate whether the process is endothermic or exothermic.

(Answers to Practice Exercises are in the back of the book.)

LO5 A calorimeter, characterized by its **calorimeter constant** (its characteristic **heat capacity**), is a device used to measure the amount of energy involved in physical and chemical processes. The enthalpy change associated with a reaction is defined by the **enthalpy of reaction** (ΔH_{rxn}). (Section 5.6)

LO6 **Hess's law** states that the enthalpy of a reaction (ΔH_{rxn}) that is the sum of two or more other reactions is equal to the sum of the ΔH_{rxn} values of the constituent reactions. It can be used to calculate enthalpy changes in reactions that are hard or impossible to measure directly. (Section 5.7)

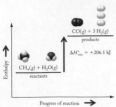

If you use the book mostly as a reference and problem-solving guide, we have a learning path for you as well. It starts with the **Summary** and a **Problem-Solving Summary** at the end of each chapter. The first is a brief synopsis of the chapter, organized by Learning Outcomes. Key figures have been added to this Summary to provide visual cues as you review. The Problem-Solving Summary organizes the chapter by problem type and summarizes relevant concepts and equations you need to solve each type of problem. The Problem-Solving Summary also points you back to the relevant Sample Exercises that model how to solve each problem and cross-references the Learning Outcomes at the beginning of the chapter.

PROBLEM-SOLVING SUMMARY

Type of Problem	Concepts and Equations		Sample Exercises
Calculating kinetic and potential energy		(5.2) (5.3)	**5.1**
Identifying endothermic and exothermic processes, and calculating internal energy change (ΔE) and P–V work	For the system: $\Delta E = q + w$ where $w = -P\Delta V$.	(5.5)	**5.2, 5.3, 5.4**
Predicting the sign of ΔH_{sys} for physical and chemical changes	Exothermic: $\Delta H_{sys} < 0$ Endothermic: $\Delta H_{sys} > 0$		**5.5, 5.6**

Following the summaries are groups of questions and problems. The first group is the **Visual Problems**. **Concept Review Questions and Problems** come next, arranged by topic in the same order as they appear in the chapter. Concept Reviews are qualitative and often ask you to explain why or how something happens. Problems are paired and can be quantitative, conceptual, or a combination of both. **Contextual problems** have a title that describes the context in which the problem is placed. **Additional Problems** can come from any section or combination of sections in the chapter. Some of them incorporate concepts from previous chapters. Problems marked with an asterisk (*) are more challenging and often require multiple steps to solve.

We want you to have confidence in using the answers in the back of the book as well as the Student's Solutions Manual, so we continue to use a rigorous triple-check accuracy program for the fifth edition. Each end-of-chapter question and problem has been solved independently by at least three PhD chemists. For the fifth edition the team included Solutions Manual author Bradley Wile and two additional chemistry educators. Brad compared his solutions to those from the two reviewers and resolved any discrepancies. This process is designed to ensure clearly written problems and accurate answers in the appendices and Solutions Manual.

No matter how you use this book, we hope it becomes a valuable tool for you and helps you not only understand the principles of chemistry but also apply them to solving global problems, such as diagnosing and treating disease or making more efficient use of Earth's natural resources.

Changes to the Fifth Edition

Dear Instructor,

As authors of a textbook we are very often asked: "Why is a fifth edition necessary? Has the science changed that much since the fourth edition?" Although chemistry is a vigorous and dynamic field, most basic concepts presented in an introductory course have not changed dramatically. However, two areas tightly intertwined in this text—pedagogy and context—have changed significantly, and those areas are the drivers of this new edition. Here are some of the most noteworthy changes we made throughout this edition:

- We welcome Stacey Lowery Bretz as our new coauthor. Stacey is a chemistry education researcher, and her insights and expertise about student misconceptions and the best way to address those misconceptions can be seen throughout the book. The most obvious examples are the new **Particulate Review** and **Particulate Preview** questions at the beginning of each chapter. The Review is a diagnostic tool that addresses important prior knowledge students must draw upon to successfully interpret molecular (particulate) images in the chapter. The Review consists of a few questions based on particulate-scale art. The Preview consists of a short series of questions about a particulate image that ask students to *extend* their prior knowledge and *speculate* about material in the chapter. The goal of the Preview is to direct students as they read, making reading more interactive.

- In addition to the Particulate Review and Preview, Stacey authored a new type of visual problem: the **visual problem matrix**. The matrix consists of macroscopic and particulate images in a grid, followed by a series of questions that ask students to identify commonalities and differences across the images based on their understanding. Versions of the Particulate Review, Preview, and the visual matrix problems are in the lecture PowerPoint presentations to use with clickers during lectures. They are also available in Smartwork5 as individual problems as well as premade assignments to use before or after class.

- We evaluated each Sample Exercise, and in simple, one-step Sample Exercises, we have streamlined the prose by combining the Collect and Organize and Analyze step. We revised numerous Sample Exercises throughout the fifth edition on the basis of reviewer and user feedback.

- The treatment of how to evaluate the precision and accuracy of experimental values in Chapter 1 has been expanded to include the identification of outliers by using standard deviations, confidence intervals, and the Grubbs test.

- We have expanded our coverage of aqueous equilibrium by adding a second chapter that doubles the number of Sample Exercises and includes Concept Tests that focus upon the molecules and ions present in titrations and buffers.

- In the fifth edition, functional groups are introduced in Chapter 2 and then seamlessly integrated into chapters as appropriate. For example, carboxylic acids and amines are introduced in Chapter 4 when students learn about acid–base reactions. This pedagogical choice enables us to weave core chemistry concepts into contexts that include a wider variety of environmental and health issues. Our hope is that it provides a stronger foundation for considering Lewis structures with a broader knowledge of the variety of molecules that are possible, as well as emphasizes the importance of structure–function from the very beginning of students' journey through chemistry.

- Given the integration of functional groups into the first 12 chapters, we now have one chapter (Chapter 20) that focuses on organic chemistry and biochemistry by discussing isomers, chirality, and the major classes of large biomolecules.

5.8. Use representations [A] through [I] in Figure P5.8 to answer questions a–f.
  a. Which processes are exothermic?
  b. Which processes have a positive ΔH?
  c. In which processes does the system gain energy?
  d. In which processes do the surroundings lose energy?
  e. Compare a flame of methane [D] at 1000°C to a flame of propane [E] at 1000°C in terms of (i) average kinetic energy and (ii) average speed of the molecules.
  f. Which substance(s) would *not* have vibrational motion or rotational motion? Why?

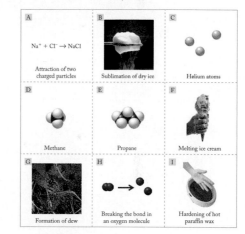

- Chapter 12, the Solids chapter, has been expanded to include polymers with a focus on biomedical applications, and band theory has been moved from the Solids chapter to the end of Chapter 9 following the discussion of molecular orbital theory.
- We took the advice of reviewers and now have two descriptive chemistry chapters at the end of the book. These chapters focus on main group chemistry and transition metals, both within the context of biological and medical applications.
- We have revised or replaced at least 10 percent of the end-of-chapter problems. We incorporated feedback from users and reviewers to address areas where we needed more problems or additional problems of varying difficulty.
- A new version of Smartwork, Smartwork5, offers more than 3600 problems in a sophisticated and user-friendly platform, and 400 new problems are designed to support the new visualization pedagogy. In addition to being tablet compatible, Smartwork5 integrates with the most common campus learning management systems.
- The nearly 100 ChemTours have been updated to better support lecture, lab, and independent student learning. The ChemTours include images, animations, and audio that demonstrate dynamic processes and help students visualize and understand chemistry at the molecular level. Forty of the ChemTours now contain greater interactivity and are assignable in Smartwork5. The ChemTours are linked directly from the ebook and are now in HTML5, which means they are tablet compatible.

Teaching and Learning Resources

Smartwork5 Online Homework for General Chemistry

digital.wwnorton.com/chem5
Smartwork5 is the most intuitive online tutorial and homework management system available for general chemistry. The many question types, including graded molecule drawing, math and chemical equations, ranking tasks, and interactive figures, help students develop and apply their understanding of fundamental concepts in chemistry.

Every problem in Smartwork5 includes response-specific feedback and general hints using the steps in COAST. Links to the ebook version of *Chemistry: The Science in Context*, Fifth Edition, take students to the specific place in the text where the concept is explained. All problems in Smartwork5 use the same language and notation as the textbook.

Smartwork5 also features Tutorial Problems. If students ask for help in a Tutorial Problem, the system breaks the problem down into smaller steps, coaching them with hints, answer-specific feedback, and probing questions within each step. At any point in a Tutorial, a student can return to and answer the original problem.

Assigning, editing, and administering homework within Smartwork5 is easy. It's tablet compatible and integrates with the most common campus learning management systems. Smartwork5 allows the instructor to search for problems by using both the text's Learning Objectives and Bloom's taxonomy. Instructors can use premade assignment sets provided by Norton authors, modify those assignments, or create their own. Instructors can also make changes in the problems at the question level. All instructors have access to our WYSIWYG (What You See Is What You Get) authoring tools—the same ones Norton authors use. Those intuitive tools make it easy to modify existing problems or to develop new content that meets the specific needs of your course.

Wherever possible, Smartwork5 makes use of algorithmic variables so that students see slightly different versions of the same problem. Assignments are graded automatically, and Smartwork5 includes sophisticated yet flexible tools for managing class data. Instructors can use the class activity report to assess students' performance on specific problems within an assignment. Instructors can also review individual students' work on problems.

Smartwork5 for *Chemistry*, Fifth Edition, features the following problem types:

- End-of-Chapter Problems. These problems, which use algorithmic variables when appropriate, all have hints and answer-specific feedback to coach students through mastering single- and multiple-concept problems based on chapter content. They make use of all of Smartwork5's answer-entry tools.
- ChemTour Problems. Forty ChemTours now contain greater interactivity and are assignable in Smartwork5.
- Visual and Graphing Problems. These problems challenge students to identify chemical phenomena and to interpret graphs. They use Smartwork5's Drag-and-Drop and Hotspot functionality.
- Reaction Visualization Problems. Based on both static art and videos of simulated reactions, these problems are designed to help students visualize what happens at the atomic level—and why it happens.
- Ranking Task Problems. These problems ask students to make comparative judgments between items in a set.
- Nomenclature Problems. New matching and multiple-choice problems help students master course vocabulary.
- Multistep Tutorials. These problems offer students who demonstrate a need for help a series of linked, step-by-step subproblems to work. They are based on the Concept Review problems at the end of each chapter.
- Math Review Problems. These problems can be used by students for practice or by instructors to diagnose the mathematical ability of their students.

Ebook

digital.wwnorton.com/chem5

An affordable and convenient alternative to the print text, Norton Ebooks let students access the entire book and much more: they can search, highlight, and take notes with ease. The Norton Ebook allows instructors to share their notes with students. And the ebook can be viewed on most devices—laptop, tablet, even a public computer—and will stay synced between devices.

The online version of *Chemistry*, Fifth Edition, also provides students with one-click access to the nearly 100 ChemTour animations.

The online ebook is available bundled with the print text and Smartwork5 at no extra cost, or it may be purchased bundled with Smartwork5 access.

Norton also offers a downloadable PDF version of the ebook.

Student's Solutions Manual

by Bradley Wile, Ohio Northern University

The Student's Solutions Manual provides students with fully worked solutions to select end-of-chapter problems using the **COAST** four-step method (**C**ollect and **O**rganize, **A**nalyze, **S**olve, and **T**hink About It). The Student's Solutions Manual contains several pieces of art for each chapter, designed to help students visualize ways to approach problems. This artwork is also used in the hints and feedback within Smartwork5.

Clickers in Action: Increasing Student Participation in General Chemistry

by Margaret Asirvatham, University of Colorado, Boulder

An instructor-oriented resource providing information on implementing clickers in general chemistry courses, *Clickers in Action* contains more than 250 class-tested, lecture-ready questions, with histograms showing student responses, as well as insights and suggestions for implementation. Question types include macroscopic observation, symbolic representation, and atomic/molecular views of processes.

Test Bank

by Chris Bradley, Mount St. Mary's University

Norton uses an innovative, evidence-based model to deliver high-quality and pedagogically effective quizzes and testing materials. Each chapter of the Test Bank is structured around an expanded list of student learning objectives and evaluates student knowledge on six distinct levels based on Bloom's Taxonomy: Remembering, Understanding, Applying, Analyzing, Evaluating, and Creating.

Questions are further classified by section and difficulty, making it easy to construct tests and quizzes that are meaningful and diagnostic, according to each instructor's needs. More than 2500 questions are divided into multiple choice and short answer.

The Test Bank is available with ExamView Test Generator software, allowing instructors to effortlessly create, administer, and manage assessments. The convenient and intuitive test-making wizard makes it easy to create customized exams with no software learning curve. Other key features include the ability to create paper exams with algorithmically generated variables and export files directly to Blackboard, Canvas, Desire2Learn, and Moodle.

Instructor's Solutions Manual

by Bradley Wile, Ohio Northern University

The Instructor's Solutions Manual provides instructors with fully worked solutions to every end-of-chapter Concept Review and Problem. Each solution uses the **COAST** four-step method (**C**ollect and **O**rganize, **A**nalyze, **S**olve, and **T**hink About It).

Instructor's Resource Manual

by Matthew Van Duzor, North Park University, and Andrea Van Duzor, Chicago State University

This complete resource manual for instructors has been revised to correspond to changes made in the fifth edition. Each chapter begins with a brief overview of the text chapter followed by suggestions for integrating the contexts featured in the book into a lecture, summaries of the textbook's Particulate Preview and Review sections, suggested sample lecture outlines, alternative contexts to use with each chapter, and instructor notes for suggested activities from the *Chem-Connections* and *Calculations in Chemistry*, Second Edition, workbooks. Suggested ChemTours and laboratory exercises round out each chapter.

Instructor's Resource Disc

This helpful classroom presentation tool features:

- Stepwise animations and classroom response questions. Developed by Jeffrey Macedone of Brigham Young University and his team, these animations, which use native PowerPoint functionality and textbook art, help instructors to "walk" students through nearly 100 chemical concepts and processes. Where appropriate, the slides contain two types of questions for students to answer in class: questions that ask them to predict what will happen next and why, and questions that ask them to apply knowledge gained from watching the animation. Self-contained notes help instructors adapt these materials to their own classrooms.
- Lecture PowerPoint (Scott Farrell, Ocean County College) slides include a suggested classroom-lecture script in an accompanying Word file. Each chapter opens with a set of multiple-choice questions based on the textbook's Particulate Review and Preview section, and concludes with another set of questions based on the textbook's visual problem matrix.
- All ChemTours.
- *Clickers in Action* clicker questions for each chapter provide instructors with class-tested questions they can integrate into their course.
- Photographs, drawn figures, and tables from the text, available in PowerPoint and JPEG format.

Downloadable Instructor's Resources

digital.wwnorton.com/chem5
This password-protected site for instructors includes:

- Stepwise animations and classroom response questions. Developed by Jeffrey Macedone of Brigham Young University and his team, these animations, which use native PowerPoint functionality and textbook art, help instructors to "walk" students through nearly 100 chemical concepts and processes. Where appropriate, the slides contain two types of questions for students to answer in class: questions that ask them to predict what will happen next and why, and questions that ask them to apply knowledge gained from watching the animation. Self-contained notes help instructors adapt these materials to their own classrooms.

- Lecture PowerPoints.
- All ChemTours.
- Test bank in PDF, Word RTF, and ExamView Assessment Suite formats.
- Solutions Manual in PDF and Word, so that instructors may edit solutions.
- All of the end-of-chapter questions and problems, available in Word along with the key equations.
- Photographs, drawn figures, and tables from the text, available in PowerPoint and JPEG format.
- *Clickers in Action* clicker questions.
- Course cartridges. Available for the most common learning management systems, course cartridges include access to the ChemTours and StepWise animations as well as links to the ebook and Smartwork5.

Acknowledgments

We begin by thanking the people who played the biggest role in getting the whole process started for the fifth edition: you, the users and the reviewers. Your suggestions, comments, critiques, and quality feedback have encouraged us to tackle the revision process to ensure the content, context, and pedagogy work better for you and maximize learning for all your students. Our deepest thanks and gratitude go to you, the users, for sharing your experiences with us. Your comments at meetings, in focus groups, in emails, and during office visits with the Norton travelers help us identify what works well and what needs to be improved pedagogically; these comments, together with sharing the new stories and current real-world examples of chemistry that capture your students' interest, provide the foundation for this revision. We are grateful to you all.

Our colleagues at W. W. Norton remain a constant source of inspiration and guidance. Their passion for providing accurate and reliable content sets a high standard that motivates all of us to create an exceptional and user-friendly set of resources for instructors and students. Our highest order of thanks must go to W. W. Norton for having enough confidence in the idea behind the first four editions to commit to the massive labor of the fifth. The people at W. W. Norton with whom we work most closely deserve much more praise than we can possibly express here. Our editor, Erik Fahlgren, continues to offer his wisdom, guidance, energy, creativity, and most impressively, an endless amount of patience, as he both simultaneously leads and pushes us to meet deadlines. Erik's leadership is the single greatest reason for this book's completion, and our greatest thanks are far too humble an offering for his unwavering vision and commitment. He is the consummate professional and a valued friend.

We are grateful beyond measure for the contributions of developmental editor Andrew Sobel. His analytical insights kept us cogent and his questions kept us focused; we are all better writers for having benefited from his mentorship. Our project editor, Carla Talmadge, brought finesse to the job of synchronizing our words and images on the page. Assistant editor Arielle Holstein kept us all organized, on track, and on time—a herculean task essential to the success of this project. Debra Morton Hoyt took our inchoate ideas and produced a spectacular new cover; Rona Tuccillo found just the right photo again and again; production manager Eric Pier-Hocking worked tirelessly behind the scenes; Julia Sammaritano managed the print supplements skillfully; Chris Rapp's vision for the new media package was imaginative and transformed the written page into interactive learning tools for instructors and students alike; and Stacy Loyal created a bold yet thoughtful strategy to ambitiously market the book and work with the Norton team in the field.

This book has benefited greatly from the care and thought that many reviewers, listed here, gave to their readings of earlier drafts. We owe an extra-special thanks to Brad Wile for his dedicated and precise work on the Solutions Manual. He, along with Timothy Chapp and Tim Brewer, are the triple-check accuracy team who solved each problem and reviewed each solution for accuracy. We are deeply grateful to David Hanson for working with us to clarify both our language and our thoughts about thermochemistry. Finally, we greatly appreciate Allen Apblett, Chuck Cornett, Joseph Emerson, Amy Johnson, Edith Kippenhan, Brian Leskiw, Steve Rathbone, Jimmy Reeves, Jason Ritchie, Mary Roslonowski, Thomas Sorensen, and David Winters for checking the accuracy of the myriad facts that form the framework of our science.

Thomas R. Gilbert
Rein V. Kirss
Natalie Foster
Stacey Lowery Bretz
Geoffrey Davies

Fifth Edition Reviewers

Kenneth Adair, *Case Western Reserve University*
Allen Apblett, *Oklahoma State University*
Mark Baillie, *University of Delaware*
Jack Barbera, *Portland State University*
Paul Benny, *Washington State University*
Simon Bott, *University of Houston*
Chris Bradley, *Mount St. Mary's University*
Vanessa Castleberry, *Baylor University*
Dale Chatfield, *San Diego State University*
Christopher Cheatum, *University of Iowa*
Gina Chiarella, *Prairie View A&M University*
Susan Collins, *California State University, Northridge*
Chuck Cornett, *University of Wisconsin, Platteville*
Mapi Cuevas, *Santa Fe College*
Vanessa dos Reis Falcao, *University of Miami*
Amanda Eckermann, *Hope College*
Emad El-Giar, *University of Louisiana at Monroe*
Joseph Emerson, *Mississippi State University*
Steffan Finnegan, *Georgia State University*
Crista Force, *Lone Star College, Montgomery*
Andrew Frazer, *University of Central Florida*
Jennifer Goodnough, *University of Minnesota, Morris*
Kayla Green, *Texas Christian University*
Alexander Grushow, *Rider University*
Joseph Hall, *Norfolk State University*
C. Alton Hassell, *Baylor University*
Andy Ho, *Lehigh University*
K. Joseph Ho, *University of New Mexico*
Donna Iannotti, *Eastern Florida State College*
Roy Jensen, *University of Alberta*
Jie Jiang, *Georgia State University*
Amy Johnson, *Eastern Michigan University*

Gregory Jursich, *University of Illinois*
Kayla Kaiser, *California State University, Northridge*
Jesudoss Kingston, *Iowa State University*
Edith Kippenhan, *University of Toledo*
Leslie Knecht, *University of Miami*
Sushilla Knottenbelt, *University of New Mexico*
Larry Kolopajlo, *Eastern Michigan University*
Richard Lahti, *Minnesota State University, Moorhead*
Patricia Lang, *Ball State University*
Neil Law, *SUNY Cobleskill*
Daniel Lawson, *University of Michigan, Dearborn*
Alistair Lees, *Binghamton University*
Brian Leskiw, *Youngstown State University*
Dewayne Logan, *Baton Rouge Community College*
Jeremy Mason, *Texas Tech University*
Lauren McMills, *Ohio University*
Zachary Mensinger, *University of Minnesota, Morris*
Drew Meyer, *Case Western Reserve University*
Rebecca Miller, *Lehigh University*
Deb Mlsna, *Mississippi State University*
Gary Mort, *Lane Community College*
Pedro Patino, *University of Central Florida*
Michael Pikaart, *Hope College*
Steve Rathbone, *Blinn College*
Jimmy Reeves, *University of North Carolina, Wilmington*
Jason Ritchie, *University of Mississippi*
Mary Roslonowski, *Eastern Florida State College*
Kresimir Rupnik, *Louisiana State University*
Omowunmi Sadik, *Binghamton University*
Akbar Salam, *Wake Forest University*
Kerri Scott, *University of Mississippi*
Fatma Selampinar, *University of Connecticut*

Trineshia Sellars, *Palm Beach State College*
Gregory Smith, *Angelo State University*
Thomas Sorensen, *University of Wisconsin, Milwaukee*
Alan Stolzenberg, *West Virginia University*
Elina Stroeva, *Georgia State University*
Wayne Suggs, *University of Georgia*
Eirin Sullivan, *Illinois State University*
Daniel Swart, *Minnesota State University, Mankato*
Brooke Taylor, *Lane Community College*
Daeri Tenery, *Valencia State College*
Mark Tinsley, *West Virginia University*
Rick Toomey, *Northwest Missouri State University*
Marie Villarba, *Seattle Central Community College*
Yan Waguespack, *University of Maryland, Eastern Shore*
John Wiginton, *University of Mississippi*
David Winters, *Tidewater Community College*
Joseph Wu, *University of Wisconsin, Platteville*
Mingming Xu, *West Virginia University*
Tatiana Zuvich, *Eastern Florida State College*

Previous Editions' Reviewers

William Acree, Jr., *University of North Texas*
R. Allendoefer, *University at Buffalo*
Thomas J. Anderson, *Francis Marion University*
Anil Banerjee, *Texas A&M University, Commerce*
Sharon Anthony, *The Evergreen State College*
Jeffrey Appling, *Clemson University*
Marsi Archer, *Missouri Southern State University*
Margaret Asirvatham, *University of Colorado, Boulder*
Robert Balahura, *University of Guelph*
Ian Balcom, *Lyndon State College of Vermont*
Sandra Banks, *Mills College*
Mikhail V. Barybin, *University of Kansas*
Mufeed Basti, *North Carolina Agricultural & Technical State University*
Shuhsien Batamo, *Houston Community College, Central Campus*
Robert Bateman, *University of Southern Mississippi*
Erin Battin, *West Virginia University*
Kevin Bennett, *Hood College*
H. Laine Berghout, *Weber State University*
Lawrence Berliner, *University of Denver*
Mark Berry, *Brandon University*
Narayan Bhat, *University of Texas, Pan American*
Eric Bittner, *University of Houston*
David Blauch, *Davidson College*
Robert Boggess, *Radford University*
Simon Bott, *University of Houston*
Ivana Bozidarevic, *DeAnza College*
Chris Bradley, *Mount St. Mary's University*
Michael Bradley, *Valparaiso University*
Richard Bretz, *Miami University*
Karen Brewer, *Hamilton College*
Timothy Brewer, *Eastern Michigan University*
Ted Bryan, *Briar Cliff University*
Donna Budzynski, *San Diego Community College*

Julia Burdge, *Florida Atlantic University*
Robert Burk, *Carleton University*
Andrew Burns, *Kent State University*
Diep Ca, *Shenandoah University*
Sharmaine Cady, *East Stroudsburg University*
Chris Cahill, *George Washington University*
Dean Campbell, *Bradley University*
Kevin Cantrell, *University of Portland*
Nancy Carpenter, *University of Minnesota, Morris*
Patrick Caruana, *SUNY Cortland*
David Cedeno, *Illinois State University*
Tim Champion, *Johnson C. Smith University*
Stephen Cheng, *University of Regina*
Tabitha Chigwada, *West Virginia University*
Allen Clabo, *Francis Marion University*
William Cleaver, *University of Vermont*
Penelope Codding, *University of Victoria*
Jeffery Coffer, *Texas Christian University*
Claire Cohen-Schmidt, *University of Toledo*
Renee Cole, *Central Missouri State University*
Patricia Coleman, *Eastern Michigan University*
Shara Compton, *Widener University*
Andrew L. Cooksy, *San Diego State University*
Elisa Cooper, *Austin Community College*
Brian Coppola, *University of Michigan*
Richard Cordell, *Heidelberg College*
Chuck Cornett, *University of Wisconsin, Platteville*
Mitchel Cottenoir, *South Plains College*
Robert Cozzens, *George Mason University*
Mark Cybulski, *Miami University*
Margaret Czrew, *Raritan Valley Community College*
Shadi Dalili, *University of Toronto*
William Davis, *Texas Lutheran University*
John Davison, *Irvine Valley College*
Laura Deakin, *University of Alberta*
Milagros Delgado, *Florida International University*
Mauro Di Renzo, *Vanier College*
Anthony Diaz, *Central Washington University*
Klaus Dichmann, *Vanier College*
Kelley J. Donaghy, *State University of New York, College of Environmental Science and Forestry*
Michelle Driessen, *University of Minnesota*
Theodore Duello, *Tennessee State University*
Dan Durfey, *Naval Academy Preparatory School*
Bill Durham, *University of Arkansas*
Stefka Eddins, *Gardner-Webb University*
Gavin Edwards, *Eastern Michigan University*
Alegra Eroy-Reveles, *San Francisco State University*
Dwaine Eubanks, *Clemson University*
Lucy Eubanks, *Clemson University*
Jordan L. Fantini, *Denison University*
Nancy Faulk, *Blinn College, Bryan*
Tricia Ferrett, *Carleton College*
Matt Fisher, *St. Vincent College*
Amy Flanagan-Johnson, *Eastern Michigan University*
George Flowers, *Darton College*

Richard Foust, *Northern Arizona University*
David Frank, *California State University, Fresno*
Andrew Frazer, *University of Central Florida*
Cynthia Friend, *Harvard University*
Karen Frindell, *Santa Rosa Junior College*
Barbara Gage, *Prince George's Community College*
Rachel Garcia, *San Jacinto College*
Simon Garrett, *California State University, Northridge*
Nancy Gerber, *San Francisco State University*
Brian Gilbert, *Linfield College*
Jack Gill, *Texas Woman's University*
Arthur Glasfeld, *Reed College*
Samantha Glazier, *St. Lawrence University*
Stephen Z. Goldberg, *Adelphi University*
Pete Golden, *Sandhills Community College*
Frank Gomez, *California State University, Los Angeles*
John Goodwin, *Coastal Carolina University*
Steve Gravelle, *Saint Vincent College*
Tom Greenbowe, *Iowa State University*
Stan Grenda, *University of Nevada*
Nathaniel Grove, *University of North Carolina, Wilmington*
Tammy Gummersheimer, *Schenectady County Community College*
Kim Gunnerson, *University of Washington*
Margaret Haak, *Oregon State University*
Christopher Hamaker, *Illinois State University*
Todd Hamilton, *Adrian College*
David Hanson, *Stony Brook University*
Robert Hanson, *St. Olaf College*
David Harris, *University of California, Santa Barbara*
Holly Ann Harris, *Creighton University*
Donald Harriss, *University of Minnesota, Duluth*
C. Alton Hassell, *Baylor University*
Dale Hawley, *Kansas State University*
Sara Hein, *Winona College*
Brad Herrick, *Colorado School of Mines*
Vicki Hess, *Indiana Wesleyan University*
Paul Higgs, *University of Tennessee, Martin*
Kimberly Hill Edwards, *Oakland University*
Andy Ho, *Lehigh University*
K. Joseph Ho, *University of New Mexico*
Donna Hobbs, *Augusta State University*
Angela Hoffman, *University of Portland*
Matthew Horn, *Utah Valley University*
Zhaoyang Huang, *Jacksonville University*
Donna Ianotti, *Brevard Community College*
Tim Jackson, *University of Kansas*
Tamera Jahnke, *Southern Missouri State University*
Shahid Jalil, *John Abbot College*
Eugenio Jaramillo, *Texas A&M International University*
David Johnson, *University of Dayton*
Kevin Johnson, *Pacific University*
Lori Jones, *University of Guelph*
Martha Joseph, *Westminster College*
David Katz, *Pima Community College*
Jason Kautz, *University of Nebraska, Lincoln*
Phillip Keller, *University of Arizona*

Resa Kelly, *San Jose State University*
Vance Kennedy, *Eastern Michigan University*
Michael Kenney, *Case Western Reserve University*
Mark Keranen, *University of Tennessee, Martin*
Robert Kerber, *Stony Brook University*
Elizabeth Kershisnik, *Oakton Community College*
Angela King, *Wake Forest University*
Edith Kippenhan, *University of Toledo*
Sushilla Knottenbelt, *University of New Mexico*
Tracy Knowles, *Bluegrass Community and Technical College*
Larry Kolopajlo, *Eastern Michigan University*
John Krenos, *Rutgers University*
C. Krishnan, *Stony Brook University*
Stephen Kuebler, *University of Central Florida*
Liina Ladon, *Towson University*
Richard Langley, *Stephen F. Austin State University*
Timothy Lash, *Illinois State University*
Sandra Laursen, *University of Colorado, Boulder*
Richard Lavrich, *College of Charleston*
David Laws, *The Lawrenceville School*
Edward Lee, *Texas Tech University*
Willem Leenstra, *University of Vermont*
Alistair Lees, *Binghamton University*
Scott Lewis, *Kennesaw State University*
Jailson de Lima, *Vanier College*
Joanne Lin, *Houston Community College, Eastside Campus*
Matthew Linford, *Brigham Young University*
George Lisensky, *Beloit College*
Jerry Lokensgard, *Lawrence University*
Boon Loo, *Towson University*
Karen Lou, *Union College*
Leslie Lyons, *Grinnell College*
Laura MacManus-Spencer, *Union College*
Roderick M. Macrae, *Marian College*
John Maguire, *Southern Methodist University*
Susan Marine, *Miami University, Ohio*
Albert Martin, *Moravian College*
Brian Martinelli, *Nevada State College*
Diana Mason, *University of North Texas*
Laura McCunn, *Marshall University*
Ryan McDonnell, *Cape Fear Community College*
Garrett McGowan, *Alfred University*
Thomas McGrath, *Baylor University*
Craig McLauchlan, *Illinois State University*
Lauren McMills, *Ohio University*
Heather Mernitz, *Tufts University*
Stephen Mezyk, *California State University, Long Beach*
Rebecca Miller, *Lehigh University*
John Milligan, *Los Angeles Valley College*
Timothy Minger, *Mesa Community College*
Ellen Mitchell, *Bridgewater College*
Robbie Montgomery, *University of Tennessee, Martin*
Joshua Moore, *Tennessee State University*
Dan Moriarty, *Siena College*
Stephanie Myers, *Augusta State College*
Richard Nafshun, *Oregon State University*

Steven Neshyba, *University of Puget Sound*

Melanie Nilsson, *McDaniel College*

Mya Norman, *University of Arkansas*

Sue Nurrenbern, *Purdue University*

Gerard Nyssen, *University of Tennessee, Knoxville*

Ken O'Connor, *Marshall University*

Jodi O'Donnell, *Siena College*

Jung Oh, *Kansas State University, Salinas*

Joshua Ojwang, *Brevard Community College*

MaryKay Orgill, *University of Nevada, Las Vegas*

Robert Orwoll, *College of William and Mary*

Gregory Oswald, *North Dakota State University*

Jason Overby, *College of Charleston*

Greg Owens, *University of Utah*

Maria Pacheco, *Buffalo State College*

Stephen R. Parker, *Montana Tech*

Jessica Parr, *University of Southern California*

Pedro Patino, *University of Central Florida*

Vicki Paulissen, *Eastern Michigan University*

Trilisa Perrine, *Ohio Northern University*

Giuseppe Petrucci, *University of Vermont*

Julie Peyton, *Portland State University*

David E. Phippen, *Shoreline Community College*

Alexander Pines, *University of California, Berkeley*

Prasad Polavarapu, *Vanderbilt University*

John Pollard, *University of Arizona*

Lisa Ponton, *Elon University*

Pete Poston, *Western Oregon University*

Gretchen Potts, *University of Tennessee, Chattanooga*

Robert Pribush, *Butler University*

Gordon Purser, *University of Tulsa*

Robert Quandt, *Illinois State University*

Karla Radke, *North Dakota State University*

Orlando Raola, *Santa Rosa College*

Casey Raymond, *State University of New York, Oswego*

Haley Redmond, *Texas Tech University*

Jimmy Reeves, *University of North Carolina, Wilmington*

Scott Reid, *Marquette University*

Paul Richardson, *Coastal Carolina University*

Alan Richardson, *Oregon State University*

Albert Rives, *Wake Forest University*

Mark Rockley, *Oklahoma State University*

Alan Rowe, *Norfolk State University*

Christopher Roy, *Duke University*

Erik Ruggles, *University of Vermont*

Beatriz Ruiz Silva, *University of California, Los Angeles*

Pam Runnels, *Germanna Community College*

Joel Russell, *Oakland University*

Jerry Sarquis, *Miami University, Ohio*

Nancy Savage, *University of New Haven*

Barbara Sawrey, *University of California, Santa Barbara*

Mark Schraf, *West Virginia University*

Truman Schwartz, *Macalester College*

Louis Scudiero, *Washington State University*

Fatma Selampinar, *University of Connecticut*

Shawn Sendlinger, *North Carolina Central University*

Susan Shadle, *Boise State University*

George Shelton, *Austin Peay State University*

Peter Sheridan, *Colgate University*

Edwin Sibert, *University of Wisconsin, Madison*

Ernest Siew, *Hudson Valley Community College*

Roberta Silerova, *John Abbot College*

Virginia Smith, *United States Naval Academy*

Sally Solomon, *Drexel University*

Xianzhi Song, *University of Pittsburgh, Johnstown*

Brock Spencer, *Beloit College*

Clayton Spencer, *Illinois College*

Estel Sprague, *University of Cincinnati*

William Steel, *York College of Pennsylvania*

Wesley Stites, *University of Arkansas*

Meredith Storms, *University of North Carolina, Pembroke*

Steven Strauss, *Colorado State University*

Katherine Stumpo, *University of Tennessee, Martin*

Mark Sulkes, *Tulane University*

Luyi Sun, *Texas State University*

Duane Swank, *Pacific Lutheran University*

Keith Symcox, *University of Tulsa*

Agnes Tenney, *University of Portland*

Charles Thomas, *University of Tennessee, Martin*

Matthew Thompson, *Trent University*

Craig Thulin, *Utah Valley University*

Edmund Tisko, *University of Nebraska, Omaha*

Brian Tissue, *Virginia Polytechnic Institute and State University*

Steve Trail, *Elgin Community College*

Laurie Tyler, *Union College*

Mike van Stipdonk, *Wichita State University*

Kris Varazo, *Francis Marion University*

William Vining, *University of Massachusetts*

Andrew Vruegdenhill, *Trent University*

Ed Walton, *California State Polytechnic University, Pomona*

Haobin Wang, *New Mexico State University*

Erik Wasinger, *California State University, Chico*

Rory Waterman, *University of Vermont*

Karen Wesenberg-Ward, *Montana Tech University*

Wayne Wesolowski, *University of Arizona*

Charles Wilkie, *Marquette University*

Ed Witten, *Northeastern University*

Karla Wohlers, *North Dakota State University*

Stephen Wood, *Brigham Young University*

Cynthia Woodbridge

Mingming Xu, *West Virginia University*

Tim Zauche, *University of Wisconsin, Platteville*

Noel Zaugg, *Brigham Young University, Idaho*

Corbin Zea, *Grand View University*

James Zimmerman, *Missouri State University*

Martin Zysmilich, *George Washington University*

Chemistry

The Science in Context

1

Particles of Matter

Measurement and the Tools of Science

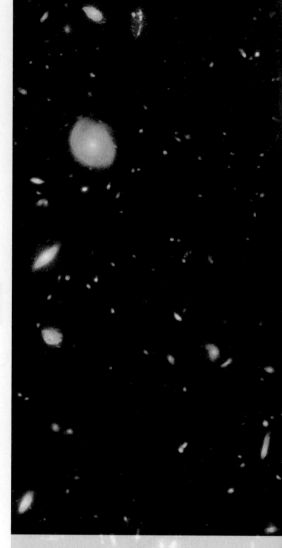

ANCIENT UNIVERSE The colors of the more than 10,000 galaxies in this image give us a glimpse into the universe as it existed about 13 billion years ago. This image was taken by NASA's Hubble Space Telescope.

PARTICULATE **REVIEW**

Atoms and Molecules: What's the Difference?

In Chapter 1 we explore how chemists classify different kinds of matter, from elements to compounds to mixtures. Hydrogen and helium were the first two elements formed after the universe began. Chemists use distinctively colored spheres to distinguish atoms of different elements in their drawings and models. For example, hydrogen is almost always depicted as white.

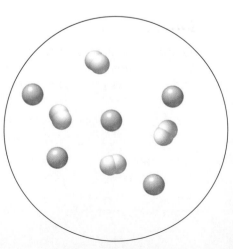

- How many of the following particles are shown in this image?
  - Hydrogen atoms?
  - Hydrogen molecules?
  - Helium atoms?
- Are molecules composed of atoms, or are atoms composed of molecules?

(Answers to Particulate Review questions are in the back of the book.)

Matter and Energy

The temperature in outer space is 2.73 K. The temperature of dry ice (carbon dioxide, CO_2) is 70 times warmer, but still cold enough to keep ice cream frozen on a hot summer day. As you read Chapter 1, look for ideas that will help you answer these questions:

- Particulate images of CO_2 as it sublimes are shown here. Which two phases of matter are involved in sublimation?

- What features of the images helped you decide which two phases were Involved?

- What is the role of energy in this transformation of matter? Must energy be added or is energy produced?

Learning Outcomes

LO1 Distinguish among pure substances, homogeneous mixtures, and heterogeneous mixtures, and between elements and compounds

LO2 Connect chemical formulas to molecular structures and vice versa

LO3 Distinguish between physical processes and chemical reactions, and between physical and chemical properties
Sample Exercise 1.1

LO4 Use a systematic approach (COAST) to problem solving

LO5 Describe the three states of matter and the transitions between them at the macroscopic and particulate levels
Sample Exercise 1.2

LO6 Describe the scientific method

LO7 Convert quantities from one system of units to another
Sample Exercises 1.3, 1.4, 1.9

LO8 Express uncertain values with the appropriate number of significant figures
Sample Exercise 1.5

LO9 Distinguish between exact and uncertain values, evaluate the precision and accuracy of experimental results, and identify outliers
Sample Exercises 1.6, 1.7, 1.8

1.1 How and Why

For thousands of years, humans have sought to better understand the world around us. For most of that time we resorted to mythological explanations of natural phenomena. Many once believed, for example, that the Sun rose in the east and set in the west because it was carried across the sky by a god driving a chariot propelled by winged horses.

In recent times we have been able to move beyond such fanciful accounts of natural phenomena to explanations based on observation and scientific reasoning. Unfortunately, this movement toward rational explanations has not always been smooth. Consider, for example, the contributions of a man whom Albert Einstein called the father of modern science, Galileo Galilei. At the dawn of the 17th century, Galileo used advanced telescopes of his own design to observe the movement of the planets and their moons. He concluded that they, like Earth, revolved around the Sun. However, this view conflicted with a belief held by many religious leaders of his time that Earth was the center of the universe. In 1633 a religious tribunal forced Galileo to disavow his conclusion that Earth orbited the Sun and banned him (or anyone) from publishing the results of studies that called into question the Earth-centered view of the universe. The ban was not completely lifted until 1835—nearly 200 years after Galileo's death.

In the last century, advances in the design and performance of telescopes have led to the astounding discovery that we live in an expanding universe that probably began 13.8 billion years ago with an enormous release of energy. In this chapter and in later ones, we examine some of the data that led to the theory of the Big Bang and that also explain the formation of the elements that make up the universe, our planet, and ourselves.

Scientific investigations into the origin of the universe have stretched the human imagination and forced scientists to develop new models and new explanations of how and why things are the way they are. Frequently these efforts have involved observing and measuring large-scale phenomena, which we refer to as *macroscopic* phenomena. We seek to explain these macroscopic phenomena through *particulate* representations that show the structure of matter on the scale of atomic and even subatomic particles. In this chapter and those that follow, you will encounter many of these macroscopic–particulate connections. The authors of this book hope that your exploration of these connections will help you better understand how and why nature is the way it is.

1.2 Macroscopic and Particulate Views of Matter

According to a formula widely used in medicine, the ideal weight for a six-foot male is 170 pounds (or 77 kilograms). On average, about 30 of these pounds are fat, with the remaining 140 pounds—including bones, organs, muscle, and blood—classified as lean body mass. These values are measures of the total *mass* of all the *matter* in the body. In general, **mass** is the quantity of matter in any object. **Matter**, in turn, is a term that applies to everything in the body (and in the universe) that has mass and occupies space. **Chemistry** is the study of the composition, structure, and properties of matter.

Classes of Matter

The different forms of matter are organized according to the classification scheme shown in Figure 1.1. We begin on the left with pure **substances**, which have a constant composition that does not vary from one sample to another. For example, the composition of pure water does not vary, no matter what its source or how much of it there is. Like all pure substances, water cannot be separated into simpler substances by any physical process. A **physical process** is a transformation of a sample of matter that does not alter the chemical identities of any of the substances in the sample, such as a change in physical state from solid to liquid.

mass the property that defines the quantity of matter in an object.

matter anything that has mass and occupies space.

chemistry the study of the composition, structure, and properties of matter, and of the energy consumed or given off when matter undergoes a change.

substance matter that has a constant composition and cannot be broken down to simpler matter by any physical process; also called *pure substance*.

physical process a transformation of a sample of matter, such as a change in its physical state, that does not alter the chemical identity of any substance in the sample.

FIGURE 1.1 The two principal classes of matter are pure substances and mixtures. A pure substance may be a compound (such as water) or an element (such as gold). A mixture is homogeneous when the substances are distributed uniformly, as they are in vinegar (a mixture of mostly acetic acid and water). A mixture is heterogeneous when the substances are not distributed uniformly—as when solids are suspended in a liquid but may settle to the bottom of the container, as they do in some salad dressings.

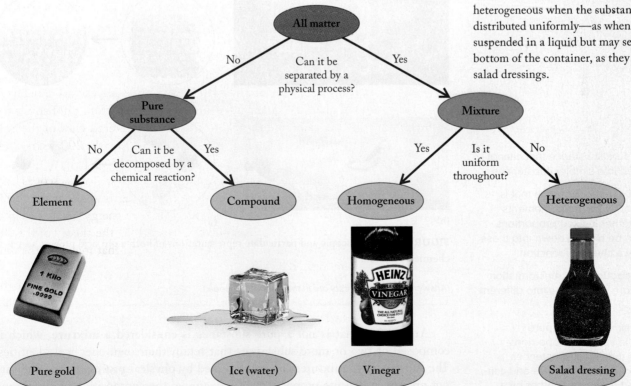

FIGURE 1.2 All matter is made up of either pure substances (of which there are relatively few in nature) or mixtures. (a) The element helium (He), the second most abundant element in the universe, is one example of a pure substance. (b) The compound carbon dioxide (CO_2), the gas used in many fire extinguishers, is also a pure substance. (c) This homogeneous mixture contains three substances: nitrogen (N_2, blue), hydrogen (H_2, white), and oxygen (O_2, red).

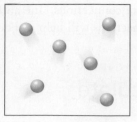
(a) Atoms of helium, an element

(b) Molecules of carbon dioxide, a compound

(c) Mixture of gases

Pure substances are subdivided into two groups: elements and compounds (Figure 1.2). An **element** is a pure substance that cannot be broken down into simpler substances. The periodic table inside the front cover shows all the known elements. Only a few of them (including gold, silver, nitrogen, oxygen, and sulfur) occur in nature uncombined with other elements. Instead, most elements in nature are found mixed with other elements in the form of **compounds**, substances whose elements can be separated from one another only by a **chemical reaction**: the transformation of one or more substances into one or more different substances. Compounds typically have properties that are very different from those of the elements of which they are composed. For example, common table salt (sodium chloride) has little in common with either sodium, which is a silver-gray metal that reacts violently when dropped in water, or chlorine, which is a toxic yellow-green gas.

CONCEPT TEST

Which photo in Figure 1.3 depicts a physical process? Which photo depicts a chemical reaction? Match each photo to its corresponding particulate representation, using what you know about the difference between a physical process and a chemical reaction.

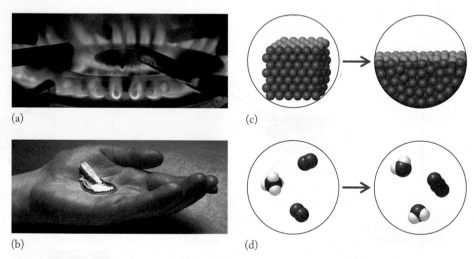

FIGURE 1.3 Macroscopic and particulate representations of both a physical process and a chemical reaction.

(Answers to Concept Tests are in the back of the book.)

element a pure substance that cannot be separated into simpler substances.

compound a pure substance that is composed of two or more elements bonded together in fixed proportions and that can be broken down into those elements by a chemical reaction.

chemical reaction the transformation of one or more substances into different substances.

mixture a combination of pure substances in variable proportions in which the individual substances retain their chemical identities and can be separated from one another by a physical process.

Any matter that is not a pure substance is considered a **mixture**, which is composed of two or more substances that retain their own chemical identities. The substances in mixtures can be separated by physical processes, and they are not present in definite proportions. For example, the composition of circulating blood in a human body is constantly changing as it delivers substances involved in energy production and cell growth to the cells and carries away the waste products of life's biochemical processes. Thus blood contains more oxygen and less carbon dioxide when it leaves our lungs than it does when it enters them.

In a **homogeneous mixture**, the substances making up the mixture are uniformly distributed. This means that the first sip you take from a bottle of water has the same composition as the last. (Keep in mind that bottled water contains small quantities of dissolved substances that either occur naturally in the water or are added prior to bottling to give it a desirable taste. Bottled drinking water is not *pure* water.) Homogeneous mixtures are also called **solutions**, a term that chemists apply to homogeneous mixtures of gases and solids as well as liquids. For example, a sample of filtered air is a solution of nitrogen, oxygen, argon, carbon dioxide, and other atmospheric gases. A "gold" ring is actually a solid solution of mostly gold plus other metals such as silver, copper, and zinc.

On the other hand, the substances in a **heterogeneous mixture** are not distributed uniformly. One way to tell that a liquid mixture is heterogeneous is to look for a boundary between the liquids in it (such as the oil and water layers in the bottle of salad dressing in Figure 1.1). Such a boundary indicates that the substances do not dissolve in one another. Another sign that a liquid may be a heterogeneous mixture is that it is not clear (transparent). Light cannot pass through such liquids because it is scattered by tiny solid particles or liquid drops that are *suspended*, but not dissolved, in the surrounding liquid. Human blood, for example, is opaque because the blood cells that are suspended in it absorb and scatter light.

A Particulate View

Given that compounds are formed from elements, the question must be asked: Do elements consist of yet smaller particles? The answer is yes. An element consists of just one type of particle, known as an **atom**. For example, elements such as gold and helium consist of individual atoms, yet the gold atoms are different from the helium atoms, as we see in Chapter 2. Atoms cannot be chemically or mechanically divided into smaller particles. Although civilizations as old as the ancient Greeks believed in atoms, people then had no evidence that atoms actually existed. Today, however, we have compelling evidence in the form of images of atoms, such as a surface of platinum (Figure 1.4) that has been magnified over 100 million times by using a device called a scanning tunneling microscope.

Some elements exist as molecules. A **molecule** is an assembly of two or more atoms that are held together in a characteristic pattern by forces called **chemical bonds**. For example, the air we breathe consists mostly of *diatomic* (two-atom)

---

homogeneous mixture a mixture in which the components are distributed uniformly throughout and have no visible boundaries or regions.

solution another name for a homogeneous mixture. Solutions are often liquids, but they may also be solids or gases.

heterogeneous mixture a mixture in which the components are not distributed uniformly, so that the mixture contains distinct regions of different compositions.

atom the smallest particle of an element that cannot be chemically or mechanically divided into smaller particles.

molecule a collection of atoms chemically bonded together in characteristic proportions.

chemical bond a force that holds two atoms or ions in a compound together.

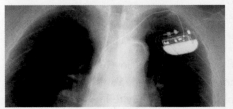

(a)

(b)

FIGURE 1.4 (a) Platinum resists oxidation and is therefore used to make expensive items such as wedding rings and pacemakers. (b) Since the 1980s, scientists have been able to image individual atoms by using an instrument called a scanning tunneling microscope (STM). In this STM image, the fuzzy hexagons (colored blue to be easier to see) are individual platinum atoms. The radius of each atom is 138 picometers (pm), or 138 trillionths of a meter.

FIGURE 1.5 The reaction between hydrogen and oxygen is depicted with molecular models (white and red spheres) and in the form of a chemical equation. Note that energy is also a product of the reaction.

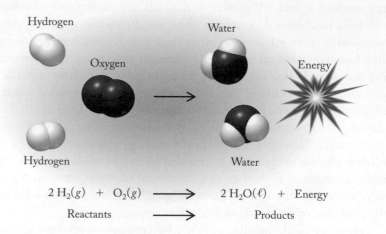

$$2\,H_2(g) \;+\; O_2(g) \;\longrightarrow\; 2\,H_2O(\ell) \;+\; Energy$$

Reactants ⟶ Products

chemical formula notation for representing elements and compounds; consists of the symbols of the constituent elements and subscripts identifying the number of atoms of each element present.

chemical equation notation in which chemical formulas express the identities and their coefficients express the quantities of substances involved in a chemical reaction.

energy the capacity to do work.

law of constant composition the principle that all samples of a particular compound contain the same elements combined in the same proportions.

ion a particle consisting of one or more atoms that has a net positive or negative electrical charge.

cation an ion with a positive charge.

anion an ion with a negative charge.

molecules of nitrogen gas, N_2, and oxygen gas, O_2. The subscripts in the **chemical formulas** of these two gases tell us that their molecules are each composed of two atoms. Other elements also exist as diatomic molecules, including H_2 and elements in column 17 of the periodic table: F_2, Cl_2, Br_2, and I_2.

Most of the molecules in the universe, however, contain atoms of more than one element, meaning that they are compounds. The chemical formula of a molecular compound tells us the number of atoms of each element in one of its molecules. For example, the formula H_2O tells us that pure water is composed of molecules that each contain two hydrogen atoms and one oxygen atom, as shown in Figure 1.5.

The 2:1 ratio of hydrogen to oxygen atoms in molecules of H_2O also reflects the proportions of H_2 gas and O_2 gas that react to form water. These gases *always* react with each other in the same proportion: two molecules of H_2 for every one molecule of O_2. This relationship is illustrated in Figure 1.5 with models of the molecules involved and the chemical equation beneath them. In a **chemical equation**, chemical formulas represent the substances involved in a chemical reaction. The arrow in the middle of a chemical equation separates the reactant(s) from the product(s). In Figure 1.5, the *phase symbol* (g) shows that the reactants are gases and (ℓ) shows that H_2O is a liquid. Solids are denoted by (s).

Note that the reaction between hydrogen and oxygen also produces **energy**, most generally defined as the capacity to do work. If we reverse the process and add enough energy to decompose water into hydrogen and oxygen (Figure 1.6), which is another example of a chemical reaction, we will always obtain two molecules of hydrogen gas for every one molecule of oxygen gas. This consistency illustrates the **law of constant composition**: every sample of a particular compound always contains the same elements combined in the same proportions.

CONCEPT TEST

A compound with the formula NO is present in the exhaust gases leaving a car's engine. As NO travels through the car's exhaust system, some of it decomposes into nitrogen gas and oxygen gas. What is the ratio of nitrogen molecules to oxygen molecules formed from the decomposition of NO?

(Answers to Concept Tests are in the back of the book.)

Chemical formulas provide information about the ratios of the elements in molecular compounds, but formulas do not tell us how the atoms of each element are bonded to one another within each molecule, nor do they tell us anything about

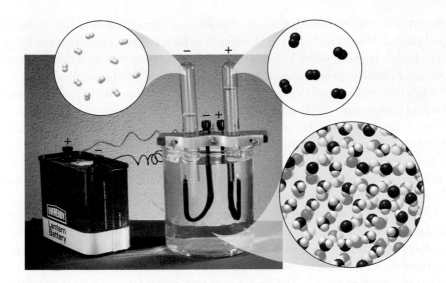

FIGURE 1.6 An electric current passed through water provides enough energy to decompose water into oxygen gas and hydrogen gas. The ratio of the gases produced is always two molecules of hydrogen for every one molecule of oxygen. These fixed proportions illustrate the law of constant composition.

the shapes of molecules. To communicate information about bonding and shape, we need to draw a *structural formula* such as the one for ethanol (C_2H_5OH) in Figure 1.7(a), which uses straight lines to represent the chemical bonds that connect the carbon (C), hydrogen (H), and oxygen (O) atoms within the molecule.

However, a structural formula does not necessarily show how atoms are arranged in three-dimensional space. *Molecular models* provide this 3-D perspective. *Ball-and-stick* molecular models (Figure 1.7b) use spheres to represent atoms and sticks to represent chemical bonds. The advantage of ball-and-stick models is that they show the correct angles between the bonds. However, there are limitations to using models to represent molecules. For example, the sizes of the spheres are not proportional to the sizes of the atoms they represent, and the atoms are spaced far enough apart to accommodate the stick bonds. (In real molecules, the atoms touch each other.) Both of these limitations are overcome with *space-filling* molecular models (Figure 1.7c), in which the spheres are drawn to scale and touch one another as atoms do in real molecules. One limitation of space-filling models is that the bond angles between atoms may be difficult to discern. An additional limitation of both the ball-and-stick and space-filling models is that atoms themselves do not have color. Representing oxygen atoms as red spheres and hydrogen atoms as white spheres is merely a convention used by chemists.

Not all compounds are molecular. Instead, some compounds consist of positively and negatively charged particles called **ions** that are electrostatically attracted to one another. For example, calcium chloride (which is used to melt snow and ice on sidewalks in winter) consists of calcium ions (Ca^{2+}) and chloride ions (Cl^-). The positive ions are called **cations**, and the negative ions are called **anions**. Ions may consist of single atoms like Ca^{2+} and Cl^-, or they may contain two or more atoms bonded together that have an overall positive or negative charge, like the hydroxide ion (OH^-).

1.3 Mixtures and How to Separate Them

As noted in Section 1.2, mixtures can be separated into their component substances by physical processes. Consider, for example, how the components of human blood can be separated. Blood is a heterogeneous mixture of hundreds of

Ethanol

H H
| |
H—C—C—O—H
| |
H H

(a) Structural formula

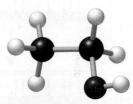

(b) Ball-and-stick model

(c) Space-filling model

FIGURE 1.7 Three ways to represent the arrangement of atoms in a molecule of ethanol: (a) structural formula; (b) ball-and-stick model, where white spheres represent hydrogen atoms, black spheres represent carbon atoms, and red spheres represent oxygen atoms; (c) space-filling model.

FIGURE 1.8 Separation of whole blood by centrifugation. The red blood cells at the bottom of the sample are separated from the upper layer of pale yellow plasma by a buff-colored layer of white blood cells and platelets.

different ions and molecules that are dissolved or suspended in a liquid called blood *plasma*. Plasma includes red blood cells (which carry oxygen from the lungs to our body's tissues), white blood cells (which play a key role in fighting infections), and platelets (which combine with clotting proteins to stop bleeding).

Each day, millions of blood samples are separated into their suspended and dissolved components for diagnostic purposes or to recover materials used to treat sick or injured patients. To begin, the blood is placed in centrifuge tubes and subjected to forces about 1000 times the force of gravity. Centrifugation causes the large, dense red blood cells to settle to the bottom of the tubes. Above them is a layer of platelets and white blood cells, and above that layer is clear, yellow-colored plasma (Figure 1.8).

Once the plasma has been separated, other techniques are used to isolate blood proteins dissolved in the plasma. Identification of these proteins helps doctors diagnose diseases. One such protein isolation technique is two-dimensional electrophoresis, in which the proteins are separated in two steps (Figure 1.9): first by charge, and then by size. The first step of this process is based on the fact that molecules of these proteins have different charges depending on their environment: they exist as anions in less acidic solutions, as cations in more acidic solutions, and as neutral molecules only at a characteristic acidity in between. In the second step, fractions from the first step are further separated by the sizes of their molecules, permitting their unique identification.

Different techniques must be used to separate solid particles from liquid and gas mixtures. Consider, for example, the use of filters to trap particles of dust suspended in the air entering an automobile engine (Figure 1.10a) or to remove bacteria from the air in hospitals to control the spread of infectious diseases (Figure 1.10b). A key parameter in designing these filters is their pore size, which must be small enough to trap the target particles or microbes, but not so small that enormous pressure would be required to force an adequate flow of air through them. Filtration is also used to separate particles suspended in liquids; it is a key step in treating the water that many of us drink (Figure 1.10c).

The components of a liquid mixture can also be separated by exploiting differences in their *volatilities*, that is, how easily they can be converted from liquids to gases. The most widely used of these techniques is distillation (Figure 1.11), which can be used to desalinate seawater because water is much more volatile than the sea salts dissolved in it. Distillation is also employed when converting fresh water into the distilled water used in laboratories and other places where highly pure water is required. When water of ultra-high purity is required, an additional separation step based on a technology called *ion exchange* is used to remove essentially all the dissolved ions from water. As the water flows through cartridges containing porous materials called ion-exchange resins, any dissolved cations in the water are replaced with H^+ ions, and any dissolved anions are replaced with OH^- ions. These two ions then combine to form additional molecules of water:

$$H^+(aq) + OH^-(aq) \rightarrow H_2O(\ell)$$

The phase symbol (*aq*) indicates that the ions are dissolved in water. Chemists call this an *aqueous* solution.

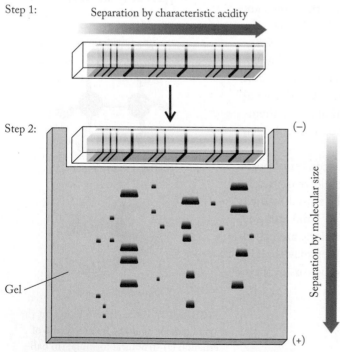

FIGURE 1.9 Separation of proteins by electrophoresis. In the first step, the proteins are separated according to the characteristic acidity at which they have no net charge. In the second step, negatively charged proteins are separated by molecular size: the lightest molecules move fastest through the gel toward a positive charge.

1.4 A Framework for Solving Problems

Throughout this book we include questions and problems that test your understanding of the material. Your success in this course depends, in part, on your ability to solve these problems. Solving chemistry problems is like playing a musical instrument: the more you practice, the better you become.

In this section we present a framework for solving problems that we follow throughout the book. Each Sample Exercise is followed by a Practice Exercise that can be solved using a similar approach. We strongly encourage you to hone your problem-solving skills by working all the Practice Exercises as you read. You should find the approach described here useful in solving the exercises in this book as well as problems you encounter in other courses and other contexts.

We use the acronym COAST (**C**ollect and **O**rganize, **A**nalyze, **S**olve, and **T**hink about the answer) to represent the four steps in this approach. As you read about it here and use it later, keep in mind that COAST is merely a *framework* for solving problems, not a recipe. Use it as a guide to develop your own approach to solving problems.

Collect and Organize The first step in solving a problem is to decide how to use the given information. You should be able to identify the key concept of the problem and the key terms used to express that concept. You may find it useful to restate the problem in your own words. Visualizing the problem may also be helpful—drawing a sketch based on molecular models or an experimental setup, or even just closing your eyes and picturing the situation. Once you have a clear understanding of what is being asked, sort through the information and separate what is relevant from what is not. Then assemble the relevant information, including chemical equations, molecular models, definitions, and the values of constants.

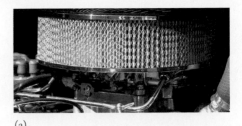

(a)

(b)

(c)

FIGURE 1.10 Filters are used to remove particles from gases such as (a) the air entering a car engine or (b) the air circulating through a hospital, and from liquids such as (c) the water being analyzed at a municipal water supply treatment plant. The pore size of the hospital filters must be very small to trap bacteria and viruses that are less than 1 micrometer (10^{-6} m) in diameter.

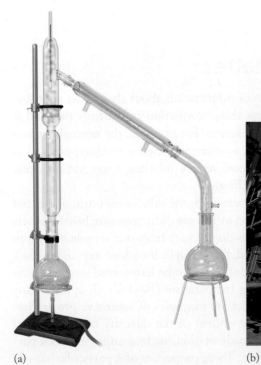

(a)

(b)

FIGURE 1.11 Distillation is used to (a) purify compounds and (b) desalinate seawater. This desalination plant is in Kuwait City.

Analyze The next step is to analyze the information you have to determine how to connect it to the answer you seek. Sometimes you can work backward to create these links: Consider the answer first and think about how you might reach it from the information provided and other sources. This approach might mean finding some intermediate quantity to use in a later step. If the problem requires a numerical answer, frequently the units of the initial values and the final answer can help you identify how they are connected and which equations may be useful. Or, this might include rearranging equations to solve for an unknown or setting up conversion factors. When possible, try to estimate the value of your answer.

Some Sample Exercises in this book involve performing a single-step calculation or answering a question based on a single concept. These exercises don't require assembling and sorting through information and equations, so we have combined the above two steps into a single Collect, Organize, and Analyze step.

Solve The solution to a problem that tests your understanding of a concept often flows directly from the Analyze step. To solve quantitative problems, you need to insert the starting values and appropriate constants into the equations and calculate the answer. Make sure that units are consistent and cancel out (use conversion factors as needed), and ensure that the answer contains the appropriate number of significant figures. (We discuss conversion factors in Section 1.9 and significant figures in Section 1.10.)

Think About It Ask yourself, "Does this answer make sense? Is this number about what I estimated it would be? Are the units correct? Is the number of significant figures appropriate?" Then ask yourself whether you could solve another problem, perhaps drawn from another context but based on the same concept.

The COAST approach should help you solve problems in a logical way and avoid pitfalls like grabbing an equation that seems to have the right variables and plugging in numbers, or resorting to trial and error. As you study each step in a Sample Exercise, be sure to ask yourself these questions: **What** is done in this step? **How** is it done? **Why** is it done? With answers to these questions in mind, you will be ready to solve the Practice Exercises and end-of-chapter Questions and Problems in a systematic way.

1.5 Properties of Matter

Our senses provide a constant stream of information about the matter that surrounds us. Subconsciously, we process this information to confirm that "all is well"—or, if it's not, to take appropriate action. For example, the normal components of the air we breathe are odorless, so when we smell something unusual as we enter a room or step outside, we know we are inhaling a gas not normally present in the air, and we respond accordingly.

Pure substances have distinctive properties. Some substances burn; others put out fires. Some solid substances are shiny; others are dull. Some are brittle; others are malleable. These and other properties do not vary from one sample of a given pure substance to the next. Consider gold, for example: its color (very distinctive), its hardness (soft for a metal), its malleability (it can be hammered into very thin sheets called gold leaf), and its melting temperature (1064°C) all apply to any sample of pure gold. These characteristics are examples of **intensive properties**: properties that characterize matter independent of the quantity of the material present. On the other hand, an object made of gold, such as an ingot, has a particular length, width, mass, and volume. These properties of a particular sample

intensive property a property that is independent of the amount of substance present.

of a pure substance, which depend on how much of the substance is present, are **extensive properties**.

CONCEPT **TEST**

Which of these properties of a sample of pure iron is intensive? (a) mass; (b) density; (c) volume

(Answers to Concept Tests are in the back of the book.)

Properties fall into two other general categories: physical and chemical. **Physical properties** are the properties of a pure substance that can be observed or measured without changing the substance into another substance. The intensive properties of gold described above are all physical properties, as is its **density (d)**—the ratio of the mass (*m*) of an object to its volume (*V*):

$$d = \frac{m}{V} \qquad (1.1)$$

As we saw in Figure 1.5, hydrogen combines with oxygen in a chemical reaction that produces water and energy. This reaction is an example of combustion (burning), and H_2 is the fuel. High flammability is a **chemical property** of hydrogen. Like any other chemical property, flammability can be determined only by reacting one substance with another substance and finding that a different material is produced. A substance's chemical reactivity, including the rates of its reactions, the identity of other substances with which it reacts, and the identity of the products formed, defines the chemical properties of the substance.

SAMPLE EXERCISE 1.1 Distinguishing Physical and   **LO3**
                                        Chemical Properties

Which of the following properties of gold are chemical and which are physical?

a. Gold metal, which is insoluble in water, can be made soluble by reacting it with a mixture of nitric and hydrochloric acids known as aqua regia.
b. Gold melts at 1064°C.
c. Gold can be hammered into sheets so thin that light passes through them.
d. Gold metal can be recovered from gold ore by treating the ore with a solution containing cyanide, which reacts with and dissolves gold.

Collect, Organize, and Analyze Chemical properties describe how a substance reacts with other substances; physical properties can be observed or measured without changing one substance into another. Properties (a) and (d) involve reactions that chemically change gold metal (which does not dissolve in water) into compounds of gold that do dissolve in water. Properties (b) and (c) describe processes in which elemental gold remains elemental gold. When it melts, gold changes its physical state, but not its chemical identity. Gold leaf is still solid, elemental gold.

Solve Properties (a) and (d) are chemical properties, and (b) and (c) are physical properties.

Think About It When possible, we fall back on our experiences and observations. We know gold jewelry does not dissolve in water. Therefore, dissolving gold metal requires a change in its chemical identity: it can no longer be elemental gold. On the other hand, physical processes such as melting do not alter the chemical identity of the gold. Gold can be melted and then cooled to produce solid gold again.

solid a form of matter that has a definite shape and volume.

liquid a form of matter that occupies a definite volume but flows to assume the shape of its container.

gas a form of matter that has neither definite volume nor shape and that expands to fill its container; also called *vapor*.

 Practice Exercise Which of the following properties of water are chemical and which are physical?

a. Water normally freezes at 0.0°C.
b. Water is useful for putting out most fires.
c. A cork floats on water, but a piece of copper sinks.
d. During digestion, starch reacts with water to form sugar.

(Answers to Practice Exercises are in the back of the book.)

1.6 States of Matter

Matter typically exists in one of three phases or physical states: solid, liquid, or gas (Figure 1.12). You are probably familiar with the characteristic properties of these states:

- A **solid** has a definite volume and shape.
- A **liquid** has a definite volume but not a definite shape.
- A **gas** (or *vapor*) has neither a definite volume nor a definite shape. Rather, it expands to occupy the entire volume and shape of its container.

Unlike a solid or a liquid, a gas is compressible, which means it can be squeezed into a smaller volume if its container is not rigid and pressure is applied to it.

FIGURE 1.12 Macroscopic and particulate views of the three states of water in this winter photograph of a hot spring in Yellowstone National Park: (a) solid (ice), (b) liquid (water: in the case of the hot spring, it is a solution of compounds in liquid water), and (c) gas (water vapor). Water vapor is invisible, but the mist above the spring consists of tiny drops of liquid water that form as water vapor condenses in the frigid air. The rich colors near the edges of the spring are produced by thermophiles, tiny organisms that thrive in extremely hot water.

(a) Solid   (b) Liquid   (c) Gas

The differences between solids, liquids, and gases can be understood if we consider the three states on the particulate level. As shown in Figure 1.12, each H_2O molecule in ice is surrounded by other H_2O molecules and is locked in place in a rigid three-dimensional array of molecules. A molecule in this structure may vibrate, but it cannot move past the other molecules that surround it. In liquid water, the H_2O molecules flow past one another, but they are still near one another, even though their nearest neighbors change over time. The H_2O molecules in water vapor are widely separated by empty space, so that the volume of the molecules themselves is much smaller than the total volume occupied by the vapor. This separation accounts for the compressibility of gases. Molecules in the gas phase move freely throughout the space occupied by the vapor.

We can transform water from one physical state to another by heating or cooling it (Figure 1.13). Ice forms on a pond when the temperature drops in the winter and the water freezes. This process is reversed when warmer temperatures return in the spring and the ice melts. The heat of the sun may vaporize liquid water during the daytime, but colder temperatures at night may cause the water vapor to condense as dew.

Some solids change directly into gases with no intervening liquid phase. For example, snow may be converted to water vapor on a cold, sunny winter day even though the air temperature remains well below freezing. This transformation of solid directly to vapor is called **sublimation**. The reverse process—in which water vapor forms a layer of frost on a cold night—is an example of a gas being transformed directly into a solid without ever being a liquid, a process called **deposition**.

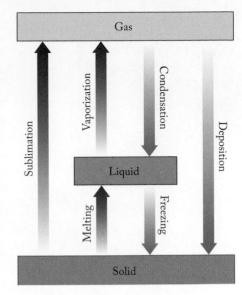

FIGURE 1.13 Matter changes from one state to another when energy is either added or removed. Arrows pointing up represent transformations that require the addition of energy; arrows pointing down represent transformations that release energy.

CONCEPT **TEST**

Energy must be added to convert solids into liquids or liquids into gases (Figure 1.13). In the case of water, energy is needed to overcome the attractions between water molecules, which are held together rigidly in ice and are packed even closer together in liquid water. Should it require (a) more energy, (b) less energy, or (c) about the same amount of energy to melt a gram of ice at its melting point as it does to boil a gram of water that has already been heated to its boiling point? Explain your selection.

(Answers to Concept Tests are in the back of the book.)

SAMPLE EXERCISE 1.2 Distinguishing between Particulate Views   **LO5**
                        of the Different States of Matter

Which physical state is represented in each box of Figure 1.14? (The particles could be atoms or molecules.) What change of state does each arrow indicate? What would the changes of state be if both arrows pointed in the opposite direction?

Collect, Organize, and Analyze We need to look for patterns among the particles in each box:

- An ordered arrangement of particles that does not fill the box or match the shape of the box represents a solid.
- A less-ordered arrangement that partially fills a box and conforms to its shape represents a liquid.
- A dispersed array of particles distributed throughout a box represents a gas.

sublimation transformation of a solid directly into a vapor (gas).

deposition transformation of a vapor (gas) directly into a solid.

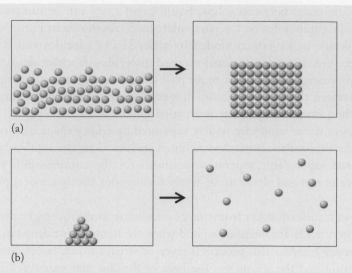

FIGURE 1.14 Changes in the physical state of matter.

Solve The particles in the left box in Figure 1.14(a) partially fill the box and adopt the shape of the container; therefore, they represent a liquid. The particles in the box on the right are ordered and form a shape different from that of the box, so they represent a solid. The arrow represents a liquid turning into a solid, and thus the physical process is freezing. An arrow in the opposite direction would represent melting. The particles in the box on the left in Figure 1.14(b) represent a solid because they are ordered and do not adopt the shape of the box. Those to the right are dispersed throughout the box, representing a gas. The arrow represents a solid turning into a gas, and thus the physical process is sublimation. The reverse process—a gas becoming a solid—would be deposition.

Think About It Notice that the particles represented in Figure 1.14(a) are very close together, which is consistent with liquids and solids being classified as condensed phases. On the other hand, the particles in gases, as shown in the right panel of Figure 1.14(b), are widely separated.

Practice Exercise (a) What physical state is represented by the particles in each box of Figure 1.15, and which change of state is represented? (b) Which change of state would be represented if the arrow pointed in the opposite direction?

FIGURE 1.15 A change in the physical state of matter.

(Answers to Practice Exercises are in the back of the book.)

scientific method an approach to acquiring knowledge based on observation of phenomena, development of a testable hypothesis, and additional experiments that test the validity of the hypothesis.

hypothesis a tentative and testable explanation for an observation or a series of observations.

scientific theory (model) a general explanation of a widely observed phenomenon that has been extensively tested and validated.

1.7 The Scientific Method: Starting Off with a Bang

In Sections 1.5 and 1.6, we examined the properties and states of matter. In this section and in Section 1.11, we examine the origins of matter and a theory of how the universe came into being.

The ancient Greeks believed that at the beginning of time there was no matter, only a vast emptiness they called Chaos, and that from that emptiness emerged the first supreme being, Gaia (also known as Mother Earth), who gave birth to Uranus (Father Sky). Other cultures and religions have described creation in similar terms of supernatural beings creating ordered worlds out of vast emptiness. The opening verses of the book of Genesis, for example, describe darkness "without form and void" from which God created the heavens and Earth. Similar stories are part of Asian, African, and Native American cultures (Figure 1.16). Today we have the technology to take a different approach to explaining how the universe was formed and what physical forces control it. Our approach is based on the **scientific method** of inquiry (Figure 1.17).

This approach evolved in the late Renaissance, during a time when economic and social stability gave people an opportunity to study nature and question old beliefs. By the early 17th century, the English philosopher Francis Bacon (1561–1626) had published his *Novum Organum* (*New Organ* or *New Instrument*) in which he described how humans can acquire knowledge and understanding of the natural world through observation, experimentation, and reflection. In the process, he described how observations of a natural phenomenon lead to a tentative explanation, or **hypothesis**, of what causes the phenomenon. Developing a hypothesis and devising experiments to test it in a decisive way are very creative aspects of science. Further testing and observation might support a hypothesis, disprove it, or perhaps require that it be modified so that it explains all of the experimental results. However, no experiment can ever "prove" a hypothesis to be correct; the best that scientists can say is that no experiment or observation has yet to contradict the hypothesis. One measure of the validity of a hypothesis is that it enables scientists to predict accurately the results of *future* experiments and observations. Ultimately, a hypothesis that withstands the tests of many experiments over time and explains the results of further observation and experimentation may be elevated to the rank of a **scientific theory** or **model**. Note that although *theory* in everyday conversation is often used as a synonym for "speculation," the word *theory* as used by scientists refers to an explanatory framework that can be used to empirically test hypotheses.

Let's examine some of the ways the scientific method helped in formulating the theory of the origin of the universe, or what has come to be called the "Big

FIGURE 1.16 For centuries, cultures have passed on creation stories in which a supreme being is responsible for creating the Sun, the Moon, and the Earth and its inhabitants. According to a creation myth of Native Americans of the Pacific Northwest, a deity in the form of a raven was responsible for releasing the Sun into the sky. This Bellacoola woman has a raven headdress and rattle.

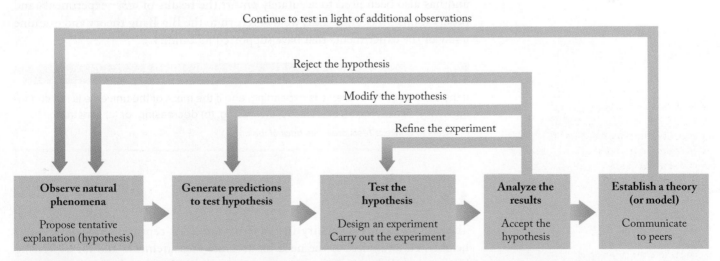

FIGURE 1.17 In the scientific method, observations lead to a tentative explanation, or hypothesis, which leads to more observations and testing, which may lead to the formulation of a succinct, comprehensive explanation called a theory. This process is rarely linear: it often involves looping back, as the results of one test lead to additional tests and a revised hypothesis. Science is a dynamic and self-correcting process.

FIGURE 1.18 The U.S. Postal Service issued a commemorative stamp in March 2008 in honor of Edwin Hubble.

Bang." In the early 20th century, American astronomer Edwin Hubble (1889–1953; Figure 1.18) and others used increasingly powerful telescopes to discover that the universe contains billions of galaxies, each containing many billions of stars. They also discovered that (1) these other galaxies are moving away from us and (2) the speeds with which these galaxies are receding are proportional to the distances from us—the farthest galaxies are moving away the fastest.

From these observations, Belgian priest and astronomer Georges-Henri Lemaître (1894–1966) proposed in 1927 that the universe we know today formed with an enormous release of energy that quickly transformed into a rapidly expanding, unimaginably hot cloud of matter in accordance with Einstein's famous formula:

$$E = mc^2$$

where E is energy, m is the equivalent mass of matter, and c is the speed of light. To understand Lemaître's proposal, consider what we would see if the motion of the matter in the universe had somehow been recorded since the beginning of time and we were able to play the recording in reverse. The other galaxies would move toward ours (and toward one another), with the ones farthest away closing in the fastest. Eventually, all the matter in the universe would approach the same point at the same time. During this compression, the universe would become tremendously dense and hot. At the very beginning, then, the universe would be squeezed into an infinitesimally small space with an inconceivably high temperature. At such a temperature, the distinction between matter and energy would become blurred. Indeed, at the very beginning there would be no matter, only energy. Now, if we were to play this recording in the forward direction, we would see an instantaneous release of an enormous quantity of energy in an event that has come to be known as the Big Bang.

Lemaître's explanation of how the universe began has been the subject of extensive testing—and controversy—since it was first proposed. Indeed, the term *Big Bang* was first used by British astronomer Sir Fred Hoyle (1915–2001) to poke fun at the idea. However, the results of many experiments and observations conducted since the 1920s support what is now described as the Big Bang theory. We use the term *theory* because this comprehensive explanation of natural phenomena has been validated by the results of extensive experimentation and observation, and has also been used to accurately *predict* the results of *other* experiments and observations. In Section 1.11, we will return to the Big Bang theory and examine a few of the experiments that have supported its validity.

CONCEPT **TEST**

If the volume of the universe is expanding, and if the mass of the universe is not changing, is the density of the universe (a) increasing, (b) decreasing, or (c) constant?

(Answers to Concept Tests are in the back of the book.)

1.8 SI Units

The rise of scientific inquiry in the 17th and 18th centuries brought about a heightened awareness of the need for accurate measurements and for expressing the results of those measurements in ways that others, including scientists in other countries, could understand. Standardization of the units of measurement was essential.

meter the standard unit of length, equivalent to 39.37 inches.

TABLE 1.1 Commonly Used Prefixes for SI Units

PREFIX		VALUE	
Name	Symbol	Numerical	Exponential
zetta	Z	1,000,000,000,000,000,000,000	10^{21}
exa	E	1,000,000,000,000,000,000	10^{18}
peta	P	1,000,000,000,000,000	10^{15}
tera	T	1,000,000,000,000	10^{12}
giga	G	1,000,000,000	10^{9}
mega	M	1,000,000	10^{6}
kilo	k	1000	10^{3}
hecto	h	100	10^{2}
deka	da	10	10^{1}
deci	d	0.1	10^{-1}
centi	c	0.01	10^{-2}
milli	m	0.001	10^{-3}
micro	μ	0.000001	10^{-6}
nano	n	0.000000001	10^{-9}
pico	p	0.000000000001	10^{-12}
femto	f	0.000000000000001	10^{-15}
atto	a	0.000000000000000001	10^{-18}
zepto	z	0.000000000000000000001	10^{-21}

In 1791 French scientists proposed a standard unit of length, which they called the **meter** (m) after the Greek *metron*, which means "measure." They based the length of the meter on 1/10,000,000 of the distance along an imaginary line running from the North Pole to the equator. By 1794, hard work by teams of surveyors had established the length of the meter that is still in use today.

These French scientists also settled on a decimal-based system for designating lengths that are multiples or fractions of a meter (Table 1.1). They chose Greek prefixes for lengths much greater than 1 meter, such as *deka-* and *kilo-* for lengths of 10 meters (1 dekameter) and 1000 meters (1 kilometer). Then they chose Latin prefixes for lengths much smaller than a meter, such as *centi-* and *milli-* for lengths of 1/100 of a meter (1 centimeter) and 1/1000 of a meter (1 millimeter).

Eventually these decimal prefixes carried over to the names of standard units for other dimensions. In July 1799 a platinum rod 1 meter long and a platinum block having a mass of 1 kilogram (1000 grams) were placed in the French National Archives to serve as the legal standards for length and mass. These objects served as references for the metric system for expressing measured quantities.

Since 1960, scientists have, by international agreement, used a modern version of the French metric system: the *Système International d'Unités*, commonly abbreviated SI. Table 1.2 contains six of the seven *SI base units*; many other SI units are derived from them. For example, a common SI unit for volume, the cubic meter (m^3), is derived from the length base unit, the meter. The common SI unit for speed, meters per second (m/s), is derived from both the length and time base units. Table 1.3 contains some of these derived units and their equivalents in the

TABLE 1.2 Six SI Base Units

Quantity or Dimension	Unit Name	Unit Abbreviation
Mass	kilogram	kg
Length	meter	m
Temperature	kelvin	K
Time	second	s
Electrical current	ampere	A
Amount of a substance	mole	mol

TABLE 1.3 SI Units and Their U.S. Customary Unit Equivalents

Quantity or Dimension	Equivalent Units
Mass	1 kg = 2.205 pounds (lb); 1 lb = 0.4536 kg = 453.6 g
	1 g = 0.03527 ounce (oz); 1 oz = 28.35 g
Length (distance)	1 m = 1.094 yards (yd); 1 yd = 0.9144 m (exactly)
	1 m = 39.37 inches (in); 1 foot (ft) = 0.3048 m (exactly)
	1 in = 2.54 cm (exactly)
	1 km = 0.6214 miles (mi); 1 mi = 1.609 km
Volume	$1 m^3 = 35.31 ft^3$; $1 ft^3 = 0.02832 m^3$
	$1 m^3 = 1000$ liters (L) (exactly)
	1 L = 0.2642 gallon (gal); 1 gal = 3.785 L
	1 L = 1.057 quarts (qt); 1 qt = 0.9464 L

TABLE 1.4 Densities of Common Materials

ITEM	DENSITY (AT 20°C)
Solids	g/cm^3
Aluminum	2.70
Copper	8.96
Lead	11.34
Zinc	7.13
Liquids	g/cm^3
Gasoline	0.78
Olive oil	0.91
Diet soda	1.00
Milk	1.03
Soda	1.03
Seawater	1.03
Whole blood (human)	1.06
Gases	g/L (at 20°C and 1 atm pressure)
Air (dry)	1.204
Ammonia	0.717
Carbon dioxide	1.842
Propane gas	1.873
Water vapor	0.804

U.S. Customary System of units. They include the volume corresponding to 1 cubic decimeter (a cube 1/10 meter on each side), which we call a liter (L).

Various combinations of base and derived SI units are used to express the densities of materials, that is, their mass-to-volume ratios. The selection of units depends on how dense the material is. For example, the densities of gases are often expressed in grams/liter (g/L) so that the density values are close to one. On the other hand, liquids are many hundreds of times denser than gases, so their densities are often expressed in g/cm^3, which is equivalent to g/mL. Solids have densities that are comparable to or greater than those of most liquids, so their values are usually expressed in g/cm^3, too (Table 1.4).

Modern science requires that the length of the meter, as well as the dimensions of other SI units, be defined by quantities that are much more constant than the length of a platinum rod in Paris. Two such quantities are the speed of light (c) and time. In 1983, 1 meter was redefined as the distance traveled in 1/299,792,458 of a second by the light emitted from a helium–neon laser. This modern definition of the meter is consistent with the one adopted in France in 1794.

1.9 Unit Conversions and Dimensional Analysis

Throughout this book we will often need to convert measurements and quantities from one unit to another. Sometimes this will require choosing the appropriate prefix, such as expressing the distance between Toronto and Montreal in kilometers instead of meters. Other times we will need to convert a value from one unit system to another, such as converting a gasoline price per liter to an equivalent price per U.S. gallon.

To do these calculations and many others, we use an approach sometimes called the *unit factor method* but more often called *dimensional analysis*. The approach makes use of **conversion factors**, which are fractions in which the numerators and denominators have different units but represent equivalent quantities. This equivalency means that multiplying a quantity by a conversion factor is equal to multiplying by 1. The intrinsic value that the quantity represents does not change; it is simply expressed in different units.

The key to using conversion factors is to set them up correctly with the appropriate units in the numerator and denominator. For example, the odometer of a Canadian car traveling between the airports serving Vancouver, BC, and Seattle, WA, records the distance as 241 km. What is that distance in miles? To find out we translate the following equivalency from Table 1.3:

$$1 \text{ km} = 0.6214 \text{ mi}$$

into this conversion factor:

$$\frac{0.6214 \text{ mi}}{1 \text{ km}}$$

and multiply the distance in kilometers by it:

$$241 \text{ km} \times \frac{0.6124 \text{ mi}}{1 \text{ km}} = 1.48 \times 10^2 \text{ mi}$$

We obtain a value with the desired units (in this case, miles) because those units are in the numerator of the conversion factor, whereas the units in the denominator (in this case, kilometers) of the conversion factor cancel out the units of the initial value itself. The general form of the equation for converting a value from one unit to another is

$$\text{initial units} \times \frac{\text{desired units}}{\text{initial units}} = \text{desired units}$$

Most conversion factors, such as the one for converting kilometers to miles, also have a "1" in the numerator or denominator. All of these ones are *exactly* one; there is no uncertainty in their value.

> **conversion factor** a fraction in which the numerator is equivalent to the denominator, even though they are expressed in different units, making the value of the fraction one.

CHEMTOUR
Dimensional Analysis

SAMPLE EXERCISE 1.3 Converting Units   **LO7**

The masses of diamonds are usually expressed in carats (1 gram = 5 carats exactly). The Star of Africa (Figure 1.19) is one of the world's largest diamonds, having a mass of 530.02 carats. What is its mass in grams? in kilograms?

Collect and Organize We have a mass value expressed in carats, and we want to convert it into grams (g) and kilograms (kg). We know there are exactly 5 carats in 1 gram and, from Table 1.1, exactly 1000 grams in 1 kilogram.

Analyze We can use the above unit equivalencies to convert carats into grams and then grams into kilograms:

$$\text{carats} \times \frac{1 \text{ g}}{5 \text{ carat}} \times \frac{1 \text{ kg}}{1000 \text{ g}}$$

Solve We multiply the diamond's mass in carats by the first conversion factor:

$$530.02 \text{ carats} \times \frac{1 \text{ g}}{5 \text{ carat}} = 106.04 \text{ g}$$

and then convert grams into kilograms:

$$106.04 \text{ g} \times \frac{1 \text{ kg}}{1000 \text{ g}} = 0.10604 \text{ kg}$$

Think About It Diamond engagement rings typically weigh about 1 carat. The Star of Africa, in comparison, is enormous—it has an estimated value of U.S. $1 billion.

FIGURE 1.19 The Star of Africa is one of the largest cut diamonds in the world. It is mounted in the handle of the Royal Sceptre in the British Crown Jewels displayed in the Tower of London.

FIGURE 1.20 A warning to Mount Washington hikers from the U.S. Forest Service.

 Practice Exercise The Eiffel Tower in Paris is 324 m tall, including a 24-m television antenna that was not there when the tower was built in 1889. What is the height of the Eiffel Tower in kilometers and in centimeters?

(Answers to Practice Exercises are in the back of the book.)

SAMPLE EXERCISE 1.4 Multistep Conversions of U.S. Customary to SI Units **LO7**

On April 12, 1934, a wind gust of 231 miles per hour was recorded at the summit of Mount Washington in New Hampshire (Figure 1.20). What is this wind speed in meters per second?

Collect and Organize Speed is given in units of miles per hour, and we need to convert it to meters per second. Table 1.3 contains the equivalency 1 km = 0.6214 mi, and we know from Table 1.1 that the *kilo-* prefix means 1000. We will also need two unit-of-time equivalencies: 1 hour = 60 minutes and 1 minute = 60 seconds.

Analyze The unit conversion patterns are

$$\text{mi} \rightarrow \text{km} \rightarrow \text{m}$$

in the numerator, and

$$\text{h} \rightarrow \text{min} \rightarrow \text{s}$$

in the denominator.

Solve We use the unit equivalencies described above as conversion factors corresponding to the conversion patterns we have identified. We arrange them so that their numerators and denominators cancel out the initial and intermediate units:

$$231 \, \frac{\text{mi}}{\text{h}} \times \frac{1 \, \text{km}}{0.6214 \, \text{mi}} \times \frac{1000 \, \text{m}}{1 \, \text{km}} \times \frac{1 \, \text{h}}{60 \, \text{min}} \times \frac{1 \, \text{min}}{60 \, \text{s}} = 103 \, \frac{\text{m}}{\text{s}}$$

Think About It The calculated value is reasonable because the product of the numerator values is 1000, and the product of the denominator values is ($\sim$2/3 $\times$ 60 $\times$ 60), or about 2400. Therefore, the calculated value should be near (1000/2400), or a little less than half the initial value, which it is.

 Practice Exercise Light travels through exactly 1 meter of outer space in 1/299,792,458 of a second. How many kilometers does it travel in one year?

(Answers to Practice Exercises are in the back of the book.)

1.10 Evaluating and Expressing Experimental Results

Although scientific measurements can be expressed in different units, they all have one thing in common: there is a limit to how accurate they can be. Nobody is perfect, and no analytical method is perfect either. Every method has an inherent limit in its capacity to produce accurate results, so we need ways to express experimental results that reflect these limits in their certainty.

Significant Figures

One way to express how well we know a measured value is by using the appropriate number of **significant figures**, which are all the certain digits in a measured value plus one estimated digit. Suppose we determine the mass of a newly minted penny on the two balances shown in Figure 1.21. The lower balance determines the masses of objects to the nearest 0.0001 gram (g); the upper balance determines the mass of objects only to the nearest 0.01 g. According to the upper balance, our penny has a mass of 2.50 g; according to the lower balance, it has a mass of 2.4987 g. The mass obtained from the top balance has three significant figures: the 2, 5, and 0 are considered *significant*, which means we are confident in their values. The mass obtained using the bottom balance has five significant figures (2, 4, 9, 8, and 7). We may conclude that the mass of the penny can be determined with greater certainty with the balance at the bottom.

Now imagine that an aspirin tablet is placed on the lower balance in Figure 1.21, and the display reads 0.0180 g. How many significant figures are in this measurement? You might be tempted to say five because that is the number of digits displayed. However, the first two zeros are not considered significant because they serve only to set the location of the decimal point. These two zeros function the way exponents do when we use scientific notation to express values. Expressing 0.0180 g in scientific notation gives us 1.80×10^{-2} g. Only three of the digits (1, 8, and 0) in the decimal part indicate how precisely we know the value; the "−2" in the exponent does not. (See Appendix 1 for a review of how to express values using scientific notation.) The following guidelines will help you handle zeros (highlighted in green) in deciding the number of significant figures in a value:

1. Zeros at the beginning of a value, as in **0.0**592, are never significant. In this example, they just set the decimal place.
2. Zeros at the end of a value and after a decimal point, as in $3.\mathbf{00} \times 10^{8}$, are always significant.
3. Zeros at the end of a value that contains no decimal point, as in 96,5**0**0, may or may not be significant. They may be there only to set the decimal place. We should use scientific notation to avoid this ambiguity. Two possible interpretations of 96,500 are 9.65×10^{4} (three significant figures) and 9.6500×10^{4} (five significant figures).
4. Zeros between nonzero digits, as in 1**0**1.3, are always significant.

CONCEPT **TEST**

How many significant figures are there in the values used as examples in guidelines 1, 2, and 4?

(Answers to Concept Tests are in the back of the book.)

Significant Figures in Calculations

Let's explore how significant figures are used to express the results of calculations involving measured quantities. Suppose we suspect that a small nugget of yellow metal is pure gold. We could test our suspicion by determining the mass and volume of the nugget and then calculating its density. If the calculated density matches that of gold (19.3 g/mL), chances are good that the nugget is pure gold because few minerals are that dense. We find that the mass of the nugget is 4.72 g and its volume is 0.25 mL. What is the density of the nugget, expressed in the appropriate number of significant figures?

FIGURE 1.21 The mass of a penny can be measured to the nearest 0.01 g with the upper balance and to the nearest 0.0001 g with the balance below.

CHEMTOUR
Significant Figures

CHEMTOUR
Scientific Notation

significant figures all the certain digits in a measured value plus one estimated digit. The greater the number of significant figures, the greater the certainty with which the value is known.

Using Equation 1.1 to calculate the density (d) from the mass (m) and volume (V) produces the following result:

$$d = \frac{m}{V} = \frac{4.72 \text{ g}}{0.25 \text{ mL}} = 18.88 \text{ g/mL}$$

This density value appears to be slightly less than that of pure gold. However, we need to answer the question, "How well do we know the result?" The mass value is known to three significant figures, but the volume value is known only to two. At this point we need to apply the *weak-link principle*, which is based on the idea that a chain is only as strong as its weakest link. In calculations involving measured values, this principle means that we can know the answer to a calculation only as well as we know the least well-known measurement. In calculations involving multiplication or division, the weak link has the fewest significant figures. In this example the weak link is the volume, 0.25 mL. Because it has only two significant figures, our final answer should be reported with only two significant figures. We must convert 18.88 to a number with two significant figures. We do this by a process called rounding off.

Rounding off a value means dropping the *insignificant digits* (all digits to the right of the first uncertain digit) and then rounding that first uncertain digit either up or down. If the first digit in the string of insignificant digits dropped is greater than 5, we round up; if it is less than 5, we round down. If it is 5 and there are nonzero digits to the right of it, we round up. For example, rounding 45.450001 to three significant figures makes it 45.5 (we rounded up because there was a nonzero digit to the right of the second 5).

If there are no nonzero digits to the right of the 5, then we need a tie-breaking rule for rounding down or up. A good rule to follow is to round to the nearest even number. For example, rounding 45.45 (or 45.450) to three significant figures makes either value 45.4, because the 4 in the tenths place is the nearest even number. However, we round off 45.55 (or 45.550) to 45.6, because the 6 in the tenths place is the nearest even number.

An important rule to remember is that you round off only at *the end of a calculation*, never on intermediate results. Otherwise, "rounding errors" can creep into calculations. A reasonable guideline to follow is to carry at least one digit to the right of the last significant digit through all intermediate steps in a multistep calculation, or simply store all intermediate values in your calculator's memory until you have the final answer.

In the case of our gold nugget, the density value resulting from our calculation, 18.88 g/mL, must be rounded to two significant figures. Because the first insignificant digit to be dropped (shown in red) is greater than 5, we round up the blue 8 to 9:

$$18.88 \text{ g/mL} = 19 \text{ g/mL}$$

The density of pure gold expressed to two significant figures is also 19 g/mL, so we conclude that the nugget could be pure gold.

The weak-link principle for significant figures also applies to calculations involving addition and subtraction. To illustrate, consider how the volume of a gold nugget might be measured. One approach is to measure the volume of water the nugget displaces. Suppose we add 50.0 mL of water to a 100-mL graduated cylinder, as shown in Figure 1.22. Note that the cylinder is graduated in milliliters. However, the space between the graduations allows us to estimate the volumes of samples to the nearest tenth of a milliliter. For example, the meniscus of the water in Figure 1.22 is aligned exactly with the 50-mL graduation. We can correctly record the volume of water as 50.0 mL because we can read the "50" part

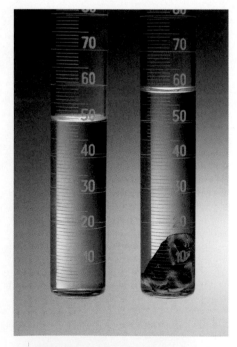

FIGURE 1.22 A nugget believed to be pure gold is placed in a graduated cylinder containing 50.0 mL of water. The volume rises to 58.5 mL, which means that the volume of the nugget is 8.5 mL.

of this value directly from the cylinder, and we can estimate that the next digit is "0." An estimated final digit in a measured value is still considered significant.

Then we place the nugget in the graduated cylinder, being careful not to splash any water out. The level of the water is now about halfway between the 58 and 59 mL graduations and so we estimate the volume of the combined sample to be 58.5 mL. Taking the difference of the two estimated values, we have the volume of our nugget:

$$
\begin{array}{r}
58.5 \text{ mL} \\
-50.0 \text{ mL} \\
\hline
8.5 \text{ mL}
\end{array}
$$

The result has only two significant figures because the initial and final volumes are known to the nearest tenth of a milliliter. Therefore, we can report the difference between them to only the nearest tenth of a milliliter. When measured numbers are added or subtracted, the result is reported with the same number of significant digits to the right of the decimal as the measured number with the fewest digits to the right of the decimal.

Consider one more example. Suppose the average mass of a recently minted penny is 2.50 g and a roll of pennies contains 50 of them. We use the upper balance from Figure 1.21 to determine the mass of a roll of pennies. Using the tare feature to correct for the mass of the wrapper, we find that the mass of pennies in the roll is 122.53 g. Dividing this value by the average mass of a penny and canceling out the grams units in the numerator and denominator, we get

$$
\frac{122.53 \text{ g}}{2.50 \text{ g/penny}} = 49.012 \text{ pennies}
$$

If we were measuring uncountable substances, rounding off to 49.0 would be appropriate (because under these circumstances, the measured value of 2.50 g/penny would be the weak link in the calculation). However, because we use only whole numbers to count items, we round to 49 and conclude that the roll contains 49 pennies. We must add one penny to make the roll complete.

Suppose the penny we add is one we know has a mass of 2.5007 g because we determined its mass on the lower balance in Figure 1.21. What is the mass of all 50 pennies to the appropriate number of significant figures? Adding the mass of the 50th penny to the mass of the other 49, we have

$$
\begin{array}{r}
122.53 \text{ g} \\
+ \quad 2.5007 \text{ g} \\
\hline
125.0307 \text{ g}
\end{array}
$$

The mass of the first 49 pennies is known to two decimal places, and the mass of the 50th is known to four decimal places. Therefore, by the weak-link principle we can know the sum of the two numbers to only two decimal places. This makes the last two digits in the sum, shown in red, not significant, so we must round off the mass to 125.03 g.

SAMPLE EXERCISE 1.5 Using Significant Figures in Calculations   **LO8**

A nugget of a shiny yellow mineral has a mass of 30.01 g. Its volume is determined by placing it in a 100-mL graduated cylinder containing 56.3 mL of water. The volume after the nugget is added is 62.6 mL. Is the nugget made of gold?

Collect and Organize We are given the mass of a nugget and water displacement data that can be used to calculate its volume. Density is mass divided by volume. We know from data given in this section that the density of gold is 19.3 g/mL.

Analyze The volume of the nugget is the difference between 62.6 and 56.3 mL, or about 6 mL. The ratio of the mass (30.01 g) to volume (~6 mL) is about 5 g/mL, which is much smaller than that of gold.

Solve

$$d = \frac{m}{V}$$

$$= \frac{30.01 \text{ g}}{(62.6 - 56.3) \text{ mL}} = \frac{30.01 \text{ g}}{6.3 \text{ mL}} = \frac{4.7635 \text{ g}}{\text{mL}}$$

Because we know the volume of the nugget (6.3 mL) to two significant figures, we can report its density to only two significant figures. So we must round off the result to 4.8 g/mL. The density of gold is 19.3 g/mL, so the nugget cannot be pure gold.

Think About It The weak link in defining the certainty in this calculation is the volume of the irregularly shaped sample calculated from water displacement data. Note that our rough estimate in the Analyze step strongly suggested that the nugget was unlikely to be pure gold.

 Practice Exercise Express the result of the following calculation to the appropriate number of significant figures:

$$\frac{0.391 \times 0.0821 \times (273 + 25)}{8.401}$$

Note: None of the starting values are exact numbers.

(Answers to Practice Exercises are in the back of the book.)

CONCEPT TEST

Lap speed in stock car racing is calculated by dividing the distance around a track (Figure 1.23) by the time it takes a car to go around the track. If lap time is measured to the nearest hundredth of a second, which dimension—distance or time—is probably the weak link in the calculation? Explain your answer.

(Answers to Concept Tests are in the back of the book.)

FIGURE 1.23 Dirt track for stock car racing.

precision agreement between the results of multiple measurements that were carried out in the same way.

accuracy agreement between one or more experimental values and the true value.

Measurements always have some degree of uncertainty, which limits the number of significant figures we can use to report any measurement. On the other hand, some values are known exactly, such as 12 eggs in a dozen and 60 seconds in a minute. Because these values are definitions, there is no uncertainty in them and they are not considered when determining significant figures in the answer to a mathematical calculation in which they appear.

SAMPLE EXERCISE 1.6 Distinguishing Exact from Uncertain Values   **LO9**

Which of the following data for the Washington Monument in Washington, DC, are exact numbers and which are not?
a. The monument is made of 36,491 white marble blocks.
b. The monument is 169 m tall.
c. There are 893 steps to the top.

d. The mass of its aluminum apex is 2.8 kg.
e. The area of its foundation is 1487 m².

Collect, Organize, and Analyze One way to distinguish exact from inexact values is to answer the question, "Which quantities can be counted?" Values that can be counted are exact.

Solve (a) The number of marble blocks and (c) the number of stairs are quantities we can count, so they are exact numbers. The other three quantities are based on measurements of (b) length, (d) mass, and (e) area and therefore are not exact.

Think About It The "you can count them" property of exact numbers works in this exercise and in most others even when it takes a long time (such as counting the hairs on a dog) or if help is needed in visualizing the objects (such as using a microscope). Actually, these examples raise a good point: Is there uncertainty in the counting process itself? Can you think of a famous example in the history of American politics of an uncertain counting process?

 Practice Exercise Which of the following statistics associated with the Golden Gate Bridge in San Francisco, CA, are exact numbers and which have some uncertainty?

a. The roadway is six lanes wide.
b. The width of the bridge is 27.4 m.
c. The bridge has a mass of 3.808×10^8 kg.
d. The length of the bridge is 2740 m.
e. The cash toll for a car traveling south increased to $7.00 on April 7, 2014.

(Answers to Practice Exercises are in the back of the book.)

Precision and Accuracy

Two terms—precision and accuracy—are used to describe how well we know a measured quantity or a value calculated from a measured quantity. **Precision** indicates how well the results of repeated measurements or analyses agree with each other, whereas **accuracy** reflects how close the results are to the true value of a quantity (Figure 1.24). The accuracy of a balance, for example, can be checked by measuring the mass of objects of known mass. Similarly, a thermometer can be calibrated by measuring the temperature at which a substance changes state. For example, ice melts to liquid water at 0.0°C, and an accurate thermometer dipped into a mixture of ice and liquid water reads 0.0°C exactly. A measurement that is validated by calibration with an accepted standard material is considered accurate.

Another way to check the accuracy of measurements is to periodically analyze *control samples*, which contain known quantities of the substance of interest. Control samples are routinely used in clinical chemistry labs to ensure the accuracy of routine analyses of blood serum and urine samples. One substance that is included in comprehensive blood tests is creatinine, a product of protein metabolism that doctors use to monitor patients' kidney function. Suppose that a control sample known to contain 0.681 milligrams of creatinine per deciliter of sample is analyzed after every ten patient samples, and that the results of five control sample analyses are 0.685, 0.676, 0.669, 0.688, and 0.692 mg/dL. How precise (repeatable) are these results, and do they mean that the analyses of the control sample (and those subsequently conducted on the real samples) are accurate?

A widely used method for expressing the precision of data such as these involves calculating the average of all the values and then calculating how much

(a)

(b)

(c)

FIGURE 1.24 (a) The three dart throws meant to hit the center of the target are both accurate and precise. (b) The three throws meant to hit the center of the target are precise but not accurate. (c) This set of throws is neither precise nor accurate.

mean (arithmetic mean, $\bar{x}$) an average calculated by summing a set of related values and dividing the sum by the number of values in the set.

standard deviation (s) a measure of the amount of variation, or dispersion, in a set of related values.

confidence interval a range of values that has a specified probability of containing the true value of a measurement.

each value deviates from the average. To calculate the average, we sum the five control sample results and divide by the number of values (5):

$$\frac{(0.685 + 0.676 + 0.669 + 0.688 + 0.692)}{5} \text{ mg/dL} = \frac{3.410}{5} \text{ mg/dL} = 0.682 \text{ mg/dL}$$

We can write a general equation to represent this calculation for any number of measurements (n) of any parameter x. Individual results are identified by the generic symbol x_i, where i can be any integer from 1 to n, the number of measurements made. The capital Greek sigma ($\sum$) with an i subscript, $\sum_i(x_i)$, represents the *sum* of all the individual x_i values. This kind of average is called the **mean** or **arithmetic mean**, and it is represented by the symbol $\bar{x}$:

$$\bar{x} = \frac{\sum_i(x_i)}{n} \tag{1.2}$$

To evaluate the precision of the five analyses of the control sample, we first calculate how much each of the values, x_i, deviates from the mean value $\bar{x}$: $(x_i - \bar{x})$. Then, to average these deviations, we apply the following procedure: First, we square each deviation $(x_i - \bar{x})^2$. Then we sum all of these squared values, divide the sum by $(n - 1)$, and finally take the square root of the quotient. The results of these calculations are shown in Table 1.5, where the final result is called the **standard deviation (s)** of the control sample values. Equation 1.3 puts the steps for calculating s in equation form:

$$s = \sqrt{\frac{\sum_i(x_i - \bar{x})^2}{n - 1}} \tag{1.3}$$

The standard deviation is sometimes reported with the mean value, using a $\pm$ sign to indicate the uncertainty in the mean value. In the case of the five control samples, this expression is 0.682 ± 0.009 mg/dL, where the s value is rounded off to match the last decimal place of the mean. The smaller the value of s, the more tightly clustered the measurements are around the mean and the more precise the data are.

Now that we have evaluated the precision of the control sample data, we should evaluate how accurate they are—that is, how well they agree with the actual creatinine value of the control sample, 0.681 mg/dL. The mean of the five values is within 0.001 mg/dL of the actual value, so the results certainly seem accurate, but there is also a way to quantitatively express the certainty that they are accurate. It involves calculating the **confidence interval**, which is a range of values around a calculated mean (0.682 mg/dL in our control sample analyses) that probably contains the true mean value, μ. Once we have calculated the size

TABLE 1.5 Calculating the Standard Deviation of the Control Sample Creatinine Values

x_i	$x_i - \bar{x}$	$(x_i - \bar{x})^2$	$\sum_i(x_i - \bar{x})^2$	$s = \sqrt{\dfrac{\sum_i(x_i - \bar{x})^2}{n - 1}}$
0.685	0.003	0.000009		
0.676	−0.006	0.000036		
0.669	−0.013	0.000169		
0.688	0.006	0.000036		
0.692	0.010	0.000100		
			0.00035	0.0094

of this range, we can determine whether the actual creatinine value is within it. If it is, then the analyses are probably accurate.

To calculate the confidence interval, we use a statistical tool called the *t*-distribution and the following equation:

$$\mu = \bar{x} \pm \frac{t\,s}{\sqrt{n}} \qquad (1.4)$$

A table of *t* values is located in Appendix 1. Table 1.6 contains a portion of it. The values are arranged based on two parameters: the number of values in a set of data (actually, $n - 1$) and the confidence level we wish to use in our decision making. A commonly used confidence level in chemical analysis is 95%. Using it means that the chances are 95% that the range we calculate using Equation 1.4 will contain the true mean value, in this case, the amount of the creatinine in our control sample. Using the 95% value for $(n - 1 = 5 - 1 = 4)$ in Table 1.6, which is 2.776, and the mean and standard deviation values calculated above:

$$\mu = \bar{x} \pm \frac{t\,s}{\sqrt{n}} = \left(0.682 \pm \frac{2.776 \times 0.0094}{\sqrt{5}}\right) \text{mg/dL} = (0.682 \pm 0.012) \text{ mg/dL}$$

Thus, we can say with 95% certainty that the true mean of our control sample data is between 0.670 and 0.694 mg/dL. Because this range includes the actual creatinine concentration, 0.681 mg/dL, of the control sample, we can infer with 95% confidence that these five analyses (and, importantly, the analyses of the patients' samples) are probably accurate.

TABLE 1.6 Values of *t*

($n - 1$)	CONFIDENCE LEVEL (%)		
	90	95	99
3	2.353	3.182	5.841
4	2.132	2.776	4.604
5	2.015	2.571	4.032
10	1.812	2.228	3.169
20	1.725	2.086	2.845
∞	1.645	1.960	2.576

CONCEPT TEST

Instead of calculating a standard deviation to express the variability in the data as we did in Table 1.5, we could have calculated a simple average deviation based on the mean of the absolute values of the deviations in the second column. What is the average deviation of the data? How does it differ from the standard deviation value? Suggest one reason for this difference.

(Answers to Concept Tests are in the back of the book.)

SAMPLE EXERCISE 1.7 Evaluating the Precision of Analytical Results   **LO9**

A group of students collects a sample of water from a river near their campus and shares it with five other groups of students. All six groups of students determine the concentration of dissolved oxygen in it. The results of their analyses are 9.2, 8.6, 9.0, 9.3, 9.1, and 8.9 mg O_2/L. Calculate the mean, standard deviation, and 95% confidence interval of these results.

Collect, Organize and Analyze We want to calculate the mean, standard deviation, and 95% confidence interval of the results of six analyses of the same sample. Equations 1.2, 1.3, and 1.4 can be used to calculate the three statistical parameters we seek. There are 6 values, so $n - 1 = 5$, and the appropriate *t* value (Table 1.6) to use in Equation 1.4 is 2.571. Mean and standard deviation functions are also included in many programmable calculators and in computer spreadsheet applications such as Microsoft Excel.

Solve
a. Calculating the mean:

$$\bar{x} = \frac{\sum_i (x_i)}{n} = \left(\frac{9.2 + 8.6 + 9.0 + 9.3 + 9.1 + 8.9}{6}\right) \text{mg/L} = \frac{54.1}{6} = 9.02 \text{ mg/L}$$

outlier a data point that is distant from the other observations.

Grubbs' test a statistical test used to detect an outlier in a set of data.

b. To calculate the standard deviation, we can set up a data table like the one in Table 1.5:

x_i	$x_i - \bar{x}$	$(x_i - \bar{x})^2$	$\sum_i (x_i - \bar{x})^2$	$\sqrt{\dfrac{\sum_i (x_i - \bar{x})^2}{n-1}}$
9.2	0.18	0.0324		
8.6	−0.42	0.1764		
9.0	−0.02	0.0004		
9.3	0.28	0.0784		
9.1	0.08	0.0064		
8.9	−0.12	0.0144		
			0.3084	0.25

Thus, the standard deviation (at the bottom of the fifth column) is 0.25.

c. Using Equation 1.4 to calculate the 95% confidence interval:

$$\mu = \bar{x} \pm \frac{t\,s}{\sqrt{n}} = \left(9.02 \pm \frac{2.571 \times 0.25}{\sqrt{6}} \right) \text{mg/L} = (9.02 \pm 0.26)\ \text{mg/L}$$

Think About It Did you notice that the six data points used in these calculations each contained two significant figures (each was known to the nearest tenth of a mg/L), yet we expressed the mean with three significant figures (to the hundredths place)? This happened because we knew the sum of the six results (54.1) to three significant figures, and dividing this value by an exact number (6 values) meant the quotient could be reported with three significant figures: 9.02. This increase in significant figures—and in the students' confidence in knowing the actual concentration of dissolved oxygen in the river—illustrates the importance of replicating analyses: doing so gives us more certainty about the true value of an experimental value than a single determination of it.

Practice Exercise Analyses of a sample of Dead Sea water produced these results for the concentration of sodium ions: 35.8, 36.6, 36.3, 36.8, and 36.4 mg/L. What are the mean, standard deviation, and 95% confidence interval of these results?

(Answers to Practice Exercises are in the back of the book.)

There is an assumption built into calculating means, standard deviations, and confidence intervals, which is that the variability in the data is random. *Random* means that data points are as likely to be above the mean value as below it and that there is a greater probability of values close to the mean than far away from it. This kind of distribution is called a *normal* distribution. Large numbers of such data produce a distribution profile called a bell curve. Figure 1.25 shows such a curve, based on a study conducted by the U.S. Centers for Disease Control and Prevention on the average weights of Americans of different ages. The data in the figure are for the weights of 19-year-old males and have a mean of 175 pounds and a standard deviation of 41 pounds (79 ± 19 kg). In randomly distributed data, 68% of the values—represented by the area under the curve highlighted in pale red—are within 1 standard deviation of the mean.

The concept of a normal distribution raises the question of how to handle **outliers**, that is, individual values that are much farther away from the mean than any of the other values. There may be a temptation to simply ignore such a value, but unless there is a valid reason for doing so, such as accidentally leaving out a step in an analytical procedure, it is unethical to disregard a value just because it is unexpected or not similar to the others.

Consider this scenario: A college freshman discovers a long-forgotten childhood piggy bank full of pennies and, being short of cash, decides to pack them into

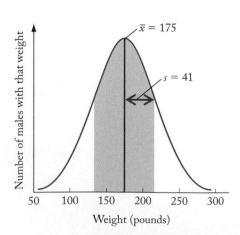

FIGURE 1.25 Weight distribution of 19-year-old American males. Source: U.S. Centers for Disease Control and Prevention (2012).

rolls of 50 to deposit them at a bank. To avoid having to count out hundreds of pennies, the student decides to weigh them out on a balance in a general chemistry lab that can weigh up to 300 grams to the nearest 0.001 g. The student weighs ten pennies from the piggy bank to determine their average mass, intending to multiply the average by 50 to calculate how many to weigh out to pack each roll. The results of the ten measurements are listed in Table 1.7 from lowest to highest values.

The results reveal that nine of the ten masses are quite close to 2.5 grams, but the tenth is considerably heavier. Was there an error in the measurement, or is there something unusual about that tenth penny? To answer questions such as these—and the broader one of whether an unusually high or low value is statistically different enough from the others in a set of data to be labeled an outlier—we analyze the data by using *Grubbs' test of a single outlier*, or **Grubbs' test**. In this test, the absolute difference between the suspected outlier and the mean of a set of data is divided by the standard deviation of the data set. The result is a statistical parameter that has the symbol Z:

$$Z = \frac{|x_i - \bar{x}|}{s} \qquad (1.5)$$

If this calculated Z value is greater than the reference Z value (Table 1.8) for a given number of data points and a particular confidence level—usually 95%—then the suspect data point is determined to be an outlier and can be discarded.

To apply Grubbs' test to the mass of the tenth and heaviest penny, we first calculate the mean and standard deviation of the masses of all ten pennies. These values are 2.562 ± 0.1915 g. We use them in Equation 1.5 to calculate Z:

$$Z = \frac{|x_i - \bar{x}|}{s} = \frac{|3.107 - 2.562|}{0.1915} = 2.846$$

Next, we check the reference Z values in Table 1.8 for $n = 10$ data points, and we find that our calculated Z value is greater than both 2.290 and 2.482—the Z values above which we can conclude with 95% and 99% confidence, respectively, that the 3.107 g data point is an outlier. Stated another way, the probability that this data point *is not* an outlier is less than 1%. Note that Grubbs' test can be used *only once* to identify *only one* outlier in a set of data.

The tenth penny probably weighed much more than the others because U.S. pennies minted since 1983 weigh 2.50 grams new, but those minted before 1983 weighed 3.11 grams new. The older pennies are 95% copper and 5% zinc, whereas the newer pennies are 97.5% zinc and are coated with a thin layer of copper. (Copper is about 25% more dense than zinc.) Thus a piggy bank containing hundreds of pennies will probably have a few that are heavier than most of the others.

TABLE 1.7 Masses in Grams of Ten Circulated Pennies

2.486
2.495
2.500
2.502
2.502
2.505
2.506
2.507
2.515
3.107

TABLE 1.8 Reference Z Values for Grubbs' Test of a Single Outlier

	CONFIDENCE LEVEL (%)	
n	95	99
3	1.155	1.155
4	1.481	1.496
5	1.715	1.764
6	1.887	1.973
7	2.020	2.139
8	2.127	2.274
9	2.215	2.387
10	2.290	2.482
11	2.355	2.564
12	2.412	2.636

CONCEPT TEST

Calculate the mean mass of one penny in a roll of pennies that contains exactly half pre-1983 pennies. One synonym for *mean* is "average." Would the mass of any penny in this roll actually equal the average mass?

(Answers to Concept Tests are in the back of the book.)

SAMPLE EXERCISE 1.8 Testing Whether a Data Point Should Be Considered an Outlier   **LO9**

Use Grubbs' test to determine whether the lowest value in the set of sodium ion concentration data from Practice Exercise 1.7 should be considered an outlier at the 95% confidence level.

Collect, Organize, and Analyze We want to determine whether the lowest value in the following set of five results is an outlier: 35.8, 36.6, 36.3, 36.8, and 36.4 mg/L. Grubbs' test (Equation 1.5) is used to determine whether a data point should be deemed an outlier. If the resulting Z value is equal to or greater than the reference Z values for $n = 5$ in Table 1.8, the suspect value is an outlier.

Solve Using the statistics functions of a programmable calculator, we find the mean and standard deviation of the data set to be 36.38 ± 0.38 mg/L. We calculate the value of Z by using Equation 1.5:

$$Z = \frac{|x_i - \bar{x}|}{s} = \frac{|35.8 - 36.38|}{0.38} = 1.5$$

This calculated Z value is less than 1.71, which is the reference Z value for $n = 5$ at the 95% confidence level (Table 1.8). Therefore, the lowest value is *not* an outlier, and it should be included with the other four values in any analysis of the data.

Think About It Although the lowest value may have appeared to be considerably lower than the others in the data set, Grubbs' test tells us that it is not *significantly* lower at the 95% confidence level.

Practice Exercise Duplicate determinations of the cholesterol concentration in a blood serum sample produce the following results: 181 and 215 mg/dL. The patient's doctor is concerned about the difference between the two results and the fact that values above 200 mg/dL are considered "borderline high," so she orders a third cholesterol test. The result was 185 mg/dL. Should the doctor ignore the 215 mg/dL value?

(Answers to Practice Exercises are in the back of the book.)

1.11 Testing a Theory: The Big Bang Revisited

Let's return to our discussion of the Big Bang. If all the matter in the universe started as a dense cloud of very hot gas accompanied by an enormous release of energy, and if the universe has been expanding ever since, then the universe must have been cooling throughout time because gases cool as they expand. This is the principle behind the operation of refrigerators and air conditioners—and behind one of the most important experiments testing the validity of the Big Bang theory.

Temperature Scales

If the universe is still expanding and cooling, some warmth must be left over from the Big Bang. By the 1950s, some scientists predicted how much leftover warmth there should be: enough to give interstellar space a temperature of 2.73 K, where K indicates a temperature value on the Kelvin scale.

Several temperature scales are in use today. In the United States the Fahrenheit scale is still the most popular. In the rest of the world and in science, temperatures are most often expressed in degrees Celsius or on the Kelvin scale. The Fahrenheit and Celsius scales differ in two ways, as shown in Figure 1.26. First, their zero points are different. Zero degrees Celsius (0°C) is the temperature at which water freezes under normal conditions, but that temperature is 32 degrees on the Fahrenheit scale (32°F). The other difference is the size of the temperature change corresponding to 1 degree. The difference between the freezing and

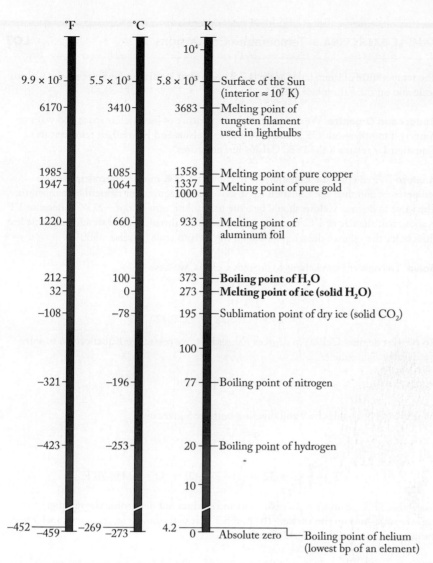

FIGURE 1.26 Three temperature scales are commonly used today, although the Fahrenheit scale is rarely used in scientific work. The freezing and boiling points of liquid water are the defining temperatures for each of the scales and their degree sizes.

boiling points of water is $212 - 32 = 180$ degrees on the Fahrenheit scale but only $100 - 0 = 100$ degrees on the Celsius scale. This difference means that a Fahrenheit degree is 100/180, or 5/9, as large as a Celsius degree.

To convert temperatures from Fahrenheit into Celsius, we need to account for the differences in zero point and in degree size. The following equation does both:

$$°C = \frac{5}{9} (°F - 32) \tag{1.6}$$

The SI unit of temperature (Table 1.2) is the **kelvin (K)**. The zero point on the Kelvin scale is not related to the freezing of a particular substance; rather, it is the coldest temperature—called **absolute zero (0 K)**—that can theoretically exist. It is equivalent to $-273.15°C$. No one has ever been able to chill matter to absolute zero, but scientists have come very close, cooling samples to less than 10^{-9} K.

The zero point on the Kelvin scale differs from that on the Celsius scale, but the size of 1 degree is the same on the two scales. For this reason, the conversion from a Celsius temperature to a Kelvin temperature is simply a matter of adding 273.15 to the Celsius value:

$$K = °C + 273.15 \tag{1.7}$$

CHEMTOUR
Temperature Conversion

kelvin (K) the SI unit of temperature.

absolute zero (0 K) the zero point on the Kelvin temperature scale; theoretically the lowest temperature possible.

FIGURE 1.27 Robert Dicke (1916–1997) predicted the existence of cosmic microwave background radiation. His prediction was confirmed by the serendipitous discovery of this radiation by Robert Wilson and Arno Penzias.

FIGURE 1.28 At normal body temperature, a human being emits radiation in the infrared region of the electromagnetic spectrum. In this thermal camera self-portrait, the red areas are the warmest and the blue regions are the coolest.

SAMPLE EXERCISE 1.9 Temperature Conversions   **LO7**

The temperature of interstellar space is 2.73 K. What is this temperature on the Celsius scale and on the Fahrenheit scale?

Collect and Organize We are given the temperature of interstellar space and want to convert it to other scales. Equation 1.6 relates Celsius and Fahrenheit temperatures; Equation 1.7 relates Kelvin and Celsius temperatures.

Analyze We can first use Equation 1.7 to convert 2.73 K into an equivalent Celsius temperature and then use Equation 1.6 to calculate an equivalent Fahrenheit temperature. The value in degrees Celsius should be close to absolute zero (about −273°C). Because 1°F is about half the size of 1°C, the temperature on the Fahrenheit scale should be a little less than twice the value of the temperature on the Celsius scale (around −500°F).

Solve To convert from kelvins to degrees Celsius, we have

$$K = °C + 273.15$$
$$°C = K - 273.15$$
$$= 2.73 - 273.15 = -270.42°C$$

To convert degrees Celsius to degrees Fahrenheit, we rearrange Equation 1.6 to solve for degrees Fahrenheit:

$$°C = \frac{5}{9}(°F - 32)$$

Multiplying both sides by 9 and dividing both by 5 gives us

$$\frac{9}{5}°C = °F - 32$$

$$°F = \frac{9}{5}°C + 32 = \frac{9}{5}(-270.42) + 32 = -454.76°F$$

The value 32°F is considered a definition and so does not determine the number of significant figures in the answer. The number that determines the accuracy to which we can know this value is −270.42°C.

Think About It The calculated Celsius value of −270.42°C makes sense because it represents a temperature only a few degrees above absolute zero, just as we estimated. The Fahrenheit value is within 10% of our estimate and so is reasonable, too.

Practice Exercise The temperature of the Moon's surface varies from −233°C at night to 123°C during the day. What are these temperatures on the Kelvin and Fahrenheit scales?

(Answers to Practice Exercises are in the back of the book.)

An Echo of the Big Bang

By the early 1960s, Princeton University physicist Robert Dicke (Figure 1.27) had suggested the presence of residual energy left over from the Big Bang, and he was eager to test this hypothesis. He proposed building an antenna that could detect microwave energy reaching Earth from outer space. Why did he pick microwaves? Even matter as cold as 2.73 K emits a "glow" (an energy signature), but not a glow you can see or feel, like the infrared rays emitted by any warm object, including humans (Figure 1.28). Dicke wanted to build a microwave antenna because the glow from a 2.73 K object takes the form of microwaves.

Dicke's microwave detector was never built because of events that occurred just a short distance from Princeton. By the early 1960s, the United States had launched *Echo* and *Telstar*, the first communication satellites. These satellites were reflective spheres designed to bounce microwave signals to receivers on Earth. An antenna designed to receive these microwave signals had been built at Bell Laboratories in Holmdel, NJ (Figure 1.29). Two Bell Labs scientists, Robert W. Wilson and Arno A. Penzias, were working to improve the antenna's reception when they encountered a problem. They found that no matter where they directed their antenna, it picked up a background microwave signal much like the hissing sound radios make when tuned between stations. They concluded that the signal was due to a flaw in the antenna or in one of the instruments connected to it. At one point they came up with another explanation: that the source of the background signal was a pair of pigeons roosting on the antenna and coating parts of it with their droppings. However, the problem persisted when the droppings were cleaned up. More testing discounted the flawed-instrument hypothesis but still left unanswered the question of where the signal was coming from.

The nuisance signal picked up by the Wilson–Penzias antenna serendipitously matched the microwave echo of the Big Bang that Dicke had predicted. When the scientists at Bell Labs learned of Dicke's prediction, they realized the significance of the strange background signal, and others did, too. Wilson and Penzias shared the Nobel Prize in Physics in 1978 for discovering the cosmic microwave background radiation of the universe. Dicke did not share in the prize, even though he had accurately predicted what Wilson and Penzias discovered by accident.

A major discovery in science usually leads to new questions. The discovery of cosmic microwaves was no exception: it supported the Big Bang theory and also raised even more questions about it. The Bell Labs antenna picked up the same microwave signal no matter where in the sky it was pointed. In other words, the cosmic microwaves appeared to be uniformly distributed throughout the universe. If the afterglow from the Big Bang really was uniform, how could galaxies have formed? Some heterogeneity had to arise in the expanding universe—some clustering of the matter in it—to allow galaxies to form.

Scientists doing work related to Dicke's proposed that, if such clustering had occurred, a record of it should exist as subtle differences in the cosmic background radiation. Unfortunately, a microwave antenna in New Jersey—or any other place on Earth—could not detect such slight variations in signals because too many other sources of microwaves interfere with the measurements. One way for scientists to determine whether these heterogeneities exist would be to take readings from a radio antenna in space.

In late 1989, the United States launched such an antenna in the form of the *Cosmic Background Explorer* (*COBE*) satellite. After many months of collecting data and many more months of analyzing it, scientists released the results in 1992. The image in Figure 1.30(a) appeared on the front pages of newspapers and magazines around the world. This was a major news story because the predicted heterogeneity had been found, providing additional support for the Big Bang theory of the origin of the universe. At a news conference, the lead scientist on the *COBE* project called the map a "fossil of creation."

COBE's measurements, and those obtained over the next two decades by a satellite with even higher resolving power (Figure 1.30b), support the theory that the universe did not expand and cool uniformly. The blobs and ripples in the images in Figure 1.30 indicate that galaxy "seed clusters" formed early in the history of the universe. Cosmologists believe these ripples are a record of the next stage after the Big Bang in the creation of matter. This stage is examined in Chapter 2, where we discuss how some elements may have formed just after the Big Bang, and how others continue to be formed by the nuclear reactions that fuel our Sun and all the stars in the universe.

FIGURE 1.29 In 1965 Robert Wilson (left) and Arno Penzias discovered the microwave echo of the Big Bang while tuning this highly sensitive "horn" antenna at Bell Labs in Holmdel, NJ.

(a)

(b)

FIGURE 1.30 (a) Map of the cosmic microwave background released in 1992. It is a 360-degree image of the sky made by collecting microwave signals for a year from the microwave telescopes of the *COBE* satellite. (b) Higher-resolution image based on measurements made in 2012 by the *WMAP* satellite. Red regions are up to 200 μK warmer than the average interstellar temperature of 2.73 K, and blue regions are up to 200 μK colder than 2.73 K.

When Lemaître first proposed how the universe might have formed following a cosmic release of energy, would it have been more appropriate to call his explanation a hypothesis or a theory? Why?

(Answers to Concept Tests are in the back of the book.)

SAMPLE EXERCISE 1.10 Integrating Concepts: Driving around Mars

In early press conferences about the *Curiosity* rover on Mars (Figure 1.31), scientists and engineers expressed distances in either the English or the metric system and temperatures in K, °C, and °F in response to questions from journalists from different countries. After an early drive, *Curiosity* stopped 8.0 ft away from an interesting football-shaped rock. *Curiosity*'s wheels are 50 cm in diameter, and the rover moves at a speed of 200 m/sol, where 1 sol = 1 Martian day = 24.65 h.

a. How many minutes would it take the rover to move to within 1.0 ft of the rock?
b. How many rotations of the wheels would be required to move that distance?
c. The rover landed in a valley about 15 miles away from the base of Mt. Sharp. How far is that in kilometers?
d. From the base to the peak of Mt. Sharp is 5.5 km; from the base to the peak of Mt. Everest on Earth is 15,000 ft. Which mountain is taller?
e. One night *Curiosity* recorded a temperature reported as "−132°" on the Fahrenheit scale. If the daytime temperature was 263 K, what was the day/night temperature range in °C?

Collect and Organize We are given a series of measurements and are asked questions that require us to convert units of time, distance, speed, and temperature.

Analyze We can use conversion factors from the chapter and from the table on the inside back cover of the book to help us estimate the answers. For part (a), the rover has to move 7 ft. The speed of the rover is 200 m/sol, which is about 600 ft/24 h. The distance moved is about 1/100 of 600 ft, so we estimate it would take about 1/100 of a day, or about 0.24 h, which is about $\frac{1}{4}$ h or 15 min. For (b), the wheels are 50 cm in diameter, or about 20 in. The formula for the circumference of a circle is $\pi \times d \approx 3 \times 20 = 60$ in, or about 5 ft. We estimate it would take more than one turn of the wheels to travel to the rock. For part (c), a mile is about 1.6 km, so we estimate that

Mt. Sharp is midway between 15 and 30 km away, or about 23 km distant. For part (d), we can estimate that 5.5 km is a bit more than 3 mi and 15,000 ft is a bit less than 3 mi, so we estimate that the heights are similar but Mt. Sharp might be taller. Finally for part (e), −132°F is (−132 − 32)°F = −164°F below the freezing point of water. One degree Celsius is about the size of 2°F, so we estimate the temperature at night was about −82°C. The temperature in Kelvin during the day was 263 K, or −10°C, so the range of temperatures on the surface was about [−10°C − (−82°C)], or about 72°C.

Solve To work out the answers more exactly:

a. $(8.0 - 1.0) \text{ ft} \times \dfrac{12 \text{ in}}{1 \text{ ft}} \times \dfrac{2.54 \text{ cm}}{1 \text{ in}} \times \dfrac{1 \text{ m}}{100 \text{ cm}}$

$\times \dfrac{1 \text{ sol}}{200 \text{ m}} \times \dfrac{24.65 \text{ h}}{1 \text{ sol}} \times \dfrac{60 \text{ min}}{1 \text{ h}} = 16 \text{ min}$

b. The diameter of the wheel is 50 cm, or

$$50 \text{ cm} \times \dfrac{1 \text{ in}}{2.54 \text{ cm}} \times \dfrac{1 \text{ ft}}{12 \text{ in}} = 1.64 \text{ ft}$$

and its circumference is

$$\pi \times d = 3.1416 \times 1.64 \text{ ft} = 5.15 \text{ ft}$$

The wheel has 7.0 ft to travel, so it requires

$$7.0 \text{ ft} \times \dfrac{1 \text{ revolution}}{5.15 \text{ ft}} = 1.4 \text{ revolutions}$$

c. $15 \text{ mi} \times \dfrac{1.6093 \text{ km}}{1 \text{ mi}} = 24 \text{ km}$

d. Mt. Sharp: $5.5 \text{ km} \times \dfrac{1 \text{ mi}}{1.6093 \text{ km}} = 3.4 \text{ mi}$

Mt. Everest: $15,000 \text{ ft} \times \dfrac{1 \text{ mi}}{5280 \text{ ft}} = 2.8 \text{ mi}$

Mt. Sharp is $(3.4 - 2.8) = 0.6$ miles higher than Mt. Everest.

e. $°C = \dfrac{5}{9}(-132°F - 32°F) = -91.1°C$

The high temperature that day was 263 K:
$°C = K - 273.15 = 263 - 273.15 = -10.15 = -10°C$
The temperature differential on the surface that day was
$$-10°C - (-91.1°C) = 81°C$$

Think About It Although none of our estimates matched the calculated values exactly, each of them was in the ball park and served as a useful accuracy check. Two of the starting values—the diameter (50 cm) of *Curiosity*'s wheels and its speed (200 m/sol)—ended with zeros and had no decimal points. In both cases we assumed the zeros were significant. This was a reasonable assumption given (1) the precision with which the components of *Curiosity* must have been fabricated and (2) the ability of the Jet Propulsion Laboratory engineers to control its speed. Moreover, the results of the calculation involving the rover's speed were eventually rounded to only two significant figures.

FIGURE 1.31 *Curiosity* takes a selfie on the surface of Mars in 2015.

SUMMARY

LO1 **Matter** exists as pure **substances**, which may be either **elements** or **compounds**, and as **mixtures**. Mixtures may be **homogeneous** (these mixtures are also called **solutions**) or **heterogeneous**. (Section 1.2)

LO2 All matter consists of **atoms**, and we use **chemical formulas** consisting of atomic symbols to express the polyatomic form of an element or the elemental composition of a compound. **Chemical equations** describe the proportions of the substances involved in a **chemical reaction**. Space-filling and ball-and-stick models are used to show the three-dimensional arrangement of atoms held together by **chemical bonds** to form a **molecule**. (Section 1.2)

LO3 Matter undergoes **physical processes**, which do not change its chemical identity, and **chemical reactions**, which transform matter into different substances. Matter is described and defined in terms of its **physical properties** and **chemical properties**. Physical properties may be used to separate mixtures into pure substances. (Sections 1.1, 1.2, 1.3, and 1.5)

LO4 The COAST framework used in this book to solve problems has four components: Collect and Organize information and ideas; Analyze the information to determine how it can be used to obtain the answer; Solve the answer to the problem (often the math-intensive step); and Think About It—consider the answer's reasonableness, including its value and units. (Section 1.4)

LO5 The differences in the states of matter and the transitions between them can be understood by viewing **solids**, **liquids**, and **gases** at both the macroscopic and the atomic levels. (Section 1.6)

LO6 The **scientific method** is the approach we use to acquire knowledge through observation, testable **hypotheses**, and experimentation. An extensively tested and well-validated hypothesis becomes accepted as a **scientific theory** or **model**. (Sections 1.7 and 1.11)

LO7 Dimensional analysis uses **conversion factors** (fractions in which the numerators and denominators have different units but represent the same quantity) to convert a value from one unit into another unit. (Section 1.9)

LO8 Accurate measurements expressed in units understandable to others are crucial in science. The limit to a measurement's **accuracy** is expressed by the number of **significant figures** in the number. (Sections 1.8 and 1.10)

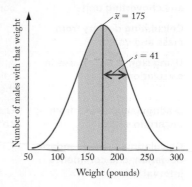

LO9 All measured values have some degree of uncertainty in their **precision**. Other values are exact—often because they are quantities that can be counted. The average value and variability of repeated measurements or analyses are determined by calculating the **arithmetic mean**, **standard deviation**, and **confidence interval**. An **outlier** in a data set may be identified based on the results of **Grubbs' test**. (Section 1.10)

PARTICULATE **PREVIEW WRAP-UP**

Sublimation is the process by which a solid changes directly into a gas. The image on the left is a solid because it consists of tightly packed and ordered molecules, whereas the image on the right is a gas because there is space between the molecules and they are spread out to fill the circle. Energy must be added to convert a solid into a gas.

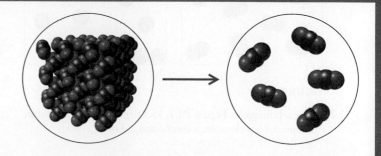

PROBLEM-SOLVING SUMMARY

Type of Problem	Concepts and Equations	Sample Exercises
Distinguishing physical properties from chemical properties	The chemical properties of a substance can be determined only by reacting it with another substance; physical properties can be determined without altering the substance's composition.	1.1
Recognizing physical states of matter	Particles in a solid are ordered; particles in a liquid are randomly arranged but close together; particles in a gas are separated by space and fill the volume of their container.	1.2
Doing dimensional analysis and converting units	Convert values from one set of units to another by multiplying by conversion factors set up so that the original units cancel.	1.3, 1.4
Calculating density from mass and volume	$$d = \frac{m}{V} \qquad (1.1)$$	1.5
Using significant figures in calculations	Apply the weak-link rule: the number of significant figures allowed in a calculated quantity involving multiplication or division can be no greater than the number of significant figures in the least-certain value used to calculate it.	1.5
Distinguishing exact from uncertain values	Quantities that can be counted are exact. Measured quantities or conversion factors that are not exact values are inherently uncertain.	1.6
Calculating mean, standard deviation, and confidence interval values	$$\bar{x} = \frac{\sum_i (x_i)}{n} \qquad (1.2)$$ $$s = \sqrt{\frac{\sum_i (x_i - \bar{x})^2}{n - 1}} \qquad (1.3)$$ $$\mu = \bar{x} \pm \frac{t\,s}{\sqrt{n}} \qquad (1.4)$$	1.7
Using Grubbs' test to identify an outlier	Calculate the value of Z for a suspected outlier x_i: $$Z = \frac{|x_i - \bar{x}|}{s} \qquad (1.5)$$ If the calculated Z value is greater than the appropriate reference Z value in Table 1.8, then x_i is an outlier.	1.8
Converting temperatures	$$°C = \frac{5}{9}(°F - 32) \qquad (1.6)$$ $$K = °C + 273.15 \qquad (1.7)$$	1.9

VISUAL PROBLEMS

(Answers to boldface end-of-chapter questions and problems are in the back of the book.)

1.1. For each image in Figure P1.1, identify what class of matter is depicted (an element, a compound, a homogeneous mixture, or a heterogeneous mixture) and identify the physical state(s).

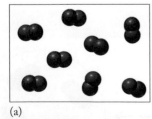
(a)

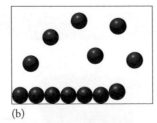
(b)

FIGURE P1.1

1.2. For each image in Figure P1.2, identify what class of matter is depicted (an element, a compound, a homogeneous

mixture, or a heterogeneous mixture) and identify the physical state(s).

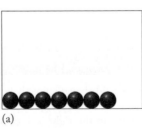
(a)

(b)

FIGURE P1.2

1.3. Which of the following statements best describes the change depicted in Figure P1.3?
a. A mixture of two gaseous elements undergoes a chemical reaction, forming a gaseous compound.
b. A mixture of two gaseous elements undergoes a chemical reaction, forming a solid compound.

c. A mixture of two gaseous elements undergoes deposition.

d. A mixture of two gaseous elements condenses.

FIGURE P1.3

1.4. Which of the following statements best describes the change depicted in Figure P1.4?

a. A mixture of two gaseous elements is cooled to a temperature at which one of them condenses.

b. A mixture of two gaseous compounds is heated to a temperature at which one of them decomposes.

c. A mixture of two gaseous elements undergoes deposition.

d. A mixture of two gaseous elements reacts together to form two compounds, one of which is a liquid.

FIGURE P1.4

1.5. A space-filling model of formic acid is shown in Figure P1.5. What is the chemical formula of formic acid?

FIGURE P1.5

1.6. A ball-and-stick model of isopropanol is shown in Figure P1.6. What is the molecular formula of isopropanol?

FIGURE P1.6

1.7. A pharmaceutical company checks the quality control process involved in manufacturing pills of one of its medicines by taking the mass of two samples of four pills each. Figure P1.7 shows graphs of the masses of the pills in each of the two samples. The pill is supposed to weigh 3.25 mg. Label each sample as both precise and accurate, precise but not accurate, accurate but not precise, or neither precise nor accurate.

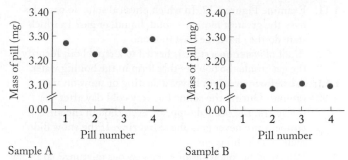

FIGURE P1.7

1.8. Use representations [A] through [I] in Figure P1.8 to answer questions a–f.

a. Which figures, if any, depict a chemical reaction?

b. Which two representations depict the same compound?

c. Which representations, if any, depict a physical process? Name the change(s).

d. List the molecules that are elements.

e. What is the formula of the compound that consists of three elements?

f. Which molecule contains the most atoms?

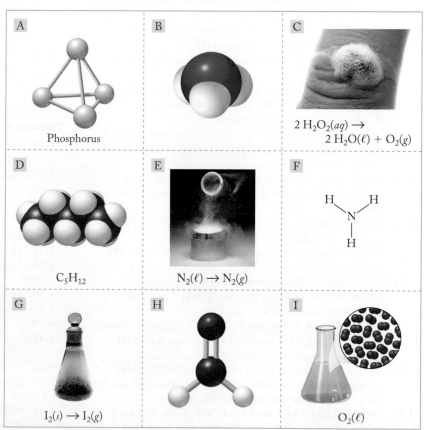

FIGURE P1.8

QUESTIONS AND PROBLEMS

Matter

Concept Review

1.9. List three differences and three similarities between a compound and an element.

1.10. What is in the space between the particles that make up a gas?

1.11. Examine Figure 1.12. In which physical state do particles have the greatest motion—solid, liquid, or gas? In which state do they have the least motion?

1.12. A pot of water on a stove is heated to a rapid boil. Identify the gas inside the bubbles that form in the boiling water.

1.13. A brief winter storm leaves a dusting of snow on the ground. During the sunny but very cold day after the storm, the snow disappears even though the air temperature never gets above freezing. If the snow didn't melt, where did it go?

1.14. Which of the following are homogeneous mixtures? (a) a gold wedding ring; (b) sweat; (c) bottled drinking water; (d) human blood; (e) compressed air in a scuba tank

1.15. Indicate whether each of the following properties is a physical or a chemical property of the element sodium:
 a. Its density is greater than that of kerosene and less than that of water.
 b. It has a lower melting point than that of most metals.
 c. It is an excellent conductor of heat and electricity.
 d. It is soft and can be easily cut with a knife.
 e. Freshly cut sodium is shiny, but it rapidly tarnishes in contact with air.
 f. It reacts very vigorously with water to form hydrogen gas (H_2) and sodium hydroxide (NaOH).

1.16. Indicate whether each of the following is a physical or a chemical property of hydrogen gas (H_2):
 a. At room temperature, its density is less than that of any other gas.
 b. It reacts vigorously with oxygen (O_2) to form water.
 c. Liquefied H_2 boils at a very low temperature ($-253°C$).
 d. H_2 gas does not conduct electricity.

1.17. Which of the following is not a pure substance? (a) air; (b) nitrogen gas; (c) oxygen gas; (d) argon gas; (e) table salt (sodium chloride)

1.18. Which of the following is a pure substance? (a) sweat; (b) blood; (c) brass (an alloy of copper and zinc); (d) sucrose (table sugar); (e) milk

1.19. Which of the following is an element? (a) Cl_2; (b) H_2O; (c) HCl; (d) NaCl

1.20. Which of the following is not an element? (a) I_2; (b) O_3; (c) ClF; (d) S_8

1.21. Which of the following is a homogeneous mixture? (a) filtered water; (b) chicken noodle soup; (c) clouds; (d) trail mix snack; (e) fruit salad

1.22. Which of the following is a heterogeneous mixture? (a) air; (b) sugar dissolved in water; (c) muddy river water; (d) brass; (e) table salt (sodium chloride)

1.23. Which of the following can be separated by filtration? (a) sugar dissolved in coffee; (b) sand and water; (c) gasoline; (d) alcohol dissolved in water; (e) damp air

*1.24. Would filtration be a suitable way to separate dissolved proteins from blood plasma? Explain why or why not.

1.25. Which of the following is an example of a chemical property of formaldehyde (CH_2O)?
 a. It has a characteristic acrid smell.
 b. It is soluble in water.
 c. It burns in air.
 d. It is a gas at room temperature.
 e. It is colorless.

1.26. Which of the following is an example of a physical property of silver (Ag)?
 a. It tarnishes over time.
 b. Tarnished silver can be cleaned to a shiny metallic finish.
 c. It reacts with chlorine to make a white solid.
 d. It sinks in water.

1.27. Can an extensive property be used to identify a substance? Explain why or why not.

1.28. Which of these properties of water are intensive and which are extensive properties?
 a. The density of water at room temperature and pressure.
 b. The temperature at which water freezes.
 c. The mass of water in your body.
 d. The mass of one molecule of water.
 e. The rate at which water is flowing over Niagara Falls.

The Scientific Method: Starting Off with a Bang

Concept Review

1.29. What kinds of information are needed to formulate a hypothesis?

1.30. How does a hypothesis become a theory?

1.31. Is it possible to disprove a scientific hypothesis?

1.32. Why was the belief that matter consists of atoms considered a philosophy in ancient Greece, but a theory by the early 1800s?

1.33. How do people use the word *theory* in normal conversation?

1.34. Can a theory be proven?

Unit Conversions and Dimensional Analysis

Concept Review

1.35. Describe in general terms how the SI and U.S. customary systems of units differ.

1.36. Suggest two reasons why SI units are not more widely used in the United States.

Problems

Note: The physical properties of the elements are listed in Appendix 3.

1.37. How many grams are there in 1.65 lbs?

1.38. How many pounds are there in 765.4 g?

1.39. How many milliliters are there in 2.44 gal?

1.40. How many gallons are there in 108 mL?

1.41. Peter is 5 ft, 11 in tall, and Paul is 176 cm tall. Who is taller?

1.42. A swimming pool is 1.12 m deep at one end and 72 in deep at the other. Which is the deeper end?

1.43. What is the mass of a magnesium block that measures 2.5 cm $\times$ 3.5 cm $\times$ 1.5 cm?

1.44. What is the mass of an osmium block that measures 6.5 cm $\times$ 9.0 cm $\times$ 3.25 cm? Do you think you could lift it with one hand?

1.45. What is the conversion factor for each of the following unit conversions? (a) picoseconds (ps) to femtoseconds (fs); (b) kilograms (kg) to milligrams (mg); (c) the mass of a block of titanium in kilograms (kg) to its volume in cubic meters (m^3)

1.46. A single strand of natural silk may be as long as 4.0×10^3 m. What is this length in miles?

*1.47. There are ten steps from the sidewalk up to the front door of a student's apartment. Each tread is 5.0 inches deep and 6.0 inches above the previous one. What is the distance diagonally from the bottom of the steps to the top in centimeters?

1.48. A swimming pool is 7.5 ft deep, 42 ft wide, and 65 ft long, and it is filled to the brim with water. What is the volume of water in the pool in cubic inches?

1.49. **Boston Marathon** To qualify to run in the 2015 Boston Marathon, a distance of 26.2 miles, an 18-year-old woman had to have completed another marathon in 3 hours and 34 minutes, or less. To qualify, what must this woman's average speed have been (a) in miles per hour and (b) in meters per second?

1.50. **Olympic Mile** An Olympic "mile" is actually 1500 m. What percentage is an Olympic mile of a U.S. mile (5280 feet)?

*1.51. If a wheelchair-marathon racer moving at 13.1 miles per hour expends energy at a rate of 665 Calories per hour, how much energy in Calories would be required to complete a marathon race (26.2 miles) at this pace?

1.52. **Nearest Star** At a distance of 4.3 light-years, Proxima Centauri is the nearest star to our solar system. What is the distance to Proxima Centauri in kilometers?

1.53. What volume of gold would be equal in mass to a piece of copper with a volume of 125 cm^3?

*1.54. A small hot-air balloon is filled with 1.00×10^6 L of air ($d = 1.20$ g/L). As the air in the balloon is heated, it expands to 1.09×10^6 L. What is the density of the heated air in the balloon?

1.55. What is the volume of 1.00 kg of mercury?

1.56. A student wonders whether a piece of jewelry is made of pure silver. She determines that its mass is 3.17 g. Then she drops it into a 10 mL graduated cylinder partially filled with water and determines that its volume is 0.3 mL. Could the jewelry be made of pure silver?

1.57. **The Density of Blood** The average density of human blood is 1.06 g/mL. What is the mass of blood in an adult with a blood volume of 5.5 L? Express your answer in (a) grams and (b) ounces.

1.58. **The Density of Earth** Earth has a mass of 5.98×10^{24} kg and a volume of 1.08×10^{12} km^3. What is the average density of our planet in units of grams per cubic centimeter?

1.59. The mass of a diamond is usually expressed in carats, where 1 carat = 0.200 g. The density of diamond is 3.51 g/cm^3. What is the volume of a 5.0-carat diamond?

*1.60. If the concentration of mercury in the water of a polluted lake is 0.33 micrograms per liter, what is the total mass of mercury in the lake, in kilograms, if the lake has a surface area of 10.0 km^2 and an average depth of 15 m?

Evaluating and Expressing Experimental Results

Concept Review

1.61. How many suspect data points can be identified from a data set by using Grubbs' test?

1.62. Which confidence interval is the largest for a given value of n: 50%, 90%, or 95%?

1.63. The concentration of ammonia in an aquarium tank is determined each day for a week. Which of these measures of the variability in the results of these analyses is greater: (a) mean ± standard deviation or (b) 95% confidence interval? Explain your selection.

1.64. If an outlier could not be identified at the 95% confidence level, (a) could it be identified at the 90% confidence level? (b) Could it be identified at the 99% confidence level?

Problems

1.65. Which of these uncertain values has the smallest number of significant figures? (a) 545; (b) 6.4×10^{-3}; (c) 6.50; (d) 1.346×10^2

1.66. Which of these uncertain values has the largest number of significant figures? (a) 545; (b) 6.4×10^{-3}; (c) 6.50; (d) 1.346×10^2

1.67. Which of these uncertain values has the smallest number of significant figures? (a) 1/545; (b) $1/6.4 \times 10^{-3}$; (c) 1/6.50; (d) $1/1.346 \times 10^2$

1.68. Which of these uncertain values has the largest number of significant figures? (a) 1/545; (b) $1/6.4 \times 10^{-3}$; (c) 1/6.50; (d) $1/1.346 \times 10^2$

1.69. Which of these uncertain values have four significant figures? (a) 0.0592; (b) 0.08206; (c) 8.314; (d) 5420; (e) 5.4×10^3

1.70. Which of these uncertain values have only three significant figures? (a) 7.02; (b) 6.452; (c) 6.02×10^{23}; (d) 302; (e) 12.77

1.71. Perform each of the following calculations, and express the answer with the correct number of significant figures (only the highlighted values are exact):
   a. $0.6274 \times 1.00 \times 10^3 / [2.205 \times (2.54)^3] =$
   b. $6 \times 10^{-18} \times (1.00 \times 10^3) \times 12 =$
   c. $(4.00 \times 58.69)/(6.02 \times 10^{23} \times 6.84) =$
   d. $[(26.0 \times 60.0)/43.53]/(1.000 \times 10^4) =$

1.72. Perform each of the following calculations, and express the answer with the correct number of significant figures (only the highlighted values are exact):
   a. $[(92 \times 60.0) + 55.3]/(5.000 \times 10^3) =$
   b. $(2.00 \times 183.9)/[(6.02 \times 10^{23}) \times (1.61 \times 10^{-8})^3] =$

c. $0.8161/[2.205 \times (2.54)^3] =$

d. $(9.00 \times 60.0) + (50.0 \times 60.0) + (3.00 \times 10^1) =$

_____

*1.73. The widths of copper lines in printed circuit boards must be close to a design value. Three manufacturers were asked to prepare circuit boards with copper lines that are 0.500 μm (micrometers) wide (1 μm = 1×10^{-6} m). Each manufacturer's quality control department reported the following line widths on five sample circuit boards (given in micrometers):

Manufacturer 1	Manufacturer 2	Manufacturer 3
0.512	0.514	0.500
0.508	0.513	0.501
0.516	0.514	0.502
0.504	0.514	0.502
0.513	0.512	0.501

a. What is the mean and standard deviation of the data provided by each manufacturer?
b. For which of the three sets of data does the 95% confidence interval include 0.500 μm?
c. Which of the data sets fit the description "precise and accurate," and which is "precise but not accurate"?

1.74. **Diabetes Test** Glucose concentrations in the blood above 110 mg/dL can be an early indication of several medical conditions, including diabetes. Suppose analyses of a series of blood samples from a patient at risk of diabetes produce these results: 106, 99, 109, 108, and 105 mg/dL.
a. What are the mean and the standard deviation of the data?
b. Patients with blood glucose levels above 120 mg/dL are considered diabetic. Is this value within the 95% confidence interval of these data?

_____

1.75. Use Grubbs' test to decide whether the value 3.41 should be considered an outlier in the following data set from analyses of portions of the same sample conducted by six groups of students: 3.15, 3.03, 3.09, 3.11, 3.12, and 3.41.

1.76. Use Grubbs' test to decide whether any one of the values in this set of replicate measurements should be considered an outlier: 61, 75, 64, 65, 64, and 66.

Testing a Theory: The Big Bang Revisited

Concept Review

1.77. Can a temperature in °C ever have the same value in °F?

1.78. What is meant by an _absolute_ temperature scale?

Problems

1.79. **Radiator Coolant** The coolant in an automobile radiator freezes at −39°C and boils at 110°C. What are these temperatures on the Fahrenheit scale?

1.80. Silver and gold melt at 962°C and 1064°C, respectively. Convert these two temperatures to the Kelvin scale.

1.81. **Critical Temperature** The discovery of new "high temperature" superconducting materials in the mid-1980s spurred a race to prepare the material with the highest superconducting temperature. The _critical temperatures_ (T_c)—the temperatures at which the material becomes superconducting—of three such materials are 93.0 K, −250.0°C, and −231.1°F. Convert these temperatures into a single temperature scale, and determine which superconductor has the highest T_c value.

1.82. As air is cooled, which gas condenses first: N_2, He, or H_2O?

_____

1.83. Liquid helium boils at 4.2 K. What is the boiling point of He in °C?

1.84. Liquid hydrogen boils at −253°C. What is the boiling point of H_2 on the Kelvin scale?

_____

1.85. A person has a fever of 102.5°F. What is this temperature in °C?

1.86. Physiological temperature, or body temperature, is considered to be 37.0°C. What is this temperature in °F?

_____

1.87. **Record Low** The lowest temperature measured on Earth is −128.6°F, recorded at Vostok, Antarctica, in July 1983. What is this temperature on the Celsius and Kelvin scales?

1.88. **Record High** The highest temperature ever recorded in the United States is 134°F at Greenland Ranch, Death Valley, CA, on July 13, 1913. What is this temperature on the Celsius and Kelvin scales?

Additional Problems

*1.89. Sodium chloride (NaCl) contains 1.54 g of Cl for every 1.00 g of Na. Which of the following mixtures would react to produce sodium chloride with no Na or Cl left over?
a. 11.0 g of Na and 17.0 g of Cl
b. 6.5 g of Na and 10.0 g of Cl
c. 6.5 g of Na and 12.0 g of Cl
d. 6.5 g of Na and 8.0 g of Cl

1.90. Your laboratory instructor has given you two shiny, light gray metal cylinders. Your assignment is to determine which one is made of aluminum ($d = 2.699$ g/mL) and which one is made of titanium ($d = 4.54$ g/mL). The mass of each cylinder was determined on a balance to five significant figures. The volume was determined by immersing the cylinders in a graduated cylinder as shown in Figure P1.90. The initial volume of water was 25.0 mL in each graduated cylinder. The following data were collected:

	Mass (g)	Height (cm)	Diameter (cm)
Cylinder A	15.560	5.1	1.2
Cylinder B	35.536	5.9	1.3

a. Calculate the volume of each cylinder by using the dimensions of the cylinder only.
b. Calculate the volume from the water displacement method.

c. Which volume measurement allows for the greater number of significant figures in the calculated densities?

d. Express the density of each cylinder to the appropriate number of significant figures.

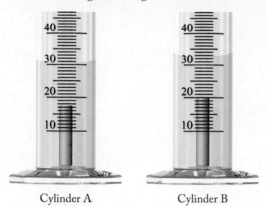

Cylinder A   Cylinder B

FIGURE P1.90

*1.91. Manufacturers of trail mix have to control the distribution of items in their products. Deviations of more than 2% outside specifications cause supply problems and downtime in the factory. A favorite trail mix is designed to contain 67% peanuts and 33% raisins. Bags of trail mix were sampled from the assembly line on different days. The bags were opened and the contents counted, with the following results:

Day	Peanuts	Raisins
1	50	32
11	56	26
21	48	34
31	52	30

a. Calculate the mean and standard deviation in the percentage of peanuts and percentage of raisins in the four samples.

b. Do the 90% confidence intervals for these percentages include the target composition values: 67% peanuts and 33% raisins?

*1.92. Gasoline and water do not mix. Regular grade (87 octane) gasoline has a lower density (0.73 g/mL) than water (1.00 g/mL). A 100 mL graduated cylinder with an inside diameter of 3.2 cm contains 34.0 g of gasoline and 34.0 g of water. What is the combined height of the two liquid layers in the cylinder? The volume of a cylinder is $\pi r^2 h$, where r is the radius and h is the height.

1.93. **Stretchy Springs** Metal springs come in many shapes and sizes. The same force is used to stretch each of two springs A and B. Spring A stretches from its natural length of 4.0 cm to a length of 5.4 cm; spring B's length increases by 15%. Which is the stronger spring, A or B?

1.94. Ms. Goodson's geology classes are popular because of their end-of-the-year field trips. Some last several days, but all involve exactly eight hours of hiking per day. On one three-day trip the class's average hiking speeds were 1.6 mi/h, 1.4 mi/h, and 1.7 mi/h each day. What was the length of their trip in miles and kilometers?

*1.95. **Toothpaste Chemistry** Most of the toothpaste sold in the United States contains about 1.00 mg of fluoride per gram of toothpaste. The fluoride compound that is most often used in toothpaste is sodium fluoride, NaF, which is 45% fluoride by mass. How many milligrams of NaF are in a typical 8.2-ounce tube of toothpaste?

*1.96. **Test for HIV** Tests called ELISAs (enzyme-linked immunosorbent assays) detect and quantify substances such as HIV antibodies in biological samples. A "sandwich" assay traps the HIV antibody between two other molecules. The trapping event causes a detector molecule to change color. To make a sandwich assay for HIV, you need the following components: one plate to which the molecules are attached; a 0.550 mg sample of the recognition molecule that "recognizes" the HIV antibody; 1.200 mg of the capture molecule that "captures" the HIV antibody in a sandwich; and 0.450 mg of the detector molecule that produces a visible color when the HIV antibody is captured. You need to make 96 plates for an assay. You are given the following quantities of material: 100.00 mg of the recognition molecule; 100.00 mg of the capture molecule; and 50.00 mg of the detector molecule.

a. Do you have sufficient material to make 96 plates?

b. If you do, how much of each material is left after 96 sandwich assays are assembled? If you do not have sufficient material to make 96 assays, how many assays can you assemble?

1.97. **Vitamin C** Some people believe that large doses of vitamin C can cure the common cold. One commercial over-the-counter product consists of 500.0 mg tablets that are 20% by mass vitamin C. How many tablets are needed for a 1.00 g dose of vitamin C?

*1.98. **Patient Data** Measurements of a patient's temperature are routinely done several times a day in hospitals. Digital thermometers are used, and it is important to evaluate new thermometers and select the best ones. The accuracy of these thermometers is checked by immersing them in liquids of known temperature. Such liquids include an ice–water mixture at 0.0°C and boiling water at 100.0°C at exactly 1 atmosphere pressure (boiling point varies with atmospheric pressure). Suppose the data shown in the following table were obtained on three available thermometers and you were asked to select the "best" one of the three.

Thermometer	Measured Temperature of Ice Water, °C	Measured Temperature of Boiling Water, °C
A	−0.8	99.9
B	0.3	99.8
C	0.3	100.3

Explain your choice of the "best" thermometer for use in the hospital.

2

Atoms, Ions, and Molecules

Matter Starts Here

BABY TEETH A young child proudly holds his tooth that has just fallen out. Calcium compounds are an important component of teeth, but a landmark 1958 study in St. Louis showed that children living hundreds, even thousands, of miles away from nuclear test sites were being exposed to and accumulating radionuclides produced during the tests.

PARTICULATE **REVIEW**

Elements versus Compounds

In Chapter 2 we explore the structure of atoms and begin the discussion of why they form the molecules and ions they do. All pure substances can be classified as either elements or compounds.

- Write the chemical formula for each substance shown here.

- Identify each substance as either an element or a compound.

   (Review Section 1.2 if you need help.)

(Answers to Particulate Review questions are in the back of the book.)

Helium

Nitrogen

Gold

Ethanol

Sodium chloride

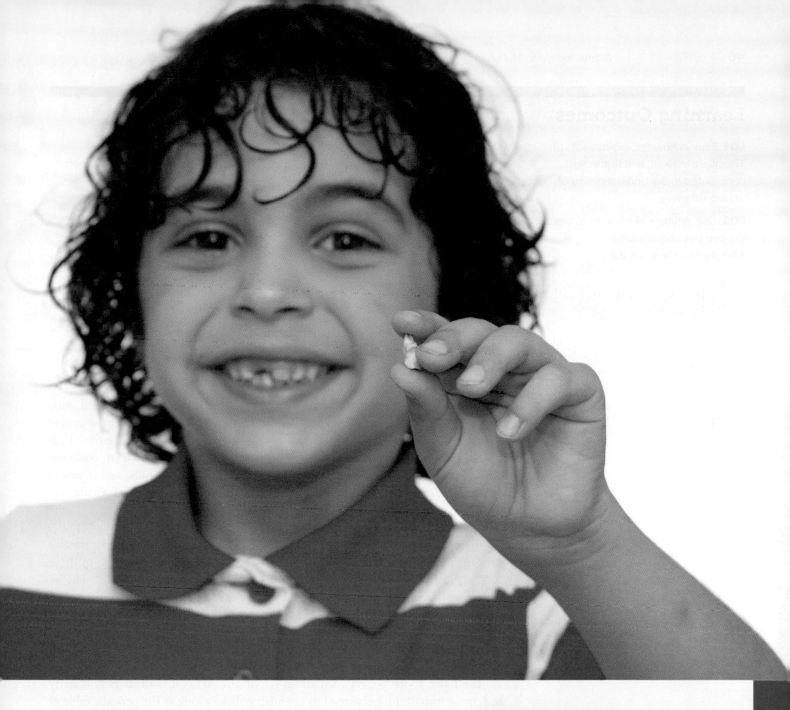

Ionic versus Covalent Compounds

Some compounds are formed when different atoms bond together; others are formed by bonds between different ions. As you read Chapter 2, look for ideas that will help you answer these questions:

- Determine if each compound represented here consists of different atoms or different ions bonded together.

- Which compound(s) are covalent? Which compound(s) are ionic?

- What features of these particulate images helped you make these distinctions?

Learning Outcomes

LO1 Explain how the experiments of Thomson, Millikan, and Rutherford contributed to our understanding of atomic structure

LO2 Use symbols of nuclides to describe the composition of atoms
Sample Exercises 2.1, 2.2

LO3 Identify isotopes and use natural abundance data to calculate average atomic masses
Sample Exercise 2.3

LO4 Use the periodic table to predict the chemical properties of elements
Sample Exercises 2.4, 2.6

LO5 Connect the law of multiple proportions to the formulas of common molecular compounds
Sample Exercise 2.5

LO6 Relate the names and formulas of common molecular and ionic compounds
Sample Exercises 2.7, 2.8, 2.9, 2.10, 2.11

LO7 Describe the origin of the elements

2.1 Atoms in Baby Teeth

In 1958 a group of physicians, scientists, and other concerned citizens in St. Louis, MO, began a project that would last for more than a decade—and alter the course of world events. Their project was known as the Baby Tooth Survey (Figure 2.1). It would eventually involve the analyses of more than 300,000 baby teeth collected from dentists' offices and young families in and around St. Louis, and it would inspire similar studies in other regions of the United States and in Canada and Germany.

The founders of the Baby Tooth Survey were worried about the effects of fallout from the hundreds of tests of nuclear weapons that had taken place in Earth's atmosphere since the end of World War II. They knew that the products of these tests included atoms exhibiting **radioactivity**, the spontaneous emission of high-energy radiation and particles. They also knew that radioactive fallout posed serious risks to human health. Their goal was to determine the impact of atmospheric nuclear testing on everyday citizens by measuring the concentration of strontium-90, a radioactive form of the element strontium that does not occur in nature, in the teeth of young children.

The Baby Tooth Survey focused on strontium-90 because of its radioactivity and, more important, because of its chemistry. Take a look at the periodic table of the elements on the inside front cover of this book and find the symbol for strontium, Sr. It's just below the one for calcium, Ca. As we discuss in this chapter, elements in the same column of the periodic table have similar chemical properties. Sometimes they can even substitute for each other in the compounds they form. This is the case for Sr^{2+} and Ca^{2+} ions, which have the same 2+ charge, as

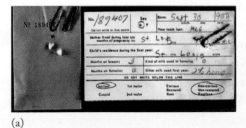

FIGURE 2.1 (a) Teeth collected in the Baby Tooth Survey and the information card used to record important data regarding the teeth; (b) a button given to children who donated their teeth.

(a)   (b)

do all the ions formed by the elements in the second column. Thus ions of strontium-90 can accumulate in the rapidly growing, calcium-rich teeth and bones of young children. The baby teeth lost by the children of St. Louis provided convenient, noninvasive biomarkers of their exposure to radioactive fallout.

The results of the survey grabbed worldwide attention. They showed that radioactivity from strontium-90 in the teeth of children born in the 1950s and 60s was 100 times greater than in baby teeth that formed before the dawn of the nuclear age in 1945, and that the amount of strontium-90 correlated with the frequency of atomic bomb testing. Early results from the Baby Tooth Survey, coupled with other data on the incidence of childhood cancer, helped persuade U.S. President John F. Kennedy to negotiate a treaty with the Soviet Union and other nuclear powers to end atmospheric testing of nuclear weapons in 1963.

In this chapter we explore the structure and composition of atoms, including the subatomic particles that cause radioactivity and the subatomic particles that are responsible for ionic charges. We begin with the discoveries made in the late 19th and early 20th centuries that led to our understanding of atomic structure and that helped explain the chemical properties of the elements, including their tendencies to form ions. We end the chapter by exploring how hydrogen and helium formed just after the Big Bang, and we will see how other elements have continued to form both in the cores of giant stars and in celestial explosions called supernovas. These are the atoms that make up the entire universe, including all of the matter in our world and in our bodies.

radioactivity the spontaneous emission of high-energy radiation and particles by materials.

cathode rays streams of electrons emitted by the cathode in a partially evacuated tube.

2.2 The Rutherford Model

By the end of the 19th century, many scientists had realized that atoms were not the smallest particles of matter, but rather consisted of even smaller *subatomic* particles. This realization came in part from the research of the British scientist Joseph John (J. J.) Thomson (1856–1940; Figure 2.2). However, not even Thomson himself recognized the fundamental importance of his work for the future of science and technology.

Electrons

Figure 2.3 shows the apparatus Thomson used in his experiments. It is called a cathode-ray tube (CRT), and it consists of a glass tube from which most of the air has been removed. Electrodes within the tube are attached to the poles of a high-voltage power supply. The electrode called the cathode is connected to the negative terminal of the power supply, and the anode is connected to the positive terminal. When these connections are made, electricity passes through the glass tube in the form of a beam of **cathode rays** emitted by the cathode. Cathode rays are invisible to the naked eye, but when the end of the CRT opposite the cathode is coated with a phosphorescent material, a glowing spot appears where the beam hits it.

Thomson found that cathode-ray beams are deflected by magnetic fields (Figure 2.3a) and by electric fields (Figure 2.3b). The directions of these deflections established that cathode rays are not rays of energy, but rather consist of negatively charged particles. By adjusting the strengths of the electric and magnetic fields, Thomson was able to balance out the deflections (Figure 2.3c). From the strengths of the two opposing fields, he calculated the mass-to-charge (*m/e*) ratio of the particles. Thomson and others observed that these particles always

FIGURE 2.2 J. J. Thomson (1856–1940) discovered electrons in 1897 by using a cathode-ray tube, but he was not sure where electrons fit into the structure of atoms.

CHEMTOUR
Cathode-Ray Tube

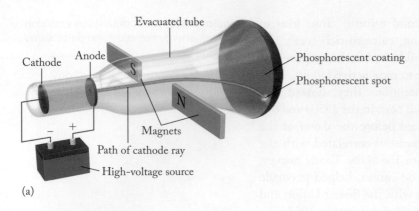

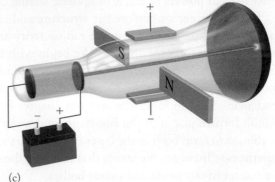

FIGURE 2.3 A cathode "ray"—actually a beam of electrons—carries electricity through this partially evacuated tube. Though invisible, the path of the beam can be inferred by the bright spot it makes in a phosphorescent coating on the end of the tube. (a) The beam is deflected in one direction by a magnetic field, (b) deflected in the opposite direction by an electric field, and (c) not deflected at all if the electric and magnetic fields are tuned to balance out the deflections.

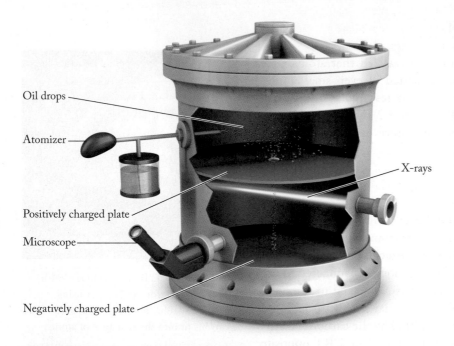

FIGURE 2.4 Millikan's oil-drop experiment.

electron a subatomic particle that has a negative charge and little mass.

beta (β) particle a radioactive emission equivalent to a high-energy electron.

alpha (α) particle a radioactive emission with a charge of 2+ and a mass equivalent to that of a helium nucleus.

behave the same way and always have the same mass-to-charge ratio, no matter what cathode material is used. This observation established that the particles in the cathode rays, which are now known as **electrons**, are fundamental particles present in all forms of matter.

In 1909 the American physicist Robert Millikan (1868–1953) advanced Thomson's work by quantifying the charge on the electron. Thomson had calculated the mass-to-charge ratio of the electron, and if Millikan could measure the charge of the electron, then he could determine its mass. Figure 2.4 illustrates Millikan's experiment: High-energy X-rays remove electrons from molecules found in air in the lower of two connected chambers. As oil drops fall from the upper chamber into the lower one, they collide with the electrons and pick up their charge. Millikan measured the mass of the drops in the absence of an electric field, when their rate of fall was governed by gravity. He then repeated the experiment in the presence of an electric field, adjusting the field strength to make the drops fall at different rates and even suspending some of them in midair. From the strength of the electric field and the rate of fall, he calculated the charge on a drop. By measuring the charges on hundreds of drops, Millikan determined that the charge on each drop was a whole-number multiple of a minimum charge. He concluded that this minimum charge had to be the charge on one electron. Millikan's value was within 1% of the modern value: -1.602×10^{-19} C. (The coulomb, abbreviated C, is the SI unit for electric charge.) Knowing Thomson's value of *m/e* for the electron, Millikan calculated the mass of the electron: 9.109×10^{-28} g.

The discovery of the electron raised the possibility of other subatomic particles. Scientists in the 1890s knew that matter was electrically neutral, but they did not know how the electrons and the positive charges were arranged at the atomic level. Thomson proposed a *plum-pudding model* in which the atom is a diffuse sphere of positive charge with negatively charged electrons embedded in the sphere, like raisins in an English plum pudding (or blueberries in a muffin; Figure 2.5). Thomson's model did not last long. Its demise was linked to another scientific discovery in the 1890s: radioactivity.

CONCEPT TEST

Why did Thomson's discovery of the electron lead to proposals that a positive particle might exist within the atom?

(Answers to Concept Tests are in the back of the book.)

Radioactivity

In 1896 the French physicist Henri Becquerel (1852–1908) discovered that a mineral then called *pitchblende* (now known as *uraninite*, Figure 2.6), the principal natural source of the element uranium, produces radiation that can be detected on photographic plates. Becquerel and his contemporaries initially thought that this radiation consisted of the X-rays that had just been discovered by the German scientist Wilhelm Conrad Röntgen (1845–1923).[1] However, additional experiments by Becquerel, by the French wife-and-husband team of Marie (1867–1934) and Pierre (1859–1906) Curie, and by the British scientist Ernest Rutherford (Figure 2.7) showed that this radiation contained particles as well as rays. The scientists had discovered radioactivity.

In studying the particles emitted by pitchblende, Rutherford found that one type, which he named **beta (β) particles**, penetrated solid materials better than the second type, which he named **alpha (α) particles**. The degree to which β particles were deflected in a magnetic field allowed their mass-to-charge ratio to be calculated, and the results matched the electron mass-to-charge ratio determined by J. J. Thomson. These data established that β particles were simply high-energy electrons.

Rutherford discovered that α particles were deflected by an electric field in the opposite direction from β particles; the same was true for the two particles in a magnetic field. Therefore, he concluded that α particles were positively charged. The β particle (electron) was assigned a relative charge of 1−. The corresponding charge assigned to an α particle was 2+. In addition, α particles have nearly the same mass as an atom of helium, meaning they are over 10^3 times more massive than β particles.

The experiment that disproved Thomson's plum-pudding model was directed by Rutherford and carried out by two of his students at Manchester University: Hans Geiger (1882–1945; for whom the Geiger counter was named) and Ernest Marsden (1889–1970). Geiger and Marsden bombarded a thin foil of gold with a beam of α particles emitted from a radioactive source (Figure 2.8a). They then

CHEMTOUR
Millikan Oil-Drop Experiment

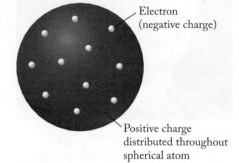

Electron (negative charge)

Positive charge distributed throughout spherical atom

FIGURE 2.5 In J. J. Thomson's plum-pudding model, atoms consist of electrons distributed throughout a massive, positively charged but very diffuse sphere. The plum-pudding model lasted only a few years before it was replaced by a model based on experiments carried out under the direction of Thomson's former student, Ernest Rutherford.

FIGURE 2.6 The mineral now known as uraninite is the chief natural source of uranium.

---

[1]Röntgen discovered X-rays in experiments with a cathode-ray tube much like the apparatus used by J. J. Thomson. After encasing the tube in a black carton, Röntgen discovered that invisible rays escaped the carton and were detected by a photographic plate. Because he knew so little about these rays, he called them X-rays.

FIGURE 2.7 Ernest Rutherford (1871–1937) was born in New Zealand and was awarded a scholarship in 1894 that enabled him to go to Trinity College in Cambridge, England. There he was a research assistant in the laboratory of J. J. Thomson. His contributions included characterizing the properties of α and β particles. By 1907, he was a professor at the University of Manchester, where his famous gold-foil experiments led to our modern view of atomic structure. He was awarded the Nobel Prize in Chemistry in 1908.

 **CHEMTOUR**
Rutherford Experiment

measured how many particles were deflected and to what extent. Rutherford's hypothesis was that if Thomson's model were correct, most of the α particles would pass straight through the diffuse positive spheres of gold atoms (each atom a "pudding" like the one shown in Figure 2.5), but a few of the particles would interact with the electrons (the "raisins") embedded in these spheres and be deflected slightly (Figure 2.8b).

Instead, Geiger and Marsden observed something quite unexpected. Although most of the α particles did indeed pass directly through the gold, about 1 in every 8000 particles was deflected from the foil through an average angle of 90 degrees. Even more surprising, a very few (about 1 out of 100,000) bounced back in the direction from which the α particles came (Figure 2.8c). Rutherford described his amazement at the result: "It is about as incredible as if you had fired a 15-inch shell at a piece of tissue paper and it came back and hit you." (The largest guns on British battleships in those days fired 15-inch-diameter shells.)

The results of the gold-foil experiments ended the short life of the plum-pudding model because the model could not account for the large angles of deflection. Rutherford concluded that those deflections occurred because the α particles occasionally encountered small, yet massive, regions of highly positive charge. Rutherford determined that the region of positive charge was only about 1/10,000 of the overall size of a gold atom. His model of the atom became the basis for our current understanding of atomic structure. It incorporates the assumption that an atom consists of a massive, but tiny, positively charged **nucleus** surrounded by a diffuse cloud of negatively charged electrons.

Protons and Neutrons

In the decade following the gold-foil experiments, Rutherford and others observed that bombarding elements with α particles could change, or *transmute*, these elements into other elements. They also discovered that hydrogen nuclei were sometimes produced during transmutation reactions. By 1920, agreement was growing

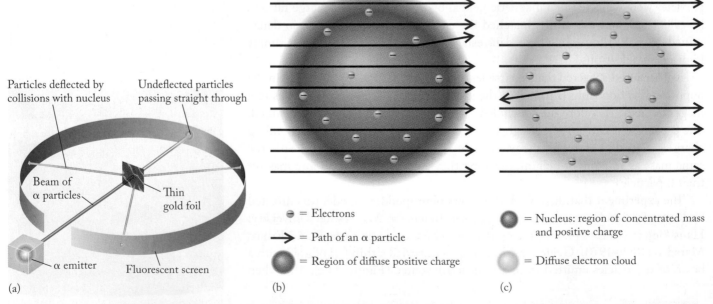

FIGURE 2.8 (a) In the Rutherford–Geiger–Marsden experiment, α particles from a radioactive source were allowed to strike a thin gold foil. A fluorescent screen surrounded the foil to detect any deflected particles. (b) If Thomson's plum-pudding model were correct, most of the α particles would pass through the gold foil and a few would be deflected slightly. (c) In fact, most particles passed straight through, but a few were scattered widely. This unexpected result led to the theory that an atom has a small, positively charged nucleus that contains most of the mass of the atom.

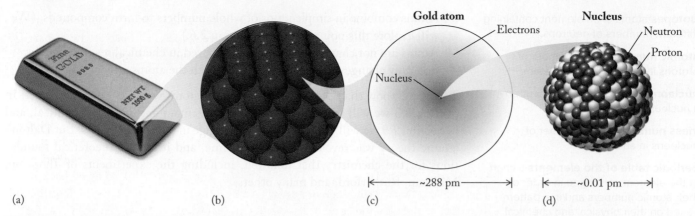

FIGURE 2.9 (a) A gold bar and (b) a particulate view of how the atoms are arranged within it. (c) The modern view of Rutherford's model of the gold atom includes electrons that surround (d) the nucleus, which is about 1/10,000 the overall size of the atom. Note that the scales have units of picometers (1 pm = 10^{-12} m). The nucleus would be too small to see if drawn to scale in (c) and (d). If an atom were the size of the Rose Bowl (an oval stadium about 212 m across in the narrow direction), the nucleus of the atom would be the size of a dime at midfield.

that hydrogen nuclei, which Rutherford called **protons** (from the Greek *protos*, meaning "first"), were part of all nuclei. For example, to account for both the mass and charge of an α particle, Rutherford assumed it was made of four protons, two of which had combined with two electrons to form two electrically neutral particles that he called **neutrons**. Repeated attempts to produce neutrons by neutralizing protons with electrons were unsuccessful. However, in 1932 one of Rutherford's students, James Chadwick (1891–1974), discovered and characterized neutrons. With the discovery of neutrons, the current model of atomic structure was complete, as illustrated for the gold atom in Figure 2.9.

Table 2.1 summarizes the properties of neutrons, protons, and electrons. For convenience, the masses of these tiny particles are expressed in **atomic mass units (amu)**. These units are also called **daltons (Da)**, or *unified atomic mass units* (u). One amu is exactly 1/12 the mass of a carbon atom that has six protons and six neutrons in its nucleus. As you can see from the data in Table 2.1, the masses of the neutron and the proton are nearly the same; hence both are assigned a mass of 1 amu. If you compare the amu values in the table with the masses of the particles in grams, you will see that these particles are tiny indeed: 1 amu is only 1.66054×10^{-24} g.

The unit name "dalton" honors the English chemist John Dalton (1766–1844), who published the first table of atomic masses in 1803. Dalton is also remembered for his *atomic theory*, published in 1808, in which he proposed that:

1. Each element consists of tiny, indestructible particles called atoms.
2. All atoms of an element are identical, and they are different from the atoms of any other element.

TABLE 2.1 Properties of Subatomic Particles

		MASS		CHARGE	
Particle	Symbol	In Atomic Mass Units (amu)	In Grams (g)	Relative Value	Charge (C^a)
Neutron	^1_0n	$1.00867 \approx 1$	1.67493×10^{-24}	0	0
Proton	^1_1p	$1.00728 \approx 1$	1.67262×10^{-24}	1+	$+1.602 \times 10^{-19}$
Electron	$^0_{-1}e$	$5.485799 \times 10^{-4} \approx 0$	9.10939×10^{-28}	1−	-1.602×10^{-19}

^{a}The *coulomb* (C) is the SI unit of electric charge. When a current of 1 ampere (see Table 1.2) passes through a conductor for 1 second, the quantity of electric charge that moves past any point in the conductor is 1 C.

nucleus the positively charged center of an atom that contains nearly all the atom's mass.

proton a positively charged subatomic particle present in the nucleus of an atom.

neutron an electrically neutral (uncharged) subatomic particle found in the nucleus of an atom.

atomic mass unit (amu) unit used to express the relative masses of atoms and subatomic particles; it is exactly 1/12 the mass of one atom of carbon with six protons and six neutrons in its nucleus.

dalton (Da) a unit of mass identical to 1 atomic mass unit.

isotopes atoms of an element containing different numbers of neutrons.

atomic number (Z) the number of protons in the nucleus of an atom.

nucleon either a proton or a neutron in a nucleus.

mass number (A) the number of nucleons in an atom.

periodic table of the elements a chart of the elements arranged in order of their atomic numbers and in a pattern based on their physical and chemical properties.

3. Atoms combine in simple ratios of whole numbers to form compounds. (We will explore this point further in Section 2.6.)
4. Atoms are not changed, created, or destroyed in chemical reactions; a reaction only changes the arrangement in which the atoms are bound.

Today, we know that Dalton's descriptions of atoms were not entirely correct: In Section 2.3 we will see that all atoms of an element are not quite identical, and in Section 2.9 we will see that atoms *can* change in nuclear reactions. But Dalton's atomic theory was revolutionary for its time, and it set the theoretical foundations for the chemistry that followed, including the experiments of Thomson, Millikan, Rutherford, and many others.

2.3 Isotopes

While Thomson investigated the properties of cathode rays in 1897, other scientists designed and built devices to produce beams of *positively* charged particles. One of Thomson's former students, Francis W. Aston (1877–1945), built modified cathode-ray tubes that were evacuated except for small quantities of *fill gases* such as neon. With these tubes he detected conventional beams of cathode rays, but he also detected secondary beams of positively charged particles. Charge was not the only thing different about the particles in these secondary beams. Whereas cathode rays are streams of electrons that all have the same mass and charge regardless of the cathode material or fill gas, the masses of the particles composing Aston's positive rays did depend on the identity of the fill gas. Aston's positively charged particles were not individual protons, but rather atoms of the fill gas that had lost electrons to form positively charged ions.

Aston used his positive-ray analyzer (Figure 2.10) to pass positively charged beams of particles through a magnetic field. Each particle in the beam was deflected along a path determined by the particle's mass: the greater the mass, the smaller the deflection. Using the purest sample of neon gas available, Aston determined that most of the particles had a mass of 20 amu, but about 1 in 10 had a mass of 22 amu.

Since the time of John Dalton, scientists had believed that each element was composed of identical atoms, each having the same mass. Aston's research contradicted this long-held idea. To explain his data, Aston proposed that neon consists of two kinds of atoms, or **isotopes**. Both isotopes of neon had the same number of protons (10) in the nucleus, but one isotope had 10 neutrons in its nucleus, giving it a mass of 20 amu, whereas the other isotope had 12 neutrons in its nucleus, giving it a mass of 22 amu.

Aston's work showed that each element is in fact composed of atoms each having the *same number of protons* in its nucleus, but not necessarily the same number of neutrons, and therefore not the same mass. The number of protons is called the **atomic number (Z)** of the element. The total number of **nucleons** (neutrons and protons) in the nucleus of an atom defines its **mass number (A)**. Isotopes of a given element thus all have the same atomic number, Z, but different mass numbers, A. A neutral atom has the same number of electrons as protons in its nucleus. The modern **periodic table of the elements** (inside the front cover of this book) displays the elements in order of atomic number.

FIGURE 2.10 Aston's positive-ray analyzer. A beam of positively charged ions of neon gas is passed through a focusing slit into a region of electric and magnetic fields. The ions are separated according to mass: those with a mass of 20 amu—90% of the sample—hit the detector at one spot, and those with a mass of 22 amu—the remaining 10%—hit the detector at a different spot. Aston's positive-ray analyzer was the forerunner of the modern mass spectrometer.

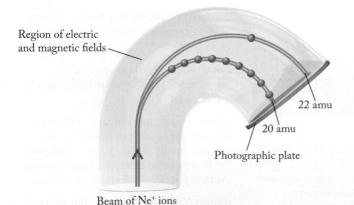

Region of electric and magnetic fields

22 amu

20 amu

Photographic plate

Beam of Ne⁺ ions

An atom with a specific combination of neutrons and protons is called a **nuclide**. The general symbol for identifying a particular nuclide is

$$^A_Z X$$

where X represents the one- or two-letter symbol for the element. For example, the two isotopes of neon identified by Aston have the symbols:

$$^{20}_{10}Ne \qquad ^{22}_{10}Ne$$

Because Z and X provide the same information—each by itself identifies the element—the subscript Z is frequently omitted: often the isotope symbol is simply written as $^A X$ (for example, ^{20}Ne and ^{22}Ne for Aston's isotopes). This same information—mass number and element name—may also be spelled out. For example, the names of the two isotopes of neon that Aston discovered may be written neon-20 and neon-22.

nuclide an atom with particular numbers of neutrons and protons in its nucleus.

CONCEPT TEST

The radioactive atoms measured in the Baby Tooth Survey were strontium-90. Only one of the nuclides below is an isotope of strontium. Which one is it and why?

$$^{87}_{38}Q \qquad ^{90}_{40}X \qquad ^{234}_{90}Z$$

(Answers to Concept Tests are in the back of the book.)

SAMPLE EXERCISE 2.1 Writing Nuclide Symbols   **LO2**

Write symbols in the form $^A_Z X$ for the nuclides that have (a) 6 protons and 6 neutrons, (b) 11 protons and 12 neutrons, and (c) 92 protons and 143 neutrons.

Collect, Organize, and Analyze We know the number of protons and neutrons in the nucleus of each nuclide. We need to write symbols in the $^A_Z X$ form, where Z is the atomic number, A is the mass number, and X is the symbol of the element. The number of protons in the nucleus of an atom defines its atomic number (Z) and defines which element it is (X). The sum of the nucleons (protons plus neutrons) is the mass number (A).

Solve

a. This nuclide has six protons, so $Z = 6$. It must be an isotope of carbon. Six protons plus six neutrons give the isotope a mass number of 12, which makes it carbon-12, $^{12}_6C$.

b. This nuclide has 11 protons, which means $Z = 11$, so it must be an isotope of sodium. Eleven protons and 12 neutrons give the isotope a mass number of 23, so the isotope is sodium-23, $^{23}_{11}Na$.

c. This nuclide has 92 protons, so $Z = 92$, which makes it an isotope of uranium. The mass number is $92 + 143 = 235$. This isotope is uranium-235, $^{235}_{92}U$.

Think About It In working through this exercise, did you use the periodic table of the elements? Once you identify the number of protons in a nucleus (its atomic number), finding a symbol and identifying the element it represents is easy because the elements in the periodic table are arranged in order of increasing atomic number.

Practice Exercise Use the format $^A X$ to write the symbols of the nuclides having (a) 26 protons and 30 neutrons, (b) 7 protons and 8 neutrons, (c) 17 protons and 20 neutrons, and (d) 19 protons and 20 neutrons.

(Answers to Practice Exercises are in the back of the book.)

average atomic mass a weighted average of the masses of all the isotopes of an element, calculated by multiplying the natural abundance of each isotope by its mass in atomic mass units and then summing these products.

natural abundance the proportion of a particular isotope, usually expressed as a percentage, relative to all the isotopes of that element in a natural sample.

SAMPLE EXERCISE 2.2 Determining the Number of Neutrons in a Nuclide   **LO2**

How many neutrons are in each of the following nuclides: (a) ^{14}N; (b) ^{32}P; (c) ^{157}Gd?

Collect, Organize, and Analyze We are given the symbols of three nuclides and asked to determine the number of neutrons in each of their nuclei. We know the value of Z from the element's symbol. Subtracting Z from A gives us the number of neutrons.

Solve
a. ^{14}N is a nuclide of nitrogen, whose atoms each have seven protons. The number of neutrons is $A - Z = 14 - 7 = 7$.
b. ^{32}P is a nuclide of phosphorus ($Z = 15$) with 32 nucleons per nucleus. The number of neutrons is $32 - 15 = 17$.
c. ^{157}Gd is a nuclide of gadolinium ($Z = 64$). The number of neutrons is $157 - 64 = 93$.

Think About It These three nuclides illustrate a trend among stable nuclei: the ratios of neutrons to protons in stable nuclei increase as atomic number increases.

Practice Exercise Determine the number of protons and neutrons in each of these radioactive nuclides: (a) ^{60}Co, used in cancer therapy; (b) ^{131}I, used in thyroid therapy; (c) ^{192}Ir, used to treat coronary disease.

(Answers to Practice Exercises are in the back of the book.)

2.4 Average Atomic Mass

Each element in the periodic table is represented by its symbol. The number above the symbol is the element's atomic number (Z), and the number below the symbol is the element's **average atomic mass**. More precisely, the number below the symbol is the *weighted average* of the masses of all the isotopes of the element.

To understand the meaning of a weighted average, consider the masses and **natural abundances** of the three isotopes of neon in the table shown here. Natural abundances are usually expressed in percentages. Thus, 90.4838% of all neon atoms are neon-20, 9.2465% are neon-22, and only 0.2696% are neon-21. The abundance of neon-21 is so small that Aston could not detect it with his positive-ray analyzer. Modern mass spectrometers, which are the source of natural abundance data such as these, are vastly more sensitive and more precise than Aston's prototype.

To determine the average atomic mass of any element, we multiply the mass of each isotope by its natural abundance (in the language of mathematics, we *weight* the isotope's mass by using natural abundance as the *weighting factor*) and then sum the three weighted masses. To simplify the calculation for neon, we convert the percent abundance values into their decimal equivalents:

Isotope	Mass (amu)	Natural Abundance (%)
Neon-20	19.9924	90.4838
Neon-21	20.9940	0.2696
Neon-22	21.9914	9.2465

$$
\begin{aligned}
\text{Average atomic mass of neon} =\quad & (19.9924 \text{ amu} \times 0.904838) = 18.08988 \\
+ & (20.9940 \text{ amu} \times 0.002696) = 0.05660 \\
+ & \underline{(21.9914 \text{ amu} \times 0.092465) = 2.03344} \\
& \phantom{(21.9914 \text{ amu} \times 0.092465) = } 20.17996 \text{ amu}
\end{aligned}
$$

or, accounting for significant figures, 20.1800 amu. No atom of neon has the average atomic mass; every atom of neon in the universe must have a mass equal to that of one of the three neon isotopes. The value we have calculated is simply the weighted average of these three isotopic masses.

This method of calculating average atomic mass works for every element. The general formula for these calculations is

$$m_X = a_1 m_1 + a_2 m_2 + a_3 m_3 + \ldots \qquad (2.1)$$

where m_X is the average atomic mass of element X, which has isotopes with masses $m_1, m_2, m_3, \ldots$, the natural abundances of which, expressed in decimal form, are $a_1, a_2, a_3, \ldots$.

SAMPLE EXERCISE 2.3 Calculating an Average Atomic Mass   **LO3**

Although strontium-90 does not occur in nature, there are four naturally occurring isotopes of strontium: ^{84}Sr, ^{86}Sr, ^{87}Sr, and ^{88}Sr. Calculate the average atomic mass of strontium ($Z = 38$), given that its stable isotopes have these natural abundances:

Symbol	Mass (amu)	Natural Abundance (%)
^{84}Sr	83.9134	0.56
^{86}Sr	85.9094	9.86
^{87}Sr	86.9089	7.00
^{88}Sr	87.9056	82.58

Collect, Organize, and Analyze We know the masses and natural abundances of each of the four isotopes of strontium, and we can combine these data by using Equation 2.1 to calculate average atomic mass.

Solve

$$
\begin{aligned}
\text{Average atomic mass} = \quad &(83.9134 \text{ amu} \times 0.0056) = 0.470 \text{ amu} \\
+ \ &(85.9094 \text{ amu} \times 0.0986) = 8.471 \text{ amu} \\
+ \ &(86.9089 \text{ amu} \times 0.0700) = 6.083 \text{ amu} \\
+ \ &(87.9056 \text{ amu} \times 0.8258) = 72.592 \text{ amu} \\
\cline{2-2}
& \phantom{(87.9056 \text{ amu} \times 0.8258) = } 87.616 \text{ amu}
\end{aligned}
$$

We have retained one more digit than is significant to avoid rounding errors in the calculation, so adjusting for the correct number of significant figures, we report the answer as 87.62 amu.

Think About It Note that the four values of natural abundances expressed as decimals should add up to 1.0000, and they do. Sometimes this is not the case (check the neon abundances earlier). Uncertainties in the last decimal place may be due to uncertainties in measured or calculated values or in rounding them off. The calculated average atomic mass of strontium is consistent with the value given inside the front cover.

Practice Exercise Silver (Ag) has two stable isotopes: silver-107 (106.905 amu) and silver-109 (108.905 amu). If the average atomic mass of silver is 107.868 amu, what is the natural abundance of each isotope? *Hint*: Let x be the natural abundance of one of the isotopes. Then $1 - x$ is the natural abundance of the other.

(Answers to Practice Exercises are in the back of the book.)

2.5 The Periodic Table of the Elements

Long before chemists knew about electrons, protons, and neutrons, they knew that groups of elements, such as Li, Na, and K, or F, Cl, and Br, had similar chemical (and sometimes physical) properties. When the elements were arranged

period a horizontal row in the periodic table.

group all the elements in the same column of the periodic table; also called *family*.

metals the elements on the left side of the periodic table that are typically shiny solids that conduct heat and electricity well and are malleable and ductile.

nonmetals elements with properties opposite those of metals, including poor conductivity of heat and electricity.

metalloids (also called **semimetals**) elements along the border between metals and nonmetals in the periodic table; they have some metallic and some nonmetallic properties.

main group elements (also called **representative elements**) the elements in groups 1, 2, and 13 through 18 of the periodic table.

transition metals the elements in groups 3 through 12 of the periodic table.

by increasing atomic mass, repeating patterns of similar properties appeared among the elements. This *periodicity* in the chemical properties of the elements inspired several 19th-century scientists to create tables of the elements in which the elements were arranged in patterns based on similarities in their chemical properties.

The most successful of these scientists was the Russian chemist Dmitri Mendeleev (1834–1907). In 1872 he published a table (Figure 2.11) that was the forerunner of the modern periodic table (Figure 2.12). In addition to organizing all the elements that were known at the time, Mendeleev realized that there might be elements in nature that were yet to be discovered, so he left empty cells in his table for those unknown elements. Doing so allowed him to align the known elements so that those in each column had similar chemical properties. On the basis of the locations of the empty cells, Mendeleev predicted the chemical properties of the missing elements that ultimately were discovered. Note that Mendeleev arranged the elements in his periodic table in order of increasing atomic mass. In modern periodic tables the elements appear in order of increasing atomic number.

CONCEPT **TEST**

Why did Mendeleev skip cells in his periodic table?

(Answers to Concept Tests are in the back of the book.)

Navigating the Modern Periodic Table

The modern periodic table (Figure 2.12) contains seven horizontal rows (also called **periods**) and 18 columns (known as **groups** or *families*) of elements. The periods are numbered at the far left of each row, and the group numbers appear at the top of each column. The periodic table inside the front cover shows a second set of column headings containing numbers followed by the letter A or B. These secondary headings were widely used in earlier versions of the table, and many scientists (and students) still find them useful.

The elements in the periodic table are also divided into three broad categories highlighted by the three colors in Figure 2.12. Elements highlighted in tan are **metals**. They tend to conduct heat and electricity well; they tend to be malleable (capable of being shaped by hammering) or ductile (capable of being drawn out in a wire), and they are shiny solids at room temperature, except for mercury (Hg), which is a liquid at room temperature (Figure 2.13a). Elements highlighted in blue are **nonmetals**. They are poor conductors of heat and electricity; the solids among them tend to be brittle, and most are gases at room temperature, except for bromine, which is a liquid with a low boiling point (Figure 2.13b). Lastly, the elements highlighted in green are called **metalloids** or **semimetals**, so named because they tend to have the physical properties of metals but the chemical properties of nonmetals (Figure 2.13c).

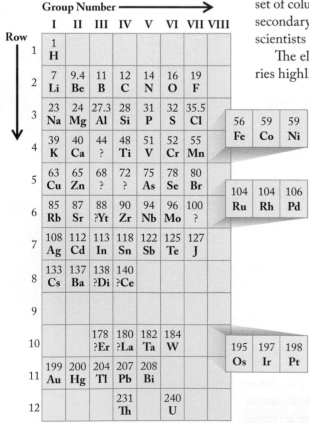

FIGURE 2.11 Mendeleev organized his periodic table on the basis of chemical and physical properties and atomic masses. He assigned three elements with similar properties to group VIII in rows 4, 6, and 10. Because he did this, the elements in the rows that followed lined up in columns with similar properties. In this way, rows 4 and 5 when combined contain spaces for 18 elements, corresponding to the 18 groups in the modern periodic table.

1																	18
1 H	2											13	14	15	16	17	2 He
3 Li	4 Be											5 B	6 C	7 N	8 O	9 F	10 Ne
11 Na	12 Mg	3	4	5	6	7	8	9	10	11	12	13 Al	14 Si	15 P	16 S	17 Cl	18 Ar
19 K	20 Ca	21 Sc	22 Ti	23 V	24 Cr	25 Mn	26 Fe	27 Co	28 Ni	29 Cu	30 Zn	31 Ga	32 Ge	33 As	34 Se	35 Br	36 Kr
37 Rb	38 Sr	39 Y	40 Zr	41 Nb	42 Mo	43 Tc	44 Ru	45 Rh	46 Pd	47 Ag	48 Cd	49 In	50 Sn	51 Sb	52 Te	53 I	54 Xe
55 Cs	56 Ba	57 La	72 Hf	73 Ta	74 W	75 Re	76 Os	77 Ir	78 Pt	79 Au	80 Hg	81 Tl	82 Pb	83 Bi	84 Po	85 At	86 Rn
87 Fr	88 Ra	89 Ac	104 Rf	105 Db	106 Sg	107 Bh	108 Hs	109 Mt	110 Ds	111 Rg	112 Cn	113 Nh	114 Fl	115 Mc	116 Lv	117 Ts	118 Og

4 —— Atomic number
Be —— Symbol for element

Period labels rows on the left; rows 1–7.

6 Lanthanides	58 Ce	59 Pr	60 Nd	61 Pm	62 Sm	63 Eu	64 Gd	65 Tb	66 Dy	67 Ho	68 Er	69 Tm	70 Yb	71 Lu
7 Actinides	90 Th	91 Pa	92 U	93 Np	94 Pu	95 Am	96 Cm	97 Bk	98 Cf	99 Es	100 Fm	101 Md	102 No	103 Lr

FIGURE 2.12 In the modern periodic table, the elements are arranged in order of atomic number (Z) and in a pattern related to their physical and chemical properties. The rows are called periods, and the columns contain groups (or families) of elements. The elements shown in tan are classified as metals; those shown in blue, nonmetals; and those shown in green, metalloids (also called semimetals).

In the modern periodic table, groups 1, 2, and 13–18 are referred to collectively as **main group elements**, or **representative elements** (Figure 2.14a). These groups include the most abundant elements in the solar system and many of the most abundant elements on Earth. Note that these are the "A" elements in the older group labeling system shown on the table inside the front cover. The elements in groups 3 through 12 are called **transition metals**; these are the "B" elements. Nearly all the elements in groups 3 through 12 exhibit the characteristic properties of metals: namely, they are hard, shiny, ductile, malleable, and excellent conductors of heat and electricity.

The first row contains only two elements—hydrogen and helium—and the second and third rows each contain only eight. Starting with the fourth row, all 18 columns are full. Actually, the sixth and seventh rows contain additional elements, which appear in the two separate rows at the bottom of the main table. Elements in the row with atomic numbers from 58 to 71 are called the lanthanides (after element

(a) Metals

(b) Nonmetals

(c) Metalloids

FIGURE 2.13 (a) Metals: a spool of copper, gold that has been hammered into a thin foil, and mercury. (b) Nonmetals: sulfur, chlorine, and bromine. (c) Metalloids: silicon, germanium, and antimony.

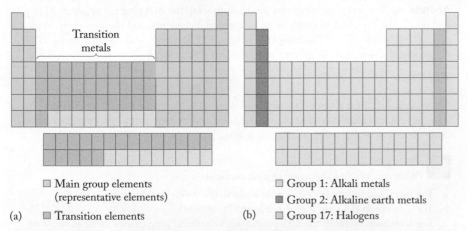

Transition metals

☐ Main group elements (representative elements)
☐ Transition elements

☐ Group 1: Alkali metals
■ Group 2: Alkaline earth metals
☐ Group 17: Halogens

(a)　(b)

FIGURE 2.14 (a) The *main group* (or *representative*) elements are in groups 1, 2, and 13–18. In between are the *transition metals* in groups 3–12. (b) The commonly used names of groups 1, 2, and 17.

halogens the elements in group 17 of the periodic table.

alkali metals the elements in group 1 of the periodic table.

alkaline earth metals the elements in group 2 of the periodic table.

noble gases the elements in group 18 of the periodic table.

law of multiple proportions the principle that, when two masses of one element react with a given mass of another element to form two different compounds, the two masses of the first element have a ratio of two small whole numbers.

57, lanthanum), and those with atomic numbers between 90 and 103 are called actinides (after element 89, actinium). All isotopes of the actinide elements are radioactive, and none of those with atomic numbers above 94 occur in nature. Therefore, they have no *natural* abundance.

Several of the groups have a name in addition to a number. The names are typically based on properties common to all the elements in that group (Figure 2.14b). For example, the elements in group 17 are called **halogens**. The word *halogen* is derived from the Greek for "salt former." Chlorine (Cl), for example, reacts with sodium (a metal in group 1) to form table salt. Other elements in group 1 (which are called **alkali metals**) and in group 2 (which are called **alkaline earth metals**) also react with members of the halogen family to form different salts. These reactivity patterns, and others, were the basis for Mendeleev's arrangement of the elements in his periodic table, and they illustrate what we mean by "similar chemical properties."

Take a moment to compare Figures 2.11 and 2.12. First note the similarity in the arrangements of the lighter (smaller atomic number) elements through calcium ($Z = 20$). All of the elements in groups 1, 2, and 13 through 17 of the modern table appear in the same order in Mendeleev's table, but group 18 (the **noble gases**) is missing from Mendeleev's table. There is a good reason for this: Helium was the first noble gas to be discovered, and that was not until 1895, many years after Mendeleev published his table. Noble gases have very limited chemical reactivity (indeed, helium and neon don't react at all) and so were elusive substances for early chemists to isolate and identify. Because Mendeleev arranged his table largely on the basis of reactivity, he had no reason to predict the existence of the noble gases.

SAMPLE EXERCISE 2.4 Navigating the Periodic Table   **LO4**

Use the periodic table on the inside front cover to determine the symbol and name of each of the following elements:

  a. The third-row element in group 14
  b. The fourth-row alkaline earth metal
  c. The halogen with fewer than 16 protons in its nucleus

Collect and Organize We need to identify each of four elements on the basis of its: (a) row number and column number, (b) row number and the common name of its group, and (c) group name and atomic number.

Analyze (a) The first element in row 3 is sodium, Na. The first element in group 14 is carbon, C. (b) The alkaline earth metals are group 2 elements, and the first element in row 4 is potassium, K. (c) The halogens are group 17 elements, and the only group 17 element with fewer than 16 protons in its nucleus is fluorine, F ($Z = 9$).

Solve (a) Si, silicon; (b) Ca, calcium; (c) F, fluorine.

Think About It Each element has a unique location in the periodic table determined by its atomic number, which defines the row it is in, and by its patterns of reactivity with other elements, which defines the group it is in.

 Practice Exercise
Write the symbol and name of each element:

  a. The metalloid in group 15 closest in mass to the noble gas krypton
  b. The element in the fourth row that is an alkali metal
  c. The transition metal in the fifth period with chemical properties similar to those of zinc ($Z = 30$)
  d. The nonmetal in the fourth period with chemical properties similar to those of sulfur

(Answers to Practice Exercises are in the back of the book.)

2.6 Trends in Compound Formation

Mendeleev used patterns of reactivity to place elements in different groups in his early periodic table. Both Dalton and Mendeleev knew that when elements combine to form compounds, they do so in characteristic ratios. These ratios are reflected in the chemical formulas of compounds. For example, the formula of carbon dioxide, CO_2, tells us that in every molecule of CO_2, one atom of carbon is combined with two atoms of oxygen.

Dalton's atomic view of compounds also explains why some elements (for example, S and O) can form more than one compound (as in SO_2 and SO_3). Dalton determined that the ratio of the different masses of oxygen that react with a given mass of sulfur to form two distinct compounds can be expressed as a ratio of two small whole numbers. This principle was observed experimentally and is known as Dalton's **law of multiple proportions**.

To see what this principle means, consider SO_2 and SO_3. We determine in an experiment that under one set of conditions, 10 g of sulfur reacts with 10 g of oxygen to form SO_2. However, under different conditions, 10 g of sulfur reacts with 15 g of oxygen to form SO_3. The ratio of the two masses of oxygen is 10:15, or 2:3, which is a ratio of two small whole numbers. This example illustrates the law of multiple proportions.

Similarly, we can confirm experimentally that the mass of oxygen that reacts with a given mass of nitrogen to form NO_2 (22.8 g of O for every 10.0 g of N) is twice as much as the mass of oxygen that reacts with the same mass of nitrogen to form NO (11.4 g of O for every 10.0 g of N). The ratio of the two oxygen masses is 22.8:11.4, or 2:1, again a ratio of small whole numbers. The law of multiple proportions and the law of constant composition (Section 1.2) were key ideas that formed the basis for Dalton's atomic theory.

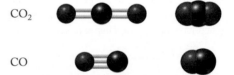

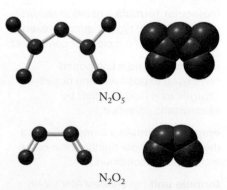

FIGURE 2.15 The "sticks" (bonds) between atoms in the ball-and-stick models indicate that two and three pairs of electrons, respectively, are shared. We discuss single (one bond, one pair of electrons shared), double (two bonds, two pairs), and triple (three bonds, three pairs) bonds in Chapter 8.

SAMPLE EXERCISE 2.5 Relating Chemical Formulas to the Law of Multiple Proportions   **LO5**

Carbon can combine with oxygen to form either CO or CO_2 (Figure 2.15), depending on reaction conditions. If 26.6 g of O_2 reacts with 10.0 g of C to make CO_2, how many grams of O_2 react with 10.0 g of C to make CO?

Collect, Organize, and Analyze We know the mass ratio of oxygen to carbon in CO_2 and need to calculate the mass ratio of oxygen to carbon in CO. The ratio of O atoms to C atoms in CO_2 is 2:1. The ratio of the O atoms to C atoms in CO is 1:1. Therefore, half as much oxygen reacts with 10.0 g of carbon to make CO as reacts with 10.0 g of carbon to make CO_2.

Solve $(26.6 \text{ g of oxygen}) \times \dfrac{1}{2} = 13.3 \text{ g of oxygen}$

Think About It The solution to this problem was based on Dalton's atomic view of these compounds as conveyed by their chemical formulas. In practice, the process runs in reverse: chemists determine the masses of the elements in a compound and use that information to determine its chemical formula.

⊛ **Practice Exercise** Predict the mass of oxygen required to react with 14.0 g of nitrogen to make N_2O_5 if 16.0 g of oxygen reacts with 14.0 g of nitrogen to make N_2O_2 (Figure 2.16).

(Answers to Practice Exercises are in the back of the book.)

N_2O_5

N_2O_2

FIGURE 2.16 Two compounds of nitrogen and oxygen.

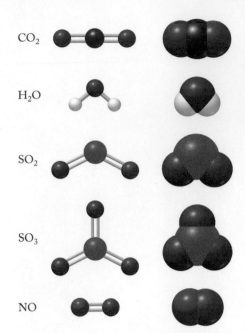

CO_2

H_2O

SO_2

SO_3

NO

FIGURE 2.17 Ball-and-stick and space-filling models of some molecular compounds found in the atmosphere, particularly near sources of automobile and industrial emissions.

 **CONNECTION** In Chapter 1 we saw that ions are particles with either a positive charge (cations) or a negative charge (anions).

CHEMTOUR
NaCl Reaction

molecular compound a compound composed of molecules that contain the atoms of two or more elements.

covalent bond a bond between two atoms created by sharing one or more pairs of electrons.

molecular formula a notation showing the number and type of atoms present in one molecule of a molecular compound.

ionic compound a compound composed of positively and negatively charged ions held together by electrostatic attraction.

empirical formula a formula showing the smallest whole-number ratio of the elements in a compound.

formula unit the smallest electrically neutral unit of an ionic compound.

Molecular Compounds

The compounds we have examined so far in this section have been **molecular compounds** formed from two nonmetals. The molecular structures of several of these compounds are shown in Figure 2.17. Other molecular compounds contain atoms of three, four, or more elements, be they nonmetals or metalloids. The building blocks of these molecules are atoms that have combined through shared pairs of electrons called **covalent bonds**.

All the compounds in Figure 2.17 are present in the air we breathe. They all contain oxygen and another nonmetal, so they are examples of *nonmetal oxides*. Each chemical formula specifies the number of atoms of each element in one molecule of the compound. Therefore, these chemical formulas are **molecular formulas**. The fact that the same two elements can form compounds with different molecular formulas (consistent with Dalton's law of multiple proportions) means that there are different ways to form covalent bonds between atoms of the same two elements.

Ionic Compounds

Now we shift our focus to binary (two-element) **ionic compounds**, which are formed by cations of metals (shown in tan in the periodic table inside the front cover) and anions of nonmetals (shown in blue). The cations and anions in all ionic compounds are held together by the strong electrostatic attraction between ions of opposite charge. Because all the cations in a binary ionic compound come from one element and all the anions come from another element, the ions are referred to as *monatomic* ions.

Let's consider how two of the most common ions in nature, Na^+ and Cl^-, might form from single atoms of sodium (a metal) and chlorine (a nonmetal). Each sodium atom loses an electron and forms a sodium cation:

$$Na \rightarrow Na^+ + e^-$$

Each chlorine atom gains an electron and forms a chloride anion:

$$Cl + e^- \rightarrow Cl^-$$

Figure 2.18(a) illustrates the loss and gain of electrons by these atoms. Notice that when the sodium atom loses an electron and forms the sodium ion, the cation is smaller than the neutral atom. When a chlorine atom gains an electron, the anion is larger than the neutral atom. We explore *why* these changes in size happen in Chapter 7.

Figure 2.18(b) shows several crystals of sodium chloride (table salt) and a particulate view of part of a crystal, revealing the three-dimensional array of the equal numbers of Na^+ ions and Cl^- ions that make up this ionic compound. Within the crystal, each sodium ion is surrounded by six chloride ions, and each chloride ion is surrounded by six sodium ions. However, the smallest whole-number ratio of sodium ions to chloride ions in the crystal is simply 1:1. Formulas based on the lowest whole-number ratio of the elements in a compound are called **empirical formulas**. The chemical formulas of ionic compounds, such as NaCl for sodium chloride, are examples of empirical formulas. The empirical formula of an ionic compound describes a **formula unit**, the smallest electrically neutral unit within the crystal.

Because the periodic table is arranged in part due to observed patterns of reactivity, we can predict the charges on the monatomic ions that elements form and thereby predict the empirical formulas of ionic compounds. For example, atoms of group 1 elements each lose one electron and form 1+ ions; atoms of group 2 elements each lose two electrons and form 2+ ions (Figure 2.19). Note that the charges on these monatomic cations match the group numbers. However, no strong correlation exists between group number and cation charge among the transition metals and the metallic elements on the right side of the periodic table, as you can see in Figure 2.19. Still, some similarities are apparent within groups. For example, the most common charge of the group 13 monatomic ions is 3+.

As metallic elements lose electrons in forming ionic compounds, nonmetals gain them so that the overall charge on the resultant compound is zero. As Figure 2.19 shows, the charge on the monatomic anions formed by the group 17 elements is 1−; the charge on the monatomic anions formed by the group 16 nonmetals is 2−; and the charge is 3− for the nonmetals in group 15.

(a)

CONCEPT **TEST**

Which of the following formulas does not represent an electrically neutral compound? *Hint*: Base your selection on the charges of the common ions in Figure 2.19. (a) KBr; (b) MgF_2; (c) CsN; (d) TiO_2; (e) AgCl

(Answers to Concept Tests are in the back of the book.)

SAMPLE EXERCISE 2.6 Classifying Compounds as Molecular or Ionic   **LO4**

Identify each of the following compounds as ionic or molecular: (a) sodium bromide (NaBr); (b) carbon dioxide (CO_2); (c) lithium iodide (LiI); (d) magnesium fluoride (MgF_2); (e) calcium chloride ($CaCl_2$).

Collect, Organize, and Analyze We need to distinguish between compounds that are composed of ions and those that are composed of molecules. Metallic and nonmetallic elements form ionic compounds when they combine, whereas binary molecular compounds form when two nonmetals or metalloids combine.

Solve NaBr, LiI, MgF_2, and $CaCl_2$ all contain a group 1 or group 2 metal and a group 17 nonmetal; therefore they are ionic compounds. Only CO_2 is composed of two nonmetals, which means it is a molecular compound.

Think About It Labeling compounds as ionic or molecular is not as clear-cut as you might think based on this sample exercise. In later chapters you will encounter covalent bonds that have a degree of ionic "character," and we will explore ways that enable us to determine how much ionic character covalent bonds have.

(b)   One formula unit

FIGURE 2.18 (a) A sodium atom forms a Na^+ cation by losing one electron. A chlorine atom forms a Cl^- anion by gaining one electron. (b) Crystals of sodium chloride. The cubic shape of the crystals mirrors the cubic array of Na^+ and Cl^- ions that make up its structure. The empirical formula NaCl describes the smallest whole-number ratio of cations to anions in the structure, which is electrically neutral.

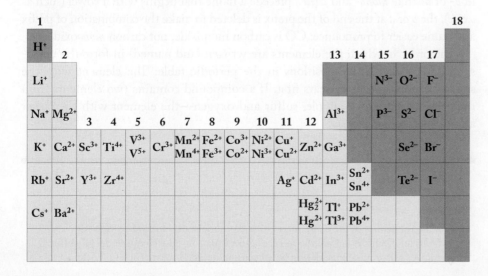

FIGURE 2.19 The most common charges on the ions of some common elements. For main group elements (groups 1, 2, and 13–18), all the ions within a group typically have the same charge. All the ions shown are monatomic except for the mercury ion Hg_2^{2+}, which consists of two mercury atoms covalently bonded to each other.

FIGURE 2.20 Hydrogen peroxide.

Practice Exercise Identify the following compounds as molecular or ionic: (a) carbon disulfide (CS_2); (b) carbon monoxide (CO); (c) ammonia (NH_3); (d) water (H_2O); (e) sodium iodide (NaI).

(Answers to Practice Exercises are in the back of the book.)

CONCEPT **TEST**

Figure 2.20 shows a space-filling model of hydrogen peroxide. What are the molecular formula and the empirical formula of hydrogen peroxide?

(Answers to Concept Tests are in the back of the book.)

2.7 Naming Compounds and Writing Formulas

At this point we need to establish some rules for naming compounds and writing their chemical formulas. These names and formulas are a foundation of the language of chemistry. The periodic table is a valuable resource for naming simple compounds, for translating names into chemical formulas, and for translating formulas into names.

Molecular Compounds

The molecular formula of a molecular compound can be translated into a two-word compound name in three steps:

1. The first word is the name of the first element in the formula.
2. For the second word, change the ending of the name of the second element to *-ide*.
3. Use prefixes (Table 2.2) to indicate the number of atoms of each element in the molecule. (Exception: do not use the prefix *mono-* with the first element in a name.)

For example, NO is nitrogen monoxide (not *mono*nitrogen monoxide), NO_2 is nitrogen dioxide, SO_2 is sulfur dioxide, and SO_3 is sulfur trioxide. When prefixes ending in *o-* or *a-* (like *mono-* and *tetra-*) precede a name that begins with a vowel (such as *oxide*), the *o* or *a* at the end of the prefix is deleted to make the combination of prefix and name easier to pronounce: CO is carbon monoxide, not carbon *mono*oxide.

The order in which the elements are written (and named) in formulas corresponds to their relative positions in the periodic table: The element with the smaller group number appears first. If a compound contains two elements from the same group—for example, sulfur and oxygen—the element with the larger atomic number goes first.

TABLE 2.2 Prefixes for Naming Molecular Compounds

one	*mono-*
two	*di-*
three	*tri-*
four	*tetra-*
five	*penta-*
six	*hexa-*
seven	*hepta-*
eight	*octa-*
nine	*nona-*
ten	*deca-*

SAMPLE EXERCISE 2.7 Relating the Formulas and Names of Molecular Compounds

LO6

What are the names of the compounds with these molecular formulas: (a) N_2O; (b) N_2O_4; (c) N_2O_5? What are the molecular formulas of these compounds: (d) sulfur trioxide; (e) sulfur monoxide; (f) diphosphorus pentoxide?

Collect and Organize We are asked to translate the formulas of three molecular compounds into names and the names of three compounds into molecular formulas. The prefixes in Table 2.2 are used in the names of compounds to indicate the number of atoms of each element in a molecule: *mono-* means 1 atom, *di-* means 2 atoms, *tri-* means 3, *tetra-* means 4, and *penta-* means 5.

Analyze In the first question, the first element in all three compounds is nitrogen, so the first word in each name is *nitrogen* with the appropriate prefix. The second element in all three compounds is oxygen, so the second word in each name is *oxide* with the appropriate prefix. In the second question, all the molecules contain oxygen because their names end in *oxide*.

Solve (a) dinitrogen monoxide; (b) dinitrogen tetroxide; (c) dinitrogen pentoxide; (d) SO_3; (e) SO; (f) P_2O_5.

Think About It Note that the prefixes *mono-*, *tetra-*, and *penta-* lost their final letter when they combined with *oxide* to make the names of the compounds easier to pronounce.

⊛ **Practice Exercise** Name these compounds: (a) P_4O_{10}; (b) CO; (c) NCl_3. Write the formulas for these compounds: (d) sulfur hexafluoride; (e) iodine monochloride; (f) dibromine monoxide.

(Answers to Practice Exercises are in the back of the book.)

Ionic Compounds

Ionic compounds also have two-word names. To name a binary ionic compound:

1. The first word is the name of the cation, which is simply the name of its parent element.
2. The second word is the name of the anion, which is the name of its parent element, except that the ending is changed to *-ide*.

Prefixes are not used in naming binary ionic compounds of representative elements because metals in groups 1 and 2 and aluminum in group 13 all have characteristic positive charges, as do the monatomic anions formed by the elements in groups 16 and 17. Ionic compounds are electrically neutral, so the negative and positive charges in an ionic compound must balance, which dictates the number of each of the ions in the formula. Therefore, the name *magnesium fluoride*, for example, is unambiguous: it can mean only MgF_2.

SAMPLE EXERCISE 2.8 Relating the Formulas and Names   **LO6**
of Ionic Compounds

Write the formulas of (a) potassium bromide, (b) calcium oxide, (c) sodium sulfide, (d) magnesium chloride, and (e) aluminum oxide.

Collect, Organize, and Analyze We need to write the formulas of five binary ionic compounds. Checking the names of the compounds against the positions of the elements in the periodic table and the charges on the monatomic ions they form (Figure 2.19), we see that all five are made up of main group elements that form these ions: K^+, Br^-, Ca^{2+}, O^{2-}, Na^+, S^{2-}, Mg^{2+}, Cl^-, and Al^{3+}.

TABLE 2.3 Names, Formulas, and Charges of Some Common Polyatomic Ions

Name	Chemical Formula
Acetate	CH_3COO^-
Carbonate	CO_3^{2-}
Hydrogen carbonate or bicarbonate	HCO_3^-
Cyanide	CN^-
Hypochlorite	ClO^-
Chlorite	ClO_2^-
Chlorate	ClO_3^-
Perchlorate	ClO_4^-
Dichromate	$Cr_2O_7^{2-}$
Chromate	CrO_4^{2-}
Permanganate	MnO_4^-
Azide	N_3^-
Ammonium	NH_4^+
Nitrite	NO_2^-
Nitrate	NO_3^-
Hydroxide	OH^-
Peroxide	O_2^{2-}
Phosphate	PO_4^{3-}
Hydrogen phosphate	HPO_4^{2-}
Dihydrogen phosphate	$H_2PO_4^-$
Disulfide	S_2^{2-}
Sulfate	SO_4^{2-}
Hydrogen sulfate or bisulfate	HSO_4^-
Sulfite	SO_3^{2-}
Hydrogen sulfite or bisulfite	HSO_3^-
Thiocyanate	SCN^-

Solve We must balance the positive and negative charges on the ions in each compound:

a. The ions in potassium bromide are K^+ and Br^-. A 1:1 ratio of the ions is required for electrical neutrality, making the formula KBr.
b. The ions in calcium oxide are Ca^{2+} and O^{2-}. A 1:1 ratio of ions balances their charges, making the formula CaO.
c. The ions in sodium sulfide are Na^+ and S^{2-}. We need twice as many Na^+ ions as S^{2-} ions to balance their positive and negative charges. This means the formula of sodium sulfide is Na_2S.
d. The ions in magnesium chloride are Mg^{2+} and Cl^-. We need twice as many Cl^- ions as Mg^{2+} ions to balance their positive and negative charges. This means the formula of magnesium chloride is $MgCl_2$.
e. The ions in aluminum oxide are Al^{3+} and O^{2-}. To balance their positive and negative charges we need two Al^{3+} ions for every three O^{2-} ions because $2 \times (3+) + 3 \times (2-) = 0$. This means the formula of aluminum oxide is Al_2O_3.

Think About It The basic principle is that the positive charge on the cations and the negative charge on the anions must balance to sum to a net charge of zero.

 Practice Exercise Write the chemical formulas of (a) strontium chloride, (b) magnesium oxide, (c) sodium fluoride, and (d) calcium bromide.

(Answers to Practice Exercises are in the back of the book.)

Compounds of Transition Metals

Many metallic elements, including most of the transition metals, form several cations carrying different charges. For example, most of the copper found in nature is present as Cu^{2+}; however, some copper compounds contain Cu^+ ions. Therefore, the name *copper chloride* is ambiguous because it does not distinguish between $CuCl_2$ and CuCl. To name these compounds, chemists historically used different names for Cu^+ and Cu^{2+} ions: they were called *cuprous* and *cupric* ions, respectively. Similarly, Fe^{2+} and Fe^{3+} ions were called *ferrous* and *ferric* ions. Note that, in both pairs of ions, the name with the lower charge ends in *-ous* and the name with the higher charge ends in *-ic*.

More recently, chemists have adopted a more systematic way of distinguishing between differently charged ions of the same transition metal. The new system, called the Stock system after the German chemist Alfred Stock (1876–1946), uses a Roman numeral to indicate the charge on the transition metal ion in a compound. Thus the modern name of $CuCl_2$ is copper(II) chloride (pronounced "copper-two chloride"), and CuCl is copper(I) chloride. Roman numerals are used to indicate the charge of transition metal cations unless the metal forms just one cation, as with Ag^+, Cd^{2+}, and Zn^{2+}.

SAMPLE EXERCISE 2.9   Relating the Formulas and Names   **LO6**
of Transition Metal Compounds

What are the chemical formulas of (a) iron(II) sulfide and (b) chromium(III) oxide? What are the names of (c) V_2O_5 and (d) $NiCl_2$?

Collect, Organize, and Analyze We are asked to write the formulas of two transition metal compounds from their names and to write the names of two others from their formulas. In the names of all four compounds, Roman numerals indicate the charges on the transition metal cations. The charges of the most common monatomic anions are shown in Figure 2.19. They include S^{2-}, O^{2-}, and Cl^-. As with all ionic compounds, the sum of the positive and negative charges of their ions must be zero.

Solve

a. The Roman numeral II in the compound's name means that the iron ions are Fe^{2+} ions. All sulfide ions are 2−. A 1:1 ratio of Fe^{2+} and S^{2-} ions balances their charges, making the formula of iron(II) sulfide FeS.

b. The Roman numeral III means that the chromium ions in the compound are Cr^{3+} ions. All oxide ions are 2−. To balance their positive and negative charges, we need two Cr^{3+} ions for every three O^{2-} ions because $2 \times (3+) + 3 \times (2-) = 0$. Therefore the formula of chromium(III) oxide is Cr_2O_3.

c. The compound with the formula V_2O_5 has five O^{2-} ions for every two vanadium ions. This means the charge on the two vanadium ions must sum to $-(5 \times (2-)) = 10+$, which means that each vanadium ion is 5+. Expressing this 5+ charge with the Roman numeral V gives us the compound name vanadium(V) oxide.

d. There are two Cl^- ions for every one nickel ion in $NiCl_2$. This means the charge on the nickel ion must be $-(2 \times (1-)) = 2+$. Expressing this 2+ charge with the Roman numeral II gives us the compound name nickel(II) chloride.

Think About It We use the Stock system for designating the charges on the ions of most transition metals and some other metallic elements, such as lead and tin, but you may encounter *-ous/-ic* nomenclature in older books and articles and in chemical catalogues.

 Practice Exercise Write the formulas of manganese(II) chloride and manganese(IV) oxide.

(Answers to Practice Exercises are in the back of the book.)

polyatomic ion a charged group of two or more atoms joined by covalent bonds.

oxoanion a polyatomic ion that contains oxygen in combination with one or more other elements.

Polyatomic Ions

Table 2.3 lists some common **polyatomic ions**, which means the ions consist of two or more atoms joined by covalent bonds. The ammonium ion (NH_4^+) is the only common polyatomic cation; all the others are anions. When writing the formula of a compound with two or more of the same polyatomic ion per formula unit, we put parentheses around the formula of the polyatomic ion to make it clear that the subscript that follows applies to the entire ion. For example, there are three sulfate ions in $Al_2(SO_4)_3$.

Polyatomic ions containing oxygen and one or more other elements are called **oxoanions**. Most oxoanions have a name based on the name of the element that appears first in the formula, with its ending changed to either *-ite* or *-ate*. The *-ate* oxoanion of an element has a greater number of oxygen atoms than its *-ite* counterpart; for example, SO_4^{2-} is the sulfate ion and SO_3^{2-} is the sulfite ion.

If an element forms more than two oxoanions, as chlorine, bromine, and iodine do, prefixes are used to distinguish among them (Table 2.4). The oxoanion with the largest number of oxygen atoms may have the prefix *per-*, and the one with the smallest number of oxygen atoms may have the prefix *hypo-* in its name. Because these rules may not enable you to predict the chemical formula of an oxoanion from its name or its name from its formula, you need to memorize the formulas, charges, and names of the polyatomic ions in Tables 2.3 and 2.4 that your instructor thinks are most important.

TABLE 2.4 Oxoanions of Bromine and Their Corresponding Acids

Ions	
BrO^-	Hypobromite
BrO_2^-	Bromite
BrO_3^-	Bromate
BrO_4^-	Perbromate
Acids	
HBrO	Hypobromous acid
$HBrO_2$	Bromous acid
$HBrO_3$	Bromic acid
$HBrO_4$	Perbromic acid

SAMPLE EXERCISE 2.10 Relating the Formulas and Names **LO6**
of Compounds Containing Oxoanions

What are the formulas of (a) sodium sulfite and (b) magnesium phosphate?
What are the names of (c) $CaCO_3$ and (d) $KClO_4$?

Collect, Organize, and Analyze We are asked to write the formulas of two compounds
whose names tell us they contain oxoanions, and to write the names of two other
compounds from formulas that contain oxoanions. Table 2.3 contains these pairs of names
and formulas: sulfite is SO_3^{2-}, phosphate is PO_4^{3-}, carbonate is CO_3^{2-}, and perchlorate is
ClO_4^-. The charges of the cations in all four compounds are shown in Figure 2.19: they
are Na^+, Mg^{2+}, Ca^{2+}, and K^+. As with all ionic compounds, the sum of the positive and
negative charges of the ions in these compounds must be zero.

Solve

a. To balance charges, we need twice as many Na^+ ions as SO_3^{2-} ions. Therefore the
   formula of sodium sulfite is Na_2SO_3.
b. To balance charges, we need three Mg^{2+} ions for every two PO_4^{3-} ions. Therefore
   the formula of magnesium phosphate is $Mg_3(PO_4)_2$.
c. Combining the names of the Ca^{2+} ions and CO_3^{2-} ions, we have calcium carbonate.
d. Combining the names of the K^+ ions and ClO_4^- ions, we have potassium perchlorate.

Think About It To complete this exercise we had to know the formulas and charges of
several monatomic ions and oxoanions. The charges on the most common monatomic
ions can be inferred from the positions of their parent elements in the periodic table,
but knowing the names and formulas of common polyatomic ions requires at least some
memorization. That said, there are patterns that can reduce how much memorization
you have to do. For example, if you learn the formulas and charges of the nitrate and
sulfate ions, then you can remember that removing one oxygen atom from each yields
the formulas and charges of the nitrite and sulfite ions.

 Practice Exercise What are the formulas of (a) strontium nitrate and
(b) potassium sulfate, and the names of (c) NaClO and (d) $KMnO_4$?

(Answers to Practice Exercises are in the back of the book.)

Acids

Some compounds have special names that highlight particular chemical properties. Among these are acids. We discuss acids in greater detail in later chapters,
but for now it is sufficient to say that acids are compounds that release hydrogen
ions (H^+) when they dissolve in water. For example, when the molecular compound hydrogen chloride (HCl) dissolves in water, it produces the solution we call
hydrochloric acid. In this aqueous solution, every molecule of HCl has separated
into a H^+ ion and a Cl^- ion. To name the aqueous solutions of acids such as HCl:

1. Affix the prefix *hydro-* to the name of the second element in the formula.
2. Replace the last syllable in the second element's name with the suffix *-ic*, and
   add *acid*.

Common acids include compounds of hydrogen and the halogens. Their aqueous
solutions are hydrofluoric, hydrochloric, hydrobromic, and hydroiodic acid.

The scheme for naming the acids of oxoanions, which are called *oxoacids*, is
illustrated in Table 2.4. If the oxoanion name ends in *-ate*, the name of the corresponding oxoacid ends in *-ic*; if the oxoanion name ends in *-ite*, the name of the
oxoacid ends in *-ous*. Thus, the acid that forms perchlor*ate* (ClO_4^-) ions in solution is perchlor*ic* acid ($HClO_4$) and the acid that forms nitr*ite* (NO_2^-) ions in
solution is nitr*ous* acid (HNO_2).

SAMPLE EXERCISE 2.11 Relating the Formulas and Names **LO6**
of Oxoacids and Oxoanions

What are the names of the oxoacids formed by the following oxoanions: (a) SO_3^{2-}; (b) ClO_4^-; (c) NO_3^-?

Collect, Organize, and Analyze We are given the formulas of three oxoanions and asked to name the oxoacids formed when they combine with H^+ ions. According to Table 2.3, the names of the oxoanions are (a) sulfite, (b) perchlorate, and (c) nitrate. When the oxoanion name ends in *-ite*, the corresponding oxoacid name ends in *-ous*. When the anion name ends in *-ate*, the oxoacid name ends in *-ic*.

Solve Making the appropriate changes to the endings of the oxoanion names and adding the word *acid*, we get (a) sulfurous acid, (b) perchloric acid, and (c) nitric acid.

Think About It Note that you cannot tell the names of the oxoanions just by looking at them. You have to remember the names associated with each family of polyatomic ions.

 Practice Exercise
Name these acids: (a) HBrO; (b) $HBrO_3$; (c) H_2CO_3.

(Answers to Practice Exercises are in the back of the book.)

2.8 Organic Compounds: A First Look

All of the compounds we have discussed so far in this chapter—those composed of molecules and those made of ions—have been *inorganic* compounds. These are the compounds that make up Earth's geosphere, including its crust (part of the lithosphere), the air we breathe (the atmosphere), and the water that covers most of Earth's surface (the hydrosphere). Still, despite their abundance in the world *around* us, these compounds constitute much less of the matter *inside* us.

We are living organisms, which means our bodies are mostly water and *organic* compounds. **Organic compounds** are composed of molecules that always contain carbon atoms, almost always contain hydrogen atoms, and frequently contain **heteroatoms**—typically oxygen, nitrogen, sulfur, phosphorus, and the halogens. The study of these compounds is called **organic chemistry**. Note that while *organic* has connotations in everyday conversation that often refer to farming and foods grown without the use of pesticides or synthetic fertilizers, the word *organic* as used by chemists refers to compounds with a particular composition and structure.

For centuries, philosophers and scientists thought that organic compounds could be synthesized only by biological processes. However, in the 1820s a young German chemist named Friedrich Wöhler (1800–1882) showed that they could also be synthesized in the laboratory from inorganic starting materials. Today, we tend to distinguish between organic compounds that are *natural* (produced by living organisms) and *synthetic* (not found in nature). Altogether, there are tens of millions of them that we know of, and more are being synthesized every day.

Hydrocarbons

Organic compounds that contain no heteroatoms are called **hydrocarbons** because their molecules contain only hydrogen and carbon atoms. One class of hydrocarbons, called **alkanes**, is composed of molecules in which each carbon

organic compound a molecule containing carbon atoms whose structure typically consists of carbon–carbon bonds and carbon–hydrogen bonds, and may include one or more heteroatoms such as oxygen, nitrogen, sulfur, phosphorus, or the halogens.

heteroatom atom of an element other than carbon and hydrogen within a molecule of an organic compound.

organic chemistry the study of organic compounds.

hydrocarbon an organic compound whose molecules are composed only of carbon and hydrogen atoms.

alkane a hydrocarbon in which all the bonds are single bonds.

alkene a hydrocarbon containing one or more carbon–carbon double bonds.

alkyne a hydrocarbon containing one or more carbon–carbon triple bonds.

alcohol an organic compound containing the OH functional group.

functional group a group of atoms in the molecular structure of an organic compound that imparts characteristic chemical and physical properties.

atom forms four single bonds to four other atoms. The names and structural formulas of the two smallest alkanes are

Methane   Ethane

Note that the name of each of these alkanes ends in *-ane*, as do the names of all alkanes. These two alkanes are important components of natural gas. Alkanes made of larger molecules make up other important fuels, including gasoline, diesel fuel, jet fuel, and heating oil. Because of their importance as fuels, we explore their molecular structures and chemical properties in detail in Chapter 5, which focuses on combustion and the energetics of chemical reactions.

Other hydrocarbons contain carbon atoms bonded to fewer than four other atoms. These compounds include **alkenes**, which contain C=C double bonds:

Ethylene (ethene)   Propylene (propene)

Like all alkenes, the names of these compounds end in *-ene*. The names in parentheses, ethene and propene, follow the naming rules established by the International Union of Pure and Applied Chemistry (IUPAC). (We followed IUPAC rules in naming the inorganic compounds in Section 2.7.) Note that the beginning letters of ethene and propene correspond to the names of the alkanes with the same number of carbon atoms (ethane and propane). This pattern is repeated for all alkenes. However, many organic compounds are better known by their common names, like ethylene and propylene, which predate the current IUPAC naming rules. We will use these common names throughout this book, providing the IUPAC names in parentheses.

Some hydrocarbons contain C≡C triple bonds. They are called **alkynes**:

Acetylene (ethyne)   Propyne

As with alkenes, the names of alkynes are linked to the names of the alkanes with the same number of carbon atoms. All of their IUPAC names end in *-yne*.

Heteroatoms and Functional Groups

Incorporation of heteroatoms into the molecular structure of hydrocarbons produces organic compounds with physical and chemical properties that are very different from those of their parent hydrocarbons. For example, replacing a hydrogen atom with an OH group in the molecular structures of methane and ethane produces molecules with these simplified structural formulas (the O–H bonds are not shown):

Methanol   Ethanol

Note that the names of these compounds use the parent alkane root, but they each end in *-ol*. The presence of the OH groups, highlighted in red, means that these compounds are **alcohols**. Unlike their parent alkanes, which are gases at room temperature and pressure, these alcohols are colorless liquids. Alkanes don't dissolve in water (at least, not much), but these alcohols dissolve freely in water (and water dissolves freely in them).

The OH group in alcohols is an example of a **functional group**, that is, a group of atoms in the molecular structure of a compound that has a significant impact on the physical and chemical properties of the compound. The common functional groups, whose chemical reactivity we explore in the chapters ahead, are listed in Table 2.5. Note that all the functional groups contain an atom of

TABLE 2.5 Classes of Organic Compounds and Their Functional Groups

Class of Compounds	Structural Formula of Functional Group	Sample Compound	Name of Compound	Common Use of Compound
Alcohols	$R-OH$	CH_3CH_2-OH	Ethanol	Alcoholic beverages
Ethers	$R-O-R'$	$H_3C-O-CH_3$	Dimethyl ether (methoxymethane)	Organic synthesis
Aldehydes	$R-\overset{\overset{\displaystyle O}{\|\|}}{C}-H$	$H_3C-\overset{\overset{\displaystyle O}{\|\|}}{C}-H$	Ethanal (acetaldehyde)	Making plastic resin
Ketones	$R-\overset{\overset{\displaystyle O}{\|\|}}{C}-R'$	$H_3C-\overset{\overset{\displaystyle O}{\|\|}}{C}-CH_3$	Acetone (propanone)	Nail polish remover
Carboxylic acids	$R-\overset{\overset{\displaystyle O}{\|\|}}{C}-OH$	$H_3C-\overset{\overset{\displaystyle O}{\|\|}}{C}-OH$	Acetic acid (ethanoic acid)	Vinegar
Esters	$R-\overset{\overset{\displaystyle O}{\|\|}}{C}-O-R'$	$H_3C-\overset{\overset{\displaystyle O}{\|\|}}{C}-O-CH_3$	Methyl acetate (methyl ethanoate)	Nail polish remover, glue
Amines	$R-NH_2$	H_3C-NH_2	Methylamine (aminomethane)	Synthesis of pharmaceuticals and industrial chemicals
Amides	$R-\overset{\overset{\displaystyle O}{\|\|}}{C}-NH_2$	$H_3C-\overset{\overset{\displaystyle O}{\|\|}}{C}-NH_2$	Acetamide (ethanamide)	Solvent, plastics additive

oxygen or nitrogen or both. The symbols R and R′ are used to represent portions of the molecule that contain only carbon and hydrogen atoms.

We will not dwell on all the rules that apply to naming compounds such as those in Table 2.5. There are many of them. (Appendix 7 contains a comprehensive summary.) For now, try to relate the names and structures of the sample compounds by using the names of the two alkanes described above and the class of compounds to which each belongs. For example, the name of the sample aldehyde in Table 2.5, *ethanal*, is derived from *ethane*, the alkane that also has two carbon atoms per molecule, and the functional group *al*dehyde.

CONCEPT **TEST**

What functional group is present in each of the three organic compounds shown in Figure 2.21? Note that these structures represent atoms by using color, as introduced in Section 1.2: black is carbon, white is hydrogen, red is oxygen, and blue is nitrogen.

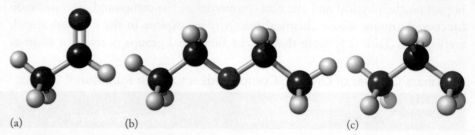

FIGURE 2.21 Organic compounds. (a) (b) (c)

(Answers to Concept Tests are in the back of the book.)

2.9 Nucleosynthesis: The Origin of the Elements

We began this chapter by discussing remarkable advances in the early 20th century that provided science with a clearer view of atomic structure and of why elements react the way they do. Later in the century, scientists increased our understanding of how the elements may have originally formed by determining the composition of the very early universe—the particles that appeared in the moments immediately following the Big Bang.

Scientists believe that much of the energy released at the instant of the Big Bang transformed into matter within a few microseconds (Figure 2.22). This matter consisted of the smallest of subatomic particles: electrons and **quarks**. Less than a millisecond later, the universe had expanded and "cooled" to a mere 10^{12} K, and quarks combined with one another to form neutrons and protons. Thus in less than a second, the matter in the universe consisted of the three types of subatomic particles that would eventually make up all atoms.

Primordial Nucleosynthesis

By about four minutes after the Big Bang, the universe had expanded and cooled to 10^9 K. In this hot, dense subatomic "soup," neutrons and protons that collided with one another began to fuse in a process called primordial **nucleosynthesis**. (We discuss fusion in more detail in Chapter 19.) In one step, protons (p) and neutrons (n) fused to form *deuterons* (d), which are nuclei of the deuterium (^2_1H) isotope of hydrogen:

$$^1_1p + ^1_0n \rightarrow ^2_1d \tag{2.2}$$

quarks elementary particles that combine to form neutrons and protons.

nucleosynthesis the natural formation of nuclei as a result of fusion and other nuclear processes.

In writing this equation, we follow the rules we described in Section 2.3 for writing nuclide symbols: the superscript is a mass number and the subscript is an atomic number. We use a similar convention for writing the symbols of subatomic particles, except that in this case a subscript represents the charge on the particle. For example, the symbol of the neutron is $_0^1n$ because a neutron has a mass number of 1 and a charge of 0.

Equation 2.2 is an example of a *nuclear equation*. It is related to the chemical equations we begin working with in Chapter 3 in that it is *balanced*. This means that the sum of the masses (superscripts) of the particles to the left of the arrow is equal to the sum of the masses of the particles to the right of the arrow. Similarly, the sum of the charges (subscripts) of the particles on the left side is equal to the sum of the charges of the particles on the right side.

Deuteron formation proceeded rapidly, consuming most of the neutrons in the universe in a matter of seconds. No sooner did deuterons form than they, too, were rapidly consumed by additional collisions and fusion reactions, which very quickly produced a universe that consisted nearly entirely of the nuclei of hydrogen (mostly $_1^1H$) and helium (mostly $_2^4He$). However, production of the other elements through continued fusion reactions did not happen. For example, collisions between $_1^1H$ and $_2^4He$ nuclei did not produce $_3^5Li$, and collisions between pairs of $_2^4He$ nuclei did not produce $_4^8Be$. Why? Because these isotopes

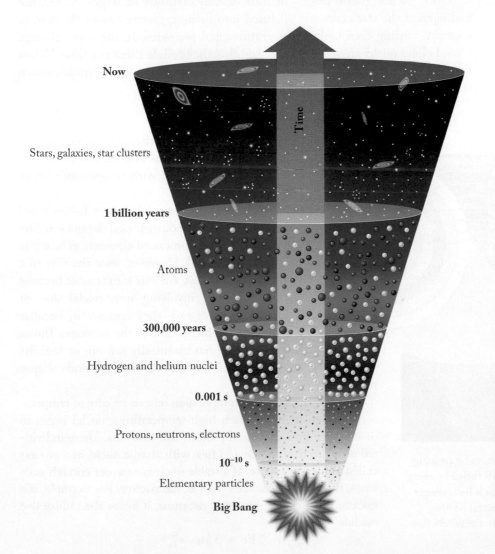

FIGURE 2.22 Timeline for energy and matter transformations believed to have occurred since the universe began. In this model, protons and neutrons were formed from quarks in the first millisecond after the Big Bang, followed by hydrogen and helium nuclei. Whole atoms of H and He did not form until after 300,000 years of expansion and cooling, and other elements did not form until the first galaxies appeared, around 1 billion years after the Big Bang. According to this model, our solar system, our planet, and all life-forms on it are composed of elements synthesized in stars that were born, burned brightly, and then disappeared millions to billions of years after the Big Bang.

neutron capture the absorption of a neutron by a nucleus.

beta (β) decay a spontaneous process by which a neutron in a radioactive nuclide is transformed into a proton and emits a high-energy electron (β particle).

CHEMTOUR
Synthesis of Elements

of lithium and beryllium are not stable. In fact, there are no nuclides with five or eight nucleons.

Therefore within minutes of the Big Bang, primordial nucleosynthesis shut down, leaving an expanding and cooling universe that consisted nearly entirely of hydrogen and helium. The fact that the universe today is still 99% hydrogen and helium, and that these elements occur in the same proportion predicted by physicists' models of primordial nucleosynthesis (about 3 parts hydrogen to 1 part helium, by mass), is strong evidence in support of the theory of the Big Bang.

Stellar Nucleosynthesis

The chemistry that occurred after the Big Bang explains the origin of hydrogen and helium. But how did the other elements in the periodic table form, including those that make up most of our planet? Scientists theorize that the synthesis of elements more massive than helium had to wait until nuclear fusion resumed in the first generation of stars. Inside the coalescing masses of hydrogen and helium that would turn into the first stars, the gases underwent enormous compressional heating. The nuclear furnaces that are the source of the energy in all stars were ignited as hydrogen nuclei began to fuse, making more helium nuclei.

Once nuclear fusion begins in stars, it may continue in stages. When the hydrogen at the star's core has all fused into helium, gravity causes the core to contract, causing even higher temperatures and pressures. In the cores of large stars, helium nuclei are so densely packed that they collide three at a time. When they do, they fuse to form a nucleus that contains 6 protons and 6 neutrons, which is the most common isotope of carbon:

$$3\,{}_{2}^{4}\text{He} \rightarrow {}_{6}^{12}\text{C} \tag{2.3}$$

With the formation of ${}^{12}\text{C}$, the barrier that had halted primordial nucleosynthesis is overcome, and the stage is set for additional fusion reactions with progressively more massive nuclei.

For billions of years, chains of fusion reactions have simultaneously fueled the nuclear furnaces of stars and produced elements as heavy as iron (Figure 2.23). However, once the core of a star turns into iron, the star is in trouble because fusion reactions involving iron nuclei do not release energy. Instead, they *consume* it, because ${}^{56}\text{Fe}$ is the most stable nuclide in the universe. Thus a star with an iron core has essentially run out of fuel. Its nuclear furnace goes out, and the star begins to cool and collapse into itself.

As the star collapses, compression reheats its core to temperatures above 10^9 K. At such high temperatures, nuclei begin to disintegrate into individual protons and neutrons. These individual neutrons may collide and fuse with atomic nuclei in a process called **neutron capture**. If a stable nucleus captures enough neutrons, it becomes unstable—that is, radioactive. For example, if a nucleus of ${}^{56}\text{Fe}$ captures three neutrons, it forms the radioactive nuclide ${}^{59}\text{Fe}$:

$$_{26}^{56}\text{Fe} + 3\,{}_{0}^{1}\text{n} \rightarrow {}_{26}^{59}\text{Fe}$$

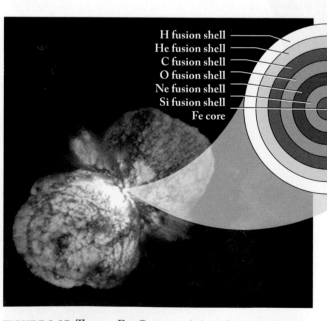

FIGURE 2.23 The star Eta Carinae is believed to be evolving toward an explosion. The outer regions are still fueled by energy released as hydrogen isotopes fuse, but the star is increasingly hotter and denser closer to the center. This central heating allows the fusion of larger nuclei and results in the production of ${}^{56}\text{Fe}$ in the core.

The neutron *richness* of ^{59}Fe means that this radioactive nuclide spontaneously undergoes a kind of radioactive decay that reduces the ratio of neutrons to protons in its nucleus. This kind of decay is called **beta (β) decay** because it involves the ejection of a high-energy electron, or β particle. The decay process is summarized in the following nuclear equation:

$$^{59}_{26}\text{Fe} \rightarrow {}^{59}_{27}\text{Co} + {}^{0}_{-1}\beta \qquad (2.4)$$

Note that the nuclide formed in the reaction has an atomic number of 27, or one more than the radioactive isotope that produced it. This increase is the result of the decomposition of a neutron into a proton, which remains in the nucleus, and an electron (β particle) that is ejected from it. The additional proton results in an increase in atomic number. The combination of repeated neutron capture and β decay events in the cores of collapsing stars produces the most massive stable nuclides in the periodic table.

Eventually, the enormous heating that occurs when a massive star collapses produces a gigantic explosion. Cosmologists call such an event a *supernova*. In addition to finishing the job of synthesizing the elemental building blocks found in the universe—up to and including the isotopes of uranium—a supernova serves as its own element-distribution system, blasting its inventory of elements throughout its galaxy (Figure 2.24). The legacies of supernovas are found in the elemental composition of later-generation stars like our Sun and in the planets that orbit these stars. Indeed, all the matter in our solar system—and in us—is essentially demolition debris from ancient exploding stars.

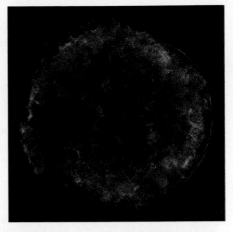

FIGURE 2.24 This colorized picture from the Chandra X-ray Observatory shows the remains of SN 1572, called "Tycho's supernova," in the constellation Cassiopeia. The expanding bubble of debris colored red and green is a cloud of hot ionized gas inside a more rapidly moving shell of extremely high-energy electrons in blue.

SAMPLE EXERCISE 2.12 Integrating Concepts: Radioactive Isotopes in Medicine

Dmitri Mendeleev predicted the existence of one of the elements that was unknown in his time—technetium ($Z = 43$). Technetium has more than 40 isotopes, with mass numbers ranging from 85 to 118. Every isotope of technetium is radioactive. Medicine has developed procedures that take advantage of this radioactivity to use ^{99}Tc to image many parts of the human body, including the brain, heart, thyroid, lungs, and kidneys (Figure 2.25). More than 20 million medical procedures using ^{99}Tc are conducted each year. One challenge, though, is how to make ^{99}Tc available to doctors precisely where and when it is needed for patients because it spontaneously undergoes radioactive decay. The solution to this problem is to transport a more stable substance, in this case, molybdenum-99, from which technetium-99 can be isolated. Most of the molybdenum-99 ($Z = 42$) on Earth is produced from ^{235}U in nuclear reactors.

a. What are the nuclide symbols for the lightest and the heaviest isotopes of technetium?
b. How many protons, neutrons, and electrons are in a neutral atom of ^{99}Tc?
c. When doctors need to administer ^{99}Tc to a patient, they obtain a sample of molybdenum-99 that then undergoes beta decay to produce ^{99}Tc. Write a balanced nuclear equation to show the beta decay of molybdenum-99 to form technetium-99.

Collect and Organize We are given the mass numbers of two technetium isotopes and are asked to write their complete nuclide symbols, and we need to translate the nuclide symbol ^{99}Tc into

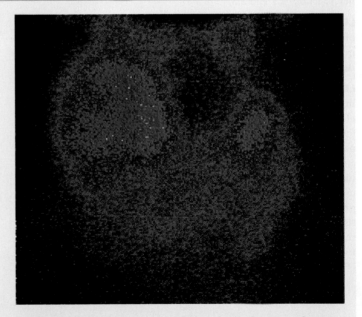

FIGURE 2.25 Technetium is used to study the function of organs such as the thyroid gland. Typically an image is taken soon after injection of a radioactive Tc compound. After a time, the thyroid is re-imaged to see how much Tc remains. The distribution of Tc should be symmetrical. Here the lobes are not the same, indicating abnormal thyroid function.

the numbers of protons, neutrons, and electrons per atom. We are also given the name and atomic number of the more stable molybdenum that decays to produce ^{99}Tc.

Analyze The atomic number (Z) of an element is the same as the number of protons and the number of electrons in each of its atoms. The mass number (A) of a nuclide is the sum of the numbers of protons and neutrons in each of its nuclei. In the symbol of a nuclide, Z is the subscript and A is the superscript that precede the element symbol. Beta decay involves the ejection of a high-energy electron, $_{-1}^{0}\beta$.

Solve

a. For the lightest isotope of Tc, $Z = 43$ and $A = 85$; for the heaviest isotope, $Z = 3$ and $A = 118$. Therefore the two symbols are $_{43}^{85}$Tc and $_{43}^{118}$Tc, respectively.

b. For ^{99}Tc: 43 protons, 56 neutrons, 43 electrons

c. $_{42}^{99}\text{Mo} \rightarrow {}_{43}^{99}\text{Tc} + {}_{-1}^{0}\beta$

Think About It Because one of its isotopes emits radiation that can be easily detected, technetium incorporated into soluble molecules can be used to observe the flow of fluids in organisms. The thyroid is a symmetrical, butterfly-shaped gland that surrounds the trachea. It appears in Figure 2.25 as the two lobes imaged after the administration of a technetium-containing agent. The gland should be symmetrical, but one lobe is clearly smaller than the other. The irregular uptake of the agent indicates a malfunctioning thyroid.

SUMMARY

LO1 Thomson's experiments with cathode rays, Millikan's oil-drop experiment, and Rutherford's gold-foil experiment, in addition to studies of **radioactivity** by Henri Becquerel and Marie and Pierre Curie, led to our understanding of atomic structure. (Sections 2.1 and 2.2)

LO2 Atoms consist of a **nucleus** containing **protons** and **neutrons** that accounts for nearly all the mass of the atom and is surrounded by **electrons**. Unique symbols for each **isotope** of every element indicate the identity of an atom by showing its **atomic number** and **mass number**. (Sections 2.2 and 2.3)

LO3 The **average atomic mass** of an element is the weighted average of all the isotopes of that element. It is calculated by multiplying the mass of each of its stable isotopes by the **natural abundance** of that isotope and then summing these products. (Section 2.4)

LO4 Elements are arranged in the **periodic table of the elements** in order of increasing atomic number and in a pattern based on their physical and chemical properties. Elements in the same vertical column are said to be in the same **group**. The periodic table consists of the **main group** (or **representative**) **elements** in groups 1–2 and 13–18, and the **transition metals** in groups 3–12. The periodic table can also be divided

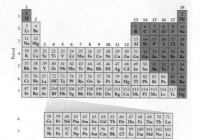

into the three categories of **metals**, **metalloids**, and **nonmetals**. (Sections 2.3 and 2.5)

LO5 The **law of multiple proportions** describes the ratio of masses of elements in simple compounds in terms of small whole numbers. Compounds may be either **molecular compounds** or **ionic compounds**. The **empirical formula** of a molecular or ionic compound gives the smallest whole-number ratio of the atoms (or ions) in it. (Section 2.6)

LO6 Standardized naming conventions allow us to translate the names of molecular and ionic compounds into their chemical formulas, and vice versa. **Organic compounds** always contain carbon atoms and almost always contain hydrogen atoms. (Sections 2.7 and 2.8)

LO7 During primordial **nucleosynthesis**, protons and neutrons fused to produce nuclei of helium. The nuclei of atoms as massive as ^{56}Fe formed when the nuclei of lighter elements fused in the cores of giant stars in a process called stellar nucleosynthesis, which continues today. Even more massive nuclei are formed by a combination of other nuclear reactions, leading to supernovas (explosions of giant stars). (Section 2.9)

PARTICULATE **PREVIEW WRAP-UP**

The middle compound is ionic because it shows ions attracted to one another in an extended, ordered array. The other two are covalent compounds that show single molecules made from attractions between atoms.

PROBLEM-SOLVING SUMMARY

Type of Problem	Concepts and Equations	Sample Exercises
Writing symbols of isotopes	To the left of the element symbol, place a superscript for the mass number (A) and (if needed) a subscript for the atomic number (Z).	**2.1, 2.2**
Calculating the average atomic mass of an element	Multiply the mass (m) of each stable isotope of the element times the natural abundance (a) of that isotope; then sum these products: $$m_X = a_1 m_1 + a_2 m_2 + a_3 m_3 \ldots \qquad (2.1)$$	**2.3**
Locating elements on the periodic table	Use the row number, group designation, and chemical properties to locate elements.	**2.4**
Predicting the composition of different compounds formed by the same two elements	Use the chemical formulas of the two compounds and the law of multiple proportions.	**2.5**
Classifying compounds as ionic or molecular	Ionic compounds contain metallic and nonmetallic elements; molecular compounds contain nonmetals or metalloids.	**2.6**
Naming binary compounds and writing their formulas	Apply the naming rules in Section 2.7.	**2.7, 2.8**
Naming transition metal compounds and writing their formulas	Use a Roman numeral to indicate the charge on the transition metal cation.	**2.9**
Naming oxoacids and compounds containing oxoanions and writing their formulas	Apply the naming rules in Section 2.7.	**2.10, 2.11**

VISUAL PROBLEMS

(Answers to boldface end-of-chapter questions and problems are in the back of the book.)

2.1. Alpha and beta particles emitted by a sample of pitchblende escape through a narrow channel in the shielding surrounding the sample and into an electrical field as shown in Figure P2.1. Which path in the figure corresponds to each form of radiation?

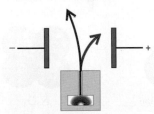

FIGURE P2.1

2.2. Does the radiation that follows the red path in Figure P2.1 penetrate solid objects better than the radiation following the green path? Explain your answer.

2.3. In Figure P2.3 the blue spheres represent nitrogen atoms and the red spheres represent oxygen atoms. The figure as a whole represents which of the following gases? (a) N_2O_3; (b) N_7O_{11}; (c) a mixture of NO_2 and NO; (d) a mixture of N_2 and O_3

FIGURE P2.3

2.4. In Figure P2.4 the black spheres represent carbon atoms and the red spheres represent oxygen atoms. Which of the following statements about the two equal-volume compartments is or are true?

FIGURE P2.4

a. The compartment on the left contains CO_2; the one on the right contains CO.
b. The compartments contain the same mass of carbon.
c. The ratio of oxygen to carbon in the gas in the left compartment is twice that of the gas in the right compartment.
d. The pressures inside the two compartments are equal. (Assume that the pressure of a gas is proportional to the number of particles in a given volume.)

2.5. Which of the highlighted elements in Figure P2.5 is (a) a reactive nonmetal, (b) a chemically inert gas, (c) a reactive metal?

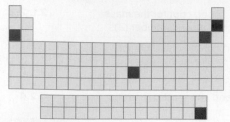

FIGURE P2.5

2.6. Which of the highlighted elements in Figure P2.6 forms monatomic ions with a charge of (a) 1+, (b) 2+, (c) 3+, (d) 1−, (e) 2−?

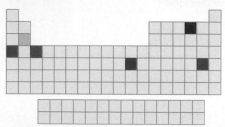

FIGURE P2.6

2.7. Which of the highlighted elements in Figure P2.7 forms an oxide with the following formula: (a) XO; (b) X_2O; (c) XO_2; (d) X_2O_3?

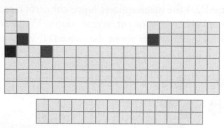

FIGURE P2.7

2.8. Which of the highlighted elements in Figure P2.8 forms an oxoanion with the following generic formula: (a) XO_4^-; (b) XO_4^{2-}; (c) XO_3^{3-}; (d) XO_3^-?

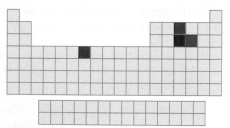

FIGURE P2.8

2.9. Which of the highlighted elements in Figure P2.9 is *not* formed by the fusion of lighter elements in the cores of giant stars?

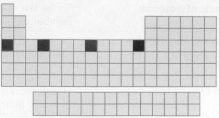

FIGURE P2.9

2.10. Use representations [A] through [I] in Figure P2.10 to answer questions a–f.
a. Which are covalent compounds?
b. Which are ionic compounds?
c. Which incorporates both covalent and ionic bonding?
d. Which two representations demonstrate the law of multiple proportions?
e. Which, if any, of the representations have the same empirical formula?
f. What are the names of the substances in [D], [F], and [H]?

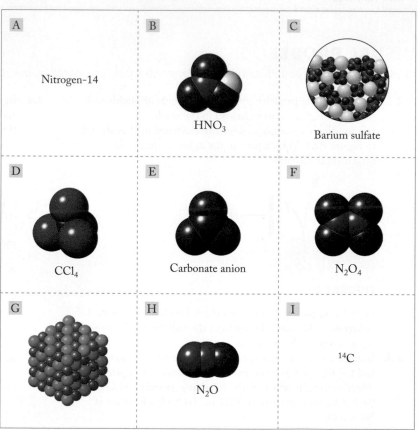

A   Nitrogen-14

B HNO_3

C   Barium sulfate

D CCl_4

E   Carbonate anion

F N_2O_4

G

H N_2O

I ^{14}C

FIGURE P2.10

QUESTIONS AND PROBLEMS
The Rutherford Model
Concept Review

2.11. Explain how the results of the gold-foil experiment led Rutherford to dismiss the plum-pudding model of the atom and create his own model based on a nucleus surrounded by electrons.

2.12. Had the plum-pudding model been valid, how would the results of the gold-foil experiment have differed from what Geiger and Marsden actually observed?

2.13. What properties of cathode rays led Thomson to conclude that they were not pure energy, but rather particles with an electric charge?

*__2.14.__ What would be observed if the charges on the plates of Millikan's apparatus (Figure 2.4) were reversed?

Isotopes and Average Atomic Mass
Concept Review

2.15. What is meant by a *weighted average*?

2.16. Explain how natural abundance percentages are related to average atomic masses.

2.17. Explain the inherent redundancy in the nuclide symbol $_Z^A X$.

2.18. How are the mass number and atomic number of a nuclide related to the number of neutrons and protons in each of its nuclei?

Problems

2.19. How many protons, neutrons, and electrons are there in the following atoms? (a) ^{14}C; (b) ^{59}Fe; (c) ^{90}Sr; (d) ^{210}Pb

2.20. How many protons, neutrons, and electrons are there in the following atoms? (a) ^{11}B; (b) ^{19}F; (c) ^{131}I; (d) ^{222}Rn

2.21. If the mass number of an isotope is more than twice the atomic number, is the neutron-to-proton ratio less than, greater than, or equal to 1?

2.22. In each of the following pairs of isotopes, which isotope has more protons and which has more neutrons? (a) ^{127}I or ^{131}I; (b) ^{188}Re or ^{188}W; (c) ^{14}N or ^{14}C

2.23. Fill in the missing information about atoms of the four nuclides in the following table.

Symbol	^{16}O	?	?	?
Number of Protons	?	26	?	79
Number of Neutrons	?	30	?	?
Number of Electrons	?	?	50	?
Mass Number	?	?	118	197

2.24. Fill in the missing information about atoms of the four nuclides in the following table.

Symbol	^{27}Al	?	?	?
Number of Protons	?	42	?	92
Number of Neutrons	?	56	?	?
Number of Electrons	?	?	60	?
Mass Number	?	?	143	238

2.25. Boron, lithium, nitrogen, and neon each have two stable isotopes. In which of the following pairs of isotopes is the heavier isotope more abundant? (a) ^{10}B or ^{11}B (average atomic mass, 10.81 amu); (b) ^6Li or ^7Li (average atomic mass, 6.941 amu); (c) ^{14}N or ^{15}N (average atomic mass, 14.01 amu); (d) ^{20}Ne or ^{22}Ne (average atomic mass, 20.18 amu)

2.26. The average atomic mass of copper is 63.546 amu. It is composed of copper-63 and copper-65 isotopes. The natural abundance of copper-63 (62.9296 amu) is 69.17%. What is the natural abundance of copper-65 (64.9278 amu)?

2.27. **Chemistry of Mars** The 1997 mission to Mars included a small robot, the *Sojourner*, that analyzed the composition of Martian rocks. Magnesium oxide from a boulder dubbed "Barnacle Bill" was analyzed and found to have the following isotopic composition:

Mass (amu)	Natural Abundance (%)
39.9872	78.70
40.9886	10.13
41.9846	11.17

If essentially all of the oxygen in the Martian MgO sample is oxygen-16 (which has an exact mass of 15.9948 amu), is the average atomic mass of magnesium on Mars the same as on Earth (24.31 amu)?

2.28. Platinum has six isotopes: ^{190}Pt, ^{192}Pt, ^{194}Pt, ^{195}Pt, ^{196}Pt, and ^{198}Pt.
   a. How many neutrons are there in each isotope?
   b. The natural abundances of the six isotopes are 0.014% ^{190}Pt (189.96 amu); 0.782% ^{192}Pt (191.96 amu); 32.967% ^{194}Pt (193.96 amu); 33.832% ^{195}Pt (194.97 amu); 25.242% ^{196}Pt (195.97 amu); and 7.163% ^{198}Pt (197.97 amu). Calculate the average atomic mass of platinum and compare it with the value in the periodic table on the inside front cover.

2.29. Use the following table of abundances and masses of five naturally occurring titanium isotopes to calculate the mass of ^{48}Ti.

Symbol	Mass (amu)	Natural Abundance (%)
^{46}Ti	45.95263	8.25
^{47}Ti	46.9518	7.44
^{48}Ti	?	73.72
^{49}Ti	48.94787	5.41
^{50}Ti	49.9448	5.18
Average	47.87	

2.30. Use the following table of abundances and masses of the three naturally occurring argon isotopes to calculate the mass of ^{40}Ar.

Symbol	Mass (amu)	Natural Abundance (%)
^{36}Ar	35.96755	0.337
^{38}Ar	37.96272	0.063
^{40}Ar	?	99.60
Average	39.948	

The Periodic Table of the Elements

Concept Review

2.31. Mendeleev ordered the elements in his version of the periodic table according to their atomic masses instead of their atomic numbers. Why?

2.32. Why did Mendeleev not include the noble gases in his version of the periodic table?

2.33. Mendeleev arranged the elements on the left side of his periodic table according to the formulas of the binary compounds they form with oxygen, and he used those formulas as column labels. For example, group 1 of the modern periodic table was labeled "R_2O" in Mendeleev's table, where "R" represented one of the elements in the group. What labels did Mendeleev use for groups 2, 3, and 4 of the modern periodic table?

2.34. Mendeleev arranged the elements on the right side of his periodic table according to the formulas of the binary compounds they form with hydrogen, and he used those formulas as column labels. Which groups of the modern periodic table were labeled "HR," "H_2R," and "H_3R" in Mendeleev's table, where "R" represented one of the elements in the group?

Problems

2.35. Which element is most likely to form a cation with a 2+ charge? (a) S; (b) P; (c) Be; (d) Al

2.36. Which element is most likely to form an anion with a 2− charge? (a) S; (b) P; (c) Be; (d) Al

2.37. Which ions have the same number of electrons as an atom of argon? (a) S^{2-}; (b) P^{3-}; (c) Be^{2+}; (d) Ca^{2+}

2.38. Which ions have the same number of electrons as an atom of krypton? (a) Se^{2-}; (b) As^{3-}; (c) Ca^{2+}; (d) K^+

2.39. Classify each of the following third-row elements as a metal, a metalloid, or a nonmetal: (a) Mg; (b) Al; (c) Si; (d) S; (e) Ar.

2.40. Classify each of the following fourth-row elements as a metal, a metalloid, or a nonmetal: (a) Ti; (b) Ni; (c) As; (d) Se; (e) Kr.

2.41. Classify each of the following fourth-row elements as an alkali metal, an alkaline earth metal, a transition metal, a halogen, or a noble gas: (a) Br; (b) Ca; (c) K; (d) Kr; (e) V.

2.42. Which element in the second row of the periodic table is (a) a halogen; (b) an alkali metal; (c) an alkaline earth metal; (d) a noble gas?

2.43. **Elements in TNT** Molecules of the explosive called TNT contain atoms of hydrogen and of the second-row elements in groups 14, 15, and 16. Which three elements are these?

2.44. **Chemical Weapons** Phosgene, a colorless, poisonous gas, was used as a chemical weapon during World War I. Despite its name, molecules of phosgene contain no atoms of phosphorus. Instead, they contain atoms of carbon, of the group 16 element in the second row of the periodic table, and of the group 17 element in the third row. What are the names and the atomic numbers of these last two elements?

2.45. **Catalytic Converters** The catalytic converters used to remove pollutants from automobile exhaust contain the compounds of several fairly expensive elements, including those described below. Which elements are they?
   a. The group 10 transition metal in the fifth row of the periodic table.
   b. The transition metal whose symbol is to the left of your answer to part (a).
   c. The transition metal whose symbol is directly below your answer to part (a).

2.46. **Swimming Pool Chemistry** Compounds containing chlorine have long been used to disinfect the water in swimming pools, but in recent years a compound of a less corrosive halogen has become a popular alternative disinfectant. What is the name of this fourth-row element?

Trends in Compound Formation

Concept Review

2.47. How does Dalton's atomic theory of matter explain the fact that when water is decomposed into hydrogen and oxygen gas, the volume of hydrogen is always twice that of oxygen?

2.48. **Pollutants in Automobile Exhaust** In the internal combustion engines that power most automobiles, nitrogen and oxygen may combine to form NO. When NO in automobile exhaust is released into the atmosphere, it reacts with more oxygen, forming NO_2, a key ingredient in smog. How do these reactions illustrate Dalton's law of multiple proportions?

2.49. Describe the types of elements that combine to form molecular compounds and the types that combine to form ionic compounds.

2.50. How do the properties of ionic compounds differ from those of molecular compounds?

Problems

2.51. Cobalt forms two sulfides: CoS and Co_2S_3. Predict the ratio of the two masses of sulfur that combine with a fixed mass of cobalt to form CoS and Co_2S_3.

2.52. Lead forms two oxides: PbO and PbO_2. Predict the ratio of the two masses of oxygen that combine with a fixed mass of lead to form PbO and PbO_2.

2.53. When 5.0 g of sulfur is combined with 5.0 g of oxygen, 10.0 g of sulfur dioxide (SO_2) is formed. What mass of oxygen would be required to convert 5.0 g of sulfur into sulfur trioxide (SO_3)?

*__2.54.__ Nitrogen monoxide (NO) is 46.7% nitrogen by mass. Use the law of multiple proportions to calculate the mass percentage of nitrogen in nitrogen dioxide (NO_2).

2.55. Fill in the missing information in the following table of monatomic ions.

Symbol	$^{35}Cl^-$	?	?	?
Number of Protons	?	11	?	82
Number of Neutrons	?	12	46	?
Number of Electrons	?	10	36	80
Mass Number	?	?	81	210

2.56. Fill in the missing information in the following table of monatomic ions.

Symbol	$^{137}Cs^+$	?	?	?
Number of Protons	?	30	?	40
Number of Neutrons	?	34	16	?
Number of Electrons	?	28	18	36
Mass Number	?	?	32	90

2.57. Which of these compounds consist of molecules, and which consist of ions? (a) P_4O_{10}; (b) $SrCl_2$; (c) MgF_2; (d) SO_3

2.58. Which of these compounds consist of molecules, and which consist of ions? (a) Mg_3N_2; (b) BaS; (c) AgCl; (d) NCl_3

2.59. Would compounds formed from each pair of elements contain covalent bonds or ionic bonds? (a) cesium and fluorine; (b) nitrogen and chlorine; (c) carbon and oxygen; (d) magnesium and oxygen

2.60. Would compounds formed from each pair of elements contain covalent bonds or ionic bonds? (a) silver and oxygen; (b) sulfur and oxygen; (c) carbon and hydrogen; (d) barium and nitrogen

2.61. Give the total number of atoms in a formula unit of these compounds: (a) LaF_3; (b) In_2S_3; (c) Na_3P; (d) Ca_3N_2.

2.62. Give the total number of atoms in a formula unit of these compounds: (a) In_2O_3; (b) CeS_2; (c) TlF_3; (d) CaO.

Naming Compounds and Writing Formulas

Concept Review

2.63. Consider a mythical element X, which forms only two oxoanions: XO_2^{2-} and XO_3^{2-}. Which of the two has a name that ends in *-ite*?

2.64. Concerning the oxoanions in Problem 2.63, would the name of either of them require a prefix such as *hypo-* or *per-*? Explain why or why not.

2.65. What is the role of Roman numerals in the names of the compounds formed by transition metals?

2.66. Why do the names of the ionic compounds formed by the alkali metals and by the alkaline earth metals not include Roman numerals?

Problems

2.67. What are the names of these compounds of nitrogen and oxygen? (a) NO_3; (b) N_2O_5; (c) N_2O_4; (d) NO_2

2.68. What are the names of these compounds of nitrogen and oxygen? (a) N_2O_3; (b) NO; (c) N_2O; (d) N_4O

2.69. What are the formulas and names of the ionic compounds containing the following pairs of elements? (a) sodium and sulfur; (b) strontium and chlorine; (c) aluminum and oxygen; (d) lithium and hydrogen

2.70. What are the formulas and names of the ionic compounds containing the following pairs of elements? (a) potassium and bromine; (b) calcium and hydrogen; (c) lithium and nitrogen; (d) aluminum and chlorine

2.71. Milk of Magnesia Milk of magnesia is a slurry of $Mg(OH)_2$ in water. What is the chemical name of this compound?

2.72. Smelling Salts Smelling salts are an antidote for fainting, made of $(NH_4)_2CO_3$, which smells of ammonia. What is the chemical name of $(NH_4)_2CO_3$?

2.73. What are the names of these sodium compounds? (a) Na_2O; (b) Na_2S; (c) Na_2SO_4; (d) $NaNO_3$; (e) $NaNO_2$

2.74. What are the names of these potassium compounds? (a) K_3PO_4; (b) K_2O; (c) K_2SO_3; (d) KNO_3; (e) KNO_2

2.75. What are the formulas of these compounds? (a) potassium sulfide; (b) potassium selenide; (c) rubidium sulfate; (d) rubidium nitrite; (e) magnesium sulfate

2.76. What are the formulas of these compounds? (a) rubidium nitride; (b) potassium selenite; (c) rubidium sulfite; (d) rubidium nitrate; (e) magnesium sulfite

2.77. What are the formulas of these compounds? (a) sodium hypobromite; (b) potassium sulfate; (c) lithium iodate; (d) magnesium nitrite

*__2.78.__ What are the formulas of these compounds? (a) potassium tellurite; (b) sodium arsenate; (c) calcium selenite; (d) potassium chlorate

2.79. Which compound is magnesium nitrite? (a) Mg_3N; (b) $Mg(NO_2)_2$; (c) $Mg(NO_3)_2$; (d) $Mg(NO)_2$

2.80. Which compound is lithium phosphate? (a) Li_3P; (b) Li_3PO_3; (c) Li_3PO_4; (d) $Li_2(PO_4)_3$

2.81. Which of these chemical names is followed by a chemical formula that does not fit the name? (a) calcium oxide, CaO; (b) lithium sulfate, $LiSO_4$; (c) barium sulfide, BaS; (d) potassium oxide, K_2O

2.82. Which of these chemical names is followed by a chemical formula that does not fit the name? (a) aluminum nitride, AlN; (b) aluminum sulfate, $Al_2(SO_4)_3$; (c) potassium chloride, KCl_2; (d) cesium sulfate, Cs_2SO_4

2.83. Ions in Blood The most abundant cations in blood plasma are Na^+, K^+, Mg^{2+}, and Ca^{2+}. Two of the principal anions are Cl^- and $H_2PO_4^-$. Write the formulas of the eight ionic compounds these cations and anions form.

2.84. Evaporation of seawater gives a mixture of ionic compounds containing sodium combined with chloride, sulfate, carbonate, bicarbonate, bromide, fluoride, and tetrahydroborate, $B(OH)_4^-$. Write the chemical formulas of all these compounds.

2.85. What are the names of these compounds? (a) Cr_2Te_3; (b) $V_2(SO_4)_3$; (c) Fe_2CrO_4; (d) MnO

2.86. What are the names of these compounds? (a) $FePO_4$; (b) $CuSO_4$; (c) Ag_2CO_3; (d) $Zn(NO_2)_2$

2.87. What are the formulas for these transition metal compounds? (a) zinc dichromate; (b) iron(III) acetate; (c) mercury(I) peroxide; (d) scandium thiocyanate

2.88. What are the formulas for these transition metal compounds? (a) mercury(II) hydroxide; (b) silver perchlorate; (c) manganese(IV) nitrate; (d) vanadium(IV) oxide

2.89. What are the formulas of the following copper minerals? (a) cuprite, copper(I) oxide; (b) chalcocite, copper(I) sulfide; (c) covellite, copper(II) sulfide

2.90. What are the names of the cobalt oxides with the following formulas? (a) CoO; (b) Co_2O_3; (c) CoO_2

2.91. What is the name of each of the following acids? (a) HF; (b) $HBrO_3$; (c) HBr; (d) HIO_4

2.92. What are the formulas of the following acids? (a) selenous acid; (b) hydrocyanic acid; (c) phosphoric acid; (d) nitrous acid

Organic Compounds: A First Look

Concept Review

2.93. What classes of organic compounds contain no heteroatoms?

2.94. What classes of organic compounds contain oxygen atoms?

2.95. What classes of organic compounds contain nitrogen atoms? Does any class of organic compounds contain both oxygen and nitrogen atoms?

2.96. Are organic compounds made of covalent or ionic bonds? Do they exist as molecules or as ionic lattices?

Problems

2.97. To which class of organic compounds does each of the following compounds belong?
a. $CH_3CH_2CH_2CH_2CH_2CH_2CH_2CH_3$ (octane—a principal component of gasoline)
b. $HC{\equiv}CH$ (ethyne, also known as acetylene—used in welding torches)

2.98. To which class of organic compounds does each of the following compounds belong?
a. CH_3CH_2—O—CH_2CH_3 (diethyl ether—once used as anesthesia for surgery)
b. $CH_3CH_2CH_2CH_2OH$ (butanol—a component of brake fluids and perfumes)

2.99. **Scent of Pears** What is the functional group present in propyl acetate (the substance responsible for the smell of pears; Figure P2.99)?

$$CH_3-\overset{\overset{\displaystyle O}{\|}}{C}-O-CH_2-CH_2-CH_3$$

FIGURE P2.99

*2.100. What two functional groups are present in glycine (the simplest amino acid; Figure P2.100)?

$$H_2N-CH_2-\overset{\overset{\displaystyle O}{\|}}{C}-OH$$

FIGURE P2.100

Nucleosynthesis: The Origin of the Elements

Concept Review

2.101. Write brief (one-sentence) definitions of *chemistry* and *cosmology*, and then give as many examples as you can of how the two sciences are related.

2.102. In the history of the universe, which of these particles formed first, and which formed last? (a) deuteron; (b) neutron; (c) proton; (d) quark

2.103. Chemists do not include quarks in the category of subatomic particles. Why?

2.104. Why did early nucleosynthesis last such a short time?

2.105. In the current cosmological model, the volume of the universe is increasing with time. How might this expansion affect the density of the universe?

2.106. **Components of Solar Wind** Most of the ions in the solar wind emitted by the Sun are hydrogen ions. The ions of which element should be next most abundant?

Additional Problems

2.107. The molecular compounds HClO, $HClO_2$, and $HClO_4$ are three of the four common oxoacids of chlorine. Give the formula and chemical name of the fourth oxoacid of chlorine.

2.108. **Fusion in Stars** One reaction in the process of carbon fusion in massive stars involves the combination of two carbon-12 nuclei to form the nucleus of a new element and an alpha particle. Write an equation that describes this process.

2.109. A process called neon fusion takes place in massive stars. In one of the reactions in this process, an alpha particle combines with a neon-21 nucleus to produce another element and a neutron. Write an equation that describes this process.

2.110. In April 1897, J. J. Thomson presented the results of his experiment with cathode-ray tubes (Figure P2.110) in which he proposed that the rays were actually beams of negatively charged particles, which he called "corpuscles."
a. What is the name we use for these particles today?
b. Why did the beam deflect when passed between electrically charged plates, as shown in Figure P2.110?
c. If the polarity of the plates was switched, how would the position of the light spot on the phosphorescent screen change?
d. If the voltage on the plates was reduced by half, how would the position of the light spot change?

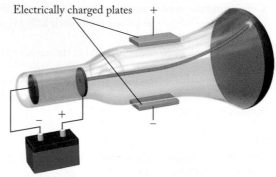

Electrically charged plates

FIGURE P2.110

*2.111. Suppose the electrically charged discs at the end of the cathode-ray tube were replaced with a radioactive source, as shown in Figure P2.111. Also suppose the radioactive material inside the source emits α and β particles. The only way for either kind of particle to escape the source is through a narrow channel drilled through a block of lead.
a. How many light spots do you expect to see on the phosphorescent screen?
b. What are their positions relative to the electrical plates, and which particle produces which spot?

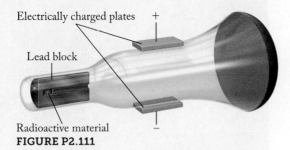

Electrically charged plates

Lead block

Radioactive material
FIGURE P2.111

*2.112. Suppose the radioactive material inside the source in the apparatus shown in Figure P2.111 emits protons and α particles, and suppose both kinds of particles have the same velocities.
  a. How many light spots do you expect to see on the phosphorescent screen?
  b. What are their positions on the screen (above, at, or below the center)? Which particle produces which spot?

2.113. Early Universe Cosmologists estimate that the matter in the early universe was 75% hydrogen-1 and 25% helium-4, by mass, when atoms first formed.
  a. Assuming these proportions are correct, what was the ratio of hydrogen to helium *atoms* in the early universe?
  b. The ratio of hydrogen to helium atoms in our solar system is slightly less than 10:1. Compare this value with the value you calculated in part (a).
  c. Propose a hypothesis that accounts for the difference in composition between the solar system and the early universe.
  d. Describe an experiment that would test your hypothesis.

2.114. Emergency Air Supply Potassium forms three compounds with oxygen: K_2O (potassium oxide), K_2O_2 (potassium peroxide), and KO_2 (potassium superoxide). Elemental potassium is rarely encountered; it reacts violently with water and is very corrosive to human tissue. Potassium superoxide is used in self-contained breathing apparatuses as a source of oxygen for use in mines, submarines, and spacecraft. Potassium peroxide binds carbon dioxide and is used to scrub (remove) toxic CO_2 from the air in submarines. Predict the ratio of the masses of oxygen that combine with a fixed mass of potassium in K_2O, K_2O_2, and KO_2.

*2.115. **Bronze Age** Historians and archaeologists often apply the term "Bronze Age" to the period in Mediterranean and Middle Eastern history when bronze was the preferred material for making weapons, tools, and other metal objects. Ancient bronze was an alloy prepared by blending molten copper (88% by mass) and tin (12% by mass). What is the ratio of copper to tin atoms in a piece of bronze with this composition?

*2.116. In his version of the periodic table, Mendeleev arranged elements according to the formulas of the compounds they formed with hydrogen and oxygen. The elements in one of his eight groups formed compounds with the generic formulas MH_3 and M_2O_5, where M was the symbol of an element in the group. Which Roman numeral did Mendeleev assign to this group?

2.117. In the Mendeleev table in Figure 2.11, there are no symbols for elements with predicted atomic masses of 44, 68, and 72.
  a. Which elements are these?
  b. Mendeleev anticipated the later discovery of these three elements and gave them the tentative names ekaaluminum, ekaboron, and ekasilicon, reflecting the probability that their properties would resemble those of aluminum, boron, and silicon, respectively. What are the modern names of ekaaluminum, ekaboron, and ekasilicon?
  c. When were these elements finally discovered? To answer this question you may wish to consult a reference such as webelements.com.

2.118. In chemical nomenclature, the prefix *thio-* is used to indicate that a sulfur atom has replaced an oxygen atom in the structure of a molecule or a polyatomic ion.
  a. With this rule in mind, write the formula for the thiosulfate ion.
  b. What is the formula of sodium thiosulfate?

2.119. There are two stable isotopes of gallium. Their masses are 68.92558 and 70.9247050 amu. If the average atomic mass of gallium is 69.7231 amu, what is the natural abundance of the lighter isotope?

2.120. There are two stable isotopes of bromine. Their masses are 78.9183 and 80.9163 amu. If the average atomic mass of bromine is 79.9091 amu, what is the natural abundance of the heavier isotope?

*2.121. Using the information in the previous question:
  a. Predict the possible masses of individual molecules of Br_2.
  b. Calculate the natural abundance of molecules with each of the masses predicted in part (a) in a sample of Br_2.

*2.122. There are three stable isotopes of magnesium. Their masses are 23.9850, 24.9858, and 25.9826 amu. If the average atomic mass of magnesium is 24.3050 amu and the natural abundance of the lightest isotope is 78.99%, what are the natural abundances of the other two isotopes?

2.123. Write the names and formulas of the following compounds from this list of elements: Li, Fe, Al, O, C, and N.
  a. A molecular substance AB_2, where A is a group 14 element and B is a group 16 element
  b. An ionic compound C_3D, where C is a group 1 element and D is a group 15 element

2.124. Write the chemical symbol of each of the following species: (a) a cation with a mass number of 24, an atomic number of 12, and a charge of 2+; (b) a member of group 15 that has a charge of 3+, 48 electrons, and 70 neutrons; (c) a noble gas atom with 48 neutrons.

*2.125. Predict some physical and chemical properties of radium (Ra). Predict the melting points of $RaCl_2$ and RaO. (*Hint*: research the melting points of the other alkaline earth element chlorides and oxides.)

2.126. From their positions in the periodic table, predict which of the following elements would be good electrical conductors: Ti, Ne, N, Ag, Tb, Br, and Mo.

2.127. Argon has a larger average atomic mass than potassium, yet it is placed before potassium in the modern periodic table. Explain.

*2.128. It takes nearly twice the energy to remove an electron from a helium atom as it does to remove an electron from a hydrogen atom. Propose an explanation for this.

3

Stoichiometry
Mass, Formulas, and Reactions

METHANE BUBBLES An ecologist pierces a large methane bubble trapped in the ice of an Alaskan lake. Burning methane is an example of a chemical reaction.

PARTICULATE **REVIEW**

Molecules and Names

In Chapter 3 we examine the composition of pure substances and explore how the quantities of reactants and products in chemical reactions are related. Unlike today's atmosphere, which is mostly elemental nitrogen and oxygen, Earth's early atmosphere consisted predominantly of the compounds shown here.

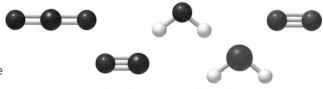

- Name these compounds.

    (Review Section 2.7 if you need help.)

(Answers to Particulate Review questions are in the back of the book.)

How Many Particles?

When large methane bubbles burn, they do so by reacting with oxygen. As you read Chapter 3, look for ideas that will help you answer these questions:

- How many oxygen molecules are necessary to completely react with five methane molecules?

- Identify the products of this reaction.

- Draw all the molecules of product formed in this reaction.

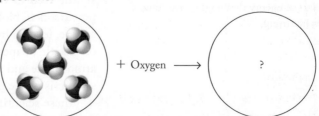

Learning Outcomes

LO1 Use Avogadro's number and the definition of the mole in calculations
Sample Exercises 3.1, 3.2, 3.3, 3.4, 3.5, 3.6, 3.7

LO2 Write balanced chemical equations that describe chemical reactions
Sample Exercises 3.8, 3.9, 3.10, 3.11

LO3 Use balanced chemical equations to relate the mass of a reactant consumed to the mass of a product formed
Sample Exercises 3.12, 3.13

LO4 Determine an empirical formula from the percent composition of a substance
Sample Exercises 3.14, 3.15, 3.16

LO5 Determine a molecular formula from the empirical formula and molar mass of a substance
Sample Exercise 3.17

LO6 Use data from combustion reactions to determine empirical formulas of substances
Sample Exercises 3.18, 3.19

LO7 Determine the limiting reactant in a chemical reaction
Sample Exercises 3.20, 3.21

LO8 Calculate the theoretical and percent yields in a chemical reaction
Sample Exercises 3.22, 3.23

3.1 Air, Life, and Molecules

FIGURE 3.1 The apparatus used by Miller (shown) and Urey to simulate the synthesis of amino acids in the atmosphere of early (prebiotic) Earth. Discharges between the tungsten electrodes were meant to provide the sort of energy that might have come from lightning.

In Chapter 2 we described how elements are synthesized in the cores of giant stars and dispersed throughout galaxies when the stars explode as supernovas. The dispersed elements become the building blocks of other stars and planets. As our solar system formed, the inner planets—Mercury, Venus, Earth, and Mars—were rich in nonvolatile elements such as iron, silicon, magnesium, and aluminum. Earth was also rich in oxygen—in the form of stable compounds with these and other elements. These compounds formed the rocks and minerals of Earth's crust and provided an early atmosphere. As Earth cooled, atmospheric water vapor condensed into torrential rains that filled up depressions in the crust, forming the first oceans.

Could the substances in lifeless rocks and air or dissolved in primordial seas have combined to become the organic building blocks of life—compounds like simple sugars, amino acids, and the molecules that form DNA? No one knows for sure, but in the early 1950s Nobel Prize winner Harold Urey (1893–1981) and his student Stanley Miller (1930–2007) tested this hypothesis by assembling a mixture of gases believed to have been present in Earth's early (prebiotic) atmosphere. Urey and Miller subjected that mixture of gases to an electric current, simulating the lightning that would have been prevalent on early Earth (Figure 3.1). The resulting chemical reactions produced several amino acids, the building blocks of proteins. Samples from additional experiments by Miller in an apparatus simulating the composition of gases from a volcanic eruption were reanalyzed in 2007, and more than 20 amino acids were detected.

Amino acids have also been found in meteorites, which have bombarded Earth since it formed. Extensive analyses of the Murchison meteorite, which landed in Australia in 1969, showed that it contains many of the same amino acids synthesized in the Miller–Urey experiments. Regardless of whether important precursors to biological compounds were synthesized throughout the early atmosphere, formed in the localized environment around volcanic eruptions, or arrived on Earth in meteors, current thinking favors an early Earth with a primitive atmosphere and an ocean of water that provided an environment conducive to the synthesis of the molecules of life.

These scientific theories about the origin of life from simple inorganic molecules and the evidence that supports them are the product of research carried out in the 20th and 21st centuries. In the early 19th century, when molecular science was in its infancy, it was believed that organic compounds could only be made by living organisms. In addition, the prevailing understanding held that chemical compounds were different only because they had different elemental compositions. A serendipitous experiment by Friedrich Wöhler (1800–1882) in 1828 led to the demise of both ideas and to their replacement with a view of matter that is a cornerstone of modern molecular science.

Wöhler attempted to synthesize the compound ammonium cyanate (NH_4NCO; Figure 3.2a,b) by using silver cyanate (AgNCO) and ammonium chloride (NH_4Cl). The product of his reaction demonstrated none of the chemical or physical properties of cyanates. After several other attempts at synthesis, he established that the white crystals produced by the reaction were urea (H_2NCONH_2; Figure 3.2c). At that time, urea was known only as a material isolated from urine, clearly the product of a living being. Wöhler famously reported to his scientific mentor Jöns Jacob Berzelius (1779–1848) that he had succeeded in making urea "without the intervention of a kidney." This result and others like it ultimately led to the modern definition of organic compounds as introduced in Chapter 2—compounds of carbon and hydrogen, often including other elements—and to the understanding that organic compounds identical to those found in living systems can be made in the laboratory.

But Wöhler's result had another very important consequence. Look carefully at the formulas for ammonium cyanate and urea. If we write them simply in a way that illustrates their composition, ammonium cyanate is CH_4N_2O, and urea is exactly the same: CH_4N_2O. The two different substances have the same atomic composition. Scientists ultimately reasoned that the *arrangement* of atoms within the molecules of each compound must be different. This idea led to the modern statement that chemical substances must be defined by not only the number and kind of atoms in their molecules but also the arrangement of those atoms. Composition and arrangement both matter. Berzelius called urea and ammonium cyanate **isomers** (literally, "same units"), which is the term still used to describe two or more different compounds with the same atomic composition.

In this chapter we begin with a molecular view of the physical processes and chemical reactions that may have occurred on early Earth. We explore how the quantities of substances produced and consumed in chemical reactions are related. Then we examine the analytical methods that reveal the composition of pure substances. Finally, we consider a class of chemical reactions that consume oxygen and a fuel, liberating energy—including the energy necessary for life. We will concentrate on the quantitative aspects of identifying chemical substances and their reactions, but keep Wöhler's result in mind: how the atoms within molecules are arranged matters, too. We will address the important issue of how atoms connect to each other and the resulting shapes of molecules in detail in Chapters 8 and 9.

Chemical Reactions and Earth's Early Atmosphere

The Earth that formed 4.6 billion years ago was a hot, molten sphere that gradually separated into distinct regions based on differences in density and melting point. The densest elements, notably iron and nickel, sank to the center of the planet. A less dense mantle, rich in compounds containing aluminum, magnesium, silicon, and oxygen, formed around the core. As time passed and Earth cooled, the mantle fractionated further, allowing a solid crust to form from the components of the mantle that were the least dense and had the highest melting points. The core also

isomers compounds with the same molecular formula but different arrangements of the atoms in their molecules.

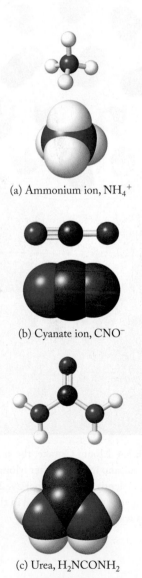

(a) Ammonium ion, NH_4^+

(b) Cyanate ion, CNO^-

(c) Urea, H_2NCONH_2

FIGURE 3.2 (a, b) Ball-and-stick and space-filling models of the ions in ammonium cyanate and (c) a molecule of urea.

FIGURE 3.3 Earth is composed of a solid inner core, consisting mostly of nickel and iron, surrounded by a molten outer core of similar composition. A rocky mantle, composed mostly of oxygen, silicon, magnesium, and iron, lies between the outer core and a relatively thin solid crust.

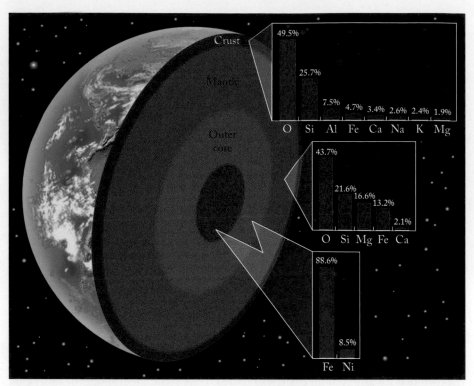

FIGURE 3.4 Mount Ontake, the second-highest volcano in Japan (after Mount Fuji), erupted violently on September 27, 2014. The most abundant gas released in this eruption was water vapor.

separated into a solid inner core and a molten outer core. Figure 3.3 shows the elemental compositions of these layers.

Earth's early crust was torn by the impact of asteroids and widespread volcanic activity. The gases released by these impacts and eruptions generated a primitive atmosphere with a chemical composition different from that of the air we breathe now. Current research favors a view of the early atmosphere of Earth being nearly devoid of molecular oxygen (O_2) yet rich in oxygen-containing compounds, including carbon dioxide (CO_2), carbon monoxide (CO), and water vapor (H_2O). Minor components of the atmosphere were hydrogen (H_2), methane (CH_4), and hydrogen sulfide (H_2S). Oxides of nitrogen and additional CO may have arisen as a result of heating of the atmosphere during bombardment by large meteorites, whereas traces of other volatile oxides, including sulfur dioxide (SO_2) and sulfur trioxide (SO_3), may have resulted from volcanic activity. Today, the most abundant gases released by volcanoes like Japan's Mount Ontake are water vapor, CO_2, and SO_2 (Figure 3.4).

Sometimes these compounds in Earth's prebiotic atmosphere combined to make substances with more elaborate molecular structures. For example, sulfur trioxide gas and water vapor are the **reactants** that combine to form liquid sulfuric acid, H_2SO_4, as a **product** (Figure 3.5). We use the formulas of these substances in a chemical equation to describe the reaction:

$$SO_3(g) + H_2O(g) \rightarrow H_2SO_4(\ell) \qquad (3.1)$$

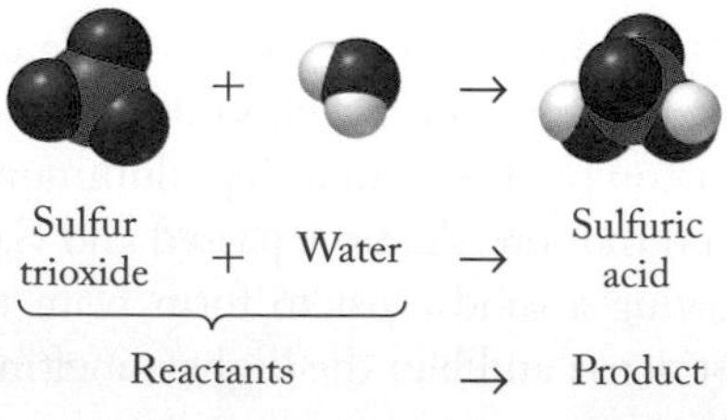

Sulfur trioxide + Water → Sulfuric acid

Reactants → Product

FIGURE 3.5 In this combination reaction, a molecule of SO_3 and a molecule of H_2O form a molecule of H_2SO_4.

The reaction between SO_3 and H_2O is an example of a **combination reaction**, a reaction where two (or more) substances combine to form one product. An important feature of any chemical equation is that it is *balanced*: every atom that is present in the reactants is also present in the products. This conservation of atoms means that there is also a conservation of mass: the sum of the masses of the reactants always equals the sum of the masses of the products.

The sulfuric acid that formed in Earth's early atmosphere eventually fell to the planet's surface as highly acidic rain. This rain landed on a crust made up mostly of metal oxides and metalloid oxides, including the mineral hematite, Fe_2O_3. When that sulfuric acid mixed with hematite, another chemical reaction took place—one that produced slightly water-soluble iron(III) sulfate, $Fe_2(SO_4)_3$, and liquid water. This reaction is described by the following chemical equation:

$$Fe_2O_3(s) + 3\,H_2SO_4(aq) \rightarrow 3\,H_2O(\ell) + Fe_2(SO_4)_3(aq) \qquad (3.2)$$

The "3"s in front of both H_2SO_4 and H_2O are coefficients that balance the number of atoms of each element on both sides of the reaction arrow, as we explore more closely in Section 3.3. The absence of a coefficient in front of Fe_2O_3 and $Fe_2(SO_4)_3$ is the same as having a coefficient of "1" in front of them. The chemical formula itself indicates the composition of one molecule (or one formula unit) of the substance. As we will see in the next section, the chemical formula also indicates the number of *moles* of each element in one *mole* of a compound.

3.2 The Mole

Equation 3.1 describes a chemical reaction between a molecule of sulfur trioxide and a molecule of water. In our macroscopic world, chemists rarely work with individual atoms or single molecules because they are too small to manipulate easily. Instead, we usually deal with quantities of reactants and products large enough to see and work with comfortably. Such measurable quantities contain enormous numbers of particles—atoms, molecules, or ions—and consequently we need a unit that can relate measurable quantities of substances to the number of particles they contain. That unit is the **mole (mol)**, the SI base unit for expressing quantities of substances (see Table 1.2).

Moles allow us to relate the mass of a pure substance to the number of particles it contains. One mole of any substance is defined as the quantity of the substance that contains the same number of particles as the number of carbon atoms in exactly 12 g of the nuclide carbon-12. This number is $6.02214154 \times 10^{23}$ carbon-12 atoms. (We will round it off to four significant figures, 6.022×10^{23}, throughout this book.) This very large value is called **Avogadro's number (N_A)** after the Italian scientist Amedeo Avogadro (1776–1856), whose research enabled other scientists to accurately determine the atomic masses of the elements. One mole of water also contains 6.022×10^{23} particles—in this case, molecules of water. To put a number of this magnitude in perspective, consider an empty sphere the size of Earth. It would require about 1 mole of basketballs to fill that sphere.

Avogadro's number is often used as part of a conversion factor between the number of particles and the number of moles of a substance. Dividing the number of particles in a sample by Avogadro's number yields the number of moles of those particles. For example, a jet airplane flying at an altitude of 11,300 m (37,000 ft)

CONNECTION In Chapter 1 we defined a *chemical reaction* as the transformation of one or more substances into different substances, and we used *chemical formulas* in a *chemical equation* to describe this transformation.

CONNECTION In Section 2.9 we balanced *nuclear equations* in terms of both mass and charge.

CHEMTOUR
Avogadro's Number

reactant a substance consumed during a chemical reaction.

product a substance formed during a chemical reaction.

combination reaction a reaction in which two (or more) substances combine to form one product.

mole (mol) an amount of material (atoms, ions, or molecules) that contains Avogadro's number ($N_A = 6.022 \times 10^{23}$) of particles.

Avogadro's number (N_A) the number of carbon atoms in exactly 12 grams of the carbon-12 isotope; $N_A = 6.022 \times 10^{23}$. It is the number of particles in one mole.

FIGURE 3.6 Converting between a number of particles and an equivalent number of moles (or vice versa) is a matter of dividing (or multiplying) by Avogadro's number. Note how the units cancel, leaving only the units we sought. Although it is sometimes useful to know the number of molecules of a substance in a sample, we more often care how many *moles* are in a particular quantity of a substance.

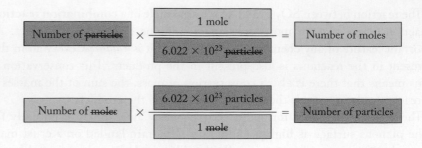

is flying through air that contains about 7.0×10^{21} particles per liter. This number is equivalent to 1.2×10^{-2} moles of particles per liter of air:

$$\frac{7.0 \times 10^{21} \text{ particles}}{1 \text{ L}} \times \frac{1 \text{ mol}}{6.022 \times 10^{23} \text{ particles}} = \frac{1.2 \times 10^{-2} \text{ mol}}{\text{L}}$$

On the other hand, multiplying a number of moles by Avogadro's number gives us the number of particles in that many moles. For example, a liter bottle of seltzer water contains 55 moles of H_2O. The number of molecules of water in that liter bottle is a very large number:

$$\frac{55 \text{ mol } H_2O}{1 \text{ L}} \times \frac{6.022 \times 10^{23} \text{ molecules } H_2O}{1 \text{ mol } H_2O} = \frac{3.3 \times 10^{25} \text{ molecules } H_2O}{1 \text{ L}}$$

The use of this conversion factor is illustrated in Figure 3.6, and the quantities of some common elements equivalent to 1 mole are illustrated in Figure 3.7.

SAMPLE EXERCISE 3.1 Converting Number of Moles   **LO1**
                                     into Number of Particles

It's not unusual for the polluted air above a large metropolitan area to contain as much as 5×10^{-10} moles of SO_2 per liter of air. What is this concentration of SO_2 in molecules per liter?

Collect and Organize The problem gives the number of moles of SO_2 per liter of air. We want to find the number of molecules of SO_2 in 1 L. There are 6.022×10^{23} particles in 1 mole of anything.

Analyze We convert the number of moles/liter into the number of molecules/liter by multiplying by 6.022×10^{23} molecules/mole:

$$\boxed{\frac{\text{mole}}{\text{L}}} \xrightarrow[\text{1 mole}]{6.022 \times 10^{23} \text{ molecules}} \boxed{\frac{\text{molecules}}{\text{L}}}$$

We can estimate the answer by considering the approximate value of Avogadro's number (about 6×10^{23}) and the number of moles of SO_2 (5×10^{-10}). The product of the two is 3×10^{14}.

Solve The number of SO_2 molecules per liter is:

$$\frac{5 \times 10^{-10} \text{ mol } SO_2}{1 \text{ L air}} \times \frac{6.022 \times 10^{23} \text{ molecules } SO_2}{1 \text{ mol } SO_2} = \frac{3 \times 10^{14} \text{ molecules } SO_2}{1 \text{ L air}}$$

Think About It The result tells us that a tiny fraction of 1 mole of a molecular substance (10^{-10}) is equivalent to a very large number of molecules (10^{14}), which makes sense given the immensity of Avogadro's number (10^{23}).

FIGURE 3.7 The quantities shown are equivalent to 1 mole of each material: 4.003 g of helium gas in the balloon and, left to right in front of the balloon, 32.06 g of solid sulfur, 63.55 g of copper metal, and 200.59 g of liquid mercury.

Practice Exercise If 1.0 mL of seawater contains about 2.5×10^{-14} moles of dissolved gold, how many atoms of gold are in that volume of seawater?

(Answers to Practice Exercises are in the back of the book.)

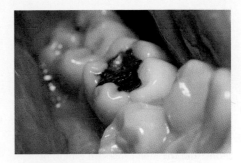

FIGURE 3.8 One of the materials used to fill dental caries is a mixture of three compounds: Ag_2Hg_3, Ag_3Sn, and Sn_8Hg. These materials are a kind of compound known as an alloy, which consists of a metal and at least one additional element mixed or combined together.

SAMPLE EXERCISE 3.2 Relating Formulas, Moles, and Particles   **LO1**

Dental fillings contain compounds with the formulas Ag_2Hg_3, Ag_3Sn, and Sn_8Hg (Figure 3.8). Which compound contains the most atoms per mole? How many atoms of silver are there in one mole of Ag_3Sn?

Collect, Organize, and Analyze We are given the formulas of three compounds: Ag_2Hg_3, Ag_3Sn, and Sn_8Hg. A chemical formula indicates the number of moles of each element in one mole of a compound. The number of atoms in a mole of *any* element is defined by Avogadro's number: $N_A = 6.022 \times 10^{23}$ atoms/mol.

Solve From the formulas of the three compounds we have:

- 1 mole of Ag_2Hg_3 contains 2 moles of silver and 3 moles of mercury, or $2 + 3 = 5$ total moles of atoms.
- 1 mole of Ag_3Sn contains 3 moles of silver and 1 mole of tin, or $3 + 1 = 4$ total moles of atoms.
- 1 mole of Sn_8Hg contains 8 moles of tin and 1 mole of mercury, or $8 + 1 = 9$ total moles of atoms.

Thus 1 mole of Sn_8Hg contains the greatest number of atoms:

$$\frac{9 \text{ total mol}}{\text{mol } Sn_8Hg} \times \frac{6.022 \times 10^{23} \text{ atoms}}{1 \text{ mol}} = \frac{5.420 \times 10^{24} \text{ atoms}}{\text{mol } Sn_8Hg}$$

The number of atoms of Ag per mole of Ag_3Sn is:

$$\frac{3 \text{ mol Ag}}{\text{mol } Ag_3Sn} \times \frac{6.022 \times 10^{23} \text{ atoms Ag}}{1 \text{ mol Ag}} = \frac{1.807 \times 10^{24} \text{ atoms Ag}}{\text{mol } Ag_3Sn}$$

Think About It Chemical formulas reveal the number of moles of each element per mole of a compound. When we compare two compounds, their chemical formulas allow us to quickly determine the relative numbers of each type of atom.

 Practice Exercise How many electrons are in a one-ounce gold coin (Figure 3.9) that contains 0.143 mol of pure gold?

(Answers to Practice Exercises are in the back of the book.)

CONCEPT TEST

How does a unit of measure such as a gross (144 units) relate to the concept of the mole?

(Answers to Concept Tests are in the back of the book.)

Molar Mass

The mole provides an important link from the values of atomic mass in the periodic table to masses of macroscopic quantities of elements and compounds. Recall from Section 2.4 that the mass value listed for each element in the periodic table is the average atomic mass of the atoms of the element, expressed in atomic mass

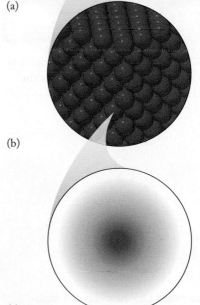

(a)

(b)

(c)

FIGURE 3.9 (a) Five 1-ounce gold coins, each containing about 0.143 mole of pure gold. (b) A lattice of gold atoms. (c) The nucleus of one gold atom, surrounded by electrons.

| 2 |
| He |
| 4.003 |

Atomic mass of He	4.003 amu/atom
Mass of 1 mol of He	4.003 g
Molar mass of He	4.003 g/mol

FIGURE 3.10 The atomic mass (in amu/atom) and the molar mass (in g/mol) of helium have the same numerical value.

FIGURE 3.11 A dozen golf balls weigh less than a dozen baseballs, but both contain the same number of balls.

molar mass (*M*) the mass of 1 mole of a substance. The molar mass of an element in grams per mole is numerically equal to that element's average atomic mass in atomic mass units.

molecular mass the mass of one molecule of a molecular compound.

units (amu) or daltons (Da). That same value is also the mass of 1 mole of atoms of that element expressed in grams. Thus the average mass of one atom of helium is 4.003 amu, and the mass of 1 mole of helium is 4.003 g. The mass in grams of 1 mole of any substance is called the **molar mass (*M*)** of the substance, and we say that the molar mass of helium is 4.003 g/mol (Figure 3.10). Note that because the mass of electrons is insignificant compared with that of protons and neutrons (Table 2.1), the mass of an ion is essentially the same as the mass of an atom, even though cations have fewer electrons than neutral atoms and anions have more.

As mentioned earlier, moles allow us to compare the number of particles in different masses of two different substances. Molar mass, in turn, can be used to convert between the mass of a sample of any substance and the number of moles in that sample. Remember that although chemical reactions take place between particles (atoms or molecules or ions), in the laboratory we almost always measure the *masses* of products and reactants.

The mole enables us to calculate the number of particles in any sample of a given substance from the known (or measured) mass of the sample. We can do this because the mole represents a fixed number of particles (Avogadro's number). One useful analogy is to compare golf balls with baseballs. A dozen golf balls and a dozen baseballs have very different masses, but each contains 12 balls (Figure 3.11). Indeed, the mole is sometimes referred to as "the chemist's dozen."

SAMPLE EXERCISE 3.3 Converting Mass into Number of Moles   **LO1**

Some antacid tablets contain 425 mg of calcium (as Ca^{2+} ions). How many moles of calcium are in each tablet?

Collect, Organize, and Analyze We are given the mass of Ca^{2+} ions in each tablet. We are asked to convert this mass into an equivalent number of moles. The number of moles of a substance is related to the mass of the substance by its molar mass. We begin by converting milligrams to grams, and then we convert grams of Ca^{2+} into moles of Ca^{2+} by using the molar mass of Ca^{2+}.

$$\boxed{mg\ Ca^{2+}} \xrightarrow{\frac{1\ g}{10^3\ mg}} \boxed{g\ Ca^{2+}} \xrightarrow{\frac{1}{molar\ mass}} \boxed{mol\ Ca^{2+}}$$

The average atomic mass of a calcium atom is 40.078 amu, which means the molar mass of calcium is 40.08 g/mol when rounded to four significant figures.[1] We can estimate the answer to be about 0.01 mol because we are dividing about 0.4 g by about 40 g/mol.

Solve

$$425\ \cancel{mg\ Ca^{2+}} \times \frac{1\ g}{10^3\ \cancel{mg}} \times \frac{1\ mol\ Ca^{2+}}{40.08\ \cancel{g\ Ca^{2+}}} = 0.0106\ mol\ Ca^{2+} = 1.06 \times 10^{-2}\ mol\ Ca^{2+}$$

Think About It People seem most comfortable dealing with numbers between 1 and 1000, which is why the mass is given in milligrams in the antacid tablet. The answer here seems appropriate based on the small number of grams of calcium in the tablet.

Practice Exercise The mass of the diamond in Figure 3.12 is 3.25 carats (1 carat = 0.200 g). Diamonds are nearly pure carbon. How many moles of carbon are in the diamond?

(Answers to Practice Exercises are in the back of the book.)

[1]In Sample Exercises involving molar masses, we will usually round off the values given in the periodic table to four significant figures, except for elements with molar masses greater than 100.

SAMPLE EXERCISE 3.4 Converting Number of Moles into Mass **LO1**

A helium balloon sold at an amusement park contains 0.462 moles of He. How many grams of He are in the balloon?

Collect, Organize, and Analyze We are asked to convert 0.462 mol of He to grams of He. We will need to use the molar mass of helium, 4.003 g/mol, to convert from moles to grams:

Because the balloon contains approximately $\frac{1}{2}$ mole of helium, we estimate the answer should be around 2 g.

Solve

$$0.462 \ \cancel{\text{mol He}} \times \frac{4.003 \ \text{g He}}{1 \ \cancel{\text{mol He}}} = 1.85 \ \text{g He}$$

Think About It Two grams is very little mass, which seems reasonable for a balloon that is lighter than air.

 Practice Exercise
How many grams of gold are there in 0.250 mole of gold?

(Answers to Practice Exercises are in the back of the book.)

CONCEPT **TEST**

Which contains more atoms: 1 gram of gold (Au) or 1 gram of silver (Ag)?

(Answers to Concept Tests are in the back of the book.)

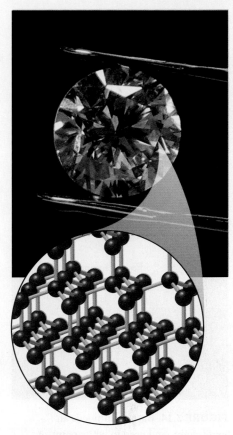

FIGURE 3.12 A 3.25-carat diamond.

Molecular Masses and Formula Masses

The **molecular mass** of a molecular compound is the sum of the atomic masses of the atoms in that molecule. Because atomic masses are given in atomic mass units, molecular masses are also reported in amu. Thus the molecular mass of sulfur trioxide, SO_3, is the sum of the masses of 1 sulfur atom and 3 oxygen atoms:

$$32.06 \ \text{amu} + (3 \times 16.00 \ \text{amu}) = 80.06 \ \text{amu}$$

Just as the molar mass of an element in grams per mole is numerically the same as the atomic mass of one atom of the element in atomic mass units (Figure 3.10), the molar mass of a molecular compound in grams per mole is numerically the same as its molecular mass in atomic mass units. Thus the molar mass of SO_3 is 80.06 g/mol (Figure 3.13).

As with atomic masses, we can use the concept of moles to scale from molecular masses to molar masses, which are quantities that are large enough to measure and manipulate in the laboratory. For example, the molar mass of SO_3 in grams per mole is the sum of the masses of 1 mole of sulfur atoms and 3 moles of oxygen atoms:

$$\mathcal{M}_{SO_3} = 32.06 \ \text{g/mol} + 3(16.00 \ \text{g/mol}) = 80.06 \ \text{g/mol} \ SO_3$$

The concept of molar mass applies to not only molecular compounds but also ionic compounds. For example, 1 mole of BaS contains 1 mole (137.33 g) of Ba^{2+} ions and 1 mole (32.06 g) of S^{2-} ions. Therefore the molar mass of BaS is

$$\mathcal{M}_{BaS} = 137.33 \ \text{g/mol} + 32.06 \ \text{g/mol} = 169.39 \ \text{g/mol} \ BaS$$

Molecular mass of SO_3	80.06 amu/molecule
Mass of 1 mol of SO_3	80.06 g
Molar mass of SO_3	80.06 g/mol

FIGURE 3.13 The molecular mass (in amu/molecule) and the molar mass (in g/mol) of sulfur trioxide have the same numerical value.

CONNECTION In Section 2.6 we defined the composition of an ionic compound in terms of the formula unit.

FIGURE 3.14 Barium sulfide is an ionic compound used in white paint. A crystal of barium sulfide consists of a three-dimensional array of Ba^{2+} ions and S^{2-} ions. The 1:1 ratio of Ba^{2+} ions to S^{2-} ions is represented in the formula unit of the crystal (enclosed in a dashed oval) and in the empirical formula of the compound (BaS).

Keep in mind that there are no discrete molecules of BaS. This compound is ionic, and its crystals consist of ordered arrays of cations and anions (Figure 3.14). The formula unit defines the smallest integer ratio of positive and negative ions, 1:1 in this case, that describes the composition of an ionic compound. The mass of one formula unit of an ionic compound is called its **formula mass**.

Figure 3.15 summarizes how to use Avogadro's number and molar masses for a given quantity of an element or compound to convert among its mass in grams, the number of moles, or the number of particles.

SAMPLE EXERCISE 3.5 Calculating the Molar Mass of a Compound   **LO1**

Calculate the molar masses of (a) H_2O and (b) H_2SO_4.

Collect and Organize We are given the formulas of two compounds and want to find their molar masses. The molar mass of a molecular compound is the sum of the molar masses of the elements in the molecular formula of the compound, each multiplied by the number of atoms of that element in the molecular formula. The molar mass of an element in grams per mole can be found in the periodic table.

Analyze
a. One mole of H_2O contains 2 moles of H atoms and 1 mole of O atoms. Therefore,

$$\boxed{\mathcal{M}_{H_2O}} = \boxed{2 \times \mathcal{M}_H} + \boxed{\mathcal{M}_O}$$

b. One mole of H_2SO_4 contains 2 moles of H atoms, 1 mole of S atoms, and 4 moles of O atoms. Therefore,

$$\boxed{\mathcal{M}_{H_2SO_4}} = \boxed{2 \times \mathcal{M}_H} + \boxed{\mathcal{M}_S} + \boxed{4 \times \mathcal{M}_O}$$

Solve
a. The molar mass of H_2O is

$$2(1.008 \text{ g/mol}) + 16.00 \text{ g/mol} = 18.02 \text{ g/mol } H_2O$$

b. The molar mass of H_2SO_4 is

$$2(1.008 \text{ g/mol}) + 32.06 \text{ g/mol} + 4(16.00 \text{ g/mol}) = 98.08 \text{ g/mol } H_2SO_4$$

Think About It The molar mass of a compound is the sum of the masses of the elements in that compound. Because H_2SO_4 contains more atoms and heavier atoms than H_2O does, we expect the molar mass of sulfuric acid to be larger than that of water.

Practice Exercise Green plants take in water (H_2O) and carbon dioxide (CO_2) and produce glucose ($C_6H_{12}O_6$) and oxygen (O_2). Calculate the molar masses of carbon dioxide, oxygen, and glucose.

(Answers to Practice Exercises are in the back of the book.)

FIGURE 3.15 The mass of a substance, the number of moles, and the number of particles are related by the molar mass ($\mathcal{M}$) of the substance and Avogadro's number (N_A).

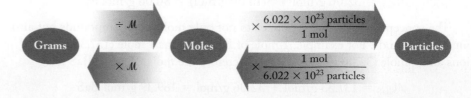

formula mass the mass of one formula unit of an ionic compound.

SAMPLE EXERCISE 3.6 Interconverting Grams, Moles, and Molecules   **LO1**

Lithium carbonate is used in the treatment of bipolar disorder. Calculate the number of moles and the number of formula units of lithium carbonate in a tablet containing 0.350 g of Li_2CO_3.

Collect and Organize We are given the mass of Li_2CO_3. We are asked to express that mass in terms of the number of moles of Li_2CO_3 and the number of formula units. Lithium carbonate is an ionic compound with the formula unit Li_2CO_3. We need to determine the formula mass of Li_2CO_3 and use Avogadro's number.

Analyze Figure 3.15 illustrates the general method we can apply. To convert a mass of Li_2CO_3 in grams into the equivalent number of formula units, we divide by its molar mass and multiply by Avogadro's number:

$$\boxed{g\ Li_2CO_3} \xrightarrow{\dfrac{1}{molar\ mass}} \boxed{mol\ Li_2CO_3} \xrightarrow{N_A} \boxed{formula\ units\ Li_2CO_3}$$

Because we have a relatively small mass of Li_2CO_3, we expect the number of moles of Li_2CO_3 to be less than one. Avogadro's number, however, is very large, so the number of formula units of Li_2CO_3 should be very large.

Solve First we need to calculate the molar mass of Li_2CO_3 by adding twice the molar mass of Li, the molar mass of C, and three times the molar mass of O:

$$2(6.941\ g/mol) + 12.01\ g/mol + 3(16.00\ g/mol) = 73.89\ g/mol\ Li_2CO_3$$

Then use this molar mass to convert from grams to moles:

$$0.350\ \cancel{g\ Li_2CO_3} \times \frac{1\ mol\ Li_2CO_3}{73.89\ \cancel{g\ Li_2CO_3}} = 0.004737\ mol\ Li_2CO_3 = 4.737 \times 10^{-3}\ mol\ Li_2CO_3$$

We report the answer with three significant figures, 4.74×10^{-3} moles, because the mass of Li_2CO_3 has only three significant figures.

Carrying on the calculation with the intermediate value and multiplying by Avogadro's number gives us

$$4.737 \times 10^{-3}\ \cancel{mol\ Li_2CO_3} \times \frac{6.022 \times 10^{23}\ formula\ units\ Li_2CO_3}{1\ \cancel{mol\ Li_2CO_3}}$$
$$= 2.852 \times 10^{21}\ formula\ units\ Li_2Co_3$$

The final answer can have only three significant figures, so the number of formula units of Li_2CO_3 is 2.85×10^{21}.

Think About It Because atoms and ions are tiny, we expect a large number of formula units in a relatively small mass of Li_2CO_3. We cannot count individual formula units of an ionic compound, but we can equate the number of formula units with a specific mass, a quantity we can readily measure in the laboratory.

 Practice Exercise How many moles of $CaCO_3$ are in an antacid tablet containing 0.500 g of $CaCO_3$? How many formula units of $CaCO_3$ are in this tablet?

(Answers to Practice Exercises are in the back of the book.)

SAMPLE EXERCISE 3.7 Calculating the Mass of an Element   **LO1**
                        in a Mixture of Its Compounds

The discovery that iron metal could be extracted from iron minerals such as Fe_2O_3 and Fe_3O_4 ushered in the Iron Age around 2000 BCE. If an ancient metalworker converts all the iron in a mixture of 341 g of Fe_2O_3 and 113 g of Fe_3O_4 to iron metal, what mass of iron does this represent?

Collect and Organize We are given the masses of two iron oxides and asked to calculate the combined mass of iron in the two oxides. The chemical formulas of the oxides tell us the number of moles of Fe per mole of each oxide. The molar mass of a compound or element can be used to interconvert its mass and the equivalent number of moles as shown in Figure 3.15.

Analyze To convert the masses of two iron oxides into equivalent masses of iron, we need to calculate the number of moles of each oxide in the mixture and then use the formulas of the two oxides to convert the number of moles of each oxide to moles of iron. That value multiplied by the molar mass of Fe will give us the mass of Fe in the oxide. So, first we need to calculate the molar masses of Fe_2O_3 and Fe_3O_4, which will serve as conversion factors for calculating the number of moles of Fe_2O_3 and Fe_3O_4 (see Figure 3.15). Next we convert these values into moles of Fe. There are 2 mol of Fe per mol of Fe_2O_3 and 3 mol of Fe per mol of Fe_3O_4, so the conversion factors in this step are 2 and 3. Then we convert mol Fe values to masses in grams by multiplying by the molar mass of Fe. Finally we sum the two masses of Fe. Summarizing all but the last of these steps for Fe_2O_3:

$$\boxed{\text{g } Fe_2O_3} \xrightarrow[\text{molar mass}]{1} \boxed{\text{mol } Fe_2O_3} \xrightarrow[\text{mol } Fe_2O_3]{2 \text{ mol Fe}} \boxed{\text{mol Fe}} \xrightarrow[\text{molar mass}]{} \boxed{\text{g Fe}}$$

Solve First we need to determine the formula masses of Fe_2O_3 and Fe_3O_4 by using the molar mass of Fe and the molar mass of O:

$$Fe_2O_3: 2(55.84 \text{ g/mol}) + 3(16.00 \text{ g/mol}) = 159.68 \text{ g/mol } Fe_2O_3$$

$$Fe_3O_4: 3(55.84 \text{ g/mol}) + 4(16.00 \text{ g/mol}) = 231.52 \text{ g/mol } Fe_3O_4$$

Converting from grams of Fe_2O_3 to moles of Fe_2O_3 and then to number of ions gives us

$$341 \text{ g } Fe_2O_3 \times \frac{1 \text{ mol } Fe_2O_3}{159.68 \text{ g } Fe_2O_3} \times \frac{2 \text{ mol Fe}}{1 \text{ mol } Fe_2O_3} \times \frac{55.84 \text{ g Fe}}{1 \text{ mol Fe}} = 238 \text{ g Fe}$$

We repeat the calculation for Fe_3O_4:

$$113 \text{ g } Fe_3O_4 \times \frac{1 \text{ mol } Fe_3O_4}{231.52 \text{ g } Fe_3O_4} \times \frac{3 \text{ mol Fe}}{1 \text{ mol } Fe_3O_4} \times \frac{55.84 \text{ g Fe}}{1 \text{ mol Fe}} = 81.8 \text{ g Fe}$$

The total mass of iron in the two samples is $238 + 81.8 = 319.8$ g.

Because the masses of Fe_2O_3 and Fe_3O_4 were each given to three significant figures, our calculations of the grams Fe in each must be reported to three significant figures. When adding the two masses together, we report the sum to the ones place: 320 g of Fe or 3.20×10^2 g of Fe.

Think About It The mass of iron recovered from a total of 454 g of iron oxides ($\approx$1 pound) represents

$$\frac{320 \text{ g Fe}}{454 \text{ g } Fe_2O_3 \text{ and } Fe_3O_4} \times 100\% = 70.5\%$$

of the mass of the oxides. This high percentage makes sense because the molar mass of Fe is more than three times the molar mass of oxygen. These iron oxides were valuable commodities in the Iron Age, as they are today.

 Practice Exercise How many grams of chromium are contained in a mixture of 125 g of sodium chromate (Na_2CrO_4) and 225 g of sodium dichromate ($Na_2Cr_2O_7$)?

(Answers to Practice Exercises are in the back of the book.)

Moles and Chemical Equations

The concepts of mole and molar mass offer three new interpretations of Equation 3.1:

$$SO_3(g) + H_2O(g) \rightarrow H_2SO_4(\ell)$$

1. The coefficients in the chemical equation tell us that 1 *mole* of SO_3 reacts with 1 *mole* of H_2O, producing 1 *mole* of H_2SO_4.
2. The molar masses of the reactants and products allow us to say that 80.06 *grams* of SO_3 react with 18.02 *grams* of H_2O, producing 98.08 *grams* of H_2SO_4.
3. Avogadro's number tells us that 6.022×10^{23} *molecules* of SO_3 react with 6.022×10^{23} *molecules* of H_2O, forming 6.022×10^{23} *molecules* of H_2SO_4. In terms of lowest whole-number ratios, 1 *molecule* of SO_3 reacts with 1 *molecule* of H_2O, forming 1 *molecule* of H_2SO_4.

The several interpretations of this equation, however, do not limit the amount of SO_3 that can react to 80.06 g. Reacting twice that amount, or 160.12 g of SO_3, is possible as long as twice the molar mass of water, 36.04 g of H_2O, is available. That is, 2 moles of SO_3 can react with 2 moles of H_2O to produce 2 moles of H_2SO_4. In fact, the equation tells us that *any* number of moles (x mol) of SO_3 react with an equal number of moles (x mol) of H_2O to produce the same quantity (x mol) of H_2SO_4. This interpretation is valid because the mole ratio (the ratio of the coefficients) of SO_3 to H_2O to H_2SO_4 in the equation is 1:1:1. The mole ratio differs when different substances combine and is therefore specific for a given equation. For example, in the reaction in Equation 3.2 between sulfuric acid and Fe_2O_3:

$$Fe_2O_3(s) + 3\ H_2SO_4(aq) \rightarrow 3\ H_2O(\ell) + Fe_2(SO_4)_3(aq)$$

the mole ratio of Fe_2O_3 to H_2SO_4 is 1:3.

From Equation 3.1, we can determine how much of one reactant is needed to completely react with any quantity of the other reactant. We can also calculate how much product can be made from any quantity of reactants by multiplying x by the appropriate molar mass.

	$SO_3(g)$	+	$H_2O(g)$	→	$H_2SO_4(\ell)$
Particle Ratios	1 molecule	+	1 molecule	→	1 molecule
Mole Ratios	1 mol	+	1 mol	→	1 mol
Mass Ratios	80.06 g	+	18.02 g	→	98.08 g
General Case (moles)	x mol	+	x mol	→	x mol
General Case (masses)	x(80.06 g)	+	x(18.02 g)	→	x(98.08 g)

This quantitative relationship between the reactants and products involved in a chemical reaction is called the **stoichiometry** of the reaction. In a reaction where all reactants are completely converted into products, the sum of the masses of the reactants equals the sum of the masses of the products (Figure 3.16). This fact illustrates a fundamental relation known as the **law of conservation of mass**, which applies to all chemical reactions. This law works in an equation representing any chemical reaction for two reasons: (1) the total number of atoms (and hence moles) of each element to the left of the reaction arrow must equal the total number of atoms (and moles) of that element to the right of the reaction arrow, and (2) the identity of atoms does not change in a chemical reaction.

stoichiometry the quantitative relation between reactants and products in a chemical reaction.

law of conservation of mass the principle that the sum of the masses of the reactants in a chemical reaction is equal to the sum of the masses of the products.

FIGURE 3.16 The law of conservation of mass states that the total mass of reactants consumed in a chemical reaction equals the total mass of products formed in the reaction. In the reaction shown here, the combined mass of SO_3 and H_2O consumed equals the mass of the H_2SO_4 that forms. The number of each kind of atom is the same on the left and right sides of the balanced equation.

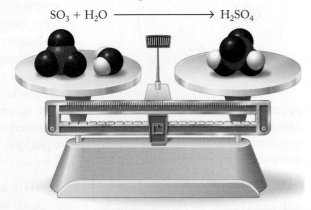

$$SO_3 + H_2O \longrightarrow H_2SO_4$$

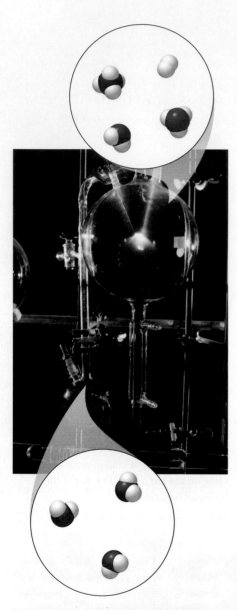

FIGURE 3.17 A molecular view of the Miller–Urey experiment. The large flask contained water, ammonia, hydrogen, and methane in ratios believed at that time to represent the composition of Earth's early atmosphere. The smaller flask was heated to add water vapor to the gas mixture in the large flask.

SAMPLE EXERCISE 3.8 Visualizing Chemical Reactions   **LO2**

Using blue spheres to represent nitrogen atoms and red spheres to represent oxygen atoms, sketch a picture describing the reaction between two molecules of N_2 and four molecules of O_2 to form NO_2.

Collect, Organize, and Analyze We are given the chemical formulas for two reactants and will use colored spheres to sketch their conversion to a product. We choose blue spheres to represent nitrogen atoms and red spheres to represent oxygen atoms. We use the atoms contained in two molecules of nitrogen and four molecules of oxygen to construct the appropriate number of NO_2 molecules.

Solve From a total of 4 nitrogen and 8 oxygen atoms, we can make four molecules of NO_2:

Think About It A chemical reaction involves rearrangements of atoms and molecules, resulting in new molecules. The number of each type of atom present before the reaction equals the number of each type of atom present after the reaction.

Practice Exercise Using colored spheres to represent nitrogen atoms (blue) and hydrogen atoms (white), sketch a picture describing the reaction between two molecules of N_2 and six molecules of H_2 to form ammonia, NH_3.

(Answers to Practice Exercises are in the back of the book.)

3.3 Writing Balanced Chemical Equations

A *balanced* chemical equation has the same number of atoms of each element on both sides of the equation. In this section we practice writing balanced chemical equations describing chemical reactions that may account for the products observed by Miller and Urey in their famous experiments on the origins of life on Earth mentioned at the beginning of this chapter.

Figure 3.17 revisits the apparatus we saw in Figure 3.1. A key molecule in the synthesis of glycine, one of the amino acids produced in the Miller–Urey experiment intended to simulate the steam-rich environment of volcanic eruptions, is hydrogen cyanide (HCN). Hydrogen cyanide can be formed from methane (CH_4) and ammonia (NH_3) in a reaction that also produces hydrogen (H_2). Let's write a balanced chemical equation describing this reaction. We know the formulas of the reactants and products, but we don't know the stoichiometry of the reaction—that is, we need to determine the coefficients that precede their formulas in a balanced chemical equation. There are several approaches to writing chemical equations. Let's apply the following four-step method, which works well if we have correctly identified all the reactants and products.

1. *Write a preliminary expression containing a single particle (atom, molecule, or formula unit) of each reactant and product with a reaction arrow separating reactants from products. Include phase symbols indicating physical states.*

Below is such an expression for the reaction of interest, including particulate models of the molecules involved.

$$CH_4(g) \;+\; NH_3(g) \;\rightarrow\; HCN(g) \;+\; H_2(g)$$

$$\text{reactants} \qquad\qquad \rightarrow \qquad\qquad \text{products}$$

CHEMTOUR
Balancing Equations

2. *Check whether the expression is balanced by counting the atoms of each element on each side of the reaction arrow.*

Adding up the atoms of each element in the reactants and products, we find that the expression above is balanced in terms of the numbers of C atoms and N atoms, but not in terms of H atoms: there are 7 on the reactant side but only 3 on the product side.

Element	Reactant Side	Product Side	Balanced?
C	1	1	✔
N	1	1	✔
H	4 + 3 = 7	1 + 2 = 3	✗

3. *Choose an element that appears in only one reactant and one product and balance it first.*

In this example, the two elements that occur only once on each side of the reaction arrow are C and N, but they are already balanced, so we can skip this step.

4. *Choose coefficients for the other substances so that the number of atoms for each element is the same on both sides of the reaction arrow.*

We need 4 more H atoms on the product side to have a balanced equation. Changing the coefficient in front of either HCN or H_2 could increase the number of H atoms, but changing the coefficient of HCN would also increase the numbers of C and N atoms, which are already balanced. Therefore the correct approach is to increase the coefficient of H_2 from 1 to 3, which increases the number of H atoms from 2 for one molecule of H_2 to 6 for three molecules of H_2.

$$CH_4(g) \;+\; NH_3(g) \;\rightarrow\; HCN(g) \;+\; \underline{\mathbf{3}}\, H_2(g)$$

A final check of the number of atoms of each element on both sides of the reaction arrow confirms that the reaction is balanced:

Element	Reactant Side	Product Side	Balanced?
C	1	1	✔
N	1	1	✔
H	4 + 3 = 7	1 + (**3** × 2) = 7	✔

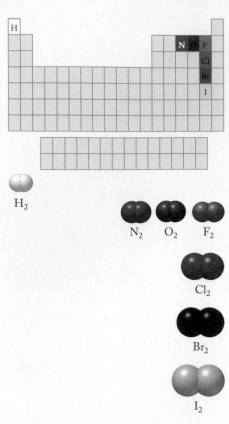

FIGURE 3.18 The diatomic elements are hydrogen, nitrogen, oxygen, and the group 17 halogens. The halogen astatine (At) at the bottom of group 17 is the rarest terrestrial element, and virtually none of its bulk physical properties are known.

Note that balance was achieved by changing a coefficient, *not* by changing any subscripts in chemical formulas. Changing subscripts would change the identities of substances, and that would change the nature of the reaction. Only coefficients may be changed in writing balanced chemical equations.

Let's get more practice writing balanced chemical equations by using another important reaction in the Miller–Urey experiment. Analysis of the vapor above the solution in the large flask revealed the presence of both carbon dioxide and carbon monoxide gas. It has been shown that the reaction between methane and water yields carbon dioxide and hydrogen.

1. *Write a preliminary expression containing a single particle of each reactant and product.* Methane has the formula CH_4, water is H_2O, carbon dioxide is CO_2, and hydrogen is H_2. Remember that we learned in Chapter 1 that all of the gaseous nonmetallic elements—except for the noble gases—exist in nature as diatomic molecules: H_2, N_2, O_2, F_2, and Cl_2. Bromine, a liquid at room temperature, and I_2, a solid, are also diatomic (Figure 3.18). Whenever these substances are involved in chemical reactions, they must be written as diatomic molecules.

$$CH_4(g) + H_2O(g) \rightarrow CO_2(g) + H_2(g)$$

2. *Check whether the expression is balanced.*

Element	Reactant Side	Product Side	Balanced?
C	1	1	✔
O	1	2	✗
H	4 + 2 = 6	2	✗

The equation as written is not yet balanced.

3. *Choose an element that appears in only one reactant and one product to balance first.* In this case we could choose either carbon or oxygen. Carbon is already balanced, so we focus on oxygen. With 1 O atom on the reactant side and 2 O atoms on the product side, we can make the number of O atoms equal on both sides of the equation by placing a coefficient of 2 in front of H_2O.

$$\_ CH_4(g) + \underline{\mathbf{2}}\, H_2O(g) \rightarrow \_ CO_2(g) + \_ H_2(g)$$

Assigning this coefficient balances the number of oxygen atoms but not the number of hydrogen atoms.

Element	Reactant Side	Product Side	Balanced?
C	1	1	✔
O	**2** × 1 = 2	2	✔
H	4 + (**2** × 2) = 8	2	✗

4. *Choose coefficients for the other substances so that the numbers of atoms of each element are the same on both sides of the equation.*

We can balance the H atoms by placing a coefficient of 4 in front of H_2, which finishes the balancing process:

$$CH_4(g) + \underline{\mathbf{2}}\,H_2O(g) \rightarrow CO_2(g) + \underline{\mathbf{4}}\,H_2(g)$$

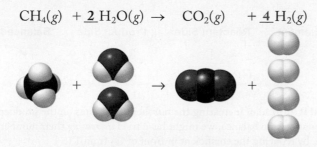

Element	Reactant Side	Product Side	Balanced?
C	1	1	✔
O	$\mathbf{2} \times 1 = 2$	2	✔
H	$4 + (\mathbf{2} \times 2) = 8$	$4 \times 2 = 8$	✔

SAMPLE EXERCISE 3.9 Writing a Balanced Chemical Equation **LO2**

Write a balanced chemical equation for the gas-phase reaction between SO_2 and O_2 that forms SO_3. This reaction, one of the causes of acid rain, occurs when fuels containing sulfur are burned. Acid rain is harmful to aquatic life and damages terrestrial plants.

Collect, Organize, and Analyze We are given the chemical formulas and physical states of the reactants and products. We need to write a balanced chemical equation such that we have the same number of atoms of each element on both sides of the equation. To balance the equation, we assign coefficients as needed to make the numbers of atoms of each element on the reactant (left) and product (right) sides of the reaction arrow equal.

Solve
1. Write a reaction expression based on single molecules of the reactants and product:

$$SO_2(g) + O_2(g) \rightarrow SO_3(g)$$

2. A check of whether the expression is balanced:

Element	Reactant Side	Product Side	Balanced?
S	1	1	✔
O	2 + 2 = 4	3	✗

discloses that it is not.
3. Sulfur is balanced, so we move on to oxygen.
4. Balancing the O atoms presents a challenge because increasing their number on the product side by increasing the SO_3 coefficient from 1 to 2:

$$SO_2(g) + O_2(g) \rightarrow \underline{\mathbf{2}}\,SO_3(g)$$

means that there are now two more O atoms and one more S atom on the product side than on the reactant side. Fortunately, both of these imbalances can be solved simultaneously by increasing the SO_2 coefficient from 1 to 2:

$$\underline{\mathbf{2}}\,SO_2(g) + O_2(g) \rightarrow \underline{\mathbf{2}}\,SO_3(g)$$

A check of the atom inventory on both sides confirms that the equation is now balanced:

Element	Reactant Side	Product Side	Balanced?
S	$2 \times 1 = 2$	$2 \times 1 = 2$	✔
O	$(2 \times 2) + 2 = 6$	$2 \times 3 = 6$	✔

Think About It Instead of increasing the number of O atoms on the product side to bring the equation into balance, we might have tried *decreasing* their number on the reactant side by reducing the coefficient in front of O_2 from 1 to $\frac{1}{2}$:

$$SO_2(g) + \tfrac{1}{2} O_2(g) \rightarrow SO_3(g)$$

This fix has the advantage of bringing the number of O atoms into balance without imbalancing the number of S atoms. It has the disadvantage of creating a fractional coefficient, but this problem can be solved by multiplying all of the coefficients by 2, which gives us the same balanced equation we obtained in the Solve section:

$$2\,SO_2(g) + O_2(g) \rightarrow 2\,SO_3(g)$$

We will use this strategy again in Sample Exercise 3.10 and in writing chemical equations for some of the combustion reactions in Section 3.4.

Practice Exercise Balance the chemical equations for (a) the reaction between elemental phosphorus, $P_4(s)$, and oxygen to make $P_4O_{10}(s)$ and (b) the reaction of water with $P_4O_{10}(s)$ to produce phosphoric acid, $H_3PO_4(aq)$.

(Answers to Practice Exercises are in the back of the book.)

CONCEPT TEST

Why can't we write a balanced equation for the reaction between SO_2 and O_2 in Sample Exercise 3.9 this way?

$$SO_2(g) + O(g) \rightarrow SO_3(g)$$

(Answers to Concept Tests are in the back of the book.)

Note that a chemical equation is analogous to a mathematical equation. The equation in Sample Exercise 3.9 is still balanced if we multiply all the coefficients by the same number, such as 2:

$$4\,SO_2(g) + 2\,O_2(g) \rightarrow 4\,SO_3(g)$$

It is also still balanced if we multiply all the coefficients by a fraction, such as $\frac{1}{2}$. However, it is conventional to report a balanced equation with the smallest whole-number coefficients.

SAMPLE EXERCISE 3.10 Balancing Chemical Equations Using Fractional Coefficients   **LO2**

Write a balanced chemical equation for the gas-phase reaction in which dinitrogen pentoxide is formed from nitrogen and oxygen.

Collect, Organize, and Analyze We are asked to write a balanced chemical equation given the names and physical states of the reactants and products. Nitrogen and oxygen gases exist in nature as diatomic molecules, so their molecular formulas are N_2 and O_2. The formula for dinitrogen pentoxide is N_2O_5. To write a balanced chemical equation, we must assign coefficients as needed to make the numbers of atoms of each element on the reactant (left) and product (right) sides of the reaction arrow equal.

Solve

1. Write an equation using correct formulas:

$$N_2(g) + O_2(g) \rightarrow N_2O_5(g)$$

2. The equation is not balanced as written:

Element	Reactant Side	Product Side	Balanced?
N	2	2	✔
O	2	5	✗

3. Two N atoms appear on each side of the equation. Oxygen is the only other element to balance in the equation. With 2 O atoms on the reactant side and 5 O atoms on the product side, there is no whole-number coefficient that we can place in front of the O_2 to make the number of O atoms equal on both sides of the equation. A coefficient of $\frac{5}{2}$ in front of O_2 would balance the number of O atoms, but we need to use whole numbers. Because $\frac{5}{2}$ times 2 is the whole number 5, we can place a coefficient of 5 in front of O_2 and a coefficient of 2 in front of N_2O_5:

$$\underline{}\, N_2(g) + \underline{5}\, O_2(g) \rightarrow \underline{2}\, N_2O_5(g)$$

However, the number of nitrogen atoms is no longer equal on both sides of the equation:

Element	Reactant Side	Product Side	Balanced?
N	2	**2** × 2 = 4	✗
O	**5** × 2 = 10	**2** × 5 = 10	✔

To balance the number of nitrogen atoms, we add a coefficient of 2 in front of N_2:

$$\underline{2}\, N_2(g) + \underline{5}\, O_2(g) \rightarrow \underline{2}\, N_2O_5(g)$$

The equation is now balanced.

Think About It In balancing a chemical equation, always check the equation after you are done to make sure that an element balanced in a previous step is still balanced at the end.

 Practice Exercise Balance the chemical equation for the reaction between carbon monoxide, $CO(g)$, and oxygen to form carbon dioxide, $CO_2(g)$.

(Answers to Practice Exercises are in the back of the book.)

3.4 Combustion Reactions

Both methane and carbon monoxide are still present in our atmosphere, although at much lower concentrations than on early Earth. Atmospheric methane (natural gas) can be traced to several sources, including wetlands, cattle ranching, rice production, and oil drilling and fracking operations. Carbon dioxide in the atmosphere arises from burning carbon-containing compounds. Natural sources of CO_2 include volcanic activity and forest fires. In our industrialized world, the greatest use of methane and petroleum-based fuels is energy production.

CONNECTION Molecules like methane (CH_4) are members of a class of organic compounds known as hydrocarbons because they are composed of only hydrogen and carbon, as we saw in Chapter 2.

FIGURE 3.19 The combustion of methane in a gas stove produces carbon dioxide and water.

Natural gas and other hydrocarbon fuels are burned to heat buildings and to warm water. Burning CH_4 is an example of an important class of chemical reactions known as **combustion reactions**, or simply *combustion* (Figure 3.19). Combustion refers to the reaction of a fuel with oxygen, such as the burning of a substance in air, which contains 21% oxygen.

When the combustion of a hydrocarbon is complete, the only products are carbon dioxide and water. In the presence of O_2, the more stable form of carbon is CO_2, not CO, so carbon monoxide also reacts with oxygen to form carbon dioxide. In the combustion reaction between methane and oxygen, the carbon becomes carbon dioxide when a sufficient supply of oxygen is available. The hydrogen in the fuel combines with oxygen to form water vapor. Let's see how to balance the equation describing the combustion of methane.

1. Our first step is to write the formulas of the reactants on the left and products on the right.

$$CH_4(g) + O_2(g) \rightarrow CO_2(g) + H_2O(g)$$

Element	Reactant Side	Product Side	Balanced?
C	1	1	✔
H	4	2	✗
O	2	2 + 1 = 3	✗

The carbon atoms are balanced but the hydrogen and oxygen atoms are not.

2. Like carbon, H appears in only one reactant and one product. Adding a coefficient of 2 in front of H_2O makes the number of H atoms equal on both sides of the equation but increases the number of O atoms on the product side:

$$CH_4(g) + \underline{}\, O_2(g) \rightarrow CO_2(g) + \underline{\mathbf{2}}\, H_2O(g)$$

Element	Reactant Side	Product Side	Balanced?
C	1	1	✔
H	4	**2** × 2 = 4	✔
O	2	2 + (**2** × 1) = 4	✗

3. This leaves only the O atoms to balance. A coefficient of 2 in front of O_2 makes the atoms of O equal on both sides of the equation:

$$CH_4(g) + \underline{\mathbf{2}}\, O_2(g) \rightarrow CO_2(g) + \underline{\mathbf{2}}\, H_2O(g)$$

Element	Reactant Side	Product Side	Balanced?
C	1	1	✔
H	4	**2** × 2 = 4	✔
O	**2** × 2 = 4	2 + (**2** × 1) = 4	✔

combustion reaction a heat-producing reaction between oxygen and another element or compound.

Balancing first C, then H, then O is a useful strategy when writing the equations for many combustion reactions involving hydrocarbons.

SAMPLE EXERCISE 3.11 Writing a Balanced Chemical Equation **LO2**
for a Combustion Reaction

Methane (CH_4) is the principal ingredient in natural gas, but significant concentrations of the hydrocarbon gases ethane (C_2H_6) and propane (C_3H_8) are also present in most natural gas samples. Write a balanced chemical equation describing the complete combustion of C_2H_6.

Collect, Organize, and Analyze We need to write a balanced chemical equation for the combustion of C_2H_6. The products of complete combustion of hydrocarbons are carbon dioxide and water.

Solve The preliminary expression with single particles of all known reactants and products describing the combustion of ethane is

$$C_2H_6(g) + O_2(g) \rightarrow CO_2(g) + H_2O(g)$$

Element	Reactant Side	Product Side	Balanced?
C	2	1	✗
H	6	2	✗
O	2	2 + 1 = 3	✗

Because this reaction is between oxygen and a hydrocarbon, we balance C first, then H, then O. Balance the carbon atoms first by giving CO_2 in the product a coefficient of 2:

$$\_\,C_2H_6(g) + \_\,O_2(g) \rightarrow \underline{2}\,CO_2(g) + \_\,H_2O(g)$$

Element	Reactant Side	Product Side	Balanced?
C	2	**2** × 1 = 2	✔
H	6	2	✗
O	2	(**2** × 2) + 1 = 5	✗

Then balance the hydrogen atoms by giving H_2O a coefficient of 3:

$$\_\,C_2H_6(g) + \_\,O_2(g) \rightarrow \underline{2}\,CO_2(g) + \underline{3}\,H_2O(g)$$

Element	Reactant Side	Product Side	Balanced?
C	2	**2** × 1 = 2	✔
H	6	**3** × 2 = 6	✔
O	2	(**2** × 2) + (**3** × 1) = 7	✗

At this stage, the oxygen atoms cannot be balanced with a simple whole-number coefficient for O_2; we need $\frac{7}{2}\,O_2$. However, if we give O_2 a coefficient of 7 and double the coefficients for ethane, carbon dioxide, and water, we can write

$$\underline{2}\,C_2H_6(g) + \underline{7}\,O_2(g) \rightarrow \underline{4}\,CO_2(g) + \underline{6}\,H_2O(g)$$

Element	Reactant Side	Product Side	Balanced?
C	**2** × 2 = 4	**4** × 1 = 4	✔
H	**2** × 6 = 12	**6** × 2 = 12	✔
O	**7** × 2 = 14	(**4** × 2) + (**6** × 1) = 14	✔

This equation is balanced.

3.5 Stoichiometric Calculations and the Carbon Cycle

Earth's atmosphere underwent a major change beginning about 2.5 billion years ago with the evolution of bacteria capable of *photosynthesis*. Photosynthesis is driven by the energy in sunlight and involves several steps, but the overall reaction is

$$6\,CO_2(g) + 6\,H_2O(\ell) \rightarrow C_6H_{12}O_6(aq) + 6\,O_2(g)$$
$$\text{Glucose}$$

Although photosynthetic bacteria are believed to have been the initial source of oxygen in our atmosphere, today O_2 comes mostly from green plants. The reverse reaction, called *respiration*, is the major source of energy for all living things on Earth:

$$C_6H_{12}O_6(aq) + 6\,O_2(g) \rightarrow 6\,CO_2(g) + 6\,H_2O(\ell)$$
$$\text{Glucose}$$

Photosynthesis and respiration are key reactions in the *carbon cycle* (Figure 3.20). The two processes are nearly, but not exactly, in balance in Earth's biosphere. If they were exactly in balance, no net change would have taken place in the concentrations of atmospheric carbon dioxide or oxygen in the past 2.5 billion years. However, about 0.01% of the decaying mass of plants and animals (called *detritus*) is incorporated into sediments and soil when organisms die. Shielded in this way from exposure to oxygen, the carbon in this mass is not converted back into CO_2. Although 0.01% may not seem like much, over hundreds of millions of years it has added up to the removal of about 10^{20} kg of carbon dioxide from the atmosphere. About 10^{15} kg of this buried carbon is in the form of fossil fuels: coal, petroleum, and natural gas.

As a result of human activity and the combustion of fossil fuels, the natural balance that limited the concentration of CO_2 in the atmosphere is being altered. Annually about 6.8 trillion (6.8×10^{12}) kg of carbon is reintroduced into the atmosphere as CO_2 as a result of the combustion of fossil fuels, and deforestation adds another 2×10^{12} kg each year. The effects of these additions on global climate have been the subject of considerable debate, and we examine them in Chapter 8. Here, though, we will use a typical reaction involving CO_2 to learn how to use a balanced chemical equation to calculate the mass of products in a chemical reaction.

If combustion of fossil fuels adds 6.8×10^{12} kg of carbon to the atmosphere each year as CO_2, what is the mass of the carbon dioxide added? The mass must be more than 6.8×10^{12} kg, because this amount is only the mass due to carbon. The balanced chemical equation for the combustion of carbon to carbon dioxide in excess O_2 is

$$C(s) + O_2(g) \rightarrow CO_2(g)$$

CHEMTOUR
Carbon Cycle

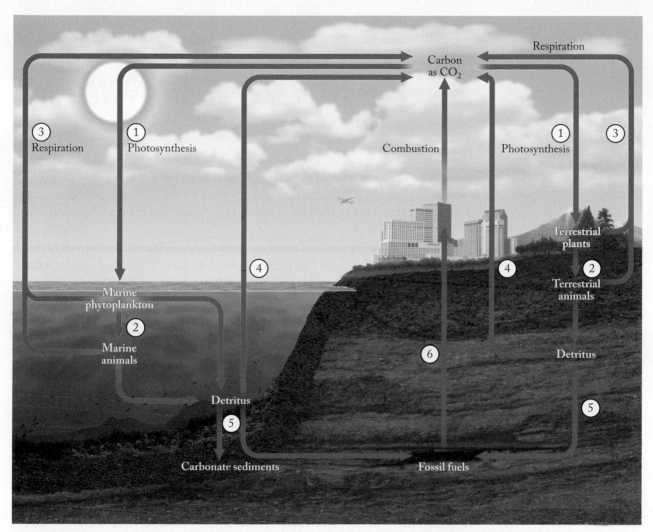

To use this equation to determine the mass of CO_2 released, we first use the molar mass of carbon, 12.01 g/mol, to convert mass into moles:

$$6.8 \times 10^{12} \ \cancel{kg \ C} \times \frac{10^3 \ g}{1 \ \cancel{kg}} \times \frac{1 \ mol \ C}{12.01 \ \cancel{g \ C}} = 5.7 \times 10^{14} \ mol \ C$$

We know from the coefficients for CO_2 and C in the balanced equation that the mole ratio of CO_2 to C is 1:1, so the amount of CO_2 produced is also 5.7×10^{14} mol:

$$5.7 \times 10^{14} \ \cancel{mol \ C} \times \frac{1 \ mol \ CO_2}{1 \ \cancel{mol \ C}} = 5.7 \times 10^{14} \ mol \ CO_2$$

To convert moles of CO_2 to mass, we first calculate the molar mass of CO_2:

$$12.01 \ g/mol + 2(16.00 \ g/mol) = 44.01 \ g/mol \ CO_2$$

Multiplying the moles of CO_2 by the molar mass of CO_2 and converting grams to kilograms gives us the mass of CO_2:

$$5.7 \times 10^{14} \ \cancel{mol \ CO_2} \times \frac{44.01 \ \cancel{g \ CO_2}}{1 \ \cancel{mol \ CO_2}} \times \frac{1 \ kg}{10^3 \ \cancel{g}} = 2.5 \times 10^{13} \ kg \ CO_2$$

FIGURE 3.20 The carbon cycle. ① Green plants and marine phytoplankton incorporate CO_2 into their biomass. ② Some of this biomass becomes the biomass of animals. ③ As photosynthetic organisms and animals respire, they release CO_2 back into the environment. ④ When they die, the decay of their tissues releases most of their carbon content as CO_2, but about 0.01% is incorporated into carbonate minerals and deposits of coal, petroleum, and natural gas (fossil fuels, ⑤). ⑥ Mining and the combustion of fossil fuels for human use are shifting the natural equilibrium that has controlled the concentration of CO_2 in the atmosphere.

Notice that the answer in each of these steps is the starting point for the next step. Therefore we can combine the three separate calculations into a single calculation:

$$6.8 \times 10^{12} \; \cancel{kg \; C} \times \frac{10^3 \; \cancel{g}}{1 \; \cancel{kg}} \times \frac{1 \; \cancel{mol \; C}}{12.01 \; \cancel{g \; C}} \times \frac{1 \; \cancel{mol \; CO_2}}{1 \; \cancel{mol \; C}} \times \frac{44.01 \; \cancel{g \; CO_2}}{1 \; \cancel{mol \; CO_2}} \times \frac{1 \; kg}{10^3 \; \cancel{g}}$$

$$= 2.5 \times 10^{13} \; kg \; CO_2$$

This procedure can be applied to determining the mass of any substance (reactant or product) involved in any chemical reaction if we know (1) the mass of another substance in the reaction and (2) the *stoichiometric relation* between the two substances, that is, their mole ratio in the balanced chemical equation.

SAMPLE EXERCISE 3.12 Calculating a Product Mass   **LO3**
from a Reactant Mass

In 2014 electric power plants in the United States consumed about 1.57×10^{11} kg of natural gas. Natural gas is mostly methane, CH_4, so we can approximate the combustion reaction generating the energy by the equation:

$$CH_4(g) + 2 \, O_2(g) \rightarrow CO_2(g) + 2 \, H_2O(g)$$

How many kilograms of CO_2 were released into the atmosphere from these power plants in 2014?

Collect, Organize, and Analyze We are asked to calculate the mass of CO_2 produced from combustion of 1.57×10^{11} kg of CH_4. The balanced equation tells us that 1 mole of carbon dioxide is produced for every 1 mole of methane consumed. We can change the mass of CH_4 in kilograms to grams and then convert into moles of CH_4 by multiplying by the molar mass of CH_4. In this reaction the moles of CH_4 consumed is equal to the number of moles of CO_2 produced, so the mole ratio is 1:1. Finally we convert moles of CO_2 into kilograms of CO_2. We can combine all of these steps into a single calculation:

$$\boxed{kg \; CH_4} \xrightarrow{\frac{10^3 \; g}{1 \; kg}} \boxed{g \; CH_4} \xrightarrow{\frac{1}{molar \; mass}} \boxed{mol \; CH_4} \xrightarrow{\frac{1 \; mol \; CO_2}{1 \; mol \; CH_4}}$$

$$\boxed{mol \; CO_2} \xrightarrow{molar \; mass} \boxed{g \; CO_2} \xrightarrow{\frac{1 \; kg}{10^3 \; g}} \boxed{kg \; CO_2}$$

Solve The chemical equation is balanced as written (you should always check to be sure). To convert between grams and moles of CH_4 and CO_2, we need to calculate the molar masses of these compounds. The molar mass of CH_4 is

$$12.01 \; g/mol + 4(1.008 \; g/mol) = 16.04 \; g/mol \; CH_4$$

and the molar mass of CO_2 is

$$12.01 \; g/mol + 2(16.00 \; g/mol) = 44.01 \; g/mol \; CO_2$$

Convert the mass of CH_4 into moles:

$$1.57 \times 10^{11} \; \cancel{kg \; CH_4} \times \frac{10^3 \; \cancel{g}}{1 \; \cancel{kg}} \times \frac{1 \; mol \; CH_4}{16.04 \; \cancel{g \; CH_4}} = 9.788 \times 10^{12} \; mol \; CH_4$$

Convert moles of CH_4 into moles of CO_2 by using the mole ratio from the balanced chemical equation:

$$9.788 \times 10^{12} \; \cancel{mol \; CH_4} \times \frac{1 \; mol \; CO_2}{1 \; \cancel{mol \; CH_4}} = 9.788 \times 10^{12} \; mol \; CO_2$$

Convert moles of CO_2 into mass of CO_2:

$$9.788 \times 10^{12} \text{ mol CO}_2 \times \frac{44.01 \text{ g CO}_2}{1 \text{ mol CO}_2} \times \frac{1 \text{ kg}}{10^3 \text{ g}} = 4.31 \times 10^{11} \text{ kg CO}_2$$

We can combine the three separate calculations into a single calculation, and in subsequent problems we may not show the individual steps:

$$1.57 \times 10^{11} \text{ kg CH}_4 \times \frac{10^3 \text{ g}}{1 \text{ kg}} \times \frac{1 \text{ mol CH}_4}{16.04 \text{ g CH}_4} \times \frac{1 \text{ mol CO}_2}{1 \text{ mol CH}_4} \times \frac{44.01 \text{ g CO}_2}{1 \text{ mol CO}_2} \times \frac{1 \text{ kg}}{10^3 \text{ g}}$$

$$= 4.31 \times 10^{11} \text{ kg CO}_2$$

Think About It The answer 4.31×10^{11} kg of CO_2 is about 1.7% of the mass of CO_2 we calculated for total annual CO_2 production by combustion of fossil fuels.

Practice Exercise Disposable lighters burn butane (C_4H_{10}) and produce CO_2 and H_2O. Balance the chemical equation for this combustion reaction, and determine how many grams of CO_2 are produced by burning 1.00 g of C_4H_{10}.

(Answers to Practice Exercises are in the back of the book.)

SAMPLE EXERCISE 3.13 Calculating the Masses of Reactants   **LO3**

Copper was among the first metals to be refined from minerals collected by ancient metalworkers. The production of copper was already an industry by 3500 BCE. Cuprite is a copper mineral commonly found near Earth's surface, making it a likely resource for Bronze Age coppersmiths. Cuprite has the formula Cu_2O and can be converted to copper metal by reacting it with charcoal:

$$2 \text{ Cu}_2\text{O}(s) + \text{C}(s) \rightarrow 4 \text{ Cu}(s) + \text{CO}_2(g)$$

How much cuprite and how much carbon are needed to prepare a 256 g copper bracelet such as the one shown in Figure 3.21?

Collect and Organize We are given a balanced chemical equation and the mass of product. We are asked to find the masses of the reactants. Because the chemical equation relates moles, not masses, of reactants and products, we need to find the molar mass of each substance in the reaction.

Analyze The balanced chemical equation tells us that for every 4 moles of copper we produce, 2 moles of Cu_2O and 1 mole of C must react. To use these mole ratios, we first need to convert the mass of copper to the equivalent number of moles of Cu by using the molar mass of Cu. We then work a separate conversion for each reactant: the first conversion uses the mole ratio of Cu_2O to Cu (2:4) from the balanced equation, and the second uses the mole ratio of C to Cu (1:4). Finally, we use the molar masses of Cu_2O and C to find the mass of each required for the reaction.

Solve Carrying out the steps described in the analysis of the problem:

$$256 \text{ g Cu} \times \frac{1 \text{ mol Cu}}{63.55 \text{ g Cu}} \times \frac{2 \text{ mol Cu}_2\text{O}}{4 \text{ mol Cu}} \times \frac{143.09 \text{ g Cu}_2\text{O}}{1 \text{ mol Cu}_2\text{O}} = 288 \text{ g Cu}_2\text{O}$$

$$256 \text{ g Cu} \times \frac{1 \text{ mol Cu}}{63.55 \text{ g Cu}} \times \frac{1 \text{ mol C}}{4 \text{ mol Cu}} \times \frac{12.01 \text{ g C}}{1 \text{ mol C}} = 12.1 \text{ g C}$$

FIGURE 3.21 Copper bracelet.

We are allowed only three significant figures in our answer, so the final masses of the reactants we need are 288 g of Cu_2O and 12.1 g of C.

Think About It To prepare a given mass of copper, we must have sufficient amounts of both copper ore and charcoal. Usually there is a limited supply of ore and more than enough charcoal.

Practice Exercise The copper mineral chalcocite, Cu_2S, can be converted to copper simply by heating in air: $Cu_2S(s) + O_2(g) \rightarrow 2\ Cu(s) + SO_2(g)$. How much Cu_2S is needed to make 256 g of Cu? How many grams of SO_2 are produced?

(Answers to Practice Exercises are in the back of the book.)

3.6 Determining Empirical Formulas from Percent Composition

Natural sources of methane such as swamps, bogs, rice paddies, and animals introduce 0.570 teragrams (1 teragram = 1 Tg = 1 million metric tons = 10^{12} g) or 5.70×10^{11} g of CH_4 into the atmosphere every day. Hydrocarbons are also emitted into the atmosphere by other sources, such as conifers like the longleaf pine shown in Figure 3.22. Conifers emit a variety of hydrocarbons into the air, including pinene, which has the chemical formula $C_{10}H_{16}$. One way to distinguish between two hydrocarbons such as CH_4 and $C_{10}H_{16}$ is in terms of **percent composition**: the composition of a compound expressed in terms of the percentages of the masses of the constituent elements with respect to the total mass of the compound.

To calculate the percent composition of both CH_4 and $C_{10}H_{16}$, we must calculate the carbon content of both compounds. Suppose we have 1 mole of CH_4, which has a molar mass of

$$4(1.008\ \text{g H/mol}) + 12.01\ \text{g C/mol} = 16.04\ \text{g/mol}$$

Of this 16.04 g, C accounts for 12.01 g. Therefore the C content of CH_4 is

$$\frac{\text{mass of C}}{\text{total mass}} = \frac{12.01\ \text{g C}}{16.04\ \text{g CH}_4} = 0.7488$$

or

$$0.7488 \times 100\% = 74.88\%\ \text{C}$$

It follows that the hydrogen content is

$$\frac{\text{mass of H}}{\text{total mass}} = \frac{4(1.008\ \text{g H})}{16.04\ \text{g CH}_4} = 0.2512\ \text{or}\ 25.12\%\ \text{H}$$

Because CH_4 contains only C and H, we could also determine the percent H by subtracting the percent C from 100%:

$$100.00\% - 74.88\% = 25.12\%$$

Thus the percent composition of CH_4 is 74.88% C and 25.12% H.

How does this compare with the percent C in pinene? If we repeat the calculation for $C_{10}H_{16}$, the molar mass is

$$16(1.008\ \text{g H/mol}) + 10(12.01\ \text{g C/mol}) = 136.23\ \text{g/mol}$$

CHEMTOUR
Percent Composition

FIGURE 3.22 The odor of the longleaf pine (*Pinus palustris*) comes from the volatile hydrocarbon pinene.

The C content of $C_{10}H_{16}$ is

$$\frac{\text{mass of C}}{\text{total mass}} = \frac{10(12.01 \text{ g C})}{136.23 \text{ g } C_{10}H_{16}} = 0.8816 = 88.16\%$$

and the hydrogen content is

$$100.00\% - 88.16\% = 11.84\%$$

Pinene contains a higher percent carbon by mass, and a correspondingly lower percent hydrogen, than methane. Actually, methane has the highest percent hydrogen of all hydrocarbons.

percent composition the composition of a compound expressed in terms of the percentage by mass of each element in the compound.

SAMPLE EXERCISE 3.14 Calculating Percent Composition **LO4**
from a Chemical Formula

The compound $C_2H_2F_4$ is a propellant in the inhalers used by asthma sufferers (Figure 3.23). What is the percent composition of $C_2H_2F_4$?

Collect, Organize, and Analyze We are asked to calculate the percent composition of $C_2H_2F_4$. We need the molar masses of each of the elements (C, H, and F) to calculate the molar mass of $C_2H_2F_4$ and the relative contribution of each element to the molar mass. To calculate the percent composition of $C_2H_2F_4$, we must determine the percentage by mass of each element in one mole of $C_2H_2F_4$. These percentages can be calculated by dividing the mass of each element in 1 mole of $C_2H_2F_4$ by the molar mass of $C_2H_2F_4$.

Solve The molar mass of $C_2H_2F_4$ is

$$2(12.01 \text{ g/mol}) + 2(1.008 \text{ g/mol}) + 4(19.00 \text{ g/mol}) = 102.04 \text{ g/mol}$$

Thus the percent composition of this compound is

$$\%C = \frac{24.02 \text{ g C}}{102.04 \text{ g}} \times 100\% = 23.54\% \text{ C}$$

$$\%H = \frac{2.016 \text{ g H}}{102.04 \text{ g}} \times 100\% = 1.98\% \text{ H}$$

$$\%F = \frac{76.00 \text{ g F}}{102.04 \text{ g}} \times 100\% = 74.48\% \text{ F}$$

Think About It Although C and F have similar molar masses, the observation that the percentage of F is nearly three times the percentage of C makes sense because there are two moles of F for every 1 mole of C in $C_2H_2F_4$. It also makes sense that the percentage of the lightest element, hydrogen, is very small.

 Practice Exercise Determine the percent composition of Fluosol-DA, $C_{10}F_{18}$, the only FDA-approved synthetic blood substitute (Figure 3.24).

(Answers to Practice Exercises are in the back of the book.)

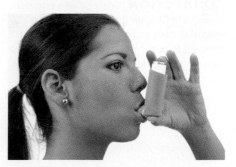

FIGURE 3.23 Patients suffering from asthma and other respiratory ailments use inhalers to relieve their discomfort. The medication is delivered using compounds such as $C_2H_2F_4$ as the propellant.

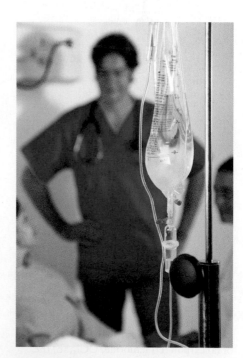

FIGURE 3.24 The fluorocarbon $C_{10}F_{18}$ is the only FDA-approved synthetic blood substitute.

Note in Sample Exercise 3.14 that the three percentages sum to 100.00%. Percent composition values should always sum to 100%, or very close to 100%, if we have accounted for all the elements that make up the total mass of the compound. The total may deviate slightly from 100% because of rounding.

We typically determine the percent composition of a substance in the laboratory by measuring the amount of each element in a given mass of the substance.

We can use these data to determine the *empirical formula* of the substance. There are four steps to deriving an empirical formula from percent composition data obtained in the laboratory:

1. Assume you have exactly 100 g of the substance, so that the percent composition values are equivalent to the values of the masses of the elements expressed in grams. As an initial check, add the mass percentages given in the problem to make sure they total 100%.
2. Convert the mass of each element into moles.
3. Compute the mole ratio by reducing one of the mole values to 1.
4. If necessary, convert the ratio from step 3 into a ratio of whole numbers.

CONNECTION We discussed empirical formulas in Chapter 2. *Empirical*, which is related to the word *experiment*, means "derived from experimental data" such as the data from experiments to determine percent composition.

Let's look at an example. Carbon tetrachloride was once widely used as a fire-extinguishing agent and as a solvent in dry cleaning. It is 7.808% C and 92.19% Cl. These two values sum to 99.998%, so we know all the elements are accounted for and the compound contains only carbon and chlorine. In exactly 100 g of this substance, there are

$$7.808\% \text{ C} = 7.808 \text{ g C in } 100 \text{ g}$$

$$92.19\% \text{ Cl} = 92.19 \text{ g Cl in } 100 \text{ g}$$

Now we determine the number of moles of C and Cl in this sample:

$$7.808 \text{ g C} \times \frac{1 \text{ mol C}}{12.01 \text{ g C}} = 0.6501 \text{ mol C}$$

$$92.19 \text{ g Cl} \times \frac{1 \text{ mol Cl}}{35.45 \text{ g Cl}} = 2.600 \text{ mol Cl}$$

The mole ratio of carbon to chlorine is 0.6501:2.600, but we need to express this in a chemical formula in terms of whole numbers. We divide both numbers by the smaller of the two; this ensures that one of the numbers in the ratio will have a value of 1 and the other will be larger than 1:

$$\frac{0.6501 \text{ mol C}}{0.6501} = 1.000 \text{ or } 1 \text{ mol C} \qquad \frac{2.600 \text{ mol Cl}}{0.6501} = 3.999 \text{ or } 4 \text{ mol Cl}$$

That means the empirical formula of carbon tetrachloride is CCl_4.

We can use these same steps for substances that contain more than two elements. Let's see how this procedure works with a commonly used refrigeration gas. Elemental analysis reveals that the percent composition of a sample is 28.58% C, 3.60% H, and 67.82% F. The sum of these percentages totals 100%, so all elements are accounted for in our analysis. Our calculations are then

$$28.58\% \text{ C} = 28.58 \text{ g C in } 100 \text{ g}$$

$$3.60\% \text{ H} = 3.60 \text{ g H in } 100 \text{ g}$$

$$67.82\% \text{ F} = 67.82 \text{ g F in } 100 \text{ g}$$

and the numbers of moles are

$$28.58 \text{ g C} \times \frac{\text{mol C}}{12.01 \text{ g C}} = 2.380 \text{ mol C} \qquad 67.82 \text{ g F} \times \frac{\text{mol F}}{19.00 \text{ g F}} = 3.569 \text{ mol F}$$

$$3.60 \text{ g H} \times \frac{\text{mol H}}{1.008 \text{ g H}} = 3.571 \text{ mol H}$$

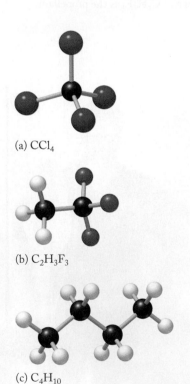

(a) CCl_4

(b) $C_2H_3F_3$

(c) C_4H_{10}

FIGURE 3.25 The molecular structures of (a) carbon tetrachloride, CCl_4; (b) the refrigerant gas, $C_2H_3F_3$; and (c) the fuel gas butane, C_4H_{10}.

The mole ratio for C:H:F is 2.380:3.569:3.571. To convert this ratio to small whole numbers, we divide all values by the smallest value:

$$\frac{2.380 \text{ mol C}}{2.380} = 1.000 \text{ or } 1 \text{ mol C} \qquad \frac{3.569 \text{ mol F}}{2.380} = 1.500 \text{ or } 1.5 \text{ mol F}$$

$$\frac{3.571 \text{ mol H}}{2.380} = 1.500 \text{ or } 1.5 \text{ mol H}$$

To change the mole ratio 1:1.5:1.5 to a ratio of whole numbers, we multiply the numbers by a factor that results in whole numbers. Multiplying 1.5 (or $\frac{3}{2}$) by 2 yields 3, so we multiply all three terms in this ratio by 2 to obtain a C:H:F ratio of 2:3:3. Thus the empirical formula is $C_2H_3F_3$.

Both CCl_4 and $C_2H_3F_3$ are molecular compounds whose empirical formulas match their molecular formulas (Figure 3.25a,b). This is not always the case. Consider another hydrocarbon found in natural gas and used as a fuel: butane. Butane has a percent composition of 82.66% C and 17.34% H. We can calculate its empirical formula:

$$82.66\% \text{ C} = 82.66 \text{ g C} \times \frac{1 \text{ mol C}}{12.01 \text{ g C}} = 6.883 \text{ mol C}$$

$$17.34\% \text{ H} = 17.34 \text{ g H} \times \frac{1 \text{ mol H}}{1.008 \text{ g H}} = 17.20 \text{ mol H}$$

Examining the mole ratio 6.883:17.20 gives us

$$\frac{6.883}{6.883} : \frac{17.20}{6.883} = 1:2.499$$

When we multiply both numbers by 2, the whole-number ratio becomes 2:5, and the empirical formula for butane is C_2H_5. The structure of a molecule of butane is shown in Figure 3.25(c); it has the *molecular* formula C_4H_{10}. Whereas the percent composition data gives us the empirical formula—the lowest whole-number ratio of atoms in the molecule—it is the molecular formula that tells us the actual numbers of atoms in the molecule.

For ionic compounds such as FeO and Fe_2O_3, the chemical formulas are empirical formulas that represent the ratio of the ions in one formula unit of the compound. However, FeO consists of a three-dimensional array of Fe^{2+} and O^{2-} ions, whereas Fe_2O_3 consists of a three-dimensional array of Fe^{3+} and O^{2-} ions. In FeO, the Fe^{2+} and O^{2-} ions are arranged as shown in Figure 3.26, often called the *halite structure* because it is also the structure seen in the mineral halite (NaCl). In an ionic compound the chemical formula, the empirical formula, and the formula unit are all the same.

CONCEPT TEST

Which pairs of the following compounds have the same empirical formula and same percent composition?

a. Ethylene (C_2H_4), a gas used to ripen bananas

b. Eicosene ($C_{20}H_{40}$), a compound used to attract Japanese beetles to traps

c. Acetylene (C_2H_2), a gas used in welding

d. Benzene (C_6H_6), a known carcinogen found in gasoline

(Answers to Concept Tests are in the back of the book.)

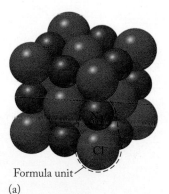

Formula unit

(a)

Formula unit

(b)

FIGURE 3.26 (a) The halite structure is the three-dimensional pattern of the Na^+ and Cl^- ions in NaCl; (b) the same pattern is found in the Fe^{2+} and O^{2-} ions in FeO.

SAMPLE EXERCISE 3.15 Deriving an Empirical Formula
from Percent Composition

LO4

Nitrous oxide, the laughing gas used in dentists' offices, is 63.65% N. What is the empirical formula of nitrous oxide?

Collect and Organize We are given the percent composition by mass of N (63.65%) and asked to determine the simplest whole-number ratio of nitrogen and oxygen in the chemical formula.

Analyze We can obtain the percent composition by mass of oxygen in nitrous oxide by subtracting the percent N from 100%. We can use the four steps described earlier to determine the empirical formula.

Solve The percent O in the sample is:

$$\%O = 100.00\% - \%N = 100.00\% - 63.65\% = 36.35\%$$

Now we have the data we need to determine the empirical formula of nitrous oxide.

1. Convert percent composition values into masses by assuming a sample size of exactly 100 g:

$$63.65\% \text{ N} = 63.65 \text{ g N in 100 g}$$

$$36.35\% \text{ O} = 36.35 \text{ g O in 100 g}$$

2. Convert masses into moles:

$$63.65 \text{ g N} \times \frac{1 \text{ mol N}}{14.01 \text{ g N}} = 4.543 \text{ mol N}$$

$$36.35 \text{ g O} \times \frac{1 \text{ mol O}}{16.00 \text{ g O}} = 2.272 \text{ mol O}$$

3. Simplify the mole ratio 4.543:2.272 by dividing by the smaller value:

$$\frac{4.543 \text{ mol N}}{2.272} = 2.000 \text{ mol N} \qquad \frac{2.272 \text{ mol O}}{2.272} = 1.000 \text{ mol O}$$

The mole ratio N:O is 2:1.
4. Since the mole ratio of N to O is already a ratio of whole numbers, we have the empirical formula of nitrous oxide: N_2O.

Think About It The molar masses of N and O are similar, so a $\approx$2:1 ratio of the mass of N to the mass of O is consistent with an empirical formula that is 2:1 nitrogen to oxygen. Note that "nitrous oxide"—whose molecular formula is also N_2O—is formally named "dinitrogen monoxide" under the rules we learned in Chapter 2.

Practice Exercise A different nitrogen oxide, containing 46.16% N, acts as a vasodilator, lowering blood pressure in the human body. What is its empirical formula?

(Answers to Practice Exercises are in the back of the book.)

SAMPLE EXERCISE 3.16 Deriving an Empirical Formula
of a Compound Containing More
than Two Elements

LO4

Rechargeable lithium batteries are essential to portable electronic devices and are increasingly important to the automotive industry in gas–electric hybrid and all-electric vehicles (Figure 3.27). Analysis of the most common lithium-containing mineral, spodumene, shows it is composed of 3.73% Li, 14.50% Al, 30.18% Si, and 51.59% O. What is the empirical formula of spodumene?

Collect and Organize We are given the percent composition by mass of spodumene and asked to determine the simplest whole-number ratio of the four elements (Li, Al, Si, and O), which defines the empirical formula of spodumene.

Analyze We follow the same steps as in Sample Exercise 3.15, except that here we must calculate more than one mole ratio.

Solve

1. Assuming an exactly 100 g sample of spodumene, we have 3.73 g of Li, 14.50 g of Al, 30.18 g of Si, and 51.59 g of O.
2. Converting the masses into moles, we have

$$3.73 \text{ g Li} \times \frac{1 \text{ mol Li}}{6.941 \text{ g Li}} = 0.5374 \text{ mol Li}$$

$$14.50 \text{ g Al} \times \frac{1 \text{ mol Al}}{26.98 \text{ g Al}} = 0.5374 \text{ mol Al}$$

$$30.18 \text{ g Si} \times \frac{1 \text{ mol Si}}{28.09 \text{ g Si}} = 1.074 \text{ mol Si}$$

$$51.59 \text{ g O} \times \frac{1 \text{ mol O}}{16.00 \text{ g O}} = 3.224 \text{ mol O}$$

3. We divide each mole value by the smallest value to obtain a simple ratio of the four elements:

$$\frac{0.5374 \text{ mol Li}}{0.5374} = 1.000 \text{ or } 1 \text{ mol Li} \qquad \frac{0.5374 \text{ mol Al}}{0.5374} = 1.000 \text{ or } 1 \text{ mol Al}$$

$$\frac{1.074 \text{ mol Si}}{0.5374} = 1.999 \text{ or } 2 \text{ mol Si} \qquad \frac{3.224 \text{ mol O}}{0.5374} = 5.999 \text{ or } 6 \text{ mol O}$$

4. All these results are either whole numbers or very close to whole numbers, so no further simplification of terms is needed. Our mole ratio is Li:Al:Si:O = 1:1:2:6, which means the empirical formula of spodumene is $LiAlSi_2O_6$.

Think About It The sum of the mass percentages should be very close to 100%; in this case, 3.73% + 14.50% + 30.18% + 51.59% = 100.00%, so we know we have accounted for all the elements. Notice that lithium is present in the lowest percentage by mass of all four elements in spodumene. Nevertheless, spodumene remains the most commercially important source of lithium for batteries.

Practice Exercise Spodumene is processed chemically to convert it to a material having the percent composition 18.79% Li, 16.25% C, and 64.96% O. What is the empirical formula of this lithium compound?

(Answers to Practice Exercises are in the back of the book.)

(a)

(b)

(c)

FIGURE 3.27 (a) Spodumene is the most important source of lithium metal for use in (b) lithium batteries used in portable electronic devices and (c) in automobiles.

3.7 Comparing Empirical and Molecular Formulas

An empirical formula tells us the simplest whole-number ratio of the elements contained in a compound, but—as we have already seen—it is not necessarily the same as the molecular formula of a molecular compound. To further examine the difference between empirical and molecular formulas, let's consider the organic molecule glycolaldehyde (Figure 3.28). About one in 20 of the meteorites that fall to Earth

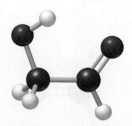

FIGURE 3.28 Glycolaldehyde is one of over 100 organic compounds detected in interstellar gases. It is also the smallest molecule among those identified as sugars.

today contains a variety of organic compounds, including glycolaldehyde. Some scientists view the presence of these compounds as evidence that the molecular building blocks of life on Earth may have come from space. Glycolaldehyde is also present in our bodies as a product of the metabolism of sugars and proteins.

Elemental analysis of glycolaldehyde provides the following percent composition data: 40.00% C, 6.71% H, and 53.28% O. We can use this information to calculate the empirical formula of the compound and then compare our result with the molecular structure in Figure 3.28. Using the method developed in Section 3.6, we obtain these results:

$$40.00 \text{ g C} \times \frac{1 \text{ mol C}}{12.01 \text{ g C}} = 3.331 \text{ mol C}$$

$$6.71 \text{ g H} \times \frac{1 \text{ mol H}}{1.008 \text{ g H}} = 6.66 \text{ mol H}$$

$$53.28 \text{ g O} \times \frac{1 \text{ mol O}}{16.00 \text{ g O}} = 3.330 \text{ mol O}$$

The mole ratio of carbon to hydrogen to oxygen is 3.331:6.66:3.330 = 1:2:1, which defines an empirical formula of CH_2O.

Comparing this empirical formula with the molecule in Figure 3.28 reveals that the empirical formula and the molecular formula are related, but not equivalent. Counting the atoms in the model, we arrive at the molecular formula $C_2H_4O_2$, which represents the actual numbers of C, H, and O atoms in one molecule of glycolaldehyde. Note that the molecular formula is a multiple of the empirical formula (this is consistent with the idea that the empirical formula represents a ratio of the atoms of the elements). In this case, we would multiply each subscript in the empirical formula by 2 to obtain the molecular formula.

CONCEPT TEST

Which of the following compounds have empirical formulas that are identical to their molecular formulas?

a. $C_2H_6O_2$, ethylene glycol, used in antifreeze

b. C_3H_8O, isopropanol, also known as rubbing alcohol

c. $C_6H_{12}O_6$, glucose, also known as "blood sugar" and the primary source of energy for our body's cells

(Answers to Concept Tests are in the back of the book.)

Occasionally, empirical and molecular formulas are identical. For example, formaldehyde—which, like glycolaldehyde, has the percent composition 40.00% C, 6.71% H, 53.28% O—has the molecular formula CH_2O, the same as its empirical formula (Figure 3.29). Many other compounds, such as glucose, $C_6H_{12}O_6$, also have the empirical formula CH_2O. Each subscript in these molecular formulas is a multiple of the corresponding subscript in the empirical formula. For example, a multiplier (n) of 6 converts the empirical formula for glucose and fructose, CH_2O, into their common molecular formula, $C_6H_{12}O_6$:

$$(CH_2O)_n = (CH_2O)_6 = C_6H_{12}O_6$$
$$\text{Glucose}$$

The key to translating empirical formulas into molecular formulas is to determine the value of n, and the key to determining the value of n is knowing the

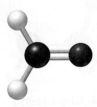

FIGURE 3.29 The empirical formula of the molecular compound formaldehyde is identical to its molecular formula, CH_2O.

molar mass. For example, suppose elemental analysis tells us that the empirical formula of a liquid hydrocarbon is C_4H_5. Using the mole ratio given by this formula and the average atomic masses of the two elements, we can calculate the molar mass corresponding to the empirical formula:

$$(4 \text{ mol} \times 12.01 \text{ g/mol}) + (5 \text{ mol} \times 1.008 \text{ g/mol}) = 53.08 \text{ g}$$

Each molecule of this hydrocarbon consists of either the number of atoms in the empirical formula or a multiple of those atoms. The empirical formula tells us the molecule has a ratio of 4 carbon atoms to 5 hydrogen atoms. Additional analysis of the compound reveals its molar mass to be 106 g. What multiple of 53.08 g (the mass of 4 carbon atoms and 5 hydrogen atoms) has a mass of 106 g? The following calculation tells us:

$$\frac{\text{molar mass}}{\text{empirical formula mass}} = \frac{106 \text{ g/mol}}{53 \text{ g/mol}} = 2$$

Thus 2 is the value of the multiplier, n, and the molecular formula of the compound is

$$(C_4H_5)_n = (C_4H_5)_2 = C_8H_{10}$$

SAMPLE EXERCISE 3.17 Deriving a Molecular Formula   **LO5**
from an Empirical Formula

Lycopene (molar mass 536.88 g/mol) and cymene (134.22 g/mol) are natural products found in tomatoes and cumin (a common spice), respectively (Figure 3.30). Lycopene in the diet has been shown to reduce the risk of prostate cancer, whereas cumin is a home remedy for stomach ailments. Both compounds are 89.49% C and 10.51% H by mass. What are the empirical and molecular formulas of lycopene and cymene?

Collect and Organize The problem gives the percent composition and molar masses of two compounds. We want to determine both the empirical and molecular formulas of lycopene and cymene.

Analyze The data on percent composition lead directly to the empirical formula by using the procedure described in Section 3.6. The molar masses of lycopene and cymene in g/mol are numerically the same as their molecular masses in amu and can be used to determine the value of the multipliers needed to convert their empirical formulas into molecular formulas.

Solve Assuming a 100 g sample, we have 89.49 g of C and 10.51 g of H, which we convert to moles:

$$89.49 \text{ g C} \times \frac{1 \text{ mol C}}{12.01 \text{ g C}} = 7.451 \text{ mol C}$$

$$10.51 \text{ g H} \times \frac{1 \text{ mol H}}{1.008 \text{ g H}} = 10.43 \text{ mol H}$$

The mole ratio of C to H is 7.451:10.43, or 1:1.4. We can convert the fractional mole ratio to a ratio of whole numbers by multiplying by 5:

$$C:H = 5 \times (1:1.4) = 5:7$$

which means the empirical formula is C_5H_7.

To determine the molecular formula, we first determine the molar mass of the empirical formula C_5H_7:

$$5 \text{ mol}(12.01 \text{ g/mol}) + 7 \text{ mol}(1.008 \text{ g/mol}) = 67.11 \text{ g}$$

(a)

(b)

FIGURE 3.30 (a) Tomatoes and (b) ground cumin (a cooking spice) both contain hydrocarbons reported to offer therapeutic benefits.

Next we divide the molar masses for lycopene and cymene by their respective empirical formula masses to determine the values of the multiplier n:

$$n = \frac{\text{molar mass lycopene}}{\text{empirical formula mass}} = \frac{536.88 \text{ g/mol}}{67.11 \text{ g/mol}} = 8$$

$$n = \frac{\text{molar mass cymene}}{\text{empirical formula mass}} = \frac{134.22 \text{ g/mol}}{67.11 \text{ g/mol}} = 2$$

The molecular formula of lycopene is

$$(C_5H_7)_n = (C_5H_7)_8 = C_{40}H_{56}$$

and the molecular formula of cymene is

$$(C_5H_7)_n = (C_5H_7)_2 = C_{10}H_{14}$$

Think About It The percent composition of a compound reveals only the empirical formula. All compounds with the same empirical formula *must* have the same percent composition.

Practice Exercise *Pheromones* are chemical substances secreted by members of a species to stimulate a response in other individuals of the same species. For example, certain pheromones are secreted by females of a species to attract males for mating. The percent composition of eicosene (280 g/mol), a compound similar to the Japanese beetle mating pheromone, is 85.63% C and 14.37% H. Determine its molecular formula.

(Answers to Practice Exercises are in the back of the book.)

CONCEPT **TEST**

Lycopene and cymene (Sample Exercise 3.17) have the same percent composition. Are they isomers?

(Answers to Concept Tests are in the back of the book.)

Molecular Mass and Mass Spectrometry

CONNECTION The positive-ray analyzer that Francis Aston built and used to separate two of the isotopes of neon (see Section 2.3) was the forerunner of today's mass spectrometers.

How do chemists who isolate an unknown molecular compound from reaction mixtures or biological systems determine the molecular mass of the compound? They often use a powerful analytical technique called *mass spectrometry*.

Mass spectrometers ionize molecules and then separate the ions by the ratio of their masses (m) to their electric charges (z). In many mass spectrometers, samples are vaporized and then bombarded with a beam of high-energy electrons (Figure 3.31). These electrons smash into gas-phase molecules with such force that they knock electrons off the molecules, forming a positively charged ion known as the **molecular ion (M^+)**:

$$M + e^- \rightarrow M^+ + 2\,e^-$$

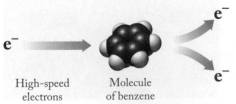
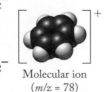

High-speed electrons   Molecule of benzene   Molecular ion ($m/z = 78$)

FIGURE 3.31 In many mass spectrometers, sample molecules are bombarded with beams of high-energy electrons, producing molecular ions with positive charges, as shown here for benzene, C_6H_6.

Sometimes the collisions are so forceful that they break molecules into fragments and ionize the fragments. The charged fragments and molecular ions are ejected into a second region of the mass spectrometer where they are separated based on their m/z ratios. Then they pass into a detector and are counted. The resulting data are displayed as a **mass spectrum** (plural *spectra*) in which the horizontal axis is m/z and the spectrum itself is a series of vertical bars at various m/z values. The heights of the bars indicate the number of ions reaching the detector with that particular m/z value. The bar with the highest m/z value in the spectrum often represents molecular ions,

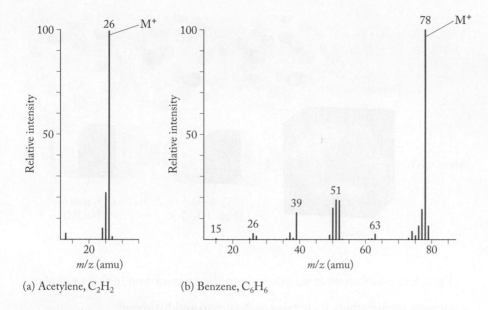

FIGURE 3.32 Mass spectra of (a) acetylene and (b) benzene. The molecular-ion bar in both spectra is labeled M$^+$. The bars at much smaller m/z values are produced when the ionizing beam also breaks apart the molecules into fragments. The molar masses of C_2H_2 and C_6H_6 are 26 and 78 g/mol, respectively.

M$^+$. If they are singly charged ions ($z = 1+$), then the m/z value of that bar is simply m, the molecular mass of the compound.

Figure 3.32 shows the mass spectra of acetylene and benzene. In both spectra, the tall bar with the greatest m/z value (at 26 amu in the acetylene spectrum and at 78 amu in the benzene spectrum) is the molecular ion. The bars at lower m/z values represent fragments. Distinctive fragmentation patterns help scientists confirm molecular structures.

> ## CONCEPT **TEST**
>
> Does the molecular ion in a mass spectrum correspond to the mass associated with the empirical formula or the molecular formula of a compound?
>
> *(Answers to Concept Tests are in the back of the book.)*

3.8 Combustion Analysis

In **combustion analysis**, the complete combustion (defined in Section 3.4) of a compound, followed by an analysis of the products, enables chemists to determine the chemical composition of that compound. To ensure that combustion is complete, the process is carried out in excess oxygen, which means more oxygen is present than the stoichiometric amount. Combusting an organic compound means converting all of the carbon in the compound to CO_2 and all of the hydrogen in the compound to H_2O:

$$C_aH_b + \text{excess } O_2(g) \rightarrow a\, CO_2(g) + b/2\, H_2O(g)$$

Consider burning a hydrocarbon of unknown composition in a chamber through which a stream of pure oxygen flows (Figure 3.33). The $CO_2(g)$ and $H_2O(g)$ produced flow first through a tube packed with $Mg(ClO_4)_2(s)$, which selectively absorbs the $H_2O(g)$, and then through a tube containing $NaOH(s)$, which absorbs the $CO_2(g)$. The masses of these tubes are measured before and after combustion. Suppose the mass of the tube that traps $CO_2(g)$ increases by 1.320 g, and the mass of the tube that traps $H_2O(g)$ increases by 0.540 g. How can we use these results to determine the empirical formula of the hydrocarbon?

mass spectrometer an instrument that separates and counts ions according to their mass-to-charge ratios.

molecular ion (M$^+$) the peak of highest mass in a mass spectrum; it has the same mass as the molecule from which it came.

mass spectrum a graph of the data from a mass spectrometer, where m/z ratios of the deflected particles are plotted against the number of particles with a particular mass.

combustion analysis a laboratory procedure for determining the composition of a substance by burning it completely in oxygen to produce known compounds whose masses are used to determine the composition of the original material.

FIGURE 3.33 A carbon/hydrogen elemental analyzer relies on the complete combustion of organic compounds in excess oxygen. The products are H_2O vapor and CO_2. Water vapor is absorbed by a $Mg(ClO_4)_2$ filter, and carbon dioxide is absorbed by a NaOH filter. The empirical formula of the compound is calculated from the masses of H_2O and CO_2 absorbed.

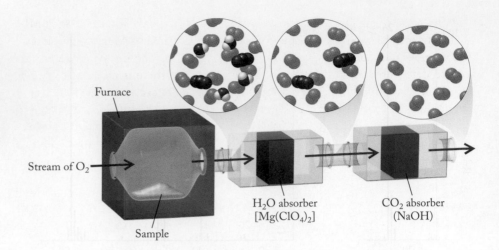

Furnace

Stream of O_2

Sample

H_2O absorber [$Mg(ClO_4)_2$]

CO_2 absorber (NaOH)

First, let's establish what we know about the hydrocarbon in this example:

1. Being a hydrocarbon, it contains only carbon and hydrogen.
2. Complete conversion of its carbon into CO_2 produces 1.320 g of CO_2.
3. Complete conversion of its hydrogen into H_2O produces 0.540 g of H_2O.

To derive an empirical formula for the hydrocarbon, we must determine the number of moles of carbon in 1.320 g of carbon dioxide and the number of moles of hydrogen in 0.540 g of water vapor. These quantities are directly related to the number of moles of carbon and hydrogen in the sample we burned. Converting the mass of carbon dioxide to moles of carbon and the mass of water to moles of hydrogen:

$$1.320 \text{ g } CO_2 \times \frac{1 \text{ mol } CO_2}{44.01 \text{ g } CO_2} \times \frac{1 \text{ mol C}}{1 \text{ mol } CO_2} = 0.02999 \text{ mol C} \approx 0.0300 \text{ mol C}$$

$$0.540 \text{ g } H_2O \times \frac{1 \text{ mol } H_2O}{18.02 \text{ g } H_2O} \times \frac{2 \text{ mol H}}{1 \text{ mol } H_2O} = 0.05993 \text{ mol H} \approx 0.0600 \text{ mol H}$$

The empirical formula is based on the mole ratio of C and H. If we divide both molar amounts by the smaller one:

$$\frac{0.0300 \text{ mol C}}{0.0300} = 1 \text{ mol C} \qquad \frac{0.0600 \text{ mol H}}{0.0300} = 2 \text{ mol H}$$

we find that the empirical formula of the hydrocarbon is CH_2.

If we want to extend this analysis to determine a molecular formula, we need to know the molecular mass of the hydrocarbon. Suppose its mass spectrum has a molecular ion with a mass of 84 amu. This means the molar mass of the compound is 84 g/mol. The empirical formula mass of CH_2 is 14 g/mol. Dividing the molar mass by the empirical formula mass, we get the multiplier, n:

$$n = \frac{84 \text{ g/mol}}{14 \text{ g/mol}} = 6$$

The molecular formula is therefore

$$(CH_2)_6 = C_6H_{12}$$

Note that in this problem we did not need to know the initial mass of the sample to determine its empirical formula. We only needed to know that the sample was a hydrocarbon and that it was completely converted into the stated amounts of CO_2 and H_2O.

What if we knew that our sample was *not* a hydrocarbon? What if we had isolated a pharmacologically promising compound from a tropical plant, and we knew that its molecules contained atoms of carbon, hydrogen, and oxygen? We would need to determine the proportion of oxygen in it, but there is no simple way of measuring that directly when the compound is burned in excess oxygen. Combustion analyses yield the percentages by mass of all atoms in a sample *except* oxygen. When given the percentages of atoms by mass from a combustion analysis, always check to see whether the percentages add up to 100%. If they do, all the elements in the sample are accounted for in the results. If they do not, the missing mass is probably due to oxygen.

SAMPLE EXERCISE 3.18 Combustion Analysis of a Hydrocarbon **LO6**

Limonene is a hydrocarbon that contributes to the odor of citrus fruits, including lemons (Figure 3.34). Combustion of 0.671 g of limonene yielded 2.168 g of CO_2 and 0.710 g of H_2O. What is the empirical formula of limonene? The molar mass of limonene is 136.24 g/mol; what is the molecular formula of limonene?

Collect and Organize We know the masses of CO_2 and H_2O produced during the combustion of a hydrocarbon sample, and we are asked to determine its empirical formula and then, from its molecular mass, its molecular formula.

Analyze First we determine the number of moles of C and H in the CO_2 and H_2O produced during combustion. These values are equal to the number of moles of C and H in the combusted sample. Then we calculate the C:H mole ratio and convert it into a ratio of small whole numbers to obtain the empirical formula. Next we calculate the mass of the empirical formula and divide this mass into the molar mass to obtain the multiplier, n, that allows us to convert the empirical formula to a molecular formula.

$$\boxed{\text{g } CO_2} \xrightarrow[\text{molar mass}]{1} \boxed{\text{mol } CO_2} \xrightarrow[\text{1 mol } CO_2]{\text{1 mol C}} \boxed{\text{mol C}}$$

$$\boxed{\text{g } H_2O} \xrightarrow[\text{molar mass}]{1} \boxed{\text{mol } H_2O} \xrightarrow[\text{1 mol } H_2O]{\text{2 mol H}} \boxed{\text{mol H}}$$

Solve The moles of C and H in the CO_2 and H_2O collected during combustion are

$$2.168 \text{ g } CO_2 \times \frac{1 \text{ mol } CO_2}{44.01 \text{ g } CO_2} \times \frac{1 \text{ mol C}}{1 \text{ mol } CO_2} = 0.04926 \text{ mol C}$$

$$0.710 \text{ g } H_2O \times \frac{1 \text{ mol } H_2O}{18.02 \text{ g } H_2O} \times \frac{2 \text{ mol H}}{1 \text{ mol } H_2O} = 0.0788 \text{ mol H}$$

The mole ratio of the two elements in the sample is

$$0.04926 \text{ mol C} : 0.0788 \text{ mol H}$$

Dividing through by the smallest value (0.04926 mol) gives a mole ratio of 1:1.6. We can convert this ratio to whole numbers by multiplying by 5, making the empirical formula of the sample C_5H_8.

The molar mass of limonene is 136.24 g/mol. The mass of the empirical formula of limonene is 5(12.01 g/mol) + 8(1.008 g/mol) = 68.11 g/mol, so the multiplier n is

$$n = \frac{\text{molar mass}}{\text{empirical formula mass}} = \frac{136.24 \text{ g/mol}}{68.11 \text{ g/mol}} = 2$$

and the molecular formula is

$$(C_5H_8)_n = (C_5H_8)_2 = C_{10}H_{16}$$

FIGURE 3.34 Citrus fruits such as lemons, limes, oranges, and grapefruit get their characteristic odor from a hydrocarbon compound, limonene.

Think About It The subscripts in our answer, 5 for C and 8 for H, make sense because the moles C $\approx 5 \times 10^{-2}$ and moles H $\approx 8 \times 10^{-2}$ are essentially at a 5:8 ratio.

Practice Exercise Cembrene A is another naturally occurring hydrocarbon, extracted from coral (Figure 3.35). Combustion of 0.0341 g of cembrene A yields 0.1101 g of CO_2 and 0.0360 g of H_2O. What is the empirical formula of cembrene A?

FIGURE 3.35 Cembrene A is a colorless oil that can be isolated from some species of coral.

(Answers to Practice Exercises are in the back of the book.)

SAMPLE EXERCISE 3.19 Combustion Analysis for Compounds   **LO6**
Containing Oxygen

Eugenol is an ingredient in several spices, including bay leaves and cloves (Figure 3.36). Eugenol contains carbon, hydrogen, and oxygen. Combustion of 21.80 mg of eugenol yields 58.5 mg of CO_2 and 14.4 mg of H_2O. What is the empirical formula of eugenol? If the mass spectrum of eugenol shows a molecular ion at 164 amu, what is the molecular formula of eugenol?

Collect and Organize We are given the masses of CO_2 (58.5 mg) and H_2O (14.4 mg) produced by the combustion of 21.80 mg of eugenol and are asked to determine its empirical formula. We also know its molar mass, and we need to determine its molecular formula.

Analyze First we determine the number of moles of C and H in the CO_2 and H_2O produced during combustion. These values are equal to the number of moles of C and H in the combusted sample. Multiplying these mole values by the atomic masses of C and H yields the mass of these elements in the original sample, which we then subtract from the total mass of the sample to obtain the mass of O in the sample. This mass can then be converted into moles of O in the sample:

$$\boxed{\text{mg sample}} - \boxed{\text{mg C}} - \boxed{\text{mg H}} = \boxed{\text{mg O}} \xrightarrow{\frac{1\,g}{10^3\,mg}} \boxed{g\ O} \xrightarrow{\frac{1}{\text{molar mass}}} \boxed{\text{mol O}}$$

(a)

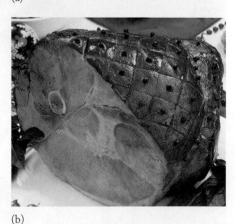

(b)

FIGURE 3.36 (a) The bay leaf is commonly used to season soups and tomato sauces, whereas (b) cloves are often added as a spice to baked hams. Both of these cooking staples contain the compound eugenol.

The mole ratio of C:H:O is then calculated and reduced to a ratio of small whole numbers to obtain the empirical formula. We divide the mass of the empirical formula into the molar mass to obtain the multiplier, n, that allows us to convert the empirical formula to a molecular formula.

Solve The moles of C and H in the CO_2 and H_2O collected during combustion are

$$58.5 \text{ mg } CO_2 \times \frac{1 \text{ g } CO_2}{10^3 \text{ mg } CO_2} \times \frac{1 \text{ mol } CO_2}{44.01 \text{ g } CO_2} \times \frac{1 \text{ mol C}}{1 \text{ mol } CO_2} = 1.329 \times 10^{-3} \text{ mol C}$$

$$14.4 \text{ mg } HO_2 \times \frac{1 \text{ g } H_2O}{10^3 \text{ mg } H_2O} \times \frac{1 \text{ mol } H_2O}{18.02 \text{ g } H_2O} \times \frac{2 \text{ mol H}}{1 \text{ mol } H_2O} = 1.598 \times 10^{-3} \text{ mol H}$$

The masses of C and H are

$$1.329 \times 10^{-3} \text{ mol C} \times \frac{12.01 \text{ g C}}{1 \text{ mol C}} = 1.596 \times 10^{-2} \text{ g C}$$

$$1.598 \times 10^{-3} \text{ mol H} \times \frac{1.008 \text{ g H}}{1 \text{ mol H}} = 1.611 \times 10^{-3} \text{ g H}$$

In order to more easily compare the masses of C and H to the original sample mass of 21.8 mg, let's convert them into milligrams as well:

$$1.596 \times 10^{-2} \text{ g C} \times \frac{10^3 \text{ mg C}}{1 \text{ g C}} = 15.96 \text{ mg C}$$

$$1.611 \times 10^{-3} \text{ g H} \times \frac{10^3 \text{ mg H}}{1 \text{ g H}} = 1.611 \text{ mg H}$$

The sum of these two masses (15.96 mg C + 1.611 mg H = 17.57 mg) is less than the mass of the sample (21.8 mg). The difference must be the mass of oxygen in the sample:

$$\text{Mass of oxygen} = 21.80 \text{ mg} - 17.57 \text{ mg} = 4.23 \text{ mg O}$$

The number of moles of O atoms in the sample is

$$4.23 \times 10^{-3} \text{ g O} \times \frac{1 \text{ mol O}}{16.00 \text{ g O}} = 2.64 \times 10^{-4} \text{ mol O}$$

The mole ratio of the three elements in the sample is

$$1.329 \times 10^{-3} \text{ mol C} : 1.598 \times 10^{-3} \text{ mol H} : 2.64 \times 10^{-4} \text{ mol O}$$

Dividing through by the smallest value (2.64×10^{-4} mol) gives a mole ratio of 5:6:1, making the empirical formula of the sample C_5H_6O. The empirical formula mass is

$$(5 \times 12.01 \text{ g/mol} + 6 \times 1.008 \text{ g/mol} + 16.00 \text{ g/mol})n = 82.10 \text{ g/mol}$$

The molecular ion identifies the molecular mass of eugenol as 164 amu, so, the molar mass of eugenol is 164 g/mol. Therefore the multiplier, n, to convert the empirical formula into a molecular formula is

$$n = \frac{\text{molar mass}}{\text{empirical formula mass}} = \frac{164 \text{ g/mol}}{82.10 \text{ g/mol}} = 2$$

and the molecular formula is

$$(C_5H_6O)_n = (C_5H_6O)_2 = C_{10}H_{12}O_2$$

Think About It Our answer makes sense because we have a simple whole-number ratio of the elements.

Practice Exercise Vanillin is a compound containing carbon, hydrogen, and oxygen that gives vanilla beans their distinctive flavor (Figure 3.37). The combustion of 30.4 mg of vanillin produces 70.4 mg of CO_2 and 14.4 mg of H_2O. The mass spectrum of vanillin shows a molecular ion at 152 amu. Use this information to determine the molecular formula of vanillin.

(Answers to Practice Exercises are in the back of the book.)

FIGURE 3.37 The distinctive flavor of vanilla ice cream comes from the seeds of the vanilla bean and the compound vanillin.

limiting reactant a reactant that is consumed completely in a chemical reaction. The amount of product formed depends on the amount of the limiting reactant available.

3.9 Limiting Reactants and Percent Yield

Let's return to photosynthesis, the process responsible for the O_2 in our present-day atmosphere and for the energy that sustains life on Earth's surface. The stoichiometry of the reaction calls for equal moles, referred to as *equimolar amounts*, of CO_2 and H_2O:

$$6\ CO_2(g) + 6\ H_2O(\ell) \rightarrow C_6H_{12}O_6(aq) + 6\ O_2(g)$$
$$\text{Glucose}$$

Because in nature there is little likelihood of having the exact mole ratio of reactants at a reaction site, let's consider what happens when more than six molecules of water are available for every six molecules of CO_2 (this is not a theoretical consideration because having water in excess is a common occurrence in biological systems). The photosynthetic production of glucose will continue until all the CO_2 is consumed, leaving the extra molecules of water unreacted. In this example, carbon dioxide is the **limiting reactant**, meaning that the extent to which the reaction proceeds is determined by the quantity of CO_2 available and not by the quantity of H_2O, which is in excess. Figure 3.38 illustrates the concept of a limiting reactant.

FIGURE 3.38 The reaction of hydrogen with oxygen produces water. A mixture containing equal numbers of hydrogen and oxygen molecules produces only as many water molecules as the number of H_2 molecules available. In this case, H_2 is the limiting reactant and O_2 is in excess.

Another way of thinking about limiting reactants is to consider the following task. You are asked to assemble bicycles containing one bike frame, one bike seat, and two wheels. You have 100 bike frames, 100 seats, and 100 wheels. The wheels are the limiting reactant in this scenario.

Calculations Involving Limiting Reactants

Suppose you are asked to calculate the mass of a product formed from given masses of reactants. How do you know whether one reactant is limiting? And how do you know which one? It is tempting to select the reactant present in the smaller amount by mass. Avoid this temptation and instead take a systematic approach based on the stoichiometry of the reaction. Several approaches are possible, two of which we present here—one that uses stoichiometric calculations and one that uses mole ratios.

In our first method, we calculate how much product would be formed if reactant A were completely consumed. Then we repeat the calculation for reactant B (and so on for any additional reactants). Let's try this approach with the reaction between sulfur trioxide gas and water vapor to form sulfuric acid:

$$SO_3(g) + H_2O(g) \rightarrow H_2SO_4(\ell) \qquad (3.1)$$

This reaction was described in Section 3.1 as contributing to the acidity of Earth's early atmosphere. Suppose we carry out this reaction in the laboratory by using 20.00 g of $SO_3(g)$ and 10.00 g of $H_2O(g)$. Which is the limiting reactant, and how many grams of H_2SO_4 are produced from these masses of reactants?

We first calculate the molar masses and then the numbers of moles of SO_3 and H_2O from the two masses:

$$\mathcal{M}_{SO_3} = 32.06\ \text{g/mol} + 3(16.00\ \text{g/mol}) = 80.06\ \text{g/mol}$$

$$20.00\ \text{g}\ SO_3 \times \frac{1\ \text{mol}\ SO_3}{80.06\ \text{g}\ SO_3} = 0.2498\ \text{mol}\ SO_3$$

CHEMTOUR
Limiting Reactant

$$10.00 \text{ g } \cancel{H_2O} \times \frac{1 \text{ mol } H_2O}{18.02 \text{ g } \cancel{H_2O}} = 0.5549 \text{ mol } H_2O$$

Then we calculate the mass of sulfuric acid produced if all of the SO_3 were consumed:

$$0.2498 \cancel{\text{ mol } SO_3} \times \frac{1 \cancel{\text{ mol } H_2SO_4}}{1 \cancel{\text{ mol } SO_3}} \times \frac{98.08 \text{ g } H_2SO_4}{1 \cancel{\text{ mol } H_2SO_4}} = 24.50 \text{ g } H_2SO_4$$

Next we carry out the same calculation to determine how much sulfuric acid is produced if the 10.00 g of H_2O were to react completely:

$$0.5549 \cancel{\text{ mol } H_2O} \times \frac{1 \cancel{\text{ mol } H_2SO_4}}{1 \cancel{\text{ mol } H_2O}} \times \frac{98.08 \text{ g } H_2SO_4}{1 \cancel{\text{ mol } H_2SO_4}} = 54.42 \text{ g } H_2SO_4$$

The SO_3 is the limiting reactant because, when it is completely consumed, a smaller amount of product is formed. Thus when 20.00 g of SO_3 and 10.00 g of H_2O react, the maximum amount of product that can form is 24.50 g of H_2SO_4. Making that amount of sulfuric acid consumes all the available SO_3 but not all the available H_2O.

The second method compares the mole ratio of the reactants to the mole ratio required by the balanced chemical equation. To use this approach, we take the following steps:

1. Convert masses of reactants A and B into moles.
2. Calculate the mole ratio of A to B.
3. Compare this mole ratio with the stoichiometric mole ratio from the balanced chemical equation. If

$$\left(\frac{\text{mol A}}{\text{mol B}} \right)_{\text{given}} > \left(\frac{\text{mol A}}{\text{mol B}} \right)_{\text{stoichiometric}} \qquad (3.3)$$

then B is the limiting reactant. If

$$\left(\frac{\text{mol A}}{\text{mol B}} \right)_{\text{given}} < \left(\frac{\text{mol A}}{\text{mol B}} \right)_{\text{stoichiometric}} \qquad (3.4)$$

then A is the limiting reactant. If the two ratios are equal, the masses are the stoichiometric amounts and both reactants are consumed completely.

For the same example using 20.00 g of SO_3 and 10.00 g of H_2O, we calculate the mole ratio of SO_3 to H_2O after calculating the moles of SO_3 and H_2O available:

$$\frac{\text{mol } SO_3}{\text{mol } H_2O} = \frac{20.00 \text{ g } \cancel{SO_3} \times \dfrac{1 \text{ mol } SO_3}{80.06 \text{ g } \cancel{SO_3}}}{10.00 \text{ g } \cancel{H_2O} \times \dfrac{1 \text{ mol } H_2O}{18.02 \text{ g } \cancel{H_2O}}} = \frac{0.2498 \text{ mol } SO_3}{0.5549 \text{ mol } H_2O} = 0.4502$$

Equation 3.1 tells us that the stoichiometric ratio of SO_3 to H_2O is $1/1 = 1$. The SO_3/H_2O mole ratio for our reaction conditions is $0.2498/0.5549 = 0.4502$. Expressing this outcome as a mathematical relationship, we get

$$\left(\frac{\text{mol } SO_3}{\text{mol } H_2O} \right)_{\text{given}} < \left(\frac{\text{mol } SO_3}{\text{mol } H_2O} \right)_{\text{stoichiometric}}$$

This inequality matches Equation 3.4, which tells us that SO_3 (reactant A in Equation 3.4) must be the limiting reactant.

These two approaches to determining a limiting reactant lead to the same conclusion and are of similar complexity. In both cases, because the masses of the reactants are given in grams, we must convert those masses into moles. The

theoretical yield the maximum amount of product possible in a chemical reaction for given quantities of reactants; also called *stoichiometric yield*.

advantage of the first method is that it not only determines the identity of the limiting reactant, it also determines the **theoretical yield**, the maximum amount of product that can be produced by the given mixture. The theoretical yield is sometimes called the *stoichiometric yield* or *100% yield*.

SAMPLE EXERCISE 3.20 Identifying Limiting Reactants **LO7**

During strenuous exercise, respiration (the reaction between glucose and oxygen, Section 3.5) is limited by the availability of oxygen to the muscles. Under *anaerobic* conditions, our bodies produce lactic acid, which accumulates in muscles and accompanies the "burning" sensation you may have felt. We can model this situation by considering the reaction between 1.32 g of $C_6H_{12}O_6$ and 1.32 g of O_2 to produce CO_2 and water. Is one of these reactants a limiting reactant, or is the mixture stoichiometric? If one reactant is limiting, which one is it?

Collect and Organize We are given quantities of two reactants ($C_6H_{12}O_6$ and O_2) and want to determine which, if either, is limiting.

Analyze We can determine the limiting reactant by calculating the ratio of $C_6H_{12}O_6$ to O_2 in the balanced chemical equation for the reaction. That means we will use the second method described in the preceding discussion. An unbalanced equation for this combustion reaction is

$$C_6H_{12}O_6(s) + O_2(g) \rightarrow CO_2(g) + H_2O(g)$$

After balancing this equation, we compare the ratio of the amounts of reactants with the stoichiometric ratio in the balanced equation to determine whether one of the reactants is limiting.

Solve First we need to balance the combustion equation:

$$\_\, C_6H_{12}O_6(s) + \_\, O_2(g) \rightarrow \_\, CO_2(g) + \_\, H_2O(g)$$

$$C_6H_{12}O_6(s) + 6\, O_2(g) \rightarrow 6\, CO_2(g) + 6\, H_2O(g)$$

This balanced equation tells us that the mole ratio of $C_6H_{12}O_6$ to O_2 is 1:6. Next we convert the given masses of reactants to moles:

$$1.32\ \text{g}\ O_2 \times \frac{1\ \text{mol}\ O_2}{32.00\ \text{g}\ O_2} = 4.13 \times 10^{-2}\ \text{mol}\ O_2$$

$$1.32\ \text{g}\ C_6H_{12}O_6 \times \frac{1\ \text{mol}\ C_6H_{12}O_6}{180.16\ \text{g}\ C_6H_{12}O_6} = 7.33 \times 10^{-3}\ \text{mol}\ C_6H_{12}O_6$$

Then we calculate the mole ratio:

$$\frac{4.13 \times 10^{-2}\ \text{mol}\ O_2}{7.33 \times 10^{-3}\ \text{mol}\ C_6H_{12}O_6} = 5.6$$

This ratio is less than the stoichiometric $C_6H_{12}O_6$ to O_2 of 1:6, so oxygen is the limiting reactant, and there is a slight excess of glucose.

Think About It The calculation indicates that when equal masses of glucose and oxygen react, oxygen is the limiting reactant.

Practice Exercise Any fuel–oxygen mixture that contains more oxygen than is needed to burn the fuel completely is called a *lean mixture*, whereas a mixture containing too little oxygen to allow complete combustion of the fuel is called a *rich mixture*. A high-performance heater that burns propane, $C_3H_8(g)$, is adjusted so that 100.0 g of $O_2(g)$ enters the system for every 26.0 g of propane. Is this mixture rich or lean?

(Answers to Practice Exercises are in the back of the book.)

SAMPLE EXERCISE 3.21 Limiting Reactants in Chemical Reactions   **LO7**

The results of the Miller–Urey experiment require a chemical reaction to account for the formation of glycine. The reaction between carbon dioxide, ammonia, and methane that produces glycine, water, and carbon monoxide is described by the following chemical equation:

$$2 CO_2(g) + NH_3(g) + CH_4(g) \rightarrow C_2H_5NO_2(s) + H_2O(\ell) + CO(g)$$

How much glycine could be expected from the reaction of 29.3 g of CO_2, 4.53 g of NH_3, and 4.27 g of CH_4?

Collect and Organize We are given a balanced chemical equation and the quantities of three reactants (CH_4, NH_3, and CO_2). We are asked how much glycine can be produced, which will require us to identify which one of the reactants is the limiting reactant. Since there are three reactants, we must use the first method discussed previously.

Analyze First we calculate the number of moles of each reactant we have available. Then, we calculate the maximum amount of glycine that would be produced from each reactant. The reactant yielding the smallest amount of glycine is the limiting reactant and determines the maximum amount of glycine that we can expect from the reaction.

Solve
1. Using the molar masses of CO_2, NH_3, and CH_4, we calculate the number of moles of each that are available:

$$29.3 \text{ g CO}_2 \times \frac{1 \text{ mol CO}_2}{44.01 \text{ g CO}_2} = 0.6658 \text{ mol CO}_2$$

$$4.53 \text{ g NH}_3 \times \frac{1 \text{ mol NH}_3}{17.03 \text{ g NH}_3} = 0.2660 \text{ mol NH}_3$$

$$4.27 \text{ g CH}_4 \times \frac{1 \text{ mol CH}_4}{16.04 \text{ g CH}_4} = 0.2662 \text{ mol CH}_4$$

2. Using the mole ratios of $C_2H_5NO_2$ for each reactant and the molar mass of $C_2H_5NO_2$, we calculate the amount of glycine that could be produced from each reactant:

$$0.6658 \text{ mol CO}_2 \times \frac{1 \text{ mol C}_2\text{H}_5\text{NO}_2}{2 \text{ mol CO}_2} \times \frac{75.07 \text{ g C}_2\text{H}_5\text{NO}_2}{1 \text{ mol C}_2\text{H}_5\text{NO}_2} = 25.0 \text{ g C}_2\text{H}_5\text{NO}_2$$

$$0.2660 \text{ mol NH}_3 \times \frac{1 \text{ mol C}_2\text{H}_5\text{NO}_2}{1 \text{ mol NH}_3} \times \frac{75.07 \text{ g C}_2\text{H}_5\text{NO}_2}{1 \text{ mol C}_2\text{H}_5\text{NO}_2} = 20.0 \text{ g C}_2\text{H}_5\text{NO}_2$$

$$0.2662 \text{ mol CH}_4 \times \frac{1 \text{ mol C}_2\text{H}_5\text{NO}_2}{1 \text{ mol CH}_4} \times \frac{75.07 \text{ g C}_2\text{H}_5\text{NO}_2}{1 \text{ mol C}_2\text{H}_5\text{NO}_2} = 20.0 \text{ g C}_2\text{H}_5\text{NO}_2$$

Of these three possible yields of $C_2H_5NO_2$, we can only make the smallest amount, 20.0 g of $C_2H_5NO_2$, so both ammonia and methane are considered limiting reactants in this reaction.

Think About It Although the mass of methane available for the reaction is the lowest of the three reactants, there is an equal number of moles of methane and ammonia. Since carbon dioxide is present in excess, both methane and ammonia are limiting reactants. With three or more reactants, one or more may turn out to be in excess.

 Practice Exercise One of the intermediates in the synthesis of glycine from ammonia, carbon dioxide, and methane is $C_2H_4N_2$, produced by this reaction:

$$3 CH_4(g) + 5 CO_2(g) + 8 NH_3(g) \rightarrow 4 C_2H_4N_2(g) + 10 H_2O(\ell)$$

How many grams of $C_2H_4N_2$ could be expected from the reaction of 14.2 g of CO_2, 2.27 g of NH_3, and 2.14 g of CH_4?

(Answers to Practice Exercises are in the back of the book.)

actual yield the amount of product obtained from a chemical reaction, which is often less than the theoretical yield.

percent yield the ratio, expressed as a percentage, of the actual yield of a chemical reaction to the theoretical yield.

Actual Yields versus Theoretical Yields

At the beginning of this section, when we calculated the mass of sulfuric acid formed by 10.00 g of water and 20.00 g of sulfur trioxide, we defined the value as a theoretical yield: the maximum amount of product possible from the given quantities of reactants. In nature, industry, or the laboratory, the **actual yield** is often less than the theoretical yield for several reasons. Sometimes reactants combine to form products other than the ones desired. Some reactions are so slow that a fraction of the reactants remain unreacted even after long reaction times. Still other reactions do not go to completion no matter how long they are allowed to run, yielding a mixture of reactants and products whose composition does not change with time. For these and other reasons, it is useful to distinguish between the theoretical yield and the actual yield of a chemical reaction and to calculate the **percent yield**:

$$\text{Percent yield} = \frac{\text{actual yield}}{\text{theoretical yield}} \times 100\% \qquad (3.5)$$

SAMPLE EXERCISE 3.22 Calculating Percent Yield   **LO8**

The industrial process for making the ammonia used in fertilizer, explosives, and many other products is based on the reaction between nitrogen and hydrogen at high temperature and pressure:

$$N_2(g) + 3\,H_2(g) \rightarrow 2\,NH_3(g)$$

If 18.20 kg of NH_3 is produced by a reaction mixture that initially contains 6.00 kg of H_2 and an excess of N_2, what is the percent yield of the reaction?

Collect and Organize We know that the actual yield of NH_3 is 18.20 kg. We also know that the reaction mixture initially contained 6.00 kg of H_2 and that H_2 must be the limiting reactant because the problem specifies the presence of excess N_2. We need to find the percent yield.

Analyze We need to use the mass of H_2 to calculate how much NH_3 could have been produced—the theoretical yield. We can then use the theoretical yield and the actual yield to calculate percent yield. We need the molar masses of H_2 and NH_3 and the stoichiometry of the reaction, which tells us that 2 moles of NH_3 are produced for every 3 moles of H_2 consumed.

$$\boxed{g\,H_2} \xrightarrow[\text{molar mass}]{1} \boxed{\text{mol}\,H_2} \xrightarrow[\text{3 mol}\,H_2]{2\,\text{mol}\,NH_3} \boxed{\text{mol}\,NH_3} \xrightarrow[\text{molar mass}]{} \boxed{g\,NH_3}$$

$$= \boxed{\text{theoretical yield}}$$

$$\boxed{\text{\% yield}} = \boxed{\frac{\text{actual yield}}{\text{theoretical yield}}} \times 100\%$$

Solve The molar masses we need are:

$$\mathcal{M}_{H_2} = 2(1.008\ \text{g/mol}) = 2.016\ \text{g/mol}$$

$$\mathcal{M}_{NH_3} = 14.01\ \text{g/mol} + 3(1.008\ \text{g/mol}) = 17.03\ \text{g/mol}$$

We calculate the theoretical yield of NH_3:

$$6.00\ \text{kg}\,H_2 \times \frac{10^3\ \text{g}\,H_2}{1\ \text{kg}\,H_2} \times \frac{1\ \text{mol}\,H_2}{2.016\ \text{g}\,H_2} \times \frac{2\ \text{mol}\,NH_3}{3\ \text{mol}\,H_2} \times \frac{17.03\ \text{g}\,NH_3}{1\ \text{mol}\,NH_3} \times \frac{1\ \text{kg}\,NH_3}{10^3\ \text{g}\,NH_3}$$

$$= 33.8\ \text{kg}\,NH_3$$

Then we divide the actual yield by this theoretical yield to determine the percent yield:

$$\frac{18.20 \text{ kg NH}_3}{33.8 \text{ kg NH}_3} \times 100\% = 53.8\%$$

Think About It A yield of about 54% may seem low, but it may be the best that can be achieved for a particular process. A great deal of chemical research goes into trying to improve the percent yield of industrial chemical reactions.

 Practice Exercise The combustion of 58.0 g of butane (C_4H_{10}) produces 158 g of CO_2. What is the percent yield of the reaction?

(Answers to Practice Exercises are in the back of the book.)

SAMPLE EXERCISE 3.23 Effect of Percent Yield on Calculating   **LO8**
the Mass of Reactant Needed

One way of producing hydrogen for use in hydrogen-powered vehicles is the water–gas shift reaction:

$$H_2O(g) + CO(g) \rightarrow H_2(g) + CO_2(g)$$

At 200°C, the reaction produces a 96% yield. How many grams of H_2O and CO are needed to generate 176 g of H_2 under these conditions?

Collect and Organize This is a stoichiometry problem like Sample Exercises 3.12 and 3.13, except that the yield of reaction is only 96%.

Analyze We need to convert the mass of H_2 we wish to make into moles of H_2, and then we use the mole ratios of H_2:CO and H_2:H_2O to find the amount of reactants we need. To produce 176 g of H_2 by using a reaction that produces 96% of the theoretical yield, we consider the theoretical yield to be 96% of the total yield we need. Then by analogy to Sample Exercise 3.13, we can calculate the amount of both reactants needed.

Solve

1. The number of moles of H_2 we wish to produce is:

$$176 \text{ g H}_2 \times \frac{1 \text{ mol H}_2}{2.016 \text{ g H}_2} = 87.3 \text{ mol H}_2$$

However, since the reaction produces only 96% of the theoretical yield, we need to base our calculation on what 100% yield would be if 87.30 moles is 96% of the theoretical yield:

$$87.30 \text{ mol} = 0.96x \qquad x = 91 \text{ mol}$$

2. Using the mole ratios of H_2:H_2O and H_2:CO, we find the number of moles of H_2O and CO:

$$91 \text{ mol H}_2 \times \frac{1 \text{ mol H}_2O}{1 \text{ mol H}_2} = 91 \text{ mol H}_2O \qquad 91 \text{ mol H}_2 \times \frac{1 \text{ mol CO}}{1 \text{ mol H}_2} = 91 \text{ mol CO}$$

3. We convert from moles of reactants to the mass of the reactants by multiplying by the molar mass:

$$91 \text{ mol H}_2O \times \frac{18.02 \text{ g H}_2O}{1 \text{ mol H}_2O} = 1.6 \times 10^3 \text{ g H}_2O$$

$$91 \text{ mol CO} \times \frac{28.01 \text{ g CO}}{1 \text{ mol CO}} = 2.5 \times 10^3 \text{ g CO}$$

As a check of our answers, let's calculate how much H_2 would be produced from 91 moles of CO if the reaction proceeded in 100% yield:

$$91 \ \text{mol CO} \times \frac{1 \ \text{mol } H_2}{1 \ \text{mol CO}} \times \frac{2.016 \ \text{g } H_2}{1 \ \text{mol } H_2} = 183 \ \text{g } H_2$$

The percent yield is

$$\frac{176 \ \text{g } H_2}{183 \ \text{g } H_2} \times 100\% = 96\%$$

Think About It On a large scale, any yield less than 100% could be a significant cost for rare and expensive reactants.

Practice Exercise How much aluminum oxide and how much carbon are needed to prepare 454 g (1 pound) of aluminum by the balanced chemical equation:

$$2 \ Al_2O_3(s) + 3 \ C(s) \rightarrow 4 \ Al(s) + 3 \ CO_2(g)$$

if the reaction proceeds to 78% yield?

(Answers to Practice Exercises are in the back of the book.)

SAMPLE EXERCISE 3.24 Integrating Concepts: Taxol™

Taxol, known generically as paclitaxel, is a molecular compound used in the treatment of cancer. It was originally isolated from the bark of the Pacific yew tree (Figure 3.39) and subsequently has been synthesized in the laboratory.

a. The yew tree provides about 100 mg of Taxol for every 1.00 kg of bark. If 1 tree has about 3 kg of bark and 9000 trees were needed to isolate enough Taxol for its first clinical trial, how many grams of Taxol were used for those initial tests?

b. One of the laboratory syntheses of Taxol begins with camphor (78.89% C; 10.59% H; $\mathcal{M}$ = 152.23 g/mol) and requires 23 different chemical reactions in sequence to convert it into Taxol (66.11% C; 6.02% H; 1.64% N; $\mathcal{M}$ = 853.88 g/mol). The overall yield of the process to produce 1 mole of Taxol, starting with 1 mole camphor, is 0.10%. (i) What are the molecular formulas of camphor and Taxol? (ii) How much camphor must one use to make the amount of Taxol provided by 9000 trees?

Collect and Organize We are given information about the amount of Taxol in bark, the percent composition and molar mass of camphor and Taxol, and the yield of the laboratory synthesis. Using dimensional analysis and other methods from this chapter, we can answer the questions posed.

Analyze The percent compositions given for both substances do not add to 100%. We may assume that both compounds contain oxygen.

Solve

a. We begin by calculating the amount of Taxol in 9000 trees:

$$9000 \ \text{trees} \times \frac{3 \ \text{kg bark}}{1 \ \text{tree}} \times \frac{0.100 \ \text{g Taxol}}{1.00 \ \text{kg bark}} = 3 \times 10^3 \ \text{g Taxol}$$

FIGURE 3.39 An ancient yew tree whose bark contains powerful cancer-fighting molecules of paclitaxel, marketed under the commercial name Taxol.

b. (i) The percent carbon in camphor added to the percent hydrogen is less than 100%:

$$78.89\% + 10.59\% = 89.48\%$$

Because the reported percentages do not add to 100%, the camphor most likely contains oxygen, the percent of which we cannot find in a combustion analysis. We can assume the percent oxygen in camphor is 100% − 89.48% = 10.52%. We can now determine the empirical formula of camphor by considering that 100.00 g of camphor would contain 78.89 g of C, 10.59 g of H, and 10.52 g of O:

$$78.89 \text{ g C} \times \frac{1 \text{ mol C}}{12.01 \text{ g C}} = 6.569 \text{ mol C}$$

$$10.59 \text{ g H} \times \frac{1 \text{ mol H}}{1.008 \text{ g H}} = 10.51 \text{ mol H}$$

$$10.52 \text{ g O} \times \frac{1 \text{ mol O}}{16.00 \text{ g O}} = 0.6575 \text{ mol O}$$

Dividing by the smallest number of moles:

$$\frac{6.569 \text{ mol C}}{0.6575} = 10 \text{ mol C}$$

$$\frac{10.51 \text{ mol H}}{0.6575} = 16 \text{ mol H}$$

$$\frac{0.6575 \text{ mol O}}{0.6575} = 1 \text{ mol O}$$

The empirical formula of camphor is $C_{10}H_{16}O$. The corresponding mass is [(10 × 12.01 g/mol) + (16 × 1.008 g/mol) + (16.00 g/mol)] = 153.23 g/mol, which matches the molar mass of camphor. Therefore the molecular formula of camphor is the same as its empirical formula.

Carrying out the same procedure for Taxol:

$$66.11\% + 6.02\% + 1.64\% = 73.77\%$$

$$100\% - 73.77\% = 26.23\% \text{ oxygen}$$

Calculating the empirical formula:

$$66.11 \text{ g C} \times \frac{1 \text{ mol C}}{12.01 \text{ g C}} = 5.505 \text{ mol C}$$

$$6.02 \text{ g H} \times \frac{1 \text{ mol H}}{1.008 \text{ g H}} = 5.97 \text{ mol H}$$

$$1.64 \text{ g N} \times \frac{1 \text{ mol N}}{14.01 \text{ g N}} = 0.117 \text{ mol N}$$

$$26.23 \text{ g O} \times \frac{1 \text{ mol O}}{16.00 \text{ g O}} = 1.639 \text{ mol O}$$

Dividing by the smallest number of moles:

$$\frac{5.505 \text{ mol C}}{0.117} = 47 \text{ mol C}$$

$$\frac{5.97 \text{ mol H}}{0.117} = 51 \text{ mol H}$$

$$\frac{0.117 \text{ mol N}}{0.117} = 1 \text{ mol N}$$

$$\frac{1.639 \text{ mol O}}{0.117} = 14 \text{ mol O}$$

The empirical formula of Taxol is $C_{47}H_{51}NO_{14}$, which has a mass of [(47 × 12.01 g/mol) + (51 × 1.008 g/mol) + (1 × 14.01 g/mol) + (14 × 16.00 g/mol)] = 853.88 g/mol. This value is the same as the molar mass of Taxol, so $C_{47}H_{51}NO_{14}$ is also the molecular formula.

(ii) We need 3×10^3 g of Taxol from a process that has a yield of 0.10%. The molar ratio in the process is 1 mole of camphor produces 1 mole of Taxol.

$$3 \times 10^3 \text{ g Taxol} = 0.10\% \text{ of the theoretical yield}$$

The theoretical yield for the process is

$$3 \times 10^3 \text{ g} = 0.0010x \qquad x = 3 \times 10^6 \text{ g}$$

Starting with that amount of Taxol as the theoretical yield:

$$3 \times 10^6 \text{ g Taxol} \times \frac{1 \text{ mol Taxol}}{853.88 \text{ g Taxol}} \times \frac{1 \text{ mol camphor}}{1 \text{ mol Taxol}}$$

$$\times \frac{152.23 \text{ g camphor}}{1 \text{ mol camphor}} = 5 \times 10^5 \text{ g camphor}$$

Thus, we need 500 kg of camphor to synthesize as much Taxol as there is in 9000 Pacific yew trees.

Think About It Camphor is quite inexpensive, but this example clearly illustrates how much material can be required to make a valuable product if the process has a low overall yield.

SUMMARY

LO1 Avogadro's number ($N_A = 6.022 \times 10^{23}$ particles) and the **mole** (the amount of a material that contains Avogadro's number of particles) can be used to convert grams of a substance to moles or to number of particles, moles of a substance to grams or to number of particles, and number of particles to moles or to number of grams. One mole of any substance has a mass equal to the sum of the masses of the constituent particles in its formula. (Section 3.2)

LO2 Balanced chemical equations are essential tools to describe chemical reactions. Correct equations indicate the quantitative relation in terms of moles between **reactants**, whose formulas appear first in the equation, and **products**, whose formulas appear after the reaction arrow. In a balanced chemical equation, the number of atoms of each element is the same on the reactant side and on the product side. (Sections 3.1–3.4)

LO3 The mole ratio from balanced chemical equations can be used to determine the masses of reactants consumed and products formed in a chemical reaction. (Section 3.5)

LO4 Writing equations requires knowing correct formulas for substances. Empirical formulas for substances can be determined from their **percent composition** by mass. (Section 3.6)

LO5 The molecular formula of a compound may or may not be the same as the empirical formula. It can be determined from the empirical formula and the molecular mass of the compound. (Section 3.7)

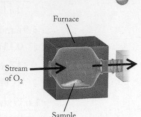

LO6 Empirical formulas for substances can be determined from the **combustion analysis** (the burning of a substance in oxygen) of compounds. In the combustion analysis of organic compounds, the production of CO_2, H_2O, and other oxides can provide the percent by mass of all elements in the compound except oxygen, which we can

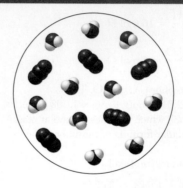

determine by applying the law of conservation of mass. (Sections 3.7 and 3.8)

LO7 Chemical reactions are not always run with exact stoichiometric amounts of reactants. Balanced equations and stoichiometric calculations can be used to determine which substances are **limiting reactants**. (Section 3.9)

LO8 The **actual yield** (often expressed as **percent yield**) of a chemical reaction is frequently less than the **theoretical yield** predicted by stoichiometry. If we know the amount of material isolated as a result of a reaction, we can calculate the percent yield as the ratio of actual yield to theoretical yield, expressed as a percentage. (Section 3.9)

PARTICULATE PREVIEW WRAP-UP

Ten oxygen molecules will completely combust the five methane molecules to produce five carbon dioxide molecules and ten water molecules.

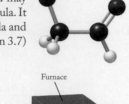

PROBLEM-SOLVING SUMMARY

Type of Problem	Concepts and Equations	Sample Exercises
Converting number of particles into number of moles (or vice versa)	Convert number of particles to moles by dividing by Avogadro's number ($N_A = 6.022 \times 10^{23}$ particles/mol). Convert number of moles to particles by multiplying by Avogadro's number ($N_A = 6.022 \times 10^{23}$ particles/mol).	**3.1, 3.2**
Converting mass of a substance into number of moles (or vice versa)	Convert mass of substance to moles by dividing by the molar mass ($\mathcal{M}$) of the substance. Convert moles of substance to mass by multiplying by the molar mass ($\mathcal{M}$) of the substance.	**3.3, 3.4, 3.6, 3.7**
Calculating molar mass of a compound	Multiply the molar mass of each element by its subscript in the compound's formula; then add the products.	**3.5, 3.6, 3.7**
Balancing a chemical reaction	Change the coefficients in the equation so that the numbers of atoms of each element are the same on both sides of the reaction arrow.	**3.8, 3.9, 3.10**
Writing and balancing a chemical equation for a combustion reaction	The C and H in organic compounds react with O_2 to form CO_2 and H_2O, for example: $$CH_4(g) + 2\,O_2(g) \rightarrow 2\,H_2O(g) + CO_2(g)$$ Balance moles of C first, then H, then O.	**3.11**
Calculating mass of a product from mass of a reactant	Use $\mathcal{M}_{\text{reactant}}$ to convert mass of reactant into moles of reactant, use reaction stoichiometry to calculate moles of product, and then use $\mathcal{M}_{\text{product}}$ to calculate mass of product.	**3.12, 3.13**

Type of Problem	Concepts and Equations	Sample Exercises
Calculating percent composition from a chemical formula	Calculate the molar mass of the compound represented by the chemical formula. Determine the mass of each element in grams from the chemical formula. Divide each element mass by the molar mass of the compound. Express each result as a percentage; percentages should add up to 100%.	**3.14**
Determining an empirical formula from percent composition	Assuming a 100.00 g sample, assign the mass (g) of each element to equal its percentage. Divide the mass of each element by its molar mass to get moles. Simplify mole ratios to lowest whole numbers, and use those numbers as subscripts in the empirical formula of the compound.	**3.15, 3.16**
Relating empirical and molecular formulas	Calculate the multiplier n by dividing the compound's molecular mass by its empirical formula mass. Multiply the subscripts in the empirical formula by this conversion factor.	**3.17**
Deriving an empirical formula from combustion analysis	For hydrocarbons, convert given masses of CO_2 and H_2O into moles of CO_2 and H_2O, and then to moles of C and H. For compounds containing O, convert moles of C and H into masses of C and H and subtract the sum of these values from the sample mass to calculate mass of O. Convert mass of O to moles of O. Simplify the mole ratio of C to H to O.	**3.18, 3.19**
Identifying the limiting reactant	Method 1: Calculate how much product each reactant could make; the reactant making the least amount of product is the limiting reactant. Method 2: Convert given masses of reactants into moles. Compare the mole ratio of reactants to the corresponding mole ratio in the stoichiometric equation. If $$\left(\frac{\text{mol A}}{\text{mol B}}\right)_{\text{given}} > \left(\frac{\text{mol A}}{\text{mol B}}\right)_{\text{stoichiometric}} \qquad (3.3)$$ then B is the limiting reactant. If $$\left(\frac{\text{mol A}}{\text{mol B}}\right)_{\text{given}} < \left(\frac{\text{mol A}}{\text{mol B}}\right)_{\text{stoichiometric}} \qquad (3.4)$$ then A is the limiting reactant.	**3.20, 3.21**
Calculating percent yield	Calculate the theoretical yield of product using mass of limiting reactant. Divide actual yield (given) by theoretical yield: $$\text{Percent yield} = \frac{\text{actual yield}}{\text{theoretical yield}} \times 100\% \qquad (3.5)$$	**3.22, 3.23**

VISUAL PROBLEMS

(Answers to boldface end-of-chapter questions and problems are in the back of the book.)

3.1. Each of the pairs of containers pictured in Figure P3.1 contains substances composed of elements X (red spheres) and Y (blue spheres). For each pair, write a balanced chemical equation describing the reaction that takes place. Be sure to indicate the physical states of the reactants and products, using the appropriate symbols in parentheses.

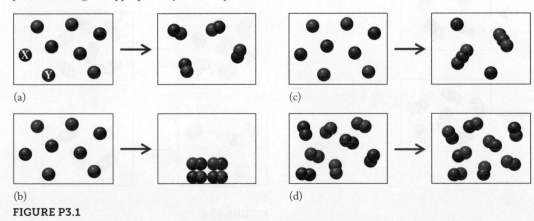

(a)

(b)

(c)

(d)

FIGURE P3.1

3.2. Identify the limiting reactant in each of the pairs of containers pictured in Figure P3.2. The red spheres represent atoms of element X, and the blue spheres represent atoms of element Y. Each question mark means that there is unreacted reactant left over.

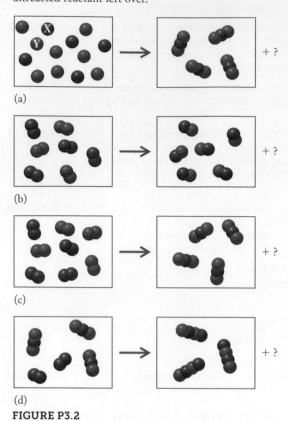

(a)

(b)

(c)

(d)

FIGURE P3.2

3.3. Which of the drawings in Figure P3.3 best illustrates the 100% reaction between N_2 and O_2 to produce N_2O? The red spheres represent atoms of oxygen and the blue spheres represent atoms of nitrogen.

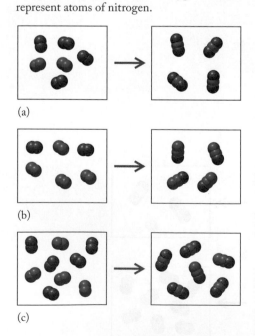

(a)

(b)

(c)

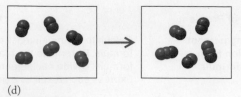

(d)

FIGURE P3.3

3.4. Is there a limiting reactant in any of the reactions depicted in Figure P3.3? If so, what is it, and how much of the excess reactant remains?

3.5. Which of the molecules in Figure P3.5 have the same empirical formulas?

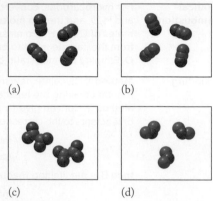

(a)   (b)

(c)   (d)

FIGURE P3.5

3.6. Which of the drawings in Figure P3.6 represent balanced chemical equations? Write balanced chemical equations for any unbalanced equations in Figure P3.6. The red spheres represent oxygen, black represent carbon, blue represent nitrogen, and yellow represent sulfur.

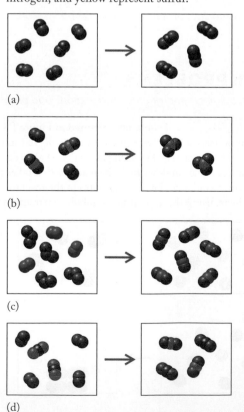

(a)

(b)

(c)

(d)

FIGURE P3.6

3.7. The two major products of combustion are CO_2 and H_2O. Figure P3.7 shows two mass spectra. Which belongs to water, and which belongs to carbon dioxide?

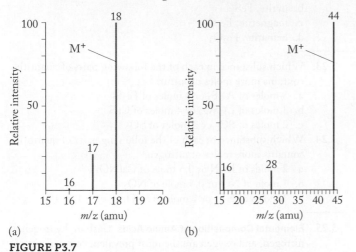

(a)   (b)

FIGURE P3.7

3.8. Figure P3.8 shows the mass spectrum of a simple hydrocarbon that can be combusted as a fuel. What is the molecular formula of this hydrocarbon?

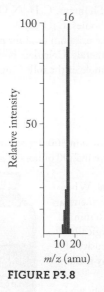

FIGURE P3.8

3.9. What is the percent yield of NH_3 for the reaction depicted in Figure P3.9? The blue spheres represent nitrogen, and the white spheres are hydrogen.

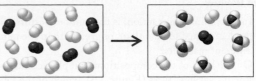

FIGURE P3.9

3.10. Use representations [A] through [I] in Figure P3.10 to answer questions a–f.
   a. Which two compounds have the same empirical formula?
   b. Which compound has a molecular mass of 180 amu?
   c. Which compound might have a molar mass of 180 g?
   d. Which compound has the largest percent oxygen by mass?
   e. The mass of one gold bar in [D] is 12.4 kg. The mass of one silver bar in [E] is 31 kg. Which contains more atoms—the gold bar or the silver bar?
   f. When completely combusted, which compound will produce more moles of carbon dioxide—benzene [A] or table sugar [C]?

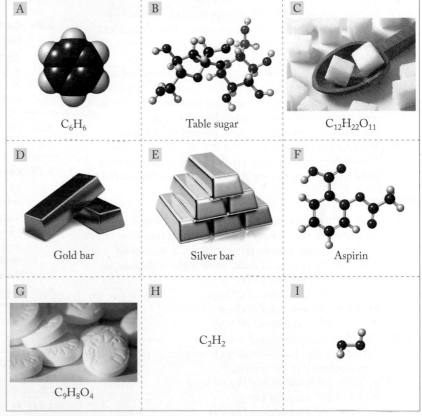

FIGURE P3.10

QUESTIONS AND PROBLEMS

Air, Life, and Molecules

Concept Review

3.11. In a combination reaction, is the number of different products equal to, less than, or greater than the number of different reactants?

3.12. On the basis of the compositions and physical states of Earth's various layers (Figure 3.3), which of the following substances has the higher melting point: Al_2O_3 or Ni?

3.13. On the basis of the distribution of the elements in Earth's layers (see Figure 3.3), which of the following substances should be the most dense? (a) $SiO_2(s)$; (b) $Al_2O_3(s)$; (c) $Fe(\ell)$

*3.14. The proportions of the elements that make up the asteroid 433 Eros are similar to those that make up Earth. Scientists believe that this similarity means that 433 Eros and Earth formed around the same time. If another asteroid formed *after* Earth formed a solid crust, and if this asteroid was the product of a collision between Earth and an even larger asteroid, how would the new asteroid's composition differ from that of 433 Eros?

The Mole

Concept Review

3.15. In principle we could use the more familiar unit *dozen* in place of mole when expressing the quantities of particles (atoms, ions, or molecules) in chemical reactions. What would be the disadvantage in doing so?

3.16. In what way are the molar mass of an ionic compound and its formula mass the same, and in what ways are they different?

3.17. Do molecular compounds containing three atoms per molecule always have a molar mass greater than that of molecular compounds containing two atoms per molecule? Explain.

3.18. Without calculating their molar masses (though you may consult the periodic table), predict which of the following oxides of nitrogen has the larger molar mass: NO_2 or N_2O.

Problems

3.19. Earth's atmosphere contains many volatile substances that are present in trace amounts. The following quantities of these trace gases were found in a 1.0 mL sample of air. Calculate the number of moles of each gas in the sample.
a. 4.4×10^{14} atoms of $Ne(g)$
b. 4.2×10^{13} molecules of $CH_4(g)$
c. 2.5×10^{12} molecules of $O_3(g)$
d. 4.9×10^{9} molecules of $NO_2(g)$

3.20. The following quantities of trace gases were found in a 1.0 mL sample of air. Calculate the number of moles of each compound in the sample.
a. 1.4×10^{13} molecules of $H_2(g)$
b. 1.5×10^{14} atoms of $He(g)$
c. 7.7×10^{12} molecules of $N_2O(g)$
d. 3.0×10^{12} molecules of $CO(g)$

3.21. How many atoms of titanium are there in 0.125 mole of each of the following?
a. ilmenite, $FeTiO_3$
b. titanium(IV) chloride
c. Ti_2O_3
d. Ti_3O_5

3.22. How many atoms of iron are there in 2.5 moles of each of the following?
a. wolframite, $FeWO_4$
b. pyrite, FeS_2
c. magnetite, Fe_3O_4
d. hematite, Fe_2O_3

3.23. Which substance in each of the following pairs of quantities contains more moles of sulfur?
a. 3 moles of Al_2S_3 or 4 moles of Fe_2S_3
b. 3 moles of Li_2SO_4 or 4 moles of CaS
c. 2 moles of SO_3 or 2 moles of SO_2

3.24. Which substance in each of the following pairs of quantities contains more moles of nitrogen?
a. 2 moles of Li_3N or 0.5 mole of $Ca(NO_3)_2$
b. 1 mole of NO or 0.4 mole of NO_2
c. 3 moles of NF_3 or 1 mole of N_2O_5

3.25. Elemental Composition of Amino Acids Carbon, hydrogen, nitrogen, and oxygen are the most prevalent components of amino acids, the fundamental building blocks of proteins in living systems. How many moles of oxygen are in 1.50 mol of the following compounds?
a. Glycine, the smallest amino acid: $C_2H_5NO_2$
b. Lysine, essential in the diet of humans: $C_6H_{14}N_2O_2$
c. Asparagine, produced in our bodies: $C_4H_8N_2O_3$

3.26. Composition of Uranium Ores The uranium used for nuclear fuel exists in nature in several minerals. Calculate how many moles of uranium are in 1 mole of the following:
a. carnotite, $K_2(UO_2)_2(VO_4)_2$
b. uranophane, $CaU_2Si_2O_{11}$
c. autunite, $Ca(UO_2)_2(PO_4)_2$

3.27. How many moles of carbon are there in 500.0 g of carbon?

3.28. How many moles of gold are there in 2.00 ounces of gold?

3.29. Cancer Therapy with Iridium Metal When iridium-192 is used in cancer treatment, a small cylindrical piece of ^{192}Ir, 0.6 mm in diameter and 3.5 mm long, is surgically inserted into the tumor. If the density of iridium is 22.42 g/cm^3, how many iridium atoms are present in the sample?

*3.30. **Gold Nanoparticles** The product shown in Figure P3.30 contains gold nanoparticles in water. The manufacturer claims that drinking this beverage improves human health. How many gold atoms are in a gold nanoparticle with a diameter of 2.00 nm ($d = 19.3$ g/mL, 1 amu = 1.66054×10^{-24} g)?

FIGURE P3.30

3.31. How many moles of iron are there in 1 mole of the following compounds? (a) FeO; (b) Fe_2O_3; (c) $Fe(OH)_3$; (d) Fe_3O_4

3.32. How many moles of sodium are there in 1 mole of the following compounds? (a) $NaCl$; (b) Na_2SO_4; (c) Na_3PO_4; (d) $NaNO_3$

3.33. Calculate the molar masses of the following atmospheric molecules: (a) SO_2; (b) O_3; (c) CO_2; (d) N_2O_5.

3.34. Determine the molar masses of the following minerals:
   a. rhodonite, $MnSiO_3$
   b. scheelite, $CaWO_4$
   c. ilmenite, $FeTiO_3$
   d. magnesite, $MgCO_3$

3.35. **Flavoring Additives** Calculate the molar masses of the following common flavors in food:
   a. vanillin, $C_8H_8O_3$
   b. oil of cloves, $C_{10}H_{12}O_2$
   c. anise oil, $C_{10}H_{12}O$
   d. oil of cinnamon, C_9H_8O

3.36. **Sweeteners** Calculate the molar masses of the following common sweeteners:
   a. sucrose, $C_{12}H_{22}O_{11}$
   b. saccharin, $C_7H_5NO_3S$
   c. aspartame, $C_{14}H_{18}N_2O_5$
   d. fructose, $C_6H_{12}O_6$

3.37. Suppose pairs of balloons are filled with 10.0 g of the following pairs of gases. Which balloon in each pair has the greater number of particles? (a) CO_2 or NO; (b) CO_2 or SO_2; (c) O_2 or Ar

3.38. If you had equal masses of the substances in the following pairs of compounds, which of the two would contain the greater number of ions? (a) NaBr or KCl; (b) NaCl or $MgCl_2$; (c) $BaCl_2$ or Li_2CO_3

3.39. How many moles of SiO_2 are there in a quartz crystal (SiO_2) that has a mass of 45.2 g?

3.40. How many moles of NaCl are there in a crystal of halite that has a mass of 6.82 g?

3.41. What is the mass of 0.122 mol $MgCO_3$?

3.42. What is the volume of 1.00 mol benzene (C_6H_6) at 20°C? The density of benzene is 0.879 g/mL.

*3.43. The density of uranium (U; 19.05 g/cm³) is more than five times as great as that of diamond (C; 3.514 g/cm³). If you have a cube (1 cm on a side) of each element, which cube contains more atoms?

*3.44. Aluminum (d = 2.70 g/mL) and strontium (d = 2.64 g/mL) have nearly the same density. If we manufacture two cubes, each containing 1 mole of one element or the other, which cube will be smaller? What are the dimensions of this cube?

Writing Balanced Chemical Equations

Concept Review

3.45. In a balanced chemical equation, does the number of moles of reactants always equal the number of moles of products?

3.46. In a balanced chemical equation, does the sum of the coefficients for the reactants always equal the sum of the coefficients for the products?

3.47. In a balanced chemical equation, must the sum of the masses of all the gaseous reactants always equal the sum of the masses of the gaseous products?

3.48. In a balanced chemical equation, must the sum of the volumes occupied by the gaseous reactants always equal the sum of the volumes occupied by the gaseous products?

Problems

3.49. Using different-colored spheres to represent C and O, sketch the reaction between five C atoms and the necessary number of O_2 molecules to produce a 50% mixture of CO and CO_2.

3.50. Using different-colored spheres to represent N and O, sketch the reaction between three molecules of N_2 and sufficient O_2 to produce a mixture containing 50% NO_2 and 50% N_2O_4.

3.51. Fluorine is a very reactive, dangerous element, as can be seen in the following unbalanced equation:

$$F_2(g) + H_2O(\ell) \rightarrow HF(aq) + O_2(g)$$

After balancing this chemical equation, what is the coefficient of HF?

3.52. Aluminum reacts with elemental oxygen at high temperatures to give pure aluminum oxide. What is the coefficient of $O_2(g)$ in the balanced chemical equation?

*3.53. **Chemical Weathering of Rocks and Minerals** Balance the following chemical reactions, which contribute to weathering of the iron–silicate minerals ferrosilite ($FeSiO_3$), fayalite (Fe_2SiO_4), and greenalite [$Fe_3Si_2O_5(OH)_4$]:
   a. $FeSiO_3(s) + H_2O(\ell) \rightarrow Fe_3Si_2O_5(OH)_4(s) + H_4SiO_4(aq)$
   b. $Fe_2SiO_4(s) + CO_2(g) + H_2O(\ell) \rightarrow$ $FeCO_3(s) + H_4SiO_4(aq)$
   c. $Fe_3Si_2O_5(OH)_4(s) + CO_2(g) + H_2O(\ell) \rightarrow$ $FeCO_3(s) + H_4SiO_4(aq)$

3.54. **Chemistry of Geothermal Vents** Some scientists believe that life on Earth may have originated near deep-ocean vents. Balance the following reactions, which are among those taking place near such vents:
   a. $CH_3SH(aq) + CO(aq) \rightarrow CH_3COSCH_3(aq) + H_2S(aq)$
   b. $H_2S(aq) + CO(aq) \rightarrow CH_3CO_2H(aq) + S_8(s)$

3.55. **Physiologically Active Nitrogen Oxides** The oxides of nitrogen are biologically reactive substances now known to be formed endogenously in the human lung: NO is a powerful agent for dilating blood vessels; N_2O is the anesthetic known as laughing gas; NO_2 has an acrid odor and is corrosive to lung tissue. Balance the following reactions for the formation of nitrogen oxides:
   a. $N_2(g) + O_2(g) \rightarrow NO(g)$
   b. $NO(g) + O_2(g) \rightarrow NO_2(g)$
   c. $NO(g) + NO_3(g) \rightarrow NO_2(g)$
   d. $N_2(g) + O_2(g) \rightarrow N_2O(g)$

3.56. **Purifying Natural Gas** If natural gas contains significant amounts of sulfur as H_2S, it is called sour natural gas. For the gas to be commercially useful as a fuel, the H_2S must be removed. Once it is separated from the natural gas, it is reacted with oxygen in two different processes to yield either elemental sulfur (S_8), a commercial material that can be sold, or sulfur dioxide (SO_2). This sulfur dioxide product can be reacted with more H_2S to make additional elemental sulfur. Balance the following reactions that describe the production of elemental sulfur.
   a. $H_2S(g) + O_2(g) \rightarrow S_8(s) + H_2O(g)$
   b. $H_2S(g) + O_2(g) \rightarrow SO_2(g) + H_2O(g)$
   *c. $H_2S(g) + SO_2(g) \rightarrow S_8(s) + H_2O(g)$

*3.57. Write a balanced chemical equation for each of the following reactions:
   a. Dinitrogen pentoxide reacts with sodium metal to produce sodium nitrate and nitrogen dioxide.
   b. A mixture of nitric acid and nitrous acid is formed when water reacts with dinitrogen tetroxide.
   c. At high pressure, nitrogen monoxide decomposes to dinitrogen monoxide and nitrogen dioxide.
   d. Acetylene, C_2H_2, burns and becomes carbon dioxide and water vapor.

3.58. Write a balanced chemical equation for each of the following reactions:
   a. Carbon dioxide reacts with carbon to form carbon monoxide.
   b. Potassium reacts with water to give potassium hydroxide and hydrogen gas.
   c. Phosphorus, P_4, burns in air to give diphosphorus pentoxide.
   d. Octane, C_8H_{18}, burns and becomes carbon dioxide and water vapor.

Combustion Reactions; Stoichiometric Calculations and the Carbon Cycle

Concept Review

3.59. Does the sum of the masses of the products always equal the sum of the masses of the reactants in a balanced chemical equation?

*3.60. There are two ways to write the equation for the combustion of ethane:

$$C_2H_6(g) + \tfrac{7}{2} O_2(g) \rightarrow 3\,H_2O(g) + 2\,CO_2(g)$$
$$2\,C_2H_6(g) + 7\,O_2(g) \rightarrow 6\,H_2O(g) + 4\,CO_2(g)$$

Do these two different ways of writing the equation affect the calculation of how much CO_2 is produced from a known quantity of C_2H_6?

Problems

3.61. When $NaHCO_3$ is heated above 270°C, it decomposes to $Na_2CO_3(s)$, $H_2O(g)$, and $CO_2(g)$.
   a. Write a balanced chemical equation for the decomposition reaction.
   b. Calculate the mass of CO_2 produced from the decomposition of 25.0 g of $NaHCO_3$.

3.62. **Egyptian Cosmetics** $Pb(OH)Cl$, one of the lead compounds used in ancient Egyptian cosmetics, was prepared from PbO according to the following recipe:

$$PbO(s) + NaCl(aq) + H_2O(\ell) \rightarrow Pb(OH)Cl(s) + NaOH(aq)$$

How many grams of PbO and how many grams of NaCl would be required to produce 10.0 g of $Pb(OH)Cl$?

3.63. The manufacture of aluminum includes the production of cryolite (Na_3AlF_6) from the following reaction:

$$6\,HF(g) + 3\,NaAlO_2(s) \rightarrow Na_3AlF_6(s) + 3\,H_2O(\ell) + Al_2O_3(s)$$

How much $NaAlO_2$ (sodium aluminate) is required to produce 1.00 kg of Na_3AlF_6?

3.64. Chromium metal can be produced from the high-temperature reaction of Cr_2O_3 [chromium(III) oxide] with silicon or aluminum by each of the following reactions:

$$Cr_2O_3(s) + 2\,Al(\ell) \rightarrow 2\,Cr(\ell) + Al_2O_3(s)$$
$$2\,Cr_2O_3(s) + 3\,Si(\ell) \rightarrow 4\,Cr(\ell) + 3\,SiO_2(s)$$

   a. Calculate the number of grams of aluminum required to prepare 400.0 g of chromium metal by the first reaction.
   b. Calculate the number of grams of silicon required to prepare 400.0 g of chromium metal by the second reaction.

*3.65. Suppose 25 metric tons of coal that is 3.0% sulfur by mass is burned at an electric power plant (1 metric ton = 10^3 kg). During combustion, the sulfur is converted into sulfur dioxide. How many tons of sulfur dioxide are produced?

*3.66. The uranium minerals found in nature must be refined and enriched in ^{235}U before the uranium can be used as a fuel in nuclear reactors. One procedure for enriching uranium relies on the reaction of UO_2 with HF to form UF_4, which is then converted into UF_6 by reaction with fluorine:

$$UO_2(g) + 4\,HF(aq) \rightarrow UF_4(g) + 2\,H_2O(\ell)$$
$$UF_4(g) + F_2(g) \rightarrow UF_6(g)$$

   a. How many kilograms of HF are needed to completely react with 5.00 kg of UO_2?
   b. How much UF_6 can be produced from 850.0 g of UO_2?

3.67. In Brazil automobiles use ethanol, C_2H_6O, as fuel, whereas in the United States we rely on gasoline. Using C_8H_{18} (octane) to represent gasoline, write balanced chemical equations for the complete combustion of ethanol and octane. Which fuel produces more CO_2 per gram of fuel?

3.68. Driving 1000 miles a month is not unusual for a short-distance commuter. If your vehicle gets 25 mpg, you would use 40 gallons ($\approx$150 L) of gasoline every month. If gasoline is approximated as C_8H_{18} ($d = 0.703$ g/mL), how much carbon dioxide does your vehicle emit every month? The *unbalanced* chemical equation for the reaction is

$$C_8H_{18}(\ell) + O_2(g) \rightarrow CO_2(g) + H_2O(g)$$

3.69. Chalcopyrite ($CuFeS_2$) is an abundant copper mineral that can be converted into elemental copper. How much Cu could be produced from 1.00 kg of $CuFeS_2$?

*3.70. **Mining for Gold** Unlike most metals, gold is found in nature as the pure element. Miners in California in 1849 searched for gold nuggets and gold dust in streambeds, where the denser gold could be easily separated from sand and gravel. However, larger deposits of gold are found in veins of rock and can be separated chemically in a two-step process:
(1) $4\,Au(s) + 8\,NaCN(aq) + O_2(g) + 2\,H_2O(\ell) \rightarrow$
$$4\,NaAu(CN)_2(aq) + 4\,NaOH(aq)$$
(2) $2\,NaAu(CN)_2(aq) + Zn(s) \rightarrow$
$$2\,Au(s) + Na_2[Zn(CN)_4](aq)$$
If a 1.0×10^3 kg sample of rock is 0.019% gold by mass, how much Zn is needed to react with the gold extracted from the rock? Assume that reactions (1) and (2) are 100% efficient.

Determining Empirical Formulas from Percent Composition; Comparing Empirical and Molecular Formulas

Concept Review

3.71. What is the difference between an empirical formula and a molecular formula?

3.72. Do the empirical and molecular formulas of a compound always have the same percent composition values? Explain your answer.

3.73. Is the element with the largest atomic mass always the element present in the highest percentage by mass in a compound? Explain your answer.

3.74. Sometimes the composition of a compound is expressed as a mole percentage, and sometimes as an atom percentage. Are the values of these parameters likely to be the same for a given compound, or different?

3.75. Among the naturally occurring hydrocarbons emitted by plants are three compounds named camphene, carene, and thujene. If all three of these compounds have the same percent composition and the same molar mass, do they have the same empirical formula? Do they have the same molecular formula? Are they isomers?

*3.76.** How might the compounds in Problem 3.75 differ from each other if they have the same molar mass and percent composition?

Problems

3.77. Gasoline consists primarily of a mixture of the hydrocarbons C_6H_{14}, C_7H_{16}, C_8H_{18}, and C_9H_{20}. What is the empirical formula of each compound?

3.78. The biosynthesis of carbohydrate includes the molecules shown below. What is the empirical formula of each compound shown in Figure P3.78?

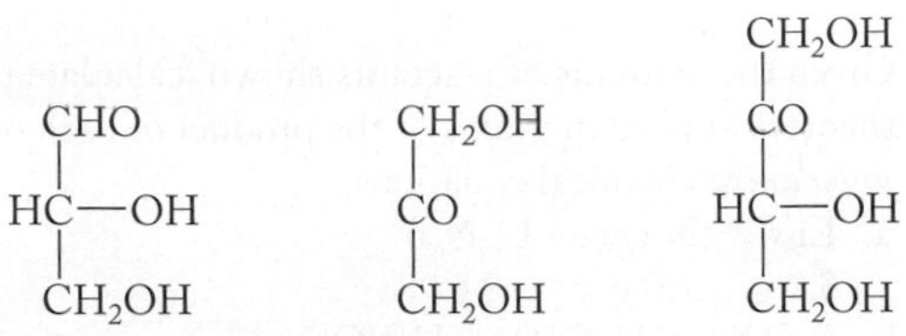

CH₂OH   CH₂OH

CHO   CH₂OH   CO

HC—OH   CO   HC—OH

CH₂OH   CH₂OH   CH₂OH

Glyceraldehyde   Dihydroxyacetone   Erythrulose

FIGURE P3.78

3.79. Calculate the percent composition of (a) Na_2O, (b) $NaOH$, (c) $NaHCO_3$, and (d) Na_2CO_3.

3.80. Calculate the percent composition of (a) sodium sulfate, (b) dinitrogen tetroxide, (c) strontium nitrate, and (d) aluminum sulfide.

3.81. **Organic Compounds in Space** The following compounds have been detected in space. Which of them contains the greatest percentage of carbon by mass? Do any two of the following compounds have the same empirical formula?
  a. naphthalene, $C_{10}H_8$
  b. chrysene, $C_{18}H_{12}$
  c. pentacene, $C_{22}H_{14}$
  d. pyrene, $C_{16}H_{10}$

3.82. Of the nitrogen oxides—N_2O, NO, N_2O_3, N_2O_2, NO_2, and N_2O_4—which are more than 50% oxygen by mass? Which, if any, have the same empirical formula?

3.83. Methane (CH_4) and tetrafluoromethane (CF_4) both contain 20% carbon per mole. Which one has the greater percent C by mass?

3.84. Silane (SiH_4) is used in the electronics industry to manufacture thin films of silicon. What is the percent Si by mass in SiH_4? Does silane have the same percent Si by mass as disilane, Si_2H_6?

3.85. **Surgical-Grade Titanium** Medical implants and high-quality jewelry items for body piercings are frequently made of a material known as G23Ti, or surgical-grade titanium. The percent composition of the material is 64.39% titanium, 24.19% aluminum, and 11.42% vanadium. What is the empirical formula for surgical-grade titanium?

3.86. A sample of an iron-containing compound is 22.0% iron, 50.2% oxygen, and 27.8% chlorine by mass. What is the empirical formula of this compound?

3.87. **Sour Candy** Tartaric acid, $C_4H_6O_6$, and citric acid, $C_6H_8O_7$, are both used commercially to give sour candies (Figure P3.87) their characteristic sour taste. Which compound has the larger percent C by mass?

FIGURE P3.87

3.88. **Chlorofluorocarbons** CFCs (chlorofluorocarbons) are molecules used as refrigerants, but they also contribute to the destruction of the ozone layer. One CFC known as Freon consists of two carbon atoms, two fluorine atoms, and four chlorine atoms. What is the empirical formula of Freon? What is its molecular formula?

3.89. **Asbestosis** Asbestosis is a lung disease caused by inhaling asbestos fibers. In addition, fiber from a form of asbestos called chrysotile is considered to be a human carcinogen by the U.S. Department of Health and Human Services. Chrysotile's composition is 26.31% magnesium, 20.20% silicon, and 1.45% hydrogen with the remainder of the mass as oxygen. Determine the empirical formula of chrysotile.

3.90. **Chemistry of Soot** A candle flame produces easily seen specks of soot near the edges of the flame, especially when the candle is moved. A piece of glass held over a candle flame will become coated with soot, which is the result of the incomplete combustion of candle wax. Elemental analysis of a compound extracted from a sample of this soot gave these results: 92.26% C and 7.74% H by mass. Calculate the empirical formula of the compound.

3.91. Making the DNA Bases Adenine (135.14 g/mol; 44.44% C, 3.73% H, and 51.84% N) was detected in mixtures of HCN, ammonia, and water under conditions that simulate early Earth. This observation suggests a possible origin for one of the bases found in DNA. What are the empirical and molecular formulas for adenine?

3.92. Making Sugars for RNA Ribose, the sugar found in RNA, has been detected in experiments designed to mimic the conditions of early Earth. If ribose contains 40.00% C, 6.71% H, and 53.28% O, with a molar mass of 150.13 g/mol, what are the empirical and molecular formulas for ribose?

Combustion Analysis

Concept Review

3.93. Explain why it is important for combustion analysis to be carried out in an excess of oxygen.

3.94. Why is the quantity of CO_2 obtained in a combustion analysis not a direct measure of the oxygen content of the starting compound?

3.95. Can the results of a combustion analysis ever give the true molecular formula of a compound?

3.96. What additional information is needed to determine a molecular formula from the results of an elemental analysis of an organic compound?

*****3.97.** If a compound containing nitrogen is subjected to combustion analysis in excess oxygen, what is the most likely molecular formula for the nitrogen-containing product?

*****3.98.** Suppose an insufficient amount of oxygen is used for a combustion analysis of a hydrocarbon, and some of the carbon is converted to CO rather than to CO_2. Will the empirical formula determined from the results of this analysis be too low or too high in carbon?

Problems

3.99. The combustion of 135.0 mg of a hydrocarbon produces 440.0 mg of CO_2 and 135.0 mg of H_2O. The molar mass of the hydrocarbon is 270 g/mol. Determine the empirical and molecular formulas of this compound.

3.100. A 0.100 g sample of a compound containing C, H, and O is burned in oxygen, producing 0.1783 g of CO_2 and 0.0734 g of H_2O. Determine the empirical formula of the compound.

3.101. GRAS List for Food Additives The compound geraniol is on the Food and Drug Administration's GRAS (generally recognized as safe) list and can be used in foods and personal care products. By itself, geraniol smells like roses but it is frequently blended with other fragrances on the GRAS list and then added to products to produce a pleasant peach- or lemon-like aroma. In an analysis, the complete combustion of 175 mg of geraniol produced 499 mg of CO_2 and 184 mg of H_2O. What is the empirical formula for geraniol?

*****3.102.** The combustion of 40.5 mg of a compound containing C, H, and O, and extracted from the bark of the sassafras tree, produces 110.0 mg of CO_2 and 22.5 mg of H_2O. The molar mass of the compound is 162 g/mol. Determine its empirical and molecular formulas.

3.103. Ethnobotany One of the ingredients in the Native American stomachache remedy derived from common chokecherry is caffeic acid. Combustion of 1.00×10^2 mg

of caffeic acid yielded 220 mg of CO_2 and 40.3 mg of H_2O. Determine the empirical formula of caffeic acid.

3.104. Coniine, a substance isolated from poison hemlock, contains only carbon, hydrogen, and nitrogen. Combustion of 5.024 mg of coniine yields 13.90 mg CO_2 and 6.048 mg of H_2O. What is the empirical formula of coniine?

Limiting Reactants and Percent Yield

Concept Review

3.105. If a reaction vessel contains equal masses of Fe and S, a mass of FeS corresponding to which of the following could theoretically be produced?
a. the sum of the masses of Fe and S
b. more than the sum of the masses of Fe and S
c. less than the sum of the masses of Fe and S

3.106. Can the percent yield of a chemical reaction ever exceed 100%?

3.107. Give two reasons why the actual yield from a chemical reaction is usually less than the theoretical yield.

3.108. A chemical reaction produces less than the expected amount of product. Is this result a violation of the law of conservation of mass?

Problems

3.109. Making Hollandaise Sauce A recipe for 1 cup of hollandaise sauce calls for $\frac{1}{2}$ cup of butter, $\frac{1}{4}$ cup of hot water, 4 egg yolks, and the juice of a medium-sized lemon. How many cups of this sauce can be made from a pound (2 cups) of butter, a dozen eggs, 4 medium lemons, and an unlimited supply of hot water?

3.110. A factory making toy wagons has 13,466 wheels, 3360 handles, and 2400 wagon beds in stock. What is the maximum number of wagons the factory can make?

3.111. Given the amounts of reactants shown, calculate the theoretical yield in grams of the product of each of these *unbalanced* chemical equations.
a. $Li(s) + N_2(g) \rightarrow Li_3N(s)$
     5.0 g   2.0 g   ? g
b. $P_2O_5(s) + H_2O(\ell) \rightarrow H_3PO_4(aq)$
     25.0 g   36.0 g   ? g
c. $SO_2(g) + O_2(g) \rightarrow SO_3(g)$
     6.4 g   4.0 g   ? g

3.112. Given the amounts of reactants shown, calculate the theoretical yield in grams of the product indicated by the question mark for each of these *unbalanced* chemical reactions.
a. $Cu_2O(s) + H_2(g) \rightarrow Cu(s) + H_2O(\ell)$
     30.0 g   12.0 g   ? g
b. $Mg(s) + HCl(g) \rightarrow MgCl_2(s) + H_2(g)$
     24.3 g   10.0 g   ? g
c. $CuCl_2(aq) + Zn(s) \rightarrow Cu(s) + ZnCl_2(aq)$
     11.6 g   10.0 g   ? g

3.113. Ammonia rapidly reacts with hydrogen chloride, making ammonium chloride. Write a balanced chemical equation for the reaction, and calculate the number of grams of excess reactant when 3.0 g of NH_3 reacts with 5.0 g of HCl.

3.114. Sulfur trioxide dissolves in water, producing H_2SO_4. How much sulfuric acid can be produced from 10.0 mL of water ($d = 1.00$ g/mL) and 25.6 g of SO_3?

*3.115. Phosgenite, a lead compound with the formula $Pb_2Cl_2CO_3$, is found in ancient Egyptian cosmetics. Phosgenite was prepared by the reaction of PbO, NaCl, H_2O, and CO_2. An unbalanced equation of the reaction is
$$PbO(s) + NaCl(aq) + H_2O(\ell) + CO_2(g) \rightarrow$$
$$Pb_2Cl_2CO_3(s) + NaOH(aq)$$
   a. Balance the equation.
   b. How many grams of phosgenite can be obtained from 10.0 g of PbO and 10.0 g of NaCl in the presence of excess water and CO_2?
   c. If 2.72 g of phosgenite is produced in the laboratory from the amounts of starting materials stated in part (b), what is the percent yield of the reaction?

3.116. Potassium superoxide, KO_2, reacts with carbon dioxide to form potassium carbonate and oxygen:
$$4\,KO_2(s) + 2\,CO_2(g) \rightarrow 2\,K_2CO_3(s) + 3\,O_2(g)$$
This reaction makes potassium superoxide useful in a self-contained breathing apparatus. How much O_2 could be produced from 2.50 g of KO_2 and 4.50 g of CO_2?

3.117. The reaction of 5.0 g of pentane (C_5H_{12}) with 5.0 g of oxygen gas produces 20.4 g of CO_2. What is the percent yield of this reaction?

3.118. Baking soda ($NaHCO_3$) can be made in large quantities by the following reaction:
$$NaCl(aq) + NH_3(aq) + CO_2(aq) + H_2O(\ell) \rightarrow$$
$$NaHCO_3(s) + NH_4Cl(aq)$$
If 10.0 g of NaCl reacts with excesses of the other reactants and 4.2 g of $NaHCO_3$ is isolated, what is the percent yield of the reaction?

3.119. Chemistry of Fermentation Yeast converts glucose ($C_6H_{12}O_6$) into ethanol ($d = 0.789$ g/mL) in a process called fermentation. An equation for the reaction can be written as follows:
$$C_6H_{12}O_6(aq) \rightarrow C_2H_5OH(\ell) + CO_2(g)$$
   a. Write a balanced chemical equation for this fermentation reaction.
   b. If 100.0 g of glucose yields 50.0 mL of ethanol, what is the percent yield for the reaction?

*3.120. Composition of Seawater A 1-liter sample of seawater contains 19.4 g of Cl^-, 10.8 g of Na^+, and 1.29 g of Mg^{2+}.
   a. How many moles of each ion are present?
   b. If we evaporated the seawater, would there be enough Cl^- present to form the chloride salts of all the sodium and magnesium present?

Additional Problems

*3.121. As a solution of copper sulfate slowly evaporates, beautiful blue crystals made of copper(II) and sulfate ions form such that water molecules are trapped inside the crystals. The overall formula of the compound is $CuSO_4 \cdot 5H_2O$.
   a. What is the percent water in this compound?
   b. At high temperatures, the water in the compound is driven off as steam. What mass percentage of the original sample of the blue solid is lost as a result?

3.122. Production of Aluminum Aluminum is mined as the mineral bauxite, which consists primarily of Al_2O_3 (alumina).
   a. How much aluminum is produced from 1 metric ton (1 metric ton = 10^3 kg) of Al_2O_3?
$$2\,Al_2O_3(s) \rightarrow 4\,Al(s) + 3\,O_2(g)$$

   b. The oxygen produced in part (a) is allowed to react with carbon to produce carbon monoxide:
$$O_2(g) + 2\,C(s) \rightarrow 2\,CO(g)$$
   Balance the following equation describing the reaction of alumina with carbon:
$$Al_2O_3(s) + C(s) \rightarrow Al(s) + CO(g)$$
   c. How much CO can be produced from the O_2 made in part (a)?

*3.123. Chemistry of Copper Production "Native," or elemental, copper can be found in nature, but most copper is mined as oxide or sulfide minerals. Chalcopyrite ($CuFeS_2$) is one copper mineral that can be converted to elemental copper in a series of chemical steps. Reacting chalcopyrite with oxygen at high temperature produces a mixture of copper sulfide and iron oxide. The iron oxide is separated from CuS by reaction with sand. CuS is converted to Cu_2S, and the Cu_2S is then burned in air to produce Cu and SO_2:
   (1) $2\,CuFeS_2(s) + 3\,O_2(g) \rightarrow$
$$2\,CuS(s) + 2\,FeO(s) + 2\,SO_2(g)$$
   (2) $FeO(s) + SiO_2(s) \rightarrow FeSiO_3(s)$
   (3) $2\,CuS(s) \rightarrow Cu_2S(s) + \frac{1}{8}\,S_8(s)$
   (4) $Cu_2S(s) + O_2(g) \rightarrow 2\,Cu(s) + SO_2(g)$
An average copper penny minted in the 1960s has a mass of about 3.0 g.
   a. How much chalcopyrite had to be mined to produce one dollar's worth of pennies?
   b. How much chalcopyrite had to be mined to produce one dollar's worth of pennies if reaction 1 above had a percent yield of 85% and reactions 2, 3, and 4 had percent yields of essentially 100%?
   c. How much chalcopyrite had to be mined to produce one dollar's worth of pennies if each reaction involving copper proceeded with an 85% yield?

*3.124. Mining for Gold Gold can be extracted from the surrounding rock by using a solution of sodium cyanide. While effective for isolating gold, toxic cyanide finds its way into watersheds, causing environmental damage and harming human health.
$$4\,Au(s) + 8\,NaCN(aq) + O_2(g) + 2\,H_2O(\ell) \rightarrow$$
$$4\,NaAu(CN)_2(aq) + 4\,NaOH(aq)$$
$$2\,NaAu(CN)_2(aq) + Zn(s) \rightarrow 2\,Au(s) + Na_2[Zn(CN)_4](aq)$$
   a. If a sample of rock contains 0.009% gold by mass, how much NaCN is needed to extract the gold from 1 metric ton (1 metric ton = 10^3 kg) of rock as $NaAu(CN)_2$?
   b. How much zinc is needed to convert the $NaAu(CN)_2$ from part (a) to metallic gold?
   c. The gold recovered in part (b) is manufactured into a gold ingot in the shape of a cube. The density of gold is 19.3 g/cm^3. How big is the cube of gold?

*3.125. Preparing Nuclear Reactor Fuel Uranium oxides used in the preparation of fuel for nuclear reactors are separated from other metals in minerals by converting the uranium to $UO_x(NO_3)_y(H_2O)_z$, where uranium has a positive charge ranging from 3+ to 6+.
   a. Roasting $UO_x(NO_3)_y(H_2O)_z$ at 400°C leads to loss of water and decomposition of the nitrate ion to nitrogen oxides, leaving behind a product with the formula U_aO_b that is 83.22% U by mass. What are the values of a and b? What is the charge on U in U_aO_b?

b. Higher temperatures produce a different uranium oxide, U_cO_d, with a higher uranium content, 84.8% U. What are the values of c and d? What is the charge on U in U_aO_b?

c. The values of x, y, and z in $UO_x(NO_3)_y(H_2O)_z$ are found by gently heating the compound to remove all of the water. In a laboratory experiment, 1.328 g of $UO_x(NO_3)_y(H_2O)_z$ produced 1.042 g of $UO_x(NO_3)_y$. Continued heating generated 0.742 g of U_nO_m. Using the information in parts (a) and (b), calculate x, y, and z.

*3.126. **Fate of Fertilizer** Large quantities of fertilizer are washed into the Mississippi River from agricultural land in the Midwest. The excess nutrients collect in the Gulf of Mexico, promoting the growth of algae and endangering other aquatic life.

a. One commonly used fertilizer is ammonium nitrate. What is the chemical formula of ammonium nitrate?

b. Corn farmers typically use 5.0×10^3 kg of ammonium nitrate per square kilometer of cornfield per year. Ammonium nitrate can be prepared by the following reaction:

$$NH_3(aq) + HNO_3(aq) \rightarrow NH_4NO_3(aq)$$

How much nitric acid would be required to make the fertilizer needed for 1 km² of cornfield per year?

c. The ammonium ions can be converted into NO_3^- by bacterial action.

$$NH_4^+(aq) + 2 O_2(g) \rightarrow NO_3^-(aq) + H_2O(\ell) + 2 H^+(aq)$$

If 10% of the ammonium component of 5.0×10^2 kg of fertilizer ends up as nitrate, how much oxygen would be consumed?

3.127. **Composition of Over-the-Counter Medicines** Calculate the number of molecules or formula units of compound in each of the following common, over-the-counter medications:

a. ibuprofen, a pain reliever and fever reducer that contains 200.0 mg of the active ingredient, $C_{13}H_{18}O_2$

b. an antacid containing 500.0 mg of calcium carbonate

c. an allergy tablet containing 4 mg of chlorpheniramine ($C_{16}H_{19}ClN_2$)

3.128. **Chemistry of Pain Relievers** The common pain relievers aspirin ($C_9H_8O_4$), acetaminophen ($C_8H_9NO_2$), and naproxen sodium ($C_{14}H_{13}O_3Na$) are all available in tablets containing 200.0 mg of the active ingredient. Which compound contains the greatest number of molecules per tablet? How many molecules of the active ingredient are present in each tablet?

3.129. Some catalytic converters in automobiles contain the manganese oxides Mn_2O_3 and MnO_2.

a. Give the names of Mn_2O_3 and MnO_2.

b. Calculate the percent manganese by mass in Mn_2O_3 and MnO_2.

c. Explain how Mn_2O_3 and MnO_2 are consistent with the law of multiple proportions.

*3.130. Several chemical reactions have been proposed for the formation of organic compounds from inorganic precursors. Here is one of them:

$$H_2S(g) + FeS(s) + CO_2(g) \rightarrow FeS_2(s) + HCO_2H(\ell)$$

a. Identify the ions in FeS and FeS_2. Give correct names for each compound.

*b. How much HCO_2H is obtained by reacting 1.00 g of FeS, 0.50 g of H_2S, and 0.50 g of CO_2 if the reaction results in a 50.0% yield?

*3.131. The formation of organic compounds by the reaction of iron(II) sulfide with carbonic acid is described by the following chemical equation:

$$2\ FeS + H_2CO_3 \rightarrow 2\ FeO + 1/n\ (CH_2O)_n + 2\ S$$

a. How much FeO is produced starting with 1.50 g of FeS and 0.525 mol of H_2CO_3 if the reaction results in a 78.5% yield?

*b. If the carbon-containing product has a molar mass of 3.00×10^2 g/mol, what is the chemical formula of the product?

*3.132. **Marine Chemistry of Iron** On the seafloor, iron(II) oxide reacts with water to form Fe_3O_4 and hydrogen in a process called serpentization.

a. Balance the following equation for serpentization:

$$FeO(s) + H_2O(\ell) \rightarrow Fe_3O_4(s) + H_2(g)$$

b. When CO_2 is present, the product is methane, not hydrogen. Balance the following chemical equation:

$$FeO(s) + H_2O(\ell) + CO_2(g) \rightarrow Fe_3O_4(s) + CH_4(g)$$

3.133. **Composition of Solar Wind** The solar wind is made up of ions, mostly protons, flowing out from the sun at about 400 km/s. Near Earth, each cubic kilometer of interplanetary space contains, on average, 6×10^{15} solar-wind ions. How many moles of ions are in a cubic kilometer of near-Earth space?

3.134. The famous Hope Diamond at the Smithsonian National Museum of Natural History has a mass of 45.52 carats (Figure P3.134). Diamond is a crystalline form of carbon.

a. How many moles of carbon are in the Hope Diamond (1 carat = 200.0 mg)?

b. How many carbon atoms are in the diamond?

FIGURE P3.134

*3.135. E-85 is an alternative fuel for automobiles and light trucks that consists of 85% (by volume) ethanol, CH_3CH_2OH, and 15% gasoline. The density of ethanol is 0.79 g/mL. How many moles of ethanol are in a gallon of E-85?

*3.136. A 100.00 g sample of white powder A is heated to 550°C. At that temperature the powder decomposes, giving off colorless gas B, which is denser than air and is neither flammable nor does it support combustion. The products also include 56 g of a second white powder C. When gas

B is bubbled through a solution of calcium hydroxide, substance A reforms. What are the identities of substances A, B, and C?

*3.137. You are given a 0.6240 g sample of a substance with the generic formula $MCl_2(H_2O)_2$. After complete drying of the sample (which means removing the 2 mol of H_2O per mole of MCl_2), the sample has a mass of 0.5471 g. What is the identity of element M?

3.138. A compound found in crude oil consists of 93.71% C and 6.29% H by mass. The molar mass of the compound is 128 g/mol. What is its molecular formula?

3.139. A reaction vessel for synthesizing ammonia by reacting nitrogen and hydrogen is charged with 6.04 kg of H_2 and excess N_2. A total of 28.0 kg of NH_3 is produced. What is the percent yield of the reaction?

3.140. If a cube of table sugar, which is made of sucrose, $C_{12}H_{22}O_{11}$, is added to concentrated sulfuric acid, the acid "dehydrates" the sugar, removing the hydrogen and oxygen from it and leaving behind a lump of carbon. What percentage of the initial mass of sugar is carbon?

*3.141. A power plant burns 1.0×10^2 metric tons of coal that contains 3.0% (by mass) sulfur (1 metric ton = 10^3 kg). The sulfur is converted to SO_2 during combustion.
   a. How many metric tons of SO_2 are produced?
   b. When SO_2 escapes into the atmosphere it may combine with O_2 and H_2O, forming sulfuric acid, H_2SO_4. Write a balanced chemical equation describing this reaction.
   c. How many metric tons of sulfuric acid, a component of acid rain, could be produced from the quantity of SO_2 calculated in part (a)?

3.142. **Reducing SO$_2$ Emissions** With respect to the previous question, one way to reduce the formation of acid rain involves trapping the SO_2 by passing smokestack gases through a spray of calcium oxide and O_2. The product of this reaction is calcium sulfate.
   a. Write a balanced chemical equation describing this reaction.
   b. How many metric tons of calcium sulfate would be produced from each ton of SO_2 that is trapped?

3.143. In the early 20th century, Londoners suffered from severe air pollution caused by burning high-sulfur coal. The sulfur dioxide that was emitted into the air mixed with London fog, forming sulfuric acid. For every gram of sulfur that was burned, how many grams of sulfuric acid could have formed?

*3.144. **Gas Grill Reaction** The burner in a gas grill mixes 24 volumes of air for every one volume of propane (C_3H_8) fuel. Like all gases, the volume that propane occupies is directly proportional to the number of moles of it at a given temperature and pressure. Air is 21% (by volume) O_2. Is the flame produced by the burner fuel-rich (excess propane in the reaction mixture), fuel-lean (not enough propane), or stoichiometric (just right)?

3.145. A common mineral in Earth's crust has the chemical composition 34.55% Mg, 19.96% Si, and 45.49% O. What is its empirical formula?

*3.146. **Ozone Generators** Some indoor air-purification systems work by converting a little of the oxygen in the air to ozone, which oxidizes mold and mildew spores and other biological air pollutants. The chemical equation for the ozone generation reaction is

$$3\ O_2(g) \rightarrow 2\ O_3(g)$$

It is claimed that one such system generates 4.0 g of O_3 per hour from dry air passing through the purifier at a flow of 5.0 L/min. If 1 liter of indoor air contains 0.28 g of O_2,
   a. what fraction of the molecules of O_2 is converted to O_3 by the air purifier?
   b. what is the percent yield of the ozone generation reaction?

*3.147. **Rebreathing Devices** In the first episode of George Lucas's *Star Wars* series, Qui-Gon Jinn and Obi-Wan Kenobi can visit the underwater world of the Gungans only by using A99 Aquata Breathers, which allow them to survive underwater for up to two hours. While the tiny devices may be from the farfetched world of science fiction, current technology exists for transforming carbon dioxide to oxygen. These self-contained rebreathers are used by a select group of underwater cave explorers and can act as self-rescue devices. The chemistry is based on the following chemical reactions, using either potassium superoxide or sodium peroxide:

$$4\ KO_2(s) + 2\ CO_2(g) \rightarrow 2\ K_2CO_3(s) + 3\ O_2(g)$$
$$2\ Na_2O_2(s) + 2\ CO_2(g) \rightarrow 2\ Na_2CO_3(s) + O_2(g)$$

   a. The respiratory rate at rest for an average, healthy adult is 12 breaths per minute. If the average breath takes in 0.500 L of O_2 ($d = 1.429$ g/L) into the lungs, how many grams of KO_2 are needed to produce enough oxygen for two hours underwater?
   b. Would you need more or less Na_2O_2 to produce an equivalent amount of oxygen?
   c. Given the densities of KO_2 (2.14 g/mL) and Na_2O_2 (2.805 g/mL), which solid material would occupy less volume in a rebreather device?

3.148. **Smelly Socks** Socks containing silver nanoparticles embedded in the fabric are currently marketed as an antidote to smelly socks. Silver is known to have antimicrobial properties, and silver ions are toxic to aquatic life. A study at Arizona State University found that much of the silver particles are lost upon laundering the socks in mild acid.
   a. Each sock in the study began with 1360 μg of silver. How many moles of silver are contained in each sock?
   b. As much as 650 μg of silver was lost after four washings. What percent of the silver was lost?

4

Reactions in Solution

Aqueous Chemistry in Nature

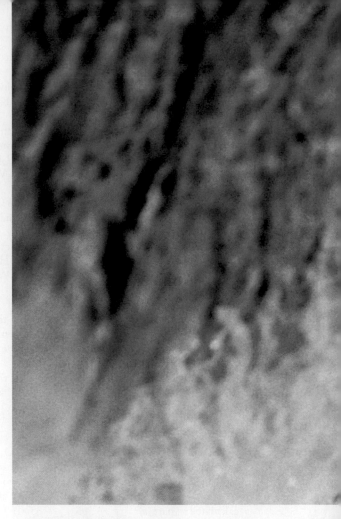

EARTH, WATER, AND LIFE Fish, like this salmon, that migrate between seawater and freshwater are exposed to water containing very different concentrations of dissolved substances.

PARTICULATE **REVIEW**

Ionic or Covalent in Solution?

In Chapter 4 we examine what happens to ionic and covalent compounds when they dissolve in water and the reactions that result from the collisions of the dissolved particles. The beaker on the left contains 0.1 moles of ethylene glycol dissolved in water; the beaker on the right contains 0.1 moles of sodium sulfate dissolved in water. Answer the questions below about these two aqueous solutions.

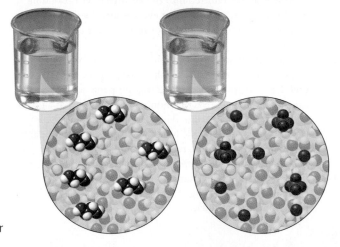

- Is ethylene glycol a covalent compound or an ionic compound? Are its bonds ionic or covalent?

- Is sodium sulfate a covalent compound or an ionic compound? Are its bonds ionic or covalent?

- What specific features of the particles help you answer these questions?

   (Review Figures 1.7, 2.17, and 2.18 if you need help.)

(Answers to Particulate Review questions are in the back of the book.)

Products and Leftover Reactants

This picture shows 200 mL of 0.1 M NH_4Cl being poured into 100 mL of 0.1 M $AgNO_3$. As you read Chapter 4, look for ideas that will help you answer these questions:

- What are the possible products? Will they dissolve in water or form a solid?

- What is the limiting reactant? Will any nitrate ions remain in solution? Why or why not?

- Sketch a drawing of the particulate contents after pouring all the NH_4Cl solution into the $AgNO_3$ solution.

Learning Outcomes

LO1 Express the concentrations of solutions in different units and convert from one set of units to another
Sample Exercise 4.1

LO2 Calculate the molar concentration (molarity) of a solution, the mass of a solute, or the volume of a stock solution required to make a solution of specified molarity
Sample Exercises 4.2, 4.3, 4.4, 4.5

LO3 Calculate the concentration of a solution from stoichiometry and titration

data, and from absorbance data and Beer's Law
Sample Exercises 4.6, 4.9, 4.10

LO4 Write molecular, overall ionic, and net ionic equations for reactions
Sample Exercises 4.7, 4.11, 4.12

LO5 Identify strong electrolytes, weak electrolytes, nonelectrolytes, and Brønsted–Lowry acids and bases from their behavior in solution
Sample Exercise 4.8

LO6 Predict the products of precipitation reactions by using solubility rules, quantify results from precipitation titrations, and explain ion exchange
Sample Exercises 4.11, 4.12, 4.13, 4.14, 4.15

LO7 Identify redox reactions, oxidizing agents, and reducing agents, and balance redox reactions
Sample Exercises 4.16, 4.17, 4.18, 4.19, 4.20, 4.21

4.1 Ions and Molecules in Oceans and Cells

We began Chapter 3 with a description of how Earth formed about 4.6 billion years ago: as a hot, molten sphere that gradually cooled and formed a solid crust surrounded by an atmosphere of gases. Further cooling allowed one of the atmosphere's principal components, water vapor, to condense. The resultant torrential rain ran down from the crust's highlands and collected in its depressions, forming the first oceans. Life on Earth may have begun in these oceans. According to one theory, life originated deep beneath the surface of the oceans at hydrothermal vents where hot, mineral-laden seawater emerging from Earth's crust mixed with cold seawater, forming mineral-rich deposits. Others have argued that only shallow ponds, not the marine environment, provided the appropriate concentrations of ions for cells to arise. Whether either theory is correct or not, water is essential for the chemistry of all biological systems now on Earth, and on early Earth it provided the medium in which dissolved molecules and ions moved about, collided with one another, and participated in chemical reactions that led to the formation of more complex molecules, and eventually to assemblies of molecules capable of reproduction.

Debate over the existence of life elsewhere in the universe hinges on the prospect of liquid water on other planets. The controversial belief that Martian

CONNECTION In Chapter 1 we saw that solutions are mixtures that are homogeneous: the substances making up the mixture are distributed uniformly.

FIGURE 4.1 (a) Meteorite ALH84001, thought to have originated from Mars, contains (b) microscopic features that some scientists believe to be fossilized bacteria. (c) An Earth-grown colony of the bacterium *Escherichia coli*, for comparison.

(a)

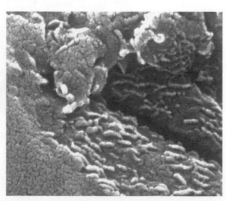
(b)

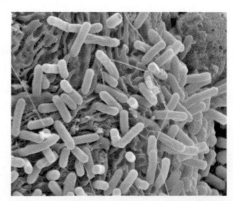
(c)

1 cm

1 cm

(a) Mars

(b) Earth

FIGURE 4.2 (a) Rocks on the Martian surface are similar in size and shape to (b) rocks on Earth in an area where a fast-moving stream of water once flowed.

meteorites collected in Antarctica contain fossilized life-forms (Figure 4.1) is consistent with evidence that water flowed on the surface of Mars in the distant past. Some Martian rocks and pebbles photographed by the *Curiosity* rover are round and smooth like streambed deposits on Earth, suggesting that they were formed by similar processes (Figure 4.2). Furthermore, analysis by *Curiosity* has revealed that some samples of Martian surface soil contain 2 percent water by weight. These and other observations support the idea that liquid water may now exist just under the surface of Mars, and where there is water, there could be life.

Understanding the chemistry of the biosphere requires understanding the principles of chemical reactions between substances dissolved in water. When one element or compound dissolves in another, a *solution* forms—a homogeneous mixture of two or more substances (Figure 4.3). The substance present in the greatest number of moles is called the **solvent**, and the other substances in the solution are called **solutes**. When the solvent is water, the solution is an *aqueous solution*, and in this chapter you may assume all of the solutions we discuss are aqueous. (It bears noting, however, that in the most general case, a solvent does not have to be a liquid. Many materials in Earth's crust, for example, are *solid solutions*, which are uniform mixtures of solid substances.)

CONCEPT TEST

Which, if any, of the following are solutions? (a) muddy river water; (b) helium gas; (c) clear cough syrup; (d) filtered dry air

(Answers to Concept Tests are in the back of the book.)

Dissolved ionic and molecular compounds are present in all of the waters on Earth's surface. The amounts of these dissolved compounds present range from very low (in the water from melting glaciers) to very high (in seawater). The liquids in the cells of all living things are also saline solutions, though not as salty as seawater. For example, the most abundant ionic solutes in our cells and in our blood plasma—Na^+, Cl^-, HCO_3^-, K^+, Ca^{2+}, HPO_4^{2-}, and Mg^{2+} ions—are only a fraction of the amount

solvent the component of a solution that is present in the largest number of moles.

solute any component in a solution other than the solvent. A solution may contain one or more solutes.

Table sugar

Water

Homogeneous solution

FIGURE 4.3 Adding table sugar (the solute) to water (the solvent) produces a solution of sugar molecules evenly distributed among water molecules. Note that the covalent bonds in the sugar molecules (shown here as space-filling models) do not break upon dissolving, and the particles of solute in the solution are neutral molecules.

present in seawater. The quantities of these ions present in the blood of healthy people are remarkably constant, as is also the case for the principal molecular solutes: CO_2, $C_6H_{12}O_6$ (glucose), and hemoglobin. This is why clinical assessments of a patient's well-being often include tests of whether these solutes are within their normal quantitative ranges. Departures from normal ranges may indicate problems such as kidney disease or drug abuse.

Dissolved particles are involved in several major classes of chemical reactions that are essential to the proper function of all living cells. Molecules of CO_2, produced when our bodies metabolize food to produce energy, combine with molecules of H_2O, forming hydrogen ions and bicarbonate ions:

$$CO_2(aq) + H_2O(\ell) \rightarrow H^+(aq) + HCO_3^-(aq)$$

The H^+ ions produced in this reaction combine with additional molecules of water, forming **hydronium ions, H_3O^+**:

$$H^+(aq) + H_2O(\ell) \rightarrow H_3O^+(aq)$$

Reactions such as these that involve the transfer of H^+ ions are called *acid–base reactions*. Acid–base reactions are a major class of aqueous-phase reactions that we explore in this chapter.

The second class of reactions involves pairs of dissolved cations and anions that collide to form insoluble ionic compounds that precipitate from (settle out of) solution. Such a *precipitation reaction* occurs when Ca^{2+} and HPO_4^{2-} ions combine with water molecules to form a mineral that has the common name hydroxyapatite, which is the principal component in tooth enamel:

$$10\ Ca^{2+}(aq) + 6\ HPO_4^{2-}(aq) + 2\ H_2O(\ell) \rightarrow Ca_{10}(PO_4)_6(OH)_2(s) + 8\ H^+(aq)$$

In this chapter we examine trends in the solubility of ionic compounds that enable you to predict whether precipitates form when solutions of different ionic compounds are mixed.

Much of a cell's energy comes from a process we discussed in Chapter 3 in which glucose reacts with O_2 (conveyed in the blood by hemoglobin), forming carbon dioxide and water:

$$C_6H_{12}O_6(aq) + 6\ O_2(g) \rightarrow 6\ CO_2(g) + 6\ H_2O(\ell)$$

This reaction is an example of a third important class of reactions, called *redox* reactions, in which elements (oxygen in this case) are *red*uced, which means that their atoms gain electrons, while others (the carbon in glucose) are *ox*idized, which means that their atoms lose electrons. We explore redox reactions in this chapter as well.

4.2 Quantifying Particles in Solution

In Chapter 3 we used the ideas of the mole and the stoichiometric relationship between reactants and products to deal with macroscopic quantities of materials involved in chemical reactions. In this section, we develop additional tools that enable us to apply the concepts of stoichiometry to reactions taking place in solutions. Figure 4.4 shows a common acid–base reaction: An antacid tablet containing sodium bicarbonate is added to stomach acid, releasing bubbles of $CO_2(g)$ and neutralizing some of the $HCl(aq)$ in gastric fluid in order to relieve an upset stomach. How does a manufacturer of antacid tablets determine the quantity of sodium bicarbonate needed to neutralize a portion of stomach acid? The answer lies in part in knowing both the balanced equation that describes the reaction and the *concentration* of the solution.

hydronium ion (H_3O^+) a H^+ ion bonded to a molecule of water, H_2O; the form in which the hydrogen ion is found in an aqueous solution.

concentration the amount of a solute in a particular amount of solvent or solution.

Concentration Units

The **concentration** of any solution, or the amount of solute in a given amount of solvent or solution, can be expressed in a variety of ways. Qualitatively, a solution can be described as being *concentrated* if it contains a much larger ratio of solute to solvent than does a *dilute* solution. A dilute solution contains a very small amount of solute compared with the amount of solvent. These are relative terms—what constitutes a large amount of solute versus a small amount?—but they are used frequently to compare solutions containing different quantities of solutes.

To be more specific, we need to be able to quantify the concentration of a solution. Some quantitative measures of concentration are based on *mass-to-volume ratios*, such as milligrams of solute per liter of solution. When clinical laboratories report the concentration of molecules in blood, the reference ranges for typical concentrations may be expressed as milligrams of substance (solute) per deciliter (dL; 1 dL = 0.10 L or 100 mL) of plasma (solution). For example, the reference concentration for total cholesterol is under 200 mg/dL. A solution of 1.0 mg per dL is the same as 10 mg per L:

$$\frac{1.0 \text{ mg}}{dL} \times \frac{1.0 \text{ dL}}{100 \text{ mL}} \times \frac{1000 \text{ mL}}{1 \text{ L}} = \frac{10 \text{ mg}}{L}$$

Most scientists interested in studying chemical processes in natural waters or laboratory solutions prefer to work with concentrations based on moles of solute.

$$NaHCO_3(s) + H_3O^+(aq) + Cl^-(aq) \rightarrow Na^+(aq) + Cl^-(aq) + 2 H_2O(\ell) + CO_2(g)$$

} = Ions in solution

FIGURE 4.4 The reaction between a base (sodium bicarbonate) and stomach acid generates bubbles of carbon dioxide and provides relief for an upset stomach.

molarity (*M*) the concentration of a solution expressed in moles of solute per liter of solution (*M* = *n*/*V*).

CHEMTOUR
Molarity

For these scientists, the preferred concentration unit is moles of solute per liter of solution. This ratio is called **molarity (*M*)**:[1]

$$\text{molarity } (M) = \frac{\text{moles } (n) \text{ of solute}}{\text{volume } (V) \text{ of solution in liters}}$$

or

$$M = \frac{n}{V} \tag{4.1}$$

If we know the volume and molarity for any solution, we can readily calculate the mass of solute in the solution. First we rearrange Equation 4.1 to

$$n = V \times M$$

Remember from Chapter 3 that the moles (*n*) of a substance can also be calculated from mass and molar mass (*M*):

$$\text{moles} = \frac{\text{mass } (g)}{\text{molar mass } (g/mol)}$$

Setting these two expressions for calculating the number of moles of solute equal to one another:

$$\frac{\text{mass } (g)}{\text{molar mass } (g/mol)} = V \times M$$

Solving for the mass of solute, m_{solute}:

$$m_{\text{solute}} = (V \times M) \times \mathcal{M} \tag{4.2}$$

Note how the units cancel out in Equation 4.2:

$$g = \left(\cancel{L} \times \frac{\cancel{mol}}{\cancel{L}} \right) \times \frac{g}{\cancel{mol}}$$

Equation 4.2 is useful when we want to calculate the mass of a solute needed to prepare a solution of a desired volume and molarity, or when we have a solution of known molarity and want to know the mass of solute in a given volume of the solution.

In many environmental and biological systems, solute concentrations are often much less than 1.0 *M*. Table 4.1 lists the major ions in seawater and in human serum and their average concentrations, using three common units. In the table, for instance, the concentration units in column 4 are not moles per liter but *millimoles* per liter, which means we are describing concentration in terms of *millimolarity* (m*M*; 1 m*M* = 10^{-3} *M*) rather than molarity. On an even smaller scale, the concentrations of minor and trace elements in seawater are often expressed in terms of *micromolarity* (μ*M*; 1 μ*M* = 10^{-6} *M*), *nanomolarity* (n*M*; 1 n*M* = 10^{-9} *M*), and even *picomolarity* (p*M*; 1 p*M* = 10^{-12} *M*). The concentration ranges of many biologically active substances in blood, urine, and other biological liquids are also so small that they are often expressed in units such as these.

CONCEPT TEST

Which of the following aqueous sodium chloride solutions is the least concentrated?
(a) 0.0053 *M* NaCl; (b) 54 m*M* NaCl; (c) 550 μ*M* NaCl; (d) 56,000 n*M* NaCl

(Answers to Concept Tests are in the back of the book.)

[1]Note that molarity is symbolized by *M*, whereas molar mass is $\mathcal{M}$ throughout this text.

TABLE 4.1 Average Concentrations of 11 Major Constituents of Seawater and Their Normal Ranges in Human Serum

	SEAWATER			HUMAN SERUM
Constituent	g/kg[a]	mmol/kg[b]	mmol/L = mM[c]	mmol/L = mM[c]
Na^+	10.781	468.96	480.57	135–145
K^+	0.399	10.21	10.46	3.5–5.0
Mg^{2+}	1.284	52.83	54.14	0.08–0.12
Ca^{2+}	0.4119	10.28	10.53	0.2–0.3
Sr^{2+}	0.00794	0.0906	0.0928	$<3 \times 10^{-4}$
Cl^-	19.353	545.88	559.40	98–108
SO_4^{2-}	2.712	28.23	28.93	0.3
HCO_3^-	0.126	2.06	2.11	22–30
Br^-	0.0673	0.844	0.865	0.04–0.06
$B(OH)_3$	0.0257	0.416	0.426	$<8 \times 10^{-4}$
F^-	0.00130	0.068	0.070	5–6

[a]g/kg = grams of solute per kilogram of solution.
[b]mmol/kg = millimoles of solute per kilogram of solution. (Oceanographers prefer this unit to one based on solution volume because the volume of a given mass of water varies with changing temperature and pressure, whereas its mass remains constant.)
[c]mmol/L = mM = millimoles of solute per liter of solution.

Some concentration units are not mass-to-volume ratios like molarity but rather are *mass-to-mass ratios*. Two such mass-to-mass concentration units are *parts per million* (ppm) and *parts per billion* (ppb). These units are convenient for expressing the very small concentrations frequently encountered in environmental samples: The Safe Drinking Water Act passed by the U.S. Congress in 1974 uses these units to identify safe maximum levels for contaminants in our drinking water, from harmful substances like lead to helpful substances like fluoride to prevent dental caries. A solution with a concentration of 1 ppm contains 1 part solute for every million parts of solution, which means 1 g of solute for every million (10^6) grams of solution. Put another way, each gram of a 1 ppm solution contains one-millionth of a gram (10^{-6} g = 1 μg) of solute. A solution of 1 ppm is also the same as 1 mg of solute per kilogram of solution:

$$1 \text{ ppm} = \frac{1 \text{ μg solute}}{1 \text{ g solution}} \times \frac{1 \text{ mg solute}}{10^3 \text{ μg solute}} \times \frac{10^3 \text{ g solution}}{1 \text{ kg solution}} = \frac{1 \text{ mg solute}}{1 \text{ kg solution}}$$

Let's express the smallest seawater concentration in Table 4.1 (0.00130 g F^-/kg seawater) in parts per million. To do so, we convert grams of F^- into milligrams of F^- because 1 mg of solute per kilogram of solution is the same as 1 ppm:

$$\frac{0.00130 \text{ g } F^-}{1 \text{ kg seawater}} \times \frac{1 \text{ mg } F^-}{0.001 \text{ g } F^-} = \frac{1.30 \text{ mg } F^-}{1 \text{ kg seawater}} = 1.30 \text{ ppm } F^-$$

As you can see, parts per million is a particularly convenient unit for expressing the concentration of F^- ions in seawater because it avoids the use of exponents or lots of zeroes to set the decimal place.

Some substances are so harmful, though, that a concentration of less than 1 ppm can still be cause for concern. For these substances, the ppb is a more appropriate unit. A concentration of 1 ppb is 1/1000 as concentrated as 1 ppm. Because 1 ppm is the same as 1 mg of solute per kilogram of solution, 1 ppb is equivalent to 1 μg of solute per kilogram of solution:

$$1 \text{ ppb} = \frac{1 \text{ ppm}}{1000} = \frac{0.001 \text{ μg solute}}{1 \text{ g solution}} \times \frac{10^3 \text{ g solution}}{1 \text{ kg solution}} = \frac{1 \text{ μg solute}}{1 \text{ kg solution}}$$

According to the Environmental Protection Agency, the maximum contaminant level for lead is 15 ppb, as opposed to that of copper, which is 1.3 ppm.

CONCEPT **TEST**

Samples of drinking water are drawn from two old homes suspected to have lead pipes. Analysis indicates the first sample contains 0.25 ppm lead, whereas the second sample contains 27 ppb lead. Would the water in either home be safe to drink?

(Answers to Concept Tests are in the back of the book.)

SAMPLE EXERCISE 4.1 Comparing Ion Concentrations in Aqueous Solutions   **LO1**

The average concentration of chloride ion in seawater is 19.353 g Cl^-/kg solution. The World Health Organization (WHO) recommends that the concentration of Cl^- ions in drinking water not exceed 250 ppm (2.50×10^2 ppm). How many times as much chloride ion is there in seawater than in the maximum concentration allowed in drinking water?

Collect and Organize We are given the concentrations of Cl^- ions in two different units, g Cl^-/kg and ppm, and we are asked to determine the ratio of Cl^- ions in seawater to the upper limit of Cl^- ions allowed in drinking water.

Analyze We know how to create conversion factors from equalities. We can convert the seawater concentration to milligrams of Cl^- per kilogram of seawater by using the conversion factor 10^3 mg/1 g and then use the fact that 1 mg solute/kg solution = 1 ppm. We expect the value in mg/kg to be much larger than the value given in g/kg.

$$\boxed{\frac{\text{g}}{\text{kg}} Cl^-} \xrightarrow{\frac{10^3 \text{ mg}}{\text{g}}} \boxed{\frac{\text{mg}}{\text{kg}} Cl^-} = \boxed{\text{ppm } Cl^-}$$

Solve

$$19.353 \frac{\text{g } Cl^-}{\text{kg seawater}} \times \frac{10^3 \text{ mg}}{\text{g}} = 19,353 \frac{\text{mg } Cl^-}{\text{kg seawater}} = 19,353 \text{ ppm } Cl^-$$

Next we take the ratio of the two concentrations:

$$\frac{19,353 \text{ ppm } Cl^- \text{ in seawater}}{250 \text{ ppm } Cl^- \text{ in drinking water}} = 77.4$$

The concentration of Cl^- ions in seawater is 77.4 times greater than would be acceptable in drinking water.

Think About It Common sense says that seawater is much saltier than drinking water. Drinking seawater induces nausea and vomiting, and it may even cause death. We could also have converted the drinking water concentration to grams of Cl^- per kilogram of drinking water and compared that number with 19.353 g Cl^-/kg seawater. It does not matter which units you choose when making a comparison as long as the units of the two numbers are the same.

Practice Exercise The WHO drinking water standard for arsenic is 10.0 μg/L. Water from some wells in Bangladesh was found to contain as much as 1.2 mg of arsenic per liter of water. How many times above the WHO standard is this level? Assume that all the samples have a density of 1 g/mL.

(Answers to Practice Exercises are in the back of the book.)

CONNECTION In Chapter 1 we discussed distilling seawater in order to separate the sea salts from the water, thereby generating fresh drinking water.

SAMPLE EXERCISE 4.2 Calculating Molarity from Mass and Volume   **LO2**

Pipes made of polyvinyl chloride (PVC; Figure 4.5) are widely used in homes and office buildings, though their use is restricted because vinyl chloride (VC) may leach from them. The maximum concentration of vinyl chloride (CH_2CHCl) allowed in drinking water in the United States is 0.002 mg CH_2CHCl/L solution. What molarity is this?

Collect, Organize, and Analyze We are given a concentration in milligrams of solute per liter of solution, and we are asked to convert it to units of molarity (moles of solute per liter of solution). The molar mass of a substance relates the mass and number of moles of a given quantity of the substance. Because the solute mass is given in milligrams, we need to convert this mass first to grams and then to moles. Because the number of grams of vinyl chloride is very small, we can expect that the molarity will be smaller still.

$$\boxed{\dfrac{mg\ VC}{L}} \xrightarrow{\frac{g}{10^3\ mg}} \boxed{\dfrac{g\ VC}{L}} \xrightarrow{\frac{1}{molar\ mass}} \boxed{\dfrac{mol\ VC}{L}}$$

Solve The conversion factors we need are the molar mass of CH_2CHCl and the equality 1 g = 10^3 mg. The molar mass of vinyl chloride is:

$$\mathcal{M} = 2(12.01\ g\ C/mol) + 3(1.008\ g\ H/mol) + 35.45\ g\ Cl/mol = 62.49\ g/mol$$

and the molarity is

$$\frac{0.002\ \cancel{mg}}{L} \times \frac{1\ g}{10^3\ \cancel{mg}} \times \frac{1\ mol}{62.49\ g} = \frac{3.2 \times 10^{-8}\ mol}{L} = 3 \times 10^{-8}\ M,\ or\ 0.03\ \mu M$$

Think About It This concentration may seem very low, but remember that the amount of vinyl chloride (0.002 mg, or 2×10^{-6} g) in 1 L of water is very small. The extremely low limit of only 0.002 mg/L or 3×10^{-8} M reflects recognition of the danger vinyl chloride poses to human health.

Practice Exercise When 1.00 L of water from the surface of the Dead Sea is evaporated, 179 g of $MgCl_2$ is recovered. What is the molarity of $MgCl_2$ in the original sample?

(Answers to Practice Exercises are in the back of the book.)

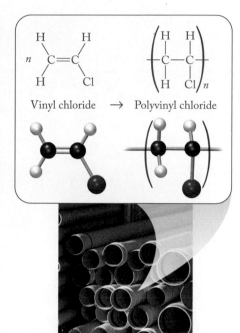

FIGURE 4.5 Vinyl chloride, a suspected carcinogen, enters drinking water by leaching from pipes made of polyvinyl chloride (PVC). Vinyl chloride is the common name of a small molecule, called a monomer (Greek for "one unit"), from which the very large molecule called a polymer ("many units") is formed. The PVC polymer usually consists of hundreds of molecules of vinyl chloride bonded together.

SAMPLE EXERCISE 4.3 Calculating Molarity from Density   **LO2**

A water sample from the Great Salt Lake in Utah contains 83.6 mg Na^+/g solution. What is the molarity of Na^+ if the density of the water is 1.160 g/mL solution?

Collect and Organize Our task is to convert concentration units from mg Na^+/g of solution to mol Na^+/L of solution. We know the lake water's density, which we will use to convert a concentration based on mass of solution to molarity, which is based on volume of solution. The molar mass of Na is 22.99 g/mol.

Analyze To convert a concentration based on mass of solute per mass of solution into molarity (moles of solute per volume of solution), we need to convert solute mass into moles (using the molar mass) and solution mass into a volume in liters (using the density). Then we can use the results of those two calculations to find molarity by dividing the moles of solute by liters of solution.

$$\boxed{\text{mg Na}^+} \xrightarrow{\dfrac{\text{g}}{10^3\ \text{mg}}} \boxed{\text{g Na}^+} \xrightarrow{\dfrac{1}{\text{molar mass}}} \boxed{\text{mol Na}^+}$$

$$\boxed{\text{g solution}} \xrightarrow{\dfrac{1}{\text{density}}} \boxed{\text{mL solution}} \xrightarrow{\dfrac{\text{L}}{10^3\ \text{mL}}} \boxed{\text{L solution}}$$

$$\boxed{\dfrac{\text{mol Na}^+}{\text{L solution}}} = M\,\text{Na}^+$$

Solve Calculating moles of Na^+ in 1 g of lake water (solution):

$$83.6\ \text{mg Na}^+ \times \frac{1\ \text{g}}{10^3\ \text{mg}} \times \frac{1\ \text{mol Na}^+}{22.99\ \text{g Na}^+} = 3.64 \times 10^{-3}\ \text{mol Na}^+$$

The volume of exactly 1 g of lake water (solution) is

$$1.000\ \text{g} \times \frac{1\ \text{mL}}{1.160\ \text{g}} \times \frac{1\ \text{L}}{10^3\ \text{mL}} = 8.621 \times 10^{-4}\ \text{L}$$

and the molarity of Na^+ ions in the lake is

$$\frac{\text{mol}}{\text{L}} = \frac{3.64 \times 10^{-3}\ \text{mol Na}^+}{8.621 \times 10^{-4}\ \text{L}} = 4.22\ M$$

Think About It According to Table 4.1, second column, the Na^+ concentration in seawater is about 10 g of Na^+ per kilogram of seawater, or about 10 mg of Na^+ per gram of seawater, which is approximately 10 mg/mL if we assume that the density of the water is about 1.0 g/mL. The concentration given for Na^+ in the Great Salt Lake, 83.6 mg/g lake water, is more than eight times the seawater concentration, so the molarity should be more than eight times the millimolar concentration given in column 4 of Table 4.1 (480.57 m$M \approx$ 0.48 M). Our answer, 4.22 M, is reasonable in light of that comparison. The Great Salt Lake is the result of evaporation of lake water over thousands of years. Evaporation of the solvent from any solution reduces the volume of solution and increases its concentration because the amount of solute (n) is unchanged.

Practice Exercise If the density of ocean water at a depth of 10,000 m is 1.071 g/mL and if 25.0 g of water at that depth contains 190 mg of potassium chloride, what is the molarity of potassium chloride in the sample?

(Answers to Practice Exercises are in the back of the book.)

CONNECTION In Chapter 1 we introduced density, the ratio of mass of a quantity of material to its volume, or $d = m/V$, as an intensive physical property.

SAMPLE EXERCISE 4.4   Calculating the Quantity of Solute   **LO2**
Needed to Prepare a Solution
of Known Concentration

An aqueous solution called *Kalkwasser* [0.0225 M Ca(OH)$_2$] may be added to saltwater aquaria to reduce acidity and add Ca^{2+} ions. How many grams of Ca(OH)$_2$ do you need to make 500.0 mL of Kalkwasser?

(a)

(b)

(c)

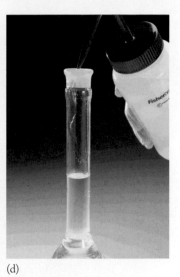
(d)

FIGURE 4.6 Preparing 500.0 mL of 0.0225 *M* Ca(OH)$_2$. (a) Weigh out the desired quantity of solid Ca(OH)$_2$; (b) transfer the Ca(OH)$_2$ to a 500 mL volumetric flask and add 200 to 300 mL of water; (c) swirl to suspend the solute; (d) dilute to exactly 500 mL. Stopper the flask and invert it several times to thoroughly mix the solution.

Collect, Organize, and Analyze We know the volume and molarity of the solution we must prepare. We also know the identity of the solute, which enables us to calculate its molar mass. We use Equation 4.2 to calculate the mass of Ca(OH)$_2$ needed. We also must convert the volume from milliliters to liters to obtain the mass, in grams, of solute required.

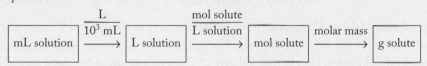

The molar mass of Ca(OH)$_2$ is

$$40.08 \text{ g/mol} + 2(16.00 \text{ g/mol}) + 2(1.008 \text{ g/mol}) = 74.10 \text{ g/mol}$$

Solve Equation 4.2 can be used to calculate the mass of Ca(OH)$_2$ needed:

$$m = V \times M \times \mathcal{M}$$

$$= 500.0 \text{ mL} \times \frac{1 \text{ L}}{10^3 \text{ mL}} \times \frac{0.0225 \text{ mol}}{1 \text{ L}} \times \frac{74.10 \text{ g}}{1 \text{ mol}} = 0.834 \text{ g}$$

Figure 4.6 shows the technique for making this solution. A key feature in the preparation is that we suspend 0.834 g of solute in 200–300 mL of water and then add additional water until the final solution has the volume specified.

Think About It To solve a problem like this, all we need to remember is the definition of molarity: moles of solute per liter of solution. The solution we want has a concentration of 0.0225 *M*, or 0.0225 mol/L. Because we need only 0.5000 L of this solution, we do not need 0.0225 mol of Ca(OH)$_2$, but only half that amount (0.0112 mol). Finally, we calculate the number of grams of calcium hydroxide in 0.0112 mol of calcium hydroxide (0.834 g). The amount of Ca(OH)$_2$ needed is less than a teaspoon.

Practice Exercise An aqueous solution known as Ringer's lactate is administered intravenously to trauma victims suffering from blood loss or severe burns. The solution contains the chloride salts of sodium, potassium, and calcium and is 4.00 m*M* in sodium lactate (NaC$_3$H$_5$O$_3$; Figure 4.7). How many grams of sodium lactate are needed to prepare 10.0 L of Ringer's lactate?

(Answers to Practice Exercises are in the back of the book.)

(a)

Sodium lactate
(b)

FIGURE 4.7 (a) 1000 mL of Ringer's lactate solution. (b) Structure of sodium lactate, NaC$_3$H$_5$O$_3$.

4.3 Dilutions

In laboratories such as those that test drinking water quality, scientists purchase **stock solutions** of substances such as pesticides and toxic metals from chemical manufacturers. To make their distribution more efficient, these solutions are packaged and sold in high concentrations—too high to be used directly in analytical procedures. So, to prepare a solution with a solute concentration that is near the substance's expected concentration in the sample of drinking water being tested, the laboratory will *dilute* the stock solution. **Dilution** is the process of lowering the concentration of a solution by adding solvent to a known volume of the initial solution. The stock solution is concentrated—it contains a much larger solute-to-solvent ratio than does the dilute solution.

CHEMTOUR
Dilution

We can use the definition of molarity given in Equation 4.1:

$$M = \frac{n}{V}$$

to determine how to dilute a stock solution to make a solution of a desired concentration. Suppose we need to prepare 250.0 mL of an aqueous solution that is $0.0100\ M\ Cu^{2+}$ by starting with a stock solution that is $0.1000\ M\ Cu^{2+}$. What volume of the stock solution do we need?

Rearranging Equation 4.1 to solve for the number of moles, n, gives the moles of Cu^{2+} needed in the dilute solution we want to make:

$$\text{Moles of solute} = n_{\text{diluted}} = V_{\text{diluted}} \times M_{\text{diluted}}$$

$$= (0.2500\ \text{L})(0.0100\ \text{mol/L}) = 2.50 \times 10^{-3}\ \text{mol}\ Cu^{2+}$$

The subscript "diluted" refers to the solution we are preparing, and V refers to the final volume of the solution in liters. Using the same equation, we can calculate the *initial* quantities, that is, the volume of the stock solution we need to deliver 2.50×10^{-3} mol of Cu^{2+}:

$$n_{\text{initial}} = V_{\text{initial}} \times M_{\text{initial}}$$

$$V_{\text{initial}} = \frac{n_{\text{initial}}}{M_{\text{initial}}} = \frac{2.50 \times 10^{-3}\ \text{mol}}{0.1000\ \text{mol/L}} = 2.50 \times 10^{-2}\ \text{L} = 25.0\ \text{mL}$$

To make the diluted solution, we add water to 25.0 mL of the stock solution to make a final volume of 250.0 mL (Figure 4.8).

Because the number of moles of solute (Cu^{2+} ions) is the same in the diluted solution and in the portion of stock solution required ($n_{\text{diluted}} = n_{\text{initial}}$), we can combine these two equations into one that applies to all situations in which we know any three of these four variables: the initial volume (V_{initial}) of stock solution, the initial solute concentration (M_{initial}), the volume (V_{diluted}) of the diluted solution, and the solute concentration (M_{diluted}) in the diluted solution:

$$V_{\text{initial}} \times M_{\text{initial}} = n_{\text{initial}} = n_{\text{diluted}} = V_{\text{diluted}} \times M_{\text{diluted}}$$

or simply

$$V_{\text{initial}} \times M_{\text{initial}} = V_{\text{diluted}} \times M_{\text{diluted}} \qquad (4.3)$$

stock solution a concentrated solution of a substance used to prepare solutions of lower concentration.

dilution the process of lowering the concentration of a solution by adding more solvent.

Equation 4.3 works because each side of the equation represents a quantity of solute that does not change because of dilution. Furthermore, this equation can be used for *any* units of volume and concentration, as long as the units used to express the initial and diluted volumes are the same and the units for the initial and diluted concentrations are the same.

Equation 4.3 applies to all systems, even those much larger than a laboratory stock solution to be diluted. For instance, in estuaries, where rivers flow into the sea, salinity is reduced because the seawater is diluted by the freshwater that enters it. Suppose a coastal bay contains 3.8×10^{12} L of ocean water mixed with 1.2×10^{12} L of river water for a total volume of 5.0×10^{12} L. We can calculate the Na^+ concentration in the bay if we know that the Na^+ concentration in the open ocean is 0.48 M and if we assume that the river water does not contribute significantly to the Na^+ content in the bay. Using these values to solve Equation 4.3 for $M_{diluted}$, the Na^+ concentration in the bay, we have

$$M_{diluted} = \frac{V_{initial} \times M_{initial}}{V_{diluted}} = \frac{(3.8 \times 10^{12} \text{ L}) \times 0.48 \ M}{5.0 \times 10^{12} \text{ L}} = 0.36 \ M$$

The bay is less salty than the ocean because it is diluted by the freshwater from the river.

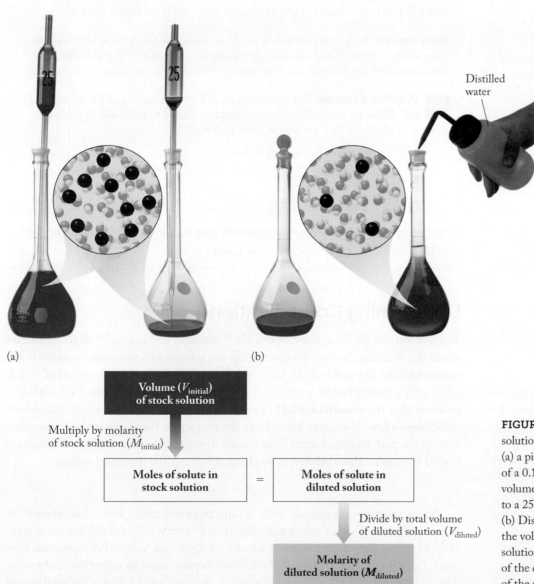

(a)

(b)

Distilled water

Volume ($V_{initial}$) of stock solution

Multiply by molarity of stock solution ($M_{initial}$)

Moles of solute in stock solution = **Moles of solute in diluted solution**

Divide by total volume of diluted solution ($V_{diluted}$)

Molarity of diluted solution ($M_{diluted}$)

FIGURE 4.8 To prepare 250.0 mL of a solution that is 0.0100 M in Cu^{2+}, (a) a pipet is used to withdraw 25.0 mL of a 0.1000 M stock solution. This volume of stock solution is transferred to a 250.0 mL volumetric flask. (b) Distilled water is added to bring the volume of the transferred stock solution to 250.0 mL. Note that the color of the diluted solution is lighter than that of the stock solution.

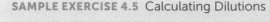

The solution used in hospitals for intravenous infusion—called *physiological saline* or *saline solution*—is 0.155 M in NaCl. It is typically prepared by diluting a stock solution, the concentration of which is 1.76 M, with water. What volume of stock solution and what volume of water are required to prepare 2.00 L of physiological saline?

Collect, Organize, and Analyze We know the volume ($V_{\text{diluted}} = 2.00$ L) and concentration ($M_{\text{diluted}} = 0.155$ M) of the diluted solution and the concentration of stock solution available ($M_{\text{initial}} = 1.76$ M). That is, we know three of the four variables in Equation 4.3. We must find V_{initial}, the volume of the stock solution required to make 2.00 L of physiological saline.

Solve Solving Equation 4.3 for V_{initial} and using the values given:

$$V_{\text{initial}} = \frac{V_{\text{diluted}} \times M_{\text{diluted}}}{M_{\text{initial}}} = \frac{2.00 \text{ L} \times 0.155 \text{ } M}{1.76 \text{ } M} = 0.176 \text{ L}$$

To make 2.00 L of the diluted solution, we need to add: 2.00 L − 0.176 L = 1.82 L of water. If we prepare the dilute solution in a volumetric flask like that in Figure 4.8, rather than add 1.82 L of water, we would add enough water to fill the flask to the mark.

Think About It Because the concentration of the stock solution is about ten times the concentration of the diluted solution, the volume of stock solution required should be about one-tenth the final volume. Our result of 0.176 L of stock solution is reasonable.

Practice Exercise The concentration of Pb^{2+} in a stock solution is 1.000 mg/mL. What volume of this solution should be diluted to 500.0 mL to produce a solution in which the Pb^{2+} concentration is 25.0 mg/L?

(Answers to Practice Exercises are in the back of the book.)

CONCEPT **TEST**

Why is a stock solution always more concentrated than the solutions made from it?

(Answers to Concept Tests are in the back of the book.)

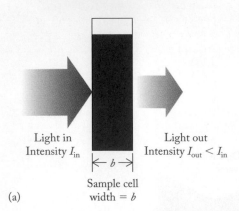

Light in
Intensity I_{in}

Light out
Intensity $I_{\text{out}} < I_{\text{in}}$

$\leftarrow b \rightarrow$

Sample cell
width = b

(a)

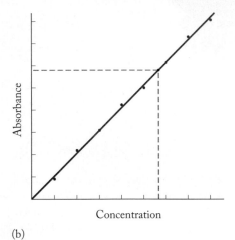

Absorbance

Concentration

(b)

FIGURE 4.9 (a) A spectrophotometer, used to measure the absorbance of solutions. The intensity of light leaving the solution (I_{out}) is less than the intensity of light entering the solution (I_{in}). (b) The red line represents a calibration curve for Beer's law, plotting absorbance versus concentration for a series of standard samples. The horizontal dashed blue line represents the absorbance of a solution of unknown concentration. By extending the line to where it meets the calibration curve and reading down, we can determine the concentration of the unknown solution.

Determining Concentration

The intensity of the blue color of the Cu^{2+} solution in Figure 4.8 decreases as we dilute the solution. Is there a way of using the intensity of the color to quantify the concentration? The color of the Cu^{2+} solution arises from the absorption of visible light, a topic we explore in more detail in Chapters 7 and 22. For now, it is sufficient to know that the **absorbance (A)** of a solution is a measure of the intensity of the color. **Beer's law** (Equation 4.4) relates absorbance to three quantities: concentration (c), the path length that the light travels through the solution (b), and a constant (called the **molar absorptivity**, or ε) characteristic of the dissolved solute.

$$A = \varepsilon b c \qquad (4.4)$$

Absorbance is measured with a spectrophotometer like that shown in Figure 4.9(a), in which a solution is placed in a sample cell of fixed length b, typically 1.00 cm. The cell is placed in a beam of light. The value of A represents how much light is absorbed by the solution. If we express the concentration of the solution, c, in molarity (M), we can calculate ε. A graph of A versus c for solutions of different Cu^{2+} concentration yields a straight line because the amount of light absorbed by a sample is *directly proportional* to the concentration of that sample. Such a graph (Figure 4.9b) is called a **calibration curve**. Once a calibration curve is generated, the concentration of any solution of Cu^{2+} can be determined by measuring its absorbance, finding that value on the *y*-axis, reading across to where it intersects the calibration curve, and then reading the value on the *x*-axis below, which is the concentration. We can also use a calibration curve to determine molar absorptivity. If $b = 1.00$ cm, then the slope of the graph is equal to ε in units of M^{-1} cm^{-1}. This method of analysis is called *spectrophotometry*. Let's see how we can apply Beer's law to find the concentration of a solution of iron ions in Sample Exercise 4.6.

absorbance (A) a measure of the quantity of light absorbed by a sample.

Beer's law the relation of the absorbance of a solution (A) to concentration (c), the light's path length (b), and the solute's molar absorptivity (ε) by the equation $A = \varepsilon b c$.

molar absorptivity (ε) a measure of how well a compound or ion absorbs light.

calibration curve a graph showing how a measurable property, such as absorbance, varies for a set of standard samples of known concentration that can later be used to identify an unknown concentration from a measured absorbance.

SAMPLE EXERCISE 4.6 Applying Beer's Law   **LO3**

One way to determine the concentration of Fe^{2+} in an environmental sample is to add a solution of phenanthroline (phen) molecules that will bond to the metal ions, creating an intensely orange-red solution (Figure 4.10). The absorbance of a 5.00×10^{-5} M solution is measured as 0.55 in a 1.00 cm cell. (a) Calculate the molar absorptivity, ε, and (b) determine the concentration of such a solution whose absorbance is 0.36.

Collect and Organize Given the concentration and absorbance of a solution, we are asked to calculate the molar absorptivity (ε).

Analyze (a) We can rearrange Beer's law to solve for ε and then use this value to find the concentration of the unknown solution. (b) Concentration and absorbance are linearly related by Beer's law, so we predict that a smaller absorbance for the unknown solution should correspond to a lower concentration. (In fact, a concentration of zero would correspond to zero absorbance.)

Solve
a. Rearranging Equation 4.4 and substituting for A, b, and c:

$$\varepsilon = \frac{A}{bc} = \frac{0.55}{(1.00 \text{ cm})(5.00 \times 10^{-5} M)} = 1.1 \times 10^4 \, M^{-1} \text{ cm}^{-1}$$

b. Solving Beer's law for concentration and substituting for A, b, and ε:

$$c = \frac{A}{\varepsilon b} = \frac{0.36}{(1.1 \times 10^4 \, M^{-1} \cdot \text{cm}^{-1})(1.00 \text{ cm})} = 3.3 \times 10^{-5} M$$

Think About It As predicted, the concentration of the unknown solution is lower than 5.00×10^{-5} M.

Practice Exercise One analytical method for determining the concentration of copper ions in a sample uses a solution of cuproine molecules to bond to the metal, similar to the bonds between Fe^{2+} ions and phenanthroline molecules. The following data were collected for four standard solutions of Cu^{2+} with cuproine:

Concentration of Copper (M)	Absorbance
1.00×10^{-4}	0.64
1.25×10^{-4}	0.80
1.50×10^{-4}	0.96
1.75×10^{-4}	1.12

Construct a calibration curve based on these data and use it to determine the concentration of a copper solution with an absorbance $A = 0.74$.

(Answers to Practice Exercises are in the back of the book.)

FIGURE 4.10 Adding a colorless solution of phenanthroline to a colorless solution of Fe^{2+} forms an orange-red solution when the phenanthroline molecules bond to the Fe^{2+} ions.

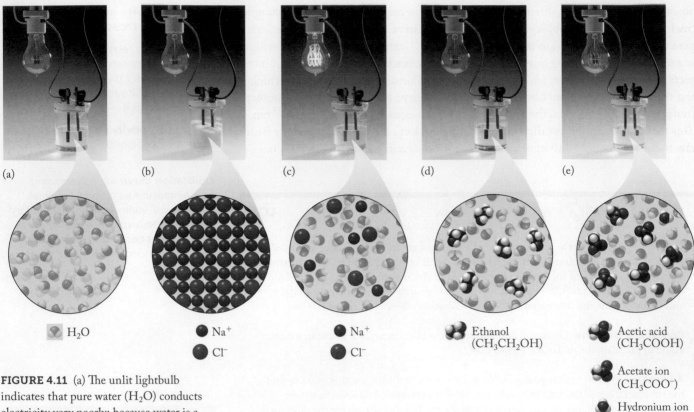

(a) (b) (c) (d) (e)

H₂O   Na⁺   Na⁺   Ethanol   Acetic acid
      Cl⁻   Cl⁻   (CH₃CH₂OH)   (CH₃COOH)

                              Acetate ion
                              (CH₃COO⁻)

                              Hydronium ion
                              (H₃O⁺)

FIGURE 4.11 (a) The unlit lightbulb indicates that pure water (H₂O) conducts electricity very poorly; because water is a molecular material, it contains very few ions. (b) The unlit lightbulb indicates that solid sodium chloride (NaCl) does not conduct electricity. (c) The brightly lit bulb indicates that a 0.50 *M* solution of NaCl conducts electricity very well. (d) The unlit bulb indicates that a 0.50 *M* solution of ethanol (CH₃CH₂OH) does not conduct electricity any better than pure water. (e) The dimly lit bulb indicates that a 0.50 *M* solution of acetic acid (CH₃COOH) conducts electricity better than pure water or the ethanol solution, but not as well as the NaCl solution.

CHEMTOUR
Ions in Solution

4.4 Electrolytes and Nonelectrolytes

The high concentrations of NaCl and other salts in seawater (Table 4.1) make it a good conductor of electricity. Suppose we immerse two *electrodes* (solid conductors) in distilled water (Figure 4.11a) and connect them to a battery and a lightbulb. For electricity to flow and the bulb to light up, the circuit must be completed by mobile charge carriers (ions) in the solution. The lightbulb does not light when the electrodes are placed in distilled water because there are very few ions in distilled water. When the electrodes are pressed into solid NaCl (Figure 4.11b), no electric current flows and the bulb does not light because the Na⁺ and Cl⁻ ions in solid NaCl are fixed in position. However, when the electrodes are immersed in aqueous 0.50 *M* NaCl (Figure 4.11c), the bulb lights because the many mobile ions in the solution act as charge carriers between the two electrodes. The saline solution conducts electricity because Na⁺ ions are attracted to and migrate toward the electrode connected to the negative terminal of the battery, and Cl⁻ ions are attracted to and migrate toward the positive electrode, completing an electric circuit.

Any solute that imparts electrical conductivity to an aqueous solution is called an **electrolyte**. Sodium chloride is considered a **strong electrolyte** because it *dissociates* completely in water, which means it completely breaks up into its component ions when it dissolves, releasing charge-carrying Na⁺ cations and Cl⁻ anions.

Other substances, however, do *not* dissociate when dissolved in water. For instance, a 0.50 *M* solution of ethanol (CH₃CH₂OH) conducts electricity no better than pure water because each CH₃CH₂OH molecule stays intact in the solution and no ions are formed (Figure 4.11d). Solutes that do not form ions when dissolved in water are called **nonelectrolytes**.

In addition to strong electrolytes (complete dissociation into ions) and non-electrolytes (no dissociation), **weak electrolytes** also exist. For a weak electrolyte, a small fraction of its molecules dissociate into ions, but most stay intact as whole, neutral molecules. For this reason, weak electrolytes are sometimes described as "partially dissociating" when dissolved in water. One example of a weak electrolyte is acetic acid, CH_3COOH. In Figure 4.11(e), the beaker contains a 0.50 M aqueous solution of acetic acid. Note that the molarity of the solution in part (e) is equivalent to that in parts (c) and (d) of the figure, meaning that the same number of moles of acetic acid were dissolved as the number of moles of NaCl or moles of CH_3CH_2OH. Because the bulb connected to the acetic acid solution glows, but not very brightly, we can conclude that there are some ions present, but not as many as are present in Figure 4.11(c) with the 0.50 M NaCl solution. The acetic acid molecules dissociate to some extent, forming hydronium ions and acetate ions, but the process does not go to completion:

$$CH_3COOH(aq) + H_2O(\ell) \rightleftharpoons CH_3COO^-(aq) + H_3O^+(aq)$$

In this case, the arrow with the top half pointing right and the bottom half pointing left indicates that all species (the reactant and both products) are simultaneously present in solution: undissociated CH_3COOH molecules, acetate ions, and hydrogen ions. This type of arrow is used to indicate that the dissociation process is not complete. It is important to understand that dissolving and dissociating are two different processes. All of the acetic acid molecules dissolve in water, but only a tiny fraction of them dissociate into ions. A balance, called a *dynamic equilibrium*, is achieved in solutions of weak electrolytes, at which point the concentrations of the reactants and the products in the dissociation reaction do not change. Solutions of weak electrolytes, such as weak acids, are characterized by dissociated ions and undissociated molecules existing together in dynamic equilibrium. For this reason, the arrow used above is called an *equilibrium arrow*.

CONCEPT TEST

Why were equivalent molar concentrations used in the solutions in Figure 4.11?

(Answers to Concept Tests are in the back of the book.)

4.5 Acid–Base Reactions: Proton Transfer

While the behavior of acetic acid molecules when dissolved in water is best described as a dynamic equilibrium, not all acids are weak electrolytes. Let's look at the behavior of another common acid: hydrochloric acid (HCl). Pure HCl is a molecular compound and a gas. However, when HCl dissolves in water it ionizes completely, producing $H_3O^+(aq)$ and $Cl^-(aq)$.

CONCEPT TEST

Which drawing in Figure 4.12 depicts the particles in an aqueous solution of an acid that is a weak electrolyte, and which drawing depicts the particles in an aqueous solution of an acid that is a strong electrolyte? (Note that the solvent water molecules are omitted.)

(Answers to Concept Tests are in the back of the book.)

electrolyte a substance that dissociates into ions when it dissolves in water, enhancing the conductivity of the solvent.

strong electrolyte a substance that dissociates completely into ions when it dissolves in water.

nonelectrolyte a substance that does not dissociate into ions and therefore does not result in conductivity when dissolved in water.

weak electrolyte a substance that dissolves in water with most molecules staying intact, but with a small fraction dissociating into ions.

(a)

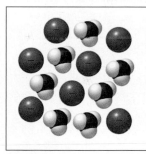

(b)

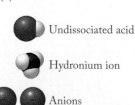

Undissociated acid

Hydronium ion

Anions

FIGURE 4.12 Aqueous solutions of acids. (Note that water molecules are omitted for clarity.)

CONNECTION We discussed the naming of acids in Section 2.7.

acid (Brønsted–Lowry acid) a proton donor.

base (Brønsted–Lowry base) a proton acceptor.

neutralization reaction a reaction that takes place when an acid reacts with a base and produces a solution of a salt in water.

salt the product of a neutralization reaction; it is made up of the cation of the base in the reaction plus the anion of the acid.

molecular equation a balanced equation describing a reaction in solution in which the reactants and products are written as undissociated compounds.

overall ionic equation a balanced equation that shows all the species, both ionic and molecular, present in a reaction occurring in aqueous solution.

net ionic equation a balanced equation that describes the actual reaction taking place in aqueous solution; it is obtained by eliminating the spectator ions from the overall ionic equation.

spectator ion an ion that is present in a reaction vessel when a chemical reaction takes place but is unchanged by the reaction; spectator ions appear in an overall ionic equation but not in a net ionic equation.

As a molecule of HCl dissolves, it donates a proton to a water molecule:

$$\underset{\substack{\text{proton donor}\\\text{(acid)}}}{HCl(g)} + \underset{\substack{\text{proton acceptor}\\\text{(base)}}}{H_2O(\ell)} \rightarrow H_3O^+(aq) + Cl^-(aq) \quad (4.5)$$

This behavior is consistent with our definition of an **acid** as a *proton donor*. Because the water molecule accepts the proton, water in this case is a *proton acceptor*, which is the corresponding definition of a **base**. These definitions were originally proposed by chemists Johannes Brønsted (1879–1947) and Thomas Lowry (1874–1936), so they are called **Brønsted–Lowry acids** and **Brønsted–Lowry bases**. We develop this concept of proton transfer more completely later, but for now it provides a convenient way to define acids and bases in aqueous solutions.

Note that when we write equations in this chapter, we often use $H^+(aq)$ to represent the cation produced in aqueous solution by an acid. Remember that hydrogen ions do not have an independent existence in water because each proton combines with a water molecule to form a hydronium ion (H_3O^+):

$$H^+(aq) + H_2O(\ell) \rightarrow H_3O^+(aq)$$

In aqueous acid–base chemistry, the terms hydrogen ion, proton, and hydronium ion all refer to the same species.

In the reaction between aqueous solutions of HCl and NaOH,

$$HCl(aq) + NaOH(aq) \rightarrow NaCl(aq) + H_2O(\ell) \quad (4.6)$$

the HCl functions as an acid by donating a proton to the hydroxide ion, and the hydroxide ion functions as a base by accepting the proton; the H^+ and OH^- ions combine to form a molecule of water. The remaining two ions—the anion Cl^- and the cation Na^+—form a salt, sodium chloride, which remains dissolved and dissociated in the aqueous solution. This reaction is called a **neutralization reaction** because the acid and the base no longer exist as a result of the reaction. The products of the neutralization of an acid like HCl with a base like NaOH in solution are always water and a salt. This provides us with the definition of a **salt** as a substance formed along with water as a product of a neutralization reaction.

Equation 4.6 is written as a **molecular equation**, meaning each reactant and product is written as a neutral compound. Molecular equations are sometimes the easiest equations to balance, so reactions involving substances that dissociate in solution are often written first as molecular equations to indicate each reactant and product and to simplify balancing the equation.

Another equation representing the reaction in Equation 4.6 is one that shows the reactants and products as the particles that exist in solution, which means that any substance that dissociates completely is written as individual ions:

$$\begin{array}{c} HCl(aq) + NaOH(aq) \rightarrow NaCl(aq) + H_2O(\ell) \\ \overline{H^+(aq) + Cl^-(aq)} + \overline{Na^+(aq) + OH^-(aq)} \rightarrow \overline{Na^+(aq) + Cl^-(aq)} + H_2O(\ell) \quad (4.7) \end{array}$$

This form, called an **overall ionic equation**, distinguishes ionic substances from molecular substances in a chemical reaction taking place in solution. Any nonelectrolytes or weak electrolytes are written as neutral molecules.

Notice that in Equation 4.7 chloride ions and sodium ions appear on both sides of the reaction arrow. If this were an algebraic equation, we would remove the terms that are the same on both sides. When we do the same thing in chemistry, the result is a **net ionic equation**, as written in Equation 4.8. The ions that are removed (in this case, sodium ion and chloride ion) are **spectator ions**, and the resulting equation shows only the species taking part in the reaction:

$$H^+(aq) + OH^-(aq) \rightarrow H_2O(\ell) \qquad (4.8)$$

The spectator ions are unchanged by the reaction and remain in solution. The chemical change in a neutralization reaction occurs when a hydrogen ion reacts with a hydroxide ion to produce a molecule of water. The net ionic equation enables us to focus on those species that actually participate in the reaction. The three ways of depicting the reaction between HCl and NaOH—molecular, overall ionic, and net ionic equations—are summarized in Figure 4.13.

(a) Molecular equation $HCl(aq) + NaOH(aq) \rightarrow NaCl(aq) + H_2O(\ell)$

(b) Overall ionic equation

$$H^+(aq) + Cl^-(aq) + Na^+(aq) + OH^-(aq) \rightarrow Na^+(aq) + Cl^-(aq) + H_2O(\ell)$$

  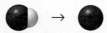

FIGURE 4.13 The neutralization reaction between aqueous hydrochloric acid and sodium hydroxide that produces sodium chloride and water can be written in three ways: (a) a molecular equation, (b) an overall ionic equation, and (c) a net ionic equation.

(c) Net ionic equation = overall ionic equation − spectator ions

$$H^+(aq) + \cancel{Cl^-(aq)} + \cancel{Na^+(aq)} + OH^-(aq) \rightarrow \cancel{Na^+(aq)} + \cancel{Cl^-(aq)} + H_2O(\ell)$$
$$H^+(aq) + OH^-(aq) \rightarrow H_2O(\ell)$$

SAMPLE EXERCISE 4.7 Writing Neutralization Reaction Equations   **LO4**

Write the balanced (a) molecular, (b) overall ionic, and (c) net ionic equations that describe the reaction that takes place when an aqueous solution of nitric acid is neutralized by an aqueous solution of calcium hydroxide.

Collect and Organize Solutions of nitric acid and calcium hydroxide take part in a neutralization reaction. Our task is to write three different equations describing the neutralization reaction.

Analyze We know that calcium hydroxide [$Ca(OH)_2$] is the base and nitric acid (HNO_3) is the acid. The products of a neutralization reaction are water and a salt. (a) All species are written as neutral compounds in the molecular equation. The products are water and calcium nitrate, the salt made from the cation of the base and the anion of the acid. (b) We can write each substance in the balanced molecular equation in ionic form to generate the overall ionic equation. (c) Removing spectator ions from the overall ionic equation gives us the net ionic equation.

Solve
a. The molecular equation is

$$HNO_3\,(aq) + Ca(OH)_2(aq) \rightarrow Ca(NO_3)_2(aq) + H_2O(\ell) \qquad \text{(unbalanced)}$$
$$2\,HNO_3\,(aq) + Ca(OH)_2(aq) \rightarrow Ca(NO_3)_2(aq) + 2\,H_2O(\ell) \qquad \text{(balanced)}$$

b. The overall ionic equation is

$$2\,H^+(aq) + 2\,NO_3^-(aq) + Ca^{2+}(aq) + 2\,OH^-(aq) \rightarrow Ca^{2+}(aq) + 2\,NO_3^-(aq) + 2\,H_2O(\ell)$$

hydrolysis the reaction of water with another material. The hydrolysis of nonmetal oxides produces acids.

strong acid an acid that completely dissociates into ions in aqueous solution.

weak acid an acid that is a weak electrolyte and so has a limited capacity to donate protons to the medium.

carboxylic acid an organic compound containing the –COOH group.

FIGURE 4.14 In Carlsbad Caverns in New Mexico, stalagmites of limestone grow up from the cavern floor and stalactites grow downward from the ceiling.

c. The spectator ions are Ca^{2+} and NO_3^-. Removing them gives us the net ionic equation:

$$2\,H^+(aq) + 2\,OH^-(aq) \rightarrow 2\,H_2O(\ell)$$

or

$$H^+(aq) + OH^-(aq) \rightarrow H_2O(\ell)$$

Think About It We treat chemical equations just like algebraic equations, canceling out spectator ions that are unchanged in the reaction and appear on both sides of the reaction arrow in the *overall* ionic equation. After canceling the spectator ions, we are left with the net ionic equation, which focuses on the species that were changed as a result of the reaction. It is also important to remember that nitrate is a polyatomic ion (see Table 2.3) that retains its identity in solution and does not separate into atoms and ions. Note that the net ionic equation is the same as in the reaction of an aqueous solution of HCl with aqueous NaOH.

Practice Exercise Write balanced (a) molecular, (b) overall ionic, and (c) net ionic equations for the reaction between an aqueous solution of acetic acid, $CH_3COOH(aq)$, and an aqueous solution of sodium hydroxide. The products are sodium acetate and water. *Note*: acetic acid is a weak electrolyte.

(Answers to Practice Exercises are in the back of the book.)

For billions of years, neutralization reactions have played key roles in the chemical transformations of Earth's crust that geologists call *chemical weathering*. One type of chemical weathering occurs when carbon dioxide in the atmosphere dissolves in rainwater to create a weakly acidic solution of carbonic acid, $H_2CO_3(aq)$. This solution reacts with calcium carbonate, which occurs as chalk, limestone, and marble and is insoluble in pure water. The reaction between carbonic acid and calcium carbonate is responsible for the formation of stalactites and stalagmites (Figure 4.14). Chemical weathering caused by even more acidic rain is responsible for the degradation of statues and the exteriors of buildings made of marble (Figure 4.15). Table 4.2 lists some common nonmetal oxides that are classified as atmospheric pollutants because they dissolve in water to produce acid rain.

FIGURE 4.15 Atmospheric sulfuric acid, made when $SO_3(g)$ dissolves in rainwater, attacks marble statues by converting the calcium carbonate to calcium sulfate, which is slowly washed away by rain and melting snow. The picture on the left was taken in 1908; the one on the right, in 1968.

Nonmetal oxides other than carbon dioxide form solutions that are much more acidic than carbonic acid. Dinitrogen pentoxide, for example, forms nitric acid in water in a **hydrolysis** reaction.

$$N_2O_5(g) + H_2O(\ell) \rightarrow 2\,HNO_3(aq)$$

Similarly, sulfur trioxide forms sulfuric acid:

$$SO_3(g) + H_2O(\ell) \rightarrow H_2SO_4(aq)$$

Nitric acid and sulfuric acid are **strong acids** because they dissociate completely in water.

$$H_2O(\ell) + HNO_3(aq) \rightarrow H_3O^+(aq) + NO_3^-(aq)$$
$$H_2O(\ell) + H_2SO_4(aq) \rightarrow H_3O^+(aq) + HSO_4^-(aq)$$

They are listed in Table 4.3 along with other common strong acids. Strong acids are strong electrolytes. The arrows in the ball-and-stick figures of the acids point to the bonds that ionize in solution to produce protons. Because one molecule of nitric acid donates one proton, nitric acid is called a *monoprotic acid*. Hydrochloric acid is another strong monoprotic acid. Sulfuric acid can donate up to two protons per molecule, which makes it a *diprotic acid*. However, in some H_2SO_4 solutions, only one of the two H atoms in each molecule ionizes. This behavior means that sulfuric acid is a strong acid in terms of donating the first proton. This dissociation results in the formation of the HSO_4^- ion (hydrogen sulfate), which is also an acid because some HSO_4^- ions in solution dissociate into H^+ and SO_4^{2-} ions. However, because not all HSO_4^- ions dissociate, the hydrogen sulfate anion is a **weak acid**, which is defined as one where only a small fraction dissociates in water.

The two equations for the dissociation of the two protons of sulfuric acid are written differently to reflect these different dissociation behaviors. First, the complete dissociation of the strong acid H_2SO_4:

$$H_2O(\ell) + H_2SO_4(aq) \rightarrow H_3O^+(aq) + HSO_4^-(aq)$$

Second, the dynamic equilibrium and partial dissociation of the weak acid HSO_4^-:

$$H_2O(\ell) + HSO_4^-(aq) \rightleftharpoons H_3O^+(aq) + SO_4^{2-}(aq)$$

Beyond the strong acids listed in Table 4.3, other acids should be considered weak acids, even if they are *polyprotic acids* such as carbonic acid (H_2CO_3) or phosphoric acid (H_3PO_4). For any diprotic acid, the HX^- anion formed when the first proton dissociates is a weaker acid than the original acid H_2X. The pattern continues for triprotic acids (H_3X) as well in terms of their strength as acids: the proton donors in a triprotic acid are in the order $H_3X > H_2X^- > HX^{2-}$.

We have already seen an example of an additional common type of acid that you can recognize based on its formula. In Section 4.4, our example of a weak electrolyte was acetic acid (CH_3COOH), which is also a weak acid. Acetic acid is a member of a large group of organic compounds called **carboxylic acids** because they all contain

TABLE 4.2 Volatile Nonmetal Oxides and Their Acids

Oxide	Acid
SO_2	H_2SO_3
SO_3	H_2SO_4
NO_2	HNO_2, HNO_3
N_2O_5	HNO_3
CO_2	H_2CO_3

TABLE 4.3 Strong Acids

Acid	Molecular Formula	Bonds Ionizing in Solution
Hydrochloric acid	HCl	
Hydrobromic acid	HBr	
Hydroiodic acid	HI	
Nitric acid[a]	HNO_3	
Sulfuric acid	H_2SO_4	
Perchloric acid	$HClO_4$	

[a]The dashed bonds in the model of nitric acid indicate bonds that are part single bond and part double bond in character. We discuss that phenomenon in detail in Chapter 8.

strong base a base that completely dissociates into ions in aqueous solution.

weak base a base that is a weak electrolyte and so has a limited capacity to accept protons.

amine an organic compound that functions as a base and has the general formula RNH_2, R_2NH, or R_3N, where R is any organic subgroup.

amphiprotic describes a substance that can behave as either a proton acceptor or a proton donor.

the –COOH group in their structures. Carboxylic acids are weak acids and hence dissociate only partially in water.

CONCEPT TEST

A molecule called EDTA has four carboxylic acid groups and hence four acidic protons. We can symbolize it as H_4E to highlight this property. Pick the more acidic member of each pair listed: (a) H_2E^{2-} or HE^{3-}; (b) H_4E or H_3E^-.

(Answers to Concept Tests are in the back of the book.)

In a classification scheme similar to that used for acids, bases are classified as **strong bases** or **weak bases**, depending on the extent to which they dissociate in aqueous solution. Strong bases are strong electrolytes; weak bases are weak electrolytes. Strong bases include the hydroxides of group 1 and group 2 metals, which dissociate completely when dissolved in water.

Although substances such as $NaOH(aq)$ and $Ca(OH)_2(aq)$ can be considered bases because they produce hydroxide ions in aqueous solution, it is important to remember the Brønsted–Lowry definition of a base as a proton acceptor. Hydroxide ions are bases because they can accept protons. This definition permits other substances that do not consist of hydroxide ions to be categorized as bases as well.

One such example is ammonia (NH_3), which is considered to be a weak base, even though it does not contain hydroxide ions. Ammonia in its molecular form is a gas at room temperature. When NH_3 dissolves in water it produces a solution that conducts electricity weakly, which means that ammonia is a weak electrolyte:

$$NH_3(aq) + H_2O(\ell) \rightleftharpoons NH_4^+(aq) + OH^-(aq) \quad (4.9)$$

$$\underset{\text{(proton acceptor)}}{\underset{\text{base}}{NH_3}} \quad \underset{\text{(proton donor)}}{\underset{\text{acid}}{H_2O}}$$

Ammonia is an inorganic molecule, but it can be viewed as the parent in a family of organic molecules that are typically weak bases. These compounds are called **amines**, and they are characterized by having one or more of the hydrogen atoms in NH_3 replaced by an organic group. We deal with the structure of ammonia in Chapters 8 and 9, but for now you need only to recognize amines as weak bases. They all behave similarly to ammonia in water by accepting a proton from the water molecule and producing a hydroxide ion in solution. The simplest member of this family is methylamine: CH_3NH_2. Because amines are weak bases, only a small fraction of the molecules accept protons in an aqueous solution; this behavior is indicated by an equilibrium arrow, just as we used with ammonia:

$$CH_3NH_2(g) + H_2O(\ell) \rightleftharpoons CH_3NH_3^+(aq) + OH^-(aq)$$

If a second or third hydrogen atom in NH_3 is replaced, whether by the same or different organic group, the compound still behaves as a weak base. For example, dimethyl amine, $(CH_3)_2NH$, contains one nitrogen atom with two CH_3 groups and one hydrogen atom bonded to it. Trimethyl amine has three CH_3 groups (and no hydrogen atoms) bonded to the central nitrogen atom: $(CH_3)_3N$. The most general way of describing amines is to use the symbol R for any organic group. Then the three general types of amines may be symbolized as RNH_2, R_2NH, and R_3N.

Finally, some substances can behave as either acids or bases, depending on what other substances are in solution. For example, note that water behaves as a base in Equation 4.5 and as an acid in Equation 4.9. Water is called an **amphiprotic** substance because it can function as a proton acceptor (Brønsted–Lowry base) in solutions containing acid solutes or as a proton donor (Brønsted–Lowry acid) in solutions containing basic solutes. In the hydrolysis of HCl, water functions as a base; in the hydrolysis of NH_3, water functions as an acid.

CONCEPT TEST

Identify the following compounds as acids or bases or neither in water. If the compound is an acid or base, identify it as weak or strong. (a) CH_3CH_2COOH; (b) CH_3OH; (c) $(CH_3CH_2)_2NH$; (d) HNO_3

(Answers to Concept Tests are in the back of the book.)

SAMPLE EXERCISE 4.8 Comparing Electrolytes, Acids, and Bases   **LO5**

Classify each of the following compounds in aqueous solution as a strong electrolyte, a weak electrolyte, or a nonelectrolyte; and then as an acid, a base, or neither an acid nor a base: (a) NaCl; (b) HCl; (c) NaOH; (d) CH_3COOH.

Collect and Organize We are given the formulas of four compounds and asked to classify them based on their behavior in an aqueous solution. We will work with the definitions of the given terms.

Analyze Strong electrolytes are substances that dissociate completely into ions when dissolved in water. Weak electrolytes dissociate only partially in water. Nonelectrolytes do not dissociate at all. A Brønsted–Lowry acid is defined as a proton donor, and a Brønsted–Lowry base is a proton acceptor.

Solve
a. NaCl is an ionic compound that dissociates completely into ions in aqueous solution, as shown in Figure 4.11(c). Therefore NaCl is a strong electrolyte. Sodium chloride has no protons to donate, and neither Na^+ nor Cl^- ions readily accept protons. Thus NaCl is neither an acid nor a base.
b. HCl dissociates completely into H^+ and Cl^- ions when dissolved in water, making HCl a strong electrolyte. As a proton donor, aqueous HCl is also a Brønsted–Lowry acid.
c. NaOH is an ionic compound that dissociates completely into Na^+ and OH^- ions in aqueous solution. NaOH is a strong electrolyte. The OH^- ion formed upon dissolution of NaOH readily accepts a proton to form water, making NaOH a Brønsted–Lowry base.
d. Some molecules of acetic acid, CH_3COOH, when dissolved in water, dissociate into H^+ and CH_3COO^- ions (Figure 4.11e), creating a weakly conducting solution; thus it is a weak electrolyte. Since acetic acid is a proton donor, it is considered a Brønsted–Lowry acid.

Think About It You should not be surprised that some strong electrolytes are also acids or bases. Acidity and basicity have different definitions from those for electrolytes, and substances can be classified as either an acid or a base while at the same time being either a strong or weak electrolyte. A weak electrolyte that donates a proton is a weak acid.

Practice Exercise Classify aqueous solutions of each of the following compounds as a strong electrolyte, weak electrolyte, or nonelectrolyte; and as an acid, a base, or neither: (a) potassium sulfate; (b) HBr; (c) NH_3; (d) ethanol (CH_3CH_2OH).

(Answers to Practice Exercises are in the back of the book.)

4.6 Titrations

We can use our understanding of the reactions of acids and bases to analyze solutions and determine the concentrations of dissolved substances. An approach called **titration** is a common analytical method based on measured volumes of reactants. Acid–base neutralization reactions are frequently the basis of analyses done by titration.

One use of titrations is to analyze water draining from abandoned coal mines. Sulfide-containing minerals called pyrites may be exposed when coal is removed from a site. The ensuing reaction between the minerals and microorganisms in the presence of oxygen produces very acidic solutions of sulfuric acid. To manage these hazardous wastes appropriately, we must know the concentrations of acid present in them. Titrations give us this information.

Suppose we have 100.0 mL of drainage water containing an unknown concentration of sulfuric acid. We can use an acid–base titration in which the sulfuric acid in the water sample is neutralized by reaction with a standard solution of 0.00100 M NaOH. The NaOH solution in this case is called the **titrant**; it is a **standard solution**, meaning its concentration is known accurately. The neutralization reaction is

$$H_2SO_4(aq) + 2\,NaOH(aq) \rightarrow Na_2SO_4(aq) + 2\,H_2O(\ell)$$

To determine the concentration of sulfuric acid in the sample, we determine the volume of titrant needed to neutralize a known volume of the sample.

Suppose 22.40 mL of the NaOH solution is required to react completely with the H_2SO_4 in the sample. Because we know the volume and molarity of the NaOH solution, we can calculate the number of moles of NaOH reacted:

$$n = V \times M$$
$$= 22.40\ \text{mL} \times \frac{1\ \text{L}}{10^3\ \text{mL}} \times \frac{1.00 \times 10^{-3}\ \text{mol NaOH}}{\text{L}}$$
$$= 2.24 \times 10^{-5}\ \text{mol NaOH}$$

We know from the stoichiometry of the reaction that 2 moles of NaOH are required to neutralize 1 mole of H_2SO_4, so the number of moles of H_2SO_4 in the 100.0 mL sample must be

$$2.24 \times 10^{-5}\ \text{mol NaOH} \times \frac{1\ \text{mol } H_2SO_4}{2\ \text{mol NaOH}} = 1.12 \times 10^{-5}\ \text{mol } H_2SO_4$$

We can combine these two steps into one mathematical expression and will do so throughout the rest of the text:

$$0.02240\ \text{L} \times \frac{1.00 \times 10^{-3}\ \text{mol NaOH}}{\text{L}} \times \frac{1\ \text{mol } H_2SO_4}{2\ \text{mol NaOH}}$$
$$= 1.12 \times 10^{-5}\ \text{mol } H_2SO_4$$

Therefore the concentration of H_2SO_4 is

$$\frac{1.12 \times 10^{-5}\ \text{mol } H_2SO_4}{0.1000\ \text{L}} = \frac{1.12 \times 10^{-4}\ \text{mol } H_2SO_4}{\text{L}} = 1.12 \times 10^{-4}\ M\ H_2SO_4$$

How did we determine in the first place that exactly 22.40 mL of the NaOH solution was needed to react with the sulfuric acid in the sample? The glassware required to perform an acid–base titration is illustrated in Figure 4.16. The standard solution of NaOH is poured into a *buret*, a narrow glass cylinder with volume

titration an analytical method for determining the concentration of a solute in a sample by reacting the solute with a standard solution of known concentration.

titrant the standard solution added to the sample in a titration.

standard solution a solution of known concentration used in titrations.

equivalence point the point in a titration at which just enough titrant has been added to completely react the substance being analyzed.

end point the point in a titration that is reached when just enough standard solution has been added to cause the indicator to change color.

markings. The solution is gradually added to the sample until the point in the titration when just enough standard solution has been added to completely react with all the solute in the sample. This is called the **equivalence point** of the titration.

How do we know when the equivalence point has been reached? Chemists add a small amount of indicator solution to the sample before it is titrated. Indicator solutions contain molecules that change color as the concentration of $H^+(aq)$ ions changes. To detect the end point in our sulfuric acid–sodium hydroxide titration, we might use the indicator *phenolphthalein*, which is colorless in acidic solutions but pink in basic solutions. If the correct indicator has been chosen, the equivalence point is very close to the **end point**, the point at which the indicator changes color. The part of the titration that requires the most skill is adding just enough NaOH solution to reach the end point, the point at which a pink color first persists in the solution being titrated. To catch this end point, one must add the standard solution no faster than one drop at a time with thorough mixing between drops.

Titration is an effective method for determining solution concentrations; however, it is not applicable in all situations. We have already seen in Section 4.3 how spectrophotometry can be used to determine concentration. We explore yet another method in Section 4.7.

(a)

(b)

SAMPLE EXERCISE 4.9 Calculating Molarity from Titration Data   **LO3**

Vinegar is an aqueous solution of acetic acid (CH_3COOH) that can be made from any source containing starch or sugar. Apple cider vinegar is made from apple juice that is fermented to produce alcohol, which then reacts with oxygen from the air in the presence of certain bacteria to produce vinegar. Commercial vinegar must contain no less than 4% acetic acid—that is, no fewer than 4 grams of acetic acid per 100 mL of vinegar. Suppose the titration of a 25.00 mL sample of vinegar requires 11.20 mL of a 5.95 M solution of NaOH. What is the molarity of the vinegar? Could this be a commercial sample of vinegar?

Collect, Organize, and Analyze We are given the volume and concentration of the titrant and the volume of the sample, and we are asked to calculate the concentration of the sample. We start with a balanced chemical equation for the neutralization reaction:

$$CH_3COOH(aq) + NaOH(aq) \rightarrow CH_3COONa(aq) + H_2O(\ell)$$

We then calculate the molarity of the vinegar so we can determine the number of grams of acetic acid in the sample and thereby determine whether the sample is commercial grade:

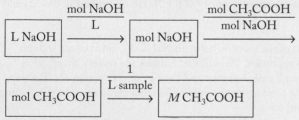

Solve The stoichiometry shows that 1 mole of base reacts with 1 mole of acid. The number of moles of acid titrated is

$$0.01120 \text{ L NaOH} \times \frac{5.95 \text{ mol NaOH}}{1 \text{ L NaOH}} \times \frac{1 \text{ mol } CH_3COOH}{1 \text{ mol NaOH}} = 0.0666 \text{ mol } CH_3COOH$$

The molarity of the vinegar is

$$\frac{0.0666 \text{ mol } CH_3COOH}{0.02500 \text{ L vinegar}} = 2.66 \, M$$

(c)

FIGURE 4.16 Determining a sulfuric acid concentration. (a) A known volume of a sample containing H_2SO_4 is placed in the flask. The buret is filled with the titrant, which is an aqueous NaOH solution of known concentration. A few drops of phenolphthalein indicator solution are added to the flask. (b) Titrant is carefully added to the flask until the indicator changes from colorless to faint pink, signaling that the acid has been neutralized and the end point has been reached. (c) If more titrant is added, the color becomes darker, signaling that more titrant than was needed to reach the end point has been added.

The number of grams of acetic acid in the sample is

$$0.0666 \text{ mol } CH_3COOH \times \frac{60.05 \text{ g } CH_3COOH}{1 \text{ mol } CH_3COOH} = 4.00 \text{ g } CH_3COOH$$

The original sample had a volume of 25.00 mL, so the number of grams of acetic acid per 100 mL of sample is

$$\frac{4.00 \text{ g } CH_3COOH}{25.00 \text{ mL vinegar}} \times 100 \text{ mL vinegar} = 16.0 \text{ g}$$

The sample has 16.0 g in 100 mL of solution, so it could certainly be a commercial vinegar.

Think About It About 11 mL of titrant was needed to neutralize the sample. That is about half the volume of the sample, so the molarity of the vinegar should be about half that of the titrant. The answer for the concentration of the vinegar seems reasonable. Vinegar this concentrated is typically used for pickling. Vinegar with 4% to 8% acetic acid is table vinegar, used in salad dressings.

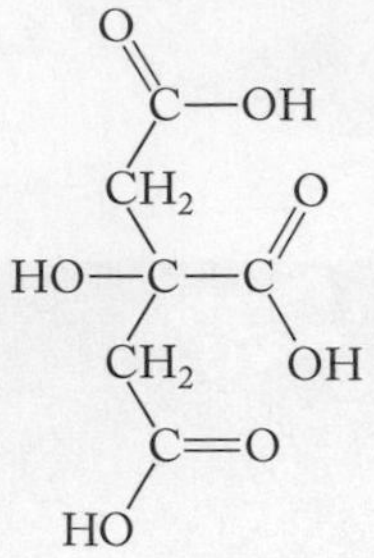

Practice Exercise Citric acid ($C_6H_8O_7$; Figure 4.17) is a weak triprotic acid found naturally in lemon juice, and it is widely used as a flavoring in beverages. What is the molarity of $C_6H_8O_7$ in commercially available lemon juice if 14.26 mL of 1.751 M NaOH is required in a titration to neutralize 20.00 mL of the juice? How many grams of citric acid are in 100.0 mL of the juice?

(Answers to Practice Exercises are in the back of the book.)

Citric acid

FIGURE 4.17 Structure of citric acid; acidic protons are shown in red.

SAMPLE EXERCISE 4.10 Calculating Molarity
of a Carbonate Solution   **LO3**

Many over-the-counter antacids, like the one shown in Figure 4.18, contain sodium bicarbonate. Sodium bicarbonate reacts with excess stomach acid (aqueous HCl) in a neutralization reaction. A 650 mg tablet is dissolved in 25.00 mL of water. Titration of this solution requires 25.27 mL of a 0.3000 M solution of HCl. (a) Write a net ionic equation for the reaction between $NaHCO_3$ and HCl. (b) What is the molarity of the antacid solution? (c) Is the tablet pure $NaHCO_3$?

Collect and Organize We are assigned three tasks in this sample exercise. We are given the names of the reactants and asked to write a balanced net ionic equation. We are given the volume and concentration of the titrant and the volume of the sample and asked to calculate the concentration of the solution. Finally, using the results of our calculations, we are asked to assess the purity of the $NaHCO_3$ present in the original sample.

Analyze (a) The products of the reaction between $NaHCO_3$ and HCl are sodium chloride, water, and carbon dioxide. We will first balance the molecular equation and then generate the overall ionic equation and the net ionic equation. (b) The steps involved in calculating the molarity of the sodium bicarbonate solution (not including liter–milliliter conversions) are

$$\boxed{\text{L HCl}} \xrightarrow{\frac{\text{mol HCl}}{\text{L}}} \boxed{\text{mol HCl}} \xrightarrow{\frac{\text{mol } NaHCO_3}{\text{mol HCl}}} \boxed{\text{mol } NaHCO_3} \xrightarrow{\frac{1}{\text{L sample}}} \boxed{M \, NaHCO_3}$$

(c) The purity of the tablet can be determined by converting the number of moles of $NaHCO_3$ calculated in part (b) into an equivalent number of grams:

$$\boxed{\text{mol } NaHCO_3} \xrightarrow{\text{molar mass}} \boxed{\text{g } NaHCO_3}$$

= Na$^+$

= HCO$_3^-$

FIGURE 4.18 Sodium bicarbonate antacid tablets.

and comparing the result with the mass of the tablet (650 mg).

Solve

a. The balanced molecular equation for the neutralization reaction is

$$NaHCO_3(aq) + HCl(aq) \rightarrow H_2O(\ell) + CO_2(g) + NaCl(aq)$$

The balanced ionic equation is

$$Na^+(aq) + HCO_3^-(aq) + H^+(aq) + Cl^-(aq) \rightarrow H_2O(\ell) + CO_2(g) + Na^+(aq) + Cl^-(aq)$$

and the net ionic equation is

$$\cancel{Na^+(aq)} + HCO_3^-(aq) + H^+(aq) + \cancel{Cl^-(aq)} \rightarrow H_2O(\ell) + CO_2(g) + \cancel{Na^+(aq)} + \cancel{Cl^-(aq)}$$

$$HCO_3^-(aq) + H^+(aq) \rightarrow H_2O(\ell) + CO_2(g)$$

b. The stoichiometry of the reaction shows that 1 mole of $NaHCO_3$ reacts with 1 mole of HCl. The number of moles of $NaHCO_3$ titrated is

$$0.02527 \text{ L HCl} \times \frac{0.3000 \text{ mol HCl}}{1 \text{ L HCl}} \times \frac{1 \text{ mol NaHCO}_3}{1 \text{ mol HCl}} = 0.007581 \text{ mol NaHCO}_3$$

The molarity of the HCO_3^- is

$$\frac{0.007581 \text{ mol NaHCO}_3}{0.02500 \text{ L solution}} = 0.3032 \text{ } M \text{ NaHCO}_3$$

c. To calculate the number of grams of sodium bicarbonate in the sample, we first need to know the molar mass of $NaHCO_3$:

$$\mathcal{M}_{NaHCO_3} = 22.99 \text{ g/mol} + 1.008 \text{ g/mol} + 12.01 \text{ g/mol} + 3(16.00 \text{ g/mol})$$
$$= 84.01 \text{ g/mol NaHCO}_3$$

Therefore, the mass of the tablet is

$$0.007581 \text{ mol NaHCO}_3 \times \frac{84.01 \text{ g NaHCO}_3}{1 \text{ mol NaHCO}_3} = 0.6369 \text{ g NaHCO}_3$$

The mass of the tablet was 650 mg. On the basis of the titration, the sample contained only 637 mg of $NaHCO_3$ and hence is not pure $NaHCO_3$.

Think About It The stoichiometry of the net ionic equation indicates a 1:1 ratio of acid to base. Because a nearly equivalent volume of aqueous HCl was needed to titrate 25 mL of antacid solution, it makes sense that the concentration of the base is similar to that of the acid.

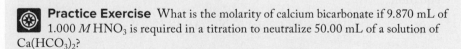 **Practice Exercise** What is the molarity of calcium bicarbonate if 9.870 mL of 1.000 M HNO_3 is required in a titration to neutralize 50.00 mL of a solution of $Ca(HCO_3)_2$?

(Answers to Practice Exercises are in the back of the book.)

4.7 Precipitation Reactions

Some of the most abundant elements in Earth's crust, including silicon (Si), aluminum (Al), and iron (Fe), are not abundant in seawater for the simple reason that most compounds containing these elements are insoluble in water. Earth's crust must be made of compounds with limited water solubility, or else they would never survive as solids in the presence of all the water on our planet. The reason why some compounds are insoluble in water whereas others are soluble involves many factors. For now we will rely on the solubility rules in Table 4.4, which result from our empirical findings and allow us to predict the solubility of

TABLE 4.4 Solubility Rules for Common Ionic Compounds in Water

All compounds containing the following ions are soluble:
- Cations: Group 1 ions (alkali metals) and NH_4^+
- Anions: NO_3^- and CH_3COO^- (acetate)

Compounds containing the following anions are soluble except as noted:
- Cl^-, Br^-, and I^-, except those of Ag^+, Cu^+, Hg_2^{2+}, and Pb^{2+}
- SO_4^{2-}, except those of Ba^{2+}, Ca^{2+}, Hg_2^{2+}, Pb^{2+}, and Sr^{2+}

Insoluble compounds include the following:
- All hydroxides except those of group 1 cations and $Ca(OH)_2$, $Sr(OH)_2$, and $Ba(OH)_2$
- All sulfides except those of group 1 cations and NH_4^+, CaS, SrS, and BaS
- All carbonates except those of group 1 cations and NH_4^+
- All phosphates except those of group 1 cations and NH_4^+
- Most fluorides, though not those of group 1 cations and NH_4^+

common ionic compounds. *Soluble* and *insoluble* are qualitative terms. In principle, all ionic compounds dissolve in water to some extent. When ionic compounds dissolve in water, they do so by dissociating into ions. In practice, we consider a compound to be insoluble in water if the maximum amount that dissolves gives a concentration of less than 0.01 *M*. At this level, a solid appears to be insoluble to the naked eye; the tiny amount that dissolves is negligible compared with the amount that remains in the solid state and in contact with the solvent.

Making Insoluble Salts

In Section 4.5, we examined what happens in neutralization reactions such as the addition of aqueous sulfuric acid to aqueous barium hydroxide. However, as Figure 4.19 illustrates, when this reaction occurs, a white solid called a **precipitate** forms at the bottom of the beaker. Such reactions are called **precipitation reactions**.

We can use Table 4.4 to determine whether a precipitate forms when solutions of ionic solutes are mixed together. For example, does a precipitate form when aqueous solutions of potassium nitrate (KNO_3) and sodium iodide (NaI) are mixed? To answer this question we need to follow two steps:

1. *Recognize that both salts are in solution, which means that they are both soluble in water and therefore have dissociated into their respective cations and anions.* Table 4.4 specifies that all compounds containing cations from group 1 in the periodic table are soluble in water. Potassium and sodium are in group 1, so salts containing them are soluble and dissociate in water into their ions:

   Solution 1 consists of potassium ions and nitrate ions: $K^+(aq)$ and $NO_3^-(aq)$
   Solution 2 consists of sodium ions and iodide ions: $Na^+(aq)$ and $I^-(aq)$

   From this information we can set up the reactant side of the overall ionic equation:

   $$K^+(aq) + NO_3^-(aq) + Na^+(aq) + I^-(aq) \rightarrow \,?$$

2. *Determine whether any of the possible combinations of ions produces an insoluble product.* In step 1 we determined that there are four kinds of ions present when the two solutions are mixed. When these ions collide with one another, do any of them form an insoluble ionic compound, that is, a precipitate? KI and $NaNO_3$ are soluble according to Table 4.4. (The other possible combinations of ions are KNO_3 and NaI, but those are the compounds we started with and we already determined them to be soluble and dissociate.) Therefore no precipitate forms when the two solutions are mixed:

   $$K^+(aq) + NO_3^-(aq) + Na^+(aq) + I^-(aq) \rightarrow \text{no reaction}$$

Table 4.5 provides a convenient summary of the list of these solubility rules. It contains the same information as Table 4.4, but organized to show trends across common cations and anions. Let's consider mixing a different solution with NaI(*aq*). Does a precipitate form when an aqueous solution of lead(II) nitrate is added to an aqueous solution of sodium iodide? As before, NaI(*aq*) exists in

$$Ba^{2+}(aq) + SO_4^{2-}(aq) \rightarrow BaSO_4(s)$$

⬤ Ba^{2+} ⬤ SO_4^{2-}

FIGURE 4.19 Addition of aqueous sulfuric acid to a solution of barium hydroxide precipitates white barium sulfate.

TABLE 4.5 Solubility Matrix for Common Ionic Compounds in Water

ANION	CATION			
	Group 1 or NH_4^+	Ca^{2+} or Sr^{2+} or Ba^{2+}	Cu^+ or Ag^+	Hg_2^{2+} or Pb^{2+}
NO_3^- or CH_3COO^-	Soluble	Soluble	Soluble	Soluble
Cl^- or Br^- or I^-	Soluble	Soluble	Insoluble	Insoluble
OH^- or S^{2-}	Soluble	Soluble	Insoluble	Insoluble
SO_4^{2-}	Soluble	Insoluble	Soluble	Insoluble
F^- or CO_3^{2-} or PO_4^{3-}	Soluble	Insoluble	Insoluble	Insoluble

precipitate a solid product formed from a reaction in solution.

precipitation reaction a reaction that produces an insoluble product upon mixing two solutions.

solution as $Na^+(aq)$ and $I^-(aq)$ ions. Because Table 4.4 (and Table 4.5) indicate that all nitrate salts are soluble, $Pb(NO_3)_2$ dissolves in water and forms ions. The ions present in the mixed solutions are

$$Pb^{2+}(aq) + 2\,NO_3^-(aq) + Na^+(aq) + I^-(aq) \rightarrow ?$$

[Note that there is a coefficient of 2 present in front of nitrate because there are 2 moles of nitrate for every mole of $Pb^{2+}(aq)$ ions.] When these ions collide, do ionic bonds form and produce an insoluble ionic compound? The solubility rules in Table 4.4 indicate that I^- ions form insoluble compounds with $Pb^{2+}(aq)$, so we predict that $PbI_2(s)$ will precipitate (Figure 4.20).

● Na^+ ○ I^-

● Pb^{2+} ❀ NO_3^-

(a)   (b)

FIGURE 4.20 (a) One beaker contains a 0.1 M solution of $Pb(NO_3)_2$, and the other contains a 0.1 M solution of NaI. Both solutions are colorless. (b) As the NaI solution is poured into the $Pb(NO_3)_2$ solution, a yellow precipitate of PbI_2 forms.

Because a reaction takes place, let's write a balanced equation that describes the process, starting with the molecular equation:

Unbalanced: $Pb(NO_3)_2(aq) + NaI(aq) \rightarrow PbI_2(s) + NaNO_3(aq)$

Balanced: $Pb(NO_3)_2(aq) + 2\,NaI(aq) \rightarrow PbI_2(s) + 2\,NaNO_3(aq)$

From the balanced molecular equation we get the overall ionic equation:

$$Pb^{2+}(aq) + 2\,NO_3^-(aq) + 2\,Na^+(aq) + 2\,I^-(aq) \rightarrow$$
$$PbI_2(s) + 2\,Na^+(aq) + 2\,NO_3^-(aq)$$

Removing the spectator ions NO_3^- and Na^+ gives us the net ionic equation:

$$Pb^{2+}(aq) + 2\,I^-(aq) \rightarrow PbI_2(s)$$

SAMPLE EXERCISE 4.11 Writing Net Ionic Equations
for Precipitation Reactions I   **LO4, LO6**

Write a balanced net ionic equation for the reaction between aqueous sulfuric acid and barium hydroxide as pictured in Figure 4.19.

Collect and Organize We are asked to write a balanced net ionic equation given the names of the reactants. Figure 4.19 shows that the reaction produces a white precipitate.

Analyze First, we need to write a balanced molecular equation for the reaction. Next we must identify which reactants and products are soluble in water. We write any soluble ionic compounds as ions, leaving any precipitates and molecular compounds unchanged. Finally, we will simplify the overall ionic equation by removing any spectator ions that appear on both sides of the equation to obtain the net ionic equation.

Note that sulfuric acid is a strong acid. However, as discussed earlier in this chapter, its first proton completely dissociates but its second proton does not:

$$H_2SO_4(aq) \rightarrow H^+(aq) + HSO_4^-(aq) \qquad HSO_4^-(aq) \rightleftharpoons H^+(aq) + SO_4^{2-}(aq)$$

Therefore the correct way to write the ionic species present in solution in an overall ionic equation when sulfuric acid is a reactant is $H^+(aq) + HSO_4^-(aq)$, *not* $2\,H^+(aq) + SO_4^{2-}(aq)$.

Solve
1. Molecular equation: $Ba(OH)_2(aq) + H_2SO_4(aq) \rightarrow BaSO_4(s) + 2\,H_2O(\ell)$
2. Overall ionic equation:

$$Ba^{2+}(aq) + 2\,OH^-(aq) + H^+(aq) + HSO_4^-(aq) \rightarrow BaSO_4(s) + 2\,H_2O(\ell)$$

3. The equation cannot be simplified any further, so the net ionic equation is the same as the overall ionic equation in this case.

Think About It The reaction between barium hydroxide and sulfuric acid is *both* a neutralization reaction *and* a precipitation reaction. The neutralization is indicated by the formation of a salt ($BaSO_4$) and water; because the salt is insoluble, the reaction is also a precipitation reaction.

 Practice Exercise Write a balanced net ionic equation for the reaction between barium hydroxide and phosphoric acid (H_3PO_4) in water.

(Answers to Practice Exercises are in the back of the book.)

SAMPLE EXERCISE 4.12 Writing Net Ionic Equations   **LO4, LO6**
                                for Precipitation Reactions II

A precipitate forms when aqueous solutions of ammonium sulfate and barium chloride are mixed. Write the net ionic equation for the reaction. Is this the same net ionic equation as in Sample Exercise 4.11?

Collect and Organize Mixing two solutions causes an insoluble compound to precipitate. Using the names of the reactants, we are asked to identify the precipitate. We then need to write the net ionic equation and compare it with the one from the previous Sample Exercise.

Analyze The two salts are in solution, so they both must dissociate in water. This is consistent with the solubility rules in Table 4.4 or Table 4.5. We need to determine which combinations of the ions produce insoluble solids. First, we need to write correct formulas of the salts. Then we can check Table 4.4 or Table 4.5 to see which of the possible combinations of ions are insoluble in water.

Solve

Solution 1 consists of ammonium ions and sulfate ions: $NH_4^+(aq)$ and $SO_4^{2-}(aq)$

Solution 2 consists of barium ions and chloride ions: $Ba^{2+}(aq)$ and $Cl^-(aq)$

The new combinations are $BaSO_4$ and NH_4Cl. Table 4.4 indicates that all ammonium salts are soluble, so the ammonium and chloride ions remain dissolved as solvated ions. Barium sulfate is insoluble, however, and that salt precipitates. The overall ionic equation describes the species in solution:

$$2\,\cancel{NH_4^+(aq)} + SO_4^{2-}(aq) + Ba^{2+}(aq) + 2\,\cancel{Cl^-(aq)} \rightarrow$$
$$2\,\cancel{NH_4^+(aq)} + 2\,\cancel{Cl^-(aq)} + BaSO_4(s)$$

We can simplify the overall ionic equation by eliminating the spectator ions to yield the net ionic equation, which describes the formation of the precipitate:

$$Ba^{2+}(aq) + SO_4^{2-}(aq) \rightarrow BaSO_4(s)$$

This net ionic equation is not the same as for the reaction between barium hydroxide and sulfuric acid in Sample Exercise 4.11, even though both reactions form the same precipitate.

Think About It To identify whether a precipitation reaction occurs when you mix two solutions containing ions, check whether an insoluble compound is formed when the ions in solution collide with one another. In this exercise, sulfate ions originally in the ammonium sulfate solution collide with barium ions originally in the barium chloride solution, forming insoluble $BaSO_4$. The other ions, NH_4^+ and Cl^-, remain in solution.

Practice Exercise Does a precipitate form when you mix aqueous solutions of (a) sodium acetate and ammonium sulfate; (b) calcium chloride and mercury(I) nitrate? (c) If you answered yes in either case, write the net ionic equation for the reaction.

(Answers to Practice Exercises are in the back of the book.)

Precipitation reactions can be used to synthesize water-insoluble salts. For example, barium sulfate, used in medical procedures to image the gastrointestinal tract, can be made by mixing an aqueous solution of any water-soluble barium salt with a solution of a soluble sulfate salt. The precipitate from such a reaction can be collected by filtration, dried, and used for whatever purpose we may have. The soluble compound remaining in solution can also be collected. The ammonium chloride left in solution in Sample Exercise 4.12, for instance, can be isolated by boiling off the water to leave a residue of solid NH_4Cl.

| CONCEPT **TEST**

Lead(II) dichromate ($PbCr_2O_7$; the dichromate ion is $Cr_2O_7^{2-}$) is a water-insoluble pigment called school bus yellow that was used for decades to paint lines on highways. Design a synthesis of school bus yellow that uses a precipitation reaction.

(Answers to Concept Tests are in the back of the book.)

Using Precipitation in Analysis

Chemists use the insolubility of ionic compounds to determine concentrations of ions in solution. For example, NaCl (in the form of *rock salt*) is used to melt ice and snow on roads during the winter. However, the resulting NaCl solutions may run off into nearby water supplies, and that water can become contaminated with high levels of sodium and chloride ions. To determine whether sources of drinking water have become contaminated with such runoff, analytical chemists can determine the concentration of chloride ion by reacting a sample of the water with a solution of silver nitrate, $AgNO_3(aq)$. Any chloride ions in the water combine with Ag^+ ions to form a precipitate of AgCl:

$$NaCl(aq) + AgNO_3(aq) \rightarrow AgCl(s) + NaNO_3(aq) \qquad (4.10)$$

This precipitate can be filtered and dried, and its mass determined. From its mass, we can calculate the concentration of chloride in the original water sample.

Equation 4.10 is the molecular equation describing the formation of AgCl. Because both reactants are soluble ionic compounds and completely dissociate in solution, the overall ionic equation is

$$Na^+(aq) + Cl^-(aq) + Ag^+(aq) + NO_3^-(aq) \rightarrow$$
$$AgCl(s) + Na^+(aq) + NO_3^-(aq) \qquad (4.11)$$

Eliminating the spectator ions $Na^+(aq)$ and $NO_3^-(aq)$ yields the net ionic equation:

$$Ag^+(aq) + Cl^-(aq) \rightarrow AgCl(s) \qquad (4.12)$$

When performing this analysis, we must add enough Ag^+ ions to ensure that all the Cl^- ions precipitate. Specifically, if we use a molar ratio of $Ag^+(aq)/Cl^-(aq)$ greater than one, then the $Ag^+(aq)$ is present *in excess*. Precipitation reactions such as this may also include indicators that produce a visual signal (like a color change) that defines when the reaction is complete.

CONNECTION In Chapter 3 we introduced the idea of limiting reactants and reactants in excess when calculating the yield of reactions.

SAMPLE EXERCISE 4.13 Calculating the Mass of a Precipitate **LO6**

When hydrofluoric acid, HF(*aq*), gets on the skin, it migrates away from the site of exposure and shuts down the microcapillaries that carry blood. The clinical management of burns on the skin from exposure to hydrofluoric acid may include the injection of a soluble calcium salt in the skin near the site of contact to stop the migration of HF(*aq*). As F^- ions from the acid move through the skin, they combine with Ca^{2+} ions and form deposits of insoluble CaF_2. If 1.00 mL of a 2.24 M aqueous solution of Ca^{2+} is injected, what is the mass of CaF_2 produced if all the calcium ions react with fluoride ions?

Collect and Organize We are given the volume and concentration of a solution of Ca^{2+} ions and asked to determine the maximum amount of CaF_2 precipitate that could be formed upon reaction with F^- ions.

Analyze The net ionic reaction for the process involved is

$$Ca^{2+}(aq) + 2\ F^-(aq) \rightarrow CaF_2(s)$$

The problem states that all the calcium ions react, so we calculate the mass of product, assuming that Ca^{2+} ions are the limiting reactant.

$$\boxed{L\ Ca^{2+}\ \text{solution}} \xrightarrow[\text{L } Ca^{2+}]{\frac{\text{mol } Ca^{2+}}{}} \boxed{\text{mol } Ca^{2+}} \xrightarrow[\text{mol } Ca^{2+}]{\frac{\text{mol } CaF_2}{}} \boxed{\text{mol } CaF_2} \xrightarrow{\text{molar mass}} \boxed{g\ CaF_2}$$

Solve We can calculate the number of moles of $Ca^{2+}(aq)$ and use the stoichiometric relationships in the net ionic equation to determine the mass of product. We will also need the molar mass of CaF_2:

$$\mathcal{M}_{CaF_2} = 40.08\ \text{g/mol} + 2(19.00\ \text{g/mol}) = 78.08\ \text{g/mol } CaF_2$$

$$1.00 \times 10^{-3}\ \text{L } Ca^{2+} \times \frac{2.24\ \text{mol } Ca^{2+}}{1\ \text{L } Ca^{2+}} \times \frac{1\ \text{mol } CaF_2}{1\ \text{mol } Ca^{2+}} \times \frac{78.08\ \text{g } CaF_2}{1\ \text{mol } CaF_2} = 0.175\ \text{g } CaF_2$$

Think About It This problem asks for the mass of CaF_2 that could be formed, assuming that Ca^{2+} ions were the limiting reactant. However, when this treatment is actually used, it is more likely that the Ca^{2+} ions will be in excess to make sure all the fluoride ions are consumed.

Practice Exercise Vermilion is a very rare and expensive solid natural pigment. Known as Chinese red, it is the pigment used to print the author's signature on works of art. It consists of mercury(II) sulfide (HgS) and is very insoluble in water. What is the maximum number of grams of vermilion you can make from the reaction of 50.00 mL of a 0.0150 M aqueous solution of mercury(II) nitrate [$Hg(NO_3)_2$] with an excess of a concentrated aqueous solution of sodium sulfide (Na_2S)?

(Answers to Practice Exercises are in the back of the book.)

SAMPLE EXERCISE 4.14 Calculating a Solute Concentration **LO6**
from Mass of a Precipitate

To determine the concentration of chloride ions in a 100.0 mL sample of groundwater, a chemist adds a large enough volume of a solution of $AgNO_3$ to the sample to precipitate all the Cl^- as AgCl. The mass of the resulting AgCl precipitate is 71.7 mg. What is the chloride concentration in milligrams of Cl^- per liter of groundwater?

Collect and Organize We are given the sample volume, 100.0 mL, and the mass of AgCl formed, 71.7 mg. Our task is to determine the chloride concentration in milligrams of Cl^- per liter of groundwater.

Analyze We need a balanced chemical equation for the reaction. Although we can use the molecular, overall ionic, or net ionic equation, we choose the net ionic equation because it contains only those species involved in the precipitation reaction. The net ionic equation is

$$Ag^+(aq) + Cl^-(aq) \rightarrow AgCl(s)$$

It shows that 1 mole of Cl^- ions reacts with 1 mole of Ag^+ ions to produce 1 mole of AgCl. The molar mass of Cl^- is 35.45 g/mol, and that of AgCl is 143.32 g/mol. We use the methods developed in Chapter 3 to determine the mass of chloride ion in 100.0 mL of groundwater and then use this mass to calculate the concentration we need.

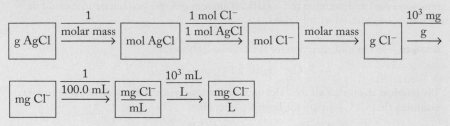

Solve Because our molar masses are in grams per mole, a logical first step is to convert the AgCl(s) mass to grams:

$$71.7 \text{ mg} \times \frac{10^{-3} \text{ g}}{1 \text{ mg}} = 0.0717 \text{ g}$$

The mass of Cl^- ions in the 100.0 mL sample of groundwater is therefore

$$0.0717 \text{ g AgCl}(s) \times \frac{1 \text{ mol AgCl}(s)}{143.32 \text{ g AgCl}(s)} \times \frac{1 \text{ mol } Cl^- (aq)}{1 \text{ mol AgCl}(s)} \times \frac{35.45 \text{ g } Cl^- (aq)}{1 \text{ mol } Cl^- (aq)}$$

$$= 0.0177 \text{ g } Cl^- (aq) = 17.7 \text{ mg } Cl^- (aq)$$

The Cl^- ion concentration in milligrams per liter of water is

$$\frac{17.7 \text{ mg } Cl^-}{100.0 \text{ mL}} \times \frac{1000 \text{ mL}}{1 \text{ L}} = 177 \text{ mg } Cl^-/\text{L groundwater}$$

An alternative way to solve this problem is to determine the mass of Cl^- ions in 1 mg of AgCl(s):

$$\frac{1 \text{ mol } Cl^-}{1 \text{ mol AgCl}} \times \frac{35.45 \text{ g } Cl^-/\text{mol } Cl^-}{143.32 \text{ g AgCl/mol AgCl}} = \frac{0.2473 \text{ g } Cl^-}{\text{g AgCl}} = \frac{0.2473 \text{ mg } Cl^-}{\text{mg AgCl}}$$

Using this factor with the mass of AgCl(s) precipitated, we get

$$71.7 \text{ mg AgCl} \times \frac{0.2743 \text{ mg } Cl^-}{1 \text{ mg AgCl}} = 17.7 \text{ mg } Cl^- \text{ precipitated as AgCl}(s)$$

From here, the problem proceeds as before.

Think About It In Sample Exercise 4.1 we noted that the chloride concentration of drinking water should not exceed 250 ppm. Since mg/L is equivalent to ppm for dilute solutions, the concentration of Cl^- in the sample for this exercise is 177 ppm, which meets the guidelines for drinking water.

Practice Exercise The concentration of $SO_4^{2-}(aq)$ in a 50.0 mL sample of river water is determined using a precipitation titration in which $Ba^{2+}(aq)$ is added to precipitate $BaSO_4(s)$. What is the $SO_4^{2-}(aq)$ concentration in molarity if 6.55 mL of a 0.00100 M $Ba^{2+}(aq)$ solution is consumed?

(Answers to Practice Exercises are in the back of the book.)

SAMPLE EXERCISE 4.15 Limiting Reactants in Solution Reactions   **LO6**

Acid rain in the form of aqueous H_2SO_4 slowly erodes marble statues, as shown in Figure 4.15. If we add 16.75 mL of 0.0100 M H_2SO_4 to 25.00 mg of $CaCO_3$, will all of the $CaCO_3$ react?

Collect and Organize Given the volume and molarity of the H_2SO_4 solution, we can calculate the mass of $CaCO_3$ that reacts.

Analyze We need the balanced chemical equation for the reaction to determine the stoichiometry of the reaction between $CaCO_3$ and H_2SO_4. The molecular equation is

$$CaCO_3(s) + H_2SO_4(aq) \rightarrow CaSO_4(s) + H_2O(\ell) + CO_2(g)$$

One mole of $H_2SO_4(aq)$ reacts with 1 mole of $CaCO_3(s)$. The steps involved in calculating the moles of H_2SO_4 available and the mass (in milligrams) of calcium carbonate that will react with it are

$$\boxed{\text{mL } H_2SO_4} \xrightarrow[10^3 \text{ mL}]{L} \boxed{\text{L } H_2SO_4} \xrightarrow[\text{L } H_2SO_4]{\text{mol } H_2SO_4} \boxed{\text{mol } H_2SO_4} \xrightarrow[1 \text{ mol } H_2SO_4]{1 \text{ mol } CaCO_3}$$

$$\boxed{\text{mol } CaCO_3} \xrightarrow{\text{molar mass}} \boxed{\text{g } CaCO_3} \xrightarrow[\text{g}]{10^3 \text{ mg}} \boxed{\text{mg } CaCO_3}$$

Solve We calculate the mass of calcium carbonate that reacts, starting with the volume of sulfuric acid available and its molarity:

$$16.75 \text{ mL } H_2SO_4 \times \frac{1 \text{ L}}{10^3 \text{ mL}} \times \frac{1.00 \times 10^{-2} \text{ mol } H_2SO_4}{1 \text{ L } H_2SO_4} \times \frac{1 \text{ mol } CaCO_3}{1 \text{ mol } H_2SO_4}$$

$$\times \frac{100.09 \text{ g } CaCO_3}{1 \text{ mol } CaCO_3} \times \frac{1000 \text{ mg } CaCO_3}{1 \text{ g } CaCO_3} = 16.8 \text{ mg } CaCO_3$$

Only 16.8 mg of the 25.00 mg sample will react to form solid calcium sulfate (Table 4.4).

Think About It The sulfuric acid is the limiting reactant, since only 67% of the $CaCO_3$ reacts under the conditions in this exercise.

Practice Exercise Heavy metal ions such as lead(II) can be precipitated from laboratory wastewater by adding sodium sulfide, Na_2S. Will all of the lead be removed from 15.00 mL of 7.52×10^{-3} M $Pb(NO_3)_2$ upon addition of 12.50 mL of 0.0115 M Na_2S?

(Answers to Practice Exercises are in the back of the book.)

> **saturated solution** a solution that contains the maximum concentration of a solute possible at a given temperature.

> **solubility** the maximum amount of a substance that dissolves in a given quantity of solvent at a given temperature.

Saturated Solutions and Supersaturation

A solution containing the maximum amount of solute that can dissolve is a **saturated solution** (Figure 4.21). The maximum amount of solute that can dissolve in a given quantity of solvent at a given temperature is called the **solubility** of that solute in the solvent. Precipitates form when the solubility of a solute is exceeded.

Solubility depends on temperature; often, the higher the temperature, the greater the solubility of solids in water. For example, more table sugar dissolves in hot water than in cold, a fact applied in the making of rock candy

FIGURE 4.21 (a) Seawater, although a concentrated aqueous solution of NaCl, is far from saturated. (b) This solution is in equilibrium with solid NaCl (the white mass at the bottom of the beaker) at 20°C, so the liquid is a saturated solution of NaCl. Crystals in the solid are constantly dissolving, while ions in the solution are constantly precipitating. Because of this dynamic equilibrium, the concentration of solute [$Na^+(aq)$ and $Cl^-(aq)$ ions] in the solvent remains constant.

$$\frac{3.3 \text{ g NaCl}}{100 \text{ mL}} \qquad \frac{35.9 \text{ g NaCl}}{100 \text{ mL}}$$

(a) Seawater   (b) Saturated NaCl

(a) *T* = close to boiling point

(b) Put in a wooden stirrer

(c) Cool to room temperature

(d) Rock candy

FIGURE 4.22 Making rock candy. (a) A large quantity of table sugar is dissolved in hot water. The solution is not saturated at this high temperature, however. (b) A wooden stirrer is added and the solution is allowed to cool. (c) As the solution cools, it reaches the temperature at which the concentration of sugar exceeds the maximum that can remain in solution, and crystals ("rocks") of solid sugar ("candy") precipitate and attach to the stirrer. (d) Rock candy.

(Figure 4.22). After a large amount of sugar is dissolved in hot water, the solution is slowly cooled. An object with a slightly rough surface (like a string or a wooden stirrer) suspended in the solution serves as a site for crystallization. As the solution cools below the temperature at which the solubility of sugar is exceeded, crystals of sugar grow on the rough surface. The formation of crystals as the temperature of a saturated solution is lowered is a form of precipitation.

Sometimes, more solute dissolves in a volume of liquid than the amount predicted by the solute's solubility in that liquid, creating a **supersaturated solution**. This can happen when the temperature of a saturated solution drops slowly or when the volume of an unsaturated solution is reduced through slow evaporation. Slow evaporation of the solvent is partly responsible for the formation of stalactites and stalagmites, as well as of the salt flats in the American Southwest. Sooner or later, usually in response to a disruption such as a change in temperature, a mechanical shock, or the addition of a *seed crystal* (a small crystal of the solute that provides a site for crystallization), the solute in a supersaturated solution rapidly comes out of solution (Figure 4.23).

4.8 Ion Exchange

We have seen how dissolved cations can be soluble in a solution containing one anion but form an insoluble precipitate when in solution with another anion. This idea of **ion exchange** is important in water purification. Water containing certain metal ions—principally Ca^{2+} and Mg^{2+}—is called *hard water*, and it causes problems in industrial water supplies and in homes. Because hard water combines with soap to form a gray scum, clothes washed in hard water appear gray and dull. Hard water forms scale (an incrustation) in boilers, pipes, and kettles, diminishing their ability to conduct heat and carry water; moreover, hard water sometimes has an unpleasant taste. In one common method of *water softening* (removing the ions responsible for the undesirable properties of hard water), hard water is passed through a system that exchanges calcium and magnesium ions for sodium ions (Figure 4.24).

The system consists of a cartridge packed with beads of a porous plastic resin (R) bonded to anions capable of binding with hard-water cations—mainly Ca^{2+}, Mg^{2+},

FIGURE 4.23 (a) A seed crystal is added to a supersaturated solution of sodium acetate. (b) The seed crystal becomes a site for rapid growth of sodium acetate crystals. (c) Crystal growth continues until the solution is no longer supersaturated but merely saturated with sodium acetate.

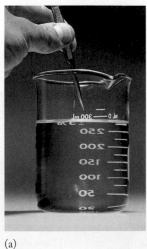

(a)

(b)

(c)

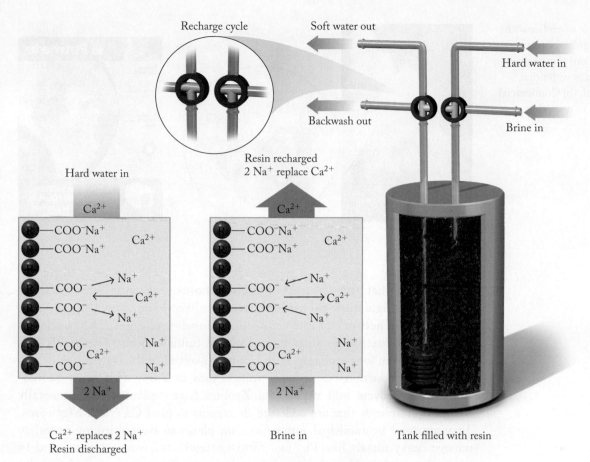

FIGURE 4.24 Residential water softeners use ion exchange to remove 2+ ions (such as Ca^{2+}) that make water hard. The ion-exchange resin contains cation-exchange sites that are initially occupied by Na^+ ions. These ions are replaced by 2+ "hardness" ions as water flows through the resin. Eventually most of the ion-exchange sites are occupied by 2+ ions, and the system loses its water-softening ability. The resin is then backwashed with a saturated solution of NaCl (*brine*), displacing the hardness ions (which wash down the drain) and restoring the resin to its Na^+ form.

and Fe^{2+}. The carboxylate anion (COO^-) is often used, and water in the ion-exchange cartridge contains Na^+ to balance the negative charges on the anions. We can therefore think of the resin as beads containing $(R–COO^-)Na^+$ units. As hard water flows through the cartridge, 2+ ions in the water exchange places with sodium ions on the resin. This exchange takes place because cations with 2+ and 3+ charges bind more strongly to the $R–COO^-$ groups on the resin than sodium cations do. The ion-exchange reaction, with calcium ion as an example, is

$$2\ (R–COO^-)Na^+(s) + Ca^{2+}(aq) \rightarrow (R–COO^-)_2Ca^{2+}(s) + 2\ Na^+(aq)$$

Hard water that has been softened in this way contains increased concentrations of sodium ions. Although this may not be a problem for healthy children and adults, people suffering from high blood pressure often must limit their intake of Na^+ and so should not drink water softened by this kind of ion-exchange reaction.

Not only synthetic materials perform ion exchange. Naturally occurring minerals called **zeolites** are extensively mined in many parts of the world and have huge commercial and technological utility as water softeners and purifiers, livestock feed additives, and odor suppressants. Zeolites, formed in nature as a result of the chemical reaction between molten lava from volcanic eruptions and salt water, are crystalline solids with tiny pores. Synthetic zeolites have been

supersaturated solution a solution that contains more than the maximum quantity of solute predicted to be soluble in a given volume of solution at a given temperature.

ion exchange a process by which one ion is displaced by another.

zeolites natural crystalline minerals or synthetic materials consisting of three-dimensional networks of channels that contain sodium or other 1+ cations.

FIGURE 4.25 (a) A sample of zeolite with a drawing showing the regular pattern of pores containing sodium ions that exchange for other cations dissolved in water that flows through the material. (b) Commercial cat litter includes zeolites to neutralize odors.

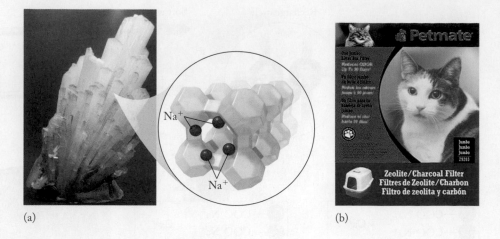

(a)   (b)

manufactured that have structures similar to those of the natural materials. All zeolites have a rigid three-dimensional structure like a honeycomb (Figure 4.25), consisting of a network of interconnecting tunnels and cages. They work just like the plastic resins: as water flows through tunnels (pores) lined with Na^+ ions, the sodium ions exchange with cations dissolved in the water.

Zeolite production reached 4.0 million tons in 2014, with many zeolite-containing products sold worldwide. Zeolites have replaced environmentally harmful phosphates that are added to detergents to bind Ca^{2+} and Mg^{2+} ions. They are used in municipal water filtration plants to treat drinking water, to remove heavy metals like $Pb^{2+}(aq)$ from contaminated waste streams, and in swimming pools to keep the water clean and clear. They are even used for odor control in barns and feedlots where animals are confined. Animal waste contains large quantities of ammonium ion [$NH_4^+(aq)$] that can be exchanged for the naturally occurring cations in zeolites. When $NH_4^+(aq)$ ions participate in acid–base reactions with basic materials (B = base), they liberate ammonia [$NH_3(g)$], which is partially responsible for the characteristic smell of cat boxes and barnyards:

$$NH_4^+(aq) + B \rightarrow NH_3(g) + HB^+(aq)$$

Zeolites reduce odors by selectively removing $NH_4^+(aq)$ through ion exchange.

Other ions can be exchanged for the sodium ion to prepare zeolites with special properties. Certain zeolites may be poured directly on wounds to stop bleeding by absorbing water from the blood, thereby concentrating clotting factors to promote coagulation. If the cations in these zeolites are exchanged for silver ions (Ag^+), the resulting product also has an antimicrobial effect and not only stems bleeding but also reduces the risk of infection.

4.9 Oxidation–Reduction Reactions: Electron Transfer

In the 15th century, Leonardo da Vinci observed: "Where flame cannot live, no animal that draws breath can live." Although oxygen was not isolated and recognized as an element until the 18th century, da Vinci captured an important property of the substance as a requirement for combustion and metabolism. Oxygen makes up about 50% by mass of Earth's crust, it is almost 89% by mass of the water that covers 70% of Earth's surface, and it is about 20% of the volume of Earth's atmosphere. It is essential to the metabolism of most creatures, just as it

CONNECTION In Chapter 3 we defined combustion as a heat-producing reaction between oxygen and another element or compound.

is essential for combustion reactions. What type of reaction is involved in metabolism and in the combustion reactions in Figure 4.26? Methane and oxygen are neither acids nor bases. No precipitates are formed in the reactions, which produce water and carbon dioxide.

Combination reactions of oxygen with nonmetals such as carbon, sulfur, and nitrogen produce volatile oxides. Combination reactions of O_2 with metals and semimetals produce solid oxides. All of these are examples of an extremely important type of reaction: oxidation–reduction reactions. *Oxidation* was first defined as a reaction that increased the oxygen content of a substance. Hence two reactions involving oxygen—with methane to produce $CO_2(g)$ and $H_2O(\ell)$, and with elemental iron to produce $Fe_2O_3(s)$—are both oxidations from the point of view of the CH_4 and the Fe because the products contain more oxygen than the starting materials:

$$CH_4(g) + 2\,O_2(g) \rightarrow CO_2(g) + 2\,H_2O(g)$$
$$4\,Fe(s) + 3\,O_2(g) \rightarrow 2\,Fe_2O_3(s)$$

Reduction reactions were originally defined in a similar fashion—reactions in which the oxygen content of a substance is reduced. A classic example is the reduction of iron ore (mostly Fe_2O_3) to metallic iron:

$$Fe_2O_3(s) + 3\,CO(g) \rightarrow 2\,Fe(s) + 3\,CO_2(g)$$

The oxygen content of the iron ore is reduced—Fe_2O_3 is converted into Fe—as a result of its reaction with $CO(g)$. Note, however, that the carbon in CO is oxidized because the oxygen content of CO increases as it is converted into CO_2.

CONCEPT TEST

The following equation describes the conversion of one iron mineral, called magnetite (Fe_3O_4), into another, called hematite (Fe_2O_3):

$$4\,Fe_3O_4(s) + O_2(g) \rightarrow 6\,Fe_2O_3(s)$$

Identify which element is oxidized and which is reduced and explain your choices.

(Answers to Concept Tests are in the back of the book.)

Oxidation Numbers

The processes of oxidation and reduction always occur together, which gives rise to the common way of referring to them as *redox reactions*. To distinguish between oxidation and reduction in redox reactions, chemists have developed a system of oxidation numbers to assign to atoms in reactants and products. If the oxidation number of an atom changes as a result of a reaction—if the oxidation number of an atom on the left-hand side of an equation is different from the oxidation number of that same atom on the right-hand side—then the reaction is a redox reaction.

The **oxidation number (O.N.)**, or **oxidation state**, of an atom in a molecule or ion may be a positive or negative number or zero. The oxidation number represents the positive or negative character an atom has or appears to have as determined by the following rules:

1. The oxidation numbers of the atoms in a neutral molecule sum to zero; those of the atoms in an ion sum to the charge on the ion.
2. Each atom in a pure element has an oxidation number of zero:

   F_2 O.N. = 0 for each F O_2 O.N. = 0 for each O
   Fe O.N. = 0   Na O.N. = 0

(a)

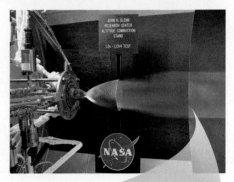

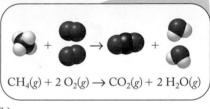

$$CH_4(g) + 2\,O_2(g) \rightarrow CO_2(g) + 2\,H_2O(g)$$

(b)

FIGURE 4.26 (a) Forest fires and (b) the combustion of methane in this rocket engine being test fired at NASA are oxidation–reduction reactions.

oxidation number (O.N.) (also called **oxidation state**) a positive or negative number based on the number of electrons an atom gains or loses when it forms an ion, or that it shares when it forms a covalent bond with another element; pure elements have an oxidation number of zero.

3. In monatomic ions, the oxidation number is the charge on the ion:

F^- O.N. = -1 O^{2-} O.N. = -2

Fe^{3+} O.N. = $+3$ Na^+ O.N. = $+1$

Note that the number comes first when we write the symbol of an ion, Fe^{3+}, but the sign comes first when we write an oxidation number, $+3$.

4. In compounds containing fluorine and one or more other elements, the oxidation number of the fluorine is *always* -1:

KF O.N. = -1 for F, which means O.N. = $+1$ for K (rule 1)

OF_2 O.N. = -1 for each F, which means O.N. = $+2$ for O (rule 1)

CF_4 O.N. = -1 for each F, which means O.N. = $+4$ for C (rule 1)

5. In most compounds, the oxidation number of hydrogen is $+1$ and that of oxygen is -2. Exceptions include hydrogen in metal hydrides (for example, LiH), where O.N. = -1 for H, and oxygen in the peroxide ion, O_2^{2-} (for example, H_2O_2), where O.N. = -1 for each O.

6. Unless combined with oxygen or fluorine, chlorine and bromine have an oxidation number of -1:

$CaCl_2$ O.N. = -1 for Cl, which means O.N. = $+2$ for Ca (rule 1)

$AlBr_3$ O.N. = -1 for Br, which means O.N. = $+3$ for Al (rule 1), *but*

ClO_4^- O.N. = -2 for O (rule 5), which means O.N. = $+7$ for Cl (rule 1)

SAMPLE EXERCISE 4.16 Determining Oxidation Numbers   **LO7**

What is the oxidation number of sulfur in (a) SO_2, (b) Na_2S, and (c) $CaSO_4$?

Collect, Organize, and Analyze To assign the oxidation number of sulfur in three of its compounds, we apply the rules for determining the oxidation numbers of the elements in each compound. SO_2 is the molecule sulfur dioxide. Na_2S is an ionic compound, and the oxidation number of an ion in an ionic compound equals its charge. Sulfur in $CaSO_4$ is part of the sulfate ion, which means its oxidation number added to those of the four O atoms must add up to the charge of the ion; we determine the oxidation number for calcium from rule 1.

Solve

a. From rule 5, O.N. = -2 for the O in SO_2. Rule 1 says that the sum of the oxidation numbers for S and the two O in the neutral molecule must be zero. Letting O.N. for S be x:

$$x + 2(-2) = 0$$
$$x = +4 \qquad \text{O.N. for S in } SO_2 = +4$$

b. Sodium forms only one ion, Na^+, which means that, according to rule 3, O.N. = $+1$. To balance the O.N. values in Na_2S (rule 1), we let y stand for the O.N. of sulfur:

$$2(+1) + y = 0$$
$$y = -2 \qquad \text{O.N. for S in } Na_2S = -2$$

c. The charge on the calcium ion is always 2+. This means that the charge on the sulfate ion must be 2−. Assigning O.N. = -2 for oxygen (rule 5) and z for sulfur:

$$z + 4(-2) = -2$$
$$z = +6 \qquad \text{O.N. for S in } CaSO_4 = +6$$

Think About It We found oxidation numbers for sulfur ranging from -2 to $+6$. Many other elements have a range of oxidation numbers in their compounds. For example, the oxidation numbers of iodine range from -1 in KI to $+7$ in KIO_4.

⊛ **Practice Exercise** Determine the oxidation number of nitrogen in (a) NO_2, (b) N_2O, and (c) HNO_3.

(Answers to Practice Exercises are in the back of the book.)

Considering Changes in Oxidation Number in Redox Reactions

Now that we know how to assign oxidation numbers to atoms in molecules and ions, we can examine how oxidation numbers change during a redox reaction. Let's look at a reaction that we balanced in Chapter 3: the combustion of methane (CH_4) to produce CO_2 and H_2O.

$$CH_4(g) + 2\,O_2(g) \rightarrow CO_2(g) + 2\,H_2O(g) \qquad (4.13)$$

First we assign oxidation numbers to all the atoms in the reactants and products:

OXIDATION NUMBERS			
Atoms in Reactants		**Atoms in Products**	
C in CH_4:	−4	C in CO_2:	+4
H in CH_4:	+1	H in H_2O:	+1
O in O_2:	0	O in CO_2:	−2
		O in H_2O:	−2

The oxidation numbers of the carbon and oxygen atoms change; such a change for any element defines a reaction as redox. One carbon atom on the left goes from O.N. = −4 in CH_4 to +4 in CO_2 while oxygen atoms go from O.N. = 0 in O_2 to −2 in CO_2 and H_2O. Because the oxidation number of carbon increases, carbon is said to be oxidized in this reaction. Likewise, because the oxidation number of oxygen decreases, oxygen is said to be reduced. In redox reactions, we can keep track of the oxidation numbers and the electrons transferred by annotating the equation in the following way:

Carbon: −4 ⎯⎯ oxidation ⎯⎯→ +4

$$CH_4(g) + 2\,O_2(g) \quad \rightarrow \quad CO_2(g) + 2\,H_2O(\ell)$$

Oxygen: 0 reduction −2   −2

Note that when we assigned oxidation numbers to the atoms in this reaction, the value for hydrogen did not change. This means that hydrogen is neither oxidized nor reduced in this reaction, and we do not need to incorporate it into our annotation above.

CONCEPT TEST

Determine whether the following are redox reactions. For the redox reactions, identify the atoms that undergo changes in oxidation number.
a. $Sn^{2+}(aq) + Br_2(aq) \rightarrow Sn^{4+}(aq) + 2\,Br^-(aq)$
b. $2\,F_2(g) + 2\,H_2O(\ell) \rightarrow 4\,HF(aq) + O_2(g)$
c. $NaHCO_3(aq) + HCl(aq) \rightarrow NaCl(aq) + CO_2(g) + H_2O(\ell)$

(Answers to Concept Tests are in the back of the book.)

oxidation a chemical change in which a species loses electrons; the oxidation number of the species increases.

reduction a chemical change in which a species gains electrons; the oxidation number of the species decreases.

oxidizing agent a substance in a redox reaction that contains the element being reduced.

reducing agent a substance in a redox reaction that contains the element being oxidized.

Considering Electron Transfer in Redox Reactions

Ultimately, the meanings of oxidation and reduction were expanded, so that **oxidation** now refers to any chemical reaction in which a substance *loses electrons* and **reduction** refers to any reaction in which a substance *gains electrons*. If one substance loses electrons, another substance must gain them; if one species is oxidized, another must be reduced. The defining event in oxidation–reduction is the *transfer* of electrons from one substance to another.

For example, when a piece of copper wire [elemental copper, $Cu(s)$ is placed in a colorless solution of silver nitrate [$Ag^+(aq) + NO_3^-(aq)$], the solution gradually turns blue, the color of a solution of $Cu^{2+}(aq)$, and branchlike structures of $Ag(s)$ form in the medium (Figure 4.27). As silver ions (Ag^+) collide with the surface of the solid Cu, electrons transfer from the copper atoms to the silver ions, resulting in a layer of silver atoms depositing on the copper solid. Two electrons transfer, one to each Ag^+ ion, from one copper atom. Simultaneously, for each two Ag^+ ions that deposit as silver atoms, one copper atom enters solution as an ion, Cu^{2+}. The process can be summarized as

$$\text{Silver:} \quad +1 \quad \xrightarrow{\text{reduction}} \quad 0$$

$$Ag^+(aq) + Cu(s) \rightarrow Ag(s) + Cu^{2+}(aq)$$

$$\text{Copper:} \quad 0 \quad \xrightarrow{\text{oxidation}} \quad +2$$

Because the silver ion is reduced as the copper atom is oxidized, the silver ion is called the **oxidizing agent** in the redox reaction. We say that "silver oxidizes copper." As an oxidizing agent, the silver ion is reduced. This is always the case: the *oxidizing* agent in any redox reaction is always *reduced* in the process. Any species that is reduced experiences a *reduction in oxidation number*, as is the case for the silver ion where it goes from O.N. = +1 to O.N. = 0. Likewise, as the copper atom is oxidized, the silver ion is reduced. Copper in this reaction is therefore called the **reducing agent**. The *reducing* agent is always *oxidized* and experiences an *increase in oxidation number;* in this reaction, the copper atom goes from O.N. = 0 to O.N. = +2. Every redox reaction has both an oxidizing agent and a reducing agent.

FIGURE 4.27 (a) When a Cu wire is immersed in a solution of $AgNO_3$, Cu metal oxidizes to Cu^{2+} ions as Ag^+ ions are reduced to Ag metal. (b) A day later, the solution has the blue color of a solution of $Cu(NO_3)_2$, and the wire is coated with Ag metal. (c) Copper atoms donate two electrons to silver ions, as silver atoms form a layer on the surface of the copper wire and copper ions enter solution.

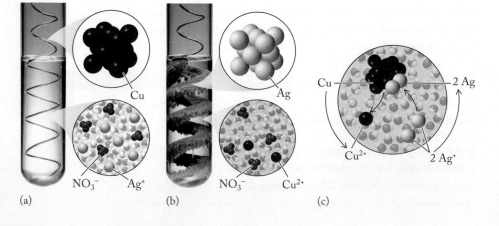

SAMPLE EXERCISE 4.17 Identifying Oxidizing Agents   **LO7**
and Reducing Agents

The reaction of oxygen with hydrazine is used to remove dissolved oxygen gas from aqueous solutions:

$$O_2(aq) + N_2H_4(aq) \rightarrow 2\,H_2O(\ell) + N_2(g)$$

and the combustion of hydrazine produces enough energy that the substance is used as a rocket fuel. Identify the species oxidized, the species reduced, the oxidizing agent, and the reducing agent in the balanced chemical equation.

Collect and Organize The reactants are the element oxygen and the compound hydrazine. The products are water and nitrogen.

Analyze The O.N. of the oxygen atoms in O_2 changes from zero to -2 in H_2O. The O.N. of nitrogen changes from -2 in N_2H_4 to zero in N_2. The O.N. $= +1$ of hydrogen is unchanged.

Solve

Nitrogen: $\quad -2 \quad\quad$ oxidation $\quad\quad 0$

$$O_2(aq) + N_2H_4(aq) \rightarrow 2\,H_2O(\ell) + N_2(g)$$

Oxygen: $\quad 0 \quad\quad$ reduction $\quad\quad -2$

We see that oxygen is the oxidizing agent because the oxidation number of nitrogen increases. This means that O_2 is reduced and N_2H_4 is the reducing agent.

Think About It We tend to identify the specific *atoms* that are oxidized or reduced, whereas whole *molecules or ions* are identified as the oxidizing agents or reducing agents. In the equation in this exercise, nitrogen is oxidized but N_2H_4 is the reducing agent.

⊛ **Practice Exercise** In the reaction between $O_2(g)$ and $SO_2(g)$ to make $SO_3(g)$, identify the species oxidized, the species reduced, the oxidizing agent, and the reducing agent.

(Answers to Practice Exercises are in the back of the book.)

Balancing Redox Reactions by Using Half-Reactions

At first glance, the equation for the redox reaction between $Ag^+(aq)$ and $Cu(s)$ may seem balanced—and in a material sense it is: the number of silver and copper atoms or ions is the same on both sides. However, the charges are not balanced: the sum of the charges is 1+ for the reactants but 2+ for the products. The reason for this imbalance is that one electron is involved in the reduction reaction but two are involved in the oxidation reaction. Equal numbers of electrons must be lost and gained, so to balance equations describing redox reactions like this one, we must develop a method that accounts for electron transfer.

The change in oxidation number reflects the change in the number of electrons associated with an atom or ion. This change in the number of electrons happens because electrons are transferred from one atom or ion to another during a redox reaction. By tracking changes in oxidation numbers, we can gain additional insight into the transfer of electrons as the redox reaction takes place.

Think of this reaction between $Cu(s)$ and $AgNO_3(aq)$ as consisting of one oxidation and one reduction. Each of these reactions represents *half* of the overall

CONNECTION In Chapter 2 we examined nuclear equations that must also be balanced in both mass and charge.

half-reaction one of the two halves of an oxidation–reduction reaction; one half-reaction is the oxidation component, and the other is the reduction component.

reaction; hence, each is called a **half-reaction**. We balance the equation by following these five steps:

1. *Write one equation for the oxidation half-reaction and a separate equation for the reduction half-reaction*:

   Oxidation: $Cu(s) \rightarrow Cu^{2+}(aq)$
   Reduction: $Ag^+(aq) \rightarrow Ag(s)$

2. *Balance the number of particles in each half-reaction.* This example requires no changes at this point because both equations are balanced in terms of mass.

3. *Balance the charge in each half-reaction* by adding electrons to the appropriate side. Always *add* electrons; never subtract them. We add 2 electrons to the product side of the Cu half-reaction:

   Oxidation: $Cu(s) \rightarrow Cu^{2+}(aq) + 2\ e^-$

   and 1 electron to the reactant side of the Ag half-reaction:

   Reduction: $1\ e^- + Ag^+(aq) \rightarrow Ag(s)$

   This gives us a total charge of 0 on both sides of both half-reactions. Note that electrons are added on the reactant side of the reduction equation (reduction is the gain of electrons) and on the product side of the oxidation equation (oxidation is the loss of electrons). This is always the case.

4. *Multiply each half-reaction by the appropriate whole number* to make the number of electrons lost in the oxidation half-reaction equal the number of electrons gained in the reduction half-reaction:

   Oxidation: $Cu(s) \rightarrow Cu^{2+}(aq) + 2\ e^-$
   Reduction: $2 \times [1\ e^- + Ag^+(aq) \rightarrow Ag(s)]$

   As a result we have the following half-reactions:

   Oxidation: $Cu(s) \rightarrow Cu^{2+}(aq) + 2\ e^-$
   Reduction: $2\ e^- + 2\ Ag^+(aq) \rightarrow 2\ Ag(s)$

5. *Add the two half-reactions to generate the equation representing the overall redox reaction*:

   Oxidation: $Cu(s) \rightarrow Cu^{2+}(aq) + \cancel{2\ e^-}$
   Reduction: $\cancel{2\ e^-} + 2\ Ag^+(aq) \rightarrow 2\ Ag(s)$
   _____
   Overall equation: $2\ Ag^+(aq) + Cu(s) \rightarrow 2\ Ag(s) + Cu^{2+}(aq)$

If we have carried out step 4 correctly, the number of electrons in the oxidation half-reaction is the same as the number of electrons in the reduction half-reaction, and they cancel out when we write the overall equation. In redox reactions, we can keep track of the oxidation numbers and the electrons transferred by annotating the equation in the following way:

Silver: $\Delta\ O.N. = 2(1) - 2(0) = +2$
2 electrons gained

$+1$ reduction 0

$2\ Ag^+(aq) + Cu(s) \rightarrow 2\ Ag(s) + Cu^{2+}(aq)$

0 oxidation $+2$

Copper: $\Delta\ O.N. = 0 - 2 = -2$
2 electrons lost

The equation is now balanced in terms of mass and charge. Note that the overall equation is actually a net ionic equation, since only the species involved in the redox reaction are shown. The nitrate ions from the silver nitrate solution (Figure 4.27) are spectator ions and are not shown in the final net ionic equation or in any of the intermediate equations we used to generate it.

We review the methods for balancing chemical equations as described here when we discuss applications of oxidation–reduction reactions in Chapter 18.

SAMPLE EXERCISE 4.18 Balancing Redox Reactions **LO7**
with Half-Reactions

Iodine is slightly soluble in water and dissolves to make a yellow-brown solution of $I_2(aq)$. When colorless $Sn^{2+}(aq)$ is dissolved in it (Figure 4.28), the solution turns colorless as $I^-(aq)$ forms and $Sn^{2+}(aq)$ is converted into $Sn^{4+}(aq)$. (a) Is this a redox reaction? (b) Balance the equation that describes the reaction.

Collect and Organize The reactants in this process are $I_2(aq)$ and $Sn^{2+}(aq)$; the products are $I^-(aq)$ and $Sn^{4+}(aq)$.

Analyze The oxidation number of tin changes from +2 in $Sn^{2+}(aq)$ to +4 in $Sn^{4+}(aq)$. The oxidation number of iodine changes from 0 in $I_2(aq)$ to −1 in $I^-(aq)$. We write tin and iodine in separate equations when we write the respective half-reactions.

Solve

a. Because the oxidation numbers change, this is a redox reaction.
b. The unbalanced equation is

$$Sn^{2+}(aq) + I_2(aq) \rightarrow I^-(aq) + Sn^{4+}(aq)$$

We balance it via our five-step procedure.
1. Separate the half-reactions.

Oxidation: $Sn^{2+}(aq) \rightarrow Sn^{4+}(aq)$
Reduction: $I_2(aq) \rightarrow I^-(aq)$

2. Balance masses.

Oxidation: $Sn^{2+}(aq) \rightarrow Sn^{4+}(aq)$
Reduction: $I_2(aq) \rightarrow 2\,I^-(aq)$

3. Balance the charges by adding electrons.

Oxidation: $Sn^{2+}(aq) \rightarrow Sn^{4+}(aq) + 2\,e^-$
Reduction: $2\,e^- + I_2(aq) \rightarrow 2\,I^-(aq)$

4. Balance the numbers of electrons. This step is not needed here because the numbers of electrons are the same in the two half-reactions.
5. Add the two half-reactions.

Oxidation: $Sn^{2+}(aq) \rightarrow Sn^{4+}(aq) + \cancel{2\,e^-}$
Reduction: $\cancel{2\,e^-} + I_2(aq) \rightarrow 2\,I^-(aq)$

Overall equation: $Sn^{2+}(aq) + I_2(aq) \rightarrow 2\,I^-(aq) + Sn^{4+}(aq)$

Check the overall equation for mass balance: 1 Sn + 2 I = 2 I + 1 Sn. Check the charge balance: reactant side 2+, product side 2(1−) + (4+) = 2+.

Think About It We can balance redox reactions by using half-reactions. Adding the half-reactions gives us the net ionic equation and tells us the number of electrons transferred from the reducing agent to the oxidizing agent. Two moles of electrons are transferred to iodine for each mole of $Sn^{2+}(aq)$ that is oxidized. A balanced redox equation must be balanced with regard to electrons transferred as well as numbers of atoms.

(a)

(b)

(c)

FIGURE 4.28 The test tubes in these photos contain an aqueous solution of iodine. (a) When drops of a solution of $SnCl_2$ are added to the tube on the right, the yellow-brown color of the iodine solution begins to fade (b) as Sn^{2+} ions reduce I_2 to colorless I^- ions. (c) The color disappears when enough Sn^{2+} has been added to reduce all the I_2.

activity series a qualitative ordering of the oxidizing ability of metals and their cations.

> ⊕ **Practice Exercise** A nail made of Fe(s) that is placed in an aqueous solution of a soluble palladium(II) salt [containing $Pd^{2+}(aq)$] gradually disappears as the iron enters the solution as $Fe^{3+}(aq)$ and palladium metal [Pd(s)] forms. (a) Is this a redox reaction? (b) Balance the equation that describes this reaction.
>
> *(Answers to Practice Exercises are in the back of the book.)*

The Activity Series for Metals

If we replaced the $AgNO_3(aq)$ solution in Figure 4.27(a) with a $Zn(NO_3)_2$ solution, no reaction would be observed between $Zn(NO_3)_2(aq)$ and Cu(s). By testing additional solutions of different metal cations with copper wire, for example, we could establish that Ag^+ and Au^{3+} ions both oxidize copper, but Zn^{2+}, Ni^{2+}, and Al^{3+} ions do not. Similarly, we could explore the oxidation of zinc metal with these same cations. We would find that all these ions except Al^{3+} (and Zn^{2+}) will oxidize zinc. Why do metal cations oxidize some metals but not others? Although an explanation for these observations will have to wait until Chapter 18, for now we can use an **activity series** that summarizes the results of experiments to investigate whether one metal ion will oxidize a different metal. A metal cation on the list in Table 4.6 will oxidize any metal above it in the activity series. Conversely, a metal will be oxidized by any cation below it.

TABLE 4.6 An Activity Series for Metals in Aqueous Solution

Metal	Oxidation Reaction
Lithium	$Li(s) \rightarrow Li^+(aq) + e^-$
Potassium	$K(s) \rightarrow K^+(aq) + e^-$
Barium	$Ba(s) \rightarrow Ba^{2+}(aq) + 2\,e^-$
Calcium	$Ca(s) \rightarrow Ca^{2+}(aq) + 2\,e^-$
Sodium	$Na(s) \rightarrow Na^+(aq) + e^-$
Magnesium	$Mg(s) \rightarrow Mg^{2+}(aq) + 2\,e^-$
Aluminum	$Al(s) \rightarrow Al^{3+}(aq) + 3\,e^-$
Manganese	$Mn(s) \rightarrow Mn^{2+}(aq) + 2\,e^-$
Zinc	$Zn(s) \rightarrow Zn^{2+}(aq) + 2\,e^-$
Chromium	$Cr(s) \rightarrow Cr^{3+}(aq) + 3\,e^-$
Iron	$Fe(s) \rightarrow Fe^{2+}(aq) + 2\,e^-$
Cobalt	$Co(s) \rightarrow Co^{2+}(aq) + 2\,e^-$
Nickel	$Ni(s) \rightarrow Ni^{2+}(aq) + 2\,e^-$
Tin	$Sn(s) \rightarrow Sn^{2+}(aq) + 2\,e^-$
Lead	$Pb(s) \rightarrow Pb^{2+}(aq) + 2\,e^-$
Hydrogen	$H_2(g) \rightarrow 2\,H^+(aq) + 2\,e^-$
Copper	$Cu(s) \rightarrow Cu^{2+}(aq) + 2\,e^-$
Silver	$Ag(s) \rightarrow Ag^+(aq) + e^-$
Mercury	$Hg(\ell) \rightarrow Hg^{2+}(aq) + 2\,e^-$
Platinum	$Pt(s) \rightarrow Pt^{2+}(aq) + 2\,e^-$
Gold	$Au(s) \rightarrow Au^{3+}(aq) + 3\,e^-$

SAMPLE EXERCISE 4.19 Using the Activity Series   **LO7**

An aluminum wire is dipped into a solution containing sodium, tin, and copper ions. Which ions will be present in the solution after the reaction is complete?

Collect and Organize We are asked to predict whether the metal ions in solution will oxidize the aluminum wire. We use the activity series in Table 4.6 to guide our answer.

Analyze The activity series is arranged in terms of increasing oxidizing ability of cations as we proceed down Table 4.6. Alternatively, the ease of oxidation of the corresponding metal decreases down the list. Any metal in Table 4.6 is oxidized by a cation below it in the activity series.

Solve Cu^{2+} and Sn^{2+} are both listed below Al metal, so they will oxidize the aluminum wire. Sodium ion is above Al in the activity series, so it will not react. Thus both Cu^{2+} and Sn^{2+} ions are removed from the solution, but Na^+ ions remain. Because electron transfer from aluminum to both Cu^{2+} and Sn^{2+} occurs, Al^{3+} ions are also present in the solution at the end of the experiment.

Think About It The activity series allows us to predict which metals are oxidized by other members of the series. Table 4.6 does not explain *why* a particular metal is a better oxidizing agent than another.

  Practice Exercise Which of the metals listed in Table 4.6 is oxidized by half of the cations in the table?

(Answers to Practice Exercises are in the back of the book.)

Why is hydrogen gas listed in Table 4.6? What does its position in the activity series tell us? Table 4.6 allows us to predict which metals will react with strong aqueous acids like HCl. Any metal below H_2 in Table 4.6 will not react with H^+ and hence will be resistant to strong acids. One reason coins and jewelry made from silver, gold, and platinum last is that they do not react with most acids.

SAMPLE EXERCISE 4.20 Writing Net Ionic Equations   **LO7**
                          for Redox Reactions

Predict the products of the reaction between zinc metal and aqueous hydrochloric acid. Write balanced molecular and net ionic equations for the reaction. Identify the oxidation and reduction half-reactions.

Collect and Organize The reactants in this process are Zn(*s*) and HCl(*aq*). The products are determined by the activity series.

Analyze Zinc metal is listed above H^+ in Table 4.6; thus it is oxidized by H^+ from hydrochloric acid, producing Zn^{2+} and hydrogen gas. After writing a molecular equation for the reaction, we will convert it to a net ionic equation, as in Sample Exercise 4.7. After assigning oxidation numbers, we can identify the half-reactions.

Solve The unbalanced molecular equation is

$$Zn(s) + HCl(aq) \rightarrow ZnCl_2(aq) + H_2(g)$$

Adding a coefficient of 2 in front of the HCl balances all the elements:

$$Zn(s) + 2\,HCl(aq) \rightarrow ZnCl_2(aq) + H_2(g)$$

FIGURE 4.29 Red-orange rock formations called hoodoos in Bryce Canyon, Utah, owe their color to high concentrations of iron(III) oxide that result from weathering of the rock.

The overall ionic equation is:

$$Zn(s) + 2\,H^+(aq) + 2\,Cl^-(aq) \rightarrow Zn^{2+}(aq) + 2\,Cl^-(aq) + H_2(g)$$

The net ionic equation becomes:

$$Zn(s) + 2\,H^+(aq) + \cancel{2\,Cl^-(aq)} \rightarrow Zn^{2+}(aq) + \cancel{2\,Cl^-(aq)} + H_2(g)$$

$$Zn(s) + 2\,H^+(aq) \rightarrow Zn^{2+}(aq) + H_2(g)$$

The oxidation numbers of Zn, Zn^{2+}, and H^+ are 0, +2, and +1, respectively. The oxidation number of H_2 gas is zero because it is the elemental form of hydrogen. Zinc is oxidized and hydrogen ion is reduced. The two half-reactions are

Oxidation: $\qquad\qquad Zn(s) \rightarrow Zn^{2+}(aq) + 2\,e^-$

Reduction: $\quad 2\,e^- + 2\,H^+(aq) \rightarrow H_2(g)$

Think About It Including the spectator ions in a redox equation makes it more difficult to identify which atoms or ions are involved in the electron transfer. That is why most redox equations in this chapter and elsewhere in this book are written as net ionic equations.

Practice Exercise Predict the products of the reaction between aluminum metal and silver nitrate. Write balanced molecular and net ionic equations for the reaction. Identify the oxidation and reduction half-reactions.

(Answers to Practice Exercises are in the back of the book.)

Redox in Nature

Redox processes play a major role in determining the character of Earth's rocks and soils. For example, the orange-red rocks of Bryce Canyon National Park in Utah contain the iron(III) oxide mineral called hematite (Figure 4.29). Rocks containing hematite are relatively weak and easily broken, causing the fascinating shapes of such deposits. Iron(III) oxide is also the form of iron known as rust—the crumbly, red-brown solid that forms on iron objects exposed to water and oxygen.

Soil color is influenced by mineral content. Wetlands, which are areas that are either saturated or flooded with water during much of the year, are protected environments in many areas of the United States. Defining a given area as a wetland relies in part on the characteristics of the soils in the area, and soil color is an important diagnostic feature (Figure 4.30).

The redox reactions of iron(II) and iron(III) compounds can be modeled in the laboratory. Figure 4.31(a) shows a solution of iron(II) ammonium sulfate, $(NH_4)_2Fe(SO_4)_2(aq)$, upon addition of $NaOH(aq)$. The pale greenish color of the liquid is characteristic of $Fe^{2+}(aq)$. Addition of $NaOH(aq)$ causes $Fe(OH)_2(s)$ to precipitate as a blue-gray solid, similar in color to the wetland soils shown in Figure 4.30. When this mixture is filtered, the iron(II) hydroxide residue immediately starts to darken (Figure 4.31b). After about 20 minutes of exposure to oxygen in the air (Figure 4.31c), the precipitate turns the orange-red color of iron(III) hydroxide. The iron(II) has been oxidized—exactly the way the iron(II) compounds in a wetland soil are oxidized when exposed to oxygen.

Many redox reactions take place either in acidic solutions or in basic solutions, and the $H^+(aq)$, $OH^-(aq)$, or even the water may play a role in the reaction. Let's look at what we must do to balance and write net ionic equations for such reactions.

A method for determining the concentration of $Fe^{2+}(aq)$ in an acidic solution involves its oxidation to $Fe^{3+}(aq)$ by the

FIGURE 4.30 (a) The soil in this wetland has a blue-gray color because of the presence of iron(II) compounds. (b) Orange-red mottling indicates the presence of iron(III) oxides formed as a result of O_2 permeation through channels made by plant roots.

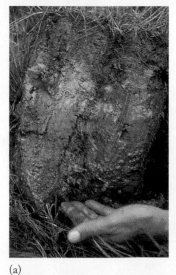

(a)   (b) Fe^{3+} Fe^{2+}

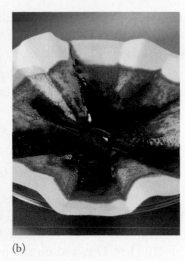

(a)   (b)   (c)

FIGURE 4.31 (a) A solution of iron(II) ammonium sulfate upon addition of NaOH(*aq*). The precipitate is iron(II) hydroxide. (b) The iron(II) hydroxide precipitate immediately after filtration. (c) The same precipitate after about 20 minutes of exposure to air. The color change indicates oxidation of iron(II) to iron(III).

intensely purple permanganate ion, $MnO_4^-(aq)$ (Figure 4.32). In the reaction, the permanganate ion is reduced to $Mn^{2+}(aq)$:

$$Fe^{2+}(aq) \quad + \quad MnO_4^-(aq) \quad \rightarrow \quad Fe^{3+}(aq) \quad + \quad Mn^{2+}(aq)$$
Appears colorless   Deep purple   Yellow-orange   Pale pink

This reaction occurs in an acidic solution, which suggests that it involves protons, and it is clearly redox because $Fe^{2+}(aq)$ is oxidized to $Fe^{3+}(aq)$. The permanganate ion must be the oxidizing agent, so it must be reduced. Indeed, the O.N. of the Mn atom in MnO_4^- is +7, and it is reduced to +2 in Mn^{2+}.

A chemical equation describing the reaction will be balanced when (a) the number of atoms of each element on both sides of the reaction is the same and (b) the total charges on each side of the reaction arrow are the same. We proceed in the same way as for the previous examples in neutral medium, but we must add a few new steps to address the role of the protons in the reaction.

1. Separate iron and manganese in their respective half-reactions.

   Oxidation: $Fe^{2+}(aq) \rightarrow Fe^{3+}(aq)$
   Reduction: $MnO_4^-(aq) \rightarrow Mn^{2+}(aq)$

2. Balance the masses in three steps to account for any role played by the aqueous acid. First we balance all the elements *except hydrogen and oxygen*:

   a. Oxidation: $Fe^{2+}(aq) \rightarrow Fe^{3+}(aq)$ (mass balanced)
      Reduction: $MnO_4^-(aq) \rightarrow Mn^{2+}(aq)$ (O mass not balanced)

   Next we balance oxygen by adding water as a product.

   b. Oxidation: $Fe^{2+}(aq) \rightarrow Fe^{3+}(aq)$ (mass balanced)
      Reduction: $MnO_4^-(aq) \rightarrow Mn^{2+}(aq) + 4\,H_2O(\ell)$
                                              (O mass balanced, H mass not)

   In acidic solutions, the only source of H to make $H_2O(\ell)$ is H^+ from a strong acid, so we balance the hydrogen by adding $H^+(aq)$:

   c. Oxidation: $Fe^{2+}(aq) \rightarrow Fe^{3+}(aq)$ (mass balanced)
      Reduction: $8\,H^+(aq) + MnO_4^-(aq) \rightarrow Mn^{2+}(aq) + 4\,H_2O(\ell)$
                                              (mass balanced)

3. Balance charge by adding electrons.

   Oxidation: $Fe^{2+}(aq) \rightarrow Fe^{3+}(aq) + 1\,e^-$
   Reduction: $5\,e^- + 8\,H^+(aq) + MnO_4^-(aq) \rightarrow Mn^{2+}(aq) + 4\,H_2O(\ell)$

FIGURE 4.32 The test tubes on the left and right initially contain the same solution of Fe^{2+} ions, which appear colorless. When drops of the purple solution of $KMnO_4$ from the middle test tube are added to the solution on the right, the mixture turns yellow-orange, which is the color of Fe^{3+} ions in solution. The purple color disappears as MnO_4^- ions are reduced to Mn^{2+} ions.

Note that the number of electrons added to the reduction half-reaction also accounts for the oxidation state of Mn changing from +7 to +2.

4. Balance numbers of electrons lost and gained.

Oxidation: $\qquad 5 \times [Fe^{2+}(aq) \rightarrow Fe^{3+}(aq) + 1\ e^-]$

Reduction: $\quad 1 \times [5\ e^- + 8\ H^+(aq) + MnO_4^-(aq) \rightarrow Mn^{2+}(aq) + 4\ H_2O(\ell)]$

5. Add the two equations.

Oxidation: $\qquad 5\ Fe^{2+}(aq) \rightarrow 5\ Fe^{3+}(aq) + \cancel{5\ e^-}$

Reduction: $\quad \cancel{5\ e^-} + 8\ H^+(aq) + MnO_4^-(aq) \rightarrow Mn^{2+}(aq) + 4\ H_2O(\ell)]$

$$8\ H^+(aq) + MnO_4^-(aq) + 5\ Fe^{2+}(aq) \rightarrow 5\ Fe^{3+}(aq) + Mn^{2+}(aq) + 4\ H_2O(\ell)$$

Always check the mass and charge balance of the final net ionic equation. The reactant side contains 8 H, 1 Mn, 4 O, and 5 Fe atoms and the product side contains 8 H, 1 Mn, 4 O, and 5 Fe atoms, so mass is balanced. On the reactant side, the charge is $(8+) + (1-) + (10+) = 17+$, and on the product side it is $(15+) + (2+) = 17+$, so charge is balanced.

This method works for writing the net ionic equation for any redox reaction taking place in an aqueous acid. The oxidation in Figure 4.31, however, was carried out in a basic medium, so now let's look at how to balance the equation for that reaction. We balance these equations in almost the same way we do for acidic media: in fact, we carry out exactly the same steps outlined in the example we just worked, as though the reaction were run in acidic medium, except we add a final step in which sufficient hydroxide ion is added to both sides of the equation to convert any $H^+(aq)$ into $H_2O(\ell)$. This step converts the medium into an aqueous basic solution, as specified by the conditions.

SAMPLE EXERCISE 4.21 Balancing Redox Equations **LO7**
That Involve Hydroxide Ions

Wetland soil is blue-gray due to $Fe(OH)_2(s)$, whereas well-aerated soils are often orange-red due to the presence of $Fe(OH)_3(s)$ (Figure 4.30). Write the balanced equation for the reaction of $O_2(g)$ with $Fe(OH)_2(s)$ in soil that produces $Fe(OH)_3(s)$ in basic solution.

Collect and Organize We know the reactants and product and can write the unbalanced equation:

$$Fe(OH)_2(s) + O_2(g) \rightarrow Fe(OH)_3(s)$$

Our task is to balance the equation with respect to both mass and charge.

Analyze This reaction takes place with both iron compounds in the solid phase in the presence of hydroxide ions in water. The oxidation half-reaction is clear: iron(II) hydroxide is oxidized to iron(III) hydroxide. For the reduction half-reaction, O_2 must be converted into the additional OH^- ion in the iron(III) hydroxide product.

Solve

1. Separate the equations.

Oxidation: $\quad Fe(OH)_2(s) \rightarrow Fe(OH)_3(s)$

Reduction: $\qquad O_2(g) \rightarrow OH^-(aq)$

a. Balance all masses except H and O. This step is not needed because the only mass not attributed to H and O is Fe, which is balanced.

b. Balance O by adding water as needed.

Oxidation: $H_2O(\ell) + Fe(OH)_2(s) \rightarrow Fe(OH)_3(s)$
Reduction: $O_2(g) \rightarrow OH^-(aq) + H_2O(\ell)$

c. Balance H by adding $H^+(aq)$ as needed.

$$H_2O(\ell) + Fe(OH)_2(s) \rightarrow Fe(OH)_3(s) + H^+(aq)$$
$$3\,H^+(aq) + O_2(g) \rightarrow OH^-(aq) + H_2O(\ell)$$

3. Balance the charges.

$$H_2O(\ell) + Fe(OH)_2(s) \rightarrow Fe(OH)_3(s) + H^+(aq) + 1\,e^-$$
$$4\,e^- + 3\,H^+(aq) + O_2(g) \rightarrow OH^-(aq) + H_2O(\ell)$$

4. Balance the numbers of electrons lost and gained.

$$4 \times [H_2O(\ell) + Fe(OH)_2(s) \rightarrow Fe(OH)_3(s) + H^+(aq) + 1\,e^-]$$
$$1 \times [4\,e^- + 3\,H^+(aq) + O_2(g) \rightarrow OH^-(aq) + H_2O(\ell)]$$

5. Add the two equations.

$$3\;4\,H_2O(\ell) + 4\,Fe(OH)_2(s) \rightarrow 4\,Fe(OH)_3(s) + 4\,H^+(aq) + 4\,e^-$$
$$4\,e^- + 3\,H^+(aq) + O_2(g) \rightarrow OH^-(aq) + H_2O(\ell)$$

This gives us the balanced equation:

$$3\,H_2O(\ell) + 4\,Fe(OH)_2(s) + O_2(g) \rightarrow 4\,Fe(OH)_3(s) + OH^-(aq) + H^+(aq)$$

which can be simplified:

$$3\,H_2O(\ell) + 4\,Fe(OH)_2(s) + O_2(g) \rightarrow 4\,Fe(OH)_3(s) + \underbrace{OH^-(aq) + H^+(aq)}_{H_2O(\ell)}$$

to

$$2\,H_2O(\ell) + 4\,Fe(OH)_2(s) + O_2(g) \rightarrow 4\,Fe(OH)_3(s)$$

Think About It This reaction occurs in neutral and basic soils as H_2O and O_2 combine to form the OH^- ions needed in the conversion of $Fe(OH)_2(s)$ to $Fe(OH)_3(s)$.

Practice Exercise The hydroperoxide ion, $HO_2^-(aq)$, reacts with permanganate ion, $MnO_4^-(aq)$, to produce $MnO_2(s)$ and oxygen gas. Balance the equation for the oxidation of hydroperoxide ion to $O_2(g)$ by permanganate ion in a basic solution.

(Answers to Practice Exercises are in the back of the book.)

Reactions in aqueous solutions are an integral part of our daily lives. On a large scale, in oceans, rivers, and rain, they shape our physical world. Many reactions that produce the substances that are part of modern life—from paint pigments to drugs—are run in water, and many analytical procedures rely on reactions in water to determine the content of aqueous solutions that we drink, swim in, and use in countless consumer products like car batteries and shampoos. On a small scale, reactions in the water within the cells of our bodies and in all living organisms make the chemical processes that are essential to life possible. On Earth, there may be water without life, but there is no life without water.

SAMPLE EXERCISE 4.22 Integrating Concepts: Shelf-Stability of Drugs

Commercial pharmaceutical agents undergo extensive analysis to establish how long they may be stored without degradation and loss of potency. A candidate drug has the following percent composition from combustion analysis: C, 62.50%; H, 4.20%. The drug is a diprotic acid. A standard tablet containing 325 mg of the drug is dissolved in 100.00 mL of water and titrated to the equivalence point with 16.45 mL of 0.2056 M NaOH(aq). After storage at 50°C under high humidity for one month, a second tablet, when dissolved in 50.00 mL of water, requires 10.10 mL of 0.1755 M NaOH(aq) to be completely neutralized.

a. What is the empirical formula of the drug?
b. What is its molar mass?
c. What is its molecular formula?
d. What is the percent of active drug substance remaining in the tablet after storage?

Collect and Organize We are given the percent composition of the drug and information from titration experiments. From these data we can find the empirical formula, the molar mass, and the amount of active drug in the stored tablet.

Analyze The percent composition data from the combustion analysis do not add to 100%. We may assume that the amount missing is due to oxygen in the molecule. The drug is a diprotic acid, so neutralizing 1 mole of the drug requires 2 moles of NaOH. The titration of the drug requires about 20 mL of 0.2 M NaOH, which is about 0.004 mol OH$^-$. That means the pure sample is about 0.002 moles of the drug; if about 0.3 grams contains 0.002 moles, the drug must have a molar mass of about 150 g. If any of the drug decomposes upon storage, we should have less than 325 mg in the second sample.

Solve

a. The percent oxygen in the drug molecule is

$$100.00\% - (62.50\% + 4.20\%) = 33.30\%$$

Calculating the moles of C, H, and O in exactly 100 g of the drug:

$$\frac{62.50 \text{ g C}}{12.01 \text{ g/mol}} = 5.20 \text{ mol C}$$

$$\frac{4.20 \text{ g H}}{1.008 \text{ g/mol}} = 4.17 \text{ mol H}$$

$$\frac{33.30 \text{ g O}}{16.00 \text{ g/mol}} = 2.08 \text{ mol O}$$

Reducing these quantities to ratios of small whole numbers:

$$\frac{5.20 \text{ mol C}}{2.08} = 2.50 \text{ mol C}$$

$$\frac{4.17 \text{ mol H}}{2.08} = 2.00 \text{ mol H}$$

$$\frac{2.08 \text{ mol O}}{2.08} = 1.00 \text{ mol O}$$

requires multiplying by 2, which gives us

$$5 \text{ mol C:4 mol H:2 mol O}$$

and the empirical formula: $C_5H_4O_2$.

b. To find the molar mass of the drug from the titration data, we first find the number of moles of drug in the sample:

$$0.01645 \text{ L NaOH} \times \frac{0.2056 \text{ mol NaOH}}{1 \text{ L NaOH}} \times \frac{1 \text{ mol drug}}{2 \text{ mol NaOH}}$$

$$= 0.001691 \text{ mol drug}$$

If 0.325 g of drug is 0.001691 mol of drug, then the mass of 1 mole is

$$\frac{0.325 \text{ g}}{0.001691 \text{ mol}} = 192.2 \text{ g/mol} = \mathcal{M}_{\text{drug}}$$

c. The mass of 1 mole of empirical formula units ($C_5H_4O_2$) is

$$(5 \times 12.01 \text{ g/mol}) + (4 \times 1.008 \text{ g/mol}) + (2 \times 16.00 \text{ g/mol})$$
$$= 96.08 \text{ g/mol}$$

The number of moles of formula units per mole of molecules is

$$\frac{192.16 \frac{\text{g}}{\text{mol}}}{96.08 \frac{\text{g}}{\text{mol}}} = 2$$

So the molecular formula of the drug is

$$(C_5H_4O_2)_2 = C_{10}H_8O_4$$

d. The stored tablet contains

$$0.01010 \text{ L NaOH} \times \frac{0.1755 \text{ mol NaOH}}{1 \text{ L NaOH}} \times \frac{1 \text{ mol drug}}{2 \text{ mol NaOH}}$$

$$= 0.0008863 \text{ mol drug} = 8.863 \times 10^{-4} \text{ mol}$$

Comparing this value with the original amount:

$$\frac{8.863 \times 10^{-4} \text{ mol}}{1.691 \times 10^{-3} \text{ mol}} \times 100\% = 52.41\%$$

of the active drug is left in the tablet.

Think About It The molar mass we calculated is close to our estimated value, so it seems reasonable. If only half the active agent is present in the tablet after storage under these conditions, the manufacturer may have to reformulate the drug or package it differently to protect it from its surroundings.

SUMMARY

LO1 The **concentration** of **solute** in a solution can be expressed many different ways: as mass of solute per mass of solution (such as grams of solute per kilogram of solution), parts per million (1 ppm = 1 μg solute/g solution = 1 mg solute/kg solution), or parts per billion (1 ppb = 1 μg solute/kg solution). Solute concentration can also be expressed as mass of solute per volume of **solvent** and as moles of solute per liter of solution, or **molarity** (*M*). One set of units can be converted into another by using dimensional analysis. (Sections 4.1 and 4.2)

LO2 Two common techniques to make solutions of desired concentration are determining the mass of solute to be dissolved in an appropriate amount of solvent to produce the quantity of solution needed and the **dilution** of a **stock solution**. (Sections 4.2 and 4.3)

LO3 We can determine the concentration of a solution by spectrophotometry and the application of **Beer's law**, which relates the **absorbance** of a solution (A) to concentration (c), path length (b) and **molar absorptivity (ε)** by the equation $A = \varepsilon b c$. We can also use **titrations** based on acid–base **neutralizations** or **precipitation reactions**. (Sections 4.3, 4.5, 4.6, and 4.7)

LO4 There are three different ways to write chemical equations. **Molecular equations** may be the easiest to balance; **overall ionic equations** show all the species present in solution; **net ionic equations** eliminate **spectator ions** and show only the species that are involved in the reaction. (Sections 4.5, 4.6, and 4.7)

LO5 Measuring the conductivity of a solution enables us to identify solutes as **strong electrolytes**, **weak electrolytes**, or **nonelectrolytes**. In aqueous solution, **Brønsted–Lowry acids** are proton donors and **Brønsted–Lowry bases** are proton acceptors; they may also be strong or weak electrolytes. (Sections 4.4 and 4.5)

LO6 **Solubility** rules define substances that dissolve readily in water and those that have very limited solubility. We can use them to determine combinations of ions in solution that result in **precipitate** formation, and we can quantify the amount of precipitate formed and the concentration of ions left in solution after reaction. We can remove ions from solution by precipitation and by **ion exchange**. (Sections 4.7 and 4.8)

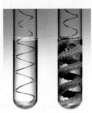

LO7 In a redox reaction, substances either gain electrons (and thereby undergo **reduction**) or lose electrons (undergo **oxidation**). A reaction is a redox reaction if the **oxidation numbers (O.N.)**, or **oxidation states**, of the atoms in the reactants change during the reaction. An **activity series** allows us to predict whether a particular metal ion will oxidize a different metal. In balancing equations for redox reactions, we must consider the number of electrons transferred as well as the number of atoms involved. (Section 4.9)

PARTICULATE **PREVIEW WRAP-UP**

The balanced equation when the contents of the two beakers are mixed is

$$NH_4Cl(aq) + AgNO_3(aq) \rightarrow NH_4NO_3(aq) + AgCl(s)$$

Therefore ammonium nitrate remains as dissociated ions dissolved in water, while silver chloride precipitates as a solid. $AgNO_3(aq)$ is the limiting reagent, and all the nitrate ions (0.01 moles) remain in solution because they are spectator ions. The starting materials are drawn showing four cations and four anions for each reactant, so they form the products drawn here.

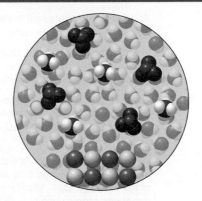

PROBLEM-SOLVING SUMMARY

Type of Problem	Concepts and Equations	Sample Exercises
Comparing concentrations in aqueous solutions	Use conversion factors to express concentrations in different units.	**4.1**
Calculating molarity from solute mass and solution volume or from solute mass, solution mass, and density	Convert the solute mass into grams and then into moles. Convert the solution mass into volume by dividing by the density of the solvent. Divide moles of solute by liters of solution: $$\text{Molarity} = \frac{\text{moles of solute}}{\text{liter of solution}}$$	**4.2, 4.3**

Type of Problem	Concepts and Equations	Sample Exercises
Calculating the mass of solute or volume of stock solution to prepare a solution	a. Multiply the known concentration (in mol/L) by the target volume (in L) to obtain the moles of solute needed. Then multiply moles of solute by solute molar mass, $\mathcal{M}$, to get mass of solute needed: $$\text{Mass of solute (g)} = (V \times M) \times \mathcal{M} \qquad (4.2)$$ b. Given three of the four variables, use $$V_{initial} \times M_{initial} = V_{diluted} \times M_{diluted} \qquad (4.3)$$ to solve for the fourth.	**4.4, 4.5**
Applying Beer's law	Substitute for A, b, and c in Equation 4.4 and solve for ε for a solution of known concentration: $$\varepsilon = \frac{A}{bc}$$ Use A, b, and ε to find c for a solution of unknown concentration: $$c = \frac{A}{\varepsilon b}$$	**4.6**
Writing neutralization reaction equations	Balance the molecular equation by balancing the moles of H^+ ions donated by the acid and accepted by the base. Next, create the overall ionic equation by writing strong electrolytes in their ionic form. Finally, create the net ionic equation by eliminating spectator ions from the overall ionic equation.	**4.7, 4.11**
Comparing electrolytes, acids, and bases	Use the definitions of a strong electrolyte, a weak electrolyte, a nonelectrolyte, an acid, and a base to classify compounds into one or more categories.	**4.8**
Calculating molarity from titration data	Use the volume and concentration of titrant used to neutralize a sample along with a balanced chemical equation to calculate the number of moles in a sample of known volume and determine its molarity.	**4.9, 4.10**
Predicting precipitation reactions	Write all the ions present in the solutions being mixed. If any cation/anion pair forms an insoluble compound, that compound will precipitate.	**4.12**
Calculating the mass of a precipitate	Find the limiting reactant. Use the stoichiometry of the net ionic equation to calculate the moles of precipitate, and then convert moles into mass of precipitate by using the precipitate's molar mass.	**4.13, 4.15**
Calculating a solute concentration from a precipitate mass	Convert precipitate mass into moles by dividing by its molar mass. Convert moles of precipitate into moles of solute. Calculate molarity of the solute in the sample by dividing the moles of solute by the volume of sample in liters.	**4.14**
Determining oxidation numbers (O.N.)	O.N. of a monatomic ion is equal to the ion's charge. O.N. of a pure element is 0. To assign O.N. in a molecule containing more than one type of atom, assign O.N. +1 to H, −2 to O, and then calculate O.N. for any remaining atoms such that all the O.N. sum to 0. For polyatomic ions, the sum of the O.N. values must equal the charge on the ion.	**4.16**
Identifying oxidizing and reducing agents and number of electrons transferred	The oxidizing agent contains an atom whose O.N. decreases during the reaction; the reducing agent contains an atom whose O.N. increases; the change in O.N. determines the number of electrons transferred.	**4.17**
Balancing redox reactions with half-reactions	Multiply one or both half-reactions by the appropriate coefficient(s) to balance the loss and gain of electrons. Combine the two half-reactions and simplify.	**4.18**
Using the activity series	Any metal in Table 4.6 will be oxidized by a cation listed below it in the activity series.	**4.19**
Balancing redox reactions that involve acidic or basic conditions	Follow the steps described on pp. 190–193.	**4.20, 4.21**

VISUAL PROBLEMS

(Answers to boldface end-of-chapter questions and problems are in the back of the book.)

4.1. In Figure P4.1, which shows a solution containing three binary acids, one of the three is a weak acid and the other two are strong acids. Which color sphere represents the anion of the dissociated weak acid?

FIGURE P4.1

4.2. Solutions of sodium chloride and silver iodide are mixed together and vigorously shaken. Which colored spheres in Figure P4.2 represent the following ions? (a) Na^+; (b) Cl^-; (c) I^-

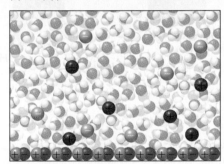

FIGURE P4.2

4.3. Which of the highlighted elements in Figure P4.3 forms an acid with the following generic formula? (a) HX; (b) H_2XO_4; (c) HXO_3; (d) H_3XO_4

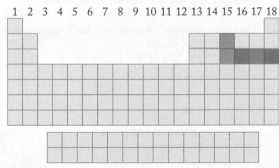

1 2 3 4 5 6 7 8 9 10 11 12 13 14 15 16 17 18

FIGURE P4.3

4.4. In which of the highlighted groups of elements in Figure P4.4 will you find an element that forms the following? (a) insoluble halides; (b) insoluble hydroxides; (c) hydroxides that are soluble; (d) binary compounds with hydrogen that are strong acids

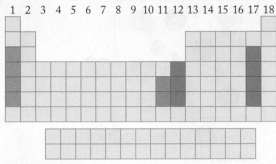

1 2 3 4 5 6 7 8 9 10 11 12 13 14 15 16 17 18

FIGURE P4.4

4.5. Which of the drawings in Figure P4.5 depicts a strong electrolyte? A weak electrolyte? A strong acid? A weak acid? A nonelectrolyte? Each drawing may fit more than one category.

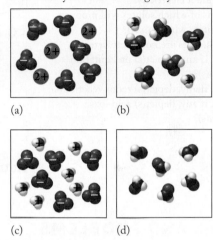

(a)   (b)

(c)   (d)

FIGURE P4.5

4.6. Which of the three half-reactions shown in Figure P4.6 depicts an oxidation? Which depicts a reduction?

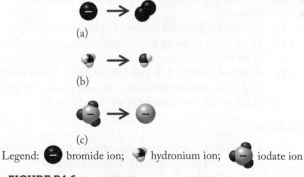

(a)

(b)

(c)

Legend: ⚫ bromide ion; ⚫ hydronium ion; ⚫ iodate ion

FIGURE P4.6

4.7. Which ions in Figure P4.7 will remain in solution?

Legend: ⚫− bromide ions; ②⁺ lead(II) ions

FIGURE P4.7

4.8. Use representations [A] through [I] in Figure P4.8 to answer questions a–f.
   a. Which solutes form aqueous solutions that conduct electricity?
   b. Solutions of which solutes will produce a precipitate when mixed?
   c. Which solutes are nonelectrolytes?
   d. Which, if any, depict(s) precipitation reaction(s)?
   e. Which, if any, depict(s) redox reaction(s)?
   f. Which, if any, depict(s) acid–base reaction(s)?

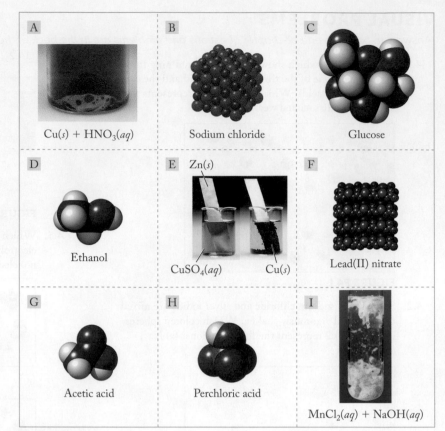

A $Cu(s) + HNO_3(aq)$

B Sodium chloride

C Glucose

D Ethanol

E $Zn(s)$ $CuSO_4(aq)$ $Cu(s)$

F Lead(II) nitrate

G Acetic acid

H Perchloric acid

I $MnCl_2(aq) + NaOH(aq)$

FIGURE P4.8

QUESTIONS AND PROBLEMS

Quantifying Particles in Solution

Concept Review

4.9. How do you decide which component in a solution is the solvent?

4.10. Can a solid ever be a solvent? Explain.

4.11. What is the molarity of a solution that contains 1.00 mmol of solute per milliliter of solution?

*4.12. A beaker contains 100 g of 1.00 M NaCl. If you transfer 50 g of the solution to another beaker, what is the molarity of the solution remaining in the first beaker?

Problems

4.13. Calculate the molarity of each of the following solutions:
   a. 0.56 mol of $BaCl_2$ in 100.0 mL of solution
   b. 0.200 mol of Na_2CO_3 in 200.0 mL of solution
   c. 0.325 mol of $C_6H_{12}O_6$ in 250.0 mL of solution
   d. 1.48 mol of KNO_3 in 250.0 mL of solution

4.14. Calculate the molarity of each of the following solutions:
   a. 0.150 mol of urea (CH_4N_2O) in 250.0 mL of solution
   b. 1.46 mol of $NaC_2H_3O_2$ in 1.000 L of solution
   c. 1.94 mol of methanol (CH_3OH) in 5.000 L of solution
   d. 0.045 mol of sucrose ($C_{12}H_{22}O_{11}$) in 50.0 mL of solution

4.15. Calculate the molarity of each of the following ions:
   a. 0.33 g Na^+ in 100.0 mL of solution
   b. 0.38 g Cl^- in 100.0 mL of solution
   c. 0.46 g SO_4^{2-} in 50.0 mL of solution
   d. 0.40 g Ca^{2+} in 50.0 mL of solution

4.16. Calculate the molarity of each of the following solutions:
   a. 64.7 g LiCl in 250.0 mL of solution
   b. 29.3 g $NiSO_4$ in 200.0 mL of solution
   c. 50.0 g KCN in 500.0 mL of solution
   d. 0.155 g $AgNO_3$ in 100.0 mL of solution

4.17. How many grams of solute are needed to prepare each of the following solutions?
   a. 1.000 L of 0.200 M NaCl
   b. 250.0 mL of 0.125 M $CuSO_4$
   c. 500.0 mL of 0.400 M CH_3OH

4.18. How many grams of solute are needed to prepare each of the following solutions?
   a. 500.0 mL of 0.250 M KBr
   b. 25.0 mL of 0.200 M $NaNO_3$
   c. 100.0 mL of 0.375 M CH_3OH

4.19. River Water The Mackenzie River in northern Canada contains, on average, 0.820 mM Ca^{2+}, 0.430 mM Mg^{2+}, 0.300 mM Na^+, 0.0200 M K^+, 0.250 mM Cl^-, 0.380 mM SO_4^{2-}, and 1.82 mM HCO_3^-. What, on average, is the total mass of these ions in 2.75 L of Mackenzie River water?

4.20. **Toxicity of Metal Ions** Zinc, copper, lead, and mercury ions are toxic to Atlantic salmon at concentrations of 6.42×10^{-2} mM, 7.16×10^{-3} mM, 0.965 mM, and

5.00×10^{-2} mM, respectively. What are the corresponding concentrations in milligrams per liter?

4.21. Calculate the number of moles of solute contained in the following volumes of aqueous solutions of four pesticides:
   a. 0.400 L of 0.024 M lindane
   b. 1.65 L of 0.473 mM dieldrin
   c. 25.8 L of 3.4 mM DDT
   d. 154 L of 27.4 mM aldrin

4.22. **Hemoglobin in Blood** A typical adult body contains 6.0 L of blood. The hemoglobin content of blood is about 15.5 g/100.0 mL of blood. The approximate molar mass of hemoglobin is 64,500 g/mol. How many moles of hemoglobin are present in a typical adult?

4.23. **DDT Affects Neurons** The pesticide DDT ($C_{14}H_9Cl_5$) kills insects such as malaria-carrying mosquitoes by opening sodium ion channels in neurons, causing them to fire spontaneously, which leads to spasms and eventual death. However, its toxicity in wildlife and humans led to the banning of its use in the United States in 1972. Analysis of DDT concentrations in groundwater samples between 1969 and 1971 in Pennsylvania yielded the following results:

Location	Sample Size	Mass of DDT
Orchard	250.0 mL	0.030 mg
Residential	1.750 L	0.035 mg
Residential after a storm	50.0 mL	0.57 mg

Express these concentrations in ppm and in millimoles per liter.

4.24. **Pesticides in the Environment** Pesticide concentrations in the Rhine River between Germany and France between 1969 and 1975 averaged 0.55 mg/L of hexachlorobenzene (C_6Cl_6), 0.06 mg/L of dieldrin ($C_{12}H_8Cl_6O$), and 1.02 mg/L of hexachlorocyclohexane ($C_6H_6Cl_6$). Express these concentrations in ppb and in millimoles per liter.

4.25. Nitrogen trifluoride, NF_3, is used in the production of flat panel displays. It is also a potent greenhouse gas. The average concentration of NF_3 in the atmosphere increased from 0.02 parts per trillion (ppt) in 1978 to 0.454 ppt in 2008. What is the concentration of NF_3 in mg per kg of air?

*4.26. **Gases Found in Air** Sulfur hexafluoride, SF_6, is used in electrical transformers. Like NF_3, it has a potential impact on climate. Between 1978 and 2012, the concentration of SF_6 increased from 0.51 parts per trillion (ppt) to 7.48 ppt. How many *more* molecules of SF_6 were found in one liter of air in 2012 than in 1978? (1 mole of gas = 22.4 L of gas.)

*4.27. The concentration of copper(II) sulfate in one brand of soluble plant fertilizer is 0.07% by mass. If a 20 g sample of this fertilizer is dissolved in 2.0 L of solution, what is the molarity of Cu^{2+}?

*4.28. For which of the following compounds is it possible to make a 1.0 M solution at 20°C?
   a. $CuSO_4$, solubility = 32.0 g/100 mL
   b. $Ba(OH)_2$, solubility = 3.9 g/100 mL
   c. $FeCl_2$, solubility = 68.5 g/100 mL
   d. $Ca(OH)_2$, solubility = 0.173 g/100 mL

Dilutions

Problems

4.29. Calculate the final concentrations of the following aqueous solutions after each has been diluted to a final volume of 25.0 mL:
   a. 3.00 mL of 0.175 M K$^+$
   b. 2.50 mL of 10.6 mM LiCl
   c. 15.00 mL of 7.24×10^{-2} mM Zn^{2+}

4.30. **Dilution of Adult-Strength Cough Syrup** A standard dose of an over-the-counter cough suppressant for adults is 20.0 mL. A portion this size contains 35 mg of the active pharmaceutical ingredient (API). Your pediatrician says you may give this medication to your 6-year-old child, but the child may take only 10.0 mL at a time and receive a maximum of 4.00 mg of the API. What is the concentration in mg/mL of the adult-strength medication, and how many millimeters of it would you need to dilute to make 100.0 mL of child-strength cough syrup?

4.31. The concentration of Na$^+$ in seawater, 0.481 M, is higher than in the cytosol, the fluid inside human cells (12 mM). How much water must be added to 1.50 mL of seawater to make the Na$^+$ concentration equal to that found in the cytosol? Assume the volumes are additive.

4.32. The concentration of chloride ion in blood, 116 mM, is less than that in the ocean, 0.559 M. Describe how you would prepare 2.50 mL of a solution of 116 mM chloride ion from seawater.

4.33. Water is allowed to evaporate from 100.0 mL of 0.24 M Na_2SO_4 until the solution volume is 60.0 mL. What is the molar concentration of the evaporated solution?

*4.34. **Mixing Fertilizer** The label on a bottle of "organic" liquid fertilizer concentrate states that it contains 8 g of phosphate per 100.0 mL and that 16 fluid ounces should be diluted with water to make 32 gallons of fertilizer to be applied to growing plants. What is the phosphate concentration in grams per liter in the diluted fertilizer? (1 gallon = 128 fluid ounces.)

4.35. If the absorbance of a solution of copper ion decreases by 45% upon dilution, how much water was added to 15.0 mL of a 1.00 M solution of Cu^{2+}?

4.36. By what percentage does the absorbance decrease if 12.25 mL of water is added to a 16.75 mL sample of 0.500 M Cr^{3+}?

*4.37. The reaction of $SnCl_2(aq)$ with $Pt^{4+}(aq)$ in aqueous HCl yields a yellow-orange solution of a 1:1 Pt–Sn compound with a molar absorptivity (ε) of 1.3×10^4 M^{-1} cm^{-1}. What is the absorbance in a cell with a path length of 1.00 cm of a solution prepared by adding 100 mL of an aqueous solution of 5.2 mg $(NH_4)_2PtCl_6$ to 100 mL of an aqueous solution of 2.2 mg $SnCl_2$?

*4.38. The reaction of $SnCl_2(aq)$ with $RhCl_3(aq)$ in aqueous HCl yields a red solution of a 1:1 Rh–Sn compound. If a solution prepared by adding 150 mL of a 0.272 mM aqueous solution of $SnCl_2$ to 50 mL of an aqueous solution of 8.5 mg $RhCl_3$ has an absorbance of 0.85, as measured in a 1.00 cm cell, what is the molar absorptivity of the red compound?

Electrolytes and Nonelectrolytes

Concept Review

4.39. A solution of table salt is a good conductor of electricity, but a solution containing an equal molar concentration of table sugar is not. Why?

4.40. **Corrosion at Sea** Metallic fixtures on the bottom of a ship corrode more quickly in seawater than in freshwater. Why?

4.41. Explain why liquid methanol, CH_3OH, cannot conduct electricity, whereas molten NaOH can.

4.42. **Fuel Cells** The electrolyte in an electricity-generating device called a *fuel cell* consists of a mixture of Li_2CO_3 and K_2CO_3 heated to 650°C. At this temperature the ionic solids melt. Explain how this mixture of molten carbonates can conduct electricity.

4.43. Rank the following solutions on the basis of their ability to conduct electricity, starting with the most conductive: (a) 1.0 M NaCl; (b) 1.2 M KCl; (c) 1.0 M Na$_2$SO$_4$; (d) 0.75 M LiCl.

4.44. Rank the conductivities of 1 M aqueous solutions of each of the following solutes, starting with the most conductive: (a) acetic acid; (b) methanol; (c) sucrose (table sugar); (d) hydrochloric acid.

Problems

4.45. Calculate the molarity of Na^+ ions in a 0.025 M aqueous solution of: (a) NaBr; (b) Na$_2$SO$_4$; (c) Na$_3$PO$_4$.

4.46. Calculate the molarity of each ion in a 0.025 M aqueous solution of: (a) KCl; (b) CuSO$_4$; (c) CaCl$_2$.

4.47. Which of the following solutions has the greatest number of particles (atoms or ions) of solute per liter? (a) 1 M NaCl; (b) 1 M CaCl$_2$; (c) 1 M ethanol; (d) 1 M acetic acid

4.48. Which of the following solutions contains the most solute particles per liter? (a) 1 M KBr; (b) 1 M Mg(NO$_3$)$_2$; (c) 4 M ethanol; (d) 4 M acetic acid

Acid–Base Reactions: Proton Transfer

Concept Review

4.49. What name is given to a proton donor?

4.50. What is the difference between a strong acid and a weak acid?

4.51. Identify each compound as either a weak acid or a strong acid in aqueous solution: (a) HNO$_3$; (b) HNO$_2$; (c) CH$_3$CH$_2$CH$_2$COOH; (d) H$_2$SO$_4$.

4.52. Why is $HSO_4^-(aq)$ a weaker acid than $H_2SO_4(aq)$?

4.53. What name is given to a proton acceptor?

4.54. What is the difference between a strong base and a weak base?

4.55. Identify each compound as either a weak base or a strong base in aqueous solution: (a) Ca(OH)$_2$; (b) NH$_3$; (c) CH$_3$CH$_2$NH$_2$; (d) NaOH.

4.56. Write the net ionic equation for the neutralization of a strong acid by a strong base.

Problems

4.57. For each of the following acid–base reactions, identify the acid and the base, and then write the overall ionic and net ionic equations.
  a. $H_2SO_4(aq) + Ca(OH)_2(aq) \rightarrow CaSO_4(s) + 2\ H_2O(\ell)$
  b. $PbCO_3(s) + H_2SO_4(aq) \rightarrow$
     $PbSO_4(s) + CO_2(g) + H_2O(\ell)$
  c. $Ca(OH)_2(s) + 2\ CH_3COOH(aq) \rightarrow$
     $Ca(CH_3COO)_2(aq) + 2\ H_2O(aq)$

4.58. Complete and balance each of the following neutralization reactions, name the products, and write the overall ionic and net ionic equations.
  a. $HBr(aq) + KOH(aq) \rightarrow$
  b. $H_3PO_4(aq) + Ba(OH)_2(aq) \rightarrow$
  c. $Al(OH)_3(s) + HCl(aq) \rightarrow$
  d. $CH_3COOH(aq) + Sr(OH)_2(aq) \rightarrow$

4.59. Write a balanced molecular equation and a net ionic equation for the following reactions:
  a. Solid magnesium hydroxide reacts with a solution of sulfuric acid.
  b. Solid magnesium carbonate reacts with a solution of hydrochloric acid.
  c. Ammonia gas reacts with hydrogen chloride gas.
  d. Gaseous sulfur trioxide is dissolved in water and reacts with a solution of sodium hydroxide.

4.60. Write a balanced molecular equation and a net ionic equation for the following reactions:
  a. Solid aluminum hydroxide reacts with a solution of hydrobromic acid.
  b. A solution of sulfuric acid reacts with solid sodium carbonate.
  c. A solution of calcium hydroxide reacts with a solution of nitric acid.
  d. Solid potassium oxide is dissolved in water and reacts with a solution of sulfuric acid.

4.61. **Toxicity of Lead Pigments** The use of lead(II) carbonate and lead(II) hydroxide as white pigments in paint was discontinued because children have been known to eat paint chips. The pigments dissolve in stomach acid, and lead ions enter the nervous system and interfere with neurotransmissions in the brain, causing neurological disorders. Using net ionic equations, show why lead(II) carbonate and lead(II) hydroxide dissolve in acidic solutions.

4.62. **Lawn Care** Many homeowners treat their lawns with $CaCO_3(s)$ to reduce the acidity of the soil. Write a net ionic equation for the reaction of $CaCO_3(s)$ with a strong acid.

Titrations

Problems

4.63. How many milliliters of 0.250 *M* NaOH are required to neutralize the following solutions?
a. 60.0 mL of 0.0750 *M* HCl
b. 35.0 mL of 0.226 *M* HNO₃
c. 75.0 mL of 0.190 *M* H₂SO₄

4.64. How many milliliters of 0.250 *M* HNO₃ are needed to neutralize the following solutions?
a. 25.0 mL of 0.395 *M* KOH
b. 78.6 mL of 0.0100 *M* Al(OH)₃
c. 65.9 mL of 0.475 *M* NaOH

*4.65. The solubility of slaked lime, Ca(OH)₂, in water at 20°C is 0.185 g/100.0 mL. What volume of 0.00100 *M* HCl is needed to neutralize 10.0 mL of a saturated Ca(OH)₂ solution?

*4.66. The solubility of magnesium hydroxide, Mg(OH)₂, in water is 9.0×10^{-4} g/100.0 mL. What volume of 0.00100 *M* HNO₃ is required to neutralize 1.00 L of saturated Mg(OH)₂ solution?

4.67. A 10.0 mL dose of the antacid in Figure P4.67 contains 830 mg of magnesium hydroxide. What volume of 0.10 *M* stomach acid (HCl) could one dose neutralize?

FIGURE P4.67

*4.68. **Exercise Physiology** The ache, or "burn," you feel in your muscles during strenuous exercise is related to the accumulation of lactic acid, which has the structure shown in Figure P4.68. Only the hydrogen atom in the −COOH group is acidic, that is, can be released as an H⁺ ion in aqueous solutions. To determine the concentration of a solution of lactic acid, a chemist titrates a 20.00 mL sample of it with 0.1010 *M* NaOH and finds that 12.77 mL of titrant is required to reach the equivalence point. What is the concentration of the lactic acid solution in moles per liter?

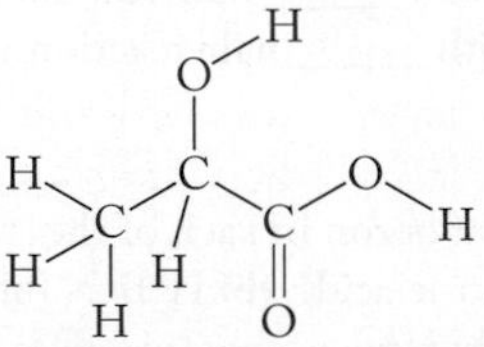

FIGURE P4.68

Precipitation Reactions

Concept Review

4.69. What is the difference between a saturated solution and a supersaturated solution?

4.70. What are common solubility units?

4.71. An aqueous solution containing Ca²⁺, Cl⁻, CO₃²⁻, and NO₃⁻ is allowed to evaporate. Which compound will precipitate first?

4.72. A precipitate may appear when two completely clear aqueous solutions are mixed. What circumstances are responsible for this event?

4.73. Is a saturated solution always a concentrated solution? Explain.

4.74. **Behavior of Honey** Honey is a concentrated solution of sugar molecules in water. Clear, viscous honey becomes cloudy after being stored for long periods. Explain how this transition illustrates supersaturation.

Problems

4.75. According to the solubility rules in Table 4.4 and Table 4.5, which of the following compounds have limited solubility in water? (a) barium sulfate; (b) barium hydroxide; (c) lanthanum nitrate; (d) sodium acetate; (e) lead hydroxide; (f) calcium phosphate

4.76. **Ocean Vents** The black "smoke" that flows out of deep ocean hydrothermal vents (Figure P4.76) is made of insoluble metal sulfides suspended in seawater. Of the following cations that are present in the water flowing up through these vents, which ones could contribute to the formation of the black smoke? Na⁺, Li⁺, Mn²⁺, Fe²⁺, Ca²⁺, Mg²⁺, Zn²⁺, Pb²⁺, Cu²⁺

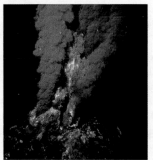

FIGURE P4.76

4.77. Complete and balance the molecular equations for the precipitation reactions, if any, between the following pairs of reactants, and write the overall and net ionic equations.
a. Pb(NO₃)₂(*aq*) + Na₂SO₄(*aq*) →
b. NiCl₂(*aq*) + NH₄NO₃(*aq*) →
c. FeCl₂(*aq*) + Na₂S(*aq*) →
d. MgSO₄(*aq*) + BaCl₂(*aq*) →

*4.78. **Wastewater Treatment** Show with appropriate net ionic equations how Cr³⁺ and Cd²⁺ can be removed from wastewater by treatment with solutions of sodium hydroxide.

4.79. Calculate the mass of MgCO₃ precipitated by mixing 10.0 mL of a 0.200 *M* Na₂CO₃ solution with 5.00 mL of 0.0500 *M* Mg(NO₃)₂ solution.

4.80. Toxic chromate can be precipitated from an aqueous solution by bubbling SO₂ through the solution. How many grams of SO_2 are required to treat 3.0×10^8 L of 0.050 mM CrO_4^-?

$$2\ CrO_4^{2-}(aq) + 3\ SO_2(g) + 4\ H^+(aq) \rightarrow$$
$$Cr_2(SO_4)_3(s) + 2\ H_2O(\ell)$$

4.81. Iron(II) can be precipitated from a slightly basic aqueous solution by bubbling oxygen through the solution, which converts soluble $Fe(OH)^+$ to insoluble $Fe(OH)_3$. How many grams of O_2 are consumed to precipitate all of the iron in 75 mL of 0.090 M iron(II)?

$$4\ Fe(OH)^+(aq) + 4\ OH^-(aq) + O_2(g) + 2\ H_2O(\ell) \rightarrow$$
$$4\ Fe(OH)_3(s)$$

4.82. Given the following equation, how many grams of $PbCO_3$ will dissolve when 1.00 L of 1.00 M H^+ is added to 5.00 g of $PbCO_3$?

$$PbCO_3(s) + 2\ H^+(aq) \rightarrow Pb^{2+}(aq) + H_2O(\ell) + CO_2(g)$$

*4.83. Treating Drinking Water Phosphate can be removed from drinking-water supplies by treating the water with $Ca(OH)_2$. How much $Ca(OH)_2$ is required to remove 90% of the PO_4^{3-} from 4.5×10^6 L of drinking water containing 25 mg/L of PO_4^{3-}?

$$5\ Ca(OH)_2(aq) + 3\ PO_4^{3-}(aq) \rightarrow Ca_5OH(PO_4)_3(s) + 9\ OH^-(aq)$$

4.84. Toxic cyanide ions can be removed from wastewater by adding hypochlorite.

$$2\ CN^-(aq) + 5\ OCl^-(aq) + H_2O(\ell) \rightarrow$$
$$N_2(g) + 2\ HCO_3^-(aq) + 5\ Cl^-(aq)$$

   a. If 1.50×10^3 L of 0.125 M OCl^- are required to remove the CN^- in 3.4×10^6 L of wastewater, what is the CN^- concentration in the water in mg/L?

   *b. How many milliliters of 0.575 M $AgNO_3$ would you need to add to a 50.00 mL aliquot of the final solution (consider the volumes simply additive) to precipitate the chloride ions formed in the reaction?

4.85. For each of the following aqueous mixtures, determine which ionic concentrations decrease and which remain the same.

   a. Sodium chloride and silver nitrate are dissolved in 100 mL of water.

   b. Equimolar amounts of sodium hydroxide and hydrochloric acid react.

   c. Ammonium sulfate and potassium bromide are dissolved in 100 mL of water.

4.86. For each of the following aqueous mixtures, determine which ionic concentrations decrease and which remain the same.

   a. Sodium chloride and iron(II) chloride are dissolved in 100 mL of water.

   b. Equimolar amounts of sodium carbonate and sulfuric acid react.

   c. Potassium sulfate and barium nitrate are dissolved in 100 mL of water.

Ion Exchange

Concept Review

4.87. Explain how a mixture of anion and cation exchangers can be used to deionize water.

4.88. Describe the process by which the ion exchanger in a home water softener is regenerated for further use.

4.89. (a) Use the solubility rules to write the balanced net ionic equation for each of the following "molecular" reactions. If there is no net reaction, write "NR." (b) Which of these three reactions give clear visual evidence of the ion exchange process?

   1. $NaCl(aq) + AgNO_3(aq) \rightarrow AgCl(s) + NaNO_3(aq)$
   2. $NaCl(aq) + KNO_3(aq) \rightarrow NaNO_3(aq) + KCl(aq)$
   3. $MgCl_2(aq) + KOH(aq) \rightarrow Mg(OH)_2(s) + KCl(aq)$

4.90. (a) Use the solubility rules to write the balanced net ionic equation for each of the following "molecular" reactions. If there is no net reaction, write "NR." (b) Which of these three reactions give clear visual evidence of the ion exchange process?

   1. $BaCl_2(aq) + Na_2CO_3(aq) \rightarrow BaCO_3(s) + NaCl(aq)$
   2. $NaCl(aq) + KOH(aq) \rightarrow NaOH(aq) + KCl(aq)$
   3. $Na_3PO_4(aq) + CaCl_2(aq) \rightarrow Ca_3(PO_4)_2(s) + NaCl(aq)$

Oxidation–Reduction Reactions: Electron Transfer

Concept Review

4.91. How are the gains or losses of electrons related to changes in oxidation numbers?

4.92. What is the sum of the oxidation numbers of the atoms in a molecule?

4.93. What is the sum of the oxidation numbers of all the atoms in each of the following polyatomic ions? (a) OH^-; (b) NH_4^+; (c) SO_4^{2-}; (d) PO_4^{3-}

4.94. Gold does not dissolve in concentrated H_2SO_4 but readily dissolves in H_2SeO_4 (selenic acid). Which acid is the stronger oxidizing agent?

4.95. Silver dissolves in sulfuric acid to form silver sulfate and H_2, but gold does not dissolve in sulfuric acid to form gold sulfate. Which of the two metals is the better reducing agent?

4.96. What is meant by a half-reaction?

4.97. What are the half-reactions that take place in the electrolysis of molten NaCl?

4.98. Electron gain is associated with _____ half-reactions, and electron loss is associated with _____ half-reactions.

Problems

4.99. Give the oxidation number of boron in each of the following: (a) HBO_2 (metaboric acid); (b) H_3BO_3 (boric acid); (c) $Na_2B_4O_7$ (sodium borate).

4.100. Give the oxidation number of nitrogen in each of the following: (a) elemental nitrogen (N_2); (b) hydrazine (N_2H_4); (c) ammonium ion (NH_4^+).

4.101. Balance the following half-reactions by adding the appropriate number of electrons. Identify the oxidation half-reactions and the reduction half-reactions.
   a. $Br_2(\ell) \rightarrow 2\ Br^-(aq)$
   b. $Pb(s) + 2\ Cl^-(aq) \rightarrow PbCl_2(s)$
   c. $O_3(g) + 2\ H^+(aq) \rightarrow O_2(g) + H_2O(\ell)$
   d. $2\ H_2SO_3(aq) + H^+(aq) \rightarrow HS_2O_4^-(aq) + 2\ H_2O(\ell)$

4.102. Balance the following half-reactions by adding the appropriate number of electrons. Which are oxidation half-reactions and which are reduction half-reactions?
   a. $Fe^{2+}(aq) \rightarrow Fe^{3+}(aq)$
   b. $AgI(s) \rightarrow Ag(s) + I^-(aq)$
   c. $VO_2^+(aq) + 2\ H^+(aq) \rightarrow VO^{2+}(aq) + H_2O(\ell)$
   d. $I_2(s) + 6\ H_2O(\ell) \rightarrow 2\ IO_3^-(aq) + 12\ H^+(aq)$

4.103. Balance the following net ionic reactions, and identify which elements are oxidized and which are reduced:
   a. $MnO_2(s) + HCl(aq) \rightarrow Mn^{2+}(aq) + Cl_2(g)$
   b. $I_2(s) + S_2O_3^{2-}(aq) \rightarrow S_4O_6^{2-}(aq) + I^-(aq)$
   c. $MnO_4^-(aq) + Fe^{2+}(aq) \rightarrow Mn^{2+}(aq) + Fe^{3+}(aq)$

4.104. Balance the following net ionic reactions, and identify which elements are oxidized and which are reduced:
   a. $MnO_4^-(aq) + S^{2-}(aq) \rightarrow MnO_2(s) + S(s)$
   b. $IO_3^-(aq) + I^-(aq) \rightarrow I_2(s)$
   c. $Mn^{2+}(aq) + BiO_3^-(aq) \rightarrow MnO_4^-(aq) + Bi^{3+}(aq)$

4.105. **Earth's Crust** The following chemical reactions have helped to shape Earth's crust. Determine the oxidation numbers of all the elements in the reactants and products, and identify which elements are oxidized and which are reduced.
   a. $3\ SiO_2(s) + 2\ Fe_3O_4(s) \rightarrow 3\ Fe_2SiO_4(s) + O_2(g)$
   b. $SiO_2(s) + 2\ Fe(s) + O_2(g) \rightarrow Fe_2SiO_4(s)$
   c. $4\ FeO(s) + O_2(g) + 6\ H_2O(\ell) \rightarrow 4\ Fe(OH)_3(s)$

4.106. Determine the oxidation numbers of each of the elements in the following reactions, and identify which of them are oxidized or reduced, if any.
   a. $SiO_2(s) + 2\ H_2O(\ell) \rightarrow H_4SiO_4(aq)$
   b. $2\ MnCO_3(s) + O_2(g) \rightarrow 2\ MnO_2(s) + 2\ CO_2(g)$
   c. $3\ NO_2(g) + H_2O(\ell) \rightarrow$
      $2\ NO_3^-(aq) + NO(g) + 2\ H^+(aq)$

4.107. Combine the half-reaction for the reduction of O_2
   $$O_2(aq) + 4\ H^+(aq) + 4\ e^- \rightarrow 2\ H_2O(\ell)$$
   with the following oxidation half-reactions (which are based on common iron minerals) to develop complete redox reactions:
   a. $2\ FeCO_3(s) + H_2O(\ell) \rightarrow$
      $Fe_2O_3(s) + 2\ CO_2(g) + 2\ H^+(aq) + 2\ e^-$
   b. $3\ FeCO_3(s) + H_2O(\ell) \rightarrow$
      $Fe_3O_4(s) + 3\ CO_2(g) + 2\ H^+(aq) + 2\ e^-$
   c. $2\ Fe_3O_4(s) + H_2O(\ell) \rightarrow 3\ Fe_2O_3(s) + 2\ H^+(aq) + 2\ e^-$

4.108. Uranium is found in Earth's crust as UO_2 and an assortment of compounds containing UO_2^{n+} cations. Add the following pairs of reduction and oxidation equations to

develop overall equations for converting soluble uranium polyatomic ions into insoluble UO_2.
   a. $6\ H^+(aq) + UO_2(CO_3)_3^{4-}(aq) + 2\ e^- \rightarrow$
      $UO_2(s) + 3\ CO_2(g) + 3\ H_2O(\ell)$
      $Fe^{2+}(aq) + 3\ H_2O(\ell) \rightarrow Fe(OH)_3(s) + 3\ H^+(aq) + e^-$
   b. $6\ H^+(aq) + UO_2(CO_3)_3^{4-}(aq) + 2\ e^- \rightarrow$
      $UO_2(s) + 3\ CO_2(g) + 3\ H_2O(\ell)$
      $HS^-(aq) + 4\ H_2O(\ell) \rightarrow SO_4^{2-}(aq) + 9\ H^+(aq) + 8\ e^-$
   c. $2\ e^- + UO_2(HPO_4)_2^{2-}(aq) \rightarrow UO_2(s) + 2\ HPO_4^{2-}(aq)$
      $3\ OH^-(aq) \rightarrow H_2O(\ell) + HO_2^-(aq) + 2\ e^-$

4.109. Nitrogen in the hydrosphere is found primarily as ammonium ions and nitrate ions. Complete and balance the following chemical equation describing the oxidation of ammonium ions to nitrate ions in acid solution:
   $$NH_4^+(aq) + O_2(g) \rightarrow NO_3^-(aq)$$

4.110. **When Soil Smells Bad** In sediments and waterlogged soil, dissolved O_2 concentrations are so low that the microorganisms living there must rely on other sources of oxygen for respiration. Some bacteria can extract the oxygen from sulfate ions, reducing the sulfur in them to hydrogen sulfide gas and giving the sediments or soil a distinctive rotten-egg odor.
   a. What is the change in oxidation state of sulfur as a result of this reaction?
   b. Write the balanced net ionic equation for the reaction, under acidic conditions, that releases O_2 from sulfate and forms hydrogen sulfide gas.

4.111. Chromium is more toxic and more soluble in natural waters as $HCrO_4^-$ than as chromium(III) ion. In the presence of H_2S, the following reaction takes place in neutral solution:
   $$HCrO_4^-(aq) + H_2S(aq) \rightarrow Cr_2O_3(s) + SO_4^{2-}$$
   a. Assign oxidation numbers to the reactants and products.
   b. Balance the equation.
   c. How many electrons are transferred for each atom of chromium that reacts?

4.112. The water-soluble uranyl cation, UO_2^+, can be removed by reaction with methane gas:
   $$UO_2^+(aq) + CH_4(g) \rightarrow UO_2(s) + HCO_3^-(aq)$$
   a. Assign oxidation numbers to the reactants and products.
   b. Balance the equation in acidic solution.
   c. How many electrons are transferred for each atom of uranium that reacts?

4.113. The solubilities of Fe and Mn in freshwater streams are affected by changes in their oxidation states. Complete and balance the following redox equation in which soluble Mn^{2+} becomes solid MnO_2:
   $$Fe(OH)_2^+(aq) + Mn^{2+}(aq) \rightarrow MnO_2(s) + Fe^{2+}(aq)$$

4.114. Bactericide and Virucide The water-soluble gas ClO_2 is known as an oxidative biocide. It destroys bacteria by oxidizing their cell walls and destroys viruses by attacking their viral envelopes. ClO_2 may be prepared for use as a decontaminating agent from several different starting materials in slightly acidic solutions. Complete and balance the following chemical equations for the synthesis of ClO_2.

a. $ClO_3^-(aq) + SO_2(g) \rightarrow ClO_2(g) + SO_4^{2-}(aq)$
b. $ClO_3^-(aq) + Cl^-(aq) \rightarrow ClO_2(g) + Cl_2(g)$
c. $ClO_3^-(aq) + Cl_2(g) \rightarrow ClO_2(g) + O_2(g)$

4.115. Refer to Table 4.6 to determine which of the following metals will reduce aqueous Fe^{2+} to iron metal: lead, copper, zinc, or aluminum.

4.116. Which ions will oxidize aluminum? Li^+; Ca^{2+}; Ag^+; Sn^{2+}

4.117. Through appropriate experiments, we could expand the activity series in Table 4.6 to include additional metals. If aluminum is oxidized by V^{3+} but aluminum does not reduce Sc^{3+}, where would you place vanadium and scandium in the activity series? Which metal would you test to firmly establish scandium's position?

4.118. If iron is oxidized by Cd^{2+} but iron does not reduce Ga^{3+}, where would you place cadmium and gallium in the activity series? Which metal would you test to firmly establish gallium's position?

4.119. Dichromate ion oxidizes Fe^{2+} ion in aqueous, acidic solution, producing Fe^{3+} and Cr^{3+} by the unbalanced chemical equation:

$$Cr_2O_7^{2-}(aq) + Fe^{2+}(aq) \rightarrow Fe^{3+} + 2\,Cr^{3+}$$

a. Balance the equation.
b. If 15.2 mL of 0.135 M $Cr_2O_7^{2-}$ is required to completely react with 100.0 mL of Fe^{2+}, what is the concentration of the Fe^{2+} solution?

*4.120. Ozone, O_3, reacts with iodide ion (I^-) in basic solution to form O_2 and I_2 by the unbalanced chemical equation:

$$O_3(aq) + I^-(aq) \rightarrow O_2(g) + I_2(aq)$$

a. Balance the equation.
b. A saturated solution of ozone in 125 mL of water at 0°C is treated with 10 mL of 2.0 M KI. After the reaction is complete, the solution is titrated with 0.100 M H^+. If 54.7 mL of acid is needed, what is the concentration of O_3 in a saturated solution?

Additional Problems

4.121. A puddle of coastal seawater, caught in a depression formed by some coastal rocks at high tide, begins to evaporate on a hot summer day as the tide goes out. If the volume of the puddle decreases to 23% of its initial volume, what is the concentration of Na^+ after evaporation if initially it was 0.449 M?

4.122. Antifreeze Ethylene glycol is the common name for the liquid used to keep the coolant in automobile cooling systems from freezing. It is 38.7% carbon, 9.7% hydrogen,

and 51.6% oxygen by mass. Its molar mass is 62.07 g/mol, and its density is 1.106 g/mL at 20°C.

a. What is the empirical formula of ethylene glycol?
b. What is the molecular formula of ethylene glycol?
c. In a solution prepared by mixing equal volumes of water and ethylene glycol, which ingredient is the solute and which is the solvent?

4.123. According to the label on a bottle of concentrated hydrochloric acid, the contents are 36.0% HCl by mass and have a density of 1.18 g/mL.

a. What is the molarity of concentrated HCl?
b. What volume of it would you need to prepare 0.250 L of 2.00 M HCl?
c. What mass of sodium hydrogen carbonate would be needed to neutralize the spill if a bottle containing 1.75 L of concentrated HCl dropped on a lab floor and broke open?

4.124. Synthesis and Toxicity of Chlorine Chlorine was first prepared in 1774 by heating a mixture of NaCl and MnO_2 in sulfuric acid:

$$NaCl(aq) + H_2SO_4(aq) + MnO_2(s) \rightarrow$$
$$Na_2SO_4(aq) + MnCl_2(aq) + H_2O(\ell) + Cl_2(g)$$

a. Assign oxidation numbers to the elements in each compound, and balance the redox reaction in acid solution.
b. Write a net ionic equation describing the reaction for formation of chlorine.
c. If chlorine gas is inhaled, it causes pulmonary edema (fluid in the lungs) because it reacts with water in the alveolar sacs of the lungs to produce the strong acid HCl and the weaker acid HOCl. Balance the equation for the conversion of Cl_2 to HCl and HOCl.

*4.125. When a solution of dithionate ions ($S_2O_4^{2-}$) is added to a solution of chromate ions (CrO_4^{2-}), the products of the reaction under basic conditions include soluble sulfite ions and solid chromium(III) hydroxide. This reaction is used to remove chromium(VI) from wastewater generated by factories that make chrome-plated metals.

a. Write the net ionic equation for this redox reaction.
b. Which element is oxidized and which is reduced?
c. Identify the oxidizing and reducing agents in this reaction.
d. How many grams of sodium dithionate would be needed to remove the chromium(VI) in 100.0 L of wastewater that contains 0.00148 M chromate ion?

4.126. An Iron Battery A prototype battery based on iron compounds with large, positive oxidation numbers was developed in 1999. In the following reactions, assign oxidation numbers to the elements in each compound, and balance the redox reactions in basic solution.

a. $FeO_4^{2-}(aq) + H_2O(\ell) \rightarrow$
$FeOOH(s) + O_2(g) + OH^-(aq)$
b. $FeO_4^{2-}(aq) + H_2O(\ell) \rightarrow Fe_2O_3(s) + O_2(g) + OH^-(aq)$

4.127. Polishing Silver Silver tarnish is the result of silver metal reacting with sulfur compounds, such as H_2S, in the air. The tarnish on silverware (Ag_2S) can be removed by soaking in a solution of $NaHCO_3$ (baking soda) in a basin lined with aluminum foil.

a. Write a balanced equation for the tarnishing of Ag to Ag_2S, and assign oxidation numbers to the reactants and products. How many electrons are transferred per mole of silver?

b. Write a balanced equation for the reaction of Ag_2S with Al metal, $NaHCO_3$, and water to produce $Al(OH)_3$, H_2S, H_2, and Ag metal.

4.128. Many nonmetal oxides react with water to form acidic solutions. Give the formula and name for the acids produced from the following reactions:

a. $P_4O_{10} + 6\ H_2O \rightarrow$
b. $SeO_2 + H_2O \rightarrow$
c. $B_2O_3 + 3\ H_2O \rightarrow$

4.129. Write overall and net ionic equations for the reactions that occur when

a. a sample of acetic acid is titrated with a solution of KOH.

b. a solution of sodium carbonate is mixed with a solution of calcium chloride.

c. calcium oxide dissolves in water.

***4.130. Fluoride Ion in Drinking Water** Sodium fluoride is added to drinking water in many municipalities to protect teeth against cavities. The target of the fluoridation is hydroxyapatite, $Ca_{10}(PO_4)_6(OH)_2$, a compound in tooth enamel. There is concern, however, that fluoride ions in water may contribute to skeletal fluorosis, an arthritis-like disease.

a. Write a net ionic equation for the reaction between hydroxyapatite and sodium fluoride that produces fluorapatite, $Ca_{10}(PO_4)_6F_2$.

b. The EPA currently restricts the concentration of F^- in drinking water to 4 mg/L. Express this concentration of F^- in molarity.

c. One study of skeletal fluorosis suggests that drinking water with a fluoride concentration of 4 mg/L for 20 years raises the fluoride content in bone to 6 mg/g, a level at which a patient may experience stiff joints and other symptoms. How much fluoride (in milligrams) is present in a 100 mg sample of bone with this fluoride concentration?

***4.131. Rocket Fuel in Drinking Water** Near Las Vegas, NV, improper disposal of perchlorates used to manufacture rocket fuel has contaminated a stream that flows into Lake Mead, the largest artificial lake in the United States and a major supply of drinking and irrigation water for the American Southwest. The EPA has proposed an advisory range for perchlorate concentrations in drinking water of 4 to 18 μg/L. The perchlorate concentration in the stream averages 700.0 μg/L, and the stream flows at an average rate of 161 million gallons per day (1 gal = 3.785 L).

a. What are the formulas of sodium perchlorate and ammonium perchlorate?

b. How many kilograms of perchlorate flow from the Las Vegas stream into Lake Mead each day?

c. What volume of perchlorate-free lake water would have to mix with the stream water each day to dilute the stream's perchlorate concentration from 700.0 to 4 μg/L?

d. Since 2003, Maryland, Massachusetts, and New Mexico have limited perchlorate concentrations in drinking water to 0.1 μg/L. Five replicate samples were analyzed for perchlorates by laboratories in each state, and the following data (μg/L) were collected:

MD	MA	NM
1.1	0.90	1.2
1.1	0.95	1.2
1.4	0.92	1.3
1.3	0.90	1.4
0.9	0.93	1.1

Which of the labs produced the most precise analytical results?

***4.132. Acidic Mine Drainage** Water draining from abandoned mines on Iron Mountain in California is extremely acidic and leaches iron, zinc, and other metals from the underlying rock (Figure P4.132). One liter of drainage contains as much as 80.0 g of dissolved iron and 6 g of zinc.

a. Calculate the molarity of iron and of zinc in the drainage.

b. One source of the dissolved iron is the reaction between water containing H_2SO_4 and solid $Fe(OH)_3$. Complete the following chemical equation, and write a net ionic equation for the process.

$$2\ Fe(OH)_3(s) + 3\ H_2SO_4(aq) \rightarrow$$

c. Sources of zinc include the mineral smithsonite, $ZnCO_3$. Write a balanced net ionic equation for the reaction between smithsonite and H_2SO_4 that produces $Zn^{2+}(aq)$.

d. One member of a class of minerals called ferrites is found to contain a mixture of zinc(II), iron(II), and iron(III) oxides. The generic formula for the mineral is $Zn_xFe_{1-x}O \cdot Fe_2O_3$. If acidic mine waste flowing through a deposit of this mineral contains 80 g of Fe and 6 g of Zn as a result of dissolution of the mineral, what is the value of x in the formula of the mineral in the deposit?

FIGURE P4.132

*4.133. Making Apple Cider Vinegar Some people who prefer natural foods make their own apple cider vinegar. They start with freshly squeezed apple juice that contains about 6% natural sugars. These sugars, which all have nearly the same empirical formula, CH_2O, are fermented with yeast in a chemical reaction that produces equal numbers of moles of ethanol (Figure P4.133a) and carbon dioxide. The product of this fermentation, called hard cider, undergoes an acid fermentation step in which ethanol and dissolved oxygen gas react to form acetic acid (Figure P4.133b) and water. This acetic acid is the principal solute in vinegar.

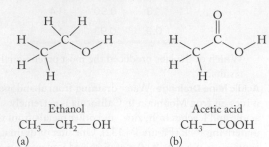

Ethanol	Acetic acid
CH_3-CH_2-OH	CH_3-COOH
(a)	(b)

FIGURE P4.133

a. Write a balanced chemical equation describing the fermentation of natural sugars to ethanol and carbon dioxide. You may use the empirical formula given in the above paragraph.
b. Write a balanced chemical equation describing the acid fermentation of ethanol to acetic acid.
c. What are the oxidation states of carbon in the reactants and products of the two fermentation reactions?
d. If a sample of apple juice contains 1.00×10^2 g of natural sugar, what is the maximum quantity of acetic acid that could be produced by the two fermentation reactions?

*4.134. A food chemist determines the concentration of acetic acid in a sample of apple cider vinegar (see Problem 4.133) by acid–base titration. The density of the sample is 1.01 g/mL. The titrant is 1.002 M NaOH. The average volume of titrant required to titrate 25.00 mL aliquots of the vinegar is 20.78 mL. What is the concentration of acetic acid in the vinegar? Express your answer the way a food chemist probably would: as percent by mass.

*4.135. One way to follow the progress of a titration and detect its equivalence point is by monitoring the conductivity of the titration reaction mixture. For example, consider the way the conductivity of a sample of sulfuric acid changes as it is titrated with a standard solution of barium hydroxide before and then after the equivalence point.
a. Write the overall ionic equation for the titration reaction.
b. Which of the four graphs in Figure P4.135 comes closest to representing the changes in conductivity during the titration? (The zero point on the y-axis of these graphs represents the conductivity of pure water; the break points on the x-axis represent the equivalence point.)

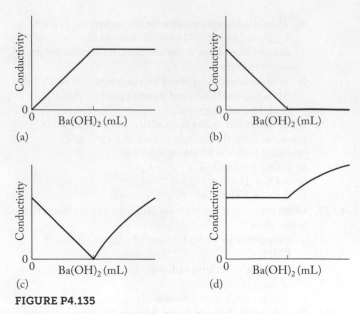

FIGURE P4.135

*4.136. Which of the graphs in Figure P4.136 best represents the changes in conductivity that occur before and after the equivalence point in each of the following titrations:
a. sample, $AgNO_3(aq)$; titrant, $KCl(aq)$
b. sample, $HCl(aq)$; titrant, $LiOH(aq)$
c. sample, $CH_3COOH(aq)$; titrant, $NaOH(aq)$

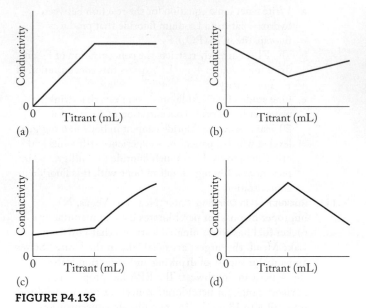

FIGURE P4.136

4.137. When electrodes connected to a lightbulb are inserted into an aqueous solution of acetic acid, the bulb glows dimly. Will the bulb become brighter, remain the same, or turn off after one equivalent of aqueous NaOH is added to the solution? Write a balanced net ionic equation that supports your answer.

*4.138. When electrodes connected to a lightbulb are inserted into a beaker containing silver carbonate and water, will the bulb not glow, glow dimly, or glow brightly? What do you think will happen after addition of one equivalent of aqueous HCl? Write a balanced net ionic equation that supports your answer.

4.139. Superoxide Dismutases Oxygen in the form of superoxide ions, O_2^-, is quite hazardous to human health. Superoxide dismutases represent a class of enzymes that convert superoxide ions to hydrogen peroxide and oxygen by the unbalanced chemical equation:

$$O_2^-(aq) + H^+(aq) \rightarrow H_2O_2(aq) + O_2(aq)$$

a. Identify the oxidation and reduction half-reactions.
b. Balance the equation.

4.140. Nitrogen-Fixing Bacteria Bacteria found among the roots of legumes perform an important biological function, converting nitrogen to ammonia in a process known as nitrogen fixation. The electrons required for this redox reaction are supplied by transition metal–containing enzymes called nitrogenases. The unbalanced chemical equation for this process is

$$N_2(g) + H^+(aq) + M^{2+}(aq) \rightarrow NH_3(aq) + H_2(g) + M^{3+}(aq)$$

where M represents a transition metal such as iron. Nitrification is a multistep process in which the nitrogen in organic and inorganic compounds is biochemically oxidized. Bacteria and fungi are responsible for a part of the *nitrification process* described by the reaction:

$$NH_4^+(aq) + M^{3+}(aq) \rightarrow NO_2^-(aq) + M^{2+}(aq)$$

a. What are the oxidation numbers of nitrogen in the reactants and products of each reaction?
b. Which compounds or ions are being reduced in each reaction?
c. Balance the equations in acidic solution.

4.141. Rocks in Caves The stalactites and stalagmites in most caves are made of limestone (calcium carbonate; see Figure 4.14). However, in the Lower Kane Cave in Wyoming they are made of gypsum (calcium sulfate). The presence of $CaSO_4$ is explained by the following sequence of reactions:

$$H_2S(aq) + 2 O_2(g) \rightarrow H_2SO_4(aq)$$

$$H_2SO_4(aq) + CaCO_3(s) \rightarrow CaSO_4(s) + H_2O(\ell) + CO_2(g)$$

a. Which (if either) of these reactions is a redox reaction? How many electrons are transferred?

b. Write a net ionic equation for the reaction of H_2SO_4 with $CaCO_3$.
c. How would the net ionic equation be different if the reaction were written as follows?

$$H_2SO_4(aq) + CaCO_3(s) \rightarrow CaSO_4(s) + H_2CO_3(aq)$$

4.142. Balance this net ionic reaction and answer the questions that follow:

$$BrO_3^-(aq) + Br^-(aq) \rightarrow Br_2(aq)$$

a. Is $BrO_3^-(aq)$ reduced?
b. What is the product of $BrO_3^-(aq)$ reduction in this reaction?
c. Is $Br^-(aq)$ oxidized?
d. What is the product of $Br^-(aq)$ oxidation in this reaction?

4.143. Which of the following reactions of calcium compounds is/are redox reactions?
a. $CaCO_3(s) \rightarrow CaO(s) + CO_2(g)$
b. $CaO(s) + SO_2(g) \rightarrow CaSO_3(s)$
c. $CaCl_2(s) \rightarrow Ca(s) + Cl_2(g)$
d. $3\ Ca(s) + N_2(g) \rightarrow Ca_3N_2(s)$

4.144. Preparation of Fluorine Gas HF is prepared by reacting CaF_2 with H_2SO_4:

$$CaF_2(s) + H_2SO_4(\ell) \rightarrow 2\ HF(g) + CaSO_4(s)$$

HF can in turn be electrolyzed when dissolved in molten KF to produce fluorine gas:

$$2\ HF(\ell) \rightarrow F_2(g) + H_2(g)$$

Fluorine is extremely reactive, so it is typically sold as a 5% mixture by volume in an inert gas such as helium. How much CaF_2 is required to produce 500.0 L of 5% F_2 in helium? Assume the density of F_2 gas is 1.70 g/L.

*4.145. A piece of Zn metal is placed in a solution containing Cu^{2+} ions. At the surface of the Zn metal, Cu^{2+} ions react with Zn atoms, forming Cu atoms and Zn^{2+} ions. Is this reaction an example of ion exchange? Explain why or why not.

5

Thermochemistry
Energy Changes in Reactions

SUNLIGHT AND LIFE The Sun is the ultimate source of energy for most forms of life on Earth.

Acid and Base

In Chapter 5 we consider the energy changes that occur during reactions such as the combustion reactions from Chapter 3 and neutralization reactions from Chapter 4.

- Here we see the key molecules and ions involved in the titration of hydrochloric acid with sodium hydroxide. Name each molecule or ion and write its formula.

- The colorless solution in the flask on the left is hydrochloric acid. The colorless solution in the buret is sodium hydroxide. On the right is a picture of the titration after all the acid has been neutralized. Which of the illustrated particles are present in the buret, the flask on the left, and the flask on the right?

  (Review Sections 4.5–4.6 if you need help.)

(Answers to Particulate Review questions are in the back of the book.)

(a)   (b)

Breaking Bonds and Energy

When ozone molecules absorb ultraviolet rays (UV rays) from the Sun, the ozone falls apart into oxygen molecules and oxygen atoms according to the chemical reaction depicted here. As you read Chapter 5, look for ideas that will help you answer these questions:

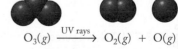

$$O_3(g) \xrightarrow{\text{UV rays}} O_2(g) + O(g)$$

- What role does energy play in breaking the bonds?

- Does bond breaking occur when energy is absorbed? Or does breaking a bond release energy?

Learning Outcomes

LO1 Explain kinetic and potential energies at the molecular level
Sample Exercise 5.1

LO2 Identify familiar endothermic and exothermic processes
Sample Exercise 5.2

LO3 Calculate changes in the internal energy of a system
Sample Exercises 5.3, 5.4

LO4 Calculate the amount of heat transferred in physical or chemical processes
Sample Exercises 5.5, 5.6, 5.7, 5.8, 5.9

LO5 Calculate thermochemical values by using data from calorimetry experiments
Sample Exercises 5.10, 5.11

LO6 Calculate enthalpies of reaction
Sample Exercises 5.12, 5.13, 5.15

LO7 Recognize and write equations for formation reactions
Sample Exercise 5.14

LO8 Calculate and compare fuel and food values and fuel densities
Sample Exercises 5.16, 5.17

5.1 Sunlight Unwinding

All physical changes of matter, such as ponds freezing in winter and thawing in spring, involve changes in energy, as do the chemical reactions we studied in Chapter 4. Energy—to power an automobile, heat a home, or support life—may seem like an abstract idea: Energy has no mass and no volume. However, we see its effects very clearly when matter changes from one state to another—as sunlight melts snow, or a gas flame boils water and converts it into steam—and when energy itself is converted from one form into another, as when the chemical energy of gasoline is converted into mechanical energy to move a car. Part of the energy in the gasoline contributes nothing to moving the vehicle and is lost to the surroundings as heat. The sum of the energy used to move the vehicle and the energy lost as heat equals the energy released during the combustion of the gasoline. In other words, energy is neither created nor destroyed during chemical reactions.

We can roast marshmallows by using energy from a campfire, but where does that energy come from? R. Buckminster Fuller (1895–1983), a 20th-century architect, inventor, and futurist, described a burning log like this: Trees gather the energy in sunlight and combine it with water and carbon dioxide to make the molecules that compose wood. When the wood is burned, the chemical products are carbon dioxide and water, and the fire is, as Fuller said, "all that sunlight unwinding." The sunlight unwinding is the net release of chemical energy as bonds break between atoms in the molecules of the wood and bonds form between atoms in the molecules of carbon dioxide and water. Through the transforming power of green plants, sunlight is the source of the chemical energy stored in all the substances we consume as food and fuel.

Nearly every chemical reaction, from combustion of fuels to neutralization reactions to the dissolution of salts, involves energy as either a product or a reactant. For example, the reaction in which hydrogen combines with oxygen to form water releases energy:

$$2\,H_2(g) + O_2(g) \rightarrow 2\,H_2O(g) + \text{energy}$$

The energy derived from the reaction between hydrogen and oxygen may be converted into motion, as when a spacecraft fueled by hydrogen lifts off. The study of energy and its transformations from one form to another is called

CONNECTION We introduced energy as the capacity to do work in Chapter 1.

thermodynamics. The part of thermodynamics that deals with changes in energy accompanying chemical reactions is known as **thermochemistry**.

Changes in energy can cause changes in the state of a material, as when a solid melts or a liquid freezes. When we put an ice cube, initially at −18°C—the typical temperature of a freezer—into room-temperature water (25°C), the ice cube melts and the water cools because energy moves from the room-temperature water into the colder ice cube. The process by which energy moves from a warmer object to a cooler object is called *heat transfer*, or more generally *energy transfer*. The difference in temperature defines the direction of energy flow when two objects come into contact: energy transfer in the form of **heat** always flows from the hotter object to the colder one (Figure 5.1). The ice cube, for example, changes state from solid to liquid as energy is transferred to it from the liquid water. The water remains in the liquid state, but its temperature drops as energy from it is transferred to the ice cube. Ultimately, these two portions of matter achieve the same temperature, which is higher than the initial temperature of the ice cube but lower than the initial temperature of the water. At this point, **thermal equilibrium** has been reached, which means the temperature is the same throughout the combined material and no further energy transfer occurs.

How do we measure the amount of energy involved in physical and chemical processes? We cannot directly measure the amount of energy in a chemical reaction, but by measuring changes in the temperature, we can relate *the change* in energy to the identities and amounts of reactants and products involved in changes of state and in chemical reactions. Studying energy changes gives us important insights into the way nature works and how human activities affect our world.

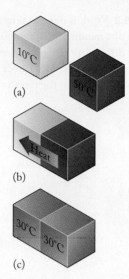

FIGURE 5.1 (a) Two identical blocks at different temperatures are brought into contact. (b) Heat is transferred from the block at higher temperature to the block at lower temperature until thermal equilibrium (equal temperature) (c) is reached.

5.2 Forms of Energy

The energy produced by a chemical reaction can be used to do **work**, can be transferred to an object to raise its temperature, or both. Energy that does work includes electrical, mechanical, light, and sound energy. Whatever the form, energy used to do work causes motion: The location or shape of an object changes when an energy source does work on the object. For example, the energy derived from the reaction between hydrogen and oxygen is converted into motion when a hydrogen-fueled spacecraft blasts off.

Work, Potential Energy, and Kinetic Energy

How does the combustion of hydrogen lead to the work done in lifting a spacecraft into orbit? Let's start with the classical view of work and energy from physics and consider a skier poised on top of a steep slope. In the physical sciences, work (w) is done whenever a force (F) moves an object through a distance (d). The amount of mechanical work done is

$$w = F \times d \qquad (5.1)$$

Consider Equation 5.1 as it relates to skiers ascending a mountain (Figure 5.2). The work (w) done by the lift on a skier equals the length of the ride (d) times the force (F) needed to overcome gravity and transport the skier up the mountain. Some of the work done is stored in the skier as **potential energy (PE)**, which is the energy an object has because of its position. The farther up the mountain the

thermodynamics the study of energy and its transformations.

thermochemistry the study of the relation between chemical reactions and changes in energy.

heat the energy transferred between objects because of a difference in their temperatures.

thermal equilibrium a condition in which temperature is uniform throughout a material and no energy flows from one point to another.

work a form of energy: the energy required to move an object through a given distance.

potential energy (PE) the energy stored in an object because of its position.

FIGURE 5.2 Work is done as skiers ascend to the top of a mountain. The amount of work may differ, depending on whether the skiers (1) ride a gondola on a direct route to the top or (2) hike to the top along a winding path.

state function a property of an entity based solely on its chemical or physical state or both, but not on how it achieved that state.

kinetic energy (KE) the energy of an object in motion due to its mass (m) and its speed (u): $KE = \frac{1}{2}mu^2$.

law of conservation of energy the principle that energy cannot be created or destroyed but can be converted from one form into another.

CHEMTOUR
State Functions and Path Functions

skier is carried, the greater her potential energy. The mathematical expression for the skier's potential energy is

$$PE = m \times g \times h$$

where m is the skier's mass, g is the acceleration due to the force of gravity, and h is the vertical distance between the skier's location on the mountain and her starting point. How she gets to that position is not important. Because the potential energy of any object does not depend on how the object gets to a particular point, potential energy is a **state function**, which means it is independent of the path followed to acquire the potential energy. Only position is important in considering potential energy (Figure 5.3). The term *state function* refers to a property of a system that is determined by the position or condition of the system; do not confuse it with the term *change of physical state*, as in a phase change.

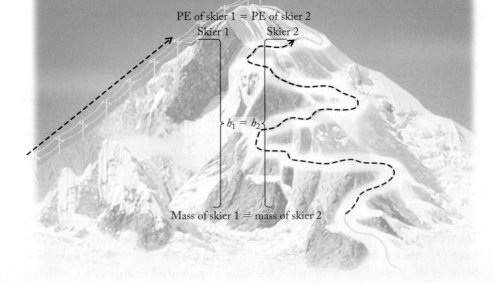

FIGURE 5.3 The potential energy of a skier depends only on the skier's mass and height above the base of the slope. If two skiers are at the same height ($h_1 = h_2$) and both skiers have the same mass, then they have the same potential energy, no matter how each skier got to that height.

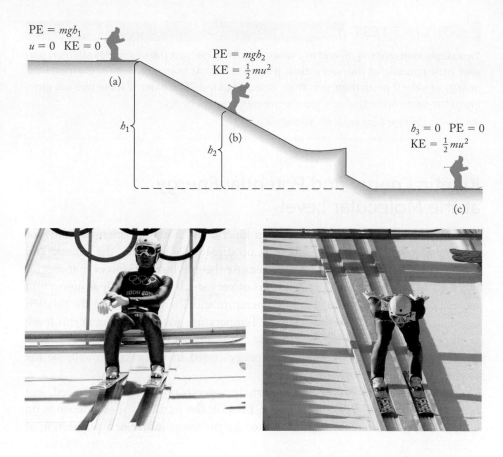

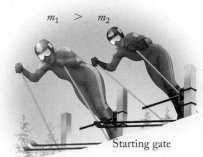

FIGURE 5.4 (a) A skier at the starting gate of a ski jump has potential energy (PE) due to his position (h_1) above the bottom of the slope, his mass (m), and the acceleration due to gravity (g): PE = mgh_1. (b) During his run, the skier's potential energy is converted into kinetic energy: KE = $\frac{1}{2}mu^2$. While he moves down the slope, he has both KE and PE. (c) At the end of the run, the skier's potential energy is 0. His KE decreases from its maximum value to 0 as he slows to a stop at some point along the flat region.

Now consider the potential energy of a skier standing still at the top of a ski jump (Figure 5.4a). At this position, his energy is all potential energy, but as he moves down the slope, his potential energy is converted into **kinetic energy (KE)**, the energy of motion (Figure 5.4b). At any moment between the start of the run and coming to a stop at the bottom of the hill, the jumper's kinetic energy is proportional to his mass (m) times the square of his speed (u):

$$\text{KE} = \tfrac{1}{2}mu^2 \qquad (5.2)$$

This equation tells us that a heavier skier (larger m) moving at the same speed as a lighter skier (smaller m) has more kinetic energy. Our intuition and experience tell us this as well. If you were standing at the bottom of the ski jump, how would the impact differ if a 136 kg (300 lb) skier ran into you rather than a 45 kg (100 lb) skier going the same speed? The difference lies in their relative kinetic energies.

According to the **law of conservation of energy**, energy cannot be created or destroyed. However, it can be converted from one form to another, as this example illustrates. Potential energy at the top of the slope becomes kinetic energy during the run. The total energy at any position on the hill is the sum of the skier's potential and kinetic energies.

CONCEPT **TEST**

Two skiers with masses m_1 and m_2, where $m_1 > m_2$, are poised at the starting gate of a downhill course (Figure 5.5a). Is the potential energy of skier 1 the same as that of skier 2? If the energies are different, which skier has more potential energy?

(Answers to Concept Tests are in the back of the book.)

FIGURE 5.5 (a) Two skiers of different mass are at the same position at the start of a race with respect to the bottom of the hill; (b) the same two skiers during the course of a race pass the same elevation at the same time.

Two skiers with masses m_1 and m_2, where $m_1 > m_2$, go past the same elevation on parallel race courses at the same time (Figure 5.5b). At that moment, is the potential energy of skier 2 more than, less than, or equal to the PE of skier 1? If the two are moving at the same speed, which has the greater kinetic energy?

(Answers to Concept Tests are in the back of the book.)

Kinetic Energy and Potential Energy at the Molecular Level

The relationship just described between kinetic and potential energies holds for atoms and molecules as well. There is no direct analogy with the kinetic and potential energies of a skier, however, because the gravitational forces that control the skier play no role in the interactions of very small objects such as atoms and molecules. At the molecular level, temperature and charge dominate the relation between kinetic and potential energies. Temperature governs motion at this level. Chemical bonds and differential electric charges cause interactions between particles that give rise to the potential energy stored in the arrangements of the atoms, ions, and molecules in matter.

The kinetic energy of an atomic particle depends on its mass and speed, just as with macroscopic objects. However, because the particle's speed depends on temperature, its kinetic energy does, too. As the temperature of a population of

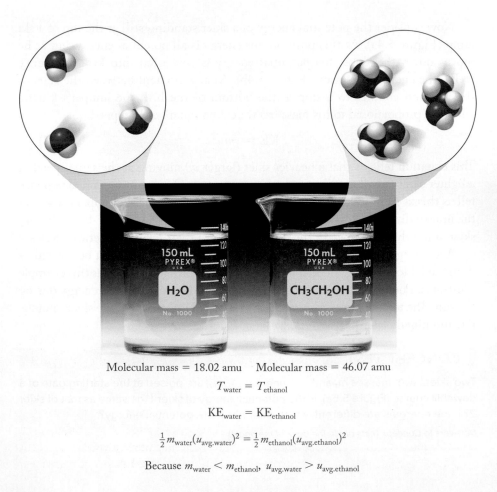

FIGURE 5.6 Two populations of gas-phase molecules have the same temperature and therefore the same average kinetic energy. Because ethanol molecules have a greater mass than water molecules, the average speed of the water molecules in the water vapor above the liquid water is greater than the average speed of the gas-phase ethanol molecules above the liquid ethanol.

Molecular mass = 18.02 amu   Molecular mass = 46.07 amu

$$T_{\text{water}} = T_{\text{ethanol}}$$

$$\text{KE}_{\text{water}} = \text{KE}_{\text{ethanol}}$$

$$\tfrac{1}{2} m_{\text{water}} (u_{\text{avg.water}})^2 = \tfrac{1}{2} m_{\text{ethanol}} (u_{\text{avg.ethanol}})^2$$

Because $m_{\text{water}} < m_{\text{ethanol}},\ u_{\text{avg.water}} > u_{\text{avg.ethanol}}$

particles increases, the average kinetic energy of the particles also increases. Consider, for example, how the molecules in the vapor phase above a liquid behave at different temperatures. We pick the gas phase because the molecules in a gas at normal pressures are widely separated and behave essentially independently of one another, and they all behave the same way regardless of their identities. If we have two samples of water vapor (molecular mass 18.02 amu) at room temperature, the two populations of H_2O molecules have the same average kinetic energy. The average speeds of the molecules are the same because their masses are identical. If we increase the temperature of one sample, that population of H_2O molecules acquires a higher average kinetic energy and the average speed of the molecules increases. An equivalent population of ethanol molecules (molecular mass 46.07 amu) in the gas phase at room temperature has the same average kinetic energy as the water molecules at room temperature, but the average speed of the ethanol molecules is lower because their mass is higher (Figure 5.6).

The kinetic energy associated with the total random motion of molecules is called **thermal energy**, and the thermal energy of a given sample of matter is proportional to the temperature of the sample. However, thermal energy also depends on the number of particles in a sample. The water in a swimming pool and in a cup of water taken from the pool have the same temperature, so their molecules have the same average kinetic energy. The water in the pool has considerably more thermal energy than the water in the cup, however, simply because there are more molecules in the pool. A large number of particles at a given temperature has a higher total thermal energy than a small number of particles at the same temperature.

> **thermal energy** the kinetic energy of atoms, ions, and molecules.
>
> **electrostatic potential energy (E_{el})** the energy a particle has because of its electrostatic charge and its position with respect to another particle; it is directly proportional to the product of the charges of the particles and inversely proportional to the distance between them.

CONCEPT **TEST**

If we heat a cup of water from a swimming pool almost to the boiling point, will its thermal energy be more than, less than, or the same as the thermal energy of all the water in the pool?

(Answers to Concept Tests are in the back of the book.)

An important form of potential energy at the atomic level arises from electrostatic interactions between charged particles. Just as the potential energy of skiers is determined by their distances above the bottom of the slope, the **electrostatic potential energy (E_{el})** of charged particles is determined by the distance between them. The magnitude of this electrostatic potential energy, also known as *coulombic interaction*, is directly proportional to the charges (Q_1 and Q_2) on the particles and is inversely proportional to the distance (d) between them:

$$E_{el} \propto \frac{Q_1 \times Q_2}{d} \tag{5.3}$$

where the symbol $\propto$ means "is proportional to." Coulombic interactions determine the potential energy of matter at the atomic level because they determine the relative positions of particles. Equation 5.3 is called Coulomb's law because it traces back to the work done by a French engineer, Charles Augustin de Coulomb (1736–1806), who first measured the interactions between charged particles.

Ions that are far enough apart do not interact (Figure 5.7a). If the two charges in Equation 5.3 are either both positive or both negative, their product is positive, E_{el} is positive, and the particles repel each other. If one particle is positive and the other negative, E_{el} is negative and the particles attract each other (Figure 5.7b and c). A lower electrostatic potential energy (a more negative value of E_{el}) corresponds to greater stability, so particles that attract each other because of their

FIGURE 5.7 Ionic interaction. (a) A positive ion and a negative ion are so far apart (*d* is large) that they do not interact at all. (b) As the ions move closer together (*d* decreases), the electrostatic potential energy between them becomes more negative. (c) At this distance, the attraction between them produces an arrangement that is the most favorable energetically because the ions have the lowest electrostatic potential energy. (d) If the ions are forced even closer together, they repel each other.

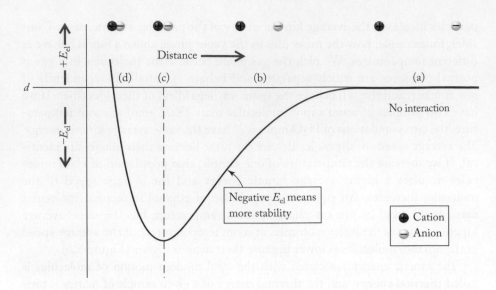

charges form an arrangement with a lower electrostatic potential energy than particles that repel each other. However, there is a limit to how close two particles can be. Recall that ions, whatever their charge, are surrounded by clouds of electrons that repel each other if they are pushed together too closely (Figure 5.7d).

Ions are not the only particles that experience coulombic interactions. Neutral species such as water molecules attract each other as well, due to an uneven electron distribution throughout the molecule. The same ideas we use to describe the behavior of oppositely charged ions apply to molecules as well. Whether we are dealing with matter composed of atoms, molecules, or ions, the total energy at the atomic level is the sum of the kinetic energy due to the random motion of particles and the potential energy due to their arrangement.

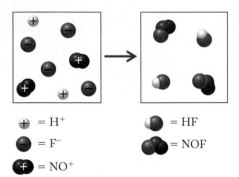

- + = H⁺
- = F⁻
- + = NO⁺
- = HF
- = NOF

FIGURE 5.8 Protons and nitrosonium ions react with fluoride ions to form HF and NOF, respectively.

SAMPLE EXERCISE 5.1 Kinetic Energy of Gas-Phase Ions **LO1**

Protons (H^+) and nitrosonium ions (NO^+) react with fluoride ions (F^-) in the gas phase under laboratory conditions (constant temperature and pressure) to form HF and NOF, respectively, as shown in Figure 5.8. (a) At a given temperature, which ion has the greatest kinetic energy? (b) Which cation is moving at the higher average speed?

Collect and Organize We are asked to predict which of three gas-phase particles has the greatest kinetic energy, and which of two positively charged particles has the higher average speed.

Analyze Particles of gas at the same temperature have the same kinetic energy (KE), which is related to their average speeds (*u*) by Equation 5.2, $KE = \frac{1}{2}mu^2$.

Solve
a. Because the temperature is constant, all the ions have the same kinetic energy.
b. Expressing this equality using Equation 5.2:

$$\tfrac{1}{2}m_{H^+} \times u_{H^+}^2 = \tfrac{1}{2}m_{NO^+} \times u_{NO^+}^2$$

Rearranging the terms and simplifying:

$$\frac{m_{NO^+}}{m_{H^+}} = \left(\frac{u_{H^+}}{u_{NO^+}}\right)^2$$

The mass of a NO^+ ion is much greater than the mass of a proton; therefore the protons in the reaction mixture have a higher average speed.

Think About It Heavier particles move more slowly than lighter particles with the same kinetic energy.

Practice Exercise Consider two positively charged particles, A^+ and B^+. If A^+ is 33% heavier than B^+, how much slower must it move to equal the kinetic energy of B^+?

(Answers to Practice Exercises are in the back of the book.)

FIGURE 5.9 Energy from the combustion of hydrogen can be used to launch rockets.

The energy given off or absorbed during a chemical reaction is equal to the difference in the energy of the reactants and products. For example, when hydrogen molecules burn in oxygen, the products are water and a considerable amount of energy (Figure 5.9). The energy given off by this reaction can be used to power rockets and is now being used to run some buses and automobiles. The fact that energy is given off in the reaction tells us that the product molecules must be at a lower potential energy than the reactant molecules. The difference between the energy of the products and the energy of the reactants is the energy released. In the case of hydrogen combustion (Figure 5.10a), this energy is sufficient to run vehicles once powered by fossil fuels (Figure 5.10b).

Where does the energy released by the reaction between hydrogen and oxygen in Figure 5.10 come from? During the combustion of H_2, the bonds in molecules of H_2 and O_2 must be broken, which requires energy, but even more energy is released when the O–H bonds in H_2O form. The result is a net release of energy.

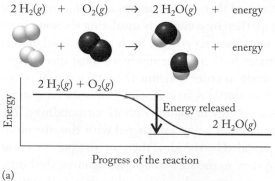

(a) (b)

FIGURE 5.10 (a) Hydrogen reacts with oxygen to produce water. Because this reaction releases energy, the product molecules are at a lower energy than those of the reactants. (b) Fueling a hydrogen-powered vehicle.

5.3 Systems, Surroundings, and Energy Transfer

In both thermochemistry and thermodynamics, the specific part of the universe we are studying is called the **system** and everything else is called the **surroundings**. Although a system can be as large as a galaxy or as small as a living cell, most of the systems we examine in this chapter fit on a laboratory bench. Typically, we limit our concern about surroundings to that part of the universe that can exchange energy and matter with the system. In evaluation of the energy gained or lost in a chemical reaction, the system may be just the particles involved in the reaction, or it may also include the contents along with the vessel in which the reaction occurs.

system the part of the universe that is the focus of a thermochemical study.

surroundings everything that is not part of the system.

(a) **Isolated system:** A thermos bottle containing hot soup with the lid screwed on tightly

(b) **Closed system:** A cup of hot soup with a lid

(c) **Open system:** An open cup of hot soup

FIGURE 5.11 Transfer of energy and matter in isolated, closed, and open systems. (a) Hot soup in a tightly sealed thermos bottle approximates an isolated system: No vapor escapes, no matter is added or removed, and no energy escapes to the surroundings. (b) Hot soup in a cup with a lid is a closed system: The soup transfers energy to the surroundings as it cools; however, no matter escapes and none is added. (c) Hot soup in a cup with no lid is an open system: It transfers both matter (steam) and energy (heat) to the surroundings as it cools. Matter in the form of pepper, grated cheese, crackers, or other matter from the surroundings may be added to the soup.

isolated system a system that exchanges neither energy nor matter with the surroundings.

closed system a system that exchanges energy but not matter with the surroundings.

open system a system that exchanges both energy and matter with the surroundings.

exothermic process a thermochemical process in which energy flows from a system into its surroundings.

endothermic process a thermochemical process in which energy flows from the surroundings into the system.

Isolated, Closed, and Open Systems

There are three types of thermodynamic systems: isolated, closed, and open (Figure 5.11). These designations are important because they define the system we are dealing with and the part of the universe that the system interacts with. In the following discussion, hot soup is the system.

Consider hot soup in an ideal, closed thermos bottle. An ideal thermos bottle takes no energy away from the soup, thereby completely insulating the soup from the rest of the universe, and the soup loses no energy. Such perfect insulation is impossible in practice, but sometimes in thermochemistry we must discuss systems that do not exist in the real world in order to define the total range of possibilities. The soup in the ideal thermos bottle is an example of an **isolated system**, which is a system that exchanges no energy or matter with its surroundings. The ideal thermos bottle prevents matter from being exchanged with the surroundings, and the thermal insulation provided by the ideal thermos also prevents heat from being transferred from the system to the surroundings, including the bottle itself. From the soup's point of view, the soup *is* the entire universe; it neither picks up nor donates any matter to the rest of the world and it loses no energy. In an isolated system, the system has no surroundings, which is why it is called *isolated*. The mass of the system does not change, and its energy content is constant; the soup stays at one constant temperature. Actually, even the best real thermos bottle cannot maintain the soup as an isolated system over time because heat will leak out, and the contents will cool. But for short periods, the soup in a good thermos bottle approximates an isolated system.

Hot soup in a cup with a lid is an example of a **closed system**, which is a system that exchanges energy but not matter with its surroundings. Because the cup has a lid, no vapor (which is matter) escapes from the soup and no matter can be added. Only energy is exchanged between the soup and its surroundings. Energy from the soup is transferred—first into the cup walls, then into the air and the tabletop—and the soup gradually cools. Many real systems are closed systems.

Soup in an open cup is an **open system**, one that can exchange both energy and matter with its surroundings. Energy from the soup is transferred to the surroundings (cup, air, tabletop), and matter in the form of water vapor leaves the system and enters the air. We may add matter to the system from the surroundings by sprinkling on a little grated cheese, some ground pepper, or a few crumbled crackers.

Most of the real systems we deal with are closed, and we may treat some of them as isolated, at which point we assume behavior that is more ideal than real. We do this because isolated and closed systems are easier to model quantitatively than open systems. However, many important systems are open—including cells, organisms, and Earth itself.

CONCEPT **TEST**

Identify the following systems as isolated, closed, or open: (a) the water in a pond; (b) a carbonated beverage in a sealed bottle; (c) a sandwich wrapped in thermally conducting plastic wrap; (d) a live chicken.

(Answers to Concept Tests are in the back of the book.)

Exothermic and Endothermic Processes

Chemists classify chemical reactions and changes of state on the basis of whether they give off or absorb energy. A thermochemical process that results in the transfer of energy from a system to its surroundings is **exothermic** from the point of view of the system (Figure 5.12a). This flow of energy can be detected because it causes an increase in the temperature of the surroundings. Combustion reactions (the reactants are the system) release energy and are examples of exothermic reactions. In contrast, a chemical reaction or change of state that absorbs energy from the surroundings is **endothermic** from the point of view of the system (Figure 5.12b). For example, ice cubes (the system) in a glass of warm water (the surroundings) absorb energy from the water, which causes the cubes to melt.

In another phase change, condensation of water vapor (the system) from humid air, forming droplets of liquid water on the outside of a glass containing an ice-cold drink (part of the surroundings), requires that energy flow from the system to the surroundings. From the point of view of the system, this process is exothermic. If we pour out the cold drink and pour hot coffee into the glass, the water droplets (the system) on the outside surface of the glass absorb energy from the coffee and vaporize in a process that is endothermic from the point of view of the system. This illustrates an important concept: a process that is exothermic in one direction (vapor → liquid; releases energy) is endothermic in the reverse direction (liquid → vapor; absorbs energy).

We use the symbol q to represent the *quantity* of energy transferred during a chemical reaction or a change of state. For historical reasons, q is called heat and represents energy transferred directly because of a difference in temperature. If the reaction or process is endothermic, q is positive, indicating that energy is *gained* by the system. If the reaction or process is exothermic, q is negative, meaning that the system *loses* energy to its surroundings. In Figure 5.13, endothermic changes of state are represented by arrows pointing upward: solid → liquid, liquid → gas, and solid → gas. The opposite changes: liquid → solid, gas → liquid, and gas → solid, represented by arrows pointing downward, are exothermic. To summarize:

$$\text{Exothermic: } q < 0 \qquad \text{Endothermic: } q > 0$$

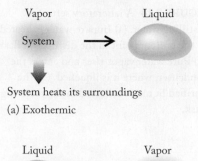

System heats its surroundings
(a) Exothermic

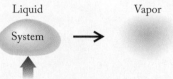

Surroundings heat the system
(b) Endothermic

FIGURE 5.12 A process that is exothermic in one direction, such as (a) the condensation of a vapor, is endothermic in the reverse direction, such as (b) the evaporation of a liquid. Reversing a process changes the direction in which energy is transferred but not the quantity of energy transferred.

CONNECTION In Chapter 3 we defined combustion as the reaction of oxygen with another element.

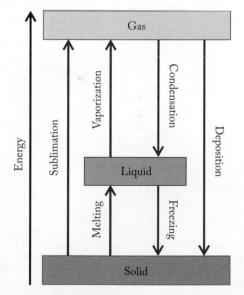

FIGURE 5.13 Matter can be transformed from one physical state to another through adding or removing energy. Red, upward-pointing arrows represent endothermic processes (energy enters the system from the surroundings; $q > 0$). Blue, downward-pointing arrows represent exothermic processes (energy leaves the system and enters the surroundings; $q < 0$).

FIGURE 5.14 A laboratory setup for distilling water. (a) Impure water is heated to the boiling point in the distillation flask. (b) Pure water vapor rises and enters the condenser, where it is liquefied. (c) The purified liquid is collected in the receiving flask.

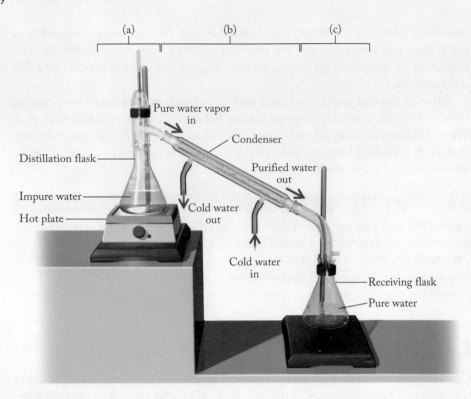

internal energy (E) the sum of all the kinetic and potential energies of all the components of a system.

SAMPLE EXERCISE 5.2 Identifying Exothermic and Endothermic Processes **LO2**

Describe the flow of energy during the purification of water by distillation (Figure 5.14), identify the steps in the process as either endothermic or exothermic, and give the sign of *q* associated with each step. Consider the water being purified to be the system.

Collect and Organize We are given that the water is the system. We must evaluate how the water gains or loses energy during distillation.

Analyze In distillation, energy flows in three steps: (1) liquid water is heated to the boiling point and (2) vaporizes. (3) The vapors are cooled and condense as they pass through the condenser.

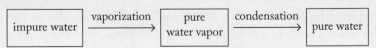

Solve Energy flows from the surroundings (hot plate) to heat the impure water (the system) to its boiling point and then to vaporize it. Therefore processes 1 and 2 are endothermic. The sign of *q* is positive for both. Because energy flows from the system (water vapor) into the surroundings (condenser walls), process 3 is exothermic. Therefore, the sign of *q* is negative.

Think About It *Endothermic* means that energy is transferred from the surroundings into the system—the water in the distillation flask. When the water vapor is cooled in the condenser, energy flows from the vapor as it is converted from a gas to a liquid; the process is exothermic.

⊛ **Practice Exercise** What is the sign of *q* as (a) a match burns, (b) drops of molten candle wax solidify, and (c) perspiration evaporates from skin? In each case, define the system and indicate whether the process is endothermic or exothermic.

(Answers to Practice Exercises are in the back of the book.)

(a) Molecules close together; same nearest neighbor over time

(b) Molecules close together but moving; exchanging nearest neighbors

(c) Molecules widely separated; moving rapidly

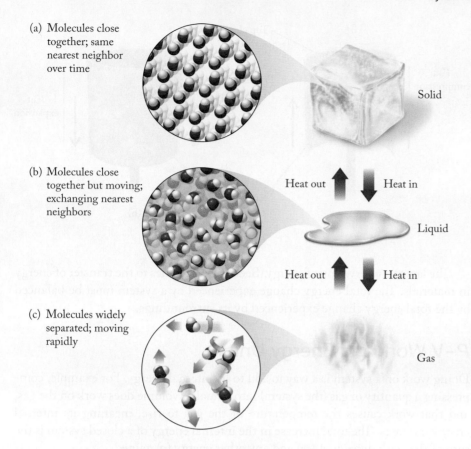

Solid

Heat out ↑ ↓ Heat in

Liquid

Heat out ↑ ↓ Heat in

Gas

FIGURE 5.15 Changes of state. (a) The molecules in solid ice are held together in a rigid three-dimensional arrangement; they have the same nearest neighbors over time. Solid ice absorbs energy and is converted to liquid water. (b) As the solid melts, the molecules in the liquid state exchange nearest neighbors and occupy many more positions relative to one another than were possible in the solid. The liquid absorbs energy and is converted to a gas. (c) The molecules in a gas are widely separated from one another and move rapidly. The reverse of these processes occurs when water in the gas state loses energy and condenses to a liquid. The liquid also loses energy when it is converted to a solid.

CHEMTOUR
Internal Energy

CONNECTION In Chapter 1 we discussed the arrangement of molecules in ice, water, and water vapor.

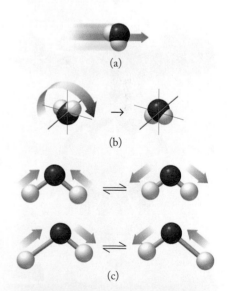

(a)

(b)

(c)

FIGURE 5.16 Some of the types of molecular motion that contribute to the overall internal energy of a system: (a) translational motion, motion from place to place along a path; (b) rotational motion, motion about a fixed axis; and (c) vibrational motion, movement back and forth from some central position.

Let's look at what happens at the molecular level as ice melts. These same types of changes occur when any substance goes through a phase change. Consider the flow of energy when an ice cube is left on a kitchen counter (Figure 5.15). Qualitatively, as the cube (the system) absorbs energy from the air and the counter (the surroundings) and starts to melt, the attractive forces that hold the water molecules in place in solid ice are overcome. The water molecules now have more positions available that they can occupy because they are not held in a rigid lattice, so they have more potential energy. After all the ice has melted into liquid water at 0°C, the temperature of the water (the system) slowly rises to room temperature. As the temperature increases, the average kinetic energy of the molecules increases and they move more rapidly. (In Section 5.4 we present a quantitative view of this process.)

The kinetic energy of a system is part of its **internal energy (E)**, defined as the sum of the kinetic and potential energies of all the components of the system (Figure 5.16). It is not possible to determine the absolute values of kinetic and potential energies, but *changes* in internal energy (ΔE) are fairly easy to measure because a change in a system's physical state or temperature is a measure of the change in its internal energy. (The capital Greek delta, Δ, is the standard way scientists symbolize change in a quantity.) The change in internal energy is the difference between the final internal energy of the system and its initial internal energy:

$$\Delta E = E_{\text{final}} - E_{\text{initial}} \tag{5.4}$$

Internal energy is a state function because ΔE depends only on the initial and final states. How the change occurs in the system does not matter.

FIGURE 5.17 Work performed by changing the volume of a gas. (a) Highest position of the piston after it compresses the gas in the cylinder, causing the fuel to ignite in a diesel engine. The piston does work on the gas in decreasing the gas volume. (b) The exploding fuel releases energy, causing the gas in the cylinder to expand and push the piston down. The gas has done work on the piston.

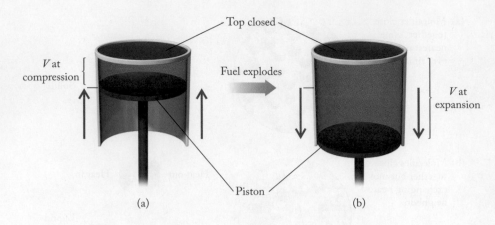

(a)   (b)

CHEMTOUR
Pressure–Volume Work

FIGURE 5.18 In a completely deflated hot-air balloon, $V = 0$. Inflating the balloon causes it to do work on the atmosphere by pushing against the air. Because this work involves a change in volume, it is called P–V work. The blue arrow represents collisions by the atmosphere against the outside of the balloon.

The law of conservation of energy (Section 5.2) applies to the transfer of energy in materials. The total energy change experienced by a system must be balanced by the total energy change experienced by its surroundings.

P–V Work and Energy Units

Doing work on a system is a way to add to its internal energy. For example, compressing a quantity of gas (the system) into a smaller volume does work on the gas, and that work causes the temperature of the gas to rise, meaning its internal energy increases. The total increase in the internal energy of a closed system is the sum of the work done on it (w) and any other energy (q) gained:

$$\Delta E = q + w \tag{5.5}$$

Equation 5.5 expresses the **first law of thermodynamics**: The energy changes ($\Delta E = q + w$) of a closed system and its surroundings are equal in magnitude but opposite in sign, so their sum is zero:

$$\Delta E_{sys} + \Delta E_{surr} = 0$$

When work is done *by* a system on its surroundings, the internal energy of the system decreases. For example, when fuel in the cylinder of a diesel engine ignites and produces hot gases, the gases (the system) expand and do work on the surroundings by pushing on the piston (Figure 5.17).

As another example of work being done by a system on its surroundings, consider a hot-air balloon (Figure 5.18), where we define the air in the balloon as the system. The air in the balloon is heated by a burner located at the balloon's bottom, which is open. Heating this air causes the balloon to expand. As the volume of the balloon increases, the balloon presses against the air outside the balloon, thus doing work on that outside air (the surroundings). This type of work, in which the pressure on a system remains constant but the volume of the system changes, is called **pressure–volume (P–V) work**.

The pressure referred to in the example of the hot-air balloon is *atmospheric pressure*. The pressure of the atmosphere on the balloon in this illustration, and indeed on all objects, is a result of Earth's gravity pulling the atmosphere toward the surface of the planet, where it exerts a force on all things because of its mass. The pressure at sea level on a dry day is about 1.00 atmosphere (atm). Atmospheric pressure varies slightly with the weather, but in an example involving P–V work like this one with the balloon, the important feature is that the pressure is constant.

Let's examine what happens to the balloon in terms of heat transferred and work done under constant pressure. First, adding hot air to the balloon increases its internal energy and causes the balloon to expand. Second, the expansion of the balloon against the pressure of its surroundings is $P–V$ work done by the system on its surroundings. The internal energy of the system decreases as it performs this $P–V$ work. We can relate this change in internal energy to the energy gained by the system by heating (q) and the work done by the system ($P\Delta V$) by writing Equation 5.5 in the form

$$\Delta E = q + (-P\Delta V) = q - P\Delta V \qquad (5.6)$$

The negative sign in front of $P\Delta V$ is appropriate in this case because, when the system expands (positive change in volume ΔV), it loses energy (negative change in internal energy ΔE) as it does work on its surroundings. Correspondingly, when the surroundings do work on the system (for example, when a gas is compressed), the quantity ($-P\Delta V$) has a positive value.

The sign of q may also be either positive or negative (Figure 5.19). If the system is heated by its surroundings, then q is positive ($q > 0$). Energy is added to the balloon, for instance, when it is being inflated, because energy flows from the surroundings (the burner) into the system (the air in the balloon). When the balloon expands, work is done by the system, w is negative ($w < 0$), and the internal energy decreases. If energy is transferred from the system into the surroundings due to a temperature difference, q is negative ($q < 0$), and the internal energy decreases. From Equation 5.6 we see that the change in internal energy of a system is positive when more energy enters the system than leaves and negative when more energy leaves the system than enters. Figure 5.19 also illustrates the first law of thermodynamics: the energy changes ($\Delta E = q + w$) of the system and the surroundings are equal in magnitude but opposite in sign, so their sum is zero.

Let's pause for a moment and consider the units needed to calculate ΔE by using Equation 5.6. Energy changes that accompany chemical reactions and changes in physical state are sometimes expressed in calories. A **calorie (cal)** is the quantity of energy required to raise the temperature of 1 g of water from 14.5°C to 15.5°C. The SI unit of energy, used throughout this text, is the **joule (J)**; 1 cal = 4.184 J. The *Calorie* (Cal; note the capital *C*) in nutrition is actually one kilocalorie (kcal): 1 Cal = 1 kcal = 1000 cal. If the SI unit for energy is a joule, but pressure is measured in atmospheres and volume in liters, how can we calculate work? A conversion factor, 1 L · atm = 101.32 J, connects joules to liter-atmospheres.

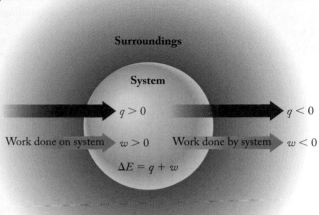

FIGURE 5.19 Energy entering a system by heating and by work done on the system by the surroundings are both positive quantities because both increase the internal energy of the system. Energy released by a system to the surroundings and work done by a system on the surroundings are both negative quantities because both decrease the internal energy of a system.

FIGURE 5.20 Hot gases expand, pushing down on the piston.

SAMPLE EXERCISE 5.3 Calculating Changes in Internal Energy   **LO3**

Figure 5.20 shows a simplified version of a piston and cylinder in an engine. Suppose combustion of fuel injected into the cylinder produces 155 J of energy. The hot gases in the cylinder expand, pushing the piston down and doing 93 J of $P–V$ work on the piston. If the system is the gases in the cylinder, what is the change in internal energy of the system?

Collect and Organize We are given q and w and asked to calculate ΔE. We will need to decide whether q and w are positive or negative according to the sign convention in Figure 5.19.

Analyze The change in internal energy is related to the work done by or on a system and the energy gained or lost by the system (Equation 5.5). The system (gases in

first law of thermodynamics the principle that the energy gained or lost by a system must equal the energy lost or gained by the surroundings.

pressure–volume (P–V) work the work associated with the expansion or compression of a gas.

calorie (cal) the amount of energy necessary to raise the temperature of 1 gram of water by 1°C, from 14.5°C to 15.5°C.

joule (J) the SI unit of energy; 4.184 J = 1 cal.

the cylinder) absorbs energy by heating, so $q > 0$, and the system does work on the surroundings (the piston), so $w < 0$.

Solve When we substitute values into Equation 5.5, the change in internal energy for the system is

$$\Delta E = q + w = (155\,\mathrm{J}) + (-93\,\mathrm{J}) = 62\,\mathrm{J}$$

Think About It More energy enters the system (155 J) than leaves it (93 J), so a positive value of ΔE is reasonable.

Practice Exercise In another event, the piston in Figure 5.20 compresses the air in the cylinder by doing 64 J of work on the gas. As a result, the air gives off 32 J of energy to the surroundings. If the system is the air in the cylinder, what is the change in its internal energy?

(Answers to Practice Exercises are in the back of the book.)

SAMPLE EXERCISE 5.4 Calculating *P–V* Work   **LO3**

A tank of compressed helium is used to inflate balloons for sale at a carnival on a day when the atmospheric pressure is 0.99 atm. If each balloon is inflated from an initial volume of 0.0 L to a final volume of 4.8 L, how much *P–V* work, in joules, is done on the surrounding atmosphere by 100 balloons when they are inflated? The atmospheric pressure remains constant during the filling process.

Collect and Organize Each of the 100 balloons goes from empty ($V = 0.0$ L) to 4.8 L, which means $\Delta V = 4.8$ L, and the atmospheric pressure P is constant at 0.99 atm. The identity of the gas used to fill the balloons does not matter because under these conditions all gases behave the same way, regardless of their identity. Our task is to determine how much *P–V* work is done by the 100 balloons on the air that surrounds them. To express our answer in units of joules, we will need the conversion factor: $1\,\mathrm{L} \cdot \mathrm{atm} = 101.32\,\mathrm{J}$.

Analyze We focus on the work done on the atmosphere (the surroundings) by the system; the balloons and the helium they contain are the system.

Solve The volume change (ΔV) as all the balloons are inflated is

$$100\ \cancel{\text{balloons}} \times 4.8\ \mathrm{L/}\cancel{\text{balloon}} = 4.8 \times 10^2\ \mathrm{L}$$

The work (w) done by our system (the balloons) as they inflate against an external pressure of 0.99 atm is

$$w = -P\Delta V = -0.99\ \mathrm{atm} \times 4.8 \times 10^2\ \mathrm{L}$$

$$= -4.8 \times 10^2\ \mathrm{L} \cdot \mathrm{atm}$$

Finally, we convert the value for work from L · atm to J by using the conversion factor:

$$w = -4.8 \times 10^2\ \cancel{\mathrm{L \cdot atm}} \times \frac{101.32\ \mathrm{J}}{\cancel{\mathrm{L \cdot atm}}} = -4.9 \times 10^4\ \mathrm{J} = -49\ \mathrm{kJ}$$

Think About It Because work is done *by* the system on its surroundings, the work is negative from the point of view of the system.

Practice Exercise The balloon *Spirit of Freedom* (Figure 5.21), flown around the world by American aviator Steve Fossett in 2002, contained 5.50×10^5 cubic feet of helium. How much *P–V* work was done by the balloon on the surrounding atmosphere while the balloon was being inflated, if we assume that the atmospheric pressure was 1.00 atm? ($1\,\mathrm{m}^3 = 1000\,\mathrm{L} = 35.3\,\mathrm{ft}^3$)

(Answers to Practice Exercises are in the back of the book.)

FIGURE 5.21 The balloon *Spirit of Freedom* was flown around the world in 2002.

5.4 Enthalpy and Enthalpy Changes

enthalpy change (ΔH) the heat absorbed by an endothermic process or given off by an exothermic process occurring at constant pressure.

enthalpy (H) the sum of the internal energy and the pressure–volume product of a system; $H = E + PV$.

Many physical and chemical changes take place at constant atmospheric pressure (P). For example, boiling water for tea or any of the chemical reactions described in Chapter 4 takes place in vessels open to the air. The thermodynamic parameter that relates the flow of energy into or out of a system during chemical reactions or physical changes at constant pressure is called the **enthalpy change (ΔH)**. We symbolize this enthalpy change at constant pressure as q_P, where the subscript P specifies a process taking place at constant pressure. We then rearrange Equation 5.6 to define ΔH as

$$\Delta H = q_P = \Delta E + P\Delta V \qquad (5.7)$$

Thus the change in enthalpy is the change in the internal energy of the system at constant pressure plus the P–V work done by the system on its surroundings.

The **enthalpy (H)** of a thermodynamic system is the sum of the internal energy and the pressure–volume product ($H = E + PV$). However, as we saw with internal energy, determining the absolute values of these parameters is difficult, whereas determining *changes* is fairly easy. We therefore concentrate on the *change* in enthalpy (ΔH) of a system or the surroundings. Equation 5.7 tells us that for a reaction run at constant pressure, the enthalpy change is equal to q_P, which is the heat gained or lost by the system during the reaction. This statement also means that the units for ΔH are the same as those for q: enthalpy has units of joules (J), and, if it is reported with respect to the quantity of a substance, J/g or J/mol.

Both ΔH and ΔE represent changes in a state function of a system. These two terms are similar, but the difference between them is important. ΔE includes *all* the energy (heat and work) exchanged by the system with the surroundings: $\Delta E = q + w$. However, ΔH is only q, the heat, exchanged at constant pressure: $\Delta H = q_P$. If a chemical reaction does not involve changes in volume, then ΔE and ΔH have very similar values. But if, for example, a reaction consumes or produces gas and the system experiences large changes in volume as a result, then ΔE and ΔH can be quite different.

When energy flows out of a system, q is negative according to our sign convention (Figure 5.19), so the enthalpy change is negative: $\Delta H < 0$. When energy flows into a system, q is positive, so $\Delta H > 0$. As we saw in Section 5.3, positive q values indicate endothermic processes, and negative q values indicate exothermic processes. Knowing that, we can relate the terms *exothermic* and *endothermic* to enthalpy changes as well.

For example, consider the flow of energy into a melting ice cube (the system; Figure 5.22) from its surroundings. This process is endothermic, meaning that q—and therefore ΔH—is positive. However, to make ice cubes in a freezer, the water (the system) must lose energy, which means the process is exothermic and both q and ΔH are negative. The enthalpy changes for the two processes—melting and freezing—have different signs but, for a given amount of water, they have the same absolute value. As Figure 5.22 illustrates, the energy required to melt a given quantity of a substance has the same magnitude but is opposite in sign to the energy given off when that same quantity of material freezes. For example, $\Delta H_{fus} = 6.01$ kJ if 1 mole of ice melts, but $\Delta H_{solid} = -6.01$ kJ if 1 mole of water freezes. We add the subscript *fus* or *solid* to ΔH to identify which process is occurring: melting (*fusion*) or freezing (*solidification*). The magnitudes and signs

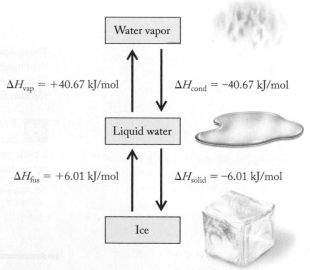

$\Delta H_{vap} = +40.67$ kJ/mol $\Delta H_{cond} = -40.67$ kJ/mol

$\Delta H_{fus} = +6.01$ kJ/mol $\Delta H_{solid} = -6.01$ kJ/mol

FIGURE 5.22 The enthalpy changes at 100°C for vaporization of liquid water, $\Delta H_{vap} = 40.67$ kJ/mol, and condensation of water vapor, $\Delta H_{cond} = -40.67$ kJ/mol, are equal in magnitude but opposite in sign. So, too, are the enthalpy changes at 0°C for melting ice, $\Delta H_{fus} = 6.01$ kJ/mol, and freezing water, $\Delta H_{solid} = -6.01$ kJ/mol.

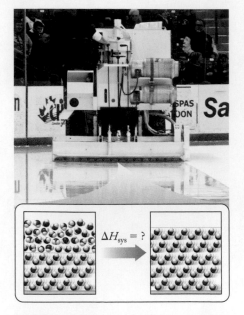

FIGURE 5.23 An ice-resurfacing machine spreads a thin layer of liquid water on top of the existing ice. The liquid freezes, providing a fresh, smooth surface for skaters. If the water represents the system, the energy change for the freezing of the water is described by ΔH_{sys}.

of enthalpies associated with other paired phase changes—vaporization and condensation, and sublimation and deposition—also work this way.

In terms of symbols, we put subscripts on ΔH to clarify not only the specific process but also the part of the universe to which the value applies. If we wish to indicate the enthalpy change associated with the system, we may write ΔH_{sys}.

SAMPLE EXERCISE 5.5 Determining the Value and Sign of ΔH **LO4**

Between periods of a hockey game, an ice-resurfacing machine (Figure 5.23) spreads 855 L of water across the surface of a hockey rink. (a) If the system is the water, what is the sign of ΔH_{sys} as the water freezes? (b) To freeze this volume of water at 0°C, what is the value of ΔH_{sys}? The density of water is 1.00 g/mL.

Collect, Organize, and Analyze We are given the volume and density of the water, and we need to determine the sign and value of ΔH_{sys}. The water from the ice-resurfacing machine is the system, so we can determine the sign of ΔH by determining whether energy transfer is into or out of the water. (a) The freezing takes place at constant pressure, so $\Delta H_{sys} = q_P$. (b) To calculate the amount of energy lost from the water as it freezes, we must convert 855 L of water into moles of water because the conversion factor between the quantity of energy removed and the quantity of water that freezes is the enthalpy of solidification of water, $\Delta H_{solid} = -6.01$ kJ/mol.

Solve
a. For the water (the system) to freeze, energy must be removed from it. Therefore, ΔH_{sys} must be a negative value.
b. We convert the volume of water into moles:

$$855 \text{ L} \times \frac{1000 \text{ mL}}{1 \text{ L}} \times \frac{1.00 \text{ g}}{1 \text{ mL}} \times \frac{1 \text{ mol}}{18.02 \text{ g}} = 4.745 \times 10^4 \text{ mol}$$

Then we calculate ΔH_{sys}:

$$\Delta H_{sys} = 4.745 \times 10^4 \text{ mol} \times \frac{-6.01 \text{ kJ}}{1 \text{ mol}} = -2.85 \times 10^5 \text{ kJ}$$

Think About It The enthalpy change ΔH varies with pressure, but the magnitude of the change is negligible for pressures near normal atmospheric pressure. The magnitude of the answer to part (b) is reasonable given the large amount of water that is frozen to refinish the rink's surface.

Practice Exercise The flame in a torch used to cut metal is produced by burning acetylene (C_2H_2) in pure oxygen. If we assume that the combustion of 1 mole of acetylene releases 1251 kJ of energy, what mass of acetylene is needed to cut through a piece of steel if the process requires 5.42×10^4 kJ of energy?

(Answers to Practice Exercises are in the back of the book.)

SAMPLE EXERCISE 5.6 The Sign of ΔH in a Chemical Reaction   **LO4**

A chemical cold pack (Figure 5.24) contains a small pouch of water inside a bag of solid ammonium nitrate. To activate the pack, you press on it to rupture the pouch, allowing the ammonium nitrate to dissolve in the water. The result is a cold, aqueous solution of ammonium nitrate. Write a balanced chemical equation describing the dissolution of solid NH_4NO_3. If we define the system as the NH_4NO_3 and the surroundings as the water, what are the signs of ΔH_{sys} and q_{surr}?

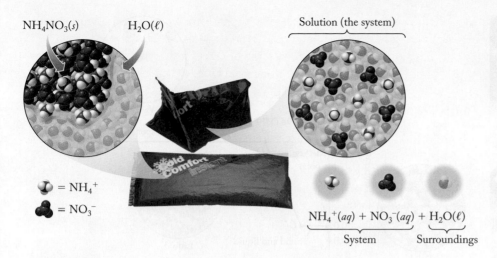

FIGURE 5.24 A chemical cold pack contains a bag of water inside a larger bag of ammonium nitrate. Breaking the inner bag and allowing the NH_4NO_3 to dissolve leads to a decrease in temperature.

Collect, Organize, and Analyze The solution's temperature drops when we dissolve NH_4NO_3 in water. We want to write a balanced chemical equation for the dissolution of ammonium nitrate and determine the sign of ΔH for the solution (the system) and q_{surr}. The first law of thermodynamics says energy must be conserved, so all the energy must be accounted for between system and surroundings.

Solve The solubility rules in Tables 4.4 and 4.5 tell us that all ammonium salts are soluble. We can write a balanced chemical equation for the dissolution of NH_4NO_3 (the system) as

$$NH_4NO_3(s) \rightarrow NH_4^+(aq) + NO_3^-(aq)$$

The low temperature of the cold pack (the system) means that heat flows into it from its warmer surroundings. Therefore the sign of ΔH_{sys} is positive ($\Delta H_{sys} > 0$), and the sign of q_{surr} is negative ($q_{surr} < 0$).

Think About It The first law of thermodynamics tells us that the signs of ΔH_{sys} and q_{surr} must be opposite.

Practice Exercise Potassium hydroxide can be used to unclog sink drains. The reaction between potassium hydroxide and water is quite exothermic. What are the signs of ΔH_{sys} and q_{surr}?

(Answers to Practice Exercises are in the back of the book.)

5.5 Heating Curves, Molar Heat Capacity, and Specific Heat

Winter hikers and high-altitude mountain climbers use portable stoves fueled by propane or butane to prepare hot meals. Ice or snow may be their only source of water. In this section, we use this scenario to examine the transfer of energy into water that begins as snow and ends up as vapor.

Hot Soup on a Cold Day

Let's consider the changes of temperature and state that water undergoes as some hikers melt snow as a first step in preparing soup from a dry soup mix. Suppose they start with a saucepan filled with snow at −18°C at constant pressure. They place the pan above the flame of a portable stove, and energy begins to flow into the snow. The

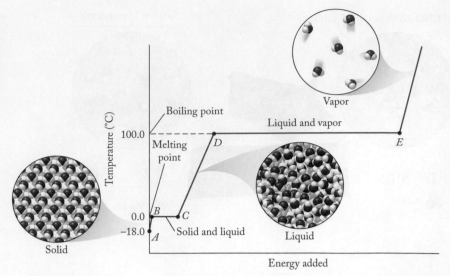

FIGURE 5.25 The energy required to melt snow and boil the resultant water is illustrated by the four line segments on the heating curve of water: heating snow to its melting point, $\overline{AB}$; melting the snow to form liquid water, $\overline{BC}$; heating the water to its boiling point, $\overline{CD}$; and boiling the water to convert it to vapor, $\overline{DE}$.

CHEMTOUR
Heating Curves

TABLE 5.1 Molar Heat Capacities of Selected Substances at 25°C

Substance	c_P [J/(mol · °C)]
Al(s)	24.4
Cu(s)	24.5
Fe(s)	25.1
C(s), graphite	8.54
$CH_3CH_2OH(\ell)$, ethanol	113.1
$H_2O(s)$	37.1
$H_2O(\ell)$	75.3
$H_2O(g)$	33.6

temperature of the snow immediately begins to rise. If the flame is steady so that energy flow is constant, the temperature of the snow changes as indicated on the graph shown in Figure 5.25. First, heating increases the temperature of the snow to its melting point, 0°C. This temperature rise and the energy transfer into the snow are represented by line segment $\overline{AB}$ in Figure 5.25. The temperature of the snow then remains steady at 0°C while the snow continues to absorb energy and melt. This state change that produces liquid water from snow takes place at constant temperature and is represented by the constant-temperature (horizontal) line segment $\overline{BC}$. Phase changes of pure materials take place at constant temperature and pressure.

When all the snow has melted, the temperature of the liquid water rises (line segment $\overline{CD}$) until it reaches 100°C. At 100°C, the temperature of the water remains steady while another state change takes place: liquid water becomes water vapor. This constant-temperature process is represented by line segment $\overline{DE}$ in Figure 5.25. If all the liquid water in the pan were converted into vapor, the temperature of the vapor would then rise as long as heat was added, as indicated by the final (slanted) line segment in Figure 5.25.

The difference in the x-axis coordinates for the beginning and end of each line segment in Figure 5.25 indicates how much energy is required in each step in this process. In the first step (line segment $\overline{AB}$), the energy required to raise the temperature of the snow from −18°C to 0°C can be calculated if we know how many moles of snow we have and how much energy is required to change the temperature of 1 mole of snow.

Molar heat capacity (c_P) is the quantity of energy required to raise the temperature of 1 *mole* of a substance by 1°C. The symbol we use for molar heat capacity is c_P, where the subscript P again indicates the value for a process taking place at *constant pressure*. Molar heat capacities for several substances are given in Table 5.1, from which we see that ice has a molar heat capacity of 37.1 J/(mol · °C) at constant pressure. From these units, we see that if we know the number of moles of snow and the temperature change it experiences as we bring it to its melting point, we can calculate q for the process:

$$q = nc_P\Delta T \tag{5.8}$$

where n is the number of moles of the substance absorbing or releasing energy and ΔT is the temperature change in degrees Celsius.

In addition to molar heat capacity, other measures exist that specify how much energy is required to increase the temperature of a substance, each referring to a specific amount of the substance. For example, some tables of thermodynamic data list values of *specific heat capacity*, or **specific heat (c_s)**, which is the energy required to raise the temperature of 1 *gram* of a substance by 1°C; c_s has units of J/(g · °C). Note the difference that specific heat is for a specific mass, whereas molar heat capacity is for 1 mole of a substance.

Let's assume the hikers decide to cook their meal with 270 g of snow. Dividing that mass by the molar mass of water (18.02 g/mol), we find that they melt 15.0 moles of snow. To calculate how much energy is needed to raise the temperature of 15.0 moles of $H_2O(s)$ from −18°C to 0°C, we use Equation 5.8:

$$q = nc_P\Delta T$$

$$= 15.0 \text{ mol} \times \frac{37.1 \text{ J}}{\text{mol} \cdot °C} \times (+18°C) = 1.0 \times 10^4 \text{ J} = 10 \text{ kJ}$$

Notice that this value is positive, which means that the system (the snow) gains energy as it warms up.

The energy absorbed as the snow melts, or *fuses* (line segment $\overline{BC}$ in Figure 5.25), can be calculated using the enthalpy change that takes place as 1 mole of snow melts. This enthalpy change is called the **molar enthalpy of fusion (ΔH_{fus})**, and the energy absorbed as *n* moles of a substance melts is given by

$$q = n\Delta H_{fus} \tag{5.9}$$

The molar enthalpy of fusion for water is 6.01 kJ/mol, and using this value and the known number of moles of snow, we get

$$q = 15.0 \text{ mol} \times \frac{6.01 \text{ kJ}}{\text{mol}} = 90.2 \text{ kJ}$$

This value is positive because energy enters the system. This energy overcomes the attractive forces between the water molecules in the solid, and the snow becomes a liquid. Note that no factor for temperature appears in this calculation because state changes of pure substances take place at constant temperature, as the two horizontal line segments in Figure 5.25 indicate. Snow at 0°C becomes liquid water at 0°C.

While the water temperature increases from 0°C to 100°C, the relation between temperature and energy absorbed is again defined by Equation 5.8, but this time c_P represents not the molar heat capacity of $H_2O(s)$ but the molar heat capacity of $H_2O(\ell)$, 75.3 J/(mol · °C) (see Table 5.1):

$$q = nc_P\Delta T$$

$$= 15.0 \text{ mol} \times \frac{75.3 \text{ J}}{\text{mol} \cdot °C} \times 100°C = 1.13 \times 10^5 \text{ J} = 113 \text{ kJ}$$

This value is positive because the system takes in energy as its temperature rises.

At this point in our story, our hiker–chefs are ready to make their soup and enjoy a hot meal. However, if they accidentally leave the boiling water unattended, it will eventually vaporize completely. How much heat is needed to boil away all the water in the pot? As line segment $\overline{DE}$ in Figure 5.25 shows, the temperature of the water remains at 100°C until all of it is vaporized. The enthalpy change associated with changing 1 mole of a liquid to a gas is the **molar enthalpy of vaporization (ΔH_{vap})**, and the quantity of energy absorbed as *n* moles of a substance vaporizes is given by

$$q = n\Delta H_{vap} \tag{5.10}$$

molar heat capacity (c_P) the quantity of energy required to raise the temperature of 1 mole of a substance by 1°C at constant pressure.

specific heat (c_s) the quantity of energy required to raise the temperature of 1 gram of a substance by 1°C at constant pressure.

molar enthalpy of fusion (ΔH_{fus}) the energy required to convert 1 mole of a solid substance at its melting point into the liquid state.

molar enthalpy of vaporization (ΔH_{vap}) the energy required to convert 1 mole of a liquid substance at its boiling point to the vapor state.

FIGURE 5.26 Macroscopic and molecule-level views of (a) ice, (b) water, and (c) water vapor.

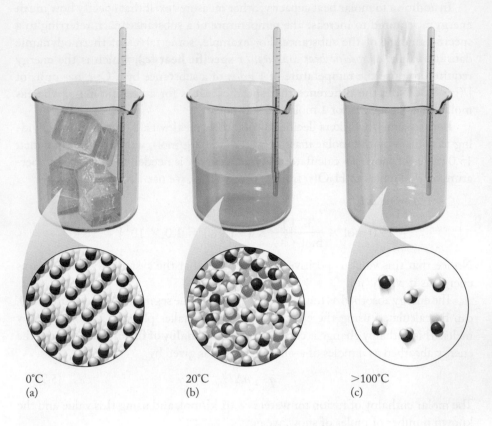

FIGURE 5.26 Macroscopic and molecule-level views of (a) ice, (b) water, and (c) water vapor.

The molar enthalpy of vaporization for water is 40.67 kJ/mol, so to completely vaporize the water we need an additional 610 kJ of energy:

$$q = (15.0 \text{ mol})(40.67 \text{ kJ/mol}) = 610 \text{ kJ}$$

This value is positive because the system takes in energy as it converts liquid into vapor. Again, no factor for temperature appears because the phase change takes place at constant temperature. Only after all the water has vaporized does its temperature increase above 100°C, along the line above point E in Figure 5.25. The molar heat capacity of steam, 33.6 J/(mol · °C), is used to calculate the energy required to heat the vapor to any temperature above 100°C.

The c_P values for $H_2O(s)$, $H_2O(\ell)$, and $H_2O(g)$ are different. Nearly all substances have different molar heat capacities in their different physical states.

Notice in Figure 5.25 that line segment $\overline{DE}$, the phase change from liquid water to water vapor, is much longer than the line segment $\overline{BC}$ representing the change from solid snow to liquid water. The relative lengths of these lines indicate that the molar enthalpy of vaporization of water (40.67 kJ/mol) is much larger than the molar enthalpy of fusion of snow (6.01 kJ/mol). Why does it take more energy to boil 1 mole of water than to melt 1 mole of snow? The answer is related to the extent to which attractive forces between molecules must be overcome in each process and to how the internal energy of the system changes as energy is added (Figure 5.26).

Attractive forces determine both the organization of the molecules in any substance and their relation to their nearest neighbors. In forms of frozen water, like ice and snow, these attractive forces are strong enough to hold the water molecules in place relative to one another. Melting the snow requires that these attractive forces be overcome. The energy added to the system at the melting point is sufficient to overcome the attractive forces and change the arrangement (and hence the potential energy).

Intermolecular attractive forces still exist in liquid water, but the energy added at the melting point increases the energy of the molecules and enables them to move with respect to their nearest neighbors. The molecules are closer together in the liquid than they are in the solid, but their relative positions constantly change; they have different nearest neighbors over time.

Once the snow has completely melted, added energy causes the temperature of the liquid water to rise, increasing its internal energy. When water vaporizes at the boiling point, the attractive forces between water molecules must be overcome to separate the molecules from one another as they enter the gas phase. Separating the molecules widely in space requires work (that is, energy), and the amount required is much larger than that for the solid-to-liquid state change. This difference is consistent with the relative lengths of $\overline{BC}$ and $\overline{DE}$ in Figure 5.25.

SAMPLE EXERCISE 5.7 Calculating the Energy Required   **LO4**
to Raise the Temperature of Water

Calculate the amount of energy required to convert 237 g of solid ice at 0.0°C to hot water at 80.0°C. The molar enthalpy of fusion (ΔH_{fus}) of ice is 6.01 kJ/mol. The molar heat capacity of liquid water is 75.3 J/(mol · °C).

Collect, Organize, and Analyze This problem refers to the process symbolized by segments $\overline{BC}$ and $\overline{CD}$ in Figure 5.25. We can calculate the temperature change (ΔT) of the water from the initial and final temperatures. Four mathematical steps are required: (1) calculate n, the number of moles of ice; (2) determine the amount of energy required to melt the ice, using $\Delta H_{fus} = 6.01$ kJ/mol in Equation 5.9; (3) calculate the amount of energy required to raise the liquid water temperature from 0.0°C to 80.0°C, using $c_P = 75.3$ J/(mol · °C) in Equation 5.8; and (4) add the results of steps (2) and (3). We can calculate the number of moles in 237 g of water by using the molar mass of water ($\mathcal{M} = 18.02$ g/mol).

Solve

1. Calculate the number of moles of water:

$$n = 237 \text{ g H}_2\text{O} \times \frac{1 \text{ mol H}_2\text{O}}{18.02 \text{ g H}_2\text{O}} = 13.2 \text{ mol H}_2\text{O}$$

2. Determine the amount of energy needed to melt the ice (Equation 5.9):

$$q_1 = n\Delta H_{fus} = 13.2 \text{ mol} \times 6.01 \text{ kJ/mol} = 79.3 \text{ kJ}$$

3. Determine the amount of energy needed to warm the water (Equation 5.8):

$$q_2 = nc_P\Delta T$$

$$= 13.2 \text{ mol} \times \frac{75.3 \text{ J}}{\text{mol} \cdot °\text{C}} \times (80.0 - 0.0)°\text{C} = 79{,}517 \text{ J} = 79.5 \text{ kJ}$$

4. Add the results of parts 2 and 3:

$$q_1 + q_2 = 79.3 \text{ kJ} + 79.5 \text{ kJ} = 158.8 \text{ kJ}$$

Think About It Using the definitions of molar enthalpy of fusion and molar heat capacity, we can also think our way through the solution to this exercise without referring to Equations 5.8 and 5.9. Molar enthalpy of fusion defines the amount of energy needed to melt 1 mole of ice. Multiplying that value (6.01 kJ/mol) by the number of moles (13.2 mol) gives us the energy required to melt the given amount of ice. By the same token, the molar heat capacity of water [75.3 J/(mol · °C)] defines the amount of energy needed to raise the temperature of 1 mole of liquid water by 1°C. We know the number of moles (13.2 mol), and we know the number of degrees by which we want to raise the temperature (80.0°C − 0.00°C = 80.0°C); multiplying those factors together gives us the energy needed to raise the temperature of the water.

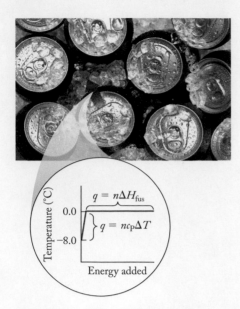

FIGURE 5.27 When beverage cans are chilled in an ice-filled cooler, energy flows from the cans to the ice until the temperature of the cans equals that of the ice.

 Practice Exercise Calculate the change in energy when 125 g of water vapor at 100.0°C condenses to liquid water and then cools to 25.0°C.

(Answers to Practice Exercises are in the back of the book.)

Water is an extraordinary substance for many reasons, one of which is its high molar heat capacity. The ability of water to absorb large quantities of thermal energy is one reason it is used as a *heat sink* both in automobile radiators and in our bodies. The term "heat sink" is often used to identify matter that can absorb energy without changing phase or significantly changing its temperature. Weather and climate changes are largely driven and regulated by cycles involving retention of energy by our planet's oceans, which serve as giant heat sinks for solar energy.

Cold Drinks on a Hot Day

Let's consider another familiar application involving energy transfer. Suppose we throw a party and plan to chill three cases (72 aluminum cans, each containing 355 mL) of beverages by placing the cans in an insulated cooler and covering them with ice cubes (Figure 5.27). If the temperature of the ice (sold in 10-pound bags) is −8.0°C and the temperature of the beverages is initially 25.0°C, how many bags of ice do we need to chill the cans and their contents to 0.0°C (as in "ice cold")?

You may already have an idea that more than one bag, but probably fewer than ten, will be needed. We can use the energy transfer relationships we have defined to predict more accurately how much ice is required. In doing so, we assume that whatever energy is absorbed by the ice is lost by the cans and the beverages in them. As the ice absorbs energy from the cans and the beverages, the temperature of the ice increases from −8.0°C to 0.0°C, and, as we saw in analyzing Figure 5.25, the resulting liquid water remains at 0.0°C until all the ice has melted. We need enough ice so that the last of it melts just as the temperature of the beverages and the cans reaches 0.0°C. Our analysis of this cooling process is a little simpler than our snow/liquid water/water vapor analysis because here there is no change of state. All the energy transferred goes only into cooling the cans and beverages to 0.0°C. The cans stay in the solid state, and the beverages stay in the liquid state.

First let's consider the energy lost in the cooling process. Two materials are to be chilled: 72 aluminum cans and 72 × 355 mL = 25,600 mL of beverages. The beverages are mostly water. The other ingredients in the beverages are present in such small concentrations that they will not affect our calculations, so we can assume that we need to reduce the temperature of 25,600 mL of water by 25.0°C. We can calculate the amount of energy lost with Equation 5.8 if we first calculate the number of moles of water in 25,600 mL, assuming a density of 1.000 g/mL:

$$25{,}600 \ \text{mL} \ \text{H}_2\text{O} \times 1.000 \ \frac{\text{g}}{\text{mL}} \times \frac{1 \ \text{mol} \ \text{H}_2\text{O}}{18.02 \ \text{g} \ \text{H}_2\text{O}} = 1420 \ \text{mol} \ \text{H}_2\text{O}$$

We can use the molar heat capacity of water and Equation 5.8 to calculate the energy lost by 1420 moles of water as its temperature decreases from 25.0°C to 0.0°C:

$$q_{\text{beverage}} = nc_{\text{p}}\Delta T$$

$$= 1420 \ \text{mol} \times \frac{75.3 \ \text{J}}{\text{mol} \cdot °\text{C}} \times (-25.0°\text{C})$$

$$= -2.67 \times 10^6 \ \text{J}$$

We must also consider the energy released in lowering the temperature of 72 aluminum cans by 25.0°C. The typical mass of a soda can is 12.5 g. The molar heat capacity of solid aluminum (Table 5.1) is 24.4 J/(mol · °C), and the molar mass of aluminum is 26.98 g/mol. Thus

$$q_{cans} = nc_P\Delta T$$

$$= 72 \text{ cans} \times \frac{12.5 \text{ g Al}}{\text{can}} \times \frac{1 \text{ mol}}{26.98 \text{ g Al}} \times \frac{24.4 \text{ J}}{\text{mol} \cdot {}^\circ\text{C}} \times (-25.0{}^\circ\text{C})$$

$$= -2.03 \times 10^4 \text{ J}$$

The total quantity of energy that must be removed from the cans and beverages is

$$q_{total\,lost} = q_{beverage} + q_{cans} = [(-2.67 \times 10^6) + (-2.03 \times 10^4)] \text{ J}$$

$$= (-2.67 - 0.0203) \times 10^6 \text{ J} = -2.69 \times 10^6 \text{ J}$$

$$= -2.69 \times 10^3 \text{ kJ}$$

This quantity of energy must be absorbed by the ice as it warms to its melting point and then melts. The calculation of the amount of ice needed is an algebra problem. Let n be the number of moles of ice needed. The energy absorbed is the sum of (1) the energy needed to raise the temperature of n moles of ice from $-8.0°$ to 0.0°C and (2) the energy needed to melt n moles of ice. These quantities can be calculated with Equation 5.8 for step 1 and Equation 5.9 for step 2. Note that the c_P values in Table 5.1 have units of joules, whereas ΔH_{fus} values given earlier have units of kilojoules. We need to have both terms in the same units, so we use 0.0371 kJ/(mol · °C) for c_P:

$$q_{total\,gained} = q_1 + q_2$$

$$= nc_P\Delta T + n\Delta H_{fus}$$

$$= n\left(\frac{0.0371 \text{ kJ}}{\text{mol} \cdot {}^\circ\text{C}}\right)(80{}^\circ\text{C}) + n(6.01 \text{ kJ/mol})$$

$$= n(6.31 \text{ kJ/mol})$$

The energy lost by the cans and the beverages balances the energy gained by the ice:

$$-q_{total\,lost} = +q_{total\,gained}$$

$$2.69 \times 10^3 \text{ kJ} = n(6.31 \text{ kJ/mol})$$

$$n = 4.26 \times 10^2 \text{ mol ice}$$

Converting 4.26×10^2 mol of ice into pounds gives us

$$(4.26 \times 10^2 \text{ mol}) \times \frac{18.02 \text{ g}}{\text{mol}} \times \frac{1 \text{ lb}}{453.6 \text{ g}} = 16.9 \text{ lb of ice}$$

Thus we need at least two 10-pound bags of ice to chill three cases of our favorite beverages.

CONCEPT TEST

The energy lost by the beverages inside the 72 cans in the preceding discussion was more than 100 times the energy lost by the aluminum cans themselves. What factors contributed to this large difference between the energy lost by the cans and the energy lost by their contents?

(Answers to Concept Tests are in the back of the book.)

SAMPLE EXERCISE 5.8 Calculating the Temperature of Iced Tea **LO4**

If you add 250.0 g of ice, initially at $-18.0°C$, to 237 g (1 cup) of freshly brewed tea, initially at $100.0°C$, and the ice melts, what is the final temperature of the tea? Assume that the mixture is an isolated system (in an ideal insulated container) and that tea has the same molar heat capacity, density, and molar mass as water.

Collect and Organize We know the mass of the tea, its initial temperature, and the molar heat capacity of tea, for which we use the c_P value for water. We know the amount of ice, its initial temperature, and the molar enthalpy of fusion of ice. Our task is to find the final temperature of the tea–ice-water mixture.

Analyze The amount of energy released when the tea is cooled will be the same as the amount of energy gained by the ice and, once it is melted, the amount gained by the water that is formed as it warms to the final temperature. If we assume that the ice melts completely, the tea loses energy through three processes:

1. q_1: raising the temperature of the ice to $0.0°C$
2. q_2: melting the ice
3. q_3: bringing the mixture to the final temperature; the temperature of the water from the melted ice rises and the temperature of the tea ($T_{initial} = 100.0°C$) falls to the final temperature of the mixture ($T_{final} = ?$)

The energy gained by the ice equals the energy lost by the tea:

$$q_{ice} = -q_{tea}$$

From our analysis, we know that

$$q_{ice} = q_1 + q_2 + q_3$$

Solve The energy lost by the tea as it cools from $100.0°C$ to T_{final} is, from Equation 5.8,

$$q_{tea} = nc_P\Delta T_{tea}$$

$$= 237 \text{ g} \times \frac{1 \text{ mol}}{18.02 \text{ g}} \times \frac{75.3 \text{ J}}{\text{mol} \cdot °C} \times (T_{final} - 100.0°C)$$

$$= (990 \text{ J/}°C)(T_{final} - 100.0°C)$$

The transfer of this energy is responsible for the changes in the ice. We can treat the energy transfer from the hot tea to the ice in terms of the three processes we identified in Analyze. In step 1, the ice is warmed from $-18.0°C$ ($T_{initial}$) to $0.0°C$ (T_{final} for this step):

$$q_1 = n_{ice}c_{P,ice}\Delta T_{ice}$$

$$= 250.0 \text{ g} \times \frac{1 \text{ mol}}{18.02 \text{ g}} \times \frac{37.1 \text{ J}}{\text{mol} \cdot °C} \times [0.0°C - (-18.0°C)]$$

$$= 9.26 \times 10^3 \text{ J} = 9.26 \text{ kJ}$$

In step 2, the ice melts, requiring the absorption of energy:

$$q_2 = n_{ice}\Delta H_{fus,ice}$$

$$= 250.0 \text{ g} \times \frac{1 \text{ mol}}{18.02 \text{ g}} \times \frac{6.01 \text{ kJ}}{\text{mol}}$$

$$= 83.4 \text{ kJ}$$

In step 3, the water from the ice, initially at $0.0°C$, warms to the final temperature (where $\Delta T = T_{final} - T_{initial} = T_{final} - 0.0°C$):

$$q_3 = n_{water}c_{P,water}\Delta T_{water}$$

$$= 250.0 \text{ g} \times \frac{1 \text{ mol}}{18.02 \text{ g}} \times \frac{75.3 \text{ J}}{\text{mol} \cdot °C} \times (T_{final} - 0.0°C)$$

$$= (1045 \text{ J/}°C)(T_{final})$$

The sum of the quantities of energy absorbed by the ice and the water from it during steps 1 through 3 must balance the energy lost by the tea:

$$q_{ice} = q_1 + q_2 + q_3 = -q_{tea}$$

$$9.26 \text{ kJ} + 83.4 \text{ kJ} + (1045 \text{ J/°C})(T_{final}) = -[(990 \text{ J/°C})(T_{final} - 100.0°C)]$$

Expressing all values in kilojoules:

$$9.26 \text{ kJ} + 83.4 \text{ kJ} + (1.045 \text{ kJ/°C})(T_{final}) = -[(0.990 \text{ kJ/°C})(T_{final} - 100.0°C)]$$

Rearranging the terms to solve for T_{final}, we have

$$(2.04 \text{ kJ/°C})T_{final} = -9.26 \text{ kJ} - 83.4 \text{ kJ} + 99.0 \text{ kJ} = 6.34 \text{ kJ}$$

$$T_{final} = 3.1°C$$

Think About It This calculation was carried out under the assumption that the system (the tea plus the ice) was isolated and the vessel was a perfect insulator. Our answer, therefore, is an "ideal" answer and reflects the coldest temperature we can expect the tea to reach. In the real world, the ice would absorb some energy from the surroundings (the container and the air) and the tea would lose some energy to the surroundings, so the final temperature of the beverage could be different from the ideal value we calculated.

 Practice Exercise Calculate the final temperature of a mixture of 350 g of ice, initially at −18.0°C, and 237 g of water, initially at 100.0°C.

(Answers to Practice Exercises are in the back of the book.)

5.6 Calorimetry: Measuring Heat Capacity and Enthalpies of Reaction

Up to this point we have discussed molar heat capacities (c_P) and the enthalpy changes ΔH_{fus} and ΔH_{vap} associated with phase changes, but we have not broached the issue of how we know the values of these parameters. The experimental method of measuring the quantities of energy associated with chemical reactions and physical changes is called **calorimetry**. The device used to measure the energy released or absorbed during a process is a **calorimeter**.

CHEMTOUR
Calorimetry

Determining Molar Heat Capacity and Specific Heat

When we determined the amount of ice needed to cool 72 aluminum beverage cans in Section 5.5, we used the molar heat capacity of aluminum to determine the quantity of energy lost by the cans. How are molar heat capacities measured?

We can apply the first law of thermodynamics and design an experiment to determine the specific heat of aluminum, from which we can calculate its molar heat capacity. Recall from Section 5.5 that specific heat c_s is defined as the quantity of energy required to raise the temperature of 1 g of a substance by 1°C. The units of specific heat, J/(g · °C), tell us exactly what we have to do. "Determine the specific heat of aluminum" means that we must find the quantity of energy (in joules) required to change the temperature of 1 g of aluminum by 1°C. If we determine how much energy is required to change the

calorimetry the measurement of the quantity of energy transferred during a physical change or chemical process.

calorimeter a device used to measure the absorption or release of energy by a physical change or chemical process.

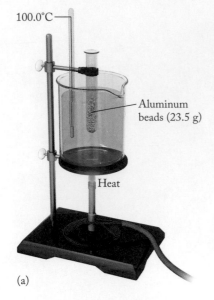

FIGURE 5.28 Experimental setup to determine the molar heat capacity of a metal. (a) Pure aluminum beads having a combined mass of 23.5 g are heated to 100.0°C in boiling water; (b) 130.0 g of water at 23.0°C is placed in a Styrofoam box; (c) the hot Al beads are dropped into the water, and the temperature at thermal equilibrium is 26.0°C.

temperature of any known mass of aluminum by any measured number of degrees, we can calculate how much energy is required to change the temperature of 1 g of aluminum by 1°C—in other words, we will have the value of the specific heat of aluminum.

Suppose we have beads of pure aluminum with a total mass of 23.5 g. We want to transfer a known amount of energy to or from the aluminum (Figure 5.28). First, we must warm the known quantity of aluminum to a known temperature. We can do this by boiling some water in a beaker containing a test tube holding the aluminum beads (the system). We wait a few minutes while the beads and the boiling water bath come to thermal equilibrium at 100.0°C (Figure 5.28a). Because water (or any pure liquid) boils at a constant temperature, we can use this step to bring the beads to a known temperature. No matter how rapidly the water is boiling, it will maintain a temperature of 100.0°C.

While the metal is warming, we place a measured mass of water (in this case, 130.0 g) in a beaker in a Styrofoam box (Figure 5.28b). We consider this box a perfect insulator that effectively isolates its contents from the rest of the universe. Through a small hole in the box lid (no energy escapes through the hole because the box is a perfect insulator), we insert a thermometer into the water, read the temperature (23.0°C), and leave the thermometer in place. When we have waited long enough for the beads to warm to 100.0°C, we remove the test tube from the boiling water, remove the lid from the insulated box, pour the beads out of the test tube into the water in the beaker, and quickly close the lid (Figure 5.28c). The temperature of the water rises because it is now in contact with the hot aluminum. We assume that all the energy leaving the beads goes into the water in the beaker.

Suppose the temperature of the aluminum–water mixture in the box rises to 26.0°C and stays at that temperature. Energy flows from the beads to the water. From the first law of thermodynamics we know

$$-q_{\text{aluminum}} = q_{\text{water}}$$

where we show a minus sign on q_{aluminum} because energy leaves the aluminum. (q_{water} is positive because energy enters the water.) To determine the amount of energy transferred from the aluminum, we need to know how much energy enters the water. We know the mass of water (130.0 g) and the temperature change the water experiences ($\Delta T_{\text{water}} = 26.0°C - 23.0°C = 3.0°C$); we need the specific heat of water to calculate q_{water}.

When we defined units of energy in Section 5.3, we defined 4.184 joules (J) as the amount of energy needed to raise the temperature of 1 g of water by 1°C; that value is the specific heat of water. We could also calculate $c_{\text{s,water}}$ by dividing the molar heat capacity of water [75.3 J/(mol · °C); Table 5.1] by its molar mass ($\mathcal{M} = 18.02$ g/mol). With that value we now have everything we need to calculate the specific heat c_{s} of aluminum.

First we determine the amount of energy gained by the water in the box. The units of specific heat tell us what to do:

$$c_{\text{s,water}} = \frac{4.184 \text{ J}}{\text{g} \cdot °C}$$

If we multiply the mass of the water in grams by the specific heat and the temperature change of the water, the result is the energy absorbed by the water:

$$q_{\text{water}} = (130.0 \text{ g H}_2\text{O})\left(\frac{4.184 \text{ J}}{\text{g H}_2\text{O} \cdot °C}\right)(3.0°C)$$
$$= 1.6 \times 10^3 \text{ J}$$

The energy gained by the water has a positive value because the water absorbs energy from the beads. This expression can be generalized to calculate energy absorbed or released from any mass of matter over any temperature change:

$$q = mc_s\Delta T \qquad (5.11)$$

where m is the mass, c_s is specific heat, and $\Delta T = T_{final} - T_{initial}$.

To find the specific heat of aluminum, we recognize that the 1600 J of energy that increased the temperature of the water came from the aluminum beads. Therefore

$$q_{aluminum} = -q_{water} = -1.6 \times 10^3 \, J = mc_s\Delta T_{aluminum}$$

where m is the mass of aluminum, c_s is the specific heat of aluminum, and $\Delta T_{aluminum} = T_{final} - T_{initial} = (26.0 - 100.0)°C = -74.0°C$. Note that ΔT for the aluminum is different from ΔT for the water. The aluminum is initially at 100°C and drops to a final temperature of 26.0°C. Thus

$$-1.6 \times 10^3 \, J = (23.5 \, g)(c_s)(-74.0°C)$$

where c_s is the unknown value we seek.

Solve for c_s:

$$c_s = \frac{-1.6 \times 10^3 \, J}{(23.5 \, g)(-74.0°C)} = \frac{0.92 \, J}{g \cdot °C}$$

From this specific heat value, we can also calculate the molar heat capacity of aluminum by multiplying c_s by the mass of 1 mole of aluminum:

$$c_P = c_s \times \mathcal{M} = \frac{0.92 \, J}{g \cdot °C} \times 26.98 \frac{g}{mol} = 25 \frac{J}{mol \cdot °C}$$

SAMPLE EXERCISE 5.9 Calculating the Final Temperature **LO4**
of a Hot Object Dropped Into Water

A blacksmith drops a 1.50 kg piece of iron heated to 525°C into 2.00 kg of water at 15.0°C. Given the molar heat capacities $c_{P,Fe} = 25.1 \, J/(mol \cdot °C)$ and $c_{P,H_2O} = 75.3 \, J/(mol \cdot °C)$, calculate the final temperature of the water.

Collect, Organize, and Analyze We are given the masses, initial temperatures, and molar heat capacities of the iron and the water; we need to calculate the final temperature of the water. After the iron is dropped into the water, the temperature of the iron decreases and the temperature of the water increases until they reach thermal equilibrium. This means that the heat lost by the iron will equal the heat gained by the water, $q_{Fe} = -q_{water}$. We use Equation 5.8 to relate q_{Fe} and q_{H_2O} to the final temperature. The masses of the iron and water involved here are similar, but the molar heat capacity of water is much greater than that of iron; therefore we expect the water's final temperature to be closer to its initial temperature than to the iron's.

Solve We start with the equation $q_{Fe} = -q_{H_2O}$ and substitute $q = nc_P\Delta T$:

$$n_{Fe}c_{P,Fe}\Delta T = -n_{H_2O}c_{P,H_2O}\Delta T$$

where $\Delta T = T_{final} - T_{initial}$ for each material. The final temperature of the water and iron must be equal,

$$T_{final,Fe} = T_{final,H_2O} = T_{final}$$

so

$$n_{Fe}c_{P,Fe}(T_{final,Fe} - T_{initial,Fe}) = -n_{H_2O}c_{P,H_2O}(T_{final,H_2O} - T_{initial,H_2O})$$

We calculate n_{Fe} and n_{H_2O} by using the appropriate molar masses:

$$n_{Fe} = 1.50 \text{ kg Fe} \times \frac{10^3 \text{ g}}{1 \text{ kg}} \times \frac{1 \text{ mol Fe}}{55.84 \text{ g Fe}} = 26.9 \text{ mol Fe}$$

$$n_{H_2O} = 2.00 \text{ kg H}_2\text{O} \times \frac{10^3 \text{ g}}{1 \text{ kg}} \times \frac{1 \text{ mol H}_2\text{O}}{18.02 \text{ g H}_2\text{O}} = 1.11 \times 10^2 \text{ mol H}_2\text{O}$$

Substitute the values for n, c_P, and $T_{initial}$:

$$26.9 \text{ mol Fe} \times \frac{25.1 \text{ J}}{\text{mol Fe} \cdot {}^\circ\text{C}} \times (T_{final} - 525{}^\circ\text{C})$$

$$= -1.11 \times 10^2 \text{ mol H}_2\text{O} \times \frac{75.3 \text{ J}}{\text{mol H}_2\text{O} \cdot {}^\circ\text{C}} (T_{final} - 15.0{}^\circ\text{C})$$

Solving, we get $T_{final} = 53.1{}^\circ\text{C}$.

Think About It The final temperature, 53.1°C, is much cooler than the initial temperature of the iron and only about 40°C warmer than the initial temperature of the water, reflecting the large difference between the molar heat capacities of Fe and water. Notice that there is not enough heat in the hot piece of iron to boil away all the water.

Practice Exercise A 255 g piece of granite, heated to 575°C in a campfire, is dropped into 1.00 L water ($d = 1.00$ g/mL) at 26°C. The molar heat capacity of water is $c_{P,H_2O} = 75.3$ J/(mol · °C) and the specific heat of the granite is $c_{s,granite} = 0.79$ J/(g · °C). Calculate the final temperature of the granite.

(Answers to Practice Exercises are in the back of the book.)

Enthalpies of Reaction

The energy transfer accompanying any chemical reaction is defined by a quantity known as the **enthalpy of reaction (ΔH_{rxn})**, also called the *heat of reaction*. The subscript may be changed to reflect a specific type of reaction being studied: for example, ΔH_{comb} for the enthalpy of a combustion reaction.

Much of the energy we use every day can be traced to combustion of a fuel such as natural gas (mostly methane, CH_4), propane (C_3H_8), or gasoline (a mixture of hydrocarbons). Our bodies obtain energy from respiration, which is essentially the combustion of glucose ($C_6H_{12}O_6$). We have already written balanced chemical equations for the combustion of CH_4, C_3H_8, and $C_6H_{12}O_6$ in Chapter 3. Now let's consider **thermochemical equations** for these processes by adding their ΔH_{rxn} values:

$$CH_4(g) + 2 \, O_2(g) \rightarrow CO_2(g) + 2 \, H_2O(g) \quad \Delta H_{rxn} = -802 \text{ kJ/mol CH}_4$$

$$C_3H_8(g) + 5 \, O_2(g) \rightarrow 3 \, CO_2(g) + 4 \, H_2O(g) \quad \Delta H_{rxn} = -2220 \text{ kJ/mol C}_3\text{H}_8$$

$$C_6H_{12}O_6(s) + 6 \, O_2(g) \rightarrow 6 \, CO_2(g) + 6 \, H_2O(g) \quad \Delta H_{rxn} = -2803 \text{ kJ/mol C}_6\text{H}_{12}\text{O}_6$$

All three of these reactions are *exothermic*, so the enthalpy change for each reaction is less than zero: $\Delta H_{rxn} < 0$.

Knowledge of ΔH_{comb} values allows us to compare fuels and to calculate how much fuel is needed to supply a desired quantity of energy. For example, we can calculate how much propane we need to melt the 270 g (15.0 mol) of ice in Figure 5.25 and raise its temperature to 100°C. In the beginning of Section 5.5,

enthalpy of reaction (ΔH_{rxn}) the energy absorbed or given off by a chemical reaction under conditions of constant pressure; also called *heat of reaction*.

thermochemical equation the chemical equation of a reaction that includes the change in enthalpy that accompanies that reaction.

we calculated that the energy required to melt this much ice is 113 kJ. If we get 2220 kJ/mol C_3H_8, then we need to burn at least 2.24 g of C_3H_8:

$$113 \text{ kJ} \times \frac{1 \text{ mol C}_3\text{H}_8}{2220 \text{ kJ}} \times \frac{44.09 \text{ g C}_3\text{H}_8}{1 \text{ mol C}_3\text{H}_8} = 2.24 \text{ g C}_3\text{H}_8$$

How much energy would we get from burning 2.24 g of glucose? Converting the mass of glucose to moles and multiplying by ΔH_{comb}:

$$2.24 \text{ g C}_6\text{H}_{12}\text{O}_6 \times \frac{1 \text{ mol C}_6\text{H}_{12}\text{O}_6}{180.16 \text{ g C}_6\text{H}_{12}\text{O}_6} \times \frac{2803 \text{ kJ}}{1 \text{ mol C}_6\text{H}_{12}\text{O}_6} = 34.9 \text{ kJ}$$

Calorimetry can be used to measure ΔH_{rxn} for chemical reactions, including the neutralization reactions discussed in Chapter 4. A simple laboratory apparatus for measuring the heat of reaction between a strong acid and a strong base is shown in Figure 5.29. A nested pair of Styrofoam coffee cups acts as our calorimeter, much like the apparatus in Figure 5.28. Note that pressure remains constant in a coffee-cup calorimeter, so q_{rxn} in this case will be the same as the enthalpy of reaction (ΔH_{rxn}).

Let's say we place 25.0 mL of 1.0 M aqueous HCl in the coffee cup and measure the temperature as 18.5°C. Rapid addition of 25.0 mL of 1.0 M NaOH causes the temperature to rise to 25.0°C. What is ΔH_{rxn} for the reaction? (Note this is the reaction depicted in the Particulate Review on the first page of this chapter.)

$$\text{HCl}(aq) + \text{NaOH}(aq) \rightarrow \text{NaCl}(aq) + \text{H}_2\text{O}(\ell)$$

The rise in temperature is the result of energy flowing from the reaction (the system) to the water (the surroundings). The amount of heat, q, gained by the surroundings is

$$q_{H_2O} = n_{H_2O}c_P\Delta T$$

First we need to calculate the moles of water, using the density of H_2O as a reasonable approximation of the density of the solution:

$$n_{H_2O} = 50.0 \text{ mL H}_2\text{O} \times \frac{1.00 \text{ g H}_2\text{O}}{1 \text{ mL H}_2\text{O}} \times \frac{1.00 \text{ mol H}_2\text{O}}{18.02 \text{ g H}_2\text{O}} = 2.77 \text{ mol H}_2\text{O}$$

Notice that the total volume of the solution is 50.0 mL because we have mixed two 25.0 mL solutions in our calorimeter. Substituting n_{H_2O} and the values of c_P and ΔT:

$$q_{H_2O} = n_{H_2O}c_P\Delta T = (2.77 \text{ mol H}_2\text{O}) \times \left(\frac{75.3 \text{ J}}{1 \text{ mol H}_2\text{O} \cdot °\text{C}}\right) \times (25.0°\text{C} - 18.5°\text{C})$$

$$= 1356 \text{ J}$$

The heat gained by the surroundings must equal the heat lost by the system,

$$q_{H_2O} = -q_{rxn} = -1356 \text{ J}$$

Because 25.0 mL of 1.0 M HCl contains 0.025 mol of HCl, ΔH_{rxn} equals

$$\Delta H_{rxn} = \frac{-1356 \text{ kJ}}{0.025 \text{ mol HCl}} = -54 \text{ kJ/mol}$$

In this experiment we made three important assumptions:

1. The coffee cup does not absorb any of the heat generated by the reaction.
2. The densities of the relatively dilute solutions are the same as that of pure water.
3. The molar heat capacity of the solution is the same as that of pure water.

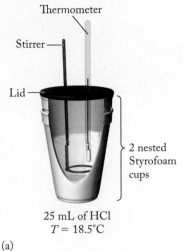

FIGURE 5.29 A cutaway image of a coffee-cup calorimeter made of two nested Styrofoam cups. A thermometer enables us to observe the temperature of the contents, which we can agitate using a simple stirrer to ensure mixing of solutions. (a) The inside cup contains 25 mL of 1.0 M HCl. (b) The rapid addition of 25 mL of 1.0 M NaOH causes the temperature to rise because of the energy given off in the ensuing neutralization reaction.

These assumptions are reasonably valid for our example, but they would not be in cases where the concentrations of the solutions are substantially higher. Calorimeters suitable for accurate determination of ΔH are commercially available.

SAMPLE EXERCISE 5.10 Coffee-Cup Calorimetry: ΔH_{soln} of NH_4NO_3 **LO5**

Dissolving 80.0 g of NH_4NO_3 in 505 g of water in a coffee-cup calorimeter causes the temperature of the water to decrease by 13.3°C. What is the enthalpy change that accompanies the dissolution process, ΔH_{soln}, expressed in kJ/mol?

Collect and Organize We are given the masses of solute and solvent used to prepare a solution of ammonium nitrate. The temperature decreases by 13.3°C. We need to determine the value of ΔH_{soln}.

Analyze The ammonium nitrate is the system and the water represents the surroundings.

a. First we use Equation 5.8 to find q_{surr}.

$$\boxed{\text{mass } H_2O} \xrightarrow{\text{molar mass}} \boxed{\text{mol } H_2O} \xrightarrow{q = nc_p\Delta T} \boxed{\text{heat lost by } H_2O}$$

b. Then we use the relationship $q_{sys} = -q_{surr}$ to obtain q_{sys}.
c. Finally, we calculate ΔH_{soln} by dividing q_{sys} by the number of moles of NH_4NO_3.

$$\boxed{\text{heat lost by } H_2O} \xrightarrow{q_{surr} = -q_{sys}} \boxed{\text{heat gained by system}} \xrightarrow{\dfrac{1}{\text{mol } NH_4NO_3}} \boxed{\Delta H_{soln}}$$

Solve
a. The heat lost by the water, q_{surr}, is

$$505 \text{ g } H_2O \times \frac{1 \text{ mol } H_2O}{18.02 \text{ g } H_2O} = 28.0 \text{ mol } H_2O$$

$$q_{surr} = n_{H_2O}c_{P,H_2O}\Delta T = 28.0 \text{ mol } H_2O \times \frac{75.3 \text{ J}}{\text{mol } H_2O \cdot °C} \times (-13.3°C) = -2.81 \times 10^4 \text{ J}$$

b. The heat gained by the system, q_{sys}, is

$$q_{sys} = -q_{surr} = -(-2.81 \times 10^4 \text{ J}) = 2.81 \times 10^4 \text{ J}$$

c. The number of moles of NH_4NO_3 is

$$80.0 \text{ g } NH_4NO_3 \times \frac{1 \text{ mol } NH_4NO_3}{80.05 \text{ g } NH_4NO_3} = 0.999 \text{ mol } NH_4NO_3$$

and the enthalpy of solution of NH_4NO_3 is

$$\Delta H_{soln} = \frac{2.81 \times 10^4 \text{ J}}{0.999 \text{ mol } NH_4NO_3} = 2.81 \times 10^4 \text{ J/mol } NH_4NO_3 = 28.1 \text{ kJ/mol } NH_4NO_3$$

Think About It The value for ΔH_{soln}, 28.1 kJ/mol, is positive, and it is consistent with the use of ammonium nitrate in chemical cold packs.

Practice Exercise Addition of 114 g of potassium fluoride to 0.600 L of water ($d = 1.00$ g/mL) causes the temperature to rise by 3.6°C. What is ΔH_{soln} for KF?

(Answers to Practice Exercises are in the back of the book.)

Determining Calorimeter Constants

Although the apparatus in Figure 5.29 is useful for measuring the enthalpy change of a reaction in solution, enthalpies of combustion are best measured with a device called a **bomb calorimeter** (Figure 5.30). To measure the enthalpy of a combustion reaction, the sample is placed in a sealed vessel (called a *bomb*) capable of withstanding high pressures and submerged in a large volume of water in a heavily insulated container. Oxygen is introduced into the bomb, and the mixture is ignited with an electric spark. As combustion occurs, energy generated by the reaction flows into the walls of the bomb and then into the water surrounding the bomb. A good bomb calorimeter keeps the system contained within the bomb and ensures that all energy generated by the reaction stays in the calorimeter. The surroundings consist of the bomb, the water, the insulated container, and minor components (stirrer, thermometer, and any other materials).

The energy produced by the reaction is determined by measuring the temperature of the water before and after the reaction. The water is at the same temperature as the parts of the calorimeter it contacts—the walls of the bomb, the thermometer, and the stirrer—so the temperature change of the water takes the entire calorimeter into account.

Measuring the change in temperature of the water is not the whole story, however. We also need to know the **heat capacity (C_P)** of the calorimeter, the quantity of energy required to increase the temperature of *a particular object* (for example, a calorimeter) by 1°C at constant pressure. Note the difference between heat capacity and two terms we have previously defined: *specific heat* [J/(g · °C)] is the energy required to raise the temperature of 1 *gram* of a substance by 1°C, and *molar heat capacity* [J/(mol · °C)] is the energy required to raise the temperature of 1 *mole* of a substance by 1°C. Because specific heat and molar heat capacity refer to quantities of substances, units for these quantities include grams and moles. Because the heat capacity is a value unique to every calorimeter, it is frequently referred to as a **calorimeter constant ($C_{\text{calorimeter}}$)** and has units of J/°C, which is understood to mean "J/°C for this specific calorimeter." If we know the value of $C_{\text{calorimeter}}$ and if we can measure the change in water temperature, we can calculate the quantity of energy that flowed from the reactants into the calorimeter ($q_{\text{calorimeter}}$) to cause the temperature change of the water:

$$q_{\text{calorimeter}} = C_{\text{calorimeter}} \, \Delta T \qquad (5.12)$$

Rearranging terms:

$$C_{\text{calorimeter}} = \frac{q_{\text{calorimeter}}}{\Delta T} \qquad (5.13)$$

This equation indicates that heat capacity is expressed in units of energy divided by temperature, usually kilojoules per degree Celsius (kJ/°C).

Equation 5.13 can be used to determine $C_{\text{calorimeter}}$ for a bomb calorimeter. To do this, we burn a quantity of material in the calorimeter that produces a known quantity of energy when it burns—in other words, a material whose ΔH_{comb} value is known. Benzoic acid ($C_7H_6O_2$) is often used for this purpose because it can be obtained in a very pure form. Once $C_{\text{calorimeter}}$ has been determined, the calorimeter can be used to determine ΔH_{comb} for other substances. The observed increases in water temperature can be used to calculate the quantities of energy produced by combustion reactions on a per-gram or per-mole basis.

Because there is no change in the volume of the reaction mixture in a bomb calorimeter, this technique is referred to as *constant-volume calorimetry*. No P–V work is done, so according to Equation 5.12 the energy gained by the calorimeter

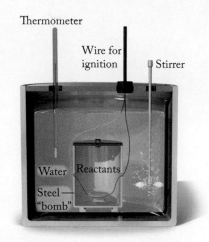

FIGURE 5.30 A bomb calorimeter.

bomb calorimeter a constant-volume device used to measure the energy released during a combustion reaction.

heat capacity (C_P) the quantity of energy needed to raise the temperature of an object by 1°C at constant pressure.

calorimeter constant ($C_{\text{calorimeter}}$) the heat capacity of a calorimeter.

during the combustion equals the internal energy lost by the reaction system during the combustion:

$$q_{calorimeter} = -\Delta E_{comb} \qquad (5.14)$$

The pressure inside a bomb calorimeter may change as a result of a combustion reaction, and in such cases ΔE_{comb} is not *exactly* the same as ΔH_{comb}. However, the pressure effects are usually so small that ΔE_{comb} is *nearly* the same as ΔH_{comb}, and we do not worry about the very small differences. Hence we discuss enthalpies of reactions (ΔH_{rxn}) throughout and apply the approximate relation

$$q_{calorimeter} = -\Delta H_{comb}$$

or even more generally

$$q_{calorimeter} = -\Delta H_{rxn} \qquad (5.15)$$

Other types of calorimeters (like the coffee-cup calorimeter in Figure 5.29) allow the volume to change while the pressure remains constant, so $q_{calorimeter}$ in those cases is exactly the same as the enthalpy of reaction (ΔH_{rxn}).

SAMPLE EXERCISE 5.11 Determining a Calorimeter Constant   **LO5**

What is the calorimeter constant of a bomb calorimeter if burning 1.000 g of benzoic acid in it causes the temperature of the calorimeter to rise by 7.248°C? The enthalpy of combustion of benzoic acid is $\Delta H_{comb} = -26.38$ kJ/g.

Collect and Organize We are asked to find the calorimeter constant of a calorimeter. We are given data describing how much the temperature of the calorimeter rises when a known amount of benzoic acid is burned in it, and we are given the enthalpy of combustion of benzoic acid.

Analyze We need to determine the amount of energy required to raise the temperature of the calorimeter by 1°C. The heat capacity of a calorimeter can be calculated using Equation 5.13 and the knowledge that the combustion of 1.000 g of benzoic acid produces 26.38 kJ of heat.

Solve

$$C_{calorimeter} = \frac{q_{calorimeter}}{\Delta T}$$

$$= \frac{26.38 \text{ kJ}}{7.248°C}$$

$$= 3.640 \text{ kJ/°C}$$

Knowing the heat capacity of this calorimeter means we could now use it to determine the enthalpy of combustion of any combustible material.

Think About It The calorimeter constant is determined for a specific calorimeter. If anything changes—if the thermometer breaks and has to be replaced, or if the calorimeter loses any of the water it contains—a new constant must be determined.

Practice Exercise When 0.500 g of a mixture of hydrocarbons is burned in the bomb calorimeter from Sample Exercise 5.11, its temperature rises by 6.76°C. How much energy (in kilojoules) is released during combustion? How much energy is released with the combustion of 1.000 g of the same mixture?

(Answers to Practice Exercises are in the back of the book.)

5.7 Hess's Law

In the previous section we were reminded of the importance of glucose as a fuel for our bodies. Most organisms rely on glucose, but not all of them produce CO_2 and water as the only products. The natural world abounds in organisms that metabolize glucose in different ways. For example, some anaerobic bacteria convert glucose to carbon dioxide, acetic acid, and hydrogen in a reaction that has an enthalpy of reaction of +90 kJ:

$$C_6H_{12}O_6(s) + 2\,H_2O(\ell) \rightarrow$$
$$2\,CH_3CO_2H(\ell) + 4\,H_2(g) + 2\,CO_2(g) \qquad \Delta H_1 = +90\text{ kJ} \quad (5.16)$$

Methanosarcina, a strain of bacteria, converts acetic acid to methane (CH_4) in a reaction that has an enthalpy of reaction of +16 kJ:

$$CH_3CO_2H(\ell) \rightarrow CH_4(g) + CO_2(g) \qquad \Delta H_2 = +16\text{ kJ} \quad (5.17)$$

In a world where fuel supplies are short, could we use these two reactions to prepare methane and hydrogen, two useful fuels? In principle, we can write an overall reaction equation that results from adding the reactions in Equations 5.16 (step 1) and 5.17 (step 2). However, doing so requires that we multiply Equation 5.17 by 2.

(1) $C_6H_{12}O_6(s) + 2\,H_2O(\ell) \rightarrow 2\,CH_3CO_2H(\ell) + 4\,H_2(g) + 2\,CO_2(g)$

(2) $\qquad\quad 2\,[CH_3CO_2H(\ell) \rightarrow CH_4(g) + CO_2(g)]$

_____

(3) $C_6H_{12}O_6(s) + 2\,H_2O(\ell) \rightarrow 4\,H_2(g) + 4\,CO_2(g) + 2\,CH_4(g)$ (5.18)

After simplifying, we obtain the reaction in Equation 5.18: the conversion of glucose to methane, hydrogen, and carbon dioxide (step 3). Just as we obtain the overall chemical equation 3 by adding equations 1 and 2, we obtain the enthalpy of reaction for the overall reaction by adding the ΔH values for reactions 1 and 2. Since we multiplied Equation 5.17 by 2 before adding it to Equation 5.16, we also need to multiply ΔH_2 by 2. The thermochemical equation for the overall reaction is therefore

$$\Delta H_1 + 2\Delta H_2 = \Delta H_3$$

$$+90\text{ kJ} + 2(16\text{ kJ}) = +122\text{ kJ}$$

This calculation for ΔH_{rxn} is an application of **Hess's law**. Also known as *Hess's law of constant heat of summation*, it states that the enthalpy of reaction ΔH_{rxn} for a process that is the sum of two or more other reactions is equal to the sum of the ΔH_{rxn} values of the constituent reactions.

Hess's law is especially useful for calculating enthalpy changes that are difficult to measure directly. For example, CO_2 is the principal product of the combustion of carbon in the form of charcoal:

Reaction A: $C(s) + O_2(g) \rightarrow CO_2(g) \qquad \Delta H_{comb} = -393.5\text{ kJ}$

When the oxygen supply is limited, however, the products include carbon monoxide:

Reaction B: $C(s) + \frac{1}{2}O_2(g) \rightarrow CO(g)$

It is difficult to measure the enthalpy of combustion of this reaction directly because, as long as any oxygen is present, some of the $CO(g)$ formed reacts with the O_2 to form $CO_2(g)$, yielding a mixture of CO and CO_2 as the product. However, we can use Hess's law to obtain this value *indirectly* by working with ΔH_{comb} values we can measure.

CHEMTOUR
Hess's Law

Because we can run reaction A with excess oxygen and thereby force it to completion, we can measure the enthalpy of combustion, which is −393.5 kJ. We can also react a sample of pure $CO(g)$ with oxygen and measure ΔH_{comb} for that reaction:

$$\text{Reaction C:} \quad CO(g) + \tfrac{1}{2} O_2(g) \rightarrow CO_2(g) \qquad \Delta H_{comb} = -283.0 \text{ kJ}$$

Hess's law gives us a way to calculate ΔH_{comb} for reaction B from the measured values for reactions A and C. To do this, we must find a way to combine the equations for reactions A and C so that the sum equals reaction B. Once we have that combination, because we know two of the ΔH_{comb} values, we can calculate the one we do not know.

One approach to this analysis is to focus on the reactants and products in the reaction whose ΔH_{comb} value is unknown—reaction B in this example. This reaction has carbon and oxygen as reactants and carbon monoxide as a product. Note that reaction C has carbon monoxide as a reactant. If we add reaction C to reaction B, the carbon monoxide cancels out, and we end up with reaction A as the sum of C and B:

$$\text{Reaction B:} \quad C(s) + \tfrac{1}{2} O_2(g) \rightarrow \cancel{CO(g)} \qquad \Delta H_{comb} = ?$$
$$\text{Reaction C:} \quad \cancel{CO(g)} + \tfrac{1}{2} O_2(g) \rightarrow CO_2(g) \qquad \Delta H_{comb} = -283.0 \text{ kJ}$$
$$\overline{\text{Reaction A:} \quad C(s) + O_2(g) \rightarrow CO_2(g) \qquad \Delta H_{comb} = -393.5 \text{ kJ}}$$

We now use algebra to find ΔH_{comb} for reaction B:

$$\Delta H_B + \Delta H_C = \Delta H_A$$
$$\Delta H_B = \Delta H_A - \Delta H_C$$
$$= -393.5 \text{ kJ} - (-283.0 \text{ kJ}) = -110.5 \text{ kJ}$$

Recall that ΔH is a state function. This means we can manipulate equations in two important ways when applying Hess's law, should the need arise. (1) We can multiply the coefficients in a balanced equation and the ΔH for the reaction by the same factor to change the quantity of material we are dealing with, as we did with Equation 5.17. (2) We can reverse a reaction (make the reactants the products and the products the reactants) if we also change the sign of ΔH.

Hess's law is a direct consequence of the fact that enthalpy is a state function. In other words, for a particular set of reactants and products, the enthalpy change of the reaction is the same whether the reaction takes place in one step or in a series of steps. The concept of enthalpy and its expression in Hess's law are very useful because the enthalpy changes associated with many reactions can be calculated from a few that have been measured.

SAMPLE EXERCISE 5.12 Applying Hess's Law in Biology **LO6**

One source of energy in our bodies is the conversion of sugars to CO_2 and H_2O (called respiration; see Section 3.5). Maltose is one of the sugars found in foods. Show how we could combine reactions A and B below to get reaction C. How is ΔH_C related to the values of ΔH_A and ΔH_B?

$$\text{Reaction A:} \quad \text{maltose}(s) + H_2O(\ell) \rightarrow 2 \text{ glucose}(s) \qquad \Delta H_A$$

$$\text{Reaction B:} \quad 6 CO_2(g) + 6 H_2O(\ell) \rightarrow \text{glucose}(s) + 6 O_2(g) \qquad \Delta H_B$$

$$\text{Reaction C:} \quad \text{maltose}(s) + 12 O_2(g) \rightarrow 12 CO_2(g) + 11 H_2O(\ell) \qquad \Delta H_C$$

Collect and Organize We are asked to combine two chemical equations to write the equation of an overall reaction and to combine two ΔH_{rxn} values to obtain an overall ΔH_{rxn} value. Hess's law allows us to combine the enthalpy changes that accompany the steps in an overall chemical reaction to calculate the value of ΔH_{rxn} of the overall reaction.

Analyze Maltose is a reactant in equations A and C, so we can leave reaction A unchanged. The products of reaction C appear as reactants in equation B, so we need to reverse B in order to get CO_2 and H_2O on the product side. This results in only 6 moles of CO_2 and H_2O to the product side, so we also need to multiply the reversed equation B by 2 before adding it to equation A. Reversing equation B and multiplying its coefficients by 2 means that we also have to change the sign of ΔH_B and multiply its value by 2.

Solve We start with reaction A as written, and we reverse and double reaction B:

$$maltose(s) + H_2O(\ell) \rightarrow 2\,glucose(s) \qquad\qquad \Delta H_A$$
$$2 \times [glucose(s) + 6\,O_2(g) \rightarrow 6\,CO_2(g) + 6\,H_2O(\ell)] \qquad 2 \times [-\Delta H_B]$$

Adding A + 2(−B):

A: $\quad maltose(s) + \cancel{H_2O(\ell)} \rightarrow \cancel{2\,glucose(s)} \qquad\qquad \Delta H_A$

−(2 × B): $\quad \cancel{2\,glucose(s)} + 12\,O_2(g) \rightarrow 12\,CO_2(g) + \overset{11}{\cancel{12}}\,H_2O(\ell) \qquad -2\Delta H_B$

C: $\quad maltose(s) + 12\,O_2(g) \rightarrow 12\,CO_2(g) + 11\,H_2O(\ell) \qquad \Delta H_C$

Hess's law allows us to add the respective enthalpies: $\Delta H_C = \Delta H_A - 2\Delta H_B$.

Think About It We can add chemical equations just like algebraic equations, multiplying them by coefficients if necessary, to obtain an equation for a new chemical reaction. Then we can use Hess's law to find the enthalpy change for the new reaction.

 Practice Exercise How would you combine reactions A–C, shown below, to obtain reaction D?

A: $2\,H_2(g) + O_2(g) \rightarrow 2\,H_2O(g)$
B: $H_3BNH_3(s) \rightarrow NH_3(g) + BH_3(g)$
C: $H_3BNH_3(s) \rightarrow 2\,H_2(g) + HBNH(s)$
D: $NH_3(g) + BH_3(g) + O_2(g) \rightarrow 2\,H_2O(g) + HBNH(s)$

(Answers to Practice Exercises are in the back of the book.)

SAMPLE EXERCISE 5.13 Calculating Enthalpies of Reaction by Using Hess's Law **LO6**

Hydrocarbons burned in a limited supply of air may not burn completely, and $CO(g)$ may be generated. One reason furnaces and hot-water heaters fueled by natural gas need to be vented is that incomplete combustion can produce toxic carbon monoxide:

Reaction A: $2\,CH_4(g) + 3\,O_2(g) \rightarrow 2\,CO(g) + 4\,H_2O(g) \qquad \Delta H_A = ?$

Use reactions B and C to calculate the ΔH_{comb} for reaction A.

Reaction B: $\quad CH_4(g) + 2\,O_2(g) \rightarrow CO_2(g) + 2\,H_2O(g) \qquad \Delta H_B = -802\ kJ$
Reaction C: $\quad 2\,CO(g) + O_2(g) \rightarrow 2\,CO_2(g) \qquad \Delta H_C = -566\ kJ$

Collect and Organize We are given two reactions (B and C) with thermochemical data and a third (A) for which we are asked to find ΔH_A. All the reactants and products of reaction A are present in reaction B or C or both.

Analyze We can manipulate the equations for reactions B and C so that they sum to give the equation for which ΔH_{comb} is unknown. Then we can calculate this unknown value by applying Hess's law.

The reaction of interest (A) has methane on the reactant side. Because reaction B also has methane as a reactant, we can use B as written. Reaction A has CO as a product. Reaction C involves CO as a reactant, so we have to reverse C in order to get CO on the product side. Once we reverse C, we must change the sign of ΔH_C. If the coefficients as given do not allow us to sum the two reactions to yield reaction A, we can multiply one or both reactions by other factors.

Solve We start with reaction B as written and add the reverse of reaction C, remembering to change the sign of its ΔH_{comb}:

$$B:\quad CH_4(g) + 2\,O_2(g) \rightarrow CO_2(g) + 2\,H_2O(g) \qquad \Delta H_B = -802\text{ kJ}$$

$$C\text{ (reversed)}:\ 2\,CO_2(g) \rightarrow 2\,CO(g) + O_2(g) \qquad -\Delta H_C = 566\text{ kJ}$$

Because methane has a coefficient of 2 in reaction A, we multiply all the terms in reaction B, including ΔH_{comb}, by 2:

$$2 \times [CH_4(g) + 2\,O_2(g) \rightarrow CO_2(g) + 2\,H_2O(g)] \qquad 2 \times [\Delta H_B = -802\text{ kJ}]$$

Because the carbon monoxide in reaction A has a coefficient of 2, we do not need to multiply reaction C by any factor. Now we add $(2 \times B)$ to the reverse of C and cancel out common terms:

$$2 \times B:\ 2\,CH_4(g) + \overset{3}{\cancel{4}}\,O_2(g) \rightarrow \cancel{2\,CO_2(g)} + 4\,H_2O(g) \qquad 2\Delta H_B = -1604\text{ kJ}$$

$$C\text{ (reversed)}:\quad \cancel{2\,CO_2(g)} \rightarrow 2\,CO(g) + \cancel{O_2(g)} \qquad -\Delta H_C = 566\text{ kJ}$$

$$\overline{A:\quad 2\,CH_4(g) + 3\,O_2(g) \rightarrow 2\,CO(g) + 4\,H_2O(g) \qquad \Delta H_A = -1038\text{ kJ}}$$

Think About It We used Hess's law to calculate the enthalpy of combustion of methane to make CO. This is actually impossible to achieve in an experiment because any CO produced will react with O_2 to give CO_2, resulting in a product mixture of CO and CO_2. The answer of -1038 kJ is less negative than the enthalpy change accompanying the complete combustion of 2 moles of CH_4 [$2 \times (-802$ kJ)], so the answer is reasonable.

Practice Exercise It does not matter how you assemble the equations in a Hess's law problem. Show that reactions A and C can be summed to give reaction B and result in the same value for ΔH_{comb}.

(Answers to Practice Exercises are in the back of the book.)

standard enthalpy of formation (ΔH_f°) the enthalpy change of a formation reaction; also called *standard heat of formation* or *heat of formation*.

formation reaction a reaction in which 1 mole of a substance is formed from its component elements in their standard states.

standard conditions in thermodynamics: a pressure of 1 bar ($\sim$1 atm) and some specified temperature, assumed to be 25°C unless otherwise stated; for solutions, a concentration of 1 M is specified.

standard state the most stable form of a substance under 1 bar pressure and some specified temperature (25°C unless otherwise stated).

standard enthalpy of reaction (ΔH_{rxn}°) the energy associated with a reaction that takes place under standard conditions; also called *standard heat of reaction*.

5.8 Standard Enthalpies of Formation and Reaction

As noted in Section 5.3, it is impossible to measure the *absolute* value of the internal energy of a substance. The same is true for the enthalpy of a substance. However, we can establish *relative* enthalpy values that are referenced to a convenient standard. This approach is similar to using the freezing point of water as the zero point on the Celsius temperature scale or to using sea level as the zero point for expressing altitude. The enthalpy value referenced to this zero point is a substance's **standard enthalpy of formation (ΔH_f°)**, or *standard heat of formation*, defined as the enthalpy change that takes place at constant pressure when 1 mole of a substance is formed from its constituent elements in their standard states. A reaction that fits this description is known as a **formation reaction**.

The adjective *standard*, indicated by the symbol ° as seen in ΔH_f°, indicates that the associated reaction occurs under **standard conditions**, which means at a constant pressure of 1 bar and at some specified temperature. The value of 1 bar of pressure is very close to 1 atm; for the level of precision used in this book, a standard pressure of 1 bar will be considered equivalent to 1 atm. There is no universal standard temperature, though many tables of thermodynamic data, including those in the appendix of this book, apply to processes occurring at 25°C.

We also use the term **standard state** to describe the most stable physical state of a substance under standard conditions. In their standard states, oxygen is a gas, water is a liquid, and carbon is solid graphite. By definition, for a pure element in its most stable form under standard conditions, $\Delta H_f^\circ = 0$. This is the zero point of enthalpy values. Table 5.2 lists standard enthalpies of formation for several substances. (A more complete list can be found in Appendix 4.[1])

The general symbol for the enthalpy change associated with a reaction that takes place under standard conditions is ΔH_{rxn}°, and the value is called either a **standard enthalpy of reaction** or a *standard heat of reaction*. Implied in our notion of standard states and standard conditions is the assumption that parameters such as ΔH change with temperature and pressure. That assumption is correct, although the changes are so small that we ignore them in the calculations in this textbook.

Because the definition of a formation reaction specifies 1 mole of product, writing balanced equations for formation reactions may require the use of something we typically avoided in Chapter 3: fractional coefficients in the final form of our balanced equations. For example, the thermochemical equation for the production of ammonia from nitrogen and hydrogen is usually written

$$N_2(g) + 3\,H_2(g) \rightarrow 2\,NH_3(g) \qquad \Delta H_{rxn}^\circ = -92.2\ kJ$$

Although all reactants and products in this equation are in their standard states, it is not a formation reaction because 2 moles of product are formed, which is why we denote the enthalpy change of this reaction by ΔH_{rxn}° rather than ΔH_f°. The formation reaction for ammonia must, by definition, show 1 mole of product being formed, so we divide each coefficient in the above equation by 2. Because energy is a stoichiometric quantity, the enthalpy of reaction is divided by 2 as well. Thus, the thermochemical equation representing the formation reaction of ammonia is

$$\tfrac{1}{2} N_2(g) + \tfrac{3}{2} H_2(g) \rightarrow NH_3(g) \qquad \Delta H_f^\circ = -46.1\ kJ$$

TABLE 5.2 Standard Enthalpies of Formation for Selected Substances at 25°C

Substance	ΔH_f° (kJ/mol)
$Br_2(\ell)$	0
C(s, graphite)	0
$CH_4(g)$, methane	−74.8
$C_2H_2(g)$, acetylene	226.7
$C_2H_4(g)$, ethylene	52.26
$C_2H_6(g)$, ethane	−84.68
$C_3H_8(g)$, propane	−103.8
$C_4H_{10}(g)$, butane	−125.6
$CH_3OH(\ell)$, methanol	−238.7
$CH_3CH_2OH(\ell)$, ethanol	−277.7
$CH_3COOH(\ell)$, acetic acid	−484.5
CO(g)	−110.5
$CO_2(g)$	−393.5
$H_2(g)$	0
$H_2O(g)$	−241.8
$H_2O(\ell)$	−285.8
$N_2(g)$	0
$NH_3(g)$, ammonia	−46.1
$N_2H_4(g)$, hydrazine	95.4
$N_2H_4(\ell)$	50.63
NO(g)	90.3
$O_2(g)$	0

SAMPLE EXERCISE 5.14 Recognizing Formation Reactions   **LO7**

Which of the following reactions are formation reactions at 25°C? For those that are not, explain why not.

a. $H_2(g) + \tfrac{1}{2} O_2(g) \rightarrow H_2O(g)$
b. C(s, graphite) + 2 H_2(g) + \tfrac{1}{2} O_2(g) \rightarrow CH_3OH(\ell)$
   (CH_3OH is methanol, a liquid in its standard state.)
c. $CH_4(g) + 2\,O_2(g) \rightarrow CO_2(g) + 2\,H_2O(\ell)$
d. $P_4(s) + 2\,O_2(g) + 6\,Cl_2(g) \rightarrow 4\,POCl_3(\ell)$
   (In its standard state, P_4 is a solid; Cl_2 is a gas, and $POCl_3$ is a liquid.)

---

[1]The website of the National Institute of Standards and Technology contains much additional data: http://webbook.nist.gov.

Collect and Organize We are given four balanced chemical equations and information about the standard states of specific reactants and products. We want to determine which of these are formation reactions.

Analyze For a reaction to be a formation reaction, it must meet the criteria that it produces 1 mole of a substance from its component elements in their standard states. Therefore we must evaluate each reaction for the quantity of product and for the state of each reactant.

Solve
a. The reaction shows 1 mole of water vapor formed from its constituent elements in their standard states. Therefore this is the formation reaction for $H_2O(g)$, and its heat of reaction is ΔH_f°.
b. The reaction shows 1 mole of liquid methanol formed from its constituent elements in their standard states. This reaction is a formation reaction.
c. This is not a formation reaction because the reactants are not elements in their standard states and because more than one product is formed.
d. This is not a formation reaction because the product is 4 moles of $POCl_3$. Note, however, that all the constituent elements are in their standard states and that only one compound is formed. We could easily convert this into a formation reaction by dividing all the coefficients by 4.

Think About It Just because we can write formation reactions for substances like methanol does not mean that anyone would ever use that reaction to make methanol. Remember that formation reactions are defined to provide a standard against which other reactions can be compared when evaluating their thermochemistry.

 Practice Exercise Write formation reactions for (a) $CaCO_3(s)$, (b) $CH_3COOH(\ell)$ (acetic acid), and (c) $KMnO_4(s)$.

(Answers to Practice Exercises are in the back of the book.)

Note that the standard enthalpies of formation of acetylene and ethylene are positive, which means that the formation reactions for these compounds are endothermic. The other hydrocarbon fuels in Table 5.2 have negative enthalpies of formation, but the values for all of them are less negative than the values for water and carbon dioxide. Recall that the products of the complete combustion of hydrocarbons are carbon dioxide and water. The more negative the enthalpy of formation (ΔH_f°) of a substance, the more stable it is. The implication of this, as stated earlier, is that the reaction of fuels with oxygen to produce CO_2 and H_2O is exothermic. The reactions produce energy, which is why hydrocarbons are useful as fuels.

Standard enthalpies of formation ΔH_f° are used to predict standard enthalpies of reaction ΔH_{rxn}°. We can calculate the standard heat of reaction for any reaction by determining the difference between the ΔH_f° values of the products and the ΔH_f° values of the reactants. To see how this approach works, consider the reaction we saw earlier as Equation 5.18:

$$C_6H_{12}O_6(s) + 2\,H_2O(\ell) \rightarrow 4\,H_2(g) + 4\,CO_2(g) + 2\,CH_4(g)$$

In this calculation, we use data from Table 5.2 in the following equation:

$$\Delta H_{rxn}^\circ = \sum n_{products}\,\Delta H_{f,products}^\circ - \sum n_{reactants}\,\Delta H_{f,reactants}^\circ \qquad (5.19)$$

where $n_{products}$ is the number of moles of each product in the balanced equation and $n_{reactants}$ is the number of moles of each reactant. Equation 5.19 states that the value of ΔH°_{rxn} for any chemical reaction equals the sum ($\sum$) of the ΔH°_{f} value for each product times the number of moles of that product in the balanced equation, minus the sum of the ΔH°_{f} value for each reactant times the number of moles of that reactant in the balanced chemical equation. Thus we multiply the value of ΔH°_{f} for H_2O in Equation 5.18 by 2 before summing with the value of ΔH°_{f} for $C_6H_{12}O_6$ because the H_2O coefficient is 2 in the balanced equation. We do the same with the ΔH°_{f} values of the products, multiplying them by 4, 4, and 2, respectively.

Once we insert ΔH°_{f} values from Table 5.2 for the reactants and the products in Equation 5.18, remembering that for H_2 gas $\Delta H^{\circ}_{f} = 0$ because it is a pure element in its most stable form, Equation 5.19 becomes

$$\Delta H^{\circ}_{rxn} = [(4 \text{ mol } H_2)(0.0 \text{ kJ/mol}) + (4 \text{ mol } CO_2)(-393.5 \text{ kJ/mol})$$
$$+ (2 \text{ mol } CH_4)(-74.8 \text{ kJ/mol})] - [(1 \text{ mol } C_6H_{12}O_6)(-1274.4 \text{ kJ/mol})$$
$$+ (2 \text{ mol } H_2O)(-285.8 \text{ kJ/mol})]$$
$$= [(0.0 \text{ kJ}) + (-1574 \text{ kJ}) + (-149.6 \text{ kJ})] - [(-1274.4 \text{ kJ}) + (-571.6 \text{ kJ})]$$
$$= -122.4 \text{ kJ}$$

Calculations of standard enthalpies of reaction ΔH°_{rxn} from standard enthalpies of formation ΔH°_{f} can be carried out for all kinds of chemical reactions, including reactions that occur in solution. Consider the reaction between methane and steam, which yields a mixture of hydrogen and carbon dioxide:

$$CH_4(g) + 2 H_2O(g) \rightarrow CO_2(g) + 4 H_2(g)$$

This reaction is used to synthesize the hydrogen used in the steel industry to remove impurities from molten iron and in the chemical industry to make hundreds of compounds, including ammonia (NH_3) for use in fertilizers and as a refrigerant. The reaction is also important in the manufacture of hydrogen fuel for fuel cells, which are used to generate electricity directly from a chemical reaction.

Inserting the appropriate values from Table 5.2 into Equation 5.19, along with the coefficients 2 for H_2O and 4 for H_2, we calculate

$$\Delta H^{\circ}_{rxn} = [(1 \text{ mol } CO_2)(-393.5 \text{ kJ/mol}) + (4 \text{ mol } H_2)(0.0 \text{ kJ/mol})]$$
$$- [(1 \text{ mol } CH_4)(-74.8 \text{ kJ/mol}) + (2 \text{ mol } H_2O)(-241.8 \text{ kJ/mol})]$$
$$= +165 \text{ kJ}$$

The positive standard enthalpy of reaction tells us that this reaction between water vapor and methane (called steam–methane reforming, or simply *steam reforming*) is endothermic. Therefore energy must be added to make the reaction take place (typically conducted at temperatures near 1000°C). Thus although hydrogen is attractive as a fuel because it burns vigorously and produces only water as a product,

$$2 H_2(g) + O_2(g) \rightarrow 2 H_2O(\ell) + \text{energy}$$

its production from methane gas does not reduce the use of fossil fuels because most of the methane currently comes from fossil fuel. (This could change if production of methane from renewable sources becomes more feasible.) Fossil fuels are also burned to generate the heat that must be added to the reaction to make it run. Using hydrogen-powered vehicles merely shifts consumption of fossil fuels from the consumer (at the filling station) to a centralized location (the hydrogen production facility).

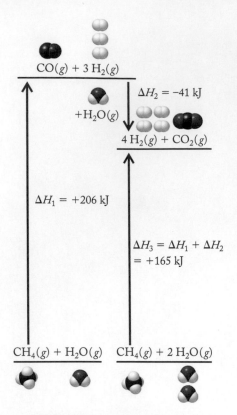

FIGURE 5.31 Hess's law predicts that the enthalpy change for the production of 4 moles of $H_2(g)$ and 1 mole of $CO_2(g)$ from 1 mole of $CH_4(g)$ and 2 moles of $H_2O(g)$. ΔH_3 is the sum of the enthalpies of two reactions: $\Delta H_3 = \Delta H_1 + \Delta H_2$.

FIGURE 5.32 (a) The reaction of methane with water to produce carbon monoxide and hydrogen is endothermic. (b) The reverse reaction, carbon monoxide plus hydrogen producing methane and water, is exothermic. The value of the enthalpy change of the two reactions has the same magnitude but is positive for the endothermic reaction and negative for the exothermic reaction.

Hydrocarbons other than methane are also used as the starting materials in the production of $H_2(g)$. However, the reactions are still usually endothermic and require the input of energy, which means fossil fuels are consumed to generate that energy. Also, CO_2 is released into the environment as a result of the production of hydrogen in any of these processes.

Enthalpy is a state function, and, as noted in Section 5.2, the value of a state function is independent of the path taken to achieve that state. Thus values of ΔH_{rxn} are independent of pathway. It does not matter what path we take to get from the reactants to the products; the enthalpy difference between them is always the same. In fact, on an industrial scale, the reaction to produce hydrogen gas from methane is carried out in two steps. In the first step, methane reacts with a limited supply of steam to produce carbon monoxide and hydrogen gas. In the second step, the carbon monoxide reacts with additional steam and more hydrogen gas:

Step 1:	$CH_4(g) + H_2O(g) \rightarrow \cancel{CO(g)} + 3\,H_2(g)$	$\Delta H_1 = +206$ kJ
Step 2:	$\cancel{CO(g)} + H_2O(g) \rightarrow CO_2(g) + H_2(g)$	$\Delta H_2 = -41$ kJ
Overall reaction:	$CH_4(g) + 2\,H_2O(g) \rightarrow CO_2(g) + 4\,H_2(g)$	$\Delta H_3 = +165$ kJ

Figure 5.31 depicts the enthalpy changes involved in these two steps, as well as the overall reaction. Note that the enthalpy change for the overall reaction has a value of +165 kJ, the same value we calculated with Equation 5.19 by using enthalpies of formation.

Just as we saw in Section 5.3 with state changes, once we know the value of the energy associated with running a reaction in one direction, we also know the value of the energy associated with running the reaction in the reverse direction because it has the same magnitude but the opposite sign. This means that, having calculated a value of $\Delta H^\circ_{rxn} = +165$ kJ for the steam reforming reaction, we can write

$$CO_2(g) + 4\,H_2(g) \rightarrow CH_4(g) + 2\,H_2O(g) = \Delta H^\circ_{rxn} -165 \text{ kJ}$$

for the exothermic reverse reaction, as illustrated in Figure 5.32.

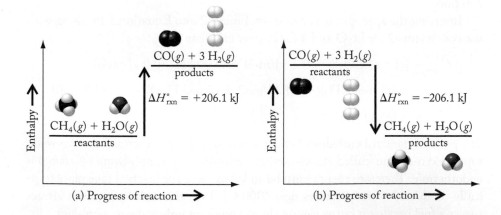

SAMPLE EXERCISE 5.15 Calculating Enthalpies of Reaction **LO6**

Using the appropriate values from Table 5.2, calculate ΔH°_{rxn} for the combustion of the fuel propane (C_3H_8) in air.

Collect and Organize The reactants are propane and elemental oxygen, O_2, and the products are carbon dioxide and water. Even though combustion is an exothermic reaction, we assume that water is produced as liquid, $H_2O(\ell)$, so that all products and reactants are in their standard states. The heats of formation of propane, carbon dioxide, and $H_2O(\ell)$ are given in Table 5.2. The heat of formation of the element O_2 in its standard state is zero. Our task is to use the balanced equation and the data from Table 5.2 to calculate the enthalpy of combustion.

Analyze Equation 5.19 defines the relation between heats of formation of reactants and products and the standard enthalpy of reaction. We also need the balanced equation for combustion of propane:

$$C_3H_8(g) + 5\,O_2(g) \rightarrow 3\,CO_2(g) + 4\,H_2O(\ell)$$

Solve Inserting ΔH_f° values for the products [$CO_2(g)$ and $H_2O(\ell)$] and reactants [$C_3H_8(g)$ and $O_2(g)$] from Table 5.2 and the coefficients in the balanced chemical equation into Equation 5.17, we get

$$\Delta H_{rxn}^\circ = [(3 \text{ mol } CO_2)(-393.5 \text{ kJ/mol}) + (4 \text{ mol } H_2O)(-285.8 \text{ kJ/mol})]$$
$$-[(1 \text{ mol } C_3H_8)(-103.8 \text{ kJ/mol}) + (5 \text{ mol } O_2)(0.0 \text{ kJ/mol})]$$
$$= -2219.9 \text{ kJ}$$

Think About It The result of the calculation has a large negative value, which means that the combustion reaction is highly exothermic, as expected for a hydrocarbon fuel.

 Practice Exercise Calculate ΔH_{rxn}° for the *water–gas shift reaction*:

$$CO(g) + H_2O(g) \rightarrow CO_2(g) + H_2(g)$$

(Answers to Practice Exercises are in the back of the book.)

Hydrocarbons such as methane, ethane, and propane are components of natural gas and are excellent fuels. However, they are currently classified as nonrenewable fuels, and their combustion produces carbon dioxide, contributing to climate change.

CONCEPT TEST

What is the enthalpy of reaction ΔH_{rxn}° for the production of 1 mole of C_3H_8 and oxygen from $CO_2(g)$ and $H_2O(\ell)$? Explain the reasoning you used to arrive at your answer. (See the information in Sample Exercise 5.15, and do not do any mathematical calculations in answering this question.)

(Answers to Concept Tests are in the back of the book.)

When we use standard enthalpies of formation (ΔH_f°) to determine standard enthalpies of reaction ΔH_{rxn}°, we are applying Hess's law. To see that this is the case, consider the reaction of ammonia with oxygen to make $NO(g)$ and liquid water (this is the first step of the industrial synthesis of nitric acid):

$$4\,NH_3(g) + 5\,O_2(g) \rightarrow 4\,NO(g) + 6\,H_2O(\ell)$$

First we write an equation for the formation reaction for each reactant and product that is not an element in its standard state (remember that $\Delta H_f^\circ = 0$ for elements in their standard states), using the ΔH_f° values in Table 5.2:

$$\tfrac{1}{2} N_2(g) + \tfrac{3}{2} H_2(g) \rightarrow NH_3(g) \qquad \Delta H_f^\circ = -46.1 \text{ kJ/mol}$$

$$\tfrac{1}{2} N_2(g) + \tfrac{1}{2} O_2(g) \rightarrow NO(g) \qquad \Delta H_f^\circ = +90.3 \text{ kJ/mol}$$

$$H_2(g) + \tfrac{1}{2} O_2(g) \rightarrow H_2O(\ell) \qquad \Delta H_f^\circ = -285.8 \text{ kJ/mol}$$

We reverse the NH_3 equation so that the NH_3 is a reactant and multiply each equation by a factor that matches the product's or reactant's coefficient in the balanced equation.

$$4[NH_3(g) \rightarrow \tfrac{1}{2} N_2(g) + \tfrac{3}{2} H_2(g)] \qquad 4 \times (\Delta H_f^\circ = +46.1 \text{ kJ/mol})$$

$$4[\tfrac{1}{2} N_2(g) + \tfrac{1}{2} O_2(g) \rightarrow NO(g)] \qquad 4 \times (\Delta H_f^\circ = +90.3 \text{ kJ/mol})$$

$$6[H_2(g) + \tfrac{1}{2} O_2(g) \rightarrow H_2O(\ell)] \qquad 6 \times (\Delta H_f^\circ = -285.8 \text{ kJ/mol})$$

Their sum yields the reaction of interest:

$$4\,NH_3(g) \rightarrow \cancel{2\,N_2(g)} + \cancel{6\,H_2(g)} \qquad \Delta H_{rxn}^\circ = +184.4 \text{ kJ}$$

$$\cancel{2\,N_2(g)} + 2\,O_2(g) \rightarrow 4\,NO(g) \qquad \Delta H_{rxn}^\circ = +361.2 \text{ kJ}$$

$$\underline{\cancel{6\,H_2(g)} + 3\,O_2(g) \rightarrow 6\,H_2O(\ell) \qquad \Delta H_{rxn}^\circ = -1714.8 \text{ kJ}}$$

$$4\,NH_3(g) + 5\,O_2(g) \rightarrow 4\,NO(g) + 6\,H_2O(\ell) \qquad \Delta H_{rxn}^\circ = -1169.2 \text{ kJ}$$

This is exactly the same mathematical operation that results when we use Equation 5.19:

$$\Delta H_{rxn}^\circ = \sum n_{products}\, \Delta H_{f,products}^\circ - \sum n_{reactants}\, \Delta H_{f,reactants}^\circ$$

$$= [4 \text{ mol } NO(+90.3 \text{ kJ/mol}) + 6 \text{ mol } H_2O(-285.8 \text{ kJ/mol})]$$

$$- [4 \text{ mol } NH_3(-46.1 \text{ kJ/mol}) + 5 \text{ mol } O_2(0 \text{ kJ/mol})] = -1169.2 \text{ kJ}$$

5.9 Fuels, Fuel Values, and Food Values

In Section 5.6 we saw that the enthalpy of reaction for 1 mole of propane (-2220 kJ) is much greater than for 1 mole of methane (-802 kJ)—that is, much more energy is released in the combustion of propane. Does this make propane an inherently better (higher-energy) fuel? Not necessarily. Expressing ΔH_{rxn} values on a per-mole basis is the only way to ensure that we are talking about the same number of molecules. However, we do not purchase fuels, or anything else for that matter, in units of moles. Depending on the fuel, we buy it either by mass (coal by the ton) or by volume (gasoline by the gallon or the liter). Before addressing the question of which fuel is superior, let's take a quick look at methane, propane, and other related *hydrocarbons*.

Alkanes

Methane, propane, and gasoline belong to the family of organic compounds called *alkanes*, hydrocarbons distinguished by the fact that each carbon atom is bonded to four other atoms. Alkanes are classified as **saturated hydrocarbons** because they contain the maximum ratio of hydrogen atoms to carbon atoms.

Alkanes are also known as *paraffins*, a name derived from Latin meaning "little affinity." This is a perfect description of the alkanes, which tend to be much less reactive than the other hydrocarbon families. *Unreactive* may not seem like the correct term to apply to compounds that are fuels, but alkanes do not react readily, even with oxygen. As evidence of this, consider that most fuels need some source of energy, such as a spark from a spark plug in an engine, to initiate combustion.

A brief word about alkane nomenclature will aid us in our discussion. The first four members of the alkane family (Figure 5.33) are methane (1 carbon atom: a C_1 alkane), ethane (2 carbon atoms: a C_2 alkane), propane (a C_3 alkane), and

CONNECTION In Chapter 2 we defined hydrocarbons as compounds composed of only carbon and hydrogen atoms, and we saw that alkanes are hydrocarbons in which each carbon atom is bonded to four other atoms.

saturated hydrocarbon an alkane; compounds containing the maximum ratio of hydrogen atoms to carbon atoms.

methylene group ($-CH_2-$) a structural unit that can make two bonds.

methyl group ($-CH_3$) a structural unit that can make only one bond.

straight-chain hydrocarbon a hydrocarbon in which the carbon atoms are bonded together in one continuous carbon chain.

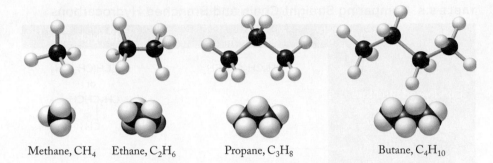

FIGURE 5.33 Ball-and-stick and space-filling models of the first four members of the alkane family: methane, ethane, propane, and butane.

Methane, CH_4 Ethane, C_2H_6 Propane, C_3H_8 Butane, C_4H_{10}

butane (a C_4 alkane). The names of alkanes with more than 4 carbon atoms are derived from the IUPAC prefix for the number of carbon atoms per molecule, followed by -ane. Table 5.3 lists the prefixes for the C_1 through C_{10} alkanes.

Alkanes all have the general formula C_nH_{2n+2}, and each alkane differs from the previous one by one $-CH_2-$ unit, which is called a **methylene group**. The terminal $-CH_3$ groups are called **methyl groups**. All the alkanes in the table are **straight-chain hydrocarbons**, which means a continuous sequence of carbon atoms with no branching. Straight-chain alkanes have a methyl group at each end with methylene groups connecting them. Table 5.4 lists a few physical properties for some straight-chain alkanes. Note that the data in Table 5.4 show similar trends in melting and boiling points: as straight-chain alkanes increase in molar mass, their melting and boiling points increase.

CONCEPT **TEST**

Alkanes have the general formula C_nH_{2n+2}. If an alkane has a molar mass of 114 g/mol, what is the value of n?

(Answers to Concept Tests are in the back of the book.)

TABLE 5.3 Prefixes for Naming Alkanes

Prefix	Condensed Structure	Name
Meth-	CH_4	Methane
Eth-	CH_3CH_3	Ethane
Prop-	$CH_3CH_2CH_3$	Propane
But-	$CH_3(CH_2)_2CH_3$	Butane
Pent-	$CH_3(CH_2)_3CH_3$	Pentane
Hex-	$CH_3(CH_2)_4CH_3$	Hexane
Hept-	$CH_3(CH_2)_5CH_3$	Heptane
Oct-	$CH_3(CH_2)_6CH_3$	Octane
Non-	$CH_3(CH_2)_7CH_3$	Nonane
Dec-	$CH_3(CH_2)_8CH_3$	Decane

TABLE 5.4 Melting Points and Boiling Points for Selected Straight-Chain Alkanes

Condensed Structure	Use	Melting Point (°C)	Normal Boiling Point (°C)
$CH_3CH_2CH_3$		−190	−42
$CH_3(CH_2)_2CH_3$	Gaseous fuels	−138	−0.5
$CH_3(CH_2)_3CH_3$		−130	36
$CH_3(CH_2)_4CH_3$		−95	69
$CH_3(CH_2)_5CH_3$	Gasoline	−91	98
$CH_3(CH_2)_6CH_3$		−57	126
$CH_3(CH_2)_7CH_3$		−54	151
$CH_3(CH_2)_{10}CH_3$ through $CH_3(CH_2)_{16}CH_3$	Diesel fuel and heating oil	−10	216
		28	316
$CH_3(CH_2)_{18}CH_3$ through $CH_3(CH_2)_{32}CH_3$	Paraffin candle wax	37	343
		72–75	na[a]
$CH_3(CH_2)_{34}CH_3$ and higher homologs	Asphalt	72–76	na

[a]na = not available; compound decomposes before boiling at 1 atm pressure.

TABLE 5.5 Comparing Straight-Chain and Branched Hydrocarbons

	Butane	2-Methylpropane
Condensed Structure	$CH_3CH_2CH_2CH_3$	$CH_3CH(CH_3)CH_3$ or CH_3CHCH_3 | CH_3
Ball-and-Stick Model		
Space-Filling Model		
Melting Point (°C)	−138	−160
Normal Boiling Point (°C)	0	−12
Density of Gas (g/L)	0.5788	0.5934

The alkane family would be huge even if it consisted of only straight-chain hydrocarbons, but another structural possibility makes the family even larger. For example, alkanes with four or more carbon atoms can also be **branched-chain hydrocarbons,** as shown for the C_4 alkanes in Table 5.5. A *branch* is a side chain attached to the main carbon chain. The molecular formula of both structures in Table 5.5 is C_4H_{10}, but they represent different compounds with different properties. In recognition of these differences, the two compounds have different formal names (butane and 2-methylpropane).

The principal source of liquid alkanes on Earth is crude oil. Natural gas is the major source for the simplest alkane, methane, as well as smaller quantities of low-molar-mass alkanes including ethane, propane, and perhaps some butanes. Methane is often associated with oil deposits, but it is also produced during bacterial decomposition of vegetable matter in the absence of air, a condition that frequently arises in swamps. Hence methane's common name is swamp gas or marsh gas (Figure 5.34).

By far the most common use of alkanes in our lives is as fuels. Combustion reactions between alkanes and oxygen provide energy to power vehicles, generate electricity, warm our homes, and cook our meals. Gasoline, kerosene, and diesel fuels are mostly mixtures of alkanes containing up to 20 carbon atoms per molecule, as shown in Table 5.4, and the enthalpies of combustion for some of the more common fuels are listed in Table 5.6. Alkanes with higher boiling points are viscous liquids used as lubricating oils. Low-melting solid alkanes (C_{20}–C_{40}) are

branched-chain hydrocarbon a hydrocarbon in which the chain of carbon atoms is not linear.

fuel value the quantity of energy released during the complete combustion of 1 gram of a substance.

fuel density the quantity of energy released during the complete combustion of 1 liter of a liquid fuel.

(a)

(b)

FIGURE 5.34 Methane (CH_4) is a renewable source of energy because it can be produced by degradation of organic matter by methanogenic bacteria. Methane is produced in swamps; hence it is commonly called swamp gas. (a) Methane bubbles trapped in a frozen pond. Methane is produced by rotting organic matter at the bottom of the pond. (b) Experiments in the bulk production of methane from natural sources, like the one shown here of a plastic tarp covering a lagoon of animal waste, are being carried out worldwide to augment the fuel supply.

used in candles and in manufacturing matches. Very heavy hydrocarbon gums and solid residues (C_{36} and up) are used for paving roads.

Fuel Value

To calculate the enthalpy change that takes place when 1 g of methane or 1 g of propane burns in air, producing CO_2 and liquid water, we divide the absolute value of ΔH°_{comb} (in kilojoules per mole) for each reaction by the molar mass of the hydrocarbon. This gives us the number of kilojoules of energy released per gram of substance:

$$CH_4: \frac{802.3 \text{ kJ}}{\text{mol}} \times \frac{1 \text{ mol}}{16.04 \text{ g}} = 50.02 \text{ kJ/g}$$

$$C_3H_8: \frac{2219.9 \text{ kJ}}{\text{mol}} \times \frac{1 \text{ mol}}{44.10 \text{ g}} = 50.34 \text{ kJ/g}$$

These quantities of energy per gram of fuel are called **fuel values**. If we carry out similar calculations for the other hydrocarbons larger than methane in Table 5.3, we can determine their fuel values. When we do this, we find that the values decrease with increasing molar mass. Why is that the case?

The answer lies in the hydrogen-to-carbon ratios in these compounds. As the number of carbon atoms per molecule increases, the hydrogen-to-carbon ratio decreases. Consider that there are 12 times as many H atoms in 1 g of hydrogen as there are C atoms in 1 g of carbon; this means that, *gram for gram*, 1 g of hydrogen can form 6 times as many moles of H_2O as 1 g of carbon can form moles of CO_2. More energy is released in a combustion reaction by the formation of 1 mole of $CO_2(g)$ (393.5 kJ) than by 1 mole of $H_2O(g)$ (241.8 kJ), but even if we take this difference into account, *gram for gram* hydrogen has many times the fuel value of carbon.

Another term frequently used to compare the energy content of liquid fuels is **fuel density**. Fuel density describes the amount of energy available per unit volume of a liquid fuel and is typically reported as energy released when 1 liter of liquid is completely burned. Both fuel value and fuel density are reported as positive numbers, and it is simply understood that these values refer to the energy released from the fuel when it is burned.

TABLE 5.6 Standard Enthalpies of Combustion

Substance	ΔH°_{comb} (kJ/mol)
$CO(g)$	−283.0
$CH_4(g)$, methane	−802.3
$C_3H_8(g)$, propane	−2219.9
$C_5H_{12}(\ell)$, pentane	−3535
$C_9H_{20}(\ell)$, avg. gasoline compound	−6160
$C_{14}H_{30}(\ell)$, avg. diesel compound	−7940

CONCEPT TEST

Without doing any calculations, predict which compound in each pair releases more energy during combustion in air: (a) 1 mole of CH_4 or 1 mole of H_2; (b) 1 g of CH_4 or 1 g of H_2.

(Answers to Concept Tests are in the back of the book.)

SAMPLE EXERCISE 5.16 Comparing Fuel Values and Fuel Densities **LO8**

Most automobiles run on either gasoline or diesel fuel. Both fuels are mixtures, but the energy content in gasoline can be approximated by a hydrocarbon with the formula C_9H_{20} ($d = 0.718$ g/mL). Diesel fuel may be considered to be $C_{14}H_{30}$ ($d = 0.763$ g/mL). Using these two formulas, compare (a) the fuel value per gram of each fuel and (b) the fuel density per liter of each fuel. Some standard enthalpies of combustion are given in Table 5.6.

Collect and Organize We are given representative chemical formulas for gasoline (C_9H_{20}) and diesel fuel ($C_{14}H_{30}$) and can calculate the molar mass for each fuel. We are also given a table of $\Delta H°_{comb}$ values that includes data for gasoline and diesel fuel. We want to use these data and the respective molar masses to obtain the fuel values. We can then use the given density of each fuel to convert the fuel value to fuel density.

Analyze The enthalpies of combustion in Table 5.6 are given in kilojoules per mole, so to answer this question in terms of grams, we need molar masses to convert moles to grams in part (a). For part (b) we need density ($d = m/V$, grams per milliliter) to convert grams to liters. Taking C_9H_{20} and $C_{14}H_{30}$ as the average molecules in regular gasoline and diesel fuel, respectively, we can determine their molar masses. Then the fuel value can be calculated from the enthalpies of combustion. Using the densities given in the problem, we can calculate fuel densities in kJ/L.

Solve

a. Fuel values

$$\text{Gasoline as } C_9H_{20}: 6160 \frac{kJ}{mol} \times \frac{1 \ mol}{128.25 \ g} = 48.0 \ kJ/g$$

$$\text{Diesel as } C_{14}H_{30}: 7940 \frac{kJ}{mol} \times \frac{1 \ mol}{198.238 g} = 40.0 \ kJ/g$$

b. Fuel densities

$$\text{Gasoline as } C_9H_{20}: 48.0 \frac{kJ}{g} \times 0.718 \frac{g}{mL} \times \frac{10^3 \ mL}{L} = 34{,}500 \ kJ/L$$

$$\text{Diesel as } C_{14}H_{30}: 40.0 \frac{kJ}{g} \times 0.763 \frac{g}{mL} \times \frac{10^3 \ mL}{L} = 30{,}500 \ kJ/L$$

Think About It We would not expect the fuel values of gasoline and diesel fuel to be very different, or there would be a strong preference for gasoline-fueled cars over diesel or vice versa. The values are similar to the fuel values of methane and propane calculated in the text, so these answers seem reasonable. In the United States, car and truck fuel is purchased by the gallon. Its consumption is rated in miles per gallon, whereas in most countries it is bought by the liter (1 U.S. gal = 3.785 L). Because of the way we buy automobile fuels, it makes sense to compare fuel densities rather than fuel values. On either basis, diesel is a slightly inferior fuel compared with gasoline. Diesel engines, however, tend to deliver greater fuel efficiency (energy converted to w rather than released as q) than gasoline engines.

Practice Exercise Kerosene, used as a fuel in high-performance aircraft and in space heaters, is a hydrocarbon intermediate in composition between gasoline and diesel fuel and may be approximated as $C_{12}H_{26}$ ($d = 0.750$ g/mL; $\Delta H°_{comb} = -7050$ kJ/mol). Estimate the fuel value and the fuel density of kerosene.

(Answers to Practice Exercises are in the back of the book.)

Food Value

Food serves the same purpose in living systems as fuel does in mechanical systems. The chemical reactions that convert food into energy resemble combustion but consist of many more steps that are much more highly controlled. Carbon dioxide and water are the ultimate products, however, and in a fundamental way, metabolism of food by a living system and combustion of fuel in an engine are the same process. The **food value** of the material we eat—the amount of energy produced when food is burned completely—can be determined using the same equipment and applying the same concepts of thermochemistry we have developed to evaluate fuels for vehicles. We can analyze the relative food value of the items we consume in the same way we analyzed fuel value: by burning material in a bomb calorimeter and measuring the quantity of energy released.

As an illustration, let's consider the food value of a serving of peanuts: about 28 grams, or 40 peanuts. To determine its heat energy content per unit of mass, we burn the peanuts in a calorimeter. A portion of the mass of the peanuts (about 1.5 g) is water, so we first prepare the sample by drying it. This is a necessary step to get an accurate value for the enthalpy of reaction ΔH_{rxn} because we need to make sure that all the sample mass is due to its carbon-containing components.

Once the peanuts are dry, suppose they have a mass of 26.5 g. We put them in a calorimeter for which $C_{calorimeter}$ = 41.8 kJ/°C and burn them completely in excess oxygen. If the temperature of the calorimeter rises by 16.0°C, what is the food value of the peanuts?

To answer this question, we use Equation 5.12 to determine the quantity of energy that flowed from the peanuts to the calorimeter:

$$q_{calorimeter} = C_{calorimeter}\Delta T = (41.8\ \text{kJ/°C})(16.0\text{°C}) = +669\ \text{kJ}$$

where the plus sign reminds us that energy flowed into the calorimeter. The energy lost by the peanuts as they burned has the same value but opposite sign:

$$-q_{peanuts} = q_{calorimeter}$$
$$= -669\ \text{kJ}$$

Just as with fuel values, however, food value is reported as a positive number, and it is understood that the energy contained in the food is released when the food is metabolized. The peanuts' food value is therefore

$$\frac{669\ \text{kJ}}{1\ \text{serving of peanuts}} \times \frac{1\ \text{serving of peanuts}}{26.5\ \text{g}} = 25.2\ \text{kJ/g}$$

Most of us still think in terms of nutritional Calories (kilocalories), and we can use the definition of Calorie from Section 5.3 to convert energy in kilojoules into the familiar unit used by nutritionists:

$$\frac{669\ \text{kJ}}{\text{serving of peanuts}} \times \frac{1\ \text{Cal}}{4.184\ \text{kJ}} = \frac{160\ \text{Cal}}{\text{serving of peanuts}}$$

A single serving of peanuts represents 6% to 8% of the Calories needed by an adult with normal activity in one day. Figure 5.35 illustrates some other foods that have a similar food value.

food value the quantity of energy produced when a material consumed by an organism for sustenance is burned completely; it is typically reported in Calories (kilocalories) per gram of food.

(a)

(b)

(c)

FIGURE 5.35 (a) A can of soda, (b) a protein bar, and (c) a stack of saltines all contain between 120 and 160 Calories, or about 6% to 8% of our daily energy needs.

SAMPLE EXERCISE 5.17 Calculating Food Value **LO8**

Glucose ($C_6H_{12}O_6$) is a simple sugar formed by photosynthesis in plants. The complete combustion of 0.5763 g of glucose in a calorimeter ($C_{calorimeter}$ = 6.20 kJ/°C) raises the temperature of the calorimeter by 1.45°C. What is the food value of glucose in Calories per gram?

Collect and Organize We are asked to determine the food value of glucose, which means the energy given off when 1 g is burned. We have the mass of glucose burned and the calorimeter constant. We can convert kilojoules to Calories by using the conversion factor 1 Cal = 4.184 kJ.

Analyze We can relate the energy given off by the glucose to the energy gained by the calorimeter (Equation 5.15) and then to the temperature change and the calorimeter constant (Equation 5.12).

Solve

$$q_{calorimeter} = C_{calorimeter}\Delta T = (6.20 \text{ kJ/°C})(1.45°C) = 8.99 \text{ kJ}$$

To convert this quantity of energy to a food value, we divide by the sample mass:

$$\frac{8.99 \text{ kJ}}{0.5763 \text{ g}} = 15.6 \text{ kJ/g}$$

We can then convert this value into Calories:

$$(15.6 \text{ kJ/g})\left(\frac{1 \text{ Cal}}{4.184 \text{ kJ}}\right) = 3.73 \text{ Cal/g}$$

Think About It One gram of peanuts has about 50% more fuel value than 1 g of glucose (25.2 kJ/g and 15.6 kJ/g, respectively), so the peanuts have a greater energy density. This is important to hikers who carry their own food.

Practice Exercise Sucrose (table sugar) has the formula $C_{12}H_{22}O_{11}$ ($\mathcal{M}$ = 342.30 g/mol) and a food value of 16.4 kJ/g. Determine the calorimeter constant of the calorimeter in which the combustion of 1.337 g of sucrose raises the temperature by 1.96°C.

(Answers to Practice Exercises are in the back of the book.)

In this chapter we have considered the flow of energy and its role in defining the behavior of physical, chemical, and biological processes. The interaction of energy with matter—the transfer of energy into and out of materials—causes phase changes and alters the temperature of matter. The magnitude of these changes and the temperature ranges over which they occur are characteristic of the quantity and identity of the matter involved. Energy is also a product or a reactant in virtually all chemical reactions and, as such, behaves stoichiometrically, which means the chemical changes that a specific quantity of material undergoes are characterized by the release or consumption of a specific amount of energy. Understanding the energy contained in fuels is particularly important because fuels provide the bulk of the power needed by the equipment and devices of modern life—from cars and airplanes to computers, air conditioners, and lightbulbs. Just as significant is the energy contained in foods that sustain life. How we deal with the needs for energy in the near future—in terms of both fuel and food—will determine the quality of all our lives and the health of our planet.

SAMPLE EXERCISE 5.18 Integrating Concepts: Recycling Aluminum

Chemical engineers frequently analyze the energy required to carry out industrial procedures to assess costs of operations and support new approaches to producing materials. Over the last century, aluminum—both alone and in combination with other metals—has replaced steel for use where high strength-to-weight ratios and corrosion resistance are paramount. The industrial process for converting Al_2O_3 (alumina) into aluminum is based on passing an electric current through a solution of alumina dissolved in molten cryolite (Na_3AlF_6). As electricity passes through the solution, aluminum ions are reduced to aluminum metal, while the positively charged carbon electrode is oxidized to carbon dioxide. The process is described by the following reaction:

$$2\ Al_2O_3 \text{ (in molten } Na_3AlF_6) + 3\ C(s, \text{graphite}) \rightarrow 4\ Al(\ell) + 3\ CO_2(g)$$

The principal energy cost of the Hall–Héroult process is the electricity needed to reduce Al_2O_3; the major cost in recycling is the energy required to melt aluminum metal. We can use thermochemistry principles to estimate the energy requirements for the Hall–Héroult process and compare it with the energy needed for recycling.

a. Assign oxidation numbers to the reactants and products in the reaction and determine how many moles of electrons are needed per mole of aluminum produced.
b. Calculate the standard enthalpy of reaction for the reduction reaction of alumina from the standard enthalpies of formation. Use Table A4.3 in Appendix 4 as needed.
c. Estimate the energy required to recycle 1.00 mole of aluminum by heating it from 25°C to its melting point (660°C) until all the aluminum melts.

Collect and Organize We are given a balanced chemical equation. We will need the melting point of aluminum, 660°C; the molar enthalpy of fusion, $\Delta H_{fus,Al} = 10.79$ kJ/mol; and the molar specific heat of aluminum, $c_{P,Al}$, 24.4 J/(mol · K). Additional thermodynamic data are available in Appendix 4 to calculate ΔH_{rxn}°. We are asked to compare the energy savings for recycling aluminum with reducing it from Al_2O_3.

Analyze The energy needed to make 1 mole of aluminum—that is, the standard enthalpy change for the reaction—is calculated using standard enthalpies of formation and Equation 5.19. The energy required to recycle 1 mole of aluminum represents the sum of the energy needed to melt a mole of aluminum (the enthalpy of fusion) and the energy needed to heat the aluminum to its melting point, which can be calculated using its molar heat capacity and Equation 5.8.

Solve
a. The balanced chemical equation is given as:

$$2\ Al_2O_3 + 3\ C(s, \text{graphite}) \rightarrow 4\ Al(\ell) + 3\ CO_2(g)$$

where we have omitted the "in molten Na_3AlF_6" because the latter is the solvent for the reaction. Applying the rules for assigning oxidation numbers from Chapter 4, we recognize that both liquid Al and solid C have an oxidation number of zero; they are pure elements. Assigning the oxidation number

for oxygen as −2 in both Al_2O_3 and CO_2, we can calculate the oxidation numbers for Al and C as follows:

$$Al: \quad 2x + 3(-2) = 0 \qquad C: \quad x + 2(-2) = 0$$
$$x = +3 \qquad\qquad\qquad x = +4$$

To determine how many moles of electrons are needed per mole of aluminum produced, we need to determine the number of electrons transferred in this process:

Carbon: $\quad\quad\quad\quad\quad \Delta \text{ O.N.} = 3(+4) - 3(0) = +12$

12 electrons lost

$$\begin{array}{ccc} 0 & \text{oxidation} & +4 \end{array}$$

$$2\ Al_2O_3 + 3\ C(s, \text{graphite}) \rightarrow 4\ Al(\ell) + 3\ CO_2(g)$$

$$\begin{array}{ccc} +3 & \text{reduction} & 0 \end{array}$$

Aluminum: $\quad\quad \Delta \text{ O.N.} = 4(0) - 4(+3) = -12$

12 electrons gained

Given that a total of 12 moles of electrons are transferred for every 4 moles of aluminum produced, then 3 moles of electrons are required to produce each mole of aluminum.
b. Using Equation 5.19, we can calculate the standard enthalpy of reaction for the reduction reaction of alumina from the standard enthalpies of formation of the reactants and products:

$$\Delta H_{rxn}^\circ = [3(\Delta H_{f,CO_2}^\circ + 4(\Delta H_{f,Al(\ell)}^\circ)] - [2(\Delta H_{f,Al_2O_3}^\circ) + 3(\Delta H_{f,C}^\circ)]$$

$$= \left[(3 \text{ mol } CO_2)\left(\frac{-393.5 \text{ kJ}}{1 \text{ mol } CO_2}\right) + (4 \text{ mol } Al)\left(\frac{10.6 \text{ kJ}}{1 \text{ mol } Al}\right) \right]$$
$$- \left[(2 \text{ mol } Al_2O_3)\left(\frac{-1675.7 \text{ kJ}}{1 \text{ mol } Al_2O_3}\right) + (3 \text{ mol } C)\left(\frac{0.0 \text{ kJ}}{1 \text{ mol } C}\right) \right]$$

$$= +2213.3 \text{ kJ}$$

Dividing this value by the 4 moles of aluminum produced in the reaction as written, we get

$$\frac{2213.3 \text{ kJ}}{4 \text{ mol } Al} = 553.3 \text{ kJ/mol}$$

This reaction is endothermic, requiring 553 kJ for every mole of aluminum produced.
c. The energy needed to heat 1 mole of aluminum from 25°C to 660°C can be calculated from its molar heat capacity [24.4 J/(mol · °C), per Table 5.1] and Equation 5.8:

$$q = n c_P \Delta T$$

$$= 1.00 \text{ mol} \times 24.4 \frac{J}{\text{mol} \cdot °C} \times (660 - 25)°C \times \frac{1 \text{ kJ}}{1000 \text{ J}}$$

$$= 15.5 \text{ kJ}$$

Once the aluminum is at its melting point, the energy required to melt 1 mole of Al is its enthalpy of fusion ($\Delta H_{fus} = 10.6$ kJ/mol):

$$10.6 \text{ kJ/mol} \times 1.00 \text{ mol} = 10.6 \text{ kJ}$$

The estimated total energy to heat and melt 1.00 mole of aluminum is

$$15.5 \text{ kJ} + 10.6 \text{ kJ} = 26.1 \text{ kJ}$$

This value represents

$$\frac{26.1 \text{ kJ}}{553 \text{ kJ}} \times 100\% = 4.75\%$$

of the energy needed electrically to produce 1 mole of aluminum from its ore. The high cost of electricity makes recycling aluminum economically attractive as well as environmentally sound.

Think About It Other energy costs arise in both the production and recycling of aluminum, and the numbers calculated here should be viewed as estimates based on ideal situations; overall, however, recycling saves aluminum manufacturers about 95% of the energy required to produce the metal from the ore. This energy savings has inspired the rapid growth of a global aluminum recycling industry, and in the United States alone, aluminum recycling is a $1 billion-per-year business. Junkyards in the United States currently recycle 85% of the aluminum in cars and more than 50% of the aluminum in food and beverage containers.

SUMMARY

LO1 **Potential energy (PE)** is the energy stored in an object because of its position and is a **state function**. **Kinetic energy (KE)** is the energy of motion. Heating a sample increases the average kinetic energy of the atoms in the sample. Energy is stored in compounds, and energy is absorbed or released when they are transformed into different compounds or when a change of state occurs. (Sections 5.1 and 5.2)

LO2 A thermochemical study defines a **system** as its focus; everything other than the system is considered the **surroundings**. In an **exothermic process**, the system loses energy to its surroundings ($q < 0$); in an **endothermic process**, the system absorbs energy ($q > 0$) from its surroundings. (Section 5.3)

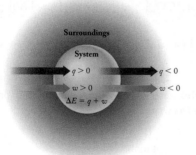

LO3 The sum of the kinetic and potential energies of a system is called its **internal energy (E)**. The internal energy of a system is increased ($\Delta E = E_{\text{final}} - E_{\text{initial}}$ is positive) when it is heated ($q > 0$) or if work is done on it ($w > 0$). When the pressure on a system is constant but the volume of the system changes, the work done is called **pressure–volume (P–V) work**. (Section 5.3)

LO4 The **enthalpy (H)** of a system is given by $H = E + PV$. The **enthalpy change (ΔH)** of a system is equal to the energy in the form of heat (q_P) added to or removed from the system at constant pressure: $\Delta H > 0$ for endothermic

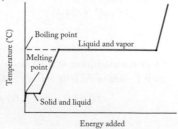

reactions and $\Delta H < 0$ for exothermic reactions. Thermodynamic values like **molar heat capacity (c_P)** and **specific heat (c_s)** can be used to quantify changes in systems involved in physical and chemical changes. (Sections 5.4 and 5.5)

LO5 A **calorimeter**, characterized by its **calorimeter constant** (its characteristic **heat capacity**), is a device used to measure the amount of energy involved in physical and chemical processes. The enthalpy change associated with a reaction is defined by the **enthalpy of reaction (ΔH_{rxn})**. (Section 5.6)

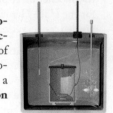

LO6 **Hess's law** states that the enthalpy of a reaction (ΔH_{rxn}) that is the sum of two or more other reactions is equal to the sum of the ΔH_{rxn} values of the constituent reactions. It can be used to calculate enthalpy changes in reactions that are hard or impossible to measure directly. (Section 5.7)

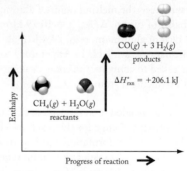

LO7 The **standard enthalpy of formation (ΔH_f°)** of a substance is the amount of energy involved in a **formation reaction**, in which 1 mole of the substance is made from its constituent elements in their **standard states** (under **standard conditions**). Enthalpy changes for physical changes and chemical reactions can be calculated from the enthalpies of formation of the reactants and products. (Section 5.8)

LO8 **Fuel value** is the amount of energy released on complete combustion of 1 g of a fuel. **Food value** is the amount of energy released when a material consumed by an organism for sustenance is burned completely; nutritionists often express food values in Calories (kilocalories) rather than in the SI unit kilojoules. (Section 5.9)

PARTICULATE **PREVIEW WRAP-UP**

Energy is absorbed by ozone to break its bonds. Bond breaking is endothermic; bond formation is exothermic.

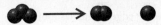

PROBLEM-SOLVING SUMMARY

Type of Problem	Concepts and Equations	Sample Exercises
Calculating kinetic and potential energy	$$KE = \frac{1}{2}mu^2 \qquad (5.2)$$ $$E_{el} \propto \frac{Q_1 \times Q_2}{d} \qquad (5.3)$$	**5.1**
Identifying endothermic and exothermic processes, and calculating internal energy change (ΔE) and P–V work	For the system: $$\Delta E = q + w \qquad (5.5)$$ where $w = -P\Delta V$.	**5.2, 5.3, 5.4**
Predicting the sign of ΔH_{sys} for physical and chemical changes	Exothermic: $\Delta H_{sys} < 0$ Endothermic: $\Delta H_{sys} > 0$	**5.5, 5.6**
Determining the flow of energy (q) associated with a change of state or with changing the temperature of a substance	Heating a substance: $$q = nc_p\Delta T \qquad (5.8)$$ or melting a solid at its melting point: $$q = n\Delta H_{fus} \qquad (5.9)$$ or vaporizing a liquid at its boiling point: $$q = n\Delta H_{vap} \qquad (5.10)$$	**5.7, 5.8, 5.9**
Measuring the enthalpy of reaction	The energy change of the system is equal in magnitude, but opposite in sign, to the energy change of the surroundings: $$q_{gained,sys} = -q_{lost,surr}$$ where the energy depends upon the amount of substance (moles), molar heat capacity, and change in temperature: $$q_{gained,sys} = n_{sys}c_{P,sys}\Delta T$$ $$q_{lost,surr} = n_{surr}c_{P,surr}\Delta T$$	**5.10**
Measuring the heat capacity (calorimeter constant) of a calorimeter	$$C_{calorimeter} = q_{calorimeter}/\Delta T \qquad (5.13)$$ where $C_{calorimeter}$ is the heat capacity of the calorimeter, $q_{calorimeter}$ is the heat released by a standard combustion reaction, and ΔT is the temperature change of the calorimeter.	**5.11**
Using Hess's law	Reorganize the information so that the reactions add together as desired. Reversing a reaction changes the sign of the reaction's ΔH_{rxn} value. Multiplying the coefficients in a reaction by a factor requires that the reaction's ΔH_{rxn} value be multiplied by the same factor.	**5.12, 5.13**
Recognizing and writing formation reactions	In a formation reaction, the reactants are elements in their standard states and the product is 1 mole of a single compound.	**5.14**
Calculating standard enthalpies of reaction from heats of formation	$$\Delta H^\circ_{rxn} = \sum n_{products}\,\Delta H^\circ_{f,products} - \sum n_{reactants}\,\Delta H^\circ_{f,reactants} \qquad (5.19)$$	**5.15**
Calculating fuel value and food value	The fuel value or food value of a substance is the energy released by the complete combustion of 1 g of the substance.	**5.16, 5.17**

VISUAL PROBLEMS

(Answers to boldface end-of-chapter questions and problems are in the back of the book.)

5.1. A brick lies perilously close to the edge of the flat roof of a building (Figure P5.1). The roof edge is 50 ft above street level, and the brick has 500 J of potential energy with respect to street level. Someone edges the brick off the roof, and it begins to fall. What is the brick's kinetic energy when it is 35 ft above street level? What is its kinetic energy the instant before it hits the street surface?

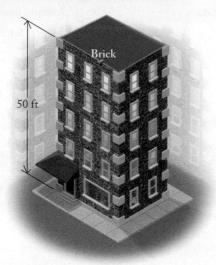

Brick

50 ft

FIGURE P5.1

5.2. Figure P5.2 shows pairs of cations and anions in contact. The energy of each interaction is proportional to Q_1Q_2/d, where Q_1 and Q_2 are the charges on the cation and anion, and d is the distance between their nuclei. Which pair has the greatest interaction energy and which has the smallest?

(a)   (b)   (c)

FIGURE P5.2

5.3. Figure P5.3 shows a sectional view of an assembly, consisting of a gas trapped inside a stainless steel cylinder with a stainless-steel piston and a block of iron on top of the piston to keep it in place.

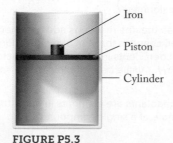

Iron

Piston

Cylinder

FIGURE P5.3

    a. Sketch the situation after the assembly has been heated for a few seconds with a blowtorch.

    b. Is the piston higher or lower in the cylinder?

    c. Has heat (q) been added to the system?

    d. Has the system done work (w) on its surroundings, or have the surroundings done work on the system?

5.4. The closed, rigid, metal box lying on the wooden kitchen table in Figure P5.4 is about to be involved in an accident.

    a. Are the contents of the box an isolated system, a closed system, or an open system?

    b. What will happen to the internal energy of the system if the table catches fire and burns?

    c. Will the system do any work on the surroundings while the fire is burning or after the table has burned away and collapsed?

FIGURE P5.4

5.5. The diagram in Figure P5.5 shows how a chemical reaction in a cylinder with a piston affects the volume of the system.

    a. In this reaction, does the system do work on the surroundings?

    b. If the reaction is endothermic, does the internal energy of the system increase or decrease when the reaction is proceeding?

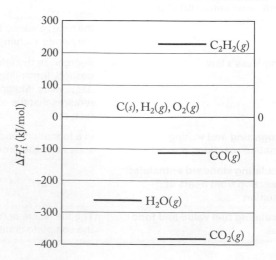

Piston

Reaction

System

FIGURE P5.5

5.6. The enthalpy diagram in Figure P5.6 indicates the enthalpies of formation of four compounds made from the elements listed on the "zero" line of the vertical axis.

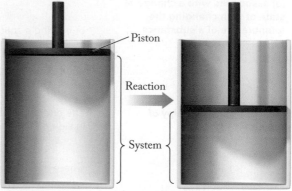

FIGURE P5.6

a. Why are the elements all put on the same horizontal line?
b. Why is $C_2H_2(g)$ sometimes called an "endothermic" compound?
c. How could the data be used to calculate the heat of the reaction that converts a stoichiometric mixture of $C_2H_2(g)$ and $O_2(g)$ to $CO(g)$ and $H_2O(g)$?

5.7. The process illustrated in Figure P5.7 takes place at constant pressure.
a. Write a balanced equation for the process.
b. Is w positive, negative, or zero for this reaction?
c. Using data from Appendix 4, calculate ΔH°_{rxn} for the formation of 1 mole of the product.

5.8. Use representations [A] through [I] in Figure P5.8 to answer questions a–f.
a. Which processes are exothermic?
b. Which processes have a positive ΔH?
c. In which processes does the system gain energy?
d. In which processes do the surroundings lose energy?
e. Compare a flame of methane [D] at 1000°C to a flame of propane [E] at 1000°C in terms of (i) average kinetic energy and (ii) average speed of the molecules.
f. Which substance(s) would *not* have vibrational motion or rotational motion? Why?

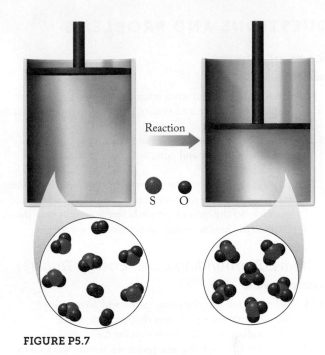

FIGURE P5.7

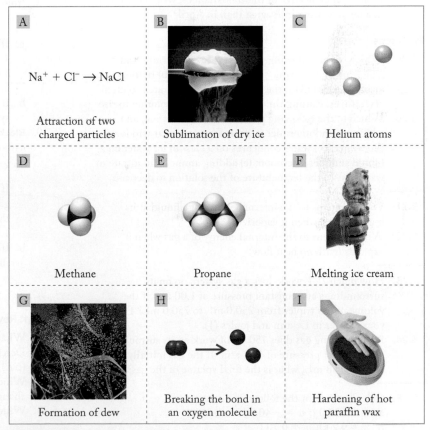

A	B	C
$Na^+ + Cl^- \rightarrow NaCl$ Attraction of two charged particles	Sublimation of dry ice	Helium atoms
D Methane	E Propane	F Melting ice cream
G Formation of dew	H Breaking the bond in an oxygen molecule	I Hardening of hot paraffin wax

FIGURE P5.8

QUESTIONS AND PROBLEMS

Forms of Energy

Concept Review

5.9. How are energy and work related?

5.10. Explain the difference between potential energy and kinetic energy.

5.11. Explain what is meant by a state function.

5.12. Are kinetic energy and potential energy both state functions?

5.13. Explain the nature of the potential energy in the following: (a) the new battery for your remote control; (b) a gallon of gasoline; (c) the crest of a wave before it crashes onto shore.

5.14. Explain the kinetic energy in a stationary ice cube.

Systems, Surroundings, and Energy Transfer

Concept Review

5.15. What is meant by the terms *system* and *surroundings*?

5.16. Do all exothermic processes that do work on the surroundings have the same sign for ΔE?

5.17. Why don't all endothermic processes that do work on the surroundings have the same sign for ΔE?

5.18. From the perspective of thermodynamics, why is the water in a pond warmer in August than in April?

Problems

5.19. Which of the following processes are exothermic, and which are endothermic? (a) a candle burns; (b) rubbing alcohol feels cold on the skin; (c) a supersaturated solution crystallizes, causing the temperature of the solution to rise

5.20. Which of the following processes are exothermic, and which are endothermic? (a) frost forms on a car window in the winter; (b) water condenses on a glass of ice water on a humid summer afternoon; (c) adding ammonium nitrate to water causes the temperature of the solution to decrease

5.21. What happens to the internal energy of a liquid at its boiling point when it vaporizes?

5.22. What happens to the internal energy of a gas when it expands (with no heat flow)?

5.23. How much $P–V$ work does a gas system do on its surroundings at a constant pressure of 1.00 atm if the volume of gas triples from 250.0 mL to 750.0 mL? Express your answer in L · atm and joules (J).

5.24. An expanding gas does 150.0 J of work on its surroundings at a constant pressure of 1.01 atm. If the gas initially occupied 68 mL, what is the final volume of the gas?

5.25. Calculate ΔE for the following situations:
   a. $q = 120.0$ J; $w = -40.0$ J
   b. $q = 9.2$ kJ; $w = 0.70$ J
   c. $q = -625$ J; $w = -315$ J

5.26. Calculate ΔE for
   a. the combustion of a gas that releases 210.0 kJ of heat to its surroundings and does 65.5 kJ of work on its surroundings.
   b. a chemical reaction that produces 90.7 kJ of heat but does no work on its surroundings.

*5.27. The following reactions take place in a cylinder equipped with a movable piston at atmospheric pressure (Figure P5.27). Which reactions will result in work being done on the surroundings? Assume the system returns to an initial temperature of 110°C. (*Hint*: The volume of a gas is proportional to number of moles, n, at constant temperature and pressure.)

FIGURE P5.27

   a. $CH_4(g) + 2\ O_2(g) \rightarrow CO_2(g) + 2\ H_2O(g)$
   b. $C_3H_8(g) + 5\ O_2(g) \rightarrow 3\ CO_2(g) + 4\ H_2O(g)$
   c. $N_2(g) + 2\ O_2(g) \rightarrow 2\ NO_2(g)$

*5.28. In which direction will the piston shown in Figure P5.27 move when the following reactions are carried out at atmospheric pressure inside the cylinder and after the system has returned to its initial temperature of 110°C? (*Hint*: The volume of a gas is proportional to number of moles, n, at constant temperature and pressure.)
   a. $N_2(g) + 3\ H_2(g) \rightarrow 2\ NH_3(g)$
   b. $C(s) + O_2(g) \rightarrow CO_2(g)$
   c. $CH_3CH_2OH(g) + 3\ O_2(g) \rightarrow 2\ CO_2(g) + 3\ H_2O(g)$

*5.29. **Filling Air Bags** Automobile air bags produce nitrogen gas from the reaction:

$$2\ NaN_3(s) \rightarrow 2\ Na(s) + 3\ N_2(g)$$

   a. If 2.25 g of NaN_3 reacts to fill an air bag, how much $P–V$ work will the N_2 do against an external pressure of 1.00 atm given that the density of nitrogen is 1.165 g/L at 20°C?
   b. If the process releases 2.34 kJ of heat, what is ΔE for the system?

*5.30. **Black Powder** Igniting gunpowder produces nitrogen and carbon dioxide gas that propels the bullet by the reaction:

$$2\ KNO_3(s) + \tfrac{1}{8}\ S_8(s) + 3\ C(s) \rightarrow K_2S(s) + N_2(g) + 3\ CO_2(g)$$

   a. If 1.00 g of KNO_3 reacts, how much $P–V$ work will the gases do against an external pressure of 1.00 atm given that the densities of nitrogen and CO_2 are 1.165 g/L and 1.830 g/L, respectively, at 20°C?
   b. If the reaction produces 21.6 kJ of heat, what is ΔE for the system?

Enthalpy and Enthalpy Changes

Concept Review

5.31. What is meant by an *enthalpy change*?

5.32. Describe the difference between an internal energy change (ΔE) and an enthalpy change (ΔH).

5.33. Which symbol, ΔH_{comb} or ΔH_{fus}, refers to a physical change?

5.34. Which symbol, ΔH_f or ΔH_{fus}, refers to a chemical change?

Problems

5.35. **A Clogged Sink** Adding Drano to a clogged sink causes the drainpipe to get warm. What is the sign of ΔH for this process?

5.36. **Hot Packs for Cold Fingers** Skiers can purchase small packets of finely divided iron that, after reaction with water, provide heat to frozen fingers and toes (Figure P5.36).

Define the system and the surroundings in this example, and indicate the sign of ΔH_{sys}.

(a)   (b)

FIGURE P5.36

5.37. Break a Bond The stable form of oxygen at room temperature and pressure is the diatomic molecule O_2. What is the sign of ΔH for the following process?

$$O_2(g) \rightarrow 2\ O(g)$$

5.38. Plaster of Paris Gypsum is the common name of calcium sulfate dihydrate ($CaSO_4 \cdot 2\ H_2O$). When gypsum is heated to 150°C, it loses most of the water in its formula and forms plaster of Paris ($CaSO_4 \cdot 0.5\ H_2O$):

$$2\ (CaSO_4 \cdot 2\ H_2O)(s) \rightarrow 2\ (CaSO_4 \cdot 0.5\ H_2O)(s) + 3\ H_2O(g)$$

What is the sign of ΔH for making plaster of Paris from gypsum?

5.39. Metallic Hydrogen A solid with metallic properties is formed when hydrogen gas is compressed under extremely high pressures. Predict the sign of the enthalpy change for the following reaction:

$$H_2(g) \rightarrow H_2(s)$$

5.40. Kitchen Chemistry A simple "kitchen chemistry" experiment requires placing some vinegar in a soda bottle. A deflated balloon containing baking soda is stretched over the mouth of the bottle. Adding the baking soda to the vinegar starts the following reaction and inflates the balloon:

$$NaHCO_3(aq) + CH_3COOH(aq) \rightarrow$$
$$CH_3COONa(aq) + CO_2(g) + H_2O(\ell)$$

If the contents of the bottle are considered the system, is work being done on the surroundings or on the system?

Heating Curves, Molar Heat Capacity, and Specific Heat

Concept Review

5.41. What is the difference between *specific heat* and *molar heat capacity*?

5.42. What happens to the molar heat capacity of a material if its mass is doubled? Is the same true for the specific heat?

5.43. Are the enthalpies of fusion and vaporization of a given substance usually the same?

5.44. An equal amount of energy is added to pieces of metal A and metal B that have the same mass. Why does the metal with the smaller specific heat reach the higher temperature?

***5.45. Cooling an Automobile Engine** Most automobile engines are cooled by water circulating through them and a radiator. However, the original Volkswagen Beetle had an air-cooled engine. Why might car designers choose water cooling over air cooling?

***5.46. Nuclear Reactor Coolants** The reactor-core cooling systems in some nuclear power plants use liquid sodium as the coolant. Sodium has a thermal conductivity of 1.42 J/ (cm · s · K), which is quite high compared with that of water [6.1×10^{-3} J/(cm · s · K)]. The respective molar heat capacities are 28.28 J/(mol · K) and 75.31 J/(mol · K). What is the advantage of using liquid sodium over water in this application?

Problems

5.47. How much energy is required to raise the temperature of 100.0 g of water from 30.0°C to 100.0°C?

5.48. Cooking in the Mountains At an elevation where the boiling point of water is 93°C, 100.0 g of water at 30°C absorbs 290.0 kJ of energy from a mountain climber's stove. Is this amount of energy sufficient to heat the water to its boiling point?

5.49. Use the following data to sketch a heating curve for 1 mole of methanol. Start the curve at −100°C and end it at 100°C.

Boiling point	65°C
Melting point	−94°C
ΔH_{vap}	37 kJ/mol
ΔH_{fus}	3.18 kJ/mol
Molar heat capacity (ℓ)	81.1 J/(mol · °C)
(g)	43.9 J/(mol · °C)
(s)	48.7 J/(mol · °C)

5.50. Use the following data to sketch a heating curve for 1 mole of octane. Start the curve at −57°C and end it at 150°C.

Boiling point	125.7°C
Melting point	−56.8°C
ΔH_{vap}	41.5 kJ/mol
ΔH_{fus}	20.7 kJ/mol
Molar heat capacity (ℓ)	254.6 J/(mol · °C)
Molar heat capacity (g)	316.9 J/(mol · °C)

5.51. Keeping an Athlete Cool During a strenuous workout, an athlete generates 2000.0 kJ of energy. What mass of water would have to evaporate from the athlete's skin to dissipate this much heat?

5.52. Hypothermia Damp clothes can prove fatal when outdoor temperatures drop (death by hypothermia).
a. If the clothes you are wearing absorb 1.00 kg of water and then dry in a cold wind on Mount Washington, how much heat would your body lose during this process? $c_{P,H_2O} = 75.3$ J/(mol · °C), and $\Delta H_{vap,H_2O} = 40.67$ kJ/mol.

*b. If the heat lost by your body is not replaced, what will the final temperature of your body be after the 1.00 kg of water has evaporated? Use your own body weight and assume that the specific heat of your body is 4.25 J/g. (Note that the specific heat is given in units of J/g.)

5.53. The same quantity of energy is added to 10.00 g pieces of gold, magnesium, and platinum, all initially at 25°C. The molar heat capacities of these three metals are 25.41 J/(mol · °C), 24.79 J/(mol · °C), and 25.95 J/(mol · °C), respectively. Which piece of metal has the highest final temperature?

5.54. Which of the following would reach the higher temperature after 10.00 g of iron [c_P = 25.1 J/(mol · °C)] at 150°C is added: 100 mL of water [d = 1.00 g/mL, c_P = 75.3 J/(mol · °C)] or 200 mL of ethanol [CH_3CH_2OH, d = 0.789 g/mL, c_P = 113.1 J/(mol · °C)]?

***5.55.** Exactly 10.0 mL of water at 25.0°C is added to a hot iron skillet. All the water is converted into steam at 100.0°C. The mass of the pan is 1.20 kg and the molar heat capacity of iron is 25.19 J/(mol · °C). What is the temperature change of the skillet?

***5.56.** A 20.0 g piece of iron and a 20.0 g piece of gold at 100.0°C were dropped into the same 1.00 L of water at 20.0°C. The molar heat capacities of iron and gold are 25.19 J/(mol · °C) and 25.41 J/(mol · °C), respectively. What is the final temperature of the water and pieces of metal?

Calorimetry: Measuring Heat Capacity and Enthalpies of Reaction

Concept Review

5.57. Why is it necessary to know the heat capacity of a calorimeter?

5.58. Could an endothermic reaction be used to measure the heat capacity of a calorimeter?

5.59. If we replace the water in a bomb calorimeter with another liquid, why do we need to redetermine the heat capacity of the calorimeter?

5.60. When measuring the enthalpy of combustion of a very small amount of material, would you prefer to use a calorimeter having a heat capacity that is small or large? Explain your reasoning.

Problems

5.61. **Designing Cold Packs** If you were designing a chemical cold pack, which of the following salts would you choose to provide the greatest drop in temperature per gram: NH_4Cl (ΔH_{soln} = 14.6 kJ/mol), NH_4NO_3 (ΔH_{soln} = 25.7 kJ/mol), or $NaNO_3$ (ΔH_{soln} = 20.4 kJ/mol)?

5.62. Dissolving calcium hydroxide (ΔH_{soln} = −16.2 kJ/mol) in water is an exothermic process. How much lithium hydroxide (ΔH_{soln} = −23.6 kJ/mol) would be required to produce the same amount of energy as dissolving 15.00 g of $Ca(OH)_2$?

5.63. The standard enthalpy of combustion of benzoic acid (molar mass 122 g/mol) is −3225 kJ/mol. Calculate the heat capacity of a bomb calorimeter if a temperature increase of 2.16°C occurs on combusting 0.500 g of benzoic acid in the presence of excess O_2.

5.64. Burning 1.43 g of methane [$CH_4(g)$, molar mass 16.0 g/mol] with excess O_2 raised the temperature of 250 g of water in a bomb calorimeter from 22°C to 98°C. Calculate the molar enthalpy of combustion of methane.

5.65. The complete combustion of 1.200 g of cinnamaldehyde (C_9H_8O, one of the compounds in cinnamon) in a bomb calorimeter ($C_{calorimeter}$ = 3.640 kJ/°C) produced an increase in temperature of 12.79°C. Calculate the molar enthalpy of combustion of cinnamaldehyde (ΔH_{comb}) in kilojoules per mole of cinnamaldehyde.

5.66. **Aromatic Spice** The aromatic hydrocarbon cymene ($C_{10}H_{14}$) is found in nearly 100 spices and fragrances, including coriander, anise, and thyme. The complete combustion of 1.608 g of cymene in a bomb calorimeter ($C_{calorimeter}$ = 3.640 kJ/°C) produced an increase in temperature of 19.35°C. Calculate the molar enthalpy of combustion of cymene (ΔH_{comb}) in kilojoules per mole of cymene.

5.67. **Hormone Mimics** Phthalates, used to make plastics flexible, are among the most abundant industrial contaminants in the environment. Several have been shown to act as hormone mimics in humans by activating the receptors for estrogen, a female sex hormone. In characterizing the compounds completely, the value of ΔH_{comb} for dimethyl phthalate ($C_{10}H_{10}O_4$) was determined to be −4685 kJ/mol. Assume that 1.00 g of dimethyl phthalate is combusted in a calorimeter whose heat capacity ($C_{calorimeter}$) is 7.854 kJ/°C at 20.215°C. What is the final temperature of the calorimeter?

5.68. **Flavorings** The flavor of anise is due to anethole, a compound with the molecular formula $C_{10}H_{12}O$. The ΔH_{comb} value for anethole is −5541 kJ/mol. Assume 0.950 g of anethole is combusted in a calorimeter whose heat capacity ($C_{calorimeter}$) is 7.854 kJ/°C at 20.611°C. What is the final temperature of the calorimeter?

5.69. Adding 2.00 g of Mg metal to 95.0 mL of 1.00 M HCl in a coffee-cup calorimeter leads to a temperature increase of 9.2°C.
  a. Write a balanced net ionic equation for the reaction.
  b. If the molar heat capacity of 1.00 M HCl is the same as that for water [c_P = 75.3 J/(mol · °C)], what is ΔH_{rxn}?

5.70. What is the ΔH_{rxn} for the precipitation of AgCl if adding 125 mL of 1.00 M $AgNO_3$ to 125 mL of 1.00 M NaCl at 18.6°C causes the temperature to increase to 26.4°C? (Assume $c_{P,soln}$ = $c_{P,water}$.)

***5.71.** How much glucose must be metabolized to completely evaporate 1.00 g of water at 37°C, given $\Delta H_{comb,glucose}$ = −2803 kJ/mol?

5.72. If all the energy obtained from burning 275 g of propane ($\Delta H_{comb,C_3H_8}$ = −2220 kJ/mol) is used to heat water, how many liters of water can be heated from 20.0°C to 100.0°C?

Hess's Law

Concept Review

5.73. How is Hess's law consistent with the law of conservation of energy?

5.74. Why is it important for Hess's law that enthalpy is a state function?

Problems

5.75. How can the first two of the following reactions be combined to obtain the third reaction?
  a. $CO(g) + NH_3(g) \rightarrow HCN(g) + H_2O(g)$
  b. $CO(g) + 3 H_2(g) \rightarrow CH_4(g) + H_2O(g)$
  c. $CH_4(g) + NH_3(g) \rightarrow HCN(g) + 3 H_2(g)$

5.76. **Cleansing the Atmosphere** The atmosphere contains the highly reactive molecule OH, which acts to remove selected pollutants. Use the values for ΔH_{rxn} given below to find the ΔH_{rxn} for the formation of OH and H from water.

$$\tfrac{1}{2} H_2(g) + \tfrac{1}{2} O_2(g) \rightarrow OH(g) \qquad \Delta H_{rxn} = 42.1 \text{ kJ}$$
$$H_2(g) \rightarrow 2 H(g) \qquad \Delta H_{rxn} = 435.9 \text{ kJ}$$
$$H_2(g) + \tfrac{1}{2} O_2(g) \rightarrow H_2O(g) \qquad \Delta H_{rxn} = -241.8 \text{ kJ}$$
$$H_2O(g) \rightarrow H(g) + OH(g) \qquad \Delta H_{rxn} = ?$$

5.77. **Ozone Layer** The destruction of the ozone layer by chlorofluorocarbons (CFCs) can be described by the following reactions:

$$ClO(g) + O_3(g) \rightarrow Cl(g) + 2 O_2(g) \qquad \Delta H_{rxn} = -29.90 \text{ kJ}$$
$$2 O_3(g) \rightarrow 3 O_2(g) \qquad \Delta H_{rxn} = 24.18 \text{ kJ}$$

Determine the value of the heat of reaction for the following:

$$Cl(g) + O_3(g) \rightarrow ClO(g) + O_2(g) \qquad \Delta H_{rxn} = ?$$

5.78. What is ΔH_{rxn} for the reaction between H_2S and O_2 that yields SO_2 and water,

$$2 H_2S(g) + 3 O_2(g) \rightarrow 2 SO_2(g) + 2 H_2O(g) \qquad \Delta H_{rxn} = ?$$

given ΔH_{rxn} for the following reactions?

$$H_2(g) + \tfrac{1}{2} O_2(g) \rightarrow H_2O(g) \qquad \Delta H_{rxn} = -241.8 \text{ kJ}$$
$$SO_2(g) + 3 H_2(g) \rightarrow H_2S(g) + 2 H_2O(g) \qquad \Delta H_{rxn} = 34.8 \text{ kJ}$$

Standard Enthalpies of Formation and Reaction

Concept Review

5.79. Explain how the use of ΔH_f° to calculate ΔH_{rxn}° is an example of Hess's law.

5.80. Why is the standard enthalpy of formation of $CO(g)$ difficult to measure experimentally?

5.81. Oxygen and ozone are both forms of elemental oxygen. Are the standard enthalpies of formation of oxygen and ozone the same? Explain.

5.82. Explain why the heats of formation of elements in their standard states are zero.

Problems

5.83. For which of the following reactions does ΔH_{rxn}° represent an enthalpy of formation?
  a. $C(s) + O_2(g) \rightarrow CO_2(g)$
  b. $CO_2(g) + C(s) \rightarrow 2 CO(g)$
  c. $CO_2(g) + H_2(g) \rightarrow H_2O(g) + CO(g)$
  d. $2 H_2(g) + C(s) \rightarrow CH_4(g)$

5.84. For which of the following reactions does ΔH_{rxn}° also represent an enthalpy of formation?
  a. $2 N_2(g) + 3 O_2(g) \rightarrow 2 NO_2(g) + 2 NO(g)$
  b. $N_2(g) + O_2(g) \rightarrow 2 NO(g)$
  c. $2 NO_2(g) \rightarrow N_2O_4(g)$
  d. $N_2(g) + 2 O_2(g) \rightarrow 2 NO_2(g)$

5.85. Use the following standard heats of formation to calculate the molar enthalpy of vaporization of liquid hydrogen peroxide: ΔH_f° of $H_2O_2(\ell)$ is -188 kJ/mol and ΔH_f° of $H_2O_2(g)$ is -136 kJ/mol.

5.86. Use the following standard heats of formation to calculate the molar enthalpy of vaporization of acetic acid: ΔH_f° of $CH_3COOH(\ell)$ is -484.5 kJ/mol and ΔH_f° of $CH_3COOH(g)$ is -432.8 kJ/mol.

5.87. Ammonium nitrate decomposes to N_2O and water vapor at temperatures between 250°C and 300°C. Write a balanced chemical reaction describing the decomposition of ammonium nitrate, and calculate the enthalpy of reaction by using the appropriate enthalpies of formation from Appendix 4.

5.88. **Military Explosives** Explosives called amatols are mixtures of ammonium nitrate and TNT introduced during World War I when TNT was in short supply. The mixtures can provide 30% more explosive power than TNT alone. Above 300°C, ammonium nitrate decomposes to N_2, O_2, and H_2O. Write a balanced chemical reaction describing the decomposition of ammonium nitrate, and determine the standard enthalpy of reaction by using the appropriate standard enthalpies of formation from Appendix 4.

5.89. **Improvised Explosives** Mixtures of fertilizer (ammonium nitrate) and fuel oil (a mixture of long-chain hydrocarbons similar to decane, $C_{10}H_{22}$) are the basis for powerful explosions. Determine the enthalpy change of the following explosive reaction by using the appropriate enthalpies of formation ($\Delta H_{f,C_{10}H_{22}}^\circ = 249.7$ kJ/mol):

$$3 NH_4NO_3(s) + C_{10}H_{22}(\ell) + 14 O_2(g) \rightarrow$$
$$3 N_2(g) + 17 H_2O(g) + 10 CO_2(g)$$

*5.90. **A Little TNT** Trinitrotoluene (TNT) is a highly explosive compound. The thermal decomposition of TNT is described by the following chemical equation:

$$2 C_7H_5N_3O_6(s) \rightarrow 12 CO(g) + 5 H_2(g) + 3 N_2(g) + 2 C(s)$$

If ΔH_{rxn} for this reaction is $-10{,}153$ kJ/mol, how much TNT is needed to equal the explosive power of 1 mole of ammonium nitrate (Problem 5.89)?

5.91. How can the standard enthalpy of formation of $CO(g)$ be calculated from the standard enthalpy of formation ΔH_f° of $CO_2(g)$ and the standard enthalpy of combustion ΔH_{comb}° of $CO(g)$?

5.92. Calculate the standard enthalpy of formation of $SO_2(g)$ from the standard enthalpy changes of the following reactions:

$$2 SO_2(g) + O_2(g) \rightarrow 2 SO_3(g) \qquad \Delta H_{rxn}^\circ = -196 \text{ kJ}$$
$$\tfrac{1}{4} S_8(s) + 3 O_2(g) \rightarrow 2 SO_3(g) \qquad \Delta H_{rxn}^\circ = -790 \text{ kJ}$$
$$\tfrac{1}{8} S_8(s) + O_2(g) \rightarrow SO_2(g) \qquad \Delta H_f^\circ = ?$$

5.93. Use the following data to calculate the enthalpy of formation of NO_2Cl from N_2, O_2, and Cl_2:

$$NO_2Cl(g) \rightarrow NO_2(g) + \tfrac{1}{2} Cl_2(g) \qquad \Delta H_{rxn}^\circ = +20.6 \text{ kJ}$$
$$\tfrac{1}{2} N_2(g) + O_2(g) \rightarrow NO_2(g) \qquad \Delta H_f^\circ = +33.2 \text{ kJ}$$
$$\tfrac{1}{2} N_2(g) + O_2(g) + \tfrac{1}{2} Cl_2(g) \rightarrow NO_2Cl(g) \qquad \Delta H_f^\circ = ?$$

5.94. Formation of CO$_2$ Baking soda decomposes on heating as follows, creating the holes in baked bread:

$$2\,NaHCO_3(s) \rightarrow Na_2CO_3(s) + CO_2(g) + H_2O(\ell)$$

Calculate the standard enthalpy of formation of $NaHCO_3(s)$ from the following information:

$$\Delta H°_{rxn} = -129.3 \text{ kJ} \qquad \Delta H°_f[Na_2CO_3(s)] = -1131 \text{ kJ/mol}$$

$$\Delta H°_f[CO_2(g)] = -394 \text{ kJ/mol} \qquad \Delta H°_f[H_2O(\ell)] = -286 \text{ kJ/mol}$$

Fuels, Fuel Values, and Food Values

Concept Review

5.95. What is meant by *fuel value*?

5.96. What are the units of fuel values?

5.97. How are fuel values calculated from molar enthalpies of combustion?

5.98. Is the fuel value of liquid propane the same as that of propane gas?

Problems

5.99. Flex Fuel Vehicles An increasing number of vehicles in the United States can run on either gasoline [$C_9H_{20}(\ell)$, $\Delta H_{comb,gasoline} = -6160$ kJ/mol] or ethanol [$CH_3CH_2OH(\ell)$, $\Delta H_{comb,ethanol} = -1367$ kJ/mol]. Which fuel has the greater fuel value?

5.100. Arctic Exploration Food contains three main categories of compounds: carbohydrate, protein, and fat. Arctic explorers often eat a high-fat diet because fats have a high food value. The average food values for carbohydrate, protein, and fat are 4 Cal/g, 4 Cal/g, and 9 Cal/g, respectively. Consider a typical fat to have the chemical formula $C_{18}H_{36}O_2$, glucose ($C_6H_{12}O_6$) to be a representative carbohydrate, and the amino acid alanine ($C_3H_6NO_2$) a building block of proteins. Express the food values given above as ΔH_{comb} in kJ/mol.

5.101. Cooking when Camping Lightweight camping stoves typically use *white gas*, a mixture of C_5 and C_6 hydrocarbons.
  a. Calculate the fuel value of C_5H_{12}, given that $\Delta H°_{comb} = -3535$ kJ/mol.
  b. How much energy is released during the combustion of 1.00 kg of C_5H_{12}?
  c. How many grams of C_5H_{12} must be burned to heat 1.00 kg of water from 20.0°C to 90.0°C? Assume that all the energy released during combustion is used to heat the water.

5.102. The heavier hydrocarbons in white gas are hexanes (C_6H_{14}).
  a. Calculate the fuel value of C_6H_{14}, given that $\Delta H°_{comb} = -4163$ kJ/mol.
  b. How much energy is released during the combustion of 1.00 kg of C_6H_{14}?
  c. How many grams of C_6H_{14} are needed to heat 1.00 kg of water from 25.0°C to 85.0°C? Assume that all the energy released during combustion is used to heat the water.
  d. Assume white gas is 25% C_5 hydrocarbons and 75% C_6 hydrocarbons; how many grams of white gas are needed to heat 1.00 kg of water from 25.0°C to 85.0°C?

Additional Problems

5.103. Diagnosing Car Trouble Mechanics sometimes use diethyl ether ($CH_3CH_2OCH_2CH_3$) to diagnose starting problems in vehicles because it has a low ignition temperature. If sprayed into the air intake, it can help determine whether the spark and ignition system of the car is functioning and the fuel delivery system is not. If the normal fuel is not entering the engine, the engine will run only until the diethyl ether vapors are completely burned. Compare the fuel value and fuel density of diethyl ether ($\Delta H°_{comb} = -2726.3$ kJ/mol; density = 0.7134 g/mL) to that of diesel fuel (determined in Sample Exercise 5.16).

5.104. The standard enthalpies of combustion of ethyne [$C_2H_2(g)$], C(s), and $H_2(g)$ are −1299.6, −393.5, and −285.9 kJ/mol, respectively. Use this information to calculate the standard enthalpy of formation of ethyne.

5.105. Carbon tetrachloride (CCl_4) was at one time used as a fire-extinguishing agent. It has a molar heat capacity c_P of 131.3 J/(mol · °C). How much energy is required to raise the temperature of 275 g of CCl_4 from room temperature (22°C) to its boiling point (77°C)?

5.106. Ethylene glycol ($HOCH_2CH_2OH$) is mixed with the water in radiators to cool car engines. How much heat will 725 g of pure ethylene glycol remove from an engine as it is warmed from 0°C to its boiling point of 196°C? The c_P of ethylene glycol is 149.5 J/(mol · °C).

5.107. Sodium may be used as a heat-storage material in some devices. The specific heat (c_s) of sodium metal is 1.23 J/(g · °C). How many moles of sodium metal are required to absorb 1.00×10^3 kJ of energy?

5.108. The water in a bomb calorimeter was replaced with the organic compound methylene chloride (CH_2Cl_2). Burning 2.23 g of glucose ($C_6H_{12}O_6$; $\Delta H_{comb} = -2801$ kJ/mol) in the calorimeter causes its temperature to rise 9.64°C. What is the heat capacity of the calorimeter?

5.109. The standard enthalpy of formation of NH_3 is −46.1 kJ/mol. What is $\Delta H°_{rxn}$ for the following reactions?
  a. $N_2(g) + 3\,H_2(g) \rightarrow 2\,NH_3(g)$
  b. $NH_3(g) \rightarrow \frac{1}{2}\,N_2(g) + \frac{3}{2}\,H_2(g)$

*5.110. Hung Out to Dry Laundry left outside to dry on a clothesline in the winter slowly dries by "ice vaporization" (sublimation). The increase in internal energy of water vapor produced by sublimation is less than the amount of heat absorbed. Explain.

5.111. Chlorofluorocarbons (CFCs) such as CF_2Cl_2 are refrigerants whose use has been phased out because of their destructive effect on Earth's ozone layer. The standard enthalpy of vaporization of CF_2Cl_2 is 17.4 kJ/mol, compared with $\Delta H°_{vap} = 40.67$ kJ/mol for liquid water. How many grams of liquid CF_2Cl_2 are needed to cool 200.0 g of water from 50.0°C to 40.0°C? The specific heat of water is 4.184 J/(g · °C).

5.112. A 100.0 mL sample of 1.0 *M* NaOH is mixed with 50.0 mL of 1.0 *M* H_2SO_4 in a large Styrofoam coffee cup; the cup is fitted with a lid through which a calibrated thermometer passes. The temperature of each solution before mixing is 22.3°C. After the NaOH solution is added to the coffee cup and the mixed solutions are stirred with the thermometer, the maximum temperature measured is 31.4°C. Assume that the density of the mixed solutions is 1.00 g/mL, the specific heat of the mixed solutions is 4.18 J/(g · °C), and no heat is lost to the surroundings.
  a. Write a balanced chemical equation for the reaction that takes place in the Styrofoam cup.

b. Is any NaOH or H_2SO_4 left in the Styrofoam cup when the reaction is over?

c. Calculate the enthalpy change per mole of H_2SO_4 in the reaction.

5.113. Varying the scenario in Problem 5.112, assume this time that 65.0 mL of 1.0 M H_2SO_4 is mixed with 100.0 mL of 1.0 M NaOH and that both solutions are initially at 25.0°C. Assume that the mixed solutions in the Styrofoam cup have the same density and specific heat as in Problem 5.112 and no heat is lost to the surroundings. What is the maximum measured temperature in the Styrofoam cup?

*5.114. An insulated container is used to hold 50.0 g of water at 25.0°C. A 7.25 g sample of copper is placed in a dry test tube and heated for 30 minutes in a boiling water bath at 100.1°C. The heated test tube is carefully removed from the water bath with laboratory tongs and inclined so that the copper slides into the water in the insulated container. Given that the specific heat of solid copper is 0.385 J/(g · °C), calculate the maximum temperature of the water in the insulated container after the copper metal is added.

5.115. The mineral magnetite (Fe_3O_4) is magnetic, whereas iron(II) oxide is not.

a. Write and balance the chemical equation for the formation of magnetite from iron(II) oxide and oxygen.

b. Given that 318 kJ of heat is released for each mole of Fe_3O_4 formed, what is the enthalpy change of the balanced reaction of formation of Fe_3O_4 from iron(II) oxide and oxygen?

5.116. Which of the following substances has a standard enthalpy of formation equal to zero? (a) Pb at 1000°C; (b) $C_3H_8(g)$ at 25.0°C and 1 atm pressure; (c) solid glucose at room temperature; (d) $N_2(g)$ at 25.0°C and 1 atm pressure

*5.117. The standard enthalpy of formation of liquid water is −285.8 kJ/mol.

a. What is the significance of the negative sign associated with this value?

b. Why is the magnitude of this value so much larger than the enthalpy of vaporization of water (ΔH°_{vap} = 40.67 kJ/mol)?

c. Calculate the amount of heat produced in making 50.0 mL of water from its elements under standard conditions.

5.118. Acetylene, C_2H_2 (ΔH°_f = 226.7 kJ/mol), and benzene, C_6H_6 (ΔH°_f = 49.0 kJ/mol), are sometimes referred to as endothermic compounds.

a. Why are C_2H_2 and C_6H_6 called endothermic compounds?

b. Calculate the standard molar enthalpy of combustion of acetylene and benzene.

*5.119. Balance the following chemical equation, name the reactants and products, and calculate the standard enthalpy change by using the data in Appendix 4.

$$FeO(s) + O_2(g) \rightarrow Fe_2O_3(s)$$

*5.120. Add reactions 1, 2, and 3, and label the resulting reaction 4. Consult Appendix 4 to find the standard enthalpy change for the balanced reaction 4.

(1) $Zn(s) + \frac{1}{8} S_8(s) \rightarrow ZnS(s)$

(2) $ZnS(s) + 2\,O_2(g) \rightarrow ZnSO_4(s)$

(3) $\frac{1}{8} S_8(s) + O_2(g) \rightarrow SO_2(g)$

5.121. The standard enthalpies of formation of benzene $C_6H_6(\ell)$, $CO_2(g)$, and $H_2O(\ell)$ are 49.0, −394, and −286 kJ/mol, respectively. Use this information to calculate the standard enthalpy of combustion of $C_6H_6(\ell)$.

5.122. The specific heat of solid copper is 0.385 J/(g · °C). What thermal energy change occurs when a 35.3 g sample of copper is cooled from 35.0°C to 15.0°C? Be sure to give your answer the proper sign. This amount of energy is used to melt solid ice at 0.0°C. The molar enthalpy of fusion of ice is 6.01 kJ/mol. How many moles of ice are melted?

*5.123. **Metabolism of Methanol** Methanol is toxic because it is metabolized in a two-step process in vivo to formic acid (HCOOH). Consider the following overall reaction under standard conditions:

$$O_2(g) + 2\,CH_3OH(\ell) \rightarrow 2\,HCOOH(\ell) + 2\,H_2O(\ell) + 1019.6\ kJ$$

a. Is this reaction endothermic or exothermic?

b. What is the value of ΔH°_{rxn} for this reaction?

c. How much energy would be absorbed or released if 60.0 g of methanol were metabolized in this reaction?

d. In the first step of metabolism, methanol is converted into formaldehyde (CH_2O), which is then converted into formic acid. Would you expect ΔH°_{rxn} for the metabolism of 1 mole of $CH_3OH(\ell)$ to give 1 mole of formaldehyde to be larger or smaller than 509.8 kJ?

5.124. Use Hess's law and the following data to calculate the standard enthalpy of formation of ethanol, $CH_3CH_2OH(\ell)$.

$CH_3CH_2OH(\ell) + 3\,O_2(g) \rightarrow$
$\qquad 2\,CO_2(g) + 3\,H_2O(\ell) \qquad \Delta H^{\circ}_{rxn} = -1368.2$ kJ/mol

$C(s) + O_2(g) \rightarrow CO_2(g) \qquad\qquad \Delta H^{\circ}_f = -393.5$ kJ/mol

$H_2(g) + \frac{1}{2} O_2(g) \rightarrow H_2O(\ell) \qquad \Delta H^{\circ}_f = -285.9$ kJ/mol

5.125. Use Hess's law and the following data to calculate the standard enthalpy of formation of $CH_4(g)$.

$C(s) + O_2(g) \rightarrow CO_2(g) \qquad\qquad \Delta H^{\circ}_f = -393.5$ kJ/mol

$H_2(g) + \frac{1}{2} O_2(g) \rightarrow H_2O(\ell) \qquad \Delta H^{\circ}_f = -285.9$ kJ/mol

$CO_2(g) + 2\,H_2O(\ell) \rightarrow$
$\qquad CH_4(g) + 2\,O_2(g) \qquad \Delta H^{\circ}_{rxn} = -890.4$ kJ/mol

5.126. The reaction of $CH_3OH(g)$ with $N_2(g)$ to give $HCN(g)$ and $NH_3(g)$ requires 164 kJ/mol of heat.

a. Write a balanced chemical equation for this reaction.

b. Should the thermal energy involved be written as a reactant or as a product?

c. What is the enthalpy change in the reaction of 60.0 g of $CH_3OH(g)$ with excess $N_2(g)$ to give $HCN(g)$ and $NH_3(g)$ in this reaction?

5.127. Calculate ΔH°_{rxn} for the reaction

$$2\,Ni(s) + \tfrac{1}{4} S_8(s) + 3\,O_2(g) \rightarrow 2\,NiSO_3(s)$$

from the following information:

(1) $NiSO_3(s) \rightarrow NiO(s) + SO_2(g) \qquad \Delta H^{\circ}_{rxn} = 156$ kJ

(2) $\frac{1}{8} S_8(s) + O_2(g) \rightarrow SO_2(g) \qquad \Delta H^{*}_{rxn} = -297$ kJ

(3) $Ni(s) + \frac{1}{2} O_2(g) \rightarrow NiO(s) \qquad \Delta H^{\circ}_{rxn} = -241$ kJ

5.128. Use the following information to calculate the enthalpy change involved in the complete reaction of 3.0 g of carbon to form $PbCO_3(s)$ in reaction 4. Be sure to give the proper sign (positive or negative) with your answer.

(1) $Pb(s) + \frac{1}{2} O_2(g) \rightarrow PbO(s)$ $\Delta H_{rxn}^\circ = -219$ kJ

(2) $C(s) + O_2(g) \rightarrow CO_2(g)$ $\Delta H_{rxn}^\circ = -394$ kJ

(3) $PbCO_3(s) \rightarrow PbO(s) + CO_2(g)$ $\Delta H_{rxn}^\circ = 86$ kJ

(4) $Pb(s) + C(s) + \frac{3}{2} O_2(g) \rightarrow PbCO_3(s)$ $\Delta H_{rxn}^\circ = ?$

*5.129. **Ethanol as Automobile Fuel** Brazilians are quite familiar with fueling their automobiles with ethanol, a fermentation product from sugarcane. Calculate the standard molar enthalpy for the complete combustion of liquid ethanol by using the standard enthalpies of formation of the reactants and products as given in Appendix 4.

5.130. Adding 1.56 g of K_2SO_4 to 6.00 mL of water at 16.2°C causes the temperature of the solution to drop by 7.70°C. How many grams of NaOH ($\Delta H_{soln} = -44.3$ kJ/mol) would you need to add to raise the temperature back to 16.2°C?

5.131. Two solids, 5.00 g of NaOH and 4.20 g of KOH, are added to 150 mL of water [$c_P = 75.3$ J/(mol · °C); $T = 23$°C] in a calorimeter. Given $\Delta H_{soln,NaOH} = -44.3$ kJ/mol and $\Delta H_{soln,KOH} = -56.0$ kJ/mol, what is the final temperature of the solution?

5.132. **The 100-Meter Dash** A 1995 article in *Discover* magazine on world-class sprinters contained the following statement: "In one race, a field of eight runners releases enough energy to boil a gallon jug of ice at 0.0°C in ten seconds!" How much "energy" do the runners release in 10 seconds? Assume that the ice has a mass of 128 ounces.

*5.133. **Specific Heats of Metals** In 1819, Pierre Dulong and Alexis Petit reported that the product of the atomic mass of a metal times its specific heat is approximately constant, an observation called the *law of Dulong and Petit*. Use the data in the table below to answer the following questions.

Element	$\mathcal{M}$ (g/mol)	c_s [J/(g · °C)]	$\mathcal{M} \times c_s$
Bismuth		0.120	
Lead	207.2	0.123	25.5
Gold	197.0	0.125	
Platinum	195.1		
Tin	118.7	0.215	
		0.233	
Zinc	65.4	0.388	
Copper	63.5	0.397	
		0.433	
Iron	55.8	0.460	
Sulfur	32.1		
		Average value:	

a. Complete each row in the table by multiplying each given molar mass and specific heat pair (one result has been entered in the table). What are the units of the resulting values in column 4?

b. Next, calculate the average of the values in column 4.

c. Use the mean value from part (b) to calculate the missing atomic masses in the table. Do you feel confident in identifying the elements from the calculated atomic masses?

d. Use the average value from part (b) to predict the missing specific heat values in the table.

*5.134. **Odor of Urine** Urine odor gets worse with time because urine contains the metabolic product urea [$CO(NH_2)_2$], a compound that is slowly converted to ammonia, which has a sharp, unpleasant odor, and carbon dioxide:

$$CO(NH_2)_2(aq) + H_2O(\ell) \rightarrow CO_2(aq) + 2\,NH_3(aq)$$

This reaction is much too slow for the enthalpy change to be measured directly by using a temperature change. Instead, the enthalpy change for the reaction may be calculated from the following data:

Compound	ΔH_f° (kJ/mol)
Urea(aq)	−319.2
$CO_2(aq)$	−412.9
$H_2O(\ell)$	−285.8
$NH_3(aq)$	−80.3

Calculate the standard molar enthalpy change for the reaction.

5.135. **Experiment with a Metal** Explain how the specific heat of a metal sample could be measured in the lab. (*Hint*: You'll need a test tube, a boiling water bath, a Styrofoam cup calorimeter containing a known mass of water, a calibrated thermometer, and a known mass of the metal.)

*5.136. **Rocket Fuels** The payload of a rocket includes a fuel and oxygen for combustion of the fuel. Reactions 1 and 2 describe the combustion of dimethylhydrazine and hydrogen, respectively. Pound for pound, which is the better rocket fuel, dimethylhydrazine or hydrogen?

(1) $(CH_3)_2NNH_2(\ell) + 4\,O_2(g) \rightarrow$
$N_2(g) + 4\,H_2O(g) + 2\,CO_2(g)$ $\Delta H_{rxn}^\circ = -1694$ kJ

(2) $H_2(g) + \frac{1}{2} O_2(g) \rightarrow H_2O(g)$ $\Delta H_{rxn}^\circ = -242$ kJ

5.137. At high temperatures, such as those in the combustion chambers of automobile engines, nitrogen and oxygen form nitrogen monoxide:

$$N_2(g) + O_2(g) \rightarrow 2\,NO(g) \quad \Delta H_{comb}^\circ = +180 \text{ kJ}$$

Any NO released into the environment is oxidized to NO_2:

$$2\,NO(g) + O_2(g) \rightarrow 2\,NO_2(g) \quad \Delta H_{comb}^\circ = -112 \text{ kJ}$$

Is the overall reaction

$$N_2(g) + 2\,O_2(g) \rightarrow 2\,NO_2(g)$$

exothermic or endothermic? What is ΔH_{comb}° for this reaction?

5.138. You are given the following data:

$\frac{1}{2} N_2(g) + \frac{1}{2} O_2(g) \rightarrow NO(g)$ $\Delta H_{rxn}^\circ = +90.3$ kJ

$NO(g) + \frac{1}{2} Cl_2(g) \rightarrow NOCl(g)$ $\Delta H_{rxn}^\circ = -38.6$ kJ

$2\,NOCl(g) \rightarrow N_2(g) + O_2(g) + Cl_2(g)$ $\Delta H_{rxn}^\circ = ?$

a. Which of the ΔH_{rxn}° values represent enthalpies of formation?

b. Determine ΔH_{rxn}° for the decomposition of NOCl.

5.139. Hydrogen as Fuel Hydrogen is attractive as a fuel because it has a high fuel value and produces no CO_2.

$$H_2(g) + \tfrac{1}{2} O_2(g) \rightarrow H_2O(g)$$

Unfortunately, the production and storage of hydrogen fuel remain problematic.

a. One problem with hydrogen is its low fuel density. What is the fuel density (kJ/L) for H_2 given that the density of hydrogen gas is 0.0899 g/L? If we could liquefy hydrogen, what would the fuel density of hydrogen be, given that the density of liquid hydrogen is 70.8 g/L?

b. One solution to the problem of hydrogen storage is to use solid, hydrogen-containing compounds that release hydrogen upon heating at low temperature. One such compound is ammonia borane, H_3NBH_3. Calculate the enthalpy change (ΔH) for the reactions shown below given $\Delta H^\circ_{f,H_3NBH_3} = -38.1$ kJ/mol, $\Delta H^\circ_{f,BH_3} = +110.2$ kJ/mol, $\Delta H^\circ_{f,NH_3} = -46.1$ kJ/mol, $\Delta H^\circ_{f,H_2NBH_2} = -66.5$ kJ/mol, and $\Delta H^\circ_{f,HNBH} = +56.9$ kJ/mol.

$$H_3NBH_3(g) \rightarrow NH_3(g) + BH_3(g)$$

$$H_3NBH_3(g) \rightarrow H_2(g) + H_2NBH_2(g)$$

$$H_2NBH_2(g) \rightarrow H_2(g) + HNBH(g)$$

c. How many kilograms of ammonia borane are needed to supply 10.0 kg of hydrogen?

*5.140. Industrial Use of Cellulose Research is being carried out on cellulose as a source of chemicals for the production of fibers, coatings, and plastics. Cellulose consists of long chains of glucose molecules ($C_6H_{12}O_6$), so for the purposes of modeling the reaction, we can consider the conversion of glucose to formaldehyde (CH_2O).

a. Is the reaction to convert glucose into formaldehyde an oxidation or a reduction?

b. Calculate the heat of reaction for the conversion of 1 mole of glucose into formaldehyde, given the following thermochemical data:

ΔH°_{comb} of formaldehyde gas $= -572.9$ kJ/mol

ΔH°_f of solid glucose $= -1274.4$ kJ/mol

$$C_6H_{12}O_6(s) \rightarrow 6 \ CH_2O(g)$$

Glucose   Formaldehyde

5.141. Smelting Iron Iron metal is obtained by reducing iron oxide with carbon. The balanced chemical equation for making iron from Fe_2O_3 is:

$$2 \ Fe_2O_3(s) + 3 \ C(s) \rightarrow 4 \ Fe(s) + 3 \ CO_2(g)$$

Iron melts at 1538°C with $\Delta H_{fus} = 19.4$ kJ/mol. The molar heat capacity of iron is $c_{P,Fe} = 25.1$ J/(mol · °C). Is the energy required to melt recycled iron less than that needed to reduce the iron in Fe_2O_3 to the free metal?

5.142. Recycling Copper How does the energy required to recycle 1.00 mole of copper compare with that required to recover copper from CuO? The balanced chemical equation for the smelting of copper is: $CuO(s) + CO(g) \rightarrow Cu(s) + CO_2(g)$. Copper melts at 1084.5°C with $\Delta H^\circ_{fus} = 13.0$ kJ/mol and a molar heat capacity $c_{P,Cu} = 24.5$ J/(mol · °C). In addition, $\Delta H^\circ_{f,CuO} = -155$ kJ/mol.

5.143. Use the following data to determine whether the conversion of diamond into graphite is exothermic or endothermic:

$C(s, \text{diamond}) + O_2(g) \rightarrow CO_2(g)$	$\Delta H^\circ = -395.4$ kJ
$2 \ CO_2(g) \rightarrow 2 \ CO(g) + O_2(g)$	$\Delta H^\circ = 566.0$ kJ
$2 \ CO(g) \rightarrow C(s, \text{graphite}) + CO_2(g)$	$\Delta H^\circ = -172.5$ kJ
$C(s, \text{diamond}) \rightarrow C(s, \text{graphite})$	$\Delta H^\circ = ?$

5.144. Converting Diamond to Graphite The standard state of carbon is graphite. $\Delta H^\circ_{f,\text{diamond}}$ is +1.896 kJ/mol. Diamond masses are normally given in carats, where 1 carat = 0.20 g. Determine the standard enthalpy of the reaction for the conversion of a 4-carat diamond into an equivalent mass of graphite. Is this reaction endothermic or exothermic?

6

Properties of Gases

The Air We Breathe

PARTICULATE **REVIEW**

Classifying Gases as Matter

In Chapter 6 we focus on the properties of gases. Gases are often the products of the reactions describing the combustion of fuels, which we examined in Chapter 5. An alternative fuel for cars is hydrogen gas. One way to produce $H_2(g)$ is to run electricity through water, causing the water to decompose to $H_2(g)$ and $O_2(g)$.

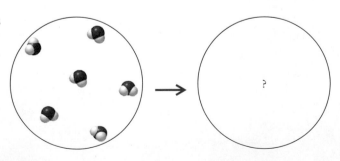

- Write a balanced chemical equation for this process.

- Given the sample of water shown here, draw the products after it undergoes electrolysis.

- Classify the products as elements, compounds, or a mixture.

    (Review Sections 1.1, 1.2, and 3.3 if you need help.)

(Answers to Particulate Review questions are in the back of the book.)

Volume and Temperature

A steel tank of helium gas has been used to fill several party balloons. As you read Chapter 6, look for ideas that will help you answer these questions.

- Draw a particulate image of the helium in the tank. Then draw a particulate image of the helium in one of the balloons. How do these drawings differ?

- Suppose you put both the tank and the helium-filled balloons in the trunk of your car on a hot summer day.
  - How would your particulate image for the helium in the balloon change?
  - How would your particulate image for the helium in the tank change?

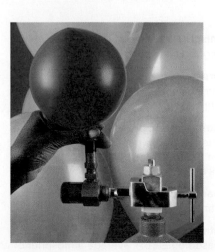

Learning Outcomes

LO1 Distinguish gases from liquids and solids

LO2 Measure pressure and convert between the different units used to quantify it
Sample Exercises 6.1, 6.2

LO3 Calculate changes in the volume, temperature, pressure, and number of moles of a gas by using the individual, combined, and ideal gas laws
Sample Exercises 6.3, 6.4, 6.5, 6.6, 6.7

LO4 Use balanced chemical equations to relate the volume of a gas-phase reactant to the amount of a product by using the stoichiometry of the reaction and the ideal gas law
Sample Exercises 6.8, 6.9

LO5 Calculate the density and molar mass of any gas
Sample Exercises 6.10, 6.11

LO6 Determine the mole fraction and the partial pressure of a gas in a mixture
Sample Exercises 6.12, 6.13, 6.14

LO7 Use kinetic molecular theory to explain the behavior of gases

LO8 Calculate the root-mean-square speed of a gas and relative rates of effusion and diffusion
Sample Exercises 6.15, 6.16

LO9 Use the van der Waals equation to correct for nonideal behavior
Sample Exercise 6.17

TABLE 6.1 Composition of Dry Air^a

Component	% (by volume)
Nitrogen	78.08
Oxygen	20.95
Argon	0.934
Carbon dioxide	0.0400^b
Neon	0.0018
Helium	0.00052
Methane	0.00018
Hydrogen	0.00011

^{a}Includes major and minor gases (with concentrations > 1 ppm by volume).
^{b}Value as of May 2015. Atmospheric CO_2 is increasing by about 2 ppm each year.

miscible capable of being mixed in any proportion (without reacting chemically).

pressure (P) the ratio of a force to the surface area over which the force is applied.

atmospheric pressure (P_{atm}) the force exerted by the gases surrounding Earth on Earth's surface and on all surfaces of all objects.

barometer an instrument that measures atmospheric pressure.

atmosphere (1 atm) a unit of pressure based on Earth's average atmospheric pressure at sea level.

6.1 Air: An Invisible Necessity

How often do you think about air? You remember how important air is when you do not have enough—for example, when you dive into a pool of water or hike to the top of a tall mountain. Otherwise you probably do not give much thought to breathing or to the mixture of gases that make up the air you breathe. You may not think much about the oxygen needed to stay alive because it is colorless, tasteless, odorless—and free. But the exchange of gases between your lungs and the surrounding atmosphere is crucial to life; air is an invisible necessity.

Having the right mixture of gases in our bodies is critical during surgery, which is why anesthesiologists in a hospital operating room constantly monitor levels of oxygen and carbon dioxide in the blood. The management of the delicate balance of gases entering and leaving a patient can mean the difference between a normal recovery and an irreversible coma.

We have seen how dissolved compounds react in aqueous solution. Chemical reactions also take place in the gas phase, and gases are intimately involved in chemical reactions in living systems as well as in the material world. Most life in our biosphere requires oxygen. Insects, birds, mammals, plants, and even underwater organisms need O_2 to metabolize nutrients.

How do gases differ from solids and liquids? Gases have neither definite volumes nor definite shapes; they expand to occupy the entire volume of their container and assume the container's shape. Under everyday conditions, other properties also distinguish gases from liquids and solids:

1. Unlike the volume occupied by a liquid or solid, the volume occupied by a gas changes significantly with pressure. If we carry an inflated balloon from sea level (0 m) to the top of a 1600-m mountain, the balloon volume increases by about 20%. The volume of a liquid or solid is unchanged under these conditions.

2. The volume of a gas changes with temperature. For example, the volume of a balloon filled with room-temperature air decreases when the balloon is taken outside on a cold winter's day. A temperature decrease from 20°C to 0°C leads to a volume decrease of about 7%, whereas the volume of a liquid or solid remains practically unchanged by this modest temperature change.

3. Gases are **miscible**, which means they can be mixed in any proportion (unless they chemically react with one another). A hospital patient experiencing respiratory difficulties may be given a mixture of nitrogen and oxygen in which the proportion of oxygen is much higher than its proportion in air. Alternatively, a scuba diver may leave the ocean surface with a tank of air containing a homogeneous mixture of 17% oxygen, 34% nitrogen, and 49% helium. In contrast, many liquids are immiscible, such as oil and water.

4. Gases are typically much less dense than liquids or solids. One indicator of this large difference is that gas densities are expressed in grams per *liter* but liquid densities are expressed in grams per *milliliter*. The density of dry air at 20°C at typical atmospheric pressure is 1.20 g/L, for example, whereas the density of liquid water under the same conditions is 1.00 g/mL—more than 800 times greater than the density of dry air.

These four observations about gases are consistent with the idea that the particles of a gas (be they molecules or atoms) are farther apart than the particles in solids and liquids. The larger spaces between the molecules in air, for example, account for the ready compressibility of air into scuba tanks. Greater distances between molecules account for both the lower densities and the miscibility of gases.

6.2 Atmospheric Pressure and Collisions

Earth is surrounded by a layer of gases 50 kilometers thick. We call this mixture of gases either *air* or *the atmosphere*. By volume it is composed primarily of nitrogen (78%) and oxygen (21%), with lesser amounts of other gases (Table 6.1). To put the thickness of the atmosphere in perspective: if Earth were the size of an apple, the atmosphere would be about as thick as the apple's skin.

Earth's atmosphere is pulled toward Earth by gravity and exerts a force that is spread across the entire surface of the planet (Figure 6.1). The molecules of N_2 and O_2 (along with the other elements and compounds in Table 6.1) are in constant motion in the atmosphere. As these molecules collide with one another and with the surface of Earth, the force of each collision creates **pressure (P)**. Pressure is defined as the ratio of force (F) to surface area (A):

$$P = \frac{F}{A} \qquad (6.1)$$

The force exerted by the atmosphere of colliding atoms and molecules on Earth's surface is called **atmospheric pressure (P_{atm})**.

Atmospheric pressure is measured with an instrument called a **barometer**. A simple but effective barometer design consists of a tube nearly 1 meter long, filled with mercury, and closed at one end (Figure 6.2). The tube is inverted with its open end immersed in a pool of mercury that is open to the atmosphere (meaning that molecules of N_2 and O_2 are in constant motion and collide against the surface of the liquid mercury). Gravity pulls the mercury in the tube downward, creating a vacuum at the top of the tube, while atmospheric pressure pushes the mercury in the pool up into the tube (due to the force of the collisions). The net effect of these opposing forces is indicated by the height of the mercury in the tube, which provides a measure of atmospheric pressure.

Atmospheric pressure varies from place to place and with changing weather conditions. Several units are used to express pressure. An **atmosphere (1 atm)**

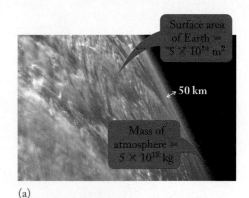

(a)

(b)

FIGURE 6.1 (a) Atmospheric pressure results from the force exerted by the atmosphere on Earth's surface. (b) If you stretch out your hand palm upward, the mass of the column of air above your palm is about 100 kg. This textbook has a mass of about 2.5 kg, so the mass of the atmosphere on your palm is equivalent to the mass of about 40 textbooks.

millimeters of mercury (mmHg) (also called **torr**) a unit of pressure where 1 atm = 760 mmHg = 760 torr.

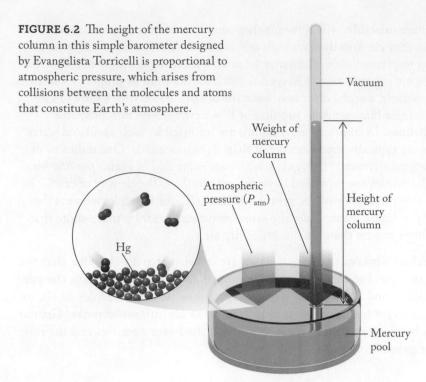

FIGURE 6.2 The height of the mercury column in this simple barometer designed by Evangelista Torricelli is proportional to atmospheric pressure, which arises from collisions between the molecules and atoms that constitute Earth's atmosphere.

Vacuum

Weight of mercury column

Atmospheric pressure (P_{atm})

Height of mercury column

Hg

Mercury pool

CONNECTION We encountered atmospheric pressure in our discussion of *P–V* work in Chapter 5.

of pressure is the pressure capable of supporting a column of mercury 760 mm high in a barometer. This column height is the average height of mercury in a barometer at sea level. Another pressure unit, **millimeters of mercury (mmHg)**, is more explicitly related to column height, where 1 atm = 760 mmHg. Pressure in mmHg is also expressed in a unit called the **torr** in honor of Evangelista

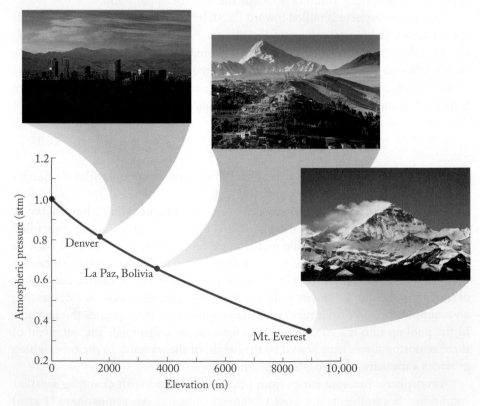

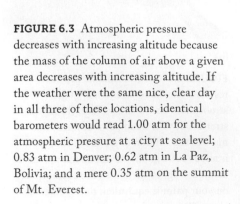

FIGURE 6.3 Atmospheric pressure decreases with increasing altitude because the mass of the column of air above a given area decreases with increasing altitude. If the weather were the same nice, clear day in all three of these locations, identical barometers would read 1.00 atm for the atmospheric pressure at a city at sea level; 0.83 atm in Denver; 0.62 atm in La Paz, Bolivia; and a mere 0.35 atm on the summit of Mt. Everest.

Denver

La Paz, Bolivia

Mt. Everest

Torricelli (1608–1647), the Italian mathematician and physicist who invented the barometer. Thus

$$1 \text{ atm} = 760 \text{ mmHg} = 760 \text{ torr}$$

The SI unit of pressure is the *pascal* (Pa), named in honor of the French mathematician and physicist Blaise Pascal (1623–1662), who was the first to propose that atmospheric pressure decreases with increasing altitude. We can explain this phenomenon by noting that the atmospheric pressure at any given location on Earth's surface is related to the mass of the column of air *above* that location (Figure 6.3). As altitude increases, the height of the column of air above a location decreases, which means the mass of the gases in the column decreases. Less mass means fewer collisions among molecules in the atmosphere and therefore a smaller force exerted by the air. According to Equation 6.1, a smaller force means less pressure.

The pascal is a derived SI unit; it is defined using the SI base units kilogram, meter, and second:

$$1 \text{ Pa} = \frac{1 \text{ kg}}{\text{m} \cdot \text{s}^2}$$

To understand the logic of this combination of units, consider the relationship in physics that states that pushing an object of mass m with a force F causes the object to accelerate at the rate a:

$$F = ma \qquad (6.2)$$

Combining Equations 6.1 and 6.2, we have

$$\frac{F}{A} = P = \frac{ma}{A} \qquad (6.3)$$

If m is in kilograms, a is in meters per second-squared (m/s^2), and A is in square meters (m^2), the units for P are

$$\frac{\text{kg} \dfrac{\text{m}}{\text{s}^2}}{\text{m}^2} = \text{kg} \frac{\text{m}}{\text{s}^2} \times \frac{1}{\text{m}^2} = \frac{\text{kg}}{\text{m} \cdot \text{s}^2}$$

The relationship between atmospheres and pascals is

$$1 \text{ atm} = 101{,}325 \text{ Pa}$$

Clearly, 1 Pa is a tiny quantity of pressure. In many applications it is more convenient to express pressure in kilopascals.

For many years, meteorologists have expressed atmospheric pressure in *millibars* (mbar). The weather maps prepared by the U.S. National Weather Service show changes in atmospheric pressure by constant-pressure contour lines, called *isobars*, spaced 4 mbar apart (Figure 6.4). There are exactly 10 mbar in 1 kPa; thus

$$1 \text{ atm} = (101.325 \text{ kPa})(10 \text{ mbar/kPa})$$
$$= 1013.25 \text{ mbar}$$

Other units of pressure are derived from masses and areas in the U.S. Customary System, such as pounds per square inch (lb/in^2, psi) for tire pressures and inches of mercury for atmospheric pressure in weather reports. The relationships between different units for pressure and 1 atm are summarized in Table 6.2.

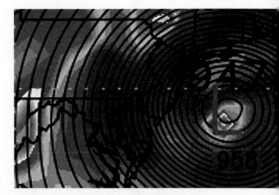

FIGURE 6.4 Small differences in atmospheric pressure are associated with major changes in weather. Adjacent isobars on this map of Hurricane Sandy differ by 4 mbar of pressure.

TABLE 6.2 Units for Expressing Pressure

Unit	Value
Atmosphere (atm)	1 atm
Millimeter of mercury (mmHg)	1 atm = 760 mmHg
Torr	1 atm = 760 torr
Pascal (Pa)	1 atm = 1.01325×10^5 Pa
Kilopascal (kPa)	1 atm = 101.325 kPa
Bar	1 atm = 1.01325 bar
Millibar (mbar)	1 atm = 1013.25 mbar
Pounds per square inch (psi)	1 atm = 14.7 psi
Inches of mercury	1 atm = 29.92 inches of Hg

manometer an instrument for measuring the pressure exerted by a gas.

SAMPLE EXERCISE 6.1 Calculating Atmospheric Pressure　　　**LO2**

The atmosphere on Titan, one of Saturn's moons, is composed almost entirely of nitrogen (98%). However, its oceans contain substances thought to be important in the evolution of life, making Titan interesting for planetary scientists. The mass of Titan's atmosphere is estimated to be 9.0×10^{18} kg. The surface area of Titan is 8.3×10^{13} m^2. The acceleration due to gravity on Titan's atmosphere is 1.35 m/s^2. From these values, calculate an average atmospheric pressure in kilopascals.

Collect, Organize, and Analyze We are given the mass of the atmosphere in kilograms, the surface area of Titan in square meters, and the acceleration due to gravity in meters per second-squared. Equation 6.3 allows us to combine these units to calculate atmospheric pressure in pascals.

Solve Substituting the values into Equation 6.3 gives us

$$P = \frac{ma}{A}$$

$$= \frac{(9.0 \times 10^{18} \text{ kg})(1.35 \text{ m/s}^2)}{8.3 \times 10^{13} \text{ m}^2}$$

$$= \frac{1.46 \times 10^5 \text{ kg}}{\text{m} \cdot \text{s}^2} = 1.5 \times 10^5 \text{ Pa}$$

$$= 1.5 \times 10^2 \text{ kPa}$$

Think About It The atmospheric pressure on Titan is similar to that on Earth. The greater mass of Titan's atmosphere and its smaller surface area compensate for the lower acceleration due to gravity.

Practice Exercise Calculate the pressure in pascals exerted on a tabletop by a sugar cube that is 1.00 cm on each side and has a mass of 4.15 g. The acceleration due to gravity on Earth is 9.8 m/s^2.

(Answers to Practice Exercises are in the back of the book.)

Scientists conducting experiments with gases may monitor the pressures exerted by the gases by using a **manometer**. Two types of manometers are illustrated in Figure 6.5. In each case, a U-shaped tube filled with mercury (or another dense liquid) is connected to a container holding the gas sample whose pressure is being measured. When the valve is opened, the gas exerts pressure on the mercury.

FIGURE 6.5 (a) A closed-end manometer is designed to measure the pressure of a gas sample that is less than atmospheric pressure. (b) The pressure of the gas sample is the difference between the heights of the mercury columns. (c) An open-end manometer is designed to measure the pressure of a gas sample that is greater than atmospheric pressure. (d) The difference in the heights of the two mercury columns is the difference between the pressure of the sample and that of the atmosphere.

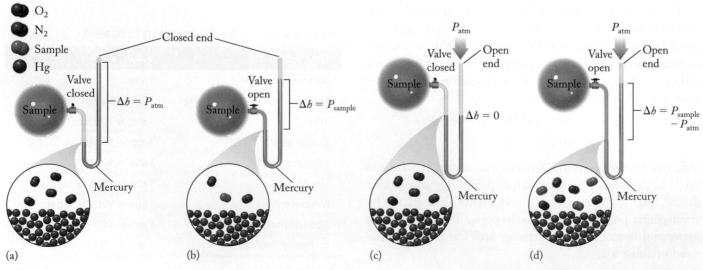

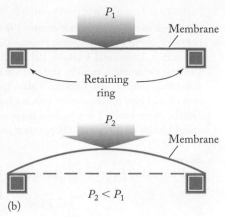

P_1

Membrane

Retaining
ring

P_2

Membrane

$P_2 < P_1$

(a)

(b)

FIGURE 6.6 (a) The pressure sensor in this barograph is a partially evacuated corrugated metal can. (b) As atmospheric pressure decreases, the lid of the can distorts outward. This motion is amplified by a series of levers and transmitted via the horizontal arm to a pen tip that records the pressure on the graph paper as the drum slowly turns. A week's worth of barometric data can be recorded in this way.

The difference between the two manometers is whether the end of the tube not connected to the flask containing the sample is closed or open to the atmosphere. In a closed-end manometer, the difference in the heights of the mercury columns in the two arms of the U-shaped tube is a direct measure of the gas pressure (Figure 6.5b). In an open-end manometer, the difference in heights represents the difference between the pressure in the flask and atmospheric pressure. When the pressure in the flask is greater than atmospheric pressure (Figure 6.5d), the level in the arm open to the atmosphere is higher than the level in the other arm; when the pressure in the flask is lower than atmospheric pressure, the level in the arm attached to the flask is higher than the level in the other arm. Torricelli's barometer (Figure 6.2), which he used to measure atmospheric pressure, is an example of a closed-end manometer.

CONCEPT **TEST**

In Figure 6.5(d), which gas has more collisions with the surface of the mercury: the gas in the flask or the atmosphere pushing down on the open end of the U-tube?

(Answers to Concept Tests are in the back of the book.)

Manometers have been largely displaced by pressure sensors based on flexible metallic or ceramic diaphragms. As the pressure on one side of the diaphragm increases, it distorts away from that side. This is the mechanism used to sense changes in atmospheric pressure in most present-day barometers, including the recording barometer, or *barograph*, shown in Figure 6.6.

SAMPLE EXERCISE 6.2 Measuring Gas Pressure with a Manometer **LO2**

Oyster shells are composed of calcium carbonate ($CaCO_3$). When the shells are roasted, they produce solid calcium oxide (CaO) and CO_2. A chemist roasts some oyster shells in an evacuated flask that is attached to a closed-end manometer (Figure 6.7a). When the roasting is complete and the system has cooled to room temperature, the difference in levels of mercury (Δh) in the arms of the manometer is 143.7 mm (Figure 6.7b). Calculate the pressure of the $CO_2(g)$ in (a) torr, (b) atmospheres, and (c) kilopascals.

Collect, Organize, and Analyze We are given the difference between the mercury levels in the arms of the manometer, a direct measure of the pressure in the flask,

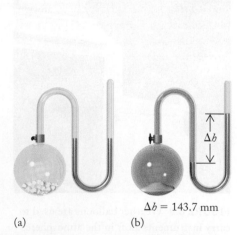

Δh

$\Delta h = 143.7$ mm

(a)

(b)

FIGURE 6.7 Roasting $CaCO_3$ in a closed-end manometer. (a) An evacuated flask containing oyster shells. (b) The same setup after the shells have been roasted, a decomposition reaction that produces CO_2 in the flask.

FIGURE 6.8 Before roasting oyster shells in an open-end manometer.

measured in millimeters of mercury. We need to express this value in three units of pressure, using the appropriate conversion factors from Table 6.2: 1 mmHg = 1 torr, 760 torr = 1 atm, and 1 atm = 101.325 kPa. We can estimate our answers quite readily. Because the units of mmHg and torr are equal, the numerical value of pressure is the same using either unit. Because 1 atmosphere equals hundreds of torr, the pressure in atm should be a much smaller number than the value of the pressure expressed in torr. On the other hand, 1 atmosphere is approximately 100 kPa, so the pressure in kPa should be about 10^2 times the pressure in atm.

Solve

a. Converting torr to atmospheres:

$$143.7 \ \text{torr} \times \frac{1 \ \text{atm}}{760 \ \text{torr}} = 0.1891 \ \text{atm}$$

b. Converting atmospheres to kilopascals:

$$0.1891 \ \text{atm} \times \frac{101.325 \ \text{kPa}}{1 \ \text{atm}} = 19.16 \ \text{kPa}$$

Think About It The calculated pressure values make sense according to our estimates: a pressure less than 760 torr should also be less than 1 atm; by the same token, a pressure less than 1 atm should also be less than 101.325 kPa.

Practice Exercise The same quantity of $CO_2(g)$ produced in Sample Exercise 6.2 is produced in a flask connected to an open-end manometer (Figure 6.8). The atmospheric pressure and the pressure inside the flask containing the oyster shells are both 760 torr at the start of the experiment. Indicate on the drawing where the mercury levels in the two arms should be at the conclusion of the experiment. What is the value of Δh in millimeters?

6.3 The Gas Laws

In Section 6.1 we summarized some of the properties of gases and described the effect of pressure and temperature on volume, mostly in qualitative terms. Our knowledge of the quantitative relationships among P, T, and V goes back more than three centuries, to a time before the field of chemistry as we now know it even existed. Some of the experiments that led to our understanding of how gases behave were driven by human interest in hot-air balloons, and today balloons are still used to study weather and atmospheric phenomena (Figure 6.9). Their successful and safe use requires an understanding of gas properties that was first gained in the 17th and 18th centuries.

Boyle's Law: Relating Pressure and Volume

Did you notice that as the diver in the photograph on the first page of this chapter exhales, the bubbles increase in size as they rise toward the surface? The reason is that as bubbles rise, the pressure exerted on them decreases—and as the pressure exerted on a gas decreases, its volume increases. The inverse is also true: as the pressure on a gas increases, its volume decreases. In other words, gases are compressible, a property that allows us to store gases (which would typically occupy large volumes under atmospheric pressure) in relatively small metal cylinders at high pressure.

The relationship between the pressure and volume of a fixed quantity of gas (constant value of n, where n is the number of moles) at constant temperature was investigated by the British chemist Robert Boyle (1627–1691), who conducted

FIGURE 6.9 Weather balloons are used to carry instruments aloft in the atmosphere.

Boyle's law the principle that the volume of a given amount of gas at constant temperature is inversely proportional to its pressure.

experiments with a J-shaped tube open at one end and closed at the other end (Figure 6.10). When a small amount of mercury was poured into the tube, a column of air became trapped at the closed end. The pressure on the trapped air was changed by varying the amount of mercury poured into the open arm. The more mercury added, the greater the force exerted by the mercury ($F = ma$), and hence the greater the force per unit area ($P = F/A$) on the column of trapped air in the closed arm. The pressure exerted on this trapped air depended on the difference in the height of the mercury in the two sides of the tube and the atmospheric pressure.

As Boyle added mercury to the open arm, the volume occupied by the trapped air decreased, even though the *amount* (number of moles) of trapped air remained constant. At the time Boyle discovered the relationship between P and V, scientists lacked a clear understanding of atoms or molecules. Given what we now know about matter, we recognize that because the molecules in the trapped gas were in constant random motion within a smaller, confined space, the number of collisions with one another and with the surface of the mercury *increased*. The more frequent the collisions, the greater the force exerted by the molecules against the surface of the mercury and thus the greater the pressure of the gas. That is to say, as the volume of the trapped gas decreased, the pressure of the trapped gas increased.

The mathematical relationship describing this inverse relationship between volume and pressure (under conditions of constant moles and constant temperature) is known as **Boyle's law**. Boyle observed that the volume of a given amount of a gas is inversely proportional to its pressure when kept at a constant temperature:

$$P \propto \frac{1}{V} \quad (T \text{ and } n \text{ constant}) \qquad (6.4)$$

In other words, as the volume of a constant amount of gas (at a constant temperature) increases, the pressure of the gas decreases because of fewer collisions, and conversely, as the volume of the gas decreases, its pressure increases because of more collisions (Figure 6.11).

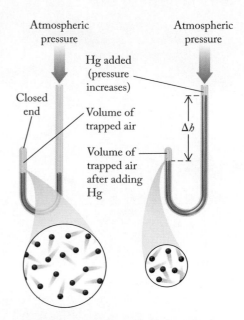

FIGURE 6.10 Boyle used a J-shaped tube for his experiments on the relationship between P and V. The volume of air trapped in the closed end of the tube decreases as the difference in height of mercury in the tube increases. The total pressure on the gas is equal to the pressure exerted by the added mercury (Δh) plus atmospheric pressure.

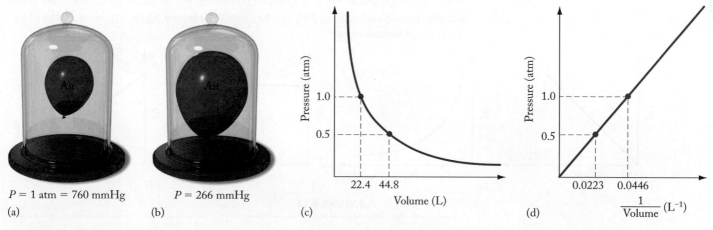

$P = 1$ atm $= 760$ mmHg
(a)

$P = 266$ mmHg
(b)

(c)

(d)

FIGURE 6.11 The change in the size of the balloon demonstrates the inverse relationship between pressure and volume. (a) The balloon inside the bell jar is at a pressure of 760 mmHg (1 atm), meaning that the pressure inside the balloon is 760 mmHg and the pressure outside the balloon (but inside the bell jar) is also 760 mmHg. (b) A slight vacuum applied to the bell jar reduces the pressure of the air outside the balloon (but inside the bell jar) to 266 mmHg. Because the pressure (number of collisions) on the outside of the balloon decreases, the collisions on the inside of the balloon cause the balloon to expand until the pressure inside the balloon equals the pressure outside the balloon; that is, the number of collisions on the outside of the balloon equals the number on the inside of the balloon. (c) At constant temperature, the pressure of a given quantity of gas is inversely proportional to the volume occupied by the gas; the graph of an inverse proportion is a hyperbola. (d) The inverse proportion between P and V means that a plot of P versus $1/V$ is a straight line.

Mathematically, we can replace the proportionality symbol with an equals sign and a constant:

$$P = (\text{constant})\frac{1}{V}$$

$$PV = \text{constant} \tag{6.5}$$

The value of the constant depends on the mass of trapped air and its temperature.

Because the value of the product PV in Equation 6.5 does not change for a given mass of trapped air at constant temperature, any two combinations of pressure and volume for a given sample of gas at constant temperature are related as follows:

$$P_1V_1 = P_2V_2 \tag{6.6}$$

This relationship, which applies to all gases, is illustrated by the dashed lines in Figure 6.11(c,d). When, for example, 44.8 L (V_1) of gas has a pressure of 0.500 atm (P_1), we have

$$P_1V_1 = (0.500 \text{ atm})(44.8 \text{ L}) = 22.4 \text{ L} \cdot \text{atm}$$

Equation 6.6 enables us to calculate the pressure if this quantity of gas were compressed (or expanded) to any other volume. For example, if we want to compress this quantity of the gas into a container half the size, then V_2 would equal 22.4 L:

$$P_1V_1 = P_2V_2$$

$$P_2 = \frac{P_1V_1}{V_2} = \frac{(0.500 \text{ atm})(44.8 \text{ L})}{22.4 \text{ L}}$$

$$= 1.00 \text{ atm}$$

Note that if the volume were to be cut in half, Boyle's law enables us to calculate that the pressure would double.

CONCEPT TEST

Which of the graphs in Figure 6.12 correctly depicts the relationship between the product of pressure and volume (PV) as a function of pressure (P) for a given quantity of gas at constant temperature?

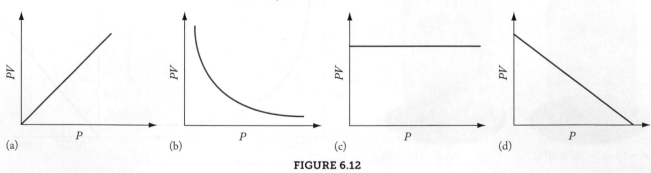

(a)   (b)   (c)   (d)

FIGURE 6.12

SAMPLE EXERCISE 6.3 Using Boyle's Law   **LO3**

A balloon is partially inflated with 5.00 L of helium at sea level, where the atmospheric pressure is 1.00 atm. The balloon ascends to an altitude of 1600 m, where the pressure is

0.83 atm. (a) What is the volume of the balloon at the higher altitude if the temperature of the helium does not change during the ascent? (b) What is the percent increase in volume?

Collect and Organize We are given the volume ($V_1 = 5.00$ L) of a gas and its pressure ($P_1 = 1.00$ atm) and asked to find the volume (V_2) of the gas when the pressure changes ($P_2 = 0.83$ atm). The problem states that the temperature of the gas does not change.

Analyze The balloon contains a fixed amount of gas and its temperature is constant, so pressure and volume are related by Boyle's law (Equation 6.6). We predict that the volume should increase because pressure decreases.

Solve

a. Rearranging Equation 6.6 to solve for V_2 and inserting the given values of P_1, P_2, and V_1 gives us

$$V_2 = \frac{P_1 V_1}{P_2}$$

$$V_2 = \frac{(1.00 \text{ atm})(5.00 \text{ L})}{0.83 \text{ atm}} = 6.0 \text{ L}$$

b. To calculate the percent increase in the volume, we must determine the difference between V_1 and V_2 and compare the result with V_1:

$$\% \text{ increase in volume} = \frac{V_2 - V_1}{V_1} \times 100\%$$

$$= \frac{6.0 \text{ L} - 5.0 \text{ L}}{5.0 \text{ L}} \times 100\% = 20\%$$

Think About It The prediction we made about the volume increasing is confirmed. The lower atmospheric pressure means fewer collisions by molecules in the atmosphere on the outside of the helium balloon. The helium balloon will rise until it reaches an altitude where the pressure inside the balloon (that is, the number of collisions by the helium atoms) equals the atmospheric pressure. Notice how Equation 6.6 can also be used to calculate the change in pressure that takes place at constant temperature when the volume of a quantity of gas changes.

⊚ **Practice Exercise** A scuba diver exhales 3.50 L of air while swimming at a depth of 20.0 m, where the sum of atmospheric pressure and water pressure is 3.00 atm. By the time the exhaled air rises to the surface, where the pressure is 1.00 atm, what is its volume?

Charles's Law: Relating Volume and Temperature

Nearly a century after Boyle's discovery of the inverse relationship between the pressure exerted by a gas and its volume, the French scientist Jacques Charles (1746–1823) documented the linear relationship between the volume and temperature of a fixed quantity of gas at constant pressure. Now known as **Charles's law**, this relationship states that, when the pressure exerted by a gas is held constant, the volume of a fixed quantity of gas is directly proportional to the **absolute temperature** (the temperature on the Kelvin scale) of the gas:

$$V \propto T \quad (P \text{ and } n \text{ constant}) \tag{6.7}$$

Charles's law the principle that the volume of a fixed quantity of gas at constant pressure is directly proportional to its absolute temperature.

absolute temperature temperature expressed in kelvins on the absolute (Kelvin) temperature scale, on which 0 K is the lowest possible temperature.

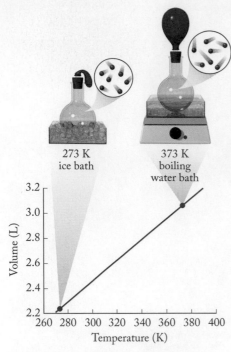

273 K
ice bath

373 K
boiling
water bath

FIGURE 6.13 A balloon attached to a flask inflates as the temperature of the gas inside the flask increases from 273 K to 373 K at constant atmospheric pressure. This behavior is described by Charles's law.

Because the lowest temperature on the absolute temperature scale is 0 K, temperatures expressed in kelvins are always positive numbers.

The effects of Charles's law can be seen in Figure 6.13, in which a balloon has been attached to a flask, trapping a fixed amount of gas in the apparatus. Heating the flask causes the gas to expand, inflating the balloon. The higher the temperature, the larger the volume occupied by the gas and the bigger the balloon.

As with Boyle's law, we can replace the proportionality symbol in Equation 6.7 with an equals sign if we include a proportionality constant:

$$V = (\text{constant})\,T$$

$$\frac{V}{T} = \text{constant} \tag{6.8}$$

The value of the constant depends on the number of moles of gas in the sample (n) and on the pressure of the gas (P). When those two parameters are held constant, the ratio V/T does not change, and any two combinations of volume and temperature are related as follows:

$$\frac{V_1}{T_1} = \frac{V_2}{T_2} \tag{6.9}$$

Charles's law allows us to predict that a decrease in temperature from 20°C to 0°C reduces volume by about 7%. Let's see how we arrive at those numbers. Consider what happens when a balloon containing 2.00 L of air at 20°C is taken outside on a day when the temperature is 0°C. The amount of gas in the balloon is fixed, and the atmospheric pressure is constant. Experience tells us that the volume of the balloon decreases; Charles's law enables us to quantify that change. We can calculate the final volume with Equation 6.9, provided that we express T_1 and T_2 in kelvins, not degrees Celsius. Solving Equation 6.9 for V_2:

$$V_2 = \frac{V_1 T_2}{T_1}$$

and substituting the known volume and temperature values:

$$V_2 = \frac{2.00\ \text{L} \times 273\ \text{K}}{293\ \text{K}} = 1.86\ \text{L}$$

As predicted, the volume of the gas decreases when the temperature of the gas decreases. The percent change in the volume of the balloon is

$$\%\ \text{decrease} = \frac{V_1 - V_2}{V_1} \times 100\% = \frac{2.00\ \text{L} - 1.86\ \text{L}}{2.00\ \text{L}} \times 100\% = 7.0\%$$

CONNECTION We learned in Chapter 1 that Kelvin and Celsius temperatures are related by the equation K = °C + 273.15.

CONCEPT TEST

Suppose you make two graphs: Graph 1 plots volume (V) as a function of temperature (T) for 1 mole of a gas at a pressure of 1.00 atm, and Graph 2 plots volume (V) as a function of T for 1 mole of a gas at 2.00 atm pressure. How do the slopes of the two graphs differ?

SAMPLE EXERCISE 6.4 Using Charles's Law   **LO3**

Charles was drawn to the study of gases in the late 18th century because of his interest in hot-air balloons. Scientists of his time might have designed an experiment to answer the question: what Celsius temperature is required to increase the volume of a sealed balloon from 2.00 L to 3.00 L if the initial temperature is 15°C and the atmospheric pressure is constant?

Collect and Organize We are given the volume ($V_1 = 2.00$ L) of a gas at an initial Celsius temperature ($T = 15°C$), and we are asked to calculate the Celsius temperature needed to increase the volume of the gas ($V_2 = 3.00$ L). The container (a balloon) is sealed, so n is constant, as is atmospheric pressure.

Analyze The relationship between volume and temperature is given by Charles's law in Equation 6.9. Charles's law tells us that for volume to increase, temperature must also increase. If the volume increases by 50%, we predict that the absolute temperature must also increase by 50%.

Solve First we rearrange Equation 6.9 to solve for T_2:

$$T_2 = \frac{V_2 T_1}{V_1}$$

We then substitute for V_1, V_2, and T_1, remembering to convert temperature to kelvins:

$$T_2 = \frac{V_2 T_1}{V_1} = \frac{(3.00 \text{ L})(288 \text{ K})}{2.00 \text{ L}} = 432 \text{ K}$$

The final step is to convert the final temperature to degrees Celsius:

$$T_2 = 432 \text{ K} - 273 = 159°C$$

Think About It To increase the volume by 1.5 times [$(1.5)(2.00$ L$) = 3.00$ L], *absolute* temperature must increase 1.5 times [$(1.5)(288$ K$) = 432$ K].

Practice Exercise Hot expanding gases can be used to perform useful work in a cylinder fitted with a movable piston, as in Figure 6.14. If the temperature of a gas confined to such a cylinder is increased from 245°C to 560°C, what is the ratio of the initial volume to the final volume if the pressure exerted on the gas remains constant?

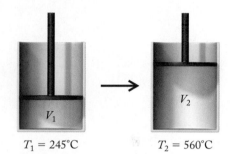

$T_1 = 245°C$ $T_2 = 560°C$

FIGURE 6.14 The volume of a gas in a cylinder changes as the temperature of the gas increases from 245°C to 560°C.

Charles's law also allowed the original determination that absolute zero (0 K) is equal to −273.15°C. Consider what happens when the volume of a fixed quantity of gas at constant pressure is plotted as a function of temperature. Some typical results for three gases are graphed in Figure 6.15, showing that the decrease in volume is linear as temperature decreases. There is a limit to how much the temperature of a gas can decrease because at some temperature the gas liquefies, at which point the gas laws no longer describe the behavior of the substance because it is no longer a gas. However, we can *extrapolate* from our measured data to the point where the volume of a gas would reach zero if condensation did not occur. The dashed lines in Figure 6.15 cross the temperature axis at −273.15°C, the temperature defined as 0 K, or absolute zero.

Avogadro's Law: Relating Volume and Quantity of Gas

Boyle's and Charles's experiments involved a constant amount of gas in a confined sample. Now consider what happens when the amount of gas is *not* held constant, such as when a balloon is inflated. As more and more gas is blown into the balloon, the balloon grows larger until it finally bursts. Clearly, the volume depends on the quantity (number of grams or moles) of the gas.

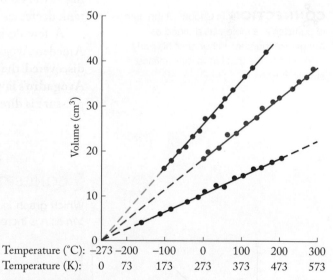

| Temperature (°C): | −273 | −200 | −100 | 0 | 100 | 200 | 300 |
| Temperature (K): | 0 | 73 | 173 | 273 | 373 | 473 | 573 |

FIGURE 6.15 The volumes of three gases (when n and P are held constant) plotted against temperature on both the Celsius and Kelvin scales. As predicted by Charles's law, volume decreases as the temperature decreases; the relationship is linear on both scales. The dashed lines show an extrapolation from the experimental data to a point corresponding to zero volume. This temperature is known as absolute zero: −273.15°C on the Celsius scale and 0 K on the Kelvin scale.

FIGURE 6.16 The balloon on the right has twice the volume of the balloon on the left because it is filled with twice the quantity of helium. This direct relationship between volume and number of moles of a gas at constant temperature and pressure demonstrates Avogadro's law.

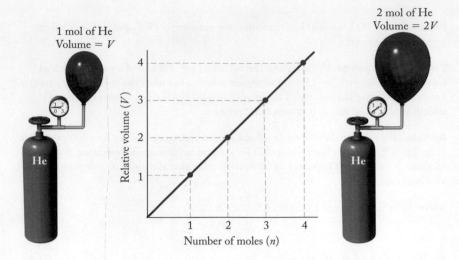

If you were to poke a hole in the balloon, the gas would escape, resulting in a smaller quantity of gas remaining inside the balloon and hence a smaller volume. The relationship between the pressure, volume, and the quantity of the gas can be illustrated in several ways.

Consider a bicycle tire that is inflated enough to hold its shape but is nevertheless too soft to ride on. We can increase the pressure inside the tire by pumping more air into it. Note, however, that the tire will not expand much because the heavy rubber is much more rigid than the material in a balloon. As another example, consider a carnival vendor who fills balloons from a helium tank fitted with a pressure gauge (Figure 6.16). Because the balloons are much more flexible than a tire, the size of the inflated balloons depends on the amount of gas the vendor dispenses: the more gas he adds to a balloon, the larger it becomes. When sales are brisk, it is possible to observe the pressure reading on the gauge decreasing as the vendor fills more and more balloons. The pressure of the helium in the tank decreases as the amount of gas in the cylinder decreases.

A few decades after Charles discovered the relationship between V and T, Amedeo Avogadro (1776–1856), whom we know from Avogadro's number (N_A), discovered the relationship between V and n. This relationship is described by **Avogadro's law**, which states that the volume of a gas at a given temperature and pressure is directly proportional to the quantity of the gas in moles:

$$V \propto n \quad \text{or} \quad \frac{V}{n} = \text{constant} \quad (P \text{ and } T \text{ constant}) \tag{6.10}$$

CONNECTION In Chapter 3 the number of particles in a mole was defined as Avogadro's number, in honor of his early work with gases that led to determining atomic masses.

CONCEPT TEST

Which graph in Figure 6.17 correctly describes the relationship between the value of V/n as n is increased (at constant P and T)?

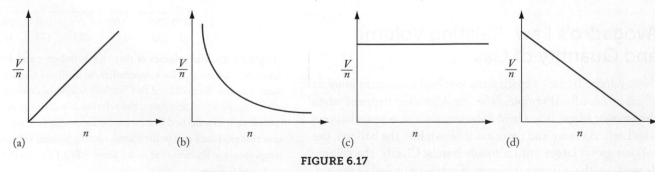

FIGURE 6.17

Amontons's Law: Relating Pressure and Temperature

Both Boyle's law (PV = constant) and Charles's law (V/T = constant) examine what happens to the volume of a gas, but what is the relationship between pressure and temperature? Experiments show that pressure is directly proportional to temperature when n and V are constant:

$$P \propto T \quad \text{or} \quad \frac{P}{T} = \text{constant} \quad (V \text{ and } n \text{ constant}) \qquad (6.11)$$

This relationship means that as the absolute temperature of a fixed amount of gas held at a constant volume increases, the pressure of the gas increases (Figure 6.18). This statement is referred to as **Amontons's law** in honor of the French physicist Guillaume Amontons (1663–1705), a contemporary of Robert Boyle's, who constructed a thermometer based on the observation that the pressure of a gas is directly proportional to its temperature.

Where do we see evidence of Amontons's law? Consider a bicycle tire inflated to a recommended pressure of 100 psi (6.8 atm) on an afternoon in late autumn when the temperature is 25°C. The next morning, after an early freeze in which the temperature dropped to 0°C, the pressure in the tire drops to about 91 psi (6.2 atm). Did the tire leak? Perhaps, but the decrease in pressure could also be explained by Amontons's law.

We can confirm that the decrease in tire pressure was consistent with the decrease in temperature by applying the logic we used to develop Equations 6.6 and 6.9. Relating initial pressure and temperature to final pressure and temperature, we can write

$$\frac{P_1}{T_1} = \frac{P_2}{T_2} \qquad (6.12)$$

Solving for P_2, we get

$$P_2 = \frac{P_1 T_2}{T_1} = \frac{(6.8 \text{ atm})(273 \text{ K})}{298 \text{ K}} = 6.2 \text{ atm}$$

Why is pressure directly proportional to temperature? Like pressure, temperature is directly related to molecular motion. The average speed at which a population of molecules moves increases with increasing temperature (see Figure 6.18). For a given number of gas molecules, increasing the temperature increases the average speed and therefore increases both the frequency and force with which the molecules collide with the walls of their container. Because pressure is related to the frequency and force of the collisions, it follows that higher temperatures produce higher pressures, as long as other factors such as volume and quantity of gas are held constant.

Avogadro's law the principle that the volume of a gas at a given temperature and pressure is proportional to the quantity of the gas.

Amontons's law the principle that the pressure of a quantity of gas at constant volume is directly proportional to its absolute temperature.

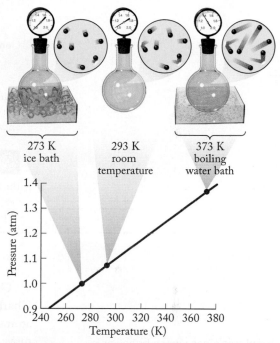

FIGURE 6.18 The pressure of a given quantity of gas is directly proportional to its absolute temperature when the moles of gas and its volume are held constant. Each flask has the same volume and contains the same number of molecules. The relative speeds of the molecules (represented by the lengths of their "tails") increase with increasing temperature, causing more frequent and more forceful collisions with the walls of the flasks and hence higher pressures.

SAMPLE EXERCISE 6.5 Using Amontons's Law   **LO3**

Labels on aerosol cans caution against incineration because the cans may explode when the pressure inside them exceeds 3.00 atm. At what temperature in degrees Celsius will an aerosol can burst if the initial pressure inside the can is 2.20 atm at 25°C?

Collect and Organize We are given the temperature (T_1 = 25°C) and pressure (P_1 = 2.20 atm) of a gas, and we are asked to determine the temperature (T_2) at which the pressure (P_2 = 3.00 atm) will cause the can to explode.

Analyze Because the gas is enclosed in an aerosol can, we know that the quantity of gas and volume of the gas are constant. We can use Amontons's law (Equation 6.12) because it describes what happens to pressure as the temperature of a gas is changed. We can estimate the answer by considering that the pressure in the can must increase by about 50% in order for the pressure to exceed 3 atm. Pressure is directly proportional to temperature, so the absolute temperature must also increase by about 50%. We should note that the initial temperature is given in degrees Celsius. Because Equation 6.12 works only with absolute temperatures, we must convert the initial temperature to kelvins. After solving for T_2, we need to convert kelvins back to degrees Celsius to report our answer.

Solve We rearrange Equation 6.12 to solve for T_2:

$$T_2 = \frac{T_1 P_2}{P_1}$$

After converting T_1 from degrees Celsius to kelvins, we have

$$T_2 = \frac{(25 + 273)\text{K} \times 3.00 \text{ atm}}{2.20 \text{ atm}} = 406 \text{ K}$$

Converting T_2 to degrees Celsius gives us

$$T_2 = 406 \text{ K} - 273 = 133°\text{C}$$

Think About It This temperature is certainly higher than the original temperature. In fact 406 K is about 1.5 times (or 50% higher than) the initial temperature, 298 K, so the answer makes sense.

Practice Exercise The air pressure in the tires of an automobile is adjusted to 28 psi at a gas station in San Diego, where the air temperature is 68°F. The air in the tires is at the same temperature as the atmosphere. The automobile is then driven east along a hot desert highway, and the temperature inside the tires reaches 140°F. What is the measured pressure in the tires at 140°F?

Gases that behave according to the linear relationships discovered by Boyle, Charles, Avogadro, and Amontons are called **ideal gases**. In an ideal gas, the atoms or molecules are assumed not to interact with one another; rather, they move independently with speeds that are related to their masses and to the temperature of the gas. Most gases exhibit ideal behavior at the pressures and temperatures typically encountered in the atmosphere. Under these conditions, the volumes occupied by the gas molecules or atoms are insignificant compared with the overall volume occupied by the gas.

6.4 The Ideal Gas Law

What would happen if we launched a weather balloon from the surface of Earth and allowed it to drift to an elevation of 10,000 m? The volume of the balloon would expand as the atmospheric pressure decreased, but the air temperature would simultaneously decrease as the balloon ascended, tending to make the volume smaller. How do we determine the final volume of the balloon when both pressure and temperature change at the same time for a given quantity of gas? Taken individually, none of the four gas laws and their accompanying mathematical relations apply. Nevertheless, we can derive a relationship that describes this situation.

Boyle's law states that volume and pressure are inversely proportional. Charles's law tells us that volume is directly proportional to temperature. Putting

ideal gas a gas whose behavior is predicted by the linear relationships defined by Boyle's, Charles's, Avogadro's, and Amontons's laws.

ideal gas equation (also called **ideal gas law**) the principle relating the pressure, volume, number of moles, and temperature of an ideal gas; expressed as $PV = nRT$, where R is the universal gas constant.

universal gas constant the constant R in the ideal gas equation; its value and units depend on the units used for the variables in the equation.

standard temperature and pressure (STP) 0°C and 1 bar as defined by IUPAC; 0°C and 1 atm are commonly used in the United States.

molar volume the volume occupied by 1 mole of an ideal gas at STP; 22.4 L.

Boyle's and Charles's laws together (Equations 6.5 and 6.8) allows us to write an equation relating P, V, and T and combining the constants:

$$PV = \text{constant} \qquad \text{Boyle's Law}$$

$$\frac{V}{T} = \text{constant} \qquad \text{Charles's Law}$$

$$\frac{PV}{T} = \text{combined constant} \qquad (6.13)$$

Avogadro's law tells us that volume is directly proportional to the number of moles of gas when T and P are constant, so we can include n in Equation 6.13 and combine its constant with the "combined constant" in that equation:

$$\frac{PV}{nT} = \text{combined constant} \quad \text{or} \quad PV = (\text{combined constant}) \times nT \quad (6.14)$$

We turn Equation 6.14 into a meaningful equality known as the **ideal gas equation** or **ideal gas law** (Equation 6.15) by inserting an appropriate constant. By convention, this constant is called the **universal gas constant** R:

$$PV = nRT \qquad (6.15)$$

As Table 6.3 shows, R has different values depending on the units used. For many calculations in chemistry, it is convenient to use $R = 0.08206 \text{ L} \cdot \text{atm}/(\text{mol} \cdot \text{K})$. As the units tell us, however, we can use this value only when the quantity of gas is expressed in moles, the volume in liters, the pressure in atmospheres, and the temperature in kelvins.

A useful reference point when studying the properties of gases is **standard temperature and pressure (STP)**, defined by the International Union of Pure and Applied Chemistry (IUPAC) as 0°C and 1 bar. The more familiar unit of 1 atm is very close to 1 bar, and at the level of accuracy of calculations in this text, this substitution makes little difference, so we consider STP to be 0°C and 1 atm. The volume of 1 mole of an ideal gas at STP is known as the **molar volume** (Figure 6.19). We can calculate the molar volume from the ideal gas equation by solving the equation for V and inserting the values of n, P, and T at STP:

$$V = \frac{(1 \text{ mol})\left(0.08206 \dfrac{\text{L} \cdot \text{atm}}{\text{mol} \cdot \text{K}}\right)(273 \text{ K})}{1 \text{ atm}} = 22.4 \text{ L}$$

Many chemical and biochemical processes take place at pressures near 1 atm and at temperatures between 0°C and 40°C. Within this range, the volume that 1 mole of gaseous reactant or product occupies is no more than about 15% greater than the molar volume. Therefore volumes can be estimated easily if molar amounts are known. An important feature of molar volume is that it applies to any ideal gas, independent of its chemical composition. In other words, at STP, 1 mole of helium occupies the same volume—22.4 L—as 1 mole of methane (CH_4), or carbon dioxide (CO_2), or even a compound like uranium hexafluoride (UF_6), which has a molar mass almost 90 times that of helium.

We can derive another relationship that is useful when a system starts in an initial state (P_1, V_1, T_1, n_1) and moves to a final state (P_2, V_2, T_2, n_2):

$$P_1 V_1 = n_1 R T_1 \quad \text{so} \quad \frac{P_1 V_1}{n_1 T_1} = R \quad \text{and}$$

$$P_2 V_2 = n_2 R T_2 \quad \text{so} \quad \frac{P_2 V_2}{n_2 T_2} = R \qquad (6.16)$$

CHEMTOUR
The Ideal Gas Law

TABLE 6.3 Values for the Universal Gas Constant (R)

Value of R	Units
0.08206	$\text{L} \cdot \text{atm}/(\text{mol} \cdot \text{K})$
8.314	$\text{kg} \cdot \text{m}^2/(\text{s}^2 \cdot \text{mol} \cdot \text{K})$
8.314	$\text{J}/(\text{mol} \cdot \text{K})$
8.314	$\text{m}^3 \cdot \text{Pa}/(\text{mol} \cdot \text{K})$
62.37	$\text{L} \cdot \text{torr}/(\text{mol} \cdot \text{K})$

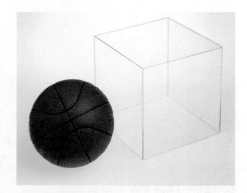

FIGURE 6.19 The box contains 1 mole of gas; the molar volume of an ideal gas is 22.4 L at 0°C and 1 atm of pressure. A basketball fits loosely into a box having this volume.

FIGURE 6.20 Breathing illustrates the relationship between P, V, and n. (a) When you inhale, your rib cage expands and your diaphragm moves down, increasing the volume of your lungs. Increased volume decreases the pressure inside your lungs, in accord with Boyle's law: $PV = $ constant. Decreased pressure allows more air to enter until the pressure inside your lungs matches atmospheric pressure. (b) Upon exhaling, lung volume decreases.

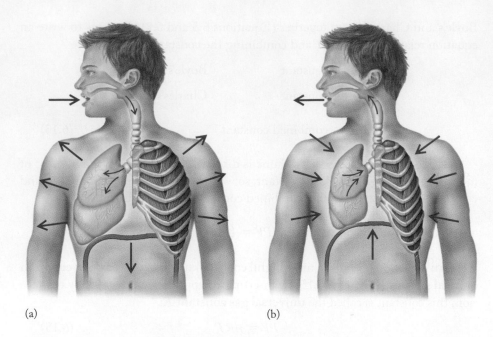

(a) (b)

Because both the initial-state and final-state expressions are equal to R, they are also equal to each other, allowing us to simplify Equation 6.16 to

$$\frac{P_1V_1}{n_1T_1} = \frac{P_2V_2}{n_2T_2} \tag{6.17}$$

Furthermore, because we frequently work with systems like weather balloons and gas canisters in which the amount of gas is constant ($n_1 = n_2$), but pressure, temperature, and volume vary, we define an especially useful form of this equation:

$$\frac{P_1V_1}{T_1} = \frac{P_2V_2}{T_2} \quad \text{for constant } n \tag{6.18}$$

Equation 6.18 is known as the **combined gas law**, or the **general gas equation**. Sample Exercise 6.6 gives an example of how to use this equation for determining the effect of changes in P and T on the volume (V) of a weather balloon. Other simplified versions of Equation 6.17 are certainly possible when variables in addition to n are fixed (Figure 6.20), but only Equation 6.18 is known as the combined gas law. Figure 6.21 summarizes the four laws expressed within the ideal gas law.

CONCEPT TEST

Which are correct variations of Equation 6.17?

a. $\dfrac{n_2T_2}{P_2} = \dfrac{n_1T_1}{P_1}$ at constant V

b. $\dfrac{n_2V_2}{P_2} = \dfrac{n_1V_1}{P_1}$ at constant T

c. $\dfrac{n_2T_2}{V_2} = \dfrac{n_1T_1}{V_1}$ at constant P

d. $\dfrac{T_1}{n_1V_1} = \dfrac{T_2}{n_2V_2}$ at constant P

combined gas law (also called **general gas equation**) the principle relating the pressure, volume, and temperature of a quantity of an ideal gas:

$$\frac{P_1V_1}{T_1} = \frac{P_2V_2}{T_2}$$

(a) **Boyle's law:** volume inversely proportional to pressure; n and T fixed

$$PV = \text{constant}$$

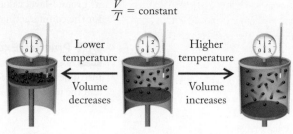

Volume decreases ← Volume increases →

Pressure increases / Pressure decreases

(b) **Charles's law:** volume directly proportional to temperature; n and P fixed

$$\frac{V}{T} = \text{constant}$$

Lower temperature ← Higher temperature →

Volume decreases / Volume increases

(c) **Avogadro's law:** volume directly proportional to number of moles; T and P fixed

$$\frac{V}{n} = \text{constant}$$

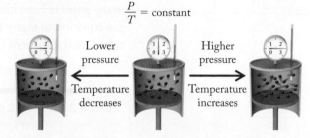

Remove gas ← Add gas →

Volume decreases / Volume increases

Gas cylinder

(d) **Amontons's law:** pressure directly proportional to temperature; n and V fixed

$$\frac{P}{T} = \text{constant}$$

Lower pressure ← Higher pressure →

Temperature decreases / Temperature increases

FIGURE 6.21 Relationships between pressure, volume, temperature, and/or moles of gas. (a) Increasing or decreasing V at constant n and T: Boyle's law. (b) Increasing or decreasing T at constant n and P: Charles's law. (c) Increasing or decreasing n at constant T and P: Avogadro's law. (d) Increasing or decreasing P at constant n and V: Amontons's law.

SAMPLE EXERCISE 6.6 Calculations Involving Changes in P, V, and T **LO3**

A weather balloon filled with 100.0 L of He is launched from ground level ($T = 20°C$, $P = 755$ torr). No gas is added or removed from the balloon during its flight. Calculate its volume at an altitude of 10 km, where the temperature and pressure in the balloon are $-52°C$ and 195 torr, respectively.

Collect, Organize, and Analyze We are given the initial temperature, pressure, and volume of a gas, and we are asked to determine the final volume after the pressure and temperature have changed. The decrease in temperature leads to a decrease in volume, but the decrease in pressure leads to an increase in volume. We need to express the temperatures in kelvins and estimate which variable dominates the change. Because the quantity of gas does not change (n is constant), we can solve for V_2 in the combined gas law.

The decrease in pressure to 195 torr (about 25% of the starting value) is considerably larger than the decrease in temperature from 20°C to −52°C (about 75% of the starting value). Therefore, we conclude that pressure is the dominant variable and that the volume of the balloon increases.

Solve First we convert the given Celsius temperatures to kelvins:

$$T_1 = 20°C + 273 = 293 \text{ K}$$

$$T_2 = -52°C + 273 = 221 \text{ K}$$

Then we use the general gas equation to solve for V_2:

$$\frac{P_1 V_1}{T_1} = \frac{P_2 V_2}{T_2}$$

$$V_2 = V_1 \times \frac{P_1}{P_2} \times \frac{T_2}{T_1}$$

$$V_2 = (100.0 \text{ L}) \times \frac{755 \text{ torr}}{195 \text{ torr}} \times \frac{221 \text{ K}}{293 \text{ K}} = 292 \text{ L}$$

Think About It The volume increases nearly threefold as the balloon ascends to 10 km. This result makes sense because atmospheric pressure decreases to about one-quarter of its ground-level value during the ascent. The volume would have increased more if not for the counteracting effect of the lower temperature at 10 km.

🔘 **Practice Exercise** The balloon in Sample Exercise 6.6 is designed to continue its ascent to an altitude of 30 km, where it bursts, releasing a package of meteorological instruments that parachute back to Earth. If the atmospheric pressure at 30 km is 28.0 torr and the temperature is −45°C, what is the volume of the balloon when it bursts?

The ideal gas law describes the relations among number of moles, pressure, volume, and temperature for any gas, provided it behaves as an ideal gas. Because many gases behave ideally at typical atmospheric pressures, we can apply the ideal gas equation to many situations outside the laboratory. In Sample Exercise 6.7, we calculate the mass of oxygen in an alpine climber's compressed-oxygen cylinder.

SAMPLE EXERCISE 6.7 Applying the Ideal Gas Law   **LO3**

Bottles of compressed O_2 carried by climbers ascending Mt. Everest have an internal volume of 5.90 L. Assume that such a bottle has been filled with O_2 to a pressure of 2025 psi at 25°C. Also assume that O_2 behaves as an ideal gas. (a) How many moles of O_2 are in the bottle? (b) What is the mass in grams of O_2 in the bottle?

Collect, Organize, and Analyze The ideal gas equation enables us to use the given values for P, V, and T to calculate n, the number of moles of O_2. Then we can use its molar mass (32.00 g/mol) to calculate the mass of O_2 in the bottle. We can estimate the answer by considering that 1 mole of gas at STP occupies 22.4 L. Our oxygen bottle has a volume of about 6 L, which is about one-quarter of the molar volume at STP, but the pressure (2025 psi) is more than 100 times greater than 1 atm (see Table 6.2). Thus we predict that the bottle contains more than 25 moles of O_2.

Solve

a. Let's start with the ideal gas equation rearranged to solve for n:

$$PV = nRT$$
$$n = \frac{PV}{RT}$$

Before using this expression for n, we need to convert pressure into atmospheres and temperature into kelvins:

$$P = (2025 \text{ psi})\left(\frac{1 \text{ atm}}{14.7 \text{ psi}}\right) = 138 \text{ atm} \quad T = 25°C + 273 = 298 \text{ K}$$

$$n = \frac{(138 \text{ atm})(5.90 \text{ L})}{\left(0.08206 \dfrac{\text{L} \cdot \text{atm}}{\text{mol} \cdot \text{K}}\right)(298 \text{ K})} = 33.3 \text{ mol}$$

b. Converting moles into grams is a matter of multiplying by the molar mass:

$$(33.3 \text{ mol}) \frac{(32.00 \text{ g})}{1 \text{ mol}} = 1.07 \times 10^3 \text{ g}$$

Think About It Our answer in part (a) is certainly reasonable based on our estimate. Most climbers require several bottles to climb Mt. Everest and return.

🔘 **Practice Exercise** Starting with the moles of O_2 calculated in Sample Exercise 6.7, calculate the volume of O_2 the bottle could deliver to a climber at an altitude where the temperature is −38°C and the atmospheric pressure is 0.35 atm.

6.5 Gases in Chemical Reactions

Gases are reactants or products in many important reactions. For example, if a commercial airliner flying at a high altitude loses cabin pressure, oxygen masks are deployed automatically for the passengers to breathe until the aircraft can descend or the cabin can be repressurized. The oxygen gas that flows in the masks is produced by the decomposition of solid sodium chlorate, $NaClO_3$. Similarly, when an air bag deploys during an automobile accident, the nitrogen gas that rapidly inflates the protective air bag is the product of the decomposition of solid sodium azide, NaN_3. Gases are also reactants, as in the combustion of charcoal in a backyard grill. Solid carbon reacts with oxygen gas in the air to produce carbon dioxide and the heat used to cook our food:

$$C(s) + O_2(g) \rightarrow CO_2(g) + \text{heat} \qquad (6.19)$$

In any chemical reaction that involves a gas as either a reactant or a product, the volume of the gas indirectly defines the amount of it in the reaction. If T and P are known, we can use the ideal gas equation to relate volume to the number of moles of gas in the system. Once we know that, we can use stoichiometric calculations to relate quantities of gas to quantities of other reactants and products, including heat. For example, consider the volume of oxygen needed to completely burn 1.00 kg (about 2 lb) of charcoal at 1.00 atm of pressure on an average summer day (25°C). Before starting the calculation we must first write a balanced chemical equation for the reaction. Equation 6.19 is indeed balanced, and 1 mole of C reacting with 1 mole of oxygen should yield 1 mole of CO_2. If we start with 1.00 kg of C, we have

$$1.00 \; \cancel{\text{kg}} \; \cancel{\text{C}} \times \frac{10^3 \; \cancel{\text{g}}}{1 \; \cancel{\text{kg}}} \times \frac{1 \; \text{mol C}}{12.01 \; \cancel{\text{g}} \; \cancel{\text{C}}} = 83.3 \; \text{mol C}$$

The stoichiometry of the reaction tells us that we need 83.3 moles of O_2 to completely react with the given amount of carbon. The volume of O_2 that corresponds to 83.3 moles of O_2 is calculated using the ideal gas equation by first rearranging the equation to solve for V and then substituting the values of n, R, T (in kelvins), and P:

$$V = \frac{nRT}{P} = \frac{(83.3 \; \cancel{\text{mol}}) \left(0.08206 \; \dfrac{\text{L} \cdot \cancel{\text{atm}}}{\cancel{\text{mol}} \cdot \cancel{\text{K}}} \right) (298 \; \cancel{\text{K}})}{1.00 \; \cancel{\text{atm}}} = 2.04 \times 10^3 \; \text{L}$$

SAMPLE EXERCISE 6.8 Combining Stoichiometry and   **LO4**
the Ideal Gas Law

Oxygen generators in some airplanes (Figure 6.22) are based on the chemical reaction between solid sodium chlorate and iron:

$$NaClO_3(s) + Fe(s) \rightarrow O_2(g) + NaCl(s) + FeO(s)$$

The resultant O_2 is blended with cabin air to provide 10–15 minutes of breathable air for passengers. How many grams of $NaClO_3$ are needed in a typical generator to produce 125 L of O_2 gas at 1.00 atm and 20.0°C?

Collect and Organize We are given the volume of O_2 we need to prepare at a particular pressure and temperature. With this information we can determine the mass of $NaClO_3$ needed based on the stoichiometric relations in the balanced chemical equation.

FIGURE 6.22 A flight attendant demonstrates how to use the oxygen mask on an airplane.

Analyze The solution requires two calculations. (1) We start by recognizing that the balanced chemical equation tells us that 1 mole of $NaClO_3$ is needed to produce 1 mole of oxygen. If we can determine how many moles of O_2 occupy a volume of 125 L at 1.00 atm pressure and 20.0°C, we can determine the number of moles of $NaClO_3$ we need. To determine the moles of oxygen (n_{O_2}), we can use the ideal gas law (Equation 6.15). (2) Then we use our calculated value of n_{O_2} and the balanced chemical equation to determine the number of moles and number of grams of $NaClO_3(s)$ required.

We can estimate the answer by comparing the volume of gas we wish to make (125 L) with the molar volume of an ideal gas at STP (22.4 L). Considering that the difference between the temperature at STP (0°C = 273 K) and the temperature in this problem (20°C = 293 K) is relatively small, we estimate that we need about 5 moles of O_2, requiring 5 moles of $NaClO_3$.

Solve We use the rearranged ideal gas equation to solve for moles of O_2:

$$n = \frac{PV}{RT}$$

$$n = \frac{(1.00 \text{ atm})(125 \text{ L})}{\left(0.08206 \dfrac{\text{L} \cdot \text{atm}}{\text{mol} \cdot \text{K}}\right)(273 + 20.0) \text{ K}} = 5.20 \text{ mol } O_2$$

To convert moles of O_2 into an equivalent mass of $NaClO_3$, we use the stoichiometry of the reaction to calculate the equivalent number of moles of $NaClO_3$, and we use molar mass to determine the number of grams of $NaClO_3$ required:

$$5.20 \text{ mol } O_2 \times \frac{1 \text{ mol } NaClO_3}{1 \text{ mol } O_2} \times \frac{106.44 \text{ g } NaClO_3}{1 \text{ mol } NaClO_3} = 553 \text{ g } NaClO_3$$

Think About It We predicted that about 5 moles of $NaClO_3$ would be needed to produce 125 L of oxygen, which is quite close to the calculated value of 5.20 moles of $NaClO_3$. Although the conditions given in this problem are not at STP, the molar volume of an ideal gas at STP is 22.4 L, so a quick estimate suggests that 125 L of oxygen would occupy the volume of approximately 5 basketballs (see Figure 6.19). Carrying that much oxygen in gas form aboard an airplane would create a storage challenge. Carrying 553 g of solid $NaClO_3$ that can readily be converted into oxygen gas is a more practical choice.

Practice Exercise Automobile air bags (Figure 6.23) inflate during a crash or sudden stop by the rapid generation of nitrogen gas from solid sodium azide, according to the reaction

$$2 NaN_3(s) \rightarrow 2 Na(s) + 3 N_2(g)$$

How many grams of sodium azide are needed to produce 50.0 L of N_2 at a pressure of 1.20 atm at 15°C?

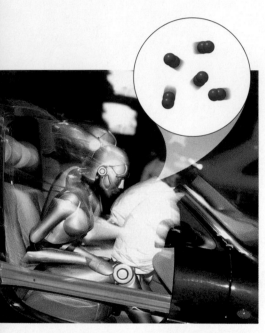

FIGURE 6.23 An automobile air bag inflates when solid NaN_3 decomposes, rapidly producing N_2 gas.

SAMPLE EXERCISE 6.9 Reactant Volumes and the Ideal Gas Law **LO4**

Reactions between methane and ammonia are of interest to scientists exploring nonbiological reactions that lead to amino acids. One such reaction produces cyanamide, H_2NCN:

$$CH_4(g) + 2 NH_3(g) \rightarrow H_2NCN(g) + 4 H_2(g)$$

If a reaction mixture contains 1.25 L of methane and 3.00 L of ammonia, what volume of hydrogen is produced at constant temperature and pressure?

Collect and Organize We are given a balanced chemical equation and the volumes of the starting materials. We are asked to calculate the volume of hydrogen produced. Temperature and pressure are constant, but their values are not specified.

Analyze The balanced chemical equation relates the numbers of moles of reactants and products. In this case, 4 moles of H_2 are produced for every 1 mole of CH_4 and every 2 moles of NH_3. We could calculate the moles of CH_4 and NH_3 by using the ideal gas law, Equation 6.15, but we lack values for P and T. However, since P and T are constant, we can use Avogadro's law instead, which states that $V \propto n$ (Equation 6.10), and use the volumes of the starting materials in place of moles.

Solve The balanced chemical equation tells us that complete reaction of 1.25 L of methane requires 2.5 L of ammonia:

$$V \propto n$$

$$1.25 \; \cancel{\text{L CH}_4} \times \frac{2 \text{ L NH}_3}{1 \; \cancel{\text{L CH}_4}} = 2.50 \text{ L NH}_3$$

Because we have 3.00 L of ammonia available, NH_3 is in excess and CH_4 is the limiting reactant. The volume of hydrogen produced by the reaction is

$$1.25 \; \cancel{\text{L CH}_4} \times \frac{4 \text{ L H}_2}{1 \; \cancel{\text{L CH}_4}} = 5.00 \text{ L H}_2$$

Think About It At a constant temperature and pressure, the ideal gas law tells us that the volume of 1.00 mol of gas is the same, regardless of its identity. This means that we can compare the volumes of gases the same way we compare moles of gases. It is sometimes easier to measure the volumes of gases rather than their masses in reactions.

Practice Exercise Dinitrogen monoxide, N_2O, commonly known as "nitrous oxide," is used as an anesthetic or "laughing gas" in many dental procedures (Figure 6.24). If 316 mL of nitrogen is combined with 178 mL of oxygen, what volume of N_2O is produced at constant temperature and pressure if the reaction proceeds to 82% yield?

$$2 N_2(g) + O_2(g) \rightarrow 2 N_2O(g)$$

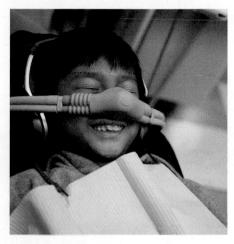

FIGURE 6.24 A child is administered "laughing gas" before a dental procedure.

6.6 Gas Density

Carbon dioxide is a relatively minor component of our atmosphere. Because CO_2 is produced by the combustion of fuels, however, the amount of it in the atmosphere has increased since the Industrial Revolution. This increase is of great concern because of its impact on climate change.

Another source of atmospheric carbon dioxide is volcanoes. Sudden releases of large quantities of $CO_2(g)$ from volcanic activity are life-threatening events. On the Dieng Plateau in Indonesia in 1979, 149 people died from asphyxiation in a valley after a massive, sudden release of carbon dioxide from one of the volcanoes around the valley (Figure 6.25). Because the density of $CO_2(g)$ is greater than that of air, the carbon dioxide settled in the valley, displacing the air containing the oxygen necessary for life.

The density of a gas at STP can be calculated by dividing its molar mass by its molar volume. Carbon dioxide, for example, has a molar mass of 44.01 g/mol. Therefore the density of CO_2 at STP is

$$\frac{44.01 \text{ g/}\cancel{\text{mol}}}{22.4 \text{ L/}\cancel{\text{mol}}} = 1.96 \text{ g/L}$$

The density of air at STP is about 1.3 g/L, so it is not surprising that the CO_2 in the Dieng Plateau disaster concentrated at the bottom of the valley. That the

FIGURE 6.25 The release of CO_2 from volcanic centers on Indonesia's Dieng Plateau in 1979 killed many people and animals as the dense gas flowed over the valley floor, effectively displacing the air (and oxygen) that the inhabitants and their livestock needed to survive.

density of CO_2 is greater than that of air also gives rise to its effectiveness in fighting fires. CO_2 blankets a fuel, separating it from the O_2 it needs to burn and thus extinguishing the fire.

Which gas has the highest density at STP: CH_4, Cl_2, Kr, or C_3H_8?

He: 4.003 g/mol, density = 0.169 g/L

N_2: 28.02 g/mol, density = 1.19 g/L

CO_2: 44.01 g/mol, density = 1.86 g/L

FIGURE 6.26 At 15°C and 1 atm, the density of the gas inside each balloon determines whether the balloon floats (helium), hovers just slightly above the benchtop (nitrogen), or sinks (carbon dioxide). Note the correlation between molar mass and density: the larger the molar mass of a gas, the greater the density.

We can calculate the density of an ideal gas at any temperature and pressure by using the ideal gas equation. Because density is the mass of a sample divided by its volume, $d = m/V$, we need to identify the variables in the ideal gas equation that represent the density. Mass can be determined from the number of moles, and of course the V in the ideal gas equation is the volume. We can rearrange Equation 6.15:

$$\frac{P}{RT} = \frac{n}{V} \tag{6.20}$$

For a given sample of gas, n equals the mass of the gas (in units of grams) divided by the molar mass of the gas (in units of grams/mol): $m/\mathcal{M}$. We can substitute this ratio (which has units of moles) for n in Equation 6.20 and solve the resulting expression for m/V (the density):

$$\frac{P}{RT} = \frac{m/\mathcal{M}}{V} = \frac{m}{\mathcal{M}V}$$

$$d = \frac{m}{V} = \frac{P\mathcal{M}}{RT} \tag{6.21}$$

If the term on the right in Equation 6.21 has P in atmospheres, $\mathcal{M}$ in grams per mole, R in $L \cdot atm/(mol \cdot K)$, and T in kelvins, then the units of density are g/L:

$$\frac{(\text{atm})\left(\dfrac{g}{\text{mol}}\right)}{\left(\dfrac{L \cdot \text{atm}}{\text{mol} \cdot K}\right)(K)} = g/L$$

Figure 6.26 illustrates the relationship between density and molar mass of three pure gases and air. Note how the balloon containing $CO_2(g)$ does not float at all but sinks to the benchtop because the density of this pure gas is greater than the density of air. In the absence of any wind currents or mixing, carbon dioxide will always fall downward through air, just as it filled the valley in the Dieng Plateau disaster, cutting the inhabitants off from their supply of oxygen.

Equation 6.21 can also be used to explain why a balloon inflated with hot air rises (Figure 6.27). Like a bicycle tire, the volume of the balloon is essentially constant once inflated. As the temperature of the air inside the balloon increases, the balloon can no longer expand. However, the balloons in Figure 6.27 represent open systems—gas can escape from the bottom of the balloon and n decreases. This effect can be seen in Equation 6.20, where increasing T at constant P and V means that n decreases. If n decreases, d will also decrease. Put another way, the density of a given quantity of gas decreases with increasing temperature, giving the balloon buoyancy.

FIGURE 6.27 Increasing the temperature of the air in the balloon leads to a decrease in density as the air expands. Hot air in a balloon is less dense than cooler air, making the balloon buoyant.

CONCEPT **TEST**

Which graph in Figure 6.28 best approximates (a) the relationship between density and pressure (n and T constant) and (b) the relationship between density and temperature (n and P constant) for an ideal gas?

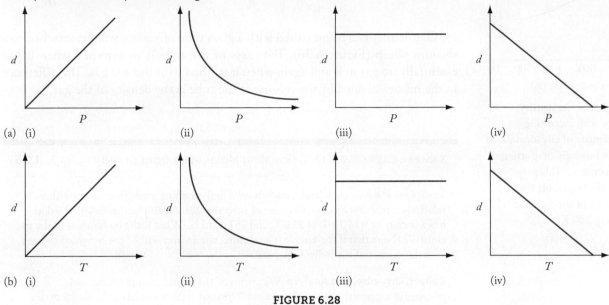

(a) (i) (ii) (iii) (iv)

(b) (i) (ii) (iii) (iv)

FIGURE 6.28

CONCEPT **TEST**

The density of methane is measured at four sets of conditions. Under which set of conditions is the density of CH_4 the greatest? (a) $T = 300$ K, $P = 1$ atm; (b) $T = 325$ K, $P = 1$ atm; (c) $T = 275$ K, $P = 2$ atm; (d) $T = 300$ K, $P = 2$ atm

---

SAMPLE EXERCISE 6.10 Calculating the Density of a Gas **LO5**

Calculate the density of air at 1.00 atm and 302 K and compare your answer with the density of air at STP (1.29 g/L). Assume that air has an average molar mass of 28.8 g/mol. (This average molar mass is the weighted average of the molar masses of the gases in air as listed in Table 6.1.)

Collect, Organize, and Analyze We are provided with the average molar mass, temperature, and atmospheric pressure of air, and we can use Equation 6.21 to calculate its density. The density of a gas is inversely proportional to temperature, and the temperature given in this problem (302 K) is higher than the temperature at STP, so we predict that the density of the air will be less than 1.29 g/L.

Solve Inserting the values of P, T, and $\mathcal{M}$ into Equation 6.21, we have

$$d = \frac{P\mathcal{M}}{RT} = \frac{(1.00\ \text{atm})\left(28.8\ \dfrac{\text{g}}{\text{mol}}\right)}{[0.08206\ \text{L} \cdot \text{atm}/(\text{mol}) \cdot \text{K}](302\ \text{K})} = 1.16\ \text{g/L}$$

Think About It The solution confirms that the density of air calculated at 302 K (1.16 g/L) is less than the value at 273 K (1.29 g/L).

Practice Exercise Air is a mixture of mostly nitrogen and oxygen. A balloon filled with only oxygen is released in a room full of air. Will it sink to the floor or float to the ceiling?

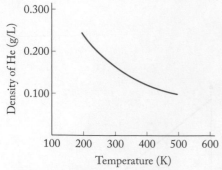

FIGURE 6.29 At 1 atm, the density of helium decreases with increasing temperature. The density of any ideal gas at constant pressure is inversely proportional to its absolute temperature. This graph shows a portion of the hyperbola that describes the behavior of the gas from approximately 200 to 500 K.

Equation 6.21 tells us that density decreases as temperature increases (Figure 6.29). We can rearrange this equation to calculate the molar mass of a gas from its density at any temperature and pressure:

$$\mathcal{M} = \frac{dRT}{P} \qquad (6.22)$$

Gas density can be measured with a glass tube of known volume attached to a vacuum pump (Figure 6.30). The mass of the tube is determined when it has essentially no gas in it and again when it is filled with the test gas. The difference in the masses divided by the volume of the tube is the density of the gas.

SAMPLE EXERCISE 6.11 Calculating Molar Mass from Density   **LO5**

Vent pipes at solid-waste landfills often emit foul-smelling gases that may be either relatively pure substances or mixtures of several gases. A sample of such an emission has a density of 0.650 g/L at 25.0°C and 757 mmHg. What is the molar mass of the gas emitted? (Note that if the sample is a mixture, the answer will be the weighted average of molar masses of the individual gases.)

Collect, Organize, and Analyze We are given the density, temperature, and pressure of a gaseous sample. We can use Equation 6.22 to calculate the molar mass, remembering to convert the temperature to kelvins and the pressure to atmospheres before using these values in the equation.

$$T = 273 + 25.0 = 298 \text{ K}$$

$$P = 757 \text{ mmHg} \times \frac{1 \text{ atm}}{760 \text{ mmHg}} = 0.996 \text{ atm}$$

Many gases have relatively low molar masses, so we expect that the molar mass should be similar to that of typical gases we have discussed in this chapter.

Solve

$$\mathcal{M} = \frac{dRT}{P} = \left(\frac{0.650 \text{ g}}{L}\right)\left(\frac{0.08206 \text{ L} \cdot \text{atm}}{\text{mol} \cdot \text{K}}\right)\left(\frac{298 \text{ K}}{0.996 \text{ atm}}\right)$$

$$= 16.0 \text{ g/mol}$$

Think About It The molar mass of the gas is 16.0 g/mol, a fairly small molar mass consistent with methane, a principal component in the mixture of gases emitted by decomposing solid waste.

FIGURE 6.30 (a) A gas collection tube with an internal volume of 235 mL is evacuated by connecting it to a vacuum pump. (b) The mass of the evacuated tube is measured. (c) Then the tube is opened to the atmosphere and refills with air. The difference in mass between the filled and evacuated tubes (0.273 g, or 273 mg) is the mass of the air inside it. The density of the air sample is 273 mg/235 mL, or 1.16 mg/mL.

(a)

(b)

(c)

6.7 Dalton's Law and Mixtures of Gases

As noted in Table 6.1, air is mostly N$_2$ and O$_2$ with smaller quantities of other gases. Each gas in air, and in any other gas mixture, exerts its own pressure, called a **partial pressure**. Atmospheric pressure is the sum of the partial pressure of each gas in the air:

$$P_{atm} = P_{N_2} + P_{O_2} + P_{Ar} + P_{CO_2} + \ldots$$

A similar expression can be written for any mixture of gases:

$$P_{total} = P_1 + P_2 + P_3 + P_4 + \ldots \tag{6.23}$$

This is a mathematical expression of **Dalton's law of partial pressures**: the total pressure of any mixture of gases equals the sum of the partial pressure of each gas in the mixture.

 **CHEMTOUR**
Dalton's Law

The most abundant gases in a mixture have the greatest partial pressures and contribute the most to the total pressure of the mixture: their particles have the largest number of collisions simply because there are more of them. The mathematical term used to express the abundance of each component, x, is its **mole fraction (X_x)**,

$$X_x = \frac{n_x}{n_{total}} \tag{6.24}$$

where n_x is the number of moles of the component and n_{total} is the sum of the number of moles of all the components of the mixture.

The mole fraction as an expression of concentration has three characteristics worth noting: (1) unlike molarity, mole fractions have no units; (2) unlike molarity, mole fractions are based on numbers of moles and can be used for any mixture or solution (solid, liquid, or gas); and (3) the mole fractions of all the components in a mixture must sum to 1.

To see how mole fractions work, consider a sample of the atmosphere (air) that contains 100.0 moles of atmospheric gases. This sample consists of 21.0 moles of O$_2$ and 78.1 moles of N$_2$. The mole fraction of O$_2$ in the air sample is

$$X_{O_2} = \frac{n_{O_2}}{n_{total}} = \frac{21.0}{100.0} = 0.210$$

and the mole fraction of N$_2$ is

$$X_{N_2} = \frac{n_{N_2}}{n_{total}} = \frac{78.1}{100.0} = 0.781$$

Note that mole fraction is an expression of concentration based on numbers of moles of material present. In this example, there are more nitrogen molecules in air than there are molecules of oxygen:

N$_2$ 78.1 mol × (6.022 × 10^{23} molecules/mol) = 4.70 × 10^{25} molecules

O$_2$ 21.0 mol × (6.022 × 10^{23} molecules/mol) = 1.26 × 10^{25} molecules

partial pressure the contribution to the total pressure made by one gas in a mixture of gases.

Dalton's law of partial pressures the principle that the total pressure of any mixture of gases equals the sum of the partial pressure of each gas in the mixture.

mole fraction (X_x) the ratio of the number of moles of a component in a mixture to the total number of moles in the mixture.

8 mol of N_2

8 mol of O_2

8 mol of gas

FIGURE 6.31 Pressure is directly proportional to number of moles of an ideal gas, independent of the identity of the gas and independent of whether the sample is a pure gas or a mixture. In three containers that have the same volume and are at the same temperature, 8 moles of nitrogen, 8 moles of oxygen, and 8 moles of a 50:50 mixture of nitrogen and oxygen all exert the same pressure, as indicated by the gauges.

CONCEPT TEST

The partial pressure of one component of a gas mixture cannot be directly measured. Why?

The partial pressure of any gas in air can be calculated by multiplying the mole fraction of the gas times the total atmospheric pressure. If the total pressure is 1.00 atm, for instance, as it often is at sea level, the partial pressures of O_2 and N_2 in the example above are

$$P_{O_2} = X_{O_2}P_{total} = (0.210)(1.00 \text{ atm}) = 0.210 \text{ atm}$$

$$P_{N_2} = X_{N_2}P_{total} = (0.781)(1.00 \text{ atm}) = 0.781 \text{ atm}$$

These two equations are specific instances of the general equation for the partial pressure of one gas in a mixture of gases:

$$P_x = X_x P_{total} \tag{6.25}$$

The pressure of a gas is directly proportional to the quantity of the gas in a given volume and does not depend on the identity of the gas or whether the gas is pure or a mixture (Figure 6.31). Because P_{total} is a measure of the total number of collisions within a mixture, it makes sense that the partial pressure of one gas would be due to some fraction of the total collisions. That fraction is the mole fraction of that gas.

SAMPLE EXERCISE 6.12 Calculating Mole Fraction **LO6**

Scuba divers who descend more than 45 m below the surface may use a gas mixture called Trimix to support their breathing. Trimix is available in several different concentrations; one is 11.7% He, 56.2% N_2, and 32.1% O_2, by mass. This mixture of gases avoids high concentrations of dissolved nitrogen in the blood at great depth, which can lead to *nitrogen narcosis*, a potentially deadly condition for divers. Calculate the mole fraction of each gas in this mixture.

Collect, Organize, and Analyze If we assume a 100.0 g sample of Trimix with a specified composition, then we can calculate the mass of each gas from the mass percentages. Next we can convert each mass to moles, from which we can determine the total number of moles in the sample and the mole fraction of each gas. Mole fraction is defined in Equation 6.24, and to calculate it we need to calculate both the total number of moles of gas in the mixture and the number of moles of each component.

We can estimate the answer by considering the relative molar masses of the three gases. The molar masses of nitrogen and oxygen are similar and seven to eight times larger, respectively, than the molar mass of helium. Because N_2 weighs less than O_2 and is a higher percentage of the mass of the mixture than is O_2, we predict that the mole fraction of nitrogen will be larger than that of oxygen. Furthermore, we predict that the mole fractions of He, N_2, and O_2 will be quite different from the mass percentages of the three gases.

Solve In 100.0 g of this mixture we would have 11.7 g of He, 56.2 g of N_2, and 32.1 g of O_2. Using molar masses to convert these masses into moles yields

$$(11.7 \text{ g He})\left(\frac{1 \text{ mol He}}{4.003 \text{ g He}}\right) = 2.92 \text{ mol He}$$

$$(56.2 \text{ g N}_2)\left(\frac{1 \text{ mol N}_2}{28.02 \text{ g N}_2}\right) = 2.01 \text{ mol N}_2$$

$$(32.1 \text{ g O}_2)\left(\frac{1 \text{ mol O}_2}{32.00 \text{ g O}_2}\right) = 1.00 \text{ mol O}_2$$

Mole fractions are calculated with Equation 6.24. The total number of moles in 100.0 g of Trimix is the sum of the number of moles of the constituent gases: He, N_2, and O_2.

$$n_{total} = 2.92 + 2.01 + 1.00 = 5.93 \text{ mol}$$

Thus, X_{He}, X_{N_2}, and X_{O_2} are

$$X_{He} = \frac{2.92 \text{ mol He}}{5.93 \text{ mol}} = 0.492$$

$$X_{N_2} = \frac{2.01 \text{ mol N}_2}{5.93 \text{ mol}} = 0.339$$

$$X_{O_2} = \frac{1.00 \text{ mol O}_2}{5.93 \text{ mol}} = 0.169$$

Alternatively, we could use Equation 6.24 for all but one of the gases in the sample and then calculate the mole fraction of the final component by subtracting the sum of the calculated mole fractions from 1.00. For example, we could use the equation to calculate the mole fractions of He and N_2 and then determine the O_2 mole fraction by difference:

$$X_{O_2} = 1 - (X_{He} + X_{N_2}) = 1 - (0.492 + 0.339) = 0.169$$

Think About It We can check the answer by summing the mole fractions; they must add up to 1.

$$X_{He} + X_{N_2} + X_{O_2} = 1$$
$$0.492 + 0.339 + 0.169 = 1.00$$

Finally, as we predicted, the mole fractions of the three gases do not reflect their mass percentages. Although nitrogen is present in the greatest amount by mass in Trimix, helium has the largest mole fraction.

Practice Exercise A gas mixture called Heliox, which is 52.17% O_2 and 47.83% He by mass, is used in scuba tanks for descents more than 65 m below the surface. Calculate the mole fractions of He and O_2 in this mixture.

CONCEPT TEST

Does each of the gases in an equimolar mixture of four gases have the same mole fraction?

Let's consider the implications of Equation 6.25 in the context of the "thin" air at high altitudes. The mole fraction of oxygen in air is 0.210, which does not change significantly with increasing altitude. However, the total pressure of the atmosphere, and therefore the partial pressure of oxygen, decreases with increasing altitude, as illustrated in the following Sample Exercise.

SAMPLE EXERCISE 6.13 Calculating Partial Pressure   **LO6**

Calculate the partial pressure of O_2 (in atm) in the air outside an airplane cruising at an altitude of 10 km, where the atmospheric pressure is 190.0 mmHg. The mole fraction of O_2 in the air is 0.210.

FIGURE 6.32 At high altitudes, atmospheric pressure is low and the partial pressure of oxygen falls below 0.21 atm, the optimum value for humans. Therefore mountaineers must carry supplemental oxygen to compensate for the "thinner" air.

Collect, Organize, and Analyze We are given both the mole fraction of oxygen and atmospheric pressure, and we are asked to calculate the partial pressure of oxygen. Equation 6.25 relates the partial pressure of a gas in a gas mixture to its mole fraction in the mixture and to the total gas pressure.

Solve

$$P_{O_2} = X_{O_2}P_{total} = (0.210)(190.0 \text{ mmHg})\left(\frac{1 \text{ atm}}{760 \text{ mmHg}}\right)$$

$$= 0.0525 \text{ atm}$$

Think About It As discussed earlier, the partial pressure of oxygen in air at sea level is 0.210 atm. At higher altitudes, the atmosphere becomes thinner (less dense). Our answer of 0.0525 atm makes sense because, at an altitude of 10 km, the air is much less dense than at sea level.

Practice Exercise Assume a scuba diver is working at a depth where the total pressure is 5.0 atm (about 50 m below the surface). What mole fraction of oxygen is necessary in the gas mixture the diver breathes for the partial pressure of oxygen to be 0.21 atm?

Sample Exercises 6.12 and 6.13 illustrate how dependent humans are on the atmosphere and the properties of gases that compose it. If we venture very far from Earth's surface into the mountains or into the oceans, we have to take the right blend of gases with us to sustain life. The scuba diver can't breathe the same mixture of gases as we do on the surface of Earth because the increased partial pressure of O_2 can be toxic below 30 m. Similarly, a higher partial pressure of N_2 could lead to nitrogen narcosis, a dangerous condition caused by a high concentration of nitrogen in the blood that leads to hallucinations. Alpine climbers on the tallest peaks on Earth must be vigilant against the opposite problem. For them, the lower atmospheric pressure leads to a lower partial pressure of oxygen, making it hard to get enough oxygen into the blood. Most mountaineers in these locations carry bottled oxygen (Figure 6.32). The lower partial pressure of oxygen at high elevations is also the reason why aircraft cabins are pressurized.

Dalton's law of partial pressures is also useful in the laboratory when we want to measure the pressure exerted by a gaseous product in a chemical reaction. For example, heating potassium chlorate ($KClO_3$) in the presence of MnO_2 causes it to decompose into $KCl(s)$ and $O_2(g)$. The oxygen gas produced by this reaction can be collected by bubbling the gas into an inverted bottle that is initially filled with water (Figure 6.33). As the reaction proceeds, $O_2(g)$ displaces the water in the bottle. When the reaction is complete, the volume of water displaced provides a measure of the volume of O_2 produced. If you know the temperature of the water and the barometric pressure at the time of the reaction, then you can use the ideal gas law to calculate the number of moles of O_2 produced.

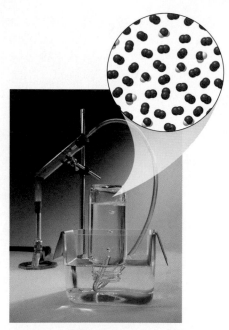

FIGURE 6.33 Collecting O_2 gas by water displacement. Oxygen is produced by the thermal decomposition of $KClO_3$. The tubing through which the oxygen passes from the reaction test tube to the jar is removed when the height of the water in the collection jar matches the level of water around it, ensuring that the pressure in the jar is equal to atmospheric pressure.

This procedure works for any gas that neither reacts with nor dissolves appreciably in water. However, one additional step is needed to apply the ideal gas law in calculating the number of moles of gas produced. At room temperature (nominally 20°C), any enclosed space above a pool of liquid water contains some $H_2O(g)$, meaning some water vapor is in the collection flask in addition to the gas produced by the reaction. Thus in the $KClO_3$ reaction in Figure 6.33, the gas collected is a mixture of both $O_2(g)$ and $H_2O(g)$. A mixture of a gas and water vapor is said to be *wet*, in contrast to a gas that contains no water vapor, which is said to be *dry*. The amount of water vapor increases as temperature increases (Table 6.4). Dalton's law of partial pressures gives us the total pressure of the mixture at the point where the water level inside the collection vessel matches the water level outside the vessel:

$$P_{total} = P_{atm} = P_{O_2} + P_{H_2O} \qquad (6.26)$$

To calculate the quantity of oxygen produced using the ideal gas law, we must know P_{O_2}, which we get by subtracting P_{H_2O} at 20°C (see Table 6.4) from P_{atm}. If we know the values of P_{O_2}, T, and V, we can calculate the number of moles (or subsequently, the number of grams) of oxygen produced.

TABLE 6.4 Partial Pressure of $H_2O(g)$ at Selected Temperatures

Temperature (°C)	Pressure (mmHg)
5	6.5
10	9.2
15	12.8
20	17.5
25	23.8
30	31.8
35	42.2
40	55.3
45	71.9
50	92.5

SAMPLE EXERCISE 6.14 Determining the Partial Pressure of a Gas Collected over Water   **LO6**

During the decomposition of $KClO_3$, 92.0 mL of gas was collected by the displacement of water at 25.0°C. If atmospheric pressure is 756 mmHg, what mass of O_2 was collected?

Collect, Organize, and Analyze We are given the atmospheric pressure, the volume of oxygen, and the temperature, and we are asked to calculate the mass of oxygen collected by water displacement. A quick estimation can give us an idea about the value of our final answer. The molar volume of oxygen is about 20 L at 25.0°C; if the oxygen were dry, about 100 mL would be collected, or (0.100 L)/(20 L/mol) = 0.005 mol of oxygen. Since oxygen has a molar mass of 32.00 g, 0.005 mol is about 0.16 g.

When a gas is collected over water, however, the collection vessel contains both water vapor and the gas of interest. According to Equation 6.26, we need to know the partial pressure of $H_2O(g)$ at the temperature of the apparatus to calculate the partial pressure due to O_2 in the collection flask. We are not given this pressure for the $H_2O(g)$, but we can find it in Table 6.4. The presence of water vapor in the gas mixture that we collect means that the partial pressure of the gas being collected is less than the total pressure in the vessel. This also means that our estimated value for the mass of oxygen, based on the entire volume of gas being O_2, is a high estimate.

Solve Table 6.4 indicates that P_{H_2O} at 25°C is 23.8 mmHg. To calculate P_{O_2} in the collected gas, we subtract this value from P_{total}:

$$P_{O_2} = P_{total} - P_{H_2O} = (756 - 23.8)\ \text{mmHg}$$
$$= 732\ \text{mmHg}$$

We can use the ideal gas equation to calculate the moles of O_2 produced; recall that we need to convert P_{O_2} to atmospheres, V to liters, and T to kelvins. The number of moles, n, is equal to the grams of O_2 collected (the unknown in this problem) divided by the molar mass of O_2 (32.00 g/mol).

$$732\ \text{mmHg} \times \frac{1\ \text{atm}}{760\ \text{mmHg}} = 0.963\ \text{atm}$$

$$92.0\ \text{mL} \times \frac{10^{-3}\ \text{L}}{1\ \text{mL}} = 0.0920\ \text{L}$$

$$25.0°C = 298\ \text{K}$$

$$n = \frac{PV}{RT} = \frac{(0.963\ \text{atm})(0.0920\ \text{L})}{[0.08206\ \text{L} \cdot \text{atm/(mol} \cdot \text{K)}](298\ \text{K})} = 0.00362\ \text{mol}$$

$$m_{O_2} = 0.00362\ \text{mol} \times \frac{32.00\ \text{g}}{1\ \text{mol}} = 0.116\ \text{g}$$

Think About It The pressure of the $O_2(g)$ is slightly less than P_{total}, as we predicted. Our answer (0.116 g) is close enough to our estimate that we can be confident our answer is reasonable.

Practice Exercise Electrical energy can be used to separate water into $O_2(g)$ and $H_2(g)$. In one demonstration of this reaction, 27 mL of H_2 was collected over water at 25°C. Atmospheric pressure is 761 mmHg. How many milligrams of H_2 were collected?

6.8 The Kinetic Molecular Theory of Gases

The fundamental discoveries of Boyle, Charles, and Amontons all took place before scientists recognized the existence of molecules. Only Avogadro, because he lived after Dalton, had the advantage of knowing Dalton's proposal that matter was composed of tiny particles called atoms. Thus Avogadro recognized that the volume of a gas was directly proportional to the number of particles (moles) of gas. Still, Avogadro's law, V/n = constant (Equation 6.10), does not explain why 1 mole of relatively small He(g) atoms at STP has the same volume as 1 mole of, for example, much larger $SF_6(g)$ molecules. Dalton's law of partial pressures (Equation 6.23) does not explain *why* each gas in a mixture contributes a partial pressure based on its mole fraction. Similarly, Boyle's, Charles's, and Amontons's laws demand explanations for why pressure and volume are inversely proportional to each other but both are directly proportional to temperature.

In this section we examine the **kinetic molecular theory** of gases, a unifying theory developed in the late 19th century that explains the relationships described by the ideal gas law and its predecessors, the laws of Boyle, Charles, Dalton, Amontons, and Avogadro.

The main assumptions of the kinetic molecular theory are:

1. Gas molecules have tiny volumes compared with the collective volume they occupy. Their individual volumes are so small as to be considered negligible, allowing particles in a gas to be treated as *point masses*—masses with essentially no volume. Gas molecules are separated by large distances; hence a gas is mostly empty space.
2. Gas molecules move constantly and randomly throughout the volume they collectively occupy.
3. The motion of these molecules is associated with an average kinetic energy that is proportional to the absolute temperature of the gas. All populations of gas molecules at the same temperature have the same average kinetic energy.
4. Gas molecules continually collide with one another and with their container walls. These collisions are *elastic*; that is, they result in no net transfer of energy to the walls. Therefore, the average kinetic energy of gas molecules is not affected by these collisions and remains constant as long as there is no change in temperature.
5. Each gas molecule acts independently of all other molecules in a sample. We assume there are no forces of attraction or repulsion between the molecules.

Explaining Boyle's, Dalton's, and Avogadro's Laws

We already have a picture in our minds that describes the origin of gas pressure. Every collision between a gas molecule and a wall of its container generates a force. The more frequent the collisions, the greater the force and thus the greater the pressure. Compressing molecules into a smaller space by reducing the volume of the container means more collisions take place per unit time, and the pressure increases (Boyle's law: $P \propto 1/V$). Because $P \propto n$, we can also increase the number of collisions by increasing the number of gas molecules (number of moles, n) in a container of fixed volume (Figure 6.34a and b). We can increase n either by adding more of the same gas to a container of fixed volume or by

kinetic molecular theory a model that describes the behavior of ideal gases; all equations defining relationships between pressure, volume, temperature, and number of moles of gases can be derived from the theory.

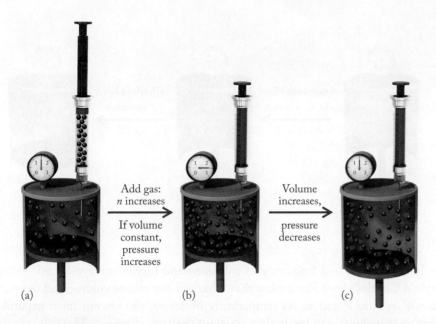

FIGURE 6.34 (a) Gas contained in a cylinder fitted with a movable piston exerts a pressure equal to atmospheric pressure ($P_{gas} = P_{atm}$). (b) When more gas is added to the cylinder (n increases), P_{gas} increases because the number of collisions with the walls of the container increases. (c) In order for the pressure to return to the same pressure as in part (a), the volume must increase.

adding some quantity of a different gas to the container. If we add two gases, such as N_2 and O_2, to a container in a 4:1 mole ratio, then the number of collisions involving nitrogen molecules and a container wall will be greater than the number involving oxygen molecules. Therefore the pressure due to nitrogen (P_{N_2}) will be greater than that due to oxygen (P_{O_2}), and the total pressure will be $P_{total} = P_{O_2} + P_{N_2}$. This statement is precisely what Dalton proposed in his law of partial pressures.

Avogadro's law ($V \propto n$) is also explained by the kinetic molecular theory. Avogadro observed that the volume of a gas *at constant pressure* is directly proportional to the number of moles of the gas. Because additional gas molecules in a given volume lead to increased numbers of collisions and an increase in pressure, the only way to reduce the pressure is to allow the gas to expand (Figure 6.34c).

CONCEPT TEST

If the collisions between molecules and the walls of the container were not elastic and energy was lost to the walls, would the pressure of the gas be higher or lower than that predicted by the ideal gas law?

Explaining Amontons's and Charles's Laws

Amontons's law ($P \propto T$) and Charles's law ($V \propto T$) both depend on temperature, and both can be explained in terms of kinetic molecular theory. Why is pressure directly proportional to temperature? Like pressure, temperature is directly related to molecular motion. The average speed at which a population of molecules moves increases with increasing temperature. For a given number of gas molecules, increasing the temperature increases the velocity of the molecules and therefore increases the frequency and average kinetic energy with which the molecules collide with one another and with the walls of their container. Because pressure is related to the frequency and force of these collisions, it follows that higher temperatures produce higher pressures, as long as volume and quantity of gas are constant. Figure 6.35(a and b) illustrates Amontons's law on a molecular level.

FIGURE 6.35 (a) As the temperature of a given amount of gas in a cylinder increases while the volume is held constant, the number and force of molecular collisions with the walls of the container increase, causing (b) the pressure to increase, as stated by Amontons's law. For the pressure to remain constant as the temperature of the gas increases, (c) the volume must increase until $P_{gas} = P_{atm}$. Hence as temperature increases at constant pressure, volume increases, as stated by Charles's law.

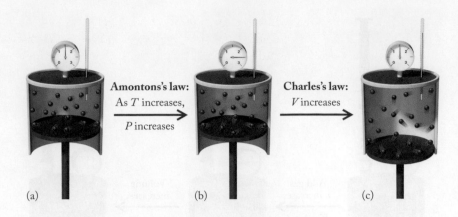

Amontons's law:
As T increases,
P increases

Charles's law:
V increases

(a)　　　　　(b)　　　　　(c)

How do increased frequency of collisions and higher average kinetic energy explain Charles's law? Remember that Charles's law relates volume and temperature at *constant P* and *n*. As temperature increases, the system must expand—increase its volume—to maintain a constant pressure (Figure 6.35c).

Molecular Speeds and Kinetic Energy

Kinetic molecular theory tells us that all populations of gas particles at a given temperature have the same *average* kinetic energy. The kinetic energy of a single molecule or atom of a gas can be calculated by using the equation

$$KE = \tfrac{1}{2}mu^2$$

where *m* is the mass of a molecule of the gas and *u* is its speed. At any given moment, however, not all gas molecules in a population are traveling at exactly the same speed. Even elastic collisions between two molecules may result in one of them moving with a greater speed than the other after the collision. One molecule might even stop moving. Thus collisions between gas molecules cause the molecules in any sample to have a range of speeds.

Figure 6.36(a) shows a typical distribution of speeds in a population of gas molecules. The peak in the curve represents the *most probable speed* (u_m) of molecules in the population. It is the speed that characterizes the largest number of molecules in the sample. Because the distribution of speeds is not symmetrical, the *average speed* (u_{avg}), which is simply the arithmetic average of all the speeds of all the molecules in the population, is a little higher than the most probable speed.

A very important value is the **root-mean-square speed (u_{rms})**; this is the speed of a molecule possessing the average kinetic energy. The absolute temperature of

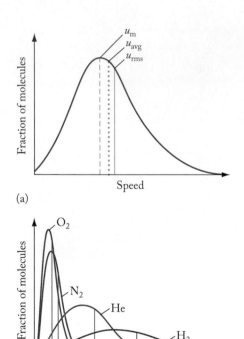

FIGURE 6.36 At any given temperature, the speeds of gas molecules cover a range of values. (a) The most probable speed (u_m, dashed line), at the highest point on the curve, is the speed of the largest fraction of molecules in the population; in other words, more molecules have the most probable speed than any other speed. The average speed (u_{avg}, dotted line), a little faster than the most probable speed, is the arithmetic average of all the speeds. The root-mean-square speed (u_{rms}, solid line), a little faster than the average speed, is directly proportional to the square root of the absolute temperature of the gas and inversely proportional to the square root of its molar mass. (b) Molecular speed distributions for samples of oxygen, nitrogen, helium, and hydrogen gas at the same temperature. On each curve, the vertical solid line is the u_{rms}. The lower the molar mass of the molecules, the higher is their root-mean-square speed (u_{rms}) and the broader the distribution of speeds. Having the lowest molar mass of the four gases, the H_2 molecules have the highest u_{rms} and the broadest curve. Having the highest molar mass in the group, the O_2 molecules have the lowest u_{rms} and the narrowest curve.

a gas is a measure of the average kinetic energy of the population of gas molecules. At a given temperature, the population of molecules in a gas has the same *average* kinetic energy as every other population of gas molecules at that same temperature, and this *average* kinetic energy is defined as

$$KE_{avg} = \frac{1}{2}m(u_{rms})^2 \qquad (6.27)$$

The root-mean-square speed (u_{rms}) of a gas with molar mass $\mathcal{M}$ at temperature T is defined as

$$u_{rms} = \sqrt{\frac{3RT}{\mathcal{M}}} \qquad (6.28)$$

Because different gases have different molar masses, this equation indicates that more massive molecules move more slowly at a given temperature than lighter molecules. Figure 6.36(b) shows the different distributions of speeds for several gases at a constant temperature.

Because u_{rms} is typically expressed in meters per second, care must be taken in choosing the units for R in Equation 6.28. One value of R (Table 6.3) that has meters in its units is

$$R = 8.314 \text{ kg} \cdot \text{m}^2/(\text{s}^2 \cdot \text{mol} \cdot \text{K})$$

and this is the most convenient value to use when working with Equation 6.28. Using this value for R requires that we express molar mass in kilograms per mole rather than the more common grams per mole.

Figure 6.37 shows how u_m increases with temperature. How fast do gas molecules move? Very rapidly at or slightly above room temperature—many hundreds of meters per second, depending on molar mass.

CHEMTOUR
Molecular Speed

CHEMTOUR
Molecular Motion

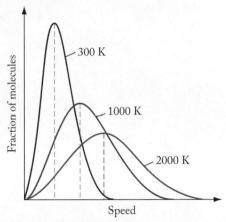

FIGURE 6.37 The most probable speed (u_m, dashed lines) increases with increasing temperature. Notice that the distributions broaden as the temperature increases: a smaller fraction of the molecules moves at any given speed, and more speeds are represented by a significant fraction of the population.

CONCEPT TEST

Rank these gases in order of increasing root-mean-square speed at 20°C, lowest speed first: H_2, CO_2, Ar, SF_6, UF_6, Kr.

SAMPLE EXERCISE 6.15 Calculating Root-Mean-Square Speeds   **LO8**

Calculate the root-mean-square speed of nitrogen molecules at 300.0 K in meters per second and miles per hour.

Collect and Organize Given only the temperature of a sample of nitrogen gas, we are asked to calculate the root-mean-square speed of the molecules in the sample. According to Equation 6.28, we need the molar mass of the gas in addition to the temperature. We also need $R = 8.314 \text{ kg} \cdot \text{m}^2/(\text{s}^2 \cdot \text{mol} \cdot \text{K})$ because it gives us a speed in meters per second.

Analyze Nitrogen gas (N_2) has a molar mass of 28.02 g/mol. Having chosen the value of R, we need to express the molar mass in kilograms. We can estimate the magnitude of the answer by considering the relative magnitudes of R, T, and molar mass. When T is expressed in kelvins, it is on the order of 10^2, whereas the correct value of R is approximately 10. The molar mass of nitrogen expressed as kilograms per mole is a small number, on the order of 10^{-2} kg/mol. When these values are substituted into Equation 6.28, the answer will be a large number, even after taking the square root: $(10 \times 10^2/10^{-2})^{1/2} \approx 10^2$ to 10^3 m/s.

root-mean-square speed (u_{rms}) the square root of the average of the squared speeds of all the molecules in a population of gas molecules; a molecule possessing the average kinetic energy moves at this speed.

Solve

$$u_{\text{rms},N_2} = \sqrt{\frac{3RT}{\mathcal{M}}} = \sqrt{\frac{3\left(8.314 \dfrac{\text{kg} \cdot \text{m}^2}{\text{s}^2 \cdot \text{mol} \cdot \text{K}}\right)(300.0 \text{ K})}{0.02802 \dfrac{\text{kg}}{\text{mol}}}}$$

$$= 516.8 \text{ m/s} = 5.168 \times 10^2 \text{ m/s}$$

$$u_{\text{rms},N_2} = \left(516.8 \frac{\text{m}}{\text{s}}\right)\left(\frac{1 \text{ mi}}{1.6093 \times 10^3 \text{ m}}\right)\left(3600 \frac{\text{s}}{\text{hr}}\right)$$

$$= 1156 \text{ mi/hr} = 1.156 \times 10^3 \text{ mi/hr}$$

Think About It The average speed of a nitrogen molecule does indeed fall between 10^2 and 10^3 m/s as predicted. In units of miles per hour, the average speed of a nitrogen molecule at this temperature is approximately twice the speed of an airplane at cruising altitude! The relatively small molar mass contributes to the large root-mean-square speed.

Practice Exercise Calculate the root-mean-square speed of helium at 300.0 K in meters per second, and compare your result with the root-mean-square speed of nitrogen calculated in Sample Exercise 6.15.

It is not immediately obvious why the pressures exerted by two different gases are the same at a given temperature when their root-mean-square speeds are quite different. The answer lies in the assumption in the kinetic molecular theory that all populations of gas molecules have the same average kinetic energy at a given temperature, independent of their molar mass. To examine this issue, let's calculate the average kinetic energy of 1 mole of N_2 and 1 mole of He at 300 K. The solution to Sample Exercise 6.15 reveals that the u_{rms} of N_2 molecules at 300 K is 517 m/s. The root-mean-square speed of He atoms is 1370 m/s (the result for the Practice Exercise of Sample Exercise 6.15). Using these values of speed and the mass of 1 mole of each gas ($m = \mathcal{M}$) in Equation 6.27 gives

$$\text{KE}_{N_2} = \tfrac{1}{2}\mathcal{M}_{N_2}\left(u_{\text{rms},N_2}\right)^2 = \tfrac{1}{2}(2.802 \times 10^{-2} \text{ kg})(517 \text{ m/s})^2$$

$$= 3.74 \times 10^3 \text{ kg} \cdot \text{m}^2/\text{s}^2$$

$$\text{KE}_{\text{He}} = \tfrac{1}{2}\mathcal{M}_{\text{He}}\left(u_{\text{rms,He}}\right)^2 = \tfrac{1}{2}(4.003 \times 10^{-3} \text{ kg})(1370 \text{ m/s})^2$$

$$= 3.76 \times 10^3 \text{ kg} \cdot \text{m}^2/\text{s}^2$$

This calculation demonstrates that the same quantities of two *different* gases have essentially the same average kinetic energy (when the correct number of significant figures is used). Therefore at the same temperature, they exert the same pressure. The slower N_2 molecules collide with the container walls less often, but because they are more massive than He atoms, the N_2 molecules exert a greater force during each collision.

Equation 6.28 enables us to compare the relative root-mean-square speeds of two gases at the same temperature. Consider an equimolar mixture of $N_2(g)$ and $He(g)$. We have just demonstrated that at any given temperature their average kinetic energies are the same:

$$\text{KE}_{N_2} = \tfrac{1}{2}m_{N_2}(u_{\text{rms},N_2})^2 = \tfrac{1}{2}m_{\text{He}}(u_{\text{rms,He}})^2 = \text{KE}_{\text{He}}$$

or

$$m_{N_2}(u_{rms,N_2})^2 = m_{He}(u_{rms,He})^2$$

Rearranging this equation to express the ratio of the root-mean-square speeds in terms of the ratio of the molar masses and then taking the square root of each side, we get

$$\frac{(u_{rms,He})^2}{(u_{rms,N_2})^2} = \frac{\mathcal{M}_{N_2}}{\mathcal{M}_{He}}$$

$$\frac{u_{rms,He}}{u_{rms,N_2}} = \sqrt{\frac{\mathcal{M}_{N_2}}{\mathcal{M}_{He}}} = \sqrt{\frac{28.02 \text{ g/mol}}{4.003 \text{ g/mol}}} = 2.646$$

The root-mean-square speed of helium atoms at 300 K is 2.646 times higher than that of nitrogen molecules. We can test this by using the values for u_{rms} we used above to calculate KE_{N_2} and KE_{He}:

$$\frac{u_{rms,He}}{u_{rms,N_2}} = \frac{1370 \text{ m/s}}{517 \text{ m/s}} = 2.65$$

The relationship between root-mean-square speeds and molar masses applies to any pair of gases x and y:

$$\frac{u_{rms,x}}{u_{rms,y}} = \sqrt{\frac{\mathcal{M}_y}{\mathcal{M}_x}} \tag{6.29}$$

This equation leads to the conclusion that the root-mean-square speed of more massive particles is lower than the root-mean-square speed of lighter particles.

CONCEPT TEST

Since root-mean-square speed depends on temperature, why doesn't the ratio of the u_{rms} of two gases change with temperature?

Graham's Law: Effusion and Diffusion

Let's consider what happens to the two balloons in Figure 6.38, one filled with nitrogen and the other with helium. The volume, temperature, and pressure of the gases are identical in the two balloons, and the pressure inside each balloon is greater than atmospheric pressure. Over time, the volume of the helium balloon decreases significantly, but the volume of the nitrogen balloon does not.

The skin of any balloon is slightly permeable; that is, it has microscopic holes that allow gas to escape, reducing the pressure inside the balloon. Why does the helium leak more quickly than the nitrogen? We now know the relationship between root-mean-square speeds and molar masses of two gases, and we have calculated that helium atoms move about 2.65 times faster than nitrogen atoms at the same temperature. The gas with the greater root-mean-square speed (He) leaks out of the balloon at a higher rate. The process of moving through a small opening from a higher-pressure region to a lower-pressure region is called **effusion**.

In the 19th century, Scottish chemist Thomas Graham (1805–1869) recognized that the effusion rate of a gas is related to its molar mass. Today this relation is known as **Graham's law of effusion**, and it states that the effusion rate of any gas is inversely proportional to the square root of its molar mass. We can derive a

effusion the process by which a gas escapes from its container through a tiny hole into a region of lower pressure.

Graham's law of effusion the principle that the rate of effusion of a gas is inversely proportional to the square root of its molar mass.

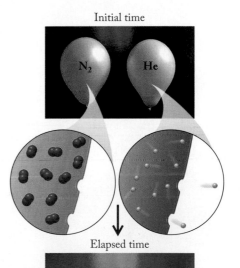

FIGURE 6.38 Two balloons at the same volume, temperature and pressure—one filled with nitrogen gas and the other filled with helium gas. Over time, the volume of the helium balloon decreases more quickly than the nitrogen balloon. The helium atoms are lighter than the nitrogen molecules and therefore have higher root-mean-square speeds. As a consequence, the helium atoms escape from the balloon faster, causing the helium-containing balloon to shrink more quickly than the nitrogen balloon.

mathematical representation of Graham's law starting from Equation 6.28 for two gases x and y:

$$u_{rms,x} = \sqrt{\frac{3RT}{\mathcal{M}_x}}$$

$$u_{rms,y} = \sqrt{\frac{3RT}{\mathcal{M}_y}}$$

We take the ratio of the root-mean-square speeds, which we designate as the effusion rates r_x and r_y:

$$\frac{r_x}{r_y} = \frac{u_{rms,x}}{u_{rms,y}} = \frac{\sqrt{\dfrac{3RT}{\mathcal{M}_x}}}{\sqrt{\dfrac{3RT}{\mathcal{M}_y}}} \tag{6.30}$$

This equation simplifies to the usual form of Graham's law:

$$\frac{r_x}{r_y} = \sqrt{\frac{\mathcal{M}_y}{\mathcal{M}_x}} \tag{6.31}$$

Using Equation 6.31 for the helium and nitrogen balloons, we see that the effusion rate of helium compared to nitrogen is:

$$\frac{r_{He}}{r_{N_2}} = \sqrt{\frac{\mathcal{M}_{N_2}}{\mathcal{M}_{He}}} = \sqrt{\frac{28.02 \; \cancel{g/mol}}{4.003 \; \cancel{g/mol}}} = 2.646$$

Notice that this expression has the same form—and hence the ratio has the same value—as the ratio of the root-mean-square speeds of the N_2 and He molecules.

Two things about Equation 6.31 are worth noting: (1) Like the root-mean-square speed in Equation 6.29, Graham's law involves a square root. (2) With labels x and y for two gases in a mixture, the left side of Equation 6.31 has the x term in the numerator and the y term in the denominator, whereas the right side of the equation has the opposite: the y term is in the numerator and x term in the denominator. Keeping the labels straight is the key to solving problems involving Graham's law.

How does the kinetic molecular theory explain Graham's law of effusion? The escape of a gas molecule from a balloon requires that the molecules encounter one of the microscopic holes in the balloon. The faster a gas molecule moves, the more likely it is to find one of these holes.

Effusion of gas molecules is related to **diffusion**, which is the spread of one substance through another. If we focus our attention on gases, then diffusion contributes to the passage of odors such as a perfume throughout a room and the smell of baking bread throughout a house. The rates of diffusion of gases depend on the average speeds of the gas molecules and on their molar masses. Like effusion, diffusion of gases is described by Graham's law.

SAMPLE EXERCISE 6.16 Applying Graham's Law   **LO8**
                                          to Diffusion of a Gas

An odorous gas emitted by a hot spring was found to diffuse 2.92 times slower than helium. What is the molar mass of the emitted gas?

Collect and Organize We are asked to determine the molar mass of an unknown gas on the basis of its rate of diffusion relative to the rate for helium.

diffusion the spread of one substance (usually a gas or liquid) through another.

Analyze Graham's law (Equation 6.31) relates the relative rate of diffusion of two gases to their molar masses. We know that helium ($\mathcal{M}_{He}$ = 4.003 g/mol) diffuses 2.92 times faster than the unidentified gas (*y*). Because helium diffuses faster than the unknown gas, we predict that the unidentified gas has a larger molar mass. In fact, it should be about nine (or 3^2) times the mass of He according to Graham's law, or about 36 g/mol, because the ratio of the molar masses will equal the square of the ratio of the diffusion rates.

Solve Rearranging Equation 6.31, we obtain an expression for $\mathcal{M}_y$:

$$\mathcal{M}_{He}\left(\frac{r_{He}}{r_y}\right)^2 = \mathcal{M}_y$$

Substituting for the ratio of the diffusion rates and the mass of 1 mole of helium gives

$$\mathcal{M}_y = \mathcal{M}_{He}\left(\frac{r_{He}}{r_y}\right)^2 = (4.003 \text{ g/mol})(2.92)^2 = 34.1 \text{ g/mol}$$

Think About It The molar mass of the unidentified gas is 34.1 g/mol, which is consistent with our prediction. One possibility for the identity of this gas is $H_2S(g)$, $\mathcal{M}$ = 34.08 g/mol, a foul-smelling and toxic gas frequently emitted from volcanoes and also responsible for the odor of rotten eggs.

 Practice Exercise Helium effuses 3.16 times as fast as which other noble gas?

6.9 Real Gases

Up to now we have treated all gases as ideal. This is acceptable because, under typical atmospheric pressures and temperatures, most gases *do* behave ideally. We have also assumed, according to kinetic molecular theory, that the volume occupied by individual gas molecules is negligible compared with the total volume occupied by the gas. In addition, we have assumed that no interactions occur between gas molecules other than random elastic collisions. These assumptions are not valid, however, when we begin to compress gases into increasingly smaller volumes.

Deviations from Ideality

Let's consider the behavior of 1.0 mole of a gas as we increase the pressure on it. From the ideal gas law, we know that $PV/RT = n$, so for 1 mole of gas, PV/RT should remain equal to 1.0 regardless of how we change the pressure. The relationship between PV/RT and P for an ideal gas is shown by the purple line in Figure 6.39. However, the curves for PV/RT versus P for CH_4, H_2, and CO_2 at pressures above 10 atm are not horizontal, straight lines like this ideal curve. Not only do the curves diverge from the ideal, but the shapes of the curves also differ for each gas, indicating that when we are dealing with real gases rather than ideal ones, the identity of the gas does matter.

Why don't real gases behave like ideal gases at high pressure? One reason is that the ideal gas law considers gas molecules to have so little volume compared with the volume of their container that they are assumed to have no volume at all. However, at high pressures, more molecules are squeezed into a given volume

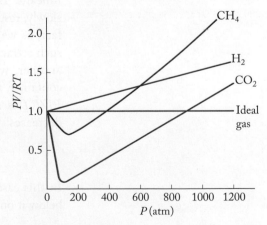

FIGURE 6.39 The effect of pressure on the behavior of real and ideal gases. The curves diverge from ideal behavior in a manner unique to each gas. When gases deviate from ideal behavior, their identity matters.

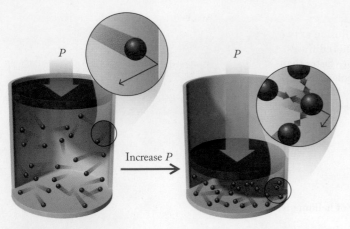

FIGURE 6.40 At high pressures, a larger fraction of the volume of a gas is occupied by the gas molecules. The larger density of the molecules also results in more interactions between them (broad red arrows in expanded view). These interactions reduce the frequency and force of collisions with the walls of the container (red arrow), thereby reducing the pressure.

(Figure 6.40). Under these conditions the volume occupied by the molecules can become significant. What is the impact of this on the graph of PV/RT versus P?

In PV/RT, the V actually refers to the *free volume* ($V_{\text{free volume}}$), the empty space not occupied by gas molecules. Because $V_{\text{free volume}}$ is difficult to measure, we instead measure V_{total}, the total volume of the container holding the gas:

$$V_{\text{total}} = V_{\text{free volume}} + V_{\text{molecules}}$$

In an ideal gas and in a real gas at low pressure, $V_{\text{total}} \approx V_{\text{free volume}}$ because $V_{\text{molecules}}$ is considered to be negligible (≈ 0) in relation to the volume of the container (V_{total}), and therefore we can use the ideal gas equation with confidence. Consider, however, what happens to a real gas in a closed but flexible container as we increase the external pressure, causing the container—and hence the volume occupied by the gas—to shrink. As the external pressure increases, the assumption that $V_{\text{molecules}} \approx 0$ becomes less valid because the portion of the container's volume occupied by the molecules themselves increases. The relationship $V_{\text{total}} = V_{\text{free volume}} + V_{\text{molecules}}$ still applies, but now $V_{\text{molecules}} \neq 0$, which means $V_{\text{free volume}}$ can no longer be approximated by V_{total}. Since $V_{\text{total}} > V_{\text{free volume}}$, the ratio PV/RT is also larger for a real gas than for an ideal gas:

$$\frac{PV_{\text{total}}}{RT} > \frac{PV_{\text{free volume}}}{RT}$$

As a consequence, the curve for a real gas in Figure 6.39 diverges upward from the line for an ideal gas.

A second factor also causes the ratio PV/RT to diverge from 1. Kinetic molecular theory assumes that molecules do not interact, but real molecules do attract one another. These attractive forces function over short distances, so the assumption that molecules behave independently is a good one as long as the molecules are far apart. However, as pressure increases on a population of gas molecules and they are pushed closer together, intermolecular attractive forces can become significant. Likewise, lowering the temperature causes molecules to move more slowly, resulting in more attractions and more deviations from ideal behavior. In fact, as we see in Chapter 10, a sufficiently cooled gas will condense because of such attractions. These attractive forces cause the molecules to associate with one another, which decreases the force of their collisions with the walls of their container, thereby decreasing the pressure exerted by the gas. If the value of P in PV/RT is smaller in the real gas than the ideal value, the value of the ratio decreases:

$$\frac{P_{\text{real}}V}{RT} < \frac{P_{\text{ideal}}V}{RT}$$

In this case, the curve for a real gas diverges from the ideal gas line by moving below it on the graph.

CONCEPT **TEST**

For which gases in Figure 6.39 does the effect of intermolecular attractive forces outweigh the effect of pressure on free volume at 200 atm?

The van der Waals Equation for Real Gases

Because the ideal gas equation does not hold at high pressures, we need another equation that can be used under nonideal conditions—one that accounts for the following facts:

1. The free volume of a real gas is less than the total volume because its molecules occupy significant space.
2. The observed pressure is less than the pressure of an ideal gas because of intermolecular attractions.

The **van der Waals equation**

$$\left(P + \frac{n^2 a}{V^2}\right)(V - nb) = nRT \qquad (6.32)$$

includes terms to correct for pressure $(n^2 a/V^2)$ and volume (nb). The values of a and b, called *van der Waals constants*, have been determined experimentally for many gases (Table 6.5). Both a and b increase with increasing molar mass and with the number of atoms in each molecule of a gas.

TABLE 6.5 Van der Waals Constants of Selected Gases

Substance	a (L$^2 \cdot$ atm/ mol^2)	b (L/mol)
H_2	0.244	0.0266
He	0.0341	0.02370
N_2	1.39	0.0391
O_2	1.36	0.0318
Ar	1.34	0.0322
CH_4	2.25	0.0428
CO	1.45	0.0395
CO_2	3.59	0.0427
H_2O	5.46	0.0305
HCl	3.67	0.04081
NO	1.34	0.02789
NO_2	5.28	0.04424
SO_2	6.71	0.05636

SAMPLE EXERCISE 6.17 Calculating Pressure with the van der Waals Equation **LO9**

Calculate the pressure of 1.00 mol of N_2 in a 1.00 L container at 300.0 K, using first the van der Waals equation and then the ideal gas equation.

Collect and Organize We are given the amount of nitrogen, its volume, and its temperature, and we need to calculate its pressure by using both the van der Waals equation and the ideal gas equation. Using the van der Waals equation means we need to know the values of the van der Waals constants for nitrogen.

Analyze Table 6.5 lists the van der Waals constants for nitrogen gas as $a = 1.39$ L$^2 \cdot$ atm/mol^2 and $b = 0.0391$ L/mol. The values of a and b represent experimentally determined corrections for the interactions between molecules and the portion of the total volume occupied by the gas molecules. One mole of ideal gas at STP occupies 22.4 L. Compressing the gas to a volume of about 1/20 of the initial volume requires a pressure of about 20 atm. Heating it to 300 K at constant volume causes a relatively minor change in the pressure. We can estimate the effect of a pressure of about 20 atm on the ideality of the nitrogen from the curves in Figure 6.39. Although we do not know which curve best describes the behavior of N_2, the scale of the x-axis is rather large. At a pressure of 20 atm, none of the curves deviate substantially from ideality, so we expect a relatively small difference in the pressure calculated for N_2 as an ideal gas or by using the van der Waals equation.

Solve We solve Equation 6.32 for pressure:

$$P = \frac{nRT}{V - nb} - \frac{n^2 a}{V^2}$$

$$P_{N_2} = \frac{(1.00\ \text{mol})\left(0.08206\ \dfrac{\text{L} \cdot \text{atm}}{\text{mol} \cdot \text{K}}\right)(300.0\ \text{K})}{1.00\ \text{L} - (1.00\ \text{mol})\left(0.0391\ \dfrac{\text{L}}{\text{mol}}\right)} - \frac{(1.00\ \text{mol})^2\left(1.39\ \dfrac{\text{L}^2 \cdot \text{atm}}{\text{mol}^2}\right)}{(1.00\ \text{L})^2}$$

$$= 24.2\ \text{atm}$$

van der Waals equation an equation that includes experimentally determined factors a and b that quantify the contributions of non-negligible molecular volume and non-negligible intermolecular interactions to the behavior of real gases with respect to changes in P, V, and T.

If nitrogen behaved as an ideal gas, we would have

$$P = \frac{nRT}{V}$$

$$P_{N_2} = \frac{(1.00 \text{ mol})\left(0.08206 \dfrac{\text{L} \cdot \text{atm}}{\text{mol} \cdot \text{K}}\right)(300.0 \text{ K})}{1.00 \text{ L}}$$

$$= 24.6 \text{ atm}$$

Think About It This pressure is about 25 times greater than normal atmospheric pressure, so we expect some difference in the calculated pressures. However, the small deviation from ideality, $0.4/24.6 = 2\%$, supports our prediction. Furthermore, because of the likely deviation from ideal behavior, we expect the pressure calculated using the ideal gas equation to be too high. In this case, we observe that P_{ideal} is greater than P_{real}.

Practice Exercise Assuming the conditions stated in Sample Exercise 6.17, use the van der Waals equation to calculate the pressure for 1.00 mol of He gas. Which gas behaves more ideally at 300 K, He or N_2?

We began this chapter with a description of our atmosphere as an invisible necessity. Looking back, we can now answer the questions we raised about gases and our atmosphere. We now understand how elevation affects the quantity of gas available, making pressurized cabins on commercial airliners a necessity for safe, comfortable travel. We have seen the chemical reactions used to supply oxygen in aircraft under emergency conditions and applied principles of stoichiometry to calculate the amount of material required to produce sufficient breathable air. Undersea explorers carry a different mixture of gases in a scuba tank from what we breathe under normal conditions because the high pressures under water increase the partial pressures of oxygen and nitrogen to unhealthy levels. Hot-air balloons rise because volume is directly proportional to temperature and because the density of air decreases as temperature rises. Understanding the physical properties of gases, especially their compressibility and their responses to changing conditions, makes their use in many everyday items, from automobile tires to weather balloons and fire extinguishers, possible because their behavior is reliable and predictable.

SAMPLE EXERCISE 6.18 Integrating Concepts: Scuba Diving

The photograph on the opening page of this chapter shows a scuba diver whose exhaled air bubbles expand as they rise to the surface. The lungs of a diver behave similarly unless certain precautions are taken, and serious medical consequences may result if divers deviate from these precautions. In the following calculations, assume that the pressure at sea level is 1.0 atm and that the pressure increases by 1.0 atm for every 10 m you descend under the surface. Assume your lungs have a volume of 5.1 L and you are breathing dry air that is 78.1% N_2, 20.9% O_2, and 1.00% Ar by volume. Because you are breathing constantly as you descend, your lungs maintain this volume throughout the dive. (a) What is the partial pressure of each gas in your lungs before the dive? (b) What is the partial pressure of each gas in your lungs when you reach a depth of 30 m? (c) Suppose you are accompanied by a breath-hold diver on this descent. She fills her lungs (which also have a volume of 5.1 L) with air at the surface, dives to 30 m with you, and ascends to the surface, all without exhaling. What is the volume of her lungs at a depth of 30 m and upon the return to the surface? (d) As you return to the surface from this depth, you hold your breath. What is the volume of your lungs when you reach the surface?

Air may be used in scuba for shallow dives. However, nitrogen narcosis (the symptoms of which include dulling of the senses, inactivity, and even unconsciousness) can result during deeper dives because of the anesthetic qualities of nitrogen under pressure.

Breathing pure oxygen is not a solution because oxygen at high levels is quite toxic to the central nervous system and other parts of the body. Therefore mixtures of N_2, O_2, and He, generically called Trimix, are used to minimize nitrogen narcosis and oxygen toxicity. For most dives, the oxygen concentration in the tank is adjusted to a maximum partial pressure of 1.5 atm at the working depth; the "equivalent narcotic depth" (END) for N_2 pressure is normally set at 4.0 atm. Helium is added as needed to achieve these values. (e) What is the best Trimix composition for a dive 225 ft below the surface?

Collect and Organize We are given initial and final conditions for mixtures of gases and desired partial pressures for gases at known total pressures. We can use Boyle's law and Dalton's law of partial pressures to answer these questions about gas mixtures in scuba tanks.

Analyze We can convert percent composition data into partial pressures at 1 atm and then into partial pressures under other conditions. The pressure at 30 m is about 4 atm, so the partial pressures of gases in the scuba diver's lungs will increase by a factor of 4. The scuba diver's lungs will have the same volume at 30 m; the breath-holding diver's lungs will shrink to 1/4 their volume on the surface. As the two ascend, both divers' lungs will expand as pressure drops, but the scuba diver's lungs start at a much higher volume than the breath-hold diver's. A dive 225 ft below the surface is around 70 m, where the pressure will be about 8 atm. Using that as the total pressure, we can estimate that the P_{He} should be about 2.5 atm ($P_{total} \approx 8 \text{ atm} = P_{N_2} + P_{O_2} + P_{He} = 4.0 + 1.5 + x$).

Solve

a. The percent composition by volume can be converted into mole fractions by recognizing, for example, that 78.1% nitrogen means 0.781 moles of nitrogen for every 1 mole of gas:

$$X_{N_2} = \frac{n_{N_2}}{n_{total}} = \frac{0.781}{1} = 0.781$$

The mole fraction of oxygen in your lungs is 0.209, and that of argon is 0.010.

b. For every 10 m of descent below the surface, the pressure increases by 1.0 atm. At 30 m, the pressure would increase by 3.0 atm, so the total pressure is 3.0 atm + 1.0 atm = 4.0 atm. The partial pressure of each gas at 4.0 atm total pressure is:

$$P_{N_2} = X_{N_2} P_{total} = 0.781 \times 4.0 \text{ atm} = 3.1 \text{ atm}$$

$$P_{O_2} = X_{O_2} P_{total} = 0.209 \times 4.0 \text{ atm} = 0.84 \text{ atm}$$

$$P_{Ar} = X_{Ar} P_{total} = 0.010 \times 4.0 \text{ atm} = 0.040 \text{ atm}$$

c. The breath-hold diver does not breathe in more air, so the quantity of gas in her lungs remains constant. As the pressure increases, Boyle's law predicts her lungs will decrease in volume:

$$V_{final} = V_{initial} \times \frac{P_{initial}}{P_{final}} = 5.1 \text{ L} \times \frac{1.0 \text{ atm}}{4.0 \text{ atm}} = 1.3 \text{ L}$$

Upon return to the surface, her lungs will re-expand to their original volume of 5.1 L.

d. Because you have been breathing air supplied by your scuba tank, your lungs retain their volume of 5.1 L at a depth of 30 m. If you hold your breath as you ascend to the surface, your lungs will expand (or try to):

$$5.1 \text{ L} \times \frac{4.0 \text{ atm}}{1.0 \text{ atm}} = 20.4 \text{ L} = 20 \text{ L}$$

e. A dive 225 ft below the surface in terms of meters is

$$225 \text{ ft} \times \frac{12 \text{ in}}{1 \text{ ft}} \times \frac{2.54 \text{ cm}}{1 \text{ in}} \times \frac{1 \text{ m}}{100 \text{ cm}} = 68.6 \text{ m}$$

The pressure would rise:

$$68.6 \text{ m} \times \frac{1.0 \text{ atm}}{10 \text{ m}} = 6.9 \text{ atm}$$

The total pressure at 225 ft is 6.9 atm + 1.0 atm = 7.9 atm. The partial pressure of helium in the tank must be

$$7.9 \text{ atm} = 4.0 \text{ atm} + 1.5 \text{ atm} + x; \quad x = 2.4 \text{ atm}$$

To achieve these partial pressures at a depth of 225 ft, the percent composition of the gas mixture in the scuba tank must be:

$$N_2: \quad \frac{P_{N_2}}{P_{total}} \times 100\% = \frac{4.0 \text{ atm}}{7.9 \text{ atm}} \times 100\% = 50.6 = 51\%$$

$$O_2: \quad \frac{P_{O_2}}{P_{total}} \times 100\% = \frac{1.5 \text{ atm}}{7.9 \text{ atm}} \times 100\% = 19.0 = 19\%$$

$$He: \quad \frac{P_{He}}{P_{total}} \times 100\% = \frac{2.4 \text{ atm}}{7.9 \text{ atm}} \times 100\% = 30.4 = 30\%$$

Think About It An important rule in scuba diving is "never hold your breath." If a scuba diver holds his breath on an ascent, the resulting expansion of his lungs could rupture his lungs. In addition, air may also be pushed into the skin, the spaces between tissues, and even into blood vessels, causing embolisms. A breath-hold diver is not immune to damage from pressure changes either. During deep dives, a condition called thoracic squeeze may develop, which can result in hemorrhage of lung tissues.

SUMMARY

LO1 Gases occupy the entire volume of their container. The volume occupied by a gas changes significantly with pressure and temperature. Gases are **miscible**, mixing in any proportion. They are much less dense than liquids or solids. (Section 6.1)

LO2 Pressure is defined as the ratio of force to surface area, and gas pressure is measured with **barometers** and **manometers**. Pressure can be expressed in a variety of units, and we can convert between them by using dimensional analysis. (Section 6.2)

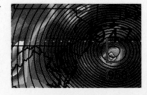

LO3 Boyle's law, Charles's law, Avogadro's law, and **Amontons's law** describe the behavior of gases under different conditions. They may be used individually or in a combined form called the **ideal gas law** ($PV = nRT$) to calculate the volume, temperature, pressure, or number of moles of a gas. (Sections 6.3 and 6.4)

LO4 The ideal gas equation and the stoichiometry of a chemical reaction can be used to calculate the volumes of gases required or produced in the reaction. (Section 6.5)

LO5 The density of a gas can be calculated for a given set of conditions of volume, temperature, and pressure. The density of a gas can also be used to calculate its molar mass. (Section 6.6)

LO6 In a gas mixture, the contribution each component gas makes to the total gas pressure is called the **partial pressure** of that gas. **Dalton's law of partial pressures** allows us to calculate the partial pressure (P_x) of any constituent gas x in a gas mixture if we know its **mole fraction** (X_x) and the total pressure. (Section 6.7)

LO7 Kinetic molecular theory describes gases as particles in constant random motion. This motion is associated with an average kinetic energy that is proportional to the absolute temperature of the gas. Pressure arises from elastic collisions between gases and the walls of their container. All relationships between P, V, T, and n can be derived from kinetic molecular theory. (Section 6.8)

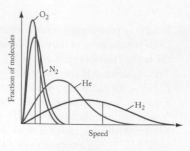

LO8 Particles move at **root-mean-square speeds** (u_{rms}) that are inversely proportional to the square root of their molar masses and directly proportional to the square root of their temperature. **Graham's law of effusion** states that the rate of **effusion** (escape through a pinhole) or **diffusion** (spreading) of a gas at a fixed temperature is inversely proportional to the square root of its molar mass. (Section 6.8)

LO9 At high pressures, the behavior of real gases deviates from the predictions of the ideal gas law. The **van der Waals equation**, a modified form of the ideal gas equation, accounts for real gas properties. (Section 6.9)

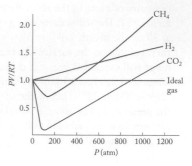

PARTICULATE **PREVIEW WRAP-UP**

The helium atoms fill both the tank and the balloon, but there are significantly more atoms in the tank than in a single balloon, as one tank purchased at a party store can fill up to 20–30 balloons. The drawing for the tank does not change after being in the trunk of the car on a hot day; the helium atoms move faster at a higher temperature, but the size of the tank and the distribution of the atoms do not change. The helium-filled balloons will expand because as the temperature rises, the kinetic energy of the atoms increases, there are more collisions with the walls of the balloons. Because the balloon expands, the same number of particles would occupy a larger space, so the circle describing the gas in the balloon would contain fewer particles.

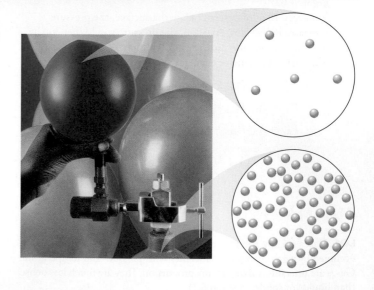

PROBLEM-SOLVING SUMMARY

Type of Problem	Concepts and Equations	Sample Exercises
Calculating pressure of any gas, calculating atmospheric pressure	Divide the force by the area over which the force is applied, using the equation $$P = \frac{F}{A} \quad (6.1)$$	**6.1, 6.2**
Calculating changes in P, V, and/or T in response to changing conditions	Rearrange $$\frac{P_1V_1}{T_1} = \frac{P_2V_2}{T_2} \quad (6.18)$$ for whichever variable is sought and then substitute given values. (T must be in kelvins, and n must be constant.)	**6.3, 6.4, 6.5, 6.6**
Determining n from P, V, and T	Rearrange $$PV = nRT \quad (6.15)$$ for n and then substitute given values of P, T, and V. (T must be in kelvins.)	**6.7, 6.8, 6.9**
Calculating the density of a gas and calculating molar mass from density	Substitute values for pressure, absolute temperature, and molar mass into the equation $$d = \frac{P\mathcal{M}}{RT} \quad (6.21)$$ Substitute values for pressure, absolute temperature, and density into the equation $$\mathcal{M} = \frac{dRT}{P} \quad (6.22)$$	**6.10, 6.11**
Calculating mole fraction for one component gas in a mixture	Divide the number of moles of the component gas by the total number of moles in the mixture: $$X_x = \frac{n_x}{n_{total}} \quad (6.24)$$	**6.12**
Calculating partial pressure of one component gas in a mixture and total pressure in the mixture	Substitute the mole fraction of the component gas and the total pressure in the equation $$P_x = X_x P_{total} \quad (6.25)$$ Solve the equation $$P_{total} = P_1 + P_2 + P_3 + P_4 + \dots \quad (6.23)$$ for the partial pressure of the component gas and then substitute given values for other partial pressures and total pressure.	**6.13, 6.14**
Calculating root-mean-square speeds	Substitute absolute temperature, molar mass, and the value $8.314 \text{ kg} \cdot \text{m}^2/(\text{s}^2 \cdot \text{mol} \cdot \text{K})$ for R in the equation $$u_{rms} = \sqrt{\frac{3RT}{\mathcal{M}}} \quad (6.28)$$	**6.15**
Calculating relative rate of effusion or diffusion from molar masses	Substitute values for molar masses into the equation $$\frac{r_x}{r_y} = \sqrt{\frac{\mathcal{M}_y}{\mathcal{M}_x}} \quad (6.31)$$	**6.16**
Calculating P for a real gas	Solve the van der Waals equation $$\left(P + \frac{n^2 a}{V^2}\right)(V - nb) = nRT \quad (6.32)$$ for P and substitute given values of n, V, and T (in kelvins), plus values of a and b from Table 6.5.	**6.17**

VISUAL PROBLEMS

(Answers to boldface end-of-chapter questions and problems are in the back of the book.)

6.1. Shown in Figure P6.1 are three barometers. The one in the center is located at sea level. Which barometer is most likely to reflect the atmospheric pressure in Denver, CO, where the elevation is approximately 1500 m? Explain your answer.

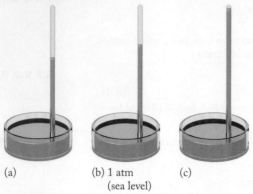

(a)   (b) 1 atm   (c)
             (sea level)

FIGURE P6.1

6.2. A rubber balloon is filled with helium gas. Which of the drawings in Figure P6.2 most accurately reflects the gas in the balloon on a molecular level? The blue spheres represent helium atoms. Explain your answer. Because atoms are in constant motion, the spheres represent their average position.

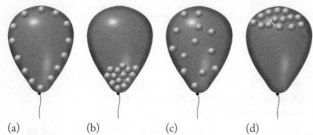

(a)   (b)   (c)   (d)
FIGURE P6.2

6.3. Which of the three changes shown in Figure P6.3 best illustrates what happens when the atmospheric pressure on a helium-filled rubber balloon is increased at constant temperature? Because atoms are in constant motion, the spheres represent their average position.

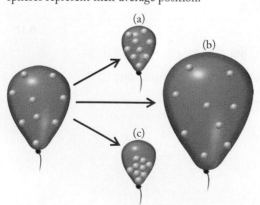

FIGURE P6.3

6.4. Which of the drawings in Figure P6.3 best illustrates what happens when the temperature of a helium-filled rubber balloon is increased at constant pressure?

6.5. Which of the three changes shown in Figure P6.5 best illustrates what happens when the amount of gas in a helium-filled rubber balloon is increased at constant temperature and pressure?

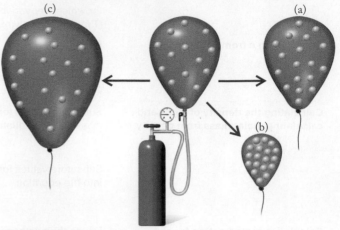

FIGURE P6.5

6.6. Which line in the plot of volume versus reciprocal pressure in Figure P6.6 corresponds to the higher temperature?

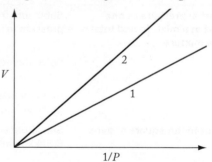

FIGURE P6.6

6.7. In Figure P6.7, which line in the plot of volume versus temperature represents a gas at higher pressure? Is the *x*-axis an absolute temperature scale?

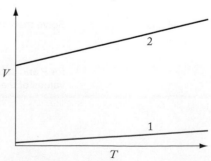

FIGURE P6.7

6.8. In Figure P6.8, which of the two plots of volume versus pressure at constant temperature is *not* consistent with the ideal gas law?

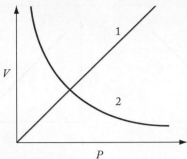

FIGURE P6.8

6.9. In Figure P6.9, which of the two plots of volume versus temperature at constant pressure is *not* consistent with the ideal gas law?

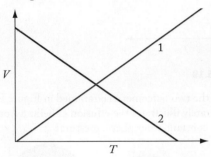

FIGURE P6.9

6.10. In Figure P6.10, which line in the plot of density versus pressure at constant temperature for methane (CH_4) and nitrogen (N_2) should be labeled *methane*?

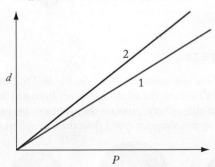

FIGURE P6.10

6.11. Add lines showing the densities of He and NO as a function of pressure to a copy of the graph in Figure P6.10.

6.12. Which of the graphs in Figure P6.12 best represents the relationship between number of collisions with the container and pressure?

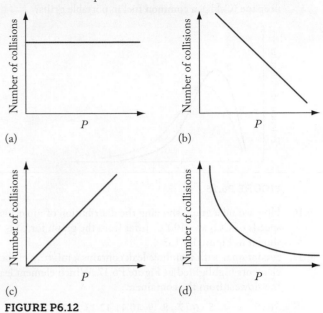

FIGURE P6.12

6.13. Which of the drawings in Figure P6.13 best depicts the arrangement of molecules in a mixture of gases? The red and blue spheres represent atoms of two different gases, such as helium and neon.

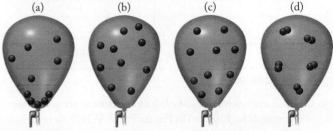

FIGURE P6.13

6.14. The drawings in Figure P6.14 illustrate four mixtures of two diatomic gases (such as N_2 and O_2) in flasks of identical volume and at the same temperature. Is the total pressure in each flask the same? Which flask has the highest partial pressure of nitrogen, depicted by the blue molecules?

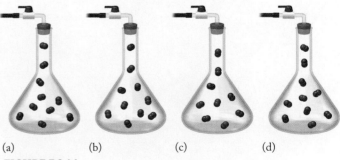

FIGURE P6.14

6.15. Figure P6.15 shows the distribution of molecular speeds of CO_2 and SO_2 molecules at 25°C. Which curve is the profile for SO_2? Which of these profiles should match that of propane (C_3H_8), a common fuel in portable grills?

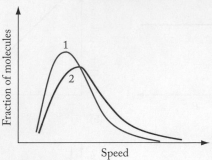

FIGURE P6.15

6.16. How would a graph showing the distribution of molecular speeds of CO_2 at −100°C differ from the graph for CO_2 shown in Figure P6.15?

6.17. A container with a pinhole leak contains a mixture of the elements highlighted in Figure P6.17. Which element leaks the slowest from the container?

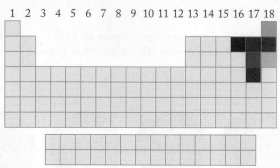

FIGURE P6.17

6.18. A container with a pinhole leak contains a mixture of the elements highlighted in Figure P6.17. Which element has the smallest root-mean-square speed?

6.19. The rate of effusion of a gas increases with temperature. Which graph in Figure P6.19 best describes the ratio of the rates of effusion of two gases (x and y) as a function of temperature?

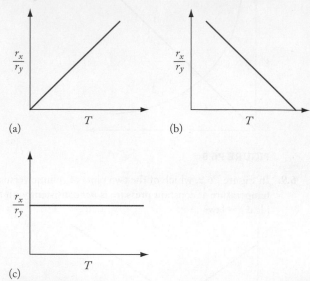

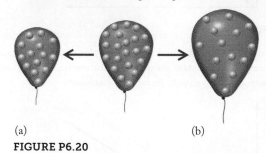

FIGURE P6.19

6.20. Which of the two outcomes diagrammed in Figure P6.20 more accurately illustrates the effusion of helium from a balloon at constant atmospheric pressure?

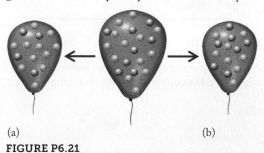

FIGURE P6.20

6.21. Which of the two outcomes shown in Figure P6.21 more accurately illustrates the effusion of gases from a balloon at constant atmospheric pressure if the red spheres have a greater root-mean-square speed than the blue spheres?

(a)   (b)
FIGURE P6.21

6.22. Use representations [A] through [I] in Figure P6.22 to answer questions a–f. The blue arrows indicate changes in pressure, temperature, volume, and/or the number of moles of gas in the cylinder.

a. Describe which parameters change and which do not for each blue arrow.

b. In which cylinder are the collisions between particles and the piston the most frequent?

c. Which cylinders have an equal frequency of collisions?

d. Which transformation(s) result(s) in more frequent collisions in the final state when compared to the initial state of [E]?

e. How does the root-mean-square speed change from [E] to [I]?

f. Are more molecules moving at the root-mean-square speed in [E] or in [I]?

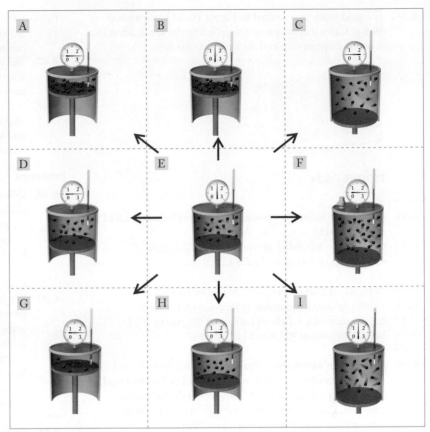

FIGURE P6.22

QUESTIONS AND PROBLEMS

Atmospheric Pressure and Collisions

Concept Review

6.23. Describe the difference between force and pressure.

6.24. Describe the role of collisions in generating pressure.

6.25. How does Torricelli's barometer measure atmospheric pressure?

6.26. What is the relationship between *torr* and *atmospheres* of pressure?

6.27. What is the relationship between *millibars* and *pascals* of pressure?

6.28. Three barometers based on Torricelli's design are constructed using water (d = 1.00 g/mL), ethanol (d = 0.789 g/mL), and mercury (d = 13.546 g/mL). Which barometer contains the tallest column of liquid?

6.29. Why does an ice skater exert more pressure on ice when wearing newly sharpened skates than when wearing skates with dull blades?

6.30. Why is it easier to travel over deep snow when wearing boots and snowshoes (Figure P6.30) rather than just boots?

6.31. Why does atmospheric pressure decrease with increasing elevation?

*6.32. Pieces of different metals have the same mass but different densities. Could these objects ever exert the same pressure?

Problems

6.33. Calculate the downward pressure due to gravity exerted by the bottom face of a 1.00 kg cube of iron that is 5.00 cm on a side.

FIGURE P6.30

6.34. The gold block represented in Figure P6.34 has a mass of 38.6 g. Calculate the pressure exerted by the block when it is on (a) a square face and (b) a rectangular face.

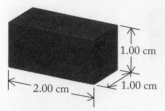

FIGURE P6.34

6.35. Convert the following pressures into atmospheres: (a) 2.0 kPa; (b) 562 mmHg.

6.36. Convert the following pressures into millimeters of mercury: (a) 0.541 atm; (b) 2.8 kPa.

6.37. Which is the larger pressure: 10 torr or 10 atm?

6.38. Which helium balloon has more collisions at room temperature: a 1 L helium balloon at 0.1 atm or a 1 L helium balloon at 455 mm Hg?

6.39. Record High Atmospheric Pressure The highest atmospheric pressure recorded on Earth was measured at Tosontsengel, Mongolia, on December 19, 2001, when the barometer read 108.6 kPa. Express this pressure in (a) millimeters of mercury, (b) atmospheres, and (c) millibars.

6.40. Record Low Atmospheric Pressure Hurricane Irene registered an atmospheric pressure of 982 mbar in August 2011. Hurricane Katrina registered an atmospheric pressure of 86.2 kPa in September 2005. What was the *difference* in pressure between the two hurricanes in (a) millimeters of mercury, (b) atmospheres, and (c) millibars?

*6.41. Pressure on Venus Venus is similar to Earth in size and distance from the Sun but has an atmosphere inhospitable for human life that consists of 96.5% CO_2 and 3.5% N_2 by mass.
   a. Calculate the atmospheric pressure on Venus given the mass of the Venusian atmosphere (4.8×10^{20} kg), the surface area (4.6×10^{14} m^2), and the acceleration due to gravity (8.87 m/s^2).
   *b. Given the similarity between Venus and Earth in surface area, why do you suppose atmospheric pressure on Venus is so much higher than on Earth?

*6.42. Pressure on Uranus Uranus is much larger than Earth, with a surface area of 8.1×10^{15} m^2 with an atmosphere composed of low-density gases: 82.5% H_2, 15.2% He, and 2.3% CH_4. The acceleration due to gravity is similar to that on Earth, 8.7 m/s^2, yet the atmospheric pressure on Uranus is estimated to exceed 10^8 Pa. What is the mass of the atmosphere on Uranus?

The Gas Laws

Concept Review

6.43. From the molecular perspective, why is pressure directly proportional to temperature at fixed volume (Amontons's law)?

6.44. How do we explain Boyle's law on a molecular basis?

6.45. A balloonist is rising too fast for her taste. Should she increase the temperature of the gas in the balloon or decrease it?

6.46. Could the pilot of the balloon in Problem 6.45 reduce her rate of ascent by allowing some gas to leak out of the balloon? Explain your answer.

6.47. The volume of a quantity of gas decreases by 50% when it cools from 20°C to 10°C. Does the pressure of the gas increase, decrease, or remain the same?

6.48. If the volume of gasoline vapor and air in an automobile engine cylinder is reduced to 1/10 of its original volume before ignition, by what factor does the pressure in the cylinder increase? (Assume there is no change in temperature.)

Problems

6.49. What is the final pressure of 1.00 mol of ammonia gas, initially at 1.00 atm, if the volume is
   a. gradually decreased from 78.0 mL to 39.0 mL at constant temperature?
   b. increased from 43.5 mL to 65.5 mL at constant temperature?
   c. decreased by 40% at constant temperature?

6.50. The behavior of 485 mL of an ideal gas in response to pressure is studied in a vessel with a movable piston. What is the final volume of the gas if the pressure on the sample is
   a. increased from 715 mmHg to 3.55 atm at constant temperature?
   b. decreased from 1.15 atm to 520 mmHg at constant temperature?
   c. increased by 26% at constant temperature?

6.51. Underwater Archeology A scuba diver releases a balloon containing 153 L of helium attached to a tray of artifacts at an underwater archaeological site (Figure P6.51). When the balloon reaches the surface, it has expanded to a volume of 352 L. The pressure at the surface is 1.00 atm; what is the pressure at the underwater site? Pressure increases by 1.0 atm for every 10 m of depth; at what depth was the diver working? Assume the temperature remains constant.

FIGURE P6.51

6.52. Breath-Hold Diving The world record for diving without supplemental air tanks (breath-hold diving) is about 125 m, a depth at which the pressure is about 12.5 atm. If a diver's lungs have a volume of 6 L at the surface of the water, what is their volume at a depth of 125 m?

6.53. Use the following data to draw a graph of the volume of 1 mole of H_2 as a function of the reciprocal of pressure at 298 K:

P (mmHg)	V (L)
100	186
120	155
240	77.5
380	48.9
500	37.2

Would the graph be the same for the same number of moles of argon?

6.54. The following data for P and V were collected for 1 mole of argon at 300 K. Draw a graph of the volume of 1 mole of Ar as a function of the reciprocal of pressure. Does this graph look like the graph you drew for Problem 6.53?

P (atm)	V (L)
0.10	246.3
0.25	98.5
0.50	49.3
0.75	32.8
1.0	24.6

6.55. Use the following data to draw a graph of the volume of He as a function of temperature for 1.0 mol of He gas at a constant pressure of 1.00 atm:

V (L)	T (K)
7.88	96
3.94	48
1.97	24
0.79	9.6
0.39	4.8

How would the graph change if the amount of gas were halved?

6.56. Use the following data to draw a graph of the volume of He as a function of temperature for 0.50 mol of He gas at a constant pressure of 1.00 atm:

V (L)	T (K)
3.94	96
1.97	48
0.79	24
0.39	9.6
0.20	4.8

Does this graph match your prediction from Problem 6.55?

6.57. A cylinder with a piston (Figure P6.57) contains a sample of gas at 25°C. The piston moves to keep the pressure constant inside the cylinder. At what gas temperature would the piston move so that the volume inside the cylinder doubled?

6.58. The temperature of the gas in Problem 6.57 is reduced to a temperature at which the volume inside the cylinder has decreased by 25% from its initial volume at 25.0°C. What is the new temperature?

FIGURE P6.57

6.59. Calculate the volume of a 2.68 L sample of gas in the cylinder shown in Figure P6.57 after it is subjected to the following changes in conditions:
 a. The gas is warmed from 250 K to a final temperature of 398 K at constant pressure.
 b. The pressure is increased by 33% at constant temperature.
 c. Ten percent of the gas leaks from the cylinder at constant temperature and pressure.

6.60. Calculate the temperature of a 5.6 L sample of gas in the cylinder in Figure P6.57 after it is subjected to the following changes in conditions:
 a. The gas, at constant pressure, is cooled from 78°C to a temperature at which its volume is 4.3 L.
 b. The external pressure is doubled at constant volume.
 c. The number of molecules in the cylinder is increased by 15% at constant pressure and volume.

*6.61. An empty balloon is filled with 1.75 mol of He at 17°C and 1.00 atm pressure. How much work is done on the system if the temperature remains constant?

*6.62. A student holding a 50.0 L balloon containing 800 mmHg of hydrogen at 19°C lets go of the balloon, allowing the gas to escape. How many moles of gas did the balloon contain and how much work was done on the surroundings if the temperature remains constant?

6.63. Which of the following actions would produce the greater increase in the volume of a gas sample: (a) lowering the pressure from 760 mmHg to 720 mmHg at constant temperature or (b) raising the temperature from 10°C to 40°C at constant pressure?

6.64. Which of the following actions would produce the greater increase in the volume of a gas sample: (a) doubling the amount of gas in the sample at constant temperature and pressure or (b) raising the temperature from 244°C to 1100°C at constant pressure?

*6.65. What happens to the volume of gas in a cylinder with a movable piston under the following conditions?
 a. Both the absolute temperature and the external pressure on the piston double.
 b. The absolute temperature is halved, and the external pressure on the piston doubles.
 c. The absolute temperature increases by 75%, and the external pressure on the piston increases by 50%.

*6.66. What happens to the pressure of a gas under the following conditions?
 a. The absolute temperature is halved and the volume doubles.
 b. Both the absolute temperature and the volume double.
 c. The absolute temperature increases by 75%, and the volume decreases by 50%.

6.67. A sample of a gas enclosed in a cylinder with a piston has a volume of 500.0 mL at 30.0°C. What is the sample volume at 100.0°C?

6.68. A tank containing 5.00 L of nitrogen at 75.0 atm pressure and 28°C is left standing in sunlight. Its temperature rises to 50.0°C. What is the gas pressure at this higher temperature?

6.69. Temperature Effects on Bicycle Tires A bicycle racer inflates his tires to 7.1 atm on a warm autumn afternoon when temperatures reached 27°C. By morning the temperature has dropped to 5.0°C. What is the pressure in the tires if we assume that the volume of the tire does not change significantly?

*6.70. A balloon vendor at a street fair is using a tank of helium to fill her balloons. The tank has a volume of 145 L and a pressure of 136 atm at 25°C. After a while she notices that the valve has not been closed properly, and the pressure has dropped to 94 atm. How many moles of gas have been lost?

The Ideal Gas Law; Gases in Chemical Reactions

Concept Review

6.71. What is meant by standard temperature and pressure (STP)? What is the volume of 1 mole of an ideal gas at STP?

6.72. Which of the following are not characteristics of an ideal gas?
a. The molecules of gas have little volume compared with the volume that they occupy.
b. Its volume is independent of temperature.
c. The density of all ideal gases is the same.
d. Gas atoms or molecules do not interact with one another.

*6.73. What does the slope represent in a graph of pressure as a function of $1/V$ at constant temperature for an ideal gas?

*6.74. How would the graph in Problem 6.73 change if we increased the temperature to a larger but constant value?

Problems

6.75. How many moles of air are there in a bicycle tire with a volume of 2.36 L if it has an internal pressure of 6.8 atm at 17.0°C?

6.76. At what temperature will 1.00 mole of an ideal gas in a 1.00 L container exert a pressure of
a. 1.00 atm?
b. 2.00 atm?
c. 0.75 atm?

6.77. Hyperbaric Oxygen Therapy Hyperbaric oxygen chambers are used to treat divers suffering from decompression sickness (the "bends") with pure oxygen at greater than atmospheric pressure. Other clinical uses include treatment of patients with thermal burns, necrotizing fasciitis, and CO poisoning. What is the pressure in a chamber with a volume of 2.36×10^3 L that contains 4635 g of $O_2(g)$ at a temperature of 298 K?

6.78. Hydrogen holds promise as an "environment friendly" fuel. How many grams of H_2 gas are present in a 50.0 L fuel tank at a pressure of 2850 lb/in² (psi) at 20°C? Assume that 1 atm = 14.7 psi.

6.79. A weather balloon with a volume of 200.0 L is launched at 20°C at sea level, where the atmospheric pressure is 1.00 atm.
a. What is the volume of the balloon at 20,000 m where atmospheric pressure is 63 mmHg and the temperature is 220 K?
b. At even higher elevations (the stratosphere, up to 50,000 m), the temperature of the atmosphere increases to 270 K but the pressure decreases to 0.80 mmHg. What is the volume of the balloon under these conditions?
c. If the balloon is designed to rupture when the volume exceeds 400.0 L, will the balloon break under either of these sets of conditions?

6.80. A sealed, flexible foil bag of potato chips, containing 0.500 L of air, is carried from Boston ($P_{atm} = 1.0$ atm) to Denver ($P_{atm} = 0.83$ atm).
a. What would the volume of the bag be upon arrival in Denver?
b. If the structural limitations of the bag allow for only a 10% expansion in the volume of the bag, what is the pressure in the bag?
c. If the bag is placed in checked luggage, the pressure and temperature in the hold will decrease considerably during flight to $T = 210$ K and $P = 126$ torr. Calculate the pressure in the bag during flight.

6.81. Miners' Lamps Before the development of reliable batteries, miners' lamps burned acetylene produced by the reaction of calcium carbide with water:

$$CaC_2(s) + H_2O(\ell) \rightarrow C_2H_2(g) + CaO(s)$$

A lamp uses 1.00 L of acetylene per hour at 1.00 atm pressure and 18°C.
a. How many moles of C_2H_2 are used per hour?
b. How many grams of calcium carbide must be in the lamp for a 4-h shift?

6.82. Acid precipitation dripping on limestone produces carbon dioxide by the following reaction:

$$CaCO_3(s) + 2 H_3O^+(aq) \rightarrow Ca^{2+}(aq) + CO_2(g) + 3 H_2O(\ell)$$

If 15.0 mL of CO_2 was produced at 25°C and 760 mmHg, then
a. how many moles of CO_2 were produced?
b. how many milligrams of $CaCO_3$ were consumed?

*6.83. Air is about 78% nitrogen by volume and 21% oxygen. Pure nitrogen can be produced by the decomposition of ammonium dichromate:

$$(NH_4)_2Cr_2O_7(s) \rightarrow N_2(g) + Cr_2O_3(s) + 4 H_2O(g)$$

Oxygen can be generated by the thermal decomposition of potassium chlorate:

$$2 KClO_3(s) \rightarrow 2 KCl(s) + 3 O_2(g)$$

How many grams of ammonium dichromate and how many grams of potassium chlorate would be needed to make 200.0 L of "air" at 0.85 atm and 273 K?

*6.84. Nitrogen can be produced from sodium metal and potassium nitrate by the reaction:

$$10 Na(s) + 2 KNO_3(s) \rightarrow K_2O(s) + 5 Na_2O(s) + N_2(g)$$

If we generate oxygen by the thermal decomposition of potassium chlorate:

$$2 KClO_3(s) \rightarrow 2 KCl(s) + 3 O_2(g)$$

how many grams of potassium nitrate and how many grams of potassium chlorate would be needed to make 200.0 L of gas containing 160.0 L of N_2 and 40.0 L of O_2 at 1.00 atm and 290 K?

6.85. **Healthy Air for Sailors** The CO_2 that builds up in the air of a submerged submarine can be removed by reacting it with sodium peroxide:

$$2\,Na_2O_2(s) + 2\,CO_2(g) \rightarrow 2\,Na_2CO_3(s) + O_2(g)$$

If a sailor exhales 150.0 mL of CO_2 per minute at 20°C and 1.02 atm, how much sodium peroxide is needed per sailor in a 24-hr period?

6.86. **Rescue Breathing Devices** Self-contained self-rescue breathing devices, like the one shown in Figure P6.86, convert CO_2 into O_2 according to the following reaction:

$$4\,KO_2(s) + 2\,CO_2(g) \rightarrow 2\,K_2CO_3(s) + 3\,O_2(g)$$

How many grams of KO_2 are needed to produce 100.0 L of O_2 at 20°C and 1.00 atm?

FIGURE P6.86

Gas Density

Concept Review

6.87. Do all gases at the same pressure and temperature have the same density? Explain your answer.

6.88. Birds and sailplanes take advantage of thermals (rising columns of warm air) to gain altitude with less effort than usual. Why does warm air rise?

6.89. How does the density of a gas sample change when (a) its pressure is increased and (b) its temperature is decreased?

6.90. Which will increase the density of a gas: doubling its temperature or doubling its pressure?

Problems

6.91. **Biological Effects of Radon Exposure** Radon is a naturally occurring radioactive gas found in the ground and in building materials. It is easily inhaled and emits α particles when it decays. Cumulative radon exposure is a significant risk factor for lung cancer.
   a. Calculate the density of radon at 298 K and 1.00 atm of pressure.
   b. Are radon concentrations likely to be greater in the basement or on the top floor of a building?

*6.92. Four empty balloons, each with a mass of 10.0 g, are inflated to a volume of 20.0 L. The first balloon contains He; the second, Ne; the third, CO_2; and the fourth, CO. If the density of air at 25°C and 1.00 atm is 0.00117 g/mL, which of the balloons float in this air?

6.93. A 150.0 mL flask contains 0.391 g of a volatile oxide of sulfur. The pressure in the flask is 750 mmHg, and the temperature is 22°C. Is the gas SO_2 or SO_3?

6.94. A 100.0 mL flask contains 0.193 g of a volatile oxide of nitrogen. The pressure in the flask is 760 mmHg at 17°C. Is the gas NO, NO_2, or N_2O_5?

6.95. A 0.375 g sample of benzene vapor has a volume of 149 mL measured at 95.0°C and 740.0 torr. Calculate the molar mass of benzene.

6.96. Calculate the density of toluene vapor (molar mass 92 g/mol) at 1.00 atm pressure and 227.0°C.

Dalton's Law and Mixtures of Gases

Concept Review

6.97. What is meant by the *partial pressure* of a gas?

6.98. Can a barometer be used to measure just the partial pressure of oxygen in the atmosphere? Why or why not?

6.99. Which gas sample has the largest volume at 25°C and 1 atm pressure? (a) 0.500 mol of dry H_2; (b) 0.500 mol of dry N_2; (c) 0.500 mol of wet H_2 (H_2 collected over water)

6.100. Two identical balloons are filled to the same volume at the same pressure and temperature. One balloon is filled with air and the other with helium. Which balloon contains more particles (atoms and molecules)?

6.101. A mixture of gases has a partial pressure of $N_2 = 0.5$ atm and a partial pressure of $O_2 = 1.0$ atm. Which molecules experience more collisions: the molecules of nitrogen or the molecules of oxygen?

6.102. Why does the partial pressure of water increase as temperature increases?

Problems

6.103. What pressure is exerted by a gas mixture containing 2.00 g of H_2 and 7.00 g of N_2 at 273°C in a 10.0 L container? What is the contribution of N_2 to the total pressure?

6.104. A gas mixture contains 7.0 g of N_2, 2.0 g of H_2, and 16.0 g of CH_4. What is the mole fraction of H_2 in the mixture? Calculate the pressure of the gas mixture and the partial pressure of each constituent gas if the mixture is in a 1.00 L vessel at 0°C.

6.105. **The X-15 and Mach 6** On November 9, 1961, the Bell X-15 test aircraft exceeded Mach 6, or six times the speed of sound. A few weeks later it reached an elevation above 300,000 feet, effectively putting it in space. The combustion of ammonia and oxygen provided the fuel for the X-15.
   a. Write a balanced chemical equation for the combustion of ammonia given that nitrogen ends up as nitrogen dioxide.
   b. What ratio of partial pressures of ammonia and oxygen is needed for this reaction?

6.106. **Rocket Fuels** Before settling on hydrogen as the fuel of choice for space vehicles, several other fuels were explored, including hydrazine, N_2H_4, and pentaborane, B_5H_{11}. Both are gases under conditions in space.
   a. Write balanced chemical equations for the combustion of hydrazine and pentaborane given that the products in addition to water are $NO_2(g)$ and $B_2O_3(s)$, respectively.

b. What ratios of partial pressures of hydrazine to oxygen and pentaborane to oxygen are needed for these reactions?

6.107. A sample of oxygen was collected over water at 25°C and 1.00 atm.
  a. If the total sample volume was 0.480 L, how many moles of O_2 were collected?
  *b. If the same volume of oxygen is collected over ethanol instead of water, does it contain the same number of moles of O_2?

6.108. Water and ethanol were removed from the O_2 samples in Problem 6.107.
  a. What is the volume of the dry O_2 gas sample at 25°C and 1.00 atm?
  b. What is the volume of the dry O_2 gas sample at 25°C and 1.00 atm if $P_{ethanol} = 50$ mmHg at 25°C ?

6.109. The following reactions were carried out in sealed containers. Will the total pressure after each reaction is complete be greater than, less than, or equal to the total pressure before the reaction? Assume all reactants and products are gases at the same temperature.
  a. $N_2O_5(g) + NO_2(g) \rightarrow 3\ NO(g) + 2\ O_2(g)$
  b. $2\ SO_2(g) + O_2(g) \rightarrow 2\ SO_3(g)$
  c. $C_3H_8(g) + 5\ O_2(g) \rightarrow 3\ CO_2(g) + 4\ H_2O(g)$
  d. $4\ NH_3(g) + 5\ O_2(g) \rightarrow 4\ NO(g) + 6\ H_2O(g)$

6.110. The following reactions were carried out in a cylinder with a piston. If the external pressure is constant, in which reactions will the volume of the cylinder increase? Assume all reactants and products are gases at the same temperature.
  a. $CH_4(g) + NH_3(g) \rightarrow HCN(g) + 3\ H_2(g)$
  b. $H_2S(g) + 2\ O_2(g) \rightarrow H_2O(g) + SO_3(g)$
  c. $H_2(g) + Cl_2(g) \rightarrow 2\ HCl(g)$
  d. $2\ NO_2(g) \rightarrow 2\ NO(g) + O_2(g)$

6.111. High-Altitude Mountaineering Climbers use pure oxygen near the summits of 8000-m peaks, where $P_{atm} = 0.35$ atm (Figure P6.111). How much more O_2 is there in a lung full of pure O_2 at this elevation than in a lung full of air at sea level?

FIGURE P6.111

6.112. Scuba Diving A scuba diver is at a depth of 50 m, where the pressure is 5.0 atm. What should be the mole fraction of O_2 in the gas mixture the diver breathes to produce the same partial pressure of oxygen as the gas mixture at sea level?

6.113. Carbon monoxide at a pressure of 680 mmHg reacts completely with O_2 at a pressure of 340 mmHg in a sealed vessel to produce CO_2. What is the final pressure in the flask?

6.114. Ozone reacts completely with NO, producing NO_2 and O_2. A 10.0 L vessel is filled with 0.280 mol of NO and 0.280 mol of O_3 at 350 K. Find the partial pressure of each product and the total pressure in the flask at the end of the reaction.

***6.115. Ammonia Production** Ammonia is produced industrially from the reaction of hydrogen with nitrogen under pressure in a sealed reactor. What is the percent decrease in pressure of a sealed reaction vessel during the reaction between 3.60×10^3 mol of H_2 and 1.20×10^3 mol of N_2 if half of the N_2 is consumed?

***6.116.** A mixture of 0.156 mol of C is reacted with 0.117 mol of O_2 in a sealed, 10.0 L vessel at 500 K, producing a mixture of CO and CO_2. The total pressure is 0.640 atm. What is the partial pressure of CO?

The Kinetic Molecular Theory of Gases

Concept Review

6.117. What is meant by the *root-mean-square speed* of gas molecules?

6.118. Why don't all molecules in a sample of air move at exactly the same speed?

6.119. How does the root-mean-square speed of the molecules in a gas vary with (a) molar mass and (b) temperature?

6.120. Does pressure affect the root-mean-square speed of the molecules in a gas? Explain your answer.

6.121. How can Graham's law of effusion be used to determine the molar mass of an unknown gas?

6.122. Is the ratio of the rates of effusion of two gases the same as the ratio of their root-mean-square speeds?

6.123. What is the difference between *diffusion* and *effusion*?

6.124. If gas X diffuses faster in air than gas Y, is gas X also likely to effuse faster than gas Y?

Problems

6.125. Rank the gases SO_2, CO_2, and NO_2 in order of increasing root-mean-square speed at 0°C.

6.126. In a mixture of CH_4, NH_3, and N_2, which gas molecules are, on average, moving fastest?

6.127. At 286 K, three gases, A, B, and C, have root-mean-square speeds of 360 m/s, 441 m/s, and 472 m/s, respectively. Which gas is O_2?

6.128. Determine the root-mean-square speed of CO_2 molecules that have an average kinetic energy of 4.2×10^{-21} J per molecule.

***6.129.** The root-mean-square speed of He gas at 300 K is 1.370×10^3 m/s. Sketch a graph of $u_{rms,He}$ versus T for $T = 300, 450, 600, 750$, and 900 K. Is the graph linear?

***6.130.** The root-mean-square speed of N_2 gas at 300 K is 516.8 m/s. Why doesn't doubling the temperature also double u_{rms,N_2}?

6.131. What is the ratio of the root-mean-square speed of D_2 to that of H_2 at constant temperature?

6.132. Enriching Uranium The two isotopes of uranium, ^{238}U and ^{235}U, can be separated by diffusion of the corresponding UF_6 gases. What is the ratio of the root-mean-square speed of $^{238}UF_6$ to that of $^{235}UF_6$ at constant temperature?

6.133. A student measured the relative rate of effusion of carbon dioxide and propane, C_3H_8, and found that they effuse at exactly the same rate. Did the student make a mistake?

*6.134. The density of Ne at 760 mmHg is exactly half its density at 2.00 atm at constant temperature. Is the root-mean-square speed of Ne at 2.00 atm half, twice, or the same as $u_{rms,Ne}$ at 760 mmHg?

6.135. An unknown pure gas X effuses at half the rate of effusion of O_2 at the same temperature and pressure. What is the molar mass of gas X?

6.136. Decay Products of Uranium Minerals Radon and helium are both by-products of the radioactive decay of uranium minerals. A fresh sample of carnotite, $K_2(UO_2)_2(VO_4)_2 \cdot 3 H_2O$, is put on display in a museum. Calculate the relative rates of diffusion of helium and radon under fixed conditions of pressure and temperature. Which gas diffuses more rapidly through the display case?

6.137. Isotope Use by Plants During photosynthesis, green plants preferentially use $^{12}CO_2$ over $^{13}CO_2$ in making sugars, and food scientists can frequently determine the source of sugars used in foods on the basis of the ratio of ^{12}C to ^{13}C in a sample.
a. Calculate the relative rates of diffusion of $^{13}CO_2$ and $^{12}CO_2$.
b. Specify which gas diffuses faster.

6.138. At a fixed temperature, how much faster does NO effuse than NO_2?

6.139. One balloon was filled with H_2, another with He. The person responsible for filling them neglected to label them. After 24 h the volumes of both balloons had decreased but by different amounts. Which balloon contained hydrogen?

6.140. Compounds sensitive to oxygen are often manipulated in *glove boxes* that contain a pure nitrogen or pure argon atmosphere. A balloon filled with carbon monoxide was placed in a glove box. After 24 h, the volume of the balloon was unchanged. Did the glove box contain N_2 or Ar?

Real Gases

Concept Review

6.141. Rearrange the van der Waals equation to solve for *P*. Why is the pressure exerted by a real gas lower than the pressure for an ideal gas at the same temperature and volume?

6.142. Under what conditions is the pressure exerted by a real gas *less* than that predicted for an ideal gas?

6.143. The van der Waals equation contains two constants, *a* and *b*, that depend on the identity of the gas. Which gas, Ne or Kr, has a higher value of constant *a*?

6.144. Explain why the constant *a* in the van der Waals equation generally increases with the molar mass of the gas.

Problems

6.145. The graphs of *PV/RT* versus *P* (see Figure 6.39) for 1 mole of CH_4 and 1 mole of H_2 differ in how they deviate from ideal behavior. For which gas is the effect of the volume occupied by the gas molecules more important than the attractive forces between molecules?

6.146. Which noble gas is expected to deviate the most from ideal behavior in a graph of *PV/RT* versus *P*?

6.147. At high pressures, real gases do not behave ideally. (a) Use the van der Waals equation and data in the text to calculate the pressure exerted by 40.0 g of H_2 at 20°C in a 1.00 L container. (b) Repeat the calculation, assuming that the gas behaves like an ideal gas.

6.148. (a) Calculate the pressure exerted by 1.00 mol of CO_2 in a 1.00 L vessel at 300 K, assuming that the gas behaves ideally. (b) Repeat the calculation using the van der Waals equation.

Additional Problems

6.149. The volume of a sample of propane gas at 12.5 atm is 10.6 L. What volume does the gas occupy if the pressure is reduced to 1.05 atm and the temperature remains constant?

6.150. A 22.4 L sample of hydrogen chloride gas is heated from 15°C to 78°C. What volume does it occupy at the higher temperature if the pressure remains constant?

6.151. A gas cylinder in the back of an open truck experiences a temperature change from 12°F to 143°F as it is driven from high in the Rocky Mountains of Colorado to Death Valley, CA. If the pressure gauge on the tank reads 2200 psi in Colorado, what will the gauge read in Death Valley to the nearest psi?

6.152. The temperature of a quantity of methane gas at a pressure of 761 mmHg is 18.6°C. Predict the temperature of the gas if the pressure is reduced to 355 mmHg while the volume remains constant.

6.153. A souvenir soccer ball is partially deflated and then put into a suitcase. At a pressure of 0.947 atm and a temperature of 27°C, the ball has a volume of 1.034 L. What volume does it occupy during an airplane flight if the pressure in the baggage compartment is 0.235 atm and the temperature is −35°C?

6.154. Asthma Therapy A gas mixture used experimentally for asthma treatments contains 17.5 mol of helium for every 0.938 mol of oxygen. What is the mole fraction of oxygen in the mixture?

6.155. Blood Pressure A typical blood pressure in a resting adult is "120 over 80," meaning 120 mmHg with each beat of the heart and 80 mmHg of pressure between heartbeats. Express these pressures in the following units: (a) torr; (b) atm; (c) bar; (d) kPa.

*6.156. Scuba Tanks** A popular scuba tank is the "aluminum 80," so named because it can deliver 80 cubic feet of air at "normal" temperature (72°F) and pressure (1.00 atm, 14.7 psi) when filled with air at a pressure of 3000 psi. A particular aluminum 80 tank has a mass of 15 kg when empty. What is its mass when filled with air at 3000 psi?

6.157. The flame produced by the burner of a gas (propane) grill is a pale blue color when enough air mixes with the propane (C_3H_8) to burn it completely. For every gram of propane that flows through the burner, what volume of air is needed to burn it completely? Assume that the temperature of the burner is 200°C, the pressure is 1.00 atm, and the mole fraction of O_2 in air is 0.21.

6.158. Which noble gas effuses at about half the effusion rate of O_2?

6.159. **Anesthesia** A common anesthesia gas is halothane, with the formula $C_2HBrClF_3$ and the structure shown in Figure P6.159. Liquid halothane boils at 50.2°C and 1.00 atm. If halothane behaved as an ideal gas, what volume would 10.0 mL of liquid halothane ($d = 1.87$ g/mL) occupy at 60°C and 1.00 atm of pressure? What is the density of halothane vapor at 55°C and 1.00 atm of pressure?

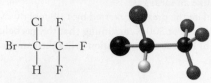

FIGURE P6.159

6.160. One cotton ball soaked in ammonia and another soaked in hydrochloric acid were placed at opposite ends of a 1.00 m glass tube (Figure P6.160). The vapors diffused toward the middle of the tube and formed a white ring of ammonium chloride where they met.
   a. Write the chemical equation for this reaction.
   b. Will the ammonium chloride ring form closer to the end of the tube with ammonia or the end with hydrochloric acid? Explain your answer.
   c. Calculate the distance from the ammonia end to the position of the ammonium chloride ring.

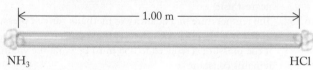

NH_3 HCl
FIGURE P6.160

*6.161. The same apparatus described in Problem 6.160 was used in another series of experiments. A cotton ball soaked in either hydrochloric acid (HCl) or acetic acid (CH_3COOH) was placed at one end. Another cotton ball soaked in one of three amines—CH_3NH_2, $(CH_3)_2NH$, or $(CH_3)_3N$—was placed in the other end (Figure P6.161).
   a. In one combination of acid and amine, a white ring was observed almost exactly halfway between the two ends. Which acid and which amine were used?
   b. Which combination of acid and amine would produce a ring closest to the amine end of the tube?
   c. Do any two of the six combinations result in the formation of product at the same position in the ring? Assume measurements can be made to the nearest centimeter.

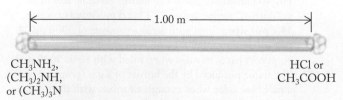

CH_3NH_2, HCl or
$(CH_3)_2NH$, CH_3COOH
or $(CH_3)_3N$
FIGURE P6.161

6.162. **Liquid Nitrogen–Powered Car** Students at the University of North Texas and the University of Washington built a car propelled by compressed nitrogen gas. The gas was obtained by boiling liquid nitrogen stored in a 182-L tank.

What volume of N_2 is released at 0.927 atm of pressure and 25°C from a tank full of liquid N_2 ($d = 0.808$ g/mL)?

6.163. The pressure in an aerosol can is 1.5 atm at 27°C. The can will withstand a pressure of about 2 atm. Will it burst if heated in a campfire to 450°C?

6.164. **Hydrogen-Powered Vehicles** Combustion of hydrogen supplies a great deal of energy per gram. One challenge in designing hydrogen-powered vehicles is storing hydrogen. The pressure limit for carbon fiber–reinforced gas cylinders is 10,000 psi (1 atm = 14.7 psi). A fuel tank capable of holding up to 10 kg of H_2 is required to meet a reasonable driving range for a vehicle. If we limit the volume of the fuel tank to 400 L, what pressure must the hydrogen be under at 300 K? Does your answer conform to the safety requirements?

6.165. A sample of 11.4 L of an ideal gas at 25.0°C and 735 torr is compressed and heated so that the volume is 7.9 L and the temperature is 72.0°C. What is the pressure in the container?

6.166. A sample of argon gas at STP occupies 15.0 L. What mass of argon is present in the container?

6.167. A sample of a gas has a mass of 2.889 g and a volume of 940 mL at 735 torr and 31°C. What is its molar mass?

6.168. Uranus has a total atmospheric pressure of 130 kPa and its atmosphere consists of the following gases: 83% H_2, 15% He, and 2% CH_4 by volume. Calculate the partial pressure of each gas in Uranus's atmosphere.

6.169. A sample of $N_2(g)$ requires 240 s to diffuse through a porous plug. It takes 530 s for an equal number of moles of an unknown gas X to diffuse through the plug under the same conditions of temperature and pressure. What is the molar mass of gas X?

6.170. The rate of effusion of an unknown gas is 0.10 m/s and the rate of effusion of $SO_3(g)$ is 0.052 m/s under identical experimental conditions. What is the molar mass of the unknown gas?

6.171. Derive an equation that describes the relationship between root-mean-square speed, u_{rms}, of a gas and its density.

6.172. Derive an equation that expresses the ratio of the densities (d_1 and d_2) of a gas under two different combinations of temperature and pressure, (T_1, P_1) and (T_2, P_2).

6.173. **Denitrification in the Environment** In some aquatic ecosystems, nitrate (NO_3^-) is converted to nitrite (NO_2^-), which then decomposes to nitrogen and water. As an example of this second reaction, consider the decomposition of ammonium nitrite:

$$NH_4NO_2(aq) \rightarrow N_2(g) + 2\,H_2O(\ell)$$

What would be the change in pressure in a sealed 10.0 L vessel due to the formation of N_2 gas when the ammonium nitrite in 1.00 L of 1.0 M NH_4NO_2 decomposes at 25°C?

*6.174. When sulfur dioxide bubbles through a solution containing nitrite, chemical reactions that produce gaseous N_2O and NO may occur.
   a. How much faster, on average, would NO molecules be moving than N_2O molecules in such a reaction mixture?
   b. If these two nitrogen oxides were to be separated based on differences in their rates of effusion, would unreacted SO_2 interfere with the separation? Explain your answer.

*6.175. Using Wetlands to Treat Agricultural Waste Wetlands can play a significant role in removing fertilizer residues from rain runoff and groundwater; one way they do this is through denitrification, which converts nitrate ions to nitrogen gas:

$$2\,NO_3^-(aq) + 5\,CO(g) + 2\,H^+(aq) \rightarrow N_2(g) + H_2O(\ell) + 5\,CO_2(g)$$

Suppose 200.0 g of NO_3^- flows into a swamp each day. What volume of N_2 would be produced at 17°C and 1.00 atm if the denitrification process were complete? What volume of CO_2 would be produced? Suppose the gas mixture produced by the decomposition reaction is trapped in a container at 17°C; what is the density of the mixture if we assume that $P_{total} = 1.00$ atm?

6.176. Ammonium nitrate decomposes on heating. The products depend on the reaction temperature:

$$NH_4NO_3(s) \xrightarrow{>300°C} N_2(g) + \tfrac{1}{2}O_2(g) + 2\,H_2O(g)$$
$$\xrightarrow{200-260°C} N_2O(g) + 2\,H_2O(g)$$

A sample of NH_4NO_3 decomposes at an unspecified temperature, and the resulting gases are collected over water at 20°C.

a. Without completing a calculation, predict whether the volume of gases collected can be used to distinguish between the two reaction pathways. Explain your answer.

*b. The gas produced during the thermal decomposition of 0.256 g of NH_4NO_3 displaces 79 mL of water at 20°C and 760 mmHg of atmospheric pressure. Is the gas N_2O or a mixture of N_2 and O_2?

6.177. Hydrogen-Producing Enzymes Generating hydrogen from water or methane is energy intensive. A non-natural enzymatic process has been developed that produces 12 moles of hydrogen per mole of glucose by the reaction:

$$C_6H_{12}O_6(aq) + 6\,H_2O(\ell) \rightarrow 12\,H_2(g) + 6\,CO_2(g)$$

What volume of hydrogen could be produced from 256 g of glucose at STP?

*6.178. Oxygen Generators Several devices are available for generating oxygen. Breathing pure oxygen for any length of time, however, can be dangerous to the human body, so a better rebreathing apparatus would be one that also produces nitrogen. One possible reaction that produces both N_2 and O_2 is the thermal decomposition of ammonium nitrate:

$$2\,NH_4NO_3(s) \rightarrow 2\,N_2(g) + O_2(g) + 4\,H_2O(g)$$

a. The respiratory rate at rest for an average, healthy adult is 12 breaths per minute. If the average breath takes in 500 mL of gas into our lungs, what mass in grams of

NH_4NO_3 is needed to satisfy these needs for 1 h if $P = 1.00$ atm and $T = 37$°C?

b. What are the partial pressures of N_2 and O_2 in this mixture?

6.179. Tropical Storms The severity of a tropical storm is related to the depressed atmospheric pressure at its center. In August 1985, Typhoon Odessa reached maximum winds of about 90 mi/h and the pressure was 40 mbar lower at the center than normal atmospheric pressure. In contrast, the central pressure of Hurricane Andrew (Figure P6.179) was 90 mbar lower than its surroundings when it hit southern Florida with winds as high as 165 mi/h. If a small weather balloon with a volume of 50.0 L at a pressure of 1.0 atmosphere was deployed above the center of Andrew, what was the volume of the balloon when it reached the surface of the ocean?

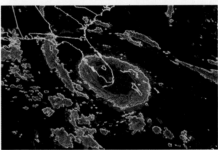

FIGURE P6.179

6.180. Manufacturing a Proper Air Bag Here is the overall reaction in an automobile air bag:

$$20\,NaN_3(s) + 6\,SiO_2(s) + 4\,KNO_3(s) \rightarrow$$
$$32\,N_2(g) + 5\,Na_4SiO_4(s) + K_4SiO_4(s)$$

Calculate how many grams of sodium azide (NaN_3) are needed to inflate a $40 \times 40 \times 20$ cm bag to a pressure of 1.25 atm at a temperature of 20°C. How much more sodium azide is needed if the air bag must produce the same pressure at 10°C?

6.181. 0.200 L of O_2 is collected over water at 25.0°C. The atmospheric pressure is 750.0 torr. The vapor pressure of water at 25.0°C is 24.0 torr. How many moles of O_2 have been collected?

6.182. Use Dalton's law of partial pressures to calculate the mole fraction of water vapor in the gas sample collected over water in Problem 6.181.

7

A Quantum Model of Atoms

Waves, Particles, and Periodic Properties

RAINBOWS When sunlight passes through water droplets in the sky, a rainbow is often the result. Scientists' efforts to explain why energy (such as sunlight) interacted with matter (such as raindrops) to produce the colors of a rainbow led to the development of quantum theory and an entirely new way of viewing matter at the atomic level.

PARTICULATE **REVIEW**

Atoms, Ions, and Their Electrons

In Chapter 7 we explore periodic trends regarding the sizes of atoms and monatomic ions, along with the energy changes involved with forming ions from neutral parent atoms.

- The spheres shown here represent the sizes and charges of the most common ions formed by potassium, aluminum, fluorine, and sulfur. Match each element to its ion.

- How many electrons are in a neutral atom of each of these elements?

- How many electrons are in each ion shown here?

    (Review Sections 2.3 and 2.6 if you need help.)

(Answers to Particulate Review questions are in the back of the book.)

Energy Changes and Ion Formation

In Chapter 5 we learned that separating two charged particles is an endothermic process. As you read Chapter 7, look for ideas that will help you answer these questions:

- Is an electron gained or lost in each of the two processes depicted here?

- Which process usually emits energy?

- Which process requires the absorption of energy?

331

Learning Outcomes

LO1 Describe the different forms of electromagnetic radiation and relate their wavelengths and frequencies
Sample Exercise 7.1

LO2 Describe quantum theory and use it to explain the photoelectric effect
Sample Exercises 7.2, 7.3

LO3 Relate the wavelengths of atomic spectral lines to the energies gained and lost during electron transitions within atoms
Sample Exercises 7.4, 7.5

LO4 Apply the concept of matter waves to electrons in atoms and other objects in motion
Sample Exercise 7.6

LO5 Describe and apply the Heisenberg uncertainty principle
Sample Exercise 7.7

LO6 Assign quantum numbers to orbitals and use their values to describe the size, energy, and orientation of orbitals
Sample Exercises 7.8, 7.9

LO7 Use the aufbau principle and Hund's rule to write electron configurations and draw orbital diagrams of atoms and monatomic ions
Sample Exercises 7.10, 7.11, 7.12, 7.13

LO8 Relate the ionization energies and electron affinities of elements and the sizes of their atoms and monatomic ions to their atomic structures and the concept of effective nuclear charge
Sample Exercises 7.14, 7.15

7.1 Rainbows of Light

FIGURE 7.1 Isaac Newton used a prism to separate sunlight into a spectrum containing all the colors of the rainbow.

Rainbows can put visual exclamation points on passing summer showers. They often appear just after the clouds that brought the shower begin to clear, allowing sunlight to illuminate the trailing edge of the receding rain. Have you ever noticed that rainbow colors are nearly always arranged in the same pattern: red on top, violet on the bottom, and all of the other colors in between? Have you ever wondered why?

The answer has to do with the ways that radiant energy (sunlight) interacts with matter (raindrops). Such interactions are a recurring theme in this chapter. Although humans have enjoyed the colors of rainbows ever since we evolved as an intelligent species, it was only a few hundred years ago that a brilliant Englishman named Isaac Newton (1642–1727) was able to explain how and why rainbows appear. He discovered that passing sunlight through a glass prism (Figure 7.1) separated the light into its component colors in much the way raindrops do. Then he showed how the colors could be recombined into white sunlight with a second prism, thereby proving that white light is a blend of all of the colors in a rainbow.

In the decades following Newton's discoveries, the quality of prisms improved dramatically, enabling scientists in Europe and Great Britain to separate sunlight into its component colors with increasing resolution. Then, in 1800 another English scientist named William Hyde Wollaston (1766–1828) made a second discovery about the sun's rainbow of colors: it was not continuous. Rather, it

Fraunhofer lines a set of dark lines in the otherwise continuous solar spectrum.

atomic emission spectrum a characteristic series of bright lines produced by high-temperature atoms.

atomic absorption spectrum a characteristic series of dark lines produced when free, gaseous atoms are illuminated by a continuous source of radiation.

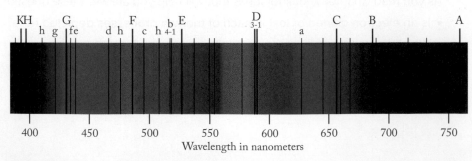

FIGURE 7.2 The spectrum of sunlight contains many narrow gaps, which appear as dark lines called Fraunhofer lines.

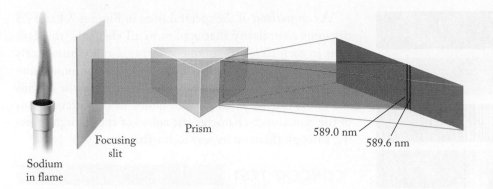

FIGURE 7.3 Sodium atoms heated in a flame produce a bright yellow-orange light that is actually due to two closely spaced emission lines at 589.0 nm and 589.6 nm.

(a) Visible emission spectrum of hydrogen

(b) Visible emission spectrum of helium

(c) Visible emission spectrum of neon

FIGURE 7.4 Atoms of elements such as (a) hydrogen, (b) helium, and (c) neon emit characteristic line spectra in gas discharge tubes.

contained dark, narrow lines. Using even better prisms, the German physicist Joseph von Fraunhofer (1787–1826) resolved more than 500 of these lines, which are now called **Fraunhofer lines** (Figure 7.2). He labeled the darkest lines with the letters of the alphabet, starting with A at the red end of the Sun's spectrum.

Labeling his lines, however, did not mean that Fraunhofer, or anyone else, knew why they existed. Discovering why would come later in the 19th century with the work of two more German scientists: the chemist Robert Wilhelm Bunsen (1811–1899) and physicist Gustav Robert Kirchhoff (1824–1887). They collaborated on extensive studies of the light emitted (given off) by elements, particularly group 1 and 2 elements, when they are vaporized and their atoms are heated by the transparent flames produced by a burner designed by Bunsen. Unlike the spectrum of sunlight, which displays narrow gaps in an otherwise continuous spectrum of all colors, the spectra produced by hot atoms in flames consist of only a few bright lines on a dark background, as shown in Figures 7.3 and 7.4. Bunsen and Kirchhoff discovered that many of the lines in these **atomic emission spectra** exactly matched colors that were missing in the Sun's visible spectrum. For example, the Fraunhofer D line corresponded to the yellow-orange light produced by hot sodium vapor (Figure 7.3).

These experiments, and others employing light sources called gas discharge tubes, showed that each element emits a characteristic line spectrum when its atoms are heated to a sufficiently high temperature (Figure 7.4). Other experiments showed that when free atoms of an element are illuminated by a continuous source of light, such as that emitted by a light bulb or the surface of the Sun, the atoms may absorb particular colors of light, producing an **atomic absorption spectrum** consisting of dark lines in an otherwise continuous spectrum (Figure 7.5).

Absorption spectrum of hydrogen

Absorption spectrum of helium

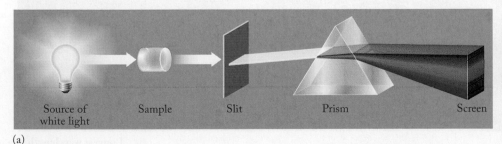

(a)

FIGURE 7.5 (a) When gaseous atoms of hydrogen, helium, and neon are illuminated by an external source of white light (containing all colors of the visible spectrum), the resultant atomic absorption spectra contain dark lines that are characteristic of each element. (b) The dark lines in the atomic absorption spectra of these elements match the bright lines in their atomic emission spectra, shown in Figure 7.4.

Absorption spectrum of neon

(b)

FIGURE 7.6 Atomic emission spectrum of mercury vapor (top) and four atomic absorption spectra (a–d).

A comparison of the spectral lines in Figures 7.4 and 7.5 discloses a similarity that applies to all elements: the dark lines in each element's atomic absorption spectrum exactly match the colors of the lines in that element's atomic emission spectrum. Atomic absorption also explains the origins of the Fraunhofer lines: gaseous atoms in the outer regions of the Sun absorb characteristic colors of the sunlight passing through them on its way to Earth.

CONCEPT **TEST**

The top spectrum in Figure 7.6 is the emission spectrum of mercury vapor. Select the absorption spectrum of mercury vapor from the other four spectra.

(Answers to Concept Tests are in the back of the book.)

In this chapter we explore why the atoms of the elements have unique and complementary atomic emission and absorption spectra. We also link the positions of the lines in these spectra to the internal structure of atoms and, in

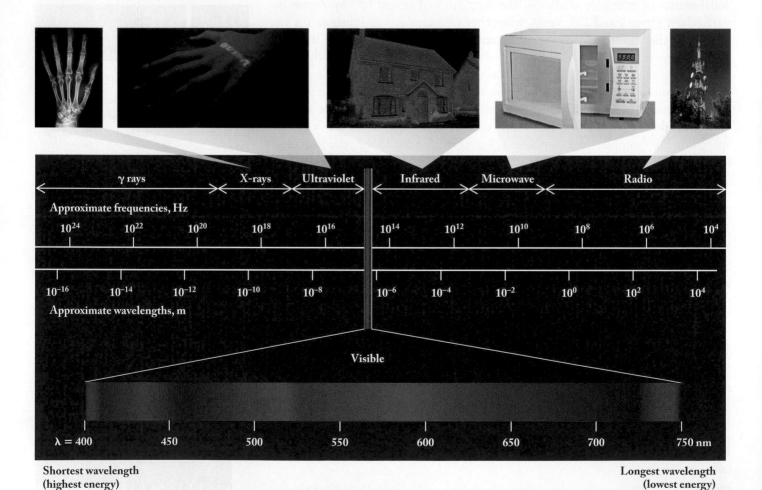

FIGURE 7.7 Visible light occupies a tiny fraction of the electromagnetic spectrum, which ranges from ultrashort-wavelength, high-frequency gamma (γ) rays to long-wavelength, low-frequency radio waves. Note that frequencies increase from right to left, and wavelengths increase from left to right.

particular, to the arrangement and motion of the electrons inside atoms. We begin this exploration of atomic structure by addressing the properties of sunlight and other invisible forms of radiant energy, and how the nature of radiant energy allows it to interact with matter.

7.2 Waves of Energy

Visible light is a small part of the **electromagnetic spectrum** (Figure 7.7). This spectrum is a continuous range of radiant energy that extends from low-energy radio waves to ultrahigh-energy gamma rays. All forms of radiant energy are examples of **electromagnetic radiation**.

The term *electromagnetic* comes from a theory, proposed by the Scottish scientist James Clerk Maxwell (1831–1879), that radiant energy moves through space (or any transparent medium) in a way that resembles waves flowing across a body of water. Unlike water waves, which oscillate only up and down, Maxwell's waves of radiant energy have two components: an oscillating electric field and an oscillating magnetic field (Figure 7.8). These two fields are perpendicular to each other and travel together through space. Maxwell derived a set of equations based on his oscillating-wave model that accurately describes nearly all the observed properties of light.

A wave of electromagnetic radiation, like any wave traveling through any medium, has a characteristic **wavelength** (**λ**, the distance from crest to crest) and **frequency** (**ν**, the number of crests that pass a stationary point of reference per second), as shown in Figure 7.9(a). Frequencies have units of **hertz (Hz)**, also called *cycles per second* (cps): 1 Hz = 1 cps = 1/s. The product of the wavelength and frequency of any electromagnetic radiation is the universal constant c, which is the symbol for the speed of light in a vacuum (2.998×10^8 m/s):

$$c = \lambda\nu \tag{7.1}$$

Thus wavelength and frequency have a reciprocal relationship: as wavelength decreases, frequency increases. Another characteristic of a wave is its **amplitude**, the height of the crest or the depth of the trough with respect to the center line of the wave (Figure 7.9b).

CHEMTOUR
Electromagnetic Radiation

CONNECTION In Chapter 1 we discussed the background microwave radiation of the universe, discovered by Penzias and Wilson, that provided evidence for the Big Bang.

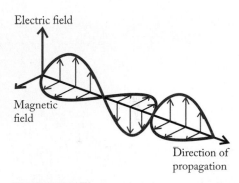

FIGURE 7.8 Electromagnetic waves consist of electric and magnetic fields that oscillate in planes oriented at right angles to each other.

electromagnetic spectrum a continuous range of radiant energy that includes radio waves, microwaves, infrared radiation, visible light, ultraviolet radiation, X-rays, and gamma rays.

electromagnetic radiation any form of radiant energy in the electromagnetic spectrum.

wavelength (λ) the distance from crest to crest or trough to trough on a wave.

frequency (ν) the number of crests of a wave that pass a stationary point of reference per second.

hertz (Hz) the SI unit of frequency, equivalent to 1 cycle per second, or simply 1/s.

amplitude the height of the crest or depth of the trough of a wave with respect to the center line of the wave.

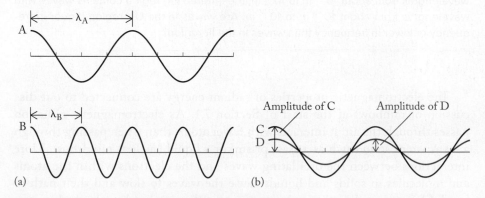

FIGURE 7.9 Every wave has a characteristic wavelength (λ), frequency (ν), and amplitude (intensity). (a) Wave A has a longer wavelength (and lower frequency) than wave B. (b) Waves C and D have the same wavelength and frequency, but the amplitude of C is greater than that of D.

SAMPLE EXERCISE 7.1 Calculating Frequency from Wavelength   **LO1**

What is the frequency of the yellow-orange light (λ = 589 nm) produced by sodium-vapor streetlights?

Collect and Organize We are asked to find the frequency of light that has a wavelength of 589 nm. Frequency and wavelength are related by Equation 7.1 ($c = \lambda\nu$), where c is the speed of light = 2.998×10^{8} m/s.

Analyze Before using Equation 7.1, we need to convert the wavelength of the light from nanometers to meters (1 nm = 10^{-9} m). The unit labels in Figure 7.7 indicate that frequencies of visible light are in the 10^{14} Hz range, so a correct answer should have a value in this range.

Solve Rearrange Equation 7.1 to solve for frequency:

$$\nu = \frac{c}{\lambda}$$

Convert wavelength from nanometers to meters:

$$589 \ \text{nm} \times \frac{10^{-9} \ \text{m}}{1 \ \text{nm}} = 589 \times 10^{-9} \ \text{m} = 5.89 \times 10^{-7} \ \text{m}$$

Use this wavelength value to calculate frequency:

$$\nu = \frac{2.998 \times 10^{8} \ \text{m/s}}{5.89 \times 10^{-7} \ \text{m}}$$
$$= 5.09 \times 10^{14} \ \text{s}^{-1} = 5.09 \times 10^{14} \ \text{Hz}$$

Think About It Our calculated frequency is in the range we expected. Remember that wavelengths and frequencies are inversely proportional to each other: as one increases, the other decreases.

 Practice Exercise The radio waves transmitted by a radio station have a frequency of 90.9 MHz. What is the wavelength of these waves in meters?

(Answers to Practice Exercises are in the back of the book.)

CONCEPT TEST

The ultraviolet (UV) region of the electromagnetic spectrum contains waves with wavelengths from about 10^{-9} m to 10^{-7} m; the infrared (IR) region contains waves with wavelengths from about 10^{-6} m to 10^{-4} m. Are waves in the UV region higher in frequency or lower in frequency than waves in the IR region?

The electromagnetic properties of radiant energy are connected to our discussion of rainbows at the start of Section 7.1. As electromagnetic radiation passes through the air, it interacts with fewer atoms than when passing through a transparent solid (such as a glass prism) or a liquid (such as raindrops). More interactions between the oscillating waves and the electrons within the atoms and molecules in solids and liquids cause the waves to slow and their path to bend. Shorter wavelengths are bent more than longer wavelengths as they pass between the air and the raindrops, causing violet to always appear on the bottom of rainbows, while red (which has longer wavelengths than violet) always appears at the top.

7.3 Particles of Energy and Quantum Theory

As studies of electromagnetic radiation progressed in the 19th century, scientists discovered limitations to the wave model when describing radiation. One limitation was the inability to account for radiation given off by very hot objects. Consider, for example, what happens when a metal rod is heated in a flame. At first the rod gives off only heat in the form of infrared radiation, which we can feel but not see. As the temperature of the metal is increased from 500 to 1000 K (Figure 7.10a), the most intense radiation is still in the infrared region of the spectrum, but a small fraction of the emitted light is in the red region of the visible spectrum, as shown in Figure 7.10(b).

quantum (plural *quanta*) the smallest discrete quantity of a particular form of energy.

quantum theory a model based on the idea that energy is absorbed and emitted in discrete quantities of energy called quanta.

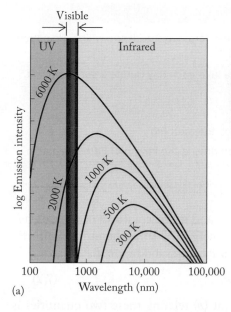

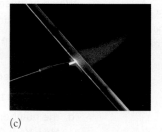

(b)

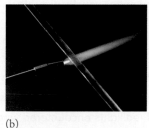

(c)

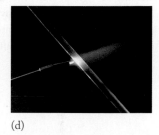

(d)

(a)

FIGURE 7.10 (a) Plots of emission intensity versus wavelength show that as the temperature of a solid increases, the intensity of the radiation that it emits increases and the wavelength of maximum intensity decreases. As a metal rod is heated, (b) it first glows red, (c) then turns orange, and (d) finally becomes white-hot as light of all colors is emitted.

With still more heating, the intensity of the red glow increases; and the color of the rod begins to shift to red-orange, orange, yellow (Figure 7.10c), and eventually to white (Figure 7.10d) as the metal emits all wavelengths of visible light. Even white-hot metal emits little radiation in the ultraviolet region and none at even shorter wavelengths.

This phenomenon was well known at the end of the 19th century. However, none of Maxwell's equations for electromagnetic radiation could account for the emission spectrum of a heated metal filament. Another explanation—indeed, another model of radiation behavior—was required.

Quantum Theory

In 1900 the German scientist Max Planck (1858–1947) proposed such a model. Planck proposed that light and all other forms of electromagnetic radiation have not only wavelike properties, as Maxwell had proposed, *but also*, at the atomic level, particle-like properties. Planck (Figure 7.11) called a particle of radiant energy a **quantum** (plural *quanta*). In his model, which we call **quantum theory**, quanta are the smallest amounts of radiant energy in nature, the sort that a single atom or molecule might absorb or emit. An object made of a large, but discrete,

FIGURE 7.11 German scientist Max Planck is considered the father of quantum physics. He won the 1918 Nobel Prize in Physics for his pioneering work on the quantized nature of electromagnetic radiation. Planck was revered by his colleagues for his personal qualities as well as his scientific accomplishments.

FIGURE 7.12 Quantized and unquantized heights. A flight of stairs exemplifies quantization: each step rises by a discrete height to the next step. In contrast, the rise on a ramp is not quantized, but rather is continuous.

number of atoms or molecules could therefore emit only a discrete number of quanta. In other words, light from such an object must be **quantized**. A quantum of light is called a **photon**, derived from the Greek *photo*, meaning "light."

To visualize the meaning of Planck's quanta, consider two ways you might get from the sidewalk to the entrance of a building (Figure 7.12). If you walked up the steps, you would be able to stand at only discrete heights above the sidewalk—each height equal to the rise of a single step. You could not stand at a height between two adjacent steps because there would be nothing to stand on at that height. If you walked up the ramp, however, you could stop at any height between the sidewalk and the entrance. The discrete height changes represented by the steps are also a model of Planck's hypothesis that energy is released (analogous to walking down the steps) or absorbed (walking up the steps) in discrete packets, or quanta, of energy.

CONCEPT TEST

Which of these quantities vary by discrete values (are quantized) and which are continuous (not quantized)?

a. the volume of water that evaporates from a lake each day during a summer heat wave

b. the number of eggs remaining in a carton

c. the time it takes you to get ready for class in the morning

d. the number of red lights encountered when driving the length of Fifth Avenue in New York City

Quantum theory also connects the wave and particle natures of electromagnetic radiation. Planck proposed that the energy (E) of a photon was directly proportional to the frequency of the wave of its corresponding radiation:

$$E = h\nu \tag{7.2}$$

The value of the proportionality constant (h) relating these two quantities is 6.626×10^{-34} J · s. It is now called the **Planck constant**. When we solve Equation 7.1, $c = \lambda\nu$, for ν and substitute into Equation 7.2, we find that the energy of a photon is inversely proportional to its wavelength:

$$E = \frac{hc}{\lambda} \tag{7.3}$$

quantized having values restricted to whole-number multiples of a specific base value.

photon a quantum of electromagnetic radiation.

Planck constant (h) the proportionality constant between the energy and frequency of electromagnetic radiation expressed in $E = h\nu$; $h = 6.626 \times 10^{-34}$ J · s.

SAMPLE EXERCISE 7.2 Calculating the Energy of a Photon   **LO2**

Chlorophyll *b* absorbs light with wavelengths of 462 nm and 647 nm. What are the energies of photons with these wavelengths? What colors of light does chlorophyll *b* absorb?

Collect, Organize, and Analyze We are asked to calculate the energies of photons with two different wavelengths. To use Equation 7.3, which relates these quantities, we need to use the value of the Planck constant (h), which is 6.626×10^{-34} J · s, and the speed of light (c), 2.998×10^{8} m/s. The photons' wavelengths are given in nanometers, but this value of c has units of meters per second, so we need to convert nanometers to meters for the distance units to cancel. The value of h is extremely small, so it is likely that the results of our calculation, even factoring in the speed of light, will be very small too.

Solve

$$E = \frac{hc}{\lambda} = \frac{(6.626 \times 10^{-34}\,\text{J}\cdot\text{s})\left(2.998 \times 10^8\,\frac{\text{m}}{\text{s}}\right)}{462\,\text{nm} \times \dfrac{10^{-9}\,\text{m}}{1\,\text{nm}}}$$

$$= 4.30 \times 10^{-19}\,\text{J}$$

$$E = \frac{hc}{\lambda} = \frac{(6.626 \times 10^{-34}\,\text{J}\cdot\text{s})\left(2.998 \times 10^8\,\frac{\text{m}}{\text{s}}\right)}{647\,\text{nm} \times \dfrac{10^{-9}\,\text{m}}{1\,\text{nm}}}$$

$$= 3.07 \times 10^{-19}\,\text{J}$$

We can use Figure 7.7 to determine which colors of light are absorbed. A wavelength of 462 nm is blue light, and $\lambda = 647$ nm is orange light.

Think About It These quantities of energy are extremely small, as they should be, because a photon is an atomic-level particle of radiant energy. The results of the calculations confirm that a photon of orange light has less energy than a photon of blue light.

Practice Exercise Leaves contain an array of different light-absorbing molecules to harvest the full spectrum of visible light. Some of these include β-carotene (the compound that makes carrots orange), which absorbs at $\lambda = 453$ nm and 482 nm. How much energy do single photons of 453-nm and 482-nm light have?

The Photoelectric Effect

Although Planck's quantum model explained the emission spectra of hot objects, there was no experimental evidence in 1900 to support the existence of quanta of energy. In 1905 Albert Einstein (1879–1955) supplied that evidence. It came from his studies of a phenomenon called the **photoelectric effect**, in which electrons are emitted from metals when they are illuminated by and absorb electromagnetic radiation. Because light releases these electrons, they are called *photoelectrons*.

Photoelectrons are emitted by a material when the frequency of incident radiation is above a minimum **threshold frequency (ν_0)** characteristic of that material (Figure 7.13). Radiation at frequencies less than the threshold value produces no photoelectrons, no matter how intense the radiation is. On the other hand, even a dim source of radiant energy produces at least a few photoelectrons when the frequencies it emits are equal to, or greater than, the threshold frequency.

Einstein used Planck's quantum theory to explain this behavior. He proposed that the threshold frequency is the frequency of the minimum quantum of absorbed energy needed to remove a single electron from the surface of a material. This minimum quantity of energy is related to the strength of the attraction between the nuclei of surface atoms and the electrons surrounding them, and it is called the material's **work function (φ)**:

$$\phi = h\nu_0 \tag{7.4}$$

If a photoelectric material is illuminated with radiation frequencies above the threshold frequency ($\nu > \nu_0$), any energy in excess of φ is imparted to each ejected electron as kinetic energy:

$$\text{KE}_{\text{electron}} = h\nu - h\nu_0 = h\nu - \phi \tag{7.5}$$

The higher the frequency of the incident light above the threshold frequency, the higher the kinetic energy, and hence the velocity, of the ejected electrons.

photoelectric effect the phenomenon of light striking a metal surface and producing an electric current (a flow of electrons).

threshold frequency (ν_0) the minimum frequency of light required to produce the photoelectric effect.

work function (φ) the amount of energy needed to remove an electron from the surface of a metal.

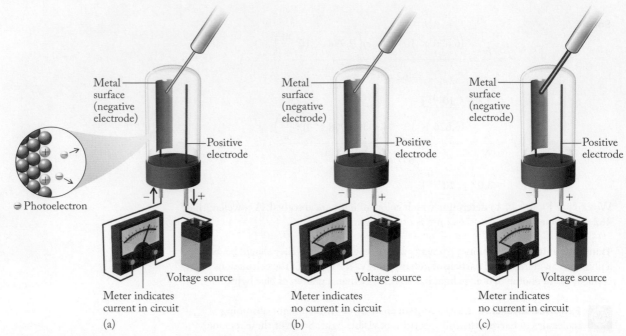

(a)   (b)   (c)

FIGURE 7.13 A phototube includes a positive electrode and a negative metal electrode. (a) If radiation of high enough frequency and energy (violet light in this illustration) illuminates the negative electrode, electrons are dislodged from the surface and flow toward the positive electrode. This flow of electrons produces an electric current that is detected by the meter. The size of the current is proportional to the intensity of the radiation—to the number of photons per unit time striking the negative electrode. (b) The frequency of the red light is below the threshold frequency, so it cannot dislodge electrons and does not produce the photoelectric effect. (c) Even if many low-frequency photons bombard the surface of the metal, no electrons are emitted, and no electrical current flows.

Wave–Particle Duality

Einstein's explanation of the photoelectric effect was of such profound significance that he was awarded the Nobel Prize for it in 1921. Why does the concept of photons have such far-reaching significance in science? Classical physics had identified the wave nature of light but could not account for light behaving as a particle. Only quantum physics recognizes the **wave–particle duality** of light: light behaves as both a wave *and* a particle.

Photosynthesis depends on the absorption of photons of a particular wavelength by chlorophyll molecules. These molecules use the energy from the light to power chemical reactions that convert carbon dioxide and water to glucose and oxygen. The photoelectric effect itself finds its way into our daily lives through technology such as TV remote control devices and motion sensors. When we press the power button on a TV remote, the remote emits a beam of infrared light that impinges on the "electric eye" in the television set. The resulting emission of electrons signals the television to turn on or off. The same principle applies to photosensors found in some motion detectors. In this case, a beam of light shines continually on a detector that emits electrons and causes a current to flow. If anything interrupts the light beam, the current drops. A door may open or an alarm may be triggered as a result of the drop in current. Other applications of the photoelectric effect include turning on an automatic faucet or hand dryer in a public restroom.

wave–particle duality the behavior of an object that exhibits the properties of both a wave and a particle.

SAMPLE EXERCISE 7.3 Applying the Photoelectric Effect   **LO2**

Can germanium be used to detect infrared radiation emitted by a remote control device that has a wavelength of 902 nm? The work function of germanium is 7.61×10^{-19} J.

Collect, Organize, and Analyze We know the work function of germanium and a wavelength of infrared radiation, and we are asked to determine whether the radiation will dislodge photoelectrons from germanium. Equation 7.4 relates a material's work function (ϕ) to its threshold frequency (ν_0). To use Equation 7.4 to solve this problem, we need to convert wavelength to frequency, which we can do using Equation 7.1.

Solve Calculate the frequency of a 902-nm photon:

$$\lambda\nu = c = 2.998 \times 10^8 \text{ m/s}$$

$$\nu = \frac{c}{\lambda} = \frac{2.998 \times 10^8 \text{ m/s}}{902 \text{ nm} \times \dfrac{10^{-9} \text{ m}}{1 \text{ nm}}} = 3.33 \times 10^{14} \text{ s}^{-1}$$

Rearrange Equation 7.4 to solve for the threshold frequency for germanium:

$$\phi = h\nu_0$$

$$\nu_0 = \frac{\phi}{h} = \frac{7.61 \times 10^{-19} \text{ J}}{6.626 \times 10^{-34} \text{ J} \cdot \text{s}} = 1.15 \times 10^{15} \text{ s}^{-1}$$

The frequency of the radiation is lower than the threshold frequency. Therefore, it will not liberate photoelectrons, and germanium cannot be used to detect it.

Think About It Another way to solve this problem would have been to use Equation 7.3 to determine the energy of a photon of 902-nm radiation. If this energy were equal to or greater than the work function, germanium could be used to detect the radiation. According to the results of the calculation above, the energy of a 902-nm photon is less than the work function of germanium.

Practice Exercise The work function of silver is 7.59×10^{-19} J. What is the longest wavelength (in nanometers) of electromagnetic radiation that can eject an electron from the surface of a piece of silver?

CONCEPT **TEST**

Consult Figure 7.7 to answer this question without making a calculation. If a photon of orange light has sufficient energy to eject a photoelectron from the surface of a metal, does a photon of green light have enough energy to do so?

7.4 The Hydrogen Spectrum and the Bohr Model

In formulating his quantum theory, Planck was influenced by the results of investigating the emission spectra produced by free (gas-phase) atoms, which led him to question whether any spectrum, even that of an incandescent light bulb, was truly continuous. Among these earlier results was a discovery made in 1885 by a Swiss mathematician and schoolteacher named Johann Balmer (1825–1898).

The Hydrogen Emission Spectrum

Balmer determined that the wavelengths of the four brightest lines in the visible region of the emission spectrum of hydrogen (Figure 7.4a) fit the simple equation

$$\lambda = 364.5 \text{ nm}\left(\frac{m^2}{m^2 - n^2}\right) \tag{7.6}$$

where n is 2 and m is a whole number greater than 2—specifically, 3 for the red line, 4 for the green, 5 for the blue, and 6 for the violet. Without having seen any other lines in hydrogen's visible emission spectrum, Balmer predicted there should be at least one more ($n = 7$) at the edge of the violet region, and indeed such a line was later discovered.

Balmer also predicted that a series of hydrogen emission lines should exist in regions outside the visible range, lines with wavelengths calculated by replacing $n = 2$ in Equation 7.6 with $n = 1, 3, 4$, and so forth. He was right. In 1908 the German physicist Friedrich Paschen (1865–1947) discovered hydrogen emission lines in the infrared region, corresponding to $n = 3$ in Balmer's equation. A few years later, Theodore Lyman (1874–1954) at Harvard University discovered hydrogen emission lines in the UV region corresponding to $n = 1$. By the 1920s, the $n = 4$ and $n = 5$ series of emission lines also had been discovered. Like the $n = 3$ lines, they are in the infrared region.

In 1888 the Swedish physicist Johannes Robert Rydberg (1854–1919) revised Balmer's equation, changing wavelength to *wave number* ($1/\lambda$), which is the number of wavelengths per unit of distance. Rydberg's equation is

$$\frac{1}{\lambda} = [1.097 \times 10^{-2}\ \text{nm}^{-1}]\left(\frac{1}{n_1^2} - \frac{1}{n_2^2}\right) \tag{7.7}$$

where n_1 is a positive whole number that remains fixed for a series of emission lines and n_2 is a whole number equal to $n_1 + 1, n_1 + 2, \ldots$, for successive lines in the series. Values of n_1 in Rydberg's equation correspond to values of n in Balmer's equation, so setting $n_1 = 2$ and $n_2 = 3, 4, 5, 6$ produces the wave numbers of the visible hydrogen emission lines.

When Balmer and Rydberg derived their equations describing the hydrogen spectrum, they did not know why the equations worked. The discrete frequencies of hydrogen's emission lines indicated that only certain levels of internal energy were available in hydrogen atoms. However, classical physics could not explain the existence of these internal energy levels. A new model that could explain these observations at the atomic level was needed.

SAMPLE EXERCISE 7.4 Calculating the Wavelength of a Line in the Hydrogen Emission Spectrum   LO3

What is the wavelength of the visible hydrogen atomic emission line corresponding to $n_2 = 3$ in Equation 7.7?

Collect, Organize, and Analyze We are asked to calculate the wavelength of a line in the hydrogen emission spectrum given an n_2 value of 3. The n_1 value of all visible hydrogen emission lines is 2. Wavelengths of visible radiation are between 400 nm and 750 nm.

Solve

$$\frac{1}{\lambda} = [1.097 \times 10^{-2}\ \text{nm}^{-1}]\left(\frac{1}{2^2} - \frac{1}{3^2}\right) = [1.097 \times 10^{-2}\ \text{nm}^{-1}]\left(\frac{1}{4} - \frac{1}{9}\right)$$

$$= [1.097 \times 10^{-2}\ \text{nm}^{-1}](0.1389) = 1.524 \times 10^{-3}\ \text{nm}^{-1}$$

$$\lambda = 656\ \text{nm}$$

Think About It The calculated wavelength is in the visible region of the electromagnetic spectrum, so the answer is reasonable.

 Practice Exercise What is the wavelength, in nanometers, of the line in the hydrogen spectrum corresponding to $n = 2$ and $m = 4$ in Equation 7.6?

The Bohr Model of Hydrogen

In the early 20th century, Ernest Rutherford established that atoms are mostly empty space occupied by negatively charged electrons surrounding a tiny nucleus containing most of the atom's mass and all of its positive charge. This model led to the question: What keeps the electrons from falling into the nucleus? Rutherford suggested that the electrons might orbit the nucleus the way planets orbit the Sun. However, classical physics predicts that negative electrons orbiting a positive nucleus should emit energy in the form of electromagnetic radiation and eventually spiral into the nucleus. If this happened, no atom would be stable.

The Danish physicist Niels Bohr (1885–1962) was familiar with the challenge of trying to explain atom stability on the basis of the Rutherford model because he had studied with Rutherford. Bohr proposed a theoretical model for the hydrogen atom that assumed its one electron travels around the nucleus in one of an array of concentric orbits. Each orbit represents an allowed energy level and is designated by the value of n as shown in Equation 7.8:

$$E = -2.178 \times 10^{-18} \text{ J}\left(\frac{1}{n^2}\right) \qquad (7.8)$$

where $n = 1, 2, 3, \ldots, \infty$. In the Bohr model, an electron in the orbit closest to the nucleus ($n = 1$) has the lowest energy:

$$E = -2.178 \times 10^{-18} \text{ J}\left(\frac{1}{1^2}\right) = -2.178 \times 10^{-18} \text{ J}$$

The next-closest orbit has an n value of 2 and the energy of an electron in it is

$$E = -2.178 \times 10^{-18} \text{ J}\left(\frac{1}{2^2}\right) = -5.445 \times 10^{-19} \text{ J}$$

Note that this value is less negative than the value for the electron in the $n = 1$ orbit. As the value of n increases, the radius of the orbit increases and so, too, does the energy of an electron in the orbit; its value becomes less negative. As n approaches ∞, E approaches zero:

$$E = -2.178 \times 10^{-18} \text{ J}\left(\frac{1}{\infty^2}\right) = 0$$

Zero energy means that there is no longer any electrostatic attraction between the electron and the positively charged nucleus; in other words, the electron is no longer part of the atom. The H atom has become two separate particles: an H^+ ion and a free electron.

An important feature of the Bohr model is that it provides a theoretical framework for explaining the experimental observations and equations developed by Balmer and by Rydberg. To see how, consider what happens when an electron moves between two allowed energy levels in Bohr's model. If we label the initial energy level (the level where the electron starts) $n_{initial}$, and we label the second level (the level where the electron ends up) n_{final}, then the change in energy of the electron is

$$\Delta E = -2.178 \times 10^{-18} \text{ J}\left(\frac{1}{n_{final}^2} - \frac{1}{n_{initial}^2}\right) \qquad (7.9)$$

If the electron moves to an orbit farther from the nucleus, then $n_{final} > n_{initial}$, and the value of the terms inside the parentheses in Equation 7.9 is negative because

$$\frac{1}{n_{final}^2} < \frac{1}{n_{initial}^2}$$

CONNECTION We discussed Rutherford's gold-foil experiment and the development of the idea of the nuclear atom in Section 2.2.

CHEMTOUR Emission Spectra and the Bohr Model of the Atom

CONNECTION We discussed the potential energy of attraction between charged particles in Section 5.2.

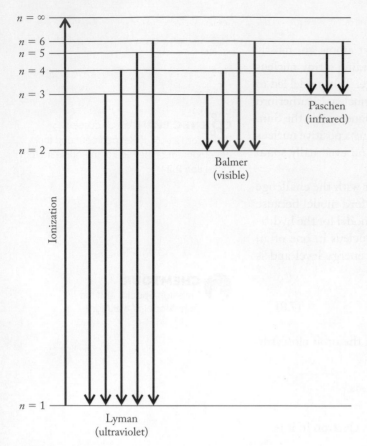

FIGURE 7.14 An energy-level diagram showing possible electron transitions for the electron in the hydrogen atom. The arrow pointing up represents ionization. Arrows pointing down represent the emission of energy that accompanies an electron falling to a lower energy level. This diagram shows several possible transitions for the single electron in hydrogen; each arrow does *not* represent a different electron in the atom.

This negative value multiplied by the negative coefficient would give us a positive ΔE and represent an increase in electron energy. On the other hand, if the electron moves from an outer orbit to one closer to the nucleus, then $n_{\text{final}} <$ n_{initial}, and the sign of ΔE is negative. This means the electron has lost energy. Equation 7.9 demonstrates that energy in the hydrogen atom is *quantized* because ΔE can have only certain values determined by n_{final} and n_{initial}.

When the electron in a hydrogen atom is in the lowest ($n = 1$) energy level, the atom is said to be in its **ground state**. If the electron in a hydrogen atom is in an energy level above $n = 1$, then the atom is said to be in an **excited state**. According to the Bohr model, a hydrogen atom's electron can move from the ground state ($n = 1$) energy level to an excited state (for example, $n = 3$) by absorbing a quantity of energy (ΔE) that exactly matches the energy difference between the two states. Similarly, an electron in an excited state can move to an even higher energy level by absorbing a quantity of energy that exactly matches the energy difference between the two excited states. An electron in an excited state can also move to a lower-energy excited state, or to the ground state, by emitting a quantity of energy that exactly matches the energy difference between those two states. Any change in electron energy that occurs by absorption or emission of energy is called an **electron transition**.

*Energy-level diagram*s show the transitions that electrons can make in atoms from one energy level to another. Figure 7.14 is such a diagram for the hydrogen atom. The black arrow pointing upward represents the absorption of energy sufficient to completely remove the electron from a hydrogen atom (ionization). The downward-pointing colored arrows represent decreases in the internal energy of the hydrogen atom that occur when photons are emitted as the electron moves from a higher energy level to a lower energy level. If the colored arrows pointed up, they would represent absorption of photons leading to increases in the internal energy of the atom. In every case, the energy of the photon matches the absolute value of ΔE.

If you compare Equation 7.9 with Equation 7.7, you will see that they are much alike. The coefficients differ only because of the different units used to express wave number and energy. In addition, Bohr was able to derive his coefficient from a combination of physical constants, including the Planck constant and the charge of an electron. It is not an arbitrary value that just happens to work; it is part of atomic structure. The key point is that the equation developed to fit the absorption and emission spectra of hydrogen has the same form as the theoretical equation developed by Bohr to explain the internal structure of the hydrogen atom. Thus atomic emission and absorption spectra reveal the energies of electrons inside atoms.

CONCEPT **TEST**

On the basis of the lengths of the arrows in Figure 7.14, rank the following electron transitions in order of greatest change in energy to smallest change:

a. $n = 4 \rightarrow n = 2$

b. $n = 3 \rightarrow n = 2$

c. $n = 2 \rightarrow n = 1$

d. $n = 4 \rightarrow n = 3$

ground state the most stable, lowest-energy state available to an atom or ion.

excited state any energy state above the ground state in an atom or ion.

electron transition movement of an electron between energy levels.

SAMPLE EXERCISE 7.5 Calculating the Energy Change of an Electron Transition **LO3**

How much energy is required to ionize a ground-state hydrogen atom?

Collect and Organize We are asked to determine the energy required to remove an electron from a hydrogen atom in its ground state. Equation 7.9 enables us to calculate the energy change associated with any electron transition in a hydrogen atom.

Analyze To use Equation 7.9, we need to identify the initial (n_{initial}) and final (n_{final}) energy levels. The ground state of a hydrogen atom corresponds to the $n = 1$ energy level. If the atom is ionized, $n = \infty$ and there is no longer any electrostatic attraction between the positively charged nucleus and the electron.

Solve

$$\Delta E = -2.178 \times 10^{-18}\,\text{J}\left(\frac{1}{n_{\text{final}}^2} - \frac{1}{n_{\text{initial}}^2}\right)$$

$$= -2.178 \times 10^{-18}\,\text{J}\left(\frac{1}{\infty^2} - \frac{1}{1^2}\right)$$

Dividing by ∞^2 yields zero, so the difference in parentheses simplifies to −1. Therefore:

$$\Delta E = 2.178 \times 10^{-18}\,\text{J}$$

Think About It This is a small amount of energy, but it is the energy change for just a single atom. The sign of ΔE is positive because energy must be added to overcome the electrostatic attraction between an electron and its positive nucleus.

Practice Exercise Calculate the energy required to ionize an excited-state hydrogen atom with an electron initially in the $n = 3$ energy level. Before doing the calculation, predict whether this energy is greater than or less than the 2.178×10^{-18} J needed to ionize a ground-state hydrogen atom.

One of the strengths of the Bohr model of the hydrogen atom is that it accurately predicts the energy needed to remove the electron. This energy is called the ionization energy of the hydrogen atom. We examine the ionization energies of other elements in Section 7.11. However, the Bohr model applies only to hydrogen atoms and to ions that have only a single electron. The model does not agree with the observed spectra of multielectron atoms and ions because it does not account for the way electrons interact with each other. Thus the picture of the atom provided by Bohr's model is extremely limited. Nevertheless, it enabled scientists to begin using quantum theory to explain the behavior of matter at the atomic level.

CONCEPT TEST

Can the Bohr model be used to explain the emission spectrum of He atoms? Why or why not?

7.5 Electron Waves

A decade after Bohr published his model of the hydrogen atom, a French graduate student named Louis de Broglie (1892–1987) provided a theoretical basis for the stability of electron orbits. His approach incorporated yet another significant advance in the way early-20th-century scientists conceived of atoms

matter wave the wave associated with any particle.

standing wave a wave confined to a given space, with a wavelength (λ) related to the length L of the space by $L = n(\lambda/2)$, where n is a whole number.

node a location in a standing wave that experiences no displacement. In the context of orbitals, nodes are locations at which electron density goes to zero.

and subatomic particles—namely, to think of them not only as particles of matter but also as waves. His work provided an answer to the question of why an electron in the hydrogen atom does not spiral into the nucleus.

De Broglie Wavelengths

De Broglie proposed that if light, which we normally think of as a wave, has particle properties, then perhaps the electron, which we normally think of as a particle, has wave properties. If that were true, an electron moving around in an atom should have a characteristic wavelength. De Broglie calculated electron wavelengths from Einstein's equations relating energy and mass, $E = mc^2$, and the energy and wavelength of a photon, $E = hc/\lambda$:

$$\lambda = \frac{hc}{E} = \frac{hc}{mc^2} = \frac{h}{mc} \tag{7.10}$$

To apply Equation 7.10 to electrons, de Broglie replaced c (the speed of light) with u, the speed of an orbiting electron in an atom:

$$\lambda = \frac{h}{mu} \tag{7.11}$$

where h is the Planck constant, m is the mass of the electron in kilograms, and u is its velocity in meters per second. The wavelength of an electron calculated in this way is often called the *de Broglie wavelength* of the electron.

De Broglie's equation is not restricted to electrons. It tells us that any moving particle has wavelike properties. In other words, the particle behaves as a **matter wave**. De Broglie predicted that moving particles much bigger than electrons, such as atomic nuclei, molecules, and even macroscopic objects like tennis balls and airplanes, should have characteristic wavelengths described by Equation 7.11. The wavelengths of such large objects are extremely small given the tiny size of the Planck constant in the numerator of Equation 7.11, and for this reason we never notice the wave nature of large objects in motion.

CHEMTOUR
De Broglie Wavelength

SAMPLE EXERCISE 7.6 Calculating the Wavelength   **LO4**
of a Particle in Motion

(a) Compare the wavelength of a 142-g baseball thrown at 44 m/s (98 mi/h) with the size of the ball, which has a diameter of 7.5 cm. (b) Compare the wavelength of an electron ($m_e = 9.109 \times 10^{-31}$ kg) moving at one-tenth the speed of light in a hydrogen atom with the size of the atom (diameter = 1.06×10^{-10} m).

Collect, Organize, and Analyze Given the masses and velocities of these two moving objects, Equation 7.11 may be used to calculate their wavelengths. Given the small value of h, it is likely that the wavelength of a pitched baseball is only a tiny fraction of the size of the baseball. However, the wavelength of an electron moving at one-tenth the speed of light may be a much greater fraction of the size of the atom that it orbits. The fraction on the right side of Equation 7.11 has units of joule-seconds in the numerator and mass and velocity in the denominator. To combine these units in a way that gives us a unit of length, we need to use the following conversion factor:

$$1\text{ J} = 1\text{ kg} \cdot \text{m}^2/\text{s}^2$$

To use this equality, we must express the mass of the baseball in kilograms: 142 g = 0.142 kg.

Solve

a. For the baseball:

$$\lambda = \frac{h}{mu} = \frac{6.626 \times 10^{-34}\,\text{J}\cdot\text{s}}{(0.142\,\text{kg})(44\,\text{m/s})} \times \frac{1\,\text{kg}\cdot\text{m}^2/\text{s}^2}{1\,\text{J}}$$

$$= 1.06 \times 10^{-34}\,\text{m}$$

The wavelength of the baseball is

$$\frac{1.06 \times 10^{-34}\,\text{m}}{0.075\,\text{m}} \times 100\% = 1.4 \times 10^{-31}\%$$

of the ball's diameter.

b. The wavelength of an electron moving at one-tenth the speed of light is

$$\lambda = \frac{h}{mu} = \frac{6.626 \times 10^{-34}\,\text{J}\cdot\text{s}}{(9.109 \times 10^{-31}\,\text{kg})(2.998 \times 10^{7}\,\text{m/s})} \times \frac{1\,\text{kg}\cdot\text{m}^2/\text{s}^2}{1\,\text{J}}$$

$$= 2.43 \times 10^{-11}\,\text{m}$$

and

$$\frac{2.43 \times 10^{-11}\,\text{m}}{1.06 \times 10^{-10}\,\text{m}} \times 100\% = 23\%$$

of the diameter of a hydrogen atom.

Think About It The matter wave of the baseball is much too small to be observed, so its character contributes nothing to the behavior of the baseball. We expected that. Our experience with objects in the world is that they behave like particles, not like waves. For the electron, however, the wavelength is a significant percentage of the size of the hydrogen atom. This is not a familiar experience to us—a particle that also has properties of waves such as frequency and wavelength. The implications of this for our understanding of the structure of the atom are discussed in the rest of this section.

 Practice Exercise The velocity of the electron in the ground state of a hydrogen atom is 2.2×10^{6} m/s. What is the wavelength of this electron in meters?

De Broglie explained the stability of the electron levels in Bohr's model of the hydrogen atom by proposing that the electron in a hydrogen atom behaves like a circular wave oscillating around the nucleus. To understand the implications of this statement, we need to examine what is required to make a stable, circular wave. Consider the motion of a vibrating violin string of length L (Figure 7.15a). Because the string is fixed at both ends, there is no vibration at the ends and the vibration is at maximum in the middle. The wave created by this combination of fixed ends and maximum vibration in the middle is a **standing wave**: a wave that oscillates back and forth within a fixed space rather than moving through space the way waves of light travel through space. On a standing wave, any points that have zero displacement, such as the two ends of the string, are called **nodes**. The sound wave produced in this way on a violin string is called the *fundamental* of the string. The wavelength of the fundamental is $2L$. It has the lowest frequency and longest wavelength possible for that string. The length of the string is one-half the wavelength of the fundamental ($L = \lambda/2$).

If the string is held down in the middle (creating a third node) and plucked halfway between the middle and one end, a new, higher-frequency wave called the *first harmonic* is produced. The wavelength of this harmonic is equal to L

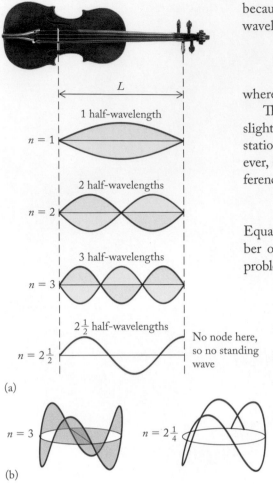

(a)

(b)

FIGURE 7.15 Linear and circular standing waves. (a) The wavelength of a standing wave in a violin string fixed at both ends is related to the distance L between the ends of the string by the equation $L = n(\lambda/2)$. In the standing waves shown, $n = 1, 2,$ and 3. An n value of $2\frac{1}{2}$ does not produce a standing wave because a standing wave can have no string motion at either end. (b) The circular standing waves proposed by de Broglie account for the stability of the energy levels in Bohr's model of the hydrogen atom. Each stable wave must have a circumference equal to $n\lambda$ with n restricted to being an integer, such as the $n = 3$ circular standing wave shown here. If the circumference is not an exact multiple of λ, as shown in the $2\frac{1}{4}$ image, no standing wave occurs.

because $L = 2(\lambda/2) = \lambda$. We can continue generating higher frequencies with wavelengths that are related to the length of the string by the equation

$$L = \frac{n\lambda}{2}$$

where n is a whole number and $L = 3(\lambda/2), 4(\lambda/2), 5(\lambda/2),$ and so on.

The standing-wave pattern for a circular wave generated by an electron differs slightly from that of a vibrating violin string in that the circular wave has no defined stationary ends. Instead, the electron vibrates in an endless series of waves. However, standing waves are only produced if, as shown in Figure 7.15(b), the circumference of the circle equals a whole-number multiple of the electron's wavelength:

$$\text{Circumference} = n\lambda \qquad (7.12)$$

Equation 7.12 offers a new interpretation of Bohr's number n: it represents the number of matter waves in a given energy level. De Broglie's work also solved the problem of a negatively charged electron spiraling into the positively charged nucleus. Since $n = 1$ represents the minimum circumference of the circular wave of a moving electron, the electron must remain a minimum distance from the nucleus at all times. Furthermore, the lowest energy possible for an electron in a hydrogen atom occurs for a standing wave with circumference λ, so the electron cannot get any closer to the nucleus than that.

De Broglie's research created a quandary for the graduate faculty at the University of Paris, where he studied. Bohr's model of electrons moving between allowed energy levels had been widely criticized as an arbitrary suspension of well-tested physical laws. De Broglie's rationalization of Bohr's model seemed even more outrageous to many scientists. Before the faculty would accept his thesis, they wanted another opinion, so they sent it to Albert Einstein for review. Einstein wrote back that he found the young man's work "quite interesting." That endorsement was good enough for the faculty: de Broglie's thesis was accepted in 1924 and immediately submitted for publication. Five years later, he was awarded the Nobel Prize.

The Heisenberg Uncertainty Principle

After de Broglie proposed that electrons exhibited both particle-like and wavelike behavior, questions arose about the impact of wave behavior on our ability to locate the electron. A wave, by its very nature, is spread out in space. The question "Where is the electron?" has one answer if we treat the electron as a particle and a very different answer if we treat it as a wave. This issue was addressed by the German physicist Werner Heisenberg (1901–1976), who proposed the following thought experiment: Watch an electron with a microscope to "see" the electron's path around an atom. The microscope (if it existed) would need to use gamma rays for illumination (rather than visible light) because gamma rays are the only part of the electromagnetic spectrum with wavelengths short enough to match the diminutive size of electrons. However, Equations 7.2 and 7.3 tell us that the short wavelengths and high frequencies of gamma rays mean that they have enormous energies—so large that any gamma ray striking an electron would knock the electron off course. The only way to not affect the electron's motion would be to use a much lower-energy, longer-wavelength source of radiation to illuminate it—but then we would not be able to see the tiny electron clearly.

This situation presents a quantum mechanical dilemma. The only means for clearly observing an electron make it impossible to know the electron's motion or, more precisely, its momentum, which is defined as an object's velocity times its mass. Therefore we can never know exactly both the position and the momentum of the electron simultaneously. This conclusion is known as the **Heisenberg uncertainty principle**, and it is mathematically expressed by

$$\Delta x \cdot m\Delta u \geq \frac{h}{4\pi} \tag{7.13}$$

where Δx is the uncertainty in the position of the electron, m is its mass, Δu is the uncertainty in its velocity, and h is the Planck constant. To Heisenberg, this uncertainty was the essence of quantum mechanics. Its message for us is that there are limits to what we can observe, measure, and therefore know.

Heisenberg uncertainty principle the principle that we cannot determine both the position and the momentum of an electron in an atom at the same time.

SAMPLE EXERCISE 7.7 Calculating (Heisenberg) Uncertainty   LO5

Use the data in Sample Exercise 7.6 to compare the uncertainty in the velocity of the baseball with the uncertainty in the velocity of the electron. Assume that the position of the baseball is known to within one wavelength of red light ($\Delta x_{baseball}$ = 680 nm) and that the position of the electron is known to within the radius of the hydrogen atom ($\Delta x_{electron} = 5.29 \times 10^{-11}$ m).

Collect and Organize We are asked to calculate the uncertainty in the velocities of two particles given the uncertainties in their positions. From Sample Exercise 7.6 we know that the mass of a baseball is 0.142 kg and the mass of an electron is 9.109×10^{-31} kg.

Analyze Equation 7.13 provides a mathematical connection between these variables. According to Equation 7.13, the uncertainty in the velocity and position of a particle is inversely proportional to its mass. Therefore, we can expect little uncertainty in the velocity of the baseball, but much greater uncertainty in the velocity of the electron. We need to rearrange the terms in the equation to solve for the uncertainty in velocity (Δu):

$$\Delta u \geq \frac{h}{4\pi \Delta x m}$$

Solve For the baseball:

$$\Delta u \geq \frac{6.626 \times 10^{-34} \text{ J} \cdot \text{s}}{4\pi (6.8 \times 10^{-7} \text{ m})(0.142 \text{ kg})} \times \frac{1 \text{ kg} \cdot \text{m}^2/\text{s}^2}{1 \text{ J}}$$

$$\geq 5.5 \times 10^{-28} \text{ m/s}$$

For the electron:

$$\Delta u \geq \frac{6.626 \times 10^{-34} \text{ J} \cdot \text{s}}{4\pi (5.29 \times 10^{-11} \text{ m})(9.109 \times 10^{-31} \text{ kg})} \times \frac{1 \text{ kg} \cdot \text{m}^2/\text{s}^2}{1 \text{ J}}$$

$$\geq 1.09 \times 10^{6} \text{ m/s}$$

Comparing the two values, we see that the uncertainty in the velocity of the baseball is extremely small (about 10^{-28} m/s), whereas that of the electron is huge (greater than 1 million m/s).

Think About It The uncertainty in the measurement of the baseball's velocity is so minuscule that it is insignificant. This result is expected for objects in the macroscopic world. However, the uncertainty in the measurement of the velocity of the electron is huge, which is what we must expect at the atomic level, where "particles" such as the electron also behave like waves.

Practice Exercise What is the uncertainty, in meters, in the position of an electron moving near a nucleus at a speed of 8×10^{7} m/s? Assume the uncertainty in the velocity of the electron is 1% of its value—that is, $\Delta u = (0.01)(8 \times 10^{7}$ m/s).

When Heisenberg proposed his uncertainty principle, he was working with Bohr at the University of Copenhagen. The two scientists had widely different views about the significance of the uncertainty principle and the idea that particles could behave like waves. To Heisenberg, uncertainty was a fundamental characteristic of nature. To Bohr, it was merely a mathematical consequence of the wave–particle duality of electrons; there was no physical meaning to an electron's position and path. The debate between these two gifted scientists was heated at times. Heisenberg later wrote about one particularly emotional debate:

> [A]t the end of the discussion I went alone for a walk in the neighboring park [and] repeated to myself again and again the question: "Can nature possibly be as absurd as it seems?"[1]

The Heisenberg uncertainty principle is fundamental to our present understanding of the atom. If we cannot know both the position and momentum of an electron in a hydrogen atom, then the electron cannot be moving in circular orbits as implied by Bohr's original model. As we see in the next section, the Heisenberg uncertainty principle limits us to knowing only the probability of finding an electron at a particular location in an atom.

7.6 Quantum Numbers and Electron Spin

Many of the leading scientists of the 1920s were unwilling to accept the dual wave–particle nature of electrons proposed by de Broglie unless the model could be used to predict the features of the hydrogen emission spectrum. Such application required the development of equations describing the behavior of electron waves. During his Christmas vacation in 1925, the Austrian physicist Erwin Schrödinger (1887–1961) did just that, developing in a few weeks the mathematical foundation for what came to be called **wave mechanics** or **quantum mechanics**.

Schrödinger's mathematical description of electron waves is called the **Schrödinger wave equation**. Although it is not discussed in detail in this book, you should know that solutions to the wave equation are called **wave functions (ψ)**: mathematical expressions that describe how the matter wave of an electron in an atom varies both with time and with the location of the electron in the atom. Wave functions define the energy levels in the hydrogen atom. They can be simple trigonometric functions, such as sine or cosine waves, or they can be very complex.

What is the physical significance of a wave function? Actually there is none. However, the *square of a wave function* (ψ^2) does have physical meaning. Initially, Schrödinger believed that a wave function depicted the "smearing" of an electron through three-dimensional space. This notion of subdividing a discrete particle was later rejected in favor of the model developed by German physicist Max Born (1882–1970), who proposed that ψ^2 defines an **orbital**: the space around the nucleus of an atom where the probability of finding an electron is high. Born later showed that his interpretation could be used to calculate the probability of a transition between two orbitals, as happens when an atom absorbs or emits a photon.

To help visualize the probabilistic meaning of ψ^2, consider what happens when we spray ink onto a flat surface (Figure 7.16). If we then draw a circle encompassing most of the ink spots, we are identifying the region of maximum probability for finding the spots.

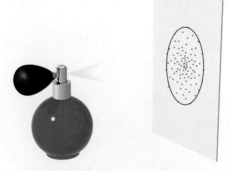

FIGURE 7.16 The probability of encountering an ink spot in the pattern produced by an ink spray decreases with increasing distance from the center of the pattern.

[1]Heisenberg, W. *Physics and Philosophy: The Revolution in Modern Science* (Harper & Row, 1958), p. 42.

It is important to understand that quantum mechanical orbitals in an atom are not two-dimensional concentric orbits, as in Bohr's model of the hydrogen atom, or even two-dimensional circles, as in Figure 7.16. Instead, as we see in detail in Section 7.7, orbitals are three-dimensional regions of space with distinctive shapes, orientations, and average distances from the nucleus. Each orbital is a solution to Schrödinger's wave equation and is identified by a unique combination of three integers, or **quantum numbers**, whose values flow directly from the mathematical solutions to the wave equation. The quantum numbers are as follows:

- The **principal quantum number** n is like Bohr's number n for the hydrogen atom in that it is a positive integer that indicates the relative size and energy of an orbital or group of orbitals in an atom. Orbitals with the same value of n are in the same *shell*. Orbitals with larger values of n are farther from the nucleus and, in the hydrogen atom, represent higher energy levels, consistent with Bohr's model of the hydrogen atom. In multielectron atoms, the relationship between energy levels and orbitals is more complex, but increasing values of n generally represent higher energy levels.

- The **angular momentum quantum number** ℓ is an integer with a value ranging from zero to $n - 1$ that defines the shape of an orbital. Orbitals with the same values of n and ℓ are in the same *subshell* and represent equal energy levels. Orbitals with a given value of ℓ are identified with a letter according to the following scheme:

Value of ℓ	0	1	2	3
Type of Orbital	*s*	*p*	*d*	*f*

- The choice of letters to designate the values of ℓ (that is, *s*, *p*, *d*, and *f*) may seem a bit odd. Before quantum mechanics was developed, scientists recording line spectra of the elements described the lines they observed as *sharp*, *principal*, *diffuse*, and *fundamental*. Designating orbitals with the letters *s*, *p*, *d*, and *f* recognizes this historical convention.

- The **magnetic quantum number** m_ℓ is an integer with a value from $-\ell$ to $+\ell$. It defines the orientation of an orbital in the space around the nucleus of an atom.

Each subshell in an atom has a two-part designation containing the appropriate value of n and a letter designation for ℓ. For example, orbitals with $n = 3$ and $\ell = 1$ are called 3*p* orbitals, and electrons in 3*p* orbitals are called 3*p* electrons. How many 3*p* orbitals are there? We can answer this question by finding all possible values of m_ℓ. Because *p* orbitals are those for which $\ell = 1$, they have m_ℓ values of -1, 0, and $+1$. These three values mean that there are three 3*p* orbitals, each with a unique combination of n, ℓ, and m_ℓ values. All the possible combinations of these three quantum numbers for the orbitals of the first four shells are given in Table 7.1.

wave mechanics (also called **quantum mechanics**) a mathematical description of the wavelike behavior of particles on the atomic level.

Schrödinger wave equation a description of how the electron matter wave varies with location and time around the nucleus of a hydrogen atom.

wave function (ψ) a solution to the Schrödinger wave equation.

orbital a region around the nucleus of an atom where the probability of finding an electron is high; each orbital is defined by the square of the wave function (ψ^2) and identified by a unique combination of three quantum numbers.

quantum number a number that specifies the energy, the probable location or orientation of an orbital, or the spin of an electron within an orbital.

principal quantum number (n) a positive integer describing the relative size and energy of an atomic orbital or group of orbitals in an atom.

angular momentum quantum number (ℓ) an integer having any value from 0 to $n - 1$ that defines the shape of an orbital.

magnetic quantum number (m_ℓ) defines the orientation of an orbital in space; an integer that may have any value from $-\ell$ to $+\ell$, where ℓ is the angular momentum quantum number.

 **CHEMTOUR**
Quantum Numbers

SAMPLE EXERCISE 7.8 Identifying the Subshells and   **LO6**
                     Orbitals in an Energy Level

(a) What are the designations of all the subshells in the $n = 4$ shell? (b) How many orbitals are in these subshells?

Collect and Organize We are asked to describe the subshells in the fourth shell and to determine how many orbitals are in all these subshells. Table 7.1 contains an inventory of all the subshells in the first four shells.

TABLE 7.1 Quantum Numbers of the Orbitals in the First Four Shells

Value of n	Allowed Value of ℓ	Subshell Label	Allowed Values of m_ℓ	NUMBER OF ORBITALS	
				In Subshell	In Shell
1	0	s	0	1	1
2	0	s	0	1	4
	1	p	−1, 0, +1	3	
3	0	s	0	1	9
	1	p	−1, 0, +1	3	
	2	d	−2, −1, 0, +1, +2	5	
4	0	s	0	1	16
	1	p	−1, 0, +1	3	
	2	d	−2, −1, 0, +1, +2	5	
	3	f	−3, −2, −1, 0, +1, +2, +3	7	

Analyze The designations of subshells are based on the possible values of quantum numbers n and ℓ. The allowed values of ℓ depend on the value of n, in that ℓ is an integer from 0 up to $n - 1$. The number of orbitals in a subshell depends on the number of possible values of m_ℓ, from $-\ell$ to $+\ell$.

Solve
a. The allowed values of ℓ for $n = 4$ range from 0 through $(n - 1)$, so they are 0, 1, 2, and 3. These ℓ values correspond to the subshell designations s, p, d, and f, respectively. The appropriate subshell names are thus 4s, 4p, 4d, and 4f.
b. The possible values of m_ℓ from $-\ell$ to $+\ell$ are as follows:

   $\ell = 0$; $m_\ell = 0$: This combination of ℓ and m_ℓ values for the $n = 4$ shell represents a single 4s orbital.

   $\ell = 1$; $m_\ell = -1$, 0, or +1: These three combinations of ℓ and m_ℓ values for the $n = 4$ shell represent the three 4p orbitals.

   $\ell = 2$; $m_\ell = -2$, −1, 0, +1, or +2: These five combinations of ℓ and m_ℓ values represent the five 4d orbitals.

   $\ell = 3$; $m_\ell = -3$, −2, −1, 0, +1, +2, or +3: These seven combinations of ℓ and m_ℓ values represent the seven 4f orbitals.

Think About It There are $1 + 3 + 5 + 7 = 16$ orbitals in the $n = 4$ shell. There would be fewer orbitals in the third shell because the $\ell = 3$ subshell would not exist in the $n = 3$ shell. Therefore, there would be $1 + 3 + 5 = 9$ orbitals in the $n = 3$ shell.

 Practice Exercise
How many orbitals are there in the $n = 5$ shell?

spin magnetic quantum number (m_s) either $+\frac{1}{2}$ or $-\frac{1}{2}$, indicating that the spin orientation of an electron is either up or down.

Pauli exclusion principle the principle that no two electrons in an atom can have the same set of four quantum numbers.

Figure 7.17 provides a visual summary of the quantum number rules. This system of quantum numbers provides a useful "shorthand" to refer to the orbitals that are solutions to Schrödinger's wave equation. Even so, the Schrödinger wave equation could account for most, but not all, aspects of atomic spectra. The emission spectrum of hydrogen, for example, has a pair of red lines at 656 nm where Balmer thought there was only one line (Figure 7.18). There are also pairs of lines in the spectra of multielectron atoms that have a single electron in their outermost shells (see Figure 7.3). The Schrödinger equation cannot explain these pairs of lines.

In 1925, two students at the University of Leiden in the Netherlands, Samuel Goudsmit (1902–1978) and George Uhlenbeck (1900–1988), proposed

Name, Symbol (Property)	Allowed Values	Quantum Numbers
Principal, n (size, energy)	Positive integers $(1, 2, 3, \ldots)$	
Angular momentum, ℓ (shape)	From 0 to $n - 1$	
Magnetic, m_ℓ (orientation)	$-\ell, \ldots, 0, \ldots, +\ell$	

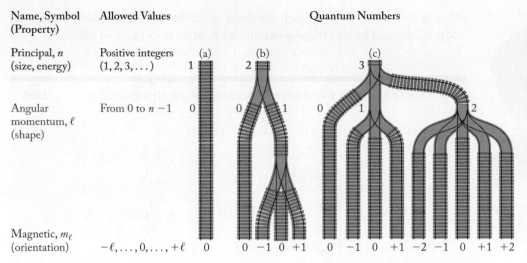

FIGURE 7.17 These three sets of train tracks provide a visual metaphor for the orbitals in the first three energy shells in an atom. The single track labeled (a) represents the first shell and its $1s$ orbital, which has ℓ and m_ℓ values of 0. The set labeled (b) has two ℓ branches, labeled 0 and 1, representing the $2s$ and $2p$ orbitals. The 1 branch further divides into branches with m_ℓ values of -1, 0, and $+1$, representing the three $2p$ orbitals. The set labeled (c) divides into three main branches with ℓ values of 0, 1, and 2, representing the $3s$, $3p$, and $3d$ subshells, respectively.

that the pairs of lines, called *doublets*, were caused by a property they called electron spin. In their model, electrons spin in one of two directions, designated "spin up" and "spin down." A moving electron (or any charged particle) creates a magnetic field by virtue of its motion. The spinning motion produces a second magnetic field oriented up or down. To account for these two spin orientations, Goudsmit and Uhlenbeck proposed a fourth quantum number, the **spin magnetic quantum number (m_s)**. The values of m_s are $+\frac{1}{2}$ for spin up and $-\frac{1}{2}$ for spin down.

Even before Goudsmit and Uhlenbeck proposed the electron-spin hypothesis, two other scientists, Otto Stern (1888–1969) and Walther Gerlach (1889–1979), had observed the effect of electron spin when they shot a beam of silver ($Z = 47$) atoms through a magnetic field (Figure 7.19). Those atoms in which the net electron spin was "up" were deflected in one direction by the field; those in which the net electron spin was "down" were deflected in the opposite direction.

In 1925 Austrian physicist Wolfgang Pauli (1900–1958; Figure 7.20) proposed that no two electrons in a multielectron atom have the same set of four quantum numbers. This idea is known as the **Pauli exclusion principle**. The three quantum numbers from Schrödinger's wave equation define the orbitals where an atom's electrons are likely to be. The two allowed values of the spin magnetic quantum number indicate that each orbital can hold two electrons, one with $m_s = +\frac{1}{2}$ and the

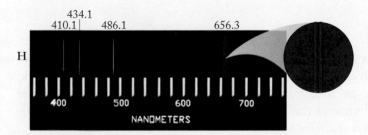

FIGURE 7.18 The Schrödinger equation does not account for the appearance of closely spaced pairs of bright lines in the emission spectra of atoms, such as the red lines at 656.272 nm and 656.285 nm in the spectrum of hydrogen.

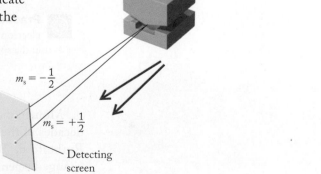

FIGURE 7.19 A narrow beam of silver atoms passed through a magnetic field is split into two beams because of the interactions between the field and the spinning electrons in the atoms. This observation led to proposing the fourth quantum number, m_s. (The blue arrows indicate the direction of the beam.)

FIGURE 7.20 Wolfgang Pauli (left) and Niels Bohr are apparently amused by the behavior of a toy called a tippe top, which, when spun on its base, tips itself over and spins on its stem. Although the toy's behavior is caused by a combination of friction and the top's angular momentum and is not quantum mechanical in origin, it provides a visual metaphor for the two spin orientations of an electron in an orbital.

other with $m_s = -\frac{1}{2}$. Thus each electron in an atom has a unique "quantum address" defined by a particular combination of n, ℓ, m_ℓ, and m_s values.

SAMPLE EXERCISE 7.9 Identifying Valid Quantum Number Sets   **LO6**

Which of these combinations of quantum numbers are valid?

	n	ℓ	m_ℓ	m_s
(a)	1	0	−1	$+\frac{1}{2}$
(b)	3	2	−2	$+\frac{1}{2}$
(c)	2	2	0	0
(d)	2	0	0	$-\frac{1}{2}$
(e)	−3	−2	−1	$-\frac{1}{2}$

Collect, Organize, and Analyze We are asked to validate five sets of quantum numbers. These rules apply: A principal quantum number (n) can be any positive integer. The values of ℓ in a given shell are integers from 0 to $(n-1)$; the values of m_ℓ in a given subshell are all integers from $-\ell$ to $+\ell$ including 0. The only two options for m_s are $+\frac{1}{2}$ or $-\frac{1}{2}$.

Solve

a. Because n is 1, the maximum (and only) value of ℓ is $(n-1) = 1 - 1 = 0$. Therefore the values of n and ℓ are valid. However, if $\ell = 0$, then m_ℓ must be 0; it cannot be −1. Therefore this set is not valid. The spin quantum number is a possible value.
b. Because n is 3, ℓ can be 2 and m_ℓ can be −2. Also, $m_s = +\frac{1}{2}$ is a valid choice for the spin magnetic quantum number. This set is valid.
c. Because n is 2, ℓ cannot be 2, making this set invalid. In addition, m_s has an invalid value (0).
d. Because n is 2, ℓ can be 0, and for that value of ℓ, m_ℓ must be 0. The value of m_s is also valid, and so is the set.
e. This set contains two impossible values, $n = -3$ and $\ell = -2$, so it is invalid.

Think About It The values of n, ℓ, and m_ℓ are related mathematically, and m_s can be either $+\frac{1}{2}$ or $-\frac{1}{2}$. The quantum numbers are the "address" of the electron in an atom, and every electron has its own unique "address"—its own unique set of four quantum numbers.

 Practice Exercise Write all the possible sets of quantum numbers for an electron in the $n = 3$ shell that has an angular momentum quantum number $\ell = 1$ and a spin quantum number $m_s = +\frac{1}{2}$.

Momentous advances in chemistry and physics were made in the first three decades of the 20th century because of the brilliant minds of Einstein, Planck, Rutherford, Bohr, de Broglie, Schrödinger, and others, who have forever changed scientists' view of the fundamental structure of matter and the universe. Figure 7.21 summarizes and connects some of these advances.

How different is the view of the interaction of matter and energy provided by quantum mechanics from the laws governing the behavior of large objects? A pebble picked up and dropped immediately falls to the ground. However, electrons remain in excited states for indeterminate (though usually short) times before falling to their ground states. This lack of determinacy bothered Einstein and many of his colleagues. Had they discovered an underlying theme of nature—that some processes cannot be described or known with certainty? Are there fundamental limits to how well we can know and understand our world and the events that change it?

The variable lifetimes of excited states are the basis for the lasers used in bar-code scanners at grocery stores, laser pointers, DVD players, and in surgery (Figure 7.22). The word *laser* is actually an acronym for **L**ight **A**mplification by **S**timulated **E**mission of **R**adiation. The phrase *stimulated emission* is linked to the fact that the atoms in lasers linger in particular excited states for unusually long times, which increases the populations of these excited states. Then, if a photon with just the right energy encounters one of these excited-state atoms, it stimulates the decay of the excited state, producing a second photon that matches the first: same wavelength, same phase, and going in the same direction. When these two photons encounter two more excited-state atoms, they could produce four in-phase photons, which could produce four more, and so on. In this way, one incident photon is rapidly amplified into an intense pulse of monochromatic light. To produce a steady beam of this light, electrical energy continuously pumps up the lasing material, repopulating the excited state. By manipulating the composition of lasers, scientists can adjust the electronic energy levels inside them and the color of the light they produce.

7.7 The Sizes and Shapes of Atomic Orbitals

As noted earlier, the orbitals that are solutions to Schrödinger's wave equation have three-dimensional shapes that are graphical representations of ψ^2. In this section we examine the shapes of atomic orbitals and how those shapes affect the energies of the electrons in them.

s Orbitals

Figure 7.23 provides several representations of the 1s orbital of the hydrogen atom. In Figure 7.23(a), electron density is plotted against distance from the nucleus and shows that density decreases with increasing distance. However, Figure 7.23(b) provides a more useful profile of electron distribution. To understand why, imagine if the hydrogen atom were like an onion, made of many concentric spherical layers, all of the same thickness. A cross section of this image of the atom is shown in Figure 7.23(c). What would be the probability of finding the electron in one of these spherical layers? A layer very close to the nucleus has a very small radius, so it accounts for only a small fraction of the total volume of the atom. A layer with a larger radius makes up a much larger fraction of the volume of the atom because the volume of the layers increases as a function of r^2. (Note that the volume of a sphere depends on r^3, but here we are discussing a spherical

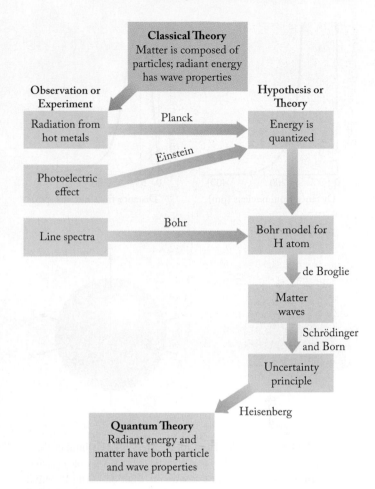

FIGURE 7.21 During the first three decades of the 20th century, quantum theory evolved from classical 19th-century theories of the nature of matter and energy. The arrows trace the development of modern quantum theory, which incorporates the assumption that radiant energy has both wavelike and particle-like properties and that mass (matter) also has both wavelike and particle-like properties.

FIGURE 7.22 Laser pointers emitting different colors are commonplace.

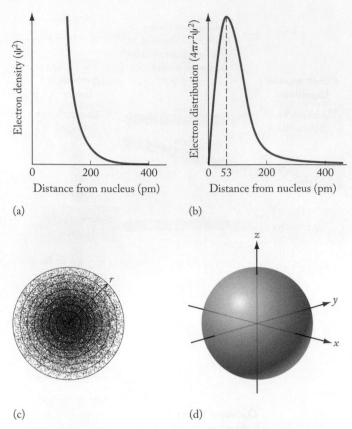

(a)

(b)

(c)

(d)

FIGURE 7.23 (a) Probable electron density in the 1s orbital of the hydrogen atom represented by a plot of electron density (ψ^2) versus distance from the nucleus. (b) Electron distribution in the 1s orbital versus distance from the nucleus. The distribution is essentially zero both for very short distances from the nucleus and for very long distances from the nucleus. The maximum probability occurs at $r = 53$ pm. (c) Cross section through the hydrogen atom, with the space surrounding the nucleus divided into an arbitrary number of thin, concentric, hollow layers. Each layer has a unique value for radius r. The probability of finding an electron in a particular layer of radius r depends on the volume of the layer and the density of electrons in the layer. (d) Boundary–surface representation of a sphere within which the probability of finding a 1s electron is 90%.

shell, the volume of which depends on r^2.) Even though electron densities are higher closer to the nucleus (as Figure 7.23a shows), the volumes of the spherical layers closest to the nucleus are so small that the chances of the electron being near the center of an atom are extremely low; this is shown in Figure 7.23(b), where the curve starts off at essentially zero for electron distribution values at distances very close to the nucleus. Farther from the nucleus, electron densities are lower but the volumes of the layers are much larger, so the probability of the electron being in one of these layers is relatively high, represented by the peak in the curve of Figure 7.23(b). At greater distances, volumes of the layers are very large but ψ^2 drops to nearly zero (see Figure 7.23a); therefore the chances of finding an electron in layers far from the nucleus are very small.

Thus Figure 7.23(b) represents a combination of two competing factors: increasing layer volume and decreasing probability of finding an electron in a given layer. This combination produces a *radial distribution profile* for the electron. Figure 7.23(b) is not a plot of ψ^2 versus distance from the nucleus as in Figure 7.23(a), but rather a plot of $4\pi r^2\psi^2$ versus distance from the nucleus. In geometry, $4\pi r^2$ is the formula for the surface area of a sphere, but here it represents the volume of one of the thin spherical layers in Figure 7.23(c).

A significant feature of the curve in Figure 7.23(b) is that its maximum value corresponds to the most likely radial distance of the electron from the nucleus. The value of r corresponding to this maximum for the 1s orbital of hydrogen is 53 pm.

Figure 7.23(d) provides a view of the spherical shape of this (or any other) s orbital. The surface of the sphere encloses the volume within which the probability of finding a 1s electron is 90%. This type of depiction, called a *boundary–surface representation*, is one of the most useful ways to view the relative sizes, shapes, and orientations of orbitals. All s orbitals are spheres, which have only one orientation and in which electron density depends only on distance from the nucleus. Boundary surfaces are a useful way to depict the shape and relative size of an orbital.

The relative sizes of 1s, 2s, and 3s orbitals are shown in Figure 7.24. Note that orbital size increases with increasing values of the principal quantum number n. Note also that the sections of the spheres above the profile curves show bands in which the density of dots is high. The dots represent the probability of an electron being in these regions of three-dimensional space, and each band is called a *local maximum* of electron density. In all three profiles, a local maximum occurs close to the nucleus. This means that electrons in s orbitals—even s orbitals with high values of n—have some probability of being close to the nucleus.

The local maxima in any s orbital are separated from other local maxima by nodes. The number of nodes in any s orbital is equal to $n - 1$. Nodes have the same meaning here as they do in one-dimensional standing waves: they are places where the wave has zero amplitude. In the context of electrons as three-dimensional matter waves, nodes are locations at which electron density goes to zero.

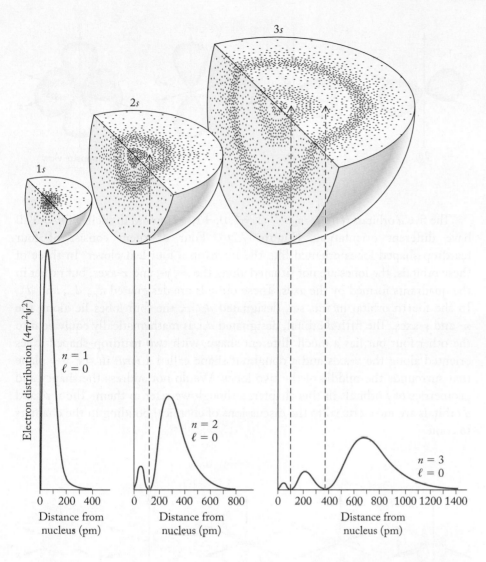

FIGURE 7.24 These radial distribution profiles of 1s, 2s, and 3s orbitals have 0, 1, and 2 nodes, respectively, identifying (with dashed arrows) locations of zero electron density. Electrons in all these s orbitals have some probability of being close to the nucleus, but 3s electrons are more likely to be farther away from the nucleus than 2s electrons, which are more likely to be farther away than 1s electrons.

CONCEPT **TEST**

How many nodes are there in the electron distribution profile of the 6s orbital?

p and d Orbitals

All shells with $n \geq 2$ have a subshell containing three p orbitals ($\ell = 1$; $m_\ell = -1$, 0, +1). Each of these orbitals has two teardrop-shaped lobes, oriented opposite each other, along one of the three perpendicular Cartesian axes x, y, z (Figure 7.25). These orbitals are designated p_x, p_y, and p_z, depending on the axis along which the lobes are situated. The two lobes of a p orbital are sometimes labeled with plus and minus signs, indicating that the sign of the wave function defining them is either $+\psi$ or $-\psi$. (These signs are not in any way connected with electric charges.) An electron in a p orbital occupies *both* lobes. Because a node of zero probability separates the two lobes, you may wonder how an electron gets from one lobe to the other. One way to think about how this happens is to remember that an electron behaves as a three-dimensional standing wave, and waves have no difficulty passing through nodes. After all, Figure 7.15 shows a node right in the middle of the $n = 2$ standing wave separating the positive and negative displacement regions in the vibrating string.

FIGURE 7.25 Boundary–surface views of the three *p* orbitals, showing their orientation along the *x*-, *y*-, and *z*-axes. The nucleus of the atom containing these orbitals is located at the origin. These shapes are an elongated version of the theoretical shapes of the orbitals. We use this version throughout the book to make it easier to see the orientation of the lobes.

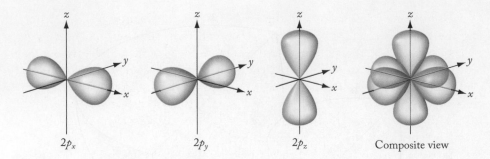

$2p_x$ $2p_y$ $2p_z$ Composite view

The five *d* orbitals ($\ell = 2$; $m_\ell = -2, -1, 0, +1, +2$) found in shells of $n \geq 3$ all have different orientations (Figure 7.26). Four of them consist of four teardrop-shaped lobes oriented like the leaves in a four-leaf clover. In three of these orbitals, the lobes are not situated along the *x*-, *y*-, and *z*-axes, but rather in the quadrants formed by the axes. These orbitals are designated d_{xy}, d_{xz}, and d_{yz}. In the fourth orbital in this set, designated $d_{x^2-y^2}$, the four lobes lie along the *x*- and *y*-axes. The fifth *d* orbital, designated d_{z^2}, is mathematically equivalent to the other four but has a much different shape, with two teardrop-shaped lobes oriented along the *z*-axis and a doughnut shape called a *torus* in the *x*–*y* plane that surrounds the middle of the two lobes. We do not address the shapes and geometries of *f* orbitals in this chapter, although we refer to them. The *s*, *p*, and *d* orbitals are most crucial to the discussions of chemical bonding in the chapters to come.

FIGURE 7.26 Boundary–surface views of the five *d* orbitals, showing their orientation relative to the *x*-, *y*-, and *z*-axes. The d_{xy}, d_{xz}, and d_{yz} orbitals are not aligned along any axis; the $d_{x^2-y^2}$ orbital lies along the *x* and *y* axes; the d_{z^2} orbital consists of two teardrop-shaped lobes along the *z*-axis with a doughnut-shaped torus ringing the point where the two lobes meet.

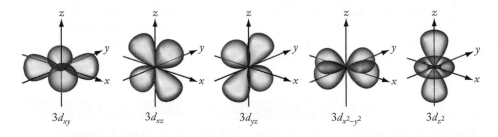

$3d_{xy}$ $3d_{xz}$ $3d_{yz}$ $3d_{x^2-y^2}$ $3d_{z^2}$

CHEMTOUR
Electron Configuration

CONNECTION We described in Section 2.5 how Mendeleev developed the first useful periodic table of the elements, not only decades before de Broglie gave us electron waves and Schrödinger pioneered quantum mechanics, but even before atoms were known to consist of electrons, protons, and neutrons.

7.8 The Periodic Table and Filling the Orbitals of Multielectron Atoms

Now that we have developed a model to describe the atomic orbitals, we can use that model to describe atoms containing more than one electron. To explore the orbital-filling sequence, let's start at the beginning of the periodic table with hydrogen and put each successive electron into the lowest-energy orbital available as we move through the table element by element. This method is based on the **aufbau principle** (German *aufbauen*, "to build up"), which states that the most stable atomic structures are those in which the electrons are in the lowest-energy orbitals available.

To decide which orbitals contain electrons, we start with two rules:

1. Electrons always go into the lowest-energy orbital available.
2. Each orbital can hold up to two electrons.

Using these rules, let's assign the single electron in a hydrogen ($Z = 1$) atom to the appropriate orbital. Actually, we have already discussed how the single electron in a ground-state atom of hydrogen is in the $1s$ orbital. We represent this arrangement with the **electron configuration** $1s^1$, where the first "1" indicates the principal quantum number (n) of the orbital, "s" indicates the type of orbital, and the superscript "1" indicates that there is *one* electron in this $1s$ orbital. Because the hydrogen atom has only one electron, $1s^1$ is the complete electron configuration for a hydrogen atom in its ground state.

To be unambiguous about the electron configuration, we use **orbital diagrams** to show how electrons, represented by single-headed arrows, are distributed among orbitals, represented by boxes. A single-headed arrow pointing upward represents an electron with spin up ($m_s = +\frac{1}{2}$), and a downward-pointing single-headed arrow represents an electron with spin down ($m_s = -\frac{1}{2}$). Because hydrogen has only one electron, its orbital diagram contains one single-headed arrow in a box labeled $1s$ (Figure 7.27a).

The atomic number of helium is 2, which tells us there are two protons and two electrons in the neutral atom. Using the aufbau principle, we simply add another electron to the $1s$ orbital. We know that each orbital can contain two electrons, but the spin quantum numbers for the two electrons cannot be the same. One must be $+\frac{1}{2}$ and the other must be $-\frac{1}{2}$. These two electrons are said to be *spin-paired*. Their presence gives helium a ground-state electron configuration of $1s^2$. With two electrons, the $1s$ orbital is now full and so is the $n = 1$ shell (Figure 7.27b).

The concept of a *filled shell* is key to understanding chemical properties: elements composed of atoms that have filled s and p sub-shells in their outermost shells are chemically stable and generally unreactive. Helium is such an element, as are all the other elements in group 18.

The location of lithium ($Z = 3$) in the periodic table—the first element in the second period—is a signal that lithium has one electron in its $n = 2$ shell. The row numbers in the periodic table correspond to the n values of the outermost shells of the elements in the rows. The second shell has four orbitals (one $2s$ and three $2p$), so it can hold up to eight electrons. Lithium's third electron occupies the lowest-energy orbital in the second shell, which is the $2s$ orbital, making its electron configuration $1s^2 2s^1$.

Now that we've begun to place electrons in the second shell, it's worth taking a closer look at the forces that a second-shell electron experiences. An electron in a $2s$ or $2p$ orbital has an electrostatic attraction to the positively charged nucleus, but it is also shielded from the nucleus by the negative charges of the two $1s$ electrons (Figure 7.28). This shielding effect reduces the electrostatic attraction, known as the **effective nuclear charge (Z_{eff})**, between a second-shell electron and the nucleus.

Returning to the ground-state lithium atom, we saw that its third electron occupies the $2s$ orbital. Perhaps you are wondering why the $2s$ orbital is lower in energy than any of the $2p$ orbitals. We can best explain this difference in energies by comparing the radial distribution profiles of these orbitals (Figure 7.29). The small peak on the $2s$ curve near the nucleus tells us that an electron in the $2s$ orbital is closer to the nucleus more of the time than an electron in a $2p$ orbital, which has no such secondary peak. This closer proximity means that a $2s$ electron experiences a greater Z_{eff} than an electron in a $2p$ orbital, even though both are

aufbau principle the method of building electron configurations of atoms by adding one electron at a time as atomic number increases across the rows of the periodic table; each electron goes into the lowest-energy orbital available.

electron configuration the distribution of electrons among the orbitals of an atom or ion.

orbital diagram depiction of the arrangement of electrons in an atom or ion, using boxes to represent orbitals.

effective nuclear charge (Z_{eff}) the attraction toward the nucleus experienced by an electron in an atom, equal to the positive charge on the nucleus reduced by the extent to which other electrons in the atom shield the electron from the nucleus.

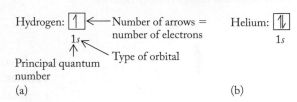

(a)

(b)

FIGURE 7.27 Orbital diagrams for hydrogen and helium indicate the orbitals in which the electrons are found and the number of electrons in each orbital. The label on the orbital, $1s$, indicates the value of the principal quantum number for the orbital ($n = 1$) and the type of orbital (s). (a) Hydrogen atoms have one electron in the $1s$ orbital, whereas (b) helium atoms have two.

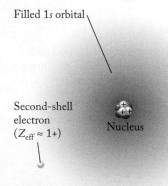

FIGURE 7.28 The effective nuclear charge (Z_{eff}) experienced by an outer-shell electron in Li equals the actual nuclear charge (3+) less the shielding effect of the negative charges of the two $1s$ electrons (2−). This shielding produces a net Z_{eff} of about 1+.

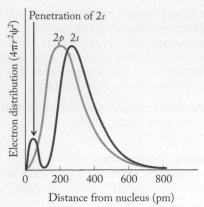

FIGURE 7.29 Radial distribution profiles of electrons in 2s and 2p orbitals. The 2s orbital is lower in energy because electrons in it penetrate more closely to the nucleus, as indicated by the local maximum in electron distribution about 50 pm from the nucleus. As a result 2s electrons experience a greater effective nuclear charge than 2p electrons.

core electrons electrons in the filled, inner shells of an atom or ion that are not involved in chemical reactions.

valence electrons electrons in the outermost occupied shell of an atom. These are the electrons that are transferred or shared in chemical reactions.

valence shell the outermost occupied shell of an atom.

degenerate (orbitals) describes orbitals of the same energy.

Hund's rule the lowest-energy electron configuration of an atom has the maximum number of unpaired electrons, all of which have the same spin, in degenerate orbitals.

shielded by the 1s electrons. Therefore, a 2s electron has lower energy than a 2p electron, which is why the 2s orbital fills first.

We can simplify the electron configuration of Li and all the elements that follow it in the periodic table. The simplified form is called a *condensed electron configuration*. In this form the symbols representing all of the electrons in orbitals that were filled in the rows above the element of interest are replaced by the symbol of the group 18 element at the end of the row above the element. For example, the condensed electron configuration of Li is [He]2s¹. Condensed electron configurations are useful because they replace the orbital notation for the **core electrons** in filled shells and subshells with the symbol of a single noble gas. Only the orbital notation for the electrons in the outermost shell is written out. These outermost electrons are called **valence electrons**. The distinction between core and valence electrons will be useful in later chapters to explain reactions and bond formation.

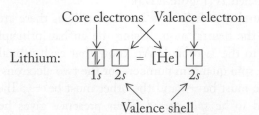

In any atom, the shell containing the valence electrons is referred to as the **valence shell**. Notice that lithium has a single electron in its valence-shell *s* orbital, as does hydrogen, the element directly above it in the periodic table; therefore both atoms have the valence-shell configuration ns^1. Here *n* represents both the number of the row in which the atom is located on the table and the principal quantum number of the valence shell in the atom.

Beryllium ($Z = 4$) is the fourth element in the periodic table and the first in group 2. The configuration for its four electrons is $1s^22s^2$, or $[He]2s^2$. The other elements in group 2 also have two spin-paired electrons in the *s* orbital of their valence shell. The second shell is not filled at this point because it also has three empty *p* orbitals that will fill as we move to the next elements in the periodic table.

Boron ($Z = 5$) is the next element and the first element in group 13. Its fifth electron is in one of its three 2p orbitals, resulting in the condensed electron configuration $[He]2s^22p^1$. Designating the 2p orbital that contains the fifth electron as $2p_x$, $2p_y$, or $2p_z$ is not important because these three orbitals all have the same energy; we say they are **degenerate** orbitals.

The next element is carbon ($Z = 6$). It has four electrons in its valence shell (the $n = 2$ shell), so its condensed electron configuration is $[He]2s^22p^2$. Note that there are two electrons in the 2p orbitals. Are they both in the same orbital? Remember that all electrons have a negative charge and repel one another. Thus they occupy orbitals that are as far away from each other as possible, which means the two 2p electrons in carbon occupy separate 2p orbitals. This distribution pattern follows **Hund's rule**, named after the German physicist Friedrich Hund (1896–1997), which states that the lowest-energy electron configuration for degenerate orbitals (such as the $2p_x$, $2p_y$, and $2p_z$ orbitals in the 2p subshell) is the one with the maximum number of unpaired valence electrons, all of which have the same spin.

Hund's rule is the third rule we apply when determining electron configurations. By convention, the first electron placed in an orbital has a positive spin.

When electrons fill a degenerate set of orbitals, Hund's rule requires that an electron with a positive spin occupy each valence orbital before any electron with a negative spin enters any orbital. Electrons do not pair in degenerate orbitals until each orbital is occupied by a single electron. To obey Hund's rule, the orbital diagram for carbon must be

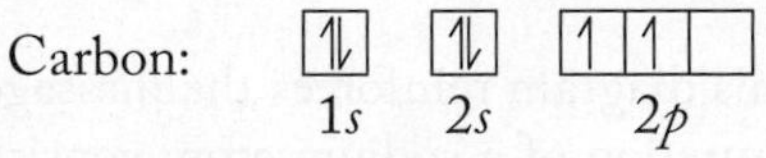

Carbon:

which shows that the two $2p$ electrons are unpaired and have the same spin.

The next element is nitrogen ($Z = 7$), represented by either $1s^2 2s^2 2p^3$ or $[He]2s^2 2p^3$. According to Hund's rule, the third $2p$ electron resides alone in the third $2p$ orbital, so that the electron distribution is

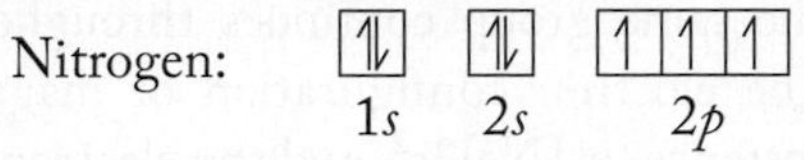

Nitrogen:

As we proceed across the second row to neon ($Z = 10$), we fill the $2p$ orbitals as shown in Figure 7.30. The last three $2p$ electrons added (in oxygen, fluorine, and neon) spin pair with the first three, so that in neon, the three $2p$ orbitals are all filled. At this point the $n = 2$ shell is full. Note that neon is directly below helium in group 18. Helium, neon, and all the other noble gases in group 18 have filled s and p orbitals in their valence shells. This same trend is seen throughout the periodic table: *main group elements in the same column of the periodic table have the same valence-shell electron configuration.* Argon, krypton, xenon, and radon also have filled s and p orbitals in their outermost occupied shells, and they also are chemically inert gases at room temperature. Thus, chemical inertness is associated with filled s and p orbitals in the valence shell.

Sodium ($Z = 11$) follows neon in the periodic table. It is the third element in group 1 and the first element in the third period. Ten of its electrons are distributed as in neon. The 11th electron is in the lowest-energy orbital available after $2p$ has been filled, which is $3s$. The condensed electron configuration of Na is $[Ne]3s^1$. Just as we write condensed electron configurations that provide orbital notation

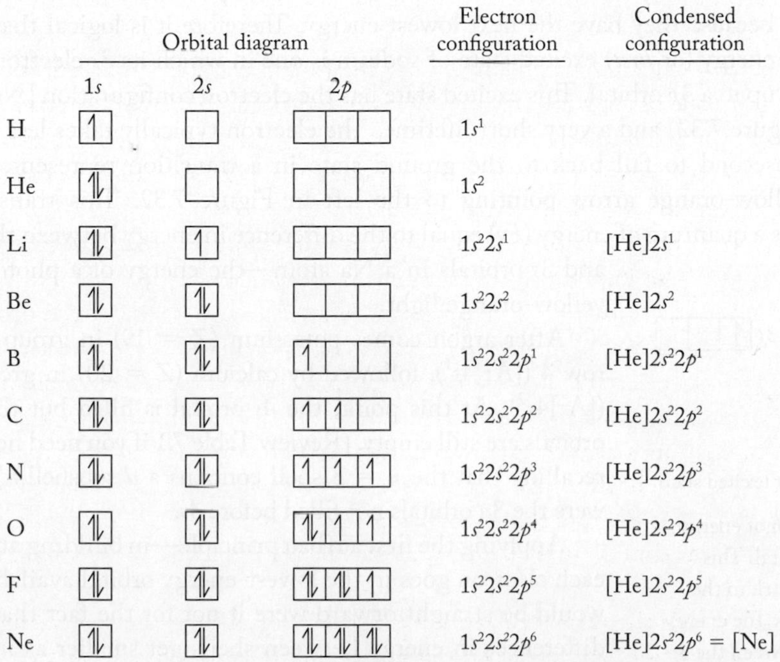

	Orbital diagram			Electron configuration	Condensed configuration
	1s	2s	2p		
H	↑			$1s^1$	
He	↑↓			$1s^2$	
Li	↑↓	↑		$1s^2 2s^1$	$[He]2s^1$
Be	↑↓	↑↓		$1s^2 2s^2$	$[He]2s^2$
B	↑↓	↑↓	↑	$1s^2 2s^2 2p^1$	$[He]2s^2 2p^1$
C	↑↓	↑↓	↑ ↑	$1s^2 2s^2 2p^2$	$[He]2s^2 2p^2$
N	↑↓	↑↓	↑ ↑ ↑	$1s^2 2s^2 2p^3$	$[He]2s^2 2p^3$
O	↑↓	↑↓	↑↓ ↑ ↑	$1s^2 2s^2 2p^4$	$[He]2s^2 2p^4$
F	↑↓	↑↓	↑↓ ↑↓ ↑	$1s^2 2s^2 2p^5$	$[He]2s^2 2p^5$
Ne	↑↓	↑↓	↑↓ ↑↓ ↑↓	$1s^2 2s^2 2p^6$	$[He]2s^2 2p^6 = [Ne]$

FIGURE 7.30 Orbital diagrams and condensed electron configurations for the first ten elements show that each orbital (indicated by a square in the orbital diagrams) holds at most two electrons and the two electrons must be of opposite spin. The orbitals are filled in order of increasing quantum numbers n and ℓ. Condensed electron configurations for all elements are given in Appendix 3.

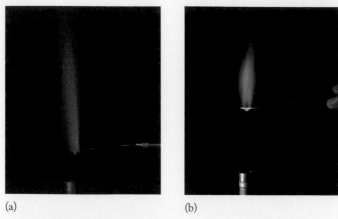

(a)   (b)

FIGURE 7.31 The alkali metals produce characteristic colors in Bunsen burner flames because the high flame temperatures produce excited-state atoms of these elements. (a) The yellow-orange glow of Na atoms. (b) The lavender color of potassium atoms.

for only the outermost occupied shell, we can also condense orbital diagrams in this same way, so that the condensed orbital diagram for sodium is

Sodium: [Ne] $\boxed{\uparrow}$
                    $3s$

This diagram reinforces the message that the electron configuration of a sodium atom consists of a neon core plus a single electron in the $3s$ orbital of the valence shell. The sodium atom has the same generic valence-shell configuration as lithium and hydrogen, namely, ns^1, where n is the period number. This pattern for main group elements in the same group continues throughout the periodic table. The electron configuration of magnesium ($Z = 12$), for instance, is $[Ne]3s^2$, and the electron configuration of every other element in group 2 consists of the immediately preceding noble gas core followed by ns^2.

The next six elements in the periodic table—aluminum, $[Ne]3s^23p^1$, to argon, $[Ne]3s^23p^6$—show a pattern of increasing numbers of $3p$ electrons, a trend that continues until all three $3p$ orbitals are filled (six electrons), which means the s and p orbitals of the $n = 3$ shell are filled (eight electrons). Thus, argon is chemically inert, as predicted by its position in group 18.

Before leaving the third row, let's revisit the condensed electron configuration of sodium, $[Ne]3s^1$. This configuration represents a ground-state sodium atom because all of the electrons, and most importantly its valence electron, occupy the lowest-energy orbitals available. Now think back to the discussion about atomic emission spectra in Section 7.1 and the distinctive yellow-orange glow that sodium makes in the flames of Bunsen burners, as shown back in Figure 7.3 and here in Figure 7.31(a). Each sodium atom absorbs a quantum of energy from the Bunsen burner that raises the valence electron from the ground state to an excited state (a transition represented by the orange arrow pointing to the right in Figure 7.32). The easiest excited state to populate is the one with the smallest energy above the ground state. We have seen that the $3p$ orbitals fill after the $3s$ orbital because they have the next lowest energy. Therefore it is logical that the lowest-energy (or *first*) excited state of sodium is one in which its $3s$ electron has moved up to a $3p$ orbital. This excited state has the electron configuration $[Ne]3p^1$ (see Figure 7.32) and a very short lifetime. The electron typically takes less than a nanosecond to fall back to the ground state in a transition represented by the yellow-orange arrow pointing to the left in Figure 7.32. This transition releases a quantum of energy ($h\nu$) equal to the difference in energy between the $3p$ and $3s$ orbitals in a Na atom—the energy of a photon of yellow-orange light.

After argon comes potassium ($Z = 19$) in group 1 of row 4 ($[Ar]4s^1$), followed by calcium ($Z = 20$) in group 2 ($[Ar]4s^2$). At this point, the $4s$ orbital is filled but the $3d$ orbitals are still empty. (Review Table 7.1 if you need help in recalling that the $n = 3$ shell contains a d subshell.) Why were the $3d$ orbitals not filled before $4s$?

Applying the first aufbau principle—in building atoms, each electron goes in the lowest-energy orbital available—would be straightforward were it not for the fact that the differences in energy between shells get smaller as n gets larger (Figure 7.33). These smaller differences result in

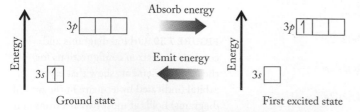

FIGURE 7.32 A ground-state Na atom absorbs a quantum of energy as its valence electron moves from the $3s$ orbital to a $3p$ orbital. This $3p$ electron in the excited-state atom spontaneously falls back to the empty $3s$ orbital, emitting a photon of yellow-orange light. The energy of the photon exactly matches the difference in energy between the $3p$ and $3s$ orbitals of Na atoms.

orbitals with large ℓ values in one shell having energies similar to those of orbitals with small ℓ values in the next higher shell. Note in Figure 7.33 that the energy of the 4s orbital is slightly lower than that of the 3d orbitals. The 4s orbitals in potassium and calcium are the lowest-energy orbitals available and are filled before any electrons go into a 3d orbital.

The element after calcium is scandium ($Z = 21$). It is the first element in the central region of the periodic table, the region populated by transition metals. Scandium has the condensed electron configuration $[Ar]3d^1 4s^2$. Note that the orbitals are listed in order of increasing principal quantum number, not necessarily in the order in which they were filled. The 3d orbitals are filled in the transition metals from scandium to zinc ($Z = 30$). This pattern of filling the d orbitals of the shell whose principal quantum number is 1 less than the period number, $(n − 1)d$, is followed throughout the periodic table: the 4d orbitals are filled in the transition metals of the fifth period, and so on (Figure 7.34).

The element after scandium is titanium ($Z = 22$), which has one more d electron than scandium, so its condensed electron configuration is $[Ar]3d^2 4s^2$. Vanadium ($Z = 23$) has the expected configuration $[Ar]3d^3 4s^2$. At this point, you may

FIGURE 7.33 The energy levels in multielectron atoms increase with increasing values of n and with increasing values of ℓ within a shell. The difference in energy between adjacent shells decreases with increasing values of n, which may result in the energies of subshells in two adjacent shells overlapping. For example, electrons in 3d orbitals have slightly higher energy than those in the 4s orbital, resulting in the order of subshell filling $4s \to 3d \to 4p$.

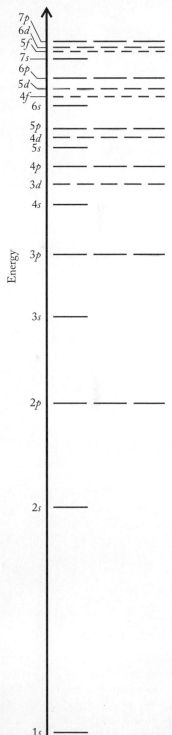

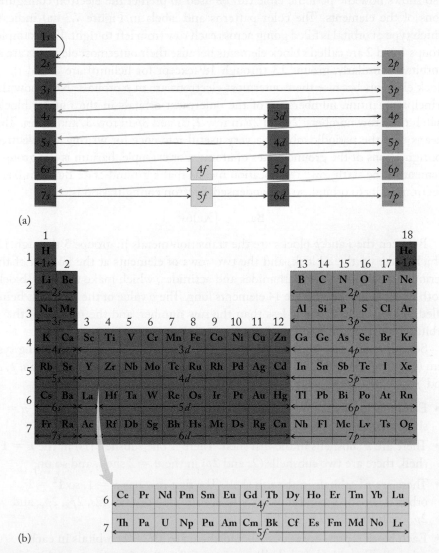

FIGURE 7.34 (a) This diagram shows the sequence in which atomic orbitals fill. (b) The same color coding in this version of the periodic table highlights the four "blocks" of elements in which valence-shell s (green), p (blue), d (orange), and f (purple) orbitals are filled as atomic number increases across a row of the table.

feel you can accurately predict the electron configurations of the remaining transition metals in the fourth period. However, because the energies of the $3d$ and $4s$ orbitals are similar, the sequence of d-orbital filling deviates in two spots from the pattern you might expect. The next element, chromium ($Z = 24$), has the configuration $[Ar]3d^5 4s^1$:

Chromium: [Ar] $\boxed{\uparrow}\boxed{\uparrow}\boxed{\uparrow}\boxed{\uparrow}\boxed{\uparrow}$ $\boxed{\uparrow}$
$ 3d^5 4s^1$

This half-filled set of d orbitals is an energetically favored configuration. Apparently, the stability of having five half-filled $3d$ orbitals compensates for the energy needed to raise a $4s$ electron to a $3d$ orbital. As a result, $[Ar]3d^5 4s^1$ is a lower-energy electron configuration than $[Ar]3d^4 4s^2$.

Another deviation from the expected filling pattern is observed near the end of a row of transition metals. Copper ($Z = 29$) has the electron configuration $[Ar]3d^{10}4s^1$ instead of the expected $[Ar]3d^9 4s^2$ because a filled set of d orbitals also represents a lower-energy electron configuration.

Figure 7.34 illustrates the overall orbital filling pattern described above. It also shows how the periodic table can be used to predict the electron configurations of the elements. The color patterns and labels in Figure 7.34(a) indicate which type of orbital is filled going across each row from left to right. For example, groups 1 and 2 are called s block elements because their outermost electrons are in s orbitals. Similarly, groups 13 through 18 (except for helium) are called the p block elements because their outermost electrons are in p orbitals. Note how the principal quantum numbers (n) of the outermost orbitals in the s and p blocks match their row numbers: $2s$ and $2p$ in row 2, $3s$ and $3p$ in row 3, and so on. This means that the periodic table is a very useful reference for writing the electron configurations of the ground states of atoms. For example, barium is the group 2 element in the sixth row. This location means that a ground-state Ba atom has 2 electrons in its $6s$ orbital, so its condensed electron configuration is

Ba: $[Xe]6s^2$

Between the s and p blocks are the transition metals in groups 3 through 12, which make up the d block, and the two rows of elements at the bottom of the periodic table, called the lanthanides and actinides, which make up the f block. Both of the f block series are 14 elements long. The n value of the d orbitals being filled in a row is always one less than the row number, and the n value of the f orbitals being filled is two less than the row number.

Several relationships are worth noting between the quantum numbering system (Figure 7.17) and the organization of the periodic table into rows and s, p, d, and f blocks (Figure 7.34b):

- Each row number corresponds to a value of n. The first row begins the $n = 1$ shell, the second row begins the $n = 2$ shell, and so on.
- There are n subshells in the nth shell. There is one subshell ($1s$) in the $n = 1$ shell, there are two subshells ($2s$ and $2p$) in the $n = 2$ shell, and so on.
- There are n^2 orbitals in the nth shell. The first row has $n = 1$, so $1^2 = 1$ orbital ($1s$). The second row has $n = 2$, so $2^2 = 4$ orbitals ($2s$, $2p_x$, $2p_y$, and $2p_z$), and so on.
- Each block represents a subshell and there are $(2\ell + 1)$ orbitals in each subshell. The s block (subshell) consists of $(2 \times 0 + 1 = 1)$ one s orbital in each row. The p block consists of $(2 \times 1 + 1 = 3)$ three p orbitals in each row. The d block consists of $(2 \times 2 + 1 = 5)$ five d orbitals in each row. Lastly, the f block consists of $(2 \times 3 + 1 = 7)$ seven d orbitals in each row.

- Because the s, p, d, and f subshells always contain 1, 3, 5, and 7 orbitals, respectively, and each orbital contains 2 electrons, the s block consists of 2 elements in one row of the periodic table, the p block consists of 6 elements, the d block consists of 10 elements, and the f block consists of 14 elements. For example, in the $n = 4$ shell, there are 2 elements in the s block (K and Ca), 6 elements in the p block (Ga through Kr), 10 elements in the d block (Y through Cd), and 14 elements in the f block (Ce through Lu).

With these patterns in mind, let's write the condensed electron configuration of lead ($Z = 82$), which is the group 14 element in the sixth row. The noble gas that most closely precedes it is Xe ($Z = 54$). The difference in atomic numbers means that we need to account for $82 - 54 = 28$ electrons in the electron configuration symbols. The location of Pb and the block labels in Figure 7.34(b) tell us that these 28 electrons are distributed as follows:

2 electrons in $6s$

14 electrons in $4f$

10 electrons in $5d$

2 electrons in $6p$

The electron configuration of ground-state lead atoms reflects this distribution:

$$\text{Pb:} \qquad [\text{Xe}]4f^{14}5d^{10}6s^26p^2$$

SAMPLE EXERCISE 7.10 Writing Electron Configurations   **LO7**

Strontium salts give off a bright red light at the high temperature of fireworks, explosions, and signal flares (Figure 7.35). What is the ground-state electron configuration of strontium atoms?

Collect, Organize, and Analyze Strontium is in the fifth row of group 2 of the periodic table. This position of Sr means that the last orbital to be filled in an atom of Sr is $5s$ (see Figure 7.34b). The sequence of filling the atom's inner-shell orbitals (Figure 7.34a) is $1s$, $2s$, $2p$, $3s$, $3p$, $4s$, $3d$, $4p$, and then $5s$. The atomic number of strontium, $Z = 38$, means that each strontium atom has 38 electrons.

Solve An s orbital can hold up to 2 electrons, a set of three p orbitals can hold 6 electrons, and a set of five d orbitals can hold 10. Therefore the distribution of the 38 electrons in an atom of Sr is $1s^22s^22p^63s^23p^63d^{10}4s^24p^65s^2$.

Think About It The location of Sr in Figure 7.34 also tells us that the condensed electron configuration of Sr is $[\text{Kr}]5s^2$.

Practice Exercise Gallium arsenide, GaAs, is used in the red lasers in bar-code readers. Write the electron configuration of a ground-state atom of gallium ($Z = 31$) and a ground-state arsenic atom ($Z = 33$).

FIGURE 7.35 Strontium nitrate is a common ingredient in road flares that give off red light.

SAMPLE EXERCISE 7.11 Writing Condensed Electron Configurations   **LO7**
                          of Transition Metal Atoms

Write the condensed electron configuration of an atom of silver.

Collect and Organize We need to write an element's condensed electron configuration, in which filled inner-shell orbitals are represented by the atomic symbol of the noble gas immediately preceding the element of interest in the periodic table. Silver ($Z = 47$) is the group 11 element in the fifth row of the periodic table.

Analyze Krypton ($Z = 36$) is the noble gas at the end of the fourth row. The difference between their atomic numbers ($47 - 36$) means that we need to assign 11 electrons to the orbitals after [Kr] in the condensed electron configuration of silver.

Solve According to the orbital filling pattern in Figure 7.34(a), 2 of the 11 electrons would be in the $5s$ orbital and the other 9 would be in $4d$ orbitals. However, a filled set of d orbitals is more stable than a partially filled set. Therefore silver, like copper just above it in the periodic table, has ten electrons in its outermost d orbital and only one electron in its outermost s orbital. The condensed electron configuration of an Ag atom is $[Kr]4d^{10}5s^1$.

Think About It We can generate a tentative electron configuration by simply moving across a row in Figure 7.34(b) until we come to the element of interest. However, we may need to move an s electron to a d orbital to take into account the special stability of half-filled and filled d orbitals.

 Practice Exercise Write the condensed electron configuration of a ground-state atom of cobalt ($Z = 27$).

7.9 Electron Configurations of Ions

In Section 2.6 we learned that metals and nonmetals can form cations or anions (Figure 2.19) by losing or gaining one or more electrons. How does the electron configuration of an ion compare with that of the neutral parent atom? To write the electron configuration of an ion, we begin with the electron configuration of its neutral parent atom. For a cation, we remove the appropriate number of electrons from the orbital(s) with the highest principal quantum number. For an anion, we add the appropriate number of electrons to one or more partially filled outer-shell orbitals.

Ions of the Main Group Elements

The s block elements (see Figure 7.34) form monatomic cations by losing all the outer-shell electrons, producing ions with the electron configurations of the noble gases immediately preceding in the periodic table. For example, an atom of sodium forms a Na$^+$ ion by losing the single $3s$ electron:

$$Na \rightarrow Na^+ + e^-$$
$$[Ne]3s^1 \rightarrow [Ne] + e^-$$

The nonmetal elements of the p block that form monatomic anions do so by gaining enough electrons to fill the outer-shell p orbitals, producing ions with the electron configurations of the noble gases at the right ends of the corresponding rows in the periodic table. For example, an atom of fluorine forms a F$^-$ ion by gaining one electron, which fills the set of three $2p$ orbitals and gives it the electron configuration of neon:

$$F + e^- \rightarrow F^-$$
$$[He]2s^22p^5 + e^- \rightarrow [He]2s^22p^6 = [Ne]$$

Note that a Na^+ ion and a F^- ion both have the same electron configuration as the neutral atom of Ne. We say that these three species, Na^+, F^-, and Ne, are **isoelectronic**, meaning they have the same electron configuration.

CONNECTION In previous chapters we learned the charges on the ions of common elements. Electron configurations help us understand why these ions have the charges they do.

SAMPLE EXERCISE 7.12 Determining Isoelectronic Species   **LO7**

(a) Write the condensed electron configurations of the ions in CsF, $MgCl_2$, CaO, and KBr. (b) Which ions in part (a) are isoelectronic?

Collect and Organize In part (a), we must determine the electron configuration of each ion in four binary ionic compounds. In part (b), our task is to identify the configurations from part (a) that are equivalent. Figure 7.34(a) provides information on the order in which orbitals are filled.

Analyze The elements in the compounds include
- two from group 1, Cs and K, which are present as 1+ cations;
- two from group 2, Mg and Ca, which are present as 2+ cations;
- one from group 16, O, which is present as a 2− anion;
- three from group 17, F, Cl, and Br, which are present as 1− anions.

Solve

a. Let's arrange these atoms and ions in a table in which we list the condensed electron configurations of the parent atoms, the formulas of the ions (derived from the parent elements' positions in the periodic table—see Figure 2.19), followed by the number of electrons lost or gained to form each ion, and finally the condensed electron configuration of the ion that is produced.

Element	Electron Configuration of Atom	Formula of Ion	Number of Electrons Lost/Gained	Electron Configuration of Ion
Cs	$[Xe]6s^1$	Cs^+	−1	$[Xe]$
K	$[Ar]4s^1$	K^+	−1	$[Ar]$
Mg	$[Ne]3s^2$	Mg^{2+}	−2	$[Ne]$
Ca	$[Ar]4s^2$	Ca^{2+}	−2	$[Ar]$
O	$[He]2s^22p^4$	O^{2-}	+2	$[He]2s^22p^6 = [Ne]$
F	$[He]2s^22p^5$	F^-	+1	$[He]2s^22p^6 = [Ne]$
Cl	$[Ne]3s^23p^5$	Cl^-	+1	$[Ne]2s^22p^6 = [Ar]$
Br	$[Ar]3d^{10}4s^24p^5$	Br^-	+1	$[Ar]3d^{10}4s^24p^6 = [Kr]$

b. Examining the ion electron configurations in the last column, we have three ions that are isoelectronic with Ne (Mg^{2+}, O^{2-}, and F^-), and three that are isoelectronic with Ar (K^+, Ca^{2+}, and Cl^-).

Think About It Each of the ions has the electron configuration of a noble gas atom. These configurations are associated with stable ions of many main group elements.

 Practice Exercise Write the electron configurations of the ions in KI, BaO, Rb_2O, and $AlCl_3$. Which of these ions are isoelectronic with Ar?

CONCEPT TEST

Potassium has one valence electron in the $4s$ orbital. Figure 7.31(b) shows the color produced from excited-state potassium atoms. Write a condensed electron configuration for an excited-state potassium atom. How does that electron configuration differ from that of a potassium ion?

isoelectronic describes atoms or ions that have identical electron configurations.

Transition Metal Cations

As with the main group elements, writing the electron configurations of transition metal cations begins by considering the metal atoms from which the cations form. Zinc atoms, like those of many transition metals, form ions with 2+ charges by losing both electrons from the *s* orbital in the outermost shell:

$$Zn \rightarrow Zn^{2+} + 2\,e^-$$
$$[Ar]3d^{10}4s^2 \rightarrow [Ar]3d^{10} + 2\,e^-$$

A few transition metals, including silver, have only one outer-shell *s* electron and may form singly charged ions:

$$Ag \rightarrow Ag^+ + e^-$$
$$[Kr]4d^{10}5s^1 \rightarrow [Kr]4d^{10} + e^-$$

We might have expected Zn and Ag atoms to lose the 3*d* and 4*d* electrons, respectively, reasoning that the last orbitals to be filled should be the first to lose electrons when an atom forms a positive ion. However, the rule that the electrons in orbitals with the highest *n* value ionize first applies to all atoms, including transition metals. Preferential loss of outer-shell *s* electrons explains why the most frequently encountered charge on transition metal ions is 2+.

Transition metal atoms may also lose one or more *d* electrons in addition to the two valence *s* electrons when forming ions with charges $\geq$ 2+. For example, an atom of scandium, $[Ar]3d^14s^2$, loses both 4*s* electrons *and* the 3*d* electron as it forms a Sc^{3+} ion. The chemistry of titanium, $[Ar]3d^24s^2$, is dominated by its atoms losing all four of the 4*s* and 3*d* electrons to form Ti^{4+} ions. In other cases, some—but not all—of the outermost *d* electrons are lost when a transition metal atom forms an ion. For example, when an atom of iron, $[Ar]3d^64s^2$, forms a Fe^{3+} ion, it loses both 4*s* electrons but only one 3*d* electron, which results in the condensed electron configuration $[Ar]3d^5$.

SAMPLE EXERCISE 7.13 Writing Condensed Electron   **LO7**
                          Configurations of Transition Metal Ions

What are the condensed electron configurations of Fe^{3+} and Ni^{2+}?

Collect and Organize We are asked to write the electron configurations of ions formed by two transition metals: iron ($Z = 26$) and nickel ($Z = 28$). Figure 7.34(a) provides information on the order in which orbitals are filled. Transition metal atoms preferentially lose their outermost *s* electrons when forming ions.

Analyze The location of iron in group 8 of the periodic table tells us that its atoms have two 4*s* electrons and six 3*d* electrons built on an argon core. Nickel is in group 10, so its atoms have two more 3*d* electrons than iron. Therefore the electron configurations of Fe and Ni atoms are

$$Fe = [Ar]3d^64s^2 \text{ and } Ni = [Ar]3d^84s^2$$

Solve Fe atoms lose their 4*s* electrons *and* one 3*d* electron when forming Fe^{3+} ions, and Ni atoms lose two 4*s* electrons to form Ni^{2+} ions, producing these condensed electron configurations:

$$Fe \rightarrow Fe^{3+} + 3\,e^-$$
$$[Ar]3d^64s^2 \rightarrow [Ar]3d^5 + 3\,e^-$$
$$Ni \rightarrow Ni^{2+} + 2\,e^-$$
$$[Ar]3d^84s^2 \rightarrow [Ar]3d^8 + 2\,e^-$$

Practice Exercise Write the electron configurations for the manganese atom and the ions Mn^{3+} and Mn^{4+}.

We have yet to consider the lanthanides (elements 58 through 71) and actinides (elements 90 through 103), represented by the two purple rows in Figure 7.34(b). The lanthanides have partly filled $4f$ orbitals, and the actinides have partly filled $5f$ orbitals. There are 14 elements in each group, reflecting the capacity of the seven orbitals in each f subshell ($\ell = 3$; $m_\ell = -3, -2, -1, 0, +1, +2,$ and $+3$). As you can see in Figure 7.34(a), the $4f$ orbitals are not filled until after the $6s$ orbital has been filled. This order of filling is due to the similar energies of $6s$ and $4f$ orbitals (Figure 7.33). Similarly, the $5f$ orbitals are filled after the $7s$ orbital is filled.

The periodic table is a useful reference for predicting the physical and chemical properties of elements. Our quantum mechanical perspectives on atomic structure provide us with a theoretical basis for explaining why particular families of elements behave similarly. The original table was based on periodic trends in observable chemical properties. Now we know that the chemical properties of an element are closely linked to the electron configurations of the atoms of the element.

7.10 The Sizes of Atoms and Ions

The sizes of atoms are usually expressed in terms of their radii. The atomic radius of an element that occurs in nature as a diatomic molecule, such as N_2 and O_2, is simply half the distance between the nuclear centers in the molecule (Figure 7.36a). The atomic radius of a metal, also called its *metallic* radius, is half the distance between the nuclear centers in the solid metal (Figure 7.36b). The values of ionic radii are derived from the distances between nuclear centers in solid ionic compounds (Figure 7.36c). The periodic trends in the relative sizes of atoms are shown in Figure 7.37. Several factors contribute to these trends.

Trends in Atom and Ion Sizes

The radial distribution profiles in Figure 7.24 show how the distance from the nucleus to the s orbitals increases as the principal quantum number n increases. This same trend is observed for p, d, and f orbitals as well. Therefore we expect the atomic radii of elements in the same group of the periodic table to increase with increasing atomic number. For example, the atomic radii of

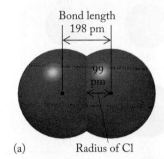

(a)   Radius of Cl

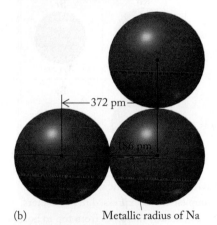

(b)   Metallic radius of Na

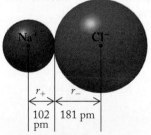

(c) Ionic radii of Na^+ and Cl^-

FIGURE 7.36 A comparison of covalent, metallic, and ionic radii. (a) A covalent radius is half the length of the bond between identical atoms in a molecule, such as the bond in Cl_2. (b) A metallic radius is based on the distance of closest approach of adjacent atoms in a crystalline metal. (c) Ionic radii are based on the distances between the centers of adjacent ions in the crystals of many ionic compounds.

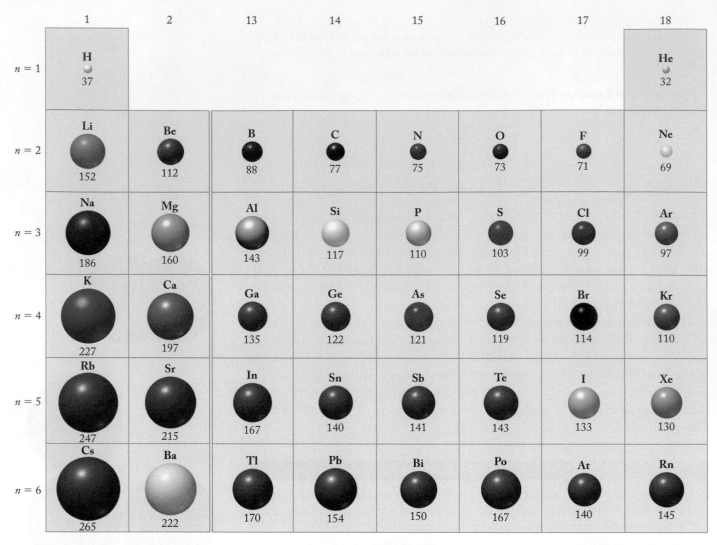

	1	2	13	14	15	16	17	18
$n = 1$	H 37							He 32
$n = 2$	Li 152	Be 112	B 88	C 77	N 75	O 73	F 71	Ne 69
$n = 3$	Na 186	Mg 160	Al 143	Si 117	P 110	S 103	Cl 99	Ar 97
$n = 4$	K 227	Ca 197	Ga 135	Ge 122	As 121	Se 119	Br 114	Kr 110
$n = 5$	Rb 247	Sr 215	In 167	Sn 140	Sb 141	Te 143	I 133	Xe 130
$n = 6$	Cs 265	Ba 222	Tl 170	Pb 154	Bi 150	Po 167	At 140	Rn 145

FIGURE 7.37 Atomic radii of the main group elements expressed in picometers. Size generally increases from top to bottom in any group and generally decreases from left to right across any period.

the halogens increase as we move from the top to the bottom of group 17 of the periodic table:

Element	F	Cl	Br	I	At
Atomic Number	9	17	35	53	85
n Value of Valence Shell	2	3	4	5	6
Atomic Radius (pm)	71	99	114	133	140

This pattern holds for other groups of elements, as shown in Figure 7.37.

As we move across a row (period) in the periodic table, each successive element has one additional electron. We might expect that the addition of electrons across a row would mean a corresponding increase in the size of the atom. Surprisingly, however, experimental data do not support this prediction. Atomic radii tend to *decrease* as the atomic number increases across a row of the periodic table. This pattern is particularly evident as we move from left to right across the main group elements (Figure 7.37). To explain these data, we need to carefully consider two competing interactions:

1. **Increasing effective nuclear charge.** Each time the atomic number increases, so does the positive charge of the nucleus. Consider, for example,

Na and Mg. An atom of sodium-23 has 11 protons, 12 neutrons, and 11 electrons, whereas an atom of magnesium-24 has 12 protons, 12 neutrons, and 12 electrons. The one valence electron of Na and the two valence electrons of Mg are both in the 3*s* orbital, meaning that the valence electron in Na is probably at the same distance from the Na nucleus as the two valence electrons in Mg are from the Mg nucleus. The two valence electrons in Mg experience a larger effective nuclear charge than the valence electron in Na: the electrostatic attraction of electrons (be it one or two) in a 3*s* orbital to 12 protons is stronger than the attraction to 11 protons because the distance is unchanged. As Z_{eff} increases, the size of atoms decreases.

2. **Increasing repulsion between valence electrons.** As atomic number increases, so does the number of valence electrons. More electrons means more electron–electron repulsions (because two negatively charged particles repel one another), which tends to increase the size of the atoms.

The experimental data tabulated in Figure 7.37 show that atomic size decreases with increasing *Z* across each row of main group elements. This means that interaction 1, the effective nuclear charge, more than offsets interaction 2, the repulsions between the increasing number of valence electrons.

Note that the cations of the main group elements are much smaller than their parent atoms, but the anions are much larger (Figure 7.38). To understand these trends, let's revisit what happens when a Na atom forms a Na$^+$ ion. The neutral atom contains 11 protons and loses one of the 11 electrons to form Na$^+$. The electrostatic attractions between the 11 protons and each of the remaining 10 electrons are now larger; that is to say, the effective nuclear charge is larger for Na$^+$ than for a neutral Na atom. On the other hand, when a Cl atom gains an electron and forms a Cl$^-$ ion, it contains one more electron (18) than the 17 protons in the nucleus. The effective nuclear charge of the 17 protons on the now 18 electrons is reduced in comparison with the electrostatic attractions between 17 protons and 17 electrons in a neutral Cl atom. A reduced Z_{eff}, along with increased electron–electron repulsions, results

CHEMTOUR
Periodic Trends

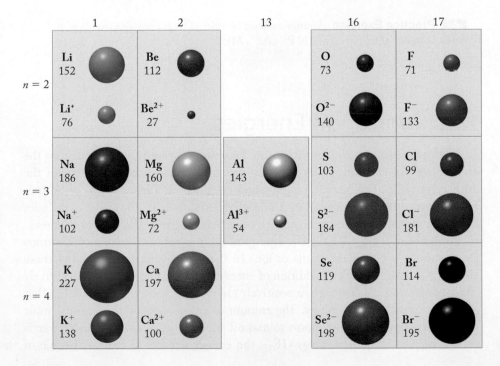

FIGURE 7.38 Comparison of atomic and ionic radii. Values are in picometers.

in an anion that is larger than its neutral parent atom. This is why Cl^- ions and all monatomic anions are larger than the atoms from which they form.

Rank the following orbitals in an atom of silver in order of decreasing effective nuclear charge experienced by the electrons in them: 1s, 2s, 3s, 4s, 2p, 3p, 4p.

SAMPLE EXERCISE 7.14 Ordering Atoms and Ions by Size **LO8**

Arrange each set by size, largest to smallest: (a) O, P, S; (b) Na^+, Na, K.

Collect and Organize We want to rank a set of three atoms according to their size and a set of two atoms and one cation of one of the atoms according to their size. The location of elements in the periodic table can be used to determine the relative sizes of the atoms.

Analyze Sizes decrease as we move from left to right across a period and increase as we move down a column. In addition, a cation is always smaller than the atom from which it is made.

Solve
a. S is below O in group 16, so in terms of atomic size, S > O. S is to the right of P, so P > S. The size order is thus P > S > O.
b. Cations are smaller than their atoms, so Na > Na^+. Size increases down a group; K is below Na in the alkali metals group, so K > Na. Therefore the size order is K > Na > Na^+.

Think About It The trend in the sizes of the atoms in set (a) reflects decreasing atomic size with increasing effective nuclear charge within a row of elements in the periodic table and increasing atomic size with increasing atomic number down a group. The relative sizes of the particles in set (b) are linked (1) to increasing atomic size as one goes down a column of elements in the periodic table and (2) to the smaller size of a cation than its parent atom.

 Practice Exercise Arrange each set in order of increasing size (smallest to largest): (a) Cl^-, F^-, Li^+; (b) P^{3-}, Al^{3+}, Mg^{2+}.

7.11 Ionization Energies

In developing electron configurations, we followed a theoretical framework for the arrangement of electrons in orbitals that was developed in the early years of the 20th century. Is there actual experimental evidence for the existence of orbitals representing different energy levels inside atoms? Yes, there is. The evidence includes the measurement of the energies needed to remove electrons from atoms and ions.

Ionization energy (IE) is the energy needed to remove 1 mole of electrons from 1 mole of gas-phase atoms or ions in their ground state. Removing these electrons always requires an addition of energy to the system because a negatively charged electron is attracted to a positively charged nucleus, and overcoming that attractive force requires energy. The amount of energy needed to remove 1 mole of electrons from 1 mole of atoms to make 1 mole of cations with a 1+ charge is called the *first ionization energy* (IE_1), the energy needed to remove 1 mole of

ionization energy (IE) the quantity of energy needed to remove 1 mole of electrons from 1 mole of ground-state atoms or ions in the gas phase.

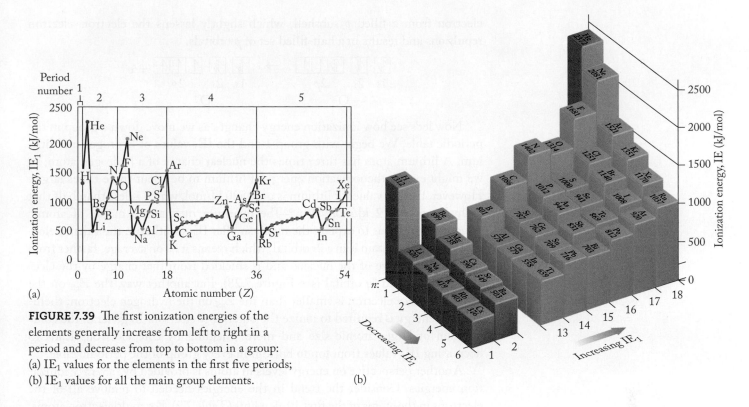

FIGURE 7.39 The first ionization energies of the elements generally increase from left to right in a period and decrease from top to bottom in a group: (a) IE_1 values for the elements in the first five periods; (b) IE_1 values for all the main group elements.

electrons from 1 mole of 1+ cations to make 1 mole of cations with a 2+ charge is the *second ionization energy* (IE_2), and so forth. For example,

$$Mg(g) \rightarrow Mg^+(g) + 1\ e^- \qquad (IE_1 = 738\ kJ/mol)$$
$$Mg^+(g) \rightarrow Mg^{2+}(g) + 1\ e^- \qquad (IE_2 = 1451\ kJ/mol)$$

The total energy required to make 1 mole of $Mg^{2+}(g)$ cations from 1 mole of $Mg(g)$ atoms is the sum of these two ionization energies:

$$Mg(g) \rightarrow Mg^{2+}(g) + 2\ e^-$$
$$Total\ IE = (738 + 1451)\ kJ/mol = 2189\ kJ/mol$$

Figure 7.39 shows how the first ionization energies of the main group elements vary. The IE_1 of hydrogen is 1312 kJ/mol. The IE_1 of helium is nearly twice as big: 2372 kJ/mol. This difference seems reasonable because He atoms have two protons per nucleus, whereas H atoms have only one. Twice the nuclear charge increases Z_{eff}, so nearly twice the ionization energy is needed to pull an electron away from the nucleus. In general, first ionization energies increase from left to right across a period. Thus the easiest element to ionize in a period is the group 1 element, and the hardest is the group 18 element. This pattern makes sense because as the charge of the nucleus increases across a row, so do Z_{eff} and the attractions between the nucleus and the valence electrons.

Two anomalies occur in the general trend of increasing IE_1 with increasing Z across a row. One shows up in the decrease in IE_1 values between the group 2 and group 13 elements in the second and third rows. Note that B and Al lose a p electron when they ionize, whereas Be and Mg lose an s electron. It is easier to remove a p electron because it experiences a smaller effective nuclear charge than an s electron in the same shell.

A second anomaly is the decrease in IE_1 values between the group 15 and 16 elements in each row. Such a decrease is associated with the removal of a single electron from a filled *p* subshell, which slightly lessens the electron–electron repulsions and results in a half-filled set of *p* orbitals.

$$\boxed{\uparrow\downarrow}\ \boxed{\uparrow\downarrow}\ \boxed{\uparrow\downarrow\ \uparrow\ \uparrow} \quad \rightarrow \quad \boxed{\uparrow\downarrow}\ \boxed{\uparrow\downarrow}\ \boxed{\uparrow\ \uparrow\ \uparrow} + e^-$$
$$\begin{array}{ccc} 1s & 2s & 2p \end{array} \qquad\qquad \begin{array}{ccc} 1s & 2s & 2p \end{array}$$
$$\text{O} \qquad\qquad\qquad \text{O}^+$$

Now let's see how ionization energy changes as we move down a group in the periodic table. We begin with group 1 and the IE_1 values of hydrogen and lithium. A lithium atom has three times the nuclear charge of a hydrogen atom, so we might expect the ionization energy of lithium to be about three times larger. However, the IE_1 value for lithium is only 520 kJ/mol, which is less than half that of hydrogen, 1312 kJ/mol. Why? The quantum mechanical model of atomic structure enables us to explain the much smaller IE_1 of lithium: the valence electron in a lithium atom is in a 2*s* orbital, which means it is, on average, farther from the positive charge of the nucleus and is shielded from that charge by the electrons in the filled 1*s* orbital (see Figure 7.28). Put another way, the Z_{eff} on the lithium valence electron is smaller than the Z_{eff} on the hydrogen electron; therefore less energy is required to ionize the outermost electron. In general, the combination of larger atomic size and more shielding by core electrons leads to decreasing IE_1 values from top to bottom in every group of the periodic table.

Another perspective on energy levels in atoms is provided by *successive* ionization energies. Consider the trend in the energies needed to remove all of the electrons in the atoms of the first 10 elements (Table 7.2). For multielectron atoms, the second ionization energy (IE_2) is always greater than the IE_1 because the second electron is being removed from an ion that already has a positive charge. Because electrons carry a negative charge, they are held more strongly in a cation (have stronger electrostatic attractions to the nucleus) than in the atom from which the cation was formed.

The energy needed to remove a third electron is greater still because it is being removed from a 2+ ion. Superimposed on this trend are much more dramatic increases in ionization energy (separated by the red line in Table 7.2) when all the valence electrons in an ionizing atom have been removed and the next electron comes from an inner shell. Core electrons experience less shielding and a greater effective nuclear charge than valence electrons, which means that much more energy is needed to ionize them.

TABLE 7.2 Successive Ionization Energies[a] of the First 10 Elements

Element	Z	IE_1	IE_2	IE_3	IE_4	IE_5	IE_6	IE_7	IE_8	IE_9
H	1	1312								
He	2	2372	5249							
Li	3	520	7296	12,040						
Be	4	900	1758	15,050	21,070					
B	5	801	2426	3660	24,682	32,508				
C	6	1086	2348	4617	6201	37,926	46,956			
N	7	1402	2860	4581	7465	9391	52,976	64,414		
O	8	1314	3383	5298	7465	10,956	13,304	71,036	84,280	
F	9	1681	3371	6020	8428	11,017	15,170	17,879	92,106	106,554
Ne	10	2081	3949	6140	9391	12,160	15,231	19,986	23,057	115,584

[a]All values in kilojoules per mole of atoms.

electron affinity (EA) the energy change that occurs when 1 mole of electrons combines with 1 mole of atoms or monatomic cations in the gas phase.

SAMPLE EXERCISE 7.15 Ranking Ionization Energies **LO8**

Arrange argon, magnesium, and phosphorus in order of increasing first ionization energy.

Collect, Organize, and Analyze We are asked to order three elements on the basis of their IE_1 values. All three are in the third row of the periodic table. First ionization energies generally increase from left to right across a row because of increasing effective nuclear charge.

Solve Ranking the elements in order of increasing first ionization energies, according to their increasing group numbers:

$$\text{Mg} < \text{P} < \text{Ar}$$

Think About It Magnesium forms stable 2+ cations, so we expect its ionization energy to be smaller than the values for phosphorus and argon, which do not form cations. Argon is a noble gas with a stable valence electron configuration, so its first ionization energy should be the highest of the three. We can check our prediction against Figure 7.39(b): Mg, 738 kJ/mol; P, 1012 kJ/mol; Ar, 1521 kJ/mol.

Practice Exercise Arrange cesium, calcium, and neon in order of decreasing first ionization energy.

CONCEPT TEST

Why does ionization energy decrease as we move down the noble gas family from helium to neon to argon?

7.12 Electron Affinities

In the preceding section we examined the periodic nature of the energy required to ionize atoms. Now we look at a complementary process and examine the change in energy when 1 mole of electrons is added to 1 mole of gas phase atoms or monatomic cations. The energies involved are called **electron affinities (EA)**.

For example, the energy associated with adding 1 mole of electrons to 1 mole of chlorine atoms in the gas phase is

$$\text{Cl}(g) + e^- \rightarrow \text{Cl}^-(g) \qquad \text{EA}_1 = -349 \text{ kJ/mol}$$

The electron affinities of many elements are negative, meaning the formation of anions with a 1− charge releases energy (Figure 7.40). However, an examination of the EA values in Figure 7.40 reveals that the trends in this property are not as regular as the trends in size and ionization energy. Electron affinity increases down a column only for the group 1 metals; other groups do not display a clear trend. In general, electron affinity becomes more negative with increasing atomic number across a row, but there are exceptions to that trend, too. The halogens of group 17 have the most negative EA values, which seems logical given that each halogen atom becomes isoelectronic with a noble gas electron configuration upon gaining one more electron.

Note that the addition of an electron to an atom of a noble gas consumes energy, which makes sense because

1	2	13	14	15	16	17	18
H −72.6							He (0.0)[a]
Li −59.6	Be >0	B −26.7	C −122	N +7	O −141	F −328	Ne (+29)[a]
Na −52.9	Mg >0	Al −42.5	Si −134	P −72.0	S −200	Cl −349	Ar (+35)[a]
K −48.4	Ca −2.4	Ga −28.9	Ge −119	As −78.2	Se −195	Br −325	Kr (+39)[a]
Rb −46.9	Sr −5.0	In −28.9	Sn −107	Sb −103	Te −190	I −295	Xe (+41)[a]
Cs −45.5	Ba −14	Tl −19.2	Pb −35.2	Bi −91.3	Po −183.3	At −270[a]	Rn (+41)[a]

[a]Calculated values.

FIGURE 7.40 Electron affinity (EA) values of main group elements are expressed in kilojoules per mole. The more negative the value, the more energy is released when 1 mole of atoms combines with 1 mole of electrons to form 1 mole of anions with a 1− charge. A greater release of energy reflects stronger electrostatic attractions between the nuclei of the elements and free electrons.

these atoms already have stable electron configurations. Beryllium and magnesium have very small, but positive, electron affinities because an additional electron occupies a valence-shell p orbital, which is significantly higher in energy than a valence-shell s orbital.

CONCEPT TEST

Describe at least one similarity and one difference in the periodic trends for first ionization energies and electron affinities among the main group elements.

SAMPLE EXERCISE 7.16 Integrating Concepts: Red Fireworks

The brilliant red color of some fireworks (Figure 7.41) is produced by the presence of lithium carbonate (Li_2CO_3) in the shells that are launched into the night sky and explode at just the right time. The explosions produce enough energy to vaporize the lithium compound and produce excited-state lithium atoms. These atoms quickly lose energy as they transition from their excited states to their ground states, emitting photons of red light that have a wavelength of 670.8 nm.

a. How much energy is in each photon of red light emitted by an excited-state lithium atom?
b. What is the frequency of the red light?
c. Lithium atoms in the lowest energy state above the ground state emit these red photons. What is the difference in energy between the ground state of a Li atom and its lowest-energy excited state?
d. What is the electron configuration of a ground-state lithium atom?
e. What is the electron configuration of a lithium atom in its lowest-energy excited state?

Collect and Organize We are asked to calculate the energy of photons of red light from their wavelength. We are also asked to relate this energy to the difference in energies between a lithium atom's lowest-energy excited state and its ground state and to write the electron configurations of these states. The energy of a photon is related to its wavelength (λ) by Equation 7.3:

$$E = \frac{hc}{\lambda}$$

Lithium is the group 1 element in the second row of the periodic table.

Analyze The energy of a photon (E) emitted by an excited-state atom is the same as the difference in energies (ΔE) between the excited state and the lower-energy state (in this case, the ground state) of the transition that produced the photon. All the group 1 elements have one electron in their valence-shell s orbital. In lithium this electron is in the $2s$ orbital, which is occupied after the $1s$ has filled. The next higher energy orbitals are the $2p$ orbitals.

Solve
a. The energy of a photon with a wavelength of 670.8 nm is

$$E = \frac{hc}{\lambda} = \frac{(6.626 \times 10^{-34}\,\text{J} \cdot \text{s})\left(2.998 \times 10^8\,\frac{\text{m}}{\text{s}}\right)}{670.8\,\text{nm}\left(\dfrac{10^{-9}\,\text{m}}{\text{nm}}\right)} = 2.961 \times 10^{-19}\,\text{J}$$

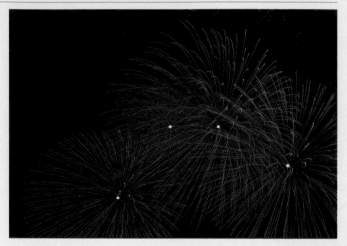

FIGURE 7.41 Lithium carbonate is sometimes used to produce vivid red colors in fireworks.

b. Frequency and wavelength are related:

$$c = \nu\lambda \qquad \text{so} \qquad \nu = \frac{c}{\lambda}$$

The frequency of red light with a wavelength of 670.8 nm is

$$\nu = \frac{2.998 \times 10^8\,\frac{\text{m}}{\text{s}}}{670.8\,\text{nm}\left(\dfrac{10^{-9}\,\text{m}}{\text{nm}}\right)} = 4.469 \times 10^{14}\,\text{s}^{-1}$$

c. The difference in energies between the ground state of a Li atom and its lowest-energy excited state must match exactly the energy of the photon emitted; therefore $\Delta E = 2.961 \times 10^{-19}$ J.
d. The electron configuration of a ground-state Li atom is $1s^2 2s^1$.
e. In the lowest-energy excited state, the valence electron has moved to the next higher energy orbital above the $2s$ orbital, which is one of the three $2p$ orbitals. Therefore the electron configuration of the excited state is $1s^2 2p^1$.

Think About It You might expect that the colors that lithium and other elements make in fireworks explosions are the same colors they produce in Bunsen burner flames. You would be right. Many of the colors are produced by transitions from lowest-energy excited states to ground states because the lowest-energy excited state is the one most easily populated by the thermal energy available in a gas flame or exploding firework shell.

SUMMARY

LO1 Light and other forms of **electromagnetic radiation** have wavelike properties described by characteristic **wavelengths** and **frequencies**. The speed of light is a constant in a vacuum but slows as it passes through liquids and solids. (Sections 7.1, 7.2)

LO2 Max Planck used **quantum theory**, which incorporates the assumption that all electromagnetic radiation consists of discrete particles of energy called **quanta**, to explain blackbody radiation. The **Planck constant** relates the energy of a quantum of light, called a **photon**, to its frequency. Einstein also used quantum theory to explain the **photoelectric effect**: a process in which a metal surface emits electrons when illuminated by electromagnetic radiation that is at or above a **threshold frequency**. (Section 7.3)

Metal surface (negative electrode)
Positive electrode
Voltage source
Meter indicates current in circuit

LO3 **Electron transitions** between energy levels in atoms may be accompanied by the absorption or emission of photons. Absorption and emission of electromagnetic radiation by free atoms produce characteristic **atomic absorption spectra** and **atomic emission spectra**. (Sections 7.1, 7.4)

LO4 De Broglie invoked the concept of **matter waves**—which incorporates the assumption that an electron in an atom, or any particle in motion, has a characteristic wavelength—to explain the stability of atoms. (Section 7.5)

LO5 The **Heisenberg uncertainty principle** states that the position and momentum of an electron cannot both be precisely known at the same time. (Section 7.5)

LO6 Solutions to the **Schrödinger wave equation** define allowed electron energy levels in atoms. **Orbitals** are the three-dimensional

regions inside an atom that describe the probability of finding an electron at a given distance from the nucleus. Orbitals have characteristic three-dimensional sizes, shapes, and orientations that are depicted by boundary–surface representations. Each orbital has a unique set of three **quantum numbers** that come directly from the solution to the Schrödinger wave equation. A fourth quantum number is necessary to explain certain characteristics of emission spectra. (Sections 7.6, 7.7)

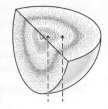

LO7 An **electron configuration** is a set of numbers and letters expressing the number of electrons in each occupied orbital in an atom or ion. All the orbitals of a given subshell are **degenerate**; they all have the same energy. In any set of degenerate orbitals, each orbital must contain one electron before any orbital in the set can accept a second electron. (Sections 7.8, 7.9)

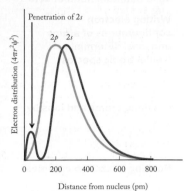

LO8 The **effective nuclear charge (Z_{eff})** is the net nuclear charge experienced by an electron that takes into account those electrons closer to the nucleus that shield valence electrons and decrease the electrostatic attraction to the nucleus. Trends in Z_{eff} explain why atomic size decreases across a period and increases down a group. **Ionization energies** generally increase with increasing effective nuclear charge across a period. **Electron affinity** values of many main group elements are negative (energy is released when they acquire electrons) but positive for nitrogen and the noble gases. (Sections 7.8, 7.10, 7.11, 7.12)

PARTICULATE **PREVIEW WRAP-UP**

Each process forms an ion from a neutral parent atom. Because cations are positively charged and smaller than their neutral parent atoms, an electron is lost in the first process. Because anions are negatively charged and larger than their neutral parent atoms, an electron is gained in the second process. Ionizing an electron from a neutral atom to form a cation requires the absorption of energy. For most elements, the second process emits energy.

PROBLEM-SOLVING SUMMARY

Type of Problem	Concepts and Equations		Sample Exercises
Calculating frequency from wavelength	$\lambda \nu = c$ where $c = 2.998 \times 10^8$ m/s	(7.1)	**7.1**
Calculating the energy of a photon	$E = \dfrac{hc}{\lambda}$	(7.3)	**7.2**
Applying the photoelectric effect	$\phi = h\nu_0$ $KE_{\text{electron}} = h\nu - \phi$	(7.4) (7.5)	**7.3**
Calculating the wavelength of a line in the hydrogen spectrum	$\dfrac{1}{\lambda} = [1.097 \times 10^{-2}\ \text{nm}^{-1}]\left(\dfrac{1}{n_1^2} - \dfrac{1}{n_2^2}\right)$	(7.7)	**7.4**

Type of Problem	Concepts and Equations	Sample Exercises
Calculating the energy needed for an electron transition	$\Delta E = -2.178 \times 10^{-18} \text{ J}\left(\dfrac{1}{n_{final}^2} - \dfrac{1}{n_{initial}^2}\right)$ (7.9)	7.5
Calculating the wavelength of particles in motion	$\lambda = \dfrac{h}{mu}$ (7.11)	7.6
Calculating uncertainty	$\Delta x \cdot m\Delta u \geq \dfrac{h}{4\pi}$ (7.13)	7.7
Identifying the subshells and orbitals in an energy level and valid quantum number sets	n is the shell number, ℓ defines both the subshell and the type of orbital; an orbital has a unique combination of allowed n, ℓ, and m_ℓ values. ℓ is any integer from 0 to $n - 1$; m_ℓ is any integer from $-\ell$ to $+\ell$, including zero.	7.8, 7.9
Writing electron configurations of atoms and ions; determining isoelectronic species	Orbitals fill up in the following sequence: $1s^2$, $2s^2$, $2p^6$, $3s^2$, $3p^6$, $4s^2$, $3d^{10}$, $4p^6$, $5s^2$, $4d^{10}$, $5p^6$, $6s^2$, $4f^{14}$, $5d^{10}$, $6p^6$ (superscripts represent maximum numbers of electrons). Arrange orbitals in electron configuration on the basis of increasing value of n. In forming transition metal ions, electrons are removed to maximize the number of d electrons; there is enhanced stability in half-filled and filled d subshells.	7.10, 7.11, 7.12, 7.13
Ordering atoms and ions by size	Effective nuclear charge and shielding explain observable periodic trends. In general, sizes decrease from left to right across a row and increase down a column of the periodic table. Cations are smaller, and anions are larger, than their parent atoms.	7.14
Ranking ionization energies	First ionization energies generally increase across a row and decrease down a column.	7.15

VISUAL PROBLEMS

(Answers to boldface end-of-chapter questions and problems are in the back of the book.)

7.1. Which of the elements highlighted in Figure P7.1 consist of ground-state atoms with:
   a. a single *s* electron in the valence shell? (More than one answer is possible.)
   b. filled sets of *s* and *p* orbitals in the valence shell?
   c. filled sets of *d* orbitals?
   d. half-filled sets of *d* orbitals?
   e. two *s* electrons in the valence shell?

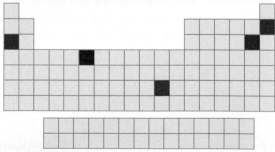

FIGURE P7.1

7.2. Which of the highlighted elements in Figure P7.1:
   a. forms a common monatomic ion that is larger than its parent atom?
   b. has the most unpaired electrons per ground-state atom?

7.3. Which of the elements highlighted in Figure P7.3 forms monatomic ions by
   a. losing an *s* electron?
   b. losing two *s* electrons?
   c. losing two *s* electrons and a *d* electron?
   d. adding an electron to a *p* orbital?
   e. adding electrons to two *p* orbitals?

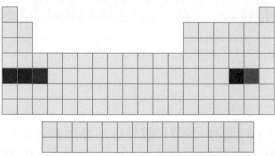

FIGURE P7.3

7.4. Which of the highlighted elements in Figure P7.3:
   a. forms common monatomic ions smaller than the parent atoms? (More than one answer is possible.)
   b. has the largest first ionization energy, IE_1?
   c. has the largest second ionization energy, IE_2?

7.5. Rank the elements highlighted in Figure P7.3 by:
   a. increasing atomic size.
   b. increasing size of the most common monatomic ions.

7.6. Which arrow in Figure P7.6 represents:
   a. emission of light with the longest wavelength?
   b. absorption of light of an atom in an excited state?
   c. a transition requiring absorption of the most energy?

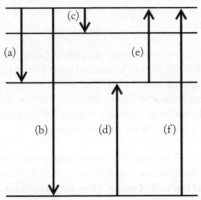

FIGURE P7.6

7.7. In Figure P7.7, which sphere represents a Na atom, Na⁺ ion, and K atom?

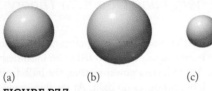

(a)   (b)   (c)

FIGURE P7.7

7.8. Which orbital diagram in Figure P7.8 represents:
   a. the ground state of an oxygen cation, O⁺?
   b. an excited state of a nitrogen atom?
   c. a violation of Hund's rule?

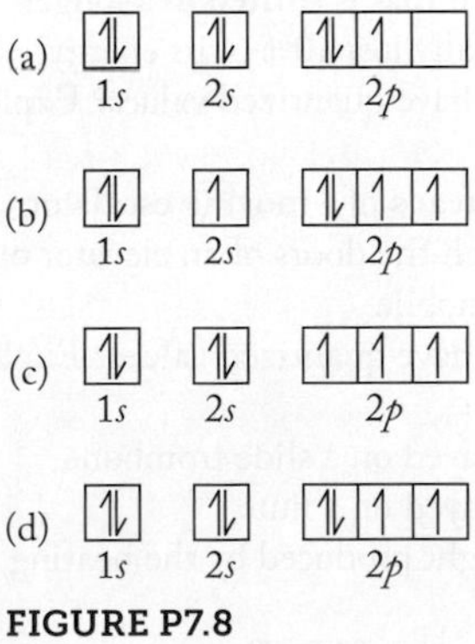

(a) $\boxed{\uparrow\downarrow}$ $\boxed{\uparrow\downarrow}$ $\boxed{\uparrow\downarrow|\uparrow}\ \ $
      1s   2s   2p

(b) $\boxed{\uparrow\downarrow}$ $\boxed{\uparrow}$ $\boxed{\uparrow\downarrow|\uparrow|\uparrow}$
      1s   2s   2p

(c) $\boxed{\uparrow\downarrow}$ $\boxed{\uparrow\downarrow}$ $\boxed{\uparrow|\uparrow|\uparrow}$
      1s   2s   2p

(d) $\boxed{\uparrow\downarrow}$ $\boxed{\uparrow\downarrow}$ $\boxed{\uparrow\downarrow|\uparrow|\uparrow}$
      1s   2s   2p

FIGURE P7.8

7.9. Which light represented by the waves in Figure P7.9 has:
   a. the highest frequency?
   b. the lowest energy?
   c. the highest energy?

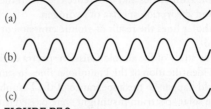

FIGURE P7.9

7.10. Use representations [A] through [I] in Figure P7.10 to answer questions a–f.
   a. Compare [A] and [H]. Which valence electron experiences the larger Z_{eff}? Which is more shielded from the positive charge of the nucleus? Which has the lower IE_1?
   b. Which representation depicts the loss of an electron?
   c. Which representations can be discussed using only a quantized model of light?
   d. Which representations are consistent with the model of an atom that discusses probability with regard to the location of electrons?
   e. What do the arrows represent in [E] and [I]?
   f. Which representation conveys that the probability of finding an electron at the nucleus is zero?

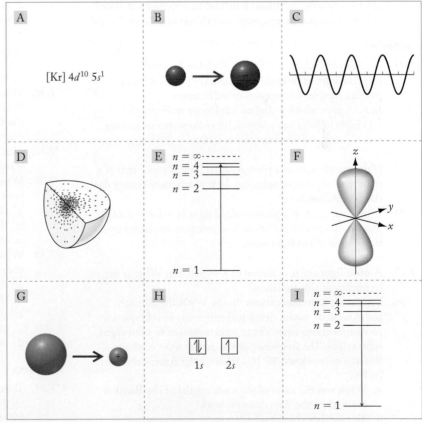

FIGURE P7.10

QUESTIONS AND PROBLEMS

Rainbows of Light; Waves of Energy

Concept Review

7.11. Describe the similarities and differences between the atomic emission and absorption spectra of hydrogen.

7.12. Are Fraunhofer lines the result of atomic emission or atomic absorption?

7.13. How did the study of the atomic emission spectra of elements lead to the identification of the Fraunhofer lines in sunlight?

*7.14. What would happen to the appearance of the Fraunhofer lines in the solar spectrum if sunlight were passed through a flame containing high-temperature calcium atoms and then analyzed?

7.15. Why are the various forms of radiant energy called *electromagnetic* radiation?

7.16. Explain with a sketch why the frequencies of long-wavelength waves of electromagnetic radiation are lower than those of short-wavelength waves.

7.17. **Dental X-rays** When X-ray images are taken of your teeth and gums in the dentist's office, your body is covered with a lead shield. Explain the need for this precaution.

7.18. **UV Radiation and Skin Cancer** Ultraviolet radiation causes skin damage that may lead to cancer, but exposure to infrared radiation does not seem to cause skin cancer. Why do you think this is so?

7.19. Gamma rays are an example of "ionizing" radiation because they have the energy to break apart molecules into molecular ions and free electrons. What other forms of electromagnetic radiation could also be ionizing radiation?

*7.20. If light consists of waves, why don't things look "wavy" to us?

Problems

7.21. In each pair, which radiation has higher frequency? (a) ultraviolet or infrared radiation; (b) visible light or gamma rays; (c) microwaves or radio waves

7.22. In each pair, which radiation has longer wavelength? (a) visible light or microwaves; (b) radio waves or gamma rays; (c) infrared or ultraviolet radiation

7.23. If the wavelength of a photon of red light is twice that of a photon of ultraviolet radiation, how much more energy does the UV photon have?

7.24. If the frequency of a photon of red light is twice that of a photon of infrared radiation, how much more energy does the photon of red light have?

7.25. A neon light emits radiation of $\lambda = 616$ nm. What is the frequency of this radiation?

7.26. **Submarine Communications** In the 1990s the Russian and American navies developed extremely low-frequency communications networks to send messages to submerged submarines. The frequency of the carrier wave of the Russian network was 82 Hz, whereas the Americans used 76 Hz.
   a. What was the ratio of the wavelengths of the Russian network to the American network?
   b. To calculate the actual underwater wavelength of the transmissions in either network, what additional information would you need?

7.27. **Broadcast Frequencies** FM radio stations broadcast at different frequencies. Calculate the wavelengths corresponding to the broadcast frequencies of the following college radio stations: (a) KCSU-FM (Fort Collins, CO), 90.5 MHz; (b) WVUD (Newark, DE), 91.3 MHz; (c) KUCR (Riverside, CA), 88.3 MHz.

7.28. Which radiation has the longer wavelength: (a) radio waves from an AM radio station broadcasting at 680 kHz or (b) infrared radiation emitted by the surface of Earth ($\lambda = 15$ μm)?

7.29. Which radiation has the lower frequency: (a) radio waves from an AM radio station broadcasting at 1030 kHz or (b) the red light ($\lambda = 633$ nm) from a helium-neon laser?

7.30. Which radiation has the higher frequency: (a) the red light on a bar-code reader at a grocery store or (b) the green light on the battery charger for a laptop computer?

7.31. **Speed of Light** How long does it take moonlight to reach Earth when the distance to the moon is 384,000 km?

7.32. **Exploration of the Solar System** How long would it take an instruction to a Martian rover to travel to Mars from a NASA site on Earth when Earth and Mars are 68.4 million km apart?

Particles of Energy and Quantum Theory

Concept Review

7.33. What is the difference between a quantum and a photon?

7.34. A variable power supply is connected to an incandescent light bulb. At the lowest power setting, the bulb feels warm to the touch but produces no light. At medium power, the light bulb filament emits a red glow. At the highest power, the light bulb emits white light. Explain this emission pattern.

*7.35. What effect does the intensity (amplitude) of a wave have on the emission of electrons from a surface, if we assume that the frequency of incident radiation is above the threshold frequency?

*7.36. Has a photon of radiation that is shifted to a longer wavelength by 10% actually lost 10% of its energy?

7.37. Which of the following have quantized values? Explain your selections.
   a. the elevation of the treads of a moving escalator
   b. the elevations at which the doors of an elevator open
   c. the speed of an automobile

7.38. Which of the following have quantized values? Explain your selections.
   a. the pitch of a note played on a slide trombone
   b. the pitch of a note played on a flute
   c. the wavelengths of light produced by the heating elements in a toaster
   d. the wind speed at the top of Mt. Everest

Problems

*7.39. Thin layers of potassium ($\phi = 3.68 \times 10^{-19}$ J) and sodium ($\phi = 4.41 \times 10^{-19}$ J) are exposed to radiation of wavelength 300 nm. Which metal emits electrons with the greater velocity? What is the velocity of these electrons?

7.40. Titanium ($\phi = 6.94 \times 10^{-19}$ J) and silicon ($\phi = 7.24 \times 10^{-19}$ J) surfaces are irradiated with UV radiation with a wavelength of 250 nm. Which surface emits electrons with the longer wavelength? What is the wavelength of the electrons emitted by the titanium surface?

7.41. **Solar Power** Photovoltaic cells convert solar energy into electricity. Could tantalum ($\phi = 6.81 \times 10^{-19}$ J) be used to convert visible light to electricity? Assume that most of the electromagnetic energy from the Sun is in the visible region near 500 nm.

7.42. With reference to Problem 7.41, could tungsten ($\phi = 7.20 \times 10^{-19}$ J) be used to construct solar cells?

7.43. The power of a red laser ($\lambda = 630$ nm) is 1.00 watt (abbreviated W, where 1 W = 1 J/s). How many photons per second does the laser emit?

*7.44. The energy density of starlight in interstellar space is 10^{-15} J/m³. If the average wavelength of starlight is 500 nm, what is the corresponding density of photons per cubic meter of space?

The Hydrogen Spectrum and the Bohr Model

Concept Review

7.45. Why should hydrogen have the simplest atomic spectrum of all the elements?

7.46. For an electron in a hydrogen atom, how is the value of n of its orbit related to its energy?

7.47. Does the electromagnetic energy emitted by an excited-state H atom depend on the individual values of n_1 and n_2, or only on the difference between them ($n_1 - n_2$)?

7.48. Explain the difference between a ground-state H atom and an excited-state H atom.

7.49. Without calculating any wavelength values, predict which of the following electron transitions in the hydrogen atom is associated with radiation having the shortest wavelength.
   a. from $n = 1$ to $n = 2$
   b. from $n = 2$ to $n = 3$
   c. from $n = 3$ to $n = 4$
   d. from $n = 4$ to $n = 5$

7.50. Without calculating any frequency values, rank the following transitions in the hydrogen atom in order of increasing frequency of the electromagnetic radiation that could produce them.
   a. from $n = 4$ to $n = 6$
   b. from $n = 6$ to $n = 8$
   c. from $n = 9$ to $n = 11$
   d. from $n = 11$ to $n = 13$

7.51. Electron transitions from $n = 2$ to $n = 3, 4, 5,$ or 6 in hydrogen atoms are responsible for some of the Fraunhofer lines in the Sun's spectrum. Are there any Fraunhofer lines due to transitions that start from $n = 3$ in hydrogen atoms?

7.52. In the visible portion of the atomic emission spectrum of hydrogen, are there any bright lines due to electron transitions to the $n = 1$ state?

7.53. Balmer observed a hydrogen emission line for the transition from $n = 6$ to $n = 2$, but not for the transition from $n = 7$ to $n = 2$. Why?

*7.54. In what ways should the emission spectra of H and He⁺ be alike, and in what ways should they be different?

Problems

7.55. What is the wavelength of the photons emitted by hydrogen atoms when they undergo transitions from $n = 4$ to $n = 3$? In which region of the electromagnetic spectrum does this radiation occur?

7.56. What is the frequency of the photons emitted by hydrogen atoms when they undergo transitions from $n = 5$ to $n = 3$? In which region of the electromagnetic spectrum does this radiation occur?

*7.57. The energies of the photons emitted by one-electron atoms and ions fit the equation

$$E = (2.178 \times 10^{-18} \text{ J})Z^2\left(\frac{1}{n_1^2} - \frac{1}{n_2^2}\right)$$

where Z is the atomic number, n_2 and n_1 are positive integers, and $n_2 > n_1$.
   a. As the value of Z increases, does the wavelength of the photon associated with the transition from $n = 2$ to $n = 1$ increase or decrease?
   b. Can the wavelength associated with the transition from $n = 2$ to $n = 1$ ever be observed in the visible region of the spectrum?

*7.58. Can transitions from higher energy states to the $n = 2$ level in He⁺ ever produce visible light? If so, for what values of n_2? (*Hint*: The equation in Problem 7.57 may be useful.)

7.59. The transition from $n = 3$ to $n = 2$ in a hydrogen atom produces a photon with a wavelength of 656 nm. What is the wavelength of the transition from $n = 3$ to $n = 2$ in a Li²⁺ ion? (Use the equation in Problem 7.57.)

*7.60. The hydrogen atomic emission spectrum includes a UV line with a wavelength of 92.3 nm.
   a. Is this line associated with a transition between different excited states or between an excited state and the ground state?
   b. What is the value of n_1 of this transition?
   c. What is the energy of the longest wavelength photon that a ground-state hydrogen atom can absorb?

Electron Waves

Concept Review

7.61. Identify the symbols in the de Broglie relation $\lambda = h/mu$, and explain how the relation links the properties of a particle to those of a wave.

7.62. How does de Broglie's hypothesis that electrons behave like waves explain the stability of the electron orbits in a hydrogen atom?

7.63. Would the density of an object have an effect on its de Broglie wavelength?

7.64. Would the shape of an object have an effect on its de Broglie wavelength?

Problems

7.65. Calculate the wavelengths of the following objects:
   a. a muon (a subatomic particle with a mass of 1.884×10^{-25} g) traveling at 325 m/s
   b. electrons ($m_e = 9.10938 \times 10^{-28}$ g) moving at 4.05×10^6 m/s in an electron microscope
   c. an 80-kg athlete running a 4-minute mile

d. Earth (mass = 6.0×10^{27} g) moving through space at 3.0×10^4 m/s

7.66. Two objects are moving at the same velocity. Which (if any) of the following statements about them is true?
   a. The de Broglie wavelength of the heavier object is longer than that of the lighter one.
   b. If one object has twice as much mass as the other, its wavelength is one-half the wavelength of the other.
   c. Doubling the velocity of one of the objects will have the same effect on its wavelength as doubling its mass.

7.67. Which (if any) of the following statements about the frequency of a particle is true?
   a. Heavy, fast-moving objects have lower frequencies than those of lighter, faster-moving objects.
   b. Only very light particles can have high frequencies.
   c. Doubling the mass of an object and halving its velocity results in no change in its frequency.

7.68. How rapidly would each of the following particles be moving if they all had the same wavelength as a photon of red light ($\lambda = 750$ nm)?
   a. an electron of mass 9.10938×10^{-28} g
   b. a proton of mass 1.67262×10^{-24} g
   c. a neutron of mass 1.67493×10^{-24} g
   d. an α particle of mass 6.64×10^{-24} g

*7.69. Kinetic molecular theory tells us that helium atoms at 500 K are in constant random motion.
   a. Calculate the root-mean-square speed of helium atoms at 500 K.
   b. What is the wavelength of a helium atom at 500 K?

*7.70. Neon atoms are expected to move slower than helium atoms at the same temperature.
   a. Using Graham's law of effusion and your answer from Problem 7.69, calculate the root-mean-square speed of neon atoms at 500 K.
   b. What is the wavelength of a neon atom at 500 K?

7.71. **Particles in a Cyclotron** The first cyclotron was built in 1930 at the University of California, Berkeley, and was used to accelerate molecular ions of hydrogen, H_2^+, to a velocity of 4×10^6 m/s. (Modern cyclotrons can accelerate particles to nearly the speed of light.) If the uncertainty in the velocity of the H_2^+ ion was 3%, what was the uncertainty of its position?

7.72. **Radiation Therapy** An effective treatment for some cancerous tumors involves irradiation with "fast" neutrons. The neutrons from one treatment source have an average velocity of 3.1×10^7 m/s. If the velocities of individual neutrons are known to within 2% of this value, what is the uncertainty in the position of one of them?

Quantum Numbers and Electron Spin; The Sizes and Shapes of Atomic Orbitals

Concept Review

7.73. How does the concept of an orbit in the Bohr model of the hydrogen atom differ from the concept of an orbital in quantum theory?

7.74. What properties of an orbital are defined by each of the three quantum numbers n, ℓ, and m_ℓ?

7.75. How many quantum numbers are needed to identify an orbital?

7.76. How many quantum numbers are needed to identify an electron in an atom?

Problems

7.77. How many orbitals are there in an atom with each of the following principal quantum numbers? (a) 1; (b) 2; (c) 3; (d) 4; (e) 5

7.78. How many orbitals are there in an atom with the following combinations of quantum numbers?
   a. $n = 3$, $\ell = 2$
   b. $n = 3$, $\ell = 1$
   c. $n = 4$, $\ell = 2$, $m_\ell = 2$

7.79. What are the possible values of quantum number ℓ when $n = 4$?

7.80. What are the possible values of m_ℓ when $\ell = 2$?

7.81. What set of orbitals corresponds to each of the following sets of quantum numbers? How many electrons could occupy these orbitals?
   a. $n = 2$, $\ell = 0$
   b. $n = 3$, $\ell = 1$
   c. $n = 4$, $\ell = 2$
   d. $n = 1$, $\ell = 0$

7.82. What set of orbitals corresponds to each of the following sets of quantum numbers? How many electrons could occupy these orbitals?
   a. $n = 2$, $\ell = 1$
   b. $n = 5$, $\ell = 3$
   c. $n = 3$, $\ell = 2$
   d. $n = 4$, $\ell = 3$

7.83. Which of the following combinations of quantum numbers are allowed?
   a. $n = 1$, $\ell = 1$, $m_\ell = 0$, $m_s = +\frac{1}{2}$
   b. $n = 3$, $\ell = 0$, $m_\ell = 0$, $m_s = -\frac{1}{2}$
   c. $n = 1$, $\ell = 0$, $m_\ell = 1$, $m_s = -\frac{1}{2}$
   d. $n = 2$, $\ell = 1$, $m_\ell = 2$, $m_s = +\frac{1}{2}$

7.84. Which of the following combinations of quantum numbers are allowed?
   a. $n = 3$, $\ell = 2$, $m_\ell = 0$, $m_s = -\frac{1}{2}$
   b. $n = 5$, $\ell = 4$, $m_\ell = 4$, $m_s = +\frac{1}{2}$
   c. $n = 3$, $\ell = 0$, $m_\ell = 1$, $m_s = +\frac{1}{2}$
   d. $n = 4$, $\ell = 4$, $m_\ell = 1$, $m_s = -\frac{1}{2}$

The Periodic Table and Filling the Orbitals of Multielectron Atoms; Electron Configurations of Ions

Concept Review

7.85. What is meant when two or more orbitals are said to be degenerate?

7.86. Explain how the electron configurations of the group 2 elements are linked to their location in the periodic table developed by Mendeleev (see Figure 2.11).

7.87. How do we know from examining the periodic table's structure that the $4s$ orbital is filled before the $3d$ orbital?

7.88. Explain why so many transition metals form ions with a 2+ charge.

7.89. Why is there only one ground-state electron configuration for an atom but many excited-state electron configurations?

7.90. Can the ground-state electron configuration for an atom ever be an excited-state electron configuration for a different atom?

Problems

7.91. List the following orbitals in order of increasing energy in a multielectron atom:
a. $n = 3, \ell = 2$
b. $n = 5, \ell = 1$
c. $n = 3, \ell = 0$
d. $n = 4, \ell = 1, m_\ell = -1$

7.92. Place the following orbitals in order of increasing energy in a multielectron atom:
a. $n = 2, \ell = 1$
b. $n = 5, \ell = 3$
c. $n = 3, \ell = 2$
d. $n = 4, \ell = 3$

7.93. What are the electron configurations of Li, Li^+, Ca, F^-, Na^+, Mg^{2+}, and Al^{3+}?

7.94. Which species listed in Problem 7.93 are isoelectronic with Ne?

7.95. What are the condensed electron configurations of K, K^+, S^{2-}, N, Ba, Ti^{4+}, and Al?

7.96. In what way are the electron configurations of H, Li, Na, K, Rb, and Cs similar?

7.97. How many unpaired electrons are there in the following ground-state atoms and ions? (a) N; (b) O; (c) P^{3-}; (d) Na^+

7.98. How many unpaired electrons are there in the following ground-state atoms and ions? (a) Sc; (b) Ag^+; (c) Cd^{2+}; (d) Zr^{4+}

7.99. Identify the atom whose electron configuration is $[Ar]3d^2 4s^2$. How many unpaired electrons are there in the ground state of this atom?

7.100. Identify the atom whose electron configuration is $[Ne]3s^2 3p^3$. How many unpaired electrons are there in the ground state of this atom?

7.101. Which monatomic ion has a charge of 1− and the electron configuration $[Ne]3s^2 3p^6$? How many unpaired electrons are there in the ground state of this ion?

7.102. Which monatomic ion has a charge of 1+ and the electron configuration $[Kr]4d^{10} 5s^2$? How many unpaired electrons are there in the ground state of this ion?

7.103. Does the ground-state electron configuration of Na^+ represent an excited-state electron configuration of Ne?

7.104. Which excited state of the hydrogen atom is likely to be of higher energy, $1s^1 2s^1$ or $1s^1 3s^1$?

7.105. Predict the charge of the monatomic ions formed by Al, N, Mg, and Cs.

7.106. Predict the charge of the monatomic ions formed by S, P, Zn, and I.

7.107. Which of the following electron configurations represent an excited state?
a. $[He]2s^1 2p^5$
b. $[Kr]4d^{10} 5s^2 5p^1$
c. $[Ar]3d^{10} 4s^2 4p^5$
d. $[Ne]3s^2 3p^2 4s^1$

7.108. Which of the following electron configurations represent an excited state?
a. $[Ne]3s^2 3p^1$
b. $[Ar]3d^{10} 4s^1 4p^2$
c. $[Kr]4d^{10} 5s^1 5p^1$
d. $[Ne]3s^2 3p^6 4s^1$

7.109. Boric acid, H_3BO_3, gives off a green color (Figure P7.109) when heated in a flame.
a. Write ground-state electron configurations for B and O.
b. Assign oxidation numbers to each of the elements in boric acid. If the oxidation number reflects the ionic charge on the element, write ground-state electron configurations for each ion.

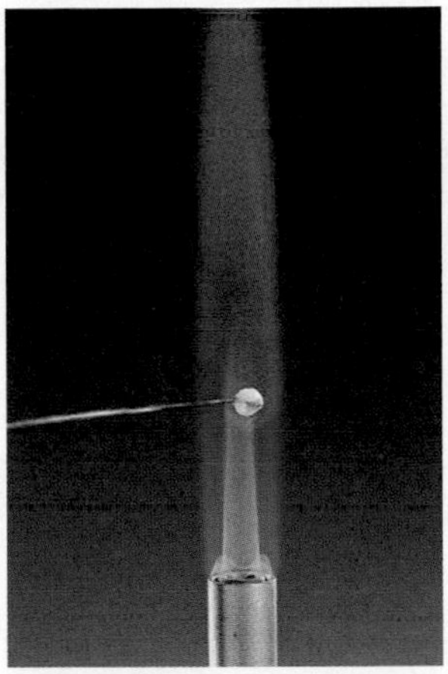

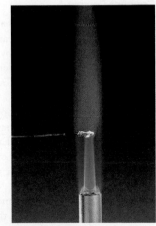

FIGURE P7.109

7.110. Introducing calcium chloride into a flame imparts an intense orange color (Figure P7.110).
a. Write ground-state electron configurations for Ca and Cl.
b. Calcium chloride contains calcium and chloride ions. Write ground-state electron configurations for each ion.

7.111. In which subshell are the highest-energy electrons in a ground-state atom of the isotope ^{131}I? Are the electron configurations of ^{131}I and ^{127}I the same?

FIGURE P7.110

7.112. Although no currently known elements contain electrons in g orbitals ($\ell = 4$), such elements may be synthesized someday. What is the minimum atomic number of an element whose ground-state atoms would have an electron in a g orbital?

The Sizes of Atoms and Ions

Concept Review

7.113. Sodium atoms are much larger than chlorine atoms, but in NaCl sodium ions are much smaller than chloride ions. Why?

7.114. Why does atomic size tend to decrease with increasing atomic number across a row of the periodic table?

7.115. Which of the following group 1 elements has the largest atoms: Li, Na, K, or Rb? Explain your selection.

7.116. Which of the following group 17 elements has the largest monatomic ions: F, Cl, Br, or I? Explain your selection.

Ionization Energies

Concept Review

7.117. How do ionization energies change with increasing atomic number (a) down a group of elements in the periodic table and (b) from left to right across a period of elements?

7.118. The first ionization energies of the main group elements are given in Figure 7.39. Explain the differences in the first ionization energy between (a) He and Li; (b) Li and Be; (c) Be and B; (d) N and O.

7.119. Explain why it is more difficult to ionize a fluorine atom than a boron atom.

7.120. Do you expect the ionization energies of anions of group 17 elements to be lower or higher than for neutral atoms of the same group?

7.121. Which of the following elements should have the smallest *second* ionization energy? Br, Kr, Rb, Sr, Y

7.122. Why is the first ionization energy (IE_1) of Al ($Z = 13$) less than the IE_1 of Mg ($Z = 12$) *and* less than the IE_1 of Si ($Z = 14$)?

Electron Affinities

Concept Review

7.123. An electron affinity (EA) value that is negative indicates that the free atoms of an element are less stable than the $1-$ anions they form by acquiring electrons. Does this mean that all of the elements with negative EA values exist in nature as anions? Give some examples to support your answer.

7.124. The electron affinities of the group 17 elements are all negative values, but the EA values of the group 18 noble gases are all positive. Explain this difference.

7.125. The electron affinities of the group 17 elements increase with increasing atomic number. Suggest a reason for this trend.

7.126. Ionization energies generally increase with increasing atomic number across the second row of the periodic table, but electron affinities generally decrease. Explain the opposing trends.

Additional Problems

7.127. **Interstellar Hydrogen** Astronomers have detected hydrogen atoms in interstellar space in the $n = 732$ excited state. Suppose an atom in this excited state undergoes a transition from $n = 732$ to $n = 731$.
  a. How much energy does the atom lose as a result of this transition?

  b. What is the wavelength of radiation corresponding to this transition?

  c. What kind of telescope would astronomers need in order to detect radiation of this wavelength? (*Hint*: It would not be one designed to capture visible light.)

*7.128. When an atom absorbs an X-ray of sufficient energy, one of its $2s$ electrons may be emitted, creating a hole that can be spontaneously filled when an electron in a higher-energy orbital—a $2p$, for example—falls into it. A photon of electromagnetic radiation with an energy that matches the energy lost in the $2p \rightarrow 2s$ transition is emitted. Predict how the wavelengths of $2p \rightarrow 2s$ photons would differ between (a) different elements in the fourth row of the periodic table and (b) different elements in the same column (for example, between the noble gases from Ne to Rn).

7.129. A green flame is observed when copper(II) chloride is heated in a flame (Figure P7.129).
  a. Write the ground-state electron configuration for copper atoms.
  b. There are two common forms of copper chloride, copper(I) chloride and copper(II) chloride. What is the difference in the ground-state electron configurations of copper in these two compounds?

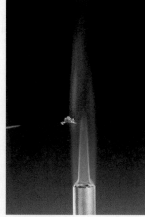
FIGURE P7.129

*7.130. Two helium ions (He^+) in the $n = 3$ excited state emit photons of radiation as they return to the ground state. One ion does so in a single transition from $n = 3$ to $n = 1$. The other does so in two steps: $n = 3$ to $n = 2$ and then $n = 2$ to $n = 1$. Which of the following statements about these two pathways is true?
  a. The sum of the energies lost in the two-step process is the same as the energy lost in the single transition from $n = 3$ to $n = 1$.
  b. The sum of the wavelengths of the two photons emitted in the two-step process is equal to the wavelength of the single photon emitted in the transition from $n = 3$ to $n = 1$.
  c. The sum of the frequencies of the two photons emitted in the two-step process is equal to the frequency of the single photon emitted in the transition from $n = 3$ to $n = 1$.
  d. The wavelength of the photon emitted by the He^+ ion in the $n = 3$ to $n = 1$ transition is shorter than the wavelength of a photon emitted by an H atom in an $n = 3$ to $n = 1$ transition.

*7.131. Use your knowledge of electron configurations to explain the following observations:
  a. Silver tends to form ions with a charge of 1+, but the element to the right of silver in the periodic table tends to form ions with 2+ charges.
  b. The heavier group 13 elements (Ga, In, Tl) tend to form ions with charges of 1+ or 3+ but not 2+.
  c. The heavier elements of group 14 (Sn, Pb) and group 4 (Ti, Zr, Hf) tend to form ions with charges of 2+ or 4+.

7.132. Trends in ionization energies of the elements as a function of the position of the elements in the periodic table are a useful test of our understanding of electronic structure.

    a. Should the same trend in the first ionization energies for elements with atomic numbers $Z = 31$ through $Z = 36$ be observed for the second ionization energies of the same elements? Explain why or why not.

    b. Which element should have the greater second ionization energy: Rb ($Z = 37$) or Kr ($Z = 36$)? Why?

7.133. Chemistry of Photo-Gray Glasses "Photo-gray" lenses for eyeglasses darken in bright sunshine because the lenses contain tiny, transparent AgCl crystals. Exposure to light removes electrons from Cl^- ions, forming a chlorine atom in an excited state (indicated below by the asterisk):

$$Cl^- + h\nu \rightarrow Cl^* + e^-$$

The electrons are transferred to Ag^+ ions, forming silver metal:

$$Ag^+ + e^- \rightarrow Ag$$

Silver metal is reflective, giving rise to the photo-gray color.

    a. Write condensed electron configurations of Cl^-, Cl, Ag, and Ag^+.

    b. What do we mean by the term *excited state*?

    c. Would more energy be needed to remove an electron from a Br^- ion or from a Cl^- ion? Explain your answer.

    *d. How might substitution of AgBr for AgCl affect the light sensitivity of photo-gray lenses?

7.134. The first ionization energy of a gas-phase atom of a particular element is 6.24×10^{-19} J. What is the maximum wavelength of electromagnetic radiation that could ionize this atom?

7.135. Tin (in group 14) forms both Sn^{2+} and Sn^{4+} ions, but magnesium (in group 2) forms only Mg^{2+} ions.

    a. Write condensed ground-state electron configurations for the ions Sn^{2+}, Sn^{4+}, and Mg^{2+}.

    b. Which neutral atoms have ground-state electron configurations identical to Sn^{2+} and Mg^{2+}?

    c. Which $2+$ ion is isoelectronic with Sn^{4+}?

7.136. Oxygen Ions in Space Between 1999 and 2007 the *Far Ultraviolet Spectroscopic Explorer* satellite analyzed the spectra of emission sources within the Milky Way. Among the satellite's findings were interplanetary clouds containing oxygen atoms that have lost five electrons.

    a. Write an electron configuration for these highly ionized oxygen atoms.

    b. Which electrons have been removed from the neutral atoms?

    c. The ionization energies corresponding to removal of the third, fourth, and fifth electrons are 4581 kJ/mol,

7465 kJ/mol, and 9391 kJ/mol, respectively. Explain why removal of each additional electron requires more energy than removal of the previous one.

    d. What is the maximum wavelength of radiation that will remove the fifth electron from an O atom?

***7.137.** Effective nuclear charge (Z_{eff}) is related to atomic number (Z) by a parameter called the shielding parameter (σ) according to the equation $Z_{eff} = Z - \sigma$.

    a. Calculate Z_{eff} for the outermost s electrons of Ne and Ar given $\sigma = 4.24$ (for Ne) and 11.24 (for Ar).

    b. Explain why the shielding parameter is much greater for Ar than for Ne.

7.138. Fog Lamp Technology Sodium fog lamps and street lamps contain gas-phase sodium atoms and sodium ions. Sodium atoms emit yellow-orange light at 589 nm. Do sodium ions emit the same yellow-orange light? Explain why or why not.

7.139. How can an electron get from the (+) lobe of a p orbital to the (−) lobe without going through the node between the lobes?

7.140. Einstein did not fully accept the uncertainty principle, remarking that "He [God] does not play dice." What do you think Einstein meant? Niels Bohr allegedly responded by saying, "Albert, stop telling God what to do." What do you think Bohr meant?

***7.141.** The wavelengths of Fraunhofer lines in galactic spectra are not exactly the same as those in sunlight: they tend to be shifted to longer wavelengths (*redshifted*), in part because of the Doppler effect. The Doppler effect is described by the equation

$$\frac{(\nu - \nu')}{\nu} = \frac{u}{c}$$

where ν is the unshifted frequency, ν' is the perceived frequency, c is the speed of light, and u is the speed at which the object is moving. If hydrogen in a galaxy that is receding from Earth at half the speed of light emits radiation with a wavelength of 656 nm, will the radiation still be in the visible part of the electromagnetic spectrum when it reaches Earth?

7.142. The work function of mercury is 7.22×10^{-19} J.

    a. What is the minimum frequency of radiation required to eject photoelectrons from a mercury surface?

    b. Could visible light produce the photoelectric effect in mercury?

8

Chemical Bonds

What Makes a Gas a Greenhouse Gas?

GREENHOUSE EFFECT The glass windows of a greenhouse allow sunlight to enter but trap the warm air inside. Greenhouse gases, such as CO_2, behave in a similar fashion. Recent increases in atmospheric CO_2 concentrations have been linked to rising global temperatures and the threat of global climate change.

PARTICULATE **REVIEW**

Breaking and Making Bonds

In Chapter 8 we learn how chemists model and represent the formation of chemical bonds between atoms.

- When methane burns in oxygen to produce carbon dioxide and water, which bonds are broken? Which bonds are formed? Space-filling models of each molecule in the reaction are shown here.

- Write a balanced equation for this reaction.

- Referring to the balanced equation, how many bonds are broken? How many bonds are formed?

  (Review Section 3.3 if you need help.)

(Answers to Particulate Review questions are in the back of the book.)

Single, Double, or Triple Bond?

These space-filling models show an oxygen molecule, a fluorine molecule, and a nitrogen molecule. As you read Chapter 8, look for ideas that will help you answer these questions:

- One of the three molecules contains a single bond, one contains a double bond, and one contains a triple bond. Does a space-filling model tell you what bonds are present?

- What is the molecular formula for each compound? Does a molecular formula tell you what bonds are present?

- What information does a ball-and-stick model contain that a space-filling model does not?

Learning Outcomes

LO1 Describe ways in which covalent, ionic, and metallic bonds are alike and ways in which they differ

LO2 Draw Lewis structures of molecular compounds and polyatomic ions, including resonance structures when appropriate
Sample Exercises 8.1, 8.2, 8.3, 8.4, 8.6, 8.7, 8.10, 8.11

LO3 Predict the polarity of covalent bonds on the basis of differences in the electronegativities of the bonded elements
Sample Exercise 8.5

LO4 Use formal charges to identify preferred resonance structures
Sample Exercises 8.8, 8.9, 8.10

LO5 Describe how bond order, bond energy, and bond length are related, and estimate enthalpy of reaction (ΔH_{rxn}) from bond energies
Sample Exercise 8.12

8.1 Types of Chemical Bonds and the Greenhouse Effect

In Chapter 7 we saw that quantum theory describes matter's interaction with energy at the atomic level. In this chapter we continue our exploration of matter and energy as we step from considering individual atoms to examining molecules, exploring how atoms are connected to one another.

As an example of how molecules interact with energy, let's look at molecules of carbon dioxide in Earth's atmosphere. Most scientists agree that recent increases in the average temperature of Earth's surface (about half a Celsius degree in the last half century) are linked to increases in the concentration of carbon dioxide and other *greenhouse gases* in the atmosphere. These gases trap Earth's heat in its atmosphere, much like panes of glass trap heat in a greenhouse or in a car on a hot summer day. Increases in atmospheric concentrations of greenhouse gases have been linked to human activity, particularly to increasing rates of fossil fuel combustion and the destruction of forests that would otherwise consume CO_2 during photosynthesis.

Were it not for the presence of atmospheric CO_2 and other greenhouse gases, Earth would be much colder than it is. Indeed, it would be too cold to be habitable. The *greenhouse effect* resulting from the presence of these gases has moderated Earth's average global temperature to within a fairly narrow range, 15°–17°C, for many thousands of years. However, during the last 60 years, atmospheric concentrations of CO_2 have increased from about 315 ppm to more than 400 ppm. The atmosphere has not contained this much CO_2 since more than half a million years ago, long before our species evolved.

Carbon dioxide allows solar radiation to pass through the atmosphere and warm Earth's surface, but it traps heat that would otherwise radiate from the warm surface back into space. Why is CO_2 so good at trapping heat? We noted in Chapter 7 that heat is associated with the infrared region of the electromagnetic spectrum. Carbon dioxide traps heat because it absorbs infrared radiation. To understand how CO_2 does this, whereas O_2 and N_2 do not, we need to learn more about the chemical bonds that hold atoms together in molecules.

Forming Bonds from Atoms

Fewer than 100 stable (nonradioactive) elements make up all the matter in our world. These elements combine to form more than 60 million compounds, and the number grows daily as chemists synthesize new ones. This multitude of compounds exists because they are more stable than the elements that react to form them. Furthermore, two atoms that are linked by a chemical bond tend to be lower in chemical energy than those same two atoms without a bond connecting them.

Why are bonded atoms more stable? We introduced the principal reason in Chapter 5 when we discussed the concept of electrostatic potential energy (E_{el}) between charged particles. We noted that E_{el} is directly proportional to the product of the charges (Q_1 and Q_2) on pairs of ions (or subatomic particles) and is inversely proportional to the distance (d) between them:

$$E_{el} \propto \frac{Q_1 \times Q_2}{d} \qquad (5.3)$$

When we apply Equation 5.3 to particles of opposite charge, such as a cation–anion pair, the product of Q_1 and Q_2 is negative and so is the value of E_{el}. Also, E_{el} becomes more negative as the particles approach each other and the distance (d) between them decreases. As the oppositely charged pair of ions moves very close together, however, their attraction to each other is offset by electrostatic repulsions between each ion's negatively charged outer-shell electrons and between their two positive nuclei. This trend was illustrated in Figure 5.7 for such a cation–anion pair. At the energy minimum in Figure 5.7, the forces of attraction and repulsion are in balance, and the ions are held together by a stable **ionic bond**.

We can apply this same concept of electrostatic potential energy to also explain why two neutral atoms may come together to form a stable molecule. Let's consider the case of two hydrogen atoms (Figure 8.1a) that are sufficiently far apart that they do not interact with each other and have zero potential energy with respect to each other. As they approach each other, the proton in the nucleus of one H atom is attracted to the electron of the other, and vice versa. This mutual attraction produces a negative E_{el} (Figure 8.1b). As the distance between the atoms decreases, the value of E_{el} continues to decrease, eventually reaching a minimum when the nuclei

ionic bond a chemical bond that results from the electrostatic attraction between a cation and an anion.

CHEMTOUR
Bonding

CONNECTION Negative values of E_{el} quantify the attraction between particles of opposite charge; positive values relate to the repulsion of particles with the same charge, as described in Chapter 5.

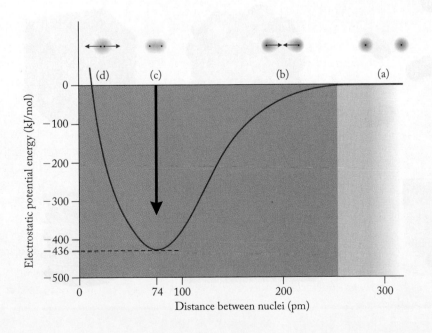

FIGURE 8.1 The electrostatic potential energy (E_{el}) profile of a H—H bond: (a) Two H atoms that are far apart do not interact, so $E_{el} = 0$. (b) As the atoms approach each other, mutual attraction between their positive nuclei and negative electrons causes E_{el} values to decrease. (c) At 74 pm E_{el} reaches a minimum as the two atoms form a H—H bond. (d) If the atoms are even closer together, repulsion between their nuclei causes E_{el} to increase and destabilizes the bond.

bond length the distance between the nuclear centers of two atoms joined in a bond.

metallic bond a chemical bond consisting of metal nuclei surrounded by a "sea" of shared electrons.

delocalized electrons electrons that are shared among more than two atoms.

CONNECTION In Chapter 1 we described a chemical bond as the force that holds two atoms in a molecule together. A covalent bond was defined in Chapter 2 as a bond between two atoms created by sharing pairs of electrons.

are 74 pm apart (Figure 8.1c). If they come any closer together, the repulsion between their two positive nuclei more than offsets the mutual proton–electron attractions, and E_{el} begins to rise rapidly with decreasing d (Figure 8.1d).

The distance between the atoms when E_{el} is at a minimum is referred to as the **bond length** of a H—H covalent bond, in this case, 74 pm. The value of E_{el} at this distance, -436 kJ/mol, is the energy released when two moles of H atoms bond to form one mole of H_2 molecules. Likewise, you would have to add this much energy to break a mole of H_2 molecules into separate H atoms. Thus, 436 kJ/mol is the H—H bond energy, or *bond strength*. We explore the significance of bond length and bond strength in more detail in Section 8.7.

We can also use electrostatic potential energy to explain the interactions between the atoms in a solid piece of metal. As with atoms in molecules, the positive nucleus of each atom in a metallic solid is attracted to the electrons of the atoms that surround it. These attractions result in the formation of **metallic bonds**, which are unlike covalent bonds in that they are not a pair of electrons shared by two atoms. Instead, the shared electrons in metallic bonds are highly **delocalized**, forming a "sea" of mobile electrons that move freely among all the atoms in a metallic solid. This electron mobility makes metals good conductors of heat and electricity. We discuss metallic bonding in greater detail in Chapter 9.

Table 8.1 summarizes some of the differences among ionic, covalent, and metallic bonds. As you consider these properties, keep in mind that many bonds do not fall exclusively into just one category, but rather have some covalent, ionic, and even metallic character.

TABLE 8.1 Types of Chemical Bonds

	Covalent	Ionic	Metallic
Elements Involved	Nonmetal and metalliods	Metals and nonmetals	Metals
Electron Distribution	Shared	Transferred	Delocalized
Particulate View			
Macroscopic View			

8.2 Lewis Structures

In 1916 American chemist Gilbert N. Lewis (1875–1946) proposed that atoms form chemical bonds by sharing electrons. He further suggested that through this sharing, each atom is surrounded by enough electrons to mimic the electron configuration of a noble gas. Today we associate such an electron configuration with filled *s* and *p* orbitals in the valence shells of atoms. Lewis's view of chemical bonding predated quantum mechanics and the notion of atomic orbitals, but it was consistent with what he called the **octet rule**: atoms of most main group elements lose, gain, or share electrons so that each atom has eight valence electrons, or an *octet* of electrons. Hydrogen is an important exception to the octet rule in that each of its atoms shares only a single pair of bonding electrons, thereby acquiring a "duet" instead of an octet. The space-filling models we used in Table 8.1 show atoms involved in different types of bonding, but they do not show how their electrons are distributed. Lewis devised a way of representing individual valence electrons in atoms so chemists could focus on how these electrons are distributed in various types of bonds.

Lewis Symbols

Lewis developed a system of symbols, called **Lewis symbols** or **Lewis dot symbols**, to explain the number of chemical bonds an atom typically forms to complete its octet. In other words, the Lewis symbol of an atom depicts its **bonding capacity**. A Lewis symbol consists of the symbol of an element surrounded by dots representing the valence electrons in one of its atoms. The dots are placed on the four sides of the symbol (top, bottom, right, and left). The order in which they are placed does not matter *as long as one dot is placed on each side before any dots are paired*. The number of *unpaired* dots in a Lewis symbol indicates the typical bonding capacity of the atom.

Figure 8.2 shows the Lewis symbols of the main group elements. Because all elements in a family have the same number of valence electrons, they all have the same arrangement of dots in their Lewis symbols. For example, the Lewis symbols of carbon and all other group 14 elements have four unpaired dots representing four unpaired electrons. Four unpaired electrons mean that a carbon atom has a bonding capacity of four: it forms four chemical bonds. In doing so, it completes its octet because these four bonds contain the carbon atom's original four valence electrons plus four more from the atoms with which it forms the bonds. Similarly, the Lewis symbols of nitrogen and all other group 15 elements have three dots representing three unpaired electrons. A nitrogen atom has a bonding capacity of three and typically forms three bonds to complete its octet.

Some important exceptions to Lewis's octet rule occur, several of which involve H, Be, and B. Compounds having fewer than eight valence electrons around an atom are referred to as *electron-deficient* compounds in Lewis theory. In another type of electron deficiency, atoms in some molecules have incomplete octets because there are odd numbers of valence electrons in those molecules. Examples include NO and NO_2. We explore electron-deficient compounds and the chemical implications of the unpaired electrons in molecules in Section 8.6.

octet rule the tendency of atoms of main group elements to make bonds by gaining, losing, or sharing electrons to achieve a valence shell containing eight electrons, or four electron pairs.

Lewis symbol (also called **Lewis dot symbol**) the chemical symbol for an atom surrounded by one or more dots representing the valence electrons.

bonding capacity the number of covalent bonds an atom forms to have an octet of electrons in its valence shell.

CONNECTION The elements in groups 1, 2, and 13 through 18 in the periodic table are called main group elements (Chapter 2). The monatomic ions of main group elements (except H^+) have filled *s* and *p* orbitals in their outermost shell and so are isoelectronic with a noble gas (Chapter 7).

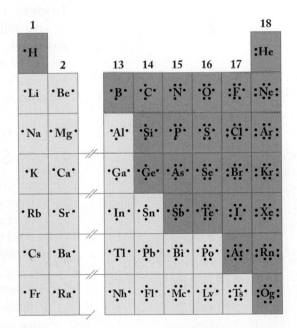

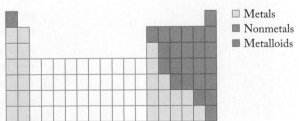

FIGURE 8.2 Lewis symbols of the main group elements of the periodic table. Because elements in a family have similar outer-shell electron configurations, they have the same number of dots in their Lewis symbols, which represent valence electrons.

Lewis structure a two-dimensional representation of the bonds and lone pairs of valence electrons in a molecule or polyatomic ion.

bonding pair a pair of electrons shared between two atoms.

single bond a chemical bond that results when two atoms share one pair of electrons.

lone pair a pair of electrons that is not shared.

 **CHEMTOUR**
Lewis Structures

Before we start to use Lewis symbols to explore how elements form compounds, we need to keep in mind that the symbols indicate the number of bonds that the atoms of an element *typically* form. We see later in this chapter that atoms may exceed their bonding capacities in some molecules and polyatomic ions, and atoms fail to reach them in others. Bonding capacity is a useful concept, but we treat it more like a guideline than a strict requirement.

Lewis Structures

The properties of molecular substances depend on how many atoms of which elements are in each of their molecules and how those atoms are bonded together. A **Lewis structure** is a two-dimensional representation of a molecule showing how the atoms in the molecule are connected. Because covalent bonds are *pairs* of shared valence electrons, Lewis structures focus on how these electron pairs, called **bonding pairs**, are distributed among the atoms in a molecule. In a Lewis structure, the pairs of electrons that are shared in chemical bonds are drawn with lines, as in the molecular structures we first saw in Chapter 1. A pair of electrons shared between two atoms is called a **single bond**. Electron pairs that are not involved in bond formation appear as pairs of dots on one atom. These unshared electron pairs are called **lone pairs**. Atoms with one or more lone pairs in their Lewis symbols frequently have that same number of lone pairs of electrons in the Lewis structures of the molecules they form.

The following guidelines describe a five-step approach to drawing Lewis structures. These initial guidelines are just a starting point in the discussion of molecular structure. We will continue to refine these guidelines as we draw structures for increasingly more challenging molecules and learn more about the distribution of electrons in bonds.

Steps to Follow When Drawing Lewis Structures

1. *Determine the number of valence electrons.* For a neutral molecule, count the valence electrons in all the atoms in the molecule. For a polyatomic ion, count the valence electrons and then add or subtract the number of electrons needed to account for the charge on the ion. This total number of valence electrons is used to form bonds and lone pairs in the Lewis structure.

2. *Arrange the symbols of the elements to show how the atoms are bonded and then connect them with single bonds (single pairs of bonding electrons).* Put the atom with the greatest bonding capacity in the center. (If two elements have the same bonding capacity, choose the one that is less electronegative, an atomic property we examine in Section 8.3.) Place the remaining atoms around the central atom and those that form the fewest bonds (such as hydrogen) around the periphery. Connect the atoms with a single bond. The resulting representation is often called a *skeletal structure*.

3. *Complete the octets or duets of the atoms bonded to the central atom by adding lone pairs of electrons.* Place lone pairs on each of the outer atoms until each outer atom has an octet or duet (including the two electrons used to connect it to the central atom in the skeletal structure).

4. *Compare the number of valence electrons in the Lewis structure with the number determined in step 1.* If valence electrons remain unused in the structure, place lone pairs on the central atom (even if doing so means giving it more than an octet of valence electrons), until all the valence electrons counted in step 1 have been included in the Lewis structure.

5. *Complete the octet on the central atom.* If there is an octet on the central atom, the structure is complete. If there is less than an octet on the central atom, create additional bonds to it by converting one or more lone pairs of electrons on outer atoms into bonding pairs.

SAMPLE EXERCISE 8.1 Drawing the Lewis Structure of Chloroform   **LO2**

Chloroform, $CHCl_3$, is a liquid with a low boiling point that was once used as an anesthetic in surgery. Draw its Lewis structure.

Collect and Organize We are given the molecular formula of chloroform, $CHCl_3$, and we can use the five-step approach described above to generate the Lewis structure.

Analyze The formula $CHCl_3$ tells us that a chloroform molecule contains one carbon atom, one hydrogen atom, and three chlorine atoms. Because carbon is a group 14 element, it has four valence electrons and needs four more to complete its octet. Hydrogen, in group 1, has one valence electron in the $1s$ orbital and needs one more to complete its duet. Chlorine, in group 17, has seven valence electrons (in the $3s$ and $3p$ atomic orbitals) and needs one more to complete its octet.

Solve

1. The number of valence electrons in the $CHCl_3$ molecule is

Element:	C	+	H	+	3 Cl	
Valence electrons:	4	+	1	+	(3×7)	= 26

2. The carbon atom has the most (four) unpaired electrons in its Lewis symbol and hence has the greatest bonding capacity. Therefore C is the central atom in the molecule. Each H atom and each Cl atom needs one more electron to achieve the electron configuration of a noble gas, and so each forms one bond to the C atom:

$$
\begin{array}{c}
\text{H} \\
| \\
\text{Cl} - \text{C} - \text{Cl} \\
| \\
\text{Cl}
\end{array}
$$

3. In this structure the hydrogen atom has a duet because it shares a bonding pair of electrons with the central carbon atom. We need to add lone pairs of electrons to complete the octets on the three chlorine atoms:

$$
\begin{array}{c}
\text{H} \\
| \\
\ddot{\text{:Cl}} - \text{C} - \ddot{\text{Cl:}} \\
| \\
\ddot{\text{:Cl:}}
\end{array}
$$

4. This structure contains four pairs of bonding electrons and nine lone pairs, for a total of

$$
(4 \times 2) + (9 \times 2) = 26 \text{ electrons}
$$

   which is the number of electrons determined in step 1.

5. The carbon atom is surrounded by four bonds, which means eight electrons, so it has a full octet. The structure is done.

Think About It In this example there was no need to change the structure during steps 4 and 5 because all the valence electrons are included and all the atoms have an octet (or duet in the case of the hydrogen atom).

 Practice Exercise
Draw the Lewis structure of methane, CH_4.

(Answers to Practice Exercises are in the back of the book.)

SAMPLE EXERCISE 8.2 Drawing the Lewis Structure of Ammonia   **LO2**

Draw the Lewis structure of ammonia, NH_3.

Collect, Organize, and Analyze The formula NH_3 tells us that each molecule contains one atom of nitrogen and three atoms of hydrogen. Nitrogen is a group 15 element with five valence electrons, three of which are unpaired, for a bonding capacity of 3. Each hydrogen atom has one valence electron in the $1s$ orbital and a bonding capacity of 1.

Solve
1. The number of valence electrons in the NH_3 molecule is

Element:	N	+	3 H
Valence electrons:	5	+	$(3 \times 1) = 8$

2. The nitrogen atom has the greater bonding capacity and is the central atom. Connecting each H atom to the nitrogen atom with a covalent bond yields

$$H\!-\!N\!-\!H$$
$$|$$
$$H$$

3. Each bonded H atom has a complete duet of electrons.
4. The three bonds in the structure represent $3 \times 2 = 6$ valence electrons, but we need 8 to match the number determined in step 1. To add 2 electrons we add a lone pair to nitrogen:

5. The N atom now has an octet of electrons, so the Lewis structure is complete.

Think About It This structure makes sense because the Lewis symbol of the nitrogen atom contains two electrons that are paired and three that are unpaired. Therefore it is reasonable that the nitrogen atom in NH_3 has one lone pair and three bonding pairs of electrons.

Practice Exercise
Draw the Lewis structure of phosphorus trichloride.

Lewis Structures of Molecules with Double and Triple Bonds

We can use Lewis structures to show the bonding in molecules in which two atoms share more than one pair of bonding electrons. A bond in which two atoms share two pairs of electrons is called a **double bond**. For example, when the two oxygen atoms in a molecule of O_2 share two pairs of electrons, they form an $O\!=\!O$ double bond. When the two nitrogen atoms in a molecule of N_2 share three pairs of electrons, they form a $N\!\equiv\!N$ **triple bond**. In Section 8.7 we discuss the characteristics of these multiple bonds and compare them with those of single bonds.

How do we know when a Lewis structure has a double or triple bond? Typically, we find out when we apply steps 3 through 5 in the guidelines. Suppose we fill the octets of all the atoms attached to the central atom in step 3, and in doing so we use all the valence electrons available. If we discover that the central atom does not have an octet, we can provide the needed electrons in step 5 by converting

double bond a chemical bond in which two atoms share two pairs of electrons.

triple bond a chemical bond in which two atoms share three pairs of electrons.

aldehyde an organic compound having a carbonyl group with a single bond to a hydrogen atom and a single bond to another atom or group of atoms, designated as R– in the general formula RCHO.

carbonyl group a carbon atom with a double bond to an oxygen atom.

one or more lone pairs of electrons from one of the other atoms into a bonding pair. The following example illustrates an application of the guidelines.

Let's draw the Lewis structure for the organic molecule formaldehyde (H_2CO).

1. The total number of valence electrons is

Element:	C	+	2 H	+	O
Valence electrons:	4	+	(2 × 1)	+	6 = 12

2. Of the three elements, carbon has the greatest bonding capacity (4). Therefore C is the central atom. Connecting it with single bonds to the other three atoms, we have:

$$H\!-\!\overset{\displaystyle |}{\underset{\displaystyle O}{C}}\!-\!H$$

3. Each H atom has a single covalent bond (2 electrons) completing its valence shell. Oxygen needs three lone pairs of electrons to complete its octet:

$$H\!-\!\overset{\displaystyle |}{\underset{\displaystyle :\ddot{O}:}{C}}\!-\!H$$

4. There are 12 valence electrons in this structure, which matches the number determined in step 1.

5. The central C atom has only 6 electrons. To provide the carbon atom with 2 more electrons so that it has an octet—without removing any electrons from oxygen, which already has an octet—we convert one of the lone pairs on the oxygen atom into a bonding pair between C and O:

$$H\!-\!\overset{|}{\underset{:\ddot{O}:}{C}}\!-\!H \quad \rightarrow \quad \overset{\displaystyle H \qquad H}{\underset{\displaystyle \cdot\ddot{O}\cdot}{\overset{\displaystyle \diagdown \;\;\diagup}{C}}}$$

It does not matter which of the three lone pairs we move because they are equivalent. The central carbon atom now has a complete octet, and the oxygen atom still does. This structure makes sense because the four covalent bonds around carbon—two single bonds and one double bond—match its bonding capacity. The double bond to oxygen makes sense because oxygen is a group 16 element with a bonding capacity of 2, and it has two bonds in this structure.

Notice that we have drawn the double bond and the two single bonds on the central carbon atom at an angle of about 120° from each other in the final structure; we have done the same thing with the lone pairs of electrons on oxygen with respect to the double bond. Electrons are negatively charged and repel each other, so we draw them as far apart from each other as possible. This logic is an important part of drawing three-dimensional structures of molecules in Chapter 9.

Formaldehyde is the smallest member of a family of organic compounds known as **aldehydes**. All aldehydes contain the functional group that consists of a carbon atom with a single bond to a hydrogen atom and a double bond to an oxygen atom (Figure 8.3). The C=O group is called a **carbonyl group**. The fourth bond on the carbon atom connects to another hydrogen atom in formaldehyde, but in all other aldehydes it connects to another carbon-containing group, giving aldehydes a general formula of RCHO. It is not necessary for you to know how to name these compounds, but it is important for you to recognize the aldehyde functional group.

CONNECTION We have already seen a few families of organic molecules. We introduced carboxylic acids (RCOOH) and amines (RNH_2, R_2NH, R_3N) in Chapter 4. We introduced alkanes (compounds of carbon atoms and hydrogen atoms linked by single bonds) in Chapter 2 and explored them further in Chapter 5. Remember that R– stands for any organic group. Table 2.5 lists many of the functional groups in organic compounds.

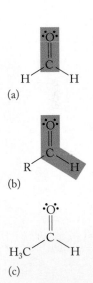

FIGURE 8.3 (a) Formaldehyde is the smallest member of the aldehyde family. The carbonyl group is highlighted. (b) All aldehydes contain a carbonyl group with a single bond to a hydrogen atom; R– represents any other organic group. The aldehyde group is highlighted. (c) Acetaldehyde is the next largest aldehyde; in acetaldehyde, R = CH_3.

The 5-step guidelines for drawing Lewis structures are particularly useful when molecules or polyatomic ions have a central atom bonded to atoms with lower bonding capacity. Many inorganic molecules, polyatomic ions, and small organic molecules (fewer than four carbon atoms) fit this description. However, in order to draw some larger Lewis structures, we may sometimes need to break the structure into subunits and consider more than one "central" atom, as in hydrogen peroxide:

$$H-\ddot{O}-\ddot{O}-H$$

In these cases we treat each atom that is bonded to two other atoms as the "central atom" within a three-atom subunit, as shown by the oxygen atoms highlighted in red:

$$-\ddot{O}-\ddot{O}-H$$

$$H-\ddot{O}-\ddot{O}-$$

SAMPLE EXERCISE 8.3 Drawing the Lewis Structure of Acetylene **LO2**

Draw the Lewis structure of acetylene, C_2H_2, the hydrocarbon fuel used in oxyacetylene torches for welding and cutting metal.

Collect, Organize, and Analyze We are asked to draw the Lewis structure of acetylene, which has the molecular formula C_2H_2. We follow the five steps in the guidelines. Each molecule contains two atoms of carbon and two atoms of hydrogen. Carbon is a group 14 element with a bonding capacity of 4.

Solve

1. The total number of valence electrons is

Element:	2 C	+	2 H
Valence electrons:	(2×4)	+	$(2 \times 1) = 10$

2. Of the two elements, carbon has the greater bonding capacity, so the two carbon atoms are placed in the center of the molecule. Connecting each carbon atom with single bonds to the other atoms, we have:

   $$H-C-C-H$$

3. Each H atom has a single covalent bond and a complete valence shell.
4. The structure has 6 valence electrons, but it should have 10. This means we have to add two pairs of electrons. One way to do that is to add two bonds between the carbon atoms. This gives the structure the right number of valence electrons, and it completes the octets of both carbon atoms (step 5):

   $$H-C\equiv C-H$$

Think About It This structure makes sense because carbon has a bonding capacity of 4, and there are four covalent bonds around each carbon atom—1 single bond and 1 triple bond—giving both atoms complete octets. Note that acetylene's triple bond makes it the smallest member of the *alkyne* family (see Section 2.8).

 Practice Exercise
Determine the Lewis structure for carbon dioxide.

Lewis Structures of Ionic Compounds

Crystals of sodium chloride are held together by the attraction between oppositely charged Na^+ and Cl^- ions. We discussed in Chapter 7 how sodium and the other group 1 and 2 elements achieve noble gas electron configurations by losing their valence-shell s electrons and forming positively charged cations. Therefore the ions of group 1 and 2 elements have no valence-shell electrons, and their Lewis structures have no dots around them. On the other hand, nonmetals such as chlorine, which acquire electrons to achieve noble gas electron configurations, have filled s and p orbitals in their valence shells. Therefore their monatomic ions have Lewis structures with four pairs of dots. Applying this notation to NaCl gives us this Lewis structure:

$$Na\cdot \quad +\ \ \cdot \ddot{\underset{\cdot\cdot}{Cl}}: \ \rightarrow\ Na^+\left[:\ddot{\underset{\cdot\cdot}{Cl}}:\right]^-$$

We place brackets around the dots to emphasize that all eight valence electrons are associated with the anion, and the charge of the ion is placed outside the bracket to indicate the overall charge of everything inside.

All alkali metal elements and alkaline earth elements tend to lose rather than share their valence electrons when bonding with nonmetals, in part because these two families of metals have low ionization energies. The cations they form by losing these electrons obey the octet rule because they have the electron configurations of the noble gases that precede the parent elements in the periodic table.

CONNECTION We defined ionization energy in Chapter 7 as the amount of energy required to remove 1 mole of electrons from 1 mole of particles in the gas phase.

SAMPLE EXERCISE 8.4 Drawing Lewis Structures of Binary Ionic Compounds   **LO2**

Draw the Lewis structure of calcium oxide.

Collect, Organize, and Analyze When a metal combines with a nonmetal, the metal forms a cation and the nonmetal forms an anion. The monatomic ions formed by calcium and oxygen are Ca^{2+} and O^{2-}, respectively.

Solve To form a cation with a 2+ charge, a Ca atom loses both its valence electrons, leaving it with none:

$$\cdot Ca\cdot \ \xrightarrow{-2\,e^-}\ Ca^{2+}$$

When an O atom accepts two electrons it has a complete octet:

$$:\ddot{\underset{\cdot}{O}}\cdot \ \xrightarrow{+2\,e^-}\ \left[:\ddot{\underset{\cdot\cdot}{O}}:\right]^{2-}$$

Combining the Lewis symbols of Ca^{2+} and O^{2-} ions, we have the Lewis structure of CaO:

$$Ca^{2+}\left[:\ddot{\underset{\cdot\cdot}{O}}:\right]^{2-}$$

Think About It The lack of dots in the symbol of the Ca^{2+} ion reinforces the fact that atoms of Ca, like all main group metal atoms, lose all the electrons in their valence shells when they form monatomic ions. In contrast, atoms of nonmetals form monatomic anions with complete octets.

 Practice Exercise
Draw the Lewis structure of magnesium fluoride.

FIGURE 8.4 Just as a battery has positive and negative poles, a polar molecule such as HCl has positive and negative ends, represented here by the arrow above the H—Cl bond. The tail of the arrow is on the hydrogen atom, which has a partial positive charge, and the arrowhead is pointed toward the Cl, which has a partial negative charge. The partial positive charge on the hydrogen atom may also be represented by a lowercase delta (δ) followed by a + sign. Correspondingly, the partial negative charge on the chlorine is represented by the $\delta-$ symbol.

CHEMTOUR
Bond Polarity and Polar Molecules

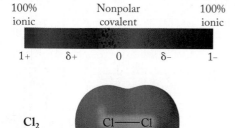

(a) Nonpolar covalent: even charge distribution

(b) Polar covalent: uneven charge distribution

(c) Ionic: complete transfer of electron

8.3 Polar Covalent Bonds

When Lewis proposed that atoms form chemical bonds by sharing electrons, he knew that electron sharing in covalent bonds did not necessarily mean *equal* sharing. For example, Lewis knew that molecules of HCl ionized to form H^+ ions and Cl^- ions when HCl dissolves in water. To explain this phenomenon, Lewis proposed that the bonding pair of electrons in a molecule of HCl is closer to the chlorine end of the bond than to the hydrogen end. This unequal sharing makes the H—Cl bond a **polar covalent bond**. As a result, when the H—Cl bond breaks, the one shared pair of electrons remains with the Cl atom to form a Cl^- ion, simultaneously changing the H atom into an H^+ ion with no electrons.

This **bond polarity** in HCl means that the bond functions as a tiny electric *dipole*, meaning that there is a slightly positive pole at the H end of the bond and a slightly negative pole at the Cl end, analogous to the positive and negative ends of a battery as shown in Figure 8.4. This figure also shows two representations we use to depict unequal sharing of bonding pairs of electrons. We use an arrow with a plus sign embedded in its tail ($\longmapsto$) to indicate the *direction of polarity*: the arrow points toward the more negative, electron-rich atom in the bond, and the position of the plus sign indicates the more positive, electron-poor atom. The other way to represent unequal sharing makes use of the lowercase Greek delta, δ, followed by a + or − sign. The deltas represent *partial* electrical charges, as opposed to full electrical charges that accompany the complete transfer of one or more electrons as atoms become ions.

Figure 8.5 shows examples of equal and unequal sharing of bonding pairs of electrons. It includes (a) Cl_2, in which the two Cl atoms share a pair of electrons equally in a **nonpolar covalent bond**; (b) HCl, which has a polar covalent bond; and (c) NaCl with its ionic bond, an extreme case of unequal sharing. In NaCl, the one valence electron of sodium has been transferred to the chlorine atom, creating a Na^+ ion and a Cl^- ion and resulting in *total*, rather than *partial*, separation of electrical charge: 1+ on Na and 1− on Cl. The degree of charge separation may be represented using color, as shown in in the color bar at the top of Figure 8.5. Yellow-green represents no charge separation (nonpolar covalent bonding), whereas violet and red represent full 1+ and 1− charges (ionic bonding), respectively. Intermediate colors indicate partial charges ($\delta+$ and $\delta-$, polar covalent bonding).

Another American chemist, Linus Pauling (1901–1994), developed the concept of **electronegativity** to explain bond polarity. Pauling assigned electronegativity values to the elements (Figure 8.6) on the basis of the idea that the bonds between atoms of different elements are neither 100% covalent nor 100% ionic, but somewhere in between. The degree of **ionic character** of a bond depends on the differences in the abilities of the two atoms to attract the electrons they share: the greater the difference, the more ionic is the bond between them.

The data in Figure 8.6 show that electronegativity is a periodic property, with values generally increasing from left to right across a row in the periodic table, and

FIGURE 8.5 Variations in valence electron density are represented using colored surfaces in these molecular models. (a) The symmetrical distribution of colors in Cl_2 indicates that the pair of bonding electrons in the molecule is shared equally. (b) Unequal sharing of the bonding pair of electrons occurs in HCl, as shown by the orange color of the Cl end of the bond and the green color around the H end. (c) In ionic NaCl, the violet color on the surface of the sodium ion indicates that it has a full 1+ charge and the red of the chloride ion reflects its charge of 1−.

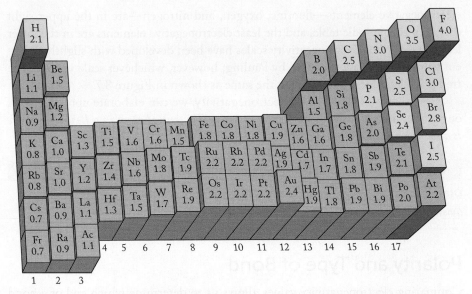

FIGURE 8.6 The Pauling electronegativity values shown below the symbols of the elements are unitless values that indicate the relative ability of an atom in a bond to attract shared electrons to itself. Electronegativity generally increases from left to right across a period and decreases from top to bottom down a group.

polar covalent bond a bond characterized by unequal sharing of bonding pairs of electrons between atoms.

bond polarity a measure of the extent to which bonding electrons are unequally shared due to differences in electronegativity of the bonded atoms.

nonpolar covalent bond a bond characterized by an even distribution of charge; electrons in the bonds are shared equally by the two atoms.

electronegativity a relative measure of the ability of an atom to attract electrons in a bond to itself.

ionic character an estimate of the magnitude of charge separation in a covalent bond.

decreasing from top to bottom down a group. The reasons for these trends are essentially the same ones that produce similar trends in first ionization energies, as we discussed in Chapter 7 (see Figure 7.39). Increasing attraction between the nuclei of atoms and their outer-shell electrons as the atomic number increases across a row produces both higher ionization energies and greater electronegativities (Figure 8.7a). Within a group of elements, the weaker attraction between nuclei and outer-shell electrons with increasing atomic number leads to lower ionization energies and smaller electronegativities (Figure 8.7b). For these two reasons, the most

FIGURE 8.7 The trends in the electronegativities of the main group elements follow those of ionization energies: both tend to (a) increase with increasing atomic number across a row and (b) decrease with increasing atomic number within a group.

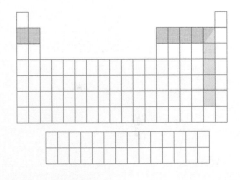

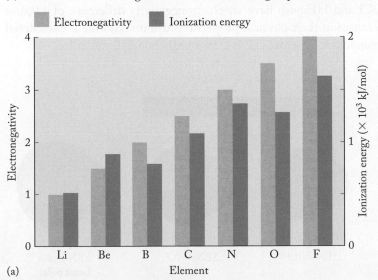

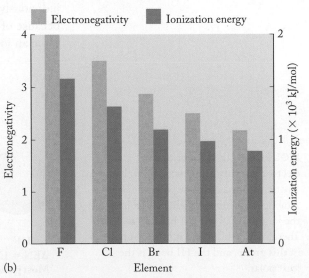

electronegative elements—fluorine, oxygen, and nitrogen—are in the upper right corner of the periodic table, and the least electronegative elements are in the lower left corner. Other electronegativity scales have been developed with slightly different values from those proposed by Pauling; however, whichever scale you use, the trends in electronegativity remain the same as shown in Figure 8.7.

Now that we have defined electronegativity, we can elaborate upon step 2 of our guidelines for drawing Lewis structures: *The central atom in a Lewis structure is often the atom with the lowest electronegativity.*

CONCEPT **TEST**

Draw arrows on the Lewis structure of CO_2 to indicate the polarity of the bonds.

(Answers to Concept Tests are in the back of the book.)

Polarity and Type of Bond

Comparing electronegativity values allows us to determine which end of a bond is relatively electron-rich and which end is relatively electron-poor. The greater the difference in electronegativity (ΔEN), the more uneven the distribution of electrons and the more polar the covalent bond. Figure 8.8 shows this trend for compounds formed between hydrogen and the halogens. For example, a H—F bond is more polar than a H—Cl bond because the ΔEN between H (2.1) and F (4.0) is 1.9, whereas the ΔEN between H (2.1) and Cl (3.0) is 0.9.

Figure 8.9 provides an approximate scale for judging the degree of polarity in a chemical bond on the basis of electronegativity differences. We saw in Figure 8.5(a) that a Cl—Cl bond is nonpolar covalent, and $\Delta EN = 0$ for this bond. On the other end of the scale, electronegativity differences greater than 2.0 frequently exist in compounds formed between metals and nonmetals. For example, calcium oxide is considered an ionic compound, and $\Delta EN = 2.5$ for Ca (1.0) and O (3.5). All other bonds, which are polar covalent, fall between nonpolar covalent and ionic, as we saw in Figure 8.8. Even small electronegativity differences, as in carbon–hydrogen bonds, where $\Delta EN = 0.4$, are considered weakly polar.

As a somewhat arbitrary guideline, we consider the bond between two atoms to be ionic rather than covalent when ΔEN values between them are equal to or greater than 2.0. Keep in mind that the scale in Figure 8.9 should be used judiciously—LiCl and HF both have an electronegativity difference of 1.9, but because of differences in their physical properties, lithium chloride is considered to be an ionic compound, whereas HF is a covalently bonded molecule.

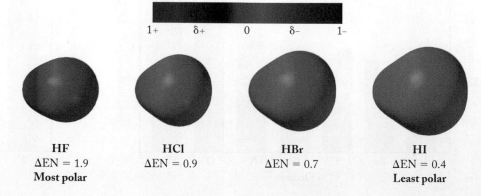

FIGURE 8.8 The greater the electronegativity difference (ΔEN) between any two atoms, the more polar the bond. Among the hydrogen halides, HF has the largest ΔEN and HI has the smallest. Therefore the HF bond is the most polar in this group, and the HI bond is the least polar.

| 1+ | δ+ | 0 | δ– | 1– |

HF	**HCl**	**HBr**	**HI**
$\Delta EN = 1.9$	$\Delta EN = 0.9$	$\Delta EN = 0.7$	$\Delta EN = 0.4$
Most polar			**Least polar**

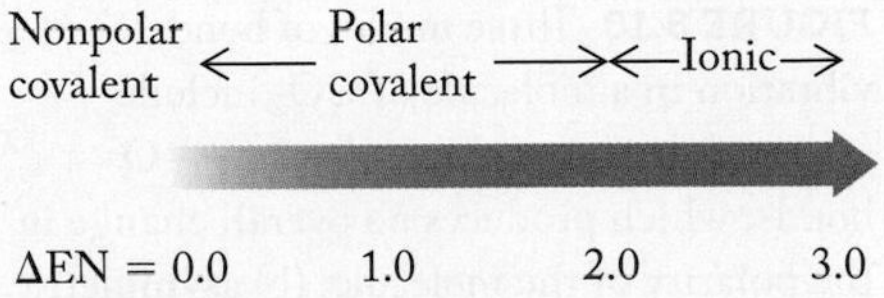

Note that the presence of a double or triple bond, such as the carbon–oxygen double bond in the carbonyl group (C=O), does not change our assessment of the overall polarity of the bond. Oxygen is more electronegative than carbon, so the carbonyl group is polarized, with the carbon atom bearing a $\delta+$ charge and the oxygen a $\delta-$ charge, as shown here in formaldehyde:

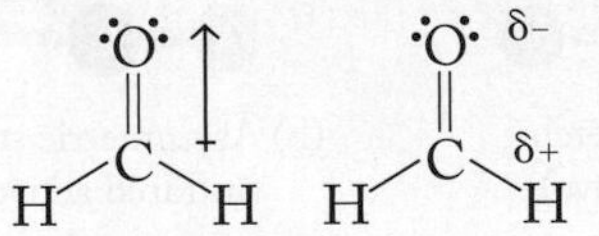

Nonpolar covalent ← Polar covalent → ←Ionic→

$\Delta EN = 0.0 \qquad 1.0 \qquad 2.0 \qquad 3.0$

FIGURE 8.9 An electronegativity difference (ΔEN) of zero is considered to be a nonpolar covalent bond. When $\Delta EN > 2.0$, the bond is ionic. When $0 < \Delta EN < 2.0$, the bond is typically considered to be polar covalent.

SAMPLE EXERCISE 8.5 Comparing the Polarity of Bonds   **LO3**

Rank, in order of increasing polarity, the bonds formed between: O and C; Cl and Ca; N and S; O and Si. Are any of these bonds considered ionic?

Collect and Organize We are given four pairs of atoms and asked to rank them according to the polarity of the bond each pair forms and to identify any ionic bonds in the set.

Analyze The polarity of a bond is related to the difference in electronegativities of the atoms in the bond. The guideline we apply is as follows: if the electronegativity difference is 2.0 or greater, the bond is considered ionic. We need to refer to the Pauling electronegativities (Figure 8.6) to judge the relative polarity.

Solve Calculate the electronegativity difference between the atoms:

O and C:	$\Delta EN = 3.5 - 2.5 = 1.0$
Cl and Ca:	$\Delta EN = 3.0 - 1.0 = 2.0$
N and S:	$\Delta EN = 3.0 - 2.5 = 0.5$
O and Si:	$\Delta EN = 3.5 - 1.8 = 1.7$

These electronegativity differences are proportional to the polarity of the bonds formed between the pairs of atoms. Therefore, ranking them in order of increasing polarity we have:

$$N—S < O—C < O—Si < Cl—Ca$$

The bond between Cl and Ca is considered ionic because $\Delta EN = 2.0$.

Think About It Calcium is a metal and chlorine is a nonmetal, so the result indicating that the bond between them is ionic is reasonable. Ionic bonds tend to be formed between metals and nonmetals. Two of the other bonds, N—S and O—C, connect pairs of nonmetals, and the O—Si bond connects a nonmetal with a metalloid. We expect these three bonds to be covalent.

 Practice Exercise Which of the following pairs forms the most polar bond: O and S; Be and Cl; N and H; C and Br? Is the bond between that pair ionic?

Vibrating Bonds and Greenhouse Gases

As we noted in Chapter 5 in our discussion of thermal energy, chemical bonds are not rigid. They vibrate a little, stretching and bending like tiny atomic-sized springs (Figure 8.10). These vibrations have natural frequencies, which match the frequencies of infrared electromagnetic radiation. This match, coupled with the unequal sharing of bonding electrons, allows some atmospheric molecules containing polar

CHEMTOUR Vibrational Modes

FIGURE 8.10 Three modes of bond vibration in a molecule of CO_2 include (a) symmetric stretching of the C=O bonds, which produces no overall change in the polarity of the molecule; (b) asymmetric stretching, which produces side-to-side fluctuations in polarity that may result in absorption of infrared radiation; (c) bending, which produces up-and-down fluctuations that also may absorb infrared radiation.

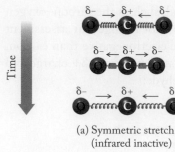
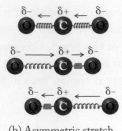
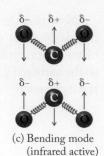

(a) Symmetric stretch (infrared inactive) (b) Asymmetric stretch (infrared active) (c) Bending mode (infrared active)

CHEMTOUR
Greenhouse Effect

covalent bonds—such as CO_2—to absorb infrared radiation emitted by Earth's surface. As a result, heat that might have dissipated into space is trapped in the atmosphere by these molecules, contributing to the atmospheric greenhouse effect as described in Section 8.1. The understanding of bond vibrations requires subtle and complex models that are beyond the scope of this book, but the basic ideas presented here provide a useful picture of the interaction between bond vibrations and infrared radiation.

Molecules with polar bonds may absorb photons, much like electrons in atoms absorb radiation and form excited states (see Section 7.4). The radiation absorbed by polar covalent bonds is typically in the infrared region of the electromagnetic spectrum (see Figure 7.7). Vibrations result in tiny fluctuating electrical fields associated with the separation of partial charges in the polar covalent bonds in the molecule. These fluctuations can alter the strengths of the fields or even create new ones, depending on the nature of the vibration. When the frequencies of the polar covalent bond vibrations match the frequencies of photons of infrared radiation, the molecules may absorb those photons. Scientists say these vibrations are *infrared active*. This is the molecular mechanism behind the greenhouse effect. The presence of fluctuating electric fields in molecules is further evidence that covalent bonds involve shared electrons.

Not all polar bond vibrations result in absorption of infrared radiation. For example, two kinds of stretching vibrations can occur in a molecule of CO_2, which has two C=O bonds. One is a *symmetric* stretching vibration (Figure 8.10a) in which the two C=O bonds stretch and then compress at the same time. In this case the two fluctuating electrical fields produced by the two C=O bonds cancel each other out, and no infrared absorption or emission is possible. This vibration is said to be *infrared inactive*. However, when the bonds stretch such that one gets shorter as the other gets longer (Figure 8.10b), the changes in charge separation do not cancel. This *asymmetric* stretch produces a fluctuating electric field that enables CO_2 to absorb infrared radiation, so this vibration is infrared active. Molecules can also bend (Figure 8.10c) to produce fluctuating electrical fields. Because the frequencies of the asymmetric stretching and bending of the bonds in CO_2 are in the same range as the frequencies of infrared radiation emitted from Earth's surface, carbon dioxide is a potent greenhouse gas.

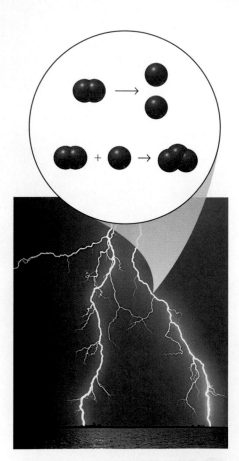

FIGURE 8.11 Lightning strikes contain sufficient energy to break O=O double bonds. The O atoms formed in this fashion collide with other O_2 molecules, forming ozone (O_3), an allotrope of oxygen.

CONCEPT **TEST**

Nitrogen and oxygen make up about 99% of the gases in the atmosphere. Could the stretching of the N≡N and O=O bonds in these molecules result in the absorption of infrared radiation? Explain why or why not.

8.4 Resonance

The atmosphere contains two types of molecular oxygen. Most of it is O_2, but trace concentrations of ozone (O_3) are also present. Ozone in the lower atmosphere is sometimes referred to as "bad ozone" because high levels damage crops, harm trees, and lead to human health problems. However, ozone is also present in the upper atmosphere, where it is considered "good ozone" because it shields life on Earth from potentially harmful ultraviolet radiation from the Sun.

Ozone is a *triatomic* (three-atom) molecule produced naturally by lightning (Figure 8.11) and accounts for the pungent odor you may have smelled after a severe thunderstorm. Ozone (O_3) and diatomic oxygen (O_2) have the same empirical formula: O. Different molecular forms of the same element, such as O_2 and O_3, are called **allotropes** of the element and have different chemical and physical properties. Ozone, for example, is an acrid, pale blue gas that is toxic even at low concentrations, whereas O_2 is a colorless, odorless gas that is essential for most life-forms.

The different molecular formulas of allotropes mean that they have different molecular structures. Let's draw the Lewis structure for ozone, O_3, following our five-step approach. Oxygen is a group 16 element and so has 6 valence electrons. Therefore the total number of valence electrons in an ozone molecule is $3 \times 6 = 18$ (step 1). Connecting the three O atoms with single bonds, we have (step 2)

$$O—O—O$$

Completing the octets of the noncentral atoms gives us (step 3)

$$:\ddot{O}—O—\ddot{O}:$$

This structure contains 16 electrons. We determined in step 1 that there are 18 valence electrons in the molecule, so we add 2 to the central oxygen atom (step 4):

$$:\ddot{O}—\ddot{O}—\ddot{O}:$$

This structure leaves the central atom two electrons short of an octet, so we convert one of the lone pairs on the O atom on the left end of the molecule into a bonding pair (step 5):

$$·\ddot{O}\!=\!\overset{\ddot{\,}}{O}\!—\ddot{O}·$$

However, we could just as well have used a lone pair from the O atom on the right, which would have given us

$$:\ddot{O}—\overset{\ddot{\,}}{O}\!=\!\ddot{O}·$$

Which structure is correct? Interestingly, experimental evidence indicates that *neither* structure is accurate. Scientists have determined that both bonds in ozone have the same length. Because a double bond between two atoms is always shorter than a single bond between the same two atoms (see Section 8.7), the structure of ozone cannot consist of one single bond and one double bond. Figure 8.12 shows that the length of the two bonds in O_3 (128 pm) is in between the length of an O—O single bond (148 pm) and an O=O double bond (121 pm). One way to explain this result is to assume that the bonding pattern of O_3 is an average of the two structures drawn above:

$$O\overset{O}{\diagdown\!\!\!-\!\!\!\diagup}O$$

allotropes different molecular forms of the same element, such as oxygen (O_2) and ozone (O_3).

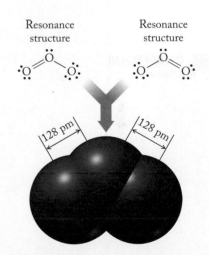

FIGURE 8.12 The molecular structure of ozone is an average of the two resonance structures shown at the top of the figure. Both bonds in ozone are 128 pm long, a value between the average length of an O—O single bond (148 pm) and the average length of an O=O double bond (121 pm). The intermediate value for the ozone bond length indicates that in an ozone molecule, the bonds are neither single bonds nor double bonds, but something in between.

In other words, each pair of oxygen atoms is held together by the equivalent of 1.5 bonds. However, this average does *not* mean that the molecule spends half its time as the left-hand structure in Figure 8.12 and half as the right-hand structure. It always has three bonding pairs spread out evenly on the two sides of the central atom.

To better understand this bonding pattern, consider what happens when the bonding electrons and the lone pair electrons in structure (a) below are rearranged as shown by the red arrows:

(a)   (b)

This rearrangement produces structure (b). The process is completely reversible, so that the electron pairs in structure (b) could just as easily be rearranged into the pattern in structure (a):

(b)   (a)

The two structures are equivalent. We use a double-headed reaction arrow between the structures to indicate that they are equivalent and that the actual structure is a blend of the two:

(a)   (b)

The ability to draw two equivalent Lewis structures for the ozone molecule illustrates an important concept in Lewis's theory called **resonance**: the existence of multiple Lewis structures, called **resonance structures** (or sometimes *resonance hybrids*), for a given arrangement of atoms. To determine whether resonance occurs in a molecule, we need to determine whether there can be alternative bond arrangements inside the molecule: arrangements in which the positions of some bonding pairs of electrons change but the positions of the atoms stay the same. One indicator of the possibility of resonance is the presence of both single and double bonds from a central atom to two or more atoms of the same element, as we just saw in ozone.

Because all resonance structures are Lewis structures by definition, the five-step guidelines still apply to drawing resonance structures. Because all the atoms in ozone are oxygen, an oxygen atom is clearly the central atom. Two additional issues, however, must be raised at this point, both of which relate to the concept of bonding capacity. First, oxygen atoms *typically* form two bonds, but Lewis theory allows both more and fewer bonds than are implied by the Lewis symbol of an atom. Remember that the resonance structures for ozone suggest that the oxygen atoms are held together by the equivalent of 1.5 bonds. Second, in previous examples we have used differences in typical bonding capacity as a criterion to select a central atom. As we deal with more molecules, the situation frequently arises that several atoms in one molecule may have the same bonding capacity. In that case, the selection of the central atom is based on electronegativity, with the least electronegative atom chosen as the central atom in the Lewis structure. Sample Exercise 8.6 explores the Lewis structure of just such a compound.

CHEMTOUR
Resonance

resonance a characteristic of electron distributions when two or more equivalent Lewis structures can be drawn for one compound.

resonance structure one of two or more Lewis structures with the same arrangement of atoms but different arrangements of bonding pairs of electrons.

SAMPLE EXERCISE 8.6 Drawing Resonance Structures of a Molecule **LO2**

Sulfur trioxide (SO_3) is a pollutant produced in the atmosphere when the SO_2 from natural and industrial sources combines with O_2. Draw all the possible resonance structures of SO_3, assuming sulfur obeys the octet rule.

Collect and Organize Each SO_3 molecule contains one atom of sulfur and three atoms of oxygen. We need to draw all resonance structures, which means at least two different Lewis structures with the same atom placement but different bonding patterns.

Analyze Both sulfur and oxygen are in group 16 with 6 valence electrons per atom and a typical bonding capacity of 2. Sulfur is less electronegative than oxygen.

Solve
1. The number of valence electrons in the SO_3 molecule is

Element:	S	+	3 O
Valence electrons:	6	+	$(3 \times 6) = 24$

2. Because sulfur is less electronegative, we select it as the central atom. Connecting it with single bonds to the three O atoms:

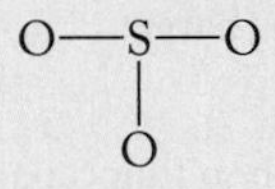

3. Each O atom has three lone pairs of electrons to complete its octet:

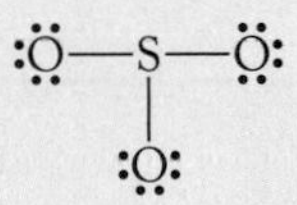

4. There are 24 valence electrons in the structure in step 3, which matches the number determined in step 1.
5. The central S atom has only 6 electrons in the structure of step 3. To complete its octet, we convert a lone pair on one of the oxygen atoms into a bonding pair:

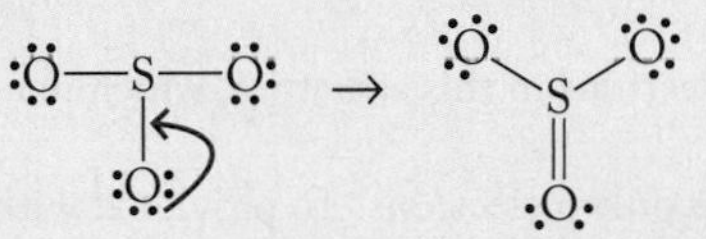

The sulfur atom now has a complete octet, and the Lewis structure is complete.
Because the three oxygen atoms are equivalent in the structure in step 3, we cannot arbitrarily choose one of them and ignore the others in forming a double bond. Therefore three resonance structures describe the bonding in SO_3:

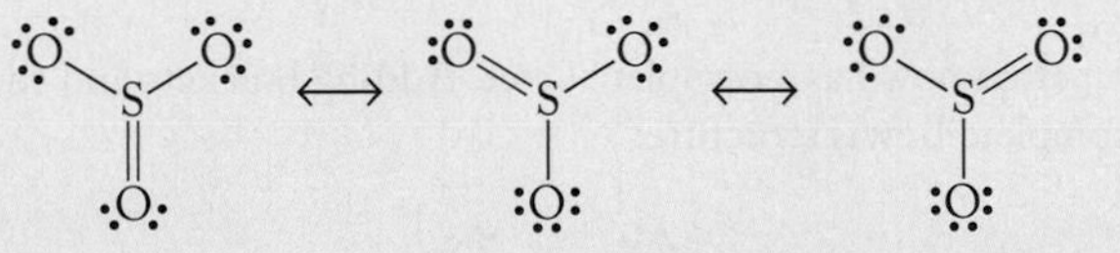

Think About It It makes sense that SO_3 has three resonance structures because three equivalent O atoms are bonded to the central S atom, and any one of the O atoms could be the one with the double bond.

Practice Exercise Draw all possible resonance structures for sulfur dioxide.

We can also draw resonance structures for polyatomic ions. In doing so, we need to account for the charge on each ion, which means adding the appropriate number of valence electrons to a polyatomic anion and subtracting the appropriate number from a polyatomic cation.

SAMPLE EXERCISE 8.7 Drawing Resonance Structures of a Polyatomic Ion **LO2**

Draw all the resonance structures for the nitrate ion, NO_3^-.

Collect and Organize We are asked to draw the resonance structures for the NO_3^- ion, which contains one nitrogen atom and three oxygen atoms. The charge of the ion is 1−.

Analyze Nitrogen is a group 15 element with 5 valence electrons per atom and a bonding capacity of 3. Oxygen is in group 16 with 6 valence electrons and a bonding capacity of 2. The 1− charge means the polyatomic ion has one additional valence electron.

Solve

1. The number of valence electrons is

Element:	N	+	3 O
Valence electrons:	5	+	$(3 \times 6) = 23$
Additional electron due to the 1− charge:			+ 1
Total valence electrons:			24

2. The nitrogen atom has the higher bonding capacity and the lower electronegativity, so N is the central atom in a NO_3^- ion. Connecting N with single bonds to the three O atoms, we have:

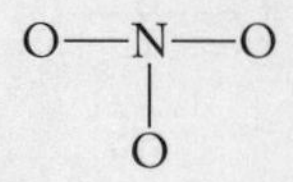

3. Each O atom has three lone pairs of electrons to complete its octet:

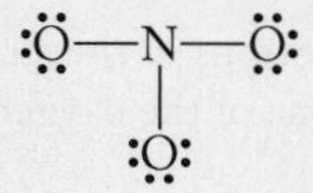

4. There are 24 valence electrons in this structure, which matches the number determined in step 1.

5. The central N atom has only 6 electrons. To provide it with 2 more to complete its octet, we convert a lone pair on one of the oxygen atoms into a bonding pair:

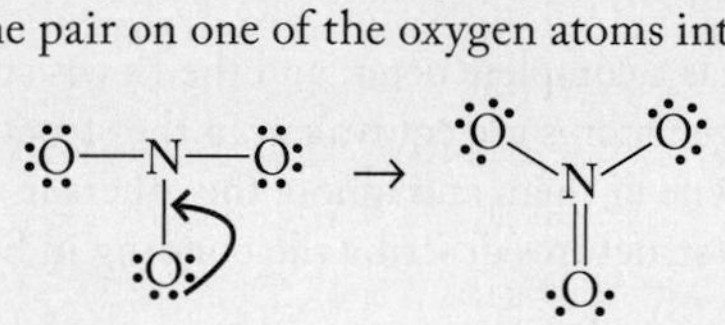

The nitrogen atom now has a complete octet. Adding brackets and the ionic charge, we have a complete Lewis structure:

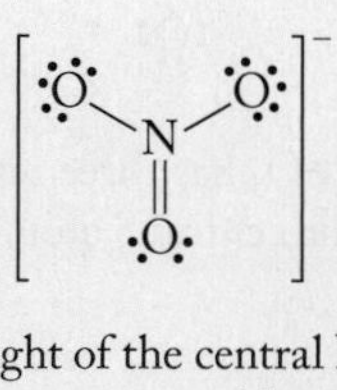

Using the O atom to the left or right of the central N atom to form the double bond, instead of the O atom below N, creates two additional resonance structures, three in all:

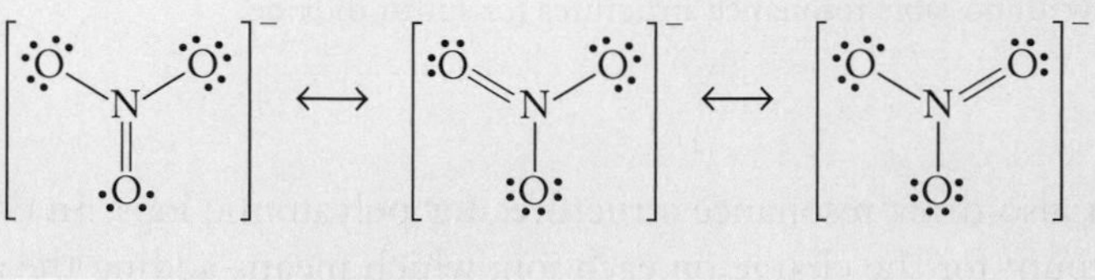

Think About It It makes sense that the NO_3^- ion has three resonance structures because three equivalent O atoms are bonded to the central N atom, and any one of the O atoms could be the one with the double bond. The additional electron from the negative charge on the ion means that NO_3^- has the same number of valence electrons as SO_3 (Sample Exercise 8.6), which also has the same number of resonance structures. Nitrogen in neutral molecules typically has a bonding capacity of 3, but in these resonance structures it forms four bonds.

 Practice Exercise Draw all the resonance forms of the azide ion, N_3^-, and the nitronium ion, NO_2^+.

Resonance also occurs in larger organic molecules that have alternating single and double bonds. Molecules of benzene (C_6H_6), for example, contain six-membered rings of carbon atoms with alternating single and double carbon–carbon bonds (Figure 8.13a). When we fix the atoms in a molecule of benzene in place, we have two equivalent ways to draw the single and double bonds. To depict their equivalency, chemists frequently draw benzene molecules with circles in the centers (Figure 8.13b). This symbol emphasizes that the six carbon–carbon bonds in the ring are all identical and intermediate in character between single and double bonds. (Note that experimental evidence confirms that all six carbon–carbon bonds in benzene are equivalent.)

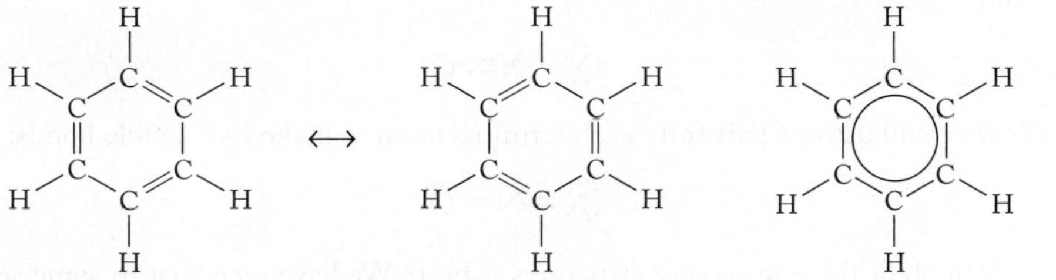

(a)   (b)

FIGURE 8.13 (a) The molecular structure of benzene is an average of two equivalent resonance structures. (b) The average is frequently represented by a circle inside the hexagonal ring of single bonds, indicating completely uniform delocalization of the electrons in the bonds around the ring.

Because of the carbon–carbon double bonds in the structure, you might think that benzene would be in the family of alkenes. However, when compared with alkenes, differences in chemical reactivity of the double bonds in benzene and other molecules where resonance influences the character of the bonds caused early chemists to consider such compounds as a distinct family: Benzene and compounds like it are classified as *aromatic compounds*. We explore these compounds in Chapter 9, where we develop models of bonding that enable us to envision the differences between alkenes like ethylene and aromatic compounds like benzene.

8.5 Formal Charge: Choosing among Lewis Structures

Now that we have explored the structure of ozone, let's turn our attention to the structure of another atmospheric pollutant, dinitrogen monoxide (N_2O), also known as nitrous oxide or, more commonly, laughing gas. Its common name is linked to the fact that people who inhale high concentrations of N_2O usually laugh spontaneously. Gaseous nitrous oxide has long been used as an anesthetic in dentistry because it has a narcotic effect.

In the lower atmosphere, nitrous oxide concentrations range between 0.1 and 1.0 ppm. It is produced naturally by bacterial action in soil and through human activities such as agriculture and sewage treatment. It is occasionally in the news because its concentration in the troposphere (the atmosphere at ground level) has been increasing. Along with CO_2 and other gases, N_2O may be contributing to global warming.

Let's draw the Lewis structure of nitrous oxide. First we count the number of valence electrons: 5 each from the two nitrogen atoms and 6 from the oxygen atom for a total of $(2 \times 5) + 6 = 16$. The central atom is a nitrogen atom because N has a higher bonding capacity than O. Connecting the atoms with single bonds, we have

$$N—N—O$$

Completing the octets of the noncentral atoms gives us a structure with 16 valence electrons, which is the number determined in step 1:

$$\ddot{\ddot{N}}—N—\ddot{\ddot{O}}:$$

However, there are only two bonds and thus only four valence electrons on the central N atom. To give it 4 more electrons, we need to convert lone pairs on the surrounding atoms to bonding pairs. Which lone pairs do we choose? We could use two lone pairs from the N atom on the left to form a $N≡N$ triple bond:

$$:N≡N—\ddot{\ddot{O}}:$$

Alternatively, we could use two lone pairs on the O atom to make a $N≡O$ triple bond:

$$:\ddot{N}—N≡O:$$

Or we could use one pair from each terminal atom to make two double bonds:

$$\ddot{:}N=N=\ddot{O}\dot{:}$$

Which of these resonance structures is best? We have seen that in some sets of resonance structures, such as those of O_3, SO_3, and NO_3^-, all the structures are equivalent, so no one of them is any more important than another in conveying the actual bonding in the molecule. This is not the case, however, with the resonance forms of N_2O. To decide which resonance form in a nonequivalent set is the most important and comes closest to representing the actual bonding pattern in a molecule, we use the concept of formal charge.

A **formal charge (FC)** is not a real charge but rather a measure of the number of electrons *formally assigned* to an atom in a molecular structure, compared with the number of electrons in the free atom. To determine a formal charge, we follow a series of steps to calculate the number of electrons formally assigned to each atom in each resonance structure.

Calculating Formal Charge of an Atom in a Resonance Structure

For each atom:

1. Determine the number of valence electrons in the free atom.
2. Count the number of lone-pair electrons on the atom in the structure.
3. Count the number of electrons in bonds on the atom and divide that number by 2.
4. Sum the results of steps 2 and 3, and subtract that sum from the number determined in step 1.

formal charge (FC) value calculated for an atom in a molecule or polyatomic ion by determining the difference between the number of valence electrons in the free atom and the sum of lone-pair electrons plus half of the electrons in the atom's bonding pairs.

Summarizing these steps in the form of an equation, we have

$$FC = \begin{pmatrix} \text{number of} \\ \text{valence } e^- \end{pmatrix} - \left[\begin{pmatrix} \text{number of} \\ \text{unshared } e^- \end{pmatrix} + \tfrac{1}{2}\begin{pmatrix} \text{number of } e^- \\ \text{in bonding pairs} \end{pmatrix} \right] \quad (8.1)$$

Once we have determined the formal charges, we then use these three criteria to select the preferred structure:

1. The preferred structure is the one with formal charges of zero.
2. If no such structure can be drawn, or if the structure is that of a polyatomic ion, then the best structure is the one where most atoms have formal charges equal to zero or closest to zero.
3. Any negative formal charges should be on the atom(s) of the most electronegative element(s).

The calculation of formal charge incorporates the assumption that each atom is formally assigned all its lone pair electrons and shares its bonding electrons equally with the atoms at the other ends of the bonds. We can confirm that the three resonance forms of N_2O are not equivalent by calculating the formal charges in each structure.

In Table 8.2 we have colored the lone pairs of electrons red and the shared pairs green to make it easier to track the quantities of these electrons in the formal charge calculation for N_2O. The numbers of valence electrons in the free atoms are shown in blue.

To illustrate one of the formal charge calculations in the table, consider the N atom at the left end of the left resonance structure. The Lewis dot symbol of nitrogen reminds us that free nitrogen atoms have five valence electrons. In this structure, the N atom has 2 electrons in a lone pair and 6 electrons in three shared (bonding) pairs. Using Equation 8.1 to calculate the formal charge on this N atom,

$$FC = 5 - [2 + \tfrac{1}{2}(6)] = 0$$

which is the first value in the bottom row of the table. The results of similar formal charge calculations for all the other atoms in the three resonance structures complete the row. Note that the sum of the formal charges on the three atoms in each structure is zero, as it should be for a neutral molecule. When we do an analysis of the formal charges of atoms in a polyatomic ion, the formal charges on its atoms must add up to the charge on the ion.

Now we must apply the three criteria for selecting the preferred resonance structure of N_2O. The first thing to note about these sets of formal charges is that in none of them are all three formal charges zero, meaning that criterion 1 is not met. The next step is to identify which structure has the most formal charge values that are the closest to zero, such as −1 or +1. On this count we have a tie between the structure on the left (0, +1, −1) and the one in the middle (−1, +1, 0).

TABLE 8.2 Formal Charge Calculations for the Resonance Structures of N_2O

Step	:N≡N—Ö:			:N̈=N=Ö:			:N̈—N≡O:		
1 Number of valence electrons	5	5	6	5	5	6	5	5	6
2 Number of electrons in lone pairs	2	0	6	4	0	4	6	0	2
3 Number of shared electrons	6	8	2	4	8	4	2	8	6
4 FC = [valence − (lone pair + ½ shared)]	0	+1	−1	−1	+1	0	−2	+1	+1

To break the tie, we invoke the third criterion and answer the question, "In which structure is the negative formal charge on the more electronegative atom?" The answer is the structure on the left, which has an oxygen atom with a formal charge of −1. We conclude that this structure is the best of the three in representing the actual bonding in a molecule of N_2O.

From experimental measurements, we know that although the N≡N—O structure contributes the most to the bonding in N_2O, the middle structure (N=N=O) also contributes to the bonding. We know this because the length of the bond between the two nitrogen atoms is between the length of a N=N bond and the length of a N≡N bond, and the nitrogen–oxygen bond is a little shorter than a typical N—O single bond.

CONCEPT **TEST**

What is the formal charge on a sulfur atom that has three lone pairs of electrons and one bonding pair?

SAMPLE EXERCISE 8.8 Selecting Resonance Structures   **LO4**
on the Basis of Formal Charges

Which of these resonance forms best describes the actual bonding in a molecule of CO_2?

$$:\ddot{O}\!-\!C\!\equiv\!O: \quad\longleftrightarrow\quad :\ddot{O}\!=\!C\!=\!\ddot{O}: \quad\longleftrightarrow\quad :O\!\equiv\!C\!-\!\ddot{O}:$$

Collect, Organize, and Analyze The preferred structure is the one in which the formal charges are closest to zero and any negative formal charges are on the more electronegative atom. In this case, oxygen is a group 16 element and is more electronegative than carbon, a group 14 element. Each free carbon atom has four valence electrons, and free oxygen atoms have six valence electrons each.

Solve We use Equation 8.1 to find the formal charge on each atom. We illustrate these results in a table:

Formal Charge Calculations for the Resonance Structures of CO_2

Step	$:\ddot{O}\!-\!C\!\equiv\!O:$			$:\ddot{O}\!=\!C\!=\!\ddot{O}:$			$:O\!\equiv\!C\!-\!\ddot{O}:$		
1 Number of valence electrons	6	4	6	6	4	6	6	4	6
2 Number of electrons in lone pairs	6	0	2	4	0	4	2	0	6
3 Number of shared electrons	2	8	6	4	8	4	6	8	2
4 FC = [valence − (lone pair + $\frac{1}{2}$ shared)]	−1	0	+1	0	0	0	+1	0	−1

The formal charges are zero on all the atoms in the middle resonance structure with the two double bonds. Therefore this structure best represents the actual bonding in a molecule of CO_2.

Think About It Notice that the sum of the formal charges is zero in all three resonance structures. Valid Lewis structures of all neutral molecules have net formal charges of zero.

 Practice Exercise Which resonance structure of the nitronium ion, NO_2^+, contributes the most to the actual bonding in each ion?

Let's take a final look at the resonance structures for N_2O. This time our purpose is to examine the link between the formal charge on an atom in a resonance structure and the bonding capacity of that atom. The Lewis symbol of nitrogen, which has three unpaired electrons, tells us that a nitrogen atom can complete its octet by forming three bonds. The two unpaired electrons in the Lewis symbol of oxygen tell us that the bonding capacity of an oxygen atom is 2. In the N_2O resonance structures in Table 8.2, the nitrogen atom with three bonds has a formal charge of zero (the left N in the first structure), and the oxygen atom with two bonds has a formal charge of zero (the O in the middle structure). As a general rule—and assuming the octet rule is obeyed—atoms have formal charges of zero in resonance structures in which the numbers of bonds they form match their bonding capacities. If an atom forms one more bond than its bonding capacity, such as an oxygen atom with three bonds (O in the third structure in the table), the formal charge is +1. If the atom forms one fewer bond than the bonding capacity, such as an oxygen atom with one bond (O in the first structure in the table), the formal charge is −1.

8.6 Exceptions to the Octet Rule

Nitric oxide (NO) is a simple and small molecule, and yet one with important roles in biology and medicine. It is crucial as a regulator of blood flow and is also a principal mediator of neurological signals for many physical and pathological processes. It is a known bioproduct in organisms ranging from bacteria to plants and higher animals, including human beings. Pharmaceuticals such as nitroglycerine, used to relieve angina (chest pain due to impaired blood flow to the heart), and Viagra, used to treat erectile dysfunction, act by producing NO *in vivo* to improve blood flow. Earth's atmosphere also contains trace concentrations of NO along with NO_2, both of which are generated in the environment by human activities like welding (Figure 8.14) and driving gasoline-burning vehicles; these molecules contribute to photochemical smog formation in urban areas. NO and NO_2 illustrate the limitations of the octet rule. Each has an odd number of valence electrons per molecule, which means that at least one atom in each molecule cannot have a complete octet.

Odd-Electron Molecules

NO is produced in the environment when high temperatures in vehicle engines or other sources lead to the reaction

$$N_2(g) + O_2(g) \rightarrow 2\,NO(g) \qquad (8.2)$$
$$\text{Nitric oxide}$$

The nitric oxide then reacts with atmospheric oxygen to produce nitrogen dioxide:

$$2\,NO(g) + O_2(g) \rightarrow 2\,NO_2(g) \qquad (8.3)$$
$$\text{Nitric oxide} \qquad\qquad \text{Nitrogen dioxide}$$

Nitric oxide is highly reactive in living systems as well as in the atmosphere because it is an odd-electron molecule. To understand the implications of this, let's draw its Lewis structure. Nitric oxide has 11 valence electrons: nitrogen contributes 5 and oxygen 6. There is no central atom, so we start with a single bond

FIGURE 8.14 The high temperatures of arc welding produce significant concentrations of nitric oxide (NO) as a result of the highly endothermic reaction $N_2(g) + O_2(g) \rightarrow 2\,NO(g)$. The U.S. Environmental Protection Agency limit on NO concentrations in the air is 25 ppm, or 0.0025% by volume.

free radical an odd-electron molecule with an unpaired electron in its Lewis structure.

between N and O and then complete the octet around O, which is the more electronegative element:

$$N\!-\!\ddot{\underset{\displaystyle ..}{O}}\!:$$

We then place the remaining 3 electrons around the N atom:

$$\dot{N}\!-\!\ddot{\underset{\displaystyle ..}{O}}\!:$$

This leaves the N atom short three valence electrons. We can increase its number by converting a lone pair on the O atom into a bonding pair:

$$\dot{\underset{\displaystyle ..}{N}}\!=\!\ddot{O}\!:$$

This change has the added advantage of creating a double-bonded O atom, which gives it a formal charge of zero. The formal charge on the N atom is also zero. The only problem with the structure is that N does not have an octet: it has only seven electrons. Because nitrogen is less electronegative than oxygen, it is reasonable that we "short-change" it when there are not enough electrons to complete the octets of both atoms. The fact that NO exists is evidence that there are exceptions to the octet rule. When that happens, the most representative Lewis structures are those that come *as close as possible* to producing zero formal charges and complete octets.

Compounds that have odd numbers of valence electrons are called **free radicals**. They are typically very reactive species because it is often energetically favorable for them to acquire an electron from another molecule or ion. This characteristic makes them excellent oxidizing agents and is responsible for the damage they may cause to materials or living tissue with which they come in contact.

SAMPLE EXERCISE 8.9 Drawing Lewis Structures   **LO4**
                          of Odd-Electron Molecules

Draw the resonance structures of nitrogen dioxide (NO_2), assign formal charges to the atoms, and suggest which structure may best represent the actual bonding in the molecule.

Collect, Organize, and Analyze Each molecule contains one atom of nitrogen (bonding capacity 3) and two atoms of oxygen (bonding capacity 2). Nitrogen dioxide is an odd-electron molecule, so we anticipate that one of the atoms will not have a complete octet. We analyze the resonance structures by assigning formal charges to select the one most representative of the actual bonding in the molecule.

Solve The number of valence electrons is

Element:	N	+	2 O
Valence electrons:	5	+	$(2 \times 6) = 17$

Nitrogen is less electronegative than oxygen and has the greater bonding capacity, so it is the central atom:

$$O\!-\!N\!-\!O$$

Completing the octets on the O atoms gives

$$\ddot{\underset{\displaystyle ..}{O}}\!-\!N\!-\!\ddot{\underset{\displaystyle ..}{O}}\!:$$

There are 16 valence electrons in this structure, but we need 17 to match the number available in the molecule. We add 1 more electron to the N atom:

$$\ddot{\underset{\displaystyle ..}{O}}\!-\!\dot{N}\!-\!\ddot{\underset{\displaystyle ..}{O}}\!:$$

There are only 5 valence electrons around the N atom, fewer than the 8 we need. We can increase this number by converting a lone pair on one of the O atoms to a bonding pair, giving the formal charges shown in red:

$$:\overset{0}{\underset{}{\ddot{O}}}=\overset{+1}{\overset{\cdot}{N}}-\overset{-1}{\underset{}{\ddot{O}}}:$$

An equivalent resonance form can be drawn with the double bond on the right side:

$$:\overset{-1}{\underset{}{\ddot{O}}}{\diagdown}\overset{+1}{\overset{\cdot}{N}}{\diagup}\overset{0}{\underset{}{\ddot{O}}}:$$

These structures are equivalent: each has a charge of 1+ on the nitrogen atom, one oxygen atom with a 1− charge, and one oxygen atom with a charge of 0 (zero).

Think About It The two Lewis structures are equivalent because the two O atoms in each structure are equivalent. The structures do not satisfy the octet rule, but the formal charges of the atoms are close to zero, and the negative formal charge is on the atom of the more electronegative element. Both O atoms have complete octets, leaving the less electronegative N atom one electron short in this odd-electron molecule.

Practice Exercise Nitrogen trioxide (NO_3) may form in polluted air when NO_2 reacts with O_3. Draw the Lewis structure(s) of NO_3 and use formal charges to select the preferred structure(s).

Atoms with More than an Octet

Another important atmospheric pollutant is sulfur hexafluoride (SF_6), which may be the most potent greenhouse gas (in terms of infrared radiation absorbed per mole) present in the atmosphere (Figure 8.15). Each of its molecules contains six sulfur–fluorine covalent bonds, which means each S atom is surrounded by 12 valence electrons, not 8.

Let's consider the Lewis structure of SF_6. Its valence electron inventory (step 1) is

Element:	S	+	6 F
Valence electrons:	6	+	$(6 \times 7) = 48$

Sulfur has a greater bonding capacity (2) than fluorine (1), so S is the central atom (step 2). However, connecting six fluorine atoms to the sulfur atom means that sulfur's bonding capacity is significantly exceeded:

$$\begin{array}{ccc} & F \quad F & \\ F & -S- & F \\ & F \quad F & \end{array}$$

Completing the octets on the fluorine atoms (step 3)

$$\begin{array}{ccc} & :\ddot{F} \quad :\ddot{F}: & \\ :\ddot{F}- & S & -\ddot{F}: \\ & :\ddot{F}. \quad .\ddot{F}: & \end{array}$$

gives us a structure with 48 valence electrons (step 4), a match with the number determined for the molecule. This structure is correct, even though it does not conform with the octet rule. It tells us that, in order for six fluorine atoms to bond

FIGURE 8.15 Electrical transformers use sulfur hexafluoride as an insulator because it is thermally stable and does not react with water. However, when SF_6 leaks into the atmosphere, it becomes a potent greenhouse gas. It absorbs much more infrared radiation than CO_2 and may remain in the atmosphere for thousands of years.

to a central sulfur atom, the sulfur atom must have an *expanded valence shell* to accommodate more than 8 valence electrons. Six S—F bonds and 12 valence electrons around S leave the SF_6 molecule with zero formal charge on each atom:

$$\text{Formal charge of S} = 6 - \left[0 + \tfrac{1}{2}(12)\right] = 0$$
$$\text{Formal charge of F} = 7 - \left[6 + \tfrac{1}{2}(2)\right] = 0$$

According to our criteria for judging Lewis structures, this is a good one, but how can a sulfur atom accommodate more than eight valence electrons?

The answer to this question is not completely clear. Some researchers believe the availability of *d* orbitals in large-*Z* atoms explains the larger valence shells, whereas other researchers think it is simply due to the larger size of the atom. We discuss this further in Chapter 9. For the moment, we note that compounds with expanded octets are observed for elements with $Z > 12$. Nevertheless, the fact that some atoms can have expanded valence shells does not mean these atoms *always* do. Instead, they tend to do so in two situations:

1. When they form compounds with strongly electronegative elements, particularly F, O, and Cl.
2. When an expanded valence shell results in smaller formal charges on the atoms in a molecule.

To illustrate a situation in which expanding a valence shell results in smaller formal charges, let's consider the sulfate ion, SO_4^{2-}. In the Lewis structure of the sulfate ion, a central S atom is bonded to four O atoms. Applying the method for drawing Lewis structures and assigning formal charges, we get

The sum of the formal charges on atoms in the ion is $1(+2) + 4(-1) = -2$. This calculation yields the correct ionic charge, but remember that the goal is to minimize formal charges in Lewis structures. We can do this by expanding the valence shell of sulfur:

Each oxygen atom still has a complete octet, but now the sulfur has an expanded valence shell to accommodate 12 electrons. In this way, the formal charges change from +2 to 0 on sulfur and from −1 to 0 on two of the four oxygen atoms. The remaining two −1 values sum to −2, which is the value required to give the structure its overall 2− charge.

We can draw the two double bonds at any locations around the sulfur atom in the Lewis structure for the sulfate ion. Consequently, this structure has six equivalent resonance forms (not shown).

If we now wanted to draw the Lewis structure for H_2SO_4, we could bond two hydrogen ions to the two negative oxygen atoms:

CHEMTOUR
Lewis Structures:
Expanded Valence Shells

$$H-\overset{\overset{\displaystyle \cdot\cdot}{\cdot O \cdot}}{\underset{\underset{\displaystyle \cdot O \cdot}{\cdot\cdot}}{S}}-\ddot{O}-H$$

Each hydrogen atom has achieved a duet of electrons, each oxygen atom has an octet, and every atom has a formal charge of zero.

H_2SO_4 is a principal component of acidic precipitation ("acid rain") in eastern North America and Europe. It forms when high-sulfur coal is burned and SO_2 is released into the atmosphere, reacting with atmospheric oxygen to form sulfur trioxide:

$$2\,SO_2(g) + O_2(g) \rightarrow 2\,SO_3(g)$$

Sulfur trioxide then reacts, in turn, with water vapor to form H_2SO_4:

$$SO_3(g) + H_2O(g) \rightarrow H_2SO_4(\ell)$$

SAMPLE EXERCISE 8.10 Drawing Lewis Structures of Ions   **LO2, LO4**
with an Expanded Valence Shell

Draw the Lewis structure for the phosphate ion (PO_4^{3-}) that minimizes the formal charges on its atoms.

Collect, Organize, and Analyze Each ion contains one atom of phosphorus and four atoms of oxygen and has an overall charge of $3-$. Phosphorus and oxygen are in groups 15 and 16 and have bonding capacities of 3 and 2, respectively. Phosphorus has an atomic number greater than 12 ($Z = 15$), so it can have an expanded octet if need be.

Solve The number of valence electrons is

Element:	P	+	4 O
Valence electrons:	5	+	$(4 \times 6) = 29$
Additional electrons due to the $3-$ charge:			$+\ 3$
Total valence electrons:			$\overline{32}$

Phosphorus has the greater bonding capacity (3) and so is the central atom:

$$\begin{array}{c} O \\ | \\ O-P-O \\ | \\ O \end{array}$$

The addition of three lone pairs of electrons to each O atom completes their octets:

$$\begin{array}{c} :\ddot{O}: \\ | \\ :\ddot{O}-P-\ddot{O}: \\ | \\ :\ddot{O}: \end{array}$$

There are 32 valence electrons in this structure, which matches the number determined for the ion. Therefore it is a complete Lewis structure of a polyatomic ion once we add the brackets and electrical charge:

$$\left[\begin{array}{c} :\ddot{O}: \\ | \\ :\ddot{O}-P-\ddot{O}: \\ | \\ :\ddot{O}: \end{array} \right]^{3-}$$

Each O has a single bond and a formal charge of −1; the four bonds around the P atom are one more than its bonding capacity, so its formal charge is +1. The sum of the formal charges, [+1 + 4(−1)], matches the charge on the ion, 3−.

We can reduce the formal charge on P by increasing the number of bonds to it, and we can do that by converting a lone pair on one of the O atoms into a bonding pair:

At the same time, we change a single-bonded O atom into a double-bonded O atom and thereby make its formal charge zero. Therefore the structure on the right, in which the P atom has ten valence electrons, is the best Lewis structure we can draw for the phosphate ion.

Think About It Note that three other equivalent resonance structures can be drawn with a double bond to one of the other O atoms.

Practice Exercise Draw the resonance structures of the selenite ion (SeO_3^{2-}) that minimize the formal charges on the atoms.

Atoms with Less than an Octet

We have seen that sulfur can combine with fluorine to form SF_6, a compound with an expanded octet. When fluorine combines with boron, a compound with the formula BF_3 is isolated. The electronegativity difference between boron and fluorine is 2.0, which puts BF_3 on the borderline between ionic and covalent compounds. If BF_3 contains three fluoride anions and a B^{3+} cation, then the ions are isoelectronic with a noble gas. Boron trifluoride, however, does not have the physical properties associated with an ionic compound; it is a gas that boils at −101°C. What does the Lewis structure for BF_3 look like?

Boron trifluoride has a total of 24 valence electrons (step 1):

Element:	B	+	3 F
Valence electrons:	3	+	$(3 \times 7) = 24$

Boron has a greater bonding capacity (3) than fluorine (1), so B is the central atom (step 2). Connecting three fluorine atoms to the boron atom and completing the octets on fluorine (step 3) gives us a structure with the requisite number of valence electrons (step 4) on each F but not on the boron atom.

Formal charges are minimized in this structure; however, boron shares only six electrons and has no electrons in lone pairs. If we share an additional pair of electrons from one of the three F atoms, as shown in the resulting three resonance forms, we can complete the octets of all the atoms:

But forming a B=F double bond results in unfavorable formal charges, particularly for the F atom. We do not expect a positive formal charge on the most electronegative of all elements.

How do we resolve this conflict? The B—F bond lengths in BF_3 are all equal, but shorter than the B—F single bonds in the BF_4^- ion:

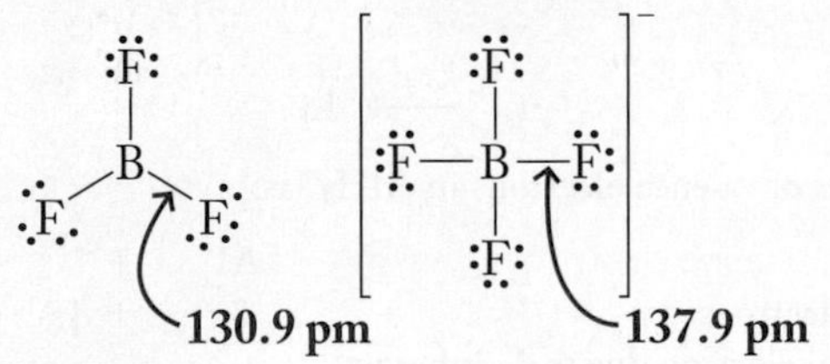

130.9 pm   137.9 pm

As we see in the next section, multiple bonds are generally shorter than single bonds. This suggests that all four of the following resonance forms contribute to the molecular structure of BF_3:

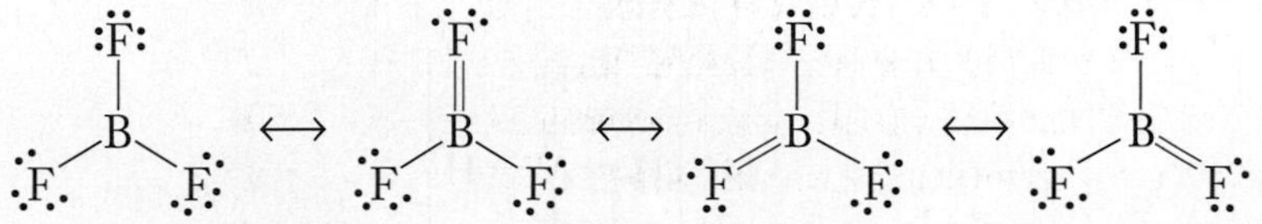

Dominant contributor

The atoms in the other group 13 halides complete their octets in different ways. For example, in aluminum chloride, the aluminum atom shares a lone pair of electrons with a chlorine atom in a second molecule, so that all the atoms in Al_2Cl_6 have a complete octet (Figure 8.16). The structure for Al_2Cl_6 is observed only in the liquid (or molten) phase. In the gas phase, aluminum chloride exists as discrete molecules of $AlCl_3$, like BF_3.

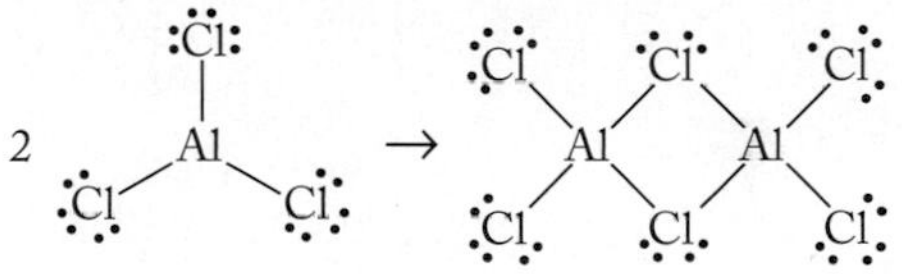

Do compounds containing elements other than group 13 ever have less than an octet around the central atom? The answer depends on which phase of the material we are considering. For example, as a solid, beryllium chloride consists of long chains of chlorine-bridged Be atoms (Figure 8.17) where all atoms have completed octets. Above 800 K, $BeCl_2$ exists in the gas phase as discrete molecules where the beryllium shares only two pairs of electrons with chlorine—much less than an octet.

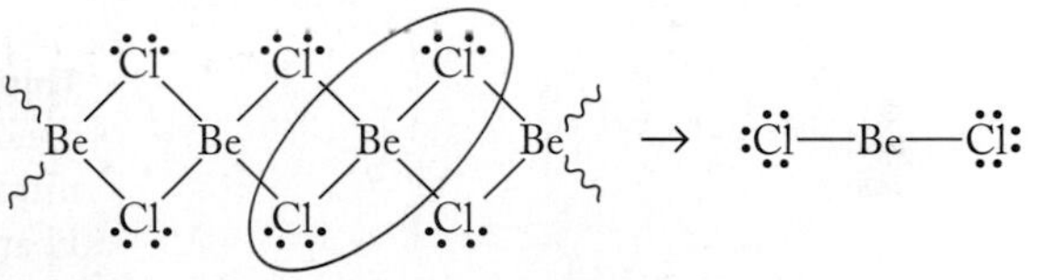

FIGURE 8.16 Aluminum chloride has a complete octet on aluminum by sharing a pair of electrons with one Cl atom from a second molecule of $AlCl_3$.

FIGURE 8.17 The chain structure of solid beryllium chloride, with bridging chlorine atoms, changes to discrete $BeCl_2$ molecules in the gas phase. Each beryllium atom in the solid has a completed octet. In the gas phase, each Be atom is surrounded by only four electrons.

SAMPLE EXERCISE 8.11   Drawing Lewis Structures of Compounds with Less than an Octet   **LO2**

Lithium aluminum hydride reacts with $[(CH_3)_3NH]^+Cl^-$ to form the products in this chemical equation:

$$Li^+[AlH_4]^-(s) + [(CH_3)_3NH]^+Cl^-(s) \rightarrow AlH_3 \cdot N(CH_3)_3(s) + LiCl(s) + H_2(g)$$

The product $AlH_3 \cdot N(CH_3)_3$ is used in the electronics industry as a source of aluminum to make thin films of aluminum metal by the thermal decomposition of AlH_3:

$$AlH_3 \cdot N(CH_3)_3(g) \rightarrow AlH_3(g) + N(CH_3)_3(g) \rightarrow Al(s) + \tfrac{3}{2} H_2(g) + N(CH_3)_3(g)$$

Draw Lewis structures for (a) $Li^+[AlH_4]^-$ and (b) $AlH_3(g)$.

Collect, Organize, and Analyze The formula of lithium aluminum hydride, $Li^+[AlH_4]^-$, tells us that it contains Li^+ cations and AlH_4^- anions. Lithium and hydrogen each have one valence electron and aluminum has three valence electrons. We use the steps outlined in Section 8.2 to draw Lewis structures for both compounds.

Aluminum is the central atom in AlH_4^- and AlH_3 because each hydrogen atom can form only one bond.

Solve

a. The lithium atom has lost its valence electron to form a Li^+ ion, so its Lewis symbol is Li^+:

$$\cdot Li \xrightarrow{-e^-} Li^+$$

The total number of valence electrons in AlH_4^- is:

Element:	Al	+	4 H
Valence electrons:	3	+	$(4 \times 1) = 7$
Additional electron due to 1− charge:			+ 1
Total valence electrons:			8

Connecting the four H atoms to the Al with single bonds results in a Lewis structure in which aluminum has a complete octet and each hydrogen atom has a complete duet.

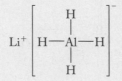

b. The total number of valence electrons in AlH_3 is:

Element:	Al	+	3 H
Valence electrons:	3	+	$(3 \times 1) = 6$

In this case we can complete the duets of each H atom, but the aluminum atom is left with only six electrons, two fewer than a complete octet.

Think About It The aluminum atom in AlH_3 must have less than an octet because there are only six total valence electrons. As a result, AlH_3 reacts readily with the lone pair on nitrogen in $N(CH_3)_3$, for example, to form $AlH_3 \cdot N(CH_3)_3$. In AlH_4^-, the additional H atom and the overall negative charge on the anion each account for one electron to complete the octet on aluminum.

Practice Exercise The boiling point of sodium chloride is approximately 1700 K ($\approx 1427°C$). In the gas phase, NaCl particles with the formula Na_2Cl_2 have been identified. Draw the Lewis structure for this particle and identify which atom does not have a complete octet.

The Limits of Bonding Models

After studying the structures of SF_6, BF_3, and other exceptions to the octet rule, you may be wondering whether the octet rule really has any validity at all. The point to remember is that the octet rule is just a *model*. The octet rule works remarkably well in many cases to predict the number of covalent bonds between main group elements and to predict the composition of ionic compounds. However, the true nature of the chemical bond is much more subtle and shows great variability among the myriad chemical compounds that have been discovered.

bond order the number of bonds between atoms: 1 for a single bond, 2 for a double bond, and 3 for a triple bond.

What about the validity of formal charges? How can F, the most electronegative atom, carry a +1 formal charge in BF_3? Like Lewis structures, formal charges represent a *model* for identifying resonance forms that contribute most to the bonding in a molecule. Sometimes our model does not fit the observed data as well as we would like. In the case of BF_3, we find that the observed B—F distances suggest that resonance forms containing B=F double bonds are important.

If aluminum chloride is a covalently bonded molecule, why do aqueous solutions of $AlCl_3$ conduct electricity? Why doesn't BF_3, with an electronegativity difference of 2.0, behave similarly? Why does $AlCl_3$ display ionic properties when ΔEN for Al—Cl bonds is only 1.5? In part, the answer lies in the partial ionic character of polar bonds. The greater difference in electronegativity between B and F (ΔEN = 2.0) suggests that B—F bonds have more ionic character than Al—Cl bonds. The fact that $AlCl_3$ does dissolve in water, forming a solution that conducts electricity, points to the limitations of the octet rule as a model. Other considerations must be taken into account: it is not only the ionic character of the bond but also bond energy and the resultant chemical reactivity of each compound that matter.

The significance of molecules with incomplete or expanded octets is that they challenge our models for bonding and push us toward better explanations for what we observe. We explore questions about bonding in more detail in Chapter 9, but before we do, let's look at some experimental data about covalent bonds: bond lengths and bond energies. As measurable quantities, these parameters allow us to test our model of the chemical bond as proposed by Gilbert Lewis.

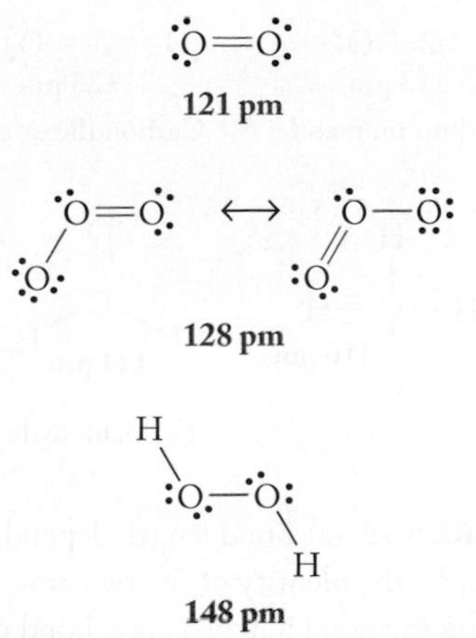

FIGURE 8.18 Resonance influences bond length. Because of resonance, the lengths of the two bonds in ozone are identical and in between the lengths of the O=O double bond in O_2 and the O—O single bond in H_2O_2.

8.7 The Lengths and Strengths of Covalent Bonds

In Section 8.4 we discussed the equivalent resonance structures of ozone. We noted that the true nature of the two oxygen–oxygen bonds in O_3 is reflected in their equal bond length, 128 pm, which is between the length of a typical O=O double bond (121 pm) and an O—O single bond (148 pm), as shown in Figure 8.18. We used these results to conclude that there are effectively 1.5 bonds between the atoms in a molecule of O_3. In this section we explore this use of bond length, and bond strength, to rationalize and validate molecular structures. We also use bond strengths to estimate the enthalpy changes that occur in chemical reactions.

Bond Length

The length of the bond between any two atoms depends on the identity of the atoms and on whether the bond is single, double, or triple (Table 8.3). The number of bonds between two atoms is called **bond order**. As bond order increases, bond length decreases, as we can see by comparing the lengths of C—C, C=C, and C≡C bonds in Table 8.3. Similarly, for carbon–oxygen bonds the C≡O

TABLE 8.3 Selected Average Covalent Bond Lengths and Bond Energies

Bond	Bond Length (pm)	Bond Energy (kJ/mol)	Bond	Bond Length (pm)	Bond Energy (kJ/mol)
C—C	154	348	N≡O	106	678
C=C	134	614	O—O	148	146
C≡C	120	839	O=O	121	495
C—N	147	293	O—H	96	463
C=N	127	615	S—O	151	265
C≡N	116	891	S=O	143	523
C—O	143	358	S—S	204	266
C=O	123	743[a]	S—H	134	347
C≡O	113	1072	H—H	75	436
C—H	110	413	H—F	92	567
C—F	133	485	H—Cl	127	431
C—Cl	177	328	H—Br	141	366
N—H	104	388	H—I	161	299
N—N	147	163	F—F	143	155
N=N	124	418	Cl—Cl	200	243
N≡N	110	941	Br—Br	228	193
N—O	136	201	I—I	266	151
N=O	122	607			

[a]The bond energy of the C=O bond in CO_2 is 799 kJ/mol.

:C≡O:
113 pm
Carbon monoxide
(a)

O=C=O
123 pm
Carbon dioxide

Methane   Formaldehyde
(b)

FIGURE 8.19 (a) Bond length depends not only upon the identity of the two atoms forming the bond but also upon bond order. (b) A bond between the same two atoms can have different lengths in different molecules. Compare the C—H length in the formaldehyde molecule with the C—H length in CH_4. Also compare the C=O length in formaldehyde with the C=O length in CO_2.

CHEMTOUR
Estimating Enthalpy Changes

triple bond in carbon monoxide is shorter than the C=O double bond in carbon dioxide (Figure 8.19a).

Measurements of the bond lengths in many molecules indicate that there are small differences in bond lengths for any given covalent bond. For example, the C—H and C=O bond lengths in formaldehyde (Figure 8.19b) are close to but not exactly the same as the C—H bond length in CH_4 and the C=O bond length in CO_2.

CONCEPT TEST

Rank the following molecules in order of decreasing lengths of their nitrogen–oxygen bonds: NO, NO_2, N_2O.

Bond Energies

The energy changes associated with chemical reactions depend on how much energy is required to break the bonds in the reactants and how much is released as the atoms recombine to form products. For example, in the methane combustion reaction

$$CH_4(g) + 2\,O_2(g) \rightarrow CO_2(g) + 2\,H_2O(g)$$

the C—H bonds in CH_4 and the O=O bonds in O_2 must be broken before the C=O bonds in CO_2 and the O—H bonds in H_2O can form. (In reality, some bond formation occurs simultaneously with bond breaking; it is not completely sequential.) Breaking bonds is endothermic (the blue "energy in" arrow in Figure 8.20), and forming bonds is exothermic (the red "energy out" arrow). If a chemical reaction is exothermic, as is methane combustion, more energy is released in forming the bonds in molecules of products than is consumed in breaking the bonds in molecules of reactants.

Bond energy, or *bond strength*, is usually expressed in terms of the enthalpy change (ΔH) that occurs when 1 mole of a particular bond in the gas phase is broken. Bond energies for some common covalent bonds are given in Table 8.3. Note that the values in Table 8.3 are all positive because bond breaking is always

FIGURE 8.20 The combustion of 1 mole of methane requires that 4 moles of C—H bonds and 2 moles of O=O bonds be broken. These processes require an enthalpy change of about +2642 kJ. In the formation of 4 moles of O—H bonds and 2 moles of C=O bonds, there is an enthalpy change of about −3450 kJ. The overall reaction is exothermic: 2642 kJ − 3450 kJ = −808 kJ.

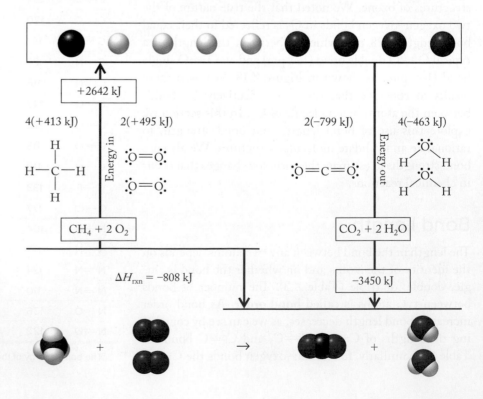

endothermic. As we noted in Section 8.1, the quantity of energy needed to break a particular bond is equal in magnitude but opposite in sign to the quantity of energy released when that same bond forms.

Like bond lengths, the bond energies in Table 8.3 are average values because bond energies vary depending on the structure of the rest of the molecule (Figure 8.19). For example, the bond energy of a $C=O$ bond in carbon dioxide is 799 kJ/mol, but $C=O$ bond energy in formaldehyde is only 743 kJ/mol.

Another view of the variability in bond energy comes from breaking the $C-H$ bonds in CH_4 in a step-by-step fashion, as shown in the following table:

Decomposition Step	Energy Needed (kJ/mol)
$CH_4 \rightarrow CH_3 + H$	435
$CH_3 \rightarrow CH_2 + H$	453
$CH_2 \rightarrow CH + H$	425
$CH \rightarrow C + H$	339
Total:	1652
Average:	413

These results tell us that the chemical environment of a bond affects the energy required to break it: breaking the first $C-H$ bond in methane is easier (requires less energy) than breaking the second but more difficult than breaking the third or fourth. The total energy needed to break all four $C-H$ bonds is 1652 kJ/mol, an average of 413 kJ/mol per bond.

The relationship between bond order and bond energy is also apparent in Table 8.3. The bond energy of the $O=O$ double bond, at 495 kJ/mol, is more than three times that of the $O-O$ single bond. This correlation between bond order and bond energy exists for other pairs of atoms: the higher the bond order, the greater the bond energy. The bond energy of the $N\equiv N$ triple bond (941 kJ/mol) is more than twice the bond energy of the $N=N$ double bond and more than five times that of the $N-N$ single bond. The large quantity of energy required to break a mole of $N\equiv N$ triple bonds is one of the reasons N_2 participates in so few chemical reactions.

In the combustion of 1 mole of CH_4 (Figure 8.20), 4 moles of $C-H$ bonds and 2 moles of $O=O$ bonds must be broken. The formation of 1 mole of CO_2 and 2 moles of H_2O requires the formation of 2 moles of $C=O$ bonds and 4 moles of $O-H$ bonds. The net change in energy resulting from breaking and forming these bonds can be estimated from average bond energies. We can take an inventory of the bond energies involved:

Bond Energies (ΔH) in Methane Combustion

Bond	Number of Bonds (mol)	Bond Energy (kJ/mol)	Bond Breaking or Forming?
$C-H$	4	413	Breaking
$O=O$	2	495	Breaking
$O-H$	4	463	Forming
$C=O$	2	799	Forming

Next we estimate the enthalpy change of the reaction by calculating the difference between the sum of the bond energies of the reactants and the sum of the bond energies of the products. Equation 8.4 can be used to do this calculation:

$$\Delta H_{rxn} = \sum \Delta H_{\text{bonds breaking}} - \sum \Delta H_{\text{bonds forming}} \tag{8.4}$$

$$= [(4 \text{ mol} \times 413 \text{ kJ/mol}) + (2 \text{ mol} \times 495 \text{ kJ/mol})]$$
$$- [(4 \text{ mol} \times 463 \text{ kJ/mol}) + (2 \text{ mol} \times 799 \text{ kJ/mol})]$$
$$= -808 \text{ kJ}$$

bond energy the energy needed to break 1 mole of a particular covalent bond in a molecule or in a polyatomic ion in the gas phase.

CONNECTION In Chapter 5 we calculated the difference between the sums of heats of formation of products and of reactants to estimate the heat of reaction.

SAMPLE EXERCISE 8.12 Estimating Heats of Reaction   **LO5**
                                   from Average Bond Energies

Use the average bond energies in Table 8.3 to estimate ΔH_{rxn} for the reaction in which $HCl(g)$ is formed from $H_2(g)$ and $Cl_2(g)$:

$$H{-}H \quad + \quad :\ddot{\underset{..}{Cl}}{-}\ddot{\underset{..}{Cl}}: \quad \rightarrow \quad 2\ H{-}\ddot{\underset{..}{Cl}}:$$

Collect and Organize We are asked to estimate the value of ΔH_{rxn} of a gas-phase reaction from the bond energies of the reactants and products.

Analyze The reaction between H_2 and Cl_2 requires that 1 mole of $H{-}H$ bonds and 1 mole of $Cl{-}Cl$ bonds be broken. Two moles of $H{-}Cl$ bonds are formed. Table 8.3 lists these average bond energies:

H—H	436 kJ/mol
Cl—Cl	243 kJ/mol
H—Cl	431 kJ/mol

Solve Using the above information in Equation 8.4:

$$\Delta H_{rxn} = \sum \Delta H_{bond\ breaking} - \sum \Delta H_{bond\ forming}$$
$$= \left[(1\ \cancel{mol} \times 436\ kJ/\cancel{mol}) + (1\ \cancel{mol} \times 243\ kJ/\cancel{mol}) \right]$$
$$\quad - \left[(2\ \cancel{mol} \times 431\ kJ/\cancel{mol}) \right]$$
$$= -183\ kJ$$

Think About It An enthalpy change of -183 kJ (note the minus sign) means that the amount of energy consumed as the bonds in the reactants are broken is 183 kJ less than the amount of energy released as the bonds in the products are formed. In other words, the reaction is exothermic. The reaction involves the formation of 2 moles of a gaseous compound from its component elements in their standard states. Therefore the result of this calculation should be close to two times the standard heat of formation (ΔH_f°) of HCl. That value (see Appendix 4) is -92.3 kJ/mol. Multiplying by 2 moles, we get -184.6 kJ, which is quite close to the estimated value.

 Practice Exercise Use average bond energies to calculate ΔH_{rxn} for the reaction of H_2 and N_2 to form ammonia:

$$:N{\equiv}N: \quad + \quad 3\ H{-}H \quad \rightarrow \quad 2\ H{-}\underset{\underset{H}{|}}{\overset{..}{N}}{-}H$$

CONCEPT TEST

Suggest a reason, based on bond energies, why O_2 is much more reactive than N_2.

In this chapter we have explored the nature of the covalent bonds that hold together molecules and polyatomic ions, observing that these bonds owe their strength to the presence of pairs of electrons shared between nuclei of atoms. Sharing does not necessarily mean equal sharing, and unequal sharing coupled with bond vibration accounts for the ability of some atmospheric gases to absorb and emit infrared radiation. As a result, these gases function as greenhouse gases.

Early in the chapter we noted that moderate concentrations of greenhouse gases are required for climate stability and to make our planet habitable. The escalating concern of many is that Earth's climate is being destabilized by too much of a good thing. Policies made by the world's governments will soon have a significant impact on the problem of global warming, one way or the other. As an informed member of the world community, you will have the opportunity to influence how those policy decisions are made.

SAMPLE EXERCISE 8.13 Integrating Concepts: Moth Balls

A compound often referred to by the abbreviation PDB is the active ingredient in most moth balls. It is also used to control mold and mildew, as a deodorant, and as a disinfectant. Tablets containing it are often stuck under the lids of garbage cans or placed in the urinals in public restrooms, producing a distinctive aroma. Molecules of PDB have the following skeletal structure:

a. Draw the Lewis structure of PDB and note any nonzero formal charges.
b. Is the structure stabilized by resonance? If so, draw all resonance structures.
c. Which, if any, of the bonds in the structure you drew are nonpolar?
d. Predict the average carbon–carbon bond length and bond strength in the structure you drew.

Collect and Organize We are given the skeletal structure of a molecule and are asked to draw its Lewis structure, including all resonance structures, and to perform a formal charge analysis. We are also asked to identify any polar bonds in the structure and to predict the length and strength of the carbon–carbon bonds. Bond polarity depends on the difference in electronegativities of the bonded atoms, which are given in Figure 8.6. Table 8.3 contains average lengths and energies (strengths) of covalent bonds.

Analyze The five-step procedure used in Sample Exercises 8.1 through 8.4 to draw Lewis structures of other small molecules should be useful in drawing the Lewis structure of PDB. Resonance structures for PDB like those for benzene (Figure 8.13) should be possible if there are alternating single and double carbon–carbon bonds in PDB's six-carbon ring.

Solve
a and b. The number of valence electrons is

Element:	6 C	+	2 Cl	+	4 H
Valence electrons:	(6×4)	+	(2×7)	+	$(4 \times 1) = 42$

Completing the octets on the Cl atoms:

gives us a structure with 36 valence electrons (12 bonding pairs and 6 lone pairs). We need 6 more electrons to reach 42. We

can achieve that number if we create three more bonds between carbon atoms by turning single C—C bonds into double bonds. We have to distribute them evenly around the ring to avoid any C atoms with five bonds. Two equivalent resonance structures can be drawn to show the bonding pattern:

Resonance stabilizes the structure of PDB. Each C atom has four bonds and each H and Cl atom has one bond, so every atom has the number of bonds that matches its bond capacity. This means that all formal charges are zero.

c. The differences in electronegativities for the bonded pairs of atoms are

C—C	$\Delta EN = 0$
Cl—C	$\Delta EN = 3.0 - 2.5 = 0.5$
C—H	$\Delta EN = 2.5 - 2.1 = 0.4$

Only the bonds between the carbon atoms are nonpolar.

d. The even distribution of a total of 9 bonding pairs of electrons among 6 C atoms means that, on average, each pair shares 1.5 pairs of bonding electrons. The corresponding bond length and bond strength should be about halfway between those of C—C single and C=C double bonds, given in Table 8.3:

Approximate bond length:

$$[(154 + 134)/2] \text{ pm} = 144 \text{ pm}$$

Approximate bond strength:

$$[(348 + 614)/2] \text{ kJ/mol} = 481 \text{ kJ/mol}$$

Think About It The resonance structures closely resemble those of benzene, which is reflected in the common name of PDB, *para*-dichlorobenzene. Note that the two polar C—Cl bonds in PDB are oriented *in opposite directions*. Thus, the unequal sharing of the bonding pair of electrons in the Cl—C bond on the left side of the molecule is offset by the unequal sharing of the bonding pair of electrons in the C—Cl bond on the right side. In Chapter 9 we explore how offsetting bond polarities in symmetrical molecules like this one explain why substances such as PDB are, overall, nonpolar.

SUMMARY

LO1 A chemical bond results when two atoms share electrons (a covalent bond) or when two ions are attracted to each other (an **ionic bond**). The atoms in metallic solids pool their electrons in forming **metallic bonds**. (Section 8.1)

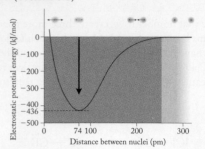

LO2 **Lewis symbols** use dots to represent paired and unpaired electrons in the ground states of atoms. A **Lewis structure** shows the bonding pattern in molecules and polyatomic ions. Two or more equivalent Lewis structures—called **resonance structures**—can sometimes be drawn for one molecule or polyatomic ion. The actual bonding pattern in a molecule is an average of equivalent resonance structures. (Sections 8.2, 8.4, and 8.6)

LO3 Unequal electron sharing between atoms of different elements results in **polar covalent bonds**. Bond polarity is a measure of how unequally the electrons in covalent bonds are shared. Greater polarity results from larger differences in the **electronegativities** of the bonded atoms. Polarity of bonds in molecules of atmospheric gases may lead to the absorption of infrared radiation, which contributes to the greenhouse effect. (Section 8.3)

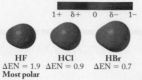

HF $\Delta EN = 1.9$ **Most polar** HCl $\Delta EN = 0.9$ HBr $\Delta EN = 0.7$ HI $\Delta EN = 0.4$ **Least polar**

LO4 The preferred resonance structure of a molecule is one in which the **formal charges** on its atoms are zero or closest to zero and any negative formal charges are on the more electronegative atoms. (Section 8.5)

LO5 **Bond order** is the number of bonding pairs in a covalent bond. **Bond energy** is the enthalpy change, ΔH, required to break 1 mole of a particular covalent bond in the gas phase; energy changes in chemical reactions depend on the energy required to break bonds in reactants and the energy released when new bonds are formed in the products. **Bond length** is the distance between the nuclear centers of two bonded atoms. (Sections 8.1, 8.7)

PARTICULATE **PREVIEW WRAP-UP**

The space-filling models look identical and do not tell us which bonds are present in these molecules. O_2, F_2, and N_2 are the molecular formulas for the oxygen, fluorine, and nitrogen molecule, respectively, from left to right. The molecular formulas also do not tell us which bonds are present in a molecule, but a ball-and-stick model would show a single bond as one stick, a double bond as two sticks, and a triple bond as three sticks.

PROBLEM-SOLVING SUMMARY

Type of Problem	Concepts and Equations	Sample Exercises
Drawing Lewis structures for molecules, monatomic ions, and ionic compounds	Connect the atoms with single covalent bonds, distribute the valence electrons to give each outer atom eight valence electrons (except two for H); use multiple bonds where necessary to complete the central atom's octet.	**8.1–8.4**
Comparing bond polarities	Calculate the difference in electronegativity (ΔEN) between the two bonded atoms; if $\Delta EN \geq 2.0$, the bond is considered ionic.	**8.5**
Drawing resonance structures of molecules and polyatomic ions	Include all possible arrangements of covalent bonds in the molecule if more than one equivalent structure can be drawn.	**8.6, 8.7**
Selecting resonance structures on the basis of formal charges	Calculate formal charge by using $$FC = \left(\begin{array}{c}\text{number of}\\\text{valence e}^-\end{array}\right) - \left[\begin{array}{c}\text{number of}\\\text{unshared e}^-\end{array} + \tfrac{1}{2}\left(\begin{array}{c}\text{number of e}^-\\\text{in bonding pairs}\end{array}\right)\right] \quad (8.1)$$ Select structures with formal charges closest to zero and with negative formal charges on the most electronegative atoms.	**8.8**
Drawing Lewis structures of odd-electron molecules	Distribute the valence electrons in the Lewis structure to leave the most electronegative atom(s) with eight valence electrons and the least electronegative atom with the odd number of electrons.	**8.9**

Type of Problem	Concepts and Equations	Sample Exercises
Drawing Lewis structures with an expanded valence shell	Distribute the valence electrons in the Lewis structure, allowing atoms of elements with $Z > 12$ to have more than eight valence electrons if more than four bonds are needed or if the structure with the expanded valence shell results in formal charges closer to zero.	**8.10**
Drawing Lewis structures with an incomplete octet	Distribute the valence electrons in the Lewis structure, allowing atoms of elements such as Be, B, and Al to have fewer than eight valence electrons if needed.	**8.11**
Estimating heats of reaction from average bond energies	Multiply bond energies by the number of bonds and calculate using $$\Delta H_{rxn} = \sum \Delta H_{bonds\ breaking} - \sum \Delta H_{bonds\ forming} \qquad (8.4)$$	**8.12**

VISUAL PROBLEMS

(Answers to boldface end-of-chapter questions and problems are in the back of the book.)

8.1. Which group highlighted in Figure P8.1 contains atoms that have the following? (a) 1 valence electron; (b) 4 valence electrons; (c) 6 valence electrons

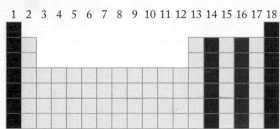

FIGURE P8.1

8.2. Which of the groups highlighted in Figure P8.2 contains atoms with the following? (a) 2 valence electrons; (b) 3 valence electrons; (c) 5 valence electrons

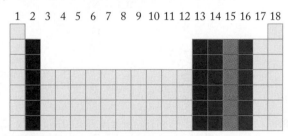

FIGURE P8.2

8.3. Which of the Lewis symbols in Figure P8.3 correctly portrays the most stable ion of magnesium?

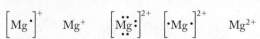

FIGURE P8.3

8.4. What changes must be made to the Lewis symbols in Figure P8.4 to make them correct?

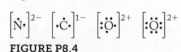

FIGURE P8.4

8.5. Which of the highlighted elements in Figure P8.5 has the greatest bonding capacity?

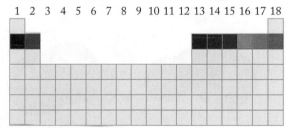

FIGURE P8.5

8.6. Which of the highlighted elements in Figure P8.6 has the greatest electronegativity?

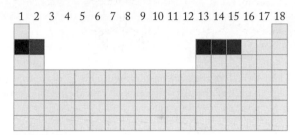

FIGURE P8.6

8.7. Which two of the highlighted elements in Figure P8.6 form the bonding pair with the most ionic character?

Note: The color scale used in Problems 8.8, 8.9, 8.12, and 8.13 is the same as in Figure 8.5, where violet is a charge of 1+, red is a charge of 1−, and green is 0. The larger the size, the greater the electron density.

8.8. Which of the drawings in Figure P8.8 is the best description of the distribution of electrical charge in ClBr?

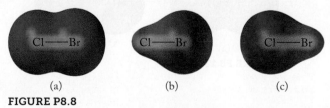

(a)   (b)   (c)

FIGURE P8.8

*8.9. Which of the drawings in Figure P8.9 best describes the distribution of electrical charge in LiF?

(a)　　　　　(b)　　　　　(c)

FIGURE P8.9

8.10. Are the three structures in Figure P8.10 resonance forms of the thiocyanate ion (SCN⁻)? Explain why or why not.

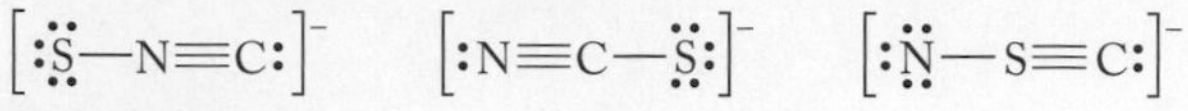

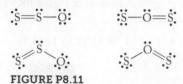

FIGURE P8.10

8.11. Why are the structures in Figure P8.11 not all resonance forms of the molecule S_2O?

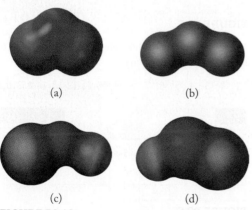

FIGURE P8.11

*8.12. Which of the drawings in Figure P8.12 most accurately describes the distribution of electrical charge in ozone? Explain your answer.

(a)　　　　　(b)

(c)　　　　　(d)

FIGURE P8.12

*8.13. Which of the drawings in Figure P8.13 most accurately describes the distribution of electrical charge in SO_2? Explain your answer.

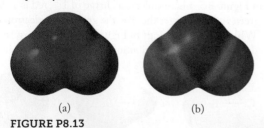

(a)　　　　　(b)

FIGURE P8.13

8.14. How many electron pairs are shared in each of the molecules and ions in Figure P8.14?

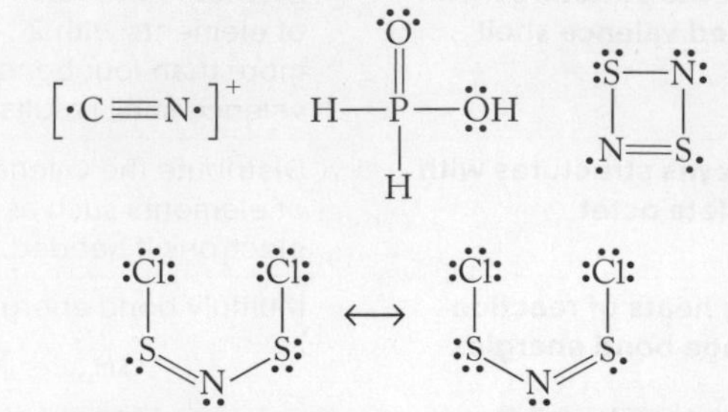

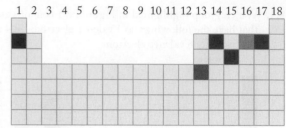

FIGURE P8.14

8.15. Krypton and xenon form compounds with only the most reactive of other elements. Which of the highlighted elements in Figure P8.15 is one of these highly reactive elements?

1 2 3 4 5 6 7 8 9 10 11 12 13 14 15 16 17 18

FIGURE P8.15

8.16. What changes must be made to the Lewis structures in Figure P8.16 to make them correct? Assume the skeletal structures shown are correct.

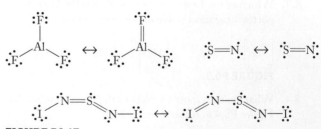

FIGURE P8.16

8.17. In each pair of resonance structures in Figure P8.17, which one contributes more to the bonding in the molecule or molecular ion?

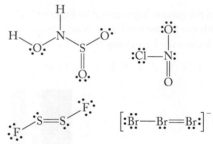

FIGURE P8.17

8.18. Use representations [A] through [I] in Figure P8.18 to answer questions a–f.

a. Which hydrocarbon has the strongest carbon–carbon bond?

b. Which hydrocarbon has the weakest carbon–carbon bond?

c. Could [B] and/or [G] absorb an infrared photon through an asymmetric stretch and contribute to the greenhouse effect?

d. Label the most polar bond in [C] and in [D], using both the arrow (↔) and δ+ and δ− notations.

e. Which process is exothermic?

f. Which representation demonstrates a key limitation of Lewis structures?

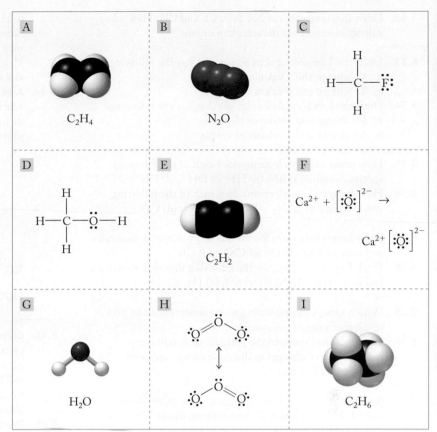

FIGURE P8.18

QUESTIONS AND PROBLEMS

Types of Chemical Bonds

Concept Review

8.19. Does the number of valence electrons in a neutral atom ever equal the atomic number?

8.20. Does the number of valence electrons in a neutral atom ever equal the group number?

8.21. Do all the elements in a group in the periodic table have the same number of valence electrons?

8.22. Distinguish between an atom's valence electrons and its total electron count.

Lewis Structures

Concept Review

8.23. Some of his critics described G. N. Lewis's approach to explaining covalent bonding as an exercise in double counting and therefore invalid. Explain the basis for this criticism.

8.24. Does the octet rule mean that a diatomic molecule must have 16 valence electrons?

8.25. Why would you not expect to find hydrogen atoms in the bonding arrangement X—H—X?

8.26. Does each atom in a pair that is covalently bonded always contribute the same number of valence electrons to form the bonds between them?

Problems

8.27. Draw Lewis symbols of atoms of lithium, magnesium, and aluminum.

8.28. Draw Lewis symbols of atoms of nitrogen, oxygen, fluorine, and chlorine.

8.29. Find the error in each of the Lewis symbols in Figure P8.29.

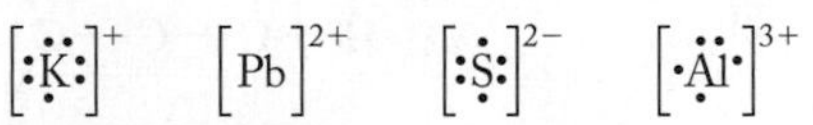

$$\left[:\ddot{K}:\right]^{+} \quad \left[Pb\right]^{2+} \quad \left[:\ddot{S}:\right]^{2-} \quad \left[\cdot\ddot{Al}\cdot\right]^{3+}$$

FIGURE P8.29

8.30. Find the error in each of the Lewis symbols in Figure P8.30.

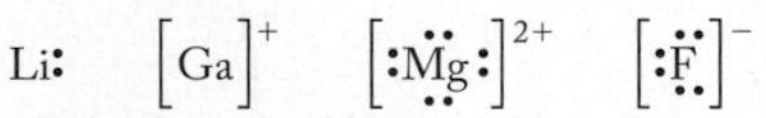

$$Li: \quad \left[Ga\right]^{+} \quad \left[:\ddot{Mg}:\right]^{2+} \quad \left[:\ddot{F}:\right]^{-}$$

FIGURE P8.30

8.31. Draw Lewis symbols for In^{+}, I^{-}, Ca^{2+}, and Sn^{2+}. Which ions have a complete valence-shell octet?

8.32. Draw Lewis symbols of Xe, Sr^{2+}, Cl, and Cl^-. How many valence electrons are in each atom or ion?

8.33. Draw the Lewis symbol of an ion that has the following:
a. 1+ charge and 1 valence electron
b. 3+ charge and 0 valence electrons

8.34. Draw the Lewis symbol of an ion that has the following:
a. 1− charge and 8 valence electrons
b. 1+ charge and 5 valence electrons

8.35. How many valence electrons does each of the following species contain? (a) BN; (b) HF; (c) OH^-; (d) CN^-

8.36. How many valence electrons does each of the following species contain? (a) N_2^+; (b) CS^+; (c) CN; (d) CO

8.37. Draw Lewis structures for the following diatomic molecules and ions: (a) CO; (b) O_2; (c) ClO^-; (d) CN^-.

8.38. Draw Lewis structures for the following diatomic molecules and ions: (a) F_2; (b) NO^+; (c) SO; (d) HI.

8.39. Which groups among main group elements have an odd number of valence electrons?

8.40. Which of the groups in the periodic table will carry negative partial charges in diatomic compounds with hydrogen, HX and H_2X?

8.41. **Greenhouse Gases** Chlorofluorocarbons (CFCs) are linked to the depletion of stratospheric ozone. They are also greenhouse gases. Draw Lewis structures for the following CFCs:
a. CF_2Cl_2 (Freon 12)
b. Cl_2FCCF_2Cl (Freon 113, containing a C—C bond)
c. C_2ClF_3 (Freon 1113, containing a C=C bond)

8.42. Draw Lewis structures for the organic compounds shown and answer the following questions. Assume the skeletal structures in Figure P8.42 are correct.
a. Which of these molecules is an alkyne?
b. Which molecule has the shortest carbon–carbon bond(s)?
c. Which of these molecules contains an aldehyde functional group?

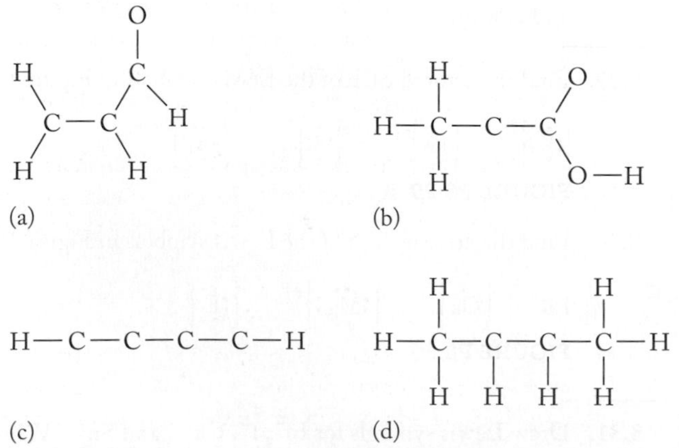

(a)
(b)
(c)
(d)

FIGURE P8.42

8.43. **Skunks and Rotten Eggs** Many sulfur-containing organic compounds have characteristically foul odors: butanethiol

$(CH_3CH_2CH_2CH_2SH)$ is responsible for the odor of skunks, and rotten eggs smell the way they do because they produce tiny amounts of pungent hydrogen sulfide, H_2S. Draw the Lewis structures for $CH_3CH_2CH_2CH_2SH$ and H_2S.

8.44. **Acid in Ants** Formic acid, HCOOH, is the smallest organic acid and was originally isolated by distilling red ants. Draw its Lewis structure given the connectivity of the atoms as shown in Figure P8.44.

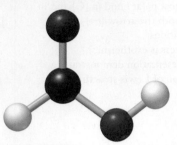

FIGURE P8.44

8.45. **Chlorine Bleach** Chlorine combines with oxygen in several proportions. Dichlorine monoxide (Cl_2O) is used in the manufacture of bleaching agents. Potassium chlorate ($KClO_3$) is used in oxygen generators aboard aircraft. Draw the Lewis structures for Cl_2O and ClO_3^- (Cl is the central atom).

8.46. **Dangers of Mixing Cleansers** Labels on household cleansers caution against mixing bleach with ammonia (Figure P8.46) because the reaction produces monochloramine (NH_2Cl) and hydrazine (N_2H_4), both of which are toxic:

$$NH_3(aq) + OCl^-(aq) \rightarrow NH_2Cl(aq) + OH^-(aq)$$

$$NH_2Cl(aq) + NH_3(aq) + OH^-(aq) \rightarrow \\ N_2H_4(aq) + Cl^-(aq) + H_2O(\ell)$$

Draw the Lewis structures for monochloramine and hydrazine.

FIGURE P8.46

Polar Covalent Bonds

Concept Review

8.47. How can we use electronegativity to predict whether a bond between two atoms is likely to be covalent or ionic?

8.48. How do the electronegativities of the elements change across a period and down a group?

8.49. Explain on the basis of atomic structure why trends in electronegativity are related to trends in atomic size.

8.50. Is the element with the most valence electrons in a period also the most electronegative? Explain.

8.51. What is meant by the term *polar covalent bond*?

8.52. What factor is responsible for the existence of polar covalent bonds?

8.53. Describe how atmospheric greenhouse gases act like the panes of glass in a greenhouse.

*8.54. Understanding the Greenhouse Effect Water vapor in the atmosphere contributes more to the greenhouse effect than carbon dioxide, yet water vapor is not considered an important factor in global warming. Propose a reason why.

8.55. Increasing concentrations of nitrous oxide in the atmosphere may be contributing to climate change. Is the ability of N_2O to absorb infrared radiation due to nitrogen–nitrogen bond stretching, nitrogen–oxygen bond stretching, or both? Explain your answer.

8.56. Is the ability of H_2O molecules to absorb photons of infrared radiation due to symmetrical stretching or asymmetrical stretching of its O—H bonds, or both? Explain your answer. (*Hint:* The angle between the two O—H bonds in H_2O is 104.5°.)

Problems

8.57. Which of the following bonds are polar: C—Se, C—O, Cl—Cl, O=O, N—H, C—H? In the bond or bonds that you selected, which atom has the greater electronegativity?

8.58. Which is the least polar bond: C—Se, C=O, Cl—Br, O=O, N—H, C—H?

_____

8.59. In which of the following binary compounds is the bond expected to have the *least* ionic character? LiCl; CsI; KBr; NaF

8.60. In which of the following compounds is the bond between the atoms expected to have the most covalent character? $AlCl_3$; $AlBr_3$; AlI_3; GaF_3

_____

8.61. Which polar bond in Figure P8.61 is correctly labeled?

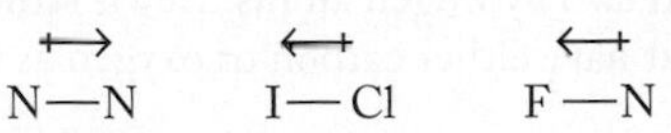

$$\overset{\longmapsto}{\text{N—N}} \qquad \overset{\longleftarrow}{\text{I—Cl}} \qquad \overset{\longleftarrow}{\text{F—N}}$$
FIGURE P8.61

8.62. Which polar bond in Figure P8.62 is correctly labeled?

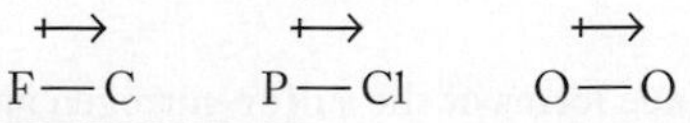

$$\overset{\longmapsto}{\text{F—C}} \qquad \overset{\longmapsto}{\text{P—Cl}} \qquad \overset{\longmapsto}{\text{O—O}}$$
FIGURE P8.62

_____

8.63. Which bond in the following is correctly labeled?

$$\overset{\delta-\ \ \delta+}{\text{O—H}} \qquad \overset{\delta-\ \ \delta+}{\text{C—Cl}} \qquad \overset{\delta-\ \ \delta+}{\text{Br—Br}}$$

8.64. Which bond in the following is correctly labeled?

$$\overset{\delta+\ \ \delta-}{\text{Na—Cl}} \qquad \overset{\delta+\ \ \delta-}{\text{Li—F}} \qquad \overset{\delta+\ \ \delta-}{\text{Si—F}}$$

8.65. Which substance has the most polar covalent bonds: PF_3, S_8, RbCl, or SF_2?

8.66. Atoms of which element are held together by nonpolar covalent bonds: lithium, phosphorus, or xenon?

Resonance

Concept Review

8.67. Explain the concept of resonance.

8.68. How does resonance influence the stability of a molecule or an ion?

8.69. What factors determine whether a molecule or ion exhibits resonance?

8.70. What structural features do all the resonance forms of a molecule or ion have in common?

8.71. Explain why NO_2 is more likely to exhibit resonance than CO_2.

8.72. Are these two skeletal structures resonance forms: X—X—O and X—O—X? Explain.

Problems

8.73. Draw Lewis structures for fulminic acid (HCNO), showing all resonance forms.

8.74. Draw Lewis structures for hydrazoic acid (HN_3), showing all resonance forms.

_____

*8.75. Oxygen and nitrogen combine to form a variety of nitrogen oxides, including the following two unstable compounds, each with two nitrogen atoms per molecule: N_2O_2 and N_2O_3. Draw Lewis structures for these molecules, showing all resonance forms.

*8.76. Oxygen and sulfur combine to form a variety of sulfur oxides. Some are stable molecules and some, including S_2O_2 and S_2O_3, decompose when they are heated. Draw Lewis structures for these two compounds, showing all resonance forms.

_____

*8.77. The oxygen–oxygen distance in F_2O_2 is about 20% shorter than in hydrogen peroxide.
   a. Draw Lewis structures for both F_2O_2 and H_2O_2.
   b. It was proposed that F_2O_2 has a resonance form of $[FO_2]^+F^-$. Draw a Lewis structure for this "ionic" form of F_2O_2 and explain how the structure is consistent with the observed O–O distance.

*8.78. The nitrogen–oxygen bond distance in NOF_3 is shorter than in $NO(CH_3)_3$.
   a. Draw Lewis structures for NOF_3 and $NO(CH_3)_3$.
   b. It was proposed that NOF_3 has a resonance form of $[NOF_2]^+F^-$. Draw a Lewis structure for this "ionic" form of NOF_3 and explain how the structure is consistent with the observed N–O distance.

_____

8.79. Chemists can use the octet rule to predict the structures of new compounds to synthesize. Draw Lewis structures showing all resonance forms for the hypothetical compound ClSeNSO, where the atoms are connected in the order they are written.

8.80. Aromatic rings can connect to make larger molecules. The skeletal structures of two such molecules are shown in Figure P8.80. Draw Lewis structures for both molecules, showing all resonance forms.

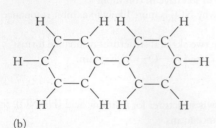

(a)

(b)

FIGURE P8.80

Formal Charge: Choosing among Lewis Structures

Concept Review

8.81. Describe how formal charges are used to choose between possible molecular structures.

8.82. How do the electronegativities of elements influence the selection of which Lewis structure is favored?

8.83. In a molecule containing S and O atoms, is a structure with a negative formal charge on sulfur more likely to contribute to bonding than an alternative structure with a negative formal charge on oxygen? Explain.

8.84. In a cation containing N and O, why do Lewis structures with a positive formal charge on nitrogen contribute more to the actual bonding in the molecule than do those structures with a positive formal charge on oxygen?

Problems

8.85. Hydrogen isocyanide (HNC) has the same elemental composition as hydrogen cyanide (HCN), but the H atom in HNC is bonded to the nitrogen atom. Draw a Lewis structure for HNC and assign formal charges to each atom. How do the formal charges on the atoms differ in the Lewis structures for HCN and HNC?

8.86. **Molecules in Interstellar Space** Hydrogen cyanide (HCN) and cyanoacetylene (HC$_3$N) have been detected in the interstellar regions of space and in comets close to Earth (Figure P8.86). Draw Lewis structures for these molecules, and assign formal charges to each atom. The hydrogen atom is bonded to the carbon atom in both cases.

FIGURE P8.86

8.87. **Origins of Life** The discovery of polyatomic organic molecules such as cyanamide (H$_2$NCN) in interstellar space has led some scientists to believe that the molecules from which life began on Earth may have come from space. Draw Lewis structures for cyanamide, and select the preferred structure on the basis of formal charges.

8.88. Nitromethane (CH$_3$NO$_2$) reacts with hydrogen cyanide to produce CNNO$_2$ and CH$_4$:

$$HCN(g) + CH_3NO_2(g) \rightarrow CNNO_2(g) + CH_4(g)$$

a. Draw Lewis structures for CH$_3$NO$_2$, showing all resonance forms.
b. Draw Lewis structures for CNNO$_2$, showing all resonance forms, based on the two possible skeletal structures for it shown in Figure P8.88. Assign formal charges, and predict which structure is more likely to exist.
c. Are the two structures of CNNO$_2$ resonance forms of each other?

$$C-N-N \begin{matrix} O \\ \\ O \end{matrix} \qquad N-C-N \begin{matrix} O \\ \\ O \end{matrix}$$

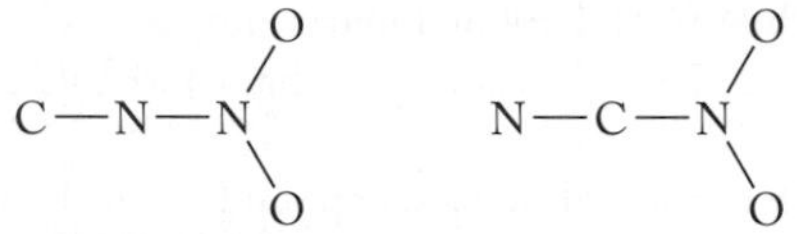

FIGURE P8.88

8.89. The formaldehyde molecule contains one carbon atom, one oxygen atom, and two hydrogen atoms. Lewis structures can be drawn that have either carbon or oxygen as the central atom.
a. Draw these structures and use formal charges to predict the more likely structure for formaldehyde.
b. Are the two structures you drew resonance forms of each other?

8.90. Draw all resonance forms of the sulfur–nitrogen anion, S$_4$N$^-$, and assign formal charges. The atoms are arranged as SSNSS.

*8.91. Nitrogen is the central atom in molecules of nitrous oxide (N$_2$O). Draw Lewis structures for another possible arrangement: N—O—N. Assign formal charges and suggest a reason why this structure is not likely to be stable.

8.92. Use formal charges to determine which resonance form of each of the following ions is preferred: CNO$^-$; NCO$^-$; CON$^-$.

Exceptions to the Octet Rule

Concept Review

8.93. Are all odd-electron molecules exceptions to the octet rule?

8.94. Describe the factors that contribute to the stability of structures in which the central atoms have more than eight valence electrons.

8.95. Why do C, N, O, and F atoms in covalently bonded molecules and ions have no more than eight valence electrons?

8.96. Do atoms with $Z > 12$ always expand their valence shell? Explain your answer.

Problems

8.97. In which of the following molecules does the sulfur atom have an expanded valence shell? (a) SF_6; (b) SF_5; (c) SF_4; (d) SF_2

8.98. In which of the following molecules does the phosphorus atom have an expanded valence shell? (a) $POCl_3$; (b) PF_5; (c) PF_3; (d) P_2F_4 (which has a P—P bond)

8.99. How many electrons are there in the covalent bonds surrounding the central atom in the following species? (a) $(CH_3)_3Al$; (b) B_2Cl_4; (c) SO_3; (d) SF_5^-

8.100. How many electrons are there in the covalent bonds surrounding the central atom in the following species? (a) $POCl_3$; (b) $InCl_5^{2-}$; (c) FBO; (d) PF_4^-

*8.101. Draw Lewis structures for NOF_3 and POF_3 in which the group 15 element is the central atom and the other atoms are bonded to it. What differences are there in the types of bonding in these molecules?

*8.102. The phosphate anion is common in minerals. The corresponding nitrogen-containing anion, NO_4^{3-}, is unstable but can be prepared by reacting sodium nitrate with sodium oxide at 300°C. Draw Lewis structures for each anion. What are the differences in bonding between these ions?

8.103. Dissolving NaF in selenium tetrafluoride (SeF_4) produces $NaSeF_5$. Draw Lewis structures for SeF_4 and SeF_5^-. In which structure does Se have more than eight valence electrons?

*8.104. Reaction between NF_3, F_2, and SbF_3 at 200°C and 100 atm pressure gives the ionic compound NF_4SbF_6:

$$NF_3(g) + 2\ F_2(g) + SbF_3(g) \rightarrow NF_4SbF_6(s)$$

Draw Lewis structures for the ions in this product.

8.105. **Ozone Depletion** The compound Cl_2O_2 may play a role in ozone depletion in the stratosphere. In the laboratory, reaction of $FClO_2$ with aluminum chloride produces Cl_2O_2 and $AlFCl_2$:

$$FClO_2(g) + AlCl_3(s) \rightarrow Cl_2O_2(g) + AlFCl_2(s)$$

Draw a Lewis structure for Cl_2O_2 based on the arrangement of atoms in Figure P8.105. Does either of the chlorine atoms in the structure have an expanded valence shell?

FIGURE P8.105

*8.106. Trimethylaluminum reacts with dimethylamine, $(CH_3)_2NH$, forming methane and $Al_2(CH_3)_4[N(CH_3)_2]_2$ by the balanced chemical equation:

$$2\ (CH_3)_3Al + 2\ HN(CH_3)_2 \rightarrow$$
$$2\ CH_4 + Al_2(CH_3)_4[N(CH_3)_2]_2$$

Draw the Lewis structures for the reactants and products. Must any of these structures contain atoms with fewer than eight valence electrons?

8.107. Which of the following chlorine oxides are odd-electron molecules? (a) Cl_2O_7; (b) Cl_2O_6; (c) ClO_4; (d) ClO_3; (e) ClO_2

8.108. Which of the following nitrogen oxides are odd-electron molecules? (a) NO; (b) NO_2; (c) NO_3; (d) N_2O_4; (e) N_2O_5

8.109. Which of the Lewis structures in Figure P8.109 contributes most to the bonding in CNO?

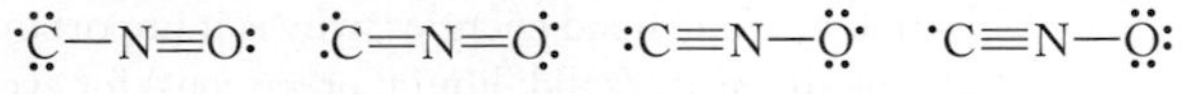

$$\overset{\bullet\bullet}{C}—N≡O: \quad :C=N=\overset{\bullet\bullet}{O}: \quad :C≡N—\overset{\bullet\bullet}{\underset{\bullet\bullet}{O}}: \quad \overset{\bullet}{C}≡N—\overset{\bullet\bullet}{\underset{\bullet\bullet}{O}}:$$
(a)   (b)   (c)   (d)

FIGURE P8.109

8.110. Why is the Lewis structure in Figure P8.110 unlikely to contribute much to the bonding in NCO?

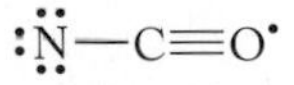

$$:\overset{\bullet\bullet}{\underset{\bullet\bullet}{N}}—C≡O\bullet$$

FIGURE P8.110

*8.111. Using Lewis structures, explain why dimethylaluminum chloride is more likely to exist as $(CH_3)_4Al_2Cl_2$ than as $(CH_3)_2AlCl$.

*8.112. When left at room temperature, mixtures of BF_3 and BCl_3 are found to contain significant amounts of BF_2Cl and $BFCl_2$. Using Lewis structures, explain the origin of the latter two compounds.

*8.113. Some have argued that SF_6 has ionic resonance forms that do not require an expanded octet for S. Draw a resonance structure consistent with this hypothesis and assign formal charges to each atom. Is this resonance form better than or the same as the one with an expanded octet?

8.114. The synthesis of an extremely unusual molecule, $[NO]_2^+[XeF_8]^{2-}$, was reported close to 50 years ago. The structure of $[XeF_8]^{2-}$ shows that all eight fluorine atoms are within bonding distance of the Xe. Draw a Lewis structure for the $[XeF_8]^{2-}$ anion.

The Lengths and Strengths of Covalent Bonds

Concept Review

8.115. Do you expect the nitrogen–oxygen bond length in the nitrate ion to be the same as in the nitrite ion? Explain.

8.116. Why is the oxygen–oxygen bond length in O_3 not the same as in O_2?

8.117. Explain why the nitrogen–oxygen bond lengths in N_2O_4 (which has a nitrogen–nitrogen bond) and N_2O are nearly identical (118 and 119 pm, respectively).

8.118. Do you expect the sulfur–oxygen bond lengths in SO_3^{2-} and SO_4^{2-} ions to be about the same? Why?

8.119. Rank the following ions in order of (a) increasing nitrogen–oxygen bond lengths and (b) increasing bond energies: NO_2^-; NO^+; NO_3^-.

8.120. Rank the following compounds and ions in order of (a) increasing carbon–oxygen bond lengths and (b) increasing bond energies: CO; CO_2; CO_3^{2-}.

*8.121. Do you expect the boron–fluorine bond energy to be the same in BF_3 and F_3BNH_3?

*8.122. The boron–oxygen distances in the BO_2^+ cation are equal. Does this mean the bond order of the B–O bond is two? Explain.

8.123. Why must the stoichiometry of a reaction be known in order to estimate the enthalpy change from bond energies?

8.124. Why must the structures of the reactants and products be known in order to estimate the enthalpy change of a reaction from bond energies?

*8.125. When calculating the enthalpy change for a chemical reaction by using bond energies, why is it important to know the phase (solid, liquid, or gaseous) for every compound in the reaction?

*8.126. If the energy needed to break 2 moles of C=O bonds is greater than the sum of the energies needed to break the O=O bonds in 1 mole of O_2 and vaporize 1 mole of carbon, why does the combustion of pure carbon release heat?

Problems

Note: Use the average bond energies in Table 8.3 to answer Problems 8.127 through 8.136.

8.127. Use average bond energies to estimate the enthalpy changes of the following reactions:
  a. $N_2(g) + 3 H_2(g) \rightarrow 2 NH_3(g)$
  b. $N_2(g) + 2 H_2(g) \rightarrow H_2NNH_2(g)$
  c. $2 N_2(g) + O_2(g) \rightarrow 2 N_2O(g)$

8.128. Use average bond energies to estimate the enthalpy changes of the following reactions:
  a. $CO_2(g) + H_2(g) \rightarrow H_2O(g) + CO(g)$
  b. $N_2(g) + O_2(g) \rightarrow 2 NO(g)$
  *c. $C(s) + CO_2(g) \rightarrow 2 CO(g)$
  (*Hint*: The enthalpy of sublimation of graphite, C(s), is 719 kJ/mol.)

8.129. The combustion of CO to CO_2 releases 283 kJ/mol. What is the bond energy of the carbon–oxygen bond in carbon monoxide?

8.130. Use average bond energies to estimate the standard enthalpy of formation of HF gas.

8.131. Estimate how much less energy is released during the incomplete combustion of 1 mole of methane to carbon monoxide and water vapor than in the complete combustion to carbon dioxide and water vapor.

8.132. Estimate how much more energy is released by the reaction

$$C(s) + O_2(g) \rightarrow CO_2(g)$$

than by the reaction

$$C(s) + \tfrac{1}{2} O_2(g) \rightarrow CO(g)$$

*8.133. Estimate ΔH_{rxn} for the following reaction:

$$4 NH_3(g) + 7 O_2(g) \rightarrow 4 NO_2(g) + 6 H_2O(g)$$

*8.134. The value of ΔH_{rxn} for the reaction

$$2 H_2S(g) + 3 O_2(g) \rightarrow 2 SO_2(g) + 2 H_2O(g)$$

is −1036 kJ. Estimate the energy of the bonds in SO_2.

8.135. A molecular view of the combustion of CS_2 is shown in Figure P8.135. If the standard enthalpy of combustion of CS_2 is −1102 kJ/mol, what is the average bond energy for the carbon–sulfur bonds in CS_2?

FIGURE P8.135

*8.136. The standard enthalpy of reaction for the decomposition of carbon oxysulfide (COS) to CO_2 and CS_2, as shown in Figure P8.136, is −1.9 kJ/mol. Are the apparent bond energies of the carbon–sulfur and carbon–oxygen bonds in COS stronger or weaker than in CS_2 and CO_2, respectively?

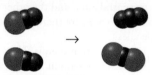

FIGURE P8.136

*8.137. Carbon and oxygen form three oxides: CO, CO_2, and carbon suboxide (C_3O_2). Draw a Lewis structure for C_3O_2 in which the three carbon atoms are bonded to each other, and predict whether the carbon–oxygen bond lengths in the carbon suboxide molecule are equal.

*8.138. Spectroscopic analysis of the linear molecule N_4O reveals that the nitrogen–oxygen bond length is 135 pm and that there are three nitrogen–nitrogen bond lengths: 148, 127, and 115 pm. Draw the Lewis structure for N_4O consistent with these observations.

Additional Problems

8.139. The unpaired dots in Lewis symbols of the elements represent valence electrons available for covalent bond formation. In Figure P8.139, which of the options for placing dots around the symbol for each element is preferred?

Be: or ·Be·	:Al· or ·Al·	·C̈· or :C̈·	He: or ·He·
(a)	(b)	(c)	(d)

FIGURE P8.139

8.140. On the basis of the Lewis symbols in Figure P8.140, predict to which group in the periodic table element X belongs.

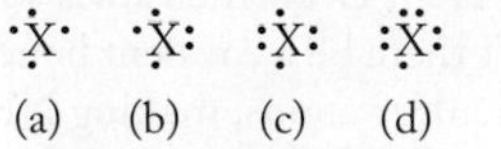

$$\cdot\dot{X} \quad \cdot\dot{X}: \quad :\dot{X}: \quad :\ddot{X}:$$

(a) (b) (c) (d)

FIGURE P8.140

8.141. Use formal charges to predict whether the atoms in carbon disulfide are arranged CSS or SCS.

8.142. The following is a family of weak acids: $HClO$, $HClO_2$, and $HClO_3$. Draw their Lewis structures, using formal charges to predict the best arrangement of atoms. Show any resonance forms of the molecules.

*8.143. Chemical Weapons Phosgene is a poisonous gas first used in chemical warfare during World War I. It has the formula $COCl_2$ (C is the central atom).
   a. Draw its Lewis structure.
   b. Phosgene kills because it reacts with water in nasal passages, in the lungs, and on the skin to produce carbon dioxide and hydrogen chloride. Write a balanced chemical equation for this process, showing the Lewis structures for reactants and products.

8.144. The dinitramide anion $[N(NO_2)_2^-]$ was first isolated in 1996. The arrangement of atoms in $N(NO_2)_2^-$ is shown in Figure P8.144.
   a. Complete the Lewis structure for $N(NO_2)_2^-$, including any resonance forms, and assign formal charges.
   b. Explain why the nitrogen–oxygen bond lengths in $N(NO_2)_2^-$ and N_2O should (or should not) be similar.
   c. $N(NO_2)_2^-$ was isolated as $[NH_4^+][N(NO_2)_2^-]$. Draw the Lewis structure for NH_4^+.

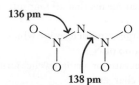

136 pm

138 pm

FIGURE P8.144

*8.145. Silver cyanate (AgOCN) is a source of the cyanate ion (OCN⁻), which reacts with several small molecules. Under certain conditions the species OCN is an anion with a charge of 1−; under other conditions it is a neutral, odd-electron molecule.
   a. Two molecules of OCN combine to form OCNNCO. Draw the Lewis structures for this molecule, including all resonance forms.
   b. The OCN⁻ ion reacts with BrNO, forming the unstable molecule OCNNO. Draw the Lewis structures for BrNO and OCNNO, including all resonance forms.
   c. The OCN⁻ ion reacts with Br_2 and NO_2 to produce N_2O, CO_2, BrNCO, and OCN(CO)NCO. Draw three of the resonance forms of OCN(CO)NCO, which has the arrangement of atoms shown in Figure P8.145.

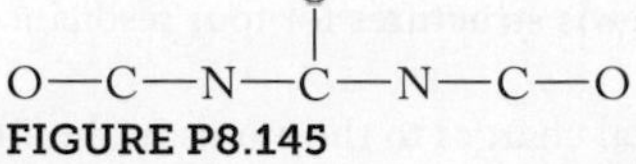

$$O-C-N-\underset{|}{\overset{\overset{\displaystyle O}{\|}}{C}}-N-C-O$$

FIGURE P8.145

*8.146. During the reaction of the cyanate ion (OCN⁻) with Br_2 and NO_2, a very unstable substance called an *intermediate* forms and then quickly falls apart. Its formula is O_2NNCO.
   a. Draw three of the resonance forms for O_2NNCO, assign formal charges, and predict which of the three contributes the most to the bonding in O_2NNCO. The connectivity of the atoms is shown in Figure P8.146(a).
   b. Which bond in O_2NNCO must break in the reaction with Br_2 to form BrNCO? What other product forms?
   c. Draw Lewis structures for a different arrangement of the N, C, and O atoms in O_2NNCO as shown in Figure P8.146(b).

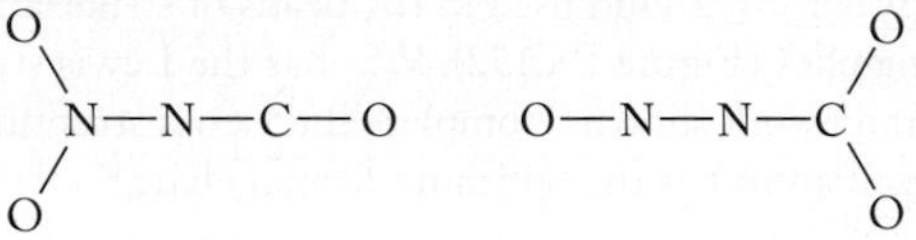

$$\underset{\overset{\displaystyle |}{\displaystyle O}}{\overset{\displaystyle O}{\diagdown}}N-N-C-O \qquad\qquad O-N-N-\underset{\displaystyle O}{\overset{\displaystyle O}{C}}$$

(a)   (b)

FIGURE P8.146

8.147. A compound with the formula Cl_2O_6 decomposes to a mixture of ClO_2 and ClO_4. Draw two Lewis structures for Cl_2O_6: one with a chlorine–chlorine bond and one with a $Cl-O-Cl$ arrangement of atoms. Draw a Lewis structure for ClO_2.

$$Cl_2O_6 \rightarrow ClO_2 + ClO_4$$

*8.148. A compound consisting of chlorine and oxygen, Cl_2O_7, decomposes by the following reaction:

$$Cl_2O_7 \rightarrow ClO_4 + ClO_3$$

   a. Draw two Lewis structures for Cl_2O_7: one with a chlorine–chlorine bond and one with a $Cl-O-Cl$ arrangement of atoms.
   b. Draw a Lewis structure for ClO_3.

*8.149. The odd-electron molecule CN dimerizes to give cyanogen (C_2N_2).
   a. Draw a Lewis structure for CN, and predict which arrangement for cyanogen is more likely: NCCN or CNNC.
   b. Cyanogen reacts slowly with water to produce oxalic acid $(H_2C_2O_4)$ and ammonia; the Lewis structure for oxalic acid is shown in Figure P8.149. Compare this structure with your answer in part (a). When the actual structures of molecules have been defined experimentally, the structures have been used to refine Lewis structures. Does this structure increase your confidence that the structure you selected in part (a) may be the better one?

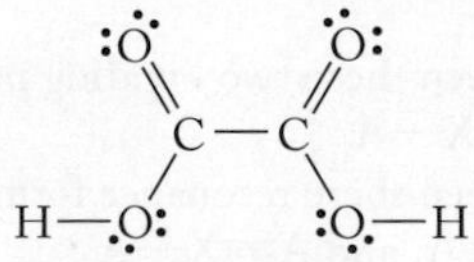

$$H-\ddot{\underset{..}{O}}:\underset{\displaystyle C}{\overset{\displaystyle :\ddot{O}:}{|}}-\underset{\displaystyle C}{\overset{\displaystyle \ddot{O}:}{|}}:\ddot{\underset{..}{O}}-H$$

FIGURE P8.149

*8.150. Draw all resonance forms for the molecules NSF and HBS where S and B are the central atoms, respectively. Include possible ionic structures for NSF.

*8.151. The molecular structure of sulfur cyanide trifluoride (SF₃CN) has been shown to have the arrangement of atoms with the indicated bond lengths in Figure P8.151. Complete the Lewis structure for SF₃CN, and assign formal charges.

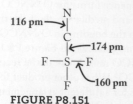

FIGURE P8.151

8.152. **Strike-Anywhere Matches** Heating phosphorus with sulfur gives P_4S_3, a solid used in the heads of strike-anywhere matches (Figure P8.152). P_4S_3 has the Lewis structure framework shown. Complete the Lewis structure so that each atom has the optimum formal charge.

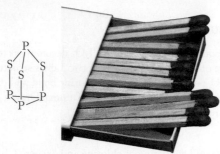

FIGURE P8.152

*8.153. The heavier group 16 elements can expand their valence shell. The $TeOF_6^{2-}$ anion was first prepared in 1993. Draw the Lewis structure for $TeOF_6^{2-}$.

*8.154. **Sulfur in the Environment** Sulfur is cycled in the environment through compounds such as dimethyl sulfide (CH_3SCH_3), hydrogen sulfide (H_2S), and sulfite and sulfate ions. Draw Lewis structures for these four species. Are expanded valence shells needed to minimize the formal charges for any of these species?

8.155. How many pairs of electrons does xenon share in the following molecules and ions? (a) XeF_2; (b) $XeOF_2$; (c) XeF^+; (d) XeF_5^+; (e) XeO_4

8.156. Consider a hypothetical structure of ozone that is cyclic (the atoms form a ring) such that its three O atoms are at the corners of a triangle. Draw the Lewis structure for this molecule.

*8.157. Bond lengths and electrostatic potential mapping provide experimental evidence about covalent bonds present in a molecule. Which of these would help characterize the bonding in a compound with the formula A_2X in terms of the following?
  a. distinguishing between these two bonding patterns:
     X—A—A and A—X—A
  b. distinguishing between these resonance forms:
     A—X≡A, A=X=A, and A≡X—A

8.158. **Explosive Cation** The highly explosive N_5^+ cation was first isolated in 1999 by reaction of N_2F^+ with HN_3:

$$N_2F^+ + HN_3 \rightarrow N_5^+ + HF$$

Draw the Lewis structures for the reactants and products, including all resonance forms.

*8.159. **Jupiter's Atmosphere** The ionic compound NH_4SH was detected in the atmosphere of Jupiter (Figure P8.159) by the *Galileo* space probe in 1995. Draw the Lewis structure for NH_4SH. Why couldn't there be a covalent bond between the nitrogen and sulfur atoms, making NH_4SH a molecular compound?

FIGURE P8.159

8.160. **Antacid Tablets** Antacids commonly contain calcium carbonate, magnesium hydroxide, or both. Draw the Lewis structures for calcium carbonate and magnesium hydroxide.

*8.161. An allotrope of nitrogen, N_4, was reported in 2002. The compound has a lifetime of 1 µs at 298 K and was prepared by adding an electron to N_4^+. Because the compound cannot be isolated, its structure is unconfirmed experimentally.
  a. Draw the Lewis structures for all the resonance forms of linear N_4. (Linear means that all four nitrogens are in a straight line.)
  b. Assign formal charges, and determine which structure is the best description of N_4.
  c. Draw a Lewis structure for a ring (cyclic) form of N_4, and assign formal charges.

*8.162. Scientists have predicted the existence of O_4 even though this compound has never been observed. However, O_4^{2-} has been detected. Draw the Lewis structures for O_4 and O_4^{2-}.

8.163. Draw a Lewis structure for $AlFCl_2$, the second product in the synthesis of Cl_2O_2 in the following reaction:

$$FClO_2(g) + AlCl_3(s) \rightarrow Cl_2O_2(g) + AlFCl_2(s)$$

*8.164. Draw Lewis structures for BF_3 and $(CH_3)_2BF$. The B—F distance in both molecules is the same (130 pm). Does this observation support the argument that all the boron–fluorine bonds in BF_3 are single bonds?

8.165. Which of the following molecules and ions contains an atom with an expanded valence shell? (a) Cl_2; (b) ClF_3; (c) ClI_3; (d) ClO^-

8.166. Which of the following molecules contains an atom with an expanded valence shell? (a) XeF_2; (b) $GaCl_3$; (c) ONF_3; (d) SeO_2F_2

*8.167. A linear nitrogen anion, N_5^-, was isolated for the first time in 1999.
  a. Draw the Lewis structures for four resonance forms of linear N_5^-.
  b. Assign formal charges to the atoms in the structures in part (a), and identify the structures that contribute the most to the bonding in N_5^-.

c. Compare the Lewis structures for N_5^- and N_3^-. In which ion do the nitrogen–nitrogen bonds have the higher average bond order?

*8.168. Carbon tetroxide (CO_4) was discovered in 2003.
   a. The four atoms in CO_4 are predicted to be arranged as shown in Figure P8.168. Complete the Lewis structure for CO_4.
   b. Are there any resonance forms for the structure you drew that have zero formal charges on all atoms?
   c. Can you draw a structure in which all four oxygen atoms in CO_4 are bonded to carbon?

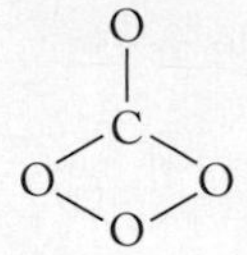

FIGURE P8.168

8.169. Plot the electronegativities of elements with $Z = 3$ to 9 (y-axis) versus their first ionization energy (x-axis). Is the plot linear? Use your graph to predict the electronegativity of neon, whose first ionization energy is 2080 kJ/mol.

8.170. Use the data in Figures 7.39 and 8.6 to plot electronegativity as a function of first ionization energy for the main group elements of the fifth row of the periodic table. From this plot estimate the electronegativity of xenon, whose first ionization energy is 1170 kJ/mol.

8.171. The cation N_2F^+ is isoelectronic with N_2O.
   a. What does it mean to be isoelectronic?
   b. Draw the Lewis structure for N_2F^+. (*Hint*: The molecule contains a nitrogen–nitrogen bond.)
   c. Which atom has the +1 formal charge in the structure you drew in part (b)?
   d. Does N_2F^+ have resonance forms?
   e. Could the middle atom in the N_2F^+ ion be a fluorine atom? Explain your answer.

8.172. **Ozone Depletion** Methyl bromide (CH_3Br) is produced naturally by fungi. Methyl bromide has also been used in agriculture as a fumigant, but this use is being phased out because the compound has been linked to ozone depletion in the upper atmosphere.
   a. Draw the Lewis structure for CH_3Br.
   b. Which bond in CH_3Br is more polar, carbon–hydrogen or carbon–bromine?

*8.173. All the carbon–fluorine bonds in tetrafluoroethylene (Figure P8.173) are polar, but experimental results show that the molecule as a whole is nonpolar. Draw arrows on each bond to show the direction of polarity, and suggest a reason why the molecule is nonpolar.

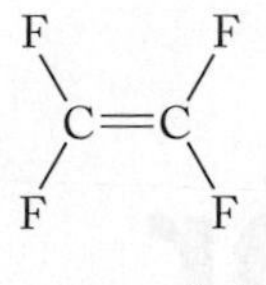

Tetrafluoroethylene

FIGURE P8.173

*8.174. Atoms of Xe have complete octets, yet the compound XeO_2 exists. How is this possible?

*8.175. Free radicals increase in stability, and hence decrease in reactivity, if they have more than one atom in their structure that can carry the unpaired electron.
   a. Draw Lewis structures for the three free radicals formed from CF_2Cl_2 in the stratosphere—ClO, Cl, and CF_2Cl—and use this principle to rank them in order of reactivity.
   b. Free radicals are "neutralized" when they react with each other, forming an electron pair (a single covalent bond) between two atoms. This is called a *termination reaction* because it shuts down any reaction that was powered by the free radical. Predict the formula of the molecules that are produced when the chlorine free radical reacts with each of the free radicals in part (a).

9

Molecular Geometry
Shape Determines Function

MOLECULES ACCELERATE RIPENING Ripening tomatoes give off the gas ethylene, which speeds up the ripening process. Ethylene molecules have a unique shape that enables them to fit into a biomolecular site that controls ripening.

PARTICULATE **REVIEW**

Smelly Molecules and Functional Groups

In Chapter 9 we explore the connection between molecular shape and properties such as taste and smell. Two molecules with very different smells are shown here. Malic acid is responsible for the smell and taste of apples. Putrescine, however, is the smell associated with rotting flesh in a cadaver.

- What is the molecular formula for malic acid?
- What is the molecular formula for putrescine?
- Identify the functional groups present in both molecules.

    (Review Section 2.8 if you need help.)

(Answers to Particulate Review questions are in the back of the book.)

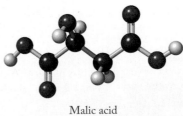

Malic acid

Putrescine

Smelly Molecules and Shapes

Your ability to detect a substance by smell depends upon both the number of molecules present and the shape of those molecules. Chemists describe shape in terms of bond angles and whether the atoms all lie within one plane, that is, whether the molecules are planar. As you read Chapter 9, look for ideas that will help you answer these questions:

- Draw a Lewis structure for both acrolein and methyl mercaptan.

- Look at the space-filling models. What are the approximate bond angles around the central carbon atom in acrolein? Around the sulfur atom in methyl mercaptan?

- Which of these molecules is planar?

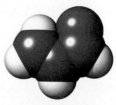

Acrolein,
present in smoky barbeque

Methyl mercaptan,
present in human "bad breath"

Learning Outcomes

LO1 Explain valence-shell electron-pair repulsion (VSEPR) theory

LO2 Use VSEPR and the concept of steric number to predict the bond angles in molecules and the shapes of molecules with one central atom
Sample Exercises 9.1, 9.2, 9.3

LO3 Predict whether a substance is polar or nonpolar on the basis of its molecular structure
Sample Exercise 9.4

LO4 Use valence bond theory to explain orbital overlap, bond angles, and molecular shapes
Sample Exercises 9.5, 9.6, 9.7, 9.8

LO5 Draw condensed and carbon-skeleton structures of organic compounds
Sample Exercise 9.9

LO6 Recognize molecular shapes that are stabilized by delocalization of π electrons

LO7 Recognize chiral molecules

LO8 Draw molecular orbital (MO) diagrams of diatomic molecules and use MO diagrams to predict magnetic properties and explain spectra
Sample Exercises 9.10, 9.11, 9.12

LO9 Use MO theory to describe metallic bonding and semiconductors
Sample Exercise 9.13

9.1 Biological Activity and Molecular Shape

Hold your hands out in front of you, palms up, fingers extended. Now rotate your wrists inward so that your thumbs point straight up. Your right hand looks the same as the image your left hand makes in a mirror. Does that mean your two hands have the same shape? If you have ever tried to put your right hand in a glove made for your left, you know that they do not have the same shape. Many other objects in our world have a "handedness" about them, from headphones to scissors to golf clubs.

This chapter focuses on the importance of shape at the molecular level. For example, the compound that produces the refreshing aroma of spearmint has the molecular formula $C_{10}H_{14}O$. The compound responsible for the musty aroma of caraway seeds has the same molecular formula *and* the same Lewis structure. To understand how two compounds can be so much alike and still have different properties, we have to consider their structures in three dimensions.

We perceive a difference in aromas in part because each molecule has a unique site where it attaches to our nasal membranes. Just as a left hand only fits a left glove, the spearmint molecule fits only the spearmint-shaped site, and the caraway molecule fits only the caraway-shaped site. This behavior, called molecular recognition, enables biomolecular structures such as nasal membranes to recognize and react when a particular molecule with a particular shape fits into a part of the structure known as an *active site*. Many of the substances we ingest, from food to pharmaceuticals, exert physiological effects because their molecules are recognized by and become bound to active sites in biomolecules.

The shape of a compound's molecules can affect many of its properties: its physical state at room temperature, its solubility in water and other solvents, its aroma, its biological activity, and its reactivity with other compounds. In Chapter 8 we drew Lewis structures to describe bonding in molecules, but Lewis structures are only two-dimensional representations of how atoms and the electron pairs that surround them are arranged in molecules. Lewis structures show how atoms are connected in molecules, but they do not show how the atoms are arranged in three dimensions, nor do they necessarily show the overall shape of the molecules.

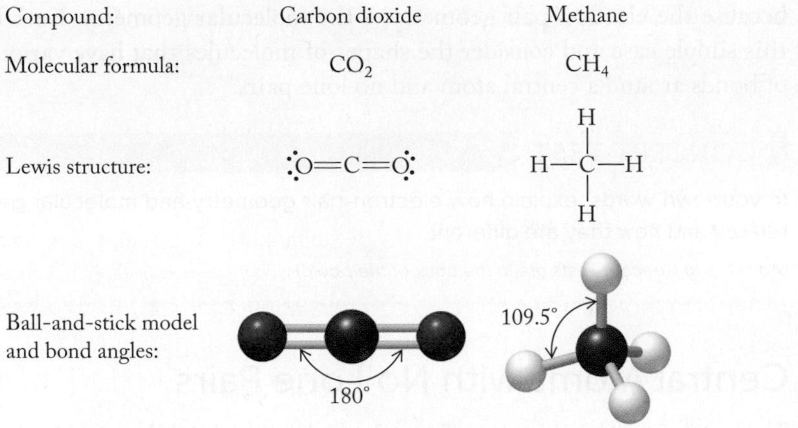

Compound:	Carbon dioxide	Methane
Molecular formula:	CO_2	CH_4

Lewis structure:

Ball-and-stick model and bond angles:

FIGURE 9.1 The Lewis structure of CO_2 matches its true molecular structure, but the Lewis structure of CH_4 does not because the C—H bonds in CH_4 extend in all three dimensions.

To illustrate this point, let's consider the Lewis structures and ball-and-stick models of two compounds: carbon dioxide and methane (Figure 9.1). The linear arrangement of atoms and bonding electrons in the Lewis structure for CO_2 corresponds to the actual linear shape of the molecule, as represented by the ball-and-stick model. The angle between the two C=O bonds is 180°, just as in the Lewis structure. On the other hand, the Lewis structure of methane does not convey the true three-dimensional orientation of the four C—H bonds in each molecule. The 90° angles between bonds in the planar Lewis structure are not close to the three-dimensional H—C—H **bond angles** of 109.5°.

Several theories of covalent bonding account for and predict the shapes of molecules. In this chapter we begin by examining small molecules with single central atoms, and we gradually apply these theories to larger molecules. We start with the shared pairs and lone pairs of electrons described by Lewis theory and then predict how those pairs should be oriented about a central atom to minimize their interactions and produce the most stable molecular structure. As we will see, each theory accurately predicts electron-pair orientations, molecular shapes, and the properties of molecular compounds.

9.2 Valence-Shell Electron-Pair Repulsion (VSEPR) Theory

Let's start with a theory based on a fundamental chemical principle: electrons have negative charges and repel each other. **Valence-shell electron-pair repulsion (VSEPR) theory** applies this principle by incorporating the assumption that pairs of valence electrons are arranged about central atoms in ways that minimize repulsions between the pairs. To predict molecular shape by using VSEPR we must consider two things:

- **Electron-pair geometry**, which defines the relative positions in three-dimensional space of all the bonding pairs and lone pairs of valence electrons on the central atom
- **Molecular geometry**, which defines the relative positions in three-dimensional space of the atoms in a molecule

The electron-pair geometry of a molecule may or may not be the same as its molecular geometry. The presence of lone pairs on the central atom makes the difference.

To accurately predict molecular geometry, we first need to know electron-pair geometry. If there are no lone pairs of electrons, then the process is simplified

bond angle the angle (in degrees) defined by lines joining the centers of two atoms to the center of a third atom to which they are chemically bonded.

valence-shell electron-pair repulsion (VSEPR) theory a model predicting that the arrangement of valence electron pairs around a central atom minimizes their mutual repulsion to produce the lowest-energy orientations.

electron-pair geometry the three-dimensional arrangement of bonding pairs and lone pairs of electrons about a central atom.

molecular geometry the three-dimensional arrangement of the atoms in a molecule.

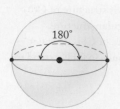

(a) Linear
SN = 2

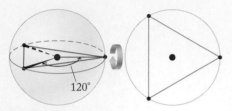

(b) Trigonal planar
SN = 3

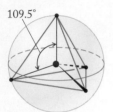

(c) Tetrahedral
SN = 4

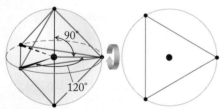

(d) Trigonal bipyramidal
SN = 5

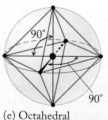

(e) Octahedral
SN = 6

FIGURE 9.2 Electron-pair geometries depend on the steric number (SN) of the central atom. In these images, there are no lone pairs of electrons on the central atoms (red dots), so the molecular geometries are the same as the electron-pair geometries. The images show bond angles for different numbers of atoms located on the surface of a sphere and bonded to an atom at the center of the sphere. The blue lines define the geometric forms that give the molecular shapes their names. Parts (b) and (d) show spheres rotated 90° so equilateral triangles in the equatorial plane are clearly visible.

because the electron-pair geometry *is* the molecular geometry. Let's begin with this simple case and consider the shapes of molecules that have various numbers of bonds around a central atom and no lone pairs.

> **CONCEPT TEST**
>
> In your own words, explain how electron-pair geometry and molecular geometry are related and how they are different.
>
> *(Answers to Concept Tests are in the back of the book.)*

Central Atoms with No Lone Pairs

To determine the geometry of a molecule, we start by drawing its Lewis structure. From the Lewis structure, we determine the **steric number (SN)** of the central atom, which is the sum of the number of atoms bonded to that atom and the number of lone pairs on it:

$$\text{SN} = \left(\begin{array}{c}\text{number of atoms} \\ \text{bonded to central atom}\end{array}\right) + \left(\begin{array}{c}\text{number of lone pairs} \\ \text{on central atom}\end{array}\right) \quad (9.1)$$

Because we are focused on molecules in which the central atom has no lone pairs, the steric number equals the number of atoms bonded to the central atom. In evaluating the shapes of these molecules, we generate five common shapes that describe both the electron-pair geometries and the molecular geometries of many covalent compounds.

Let's start by thinking of the central atom as the center of a sphere, with all the other atoms in the molecule placed on the surface of the sphere and linked to the central atom by covalent bonds. If the central atom is bonded covalently to only two other atoms, then SN = 2. How do the electron pairs in the two bonds arrange themselves to minimize their mutual repulsion? The answer is that they are as far from each other as possible: on opposite sides of the sphere (Figure 9.2a). This gives a **linear** electron-pair geometry and a linear molecular geometry. The three atoms in the molecule are arranged in a straight line, and the bond angle is 180°.

If three atoms are bonded to a central atom with no lone pairs, then SN = 3. The three bonding pairs are as far apart as possible when they are located at the three corners of an equilateral triangle. The angle between each pair of bonds is the same: 120° (Figure 9.2b). The name of this electron-pair and molecular geometry is **trigonal planar**.

With four atoms around the central atom and no lone pairs (SN = 4), we have the first case in which the atoms are not all in the same plane. Instead, the atoms bonded to the central atom occupy the four vertices of a tetrahedron, which is a four-sided pyramid (*tetra* is Greek, meaning "four"). The bonding pairs form bond angles of 109.5° with each other, as shown in Figure 9.2(c). The electron-pair geometry and molecular geometry are both **tetrahedral**.

When five atoms are bonded to a central atom with no lone pairs, SN = 5 and the atoms occupy the five corners of two triangular pyramids that share the same base. The central atom of the molecule is in the center of the sphere at the common center of the two bases, as shown in Figure 9.2(d). One bonding pair points to the tip of the top pyramid, one points to the tip of the bottom pyramid, and the other three

SN = 2   3   4   5   6

FIGURE 9.3 Balloons charged with static electricity repel each other and align themselves as far apart as possible when tied together. In doing so, they mimic the locations of electron pairs about a central atom. Each balloon represents one electron pair. Note the similarities between the balloons and the diagrams in Figure 9.2.

point to the three vertices of the shared triangular base. These three vertices lie along the equator of the sphere, so the atoms that occupy these sites and the bonds that connect them to the central atom are called *equatorial* atoms and bonds. The bond angles between the three equatorial bonds are 120° (just as in the triangle of the trigonal planar geometry in Figure 9.2b). The bond angle between an equatorial bond and either vertical, or *axial*, bond is 90°, and the angle between the two axial bonds is 180°. A molecule in which the atoms are arranged this way is said to have a **trigonal bipyramidal** electron-pair and molecular geometry.

For SN = 6, picture two pyramids that have a square base (Figure 9.2e). Put them together, base to base, and you form a shape in which all six positions around the sphere are equivalent. We can think of the six bonding pairs of electrons as three sets of two electron pairs each. The two pairs in each set are oriented at 180° to each other and at 90° to the other two pairs, just like the axes of an x–y–z coordinate system. Four equatorial atoms lie at the four vertices of the common square base and are 90° apart, another atom lies at the tip of the top pyramid, and a sixth lies at the tip of the bottom pyramid. This arrangement defines **octahedral** electron-pair and molecular geometries.

To demonstrate a real-world analogy to the above bond orientations, we start with a batch of fully inflated yellow balloons. We tie them together in clusters of two, three, four, five, and six balloons (Figure 9.3). Let us assume that all the balloons have acquired a static electrical charge, so they repel each other. If the tie points of the clusters represent the central atom in our balloon models, then the opposite ends of the balloons represent the atoms that are bonded to the central atom. Note how our clusters of two, three, four, five, and six balloons produce the same orientations that resulted from selecting points on a sphere that were as far apart as possible. The long axes of the balloons in Figure 9.3 provide an accurate representation of the bond directions in Figure 9.2.

Now let's look at five simple molecules that have no lone pairs about the central atom and apply VSEPR and the concept of steric number to predict their electron-pair and molecular geometries. To do so, we follow these three steps:

1. Draw the Lewis structure for the molecule.
2. Determine the steric number of the central atom.
3. Use the steric number to predict the electron-pair and molecular geometries by using the images in Figure 9.2.

Example I: Carbon Dioxide, CO_2

1. Lewis structure:

$$\ddot{\text{O}}\!=\!\text{C}\!=\!\ddot{\text{O}}$$

2. The central carbon atom has two atoms bonded to it and no lone pairs, so the steric number is 2.
3. Because SN = 2, the O—C—O bond angle is 180° (see Figure 9.2a) and the electron-pair and molecular geometries are both linear.

steric number (SN) the sum of both the number of atoms bonded to a central atom and the number of lone pairs of electrons on the central atom.

linear molecular geometry about a central atom with a steric number of 2 and no lone pairs of electrons; the bond angle is 180°.

trigonal planar molecular geometry about a central atom with a steric number of 3 and no lone pairs of electrons; the bond angles are all 120°.

tetrahedral molecular geometry about a central atom with a steric number of 4 and no lone pairs of electrons; the bond angles are all 109.5°.

trigonal bipyramidal molecular geometry about a central atom with a steric number of 5 and no lone pairs of electrons; three atoms occupy equatorial sites (bond angles 120°) and two other atoms occupy axial sites (bond angle 180°) above and below the equatorial plane; the bond angle between the axial and equatorial bonds is 90°.

octahedral molecular geometry about a central atom with a steric number of 6 and no lone pairs of electrons, in which all six sites are equivalent; four equatorial sites are 90° apart; two axial sites are 180° apart.

CONNECTION We learned in Chapter 8 that some molecules have central atoms that do not have an octet of electrons. The boron atom in BF_3 shares only three pairs of electrons, but because this distribution results in formal charges of zero on all the atoms, it is the preferred structure.

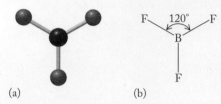

(a)   (b)

FIGURE 9.4 (a) The ball-and-stick model of BF_3 shows the orientation of the atoms. (b) All F—B—F bond angles are 120° in this trigonal planar molecular geometry.

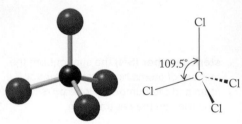

FIGURE 9.5 The ball-and-stick model shows the actual orientation of the atoms in carbon tetrachloride in three dimensions. All Cl—C—Cl bond angles are 109.5° in this tetrahedral molecular geometry.

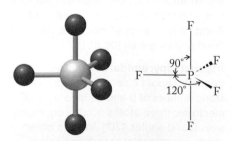

FIGURE 9.6 The ball-and-stick model shows the actual orientation of the atoms in phosphorus pentafluoride. The P—F bonds located around the equator of the imaginary sphere are 120° apart. Each of them is 90° from the two bonds connecting F atoms in the axial positions (at the north and south poles). The resulting molecular geometry is trigonal bipyramidal.

Example II: Boron Trifluoride, BF_3

1. Lewis structure:

2. Three fluorine atoms are bonded to the central boron atom, and there are no lone pairs on boron. Therefore SN = 3.
3. Because SN = 3, all four atoms lie in the same plane, forming a triangle with F—B—F bond angles of 120°. The electron-pair and molecular geometries are trigonal planar (Figure 9.4).

Example III: Carbon Tetrachloride, CCl_4

1. Lewis structure:

2. Four chlorine atoms are bonded to the central carbon atom, which gives carbon a full octet. Therefore, SN = 4.
3. Because SN = 4, the chlorine atoms are located at the vertices of a tetrahedron and all four Cl—C—Cl bond angles are 109.5°, producing tetrahedral electron-pair and molecular geometries (Figure 9.5).

Before we explore any more molecular geometries, we should explain the conventions used in drawing three-dimensional structures in two dimensions. To convey the three-dimensional structure of a molecule, we use a solid wedge (—) to indicate a bond that comes out of the paper toward the viewer. The solid wedge in the CCl_4 structure in Figure 9.5, for example, means that the chlorine in that position projects out of the plane of the paper and toward the viewer at a downward angle. A dashed wedge (⸺) indicates a bond that goes behind the plane of the paper and away from the viewer at a downward angle. Solid lines are used for bonds that lie in the plane of the paper. We typically orient a structure so that the maximum number of bonds lie in the plane of the paper.

Example IV: Phosphorus Pentafluoride, PF_5

1. Lewis structure:

2. Five fluorine atoms are bonded to the central phosphorus atom, which has no lone pairs. Therefore SN = 5.
3. Because SN = 5, the fluorine atoms are located at the five vertices of a trigonal bipyramid, and the electron-pair geometry and molecular geometry are trigonal bipyramidal (Figure 9.6).

Example V: Sulfur Hexafluoride, SF₆

1. Lewis structure:

2. Six fluorine atoms are bonded to the central sulfur atom, which has no lone pairs. Therefore SN = 6.
3. Because SN = 6, the fluorine atoms are located at the vertices of an octahedron and the electron-pair and molecular geometries are octahedral (Figure 9.7).

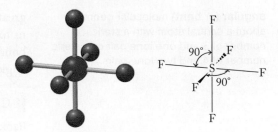

FIGURE 9.7 The ball-and-stick model of SF₆ shows how the six fluorine atoms are oriented in three dimensions about the sulfur atom. All the F—S—F bond angles are 90° in this octahedral molecular geometry.

SAMPLE EXERCISE 9.1 Using VSEPR to Predict Geometry I   **LO2**

Formaldehyde, CH₂O, is a gas at room temperature (boiling point −21°C). Aqueous solutions of formaldehyde are used to preserve biological samples. It is also a product of the incomplete combustion of hydrocarbons. Use VSEPR to predict the molecular geometry of formaldehyde.

Collect and Organize We are given the molecular formula of formaldehyde. The solution requires (1) drawing the Lewis structure, (2) determining the steric number, and (3) identifying the molecular geometry by using Figure 9.2.

Analyze Carbon is the likely central atom of the molecule because it has a bonding capacity of 4 and is less electronegative than oxygen, whose bonding capacity is 2. If the three atoms bonded to C are as far from each other as possible, the probable molecular geometry is trigonal planar.

Solve Following the procedure developed in Chapter 8 for drawing Lewis structures, we draw this one for formaldehyde:

As predicted, SN = 3 and there are no lone pairs of electrons on the central atom. According to Figure 9.2, the molecular geometry is trigonal planar.

Think About It The key to predicting the correct molecular geometry of a molecule with no lone pairs on its central atom is to determine the steric number, which is simply a matter of counting the number of atoms bonded to the central atom.

Practice Exercise Use VSEPR to determine the molecular geometry of the chloroform molecule, CHCl₃, and draw the molecule by using the solid-wedge, dashed-wedge convention.

(Answers to Practice Exercises are in the back of the book.)

Measurements of the bond angles in formaldehyde show that the H—C—H bond angle is slightly smaller than the 120° predicted for trigonal planar geometry (Figure 9.2b). The C=O double bond consists of two pairs of bonding electrons (four electrons) and exerts greater repulsion than a single bond would. This

angular (or **bent**) molecular geometry about a central atom with a steric number of 3 and one lone pair or a steric number of 4 and two lone pairs.

greater repulsion decreases the H—C—H bond angle. VSEPR does not enable us to predict the actual values of the bond angles in molecules containing double bonds to the central atom, but it does allow us to correctly predict how a bond angle deviates from the ideal value owing to the presence of a double bond.

CONCEPT **TEST**

Rank the following bond angles from largest to smallest:

a. The F—B—F bond angles in BF_3

b. The O—C—H bond angles in CH_2O

c. The H—C—H bond angle in CH_2O

Central Atoms with Lone Pairs

We have established the electron-pair and molecular geometries of molecules whose central atoms have no lone pairs of electrons. Now let's explore what happens when a central atom has one or more lone pairs. For SN = 2, the only bonding pattern possible is two atoms bound to a central atom. If we replace one bonding pair with a lone pair, we have a molecule with only two atoms, which means there is no central atom and no bond angle; it takes three atoms to define a bond angle. Such diatomic molecules must have a linear geometry. Therefore we begin the discussion of molecules containing lone pairs with SN = 3.

Consider sulfur dioxide, one of the gases produced when high-sulfur coal is burned. The resonance structures showing the distribution of electrons are

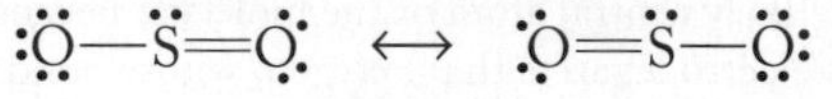

$$:\ddot{O}—\ddot{S}=\ddot{O}: \longleftrightarrow :\ddot{O}=\ddot{S}—\ddot{O}:$$

To calculate the steric number of the central S atom we need to add up the number of atoms and lone pairs of electrons that surround it. In both resonance structures, the central S atom is bonded to two atoms and has one lone pair of electrons. Therefore its SN value in either resonance structure is 2 + 1 = 3. When SN = 3, the arrangement of atoms and lone pairs about the sulfur atom is trigonal planar (Figure 9.2b). This means that the electron-pair geometry, which takes into account both the atoms and the nonbonding pairs of electrons around the central atom, is trigonal planar (as shown in Figure 9.8a). Remember that the molecular geometry describes the relative position of only the atoms in the molecule. With SN = 3 and two atoms attached to the central S atom, the SO_2 molecule has an **angular** (or **bent**) molecular geometry shown in Figure 9.8(b).

Experimental measurements establish that SO_2 is indeed a bent molecule, but the O—S—O bond angle is actually a little smaller than 120°. We explain the smaller angle with VSEPR by comparing the space occupied by bonding electrons with the space occupied by electrons in a lone pair. Because the bonding electrons are attracted to two nuclei, they have a high probability of being located

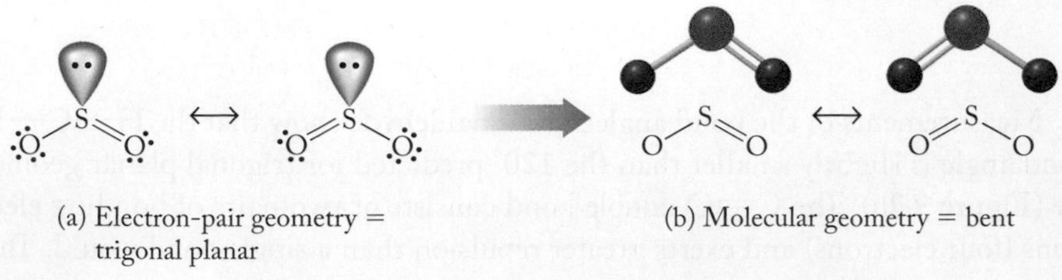

FIGURE 9.8 (a) The electron-pair geometry of SO_2 is trigonal planar because the steric number of the S atom is 3; it is bonded to two atoms and has one lone pair of electrons. (b) The molecular geometry is bent because the central sulfur atom in the structure has no bonded atom in the third position, only a lone pair of electrons.

(a) Electron-pair geometry = trigonal planar

(b) Molecular geometry = bent

between the two atomic centers that share them. In contrast, the lone pair is not shared with a second atom and is spread out around the sulfur atom, as shown in Figure 9.9. This puts the lone pair of electrons closer to the bonding pairs and produces greater repulsion. As a result, the lone pair pushes the bonding pairs closer together, thereby reducing the bond angle. In general:

- Repulsion between lone pairs and bonding pairs is greater than the repulsion between bonding pairs.
- Repulsion caused by a lone pair is greater than that caused by a double bond.
- Repulsion caused by a double bond is greater than that caused by a single bond.
- Two lone pairs of electrons on a central atom exert a greater repulsive force on the atom's bonding pairs than does one lone pair.

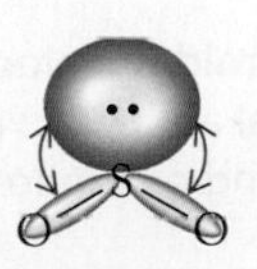

O—S—O bond angle < 120°

FIGURE 9.9 The lone pair of electrons on the central sulfur atom in SO_2 occupies more space (larger purple region) than the electron pairs in the S—O bonds (smaller purple regions). The increased repulsion (represented by the double-headed arrows), resulting from the larger volume occupied by the lone pair, forces the oxygen atoms closer together, making the O—S—O bond angle slightly less than the ideal value of 120°.

SAMPLE EXERCISE 9.2 Predicting Relative Sizes of Bond Angles   **LO2**

Rank NH_3, CH_4, and H_2O in order of decreasing bond angles in their molecular structures.

Collect and Organize We are asked to predict the relative sizes of the bond angles in three molecules, given their molecular formulas. The information in Figure 9.2 links steric numbers to electron-pair geometries and bond angles. Bond angles are also influenced by the presence of double bonds and lone pairs of electrons on the central atom.

Analyze We need to determine the steric numbers of the central atoms in these three molecules. To do that we first need to translate the molecular formulas into Lewis structures and determine how many lone pairs or double bonds they have.

Solve Using the method for drawing Lewis structures from Chapter 8 we obtain these results:

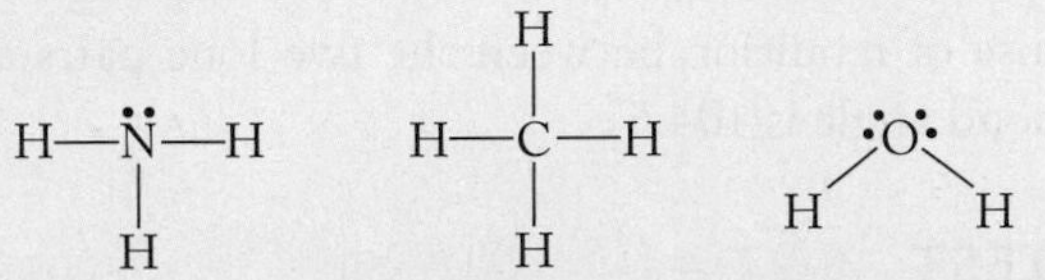

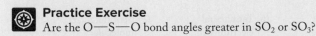

In each of these structures the central atom is surrounded by a total of four atoms or lone pairs of electrons, which means a steric number of 4. There are no double bonds.

According to Figure 9.2, the electron-pair geometry for SN = 4 is tetrahedral. In a molecule such as CH_4, in which all four tetrahedral electron pairs are equivalent bonding pairs, all bond angles are 109.5°. However, repulsion from the lone pairs of electrons in NH_3 and H_2O squeezes their bonds together, reducing the bond angles. The two lone pairs on the O atom in H_2O exert a greater repulsive force on its bonding pairs than the single lone pair on the N atom exerts on the bonding pairs in NH_3. Therefore the bond angle in H_2O should be less than the bond angles in NH_3. Ranking the three molecules in order of decreasing bond angle, we have:

$$CH_4 > NH_3 > H_2O$$

Think About It The logic used in answering this problem is supported by experimental evidence: the bond angles in molecules of CH_4, NH_3, and H_2O are 109.5°, 107.0°, and 104.5°, respectively.

Practice Exercise
Are the O—S—O bond angles greater in SO_2 or SO_3?

trigonal pyramidal molecular geometry about a central atom with a steric number of 4 and one lone pair of electrons.

As we saw in Sample Exercise 9.2, three combinations of atoms and lone pairs are possible about a central atom with a steric number of 4: four atoms and no lone pairs, three atoms and one lone pair, and two atoms and two lone pairs. The first case is illustrated by the molecular structure of methane, CH_4, in which a tetrahedral electron-pair geometry translates into a tetrahedral molecular geometry. A molecule of ammonia, NH_3, has only three bonding pairs and one lone pair. This means that the Lewis structure in Figure 9.10(a) translates into a tetrahedral electron-pair geometry in which one of the vertices is the lone pair on the N atom (Figure 9.10b). The resulting molecular geometry, which is essentially a flattened tetrahedron, is called **trigonal pyramidal** (Figure 9.10c). As we discussed in Sample Exercise 9.2, the strong repulsion produced by the diffuse lone pair of electrons on the N atom pushes the N—H bonds closer together in NH_3 and reduces the angles between them in comparison with the H—C—H bond angles in CH_4. Thus the H—N—H bond angles in ammonia are only 107.0°.

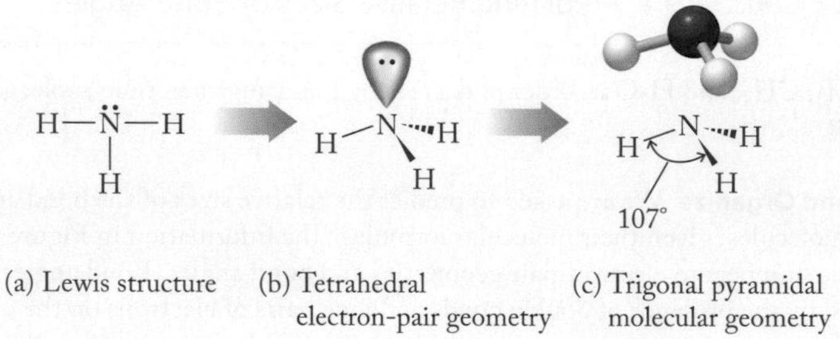

FIGURE 9.10 (a) The steric number of the N atom in NH_3 is 4, (b) so its electron-pair geometry is tetrahedral. However, one of the vertices of the tetrahedron is occupied by a lone pair of electrons, not an atom, (c) so the molecular geometry is trigonal pyramidal.

(a) Lewis structure   (b) Tetrahedral electron-pair geometry   (c) Trigonal pyramidal molecular geometry

The central O atom in a molecule of H_2O has a steric number of 4 because it is bonded to two atoms and has two lone pairs of electrons. The result is a tetrahedral electron-pair geometry (Figure 9.11). However, the presence of the two lone pairs means that two of the four tetrahedral vertices are not occupied by atoms, which leaves us with a bent (or angular) molecular geometry. As we also discussed in Sample Exercise 9.2, the H—O—H bond angle in water is smaller than 109.5° because of repulsion between the two lone pairs and each bonding pair. The actual bond angle is 104.5°.

███ CONCEPT **TEST** ████████████████████████████

The bond angles in silane (SiH_4), phosphine (PH_3), and hydrogen sulfide (H_2S) are 109.5°, 93.6°, and 92.1°, respectively. Use VSEPR to explain this trend.

Molecules with trigonal bipyramidal electron-pair geometry have four possible molecular geometries, depending on the number of lone pairs per molecule (Table 9.1). These options are possible because of the different axial and equatorial vertices in a trigonal bipyramid (Figure 9.2d). VSEPR enables us to predict which

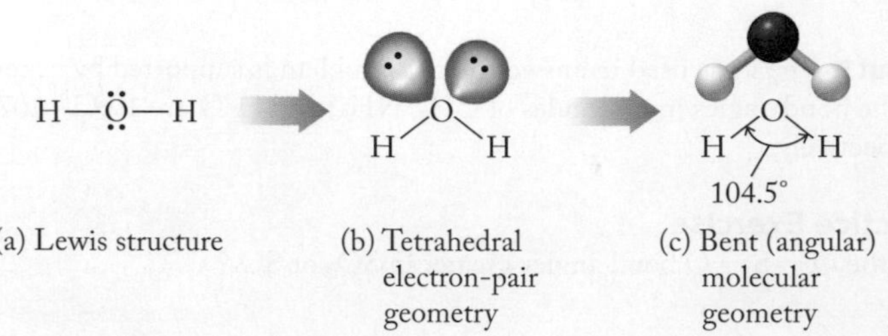

FIGURE 9.11 (a) The steric number of the O atom in H_2O is 4 (it is bonded to two atoms and has two lone pairs of electrons). Therefore (b) its electron-pair geometry is tetrahedral. However, two of the vertices of the tetrahedron are occupied by lone pairs of electrons, meaning (c) only three atoms define the molecular geometry, which is bent.

(a) Lewis structure   (b) Tetrahedral electron-pair geometry   (c) Bent (angular) molecular geometry

TABLE 9.1 Electron-Pair Geometries and Molecular Geometries

SN = 3	Electron-Pair Geometry	Number of Bonded Atoms	Number of Lone Pairs	Molecular Geometry	Structure	Theoretical Bond Angles
	Trigonal planar	3	0	Trigonal planar		120°
	Trigonal planar	2	1	Bent (angular)		<120°
SN = 4						
	Tetrahedral	4	0	Tetrahedral		109.5°
	Tetrahedral	3	1	Trigonal pyramidal		<109.5°
	Tetrahedral	2	2	Bent (angular)		<109.5°
SN = 5						
	Trigonal bipyramidal	5	0	Trigonal bipyramidal		90°, 120°
	Trigonal bipyramidal	4	1	Seesaw	=	<90°, <120°
	Trigonal bipyramidal	3	2	T-shaped	=	<90°
	Trigonal bipyramidal	2	3	Linear	=	180°
SN = 6						
	Octahedral	6	0	Octahedral		90°
	Octahedral	5	1	Square pyramidal		<90°
	Octahedral	4	2	Square planar		90°
	Octahedral	3	3	Although these geometries are possible, we will not encounter any molecules with them		
	Octahedral	2	4			

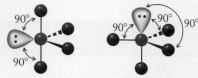

(a) Equatorial lone pair   (b) Axial lone pair

FIGURE 9.12 (a) A lone pair of electrons in an equatorial position of a molecule with trigonal bipyramidal electron-pair geometry interacts through 90° with two other electron pairs. (b) A lone pair in an axial position interacts through 90° with *three* other electron pairs. Fewer 90° interactions reduce internal electron-pair repulsion and lead to greater stability.

FIGURE 9.13 A single lone pair of electrons in an equatorial position of a trigonal bipyramidal electron-pair geometry produces a seesaw molecular geometry.

FIGURE 9.14 Two lone pairs of electrons in equatorial positions of a trigonal bipyramidal electron-pair geometry produce a T-shaped molecular geometry.

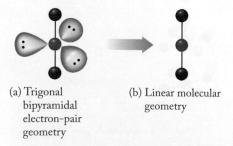

(a) Trigonal bipyramidal electron-pair geometry

(b) Linear molecular geometry

FIGURE 9.15 Three lone pairs of electrons in equatorial positions of (a) a trigonal bipyramidal electron-pair geometry produce (b) a linear molecular geometry.

vertices are occupied by atoms and which are occupied by lone pairs. The key to these predictions is the fact that the repulsions between pairs of electrons decrease as the angle between them increases: two electron pairs at 90° experience a greater mutual repulsion than two at 120°, which have a greater repulsion than two at 180°. To minimize repulsions involving lone pairs, VSEPR predicts that they preferentially occupy equatorial rather than axial vertices. Why? Because an equatorial lone pair has *two* 90° repulsions with the two axial electron pairs as shown in Figure 9.12(a), but an axial lone pair has *three* 90° repulsions (with three equatorial electron pairs), as shown in Figure 9.12(b).

When we assign one, two, or three lone pairs of valence electrons to equatorial vertices, we get three of the molecular geometries for SN = 5 in Table 9.1. When a single lone pair occupies an equatorial site, we get a molecular geometry called **seesaw** (Figure 9.13) because its shape, when rotated 90° clockwise, resembles a playground seesaw. (The formal name for this shape is *disphenoidal*.)

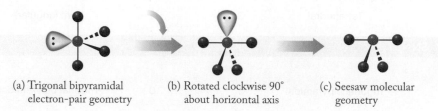

(a) Trigonal bipyramidal electron-pair geometry

(b) Rotated clockwise 90° about horizontal axis

(c) Seesaw molecular geometry

When two lone pairs occupy equatorial sites (Figure 9.14a), the molecular geometry that results is called **T-shaped**. This designation becomes more apparent when the structure in Figure 9.14(a) is rotated 90° counterclockwise. Lone-pair repulsions result in bond angles in T-shaped molecules that are slightly less than the 90° and 180° angles we would expect from a perfectly shaped "T" geometry.

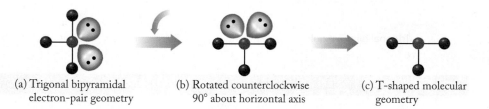

(a) Trigonal bipyramidal electron-pair geometry

(b) Rotated counterclockwise 90° about horizontal axis

(c) T-shaped molecular geometry

Finally, a SN = 5 molecule with three lone pairs and two bonded atoms has a linear geometry because all three lone pairs occupy equatorial sites, and the bonding pairs are in the two axial positions (Figure 9.15).

CONCEPT TEST

The bond angles in a trigonal bipyramidal molecular structure are 90°, 120°, or 180°. Are the corresponding bond angles in a seesaw structure likely to be larger than, the same as, or smaller than these values?

Molecules that have a central atom with a steric number of 6 have an octahedral electron-pair geometry (Figure 9.2e) and the molecular geometries listed in Table 9.1. With one lone pair of electrons, there is only one possible molecular geometry because all the sites in an octahedron are equivalent (Figure 9.16). That one molecular geometry is called **square pyramidal**: a pyramid with a square base, four triangular sides, and the central atom "embedded" in the base.

Because of stronger repulsion between lone pairs and bonding pairs, we predict the bond angles in a square pyramidal molecule to be slightly less than the ideal angle of 90°. The molecule BrF_5, for example, has a square pyramidal molecular geometry, and the angles between its equatorial and axial bonds are 85°.

When two lone pairs are present at the central atom in a molecule with octahedral electron-pair geometry, they occupy vertices on opposite sides of the octahedron to minimize the interactions between them. The resultant molecular geometry is called **square planar** because the molecule is shaped like a square and all five atoms reside in the same plane (Figure 9.17). Because the two lone pairs are on opposite sides of the bonding pairs, the presence of the lone pairs does not distort the bond angles: they are all 90°.

SAMPLE EXERCISE 9.3 Using VSEPR to Predict Geometry II   **LO2**

The Lewis structure of sulfur tetrafluoride (SF_4) is

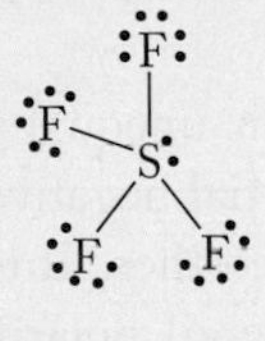

$$\ddot{F}-\underset{\displaystyle \ddot{F}\;\;\ddot{F}}{\overset{\displaystyle \ddot{F}}{S}}$$

What is its molecular geometry and what are the angles between the S—F bonds?

Collect and Organize We are given the Lewis structure of SF_4 and can determine from it the steric number of the central atom in the molecule and its electron-pair geometry.

Analyze Four atoms are bonded to the central S atom, which also has one lone pair of electrons. This means that SN = 5.

Solve With a steric number of 5 for its central atom, the electron-pair geometry of SF_4 is trigonal bipyramidal. The presence of one lone pair of electrons on the S atom means that its molecular geometry is not the same as its electron-pair geometry. Table 9.1 tells us that a seesaw molecular geometry results when a lone pair occupies one of the three equatorial positions in a trigonal bipyramidal electron-pair geometry:

$$F-\underset{\displaystyle F}{\overset{\displaystyle F}{S}}{\cdots}F \quad \Longrightarrow \quad \underset{\displaystyle F}{F-S-F}\;F \quad = \quad \underset{\displaystyle F}{\overset{\displaystyle F}{F-S}}\;F$$

| Electron-pair geometry trigonal bipyramidal | = | Molecular geometry seesaw | = | Frequently drawn from this perspective as well |

The equatorial lone pair slightly reduces the bond angles from their normal values of 90° between the axial and equatorial bonds, 120° between the two equatorial bonds, and 180° between the two axial bonds.

Think About It Any molecule with SN = 5 and one lone pair should have the same geometry as SF_4.

⊛ **Practice Exercise** Determine the molecular geometry and the bond angles of SO_2Cl_2.

seesaw molecular geometry about a central atom with a steric number of 5 and one lone pair of electrons in an equatorial position.

T-shaped molecular geometry about a central atom with a steric number of 5 and two lone pairs of electrons that occupy equatorial positions; the outer atoms occupy two axial sites and one equatorial site.

square pyramidal molecular geometry about a central atom with a steric number of 6 and one lone pair of electrons; as typically drawn, the atoms occupy four equatorial and one axial site.

square planar molecular geometry about a central atom with a steric number of 6 and two lone pairs of electrons that occupy axial sites; the atoms occupy four equatorial positions.

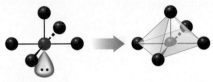

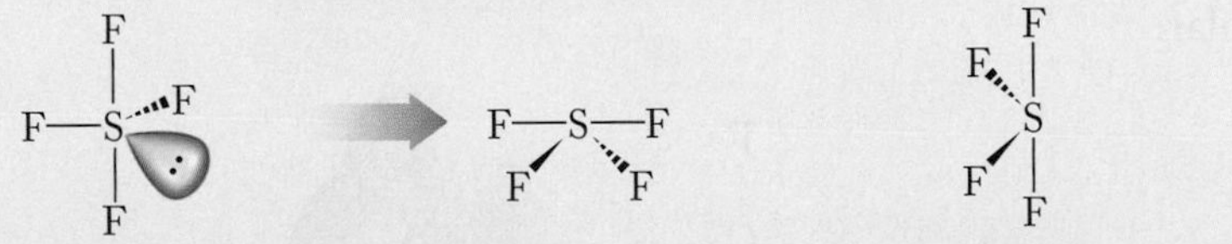

(a) Octahedral electron-pair geometry

(b) Square pyramidal molecular geometry

FIGURE 9.16 (a) A lone pair of electrons in an octahedral electron-pair geometry produces (b) a square pyramidal molecular geometry.

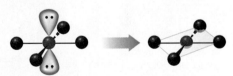

(a) Octahedral electron-pair geometry

(b) Square planar molecular geometry

FIGURE 9.17 (a) Two lone pairs of electrons on opposite sides of an octahedral electron-pair geometry produce (b) a square planar molecular geometry.

9.3 Polar Bonds and Polar Molecules

Of all the naturally occurring gases in our atmosphere, only water condenses to become a liquid at standard temperature and pressure. Why does H_2O condense to become a liquid when all other compounds of comparable molar mass such as N_2, O_2, and CO_2—and even the much heavier compound CF_4—remain gases at standard temperature and pressure? The answer, in part, is explained by the fact that water is a polar molecule.

One of the reasons why H_2O is a polar molecule is that its O—H bonds are polar bonds. However, some nonpolar molecules also contain polar bonds. For example, the C=O double bonds in CO_2 are polar because carbon and oxygen have different electronegativities. The carbon atom in the molecule has a partial positive charge and both oxygen atoms have a partial negative charge:

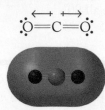

See Figures 8.4 and 8.5 to review the use of these symbols and electron-density surfaces to indicate bond polarity.

As we noted in Chapter 8, an unequal distribution of bonding electrons between two atoms produces a partial negative charge in one region of the bond and a partial positive charge in the other. This charge separation creates a **bond dipole**. We can determine the overall polarity of a molecule by summing the polarities of all the bond dipoles in the molecule. However, this summing must take into account both the strengths of the individual bond dipoles and their orientations with respect to one another. In CO_2, for example, the two dipoles are equivalent in strength because they involve the same atoms and the same kind of bond (C=O). In addition, the linear shape of the molecule means that the direction of one C=O bond dipole is exactly opposite the direction of the other. Thus, the two dipoles exactly offset so that, overall, CO_2 is a nonpolar substance.

Similarly, CF_4 is a nonpolar substance even though its molecules contain polar C—F bonds. The four C—F bond dipoles are all the same because the same atoms are involved. In addition, the tetrahedral molecular geometry means that the bond dipoles offset one another so that, overall, the CF_4 molecule is nonpolar:

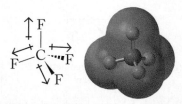

Water molecules are bent, not linear like molecules of CO_2. Therefore the dipoles of the two O—H bonds in a molecule of H_2O do not offset each other. Instead, the molecule has a net dipole with the negative end directed toward the O atom:

CONNECTION In Chapter 8 we introduced electronegativity and the unequal distribution of electrons in polar covalent bonds.

bond dipole separation of electrical charge created when atoms with different electronegativities form a covalent bond.

dipole moment (μ) a measure of the degree to which a molecule aligns itself in a strong electric field; a quantitative expression of the polarity of a molecule.

The presence of this net dipole means that water is a polar molecule. This polarity leads to interactions between H and O atoms on adjacent H_2O molecules in liquid water and solid ice. Other polar molecules exhibit similar interactions. We discuss these interactions in Chapter 10.

CONNECTION In Chapter 8 we discussed how bond polarity affects the physical properties of compounds like the hydrogen halides.

CONCEPT TEST

SCl_4 and $XeCl_4$ both contain polar bonds—yet SCl_4 is a polar molecule, whereas $XeCl_4$ is a nonpolar molecule. Explain how this is possible.

We have seen how the overall polarity of a molecule depends on the differences in the electronegativities of the bonded pairs of atoms in its molecular structure and on the arrangement of those bonded atoms. Experimentally, the polarity of a molecule can be determined by measuring its permanent **dipole moment (μ)**. The value of μ expresses the extent of the overall separation of positive and negative charge in the molecules of a substance and is determined by measuring the degree to which the molecules align with a strong electric field (Figure 9.18). Polar molecules align with the field so that their negative regions are oriented toward the positive plate and their positive regions are oriented toward the negative plate. The more polar the molecules are, the more strongly they align with the field. Dipole moments are usually expressed in units of *debyes* (D), where $1\ D = 3.34 \times 10^{-30}$ coulomb-meter.

The dipole moments of several polar substances are shown in Table 9.2. Note the structures in the last two rows of the table. Chloroform ($CHCl_3$) has a relatively strong dipole moment because its bond dipoles differ both in the *degree* of polarity

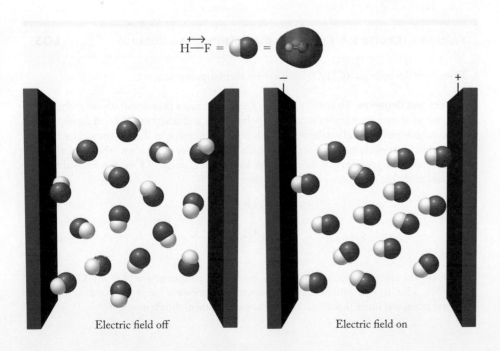

FIGURE 9.18 Gaseous HF molecules are randomly oriented in the absence of an electric field but align themselves when an electric field is applied to two metal plates. The negative (fluorine) end of each molecule is directed toward the positively charged plate; the positive (hydrogen) end, toward the negative plate.

Electric field off

Electric field on

TABLE 9.2 Permanent Dipole Moments of Several Polar Molecules

Formula	Bond Dipole(s)	Overall Dipole	Dipole Moment (debyes)
HF	H⟶F	⟶	1.82
H_2O		↑	1.85
NH_3		↑	1.47
$CHCl_3$		↓	1.01
CCl_3F		↕	0.45

and in the *direction* of the polarity (Figure 9.19a). Chlorine is more electronegative than carbon, and the two electrons in each C—Cl bond are pulled away from the carbon atom and toward the chlorine atom. Because hydrogen is less electronegative than carbon, the two electrons in the H—C bond are pulled away from the hydrogen atom and toward the carbon atom. Consequently the electron distribution is away from the top part of the molecule as drawn and toward the bottom.

In trichlorofluoromethane (CCl_3F) all four bond dipoles point away from the central carbon atom because fluorine and chlorine are more electronegative than carbon (Figure 9.19b). However, a C—F bond dipole is stronger than a C—Cl bond dipole because fluorine is more electronegative than chlorine. As a result, bonding electrons are pulled more toward the C—F (top) side of the molecule than toward the C—Cl (bottom) side. The direction of the overall dipole is upward.

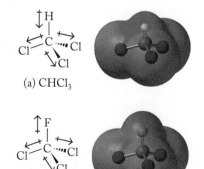

(a) $CHCl_3$

(b) CCl_3F

FIGURE 9.19 (a) The four bonds in chloroform ($CHCl_3$) are not equivalent in terms of direction or degree of polarity. The molecule has a dipole moment. (b) The magnitude of the C—F bond dipole in CCl_3F is larger than the C—Cl bond dipoles and is not offset by them. This molecule also has a dipole moment, but not as large as chloroform's.

SAMPLE EXERCISE 9.4 Predicting the Polarity of a Substance   **LO3**

Does formaldehyde gas (CH_2O) have a permanent dipole moment?

Collect and Organize To predict whether a molecule has a permanent dipole moment, we need to determine whether it contains polar bonds and whether the bond dipoles offset each other. A bond's polarity depends on the difference in the electronegativity of the bonded pair of atoms. The extent to which bond dipoles offset each other depends on the molecular geometry of the molecule. In Sample Exercise 9.1 we determined that formaldehyde has a trigonal planar structure:

Analyze The electronegativities of the elements in the compound are H = 2.1, C = 2.5, and O = 3.5. Given the differences in the electronegativities of the atoms bonded to the central atom, it is likely that formaldehyde has a permanent dipole moment.

Solve The hydrogen atoms are the least electronegative atoms in the molecule and the oxygen atom the most, so each of the bonds in the molecule has a bond dipole directed toward the oxygen atom:

Thus the formaldehyde molecule has a permanent dipole moment.

Think About It The presence of different atoms bonded to a central atom makes it highly likely that the molecule has a permanent dipole moment.

Practice Exercise Does carbon disulfide (CS_2), a gas present in small amounts in crude petroleum, have a dipole moment? Explain your answer.

CONCEPT **TEST**

Water and hydrogen sulfide both have a bent molecular geometry with dipole moments of 1.85 D and 0.98 D, respectively. Why is the dipole moment of H_2S less than that of H_2O?

9.4 Valence Bond Theory

Drawing Lewis structures and applying VSEPR theory enable us to predict the geometry of many molecules reliably. However, we have yet to make a connection between molecular geometry and the arrangement of electrons in atoms in atomic orbitals as described in Chapter 7. The search for a connection between the electronic structure of atoms and the electronic structure of molecules led to the development of bonding theories that use atomic orbitals to account for the molecular geometries predicted by VSEPR. We explore one of these theories next.

Bonds from Orbital Overlap

Valence bond theory arose in the late 1920s, largely as a result of the genius and efforts of Linus Pauling, who developed this theory of molecular bonding based on quantum mechanics. Valence bond theory incorporates the assumption that (a) a chemical bond between two atoms results from **overlap** of the atoms' atomic orbitals, and (b) the greater the overlap, the stronger and more stable the bond.

Shared electrons in a chemical bond are located between the nuclei of two atoms and are attracted to both. This attraction leads to lower potential energy and greater chemical stability than if the atoms were completely independent of one another (see Figure 8.1). This view of chemical bonding is especially useful for analyzing the physical and chemical properties of covalent substances because it provides a model of where the electrons are in a molecule. Just as the locations of electrons in atoms define atomic properties, the locations of electrons in molecules define the properties of those molecules.

Let's begin by applying Pauling's valence bond theory to the simplest of diatomic molecules: H_2. A single H atom has one electron in a $1s$ atomic orbital. According to valence bond theory, the overlap between two H atoms, each with a

valence bond theory a quantum mechanics–based theory of bonding incorporating the assumption that covalent bonds form when half-filled orbitals on different atoms overlap or occupy the same region in space.

overlap a term in valence bond theory describing bonds arising from two orbitals on different atoms that occupy the same region of space.

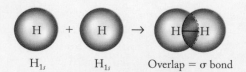

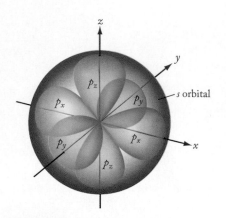

FIGURE 9.20 The overlap of the 1s orbitals on two hydrogen atoms produces a single σ bond that holds the two hydrogen atoms together in a H_2 molecule.

FIGURE 9.21 Relative orientation in space of one 2s orbital (spherical) and three 2p orbitals oriented at 90° to one another along the x-, y-, and z-axes of a coordinate system.

sigma (σ) bond a covalent bond in which the highest electron density lies between the two atoms along the bond axis.

hybridization in valence bond theory, the mixing of atomic orbitals to generate new sets of orbitals that are then available to form covalent bonds with other atoms.

hybrid atomic orbital in valence bond theory, one of a set of equivalent orbitals about an atom created when specific atomic orbitals are mixed.

sp^3 hybrid orbitals a set of four hybrid orbitals with a tetrahedral orientation, formed by mixing one s and three p atomic orbitals.

half-filled 1s orbital, produces a single H—H bond (Figure 9.20). Overlapping two orbitals from two atoms in this way increases electron density along the axis connecting the two nuclei. Whenever the region of high electron density lies along the bond axis, the resulting covalent bond is called a **sigma (σ) bond**.

SAMPLE EXERCISE 9.5 Identifying Overlapping Orbitals in a Molecule **LO4**

Identify the atomic orbitals responsible for the covalent bond in gaseous HCl.

Collect, Organize, and Analyze First we must determine the number of valence electrons in each atom and then draw Lewis structures for the molecule. Next we identify which atomic orbitals are half-filled; these are the orbitals that form the covalent bond between H and Cl.

Solve Atoms from group 17 (halogens) all have seven valence electrons, and hydrogen has one valence electron. Thus the Lewis structure for HCl is

The electron configurations of H and Cl atoms are

$$H: 1s^1 \quad \text{and} \quad Cl: [Ne]3s^23p^5$$

The covalent bond between H and Cl in HCl results from overlap between the half-filled hydrogen 1s and chlorine 3p atomic orbitals.

Think About It The σ bond between hydrogen and chlorine in HCl that arises from overlap of two atomic orbitals is consistent with the prediction from its Lewis structure.

Practice Exercise The halogens (group 17 elements) form a series of interhalogen compounds such as IBr. Identify the atomic orbitals that overlap to form the σ bond in IBr.

Hybridization

In molecules other than simple ones like H_2, an inconsistency arises between the atomic orbital shapes and orientations described in Chapter 7 and the molecular geometries we predict from VSEPR theory that are supported by experimental data. Methane, for example, has a carbon atom at its center. We established in Chapter 7 that carbon atoms have the electron configuration $[He]2s^22p^2$; that the 2s orbital is spherical; and the three 2p orbitals, p_x, p_y, and p_z, are oriented at 90° to one another (Figure 9.21). We also learned in Chapter 8 that carbon atoms form four covalent bonds. How can four equivalent bonds oriented at 109.5° to one another in a tetrahedral arrangement form from the overlap of a filled 2s orbital and two partly filled 2p orbitals? Furthermore, how can two double bonds 180° apart form around the central carbon atom in CO_2, or one double bond and two single bonds 120° apart form around the central carbon in CH_2O?

To account for the geometry of methane, carbon dioxide, formaldehyde, and many other molecules, valence bond theory describes atomic orbitals of different shapes and energies as being mixed in a process called **hybridization** to form **hybrid atomic orbitals**. Covalent bonds then result either from overlap of a hybrid orbital on one atom with an unhybridized orbital on another atom, or from overlap of two hybrid orbitals on two atoms. Let's examine the types of hybrid orbitals that form on carbon and other elements and how these orbitals account for observed molecular geometries.

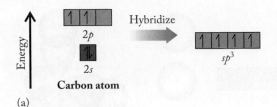

Hybridize

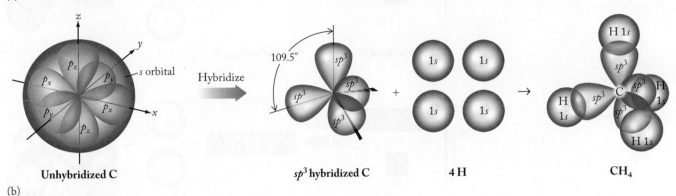

FIGURE 9.22 (a) When the 2s and 2p atomic orbitals of carbon are hybridized, one electron is promoted from the filled 2s orbital to an unoccupied 2p orbital. The four orbitals are then mixed to create four sp³ hybrid orbitals. (b) The hybrid orbitals are oriented 109.5° from one another, exactly the orientation needed to form a tetrahedral methane molecule.

Tetrahedral Geometry: sp³ Hybrid Orbitals

We have noted that methane has a tetrahedral molecular geometry with H—C—H bond angles of 109.5°. To account for this geometry by using orbital hybridization, we start with a single carbon atom with its $2s^2 2p^2$ valence-shell electron configuration (Figure 9.22a). Then we promote one electron from the 2s orbital to the empty 2p orbital. Promotion raises the energy of the 2s electron; however, this gain in energy is balanced by a slight decrease in the energies of the three p orbitals. We now have four orbitals of equal energy on the carbon atom, each containing one electron.

These four orbitals have been *hybridized* (mixed and averaged) to make a set of four equivalent hybrid orbitals, each containing one electron. An atom's steric number always indicates how many of its orbitals must be mixed to generate the hybrid set, and the number of hybrid orbitals in a set is always equal to the number of atomic orbitals mixed. The carbon atom in methane has SN = 4, so four atomic orbitals on carbon are mixed to form four hybrid orbitals.

The four orbitals are called **sp³ hybrid orbitals** because they result from the mixing of one s and three p orbitals. Their energy level is the weighted average of the energies of the s and p orbitals that formed them. They are oriented 109.5° from one another and point toward the vertices of a tetrahedron (Figure 9.22b). Now four hydrogen atoms can form σ bonds with the sp³ hybridized orbitals on carbon and form a methane molecule that has the observed tetrahedral geometry.

The sp³ hybrid orbitals in Figure 9.22(b) are shown as having one lobe each, but in fact every sp³ hybrid orbital consists of a major lobe and a minor lobe (Figure 9.23). Because the minor lobe is not involved in bonding, we ignore it in this chapter. It becomes important in the discussion of chemical reactions that are beyond the scope of this book. Also, to better show the orientation of bonds formed by hybrid orbitals, we elongate the major lobe in illustrations.

According to valence bond theory, any atom with a set of four equivalent sp³ hybrid orbitals has a tetrahedral orientation of its valence electrons. This includes atoms in which one or more hybrid orbitals are filled before any bonding takes place. For example, sp³ hybridization of the valence electrons on the nitrogen

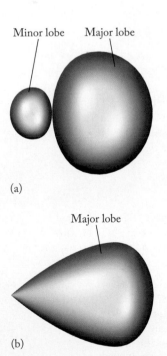

FIGURE 9.23 (a) The sp³ hybrid orbitals each consist of a major and minor lobe. (b) Because the minor lobes are not involved in orbital overlap leading to bond formation, they have been omitted from the orbital models in this book. The major lobes have also been elongated in illustrations to better show bond angles and orientations.

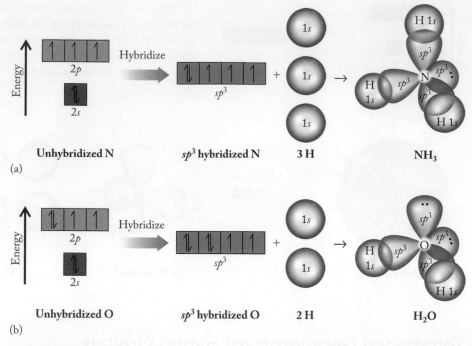

FIGURE 9.24 A set of four sp^3 hybrid orbitals point toward the vertices of a tetrahedron. (a) The nonbonding electron pair of the N in NH_3 occupies one of the four hybrid orbitals, giving the molecule a trigonal pyramidal molecular geometry. (b) The O atom in H_2O is also sp^3 hybridized. Only two of the four orbitals contain bonding pairs of electrons, giving H_2O a bent molecular geometry.

atom in ammonia produces three hybrid orbitals that are half-filled and one hybrid orbital that is filled (Figure 9.24a). The three half-filled orbitals form the three σ bonds in ammonia by overlapping with the $1s$ orbitals of three hydrogen atoms. The one filled hybrid orbital of N contains the lone pair of electrons. Similarly, the oxygen atom in water has four sp^3 hybrid orbitals in its valence shell, two filled before any bonding takes place and two half-filled and available for bond formation (Figure 9.24b). The two half-filled orbitals overlap with $1s$ orbitals from hydrogen atoms and form the two σ bonds in H_2O. Thus a carbon atom forming four σ bonds, a nitrogen atom with three σ bonds and one lone pair of electrons, and an oxygen atom with two σ bonds and two lone pairs of electrons all have SN = 4 and are all sp^3 hybridized atoms. As we will see, the steric number of an atom and its hybridization are closely related in other electron-pair geometries as well.

Trigonal Planar Geometry: sp^2 Hybrid Orbitals

We saw in Sample Exercise 9.1 that formaldehyde has a trigonal planar molecular geometry (Figure 9.25). A hybridization scheme other than sp^3 must be used to generate an orbital array with this molecular geometry. The VSEPR model defines SN = 3 for the carbon atom in a molecule of formaldehyde, so three atomic orbitals must be mixed to make three hybrid orbitals. To produce a trigonal planar geometry, the $2s$ orbital on carbon is mixed with two of the carbon $2p$ orbitals, and the third $2p$ orbital is left unhybridized (Figure 9.26a). This is the common hybridization scheme for all carbonyl (C=O) groups in organic molecules.

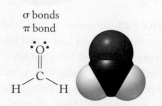

σ bonds
π bond

FIGURE 9.25 Formaldehyde has three σ bonds and one π bond.

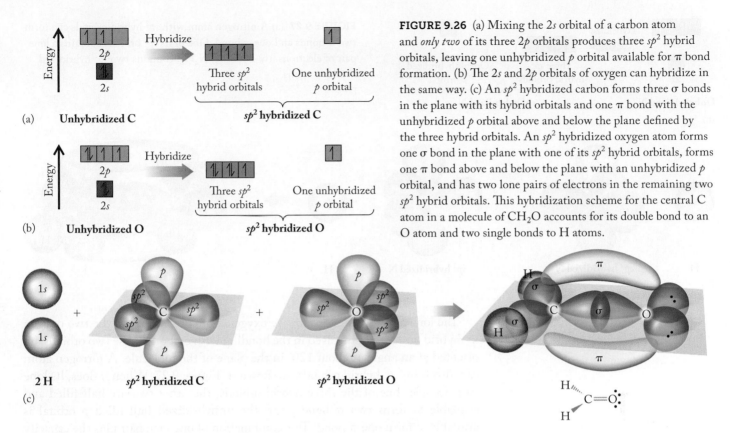

FIGURE 9.26 (a) Mixing the $2s$ orbital of a carbon atom and *only two* of its three $2p$ orbitals produces three sp^2 hybrid orbitals, leaving one unhybridized p orbital available for π bond formation. (b) The $2s$ and $2p$ orbitals of oxygen can hybridize in the same way. (c) An sp^2 hybridized carbon forms three σ bonds in the plane with its hybrid orbitals and one π bond with the unhybridized p orbital above and below the plane defined by the three hybrid orbitals. An sp^2 hybridized oxygen atom forms one σ bond in the plane with one of its sp^2 hybrid orbitals, forms one π bond above and below the plane with an unhybridized p orbital, and has two lone pairs of electrons in the remaining two sp^2 hybrid orbitals. This hybridization scheme for the central C atom in a molecule of CH_2O accounts for its double bond to an O atom and two single bonds to H atoms.

Mixing and averaging one s and two p orbitals generates three hybrid orbitals called sp^2 **hybrid orbitals**. The energy level of sp^2 orbitals is slightly lower than that of sp^3 orbitals because only two p orbitals are mixed with one s orbital in a set of sp^2 orbitals. The orbitals in an sp^2 hybridized atom all lie in the same plane and are 120° apart. The two lobes of the unhybridized p orbital lie above and below the plane of the triangle defined by the sp^2 hybrid orbitals. An sp^2 hybridized carbon atom forms three σ bonds with its three hybridized orbitals.

Valence electrons in s and p orbitals in other atoms can also be sp^2 hybridized. To complete the valence bond picture of formaldehyde or any carbonyl-containing compound, the oxygen atom must be sp^2 hybridized as well (Figure 9.26b). That both the carbon and the oxygen atoms in formaldehyde are hybridized points out another difference between VSEPR, which considers only the central atom in a molecule, and valence bond theory, which considers all the atoms in the molecule.

The valence bond view of the bonding in formaldehyde shows the $1s$ orbitals of two hydrogen atoms overlapping with two carbon sp^2 hybrid orbitals and the third carbon sp^2 hybrid orbital overlapping with one oxygen sp^2 hybrid orbital (Figure 9.26c). These overlapping orbitals constitute the σ-bonding framework of the molecule. The unhybridized carbon $2p$ orbital is parallel to the unhybridized oxygen $2p$ orbital, and the overlap of these two orbitals above and below the plane of the σ-bonding framework forms a **pi (π) bond**. The widths of the lobes on the p orbitals and the distances between atoms are not drawn to scale in this and similar figures. The lobes are actually much closer together than they appear, and they do overlap.

Pi bonds have their greatest electron density above and below the internuclear axis (or in front of and in back of the internuclear axis). They can be formed only by the overlap of partially filled p orbitals that are parallel to each other and perpendicular to the σ bond joining the atoms.

sp^2 hybrid orbitals three hybrid orbitals in a trigonal planar orientation, formed by mixing one s and two p atomic orbitals.

pi (π) bond a covalent bond in which electron density is greatest around—not along—the bonding axis.

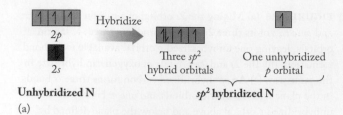

Unhybridized N

(a)

FIGURE 9.27 (a) A nitrogen atom with sp^2 hybrid orbitals can form two σ bonds and one π bond. One of its sp^2 orbitals contains a lone pair of electrons. (b) Diazene, N_2H_2, contains two sp^2 hybridized nitrogen atoms.

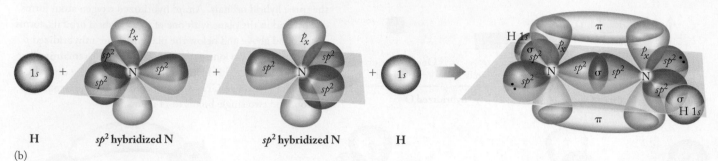

(b)

The lone pairs of electrons on the oxygen atom are located in the two oxygen sp^2 hybrid orbitals not involved in the bonding with carbon. These two orbitals are oriented at an angle of about 120° in the plane of the molecule. A nitrogen atom can also form sp^2 hybrid orbitals, as shown in Figure 9.27. When it does, its lone pair occupies one of the three hybrid orbitals, the other two are half-filled and available to form two σ bonds, and the unhybridized half-filled p orbital is available to form one π bond. This combination of one lone pair plus the capacity to form two σ bonds to two other atoms gives N a steric number of 3, the same SN value as sp^2 hybridized C and O atoms. This link to SN = 3 means that central atoms with trigonal planar electron-pair geometry are sp^2 hybridized.

Linear Geometry: *sp* Hybrid Orbitals

FIGURE 9.28 (a) Mixing one $2s$ orbital and one $2p$ orbital on a carbon atom creates two sp hybrid orbitals and leaves the carbon atom with two unhybridized $2p$ orbitals. (b) Two sp hybridized carbon atoms each bond to one hydrogen atom and to the other carbon atom via σ bonds to form the linear molecule C_2H_2. The two unhybridized p orbitals overlap above and below the plane to form one π bond and in front of and in back of the plane to form a second π bond.

We have seen hybrid orbitals generated by mixing one s orbital with three p orbitals and by mixing one s orbital with two p orbitals. The remaining possibility, one s orbital with one p orbital, forms the final set of important hybrid orbitals made from s and p atomic orbitals.

In the linear alkyne acetylene, C_2H_2, both carbon atoms have SN = 2 according to VSEPR theory. Because steric number indicates the number of hybrid orbitals that must be created, two atomic orbitals mix to make two hybrid orbitals on each carbon atom.

The results of mixing one $2s$ and one $2p$ orbital on carbon and leaving the other two $2p$ orbitals

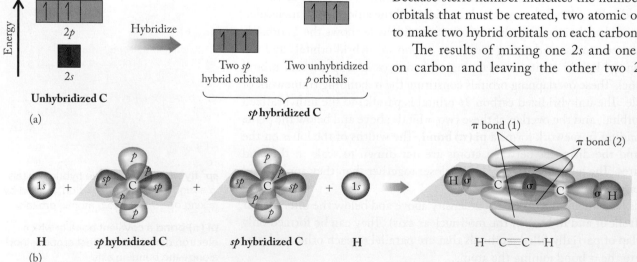

(b)

unhybridized are shown in Figure 9.28(a). The two *sp* **hybrid orbitals** have major lobes that are on opposite sides of the carbon atom. One set of the unhybridized *p* orbitals forms lobes above and below the axis of the *sp* hybrid orbitals; the second set is in front and in back of the plane of the hybrid orbitals.

The valence bond view of C_2H_2 shows a σ-bonding framework in which a hydrogen 1*s* orbital overlaps with one *sp* hybrid orbital on each carbon atom. The atom's other *sp* hybrid orbital overlaps with one *sp* hybrid orbital on the other carbon atom (Figure 9.28b). This arrangement brings the two sets of unhybridized *p* orbitals on the two carbon atoms into parallel alignment with each other. They then overlap to form two π bonds between the two carbon atoms, so that the one σ bond and the two π bonds form the triple bond of HC≡CH. The steric number of the *sp* hybridized carbon atom in acetylene is 2 (bonds to two atoms and no lone pairs).

Figure 9.29 summarizes the differences in electron distributions for single, double, and triple bonds for the carbon atoms in alkanes, alkenes, and alkynes as described by valence bond theory. The structures of these molecules determine their physical and chemical properties, but "structure" in this sense refers to more than just the location of the atoms. The distribution of electrons in three-dimensional space also greatly influences these properties. Remember: for reactions to take place, particles must interact, and the location of the valence electrons in molecules influences which bonds will break in reactions and which new bonds will form. According to valence bond theory, the electron density in the carbon–carbon σ bonds in alkanes lies on the axis directly between the two atomic centers (Figure 9.29a). In contrast, the electron density in the π bonds in alkenes and alkynes is largest not between the two carbon atoms, but rather above, below, and beside the bonding axis (Figure 9.29b,c). These electrons in π bonds are one reason why alkenes and alkynes are more reactive than alkanes with only σ bonds.

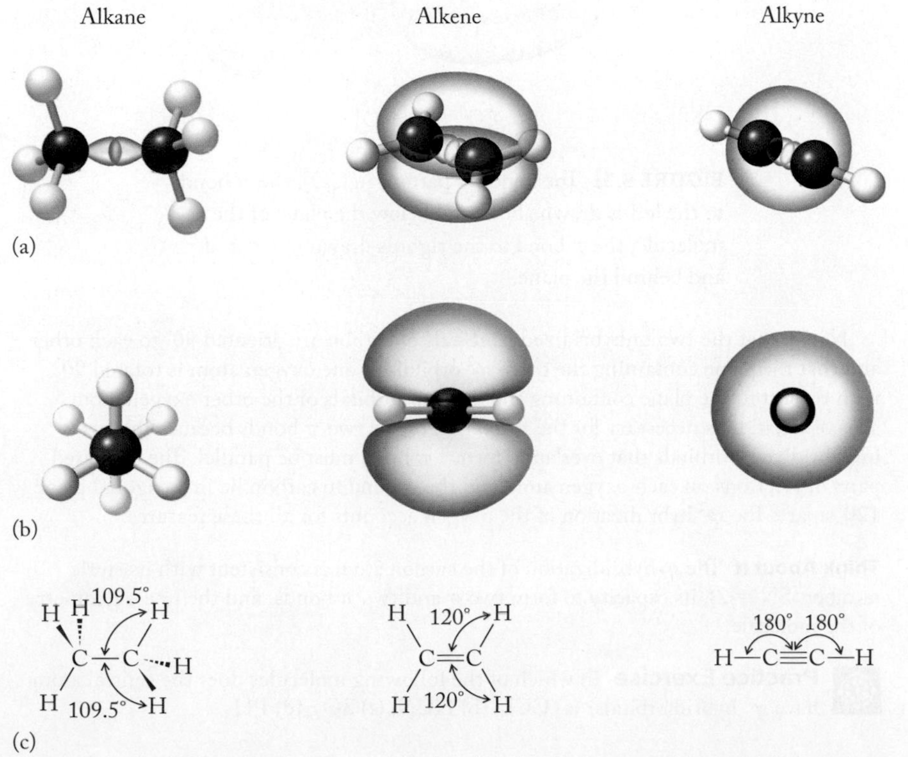

sp hybrid orbitals two hybrid orbitals oriented 180° from one another, formed by mixing one *s* and one *p* orbital.

FIGURE 9.29 Carbon–carbon bonding in alkanes, alkenes, and alkynes. The electrons in the carbon–carbon bonds in alkanes are in σ bonds; the characteristic structural feature of alkenes and alkynes is the presence of additional electrons in π bonds. To simplify these drawings, orbitals are drawn only for the carbon–carbon bonds. (a) The electron distribution in a C—C single bond, a C=C double bond, and a C≡C triple bond. (b) View of these electron distributions looking down the carbon–carbon bond axis. (c) Idealized H–C–C bond angles for single, double, and triple bonds.

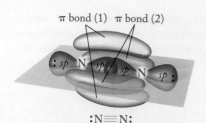

FIGURE 9.30 Valence bond representation of a N_2 molecule and its sp hybrid N atoms.

The link between SN = 2 and sp hybridization applies to atoms other than carbon—for example, the nitrogen atoms in N_2. A lone pair occupies one of the two sp orbitals on each nitrogen atom and the other sp hybrid orbital forms the σ bond (Figure 9.30).

SAMPLE EXERCISE 9.6 Describing Bonding in a Molecule   **LO4**

Use valence bond theory to account for the linear molecular geometry of CO_2, to determine the hybridization of carbon and oxygen in this molecule, and to describe the orbitals that overlap to form the bonds. Draw the molecule, showing the orbitals that overlap to form the bonds.

Collect and Organize We know that CO_2 is linear and that the Lewis structure of the molecule is

$$:\!\ddot{O}\!=\!C\!=\!\ddot{O}\!:$$

Analyze Each double bond is composed of one σ bond and one π bond. Therefore the carbon atom forms two σ bonds and two π bonds. It has no lone pairs and so SN = 2. Each oxygen atom forms one σ bond and one π bond.

Solve The carbon atom must be sp hybridized because half-filled sp hybrid orbitals have the capacity to form two σ bonds, leaving two half-filled unhybridized p orbitals available to form two π bonds (Figure 9.31). The lobes of sp hybrid orbitals are at an angle of 180°, which is consistent with the linear molecular geometry of CO_2. The oxygen atoms must be sp^2 hybridized because that hybridization leaves them with one unhybridized p orbital to form a π bond. The σ bonds form when each sp orbital on the carbon atom overlaps with one sp^2 orbital on an oxygen atom.

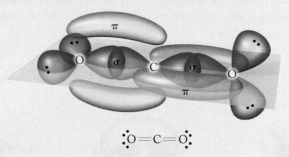

$$:\!\ddot{O}\!=\!C\!=\!\ddot{O}\!:$$

FIGURE 9.31 The bonding pattern in CO_2: the π bond to the left is drawn above and below the plane of the molecule; the π bond to the right is drawn in front of and behind the plane.

Notice that the two unhybridized p orbitals of carbon are oriented 90° to each other and that the plane containing the three sp^2 orbitals of one oxygen atom is rotated 90° with respect to the plane containing the three sp^2 orbitals of the other oxygen atom. This orientation is necessary for the formation of the two π bonds because the two unhybridized p orbitals that overlap to form a π bond must be parallel. The unshared pairs of electrons on each oxygen atom and the σ bond to carbon lie in a trigonal plane 120° apart. The sp^2 hybridization of the oxygen accounts for all these features.

Think About It The sp hybridization of the carbon atom is consistent with its steric number (SN = 2), its capacity to form two σ and two π bonds, and the linear geometry of the molecule.

 Practice Exercise In which of the following molecules does the central atom have sp^3 hybrid orbitals? (a) CCl_4; (b) HCN; (c) SO_2; (d) PH_3

sp^3d^2 hybrid orbitals six equivalent hybrid orbitals with lobes pointing toward the vertices of an octahedron, formed by mixing one s orbital, three p orbitals, and two d orbitals from the same shell.

sp^3d hybrid orbitals five equivalent hybrid orbitals with lobes pointing toward the vertices of a trigonal bipyramid, formed by mixing one s orbital, three p orbitals, and one d orbital from the same shell.

CONCEPT TEST

Why can't a carbon atom with sp^3 hybrid orbitals form π bonds?

Octahedral and Trigonal Bipyramidal Geometries: sp^3d^2 and sp^3d Hybrid Orbitals

Valence bond theory can also account for the molecular geometries of molecules with central atoms that have more than eight valence electrons. For example, the central sulfur atom in a molecule of SF_6 bonds to six fluorine atoms. As we discussed in Chapter 8, one way to explain expanded octets in atoms with Z > 12 is to assume that valence-shell d orbitals are involved in bond formation. Valence bond theory provides a way to describe an expanded octet by including these d orbitals in hybridization schemes. To form six σ bonds to six fluorine atoms, six atomic orbitals on sulfur—one 3s orbital, three 3p orbitals, and two 3d orbitals—hybridize, producing six equivalent **sp^3d^2 hybrid orbitals** and an octahedral molecular geometry (Figure 9.32a).

Other molecules, such as phosphorus pentafluoride (PF_5), have a trigonal bipyramidal geometry, and the central phosphorus atom shares 10 valence electrons. Five atomic orbitals—one 3s orbital, three 3p orbitals, and one 3d orbital—can be hybridized to produce five equivalent **sp^3d hybrid orbitals** with lobes that point toward the vertices of a trigonal bipyramid (Figure 9.32b). A summary of the shapes associated with all of the hybridization schemes we have discussed so far is given in Table 9.3.

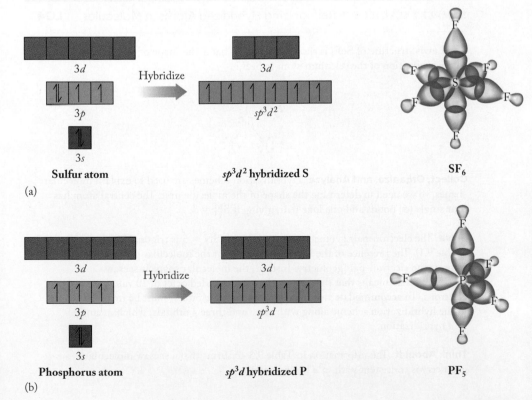

FIGURE 9.32 Hybrid orbitals can be generated by combining d orbitals with s and p orbitals in atoms with expanded octets. (a) One 3s orbital, three 3p orbitals, and two 3d orbitals mix on the central sulfur atom in SF_6 to form six sp^3d^2 hybrid orbitals. (b) One 3s, three 3p orbitals, and one 3d orbital mix on the central phosphorus atom in PF_5 to form five sp^3d hybrid orbitals. For simplicity, only the fluorine orbitals that overlap with the hybrid orbitals are shown.

TABLE 9.3 Hybrid Orbitals Based on Steric Number

Steric Number	Hybridization	Orientation of Hybrid Orbitals (purple)	Number of σ Bonds	Molecular Geometries	Theoretical Angles between Hybrid Orbitals
2	sp		2	Linear	180°
3	sp^2		3 2	Trigonal planar Bent	120°
4	sp^3		4 3 2	Tetrahedral Trigonal pyramidal Bent	109.5°
5	sp^3d		5 4 3 2	Trigonal bipyramidal Seesaw T-shaped Linear	90°, 120°, 180°
6	sp^3d^2		6 5 4	Octahedral Square pyramidal Square planar	90°, 180°

CHEMTOUR
Hybridization

SAMPLE EXERCISE 9.7 Recognizing Hybridized Atoms in Molecules   **LO4**

The Lewis structure of SeF_4 is shown at right. What is the shape of the molecule and the hybridization of the selenium atom in SeF_4?

Collect, Organize, and Analyze Hybridization schemes are used to explain molecular shapes, so we need to determine the shape of the molecule first. The central atom has four single (σ) bonds and one lone pair, giving it SN = 5.

Solve The electron-pair geometry associated with SN = 5 is trigonal bipyramidal (Table 9.1). The presence of the lone pair means that the molecular geometry is not the same as the electron-pair geometry. Instead, the molecular shape is seesaw.

  SN = 5 also means that the Se atom has an expanded octet of 10 valence electrons. To accommodate this many electrons, one d orbital must be involved in the hybridization scheme along with one s and three p orbitals, which means sp^3d hybridization.

Think About It The information in Table 9.3 confirms that a seesaw molecular geometry is consistent with sp^3d hybridization.

Practice Exercise What is the hybridization of the iodine atom in IF₅ that is consistent with the Lewis structure shown here?

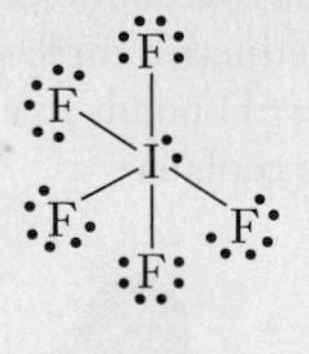

molecular recognition the process by which molecules interact with other molecules in living tissues to produce a biological effect.

9.5 Shape and Interactions with Large Molecules

Most of the molecules we have considered so far have small molar masses or a single central atom bonded to two or more other atoms. It is important to learn to apply the skills we have developed to larger molecules, however, because molecular shape is an important factor in determining the physical, chemical, and biological properties of all substances.

Living things respond to molecules that interact with regions in their tissues called *receptors* or *active sites*. The process by which these molecules and sites interact is known as **molecular recognition**. This recognition does not usually involve covalent bond formation. Instead, these noncovalent interactions require that the biologically active molecules and the receptors that respond to them fit tightly together, which means that they must have complementary three-dimensional shapes. An example of a biological effect caused by molecular recognition is the process by which produce such as green tomatoes ripens. Tomatoes ripen faster when stored in a paper or plastic bag instead of sitting on a kitchen counter because they give off ethylene gas as they ripen, and this gas accelerates the ripening process when it is trapped in a bag with the tomatoes.

The molecular structure of ethylene is shown in Figure 9.33. Both carbon atoms are bonded to three atoms and have no lone pairs of electrons. Therefore SN = 3. This means that the geometry around each carbon atom is trigonal planar and that the carbon atoms are both sp^2 hybridized (see Table 9.3). Two of these orbitals form σ bonds with hydrogen 1s orbitals. The third forms the C=C σ bond. The C=C π bond is formed by overlap of the unhybridized $2p$ orbitals on the carbon atoms. Taken together, the two trigonal planar carbon atoms produce an overall planar geometry for ethylene, which means that all six atoms lie in the same plane.

FIGURE 9.33 Ethylene molecules contain two carbon atoms that are at the centers of two overlapping triangular planes that are also coplanar. This combination means that all of the atoms are in the same plane.

SAMPLE EXERCISE 9.8 Comparing Structures by Using Valence Bond Theory **LO4**

Use valence bond theory to describe the bonding and molecular geometry of ethane, CH₃—CH₃, and compare them with the bonding and molecular geometry of ethylene.

Collect, Organize, and Analyze We are given the molecular formula of ethane. We can use its Lewis structure and VSEPR to determine the molecular geometry and hybridization of its carbon atoms. We can then compare ethane's atom locations and electron distributions to ethylene, which we saw in Figure 9.33.

Solve Each carbon atom in ethane has four single bonds and no lone pairs, which means SN = 4 for both carbon atoms, tetrahedral geometry around both, and sp^3 hybridization of both. Comparing the overall molecular structures of ethane and ethylene (Figure 9.34), we see the distinctive three-dimensionality of tetrahedral environments: only two of the six C—H bonds in ethane are in the plane of the page, whereas all the bonds in ethylene are coplanar.

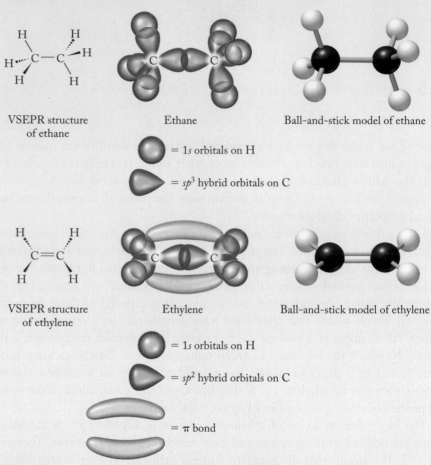

FIGURE 9.34 Molecular structures of ethane and ethylene.

Think About It Remember the analogy at the beginning of the chapter about how molecules fit receptors like hands fit gloves. It is clear from the structures that ethylene and ethane would fit into very different gloves.

Practice Exercise Diazene (N_2H_2) and hydrazine (NH_2NH_2) are reactive nitrogen compounds. Use valence bond theory to compare the bonding in these two molecules and to describe the differences in their molecular structures.

As Sample Exercise 9.8 illustrates, ethane (an alkane) and ethylene (an alkene) have similar formulas and molar masses, but very different molecular geometries. Ethane does not trigger the ripening process because its shape does not allow it to fit the receptor in plant tissue that binds the planar molecule ethylene. This structural difference is important in larger molecules, where planar regions caused by the presence of sp^2 hybridized carbon atoms and double bonds produce an overall molecular shape that is very different from the shapes associated with sp^3 hybridized carbon atoms.

Drawing Larger Molecules

Because we will see more examples of larger molecules (that is, molecules with more than just one central atom) in this chapter and throughout the text, we need to learn some of the ways that chemists have developed to represent such structures in addition to Lewis structures. Figure 9.35(a) shows the Lewis structures for both pentane and butanone. Pentane is a member of the alkane family that we discussed in Chapter 5, whereas butanone contains the functional group known as a **ketone** because the carbonyl group is bonded to a carbon atom on each side of the carbonyl carbon. A Lewis structure shows all the bonds in the molecule as well as all the lone pairs on the atoms, but it does not convey the three-dimensional shape of the molecule. Structural formulas of organic molecules that show all the bonds with lines, but omit lone pairs, are called *Kekulé structures* (Figure 9.35b) after August Kekulé (1829–1896), who first used this method for illustrating molecules. Because alkanes do not have lone pairs on any atoms, there is no difference between the Lewis structure and the Kekulé structure for pentane, but there is for butanone because the lone pairs on the oxygen atom are omitted in the Kekulé structure for butanone.

As you might imagine, it soon gets tedious to write Lewis or Kekulé structures for larger organic molecules. For this reason, chemists use even shorter notations called *condensed structures* (Figure 9.35c) to represent molecules. Condensed structures do not show the individual bonds between atoms as Lewis structures do. They use subscripts to indicate the number of times a particular subgroup is repeated. For example, the condensed structure of pentane can be written as $CH_3CH_2CH_2CH_2CH_3$, or as $CH_3(CH_2)_3CH_3$. The numerical subscript after the CH_2 group in parentheses means that three of these groups connect the two terminal $-CH_3$ groups in this compound. Note that the subscript indicating the number of CH_2 groups comes after the closing parenthesis. The subscript 2 inside the parentheses is for the two H atoms bonded to each C atom of each CH_2 group.

The most minimal notation, shown in Figure 9.35(d), is the *carbon-skeleton structure*, which has no element symbols for carbon and hydrogen atoms. Atoms other

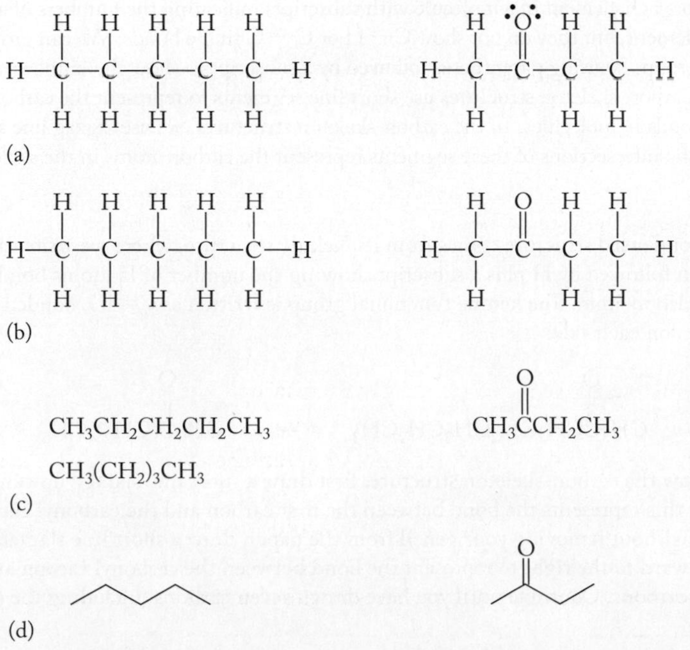

ketone organic molecule containing a carbonyl group bonded to a carbon atom on each side of the carbonyl carbon.

CONNECTION Kekulé structures are the same as the structural formulas we have been drawing throughout the text since Chapter 1.

CONNECTION We saw methyl groups ($-CH_3$) and methylene groups ($-CH_2-$) when we investigated alkanes in Section 5.9. We also classified alkanes as straight-chain or branched-chain hydrocarbons.

(a)

(b)

(c) $CH_3CH_2CH_2CH_2CH_3$

$CH_3(CH_2)_3CH_3$

$CH_3\overset{\text{O}}{\overset{\|}{C}}CH_2CH_3$

(d)

FIGURE 9.35 Several representations of the molecular structures of pentane and butanone: (a) Lewis structures; (b) Kekulé structures; (c) condensed structures; (d) carbon-skeleton structures.

CH₃CH₂CH₂CH₂CH₂CH₂CH₂CH₃

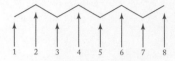

or CH₃(CH₂)₆CH₃

(a) Condensed structural formula

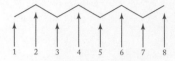

(b) Carbon-skeleton structure

FIGURE 9.36 When converting from (a) a condensed structure to (b) a carbon-skeleton structure, the carbon atoms are represented by junctions between lines, and each junction is assumed to have enough H atoms to give that carbon atom four bonds.

than C and H are shown in carbon-skeleton structures: for example, the oxygen atom that is part of the ketone function group is drawn in butanone. Figure 9.36 shows how to create a carbon-skeleton structure for octane, C_8H_{18}. Short line segments are drawn at angles to one another in a "zigzag" manner, and each line segment symbolizes one carbon–carbon bond in the molecule. The angles represent the bond angle between the two carbon atoms (109.5° for sp^3 hybridized carbon atoms in alkanes). Each end of the zigzag line is a –CH_3 group, and each vertex formed by the intersection of two line segments is either a –CH_2– group (as is the case in pentane) or a functional group (as is the case for butanone). Carbon-skeleton structures are assumed to contain the appropriate number of hydrogen atoms so that each carbon atom has a steric number of four. Hydrogen atoms are not shown in a carbon-skeleton structure because all carbon atoms are known to make four bonds, and any bonds not shown are understood to be C—H bonds.

SAMPLE EXERCISE 9.9 Drawing Condensed and Carbon-Skeleton Structures   **LO5**

Our noses are molecular detectors in that our sense of smell is due to molecular recognition by receptors in our nasal tissues. The Kekulé structure of one compound in Gorgonzola cheese is shown in Figure 9.37. (a) Write a condensed structure and (b) draw the carbon-skeleton structure for this ketone.

FIGURE 9.37 Heptane-2-one, a compound found in blue cheese.

Collect, Organize, and Analyze Figure 9.35 summarizes the differences between Kekulé, condensed, and carbon-skeleton structures. Condensed structures show the symbol for each element in a molecule with subscripts indicating the numbers of atoms of each element, but they do not show C—H or C—C single bonds. We can group all –CH_2– groups by using parentheses followed by a subscript to show the number of –CH_2– groups. Carbon-skeleton structures use short line segments to represent the carbon–carbon bonds in molecules. In the carbon-skeleton structures, we use zigzag line segments because the intersections of these segments represent the carbon atoms in the skeleton.

Solve

a. The condensed structure comes from its Kekulé structure, where we write each carbon followed by H plus a subscript showing the number of H atoms bonded to that carbon atom. The ketone functional group is written as C═O, bonded to a carbon on each side:

$$CH_3\overset{O}{\overset{\|}{C}}CH_2CH_2CH_2CH_2CH_3 \quad \text{or} \quad CH_3\overset{O}{\overset{\|}{C}}(CH_2)_4CH_3$$

b. To draw the carbon-skeleton structure, first draw a short line slanted upward to the right; this represents the bond between the first carbon and the carbonyl carbon. Then without removing your pencil from the paper, draw a short line slanted downward to the right to represent the bond between the carbonyl carbon and the third carbons. Continue until you have drawn seven carbons (including the carbonyl carbon). There must be two $CH_3–$ groups at the ends with a ketone functional group and four $–CH_2–$ groups between them:

Think About It Condensed structures and carbon-skeleton structures must both contain the same number of atoms and the same number of bonds, even though the bonds are not shown in the condensed structure and not all of the atoms are shown in the carbon-skeleton structure.

 Practice Exercise Draw the carbon-skeleton structure of hexane, $CH_3(CH_2)_4CH_3$, and a condensed structure of heptane:

Heptane

Molecules with More than One Functional Group

Now let's consider a biologically active molecule with three "central" atoms and more than one functional group. Acrolein is one of the components of barbeque smoke that contributes to the distinctive odor of a cookout. It is also a possible cancer-causing compound. The Lewis structure of acrolein is

Each molecule contains a C=O double bond and a C=C double bond, so it has both alkene and aldehyde functional groups. Both double bonds are formed by trigonal planar, sp^2 hybridized carbon atoms.

The pattern of alternating single and double bonds means that the electrons in the π bonds can be delocalized over the three carbon atoms and the oxygen atom. Delocalization can occur both when the atoms involved are all carbon atoms and when atoms of different elements are involved, as in acrolein.

Benzene, C_6H_6, is another molecule in barbeque smoke. As we saw in Chapter 8, each benzene molecule is a hexagon of six carbon atoms, each bonded to one hydrogen atom. Benzene also has three C=C bonds. Resonance structures for benzene are shown in Figure 9.38, along with a view of the π bonds located above and below the plane of the ring. Each carbon in benzene has a trigonal planar geometry and forms sp^2 hybrid orbitals. The carbon ring is made of σ bonds formed by overlapping sp^2 hybrid orbitals on adjacent carbon atoms. The C—H bonds are formed by the overlap of carbon sp^2 hybrid orbitals with hydrogen $1s$ orbitals.

The two resonance forms of benzene shown in Figure 9.38(a) correspond to shifts in the locations of the π bonds formed by the overlap of carbon $2p$ orbitals. However, because all the carbon $2p$ orbitals are identical, they are all equally likely to overlap with their neighbors, so the electrons in the π bonds are actually delocalized over all six carbon atoms rather than fixed between alternating pairs of carbon atoms,

 **CONNECTION** We defined delocalized electrons in Chapter 8 with respect to metallic bonds, but delocalization occurs in molecules and ions as well.

CHEMTOUR Structure of Benzene

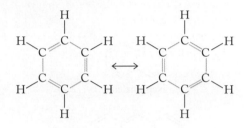

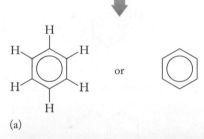

(a)

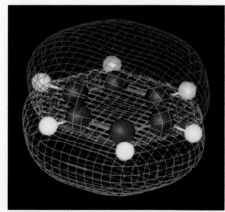

(b)

FIGURE 9.38 (a) Resonance between the two Lewis structures of benzene leads to complete delocalization of the π bonds around the benzene ring. (b) A computer-generated view of benzene's σ bonds in the plane of the ring and its delocalized π bonds above and below the ring.

aromatic compound a cyclic, planar compound with delocalized π (pi) electrons above and below the plane of the molecule.

chirality property of a molecule that is not superimposable on its mirror image.

 CONNECTION We mentioned aromatic compounds in Chapter 8 when we discussed resonance in Lewis structures.

as depicted in the computer-generated view in Figure 9.38(b). As noted in Section 8.4, the presence of these delocalized π electrons is often represented by a circle drawn in the middle of the hexagon of carbon atoms in the structural formula, corresponding to continuous rings of π electrons above and below the plane of the molecule. The carbon-skeleton structure of benzene is a simple hexagon with a circle in it.

CONCEPT TEST

In which of the molecules and polyatomic ions in Figure 9.39 are the electrons in the π bonds delocalized?

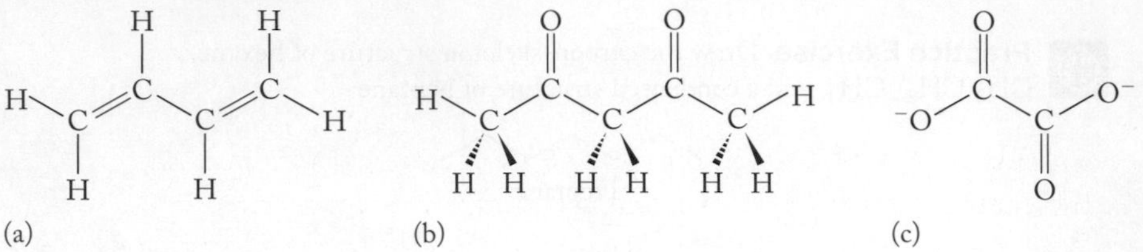

FIGURE 9.39 Organic compounds with double bonds: (a) butadiene, used to make polymers; (b) 2,4-pentanedione, a reactive organic compound used in synthesis; and (c) the oxalate ion, present in spinach.

Naphthalene

Anthracene

Phenanthrene

Benzo[*a*]pyrene

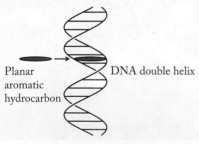

Planar aromatic hydrocarbon → DNA double helix

Intercalation of PAH in DNA

FIGURE 9.40 The molecules of these polycyclic aromatic hydrocarbons consist of fused benzene rings whose π bonds are delocalized over all the rings in each molecule. Molecules with this shape can slip between the strands of DNA and disrupt cell replication, which can lead to cell death or induce malignancy.

The molecular structures of many other compounds in addition to benzene contain carbon rings with delocalized π electrons above and below the plane of the ring. They are called **aromatic compounds**. An important class of these compounds, known as *polycyclic aromatic hydrocarbons* (PAHs), consists of molecules containing several benzene rings joined together (Figure 9.40). PAHs are formed any time coal, oil, gas, and most hydrocarbon fuels are burned, and they are also found in cigarette smoke. In 2004 they were discovered in interstellar space.

The shape of PAH molecules gives rise to a particular health hazard. After we inhale or ingest them, some PAHs may bind to the DNA in our cells in a process called *intercalation*. Because PAHs are flat, they can slide into the double helix that DNA forms, as shown in Figure 9.40. Once there, they may alter or prevent DNA replication and thereby damage or kill cells. Intercalation in DNA is one step in the process by which PAHs induce cancer.

The planarity of ethylene and aromatic compounds as described by VSEPR and valence bond theories plays a key role in determining the behavior of such molecules in biological systems. The shape of these molecules matters because it influences their interaction with other molecules, which in turn determines their biological activity.

9.6 Chirality and Molecular Recognition

Before ending our exploration of molecular shapes, we need to address the subject of handedness introduced at the beginning of this chapter. There we described how two molecules with the same molecular formula and Lewis structure can interact differently with receptors in our nasal membranes. As a result, one produces the smell of caraway seeds, and the other, spearmint leaves. You may find these different odors surprising when you consider the molecular structures of these two compounds (Figure 9.41). At first glance they may seem identical, but look closely. Notice in particular the bonding pattern around the carbon atom at the bottom of the ring. The H atom bonded directly to the bottom carbon atom is

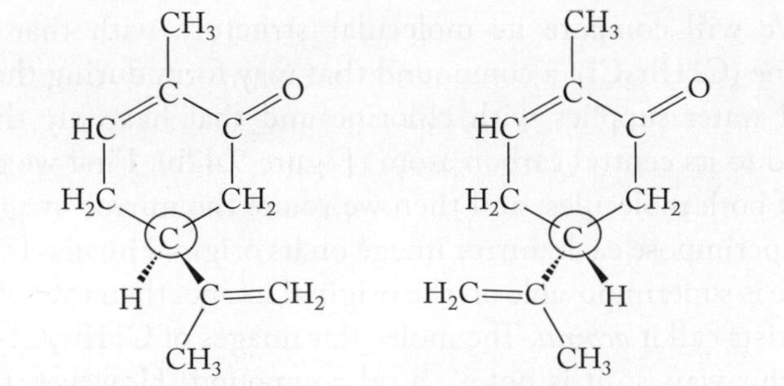

CH₃
C
HC   C═O
H₂C   CH₂
C
H   C═CH₂
CH₃

(a) (+)-Carvone
(caraway)

CH₃
C
HC   C═O
H₂C   CH₂
C
H₂C═C   H
CH₃

(b) (−)-Carvone
(spearmint)

FIGURE 9.41 The distinctive aromas of (a) caraway and (b) spearmint are primarily due to two compounds with nearly identical molecular structures. The only difference between the two is the orientation of the two groups attached to the carbon atoms highlighted with the red circles. Note that the hydrogen atom is in back of the plane of the paper and down in caraway (a) but is in front of the plane of the paper and down in spearmint (b).

on the front side of the ring in the spearmint compound, but it is on the back side in the caraway compound. This minor difference in bond orientation creates a difference in molecular shape that is easily recognized by receptors in our noses.

The structures in Figure 9.41 are called *optical isomers*. The term "optical" refers to the ways these compounds interact with a special kind of light called *plane-polarized* light. We explore this topic in more detail in Chapter 20, but it is enough now for you to know that when plane-polarized light passes through a solution of the caraway compound, the light twists in one direction. However, when the light passes through a solution of the spearmint compound, it twists in the opposite direction.

Optical isomerism has another name: **chirality**. Many molecules of biological importance, including the proteins and carbohydrates that we consume each day, are composed of chiral compounds. Chirality comes from the Greek word *chier*, meaning "hand," and is quite correctly called "handedness." Although several features within molecules can lead to chirality, the most common is the presence of a carbon atom that has four different atoms or groups of atoms attached to it.

To see how chirality works, let's look at a compound that contains a central carbon atom bonded to four different atoms. The compound is bromochloro-fluoromethane (CHBrClF; Figure 9.42a), which is used in fire extinguishers on

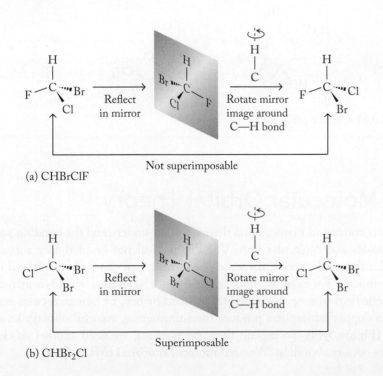

(a) CHBrClF — Reflect in mirror — Not superimposable — Rotate mirror image around C—H bond

(b) CHBr₂Cl — Reflect in mirror — Superimposable — Rotate mirror image around C—H bond

FIGURE 9.42 (a) A molecule of a chiral compound, such as CHBrClF, is not superimposable on its mirror image. (b) A molecule of a compound that is not chiral, such as CHBr₂Cl, is superimposable on its mirror image.

airplanes. We will compare its molecular structure with that of dibromo-chloromethane ($CHBr_2Cl$), a compound that may form during the purification of municipal water supplies with chlorine and that has only three different atoms bonded to its central carbon atom (Figure 9.42b). First we generate mirror images of both molecules, and then we rotate the mirror images 180° in an attempt to superimpose each mirror image on its original image. If the reflected, rotated image is superimposable on the original image, then the substance is not chiral. Scientists call it *achiral*. The molecular images of $CHBr_2Cl$ can be superimposed in this way, so it is not a chiral compound. However, the images of CHBrClF cannot be superimposed. For example, when we superimpose the F, C, and H atoms of the two images in Figure 9.42(a), the Br and Cl atoms are not aligned. This means that CHBrClF *is* a chiral compound.

Any structure like CHBrClF that has four different groups attached to an sp^3 hybridized carbon atom is chiral. It has two optical isomers that are mirror images of each other and are distinctly different compounds. You may be wondering where the chiral carbon atom is in the molecules of the caraway and spearmint compounds. If you guessed the circled carbon atoms, you were right. Two of the four different groups are easy to see: a hydrogen atom and the

$$\begin{array}{c} \quad\quad CH_2 \\ \quad\quad \| \\ -C \\ \quad\quad \backslash \\ \quad\quad CH_3 \end{array}$$

group. The other two "groups" on the bottom carbon atom are really the two sides of the ring. The left side contains a C=C double bond, and the right side contains a C=O double bond. These differences mean that the two sides are not equivalent, so the carbon atom is attached to four different groups. This makes it a chiral carbon atom, and the two compounds are optical isomers of each other.

CONCEPT TEST

Identify the molecules in Figure 9.43 that are chiral.

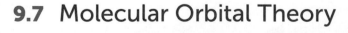

FIGURE 9.43 Organic compounds.

9.7 Molecular Orbital Theory

Lewis structures and valence bond theory help us understand the bonding patterns in molecules and molecular ions; VSEPR and valence bond theory account for molecular shapes. However, there are phenomena that cannot be explained by any of these models. For example, none of these models explains why O_2 is attracted to a magnetic field whereas N_2 is slightly repelled by one, or how molecules and ions in Earth's upper atmosphere produce the shimmering, colorful displays known as *auroras* (Figure 9.44). To explain these phenomena, we need another model that describes covalent bonding. We need **molecular orbital (MO) theory**.

FIGURE 9.44 Auroras are spectacular displays of color produced when the solar wind collides with Earth's upper atmosphere, producing excited-state atoms, ions, and molecules.

Like valence bond theory, molecular orbital theory invokes the mixing of atomic orbitals. In valence bond theory, the mixing results in hybrid atomic orbitals, whereas molecular orbital theory is based on the formation of **molecular orbitals**. A key difference between the two types of orbitals is that hybrid atomic orbitals are associated with a particular atom in a molecule, but molecular orbitals are spread out over two or more atoms in a molecule.

Molecular orbitals represent discrete energy states in molecules, just as atomic orbitals represent allowed energy states in single atoms. As with atomic orbitals, electrons fill the lowest-energy MOs first; higher-energy MOs are filled as more electrons are added. Electrons in molecules can move to higher-energy MOs when molecules absorb quanta of electromagnetic radiation. When the electrons return to lower-energy MOs, distinctive wavelengths of UV and visible radiation are emitted, including some of the shimmering colors in an aurora. The colors of an aurora are caused by collisions between atoms, molecules, and molecular ions in the atmosphere with electrons and positive particles in the solar wind that are attracted to Earth's magnetic poles. These collisions produce excited-state species. As they return to their ground states, these species emit characteristic colors of light. We discussed in Chapter 7 how excited-state atoms emit characteristic atomic spectra. In this section we use MO theory to explore how excited-state molecules and molecular ions, such as N_2 and N_2^+, do the same thing.

Unlike atomic orbitals, including hybrid atomic orbitals, MOs are not linked to single atoms but rather belong to all the atoms in a molecule. Thus any electron in a molecule could be anywhere in the molecule. This delocalized view of covalent bonding is particularly effective at describing the bonding in molecules such as benzene, in which the p orbital on each sp^2 hybridized carbon atom allows π bonds to spread out over more than one pair of atoms. The availability of electrons from other atoms also means that we do not have to involve a particular atom's d orbitals to explain expanded octets, as we sometimes need to in valence bond theory.

In MO theory, molecular orbitals are formed by combining atomic orbitals from each of the atoms in the molecule. In this book, we limit our discussion of MO theory to simple molecules in which MOs are formed from the atomic orbitals of only a few atoms. Some MOs have lobes of high electron density that lie *between* bonded pairs of atoms; they are called **bonding orbitals**. The energies of bonding MOs are lower than the energies of the atomic orbitals that combined to form them, so populating them with electrons (each MO can hold two electrons, just like an atomic orbital) stabilizes the molecule and contributes to the strength of the bonds holding its atoms together.

There are other MOs with lobes of high electron density that are not located between the bonded atoms. They are **antibonding orbitals** and have energies that are higher than the atomic orbitals that combined to form them. When electrons are in antibonding orbitals, they destabilize the molecule. As you might guess, if a molecule were to have the same number of electrons in its bonding and antibonding orbitals, then there would be no net energy holding the molecule together, and it would never have formed.

Another key point is that the *total number of molecular orbitals must match the number of atomic orbitals involved in forming them*. For example, combining two atomic orbitals, one from each of two atoms, produces one low-energy bonding orbital and one high-energy antibonding orbital. To further explore the distinction between bonding and antibonding molecular orbitals, let's apply MO theory to the simplest molecular compounds: hydrogen and helium.

CHEMTOUR
Molecular Orbitals

CONNECTION In Chapter 7 we discussed the role of atomic emission and absorption spectra in the development of quantum mechanics.

molecular orbital (MO) theory a bonding theory based on the mixing of atomic orbitals of similar shapes and energies to form molecular orbitals that extend across two or more atoms.

molecular orbital a region of characteristic shape and energy where electrons in a molecule are delocalized over two or more atoms in a molecule.

bonding orbital term in MO theory describing regions of increased electron density between nuclear centers that serve to hold atoms together in molecules.

antibonding orbital term in MO theory describing regions of electron density in a molecule that destabilize the molecule because they decrease the electron density between nuclear centers.

sigma (σ) molecular orbital in MO theory, a molecular orbital in which the greatest electron density is concentrated along an imaginary line drawn through the bonded atom centers.

molecular orbital diagram in MO theory, an energy-level diagram showing the relative energies and electron occupancy of the molecular orbitals for a molecule.

Molecular Orbitals of Hydrogen and Helium

According to MO theory, a hydrogen molecule is formed when the $1s$ atomic orbitals on two hydrogen atoms combine to form two molecular orbitals, as shown in Figure 9.45(a). Molecular orbital theory stipulates that mixing two atomic orbitals creates two molecular orbitals and that these two orbitals represent two different energy states (Figure 9.45b).

When two $1s$ atomic orbitals combine, the lower-energy bonding molecular orbital formed is oval and spans the two atomic centers. Its shape corresponds to enhanced electron density between the two atoms that donated their atomic orbitals. This enhanced electron density is a covalent bond. When two electrons occupy a bonding MO, a single bond is formed. When the region of highest density lies along the bond axis, as it does in the bonding MO in H_2, the MO is designated a **sigma (σ) molecular orbital** and the covalent bond is a σ bond. The bonding molecular orbital in H_2 is labeled σ_{1s} in Figure 9.45(a) because it is formed by mixing two $1s$ atomic orbitals.

The higher-energy (less stable) antibonding molecular orbital formed from two hydrogen atomic orbitals is designated σ_{1s}^* (pronounced "sigma star"). This antibonding orbital has two separate lobes of electron density and a region of zero electron density (a node) between the two hydrogen atoms, as shown in Figure 9.45(a).

Figure 9.45(b) is a **molecular orbital diagram**, analogous to an energy-level diagram for atomic orbitals. It shows that the σ_{1s} MO is lower in energy and therefore more stable than the $1s$ atomic orbitals by nearly the same amount that the σ_{1s}^* MO is higher in energy than the $1s$ atomic orbitals. Therefore the formation of the two MOs does not significantly change the total energy of the system. A hydrogen molecule has two valence electrons, one from each H atom, both residing in the lower-energy σ_{1s} orbital because that is the lowest-energy orbital available. As in atomic orbitals, these two σ_{1s} electrons must have opposite spins. The electron configuration that corresponds to the molecular orbital diagram in Figure 9.45 is written $(\sigma_{1s})^2$, where the superscript indicates that there are two electrons in the σ_{1s} molecular orbital. Because the energy of the electrons in a $(\sigma_{1s})^2$ configuration is lower than the energy of the electrons in two isolated hydrogen atoms, MO theory explains why hydrogen is a diatomic gas: H_2 molecules are lower in energy and so are more stable than H atoms.

FIGURE 9.45 Mixing the $1s$ orbitals of two hydrogen atoms creates two molecular orbitals: a filled bonding σ_{1s} orbital containing two electrons and an empty antibonding σ_{1s}^* orbital. (a) The lower red oval is the bonding orbital. The two red ovals at the top together make up the antibonding orbital. *Note:* The two top ovals represent only *one* molecular orbital with a node of zero electron density between. Dots show the locations of hydrogen nuclei. (b) A molecular orbital diagram shows the relative energies of bonding and antibonding molecular orbitals and of the atomic orbitals that formed them.

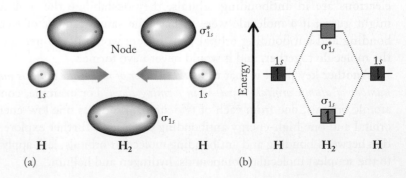

Hydrogen is a diatomic gas, but helium exists as free atoms and not as molecular He_2. MO theory explains why. One helium atom has two valence electrons in a $1s$ atomic orbital. Mixing two He $1s$ orbitals yields the same set of molecular orbitals we generated for H_2, as shown in Figure 9.46. Unlike H_2, the two helium atoms have a total of four valence electrons. Each molecular orbital in Figure 9.46—the bonding orbital and the antibonding orbital—has a maximum capacity of two electrons. Adding four valence electrons to the orbitals in Figure 9.46 means filling both orbitals. The presence of two electrons in the σ_{1s}^* orbital cancels the stability gained from having two electrons in the σ_{1s} orbital. Because there is no net gain in stability, He_2, $(\sigma_{1s})^2(\sigma_{1s}^*)^2$, does not exist.

Another way to compare the bonding in H_2 and He_2 is to look at the *bond order* in each. We previously defined bond order as the number of bonds between two atoms: a bond order of 1 for X—X, 2 for X=X, and 3 for X≡X. In MO theory we define bond order as

$$\text{Bond order} = \tfrac{1}{2}\left(\begin{array}{c}\text{number of}\\ \text{bonding electrons}\end{array} - \begin{array}{c}\text{number of}\\ \text{antibonding electrons}\end{array}\right) \quad (9.2)$$

A molecule of H_2 has two electrons in the bonding MO and none in the antibonding MO, so

$$\text{Bond order in } H_2 = \tfrac{1}{2}(2 - 0) = 1$$

For He_2, the bond order is 0 because an equal number of electrons reside in bonding and antibonding orbitals:

$$\text{Bond order in } He_2 = \tfrac{1}{2}(2 - 2) = 0$$

A bond order of 0 means that He_2 is not a stable molecule. In general, the greater the bond order, the stronger the bond and the more stable the molecule.

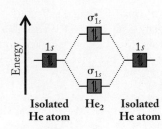

FIGURE 9.46 The molecular orbital diagram for the hypothetical molecule He_2 indicates that the same number of electrons occupy the antibonding orbital and the bonding orbital. Therefore the bond order is 0; the molecule is not stable.

C⚛NNECTION We first defined bond order in Chapter 8 when we related the length and strength of bonds to the number of electron pairs shared by two atoms.

SAMPLE EXERCISE 9.10 Using MO Diagrams to Predict Bond Order I   **LO8**

Draw the MO diagram for the molecular ion H_2^-, determine the bond order of the ion, and predict whether the ion is stable.

Collect and Organize We want to draw the MO diagram for the molecular ion H_2^- and then determine bond order by using Equation 9.2. If the value of the bond order is greater than 0, the ion may be stable.

Analyze We should be able to base the MO diagram for H_2^- on the MO diagram for H_2 (Figure 9.45) because the H_2^- ion has only one more electron than H_2 and the empty σ_{1s}^* orbital in H_2 can accommodate up to two more electrons.

Solve The σ_{1s} orbital is filled in H_2. The third electron goes into the σ_{1s}^* orbital, so the MO diagram is as shown in Figure 9.47. The notation for this electron configuration is $(\sigma_{1s})^2(\sigma_{1s}^*)^1$ (listing the molecular orbitals in order of increasing energy). The bond order is

$$\text{Bond order} = \tfrac{1}{2}(2 - 1) = 0.5$$

We predict that the H_2^- ion is less stable than H_2 but more stable than He_2.

Think About It We encountered the idea of fractional bonds in Chapter 8 in the discussion of resonance, and we encounter it again here in the MO treatment of H_2^-.

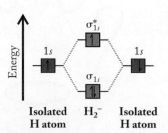

FIGURE 9.47 The molecular orbital diagram of H_2^-.

A bond order of 0.5 in MO theory means that the bond between the two atoms in H_2^- is weaker than the single bond in H_2, making H_2^- a less stable species.

 Practice Exercise
Use MO theory to predict whether the H_2^+ ion can exist.

Molecular Orbitals of Homonuclear Diatomic Molecules

Molecular orbital diagrams for homonuclear (same atom) diatomic molecules such as N_2 and O_2 are more complex than that of H_2 because of the greater number and variety of atomic orbitals in N_2 and O_2. Not all combinations of atomic orbitals result in effective bonding, but there are some general guidelines for constructing the molecular orbital diagram for any molecule:

1. The number of molecular orbitals equals the number of atomic orbitals used to create them.

FIGURE 9.48 Two atoms come together and their p atomic orbitals mix to form six molecular orbitals. (a) The $2p_z$ atomic orbitals create a σ_{2p} bonding orbital and a σ_{2p}^* antibonding orbital. (b) The $2p_x$ and $2p_y$ atomic orbitals mix to form two π_{2p} bonding molecular orbitals and two π_{2p}^* antibonding molecular orbitals.

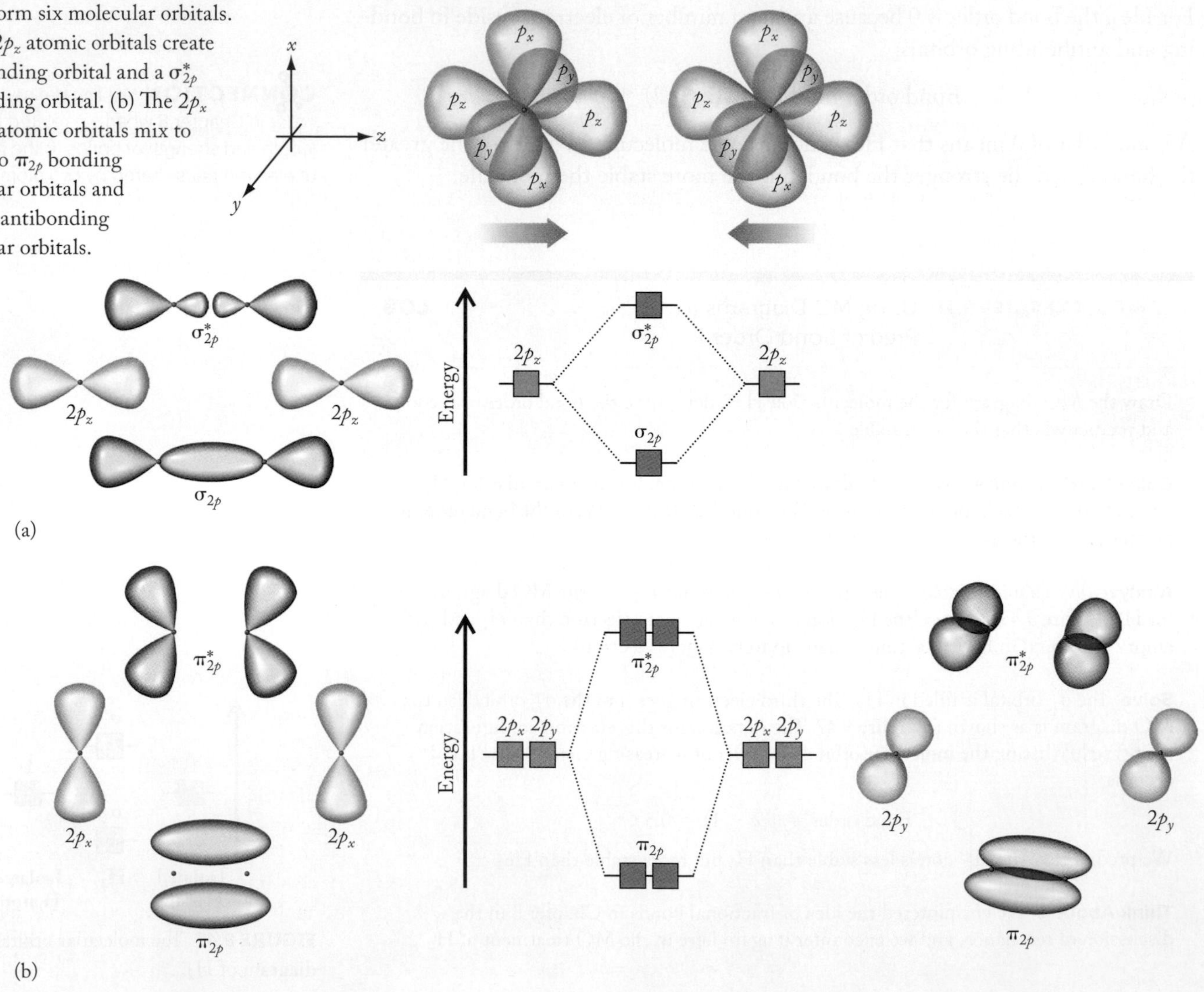

2. Atomic orbitals with similar energy and orientation mix more effectively than do those that have different energies and orientations. For example, an *s* atomic orbital mixes more effectively with another *s* atomic orbital than with a *p* orbital. The 1*s* and 2*s* orbitals have different sizes and energies, resulting in less effective mixing than two 1*s* or two 2*s* orbitals.

3. Better mixing leads to a larger energy difference between bonding and antibonding orbitals and thus greater stabilization of the bonding MOs.

4. A molecular orbital can accommodate a maximum of two electrons; two electrons in the same MO have opposite spins.

5. Electrons in ground-state molecules occupy the lowest-energy molecular orbitals available, following the aufbau principle and Hund's rule.

In mixing atomic orbitals to create molecular orbitals, we consider *only the valence electrons* on the atoms because core electrons do not participate in bonding. Focusing on N_2 and O_2 as examples, we first mix their 2*s* orbitals. The mixing process is analogous to the one we used for H_2, except that the resulting MOs are designated σ_{2s} and σ_{2s}^*.

Next we mix the three pairs of 2*p* orbitals, producing a total of six MOs. The different spatial orientations of the $2p_x$, $2p_y$, and $2p_z$ atomic orbitals result in different kinds of MOs (Figure 9.48). The $2p_z$ atomic orbitals point toward each other. When they mix, two molecular orbitals form, a σ_{2p} bonding orbital and a σ_{2p}^* antibonding orbital. The lobes of the $2p_y$ and $2p_x$ atomic orbitals are oriented at 90° to the bonding axis and at 90° to each other. When the $2p_x$ orbitals mix, and when the $2p_y$ orbitals mix, they do so around the bonding axis instead of along it. This mixing produces two **pi (π) molecular orbitals** and two π* molecular orbitals. When electrons occupy a π orbital, they form a π bond.

The relative energies of σ and π molecular orbitals for N_2 and O_2 are shown in Figure 9.49. In each molecule, the energies of the MOs derived from two

pi (π) molecular orbitals in MO theory, molecular orbitals formed by the mixing of atomic orbitals oriented above and below, or in front of and behind, the bonding axis

CONNECTION In Chapter 7 we discussed Hund's rule, which states that the lowest-energy electron configuration of a set of degenerate orbitals is the one with the maximum number of unpaired electrons, all having the same spin.

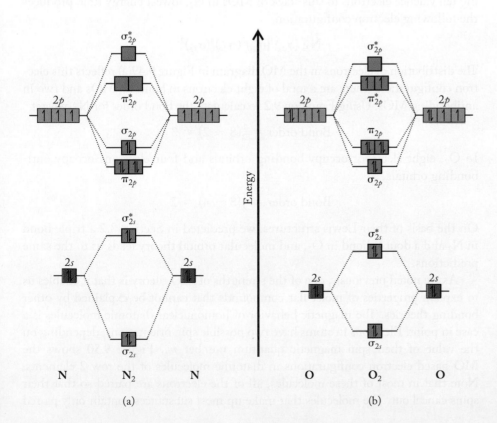

(a)   (b)

FIGURE 9.49 Molecular orbital diagrams of (a) N_2 and (b) O_2. The vertical sequence of orbitals for N_2 applies to MOs of the homonuclear diatomic molecules of elements with $Z \leq 7$. The O_2 sequence applies to homonuclear diatomic molecules of all elements beyond oxygen ($Z \geq 8$), including the halogens.

$2s$ atomic orbitals (σ_{2s} and σ_{2s}^*) are lower than the energy of the σ_{2p} MO for the same reason that a $2s$ atomic orbital is lower in energy than a $2p$ atomic orbital.

Now let's consider the relative energies of the MOs formed by mixing the $2p$ orbitals in N_2 and O_2. We begin with O_2 (Figure 9.49b) because it is representative of most homonuclear diatomic molecules, including all the halogens. In order of increasing energy, the MOs are σ_{2p}, π_{2p}, π_{2p}^*, and σ_{2p}^*. Keep in mind that there are groups of two π_{2p} and two π_{2p}^* orbitals. This means that each group of two can hold four electrons. Adding 12 valence electrons (six from each O atom) in the O_2 molecule into these MOs, starting with the lowest-energy MO first and working our way up, we get the following valence electron configuration:

$$O_2: (\sigma_{2s})^2(\sigma_{2s}^*)^2(\sigma_{2p})^2(\pi_{2p})^4(\pi_{2p}^*)^2$$

The distribution of electrons in the MO diagram follows this sequence. Note that the two π_{2p}^* orbitals are degenerate (equivalent in energy), so each contains a single electron, in accordance with Hund's rule.

The MO diagram tells us that there are two unpaired electrons in a molecule of O_2. This finding contradicts the representation of bonding obtained from the Lewis structure of O_2, from VSEPR, and from valence bond theory, all of which predict that the valence electrons are paired. We will return to this point shortly, but for now let's consider the MO diagram for N_2.

When we compare the MO diagrams of N_2 and O_2 in Figure 9.49, we see a difference in the relative energies of two of their MOs. The π_{2p} molecular orbital is lower in energy than the σ_{2p} orbital in N_2. This switch of energy levels from their relative positions in O_2 and many other diatomic molecules is thought to be due to the stability (and lower energy) of the three half-filled $2p$ orbitals in N atoms ($2s^2 2p^3$), which brings their energy closer to that of the $2s$ orbital. This proximity of orbitals has the effect of lowering the energy of the σ_{2s} molecular orbital and raising the energy of the σ_{2p} orbital—enough to put it above π_{2p}. Adding ten valence electrons to this stack of MOs in N_2, lowest energy first, produces the following electron configuration:

$$N_2: (\sigma_{2s})^2(\sigma_{2s}^*)^2(\pi_{2p})^4(\sigma_{2p})^2$$

The distribution of electrons in the MO diagram in Figure 9.49(a) reflects this electron configuration. There are a total of eight electrons in bonding MOs and two in antibonding MOs. Using Equation 9.2 to calculate the bond order for N_2, we get

$$\text{Bond order} = \tfrac{1}{2}(8 - 2) = 3$$

In O_2, eight electrons occupy bonding orbitals and four electrons occupy antibonding orbitals, so

$$\text{Bond order} = \tfrac{1}{2}(8 - 4) = 2$$

On the basis of their Lewis structures, we predicted in Section 8.2 a triple bond in N_2 and a double bond in O_2, and molecular orbital theory leads us to the same predictions.

As we noted previously, one of the strengths of MO theory is that it enables us to explain properties of molecular compounds that cannot be explained by other bonding theories. The magnetic behavior of homonuclear diatomic molecules is a case in point. Electrons in atoms have two possible spin orientations, depending on the value of their spin magnetic quantum number m_s. Figure 9.50 shows the MO-based electron configurations in diatomic molecules of the row 2 elements. Note that in most of these molecules, all of the electrons are paired so that their spins cancel out. The molecules that make up most substances contain only paired

CONNECTION In Chapter 7 we introduced the spin magnetic quantum number (m_s), which has a value of either $+\tfrac{1}{2}$ or $-\tfrac{1}{2}$ and defines the orientation of an electron in a magnetic field.

diamagnetic describes a substance with no unpaired electrons that is weakly repelled by a magnetic field.

paramagnetic describes a substance with unpaired electrons that is attracted to a magnetic field.

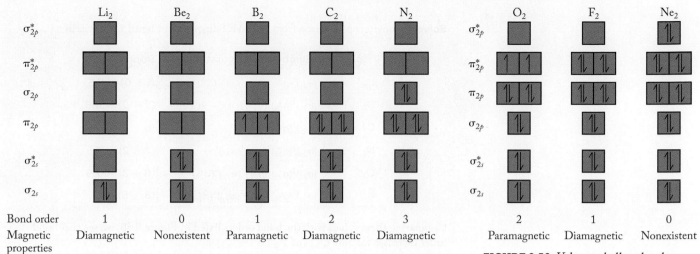

	Li₂	Be₂	B₂	C₂	N₂
Bond order	1	0	1	2	3
Magnetic properties	Diamagnetic	Nonexistent	Paramagnetic	Diamagnetic	Diamagnetic

	O₂	F₂	Ne₂
Bond order	2	1	0
Magnetic properties	Paramagnetic	Diamagnetic	Nonexistent

FIGURE 9.50 Valence-shell molecular orbital diagrams and magnetic properties of the homonuclear diatomic molecules of the second-row elements.

electrons. This complete electron pairing means that these substances are repelled slightly by a magnetic field. These substances are said to be **diamagnetic**. If a substance's molecules contain unpaired electrons, as do those of O_2, then it is attracted by a magnetic field, as shown in Figure 9.51, and is **paramagnetic**. The more unpaired electrons in a molecule, the greater its paramagnetism. Only MO theory accounts for the magnetic behavior of oxygen and of many other substances as well.

CONCEPT TEST

Can liquid N_2 and O_2 be separated from each other with a magnet?

Figure 9.50 enables us to make several predictions about diatomic molecules of the second-row elements. First, we predict that Be_2 and Ne_2 do not exist for the same reason that He_2 does not exist: both Be_2 and Ne_2 have as many antibonding electrons as they have bonding electrons and therefore have a net bond order of 0. Second, Li_2, B_2, and F_2 have a bond order of 1, whereas C_2 has a bond order of 2. Like O_2, B_2 is paramagnetic, whereas Li_2, C_2, N_2, and F_2 are diamagnetic. (We have not seen C_2 molecules before, but they actually do exist in electric arcs and in comets, and they are responsible for the blue light of candle flames.)

SAMPLE EXERCISE 9.11 Using MO Diagrams to Predict Bond Order II **LO8**

In which molecules in Figure 9.50 does the bond order increase when one electron is removed from the molecule? You may exclude the nonexistent Be_2 and Ne_2 from this analysis.

Collect, Organize, and Analyze Equation 9.2 relates bond order to the difference in the numbers of electrons in bonding and antibonding orbitals. Removing an electron from Li_2, B_2, C_2, N_2, O_2, and F_2 results in the molecular ions Li_2^+, B_2^+, C_2^+, N_2^+, O_2^+, and F_2^+, respectively. We may assume that the molecular ions have MO diagrams with orbital energies in the same order as their parent molecules. Thus the MO diagrams for the molecular ions are the same as those in Figure 9.50, but with one electron removed from the highest-energy orbital.

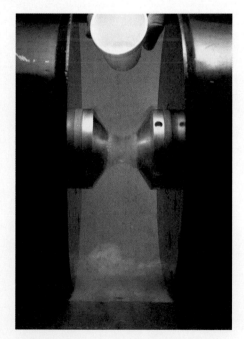

FIGURE 9.51 Liquid O_2 poured from a Styrofoam cup is suspended in the space between the poles of this magnet because the unpaired electrons in its molecules make O_2 paramagnetic. Paramagnetic substances are attracted to magnetic fields.

Solve Removing one electron from each MO diagram in Figure 9.50 gives us

Ion	Electron Configuration	Bond Order
Li_2^+	$(\sigma_{2s})^1$	$\frac{1}{2}(1-0) = 0.5$
B_2^+	$(\sigma_{2s})^2(\sigma_{2s}^*)^2(\pi_{2p})^1$	$\frac{1}{2}(3-2) = 0.5$
C_2^+	$(\sigma_{2s})^2(\sigma_{2s}^*)^2(\pi_{2p})^3$	$\frac{1}{2}(5-2) = 1.5$
N_2^+	$(\sigma_{2s})^2(\sigma_{2s}^*)^2(\pi_{2p})^4(\sigma_{2p})^1$	$\frac{1}{2}(7-2) = 2.5$
O_2^+	$(\sigma_{2s})^2(\sigma_{2s}^*)^2(\sigma_{2p})^2(\pi_{2p})^4(\pi_{2p}^*)^1$	$\frac{1}{2}(8-3) = 2.5$
F_2^+	$(\sigma_{2s})^2(\sigma_{2s}^*)^2(\sigma_{2p})^2(\pi_{2p})^4(\pi_{2p}^*)^3$	$\frac{1}{2}(8-5) = 1.5$

Comparing these values with the bond orders listed in Figure 9.50, we see that bond order increases for only O_2^+ and F_2^+.

Think About It Removing an electron from O_2 or F_2 reduces the number of electrons in antibonding molecular orbitals while leaving the number of electrons in bonding molecular orbitals unchanged. The result is an increase in bond order. In the other four homonuclear diatomic molecules, removing an electron reduces the number of electrons in bonding molecular orbitals while leaving the number of electrons in antibonding orbitals unchanged. The result is a reduction in bond order for Li_2^+, B_2^+, C_2^+, and N_2^+.

Practice Exercise In which molecules in Figure 9.50 does the bond order increase when one electron is added to the molecule?

Molecular Orbitals of Heteronuclear Diatomic Molecules

Molecular orbital theory also enables us to account for the bonding in *heteronuclear* diatomic molecules, which are molecules containing two different atoms. The bonding in some of these molecules is difficult to explain using other bonding theories. For example, it is often difficult to draw a single Lewis structure for an odd-electron molecule such as nitrogen monoxide (NO). In Chapter 8 we considered several arrangements of its valence electrons, such as

We predicted that oxygen was more likely to have a complete octet of valence electrons because it is the more electronegative element. In addition, experimental evidence allowed us to rule out structures with unpaired electrons on the oxygen atom. Our preferred structure was therefore the one shown in red. However, the bond length in NO (115 pm) is considerably shorter than the value in Table 8.3 for an average N=O double bond (122 pm). Molecular orbital theory is useful for explaining both the bonding in NO and the deviation from the expected bond length.

Let's look at the bonding first. The MO diagram for NO is different from the diagrams of homonuclear diatomic gases. Nitrogen and oxygen atoms have different numbers of protons and electrons, and the difference in effective nuclear charge in N and O atoms means that their atomic orbitals have different energies, as Figure 9.52 shows, for the $2s$ and $2p$ orbitals.

For constructing the MO diagram for NO, the guidelines described previously still apply. The number of MOs formed must equal the number of atomic orbitals combined, and the energy and orientation of the atomic orbitals being mixed must be considered. One additional factor influences the energies of the MOs in heteronuclear diatomic molecules: *bonding* MOs tend to be closer in energy to the atomic orbitals of the more electronegative atom and *antibonding* MOs tend to be closer in energy to the atomic orbitals of the less electronegative atom. The MO diagram for NO in Figure 9.52 illustrates this phenomenon. Note how the energy of the bonding σ_{2s} orbital is closer to that of the 2s orbital of the O atom, and the energy of the antibonding σ_{2s}^* orbital is closer to that of the 2s orbital of the N atom. Similarly, the π_{2p} MOs in NO are closer in energy to the 2p orbitals of oxygen, and the π_{2p}^* MOs are closer in energy to the 2p orbitals of nitrogen. The proximity of the nitrogen 2p atomic orbitals to the π_{2p}^* MOs means that the single electron in the π_{2p}^* MO is more likely to be on nitrogen than on oxygen. This prediction is consistent with our Lewis structure in which the odd electron in NO is on the nitrogen atom.

Molecular orbital theory also enables us to rationalize the relatively short bond length in NO. Equation 9.2 tells us that the bond order is $\frac{1}{2}(8 - 3) = 2.5$, halfway between the bond orders for N=O and N≡O and consistent with a bond length of 115 pm, halfway between the lengths of the N=O bond (122 pm) and the N≡O bond (106 pm).

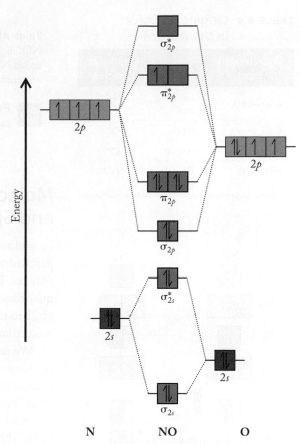

FIGURE 9.52 The molecular orbital diagram for NO shows that the unpaired electron occupies a π_{2p}^* antibonding orbital, which is closer in energy to the 2p atomic orbitals of nitrogen than to the 2p atomic orbitals of oxygen. As a result of this proximity, the electron is more likely to be located on the nitrogen atom than on the oxygen atom.

SAMPLE EXERCISE 9.12 Using MO Diagrams for Heteronuclear Diatomic Molecules **LO8**

Nitrogen monoxide reacts with many transition metals, including the iron in our blood. In these compounds, NO is sometimes present as NO^+ and at other times as NO^-. Use Figure 9.52 to predict the bond order of NO^+ and NO^-.

Collect and Organize We are asked to predict the bond order of two diatomic ions on the basis of the MO diagram of their parent molecule (Figure 9.52). Equation 9.2 relates bond order to the numbers of electrons in bonding and antibonding orbitals.

Analyze NO has 11 valence electrons. We can remove one electron from the MO diagram for NO to get the diagram for NO^+ and add one electron to get the diagram for NO^-.

Solve To generate NO^+, we remove the highest-energy electron in NO, which is the one in the π_{2p}^* molecular orbital. This gives NO^+ the electron configuration

$$NO^+: (\sigma_{2s})^2(\sigma_{2s}^*)^2(\sigma_{2p})^2(\pi_{2p})^4$$

Adding an electron to the lowest-energy MO available in NO (also π_{2p}^*) yields NO^- with the valence electron configuration

$$NO^-: (\sigma_{2s})^2(\sigma_{2s}^*)^2(\sigma_{2p})^2(\pi_{2p})^4(\pi_{2p}^*)^2$$

The bond orders of the two ions are

$$NO^+: \text{bond order} = \frac{1}{2}(8 - 2) = 3$$

$$NO^-: \text{bond order} = \frac{1}{2}(8 - 4) = 2$$

TABLE 9.4 Origins of Colors in the Aurora

Wavelength (nm)	Color	Chemical Species
650–680	Deep red	N_2^*
630	Red	O^*
558	Green	O^*
391–470	Blue-violet	N_2^{+*}

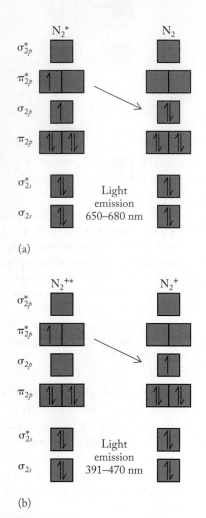

(a)

(b)

FIGURE 9.53 Molecular orbital diagrams for (a) N_2 and (b) N_2^+ show electronic transitions that result in the emission of visible light. Collisions with ions in the solar wind result in the promotion of electrons from the σ_{2p} orbitals in N_2 molecules to π_{2p}^* orbitals. Higher-energy collisions create N_2^{+*} molecular ions, which also have electrons in π_{2p}^* orbitals. When these electrons return to the ground state, red and blue-violet light are emitted.

Think About It The bond orders in N_2 and O_2 are 3 and 2, respectively. The cation NO^+ is isoelectronic with N_2, so our calculated bond order for it makes sense. The anion NO^- is isoelectronic with O_2, and so our calculated bond order for this ion is also reasonable.

 Practice Exercise Using Figure 9.52 as a guide, draw the MO diagram for carbon monoxide, and determine the bond order for the carbon–oxygen bond.

Molecular Orbitals of N_2^+ and Spectra of Auroras

In addition to predicting the magnetic properties of molecules, MO theory is particularly useful for predicting their spectroscopic properties—and the colors of auroras. In Chapter 7 we learned that the light emitted by excited free atoms is quantized and can be related to the movement of electrons between atomic orbitals. Broadly speaking, the same is true in molecules: electrons can move from one molecular orbital to another by absorbing or emitting light.

We can use this information to look again at the phenomenon described at the opening of this section: how the colors of the aurora are produced. The principal chemical species involved are listed in Table 9.4. An asterisk indicates a molecule or molecular ion in an excited state. Excited N_2 molecules produce deep crimson red (650–680 nm) light, and excited N_2^+ ions produce blue-violet (391–470 nm) light. The MO diagrams for these species are shown in Figure 9.53. Comparing the MO diagrams of N_2^* and N_2 in Figure 9.53(a), we find that one of the two electrons originally in the σ_{2p} MO in N_2 has been raised to a π_{2p}^* orbital in N_2^*, leaving an unpaired σ_{2p} electron behind. Figure 9.53(b) shows us that N_2^{+*} also has one electron in a π_{2p}^* orbital, but its σ_{2p} orbital is empty because the other σ_{2p} electron originally in the N_2 molecule was lost when the molecule was ionized. As π_{2p}^* electrons return from their antibonding, excited-state orbitals to the bonding σ_{2p} orbital in the ground state, the distinctive blue-violet and crimson emissions of N_2^+ and N_2 appear, as shown in the photograph in Figure 9.44.

CONCEPT TEST

Are the bond orders of the excited-state species in Figure 9.53 the same as the ground-state species?

Metallic Bonds and Conduction Bands

We can use the ideas in molecular orbital theory to discuss on additional type of bond we mentioned only briefly in Chapter 8: *metallic bonding*. Up to this point, our focus in the chapter has been on covalent bonding in molecules. According to valence bond theory, a covalent bond forms between two atoms when partially filled atomic orbitals—one from each atom—overlap. Using concepts from MO theory, we can now explore the bonds that form between the densely packed atoms in metallic solids. Dense packing means that the valence orbitals of these atoms overlap with the orbitals of many other nearby atoms. This large number of interactions makes metals strong. At the same time, sharing a limited number of valence electrons with many bonding partners makes the bond linking any two metal atoms relatively weak.

To understand this point, consider the bond that would result if two Cu atoms came together and formed a molecule of Cu$_2$. Copper atoms have the electron configuration [Ar]$3d^{10}4s^1$. When the partially filled 4s orbitals of the two atoms overlap, they form a diatomic molecule held together by a single covalent bond (Figure 9.54a). However, the Cu atoms in solid copper are each surrounded by a total of 12 other Cu atoms and each is bonded to all 12 of its neighbors (Figure 9.54b). This means that each Cu atom must share its 4s electron with 12 other atoms, not just one other atom. Inevitably, this dispersion of bonding electrons weakens the Cu—Cu bond between each pair of Cu atoms.

Adding to the diffuse nature of this bonding is the fact that metallic elements have lower electronegativities than nonmetals. Recall from Chapter 8 that differences in electronegativity are important for defining how bonding electrons are distributed between pairs of atoms. Atoms in a pure metal have no such differences in electronegativity. However, the low ionization energies of metals mean that the bonding electrons are not held tightly to the nuclei of individual atoms.

In Chapter 8 we described metal atoms "floating" in seas of mobile bonding electrons where the electrons are shared by all of the nuclei in the sample. The diffuse nature of metallic bonding described in the preceding paragraphs certainly fits the sea-of-electrons model; however, a more sophisticated approach, called **band theory**, better explains the bonding in metals and other solids. Let's apply band theory, which is an extension of molecular orbital theory, to explain the bonding between the atoms in solid copper.

When the 4s atomic orbitals on two Cu atoms overlap to form Cu$_2$, the atomic orbitals combine to form two molecular orbitals with different energies, equally spaced above and below the initial energy value (Figure 9.55a). This is analogous to the formation of low-energy bonding and high-energy antibonding molecular orbitals. If another two Cu atoms join the first two to form a molecule of Cu$_4$, the 4s atomic orbitals of four Cu atoms combine to form four molecular orbitals. If we add another four atoms to make Cu$_8$, a total of eight copper 4s atomic orbitals combine to form eight molecular orbitals. In all these molecules the lower-energy orbitals are filled with the available 4s electrons and the upper orbitals are empty (Figure 9.55b). If we apply this model to the enormous number of atoms in a piece of copper wire, an equally enormous number of molecular orbitals is created. The lower-energy half of them are occupied by electrons; the higher-energy half are empty. There are so many of these orbitals that they form a continuous *band* of energies with no gap between the occupied lower half and the empty upper half. Because this band of MOs was formed by combining valence-shell orbitals, it is called a **valence band**.

Band theory explains the conductivity of copper and many other metals by incorporating the assumption that essentially no gap exists between the energy of the occupied lower portion of the valence band and the empty upper portion. Therefore valence electrons can move easily from the filled lower portion to the empty upper portion, where they are free to move from one empty orbital to the next and thus flow throughout the solid.

band theory an extension of molecular orbital theory that describes bonding in solids.

valence band a band of orbitals that are filled or partially filled by valence electrons.

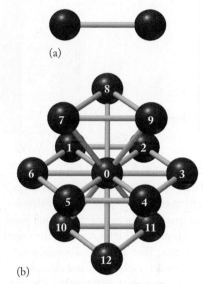

FIGURE 9.54 Covalent bonds differ from metallic bonds. (a) A ball-and-stick model of the molecule Cu$_2$ is based on the assumption that the two atoms share their 4s electrons to form a covalent bond. (b) In solid copper, the atom labeled 0 shares its 4s electron with 12 other atoms. As a result, the bonds in copper and other metals are much more diffuse than the covalent bonds in molecules.

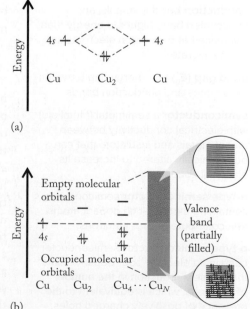

FIGURE 9.55 (a) Two 4s atomic orbitals combine to form two molecular orbitals in the molecule Cu$_2$. (b) As the half-filled 4s atomic orbitals of an increasing number of Cu atoms overlap, more and more molecular orbitals are formed; half of them are occupied, the other half are empty. As more MOs form, their energies get closer together until a continuous energy band forms—a valence band that is only half-filled with electrons. Electrons can move from the filled half (purple) to the slightly higher-energy upper half (orange), where they are free to migrate through delocalized empty orbitals throughout the solid.

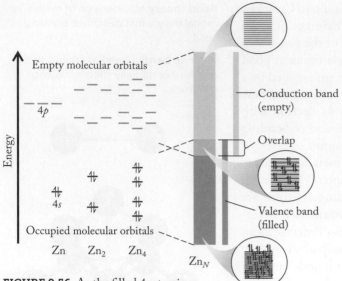

FIGURE 9.56 As the filled 4s atomic orbitals of an increasing number of Zn atoms overlap, they form a filled valence band (purple). An empty conduction band (gray) is produced by combining the empty 4p orbitals. The valence and conduction bands overlap each other, and electrons move easily from the filled valence band to the empty conduction band.

The model of a partially filled valence shell explains the conductivity of many metals, but not all. Consider, for example, zinc, copper's neighbor in the periodic table. Its electron configuration, $[Ar]3d^{10}4s^2$, tells us that all its valence-shell electrons reside in filled orbitals, which means the valence band in solid zinc is filled (Figure 9.56). With no empty space in the valence band to accommodate additional electrons, it might seem that the valence-shell electrons in Zn would be immobile, making Zn a poor electrical conductor. However, electrons in the valence band of Zn *do* migrate through the solid; zinc is a good conductor of electricity, and band theory explains why. The theory incorporates the assumption that *all* atomic orbitals of comparable shape and energy, including the empty 4p orbitals on zinc, can combine to form additional energy bands. The energy band produced by combining empty 4p orbitals, called a **conduction band**, is also empty and is broad enough to overlap the valence band. This overlap means that electrons from the valence band can move to the conduction band, where they are free to migrate from atom to atom in solid zinc, thereby conducting electricity.

CONCEPT **TEST**

Is the electrical conductivity of magnesium metal best explained in terms of overlapping conduction and valence bands or in terms of a partially filled valence band? Explain your answer.

Semiconductors

To the right of the metals in the periodic table is a "staircase" of elements that tend to have the physical properties of metals and the chemical properties of nonmetals. These semimetals (or metalloids) are not as good at conducting electricity as metals, but they are much better at it than nonmetals. We can use band theory to explain this intermediate behavior. In semimetals, conduction and valence bands do not overlap but instead are separated by an energy gap. In silicon, the most abundant semimetal, band theory predicts an energy gap, or **band gap** (E_g), of 107 kJ/mol (Figure 9.57a).

Generally, only a few valence-band electrons in Si have sufficient energy to move to the conduction band, which limits silicon's ability to conduct electricity and makes it a **semiconductor**. However, we can enhance the conductivity of solid Si, or of any other elemental semimetal, by replacing some of the Si atoms with atoms of an element of similar atomic radius but with a different number of valence electrons. The replacement process is called *doping*, and the added element is called a *dopant*. Suppose the dopant is a group 15 element such as phosphorus. Each P atom has one more electron than the atom of Si that it replaced. The energy of these additional electrons is different from the energy of the silicon electrons. They populate a narrow band located in the silicon band gap (Figure 9.57b). This arrangement effectively reduces the size of the energy gap and increases electrical conductivity because electrons can move more easily across the remaining gaps between the valence and conduction bands. Phosphorus-doped silicon is an example of an **n-type semiconductor** because the dopant contributes extra negative charges (electrons) to the structure of the host element.

conduction band in metals, an unoccupied band higher in energy than a valence band, in which electrons are free to migrate.

band gap (E_g) the energy gap between the valence and conduction bands.

semiconductor a semimetal (metalloid) with electrical conductivity between that of metals and insulators that can be chemically altered to increase its electrical conductivity.

n-type semiconductor semiconductor containing electron-rich dopant atoms that contribute excess electrons.

p-type semiconductor semiconductor containing electron-poor dopant atoms that cause a reduction in the number of electrons, which is equivalent to the presence of positively charged holes.

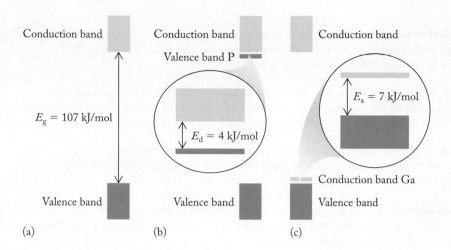

(a)   (b)   (c)

FIGURE 9.57 The electrical conductivity of semiconductors can be greatly enhanced by doping. (a) Pure Si has a band gap, E_g, of 107 kJ/mol. (b) Adding a few atoms of P, which has more valence electrons than Si, creates an n-type semiconductor with a narrow, filled valence band called the donor level, E_d, 4 kJ/mol below the conduction band of Si. (c) Adding a few atoms of Ga, which has fewer valence electrons than Si, creates an empty conduction band in the band gap of a p-type semiconductor called the acceptor level, E_a, 7 kJ/mol above the valence band of Si.

The conductivity of solid silicon can also be enhanced by replacing some Si atoms with atoms of a group 13 element such as gallium (Figure 9.57c). Because Ga atoms have one fewer valence electron than Si atoms, substituting them in the Si structure means fewer valence electrons in the solid. The result is the creation of a narrow Ga conduction band (*acceptor band*, Figure 9.58a) in the Si band gap. Because the gap between the Si valence band and the acceptor band is smaller than the band gap in pure Si, electrons from the valence band move more easily to the acceptor band, increasing electrical conductivity. This array of bands makes Si doped with Ga a **p-type semiconductor** because a reduction in the number of negatively charged electrons in the valence band is equivalent to the presence of positively charged "holes" (Figure 9.58b). The semiconductors used in solid-state electronics are combinations of n-type and p-type.

Doping is not the only way to change the conductivity of semimetals. Compounds prepared from combinations of group 13 and group 15 elements also may behave as semiconductors. For example, gallium arsenide (GaAs) is a semiconductor that emits infrared radiation ($\lambda = 874$ nm) when connected to an electrical circuit. This emission is used in devices such as bar-code readers and DVD players. Like silicon, solid gallium arsenide has both a valence band and a conduction band separated by a characteristic band gap. The energy of each photon of light corresponds to the energy gap between the valence band and the conduction band. When electrical energy is applied to the material, electrons are raised to the conduction band. When they fall back to the valence band, they emit radiation. If some aluminum is substituted for gallium in GaAs, the band gap increases, and predictably the wavelength of emitted light decreases. For example, a material with the empirical formula $AlGaAs_2$ emits orange-red light ($\lambda = 620$ nm). Many of the multicolored indicator lights in electronic devices use $AlGaAs_2$ semiconductors.

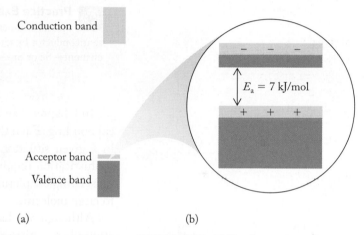

(a)   (b)

FIGURE 9.58 (a) Band structure of a p-type semiconductor with no electrons in the conduction band of the dopant. The valence band is filled. (b) In a p-type semiconductor, some electrons have enough energy to move from the valence band to the empty conduction band of the dopant, leaving behind positively charged vacancies, or "holes." The presence of holes makes the valence band partially filled, increasing electrical conductivity.

CONNECTION We introduced the semimetals (metalloids) in Section 2.5 when we described the structure of the periodic table.

CONNECTION Emission spectra obtained from gas discharge tubes were discussed in Chapter 7.

SAMPLE EXERCISE 9.13 Distinguishing p- and n-Type Semiconductors   **LO9**

Which kind of semiconductor—n-type or p-type—does doping germanium with arsenic create?

Collect and Organize We are asked to determine the type of semiconductor formed when germanium (Ge, group 14) is doped with arsenic (As, group 15). Figure 9.57 helps us distinguish between n- and p-type semiconductors.

Analyze When the dopant has more valence electrons than the host semimetal, the two form an n-type semiconductor, as in Figure 9.57(b). If the dopant has fewer valence electrons than the host, the result is a p-type semiconductor (Figure 9.57c).

Solve Arsenic is in group 15, which means that its atoms have one more valence electron than do the atoms of Ge, a group 14 element. This makes As-doped Ge an n-type semiconductor.

Think About It Arsenic is a good candidate for a dopant to make an n-type semiconductor with germanium because As atoms are nearly the same size as Ge atoms and fit easily into the structure of solid Ge.

Practice Exercise Gallium arsenide (GaAs) is a semiconductor used in optical scanners in retail stores. GaAs can be made an n-type or a p-type semiconductor by replacing some of the As atoms with another element. Which element—Se or Sn—would form an n-type semiconductor with GaAs?

In Chapter 8 and in this chapter we have presented several theories of chemical bonding. Each theory has its strengths and weaknesses. The best one to apply in a given situation depends on the question being asked and on the level of sophistication required in the answer. Molecular orbital theory may provide the most complete picture of covalent bonding, but it is also the most difficult to apply to large molecules.

Although we have focused primarily on the small molecules found in the atmosphere—nitrogen, oxygen, water, carbon dioxide, methane, ozone—and somewhat larger molecules found in living systems, it is important to realize that the principles described in this chapter apply to larger and more complex molecules and ions. We return to the importance of molecular shape, particularly in defining the biological activity of both large and small molecules, in later chapters of this book.

SAMPLE EXERCISE 9.14 Integrating Concepts: Insect Pheromones

Social insects such as bees, wasps, and ants communicate with one another by secreting and detecting molecules called pheromones.

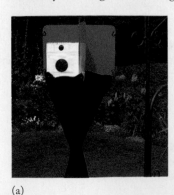

(a)   (b)

FIGURE 9.59 (a) Japanese beetle trap with pheromone-containing lure and (b) Japanese beetle on a daisy.

Small differences in the structure of these compounds, including chiral carbon atoms, lead to differences in shape that signal for either finding a mate or avoiding a predator. This molecular recognition is the basis of traps used to capture destructive pests such as the Japanese beetle (Figure 9.59).

The sex pheromone for the Japanese beetle used in these traps is called japonilure. The condensed structure for this compound is

$$CH=CH(CH_2)_7CH_3$$

a. Draw the carbon-skeleton structure of japonilure.
b. Determine the molecular geometry at the carbonyl carbon atom in the ring. What hybridization best explains this geometry? Identify two other atoms in japonilure that also have this hybridization.

c. How many π bonds are in this molecule? Do these contain delocalized π electrons?

d. Identify the chiral carbon atom in japonilure.

Collect and Organize We are given the condensed structure of japonilure. We are asked four questions about the bonding in the molecule.

Analyze The molecule consists of a 5-atom ring and a long hydrocarbon chain. Section 9.5 provides guidelines on how to draw carbon-skeleton structures and their relationship to condensed structures. We also learned in Section 9.5 that the atoms at each vertex in the ring are carbon atoms. Sections 9.2 and 9.4 provide guidelines for determining the molecular geometry of an atom and for assessing the hybrid orbitals that are consistent with those geometries, including the formation of σ and π bonds. Section 9.6 explains why carbon atoms bonded to four different groups are considered chiral.

Solve

a. The hydrocarbon chain attached to the ring can be drawn using the zigzag lines of a carbon-skeleton structure. There are 10 carbon atoms in the chain. There is a double bond between the first and second carbon atom in the chain, followed by 7 methylene groups and ending with a methyl group:

b. The carbon atom in the C=O carbonyl group has a steric number of 3 because it has single bonds to a carbon atom and a second oxygen atom. Table 9.1 tells us the molecular geometry at the carbonyl carbon atom is trigonal planar. The carbonyl carbon is sp^2 hybridized. It forms three σ bonds: one to the adjacent carbon and one to each of the two oxygen atoms. The

remaining p orbital is used to form a π bond to the oxygen atom that is part of the carbonyl group. Two other carbon atoms are also sp^2 hybridized: the two carbon atoms in the hydrocarbon chain connected by a double bond.

c. As discussed in the solution to part (b), the carbonyl group contains a π bond. The double bond between the two carbon atoms in the hydrocarbon chain consists of one σ bond and one π bond. Therefore there are two π bonds in the molecule, neither of which contain delocalized electrons because they are separated by more than one carbon atom.

d. A chiral carbon atom is bonded to four different groups. By definition, a carbon atom in a methylene ($—CH_2—$) or methyl ($—CH_3$) group cannot be chiral because it is bonded to multiple hydrogen atoms. Therefore none of the carbon atoms in the hydrocarbon chain can be the chiral atom in japonilure. That leaves us with the four carbon atoms in the ring. The carbonyl carbon has SN = 3, so it is not bonded to four groups and therefore cannot be chiral. The carbon atom adjacent to the carbonyl atom (and the next carbon atom) are both $—CH_2—$ groups. (Remember that in Kekulé and carbon-skeleton structures, hydrogen atoms are not explicitly drawn. Carbon atoms are assumed to be bonded to enough hydrogen atoms to have an octet.) That leaves us with the carbon atom in the ring to which the hydrocarbon chain is attached; it is marked with a *:

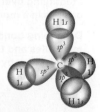

Think About It Japonilure contains one chiral carbon atom, which means that two optical isomers exist. One isomer is a sex attractant for a mate, whereas the other detects the sex attractant released by a predator. This important chiral carbon atom literally makes the difference between life and death for a Japanese beetle.

SUMMARY

LO1 Minimizing repulsion between pairs of valence electrons (the **VSEPR** model) results in the lowest-energy orientations of bonding and nonbonding electron pairs and accounts for the observed **molecular geometries** of molecules. (Section 9.2)

LO2 The shape of a molecule reflects the arrangement of the atoms in three-dimensional space and is determined largely by characteristic **bond angles**. (Sections 9.1 and 9.2)

LO3 Two covalently bonded atoms with different electronegativities have partial electrical charges of opposite sign, creating a **bond dipole**. If the individual bond dipoles in a molecule do not offset each other, the molecule is polar. If they do offset each other, the molecule is nonpolar. (Section 9.3)

Electric field on

LO4 In **valence bond theory**, the **overlap** of atomic orbitals results in covalent bonds between pairs of atoms in molecules. Molecular geometry is explained by the mixing, or **hybridizing**, of atomic orbitals to create **hybrid atomic orbitals**. (Section 9.4)

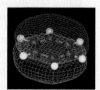

CH₄

LO5 Chemists use multiple representations to draw larger organic molecules, including condensed structures and carbon-skeleton structures. (Section 9.5)

LO6 Molecules with alternating single and double bonds are stabilized by electron delocalization over the system. (Section 9.5)

LO7 **Chiral** molecules exist in left- and right-handed forms that have different properties. Many contain an sp^3 hybridized carbon atom with four different groups attached. (Section 9.6)

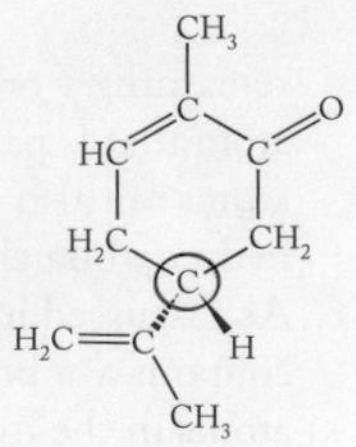

LO8 **Molecular orbital theory** is based on the formation of **molecular orbitals**, which are orbitals delocalized over two or more atoms in a

molecule. MO theory explains the magnetic and spectroscopic properties of molecules but does not explain their shapes. A **molecular orbital diagram** shows relative energies of the molecular orbitals in a molecule. (Section 9.7)

LO9 MO theory is useful to describe metallic bonding and **semiconductors**. (Section 9.7)

PARTICULATE **PREVIEW WRAP-UP**

The Lewis structures for acrolein and methyl mercaptan are shown here. Lewis structures do not include shape information (bond angles). In the space-filling models, the angle around the middle carbon atom in acrolein appears to be about 120°. From the model, the angle around the sulfur atom is bigger than 90° but less than 120°. VSEPR tells us the ideal angle is 109.5°, and we would predict the angle to be less than that value. Because all three carbon atoms in acrolein are sp^2 hybridized, this molecule is planar. The carbon atom in methyl mercaptan is sp^3 hybridized, so the molecule is not planar.

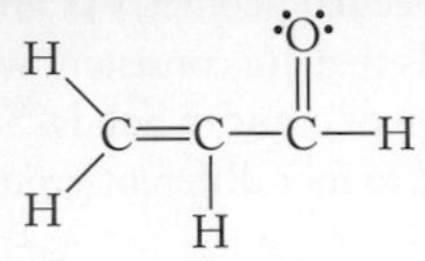

Acrolein 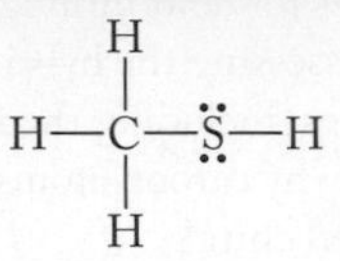   Methyl mercaptan

PROBLEM-SOLVING SUMMARY

Type of Problem	Concepts and Equations	Sample Exercises
Predicting molecular geometry	Draw a Lewis structure for the molecule. Determine the steric number (SN) of the central atom, where $$SN = \begin{pmatrix} \text{number of atoms} \\ \text{bonded to central atom} \end{pmatrix} + \begin{pmatrix} \text{number of lone pairs} \\ \text{on central atom} \end{pmatrix} \quad (9.1)$$ Choose a geometry that minimizes repulsion between electron pairs.	**9.1, 9.3**
Predicting relative sizes of bond angles	Lone pairs on a central atom push bonded atoms closer together, decreasing bond angles.	**9.2**
Predicting polarity of a substance	Assign the direction of polarity to each bond dipole and use molecular geometry to determine whether the dipoles offset each other.	**9.4**
Identifying overlapping orbitals in a molecule	Identify the partially filled atomic orbitals on the atoms.	**9.5**
Describing bonding in molecules and the shape of molecules by using hybrid orbitals	Identify the hybrid orbitals in molecules that result from mixing different numbers of s, p, and d orbitals that result in the observed molecular geometry: $s + p =$ two sp hybrid orbitals $s +$ two $p =$ three sp^2 hybrid orbitals $s +$ three $p =$ four sp^3 hybrid orbitals $s +$ three $p + d =$ five sp^3d hybrid orbitals $s +$ three $p +$ two $d =$ six sp^3d^2 hybrid orbitals	**9.6, 9.7, 9.8**
Drawing larger molecules	Write condensed structures and draw carbon-skeleton structures.	**9.9**
Using MO diagrams to predict bond order	$$\text{Bond order} = \frac{1}{2}\begin{pmatrix} \text{number of} \\ \text{bonding electrons} \end{pmatrix} - \begin{pmatrix} \text{number of} \\ \text{antibonding electrons} \end{pmatrix} \quad (9.2)$$	**9.10, 9.11, 9.12**
Describing n- and p-type semiconductors	Adding electron-rich dopants creates an n-type semiconductor. Adding electron-poor dopants creates a p-type semiconductor.	**9.13**

VISUAL PROBLEMS

(Answers to boldface end-of-chapter questions and problems are in the back of the book.)

9.1. Two compounds with the same formula, S_2F_2, have been isolated. The structures in Figure P9.1 show the arrangements of the atoms in these different compounds. Can these two compounds be distinguished by their dipole moments?

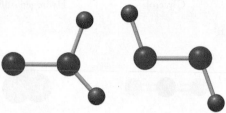

FIGURE P9.1

9.2. Could you distinguish between the two structures of N_2H_2 shown in Figure P9.2 by the magnitude of their dipole moments?

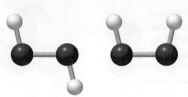

FIGURE P9.2

9.3. Which of the molecules shown in Figure P9.3 are planar; that is, all atoms are in a single plane? Are there delocalized π electrons in any of these molecules?

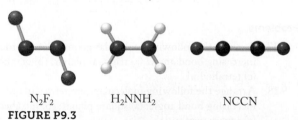

N_2F_2 $\qquad$ H_2NNH_2 $\qquad$ NCCN

FIGURE P9.3

9.4. Which of the molecules shown in Figure P9.4 is *not* planar? Are there delocalized π electrons in any of these molecules?

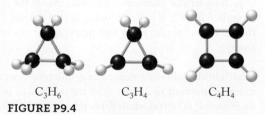

C_3H_6 $\qquad$ C_3H_4 $\qquad$ C_4H_4

FIGURE P9.4

9.5. Use the MO diagram in Figure P9.5 to predict whether O_2^+ has more or fewer electrons in antibonding molecular orbitals than O_2^{2+}.

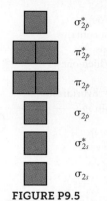

FIGURE P9.5

9.6. Under appropriate conditions, I_2 can be oxidized to I_2^+, which is bright blue. The corresponding anion, I_2^-, is not known. Use the molecular orbital diagram in Figure P9.6 to explain why I_2^+ is more stable than I_2^-.

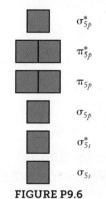

FIGURE P9.6

9.7. The molecular geometry of ReF_7 is an uncommon structure called a pentagonal bipyramid, which is shown in Figure P9.7. What are the bond angles in a pentagonal bipyramid?

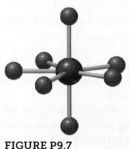

FIGURE P9.7

9.8. Use representations [A] through [I] in Figure P9.8 to answer questions a–f.

a. Which representations depict orbital overlap?

b. Which molecules contain π electrons?

c. Which molecules contain *delocalized* π electrons?

d. Identify all the functional groups present in each molecule that contains one or more functional groups.

e. Look at [C], [F], and [I]. Which ones, if any, are polar?

f. Which molecules contain sp^2 hybridized oxygen atoms? Which contain sp^3 hybridized oxygen atoms?

A	B	C
Vanillin	Glycerol	Hydrogen sulfide
D	E	F
Chlorine	Carbon dioxide	Ozone
G	H	I
Carvone	Ethylene	Acetone

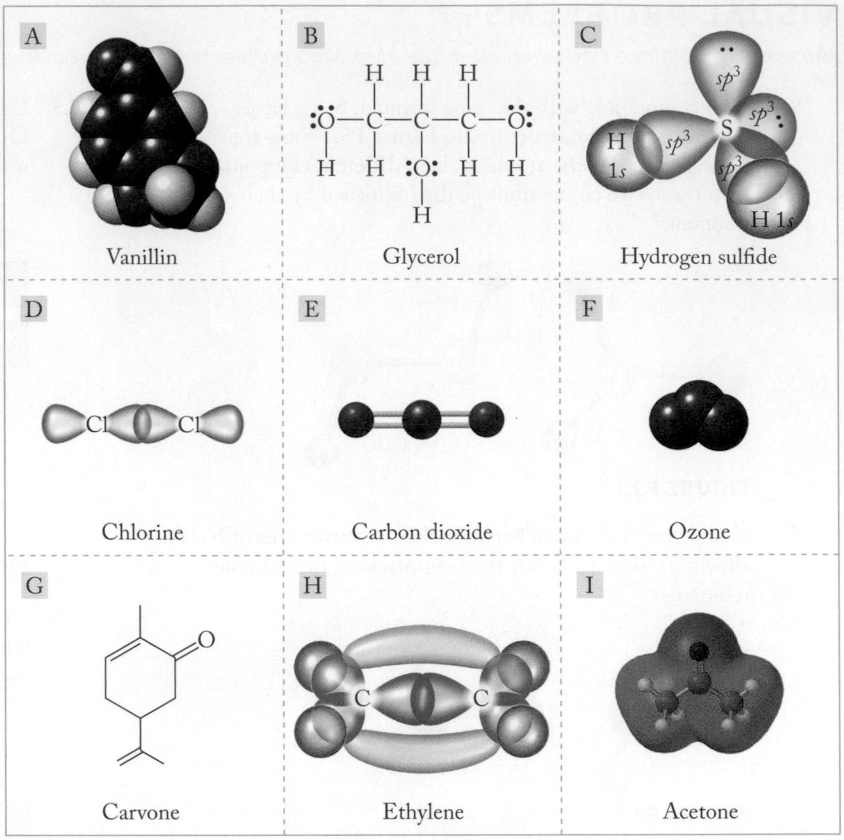

FIGURE P9.8

QUESTIONS AND PROBLEMS

Biological Activity and Molecular Shape; Valence-Shell Electron-Pair Repulsion (VSEPR) Theory

Concept Review

9.9. Why is the shape of a molecule determined by repulsions between electron pairs and not by repulsions between nuclei?

9.10. Do all resonance forms of a molecule have the same molecular geometry? Explain your answer.

9.11. How can SO_3 and BF_3 have different numbers of bonds but the same trigonal planar geometry?

9.12. Account for the range of bond angles from less than 100° to 180° in triatomic molecules.

9.13. In a molecule of ammonia, why is the repulsion between the lone pair and a bonding pair of electrons on nitrogen greater than the repulsion between two N—H bonding pairs?

9.14. Why is it important to draw a correct Lewis structure for a molecule before predicting its geometry?

9.15. Why does the seesaw structure have lower energy than a trigonal pyramidal structure derived by removing an axial atom from a trigonal bipyramidal AB_5 molecule?

*9.16. Which geometry do you predict will have lower energy: a square pyramid or a trigonal bipyramid? Why?

Problems

9.17. Arrange the following molecular geometries in order of increasing bond angle: (a) trigonal planar; (b) octahedral; (c) tetrahedral.

9.18. Arrange the following molecular geometries in order of increasing bond angle: (a) square planar; (b) tetrahedral; (c) square pyramidal.

_____

9.19. Which of the molecular geometries discussed in this chapter have more than one characteristic bond angle?

*9.20. Which molecular geometries for molecules of the general formula AB_x ($x = 2$ to 6) discussed in this chapter have the same bond angles when lone pairs replace one or more atoms?

_____

9.21. Which of the following molecular geometries does not lead to linear triatomic molecules after the removal of one or more atoms? (a) tetrahedral; (b) octahedral; (c) T-shaped

9.22. Which of the following molecular geometries does not lead to linear triatomic molecules after the removal of one or more atoms? (a) trigonal bipyramidal; (b) seesaw; (c) trigonal planar

_____

*9.23. Describe the molecular geometries that result from replacing one atom with a lone pair of electrons in an AB_7 molecule with a pentagonal bipyramidal geometry. (See Figure P9.7 for the shape of a pentagonal bipyramid.)

*9.24. Which atoms would you have to remove from the cubic AB₈ molecule shown in Figure P9.24 to create a geometry that approximates an octahedron?

FIGURE P9.24

9.25. Determine the molecular geometries of the following molecules: (a) GeH_4; (b) PH_3; (c) H_2S; (d) $CHCl_3$.

9.26. Determine the molecular geometries of the following molecules and ions: (a) NO_3^-; (b) NO_4^{3-}; (c) S_2O; (d) NF_3.

9.27. Determine the bond angles in the following ions: (a) NH_4^+; (b) SO_3^{2-}; (c) NO_2^-; (d) XeF_5^+.

9.28. Determine the bond angles in the following ions: (a) SCN^-; (b) BF_2^+; (c) ICl_2^-; (d) PO_3^{3-}.

9.29. Determine the geometries of the following ions: (a) $S_2O_3^{2-}$; (b) PO_4^{3-}; (c) NO_3^-; (d) NCO^-.

9.30. Determine the geometries of the following molecules: (a) ClO_2; (b) ClO_3; (c) IF_3; (d) SF_4.

9.31. Which of the following triatomic molecules, O_3, SO_2, N_2O, S_2O, and CO_2, have the same molecular geometry?

9.32. Which of the following species, N_3^-, O_3, CO_2, SCN^-, CNO^-, and NO_2^-, have the same molecular geometry?

9.33. The anion $C(CN)_3^-$ has a trigonal planar geometry about the central carbon atom. Draw Lewis structures for $C(CN)_3^-$, including resonance forms, and determine which structure contributes the most to the bonding.

9.34. The anion $C(NO_2)_3^-$ has a trigonal planar geometry about the carbon atom. Draw Lewis structures for $C(NO_2)_3^-$, including resonance forms, and determine which structure contributes the most to the bonding.

*9.35. The C—N—C bond angles in tri(methyl)amine, $N(CH_3)_3$, are approximately 109°, whereas the Si—N—Si bond angles in tri(silyl)amine, $N(SiH_3)_3$, are 120°, as shown in Figure P9.35. Explain the change in geometry when Si substitutes for C in this amine.

FIGURE P9.35

*9.36. The geometry about nitrogen in $N(CF_3)_3$ and $N(SCF_3)_3$ is trigonal planar for both complexes, as shown in Figure P9.36. Draw Lewis structures for each that are consistent with the observed geometry. (*Hint*: For $N(CF_3)_3$ consider an ionic form $[(CF_3)_2NCF_2]^+[F]^-$.)

FIGURE P9.36

*9.37. For many years, it was believed that the noble gases could not form covalently bonded compounds. However, xenon reacts with fluorine and oxygen. Reaction between xenon tetrafluoride and fluoride ions produces the pentafluoroxenate anion:

$$XeF_4 + F^- \rightarrow XeF_5^-$$

Draw Lewis structures for XeF_4 and XeF_5^-, and predict the geometry around xenon in XeF_4. The crystal structure of XeF_5^- compounds indicates a pentagonal bipyramidal orientation of valence pairs around Xe. Sketch the structure for XeF_5^-.

*9.38. The first compound containing a xenon–sulfur bond was isolated in 1998. Draw a Lewis structure for HXeSH and determine its molecular geometry at Xe.

*9.39. The Cl–O distances in ClO_2^+, ClO_2, and ClO_2^- are found to be 131 pm, 147 pm, and 156 pm, respectively. The corresponding O—Cl—O bond angles are 122°, 118°, and 110°. Draw Lewis structures consistent with these data.

9.40. Complete the Lewis structures of $SCNCl_3$ in Figure P9.40. Is the geometry around nitrogen the same in both molecules?

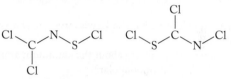

FIGURE P9.40

Polar Bonds and Polar Molecules

Concept Review

9.41. Explain the difference between a polar bond and a polar molecule.

9.42. Must a polar molecule contain polar covalent bonds? Why or why not?

9.43. Can a nonpolar molecule contain polar covalent bonds?

9.44. Compare the dipole moments of CO_2 and OCS.

Problems

9.45. Consider the following molecules: (a) CCl_4; (b) $CHCl_3$; (c) CO_2; (d) H_2S; (e) SO_2.
   a. Which of them contain polar bonds?
   b. Which are polar molecules?
   c. Which are nonpolar molecules?

9.46. **Molecules in Space** Simple diatomic molecules detected in interstellar space include CO, CS, SiO, SiS, SO, and NO. Arrange these molecules in order of increasing dipole moment on the basis of the location of the constituent elements in the periodic table, and then calculate the electronegativity differences from the data in Figure 8.6.

9.47. **Freon Ban** Compounds containing carbon, chlorine, and fluorine are known as Freons or chlorofluorocarbons (CFCs). Widespread use of these substances was banned because of their effect on the ozone layer in the upper atmosphere. Which of the following CFCs are polar and which are nonpolar? (a) Freon 11 ($CFCl_3$); (b) Freon 12 (CF_2Cl_2); (c) Freon 113 (Cl_2FCCF_2Cl)

9.48. Which of the following chlorofluorocarbons (CFCs) are polar and which are nonpolar? (a) Freon C318 (C_4F_8, cyclic structure); (b) Freon 1113 (C_2ClF_3); (c) $Cl_2HCCClF_2$

9.49. Which molecule in each of the pairs in Figure P9.49 has the larger dipole moment?

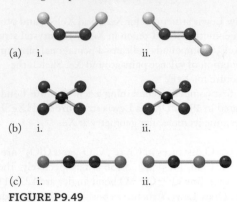

(a)   i.   ii.

(b)   i.   ii.

(c)   i.   ii.

FIGURE P9.49

9.50. Which molecule in each of the following pairs has the larger dipole moment? (a) BF_3 or BCl_3; (b) BCl_2F or $BClF_2$

9.51. Aluminum chloride has the Lewis structure shown in Figure P9.51.
  a. What is the geometry about the aluminum atom?
  b. Is Al_2Cl_6 polar or nonpolar?

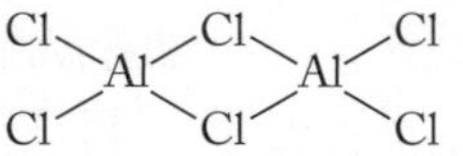

$$\begin{array}{ccc} Cl & Cl & Cl \\ \diagdown Al \diagdown & Al \diagdown & \\ Cl & Cl & Cl \end{array}$$

FIGURE P9.51

9.52. **Cleaning Silicon Chips** Nitrogen trifluoride, NF_3, is used in the electronics industry to clean surfaces. NF_3 is also a potent greenhouse gas.
  a. Draw the Lewis structure of NF_3 and determine its molecular geometry.
  b. BF_3 and NF_3 both have three covalently bonded fluorine atoms around a central atom. Do they have the same dipole moment?
  c. Could BF_3 also behave as a greenhouse gas?

Valence Bond Theory

Concept Review

9.53. Describe in your own words the differences between sigma and pi bonds.

9.54. Why aren't the orbitals on isolated atoms hybridized?

9.55. The bond angles in PF_3 are 97.8°.
  a. Explain the size of this angle by using hybrid orbitals.
  *b. Explain the size of this angle without using hybrid orbitals.

*9.56. What combination of s, p, and d orbitals would we need to form four σ and two π bonds to a sulfur atom?

Problems

9.57. What is the hybridization of nitrogen in each of the following ions and molecules? (a) NO_2^+; (b) NO_2^-; (c) N_2O; (d) N_2O_5; (e) N_2O_3

9.58. Identify the hybridization of the carbon atoms indicated by the arrows in the structures shown in Figure P9.58.

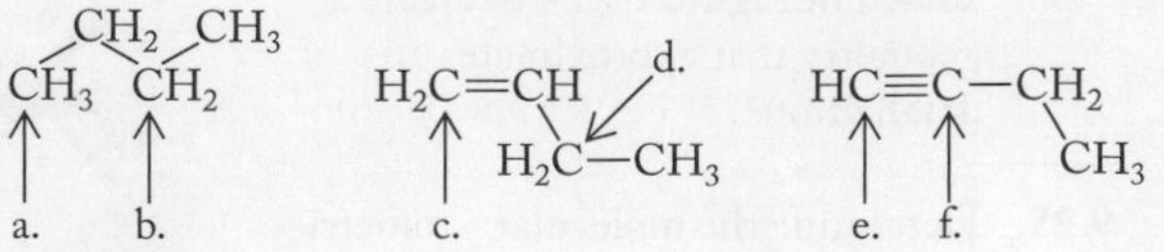

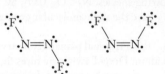

a.   b.   c.   e.   f.

FIGURE P9.58

9.59. **Airbags** Azides such as sodium azide, NaN_3, are used in automobile airbags as a source of nitrogen gas. Another compound with three nitrogen atoms bonded together is N_3F. What differences are there in the arrangement of the electrons around the nitrogen atoms in the azide ion (N_3^-) and N_3F? Is there a difference in the hybridization of the central nitrogen atom?

9.60. N_3F decomposes to nitrogen and N_2F_2 by the following reaction:

$$2\ N_3F \rightarrow 2\ N_2 + N_2F_2$$

N_2F_2 has two possible structures, as shown in Figure P9.60. Are the differences between these structures related to differences in the hybridization of nitrogen in N_2F_2? Identify the hybrid orbitals that account for the bonding in N_2F_2. Are they the same as those in acetylene, C_2H_2?

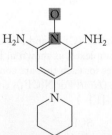

FIGURE P9.60

9.61. How does the hybridization of the sulfur atom change in the series SF_2, SF_4, and SF_6?

9.62. How does the hybridization of the central atom change in the series CO_2, NO_2, O_3, and ClO_2?

9.63. **Minoxidil** The drug minoxidil was originally developed for treating high blood pressure but is now used primarily for treating hair loss. The Lewis structure of minoxidil is shown in Figure P9.63. Complete the Lewis structure by adding lone pairs where needed. Assign formal charges to the nitrogen and oxygen highlighted in red. Describe the bonding around the nitrogen in the N–O group.

$$H_2N \overset{\displaystyle O}{\underset{\displaystyle}{\overset{\displaystyle |}{N}}} NH_2$$

FIGURE P9.63

*9.64. Draw the Lewis structure of the chlorite ion, ClO_2^-, which is used as a bleaching agent. Include all resonance structures in which formal charges are closest to zero.

What is the shape of the ion? Suggest a hybridization scheme for the central chlorine atom that accounts for the structures you have drawn.

*9.65. **Perchlorate Ion and Human Health** Perchlorate ion adversely affects human health by interfering with the uptake of iodine in the thyroid gland, but because of this behavior, it also provides a useful medical treatment for hyperthyroidism, or overactive thyroid. Draw the Lewis structure of the perchlorate ion, ClO_4^-. Include all resonance structures in which formal charges are closest to zero. What is the shape of the ion? Suggest a hybridization scheme for the central chlorine atom that accounts for this shape.

9.66. Draw a Lewis structure for CF_3PCF_2 where the fluorine atoms are all bonded to carbon atoms. Determine its molecular geometry at P and the hybridization of the phosphorus atom.

9.67. Synthesis of the first compound of argon was reported in 2000. HArF was made by reacting Ar with HF. Draw a Lewis structure for HArF, and determine the hybridization of Ar in this molecule.

9.68. The Lewis structure of N_4O, with the skeletal structure O—N—N—N—N, contains one N—N single bond, one N=N double bond, and a N≡N triple bond. Is the hybridization of all the nitrogen atoms the same?

*9.69. The trifluorosulfate anion was isolated in 1999 as the tetramethylammonium salt $[(CH_3)_4N]^+[SO_2F_3]^-$.
   a. Determine the geometry around the nitrogen atom in the cation and describe the C—N bonding according to valence bond theory.
   b. The S—O bond lengths in the anion are both 143 pm. Draw the Lewis structure that is consistent with this bond length.
   c. What is the molecular geometry of the anion?

9.70. **Treating Diabetes** The drug metformin (Figure P9.70) has been used to treat type 2 diabetes for a half-century by suppressing glucose production. Metformin contains five nitrogen atoms. Determine the geometry around each nitrogen atom, and describe the bonding according to valence bond theory.

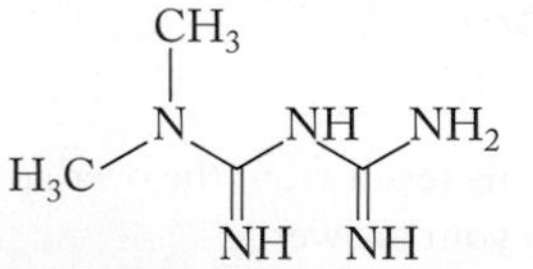

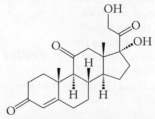

FIGURE P9.70

Shape and Interactions with Large Molecules; Chirality and Molecular Recognition

Concept Review

9.71. What features do Lewis structures, Kekulé structures, condensed structures, and carbon-skeleton structures share in common? What features differentiate these four kinds of structures?

9.72. Explain why alkanes don't have optical isomers.

9.73. Can molecules with more than one central atom have resonance forms? Explain your answer.

*9.74. Can hybrid orbitals be associated with more than one atom? Explain your answer.

*9.75. Are resonance structures examples of electron delocalization? Explain your answer.

9.76. Can sp^2 and sp hybridized carbon atoms be chiral centers? Explain your answer.

9.77. Which of the following objects are chiral? (a) a baseball bat with no lettering on it; (b) a pair of scissors; (c) a boot; (d) a fork

9.78. Why is it difficult to assign a single geometry to a molecule with more than one central atom?

Problems

9.79. Bombykol is the compound synthesized by female silkworm moths to attract mates. Convert the carbon-skeleton structure in Figure P9.79 to a condensed structure.

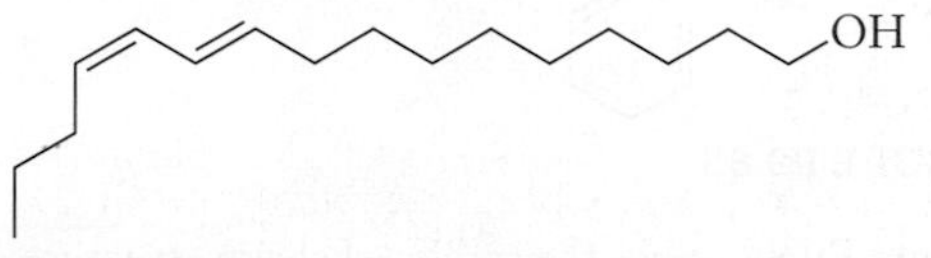

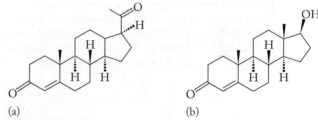

FIGURE P9.79

9.80. Fucoserratene is the compound synthesized by a brown alga to reproduce. Convert the carbon-skeleton structure in Figure P9.80 to a condensed structure.

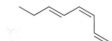

FIGURE P9.80

9.81. **Sex Hormones** Progesterone (Figure P9.81a) is a hormone involved in regulating menstrual cycles and pregnancy, and testosterone (Figure 9.81b) is the primary male sex hormone. What functional groups do these two hormones have in common?

(a)   (b)

FIGURE P9.81

9.82. **Steroid Hormones** Cortisone (Figure P9.82) is a steroid produced by the body in response to stress. It suppresses the immune system and can be administered as a drug to reduce inflammation, pain, and swelling. Identify all the functional groups in cortisone.

FIGURE P9.82

9.83. **Unsaturated Fats** Oleic acid (Figure P9.83) is a fat derived from olive oil. Write a condensed structure for oleic acid.

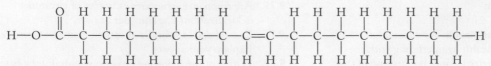

FIGURE P9.83

9.84. Draw a carbon-skeleton structure for oleic acid (Figure P9.83).

9.85. **Prozac** Fluoxetine (Figure P9.85) is an antidepressant medication sold commercially as Prozac. Circle the delocalized π electrons in fluoxetine.

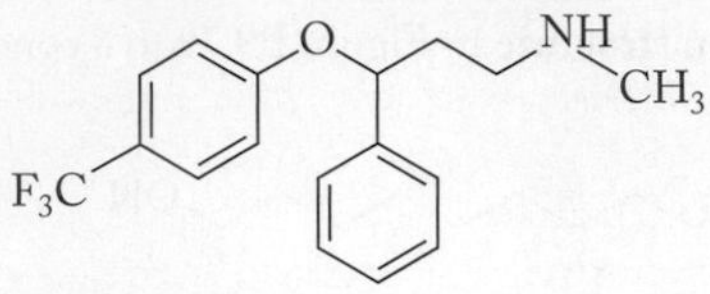

FIGURE P9.85

*9.86. Figure P9.86 shows the carbon-skeleton structure of the antifungal compound capillin. Are there delocalized electrons in capillin? If so, identify them.

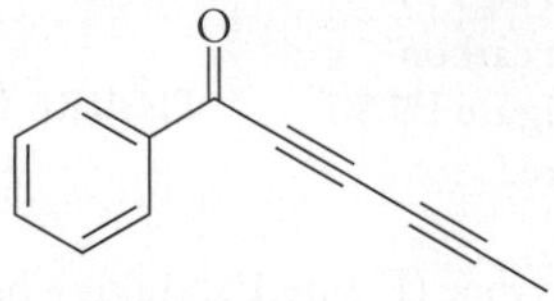

FIGURE P9.86

9.87. **Artificial Sweeteners** Acesulfame potassium is one of many artificial sweeteners used in food. It is 200 times sweeter than sugar and has the structure shown in Figure P9.87. What is the geometry at each of the atoms in the six-membered ring? Which atomic or hybrid orbitals overlap to form the C—O and C—N bonds? In which atomic or hybrid orbital is the extra electron on N located?

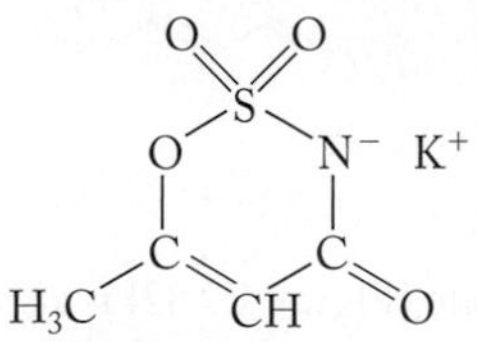

FIGURE P9.87

*9.88. **First Artificial Sweetner** Saccharin (Figure P9.88) was the first artificial sweetener, discovered in 1879. Like acesulfame potassium, it contains a sulfur atom adjacent to a nitrogen atom. Why don't all the atoms in the five-membered ring lie in the same plane? Why is it difficult to explain the bonding between S and O by using the hybrid orbitals described in Section 9.4?

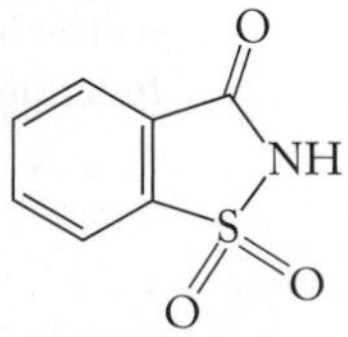

FIGURE P9.88

9.89. Which molecules in Figure P9.89 are chiral?

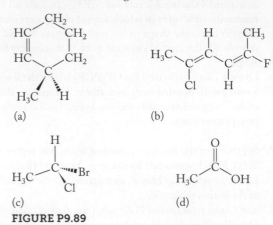

FIGURE P9.89

9.90. Which molecules in Figure P9.90 are chiral?

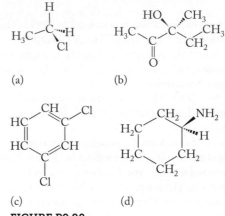

FIGURE P9.90

Molecular Orbital Theory

Concept Review

9.91. Do all σ molecular orbitals result from the overlap of *s* atomic orbitals? Explain your answer.

9.92. Do all π molecular orbitals result from the overlap of *p* atomic orbitals? Explain your answer.

9.93. Are *s* atomic orbitals with different principal quantum numbers (*n*) as likely to overlap and form MOs as *s* atomic orbitals with the same value of *n*? Explain your answer.

9.94. Which atomic orbitals are more likely to mix to form a set of molecular orbitals—a 2*s* and a 3*p* orbital or a 4*s* and a 5*p* orbital?

9.95. Why might some molecules with even numbers of valence electrons be paramagnetic?

9.96. How does the molecular orbital diagram for a homonuclear diatomic species differ from that of a heteronuclear diatomic species?

9.97. How does the sea-of-electrons model (Chapter 8) explain the high electrical conductivity of gold? How does band theory explain this?

9.98. Some scientists believe that the solid hydrogen that forms at very low temperatures and high pressures may conduct electricity. Is this hypothesis supported by band theory?

9.99. Describe in general terms the differences in composition and conduction between n-type and p-type semiconductors.

9.100. How might doping of silicon with germanium affect the conductivity of silicon?

Problems

9.101. Make a sketch showing how two $1s$ orbitals overlap to form a σ_{1s} bonding molecular orbital and a σ_{1s}^* antibonding molecular orbital.

9.102. Make a sketch showing how two $2p_y$ orbitals overlap "sideways" to form a π_{2p} bonding molecular orbital and a π_{2p}^* antibonding molecular orbital.

9.103. Consider the following molecular ions: N_2^+, O_2^+, C_2^+, and Br_2^{2-}. Using MO theory, (a) write their orbital electron configuration; (b) predict their bond orders; (c) state whether you expect any of these species to exist.

9.104. Diatomic noble gas molecules, such as He_2 and Ne_2, do not exist.
   a. Write their orbital electron configurations.
   b. Does removing one electron from each of these molecules create molecular ions (He_2^+ and Ne_2^+) that are more stable than He_2 and Ne_2?

9.105. Which of the following molecular ions is expected to have one or more unpaired electrons? (a) N_2^+; (b) O_2^+; (c) C_2^{2+}; (d) Br_2^{2-}; (e) O_2^-; (f) O_2^{2-}; (g) N_2^{2-}; (h) F_2^+

9.106. Which of the following molecular ions have electrons in π antibonding orbitals? (a) O_2^-; (b) O_2^{2-}; (c) N_2^{2-}; (d) F_2^+; (e) N_2^+; (f) O_2^+; (g) C_2^{2+}; (h) Br_2^{2+}

9.107. The odd-electron molecule ClO affects the atmospheric chemistry of chlorofluorocarbons as illustrated by the reaction (where the * indicates an excited-state oxygen atom):

$$CF_2Cl_2 + O^* \rightarrow ClO + CF_2Cl$$

Draw a molecular orbital diagram for ClO. Is the odd electron in a bonding or antibonding orbital?

9.108. The elusive molecule boron monoxide, BO, can be stabilized by bonding to platinum. Draw a molecular orbital diagram for BO. Is the odd electron in a bonding or antibonding orbital?

9.109. For which of the following diatomic molecules does the bond order increase with the gain of two electrons, forming the corresponding anion with a 2− charge?
   a. $B_2 + 2\,e^- \rightarrow B_2^{2-}$ c. $N_2 + 2\,e^- \rightarrow N_2^{2-}$
   b. $C_2 + 2\,e^- \rightarrow C_2^{2-}$ d. $O_2 + 2\,e^- \rightarrow O_2^{2-}$

9.110. For which of the following diatomic molecules does the bond order increase with the loss of two electrons, forming the corresponding cation with a 2+ charge?
   a. $B_2 \rightarrow B_2^{2+} + 2\,e^-$ c. $N_2 \rightarrow N_2^{2+} + 2\,e^-$
   b. $C_2 \rightarrow C_2^{2+} + 2\,e^-$ d. $O_2 \rightarrow O_2^{2+} + 2\,e^-$

9.111. Do the 1+ cations of homonuclear diatomic molecules of the second-row elements always have shorter bond lengths than the corresponding neutral molecules?

9.112. Do any of the anions of the homonuclear diatomic molecules formed by B, C, N, O, and F have shorter bond lengths than those of the corresponding neutral molecules? Consider only the anions with 1− or 2− charge.

9.113. Thin films of doped diamond hold promise as semiconductor materials. Trace amounts of nitrogen impart a yellow color to otherwise colorless pure diamonds.
   a. Are nitrogen-doped diamonds examples of semiconductors that are p-type or n-type?
   b. Draw a picture of the band structure of diamond to indicate the difference between pure diamond and N-doped (nitrogen-doped) diamond.
   *c. N-doped diamonds absorb violet light at about 425 nm. What is the magnitude of E_g that corresponds to this wavelength?

9.114. **Hope Diamond** Trace amounts of boron give diamonds (including the Smithsonian's Hope Diamond) a blue color (Figure P9.114).
   a. Are boron-doped diamonds examples of semiconductors that are p-type or n-type?
   b. Draw a picture of the band structure of diamond to indicate the difference between pure diamond and B-doped diamond.
   *c. What is the band gap in energy if blue diamonds absorb red-orange light with a wavelength of 675 nm?

FIGURE P9.114

Additional Problems

9.115. Draw the Lewis structure for the two ions in ammonium perchlorate (NH_4ClO_4), which is used as a propellant in solid fuel rockets, and determine the molecular geometries of the two polyatomic ions.

9.116. **Arsenic-Based DNA?** The waters of Mono Lake in the eastern Sierra Mountains of California (Figure P9.116) are rich in arsenate ion (AsO_4^{3-}). Some biochemists have proposed that microorganisms in this environment actually incorporate arsenate into their DNA in place of the phosphate ion (PO_4^{3-}). Draw the Lewis structure of the arsenate ion that yields the most favorable formal charges.

Predict the angles between the arsenic–oxygen bonds in the arsenate anion.

FIGURE P9.116

9.117. Consider the molecular structure of the amino acid glycine shown in Figure P9.117. What is the angle formed by the N—C—C bonds in this structure? What are the O—C=O and C—O—H bond angles?

FIGURE P9.117

9.118. Cl_2O_2 may play a role in ozone depletion in the stratosphere. In the laboratory, a reaction between ClO_2F and $AlCl_3$ produces Cl_2O_2 and $AlCl_2F$. Draw the Lewis structure for Cl_2O_2 on the basis of the skeletal structure in Figure P9.118. What is the geometry about the central chlorine atom?

FIGURE P9.118

9.119. Bombardment of Cl_2O_2 molecules (Figure P9.119) with intense radiation is thought to produce the two compounds with the skeletal structures shown in Figure P9.119.
a. Do both of these molecules have linear geometry? Explain your answer.
b. Do they have the same dipole moment? Explain your answer.

FIGURE P9.119

*** 9.120.** Complete the Lewis structure for the cyclic structure of Cl_2O_2 shown in Figure P9.120.
a. Is the cyclic Cl_2O_2 molecule planar?
b. Is the molecule polar or nonpolar?

FIGURE P9.120

9.121. In 1999 the ClO^+ ion, a potential contributor to stratospheric ozone depletion, was isolated in the laboratory.
a. Draw the Lewis structure for ClO^+.
b. Using the molecular orbital diagram in Figure P9.121, determine the order of the Cl—O bond in ClO^+.

9.122. The molecule trinitramide, $N(NO_2)_3$, was first prepared in late 2010 by chemists in Sweden. Draw Lewis structures for trinitramide and predict the geometry

σ^*_{3p}

π^*_{3p}

π_{3p}

σ_{3p}

σ^*_{3s}

σ_{3s}

FIGURE P9.121

about the central nitrogen atom. Do all resonance forms of $N(NO_2)_3$ have the same geometry?

9.123. Cola Beverages Phosphoric acid imparts a tart flavor to cola beverages. The skeletal structure of phosphoric acid is shown in Figure P9.123. Complete the Lewis structure for phosphoric acid in which formal charges are closest to zero. What is the molecular geometry around the phosphorus atom in your structure?

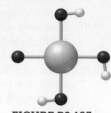

FIGURE P9.123

9.124. The fluoroaluminate anions AlF_4^- and AlF_6^{3-} have been known for more than a century, but the structure of the pentafluoroaluminate ion, AlF_5^{2-}, was not determined until 2003. Draw the Lewis structures for AlF_3, AlF_4^-, AlF_5^{2-}, and AlF_6^{3-}. Determine the molecular geometry of each molecule or ion. Describe the bonding in AlF_3, AlF_4^-, AlF_5^{2-}, and AlF_6^{3-} by using valence bond theory.

*** 9.125.** Thermally unstable compounds can sometimes be synthesized using matrix isolation methods in which the compounds are isolated in a nonreactive medium such as frozen argon. The reaction of boron with carbon monoxide produces compounds with these skeletal structures: B—B—C—O and O—C—B—B—C—O. For each of these compounds, draw the Lewis structure that minimizes formal charges. Do any of your structures contain atoms with incomplete octets? Predict the molecular geometries of BBCO and OCBBCO.

*** 9.126.** The products of the reaction between boron and NO can be trapped in solid argon matrices. Among the products is BNO. Draw the Lewis structure for BNO, including any resonance forms. Assign formal charges and predict which structure provides the best description of the bonding in this molecule. Do any of your structures contain atoms without complete octets? Predict the molecular geometry of BNO.

9.127. Compounds May Help Prevent Cancer Broccoli, cabbage, and kale contain compounds that break down in the human body to form isothiocyanates, whose presence may reduce the risk of certain types of cancer. The simplest isothiocyanate is methyl isothiocyanate, CH_3NCS. Draw the Lewis structure for CH_3NCS, including all resonance forms. Assign formal charges and determine which structure is likely to contribute the most to bonding. Predict the molecular geometry of the molecule at both carbon atoms.

9.128. Toxic to Insects and People Methyl thiocyanate (CH_3SCN) is used as an agricultural pesticide and fumigant. It is slightly water soluble and is readily absorbed through the skin; it is highly toxic if ingested. Its toxicity stems in part from its metabolism to cyanide ion. Draw three resonance structures for methyl thiocyanate. Assign formal charges and predict which structure would be the most stable. Predict the molecular geometry of the molecule at both carbon atoms.

9.129. Skunks The pungent smell of skunk spray is detected by receptors in the nose when the skunk secretes butanethiol, $CH_3(CH_2)_3SH$. Draw a carbon-skeleton structure for butanethiol.

9.130. Grapefruit Not all sulfur-containing compounds have unpleasant aromas. Figure P9.130 shows the structure of the compound primarily responsible for the aroma of grapefruit. Identify the chiral carbon in this structure.

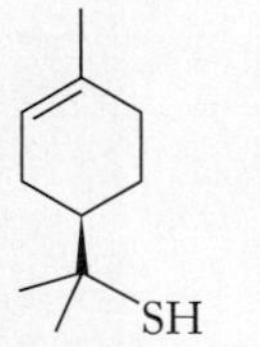

FIGURE P9.130

9.131. Borazine, $B_3N_3H_6$ (a cyclic compound with alternating B and N atoms in the ring), is isoelectronic with benzene (C_6H_6). Are there delocalized π electrons in borazine?

9.132. Unlike O_2, sulfur monoxide (SO) is highly unstable, decomposing to a mixture of S_2O and O_2 in less than 1 second. Using the O_{2s}, O_{2p}, S_{3s}, and S_{3p} atomic orbitals, construct an approximate molecular orbital diagram for SO. Is SO diamagnetic or paramagnetic?

*9.133. Some chemists think HArF consists of H^+ ions and Ar F^- ions. Using an appropriate MO diagram, determine the bond order of the Ar—F bond in ArF$^-$.

*9.134. Assume that HArF is a molecular compound.
   a. Draw its Lewis structure.
   b. What are the formal charges on Ar and F in the structure you drew?
   c. What is the shape of the molecule?
   d. Is HArF polar?

9.135. Which of the following unstable nitrogen oxides, N_2O_2, N_2O_5, and N_2O_3, are polar molecules? (N_2O_2 and N_2O_3 have N—N bonds; N_2O_5 does not.)

9.136. Explain why O_2 is paramagnetic.

9.137. Using an appropriate molecular orbital diagram, show that the bond order in the disulfide anion S_2^{2-} is equal to 1. Is S_2^{2-} diamagnetic or paramagnetic?

9.138. Use molecular orbital diagrams to determine the bond order of the peroxide (O_2^{2-}) and superoxide (O_2^-) ions. Are these bond order values consistent with those predicted from Lewis structures?

9.139. Elemental sulfur has several allotropic forms, including cyclic S_8 molecules. What is the orbital hybridization of sulfur atoms in this allotrope? The bond angles are about 108°.

*9.140. Which $3d$ atomic orbitals have the proper orientation to overlap with a $4p_z$ atomic orbital?

*9.141. Ozone (O_3) has a dipole moment (0.54 D). How can a molecule with only one kind of atom have a dipole moment?

*9.142. The bond angle in H_2O is 104.5°; the bond angles in H_2S, H_2Se, and H_2Te are very close to 90°. Which theory would you apply to describe the geometry in H_2S, H_2Se, and H_2Te: VSEPR? Valence bond without invoking hybrid orbitals? Valence bond theory, using hybrid orbitals? Why?

9.143. Garlic Garlic contains the molecule alliin (Figure P9.143). When garlic is crushed or chopped, a reaction occurs that converts alliin into the molecule allicin, which is primarily responsible for the aroma we associate with garlic.
   a. Describe the molecular geometry about the sulfur atoms in both compounds.
   b. Do any of the sulfur atoms in allicin have the same geometry as the sulfur atoms in the volatile sulfur compounds that cause bad breath (H_2S, CH_3—SH, and CH_3—S—CH_3)?

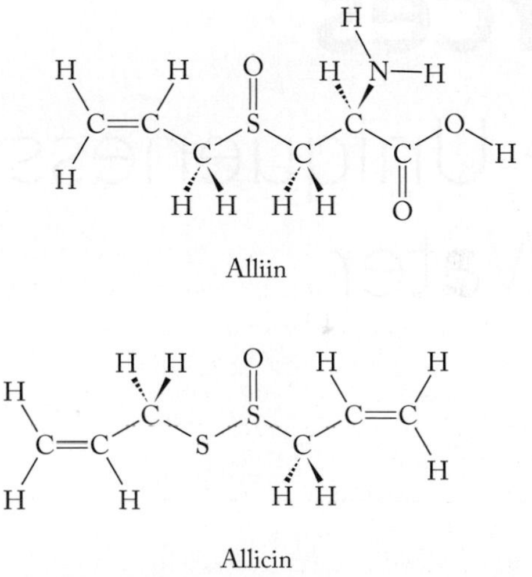

Alliin

Allicin

FIGURE P9.143

10

Intermolecular Forces

The Uniqueness of Water

WATER AND LIFE All life on Earth depends on water—the substance itself, the substances dissolved in it, and its unique physical properties.

PARTICULATE **REVIEW**

Polar Bonds versus Polar Molecules

In Chapter 10 we explore the connections between molecular polarity and attractive forces between molecules. Here are representations of four molecules: carbon dioxide, oxygen, water, and ozone.

- Label the polar covalent bonds in these molecules with partial charges ($\delta+$ and $\delta-$).

- Which molecule is nonpolar despite containing polar bonds?

- Which molecule is polar and contains polar bonds?

- Which molecule is polar even though it contains nonpolar bonds?

   (Review Section 9.3 if you need help.)

(Answers to Particulate Review questions are in the back of the book.)

One Formula, Two Structures

The physical and chemical properties of a compound
depend upon its structure. Here are ball-and-stick models
of two compounds that have the same molecular formula
but different structures and therefore different properties.
One compound is a liquid at room temperature, whereas
the other is a gas. As you read this chapter, look for ideas
that will help you answer these questions:

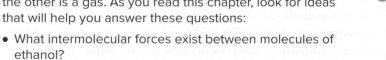

Ethanol   Dimethyl ether

- What intermolecular forces exist between molecules of
  ethanol?

- What intermolecular forces exist between molecules of dimethyl ether?

- Compare the relative strength of these intermolecular forces to determine which
  substance is a liquid at room temperature and which is a gas.

497

Learning Outcomes

LO1 Explain the origins of dispersion forces, ion–dipole forces, dipole–dipole forces, dipole–induced dipole forces, and hydrogen bonds

LO2 Explain the effect of intermolecular forces on the boiling points of compounds
Sample Exercises 10.1, 10.2

LO3 Explain the effect of intermolecular forces on the solubilities of compounds in water and other solvents
Sample Exercises 10.3, 10.4

LO4 Calculate the solubility of gases in water by using Henry's Law
Sample Exercise 10.5

LO5 Calculate the vapor pressure of a pure liquid
Sample Exercise 10.6

LO6 Identify the regions of a phase diagram and explain the effect of temperature and pressure on phase changes
Sample Exercise 10.7

LO7 Describe the role of hydrogen bonding in determining the unique properties of water

10.1 Intramolecular Forces versus Intermolecular Forces

In the past few chapters we have examined bonding in molecules and have shown how the combination of attractive and repulsive electrical forces between pairs of electrons determines molecular geometry. We refer to these interactions *within* a molecule as *intramolecular forces*. In this chapter we begin the study of the forces that act *between* molecules and *between* molecules and ions. These *intermolecular forces* are also electrostatic in nature, but they are weaker than intramolecular forces, and they typically act over longer distances. Intermolecular forces have considerable influence on the physical properties of all substances. For example, interactions between solvent and solute particles in solutions affect the solubility of one substance in another.

We often take water for granted because it is such a familiar substance. However, water's physical properties are truly remarkable and so essential to life as we know it that astronomers searching for life on other planets look first for the presence of liquid water. Water molecules are polar, which makes water capable of dissolving many substances because of favorable interactions between water molecules and the particles of these substances. Seawater contains both dissolved ionic compounds such as sodium chloride and dissolved molecular species such as oxygen gas. Aquatic life—indeed, all life on Earth—relies on the presence of these and many other substances in salt water, fresh water, and water-based biological fluids, such as blood in animals and sap in plants. The ability of water to dissolve and transport substances is governed by the strength and number of intermolecular interactions between water molecules and other ions and molecules.

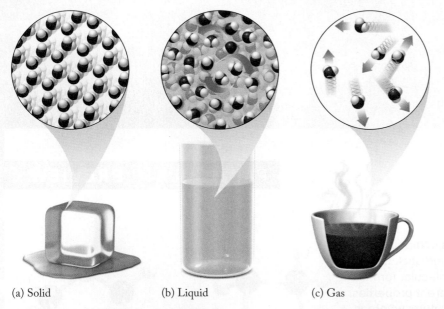

(a) Solid   (b) Liquid   (c) Gas

FIGURE 10.1 (a) The molecules of H_2O in solid ice are locked in place by the strength of intermolecular attractions. (b) In liquid water, molecules of H_2O have more energy and flow past one another. (c) In water vapor, the molecules have enough energy to overcome nearly all intermolecular attractions and move freely throughout the space they occupy.

Interactions between water molecules influence the phase of water and the conditions under which phase transitions occur. We know that ice cubes float in a glass of water and that ice floats on a pond or river. Sometimes water exists simultaneously in all three phases, such as on an early spring day when solid ice melts to liquid water in the sunlight while water vapor condenses into white clouds that dot the sky. The polarity of water influences its physical properties, such as its boiling point and melting point.

In a more general case, let's think about the three common states of matter (Figure 10.1) and how they differ based on the kinetic energy of the particles in them and their ability to overcome the attractive forces between particles. In a solid, the average kinetic energy of the particles is insufficient to overcome these forces of attraction. Consequently the particles have the same nearest neighbors over time and do not move much. In a liquid, the average kinetic energy of the particles is sufficient to overcome some of the attractive forces; particles in a liquid experience more freedom of motion and can move past one another. In gases, the average kinetic energy of the particles is sufficient to overcome essentially all of the attractive forces between them, imparting nearly complete freedom of motion to the widely separated particles.

The stronger the attractive forces among the particles in a solid, the greater the amount of energy needed to overcome those forces to cause melting or vaporization. Thus, a substance made of particles that interact relatively strongly has high melting and boiling points, which means the substance is likely to be a solid at room temperature and normal pressure. Under the same conditions, a substance with somewhat weaker particle–particle interactions has a lower melting point and is more likely to be a liquid. A substance with very weak particle–particle interactions has even lower melting and boiling points and is more likely to be a gas. Let's begin our study of intermolecular forces by examining the different types of interactions we observe between atoms and nonpolar molecules.

10.2   Dispersion Forces

The macroscopic properties of substances such as melting points and boiling points, which we can observe and measure, arise from the microscopic interactions between the particles that make up those substances. Let's see how this happens by looking at a group of elements that exist as single atoms: the noble gases. Table 10.1 lists their atomic numbers and boiling points. Note the correlation between them: boiling points increase as atomic numbers increase. To understand why this correlation exists, let's think about what happens at the particle level when a liquid vaporizes. As we discussed above, the particles in liquids (and solids) are in direct contact with each other. However, when a liquid vaporizes, the contacts are broken: the gas-phase particles become essentially independent. Separating liquid-phase particles that are attracted to each other requires energy, and the stronger the particles' attractions for each other, the greater the amount of energy needed to separate them: the greater the energy required to separate these particles, the higher the boiling points.

Why do atoms with greater atomic numbers interact more strongly? For that matter, *how* do atoms not bonded to each other interact? German-American physicist Fritz London (1900–1954) proposed one explanation for these interactions in 1930. It was based on the notion that when atoms approach each other (Figure 10.2a), they interact in ways that are similar to the electrostatic interactions involved in covalent bond formation. In other words, one atom's positive

CHEMTOUR
Intermolecular Forces

CONNECTION In Chapter 5 we looked at the flow of energy accompanying phase changes of water (solid ice ⇌ liquid water ⇌ water vapor) in terms of the changes in kinetic and potential energies of the molecules.

TABLE 10.1 Boiling Points of Noble Gases

Noble Gas	Atomic View	Z	Boiling Point (K)
He		2	4
Ne		10	27
Ar		18	87
Kr		36	120
Xe		54	165
Rn		86	211

CONNECTION Polar bond formation and the color scales and symbols used to represent it were introduced in Chapter 8.

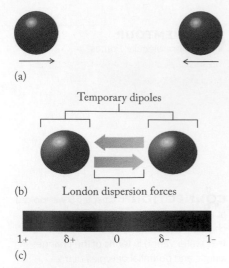

FIGURE 10.2 (a) Two atoms, each with a symmetrical distribution of electrons, approach each other and (b) create two temporary dipoles as their nuclei and electron clouds interact. (c) The strengths of temporary dipoles are shown using the same color scale used in Chapter 8 to represent the strengths of permanent dipoles in bonds.

CONNECTION The concept of screening of outer electrons by inner electrons was presented in Chapter 7 when we discussed trends in the sizes of atoms and ions.

nucleus is attracted to the other atom's negative electrons, and vice versa, even as their electron clouds repel each other. These competing interactions can cause the electrons around each atom to be distributed unevenly, producing temporary **induced dipoles** of partial electrical charge (Figure 10.2b) that are attracted to regions of opposite partial charge on the adjacent atom. (This attraction is represented by the arrows in Figure 10.2b.) In Chapter 8 we used different colors to represent the partial electrical charges created by uneven sharing of bonding pairs of electrons. In this chapter we use those same colors (Figure 10.2c) to show partial electrical charges caused by uneven distributions of electrons in neutral atoms and over entire molecules.

The presence of temporary dipoles in atoms and molecules creates a way for them to interact with other atoms and molecules. In a molecule, though, the partial charges in temporary dipoles are likely to be distributed over considerably greater distances than in single atoms because the atomic nuclei and the clouds of electrons shared by groups of atoms within the molecule (or even the entire molecule) interact with the electrons and nuclei in neighboring atoms or molecules.

Interactions based on the presence of temporary dipoles are called **dispersion forces**, also referred to as **London forces** in honor of Fritz London's pioneering work. The strengths of the interactions increase as the numbers of electrons in atoms and molecules increase. Because all atoms and molecules have electrons, all atoms and molecules experience London forces to some degree. The larger the cloud of electrons surrounding a nucleus in an atom or several nuclei in a molecule, the more likely those electrons are to be distributed unevenly, or *polarized*. Electrons in larger atoms are held less tightly by the nucleus because of both their greater average distance from the nucleus and the screening of the nuclear charge by electrons in lower-energy orbitals. Consequently, they are more easily polarized than electrons in smaller atoms or molecules. Greater **polarizability** leads to stronger temporary dipoles and stronger intermolecular interactions, so London dispersion forces become stronger as atoms and molecules become larger. This trend in polarizability accounts for the correlation between the boiling points and atomic numbers of the noble gases.

CONCEPT TEST

Rank the following atoms in order of increasing polarizability: argon, hydrogen, krypton, and neon.

(Answers to Concept Tests are in the back of the book.)

Polarizability also explains similar trends in the boiling points of the halogens (Table 10.2). These elements exist as diatomic molecules in which equal sharing of the bonding pairs of electrons by identical atoms means that the molecules have no permanent dipoles. Comparing the boiling points of the halogens with their molar masses (as a measure of particle size) reveals a trend that mimics the one we saw with the noble gases: boiling point increases as particle size increases. A similar trend is also observed in the boiling points of a series of straight-chain hydrocarbons (Figure 10.3). London's explanation of these trends was also the same: larger clouds of increasing numbers of

TABLE 10.2 Boiling Points of the Halogens

Halogen	Molecular View	Molar Mass (g/mol)	Boiling Point (K)
F_2		38	85
Cl_2		71	239
Br_2		160	332
I_2		254	457
At_2		420	610

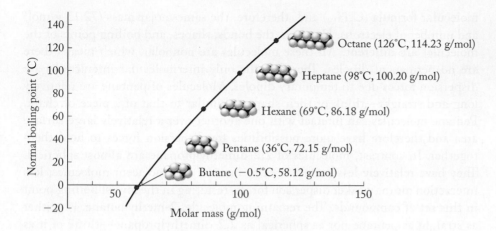

FIGURE 10.3 The boiling points of straight-chain alkanes increase as masses increase and the magnitude of the dispersion forces between molecules increases.

electrons per molecule are more polarizable. Greater polarizability means they are more likely to form temporary dipoles that attract molecules to each other in the liquid phase and inhibit their vaporization.

CONNECTION In Section 5.9 we considered the structure of straight-chain hydrocarbons.

CONCEPT **TEST**

Explain why CF_4 is a gas at room temperature but CCl_4 is a liquid.

The Importance of Shape

Molecular shape, as well as size, plays a role in determining the strength of London dispersion forces. Consider the molecular structures and boiling points of the three hydrocarbons in Figure 10.4. These compounds all have the same

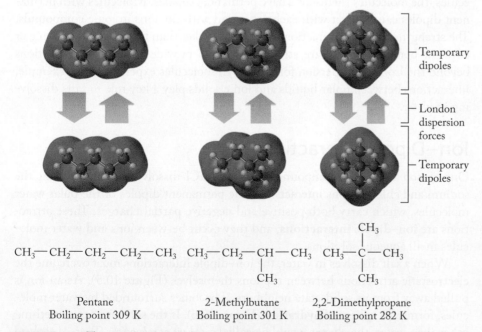

$CH_3-CH_2-CH_2-CH_2-CH_3$

Pentane
Boiling point 309 K

$CH_3-CH_2-CH-CH_3$
        |
        CH_3

2-Methylbutane
Boiling point 301 K

$CH_3-\overset{\overset{\displaystyle CH_3}{|}}{\underset{\underset{\displaystyle CH_3}{|}}{C}}-CH_3$

2,2-Dimethylpropane
Boiling point 282 K

FIGURE 10.4 Three molecular shapes are possible for C_5H_{12}. The more spread out the molecule, the greater its polarizability, the greater the opportunity for dispersion forces between molecules, and the higher the boiling point.

induced dipole the separation of charge produced in an atom or molecule by a momentary uneven distribution of electrons.

dispersion force (also called **London force**) an intermolecular force between nonpolar molecules caused by the presence of temporary dipoles in the molecules.

polarizability the relative ease with which the electron cloud in a molecule, ion, or atom can be distorted, inducing a temporary dipole.

molecular formula (C_5H_{12}) and, therefore, the same molar mass (72.15 g/mol) and number of electrons. However, the bonds, shapes, and boiling points of the molecules are different. All these molecules are nonpolar, which means there are no permanent dipoles. Therefore the only intermolecular interactions are dispersion forces due to temporary dipoles. Molecules of pentane are relatively long and straight—think of their shape as similar to that of a piece of chalk. Pentane molecules can interact with one another over a relatively large surface area and therefore have more possibilities for dispersion forces to hold them together. In contrast, molecules of 2,2-dimethylpropane are almost spherical. They have relatively less surface area to interact with adjacent molecules; less interaction means weaker dispersion forces, resulting in the lowest boiling point in this set of compounds. The remaining molecule, 2-methylbutane, is neither as straight as pentane nor as spherical as 2,2-dimethylpropane—think of it as shaped like a short, forked stick. It should come as no surprise that this compound boils at a temperature lower than pentane but higher than 2,2-dimethylpropane. In general, molecules with more branching in their structures have lower boiling points.

Dispersion forces between small atoms or molecules are weak, but large molecules have a large number of these interactions that can sometimes dominate a much smaller number of strong interactions in a system, as we see in the discussion of polarity and solubility in Section 10.4.

10.3 Interactions among Polar Molecules

In Section 8.3 we saw how the unequal distributions of bonding pairs of electrons result in partial negative charges on some bonded atoms and partial positive charges on others. When bond dipoles are arranged asymmetrically within molecules, the molecules themselves have permanent dipoles. Molecules with permanent dipoles can interact with each other and with the ions in ionic compounds. The strengths of these interactions are much weaker than the strengths of ionic or covalent bonds, but they are strong enough to provide additional interactions beyond the London dispersion forces that all molecules experience. For example, interactions between polar liquids and ionic solids play a key role in salts dissolving in water.

Ion–Dipole Interactions

One reason why ionic compounds such as NaCl dissolve in water is that the sodium and chloride ions interact with the permanent dipoles of the polar water molecules, which carry both positive and negative partial charges. These attractions are **ion–dipole interactions**, and they occur between ions and water molecules in all aqueous solutions.

When a salt dissolves in water, the ion–dipole interactions must overcome the electrostatic attractions between the ions themselves (Figure 10.5). As an ion is pulled away from its solid-state neighbors, it becomes surrounded by water molecules, forming a **sphere of hydration** (Figure 10.6). If the solvent were something other than water, the cluster would be called a *sphere of solvation*. These dissolved ions are said to be *hydrated* or, for other solvents, *solvated*.

Within a sphere of hydration, the water molecules closest to the ion are oriented so that their oxygen atoms (negative poles) are directed toward a cation or

Hydrated Na$^+$ ion   Hydrated Cl$^-$ ion

Solid NaCl

FIGURE 10.5 The hydrogen atoms (positive poles) of H_2O molecules are attracted to the Cl$^-$ ions of NaCl, and the O atoms (negative poles) are attracted to the Na$^+$ ions. Multiple ion–dipole interactions overcome the attractive forces holding ions at the surface of the solid NaCl, causing the NaCl to dissolve.

CONNECTION Bond energies of covalent bonds are discussed in Section 8.7.

CONNECTION In Chapter 4 we learned that some, but not all, ionic compounds are soluble in water.

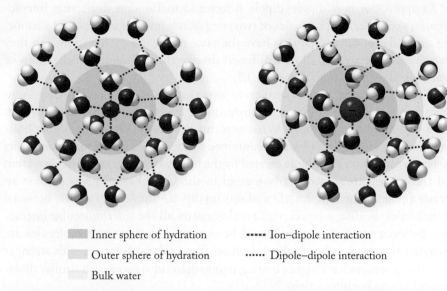

■ Inner sphere of hydration	····· Ion–dipole interaction
■ Outer sphere of hydration	····· Dipole–dipole interaction
■ Bulk water	

FIGURE 10.6 Each hydrated Na^+ ion and hydrated Cl^- ion is surrounded by six water molecules that create an inner hydration sphere. Water molecules in an outer hydration sphere surround the inner sphere. The outer sphere is the result of dipole–dipole interactions between the rest of the water (known as bulk water) and the water molecules of the inner sphere. Beyond the outer hydration sphere, dipole–dipole interactions also occur between outer-sphere water molecules and molecules in bulk water.

ion–dipole interaction an attractive force between an ion and a molecule that has a permanent dipole moment.

sphere of hydration the cluster of water molecules surrounding an ion in aqueous solution; the general term applied to such a cluster forming in any solvent is *sphere of solvation*.

dipole–dipole interaction an attractive force between polar molecules.

their hydrogen atoms (positive poles) are directed toward an anion (Figure 10.6). The number of water molecules oriented in this way depends on the size of the ion. Typically, six water molecules hydrate an ion, but the number can range from four to nine. As Figure 10.6 shows, six water molecules surround each Na^+ ion and each Cl^- ion in an aqueous solution of NaCl.

Dipole–Dipole Interactions

The water molecules further from the ions in Figure 10.6 are more randomly oriented than those in the inner hydration sphere. These molecules experience another intermolecular force, **dipole–dipole interaction**, which describes the electrostatic attractions between molecules that have permanent dipole moments—in other words, between polar molecules. In water molecules, the partial negative charges on oxygen atoms and the partial positive charges on hydrogen atoms result in attractions between a hydrogen atom of one molecule and an oxygen atom of another. Dipole–dipole interactions are not as strong as ion–dipole interactions because dipole–dipole interactions involve only partial charges, caused by unequal sharing of electrons within the molecule. In contrast, the ion involved in an ion–dipole interaction has lost or gained one or more electrons, and it has a full positive or negative charge.

CONNECTION In previous chapters we have indicated the presence of a hydration sphere around an ion in aqueous solution by placing (*aq*) after its symbol or formula.

CONNECTION In Chapter 9 we learned that permanent dipole moments are experimentally measured values, expressed in units of debyes, that define the polarity of molecules.

| **CONCEPT TEST** |

Dimethyl ether, CH_3OCH_3, and acetone, $CH_3C(O)CH_3$ (Figure 10.7), have similar formulas and molar masses. However, their dipole moments are quite different: 1.30 D for dimethyl ether and 2.88 D for acetone. Predict which compound has the higher boiling point and explain why.

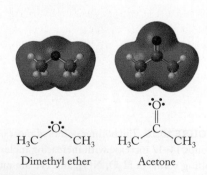

FIGURE 10.7 Molecular structures of dimethyl ether and acetone.

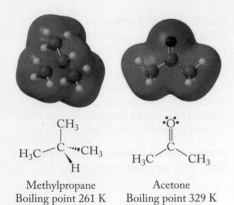

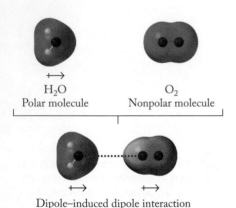

Methylpropane
Boiling point 261 K

Acetone
Boiling point 329 K

FIGURE 10.8 Molecular structures and boiling points of methylpropane and acetone.

H_2O
Polar molecule

O_2
Nonpolar molecule

Dipole–induced dipole interaction

FIGURE 10.9 The approach of a polar water molecule induces a temporary dipole in an initially nonpolar oxygen molecule by distorting the distribution of the electrons.

To appreciate how dipole–dipole forces add to London dispersion interactions, let's consider the properties of two compounds: methylpropane and acetone. The molecules of the compounds have the same molar mass (58 g/mol), and they have shapes that are not all that different despite the different hybridizations of their central carbon atoms (Figure 10.8).

We might expect the molecules to experience similar London dispersion forces, yet the boiling points of methylpropane and acetone are quite different: 261 K and 329 K, respectively. What explains the nearly 70-degree-higher boiling point of acetone? The answer is contained in the ketone functional group in its molecular structure. The dipole created by the highly electronegative oxygen atom and the less electronegative carbon atom in the C═O bond gives acetone an overall dipole moment of 2.88 D and sets up dipole–dipole interactions between its molecules. Boiling a liquid requires that nearly all the intermolecular interactions between molecules in the liquid be overcome so that these molecules are separate from one another in the gas phase. Thus polar substances with stronger interactions tend to have higher boiling points than substances with similar molar masses but weaker interactions.

A molecule with a permanent dipole can create a **dipole–induced dipole interaction** when inducing a temporary dipole in a nonpolar molecule by perturbing the electron distribution in the nonpolar molecule (Figure 10.9). The intermolecular interaction in this case is weaker than the dipole–dipole force between two polar molecules and is of the same order of magnitude as the dispersion forces between temporary dipoles in nonpolar molecules. The magnitude of an induced dipole depends on the polarizability of the electrons in a molecule, ion, or atom.

Hydrogen Bonds

In polar molecules containing O—H, N—H, or F—H bonds, the hydrogen atoms are bonded to small, highly electronegative atoms. This leads to relatively large dipole moments and stronger dipole–dipole interactions than for molecules that are approximately the same size but do not contain such bonds. Let's consider the boiling points of compounds containing one atom of a group 14, 15, 16, or 17 element bonded to enough hydrogen atoms to have a complete valence-shell octet. Most of the data, plotted in Figure 10.10, follow a familiar trend: boiling points

FIGURE 10.10 The boiling points of most of the binary hydrides in groups 14–17 increase with increasing molar mass, but not all. The boiling points of H_2O, NH_3, and HF are much higher than one might expect because of hydrogen bonding between their molecules.

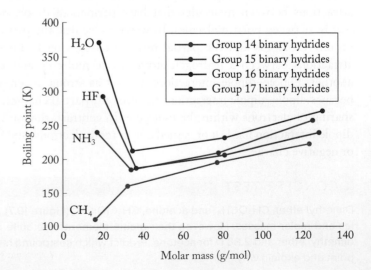

increase with increasing molar masses, as we saw with nonpolar hydrocarbons in Figure 10.3. Actually, all of the data points for the group 14 compounds fit this trend. However, the boiling points of the other three groups' compounds with the lowest molecular mass—NH_3, H_2O, and HF—are unusually high compared with the others in their series. To understand why, we need to focus on the polar bonds formed between H atoms and N, O, and F atoms. The H atoms share just two electrons, and when they are shared with a highly electronegative atom of N, O, or F, the electron density on the surface of the H atom is significantly reduced such that the H atoms interact strongly with electronegative N, O, and F atoms on neighboring molecules.

Because of its strength, this particular dipole–dipole interaction—involving one of these H atoms on one molecule and an O, N, or F atom on an adjacent molecule—merits special distinction: it is called a **hydrogen bond** (Figure 10.11). Hydrogen bonds in water play a key role in defining the remarkable behavior of H_2O. The dipole–dipole interactions between water molecules in Figure 10.6 are hydrogen bonds. Hydrogen bonds are the strongest dipole–dipole interactions and can be nearly one-tenth the strength of some covalent bonds. Table 10.3 summarizes the relative strengths of the intermolecular forces discussed here in Chapter 10.

dipole–induced dipole interaction an attraction between a polar molecule and the oppositely charged pole it temporarily induces in another molecule.

hydrogen bond the strongest dipole–dipole interaction. It occurs between a hydrogen atom bonded to a small, highly electronegative element (O, N, F) and an atom of oxygen or nitrogen in another molecule. Molecules of HF also form hydrogen bonds.

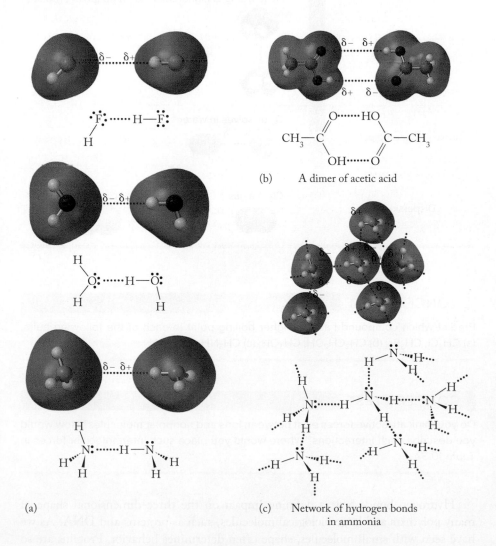

(b) A dimer of acetic acid

(a)

(c) Network of hydrogen bonds in ammonia

FIGURE 10.11 (a) Hydrogen bonds (blue dotted lines) occur between hydrogen atoms bonded to O, N, or F atoms and O, N, or F atoms in adjacent molecules. (b) Hydrogen bonding interactions are so strong in a carboxylic acid like acetic acid that two molecules stay together as a unit called a dimer in the liquid phase. (c) Hydrogen bonds between nitrogen and hydrogen in ammonia form extensive three-dimensional networks.

TABLE 10.3 Relative Strengths of Intermolecular Forces and Some Phenomena They Explain

Type of Force	Relative Strength	Phenomenon
Ion–dipole		NaCl dissolves in water
Hydrogen bonding		Water expands when it freezes
Dipole–dipole		The boiling point of dimethyl ether ($\mu = 1.30$ D, on the left) is 19°C higher than that of nonpolar propane
Dipole–induced dipole		O_2 dissolves in water
Dispersion		At 298 K: Cl_2 is a gas Br_2 is a liquid I_2 is a solid

CONCEPT TEST

Predict which compound has the higher boiling point in each of the following pairs: (a) CH_3Cl, CH_3Br; (b) CH_3CH_2OH, CH_3OH; (c) CH_3NH_2, $(CH_3)_3N$.

CONCEPT TEST

Do you think attractive forces exist between ions and nonpolar molecules? How would you describe such interactions? Where would you place such intermolecular forces in Table 10.3?

Hydrogen bonds have a defining impact on the three-dimensional shape of many polymers and large biological molecules, such as proteins and DNA. As we have seen with small molecules, shape often determines behavior. Proteins are so

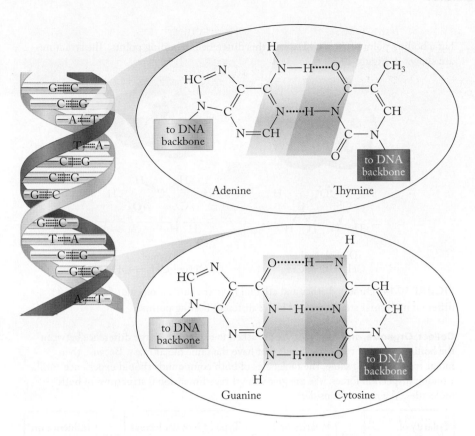

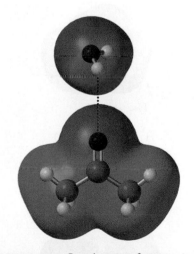

FIGURE 10.12 Hydrogen bonds (blue dotted lines) occur between hydrogen and nitrogen or oxygen in adjacent strands of DNA, contributing to its double-helix structure. The two detailed views of the DNA double helix contain the names of four of the building blocks of DNA: guanine, cytosine, adenine, and thymine. Note that there are three hydrogen bonding sites between pairs of guanine and cytosine but only two between adenine and thymine.

large that their long chains of atoms tend to fold back and wrap around themselves, such that atoms may form hydrogen bonds with other atoms in an adjacent chain. The double strands of DNA form a three-dimensional shape, called a double helix, in which pairs of the molecular building blocks of DNA, called nucleotides, form hydrogen bonds that keep the strands linked together. Pairs of two nucleotides named guanine (G) and cytosine (C) on adjacent DNA strands form three hydrogen bonds, and pairs of adenine (A) and thymine (T) form two hydrogen bonds (Figure 10.12). Because hydrogen bonds are weaker than covalent bonds, the DNA strands can be pulled apart during DNA replication.

Hydrogen bonds can also form between molecules of different substances, even when one of them has no H atoms bonded to N, O, or F atoms. For example, when acetone dissolves in water, hydrogen bonds form between the H atoms of water molecules and the O atoms of acetone molecules, as shown in Figure 10.13. These interactions happen even though the H atoms in acetone are bonded to C atoms and cannot form hydrogen bonds. The key to hydrogen bonding between acetone and water molecules is the negative partial charge of the O atoms in molecules of acetone, as indicated by the red color of the electrostatic potential map of acetone in Figure 10.13, and the blue-green color of the electron-deprived H atoms in molecules of H_2O.

FIGURE 10.13 In solutions of acetone in water, hydrogen bonds form between the hydrogen atoms of water and the oxygen atoms of acetone.

SAMPLE EXERCISE 10.1 Explaining Differences in Boiling Points   **LO2**

Dimethyl ether (C_2H_6O) has a molar mass of 46.07 g/mol and a boiling point of −24.9°C. Ethanol (C_2H_6O) has the same formula and therefore the same molar mass,

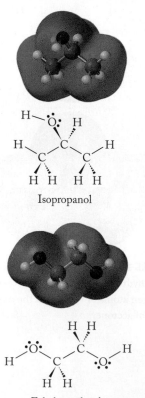

FIGURE 10.15 The polar −OH group in ethanol leads to hydrogen bonding (blue dotted lines) between molecules of ethanol. Dimethyl ether does not form hydrogen bonds but does experience weaker dipole–dipole interactions between the δ+ and δ− regions of its molecules.

FIGURE 10.16 Electrostatic potential surfaces and structures of isopropanol and ethylene glycol.

but a boiling point of 78.5°C. Explain this difference in boiling points. The structures are shown in Figure 10.14.

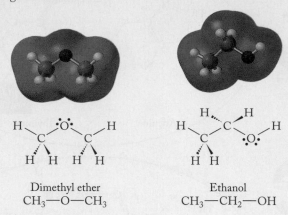

FIGURE 10.14 Dimethyl ether and ethanol have the same molecular formula but different molecular structures and hence different boiling points.

Collect, Organize, and Analyze We are asked to explain the large difference between the boiling points of two compounds that have the same molar mass. Because their molar masses are the same, the molecules of both compounds should experience similar London dispersion forces. We are given the three-dimensional structures of both molecules. We need to consider:

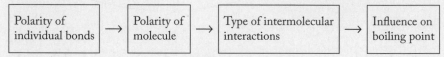

Molecules of both compounds contain O atoms bonded to atoms of less electronegative elements. This means the molecules contain bond dipoles. The shapes of the molecules show that the bond dipoles do not cancel, so both molecules are polar overall, as shown in Figure 10.15. Their polarities mean that both experience dipole–dipole interactions. Ethanol molecules contain −OH groups, which means their dipole–dipole interactions are hydrogen bonds.

Solve Hydrogen bonds are stronger than other types of dipole–dipole interactions. Therefore the intermolecular interactions in ethanol are stronger than those in dimethyl ether. More energy is required to overcome these stronger interactions, and that is why ethanol has a higher boiling point.

Think About It Dimethyl ether and ethanol have the same molecular formula and similar molecular structures, yet their boiling points differ by more than 100°C. This difference is directly linked to the presence of −OH groups capable of hydrogen bonding in ethanol.

Practice Exercise Isopropanol (molar mass 60.10 g/mol), the familiar rubbing alcohol in your medicine cabinet, boils at 82°C. Ethylene glycol, used as automotive antifreeze, has almost the same molar mass (62.07 g/mol) but boils at 196°C. Why do these substances (Figure 10.16) have such different boiling points?

(Answers to Practice Exercises are in the back of the book.)

SAMPLE EXERCISE 10.2 Explaining Trends in Boiling Points **LO2**

Table 10.4 gives the formulas, molar masses, and boiling points of several hydrocarbons and alcohols that have comparable molar masses. Figure 10.17 shows the molecular

TABLE 10.4 Boiling Point Data for Sample Exercise 10.2

HYDROCARBON			ALCOHOL		
Molecular Formula	$\mathcal{M}$ (g/mol)	Boiling Point (°C)	Molecular Formula	$\mathcal{M}$ (g/mol)	Boiling Point (°C)
1. CH_4	16.04	−161.5			
2. CH_3CH_3	30.07	−88	CH_3OH	32.04	64.5
3. $CH_3CH_2CH_3$	44.10	−42	CH_3CH_2OH	46.07	78.5
4. $CH_3CH(CH_3)CH_3$	58.12	−11.7	$CH_3CH(OH)CH_3$	60.10	82
5. $CH_3CH_2CH_2CH_3$	58.12	−0.5	$CH_3CH_2CH_2OH$	60.10	97

structures of the compounds in rows 4 and 5 of the table. (a) Explain the trend in boiling points of the five hydrocarbons and the four alcohols. (b) Explain the difference in boiling point for each hydrocarbon and the alcohol of comparable molar mass.

Collect and Organize Interpretation of trends requires evaluation of the types and relative strengths of intermolecular forces between molecules. The stronger the interactions between the molecules of a compound, the greater its boiling point. Several types of intermolecular forces have been presented in this section and are summarized in Table 10.3.

Analyze All of the compounds are molecular, so we do not need to consider ion–ion or ion–dipole forces. Alcohols contain –OH groups, whereas hydrocarbons do not. The presence of the –OH group means that alcohols can form hydrogen bonds, which hydrocarbons cannot. For both sets of compounds we need to consider:

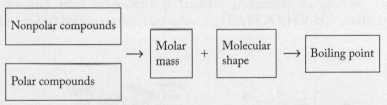

Solve

a. The trend in boiling point for both series of compounds in Table 10.4 reflects an increase in boiling point as molar mass increases. Larger molecules have more electrons distributed over larger volumes, leading to larger induced dipoles and greater dispersion forces between molecules. On the basis of the structures shown, the molecules in row 4 (Figure 10.17) are more spherical and have smaller surface areas than the molecules in row 5. The molecules in row 4 experience weaker dispersion forces. Consequently, both compounds in row 4 boil at lower temperatures than their counterparts in row 5.

b. The molar masses of ethane and methanol (row 2) are very similar, as are the molar masses of each hydrocarbon–alcohol pair; however, the alcohols all have much higher boiling points than their hydrocarbon counterparts. The presence of –OH groups means that the alcohols are capable of hydrogen bonding. Hydrogen bonds are particularly strong interactions, so hydrogen bonding among the alcohol molecules accounts for their higher boiling points. In contrast, all of the hydrocarbons interact via dispersion forces only.

Think About It Identifying the types of intermolecular interactions between molecules of different substances helps us explain trends in their boiling points. Molecular size and shape is especially important in evaluating the contribution of dispersion forces to molecular interactions.

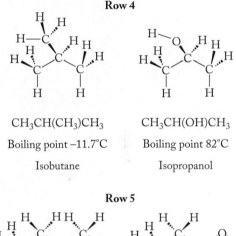

Row 4

$CH_3CH(CH_3)CH_3$
Boiling point −11.7°C
Isobutane

$CH_3CH(OH)CH_3$
Boiling point 82°C
Isopropanol

Row 5

$CH_3CH_2CH_2CH_3$
Boiling point −0.5°C
Butane

$CH_3CH_2CH_2OH$
Boiling point 97°C
Propanol

FIGURE 10.17 Structures of isobutane (a hydrocarbon) and isopropanol (an alcohol) from row 4 and butane (a hydrocarbon) and propanol (an alcohol) from row 5 in Table 10.4.

10.4 Polarity and Solubility

In Section 10.3 we explained the process by which an ionic salt dissolves in water in terms of ion–dipole interactions. The dissolving of one liquid in another or the dissolving of a gas in a liquid can also be explained in terms of intermolecular forces. We examine these processes by considering the situation where the solvent is water or some other liquid and the solute is molecular. To predict whether a given solute is soluble in a given solvent, we look at the balance between solute–solute interactions (that is, interactions between solute molecules), solvent–solvent interactions, and solvent–solute interactions.

Just as ionic compounds dissolve in polar solvents because of strong ion–dipole interactions, polar solutes tend to dissolve in polar solvents because of dipole–dipole interactions between solute and solvent molecules (Figure 10.18a). Nonpolar solutes tend not to dissolve in polar solvents because the solvent–solute interactions that promote dissolution are weaker than the solute–solute interactions that keep solute molecules together and the solvent–solvent interactions that keep solvent molecules together (Figure 10.18b). This observation is the source of a common phrase used to describe solubility: *like dissolves like.*

Let's consider an interesting medical application of "like dissolves like." Diethyl ether, $CH_3CH_2OCH_2CH_3$, a substance simply called "ether," has had

FIGURE 10.18 (a) A polar solvent such as water dissolves polar materials such as methanol because of favorable dipole–dipole interactions. In this specific case, both molecules are capable of hydrogen bonding, and solute–solvent interactions are of the same order of magnitude as solvent–solvent or solute–solute interactions. Methanol dissolves in all proportions in water. (b) Dispersion forces (indicated by the arrows between the octane molecules) between long hydrocarbon chains are the attractive forces that hold octane molecules together in the liquid phase. The highly polar water molecules interact preferentially with each other, and any solvent–solute interactions are too weak to compete with the hydrogen-bonding network within the solvent. As a result, octane has little solubility in water.

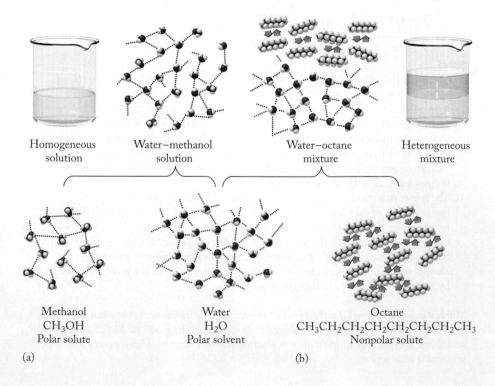

Homogeneous solution   Water–methanol solution   Water–octane mixture   Heterogeneous mixture

Methanol CH_3OH Polar solute

Water H_2O Polar solvent

Octane $CH_3CH_2CH_2CH_2CH_2CH_2CH_2CH_3$ Nonpolar solute

(a)   (b)

wide use in medicine as an anesthetic since 1842. **Ethers** have the general formula R—O—R', where R and R' are straight-chain or branched hydrocarbons or aromatic rings (R and R' may be the same or different organic groups). Because the C—O—C bond angle is close to the tetrahedral bond angle of 109.5°, the dipole moments of the two C—O bonds in an ether do not cancel, which means that ethers are polar molecules. This structural feature gives rise to the properties of typical ethers: their water solubility is comparable to that of alcohols of similar molar mass, but their boiling points are about the same as alkanes of comparable molar mass (see Table 10.5).

Although exactly how an anesthetic like diethyl ether dulls nerves and puts patients to sleep is still unknown, certain properties of diethyl ether play a role in determining its behavior as a medicinal agent. Because diethyl ether has a low boiling point, 35°C, it vaporizes easily, and a patient can inhale it. Because diethyl ether has a significant solubility in water, it is soluble in blood, which means that once inhaled, it can be easily transported throughout the body. Its low polarity and short saturated hydrocarbon chains combine to make it soluble in cell membranes, where it blocks stimuli coming into nerves. Ether has the unfortunate side effect of inducing nausea and headaches and has been replaced by new anesthetics in modern hospitals, but for many years, ether was the anesthetic of choice for surgical procedures.

Factors other than polarity also influence solubility: temperature and pressure are very important. Nevertheless, we can use the attractive forces present in any mixture of molecules as a guide to predict the relative solubilities of solutes in a given solvent.

ether organic compound with the general formula R—O—R', where R is any alkyl group or aromatic ring; the two R groups may be different.

TABLE 10.5 Functional Groups Affect Physical Properties

	Molar Mass (g/mol)	Normal Boiling Point (°C)	Solubility in Water (g/100 mL at 20°C)
CH_3CH_2—O—CH_2CH_3 Diethyl ether	74	35	6.9
$CH_3CH_2CH_2CH_2CH_3$ Pentane	72	36	0.0038
$CH_3CH_2CH_2CH_2OH$ Butanol	74	117	7.9

SAMPLE EXERCISE 10.3 Predicting Solubility in Water · · · · · · · · · · · · · · **LO3**

Considering the compounds carbon tetrachloride (CCl_4), ammonia (NH_3), hydrogen fluoride (HF), and oxygen (O_2), which should be very soluble in water and which should have limited solubility in water?

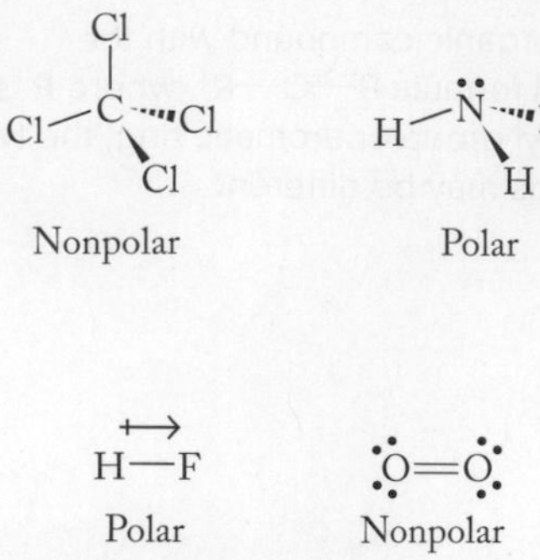

Cl
|
Cl — C ⋯ Cl
|
Cl

Nonpolar

H — N ⋯ H ↑
      |
      H

Polar

H — F

Polar

Ö = Ö

Nonpolar

FIGURE 10.19 Structures and directions of dipole moments of carbon tetrachloride, ammonia, hydrogen fluoride, and diatomic oxygen.

Collect and Organize We are asked to predict the solubilities of four substances in water. We can make our predictions based on polarity. Because water is polar, it is likely that compounds soluble in water are also polar.

Analyze To judge which of the four substances is polar, we need to draw all of their molecular structures and determine (1) which of them have polar bonds and (2) whether the arrangement of these bonds gives these molecules permanent dipoles.

Solve Applying concepts from Chapter 9, we draw the molecular structures of the four substances and determine that NH_3 and HF are polar, but CCl_4 and O_2 are not (Figure 10.19). On the premise that like dissolves like, we predict that the polar molecules (NH_3 and HF) are soluble in water, and the nonpolar molecules (CCl_4 and O_2) have only limited solubility in water. Ammonia and hydrogen fluoride also form hydrogen bonds, another property they share with water, making them even more likely to be water soluble.

Think About It The three-dimensional structure of a molecule and the electronegativity difference between the atoms in its bonds determine the polarity of a molecule. The polarity serves as a guide to solubility.

Practice Exercise In Chapter 6 we mentioned that some deep-sea divers breathe a mixture of gases rich in helium because the solubility of helium gas in blood is lower than the solubility of nitrogen gas in blood. Assuming that blood behaves like water in terms of dissolving substances, why should helium be less soluble in water than nitrogen?

FIGURE 10.20 Household cleaners such as this one are effective at removing grease stains and other nonpolar materials from surfaces and fabrics because the principal ingredient in them is a mixture of nonpolar hydrocarbons.

In Sample Exercise 10.3 we predicted that ammonia and hydrogen fluoride are water soluble, whereas carbon tetrachloride and oxygen are much less soluble. These predictions are correct. Ammonia and hydrogen fluoride are both very soluble in water. In fact, they not only dissolve in water, *they react with it*. The hydrogen bonds between NH_3 molecules and H_2O molecules are strong enough to ionize some water molecules, producing OH^- ions and H^+ ions that attach to NH_3 molecules, forming ammonium (NH_4^+) ions. The hydrogen bonds between molecules of HF and molecules of H_2O are so strong they ionize some HF molecules, producing F^- ions and H^+ ions that attach to H_2O molecules, forming hydronium (H_3O^+) ions. We will return to these reactions in our discussion of acids and bases in Chapter 15.

In contrast, the nonpolar molecules carbon tetrachloride and oxygen have very limited solubility in water because the solvent–solute interactions are very weak. The large permanent dipole of a water molecule interacts more favorably with other water dipoles than it does with the weaker, induced dipoles in the nonpolar materials. Solvent–solute interactions between water and nonpolar molecules cannot compete with the much stronger solvent–solvent interactions in these cases.

We have established that polar solutes tend to dissolve in polar solvents and that nonpolar solutes do not, at least not very much. Given the solubility of polar solutes in polar solvents, it seems reasonable that nonpolar solutes would tend to dissolve in nonpolar solvents. Indeed, they do. For example, the principal ingredient in some household cleaners (Figure 10.20) that remove grease, crayon wax, label adhesives, and other nonpolar materials from surfaces and fabrics is often labeled "petroleum distillate," which means a mixture of hydrocarbons derived from crude oil. They are effective in dissolving nonpolar substances because hydrocarbons are also nonpolar.

Nonpolar molecules are very sparingly soluble in polar solvents but can be quite soluble in nonpolar solvents as a result of compatible solute–solvent interactions.

Combinations of Intermolecular Forces

When larger molecules dissolve in a liquid solvent, we may need to take account of more than one type of intermolecular force. Consider the solubilities in water of the three organic alcohols with the structures shown in Figure 10.21. All three alcohols are liquids at room temperature, so we are looking at the solubility of one liquid in another. All three compounds contain a –OH group, which means they are like water in that they form hydrogen bonds. Considering that they are all polar molecules, why do their solubilities differ so greatly in the polar solvent water?

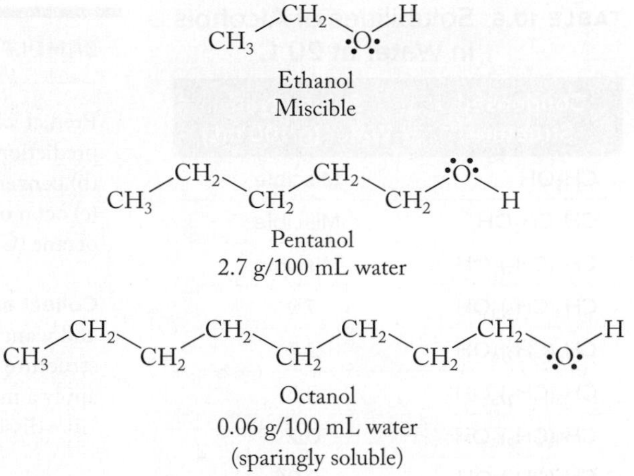

Ethanol is miscible with water, which means it is soluble in all proportions. Pentanol has a finite solubility, which means that once a solution of pentanol in water is saturated, adding more pentanol results in a heterogeneous mixture in which two layers of liquid are visible. Octanol is sparingly soluble in water, and almost any measurable amount of octanol added to water results in a heterogeneous mixture.

The feature that differentiates these molecules from one another is the number of –CH$_2$– groups. The long series of –CH$_2$– groups in octanol makes the non-OH part of the molecule nonpolar and thus unlike water. Because there are many dispersion forces between the long –CH$_2$– chains of adjacent octanol molecules, they sum to produce quite strong interactions (Figure 10.22), and these attractive forces keep the octanol molecules together, even though the polar ends of the molecules are attracted to water molecules. The minuscule solubility of octanol in water illustrates a situation where multiple dispersion forces acting together (even though they are the weakest force in Table 10.3) contribute more to intermolecular interactions than do dipole–dipole forces and hydrogen bonding, localized in one small region of a molecule.

In ethanol, the polarity and hydrogen-bonding ability of the –OH group dominate the interaction of ethanol with water because the hydrocarbon portion of the molecule is too short (only two carbon atoms long) to produce significant dispersion forces. The solubility of pentanol is less than that of ethanol, but much greater than that of octanol, because of the differences in lengths of the –CH$_2$– chains. The competing dispersion forces that limit pentanol solubility and the hydrogen-bonding interactions that promote it combine to produce moderate solubility for this alcohol. Table 10.6 lists the water solubilities for a series of alcohols. Note that the solubility in water decreases until about C$_8$, beyond which the solubility is comparable to that of the corresponding hydrocarbons.

Nonpolar interactions like the dispersion forces between hydrocarbon chains are called **hydrophobic** (literally, "water-fearing") interactions, whereas interactions that promote solubility in water are called **hydrophilic** ("water-loving") interactions. For molecules that contain both polar and nonpolar groups, as these alcohols do, solubility in water is due to the balance between hydrophilic and hydrophobic interactions. As the hydrophobic portion of the molecule increases in size, the entire molecule becomes more hydrophobic and solubility in water decreases.

FIGURE 10.21 Ethanol, pentanol, and octanol have different solubilities in water.

CONNECTION In Section 4.7 we described a solution containing the maximum amount of solute that can dissolve as being saturated.

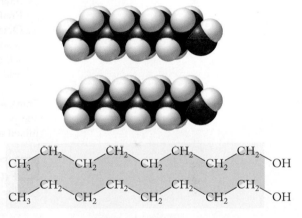

Dispersion forces

FIGURE 10.22 Even though the interaction between two –CH$_2$– units is weak, the interactions add up along the long hydrocarbon chains of octanol molecules and give rise to an attractive force that keeps the octanol molecules together in water despite the presence of the polar –OH group in the alcohol.

hydrophobic a "water-fearing," or repulsive, interaction between a solute and water that diminishes water solubility.

hydrophilic a "water-loving," or attractive, interaction between a solute and water that promotes water solubility.

TABLE 10.6 Solubilities of Alcohols in Water at 20°C

Condensed Structure	Solubility in Water (g/100 mL)
CH_3OH	Miscible
CH_3CH_2OH	Miscible
$CH_3(CH_2)_2OH$	Miscible
$CH_3(CH_2)_3OH$	7.9
$CH_3(CH_2)_4OH$	2.7
$CH_3(CH_2)_5OH$	0.6
$CH_3(CH_2)_6OH$	0.2
$CH_3(CH_2)_7OH$	0.06

SAMPLE EXERCISE 10.4 Predicting Miscibility of Liquids **LO3**

Predict which pairs of liquids are miscible. Explain your predictions. (a) Ethylene glycol ($HOCH_2CH_2OH$) and water; (b) benzene (C_6H_6) and pentane ($CH_3CH_2CH_2CH_2CH_3$); (c) octanol ($CH_3CH_2CH_2CH_2CH_2CH_2CH_2CH_2OH$) and octane ($CH_3CH_2CH_2CH_2CH_2CH_2CH_2CH_3$)

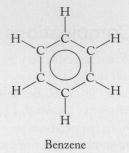

Benzene

Collect and Organize We are given formulas of liquids and want to determine their miscibility. We discussed the structure of benzene (shown here) in Section 8.4. We can apply a molecular-level understanding of the common phrase "like dissolves like" to predict miscibility.

Analyze To predict mutual solubilities and explain our predictions, we must consider:

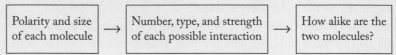

| Polarity and size of each molecule | → | Number, type, and strength of each possible interaction | → | How alike are the two molecules? |

Solve

a. Ethylene glycol is an organic liquid with two –OH groups per molecule that can form hydrogen bonds to water molecules. Prediction: the two liquids are miscible.

b. Benzene and pentane are both nonpolar hydrocarbons. Both molecules would interact via dispersion forces. They are both hydrophobic and hence very alike. Prediction: the two are miscible.

c. Octanol has a polar –OH group capable of forming hydrogen bonds, whereas octane is nonpolar. However, octanol and octane both have long hydrocarbon chains of the same length. The dispersion forces associated with these chains interacting might lead to miscibility of the two substances. Prediction: the two are miscible.

Think About It Ethylene glycol and water are miscible in all proportions, as are benzene and pentane. Octanol and octane are also miscible and are actually used together as a mixed solvent system to separate mixtures of compounds containing metal ions.

Practice Exercise Predict whether methanol (CH_3OH) or hexanol ($CH_3CH_2CH_2CH_2CH_2CH_2OH$) would be more soluble in acetone (Figure 10.7). Explain your answer.

10.5 Solubility of Gases in Water

In Sample Exercise 10.3 we considered the limited solubility of $O_2(g)$ in water. And yet, the ability of water to dissolve gases as well as nutrients is essential to both terrestrial and aquatic life. Even though oxygen is only slightly soluble in water, its limited solubility of about 10 mg/L at atmospheric pressure is usually sufficient to sustain aquatic life. Relatively weak dipole–induced dipole interactions between water molecules and the nonpolar molecules of oxygen gas account for the solubility of O_2—or, indeed, any sparingly soluble gas—in water.

However, you may have noticed fish in lakes and rivers rising to the water's surface in very warm weather. This is because the solubility of O_2 and of most other gases in water decreases with increasing temperature; therefore, warmer water lacks sufficient dissolved oxygen, and the fish gasp for air at the water's surface. Likewise, the solubility of gases in water decreases with decreasing pressure, as demonstrated by the rush of escaping gas that can be easily heard when opening a bottle containing a carbonated beverage.

CONNECTION In Chapter 6 we introduced kinetic molecular theory as a model to explain the behavior of gases. We also defined the partial pressure of gases.

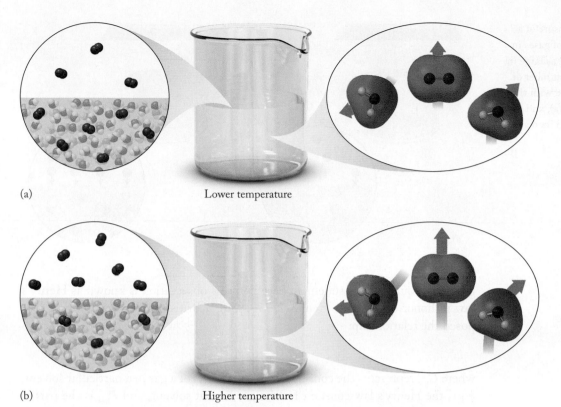

(a) Lower temperature

(b) Higher temperature

FIGURE 10.23 (a) At lower temperatures, molecules of gas dissolved in water have less kinetic energy, and dipole–induced dipole interactions between solute and solvent molecules keep more gas dissolved. (b) At higher temperatures, molecules have more kinetic energy, which overcomes solute–solvent interactions and reduces solubility.

We can visualize the influence of temperature on the solubility of gases in terms of kinetic molecular theory. Think of a population of gas molecules dissolved in a liquid (Figure 10.23a). As temperature increases, the kinetic energy of the molecules increases, and more energy is available to disrupt intermolecular attractions, reducing solubility (Figure 10.23b). More gas molecules have the necessary kinetic energy to overcome the dipole–induced dipole interactions between O_2 and H_2O molecules, and the dissolved gas escapes from the solution.

The solubility of a gas in a liquid such as water also depends on the partial pressure of the gas in the air above the surface of the liquid. For example, the partial pressure of O_2 at sea level remains fairly constant at about 0.21 atm, but the lower partial pressure of oxygen at higher altitudes can result in lower than normal concentrations of oxygen in liquids, including the blood and tissues of humans and animals. When people travel to regions of low atmospheric pressure, such as the tops of mountains, they may need supplemental oxygen to function normally because of the decreased partial pressure of oxygen. Climbers in the Himalayas may become weak and unable to think clearly because of lack of oxygen to the brain, a condition known as *hypoxia*.

We can also use kinetic molecular theory to visualize the relationship between pressure and solubility of a gas in a liquid. Think of a population of gas molecules above the surface of a liquid (Figure 10.24). The amount of gas that dissolves in a liquid depends on the frequency and number of collisions that the gas molecules have with the surface of the liquid: the more collisions, the more gas molecules that may form intermolecular attractions with the solvent (Figure 10.24b). Conversely, when the pressure above a solution decreases, the solubility of the gas decreases (Figure 10.24a). This is what happens when opening carbonated beverages; as the lid opens, the pressure drops and the CO_2 gas escapes from the fluid, resulting in bubbles and fizz.

The relationship between gas solubility in a liquid and the partial pressure of the gas in the environment surrounding the liquid applies to all sparingly soluble

FIGURE 10.24 The partial pressure of a gas (red particles) in a mixture of gases in the space above a liquid is twice as large in (b) as in (a). Consequently, the number of collisions the gas molecules have with the surface is greater in (b) than in (a), and the solubility of the gas in the liquid increases.

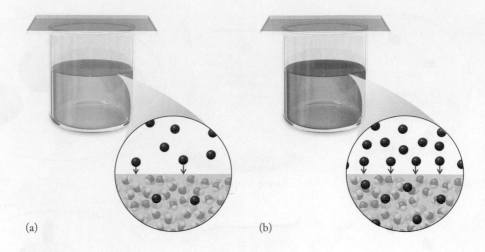

(a)   (b)

CHEMTOUR
Henry's Law

TABLE 10.7 Henry's Law Constants for Gas Solubility in Water at 20°C

Gas	k_H [mol/(L · atm)]
He	3.5×10^{-4}
O_2	1.3×10^{-3}
N_2	6.7×10^{-4}

gases. The quantitative statement based on this observation is known as **Henry's law** in honor of William Henry (1775–1836), a British physician who first proposed the relationship:

$$C_{gas} = k_H P_{gas} \qquad (10.1)$$

where C_{gas} represents the concentration (solubility) of a gas in a particular solvent, k_H is the Henry's law constant for the gas in that solvent, and P_{gas} is the partial pressure of the gas in the environment surrounding the solvent. When C_{gas} is expressed in molarity and P_{gas} in atmospheres, the units of the Henry's law constant are moles per liter-atmosphere, mol/(L · atm). Table 10.7 lists k_H values for several common gases in water.

Henry's law explains why the concentration of dissolved oxygen in blood is proportional to the partial pressure of oxygen in the air we inhale and thus is proportional to atmospheric pressure. Although this is an accurate statement of Henry's law, clearly the residents of Denver, Colorado, or Kimberley, Canada (average atmospheric pressure 0.85 atm), do not live with less oxygen in their blood than the residents of New York City or Rome, Italy (average atmospheric pressure 1.00 atm). The oxygen transport system in our bodies must adjust to accommodate local atmospheric conditions.

The amount of oxygen in the blood is related to both the concentration of hemoglobin and the fraction of the hemoglobin sites that contain oxygen as the blood leaves the lungs. This saturation of binding sites depends on the partial pressure of oxygen, as well as on proper lung function. For most people, breathing air with $P_{O_2} > 0.11$ atm results in nearly 100% saturation of hemoglobin binding sites. If P_{O_2} decreases to about 0.066 atm (as it does on high mountains), the percent saturation decreases to about 80%. Over several weeks, the body responds to lower oxygen partial pressures by producing more red blood cells and more hemoglobin. This increase in the concentration of hemoglobin, and therefore in the number of O_2 binding sites, compensates for the lower partial pressure of O_2. Even though less than 100% of the oxygen binding sites are saturated, the actual number of oxygen binding sites has increased because of the production of more hemoglobin molecules, and therefore the same amount of O_2 is delivered to tissues. Some endurance athletes try to capitalize on these physiological effects by using a technique known as "live high, train low." They acclimate to higher altitudes, typically defined as any elevation above 1500 meters (5000 ft), to bring about the physiological changes, but they continue to train at lower elevations.

Henry's law the principle that the concentration of a sparingly soluble, chemically unreactive gas in a liquid is proportional to the partial pressure of the gas.

SAMPLE EXERCISE 10.5 Calculating Gas Solubility **LO4**
by Using Henry's Law

Calculate the solubility of oxygen in water in moles per liter at 1.00 atm pressure and 20°C. The mole fraction of O_2 in air is 0.209. (Remember that the sum of the mole fractions of all the gases in a mixture equals 1.)

Collect and Organize We are given the mole fraction of oxygen in air and the total (atmospheric) pressure. Henry's law (Equation 10.1) relates solubility to partial pressure. Table 10.7 gives the Henry's law constant for oxygen at 20°C as 1.3×10^{-3} mol/(L · atm).

Analyze We need the partial pressure of oxygen for Henry's law:

$$\boxed{\text{Mole fraction}} \xrightarrow{\times P_{\text{total}}} \boxed{\text{Partial pressure}} \xrightarrow{\times k_{\text{H}}} \boxed{\text{Solubility}}$$

The product of the mole fraction (X_{O_2}) of O_2 times its total pressure gives us its partial pressure, which we use in Equation 10.1 to calculate the solubility of oxygen. We predict that oxygen is not very soluble in water because the interactions involved are weak dipole–induced dipole forces.

Solve We calculate the partial pressure of oxygen by using Equation 6.25:

$$P_{O_2} = X_{O_2}P_{\text{total}} = (0.209)(1.00 \text{ atm}) = 0.209 \text{ atm}$$

Substituting this value for P_{O_2} and k_{H} for O_2 in water in Equation 10.1 gives

$$C_{O_2} = k_{\text{H}}P_{O_2} = \left(\frac{1.3 \times 10^{-3} \text{ mol}}{L \cdot \text{atm}}\right)(0.209 \text{ atm}) = 2.7 \times 10^{-4} \text{ mol/L}$$

Think About It We predicted that oxygen is not very soluble in water because O_2 is a nonpolar solute and water is a polar solvent, and the answer agrees with that prediction.

Practice Exercise Calculate the solubility of oxygen in water at the top of Mt. Everest, where atmospheric pressure is 0.35 atm. (Assume that the Henry's law constant for oxygen at 20°C is unchanged at the top of Mt. Everest.)

10.6 Vapor Pressure of Pure Liquids

Water in a glass left on a countertop slowly disappears. Of course, it does not actually disappear—it just enters the gas phase and we can no longer see it. Molecules on the surface of the liquid vaporize, or evaporate, over time; they escape from the surface of the liquid and enter the gas phase. The rate at which the molecules make this transformation depends on the following factors:

1. Temperature: The higher the temperature, the greater the number of molecules with sufficient kinetic energy to break the attractive forces that hold them together in the liquid and to enter the gas phase.
2. Surface area: The larger the surface area of the liquid, the greater the number of molecules on the surface in a position to enter the gas phase.
3. Intermolecular forces: The stronger the intermolecular forces, the greater the kinetic energy needed for a molecule to escape the surface, and the smaller the number of molecules in the population that have this energy.

If a glass of water is covered (Figure 10.25), however, a different situation arises. Molecules at the surface still evaporate, but now they are confined to the space above the water. Some of them condense at the liquid surface and return to the liquid phase. In a short time, the rates of the two processes equalize, meaning that

↑ Evaporation ↓ Condensation

FIGURE 10.25 A covered glass of water achieves a dynamic equilibrium when the rate at which the liquid water is lost to evaporation equals the rate at which liquid water is gained by condensation.

FIGURE 10.26 A graph of vapor pressure versus temperature for six liquids shows that vapor pressure increases with increasing temperature. The temperature at which the vapor pressure equals 1 atm is the normal boiling point of the liquid.

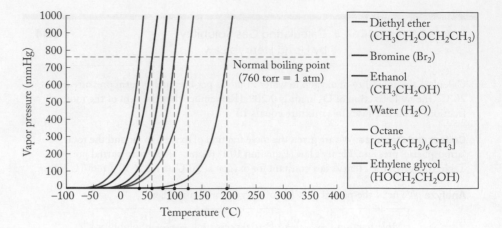

as many molecules leave the surface as reenter it. At this point of dynamic equilibrium, no further change takes place in the level of the liquid in the glass, even though molecules continue to evaporate and condense constantly. In such a situation at constant temperature, the pressure exerted by the gas in equilibrium with its liquid is called the **vapor pressure** of the liquid.

Vapor Pressure and Temperature

Figure 10.26 is a graph of the vapor pressures of several liquids as a function of temperature, showing that vapor pressure increases with increasing temperature. At some point as temperature increases, the vapor pressure of any substance reaches 1 atm; the temperature at which this occurs is the **normal boiling point** of the substance. For water, to take one example from Figure 10.26, the vapor pressure reaches 1 atm at 100°C, which we recognize as the normal boiling point of water. The reason for calling it the *normal* boiling point arises from the observation that most chemical reactions in nature, in our bodies, and in the laboratory take place at or near an atmospheric pressure of 1 atm.

Notice the order of the normal boiling points of the liquids in Figure 10.26 from lowest to highest: diethyl ether (74.12 g/mol) < bromine (159.81 g/mol) < ethanol (46.07 g/mol) < water (18.02 g/mol) < octane (114.23 g/mol) < ethylene glycol (62.07 g/mol). It would be challenging to predict this order correctly because it would be difficult to predict the relative strengths of the different types of intermolecular interactions. Take only the organic molecules as an example; the molar mass of diethyl ether is more than 1.5 times that of ethanol, but ethanol molecules are capable of hydrogen bonding, whereas diethyl ether molecules are not. Consequently, ethanol's boiling point (78.5°C) is more than 40 degrees higher than diethyl ether's (34.6°C). Octane (126°C) has only dispersion forces between its molecules, but it has a long chain of eight carbon atoms over which the molecules interact. Its molar mass is almost 2.5 times that of ethanol, and its boiling point is 48 degrees higher. Finally, ethylene glycol has a smaller molar mass than that of diethyl ether and just a little more than that of ethanol, but it has two groups capable of hydrogen bonding; its boiling point (197°C) is almost 120 degrees higher than that of ethanol.

vapor pressure the pressure exerted by a gas at a given temperature in equilibrium with its liquid phase.

normal boiling point the temperature at which the vapor pressure of a liquid equals 1 atm (760 torr).

Clausius–Clapeyron equation a relationship between the vapor pressure of a substance at two temperatures and its heat of vaporization.

CONCEPT TEST

Diesel fuel is made of hydrocarbons with an average of 13 carbon atoms per molecule, and gasoline is made of hydrocarbons with an average of 7 carbon atoms per molecule. Which fuel has the higher vapor pressure at room temperature?

Volatility and the Clausius–Clapeyron Equation

We describe substances as *volatile* if they evaporate readily at normal temperatures and pressures. At a given temperature, a more volatile substance has a higher vapor pressure than a less volatile substance, and as temperature increases, vapor pressure increases. The relationship between the vapor pressure of a pure substance and absolute temperature is not linear, as shown by the curves in Figure 10.26. However, if we graph the natural logarithm of the vapor pressure versus $1/T$, where T is the absolute temperature in kelvins, we get a straight line (Figure 10.27) described by the equation

$$\ln(P_{vap}) = -\frac{\Delta H_{vap}}{R}\left(\frac{1}{T}\right) + C$$

where ΔH_{vap} is the enthalpy of vaporization, R is the gas constant, T is in kelvin, and C is a constant that depends on the identity of the liquid. We can solve this equation for C and write it in terms of two temperatures:

$$\ln(P_{vap, T_1}) + \left(\frac{\Delta H_{vap}}{RT_1}\right) = C = \ln(P_{vap, T_2}) + \frac{\Delta H_{vap}}{RT_2}$$

Rearranging the right and left sides of this equation, we get

$$\ln\left(\frac{P_{vap, T_1}}{P_{vap, T_2}}\right) = \frac{\Delta H_{vap}}{R}\left(\frac{1}{T_2} - \frac{1}{T_1}\right) \qquad (10.2)$$

We can use this expression, called the **Clausius–Clapeyron equation**, to calculate ΔH_{vap} if the vapor pressures at two temperatures are known. We can also calculate either the vapor pressure P_{vap, T_2} at any given temperature T_2, or the temperature T_2 at any given vapor pressure P_{vap, T_2}, if ΔH_{vap}, P_{vap, T_1} and T_1 are known. Because the units of ΔH_{vap} are typically joules (J) or kilojoules (kJ) per mole, the value used for R in the Clausius–Clapeyron equation is 8.314 J/(mol · K).

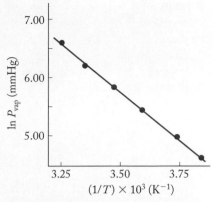

FIGURE 10.27 Plotting the natural logarithm of the vapor pressure versus the reciprocal of the absolute temperature gives a straight line described by the Clausius–Clapeyron equation. This graph shows the plot for *n*-pentane.

SAMPLE EXERCISE 10.6 Calculating Vapor Pressure **LO5**

At its normal boiling point of 126°C, octane, C_8H_{18}, has a vapor pressure of 760 torr. What is its vapor pressure at 25°C? The enthalpy of vaporization of octane is 39.07 kJ/mol.

Collect, Organize, and Analyze The vapor pressure of octane at any temperature can be calculated using Equation 10.2, the Clausius–Clapeyron equation, given octane's enthalpy of vaporization and normal boiling point. To use this equation, we must convert the given temperatures to kelvins. Because we will use 8.314 J/(mol · K) for R, we must also convert ΔH_{vap} to joules per mole. Vapor pressure decreases as temperature decreases, so we expect the vapor pressure of octane to be significantly lower than 760 torr at 25°C.

Solve The two temperatures are

$$T_1 = 126°C + 273 = 399 \text{ K} \quad \text{and} \quad T_2 = 25°C + 273 = 298 \text{ K}$$

and ΔH_{vap} in joules is

$$39.07 \frac{kJ}{mol} \times \frac{1000 \text{ J}}{1 \text{ kJ}} = 3.907 \times 10^4 \frac{J}{mol}$$

Inserting these values in Equation 10.2:

$$\ln\left(\frac{P_{vap,T_1}}{P_{vap,T_2}}\right) = \frac{\Delta H_{vap}}{R}\left(\frac{1}{T_2} - \frac{1}{T_1}\right)$$

$$\ln\left(\frac{760 \text{ torr}}{P_{vap,T_2}}\right) = \frac{3.907 \times 10^4 \dfrac{J}{mol}}{8.314 \dfrac{J}{mol \cdot K}}\left(\frac{1}{298 \text{ K}} - \frac{1}{399 \text{ K}}\right)$$

$$P_{vap,T_2} = 14.0 \text{ torr}$$

Think About It We expect octane to have a low vapor pressure at a temperature much lower than its boiling point, so this number seems reasonable.

 Practice Exercise Pentane, C_5H_{12}, boils at 36°C at $P = 1$ atm. What is its molar heat of vaporization in kilojoules per mole if its vapor pressure at 25°C is 505 torr?

10.7 Phase Diagrams: Intermolecular Forces at Work

The strengths of the attractive forces between particles determine whether a substance is a solid, liquid, or gas at a given temperature and pressure. To understand why temperature and pressure have these effects, let's think for a moment about gases. High pressures push molecules closer together; low temperatures reduce their average speed. Both of these conditions favor intermolecular interactions because molecules that are closer together collide and interact with each other more frequently, and low temperatures slow molecules down so they have more time to interact. Hence both temperature and pressure influence the degree to which molecules interact. Scientists use **phase diagrams** to show which phase of a substance is stable at a given temperature and pressure.

Phases and Phase Transformations

The phase diagram for water, like that for many other pure substances, has three regions corresponding to the three phases of matter, plus a fourth region called a *supercritical region* (Figure 10.28). The lines separating the regions are called *equilibrium lines* because the two states bordering them are at equilibrium at the temperature and pressure combinations along such a line. The blue equilibrium line separating the solid and liquid regions represents a series of freezing (or melting) points; the points on this line are combinations of temperature and pressure at which the solid and liquid states coexist. The red line separating the liquid and gas regions of Figure 10.28 represents a series of boiling points or condensation points; the points on this line are combinations of temperature and pressure at which the liquid and gaseous states coexist. The green line separating the solid and gaseous states represents a series of sublimation points (solid turning to gas) or deposition points (gas turning to solid); the points on this line are combinations of temperature and pressure at which the solid and gaseous states coexist.

Notice that the red line curves from the lower left to the upper right, separating the liquid and gas regions of the phase diagram. This line represents the

 **CONNECTION** In Section 6.9 we considered how real gases deviate from the ideal gas law, in part due to intermolecular attractions.

CHEMTOUR
Phase Diagrams

phase diagram a graphical representation of how the stabilities of the physical states of a substance depend on temperature and pressure.

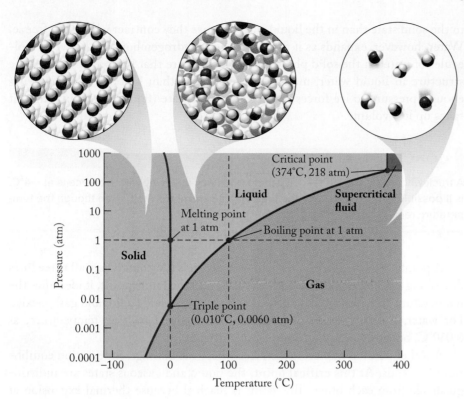

FIGURE 10.28 The phase diagram for water indicates in which phase water exists at various combinations of pressure and temperature.

changing boiling point of the liquid as a function of pressure. Its shape makes sense because when the pressure above a liquid increases, the temperature required for liquid molecules to overcome intermolecular attractive forces and enter the gas phase (that is, the boiling point) increases. The shape of the solid–gas curve also makes sense: higher pressures make it more difficult for molecules of ice to overcome intermolecular forces and sublime to water vapor.

The trends in the boiling/condensation equilibrium line and the sublimation/deposition equilibrium line in Figure 10.28 show that both the boiling point (liquid $\rightleftharpoons$ gas) and the sublimation point (solid $\rightleftharpoons$ gas) of water decrease as pressure decreases. In both cases, a phase transition from a denser phase to the gas phase occurs at a lower temperature when the pressure is lower. Lower pressure allows molecules to be farther apart, weakening the effect of intermolecular interactions and offsetting the effects of a lower temperature. Typical solid and liquid phases of a compound, like ice and water, are similar in density and many times denser than the vapor phase of the material. A decrease in pressure favors the phase change to the much less dense gas phase. Conversely, applying pressure to a gas forces its molecules closer together, which in turn increases the effect of their intermolecular interactions and favors the denser liquid and solid phases. At some point, the pressure may change the gas into a liquid or solid, which takes up much less volume.

The trend in the melting/freezing equilibrium line for water, however, shows a decrease in the melting point of ice as pressure increases. This trend in the melting/freezing points for water is opposite the trend observed for almost all other substances (see, for example, the slope direction of the blue line for CO_2 in Figure 10.29). The reason for water's unusual melting/freezing behavior is that water expands when it freezes. Most other substances are denser

FIGURE 10.29 Phase diagram for carbon dioxide.

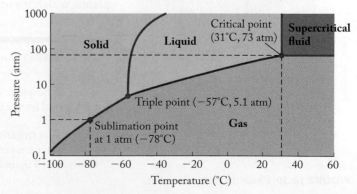

triple point the temperature and pressure at which all three phases of a substance coexist. Under these conditions, freezing and melting, boiling and condensation, and sublimation and deposition all proceed at the same rate.

critical point a specific temperature and pressure at which the liquid and gas phases of a substance have the same density and are indistinguishable from each other.

supercritical fluid a substance at conditions above its critical temperature and pressure, where the liquid and vapor phases are indistinguishable and have some characteristics of both a liquid and a gas.

in the solid state than in the liquid state because they contract when they freeze. Water, however, expands as it freezes because hydrogen bonding between molecules of water in the solid phase creates a structure that is more open than the structure in liquid water, making ice less dense than liquid water. Applying enough pressure to ice forces it into a physical state (liquid water) in which it takes up less volume.

CONCEPT TEST

A truck with a mass of 2000 kg is parked on an icy driveway where the ice is at −4°C. Is it possible that the ice under the tires of this vehicle will melt, even though the temperature remains constant? Explain your answer.

A point of special interest on a phase diagram is the point where all three lines describing the phase transitions meet. Known as the **triple point**, it identifies the temperature and pressure at which all three states (liquid, solid, and gas) coexist. For water, the triple point is just above the normal melting temperature, at 0.010°C, but at a very low pressure of 0.0060 atm.

Another point of interest is the place where the boiling/condensation equilibrium line ends. At this **critical point**, the liquid and gaseous states are indistinguishable from each other. This point is reached because thermal expansion at this high temperature causes the liquid to become less dense, while the high pressure compresses the gas into a small volume, increasing its density. At the critical point, the densities of the liquid and gaseous states are equal, so one cannot be distinguished from the other.

At temperature–pressure combinations above its critical point, a substance exists as a **supercritical fluid**. A supercritical fluid can penetrate materials like a gas but also dissolve substances in those materials like a liquid. Supercritical carbon dioxide is used in the food-processing industry to decaffeinate coffee and remove fat from potato chips. Supercritical carbon dioxide and water are sometimes mixed to generate fluids that can selectively dissolve specified materials from mixtures while leaving other components untouched.

The phase diagram of CO_2 is shown in Figure 10.29. The dashed line at $P = 1$ atm defines the phases that exist at 1 atm pressure, but note that the blue region representing liquid CO_2 does not extend below 5.1 atm. This means that solid CO_2 does not melt into a liquid at normal temperatures and pressures. Rather, it sublimes directly to CO_2 gas. This behavior gives rise to the common name for solid CO_2, *dry ice*; it is a solid that keeps things cool, as ice does, but it forms no "wet" liquid. The critical point of CO_2 is at 31°C and 73 atm, a pressure easily achieved with compressors in laboratories, factories, and food-processing plants, which means CO_2 is readily available for use as a supercritical fluid.

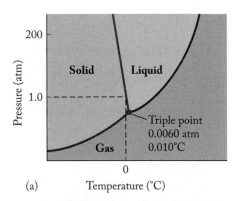

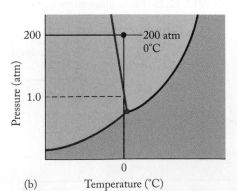

FIGURE 10.30 Phase diagrams for water.

SAMPLE EXERCISE 10.7 Reading a Phase Diagram **LO6**

Describe the phase changes that take place as the pressure on a sample of water at 0°C is increased from 0.0001 atm to 200 atm.

Collect and Organize We are asked to describe the phase changes water undergoes at a constant temperature as the pressure is increased. We can use the phase diagram for water in Figure 10.30(a). To read a phase diagram, we need to remember that every point is characterized by a temperature and a pressure and that every time we cross an equilibrium line, the phase changes.

Analyze The changes we must describe take place along the vertical line that intersects the temperature axis at 0°C.

Solve Starting at the bottom of the phase diagram with the red line (Figure 10.30b) and following the dashed line at 0°C upward as pressure increases, we approximate the location of the starting pressure, 0.0001 atm, a little below and to the left of the triple point ($T = 0.010$°C and $P = 0.0060$ atm). Water at 0°C and 0.0001 atm is a gas, as indicated by the phase diagram. As we follow the 0°C dashed line up from the temperature axis, the point where this line intersects the green equilibrium line indicates that the gas solidifies to ice at a pressure below 1 atm. Increasing the pressure toward 1 atm, we see the dashed line intersect the blue equilibrium line at that pressure. This intersection means that at this pressure the ice melts to liquid water. Extending the 0°C dashed line farther upward tells us that the water remains a liquid at pressures up to and beyond 200 atm.

Think About It The phase changes in water, from gas to solid to liquid with increasing pressure at 0°C, make sense from a molecular point of view because higher pressures favor the densest phase. Liquid water is denser than ice, so the transitions from gas to solid to liquid are transitions from the least dense to the densest phase.

Practice Exercise Describe the phase changes that occur when the temperature of CO_2 (Figure 10.29) is increased from −100°C to 50°C at a pressure of 25 atm.

CONCEPT **TEST**

Look at the phase diagram in Figure 10.29. Does solid CO_2 float on liquid CO_2 under conditions described by any point along the blue line?

10.8   Some Remarkable Properties of Water

Water has many remarkable properties. Its melting and boiling points are much higher than those of all other molecular substances with similar molar masses (Table 10.8), and its solid form (ice) is less dense than liquid water. These phenomena and many others are related to the strength of the hydrogen bonds that attract water molecules to one another. Let's look at some other unique properties resulting from the hydrogen bonding between water molecules.

CHEMTOUR
Capillary Action

TABLE 10.8   Melting Points and Boiling Points of Four Compounds of Similar Molar Mass

Substance	$\mathcal{M}$ (g/mol)	Melting Point (°C)	Boiling Point (°C)
H_2O	18.02	0	100
HF	20.01	−83	19.5
NH_3	17.03	−78	−33
CH_4	16.04	−182	−164

surface tension the energy needed to separate the molecules at the surface of a liquid.

meniscus the concave or convex surface of a liquid in a small-diameter tube.

Surface Tension, Capillary Action, and Viscosity

Surface tension is the resistance of a liquid to any increase in its surface area. Surface tension represents the energy required to move molecules apart so that an object can break through the surface.

Hydrogen bonding in water creates such a high surface tension, 7.29×10^{22} J/m^2 at 25°C, that a carefully placed steel needle floats on water and insects called water striders can walk on it (Figure 10.31). The same needle and insect would sink in oil or gasoline.

Another illustration of the intermolecular forces acting in liquids is seen in the shape of the surface of the liquid when it is in a graduated cylinder or other small-diameter tube (Figure 10.32). The surface of water is concave in such a tube, but the surface of liquid mercury is convex. Either curved surface, concave or convex, is called a **meniscus**. In both liquids, the meniscus is the result of two competing forces: *cohesive forces*, which are interactions between like particles, and *adhesive forces*, which are interactions between unlike particles. In the water sample in Figure 10.32(a), the cohesive forces are hydrogen bonds between water molecules and the adhesive forces are dipole–dipole interactions between water molecules and polar Si—O—Si groups on the surface of the glass. The adhesive forces are strong enough to cause the water to climb upward on the glass, creating the concave meniscus. The adhesive forces are greater than the cohesive forces in this case because surface water molecules next to the glass have less contact with other water molecules and therefore experience smaller cohesive forces.

In the mercury sample (Figure 10.32b), the cohesive forces are metallic bonds between mercury atoms, and the adhesive forces are interactions between induced dipoles in the mercury atoms and the polar Si—O—Si groups on the glass surface.

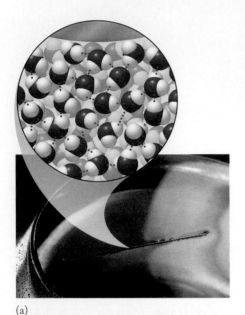

(a)

(b)

FIGURE 10.31 (a) Intermolecular forces, including hydrogen bonding, are exerted equally in all directions in the interior of a liquid. Because forces in opposite directions cancel, a molecule in the interior does not experience surface tension. However, there is no liquid water above a surface to exert attractive intermolecular forces. The resulting imbalance causes the surface water molecules to adhere tightly to one another, creating surface tension. When the magnitude of the surface tension exceeds the downward force exerted on the surface water molecules by an object, such as a needle, the object cannot break through and floats on the surface. (b) Surface tension allows a water strider to rest on top of water without sinking.

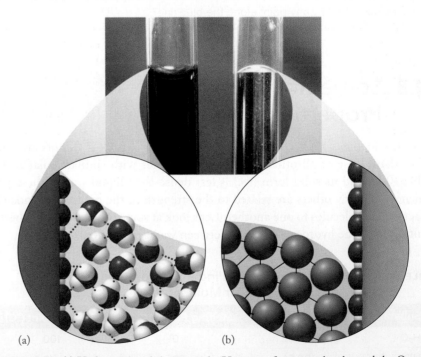

(a)   (b)

FIGURE 10.32 (a) Hydrogen bonds between the H atoms of water molecules and the O atoms of the silicon dioxide that makes up the glass are an adhesive force that causes water molecules to adhere to the glass surface and form a concave meniscus. (b) Because mercury atoms have no such attraction to the silicon dioxide in the glass, mercury forms a convex meniscus.

In this case, the adhesive forces are much weaker than the cohesive forces, and mercury atoms are more attracted to one another than to the glass. As a result, they do not adhere to the wall, forming a convex meniscus (Figure 10.32b).

The observation that adhesive forces outweigh cohesive forces on the surface of water has consequences in other situations. In a narrow tube, adhesion to the tube wall draws the outer ring of water molecules upward. At the same time, cohesive forces between the outer ring molecules and those adjacent to them draw the molecules upward (Figure 10.33). If the tube is narrow enough (a capillary tube), this combination of adhesion and cohesion draws a column of water up the tube in a phenomenon known as **capillary action**. The water column reaches its maximum height when the downward force of gravity balances the upward adhesive and cohesive forces.

In a test tube or pipette, water molecules in contact with the glass move only a very small distance up the glass before the force of gravity balances the adhesive and cohesive forces. However, when a nurse takes a blood sample after a pinprick in your finger, the blood (essentially an aqueous solution) spontaneously moves into a capillary tube because of capillary action. Capillary action also contributes to the process by which water rises 100 meters or more up the trunks of tall trees.

(a) (b)

FIGURE 10.33 Because of capillary action, colored water rises (a) in a capillary tube and (b) in a carnation.

CONCEPT **TEST**

Would mercury be spontaneously drawn into a glass capillary tube?

Viscosity, or resistance to flow, is another property of liquids related to the strength of intermolecular forces (Figure 10.34). In nonpolar compounds—for example, petroleum products—viscosity increases with increasing molar mass. Lubricating oil, for example, is much more viscous than gasoline because the hydrocarbons in a typical lubricating oil have molar masses that are two to three times those of the hydrocarbons in gasoline. Larger molar masses mean stronger dispersion forces between molecules. Because of the stronger interactions, molecules of lubricating oil do not slide past one another as easily as the shorter-chain hydrocarbons in gasoline, and bulk quantities of the larger molecules do not flow as easily when poured.

The viscosities of polar compounds are influenced by both dispersion forces and dipole–dipole interactions. Considering two of the polar alcohols in Figure 10.21, we predict that octanol has a higher viscosity than ethanol, reflecting the greater dispersion forces in octanol. Water is more viscous than gasoline even though water molecules are much smaller than the nonpolar molecules in gasoline. The remarkable viscosity of water is another property directly related to the hydrogen bonds between water molecules.

Another property unique to water is how its density changes with temperature. The density of water increases as it is cooled to 4°C (Figure 10.35), a pattern observed for most liquids and solids and for all gases. However, as water is cooled from 4°C to 0°C, it expands, and its density decreases as the pattern of its hydrogen bonds changes with temperature. As water freezes at 0°C, its density drops even more, to about 0.92 g/mL for ice, causing ice to float on liquid water (Figure 10.36). This unusual behavior is caused by the formation of a network of hydrogen bonds in ice. With each oxygen atom covalently bonded to two hydrogen atoms and

FIGURE 10.34 High viscosity results from strong intermolecular forces between large polar molecules. The white liquid is more viscous than the red, which pours freely.

capillary action the rise of a liquid in a narrow tube as a result of adhesive forces between the liquid and the tube and cohesive forces within the liquid.

viscosity the resistance to flow of a liquid.

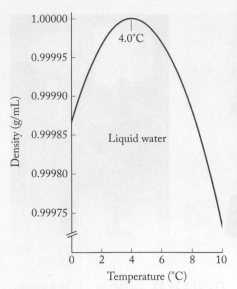

FIGURE 10.35 As water is cooled, its density increases until its temperature reaches 4°C. At this temperature the density has its maximum value. As the water cools from 4°C to its freezing point at 0°C, its density decreases.

FIGURE 10.36 Because of the changes in the density of water near the freezing point and because ice has a lower density (0.92 g/mL) than water, ice and ice-cold water float on top of warmer water (4°C) in lakes and rivers. In ice, each oxygen atom is linked to four hydrogen atoms: two by covalent bonds and two by hydrogen bonds.

hydrogen-bonded to two other hydrogen atoms, the molecules form an extensive and open hexagonal network in the solid phase. Because of the space between the molecules in the network, the same number of molecules occupies more volume in ice than in liquid water. When ice melts, some of the hydrogen bonds in the rigid array break, allowing the molecules in the liquid to be arranged more compactly, up to 4°C, whereupon the density begins to decrease once more.

Water and Aquatic Life

The expanded structure of ice plays a crucial ecological role in temperate and polar climates. The lower density of ice means that lakes, rivers, and polar oceans freeze from the top down, allowing fish and other aquatic life to survive in the liquid water below. Each fall, surface water cools first, and its density increases until its temperature reaches 4°C, the peak in the graph of Figure 10.35. This maximum-density water at 4°C sinks to the bottom, bringing warmer water to the surface, which in turn cools to 4°C and sinks, pushing the next layer of warm water to the surface in a continuing cycle (Figure 10.37). This autumnal turnover stirs up dissolved nutrients, making them available for life in the sunlit surface during the next growing season. When all of the water has reached 4°C and the surface water cools further, it becomes less dense and ice may eventually form. The layer of ice insulates the 4°C water beneath it, allowing aquatic life to survive.

In spring, the ice melts and the surface water warms to 4°C. At this temperature, the entire column of water has nearly the same temperature and density; dissolved nutrients for plant growth are evenly distributed, and the stage is set for a burst of photosynthesis and biological activity called the spring bloom.

Further warming of the surface water creates a warm upper layer separated from colder, denser water by a *thermocline*, which is a sharp change in temperature between the two layers. Biological activity depletes the pool of nutrients above the thermocline as decaying biomass settles to the bottom. Consequently, photosynthetic activity drops from its spring maximum during the summer, even though there is much more energy available from the Sun. The thermocline persists until the autumn turnover mixes the water column and the cycle begins anew.

(a) Autumn

(b) Winter

FIGURE 10.37 (a) As the surface water of a pond cools to 4°C, its density increases and it sinks to the bottom, bringing the warmer, less dense water to the surface. (b) Continued cooling of the surface water below 4°C produces a less dense layer that may eventually freeze while the more dense water deeper in the pond remains at 4°C.

SAMPLE EXERCISE 10.8 Integrating Concepts: Testing Drug Candidates

Candidate pharmaceutical agents may initially be evaluated for their partition coefficient, which is the ratio between their solubility in octanol (a mimic for cell membranes) and their solubility in water (a mimic for body fluids). A drug must be sufficiently hydrophilic to be carried in the blood and sufficiently hydrophobic to move across cell membranes. One way to determine this ratio is to put a weighed sample of the candidate agent in a vessel containing octanol and water. The container is vigorously shaken, and the amount of substance that ends up in the water and the amount in the octanol are measured. The ratio is calculated from these two numbers:

$$\text{Partition coefficient} = \frac{\text{g substance in octanol}}{\text{g substance in water}}$$

a. Draw a rough sketch of a tube containing 10 mL of water (density 1.00 g/mL) and 10 mL of octanol (density 0.83 g/mL). Indicate each fluid clearly and suggest what the meniscus would look like.

b. Two anticancer drugs, 5-fluorouracil (5-FU) and 1,3-bis(2-chloroethyl)-1-nitrosourea (BCNU), are shown in Figure 10.38. Predict which has the higher solubility in octanol and which has the higher solubility in water. Explain your prediction.

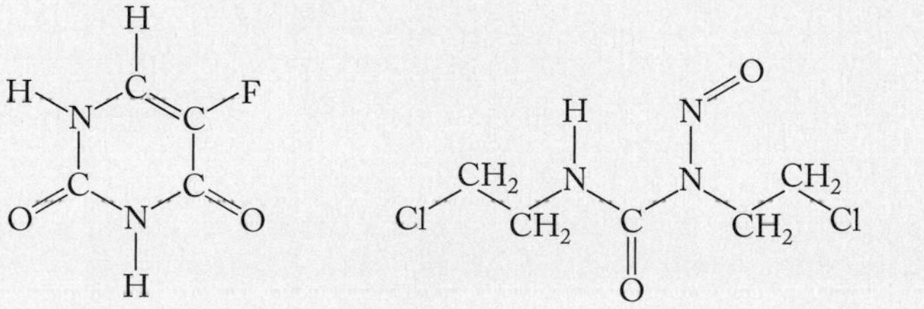

5-Fluorouracil
130.08 g/mol

1,3-Bis(2-chloroethyl)-1-nitrosourea (BCNU)
214.05 g/mol

FIGURE 10.38 Structures and molar masses of 5-FU and BCNU.

c. 5-FU has an octanol–water partition coefficient of 0.112. When 0.100 g of 5-FU was tested by this method, how much drug was found in the water and how much in the octanol?

Collect and Organize To run the experiment described, we have octanol and water together in a test tube. We are given the structures of two anticancer drugs and the value of one of their octanol–water partition coefficients.

Analyze We are first asked to show what the system looks like, so we need to think about the miscibility of water and octanol, and if they are immiscible, we need their densities to determine which would float on the other. We also need to think about how the top fluid would interact with the glass in forming a meniscus (Figure 10.39).
We need to look at the types of solvent–solute interactions the anticancer drugs have, based on their structures, and decide which is more water-like and

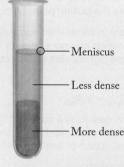

FIGURE 10.39 Test tube containing two liquids.

which more octanol-like. From molecular masses alone, we predict that 5-fluorouracil is more soluble in water than is BCNU.

Solve

a. As we discussed in Section 10.4, both water and octanol can form hydrogen bonds, but the strong dispersion forces between the hydrocarbon chains in octanol make it only slightly soluble in water. Octanol is less dense than water, so the octanol layer floats on the water layer (Figure 10.40).

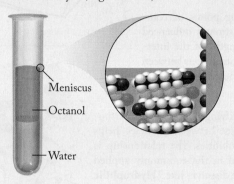

FIGURE 10.40 Test tube containing water and octanol.

Regarding the meniscus, octanol is very oil-like, despite the hydrogen-bonding –OH groups. Its meniscus would be much less concave than the meniscus of water.

b. 5-Fluorouracil has two –NH groups and two polar C=O groups capable of hydrogen bonding with water. BCNU has only one –NH group and one C=O group; it also has two other nitrogen atoms, but one is sp^2 hybridized and attached to a more electronegative oxygen atom, so it would not have electrons readily available to bond with a hydrogen atom in water. In addition, the molar mass of BCNU is almost twice that of 5-fluorouracil, which also indicates a lower water solubility. 5-Fluorouracil is most likely the more water-soluble of the two drugs and would therefore have the smaller octanol–water partition coefficient.

c. For 5-fluorouracil:

$$0.112 = \frac{\text{g drug in octanol}}{\text{g drug in water}}$$

If the total amount of drug added to the test tube is 0.100 g, we can call the amount dissolved in water x and the amount dissolved in octanol $0.100 - x$. Substituting these expressions into the ratio, we can calculate x:

$$0.112 = \frac{0.100 - x}{x} \qquad 0.112x = 0.100 - x$$
$$x + 0.112x = 0.100 \qquad x = 0.0899 \text{ g}$$

Therefore the amount dissolved in water is 0.0899 g and the amount in octanol is 0.0101 g.

Think About It The known partition coefficient of BCNU is 34.7, so it is indeed less soluble in water and more soluble in octanol than 5-fluorouracil. This evaluation is usually carried out with an aqueous solution of sodium chloride rather than pure water to even more closely mimic the behavior of drugs in blood.

SUMMARY

LO1 Ions interact with water through **ion–dipole interactions**. Dipole–dipole interactions take place between other polar molecules. The strongest dipole–dipole interactions are **hydrogen bonds**. **Dispersion (London) forces** are due to the **polarizability** of atoms and molecules and the existence of **temporary (induced) dipoles**. (Sections 10.2 and 10.3)

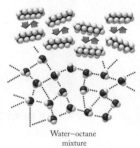

LO2 Boiling points of compounds are strongly influenced by the strengths of the intermolecular interactions between particles. (Sections 10.3 and 10.4)

LO3 Identifying the types of intermolecular forces between particles helps explain solubilities. The relationship is summarized in the commonly applied phrase "like dissolves like." **Hydrophilic** substances are more soluble in water than are **hydrophobic** substances. (Section 10.4)

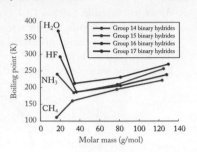

Water–octane mixture

LO4 The solubility of a gas in water depends on the temperature and pressure. **Henry's law** gives the maximum concentration (the solubility) of a sparingly soluble gas in a liquid solvent. (Section 10.5)

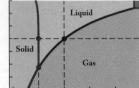

LO5 Relative volatility of substances depends on the strength of intermolecular interactions between particles. The **Clausius–Clapeyron equation** relates **vapor pressure**, the pressure exerted by a gas in equilibrium with its liquid phase, to absolute temperature. (Section 10.6)

LO6 The **phase diagram** of a substance indicates whether it exists as a solid, liquid, gas, or **supercritical fluid** at a particular pressure and temperature. (Section 10.7)

LO7 The remarkable behavior of water, including its high melting and boiling points, its **surface tension**, its **capillary action**, and its **viscosity**, results from the strength of intermolecular hydrogen bonds. (Section 10.8)

PARTICULATE **PREVIEW WRAP-UP**

Hydrogen bonding is the predominant intermolecular force for ethanol, whereas dimethyl ether molecules are held together by dipole–dipole interactions. Given that hydrogen bonding is stronger than dipole–dipole interactions, and that both molecules have the same molar mass, ethanol is the liquid and dimethyl ether is the gas.

PROBLEM-SOLVING SUMMARY

Type of Problem	Concepts and Equations	Sample Exercises
Explaining differences in boiling points of liquids and trends in boiling points of pure substances	Large molecules usually have higher boiling points than smaller molecules, with notable exceptions. The presence of polar –OH and –NH groups in molecules of a liquid leads to intermolecular hydrogen bonding that markedly increases the boiling point of the liquid.	**10.1, 10.2**
Predicting solubility in water and miscibility of liquids	Polar molecules are more soluble in water than nonpolar molecules. Molecules that form hydrogen bonds are more soluble in water than molecules that cannot form these bonds. Like dissolves like.	**10.3, 10.4**
Calculating the solubility of a gas by using Henry's law	Use Henry's law: $$C_{gas} = k_H P_{gas} \qquad (10.1)$$ where C_{gas} is the solubility of the gas; k_H is a constant that depends on the gas, the solvent, and the temperature; and P_{gas} is the pressure of the gas (or partial pressure if the gas is part of a mixture of gases).	**10.5**
Calculating the vapor pressure, enthalpy of vaporization, or temperature of a pure liquid	Use the Clausius–Clapeyron equation: $$\ln\!\left(\frac{P_{vap,T_1}}{P_{vap,T_2}}\right) = \frac{\Delta H_{vap}}{R}\left(\frac{1}{T_2} - \frac{1}{T_1}\right) \qquad (10.2)$$ and solve for any unknown value.	**10.6**

Type of Problem	Concepts and Equations	Sample Exercises
Reading a phase diagram	Locate the point (*T* and/or *P*) specified by the question as the starting point for the exercise. Temperature is on the horizontal axis; pressure is on the vertical axis. When you draw a line either horizontally rightward from the pressure axis or vertically up from the temperature axis, a phase change occurs wherever the line crosses an equilibrium line.	**10.7**

VISUAL PROBLEMS

(Answers to boldface end-of-chapter questions and problems are in the back of the book.)

10.1. Look at the pairs of ions in the structures of KF and KI represented in Figure P10.1. Which substance has the stronger cation–anion attractive forces and the higher melting point? Explain your answer.

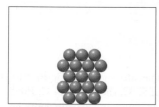

FIGURE P10.1

10.2. In Figure P10.2, identify the physical state (solid, liquid, or gas) of xenon and classify the attractive forces between the xenon atoms.

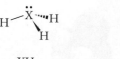

FIGURE P10.2

10.3. Figure P10.3 depicts molecules of XH_3 and YH_3 (not shown to scale) and the boiling points of XH_3 and YH_3 at 1 atm pressure. One substance is phosphine (PH_3) and the other substance is ammonia (NH_3). Which molecule is phosphine? Explain your answer.

XH_3	YH_3
Boiling point −88°C	Boiling point −33°C

FIGURE P10.3

10.4. Figure P10.4 shows representations of the molecules pentane, C_5H_{12}, and decane, $C_{10}H_{22}$. Which substance has the lower freezing point? Explain your answer.

Pentane	Decane

FIGURE P10.4

10.5. The graphs in Figure P10.5 have the same scales and describe the change in $\ln(P_{vap})$ of two pure liquids as a function of temperature. Which liquid has the stronger intermolecular attractive forces? Explain your answer.

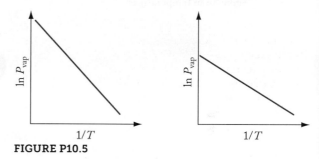

FIGURE P10.5

10.6. Which of the drawings in Figure P10.6, both of which are at constant temperature, most likely illustrates the pure liquid with the lower normal boiling point? Explain your choice.

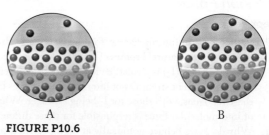

A	B

FIGURE P10.6

10.7. Examine the phase diagram of substance Z in Figure P10.7. (a) Does the freezing point of the substance increase or decrease with increasing pressure? (b) Do you predict the solid phase of substance Z would float on the liquid phase? Why?

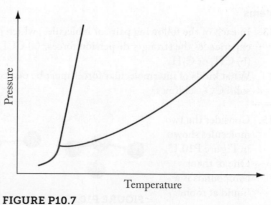

FIGURE P10.7

10.8. Use representations [A] through [I] in Figure P10.8 to answer questions a–f.

a. What are the predominant intermolecular forces in each of the substances in the electrostatic potential maps?

b. Which substance requires the lowest temperature to condense?

c. Which representations are isomers of one another?

d. Consider ethylene glycol and iodine. Which will dissolve in ethanol? Which will dissolve in carbon tetrachloride? Name the intermolecular forces responsible.

e. Which process(es) show(s) an increase in partial pressure?

f. Which process would occur due to an increase in temperature?

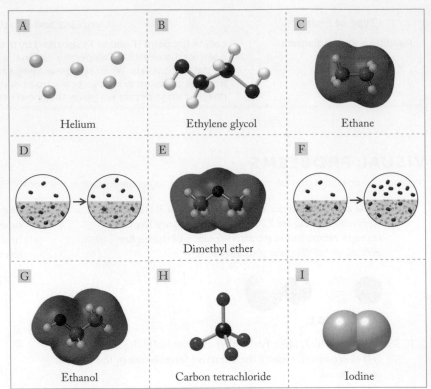

A — Helium

B — Ethylene glycol

C — Ethane

D

E — Dimethyl ether

F

G — Ethanol

H — Carbon tetrachloride

I — Iodine

FIGURE P10.8

QUESTIONS AND PROBLEMS

Dispersion Forces

Concept Review

10.9. Which type of intermolecular force exists in all substances?

10.10. At room temperature, bromine (Br_2) is a corrosive red liquid, whereas iodine (I_2) is a volatile violet solid. The differences point to different strengths of intermolecular forces between these halogens, with those for I_2 being stronger. What kind of intermolecular force is responsible for these differences?

10.11. Why do gases behave nonideally at high pressures and low temperatures?

10.12. Why are normal boiling points generally lower for branched hydrocarbons than for straight-chain hydrocarbons of the same molecular mass?

Problems

10.13. In each of the following pairs of molecules, which one experiences the stronger dispersion forces? (a) CCl_4 or CF_4; (b) CH_4 or C_3H_8

10.14. What kinds of intermolecular forces must be overcome as solid CO_2 sublimes?

*10.15. Consider the two molecules shown in Figure P10.15. One of these compounds is a liquid at room temperature; the other is a solid. Which is which? Explain why.

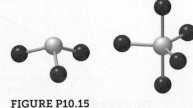

FIGURE P10.15

*10.16. Consider the two molecules shown in Figure P10.16. One has a boiling point of 57°C; the other, 84°C. Predict which has the higher boiling point, and explain your choice.

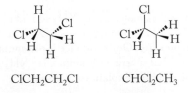

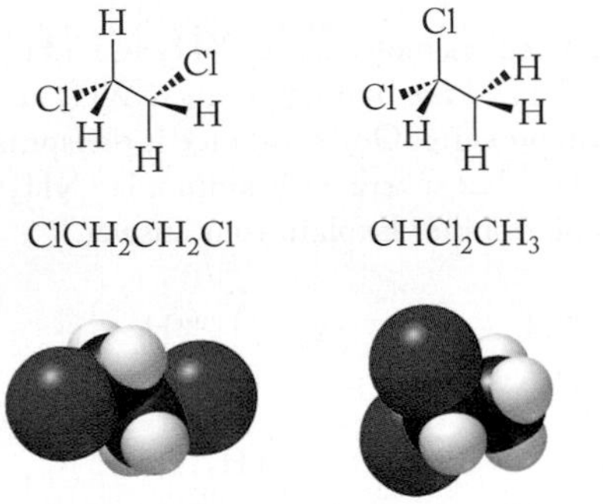

$ClCH_2CH_2Cl$ $\qquad$ $CHCl_2CH_3$

FIGURE P10.16

Interactions among Polar Molecules

Concept Review

10.17. How are the water molecules preferentially oriented around the anion in an aqueous solution of sodium chloride?

10.18. How are the water molecules preferentially oriented around the cation in an aqueous solution of potassium bromide?

10.19. Why are dipole–dipole interactions generally weaker than ion–dipole interactions?

10.20. Two liquids—one polar, one nonpolar—have the same molar mass. Which one is likely to have the higher boiling point? Explain your answer.

10.21. Why are hydrogen bonds considered a special class of dipole–dipole interactions?

10.22. Can all polar hydrogen-containing molecules form hydrogen bonds? Why or why not?

Problems

10.23. The permanent dipole moment of CH_2F_2 (1.93 D) is larger than that of CH_2Cl_2 (1.60 D), yet the boiling point of CH_2Cl_2 (40°C) is much higher than that of CH_2F_2 (−52°C). Why?

10.24. How is it that the permanent dipole moment of HCl (1.08 D) is larger than the permanent dipole moment of HBr (0.82 D), yet HBr boils at a higher temperature?

10.25. Hydrogen peroxide (H_2O_2) and water (H_2O) are both liquids at room temperature, but their standard heats of vaporization ΔH°_{vap} are different: 52 kJ/mol and 41 kJ/mol, respectively. Which substance has the stronger intermolecular forces? Can you suggest why?

10.26. From the thermochemical data for H_2O_2 and H_2O given in Problem 10.25, which liquid has the higher boiling point?

10.27. Explain why the melting point of methyl fluoride, CH_3F (−142°C), is higher than the melting point of methane, CH_4 (−182°C).

10.28. Explain why the boiling point of Br_2 (59°C) is lower than that of iodine monochloride, ICl (97°C), even though they have nearly the same molar mass.

10.29. Why doesn't fluoromethane (CH_3F) exhibit hydrogen bonding, whereas hydrogen fluoride, HF, does?

10.30. The boiling point of phosphine, PH_3 (−88°C), is lower than that of ammonia, NH_3 (−33°C), even though PH_3 has twice the molar mass of NH_3. Why?

10.31. In which of the following compounds do the molecules experience the strongest dipole–dipole attractions? (a) CF_4; (b) CF_2Cl_2; (c) CCl_4

10.32. Which of the following compounds, CO_2, NO_2, SO_2, or H_2S, is expected to have the weakest interactions between its molecules?

10.33. Which of the following molecules can form hydrogen bonds among themselves in pure samples of bulk material? (a) methanol (CH_3OH); (b) ethane (CH_3CH_3); (c) dimethyl ether (CH_3OCH_3); (d) acetic acid (CH_3COOH)

10.34. Which of the following molecules can form hydrogen bonds with molecules of water? (a) methanol (CH_3OH); (b) ethane (CH_3CH_3); (c) dimethyl ether (CH_3OCH_3); (d) acetic acid (CH_3COOH)

Polarity and Solubility

Concept Review

10.35. What is the difference between the terms *miscible* and *insoluble*?

10.36. What properties of water molecules enable them to hydrate and separate cations and anions in aqueous solutions?

10.37. One of the compounds in Figure P10.16 is insoluble in water; the other has a water solubility of 0.87 g/100 mL at 20°C. Identify which is which, and explain your reasoning.

10.38. Explain why the solubility of the series of alcohols with the formula $C_nH_{2n+2}OH$ decreases with increasing n.

10.39. In what context do the terms *hydrophobic* and *hydrophilic* relate to the solubilities of substances in water?

10.40. How does the presence of increasingly longer hydrocarbon chains in the structure affect the solubility of a series of structurally related molecules in water?

Problems

10.41. In each of the following pairs of compounds, which compound is likely to be more soluble in water?
   a. CCl_4 or $CHCl_3$
   b. CH_3OH or $C_6H_{11}OH$
   c. NaF or MgO
   d. CaF_2 or BaF_2

10.42. In each of the following pairs of compounds, which compound is likely to be more soluble in CCl_4?
   a. Br_2 or NaBr
   b. CH_3CH_2OH or CH_3OCH_3
   c. CS_2 or KOH
   d. I_2 or CaF_2

10.43. Which of these pairs of substances is likely to be miscible?
   a. Br_2 and benzene (C_6H_6)
   b. $CH_3CH_2OCH_2CH_3$ (diethyl ether) and CH_3COOH (acetic acid)
   c. C_6H_{12} (cyclohexane) and hexane ($CH_3CH_2CH_2CH_2CH_2CH_3$)
   d. CS_2 (carbon disulfide) and CCl_4 (carbon tetrachloride)

10.44. Which of these pairs of substances is likely to be miscible?
   a. CH_3CH_2OH (ethanol) and $CH_3CH_2OCH_2CH_3$ (diethyl ether)
   b. CH_3OH (methanol) and methyl amine (CH_3NH_2)
   c. CH_3CN (acetonitrile) and acetone (CH_3COCH_3)
   d. CF_3CHF_2 (a Freon replacement) and $CH_3CH_2CH_2CH_2CH_3$ (pentane)

10.45. Which of the following compounds is likely to be the most soluble in water? (a) NaCl; (b) KI; (c) $Ca(OH)_2$; (d) CaO

10.46. From the data in Figure P10.46, which has a greater effect on the solubility of oxygen in water: (a) decreasing the temperature from 20°C to 10°C or (b) raising the pressure from 1.00 atm to 1.25 atm?

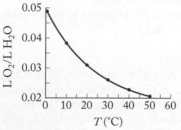

FIGURE P10.46

10.47. Which of these substances is the least soluble in water?
   a. $CH_3(CH_2)_2CH_2OH$
   b. $CH_3(CH_2)_4CH_2OH$
   c. $CH_3(CH_2)_6CH_2OH$
   d. $CH_3(CH_2)_8CH_2OH$

10.48. Which of these substances is the most soluble in water?
  a. $CH_3(CH_2)_2CH_2NH_2$
  b. $CH_3(CH_2)_4CH_2Cl$
  c. $CH_3(CH_2)_6CH_2Br$
  d. $CH_3(CH_2)_8CH_2I$

10.49. Which of the compounds in Figure P10.49 are alcohols and which ones are ethers? Place them in order of increasing boiling point.

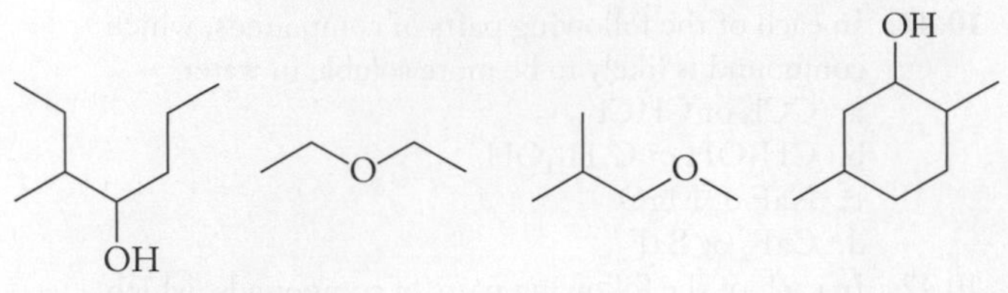

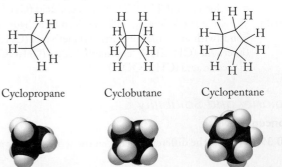

(a)   (b)   (c)   (d)

FIGURE P10.49

10.50. Which of the compounds in Figure P10.50 are alcohols and which ones are ethers? Place them in order of increasing vapor pressure at 25°C.

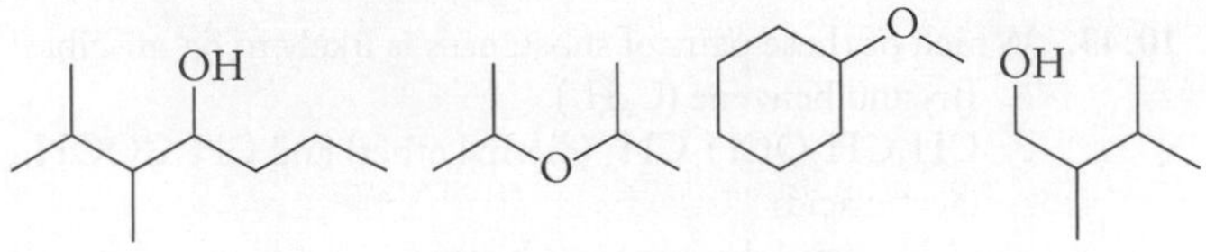

(a)   (b)   (c)   (d)

FIGURE P10.50

Solubility of Gases in Water

Concept Review

10.51. Why does the solubility of most gases in most liquids increase with decreasing temperature?

10.52. Which term, k_H or P, in Henry's law is affected by temperature?

10.53. Air is primarily a mixture of nitrogen and oxygen. Is the Henry's law constant for the solubility of air in water the sum of k_H for N_2 and k_H for O_2? Explain why or why not.

10.54. Why is the Henry's law constant for CO_2 so much larger than those for N_2 and O_2 at the same temperature?

10.55. Which sulfur oxide would you predict to be more soluble in nonpolar solvents, SO_2 or SO_3?

10.56. Which oxide would you predict to be more soluble in polar solvents, CO_2 or NO_2?

Problems

10.57. **Arterial Blood** Arterial blood contains about 0.25 g of oxygen per liter at 37°C and standard atmospheric pressure. What is the Henry's law constant [mol/(L · atm)] for O_2 dissolution in blood? The mole fraction of O_2 in air is 0.209.

10.58. The solubility of O_2 in water is 6.5 mg/L at an atmospheric pressure of 1 atm and temperature of 40°C. Calculate the Henry's law constant of O_2 at 40°C. The mole fraction of O_2 in air is 0.209.

*10.59. **Oxygen for Climbers and Divers** Use the Henry's law constant for O_2 dissolved in arterial blood from

Problem 10.57 to calculate the solubility of O_2 in the blood of (a) a climber on Mt. Everest (P_{atm} = 0.35 atm) and (b) a scuba diver at 100 feet ($P \approx 3$ atm).

*10.60. The solubility of air in water is approximately $7.9 \times 10^{-4}\ M$ at 20°C and 1.0 atm. Calculate the Henry's law constant for air. Is the k_H value of air approximately equal to the sum of the k_H values for N_2 and O_2 because these two gases make up 99% of the gases in air?

10.61. **Carbonated Sodas** Manufacturers of carbonated beverages dissolve $CO_2(g)$ in water under pressure to produce sodas. If a manufacturer uses a pressure of 4.16 atm at 25°C to carbonate the water, how many grams of CO_2 are in the average can of soda, the volume of which is 355 mL? The Henry's law constant for CO_2 at 25°C is 0.03360 mol/(L · atm).

10.62. What would be the effect on the results of the process described in Problem 10.61 (a) if the temperature of the carbonation process was changed from 25°C to 10°C, or (b) if the temperature were held constant but the pressure was doubled?

Vapor Pressure of Pure Liquids

Concept Review

10.63. Why does the vapor pressure of a liquid increase as temperature increases?

10.64. Which of the following factors influences the vapor pressure of a pure liquid?
  a. the volume of liquid present in a container
  b. the temperature of the liquid
  c. the surface area of the liquid

10.65. Is vapor pressure an intensive or extensive property of a liquid?

10.66. A chef observes bubbles while heating a pot of water to 60°C to poach vegetables. What are the gases in the bubbles and where did they come from?

Problems

10.67. Rank the following compounds in order of increasing vapor pressure at 298 K. (a) CH_3CH_2OH; (b) CH_3OCH_3; (c) $CH_3CH_2CH_3$

10.68. Rank the compounds in Figure P10.68 in order of increasing vapor pressure at 298 K.

Cyclopropane   Cyclobutane   Cyclopentane

FIGURE P10.68

10.69. Pine Oil The smell of fresh cut pine is due in part to the cyclic alkene pinene, whose carbon-skeleton structure is shown in Figure P10.69.

a. Use the data in the table to calculate the heat of vaporization, ΔH_{vap}, of pinene.

b. Use the value of ΔH_{vap} to calculate the vapor pressure of pinene at room temperature (23°C).

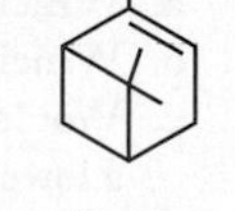

Pinene
FIGURE P10.69

Vapor Pressure (torr)	Temperature (K)
760	429
515	415
340	401
218	387
135	373

10.70 Almonds and Cherries Almonds and almond extracts are common ingredients in baked goods. Almonds contain the compound benzaldehyde (shown in Figure P10.70), which accounts for the odor of the nut. Benzaldehyde is also responsible for the aroma of cherries.

a. Use the data in the table to calculate the heat of vaporization, ΔH_{vap}, of benzaldehyde.

b. Use the value of ΔH_{vap} to calculate the vapor pressure of benzaldehyde at room temperature (23°C).

Benzaldehyde

FIGURE P10.70

Vapor Pressure (torr)	Temperature (K)
50	373
111	393
230	413
442	433
805	453

Phase Diagrams: Intermolecular Forces at Work

Concept Review

10.71. Explain the difference between sublimation and evaporation.

10.72. Can ice be melted merely by applying pressure? Explain your answer.

10.73. What phases of a substance are present (a) at its triple point and (b) at its critical point?

10.74. Explain how the solid–liquid line in the phase diagram of water differs in character from the solid–liquid line in the phase diagrams of most other substances, such as CO_2.

10.75. Which phase of a substance (gas, liquid, or solid) is most likely to be the stable phase: (a) at low temperatures and high pressures; (b) at high temperatures and low pressures?

10.76. At what temperatures and pressures does a substance behave as a supercritical fluid?

10.77. Preserving Food Freeze-drying is used to preserve food at low temperature with minimal loss of flavor. Freeze-drying works by freezing the food and then lowering the pressure with a vacuum pump to sublime the ice. Must the pressure be lower than the pressure at the triple point of H_2O? Why or why not?

10.78. Solid helium cannot be converted directly into the vapor phase. Does the phase diagram of helium have a triple point?

Problems

For help in answering Problems 10.79 through 10.90, consult Figures 10.26, 10.28, and 10.29.

10.79. What is the normal boiling point of bromine?

10.80. If water boils at 50°C, what is the pressure?

10.81. Which molecules have stronger intermolecular attractive forces: ethylene glycol or ethanol? Explain your choice.

10.82. What is the boiling point of diethyl ether at 300 mmHg of pressure?

10.83. List the steps you would take to convert a 10.0-g sample of water at 25°C and 1 atm of pressure to water at its triple point.

10.84. List the steps you would take to convert a 10.0-g sample of water at 25°C and 2 atm pressure to ice at 1 atm of pressure. At what temperature would the water freeze?

10.85. What phase changes, if any, does liquid water at 100°C undergo if the initial pressure of 5.0 atm is reduced to 0.5 atm at constant temperature?

10.86. What phase changes, if any, occur if CO_2 initially at −80°C and 8.0 atm is allowed to warm to −25°C at 5.0 atm?

10.87. Below what temperature can solid CO_2 (dry ice) be converted into CO_2 gas simply by lowering the pressure?

10.88. What is the maximum pressure at which solid CO_2 (dry ice) can be converted into CO_2 gas without melting?

10.89. Predict the phase of water that exists under the following conditions.
a. 2 atm of pressure and 110°C
b. 200 atm of pressure and 380°C
c. 6.0×10^{-3} atm of pressure and 0°C

10.90. Which phase or phases of water exist under the following conditions?
a. 2.0 atm and 50°C
b. 0.10 atm and 300°C
c. 1 atm and 0°C

Some Remarkable Properties of Water

Concept Review

10.91. Explain why a needle floats on the surface of water but sinks in a container of methanol (CH_3OH).

10.92. Explain why different liquids do not reach the same height in capillary tubes of the same diameter.

10.93. Explain why pipes filled with water are in danger of bursting when the temperature drops below 0°C.

10.94. A hot needle sinks when put on the surface of cold water. Will a cold needle float in hot water? Explain your answer.

10.95. The meniscus of water in a glass tube is concave, but that of mercury (Figure P10.95) is convex. Explain why.

*10.96. The mercury level in a capillary tube placed in a dish of mercury is below the surface of the mercury in the dish. Explain why.

10.97. Describe the origin of surface tension at the molecular level.

10.98. What is the cause of the high viscosity of molasses?

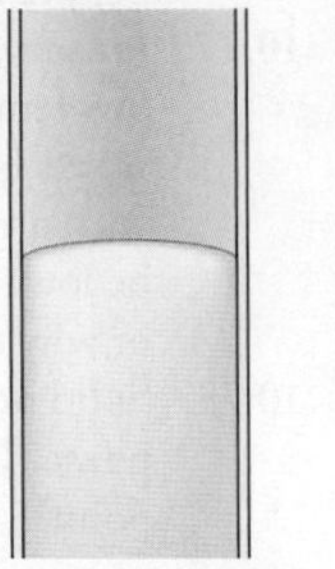

FIGURE P10.95

10.99. Describe how the surface tension and viscosity of a liquid are affected by increasing temperature.

10.100. Explain how strong intermolecular forces are expected to result in a relatively high surface tension and viscosity of a liquid.

Problems

10.101. One of two glass capillary tubes of the same diameter is placed in a dish of water and the other in a dish of ethanol (CH_3CH_2OH). Which liquid will rise higher in its tube? Explain your answer.

10.102. Would you expect water to rise to the same height in a tube made of a polyethylene plastic as it does in a glass capillary tube of the same diameter? Why or why not? The molecular structure of polyethylene is shown in Figure P10.102.

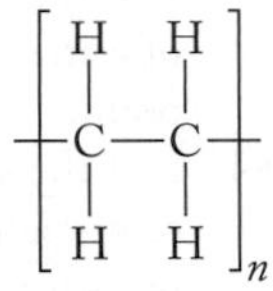

FIGURE P10.102

10.103. The normal boiling points of liquids A and B are 75.0°C and 151°C, respectively. Which of these liquids would you expect to have the higher surface tension and viscosity at 25°C? Explain your answer.

10.104. A simple viscometer consists of a thick-walled glass tube with a 0.5-mm bore. The tube has etched marks at one-quarter and three-quarters of its height. The tube is clamped with its lower end dipped in a container of the liquid to be tested. A pipette filler is used to draw liquid up the bore past the upper mark. The pipette filler is removed from the tube, and the time taken for the liquid meniscus to drain between the upper and lower viscometer marks is measured with a stopwatch. Using this viscometer to measure the drain times for two pure liquids A and B at the same temperature gives 3.45 seconds for liquid A and 4.64 seconds for liquid B.

a. Which liquid is more viscous?

b. Which liquid has weaker intermolecular forces?

c. Would the measured drain times be longer or shorter at a lower experimental liquid temperature?

Additional Problems

10.105. Why do ethers typically boil at lower temperatures than alcohols with the same molecular formula?

*10.106. **Dry Gas** During the winter months in cold climates, water condensing in a vehicle's gas tank reduces engine performance. An auto mechanic recommends adding "dry gas" to the tank during your next fill-up. Dry gas is typically an alcohol that dissolves in gasoline and absorbs water. From the structures shown in Figure P10.106, which product do you predict would do a better job—methanol or 2-propanol? Why?

CH_3OH

Methanol   2-Propanol

FIGURE P10.106

10.107. Does the sublimation point of ice increase or decrease with increasing pressure? Explain why.

10.108. Does the sublimation point of CO_2 increase or decrease with increasing pressure? Explain why.

10.109. Liquid substances are often compared for their physical properties in different applications. Comparison of two liquids A and B at constant temperature and atmospheric pressure shows that liquid A has higher viscosity and surface tension, a higher boiling point, and lower vapor pressure than liquid B. Are these data all consistent with stronger intermolecular forces in liquid A than in liquid B?

10.110. Why is methanol (CH_3OH) miscible with water, whereas CH_4 is almost completely insoluble in water?

10.111. Sketch a phase diagram for element X, which has a triple point at 152 K and a pressure of 0.371 atm, a boiling point of 166 K at a pressure of 1.00 atm, and a normal melting point of 161 K.

*10.112. The melting point of hydrogen is 14.96 K at 1.00 atm of pressure. The temperature at its triple point is 13.81 K. Does H_2 expand or contract when it freezes?

10.113. Explain why water climbs higher in a capillary tube than in a test tube.

10.114. Explain why ice floats on water.

10.115. **Fish Dying in Summer Heat** Explain why fish in a pond die if water becomes too warm.

*10.116. **Evaluation of Pharmaceuticals** A test done on new pharmaceutical agents early in the development process required the observation of their relative solubilities in octanol and water. A drug had to be sufficiently soluble in water (hydrophilic) to be carried in the bloodstream but also sufficiently hydrophobic (octanol soluble) to move across cell membranes. Pick the molecule from Figure P10.116 that you predict might have comparable solubility in both solvents. Explain your choice.

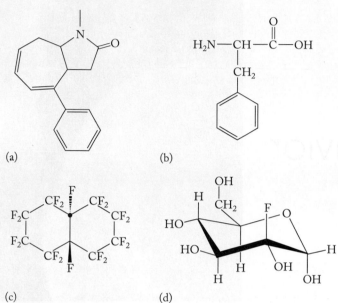

(a)   (b)

(c)   (d)

FIGURE P10.116

10.117. **First Aid for Bruises** Compounds with low boiling points may be sprayed on skin as a topical anesthetic—they chill

it as they evaporate, providing short-term relief from injuries. Predict which compound among those in Figure P10.117 has the lowest boiling point.

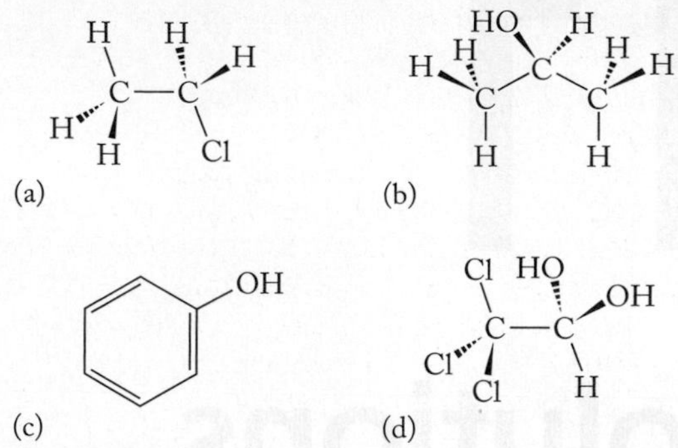

(a)   (b)

(c)   (d)

FIGURE P10.117

10.118. **Refrigerators** Refrigerators have a unit called a compressor that liquefies a gas. The refrigerator is cooled by a continuous cycle of compression of the gas to produce the liquid, followed by evaporation of the liquid to provide the cooling. Ammonia (NH_3) and sulfur dioxide (SO_2) were the gases used originally, and hexafluoroethane (C_2F_6) has been used since the 1990s. What are the intermolecular interactions that characterize these substances?

*10.119. **Disposable Wipes** Disposable wipes used to clean the skin before receiving an immunization shot contain ethanol. After the nurse wipes your arm, your skin feels cold. Why?

11

Solutions
Properties and Behavior

HEALTHY RED BLOOD CELLS
Blood is a complex mixture of dissolved and suspended components. The latter include red blood cells, which contain hemoglobin, a substance that binds oxygen and transports it from the lungs to tissues.

PARTICULATE **REVIEW**

Size Trends: Neutral Atoms versus Ions

In Chapter 11 we explore the properties of solutions containing dissolved solutes, both molecular and ionic. Neutral atoms of potassium, sodium, sulfur, and fluorine, along with their most common ions, are shown here. In each case, the neutral atom is above the ion. Answer these questions based on the periodic trends for atomic size.

- Which pair is the potassium atom and its ion?

- Which pair is the fluorine atom and its ion?

- Which of the remaining pairs is sulfur and the sulfide ion? Which pair is sodium and its ion?

   (Review Section 7.10 if you need help.)

(Answers to Particulate Review questions are in the back of the book.)

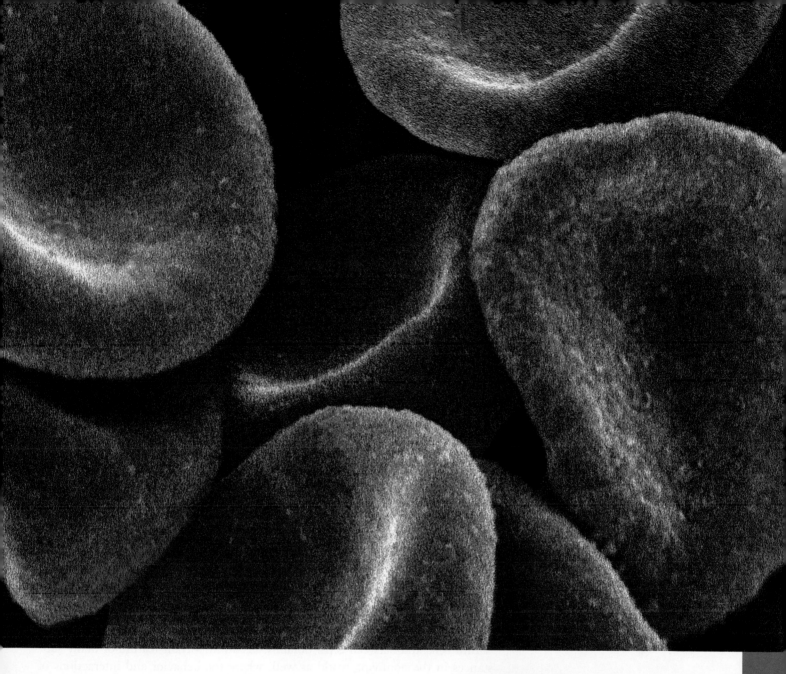

Ionic Charges, Distances, and Melting Points

Ions of cesium, barium, oxygen, and iodine are shown here from left to right. As you read Chapter 11, look for ideas that will help you answer these questions:

- Which cation and anion pair has the longer distance between their nuclei?

- Which cation and anion pair has the larger coulombic interaction?

- Which compound has a higher melting point, cesium iodide or barium oxide?

Learning Outcomes

LO1 Estimate the relative strengths of ion–ion interactions
Sample Exercise 11.1

LO2 Explain the energy changes that accompany the formation and dissolution of an ionic compound
Sample Exercises 11.2, 11.3, 11.4

LO3 Calculate the vapor pressure of a solution containing a nonvolatile solute by using Raoult's law
Sample Exercise 11.5

LO4 Describe the process of fractional distillation and interpret graphs of temperature versus volume of distillate
Sample Exercises 11.6, 11.7

LO5 Express the concentration of a solution in molality
Sample Exercise 11.8

LO6 Calculate the freezing point and boiling point of a solution of a nonvolatile solute
Sample Exercises 11.9, 11.10

LO7 Explain the significance of the van 't Hoff factor
Sample Exercises 11.11, 11.12

LO8 Predict the direction of solvent flow in osmosis and calculate osmotic pressure
Sample Exercises 11.13, 11.14, 11.15

LO9 Apply colligative properties to the determination of molar mass
Sample Exercises 11.16, 11.17

11.1 Interactions between Ions

Given that life in air and water depends on relatively narrow ranges of amounts of solutes (O_2, salts) dissolved in fluids (lakes, blood), the concentration of solute in a solution can be of critical importance. For example, the fluids dispensed intravenously in hospitals, which deliver medications to patients and restore fluid levels in trauma victims, must contain specific concentrations of solutes. Too much solute in the intravenous fluid causes dehydration, whereas too little causes retention of excess water by the body.

In this chapter we focus on the interactions between ions in solids and the ability of water and other liquids to dissolve solids, as well as on the properties of their solutions. These properties are of prime importance to all living things, which depend on aqueous solutions to contain and transport molecules and ions within cells and between cells. Solubility is an important feature of many substances in the nonliving world as well, where the behavior and interactions of substances often depend on the identity and concentration of solute dissolved in a solvent.

Solutions differ from pure substances in several important respects. For example, solutions of 50–80 percent ethylene glycol in water, known as antifreeze, are used in the radiators of vehicles because these solutions freeze at much lower temperatures than pure water or pure ethylene glycol. Antifreeze solutions also boil at higher temperatures than pure water, thereby increasing the temperature range over which the fluid can transfer heat away from the engine. The temperatures at which an antifreeze solution boils or freezes depend on the concentration of the solute (ethylene glycol) in the solvent (water).

Solutions and pure liquids behave differently because of the intermolecular forces between solvent and solute particles. Let's start our study of solutions by considering what happens when ocean waves crashing on a rocky shore create plumes of sea spray, carrying small drops of seawater into the atmosphere. As some of the water in these drops evaporates, their concentrations of dissolved ions such as Cl^-, Na^+, Mg^{2+}, Br^-, Ca^{2+}, and SO_4^{2-} increase. Eventually, the decreased volume of the drops produces supersaturated solutions of the salts, which begin to precipitate. Among the first solids to form is $CaSO_4$. Among the last is NaCl,

CONNECTION We introduced solutions and several units of concentration in Section 4.2 when discussing how much solute is dissolved in the solvent.

which does not precipitate until 90% of the seawater in a drop has evaporated. This sequence takes place even though the concentrations of Ca^{2+} and SO_4^{2-} ions are much lower than the concentrations of Na^+ and Cl^- ions. Why does $CaSO_4$ precipitate before NaCl? Put another way: why is NaCl more soluble in water than $CaSO_4$?

The lower solubility of $CaSO_4$ correlates with greater ion–ion interactions between Ca^{2+} and SO_4^{2-}, which result from higher charges on these ions than on Na^+ and Cl^- ions. In fact, ion–ion interaction is the strongest kind of interactive force between particles; the interaction between ions of opposite charge results in the formation of ionic bonds. The strengths of ion–ion interactions are defined by *coulombic interaction*. The corresponding electrostatic potential energy, E, between two particles with charges Q_1 and Q_2, separated by a distance d, is defined by a relationship we have seen before:

$$E \propto \frac{(Q_1 Q_2)}{d} \qquad (5.3)$$

The value of E in Equation 5.3 is negative when Q_1 and Q_2 have opposite signs, which means E is negative for any salt because the dominant interaction is between cations $(+Q)$ and anions $(-Q)$. Oppositely charged ions attract each other to form an arrangement characterized by a lower potential energy than the potential energy of the separated ions. When the ions are far apart, the d term in Equation 5.3 is large and E is a small negative number. When the ions are close together in a salt, d is small and E is a large negative number, which corresponds to lower energy (see Figure 5.7).

The strength of ion–ion interactions means that ionic compounds are most likely to be solids at room temperature. Whenever cations and anions in the gas phase combine to make a solid ionic compound, energy is released and the process is exothermic. The same amount of energy must be absorbed to separate the compound into its cations and anions. We can use Equation 5.3 to compare the relative strengths of interactive forces between ions in salts if we know the charges on the ions and their radii; the attractive force between two ions increases as ionic charge increases and as ionic size decreases.

The inter-ion distance d for ionic compounds is the sum of the ionic radii. Recall that the sizes of atoms and ions are a periodic property. In Chapter 7 we saw that cations are always smaller than the atoms from which they form and anions are always larger (Figure 11.1). Ions in a group of the periodic table, having the same charge, increase in size as you move down the group.

CONCEPT TEST

Rank the following sets of ions in order of increasing distance (d) between their nuclear centers: (a) KF, LiF, NaF; (b) $CaBr_2$, $CaCl_2$, CaF_2.

(Answers to Concept Tests are in the back of the book.)

To address the question of why $CaSO_4$ precipitates from sea spray before NaCl, we can use Equation 5.3 to compare the relative strengths of their ionic interactions. For NaCl the numerator is $(+1)(-1) = -1$, and for $CaSO_4$ the numerator is $(+2)(-2) = -4$, four times as large as for NaCl. For NaCl, the ionic radii are 102 pm for Na^+ and 181 pm for Cl^-, so d is $102 + 181 = 283$ pm. For $CaSO_4$, the radii are 100 pm for Ca^{2+} and 230 pm for SO_4^{2-} (Table 11.1), making $d = 330$ pm. The denominator for $CaSO_4$ is about 1.2 times the size of the NaCl denominator, so the factor-of-four difference in the numerators dominates this

CONNECTION We introduced coulombic interactions in Chapter 5 when discussing the electrostatic interaction between particles in an ionic bond. Later in that chapter we learned that the energy associated with a process has the same absolute value but the opposite sign of the energy associated with the reverse process.

CONNECTION In Chapter 7 we discussed the use of ionic radii to calculate the distance between the nuclear centers in an ionic bond.

TABLE 11.1 Estimated Radii of Polyatomic Ions

Polyatomic Ion	Radius (pm)
CO_3^{2-}	185
NO_3^-	189
SO_4^{2-}	230
PO_4^{3-}	238

Values from R. B. Heslop and K. Jones. *Inorganic Chemistry: A Guide to Advanced Study* (Elsevier, 1976), p. 123.

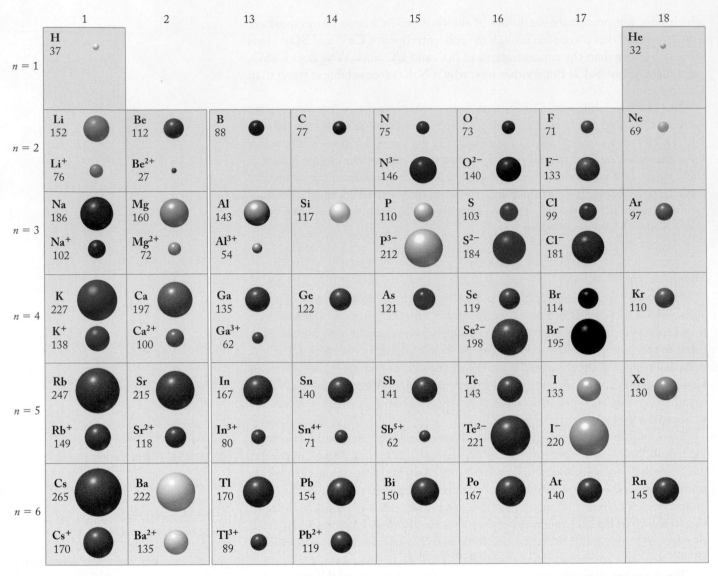

FIGURE 11.1 The radii of anions formed by the main group elements are larger than the radii of their parent atoms. The radii of cations are smaller than the radii of their parent atoms. All values are in picometers. Values from N. N. Greenwood and A. Earnshaw, *Chemistry of the Elements*, 2nd ed. (Boston: Butterworth-Heinemann, 1997); C. E. Housecroft and A. G. Sharpe, *Inorganic Chemistry*, 3rd ed. (Upper Saddle River, NJ: Pearson Prentice Hall, 2008).

comparison. Thus the attraction between Ca^{2+} ions and SO_4^{2-} ions is stronger than the attraction between Na^+ ions and Cl^- ions. The stronger attraction between its ions is not the only reason $CaSO_4$ precipitates first from sea spray, but it is a major factor.

SAMPLE EXERCISE 11.1 Predicting Relative Strengths of Ion–Ion Interactions **LO1**

Ionic compounds such as strontium sulfate and barium sulfate are major ingredients of the exoskeletons of some single-celled organisms. About 10% of kidney stones are composed primarily of calcium phosphate. List the ionic compounds $Ca_3(PO_4)_2$, $SrSO_4$, and $BaSO_4$ in order of decreasing strength of the attraction between their ions.

Collect and Organize We are given three ionic compounds and want to find the relative strength of the ion–ion interaction in each one. Equation 5.3 shows that the strength of an ion–ion attractive interaction depends on the charges on the ions and the distance between them. We know the charges on the ions in binary ionic compounds from Chapter 2. Inter-ionic distances are determined from ionic radii in Figure 11.1 and Table 11.1.

Analyze We need to compare the charges on the calcium, strontium, barium, sulfate, and phosphate ions along with the distances between the ions in $Ca_3(PO_4)_2$, $SrSO_4$, and $BaSO_4$. The distance between the ions equals the sum of their ionic radii. The group 2 cations all have a 2+ charge, and the sulfate and phosphate anions have 2− and 3− charges, respectively. Figure 11.1 indicates that the ionic radii of Ca^{2+}, Sr^{2+} and Ba^{2+} increase down the group, and Table 11.1 indicates that the phosphate ion is slightly larger than the sulfate ion. The strength of an ion–ion interaction is directly proportional to the product of the charges on the ions and inversely proportional to the distance between them. The charges of the ions are greatest in $Ca_3(PO_4)_2$, so we predict that calcium phosphate will have the greatest attraction between the ions. Strontium ions are smaller than barium ions, so the inter-ionic distance in $SrSO_4$ will be shorter than in $BaSO_4$ and, according to Equation 5.3, we can predict that the strength of the ion–ion interaction will be greater in $SrSO_4$ than in $BaSO_4$.

Solve The Q_1Q_2 values for each compound are

$$Ca^{2+} \quad PO_4^{3-} \quad Q_1Q_2 = (+2)(-3) = -6$$
$$Sr^{2+} \quad SO_4^{2-} \quad Q_1Q_2 = (+2)(-2) = -4$$
$$Ba^{2+} \quad SO_4^{2-} \quad Q_1Q_2 = (+2)(-2) = -4$$

The values of d for each compound are

$$d_{Ca_3(PO_4)_2} = 100 \text{ pm} + 238 \text{ pm} = 338 \text{ pm}$$
$$d_{SrSO_4} = 118 \text{ pm} + 230 \text{ pm} = 330 \text{ pm}$$
$$d_{BaSO_4} = 135 \text{ pm} + 230 \text{ pm} = 365 \text{ pm}$$

Substituting the values of d and Q_1Q_2 into Equation 5.3 for each compound:

$$E_{Ca_3(PO_4)_2} \propto (-6)/338 = -0.0178$$
$$E_{SrSO_4} \propto (-4)/330 = -0.0121$$
$$E_{BaSO_4} \propto (-4)/365 = -0.0110$$

Thus the predicted order of decreasing strength of ionic interactions is $Ca_3(PO_4)_2 > SrSO_4 > BaSO_4$.

Think About It The ion–ion distances are nearly the same for all three compounds, which means that the product of the ionic charges is the predominant factor influencing the relative interaction strengths. Note, however, that in the case of $SrSO_4$ and $BaSO_4$ where the ionic charges are equivalent, the size of the cation determines the relative interaction strengths of the two sulfate compounds.

 Practice Exercise Arrange the ionic compounds $CaCl_2$, BaO, and KCl in order of decreasing strength of the interaction between their ions.

(Answers to Practice Exercises are in the back of the book.)

Although ion size does affect the strength of ion–ion interactions, the product of the charges (Q_1Q_2) is typically the dominant factor in determining their relative strengths. The ionic radii of cations in Figure 11.1 range from 27 pm (Be^{2+}) to 170 pm (Cs^+), and those of monatomic anions and polyatomic anions (Figure 11.1 and Table 11.1) range from 133 pm (F^-) to 238 pm (PO_4^{3-}). If we put together the smallest cation and anion (BeF_2) and the largest cation and anion (Cs_3PO_4), the range of nucleus-to-nucleus distances is 160 to 408 pm, which is a factor of 2.6. Moreover, the radii of common ions cover an even smaller size range. Therefore, for common ionic compounds, the charge product is nearly always responsible for large differences in the strengths of ionic interactions.

11.2 Energy Changes during Formation and Dissolution of Ionic Compounds

In Section 10.2 we discussed the role of ion–dipole interactions in making salts soluble in water, and we described qualitatively how the combined effect of many ion–dipole interactions could overcome ion–ion interactions in a crystal of solute to produce hydrated ions in solution. Now we can take the next step and quantify the actual changes in energy associated with these interactions. We can then use these quantitative observations to understand further what happens at the particle level when solutes dissolve in solvents.

We can apply the qualitative picture of the dissolution process we developed in Chapter 10 to describe the factors that contribute to the overall enthalpy change that accompanies dissolution of an ionic compound in a polar solvent. The ions must be separated from one another (Figure 11.2a), a process that requires energy to break the ion–ion interactions that hold the particles in the crystal lattice ($\Delta H_{ion-ion}$). The solvent molecules must also be separated from one another so they can bind to the ions (Figure 11.2b); this process requires energy to break the dipole–dipole interactions—in water, these are hydrogen bonds—between solvent molecules ($\Delta H_{dipole-dipole}$). Last, energy is released when the solvent molecules associate with the solute ions via ion–dipole interactions ($\Delta H_{ion-dipole}$) and form hydrated ions (Figure 11.2c). The sum of the enthalpies of these interactions is the **enthalpy of solution ($\Delta H_{solution}$)**, which defines the overall change in enthalpy when an ionic solute is dissolved in a polar solvent.

$$\Delta H_{solution} = \Delta H_{ion-ion} + \Delta H_{dipole-dipole} + \Delta H_{ion-dipole} \qquad (11.1)$$

 **CONNECTION** In Figure 10.5 we illustrated the dissociation of NaCl into Na^+ and Cl^- ions, each surrounded by a hydration sphere.

CHEMTOUR Dissolution of Ammonium Nitrate

FIGURE 11.2 An ideal solution of an ionic compound in water consists of hydrated ions distributed throughout the solvent. We can account for the energy associated with establishing a solution by considering the energy (a) required to separate the ions from their lattice ($\Delta H_{ion-ion} = -U$), (b) required to overcome dipole–dipole interactions in the solvent ($\Delta H_{dipole-dipole}$), and (c) released when ion–dipole bonds form between solvent and solute particles ($\Delta H_{ion-dipole}$). The sum of (b) and (c) is $\Delta H_{hydration}$, the enthalpy of hydration. (d) The sum of all three processes is $\Delta H_{solution}$, the enthalpy of solution.

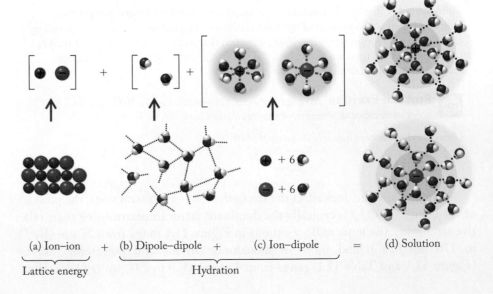

(a) Ion–ion + (b) Dipole–dipole + (c) Ion–dipole = (d) Solution

Lattice energy   Hydration

Of the quantities in Equation 11.1, we can readily measure $\Delta H_{\text{solution}}$ by using calorimetric methods. Measuring the remaining three terms independently is more difficult. The energy associated with ion–ion attraction is described by Equation 5.3; as we saw in the previous section, this energy is proportional to the product of the charges on the ions and inversely proportional to the distance between them.

We can combine the strengths of dipole–dipole and ion–dipole interactions into a single term called the **enthalpy of hydration ($\Delta H_{\text{hydration}}$)** for the compound:

$$\Delta H_{\text{hydration}} = \Delta H_{\text{dipole-dipole}} + \Delta H_{\text{ion-dipole}} \qquad (11.2)$$

Here *hydration* refers to the formation of solvated ions. Combining Equations 11.1 and 11.2, we have

$$\Delta H_{\text{solution}} = \Delta H_{\text{ion-ion}} + \Delta H_{\text{hydration}} \qquad (11.3)$$

Let's examine each of these terms in more detail.

The strength of ion–ion interactions is described by the **lattice energy (U)** of an ionic compound, which is the change in energy when free ions in the gas phase combine to form 1 mole of a solid ionic compound. The lattice energies of some common binary ionic compounds are given in Table 11.2. The formula for lattice energy is

$$U = \frac{k(Q_1 Q_2)}{d} \qquad (11.4)$$

This formula resembles Equation 5.3 except that it includes a proportionality constant, k, the value of which depends on the structure of the ionic solid. We examine the structures of solids in Chapter 12, at which point we will learn about the different arrangements possible for ions in an ionic solid. The key point here is that the same value of k is used for all compounds that have the same or nearly the same arrangement of ions. We can then substitute the lattice energy, U, of an ionic compound in Equation 11.1 for $\Delta H_{\text{ion-ion}}$:

$$\Delta H_{\text{solution}} = -U + \Delta H_{\text{hydration}}$$

or

$$\Delta H_{\text{solution}} = \Delta H_{\text{hydration}} - U \qquad (11.5)$$

There is a minus sign in front of the lattice energy term in Equation 11.5 because U is defined as the enthalpy change when gas-phase ions combine to form an ionic solid. In Equation 11.5 we are calculating the enthalpy change associated with separating the ions in an ionic compound. Recall from Chapter 5 that the enthalpy change for a process in one direction has the same magnitude but opposite sign of the process in the reverse direction.

The lattice energy of an ionic solid affects not only its solubility in water but also its melting point. Melting an ionic structure in which the ions are held together tightly should require more thermal energy (a higher temperature) than melting a structure in which the ions are held together less tightly—a trend we observe experimentally. Consider two ionic compounds: LiF ($U = -1047$ kJ/mol) and MgO ($U = -3791$ kJ/mol). The greater lattice energy of MgO, nearly four times that of LiF, is reflected in its higher melting point, 2825°C, versus 848°C for LiF, and its higher boiling point, 3600°C for MgO, versus 1673°C for LiF.

CONNECTION Calorimetry as a method for measuring enthalpy changes was described in Chapter 5.

TABLE 11.2 Lattice Energies (U) of Common Binary Ionic Compounds

Compound	U (kJ/mol)
LiF	−1047
LiCl	−864
NaCl	−786
KCl	−720
KBr	−691
$MgCl_2$	−2540
MgO	−3791

enthalpy of solution ($\Delta H_{\text{solution}}$) the overall energy change when a solute is dissolved in a solvent.

enthalpy of hydration ($\Delta H_{\text{hydration}}$) the energy change when gas-phase ions dissolve in a solvent.

lattice energy (U) the enthalpy change that occurs when 1 mole of an ionic compound forms from its free ions in the gas phase.

Na$^+$ = 102 pm
F$^-$ = 133 pm

K$^+$ = 138 pm
F$^-$ = 133 pm

Rb$^+$ = 149 pm
F$^-$ = 133 pm

FIGURE 11.3 The distance, d, between the nuclei of ions in NaF, KF, and RbF is in the order $d_{NaF} < d_{KF} < d_{RbF}$.

SAMPLE EXERCISE 11.2 Ranking Lattice Energies and Melting Points **LO2**

Rank these three ionic compounds in order of (a) increasing lattice energy and (b) increasing melting point: NaF, KF, and RbF. Assume that these compounds have the same solid structure, which means they have the same value of k in Equation 11.4.

Collect and Organize We can determine relative lattice energies from Equation 11.4. To do so, we need to establish the charges and radii of the ions in each compound, using Figure 11.1.

Analyze All the cations are alkali metal cations and have a charge of 1+; all the anions are fluoride ions, with a charge of 1−. We are told that k is the same for all three solids. Therefore any differences in lattice energy must be related to differences in the nucleus-to-nucleus distance, d, between ions. Because the fluoride ion is common to all the salts, variations in d depend only on the size of the cation.

Solve Periodic trends in size predict that the cation sizes are Na$^+$ < K$^+$ < Rb$^+$ (Figure 11.3). Therefore the compounds in order of increasing value of d are NaF < KF < RbF.

a. As d increases, lattice energy decreases, so we predict RbF has the lowest lattice energy, followed by KF, followed by NaF with the highest.
b. The same trend occurs in melting points: RbF < KF < NaF.

Think About It We predicted the order of lattice energies and melting points for three alkali metal fluorides on the basis of the radii of the ions. The predicted orders are confirmed by the experimentally measured melting points: 775°C for RbF, 846°C for KF, and 988°C for NaF.

⊛ **Practice Exercise** Predict which compound has the highest melting point: CaCl$_2$, PbBr$_2$, or TiO$_2$. All three compounds have nearly the same structure and therefore the same value of k in Equation 11.4. The radius of Ti^{4+} is 60.5 pm.

(a)

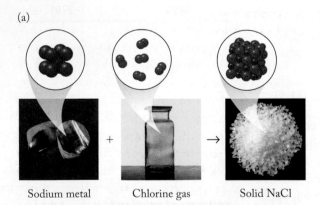

Sodium metal   Chlorine gas   Solid NaCl

FIGURE 11.4 (a) The reaction between sodium metal and chlorine gas releases more than 400 kJ of energy per mole of NaCl produced. (b) The Born–Haber cycle shows that the most exothermic step is the combination of free sodium ions Na$^+$(g) and free chloride ions Cl$^-$(g) to form NaCl(s). Although the particulate representations indicate whether each element or compound exists as atoms, molecules, or ions, they are not intended to represent the exact number of particles involved.

(b)

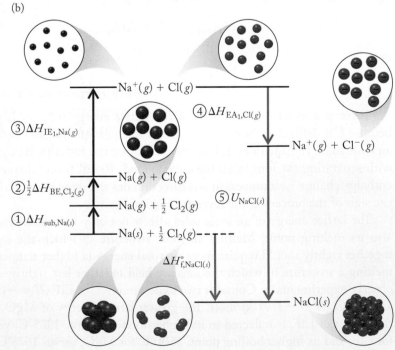

Calculating Lattice Energies by Using the Born–Haber Cycle

We predicted the relative order of lattice energies by considering atomic radii and bond distances, but we can also calculate values for lattice energies. Such calculations are necessary because measuring lattice energies directly is difficult. We can calculate the lattice energy of a binary ionic compound from its standard enthalpy of formation (ΔH_f°). We use Hess's law to determine the enthalpies of reaction, ΔH_{rxn}, associated with a series of reactions that take the constituent elements from their standard states to ions in the gas phase and then to ions in the ionic solid.

Consider the formation of NaCl:

$$Na(s) + \tfrac{1}{2}Cl_2(g) \rightarrow NaCl(s) \qquad \Delta H_f^\circ = -411.2 \text{ kJ}$$

The reaction is exothermic, and the heat produced, 411.2 kJ per mole of NaCl formed, has been measured with a calorimeter and is among those tabulated in Table A4.3 in the Appendix. In terms of Hess's law, this enthalpy of formation is the algebraic sum of all the enthalpy changes associated with five reactions that together form a **Born–Haber cycle** (Figure 11.4):

CONNECTION We used Hess's law in Chapter 5 to determine heats of reactions that are difficult to measure experimentally. We also defined a formation reaction as a reaction in which 1 mole of a substance is produced from its constituent elements in their standard states.

1. Sublimation of 1 mole of Na(s) atoms into 1 mole of Na(g) atoms: ΔH_{sub}, the molar heat of sublimation of sodium.
2. Breaking covalent bonds in $\tfrac{1}{2}$ mole of $Cl_2(g)$ molecules to make 1 mole of Cl(g) atoms: $\tfrac{1}{2}\Delta H_{BE}$, where ΔH_{BE} is the enthalpy change needed to break 1 mole of $Cl_2(g)$ bonds.
3. Ionization of 1 mole of Na(g) atoms to 1 mole of $Na^+(g)$ ions and 1 mole of electrons: IE_1, the first ionization energy of sodium.
4. Combination of 1 mole of Cl(g) atoms with 1 mole of electrons to form 1 mole of $Cl^-(g)$ ions: EA_1, the first electron affinity of chlorine.
5. Formation of 1 mole of NaCl(s) from 1 mole of $Na^+(g)$ ions and 1 mole of $Cl^-(g)$ ions: $U = \Delta H_{lattice}$, the lattice energy of NaCl.

This reaction sequence is summarized in Table 11.3. To use the Born–Haber cycle to calculate the lattice energy of NaCl(s), we start with an equation relating the

TABLE 11.3 Born–Haber Cycle for Formation of NaCl(s)

Step	Description	PROCESS Chemical Equation	Enthalpy Change (kJ)
1	Sublime 1 mol Na(s)	$Na(s) \rightarrow Na(g)$	$\Delta H_{sub,Na} = +109$
2	Break $\tfrac{1}{2}$ mol Cl—Cl bonds	$\tfrac{1}{2}Cl_2(g) \rightarrow Cl(g)$	$\tfrac{1}{2}\Delta H_{BE,Cl_2} = \tfrac{1}{2}(240) = +120$
3	Ionize 1 mol Na(g) atoms forming 1 mol $Na^+(g)$ ions	$Na(g) \rightarrow Na^+(g) + e^-$	$\Delta H_{IE_1,Na} = +495$
4	1 mol Cl(g) atoms acquires 1 mol electrons forming 1 mol $Cl^-(g)$ ions	$Cl(g) + e^- \rightarrow Cl^-(g)$	$\Delta H_{EA_1,Cl} = -349$
5	1 mol $Na^+(g)$ ions combine with 1 mol $Cl^-(g)$ ions forming 1 mol NaCl(s)	$Na^+(g) + Cl^-(g) \rightarrow NaCl(s)$	$\Delta H_{lattice} = U$

value of ΔH_f° for NaCl to the sum of the enthalpy changes of the five reactions in Table 11.3:

$$\Delta H_f^\circ = \Delta H_{\text{step 1}} + \Delta H_{\text{step 2}} + \Delta H_{\text{step 3}} + \Delta H_{\text{step 4}} + \Delta HL_{\text{step 5}}$$

$$= \Delta H_{\text{sub,Na}} + \tfrac{1}{2}\Delta H_{\text{BE,Cl}_2} + \Delta H_{\text{IE}_1,\text{Na}} + \Delta H_{\text{EA}_1,\text{Cl}} + U_{\text{NaCl}}$$

Inserting the values from Table 11.3 and solving for U:

$$-411.2 \text{ kJ} = (+109 \text{ kJ}) + \tfrac{1}{2}(+240 \text{ kJ}) + (+495 \text{ kJ}) + (-349 \text{ kJ}) + U$$

$$U = (-411.2 \text{ kJ}) - (+109 \text{ kJ}) - (+120 \text{ kJ}) - (+495 \text{ kJ}) - (-349 \text{ kJ}) = -786 \text{ kJ}$$

In addition to its usefulness for calculating lattice energies, the Born–Haber cycle can also be used to calculate other values. For example, electron affinities can be very difficult to measure, and if the thermochemical values are known for all other steps in the cycle, the Born–Haber cycle can be used to calculate electron affinities.

SAMPLE EXERCISE 11.3 Calculating Lattice Energy   **LO2**

From the relative charges on the ions, we predict that the ion–ion attractions in CaF_2 are greater than in NaF. Confirm this prediction by calculating the lattice energies of (a) NaF and (b) CaF_2, given the following information:

$$\Delta H_{\text{sub}} \text{ Na}(s) = +109 \text{ kJ/mol} \qquad \Delta H_{\text{sub}} \text{ Ca}(s) = +154 \text{ kJ/mol}$$

$$\Delta H_{\text{BE}} \text{ F}_2(g) = +154 \text{ kJ/mol} \qquad \Delta H_{\text{IE}_1} \text{ Ca}(g) = +590 \text{ kJ/mol}$$

$$\Delta H_{\text{EA}_1} \text{ F}(g) = -328 \text{ kJ/mol} \qquad \Delta H_{\text{IE}_2} \text{ Ca}(g) = +1145 \text{ kJ/mol}$$

$$\Delta H_{\text{IE}_1} \text{ Na}(g) = +495 \text{ kJ/mol}$$

Collect and Organize We are asked to calculate the lattice energy of two ionic compounds, using enthalpy values for processes that can be summed to describe an overall process that produces a salt from its constituent elements. We can use a Born–Haber cycle to calculate the unknown lattice energy. Table A4.3 in the Appendix contains standard enthalpy of formation values for NaF and CaF_2: −569.0 kJ/mol and −1228.0 kJ/mol, respectively. These enthalpy changes apply to the formation of 1 mole of the two compounds by combining their component elements in their standard states.

Analyze Figure 11.5 summarizes the Born–Haber cycles for calculating the lattice energies of NaF and CaF_2. The lattice energy U is the only unknown value in both cycles. For calcium, we must include both the first and second ionization energies because Ca loses two electrons when it forms Ca^{2+} ions. Because two fluorine atoms are needed to react with a single calcium atom, we do not need the factor of $\tfrac{1}{2}$ in front of the term for energy to break a mole of F—F bonds. However, we need to multiply the electron affinity of F by 2 because 2 moles of fluorine atoms gain 2 moles of electrons to form 2 moles of fluoride ions. From the charges on the ions, their ionic radii, and Equation 11.4, we predict that the lattice energy of CaF_2 should be greater than the lattice energy of NaF.

Solve
a. The Born–Haber cycle for the formation of NaF(s) from Na(s) and $F_2(g)$ is illustrated in Figure 11.5(a). The overall enthalpy change in the reaction producing NaF is

$$\Delta H_f^\circ = \Delta H_{\text{sub,Na}(s)} + \tfrac{1}{2}\Delta H_{\text{BE,F}_2(g)} + \Delta H_{\text{IE}_1,\text{Na}(g)} + \Delta H_{\text{EA}_1,\text{F}(g)} + U_{\text{NaF}(s)}$$

Substituting the values given:

$$-569.0 \text{ kJ} = (+109 \text{ kJ}) + \tfrac{1}{2}(+154 \text{ kJ}) + (+495 \text{ kJ}) + (-328 \text{ kJ}) + U$$

(a)

(b)

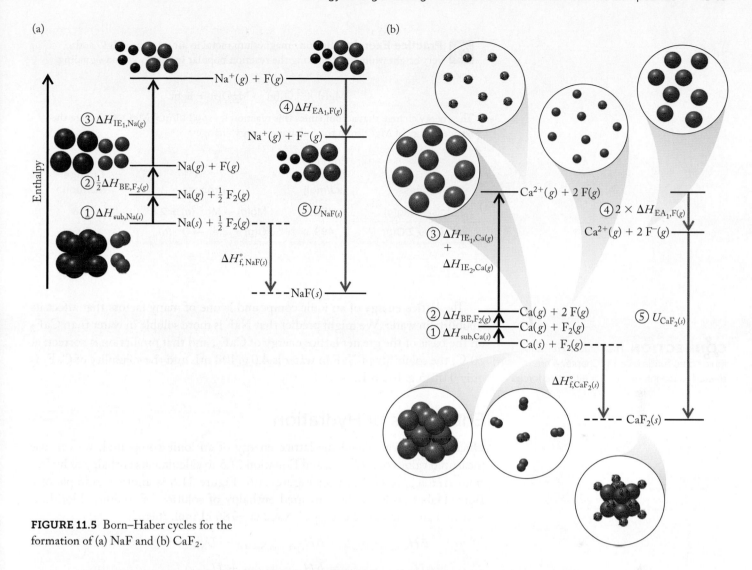

FIGURE 11.5 Born–Haber cycles for the formation of (a) NaF and (b) CaF_2.

Solving for U:

$$U = (-569 \text{ kJ}) - (+109 \text{ kJ} + 77 \text{ kJ} + 495 \text{ kJ} - 328 \text{ kJ})$$

$$= -922 \text{ kJ/mol NaF}(s) \text{ formed}$$

b. The Born–Haber cycle for the formation of $CaF_2(s)$ from $Ca(s)$ and $F_2(g)$ is illustrated in Figure 11.5(b). The overall enthalpy change in the reaction producing CaF_2 is

$$\Delta H_f^\circ = \Delta H_{\text{sub,Ca}(s)} + \Delta H_{\text{BE,F}_2(g)} + [\Delta H_{\text{IE}_1,\text{Ca}(g)} + \Delta H_{\text{IE}_2,\text{Ca}(g)}] + 2\Delta H_{\text{EA}_1,\text{F}(g)} + U_{CaF_2(s)}$$

Substituting the values given:

$$-1228.0 \text{ kJ} = (+154 \text{ kJ}) + (+154 \text{ kJ}) + [(+590 \text{ kJ}) + (+1145 \text{ kJ})] + 2(-328 \text{ kJ}) + U$$

Solving for U:

$$U = (-1228.0 \text{ kJ}) - [+154 \text{ kJ} + 154 \text{ kJ} + (590 \text{ kJ} + 1145 \text{ kJ}) - 656 \text{ kJ}]$$

$$= -2615 \text{ kJ/mol } CaF_2(s) \text{ formed}$$

Think About It From Equation 11.4, we predicted that the lattice energy of CaF_2 would be larger than that of NaF. The calculated values of U confirm that U_{CaF_2} is more negative than U_{NaF}.

Practice Exercise Burning magnesium metal in air produces MgO and a very bright white light, making the reaction popular in fireworks and signaling devices:

$$Mg(s) + \tfrac{1}{2}O_2(g) \rightarrow MgO(s) + light$$

The energy change that accompanies this reaction is -602 kJ/mol MgO. Calculate the lattice energy of MgO from the following energy changes:

Process	Enthalpy Change (kJ/mol)	Process	Enthalpy Change (kJ/mol)
$Mg(s) \rightarrow Mg(g)$	150	$Mg(g) \rightarrow Mg^{2+}(g) + 2\,e^-$	2188
$O_2(g) \rightarrow 2\,O(g)$	499	$O(g) + 2\,e^- \rightarrow O^{2-}(g)$	603

The lattice energy of an ionic compound is one of many factors that affect its solubility in water. We might predict that NaF is more soluble in water than CaF_2 on the basis of the greater lattice energy of CaF_2, and that prediction is correct: at 20°C, the solubility of NaF in water is 4.0 g/100 mL and the solubility of CaF_2 is only 0.0015 g/100 mL.

CONNECTION The solubility rules for ionic compounds given in Chapter 4 are related to the intermolecular attractive forces.

Enthalpies of Hydration

Once we have calculated the lattice energy of an ionic compound, we can use measured values of $\Delta H_{solution}$ and Equation 11.5 to calculate its enthalpy of hydration, $\Delta H_{hydration}$, as shown in Figure 11.6. Figure 11.6 is another example of a Born–Haber cycle. If the measured enthalpy of solution of sodium chloride is 4 kJ/mol and the lattice energy of NaCl is -786 kJ/mol, then:

$$\Delta H_{solution,NaCl(aq)} = \Delta H_{hydration,NaCl(g)} - U_{NaCl(s)}$$

$$\Delta H_{hydration,NaCl(aq)} = \Delta H_{solution,NaCl(g)} + U_{NaCl(s)}$$

$$= 4 \text{ kJ/mol} + (-786 \text{ kJ/mol}) = -782 \text{ kJ/mol}$$

The value for $\Delta H_{hydration,NaCl(g)}$ is negative, indicating that when Na^+ and Cl^- ions in the gas phase dissolve in water, the reaction is exothermic. The sign of $\Delta H_{hydration,NaCl(g)}$ also tells us something about the relative strengths of the hydrogen bonding interactions in water and the ion–dipole interactions in aqueous NaCl. If $\Delta H_{hydration,NaCl(g)}$ is less than zero, then according to Equation 11.5 the formation of ion–dipole interactions must release more energy than is needed to disrupt the dipole–dipole interactions (hydrogen bonds) in water.

With careful measurements and calculations, we can even separate $\Delta H_{hydration,NaCl(g)}$ into enthalpies of hydration for the individual ions:

$$\Delta H_{hydration,NaCl(g)} = \Delta H_{hydration,Na^+(g)} + \Delta H_{hydration,Cl^-(g)}$$

The values for $\Delta H_{hydration}$ for selected cations and anions are listed in Table 11.4 and can be used to calculate $\Delta H_{solution}$ for an ionic compound if we know the lattice energy, or to calculate the lattice energy if we measure the enthalpy of solution. The values in Table 11.4 are the results of several experiments and include an inherent uncertainty. The sum of the values for Na^+ and Cl^- is -786 kJ/mol; that is the same as the value calculated from the measured enthalpy of solution (Table 11.2).

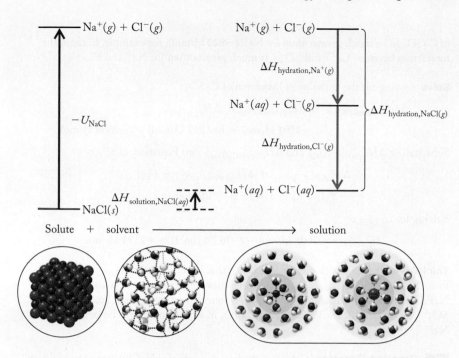

FIGURE 11.6 Born–Haber cycle for the dissolution of an ionic compound in water. The enthalpy of solution ($\Delta H_{solution}$) is the enthalpy of hydration ($\Delta H_{hydration}$) minus the lattice energy (U). The size of $\Delta H_{solution}$ is not drawn to scale; it is very small in relation to the other values.

TABLE 11.4 Enthalpies of Hydration for Selected Cations and Anions^a

Cation	$\Delta H_{hydration}$ (kJ/mol)	Anion	$\Delta H_{hydration}$ (kJ/mol)
Li^+	−536	F^-	−502
Na^+	−418	Cl^-	−368
K^+	−335	Br^-	−335
Rb^+	−305	I^-	−293
Cs^+	−289	ClO_4^-	−238
Mg^{2+}	−1903	NO_3^-	−301
Ca^{2+}	−1591	SO_4^{2-}	−1017

^{a}Based on enthalpy of hydration of H^+ as −1105 kJ/mol.

SAMPLE EXERCISE 11.4 Calculating Lattice Energy by Using
Enthalpies of Hydration   **LO2**

Calcium sulfate is used to prepare "plaster of Paris" for casting broken limbs. Use the appropriate enthalpy of hydration values in Table 11.4 and $\Delta H_{solution} = -16.7$ kJ/mol for $CaSO_4$ to calculate the lattice energy of $CaSO_4$.

Collect and Organize The enthalpies of hydration for Ca^{2+} and SO_4^{2-} are −1591 kJ/mol and −1017 kJ/mol, respectively. We can use the enthalpy of solution and Equation 11.5 to calculate the lattice energy.

Analyze To calculate U_{CaSO_4}, we must obtain a value for $\Delta H_{hydration}$ for $CaSO_4$:

$\Delta H_{hydration}$ of cation	+	$\Delta H_{hydration}$ of anion	$\longrightarrow$	$\Delta H_{hydration}$ of ionic compound

From our calculations in Sample Exercises 11.1 and 11.3, we know that ionic charges greatly influence the strength of interactions. Therefore we expect the lattice energy of CaSO$_4$ to be much greater than for NaF (-922 kJ/mol), for example, because the interaction between Ca^{2+} and SO$_4^{2-}$ is much greater than for Na$^+$ and F$^-$.

Solve Solving for the enthalpy of hydration of CaSO$_4$:

$$\Delta H_{\text{hydration,CaSO}_4(g)} = \Delta H_{\text{hydration,Ca}^{2+}(g)} + \Delta H_{\text{hydration,SO}_4^{2-}(g)}$$

$$= -1591 \text{ kJ/mol} + (-1017 \text{ kJ/mol}) = -2608 \text{ kJ/mol}$$

Substituting $\Delta H_{\text{hydration,CaSO}_4(g)}$ and $\Delta H_{\text{solution,CaSO}_4(aq)}$ into Equation 11.5:

$$\Delta H_{\text{solution,CaSO}_4(aq)} = \Delta H_{\text{hydration,CaSO}_4(g)} - U_{\text{CaSO}_4(s)}$$

$$-16.7 \text{ kJ/mol} = (-2608 \text{ kJ/mol}) - U_{\text{CaSO}_4(s)}$$

Solving for $U_{\text{CaSO}_4(s)}$:

$$U_{\text{CaSO}_4(s)} = -2608 \text{ kJ/mol} - (-16.7 \text{ kJ/mol}) = -2591 \text{ kJ/mol}$$

Think About It The lattice energy of CaSO$_4$ [that is, the value of $U_{\text{CaSO}_4(s)}$ calculated from $\Delta H_{\text{solution}}$ and $\Delta H_{\text{hydration}}$] is about three times larger than the lattice energy of NaF. The product of the charges in CaSO$_4$ is four times that in NaF, but the larger SO$_4^{2-}$ ions mean that the value of d for CaSO$_4$ in Equation 5.3 will be larger than for NaF.

 Practice Exercise Calculate the lattice energy for NaClO$_4$ by using the data in Table 11.4 and $\Delta H_{\text{solution,NaClO}_4(aq)} = 14$ kJ/mol.

11.3 Vapor Pressure of Solutions

CONNECTION In Chapter 10 we explored vapor pressure (the pressure of a gas in equilibrium with its liquid) and the normal boiling point (the temperature at which the vapor pressure equals 1 atm) of pure liquids.

In Section 10.6 we discussed how intermolecular interactions affect the vapor pressure and the normal boiling point of pure liquids. Let's see what happens to vapor pressure when nonvolatile solutes such as salts are dissolved in water.

When adjoining compartments of seawater (water that contains many dissolved nonvolatile solutes) and pure water are sealed in a chamber (Figure 11.7), the volume of fluid on the seawater side increases over time, while the volume on the side of pure water decreases at the same rate. Eventually, nearly all the water ends up in the seawater compartment. The transfer of the pure water is due to the dissolved solutes in the seawater.

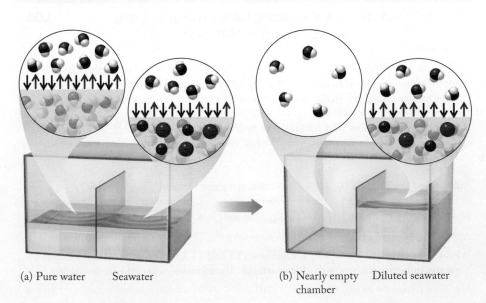

FIGURE 11.7 (a) Adjoining compartments are partially filled with pure water and seawater. (b) The slightly higher vapor pressure of the pure water leads to a net transfer of water from the pure-water compartment to the seawater compartment.

(a) Pure water   Seawater

(b) Nearly empty   Diluted seawater
    chamber

As the water in both compartments evaporates, the concentration of water vapor in the air space of the sealed chamber increases. As the concentration of water vapor increases, the pressure the water vapor exerts on the two liquid surfaces increases. At constant temperature, this pressure eventually stabilizes at a value equal to the vapor pressure of water at that temperature. At this point, the rate of evaporation from the compartments is equal to the rate of condensation.

If the evaporation and condensation rates for the pure water and the seawater were the same, the liquid levels in the compartments would not change over time. However, the liquid levels do change. The $H_2O(g)$ molecules in the sealed chamber are free to condense into either compartment, so the rate of condensation into both compartments is the same—but the rates of evaporation are not: The presence of dissolved solutes in the seawater affects its rate of evaporation. Given that most of the pure water ends up in the seawater compartment, the pure water must have a higher rate of evaporation than the seawater. This conclusion leads to another: if the seawater evaporates at a lower rate, it must have a lower vapor pressure than the pure water at the same temperature.

More water vapor enters the air in the chamber from the pure-water compartment than from the seawater compartment because the vapor pressure of the pure water is greater than the vapor pressure of the seawater. Because the condensation rates are the same, the pure water loses more water over time than is restored to it by condensation, while the seawater gains more water than it loses by evaporation. The process depicted in Figure 11.7 is an illustration of a more general observation: at a given temperature, the vapor pressure of the solvent in a solution containing nonvolatile solutes is less than the vapor pressure of the pure solvent.

Raoult's Law

The connection between the vapor pressure of a solution and the concentration of nonvolatile solutes dissolved in the solvent was studied extensively by the French chemist François Marie Raoult (1830–1901). He discovered that the relation between the vapor pressure of a solution, $P_{solution}$, and that of the pure solvent, $P°_{solvent}$, is

$$P_{solution} = X_{solvent} \, P°_{solvent} \qquad (11.6)$$

where $X_{solvent}$ is the mole fraction of solvent. This relationship is now known as **Raoult's law**.

Let's take another look at the two compartments in Figure 11.7 when the temperature in the chamber is 20°C. At 20°C, the vapor pressure of pure water is 0.0231 atm. What is the vapor pressure produced by evaporation of water from seawater if the mole fraction of water in the sample is 0.980? We can use Raoult's law as expressed by Equation 11.6 to answer this question:

$$P_{solution} = (0.980)(0.0231 \text{ atm}) = 0.0226 \text{ atm}$$

Because the vapor pressure of seawater at 20°C is slightly lower than the vapor pressure of pure water at 20°C, seawater evaporates more slowly than pure water.

The lower vapor pressure of a solution in relation to the vapor pressure of pure solvent depends only on the concentration of solute particles, not on their identity. Properties of solutions that depend only on the concentration of particles and not on their identity are called colligative properties. We examine them in detail in Section 11.5.

Solutions that obey Raoult's law are called **ideal solutions**. In ideal solutions, the solute and solvent experience similar intermolecular forces. For the most part,

CHEMTOUR
Raoult's Law

CONNECTION We defined *mole fraction* in Chapter 6 in the discussion of partial pressures of gases.

Raoult's law the principle that the vapor pressure of the solvent in a solution is equal to the vapor pressure of the pure solvent multiplied by the mole fraction of the solvent in the solution.

ideal solution a solution that obeys Raoult's law.

CONNECTION As we discussed in Chapter 1, intensive properties of matter are independent of the amount of material present, whereas extensive properties vary with the quantity of substance present.

we treat solutions as ideal systems, but you should be aware that deviations from ideal behavior exist in liquids just as they do in gases. Deviations from ideal behavior typically occur when solute–solvent interactions are much stronger than solvent–solvent interactions. We address such situations in Section 11.4 in the discussion of the distillation of crude oil to produce gasoline.

CONCEPT TEST

Is the vapor pressure of a pure solvent an intensive or an extensive property? Is the vapor pressure of a solution an intensive or an extensive property?

SAMPLE EXERCISE 11.5 Calculating the Vapor Pressure of a Solution **LO3**

The liquid used in automobile cooling systems is prepared by dissolving ethylene glycol ($HOCH_2CH_2OH$, molar mass 62.07 g/mol) in water. What is the vapor pressure of a solution prepared by mixing 1.000 L of ethylene glycol (density 1.114 g/mL) with 1.000 L of water (density 1.000 g/mL) at 100.0°C? Assume that the mixture obeys Raoult's law.

Collect and Organize The vapor pressure of a solution is a colligative property that depends on the number of solute particles and hence the concentration. We have the volume and density of the components and can use them to determine the concentration of ethylene glycol. We may treat ethylene glycol as a nonvolatile solute whose solutions obey Raoult's law. The vapor pressure curves in Figure 10.26 show that the vapor pressure of pure water at 100°C (its normal boiling point) is 1.00 atm, whereas that of ethylene glycol is less than 0.05 atm.

Analyze We have a mixture of equal volumes of two liquids. The solvent is the one present in the greater number of moles (Figure 11.8).
   We can determine the numbers of moles of each by the following calculation:

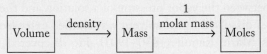

From these values we can decide which liquid is the solvent and calculate its mole fraction in the mixture. Water and ethylene glycol are both capable of hydrogen bonding, so their intermolecular interactions are similar, and we may treat the solution as ideal.

Solve For ethylene glycol:

$$1000 \text{ mL} \times \frac{1.114 \text{ g}}{1 \text{ mL}} \times \frac{1 \text{ mol}}{62.07 \text{ g}} = 17.95 \text{ mol}$$

For water:

$$1000 \text{ mL} \times \frac{1.000 \text{ g}}{1 \text{ mL}} \times \frac{1 \text{ mol}}{18.02 \text{ g}} = 55.49 \text{ mol}$$

The mole fraction of water is

$$X_{\text{water}} = \frac{55.49 \text{ mol}}{55.49 \text{ mol} + 17.95 \text{ mol}} = 0.7556$$

The number of moles of water is greater than the number of moles of ethylene glycol, so water is the solvent. Ethylene glycol is essentially nonvolatile, so the vapor pressure of the solution is due only to the solvent, and $P_{\text{solvent}} = P^\circ_{\text{H}_2\text{O}} = 1.00$ atm. Using this value and the calculated mole fraction of water in the mixture yields

$$P_{\text{solution}} = X_{\text{H}_2\text{O}} \times P^\circ_{\text{H}_2\text{O}} = (0.756)(1.00 \text{ atm}) = 0.756 \text{ atm}$$

Think About It The presence of a nonvolatile solute causes the vapor pressure of the solution to be less than the vapor pressure of a pure solvent (1 atm, in this case), giving us confidence in our result.

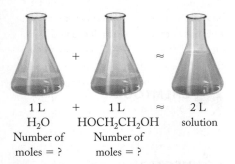

FIGURE 11.8 Forming a solution from two liquids.

Practice Exercise Glycerol [HOCH$_2$CH(OH)CH$_2$OH] is considered a nonvolatile, water-soluble liquid. Its density is 1.25 g/mL. Predict the vapor pressure of a solution of 275 mL of glycerol in 375 mL of water at the normal boiling point of water.

To boil the solution described in Sample Exercise 11.5, we must heat it to above 100°C—to a temperature at which the vapor pressure of the solution is 1 atm. This is why antifreeze works in an automobile engine: it not only lowers the freezing point of water, protecting the radiator, but also raises the temperature of the coolant above the boiling point of H$_2$O, making it possible for the coolant to remove more heat from the engine while staying in the liquid state.

11.4 Mixtures of Volatile Solutes

Thus far we have considered only the effect of essentially nonvolatile solutes on the vapor pressure of a solvent. However, in our everyday lives we encounter many solutions containing volatile solutes. For example, the natural gas used for heating and the gasoline we use to power vehicles are mixtures of hydrocarbons. Gasoline comes from crude oil, a complex mixture of compounds composed mostly of carbon and hydrogen. Depending on its source, crude oil contains varying concentrations of hydrocarbons with five or more carbon atoms in their molecular structures. The hydrocarbons with one to four carbon atoms are usually found in deposits of natural gas, although they are also dissolved in crude oil. Most hydrocarbons in gasoline have five to nine carbon atoms per molecule. Each of these compounds is volatile and has a measurable vapor pressure at 25°C. Because of this volatility, we can separate gasoline from crude oil by distillation. In this section we explore distillation on a molecular level and consider the effects of volatile solutes on the vapor pressure of a solution.

Vapor Pressures of Mixtures of Volatile Solutes

In Chapter 5 we discussed the process of distillation as a way to purify liquids, and in Chapter 10 we discussed the vaporization of pure liquids and the role of intermolecular forces in determining normal boiling point. In addition, we saw that the temperature at which a pure liquid boils remains constant throughout the process. Another type of distillation, called **fractional distillation** (Figure 11.9), is used to separate the volatile components of a mixture. This method of purifying the components of a liquid mixture is based on the observation that the boiling point of a mixture changes as the mixture is distilled.

As a mixture of volatile liquids is heated, the vapor that rises and fills the space above the liquid has a different composition from the composition of the mixture: the concentration of the component with the lowest boiling point is higher in the vapor than in the liquid. If this enriched vapor is collected, condensed, and redistilled, the vapor this time is even richer in the component with the lowest boiling point. In a fractional distillation apparatus, repeated distillation steps allow components with only slightly different boiling points to be separated from one another.

To see how a mixture behaves when boiled and how fractional distillation uses that behavior to separate the components, let's examine the

fractional distillation a method of separating a mixture of compounds on the basis of their different boiling points.

C⚛NNECTION We discussed simple distillation in Chapter 1 as a way to make drinking water from seawater.

FIGURE 11.9 Fractional distillation separates mixtures of volatile components. Vapors rise through a fractionating column, where they repeatedly condense and vaporize. The most volatile component distills first and is first to pass through the condenser and into the collecting flask. Increasingly less volatile, higher-boiling components are distilled in turn. The progress of the distillation process is monitored using the thermometer at the top of the fractionating column.

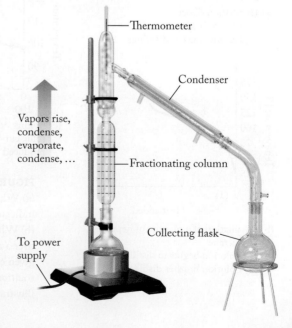

Thermometer

Condenser

Vapors rise, condense, evaporate, condense, …

Fractionating column

Collecting flask

To power supply

heating curves of a pure substance and a solution of two volatile liquids. Figure 11.10(a) is the heating curve for pure octane (C_8H_{18}, bp 126°C). As in the heating curves of Section 5.5, the phase change from liquid to vapor takes place at a constant temperature of 126°C. Figure 11.10(b) shows the heating curve for a mixture of heptane (C_7H_{16}, bp 98°C) and octane. Here the portion of the curve from point 1 to point 2 again represents a vapor–liquid phase change, but in this case the temperature is not constant during the phase change. Instead, the two components co-distill over a range of temperatures.

Now let's analyze the graphs describing this process to understand how fractional distillation works. Figure 11.10(c) shows the composition of the vapor above a series of boiling solutions, and herein lies the key to how distillation separates mixtures. In the vapor in equilibrium with a solution of two volatile substances, the concentration of the lower-boiling (more volatile) component is always greater in the vapor than in the liquid. Suppose we use the distillation apparatus from Figure 11.9 to heat 100 mL of a solution that is 50% by volume heptane and 50% octane (50:50). The blue curve in Figure 11.10(c) gives the temperature of the boiling solution, and point 1 on the curve tells us that the boiling point of the 50:50 solution is about 108°C.

The red line on the graph gives the composition of the vapor above the solution at a given temperature. To find the composition of the vapor at 108°C, we move horizontally to the left along the dashed line between points 1 and 2. This horizontal line corresponds to a temperature of 108°C on the vertical (temperature) axis. Point 2 corresponds to a different composition (on the *x*-axis) for the

CHEMTOUR
Fractional Distillation

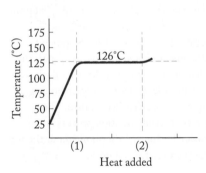

(a) Distillation of octane
  (1) Octane begins to distill.
  (2) Octane finishes distilling.

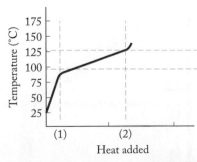

(b) Distillation of a mixture of heptane and octane
  (1) Solution begins to distill.
  (2) Solution finishes distilling.

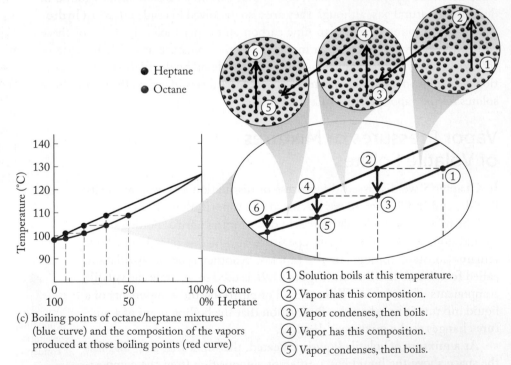

(c) Boiling points of octane/heptane mixtures (blue curve) and the composition of the vapors produced at those boiling points (red curve)

① Solution boils at this temperature.
② Vapor has this composition.
③ Vapor condenses, then boils.
④ Vapor has this composition.
⑤ Vapor condenses, then boils.

FIGURE 11.10 Heating curves describe the vaporization and distillation of volatile liquids. (a) When pure octane is distilled, the distillation takes place at a constant temperature, indicated by the horizontal line at $T = 126$°C. The liquid in the collecting flask is pure octane. (b) When a mixture of heptane (bp 98°C) and octane is distilled, the distillation takes place over a range of temperatures. (c) The blue line shows the temperatures where solutions of a given composition boil. The red line shows the composition of the vapor arising from those solutions. The stair-step line illustrates what happens in a fractionating column. The balloons illustrate the relative composition of the two phases as the fractional distillation proceeds.

vapor than for the solution at point 1. The vapor at point 2 is enriched in the lower-boiling component: it is about 65% heptane and only 35% octane.

Suppose this 65:35 vapor rises up in the distillation column, cools, and condenses in the next region of the column, a process represented by the red arrow from point 2 to point 3. The condensed liquid in this flask is 65% heptane and 35% octane. Continued heating of the column warms this liquid, and it begins boiling at about 104°C, which is the temperature at point 3. To find the concentration of the vapor above this boiling liquid, we move left from point 3 until we intersect the red curve (point 4). Reading down from point 4 to the concentration axis, we see that the vapor concentration is now about 80% heptane and only 20% octane. This vapor with 80:20 composition rises up, where it cools and condenses, and the distillation is repeated.

If we continue this process of redistilling mixtures with increasing concentrations of heptane and then cooling and condensing the vapors, we eventually obtain a condensate that is pure heptane. If we monitor the temperature at which vapors condense at the very top of our distillation column, we will see a profile of temperature versus volume of distillate produced (Figure 11.11). The first liquid to be produced is pure heptane, which has a boiling point of 98°C. Ideally, over time, all 50 mL of heptane in the original sample is recovered. As we continue to add heat at the bottom of the column, the temperature at the top rises to 126°C, signaling that the second component (pure octane, bp 126°C) is being collected. In the real world, the separation may not be as complete as described in this idealized presentation.

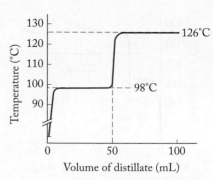

FIGURE 11.11 Fractional distillation of a mixture of 50 mL of heptane and 50 mL of octane produces two plateaus at the boiling points of the two components. If fractionation is perfect, the first 50 mL of distillate is pure heptane, and the second 50 mL is pure octane.

CONCEPT TEST

Dimethyl ether, CH_3OCH_3, and acetone, $CH_3C(O)CH_3$, shown in Figure 11.12, have similar molar masses but different dipole moments: 1.30 D and 2.88 D, respectively. Which compound would you expect to distill first from a mixture of acetone and dimethyl ether? Why?

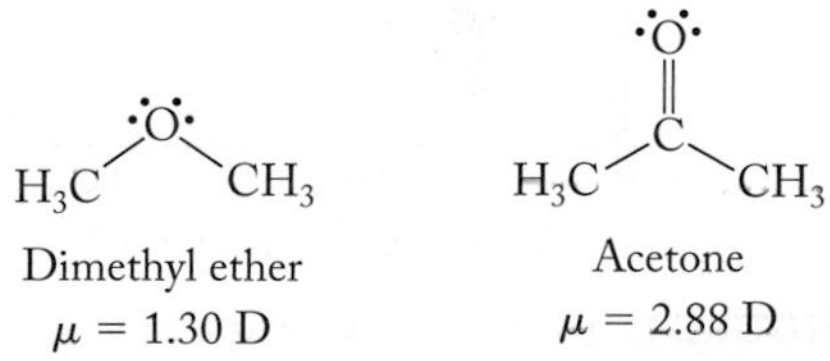

Dimethyl ether
$\mu = 1.30$ D

Acetone
$\mu = 2.88$ D

FIGURE 11.12 Lewis structures for dimethyl ether and acetone.

SAMPLE EXERCISE 11.6 Predicting the Results   **LO4**
                          of Fractional Distillation

How would the graph in Figure 11.11 be different if (a) we started with a mixture of 75 mL of octane (C_8H_{18}) and 25 mL of heptane (C_7H_{16}) and (b) we started with a mixture of 75 mL of octane (C_8H_{18}) and 25 mL of nonane (C_9H_{20})? The normal boiling points of heptane, octane, and nonane are 98°C, 126°C, and 151°C, respectively.

Collect and Organize The graph in Figure 11.11 is an idealized plot of the temperature of the solution as a function of the volume of distillate for a 50:50 mixture of C_7H_{16}:C_8H_{18}. We want to describe the changes in the graph if we change the ratio of the components and if we change the identity of one of the components. We are given the normal boiling points of all compounds.

Analyze Fractional distillation separates solutions of volatile liquids into pure substances on the basis of their different boiling points. The components distill in the order of their boiling points, with the lowest-boiling component distilling first. Ideally, the total volume of distillate equals the total volume of the solution, and the volume of each fraction reflects the volume of the component in the original solution.

Solve
a. The first component that distills is the lower-boiling heptane. If our system were perfect, 25 mL of heptane would distill at 98°C. When the heptane was removed from the solution, the only remaining component (75 mL of octane) would distill

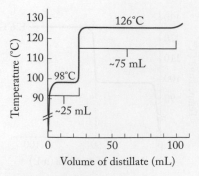

FIGURE 11.13 Temperature as a function of distillate volume for a mixture of 25 mL of C_7H_{16} and 75 mL of C_8H_{18}.

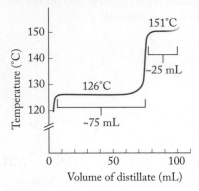

FIGURE 11.14 Temperature as a function of distillate volume for a mixture of 75 mL of C_8H_{18} and 25 mL of C_9H_{20}.

at 126°C. The idealized graph (Figure 11.13) shows the boiling points of the two substances along the y-axis and the volume of distillate along the x-axis. The lengths of the horizontal lines at 98°C and 126°C are in a 1:3 ratio, reflecting the composition of the mixture, 25 mL of C_7H_{16} and 75 mL of C_8H_{18}.

b. The first component that distills in the second mixture is the lower-boiling octane. If our system were perfect, 75 mL of octane would distill at 126°C. Once the octane is removed from the solution, the only remaining component (25 mL of nonane) would distill at 151°C. The idealized graph (Figure 11.14) shows the boiling points of the two substances along the y-axis and the volume of distillate along the x-axis.

Think About It Our graphs show the best results we could achieve. Actually, a small fraction that is a mixture of two hydrocarbons distills at temperatures between the boiling points of the two pure components.

Practice Exercise Draw an idealized graph of the temperature versus volume of distillate collected when a solution consisting of 30 mL of hexane (boiling point 69°C), 50 mL of heptane (boiling point 98°C), and 20 mL of nonane (boiling point 151°C) is fractionally distilled.

Now that we know how fractional distillation works, the next step is to understand *why* it works. We return to Raoult's law, which we used in Section 11.3 to describe the influence of nonvolatile solutes on the boiling point of a pure solvent. Raoult's law also applies to homogeneous mixtures of volatile compounds, such as crude oil. Because the solutes in a solution of crude oil are volatile, they contribute to the solution's overall vapor pressure. The total vapor pressure equals the sum of the vapor pressures of each component (P_x° is the equilibrium vapor pressure of the pure component at the temperature of interest) multiplied by the mole fraction of that component in the solution (X_x):

$$P_{total} = X_1 P_1^\circ + X_2 P_2^\circ + X_3 P_3^\circ + \ldots \qquad (11.7)$$

SAMPLE EXERCISE 11.7 Calculating the Vapor Pressure of a Solution of Volatile Substances **LO4**

Ethanol, CH_3CH_2OH, is used to disinfect the skin prior to getting a shot, such as a flu vaccine. The solution quickly evaporates at 37°C, roughly the body temperature of a healthy human. Calculate the vapor pressure of a solution prepared by dissolving 5.0 g of water in 78.0 g of ethanol (CH_3CH_2OH) at 37°C. By what factor does the concentration of the more volatile component in the vapor exceed the concentration of this component in the liquid? The vapor pressures of ethanol and water at 37°C are 115 torr and 47 torr, respectively.

Collect and Organize We can use Equation 11.7 to determine the total vapor pressure of the solution from the vapor pressures of the components after we determine the composition of the solution in terms of mole fractions.

Analyze To calculate the mole fraction of each component, we need the molar masses of water and ethanol. The mole fraction is then equal to the number of moles of each component divided by the total number of moles of material in the solution.

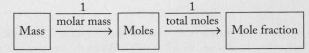

The vapor pressure of the solution must lie between the vapor pressures of the pure compounds.

Solve The number of moles of each component is

$$78.0 \text{ g C}_2\text{H}_5\text{OH} \times \frac{1 \text{ mol CH}_3\text{CH}_2\text{OH}}{46.1 \text{ g CH}_3\text{CH}_2\text{OH}} = 1.69 \text{ mol CH}_3\text{CH}_2\text{OH}$$

$$5.0 \text{ g H}_2\text{O} \times \frac{1 \text{ mol H}_2\text{O}}{18.0 \text{ g H}_2\text{O}} = 0.28 \text{ mol H}_2\text{O}$$

The total number of moles of material is (1.69 mol + 0.28 mol) = 1.97 mol. Thus the mole fraction of each component in the mixture is

$$X_{\text{ethanol}} = \frac{1.69 \text{ mol}}{1.97 \text{ mol}} = 0.858$$

$$X_{\text{water}} = 1 - X_{\text{ethanol}} = 0.142$$

Using these values and the vapor pressures of the two substances in Equation 11.7, we have

$$P_{\text{total}} = X_{\text{water}}P°_{\text{water}} + X_{\text{ethanol}}P°_{\text{ethanol}}$$

$$= 0.142(47 \text{ torr}) + 0.858(115 \text{ torr})$$

$$= 6.7 \text{ torr} + 98.7 \text{ torr} = 105.4 \text{ torr}$$

To calculate how enriched the vapor phase is in the more volatile component (the component with the greater vapor pressure, ethanol), we need to recall Dalton's law of partial pressures (Section 6.7) and the concept that the partial pressure of a gas in a mixture of gases is proportional to its mole fraction in the mixture. Therefore the ratio of the mole fraction of ethanol to that of water is the ratio of their two vapor pressures:

$$\frac{98.7 \text{ torr}}{6.7 \text{ torr}} = 14.7$$

The mole ratio of ethanol to water in the liquid mixture is

$$\frac{1.69 \text{ mol}}{0.28 \text{ mol}} = 6.0$$

Therefore the vapor phase is enriched in ethanol by a factor of

$$\frac{14.7}{6.0} = 2.4$$

Think About It As expected, the vapor pressure of the mixture (106 torr) is between the vapor pressures of the separate components. This result illustrates how fractional distillation works. The vapor is enriched in the lower-boiling component.

Practice Exercise Benzene (C_6H_6) is a trace component of gasoline. What is the mole ratio of benzene to octane in the vapor above a solution of 10% benzene and 90% octane by mass at 25°C? The vapor pressures of octane and benzene at 25°C are 11 torr and 95 torr, respectively.

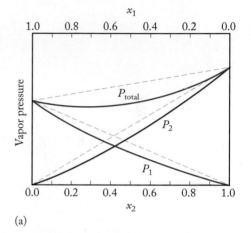

(a)

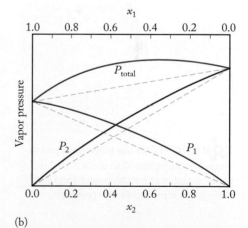

(b)

FIGURE 11.15 In a mixture of two volatile substances x_1 and x_2, the vapor pressures P_1 and P_2 may deviate from the ideal behavior predicted by Raoult's law and described by the dashed lines. (a) If solute–solvent interactions are stronger than solvent–solvent or solute–solute interactions, the deviations from Raoult's law are negative. (b) If solute–solvent interactions are weaker than solvent–solvent or solute–solute interactions, the deviations are positive.

Solutions such as the hydrocarbons in crude oil obey Raoult's law when the strengths of solvent–solvent, solute–solute, and solute–solvent interactions are similar. Under these conditions, a solution behaves ideally. A mixture of hydrocarbons is expected to behave like an ideal solution because intermolecular interactions between the components are all London forces acting on molecules of similar structure and size.

If the solute–solvent interactions are stronger than solvent–solvent or solute–solute interactions, the solute inhibits the solvent from vaporizing and the solvent inhibits the solute from vaporizing. This situation produces negative deviations from the vapor pressures predicted by Raoult's law, as shown in Figure 11.15(a). In such a solution, the rate of evaporation is slower because the vapor pressure of the mixture is lower than predicted. Because solvent and solute molecules are held at the surface by solute–solvent attractive interactions, more energy is required to separate them from the surface, and fewer vaporize at a given temperature.

If solute–solvent interactions are much weaker than solvent–solvent interactions, less energy is required to separate solute molecules from the surface, and more solute molecules vaporize. In this case the vapor pressure is greater than the value predicted by Raoult's law, as shown in Figure 11.15(b).

CONCEPT TEST

Which of the following solutions is least likely to follow Raoult's law: (a) acetone/ethanol, (b) pentane/hexane, or (c) pentanol/water? The structures of the compounds are given in Figure 11.16.

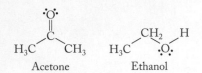

FIGURE 11.16 Lewis structures for acetone, ethanol, pentane, hexane, pentanol, and water.

11.5 Colligative Properties of Solutions

We saw in Section 11.3 that the vapor pressure of a solution containing a nonvolatile solute is lower than the vapor pressure of the pure solvent. In this section we see that many other physical properties of solvents are changed when a nonvolatile solute is added to it. These properties of solutions that depend only on the concentration of solute particles, but not on the identity of the solute, are called **colligative properties**. Solutions generally have greater densities than the solvent alone, and an aqueous solution containing a nonvolatile solute has a higher boiling point and a lower freezing point than pure water. As we saw earlier, antifreeze, an aqueous solution of ethylene glycol used to cool automobile engines, boils at a temperature above 100°C. It also has a lower freezing point than pure water.

Figure 11.17 shows the combined phase diagram of water and an aqueous solution of a nonvolatile solute. Note that the red line representing boiling/condensation points of the solution lies below the orange line for pure water. The blue melting/freezing line for the solution lies to the left of the light blue line for pure water. Following the dashed line from left to right at $P = 1$ atm, we see that the solution freezes at a lower temperature than pure water and boils at a higher temperature than pure water. Furthermore, the higher boiling point for the solution indicates that the solution has a lower vapor pressure than pure water.

Boiling point elevation and freezing point depression are both colligative properties. As noted in Section 11.3, only the number of particles in solution determines the impact of the solute on the colligative properties of the solvent. However, before we can quantify these properties, we must introduce a new concentration unit.

Molality

In Chapter 4 we introduced the concentration unit of molarity. Molarity is the concentration unit of choice when we run reactions in solutions, usually at constant temperature, so that we can measure volumes of solutions and know the number of moles of reactants we are dealing with. When we talk about colligative properties, however, we frequently deal with the properties of systems as a function of

CONNECTION We introduced phase diagrams in Chapter 10 to show how the physical states of substances change with changing temperatures and pressures.

colligative properties characteristics of solutions that depend on the concentration and not the identity of particles dissolved in the solvent.

molality (*m*) concentration expressed as the number of moles of solute per kilogram of solvent.

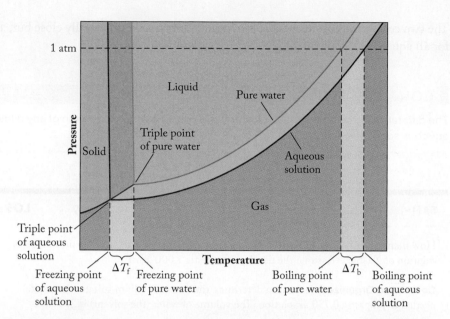

FIGURE 11.17 Combined phase diagram for pure water and a solution of a nonvolatile solute in water. Notice that the solution's boiling point is higher than the boiling point of water (the pure solvent) and that the solution's freezing point is lower than the freezing point of water.

temperature. Because volumes change with temperature, molarity changes with temperature. If we want to quantify certain colligative properties, therefore, the approach is to use a different concentration unit, called **molality (*m*)**, defined as the number of moles of solute (n_{solute}) per kilogram of solvent:

$$m = \frac{n_{\text{solute}}}{\text{kg solvent}} \qquad (11.8)$$

The difference between these two similar-sounding concentration units is that molarity is the number of moles of solute *per liter of solution*, whereas molality is the number of moles of solute *per kilogram of solvent*. Because solvent mass does not change with temperature, a concentration expressed in molality does not change with temperature.

To illustrate the difference between the molarity and molality of a solution, let's use the procedure outlined in Figure 11.18 to calculate the molality of 1 L of an aqueous solution of sodium chloride that is 0.558 *M* in NaCl (approximately the NaCl concentration in seawater) at 25°C. To calculate molality, we need the density of the solution so that we can convert liters of solution into kilograms of solvent. The density of the solution is 1.022 g/mL at 25°C, so 1 L of solution has a mass of

$$\frac{1.022 \text{ g}}{1 \text{ mL}} \times 1000 \text{ mL} = 1022 \text{ g}$$

Of this 1022 g of solution, the mass of dissolved NaCl is

$$\frac{0.558 \text{ mol NaCl}}{1 \text{ L solution}} \times \frac{58.44 \text{ g NaCl}}{1 \text{ mol NaCl}} = 32.6 \text{ g NaCl in 1 L of solution}$$

If the mass of 1 L of solution is 1022 g, and 32.6 g is due to the dissolved NaCl, then the mass of the solvent is (1022 g − 32.6 g) = 989 g = 0.989 kg. The molality of the solution (number of moles of solute per kilogram of solvent) is

$$\frac{0.558 \text{ mol NaCl}}{0.989 \text{ kg solvent}} = 0.564 \text{ } m$$

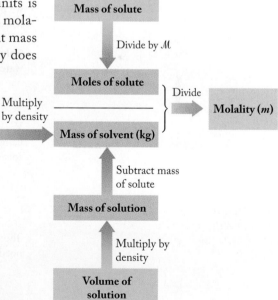

FIGURE 11.18 Flow diagram for calculating molality.

The two concentration values, 0.558 M and 0.564 m, are numerically close but, as for all aqueous solutions, molality is greater than molarity.

CONCEPT TEST

The difference between the molar concentration and molal concentration of any dilute aqueous solution is small. Why?

SAMPLE EXERCISE 11.8 Preparing a Solution of Known Molality   **LO5**

How many grams of Na_2SO_4 should be added to 275 mL of water to prepare a 0.750 m solution of Na_2SO_4? Assume the density of water is 1.000 g/mL.

Collect and Organize We want to determine the mass of sodium sulfate (the solute) needed to prepare a 0.750 m solution. The volume of water (the solvent) is 275 mL.

Analyze Following the procedure shown in Figure 11.18:

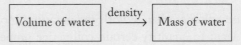

We can then work backward from molality to mass of solute:

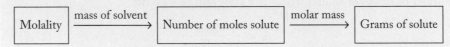

Our goal is to prepare a relatively small volume ($\sim\frac{1}{4}$ L) of a dilute solution ($< 1\ m$), so we predict that the mass of solute needed is probably less than $\frac{1}{4}$ mole of solute, which has a molar mass of about 142 g/mol, or about 35 g.

Solve Multiplying 275 mL of water by the density of water, we find that the mass of water we start with is

$$275\ \text{mL water} \times \frac{1.000\ \text{kg water}}{1000\ \text{mL water}} = 0.275\ \text{kg water}$$

We can rearrange Equation 11.8 to determine the number of moles of solute needed, using the fact that a molality of 0.750 is equivalent to 0.750 mol of solute in 1 kg of solvent:

$$n_{\text{solute}} = (m)(\text{kg solvent})$$

$$n_{Na_2SO_4} = \frac{0.750\ \text{mol}\ Na_2SO_4}{1\ \text{kg water}} \times 0.275\ \text{kg water} = 0.206\ \text{mol}\ Na_2SO_4$$

The molar mass of Na_2SO_4 is 142.04 g/mol, so the number of grams of Na_2SO_4 needed is

$$0.206\ \text{mol}\ Na_2SO_4 \times \frac{142.04\ \text{g}}{1\ \text{mol}} = 29.3\ \text{g}\ Na_2SO_4$$

Dissolving 29.3 g of Na_2SO_4 in 275 mL of water produces a 0.750 m solution.

Think About It The calculated value of 29 g of Na_2SO_4 is consistent with our prediction that about 35 g of solute would be required.

Practice Exercise What is the molality of a solution prepared by dissolving 78.2 g of ethylene glycol, $HOCH_2CH_2OH$, in 1.50 L of water? Assume the density of water is 1.000 g/mL.

Boiling Point Elevation

In Section 11.3 we discussed how the vapor pressure of a solution is reduced and the boiling point is elevated with respect to pure solvent, and at the beginning of this section we examined the phase diagrams of solutions. Now we can look quantitatively at how much the boiling points and freezing points of liquids change when solutes are present.

Boiling point elevation is a colligative property of the solvent. It is described in equation form as

$$\Delta T_b = K_b m \tag{11.9}$$

where ΔT_b is the increase in temperature above the boiling point of the pure solvent, K_b is the *boiling-point-elevation constant* of the solvent, and m is the molality of the solution. The units of K_b are $°C/m$, so the concentration of particles in solution must also have units of molality (m):

$$\Delta T_b = K_b m = \frac{°C}{m} \times m$$

The K_b of water is $0.52°C/m$, which means that for every mole of particles that dissolves in 1 kg of water (1 m), the boiling point of the solution rises by $0.52°C$:

$$\Delta T_b = K_b m = \frac{0.52°C}{m} \times 1\ m = 0.52°C$$

Therefore the boiling point (T_b) of the solution is

$$T_b = T_b° + \Delta T_b = 100.00°C + 0.52°C = 100.52°C$$

SAMPLE EXERCISE 11.9 Calculating the Boiling Point Elevation   **LO6**
of an Aqueous Solution

Hospitals use a 0.159 m NaCl solution or a 0.304 m glucose solution to deliver medications intravenously. What are the boiling points of these solutions?

Collect and Organize We are asked to compare the boiling point of an aqueous NaCl solution with a molality of 0.159 m to the boiling point of a glucose solution that is approximately twice as concentrated, 0.304 m. The boiling-point-elevation constant of water is $K_b = 0.52°C/m$, and the normal boiling point of water is 100.0°C.

Analyze We can calculate the boiling point elevation by using Equation 11.9, the boiling-point-elevation constant of water, and the solute concentration. The two solutions differ not only in concentration but also in the number of particles in solution. Specifically, sodium chloride is a strong electrolyte that dissociates into Na^+ and Cl^- ions when it dissolves. Therefore 1 mole of NaCl should in theory produce *2* moles of dissolved particles, making the *total* concentration of ions (2 × 0.159) m, or 0.318 m. Because both concentrations are close to 0.3 m, we predict that they will have similar boiling points, with boiling point elevations close to (0.3 × 0.52)°C, or about 0.15°C, for each.

Solve:
For 0.159 m NaCl:

$$\Delta T_b = K_b m = \frac{0.52°C}{m} \times 0.318\ m = 0.165°C$$

For 0.308 m glucose:

$$\Delta T_b = K_b m = \frac{0.52°C}{m} \times 0.304\ m = 0.158°C$$

The temperatures at which these solutions boil are between 0.16°C and 0.17°C higher than the normal boiling point of pure water: 100.0°C + 0.165°C = 100.17°C for 0.159 *m* NaCl, and 100.0°C + 0.158°C = 100.16°C

Think About It As predicted, the boiling points of the two solutions are essentially equal and only slightly higher than the boiling point of the pure solvent.

Practice Exercise Crude oil pumped out of the ground may be accompanied by *formation water*, a solution that contains high concentrations of NaCl and other salts. If the boiling point of a sample of formation water is 2.3°C above the boiling point of pure water, what is the molality of dissolved particles in the sample?

Freezing Point Depression

As we mentioned when discussing Figure 11.17, *freezing point depression* is a colligative property that is put to good use in car radiators to ensure that their fluid does not freeze in cold weather. The magnitude of the freezing point depression is directly proportional to the molal concentration of dissolved solute:

$$\Delta T_f = K_f m \qquad (11.10)$$

where ΔT_f is the change in the freezing temperature of the solvent, K_f is the *freezing-point-depression constant* of the solvent, and *m* is the molality of the solution. The freezing point (T_f) of a solution is

$$T_f = T_f^\circ - \Delta T_f$$

Figure 11.19 summarizes calculations of freezing point depression and boiling point elevation.

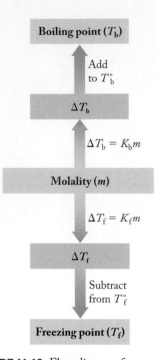

FIGURE 11.19 Flow diagram for calculating freezing points and boiling points of solutions of known molality (T_b° = normal boiling point of pure solvent; T_f° = normal freezing point of pure solvent).

SAMPLE EXERCISE 11.10 Calculating the Freezing Point of a Solution **LO6**

What is the freezing point of radiator fluid prepared by mixing 1.00 L of ethylene glycol (HOCH$_2$CH$_2$OH, density 1.114 g/mL) with 1.00 L of water (density 1.000 g/mL)? The freezing-point-depression constant of water, K_f, is 1.86°C/*m*.

Collect, Organize, and Analyze We are asked to determine the freezing point of a solution of ethylene glycol in water from the volumes of the two liquids, their densities, and K_f for water. To do so, we need to calculate ΔT_f for the solution and then subtract that value from water's normal freezing point. Because using Equation 11.10 requires us to know *m*, the molality of the solution, we need to convert volumes of solute and solvent into moles of solute and kilograms of solvent.

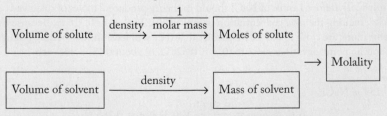

Knowing their densities allows us to make both conversions, and we use ethylene glycol's formula to calculate its molar mass. We have equal volumes of each substance, but it is logical for us to choose water as the solvent because ethylene glycol has a much higher molar mass and a density very similar to that of water, which means that 1 L of ethylene glycol contains fewer moles of material than 1 L of water. One liter of water has a mass of one kilogram (because a density of 1.000 g/mL is the same as 1.000 kg/L). The mass of one liter of ethylene glycol is 1114 g. With a solute molar mass of 62 g/mol, the molality of the solution should be between 10 and 20 m. Considering $K_f = 1.86°C/m$, we predict ΔT_f to be in the range of 20° to 40°C, and therefore the freezing point should be between −20° and −40°C.

Solve The solvent mass is

$$1.00 \text{ L} \times \frac{1000 \text{ mL}}{1 \text{ L}} \times \frac{1.000 \text{ g}}{1 \text{ mL}} \times \frac{0.001 \text{ kg}}{1 \text{ g}} = 1.00 \text{ kg solvent}$$

After calculating the ethylene glycol molar mass to be 62.07 g/mol, we have for the solute

$$1.00 \text{ L solute} \times \frac{1000 \text{ mL}}{1 \text{ L}} \times \frac{1.114 \text{ g}}{1 \text{ mL}} \times \frac{1 \text{ mol}}{62.07 \text{ g}} = 17.9 \text{ mol solute}$$

Therefore the molal concentration is

$$m = \frac{17.9 \text{ mol solute}}{1.00 \text{ kg solvent}} = 17.9 \ m$$

Using Equation 11.10 gives

$$\Delta T_f = K_f m$$

$$= \frac{1.86°C}{m} \times 17.9 \ m = 33.3°C$$

Subtracting this temperature change from the normal freezing point of water, 0.0°C, we have

$$\text{Freezing point of radiator fluid} = 0.0°C - 33.3°C = -33.3°C$$

Think About It The answer makes sense because the freezing point of radiator fluid should be below the coldest expected temperatures, and −33°C (27°F) is colder than most places in the United States ever get. In much of Alaska and Canada, however, 60%–70% by volume solutions of ethylene glycol would be advisable to prevent engines from freezing up.

Practice Exercise Concentrated solutions of ions are used in biochemistry laboratories to disrupt the structure of proteins. What is the freezing point of a solution whose ion concentration is 8.15 m?

Figure 11.20 provides a molecular view of the influence of dissolved sea salt on the freezing point of seawater. At the freezing point of pure water (0°C, Figure 11.20a), ice and liquid water coexist: molecules of water that collide with the ice surface are captured (depicted by the blue arrows), while solid particles become detached from the ice layer and enter the liquid phase (red arrows). When these two processes occur at the same rate, there is no change in the masses of ice and liquid water.

Now consider the same equilibrium in seawater (Figure 11.20b). The presence of sea salt ions appears to prevent some water molecules from reaching the ice surface, which reduces the number of freezing events (blue arrows). With no reduction in the number of melting events (red arrows), the ice melts. If the mixture is cooled, the rate of freezing increases and balance is restored (Figure 11.20c). This lower temperature is now the freezing point of seawater (−2°C).

The notion that solute particles prevent solvent particles from freezing at their normal freezing point is an appealing one for explaining freezing point depression. It is also tempting to use it to explain the lower vapor pressure (see Figure 11.7) and higher boiling points of solutions than those of pure solvents. It is important to note, however, that solute particles do not physically "block" solvent molecules

CHEMTOUR
Boiling and Freezing Points

FIGURE 11.20 The effect of solute particles on the freezing point of water. (a) A layer of ice floats on pure water in an insulated container at 0.0°C. The two phases are in equilibrium: the equal numbers of red and blue arrows indicate that water is freezing (blue arrows) just as rapidly as ice is melting (red arrows). (b) The thickness of a layer of ice floating on seawater gets smaller at 0.0°C because the rate at which the water in seawater freezes is slower than the rate at which ice melts. (c) Ice and seawater coexist at −1.9°C because the rates at which ice melts and seawater freezes are the same at this temperature.

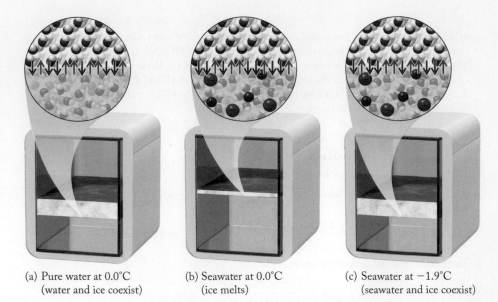

(a) Pure water at 0.0°C
(water and ice coexist)

(b) Seawater at 0.0°C
(ice melts)

(c) Seawater at −1.9°C
(seawater and ice coexist)

from reaching the surface. If that were true, bigger solute particles would be better "blockers" than little ones. However, colligative properties such as freezing point depression depend only on the number of particles, not on their size. Thus a particle-interference model is not supported by observation and experimental results. A model that is supported by the data is based on greater dispersion of the internal energy of a solution owing to the presence of solute particles. We explore this concept in detail in Chapter 17.

The van 't Hoff Factor

Recall from Section 4.4 that compounds called *electrolytes* dissociate into ions when in solution, whereas *nonelectrolytes* remain intact. We saw in Sample Exercise 11.9 that this dissociation affects how we count the number of dissolved particles that various solutes produce upon dissolution. If a solute is a nonelectrolyte, then every mole of solute yields 1 mole of particles. If, however, the solute is an electrolyte, then the number of moles of particles depends on the ratio of cations to anions that bond to form the neutral compound. For example, if we approximate seawater as 0.574 m sodium chloride, then each kilogram of water contains 1.148 moles of particles because NaCl forms Na^+ and Cl^- ions when it dissolves.

Because freezing point depression and boiling point elevation are colligative properties, the dissolution of 1 mole of a strong electrolyte such as NaCl in a given quantity of water produces the same changes in freezing point and boiling point as 1 mole of the strong electrolyte KNO_3, even though the latter has a much higher molar mass. Each of these salts adds 2 moles of particles (1 mole of cations and 1 mole of anions) to the water for every 1 mole of salt that dissolves. However, when a nonelectrolyte such as ethylene glycol is dissolved in water, 1 mole of the solute produces only 1 mole of particles (1 mole of molecules) in the solution. Therefore we would need to have a 2 m ethylene glycol solution to achieve the same boiling point and freezing point as solutions of 1 m NaCl or 1 m KNO_3.

Dutch chemist Jacobus van 't Hoff (1852–1911) studied colligative properties and defined a term i, now called the **van 't Hoff factor** (or i **factor**), which

CONNECTION In Chapter 4 we defined an electrolyte as a substance that dissociates into ions when it dissolves.

van 't Hoff factor (also called i **factor**) the ratio of the experimentally measured value of a colligative property to the theoretical value expected for that property if the solute were a nonelectrolyte.

is the ratio of the experimentally measured value of a colligative property to the value expected if the solute were a nonelectrolyte (that is, no dissociation into ions). For example, suppose we make a solution that is 0.010 m in NaCl by dissolving 0.5844 g of NaCl in 1 kg of water. If we treat the NaCl as a nonelectrolyte, then according to Equation 11.10, the freezing point of the solution should be lower than that of pure water by

$$\Delta T_f = K_f m = \frac{1.86°C}{1 \text{ mol NaCl}/1 \text{ kg water}} \times \frac{0.0100 \text{ mol NaCl}}{1 \text{ kg water}} = 0.0186°C$$

which leads to a new freezing point of

$$T_f = 0.0000° - 0.0186° = -0.0186°C$$

However, the freezing point of the solution measured in the laboratory is −0.0372°C, which is twice as low as the value predicted for a nonelectrolyte. The ratio of the experimentally measured value to the predicted value, which is the $K_f m$ term in Equation 11.10, is

$$\frac{\Delta T_{f,\text{measured}}}{K_f m} = \frac{0.0372°C}{0.0186°C} = 2 = i$$

This is the van 't Hoff factor for sodium chloride. This makes sense because sodium chloride is a strong electrolyte and forms 2 moles of ions from each mole of NaCl.

The significance of the i factor is based on the definition of colligative properties: changes due to the *total concentration of dissolved particles* present in solution. When 1 mole of a strong electrolyte such as solid sodium chloride dissolves, it produces 2 moles of particles. Thus we should expect that the freezing point depression (or boiling point elevation) that results from the dissolution of a given number of moles of NaCl should be twice as great as that produced when the same number of moles of a nonelectrolyte dissolves.

Equations 11.9 and 11.10 can be modified to include the i factor:

$$\Delta T_b = iK_b m \tag{11.11}$$

$$\Delta T_f = iK_f m \tag{11.12}$$

If the solute is molecular (such as ethylene glycol) and therefore a nonelectrolyte, then $i = 1$ because each mole of solute produces 1 mole of dissolved particles. If the solute is a strong electrolyte, then $i =$ the number of ions in one formula unit. For NaCl, therefore, $i = 2$; for Na_2SO_4, $i = 3$ (two Na^+ ions and one SO_4^{2-} ion). Note that we do *not* break a polyatomic ion such as SO_4^{2-} into its atoms when determining an i factor; polyatomic ions stay intact.

SAMPLE EXERCISE 11.11 Using the van 't Hoff Factor   **LO7**

The salt lithium perchlorate ($LiClO_4$) is one of the most water-soluble salts known. At what temperature does a 0.130 m solution of $LiClO_4$ freeze? The K_f of water is 1.86°C/m; assume $i = 2$ for $LiClO_4$, and the freezing point of pure water is 0.00°C.

Collect and Organize We are asked to determine the freezing point of a salt solution. We know the formula of the solute, its molal concentration, and K_f of the solvent. We know that the freezing point of the pure solvent is 0.00°C. We are given the value of the van 't Hoff factor as $i = 2$.

ion pair a cluster formed when a cation and an anion associate with each other in solution.

Analyze We use Equation 11.12 to solve for the freezing point depression. The value of K_f is close to 2, the value of i is 2, and the concentration of solute is close to 0.1 m, so we predict that the freezing point of the solution will be about 0.4°C lower than that of pure water.

Solve

$$\Delta T_f = iK_f m = (2)(1.86°C/m)(0.130\ m) = 0.48°C$$

The freezing point of the solution is $(0.00 - 0.48)°C = -0.48°C$.

Think About It Lithium perchlorate dissolves in aqueous solution to form 2 moles of ions for every mole of solute that dissolves: 1 mole of Li^+ cations and 1 mole of ClO_4^- anions, consistent with $i = 2$. The calculated value is consistent with our prediction.

 Practice Exercise Determine the value of the van 't Hoff factor and calculate the boiling point of a 1.75 m aqueous solution of barium nitrate, $Ba(NO_3)_2$. The K_b of water is 0.52°C/m.

FIGURE 11.21 (a) The experimentally measured freezing point of a 0.010 m solution of NaCl is the same as the theoretical value obtained with Equation 11.12. This agreement means that the solution behaves ideally and little or no ion pairing takes place. (b) The experimentally measured freezing point of a 0.10 m solution is about 0.04°C higher than the theoretical value because some of the Na^+ and Cl^- ions form ion pairs, as shown inside the red ovals. The formation of ion pairs causes the concentration of solute particles to be less than the theoretical number, and as a result the van 't Hoff factor for the solution is less than the theoretical value of 2, and the decrease in the freezing point is less than expected.

CONCEPT TEST

Which aqueous solution has the lowest freezing point: (a) 3 m glucose ($C_6H_{12}O_6$), (b) 2 m potassium iodide (KI), or (c) 1 m sodium sulfate (Na_2SO_4)?

Using Equations 11.11 and 11.12 to calculate freezing point depressions and boiling point elevations for concentrated solutions of strong electrolytes often gives larger values than the experimentally measured values. The reason is that the cations and anions produced when strong electrolytes dissolve may not be totally independent of one another. As concentration increases, cations and anions may form ionic clusters. The simplest cluster, an **ion pair**, consists of a cation and

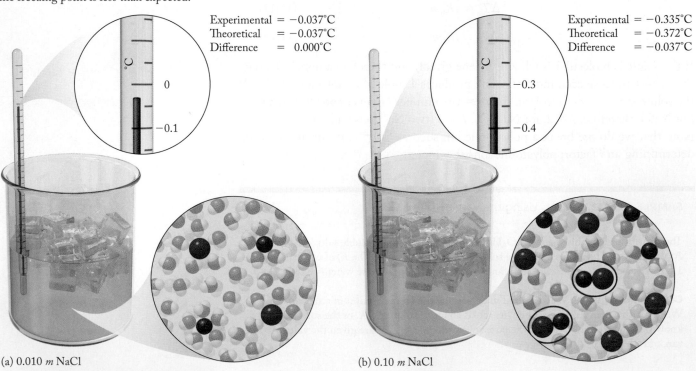

Experimental = $-0.037°C$
Theoretical = $-0.037°C$
Difference   = $0.000°C$

Experimental = $-0.335°C$
Theoretical = $-0.372°C$
Difference   = $-0.037°C$

(a) 0.010 m NaCl

(b) 0.10 m NaCl

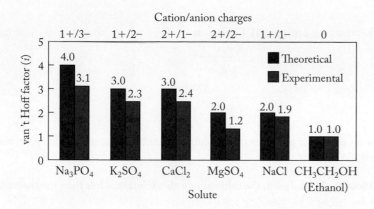

FIGURE 11.22 Theoretical and experimentally measured values for the van 't Hoff factors for 0.1 m solutions of several electrolytes and the nonelectrolyte ethanol. The higher the charge on the ions, the greater the difference between theoretical and experimentally measured values.

an anion that associate in solution, acting as a single particle. Thus the overall concentration of particles is reduced when ion pairs form, and experimentally measured freezing point depressions and boiling point elevations are smaller than the theoretical values obtained with Equations 11.11 and 11.12 (Figure 11.21).

The extent to which free ions form when a strong electrolyte dissolves is expressed by the van 't Hoff factor. The van 't Hoff factor for NaCl in water is 2 if the solution behaves ideally, because ideally 2 moles of ions are produced for each mole of NaCl that dissolves. The value of i is 2 for 0.010 m NaCl, but a little less than 2 for 0.10 m NaCl. Whenever a calculation for i gives a noninteger value, solute particles are associating in solution, and the behavior is nonideal. Figure 11.22 gives some theoretical and experimentally measured values of the van 't Hoff factor for several substances.

CONCEPT TEST

The van 't Hoff factor for an aqueous solution of an ionic compound is 2. Which of the following is (are) *not* a possible explanation?

a. The solute is a 1:1 salt behaving as an ideal solution.

b. The solute is a nonelectrolyte.

c. The solute is a 2:1 electrolyte behaving in a nonideal fashion.

d. The solute is a 1:1 salt of a weak electrolyte.

SAMPLE EXERCISE 11.12 Assessing Particle Interactions in Solution **LO7**

The experimentally measured freezing point of a 1.90 m aqueous solution of NaCl is −6.57°C. What is the value of the van 't Hoff factor for this solution? Is the solution behaving ideally, or is there evidence that solute particles are interacting with one another? The freezing-point-depression constant of water is $K_f = 1.86°C/m$, and the freezing point of pure water is 0.00°C.

Collect and Organize We are asked to calculate the i factor for a solution of known molality. We are given the measured freezing point and the K_f value. We are also asked whether the solution is behaving ideally.

Analyze Equation 11.12 tells us that $\Delta T_f = iK_f m$. We can rearrange the equation to solve for i before substituting the values for ΔT_f, K_f, and m. If the solution behaves ideally, we would expect $i = 2$ for the strong electrolyte NaCl; however, given the high concentration of NaCl (1.90 m), we predict a value less than 2.

Solve Rearranging Equation 11.12 gives us

$$i = \frac{\Delta T_{f,measured}}{K_f m}$$

$$i = \frac{6.57°C}{\left(\frac{1.86°C}{m}\right)1.90\ m} = 1.86$$

That i is not an integer tells us that the solution is not behaving ideally. Ion pairs must be forming in solution.

Think About It As predicted, the value of i for this solution is less than the theoretical value of 2.

 Practice Exercise The van 't Hoff factor for a 0.050 m aqueous solution of magnesium sulfate is 1.3. What is the freezing point of the solution?

The extent of ion pairing in a solution of a strong electrolyte generally increases with solute concentration as the solution "runs out of water" needed to form spheres of hydration around the ions. For any salt, the theoretical value of i obtained with Equation 11.11 or 11.12 is an upper limit of possible values. If ion pairing occurs, the experimentally measured value must be smaller than the theoretical value.

Osmosis and Osmotic Pressure

The final colligative property we look at in this chapter is *osmotic pressure*, the result of the process called **osmosis**—the movement of a solvent through a semipermeable membrane from a region of lower solute concentration to a region of higher solute concentration. A *semipermeable membrane* allows particles of solvent to pass through it but not particles of solute.

As an example of the importance of osmosis, we start with the observation that we all need water to survive. More than 97% of the water on Earth is seawater, and none of it is fit to drink. Let's look at the reason on a cellular level. The liquid inside each cell in the body is a complex solution of many solutes, with the average concentration of these solutes being about one-third the concentration of solutes in seawater. When cells are exposed to seawater, this substantial difference in solute concentration is the driving force for osmosis, with the cell membrane acting as the necessary semipermeable membrane (Figure 11.23). Because the solute concentration is higher outside the cell, water from inside the cell crosses the cell membrane and enters the seawater, moving from the low-solute-concentration side of the membrane to the high-solute-concentration side. As water leaves the cell, the cell shrivels and ultimately ceases to function.

Water molecules migrate through a cell membrane, or through any other semipermeable membrane, because a force makes it happen—a force caused by the different concentrations of solutes on the two sides of the membrane. When we divide the magnitude of this force, F, by the surface area, A, of the membrane, we get pressure: $P = F/A$. **Osmotic pressure (Π)** is the pressure required to halt the flow of solvent from a solution through a semipermeable membrane into pure solvent (Figure 11.24). Osmotic pressure exactly balances the pressure (F/A) driving solvent through the membrane so that no net flow of solvent takes place.

CHEMTOUR
Osmotic Pressure

osmosis the flow of a fluid through a semipermeable membrane to balance the concentration of solutes in solutions on the two sides of the membrane. The solvent particles' flow proceeds from the more dilute solution into the more concentrated one.

osmotic pressure (Π) the pressure applied across a semipermeable membrane to stop the flow of solvent from the compartment containing pure solvent or a less concentrated solution to the compartment containing a more concentrated solution. The osmotic pressure of a solution increases with solute concentration, M, and with solution temperature, T.

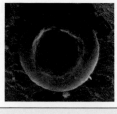

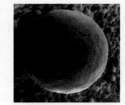

(a) Isotonic solution: total solute concentration in the solution matches that inside the cell

(b) Hypertonic solution: total solute concentration in the solution is greater than that inside the cell; water leaves cells, cells shrink

(c) Hypotonic solution: total solute concentration in the solution is less than that inside the cell; water enters cells, cells expand

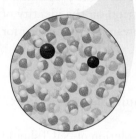
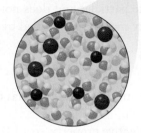
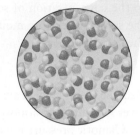

FIGURE 11.23 The membrane of a red blood cell is semipermeable, which means that water easily flows by osmosis into and out of the cell to equalize the solute concentrations on the two sides of the membrane. (a) When a cell is immersed in a solution in which the solute concentration equals the solute concentration inside the cell (isotonic conditions), the flow of water into the cell is exactly balanced by the flow of water out of the cell, and the cell size does not change. (b) When the cell is immersed in a solution in which the solute concentration is higher than the solute concentration inside the cell (hypertonic conditions), water flows by osmosis from the region of lower solute concentration to the region of higher solute concentration—in other words, out of the cell—and the cell shrinks. (c) When the cell is immersed in pure water, the solute concentration is higher inside the cell than outside (hypotonic conditions), and water flows by osmosis from the region of zero solute concentration to the region of high solute concentration—into the cell—and the cell expands.

The Greek letter pi (Π) is used as the symbol for osmotic pressure to distinguish it from the pressure exerted by gases. In Figure 11.24(a), the solution on the right has a lower concentration and lower osmotic pressure (Π_{NaCl}) than the solution on the left ($\Pi_{seawater}$): $\Pi_{NaCl} < \Pi_{seawater}$. The result is a net flow of solvent (water) from right to left until the pressure on both sides is equal. Figure 11.24(b) shows how the volumes of both solutions change: the difference in volume reflects the difference in osmotic pressures, $\Delta\Pi = \Pi_{seawater} - \Pi_{NaCl}$, in the original solutions. This process is profoundly important in living systems, where the solution

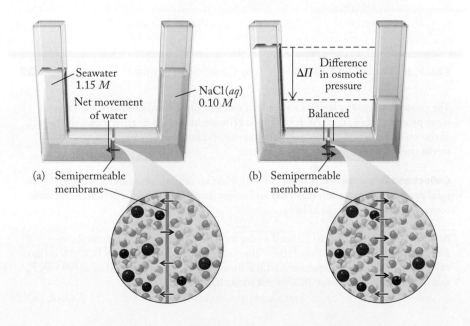

(a) Semipermeable membrane

Seawater 1.15 *M*
Net movement of water
NaCl(*aq*) 0.10 *M*

(b) Semipermeable membrane

$\Delta\Pi$ Difference in osmotic pressure
Balanced

FIGURE 11.24 (a) The solute concentration of seawater is approximately 1.15 *M*. When equal volumes of seawater and a 0.10 *M* solution of NaCl are separated by a semipermeable membrane, water moves by osmosis from 0.10 *M* NaCl (low solute concentration) to the seawater (high solute concentration) side. (b) The volume on the NaCl side decreases until the osmotic pressure of the solution equals the osmotic pressure (and concentration) of the diluted seawater solution, resulting in a difference in the heights of the liquid levels. This difference in height is proportional to the original difference in osmotic pressures ($\Delta\Pi$) between the two solutions.

inside each cell exerts an osmotic pressure on the cell membrane, a pressure pushing toward the outside of the cell. At the same time, blood or other liquid outside the cell exerts an osmotic pressure on the cell membrane, and this pressure pushes toward the cell's interior.

The osmotic pressure Π depends on the solute concentration, the absolute temperature, and the constant R, 0.0821 L · atm/(mol · K):

$$\Pi = MRT$$

where M is the molarity of the solute. Molarity is used to express concentration in calculating Π because the expression for Π can be derived from an equation similar to the ideal gas equation:

$$\Pi = P = \left(\frac{n}{V}\right)RT = MRT$$

Note that the term n/V has units of moles per liter, which matches the definition of molarity. Osmotic pressure is a colligative property because Π is proportional to the concentration of solute particles and does not depend on the identity of the solute. Therefore molarity must be multiplied by the van 't Hoff factor i for the solute:

$$\Pi = iMRT \tag{11.13}$$

Because the units of R are L · atm/(mol · K) and i has no units, the product $iMRT$ gives Π in units of atmospheres.

Osmotic pressure is a colligative property, so a 1.0 M solution of NaCl should produce the same osmotic pressure as 1.0 M KCl or 1.0 M NaNO$_3$ because all three solutions are nearly 2.0 M in total ions ($i = 2$). These solutions should have nearly twice the osmotic pressure of a 1.0 M solution of glucose at a given temperature because glucose is a nonelectrolyte and produces a solution that is only 1.0 M in dissolved particles (glucose molecules).

CONCEPT **TEST**

If a 1.0 M glucose solution is on one side of a semipermeable membrane and a 1.0 M KCl solution is on the other side, in which direction does the water flow?

SAMPLE EXERCISE 11.13 Calculating Osmotic Pressure I **LO8**

The concentration of solutes in a red blood cell is about a third of that of seawater—more precisely, about 0.30 M. If red blood cells are immersed in pure water, they swell, as shown in Figure 11.23(c). Calculate the osmotic pressure at 25°C of red blood cells across the cell membrane from pure water.

Collect and Organize We are asked to calculate an osmotic pressure. We are given the total particle concentration (0.30 M) and temperature (25°C) of the solution. We must convert the temperature to kelvins.

Analyze Osmotic pressure is related to the total concentration of all particles in solution and the absolute temperature. The total particle concentration, 0.30 M, is the value of the iM term in Equation 11.13. Estimating the answer by taking $iM = 0.3$, R at about 0.1, and T at about 300, we get about 10 atm.

Solve First we need to convert the temperature from degrees Celsius to kelvins: $T(°C) + 273 = T(K)$. Inserting the values of iM, R, and T in Equation 11.13 we have

$$\Pi = iMRT = \frac{0.30 \text{ mol}}{L} \times \frac{0.0821 \text{ L} \cdot \text{atm}}{\text{mol} \cdot \text{K}} \times (25 + 273)\text{K} = 7.3 \text{ atm}$$

Think About It Our answer is very much in line with our estimation. The calculated pressure across the membranes of red blood cells in pure water is more than 7 atmospheres: this is enough to rupture the membranes.

Practice Exercise Calculate the osmotic pressure across a semipermeable membrane separating pure water from seawater at 25°C. The total concentration of all the ions in seawater is 1.15 M.

CONCEPT **TEST**

Which has the greater effect on the osmotic pressure of a 1.0 M solution: increasing the temperature from 10°C to 20°C or adding enough solute to raise the concentration to 2.0 M?

Calculating the osmotic pressure of a solution in relation to pure solvent is straightforward. In Figure 11.24(a), however, the two solutions have different osmotic pressures, Π_{NaCl} and $\Pi_{seawater}$. Let's derive an equation for the difference between the osmotic pressures, $\Delta\Pi$, in terms of M, R, and T for the situation depicted in Figure 11.24.

Earlier in this section we said that $\Delta\Pi = \Pi_{seawater} - \Pi_{NaCl}$ and that the osmotic pressures of both solutions are described by Equation 11.13:

$$\Pi_{seawater} = iM_{seawater}RT \quad \text{and} \quad \Pi_{NaCl} = iM_{NaCl}RT$$

Taking the difference in Π expressions:

$$\Delta\Pi = \Pi_{seawater} - \Pi_{NaCl} = iM_{seawater}RT - iM_{NaCl}RT$$

$$\Delta\Pi = (iM_{seawater} - iM_{NaCl})RT \tag{11.14}$$

The total ion concentration in seawater is about 1.15 M; therefore $iM_{seawater} = 1.15$ M. If a solution of that concentration is put on one side of a semipermeable membrane and a solution of 0.10 M NaCl on the other side, then $iM_{NaCl} = (2 \times 0.10$ $M)$, and the osmotic pressure difference between the two solutions at 25°C is

$$\Delta\Pi = (iM_{seawater} - iM_{NaCl})RT$$

$$= \left(1.15 \frac{\text{mol}}{L} - 2 \times 0.10 \frac{\text{mol}}{L}\right)\left(0.0821 \frac{\text{L} \cdot \text{atm}}{\text{mol} \cdot \text{K}}\right)(298 \text{ K})$$

$$= 23 \text{ atm}$$

SAMPLE EXERCISE 11.14 Calculating Osmotic Pressure II   **LO8**

Red blood cells placed in seawater shrivel, as shown in Figure 11.23(b). Calculate the pressure across the semipermeable cell membrane separating the solution inside a red blood cell from seawater at 25°C if the total concentration of all the particles inside the cell is 0.30 *M* and the total concentration of ions in seawater is 1.15 *M*. Compare the result with the answer from Sample Exercise 11.13.

Collect and Organize We can use Equation 11.14 to calculate the difference in osmotic pressure between the two solutions. As in Sample Exercise 11.13, we must convert the temperature to units of kelvins.

Analyze We know that the concentration of particles in seawater is greater than inside a red blood cell, so solvent will flow from the cell to the seawater. Thus we predict the osmotic pressure in the cell (Π_{cell}) is less than the osmotic pressure of the seawater ($\Pi_{seawater}$). In addition, the osmotic pressure across the membrane is the difference between $\Pi_{seawater}$ and Π_{cell}, or $\Delta\Pi$. The value of $\Delta\Pi$ should be greater than the Π calculated in Sample Exercise 11.13, where red blood cells were immersed in pure water, because the difference in solute concentrations is greater.

Solve First we need to convert the temperature from degrees Celsius to kelvins: $T(°C) + 273 = T(K)$. Inserting the values of iM, R, and T in Equation 11.14 for seawater and cells we have

$$\Delta\Pi = (iM_{seawater} - iM_{cell})RT$$
$$= \left(1.15 \frac{\text{mol}}{\text{L}} - 0.30 \frac{\text{mol}}{\text{L}}\right)\left(0.0821 \frac{\text{L} \cdot \text{atm}}{\text{mol} \cdot \text{K}}\right)(298 \text{ K})$$
$$= 21 \text{ atm}$$

The pressure across a semipermeable membrane separating seawater from red blood cells, 21 atm, is greater than the pressure across the membrane separating red blood cells from pure water by almost a factor of three.

Think About It The values compare as we predicted. The effect of shrinking cells in the presence of hypertonic solutions is one of the reasons salt and sugar can be used to preserve foods. Salt rubbed on meat, for example, makes a salty solution; any bacteria exposed to this solution shrivel and die because of osmosis.

Practice Exercise Calculate the osmotic pressure at 25°C across a semipermeable membrane separating seawater (1.15 *M* total particles) from a 0.50 *M* solution of aqueous NaCl.

FIGURE 11.25 A solution of physiological saline has a concentration of 0.92 g of NaCl per 100 g of solution. The concentration of ions in this solution is equal to the concentration of ions in blood plasma. This solution is *isotonic* with blood plasma.

The osmotic pressures across cell membranes in Sample Exercises 11.13 and 11.14 are very large. The pressure of more than 7 atm calculated in Sample Exercise 11.13 for red blood cells immersed in pure water is more than three times the air pressure in a typical automobile tire and about the same as the water pressure experienced by a diver at a depth of 80 m. A pressure of 21 atm across the walls of a steel-reinforced concrete building is sufficient to cause major structural damage or even collapse the building. The possibility of serious structural damage to cells as a result of such huge pressure differentials across membranes is one reason why solutions dispensed intravenously (IV) or intramuscularly (IM) must be carefully constituted.

During a medical emergency, medication may need to be administered to a patient intravenously (Figure 11.25), and it is crucial that the osmotic pressure exerted by the intravenous solution on the body's cells be identical to the osmotic pressure exerted by the solution inside the cells. The solute concentrations in such solutions are said to be *isotonic* because they exert the same osmotic pressure as the blood exerts. As we saw in Figure 11.23, solutions with higher (*hypertonic*) or lower (*hypotonic*) solute concentrations cause the body's cells to either shrink as water leaves or swell as water enters the cell.

Two solutions are widely used to administer intravenous medications, depending on the clinical situation. One is physiological saline, which contains 0.92% NaCl by mass: 0.92 g of NaCl for every 100 g of solution. The density of dilute aqueous solutions is close to 1.00 g/mL, so 100 g of this solution has a volume of 100 mL, which means a concentration of 0.92 g/100 g is nearly the same as 0.92 g/100 mL = 9.2 g/L.

Another common IV solution is called D5W. The acronym stands for a 5.5% solution by mass of dextrose (another name for glucose; molar mass 180.16 g/mol) in water. Because this solution must be isotonic with blood and therefore isotonic with physiological saline solution, it must contain about the same concentration of solute particles as saline solution. A 5.5% by mass concentration means 5.5 g of dextrose/100 g of water, or 5.5 g of dextrose/100 mL of water, or 55 g of dextrose/L.

To compare the solute levels of these two solutions, we compare molarities:

$$\text{Physiological saline:} \quad \frac{9.2 \text{ g}}{1 \text{ L}} \times \frac{1 \text{ mol NaCl}}{58.44 \text{ g NaCl}} = 0.16 \, M$$

$$\text{D5W:} \quad \frac{55 \text{ g}}{1 \text{ L}} \times \frac{1 \text{ mol D}}{180.16 \text{ g D}} = 0.31 \, M$$

With the molar concentration of dextrose about twice that of the saline solution, how can both solutions match the solute level in blood? The answer comes again from the definition of a colligative property and the role of the i factor. We must look at the number of particles in solution in both cases. Sodium chloride is a salt with $i = 2$. Dextrose is a sugar, a molecular material, and its molar concentration directly reflects the number of sugar molecules in solution; $i = 1$. The total particle concentration in the NaCl solution is two times the molarity, or about 0.32 M, which is very close to the molar concentration of particles (molecules) in the dextrose solution. Therefore the solute concentration in 0.92% NaCl is the same as the solute concentration in 5.5% D5W, which means that the two solutions exert the same osmotic pressure.

The values used in these examples are for normal conditions with a healthy patient. Depending on the clinical condition presented, doctors may need to use solutions with different concentrations of solutes to respond most effectively to the patient's needs.

Reverse Osmosis

Many people in the world suffer from a lack of fresh water. A recent report from the United Nations (UN-Water Policy Brief on Water Quality, UN-Water 2011) suggested that by 2025, two out of three people in the world could be living in water-stressed areas—places without enough fresh water to drink or grow crops. The sea is already the source of drinking water in several desert countries bordering the Persian Gulf; to make the seawater drinkable, it must be *desalinated*, which means the salts must be removed. One way to desalinate saltwater is to distill it, but distillation requires a great deal of energy to heat seawater to its boiling point and convert it to steam.

The process of osmosis can also be applied to the desalination of seawater. As we have seen, osmosis is the movement of water from a region of low solute concentration to a region of high solute concentration. However, if a sufficiently high

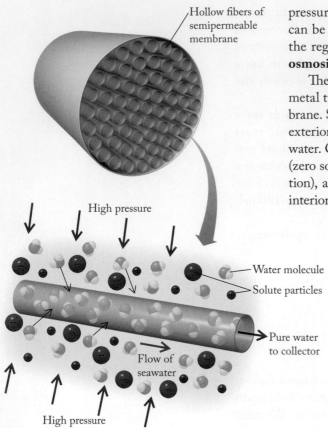

Hollow fibers of semipermeable membrane

High pressure

Water molecule

Solute particles

Pure water to collector

Flow of seawater

High pressure

FIGURE 11.26 Seawater being desalinated by reverse osmosis flows at a pressure greater than its osmotic pressure around bundles of tubes with semipermeable walls. Water molecules pass from the seawater into the tubes and flow through the tubes to a collection vessel.

pressure is applied to the region of high solute concentration, the water can be forced to move from the region of high solute concentration to the region of low solute concentration. This technique, called **reverse osmosis**, is another way of desalinating seawater to make it drinkable.

The desalination apparatus shown in Figure 11.26 consists of an outer metal tube containing many inner tubes made of a semipermeable membrane. Seawater is forced through the outer tube so that it washes over the exterior of all the inner tubes, which are initially filled with flowing pure water. Ordinarily, the direction of osmosis would be from the inner tubes (zero solute concentration) into the seawater (very high solute concentration), and indeed, the pure water does exert an osmotic pressure on the interior of the tube membranes. However, when an external pressure—a *reverse osmotic pressure*—greater than the osmotic pressure is exerted on the seawater side of the membranes, water molecules in the seawater move across the membranes into the inner tubes. The desalinated water entering the inner tubes flows into a collector and is ready for use.

Some municipal water-supply systems use reverse osmosis to make saline (brackish) water fit to drink. Some industries use this method to purify conventional tap water, and it is a common way for ships at sea to desalinate ocean water. However, very tough semipermeable membranes are necessary because reverse osmosis systems operate at very high pressures. The continued development of technologies using reverse osmosis has resulted in millions of people being supplied with sanitary drinking water. Figure 11.27 shows the range of reverse osmosis facilities available, from the world's largest desalination plant to a portable truck-mounted unit to a small device for a home water supply.

SAMPLE EXERCISE 11.15 Calculating Pressure for Reverse Osmosis **LO8**

What is the reverse osmotic pressure required at 20°C to purify brackish well water containing 0.355 M dissolved particles if the purified water is to contain no more than 87 mg of dissolved solids (measured as NaCl equivalents) per liter?

Collect and Organize We are asked to calculate the reverse osmotic pressure needed to purify water to a stated solute concentration. We are given the temperature, the solute concentration in the water to be purified, and the amount of solute (expressed as NaCl) tolerable in the product water. We can adapt Equation 11.14 to calculate the difference in osmotic pressure between the two solutions.

Analyze We need the molarity of the less concentrated solution (the drinkable water) to calculate the osmotic pressure exerted by the drinkable water, and we can calculate that from the given information. We can also calculate the osmotic pressure once we know the difference in the molarities of the solutions on the two sides of the semipermeable membrane.

Solve First we convert 87 mg of NaCl/L to molarity:

$$\frac{87 \text{ mg NaCl}}{1 \text{ L}} \times \frac{1 \text{ g}}{1000 \text{ mg}} \times \frac{1 \text{ mol NaCl}}{58.44 \text{ g NaCl}} = \frac{1.5 \times 10^{-3} \text{ mol NaCl}}{1 \text{ L}}$$

$$= 1.5 \times 10^{-3} \ M \text{ NaCl}$$

reverse osmosis a process in which solvent is forced through semipermeable membranes, leaving a more concentrated solution behind.

The total ion concentration for a $1.5 \times 10^{-3}\ M$ NaCl solution is $2(1.5 \times 10^{-3}\ M) = 0.0030\ M$. Using Equation 11.14:

$$\Delta \Pi = \Pi_{\text{brackish water}} - \Pi_{\text{drinkable water}}$$

$$= (iM_{\text{brackish water}} - iM_{\text{drinkable water}})RT$$

$$= (0.355\ M - 0.0030\ M)\left(0.0821\ \frac{\text{L} \cdot \text{atm}}{\text{mol} \cdot \text{K}}\right)(293\ \text{K})$$

$$= 8.47\ \text{atm}$$

If we maintain an osmotic pressure of exactly 8.47 atm on the side of the membrane with the brackish well water, no net flow of solvent will occur between the two solutions. Pressures greater than 8.47 atm will force water molecules from the well water through the membrane, producing water containing less than 87 mg of NaCl per liter. (We can never obtain absolutely pure water in this process; in practice, some Na^+ and Cl^- ions inevitably pass through the membrane from the well water to the drinkable water.)

Think About It In the absence of any external pressure, solvent flows from the product water to the well water, driven by the concentration difference between the two solutions as shown in Figure 11.24. However, if we apply sufficient external pressure we can reverse the flow of water. The value of the external pressure, about 8.5 atm, represents the minimum external pressure that must be applied to make this device function.

 **Practice Exercise** Calculate the minimum external pressure that must be applied in a reverse osmosis system to seawater with a total ion concentration of $1.15\ M$ at 20°C if the maximum concentration allowed in the product water is 174 mg of NaCl per liter.

CONCEPT **TEST**

If you were trying to purify water by reverse osmosis, using the apparatus in Figure 11.26, what advantage might there be in running the system at 50°C rather than 20°C?

11.6   Measuring the Molar Mass of a Solute by Using Colligative Properties

In principle, we can determine the molar mass of any solute by dissolving a known quantity of the solute in a known quantity of solvent and then measuring the effect the dissolved solute has on any colligative property of the solvent. In practice, this method works only for nonelectrolytes, which have a van 't Hoff factor of 1. Freezing-point-depression measurements, for example, can be used to find the molar mass of a molecular compound if it is sufficiently soluble in a solvent whose K_f value is known, as illustrated in Sample Exercise 11.16.

SAMPLE EXERCISE 11.16   Using Freezing Point Depression   **LO9**
to Determine Molar Mass

Eicosene is a molecular compound and nonelectrolyte with the empirical formula CH_2. It is used in water-resistant sunscreens. The freezing point of a solution prepared

(a)

(b)

(c)

FIGURE 11.27 Reverse osmosis can supply water for personal, industrial, and commercial use. (a) The world's largest reverse osmosis desalination plant at Hadera, Israel, produces up to 120 million gallons of water daily. (b) The interior of a truck carrying a portable reverse osmosis unit that can purify 500 gallons of brackish water per day. (c) A small unit attached to a kitchen sink purifies a 16-oz glass of water in about 5 minutes.

CONNECTION In Chapter 3 we used molar masses determined by mass spectrometry to convert empirical formulas derived from elemental analyses to molecular formulas. In Chapter 6 we used density measurements and the ideal gas law to calculate molar masses of gases.

by dissolving 100 mg of eicosene in 1.00 g of benzene was 1.75°C lower than the freezing point of pure benzene. What is the molar mass of eicosene? (K_f for benzene is 4.90°C/m.)

Collect and Organize We are asked to determine the molar mass of a compound. We are given the mass of the compound that lowers the freezing point of a solvent by a known amount, and we have the K_f of the solvent. Because the solute is a nonelectrolyte, $i = 1$. Equation 11.12 relates concentration to the change in freezing point.

Analyze The molar mass of a compound is expressed in units of grams per mole. Our sample consists of 100 mg, or 0.100 g. To find the molar mass we need to determine how many moles of eicosene are contained in the sample. The concentration term in Equation 11.12 is molality, or moles of solute per kilogram of solvent, so we can use the experimental data to first calculate molality. Once we know the molality of the eicosene, we know the number of moles of eicosene in 1 kg of benzene; we are given the mass of benzene used, so we can calculate the number of moles of eicosene in the sample, from which we may calculate the molar mass of eicosene. The empirical formula of eicosene enables us to check the validity of our answer because the molar mass of eicosene must be a whole-number multiple of the mass of CH_2 (14 g/mol).

Solve The molality of the eicosene solution is

$$\Delta T_f = iK_f m = K_f m$$

$$m = \frac{\Delta T_f}{K_f} = \frac{1.75°\text{C}}{4.90°\text{C}/m} = 0.357 \ m \ \text{eicosene}$$

This means that 0.357 mole of eicosene is dissolved per kilogram of solvent. Only 1.00 g (1.00×10^{-3} kg) was used in this sample. Calculating the moles of eicosene in the sample:

$$m = \frac{\text{moles of solute}}{\text{kilograms of solvent}}$$

$$0.357 \ m = \frac{\text{moles of eicosene}}{1.00 \times 10^{-3} \ \text{kg benzene}}$$

$$\text{Moles of eicosene} = \frac{0.357 \ \text{mol eicosene}}{1 \ \text{kg benzene}} \times 1.00 \times 10^{-3} \ \text{kg benzene}$$

$$= 3.57 \times 10^{-4} \ \text{mol eicosene}$$

Because the molar mass is the mass of 1 mole of eicosene, and 100 mg of eicosene was used to prepare the solution,

$$\text{Molar mass} = \frac{\text{mass of eicosene}}{\text{moles of eicosene}} = \frac{0.100 \ \text{g eicosene}}{3.57 \times 10^{-4} \ \text{mol}} = 280 \ \text{g/mol}$$

Think About It The molar mass of eicosene, 280 g/mol, is reasonable because it corresponds to $(CH_2)_n$, where $n = 20$. The molecular formula of eicosene is therefore $C_{20}H_{40}$.

Practice Exercise A solution prepared by dissolving 360 mg of a sugar (a molecular compound and a nonelectrolyte) in 1.00 g of water froze at −3.72°C. What is the molar mass of this sugar? The value of K_f of water is 1.86°C/m.

For determining the molar mass of water-soluble substances, measuring osmotic pressure is a better choice than measuring either boiling point elevation or freezing point depression for several reasons. First, the K_f and K_b values for water are much smaller than those of other solvents (Table 11.5). Thus to have a

TABLE 11.5 Molal Freezing-Point-Depression and
        Boiling-Point-Elevation Constants for Selected Solvents

Solvent	Freezing Point (°C)	K_f (°C/m)	Boiling Point (°C)	K_b (°C/m)
Water (H_2O)	0.0	1.86	100.0	0.52
Benzene (C_6H_6)	5.5	4.90	80.1	2.53
Ethanol (CH_3CH_2OH)	−114.6	1.99	78.4	1.22
Carbon tetrachloride (CCl_4)	−22.3	29.8	76.8	5.02

solution that gives a measurable boiling point or freezing point change for an aqueous solution, the concentration of the solution has to be much higher than is readily achievable. Second, biomaterials such as proteins and carbohydrates are nearly always available only in small quantities, and often they are not very soluble in nonaqueous solvents that have larger K_f or K_b values. Furthermore, these biomaterials often have high molar masses, which means that large quantities are needed to give high enough molal concentrations for reliable ΔT_f or ΔT_b measurements. Third, a solute might need to be recovered unchanged for other uses, which rules out boiling-point-elevation measurements for heat-sensitive solutes.

Perhaps the most compelling reasons for using osmotic pressure are that very small osmotic pressures can be measured precisely, the measurement equipment can be miniaturized so that only minute quantities of solute are needed, and the measurements can be made at room temperature.

SAMPLE EXERCISE 11.17   Using Osmotic Pressure to   **LO9**
                            Determine Molar Mass

A molecular compound that is a nonelectrolyte was isolated from a South African tree. A 47-mg sample was dissolved in water to make 2.50 mL of solution at 25°C, and the osmotic pressure of the solution was 0.489 atm. Calculate the molar mass of the compound.

Collect and Organize We are given the mass of a substance, the volume of its aqueous solution, the temperature, and its osmotic pressure. We can relate these parameters with Equation 11.13, using the value $i = 1$ because this is a nonelectrolyte.

Analyze We can calculate the molar concentration of the solution from the information given and Equation 11.13. Because we know the solution volume, we can calculate the number of moles of the solute from the molarity. Because we know the mass of this number of moles, we can calculate the molar mass of the solute.

Solve Rearranging Equation 11.13 to isolate M and substituting the given values:

$$M = \Pi/iRT = \frac{0.489 \text{ atm}}{(1)[0.0821 \text{ L} \cdot \text{atm}/(\text{mol} \cdot \text{K})](298 \text{ K})}$$

$$= \frac{2.00 \times 10^{-2} \text{ mol}}{\text{L}} = 2.00 \times 10^{-2} M$$

Next we solve the defining equation for molarity, $M = n/V$, for n, the number of moles of solute:

$$n = MV = \frac{2.00 \times 10^{-2} \text{ mol}}{1 \text{ L}} \times 2.50 \times 10^{-3} \text{ L} = 5.00 \times 10^{-5} \text{ mol}$$

We know that this number of moles of solute has a mass of 47 mg. The molar mass of the solute is therefore

$$\text{Molar mass} = \frac{\text{g solute}}{\text{moles of solute}}$$

$$= \frac{47 \times 10^{-3}\,\text{g}}{5.00 \times 10^{-5}\,\text{mol}} = 9.4 \times 10^{2}\,\text{g/mol}$$

Think About It One of the advantages of determining molar mass by osmotic pressure is that only a small amount of material is required. With only a 47-mg sample of a rather large molecule (molar mass 940 g/mol), the osmotic pressure is sufficiently large (0.489 atm) to enable us to calculate the molar mass accurately.

Practice Exercise A solution was made by dissolving 5.00 mg of a polysaccharide (a polymer made of sugar molecules) in water to give a final volume of 1.00 mL. The osmotic pressure of this solution was 1.91×10^{-3} atm at 25°C. Calculate the molar mass of the polysaccharide, which is a nonelectrolyte.

We very rarely deal with pure liquids in the real world. Milk, gasoline, tap water, shampoo, olive oil, cough syrup, and countless other fluids that we use in our daily lives are all solutions whose physical properties depend on the concentrations of solutes dissolved in them. By understanding the role solutes play in determining the properties of solutions, we can prepare solutions that have exactly the properties we need. The boiling and freezing points of water can be extended above 100°C and below 0°C to make antifreeze that takes heat away from an operating engine and protects the fluid from freezing in the winter. Adding the right amount of dextrose to water for an IV drip keeps trauma patients hydrated. Having too much or too little solute in a solution can have catastrophic effects: engine blocks can crack if insufficient antifreeze is dissolved in the water in a radiator to protect the fluid from freezing; a patient can die if the concentration of sodium chloride in an IV drip is too low or too high. It is important to remember that changes in boiling point elevation, freezing point depression, and osmotic pressure—as well as other properties we have discussed in this chapter—depend solely on the concentration of solute particles present in the solutions and not on their identity.

SAMPLE EXERCISE 11.18 Integrating Concepts: Fun with Eggs

The shell of a chicken's egg is mostly calcium carbonate, and it is lined with a semipermeable membrane. Explain what happens during the following series of steps, all carried out at room temperature (21°C), and answer all the associated questions.

1. You put a chicken egg in a beaker and pour enough pickling vinegar {aqueous acetic acid [CH₃COOH(aq)]} over it to submerge it. Gas bubbles out from the shell, and after several days the hard shell dissolves, leaving the sac-like semipermeable membrane containing the white and the yolk

intact. (a) Identify the gas given off and write a balanced chemical equation for the reaction. (b) The beaker is open to the air during this period. Does most of the gas dissolve in the fluid, or does most of it escape? Explain your answer.

2. At this point, you weigh the egg; its mass is 100 g.

3. You submerge the egg in a solution of corn syrup (75% by weight sucrose [C₁₂H₂₂O₁₁]; 25% water; density = 1.38 g/mL) and leave it there for 8 hours. The egg visibly shrivels. You remove it from the solution and weigh it: its mass is now 58 g. (c) Why does the egg shrivel?

4. You now submerge the egg in distilled water for 8 hours. The egg expands and its mass is now 103 g. (d) Why does the egg expand?

5. As a final step, you submerge the egg in a 0.75 M solution of table salt. (e) Predict what happens. (f) If you were to start with two fresh eggs and put one in the syrup and the other in the salt solution, which would change more in volume?

Collect and Organize We are given a series of changes an egg goes through when exposed to several solutions. We have the starting materials (calcium carbonate and acetic acid) for a chemical reaction involving the egg's shell and are asked to identify the products. We can determine the concentration of the sugar and salt solutions that cause the egg to shrink or expand.

Analyze From Chapter 4 we recall that the reaction between a carbonate and an acid produces carbon dioxide gas. We can use Henry's law (Section 10.5) to calculate the solubility of CO_2 in water. The other changes must involve osmosis across a semipermeable membrane (Section 11.5). Equation 11.13 relates solute concentration to osmotic pressure, and we can use it to compare the effects of the sugar and salt solutions.

Solve

a. $CaCO_3(s) + 2\,CH_3COOH(aq) \rightarrow$
   $Ca(CH_3COO)_2(aq) + H_2O(\ell) + CO_2(g)$
   The gas given off as the eggshell dissolves is carbon dioxide (CO_2).

b. CO_2 is nonpolar, so it would be held in the aqueous phase by relatively weak dipole–induced dipole interactions. Its Henry's law constant k_H, listed in Table 10.7, is about ten times that of oxygen, but that is still small, so we would not expect CO_2 to be very water soluble. In addition, the description states that the beaker is open to the atmosphere, so we can expect most of the CO_2 to escape into the air, just as it does if we leave a carbonated beverage open.

c. The egg shrivels (Figure 11.28) because the concentration of the syrup solution is higher than the concentration of the solution within the egg sac. Osmosis causes water to leave the egg sac and enter the surrounding solution to equalize the osmotic pressure.

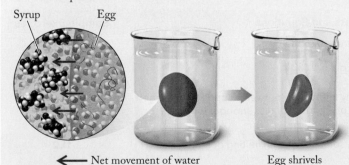

Syrup   Egg

⟵ Net movement of water   Egg shrivels

FIGURE 11.28 The egg shrivels.

d. The egg expands (Figure 11.29) when put into distilled water (containing no solute). Osmosis causes water to flow through the sac and enter the egg.

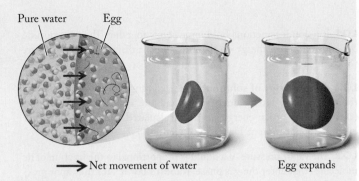

Pure water   Egg

⟶ Net movement of water   Egg expands

FIGURE 11.29 The egg expands.

e. If the egg is then placed in a salt solution, as long as the concentration of salt is higher than the concentration of the solution inside the egg, the egg will shrivel again (Figure 11.30) as water flows from inside the sac to the outside solution.

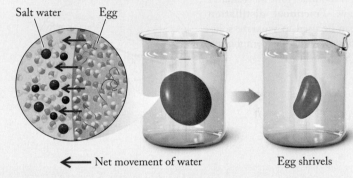

Salt water   Egg

⟵ Net movement of water   Egg shrivels

FIGURE 11.30 The egg shrivels again.

f. The solution that causes the greatest osmotic pressure difference across the semipermeable membrane will cause the greatest change in the volume of the egg. The sugar solution is 75% by weight sucrose. That means 75 g of sucrose is dissolved in 100 g of total solution. We have the density of the solution, so we can convert this concentration into molarity:

$$\frac{75\ \text{g sucrose}}{100\ \text{g solution}} \times \frac{1\ \text{mol}}{342.30\ \text{g}} \times \frac{1.380\ \text{g solution}}{1\ \text{mL solution}} \times \frac{1000\ \text{mL}}{1\ \text{L}} = 3.0\ M$$

The concentration of the salt solution is 0.75 M NaCl. We can compare the two solutions by comparing their osmotic pressures:

$$\begin{aligned} \Pi_{\text{syrup}} &= iM_{\text{syrup}}RT \\ &= (1)(3.0\ \text{mol/L})[0.0821\ \text{L} \cdot \text{atm}/(\text{mol} \cdot \text{K})](294\ \text{K}) \\ &= 72\ \text{atm} \end{aligned}$$

$$\begin{aligned} \Pi_{\text{salt water}} &= iM_{\text{salt water}}RT \\ &= (2)(0.75\ \text{mol/L})[0.0821\ \text{L} \cdot \text{atm}/(\text{mol} \cdot \text{K})](294\ \text{K}) \\ &= 36\ \text{atm} \end{aligned}$$

The syrup exerts a much higher osmotic pressure than the salt solution, so we predict that an egg placed in the syrup would shrivel more than an egg placed in the salt water.

Think About It We have to treat the solutions ideally because we have no indication about the actual value of the i factor. Because $i_{\text{salt water}} \leq 2$, Π_{syrup} has to be greater than $\Pi_{\text{salt water}}$ here.

SUMMARY

LO1 Ion–ion interactions hold ionic solids together. Their magnitude depends upon the charge on the ions and the distance between them. (Section 11.1)

LO2 The **enthalpy of solution** ($\Delta H_{solution}$) for an ionic compound is the sum of the **lattice energy** (U) and the **enthalpy of hydration** ($\Delta H_{hydration}$). Lattice energies can be calculated with a **Born–Haber cycle**, an application of Hess's law. (Section 11.2)

LO3 The vapor pressure of a liquid is proportional to the fraction of its molecules that enter the gas phase. **Raoult's law** relates the vapor pressure of a solution to its composition and to the vapor pressure of the solvent. (Section 11.3)

LO4 The vapor pressure of an ideal solution of volatile compounds follows Raoult's law. **Fractional distillation** can be used to separate solutions of volatile compounds. (Section 11.4)

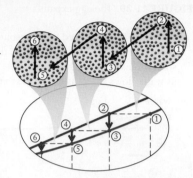

LO5 The concentration units used for **colligative property** measurements include molarity (M) and **molality** (m). (Section 11.5)

LO6 Solutes in solution elevate the solvent's boiling point and depress its freezing point. (Section 11.5)

LO7 The **van 't Hoff factor** accounts for the colligative properties of electrolytes and the formation of solute **ion pairs** in concentrated solutions. (Section 11.5)

LO8 In **osmosis**, solvent flows through a semipermeable membrane. **Osmotic pressure (Π)** is the pressure required to halt the flow of solvent from the more dilute solution across the membrane. (Section 11.5)

LO9 The molar mass of a compound can be determined by measuring the freezing point depression, boiling point elevation, or osmotic pressure of a solution of the compound. (Section 11.6)

PARTICULATE **PREVIEW WRAP-UP**

Using Equation 5.3 and Figure 11.1, we know that larger ions have larger distances between their nuclei. Given that Cs^+ is larger than Ba^{2+}, and that I^- is larger than O^{2-}, the cation and anion pair with the longer distance between their nuclei would be Cs^+ and I^-. The larger coulombic attraction would be between the ions with the larger charges, namely, Ba^{2+} and O^{2-}. BaO has a higher melting point than CsI because it contains ions with larger charges.

PROBLEM-SOLVING SUMMARY

Type of Problem	Concepts and Equations	Sample Exercises
Predicting relative strengths of ion–ion interactions	To predict relative interaction strengths, use $$E \propto \frac{(Q_1 Q_2)}{d} \quad (5.3)$$ The value of E is negative for any interaction between oppositely charged particles. More negative values of E correspond to stronger ion–ion attractions.	**11.1**
Ranking lattice energies, solubility, and melting points	Predict relative lattice energies, solubilities, and melting points by using $$U = \frac{k(Q_1 Q_2)}{d} \quad (11.4)$$	**11.2**

Type of Problem	Concepts and Equations	Sample Exercises
Calculating lattice energy with a Born–Haber cycle	The enthalpy of reaction, ΔH_{rxn}, when an ionic solid forms from its constituent elements is equal to the sum of the enthalpies of reaction for every step in the process. Solve for lattice energy U or any one unknown enthalpy change term.	**11.3**
Calculating lattice energy by using enthalpies of hydration	Use enthalpies of hydration ($\Delta H_{hydration}$) and enthalpies of solution ($\Delta H_{solution}$) to calculate lattice energy: $$\Delta H_{solution} = \Delta H_{hydration} - U \qquad (11.5)$$ where $$\Delta H_{hydration} = \Delta H_{hydration,cation} + \Delta H_{hydration,anion}$$	**11.4**
Calculating vapor pressure of a solution	Use Raoult's law, $$P_{solution} = X_{solvent}P^\circ_{solvent} \qquad (11.6)$$ where $P_{solution}$ is the vapor pressure of the solution at a given temperature, $X_{solvent}$ is the mole fraction of the solvent in the solution, and $P^\circ_{solvent}$ is the vapor pressure of the pure solvent at the same temperature. $$P_{total} = X_1P^\circ_1 + X_2P^\circ_2 + X_3P^\circ_3 \ldots \qquad (11.7)$$	**11.5, 11.6, 11.7**
Calculating molal concentrations	Molality is defined as $$m = \frac{n_{solute}}{\text{kg solvent}} \qquad (11.8)$$ where n_{solute} is the number of moles of solute.	**11.8**
Calculating boiling point or freezing point of a solution	For nonelectrolyte solutes, use $$\Delta T_b = K_b m \qquad (11.9)$$ where ΔT_b is the elevation in the boiling point of the solvent, K_b is a constant that depends only on the solvent, and m is the molality of the solution. For the freezing point, use $$\Delta T_f = K_f m \qquad (11.10)$$ where ΔT_f is the depression in the freezing point of the solvent, K_f is a constant that depends only on the solvent, and m is the molality of the solution.	**11.9, 11.10**
Assessing interactions among particles in solution by comparing the theoretical value of the van 't Hoff factor i with the experimentally measured value	For electrolytes, use $$\Delta T_b = iK_b m \qquad (11.11)$$ and $$\Delta T_f = iK_f m \qquad (11.12)$$ where the theoretical value of i is the number of particles created when an electrolyte dissociates completely: $i = 2$ for NaCl, 3 for $CaCl_2$, and 4 for Na_3PO_4. For real solutions of electrolytes where $\Delta T_{b/f}$, $K_{b/f}$, and m are known, rearrange $\Delta T_b = iK_b m$ and $\Delta T_f = iK_f m$ to calculate the value of i and compare the result with the theoretical value.	**11.11, 11.12**
Calculating osmotic pressure and reverse osmotic pressure	Use $$\Pi = iMRT \qquad (11.13)$$ where Π is the osmotic pressure, i is the van 't Hoff factor, M is the molar concentration of the solution, R is the ideal gas constant, and T is the absolute temperature of the solution.	**11.13, 11.14, 11.15**
Determining molar mass from boiling point elevation, freezing point depression, or osmotic pressure	Use $\Delta T_b = K_b m$, $\Delta T_f = K_f m$, or $\Pi = MRT$ to calculate the molality m or molarity M of the solution. Then use m or M to determine the moles of solute and molar mass of the solute.	**11.16, 11.17**

VISUAL PROBLEMS

(Answers to boldface end-of-chapter questions and problems are in the back of the book.)

11.1. Figure P11.1 shows a particle-level view of a sealed container partially filled with a solution that has two components: X (blue spheres) and Y (red spheres). Which of the following statements about substances X and Y are true?

a. X is the solvent in this solution.

b. Pure Y is a volatile liquid.

c. If Y were not present, there would be fewer X particles in the gas above the liquid solution.

d. The presence of Y increases the vapor pressure of X.

FIGURE P11.1

11.2. Figure P11.2 shows a particle-level view of a sealed container partially filled with a solution of two miscible liquids: X (blue spheres) and Y (red spheres). Which of the following statements about substances X and Y are true?

a. Y is the solvent in this solution.

b. Pure Y has a higher vapor pressure than pure X.

c. The presence of Y in the solution lowers the vapor pressure of X.

d. If Y were not present, there would be fewer total particles in the gas above the liquid solution.

FIGURE P11.2

11.3. Figure P11.3 shows particle-level views of 0.001 *M* aqueous solutions of the following four solutes: $C_6H_{12}O_6$, NaCl, $MgCl_2$, and K_3PO_4. The blue spheres represent particles of solute.

a. Which compounds are represented in images (I)–(IV)?

b. Which of the four solutions in Figure P11.3 has the highest (i) vapor pressure; (ii) boiling point; (iii) freezing point; (iv) osmotic pressure?

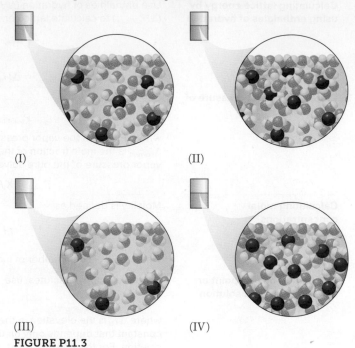

(I)   (II)

(III)   (IV)

FIGURE P11.3

11.4. The graph in Figure P11.4 describes the volume of distillate collected during the fractional distillation of a liquid. Answer the following questions about the process: (a) Is the sample a pure liquid or a mixture? (b) If it is a mixture: (i) how many components are in the mixture? (ii) What are the relative ratios of the volumes in the mixture? (iii) What are their approximate boiling points?

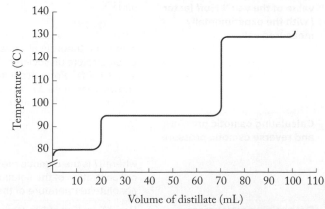

FIGURE P11.4

11.5. The graph in Figure P11.5 shows the decrease in the freezing point of water ΔT_f for solutions of two different substances, A (triangles) and B (circles), in water. Explain how you can reasonably conclude that (a) A and B are nonelectrolytes and (b) the freezing point depression constant K_f of water is independent of the solute's identity.

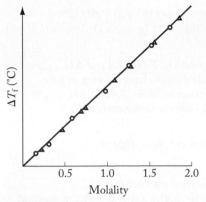

FIGURE P11.5

11.6. The arrow in Figure P11.6 indicates the direction of solvent flow through a semipermeable membrane in equipment designed to measure osmotic pressure. Which solution, A or B, is more concentrated? Explain your answer.

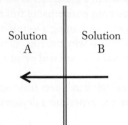

Solution A   Solution B

FIGURE P11.6

11.7. Kidney Dialysis Semipermeable membranes of the sort used in kidney dialysis do not allow large molecules and cells to pass but do allow small ions and water to pass. Figure P11.7 shows such a membrane separating fluids of various compositions.
   a. In which direction does the water flow in each apparatus?
   b. In which direction do sodium ions flow in each apparatus?
   c. In which direction do the potassium ions flow in each apparatus?

Pure H_2O	0.2 *M* KCl in H_2O
Membrane	Membrane
Proteins 0.5 *M* NaCl 0.1 *M* KCl in H_2O	Proteins 0.5 *M* NaCl 0.1 *M* KCl in H_2O
(i)	(ii)

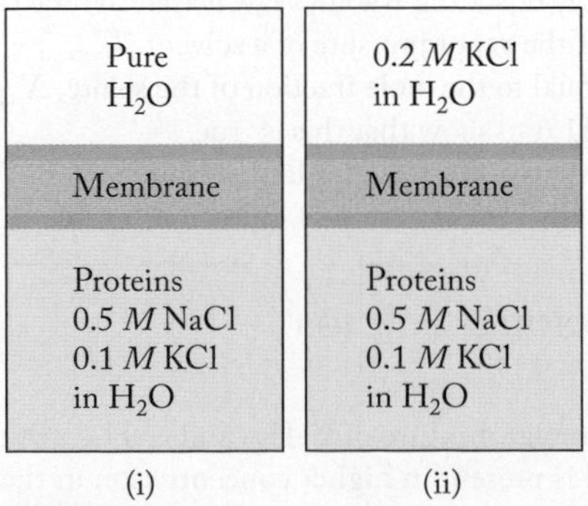

FIGURE P11.7

11.8. Use representations [A] through [I] in Figure P11.8 to answer questions a–f about the formation of an aqueous solution of potassium chloride from potassium [K(*s*), purple spheres] and chlorine [Cl$_2$(*g*), green spheres].
   a. Which process depicts the formation of a compound from its elements?
   b. Which processes require the breaking of bonds?
   c. Which processes depict transfer of electrons?
   d. Which representation illustrates ion-dipole interactions?
   e. Which processes require the input of energy to overcome intermolecular forces?
   f. Which processes are exothermic?

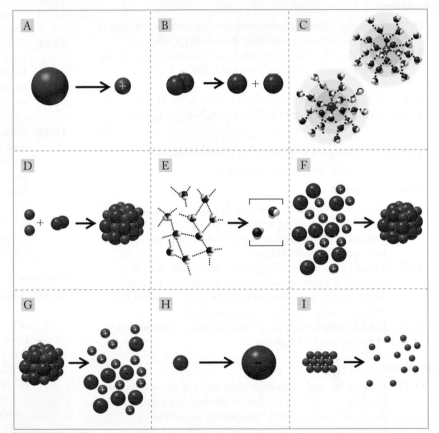

FIGURE P11.8

QUESTIONS AND PROBLEMS

Interactions between Ions

Concept Review

11.9. Indicate the substance that contains the largest anion.
(a) $BaCl_2$; (b) AlF_3; (c) KI; (d) $SrBr_2$

11.10. Indicate the substance that contains the smallest cation.
(a) $BaCl_2$; (b) Al_2O_3; (c) Mg_3N_2; (d) SrS

11.11. Why is $CaSO_4$ less soluble in water than NaCl?

11.12. Does the strength of an ion–ion attraction depend on the number of ions in the compound? Explain your answer.

Problems

11.13. Rank the following ionic compounds in order of increasing attraction between their ions: KBr, $SrBr_2$, CsBr.

11.14. Rank the following ionic compounds in order of increasing attraction between their ions: BaO, $BaCl_2$, CaO.

11.15. Which has the higher melting point, LiF or MgO?

11.16. Which has the higher boiling point, NaCl or $CaCl_2$?

Energy Changes during Formation and Dissolution of Ionic Compounds

Concept Review

11.17. Define the terms in the relation $\Delta H_{solution} = \Delta H_{hydration} - U$.

11.18. Explain why trends in lattice energies for ionic compounds parallel trends in melting points and are opposite to the trends in water solubility.

11.19. In the formula $U = kQ_1Q_2/d$ for the different lattice energies of a pair of ionic compounds such as LiCl and MgO, which factor dominates—the charge product Q_1Q_2 or the distance d between the ion nuclei?

11.20. Explain why it might be difficult to measure the enthalpy of hydration of a single ion.

Problems

11.21. How do the melting points of the series of sodium halides NaX (X = F, Cl, Br, I) relate to the atomic number of X?

11.22. Rank the following ionic compounds in order of
(a) increasing melting point and (b) increasing water solubility: BaF_2, $CaCl_2$, $MgBr_2$, and SrI_2.

11.23. Which substance has the least negative lattice energy?
(a) MgI_2; (b) $MgBr_2$; (c) $MgCl_2$; (d) MgF_2

11.24. Rank the following from lowest to highest lattice energy: NaBr, $MgBr_2$, $CaBr_2$, and KBr.

11.25. Use a Born–Haber cycle to calculate the lattice energy of potassium chloride (KCl) from the following data:

Ionization energy of $K(g)$ = 425 kJ/mol
Electron affinity of $Cl(g)$ = −349 kJ/mol
Energy to sublime $K(s)$ = 89 kJ/mol
Bond energy of $Cl_2(g)$ = 240 kJ/mol

ΔH_f for $K(s) + \frac{1}{2}Cl_2(g) \rightarrow KCl(s)$ = −438 kJ/mol

11.26. Calculate the lattice energy of sodium oxide (Na_2O) from the following data:

Ionization energy of $Na(g)$ = 495 kJ/mol
Electron affinity of $O(g)$ for 2 electrons = 603 kJ/mol
Energy to sublime $Na(s)$ = 109 kJ/mol

Bond energy of $O_2(g)$ = 499 kJ/mol
ΔH_f for $2 Na(s) + \frac{1}{2}O_2(g) \rightarrow Na_2O(s)$ = −416 kJ/mol

11.27. Using the values in Table 11.4 and $\Delta H_{solution}$ = 19.9 kJ/mol for KBr, calculate the lattice energy of KBr.

11.28. Using the values in Table 11.4 and $\Delta H_{solution}$ = −17.7 kJ/mol for KF, calculate the lattice energy of KF.

Vapor Pressure of Solutions

Concept Review

11.29. Explain the term *nonvolatile solute*.

11.30. Which has the higher vapor pressure at constant temperature, pure water or seawater? Explain your answer.

11.31. Why does the vapor pressure of a liquid increase with increasing temperature?

11.32. In the experiment shown in Figure 11.7, the vapor pressure of one of the solutions remains constant throughout the experiment, whereas the vapor pressure of the other solution changes. Which solution is which?

11.33. An experiment like that shown in Figure 11.7 is set up with the compartment containing pure ethanol full to the brim and the compartment containing a solution of sugar in ethanol half-full. Explain why the compartment that contains the ethanol–sugar solution will eventually overflow.

11.34. Explain to a nonscientist how the water gets from one compartment to the other in the experiment depicted in Figure 11.7.

Problems

11.35. A solution contains 3.5 moles of water and 1.5 moles of nonvolatile glucose ($C_6H_{12}O_6$). What is the mole fraction of water in this solution? What is the vapor pressure of the solution at 25°C, given that the vapor pressure of pure water at 25°C is 23.8 torr?

11.36. A solution contains 4.5 moles of water, 0.3 moles of sucrose ($C_{12}H_{22}O_{11}$), and 0.2 moles of glucose. Sucrose and glucose are nonvolatile. What is the mole fraction of water in this solution? What is the vapor pressure of the solution at 35°C, given that the vapor pressure of pure water at 35°C is 42.2 torr?

11.37. Another way of stating Raoult's law is that the fractional lowering of the vapor pressure of a solvent ($P°_{solvent} - P_{solvent}$)/$P°_{solvent}$ is equal to the mole fraction of the solute, X_{solute}. Use Equation 11.6 to show that this is true.

11.38. Use the statement of Raoult's law in Problem 11.37 to determine the mole fraction of glucose in Problem 11.35.

Mixtures of Volatile Solutes

Concept Review

11.39. In an equimolar mixture of C_5H_{12} and C_7H_{16}, which compound is present in higher concentration in the vapor above the solution?

11.40. Why does the boiling point of a mixture of volatile hydrocarbons increase over time during a simple distillation?

11.41. Would you expect a solution of cyclohexane, C_6H_{12}, in benzene, C_6H_6, to behave ideally? Explain your answer.

11.42. Explain the origin of negative deviations from Raoult's law for the predicted vapor pressure of a solution of two volatile liquids.

Problems

11.43. At 20°C, the vapor pressure of ethanol is 45 torr and the vapor pressure of methanol is 92 torr. What is the vapor pressure at 20°C of a solution prepared by mixing 25 g of methanol and 75 g of ethanol?

*11.44. At 90°C, the vapor pressure of styrene (C_8H_8) is 134 torr and that of ethylbenzene (C_8H_{10}) is 183 torr. What is the vapor pressure of a solution of 38% by weight styrene and 62% by weight ethylbenzene at 90°C?

*11.45. The mixture described in Problem 11.44 is separated by fractional distillation at reduced pressure so that it begins to boil when the solution in the distillation flask reaches 90°C.
 a. What is the ratio of ethylbenzene to styrene in the vapor phase as the mixture first begins to boil?
 b. What will be the temperature of the first distillate that comes off the top of the column? (i) lower than 90°C; (ii) 90°C; (iii) higher than 90°C

11.46. A bottle is half-filled with a 50:50 (mole-to-mole) mixture of heptane (C_7H_{16}) and octane (C_8H_{18}) at 25°C. What is the mole ratio of heptane vapor to octane vapor in the air space above the liquid in the bottle? The vapor pressures of heptane and octane at 25°C are 31 torr and 11 torr, respectively.

Colligative Properties of Solutions

Concept Review

11.47. Explain why freezing point depression, boiling point elevation, and the osmotic pressure of ionic solutes cannot be used to measure their formula masses.

11.48. What is the definition of the concentration scale called *molality* that is used to determine the molar mass of a nonvolatile solute by measuring the solute's effect on the freezing and boiling points of the solvent? Why must a mass-based concentration scale and not *molarity* be used?

11.49. As a solution of NaCl is heated from 5°C to 90°C, does the difference between its molarity and its molality increase or decrease? Explain your answer.

*11.50. **Diet Soft Drinks** The thermostat in a refrigerator filled with cans of soft drinks malfunctions and the temperature of the refrigerator drops below 0°C. The contents of the cans of diet soft drinks freeze, rupturing many of the cans and causing an awful mess. However, none of the cans containing regular, nondiet soft drinks rupture. Why?

11.51. Why is it important to know if a substance is a molecular compound or an ionic compound before predicting its effect on the boiling and freezing points of a solvent?

11.52. Refer to the phase diagram in Figure 11.17 and explain in your own words why the change in the vapor pressure of a solution caused by a nonvolatile solute results in a higher boiling point and a lower melting point.

11.53. Explain how the theoretical value of the van 't Hoff factor i for substances such as CH_3OH, NaBr, and K_2SO_4 can be predicted from their formulas.

11.54. Is it possible for an experimentally measured value of a van 't Hoff factor to be greater than the theoretical value? Explain your answer.

11.55. What is a semipermeable membrane?

11.56. A pure solvent is separated from a solution containing the same solvent by a semipermeable membrane. In which direction does the solvent flow across the membrane, and why?

11.57. A dilute solution is separated from a more concentrated solution containing the same solvent by a semipermeable membrane. In which direction does the solvent tend to flow across the membrane, and why?

11.58. How is the osmotic pressure of a solution related to its molar concentration and its temperature?

11.59. Explain the principle of reverse osmosis.

11.60. Explain how the minimum pressure for purification of seawater by reverse osmosis can be estimated from its composition.

11.61. Aqueous solutions of physiological saline (NaCl) and dextrose (glucose) are used to deliver intravenous medications. Why must their molar concentrations differ by a factor of two?

11.62. Why do red blood cells undergo hemolysis when they are placed in pure water?

Problems

11.63. Calculate the molality of each of the following solutions:
 a. 0.433 mol of sucrose ($C_{12}H_{22}O_{11}$) in 2.1 kg of water
 b. 71.5 mmol of acetic acid (CH_3COOH) in 125 g of water
 c. 0.165 mol of baking soda ($NaHCO_3$) in 375.0 g of water

*11.64. Table 4.1 lists molarities of major ions in seawater. Using a density of 1.022 g/mL for seawater, convert the concentrations into molalities.

11.65. What mass of the following solutions contains 0.100 mol of solute? (a) 0.135 m NH_4NO_3; (b) 3.92 m ethylene glycol, $HOCH_2CH_2OH$; (c) 1.07 m $CaCl_2$

11.66. How many moles of solute are there in the following solutions?
 a. 0.750 m glucose solution made by dissolving the glucose in 10.0 kg of water
 b. 0.183 m Na_2CrO_4 solution made by dissolving the Na_2CrO_4 in 900.0 g of water
 c. 1.425 m urea solution made by dissolving the urea in 750.0 g of water

11.67. **Fish Kills** High concentrations of ammonia (NH_3), nitrite ion, and nitrate ion in water can kill fish. Lethal concentrations of these species for rainbow trout are 1.1 mg/L, 0.40 mg/L, and 1361 mg/L, respectively. Express these concentrations in molality units, assuming a solution density of 1.00 g/mL.

11.68. The concentrations of six important elements in a sample of river water are 0.050 mg/kg of Al^{3+}, 0.040 mg/kg of Fe^{3+}, 13.4 mg/kg of Ca^{2+}, 5.2 mg/kg of Na^+, 1.3 mg/kg of K^+, and 3.4 mg/kg of Mg^{2+}. Express each of these concentrations in molality units.

11.69. **Cinnamon** Cinnamon owes its flavor and odor to cinnamaldehyde (C_9H_8O). Determine the boiling point elevation of a solution of 100 mg of cinnamaldehyde dissolved in 1.00 g of carbon tetrachloride ($K_b = 5.02°C/m$).

11.70. **Spearmint** Determine the boiling point elevation of a solution of 125 mg of carvone ($C_{10}H_{14}O$, oil of spearmint) dissolved in 1.50 g of carbon disulfide ($K_b = 2.34°C/m$).

11.71. What molality of a nonvolatile, nonelectrolyte solute is needed to lower the melting point of camphor by 1.000°C ($K_f = 39.7°C/m$)?

11.72. What molality of a nonvolatile, nonelectrolyte solute is needed to raise the boiling point of water by 7.60°C ($K_b = 0.52°C/m$)?

11.73. **Saccharin** Determine the melting point of an aqueous solution made by adding 186 mg of saccharin ($C_7H_5O_3NS$) to 1.00 mL of water (density = 1.00 g/mL, $K_f = 1.86°C/m$).

11.74. Determine the boiling point of an aqueous solution that is 2.50 m ethylene glycol ($HOCH_2CH_2OH$); K_b for water is 0.52°C/m. Assume that the boiling point of pure water is 100.00°C.

11.75. Which aqueous solution has the lowest freezing point: 0.5 m glucose, 0.5 m NaCl, or 0.5 m $CaCl_2$?

11.76. Which aqueous solution has the highest boiling point: 0.5 m methanol (CH_3OH), 0.5 m KI, or 0.5 m Na_2SO_4?

11.77. Which of the following aqueous solutions should have the highest boiling point: 0.0200 m ethanol (CH_3CH_2OH), 0.0125 m $LiClO_4$, or 0.0100 m $Mg(NO_3)_2$?

11.78. Which of the following aqueous solutions should have the lowest freezing point: 0.0500 m $C_6H_{12}O_6$, 0.0300 m KBr, or 0.0150 m Na_2SO_4?

11.79. Arrange the following aqueous solutions in order of increasing boiling point:
a. 0.06 m $FeCl_3$ ($i = 3.4$)
b. 0.10 m $MgCl_2$ ($i = 2.7$)
c. 0.20 m KCl ($i = 1.9$)

11.80. Arrange the following solutions in order of increasing freezing point depression:
a. 0.10 m $MgCl_2$ in water, $i = 2.7$, $K_f = 1.86°C/m$
b. 0.20 m toluene in diethyl ether, $i = 1.00$, $K_f = 1.79°C/m$
c. 0.20 m ethylene glycol in ethanol, $i = 1.00$, $K_f = 1.99°C/m$

11.81. The following pairs of aqueous solutions are separated by a semipermeable membrane. In which direction will the solvent flow?
a. A = 1.25 M NaCl; B = 1.50 M KCl
b. A = 3.45 M $CaCl_2$; B = 3.45 M NaBr
c. A = 4.68 M glucose; B = 3.00 M NaCl

11.82. The following pairs of aqueous solutions are separated by a semipermeable membrane. In which direction will the solvent flow?
a. A = 1.00 L of 0.48 M NaCl; B = 55.85 g of NaCl dissolved in 1.00 L of solution
b. A = 100 mL of 0.982 M $CaCl_2$; B = 16 g of NaCl in 100 mL of solution
c. A = 100 mL of 6.56 mM $MgSO_4$; B = 5.24 g of $MgCl_2$ in 250 mL of solution

11.83. Calculate the osmotic pressure of each of the following aqueous solutions at 20°C:
a. 2.39 M methanol (CH_3OH)
b. 9.45 mM $MgCl_2$
c. 40.0 mL of glycerol ($C_3H_8O_3$) in 250.0 mL of aqueous solution (density of glycerol = 1.265 g/mL)
d. 25 g of $CaCl_2$ in 350 mL of solution

11.84. Calculate the osmotic pressure of each of the following aqueous solutions at 27°C:
a. 10.0 g of NaCl in 1.50 L of solution
b. 10.0 mg/L of $LiNO_3$
c. 0.222 M glucose
d. 0.00764 M K_2SO_4

11.85. Determine the molarity of each of the following solutions from its osmotic pressure at 25°C. Include the van 't Hoff factor for the solution when the factor is given.
a. $\Pi = 0.674$ atm for a solution of ethanol (CH_3CH_2OH)
b. $\Pi = 0.0271$ atm for a solution of aspirin ($C_9H_8O_4$)
c. $\Pi = 0.605$ atm for a solution of $CaCl_2$, $i = 2.47$

11.86. Determine the molarity of each of the following solutions from its osmotic pressure at 25°C. Include the van 't Hoff factor for the solution when the factor is given.
a. $\Pi = 0.0259$ atm for a solution of urea [$CO(NH_2)_2$]
b. $\Pi = 1.56$ atm for a solution of sucrose ($C_{12}H_{22}O_{11}$)
c. $\Pi = 0.697$ atm for a solution of KI, $i = 1.90$

11.87. Is the following statement true or false? For solutions of the same reverse osmotic pressure at the same temperature, the molarity of a solution of NaCl will always be less than the molarity of a solution of $CaCl_2$. Explain your answer.

11.88. Suppose you have 1.00 M aqueous solutions of each of the following solutes: glucose ($C_6H_{12}O_6$), NaCl, and acetic acid (CH_3COOH). Which solution has the highest pressure requirement for reverse osmosis?

Measuring the Molar Mass of a Solute by Using Colligative Properties

Concept Review

11.89. What effect does dissolving a solute have on the following properties of a solvent? (a) its osmotic pressure; (b) its freezing point; (c) its boiling point

11.90. How can measurements of osmotic pressure, freezing point depression, and boiling point elevation be used to find the molar mass of a solute? Why are such determinations usually carried out on molecular substances as opposed to ionic ones?

Problems

11.91. **Throat Lozenges** A 188-mg sample of a nonelectrolyte isolated from throat lozenges was dissolved in enough water to make 10.0 mL of solution at 25°C. The osmotic pressure of the resulting solution was 4.89 atm. Calculate the molar mass of the compound.

*__11.92.__ **Antibiotic** An unknown compound (27.40 mg) with antibiotic properties was dissolved in water to make 100.0 mL of solution. The solution did not conduct electricity and had an osmotic pressure of 9.94 torr at 23.6°C. Elemental analysis revealed the substance to be 42.34% C, 5.92% H, and 32.93% N. Determine the molecular formula of this compound.

*11.93. **Cloves** Eugenol is one of the compounds responsible for the flavor of cloves. A 111-mg sample of eugenol was dissolved in 1.00 g of chloroform (K_b 53.63°C/m), increasing the boiling point of the chloroform by 2.45°C. Calculate eugenol's molar mass. Eugenol is 73.17% C, 7.32% H, and 19.51% O by mass. What is the molecular formula of eugenol?

*11.94. **Caffeine** The freezing point of a solution prepared by dissolving 150 mg of caffeine in 10.0 g of camphor is lower than that of pure camphor (K_f = 39.7°C/m) by 3.07°C. What is the molar mass of caffeine? Elemental analysis of caffeine yields the following results: 49.49% C, 5.15% H, 28.87% N, and the remainder O. What is the molecular formula of caffeine?

Additional Problems

11.95. Which substance has the least negative lattice energy? (a) SrI_2; (b) $CaBr_2$; (c) $CaCl_2$; (d) MgF_2

11.96. Explain why the boiling point of pure sodium chloride is much higher than the boiling point of an aqueous solution of sodium chloride.

*11.97. **Melting Ice** $CaCl_2$ is often used to melt ice on sidewalks. Could $CaCl_2$ melt ice at −20°C? Assume that the solubility of $CaCl_2$ at this temperature is 70.1 g of $CaCl_2$/100.0 g of H_2O and that the van 't Hoff factor for a saturated solution of $CaCl_2$ is 2.5.

*11.98. **Making Ice Cream** A mixture of table salt and ice is used to chill the contents of hand-operated ice-cream makers. What is the melting point of a mixture of 2.00 lb of NaCl and 12.00 lb of ice if exactly half of the ice melts? Assume that all the NaCl dissolves in the melted ice and that the van 't Hoff factor for the resulting solution is 1.44.

11.99. The freezing points of 0.0935 m ammonium chloride and 0.0378 m ammonium sulfate in water were found to be −0.322°C and −0.173°C, respectively. What are the values of the van 't Hoff factors for these salts?

11.100. The following data were collected for three compounds in aqueous solution. Determine the value of the van 't Hoff factor for each salt (K_f for water = 1.86°C/m).

Compound	Concentration	Experimentally Measured ΔT_f
LiCl	5.0 g/kg	0.410°C
HCl	5.0 g/kg	0.486°C
NaCl	5.0 g/kg	0.299°C

11.101. **Physiological Saline** One hundred milliliters of a solution of physiological saline (0.92% NaCl by mass) is diluted by the addition of 250.0 mL of water. What is the osmotic pressure of the final solution at 37°C? Assume that NaCl dissociates completely into $Na^+(aq)$ and $Cl^-(aq)$.

11.102. One hundred milliliters of 2.50 mM NaCl is mixed with 80.0 mL of 3.60 mM $MgCl_2$ at 20°C. Calculate the osmotic pressure of each starting solution and that of the mixture, assuming that the volumes are additive and that both salts dissociate completely into their component ions.

11.103. A solution of 7.50 mg of a small protein in 5.00 mL of aqueous solution has an osmotic pressure of 6.50 torr at 23.1°C. What is the molar mass of the protein?

11.104. **Kidney Dialysis** Hemodialysis, a method of removing waste products from the blood if the kidneys have failed, uses a tube made of a cellulose membrane that is immersed in a large volume of aqueous solution. Blood is pumped through the tube and is then returned to the patient's vein. The membrane does not allow passage of large protein molecules and cells but does allow small ions, urea, and water to pass through it. Assume that a physician wants to decrease the concentration of sodium ion and urea in a patient's blood while maintaining the concentration of potassium ion and chloride ion in the blood. What materials must be dissolved in the aqueous solution in which the dialysis tube is immersed? How must the concentrations of ions in the immersion fluid compare with those in blood?

11.105. **IV Solution** Another solution used clinically in the hospital setting for IV administration is Ringer's lactate, a solution of sodium, potassium, and calcium cations and chloride and lactate anions. This solution is isotonic with 0.9% saline and D5W described in Section 11.5. Write a mathematical statement that indicates the relationship between the concentrations of cations and anions in this solution compared with 0.9% saline.

11.106. **Injections** The injection of pharmaceutical solutions that are hypertonic in relation to human plasma can cause considerable pain at the site of injection. Why?

12

Solids

Crystals, Alloys, and Polymers

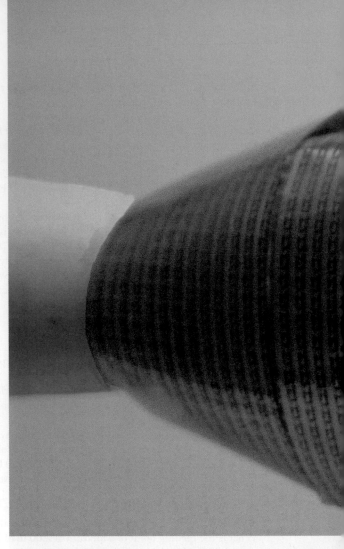

BIONIC HANDSHAKE Scientists have used shape memory alloys and synthetic polymers to create a robotic hand that is responsive to its environment, returning the correct pressure for a comfortable handshake.

PARTICULATE **REVIEW**

Ionic Compounds and Metals

In Chapter 12 we explore the structure and properties of solids. Most metals and ionic compounds are solids at room temperature, including potassium iodide and platinum, whose structures are depicted here.

- What is the role of electrons in each solid?

- Which particles experience electrostatic interactions in potassium iodide?

- Which particles experience electrostatic interactions in platinum?

(Review Section 8.1 if you need help.)

(Answers to Particulate Review questions are in the back of the book.)

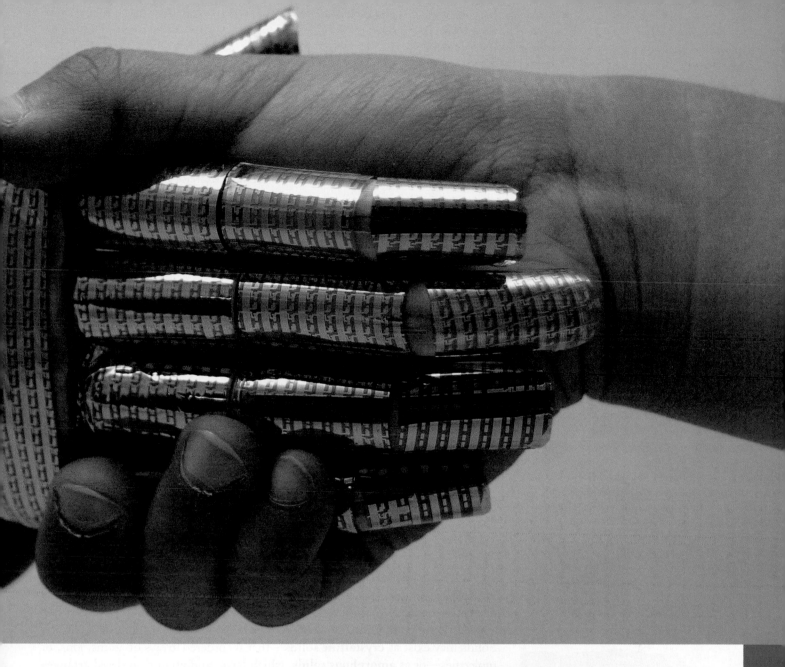

Three-Dimensional Networks: Structure and Properties

Two allotropes of carbon are depicted, both as three-dimensional networks. As you read Chapter 12, look for ideas that will help you answer these questions:

- Which allotrope has a repeating pattern that is planar? Which allotrope has a repeating pattern that is not planar?

- Which allotrope consists of carbon atoms covalently bonded to one another? Which allotrope has *both* covalent bonds *and* weak interactions between layers?

- One allotrope is used as a lubricant due to its slipperiness. Which one?

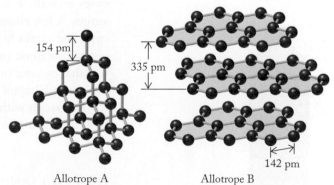

154 pm

335 pm

142 pm

Allotrope A   Allotrope B

Learning Outcomes

LO1 Identify the differences in packing schemes for atoms in the solid state, including face-centered, body-centered, and simple cubic unit cells
Sample Exercise 12.1

LO2 Describe the differences between substitutional and interstitial alloys
Sample Exercise 12.2

LO3 Describe the crystalline structures of molecular and ionic compounds
Sample Exercise 12.3

LO4 Identify monomers given the structure of a polymer, and the structure

of a polymer given the structure of a monomer
Sample Exercises 12.4, 12.6, 12.8

LO5 Analyze the properties of polymers
Sample Exercises 12.5, 12.7

12.1 The Solid State

People have placed a high value on gold for thousands of years and put gold to many different uses, including jewelry, coinage, and medicine (Figure 12.1). Today, research into solid materials is yielding applications that would have been unimaginable as recently as 50 years ago. **Nanoparticles** of gold (small particles with diameters less than 10^{-7} m) are used in medicine to treat rheumatoid arthritis and to bind drugs (using intermolecular forces) for delivery to specific target cells. Research on other solid elements has resulted in the breakthroughs responsible for miniaturized electronic devices and for the high-molar-mass organic compounds used as sutures and in knee or hip replacement surgery.

Gold is one of the few elements that exists in nature as a pure metal. Because metals are typically shiny, malleable (easily shaped), ductile (easily drawn out), and able to conduct electricity, they have many applications that take advantage of these properties. All metals except mercury exist in the solid state under standard conditions.

Many compounds, be they covalent or ionic, are solids at room temperature. Solids may exist as **crystalline solids**—that is, ordered arrays of atoms, ions, or molecules—or as **amorphous solids**, which have random or disordered arrangements of particles. There are several types of crystalline solids. Metallic elements generally crystallize as **metallic solids** consisting of ordered arrays of atoms, whereas most nonmetals crystallize as **molecular solids** consisting of neutral, covalently bonded molecules held together by intermolecular forces. The noble gases crystallize as **atomic solids** with only weak dispersion forces between the atoms. A few elements and compounds crystallize as **covalent network solids** in extended arrays held together by covalent bonds. Lastly, ionic compounds form crystalline **ionic solids** in which ions, either monatomic or polyatomic, are held together by ionic bonds. Table 12.1 summarizes the major classes of solids that we explore in Chapter 12.

Two solids with different compositions may have very different properties due to different intermolecular forces between their molecules or different bonds

(a)

(b)

(c)

FIGURE 12.1 Gold is familiar to us as (a) nuggets found naturally, (b) gold coins, and (c) tiny particles suspended in water touted as elixirs for good health. The gold in each of these forms is the same—an ordered arrangement of gold atoms—even if the colors of gold coins and small gold particles are different.

TABLE 12.1 Solid Characteristics and Classification by Particles and Electrostatic Attractions

Class	Particles	Electrostatic Attractions	Physical Properties	Examples
CRYSTALLINE SOLIDS				
Atomic	Atoms	London forces	Very low melting point, poor conductor, soft	Xe
Metallic	Atoms	Metallic bonds	Broad range of melting points, excellent conductor, malleable, ductile, soft to hard	Ag
Covalent network	Atoms	Covalent bonds	Very high melting point, poor conductor, very hard	C (diamond)
Ionic	Cations and anions	Ionic bonds	High melting point, good conductor (in molten state), hard and brittle	NaCl
Molecular	Molecules	London forces	Low melting point, poor conductor, soft	S_8 CO_2
		Dipole–dipole forces		$CHCl_3$
		Hydrogen bonds		H_2O
AMORPHOUS SOLIDS				
	Molecules	Covalent bonds	Properties vary based on composition	SiO_2 (obsidian/volcano glass)

nanoparticle an approximately spherical sample of matter with dimensions smaller than 100 nanometers (1×10^{-7} m).

crystalline solid a solid made of an ordered array of atoms, ions, or molecules.

amorphous solid a solid that lacks long-range order for the atoms, ions, or molecules in its structure.

metallic solid a solid formed by metallic bonds among atoms of metallic elements.

molecular solid a solid formed by intermolecular attractive forces among neutral, covalently bonded molecules.

atomic solid a solid formed by weak attractions between noble gas atoms.

covalent network solid a solid formed by covalent bonds among nonmetal atoms in an extended array.

ionic solid a solid formed by ionic bonds among monatomic and/or polyatomic ions.

crystal lattice a three-dimensional repeating array of particles (atoms, ions, or molecules) in a crystalline solid.

crystal structure a particular arrangement in three-dimensional space that specifies the positions of the particles (atoms, ions, or molecules) in relation to one another in a crystalline solid.

hexagonal closest-packed (hcp) a crystal lattice in which the layers of atoms or ions in hexagonal unit cells have an *ababab . . .* stacking pattern.

CONNECTION Intermolecular forces were described in Chapter 10.

CONNECTION Some physical properties of metals were described in Chapter 2.

between their atoms or ions. Molecular solids often have lower melting points than ionic solids because they are held together by relatively weak intermolecular forces compared with the strength of ionic bonds. The melting points of metallic solids cover a broad range, from gallium at 29.76°C (just above room temperature) to tungsten, which melts at 3422°C. (Mercury is the one metal that exists as a liquid at room temperature.) Because metallic solids are malleable, they can be bent into a variety of shapes. In contrast, ionic solids tend to be hard but brittle.

Mixtures of a host metal and one or more other elements are called *alloys*, and more than 500,000 alloys in the solid state have been made and characterized. Some are solid solutions, which means they are homogeneous at the atomic level. Other alloys are heterogeneous mixtures. The purpose of making alloys is to manipulate their properties by varying the proportions of their constituent metals. Alloys of nickel and titanium can be shaped into stents used to prop open arteries in heart patients. Brass, which is an alloy of copper and zinc, is used in hospitals and large kitchens because of its antibacterial properties. The transportation industry uses aluminum alloys that are ideal for making aircraft because they are strong and lighter than steel, resulting in less fuel consumption.

Large solid organic compounds with molar masses up to and exceeding 1,000,000 g/mol are called *polymers*. Like metals and alloys, different polymer compositions affect the physical properties of the material. We have seen examples of polymers in previous chapters, including polyvinyl chloride, used to make drainpipes and other building materials; and Teflon, used in nonstick cookware. Abrasion-resistant, ultrahigh-molecular-weight polyethylene is used as a coating on some artificial joints used in hip replacement. A polymer formed by the reaction of glycolic acid and lactic acid is used in dissolving sutures and to support the growth of skin cells for burn victims.

In this chapter we examine the major classes of solid materials and explore the links between their physical properties at the macroscopic level and their structures at the atomic level. Materials science is a fast-changing field, and we are able to discuss only a few of its many applications. The search for new materials and new applications is constant and constantly surprising. We start our exploration of the solid state by looking at the structures of some familiar metals.

12.2 Structures of Metals

When a metal is heated above its melting point and then slowly allowed to cool, it solidifies into a crystalline solid in which the atoms are arranged in an ordered, repeating, three-dimensional array called a **crystal lattice**. Think of a crystal lattice as stacked layers (designated *a*, *b*, *c*, . . .) of particles packed together as tightly as possible. Each atom in layer *a* touches six others in that layer, as shown in Figure 12.2(a). The atoms in layer *b* nestle into some of the spaces created by the atoms of layer *a* (Figure 12.2b), just like oranges in a fruit-stand display or cannonballs at a 16th-century fort (Figure 12.3). Similarly, the atoms in a third layer, *c*, nestle among those in layer *b*. However, two different alignments are possible for the atoms in layer *c*. They can align directly above the atoms in layer *a* (Figure 12.2c), in which case we label them as layer *a* again, or they can nestle into the atoms of layer *b* in such a way that they are not aligned directly

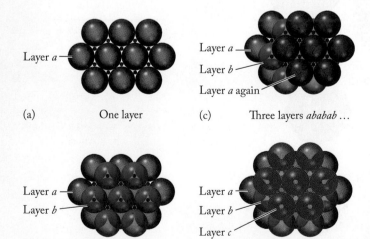

(a)   One layer

(c)   Three layers *ababab* . . .

(b)   Two layers *ab*

(d)   Three layers *abcabc* . . .

FIGURE 12.2 Two equally efficient ways to stack layers of atoms (or any particles of equal size). In both stacking patterns the atoms in all layers (shown in different colors to distinguish one layer from another) are packed as closely together as possible. Layers (a) and (b) are the same in both patterns. (c) In the *ababab . . .* pattern, atoms in the third layer are directly above the atoms in the first layer. (d) In the *abcabcabc . . .* pattern, atoms in the third layer are directly above spaces (marked by red dots) between the atoms in the first layer.

above the layer *a* atoms (Figure 12.2d). When a fourth layer is nestled into the spaces of layer *c*, the fourth layer atoms lie directly above the layer *a* atoms. In Figure 12.2(c), we have a stacking pattern *ababab* . . . throughout the crystal, and in Figure 12.2(d) we show the stacking pattern *abcabc* . . . .

Stacking Patterns and Unit Cells

Which of these two patterns do gold atoms adopt? Whether the atoms in a metal like gold are stacked in an *ababab* or an *abcabc* pattern determines the shape of the crystals the metal forms when it slowly cools and solidifies from the molten state. The **crystal structure** of an element or compound specifies the location of the atoms in three-dimensional space in relation to one another. To see how crystal structures are linked to stacking patterns, let's take a closer look at a cluster of atoms in the *ababab* . . . stacking pattern (Figure 12.4a). This cluster forms a *hexagonal* (six-sided) prism of closely packed atoms. In fact, they are as tightly packed as they can be, so the crystal lattice is called **hexagonal closest-packed (hcp)**. Titanium metal used in surgical repair of fractures is one element that crystallizes in an hcp lattice, forming hexagonal crystals (Figure 12.5). As Figure 12.6 shows, the atoms in 16 metallic elements have an hcp crystal lattice. In these metals, the cluster of atoms in Figure 12.4(a) serves

(a)　　　　　　　　　　　　　　　　(b)

FIGURE 12.3 (a) Stacks of oranges in a grocery store and (b) cannonballs at a 16th-century fort illustrate closest-packed arrays of spherical objects. Both (a) and (b) exhibit an *abcabc* . . . stacking pattern.

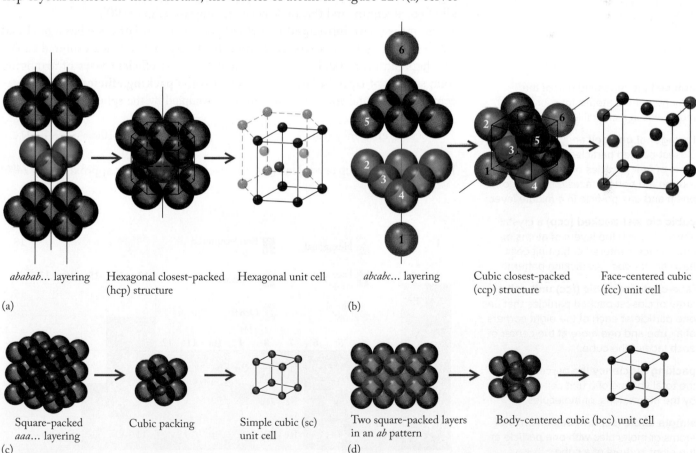

ababab... layering　　Hexagonal closest-packed (hcp) structure　　Hexagonal unit cell

(a)

abcabc... layering　　Cubic closest-packed (ccp) structure　　Face-centered cubic (fcc) unit cell

(b)

Square-packed *aaa*... layering　　Cubic packing　　Simple cubic (sc) unit cell

(c)

Two square-packed layers in an *ab* pattern　　Body-centered cubic (bcc) unit cell

(d)

FIGURE 12.4 (a) A hexagonal closest-packed (hcp) crystal structure and its hexagonal unit cell. (b) The stacking pattern *abcabc* . . . produces a face-centered cubic (fcc) unit cell. (c) In *cubic packing*, the atoms in all layers are directly above those in the *a* layer. The repeating unit of this pattern is called a *simple cubic* (sc) unit cell. (d) The atoms (blue spheres) in the second *b* layer nestle into the spaces between the square-packed atoms (red spheres) in the *a* layer. Atoms in the third layer are directly above those in the first, producing an *ababab* . . . stacking pattern and a *body-centered cubic* (bcc) unit cell.

FIGURE 12.5 Titanium crystallizes in an hcp pattern with a hexagonal unit cell. This photo shows quartz crystals coated with titanium metal.

unit cell the repeating unit of the arrangement of atoms, ions, or molecules in a crystal lattice.

hexagonal unit cell an array of closest-packed particles that includes parts of four particles on the top and four on the bottom faces of a hexagonal prism and one particle in a middle layer.

cubic closest-packed (ccp) a crystal lattice in which the layers of atoms or ions in face-centered cubic unit cells have an *abcabc* . . . stacking pattern.

face-centered cubic (fcc) unit cell an array of closest-packed particles that has one particle at each of the eight corners of a cube and one more at the center of each face of the cube.

packing efficiency the percentage of the total volume of a unit cell occupied by the atoms, ions, or molecules.

simple cubic (sc) unit cell an array of atoms or molecules with one particle at the eight corners of a cube.

body-centered cubic (bcc) unit cell an array of atoms or molecules with one particle at each of the eight corners of a cube and one at the center of the cell.

as an atomic-scale building block—a pattern of atoms repeated over and over again in all three dimensions in the metal.

We call each of these building blocks a **unit cell**. The example in Figure 12.4a shows a **hexagonal unit cell**. A unit cell represents the minimum repeating pattern that describes the three-dimensional array of atoms forming the crystal lattice of any crystalline solid, including metals. Think of unit cells as three-dimensional microscopic analogs of the two-dimensional repeating pattern in fabrics, wrapping paper, or even a checkerboard. Look carefully at Figure 12.7 to confirm that the outlined portion represents the minimum repeating pattern in the checkerboard and in the paper. A unit cell plays the same role in the crystal structure of a solid.

Atoms in solid gold and 11 other metals adopt the *abcabc* . . . stacking pattern when they solidify (Figures 12.1 and 12.6). What is the unit cell in this stacking pattern? Consider what happens when we take the cluster of atoms on the left in Figure 12.4(b), compact it, rotate the cluster, and tip it 45° to get the orientation shown in the center of Figure 12.4(b). The black outline shows that the atoms form a cube: one atom at each of the eight corners of the cube and one at the center of each of the six faces. Note that atoms at adjacent corners do not touch each other, but the three atoms along the diagonal of any face of the cube—atoms 2, 3, 4—do touch each other. Because the atoms are stacked together as closely as possible, this crystal lattice is called **cubic closest-packed (ccp)**, and the corresponding unit cell is called a **face-centered cubic (fcc) unit cell**. The unit cell edges are all of equal length, and the angle between any two edges is 90°.

So far we have introduced two *closest-packed* crystal lattices—hexagonal and cubic—along with their associated unit cells (hexagonal and face-centered cubic). The hcp and ccp crystal lattices represent the most efficient ways of arranging solid spheres of equal radius. We can express the **packing efficiency** as the percentage of the total volume of the unit cell occupied by the spheres:

$$\text{Packing efficiency (\%)} = \frac{\text{volume occupied by spheres}}{\text{volume of unit cell}} \times 100\% \quad (12.1)$$

For both hcp and ccp crystal lattices, the packing efficiency is approximately 74%.

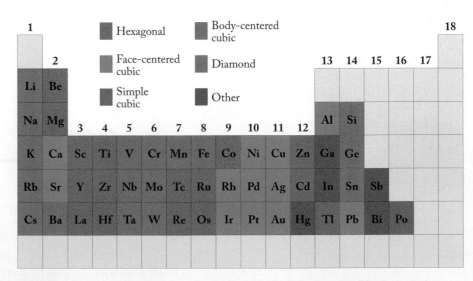

FIGURE 12.6 Unit cells of metals and metalloids in the periodic table. The five metals designated "Other" have unit cells more complicated than can easily be described in this book.

We will come back to this calculation at the end of the next subsection, but first let's examine two other packing arrangements with lower packing efficiencies.

Other crystal lattices have stacking patterns in which the atoms are arranged close together, but not as efficiently as in hcp and ccp lattices. Two of these are shown in Figure 12.4(c) and (d). The atoms are arranged in an *a* layer so that each atom touches just four adjacent atoms in that layer, an arrangement called *square packing* (Figure 12.4c). If we add a second layer of spheres directly above the first, we create the *aaa . . .* stacking pattern, which is called *cubic packing*. The three-dimensional repeating pattern of this arrangement is called a **simple cubic (sc) unit cell**. It is the least efficiently packed of the cubic unit cells and is quite rare among metals: only radioactive polonium (Po) forms a simple cubic unit cell.

For crystal lattices in which each atom in a second layer is nestled in the space created by four atoms in a square-packed *a* layer (Figure 12.4d), we have two layers in an *ab* stacking pattern. If the atoms in the third layer are directly above those in the first, then we have an *ababab . . .* stacking pattern based on layers of square-packed atoms. The simplest three-dimensional repeating unit of this pattern is called a **body-centered cubic (bcc) unit cell**. It consists of portions of nine atoms, one at each of the eight corners of a cube and one in the middle of the cube. All the group 1 metals and many transition metals have bcc unit cells (Figure 12.6). Included among the metals with a bcc unit cell is tantalum, a metal used to coat artificial joints (Figure 12.8). Table 12.2 summarizes the different stacking patterns, packing efficiencies, and unit cells described in this section.

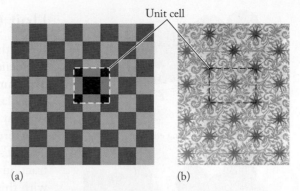

FIGURE 12.7 (a) The highlighted "unit cell" of a checkerboard is the smallest set of squares that defines the pattern repeated over the entire board. (b) This wrapping paper has a more complex pattern. One unit cell is highlighted. Can you outline another?

CONCEPT TEST

What is the difference between a crystal lattice and a unit cell?

(Answers to Concept Tests are in the back of the book.)

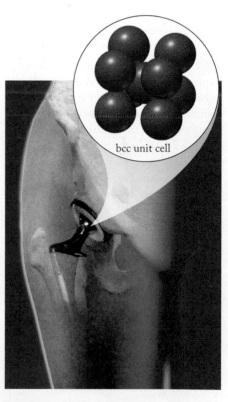

FIGURE 12.8 Tantalum metal is used to coat artificial joints such as this hip joint because tantalum metal is wear-resistant and does not react with body fluids. The crystal lattice of tantalum is body-centered cubic.

TABLE 12.2 Summary of Unit Cells, Stacking Patterns, and Packing Efficiencies for Solid Spheres

Lattice Name	Unit Cell	Type of Packing	Stacking Pattern	Number of Nearest Neighbors	Packing Efficiency
Hexagonal closest-packed (hcp)	Hexagonal	Close packing	*ababab . . .*	12	74%
Cubic closest-packed (ccp)	Face-centered cubic (fcc)	Close packing	*abcabc . . .*	12	74%
Body-centered cubic packing	Body-centered cubic (bcc)	Square packing	*ababab . . .*	8	68%
Cubic packing	Simple cubic (sc)	Square packing	*aaa . . .*	6	52%

(a) Simple cubic:
Atoms touch along edge

(b) Face-centered cubic:
Atoms touch along face diagonal

(c) Body-centered cubic:
Atoms touch along body diagonal

FIGURE 12.9 Whole-atom and cutaway views of cubic unit cells. (a) In a simple cubic unit cell, each corner atom of the unit cell is part of eight unit cells. Atoms along each edge touch with an edge length ℓ. (b) In a face-centered cubic unit cell, the face atoms are part of two unit cells. Atoms along the face diagonal touch. (c) In a body-centered cubic unit cell, one atom in the center lies entirely in one unit cell. The atoms along the body diagonal touch.

Unit Cell Dimensions

Of all the metallic elements, only Li, Na, and K have densities less than 1 g/cm³. The densities of the transition metals range from 3 g/cm³ (Sc) to 22.5 g/cm³ (Os). The number of atoms in the unit cell of a metallic element, its atomic radius, and its molar mass all contribute to its density. Figure 12.9 shows whole-atom and cutaway views of sc, fcc, and bcc unit cells. These views also provide us with a way to determine how many atoms are in each type of cubic unit cell.

Let's start with the simple cubic unit cell (Figure 12.9a). Note how only a fraction of each corner atom is inside the unit cell boundary. In a crystal lattice with this unit cell, each atom is a corner atom in eight unit cells (Figure 12.10a). Thus each atom contributes the equivalent of one-eighth of an atom to the unit cell. A cube has eight corners, so there is a total of

$$\tfrac{1}{8} \text{ corner atom/} \cancel{\text{corner}} \times 8 \ \cancel{\text{corners}}\text{/unit cell} = 1 \text{ corner atom/unit cell}$$

This calculation applies to the corner atoms in any type of cubic unit cell. Note that the two corner atoms along each edge in Figure 12.9(a) touch each other. Therefore the edge length ℓ in the simple cubic unit cell is equal to twice the atomic radius:

$$\ell = 2r$$

So, if we can measure the length of the unit cell (ℓ), we can easily calculate the atomic radius of an element with this unit cell.

An fcc unit cell (Figure 12.9b) has eight corner atoms and one atom in the center of each of the six faces (Figure 12.10b). Each face atom is shared by the two unit cells that abut each other at that face. Therefore each unit cell "owns" half of each face atom, making a total of

$$\tfrac{1}{2} \text{ face atom/} \cancel{\text{face}} \times 6 \ \cancel{\text{faces}}\text{/unit cell} = 3 \text{ face atoms/unit cell}$$

As just noted, every cubic unit cell owns the equivalent of one corner atom. Therefore an fcc unit cell consists of

$$1 \text{ corner atom} + 3 \text{ face atoms} = 4 \text{ atoms per fcc unit cell}$$

To relate the size of these atoms to the dimensions of the fcc unit cell, note in the cutaway view of Figure 12.9(b) that the corner atoms do not touch one another, but adjacent atoms along the face diagonal do touch each other. Therefore a face diagonal spans the radius r of two corner atoms and the diameter (2 radii = $2r$) of a face atom, so the length of a face diagonal is $1 + 2 + 1 = 4$ atomic radii = $4r$. A face diagonal connects the ends of two edges and forms a right triangle with

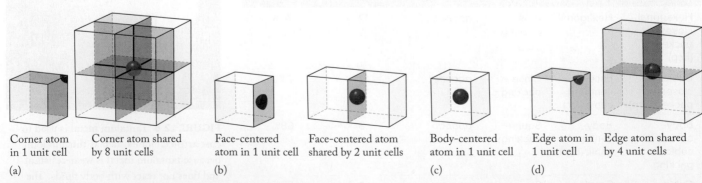

Corner atom in 1 unit cell	Corner atom shared by 8 unit cells	Face-centered atom in 1 unit cell	Face-centered atom shared by 2 unit cells	Body-centered atom in 1 unit cell	Edge atom in 1 unit cell	Edge atom shared by 4 unit cells
(a)		(b)		(c)	(d)	

FIGURE 12.10 Crystal lattices illustrating (a) corner atoms shared by eight unit cells, (b) face atoms shared by two unit cells, (c) center atoms entirely in one unit cell, and (d) edge atoms shared by four unit cells.

TABLE 12.3 Contributions of Atoms to Cubic Unit Cells

Atom Position	Contribution to Unit Cell	Unit Cell Type
Center	1 atom	bcc
Face	$\frac{1}{2}$ atom	fcc
Corner	$\frac{1}{8}$ atom	bcc, fcc, sc

TABLE 12.4 Summary of Unit Cells, Equivalent Atoms, and the Relationship between Radius of Atoms and Edge Length of Cubic Unit Cells

Unit Cell	Number of Equivalent Atoms per Unit Cell	Relationship between r and ℓ
Simple cubic	1	$r = \dfrac{\ell}{2} = 0.5\ell$
Body-centered cubic	2	$r = \dfrac{\ell\sqrt{3}}{4} = 0.4330\ell$
Face-centered cubic	4	$r = \dfrac{\ell\sqrt{2}}{4} = 0.3536\ell$

those two edges, each of length ℓ (Figure 12.9b). Therefore, according to the Pythagorean theorem, the length of a face diagonal is

$$\text{Face diagonal} = 4r = \sqrt{\ell^2 + \ell^2} = \sqrt{2\ell^2} = \ell\sqrt{2}$$

$$r = \frac{\ell\sqrt{2}}{4} = 0.3536\ell \qquad (12.2)$$

As in the case of the simple cubic unit cell, if we know the value of ℓ, we can calculate r by using Equation 12.2.

Now let's focus on the bcc unit cell in Figure 12.9(c). In addition to the one atom from one-eighth of an atom at each of the eight corners, a bcc cell also has one atom in the center of the cell that is entirely within the cell (Figure 12.10c). This means a bcc unit cell consists of

1 corner atom + 1 center atom = 2 atoms per bcc unit cell

CHEMTOUR
Unit Cell

Relating unit cell edge length to atomic radius in a bcc cell is complicated by the fact that, in addition to not touching along the edges, adjacent atoms along any face diagonal do not touch each other. However, each corner atom does touch the atom in the center of the cell, which means that the atoms touch along a *body diagonal*, which runs between opposite corners through the center of the cube. In the cutaway view in Figure 12.9(c), the body diagonal runs from the bottom left corner of the front face to the top right corner of the rear face. It spans (1) the radius of the front-face bottom left corner atom, (2) the diameter (2 radii) of the central atom, and (3) the radius of the rear-face top right atom, making the length of the body diagonal equivalent to $4r$.

We can again use the Pythagorean theorem to determine the relationship between ℓ and r. A right triangle is formed by an edge, a face diagonal, and a body diagonal serving as the hypotenuse of the triangle (Figure 12.9c). Using the face-diagonal value $\ell\sqrt{2}$, we get

$$\text{Body diagonal} = 4r = \sqrt{(\text{edge length})^2 + (\text{face diagonal})^2}$$
$$= \sqrt{\ell^2 + (\ell\sqrt{2})^2} = \sqrt{\ell^2 + \ell^2(2)} = \sqrt{3\ell^2} = \ell\sqrt{3}$$

so

$$r = \frac{\ell\sqrt{3}}{4} = 0.4330\ell \qquad (12.3)$$

Table 12.3 summarizes how atoms in different locations in sc, bcc, and fcc unit cells contribute to the total number of atoms in each unit cell. Table 12.4

summarizes the number of equivalent atoms and the relationship between r and ℓ for the three cubic unit cells.

SAMPLE EXERCISE 12.1   Calculating Atomic Radius and Density   **LO1**
from Unit Cell Dimensions

Tantalum crystallizes with a bcc unit cell as shown in Figure 12.8. Its unit cell has an edge length of 330.3 pm. (a) Calculate the radius in picometers of the tantalum atoms. Check your answer against the data in Appendix 3. (b) Calculate the density of tantalum in grams per cubic centimeter at 25°C.

Collect and Organize We are given the unit cell (bcc) and edge length (ℓ = 330.3 pm) of tantalum. Together with the atomic mass of tantalum (180.95 g/mol), we can calculate both the atomic radius of a Ta atom and the density of Ta metal.

Analyze The tantalum atoms do not touch along the unit cell edges or along any face diagonal, but they do touch along the body diagonals. Table 12.4 gives the relationship between r and ℓ for a bcc unit cell. We know that the radius is a little less than half the edge length ($\ell/2$), so we expect a value less than 165 pm for r.

We assume that the density of the unit cell is the same as the density of solid Ta. The density of the Ta bcc unit cell is the mass of two Ta atoms divided by the volume of the cell. We can calculate the mass of two Ta atoms from the molar mass, which is 180.95 g/mol. The conversion includes dividing by Avogadro's number to calculate the mass of each Ta atom in grams. The formula for the volume of a cube of edge length ℓ is $V = \ell^3$. A piece of tantalum sinks when immersed in water, so we predict that the density of tantalum should be greater than 1.0 g/cm^3.

Solve

a. Substituting the edge length into the formula for r from Table 12.4:

$$r = 0.4330 \times 330.3 \text{ pm} = 143.0 \text{ pm}$$

This is close to the reference value of 146 pm for Ta (Table A3.1).

b. We first calculate the mass m of two Ta atoms:

$$m = \frac{180.95 \text{ g Ta}}{1 \text{ mol Ta}} \times \frac{1 \text{ mol Ta}}{6.022 \times 10^{23} \text{ atoms Ta}} \times 2 \text{ atoms Ta} = 6.010 \times 10^{-22} \text{ g Ta}$$

The volume of the cell in cubic centimeters is

$$V = \ell^3 = (330.3 \text{ pm})^3 \times \frac{(10^{-10} \text{ cm})^3}{1 \text{ pm}^3} = 3.604 \times 10^{-23} \text{ cm}^3$$

The density is

$$d = \frac{m}{V} = \frac{6.010 \times 10^{-22} \text{ g}}{3.604 \times 10^{-23} \text{ cm}^3} = 16.68 \text{ g/cm}^3$$

Think About It The value of r is indeed less than half the value of the edge length, as predicted. According to the data in Table A3.2, the density of tantalum is 16.65 g/mL, so the result of that calculation is also reasonable and reflects the prediction we made. The density of the unit cell should equal the density of a bulk sample because the sample is composed of a crystal lattice of bcc unit cells.

Practice Exercise Silver and gold both crystallize in face-centered cubic unit cells with edge lengths of 407.7 and 407.0 pm, respectively. Calculate the atomic radius and density of each metal, and compare your answers with the data listed in Appendix 3.

(Answers to Practice Exercises are in the back of the book.)

(a)   (b)

Earlier in this section we mentioned that ccp and hcp are the most efficient packing schemes for spheres. Now that we know more about the fcc unit cell, let's take another look at the calculation of packing efficiency by using Equation 12.1. The unit cell for a ccp arrangement of solid spheres of radius r is an fcc unit cell that contains four equivalent spheres. The volume occupied by these spheres is

$$V_{spheres} = 4\left(\tfrac{4}{3}\pi r^3\right) = \tfrac{16}{3}\pi r^3$$

From Table 12.4, $r = (\ell\sqrt{2})/4$, and the unit cell edge (ℓ) is

$$\ell = \frac{4r}{\sqrt{2}} = 2.828r$$

This equation enables us to express the volume of the unit cell in terms of r:

$$V_{unit\ cell} = \ell^3 = (2.828r)^3 = 22.62r^3$$

Substituting $V_{spheres}$ and $V_{unit\ cell}$ into Equation 12.1:

$$\text{Packing efficiency (\%)} = \frac{V_{spheres}}{V_{unit\ cell}} \times 100\% = \frac{\tfrac{16}{3}\pi r^3}{22.62r^3} \times 100\% = 74.1\%$$

This means that the most efficient packing of spheres results in about 26% of the total volume being empty.

12.3 Alloys and Medicine

The antibacterial properties of copper metal are attractive for coating surfaces in hospitals and in food service kitchens where an infection can prove deadly (Figure 12.11). However, pure copper has two disadvantages: it is both relatively soft and very malleable, which means that pure copper objects are easily bent and damaged. We can explain the malleability of Au, Cu, and other metals in terms of the relatively weak bonds between the atoms in their cubic closest-packed crystal structure. This arrangement gives the atoms in one layer the ability, under stress, to slip past atoms in an adjacent layer (Figure 12.12), but the overall crystal structure is still cubic closest-packed. The ease with which copper atoms slip past each other makes it easy to bend copper pipes used in plumbing, but it also makes them susceptible to damage. Additionally, copper reacts with air to produce blue-green copper hydroxides and carbonates.

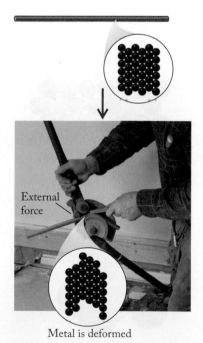

External force

Metal is deformed

FIGURE 12.12 Copper and other metals are malleable because their atoms are stacked in layers that can slip past each other under stress. Slippage is possible because of the diffuse nature of metallic bonds and the relatively weak interactions between pairs of atoms in adjoining layers.

alloy a blend of a host metal and one or more other elements, which may or may not be metals, that are added to change the properties of the host metal.

substitutional alloy an alloy in which atoms of the nonhost metal replace host atoms in the crystal lattice.

interstitial alloy an alloy in which atoms of the added element occupy the spaces between atoms of the host.

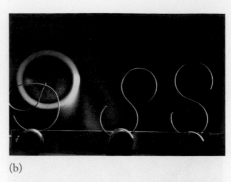

(a)　　　　(b)

FIGURE 12.13 Shape memory alloys are used in stents for heart patients. This figure illustrates how shape memory alloys work. (a) The S shape on the right is the desired final shape of the wire. The left and middle wires have been coiled to take up less space. The white circle in the background is a hair dryer heating the wires. (b) When the wire in the middle is heated, it returns to its original shape, while the wire on the left is in the process of unraveling.

CONNECTION The classification of homogeneous and heterogeneous mixtures and the difference between compounds and mixtures were described in Chapter 1.

(a)

(b)

FIGURE 12.14 Two atomic-scale views of one type of bronze, a substitutional alloy. (a) A layer of close-packed copper (Cu) atoms interspersed with a few atoms of tin (Sn). (b) One possible unit cell for bronze. In this case tin atoms have replaced one corner Cu atom and one face Cu atom.

The physical and chemical properties of copper change when mixed with zinc or tin to produce mixtures known as brass and bronze, respectively. Brass and bronze are stronger than pure copper and have improved resistance to reactions with air or moisture while retaining the antibacterial properties of pure copper. A mixture of two metals represents an **alloy**: a metallic material made when a host metal is blended with one or more other elements, which may or may not be metals, thereby changing the properties of the host metal. Like the mixtures discussed in Chapter 1, alloys can be classified according to their composition as homogeneous or heterogeneous mixtures. Brass and bronze are both *homogeneous alloys*, solid solutions in which the atoms of the added elements (in this case tin or zinc) are randomly but uniformly distributed among the atoms of the host (copper). *Heterogeneous alloys* consist of matrices of atoms of host metals interspersed with small "islands" made up of individual atoms of other elements. In both cases, the compositions may vary over a limited range.

Every year thousands of patients suffering from cardiovascular disease undergo a procedure called balloon angioplasty to open clogged arteries. This surgery involves the insertion of small metal supports, or stents, to prop open the artery. Increasingly, stents are manufactured from an alloy containing nickel and titanium. The alloy NiTi (nitinol) is an example of an *intermetallic compound*, a substance that has a reproducible stoichiometry and constant composition (just like chemical compounds). Intermetallic compounds are still commonly referred to as alloys and are considered to be a subgroup within homogeneous alloys. What makes nickel–titanium alloys particularly useful is that they demonstrate shape memory. Figure 12.13 illustrates how memory alloys work. A piece of NiTi wire is heated and formed into the desired shape for the stent. On cooling, the NiTi undergoes a phase change to a different structure. The structural change allows the stent to be easily straightened for insertion into an artery. After insertion, the coiled stent is heated, and it springs back into its original shape as it rapidly reverts to the high-temperature form.

Substitutional Alloys

If alloys are mixtures of metals, and pure metals have crystal lattices as described in Section 12.2, how are the metal atoms arranged in the crystal lattices and unit cells of alloys? How does the structure of NiTi memory metal change as a function of temperature? The answer to these questions gives rise to another classification system: a **substitutional alloy** is one in which atoms of the nonhost metal replace host atoms in the crystal lattice. Bronze and nitinol are both examples of *homogeneous, substitutional alloys*. Substitutional alloys may form between metals that have the same crystal lattice and atomic radii that are within about 15% of each other.

FIGURE 12.15 The larger Sn atoms in bronze disturb the Cu crystal lattice, producing atomic-scale bumps in the slip plane (wavy line) between layers of Cu atoms. These bumps make it more difficult for Cu atoms to slide past each other when an external force is applied.

CONCEPT TEST

Is it accurate to call bronze a *solution* of tin *dissolved* in copper? Explain why or why not.

Why is an alloy more difficult to bend than a pure metal? Figure 12.14 illustrates one layer of the crystal lattice of bronze. The radii of copper and tin atoms are similar—128 pm and 140 pm, respectively. Inserting the slightly larger Sn atoms in the cubic closest-packed Cu crystal lattice disturbs the structure a little, making the planes of copper atoms "bumpy" instead of uniform (Figure 12.15), and making it more difficult for the copper atoms to slip past one another. Less slippage makes bronze less malleable than copper, but being less malleable also means that bronze is harder and stronger.

Alloys with good strength and corrosion resistance are essential for the manufacture of surgical tools and biomedical implants. One such alloy is known as stainless steel, and its high-grade medical version is considered "surgical steel." Steel is an alloy of iron, and stainless steel owes its useful properties to chromium; surgical steel contains 13–16.5% chromium and 0.4–0.6% carbon by mass. When atoms of Cr on the surface of a piece of stainless steel combine with oxygen, they form a layer of Cr_2O_3 that tightly bonds to the surface and protects the metallic material beneath from further oxidation. This resistance to surface discoloration due to corrosion means that the surfaces of these alloys "stain less" than a surface of pure iron.

FIGURE 12.16 Carbon steel is an interstitial alloy of carbon in iron. The fcc form of iron (austenite) that forms at high temperatures can accommodate carbon atoms in its octahedral holes.

Interstitial Alloys

The major difference between iron and any kind of steel is the presence of carbon in steel. Molten iron produced in a blast furnace to which hot carbon has been added may contain up to 5% carbon. When the molten iron cools to its melting point of 1538°C, it crystallizes in a body-centered cubic structure before undergoing a phase transition at around 1390°C to austenite, a form of solid iron made up of face-centered cubic unit cells. The spaces, or *holes*, between iron atoms in austenite can accommodate carbon atoms, forming an **interstitial alloy**, so named because the carbon atoms occupy spaces, or *interstices*, between the iron atoms (Figure 12.16).

FIGURE 12.17 Close-packed atoms in adjacent layers of a crystal lattice produce octahedral holes surrounded by six host atoms and tetrahedral holes surrounded by four host atoms. Octahedral holes are larger than tetrahedral holes and so can accommodate larger nonhost atoms in interstitial alloys. Note that all the atoms are identical here; the colors are only to distinguish one layer of atoms from another.

CONCEPT TEST

Is a substitutional alloy a homogeneous or a heterogeneous alloy, or could it be both? Is an interstitial alloy a homogeneous or a heterogeneous alloy, or could it be both?

Not all interstices in a crystal lattice are equivalent, however. Indeed, holes of two different sizes occur between the atoms in any closest-packed crystal lattice (Figure 12.17). The larger holes are surrounded by clusters of six host atoms in the shape of an octahedron and are called *octahedral holes*. The

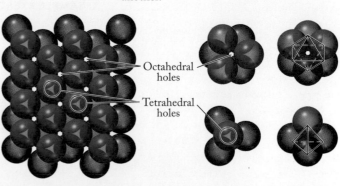

Octahedral holes

Tetrahedral holes

TABLE 12.5 Atomic Radius Ratios and Location of Nonhost Atoms in Unit Cells of Interstitial Alloys

Lattice Name	Hole Type	Atomic Radius Ratio $r_{nonhost}/r_{host}$[a]
hcp or ccp	Tetrahedral	0.22–0.41
hcp or ccp	Octahedral	0.41–0.73
Cubic packing	Cubic	0.73–1.00

[a]Radius ratios as predictors of the crystal lattice in crystalline solids are of limited value because atoms are not truly solid spheres with a constant radius; given that the radius of an atom may differ in different compounds, these ranges are approximate.

smaller holes are located between clusters of four host atoms and are called *tetrahedral holes*. The data in Table 12.5 show which holes are more likely to be occupied based on the relative sizes of nonhost and host atoms. According to Appendix 3, the atomic radii of C and Fe are 77 and 126 pm, respectively. According to Table 12.5, the ratio 77/126 = 0.61 means that C atoms should fit in the octahedral holes of austenite, as shown in Figure 12.17, but not in the tetrahedral holes.

As austenite with its fcc unit cell cools to room temperature, it changes into the crystalline solid form of iron called ferrite, which is body-centered cubic. The octahedral holes in ferrite are smaller than those in austenite, and they are too small to accommodate carbon atoms. As a result, much of the carbon precipitates as clusters of carbon atoms, or it may react with iron to form iron carbide, Fe_3C. The clusters of carbon and Fe_3C disrupt ferrite's body-centered cubic lattice and inhibit the host iron atoms from slipping past each other when a stress is applied. This resistance to slippage, which is much like that experienced by the copper atoms in bronze (Figure 12.16), makes steel much harder and stronger than pure iron. In general, the greater the carbon concentration, the stronger the steel. However, there is a trade-off in this relationship; increased strength and hardness come at the cost of increased brittleness (Table 12.6).

TABLE 12.6 Effect of Carbon Content on the Properties of Steel

Carbon Content (%)	Designation	Properties	Used to Make
0.05–0.19	Low carbon	Malleable, ductile	Nails, cables
0.20–0.49	Medium carbon	High strength	Construction girders
0.5–3.0	High carbon	Hard but brittle	Cutting tools

SAMPLE EXERCISE 12.2 Predicting the Crystal Structure of a Two-Element Alloy **LO2**

Gold nanoparticles containing 10–50% platinum show enhanced antibiotic properties in comparison with pure gold or pure platinum nanoparticles. Both elements form cubic closest-packed crystal lattices with face-centered cubic unit cells. Are these gold–platinum alloys substitutional or interstitial alloys?

Collect, Organize, and Analyze Atoms of two or more elements form a substitutional alloy when all the atoms are of similar size (within 15%). The highest $r_{nonhost}/r_{host}$ ratio in Table 12.5 for a ccp lattice, 0.73, means that an interstitial alloy can form only when the radius of the nonhost atoms is less than 73% of the radius of the host atoms. The atomic radii given in Appendix 3 are 144 pm for Au and 139 pm for Pt.

Solve The ratio of the atomic radii of Pt to Au is

$$r_{\text{nonhost}}/r_{\text{host}} = \frac{139 \text{ pm}}{144 \text{ pm}} = 0.965$$

The $r_{\text{nonhost}}/r_{\text{host}}$ ratio indicates that the platinum atoms are too big to fit into interstices in the lattice of gold atoms, regardless of whether the holes are tetrahedral or octahedral, so an interstitial alloy is impossible. Atoms of Pt can substitute for atoms of Au in the Au ccp lattice, however, with some room to spare. Figure 12.6 tells us that both elements form fcc unit cells, so little disruption to the Au lattice should result from incorporating Pt atoms. Thus gold and platinum form a substitutional alloy.

Think About It The guidelines for two metals forming a substitutional alloy are that they have the same type of crystal lattice and that their atomic radii are within 15% of each other. The radii of Au and Pt differ by only 3.5%, so we would expect gold and platinum to form a substitutional alloy.

Practice Exercise The potential for allergic reactions to nickel in nitinol shape memory alloys has spurred the exploration of titanium and niobium alloys. Would you expect niobium (atomic radius 146 pm) to form a substitutional alloy with titanium (atomic radius 147 pm)? With molybdenum (atomic radius 139 pm)?

12.4 Ionic Solids and Salt Crystals

Most of Earth's crust is composed of ionic solids, which consist of monatomic or polyatomic ions held together by ionic bonds. Most of these solids are crystalline. The simplest crystal structures are those of binary salts, such as NaCl (Figure 12.18). The cubic shape of large NaCl crystals is due to the cubic shape of the NaCl unit cell. We can describe the unit cell of NaCl (Figure 12.18b,c) as a face-centered cubic arrangement of Cl^- ions at the corners and in the center of each face, with the smaller Na^+ ions occupying the 12 octahedral holes along the edges of the unit cell and the single octahedral hole in the middle of the cell. The ordered arrangement in Figure 12.18 maximizes the coulombic forces of attraction between oppositely charged particles and minimizes the repulsions between similarly charged ions. As a result, the potential energy of the crystal structure is minimized.

In Section 12.3 we used atomic radius ratios to determine the location of atoms in a crystal lattice. The same ratios that applied to interstitial alloys guide us in understanding the structures of binary ionic compounds. Rather than examine the ratio of nonhost radii to host radii, however, we consider the ratio of the smaller ion to the larger ion. For example, in the unit cell of NaCl in Figure 12.18(b), the smaller Na^+ cations occupy the interstices in the cubic closest-packed lattice of Cl^- anions. The Na^+ ions fit better into the octahedral holes because the radius ratio of Na^+ to Cl^- is

$$\frac{r_{\text{cation}}}{r_{\text{anion}}} = \frac{102 \text{ pm}}{181 \text{ pm}} = 0.564$$

The radius ratio 0.564 is too large for Na^+ to occupy a tetrahedral hole but well within the range for occupying an octahedral hole.

CONNECTION Coulombic forces and potential energy in ionic compounds were discussed in Section 8.1.

FIGURE 12.18 (a) Sodium chloride forms cubic crystals that grow to various sizes. (b) A NaCl crystal lattice is based on cubic closest packing: an fcc unit cell made of Cl^- ions, showing the Na^+ ions in the octahedral holes. (c) Cutaway view of the unit cell in part (b).

(a)

Na$^+$

Cl$^-$

(b)

(c)

CONNECTION Periodic trends in atomic radii were discussed in Chapter 7, and the sizes of common cations and anions in comparison with their neutral parent atoms is shown in Figure 11.1.

Let's take an inventory of the ions in a unit cell of NaCl. Like the metal atoms in the fcc unit cell in Figure 12.9(b), an isolated unit cell contains portions of 14 Cl^- ions: one at each corner and one in each of the six faces of the cube. As in the analysis of Figure 12.9, accounting for partial ions gives us a total of four Cl^- ions in the unit cell in Figure 12.18(c). To count the Na^+ ions, note that one Na^+ ion occupies the central octahedral hole, and one Na^+ ion fits into each of the 12 octahedral holes along the edges of the cell. Because each Na^+ ion along an edge is shared by four unit cells, only one-fourth of each Na^+ ion on an edge is in each cell (see Figure 12.10d). Only the Na^+ in the center belongs completely to the unit cell. Therefore the total number of Na^+ ions in the unit cell is

$$(12 \times \tfrac{1}{4}) + 1 = 4 \, Na^+ \text{ ions}$$

The ratio of Na^+ to Cl^- ions in the unit cell is 4:4, consistent with the chemical formula NaCl. Because the four Na^+ ions occupy all the octahedral holes in the unit cell, the result of this calculation also means that each fcc unit cell contains the equivalent of four octahedral holes.

Note in Figure 12.18 that adjacent Cl^- ions along any face diagonal do not touch each other the way they do in Figure 12.9(b) because the Cl^- ions have to spread out a little to accommodate the Na^+ ions in the octahedral holes. Sodium ions and chloride ions touch along each edge of the unit cell, however, which means that each Na^+ ion touches six Cl^- ions and each Cl^- ion touches six Na^+ ions. This arrangement of positive and negative ions is common enough among binary ionic compounds to be assigned its own name: the *rock salt structure*.

In other binary ionic solids, the smaller ion is small enough to fit into the tetrahedral holes formed by the larger ions. For example, in the unit cell of the mineral sphalerite (zinc sulfide), the S^{2-} anions (ionic radius 184 pm) are arranged in an fcc unit cell (Figure 12.19), and half of the eight tetrahedral holes inside the cell are occupied by Zn^{2+} cations (74 pm). Therefore the unit cell contains four Zn^{2+} ions that balance the charges on the four S^{2-} ions. This pattern of half-filled tetrahedral holes in an fcc unit cell is sometimes called the *sphalerite structure*. Compounds containing zinc, cadmium, or mercury cations with a 2+ charge and heavier anions of the group 16 elements with a 2− charge, (S^{2-}, Se^{2-}, and Te^{2-}) also have a sphalerite structure.

The crystal structure of the mineral fluorite (CaF_2) is based on an fcc unit cell of smaller Ca^{2+} ions at the eight cube corners and six face centers, with all eight tetrahedral holes formed by neighboring Ca^{2+} ions filled by larger F^- ions. Because the unit cell has a total of four Ca^{2+} ions and the eight F^- ions are all completely inside the cell, this arrangement satisfies the 1:2 mole ratio of Ca^{2+} ions to F^- ions. This structure is so common that it too has its own name: the *fluorite structure* (Figure 12.20). Other compounds having this structure are SrF_2, $BaCl_2$, and PbF_2.

Some compounds in which the cation-to-anion mole ratio is 2:1 have an *antifluorite structure*. In the crystal lattices of these compounds, which include Li_2O and K_2S, the smaller cations occupy the tetrahedral holes in an fcc unit cell formed by cubic closest packing of the larger anions.

FIGURE 12.19 Many crystals of the mineral sphalerite (ZnS), like the largest ones in this photograph, have a tetrahedral shape. The crystal lattice of sphalerite is based on an fcc unit cell of S^{2-} ions with Zn^{2+} ions in four of the eight tetrahedral holes. In the expanded view of the sphalerite unit cell, each Zn^{2+} ion is in a tetrahedral hole formed by one corner S^{2-} ion and three face-centered S^{2-} ions.

SAMPLE EXERCISE 12.3 Calculating an Ionic Radius and Density   **LO3**
from a Unit Cell Dimension

The unit cell of lithium chloride (LiCl) contains an fcc arrangement of Cl^- ions
(Figure 12.21). In LiCl, the Li^+ cations (radius 76 pm) are small enough to allow
adjacent Cl^- ions to touch along any face diagonal.

a. If the edge length of the LiCl fcc cell is 513 pm, what is the radius of the
Cl^- ion?
b. Use that value to predict the type of hole the Li^+ ion occupies.
c. What is the density of LiCl at 25°C?

Collect and Organize We are given the edge length (513 pm), the radius of the Li^+
cation (76 pm), and the type of unit cell for LiCl (fcc). We are asked to calculate the
ionic radius of the Cl^- ion, predict in which type of hole the Li^+ is found, and calculate
the density of LiCl.

Analyze (a) The unit cell is an fcc array of Cl^- ions that touch along the face diagonal.
Equation 12.2 relates the edge length (ℓ) of an fcc cell to the radius (r) of the atoms or
ions that touch along its diagonals: $r = 0.3536\ell$. (b) Once we know the radius of the
Cl^- ion, the radius ratio of Li^+ to Cl^- ions and Table 12.5 allow us to predict which
type of hole Li^+ occupies. (c) The fact that LiCl has an fcc unit cell means that there are
four Cl^- ions and four Li^+ ions in the cell. The molar masses of these ions are 35.45 and
6.941 g/mol, respectively. As in Sample Exercise 12.1, we need to convert from molar
masses to the masses of individual particles by dividing by Avogadro's number. Density
is the ratio of mass to volume, and the volume of a cubic cell is the cube of its edge
length: $V = \ell^3$.

Solve
a. Substituting the edge length into Equation 12.2, we calculate the radius of Cl^-:

$$r = 0.3536 \times 513 \text{ pm} = 181 \text{ pm}$$

b. The ratio of the radius of a Li^+ ion to the radius of a Cl^- ion is

$$76 \text{ pm}/181 \text{ pm} = 0.42$$

According to Table 12.5, the Li^+ cations occupy octahedral holes in the lattice
formed by the larger Cl^- anions.
c. Calculating the mass of four Cl^- ions:

$$m = \frac{35.45 \text{ g Cl}^-}{1 \text{ mol Cl}^-} \times \frac{1 \text{ mol Cl}^-}{6.022 \times 10^{23} \text{ ions Cl}^-} \times 4 \text{ ions Cl}^-$$
$$= 2.355 \times 10^{-22} \text{ g Cl}^-$$

The mass of four Li^+ ions is

$$m = \frac{6.941 \text{ g Li}^+}{1 \text{ mol Li}^+} \times \frac{1 \text{ mol Li}^+}{6.022 \times 10^{23} \text{ ions Li}^+} \times 4 \text{ ions Li}^+$$
$$= 0.4610 \times 10^{-22} \text{ g Li}^+$$

Combining the masses of the two kinds of ions in the unit cell:

$$2.355 \times 10^{-22} \text{ g} + 0.4610 \times 10^{-22} \text{ g} = 2.816 \times 10^{-22} \text{ g}$$

The volume of the cell in cubic centimeters is

$$V = \ell^3 = (513 \text{ pm})^3 \times \frac{(10^{-10} \text{ cm})^3}{(1 \text{ pm})^3} = 1.35 \times 10^{-22} \text{ cm}^3$$

Taking the ratio of mass to volume, we have

$$d = \frac{m}{V} = \frac{2.816 \times 10^{-22} \text{ g}}{1.35 \times 10^{-22} \text{ cm}^3} = 2.09 \text{ g/cm}^3$$

FIGURE 12.20 The mineral fluorite (CaF_2)
forms cubic crystals. The crystal lattice
of CaF_2 is based on an fcc array of Ca^{2+}
ions, with F^- ions occupying all eight
tetrahedral holes. Because they are bigger
than Ca^{2+} ions, the F^- ions do not fit in the
tetrahedral holes of a cubic closest-packed
array of Ca^{2+} ions. Instead the Ca^{2+}
ions, while maintaining the fcc unit cell
arrangement, spread out to accommodate
the larger F^- ions. Note how adjacent Ca^{2+}
ions along any face diagonal do not touch
each other the way they do in the ideal fcc
unit cell in Figure 12.9(b).

FIGURE 12.21 The face-centered cubic
unit cell of LiCl.

12.5 Allotropes of Carbon

CONNECTION Allotropes are structurally different forms of the same physical state of an element, as explained in Chapter 8.

CONNECTION Hybridization and σ bonds are described in Section 9.4.

In Figure 12.6 the color key identifies the unit cells of silicon, germanium, and tin as "diamond." Carbon, of course, is a nonmetal, but the specific arrangement of carbon atoms in diamond (Figure 12.22a) is also found in some metallic materials. In addition, some diamonds may include metal impurities in the crystal lattice that give rise to distinctive colors.

Diamond is one of three allotropes of carbon, the other two being graphite and fullerenes (Figures 12.22b and 12.23). Diamond is classified as a crystalline covalent network solid because it consists of an extended three-dimensional network of atoms joined by covalent bonds. Each carbon atom in diamond bonds by overlapping one of its sp^3 hybrid orbitals with an sp^3 hybrid orbital in each of four neighboring carbon atoms, creating a network of carbon tetrahedra. The atoms in these tetrahedra are connected by localized σ bonds, making diamond a poor electrical conductor. The sigma-bond network is extremely rigid, making diamond the hardest natural material known. The atoms of other group 14 elements, particularly silicon, germanium, and tin, also form covalent network solids based on the diamond crystal lattice.

Natural diamond forms from graphite under intense heat (>1700 K) and pressure (>50,000 atm) deep in Earth. Industrial diamonds are synthesized at high temperatures and pressures from graphite or any other source rich in carbon. Synthetic diamonds are used as abrasives and for coating the tips and edges of cutting

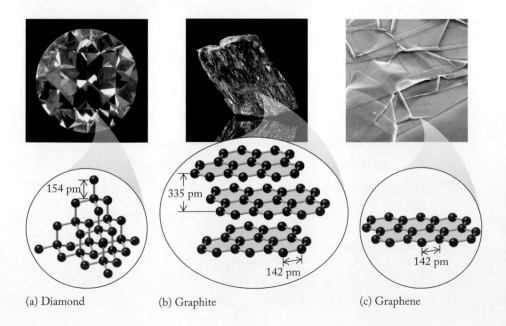

FIGURE 12.22 Two of carbon's three allotropes. (a) Diamond is a three-dimensional covalent network solid made of carbon atoms, each connected by σ bonds to four adjacent carbon atoms. (b) Graphite is a collection of layers of carbon atoms connected by σ bonds and delocalized π bonds. (c) Graphene is a single layer of C atoms arranged as the atoms are in a layer of graphite.

(a) Diamond   (b) Graphite   (c) Graphene

tools. Diamond has the highest thermal conductivity of any natural substance (five times higher than copper and silver, the most thermally conductive metals), so tools made from diamond do not become overheated. Thin diamond films grown from methane (CH_4) and hydrogen have been used to reduce wear and extend the lifetime of prosthetic implants.

By far the most abundant allotrope of carbon is graphite, another covalent network solid, frequently the principal ingredient in soot and smoke and used to make pencils, lubricants, and gunpowder. Graphite contains sheets of carbon atoms in which each atom is connected by sp^2 orbitals to a like orbital in each of three neighboring carbon atoms, forming a two-dimensional covalent network of six-membered rings (Figure 12.22b). Each carbon–carbon σ bond is 142 pm, which is shorter than the C—C σ bond in diamond (154 pm). Overlapping unhybridized p orbitals on the carbon atoms form a network of π bonds that are delocalized across the plane defined by the rings. The mobility of these delocalized electrons makes graphite a conductor of electricity.

As shown in Figure 12.22(b), the two-dimensional sheets in graphite are 335 pm apart. This distance is much too long to be a covalent bonding distance. Instead, the sheets are held together only by dispersion forces. These relatively weak interactions allow adjacent sheets to slide past each other, making graphite soft, flexible, and a good lubricant. In 2004 researchers in the United Kingdom and Russia found that they could isolate just a single layer of carbon atoms from graphite with a piece of adhesive tape. This material is called graphene (Figure 12.22c) and behaves as a semiconductor.

CONCEPT **TEST**

The diamond form of carbon is a semiconductor with a much larger band gap than silicon. With what elements might you choose to dope diamonds to form an n-type semiconductor?

The third allotrope of carbon was discovered in the 1980s. Networks of five- and six-atom carbon rings form molecules of 60, 70, or more carbon atoms that look like miniature soccer balls (Figure 12.23a). They are called *fullerenes* because their shape resembles the geodesic domes designed by the American architect R. Buckminster Fuller (1895–1983). Many chemists call them *buckyballs* for the same reason. In terms of both size and properties, buckyballs are too small to be classified as covalent network solids but too large to be molecular solids (discussed below). They fall in an ambiguous zone between small molecules and large networks and are classified as *clusters*. The 70–100 pm diameter of fullerenes means that they can fit into the three-dimensional structure of enzymes like HIV protease, inhibiting its activity. The surface of fullerenes can also be chemically modified to transport drug molecules into cells. Some fullerenes stretch the meaning of the word cluster. These include structures known as nanotubes, so named beause they are 1–2 nanometers in diameter (Figure 12.23b). Nanotubes have been fabricated that are nearly 1 meter long. Like graphene, carbon nanotubes exhibit semiconductor properties, but they are also extremely strong and have been used to strengthen fibers in sports gear.

12.6 Polymers

As part of our studies on covalent bonding in Chapters 8 and 9, we studied small organic compounds containing carbon, hydrogen, nitrogen, and oxygen. These

CHEMTOUR
Allotropes of Carbon

CONNECTION We introduced dispersion forces between molecules in Chapter 10. We discussed semiconductors in Chapter 9.

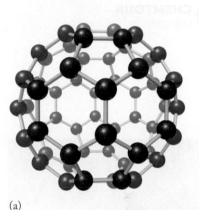

(a)

(b)

FIGURE 12.23 Some solids are described as clusters, a category between covalent network solids and molecular solids. Fullerenes, the third allotrope of carbon, are clusters. (a) One of them, buckminsterfullerene (C_{60}), is made up of 60 sp^2-hybridized carbon atoms. Both five- and six-membered rings are required to construct the nearly spherical molecule. (b) Nanotubes have diameters approximately equal to the diameter of a buckyball.

CONNECTION We introduced many organic molecules in Chapters 4 through 11. These included acetic acid; methane, ethane, and other hydrocarbon fuels; acetylene used in welding; and ethanol and other alcohols.

CHEMTOUR
Polymers

compounds had molar masses less than 1000 g/mol. Much larger organic compounds, with molar masses up to and exceeding 1,000,000 g/mol, represent a category of materials called **polymers**. In between small molecules and polymers (also called **macromolecules**) are midsize molecules called *oligomers*. The root word *meros* is Greek for "part" or "unit," so *polymer* literally means "many units" and *oligomer* means "a few units."

We have seen examples of polymers in previous chapters, including polyvinyl chloride, used to make drain pipes and other building materials; and Teflon, used in nonstick cookware. Polymers are composed of small structural units called **monomers**. Polymers may be formed from more than one type of monomer unit, they may have more than one functional group, and they may have shapes other than long chains. However, the single feature that distinguishes them as a class is their large size. In this section we explore the preparation and properties of polymers, focusing on materials with biomedical applications.

Small Molecules versus Polymers: Physical Properties

Before looking at how the composition and structure of polymers influence their behavior, let's first discuss the differences between the physical properties of small organic molecules and the physical properties of polymers. For now, we restrict our discussion to synthetic polymers made in the laboratory from small organic molecules (Figure 12.24).

All small organic compounds have well-defined properties. They have constant composition, for example, and their phase transitions take place at well-defined temperatures. In contrast, many synthetic polymers have neither constant composition nor well-defined properties because they are mixtures of large molecules that are similar but not identical: they differ in the number of monomers and the arrangement of monomers in the polymer chain. Their physical properties depend on the range and distribution of molecular masses in the sample.

Polymers typically do not have well-defined melting points and boiling points. Indeed, polymer molecules are so large that they cannot acquire sufficient kinetic energy to enter the gas phase. Furthermore, intermolecular forces between long chains in polymers can lead to both highly ordered, crystalline regions and less ordered, amorphous regions. When a polymer is heated, rearrangements of the molecules contribute to a broad range of melting temperatures.

As temperature increases, some polymers gradually soften and become more malleable. Others may eventually melt and become very viscous liquids. However, some polymers remain solids even as temperature increases—until their covalent bonds break, causing the material to decompose. For example, plastic grocery bags are made of the polymer polyethylene, the monomer unit of which is ethylene (CH_2=CH_2). The melting point of ethylene is −169°C; its boiling point is −104°C. Polyethylene, in contrast, is a tough and flexible solid at room temperature. It softens gradually over a range of temperatures from 85°C to 110°C, and at higher temperatures it tends to decompose into molecules of low-molar-mass organic gases.

FIGURE 12.24 Items made of synthetic polymers are ubiquitous in modern society.

CONNECTION The molecular structure and bonding in ethylene, CH_2=CH_2, was described in Section 9.4.

CONCEPT **TEST**

Which types of intermolecular forces are most likely to dominate between large molecules (polymers) containing long chains of carbon atoms bound to hydrogen atoms?

Polymers of Alkenes

Some widely used polymeric alkanes are produced industrially from small alkenes. The alkene polymer with the simplest structure is linear polyethylene (PE), produced from ethylene at high temperature and pressure:

$$n\ CH_2{=}CH_2 \rightarrow {+\!}CH_2{-}CH_2{\!+}_n \qquad (12.4)$$

Polyethylene is a **homopolymer**, which means it is composed of only one type of monomer. Its condensed structure is $CH_3(CH_2)_nCH_3$, but there are so many more methylene groups than methyl groups that the structure is frequently written ${+\!}CH_2CH_2{\!+}_n$ to highlight the structure and composition of the monomer. Polyethylene is also an example of an **addition polymer** because it is synthesized by adding many molecules together to form the polymer chain without the loss of any atoms or molecules. Correspondingly, the individual reactions that create an addition polymer are called **addition reactions**, in which two reactants combine to form one product.

In most products made of polyethylene, n is a very large number, ranging from 1000 to almost 1 million. Polyethylene has a wide range of properties that depend on the value of n and on whether the polymer chains are straight or branched. In low-density PE (LDPE)—a stretchable, soft plastic used in films and wrappers—n ranges from 350 to 3500 and the chains are branched. When the bagger at the grocery store asks, "Paper or plastic?" the plastic in question is LDPE.

When the molar mass of a PE polymer is between 100,000 and 500,000 and the chains are straight, the polymer has physical properties different from those of grocery bags. This straight-chain polymer—a rigid, translucent solid called high-density polyethylene (HDPE)—is used in milk containers, electrical insulation, and toys.

Why are the properties of straight-chain HDPE (rigid, tough) so different from those of branched-chain LDPE (stretchable, soft)? Think of the branched polymer (Figure 12.25) as a tree branch with lots of smaller branches attached to it. The polymer has three dimensions, so it can have branches that come out of the plane of the paper. In contrast, the straight-chain polymer is like a long, straight pole. Suppose you had a pile of 100 tree branches and a pile of

polymer (or **macromolecule**) a very large molecule with high molar mass formed by bonding many small molecules of low molecular mass.

monomer a small molecule that bonds with others like it to form polymers.

homopolymer a polymer composed of only one kind of monomer.

addition polymer a polymer formed by an addition reaction.

addition reaction a reaction in which two reactants couple to form one product without the loss of any atoms or molecules.

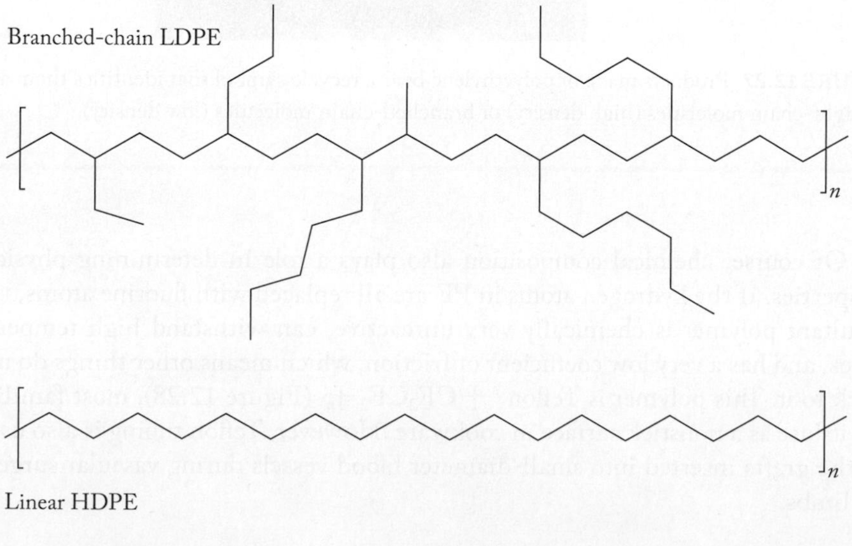

FIGURE 12.25 Low-density polyethylene consists of branched chains, but high-density polyethylene consists of straight chains. Efficient stacking—and thus high density—is possible in large polymer molecules of HDPE but not in large polymer molecules of LDPE.

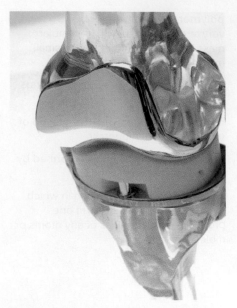

FIGURE 12.26 Devices used in knee-replacement surgery often incorporate a durable pad made of a relatively new polymer called ultrahigh molecular weight PE (UHMWPE), which is a very tough material and highly resistant to wear to cushion the joint. UHMWPE consists of linear molecules with an average molar mass over 3 million.

100 poles, and your task was to stack each pile into the smallest possible volume to fit into a truck. You can certainly pile the branches on top of one another, but they will not fit together neatly, and probably the best you can do is to make the pile a bit more compact. In contrast, you can stack the poles into a very compact pile.

The same situation arises with the branched and linear molecules of polyethylene. Branched-chain LDPE has a low density because the molecules do not line up neatly and so they stack less efficiently than do the molecules of HDPE. This makes LDPE more deformable and softer; HDPE is more rigid and even has some regions that are crystalline because the packing is so uniform. Ultrahigh molecular weight PE (UHMWPE; $n > 100{,}000$) is an even tougher material, because not only are the molecules straight, but they are also significantly larger than the molecules in HDPE. UHMWPE is used as a coating on some artificial ball-and-socket joints (Figure 12.26) because it is extremely resistant to abrasion and makes the joints last longer. The different forms of polyethylene illustrate how the size and shape of its molecules affect the physical properties of a material.

CONCEPT **TEST**

During recycling, articles made from LDPE are separated from those made from HDPE (Figure 12.27). Why?

$$-\!\!\!\left[CH_2-CH_2\right]_n$$
Polyethylene

FIGURE 12.27 Products made of polyethylene bear a recycle symbol that identifies them as straight-chain molecules (high density) or branched-chain molecules (low density).

Of course, chemical composition also plays a role in determining physical properties. If the hydrogen atoms in PE are all replaced with fluorine atoms, the resultant polymer is chemically very unreactive, can withstand high temperatures, and has a very low coefficient of friction, which means other things do not stick to it. This polymer is Teflon, $-\!\!\!\left[CF_2CF_2\right]_n$ (Figure 12.28), most familiar for its use as a nonstick surface in cookware. However, Teflon tubing is also used in the grafts inserted into small-diameter blood vessels during vascular surgery on limbs.

$$-\!\left(CF_2\!-\!CF_2\right)\!_n$$

Teflon

FIGURE 12.28 Repeating unit of polytetrafluoroethylene (Teflon).

SAMPLE EXERCISE 12.4 Identifying Monomers   **LO4**

Polypropylene, $-\!\left(CH_2CH(CH_3)\right)\!_n$, is an addition polymer used in the manufacture of fabrics, ropes, and other materials (Figure 12.29). Draw condensed and carbon-skeleton structures of the monomer used to prepare polypropylene.

Collect and Organize We are given a condensed structure of a polymer and asked to identify the monomer used in its preparation. We know that polypropylene is an addition polymer, so the monomer must be an alkene.

Analyze To understand the relationship between the polymer and the monomer from which it is made, let's look at Equation 12.4 in the reverse direction (Equation 12.5):

$$-\!\left(CH_2\!-\!CH_2\right)\!_n \rightarrow n\,CH_2\!=\!CH_2 \qquad (12.5)$$

Breaking the blue bonds in Equation 12.5 and making the red carbon–carbon single bond a double bond illustrates the relationship between polyethylene and ethylene, the alkene monomer. We need to apply a similar analysis to polypropylene.

Solve The relationship between polypropylene and its monomer is illustrated by Equation 12.6:

$$-\!\left(CH_2\!-\!\underset{\underset{\displaystyle CH_3}{|}}{CH}\right)\!_n \rightarrow n\,CH_2\!=\!CH(CH_3) \qquad (12.6)$$

Breaking the two blue bonds and making the red C—C bond a C=C double bond yields the alkene shown in Figure 12.30. This three-carbon alkene has the name propene.

Think About It The structural difference between propene and ethylene is the presence of a –CH_3 group bonded to one of the two sp^2 C atoms in propene instead of an H atom in ethylene.

Practice Exercise Draw the condensed and carbon-skeleton structures of the monomer used to make poly(methyl methacrylate), PMMA, a polymer used in shatterproof transparent plastic that can be used in place of glass:

FIGURE 12.29 Polypropylene is a common material for furniture, containers, clothing, lighting fixtures, and even objects of art. In addition to being moldable into many shapes, polypropylene is a good thermal insulator and does not absorb water easily.

$$CH_3CH\!=\!CH_2$$

FIGURE 12.30 The monomer propene is polymerized to make polypropylene.

vinyl group the subgroup CH_2=$CH-$.

copolymer a polymer formed from the chemical reaction of two different monomers.

heteropolymer a polymer made of three or more different monomer units.

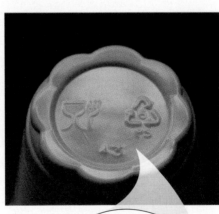

Styrene → Polystyrene

FIGURE 12.31 The monomer styrene is polymerized to make polystyrene.

C⚛NNECTION The molecular structure and bonding in benzene (C_6H_6) was introduced in Chapter 8. We explored its shape and defined it as an aromatic compound in Chapter 9.

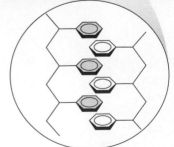

FIGURE 12.32 The aromatic rings on neighboring chains in polystyrene stack together and provide strength to the material.

Polymers Containing Aromatic Rings

As we saw in Chapter 9, benzene rings are flat molecules. Replacing one hydrogen atom in benzene with a **vinyl group**, the CH_2=$CH-$ subunit, produces the monomer styrene. The polymer polystyrene (PS) is synthesized from the styrene monomer (Figure 12.31). Notice how the polystyrene molecules tend to stack neatly (Figure 12.32), and this tendency gives rise to useful properties in materials that incorporate benzene rings and other aromatic compounds.

Solid PS is a transparent, colorless, hard, inflexible plastic. In this form it is used for the handles of some disposable razors and plastic cutlery. A more common form of PS, however, is the *expanded solid* made by blowing carbon dioxide gas or pentane gas into molten polystyrene, which then expands and retains voids in its structure when it solidifies. One form of this expanded PS is Styrofoam, the familiar material of coffee cups and take-out food containers (Figure 12.33).

The difference in properties between transparent, colorless, inflexible nonexpanded PS and opaque, white, pliable Styrofoam can be explained by considering the role of the aromatic ring in aligning the polymer chains. Branches in the chains have the same effect that we saw with polyethylene, but the aromatic rings and their tendency to stack (Figure 12.32) provide additional interactions that make chain alignment more favorable energetically. The aromatic rings along two neighboring chains can stack, and this stacking makes nonexpanded PS rigid. When the chains are blown apart by a gas, the stacking is disrupted and the chains open to form cavities that fill with air, making expanded PS a good thermal insulator and packing material. The presence of air in Styrofoam is illustrated in Figure 12.34.

Polymers of Alcohols and Ethers

More than 400,000 tons of the addition polymer poly(vinyl alcohol) (PVAL) are produced annually in the United States. It is used in fibers, adhesives, and materials known as sizing, which changes the surface properties of textiles and paper to make them smooth, less porous, and less able to absorb liquids. Because the polymer chains in PVAL are studded with −OH groups (Figure 12.35a), the surface

FIGURE 12.33 Polystyrene can be made into either rigid or foamed products. In its nonexpanded form, it is rigid and strong, suitable for making such products as plastic knives, forks, and spoons. In its expanded form, it is Styrofoam, used in carry-out food containers, packing materials, and thermal insulation in buildings.

of the polymer chains in PVAL is very polar and very water-like, and hydrocarbon solvents that are not soluble in water do not penetrate PVAL barriers. PVAL is the material of choice for laboratory gloves that are resistant to organic solvents.

PVAL is also impenetrable to carbon dioxide, and this property has led to its use in soda bottles, in which it is blended with the polymer poly(ethylene terephthalate) (PETE, Figure 12.35b). The two polymers do not mix but rather separate into layers (Figure 12.35c). The PETE makes the bottle strong enough to bear pressure changes due to temperature changes and survive the impact of falling off tables. Also, CO_2, the dissolved gas that makes soda fizz, passes readily through PETE but not through the PVAL layers, with the result that the soda does not go flat. Polymers with different properties are frequently combined like this to create new materials with desired properties.

The blend of PVAL and PETE in early soda bottles was a physical mixture of the two polymers. New materials can also be made by reacting different monomer units in one polymer molecule. This type of molecule is called a **copolymer** when two different monomers are reacted and a **heteropolymer** when three or more different monomers are reacted. One example of an addition copolymer is a material called EVAL, made from ethylene and vinyl acetate (Figure 12.36). EVAL is used in food wrappings when preservation of aroma and flavor are required. Food usually deteriorates in the presence of oxygen, and packages made of EVAL provide an excellent barrier to the entry of oxygen while retaining the flavor and fragrance of the packaged food.

Monomers forming heteropolymers or copolymers can combine in different ways. If we represent the monomer units making up a copolymer with the letters A and B, one possible way they can combine is

$$+A—B—A—B—A—B—A—B+$$

an arrangement called an *alternating copolymer*. Another possibility is

$$+A—A—A—A—B—B—B—B—$$
$$A—A—A—A—B—B—B—B+$$

called a *block copolymer*. A third possibility is a *random copolymer*:

$$+A—A—B—A—B—B—A—B—$$
$$A—B—B—A—A—A—A—B+$$

and this is the arrangement we see in the copolymer EVAL: it is a random copolymer of the monomers A = ethylene and B = vinyl acetate.

Commercially important polymers made from ethers include poly(ethylene glycol) (PEG) and poly(ethylene oxide) (PEO), which contain the same subunit

FIGURE 12.34 When the Styrofoam coffee cup on the left is placed under pressure, some of the air between the polystyrene chains is forced out. The cup shrinks to the size on the right but retains its overall shape.

CONNECTION We introduced and defined alcohols in Chapter 2 as organic molecules with the general formula R—OH. We further investigated alcohols in Chapter 10.

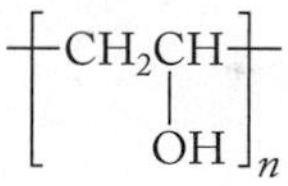

$$\left[CH_2CH \atop \quad OH \right]_n$$

PVAL

(a)

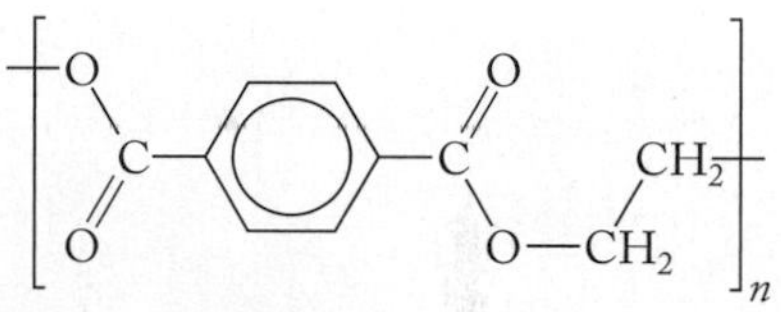

Repeating unit in PETE

(b)

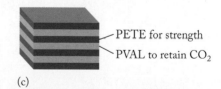

PETE for strength

PVAL to retain CO_2

(c)

FIGURE 12.35 (a) Poly(vinyl alcohol). (b) The repeating unit in PETE. (c) Layers of the polymers PVAL and PETE are used to make soda bottles. PETE makes the bottle strong, and PVAL keeps the carbon dioxide from leaking out.

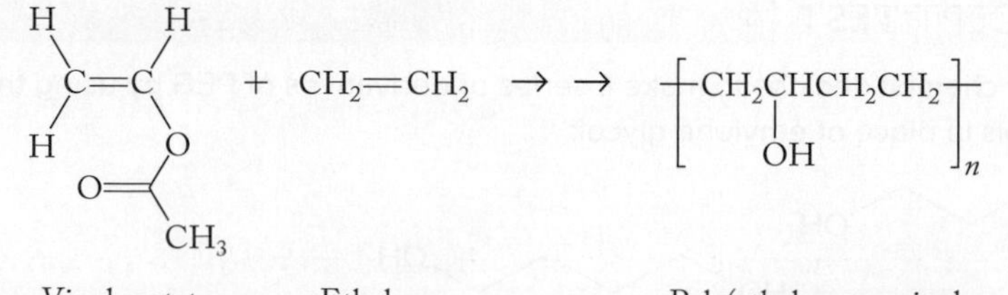

Vinyl acetate   Ethylene   Poly(ethylene-co-vinyl alcohol) = EVAL

FIGURE 12.36 EVAL is a copolymer of ethylene and vinyl acetate.

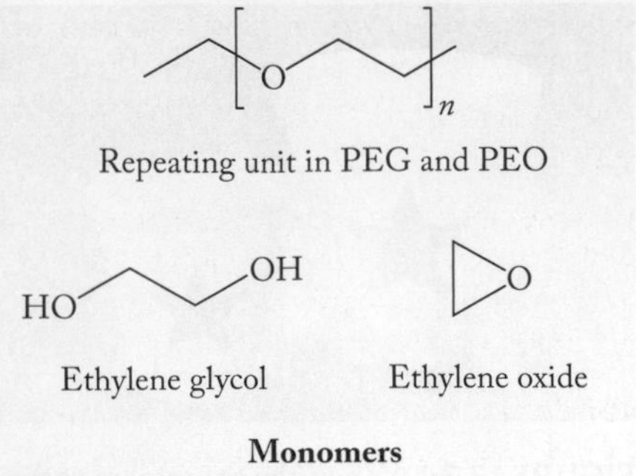

Repeating unit in PEG and PEO

Ethylene glycol   Ethylene oxide

Monomers

FIGURE 12.37 Poly(ethylene glycol) (PEG) and poly(ethylene oxide) (PEO) have the same repeating unit. The two polymers differ only in their molar masses. PEG is typically made from ethylene glycol, and ethylene oxide is the monomer of choice for making PEO.

(Figure 12.37). PEG is a low-molar-mass liquid made from ethylene glycol, and PEO is a higher-molar-mass solid made from ethylene oxide. As a polyether, PEG has properties closely related to those of diethyl ether, in that it is soluble in both polar and nonpolar liquids. It is a common component in toothpaste because it interacts both with water and with the water-insoluble materials in the paste, and keeps the toothpaste uniform both in the tube and during use. PEGs of many lengths are finding increasing use as attachments to pharmaceutical agents to improve their solubility and biodistribution: the enhanced circulation in the blood due to increased solubility.

SAMPLE EXERCISE 12.5 Analyzing Properties of Polymers   **LO5**

The polymer PEG (Figure 12.37) is used to blend materials that are typically insoluble in one another. PEG is soluble both in water and in benzene, a nonpolar solvent. Describe the structural features of PEG that make it soluble in these two liquids of very different polarities.

Collect, Organize, and Analyze The relationship between structure and solubility of compounds was discussed in Chapter 10. We learned that "like dissolves like," meaning that polar substances will dissolve in polar solvents because they both experience dipole–dipole interactions. Likewise, nonpolar substances dissolve in nonpolar solvents as a result of dispersion forces. We need to describe the intermolecular forces in PEG and see whether one part is compatible with water and another part with benzene. Water is a polar molecule and benzene is a nonpolar molecule, so we predict that PEG must contain both polar and nonpolar regions in order to be soluble in both substances.

Solve The structure of PEG consists of $-CH_2CH_2-$ groups connected by oxygen atoms. The oxygen atoms can form hydrogen bonds with water molecules, so the attractive force between PEG and water is due to hydrogen bonding. Benzene is nonpolar and is attracted to the $-CH_2CH_2-$ groups in the polymer. Nonpolar groups interact via dispersion forces, so those forces must be responsible for the solubility of PEG in nonpolar benzene.

Think About It As predicted, PEG contains both polar and nonpolar regions, allowing it to be solvated by both polar solvents such as water and nonpolar solvents such as benzene.

Practice Exercise When PEG is added to soft drinks it keeps CO_2, responsible for the fizz in soda, in solution longer when the soda is poured. What intermolecular attractive forces between PEG and CO_2 might make this application possible?

CONCEPT TEST

A polymer chemist decides to make a series of derivatives of PEG by using the following alcohols in place of ethylene glycol:

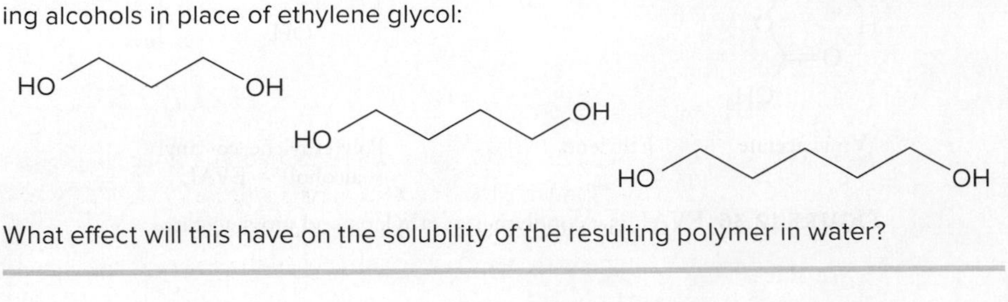

What effect will this have on the solubility of the resulting polymer in water?

Polyesters and Polyamides

Prior to this point, the polymers we have examined have been prepared from *monofunctional* monomers, in which the monomers have only one functional group. A wide range of polymers are derived from *difunctional* molecules, which contain two functional groups. Several classes of polymers are derived from difunctional carboxylic acids. We consider only two of them here: polyesters and polyamides.

An **ester** is prepared by the *esterification* of a carboxylic acid with an alcohol. In esters, the –OH group of the acid is replaced by –OR, where R can be any organic group (Figure 12.38). Esterification is not an addition reaction but rather a **condensation reaction**: two molecules react to create a larger molecule while a small molecule (typically water, hence "condensation reaction") is also formed. In Figure 12.38, the –COOH group of butyric acid reacts with the –OH group of ethanol to form a carbon–oxygen single bond in the resulting ester, ethyl butyrate, and release a molecule of water.

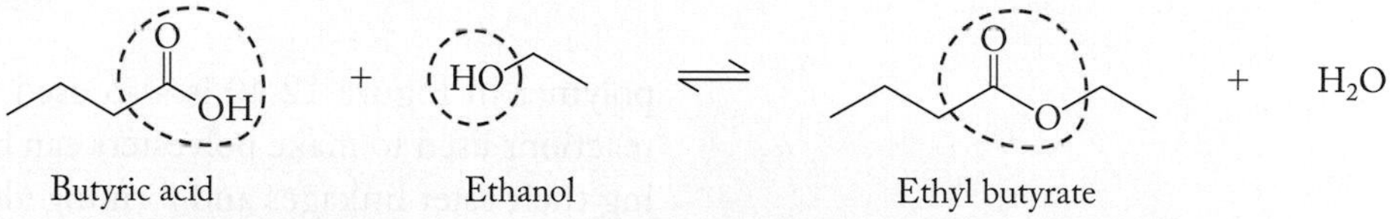

Now think about what happens at the molecular level when we have a single compound that contains a carboxylic acid functional group at one end and an alcohol functional group at the other (Figure 12.39). The carboxylic acid group of one molecule can react with the alcohol group of another molecule in a condensation reaction to generate a molecule that has a carboxylic acid group at one end, an alcohol at the other end, and an ester linkage between them. If this reaction happens repeatedly, a monomer containing one carboxylic acid and one hydroxyl group (a hydroxy acid) polymerizes, as shown in Figure 12.39, to form a *polyester*, a **condensation polymer**. We have already encountered one condensation polymer, PETE. In general, condensation polymers are formed by the reaction of monomers yielding a polymer and water as a by-product of the reaction. In addition to its use in plastic soda bottles, PETE is used extensively in medicine. Artificial heart valves and grafts for arteries are also made from PETE.

A copolymer formed by the reaction of glycolic acid and lactic acid (Figure 12.40) is used to support the growth of skin cells for burn victims. The

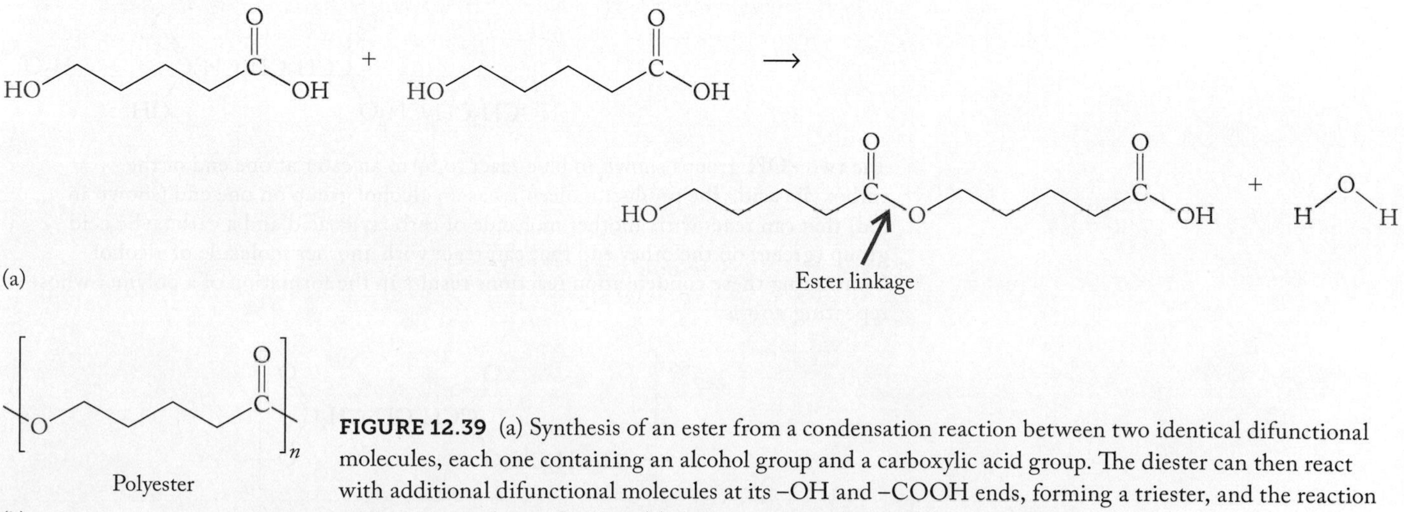

ester an organic compound in which the –OH of a carboxylic acid group is replaced by –OR, where R can be any organic group.

condensation reaction a reaction in which two molecules combine to form a larger molecule and a small molecule, typically water.

condensation polymer a polymer formed by a condensation reaction.

FIGURE 12.38 Condensation reactions between carboxylic acids and alcohols produce esters and water. Here butyric acid reacts with ethanol, forming ethyl butyrate. The functional groups (carboxylic acid, alcohol, and ester) are marked by the red, dashed circles.

CONNECTION The carboxylic acid functional group (–COOH) was described in Chapter 4.

FIGURE 12.39 (a) Synthesis of an ester from a condensation reaction between two identical difunctional molecules, each one containing an alcohol group and a carboxylic acid group. The diester can then react with additional difunctional molecules at its –OH and –COOH ends, forming a triester, and the reaction repeats over and over, forming (b) the polyester made up of the repeating monomer unit shown.

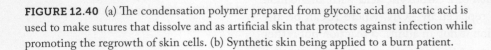

(a)

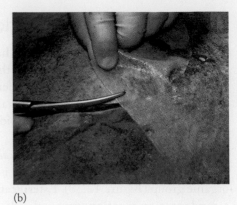

Glycolic acid   Lactic acid   A polyester

FIGURE 12.40 (a) The condensation polymer prepared from glycolic acid and lactic acid is used to make sutures that dissolve and as artificial skin that protects against infection while promoting the regrowth of skin cells. (b) Synthetic skin being applied to a burn patient.   (b)

polymer in Figure 12.40 is also used to make dissolving sutures. Esterification reactions used to make polyesters can be reversed by the addition of water, breaking their ester linkages and forming alcohol and acid functional groups.

SAMPLE EXERCISE 12.6 Making a Polyester   **LO4**

Show how a polyester can be synthesized from the difunctional alcohol $HO(CH_2)_3OH$ and the difunctional carboxylic acid $HOOC(CH_2)_3COOH$.

Collect, Organize, and Analyze An ester is the product of a reaction between a carboxylic acid and an alcohol. A polyester is a polymer with a repeating unit containing an ester functional group. We are given an alcohol and a carboxylic acid to react to make the ester repeating monomer unit. Because the starting materials are difunctional, the alcohol can react with two molecules of carboxylic acid, and the acid can react with two molecules of alcohol.

Solve The reaction is

$$HOCH_2CH_2CH_2OH \quad + \quad \overset{O}{\underset{HO}{\|}}CCH_2CH_2CH_2\overset{O}{\underset{OH}{C\|}} \quad \rightarrow$$

$$\overset{O}{\|}CCH_2CH_2CH_2\overset{O}{C\|} \quad + \quad H_2O$$
$$HOCH_2CH_2CH_2O \qquad\qquad OH$$

The two −OH groups shown in blue react to form an ester at one end of the carboxylic acid. The product molecule has an alcohol group on one end (shown in red) that can react with another molecule of carboxylic acid and a carboxylic acid group (green) on the other end that can react with another molecule of alcohol. Continuing these condensation reactions results in the formation of a polymer whose repeating unit is:

$$\left[\overset{O}{\|}CCH_2CH_2CH_2\overset{O}{C\|} \atop -CH_2CH_2CH_2O \qquad O- \right]_n$$

Think About It The repeating unit in the polyester contains one section that came from the alcohol and a second section that came from the carboxylic acid because esters are formed from alcohols and acids.

 Practice Exercise A difunctional molecule may contain two different functional groups, such as this one with an alcohol group and a carboxylic acid group:

$$HO-CH_2CH_2CH_2C\overset{\displaystyle O}{\underset{\displaystyle OH}{\Vert}}$$

Draw the repeating unit of the polyester made from this molecule.

SAMPLE EXERCISE 12.7 Comparing Properties of Polymers   **LO5**

Clothes made from the polyester fabric known as Dacron can be less comfortable in hot weather than clothes made of cotton (also a polymer) because Dacron does not absorb perspiration as effectively as cotton. On the basis of the repeating units of these two polymers (Figure 12.41), suggest a structural reason why cotton absorbs perspiration (water) better than Dacron.

Collect, Organize, and Analyze Cotton and Dacron are polymers that differ in the functional groups in their repeating units. The absorption of water by a polymer depends on the intermolecular forces present. Water is a polar molecule and is attracted to polar groups. We need to compare the different functional groups in each monomer to see which has the greatest number of polar functional groups or atoms that can form hydrogen bonds. This will be the polymer more likely to absorb water.

Solve Each monomer unit in cotton has three –OH groups attached to it, all polar and capable of hydrogen bond formation, which means they are likely to interact with the water in perspiration and thereby draw it away from the body. The monomer unit in Dacron has oxygen atoms in it, but no –OH groups. Although regions in the Dacron monomer are polar, they are not nearly as polar as the –OH groups in the cotton monomer unit.

Think About It The principle of like interacting with like works for polymers just as it does for small molecules.

 Practice Exercise Gloves made of a woven blend of cotton and polyester fibers protect the hands from exposure to oil and grease and are comfortable to wear because they "breathe"—they allow perspiration to evaporate and pass through them, thereby cooling the skin. Suggest how these gloves work at the molecular level.

FIGURE 12.41 On the basis of the molecular structures of (a) Dacron and (b) cotton, why is cotton the better material for making tee shirts worn during strenuous exercise?

Repeating unit in Dacron

(a)   (b)

Repeating unit in cotton: $n \approx 10,000$

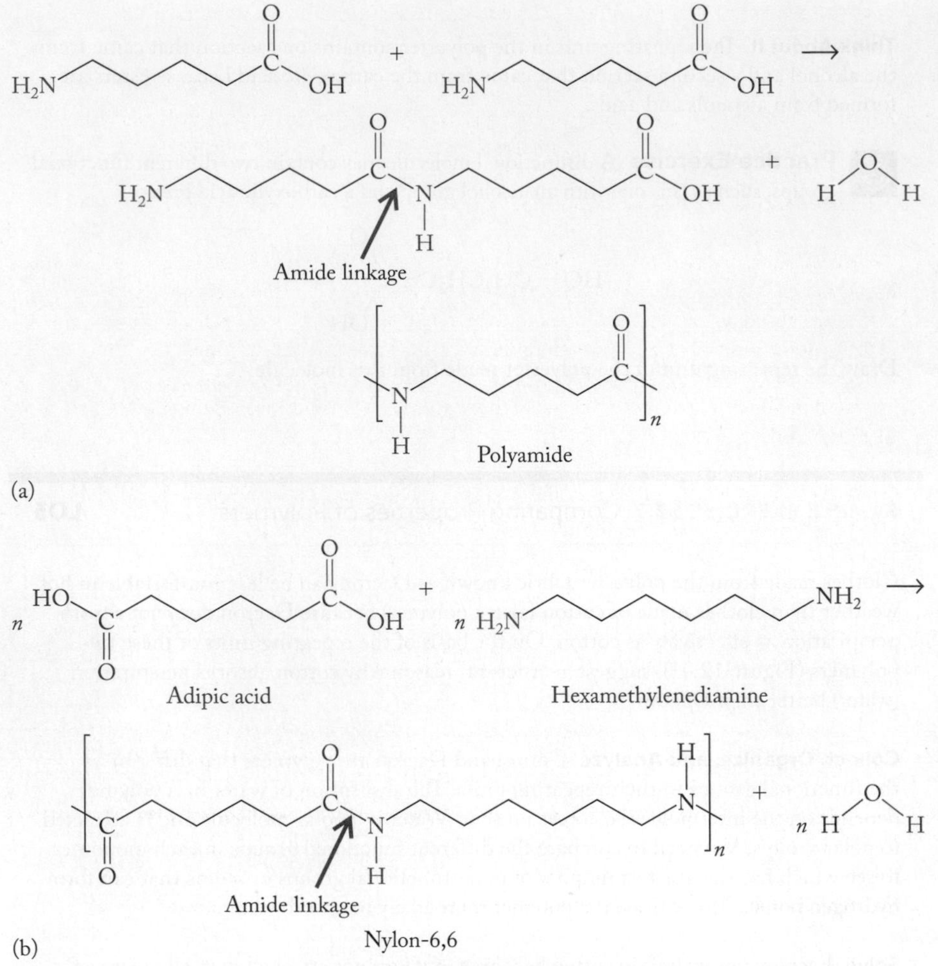

FIGURE 12.42 (a) Synthesis of a polyamide from two identical monomers, each containing a carboxylic acid functional group and an amine functional group. (b) Synthesis of the polyamide nylon-6,6 from nonidentical monomers: adipic acid (a dicarboxylic acid) and hexamethylenediamine (a diamine).

Amide linkage

Polyamide

(a)

Adipic acid   Hexamethylenediamine

Amide linkage

Nylon-6,6

(b)

Combining difunctional molecules in a condensation reaction can be used to make *polyamides*, another class of very useful synthetic polymers. The functional groups are a carboxylic acid and an amine. The difunctional monomer units can be identical (Figure 12.42a), each containing one carboxylic acid group and one amine group, or they can be different (Figure 12.42b), with one monomer containing two carboxylic acid groups (a dicarboxylic acid) and the other containing two amine groups (a diamine).

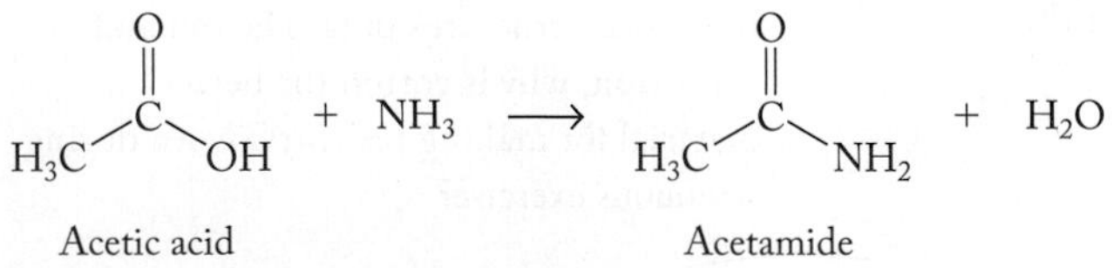

In **amides**, the –OH group of a carboxylic acid is replaced by an amine group—which can be $-NH_2$, $-NHR$, or $-NR_2$ (Figure 12.43). Amides are also polar and can form intermolecular hydrogen bonds. The hydrogen atoms on the $-NH_2$ group can hydrogen-bond with the oxygen atom in the carbonyl group of an adjacent molecule. This causes the boiling points of amides to be considerably higher than those of esters of comparable size. Although amines are bases, amides are actually weakly acidic. The carbonyl group pulls electron density away from the nitrogen, making it less favorable for the nitrogen to behave like a Brønsted–Lowry base and pick up a proton. It is actually more favorable for an amide to donate a proton and function as an acid.

Probably the most familiar polyamide is nylon-6,6, made from the monomers shown in Figure 12.42(b). Each monomer contains six carbon atoms, which is what the digits in the name represent.

Acetic acid   Acetamide

FIGURE 12.43 Condensation reactions between carboxylic acids and ammonia (or amines) produce amides. Here acetic acid reacts with ammonia, forming acetamide.

amide an organic compound in which the same carbon atom is single bonded to a nitrogen atom and double bonded to an oxygen atom.

SAMPLE EXERCISE 12.8 Identifying Monomers **LO4**

Another form of nylon is nylon-6, with the single 6 indicating that the polymer is made from the reaction of a series of identical six-carbon monomers. By analogy with the polyester in Sample Exercise 12.6 and the accompanying Practice Exercise, draw the condensed structure of a compound that could polymerize to make nylon-6.

Collect, Organize, and Analyze We are asked to draw the condensed structure for a monomer that contains both functional groups that react to form an amide and that could react with identical monomers to form a polyamide with a repeating unit six carbon atoms long. By analogy to Sample Exercise 12.6 and its accompanying Practice Exercise, we should suggest a difunctional molecule that has a carboxylic acid on one end and an amine on the other because these are the two functional groups that react to form the amide linkage.

Solve We can build the required monomer by starting with one of the functional groups. It doesn't matter which one, so let's begin with the amine:

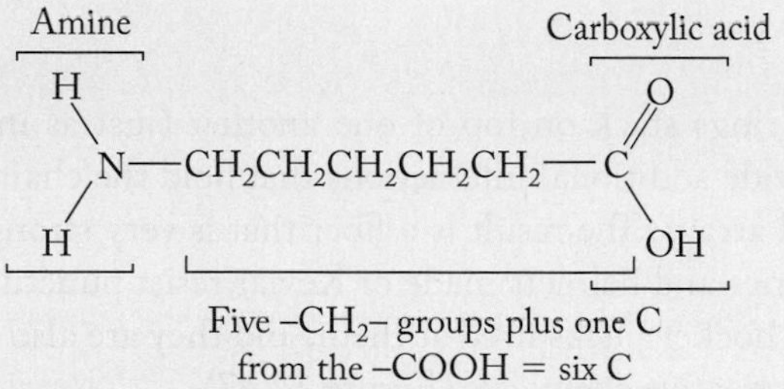

Amine   Carboxylic acid

$$\underset{H}{\overset{H}{N}}-CH_2CH_2CH_2CH_2CH_2-C\underset{OH}{\overset{O}{\Vert}}$$

Five –CH_2– groups plus one C
from the –COOH = six C

We then add a chain of five –CH_2– units because the name nylon-6 indicates that six carbon atoms separate the ends of the repeating unit. Finally, we add the carboxylic acid functional group as the second functional group and the sixth carbon atom in the chain.

Think About It Many nylons with different properties can be made by varying the length of the carbon chain in a monomer like the one in this exercise or by varying the lengths of the chains in both the difunctional acid and difunctional amine in Figure 12.42(b).

🟐 **Practice Exercise** Draw the carbon-skeleton structures of two monomers that could react with each other to make nylon-5,4. Draw the carbon-skeleton structure of the repeating unit in the polymer. (*Note:* The first number refers to the carboxylic acid monomer; the second refers to the amine monomer.)

FIGURE 12.44 Stephanie Kwolek (1923–2014), a chemist at DuPont who created Kevlar, the first member of a family of exceptionally strong, stiff synthetic fibers.

Benzene-1,4-dicarboxylic acid
(terephthalic acid)

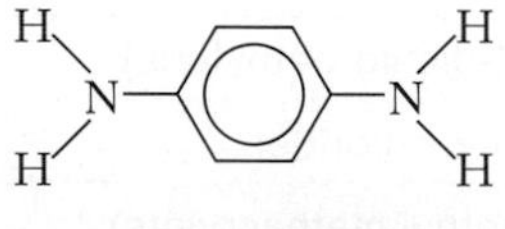

1,4-Diaminobenzene

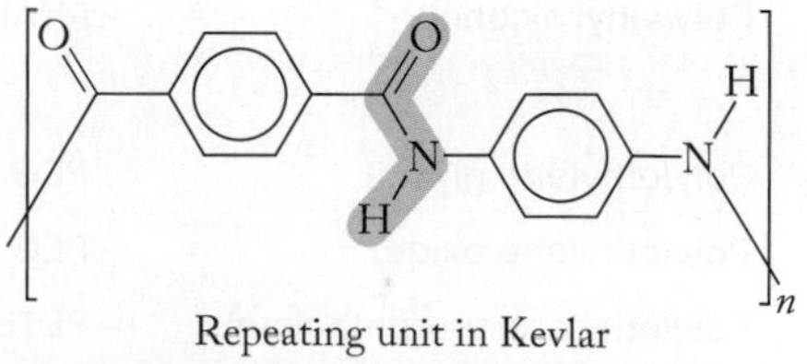

Repeating unit in Kevlar

FIGURE 12.45 The monomers used to make Kevlar are a dicarboxylic acid and a diamine. The amide bond in the repeating unit is highlighted.

Polymers of nylon make long, straight fibers that are quite strong and excellent for weaving into fabrics. Nylon is flexible and stretchable because the hydrocarbon chains can bend and curl, much like a telephone cord or a Slinky spring toy. To produce an even stronger nylon, researchers recognized they had to find some way to reduce the ability of the chains to form coils. They discovered this could be done by using monomer units containing functional groups that made it difficult for the chains to bend. One product of this work was a polyamide called Kevlar, invented by Stephanie Kwolek (Figure 12.44) of DuPont in 1965. Kevlar is formed from a dicarboxylic acid of benzene and a diamine of benzene (Figure 12.45). When these two monomers polymerize, the flat, rigid aromatic rings keep the chains straight.

Two additional intermolecular interactions orient the chains and hold them together very tightly (Figure 12.46). First, the –NH hydrogen atoms form hydrogen bonds with the oxygen atoms of carbonyl groups on adjacent chains. Second,

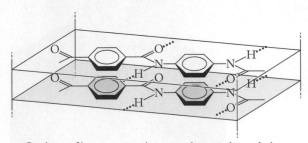

Hydrogen bonding between chains in the same plane

Stacking of benzene rings between chains in layered planes

FIGURE 12.46 Interactions between groups in Kevlar.

FIGURE 12.47 A bullet fired point-blank at a sheet of Kevlar does not puncture the fabric.

the rings stack on top of one another (just as in polystyrene) and provide additional interactions that hold the chains together in parallel arrays. The result is a fiber that is very strong but still flexible. Fabrics and helmets made of Kevlar resist puncture, even by bullets and hockey pucks fired at them, and they are also resistant to flames and reactive chemicals (Figure 12.47).

In this section we have discussed many polymers and mentioned their uses in common products. The classification of these polymeric materials as addition or condensation polymers is summarized in Table 12.7. Table 12.7 also identifies the polymers by their distinctive functional groups and mentions some common uses in our lives.

TABLE 12.7 Summary of Common Polymers and Their Uses

Name	Abbreviation	Functional Group	Use
ADDITION POLYMERS			
Polyethylene	PE	Alkane	Plastic bags and films
Polytetrafluoroethylene	Teflon	Fluoroalkane	Nonstick coatings
Poly(1,1-dichloroethylene)	Saran	Chloroalkane	Plastic wrap
Poly(vinyl chloride)	PVC	Chloroalkane	Drain pipes
Poly(methyl methacrylate)	PMMA	Alkane and ester	Shatter-resistant glass, e.g., Plexiglas, Lucite
Polystyrene	PS	Aromatic hydrocarbon	Cups, dishes, insulation
Poly(vinyl alcohol)	PVAL	Alcohol	Gloves, bottles
CONDENSATION POLYMERS			
Poly(ethylene glycol)	PEG	Ether	Pharmaceuticals, consumer products
Poly(ethylene oxide)	PEO	Ether	Same as for PEG
Poly(ethylene terephthalate)	PETE	Ester	Plastic bottles
Nylon		Amide	Clothing
Kevlar		Amide	Protective equipment
Dacron		Ester	Clothing

SAMPLE EXERCISE 12.9 Integrating Concepts: Glowing Lantern Mantles

The mantles in camping lanterns contain the mineral thorite, an oxide of thorium. Thorite crystallizes with the fluorite structure and contains 87.88% thorium and 12.12% oxygen.
a. Determine the empirical formula of thorite.
b. Calculate the density of thorite. Does the calculated density match the measured density of 9.86 g/cm³?
c. The ionic radii of Th⁴⁺ and O²⁻ ions are 102 pm and 140 pm, respectively. If thorite were to crystallize with the rock salt structure, what type of hole would the Th⁴⁺ ions most likely occupy?
d. Calculate the density of a rock salt polymorph of thorite by using the same unit cell edge length as in part (b). Compare the value with the observed and calculated values in part (b).

Collect and Organize We are given the percent composition and density of a thorium mineral. We are asked to determine the empirical formula of thorite and to calculate the density of thorite in two possible crystal structures—the fluorite (observed) and rock salt (hypothetical) structures. Methods for determining empirical formulas were described in Chapter 3. We will also make use of the information presented in Section 12.4.

Analyze We recognize that thorite contains only Th and O with the general formula Th_xO_y because the sum of the %Th and %O equals 100%, and thus we know the masses of each element in thorite. We can convert the masses to the equivalent number of moles and determine the simplest whole-number ratio of the moles of the elements. In the fluorite structure of thorite, the unit cell consists of a face-centered arrangement of Th ions with oxide ions in all of the tetrahedral holes. Following the procedure in Sample Exercise 12.3, we can calculate the density of thorite. If thorite were to crystallize in a ccp rock salt structure, then Th ions would be found in the octahedral holes of an fcc unit cell of oxide ions.

Solve
a. A 100.00-g sample of thorite contains 87.88 g of Th and 12.12 g of O. Converting these masses to moles:

$$87.88 \text{ g Th} \times \frac{1 \text{ mol Th}}{232.04 \text{ g Th}} = 0.3787 \text{ mol Th}$$

$$12.12 \text{ g O} \times \frac{1 \text{ mol O}}{16.00 \text{ g O}} = 0.7575 \text{ mol O}$$

Dividing by the smallest number of moles gives us the empirical formula: ThO_2.
b. In the fluorite structure of ThO_2, the Th⁴⁺ ions adopt an fcc unit cell with O²⁻ ions in all the tetrahedral holes. The equivalent of four Th⁴⁺ and eight O²⁻ are in the unit cell. The mass of the ions in the unit cell is

$$m = \frac{232.04 \text{ g Th}^{4+}}{1 \text{ mol Th}^{4+}} \times \frac{1 \text{ mol Th}^{4+}}{6.022 \times 10^{23} \text{ ions Th}^{4+}} \times 4 \text{ ions Th}^{4+}$$
$$= 1.541 \times 10^{-21} \text{ g Th}^{4+}$$

$$m = \frac{16.00 \text{ g O}^{2-}}{1 \text{ mol O}^{2-}} \times \frac{1 \text{ mol O}^{2-}}{6.022 \times 10^{23} \text{ ions O}^{2-}} \times 8 \text{ ions O}^{2-}$$
$$= 0.2126 \times 10^{-21} \text{ g O}^{2-}$$

Total mass of ions $= 1.541 \times 10^{-21} \text{ g Th}^{4+} + 0.2126 \times 10^{-21} \text{ g O}^{2-}$
$$= 1.754 \times 10^{-21} \text{ g}$$

The volume of the unit cell is

$$V = \ell^3 = (560 \text{ pm})^3 \times \frac{(10^{-10} \text{ cm})^3}{(1 \text{ pm})^3} = 1.756 \times 10^{-22} \text{ cm}^3$$

and the calculated density of ThO_2 is

$$d = \frac{m}{V} = \frac{1.754 \times 10^{-21} \text{ g}}{1.756 \times 10^{-22} \text{ cm}^3} = 9.99 \text{ g/cm}^3$$

The calculated density differs from the observed density of 9.86 g/cm³ by less than 2%.
c. In the hypothetical rock salt structure of ThO_2, the lattice is composed of fcc O²⁻ ions with Th⁴⁺ in holes. The radius ratio of Th⁴⁺ to O²⁻ is

$$\frac{r_{cation}}{r_{anion}} = \frac{102 \text{ pm}}{140 \text{ pm}} = 0.729$$

From data in Table 12.5, the Th⁴⁺ ions would occupy one-half of the octahedral holes.
d. By analogy to part (b), the density of ThO_2 in the hypothetical rock salt structure would be

$$m = \frac{232.04 \text{ g Th}^{4+}}{1 \text{ mol Th}^{4+}} \times \frac{1 \text{ mol Th}^{4+}}{6.022 \times 10^{23} \text{ ions Th}^{4+}} \times 2 \text{ ions Th}^{4+}$$
$$= 7.706 \times 10^{-22} \text{ g Th}^{4+}$$

$$m = \frac{16.00 \text{ g O}^{2-}}{1 \text{ mol O}^{2-}} \times \frac{1 \text{ mol O}^{2-}}{6.022 \times 10^{23} \text{ ions O}^{2-}} \times 4 \text{ ions O}^{2-}$$
$$= 1.063 \times 10^{-22} \text{ g O}^{2-}$$

$$V = \ell^3 = (560 \text{ pm})^3 \times \frac{(10^{-10} \text{ cm})^3}{(1 \text{ pm})^3} = 1.756 \times 10^{-22} \text{ cm}^3$$

$$d = \frac{m}{V} = \frac{8.769 \times 10^{-22} \text{ g}}{1.756 \times 10^{-22} \text{ cm}^3} = 4.99 \text{ g/cm}^3$$

The calculated density for ThO_2 in a rock salt structure is about half of the observed value, and it is about half of the calculated value based on the fluorite structure.

Think About It By analogy to the fluorite structure shown in Figure 12.20, the larger oxide ions occupy the tetrahedral holes. The calculated value for the density of ThO_2 is very close to the observed value, supporting the assignment of the fluorite structure for ThO_2. The rock salt structure, where the smaller Th⁴⁺ ions occupy octahedral holes in the fcc lattice of the larger O²⁻ ions, can be ruled out because the calculated density is very different from the observed density.

SUMMARY

LO1 Many metallic crystals are based on **crystal lattices** of the **cubic closest-packed (ccp)** and **hexagonal closest-packed (hcp)** types, which are the two most efficient ways of packing atoms in a solid. **Crystalline solids** contain repeating **unit cells**, which can be **simple cubic (sc)**, **body-centered cubic (bcc)**, or **face-centered cubic (fcc)**. The dimensions of the unit cell in a crystalline lattice can be used to determine the radius of the atoms or ions and to predict density. (Sections 12.1 and 12.2)

LO2 **Alloys** are blends of a host metal and one or more other elements (which may or may not be metals) added to enhance the properties of the host, including strength, hardness, and corrosion resistance. In **substitutional alloys**, atoms of the added elements replace atoms of the host metal in the crystal lattice. In **interstitial alloys**, atoms of added elements are located in the tetrahedral and/or octahedral holes between atoms of the host metal. (Section 12.3)

LO3 Many **ionic solids** consist of crystals with some number of either cations or anions forming the unit cell with the opposite ion occupying octahedral and tetrahedral holes in the unit cell. The unit cell edge lengths of ionic solids can be used to calculate ionic radii and to predict densities. Two allotropes of carbon are the **covalent network solids** graphite and diamond. (Sections 12.1, 12.4, and 12.5)

LO4 Alkenes undergo **addition reactions** to form **homopolymers** known as **addition polymers**. Monomer units derived from alcohols or ethers can be chemically reacted to make **heteropolymers** or **copolymers**. **Condensation reactions** of carboxylic acids and amines are used to prepare **condensation polymers** such as polyesters and polyamides. (Section 12.6)

PETE for strength

PVAL to retain CO₂

LO5 Properties of polymers depend on the identity and arrangement of the monomer units in the polymer. (Section 12.6)

PARTICULATE **PREVIEW WRAP-UP**

The pattern in Allotrope A is nonplanar, as each carbon atom is covalently bonded to four other carbon atoms in a tetrahedral geometry. The pattern in Allotope B is planar, as each carbon atom is covalently bonded to three other carbon atoms in a trigonal planar geometry and is attracted to the planar layers above and below through dispersion forces. Allotrope A (diamond) is very hard and resists compression, whereas Allotrope B (graphite) can be used as a lubricant because the planar layers can slide past one another.

PROBLEM-SOLVING SUMMARY

Type of Problem	Concepts and Equations	Sample Exercises
Calculating atomic or ionic radii and density from unit cell dimensions	Determine the length of a unit cell edge, face diagonal, or body diagonal along which adjacent atoms touch. Use the relationship between edge length ℓ and the atomic radius r to calculate the value of r: $r = 0.5\,\ell$ for sc unit cells, $r = 0.3536\ell$ for fcc unit cells, and $r = 0.4330\ell$ for bcc unit cells.	12.1, 12.3
	Determine the mass of the atoms in the unit cell from the molar mass and the volume of the unit cell from the unit cell edge length; then calculate the density: $$d = \frac{m}{V}$$	
Predicting the crystal structure of two-element alloys	Compare the radii of the alloying elements. Similarly sized radii (within 15%) predict a substitutional alloy. When the radius of the smaller atom in an hcp or ccp lattice is <73% of the radius of the larger atom, an interstitial alloy forms.	12.2
Predicting the location of ions in ionic compounds	Determine the ratio of the radii of the guest and the host to predict whether the ion is more likely to reside in an octahedral or tetrahedral hole.	12.3
Identifying monomers in polymers	Find the smallest portion of the polymer that is repeated.	12.4, 12.8
Analyzing properties of polymers	Evaluate the polarity of the functional groups in the polymer, and assess the relative importance of all types of intermolecular forces possible in the molecules.	12.5, 12.7
Making a polyester	Combine monomers with alcohol functional groups (ROH) and carboxylic acid functional groups (RCOOH) to form water (H_2O) and ester groups (RCOOR).	12.6

VISUAL PROBLEMS

(Answers to boldface end-of-chapter questions and problems are in the back of the book.)

12.1. In Figure P12.1, which drawings are analogous to crystalline solids, and which are analogous to amorphous solids?

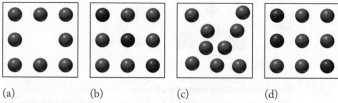

(a)   (b)   (c)   (d)

FIGURE P12.1

12.2. The unit cells in Figure P12.2 continue infinitely in two dimensions. Draw a box around the unit cell in each pattern. How many light squares and how many dark squares are in each unit cell?

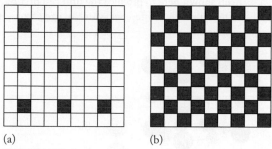

(a)   (b)

FIGURE P12.2

12.3. The pattern in Figure P12.3 continues indefinitely in three dimensions. Draw a box around the unit cell. If the red circles represent element A and the blue circles represent element B, what is the chemical formula of the compound?

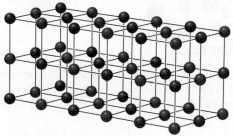

FIGURE P12.3

12.4. What is the chemical formula of the ionic compound a portion of whose unit cell is shown in Figure P12.4? (A and B are cations; X is an anion.)

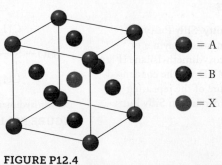

= A
= B
= X

FIGURE P12.4

12.5. What is the formula of the compound that crystallizes in a cubic closest-packed arrangement of A atoms (silver) with B atoms (red) occupying half of the octahedral holes and C atoms (blue) occupying one-eighth of the tetrahedral holes, as shown in Figure P12.5?

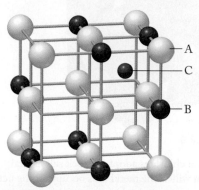

A
C
B

FIGURE P12.5

12.6. What is the formula of the compound that crystallizes with copper ions (the small spheres in Figure P12.6) occupying half of the tetrahedral holes in a cubic closest-packed arrangement of chloride ions?

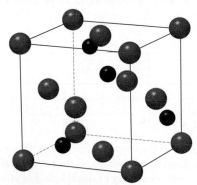

FIGURE P12.6

12.7. What is the formula of the compound that crystallizes with lithium ions occupying all of the tetrahedral holes in a cubic closest-packed arrangement of sulfide ions? See Figure P12.7.

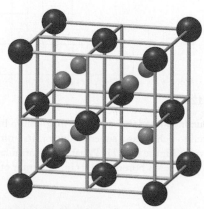

FIGURE P12.7

12.8. Figure P12.8 shows the unit cell of CsCl. From the information shown and given the radius of the chloride (corner) ions of 181 pm, calculate the radius of Cs$^+$ ions.

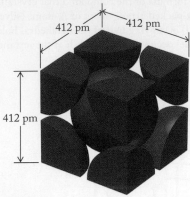

412 pm   412 pm

412 pm

FIGURE P12.8

12.9. Several metal chlorides adopt the rock salt crystal structure, in which the metal ions occupy all the octahedral holes in a face-centered cubic array of chloride ions. For at least one (maybe more) of the metallic elements highlighted in Figure P12.9, this crystal structure is not possible. Which one(s) cannot form a rock salt structure with Cl$^-$ ions?

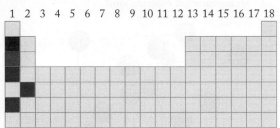

1 2 3 4 5 6 7 8 9 10 11 12 13 14 15 16 17 18

FIGURE P12.9

12.10. Several metal fluorides adopt the fluorite crystal structure, in which the fluoride ions occupy all the tetrahedral holes in a face-centered cubic array of metal ions. For at least one (maybe more) of the highlighted metallic elements in Figure P12.10, this crystal structure is not possible. Which one(s) are they?

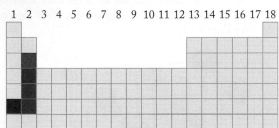

1 2 3 4 5 6 7 8 9 10 11 12 13 14 15 16 17 18

FIGURE P12.10

*12.11. Superconducting Materials I In 2000, magnesium boride was observed to behave as a superconductor. Its unit cell is shown in Figure P12.11. What is the formula of magnesium boride? A boron atom is in the center of the unit cell (on the left), which is part of the hexagonal closest-packed crystal structure (right).

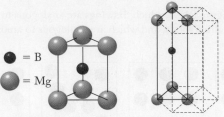

= B
= Mg

FIGURE P12.11

*12.12. Superconducting Materials II The 1987 Nobel Prize in Physics was awarded to J. G. Bednorz and K. A. Müller for their discovery of superconducting ceramic materials such as YBa$_2$Cu$_3$O$_7$. Figure P12.12 shows the unit cell of another yttrium–barium–copper oxide.
   a. What is the chemical formula of this compound?
   b. Eight oxygen atoms must be removed from the unit cell shown here to produce the unit cell of YBa$_2$Cu$_3$O$_7$. Does it make a difference which oxygen atoms are removed?

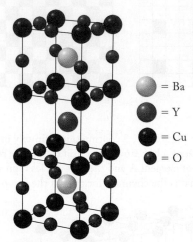

= Ba
= Y
= Cu
= O

FIGURE P12.12

*12.13. The three polymers shown in Figure P12.13 are widely used in the plastics industry. In which of them are the intermolecular forces per mole of monomer the strongest?

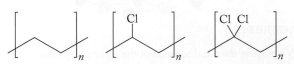

(a) Polyethylene   (b) Poly(vinyl chloride)   (c) Poly(1,1-dichloroethylene)
FIGURE P12.13

*12.14. Silly Putty Silly Putty is a condensation polymer of dihydroxydimethylsilane (Figure P12.14). Draw the condensed structure of the repeating monomer unit in Silly Putty.

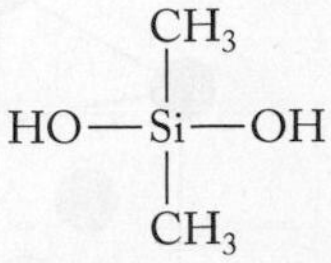

CH$_3$

HO—Si—OH

CH$_3$

Dihydroxydimethylsilane
FIGURE P12.14

12.15. **Spider Silk** Polymer chemists have long sought to mimic the physical properties of spider silk, a remarkably strong but flexible material. The structure of spider silk is illustrated in Figure P12.15.

a. What are the intermolecular forces between the strands?

b. The strands are formed during a condensation reaction. What functional group is present along each strand? What two functional groups reacted to form this functional group?

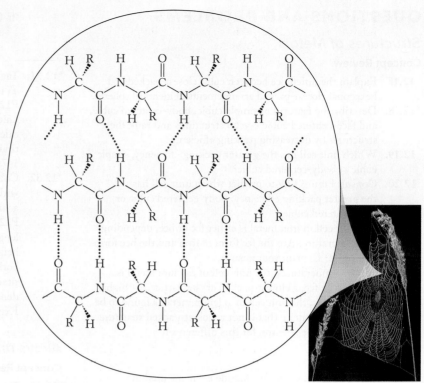

FIGURE P12.15

12.16. Use representations [A] through [I] in Figure P12.16 to answer questions a–f.

a. Two forms of iron are ferrite and austentite. One has a body-centered cubic structure, and the other takes a face-centered cubic structure. Which representations depict these common lattices for iron?

b. Which representation depicts an ionic compound?

c. Which representation depicts an interstitial alloy?

d. Which polymers were formed by an addition polymerization process? Which polymers were formed by a condensation polymerization process?

e. Which polymer structures are likely to result in rigid, hard materials?

f. Which polymer structures are likely to result in flexible, stretchable, soft materials?

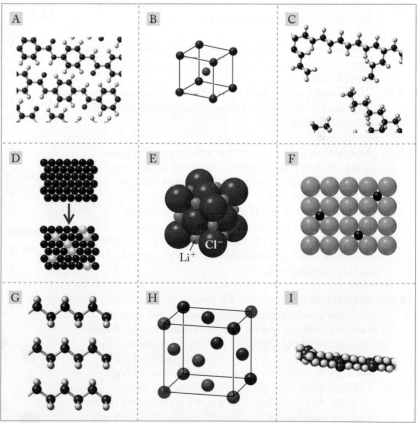

FIGURE P12.16

QUESTIONS AND PROBLEMS

Structures of Metals

Concept Review

12.17. Explain the difference between cubic closest-packed and hexagonal closest-packed arrangements of identical spheres.

*12.18. Describe the features of simple cubic, body-centered cubic, and face-centered cubic crystal structures and rank these structures by decreasing packing efficiency.

12.19. Which unit cell has the greater packing efficiency, simple cubic or body-centered cubic?

12.20. Consult Figure 12.9 to predict which unit cell has the greater packing efficiency: body-centered cubic or face-centered cubic.

12.21. The unit cell in iron metal is either fcc or bcc, depending on temperature. Are the fcc form of iron and the bcc form allotropes? Explain your answer.

*12.22. At low temperatures, the unit cell of calcium metal is found to be fcc, a closest-packed crystal lattice. At higher temperatures, the unit cell of calcium metal is found to be bcc, a crystal lattice that is not a closest-packed structure. What might be a reason for this difference?

Problems

12.23. Europium, one of the lanthanide elements used in television screens, crystallizes in a crystal lattice built on bcc unit cells, with a unit cell edge of 240.6 pm. Calculate the radius of a europium atom.

12.24. Nickel has an fcc unit cell with an edge length of 350.7 pm. Calculate the radius of a nickel atom.

12.25. What is the length of an edge of the unit cell when barium (atomic radius 222 pm) crystallizes in a crystal lattice of bcc unit cells?

12.26. What is the length of an edge of the unit cell when aluminum (atomic radius 143 pm) crystallizes in a crystal lattice of fcc unit cells?

12.27. A crystalline form of copper has a density of 8.95 g/cm^3. If the radius of copper atoms is 127.8 pm, is the copper unit cell (a) simple cubic; (b) body-centered cubic; or (c) face-centered cubic?

12.28. A crystalline form of molybdenum has a density of 10.28 g/cm^3 at a temperature at which the radius of a molybdenum atom is 139 pm. Which unit cell is consistent with these data? (a) simple cubic; (b) body-centered cubic; (c) face-centered cubic

12.29. Sodium metal crystallizes with a body-centered cubic structure at normal atmospheric pressure. The atomic radius of a sodium atom is 186 pm and the density of Na is 0.971 g/cm^3. At 613,000 atm (9 million psi) the structure of Na metal becomes fcc. Assuming the radius of Na is unchanged, what is the density of the fcc form of Na?

12.30. Calcium metal undergoes two phase changes under pressure. At atmospheric pressure, Ca crystallizes in an fcc unit cell that changes to a bcc unit cell above 20 GPa and to a simple cubic structure above 32 GPa.

a. If the radius of Ca remains unchanged, calculate the density of the simple cubic phase.

*b. Calculate the radius of a Ca atom in the simple cubic structure if the density of the sc and fcc phases is the same.

*12.31. The unit cell for hcp Ti is shown in Figure P12.31. Given the unit cell dimensions, calculate the density of titanium (atomic radius 147 pm).

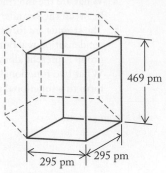

FIGURE P12.31

*12.32. Cobalt crystallizes with the same unit cell as titanium (Figure P12.31). The edge length (251 pm) and height (407 pm) are both shorter than for titanium. Is cobalt denser than titanium? Explain your answer.

Alloys and Medicine

Concept Review

12.33. Describe the structural differences between substitutional and interstitial alloys and give an example of each alloy.

12.34. Describe how homogeneous alloys and intermetallic compounds are similar and how they are different.

12.35. What effect does the substitution of Ni for Ti in the center of a bcc unit cell of Ti have on the unit cell edge length?

12.36. White gold was originally developed to give the appearance of platinum. One formulation of white gold contains 25% nickel and 75% gold. Which is more malleable, white gold or pure gold? Explain your answer.

12.37. Magnesium and hafnium have nearly the same atomic radius (within 1 pm). Does substitution of 25% of the magnesium by hafnium in an alloy increase or decrease the density of the alloy in comparison with pure magnesium? Explain your answer.

12.38. Why are the alloys that second-row nonmetals—such as B, C, and N—form with transition metals more likely to be interstitial than substitutional?

Problems

12.39. The unit cell of NiTi in Figure P12.39(a) shows a bcc arrangement of Ti atoms with Ni in the middle. Could we also draw the unit cell of NiTi as a bcc arrangement of Ni atoms with Ti in the middle as shown in Figure P12.39(b)? Explain why or why not.

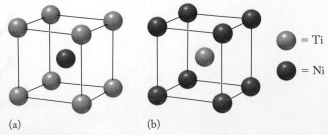

(a)   (b)   ● = Ti
                                    ● = Ni

FIGURE P12.39

12.40. Can both of the unit cells in Figure P12.40 represent the same substitutional alloy? Explain why or why not.

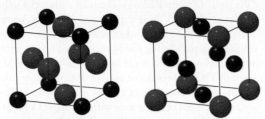

FIGURE P12.40

12.41. **Hydrogen Storage** Hydrogen is an attractive alternative fuel to hydrocarbons because it has a high fuel value and does not produce carbon dioxide when burned. One challenge to developing a hydrogen economy is storing hydrogen. Many transition metals absorb hydrogen by breaking the H—H bond and storing H atoms in the interstices between the metal atoms.

    a. What is the minimum radius for the host metal for a H atom (radius 37 pm) to fit in a tetrahedral hole of an fcc unit cell?

    b. If the H is present as a hydride ion, H^- (ionic radius = 146 pm), is the hydrogen more likely to be found as an interstitial alloy in a tetrahedral or octahedral hole or as a substitutional alloy? *Hint*: Consult Appendix 3 for atomic radii of metals.

12.42. Metal borides MB_x exist for many transition metals.

    a. What is the minimum radius for the host metal for a B atom (radius 88 pm) to fit in an octahedral hole of an fcc unit cell?

    b. What is the value of x if an average of one octahedral hole is occupied per unit cell?

12.43. **Dental Fillings** Dental fillings are mixtures of several alloys, including one with the formula Ag_3Sn. Silver (radius 144 pm) and tin (140 pm) both crystallize in an fcc unit cell. Is this alloy likely to be a substitutional alloy or an interstitial alloy?

12.44. An alloy used in dental fillings has the formula Sn_3Hg. The radii of tin and mercury atoms are 140 pm and 151 pm, respectively. Which alloy has a smaller mismatch (percent difference in atomic radii), Sn_3Hg or bronze (Cu/Sn alloys)?

12.45. What is the formula of vanadium carbide if the vanadium atoms have a cubic closest-packed structure and two of the octahedral holes are occupied by C atoms?

12.46. What is the formula of manganese nitride if the manganese atoms have an fcc unit cell and one of the tetrahedral holes is occupied by a N atom?

12.47. Calculate the density of nitinol, the memory alloy with the composition NiTi and the unit cell shown in Figure P12.47. The atomic radii of Ni and Ti are 124 and 147 pm, respectively.

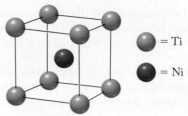

FIGURE P12.47

12.48. Calculate the density of a rose gold alloy composed of 25% Cu and 75% Au given the unit cell in Figure P12.48. The atomic radii of Cu and Au are 128 and 144 pm, respectively.

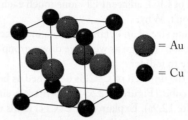

FIGURE P12.48

12.49. One of the many alloys of copper and zinc has the unit cell shown in Figure P12.49.

    a. What type of crystal structure does this alloy have?

    b. What are the proportions of copper and zinc in the alloy?

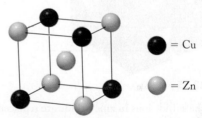

FIGURE P12.49

*12.50. The unit cell of an iron–aluminum alloy is shown in Figure P12.50. (*Hint*: Consult Figure 12.6 for help in identifying the atoms.)

    a. Is this alloy a substitutional or interstitial alloy?

    b. What is the composition of this alloy?

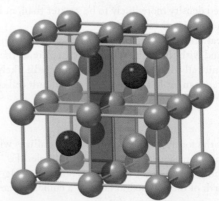

FIGURE P12.50

12.51. If the unit cell of a substitutional alloy of copper and tin has the same unit cell edge as the unit cell of copper, will the alloy have a greater density than copper? Explain your answer.

12.52. If the unit cell of an interstitial alloy of vanadium and carbon has the same unit cell edge as the unit cell of vanadium, will the alloy have a greater density than vanadium? Explain your answer.

Ionic Solids and Salt Crystals

Concept Review

12.53. Crystals of both LiCl and KCl have the rock salt structure. In the unit cell of LiCl, adjacent Cl^- ions touch each other. In KCl, they don't. Why?

12.54. Does the absolute size of an octahedral hole in an fcc lattice of halide ions change as we move down group 17 from fluoride to iodide?

*12.55. In some books the unit cell of CsCl is described as being body-centered cubic (Figure P12.55); in others, as simple cubic (see Figure 12.9a). Explain how CsCl crystals might be described by either unit cell type.

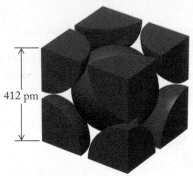

412 pm

FIGURE P12.55

12.56. If some of the sulfide ions in zinc sulfide are replaced by selenide ions, will the selenide ions occupy the same sites as the sulfide ions?

12.57. Why can't calcium chloride have a rock salt structure?

12.58. Why can't sodium fluoride have exactly the same structure as calcium fluoride?

*12.59. As the cation–anion radius ratio increases for an ionic compound with the rock salt crystal structure, is the calculated density more likely to be greater than, or less than, the measured value?

*12.60. As the cation–anion radius ratio increases for an ionic compound with the rock salt crystal structure, is the length of the unit cell edge calculated from ionic radii likely to be greater than, or less than, the observed unit cell edge length?

Problems

12.61. What is the formula of the oxide that crystallizes with Fe^{3+} ions in one-fourth of the octahedral holes, Fe^{3+} ions in one-eighth of the tetrahedral holes, and Mg^{2+} in one-fourth of the octahedral holes of a cubic closest-packed arrangement of oxide ions (O^{2-})?

12.62. What is the chemical formula of the compound that crystallizes in a simple cubic arrangement of fluoride ions with Ba^{2+} ions occupying half of the cubic holes?

12.63. **The Vinland Map** At Yale University there is a map, believed to date from the 1400s, of a landmass labeled "Vinland" (Figure P12.63). The map is thought to be evidence of early Viking exploration of North America. Debate over the map's authenticity centers on yellow stains on the map paralleling the black ink lines. One analysis suggests the yellow color is from the mineral anatase, a form of TiO_2 that was not used in 15th-century inks.
a. The crystal structure of anatase is approximated by a ccp arrangement of oxide ions with titanium(IV) ions in holes. Which type of hole are Ti^{4+} ions likely to occupy? (The radius of Ti^{4+} is 60.5 pm.)
b. What fraction of these holes is likely to be occupied?

FIGURE P12.63

*12.64. The crystal structure of olivine—M_2SiO_4 (where M = Mg or Fe)—can be viewed as a ccp arrangement of oxide ions with silicon(IV) in tetrahedral holes and the metal ions in octahedral holes.
a. What fraction of each type of hole is occupied?
b. The unit cell volumes of Mg_2SiO_4 and Fe_2SiO_4 are 2.91×10^{-26} cm³ and 3.08×10^{-26} cm³, respectively. Why is the unit cell volume of Fe_2SiO_4 larger?

12.65. The rock salt structure (Figure 12.18) of magnesium selenide (MgSe) at atmospheric pressure changes to a cesium chloride structure (Figure P12.55) under pressure.
a. How do these structures differ?
b. Explain how both structures are consistent with the observed stoichiometry of MgSe.

12.66. The sphalerite structure of ZnS changes to a rock salt structure above 15 GPa.
a. Describe the differences between these two structures.
b. Explain how both structures are consistent with the observed stoichiometry of ZnS.

12.67. The unit cell of rhenium trioxide (ReO_3) consists of a cube with rhenium atoms at the corners and an oxygen atom on each of the 12 edges. The atoms touch along the edge of the unit cell. The radii of Re and O atoms in ReO_3 are 137 and 73 pm, respectively. Calculate the density of ReO_3.

12.68. With reference to Figure P12.55, calculate the density of simple cubic CsCl.

12.69. Magnesium oxide crystallizes in the rock salt structure. Its density is 3.60 g/cm³. What is the edge length of the fcc unit cell of MgO?

12.70. Crystalline potassium bromide (KBr) has a rock salt structure and a density of 2.75 g/cm³. Calculate its unit cell edge length.

Allotropes of Carbon

Concept Review

12.71. When amorphous red phosphorus is heated at high pressure, it is transformed into the allotrope black phosphorus, which can exist in one of several forms. One form consists of six-membered rings of phosphorus atoms (Figure P12.71a). Why are the six-atom rings in black phosphorus puckered, whereas the six-atom rings in graphite (Figure P12.71b) are planar?

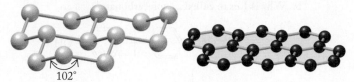

102°

(a) Black phosphorus   (b) Graphite
FIGURE P12.71

*12.72. If the carbon atoms in graphite are replaced by alternating B and N atoms, would the resulting structure contain puckered rings like black phosphorus or flat ones like graphite (see Figure P12.71)?

Problems

12.73. In the fullerene known as buckminsterfullerene, C_{60}, molecules of C_{60} form a cubic closest-packed array of spheres with a unit cell edge length of 1410 pm.
   a. What is the density of crystalline C_{60}?
   b. If we treat each C_{60} molecule as a sphere of 60 carbon atoms, what is the radius of the C_{60} molecule?

12.74. C_{60} reacts with alkali metals to form M_3C_{60} (where M = Na or K). The crystal structure of M_3C_{60} contains cubic closest-packed spheres of C_{60} with metal ions in holes. (Use the radius for C_{60} that you calculated in Problem 12.73.)
   a. If the radius of a K^+ ion is 138 pm, which type of hole is a K^+ ion likely to occupy? What fraction of the holes will be occupied?
   b. Under certain conditions, a different substance, K_6C_{60}, can be formed in which the C_{60} molecules have a bcc unit cell. Calculate the density of a crystal of K_6C_{60}.

12.75. The distance between atoms in a cubic form of phosphorus is 238 pm (Figure P12.75). Calculate the density of this form of phosphorus.

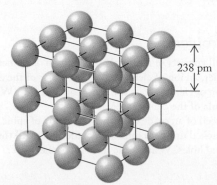

238 pm

FIGURE P12.75

12.76. **Ice under Pressure** Kurt Vonnegut's novel *Cat's Cradle* describes an imaginary, high-pressure form of ice called "ice-nine." With the assumption that ice-nine has a cubic closest-packed arrangement of oxygen atoms with hydrogen atoms in the appropriate holes, what type of hole will accommodate the H atoms?

Polymers

Concept Review

12.77. Can a polymer be composed of more than one type of monomer? Explain why or why not.

12.78. Compare the large hydrocarbon $C_{24}H_{50}$ with polyethylene. What structural feature(s) do they share in common? How do the structures of these compounds differ?

Problems

12.79. Polyethylene is prepared from the monomer ethylene, C_2H_4. About how many monomers are needed to make a polymer with a molar mass of 100,000 g/mol?

12.80. Synthetic rubber is prepared from butadiene, C_4H_6. About how many monomers are needed to make a polymer with a molar mass of 100,000 g/mol?

12.81. **Making Glue** Wood glue, or "carpenter's glue," is made of poly(vinyl acetate). Draw the carbon-skeleton structure of this polymer. The monomer is shown in Figure P12.81.

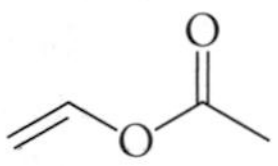

Vinyl acetate
FIGURE P12.81

12.82. The 2000 Nobel Prize in Chemistry was awarded for research on the electrically conductive polymer polyacetylene. Draw the carbon-skeleton structure of three monomeric units of the addition polymer that results from polymerization of acetylene, HC≡CH.

12.83. **Soap** One of the first chemical syntheses ever carried out was the making of soap. This same reaction is still carried out by modern-day manufacturers. The reaction is called saponification, and it is the reverse of esterification: an ester is taken apart to produce its component acid and alcohol. A compound called a glyceride, which is a fat, is reacted with an aqueous base (usually NaOH). The ester bonds are broken to yield glycerol (an alcohol) and the sodium salt of a fatty acid, which is then used as soap. Given the glyceride shown in Figure P12.83, draw the structures of the alcohol and the fatty acids that result from its saponification.

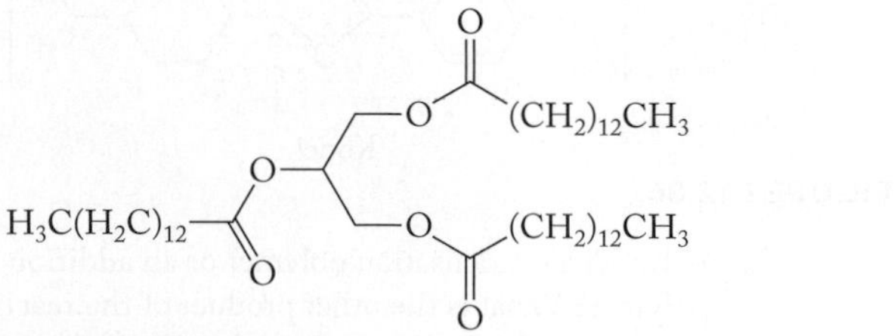

$H_3C(H_2C)_{12}$ $(CH_2)_{12}CH_3$

$(CH_2)_{12}CH_3$

FIGURE P12.83

12.84. Reactions between 1,6-diaminohexane, $H_2N(CH_2)_6NH_2$, and different dicarboxylic acids, $HOOC(CH_2)_nCOOH$, are used to prepare polymers that have a structure similar to that of nylon. How many carbon atoms (*n*) were in the dicarboxylic acids used to prepare the polymers with the repeating units shown in Figure P12.84?

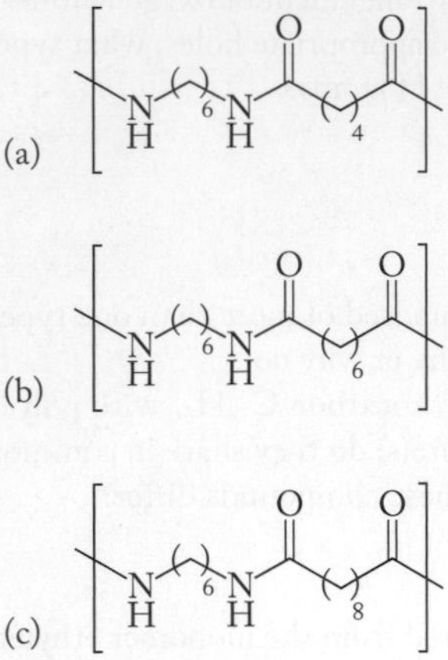

(a)

(b)

(c)

FIGURE P12.84

12.85. The two polymers in Figure P12.85 have the same empirical formula.
   a. What pairs of monomers could be used to make each of them?
   b. How might the physical properties of these two polymers differ?

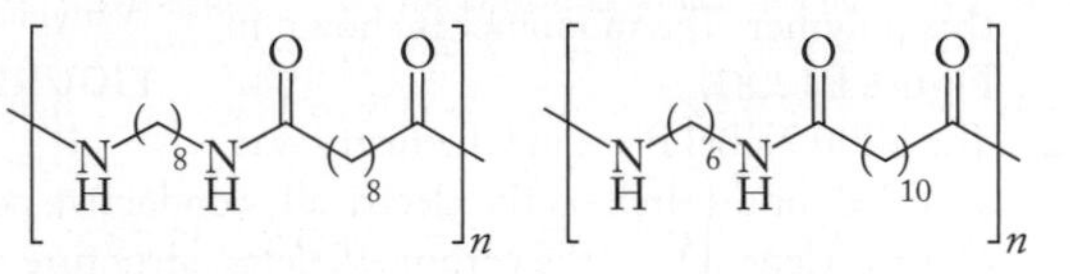

Polymer I   Polymer II

FIGURE P12.85

12.86. The polyester called Kodel is made with polymeric strands prepared by the reaction of dimethyl terephthalate with 1,4-di(hydroxymethyl)cyclohexane (Figure P12.86).

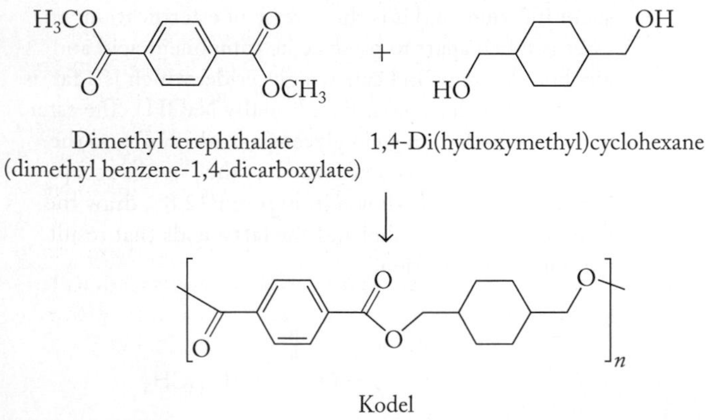

Dimethyl terephthalate
(dimethyl benzene-1,4-dicarboxylate)   1,4-Di(hydroxymethyl)cyclohexane

Kodel

FIGURE P12.86

   a. Is Kodel a condensation polymer or an addition polymer? What is the other product of the reaction?

*b. Dacron (see Sample Exercise 12.7) is made from dimethyl terephthalate and ethylene glycol. What properties of Kodel fibers might make them better than Dacron as a clothing material?

12.87. Lexan is a polymer belonging to the class of materials called polycarbonates. Figure P12.87 shows the polymerization reaction for Lexan.
   a. What other compound is formed in the polymerization reaction?
   *b. Why is Lexan called a "polycarbonate"?

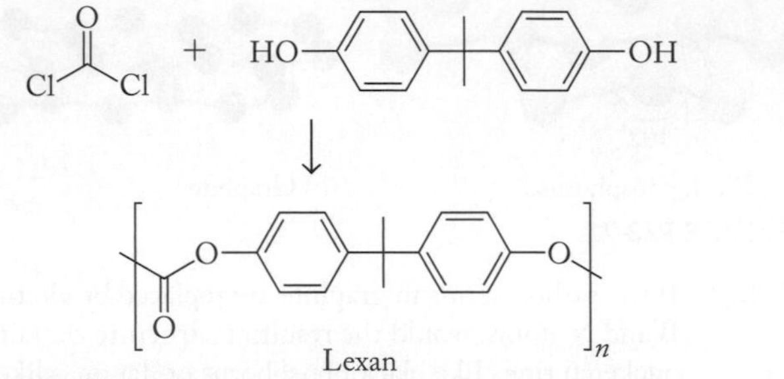

Lexan

FIGURE P12.87

12.88. **Non-Biodegradable Polymers** Biodegradable polymers are important in applications such as dissolving sutures, but biocompatible polymers that resist degradation can serve as drug delivery systems in implanted devices. Poly(caprolactone) (PCL, Figure P12.88) degrades slowly in the human body but is very permeable to the contraceptive levonorgestrel. This combination is used in implantable contraceptive devices that are effective for many years.
   *a. Why is levonorgestrel soluble in PCL?
   b. PCL does eventually degrade by reacting with water. What are the products of biodegradation?
   c. Is PCL an addition or a condensation polymer?

PCL   Levonorgestrel

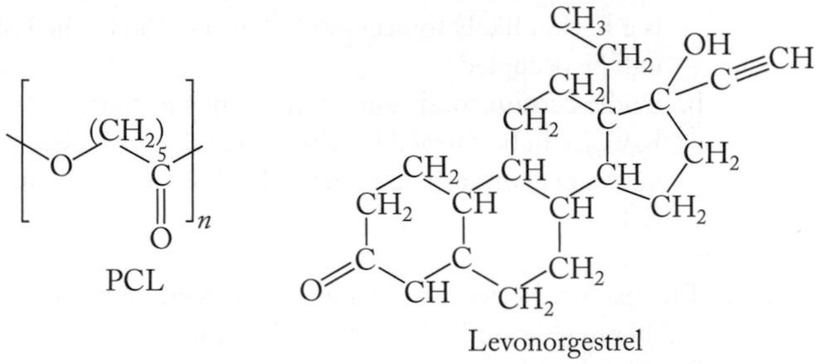

FIGURE P12.88

Additional Problems

12.89. A unit cell consists of a cube that has an ion of element X at each corner, an ion of element Y at the center of the cube, and an ion of element Z at the center of each face. What is the formula of the compound?

12.90. The unit cell of an oxide of uranium consists of cubic closest-packed uranium ions with oxide ions in all the tetrahedral holes. What is the formula of the oxide?

12.91. The phase diagram for titanium is shown in Figure P12.91.
  a. Which structure does Ti metal have at 1500 K and 6 GPa of pressure?
  b. How many phase changes does Ti metal undergo as pressure is increased at 725°C?

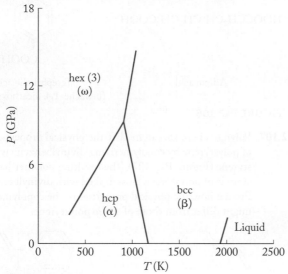

FIGURE P12.91

12.92. The phase diagram of thallium metal is shown in Figure P12.92.
  a. At what temperature and pressure are all three solid phases of thallium metal present?
  b. Does the melting point of thallium increase or decrease with increasing pressure?

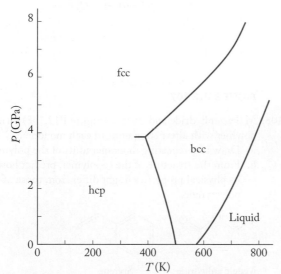

FIGURE P12.92

12.93. Silver nanoparticles are embedded in clothing fabric to kill odor-causing bacteria. Silver crystallizes in an fcc unit cell.
  a. How many unit cells are present in a cubic silver particle with an edge length of 25 nm?
  *b. How many silver atoms are there in this particle?

  c. If there are 1360 μg of silver per sock, how many of these particles does this mass correspond to?

12.94. In 2011, researchers at Duke University reported on the use of gold "nanorods," cylindrical gold particles to image brain tumors (Figure P12.94).
  a. If the nanorods have a diameter of 32 nm and a height of 67 nm, approximately how many gold atoms are in each nanorod?
  b. How many unit cells does this correspond to?
  c. What is the mass of each nanorod?

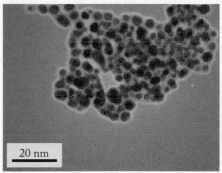

FIGURE P12.94

12.95. The center of Earth is composed of a solid iron core within a molten iron outer core. When molten iron cools, it crystallizes in different ways depending on pressure—in a bcc unit cell at low pressure and in a hexagonal unit cell at high pressures like those at Earth's center.
  a. Calculate the density of bcc iron given that the radius of an iron atom is 126 pm.
  b. Calculate the density of hexagonal iron given a unit cell volume of 5.414×10^{-23} cm^3.
  *c. Seismic studies suggest that the density of Earth's solid core is only about 90% of that of hexagonal Fe. Laboratory studies have shown that up to 4% by mass of Si can be substituted for Fe without changing the hcp crystal structure built on hexagonal unit cells. Calculate the density of such a crystal.

12.96. The unit cell of an alloy with a 1:1 ratio of magnesium and strontium is identical to the unit cell of CsCl. The unit cell edge of MgSr is 390 pm. What is the density of MgSr?

12.97. Gold and silver can be separately alloyed with zinc to form AuZn (unit cell edge 319 pm) and AgZn (unit cell edge 316 pm). The two alloys have the same unit cell. Which alloy is more dense?

12.98. Figure P12.98 shows four identical layers of atoms in an *aaaa* . . . pattern. Using the yellow triangle as a guide, how many nearest-neighbor atoms does any one hole between the layers have? Are there any other kinds of holes in this structure?

FIGURE P12.98

12.99. Manganese steels are a mixture of iron, manganese, and carbon. Is the manganese likely to occupy holes in the austenite fcc unit cell, or are manganese steels substitutional alloys?

12.100. Aluminum forms alloys with lithium (LiAl), gold (AuAl₂), and titanium (Al₃Ti). On the basis of their crystal lattices, each of these alloys is considered a substitutional alloy.
  a. Do these alloys fit the general size requirements for substitutional alloys? The atomic radii for Li, Al, Au, and Ti are 152, 143, 144, and 147 pm, respectively.
  b. If the unit cell of LiAl is bcc, what is the density of LiAl?

12.101. The aluminum alloy Cu_3Al crystallizes in a bcc unit cell. Propose a way that the Cu and Al atoms could be allocated between bcc unit cells that is consistent with the formula of the alloy.

12.102. Light-Emitting Diodes The colored lights on many electronic devices are light-emitting diodes (LEDs). One of the compounds used to make them is aluminum phosphide (AlP), which crystallizes in a sphalerite crystal structure.
  a. If AlP were an ionic compound, would the ionic radii of Al^{3+} and P^{3-} be consistent with the size requirements of the ions in a sphalerite crystal structure?
  b. If AlP were a covalent compound, would the atomic radii of Al and P be consistent with the size requirements of atoms in a sphalerite crystal structure?

12.103. Identify the reactants in the polymerization reactions that produce the polymers shown in Figure P12.103.

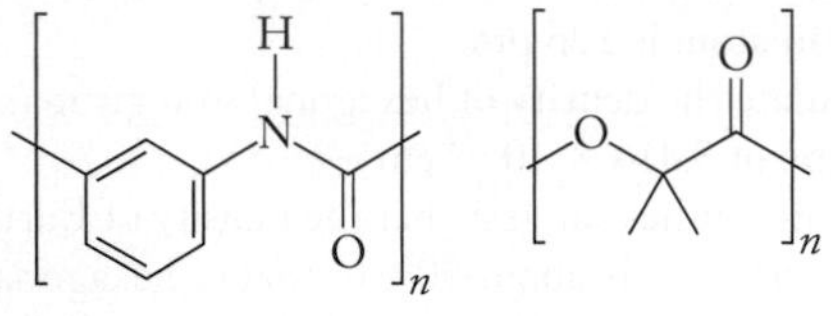

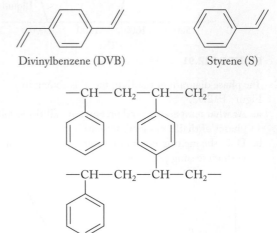

(a)   (b)

FIGURE P12.103

12.104. Raincoats "Waterproof" nylon garments have a coating to prevent water from penetrating the hydrophilic fibers. Which functional groups in the nylon molecule make it hydrophilic?

12.105. Draw the carbon-skeleton structure of the condensation polymer of $H_2N(CH_2)_6COOH$. How does this polymer compare with nylon-6?

12.106. Putrescine, $H_2N(CH_2)_4NH_2$, is one of the compounds that form in rotting meat.
  a. Draw the carbon-skeleton structures of all the trimers (a molecule formed from three monomers) that can be formed from putrescine, adipic acid, and terephthalic acid (Figure P12.106). The three monomers forming the trimer do not have to be different from one another.

  b. A chemist wishes to make a putrescine polymer containing a 1:1 ratio of adipic acid to terephthalic acid. What should be the mole ratio of the three reactants?

HOOCCH₂CH₂CH₂CH₂COOH

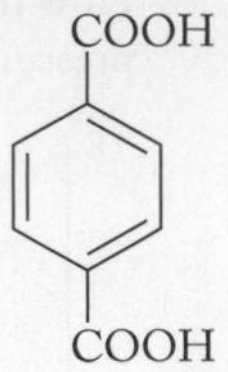

Adipic acid

Terephthalic acid
(benzene-1,4-dicarboxylic acid)

FIGURE P12.106

12.107. Polymer chemists can modify the physical properties of polystyrene by copolymerizing divinylbenzene with styrene (Figure P12.107). The resulting polymer has strands of polystyrene cross-linked with divinylbenzene. Predict how the physical properties of the copolymer might differ from those of 100% polystyrene.

Divinylbenzene (DVB)   Styrene (S)

S cross-linked with DVB

FIGURE P12.107

12.108. Maleic anhydride and styrene (Figure P12.108) form a polymer with alternating units of each monomer.
  a. Draw two repeating monomer units of the polymer.
  b. From the structure of the copolymer, predict how its physical properties might differ from those of polystyrene.

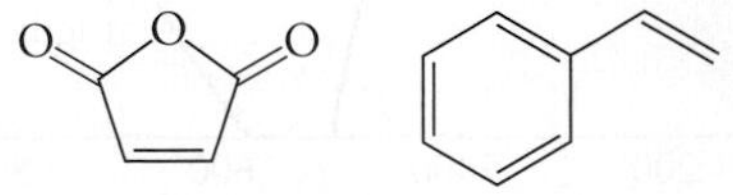

Maleic anhydride   Styrene

FIGURE P12.108

12.109. Superglue The active ingredient in superglue is methyl 2-cyanoacrylate (Figure P12.109). The liquid glue hardens rapidly when methyl 2-cyanoacrylate polymerizes. This happens when it contacts a surface containing traces of water or other compounds containing –OH or –NH– groups. Draw the carbon-skeleton structure of two repeating monomer units of poly(methyl 2-cyanoacrylate).

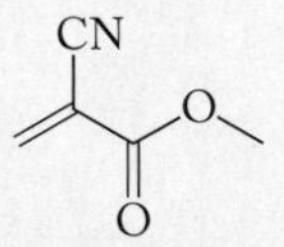

Methyl 2-cyanoacrylate
FIGURE P12.109

*12.110. Silicones are polymeric materials with the formula $[R_2SiO]_n$ (Figure P12.110). They are prepared by reaction of R_2SiCl_2 with water, yielding the polymer and aqueous HCl. Consider this reaction as taking place in two steps: (1) water reacts with 1 mole of R_2SiCl_2 to produce a new monomer and 1 mole of HCl(aq); (2) one new monomer

molecule reacts with another new monomer molecule to eliminate one molecule of H_2O and make a dimer with a Si—O—Si bond.

a. Suggest two balanced equations describing these reactions that occur over and over again to produce a silicone polymer.
b. Why are silicones water repellent?

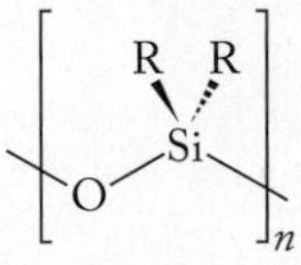

Silicone
FIGURE P12.110

13

Chemical Kinetics

Reactions in the Atmosphere

AIR QUALITY A layer of brown photochemical smog hangs over Tokyo.

PARTICULATE **REVIEW**

Concentration and Solutions

In Chapter 13 we explore how the rates of reactions depend on variables such as concentration and temperature. The image shows two graduated cylinders that contain aqueous solutions of copper(II) nitrate. Cylinder 1 has a concentration of 0.01 *M*.

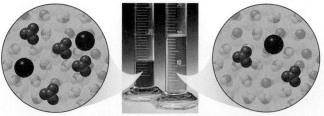

Cylinder 1   Cylinder 2

- Which cylinder contains the more concentrated solution?

- What is the concentration of the solution in cylinder 2?

- Which cylinder contains more dissolved solute?

    (Review Section 4.2 if you need help.)

(Answers to Particulate Review questions are in the back of the book.)

Collisions and Reactions

The images represent three samples of nitrogen dioxide, which is the pollutant responsible for the brown color of the atmosphere in the cover photograph of this chapter. At high temperatures, nitrogen dioxide decomposes to nitrogen monoxide and oxygen gas. As you read Chapter 13, look for ideas that will help you answer these questions:

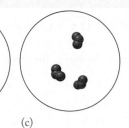

(a)   (b)   (c)

- Write a balanced equation describing the thermal decomposition of NO_2.

- In which of the NO_2 samples represented by the images do the NO_2 molecules collide most frequently, assuming the temperature of all three is the same?

- Assuming the rate of the decomposition reaction depends on the frequency of the collisions between NO_2 molecules, in which sample does the reaction proceed four times as rapidly as it does in sample (c)?

Learning Outcomes

LO1 Relate the rates of change in the concentrations of reactants and products to each other and to reaction rates
Sample Exercises 13.1, 13.2

LO2 Determine average and instantaneous reaction rates from experimental data
Sample Exercise 13.3

LO3 Derive rate laws from initial reaction rate data
Sample Exercise 13.4

LO4 Use integrated rate laws to identify zero-, first-, and second-order reactions
Sample Exercises 13.5, 13.8

LO5 Calculate changes in reactant concentration with time
Sample Exercise 13.6

LO6 Calculate half-lives of zero-, first-, and second-order reactions
Sample Exercise 13.7

LO7 Calculate the activation energy of a reaction and the effect of temperature on rate constants
Sample Exercises 13.9, 13.10

LO8 Link mechanisms to rate laws and identify catalysts and catalysis
Sample Exercises 13.11, 13.12, 13.13

FIGURE 13.1 A London policeman uses a torch so that drivers will see his traffic directions during the Great Smog of 1952.

photochemical smog a mixture of gases formed in the lower atmosphere when sunlight interacts with compounds produced in internal combustion engines and other pollutants.

13.1 Cars, Trucks, and Air Quality

Consider the brown haze hanging over Tokyo in the chapter-opening photograph. Perhaps you have seen similar conditions over a large urban area. The common name for this kind of air pollution is *smog*, a name first used over a century ago to describe the pollution that blanketed some cities during the winter when *smoke* from coal-fired stoves and furnaces combined with natural *fog* to produce air that had the appearance of pea soup. This greenish-yellow form of smog contained high concentrations of volatile sulfur compounds that combined with atmospheric moisture, forming highly acidic aerosols (suspended particles). Breathing in these pollutants can cause severe respiratory distress. During one particularly intense smog event in London, England, in December of 1952 (Figure 13.1), more than 100,000 people were sickened by the smog and as many as 12,000 people—mostly the elderly and very young children—died from impaired lung function caused by the smog. In response to this environmental disaster, Parliament passed the Clean Air Act of 1956, which banned the use of high-sulfur coal for heating homes in cities and towns across the United Kingdom.

The smog in the chapter-opening photograph is different from the pea-soup variety that covered London. It is called **photochemical smog**, and its formation begins with the combustion of the gasoline that powers most cars and trucks. The high temperatures inside the engines in these vehicles allow N_2 and O_2 to combine, producing nitrogen monoxide:

$$N_2(g) + O_2(g) \rightarrow 2\,NO(g) \tag{13.1}$$

When the NO in engine exhaust enters the atmosphere, it slowly reacts with more oxygen, producing nitrogen dioxide:

$$2\,NO(g) + O_2(g) \rightarrow 2\,NO_2(g) \tag{13.2}$$

which is the gas that gives photochemical smog its brown color. Sunlight can break the bonds in NO_2, producing NO and very reactive oxygen atoms:

$$NO_2(g) \xrightarrow{\text{sunlight}} NO(g) + O(g) \tag{13.3}$$

These atoms of oxygen may react with molecules of water vapor, producing hydroxyl radicals:

$$O(g) + H_2O(g) \rightarrow 2\,OH(g) \tag{13.4}$$

or they may combine with molecules of O_2, forming ozone.

$$O(g) + O_2(g) \rightarrow O_3(g) \tag{13.5}$$

Even trace concentrations of O_3 in the air can irritate your eyes and respiratory system. In addition, O_3 and OH radicals react with volatile organic compounds (VOCs) in the atmosphere, converting them to compounds containing C=O double bonds, such as acetaldehyde (Figure 13.2).

Acetaldehyde undergoes further oxidation and combines with nitrogen dioxide, forming a compound with the common name peroxyacetyl nitrate (PAN); Figure 13.3). PAN is an extremely potent eye and respiratory irritant and a principal contributor to the watery eyes, runny noses, sore throats, and difficulty breathing that many people experience during photochemical smog events.

The chemical reactions in Equations 13.1 to 13.5 are linked: the products of the early reactions in the sequence are reactants in later reactions. This linkage is the reason why the atmospheric concentrations of different smog components reach their maximum value at different times during the day (Figure 13.4). Also, the *rates* of the reactions—that is, the rates at which reactants are consumed and products are formed—control the atmospheric concentrations of the components, including when they form and how long they persist.

Consider, for example, the formation of NO from N_2 and O_2 (Equation 13.1). This endothermic reaction ($\Delta H° = 180.6$ kJ/mol) proceeds only at extremely high temperatures. However, the gases in the combustion chamber of an automobile engine experience such temperatures for only a tiny fraction of a second before they flow into the car's exhaust system. Therefore, NO forms only if the rate of its formation reaction is very rapid. Unfortunately, the reaction is quite rapid.

To further explore the impact of reaction rates on smog formation, let's return to Figure 13.4 and see how NO emitted by vehicles during a morning rush hour affects air quality on a sunny day with little wind over a large urban area. Heavy traffic and exhaust emissions increase the concentration of NO, which reaches a maximum in the atmosphere early in the morning—near the height of traffic. By mid-morning, NO levels start to fall and NO_2 levels start to rise as NO combines with atmospheric O_2. By noon, the concentration of NO_2 is falling as it undergoes photodecomposition, releasing O atoms that combine with O_2, forming O_3. Ozone concentrations reach their maximum in mid-afternoon and then fall as this highly reactive allotrope of oxygen combines with VOCs and nitrogen oxides, forming PAN and other volatile organic pollutants.

Smog conditions and ozone concentrations that posed a threat to human health were a worsening environmental crisis during the 1950s and 60s in many American cities, particularly in metropolitan Los Angeles. In fact, the expression "brown L.A. haze" became so common it found its way into the lyrics of a popular song.[1] To combat smog pollution, exhaust systems of automobiles sold in the United States since the 1975 model year have been equipped with devices called *catalytic converters*, which reduce the concentrations of NO in the exhaust gases.

In this chapter we explore how catalytic converters accelerate the rates of chemical reactions that turn NO and other pollutants into nonpolluting products.

[1]The chorus of "Come Monday" contains the line "I spent four lonely days in a brown L.A. haze." The song was written and recorded by Jimmy Buffett in 1974, the year that automobiles equipped with catalytic converters first went on sale.

CONNECTION A free radical, also called just a "radical," is a molecule with an odd number of electrons, which makes it chemically reactive, as we discussed in Chapter 8.

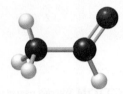

FIGURE 13.2 Acetaldehyde.

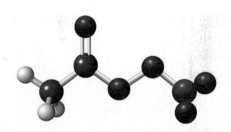

FIGURE 13.3 Peroxyacetyl nitrate (PAN).

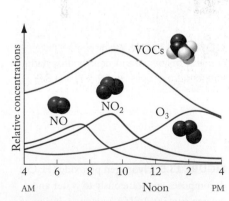

FIGURE 13.4 The concentrations of these components of photochemical smog reach their maxima in the atmosphere at different times as they are produced and then consumed in the many reactions that contribute to smog formation.

chemical kinetics the study of the rates of change of concentrations of substances involved in chemical reactions.

To understand this process, we need to investigate the factors that influence the rates of chemical reactions, and before we do that, we need to learn how to measure reaction rates and to express the results of those measurements. All of these concepts are part of **chemical kinetics**, the study of the rates of chemical reactions and the factors that influence those rates. This knowledge is used by chemists to determine how reactions happen at the molecular level—that is, the *mechanisms* of reactions. With this knowledge, chemists can manipulate reaction conditions to increase reaction rates and, ultimately, product yields. This fundamental branch of chemistry is the focus of this chapter.

CONNECTION We discussed kinetic molecular theory, a model that describes the behavior of gases, in Chapter 6.

13.2 Reaction Rates

Previous chapters in this book provide a foundation on which we can build an understanding of chemical kinetics. In Chapter 6, for example, we explored the kinetic molecular theory of gases and the relationship between temperature and the average speeds of molecules in the gas phase. Faster speeds mean more frequent and forceful collisions between molecules. As we see in this chapter, more frequent collisions can mean more rapid reactions. Reaction rates vary over a wide range (Figure 13.5), from so slow that we barely perceive they are occurring to explosively fast. Let's consider four factors that influence reaction rates:

1. *Physical states of the reactants.* The more frequently that particles interact—that is, the more often they collide with one another—the faster they can react with each other. Particles in solids have the same nearest neighbors over time, whereas the particles in liquids move in relation to one another. The particles in gases are widely separated but move very rapidly. These differences in motion are one reason why reactions in the solid phase tend to be much slower than reactions in liquids and gases.

2. *Concentration of reactants.* The rate of a reaction depends on the quantities of reactants present in a given volume. Greater quantities often produce more rapid reaction rates.

3. *Temperature.* As temperature increases, the rates of chemical reactions tend to increase. The average kinetic energy (KE) of particles increases as temperature rises, and as KE increases, the probability of collisions increases.

4. *Catalysts.* Catalytic converters contain materials that increase the rates of reactions that consume pollutants in automobile exhaust. In general, catalysts

CONNECTION We discussed translational (and vibrational and rotational) motion in Section 5.3.

CONNECTION We discussed the connections between kinetic energy and temperature in Sections 5.1, 5.2, and 6.8.

FIGURE 13.5 Hydrogen peroxide (H_2O_2) decomposes spontaneously to water and oxygen gas. (a) A 30% aqueous solution of H_2O_2 decomposes so slowly that no change is observable. (b) Pouring the solution on a wedge of potato, which contains the enzyme catalase, causes the reaction to proceed at a faster rate, clearly visible by the bubbles of O_2 gas given off. (c) The addition of a small amount of MnO_2 causes the reaction to proceed so rapidly that the heat liberated causes the solution to boil.

(a)

(b)

(c)

are materials or substances that accelerate reactions without themselves being consumed in the process. Catalysts also play a key role in biochemistry, where proteins called enzymes catalyze most reactions in living systems.

With these four ideas in mind, let's begin our investigation of reaction rates by returning to the formation of NO in internal combustion engines:

$$N_2(g) + O_2(g) \rightarrow 2\ NO(g) \qquad (13.1)$$

We can express the rate of this reaction in terms of the rate of change in the concentration of the product (NO) or the rate of change in the concentration of either reactant (N_2 or O_2). In equations describing concentration changes, we use square brackets around the formulas of substances to represent their concentrations, which are usually expressed in moles per liter. If the concentration of NO increases from $[NO]_{initial}$ to $[NO]_{final}$ during the time between $t_{initial}$ and t_{final}, then the average rate of change in the concentration of NO during this time is

CHEMTOUR
Reaction Rate

$$\frac{\Delta[NO]}{\Delta t} = \frac{[NO]_{final} - [NO]_{initial}}{t_{final} - t_{initial}} \qquad (13.6)$$

Similarly, the rates of change in the concentrations of N_2 or O_2 over the same interval are

$$\frac{\Delta[N_2]}{\Delta t} = \frac{[N_2]_{final} - [N_2]_{initial}}{t_{final} - t_{initial}} \qquad (13.7)$$

and

$$\frac{\Delta[O_2]}{\Delta t} = \frac{[O_2]_{final} - [O_2]_{initial}}{t_{final} - t_{initial}} \qquad (13.8)$$

Now let's derive an equation that relates the three rate-of-change expressions in Equations 13.6–13.8. The last two quantities, $\Delta[N_2]/\Delta t$ and $\Delta[O_2]/\Delta t$, are equal: because N_2 and O_2 are both reactants with the same coefficient (1) in Equation 13.1, their concentrations decrease at the same rate as the reaction proceeds. However, the value of $\Delta[NO]/\Delta t$ during the same interval is different because the coefficient of 2 for NO in Equation 13.1 tells us that 2 moles of NO form for every 1 mole of N_2 or O_2 that reacts. Therefore the rate of increase in the concentration of NO is *twice* the rate of decrease in the concentrations of N_2 and O_2. We can express these relative rates of change by using an equation:

$$\frac{\Delta[NO]}{\Delta t} = -2\frac{\Delta[O_2]}{\Delta t} = -2\frac{\Delta[N_2]}{\Delta t} \qquad (13.9)$$

Put another way, the rate at which either N_2 or O_2 is consumed is *half* the rate at which NO is produced, as we see if we divide Equation 13.9 through by 2:

$$-\frac{\Delta[N_2]}{\Delta t} = -\frac{\Delta[O_2]}{\Delta t} = \frac{1}{2}\frac{\Delta[NO]}{\Delta t} \qquad (13.10)$$

You may have noticed the minus signs on the terms describing the rates of consumption of N_2 and O_2. The rate of change in the concentration of a reactant ($\Delta[reactant]/\Delta t$) has a negative value because reactant concentrations decrease as a reaction proceeds; for example, in Equation 13.7, $[N_2]_{initial}$ is greater than $[N_2]_{final}$, which means that the numerator of the equation is a negative number. On the other hand, the rate of change in the concentration of a product is always positive because its value increases as the reaction proceeds. The measured rate of any reaction is defined as a positive quantity, so a minus sign is used with $\Delta[reactant]/\Delta t$ values to obtain a positive value for the reaction rate.

Also note that the number 2 in the denominator of the $\Delta[NO]/\Delta t$ term in Equation 13.10 matches the coefficient of NO in the balanced chemical equation. This pattern applies to all chemical reactions: the coefficient of each species in the balanced chemical equation appears in the *denominator* of its term in the relative rate expression if the numerator values are all 1.

SAMPLE EXERCISE 13.1 Predicting a Relative Reaction Rate   **LO1**

The synthesis of ammonia:

$$N_2(g) + 3\,H_2(g) \rightarrow 2\,NH_3(g)$$

is an important reaction in the production of agricultural fertilizers. Derive an equation that relates the rates of change in the concentrations of N_2, H_2, and NH_3.

Collect and Organize We are given the balanced chemical equation describing the synthesis of NH_3 from N_2 and H_2, and we need to relate the rates of change in the concentrations of its reactants and product: $\Delta[N_2]/\Delta t$, $\Delta[H_2]/\Delta t$, and $\Delta[NH_3]/\Delta t$.

Analyze Nitrogen and hydrogen are consumed in the reaction, which means the values of $\Delta[N_2]/\Delta t$ and $\Delta[H_2]/\Delta t$ are negative. Therefore we need to use minus signs in front of these terms in an equation that includes the positive $\Delta[NH_3]/\Delta t$ term. The coefficients of N_2, H_2, and NH_3 are 1, 3, and 2, respectively, so the coefficients in their rate of change terms are 1, $\frac{1}{3}$, and $\frac{1}{2}$.

Solve The relative rates are related as follows:

$$-\frac{\Delta[N_2]}{\Delta t} = -\frac{1}{3}\frac{\Delta[H_2]}{\Delta t} = \frac{1}{2}\frac{\Delta[NH_3]}{\Delta t}$$

Think About It The solution to the problem tells us that the rate of consumption of N_2 is one-third the rate of consumption of H_2, which makes sense given that three molecules of H_2 react for every one molecule of N_2. Likewise, the rate of consumption of N_2 is one-half the rate of production of NH_3 because two molecules of NH_3 are produced for every molecule of N_2 that reacts.

Practice Exercise How is the rate of change in the concentration of CO_2 related to the rate of change in the concentration of O_2 during the oxidation of carbon monoxide? The equation is:

$$2\,CO(g) + O_2(g) \rightarrow 2\,CO_2(g)$$

(Answers to Practice Exercises are in the back of the book.)

Experimentally Determined Reaction Rates

Rates of product formation or reactant consumption are ratios of changes in concentration divided by changes in time, and therefore they have units of concentration per unit time, such as molarity per second (M/s). Reaction rates cannot be determined by inspection of a balanced equation, but in fact can only be determined experimentally. Once we measure the rate of a reaction, we can then use the stoichiometric information from the balanced equation to calculate changes in other reactant or product concentrations.

For example, suppose we measure the rate of production of ammonia in the reaction in Sample Exercise 13.1 to be 0.472 M/s. How rapidly, then, is N_2 consumed when ammonia is produced at this rate? To answer this question, we insert the known value of $\Delta[NH_3]/\Delta t$ into the equation we derived in Sample Exercise 13.1 and solve for $\Delta[N_2]/\Delta t$:

$$-\frac{\Delta[N_2]}{\Delta t} = \frac{1}{2}\frac{\Delta[NH_3]}{\Delta t} = \frac{1}{2}(0.472\ M/s) = 0.236\ M/s$$

$$\frac{\Delta[N_2]}{\Delta t} = -0.236\ M/s$$

This result makes sense because it is negative, reflecting the decreasing concentration of a reactant, and because its absolute value is half the rate of production of ammonia, which fits the stoichiometry of the reaction equation.

So, which, if either, of these two values—$\Delta[NH_3]/\Delta t = 0.472\ M/s$, or $\Delta[N_2]/\Delta t = -0.236\ M/s$—should we use to express the rate of the reaction? The answer is that we base the reaction rate on the rate of change in the concentration of the reactant or product with a coefficient of 1 in the reaction equation. If none of the coefficients is 1, divide the rate of change in the concentration of any of the products by its coefficient to get the reaction rate. Furthermore, the reaction rate is always expressed as a positive value: therefore the rate of the reaction is 0.236 M/s.

We have just seen how a balanced chemical equation enables us to predict the *relative* rates at which reactants are consumed and products are formed in a particular reaction. However, it provides no information on the numerical values of the rates. Although computational tools can be used to predict the rates of very simple reactions, we limit our discussion of actual reaction rates in this chapter to those based on experiments.

SAMPLE EXERCISE 13.2 Converting Reaction Rates   **LO1**

Suppose that during the reaction between NO and O_2 to form NO_2,

$$2\ NO(g) + O_2(g) \rightarrow 2\ NO_2(g)$$

the rate of change in the concentration of O_2 is $-0.033\ M/s$. What is the rate of formation of NO_2?

Collect and Organize We need to determine the rate of formation of a product in a reaction, knowing the balanced chemical equation and the rate of change in concentration of one of the reactants. The rate of formation of NO_2 is related to the rates of consumption of NO and O_2 by the balanced chemical equation. The coefficients of O_2 and NO_2 are 1 and 2, respectively.

Analyze We can write an equation that expresses the relative rates of change in $[O_2]$ and $[NO_2]$ from the coefficients in the balanced chemical equation:

$$-\frac{\Delta[O_2]}{\Delta t} = \frac{1}{2}\frac{\Delta[NO_2]}{\Delta t}$$

The negative sign is needed because the concentration of O_2 decreases as the concentration of NO_2 increases.

Solve Solving for the rate of change of $[NO_2]$, we get

$$\frac{\Delta[NO_2]}{\Delta t} = -2\frac{\Delta[O_2]}{\Delta t} = -2(-0.033\ M/s) = 0.066\ M/s$$

Think About It This result is twice the magnitude of $\Delta[O_2]/\Delta t$, which is consistent with the stoichiometry of the reaction: 2 moles of NO_2 are produced for every mole of O_2 that reacts.

Average Reaction Rates

Suppose we run an experiment to determine the rate of formation of NO in an automobile engine. In the laboratory, we use a reaction vessel as hot as the combustion chambers in the engine, and we obtain the data listed in Table 13.1 and plotted in Figure 13.6. We can use the data to calculate the reaction rate based on the change in the concentration of any participant in the reaction over a particular interval. Such calculations of reaction rates based on $-\Delta[\text{reactant}]/\Delta t$ or $\Delta[\text{product}]/\Delta t$ are *average* reaction rates. Suppose, for example, we wish to calculate the average rate of change in [NO] between 5.0 and 10.0 μs:

$$\frac{\Delta[NO]}{\Delta t} = \frac{[NO]_{10.0 \ \mu s} - [NO]_{5.0 \ \mu s}}{t_{10.0 \ \mu s} - t_{5.0 \ \mu s}} = \frac{(14.8 - 7.8)\mu M}{(10.0 - 5.0)\mu s}$$

$$= 1.4 \ M/s$$

During the same interval, the average rate of change in the concentration of N_2 (or O_2) is

$$\frac{\Delta[N_2]}{\Delta t} = \frac{[N_2]_{10.0 \ \mu s} - [N_2]_{5.0 \ \mu s}}{t_{10.0 \ \mu s} - t_{5.0 \ \mu s}} = \frac{(9.6 - 13.1)\mu M}{(10.0 - 5.0)\mu s}$$

$$= -0.70 \ M/s$$

These results give us two $\Delta[X]/\Delta t$ values for expressing the rate of the reaction. Recall that, by convention, we choose the value associated with the substance having a coefficient of 1 in the balanced equation, though we have to change the value's sign if that substance is a reactant. This means we can choose the rate of change of either $[N_2]$ or $[O_2]$. Therefore the average rate of this reaction is

$$\text{Rate} = -\frac{\Delta[N_2]}{\Delta t} = 0.70 \ M/s$$

The curved lines in Figure 13.6 tell us that the numerical value of this reaction rate applies only to the interval from $t = 5.0$ μs to $t = 10.0$ μs. Any other 5.0-μs interval would have a different average rate because the reaction slows as more time passes.

Instantaneous Reaction Rates

We can also determine the *instantaneous* rate of a reaction, that is, its rate at a particular instant during the reaction. To see how, let's revisit the oxidation of NO to NO_2 in the atmosphere:

$$2 NO(g) + O_2(g) \rightarrow 2 NO_2(g) \tag{13.2}$$

Suppose that chemical analyses of a reaction mixture of these three gases yield the concentration data listed in Table 13.2. Among the three gases, only O_2 has a

TABLE 13.1 Changing Concentrations of Reactants and Product during the Reaction $N_2(g) + O_2(g) \rightarrow 2 NO(g)$

Time (μs)	$[N_2]$, $[O_2]$ (μM)	[NO] (μM)
0.0	17.0	0.0
5.0	13.1	7.8
10.0	9.6	14.8
15.0	7.6	18.6
20.0	5.8	22.2
25.0	4.5	24.8
30.0	3.6	26.7

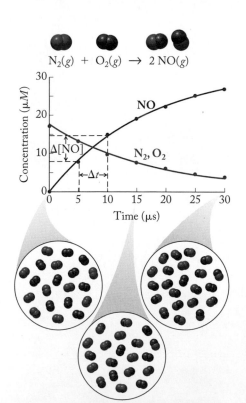

FIGURE 13.6 Concentrations of N_2, O_2, and NO over 30.0 μs for the reaction $N_2(g) + O_2(g) \rightarrow 2 NO(g)$, plotted from data in Table 13.1.

TABLE 13.2 Changing Concentrations of Reactants and Product during the Reaction $2 NO(g) + O_2(g) \rightarrow 2 NO_2(g)$ at 25°C

Time (s)	[NO] (M)	[O$_2$] (M)	[NO$_2$] (M)
0.0	0.0100	0.0100	0.0000
285.0	0.0090	0.0095	0.0010
660.0	0.0080	0.0090	0.0020
1175.0	0.0070	0.0085	0.0030
1895.0	0.0060	0.0080	0.0040
2975.0	0.0050	0.0075	0.0050
4700.0	0.0040	0.0070	0.0060
7800.0	0.0030	0.0065	0.0070

coefficient of 1 in the balanced chemical equation, so we will use the rate of change of $[O_2]$ to calculate the rate of the reaction. Now, let's determine the instantaneous rate of the reaction at $t = 2000.0$ s. We can plot the $[O_2]$ data from the table versus time—this is the green curve in Figure 13.7—and then draw a tangent to the curve at $t = 2000.0$ s. Next, we select two convenient points along the tangent—for example, at $t = 1000.0$ s and $t = 3000.0$ s—to use in calculating the slope of the tangent. The slope is a measure of the instantaneous rate of change in $[O_2]$ at $t = 2000.0$ s, and the negative of the slope is the instantaneous rate of the reaction at that time.[2]

$$\text{Slope} = \frac{\Delta[O_2]}{\Delta t} = \frac{(0.0072 - 0.0084)\ M}{(3000.0 - 1000.0)\ s} = -6.0 \times 10^{-7}\ M/s$$

$$= -6.0 \times 10^{-7}\ M\,s^{-1}$$

$$\text{Rate} = -\frac{\Delta[O_2]}{\Delta t} = -(-6.0 \times 10^{-7}\ M\,s^{-1}) = 6.0 \times 10^{-7}\ M\,s^{-1}$$

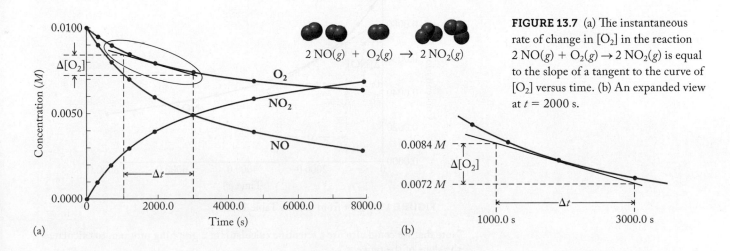

(a)

FIGURE 13.7 (a) The instantaneous rate of change in $[O_2]$ in the reaction $2 NO(g) + O_2(g) \rightarrow 2 NO_2(g)$ is equal to the slope of a tangent to the curve of $[O_2]$ versus time. (b) An expanded view at $t = 2000$ s.

[2]If you have studied calculus, you may know that the average rate of a reaction approaches the instantaneous rate as Δt approaches zero. The slope of the tangent to a curve at a given point is the derivative of the curve at that point and can be calculated using a scientific calculator. Appendix 1 discusses how to determine the slope and intercept of a line.

CONCEPT TEST

Which of the following statements is/are true about the instantaneous rate of the chemical reaction A → B as the reaction proceeds?

a. $-\Delta[A]/\Delta t$ increases, $\Delta[B]/\Delta t$ decreases
b. $-\Delta[A]/\Delta t$ decreases, $\Delta[B]/\Delta t$ increases
c. $-\Delta[A]/\Delta t$ and $\Delta[B]/\Delta t$ both increase
d. $-\Delta[A]/\Delta t$ and $\Delta[B]/\Delta t$ both decrease

(Answers to Concept Tests are in the back of the book.)

SAMPLE EXERCISE 13.3 Determining an Instantaneous Reaction Rate **LO2**

(a) What is the instantaneous rate of change of [NO] at $t = 2000.0$ s in the experiment that produced the data in Table 13.2? (b) What is the rate of the reaction based on your result in part (a)?

Collect and Organize We are asked to determine the instantaneous rate of change of [NO] and the corresponding reaction rate at $t = 2000.0$ s. The coefficient of NO is 2 in the balanced chemical equation:

$$2\,NO(g) + O_2(g) \rightarrow 2\,NO_2(g)$$

Analyze The instantaneous rate of change in [NO] at 2000.0 s can be determined from a graph of [NO] versus time and the slope of the curve at 2000.0 s. The corresponding reaction rate will be the slope value multiplied by $-\frac{1}{2}$.

Solve

a. First we plot [NO] versus time and draw a tangent to the curve at the point $t = 2000.0$ s (Figure 13.8). We then choose two points along the tangent, $t = 1000.0$ and 3000.0 s, and determine the concentrations corresponding to those times along the vertical axis. By using those values, we calculate the slope of the line:

$$\frac{\Delta[NO]}{\Delta t} = \frac{(0.0046 - 0.0070)\,M}{(3000.0 - 1000.0)\,s} = -1.2 \times 10^{-6}\,M\,s^{-1}$$

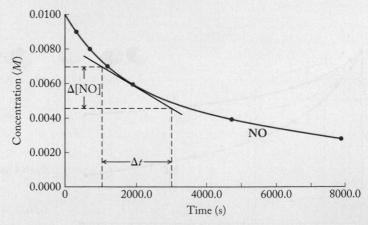

FIGURE 13.8 Plot from data in Table 13.2.

Note that we could also use a scientific calculator or a graphing program to calculate the slope of the tangent.

b. The corresponding reaction rate is

$$Rate = -\frac{1}{2}\frac{\Delta[NO]}{\Delta t} = -\frac{1}{2}(-1.2 \times 10^{-6}\,M\,s^{-1}) = 6.0 \times 10^{-7}\,M\,s^{-1}$$

Think About It The sign of Δ[NO]/Δt is negative because NO is a reactant whose concentration decreases with time. However, the rates of chemical reactions have positive values. Therefore we needed a minus sign in front of the Δ[NO]/Δt term in the solution to part (b). Note that the instantaneous reaction rate calculated in this exercise is the same as the one we calculated earlier based on the rate of change of [O₂].

 Practice Exercise What is the instantaneous rate of change in [NO₂] at t = 2000.0 s in the experiment that produced the data in Table 13.2?

13.3 Effect of Concentration on Reaction Rate

Figure 13.9 shows a typical plot of reactant concentration as a function of time. Tangents have been drawn to the curve at three points: (a) at the instant the reaction begins (t = 0), (b) when the reaction is about halfway to completion, and (c) when the reaction is nearly over. The slope of the tangent at point (a) can be used to calculate the **initial rate** of the reaction.

As with nearly all plots of reactant or product concentration versus time, the most rapid changes in concentration take place at the beginning of the reaction. As the reaction proceeds, the slopes of tangents to the curve decrease, approaching zero. When the slope of the tangent to the curve reaches zero, there are no more changes in the concentrations of the product(s) or any remaining reactant(s).

Kinetic molecular theory explains why reaction rates change over time. If we assume that most reactions take place as a result of collisions between reactant molecules, then the more reactant molecules there are in a given volume, the more collisions per unit time there are, and there are more opportunities for reactants to turn into products. As reactants collide and change into products, fewer reactant molecules are present in the system (that is, the concentrations of the reactants decrease), so the frequency of collisions decreases and the rate at which reactants change into products slows down.

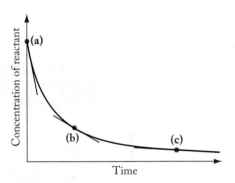

FIGURE 13.9 Typical plot of reactant concentration as a function of time with tangents drawn (a) at t = 0; (b) at an intermediate time; (c) when the reaction is nearly over.

Reaction Order and Rate Constants

Experimental observations and theoretical considerations tell us that reaction rates depend on reactant concentrations. However, they do not tell us *to what extent* rates depend on reactant concentrations. For example, if the concentration of a reactant doubles, does the reaction rate also double? The answer to this question comes from experimentally determining the **reaction order**, a parameter that tells us how reaction rate depends on reactant concentrations. Knowing the order of a reaction also provides insights into how the reaction takes place—which molecules collide with which other molecules as bonds break, new bonds form, and reactants are converted into products.

Let's look at one way in which reaction order is determined by revisiting the reaction between nitrogen monoxide and oxygen:

$$2\,NO(g) + O_2(g) \rightarrow 2\,NO_2(g) \qquad (13.2)$$

To evaluate the kinetics of this reaction, multiple experiments can be conducted, each with different initial concentrations of NO and O₂ introduced into a reaction vessel at 25°C, and the initial reaction rate is determined in each case. Table 13.3 shows the results of three such determinations.

 **CHEMTOUR**
Reaction Order

initial rate the rate of a reaction at t = 0, immediately after the reactants are mixed.

reaction order an experimentally determined number defining the dependence of the reaction rate on the concentration of a reactant.

TABLE 13.3 Effect of Reactant Concentrations on Initial Reaction Rates at 25°C for the Reaction 2 NO(g) + O$_2$(g) → 2 NO$_2$(g)

Experiment	[NO]$_0$ (M)	[O$_2$]$_0$ (M)	Initial Reaction Rate (M/s)
1	0.0100	0.0100	1.0×10^{-6}
2	0.0100	0.0050	0.5×10^{-6}
3	0.0050	0.0100	2.5×10^{-7}

To interpret the data in Table 13.3, we select pairs of experiments in which the concentrations of one reactant differ but the concentrations of the other are the same. For example, [NO]$_0$ is the same in experiments 1 and 2, but [O$_2$]$_0$ in experiment 1 is twice [O$_2$]$_0$ in experiment 2. (The zero subscripts indicate that these are the concentrations of NO and O$_2$ at the start of the experiments, when $t = 0$.) The initial reaction rate in experiment 1 is twice that in experiment 2, allowing us to state that doubling the concentration of O$_2$ while holding the NO concentration constant doubles the initial reaction rate. We conclude that the initial reaction rate is directly proportional to the O$_2$ concentration:

$$\text{Rate} \propto [\text{O}_2]$$

In experiments 1 and 3, [O$_2$]$_0$ is the same, but [NO]$_0$ in experiment 1 is twice [NO]$_0$ in experiment 3. Comparing the initial reaction rates for these two experiments, we find that the rate in experiment 1 is four times the rate in experiment 3. Thus, the reaction rate is proportional to the square of the NO concentration:

$$\text{Rate} \propto [\text{NO}]^2$$

We can combine these two rate expressions to get an overall rate expression for the reaction by multiplying the right sides of the rate expressions for the two reactants:

$$\text{Rate} \propto [\text{NO}]^2[\text{O}_2]$$

To understand why we multiply the concentration terms together, let's consider another reaction between oxygen and nitrogen monoxide. This one involves the reaction of ozone (O$_3$) with NO to produce NO$_2$ and O$_2$:

$$\text{NO}(g) + \text{O}_3(g) \rightarrow \text{NO}_2(g) + \text{O}_2(g)$$

Figure 13.10 shows how different numbers of NO and O$_3$ molecules in a reaction vessel might collide and react with each other. Note how increasing the numbers of molecules in the containers in Figure 13.10 produces increasing numbers of collisions that are proportional to the product of the number of molecules of each reactant. Increasing the numbers of molecules of each type in the vessels is equivalent to increasing the concentrations of the two gases. Therefore the rate of the reaction should (and does) depend on the product of the concentrations of NO and O$_3$:

$$\text{Rate} \propto [\text{NO}][\text{O}_3] \tag{13.11}$$

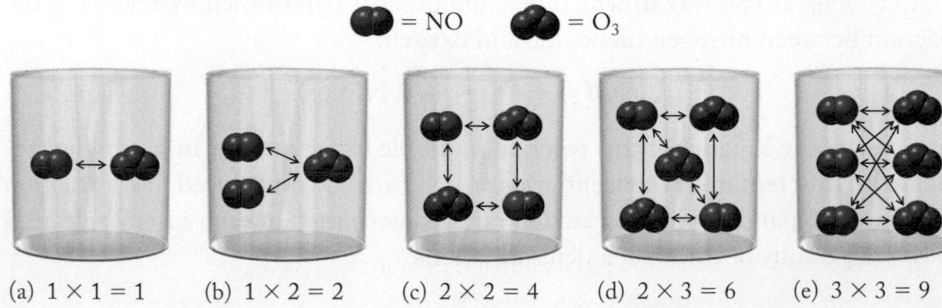

FIGURE 13.10 Increasing the concentration increases the number of possible collisions (double-headed arrows) and therefore the number of potential reaction events. Reaction rate depends on the number of collisions, which are shown for the reaction between NO and ozone (O$_3$) that produces NO$_2$ and O$_2$. With only one NO molecule and one O$_3$ molecule, as in (a), each molecule can collide only with the other, giving a relative reaction rate of $1 \times 1 = 1$. In (e), three molecules of NO can collide with three molecules of O$_3$ for a relative reaction rate of $3 \times 3 = 9$ times the rate in (a).

● = NO   ●●● = O$_3$

(a) $1 \times 1 = 1$ (b) $1 \times 2 = 2$ (c) $2 \times 2 = 4$ (d) $2 \times 3 = 6$ (e) $3 \times 3 = 9$

Similar patterns occur with all chemical reactions whose rate depends on the concentration of more than one reactant. We can modify Equation 13.11 to create the **rate law**, an equation that defines the mathematical relationship between reactant concentrations and reaction rate. We convert the proportionality to an equation by inserting a proportionality constant k, called the reaction's **rate constant**:

$$\text{Rate} = k[\text{NO}][\text{O}_3] \tag{13.12}$$

Now let's consider a generic chemical reaction with two reactants, A and B:

$$\text{A} + \text{B} \rightarrow \text{C}$$

The rate law expression for this reaction may be written

$$\text{Rate} = k[\text{A}]^m[\text{B}]^n \tag{13.13}$$

where m is the reaction order with respect to A, and n is the reaction order with respect to B. We experimentally determine these reaction order values by comparing differences in reaction rates with differences in reactant concentrations. Sometimes these comparisons are easy to make, as with the data for the reaction between NO and O_2 in Table 13.3. Other times the comparisons are not so easy. On those occasions we can use a different mathematical approach. For example, suppose we conduct three experiments to solve for the values of m and n in Equation 13.13. In experiments 1 and 2 the value of [A] is the same, but the values of [B] are different. In experiments 2 and 3 we keep [B] constant and vary the concentrations of [A]. The ratio of the reaction rates in experiments 1 (Rate_1) and 2 (Rate_2) is related to the ratio of the concentrations of B:

$$\frac{\text{Rate}_1}{\text{Rate}_2} = \left(\frac{[\text{B}]_1}{[\text{B}]_2}\right)^n$$

We can solve this equation for n by taking the logarithm of both sides:

$$\log\left(\frac{\text{Rate}_1}{\text{Rate}_2}\right) = n \log\left(\frac{[\text{B}]_1}{[\text{B}]_2}\right)$$

Now rearrange the terms to solve for n:

$$n = \frac{\log\left(\dfrac{\text{Rate}_1}{\text{Rate}_2}\right)}{\log\left(\dfrac{[\text{B}]_1}{[\text{B}]_2}\right)}$$

An equation with this format can also be used to calculate the value of m from the results of experiments 2 and 3, or to calculate the order of any reaction with respect to a reactant (X) whose concentration differs in a pair of reaction rate experiments:

$$n = \frac{\log\left(\dfrac{\text{Rate}_1}{\text{Rate}_2}\right)}{\log\left(\dfrac{[\text{X}]_1}{[\text{X}]_2}\right)} \tag{13.14}$$

rate law an equation that defines the experimentally determined relation between the concentrations of reactants in a chemical reaction and the rate of that reaction.

rate constant the proportionality constant that relates the rate of a reaction to the concentrations of reactants.

CONCEPT TEST

In the reaction A → B, the rate of the reaction triples when [A] is tripled.

a. What is the correct value of m in the rate law: rate = $k[\text{A}]^m$?

b. What is the value of m if the rate is unchanged when [A] is tripled?

overall reaction order the sum of the exponents of the concentration terms in the rate law.

The power to which a concentration term is raised (that is, m or n in Equation 13.13) is the order of the reaction in terms of that reactant. In our example from Table 13.3, the reaction of NO and O_2 is *second order* in NO and *first order* in O_2. The **overall reaction order** for a reaction is the sum of the powers in the rate equation, so the reaction of NO and O_2 is *third order* overall.

An exponent in a rate law may be a whole number, a fraction, zero, or, in rare cases, negative. It is important to remember that rate laws and reaction orders must be determined experimentally. They cannot be predicted from the coefficients in a balanced chemical equation, even though they sometimes do match the coefficients. The significance of reaction order in describing how a reaction takes place is addressed in Section 13.5.

The value of the rate constant k is unique to each particular reaction at a given temperature, and right now we will consider k simply as a proportionality constant. It does not change with concentration; in other words, *reaction rate depends on the concentration of the reactants, but the rate constant does not.* The rate constant changes only with changing temperature or in the presence of a catalyst.

We can calculate k for a reaction run at some specified temperature from initial reaction rate data. Let's do so for the reaction of O_2 and NO by selecting the results of one experiment in Table 13.3. Which experiment we use does not matter; as long as the temperature is the same, the value of k will be the same. The proportionality we derived for this reaction's overall rate expression was

$$\text{Rate} \propto [NO]^2[O_2]$$

When we convert this to an equation, we obtain

$$\text{Rate} = k[NO]^2[O_2]$$

We can insert the data from Experiment 1 into this equation and solve for k:

$$1.0 \times 10^{-6} \, M/s = k(0.0100 \, M)^2(0.0100 \, M)$$

$$k = \frac{1.0 \times 10^{-6} \, M/s}{(0.0100 \, M)^2(0.0100 \, M)} = 1.0 \, M^{-2} \, s^{-1} \qquad (13.15)$$

This value, like any rate constant calculated from experimental data, is valid only at the temperature at which the experiments were carried out.

The units of k look a bit unusual in comparison with units we have seen before. Remember that reaction rates themselves have units of concentration per unit time, such as M/s. Therefore in a first-order reaction in which concentration is expressed in molarity and time in seconds, the units of k must be per second (s^{-1}):

$$\text{Rate} = k[X]^1$$

$$k = \frac{\text{rate}}{[X]} = \frac{M/s}{M} = \frac{1}{s} = s^{-1}$$

The units of the rate constant for a reaction that is second order overall can be derived in a similar way. If the reaction is first order in reactants X and Y, we have

$$\text{Rate} = k[X][Y]$$

$$k = \frac{\text{rate}}{[X][Y]} = \frac{M/s}{M \, M} = M^{-1} \, s^{-1}$$

The units of k for a reaction that is third order overall are $M^{-2} \, s^{-1}$, as we determined in Equation 13.15. Therefore the units of k for each reaction ultimately depend on its overall reaction order.

CONCEPT **TEST**

The reaction A + 2 B → C is found to be third order overall. How many possible rate laws of the form

$$\text{Rate} = k[\text{A}]^m[\text{B}]^n$$

could we write, assuming that m and n are non-negative integers?

SAMPLE EXERCISE 13.4 Deriving a Rate Law from Initial Reaction Rate Data **LO3**

Write the rate law for the reaction of N_2 with O_2

$$N_2(g) + O_2(g) \rightarrow 2\,NO(g)$$

using the data in Table 13.4. Determine the overall reaction order and the value of the rate constant.

TABLE 13.4 Initial Reaction Rates at Constant Temperature for the Reaction $N_2(g) + O_2(g) \rightarrow 2\,NO(g)$

Experiment	$[N_2]_0$ (M)	$[O_2]_0$ (M)	Initial Reaction Rate (M/s)
1	0.040	0.020	707
2	0.040	0.010	500
3	0.010	0.010	125

Collect and Organize We need to determine the rate law and rate constant for a reaction, given the initial reaction rate (note the subscript zero on the concentration terms in Table 13.4) for each of three sets of initial concentrations.

Analyze The general form of the rate law for any reaction between reactants A and B is

$$\text{Rate} = k[\text{A}]^m[\text{B}]^n$$

We can use the experimental data given to find the values of k, m, and n for the reaction in which [A] = $[N_2]$ and [B] = $[O_2]$. The overall order of the reaction is the sum of the reaction orders of the individual reactants. Once we have established the rate law, we can calculate the rate constant by using concentrations of reactants from any row in Table 13.4. The rate constant must have units that express the reaction rate in $M\,s^{-1}$.

Solve To determine the order of the reaction (m) with respect to N_2, we use the data from experiments 2 and 3 because in these two experiments the $[N_2]$ values are different, but the $[O_2]$ values are the same. When the concentration of N_2 is increased by a factor of 4, the rate increases by a factor of 4. Thus the reaction rate is proportional to $[N_2]$, which means that $m = 1$. There are different values of $[O_2]$ in experiments 1 and 2, but $[N_2]$ is the same. Therefore we can use these data to calculate the value of n. The ratio of the reaction rates (707/500) is not a whole number, so let's apply Equation 13.14:

$$n = \frac{\log\left(\dfrac{\text{Rate}_1}{\text{Rate}_2}\right)}{\log\left(\dfrac{[O_2]_1}{[O_2]_2}\right)} = \frac{\log\left(\dfrac{707\ M\!/\!s}{500\ M\!/\!s}\right)}{\log\left(\dfrac{0.020\ M}{0.010\ M}\right)} = 0.50$$

Thus the reaction is $\frac{1}{2}$ order with respect to O_2, first order with respect to N_2, and $(\frac{1}{2} + 1) = \frac{3}{2}$ order overall:

$$\text{Rate} = k[N_2][O_2]^{1/2}$$

We can use the data from any experiment to obtain the value of k. Let's use experiment 1:

$$707 \ M/s = k(0.040 \ M)(0.020 \ M)^{1/2}$$

$$k = 1.2 \times 10^5 \ M^{-1/2} \ s^{-1}$$

Think About It The units of the rate constant, $M^{-1/2} \ s^{-1}$, are appropriate because when we substitute the units for each term into the rate law, we end up with the correct units for reaction rate: $(M^{-1/2} \ s^{-1})(M)(M^{1/2}) = M \ s^{-1}$. Note that the reaction order of $\frac{1}{2}$ for oxygen is determined from experimental data, not from the balanced chemical equation.

 Practice Exercise Nitrogen monoxide reacts rapidly with unstable nitrogen trioxide (NO_3) to form NO_2:

$$NO(g) + NO_3(g) \rightarrow 2 \ NO_2(g)$$

Determine the rate law for the reaction and calculate the rate constant from the data in Table 13.5.

TABLE 13.5 Initial Reaction Rates at 25°C for the Reaction $NO(g) + NO_3(g) \rightarrow 2 \ NO_2(g)$

Experiment	$[NO]_0$ (M)	$[NO_3]_0$ (M)	Initial Reaction Rate (M/s)
1	1.25×10^{-3}	1.25×10^{-3}	2.45×10^4
2	2.50×10^{-3}	1.25×10^{-3}	4.90×10^4
3	2.50×10^{-3}	2.50×10^{-3}	9.80×10^4

Integrated Rate Laws: First-Order Reactions

Determining a rate law by using initial reaction rate data means that several experiments must be performed with different concentrations of reactants, and that these concentrations must be manipulated in a systematic fashion. We also must accurately determine the reaction rate at the instant the reaction begins. It would be much better if we could determine the rate law and calculate the rate constant of a reaction by using data from only a single experiment. In fact, we can do this easily for reactions in which the reaction rate depends on the concentration of only one substance. One such reaction is the photochemical decomposition of ozone in the stratosphere:

$$O_3(g) \xrightarrow{\text{sunlight}} O_2(g) + O(g)$$

TABLE 13.6 Concentrations for Photochemical Decomposition of Ozone

Time (s)	$[O_3]$ (M)	$\ln[O_3]$
0.0	1.000×10^{-4}	-9.210
100.0	0.896×10^{-4}	-9.320
200.0	0.803×10^{-4}	-9.430
300.0	0.719×10^{-4}	-9.540
400.0	0.644×10^{-4}	-9.650
500.0	0.577×10^{-4}	-9.760
600.0	0.517×10^{-4}	-9.870

This decomposition reaction can be studied in the laboratory by using high-intensity ultraviolet lamps to simulate solar radiation. One such study yielded the results listed in Table 13.6 and plotted in Figure 13.11(a). Because ozone is the only reactant, the rate law for the reaction should depend only on the ozone concentration. In our interpretation of the data in Table 13.6, we start with the assumption that the reaction is first order in O_3, which means that the rate law can be written

$$\text{Rate} = k[O_3]$$

The O_3 consumption rate $-\Delta[O_3]/\Delta t$ equals the reaction rate, and we can write

$$\text{Rate} = -\frac{\Delta[O_3]}{\Delta t} = k[O_3]$$

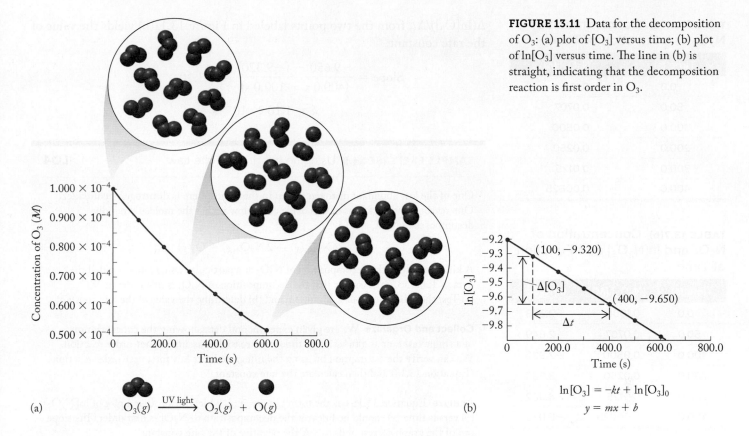

This rate law can be transformed into an expression that relates the concentration of ozone $[O_3]$ at any instant during the reaction to the initial concentration $[O_3]_0$:

$$\ln\frac{[O_3]}{[O_3]_0} = -kt \qquad (13.16)$$

This version of the rate law is called an **integrated rate law** because integral calculus is used to derive it, and it describes the change in reactant concentration with time. The general integrated rate law for any reaction that is first order in reactant X is

$$\ln\frac{[X]}{[X]_0} = -kt \qquad (13.17)$$

Using the identity $\ln(a/b) = \ln a - \ln b$, we can rearrange this equation to

$$\ln[X] = -kt + \ln[X]_0 \qquad (13.18)$$

This is the equation of a straight line of the form

$$y = mx + b$$

where $\ln[X]$ is the y variable and t is the x variable. The slope of the line (m) is $-k$, and the y intercept (b) is $\ln[X]_0$. Rearranging Equation 13.16 to fit the format of Equation 13.18 gives

$$\ln[O_3] = -kt + \ln[O_3]_0$$

Note in Figure 13.11(b) that a graph of the natural logarithm of $[O_3]$ versus time is indeed a straight line. This linearity means that our assumption was correct and the reaction is first order in O_3. Calculating the slope of the line,

integrated rate law a mathematical expression that describes the change in concentration of a reactant in a chemical reaction with time.

TABLE 13.7(a) Concentration of N_2O_5 as a Function of Time

Time (s)	$[N_2O_5]$ (M)
0.0	0.1000
50.0	0.0707
100.0	0.0500
200.0	0.0250
300.0	0.0125
400.0	0.00625

TABLE 13.7(b) Concentration of N_2O_5 and $\ln[N_2O_5]$ as a Function of Time

Time (s)	$[N_2O_5]$ (M)	$\ln[N_2O_5]$
0.0	0.1000	−2.303
50.0	0.0707	−2.649
100.0	0.0500	−2.996
200.0	0.0250	−3.689
300.0	0.0125	−4.382
400.0	0.00625	−5.075

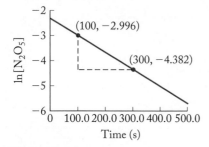

FIGURE 13.12 The slope of the line equals $-k$. Therefore the rate constant $k = 0.00693 \text{ s}^{-1}$. Note that we could also use a graphing calculator to plot $\ln[N_2O_5]$ versus t and have the calculator determine the equation of the straight line that best fits the data points.

TABLE 13.8 Concentration of Hydrogen Peroxide as a Function of Time

Time (s)	$[H_2O_2]$ (M)
0.0	0.500
100.0	0.460
200.0	0.424
500.0	0.330
1000.0	0.218
1500.0	0.144

$\Delta(\ln[O_3])/\Delta t$, from the two points labeled in Figure 13.11(b) yields the value of the rate constant:

$$\text{Slope} = \frac{-9.650 - (-9.320)}{(400.0 \text{ s} - 100.0 \text{ s})} = -1.10 \times 10^{-3} \text{ s}^{-1}$$

$$k = 1.10 \times 10^{-3} \text{ s}^{-1}$$

SAMPLE EXERCISE 13.5 Using an Integrated Rate Law   **LO4**

One of the less abundant nitrogen oxides in the atmosphere is dinitrogen pentoxide. One reason concentrations of this oxide are so low is that the molecule rapidly decomposes to N_2O_4 and O_2:

$$2 \text{ N}_2\text{O}_5(g) \rightarrow 2 \text{ N}_2\text{O}_4(g) + \text{O}_2(g)$$

A kinetic study of the decomposition of N_2O_5 at a particular temperature yielded the data in Table 13.7(a). Assume that the decomposition of N_2O_5 is first order in N_2O_5. (a) Test the validity of your assumption, and (b) determine the value of the rate constant.

Collect and Organize We are given experimental data showing the concentration of a single reactant as a function of time and are told to assume a first-order reaction. We can verify the assumption by using the integrated rate law for a first-order reaction (Equation 13.18) and then calculate the rate constant.

Analyze Equation 13.18 has the form $y = mx + b$, which means that a plot of $\ln[N_2O_5]$ (y) versus time (x) should be linear if the decomposition of N_2O_5 is first order. The slope (m) of the graph corresponds to $-k$, the negative of the rate constant.

Solve Our first step is to determine $\ln[N_2O_5]$ values (Table 13.7b).
a. The plot of $\ln[N_2O_5]$ versus t is shown in Figure 13.12. The fact that the curve in Figure 13.12 is a straight line indicates that the reaction is first order in N_2O_5.
b. Arbitrarily choosing $t = 100.0$ s and $t = 300.0$ s as two points for calculating the slope:

$$\text{Slope} = \frac{\Delta y}{\Delta x} = \frac{-4.382 - (-2.996)}{(300.0 \text{ s} - 100.0 \text{ s})} = \frac{-1.386}{200.0 \text{ s}}$$

$$= -0.00693 \text{ s}^{-1}$$

This slope corresponds to $-k$; therefore the rate constant $k = 0.00693 \text{ s}^{-1}$. As an alternative method for finding this slope, we could use a graphing calculator to plot $\ln[N_2O_5]$ versus t and to determine the best fit of the data to a straight line.

Think About It We can test whether any reaction with a single reactant (X) is first order in X by determining whether a plot of $\ln[X]$ versus t is linear.

 Practice Exercise Hydrogen peroxide (H_2O_2) decomposes into water and oxygen:

$$\text{H}_2\text{O}_2(\ell) \rightarrow \text{H}_2\text{O}(\ell) + \tfrac{1}{2}\text{O}_2(g)$$

Use the data in Table 13.8 to determine whether the decomposition of H_2O_2 is first order in H_2O_2, and calculate the value of the rate constant at the temperature of the experiment that produced the data.

SAMPLE EXERCISE 13.6 Calculating the Concentration of a Reactant   **LO5**
from an Integrated Rate Law

In Sample Exercise 13.5 we confirmed that the decomposition of dinitrogen pentoxide

$$2 \text{ N}_2\text{O}_5(g) \rightarrow 2 \text{ N}_2\text{O}_4(g) + \text{O}_2(g)$$

is first order in N_2O_5 and has a rate constant $k = 0.00693$ s^{-1}. If this reaction is run in the laboratory under the same conditions as in Sample Exercise 13.5 and the initial concentration of N_2O_5 in the reaction vessel is 0.375 M, what is the concentration of N_2O_5 after exactly 3 minutes?

Collect and Organize We are given the rate constant for a first-order reaction, the initial concentration of reactant, and the reaction time. We are asked to calculate the concentration of reactant remaining at that time. We also have the general rate equation for a first-order reaction (Equation 13.18).

Analyze We know all the terms in the rate equation

$$\ln[X] = -kt + \ln[X]_0$$

except $[X]$, which is what we want to calculate. The rate constant k is given in terms of seconds, so we must convert $t = 3$ min into $t = 180$ s. In the data set for Sample Exercise 13.5, $[N_2O_5]$ dropped to about $\frac{1}{4}[N_2O_5]_0$ in 200 s, so we estimate that the concentration of N_2O_5 in this example will drop to about one-quarter of its initial value, or around 0.09 M.

Solve Substituting the given values into Equation 13.18:

$$\ln[X] = -kt + \ln[X]_0$$
$$\ln[N_2O_5] = -(0.00693 \text{ s}^{-1})(180 \text{ s}) + \ln(0.375)$$
$$\ln[N_2O_5] = -1.247 + (-0.981) = -2.228$$
$$[N_2O_5] = e^{-2.228} = 0.108 \ M$$

Think About It The answer is less than the starting concentration of N_2O_5, as we would expect, and close to our estimated value.

 Practice Exercise Under a different set of conditions from those in Sample Exercise 13.5, dinitrogen pentoxide decomposes into nitrogen dioxide and O_2:

$$2 \ N_2O_5(g) \rightarrow 4 \ NO_2(g) + O_2(g)$$

This reaction is also first order in N_2O_5 and has a rate constant $k = 9.55 \times 10^{-4}$ s^{-1}. What is the concentration of N_2O_5 in this system after exactly 10 min if $[N_2O_5]_0 = 0.763$ M?

Reaction Half-Lives

A parameter frequently cited in kinetic studies is the **half-life ($t_{1/2}$)** of a reaction, which is the interval during which the concentration of a reactant decreases by half. Half-life is inversely related to the rate constant of a reaction: the higher the reaction rate, the shorter the half-life.

Let's consider reaction half-life in the context of another nitrogen oxide found in the atmosphere: dinitrogen monoxide, also called nitrous oxide or laughing gas, an anesthetic sometimes used by dentists. Atmospheric concentrations of this potent greenhouse gas have been increasing in recent years, although the principal source is not automotive emissions but rather bacterial degradation of nitrogen compounds in soil. Dinitrogen monoxide is not produced in internal combustion engines because at high temperatures, N_2O rapidly decomposes (Figure 13.13) into nitrogen and oxygen:

$$N_2O(g) \rightarrow N_2(g) + \tfrac{1}{2} O_2(g)$$

We can derive a mathematical relation between half-life $t_{1/2}$ and rate constant k for this or any other first-order reaction by starting with Equation 13.17:

$$\ln\frac{[X]}{[X]_0} = -kt$$

half-life ($t_{1/2}$) the time in the course of a chemical reaction during which the concentration of a reactant decreases by half.

FIGURE 13.13 The decomposition of $N_2O(g)$ is first order in N_2O. At a particular temperature the half-life of the reaction is 1.0 s, which means that, on average, half of a population of 16 N_2O molecules decomposes in 1.0 s, half of the remaining 8 molecules decomposes in the next 1.0 s, and so on.

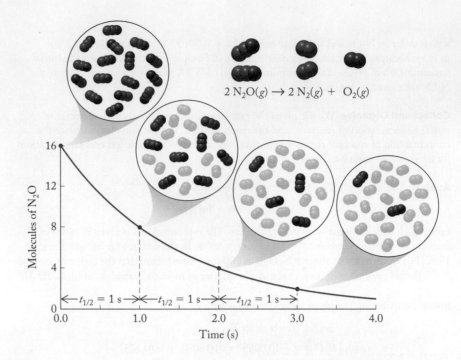

$$2\,N_2O(g) \rightarrow 2\,N_2(g) + O_2(g)$$

After one half-life has passed, $t = t_{1/2}$, and the concentration of X is half its original value: $[X] = \frac{1}{2}[X]_0$. Inserting these values for $[X]$ and t into the equation yields

$$\ln \frac{\frac{1}{2}[X]_0}{[X]_0} = -kt_{1/2}$$

$$\ln\left(\tfrac{1}{2}\right) = -kt_{1/2}$$

The natural log of $\frac{1}{2}$ is -0.693, so

$$-0.693 = -kt_{1/2}$$

or

$$t_{1/2} = \frac{0.693}{k} \tag{13.19}$$

Thus the half-life of a first-order reaction is inversely proportional to the rate constant, as noted at the beginning of this discussion. The absence of any concentration term in Equation 13.19 means that the half-life of a first-order reaction is constant throughout the reaction and independent of concentration: no matter the initial concentration of reactant, half of it is consumed in one half-life.

SAMPLE EXERCISE 13.7 Calculating the Half-Life of a First-Order Reaction **LO6**

The rate constant for the decomposition of N_2O_5 at a particular temperature is 7.8×10^{-3} s^{-1}. What is the half-life of N_2O_5 at that temperature?

Collect, Organize, and Analyze We are asked to determine the half-life of N_2O_5 from the rate constant of its decomposition reaction. We know from Sample Exercise 13.5 that the decomposition of N_2O_5 is a first-order process, so the values of $t_{1/2}$ and k are related by Equation 13.19.

Solve

$$t_{1/2} = \frac{0.693}{k} = \frac{0.693}{7.8 \times 10^{-3}\,s^{-1}} = 89\,s$$

Think About It The calculated value makes sense because the k value is small—a little less than 0.01—so dividing 0.693 by it should produce a $t_{1/2}$ value a little larger than 69 s. Remember that Equation 13.19 is valid only for first-order reactions.

Practice Exercise Environmental scientists calculating half-lives of pollutants often define a *transport rate constant* that is analogous to a reaction rate constant and describes how rapidly a pollutant washes out of an ecosystem. In a study of the gasoline additive MTBE in Donner Lake, California, scientists from the University of California, Davis, found that in the summer the half-life of MTBE in the lake was 28 days. Assuming that the transport process is first order, what was the transport rate constant of MTBE out of Donner Lake during the study? Express your answer in reciprocal days.

CONCEPT TEST

Which has a shorter half-life, a fast reaction or a slow reaction?

Integrated Rate Laws: Second-Order Reactions

In Section 13.1 we described how NO_2 exposed to UV rays from the Sun decomposes to NO and atomic oxygen, O. Nitrogen dioxide may also undergo thermal decomposition, producing NO and molecular oxygen, O_2:

$$2\,NO_2(g) \rightarrow 2\,NO(g) + O_2(g) \tag{13.20}$$

The data in Table 13.9 describe the rate of the thermal decomposition reaction. In this case, the plot of $\ln[NO_2]$ versus time (Figure 13.14a) is *not* linear, which tells us that the thermal decomposition of NO_2 is *not* first order.

So what is the order of the reaction? The answer to this question is related to how the reaction takes place. If each NO_2 molecule simply fell apart, the reaction would be first order, much like the decomposition of N_2O_5. However, if the reaction happens as a result of collisions between pairs of NO_2 molecules, the reaction

TABLE 13.9 Decomposition of NO_2 as a Function of Time

Time (s)	$[NO_2]$ (M)	$\ln[NO_2]$	$1/[NO_2]$ (1/M)
0.0	1.00×10^{-2}	−4.605	100
100.0	6.48×10^{-3}	−5.039	154
200.0	4.79×10^{-3}	−5.341	209
300.0	3.80×10^{-3}	−5.573	263
400.0	3.15×10^{-3}	−5.760	317
500.0	2.69×10^{-3}	−5.918	372
600.0	2.35×10^{-3}	−6.057	426

FIGURE 13.14 At high temperatures, NO_2 slowly decomposes into NO and O_2. (a) The plot of $\ln[NO_2]$ versus time is not linear, indicating that the reaction is not first order. (b) The plot of $1/[NO_2]$ versus time is linear, indicating that the reaction is second order in NO_2. The slope of the line in this graph equals the rate constant.

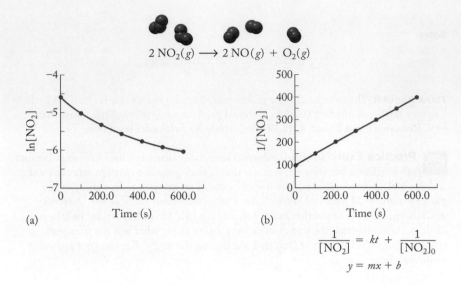

$$\frac{1}{[NO_2]} = kt + \frac{1}{[NO_2]_0}$$
$$y = mx + b$$

would be first order in each one of them and *second order* overall. Therefore the rate law expression would be

$$\text{Rate} = k[NO_2]^2 \qquad (13.21)$$

How can we determine whether this decomposition is really second order? One way is to assume that it is and then test that assumption. The test entails transforming the rate law in Equation 13.21 into the integrated rate law for a second-order reaction, again using calculus. The result of the transformation is

$$\frac{1}{[NO_2]} = kt + \frac{1}{[NO_2]_0} \qquad (13.22)$$

Like Equation 13.18, this has the form $y = mx + b$ and is the equation of a straight line, this time with $1/[NO_2]$ as the y variable and t as the x variable.

The graph obtained using data from columns 1 and 4 of Table 13.9 is shown in Figure 13.14(b). The curve is linear, which tells us that decomposition of NO_2 is second order. The slope of the line provides a direct measure of k, which is $0.544\ M^{-1}\ s^{-1}$.

A general form of Equation 13.22 that applies to any reaction that is second order in a single reactant (X) is

$$\frac{1}{[X]} = kt + \frac{1}{[X]_0} \qquad (13.23)$$

TABLE 13.10(a) Concentration of Chlorine Monoxide as a Function of Time

Time (ms)	[ClO] (M)
0.0	1.50×10^{-8}
10.0	7.19×10^{-9}
20.0	4.74×10^{-9}
30.0	3.52×10^{-9}
40.0	2.81×10^{-9}
100.0	1.27×10^{-9}
200.0	0.66×10^{-9}

SAMPLE EXERCISE 13.8 Distinguishing between First- and Second-Order Reactions   **LO4**

Chlorine monoxide accumulates in the stratosphere above Antarctica each winter and plays a key role in the formation of the ozone hole above the South Pole each spring. Eventually, ClO decomposes according to the equation

$$2\ ClO(g) \rightarrow Cl_2(g) + O_2(g)$$

The kinetics of this reaction were studied in a laboratory experiment at 298 K, and the data are shown in Table 13.10(a). Determine the order of the reaction, the rate law, and the value of k at 298 K.

Collect and Organize We are given experimental data describing the variation in concentration of ClO with time at 298 K, and we are asked to determine the order of the decomposition reaction of ClO. Two of the choices we have are first and second order.

Analyze To distinguish between first and second order for a reaction in which ClO is the single reactant, we need to calculate $\ln[ClO]$ and $1/[ClO]$ values and plot them versus time. If the $\ln[ClO]$ plot is linear, the reaction is first order; if the $1/[ClO]$ plot is linear, the reaction is second order. The rate law has the form

$$Rate = k[ClO]^m$$

where m equals 1 or 2 and, because there is only one reactant, is also the overall order of the reaction. We determine the rate constant from the slope of whichever plot is linear.

Solve Table 13.10(b) contains the needed $\ln[ClO]$ and $1/[ClO]$ values.

TABLE 13.10(b) Concentration of Chlorine Monoxide, $\ln[ClO]$, and $1/[ClO]$ as a Function of Time

Time (ms)	[ClO] (M)	ln[ClO]	1/[ClO] (1/M)
0.0	1.50×10^{-8}	−18.015	6.67×10^7
10.0	7.19×10^{-9}	−18.751	1.39×10^8
20.0	4.74×10^{-9}	−19.167	2.11×10^8
30.0	3.52×10^{-9}	−19.465	2.84×10^8
40.0	2.81×10^{-9}	−19.690	3.56×10^8
100.0	1.27×10^{-9}	−20.484	7.89×10^8
200.0	0.66×10^{-9}	−21.139	1.51×10^9

Plots of these values versus time are shown in Figure 13.15. The $\ln[ClO]$ plot is not linear, but the $1/[ClO]$ plot is, which means that the reaction is second order in ClO and second order overall. Thus, $m = 2$ and the rate law is

$$Rate = k[ClO]^2$$

Substituting [ClO] into the generic integrated rate law (Equation 13.23) gives

$$\frac{1}{[ClO]} = kt + \frac{1}{[ClO]_0}$$

From the general equation $y = mx + b$, we know that k is the slope of the graph of $1/[ClO]$ versus time. Arbitrarily choosing two convenient data points (at 0 and 100 ms), we can calculate k:

$$k = \text{slope} = \frac{\Delta y}{\Delta x} = \frac{\Delta\left(\dfrac{1}{[ClO]}\right)}{\Delta t}$$

$$= \frac{(7.89 - 0.667) \times 10^8 \ M^{-1}}{(100.0 - 0.0) \times 10^{-3} \ s}$$

$$= 7.22 \times 10^9 \ M^{-1} \ s^{-1}$$

Think About It We can distinguish between first- and second-order reactions involving a single reactant by transforming the concentration data and examining which gives a linear fit with the integrated rate law.

 Practice Exercise Experimental evidence shows that in the reaction

$$NO_2(g) + CO(g) \rightarrow NO(g) + CO_2(g)$$

the reaction rate depends only on the concentration of NO_2. Determine whether the reaction is first or second order in NO_2, and calculate the rate constant from the data in Table 13.11, which were obtained at 488 K.

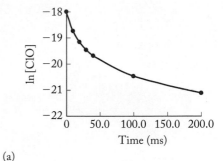

FIGURE 13.15 Plots using data from Table 13.10(b).

TABLE 13.11 Concentration of NO_2 as a Function of Time

Time (h)	[NO₂] (M)
0.00	0.250
1.39	0.198
3.06	0.159
4.72	0.132
6.39	0.114
8.06	0.099
9.72	0.088
11.39	0.080

The concept of half-life can also be applied to second-order reactions. The relationship between rate constant and half-life for the decomposition of NO_2 can be derived from Equation 13.23 if we first rearrange the terms to solve for kt:

$$kt = \frac{1}{[X]} - \frac{1}{[X]_0}$$

After one half-life has elapsed ($t = t_{1/2}$), $[X]$ has decreased to half its initial concentration. Substituting this information into the preceding equation, we have

$$kt_{1/2} = \frac{1}{\frac{1}{2}[X]_0} - \frac{1}{[X]_0}$$

$$= \frac{2}{[X]_0} - \frac{1}{[X]_0} = \frac{1}{[X]_0}$$

or

$$t_{1/2} = \frac{1}{k[X]_0} \tag{13.24}$$

Note that this value of $t_{1/2}$ is inversely proportional to the initial concentration of X. This dependence on concentration is unlike the $t_{1/2}$ values of first-order reactions, which are independent of concentration.

Zero-Order Reactions

In the Practice Exercise accompanying Sample Exercise 13.8, we introduced the reaction

$$NO_2(g) + CO(g) \rightarrow NO(g) + CO_2(g)$$

The rate law for the reaction is

$$\text{Rate} = k[NO_2]^2 \tag{13.25}$$

Because there is no concentration term for CO in the rate law, the rate of the reaction does not depend upon the concentration of CO. That is, the rate of the reaction does not change with changing [CO], even when the concentrations of CO and NO_2 are comparable.

One interpretation of Equation 13.25 is that it contains a [CO] term to the zeroth power, making the reaction *zero order* in that reactant. Because any value raised to the zeroth power equals 1, we have

$$\text{Rate} = k[NO_2]^2[CO]^0 = k[NO_2]^2(1) = k[NO_2]^2$$

Reactions with a true zero-order rate law are rare, but let's consider a hypothetical zero-order reaction involving a single reactant X that forms product Y:

$$X \rightarrow Y$$

If the reaction is zero-order in X, then the rate law is

$$\text{Rate} = -\Delta[X]/\Delta t = k[X]^0 = k$$

and the integrated rate law is

$$[X] = -kt + [X]_0 \tag{13.26}$$

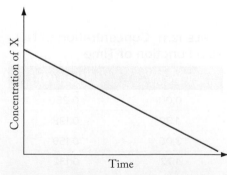

FIGURE 13.16 The change in concentration of the reactant X in the zero-order reaction $X \rightarrow Y$ is constant over time.

The slope of a plot of reactant concentration versus time (Figure 13.16) is the negative of the zero-order rate constant k.

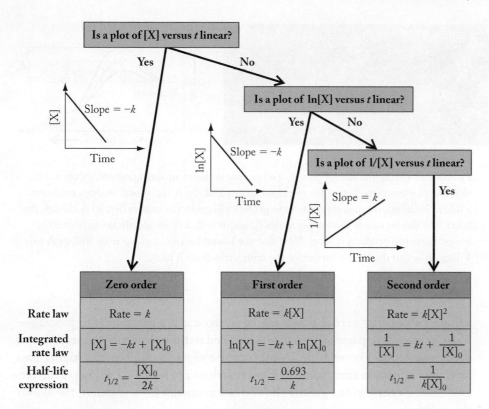

FIGURE 13.17 Summary of how to distinguish between zero-order, first-order, and second-order kinetics for reactions involving a single reactant (X).

	Zero order	First order	Second order
Rate law	Rate = k	Rate = $k[X]$	Rate = $k[X]^2$
Integrated rate law	$[X] = -kt + [X]_0$	$\ln[X] = -kt + \ln[X]_0$	$\dfrac{1}{[X]} = kt + \dfrac{1}{[X]_0}$
Half-life expression	$t_{1/2} = \dfrac{[X]_0}{2k}$	$t_{1/2} = \dfrac{0.693}{k}$	$t_{1/2} = \dfrac{1}{k[X]_0}$

We can calculate the half-life of a zero-order reaction by substituting $t = t_{1/2}$ and $[X] = [X]_0/2$ into the integrated rate law:

$$[X]_0/2 = -kt_{1/2} + [X]_0$$

$$kt_{1/2} = [X]_0 - [X]_0/2 = [X]_0/2$$

$$t_{1/2} = [X]_0/2k \qquad (13.27)$$

Figure 13.17 summarizes how to determine whether a reaction in which reactant X forms one or more products is zero, first, or second order. The figure also includes the rate laws and integrated rate laws for these reactions and the equations used to calculate the half-lives of X. The decision-making process shown in Figure 13.17 can also be used to determine the kinetics of reactions involving two reactants (X and Y) if the reaction is zero order in Y. In such cases, we focus on how the concentration of X changes with time. For now we leave the discussion of zero-order reactions with this purely mathematical treatment. We return to these reactions and examine their meaning at the molecular level in Section 13.5.

13.4 Reaction Rates, Temperature, and the Arrhenius Equation

Chemical reactions take place when molecules collide with sufficient energy to break bonds in reactants and allow bonds to form in products. The minimum amount of energy that enables this to happen is called the **activation energy** **(E_a)**. Every chemical reaction has a characteristic activation energy, usually expressed in kilojoules per mole. Activation energy, which is always a positive

activation energy (E_a) the minimum energy molecules need to react when they collide.

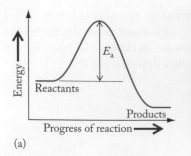

(a)

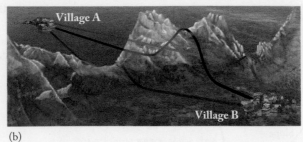

(b)

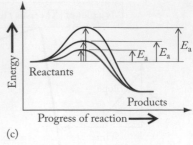

(c)

FIGURE 13.18 (a) The energy profile of a reaction includes an activation energy barrier E_a that must be overcome before the reaction can proceed. (b) A real-world analogy confronts a hiker climbing over mountain passes to get to a village in the next valley. (c) Although the hiker may choose one of the steeper routes (black or red), a molecule always goes over the lowest barrier to products (green). Note that the lowest barrier in going from Village A to Village B is also the lowest barrier for the return trip from B to A.

CONNECTION We discussed bond breaking as an endothermic process and bond formation as an exothermic process in Section 5.3.

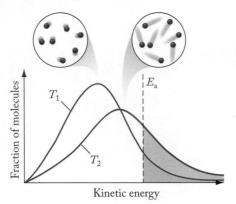

Fraction of molecules in sample with sufficient energy to react at T_1

Increase in number of molecules in sample with sufficient energy to react at T_2; $T_2 > T_1$

(a)

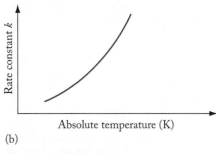

(b)

FIGURE 13.19 (a) According to kinetic molecular theory, a fraction of reactant molecules have kinetic energies equal to or greater than the activation energy (E_a) of the reaction. As temperature increases from T_1 to T_2, the number of molecules with energies exceeding E_a increases, leading to an increase in reaction rate. (b) The rate constant for any reaction increases with increasing temperature.

value, is an energy barrier that must be overcome if a reaction is to proceed—like the mountain passes that must be climbed to hike the trails connecting two valleys shown in Figure 13.18. A hiker may choose to climb over a higher pass (a higher activation energy barrier), but reactions proceed via the lowest barriers available. Generally, the greater the activation energy for a reaction, the slower the reaction.

According to kinetic molecular theory, molecules have higher average kinetic energies at higher temperatures. Put another way, raising the temperature of a collection of molecules means that more of them have the minimum amount of energy needed to overcome the activation energy barrier of a reaction, and react with each other. This principle is illustrated in Figure 13.19(a) where the vertical dashed line represents activation energy, and the shaded areas to the right of the line represent the fraction of reactant molecules with enough energy to react with each other. Notice that the size of the shaded area is larger at higher temperature. This is why the rates of chemical reactions tend to increase with increasing temperature, as shown in Figure 13.19(b).

CONCEPT **TEST**

Why does increasing temperature increase the frequency of collisions between molecules in the gas phase?

In the late 19th century, experiments carried out in the laboratories of Jacobus van 't Hoff and the Swedish chemist Svante Arrhenius (1859–1927) led to a fundamental advance in understanding how temperature affects the rates of chemical reactions. The mathematical connection between temperature, the rate constant k for a reaction, and its activation energy is given by the **Arrhenius equation**:

$$k = Ae^{-E_a/RT} \tag{13.28}$$

where R is the gas constant in J/(mol · K) and T is the reaction temperature in kelvin. The factor A, called the **frequency factor**, is the product of collision frequency and a term that accounts for the fact that not every collision results in a chemical reaction.

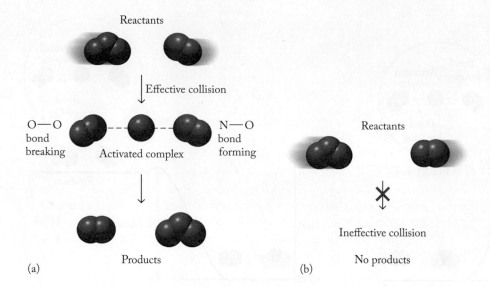

Reactants

Effective collision

O—O
bond
breaking

Activated complex

N—O
bond
forming

Products

(a)

Reactants

Ineffective collision

No products

(b)

FIGURE 13.20 The effect of molecular orientation on reaction rate. (a) When an O_3 and a NO molecule are oriented such that the collision is between an O_3 oxygen and the NO nitrogen, the collision is effective and an activated complex forms, which then yields the two product molecules NO_2 and O_2. (b) When the reactant molecules are oriented such that the collision is between an O_3 oxygen and the NO oxygen, no activated complex forms and no reaction occurs.

Some collisions do not lead to products because the colliding molecules are not oriented with respect to each other in the right way. To examine the importance of molecular orientation during collisions, consider the reaction between O_3 and NO:

$$O_3(g) + NO(g) \rightarrow O_2(g) + NO_2(g)$$

Two ways in which ozone and nitric oxide molecules might approach each other are shown in Figure 13.20. Only one of these orientations, the one in which an O_3 molecule collides with the nitrogen atom of NO, leads to a chemical reaction between the two molecules.

A collision between O_3 and NO molecules with the correct orientation and enough kinetic energy may result in the formation of the **activated complex** shown in Figure 13.20(a). In this species, one of the O—O bonds in the O_3 molecule has started to break and the new N—O bond is beginning to form. Activated complexes have extremely brief lifetimes and fall apart rapidly, either forming products or re-forming reactants. Activated complexes are formed by reacting species that have acquired enough energy to react with each other. The internal energy of an activated complex represents a high-energy **transition state** of the reaction. In fact, the energy of an activated complex for a reaction defines the height of the activation energy barrier for the reaction. The magnitudes of activation energies can vary from a few kilojoules to hundreds of kilojoules per mole.

We can draw an **energy profile** for a chemical reaction that shows the changes in energy for the reaction as a function of the progress of the reaction from reactants to products. We have already seen one energy profile in Figure 13.18(a). Now let's consider the energy profile for the reaction between nitric oxide and ozone, shown in Figure 13.21(a). The x-axis represents the progress of the reaction and the y-axis represents energy. The activation energy is equivalent to the difference in energy between the transition state and the reactants. The size of the activation energy barrier depends on the direction from which it is approached. In the forward direction (NO + O_3 → NO_2 + O_2, Figure 13.21a), E_a is smaller than in the reverse direction (NO_2 + O_2 → NO + O_3, Figure 13.21b). A smaller activation energy barrier means that the forward

 **CHEMTOUR**
Arrhenius Equation

 **CHEMTOUR**
Collision Theory

CONNECTION The van 't Hoff factor in Chapter 11 is named after the same Jacobus van 't Hoff who studied the temperature dependence of reaction rates.

Arrhenius equation an equation relating the rate constant of a reaction to absolute temperature (T), the activation energy of the reaction (E_a), and the frequency factor (A).

frequency factor (A) the product of the frequency of molecular collisions and a factor that expresses the probability that the orientation of the molecules is appropriate for a reaction to occur.

activated complex a species formed in a chemical reaction when molecules have both the proper orientation and enough energy to react with each other.

transition state a high-energy state between reactants and products in a chemical reaction.

energy profile a graph showing the changes in energy for a reaction as a function of the progress of the reaction from reactants to products.

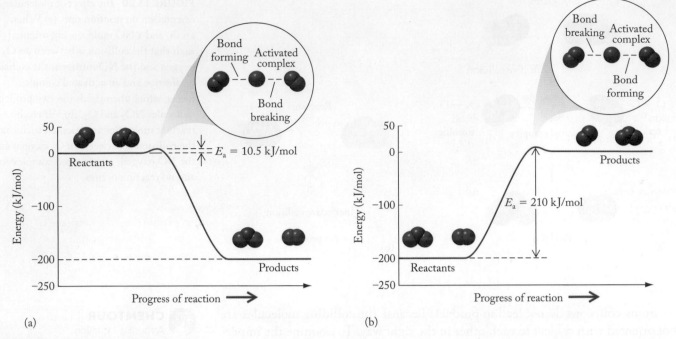

(a)

(b)

FIGURE 13.21 (a) The energy profile for the reaction $NO(g) + O_3(g) \rightarrow NO_2(g) + O_2(g)$ includes an activation energy barrier of 10.5 kJ/mol. (b) The reverse reaction has a much larger activation energy of 210 kJ/mol.

reaction proceeds at a higher rate than the reverse reaction if we have equal concentrations of reactants and products.

One of the many uses of the Arrhenius equation is to calculate the value of E_a for a chemical reaction. When we take the natural logarithm of both sides of Equation 13.28,

$$\ln k = -\frac{E_a}{R}\left(\frac{1}{T}\right) + \ln A \qquad (13.29)$$

the result fits the general equation of a straight line ($y = mx + b$) if we make ($\ln k$) the y variable and ($1/T$) the x variable. We can calculate E_a by determining the rate constant k for the reaction at several temperatures. Plotting $\ln k$ versus $1/T$ should give a straight line, the slope of which is $-E_a/R$. Table 13.12 and Figure 13.22 show data for the reaction between NO and O_3 at six temperatures. Arbitrarily picking two points on the line in Figure 13.22, we calculate its slope:

$$\text{Slope} = \frac{\Delta y}{\Delta x} = \frac{(23.814 - 24.264)}{(2.86 \times 10^{-3} - 2.50 \times 10^{-3})\ \text{K}^{-1}} = -1.25 \times 10^3\ \text{K}$$

TABLE 13.12   Temperature Dependence of the Rate of Reaction $NO(g) + O_3(g) \rightarrow NO_2(g) + O_2(g)$

T (K)	k (M⁻¹ s⁻¹)	ln k	1/T (K⁻¹)
300	1.21×10^{10}	23.216	3.33×10^{-3}
325	1.67×10^{10}	23.539	3.08×10^{-3}
350	2.20×10^{10}	23.814	2.86×10^{-3}
375	2.79×10^{10}	24.052	2.67×10^{-3}
400	3.45×10^{10}	24.264	2.50×10^{-3}
425	4.15×10^{10}	24.449	2.35×10^{-3}

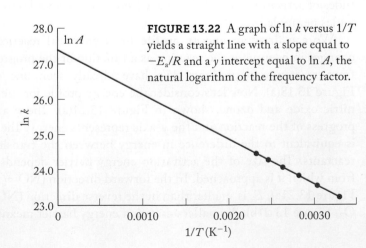

FIGURE 13.22 A graph of $\ln k$ versus $1/T$ yields a straight line with a slope equal to $-E_a/R$ and a y intercept equal to $\ln A$, the natural logarithm of the frequency factor.

The slope equals $-E_a/R$, so

$$E_a = -\text{slope} \times R$$

$$= -(-1.25 \times 10^3 \text{ K}) \times \left(\frac{8.314 \text{ J}}{\text{mol} \cdot \text{K}}\right) = 1.04 \times 10^4 \text{ J/mol}$$

$$= 10.4 \text{ kJ/mol}$$

The y intercept $(1/T = 0)$ in Figure 13.22 is 27.41. From Equation 13.29, we know that this value represents $\ln A$, which means

$$A = e^{27.41} = 8.0 \times 10^{11}$$

Now we can use the values of E_a and A to calculate k at any temperature. For example, at $T = 250$ K:

$$k = Ae^{-E_a/RT}$$

$$= (8.0 \times 10^{11})\, e^{-\left(\dfrac{1.04 \times 10^4 \text{ J/mol}}{8.314 \dfrac{\text{J}}{\text{mol} \cdot \text{K}} \cdot 250 \text{ K}}\right)}$$

$$= 5.4 \times 10^9\, M^{-1}\,\text{s}^{-1}$$

SAMPLE EXERCISE 13.9 Calculating an Activation Energy   **LO7**
from Rate Constants

The data in Table 13.13(a) were collected in a study of the effect of temperature on the rate of the decomposition reaction

$$2\, ClO(g) \rightarrow Cl_2(g) + O_2(g)$$

Determine the activation energy for the reaction.

Collect and Organize We can calculate activation energy by using the Arrhenius equation. We are given values of the rate constant k as a function of absolute temperature. According to Equation 13.29, the slope of a plot of $\ln k$ against $1/T$ is equal to $-E_a/R$, where E_a is the activation energy and R is the ideal gas constant with units of J/(mol · K).

Analyze First, we need to convert the T and k values from Table 13.13(a) to $1/T$ and $\ln k$, respectively. We predict a positive value for E_a because activation energy represents a barrier that costs energy to overcome. The rate constant for the reaction is fairly large, about $10^9\, M^{-1}\,\text{s}^{-1}$, so we predict that the activation energy barrier will be relatively low.

Solve Expanding the data table to include columns for $1/T$ and $\ln k$ yields Table 13.13(b). A plot of $\ln k$ versus $1/T$ gives us a straight line, the slope of which is -1590 K (Figure 13.23).

TABLE 13.13(b) Summary of T, $1/T$, k, and $\ln k$ for the Decomposition of ClO

T (K)	$1/T$ (K^{-1})	k (M^{-1} s^{-1})	$\ln k$
238	4.20×10^{-3}	1.9×10^9	21.365
258	3.88×10^{-3}	3.1×10^9	21.855
278	3.60×10^{-3}	4.9×10^9	22.313
298	3.36×10^{-3}	7.2×10^9	22.697

TABLE 13.13(a) Rate Constant as a Function of Temperature for the Decomposition of ClO

k (M^{-1} s^{-1})	T (K)
1.9×10^9	238
3.1×10^9	258
4.9×10^9	278
7.2×10^9	298

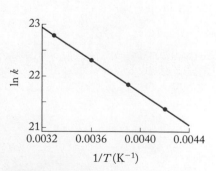

FIGURE 13.23 Plot of data in Table 13.13(b).

TABLE 13.14 Rate Constant as a Function of Temperature for the Reaction of Br with O_3

T (K)	k [cm³/(molecule · s)]
238	5.9×10^{-13}
258	7.7×10^{-13}
278	9.6×10^{-13}
298	1.2×10^{-12}

We use the values of the slope and R [8.314 J/(mol · K)] to calculate the value of E_a:

$$E_a = -\text{slope} \times R$$

$$= -(-1590 \text{ K}) \times \left(\frac{8.314 \text{ J}}{\text{mol} \cdot \text{K}}\right) = 1.32 \times 10^4 \text{ J/mol}$$

$$= 13.2 \text{ kJ/mol}$$

Think About It The activation energy is the height of a barrier that has to be overcome before a reaction can proceed. As predicted, the activation energy for ClO decomposition has a small positive value.

 Practice Exercise The rate constant for the reaction

$$\text{Br}(g) + O_3(g) \rightarrow \text{BrO}(g) + O_2(g)$$

was determined at the four temperatures shown in Table 13.14. Calculate the activation energy for this reaction.

Calculating activation energies by using the graphical method generally requires measurement of the rate constant at a minimum of three different temperatures to verify that the plot of ln k versus $1/T$ is a straight line. Once we know the value of E_a, we can use it and the value of the rate constant (k_1) of a reaction at one temperature (T_1) to calculate the value of the rate constant (k_2) at another temperature (T_2). We start by substituting k_1, k_2, T_1, and T_2 into Equation 13.29:

$$\ln k_1 = -\frac{E_a}{R}\left(\frac{1}{T_1}\right) + \ln A \qquad \ln k_2 = -\frac{E_a}{R}\left(\frac{1}{T_2}\right) + \ln A$$

Subtracting these two expressions, ln k_1 − ln k_2, gives

$$\ln k_1 - \ln k_2 = \left[-\frac{E_a}{R}\left(\frac{1}{T_1}\right) + \ln A\right] - \left[-\frac{E_a}{R}\left(\frac{1}{T_2}\right) + \ln A\right]$$

Using the mathematical properties of logarithms, we can rearrange the terms to obtain this equation:

$$\ln \frac{k_1}{k_2} = \frac{E_a}{R}\left(\frac{1}{T_2}\right) - \frac{E_a}{R}\left(\frac{1}{T_1}\right) \qquad (13.30)$$

$$= \frac{E_a}{R}\left(\frac{1}{T_2} - \frac{1}{T_1}\right)$$

SAMPLE EXERCISE 13.10 Determining the Effect of Temperature on Rate Constants   **LO7**

The kinetics of decomposition of a potential drug have been tested in experiments at room temperature (25°C) where $k = 6.45 \times 10^{-6} \ M^{-1} \text{ s}^{-1}$. If the activation energy for this reaction is 67.1 kJ, what will happen to the reaction rate when the same kinetics experiments are conducted at body temperature (37°C)?

Collect and Organize We are asked to calculate what happens to the rate of the reaction when kinetics experiments are run at a higher temperature. We are given values of the activation energy E_a, one rate constant k, and two temperatures. We can use

Equation 13.30 to solve for the second rate constant where R is the ideal gas constant with units of $J/(mol \cdot K)$.

Analyze First, we need to convert the temperature values from Celsius degrees to kelvin. We have one (k, T) pairing—$(6.45 \times 10^{-6} \, M^{-1} \, s^{-1}, 298 \, K)$—and we need to find the k value for the second temperature, 310 K. It does not matter whether we assign the given k and temperature to k_1 and T_1 or to k_2 and T_2 in Equation 13.30, as long as we keep them paired. We predict that as the temperature increases from room temperature to body temperature, the reaction will go faster and k will increase.

Solve Substituting the given values into Equation 13.30 gives us:

$$\ln \frac{k_1}{k_2} = \frac{E_a}{R}\left(\frac{1}{T_2} - \frac{1}{T_1}\right)$$

$$\ln \frac{k_1}{6.45 \times 10^{-6} \, M^{-1} \, s^{-1}} = \frac{\left(\dfrac{67.1 \, \cancel{kJ}}{\cancel{mol}}\right)\left(\dfrac{1000 \, J}{1 \, \cancel{kJ}}\right)}{\dfrac{8.314 \, J}{\cancel{mol} \cdot K}}\left(\frac{1}{298 \, K} - \frac{1}{310 \, K}\right)$$

$$k_1 = 1.84 \times 10^{-5} \, M^{-1} \, s^{-1}$$

Think About It As predicted, k increases as T increases; in fact, it nearly triples. A higher temperature increases kinetic energy, which leads to more collisions of molecules with sufficient energy to exceed the threshold value of the activation energy.

 Practice Exercise What temperature would be required to double the reaction rate at room temperature for the potential drug tested in this Sample Exercise?

CONCEPT **TEST**

Which of the following statements is/are true about activation energies?

a. Exothermic reactions have negative activation energies.

b. Fast reactions have large rate constants *and* large activation energies.

c. The forward reaction sometimes has a lower activation energy than the reverse reaction.

d. Endothermic reactions always have large activation energies.

13.5 Reaction Mechanisms

Up to this point we have described reactions in terms of macroscopically observable quantities such as pressure, temperature, volume, and numbers of moles of substances. In this section we explore how reactions proceed at the molecular level. Being able to suggest what goes on at the molecular level by observing macroscopic properties is a remarkable achievement. We will connect the minimum amount of energy that is needed to break bonds in the reactants and form bonds in the products to an exact determination of which bonds break and which bonds form. The evaluation of the stepwise behavior of molecules—the mechanism of a reaction—provides valuable insights into reactions, ranging from the subtlest biochemical processes taking place in cells to the chemistry of industrial processes that yield tons of product.

Let's start by revisiting the thermal decomposition of NO_2:

$$2 \, NO_2(g) \rightarrow 2 \, NO(g) + O_2(g)$$

CHEMTOUR
Reaction Mechanisms

reaction mechanism a set of steps that describe how a reaction occurs at the molecular level; the mechanism must be consistent with the experimentally determined rate law for the reaction.

intermediate a species produced in one step of a reaction and consumed in a subsequent step.

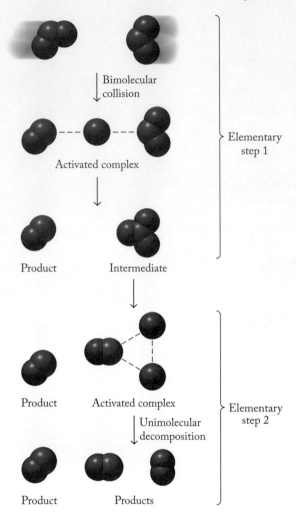

FIGURE 13.24 The decomposition of NO_2 begins when two NO_2 molecules collide, producing NO and NO_3 (elementary step 1). The NO_3 then rapidly decomposes into NO and O_2 (elementary step 2).

We noted in Section 13.3 that this reaction is second order in NO_2 because it takes place as a result of collisions between two NO_2 molecules at a time. How do the atoms in two colliding NO_2 molecules rearrange themselves to form two NO molecules and one O_2 molecule? The answer to this question is contained in the mechanism of the reaction. A **reaction mechanism** describes the stepwise manner in which the bonds in reactant molecules break and the bonds in product molecules form. Chemists use the results of reaction rate measurements to develop explanations of how reactions actually happen. Sometimes they can test for the presence of the products formed in preliminary steps, but, until recently, this was not possible for most reactions. Today, technological advances allow chemists to follow the transformation of reactants to products in the time that it takes for individual covalent bonds to break and new ones to form. These processes occur as rapidly as the bonds vibrate, that is, in femtoseconds (10^{-15} s), so the area of research based on monitoring these ultrafast processes is called *femtochemistry*.[3]

Elementary Steps

A reaction mechanism proposed for the decomposition of NO_2 is shown in Figure 13.24. In the first step of the mechanism, a collision between two NO_2 molecules produces a molecule of NO and a molecule of NO_3. In a second step, the NO_3 decomposes to NO and O_2. Both steps in the mechanism involve very short-lived activated complexes. In the activated complex of the first step, two molecules share an oxygen atom. The bonds in the activated complex of the second step rearrange so that two oxygen atoms bond together, forming a molecule of O_2 and leaving behind a molecule of NO. The molecule NO_3 is an **intermediate** in this mechanism because it is produced in one step and consumed in the next. Intermediates are not considered reactants or products and do not appear in the equation describing a reaction. In some cases, intermediates in chemical reactions are sufficiently long-lived to be isolated. In contrast, activated complexes have never been isolated, although they have been detected.

This reaction mechanism is a combination of two **elementary steps**. An elementary step that involves a single molecule is called **unimolecular**, and one that involves a collision between two molecules is **bimolecular**. Bimolecular elementary steps are much more common than **termolecular** (three-molecule) elementary steps because the chance of three molecules colliding at exactly the same time is much smaller. The terms *uni-*, *bi-*, and *termolecular* are used by chemists to describe the **molecularity** of an elementary step, which refers to the number of atoms, ions, or molecules involved in that step.

A valid reaction mechanism must be consistent with the stoichiometry of the reaction. In other words, the sum of the elementary steps in Figure 13.24 must be consistent with the observed proportions

---

[3]In 1999 Ahmed H. Zewail received the Nobel Prize in Chemistry for his pioneering research in the development of femtochemistry.

of reactants and products as defined in the balanced chemical equation. In this case, the sum matches the overall stoichiometry:

Elementary step 1 $2\,NO_2(g) \rightarrow NO(g) + NO_3(g)$

Elementary step 2 $NO_3(g) \rightarrow NO(g) + O_2(g)$

Summing the two elementary steps and simplifying by cancelling out the intermediate (NO_3) terms:

$$2\,NO_2(g) + \cancel{NO_3(g)} \rightarrow 2\,NO(g) + \cancel{NO_3(g)} + O_2(g)$$

we get the overall reaction

$$2\,NO_2(g) \rightarrow 2\,NO(g) + O_2(g)$$

Before ending this discussion, let's consider how activation energy applies to a two-step reaction such as this one. The two elementary steps produce an energy profile with two maxima. In elementary step 1, collisions between pairs of NO_2 molecules result in the formation of an activated complex associated with the first transition state in Figure 13.25. As this activated complex transforms into NO and NO_3, the energy of the system drops to the bottom of the trough between the two maxima. In elementary step 2, NO_3 forms the activated complex associated with the second transition state. As this complex transforms into the final products NO and O_2, the energy of the system drops to its final level.

Figure 13.25 shows that the energy barrier for elementary step 1 is much greater than that for elementary step 2. This difference is consistent with the relative rates of the two steps: step 1 is slower than step 2. If the reaction were to proceed in the reverse direction (as NO and O_2 react, forming NO_2), the first energy barrier would be the smaller of the two, and the first elementary step would be the more rapid one. Experimental evidence supports these expectations.

CONCEPT TEST

For a reaction mechanism with *n* elementary steps, how many activation energies does the overall reaction have?

Rate Laws and Reaction Mechanisms

Any mechanism proposed for a reaction must be consistent with the rate law derived from experimental data. Is the mechanism proposed in Figure 13.24 consistent with the reaction's second-order rate law? For a reaction mechanism to be consistent with a rate law, the molecularity of one of the elementary steps in the mechanism must be the same as the reaction order expressed in the rate law.

As noted in Section 13.3, the thermal decomposition of NO_2 is second order in NO_2. Recall that the balanced chemical equation for the overall reaction does not give us enough information to write the rate law for the reaction.

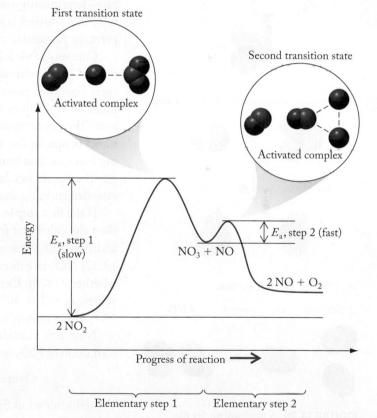

FIGURE 13.25 The energy profile for the decomposition of NO_2 to NO and O_2 shows activation energy barriers for both elementary steps. The activation energy of the first step is larger than that of the second step, so the first step is the slower of the two.

rate-determining step the slowest step in a multistep chemical reaction.

However, for any *elementary step* in a reaction mechanism, we can use the balanced chemical equation to write a rate law for that step. For the thermal decomposition of NO_2, the rate law for the first elementary step, $2 NO_2(g) \rightarrow NO_3(g) + NO(g)$, is

$$\text{Rate}_1 = k_1[NO_2]^2 \qquad (13.31)$$

We obtain this rate law by writing the concentration for each reactant raised to a power equal to the coefficient of that reactant in the balanced equation. For the second elementary step in the NO_2 decomposition, $NO_3(g) \rightarrow NO(g) + O_2(g)$, the sole reactant, NO_3, has a coefficient of 1, so the rate law is

$$\text{Rate}_2 = k_2[NO_3] \qquad (13.32)$$

How do we use the rate laws in Equations 13.31 and 13.32 to determine the validity of the mechanism, that is, to show that they conform to the observed second-order rate law for the decomposition of NO_2?

The two steps in the mechanism proceed at different rates, with different activation energies. One step is slower than the other, and this slower step is the **rate-determining step** in the reaction. The rate-determining step is the slowest elementary step in a chemical reaction, and the rate of this step controls the overall reaction rate. We can use the rate laws for the two steps to identify the rate-determining step. Because the rate law for step 1 matches the experimentally determined rate law, we may assume that step 1 defines how rapidly the reaction proceeds.

One way of visualizing the concept of a rate-determining step is to analyze the flow of people through a busy airport. Many travelers arrive at the airport with their boarding passes in hand or print them from convenient kiosks that allow them to avoid lines at ticket counters and move quickly toward their departure gate. The next step in the process is passing through security. Typically, the number of people in line outside security points greatly exceeds the number of screening machines, so that the time required to reach a departure gate depends mostly on the time needed to pass through security. Security screening is the rate-determining step on the way through an airport.

If the first step in the mechanism in Figure 13.24 is the rate-determining step, then the value of k for the overall reaction is equal to k_1 from Equation 13.31. In addition, the value of k_1 must be smaller than the value of k_2 from Equation 13.32. NO_3 is sufficiently stable that we can make small amounts of it and test whether $k_1 < k_2$. Experiments run at 300 K starting with NO_2 or NO_3 yield these values: $k_1 \approx 1 \times 10^{-10}\ M^{-1}\ s^{-1}$ and $k_2 \approx 6.3 \times 10^4\ s^{-1}$. Therefore as soon as any NO_3 forms in step 1, it rapidly falls apart to NO and O_2 in step 2.

Now let's consider the reverse reaction of NO_2 decomposition, namely, the formation of NO_2 from NO and O_2:

$$\text{Overall reaction} \qquad 2 NO(g) + O_2(g) \rightarrow 2 NO_2(g)$$

We determined in Section 13.3 that this reaction is second order in NO and first order in O_2:

$$\text{Rate} = k[NO]^2[O_2] \qquad (13.33)$$

One proposed mechanism is shown in Figure 13.26. It has two elementary steps:

Step 1 $\quad NO(g) + O_2(g) \rightarrow NO_3(g) \qquad \text{Rate} = k_1[NO][O_2] \qquad (13.34)$

Step 2 $\quad NO_3(g) + NO(g) \rightarrow 2 NO_2(g) \qquad \text{Rate} = k_2[NO_3][NO] \qquad (13.35)$

(1) Formation of intermediate: elementary step 1

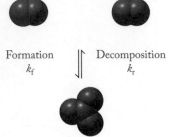

Formation k_f | Decomposition k_r

(2) Reaction of intermediate: elementary step 2

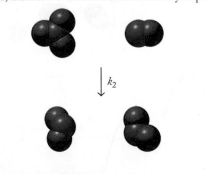

k_2

Overall reaction:

$2\ NO\ +\ O_2\ \longrightarrow\ 2\ NO_2$

FIGURE 13.26 A mechanism for the formation of NO_2 from NO and O_2 has two elementary steps: (1) a fast, reversible bimolecular reaction in which NO and O_2 form NO_3 and (2) a slower, rate-determining bimolecular reaction in which NO_3 reacts with a molecule of NO to form two molecules of NO_2.

We obtained the rate laws in Equations 13.34 and 13.35 by writing the concentration for each reactant raised to a power equal to the reactant's coefficient in the balanced equation. If step 1 were the rate-determining step, the reaction would be first order in NO and O_2, but that does not match the experimentally determined rate law. If step 2 were the rate-determining step, the reaction would be first order in NO and NO_3, but that does not match the rate law either. So, how can we account for the experimental rate law?

Consider what happens if step 2 is slow while step 1 is fast *and reversible*, which means NO_3 forms rapidly from NO and O_2 but decomposes just as rapidly back into NO and O_2. Expressing these rates in equation form:

$$\text{Rate of forward reaction} = k_f[\text{NO}][\text{O}_2] = \text{fast}$$

$$\text{Rate of reverse reaction} = k_r[\text{NO}_3] = \text{equally fast}$$

The subscripts "f" and "r" refer to the forward and reverse reactions, respectively.

Combining these two expressions, we have

$$k_f[\text{NO}][\text{O}_2] = k_r[\text{NO}_3]$$

$$[\text{NO}_3] = \frac{k_f}{k_r}[\text{NO}][\text{O}_2] \tag{13.36}$$

Now, if we replace the $[NO_3]$ term in the rate law of Equation 13.35 (which we select because step 2 is the rate-determining step) with the right side of Equation 13.36, we get

$$\text{Rate} = k_2\frac{k_f}{k_r}[\text{NO}]^2[\text{O}_2]$$

The three rate constants can be combined,

$$k_{\text{overall}} = k_2\frac{k_f}{k_r}$$

and the rate law for the overall reaction becomes

$$\text{Rate} = k_{\text{overall}}[\text{NO}]^2[\text{O}_2]$$

This expression matches the overall rate law in Equation 13.33, so the proposed mechanism—a fast and reversible step 1 followed by a slow step 2—may be valid.

As a final point regarding reaction mechanisms, even though the proposed reaction mechanism is consistent with the overall stoichiometry of the reaction and with the experimentally derived rate law, these consistencies do not *prove* that the proposed mechanism is correct. Other mechanisms could be consistent with the rate law, too. On the other hand, *not* finding a reactive (and transient) intermediate would not necessarily disprove a reaction mechanism because we could be limited by our ability to detect such a short-lived species.

CONCEPT TEST

Could the products of an elementary step in a chemical reaction include activated complexes?

SAMPLE EXERCISE 13.11 Linking Reaction Mechanisms **LO8**
to Experimental Rate Laws

The experimentally determined rate law for the reaction

$$2 NO(g) + 2 H_2(g) \rightarrow N_2(g) + 2 H_2O(g)$$

is

$$Rate = k[NO]^2[H_2]$$

A proposed mechanism for the reaction is

Elementary step 1 $2 NO(g) + H_2(g) \rightarrow N_2O(g) + H_2O(g)$
Elementary step 2 $N_2O(g) + H_2(g) \rightarrow N_2(g) + H_2O(g)$

Is this reaction mechanism consistent with the stoichiometry of the reaction and with the rate law? If so, which is the rate-determining step?

Collect and Organize We need to determine whether the proposed two-step reaction mechanism is consistent with the overall reaction stoichiometry and the rate law of the overall reaction. We know that a rate law derived from the stoichiometry of the rate-determining step in a reaction mechanism should match the rate law of the overall reaction.

Analyze To answer the questions posed in this exercise, we need to

1. Determine whether the chemical equations of the elementary steps add up to the overall reaction equation.
2. Write rate laws for each elementary step.
3. Compare the rate laws of the elementary steps to the experimental rate law of the overall reaction to assess the validity of the mechanism.
4. Decide which step is rate determining by matching its rate law to the observed rate law.

Solve Let's test whether the elementary steps add up to the overall reaction:

(1) $2 NO(g) + H_2(g) \rightarrow \cancel{N_2O(g)} + H_2O(g)$
(2) $\cancel{N_2O(g)} + H_2(g) \rightarrow N_2(g) + H_2O(g)$

$$2 NO(g) + 2 H_2(g) \rightarrow N_2(g) + 2 H_2O(g)$$

This is indeed the equation of the overall reaction. The elementary steps are consistent with the stoichiometry of the overall reaction.

Next we need to focus on the reaction mechanism. Elementary step 1 involves two molecules of NO colliding with one molecule of H_2 in a termolecular reaction. Elementary step 2 is a bimolecular reaction between the N_2O produced in step 1 and another molecule of H_2. We apply the fact that the rate law for any elementary step can be written directly from the balanced equations, using the equation coefficients as exponents in the rate law:

Step 1 $2 NO(g) + H_2(g) \rightarrow N_2O(g) + H_2O(g)$ $Rate = k_1[H_2][NO]^2$
Step 2 $N_2O(g) + H_2(g) \rightarrow N_2(g) + H_2O(g)$ $Rate = k_2[H_2][N_2O]$

The rate law of step 1 matches the observed rate law of the overall reaction. Therefore the proposed two-step mechanism is consistent with the experimental rate law, and step 1 is the rate-determining step.

Think About It We do not have definitive proof that the proposed mechanism is the correct one, but we can say that it is consistent with the available data.

⊕ **Practice Exercise** The following mechanism is proposed for a reaction between compounds A and B:

Step 1 $2 A(g) + B(g) \rightleftharpoons C(g)$ fast and reversible
Step 2 $B(g) + C(g) \rightarrow D(g)$ slow
Overall $2 A(g) + 2 B(g) \rightarrow D(g)$

What is the rate law for the overall reaction based on the proposed mechanism?

SAMPLE EXERCISE 13.12 Testing a Proposed Reaction Mechanism **LO8**

One proposed mechanism for the decomposition of N_2O_5 to NO_2 involves three elementary steps:

Step 1	$2 N_2O_5(g) \rightleftharpoons N_4O_{10}(g)$	fast and reversible
Step 2	$N_4O_{10}(g) \rightarrow N_2O_3(g) + 2 NO_2(g) + O_3(g)$	slow
Step 3	$N_2O_3(g) + O_3(g) \rightarrow 2 NO_2(g) + O_2(g)$	fast
Overall	$2 N_2O_5(g) \rightarrow 4 NO_2(g) + O_2(g)$	

What is the rate law for the overall reaction based on the proposed mechanism?

Collect and Organize The rate law for the overall reaction reflects the rate laws for the elementary reactions preceding and including the rate-determining (slowest) step in the mechanism. We are given three elementary steps and their relative rates.

Analyze The rate law for an elementary step can be written directly from the balanced chemical equation for that step. We are told that step 1 is fast *and reversible*, which means that as the reaction proceeds and $[N_4O_{10}]$ increases, eventually the rate of step 1 in the forward direction is matched by the rate of step 1 in the reverse direction. Only elementary steps 1 and 2 contribute to the rate law for the overall reaction because step 2 is the rate-determining step.

Solve The rate laws for step 1 in the forward and reverse directions are

$$\text{Rate of forward step 1} = k_f[N_2O_5]^2$$

$$\text{Rate of reverse step 1} = k_r[N_4O_{10}]$$

The rate law for step 2 is

$$\text{Rate of step 2} = k_2[N_4O_{10}]$$

To find the overall rate law, we need to express $[N_4O_{10}]$ in terms of $[N_2O_5]$ by making the step 1 rates equal to each other and rearranging the terms:

$$[N_4O_{10}] = \frac{k_f}{k_r}[N_2O_5]^2$$

Substituting this expression for $[N_4O_{10}]$ into the rate law for step 2:

$$\text{Rate} = \frac{k_f k_2}{k_r}[N_2O_5]^2$$

Think About It The rate law for the overall reaction must depend on the concentration of N_2O_5. The order of the reaction in N_2O_5 depends on the proposed mechanism for the reaction. Notice that the rate of step 3 has no impact on the rate law for this mechanism because it follows the slowest step. In addition, note that the rate law for the mechanism in this Sample Exercise does not match this reaction's experimentally determined rate law from Sample Exercise 13.5:

$$\text{Rate} = k[N_2O_5]$$

Therefore the mechanism that starts with the dimerization of N_2O_5 cannot be correct.

 Practice Exercise Here is another proposed mechanism for the reaction of NO with H_2 in Sample Exercise 13.11:

Elementary step 1	$H_2(g) + NO(g) \rightarrow N(g) + H_2O(g)$
Elementary step 2	$N(g) + NO(g) \rightarrow N_2(g) + O(g)$
Elementary step 3	$H_2(g) + O(g) \rightarrow H_2O(g)$

Is this a valid mechanism? Explain why or why not.

Mechanisms and Zero-Order Reactions

Before we leave reaction mechanisms, let's revisit the reaction between NO_2 and CO, which has an experimentally determined rate law that is zero order in CO, second order in NO_2, and second order overall:

$$NO_2(g) + CO(g) \rightarrow NO(g) + CO_2(g) \qquad \text{Rate} = k[NO_2]^2$$

What does this overall rate law tell us about how the reaction happens at the molecular level? Remember that the overall rate depends on the concentrations of the reactants in the rate-determining step, which means that CO is not a reactant in the rate-determining step. Carbon monoxide is clearly involved in the reaction—it is converted into CO_2—but whatever step involves CO must occur after the rate-determining step. This leads us to conclude that the reaction must have at least two elementary steps, one rate-determining and one not.

It has been proposed that the reaction mechanism is

(1) $\qquad\qquad 2\,NO_2(g) \rightarrow NO_3(g) + NO(g) \qquad \text{Rate} = k_1[NO_2]^2$

(2) $\qquad NO_3(g) + CO(g) \rightarrow NO_2(g) + CO_2(g) \qquad \text{Rate} = k_2[NO_3][CO]$

The experimentally determined overall rate law matches the rate law for the first step, which must be the slower, rate-determining step. The overall reaction is zero order in CO because CO is not a reactant in that step.

CONCEPT TEST

The rate law for the reaction $XO_2(g) + M(g) \rightarrow XO(g) + MO(g)$, where X and M represent metallic elements, is second order in $[XO_2]$ and independent of [M] for a wide variety of compounds XO_2 and M. Why can we conclude that these reactions probably proceed by the same mechanism?

13.6 Catalysts

We noted in Section 13.4 that slow reactions often have high activation energies. Suppose we wanted to increase the rate of such a reaction. How could we do it? One way is to increase the temperature of the reaction mixture. However, in some chemical reactions, elevated temperatures can lead to undesired products or to lower yields. Another way is to add a **catalyst**, a substance that increases the rate of a reaction but is not consumed in the process.

Catalysts and the Ozone Layer

As we discussed in Chapter 8, the way we think about ozone depends on where the ozone is located. Ozone in the stratosphere between 10 and 40 km above Earth's surface is necessary to protect us from UV radiation, but ozone at ground level is hazardous to our health. In this section we discuss the role of catalysis in the loss of stratospheric ozone that has led to the annual formation of ozone holes over Antarctica.

The natural photodecomposition of ozone in the stratosphere occurs through the reaction

$$2\,O_3(g) \rightarrow 3\,O_2(g)$$

catalyst a substance added to a reaction that increases the rate of the reaction but is not consumed in the process.

It begins with the absorption of UV radiation from the sun and the generation of atomic oxygen:

$$O_3(g) \rightarrow O_2(g) + O(g)$$

The oxygen atom may react with another ozone molecule to form two more molecules of oxygen:

$$O_3(g) + O(g) \rightarrow 2\,O_2(g)$$

The rate of the second elementary step is low because its activation energy is relatively high: 17.7 kJ/mol.

In 1974 two American scientists, Sherwood Rowland and Mario Molina, predicted significant depletion of stratospheric ozone because of the release of a class of volatile compounds called chlorofluorocarbons (CFCs) into the atmosphere at ground level, which ultimately enter the stratosphere. This prediction was later supported by experimental evidence of a thinning of the ozone layer and formation of annual ozone holes over Antarctica. By 2000 stratospheric ozone concentrations over Antarctica were less than half of what they were in 1980, and the ozone hole covered nearly all of Antarctica and the tip of South America (Figure 13.27). Less severe thinning of stratospheric ozone was observed in the Northern Hemisphere.

In 1989 an international agreement known as the Montreal Protocol, which called for an end to the production of ozone-depleting CFCs, went into effect. It has had a dramatic effect on CFC production and emission into the atmosphere, and as the trend line in Figure 13.27 indicates, the ozone layer over Antarctica appears to be slowly recovering. However, these compounds last for many years in the atmosphere, and full recovery of the ozone layer may take most of the 21st century.

How do CFCs contribute to the destruction of ozone? Three of the more widely used CFCs were CCl_2F_2, CCl_3F, and $CClF_3$. In the stratosphere, these

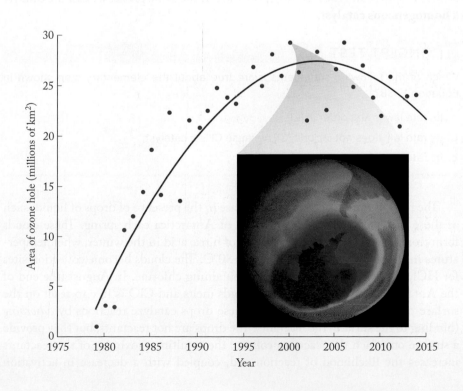

FIGURE 13.27 A 35-year trend in stratospheric ozone depletion over the South Pole. The data points represent the size of the Antarctic ozone hole observed each year. The deep blue and violet colors in the satellite image represent ozone concentrations that are less than half their normal values.

molecules encounter UV radiation with enough energy to break C—Cl bonds, releasing chlorine atoms. For example,

$$CCl_3F(g) \xrightarrow{\text{sunlight}} CCl_2F(g) + Cl(g) \qquad (13.37)$$

Free chlorine atoms react with ozone, forming chlorine monoxide:

$$Cl(g) + O_3(g) \rightarrow ClO(g) + O_2(g) \qquad (13.38)$$

Chlorine monoxide then reacts with more ozone, producing oxygen and regenerating atomic chlorine:

$$ClO(g) + O_3(g) \rightarrow Cl(g) + 2\,O_2(g) \qquad (13.39)$$

If we add Equations 13.38 and 13.39 and cancel species as needed, we get

$$2\,O_3(g) \rightarrow 3\,O_2(g)$$

The overall reaction is exactly the same as the natural photodecomposition of ozone. The difference is the presence of chlorine atoms in Equations 13.38 and 13.39. Chlorine atoms act as a catalyst for the destruction of ozone because they speed up the reaction but are not consumed by it. Rather, they are consumed in one elementary step but then regenerated in a later elementary step. Chlorine is a catalyst, not an intermediate, because in a reaction mechanism, a catalyst is consumed in an early step and then regenerated in a later one, whereas an intermediate is produced before it is consumed. A single chlorine atom can catalyze the destruction of hundreds to thousands of stratospheric O_3 molecules before it combines with other atoms and forms a less reactive molecule.

The activation energy for the O_3 decomposition following the Cl-catalyzed reaction is only 2.2 kJ/mol, whereas the activation energy for the uncatalyzed reaction pathway described in Equations 13.38 and 13.39 is 17.7 kJ/mol (Figure 13.28). Its smaller E_a value means that the catalyzed destruction of ozone is faster than the natural photodecomposition process. In the reaction describing the destruction of ozone, the catalyst, $Cl(g)$, and reactant, $O_3(g)$, exist in the same physical phase. When a catalyst and the reacting species are in the same phase, we call the catalyst a **homogeneous catalyst**.

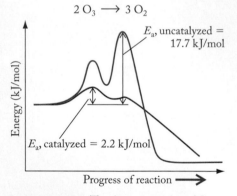

FIGURE 13.28 The decomposition of O_3 in the presence of chlorine atoms has a smaller activation energy (2.2 kJ/mol) than the naturally occurring photodecomposition of O_3 to O_2 (17.7 kJ/mol). The catalytic effect of chlorine is a key factor in the depletion of stratospheric ozone and the formation of an ozone hole over the South Pole.

CONCEPT TEST

Which of the following statements is/are true about the elementary step shown in Equation 13.38?

a. Its rate law is first order in [Cl].

b. Its rate law does not include [Cl] because Cl is a catalyst.

c. Its rate is independent of [Cl].

The rates of the above reactions increase in the presence of drops of liquid, such as those in the clouds that cover much of Antarctica each spring. These clouds form from crystals of ice and tiny drops of nitric acid in the winter, when temperatures in the lower stratosphere dip to −80°C. The clouds become collection sites for HCl, ClO, and other compounds containing chlorine. In August (the end of the Antarctic winter), the ice in the clouds melts and ClO is free to react on the surface of the drops of liquid water. These drops catalyze reactions by *adsorbing* (binding to the surface) the reactants. The drops are not reactants, but they provide a surface on which the reactants collect. The resulting proximity of the reactants increases the likelihood of reaction and, coupled with a decrease in activation energy, increases the reaction rate. When the catalyst is a liquid drop on which gas molecules adsorb, the drop is in a different phase than the reacting species and is called a **heterogeneous catalyst**. Both homogeneous and heterogeneous catalysts play a role in the reactions that diminish the amount of ozone in the stratosphere.

homogeneous catalyst a catalyst in the same phase as the reactants.

heterogeneous catalyst a catalyst in a different phase from the reactants.

CONCEPT **TEST**

Is the ClO produced in Equation 13.38 a catalyst or an intermediate?

SAMPLE EXERCISE 13.13 Identifying Catalysts in Reaction Mechanisms   **LO8**

A reaction mechanism proposed for the decomposition of ozone in the presence of NO at high temperatures consists of three elementary steps:

$$(1) \quad O_3(g) + NO(g) \rightarrow O_2(g) + NO_2(g)$$

$$(2) \quad NO_2(g) \rightarrow NO(g) + O(g)$$

$$(3) \quad O(g) + O_3(g) \rightarrow 2\,O_2(g)$$

If the rate of the overall reaction is higher than the rate of the uncatalyzed decomposition of ozone to oxygen, $2\,O_3(g) \rightarrow 3\,O_2(g)$, is NO a catalyst in the reaction?

Collect and Organize We are asked to determine whether NO is a catalyst in a reaction. A catalyst increases the rate of a reaction and is not consumed by the overall reaction. We are given the elementary steps of the reaction and are told that the reaction is more rapid in the presence of NO.

Analyze We can sum the reactions to determine the overall reaction. If NO is consumed in an early step before it is regenerated in a later one, and is not consumed in the overall process, it is a catalyst.

Solve Summing the three elementary steps:

$$O_3(g) + \cancel{NO(g)} + \cancel{NO_2(g)} + \cancel{O(g)} + O_3(g) \rightarrow$$
$$O_2(g) + \cancel{NO_2(g)} + \cancel{NO(g)} + \cancel{O(g)} + 2\,O_2(g)$$

gives the overall reaction:

$$2\,O_3(g) \rightarrow 3\,O_2(g)$$

This equation does not include NO, and the rate of the reaction is higher when NO is present. Thus NO fulfills both requirements for being a catalyst.

Think About It NO behaves much like the Cl atoms in Equations 13.38 and 13.39. NO is not an intermediate because it is used in the reaction and then regenerated in a subsequent step.

 Practice Exercise The combustion of fossil fuels results in the release of SO_2 into the atmosphere, where it reacts with oxygen to form SO_3:

$$2\,SO_2(g) + O_2(g) \rightarrow 2\,SO_3(g)$$

In the atmosphere, SO_2 may react with NO_2, forming SO_3 and NO:

$$NO_2(g) + SO_2(g) \rightarrow NO(g) + SO_3(g)$$

The rate of reaction of SO_2 with NO_2 is faster than the rate of reaction of SO_2 with oxygen. If the NO produced in the reaction of NO_2 and SO_2 is then oxidized to NO_2,

$$2\,NO(g) + O_2(g) \rightarrow 2\,NO_2(g)$$

is NO_2 a catalyst in the reaction of SO_2 with O_2?

Catalysts and Catalytic Converters

We started this chapter discussing air pollution caused by vehicles and the technology that has been developed to clean the air. Figure 13.29 shows a catalytic converter in a car's exhaust system and the reactions that take place in the converter to remove one representative pollutant, NO, from the engine exhaust. Hot exhaust gases flowing through the converter pass through a fine honeycomb mesh coated with one or more of the transition metals palladium, platinum, and rhodium. These metals are the catalysts, and they have two roles: (1) to speed up oxidation of carbon monoxide to CO_2 and of unburned hydrocarbons to CO_2 and water vapor; and (2) to convert NO and NO_2 into N_2 and O_2.

The metals used in making catalytic converters are dissolved as metal salts and dispersed on the mesh, and they are then reduced to clusters 2 to 10 nm in diameter. The large surface area of the metal clusters provides sites where the oxidation and reduction of the gases take place.

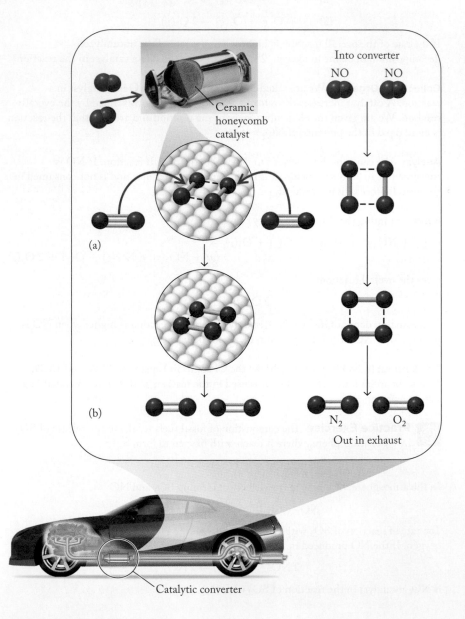

FIGURE 13.29 Catalytic converters in automobiles reduce emissions of NO by lowering the activation energy of its decomposition into N_2 and O_2. Metal catalysts are supported on a porous ceramic honeycomb. (a) NO molecules are adsorbed onto the surface of metal clusters where their NO bonds are broken, and (b) pairs of O atoms and N atoms form O_2 and N_2. The O_2 and N_2 desorb from the surface and are released to the atmosphere.

Catalysts not only speed up reactions but also allow them to take place at lower temperatures. For example, CO reacts rapidly with O_2 above 700°C, but the presence of a Pd or Pt/Rh catalyst enables this reaction to take place rapidly at the much lower temperature of automobile exhaust, about 250°C. The catalysts have similar effects on the reduction reactions taking place in the converters.

The catalysts are selective in terms of the molecules they interact with and specific in the reactions they promote. What makes the catalysts selective is that several reactions are possible for each pollutant, but one reaction proceeds more rapidly than the others. For example, the preferred reduction of the nitrogen in NO is to N_2, as shown in Figure 13.29, rather than to N_2O or to NH_3. However, carbon monoxide and hydrocarbons are also potential reducing agents for this reaction. Recall that oxidizing CO and hydrocarbons is one of the two primary goals of a catalytic converter. If these compounds are oxidized, no additional NO reduction can take place via this process. Fortunately, the reduction of NO by CO and hydrocarbons is much faster than the reactions of CO and hydrocarbons with O_2. As a result, additional NO reduction by CO and hydrocarbons is essentially complete before any of the necessary reducing agents are consumed by reaction with oxygen.

Enzymes: Biological Catalysts

Large biomolecules called **enzymes** are highly selective catalysts; they mediate very specific reactions in biological systems. For example, the chemical reactions involved in metabolism, which includes both the breaking down of molecules and the synthesis of complex substances from simpler precursors, are catalyzed in large part by enzymes. Sequences of reactions called *metabolic pathways* consist of steps, each of which is catalyzed by a specific enzyme. For example, carbonic anhydrase is an enzyme that speeds up the hydrolysis of CO_2:

$$CO_2(aq) + H_2O(\ell) \rightleftharpoons HCO_3^-(aq) + H^+(aq) \qquad (13.40)$$

In the presence of carbonic anhydrase, this reaction proceeds about 10 million times faster than in its absence. Without carbonic anhydrase, we would not be able to expel carbon dioxide fast enough to survive (as the reaction in Equation 13.40 runs in reverse). One molecule of carbonic anhydrase can hydrolyze from 10^4 to 10^6 molecules of CO_2 in 1 s. This value is called the *turnover number* for the enzyme; in general, the higher the turnover number, the faster the enzyme-catalyzed reaction proceeds. Turnover numbers for enzymes typically range from 10^3 to 10^7. The higher the turnover number, the lower the activation energy of the catalyzed reaction. As we learned earlier in this chapter, for reactions to proceed, molecules must collide with the proper orientation. Carbonic anhydrase is essentially a perfect enzyme because it catalyzes the hydrolysis reaction nearly every time it collides with CO_2.

Enzymes are specific because the idea of an effective collision has a different connotation in biochemistry from what we have depicted for other reactions and catalysts. A simplified approach to understanding and quantifying how enzymes work involves consideration of the interaction between the enzyme (E) and its *substrate* (S), the reactant molecule (Figure 13.30). We refine these ideas in Chapter 20 when we discuss specific enzymes, but for now it is sufficient to envision the substrate fitting into an *active site* in the enzyme, very much like a hand fits into a glove. Once in the active site, the substrate is converted into product (P) via a

enzyme a protein that catalyzes a reaction.

FIGURE 13.30 In this simplified view, the substrate fits into the active site of the enzyme that catalyzes the conversion of substrate into product.

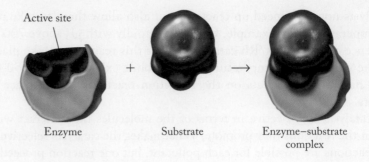

Active site

Enzyme   +   Substrate   →   Enzyme–substrate complex

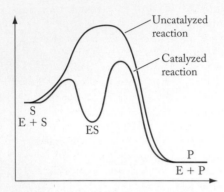

FIGURE 13.31 These superimposed reaction profiles show (black) the uncatalyzed reaction of S → P and (red) the enzyme-catalyzed reaction (E + S → E + P). The enzyme-catalyzed reaction takes place in two steps, in which the second step (ES → E + P) is rate determining.

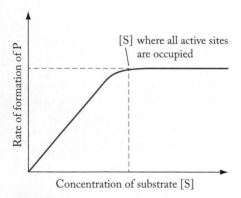

FIGURE 13.32 Plot of the rate of formation of product P versus concentration of substrate S in an enzyme-catalyzed reaction. The dotted line identifies the concentration of substrate at which all active sites are occupied.

biocatalysis the use of enzymes to catalyze reactions on a large scale; it is becoming especially important in processes that involve chiral materials.

pathway that has a lower-energy transition state (the *enzyme–substrate complex* ES), which is absent without the enzyme.

Even in the simplest organism, hundreds of enzyme-catalyzed chemical reactions are constantly taking place. Most of these enzymes are effective only under limited reaction conditions: in aqueous media and at temperatures between 4°C and 37°C. These hundreds of catalyzed reactions require hundreds of different enzymes, many operating with exquisite efficiency and producing one pure product.

In the pharmaceutical industry, research is focused on using enzymes outside living systems to produce high yields of specific products. This research area, called **biocatalysis**, uses enzymes to catalyze chemical reactions run in industrial-sized reactors. Because it deals with both isolated enzymes and microorganisms, it is considered a special type of heterogeneous catalysis.

The mathematical treatment of enzyme kinetics can be quite complex, but determining the rate of enzyme-catalyzed reactions is an important part of biochemical and medical research. We can think of the following elementary steps in the process:

$$\text{Step 1} \qquad \text{E} + \text{S} \underset{k_{-1}}{\overset{k_1}{\rightleftharpoons}} \text{ES}$$

$$\text{Step 2} \qquad \text{ES} \xrightarrow{k_2} \text{E} + \text{P}$$

Biochemists commonly assume that the formation of ES and its decomposition are both rapid and that step 2 is rate determining. A typical reaction profile for a system that meets these criteria is shown in Figure 13.31, and the rate of such a reaction is given by

$$\text{Rate} = \frac{\Delta[\text{P}]}{\Delta t} = k[\text{ES}]$$

Initially, the rate of reaction increases rapidly as the concentration of S increases (Figure 13.32). The rate of reaction is proportional to [S], so the reaction is first order and rate = k[S]. At some specific concentration of S, all of the active sites in available enzymes are occupied, and the rate of the reaction becomes constant. At this stage, the rate is zero order in S (rate = k); adding more S has no effect because all the active sites are full.

In this chapter we have seen how studying the rates of chemical reactions can lead to an understanding of how reactions happen. Determinations of the rates of chemical reactions are crucial for us to understand processes on a molecular level. The mechanisms of the reactions of volatile oxides produced during combustion and by other natural events had to be thoroughly understood so that these substances in the environment could be effectively managed. The continued development of catalytic converters for vehicles to diminish the problems caused by burning fossil fuels and to clean our air requires a thorough knowledge of the kinetics and mechanisms of many of the reactions discussed in this chapter. Many biological catalysts are responsible for myriad reactions in living systems, and their behavior may be understood by applying the same methods of study we used for inorganic catalysts in reactions of small molecules.

SAMPLE EXERCISE 13.14 Integrating Concepts: Simulations of Smog

As we have seen throughout this chapter, reactions in the atmosphere are complex. Figure 13.33 shows three graphs of data from programs that simulated atmospheric conditions arising from known reactions when the hydrocarbon propylene was introduced into air. Propylene enters the atmosphere from the evaporation of unburned petroleum fuels. The three graphs show simulated concentration versus time profiles on (a) a sunny day, (b) a cloudy day, and (c) a sunny day when the concentration of propylene is twice the amount in (a).

a. Both alkanes and alkenes are present as dissolved gases in unburned fuels. In studies of both propylene (CH_2=$CHCH_3$) and butane ($CH_3CH_2CH_2CH_3$), the alkene was shown to react more rapidly than the alkane. Suggest a reason for this observation.

b. In Figure 13.33(a), which substances are reactants and which are products? Are there any intermediates?

c. Compare Figures 13.33(a) and 13.33(c). How does doubling the concentration of propylene affect the level of air pollutants such as ozone and peroxyacetyl nitrate (PAN)?

d. What are the major differences between the reactions on a sunny day (Figure 13.33a) and a cloudy day (Figure 13.33b)?

Collect and Organize We have three graphs for different conditions, and we can use them to get relative quantitative information.

Analyze The graphs show differences in the appearance and disappearance of reactants, intermediates, and products of reactions involving an alkene, NO_x, ozone, and PAN.

Solve

a. Alkenes react more rapidly than alkanes because of the presence of a π bond. The electrons in a π bond are much more accessible to reactants than are those in a σ bond, and the π bond is also weaker than the σ bond.

b. Propylene and NO are reactants. Their concentrations start out high and drop continuously throughout the time of observation. PAN and O_3 are products. Their concentration starts out at zero and gradually builds. NO_2 is an intermediate. Its concentration begins to increase at $t = 0$, reaches a peak around 75 min, and then decreases as PAN and O_3 form and accumulate. The decrease in propylene concentration begins immediately and accelerates as O_3 concentration builds. Some organic substances other than PAN must be forming to account for the drop; their concentrations are not shown on the graph.

c. Doubling the initial concentration of propylene causes it to disappear at a faster rate; its drop-off occurs at 25 min in (c), whereas it is at 50 min in (a). NO concentration also decreases more rapidly in (c). NO_2 level peaks in (c) at 3.0 ppm, which is twice the concentration it attains at its peak in (a), and then drops off almost twice as fast; it peaks at 75 min in (a) and 40 min in (c). At higher propylene concentrations in (c), O_3 and PAN appear earlier, at 35 and 45 min versus 50 and 80 min in (a). Also, O_3 appears at higher concentrations in (c) and gradually drops at later times, after about 100 min, compared with (a), where it continues to rise at the 200-min mark. PAN is almost 4 times as abundant at 200 min in (c) compared with its value at that same time in (a). Doubling the hydrocarbon nearly quadruples the level of PAN and increases the level of ozone.

d. PAN and O_3 are not evident at all in (b) on the cloudy day, and propylene and NO disappear more slowly than on a sunny day.

Think About It Photochemical smog is aptly named. In the absence of direct sunlight, it does not seem to form.

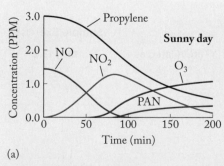

(a)

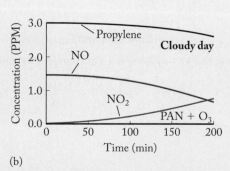

(b)

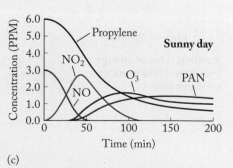

(c)

FIGURE 13.33 Graphs based on B. J. Hubert, *J. Chem. Educ.* 1974, *51*, 644–645; additional information from A. C. Baldwin, J. R. Barker, D. M. Golden, and D. G. Hendry, *J. Phys. Chem.* 1977, *81*(25), 2483–2492.

SUMMARY

LO1 The relative rates of disappearance of reactants and appearance of products are related by the stoichiometry of the reaction. **Chemical kinetics** is the study of these changing rates. (Sections 13.1 and 13.2)

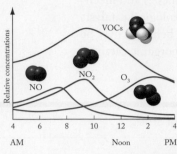

LO2 The overall rate of a reaction is typically determined from experimental measurements. It can be expressed as an average rate over a specified period or as an instantaneous rate at a particular instant. (Section 13.2)

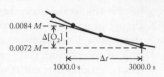

LO3 The dependence of the rate of a reaction $A + B \rightarrow C$ on reactant concentrations is expressed in the **rate law** for the reaction:

$$\text{Rate} = k[A]^m[B]^n$$

where m and n are the **reaction order** with respect to reactants A and B, respectively, and k is the **rate constant**. The order of a reaction and the rate law for the reaction can be determined from differences in the **initial rates** of reaction. (Section 13.3)

LO4 The units of a rate constant depend on the **overall reaction order**, which is the sum of the reaction orders with respect to individual reactants, observed with different concentrations of reactants or derived from the results of single kinetics experiments by using **integrated rate laws**. (Section 13.3)

LO5 An integrated rate law describes the change in concentration of a reactant over time. (Section 13.3)

LO6 The **half-life** of a reaction is the time required for the concentration of a reactant to decrease to one-half its starting concentration: the higher the reaction rate, the shorter the half-life. (Section 13.3)

LO7 Increasing the temperature of a chemical reaction increases its rate. The **activation energy** of a reaction is the minimal energy required for molecules to react when they collide. (Section 13.4)

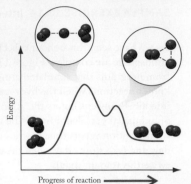

LO8 Rate studies give insight into **reaction mechanisms**, which describe what is happening at a molecular level. A reaction mechanism consists of one or more **elementary steps** that describe how the reaction takes place on a molecular level. The proposed mechanism for any reaction must be consistent with the observed rate law and with the stoichiometry of the overall reaction. **Catalysts** increase reaction rates by changing the mechanism of a reaction and decreasing activation energies. **Enzymes** are catalysts in living systems. (Sections 13.5 and 13.6)

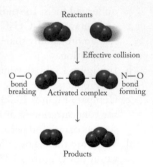

PARTICULATE **PREVIEW WRAP-UP**

The balanced equation for the decomposition of nitrogen dioxide is $2\,NO_2(g) \rightarrow 2\,NO(g) + O_2(g)$. The NO_2 molecules collide most frequently in sample (a) owing to the highest concentration of the three samples. The decomposition reaction proceeds four times as fast in sample (b) as it does in sample (c) because the reaction is second order in NO_2.

PROBLEM-SOLVING SUMMARY

Type of Problem	Concepts and Equations	Sample Exercises
Relating rates of change in reactant and product concentrations	For the reaction $xX \rightarrow yY$, the rates of change of [X] and [Y] are related as follows: $$-\frac{1}{x}\frac{\Delta[X]}{\Delta t} = \frac{1}{y}\frac{\Delta[Y]}{\Delta t}$$	**13.1, 13.2**

Type of Problem	Concepts and Equations	Sample Exercises
Determining an instantaneous rate	Determine the slope of a line tangent to a point on the plot of concentration versus time.	**13.3**
Deriving a rate law from initial reaction rate data	Compare the change in rate when the concentration of one reactant is changed (while the concentrations of other reactants are kept constant) to determine the reaction order (usually whole numbers) with respect to that reactant: $$n = \dfrac{\log\left(\dfrac{\text{Rate}_1}{\text{Rate}_2}\right)}{\log\left(\dfrac{[X]_1}{[X]_2}\right)} \qquad (13.14)$$	**13.4**
Using integrated rate laws to distinguish among zero-, first-, and second-order reactions and to calculate the value of k.	A linear plot of concentration versus time indicates a zero-order reaction with a slope of $-k$, whereas a linear plot of the natural logarithm of concentration versus time indicates a first-order reaction with a slope of $-k$, and a linear plot of the reciprocal of reactant concentration $1/[X]$ versus time indicates a second-order reaction.	**13.5, 13.8**
Calculating remaining concentration of a reactant in a zero-, first-, or second-order reaction	Use the integrated rate law Zero order: $[X] = -kt + [X]_0$ (13.26) First order: $\ln[X] = -kt + \ln[X]_0$ (13.18) Second order: $\dfrac{1}{[X]} = kt + \dfrac{1}{[X]_0}$ (13.23) to calculate concentrations of reactant X at any given time t.	**13.6**
Calculating the half-life of a zero-, first-, or second-order reaction	In a first-order reaction, Zero order: $t_{1/2} = \dfrac{[X]_0}{2k}$ (13.27) First order: $t_{1/2} = \dfrac{0.693}{k}$ (13.19) Second order: $t_{1/2} = \dfrac{1}{k[X]_0}$ (13.24)	**13.7**
Calculating an activation energy from rate constants	Using the logarithmic form of the Arrhenius equation, $$\ln k = -\dfrac{E_a}{R}\left(\dfrac{1}{T}\right) + \ln A \qquad (13.29)$$ plot $\ln k$ versus $1/T$. The slope is $-E_a/R$. The rates of a reaction at two temperatures can be found using $$\ln \dfrac{k_1}{k_2} = \dfrac{E_a}{R}\left(\dfrac{1}{T_2} - \dfrac{1}{T_1}\right) \qquad (13.30)$$	**13.9, 13.10**
Linking reaction mechanisms to experimental rate laws and testing a proposed reaction mechanism	The order of each reactant in an elementary step equals its coefficient in that step. The rate law for the mechanism must be the same as the observed rate law and does not include intermediates.	**13.11, 13.12**
Identifying catalysts in reaction mechanisms	Determine whether a potential catalyst is present by summing the elementary-step reactions to get the overall reaction. If that procedure reveals a potential catalyst, determine whether it increases the rate of reaction and whether it is initially consumed and then regenerated in the process.	**13.13**

VISUAL PROBLEMS

(Answers to boldface end-of-chapter questions and problems are in the back of the book.)

13.1. Nitrous oxide decomposes to nitrogen and oxygen in the following reaction:

$$2 N_2O(g) \rightarrow 2 N_2(g) + O_2(g)$$

In Figure P13.1, which curve represents [N₂O] and which curve represents [O₂]?

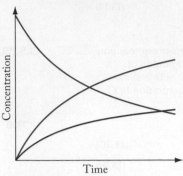

FIGURE P13.1

13.2. Sulfur trioxide is formed in the reaction

$$SO_2(g) + \tfrac{1}{2} O_2(g) \rightarrow SO_3(g)$$

In Figure P13.2, which curve represents [SO₂] and which curve represents [O₂]? All three gases are present initially.

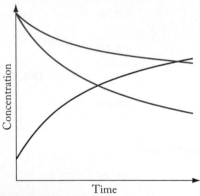

FIGURE P13.2

13.3. The rate law for the reaction 2 A → B is second order in A. Figure P13.3 represents samples with different concentrations of A; the red spheres represent molecules of A. In which sample will the reaction A → B proceed most rapidly?

 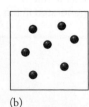 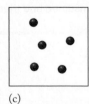

(a)   (b)   (c)

FIGURE P13.3

13.4. The rate law for the reaction A + B → C is first order in both A and B. Figure P13.4 represents samples with different concentrations of A (red spheres) and B (blue spheres). In which sample will the reaction A + B → C proceed most rapidly?

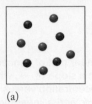 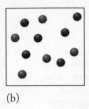 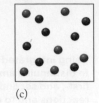

(a)   (b)   (c)

FIGURE P13.4

13.5. Figure P13.5 shows plots of reactant concentrations versus time for four reactions. Which one has the greatest initial rate?

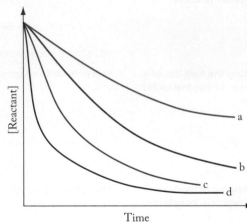

FIGURE P13.5

13.6. For a given temperature, which of the reaction profiles in Figure P13.6 represents (a) the slowest reaction? (b) the fastest reaction?

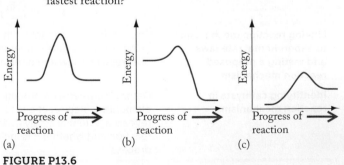

(a)   (b)   (c)

FIGURE P13.6

13.7. Which of the following mechanisms is consistent with the reaction profile shown in Figure P13.7?

a. $2A \xrightarrow{\text{slow}} B$
   $B \xrightarrow{\text{fast}} C$
b. $A + B \rightarrow C$
c. $2A \xrightleftharpoons{\text{fast}} B$
   $B \xrightarrow{\text{slow}} C$

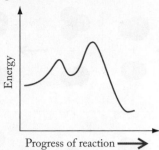

FIGURE P13.7

13.8. Which of the following mechanisms is consistent with the reaction profile shown in Figure P13.8?

a. $A + B \xrightarrow{\text{slow}} C$
   $C \xrightarrow{\text{fast}} D$
b. $A + B \rightarrow C$
c. $2A \xrightarrow{\text{fast}} B$
   $B + C \xrightarrow{\text{slow}} D$

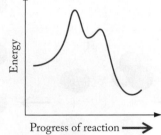

FIGURE P13.8

13.9. Refer to Figure P13.9 to answer the following questions:
   a. Which asterisk identifies a transition state?
   b. Which arrow identifies the activation energy of the reaction in the forward direction?
   c. Which arrow identifies the activation energy of the reaction in the reverse direction?
   d. Which arrow identifies the change in energy that accompanies the reaction, that is, the energy of the products less the energy of the reactants?

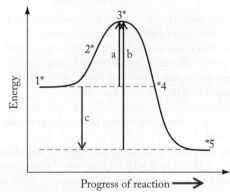

FIGURE P13.9

13.10. Refer to Figure P13.10 to answer the following questions:
   a. Which arrow identifies the activation energy of the reaction in the forward direction?
   b. Which arrow identifies the activation energy of the reaction in the reverse direction?
   c. Which arrow identifies the change in energy that accompanies the reaction, that is, the energy of the products less the energy of the reactants?

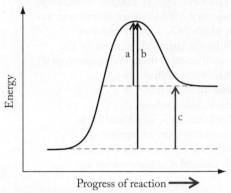

FIGURE P13.10

*13.11. The curves in Figure P13.11 show how the concentrations of four components of photochemical smog change during a sunny day. Describe how these curves would differ during a cloudy day.

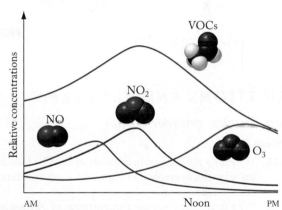

FIGURE P13.11

13.12. Use representations [A] through [I] in Figure P13.12 to answer questions a–f.
   a. Which two images describe elementary steps which, when combined, depict the chlorine-catalyzed destruction of ozone?
   b. Which representation is the overall reaction for the chlorine-catalyzed destruction of ozone?
   c. Which two images describe elementary steps that combine in an overall reaction in which NO_3 is an intermediate?
   d. Write the chemical equation for the overall reaction described in question (c).
   e. Which image shows the photodecomposition of a chlorofluorocarbon?
   f. Rank images [C], [F], and [I] in decreasing order of number of collisions between Cl atoms and O_3 molecules in the stratosphere.

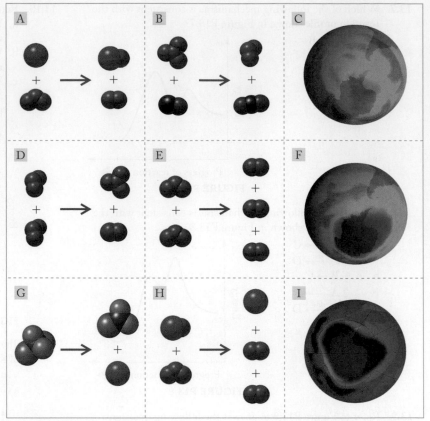

FIGURE P13.12

QUESTIONS AND PROBLEMS

Cars, Trucks, and Air Quality

Concept Review

13.13. Why does the maximum concentration of NO_2 on a smoggy day, as graphed in Figure 13.4, occur several hours after the maximum concentration of NO?

13.14. Why does the maximum concentration of ozone graphed in Figure 13.4 occur much later in the day than the maximum concentrations of NO and NO_2?

Problems

13.15. Which are more reactive: O atoms or O_2 molecules? Why?

13.16. Why is gaseous OH so much more reactive than H_2O vapor?

Reaction Rates

Concept Review

13.17. Explain the difference between the rate of a reaction at 25°C and its rate constant at 25°C.

13.18. Explain the difference between the average rate and the instantaneous rate of a chemical reaction.

13.19. Suggest three possible ways of monitoring the rate of the following reaction. Would the rate data (changing concentration with time) be the same for all the alternatives?

$$CH_3CHO(g) \rightarrow CH_4(g) + CO(g)$$

13.20. Suggest two ways to monitor the rate of the following reaction. Would the rate data (changing concentration with time) be the same for either method?

$$2\,H_2O_2(aq) \rightarrow 2\,H_2O(\ell) + O_2(g)$$

13.21. If baking soda ($NaHCO_3$) and sidewalk deicer ($CaCl_2$) are mixed, there is no sign of a chemical reaction. However, if these two solids are dissolved in water and mixed, they rapidly react. Explain the difference in reaction rate.

13.22. Any gas-phase reaction occurs more rapidly as the temperature of the gas increases. Why?

13.23. In the decomposition reaction A → B + C, how is the rate at which A is consumed related to the rate at which B is produced?

13.24. During the Haber process for synthesizing ammonia, $N_2(g) + 3\,H_2(g) \rightarrow 2\,NH_3(g)$, the rate of formation of ammonia is twice the rate at which nitrogen is consumed. Does this mean that the mass of the reaction mixture increases as the reaction proceeds—seemingly defying the law of conservation of mass? Explain why or why not.

13.25. If the rate of change in the concentration of a reactant increases (becomes less negative) with time, does the rate of change in the concentration of a product of the same reaction increase or decrease?

13.26. During a reaction, can there be a time when the instantaneous rate of the reaction does not change? If you think so, describe such a time.

Problems

13.27. Catalytic Converters in Automobiles (I) Catalytic converters combat air pollution by converting NO into N_2 and O_2.
a. How is the rate of formation of O_2 related to the rate of formation of N_2?
b. How is the rate of change in $[N_2]$ related to the rate of change in $[NO]$?

13.28. Catalytic Converters in Automobiles (II) Catalytic converters also combat air pollution by promoting the reaction between CO and O_2 that produces CO_2.
a. How is the rate of change in $[CO_2]$ related to the rate of change in $[O_2]$?
b. How is the rate of change in $[CO_2]$ related to the rate of change in $[CO]$?

13.29. Write an equation relating the rates of change in the concentrations of the products and reactants in each of the following reactions:
a. $F_2(g) + H_2O(\ell) \rightarrow HOF(g) + HF(g)$
b. $Si(s) + 3\ HCl(g) \rightarrow SiHCl_3(\ell) + H_2(g)$
c. $4\ NH_3(g) + 3\ O_2(g) \rightarrow 2\ N_2(g) + 6\ H_2O(g)$

13.30. Write an equation relating the rates of change in the concentrations of the products and reactants in each of the following reactions:
a. $SOF_2(g) + 2\ F_2(g) \rightarrow F_5SOF(g)$
b. $B_2H_6(g) + 6\ Cl_2(g) \rightarrow 2\ BCl_3(g) + 6\ HCl(g)$
c. $N_2H_4(g) + 2\ NH_2Cl(g) \rightarrow 2\ NH_4Cl(s) + N_2(g)$

13.31. In a study to determine the rate of the following reaction:

$$2\ NO(g) + O_2(g) \rightarrow 2\ NO_2(g)$$

the concentration of NO was 0.0300 M at $t = 5.0$ s and 0.0225 M at $t = 650.0$ s. What is the average rate of the reaction during this period?

13.32. In a study of the thermal decomposition of ammonia into nitrogen and hydrogen:

$$2\ NH_3(g) \rightarrow N_2(g) + 3\ H_2(g)$$

the average rate of change in the concentration of ammonia is -0.38 M/s.
a. What is the average rate of change in $[H_2]$?
b. What is the average rate of change in $[N_2]$?
c. What is the average rate of the reaction?

13.33. Power Plant Emissions Sulfur dioxide emissions in stack gases at power plants may react with carbon monoxide as follows:

$$SO_2(g) + 3\ CO(g) \rightarrow 2\ CO_2(g) + COS(g)$$

Write an equation relating the rates for each of the following:
a. The rate of formation of CO_2 to the rate of consumption of CO
b. The rate of formation of COS to the rate of consumption of SO_2
c. The rate of consumption of CO to the rate of consumption of SO_2

13.34. Reducing Power Plant Emissions Nitrogen monoxide can be removed from gas-fired power plant emissions by reaction with methane as follows:

$$CH_4(g) + 4\ NO(g) \rightarrow 2\ N_2(g) + CO_2(g) + 2\ H_2O(g)$$

Write an equation relating the rates for each of the following:
a. The rate of formation of N_2 to the rate of formation of CO_2
b. The rate of formation of CO_2 to the rate of consumption of NO
c. The rate of consumption of CH_4 to the rate of formation of H_2O

13.35. Stratospheric Ozone Depletion Chlorine monoxide (ClO) plays a major role in the creation of the ozone holes in the stratosphere over Earth's polar regions.
a. If $\Delta[ClO]/\Delta t$ at 298 K is -2.3×10^7 M/s, what is the rate of change in $[Cl_2]$ and $[O_2]$ in the following reaction?

$$2\ ClO(g) \rightarrow Cl_2(g) + O_2(g)$$

b. If $\Delta[ClO]/\Delta t$ is -2.9×10^4 M/s, what is the rate of formation of oxygen and ClO_2 in the following reaction?

$$ClO(g) + O_3(g) \rightarrow O_2(g) + ClO_2(g)$$

13.36. The chemistry of smog formation includes NO_3 as an intermediate in several reactions.
a. If $\Delta[NO_3]/\Delta t$ is -2.2×10^5 mM/min in the following reaction, what is the rate of formation of NO_2?

$$NO_3(g) + NO(g) \rightarrow 2\ NO_2(g)$$

b. What is the rate of change of $[NO_2]$ in the following reaction if $\Delta[NO_3]/\Delta t$ is -2.3 mM/min?

$$2\ NO_3(g) \rightarrow 2\ NO_2(g) + O_2(g)$$

13.37. Nitrite ion reacts with ozone in aqueous solution, producing nitrate ion and oxygen:

$$NO_2^-(aq) + O_3(g) \rightarrow NO_3^-(aq) + O_2(g)$$

The following data were collected for this reaction at 298 K. Calculate the average reaction rate between 0.0 and 100.0 μs (microseconds) and between 200.0 and 300.0 μs.

Time (μs)	$[O_3]$ (M)
0.0	1.13×10^{-2}
100.0	9.93×10^{-3}
200.0	8.70×10^{-3}
300.0	8.15×10^{-3}

13.38. Dinitrogen pentoxide (N_2O_5) decomposes as follows to nitrogen dioxide and nitrogen trioxide:

$$N_2O_5(g) \rightarrow NO_2(g) + NO_3(g)$$

Calculate the average rate of this reaction between consecutive measurement times in the following table.

Time (s)	$[N_2O_5]$ (molecules/cm^3)
0.00	1.500×10^{12}
1.45	1.357×10^{12}
2.90	1.228×10^{12}
4.35	1.111×10^{12}
5.80	1.005×10^{12}

13.39. The following data were collected for the dimerization of ClO to Cl_2O_2 at 298 K.

Time (s)	[ClO] (molecules/cm³)
0.0	2.60×10^{11}
1.0	1.08×10^{11}
2.0	6.83×10^{10}
3.0	4.99×10^{10}
4.0	3.93×10^{10}
5.0	3.24×10^{10}
6.0	2.76×10^{10}

Plot [ClO] and $[Cl_2O_2]$ as a function of time and determine the instantaneous rates of change in both at 1 s.

13.40. **Tropospheric Ozone** Tropospheric (lower atmosphere) ozone is rapidly consumed in many reactions, including

$$O_3(g) + NO(g) \rightarrow NO_2(g) + O_2(g)$$

Use the following data to calculate the instantaneous rate of the preceding reaction at $t = 0.000$ s and $t = 0.052$ s.

Time (s)	[NO] (M)
0.000	2.0×10^{-8}
0.011	1.8×10^{-8}
0.027	1.6×10^{-8}
0.052	1.4×10^{-8}
0.102	1.2×10^{-8}

Effect of Concentration on Reaction Rate

Concept Review

13.41. Why do the rates of nearly all reactions decrease as reactants form products?

13.42. Why are the units of the rate constants different for reactions of different order?

13.43. A colleague in New Zealand has sent you word that the rate constant of the reaction you are studying is 2.5×10^2 M/s at 25.0°C. What is the order of the reaction?

13.44. The rate of reaction does *not* increase with time for reactions with which order: zero, first, or second?

13.45. What effect does doubling the initial concentration of a reactant have on the half-life of a reaction that is second order in the reactant?

13.46. Suppose the decomposition reactions A → B + C and X → Y + Z are second order in A and X, respectively, and both have the same rate constant. Under what conditions do the two reactions also have the same half-life?

Problems

13.47. For each of the following rate laws, determine the order with respect to each reactant and the overall reaction order.
   a. Rate = $k[A][B]$
   b. Rate = $k[A]^2[B]$
   c. Rate = $k[A][B]^3$

13.48. Determine the overall order of the following rate laws and the order with respect to each reactant.
   a. Rate = $k[A]^2[B]^{1/2}$
   b. Rate = $k[A]^2[B][C]$
   c. Rate = $k[A][B]^3[C]^{1/2}$

13.49. Write rate laws and determine the units of the rate constant (using the units M for concentration and s for time) for the following reactions:
   a. The reaction of oxygen atoms with NO_2 is first order in both reactants.
   b. The reaction between NO and Cl_2 is second order in NO and first order in Cl_2.
   c. The reaction between Cl_2 and chloroform ($CHCl_3$) is first order in $CHCl_3$ and one-half order in Cl_2.
   *d. The decomposition of ozone (O_3) to O_2 is second order in O_3 and an order of -1 in O atoms.

13.50. Compounds A and B react to give a single product, C. Write the rate law for each of the following cases and determine the units of the rate constant by using the units M for concentration and s for time:
   a. The reaction is first order in A and second order in B.
   b. The reaction is first order in A and second order overall.
   c. The reaction is independent of the concentration of A and second order overall.
   d. The reaction is second order in both A and B.

13.51. Predict the rate law for the reaction 2 BrO(g) → $Br_2(g)$ + $O_2(g)$ if:
   a. The rate doubles when [BrO] doubles.
   b. The rate quadruples when [BrO] doubles.
   c. The rate is halved when [BrO] is halved.
   d. The rate is unchanged when [BrO] is doubled.

13.52. Predict the rate law for the reaction NO(g) + $Br_2(g)$ → $NOBr_2(g)$ if:
   a. The rate doubles when [NO] is doubled and $[Br_2]$ remains constant.
   b. The rate doubles when $[Br_2]$ is doubled and [NO] remains constant.
   c. The rate increases by 1.56 times when [NO] is increased 1.25 times and $[Br_2]$ remains constant.
   d. The rate is halved when [NO] is doubled and $[Br_2]$ remains constant.

13.53. The rate of the reaction:

$$NO(g) + O_3(g) \rightarrow NO_2(g) + O_2(g)$$

quadruples when the concentrations of NO and O_3 are doubled. Does this prove that the reaction is first order in both reactants? Why or why not?

13.54. The reaction between chlorine monoxide and nitrogen dioxide is second order overall and

$$ClO(g) + NO_2(g) + M(g) \rightarrow ClONO_2(g) + M(g)$$

produces chlorine nitrate ($ClONO_2$). A third molecule (M) takes part in the reaction but is unchanged by it. The reaction is first order in NO_2 and in ClO.
   a. Write the rate law for this reaction.
   b. What is the reaction order with respect to M?

13.55. Rate Laws for Destruction of Tropospheric Ozone The reaction of NO_2 with ozone produces NO_3 in a second-order reaction overall:

$$NO_2(g) + O_3(g) \rightarrow NO_3(g) + O_2(g)$$

a. Write the rate law for the reaction if the reaction is first order in each reactant.

b. The rate constant for the reaction is $1.93 \times 10^4\ M^{-1}\ s^{-1}$ at 298 K. What is the rate of the reaction when $[NO_2] = 1.8 \times 10^{-8}\ M$ and $[O_3] = 1.4 \times 10^{-7}\ M$?

c. What is the rate of formation of NO_3 under these conditions?

d. What happens to the rate of the reaction if the concentration of $O_3(g)$ is doubled?

13.56. Sources of Nitric Acid in the Atmosphere The reaction between N_2O_5 and water,

$$N_2O_5(g) + H_2O(g) \rightarrow 2\ HNO_3(g)$$

is a source of nitric acid in the atmosphere.

a. The reaction is first order in each reactant. Write the rate law for the reaction.

b. When $[N_2O_5]$ is 0.132 mM and $[H_2O]$ is 230 mM, the rate of the reaction is 4.55×10^{-4} mM^{-1} min^{-1}. What is the rate constant for the reaction?

13.57. Each of the following reactions is first order in the reactants and second order overall. Which reaction is fastest if the initial concentrations of the reactants are the same? All reactions are at 298 K.

a. $ClO_2(g) + O_3(g) \rightarrow ClO_3(g) + O_2(g)$
$k = 3.0 \times 10^{-19}\ cm^3/(molecule \cdot s)$

b. $ClO_2(g) + NO(g) \rightarrow NO_2(g) + ClO(g)$
$k = 3.4 \times 10^{-13}\ cm^3/(molecule \cdot s)$

c. $ClO(g) + NO(g) \rightarrow Cl(g) + NO_2(g)$
$k = 1.7 \times 10^{-11}\ cm^3/(molecule \cdot s)$

d. $ClO(g) + O_3(g) \rightarrow ClO_2(g) + O_2(g)$
$k = 1.5 \times 10^{-17}\ cm^3/(molecule \cdot s)$

13.58. Two reactions in which there is a single reactant have nearly the same magnitude rate constant. One is first order; the other is second order.

a. If the initial concentrations of the reactants are both 1.0 mM, which reaction will proceed at the higher rate?

b. If the initial concentrations of the reactants are both 2.0 M, which reaction will proceed at the higher rate?

13.59. The rate constant for the decomposition of N_2O_5 to NO_2 and O_2,

$$2\ N_2O_5(g) \rightarrow 4\ NO_2(g) + O_2(g)$$

is $3.4 \times 10^{-5}\ s^{-1}$ at 298 K. What is the rate law expression for the reaction at 298 K?

13.60. Hydroperoxyl Radicals in the Atmosphere During a smog event, trace amounts of many highly reactive substances are present in the atmosphere. One of these is the hydroperoxyl radical, HO_2, which reacts with sulfur trioxide, SO_3. The rate constant for the reaction

$$2\ HO_2(g) + SO_3(g) \rightarrow H_2SO_3(g) + 2\ O_2(g)$$

at 298 K is $2.6 \times 10^{11}\ M^{-1}\ s^{-1}$. The initial rate of the reaction doubles when the concentration of SO_3 or HO_2 is doubled. What is the rate law for the reaction?

13.61. Disinfecting Municipal Water Supplies Chlorine dioxide (ClO_2) is a disinfectant used in municipal water treatment plants (Figure P13.61). It dissolves in basic solution, producing ClO_3^- and ClO_2^-:

$$2\ ClO_2(g) + 2\ OH^-(aq) \rightarrow ClO_3^-(aq) + ClO_2^-(aq) + H_2O\ (\ell)$$

FIGURE P13.61

The following kinetic data were obtained at 298 K for the reaction:

Experiment	$[ClO_2]_0$ (M)	$[OH^-]_0$ (M)	Initial Rate (M/s)
1	0.060	0.030	0.0248
2	0.020	0.030	0.00827
3	0.020	0.090	0.0247

Determine the rate law and the rate constant for this reaction at 298 K.

13.62. The following kinetic data were collected at 298 K for the reaction of ozone with nitrite ion, producing nitrate and oxygen:

$$NO_2^-(aq) + O_3(g) \rightarrow NO_3^-(aq) + O_2(g)$$

Experiment	$[NO_2^-]_0$ (M)	$[O_3]_0$ (M)	Initial Rate (M/s)
1	0.0100	0.0050	25
2	0.0150	0.0050	37.5
3	0.0200	0.0050	50.0
4	0.0200	0.0200	200.0

Determine the rate law for the reaction and the value of the rate constant.

13.63. Hydrogen gas reduces NO to N_2 in the following reaction:

$$2\ H_2(g) + 2\ NO(g) \rightarrow 2\ H_2O(g) + N_2(g)$$

The initial reaction rates of four mixtures of H_2 and NO were measured at 900°C with the following results:

Experiment	$[H_2]_0$ (M)	$[NO]_0$ (M)	Initial Rate (M/s)
1	0.212	0.136	0.0248
2	0.212	0.272	0.0991
3	0.424	0.544	0.793
4	0.848	0.544	1.59

Determine the rate law and the rate constant for the reaction at 900°C.

13.64. The rate of the reaction

$$NO_2(g) + CO(g) \rightarrow NO(g) + CO_2(g)$$

was determined in three experiments at 225°C. The results are given in the following table.

Experiment	$[NO_2]_0$ (M)	$[CO]_0$ (M)	Initial Rate, $-\Delta[NO_2]/\Delta t$ (M/s)
1	0.263	0.826	1.44×10^{-5}
2	0.263	0.413	1.44×10^{-5}
3	0.526	0.413	5.76×10^{-5}

a. Determine the rate law for the reaction.
b. Calculate the value of the rate constant at 225°C.
c. Calculate the rate of formation of CO_2 when $[NO_2] = [CO] = 0.500\ M$.

13.65. The reaction between propionaldehyde (CH_3CH_2CHO) and hydrocyanic acid (HCN) has been studied in aqueous solution at 25°C. Concentrations of reactants as a function of time are shown in the following table.
a. What is the average rate of consumption of HCN from 11.12 min to 40.35 min?
b. What is the average rate of consumption of propionaldehyde over that same period?

Time (min)	$[CH_3CH_2CHO]$ (M)	[HCN] (M)
3.28	0.0384	0.0657
11.12	0.0346	0.0619
24.43	0.0296	0.0569
40.35	0.0242	0.0515
67.22	0.0190	0.0463

13.66. Two structural isomers of ClO_2 are shown in Figure P13.66. The isomer with the Cl–O–O skeletal arrangement is unstable and rapidly decomposes according to the reaction $2\ ClOO(g) \rightarrow Cl_2(g) + 2\ O_2(g)$. The following data were collected for the decomposition of ClOO at 298 K:

Time (μs)	[ClOO] (M)
0.0	1.76×10^{-6}
0.7	2.36×10^{-7}
1.3	3.56×10^{-8}
2.1	3.23×10^{-9}
2.8	3.96×10^{-10}

FIGURE P13.66

Determine the rate law for the reaction and the value of the rate constant at 298 K.

13.67. Hydrogen peroxide decomposes spontaneously into water and oxygen gas via a first-order reaction:

$$2\ H_2O_2\ (aq) \rightarrow 2\ H_2O(g) + O_2(g)$$

but in the absence of catalysts this reaction proceeds very slowly. If a small amount of a salt containing the Fe^{3+} ion is added to a 0.437 M solution of H_2O_2 in water, the reaction proceeds with a half-life of 17.3 min. What is the concentration of the solution after 10.0 min under these conditions?

13.68. **Yogurt Expiration Date** Labels of many food products have expiration dates, at which point they are typically removed from supermarket shelves. A particular natural yogurt degrades with a half-life of 45 days. The manufacturer of the yogurt wants unsold product pulled from the shelves when it degrades to no more than 80% of its original quality. Assume the degradation process is first order. What should be the "best if used before" date on the container with respect to the date the yogurt was packaged?

13.69. Acetoacetic acid, CH_3COCH_2COOH, decomposes in aqueous acidic solution to form acetone and carbon dioxide:

$$CH_3COCH_2COOH(aq) \rightarrow CH_3COCH_3(aq) + CO_2(g)$$

The reaction is first order. At room temperature, the half-life of the reactant is 139 min.
a. What is the rate constant of the decomposition reaction?
b. If the initial concentration of acetoacetic acid is 2.75 M, what is its concentration after 5 hours?

13.70. *p*-Toluenesulfinic acid undergoes a second-order redox reaction at room temperature (Figure P13.70).
a. The value of the rate constant k for the reaction is 0.141 L mol^{-1} min^{-1}. What is the half-life of *p*-toluenesulfinic acid?
b. If the initial concentration of *p*-toluenesulfinic acid is 0.355 M, at what time will its concentration be 0.0355 M?

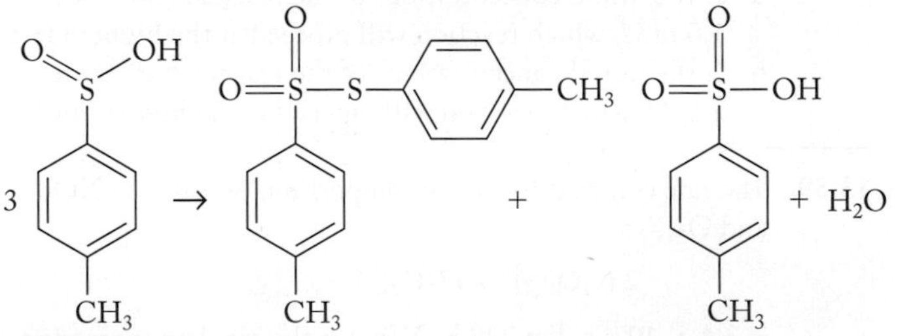

FIGURE P13.70

13.71. **Laughing Gas** Nitrous oxide (N_2O) is used as an anesthetic (laughing gas) and in aerosol cans to produce whipped cream. It is a potent greenhouse gas and decomposes slowly to N_2 and O_2:

$$2\ N_2O(g) \rightarrow 2\ N_2(g) + O_2(g)$$

a. If the plot of $\ln[N_2O]$ as a function of time is linear, what is the rate law for the reaction?

b. How many half-lives will it take for the concentration of the N_2O to reach 6.25% of its original concentration? [*Hint*: The amount of reactant remaining after time t (A_t) is related to the amount initially present (A_0) by the equation $A_t/A_0 = (0.5)^n$, where n is the number of half-lives in time t.]

13.72. The unsaturated hydrocarbon butadiene (C_4H_6) dimerizes to 4-vinylcyclohexene (C_8H_{12}). When data collected in studies of the kinetics of this reaction were plotted against reaction time, plots of $[C_4H_6]$ or $\ln[C_4H_6]$ produced curved lines, but the plot of $1/[C_4H_6]$ was linear.

a. What is the rate law for the reaction?

b. How many half-lives will it take for the $[C_4H_6]$ to decrease to 3.1% of its original concentration?

13.73. **Tracing Phosphorus in Organisms** Radioactive isotopes such as ^{32}P are used to follow biological processes. The following radioactivity data (in relative radioactivity values) were collected for a sample containing ^{32}P:

Time (days)	Radioactivity (relative radioactivity values)
0	10.0
1	9.53
2	9.08
5	7.85
10	6.16
20	3.79

a. Write the rate law for the decay of ^{32}P.

b. Determine the value of the first-order rate constant.

c. Determine the half-life of ^{32}P.

13.74. Nitrous acid slowly decomposes to NO, NO_2, and water in the following second-order reaction:

$$2\,HNO_2(aq) \rightarrow NO(g) + NO_2(g) + H_2O(\ell)$$

a. Use the data in the table to determine the rate constant for this reaction at 298 K.

Time (min)	$[HNO_2]$ (μM)
0.0	0.1560
1000.0	0.1466
1500.0	0.1424
2000.0	0.1383
2500.0	0.1345
3000.0	0.1309

b. Determine the half-life for the decomposition of HNO_2.

13.75. The dimerization of ClO,

$$2\,ClO(g) \rightarrow Cl_2O_2(g)$$

is second order in ClO. Use the following data to determine the value of k at 298 K:

Time (s)	[ClO] (molecules/cm^3)
0.0	2.60×10^{11}
1.0	1.08×10^{11}
2.0	6.83×10^{10}
3.0	4.99×10^{10}
4.0	3.93×10^{10}

Determine the half-life for the dimerization of ClO.

13.76. Kinetic data for the reaction $Cl_2O_2(g) \rightarrow 2\,ClO(g)$ are summarized in the following table. Determine the value of the first-order rate constant.

Time (μs)	$[Cl_2O_2]$ (*M*)
0.0	6.60×10^{-8}
172.0	5.68×10^{-8}
345.0	4.89×10^{-8}
517.0	4.21×10^{-8}
690.0	3.62×10^{-8}
862.0	3.12×10^{-8}

Determine the half-life for the decomposition of Cl_2O_2.

Reaction Rates, Temperature, and the Arrhenius Equation

Concept Review

13.77. In many familiar reactions, high-energy reactants form lower-energy products. In such a reaction, is the activation energy barrier higher in the forward or in the reverse direction?

13.78. Explain why gas-phase reactions go faster at a higher temperature, yet rate laws such as rate $= k[A]$ do not include T.

13.79. The order of a reaction is independent of temperature, but the value of the rate constant varies with temperature. Why?

13.80. According to the Arrhenius equation, does the activation energy of a chemical reaction depend on temperature? Explain your answer.

*13.81. Two first-order reactions have activation energies of 15 and 150 kJ/mol. Which reaction will show the larger increase in rate as temperature is increased?

*13.82. The activation energy for a particular reaction is nearly zero. Is its rate constant very sensitive to temperature changes? Explain why.

13.83. Figure P13.83 shows a plot of ln k vs. $1/T$ for two reactions with different activation energies. Which reaction has a higher activation energy?

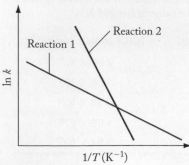

FIGURE P13.83

13.84. The rate of which reaction in Figure P13.83 is more sensitive to temperature changes?

Problems

13.85. The rate constant for the reaction of ozone with oxygen atoms was determined at four temperatures. Calculate the activation energy and frequency factor A for the reaction

$$O(g) + O_3(g) \rightarrow 2\, O_2(g)$$

given the following data:

T (K)	k [cm³/(molecule · s)]
250	2.64×10^{-4}
275	5.58×10^{-4}
300	1.04×10^{-3}
325	1.77×10^{-3}

13.86. The rate constant for the reaction

$$NO_2(g) + O_3(g) \rightarrow NO_3(g) + O_2(g)$$

was determined over a temperature range of 40 K, with the following results:

T (K)	k ($M^{-1}\,s^{-1}$)
203	4.14×10^5
213	7.30×10^5
223	1.22×10^6
233	1.96×10^6
243	3.02×10^6

a. Determine the activation energy for the reaction.
b. Calculate the rate constant of the reaction at 300 K.

13.87. **Activation Energy of a Smog-Forming Reaction** The initial step in the formation of smog is the reaction between nitrogen and oxygen. The activation energy of the reaction can be determined from the temperature dependence of the rate constants. At the temperatures indicated, values of the rate constant of the reaction

$$N_2(g) + O_2(g) \rightarrow 2\, NO(g)$$

are as follows:

T (K)	k ($M^{-1/2}\,s^{-1}$)
2000	318
2100	782
2200	1770
2300	3733
2400	7396

a. Calculate the activation energy of the reaction.
b. Calculate the frequency factor for the reaction.
c. Calculate the value of the rate constant at ambient temperature, $T = 300$ K.

13.88. Values of the rate constant for the decomposition of N_2O_5 gas at four temperatures are as follows:

T (K)	k (s^{-1})
658	2.14×10^5
673	3.23×10^5
688	4.81×10^5
703	7.03×10^5

a. Determine the activation energy of the decomposition reaction.
b. Calculate the value of the rate constant at 300 K.

13.89. **Activation Energy of Stratospheric Ozone Destruction** The value of the rate constant for the reaction between chlorine dioxide and ozone was measured at four temperatures between 193 and 208 K. The results are as follows:

T (K)	k ($M^{-1}\,s^{-1}$)
193	34.0
198	62.8
203	112.8
208	196.7

Calculate the values of the activation energy and the frequency factor for the reaction.

13.90. Chlorine atoms react with methane, forming HCl and CH_3. The rate constant for the reaction is $6.0 \times 10^7\, M^{-1}\,s^{-1}$ at 298 K. When the experiment was repeated at three other temperatures, the following data were collected:

T (K)	k ($M^{-1}\,s^{-1}$)
303	6.5×10^7
308	7.0×10^7
313	7.5×10^7

Calculate the values of the activation energy and the frequency factor for the reaction.

Reaction Mechanisms

Concept Review

13.91. The reaction between NO and Cl_2 is first order in each reactant. Does this mean that the reaction could occur in just one step? Explain your answer.

13.92. The reaction between NO and H_2 is second order in NO. Does this mean that the reaction could occur in just one step?

*13.93.** If the reaction A → B is first order in A and first order overall, does it occur in just one step?

*13.94.** If a reaction is zero order in a reactant, does that mean the reactant is never involved in collisions with other reactants? Explain your answer.

Problems

13.95. Substance A decomposes slowly into substance B, which then rapidly decomposes into substances C and D. Sketch a reaction profile for the reaction A → C + D, adding labels in the appropriate locations for the four substances involved.

13.96. How would the reaction profile in Problem 13.95 change if the first step were fast and the second step slow?

13.97. Write the rate laws for the following elementary steps and identify them as uni-, bi-, or termolecular steps:
a. $SO_2Cl_2(g) \rightarrow SO_2(g) + Cl_2(g)$
b. $NO_2(g) + CO(g) \rightarrow NO(g) + CO_2(g)$
c. $2 NO_2(g) \rightarrow NO_3(g) + NO(g)$

13.98. Write the rate laws for the following elementary steps and identify them as uni-, bi-, or termolecular steps:
a. $Cl(g) + O_3(g) \rightarrow ClO(g) + O_2(g)$
b. $2 NO_2(g) \rightarrow N_2O_4(g)$
*c.** $^{14}_{6}C \rightarrow {}^{14}_{7}N + {}^{0}_{-1}\beta$

13.99. Write the overall reaction that consists of the following elementary steps:

(1)	$N_2O_5(g) \rightarrow NO_3(g) + NO_2(g)$
(2)	$NO_3(g) \rightarrow NO_2(g) + O(g)$
(3)	$2 O(g) \rightarrow O_2(g)$

13.100. What overall reaction consists of the following elementary steps?

(1)	$ClO^-(aq) + H_2O(\ell) \rightarrow HClO(aq) + OH^-(aq)$
(2)	$I^-(aq) + HClO(aq) \rightarrow HIO(aq) + Cl^-(aq)$
(3)	$OH^-(aq) + HIO(aq) \rightarrow H_2O(\ell) + IO^-(aq)$

*13.101.** In the following mechanism for NO formation, oxygen atoms are produced by breaking O=O bonds at high temperature in a fast, reversible reaction. If $\Delta[NO]/\Delta t = k[N_2][O_2]^{1/2}$, which step in the mechanism is the rate-determining step?

(1)	$O_2(g) \rightleftharpoons 2O(g)$
(2)	$O(g) + N_2(g) \rightarrow NO(g) + N(g)$
(3)	$N(g) + O(g) \rightarrow NO(g)$
Overall	$N_2(g) + O_2(g) \rightarrow 2 NO(g)$

13.102. A proposed mechanism for the decomposition of hydrogen peroxide consists of three elementary steps:

(1)	$H_2O_2(g) \rightarrow 2 OH(g)$
(2)	$H_2O_2(g) + OH(g) \rightarrow H_2O(g) + HO_2(g)$
(3)	$HO_2(g) + OH(g) \rightarrow H_2O(g) + O_2(g)$

If the rate law for the reaction is first order in H_2O_2, which step in the mechanism is the rate-determining step?

13.103. At a given temperature, the rate of the reaction between NO and Cl_2 is proportional to the product of the concentrations of the two gases: $[NO][Cl_2]$. The following two-step mechanism was proposed for the reaction:

(1)	$NO(g) + Cl_2(g) \rightarrow NOCl_2(g)$
(2)	$NOCl_2(g) + NO(g) \rightarrow 2 NOCl(g)$
Overall	$2 NO(g) + Cl_2(g) \rightarrow 2 NOCl(g)$

Which step must be the rate-determining step if this mechanism is correct?

13.104. **Mechanism of Ozone Destruction** Ozone decomposes thermally to oxygen in the following reaction:

$$2 O_3(g) \rightarrow 3 O_2(g)$$

The following mechanism has been proposed:

$$O_3(g) \rightarrow O(g) + O_2(g)$$
$$O(g) + O_3(g) \rightarrow 2 O_2(g)$$

The reaction is second order in ozone. What properties of the two elementary steps (specifically, relative rate and reversibility) are consistent with this mechanism?

13.105. **Mechanism of NO_2 Destruction** Which of the following mechanisms are possible for the thermal decomposition of NO_2, given that the rate = $k[NO_2]^2$?
a. $NO_2(g) \xrightarrow{slow} NO(g) + O(g)$
$\quad O(g) + NO_2(g) \xrightarrow{fast} NO(g) + O_2(g)$
b. $NO_2(g) + NO_2(g) \xrightarrow{fast} N_2O_4(g)$
$\quad N_2O_4(g) \xrightarrow{slow} NO(g) + NO_3(g)$
$\quad NO_3(g) \xrightarrow{fast} NO(g) + O_2(g)$
c. $NO_2(g) + NO_2(g) \xrightarrow{slow} NO(g) + NO_3(g)$
$\quad NO_3(g) \xrightarrow{fast} NO(g) + O_2(g)$

13.106. The rate laws for the thermal and photochemical decomposition of NO_2 are different. Which of the following mechanisms are possible for the photochemical decomposition of NO_2 given that the rate = $k[NO_2]$?
a. $NO_2(g) + NO_2(g) \xrightarrow{slow} N_2O_4(g)$
$\quad N_2O_4(g) \xrightarrow{fast} N_2O_3(g) + O(g)$
$\quad N_2O_3(g) + O(g) \xrightarrow{fast} N_2O_2(g) + O_2(g)$
$\quad N_2O_2(g) \xrightarrow{fast} 2 NO(g)$
b. $NO_2(g) + NO_2(g) \xrightarrow{slow} NO(g) + NO_3(g)$
$\quad NO_3(g) \xrightarrow{fast} NO(g) + O_2(g)$
c. $NO_2(g) \xrightarrow{slow} N(g) + O_2(g)$
$\quad N(g) + NO_2(g) \xrightarrow{fast} N_2O_2(g)$
$\quad N_2O_2(g) \xrightarrow{slow} 2 NO(g)$

Catalysts

Concept Review

13.107. Does a catalyst affect both the rate and the rate constant of a reaction? Explain your answer.

13.108. Is the rate law for a catalyzed reaction the same as that for the uncatalyzed reaction?

13.109. Does a substance that increases the rate of a reaction also increase the rate of the reverse reaction?

13.110. The rate of the reaction between NO_2 and CO is independent of [CO]. Does this mean that CO is a catalyst for the reaction?

13.111. Does the concentration of a homogeneous catalyst appear in the rate law for the reaction it catalyzes?

*13.112. The rate of a chemical reaction is too slow to measure at room temperature. We could either raise the temperature or add a catalyst. Which would be a better solution for making an accurate determination of the rate constant?

Problems

13.113. Is NO a catalyst for the decomposition of N_2O in the following two-step reaction mechanism, or is N_2O a catalyst for the conversion of NO to NO_2?

$$\begin{aligned}(1) \quad & NO(g) + N_2O(g) \rightarrow N_2(g) + NO_2(g) \\ (2) \quad & 2\,NO_2(g) \rightarrow 2\,NO(g) + O_2(g)\end{aligned}$$

13.114. NO as a Catalyst for Ozone Destruction Explain why NO is a catalyst in the following two-step process that results in the depletion of ozone in the stratosphere:

$$\begin{aligned}(1) \quad & NO(g) + O_3(g) \rightarrow NO_2(g) + O_2(g) \\ (2) \quad & O(g) + NO_2(g) \rightarrow NO(g) + O_2(g) \\ \text{Overall} \quad & O(g) + O_3(g) \rightarrow 2\,O_2(g)\end{aligned}$$

13.115. On the basis of the frequency factors and activation energy values of the following two reactions, determine which one will have the larger rate constant at room temperature (298 K).

$$O_3(g) + O(g) \rightarrow O_2(g) + O_2(g)$$
$A = 8.0 \times 10^{-12}\ cm^3/(molecules \cdot s) \qquad E_a = 17.1\ kJ/mol$

$$O_3(g) + Cl(g) \rightarrow ClO(g) + O_2(g)$$
$A = 2.9 \times 10^{-11}\ cm^3/(molecules \cdot s) \qquad E_a = 2.16\ kJ/mol$

13.116. On the basis of the frequency factors and activation energy values of the following two reactions, determine which one will have the larger rate constant at room temperature (298 K).

$$O_3(g) + Cl(g) \rightarrow ClO(g) + O_2(g)$$
$A = 2.9 \times 10^{-11}\ cm^3/(molecules \cdot s) \qquad E_a = 2.16\ kJ/mol$

$$O_3(g) + NO(g) \rightarrow NO_2(g) + O_2(g)$$
$A = 2.0 \times 10^{-12}\ cm^3/(molecules \cdot s) \qquad E_a = 11.6\ kJ/mol$

Additional Problems

13.117. A student inserts a glowing wood splint into a test tube filled with O_2. The splint quickly catches fire (Figure P13.117). Why does the splint burn so much faster in pure O_2 than in air?

FIGURE P13.117

*13.118. A backyard chef turns on the propane gas to a barbecue grill. Even though the reaction between propane and oxygen is spontaneous, the gas does not begin to burn until the chef pushes an igniter button to produce a spark. Why is the spark needed?

*13.119. On average, someone who falls through the ice covering a frozen lake is less likely to experience anoxia (lack of oxygen) than someone who falls into a warm pool and is underwater for the same length of time. Why?

13.120. Ethanol Metabolism The half-life of ethanol in the human body is $t_{1/2} = [ethanol]_0/2k$, where $[ethanol]_0$ is the initial concentration of ethanol and k is the zero-order rate constant. Write the rate law for the metabolism of ethanol and sketch the concentration–time curve.

*13.121. If the rate of the reverse reaction is much slower than the rate of the forward reaction, does the method used to determine a rate law from initial concentrations and initial rates also work at some other time t? What concentrations would we use in the case where we use the rate when $t \neq 0$?

13.122. What is wrong with the following statement: The reaction rate and the rate constant for a reaction both depend on the number of collisions and on the concentrations of the reactants.

13.123. Why can't an elementary step in a mechanism have a rate law that is zero order in a reactant?

13.124. Testing for a Banned Herbicide Sodium chlorate was used in weed-control preparations, but its sale has been banned in EU countries since 2009. A simple colorimetric test for the presence of the chlorate ion in a solution of herbicide relies on the following reaction:

$$\begin{aligned}2\,MnO_4^-(aq) + 5\,ClO_3^-(aq) + 6\,H^+(aq) \rightarrow \\ 2\,Mn^{2+}(aq) + 5\,ClO_4^-(aq) + 3\,H_2O(\ell)\end{aligned}$$

The table contains rate data for this reaction.

Experiment	$[MnO_4^-]_0$ (M)	$[ClO_3^-]_0$ (M)	$[H^+]_0$ (M)	Initial Rate (M/s)
1	0.10	0.10	0.10	5.2×10^{-3}
2	0.25	0.10	0.10	3.3×10^{-2}
3	0.10	0.30	0.10	1.6×10^{-2}
4	0.10	0.10	0.20	7.4×10^{-3}

Determine the rate law and the rate constant for this reaction.

13.125. The table contains reaction rate data for the reaction

$$2\,NO(g) + Cl_2(g) \rightarrow 2\,NOCl(g)$$

Experiment	$[NO]_0$ (M)	$[Cl_2]_0$ (M)	Initial Rate (M/s)
1	0.20	0.10	0.63
2	0.20	0.30	5.70
3	0.80	0.10	2.58
4	0.40	0.20	?

Predict the initial rate of reaction in experiment 4.

13.126. An important reaction in the formation of photochemical smog is the reaction between ozone and NO:

$$NO(g) + O_3(g) \rightarrow NO_2(g) + O_2(g)$$

The reaction is first order in NO and O_3. The rate constant of the reaction is 80 M^{-1} s^{-1} at 25°C and 3000 M^{-1} s^{-1} at 75°C.

a. If this reaction were to occur in a single step, would the rate law be consistent with the observed order of the reaction for NO and O_3?

b. What is the value of the activation energy of the reaction?

c. What is the rate of the reaction at 25°C when [NO] = 3×10^{-6} M and [O_3] = 5×10^{-9} M?

d. Predict the values of the rate constant at 10°C and 35°C.

13.127. Ammonia reacts with nitrous acid to form an intermediate, ammonium nitrite (NH_4NO_2), which decomposes to N_2 and H_2O:

$$NH_3(g) + HNO_2(aq) \rightarrow NH_4NO_2(aq) \rightarrow N_2(g) + 2\,H_2O(\ell)$$

a. The reaction is first order in ammonia and second order in nitrous acid. What is the rate law for the reaction? What are the units of the rate constant if concentrations are expressed in molarity and time in seconds?

b. The rate law for the reaction has also been written as

$$\text{Rate} = k[NH_4^+][NO_2^-][HNO_2]$$

Is this expression equivalent to the one you wrote in part (a)?

c. With the data in Appendix 4, calculate the value of ΔH°_{rxn} for the overall reaction (ΔH°_f, HNO_2 = -43.1 kJ/mol).

d. Draw an energy profile for the process with the assumption that E_a of the first step is lower than E_a of the second step.

13.128. Lachrymators in Smog The combination of ozone, volatile hydrocarbons, nitrogen oxide, and sunlight in urban environments produces peroxyacetyl nitrate (PAN), a potent lachrymator (eye irritant). PAN decomposes to peroxyacetyl radicals and nitrogen dioxide in a process that is second order in PAN, as shown in Figure P13.128:

a. The half-life of the reaction, at 23°C and $P_{CH_3CO_3NO_2}$ = 10.5 torr, is 100 h. Calculate the rate constant for the reaction.

b. Determine the rate of the reaction at 23°C and $P_{CH_3CO_3NO_2}$ = 10.5 torr.

c. Draw a graph showing P_{PAN} as a function of time from 0 to 200 h starting with $P_{CH_3CO_3NO_2}$ = 10.5 torr.

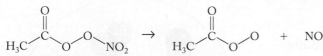

FIGURE P13.128

13.129. Nitrogen Oxide in the Human Body Nitrogen oxide is a free radical that plays many biological roles, including regulating neurotransmission and the human immune system. One of its many reactions involves the peroxynitrite ion ($ONOO^-$):

$$NO(aq) + ONOO^-(aq) \rightarrow NO_2(aq) + NO_2^-(aq)$$

a. Use the following data to determine the rate law and rate constant of the reaction at the experimental temperature at which these data were generated.

Experiment	[NO]$_0$ (M)	[ONOO$^-$]$_0$ (M)	Rate (M/s)
1	1.25×10^{-4}	1.25×10^{-4}	2.03×10^{-11}
2	1.25×10^{-4}	0.625×10^{-4}	1.02×10^{-11}
3	0.625×10^{-4}	2.50×10^{-4}	2.03×10^{-11}
4	0.625×10^{-4}	3.75×10^{-4}	3.05×10^{-11}

b. Draw the Lewis structure of peroxynitrite ion (including all resonance forms) and assign formal charges. Note which form is preferred.

c. Use the average bond energies in Table A4.1 to estimate the value of ΔH°_{rxn} by using the preferred structure from part (b).

13.130. Reducing NO Emissions Adding NH_3 to the stack gases at an electric-power-generating plant can reduce NO_x emissions. This selective noncatalytic reduction process depends on the reaction between NH_2 (an odd-electron molecule) and NO:

$$NH_2(g) + NO(g) \rightarrow N_2(g) + H_2O(g)$$

The following kinetic data were collected at 1200 K.

Experiment	[NH$_2$]$_0$ (M)	[NO]$_0$ (M)	Rate (M/s)
1	1.00×10^{-5}	1.00×10^{-5}	0.12
2	2.00×10^{-5}	1.00×10^{-5}	0.24
3	2.00×10^{-5}	1.50×10^{-5}	0.36
4	2.50×10^{-5}	1.50×10^{-5}	0.45

a. What is the rate law for the reaction?

b. What is the value of the rate constant at 1200 K?

14

Chemical Equilibrium

How Much Product Does a Reaction Really Make?

AMMONIA, FERTILIZERS, AND FOOD Nearly half of the world's food production can be attributed to the use of fertilizers. Ammonia is a key reactant in the synthesis of nitrogen fertilizers such as urea and ammonium nitrate.

PARTICULATE **REVIEW**

Simultaneous Reactions

In Chapter 14 we explore reactions in which the products can reform the reactants. Two reactions can take place simultaneously among the NO, O_2, and NO_2 molecules shown here.

- Which molecules combine to produce nitrogen dioxide? What two molecules could be produced when two molecules of nitrogen dioxide collide?

- Write one chemical equation showing that these two reactions take place simultaneously. Be sure to use the appropriate arrows.

- The forward reaction proceeds slowly at room temperature. How could the number of collisions be increased in order to increase the reaction rate?

    (Review Section 4.5 and Sections 13.2–13.4 if you need help.)

(Answers to Particulate Review questions are in the back of the book.)

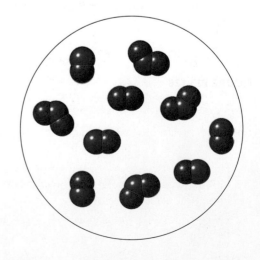

694

Equal versus Equilibrium

The images show two different mixtures of steam and carbon monoxide reacting to reversibly produce carbon dioxide and hydrogen gas. Each sample is shown both in its initial state and after 5 seconds of reaction time. As you read Chapter 14, look for ideas that will help you answer these questions:

- Write one balanced chemical equation for the two simultaneous reactions taking place in each mixture.

- How does the concentration of each reactant change in Sample A? In Sample B? How does the concentration of each product change in Sample A? In Sample B?

- Which sample is more likely to be at equilibrium?

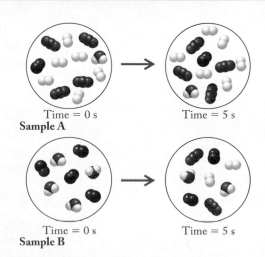

Time = 0 s
Sample A

Time = 5 s

Time = 0 s
Sample B

Time = 5 s

695

Learning Outcomes

LO1 Write mass action or equilibrium constant expressions for reversible reactions, including those involving heterogeneous equilibria
Sample Exercises 14.1, 14.9

LO2 Calculate the value of an equilibrium constant or a reaction quotient and use it to predict the direction of a reversible chemical reaction
Sample Exercises 14.2, 14.3, 14.8

LO3 Interconvert the K_c and K_p values of gas-phase reactions
Sample Exercise 14.4

LO4 Calculate the value of K for a reverse reaction, for a reaction with different coefficients, and for combined reactions
Sample Exercises 14.5, 14.6, 14.7

LO5 Predict how a reaction at equilibrium responds to changes in conditions
Sample Exercises 14.10, 14.11, 14.12

LO6 Calculate the concentrations or partial pressures of reactants and products in a reaction mixture at equilibrium from their initial values and the value of K
Sample Exercises 14.13, 14.14, 14.15

14.1 The Dynamics of Chemical Equilibrium

As the world's population continues to grow, it has become increasingly important to be able to achieve sustainable increases in food production. One strategic method for increasing both the yield and quality of crops such as corn, wheat, and soy is the use of fertilizers. Gaseous ammonia is a critical precursor to the production of nitrogen-based fertilizers such as urea and ammonium nitrate. The U.S. Geological Survey estimates that 146 million metric tons of ammonia were produced worldwide in 2015, much of it as fertilizer feedstock. Given that corn production accounts for half of fertilizer use in the United States, that almost 89 million acres of corn were planted in the United States in 2015, and that growing one bushel of corn typically requires 1.6 pounds of fertilizer, the synthesis of ammonia is essential to crop production. The manufacturing of ammonia begins with the production of hydrogen gas, often via a reaction known as the steam–methane reforming reaction. The hydrogen gas is subsequently reacted with nitrogen via the Haber–Bosch process. We explore both of these reactions in this chapter.

In the United States, about 95% of hydrogen gas needed by industry is generated in a two-step process that begins by reacting methane, the primary component of natural gas, with high-temperature steam:

$$CH_4(g) + H_2O(g) \rightleftharpoons CO(g) + 3\,H_2(g) \qquad \Delta H° = 206 \text{ kJ/mol} \qquad (14.1)$$

This is the steam–methane reforming reaction. It is highly endothermic, which is why it is typically run at temperatures between 700°C and 1000°C. To increase the rate of the reaction, nickel or nickel-containing alloys are used as catalysts, and the reaction is run at pressures as high as 25 atm.

CONCEPT TEST

Why does raising the temperature or increasing the pressure of a gas-phase reaction help increase the rate of the reaction?

(Answers to Concept Tests are in the back of the book.)

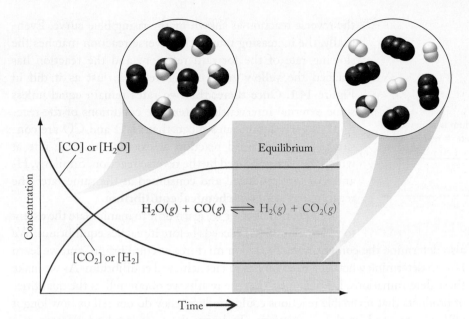

FIGURE 14.1 Concentrations of reactants (red curve) and products (blue curve) in the water–gas shift reaction change over time until equilibrium is reached (the yellow zone).

In the second step, carbon monoxide produced in the steam–methane reforming reaction reacts with more steam to produce more hydrogen gas:

$$CO(g) + H_2O(g) \rightleftharpoons CO_2(g) + H_2(g) \qquad \Delta H° = -41 \text{ kJ/mol} \qquad (14.2)$$

This reaction, called the water–gas shift reaction, also requires a catalyst and reaction temperatures greater than 350°C to achieve respectable reaction rates.

Let's focus on the double arrow that connects the reactants and products in Equations 14.1 and 14.2. This pair of arrows tells us that both the steam–methane reforming reaction and the water–gas shift reaction are reversible. Reversible reactions can proceed in both the forward and the reverse directions. In fact, they can proceed in both directions at the same time.

To better understand how reactants become products and products become reactants, consider what happens if we run the water–gas shift reaction in a sealed vessel that initially contains equal numbers of moles of steam and carbon monoxide, but no products. As the reaction proceeds, the concentrations of H_2O and CO decrease, as shown by the red curve in Figure 14.1. The concentrations of H_2 and CO_2 start at zero and increase over time, as shown by the blue curve. Because H_2O and CO react in a 1:1 stoichiometric ratio, their concentrations decrease at the same rate. Similarly, H_2 and CO_2 are formed in a 1:1 ratio, so their concentrations increase from zero at the same rate.

Eventually, both the red and blue curves level off (in the yellow region of Figure 14.1), which means there is no net change in concentrations of either the reactants of products with time. Note that the unchanging concentrations of H_2O and CO are significantly above zero. This means the reaction mixture will contain some reactants no matter how long we let it run.

To obtain another perspective on why the concentration curves in Figure 14.1 level off before all the reactants are consumed, let's revisit a key concept from Chapter 13: the rates of most chemical reactions depend on the concentrations of their reactant(s). Therefore we expect the rate of the forward reaction to decrease as the concentrations of the reactants, H_2O and CO, decrease, which is shown by the downward trend of the red curve in Figure 14.2. As the reaction proceeds, the concentrations of H_2 and CO increase, which results in an increase in the rate of

CONNECTION In Chapter 4 we introduced the double arrow when discussing the dissociation of weak electrolytes.

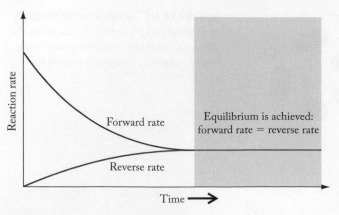

FIGURE 14.2 Equilibrium is achieved when the rates of a reversible reaction in the forward direction (red curve) and reverse direction (blue curve) become equal.

CONNECTION In Chapter 13 we discussed how reactions occur when molecules collide; the higher the concentrations of reactants, the more frequently molecules collide, and the faster a reaction proceeds.

CHEMTOUR Equilibrium

the reverse reaction as shown by the rising blue curve. Eventually, the increasing rate of the reverse reaction matches the slowing rate of the forward reaction, and the reaction has reached the yellow zone in Figure 14.2, just as it did in Figure 14.1. Once there, the two rates remain equal unless some external intervention alters the conditions of the reaction. Equal reaction rates mean that H_2O and CO are consumed in the forward reaction at exactly the same rate at which they are produced by the reverse reaction. Similarly, H_2 and CO_2 are produced and consumed at the same rate. The reaction has achieved **chemical equilibrium**.

Later in this chapter we learn how to manipulate the extent to which a reaction proceeds before it reaches equilibrium. We also determine the composition of reaction mixtures at equilibrium, and we learn how to determine whether a reaction has in fact achieved equilibrium. As we make these determinations, keep in mind that the results we obtain tell us the quantities of products that reversible reactions could produce; they do not tell us how long it will take to produce those quantities. To know the time involved also requires knowledge of the rate of the reaction, which is why this chapter on chemical equilibrium immediately follows the one on chemical kinetics.

A final key point about chemical equilibria: they are *dynamic*. As already discussed, unchanging concentrations of reactants and products do not mean that a reaction has stopped. The reaction still proceeds, but it does so equally rapidly in both the forward and reverse directions simultaneously.

14.2 The Equilibrium Constant

In this section we take a quantitative look at the concentrations of products and reactants that are present when chemical reactions come to equilibrium. Let's return to the water–gas shift reaction:

$$CO(g) + H_2O(g) \rightleftharpoons CO_2(g) + H_2(g)$$

The data in Table 14.1 describe the results of four experiments in which different quantities of H_2O gas, CO, H_2, and CO_2 are injected into a sealed reaction vessel heated to 350°C. The four gases are allowed to react and the final concentrations of all four are determined when the reaction has reached equilibrium.

In Experiment 1 the reaction vessel initially contains equimolar concentrations of H_2O and CO but no H_2 or CO_2. In Experiment 2 the vessel initially contains equimolar concentrations of H_2 and CO_2 but no H_2O or CO. The data

TABLE 14.1 Initial and Equilibrium Concentrations of the Reactants and Products in the Water–Gas Shift Reaction at 350°C

Experiment	INITIAL CONCENTRATION (*M*)				EQUILIBRIUM CONCENTRATION (*M*)			
	[H_2O]	[CO]	[H_2]	[CO_2]	[H_2O]	[CO]	[H_2]	[CO_2]
1	0.0200	0.0200	0	0	0.0036	0.0036	0.0164	0.0164
2	0	0	0.0200	0.0200	0.0036	0.0036	0.0164	0.0164
3	0.0200	0.0100	0.0200	0.0100	0.0120	0.0020	0.0280	0.0180
4	0.0100	0.0200	0.0300	0.0400	0.0050	0.0150	0.0350	0.0450

in Table 14.1 indicate that when the reaction mixtures in Experiments 1 and 2 achieve chemical equilibrium, the concentrations of H_2O and CO are the same (0.0036 M) in both experiments and so, too, are the concentrations of H_2 and CO_2 (0.0164 M). The identical equilibrium concentrations of the reactants and products in Experiments 1 and 2 confirm that the composition of a reaction mixture at equilibrium is independent of the direction in which a particular reaction proceeds to achieve equilibrium.

The data from Experiment 3 show that the concentrations of products increase and concentrations of reactants decrease when the concentrations of the products are initially the same as the concentrations of the reactants. The data from Experiment 4 show that the concentrations of products increase and concentrations of reactants decrease even when the concentrations of the products are initially *greater than* the concentrations of the reactants.

The significance of the data from the four experiments in Table 14.1 can be appreciated if we do the following math: multiply the equilibrium concentrations of the products (H_2 and CO_2) together and divide that product by the product of the equilibrium concentrations of the reactants (H_2O and CO):

$$\text{Experiments 1 and 2:} \quad \frac{[H_2][CO_2]}{[H_2O][CO]} = \frac{(0.0164)(0.0164)}{(0.0036)(0.0036)} = 21$$

$$\text{Experiment 3:} \quad \frac{[H_2][CO_2]}{[H_2O][CO]} = \frac{(0.0280)(0.0180)}{(0.0120)(0.0020)} = 21$$

$$\text{Experiment 4:} \quad \frac{[H_2][CO_2]}{[H_2O][CO]} = \frac{(0.0350)(0.0450)}{(0.0050)(0.0150)} = 21$$

Note that the ratio of the concentrations of products to reactants at equilibrium is the same in every experiment. As you might guess, we would get the same ratio from *any* combination of initial concentrations of these four gases at 350°C. This constancy applies to other reversible reactions and has been known since the mid-19th century when Norwegian chemists Cato Guldberg (1836–1902) and Peter Waage (1833–1900) discovered that any reversible reaction eventually reaches a state in which the ratio of the concentrations of products to reactants, with each value raised to a power corresponding to the coefficient for that substance in the balanced chemical equation for the reaction, has a characteristic value at a given temperature.

They called this phenomenon the **law of mass action** and the ratio of concentration terms the **mass action expression** for the reaction. This ratio is more commonly called the **equilibrium constant expression**, and its numerical value—which depends on temperature—is the **equilibrium constant (K)** of the reaction. Because K is a ratio of positive values, it is always positive, too. K values greater than 1 are associated with reaction mixtures that consist of mostly products at equilibrium. K values between 0 and 1 are associated with reaction mixtures that contain few products at equilibrium.

For the water–gas shift reaction, the equilibrium constant expression and the value of K at 350°C are

$$K = \frac{[H_2][CO_2]}{[H_2O][CO]} = 21$$

Note that none of the concentration terms has an exponent because all of their coefficients are "1" in the balanced chemical equation, and a substance's coefficient *always* becomes its exponent in the equilibrium constant expression for a reaction. This is in contrast to rate law expressions, where the exponents are

chemical equilibrium a dynamic process in which the concentrations of reactants and products remain constant over time and the rate of the reaction in the forward direction matches its rate in the reverse direction.

law of mass action the principle relating the balanced chemical equation of a reversible reaction to its mass action expression (or equilibrium constant expression).

mass action expression an expression equivalent in form to the equilibrium constant expression, but applied to reaction mixtures that may or may not be at equilibrium.

equilibrium constant expression the ratio of the concentrations or partial pressures of products to reactants at equilibrium, with each term raised to a power equal to the coefficient of that substance in the balanced chemical equation for the reaction.

equilibrium constant (K) the numerical value of the equilibrium constant expression of a reversible chemical reaction at a particular temperature.

frequently not the same as the coefficients in the balanced equation because the exponents reflect the stoichiometry of the rate-determining step, not necessarily the overall reaction.

CONCEPT TEST

Suppose equal numbers of moles of water vapor, carbon dioxide, carbon monoxide, and hydrogen gas are injected into a rigid, sealed reaction vessel and heated to 350°C. Which expression about the composition of the reaction mixture at equilibrium is true?

a. $[CO] = [H_2] = [CO_2] = [H_2O]$

b. $[CO_2] = [H_2] > [CO] = [H_2O]$

c. $[CO_2] = [H_2] < [CO] = [H_2O]$

d. $[H_2O] = [H_2] > [CO_2] = [CO]$

e. $[H_2O] = [H_2] < [CO_2] = [CO]$

Now let's generalize what we have learned about equilibrium constant expressions. For the generic reaction in which a moles of reactant A react with b moles of B to form c moles of substance C and d moles of D:

$$aA + bB \rightleftharpoons cC + dD$$

the equilibrium constant expression is

$$K_c = \frac{[C]^c[D]^d}{[A]^a[B]^b} \tag{14.3}$$

Here we have added the subscript "c" to the equilibrium constant to indicate that it is based on a ratio of concentration values. If substances A, B, C, and D are gases, then the equilibrium constant may also be expressed in terms of their partial pressures, denoted by the subscript "p":

$$K_p = \frac{(P_C)^c(P_D)^d}{(P_A)^a(P_B)^b} \tag{14.4}$$

 **CHEMTOUR**
Equilibrium in the Gas Phase

As we see later in this chapter, the values of K_c and K_p for a given gas-phase reaction at a given temperature may or may not be the same. It depends on whether the number of moles of gaseous reactants is the same as the number of moles of gaseous products. Also, K values do not have units even when there are different numbers of moles of reactants and products. Why? It is because the concentrations and partial pressures in K_c or K_p expressions are (theoretically) ratios of concentrations or partial pressures in a reaction mixture to an ideal standard concentration (1.000 M) or partial pressure (1.000 atm). These ratios take into account the nonideal behavior of substances. Because the terms in equilibrium constant expressions are, strictly speaking, ratios of values having the same units, their units cancel out, leaving unitless equilibrium constants. In practice, we use concentration and partial pressure values directly in equilibrium calculations. When we do, we are assuming that all the reactants and products are behaving ideally.

SAMPLE EXERCISE 14.1 Writing Equilibrium Constant Expressions   **LO1**

A key reaction in the formation of photochemical smog involves the reversible combination of NO and O_2 in the atmosphere, producing NO_2:

$$2\,NO(g) + O_2(g) \rightleftharpoons 2\,NO_2(g)$$

Write the K_c and K_p expressions for this reaction.

Collect, Organize, and Analyze We are given the balanced chemical equation for a reaction and asked to write K_c and K_p expressions for it. Equilibrium constant expressions are ratios of the concentrations (K_c) or partial pressures (K_p) of products to reactants, with each term raised to the power equal to its coefficient in the balanced chemical equation of the reaction.

Solve The coefficients of NO and NO_2 in the chemical equation are both 2, so the NO and NO_2 terms in the K_c and K_p expressions must be squared:

$$K_c = \frac{[NO_2]^2}{[NO]^2[O_2]}$$

$$K_p = \frac{(P_{NO_2})^2}{(P_{NO})^2(P_{O_2})}$$

Think About It The numerators and denominators of the K_c and K_p expressions contain terms for the same products and reactants, each raised to the same power. The difference between them is the nature of the terms: molar concentrations in the K_c expression and partial pressures in the K_p expression.

 Practice Exercise Write K_c and K_p equilibrium constant expressions for the steam–methane reforming reaction:

$$CH_4(g) + H_2O(g) \rightleftharpoons CO(g) + 3\,H_2(g)$$

(Answers to Practice Exercises are in the back of the book.)

SAMPLE EXERCISE 14.2 Calculating the Value of K_c from Equilibrium Concentrations   **LO2**

Table 14.2 contains data from four experiments on the dimerization of NO_2:

$$2\,NO_2(g) \rightleftharpoons N_2O_4(g)$$

The experiments were run at 100°C in a rigid, closed container. Use the data from each experiment to calculate a value of the equilibrium constant K_c for the reaction.

Collect and Organize We are given four sets of data that contain initial and equilibrium concentrations of a reactant and product. We are asked to determine the value of the equilibrium constant K_c in each experiment. The K_c expression for this reaction is the ratio of the equilibrium concentration of the product over that of the reactant, with each term raised to a power equal to its coefficient in the chemical equation.

Analyze To calculate the value of K_c, we insert the pairs of equilibrium concentrations from each experiment into the K_c expression and calculate the ratio. The $[NO_2]$ values

---

TABLE 14.2 Data for the Dimerization of NO_2 at 100°C

Experiment	INITIAL CONCENTRATION (M)		EQUILIBRIUM CONCENTRATION (M)	
	$[NO_2]$	$[N_2O_4]$	$[NO_2]$	$[N_2O_4]$
1	0.0200	0.0000	0.0172	0.00139
2	0.0300	0.0000	0.0244	0.00280
3	0.0400	0.0000	0.0310	0.00452
4	0.0000	0.0200	0.0310	0.00452

are greater than the $[N_2O_4]$ in every experiment, so the equilibrium reaction mixtures contain more reactant than product. Therefore we might expect the value of K_c to be less than 1. However, all the gas concentrations are much less than 0.1 M, and the $[NO_2]$ values in the denominator are squared, which makes the value of the denominator even smaller. Therefore, the value K_c may actually be *greater* than 1.

Solve The K_c expression is:

$$K_c = \frac{[N_2O_4]}{[NO_2]^2}$$

Inserting the data from the four experiments into this expression and doing the math:

$$\text{Experiment 1: } K_c = \frac{[N_2O_4]}{[NO_2]^2} = \frac{0.00139}{(0.0172)^2} = 4.70$$

$$\text{Experiment 2: } K_c = \frac{0.00280}{(0.0244)^2} = 4.70$$

$$\text{Experiment 3: } K_c = \frac{0.00452}{(0.0310)^2} = 4.70$$

$$\text{Experiment 4: } K_c = \frac{0.00452}{(0.0310)^2} = 4.70$$

Think About It The values calculated for K_c are the same, as they should be for the same reaction at the same temperature. As we thought they might be, they are greater than 1.

Practice Exercise A mixture of gaseous CO and H_2, called *synthesis gas*, is used commercially to prepare methanol (CH_3OH), a compound considered an alternative fuel to gasoline. Under equilibrium conditions at 700 K, $[H_2] = 0.074$ M, $[CO] = 0.025$ M, and $[CH_3OH] = 0.040$ M. What is the value of K_c for this reaction at 700 K?

SAMPLE EXERCISE 14.3 Calculating the Value of K_p from Equilibrium Partial Pressures **LO2**

A sealed chamber contains an equilibrium mixture of NO_2 and N_2O_4 at 300°C. The partial pressures of NO_2 and N_2O_4 are 1.8 and 24.1 atm, respectively. What is the value of K_p for the dimerization of NO_2: $2\ NO_2(g) \rightleftharpoons N_2O_4(g)$, at 300°C?

Collect, Organize, and Analyze We are asked to use the partial pressures of NO_2 and N_2O_4 at equilibrium to determine the value of K_p for the dimerization of NO_2. The K_p expression should have the same format as the K_c expression for this reaction that was used in Sample Exercise 14.2. The partial pressure of N_2O_4 is greater than that of NO_2, and both values are greater than 1. Therefore the value of K_p should be greater than 1.

Solve

$$K_p = \frac{P_{N_2O_4}}{(P_{NO_2})^2}$$

$$K_p = \frac{24.1}{(1.8)^2} = 7.4$$

Think About It As expected, the value of K_p is greater than 1. However, it is not the same as the K_c value for this reaction that was calculated in Sample Exercise 14.2. This is not surprising because the experiments in the two exercises were conducted at different temperatures, and K values change with changing temperature. Moreover, the K_p and K_c values of gas-phase reactions may differ, as noted earlier in this section, depending on the number of moles of gaseous reactants and products.

> **Practice Exercise** A reaction vessel contains an equilibrium mixture of SO_2, O_2, and SO_3. The partial pressures of the three gases are 0.0018 atm, 0.0032 atm, and 0.0166 atm, respectively. What is the value of K_p for the following reaction at the temperature inside the vessel?
>
> $$2\ SO_2(g) + O_2(g) \rightleftharpoons 2\ SO_3(g)$$

As we end this section, let's more closely examine the connection between K values of reactions and the composition of their reaction mixtures at equilibrium. The values we have seen thus far—$K_c = 21$ for the water–gas shift reaction at 350°C, and $K_p = 7.4$ and $K_c = 4.70$ for the dimerization of NO_2 at 300°C and 100°C, respectively—are considered intermediate values. Because the range of K values is so large ($0 < K < \infty$), all three of these values are considered *close* to 1, which means that comparable concentrations of reactants and products are likely to be present at equilibrium.

On the other hand, many reactions reach equilibrium only after essentially all the reactants have formed products. We say that these equilibria *lie far to the right* (the direction of the forward reaction arrow). For example, the combustion of H_2 gas proceeds very rapidly (Figure 14.3) until H_2 is, for all practical purposes, completely consumed. This observation is consistent with an enormous value of K for the combustion reaction at 25°C:

$$2\ H_2(g) + O_2(g) \rightleftharpoons 2\ H_2O(g) \qquad K_c = 3 \times 10^{81}$$

In other reactions, little product is formed before equilibrium is reached. These equilibria are said to favor reactants and to *lie far to the left* (the direction of the reverse reaction arrow). For example, the decomposition of CO_2 to CO and O_2 at 25°C essentially doesn't happen, as predicted by our understanding of the stability of CO_2 and the minuscule K value for the decomposition reaction:

$$2\ CO_2(g) \rightleftharpoons 2\ CO(g) + O_2(g) \qquad K_c = 3 \times 10^{-92}$$

In the rest of this chapter, and in Chapters 15 and 16, we focus mainly on equilibria that do not lie far to the left or right but are more likely to be a little to the left or right. These reaction systems tend to have relatively small concentrations of either reactants or products at equilibrium, but not so small that they can be ignored.

FIGURE 14.3 The German passenger airship *Hindenburg* was filled with hydrogen gas to make it less dense than air. The flammable gas ignited as the airship docked in Lakehurst, New Jersey on May 6, 1937. Of the 97 people aboard, 35 were killed.

14.3 Relationships between K_c and K_p Values

As we noted in Section 14.2, the values of K_c and K_p for a given reaction and temperature may or may not be the same, depending on the numbers of moles of gaseous reactants and products. To better understand this relationship, we begin with the ideal gas law:

$$PV = nRT$$

If we solve for P and express volume in liters, then n/V has units of moles per liter, which is the same as molarity (M):

$$P = \frac{n}{V}RT$$

$$P = MRT \qquad\qquad (14.5)$$

Let's apply Equation 14.5 to the gases in the NO_2/N_2O_4 equilibrium from Sample Exercise 14.3:

$$P_{NO_2} = \frac{n_{NO_2}}{V} RT = [NO_2]RT$$

$$P_{N_2O_4} = \frac{n_{N_2O_4}}{V} RT = [N_2O_4]RT$$

Substituting these values into the expression for K_p from Sample Exercise 14.3, we get

$$K_p = \frac{(P_{N_2O_4})}{(P_{NO_2})^2} = \frac{[N_2O_4]RT}{([NO_2]RT)^2} = \frac{[N_2O_4]RT}{[NO_2]^2(RT)^2}$$

The ratio of concentration terms in the expression on the right, $[N_2O_4]/[NO_2]^2$, is the same as the K_c expression for this reaction. Substituting K_c for those terms and simplifying the RT terms:

$$K_p = K_c \frac{1}{RT}$$

This last equation defines the relationship between K_c and K_p for this specific reaction. A more general expression can be derived for the generic reaction of gases A and B forming gases C and D:

$$aA + bB \rightleftharpoons cC + dD$$

The K_p expression for this reaction is

$$K_p = \frac{(P_C)^c(P_D)^d}{(P_A)^a(P_B)^b} \tag{14.4}$$

Replacing each partial pressure term in Equation 14.4 with the corresponding $[X]RT$ term (Equation 14.5) gives us the general expression

$$K_p = \frac{([C]RT)^c([D]RT)^d}{([A]RT)^a([B]RT)^b} \tag{14.6}$$

Combining the RT terms, we get

$$K_p = \frac{[C]^c[D]^d}{[A]^a[B]^b} \times (RT)^{[(c+d)-(a+b)]}$$

We can simplify the complex expression on the right side of this equation. First, we can replace the concentration ratio $[C]^c[D]^d/[A]^a[B]^b$ with K_c (see Equation 14.3). Then we can simplify the exponent on RT by noting that $(c + d)$ is the sum of the coefficients of the gaseous products in the reaction—that is, the sum of the number of moles of gases produced. Similarly, $(a + b)$ is the sum of the number of moles of gaseous reactants consumed. The difference between the two sums, $(c + d) - (a + b)$, represents the *change in the number of moles of gases* (Δn) between the product and reactant sides of the balanced chemical equation. We use the symbol Δn to represent this change. Substituting Δn for $(c + d) - (a + b)$ in Equation 14.6 gives us

$$K_p = K_c(RT)^{\Delta n} \tag{14.7}$$

In reactions in which the number of moles of gas on both sides of the reaction arrow is the same, such as the water–gas shift reaction:

$$CO(g) + H_2O(g) \rightleftharpoons CO_2(g) + H_2(g)$$

$\Delta n = 0$ and $K_p = K_c$. However, in the steam–methane reforming reaction:

$$CH_4(g) + H_2O(g) \rightleftharpoons CO(g) + 3\,H_2(g)$$

Two moles of gaseous reactants form 4 moles of gaseous products. Therefore

$$\Delta n = 4\text{ mol} - 2\text{ mol} = 2\text{ mol}$$

Inserting this value for Δn in Equation 14.7 gives us this reaction's relationship between K_p and K_c:

$$K_p = K_c(RT)^2$$

A final point about the relative sizes of K_p and K_c values: One mole of an ideal gas at STP occupies a volume of 22.4 liters. Therefore the molar concentration of an ideal gas is 1 mol/22.4 L = 0.0446 M at STP. The RT term in Equation 14.7 serves as a *conversion factor* for changing molar concentrations back into partial pressures because the value of RT at STP is

$$0.08206\,\frac{\text{L} \cdot \text{atm}}{\text{mol} \cdot \text{K}} \times 273\text{ K} = 22.4\,\frac{\text{L} \cdot \text{atm}}{\text{mol}}$$

CONNECTION We determined the molar volume occupied by an ideal gas at standard temperature and pressure in Chapter 6.

SAMPLE EXERCISE 14.4 Calculating K_c from K_p **LO3**

In Sample Exercise 14.3 we determined that $K_p = 7.4$ at 300°C for 2 $NO_2(g) \rightleftharpoons$ $N_2O_4(g)$. What is the value of K_c for this reaction at 300°C?

Collect, Organize, and Analyze We are given a K_p value and asked to calculate the corresponding K_c value for the same reaction at the same temperature. Equation 14.7 relates K_p and K_c values:

$$K_p = K_c(RT)^{\Delta n}$$

where Δn represents the change in the number of moles of gas when going from reactant to product.

Solve Two moles of gaseous reactants yield 1 mole of gaseous product. Therefore

$$\Delta n = 1\text{ mol} - 2\text{ mol} = -1\text{ mol}$$

Inserting this value, the given value of K_p, and temperature into Equation 14.7:

$$K_p = K_c(RT)^{\Delta n}$$
$$7.4 = K_c[0.08206 \times (273 + 300)]^{-1}$$
$$K_c = (7.4)(0.08206)(573) = 3.5 \times 10^2$$

Think About It The value of K_c differs from the value of K_p because 2 moles of gaseous reactants produce only 1 mole of gaseous product. The value of K_c is nearly 50 times larger than that of K_p. This makes sense because the absolute temperature of the reaction is more than twice the temperature at STP.

Practice Exercise An important industrial process for synthesizing the ammonia used in agricultural fertilizers involves the combination of N_2 and H_2:

$$N_2(g) + 3\,H_2(g) \rightleftharpoons 2\,NH_3(g) \qquad K_p = 6.1 \times 10^5 \text{ at } 25°C$$

What is the value of K_c of this reaction at 25°C?

14.4 Manipulating Equilibrium Constant Expressions

In this section we learn how to write equilibrium constant expressions for the reverse direction of reversible reactions, for chemical equations that have been multiplied or divided by a value so that a key component has a coefficient of 1, and for overall reactions that are combinations of other reactions.

K for Reverse Reactions

The K values for the forward and reverse directions of a reversible reaction are related, just as the reactions are. For example, in Sample Exercise 14.2 we wrote the K_c expression for the dimerization of NO_2:

$$2\,NO_2(g) \rightleftharpoons N_2O_4(g) \qquad K_c = \frac{[N_2O_4]}{[NO_2]^2}$$

Now let's write the K_c expression for the reverse reaction, that is, the decomposition of N_2O_4:

$$N_2O_4(g) \rightleftharpoons 2\,NO_2(g) \qquad K_c = \frac{[NO_2]^2}{[N_2O_4]}$$

Note how reversing the reaction turns the K_c expression upside down as reactants and products switch roles. Equation 14.8 extends this reciprocal relation between the forward ($K_{forward}$) and reverse ($K_{reverse}$) equilibrium constants to any reversible reaction:

$$K_{forward} = \frac{1}{K_{reverse}} \tag{14.8}$$

Equation 14.8 makes sense if we consider a generic reversible reaction in which the equilibrium favors the formation of product B over reactant A, which is reflected in its large K value:

$$A \rightleftharpoons B \qquad K > 1$$

If we write the reaction in reverse, B is still favored over A, but now B is the reactant and A is the product. Therefore the value of K for the reverse reaction should be small because the reciprocal of a large (K) value is a small one:

$$B \rightleftharpoons A \qquad K < 1$$

SAMPLE EXERCISE 14.5 Calculating the Value of K **LO4**
for a Reverse Reaction

Atmospheric NO combines with O_2 to form NO_2. The reverse reaction is the decomposition of NO_2 to NO and O_2. At 184°C, the value of K_c for the forward reaction is 1.48×10^4. Write the equilibrium constant expressions for both reactions and calculate the value of K_c for the decomposition of NO_2 at 184°C.

Collect and Organize We are asked to write equilibrium constant expressions for the formation and decomposition of NO_2 and to calculate the value of K_c of the decomposition reaction from the K_c value of the formation reaction. Each of these reactions is the reverse of the other, so their equilibrium constant expressions and K_c values are reciprocals of each other.

Analyze To write the K_c expressions for the two reactions, we first need to write a chemical equation describing one of them. That equation will provide the stoichiometric coefficients we need to use as exponents for the concentration terms in the K_c expressions for both reactions, with each expression being the reciprocal of the other.

Solve Writing a chemical equation for the formation of NO_2 from NO and O_2:

$$2\,NO(g) + O_2(g) \rightleftharpoons 2\,NO_2(g)$$

The K_c expression for this formation reaction is

$$K_{formation} = \frac{[NO_2]^2}{[NO]^2[O_2]}$$

where the "2" coefficients in front of NO and NO_2 in the chemical equation become the exponents of the NO and NO_2 concentration terms in the K_c expression.

The K_c expression for the decomposition (reverse) reaction is

$$K_{decomposition} = \frac{[NO]^2[O_2]}{[NO_2]^2}$$

and its value is: $K_{decomposition} = 1/K_{formation} = 1/1.48 \times 10^4 = 6.76 \times 10^{-5}$

Think About It The value of $K_{formation}$ is large, so it makes sense that its reciprocal, $K_{decomposition}$, is small.

Practice Exercise At 300°C, the value of K_p for the combination reaction of N_2 and H_2 that produces NH_3 gas is 4.3×10^{-3}. What is the value of K_p at the same temperature for the decomposition reaction of NH_3 to produce N_2 and H_2?

K for an Equation Multiplied or Divided by a Number

As we saw in Chapter 5, sometimes we must represent a reaction by an equation containing fractional coefficients so that, for example, there is exactly 1 mole of a particular product. Suppose we want to rewrite the formation of NO_2 from NO and O_2:

$$2\,NO(g) + O_2(g) \rightleftharpoons 2\,NO_2(g) \qquad K_c = \frac{[NO_2]^2}{[NO]^2[O_2]}$$

to describe the formation of only 1 mole of NO_2. We do this by dividing all the coefficients by 2:

$$NO(g) + \tfrac{1}{2}\,O_2(g) \rightleftharpoons NO_2(g)$$

The corresponding K_c expression is

$$K_c = \frac{[NO_2]}{[NO][O_2]^{1/2}}$$

Comparing the original K_c expression with this one, we see that the concentration terms in the original expression correspond to those in the second expression, but *squared*. Put another way, dividing all the coefficients in the original reaction equation by 2 produces a K_c expression (and a K value) that is the *square root* of the original. We can extend this pattern to all chemical equilibria with the following rule: if the balanced chemical equation of a reaction is multiplied by some factor n, then the value of K is raised to the nth power. Similarly, dividing all coefficients by n raises the value of K to the $1/n$ power.

SAMPLE EXERCISE 14.6 Calculating K for Different Coefficients   **LO4**

An important reaction in the industrial production of sulfuric acid is the oxidation of SO_2 to SO_3. One way to write a chemical equation for the oxidation reaction is

$$SO_2(g) + \tfrac{1}{2} O_2(g) \rightleftharpoons SO_3(g)$$

If the value of K_c for this reaction at 298 K is 2.8×10^{12}, what is the value of K_c at 298 K for

$$2\, SO_2(g) + O_2(g) \rightleftharpoons 2\, SO_3(g)$$

Collect, Organize, and Analyze We are given the value of K_c for a reaction that describes the production of 1 mole of SO_3. We are asked to recalculate this K_c value for the same reaction, but based on the production of 2 moles of SO_3. Multiplying the coefficients of a chemical equation by n raises the value of its equilibrium constant to the nth power.

Solve Multiplying the original chemical equation by 2 raises its K_c to the second power:

$$K_c = (2.8 \times 10^{12})^2 = 7.8 \times 10^{24}$$

Think About It Doubling the coefficients made the large original K_c value *really* large. However, if the original K_c value had been small (<1), doubling the coefficients and squaring the K_c would have made its value even smaller.

 Practice Exercise The industrial production of ammonia for fertilizers involves the reaction of nitrogen and hydrogen. If $K_c = 2.4 \times 10^{-3}$ at 1000 K for the reaction

$$N_2(g) + 3\, H_2(g) \rightleftharpoons 2\, NH_3(g)$$

what is K_c at 1000 K for this reaction?

$$\tfrac{1}{3} N_2(g) + H_2(g) \rightleftharpoons \tfrac{2}{3} NH_3(g)$$

In Sample Exercise 14.6 we saw that two equivalent ways to balance a chemical equation for the same reaction produced two different K_c expressions and two different K_c values. Surely the same ingredients should be present in the same proportions at equilibrium, no matter how we choose to write a balanced equation describing their reaction. In fact, they are. The difference between the K values is not chemical; it is only mathematical. It does not affect the composition of a reaction mixture at equilibrium—only the arithmetic we use to predict it.

Combining K Values

In Chapter 5 we applied Hess's law to calculate the enthalpies of combined reactions. We can also combine K values to obtain an overall K for a reaction that is the sum of two or more other reactions.

Consider two reactions, labeled (1) and (2) below, that were introduced in Chapter 13 and are involved in the formation of photochemical smog, wherein the NO produced in a car's engine at high temperature is oxidized to NO_2 in the atmosphere:

$$(1) \qquad N_2(g) + O_2(g) \rightleftharpoons 2\,\cancel{NO(g)}$$

$$(2) \qquad 2\,\cancel{NO(g)} + O_2(g) \rightleftharpoons 2\, NO_2(g)$$

$$\text{Overall:} \qquad N_2(g) + 2\, O_2(g) \rightleftharpoons 2\, NO_2(g)$$

The K_c expressions for the reactions (1) and (2) are:

$$K_1 = \frac{[NO]^2}{[N_2][O_2]} \quad \text{and} \quad K_2 = \frac{[NO_2]^2}{[NO]^2[O_2]}$$

How are these two individual equilibrium expressions related to the K_c expression for the overall reaction? We can derive the overall expression by multiplying K_1 and K_2 together:

$$K_1 \times K_2 = \frac{\cancel{[NO]^2}}{[N_2][O_2]} \times \frac{[NO_2]^2}{\cancel{[NO]^2}[O_2]} = \frac{[NO_2]^2}{[N_2][O_2]^2} = K_{overall}$$

This approach works for all series of reactions, and as a general rule

$$K_{overall} = K_1 \times K_2 \times K_3 \times K_4 \times \ldots \times K_n \qquad (14.9)$$

The overall equilibrium constant for a sum of two or more reactions is the product of the equilibrium constants of the individual reactions. Thus the value of K_c for the overall reaction for the formation of NO_2 from N_2 and O_2 at 1000 K is the product of the equilibrium constants for reaction 1:

$$K_1 = \frac{[NO]^2}{[N_2][O_2]} = 7.2 \times 10^{-9}$$

and reaction 2:

$$K_2 = \frac{[NO_2]^2}{[NO]^2[O_2]} = 0.020$$

Using Equation 14.9:

$$K_{overall} = K_1 \times K_2 = 7.2 \times 10^{-9} \times 0.020 = 1.4 \times 10^{-10}$$

The equilibrium constant expression for the overall reaction must contain the appropriate terms for the products and reactants of that reaction. Just as with Hess's law in thermochemical calculations, we may need to reverse an equation or multiply an equation by a factor when we combine it with another to create the right overall equation. Keep in mind that if we reverse a reaction, we must take the reciprocal of its K, and if we multiply or divide a reaction by n, we must raise its K to the n or $1/n$ power.

SAMPLE EXERCISE 14.7 Calculating an Overall K Value   **LO4**

At 1000 K, the K_c value of the following reaction is 1.5×10^6:

$$(1) \quad N_2O_4(g) \rightleftharpoons 2\,NO_2(g)$$

The K_c value at 1000 K of the following reaction is 1.4×10^{-10}:

$$(2) \quad N_2(g) + 2\,O_2(g) \rightleftharpoons 2\,NO_2(g)$$

What is the K_c value at 1000 K of the following reaction?

$$N_2(g) + 2\,O_2(g) \rightleftharpoons N_2O_4(g)$$

Collect and Organize We are given two reactions and their K_c values, and we need to combine the two reaction equations in such a way that they add up to a third reaction equation. When two reaction equations are added, their K_c values are multiplied together to obtain $K_{overall}$.

Analyze The product of the overall reaction, N_2O_4, is a reactant in reaction (1), so we need to reverse reaction (1) before combining it with reaction (2):

Reaction (1) reversed	$\cancel{2\,NO_2(g)} \rightleftharpoons N_2O_4(g)$
+ Reaction (2)	$N_2(g) + 2\,O_2(g) \rightleftharpoons \cancel{2\,NO_2(g)}$
Overall:	$N_2(g) + 2\,O_2(g) \rightleftharpoons N_2O_4(g)$

The K_c value of any reaction running in reverse is the reciprocal of the K_c value of the forward reaction. The K_c value of reaction (1) is large ($>10^6$), which means its reciprocal is small ($<10^{-6}$). The product of this small value and the smaller one for reaction (2), $\sim10^{-10}$, should be even smaller still—about 10^{-16}.

Solve

$$K_{overall} = \frac{1}{K_1} \times K_2 = \frac{1}{1.5 \times 10^6} \times 1.4 \times 10^{-10} = 9.3 \times 10^{-17}$$

Think About It The calculated $K_{overall}$ value is close to our estimate and indeed very small. Therefore the overall equilibrium lies far to the left, meaning very little N_2O_4 forms from N_2 and O_2 at 1000 K.

 Practice Exercise Calculate the value of K_c for the hypothetical reaction

$$W(g) + X(g) \rightleftharpoons M(g)$$

from the following information:

$2\,M(g) \rightleftharpoons Z(g)$	$K_c = 6.2 \times 10^{-4}$
$Z(g) \rightleftharpoons 2\,W(g) + 2\,X(g)$	$K_c = 5.6 \times 10^{-2}$

To summarize the key points for manipulating equilibrium constants:

- The K value of a reaction running in reverse is the reciprocal of the K of the forward reaction.
- If the original chemical equation describing a reversible reaction is multiplied by n, the value of K is the value of the original K raised to the nth power.
- If the original chemical equation describing a reversible reaction is divided by n, the value of K is the value of the original K raised to the $(1/n)$ power.
- If an overall chemical reaction is the sum of two or more other reactions, the overall value of K is the product of the K values of the other reactions.

14.5 Equilibrium Constants and Reaction Quotients

In Section 14.2 we introduced two key terms: *equilibrium constant expression* and *mass action expression*. Until now we have used the first term almost exclusively because we have been dealing with chemical reactions that have achieved equilibrium. Now we want to reintroduce the concept of a mass action expression because we can apply it not only to concentrations (or partial pressures) of products and reactants in reaction mixtures that have reached equilibrium, but also to reaction mixtures that are on their way to equilibrium but are not yet there.

Even if a reversible chemical reaction has not reached equilibrium, we can still insert reactant and product concentrations (or partial pressures) into its mass action expression. The mathematical result is not a K value because the reaction is not yet at equilibrium. Instead it is a Q value, where Q stands for **reaction quotient**. The value of Q provides us with a status report on how a pair of reversible reactions are proceeding.

To see how this works, let's once again consider the water–gas shift reaction and the concentration data from Experiment 4 in Table 14.1:

$$CO(g) + H_2O(g) \rightleftharpoons CO_2(g) + H_2(g) \qquad K_c = 21 \text{ at } 350°C$$

Inserting the initial concentration values into the mass action expression for the reaction yields a value for Q based on concentration; that is, Q_c:

$$Q_c = \frac{[H_2][CO_2]}{[H_2O][CO]} = \frac{(0.0300)(0.0400)}{(0.0100)(0.0200)} = 6.00$$

Let's compare this value of Q_c with the reaction's K_c value, which is 21. Clearly Q_c is less than K_c, which means there are smaller concentrations of products and larger concentrations of reactants in the initial reaction mixture than there will be at equilibrium. To achieve equilibrium, some of the reactants must form products. Mathematically, that will increase the numerator and decrease the denominator in Q_c, thereby increasing the value of Q_c until it matches the value of K_c and the equilibrium concentrations in Experiment 4 are present in the reaction vessel.

To put the results from the data in Table 14.1 in context, let's consider the curves in Figure 14.4. Starting at the left side of the graph in zone (a), we have the initial conditions of Experiment 1 from Table 14.1: equal concentrations of reactants are present, but no products. Over time, reactant concentrations (the red curve) decrease as product concentrations (the blue curve) increase. At the right side of the graph in zone (c), we have the initial conditions of Experiment 2: products are present, but no reactants. Over time, reactant concentrations increase as

> **reaction quotient (Q)** the numerical value of the mass action expression based on the concentrations or partial pressures of the reactants and products present at any time during a reaction. At equilibrium, $Q = K$.

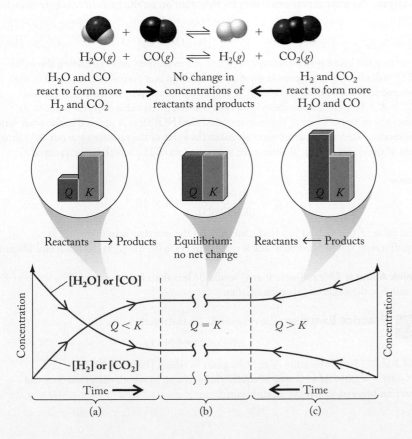

FIGURE 14.4 The value of the reaction quotient Q in relation to the equilibrium constant K for the water–gas shift reaction. (a) Reactant concentrations (red) are higher than they are once equilibrium is reached, and product concentrations (blue) are lower than at equilibrium; $Q < K$, and the reactants form more products. (b) Equilibrium concentrations are achieved; $Q = K$, and no net change in concentrations takes place. (c) Product concentrations are higher than what they are at equilibrium, and reactant concentrations are lower than at equilibrium; $Q > K$, and products form more reactants as the reaction runs in reverse.

TABLE 14.3 Comparison of Q and K Values

Value of Q	What It Means
$Q < K$	Reaction as written proceeds in forward direction ($\rightarrow$)
$Q = K$	Reaction is at equilibrium ($\rightleftharpoons$)
$Q > K$	Reaction as written proceeds in reverse direction ($\leftarrow$)

product concentrations decrease. In zone (b) in the middle of the graph, no net change occurs in the composition because the reaction is at equilibrium.

We can also characterize the three zones in Figure 14.4 on the basis of the value of Q compared with K. In zone (a), Q values are less than K and there is a net conversion of reactants into products as the forward reaction dominates. In zone (c), Q values are greater than K and a net conversion of products into reactants takes place as the reverse reaction dominates. In the middle zone (b), $Q = K$ and no change in the composition of the reaction mixture occurs over time. The relative values of Q and K and their consequences are summarized in Table 14.3.

CONCEPT TEST

In which of the three zones in Figure 14.4 do the initial reaction conditions of Experiments 3 and 4 from Table 14.1 fall?

SAMPLE EXERCISE 14.8 Using Q and K Values to Predict the Direction of a Reaction   **LO2**

At 2300 K the value of K_c of the following reaction is 1.5×10^{-3}:

$$N_2(g) + O_2(g) \rightleftharpoons 2\,NO(g)$$

At the instant when a reaction vessel at 2300 K contains 0.50 M N$_2$, 0.25 M O$_2$, and 0.0042 M NO, is the reaction mixture at equilibrium? If not, in which direction will the reaction proceed to reach equilibrium?

Collect and Organize We are asked whether a reaction mixture is at equilibrium. We are given the value of K and the concentrations of reactants and product.

Analyze The mass action expression for this reaction on the basis of concentrations is

$$Q_c = \frac{[NO]^2}{[N_2][O_2]}$$

Inserting the given concentration values into this expression and calculating the value of Q_c will allow us to determine whether the reaction is at equilibrium ($Q_c = K_c$), will proceed in the forward direction ($Q_c < K_c$), or will proceed in the reverse direction ($Q_c > K_c$). To estimate our answer, we note that the given value of [NO] is about 10^{-2} times the concentrations of the reactants, and the [NO] term is squared in the mass action expression. These two factors together make the value of the numerator about 10^{-4} times that of the denominator. Therefore the value of Q should be less than the value of K.

Solve

$$Q = \frac{(0.0042)^2}{(0.50)(0.25)} = 1.4 \times 10^{-4}$$

The value of Q is indeed less than that of K, so the reaction mixture is not at equilibrium. It will proceed in the forward direction (to the right) to reach equilibrium.

Think About It Our estimate that Q would be less than K was correct. The value of K is small, but the value of Q is even smaller.

 Practice Exercise The value of K_c for the reaction

$$2\,NO_2(g) \rightleftharpoons N_2O_4(g)$$

is 4.7 at 373 K. Is a mixture of the two gases in which [NO$_2$] = 0.025 M and [N$_2$O$_4$] = 0.0014 M in chemical equilibrium? If not, in which direction does the reaction proceed to achieve equilibrium?

homogeneous equilibria equilibria that involve reactants and products in the same phase.

heterogeneous equilibria equilibria that involve reactants and products in more than one phase.

14.6 Heterogeneous Equilibria

So far in this chapter we have focused on reactions in the gas phase. However, the principles of chemical equilibrium also apply to reactions that contain solids and liquids, particularly reactions in solution. Equilibria in which products and reactants are all in the same phase are called **homogeneous equilibria**. Equilibria in which reactants and products are in different phases are **heterogeneous equilibria**.

Let's consider a heterogeneous reaction that plays a key role in preventing SO_2 produced by coal-burning power plants from escaping into the atmosphere. The reaction occurs when tiny particles of solid calcium oxide are sprayed into exhaust gases that contain SO_2 gas:

$$CaO(s) + SO_2(g) \rightleftharpoons CaSO_3(s)$$

The large quantities of lime (CaO) needed for this reaction and for many other industrial and agricultural uses come from heating pulverized limestone, which is mostly $CaCO_3$, in kilns (Figure 14.5) operated at temperatures near 900°C. At these temperatures, $CaCO_3$ decomposes into solid lime and CO_2 gas in another reversible two-phase reaction:

$$CaCO_3(s) \rightleftharpoons CaO(s) + CO_2(g)$$

We might be tempted to write the concentration-based equilibrium constant expression for this reaction as follows:

$$K_c = \frac{[CaO][CO_2]}{[CaCO_3]}$$

However, this expression contains concentration terms for two solids: CaO and $CaCO_3$. Any pure solid has a constant concentration because it has a characteristic density, which means a constant mass (and number of moles) per unit volume. As long as there is any CaO or $CaCO_3$ present, there is no change in the "concentration" of either substance. Consequently, we remove them from the equilibrium constant expression, leaving us with

$$K_c = [CO_2]$$

This expression means that, as long as some CaO and $CaCO_3$ are present, the concentration of CO_2 gas does not vary at a given temperature, as shown in Figure 14.6. Instead, the concentration of CO_2 is the same as the value of K_c at that temperature.

(a)

(b)

FIGURE 14.5 (a) For centuries, primitive kilns were used to decompose limestone ($CaCO_3$) into lime (CaO) and CO_2. These kilns were charged with layers of fuel (originally wood, later coal) and crushed limestone, and the fuel was ignited. Farmers use lime to "sweeten" (neutralize the acidity of) soil. It is also used to make mortar and cement. (b) A modern plant for converting limestone into lime.

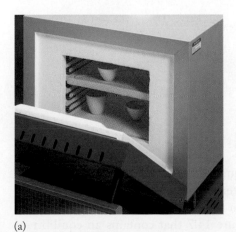

(a)

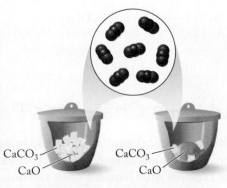

$CaCO_3$
CaO

$CaCO_3$
CaO

(b)

FIGURE 14.6 Two views of the thermal decomposition of $CaCO_3$ to CaO and CO_2. (a) Muffle furnace for heating crucibles containing $CaCO_3$ samples. (b) After heating and decomposition, the two crucibles now contain both $CaCO_3$ and CaO. Note that different crucibles with different amounts of $CaCO_3$ and CaO still have the same equilibrium concentration of CO_2 gas.

The concept of constant concentration also applies to pure liquids involved in reversible chemical reactions. As long as the liquid is present, its "concentration" is considered constant during the reaction and does not appear in the equilibrium constant expression. Similarly, most K expressions for reactions in aqueous solutions do not include a term for $[H_2O]$, even when water is a reactant or product, because its concentration does not change significantly. In summary, in writing equilibrium constant expressions for heterogeneous equilibria, we follow the rules we learned earlier, with the additional rule that pure solids and liquids do not appear in the expression.

SAMPLE EXERCISE 14.9 Writing Equilibrium Constant Expressions for Heterogeneous Equilibria **LO1**

Write K_c expressions for the following reactions:

a. $CaO(s) + SO_2(g) \rightleftharpoons CaSO_3(s)$
b. $CO_2(g) + H_2O(\ell) \rightleftharpoons H_2CO_3(aq)$

Collect, Organize, and Analyze We are asked to write equilibrium constant expressions for equilibria involving reactants and products in more than one phase. We do not include terms for liquids and solids whose concentrations do not change significantly. The first equilibrium involves two solids: CaO and $CaSO_3$. The second involves liquid H_2O.

Solve

a. An expression with terms for all reactants and products is

$$K_c = \frac{[CaSO_3]}{[CaO][SO_2]}$$

After removing the terms representing solids, we have

$$K_c = \frac{1}{[SO_2]}$$

b. In the equilibrium constant expression for reaction b, $[H_2O]$ is nearly constant and is not included, leaving

$$K_c = \frac{[H_2CO_3]}{[CO_2]}$$

Think About It When CO_2 dissolves in water, the water is, technically, no longer a pure liquid. However, its decrease in "concentration" is much too small to be significant, so there is no $[H_2O]$ term in the K_c expression in part (b).

Practice Exercise Write K_p expressions for the following reactions:

a. $C(s) + CO_2(g) \rightleftharpoons 2\,CO(g)$
b. $CO_2(g) + H_2(g) \rightleftharpoons CO(g) + H_2O(\ell)$

14.7 Le Châtelier's Principle

Once a chemical reaction has come to equilibrium, the composition of the system remains unchanged as long as no external forces perturb it. In this section we examine what happens when systems at equilibrium are perturbed. Consider, for example, the flask on the left in Figure 14.7 that contains an equilibrium mixture of two gases: NO_2 and its dimer, N_2O_4, at 25°C. NO_2 is a brown gas;

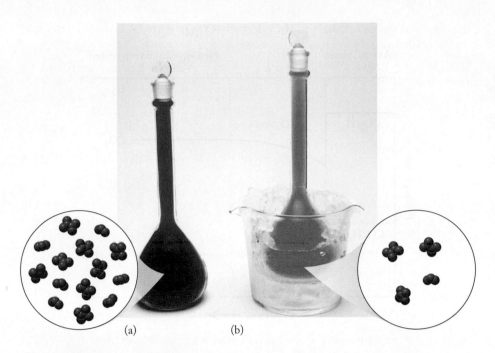

(a)   (b)

N_2O_4 is colorless. The corresponding dimerization reaction and its equilibrium constant are:

$$2\ NO_2(g) \rightleftharpoons N_2O_4(g) \qquad K_p = 6.9$$

The flask on the right in Figure 14.7 initially contained the same mixture of NO_2 and N_2O_4 as the one on the left, but was then placed in an ice bath at 0°C. Why did the temperature change cause a color change in the flask on the right? What other factors can perturb an equilibrium, and how do we explain any changes in the system?

One of the first scientists to study and then successfully predict how chemical equilibria respond to such perturbations was French chemist Henri Louis Le Châtelier (1850–1936). He articulated **Le Châtelier's principle**, which states that, if a system at equilibrium is perturbed (or subjected to an external *stress*), the position of the equilibrium shifts in either the forward or reverse direction as a response to reduce that stress. We now consider the effects of several stresses, including changes in concentration, pressure, volume, and temperature.

Effects of Adding or Removing Reactants or Products

When a reactant or product is added or removed, a system at chemical equilibrium is perturbed. Following Le Châtelier's principle, the system responds in such a way as to restore equilibrium. Through the years, chemists have used Le Châtelier's principle to increase the yields of chemical reactions that would otherwise have provided very little of a desired product.

To explore how industrial chemists exploit Le Châtelier's principle, let's look again at the water–gas shift reaction for making hydrogen:

$$CO(g) + H_2O(g) \rightleftharpoons CO_2(g) + H_2(g) \qquad (14.2)$$

To perturb the equilibrium and shift the system toward the production of more H_2 (Figure 14.8), chemists pass the reaction mixture through a scrubber containing a concentrated aqueous solution of K_2CO_3. Doing this removes CO_2 from the gaseous mixture because of the following reaction:

$$CO_2(g) + H_2O(\ell) + K_2CO_3(aq) \rightleftharpoons 2\ KHCO_3(s)$$

CHEMTOUR
Le Châtelier's Principle

Le Châtelier's principle the principle that a system at equilibrium responds to a stress in such a way that it relieves that stress.

FIGURE 14.8 Removing CO_2 from an equilibrium mixture of the water–gas shift reaction results in fewer collisions between CO_2 and H_2 (slows the rate of the reverse reaction) while the forward reaction continues at the same rate and generates additional CO_2 and H_2 until the mixture establishes a new equilibrium.

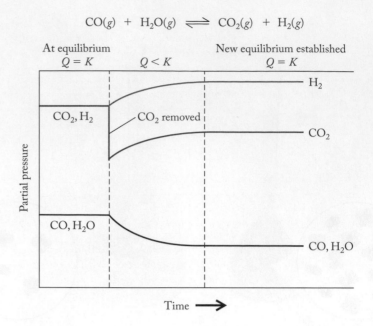

$$CO(g) \;+\; H_2O(g) \;\rightleftharpoons\; CO_2(g) \;+\; H_2(g)$$

Removing CO_2 means fewer molecules of it are available to collide with molecules of H_2 in the reverse reaction in Equation 14.2. As a result, the rate of the reverse reaction becomes slower than the rate of the forward reaction. This means the system is no longer in equilibrium. The same conclusion is reached by calculating the value of the reaction quotient, Q_p:

$$Q_p = \frac{(P_{H_2})(P_{CO_2})}{(P_{CO})(P_{H_2O})}$$

after much of the CO_2 has been removed. The depleted P_{CO_2} term in the numerator results in a Q_p value that is less than K_p, which means that the reaction proceeds in the forward direction to restore equilibrium. Chemists say the reaction *shifts to the right*, making more CO_2 to replace some of what was removed in the reaction with K_2CO_3, and in the process making more H_2 gas.

Another way to shift a reaction mixture at equilibrium is to add more reactant or product. If the goal is to form more products, then adding more reactants increases their concentration in the reaction mixture, which increases the rate of the forward reaction. Some of the added reactants are converted into additional products. This approach works even when only one of multiple reactants is added, as long as there are sufficient quantities of all the other reactants. This makes sense mathematically because adding one reactant increases the value of one of the terms in the denominator of the mass action expression. Therefore Q is reduced to a value less than K and the reaction proceeds in the forward reaction, forming more products until equilibrium is restored.

SAMPLE EXERCISE 14.10 Adding or Removing Reactants or Products to Stress an Equilibrium **LO5**

Suggest three ways the production of ammonia via the reaction

$$N_2(g) + 3\,H_2(g) \rightleftharpoons 2\,NH_3(g)$$

can be increased without changing the reaction temperature.

Collect and Organize We are given the balanced chemical equation of a reversible reaction. We are asked to suggest three ways to increase its yield—that is, to shift its equilibrium to the right.

Analyze The mass action expression for this reaction is

$$Q_p = \frac{(P_{NH_3})^2}{(P_{H_2})^3(P_{N_2})}$$

Changes in the reaction system that reduce the value of Q_p, so that $Q_p < K_p$, will result in the formation of more NH_3.

Solve Interventions that (1) increase the partial pressure of N_2, or (2) increase the partial pressure of H_2, or (3) remove NH_3 from the system will all shift the equilibrium to the right and increase the production of ammonia.

Think About It The industrial synthesis of ammonia relies on shifting the equilibrium to the right by (1) running the reaction at high partial pressures of the reactants and (2) removing the product NH_3 by passing the reaction mixture through chilled condensers. Chilling the mixture removes ammonia because it condenses at a higher temperature than N_2 or H_2.

 Practice Exercise Describe the changes that occur in a gas-phase equilibrium according to the reaction

$$2\,H_2S(g) + 3\,O_2(g) \rightleftharpoons 2\,SO_2(g) + 2\,H_2O(g)$$

if (a) the reaction mixture is cooled and water vapor condenses; (b) SO_2 gas dissolves in liquid water as it condenses; or (c) more O_2 is added.

Effects of Pressure and Volume Changes

A reaction involving gaseous reactants or products may be perturbed by altering the volume of the system, thereby changing the partial pressures of the reactants and products in accordance with Boyle's law. To see how, let's revisit the equilibrium between NO_2 and its dimer, N_2O_4:

$$2\,NO_2(g) \rightleftharpoons N_2O_4(g)$$

This time, let's examine a syringe that is filled with an equilibrium mixture of the two gases, as shown in Figure 14.9(a). The plunger on the syringe is then pushed downward until the volume of the reaction mixture is half what it was initially. If the composition of the mixture did not change, then the partial pressures of the gases would both double because the same number of molecules would be colliding in half the volume; this is consistent with Boyle's law, which states that the pressure of an ideal gas is inversely proportional to its volume. However, the composition of the mixture does change because doubling the partial pressures of both gases perturbs the equilibrium of the reaction between them. To understand why, let's assume the initial equilibrium partial pressures of NO_2 and N_2O_4 are x and y, respectively. Then the value of K_p is

$$K_p = \frac{(P_{N_2O_4})}{(P_{NO_2})^2} = \frac{y}{x^2}$$

Now consider what happens when the volume of the gases is halved and their partial pressures double to $P_{NO_2} = 2x$ and $P_{N_2O_4} = 2y$.

FIGURE 14.9 Changing pressure affects equilibrium in a gas-phase reaction. (a) A gastight syringe contains an equilibrium reaction mixture of brown $NO_2(g)$ and colorless $N_2O_4(g)$. (b) The plunger is rapidly pushed in, decreasing the sample volume by half. The color of the mixture should be darker brown as the pressure of NO_2 molecules doubles. However, there is little color change because brown NO_2 rapidly forms colorless N_2O_4, restoring chemical equilibrium.

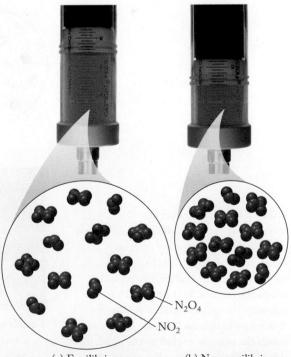

N$_2$O$_4$

NO$_2$

(a) Equilibrium   (b) New equilibrium

Inserting these pressures into the mass action expression gives us a reaction quotient that is half the value of K_p:

$$Q_p = \frac{2y}{(2x)^2} = \frac{2y}{4x^2} = \frac{1}{2} K_p$$

Because $Q_p < K_p$, the reaction proceeds in the forward direction, consuming reactants and forming products. The fact that the colors of the two samples are nearly the same confirms that most of the increase in P_{NO_2} that occurred when the sample was squeezed into half its initial volume was lost as the reaction proceeded in the forward direction.

Did you notice that converting NO_2 into N_2O_4 reduces the total number of moles of gas in the reaction mixture because 2 moles of NO_2 are consumed for every 1 mole of N_2O_4 produced? Fewer moles of gas results in fewer collisions, which ultimately reduces pressure. Whenever a reaction mixture at equilibrium is compressed—and the partial pressures of its gas-phase reactants and products increase—the equilibrium shifts toward the side of the reaction equation with fewer moles of gases. In doing so, the reaction mixture relieves some of the stress induced by the increase in pressure due to compression. Conversely, increasing the volume of a system at equilibrium at constant temperature results in fewer collisions, lowers the partial pressures of all the gaseous reactants and products, and shifts the equilibrium toward the side of the reaction equation with more moles of gases (Figure 14.10).

CONCEPT TEST

Changing the overall volume of the reaction of the water–gas shift reaction at equilibrium:

$$CO(g) + H_2O(g) \rightleftharpoons CO_2(g) + H_2(g)$$

does not shift the equilibrium. Why?

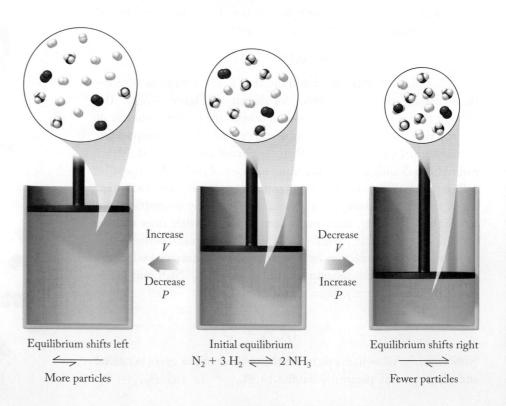

FIGURE 14.10 The effect of compressing and expanding an equilibrium mixture of the gases in the reaction $N_2(g) + 3\,H_2(g) \rightleftharpoons 2\,NH_3(g)$. An increase in volume (decrease in pressure) causes the reaction to shift to the left: the side with the greater number of particles. A decrease in volume (increase in pressure) causes the reaction to shift to the right: the side with the smaller number of particles.

Increase
V

Decrease
P

Decrease
V

Increase
P

Equilibrium shifts left
$\rightleftharpoons$
More particles

Initial equilibrium
$N_2 + 3\,H_2 \rightleftharpoons 2\,NH_3$

Equilibrium shifts right
$\rightleftharpoons$
Fewer particles

SAMPLE EXERCISE 14.11 Assessing Volume Effects   **LO5**
on Gas-Phase Equilibria

In which of the following equilibria would compressing an equilibrium reaction mixture promote the formation of more product(s)?

a. $N_2(g) + O_2(g) \rightleftharpoons 2\,NO(g)$
b. $2\,NO(g) + O_2(g) \rightleftharpoons 2\,NO_2(g)$
c. $H_2O(\ell) + CO_2(g) \rightleftharpoons H_2CO_3(aq)$
d. $CaCO_3(s) \rightleftharpoons CaCO(s) + CO_2(g)$

Collect, Organize, and Analyze We are asked to identify the reactions for which higher pressure causes an increase in product formation. We know that compression shifts a chemical equilibrium involving gases toward the side of the reaction with fewer moles of gas.

Solve Summing the number of moles of gas on the reactant side and product side in each reaction, we have:

Reaction	Moles of Gaseous Reactants	Moles of Gaseous Products
a	2	2
b	3	2
c	1	0
d	0	1

The only two reactions with fewer moles of gaseous products than reactants are reactions (b) and (c). Therefore they are the only two in which compression increases product formation.

Think About It In reaction (a), the number of moles of gas on the reactant side is the same as the number of moles on the product side, so changing pressure in either direction would not cause a shift in equilibrium. In reaction (d), the number of moles on the product side is greater than on the reactant side, so compression favors the reverse of the reaction as written and decreases product formation.

 Practice Exercise How does compressing a reaction mixture of CO, Cl_2, and $COCl_2$ affect the following equilibrium?

$$CO(g) + Cl_2(g) \rightleftharpoons COCl_2(g)$$

Effect of Temperature Changes

Let's return now to Figure 14.7 and reconsider what happens when a volumetric flask containing an equilibrium mixture of NO_2 and N_2O_4 at 25°C is placed into an ice bath. The color of the gas mixture in Figure 14.7(b) becomes much lighter, indicating that the concentration of NO_2 has decreased. Why? For one thing, the boiling point of N_2O_4 is 21°C. Therefore lowering the temperature from 25°C to 0°C removed most of the dimer from the gas phase, creating a pool of colorless N_2O_4 liquid in the bottom of the flask. The resulting decrease in the concentration and partial pressure of N_2O_4 makes the value of Q_p smaller than K_p, and the reaction shifts toward the formation of additional N_2O_4. This shift drives down the concentration of NO_2 and makes the gas phase in Figure 14.7(b) a lighter color.

CONNECTION In Chapter 6 we discussed Amontons's law, which states that the pressure of an ideal gas is proportional to its absolute temperature.

At the beginning of this chapter we noted how important the production of ammonia is for the world's food supply. Let's consider the effect of decreasing the temperature upon the synthesis of ammonia:

$$N_2(g) + 3\,H_2(g) \rightleftharpoons 2\,NH_3(g) \qquad \Delta H^\circ_{rxn} = -92.2\ kJ/mol$$

If we think of energy as a product of the forward exothermic reaction, then lowering the temperature of the reaction mixture favors the forward reaction, whereas increasing the temperature favors the reverse reaction. The situation is reversed for endothermic reactions, where raising the temperature would favor the forward reaction.

There is an important difference between applying Le Châtelier's principle to explain shifts in equilibrium position when changing concentrations or partial pressures of reactants or products versus applying Le Châtelier's principle to explain temperature changes: *changes in temperature change the value of K*. Increasing temperature reduces the yield of ammonia synthesis because increasing temperature reduces the value of *K*.

In general, the value of *K* decreases as temperature increases for exothermic reactions and increases as temperature increases for endothermic reactions. We revisit the influence of temperature on *K* values in Chapter 17, but for now this general analysis enables us to predict the direction of shifts in equilibria with changing temperature.

Temperature = 5°C   Temperature = 75°C

$$\underbrace{Co(H_2O)_6{}^{2+}(aq)}_{\text{Pink}} + 4\,Cl^-(aq) \rightleftharpoons \underbrace{CoCl_4{}^{2-}(aq)}_{\text{Royal blue}} + 6\,H_2O(\ell)$$

FIGURE 14.11 Two cobalt(II) species, one pink and one blue, are in equilibrium in aqueous hydrochloric acid solution. The equilibrium shifts in favor of the blue species as temperature increases.

SAMPLE EXERCISE 14.12 Predicting Changes in Equilibrium   **LO5**
with Changing Temperature

The color of cobalt(II) chloride dissolved in dilute hydrochloric acid depends on temperature, as shown in Figure 14.11. The solution is pink at 0°C, magenta at 25°C, and dark blue at 75°C. Is the reaction that produces the pink-to-blue color change exothermic or endothermic?

Collect, Organize, and Analyze We are asked to determine whether the reaction in Figure 14.11 is exothermic or endothermic. If it is exothermic, then increasing its temperature should shift the reaction toward the formation of the pink reactant. If the reaction is endothermic, then increasing temperature should shift the reaction toward the formation of the blue product.

Solve The blue product is favored at higher temperatures, so the reaction as written must be endothermic.

Think About It The fact that the reaction mixture in this exercise is magenta at room temperature tells us that both the pink and blue forms are present. This inference suggests that the value of *K* at room temperature is close to 1, but the concentration of $Cl^-(aq)$ is also an important factor because it is raised to the fourth power in the equilibrium constant expression.

 Practice Exercise Does the value of the equilibrium constant of the reaction

$$N_2(g) + O_2(g) \rightleftharpoons 2\,NO(g) \qquad \Delta H^\circ_{rxn} = 180.6\ kJ/mol$$

increase or decrease with increasing temperature?

Table 14.4 summarizes how a generic gas-phase reversible reaction in which 2 moles of reactants form 1 mole of product responds to various stresses.

TABLE 14.4 Response of the Reaction 2 A(*g*) ⇌ B(*g*) to Different Kinds of Stress

Kind of Stress	How System Responds	Direction of Shift
Add A	Consume A	To the right
Remove A	Produce A	To the left
Remove B	Produce B	To the right
Add B	Consume B	To the left
Compress the reaction mixture	Reduce moles of gas	To the right
Increase the volume of the reaction mixture	Increase moles of gas	To the left

CONNECTION In Chapter 13 we learned that a catalyst simultaneously increases the rate of a reaction in both the forward and reverse directions, when we discussed the effect of temperature on the rates of reactions occurring in a catalytic converter.

Catalysts and Equilibrium

The industrial production of ammonia was developed by the German chemists Fritz Haber (1868–1934) and Carl Bosch (1874–1940) in the early 20th century and is still widely referred to as the Haber–Bosch process. This process is commercially feasible because of the use of iron-based catalysts. As discussed in Chapter 13, a catalyst increases the rate of a chemical reaction by lowering its activation energy. Knowing that the synthesis of ammonia involves both equilibrium and a catalyst raises this question: if a catalyst increases the rate of a reaction, does that catalyst affect the equilibrium constant of the reaction?

To answer this question, consider the energy profiles of the catalyzed and uncatalyzed reaction in Figure 14.12. The catalyst increases the rate of the reaction by decreasing the height of the energy barrier. However, the barrier height is reduced by the same amount whether the reaction proceeds in the forward direction or in reverse. As a result, the increase in reaction rate produced by the catalyst is the same in both directions. Therefore a catalyst has no effect on the equilibrium constant of a reaction or on the composition of an equilibrium reaction mixture. A catalyst does, however, shorten the time needed for a system to reach equilibrium.

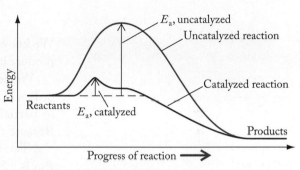

FIGURE 14.12 The effect of a catalyst on a reaction. A catalyst lowers the activation energy barrier, and as a result the rate of the reaction increases. However, because both the forward reaction and the reverse reaction occur more rapidly, the position of equilibrium (that is, the value of *K*) does not change. The system comes to equilibrium more rapidly, but the relative amounts of product and reactant present at equilibrium do not change.

14.8 Calculations Based on *K*

Reference books and the tables in Appendix 5 of this book contain lists of equilibrium constants for chemical reactions. We can use these values in several kinds of calculations, including those in which:

1. We want to determine whether a reaction mixture has reached equilibrium (Sample Exercise 14.8).
2. We know the value of *K* and the starting concentrations or partial pressures of reactants and/or products, and we want to calculate their equilibrium concentrations or pressures.

In this section we focus on the second type of calculation and introduce a useful way of handling such problems: a table of reactant and product concentrations (or partial pressures) called a *RICE table*. The acronym RICE means that the table starts with the balanced chemical equation describing the **R**eaction followed by

CHEMTOUR
Solving Equilibrium Problems

rows that contain **I**nitial concentration values, **C**hanges in those initial values as the reaction proceeds toward equilibrium, and **E**quilibrium values.

In our first example, we calculate how much hydrogen iodide forms from the reaction of hydrogen gas and iodine gas at 445°C. At this temperature, the K_p value of the reaction is 50.2:

$$H_2(g) + I_2(g) \rightleftharpoons 2\,HI(g)$$

The initial partial pressures are $P_{H_2} = 1.00$ atm and $P_{I_2} = 1.03$ atm, and no HI is present.

We start by writing the given (initial) information in our RICE table:

Reaction (R)	$H_2(g)$ +	$I_2(g)$ $\rightleftharpoons$	$2\,HI(g)$
	P_{H_2} (atm)	P_{I_2} (atm)	P_{HI} (atm)
Initial (I)	1.00	1.03	0
Change (C)			
Equilibrium (E)			

We know the reaction will proceed in the forward direction because there is no product initially present, so $Q_p = 0 < K_p$.

We need to use algebra to fill in rows C and E. We don't know how much H_2 or I_2 will be consumed or how much HI will be made. We can define the change in partial pressure of H_2 as $-x$ because H_2 is consumed during the reaction. Because the mole ratio of H_2 to I_2 in the reaction is 1:1, the change in P_{I_2} is also $-x$. Two moles of HI are produced from each mole of H_2 and I_2, so the change in P_{HI} is $+2x$. Inserting these values in the C row, we have

Reaction (R)	$H_2(g)$ +	$I_2(g)$ $\rightleftharpoons$	$2\,HI(g)$
	P_{H_2} (atm)	P_{I_2} (atm)	P_{HI} (atm)
Initial (I)	1.00	1.03	0
Change (C)	$-x$	$-x$	$+2x$
Equilibrium (E)			

Combining the values in the I and C rows, we obtain the three partial pressures at equilibrium:

Reaction (R)	$H_2(g)$ +	$I_2(g)$ $\rightleftharpoons$	$2\,HI(g)$
	P_{H_2} (atm)	P_{I_2} (atm)	P_{HI} (atm)
Initial (I)	1.00	1.03	0
Change (C)	$-x$	$-x$	$+2x$
Equilibrium (E)	$1.00 - x$	$1.03 - x$	$2x$

The next step is to substitute the terms from the E row into the K_p expression for the reaction:

$$K_p = \frac{(P_{HI})^2}{(P_{H_2})(P_{I_2})}$$

$$= \frac{(2x)^2}{(1.00 - x)(1.03 - x)}$$

Expanding the terms in the numerator and denominator gives

$$K_p = \frac{4x^2}{1.03 - 2.03 \, x + x^2} = 50.2$$

Cross-multiplying, we get

$$1.03 \times 50.2 - (2.03 \times 50.2)x + 50.2x^2 = 4x^2$$

Combining the x^2 terms and rearranging, we have

$$46.2x^2 - 101.9x + 51.7 = 0$$

You may recognize this equation as one that fits the general form of a quadratic equation:

$$ax^2 + bx + c = 0$$

We can solve for x either with a scientific calculator or by using the quadratic formula

$$x = \frac{-b \pm \sqrt{b^2 - 4ac}}{2a}$$

Two values are possible for x: 0.791 and 1.415. We focus on 0.791 atm because it is the only one of the two values that is physically possible. (Note that we start with a partial pressure of iodine gas of 1.03 atm. If we were to subtract 1.415 atm from that value, the answer would be a negative number. Because we cannot have a negative pressure, that number is a physically impossible value for x.) Using $x = 0.791$ atm to calculate the equilibrium partial pressures gives:

$$P_{H_2} = 1.00 - x$$
$$= 1.00 - 0.791 = 0.21 \text{ atm}$$
$$P_{I_2} = 1.03 - x$$
$$= 1.03 - 0.791 = 0.24 \text{ atm}$$
$$P_{HI} = 2x = 2(0.791) = 1.582 = 1.58 \text{ atm}$$

The value of x makes sense because it produces pressures of H_2 and I_2 that are less than their initial values but positive. The sum of the partial pressures of the components in the system is $(0.21 + 0.24 + 1.58) = 2.03$ atm, which is the same as the starting pressure. This result is expected, because we have 2 moles of gas on the reactant side and 2 moles of gas on the product side, so we do not expect the pressure to change as the reaction proceeds. As a final check, we can use the calculated partial pressures in the expression for K_p and calculate its value. When we do so, we calculate $K_p = (1.58)^2/[(0.21)(0.24)] = 50$, which is not significantly different from the value given of 50.2.

SAMPLE EXERCISE 14.13 Calculating an Equilibrium   **LO6**
                          Partial Pressure I

The value of K_p for the production of PCl_5 from PCl_3 and Cl_2 gas is 24.2 at 250°C. If the initial partial pressures of PCl_3 and Cl_2 are 0.43 and 0.87 atm, respectively, and no PCl_5 is initially present, what are the partial pressures of the three gases when equilibrium is achieved?

Collect and Organize We are asked to determine the equilibrium partial pressures of the three gases from the initial partial pressures of the reactants and the equilibrium constant. The chemical equation describing the reaction is

$$PCl_3(g) + Cl_2(g) \rightleftharpoons PCl_5(g) \qquad K_p = 24.2$$

Analyze The system initially contains no product, so $Q < K_p$ and the reaction will proceed in the forward direction as written. We can construct a RICE table from the information given. If $+x$ is the increase in the partial pressure of the product formed, then, given the 1:1:1 stoichiometry of the reactants and product, the decreases in both reactants will be $-x$.

Solve Setting up the RICE table in which the equilibrium (E) terms are the sum of I + C:

Reaction (R)	$PCl_3(g)$ +	$Cl_2(g)$ $\rightleftharpoons$	$PCl_5(g)$
	P_{PCl_3} (atm)	P_{Cl_2} (atm)	P_{PCl_5} (atm)
Initial (I)	0.43	0.87	0
Change (C)	$-x$	$-x$	$+x$
Equilibrium (E)	$0.43 - x$	$0.87 - x$	x

Inserting the terms from row E into the K_p expression for the reaction:

$$K_p = \frac{(P_{PCl_5})}{(P_{PCl_3})(P_{Cl_2})} = \frac{x}{(0.43 - x)(0.87 - x)}$$

Expanding the denominator and cross-multiplying:

$$K_p = \frac{x}{0.3741 - 1.30x + x^2} = 24.2$$

$$0.3741 \times 24.2 - (1.30 \times 24.2)x + 24.2x^2 = x$$

Combining the x terms and rearranging, we have

$$24.2x^2 - 32.46x + 9.05322 = 0$$

Using the quadratic equation or a scientific calculator to solve for x yields two roots: 0.3955 and 0.9458. Only 0.3955 gives positive partial pressure values for all three gases, which are

$$P_{PCl_3} = 0.43 - 0.3955 = 0.03 \text{ atm}$$

$$P_{Cl_2} = 0.87 - 0.3955 = 0.47 \text{ atm}$$

$$P_{PCl_5} = 0.3955 = 0.40 \text{ atm}$$

Think About It The values make sense because the partial pressures of both reactants have decreased and that of the product has increased. When we check our answers by substituting them into the equation for K_p:

$$K_p = \frac{(P_{PCl_5})}{(P_{PCl_3})(P_{Cl_2})} = \frac{0.40}{(0.03)(0.47)} = 28$$

we get a value that is reasonably close to the K_p value used in the calculation (24.2), given that the partial pressure of PCl_3 was rounded off to only one significant figure.

Practice Exercise Consider the reverse reaction in Sample Exercise 14.13 at the same temperature. If the vessel initially contains 1.35 atm PCl_5, what are the partial pressures of the three gases when equilibrium is achieved?

We should be alert to opportunities to simplify calculations of equilibrium concentrations of partial pressures. The following sample exercise illustrates this point.

SAMPLE EXERCISE 14.14 Calculating an Equilibrium   **LO6**
                                 Partial Pressure II

Most of the H_2 used in the Haber–Bosch process is produced by the steam–gas reforming reaction followed by the water–gas shift reaction:

$$CO(g) + H_2O(g) \rightleftharpoons CO_2(g) + H_2(g)$$

If a reaction vessel at 675 K is filled with an equimolar mixture of CO and steam, such that the partial pressure of each gas is 2.00 atm, what is the partial pressure of H_2 at equilibrium if the value of $K_p = 11.8$ at 675 K?

Collect, Organize, and Analyze We are asked to find the partial pressure of a gaseous product in an equilibrium mixture, given the initial partial pressures of reactants and the value of K_p. The system initially contains no product, which means that the reaction quotient Q_p is equal to zero. A K_p value of 11.8 means that reactants should be converted into products, but there should be significant partial pressures of both reactants and products at equilibrium. The coefficients of all the ingredients in chemical equation describing the reaction are all one, so in our RICE table we let $+x$ be the increase in the partial pressures of H_2 and CO_2 and $-x$ be the decreases in both reactants.

Solve

Reaction (R)	CO(*g*)	+	H$_2$O(*g*)	$\rightleftharpoons$	CO$_2$(*g*)	+	H$_2$(*g*)
	P_{CO} (atm)		P_{H_2O} (atm)		P_{CO_2} (atm)		P_{H_2} (atm)
Initial (I)	2.00		2.00		0.00		0.00
Change (C)	$-x$		$-x$		$+x$		$+x$
Equilibrium (E)	$2.00 - x$		$2.00 - x$		x		x

Inserting these equilibrium terms into the K_p expression:

$$K_p = \frac{(P_{CO_2})(P_{H_2})}{(P_{CO})(P_{H_2O})}$$

$$= \frac{(x)(x)}{(2.00 - x)(2.00 - x)}$$

$$= \left(\frac{x}{2.00 - x}\right)^2 = 11.8$$

Note that the partial pressure terms for both products and for both reactants are the same because in both pairs their initial pressures were the same and their changes in pressure were the same. Therefore we can simplify this equation by taking the square root of both sides:

$$\frac{x}{2.00 - x} = \sqrt{11.8} = 3.43$$

Cross-multiplying and solving for x yields $x = 1.55$ atm, which is the equilibrium partial pressure of H_2 (and CO_2).

Think About It Our prediction that significant quantities of both reactants and products would be present at equilibrium is confirmed. Let's check the accuracy of our solution by inserting the equilibrium partial pressures of H_2 and CO_2 (1.55 atm) and of steam and CO ($2.00 - 1.55 = 0.45$ atm) into the equilibrium constant expression:

$$K_p = \frac{(P_{CO_2})(P_{H_2})}{(P_{CO})(P_{H_2O})} = \frac{(1.55 \text{ atm})(1.55 \text{ atm})}{(0.45 \text{ atm})(0.45 \text{ atm})} = 11.8$$

This result matches the K_p value we started with, confirming that our calculation is correct.

> ⊛ **Practice Exercise** The chemical equation for the reaction of chlorine and bromine to produce BrCl is
>
> $$Cl_2(g) + Br_2(g) \rightleftharpoons 2\,BrCl(g)$$
>
> If the value of K_p for the reaction at a given temperature is 4.7×10^{-2}, what is the partial pressure of each reactant and product in a sealed reaction vessel if the initial partial pressures of Cl_2 and Br_2 are both 0.100 atm and no BrCl is present?

All the equilibrium calculations we have done so far have been based on K_p values in the 10^{-2} to 10^2 range. As we noted in Section 14.2, reactions with these K_p values produce reaction mixtures with significant quantities of both reactants and products at equilibrium. When equilibrium constants are very small, however, we can frequently make an approximation at the outset that enables us to simplify our mathematical operations by considering the effect that the size of an equilibrium constant has on the values in our equation. For example, the K_p value for the following reaction at 1500 K is small:

$$N_2(g) + O_2(g) \rightleftharpoons 2\,NO(g) \qquad K_p = 1.0 \times 10^{-5}$$

If the initial partial pressures of N_2 and O_2 are 0.79 and 0.21 atm, respectively, and no NO is present, we can set up a RICE table and calculate the partial pressure of NO at equilibrium as we have done previously:

Reaction (R)	$N_2(g)$	$+$	$O_2(g)$	$\rightleftharpoons$	$2\,NO(g)$
	P_{N_2} (atm)		P_{O_2} (atm)		P_{NO} (atm)
Initial (I)	0.79		0.21		0
Change (C)	$-x$		$-x$		$+2x$
Equilibrium (E)	$0.79 - x$		$0.21 - x$		$2x$

Inserting the equilibrium partial pressure terms in the K_p expression:

$$K_p = 1.0 \times 10^{-5} = \frac{(P_{NO})^2}{(P_{N_2})(P_{O_2})}$$

$$= \frac{(2x)^2}{(0.79 - x)(0.21 - x)} = \frac{4x^2}{(0.79 - x)(0.21 - x)}$$

Before we take the next usual step of expanding the equation and solving for x, let's think about the size of x. The equilibrium constant is very small, which means very little product will form. Therefore the value of x will probably be much smaller than the values of the initial partial pressures, so that $(0.79 - x)$ and $(0.21 - x)$ will be approximately the same as 0.79 and 0.21. If we assume that the $-x$ terms in the denominator are negligible and simply use the initial values for P_{N_2} and P_{O_2} instead, we obtain an equation that is much easier to solve:

$$\frac{4x^2}{(0.79)(0.21)} = 1.0 \times 10^{-5}$$

$$4x^2 = (0.79)(0.21)(1.0 \times 10^{-5}) = 1.659 \times 10^{-6}$$

$$x^2 = 4.148 \times 10^{-7}$$

$$x = 6.4 \times 10^{-4}\ \text{atm}$$

This value is in fact the same as that obtained by solving for x without neglecting the $-x$ term, at least to two significant figures, which is all we are allowed in this problem.

Generally speaking, we can ignore the $-x$ or $+x$ component of an equilibrium concentration or partial pressure term if the value of x is less than 5% of the initial value. This frequently happens when K values are small ($<\sim 3 \times 10^{-5}$) and initial concentration or partial pressure values are greater than about 0.03.

SAMPLE EXERCISE 14.15 Calculating an Equilibrium   **LO6**
                        Partial Pressure III

The value of K_p is 2.2×10^{-10} at 100°C for the decomposition of phosgene ($COCl_2$) gas:

$$COCl_2(g) \rightleftharpoons CO(g) + Cl_2(g)$$

If a sealed vessel contains only phosgene at a partial pressure of 2.75 atm, what are the partial pressures of the reactant and its decomposition products when the system comes to equilibrium?

Collect and Organize We are asked to find the partial pressures of $COCl_2$ and its decomposition products in a gas-phase equilibrium mixture. We know the initial partial pressure of $COCl_2$ and the value of K_p.

Analyze The value of K_p is very small, so we should be able to assume that the $-x$ component of the partial pressure of $COCl_2$ is negligible at equilibrium.

Solve Setting up a RICE table for this reaction system:

Reaction (R)	$COCl_2(g)$ $\rightleftharpoons$	$CO(g)$ +	$Cl_2(g)$
	P_{COCl_2} (atm)	P_{CO} (atm)	P_{Cl_2} (atm)
Initial (I)	2.75	0	0
Change (C)	$-x$	$+x$	$+x$
Equilibrium (E)	$2.75 - x$	x	x

Using the equilibrium values in the mass action expression and then simplifying and solving for x:

$$K_p = \frac{(P_{CO})(P_{Cl_2})}{(P_{COCl_2})} = 2.2 \times 10^{-10} = \frac{(x)(x)}{(2.75 - x)} \approx \frac{x^2}{(2.75)} = 2.2 \times 10^{-10}$$

$$x^2 = 2.75 \times (2.2 \times 10^{-10}) = 6.05 \times 10^{-10}$$

$$x = 2.5 \times 10^{-5}$$

Therefore at equilibrium $P_{COCl_2} = 2.75$ atm and $P_{CO} = P_{Cl_2} = 2.5 \times 10^{-5}$ atm.

Think About It The value of x is less than 10^{-5} times the initial partial pressure of $COCl_2$, so our simplifying assumption was valid.

Practice Exercise What effect will doubling the partial pressure of phosgene in the Sample Exercise have on the partial pressures of CO and Cl_2 at equilibrium?

We have seen throughout this chapter that a reversible chemical reaction has a particular equilibrium constant at a specific temperature. If we know the value of that equilibrium constant, we can calculate the concentrations of reactants and products at equilibrium from any combination of initial concentrations. Chemists use the results of such calculations to optimize the yields of chemical processes by manipulating equilibrium conditions: adjusting temperature, adding reactants as they are consumed, and harvesting products as they form.

SAMPLE EXERCISE 14.16 Integrating Concepts: Making Nitric Acid

We have mentioned the Haber–Bosch process for synthesizing ammonia, much of which is used for fertilizer. About 15% of the ammonia produced industrially, however, is later converted into nitric acid by the Ostwald process, named after the German chemist who developed it. In the Ostwald process, ammonia is burned in air at 900°C in the presence of a platinum–rhodium catalyst:

Step 1: $4\,NH_3(g) + 5\,O_2(g) \rightleftharpoons 4\,NO(g) + 6\,H_2O(g)$

Under these conditions, about 90% of the starting ammonia is converted into NO.

a. Air is 21% O_2 by volume. Typical instructions for step 1 specify that the volume of air should be 10 times the volume of ammonia reacted. (i) Using this volume ratio, which is the limiting reactant? (ii) Write the K_p expression for step 1 and explain how the partial pressure of oxygen influences the position of the equilibrium.

b. In the second step of the process, the temperature is lowered and more air is mixed with the products of step 1.

Step 2: $2\,NO(g) + O_2(g) \rightleftharpoons 2\,NO_2(g)$
$$\Delta H^\circ_{rxn} = -114.0 \text{ kJ/mol}$$

Predict what would happen to the yield in step 2 if the temperature were increased instead of lowered.

c. In the third and final step, NO_2 is passed through liquid water to form a solution of nitric acid (HNO_3); nitrogen monoxide gas is also produced. Write the K_c expression for this reaction. (*Hint*: The balanced equation reflects the fact that this is a redox reaction in acidic solution.)

Collect and Organize In part (a), we are provided with the balanced chemical equation describing a reversible gas-phase reaction and the volume ratio of the two reactants. We are asked to identify the limiting reactant, to write the K_p expression for the reaction, and to describe how altering the partial pressure of one of the reactants influences the equilibrium's position. In part (b), we are asked to predict the effect of raising reaction temperature on the yield of an exothermic reaction. In part (c), we need to write the K_c expression for a reaction happening in solution. We know at least some of the reactants and products, but we are not provided with a balanced chemical equation for this final step.

Analyze If we assume that both reactants in part (a) behave as ideal gases, then their volume ratio is the same as their mole ratio. Increasing the partial pressure of a reactant (O_2) shifts the equilibrium to the right (formation of more product). The reaction in part (b) is exothermic ($\Delta H^\circ < 0$), so the value of K_p and the yield of the reaction should decrease as temperature increases. The reaction in step 3 is a redox reaction because the oxidation state of nitrogen is initially +4 in NO_2, but it is +5 in HNO_3 and +2 in NO. Thus nitrogen is both oxidized and reduced. To write a K_c expression we first need to write a balanced chemical equation, and that requires balancing the loss and gain of electrons by the N atoms in molecules of NO_2.

Solve

a. (i) A volume ratio of 10:1 for air:ammonia translates into a mole ratio of $(10 \times 0.21) = 2.1 \text{ mol } O_2/\text{mol } NH_3$. The stoichiometric ratio is 5 moles of O_2 for every 4 moles of NH_3, or 1.25 mol O_2/mol NH_3. Therefore the reaction mixture contains excess O_2, making NH_3 the limiting reactant. (ii) The K_p expression (partial pressures of products over reactants each raised to a power equal to its stoichiometric coefficient) is

$$K_p = \frac{(P_{NO})^4 (P_{H_2O})^6}{(P_{NH_3})^4 (P_{O_2})^5}$$

The presence of excess O_2 at equilibrium shifts the position of the equilibrium toward the production of more ammonia.

b. The reaction is exothermic, so heat is a product of the reaction. If the temperature were increased, the reaction would shift in the direction of more reactants, and less NO_2 would be present at equilibrium.

c. The known reactant and products of the final reaction step are

$$NO_2(g) \rightleftharpoons NO(g) + HNO_3(aq)$$

During the reaction, the oxidation state of nitrogen simultaneously changes by +1 (from +4 in NO_2 to +5 in HNO_3) and by −2 (from +4 in NO_2 to +2 in NO). To balance this increase and decrease, molecules of NO_2 must form twice as many molecules of HNO_3 as molecules of NO. Incorporating this 2:1 ratio of products into the reaction expression above:

$$NO_2(g) \rightleftharpoons NO(g) + 2\,HNO_3(aq)$$

To write a balanced chemical equation we first need to balance the number of N atoms:

$$\mathbf{3}\,NO_2(g) \rightleftharpoons NO(g) + 2\,HNO_3(aq)$$

Next we balance the numbers of H and O atoms by adding the appropriate number of water molecules (one, in this case, to the reactant side):

$$3\,NO_2(g) + H_2O(\ell) \rightleftharpoons NO(g) + 2\,HNO_3(aq)$$

Now we have a balanced chemical equation. Because the concentration of H_2O is unlikely to change during the reaction, there is no [H_2O] term in the K_c expression:

$$K_c = \frac{[NO][HNO_3]^2}{[NO_2]^3}$$

Think About It The industrial production of nitric acid, like the industrial production of ammonia, requires balancing several competing effects. Temperatures and pressures must be adjusted so that materials can combine rapidly enough to produce product in reasonable time, while also ensuring that equilibria are stressed in the direction of product formation.

SUMMARY

LO1 According to the **law of mass action**, the ratio of the concentrations of the products of a reversible reaction divided by the concentrations of the reactants, with each raised to its stoichiometric coefficient from the balanced equation— a quantity called the **mass action expression**, or the **equilibrium constant expression**—has a characteristic value at a given temperature. At equilibrium, this value is called the **equilibrium constant**, K_c. Reversible gas–phase reactions also have characteristic K_p values based on the partial pressures of their products and reactants. The concentrations of pure liquids and solids that do not significantly change during a reaction are omitted from equilibrium constant expressions. (Sections 14.1, 14.2, 14.6)

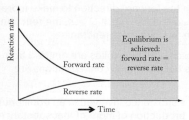

LO2 The **reaction quotient** (Q) is the value of the mass action expression at any instant during a reaction. At equilibrium, $Q = K$, but for nonequilibrium conditions, Q indicates how a reaction is proceeding. (Sections 14.2, 14.5)

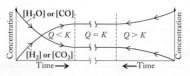

LO3 The relationship between K_c and K_p for a reversible gas-phase reaction depends on the number of moles of gaseous reactants and products in the balanced chemical equation. (Section 14.3)

LO4 The reverse of a reaction has an equilibrium constant that is the reciprocal of K for the forward reaction. If the balanced equation for a reaction is multiplied by some factor n, the value of K for that reaction is raised to the nth power. If reactions are summed to give an overall reaction, their equilibrium constants are multiplied together to obtain an overall equilibrium constant. (Section 14.4)

LO5 According to **Le Châtelier's principle**, chemical reactions at equilibrium respond to stress by shifting the position of the equilibrium to relieve the stress. A catalyst decreases the time it takes a system to achieve equilibrium but does not change the value of the equilibrium constant. (Section 14.7)

LO6 Equilibrium concentrations or partial pressures of reactants and products can be calculated from initial concentrations or pressures, the reaction stoichiometry, and the value of the equilibrium constant. (Section 14.8)

PARTICULATE **PREVIEW WRAP-UP**

The chemical equation for both samples is

$$H_2O(g) + CO(g) \rightleftharpoons H_2(g) + CO_2(g)$$

In Sample A, the concentrations of the reactants are constant; in Sample B, the concentrations of the reactants decrease. In Sample A, the concentrations of the products are constant; in Sample B, the concentrations of the products increase. Sample A is probably at equilibrium.

PROBLEM-SOLVING SUMMARY

Type of Problem	Concepts and Equations	Sample Exercises
Writing equilibrium constant expressions	For the reaction $$aA + bB \rightleftharpoons cC + dD$$ $$K_c = \frac{[C]^c[D]^d}{[A]^a[B]^b} \qquad (14.3)$$ and $$K_p = \frac{(P_C)^c(P_D)^d}{(P_A)^a(P_B)^b} \qquad (14.4)$$	**14.1**
Calculating and interconverting K_c and K_p	Insert equilibrium molar concentrations or partial pressures into the equilibrium constant expression and use the relation $$K_p = K_c(RT)^{\Delta n} \qquad (14.7)$$ where $R = 0.08206$ L · atm/ (mol · K), T is the absolute temperature, and Δn is the number of moles of product gas minus the number of moles of reactant gas in the balanced chemical equation.	**14.2, 14.3, 14.4**
Calculating K values of related reactions	$K_{reverse} = 1/K_{forward}$ for the forward and reverse reactions in an equilibrium system. If all the coefficients in a chemical equation are multiplied by n, the value of K increases by the power of n. If reactions are summed to give an overall reaction, their equilibrium constants are multiplied together to obtain an overall K.	**14.5, 14.6, 14.7**

Type of Problem	Concepts and Equations	Sample Exercises
Using Q and K values to predict the direction of a reaction	If $Q < K$, the reaction proceeds in the forward direction to make more products; if $Q = K$, the reaction is at equilibrium; if $Q > K$, the reaction proceeds in the reverse direction to make more reactants.	14.8
Writing equilibrium constant expressions for heterogeneous equilibria	Molar concentrations of pure liquids and pure solids are omitted from equilibrium constant expressions because such concentrations are constant.	14.9
Adding or removing reactants or products to stress an equilibrium	Decreasing the concentration of a substance involved in an equilibrium shifts the equilibrium toward the production of more of that substance. Increasing the concentration of a substance shifts the equilibrium to react some of that substance.	14.10
Predicting the effect of changing volume on gas-phase equilibria	Equilibria involving different numbers of moles of gaseous reactants and products shift in response to an increase (or decrease) in volume caused by a decrease (or increase) in pressure toward the side with more (or fewer) moles of gases.	14.11
Predicting changes in equilibrium with temperature	The value of K for an endothermic reaction increases with increasing temperature; the value of K for an exothermic reaction decreases with increasing temperature.	14.12
Calculating concentrations or partial pressures of reactants and products at equilibrium	Use a RICE table to develop algebraic terms for each reactant's and product's partial pressure or concentration at equilibrium. Substitute these terms into the expression for K and solve.	14.13, 14.14, 14.15

VISUAL PROBLEMS

(Answers to boldface end-of-chapter questions and problems are in the back of the book.)

14.1. Consider the graph of concentration versus time in Figure P14.1.
a. What is the mass action expression for the reaction?
b. What is the value of K_c?

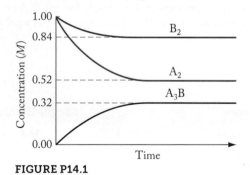

FIGURE P14.1

14.2. In Figure P14.2, the red spheres represent reactant A and the blue spheres represent product B in equilibrium with A.
a. Write a chemical equation that describes the equilibrium.
b. What is the value of the equilibrium constant K_c?

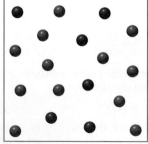

FIGURE P14.2

14.3. The equilibrium constant K_c for the reaction

A (red spheres) + B (blue spheres) $\rightleftharpoons$ AB

is 3.0 at 300.0 K. Does the situation depicted in Figure P14.3 correspond to equilibrium? If not, in what direction (to the left or to the right) will the system shift to attain equilibrium?

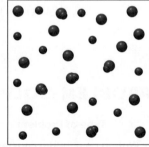

FIGURE P14.3

14.4. The diagrams in Figure P14.4 represent equilibrium states of the reaction

A (red spheres) + B (blue spheres) $\rightleftharpoons$ AB

at 300 K and 400 K, respectively. Is this reaction endothermic or exothermic? Explain.

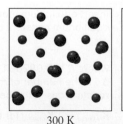 

300 K   400 K

FIGURE P14.4

14.5. Does the reaction A → 2 B represented in Figure P14.5 reach equilibrium in 20 μs? Explain your answer.

14.6. Use representations [A] through [I] in Figure P14.6 to answer questions a–f. The center cell [E] represents an equilibrium mixture of $NO_2(g)$ and $N_2O_4(g)$. Assume that temperature remains constant.

a. Write a balanced chemical equation describing the reversible reaction (NO_2 is the reactant).

b. Identify one stress to the system that would transform [E] to [C].

c. Although [E] is an equilibrium mixture, the reaction mixture in [C] is not yet at equilibrium. Which representation depicts how [C] transforms to reestablish equilibrium?

d. Are [G] and [I] at equilibrium or not? How can you tell?

e. Given that N_2O_4 is colorless and NO_2 is brown, match the pictures of the syringes to their respective particulate images.

f. Does the syringe in [H] contain any N_2O_4?

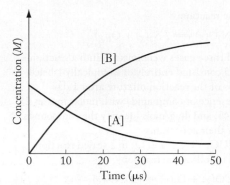

FIGURE P14.5

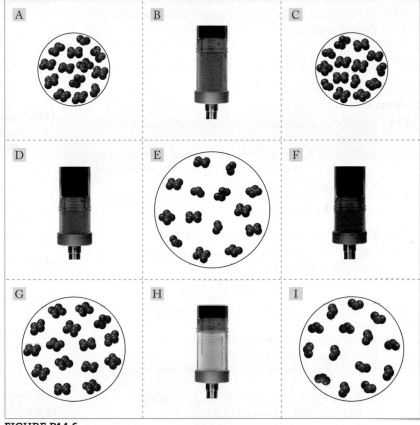

FIGURE P14.6

QUESTIONS AND PROBLEMS

The Dynamics of Chemical Equilibrium

Concept Review

14.7. How are forward and reverse reaction rates related in a system at chemical equilibrium?

14.8. Are ice cubes floating in 0°C water in an insulated container an example of a system in dynamic equilibrium? Explain why or why not.

14.9. Do rapid reversible reactions always have greater yields of product than slow reversible reactions? Explain why or why not.

14.10. At equilibrium, is the sum of the concentrations of all the reactants always equal to the sum of the concentrations of the products? Explain why or why not.

Problems

14.11. Suppose the rate constant of the forward reaction $A(g) \rightleftharpoons B(g)$ is greater than the rate constant of the reverse reaction. Do equilibrium mixtures of A and B have more A or more B in them? Explain your answer.

14.12. Suppose at 298 K the reaction $C(g) \rightleftharpoons D(g)$ has a forward rate constant of 5/s and a reverse rate constant of 10/s. What is the value of the equilibrium constant of the reaction at 298 K?

*14.13. In a study of the reaction

$$2 N_2O(g) \rightleftharpoons 2 N_2(g) + O_2(g)$$

quantities of all three gases were injected into a reaction vessel. The N_2O consisted entirely of isotopically labeled $^{15}N_2O$. Analysis of the reaction mixture after 1 day revealed the presence of compounds with molar masses 28, 29, 30, 32, 44, 45, and 46 g/mol. Identify the compounds and account for their appearance.

*14.14. A mixture of ^{13}CO, $^{12}CO_2$, and O_2 in a sealed reaction vessel was used to follow the reaction

$$2 CO(g) + O_2(g) \rightleftharpoons 2 CO_2(g)$$

Analysis of the reaction mixture after 1 day revealed the presence of compounds with molar masses 28, 29, 32, 44, and 45 g/mol. Identify the compounds and account for their appearance.

The Equilibrium Constant; Relationships between K_c and K_p Values

Concept Review

14.15. Under what conditions are the numerical values of K_c and K_p equal?

14.16. At 298 K, is K_p greater than or less than K_c if there is a net increase in the number of moles of gas in the reaction and if $K_c > 1$? Explain your answer.

14.17. What is the equilibrium constant (K_c) expression for the following reversible reaction?

$$2 A(g) + B(g) \rightleftharpoons 2 C(g)$$

14.18. What is the equilibrium constant (K_p) expression for the following reversible reaction?

$$2 A(g) + 3 B(g) \rightleftharpoons 2 C(g)$$

14.19. Write K_c and K_p expressions for the following gas-phase reactions.
   a. $C_2H_4(g) + H_2(g) \rightleftharpoons C_2H_6(g)$
   b. $2 SO_2(g) + O_2(g) \rightleftharpoons 2 SO_3(g)$

14.20. Write K_c and K_p expressions for the following reactions at 500 K.
   a. $NH_2Cl(g) + NH_3(g) \rightleftharpoons N_2H_4(g) + HCl(g)$
   b. $CO(g) + 2 H_2(g) \rightleftharpoons CH_3OH(g)$

Problems

14.21. At 1200 K the partial pressures of an equilibrium mixture of H_2S, H_2, and S are 0.020, 0.045, and 0.030 atm, respectively. Calculate the value of the following equilibrium constant at 1200 K.

$$H_2S(g) \rightleftharpoons H_2(g) + S(g) \qquad K_p = ?$$

14.22. At 1045 K the partial pressures of an equilibrium mixture of H_2O, H_2, and O_2 are 0.040, 0.0045, and 0.0030 atm, respectively. Calculate the value of the equilibrium constant K_p at 1045 K.

$$2 H_2O(g) \rightleftharpoons 2 H_2(g) + O_2(g)$$

14.23. At equilibrium, the concentrations of gaseous N_2, O_2, and NO in a sealed reaction vessel are $[N_2] = 3.3 \times 10^{-3}$ M,

$[O_2] = 5.8 \times 10^{-3}$ M, and $[NO] = 3.1 \times 10^{-3}$ M. What is the value of K_c for the reaction

$$N_2(g) + O_2(g) \rightleftharpoons 2 NO(g)$$

at the temperature of the reaction mixture?

14.24. Exactly 2 moles of ammonia are heated in a sealed 1.00-L container to 650°C. At this temperature, ammonia decomposes to nitrogen and hydrogen gas:

$$2 NH_3(g) \rightleftharpoons N_2(g) + 3 H_2(g)$$

At equilibrium, the concentration of ammonia in the container is 1.00 M. What is the value of K_c for the decomposition reaction at 650°C?

14.25. **Synthesis of Hydrogen (Step 1)** Hydrogen gas production often begins with the steam–methane reforming reaction

$$CH_4(g) + H_2O(g) \rightleftharpoons CO(g) + 3 H_2(g)$$

At 1000 K the partial pressures of the gases in an equilibrium mixture are 0.71 atm CH_4, 1.41 atm H_2O, 1.00 atm CO, and 3.00 atm H_2. What is the value of K_p for the reaction at 1000 K?

14.26. **Synthesis of Hydrogen (Step 2)** When the CO produced in the steam–methane reforming reaction is reacted with more steam at 450 K, the water–gas shift reaction:

$$CO(g) + H_2O(g) \rightleftharpoons CO_2(g) + H_2(g)$$

produces more hydrogen. If the equilibrium partial pressures of the gases in the reactor are 0.35 atm H_2O, 0.24 atm CO, 4.47 atm H_2, and 4.36 atm CO_2, what is the value of K_p?

14.27. At 500°C, the equilibrium constant K_p for the synthesis of ammonia

$$N_2(g) + 3 H_2(g) \rightleftharpoons 2 NH_3(g)$$

is 1.45×10^{-5}. What is the value of K_c?

14.28. If the value of the equilibrium constant K_c for the following reaction is 5×10^5 at 298 K, what is the value of K_p at 298 K?

$$2 CO(g) + O_2(g) \rightleftharpoons 2 CO_2(g)$$

14.29. For which of the following reactions are the values of K_p and K_c the same?
   a. $2 NH_3(g) + 2 O_2(g) \rightleftharpoons N_2O(g) + 3 H_2O(g)$
   b. $2 N_2H_4(\ell) + 2 NO_2(g) \rightleftharpoons 3 N_2(g) + 4 H_2O(\ell)$
   c. $NH_2F(g) + CaCl_2(s) \rightleftharpoons NH_2Cl(g) + CaClF(s)$

14.30. For which of the following reactions are the values of K_p and K_c different?
   a. $2 O_3(g) \rightleftharpoons 3 O_2(g)$
   b. $NH_4NO_3(s) \rightleftharpoons 2 H_2O(g) + N_2O(g)$
   c. $4 HCl(aq) + MnO_2(s) \rightleftharpoons$
      $MnCl_2(aq) + 2 H_2O(\ell) + Cl_2(g)$

14.31. **Bulletproof Glass** Phosgene ($COCl_2$) is used in the manufacture of foam rubber and bulletproof glass. It is formed from carbon monoxide and chlorine in the following reaction:

$$Cl_2(g) + CO(g) \rightleftharpoons COCl_2(g)$$

The value of K_c for the reaction is 5.0 at 327°C. What is the value of K_p at 327°C?

14.32. If the value of K_p for the following reaction

$$SO_2(g) + NO_2(g) \rightleftharpoons NO(g) + SO_3(g)$$

is 3.45 at 298 K, what is the value of K_c for the reverse reaction?

Manipulating Equilibrium Constant Expressions

Concept Review

14.33. Explain why representing the same reaction with different chemical equations, like this:

(1) $N_2(g) + 2 O_2(g) \rightleftharpoons 2 NO_2(g)$ $K_{(1)}$

(2) $\frac{1}{2} N_2(g) + O_2(g) \rightleftharpoons NO_2(g)$ $K_{(2)}$

results in different equilibrium constant values ($K_{(1)} \neq K_{(2)}$).

14.34. If the value of K_c for the reaction $A(g) \rightleftharpoons B(g)$ is 10, what is the value of K_c for the reaction $B(g) \rightleftharpoons A(g)$?

Problems

14.35. The equilibrium constant K_c for the reaction

$$I_2(g) + Br_2(g) \rightleftharpoons 2 IBr(g)$$

is 120 at 425 K. What is the value of K_c at 425 K for the following equilibrium?

$$\tfrac{1}{2} I_2(g) + \tfrac{1}{2} Br_2(g) \rightleftharpoons IBr(g)$$

14.36. The equilibrium constant K_p for the synthesis of ammonia

$$N_2(g) + 3 H_2(g) \rightleftharpoons 2 NH_3(g)$$

is 4.3×10^{-3} at 300°C. What is the value of K_p at 300°C for the following equilibrium?

$$\tfrac{1}{2} N_2(g) + \tfrac{3}{2} H_2(g) \rightleftharpoons NH_3(g)$$

14.37. The following reaction is one of the elementary steps in the oxidation of NO:

$$NO(g) + NO_3(g) \rightleftharpoons 2 NO_2(g)$$

Write an expression for the equilibrium constant K_c for this reaction and for the reverse reaction:

$$2 NO_2(g) \rightleftharpoons NO(g) + NO_3(g)$$

How are the two K_c expressions related?

14.38. **Producing Ammonia** The value of the equilibrium constant K_p for the formation of ammonia

$$N_2(g) + 3 H_2(g) \rightleftharpoons 2 NH_3(g)$$

is 4.5×10^{-5} at 450°C. What is the value of K_p at 450°C for the following reaction?

$$2 NH_3(g) \rightleftharpoons N_2(g) + 3 H_2(g)$$

14.39. The K_c value for the reaction

$$2 NOBr(g) \rightleftharpoons 2 NO(g) + Br_2(g)$$

is 3.0×10^{-4} at 298 K. What is the value of K_c at 298 K for the following reaction?

$$NOBr(g) \rightleftharpoons NO(g) + \tfrac{1}{2} Br_2(g)$$

14.40. How is the value of the equilibrium constant K_p for the reaction

$$2 H_2O(g) + N_2(g) \rightleftharpoons 2 H_2(g) + 2 NO(g)$$

related to the value of K_p for this reaction at the same temperature?

$$H_2O(g) + \tfrac{1}{2} N_2(g) \rightleftharpoons H_2(g) + NO(g)$$

14.41. At a given temperature, the equilibrium constant K_c for the reaction

$$2 SO_2(g) + O_2(g) \rightleftharpoons 2 SO_3(g)$$

is 2.4×10^{-3}. What is the value of the equilibrium constant K_c for each of the following reactions at the same temperature?

a. $SO_2(g) + \tfrac{1}{2} O_2(g) \rightleftharpoons SO_3(g)$
b. $2 SO_3(g) \rightleftharpoons 2 SO_2(g) + O_2(g)$
c. $SO_3(g) \rightleftharpoons SO_2(g) + \tfrac{1}{2} O_2(g)$

14.42. If the equilibrium constant K_c for the reaction

$$2 NO(g) + O_2(g) \rightleftharpoons 2 NO_2(g)$$

is 5×10^{12} at a given temperature, what is the value of the equilibrium constant K_c for each of the following reactions at the same temperature?

a. $NO(g) + \tfrac{1}{2} O_2(g) \rightleftharpoons NO_2(g)$
b. $2 NO_2(g) \rightleftharpoons 2 NO(g) + O_2(g)$
c. $NO_2(g) \rightleftharpoons NO(g) + \tfrac{1}{2} O_2(g)$

14.43. Calculate the value of the equilibrium constant K_p at 298 K for the reaction

$$N_2(g) + 2 O_2(g) \rightleftharpoons 2 NO_2(g)$$

from the following K_p values at 298 K:

$N_2(g) + O_2(g) \rightleftharpoons 2 NO(g)$ $K_p = 4.4 \times 10^{-31}$
$2 NO(g) + O_2(g) \rightleftharpoons 2 NO_2(g)$ $K_p = 2.4 \times 10^{12}$

14.44. Calculate the value of the equilibrium constant K_p at 298 K for the reaction

$$\tfrac{1}{4} S_8(s) + 3 O_2(g) \rightleftharpoons 2 SO_3(g)$$

from the following K_p values at 298 K:

$\tfrac{1}{8} S_8(s) + O_2(g) \rightleftharpoons SO_2(g)$ $K_p = 4.0 \times 10^{52}$
$2 SO_2(g) + O_2(g) \rightleftharpoons 2 SO_3(g)$ $K_p = 7.8 \times 10^{24}$

Equilibrium Constants and Reaction Quotients

Concept Review

14.45. How is an equilibrium constant different from a reaction quotient?

14.46. Explain how comparing the values of reaction quotient Q and equilibrium constant K for a given reaction and temperature enables chemists to predict whether a reversible reaction will proceed in the forward direction, in the reverse direction, or in neither direction.

Problems

14.47. If the equilibrium constant K_c for the hypothetical reaction $A(g) \rightleftharpoons B(g)$ is 22 at a given temperature, and if $[A] = 0.10\ M$ and $[B] = 2.0\ M$ in a reaction mixture at that temperature, is the reaction at chemical equilibrium? If not, in which direction will the reaction proceed to reach equilibrium?

14.48. The equilibrium constant K_c for the hypothetical reaction

$$2\ C(g) \rightleftharpoons D(g) + E(g)$$

is 3×10^{-3}. At a particular time, the composition of the reaction mixture is $[C] = [D] = [E] = 5 \times 10^{-4}\ M$. In which direction will the reaction proceed to reach equilibrium?

14.49. If the equilibrium constant K_c for the reaction

$$N_2(g) + O_2(g) \rightleftharpoons 2\ NO(g)$$

is 1.5×10^{-3}, in which direction will the reaction proceed if the partial pressures of the three gases are all 1.00×10^{-3} atm?

14.50. At 650 K, the value of the equilibrium constant K_p for the ammonia synthesis reaction

$$N_2(g) + 3\ H_2(g) \rightleftharpoons 2\ NH_3(g)$$

is 4.3×10^{-4}. If a vessel at 650 K contains a reaction mixture in which $[N_2] = 0.010\ M$, $[H_2] = 0.030\ M$, and $[NH_3] = 0.00020\ M$, will more ammonia form?

14.51. Use the information below to determine whether a reaction mixture in which the partial pressures of PCl_3, Cl_2, and PCl_5 are 0.20, 0.40, and 0.60 atm, respectively, is at equilibrium at 450 K.

$$PCl_3(g) + Cl_2(g) \rightleftharpoons PCl_5(g) \qquad K_p = 3.8 \text{ at } 450 \text{ K}$$

If the reaction mixture is not at equilibrium, in which direction does the reaction proceed to achieve equilibrium?

14.52. If enough PCl_3 were injected into the reaction mixture in the previous problem to double its partial pressure, would the mixture be closer to equilibrium? Explain why or why not.

Heterogeneous Equilibria

Concept Review

14.53. **SO$_2$ in the Air** Combustion of fossil fuels that contain sulfur is an important source of sulfur dioxide in the atmosphere. Write the K_p expression for the combustion of elemental sulfur:

$$\tfrac{1}{8}\ S_8(s) + O_2(g) \rightleftharpoons SO_2(g)$$

14.54. Write the K_c expression for the oxidation of calcium sulfite to gypsum (calcium sulfate):

$$2\ CaSO_3(s) + O_2(g) \rightleftharpoons 2\ CaSO_4(s)$$

14.55. Mixing aqueous solutions of sodium bicarbonate and calcium chloride results in this reaction:

$$2\ NaHCO_3(aq) + CaCl_2(aq) \rightleftharpoons$$
$$2\ NaCl(aq) + CO_2(g) + CaCO_3(s) + H_2O(\ell)$$

Write the K_c expression for the reaction.

14.56. Write K_p expressions for the reactions below that take place during the thermal decomposition of the mineral dolomite (a mixture of calcium and magnesium carbonates):

$$CaCO_3(s) \rightleftharpoons CaO(s) + CO_2(g)$$
$$MgCO_3(s) \rightleftharpoons MgO(s) + CO_2(g)$$

14.57. The brown residues that form on the surfaces of plumbing fixtures, such as inside toilet tanks, are the result of the oxidation of more soluble iron(II) compounds with dissolved oxygen, forming less soluble iron(III) compounds. Write the K_c expression for one such reaction:

$$4\ Fe(OH)_2(aq) + O_2(aq) + 2\ H_2O(\ell) \rightleftharpoons 4\ Fe(OH)_3(s)$$

*14.58. **Testing Minerals** Write the K_c expression for this reaction, which is used to test whether a white, shiny mineral is marble (calcium carbonate):

$$2\ HCl(aq) + CaCO_3(s) \rightleftharpoons CaCl_2(aq) + H_2O(\ell) + CO_2(g)$$

Le Châtelier's Principle

Concept Review

14.59. Does adding reactants to a system at equilibrium increase the value of the equilibrium constant? Why or why not?

14.60. Increasing the concentration of a reactant shifts the position of chemical equilibrium toward formation of more products. What effect does adding a reactant have on the rates of the forward and reverse reactions?

14.61. **Carbon Monoxide Poisoning** Patients suffering from carbon monoxide poisoning are treated with pure oxygen to remove CO from the hemoglobin (Hb) in their blood. The two relevant equilibria are

$$Hb(aq) + 4\ CO(g) \rightleftharpoons Hb(CO)_4(aq)$$
$$Hb(aq) + 4\ O_2(g) \rightleftharpoons Hb(O_2)_4(aq)$$

The value of the equilibrium constant for CO binding to Hb is greater than that for O_2. How, then, does this treatment work?

14.62. Is the equilibrium constant K_p for the reaction

$$2\ NO_2(g) \rightleftharpoons N_2O_4(g)$$

in air the same in Los Angeles as in Denver if the atmospheric pressure in Denver is lower but the temperature is the same? Explain your answer.

14.63. Henry's law predicts that the solubility of a gas in a liquid increases with its partial pressure. Explain Henry's law in relation to Le Châtelier's principle.

*14.64. Why does adding an inert gas such as argon to an equilibrium mixture of CO, O_2, and CO_2 in a sealed vessel increase the total pressure of the system but not shift the following equilibrium?

$$2\ CO(g) + O_2(g) \rightleftharpoons 2\ CO_2(g)$$

Problems

14.65. Which of the following equilibria will shift toward formation of more products if an equilibrium mixture is compressed into half its volume?
  a. $2\ N_2O(g) \rightleftharpoons 2\ N_2(g) + O_2(g)$
  b. $2\ CO(g) + O_2(g) \rightleftharpoons 2\ CO_2(g)$

c. $N_2(g) + O_2(g) \rightleftharpoons 2\,NO(g)$

d. $2\,NO(g) + O_2(g) \rightleftharpoons 2\,NO_2(g)$

14.66. Which of the following equilibria will shift toward formation of more products if the volume of a reaction mixture at equilibrium increases by a factor of 2?

a. $2\,SO_2(g) + O_2(g) \rightleftharpoons 2\,SO_3(g)$

b. $NO(g) + O_3(g) \rightleftharpoons NO_2(g) + O_2(g)$

c. $2\,N_2O_5(g) \rightleftharpoons 4\,NO_2(g) + O_2(g)$

d. $N_2O_4(g) \rightleftharpoons 2\,NO_2(g)$

14.67. How will the changes listed affect the equilibrium concentrations of reactants and products in the following reaction?

$$2\,O_3(g) \rightleftharpoons 3\,O_2(g)$$

a. O_3 is added to the system.

b. O_2 is added to the system.

c. The mixture is compressed to one-tenth its initial volume.

14.68. How will the changes listed affect the position of the following equilibrium?

$$2\,NO_2(g) \rightleftharpoons NO(g) + NO_3(g)$$

a. The concentration of NO is increased.

b. The concentration of NO_2 is increased.

c. The volume of the system is allowed to expand to 5 times its initial value.

14.69. How will reducing the partial pressure of O_2 affect the position of the equilibrium in the following reaction?

$$2\,SO_2(g) + O_2(g) \rightleftharpoons 2\,SO_3(g)$$

*__14.70.__ Ammonia is added to a gaseous reaction mixture containing H_2, Cl_2, and HCl that is at chemical equilibrium. How will the addition of ammonia affect the relative concentrations of H_2, Cl_2, and HCl if the equilibrium constant of reaction 2 is much greater than the equilibrium constant of reaction 1?

(1) $H_2(g) + Cl_2(g) \rightleftharpoons 2\,HCl(g)$

(2) $HCl(g) + NH_3(g) \rightleftharpoons NH_4Cl(s)$

14.71. In which of the following equilibria does an increase in temperature produce a shift toward the formation of more product?

a. $PCl_3(g) + Cl_2(g) \rightleftharpoons PCl_5(g)$ $\Delta H° < 0$

b. $CH_4(g) + H_2O(g) \rightleftharpoons CO(g) + 3\,H_2(g)$ $\Delta H° > 0$

c. $CO(g) + H_2O(g) \rightleftharpoons CO_2(g) + H_2(g)$ $\Delta H° < 0$

14.72. In which of the following equilibria does an increase in temperature produce a shift toward the formation of more product?

a. $N_2(g) + 3\,H_2(g) \rightleftharpoons 2\,NH_3(g)$ $\Delta H° < 0$

b. $2\,NO_2(g) \rightleftharpoons 2\,NO(g) + O_2(g)$ $\Delta H° > 0$

c. $C_2H_4(g) + H_2(g) \rightleftharpoons C_2H_6(g)$ $\Delta H° < 0$

Calculations Based on K

Concept Review

14.73. Why are calculations based on K often simpler when the value of K is very small?

14.74. The following reaction is carried out in a sealed, rigid vessel at constant temperature.

$$2\,NO(g) + O_2(g) \rightarrow 2\,NO_2(g)$$

a. If the change in the partial pressure of O_2 is $-x$, what are the changes in the partial pressures of NO and NO_2?

b. As the reaction proceeds, what happens to the total pressure in the reaction vessel?

Problems

14.75. For the reaction

$$PCl_5(g) \rightleftharpoons PCl_3(g) + Cl_2(g) \qquad K_p = 23.6 \text{ at } 500 \text{ K}$$

a. Calculate the equilibrium partial pressures of the reactants and products at 500 K if the initial pressures are $P_{PCl_5} = 0.560$ atm and $P_{PCl_3} = 0.500$ atm.

b. If more chlorine is added after equilibrium is reached, how will the concentrations of PCl_5 and PCl_3 change?

14.76. Enough NO_2 gas is injected into a cylindrical vessel to produce a partial pressure, P_{NO_2}, of 0.900 atm at 298 K. Calculate the equilibrium partial pressures of NO_2 and N_2O_4, given

$$2\,NO_2(g) \rightleftharpoons N_2O_4(g) \qquad K_p = 4 \text{ at } 298 \text{ K}$$

14.77. The value of K_c for the reaction between water vapor and dichlorine monoxide,

$$H_2O(g) + Cl_2O(g) \rightleftharpoons 2\,HOCl(g)$$

is 0.0900 at 25°C. Determine the equilibrium concentrations of all three compounds at 25°C if the starting concentrations of both reactants are 0.00432 M and no HOCl is present.

14.78. The value of K_p for the reaction

$$3\,H_2(g) + N_2(g) \rightleftharpoons 2\,NH_3(g)$$

is 4.3×10^{-4} at 648 K. Determine the equilibrium partial pressure of NH_3 in a reaction vessel that initially contained 0.900 atm N_2 and 0.500 atm H_2 at 648 K.

14.79. The value of K_p for the reaction

$$NO(g) + \tfrac{1}{2}O_2(g) \rightleftharpoons NO_2(g)$$

is 1.5×10^6 at 25°C. At equilibrium, what is the ratio of P_{NO_2} to P_{NO} in air at 25°C? Assume that $P_{O_2} = 0.21$ atm and does not change.

*__14.80.__ **The Water–Gas Reaction** Passing steam over hot carbon produces a mixture of carbon monoxide and hydrogen known as water gas:

$$H_2O(g) + C(s) \rightleftharpoons CO(g) + H_2(g)$$

The value of K_c for the reaction at 1000°C is 3.0×10^{-2}.

a. Calculate the equilibrium partial pressures of the products and reactants at 1000°C if $P_{H_2O} = 0.442$ atm and $P_{CO} = 5.0$ atm at the start of the reaction. Assume that the carbon is in excess.

b. Determine the equilibrium partial pressures of the reactants and products after sufficient CO and H_2 are added to the equilibrium mixture in part (a) to initially increase the partial pressures of both gases by 0.075 atm.

14.81. The value of K_p for the reaction

$$CO_2(g) + C(s) \rightleftharpoons 2\,CO(g)$$

is 1.5 at 700°C. Calculate the equilibrium partial pressures of CO and CO_2 at 700°C if initially $P_{CO_2} = 5.0$ atm and CO is not present. Pure graphite is present initially and when equilibrium is achieved.

14.82. **Composition of the Jovian Atmosphere** Ammonium hydrogen sulfide (NH_4SH) has been detected in the atmosphere of Jupiter. The equilibrium between ammonia, hydrogen sulfide, and NH_4SH is described by the following equation:

$$NH_4SH(s) \rightleftharpoons NH_3(g) + H_2S(g)$$

The value of K_p for the reaction at 24°C is 0.126. Suppose a sealed flask contains an equilibrium mixture of NH_4SH, NH_3, and H_2S at 24°C. At equilibrium, the partial pressure of H_2S is 0.355 atm. What is the partial pressure of NH_3?

*14.83. A flask containing pure NO_2 was heated to 1000 K, a temperature at which the value of K_p for the decomposition of NO_2 is 158.

$$2\,NO_2(g) \rightleftharpoons 2\,NO(g) + O_2(g)$$

The partial pressure of O_2 at equilibrium is 0.136 atm.
a. Calculate the partial pressures of NO and NO_2.
b. Calculate the total pressure in the flask at equilibrium.

14.84. The equilibrium constant K_p of the reaction

$$2\,SO_3(g) \rightleftharpoons 2\,SO_2(g) + O_2(g)$$

is 7.69 at 830°C. If a vessel at this temperature initially contains pure SO_3 and if the partial pressure of SO_3 at equilibrium is 0.100 atm, what is the partial pressure of O_2 in the flask at equilibrium?

*14.85. **NO_x Pollution** In a study of the formation of NO_x in air pollution, a chamber heated to 2200°C was filled with air (0.79 atm N_2, 0.21 atm O_2). What are the equilibrium partial pressures of N_2, O_2, and NO if $K_p = 0.050$ for the following reaction at 2200°C?

$$N_2(g) + O_2(g) \rightleftharpoons 2\,NO(g)$$

*14.86. The equilibrium constant K_p for the thermal decomposition of NO_2

$$2\,NO_2(g) \rightleftharpoons 2\,NO(g) + O_2(g)$$

is 6.5×10^{-6} at 450°C. If a reaction vessel at this temperature initially contains only 0.500 atm NO_2, what will be the partial pressures of NO_2, NO, and O_2 in the vessel when equilibrium has been attained?

14.87. The value of K_c for the thermal decomposition of hydrogen sulfide

$$2\,H_2S(g) \rightleftharpoons 2\,H_2(g) + S_2(g)$$

is 2.2×10^{-4} at 1400 K. A sample of gas in which [H_2S] = 6.00 M is heated to 1400 K in a sealed high-pressure

vessel. After chemical equilibrium has been achieved, what is the value of [H_2S]? Assume that no H_2 or S_2 was present in the original sample.

14.88. **Urban Air** On a very smoggy day, the equilibrium concentration of NO_2 in the air over an urban area reaches 2.2×10^{-7} M. If the temperature of the air is 25°C, what is the concentration of the dimer N_2O_4 in the air? Given:

$$N_2O_4(g) \rightleftharpoons 2\,NO_2(g) \qquad K_c = 6.1 \times 10^{-3}$$

*14.89. **Chemical Weapon** Phosgene, $COCl_2$, gained notoriety as a chemical weapon in World War I. Phosgene is produced by the reaction of carbon monoxide with chlorine

$$CO(g) + Cl_2(g) \rightleftharpoons COCl_2(g)$$

The value of K_c for this reaction is 5.0 at 600 K. What are the equilibrium partial pressures of the three gases if a reaction vessel initially contains a mixture of the reactants in which $P_{CO} = P_{Cl_2} = 0.265$ atm and there is no $COCl_2$?

*14.90. At 2000°C, the value of K_c for the reaction

$$2\,CO(g) + O_2(g) \rightleftharpoons 2\,CO_2(g)$$

is 1.0. What is the ratio of [CO] to [CO_2] at 2000°C in an atmosphere in which [O_2] = 0.0045 M at equilibrium?

*14.91. The water–gas shift reaction is an important source of hydrogen. The value of K_c for the reaction

$$CO(g) + H_2O(g) \rightleftharpoons CO_2(g) + H_2(g)$$

at 700 K is 5.1. Calculate the equilibrium concentrations of the four gases at 700 K if the initial concentration of each of them is 0.050 M.

*14.92. Sulfur dioxide reacts with NO_2, forming SO_3 and NO:

$$SO_2(g) + NO_2(g) \rightleftharpoons SO_3(g) + NO(g)$$

If the value of K_c for the reaction is 2.50 at a given temperature, what are the equilibrium concentrations of the products if the reaction mixture was initially 0.50 M SO_2, 0.50 M NO_2, 0.0050 M SO_3, and 0.0050 M NO?

Additional Problems

*14.93. **CO as a Fuel** Is carbon dioxide a viable source of the fuel CO? Pure carbon dioxide ($P_{CO_2} = 1$ atm) decomposes at high temperatures. For the system

$$2\,CO_2(g) \rightleftharpoons 2\,CO(g) + O_2(g)$$

the percentage of decomposition of $CO_2(g)$ changes with temperature as follows:

Temperature (K)	Decomposition (%)
1500	0.048
2500	17.6
3000	54.8

Is the reaction endothermic? Calculate the value of K_p at each temperature and discuss the results. Is the decomposition of CO_2 an antidote for global warming?

14.94. Ammonia decomposes at high temperatures. In an experiment to explore this behavior, 2.00 moles of gaseous NH_3 is sealed in a rigid 1-liter vessel. The vessel is heated at 800 K and some of the NH_3 decomposes in the following reaction:

$$2 NH_3(g) \rightleftharpoons N_2(g) + 3 H_2(g)$$

The system eventually reaches equilibrium and is found to contain 1.74 moles of NH_3. What are the values of K_p and K_c for the decomposition reaction at 800 K?

*14.95. Elements of group 16 form hydrides with the generic formula H_2X. At a certain temperature, when gaseous H_2X is bubbled through a solution containing 0.3 M hydrochloric acid, the solution becomes saturated and $[H_2X] = 0.1\ M$. The following equilibria exist in this solution:

$$H_2X(aq) + H_2O(\ell) \rightleftharpoons HX^-(aq) + H_3O^+(aq) \qquad K_1 = 8.3 \times 10^{-8}$$

$$HX^-(aq) + H_2O(\ell) \rightleftharpoons X^{2-}(aq) + H_3O^+(aq) \qquad K_2 = 1 \times 10^{-14}$$

Calculate the concentration of X^{2-} in the solution.

*14.96. Nitrogen dioxide reacts with SO_2 to form SO_3 and NO:

$$NO_2(g) + SO_2(g) \rightleftharpoons NO(g) + SO_3(g)$$

An equilibrium mixture is analyzed at a certain temperature and found to contain $[NO_2] = 0.100\ M$, $[SO_2] = 0.300\ M$, $[NO] = 2.00\ M$, and $[SO_3] = 0.600\ M$. At the same temperature, extra $SO_2(g)$ is added to make $[SO_2] = 0.800\ M$. Calculate the composition of the mixture when equilibrium has been reestablished.

*14.97. **Controlling Air Pollution** Calcium oxide is used to remove the pollutant SO_2 from smokestack gases. The overall reaction is

$$CaO(s) + SO_2(g) + \tfrac{1}{2} O_2(g) \rightleftharpoons CaSO_4(s)$$

The K_p for this reaction is 2.38×10^{73}.
a. What is P_{SO_2} in equilibrium with air and solid CaO if P_{O_2} in air is 0.21 atm?
b. Consider a sample of gas that contains 100.0 moles of gas. How many molecules of SO_2 are in that sample?

*14.98. **Volcanic Eruptions** During volcanic eruptions, gases as hot as 700°C and rich in SO_2 are released into the atmosphere. As air mixes with these gases, the following reaction converts some of the SO_2 into SO_3:

$$2 SO_2(g) + O_2(g) \rightleftharpoons 2 SO_3(g)$$

Calculate the value of K_p for this reaction at 700°C. What is the ratio of P_{SO_2} to P_{SO_3} in equilibrium with $P_{O_2} = 0.21$ atm?

*14.99. A 100-mL reaction vessel initially contains 2.60×10^{-2} mol of NO and 1.30×10^{-2} mol of H_2. At equilibrium, the concentration of NO in the vessel is 0.161 M. The vessel also contains N_2, H_2O, and H_2 at equilibrium. What is the value of the equilibrium constant K_c for the following reaction?

$$2 H_2(g) + 2 NO(g) \rightleftharpoons 2 H_2O(g) + N_2(g)$$

15

Acid–Base Equilibria

Proton Transfer in Biological Systems

SHADES OF PINK AND BLUE AND IN BETWEEN The color of hydrangea blossoms depends on the acidity of the soil in which they grow.

PARTICULATE **REVIEW**

Donating H⁺ Ions

In Chapter 15 we explore the structure and properties of acids and bases. Here we see models of four molecules that each contain at least one hydrogen atom.

- Which molecules produce H^+ ions when dissolved in water? Identify the acidic hydrogen atoms.

- Which molecule dissociates completely in water, meaning that each molecule in a sample produces a H^+ ion?

- Which molecule partially dissociates in water, meaning that only a small fraction of molecules in a sample releases H^+ ions, whereas most do not?

(Review Section 4.5 if you need help.)

(Answers to Particulate Review questions are in the back of the book.)

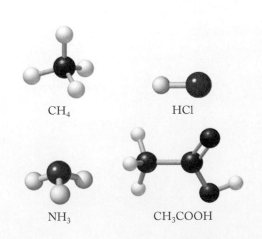

CH_4 HCl

NH_3 CH_3COOH

Accepting H⁺ Ions

Bases form covalent bonds as they accept H⁺ ions donated by acids. As you read Chapter 15, look for ideas that will help you answer these questions:

ClO⁻

NH₃

H₂O

• Which of the three species depicted form covalent bonds as they accept H⁺ ions?

• When each species accepts a proton, does it produce a cation? An anion? A neutral molecule?

• What feature do these particles have in common to form bonds to H⁺ ions?

Learning Outcomes

LO1 Relate the strengths of acids and bases to their K_a and K_b values
Sample Exercise 15.1

LO2 Recognize conjugate acid–base pairs and their complementary strengths
Sample Exercises 15.2, 15.3

LO3 Interconvert [H_3O^+], pH, pOH, and [OH^-]
Sample Exercises 15.4, 15.5, 15.6

LO4 Relate pH to percent ionization, K_a values, and K_b values
Sample Exercises 15.7, 15.8

LO5 Calculate the pH of solutions of strong and weak acids and bases
Sample Exercises 15.9, 15.10, 15.11

LO6 Calculate the pH of solutions involving autoionization and diprotic acids
Sample Exercises 15.12, 15.13, 15.14

LO7 Relate the strengths of acids to their molecular structures
Sample Exercise 15.15

LO8 Predict whether a salt is acidic, basic, or neutral, and calculate the pH of a salt solution
Sample Exercises 15.16, 15.17, 15.18

15.1 Acids and Bases: A Balancing Act

Consider the pink and blue colors of the blossoms in the chapter-opening photograph. They are produced by the same species of plant (a hydrangea) grown in the same garden, but in soils with slightly different composition. It turns out that the colors of hydrangea blossoms are controlled by the availability of aluminum ions (Al^{3+}) in the soil in which they grow. Aluminum ions are soluble in acidic soils, and hydrangeas grown in acidic soils form blue blossoms. However, in soils that are neutral or slightly basic, aluminum ions precipitate as aluminum hydroxide and are not available. As a result, hydrangea blossoms grown in those soils lack blue pigment and are pink.

For many biological systems, including the human body, acid–base balance is vital to normal function and good health. Our blood, for example, is normally slightly basic, regulated by the proportions of CO_2 gas and bicarbonate (HCO_3^-) ions dissolved in it. This regulation happens because of the following reversible chemical reactions:

$$CO_2(g) + H_2O(\ell) \rightleftharpoons H_2CO_3(aq) \tag{15.1}$$

$$H_2CO_3(aq) + H_2O(\ell) \rightleftharpoons HCO_3^-(aq) + H_3O^+(aq) \tag{15.2}$$

Note how molecules of CO_2 and H_2O combine, forming molecules of carbonic acid (H_2CO_3), which react with more water molecules, donating H^+ ions to them and forming bicarbonate and hydronium (H_3O^+) ions.

Compounds that produce hydronium ions when dissolved in water are acids, whereas compounds that produce hydroxide ions (OH^-) in aqueous solution are bases. More precisely, such compounds are referred to respectively as **Arrhenius acids** and **Arrhenius bases** in honor of Svante Arrhenius, whose research on the behavior of electrolytic solutions was recognized with the 1903 Nobel Prize in Chemistry.

If we apply Le Châtelier's principle to Equation 15.2, we can predict that the concentration of H_3O^+ ions in a solution that contains carbonic acid and bicarbonate ions will increase (as will the acidity of the solution) if the concentration of H_2CO_3 increases. The equilibrium described in Equation 15.1 tells us that the concentration of H_2CO_3 increases when the concentration of dissolved CO_2

CONNECTION In Chapter 4 we noted that, in describing aqueous solutions, the terms *hydrogen ion*, *proton*, and *hydronium ion* all refer to the same species.

Arrhenius acid a compound that produces H_3O^+ ions in aqueous solution

Arrhenius base a compound that produces OH^- ions in aqueous solution

increases. Therefore an increase in the concentration of dissolved CO_2 shifts the equilibria in *both* Equations 15.1 and 15.2 to the right, increasing the concentration of H_3O^+ ions and making the solution more acidic. On the other hand, Le Châtelier's principle also tells us that an increase in the concentration of HCO_3^- ions shifts both equilibria to the left, decreasing the concentration of H_3O^+ ions and reducing the acidity of the solution.

The equilibria in Equations 15.1 and 15.2 play a key role in human health. Normally, the concentration of HCO_3^- ions in our blood is about 20 times the dissolved concentration of CO_2, because the dissociation of carbonic acid is not our blood's only source of HCO_3^-. The abundance of HCO_3^- ions drives down the concentration of H_3O^+ ions in our blood below 10^{-7} *M*. As we discuss later in this chapter, a $[H_3O^+]$ value this small means that our blood is actually slightly basic.

Unfortunately, some people suffer from medical problems that alter the normal balance of CO_2 and HCO_3^- in their blood. For example, chronic lung disease impairs not only the ability to inhale O_2 but also the ability to exhale CO_2. As a result, CO_2 builds up in the blood and in the tissues where it is produced. The result is a condition called *respiratory acidosis*. Its symptoms include fatigue, lethargy, and shortness of breath. In severe cases, the disease can be fatal.

> **CONNECTION** In Chapter 4 we learned that the greater the concentration of H_3O^+ ions in a solution, the more acidic the solution.

CONCEPT **TEST**

When people hyperventilate, they breathe too rapidly. In fact, hyperventilation results in carbon dioxide being removed from the bloodstream more quickly than the body produces it through metabolism. What effect does this have on the equilibria in Equations 15.1 and 15.2?

(Answers to Concept Tests are in the back of the book.)

In this chapter we examine acid–base balance, both in environmental waters and in our bodies. We study why changes in this balance occur, some of the consequences of these changes, and what we can do to control them.

15.2 Strong and Weak Acids and Bases

Although the Arrhenius model of acids and bases is useful for understanding the properties of many compounds in aqueous solution, such as HCl, HNO_3, NaOH, or $Ca(OH)_2$, the model has limitations. First, although Arrhenius acids produce hydronium ions in aqueous solution, the compounds do not contain H_3O^+ ions (or even H^+ ions). Arrhenius acids always contain covalently bonded H atoms that ionize when the compound dissolves in water. Second, chemists' early classification of substances was based on their properties, not structure. Substances such as ammonia share the same properties as bases like NaOH, such as feeling slippery and tasting bitter. Such observations led to the development of a new model to explain acid–base behavior.

In Chapter 4 we introduced the *Brønsted–Lowry* model of acids and bases, in which acids are defined as substances that *donate* H^+ ions, and bases are defined as substances that *accept* H^+ ions. We also learned in Chapter 4 that acids in aqueous solutions are classified as strong or weak, depending on the degree to which they ionize and donate H^+ ions to molecules of water producing H_3O^+ ions. In this section we use the concept of chemical equilibrium to further explore what it means to be a weak or strong acid or base.

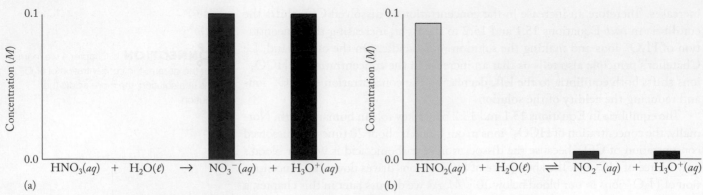

(a)

$$HNO_3(aq) + H_2O(\ell) \rightarrow NO_3^-(aq) + H_3O^+(aq)$$

(b)

$$HNO_2(aq) + H_2O(\ell) \rightleftharpoons NO_2^-(aq) + H_3O^+(aq)$$

FIGURE 15.1 Degrees of ionization of a strong acid versus a weak acid. (a) The strong acid HNO_3 is ionized completely, existing only as H_3O^+ and NO_3^- ions. (b) The weak acid HNO_2 ionizes very little. (*Note*: Although $H_2O(\ell)$ is part of the balanced chemical equation in both (a) and (b), its concentration is quite large relative to the other species and has been intentionally omitted from these graphs.)

Strong and Weak Acids

The images in Figure 15.1 show the degree to which solutions of two acids with similar chemical structures ionize when they dissolve in water. Nitric acid is a strong acid and ionizes completely in water, as shown by the bar graph in Figure 15.1(a). Note that there are essentially no intact molecules of HNO_3 in this solution; they have all donated H^+ ions to molecules of H_2O and have themselves become NO_3^- ions:

$$HNO_3(aq) + H_2O(\ell) \rightarrow NO_3^-(aq) + H_3O^+(aq) \qquad (15.3)$$

On the other hand, nitrous acid (HNO_2) is a weak acid and only a small fraction of the HNO_2 molecules ionize in water, as shown in Figure 15.1(b) and as described by the following chemical equation:

$$HNO_2(aq) + H_2O(\ell) \rightleftharpoons NO_2^-(aq) + H_3O^+(aq) \qquad (15.4)$$

TABLE 15.1 Strong Acids and Their Ionization Reactions in Water

Strong Acid	Molecular Structure	Reaction in Water
Hydrobromic	H—Br	$HBr(aq) + H_2O(\ell) \rightarrow Br^-(aq) + H_3O^+(aq)$
Hydrochloric	H—Cl	$HCl(aq) + H_2O(\ell) \rightarrow Cl^-(aq) + H_3O^+(aq)$
Hydroiodic	H—I	$HI(aq) + H_2O(\ell) \rightarrow I^-(aq) + H_3O^+(aq)$
Nitric		$HNO_3(aq) + H_2O(\ell) \rightarrow NO_3^-(aq) + H_3O^+(aq)$
Perchloric		$HClO_4(aq) + H_2O(\ell) \rightarrow ClO_4^-(aq) + H_3O^+(aq)$
Sulfuric		$H_2SO_4(aq) + H_2O(\ell) \rightarrow HSO_4^-(aq) + H_3O^+(aq)$ $HSO_4^-(aq) + H_2O(\ell) \rightleftharpoons SO_4^{2-}(aq) + H_3O^+(aq)$

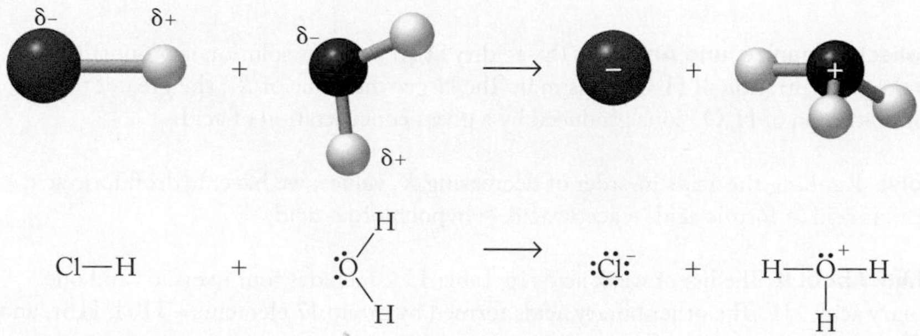

FIGURE 15.2 When hydrogen chloride dissolves in water, each molecule of HCl ionizes by reacting with a molecule of H_2O to form Cl^- and H_3O^+ ions.

Cl—H   +   [structure of water molecule] $\longrightarrow$ $:\ddot{\underset{..}{Cl}}:^-$ +   [structure of H_3O^+]

HCl   + H_2O $\longrightarrow$ Cl^- + H_3O^+
(H^+ ion donor)   (H^+ ion acceptor)

At equilibrium, molecules of HNO_2 donate H^+ ions to molecules of H_2O at the same rate that H_3O^+ ions donate H^+ ions to nitrite (NO_2^-) ions, reforming HNO_2 and H_2O. Later in this chapter (Section 15.7) we explore why two substances with such similar formulas and molecular structures as HNO_2 and HNO_3 have such different acid strengths.

Which intermolecular interactions are involved when strong acids in aqueous solutions donate H^+ ions to molecules of water? Table 15.1 contains the names and molecular structures of common strong acids. In each of these strong acids, the large difference in electronegativity between the hydrogen atom and the Br, Cl, I, or O atom that the H atom is covalently bonded to means that these bonds are polar covalent, resulting in strong dipole–dipole interactions between their H atoms and the O atoms of H_2O molecules. In aqueous solutions of HCl, for example, the strength of these interactions results in every H—Cl bond breaking to create H^+ and Cl^- ions. As the H^+ ion separates from the Cl^- ion, it covalently bonds with a molecule of water, forming a H_3O^+ ion (Figure 15.2).

As with the gas-phase equilibria we discussed in Chapter 14, the extent to which a reaction takes place is characterized by the value of its equilibrium constant, K: the greater the value of K, the higher the concentrations of products and the lower the concentrations of reactants at equilibrium. The equilibrium constants that describe the ionization of acids are often written as K_a, where the subscript "a" indicates that the equilibrium involves the ionization of an *a*cid. The large K_a values of the strong acids in Table 15.1 are often expressed "$K_a \gg 1$" to indicate that ionization of the acid is essentially complete in aqueous solutions.

Table 15.2 lists nitrous acid and several other common weak acids along with their molecular structures (in which the ionizable hydrogen atoms are highlighted in red), their ionization reactions in water, and their K_a values at 25°C. Note that the K_a values of these weak acids are much smaller than 1, and remember that the smaller the K_a for a weak acid, the fewer the number of molecules that dissociate to produce hydronium ions. In Table 15.1 and throughout this chapter, unless otherwise stated, values of K are determined at 25°C.

CHEMTOUR
Acid–Base Ionization

SAMPLE EXERCISE 15.1 Relating K_a and Acid Strength   **LO1**

Suppose 0.100 *M* aqueous solutions of each of the acids in Table 15.2 are prepared. Rank them from most acidic to least acidic.

Collect, Organize, and Analyze The acidity of an aqueous solution is proportional to the concentration of H_3O^+ ions in it. The larger the value of K_a, the greater the concentration of H_3O^+ ions produced by a given concentration of acid.

Solve Ranking the acids in order of decreasing K_a values, we have: hydrofluoric acid > nitrous acid > formic acid > acetic acid > hypochlorous acid.

Think About It The list of weak acids in Table 15.2 includes four oxoacids and one binary acid, HF. The other binary acids formed by group 17 elements—HCl, HBr, and HI—are all strong acids.

 Practice Exercise Rank the following acids in order of decreasing acid strength.

Weak Acid	K_a
HN_3	1.9×10^{-5}
$CH_2BrCOOH$	2.0×10^{-3}
$HClO_2$	1.1×10^{-2}
Lactic	1.4×10^{-4}
C_6H_5COOH	6.5×10^{-5}

(Answers to Practice Exercises are in the back of the book.)

TABLE 15.2 Some Common Weak Acids and Their Ionization Reactions in Water

Weak Acid	Molecular Structure	Reaction in Water	K_a
Acetic		$CH_3COOH(aq) + H_2O(\ell) \rightleftharpoons CH_3COO^-(aq) + H_3O^+(aq)$	1.76×10^{-5}
Formic		$HCOOH(aq) + H_2O(\ell) \rightleftharpoons HCOO^-(aq) + H_3O^+(aq)$	1.77×10^{-4}
Hydrofluoric	$F—H$	$HF(aq) + H_2O(\ell) \rightleftharpoons F^-(aq) + H_3O^+(aq)$	6.8×10^{-4}
Hypochlorous	$Cl—O—H$	$HClO(aq) + H_2O(\ell) \rightleftharpoons ClO^-(aq) + H_3O^+(aq)$	2.9×10^{-8}
Nitrous		$HNO_2(aq) + H_2O(\ell) \rightleftharpoons NO_2^-(aq) + H_3O^+(aq)$	4.0×10^{-4}

Strong and Weak Bases

The most common strong bases are hydroxides of the group 1 and 2 elements. Table 15.3 lists the ions present when these ionic compounds dissociate as they dissolve in water. Note that 1 mole of a group 1 hydroxide produces 1 mole of OH^- ions when it dissolves in water, but 1 mole of a group 2 hydroxide produces 2 moles of OH^- ions. Note that the equilibrium constants for these Arrhenius bases, which are designated K_b to indicate that the reactant is a *base*, all have values much greater than 1, that is, $K_b \gg 1$.

TABLE 15.3 Strong Bases and Their Dissociation in Water

Strong Base	Solid	Solvated Ions
Lithium hydroxide	$LiOH(s)$	$Li^+(aq) + OH^-(aq)$
Sodium hydroxide	$NaOH(s)$	$Na^+(aq) + OH^-(aq)$
Potassium hydroxide	$KOH(s)$	$K^+(aq) + OH^-(aq)$
Calcium hydroxide	$Ca(OH)_2(s)$	$Ca^{2+}(aq) + 2\,OH^-(aq)$
Barium hydroxide	$Ba(OH)_2(s)$	$Ba^{2+}(aq) + 2\,OH^-(aq)$
Strontium hydroxide	$Sr(OH)_2(s)$	$Sr^{2+}(aq) + 2\,OH^-(aq)$

Even though the weak bases listed in Table 15.4 are compounds that do not contain hydroxide ions, they produce OH^- ions when dissolved in aqueous solution. To understand how this happens, let's focus on the behavior of ammonia. In aqueous solution, some collisions between water molecules and ammonia molecules result in the transfer of a proton from a water molecule (functioning as a Brønsted–Lowry acid) to an ammonia molecule (a Brønsted–Lowry base). The transferred proton bonds to the lone pair on the nitrogen atom, resulting in NH_4^+ and OH^- ions (Figure 15.3). Many other weak bases also contain

TABLE 15.4 Some Common Weak Bases and Their Ionization Reactions in Water

Weak Base	Reaction in Water	K_b
Ammonia	$NH_3(aq) + H_2O(\ell) \rightleftharpoons NH_4^+(aq) + OH^-(aq)$	1.8×10^{-5}
Aniline	$C_6H_5NH_2(aq) + H_2O(\ell) \rightleftharpoons C_6H_5NH_3^+(aq) + OH^-(aq)$	4.0×10^{-10}
Dimethylamine	$(CH_3)_2NH(aq) + H_2O(\ell) \rightleftharpoons (CH_3)_2NH_2^+(aq) + OH^-(aq)$	5.9×10^{-4}
Methylamine	$CH_3NH_2(aq) + H_2O(\ell) \rightleftharpoons CH_3NH_3^+(aq) + OH^-(aq)$	4.4×10^{-4}
Pyridine	$C_5H_5N(aq) + H_2O(\ell) \rightleftharpoons C_5H_5NH^+(aq) + OH^-(aq)$	1.7×10^{-9}

FIGURE 15.3 (a) Molecules of ammonia can react with molecules of water, producing NH_4^+ and OH^- ions. (b) An aqueous solution of ammonia consists mostly of dissolved ammonia molecules and a small amount of ammonium ions and hydroxide ions, indicating the small extent to which the forward reaction proceeds. (*Note*: The large concentration of H_2O has been omitted.)

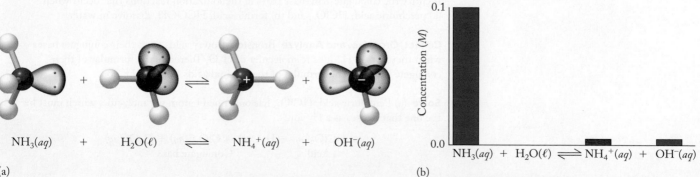

$$NH_3(aq) + H_2O(\ell) \rightleftharpoons NH_4^+(aq) + OH^-(aq)$$

(a)

(b)

a nitrogen atom with a lone pair of electrons that can bond to H^+ ions (see Table A5.3 in Appendix 5).

Conjugate Pairs

Let's revisit the nitrous acid equilibrium in Equation 15.4. In the forward direction, HNO_2 functions as a Brønsted–Lowry acid by donating H^+ ions to molecules of H_2O (Figure 15.4a). In the reverse reaction, NO_2^- ions accept H^+ ions and function as a Brønsted–Lowry base. Structurally, the difference between HNO_2 and NO_2^- is the H^+ ion that a molecule of HNO_2 donates, resulting in the formation of a NO_2^- ion, and that a NO_2^- ion accepts in the reverse reaction to reform a molecule of HNO_2. An acid and a base that are related in this way are a **conjugate acid–base pair**, or *conjugate pair*. An acid forms its **conjugate base** when it donates a H^+ ion, and a base forms its **conjugate acid** when it accepts a H^+ ion.

Figure 15.4(b) depicts the base ammonia and its conjugate acid. As we saw in Figure 15.3, when ammonia is added to water, some molecules of NH_3 accept H^+ ions from molecules of H_2O. In this way NH_3 functions as a Brønsted–Lowry base. In the reverse reaction, NH_4^+ ions donate H^+ ions and function as Brønsted–Lowry acids. Because the structure of NH_3 molecules and the structure of NH_4^+ ions differ only by the H^+ ion that a molecule of NH_3 accepts and that an NH_4^+ ion donates, NH_3 and NH_4^+ are also a conjugate acid–base pair.

What can we say about the acid–base behavior of the water molecules in the reactions in Figure 15.4? Note that H_2O functions as a base in the nitrous acid reaction, accepting a H^+ ion and forming its conjugate acid, H_3O^+, making H_2O and H_3O^+ a conjugate pair. In the ammonia reaction, however, H_2O functions as an acid by donating a H^+ ion to form its conjugate base, the OH^- ion. These two reactions illustrate that water can act as an acid or a base, depending on the acid–base properties of the substance dissolved in it. We will revisit this acid–base duality of water later in this chapter.

To summarize our discussion of conjugate acid–base pairs, let's write generic chemical equations that describe the relationships between conjugate pairs in aqueous solutions. Note how the difference within each conjugate acid–base pair is the H^+ ion that is donated by the acid to form its conjugate base and that a base accepts to form its conjugate acid.

$$\text{Acid}(aq) + H_2O(\ell) \rightleftharpoons \text{conjugate base}(aq) + H_3O^+(aq)$$

$$\text{Base}(aq) + H_2O(\ell) \rightleftharpoons \text{conjugate acid}(aq) + OH^-(aq)$$

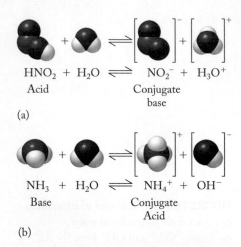

$$HNO_2 + H_2O \rightleftharpoons NO_2^- + H_3O^+$$
Acid $\qquad$ Conjugate base

(a)

$$NH_3 + H_2O \rightleftharpoons NH_4^+ + OH^-$$
Base $\qquad$ Conjugate Acid

(b)

FIGURE 15.4 Two acid–base conjugate pairs.

SAMPLE EXERCISE 15.2 Identifying Conjugate Acid–Base Pairs $\qquad$ **LO2**

Identify the conjugate acid–base pairs in the ionization reactions that occur when (a) perchloric acid, $HClO_4$, and (b) formic acid, $HCOOH$, dissolve in water.

Collect, Organize, and Analyze Brønsted–Lowry acids form their conjugate bases when they donate H^+ ions to molecules of H_2O. Therefore the formulas of their conjugate bases are the formulas of the original acids, minus a H^+ ion.

Solve (a) Perchloric acid, $HClO_4$, has only one H atom per molecule, which must be the one that it loses as a H^+ ion:

$$HClO_4(aq) + H_2O(\ell) \rightarrow ClO_4^-(aq) + H_3O^+(aq)$$
Acid $\qquad\qquad$ Conjugate base

(b) The formula of formic acid, HCOOH, tells us that it has two H atoms per molecule. However, only the one bonded to an O atom in the carboxylic acid group (see Table 15.2) is ionizable in water:

$$HCOOH(aq) + H_2O(\ell) \rightleftharpoons \underset{\text{Conjugate base}}{HCOO^-(aq)} + H_3O^+(aq)$$
$$\underset{\text{Acid}}{}$$

In both reactions, H_3O^+ and H_2O are also a conjugate acid–base pair: H_2O is the base and H_3O^+ is its conjugate acid.

Think About It Perchloric acid is a strong acid that completely dissociates in aqueous solution. Therefore ClO_4^- is a very, very weak base in that the reverse reaction in which ClO_4^- might add a H^+ ion essentially never occurs, and a single arrow is used to write this reaction. Equilibrium arrows are used for HCOOH because it is a weak acid.

 Practice Exercise Identify the conjugate acid–base pairs in the reaction that occurs when acetic acid (CH_3COOH) dissolves in water.

conjugate acid–base pair a Brønsted–Lowry acid and base that differ from each other only by a H^+ ion: acid $\rightleftharpoons$ conjugate base + H^+.

conjugate base the base formed when a Brønsted–Lowry acid donates a H^+ ion.

conjugate acid the acid formed when a Brønsted–Lowry base accepts a H^+ ion.

leveling effect the observation that all strong acids have the same strength in water and are completely converted into solutions of H_3O^+ ions; strong bases are likewise leveled in water and are completely converted into solutions of OH^- ions.

Relative Strengths of Conjugate Acids and Bases

As noted in Table 15.1, HCl is a strong acid, and just like $HClO_4$ in Sample Exercise 15.2, its acid ionization reaction goes to completion:

$$HCl(aq) + H_2O(\ell) \rightarrow Cl^-(aq) + H_3O^+(aq) \qquad (15.5)$$

Therefore the reverse reaction essentially does not happen at all, and the Cl^- ion (the conjugate base of HCl) must be a very weak base. This contrast in relative strengths applies to all conjugate pairs: strong acids have very weak conjugate bases and strong bases have very weak conjugate acids, as shown in Figure 15.5.

Between these extremes exist many weak acids with weak conjugate bases. For example, HNO_2 is a weak acid ($K_a = 4.0 \times 10^{-4}$), which means that its conjugate base, NO_2^-, is a weak base. This pairing of weak acids and weak bases applies to the vast majority of the conjugate acids and bases that occur in nature and in ourselves.

All the strong acids in Figure 15.5 ionize completely in water; that is, every molecule transfers a H^+ ion to a molecule of H_2O. In this context, water is said to *level* the strengths of these acids; they all are equally strong because they cannot be more than 100% ionized. This **leveling effect** means that the conjugate acid of H_2O, H_3O^+, is the strongest acid that can exist in water. An acid that is stronger than H_3O^+ simply donates all its ionizable H atoms to water molecules, forming H_3O^+ ions, when it dissolves in water. On the other hand, weak acids are differentiated by the fact that only a small fraction of the molecules donate their ionizable H atoms to water molecules. The weak acids higher on the list in Figure 15.5 form more acidic aqueous solutions than acids lower on the list because the larger the K_a, the greater the fraction of molecules that dissociate at a given concentration.

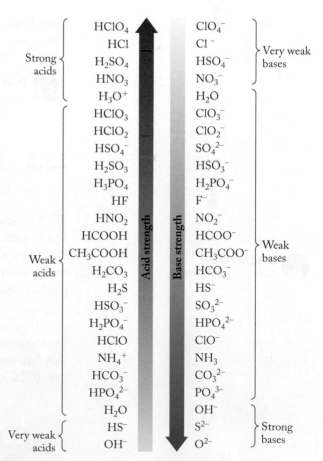

FIGURE 15.5 Opposing trends characterize the relative strengths of acids and their conjugate bases: the stronger the acid, the weaker its conjugate base. The same is true for bases: the stronger the base, the weaker its conjugate acid.

A similar leveling effect exists for strong bases. The strongest base that can exist in water is the conjugate base of H_2O: the OH^- ion. Any base that is stronger than OH^- hydrolyzes in water to produce OH^- ions. The oxide ion (O^{2-}), for example, is a very strong base, and reacts with water to produce two OH^- ions:

$$O^{2-}(aq) + H_2O(\ell) \rightarrow 2\,OH^-(aq)$$

The strengths of bases weaker than OH^- ions can be differentiated by the fraction of their molecules that accept H^+ ions from water molecules in aqueous solutions. Weaker bases are higher on the list in Figure 15.5; stronger bases are lower on the list.

SAMPLE EXERCISE 15.3 Relating the Strengths of Conjugate Pairs   **LO2**

List the following anions in order of decreasing strength as Brønsted–Lowry bases: F^-, Cl^-, OH^-, $HCOO^-$, and NO_2^-.

Collect, Organize, and Analyze All five anions are listed among the bases in Figure 15.5. Therefore all we have to do is rank them according to their location in the figure: the strongest one will be the closest to the bottom and the weakest will be the closest to the top.

Solve The anions in decreasing order of strength as bases (H^+ ion acceptors) are OH^-, $HCOO^-$, NO_2^-, F^-, and Cl^-.

Think About It A check of Figure 15.5 discloses that we have also sorted the anions in *increasing* order for the strength of their conjugate acids. The complementary sorting makes sense because the stronger the acid, the weaker its conjugate base, and vice versa.

 Practice Exercise List the following anions in order of decreasing strength as Brønsted–Lowry bases: Br^-, S^{2-}, ClO^-, CH_3COO^-, and HSO_3^-.

15.3 pH and the Autoionization of Water

We have seen that the acidity of a solution is directly related to its concentration of H_3O^+ ions. In this section we examine another way to quantify acidity. To understand this alternative approach, we first need to consider the **autoionization** of water, which is the process that can happen when two water molecules collide to produce equal and very small concentrations of H_3O^+ and OH^- ions in pure water:

$$H_2O(\ell) + H_2O(\ell) \rightleftharpoons H_3O^-(aq) + OH^-(aq) \qquad (15.6)$$

One water molecule, acting as an acid, donates a H^+ ion to another water molecule that functions as a base (Figure 15.6). The donor H_2O molecule forms its conjugate base (OH^-), and the acceptor H_2O molecule forms its conjugate acid (H_3O^+). We

CHEMTOUR
Autoionization of Water

FIGURE 15.6 The autoionization of water takes place when a proton is transferred from one water molecule to another. Both molecules are converted to ions.

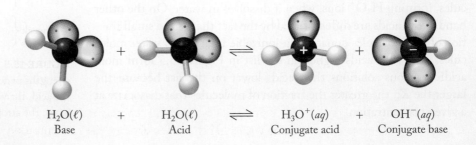

| $H_2O(\ell)$ | + | $H_2O(\ell)$ | $\rightleftharpoons$ | $H_3O^+(aq)$ | + | $OH^-(aq)$ |
| Base | | Acid | | Conjugate acid | | Conjugate base |

have already encountered this dual nature of water: molecules of H_2O act as H^+ ion acceptors in solutions of acidic solutes, and as H^+ ion donors in solutions of basic solutes. The autoionization of water is an example of amphiprotic behavior.

The equilibrium constant expression for the autoionization of water might be written

$$K_c = \frac{[H_3O^+][OH^-]}{[H_2O][H_2O]} \qquad (15.7)$$

However, water is a pure liquid, which means we do not include its concentration in equilibrium constant expressions. This reduces Equation 15.7 to an equilibrium constant expression that is given the symbol K_w:

$$K_w = [H_3O^+][OH^-] \qquad (15.8)$$

In pure water at 25°C, $[H_3O^+] = [OH^-] = 1.0 \times 10^{-7}\ M$. Inserting these values into Equation 15.8 gives us

$$K_w = [H_3O^+][OH^-] = (1.0 \times 10^{-7})(1.0 \times 10^{-7}) = 1.0 \times 10^{-14} \qquad (15.9)$$

This tiny K_w value confirms that a very small fraction of water molecules undergo autoionization at room temperature. The reverse of autoionization—the reaction between $[H_3O^+]$ and $[OH^-]$ to produce H_2O—has an equilibrium constant of $1/K_w = 1.0 \times 10^{14}$ at 25°C and essentially goes to completion:

$$H_3O^+(aq) + OH^-(aq) \rightleftharpoons H_2O(\ell) + H_2O(\ell) \qquad K = 1/K_w = 1.0 \times 10^{14}$$

Equation 15.9 means that an inverse relationship exists between $[H_3O^+]$ and $[OH^-]$ in any aqueous sample: as the concentration of one increases, the concentration of the other must decrease so that the product of the two is always 1.0×10^{-14}. A solution in which $[H_3O^+] > [OH^-]$ is acidic, a solution in which $[H_3O^+] < [OH^-]$ is basic, and a solution in which $[H_3O^+] = [OH^-] = 1.0 \times 10^{-7}\ M$ is neutral (neither acidic nor basic).

The tiny value of K_w means that autoionization of water does not contribute significantly to $[H_3O^+]$ in solutions of most acids or to $[OH^-]$ in solutions of most bases, so we can ignore the contribution of autoionization in most calculations of acid or base strength. However, if acids or bases are extremely weak, or if their concentrations are extremely low, H_2O autoionization may need to be considered.

The pH Scale

In the early 1900s scientists developed a device called the *hydrogen electrode* to determine the concentrations of hydronium ions in solutions. The electrical voltage, or *potential*, produced by the hydrogen electrode is a linear function of the logarithm of $[H_3O^+]$. This relationship led Danish biochemist Søren Sørensen (1868–1939) to propose a scale for expressing acidity and basicity on the basis of what he termed "the potential of the hydrogen ion," abbreviated **pH**. Mathematically, we define pH as the negative logarithm of $[H_3O^+]$:

$$pH = -\log[H_3O^+] \qquad (15.10)$$

For example, the pH of a solution in which $[H_3O^+] = 5.0 \times 10^{-3}\ M$ is

$$pH = -\log(5.0 \times 10^{-3}) = -(-2.30) = 2.30$$

Sørensen's pH scale has several attractive features. Because it is logarithmic, there are no exponents, as are commonly encountered in values of $[H_3O^+]$. The logarithmic scale also means that a change of one pH unit corresponds to a 10-fold change in $[H_3O^+]$, so that a solution with a pH of 5.0 has 10 times the concentration

CONNECTION We defined *amphiprotic* compounds in Chapter 4 as having both acidic and basic properties.

 **CHEMTOUR** pH Scale

autoionization the process that produces equal and very small concentrations of H_3O^+ and OH^- ions in pure water.

pH the negative logarithm of the hydronium ion concentration in an aqueous solution.

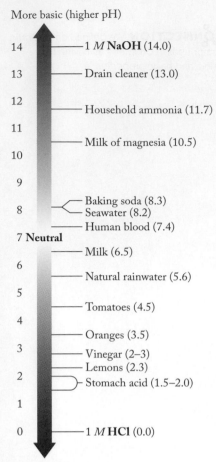

More basic (higher pH)

- 14 —— 1 M **NaOH** (14.0)
- 13 —— Drain cleaner (13.0)
- 12 —— Household ammonia (11.7)
- 11
- 10 —— Milk of magnesia (10.5)
- 9
- 8 —— Baking soda (8.3)
      —— Seawater (8.2)
      —— Human blood (7.4)
- **7 Neutral**
      —— Milk (6.5)
- 6
- 5 —— Natural rainwater (5.6)
- 4 —— Tomatoes (4.5)
- 3 —— Oranges (3.5)
      —— Vinegar (2–3)
      —— Lemons (2.3)
- 2 —— Stomach acid (1.5–2.0)
- 1
- 0 —— 1 M **HCl** (0.0)

More acidic (lower pH)

FIGURE 15.7 The pH scale is a convenient way to express the range of acidic or basic properties of some common materials.

of H_3O^+ ions and is 10 times as acidic as a solution with a pH of 6.0. Similarly, the concentration of H_3O^+ ions in a solution with a pH of 12.0 is 1/10 that of a solution with a pH of 11.0. Conversely, the concentration of OH^- ions in a solution with a pH of 12.0 is 10 times that of a solution with a pH of 11.0.

The negative sign in front of the logarithmic term in Equation 15.10 means that pH values of most aqueous solutions, except for very concentrated solutions of strong acids or bases, are positive numbers between 0 and 14. It also means that *large* pH values correspond to *small* values of $[H_3O^+]$. Acidic solutions have pH values less than 7.00 ($[H_3O^+] > 1.0 \times 10^{-7}$ M), and basic solutions have pH values greater than 7.00 ($[H_3O^+] < 1.0 \times 10^{-7}$ M). A solution with a pH of exactly 7.00 is neutral. The pH values for some common aqueous solutions are shown in Figure 15.7.

CONCEPT **TEST**

Match the pH values on the left with the descriptors on the right.

13.77	strongly acidic
10.03	weakly acidic
7.00	weakly basic
4.37	strongly basic
0.22	neutral

SAMPLE EXERCISE 15.4 Relating pH and $[H_3O^+]$ **LO3**

Suppose the pH of solution A is 8.0 and the pH of solution B is 4.0. Is each of the following statements about the two solutions true or false?

a. Solution A is 2 times as acidic as solution B.
b. Solution B is 10,000 times as acidic as solution A.
c. The concentration of OH^- ions in solution A is half their concentration in solution B.
d. The values of $[OH^-]$ and $[H_3O^+]$ are closer to each other in solution A than in solution B.

Collect, Organize, and Analyze The pH scale is logarithmic: a decrease (or increase) of one pH unit corresponds to a 10-fold increase (or a 90% decrease) in $[H_3O^+]$. The negative sign in the pH formula (pH = $-\log[H_3O^+]$) means that a higher pH value corresponds to a (much) lower $[H_3O^+]$.

Solve

a. False, because pH is a log scale, and because higher pH values mean lower $[H_3O^+]$.
b. True; a decrease of 4 pH units corresponds to a 10^4 increase in $[H_3O^+]$ and acidity.
c. False, because pH is a log scale. Actually, $[OH^-]$ in solution A is 10^4 times that of solution B.
d. True, because a pH of 8.0 is only one unit away from the pH (7.0) where $[H_3O^+] = [OH^-]$. A pH of 4.0 is 3 units below 7.0, so $[H_3O^+] = 1000 \times [OH^-]$ in solution B.

Think About It You may find it challenging at first to do log math in your head. It helps to understand that an *increase* of one pH unit corresponds to a *decrease* in $[H_3O^+]$ to only 1/10 or 10% of its initial value.

✦ **Practice Exercise** Which of the following changes in pH corresponds to the greatest percent increase in $[H_3O^+]$? Which change corresponds to the greatest percent decrease in $[H_3O^+]$? (a) pH 1 → pH 3; (b) pH 7 → pH 4; (c) pH 12 → pH 14; (d) pH 5 → pH 9; (e) pH 2 → pH 0

When we interconvert $[H_3O^+]$ and pH values, we need to be clear on how to express pH values with the appropriate number of significant digits. Because a pH value is the negative logarithm of a hydronium ion concentration, the digit or digits before the decimal point in the pH value define the location of the decimal point in the $[H_3O^+]$ value. They do not define how precisely we know the concentration value, so they are not considered significant digits. For example, $[H_3O^+]$ = 2.7×10^{-4} has two significant digits: the 2 and 7. The corresponding pH value, 3.57, should also have two significant digits after the decimal point and it does: the 5 and 7. The 3 in pH 3.57 is not a significant digit because its function is to tell us that the corresponding $[H_3O^+]$ value is between 10^{-3} and 10^{-4} M.

CONNECTION The general rules for the use of significant figures in calculations involving measured quantities were discussed in Section 1.10.

SAMPLE EXERCISE 15.5 Interconverting pH and $[H_3O^+]$ **LO3**

Cells located in the upper region of the human stomach called the *fundus* secrete gastric acid (about 0.16 M HCl) during digestion. After this acid mixes with the food being digested, the pH of the contents of the stomach is often between 3 and 4.

a. How much more acidic is 0.16 M HCl than stomach contents with a pH of 3.20?
b. Which of the two samples in part (a) has the lower $[H_3O^+]$?

Collect and Organize We are asked to compare the acidities (the ratio of the $[H_3O^+]$ values) of two solutions that both contain hydrochloric acid. We know the pH of one and the HCl concentration of the other, and we must also determine which sample has the lower $[H_3O^+]$.

Analyze Hydrochloric acid is a strong acid, so the $[H_3O^+]$ value of a solution of HCl is the same as the concentration of the acid. We can use Equation 15.10 to calculate the pH of 0.16 M HCl and to convert pH 3.20 into its corresponding $[H_3O^+]$ value. The result of the second calculation, when divided into 0.16 M, will give us the desired ratio of acidities.

Solve
a. To convert pH to $[H_3O^+]$, we need to solve Equation 15.10 for $[H_3O^+]$:

$$pH = -\log[H_3O^+]$$

or,

$$\log[H_3O^+] = -pH$$

Taking the antilog (10^x) of both sides:

$$[H_3O^+] = 10^{-pH}$$

Inserting pH = 3.20:

$$[H_3O^+] = 10^{-3.20} = 6.3 \times 10^{-4}\ M$$

Calculating the ratio of acidities ($[H_3O^+]$ values):

$$\frac{[H_3O^+](\text{gastric acid})}{[H_3O^+](\text{stomach content})} = \frac{0.16\ M}{6.3 \times 10^{-4}\ M} = 2.5 \times 10^2$$

Thus the acidity of secreted gastric acid is 2.5×10^2 times the acidity of the digestion mixture in the stomach.
b. Using the values calculated in part (a), the stomach contents have a lower $[H_3O^+]$.

Think About It The results of the two parts to this sample exercise are consistent: 0.16 M HCl is 2.5×10^2 times more acidic than the stomach content sample, and its pH is $3.20 - 0.8 = 2.4$ units lower. Note how this seemingly small change in pH corresponds to a more than 200-fold difference in acidity.

⬡ **Practice Exercise** The results of several oceanographic studies indicate that the pH of Earth's oceans is now 8.07. The hydrogen ion concentration of the oceans at the start of the industrial revolution was 6.61×10^{-9} M. How much more acidic is the ocean now than then? Does the ocean now have a lower pH than the ocean previously?

CONCEPT TEST

Is the pH of a 1.00 M solution of a weak acid higher or lower than the pH of a 1.00 M solution of a strong acid?

pOH, pK_a, and pK_b Values

The letter p as used in pH is also used with other symbols to mean *the negative logarithm* of the variable that follows it. For example, just as every aqueous solution has a pH value, it also has a **pOH** value, defined as

$$pOH = -\log[OH^-] \tag{15.11}$$

We can use the K_w expression, Equation 15.9, to relate pOH to pH. We start by taking the negative logarithm of both sides of the equation and then rearrange the terms:

$$K_w = [H_3O^+][OH^-] = 1.0 \times 10^{-14}$$

$$-\log K_w = -\log([H_3O^+][OH^-]) = -\log(1.0 \times 10^{-14})$$

$$pK_w = -\log[H_3O^+] + -\log[OH^-] = -(-14.00)$$

$$pK_w = pH + pOH = 14.00 \tag{15.12}$$

Many tables of equilibrium constants list pK values rather than K values because doing so does not require the use of exponential notation and saves space. The tables in Appendix 5 of this book contain both sets of values. Use them whenever you need a K or pK value that is not provided in a problem. For example, we can define both pK_a and pK_b values for acetic acid and methylamine, respectively, as well as the other acids and bases in Tables 15.2 and 15.4:

$$pK_a = -\log K_a \qquad\qquad pK_b = -\log K_b$$
$$= -\log(1.8 \times 10^{-5}) \qquad = -\log(4.4 \times 10^{-4})$$
$$= 4.74 \qquad\qquad\qquad = 3.36$$

SAMPLE EXERCISE 15.6 Relating [H_3O^+], [OH^-], pH, and pOH   **LO3**

Digested food travels from your stomach, where the pH is 4.50, and moves to your small intestine, where the pH increases to 7.50. Convert these pH values to pOH, [H_3O^+], and [OH^-] values.

Collect, Organize, and Analyze We are given two pH values, and we want to determine the corresponding pOH, [H_3O^+] and [OH^-] values.

- We can use Equation 15.10 to convert from pH to [H_3O^+], as we did in Sample Exercise 15.5.
- We can use Equation 15.12 to convert from pH to pOH.
- We can use Equation 15.11 to convert from pOH to [OH^-].

pOH the negative logarithm of the hydroxide ion concentration in an aqueous solution.

The [H$_3$O$^+$] value we calculate for pH 4.50 should be 10^3, or 1000, times the [H$_3$O$^+$] value we calculate for pH 7.50 (because 7.50 − 4.50 = 3.00). Correspondingly, the [OH$^-$] value we calculate for pH 7.50 should be 10^3, or 1000, times the [OH$^-$] value we calculate for pH 4.50. The inverse relationship between pH and pOH means that the pOH value we calculate for pH 4.50 should be 3 units larger than the one we get for pH 7.50.

Solve

1. Converting pH to [H$_3$O$^+$]:

$$pH = -\log[H_3O^+] \tag{15.10}$$

or

$$[H_3O^+] = 10^{-pH}$$

Stomach: $[H_3O^+] = 10^{-4.50} = 3.2 \times 10^{-5}\ M$

Small intestine: $[H_3O^+] = 10^{-7.50} = 3.2 \times 10^{-8}\ M$

2. Converting pH to pOH:

$$pH + pOH = 14.00 \tag{15.12}$$

$$pOH = 14.00 - pH$$

Stomach: $pOH = 14.00 - 4.50 = 9.50$

Small intestine: $pOH = 14.00 - 7.50 = 6.50$

3. Converting pOH to [OH$^-$]:

$$pOH = -\log[OH^-] \tag{15.11}$$

or

$$[OH^-] = 10^{-pOH}$$

Stomach: $[OH^-] = 10^{-9.50} = 3.2 \times 10^{-10}\ M$

Small intestine: $[OH^-] = 10^{-6.50} = 3.2 \times 10^{-7}\ M$

Think About It As predicted, the [H$_3$O$^+$] in the stomach is 1000 times that of the small intestine. The differences in pOH values are also as predicted, as are the differences in [OH$^-$]. All calculated concentration values were rounded off to two significant digits because each starting pH value had only two digits after the decimal point.

 Practice Exercise What are the values of [H$_3$O$^+$] and [OH$^-$] in household ammonia, an aqueous solution of NH$_3$ that has a pH of 11.70?

15.4 K_a, K_b, and the Ionization of Weak Acids and Bases

The vast majority of the acids and bases in nature are weak acids and bases. In this section and the one that follows, we explore how to quantify the degree to which these molecules react with molecules of water to form either H$_3$O$^+$ or OH$^-$ ions.

Weak Acids

We saw in Section 15.2 that nitrous acid, HNO$_2$, is a weak acid, which means it is only partially ionized when it dissolves in water:

$$HNO_2(aq) + H_2O(\ell) \rightleftharpoons NO_2^-(aq) + H_3O^+(aq) \tag{15.4}$$

Let's explore the degree to which this reaction proceeds before it reaches equilibrium. We begin by translating Equation 15.4 into an equilibrium constant expression, with products in the numerator and reactants in the denominator:

$$K_a = \frac{[NO_2^-][H_3O^+]}{[HNO_2]}$$

Remember that there is no concentration term for water in the denominator because it is the solvent in this reaction and the concentration of water changes very little during the reaction.

We can generalize this observation. The ionization of any weak acid (HA), as expressed by the reaction

$$HA(aq) + H_2O(\ell) \rightleftharpoons A^-(aq) + H_3O^+(aq) \qquad (15.13)$$

has a corresponding K_a expression that looks like:

$$K_a = \frac{[A^-][H_3O^+]}{[HA]} \qquad (15.14)$$

We can use Equation 15.14 to calculate the K_a value of an unknown weak acid if we know both the concentration of a solution of the acid and its pH. Suppose, for example, we determine that a 0.100 M solution of HA has a pH of 2.20. To calculate K_a, we first convert the pH value to $[H_3O^+]$:

$$[H_3O^+] = 10^{-2.20} = 6.3 \times 10^{-3} \ M$$

Next we assume that the only source of H_3O^+ ions is the ionization of HA. If so, then $[A^-]$ must also be $6.3 \times 10^{-3} \ M$, because 1 mole of HA ionizes to produce 1 mole of H_3O^+ and 1 mole of A^-. Also, if $[H_3O^+]$ increases by $6.3 \times 10^{-3} \ M$, then [HA] must have decreased by $6.3 \times 10^{-3} \ M$. Therefore the concentration of HA at equilibrium is

$$[HA] = (0.100 - 6.3 \times 10^{-3}) \ M = 0.094 \ M$$

Using the three calculated equilibrium concentrations in the K_a expression gives us:

$$K_a = \frac{[A^-][H_3O^+]}{[HA]} = \frac{(6.3 \times 10^{-3})(6.3 \times 10^{-3})}{(0.094)} = 4.2 \times 10^{-4}$$

The small value of K_a confirms that HA is a weak acid.

The ratio of the equilibrium $[H_3O^+]$ to the initial [HA], which equals the ratio of the equilibrium $[A^-]$ to the initial [HA], represents the **degree of ionization** of HA, which is usually expressed as a percentage of the initial acid concentration. For this reason it is also called **percent ionization**. In equation form this relationship is

$$\text{Percent ionization} = \frac{[H_3O^+]_{\text{equilibrium}}}{[HA]_{\text{initial}}} \times 100\% \qquad (15.15)$$

Inserting the data from the previous calculation into Equation 15.15:

$$\text{Percent ionization} = \frac{6.3 \times 10^{-3} \ M}{0.100 \ M} \times 100\% = 6.3\%$$

degree of ionization the ratio of the quantity of a substance that is ionized to the concentration of the substance before ionization; when expressed as a percentage, called **percent ionization**.

■ CONCEPT **TEST**

Describe how the percent ionization of a weak acid is related to its K_a value.

Now let's examine how the degree of ionization of a weak acid depends on its concentration in solution. A plot of percent ionization as a function of the initial concentration of nitrous acid is shown in Figure 15.8. Note that the degree of ionization increases as the concentration of the acid decreases. This same pattern is observed for all weak acids.

We can explain this trend by using Le Châtelier's principle. Suppose, for example, we have a 1.0 M solution of a generic weak acid, HA, in which the concentrations of A^- and H_3O^+ ions at equilibrium are both 0.0010 M. Using these concentrations in Equation 15.14 to solve for K_a, we have:

$$K_a = \frac{[A^-][H_3O^+]}{[HA]} = \frac{(0.0010)(0.0010)}{(1.0 - 0.0010)} = 1.0 \times 10^{-6}$$

Now suppose that more water is added to the HA solution, increasing its volume by a factor of 10. If no change in the degree of ionization occurred, then [HA] would be 0.10 M, and the concentrations of A^- and H_3O^+ ions would both be 1.0×10^{-4} M. However, if we insert these values into the equilibrium constant expression, we get a reaction quotient (Q) value of:

$$Q = \frac{(1.0 \times 10^{-4})(1.0 \times 10^{-4})}{(0.10 - 1.0 \times 10^{-4})} = 1.0 \times 10^{-7}$$

This Q value is only 1/10 the value of K_a, which tells us the acid ionization reaction is no longer at equilibrium. Because the value of Q is less than the value of K_a, the reaction proceeds in the forward direction. This means the concentrations of A^- and H_3O^+ ions increase, and so does the percentage of HA molecules that ionize. Further dilutions would produce even greater percent ionization values, following the trend we see for nitrous acid in Figure 15.8.

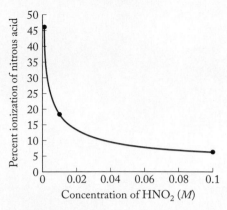

FIGURE 15.8 The degree of ionization of a weak acid increases with decreasing acid concentration. Here the degree of ionization of nitrous acid increases from about 6% in a 0.100 M solution to 18% in a 0.010 M solution to 46% in a 0.001 M solution.

SAMPLE EXERCISE 15.7 Relating pH, K_a, and Percent Ionization of a Weak Acid   **LO4**

The pH of a 1.00 M solution of formic acid (HCOOH) is 1.88.

a. What is the percent ionization of 1.00 M HCOOH?
b. What is the K_a value of formic acid?

Collect, Organize, and Analyze We are asked to determine the K_a value of formic acid and its percent ionization in a solution of known concentration and pH. There are two H atoms in a molecule of formic acid, but only one is in a carboxylic acid group and can ionize in an aqueous solution. Therefore we can write the acid ionization reaction this way:

$$HCOOH(aq) + H_2O(\ell) \rightleftharpoons HCOO^-(aq) + H_3O^+(aq)$$

The 1:1:1 stoichiometry of the ionization reaction tells us that in a solution of HCOOH at equilibrium, $[HCOO^-] = [H_3O^+]$, so using Equation 15.10, pH $= -\log[H_3O^+]$, will give us the values of $[H_3O^+]$ and $[HCOO^-]$ at equilibrium and will allow us to calculate the degree of ionization by using Equation 15.15. At equilibrium, [HCOOH] is equal to its initial concentration minus the portion that ionized, or

$$[HCOOH]_{equilibrium} = [HCOOH]_{initial} - [H_3O^+]_{equilibrium}$$

so we can calculate the value of K_a by using this equilibrium constant expression:

$$K_a = \frac{[HCOO^-][H_3O^+]}{[HCOOH]}$$

The pH of the 1.00 M solution is close to 2, which corresponds to $[H_3O^+] = 10^{-2}$ M. Therefore the percent ionization of formic acid in this solution should be about 1%.

Solve

a. The pH of the 1.00 M solution is 1.88. The corresponding $[H_3O^+]$ is

$$[H_3O^+] = 10^{-1.88} = 1.32 \times 10^{-2} \, M$$

Inserting this value and the initial concentration of HCOOH in Equation 15.15:

$$\text{Percent ionization} = \frac{[H_3O^+]_{\text{equilibrium}}}{[\text{HCOOH}]_{\text{initial}}} \times 100\%$$

$$= \frac{1.32 \times 10^{-2} \, M}{1.00 \, M} \times 100\% = 1.32\%$$

b. At equilibrium, $[\text{HCOO}^-] = [H_3O^+] = 1.32 \times 10^{-2} \, M$, and the equilibrium concentration of HCOOH is

$$(1.00 - 1.32 \times 10^{-2}) \, M = 0.99 \, M$$

Inserting these values in the expression for K_a:

$$K_a = \frac{[\text{HCOO}^-][H_3O^+]}{[\text{HCOOH}]} = \frac{(1.32 \times 10^{-2})(1.32 \times 10^{-2})}{(0.99)} = 1.8 \times 10^{-4}$$

Think About It The result of the percent ionization calculation is close to our estimate of about 1%, and the calculated K_a value is very close to the K_a value for formic acid listed in Table 15.2 (1.77×10^{-4}).

Practice Exercise The value of $[H_3O^+]$ in a 0.050 M solution of an organic acid is $5.9 \times 10^{-3} \, M$. What is the pH of the solution, the percent ionization of the acid, and its K_a value?

CONCEPT TEST

Three weak acids have these K_a values:

Acid	K_a
A	3.6×10^{-5}
B	4.9×10^{-4}
C	9.2×10^{-4}

Which acid is the most extensively ionized in a 0.10 M solution of the acid? Which acid has the lowest percent ionization in a 1.00 M solution of the acid?

Weak Bases

Now let's examine the degree of ionization of weakly basic compounds when they dissolve in water. For example, ammonia bonds to hydrogen ions transferred from water as described by the chemical equation

$$NH_3(aq) + H_2O(\ell) \rightleftharpoons NH_4^+(aq) + OH^-(aq)$$

The limited strength of ammonia as a base is reflected in its small K_b value at 25°C:

$$K_b = \frac{[NH_4^+][OH^-]}{[NH_3]} = 1.76 \times 10^{-5}$$

As with K_a expressions, there is no $[H_2O]$ term in this K_b expression because the concentration of water does not change significantly during the reaction.

More generally, the transfer of a proton to any Brønsted–Lowry weak base in aqueous solution can be written as:

$$B(aq) + H_2O(\ell) \rightleftharpoons HB^+(aq) + OH^-(aq)$$

with a corresponding equilibrium constant expression:

$$K_b = \frac{[HB^+][OH^-]}{[B]} \qquad (15.16)$$

Just as with weak acids, we can calculate the degree of ionization of a weak base and its K_b value if we know the pH and initial concentration of a solution of the base:

$$\text{Percent ionization} = \frac{[OH^-]_{equilibrium}}{[B]_{initial}} \times 100\% \qquad (15.17)$$

However, there are more steps involved than in the corresponding calculation for a solution of a weak acid, as in Sample Exercise 15.7. Because of the $[OH^-]$ term in the K_b expression of weak bases, we must convert pH to pOH and then calculate $[OH^-]$, as illustrated in the following Sample Exercise.

SAMPLE EXERCISE 15.8 Relating pH, K_b, and Percent   **LO4**
                         Ionization of a Weak Base

Trimethylamine, $(CH_3)_3N$, is a particularly foul-smelling volatile organic compound that forms during the decay of plant and animal matter. It is soluble in water, and it reacts with water molecules as described by this chemical equation:

$$(CH_3)_3N(aq) + H_2O(\ell) \rightleftharpoons (CH_3)_3NH^+(aq) + OH^-(aq)$$

The pH of a 0.50 M solution of trimethylamine is 11.75 at 25°C.

a. What is the percent ionization of trimethylamine in a 0.50 M solution?
b. What is the K_b value of trimethylamine?

Collect, Organize, and Analyze We are asked to determine the K_b value of trimethylamine and its percent ionization in a solution of known concentration and pH. The stoichiometry of the ionization reaction tells us that $[(CH_3)_3NH^+] = [OH^-]$ in an aqueous solution of $(CH_3)_3N$ at equilibrium. We also know that $[(CH_3)_3N]$ is equal to the initial concentration of the acid minus the portion of it that reacts with water to form OH^- ions:

$$[(CH_3)_3N]_{equilibrium} = [(CH_3)_3N]_{initial} - [OH^-]_{equilibrium}$$

Converting pH to pOH by using Equation 15.12 and then calculating $[OH^-]$ by using Equation 15.11 will give us $[OH^-]$ and $[(CH_3)_3NH^+]$ at equilibrium. We can then calculate the degree of ionization by using Equation 15.17. To calculate the value of K_b, the subtraction described above will give us the values we need for its equilibrium constant expression:

$$K_b = \frac{[(CH_3)_3NH^+][OH^-]}{[(CH_3)_3N]}$$

The pH of the 0.50 M solution is a little less than 12, which corresponds to a pOH that is a little more than 2 and a value for $[OH^-]$ that is a little less than 10^{-2} M. Therefore the percent ionization of trimethylamine in this solution should be about 1–2%.

Solve
a. The pH of the 0.50 M solution is 11.75. Calculating the corresponding pOH:

$$\text{pH} + \text{pOH} = 14.00$$

or

$$pOH = 14.00 - pH$$

$$= 14.00 - 11.75 = 2.25$$

Calculating $[OH^-]$:

$$[OH^-] = 10^{-pOH}$$

$$= 10^{-2.25} = 5.6 \times 10^{-3} \, M$$

Using this value and the initial concentration of trimethylamine to calculate percent ionization:

$$\text{Percent ionization} = \frac{[OH^-]_{equilibrium}}{[B]_{initial}} \times 100\%$$

$$= \frac{5.6 \times 10^{-3} \, M}{0.50 \, M} \times 100\% = 1.1\%$$

b. At equilibrium, $[(CH_3)_3NH^+] = [OH^-] = 5.6 \times 10^{-3} \, M$, and $[(CH_3)_3N]$ is

$$(0.50 - 5.6 \times 10^{-3}) \, M = 0.49 \, M$$

Inserting these values in the K_b expression:

$$K_b = \frac{[(CH_3)_3NH^+][OH^-]}{[(CH_3)_3N]} = \frac{(5.6 \times 10^{-3})(5.6 \times 10^{-3})}{0.49}$$

$$= 6.4 \times 10^{-5}$$

Think about It The result of the percent ionization calculation is in our estimated range of 1–2%, which reflects the fact that although trimethylamine may have a strong odor, it is only a weak base. The latter point is reinforced by the small K_b value.

Practice Exercise The pH of a 0.100 M aqueous solution of ethylamine, $C_2H_5NH_2$, is 11.86 at 25°C. What percentage of the $C_2H_5NH_2$ molecules in this solution are ionized as described in the chemical equation below, and what is the K_b value of ethylamine?

$$C_2H_5NH_2(aq) + H_2O(\ell) \rightleftharpoons C_2H_5NH_3^+(aq) + OH^-(aq)$$

As we end this section, consider the visual summary (Figure 15.9) of the ways we interconvert $[H_3O^+]$, $[OH^-]$, pH, and pOH. The diagram may leave you wondering which path to follow to make cross-corner (two-step) conversions, such as $[OH^-]$ to pH. Many students find the math simpler if they travel the "northern" route—that is, include a pH-to-pOH conversion in their calculation.

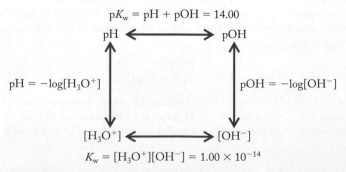

FIGURE 15.9 Pathways for interconverting $[H_3O^+]$, $[OH^-]$, pH, and pOH values.

15.5 Calculating the pH of Acidic and Basic Solutions

In this section we calculate the pH values of solutions of acids and bases. We begin with solutions of strong acids and strong bases.

Strong Acids and Strong Bases

We have seen that strong acids have large K_a values and ionize essentially completely in aqueous solutions. This means that in nearly all of their solutions, the concentration of H_3O^+ ions is the same as the initial concentration of the strong acid. For example, the pH of muriatic acid (7.4 M HCl), which is sold in hardware and building supply stores to clean concrete surfaces, is

$$pH = -\log[H_3O^+] = -\log[HCl] = -\log(7.4) = -0.87$$

This pH value is negative because $[H_3O^+]$ is greater than one, which means the value of the negative log of the concentration is less than zero. The negative sign in the pH equation gives us a negative pH value.

Like strong acids, strong bases ionize completely in aqueous solutions, so for a group 1 hydroxide such as NaOH, $[OH^-]$ will be the same as the strong base's initial concentration. For a group 2 hydroxide such as $Ca(OH)_2$, $[OH^-]$ will be twice the strong base's initial concentration. To calculate the pH of a solution of a strong base, therefore, we need to calculate the concentration of OH^- ions given the initial concentration of the base and then convert $[OH^-]$ to pOH, and then to pH. The following Sample Exercise illustrates the steps involved.

SAMPLE EXERCISE 15.9 Calculating the pH of a Solution of a Strong Base   **LO5**

Liquids such as the one in the bottle in Figure 15.10 are used to remove clogs from bathroom and kitchen drains. They contain sodium hydroxide at concentrations as high as 1.2 M. What is the pH of 1.2 M NaOH?

Collect, Organize, and Analyze NaOH is a strong base that produces 1 mole of hydroxide ions per mole of NaOH in solution, so $[OH^-]$ is the same as the initial concentration of the base. Therefore $[OH^-] = 1.2\ M$. We can convert this value to pOH and then to pH following the steps shown in Figure 15.9.

Solve

Calculating pOH: $pOH = -\log[OH^-] = -\log(1.2) = -0.079$

Calculating pH: $pH = 14.00 - pOH = 14.00 - (-0.079) = 14.08$

Think About It The calculated pH is slightly above 14 because $[OH^-]$ is slightly greater than one molar.

Practice Exercise *Kalkwasser* is the German word for "lime water," which is the common name for saturated solutions of $Ca(OH)_2$. It has many uses, including being added to water in aquarium tanks to adjust their pH (and to provide Ca^{2+} ions for the plants and animals living in the tank). What is the pH of Kalkwasser? Assume the concentration of such a solution is 0.0225 M $Ca(OH)_2$.

FIGURE 15.10 Cleaners used to remove grease and hair clogs from water pipes typically contain large amounts of sodium hydroxide.

Weak Acids and Weak Bases

In Section 15.4 we learned how pH and concentration can provide us with information about percent ionization and permit us to calculate the K_a or K_b of a weak acid or weak base, respectively. This raises the question: if we know the concentration of a weak acid solution and its K_a value (or a weak base and its K_b value), can we determine the pH of that solution? Sample Exercises 15.10 and 15.11 take us through these calculations, but in order to do so, we need to use a tool first introduced in Chapter 14—RICE tables.

CONNECTION In Chapter 14 RICE tables were a tool to help us calculate the changes (C) from initial (I) to equilibrium (E) concentrations for a given reaction (R).

FIGURE 15.11 The carabid beetle secretes formic acid as a defense against predators.

SAMPLE EXERCISE 15.10 Calculating the pH of a Solution of a Weak Acid **LO5**

The vast majority of the acids on our planet and in our bodies are weak acids, including all carboxylic acids. Formic acid is one of the more common carboxylic acids and the one with the simplest molecular structure. Secretions of the carabid beetle shown in Figure 15.11 contain 0.040 M formic acid, HCOOH. What is the pH of 0.040 M formic acid? The K_a value for formic acid is 1.77×10^{-4}.

Collect and Organize We are asked to determine the pH of a known concentration of a weak acid. We also know its K_a value. The molecular structure of formic acid in Table 15.2 indicates that only one H atom per molecule is ionizable in aqueous solution.

Analyze The acid ionization reaction is

$$HCOOH(aq) + H_2O(\ell) \rightleftharpoons HCOO^-(aq) + H_3O^+(aq)$$

and the corresponding equilibrium constant expression is

$$K_a = \frac{[HCOO^-][H_3O^+]}{[HCOOH]} = 1.77 \times 10^{-4}$$

Calculating pH involves first solving for $[H_3O^+]$, which we can do by setting up a RICE table based on the above reaction, as we did for several equilibrium calculations in Chapter 14.

Solve First we set up a RICE table with columns for the molar concentrations of the reactant and two products. The object of the table is to solve for $[H_3O^+]$ at equilibrium, so we give it the symbol x. Filling in the other cells in the table according to the 1:1:1 stoichiometry of the reaction gives us

Reaction	HCOOH(aq) + H₂O(ℓ) $\rightleftharpoons$	HCOO⁻(aq) +	H₃O⁺(aq)
	[HCOOH] (*M*)	[HCOO⁻] (*M*)	[H₃O⁺] (*M*)
Initial (I)	0.040	0	0
Change (C)	−x	+x	+x
Equilibrium (E)	0.040 − x	x	x

Inserting the equilibrium terms into the K_a expression,

$$K_a = \frac{[HCOO^-][H_3O^+]}{[HCOOH]} = \frac{(x)(x)}{(0.040 - x)} = 1.77 \times 10^{-4}$$

Solving for x by first cross multiplying:

$$x^2 = 7.08 \times 10^{-6} - (1.77 \times 10^{-4})x$$

and rearranging the terms:

$$x^2 + (1.77 \times 10^{-4})x - 7.08 \times 10^{-6} = 0$$

gives us a quadratic equation with two solutions:

$$x = -0.00275 \quad \text{and} \quad x = 0.00257$$

The negative x value has no physical meaning because it results in negative concentration values, so we conclude that $[H_3O^+] = 0.00257$.

Solving for pH: $pH = -\log[H_3O^+] = -\log(0.00257) = 2.59$

Think About It We did not attempt to simplify the calculation of $[H_3O^+]$ by eliminating "$-x$" from the denominator of the equilibrium constant expression. This was a sensible decision because the value of x is 0.00257/0.040, or 6.4%, of the initial concentration of formic acid. Put another way, the formic acid in the sample was 6.4% ionized.

Practice Exercise Acetic acid, the main ingredient in vinegar, is also a carboxylic acid. What is the pH of a 0.035 M solution of acetic acid, given that its $K_a = 1.76 \times 10^{-5}$?

SAMPLE EXERCISE 15.11 Calculating the pH of a Solution   **LO5**
of a Weak Base

The concentration of NH_3 in the household ammonia used to clean windows ranges between 50 and 100 g/L, or from about 3 M to almost 6 M. What is the pH of 3.0 M NH_3? The K_b value for ammonia is 1.76×10^{-5}.

Collect and Organize We are asked to determine the pH of a 3.0 M solution of ammonia. Figure 15.4(b) described the basic behavior of ammonia in aqueous solutions.

Analyze The hydrolysis of ammonia produces OH^- ions. We can calculate the equilibrium concentration of OH^- ions by using the K_b value and expression and then convert $[OH^-]$ to pOH and finally to pH. Given the 1:1:1 stoichiometry of NH_3, NH_4^+, and OH^-, we know that $[NH_4^+] = [OH^-]$ at equilibrium. If that equilibrium value is x, then the change in $[NH_3]$ during the reaction is $-x$. The pH value of a fairly concentrated solution of a base with a K_b value near 10^{-5} should be well above 7 but below 14.

Solve We begin by setting up a RICE table in which we let $[NH_4^+] = [OH^-] = x$ at equilibrium:

Reaction	$NH_3(aq) + H_2O(\ell)$	$\rightleftharpoons$	$NH_4^+(aq)$	$+$	$OH^-(aq)$
	$[NH_3]$ (M)		$[NH_4^+]$ (M)		$[OH^-]$ (M)
Initial	3.0		0		0
Change	$-x$		$+x$		$+x$
Equilibrium	$3.0 - x$		x		x

Because K_b is small (1.76×10^{-5}) in comparison with the initial concentration of base (3.0 M), we can make the simplifying assumption that x will be small compared with 3.0 M, and so $3.0 - x \approx 3.0$. With this assumption, our equilibrium constant expression is

$$K_b = \frac{[NH_4^+][OH^-]}{[NH_3]} = \frac{(x)(x)}{3.0} = \frac{x^2}{3.0} = 1.76 \times 10^{-5}$$

so

$$x^2 = 5.28 \times 10^{-5}$$

Solving for x gives us

$$x = [OH^-] = \sqrt{5.28 \times 10^{-5}} = 7.3 \times 10^{-3}\ M$$

Taking the negative logarithm of $[OH^-]$ to calculate pOH:

$$pOH = -\log[OH^-] = -\log(7.3 \times 10^{-3}\ M) = 2.14$$

Then we subtract this value from 14.00 to obtain the pH:

$$pH = 14.00 - pOH = 14.00 - 2.14 = 11.86$$

Think About It The calculated pH value falls in the range we predicted given the small K_b value but relatively high initial concentration of ammonia. To check our simplifying assumption, let's compare the value of x with the initial $[NH_3]$ value of 3.0 M:

$$\frac{7.3 \times 10^{-3}\ M}{3.0\ M} = 0.0024 \times 100\% = 0.24\%$$

This small percentage is acceptable, which means our simplifying assumption was justified.

 Practice Exercise What is the pH of a 0.200 M solution of methylamine (CH_3NH_2, $K_b = 4.4 \times 10^{-4}$)?

pH of Very Dilute Solutions of Strong Acids

Earlier in this chapter we discussed the autoionization of water, including the fact that water typically contributes very little to equilibrium concentrations of H_3O^+ or OH^- ions because of the small value of K_w. Now that we have developed the concept of pH and explored how it helps us quantify the acidity of a solution, let's consider a situation where the concentration of hydronium ions produced by the autoionization of water is actually greater than the concentration of hydronium ions produced by an acid. Such is the case in a very dilute solution of acid.

SAMPLE EXERCISE 15.12 pH Calculations Involving the
Autoionization of Water   **LO6**

What is the pH of $1.0 \times 10^{-8}\ M$ HCl?

Collect and Organize We are asked to calculate the pH of a very dilute solution of HCl, which is a strong acid and ionizes completely. The autoionization of pure water at 25°C produces an equilibrium $[H_3O^+] = 1.0 \times 10^{-7}\ M$.

Analyze In a $1.0 \times 10^{-8}\ M$ HCl solution, $[H_3O^+] = 1.0 \times 10^{-8}\ M$, and the pH should be:

$$pH = -\log[H_3O^+] = -\log(1.0 \times 10^{-8}) = 8.00$$

This answer is not reasonable because a solution of a strong acid, no matter how dilute it is, cannot be basic (pH > 7). To calculate pH we must also take into account the autoionization of water. Let's set up a RICE table based on the autoionization equilibrium in which x represents the increase in $[H_3O^+]$ resulting from autoionization and in which the initial value of $[H_3O^+]$ is 1.0×10^{-8} M.

Solve Setting up the RICE table and filling in the rows as described above:

Reaction	$H_2O(\ell) + H_2O(\ell)$	$\rightleftharpoons$	$H_3O^+(aq)$	$+$	$OH^-(aq)$
			$[H_3O^+]$ (M)		$[OH^-]$ (M)
Initial			1.0×10^{-8}		0
Change			$+x$		$+x$
Equilibrium			$(1.0 \times 10^{-8}) + x$		x

Substituting equilibrium values into Equation 15.8, and solving for x:

$$K_w = [H_3O^+][OH^-]$$
$$1.0 \times 10^{-14} = (1.0 \times 10^{-8} + x)(x)$$

Rearranging this equation to solve for x gives:

$$x^2 + (1.0 \times 10^{-8})x - 1.0 \times 10^{-14} = 0$$

with two solutions to this quadratic equation:

$$x = 9.5 \times 10^{-8}\ M \text{ or } x = -1.1 \times 10^{-7}\ M$$

Only the first of these numbers makes physical sense; a negative value for x would mean a negative concentration of hydroxide ions at equilibrium. Thus the total concentration of hydrogen ions in the solution is

$$[H_3O^+] = (9.5 \times 10^{-8}\ M) + (1.0 \times 10^{-8}\ M) = 10.5 \times 10^{-8}\ M = 1.05 \times 10^{-7}\ M$$

and pH is

$$pH = -\log(1.05 \times 10^{-7}\ M) = 6.98$$

Think About It This value agrees with our prediction that the solution should be (very) slightly acidic.

 Practice Exercise
What is the pH of $1.5 \times 10^{-7}\ M$ $Ca(OH)_2$?

CONCEPT TEST

In the pH calculations in this chapter, we routinely ignore the concentrations of H_3O^+ and OH^- ions produced by the autoionization of water. Suppose calculations of the pH of six different solutions produced the results shown in the following table. Which, if any, of the calculations should have taken into account the autoionization of water to obtain an accurate result?

Solution	pH
A	2.66
B	4.12
C	6.39
D	7.27
E	9.10
F	12.88

monoprotic acid an acid that has one ionizable hydrogen atom per molecule.

polyprotic acid an acid that has two or more ionizable hydrogen atoms per molecule.

CHEMTOUR
Acid Rain

15.6 Polyprotic Acids

Up to this point we have dealt with **monoprotic acids**, which have only one ionizable hydrogen atom per molecule. Acids that contain more than one ionizable hydrogen atom—such as sulfuric acid (H_2SO_4) and phosphoric acid (H_3PO_4)—are called **polyprotic acids**. For molecules with two and three ionizable hydrogen atoms, we use the more specific terms *diprotic acids* and *triprotic acids*, respectively. Let's consider the acidic properties of a strong diprotic acid, sulfuric acid.

Acid Rain

Coal naturally contains sulfur impurities, which combustion releases into the atmosphere as SO_2. For this reason, burning coal for energy production has significant environmental consequences: in the atmosphere, some SO_2 is oxidized to SO_3, which then combines with water vapor to form particles of sulfuric acid (H_2SO_4), a principal component of acid rain in many parts of the world. Sulfuric acid is a strong acid (see Table 15.1) because it is completely ionized in water:

$$H_2SO_4(aq) + H_2O(\ell) \rightarrow HSO_4^-(aq) + H_3O^+(aq) \qquad K_{a_1} \gg 1$$

However, the second ionization step may not be complete:

$$HSO_4^-(aq) + H_2O(\ell) \rightleftharpoons SO_4^{2-}(aq) + H_3O^+(aq) \qquad K_{a_2} = 0.012$$

Note that these K_a equilibrium constant symbols have an additional subscript number, with K_{a_1} and K_{a_2} corresponding to the donation of first one and then a second H^+ ion per molecule.

The combination of one complete and one incomplete ionization reaction means that many solutions of H_2SO_4 contain more than 1 mole but less than 2 moles of H_3O^+ ions for every mole of H_2SO_4 dissolved. One such solution is the subject of the following Sample Exercise.

SAMPLE EXERCISE 15.13 Calculating the pH of a Solution   **LO6**
                                     of a Strong Diprotic Acid

What is the pH of a 0.100 M solution of H_2SO_4?

Collect and Organize We are given the concentration of a solution of sulfuric acid and asked to calculate its pH. Sulfuric acid is a strong diprotic acid in that one hydrogen atom ionizes completely ($K_{a_1} \gg 1$), but a second hydrogen atom may not be completely ionized for every molecule ($K_{a_2} = 0.012$).

Analyze We start with the assumption that ionization of H_2SO_4 is complete:

$$H_2SO_4(aq) + H_2O(\ell) \rightarrow HSO_4^-(aq) + H_3O^+(aq)$$

Therefore as the second ionization step begins, $[HSO_4^-] = [H_3O^+] = 0.100\ M$. Ionization of HSO_4^- then produces additional H_3O^+ ions:

$$HSO_4^-(aq) + H_2O(\ell) \rightleftharpoons SO_4^{2-}(aq) + H_3O^+(aq)$$

We can use a RICE table to calculate how many additional H_3O^+ ions (call it x) the second ionization adds to the total $[H_3O^+]$ value. This increase in $[H_3O^+]$ will have a value between 0 and 0.1 M, which means the total $[H_3O^+]$ value will be between 0.1 and 0.2 M and the pH of the solution at equilibrium should be a little less than 1.

Solve We begin by setting up a RICE table in which initially $[HSO_4^-] = [H_3O^+] = 0.100\ M$. We let the change in $[H_3O^+]$ during the second ionization step be $+x$. Filling in the other cells of the table according to the stoichiometry of the second ionization step:

Reaction	$HSO_4^-(aq) + H_2O(\ell) \rightleftharpoons$	$SO_4^{2-}(aq)$ +	$H_3O^+(aq)$
	$[HSO_4^-]$ (M)	$[SO_4^{2-}]$ (M)	$[H_3O^+]$ (M)
Initial	0.100	0	0.100
Change	$-x$	$+x$	$+x$
Equilibrium	$0.100 - x$	x	$0.100 + x$

Inserting the equilibrium concentrations in the equilibrium constant expression for K_{a_2}:

$$K_{a_2} = \frac{[H_3O^+][SO_4^{2-}]}{[HSO_4^-]} = \frac{(0.100 + x)(x)}{(0.100 - x)} = 1.2 \times 10^{-2}$$

Cross-multiplying and rearranging the terms:

$$x^2 + 0.112x - 1.2 \times 10^{-3} = 0$$

Solving this quadratic equation for x yields these values:

$$x = 0.00985 \quad \text{and} \quad x = -0.122$$

The negative value for x has no physical meaning because it gives us a negative $[SO_4^{2-}]$ value. Therefore we use only the positive x value to calculate the total $[H_3O^+]$ at equilibrium:

$$[H_3O^+] = (0.100 + x) = (0.100 + 0.00985) = 0.10985 = 0.110\ M$$

The corresponding pH is

$$pH = -\log[H_3O^+] = -\log(0.110) = 0.96$$

Think About It As predicted, the value of $[H_3O^+]$ is between 0.1 and 0.2 M and the pH of the solution is a little less than one. The degree of ionization of HSO_4^- is

$$\frac{[SO_4^{2-}]_{\text{equilibrium}}}{[HSO_4^-]_{\text{initial}}} = \frac{0.00985\ M}{0.100\ M} = 0.0985 = 9.8\%$$

This means that ignoring the decrease in HSO_4^- to avoid solving a quadratic equation would have been a bad idea.

Practice Exercise What is the pH of a 0.200 M solution of H_2SO_4? How did doubling the concentration of H_2SO_4 affect the pH when compared with the 0.100 M solution of H_2SO_4 in the Sample Exercise above?

CONCEPT TEST

Identify all the species present in an aqueous solution of phosphoric acid, H_3PO_4.

Normal Rain

Did you know that rainwater falling from the sky is naturally acidic? The fourth-most-abundant gas in the atmosphere is CO_2, which dissolves in water to form a small amount of carbonic acid, as we discussed in Section 15.1:

$$CO_2(g) + H_2O(\ell) \rightleftharpoons H_2CO_3(aq) \tag{15.1}$$

The equilibrium constant for this reaction is only about 10^{-3}, so most of the dissolved carbon dioxide remains in the form of hydrated molecules of CO_2. However, some molecules of H_2CO_3 then go on to release H^+ ions:

$$H_2CO_3(aq) + H_2O(\ell) \rightleftharpoons HCO_3^-(aq) + H_3O^+(aq) \qquad K_{a_1} = 4.3 \times 10^{-7}$$

When using this K_{a_1} value, we need to keep in mind that its value is based on the total concentration of dissolved CO_2 present in a sample. To simplify these calculations, $[H_2CO_3]$ is used to represent the total concentration of CO_2 dissolved in the sample. The second ionization constant, K_{a_2}, has an even smaller value:

$$HCO_3^-(aq) + H_2O(\ell) \rightleftharpoons CO_3^{2-}(aq) + H_3O^+(aq) \qquad K_{a_2} = 4.7 \times 10^{-11}$$

It turns out that calculating the pH of a solution of carbonic acid is simpler than the pH calculation for sulfuric acid. The reason why is linked to the fact that K_{a_1}, while small, is still much larger than K_{a_2}. We can explain this difference on the basis of electrostatic attractions between oppositely charged ions. The first ionization step produces a negatively charged oxoanion, HCO_3^-. The second requires that a H^+ ion dissociate from HCO_3^- to produce an even more negative oxoanion, CO_3^{2-}. Separating oppositely charged ions that are naturally attracted to each other is not a process that we would expect to be favored, and the much smaller value of K_{a_2} confirms our expectation.

In general, the K_{a_2} value of any diprotic acid is smaller—often much smaller—than its K_{a_1} value. A consequence of these large differences is that essentially all the limited strength of many weak diprotic acids is derived from the first ionization reaction. For acids where $K_{a_1} > 10^3 \times K_{a_2}$, we can ignore the second ionization reaction.

SAMPLE EXERCISE 15.14 Calculating the pH of a Solution of a Weak Diprotic Acid **LO6**

What is the pH of rainwater at 25°C in equilibrium with atmospheric CO_2, producing a total dissolved CO_2 concentration of 1.4×10^{-5} M?

Collect and Organize We are asked to determine the pH of a dilute solution of dissolved CO_2. There are two ionizable H atoms in H_2CO_3. The K_{a_1} and K_{a_2} values are given in the text above. Any H_2CO_3 that dissociates to produce bicarbonate and hydronium ions will be replaced by more CO_2 dissolving from the atmosphere, so the value of $[H_2CO_3]$ in the denominator in the K_{a_1} expression will be constant at 1.4×10^{-5} M.

Analyze The large difference between the K_{a_1} and K_{a_2} values indicates that the pH of the solution is controlled by the first ionization equilibrium:

$$H_2CO_3(aq) + H_2O(\ell) \rightleftharpoons HCO_3^-(aq) + H_3O^+(aq) \qquad K_{a_1} = 4.3 \times 10^{-7}$$

Because of the small value of K_{a_1} and the small concentration of dissolved CO_2, we should obtain a pH value that is less than 7, but a lot closer to 7 than 0.

Solve First we set up a RICE table in which $x = [H_3O^+] = [HCO_3^-]$ at equilibrium and the value of $[H_2CO_3]$ is a constant 1.4×10^{-5} M.

Reaction	$H_2CO_3(aq) + H_2O(\ell)$	$\rightleftharpoons$	$HCO_3^-(aq)$	+	$H_3O^+(aq)$
	$[H_2CO_3]$ (M)		$[HCO_3^-]$ (M)		$[H_3O^+]$ (M)
Initial	1.4×10^{-5}		0		0
Change	0		$+x$		$+x$
Equilibrium	1.4×10^{-5}		x		x

$$K_{a_1} = \frac{[HCO_3^-][H_3O^+]}{[H_2CO_3]} = \frac{(x)(x)}{1.4 \times 10^{-5}} = 4.3 \times 10^{-7}$$

so

$$x^2 = 6.0 \times 10^{-12}$$

Therefore

$$x = [H_3O^+] = 2.4 \times 10^{-6} \, M$$

Taking the negative logarithm of $[H_3O^+]$ to calculate pH:

$$pH = -\log[H_3O^+] = -\log(2.4 \times 10^{-6} \, M) = 5.62$$

Think About It Carbonic acid is a weak acid, and its concentration here is small, so obtaining a pH value that is only about 1.4 units below neutral pH is reasonable.

Practice Exercise The proximity of the calculated pH value to 7.00 raises the question of whether the autoionization of water contributes significantly to $[H_3O^+]$ in the rainwater sample. Recalculate the pH of the rainwater sample in the Sample Exercise, assuming that the initial concentration of $H_3O^+ = 1.00 \times 10^{-7} \, M$.

Some acids have three ionizable H atoms per molecule. Two important triprotic acids are phosphoric acid, H_3PO_4, and citric acid, the acid responsible for the tart flavor of citrus fruits. Note in Table 15.5 how $K_{a_1} > K_{a_2} > K_{a_3}$ for both acids. This pattern is much like that for the $K_{a_1} > K_{a_2}$ values of diprotic acids and

TABLE 15.5 Ionization Equilibria of Two Triprotic Acids

for the same reason: it is more difficult to remove a second H^+ ion from the negatively charged ion formed after the first H^+ ion is removed, and it is even more difficult to remove a third H^+ ion from an ion with a 2− charge.

CONCEPT **TEST**

Do you expect the second or third acid ionization steps in phosphoric acid and citric acid to influence the pH of 0.100 *M* solutions of either acid?

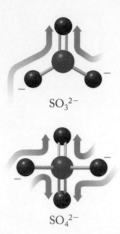

SO_3^{2-}

SO_4^{2-}

FIGURE 15.12 Sulfuric acid (H_2SO_4) is a stronger acid than sulfurous acid (H_2SO_3) because of the greater stability that comes with delocalizing the negative charge of a SO_4^{2-} ion over more atoms (shown by the curved blue arrows).

15.7 Acid Strength and Molecular Structure

In Section 15.2 we noted that nitric acid (HNO_3) is a strong acid, but nitrous acid (HNO_2) is weak. Similarly, sulfuric acid (H_2SO_4) is a strong acid, but sulfurous acid (H_2SO_3) is weak. These differences in strength are due to an important difference in these acids' molecular structures: the two strong acids have one more O atom bonded to each of their central atoms than do the two weak acids. Let's explore why this difference is important.

Recall from Section 8.3 that oxygen is the second-most-electronegative element (after fluorine). This means that oxygen atoms bonded to the central atom of an oxoacid attract electron density toward themselves. The more electron density that is drawn away from the O—H groups, the more spread out (or *delocalized*) is the negative charge on the anion that is formed when the O—H bond breaks and a H^+ ion is lost (Figure 15.12). Spreading out charge over more atoms stabilizes the anion. Thus, SO_4^{2-} ions are more stable than SO_3^{2-} ions, making H_2SO_4 a stronger acid than H_2SO_3. Similarly NO_3^- ions are more stable than NO_2^- ions, which helps make nitric acid stronger than nitrous acid. This trend of increasing acid strength with increasing numbers of oxygen atoms bonded to the central

FIGURE 15.13 In the oxoacids of chlorine, acid strength increases with increasing Cl oxidation number. The higher the oxidation number, the greater the number of O atoms bonded to the Cl atom. The greater the number of O atoms bonded to Cl, the greater the ability to delocalize the negative charge on the anion created when each acid loses its H.

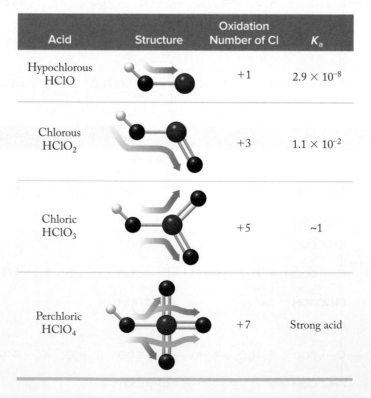

Acid	Structure	Oxidation Number of Cl	K_a
Hypochlorous HClO		+1	2.9×10^{-8}
Chlorous $HClO_2$		+3	1.1×10^{-2}
Chloric $HClO_3$		+5	~1
Perchloric $HClO_4$		+7	Strong acid

Acid	Structure	Electronegativity of Halogen Atom	K_a
Hypochlorous HClO	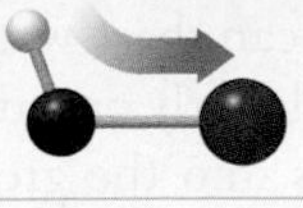 	3.0	2.9×10^{-8}
Hypobromous HBrO	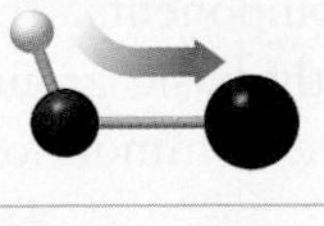	2.8	2.3×10^{-9}
Hypoiodous HIO	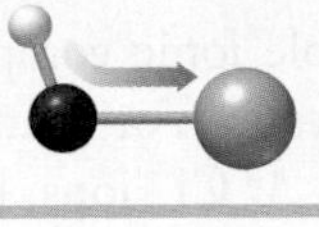	2.5	2.3×10^{-11}

FIGURE 15.14 The strengths of these three hypohalous acids are related to the electronegativities of their halogen atoms. The more electronegative the halogen atom, the more it pulls electron density (blue arrows) away from the hydrogen end of the molecule. The less electron density at the H atom, the more easily the H ionizes and so the stronger the acid.

atom (that is, with increasing oxidation number of the central atom) is true for all groups of oxoacids with the same central atom, as illustrated by the strengths of the oxoacids of chlorine (Figure 15.13).

The strength of an oxoacid is also related to the electron-withdrawing power of the central atom. Consider, for example, the relative strengths of the three hypohalous acids in Figure 15.14. The most electronegative of the three halogen atoms (Cl) has the greatest attraction for the pair of electrons it shares with oxygen. This attraction draws electron density away from hydrogen toward chlorine and toward the oxygen end of the already polar O—H bond. These shifts in electron density make the hypochlorite (ClO^-) ions better able to bear a negative charge because the charge is more delocalized. Thus HClO(aq) is the strongest of the three acids, followed by hypobromous acid [HBrO(aq)] and hypoiodous acid [HIO(aq)].

CONNECTION Oxidation numbers were introduced in Section 4.9.

CONNECTION Figure 8.6 depicts the electronegativities of the elements.

CHEMTOUR Acid Strength and Molecular Structure

SAMPLE EXERCISE 15.15 Ranking Oxoacid Strength   **LO7**

Rank the following compounds in order of decreasing acid strength: H_3PO_4, H_3AsO_4, H_3SbO_4, and H_3BiO_4.

Collect and Organize Each of these four oxoacids contains a different group 15 element. In each acid, this central atom is bonded to the same number of O atoms and has the same oxidation number (+5). The electronegativities of the elements decrease with increasing atomic number down a column in the periodic table.

Analyze In oxoacids, more electronegative central atoms draw electron density away from the H ends of the O—H bonds and help disperse the negative charges that form when oxoacids release H^+ ions to form oxoanions.

Solve Ranking the oxoacids in order of decreasing acid strength is a matter of ranking them in order of decreasing electronegativity of the central atom, that is, in order of increasing atomic number (and row number) in the periodic table: $H_3PO_4 > H_3AsO_4 > H_3SbO_4 > H_3BiO_4$.

Think About It This trend of decreasing oxoacid strength with increasing atomic number holds for each of the other groups of nonmetals—as long as the central atoms are in the same oxidation state.

 Practice Exercise Of the following compounds, which is the strongest acid and which is the weakest? H_2SeO_4, H_2SO_4, H_2SeO_3, H_2SO_3

15.8 Acidic and Basic Salts

Seawater and the freshwater in many rivers and lakes have pH values that range from weakly basic to weakly acidic. How can these waters be more basic than the acidic rainwater (pH ≤ 5.6) that serves, directly or indirectly, as their water supply? The answer is that, when rain soaks into the ground, its pH changes as it flows through soils that contain basic components. To understand the chemical processes that produce neutral or slightly basic groundwater, we first need to examine the acid–base properties of some common ionic compounds present in these waters.

As we discussed in Chapter 4, soluble ionic compounds separate into their component ions when they dissolve in water. For example, a 0.01 M solution of NaCl contains 0.01 M Na^+ ions and 0.01 M Cl^- ions. It is also a neutral solution. Neither Na^+ ions nor Cl^- ions hydrolyze (react with water) to form H_3O^+ or OH^- ions when they dissolve in water. Recall that the Cl^- ion is the conjugate base of a strong acid (HCl); therefore, the strength of the Cl^- ion can be considered negligible as a base.

Now let's consider another sodium salt, NaF. When it dissolves in water it produces F^- ions, which are the conjugate base of the *weak* acid, HF. Therefore F^- ions are weakly basic, producing at least some OH^- ions when they dissolve in water:

$$F^-(aq) + H_2O(\ell) \rightleftharpoons HF(aq) + OH^-(aq)$$

Because Na^+ ions do not influence pH, solutions of NaF are weakly basic.

If salts that contain the conjugate bases of weak acids can be basic, then it is logical that salts that contain the conjugate acids of weak bases can be acidic. An example of such a salt is NH_4Cl. We have seen that the Cl^- ions that are produced when NH_4Cl dissolves have negligible strengths as Brønsted–Lowry bases. However, NH_4^+ ions are the conjugate acid of a weak base, NH_3. Therefore NH_4^+ ions are weakly acidic, producing at least some H_3O^+ ions:

$$NH_4^+(aq) + H_2O(\ell) \rightleftharpoons NH_3(aq) + H_3O^+(aq)$$

Consequently, solutions of NH_4Cl are weakly acidic.

Table 15.6 summarizes how salts can be acidic, basic, or neutral depending on whether they include cations that are the conjugate acids of weak bases, anions that are the conjugate bases of weak acids, or both. Note that salts that contain both the conjugate base of a weak acid *and* the conjugate acid of a weak base may

TABLE 15.6 Acid–Base Properties of Salts

Anion of a	Cation of a	Aqueous Solutions Are	Example
Strong acid	Strong base	Neutral	NaCl
Strong acid	Weak base	Acidic	NH_4Cl
Weak acid	Strong base	Basic	NaF
Weak acid	Weak base	Neutral,[a] Acidic,[b] or Basic[c]	NH_4CH_3COO NH_4F NH_4HCO_3

[a] If K_a (of weak acid) = K_b (of weak base)

[b] If K_a (of weak acid) > K_b (of weak base)

[c] If K_a (of weak acid) < K_b (of weak base)

be acidic, basic, or neutral, depending on the relative strengths of the acid and base. Ammonium acetate represents the rare example of a salt in which the strengths of the conjugate acid (acetic acid) and the base (ammonia) happen to be *exactly the same*: the K_a of acetic acid = K_b of ammonia = 1.76×10^{-5}. As a result, ammonium acetate is a neutral salt.

SAMPLE EXERCISE 15.16 Predicting Whether a Salt Is   **LO8**
                                        Acidic, Basic, or Neutral

NaClO is the active ingredient in chlorine bleach. Is an aqueous solution of NaClO acidic, basic, or neutral?

Collect, Organize, and Analyze Sodium ions do not hydrolyze and do not affect the pH of aqueous solutions. However, ClO^- ions are the conjugate base of HClO, which is a weak acid (Table 15.2). Therefore, ClO^- ions are weak bases that partially hydrolyze, forming OH^- ions:

$$ClO^-(aq) + H_2O(\ell) \rightleftharpoons HClO(aq) + OH^-(aq)$$

Solve Because hydrolysis of ClO^- ions produces OH^- ions, solutions of NaClO are weakly basic.

Think About It Any sodium salt that contains an anion that is the conjugate base of a weak acid produces weakly basic aqueous solutions.

 Practice Exercise Write a chemical equation for the hydrolysis reaction that explains why an aqueous solution of K_2SO_4 is basic.

Having established that salts can be acidic, basic, or neutral, let's now address how to calculate the pH of their aqueous solutions. Our approach is much like the one we use to calculate the pH of solutions of weak acids and bases, but there is an extra step. For example, if we want to calculate the pH of a solution of ammonium chloride, we need to know the equilibrium constant of the reaction in which NH_4^+ ion functions as a Brønsted–Lowry acid:

$$NH_4^+(aq) + H_2O(\ell) \rightleftharpoons NH_3(aq) + H_3O^+(aq)$$

K_a values are typically not listed for conjugate acids of weak bases. Instead, the K_b values for the weak bases are listed, as is the case for ammonia in Appendix 5:

$$NH_3(aq) + H_2O(\ell) \rightleftharpoons NH_4^+(aq) + OH^-(aq) \qquad K_b = 1.76 \times 10^{-5}$$

We know that the strengths of conjugate acids and bases are complementary: the stronger one is, the weaker the other. Therefore we can derive the K_a value for ammonium ions from the K_b value for ammonia. To see how, let's write the K_a and K_b equilibrium constant expressions for NH_4^+ and NH_3:

$$K_a = \frac{[NH_3][H_3O^+]}{[NH_4^+]} \qquad K_b = \frac{[NH_4^+][OH^-]}{[NH_3]}$$

These expressions are similar, although any shared terms that appear in the numerator of one expression are found in the denominator of the other.

Let's take advantage of this situation by multiplying the two expressions together:

$$K_a \times K_b = \frac{[\cancel{NH_3}][H_3O^+]}{[\cancel{NH_4^+}]} \times \frac{[\cancel{NH_4^+}][OH^-]}{[\cancel{NH_3}]} = [H_3O^+][OH^-]$$

The simplified product is the familiar K_w expression for the autoionization of water. Thus

$$K_a \times K_b = K_w \qquad (15.18)$$

This is a very handy equation because (1) it works for all conjugate acid–base pairs, and (2) it allows us to calculate either the K_b of the anion in a basic salt from the K_a of its conjugate acid, or the K_a of the cation in an acidic salt from the K_b of its conjugate base.

FIGURE 15.15 Clorox is a common household cleaner.

SAMPLE EXERCISE 15.17   Calculating the pH of a Solution   **LO8**
of a Basic Salt

The bottle of chlorine bleach shown in Figure 15.15 holds an aqueous solution that contains 82.5 g/L of NaClO. What is the pH of this solution?

Collect and Organize We know the concentration of an aqueous solution of hypochlorite, NaClO, and we are asked to calculate its pH. We determined in Sample Exercise 15.16 that NaClO is a basic salt because the ClO^- ion is the conjugate base of a weak acid, HClO. The Na^+ ion plays no role in the acid–base properties of a salt. The K_a value for HClO is 2.9×10^{-8} (per Figure 15.14 and Appendix 5).

Analyze We need to first convert the K_a value for HClO into the K_b value for the ClO^- ion by using Equation 15.18: $K_a \times K_b = K_w$. We also need to convert the concentration of NaClO from g/L to molarity. That value will be the initial value of the reactant in a RICE table on the basis of its hydrolysis reaction:

$$ClO^-(aq) + H_2O(\ell) \rightleftharpoons HClO(aq) + OH^-(aq)$$

We will solve for $[OH^-]$, then pOH, and then pH. The concentration of the solution should be about 1 M, and K_b should be a little more than 10^{-7}. Therefore $[OH^-]$ of a ~1 M solution should be near the square root of that value, or ~10^{-3}, making pOH ≈ 3 and pH ≈ 11.

Solve Calculate K_b:

$$K_b = \frac{K_w}{K_a} = \frac{1.0 \times 10^{-14}}{2.9 \times 10^{-8}} = 3.4 \times 10^{-7}$$

and then calculate the molar concentration of NaClO to use in the following RICE table:

$$\frac{82.5\ g}{L} \times \frac{1\ mol}{74.44\ g} = 1.11\ mol/L$$

Reaction	$ClO^-(aq) + H_2O(\ell)$	$\rightleftharpoons$	$HClO(aq)$	+	$OH^-(aq)$
	$[ClO^-]\ (M)$		$[HClO]\ (M)$		$[OH^-]\ (M)$
Initial	1.11		0		0
Change	$-x$		$+x$		$+x$
Equilibrium	$1.11 - x$		x		x

To solve for x, we make the simplifying assumption that x will be small compared with 1.11 M because the value of K_b is so small ($<<10^{-5}$):

$$K_b = \frac{[HClO][OH^-]}{[ClO^-]} = \frac{(x)(x)}{1.11 - x} \approx \frac{x^2}{1.11} = 3.4 \times 10^{-7}$$

$$x = [OH^-] = 6.2 \times 10^{-4}$$

Calculating pOH:

$$pOH = -\log[OH^-] = -\log(6.2 \times 10^{-4}) = 3.21$$

and pH:

$$pH = 14.00 - pOH = 14.00 - 3.21 = 10.79$$

Think About It The calculated pH (10.79) is in the same ballpark as our estimate of 11. Also, the calculated $[OH^-]$ value is much less than 5% of the initial $[ClO^-]$ value, so our assumption that we could ignore the $-x$ term in the denominator of the K_b expression was justified.

Practice Exercise The pH of swimming pools is made slightly basic by spreading solid Na_2CO_3 across the surface. Another approach involves preparing concentrated solutions of Na_2CO_3 and slowly adding them to the water circulating through the pool pump and filter. What is the pH of an aqueous solution of 0.100 M Na_2CO_3?

SAMPLE EXERCISE 15.18 Calculating the pH of a Solution of an Acidic Salt   **LO8**

What is the pH of 0.25 M NH_4Cl?

Collect and Organize We are asked to calculate the pH of a solution of NH_4Cl. When NH_4Cl dissolves in water, NH_4^+ and Cl^- ions are released into solution. The NH_4^+ ion is the conjugate acid of NH_3, which is a weak base. The Cl^- ion is the conjugate base of HCl, which is a strong acid.

Analyze We have seen that the Cl^- ion has negligible strength as a Brønsted–Lowry base, so it does not contribute to the acid–base properties of NH_4Cl. Ammonia is a weak base ($K_b = 1.76 \times 10^{-5}$), which means that its conjugate acid, NH_4^+, is a weak acid, and the pH of the solution will be controlled by its hydrolysis:

$$NH_4^+(aq) + H_2O(\ell) \rightleftharpoons NH_3(aq) + H_3O^+(aq)$$

The K_a value of NH_4^+ can be calculated by dividing K_w by the K_b of ammonia. The K_b value of ammonia is close to 10^{-5}, so the K_a value of the ammonium ion will be close to 10^{-9}. Therefore $[H_3O^+]$ of a 0.25 M solution should be the square root of $\sim 10^{-10}$, or $\sim 10^{-5}$, giving a pH ≈ 5.

Solve The simplified K_a expression for the NH_4^+ ion is

$$K_a = \frac{[NH_3][H_3O^+]}{[NH_4^+]}$$

Rearranging Equation 15.18 to solve for K_a:

$$K_a = \frac{K_w}{K_b} = \frac{1.0 \times 10^{-14}}{1.76 \times 10^{-5}} = 5.68 \times 10^{-10} = \frac{[NH_3][H_3O^+]}{[NH_4^+]}$$

We set up a RICE table in which we make the usual assumptions that the reaction is the only significant source of H^+ and that $x = [H_3O^+] = [NH_3]$ at equilibrium:

Reaction	$NH_4^+(aq) + H_2O(\ell)$	$\rightleftharpoons$	$NH_3(aq)$	$+$	$H_3O^+(aq)$
	$[NH_4^+]$ (M)		$[NH_3]$ (M)		$[H_3O^+]$ (M)
Initial	0.25		0		0
Change	$-x$		$+x$		$+x$
Equilibrium	$0.25 - x$		x		x

$$K_a = 5.68 \times 10^{-10} = \frac{[NH_3][H_3O^+]}{[NH_4^+]} = \frac{(x)(x)}{0.25 - x}$$

Given the very small value of K_a, we can make the simplifying assumption that $(0.25\ M - x) \approx 0.25\ M$, which gives us

$$\frac{x^2}{0.25} = 5.68 \times 10^{-10}$$

$$x^2 = 1.42 \times 10^{-10}$$

$$x = 1.19 \times 10^{-5} = [H_3O^+]$$

$$pH = -\log[H_3O^+] = -\log(1.19 \times 10^{-5}) = 4.92$$

Think About It This result matches our prediction quite closely. The calculated $[H_3O^+]$ is much less than 5% of the initial concentration of NH_4^+, so our simplifying assumption was valid.

 Practice Exercise What is the pH of a 0.25 M solution of methylamine hydrochloride, CH_3NH_3Cl?

SAMPLE EXERCISE 15.19 Integrating Concepts: The pH of Human Blood

We began this chapter by noting how essential it is that our circulation and respiration systems efficiently remove from our bodies the CO_2 produced in our cells. An enzyme, carbonic anhydrase, plays a key role in this process by speeding up the rate at which CO_2 hydrolyzes:

$$CO_2(g) + H_2O(\ell) \rightleftharpoons H_2CO_3(aq)$$

and the rate at which H_2CO_3 ionizes:

$$H_2CO_3(aq) + H_2O(\ell) \rightleftharpoons HCO_3^-(aq) + H_3O^+(aq)$$

a. Does carbonic anhydrase increase the acid strength of dissolved CO_2, that is, the K_{a_1} of carbonic acid based on the total concentration of CO_2 in solution?

b. Suppose that, during strenuous exercise, the total concentration of dissolved CO_2 in the blood flowing through muscle tissues is $2.7 \times 10^{-3}\ M$. If the equilibrium constant of the hydrolysis reaction is 3.4×10^{-2}, what is the equilibrium concentration of carbonic acid in the blood?

c. Suppose the concentration of HCO_3^- ions in the blood in part (b) is 0.028 M. What is the pH of the blood?

Collect and Organize We are asked whether the presence of an enzyme that increases the rates at which CO_2 is hydrolyzed and undergoes acid ionization makes H_2CO_3 a stronger acid. We then need to calculate the equilibrium value of $[H_2CO_3]$ from an initial value of $[CO_2]$; we are also given the K value for the hydrolysis reaction and the pH of blood in which $[HCO_3^-]$ and $[CO_2]$ are known. Carbonic acid is a weak diprotic acid; its K_{a_1}, based on the total concentration of CO_2 in solution, is 4.3×10^{-7}.

Analyze (a) We learned in Chapter 14 that catalysts speed up reactions but do not alter their equilibrium constants. (b) Calculating $[H_2CO_3]$ will involve solving for the numerator of the equilibrium constant expression for the hydrolysis of CO_2, knowing the $[CO_2]$ value in the denominator. Assuming that continued exercise maintains $[CO_2]$ at a constant value, there will be no $-x$ term in the denominator. With a K value of 0.034 and $[CO_2] \approx 2 \times 10^{-3}\ M$, $[H_2CO_3]$ should be about $7 \times 10^{-5}\ M$. (c) Calculating the pH of a carbonic acid solution will require a

RICE table based on K_{a_1} in which $[H_3O^+] = x$ at equilibrium but in which the initial concentration of HCO_3^- is 0.028 M, not zero. The normal pH of human blood is close to 7.4, so the calculated pH should be close to that.

Solve (a) Carbonic anhydrase should not affect the value of K_{a_1}, though it should produce increases in both $[H_2CO_3]$ and $[HCO_3^-]$. These increases in the numerator and denominator of the K_{a_1} expression offset each other and do not affect pH.

(b) Inserting the known values of K and $[CO_2]$ in the hydrolysis equilibrium constant expression:

$$K = \frac{[H_2CO_3]}{[CO_2]} = \frac{x}{2.7 \times 10^{-3}} = 3.4 \times 10^{-2}$$

and solving for x:

$$x = [H_2CO_3] = 9.2 \times 10^{-5}\ M$$

(c) Setting up a RICE table based on the K_{a_1} reaction where the values in the $[H_2CO_3]$ column represent a constant total concentration of $CO_2(aq) + H_2CO_3(aq)$:

Reaction	$H_2CO_3(aq) + H_2O(\ell) \rightleftharpoons HCO_3^-(aq) + H_3O^+(aq)$		
	$[H_2CO_3]$ (M)	$[HCO_3^-]$ (M)	$[H_3O^+]$ (M)
Initial	2.7×10^{-3}	0.028	0
Change	0	$+x$	$+x$
Equilibrium	2.7×10^{-3}	$0.028 + x$	x

In solving for x, we make the simplifying assumption that its value will be much less than 0.028 M (as it must be, if the calculated pH is close to 7.4, which means $[H_3O^+] \approx 10^{-7}$), and we ignore the very small x^2 term in the numerator of the following calculation:

$$K_{a_1} = \frac{[HCO_3^-][H_3O^+]}{[H_2CO_3]} = \frac{(0.028 + x)(x)}{2.7 \times 10^{-3}} \approx \frac{0.028x}{2.7 \times 10^{-3}}$$

$$= 4.3 \times 10^{-7}$$

$$x = 4.15 \times 10^{-8}$$

Therefore the pH of the blood is

$$pH = -\log[H_3O^+] = -\log(4.15 \times 10^{-8}) = 7.38$$

SUMMARY

LO1 The strengths of acids and bases are related to the values of their acid and base ionization constants, K_a and K_b. Most acids are weak, which means their K_a is much less than 1 and they ionize only partially in water. (Section 15.2)

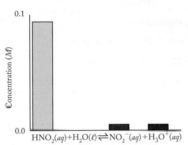

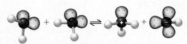
$HNO_2(aq) + H_2O(\ell) \rightleftharpoons NO_2^-(aq) + H_3O^+(aq)$

LO2 When an acid (HA) ionizes, it forms its **conjugate base**, A^-. When base B bonds to a H^+ ion, it forms its **conjugate acid**, HB^+. (Section 15.2)

LO3 In a neutral solution, $[H_3O^+] = [OH^-] = 1.0 \times 10^{-7}$. The **pH** scale is a logarithmic scale for expressing the acidic or basic strength of solutions. Acidic solutions have pH values less than 7; basic solutions have pH values greater than 7. Because pH is the negative logarithm of H_3O^+ concentration, the higher the pH, the lower the H_3O^+ concentration. An increase in one pH unit represents a decrease in $[H_3O^+]$ to 1/10 of its initial value. Likewise, **pOH** is the negative logarithm of OH^- concentration. The sum of pH and pOH equals 14 in an aqueous solution at 25°C. (Section 15.3)

LO4 The K_a and K_b values of acids and bases can be used to calculate the extent to which their solutions are ionized—that is, their **degree of ionization** or **percent ionization**—and vice versa. (Section 15.4)

LO5 To calculate the pH of a weak acid or base, use a RICE table based on the acid or base ionization reaction to determine the equilibrium value of $[H_3O^+]$ or $[OH^-]$ in a solution of a weak acid or base. (Section 15.5)

LO6 **Polyprotic acids** can undergo more than one acid ionization reaction, but for most, the first ionization reaction is the one that controls pH. (Section 15.6)

LO7 The strengths of acids are related to both the electronegativity of the atom adjacent to the hydrogen atom and the stability of the anions they form when they release H^+ ions. The stability of oxoanions is enhanced by multiple oxygen atoms bonded to the central atom, which disperse the negative charge(s) over the anion. (Section 15.7)

LO8 A salt solution is acidic if the cation in the salt is the conjugate acid of a weak base and the anion is the conjugate base of a strong acid. A salt solution is basic if the anion in the salt is the conjugate base of a weak acid and the cation is the conjugate acid of a strong base. (Section 15.8)

PARTICULATE **PREVIEW WRAP-UP**

All three species form covalent bonds when they accept donated protons. ClO^- is changed into a neutral molecule (HClO), whereas NH_3 and H_2O become cations (NH_4^+ and H_3O^+, respectively). All three species have lone pairs on an atom (O in ClO^- and H_2O, and N in NH_3) that can bond to a H^+ ion.

PROBLEM-SOLVING SUMMARY

Type of Problem	Concepts and Equations	Sample Exercises
Rank acids and bases from strongest to weakest	The strengths of acids and bases are proportional to their K_a and K_b values.	15.1, 15.3, 15.15
Identifying acid–base conjugate pairs	The formula of the base in a conjugate pair is the formula of the acid less one H^+ ion.	15.2
Interconverting $[H_3O^+]$, $[OH^-]$, pH, and pOH	Use the following: $$pH = -\log[H_3O^+] \quad (15.10)$$ $$pOH = -\log[OH^-] \quad (15.11)$$ $$pK_w = pH + pOH = 14.00 \quad (15.12)$$	15.4, 15.5, 15.6, 15.9
Determining percent ionization and K_a given the pH of a weak acid	Use the following: $$[H_3O^+] = 10^{-pH}$$ $$K_a = \frac{[A^-][H_3O^+]}{[HA]} \quad (15.14)$$ $$\text{Percent ionization} = \frac{[H_3O^+]_{equilibrium}}{[HA]_{initial}} \times 100\% \quad (15.15)$$	15.7
Determining percent ionization and K_b given the pH of a weak base	Use the following: $$[OH^-] = 10^{-pOH}$$ $$K_b = \frac{[HB^+][OH^-]}{[B]} \quad (15.16)$$ $$\text{Percent ionization} = \frac{[OH^-]_{equilibrium}}{[B]_{initial}} \times 100\% \quad (15.17)$$	15.8
Calculating the pH of a solution of weak acid HA	Set up a RICE table based on the K_a equilibrium $$HA(aq) + H_2O(\ell) \rightleftharpoons H_3O^+(aq) + A^-(aq)$$ Let $x = [H_3O^+] = [A^-]$ at equilibrium. Calculate x by using $$K_a = \frac{[A^-][H_3O^+]}{[HA]}$$ Then calculate $pH = -\log[H_3O^+]$.	15.10
Calculating the pH of a solution of weak base B	Set up a RICE table based on the equilibrium $$B(aq) + H_2O(\ell) \rightleftharpoons HB^+(aq) + OH^-(aq)$$ Let $x = [OH^-] = [HB^+]$ at equilibrium. Calculate x by using $$K_b = \frac{[HB^+][OH^-]}{[B]}$$ Then use $$pOH = -\log[OH^-]$$ $$pH = 14.00 - pOH$$	15.11
Calculating the pH of a solution of very dilute acidic (or very dilute basic) solution while considering the autoionization of water	Set up a RICE table based on the equilibrium $$H_2O(\ell) + H_2O(\ell) \rightleftharpoons H_3O^+(aq) + OH^-(aq)$$ Let $x = [H_3O^+] = [OH^-]$ because of autoionization. Add x to the $[H_3O^+]$ because of the dilute acid (or to the $[OH^-]$ because of the weak base). Solve for x by using $$K_w = [H_3O^+][OH^-] \quad (15.8)$$ Then calculate $pH = -\log[H_3O^+]$.	15.12

Type of Problem	Concepts and Equations	Sample Exercises
Calculating the pH of a solution of a strong diprotic acid	Assume $$H_2A(aq) + H_2O(\ell) \rightarrow H_3O^+(aq) + HA^-(aq)$$ is complete. Set up a RICE table based on the K_{a_2} equilibrium $$HA^-(aq) + H_2O(\ell) \rightleftharpoons H_3O^+(aq) + A^{2-}(aq)$$ Let $x =$ additional $[H^+]$ from second ionization step; $[HA^-]_{\text{initial (2nd step)}} = [H_2A]_{\text{initial}}$. Calculate x by using $$K_{a_2} = \frac{[H_3O^+][A^{2-}]}{[HA^-]}$$ Then calculate pH $= -\log[H_3O^+]$.	15.13
Calculating the pH of a solution of a weak diprotic acid	Set up a RICE table based on the K_{a_1} equilibrium $$H_2A(aq) + H_2O(\ell) \rightleftharpoons H_3O^+(aq) + HA^-(aq)$$ Let $x = [H_3O^+] = [HA^-]$ at equilibrium. Calculate x by using $$K_{a_1} = \frac{x^2}{[H_2A] - x}$$ Then calculate pH $= -\log[H_3O^+]$.	15.14
Distinguishing acidic, basic, and neutral salts	The cations in acidic salts are the conjugate acids of weak bases. The anions in basic salts are the conjugate bases of weak acids.	15.16
Calculating the pH of a solution of a basic salt	Assume the salt (MX) completely dissociates into M^+ and X^-. Set up a RICE table for the equilibrium $$X^-(aq) + H_2O(\ell) \rightleftharpoons HX(aq) + OH^-(aq)$$ Let $x = [HX] = [OH^-] =$ at equilibrium. Calculate x by using $$K_b = K_w/K_{a \text{ (of conjugate acid,HA)}} \qquad (15.18)$$ $$K_b = \frac{K_w}{K_a} = \frac{x^2}{[A^-] - x}$$ Convert $[OH^-]$ to pOH, and then pH.	15.17
Calculating the pH of a solution of an acidic salt	Assume the salt (HBX) completely dissociates into HB^+ and X^-. Set up a RICE table for the equilibrium $$HB^+(aq) + H_2O(\ell) \rightleftharpoons B(aq) + H_3O^+(aq)$$ Let $x = [H_3O^+] = [B]$ at equilibrium. Calculate x by using $$K_a = K_w/K_{b \text{ (of conjugate base,B)}}$$ $$K_a = \frac{K_w}{K_b} = \frac{x^2}{[HB^+] - x}$$ Then calculate pH.	15.18

VISUAL PROBLEMS

(Answers to boldface end-of-chapter questions and problems are in the back of the book.)

15.1. Which of the lines in Figure P15.1 best represents the dependence of the degree of ionization of acetic acid on its concentration in aqueous solution?

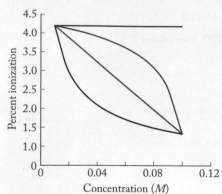

FIGURE P15.1

15.2. The graph in Figure P15.2 shows the percent ionization of two acids as a function of concentration in water. Which line describes the behavior of HNO_3, and which line describes the behavior of acetic acid (CH_3COOH)?

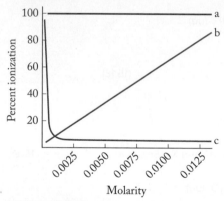

FIGURE P15.2

15.3. The bar graph in Figure P15.3 shows the degree of ionization of 1×10^{-3} M solutions of three hypohalous acids: HClO, HBrO, and HIO. Which bar corresponds to HIO?

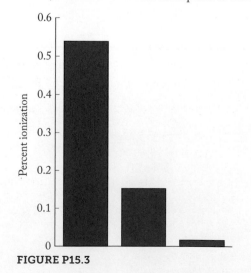

FIGURE P15.3

15.4. The bar graph in Figure P15.4 shows the degree of ionization of 1 M solutions of HClO, $HClO_2$, and $HClO_3$. Which bar corresponds to HClO?

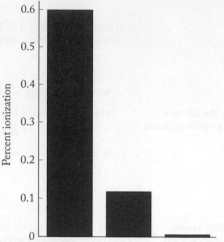

FIGURE P15.4

15.5. Figure P15.5 shows a condensed molecular structure of pseudoephedrine, a widely used decongestant and stimulant.
   a. Is pseudoephedrine an acidic, basic, or neutral compound?
   b. Which functional group in its structure gives it the property you selected in part (a)?

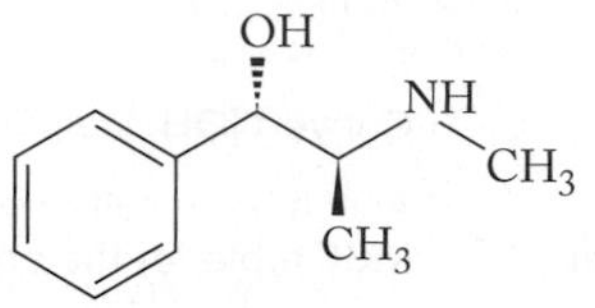

OH

NH

CH₃

CH₃

FIGURE P15.5

15.6. Figure P15.6 shows the molecular structure of alanine. Is an aqueous solution of alanine acidic, basic, or neutral? Explain your selection, or explain what additional information you need to make a more informed selection.

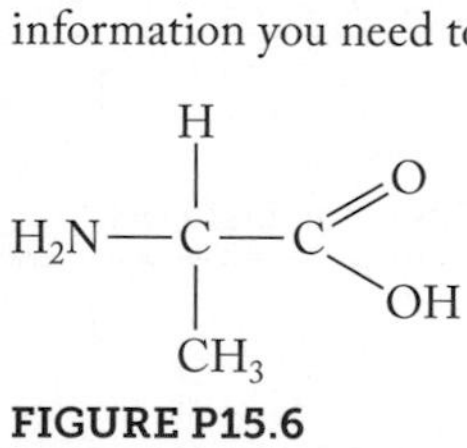

FIGURE P15.6

*15.7. Figure P15.7 contains the condensed molecular structures of piperidine and morpholine. Which is the stronger base? Explain your selection.

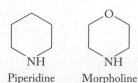

Piperidine   Morpholine

FIGURE P15.7

15.8. On the basis of the skeletal structures shown in Figure P15.8, which is the stronger base, ethanolamine or ethylamine? Explain why you think so.

Ethanolamine   Ethylamine

FIGURE P15.8

15.9. Figure P15.9 contains molecular models of acetic acid and trichloroacetic acid. Use these two models to explain why the K_a of trichloroacetic acid is about 10^4 times that of acetic acid.

Acetic acid   Trichloroacetic acid

FIGURE P15.9

15.10. Use representations [A] through [I] in Figure P15.10 to answer questions a–f. Note that water has been omitted from the solutions in [D], [E], [F], and [G] for ease of viewing.
  a. Which molecules are weak acids? Which are weak bases?
  b. Which pairs of molecules represent conjugate acid–base pairs? (Give the letter of the acid first.)
  c. Write net ionic equations describing the equilibria involving the conjugate pairs identified in part (b).
  d. Which images represent acidic solutions?
  e. Which images represent basic solutions?
  f. Which image represents a neutral solution?

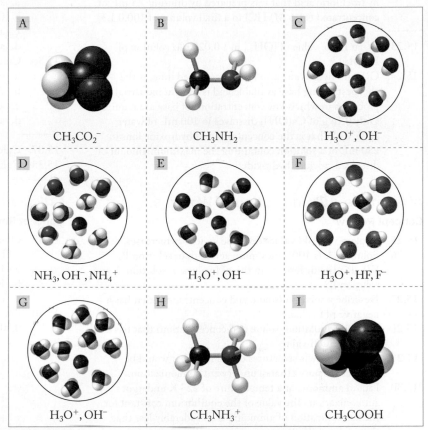

FIGURE P15.10

QUESTIONS AND PROBLEMS

Strong and Weak Acids and Bases

Concept Review

15.11. In an aqueous solution of HF, which compound acts as a Brønsted–Lowry acid and which is the Brønsted–Lowry base?

15.12. In an aqueous solution of HNO_3, which compound acts as a Brønsted–Lowry acid and which is the Brønsted–Lowry base?

15.13. In an aqueous solution of NH_3, which species acts as a Brønsted–Lowry acid and which is the Brønsted–Lowry base?

15.14. Both KOH and $Ba(OH)_2$ are strong bases. Does this mean that solutions of the two compounds with the same molarity have the same ability to accept hydrogen ions? Why or why not?

15.15. Identify the acids and bases in the following reactions:
  a. $HCl(aq) + NaOH(aq) \rightarrow NaCl(aq) + H_2O(\ell)$
  b. $MgCO_3(s) + 2\ HCl(aq) \rightarrow$
     $MgCl_2(aq) + CO_2(g) + H_2O(\ell)$
  c. $2\ NH_3(aq) + H_2SO_4(aq) \rightarrow (NH_4)_2SO_4(aq)$

15.16. Identify the acids and bases in the following reactions:
  a. $(CH_3)_3N(aq) + H_2O(\ell) \rightleftharpoons (CH_3)_3NH^+(aq) + OH^-(aq)$
  b. $CO_2(aq) + H_2O(\ell) \rightleftharpoons HCO_3^-(aq) + H_3O^+(aq)$
  c. $(CH_3)_3COH(aq) + H_3O^+(aq) \rightleftharpoons$
     $(CH_3)_3COH_2^+(aq) + H_2O(\ell)$

15.17. Identify the conjugate base of each of the following compounds: HNO_2, $HClO$, H_3PO_4, and NH_3.

15.18. Identify the conjugate acid of each of the following species: $(CH_3)_3N$, CH_3COO^-, and OH^-.

15.19. What is the conjugate acid of the bisulfate ion, HSO_4^-, and what is its conjugate base?

*15.20. Compounds that do not ionize in water have been known to ionize in nonaqueous solvents. In such a solvent, what would be the conjugate acid and conjugate base of methanol, CH_3OH?

Problems

15.21. What is the concentration of H_3O^+ ions in a 0.65 M solution of HNO_3?

15.22. What is the concentration of H_3O^+ ions in a solution of hydrochloric acid that was prepared by diluting 7.5 mL of concentrated (11.6 M) HCl to a final volume of 100.0 L?

15.23. What is the value of $[OH^-]$ in a 0.0205 M solution of $Ba(OH)_2$?

15.24. Calcium hydroxide, also known as slaked lime, is the cheapest strong base available and is used in industrial processes in which low concentrations of base are required. Only 0.16 g of $Ca(OH)_2$ dissolves in 100 mL of water at 25°C. What is the concentration of hydroxide ions in 250 mL of a solution containing the maximum amount of dissolved calcium hydroxide?

pH and the Autoionization of Water

Concept Review

15.25. Explain why pH values decrease as acidity increases.

15.26. Solution A is 100 times more acidic than solution B. What is the difference in the pH values of solution A and solution B?

15.27. Describe a solution (solute and concentration) that has a negative pH value.

15.28. Describe a solution (solute and concentration) that has a negative pOH value.

*15.29. Draw the Lewis structures of the ions that would be produced if pure ethanol underwent autoionization.

*15.30. Liquid ammonia at a temperature of 223 K undergoes autoionization. The value of the equilibrium constant for the autoionization of ammonia is considerably less than that of water. Write an equation for the autoionization of ammonia and suggest a reason why the value of K for the process is less than that of water.

Problems

15.31. Calculate the pH and pOH of solutions with the following $[H_3O^+]$ or $[OH^-]$ values. Indicate which solutions are acidic, basic, or neutral.
a. $[H_3O^+] = 5.3 \times 10^{-3}\ M$
b. $[H_3O^+] = 3.8 \times 10^{-9}\ M$
c. $[H_3O^+] = 7.2 \times 10^{-6}\ M$
d. $[OH^-] = 1.0 \times 10^{-14}\ M$

15.32. Calculate the pH and pOH of the solutions with the following hydrogen ion or hydroxide ion concentrations. Indicate which solutions are acidic, basic, or neutral.
a. $[OH^-] = 8.2 \times 10^{-11}\ M$
b. $[OH^-] = 7.7 \times 10^{-6}\ M$
c. $[H_3O^+] = 3.2 \times 10^{-4}\ M$
d. $[H_3O^+] = 1.0 \times 10^{-7}\ M$

15.33. Calculate the concentration of the following ions in the solution described:
a. $[H_3O^+]$ in $8.4 \times 10^{-4}\ M$ NaOH
b. $[H_3O^+]$ in $6.6 \times 10^{-5}\ M$ $Ca(OH)_2$
c. $[OH^-]$ in $4.5 \times 10^{-3}\ M$ HCl
d. $[OH^-]$ in $2.9 \times 10^{-5}\ M$ HCl

15.34. Determine the indicated pH or pOH values:
a. pH of a solution whose pOH = 5.5
b. pH of a solution whose pOH = 6.8
c. pOH of a solution whose pH = 9.7
d. pOH of a solution whose pH = 4.4

15.35. Calculate the pH and pOH of the following solutions:
a. stomach acid in which [HCl] = 0.155 M
b. 0.00500 M HNO_3
c. a 2:1 mixture of 0.0125 M HCl and 0.0125 M NaOH
d. a 3:1 mixture of 0.0125 M H_2SO_4 and 0.0125 M KOH

15.36. Calculate the pH and pOH of the following solutions:
a. 0.0450 M NaOH
b. 0.160 M $Ca(OH)_2$
c. a 1:1 mixture of 0.0125 M HCl and 0.0125 M $Ca(OH)_2$
d. a 2:3 mixture of 0.0125 M HNO_3 and 0.0125 M KOH

15.37. Calculate the pH of a $1.33 \times 10^{-9}\ M$ solution of LiOH.

15.38. Calculate the pH of a $6.9 \times 10^{-8}\ M$ solution of HBr.

Calculations Involving pH, K_a, and K_b

Concept Review

15.39. One-molar solutions of the following acids are prepared: CH_3COOH, HNO_2, $HClO$, and HCl.
a. Rank them in order of decreasing $[H_3O^+]$.
b. Rank them in order of increasing strength as acids (weakest to strongest).

15.40. On the basis of the following degree-of-ionization data for 0.100 M solutions, select which acid has the largest K_a.

Acid	Degree of Ionization (%)
C_6H_5COOH	2.5
HF	8.5
HN_3	1.4
CH_3COOH	1.3

15.41. A 1.0 M aqueous solution of HNO_3 is a much better conductor of electricity than is a 1.0 M solution of HNO_2. Explain why.

15.42. Hydrogen chloride and water are molecular compounds, yet a solution of HCl dissolved in H_2O is an excellent conductor of electricity. Explain why.

15.43. Hydrofluoric acid is a weak acid. Write the mass action expression for its acid ionization reaction.

15.44. **Early Antiseptic** The use of phenol, also known as carbolic acid, was pioneered in the 19th century by Sir Joseph Lister (after whom Listerine was named) as an antiseptic in surgery. Its formula is C_6H_5OH; the red hydrogen atom is ionizable. Write the mass action expression for the acid ionization equilibrium of phenol.

*15.45. The K_a values of weak acids depend on the solvent in which they dissolve. For example, the K_a of alanine in aqueous ethanol is less than its K_a in water.
a. In which solvent does alanine ionize more?

b. Which is the stronger Brønsted–Lowry base: water or ethanol?

*15.46. The K_a of proline is 2.5×10^{-11} in water, 2.8×10^{-11} in an aqueous solution that is 28% ethanol, and 1.66×10^{-8} in aqueous formaldehyde at 25°C.
   a. In which solvent is proline the strongest acid?
   b. Rank these compounds on the basis of their strengths as Brønsted–Lowry bases: water, ethanol, and formaldehyde.

15.47. When methylamine, CH_3NH_2, dissolves in water, the resulting solution is slightly basic. Which compound is the Brønsted–Lowry acid and which is the base?

*15.48. When 1,2-diaminoethane, $H_2NCH_2CH_2NH_2$, dissolves in water, the resulting solution is basic. Write the formula of the ionic compound that is formed when hydrochloric acid is added to a solution of 1,2-diaminoethane.

Problems

15.49. **Muscle Physiology** During strenuous exercise, lactic acid builds up in muscle tissues. In a 1.00 M aqueous solution, 2.94% of lactic acid is ionized. What is the value of its K_a?

15.50. **Rancid Butter** The odor of spoiled butter is due in part to butanoic acid, which results from the chemical breakdown of butterfat. A 0.100 M solution of butanoic acid is 1.23% ionized. Calculate the value of K_a for butanoic acid.

15.51. At equilibrium, the value of $[H_3O^+]$ in a 0.125 M solution of an unknown acid is 4.07×10^{-3} M. Determine the degree of ionization and the K_a of this acid.

15.52. Nitric acid (HNO_3) is a strong acid that is essentially completely ionized in aqueous solutions of concentrations ranging from 1% to 10% (1.5 M). However, in more concentrated solutions, part of the nitric acid is present as un-ionized molecules of HNO_3. For example, in a 50% solution (7.5 M) at 25°C, only 33% of the molecules of HNO_3 dissociate into H^+ and NO_3^-. What is the K_a value of HNO_3?

15.53. **Ant Stings** The venom of stinging ants contains formic acid, HCOOH, $K_a = 1.8 \times 10^{-4}$ at 25°C. Calculate the pH of a 0.055 M solution of formic acid.

15.54. **Poisonous Plant** Gifblaar is a small South African shrub and one of the most poisonous plants known because it contains fluoroacetic acid. If a 0.480 M solution of fluoroacetic acid has a pH of 1.44, what is the K_a of the acid?

15.55. **Acid Rain I** A weather system moving through the American Midwest produced rain with an average pH of 5.02. By the time the system reached New England, the rain it produced had an average pH of 4.66. How much more acidic was the rain falling in New England?

15.56. **Acid Rain II** A newspaper reported that the "level of acidity" in a sample taken from an extensively studied watershed in New Hampshire in 1998 was "an astounding 200 times lower than the worst measurement" taken in the preceding 23 years. What is this difference expressed in units of pH?

15.57. The K_b of aminoethanol, $HOCH_2CH_2NH_2$, is 3.1×10^{-5}.
   a. Is aminoethanol a stronger or weaker base than ethylamine, $pK_b = 3.36$?
   b. Calculate the pH of a 1.67×10^{-2} M solution of aminoethanol.

   c. Calculate the $[OH^-]$ concentration of a 4.25×10^{-4} M solution of aminoethanol.

15.58. **Food Dye** Quinoline is a weakly basic liquid used in the manufacture of quinolone yellow, a greenish-yellow dye for foods, and in the production of niacin. Its pK_b is 9.15.
   a. What is the pH of a 0.0752 M solution of quinolone?
   b. What is the hydroxide ion concentration of the solution in part (a)?

15.59. **Painkillers** Morphine is an effective painkiller but is also highly addictive. Codeine is a popular prescription painkiller because it is much less addictive than morphine. Codeine contains a basic nitrogen atom that can be protonated to give the conjugate acid of codeine.
   a. Calculate the pH of a 1.8×10^{-3} M solution of morphine if its $pK_b = 5.79$.
   b. Calculate the pH of a 2.7×10^{-4} M solution of codeine if the pK_a of the conjugate acid is 8.21.

15.60. The awful odor of dead fish is due mostly to trimethylamine, $(CH_3)_3N$, one of three compounds related to ammonia in which methyl groups replace 1, 2, or all 3 of the H atoms in ammonia.
   a. The K_b of trimethylamine $[(CH_3)_3N]$ is 6.5×10^{-5} at 25°C. Calculate the pH of a 3.00×10^{-4} M solution of trimethylamine.
   b. The K_b of methylamine $[(CH_3)NH_2]$ is 4.4×10^{-4} at 25°C. Calculate the pH of a 2.88×10^{-3} M solution of methylamine.
   *c. The K_b of dimethylamine $[(CH_3)_2NH]$ is 5.9×10^{-4} at 25°C. What concentration of dimethylamine is needed for the solution to have the same pH as the solution in part (b)?

Polyprotic Acids

Concept Review

15.61. Why is the K_{a_2} value of phosphoric acid less than its K_{a_1} value but greater than its K_{a_3} value?

15.62. In calculating the pH of a 1.0 M solution of H_2SO_3, we can ignore the H^+ ions produced by the ionization of the bisulfite (HSO_3^-) ion; however, in calculating the pH of a 1.0 M solution of sulfuric acid, we cannot ignore the H^+ ions produced by the ionization of the bisulfate ion. Why?

15.63. Carbonic acid, H_2CO_3, is a very weak diprotic acid ($K_{a_1} = 4.3 \times 10^{-7}$), but germanic acid, H_2GeO_3, is even weaker ($K_{a_1} = 9.8 \times 10^{-10}$). Suggest a reason why.

15.64. Figure P15.64 contains skeletal structures of two dicarboxylic acids: malonic acid on the left and oxalic acid on the right. The K_{a_1} of malonic acid is about 10^4 times as large as its K_{a_2}, whereas the K_{a_1} of oxalic acid is about 10^3 times as large as its K_{a_2}. Suggest a reason why the separation in K_a values is greater for malonic acid.

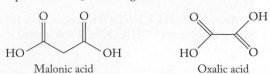

Malonic acid   Oxalic acid

FIGURE P15.64

Problems

15.65. What is the pH of a 0.75 M solution of H_2SO_4?
15.66. What is the pH of a 0.250 M solution of sulfurous acid?

15.67. Ascorbic acid (vitamin C) is a weak diprotic acid. What is the pH of a 0.250 M solution of ascorbic acid?

15.68. Rhubarb Pie The leaves of the rhubarb plant contain high concentrations of diprotic oxalic acid (HOOCCOOH) and must be removed before the stems are used to make rhubarb pie. What is the pH of a 0.0288 M solution of oxalic acid?

15.69. Nicotine Addiction Nicotine is responsible for the addictive properties of tobacco. What is the pH of a 1.00×10^{-3} M solution of nicotine?

15.70. Pseudoephedrine hydrochloride (Figure P15.70) is a common ingredient in cough syrups and decongestants. Its $pK_a = 9.22$. What is the pH of a solution that is 0.0295 M in pseudoephedrine hydrochloride?

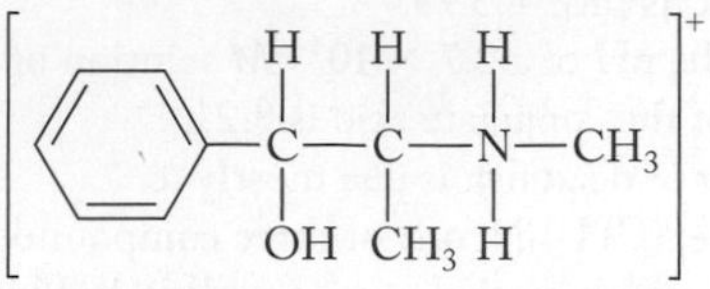

FIGURE P15.70

15.71. Malaria Treatment Quinine occurs naturally in the bark of the cinchona tree. For centuries it was the only treatment for malaria. Calculate the pH of a 0.01050 M solution of quinine in water.

15.72. Dozens of pharmaceuticals, including cyclizine for motion sickness to Viagra for impotence, are derived from the organic compound piperazine, whose structure is shown in Figure P15.72.

  a. Solutions of piperazine are basic ($K_{b_1} = 5.38 \times 10^{-5}$; $K_{b_2} = 2.15 \times 10^{-9}$). What is the pH of a 0.0125 M solution of piperazine?

  *b. Draw the structure of the ionic form of piperazine that would be present in stomach acid (about 0.16 M HCl).

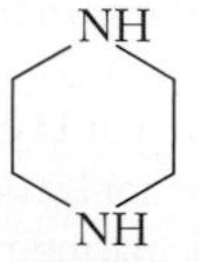

FIGURE P15.72

Acid Strength and Molecular Structure

Concept Review

15.73. Explain why the K_{a_1} of H_2SO_4 is much greater than the K_{a_1} of H_2SeO_4.

15.74. Explain why the K_{a_1} of H_2SO_4 is much greater than the K_{a_1} of H_2SO_3.

15.75. Predict which acid in the following pairs of acids is the stronger acid: (a) H_2SO_3 or H_2SeO_3; (b) H_2SeO_4 or H_2SeO_3.

15.76. Trifluoroacetic acid, CF_3COOH, is more than 10^4 times as strong as acetic acid, CH_3COOH. Explain why.

Acidic and Basic Salts

Concept Review

*15.77. The pK_a values of the conjugate acids of pyridine derivatives shown in Figure P15.77 increase as more methyl groups are added. Do more methyl groups increase or decrease the strength of the parent pyridine bases?

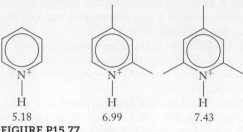

FIGURE P15.77

15.78. Why is it unnecessary to publish tables of K_b values of the conjugate bases of weak acids whose K_a values are known?

15.79. Which of the following salts produces an acidic solution in water: ammonium acetate, ammonium nitrate, or sodium formate?

15.80. Which of the following salts produces a basic solution in water: NaF, KCl, NH₄Cl?

15.81. Neutralizing the Smell of Fish Trimethylamine, $(CH_3)_3N$, $K_b = 6.5 \times 10^{-5}$ at 25°C, is a contributor to the "fishy" odor of not-so-fresh seafood. Some people squeeze fresh lemon juice (which contains a high concentration of citric acid) on cooked fish to reduce the fishy odor. Why is this practice effective?

*15.82. **Nutritional Value of Beets** Beets contain high concentrations of the calcium salt of malonic acid (see Figure P15.64). Could the presence of the calcium salt of malonic acid affect the pH balance of beets? If so, in which direction? Explain.

Problems

15.83. The K_a of the conjugate acid of the artificial sweetener saccharin is 2.1×10^{-11}. What is the pK_b for saccharin?

15.84. The K_{a_1} value for oxalic acid (HOOCCOOH) is 5.9×10^{-2}, and the K_{a_2} value is 6.4×10^{-5}. What are the values of K_{b_1} and K_{b_2} of the oxalate anion ($^-$OOCCOO$^-$)?

15.85. Dental Health Sodium fluoride is added to many municipal water supplies to reduce tooth decay. Calculate the pH of a 0.00339 M solution of NaF at 25°C.

15.86. Calculate the pH of a 1.25×10^{-2} M solution of the decongestant ephedrine hydrochloride if the pK_b of ephedrine (its conjugate base) is 3.86.

Additional Problems

15.87. Consider the following compounds: CH_3NH_2, CH_3COOH, $Ca(OH)_2$, and $HClO_4$.

  a. Identify the Arrhenius acid(s).
  b. Identify the Arrhenius base(s).
  c. Identify the Brønsted–Lowry acid(s).
  d. Identify the Brønsted–Lowry base(s).

15.88. Are all Arrhenius acids also Brønsted–Lowry acids? Are all Brønsted–Lowry acids also Arrhenius acids? If yes, explain why. If not, give a specific example to demonstrate the difference.

15.89. Are all Arrhenius bases also Brønsted–Lowry bases? Are all Brønsted–Lowry bases also Arrhenius bases? If yes, explain why. If not, give a specific example to demonstrate the difference.

15.90. Describe the intermolecular forces and changes in bonding that lead to the formation of a basic solution when methylamine (CH_3NH_2) dissolves in water.

*15.91. Describe the chemical reactions of sulfur that begin with the burning of high-sulfur fossil fuel and that end with the reaction between acid rain and building exteriors made of marble ($CaCO_3$).

15.92. The K_{a_1} of phosphorous acid, H_3PO_3, is nearly the same as the K_{a_1} of phosphoric acid, H_3PO_4.
  a. Draw the Lewis structure of phosphorous acid.
  b. Identify the ionizable hydrogen atoms in the structure.
  c. Explain why the K_{a_1} values of phosphoric and phosphorous acid are similar.

*15.93. **pH of Natural Waters** In a 1985 study of Little Rock Lake in Wisconsin, 400 gallons of 18 M sulfuric acid were added to the lake over six years. The initial pH of the lake was 6.1 and the final pH was 4.7. If none of the acid was consumed in chemical reactions, estimate the volume of the lake.

15.94. **Acid–Base Properties of Pharmaceuticals I** Zoloft is a prescription drug for the treatment of depression. It is sold as its hydrochloride salt, which is produced as shown in Figure P15.94. When the hydrochloride dissolves in water, will the resulting solution be acidic or basic?

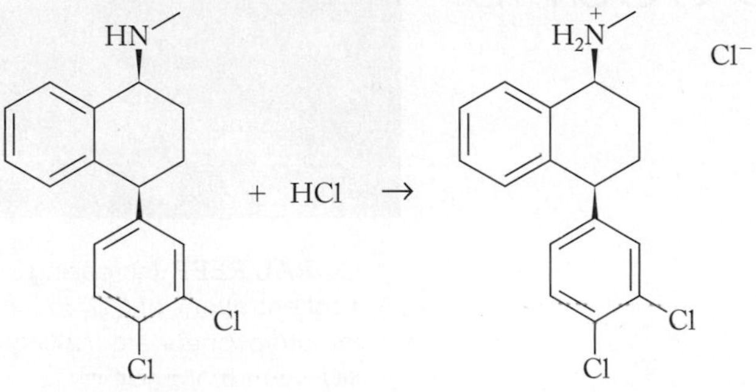

Zoloft
FIGURE P15.94

15.95. **Acid–Base Properties of Pharmaceuticals II** Prozac is a popular antidepressant drug. Its structure is given in Figure P15.95.
  a. Is a solution of Prozac in water likely to be acidic or basic? Explain your answer.
  b. Prozac is also sold as a hydrochloride salt. Which functional group is most likely to react with HCl?
  c. Prozac is sold as its hydrochloride because the solubility of the salt in water is higher than that of unreacted Prozac. Why is the salt more soluble?

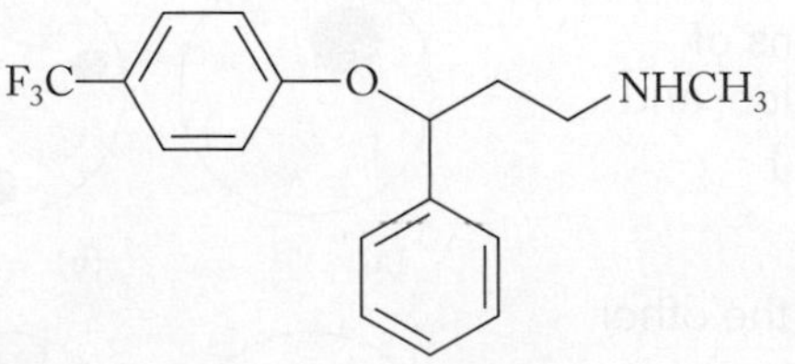

Prozac
FIGURE P15.95

15.96. Naproxen (also known as Aleve) is an anti-inflammatory drug used to reduce pain, fever, inflammation, and stiffness caused by conditions such as osteoarthritis and rheumatoid arthritis. Naproxen is an organic acid; its structure is shown in Figure P15.96. Naproxen has limited solubility in water, so it is sold as its sodium salt.
  a. Draw the molecular structure of the sodium salt.
  b. Is an aqueous solution of the salt acidic or basic? Explain why.
  c. Explain why the salt is more soluble in water than naproxen itself.

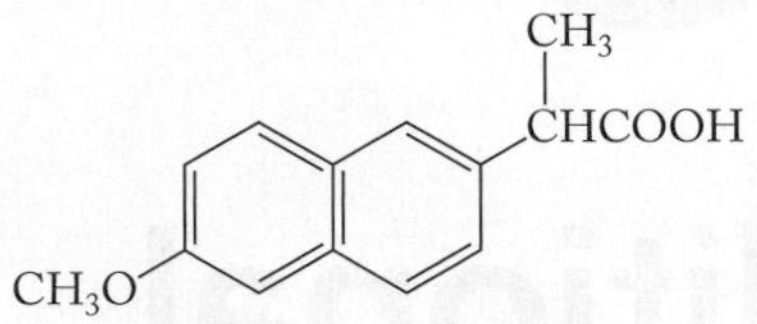

FIGURE P15.96

*15.97. Pentafluorocyclopentadiene, which has the structure shown in Figure P15.97, is a strong acid.
  a. Draw the conjugate base of C_5F_5H.
  b. Why is the compound so acidic, whereas most organic acids are weak?

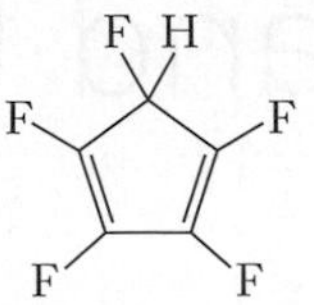

FIGURE P15.97

15.98. **Ocean Acidification** Some climate models predict a decrease in the pH of the oceans of 0.3 to 0.5 pH units by 2100 because of increases in atmospheric carbon dioxide.
  a. Explain, by using the appropriate chemical reactions and equilibria, how an increase in atmospheric CO_2 could produce a decrease in oceanic pH.
  b. How much more acidic would the oceans be if their pH dropped this much?
  c. Oceanographers are concerned about how a drop in oceanic pH could affect the survival of coral reefs. Why?

*15.99. Sulfuric acid reacts with nitric acid as follows:
$$HNO_3(aq) + 2\,H_2SO_4(aq) \rightarrow NO_2^+(aq) + H_3O^+(aq) + 2\,HSO_4^-(aq)$$
  a. Is the reaction a redox process?
  b. Identify the acid, base, conjugate acid, and conjugate base in the reaction. (*Hint*: Draw the Lewis structures for each.)

15.100. Thiosulfuric acid, $H_2S_2O_3$, can be prepared by the reaction of H_2S with HSO_3Cl:
$$HSO_3Cl(\ell) + H_2S(g) \rightarrow HCl(g) + H_2S_2O_3(\ell)$$
  a. Draw a Lewis structure for $H_2S_2O_3$, given that it is isostructural with H_2SO_4.
  b. Do you expect $H_2S_2O_3$ to be a stronger or weaker acid than H_2SO_4? Explain your answer.

16

Additional Aqueous Equilibria
Chemistry and the Oceans

CORAL REEF Increasing concentrations of CO_2 in the atmosphere are making seawater more acidic, which threatens corals and other marine life that form exoskeletons made of calcium carbonate.

PARTICULATE **REVIEW**

Soluble or Insoluble?

In Chapter 16 we revisit the limited solubility of some ionic compounds in aqueous solutions. Here we see images representing aqueous solutions of four ionic compounds: calcium fluoride, potassium fluoride, lithium sulfide, and lead(II) sulfide. (To simplify the images, water molecules are not shown.)

• What is the chemical formula of each compound?

• Which two of these compounds are much less soluble in water than the other two?

• Match each particulate representation to one of the four compounds.
   (Review Section 4.7 if you need help.)

(Answers to Particulate Review questions are in the back of the book.)

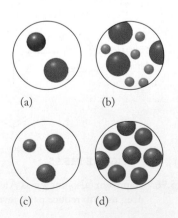

pH: To Change or Not to Change

Here we see two representations of solutions that contain (a) HF(aq) and (b) equal concentrations of both HF(aq) and NaF(aq). (To simplify the images, the solvent water molecules are not shown.) As you read Chapter 16, look for ideas that will help you answer these questions:

- Write a chemical equation describing the proton transfer equilibrium that occurs in each solution.

- Which solution(s) resist a sharp rise in pH upon the addition of OH⁻ ions?

- Which solution(s) resist a sharp drop in pH upon the addition of H_3O^+ ions?

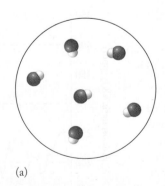

(a)

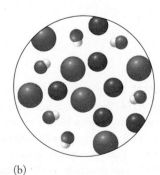

(b)

785

Learning Outcomes

LO1 Calculate the pH of a solution containing a weak acid or base and its conjugate base or acid
Sample Exercises 16.1, 16.2

LO2 Prepare a buffer with a desired pH
Sample Exercises 16.3, 16.4, 16.5

LO3 Evaluate the capacity of a buffer to resist changes in its pH
Sample Exercises 16.6, 16.7, 16.8

LO4 Calculate and interpret the results of an acid–base titration
Sample Exercises 16.9, 16.10, 16.11, 16.12

LO5 Identify a compound as a Lewis acid or Lewis base in a reaction
Sample Exercise 16.13

LO6 Use formation constants to calculate the concentrations of free and complexed metal ions in solution
Sample Exercise 16.14

LO7 Relate the acid strength of hydrated metal ions to the charge on the ions

LO8 Relate the solubility of an ionic compound to its solubility product
Sample Exercises 16.15, 16.16, 16.17, 16.18, 16.19

LO9 Separate mixtures of ionic compounds by selective precipitation reactions
Sample Exercise 16.19

16.1 Ocean Acidification: Equilibrium under Stress

The combustion of fossil fuels is increasing the concentration of carbon dioxide in the atmosphere. We know this from historical records of atmospheric CO_2 concentrations. Part of this record is based on analyzing bubbles of air trapped in layers of glacial ice, which show that CO_2 concentrations cycled between about 175 and 300 parts per million (ppm) by volume for at least 800,000 years (Figure 16.1). However, about 250 years ago, when CO_2 levels were already at the top of one of these cycles, they began to rise even more. Since the late 1950s, the increase in concentration has been the most rapid, as shown in Figure 16.2. By 2016, CO_2 levels had reached 400 ppm with no sign of slowing down.

In Chapter 8 we discussed the impact of increasing concentrations of CO_2 in the atmosphere on global warming. It turns out that more atmospheric CO_2 poses another environmental problem: ocean acidification. As we saw in Chapter 15, CO_2 gas dissolves in water, where it can form carbonic acid:

$$CO_2(g) + H_2O(\ell) \rightleftharpoons H_2CO_3(aq) \tag{16.1}$$

FIGURE 16.1 Analyses of bubbles of air trapped in Antarctic glaciers have enabled scientists to create a history of atmospheric CO_2 levels over the past 800,000 years, showing that atmospheric concentrations of CO_2 have generally oscillated between about 160 and 280 ppm. In recent years, however, CO_2 concentrations have sharply increased.

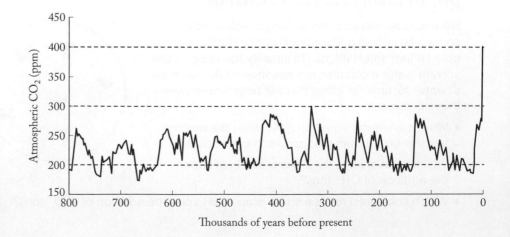

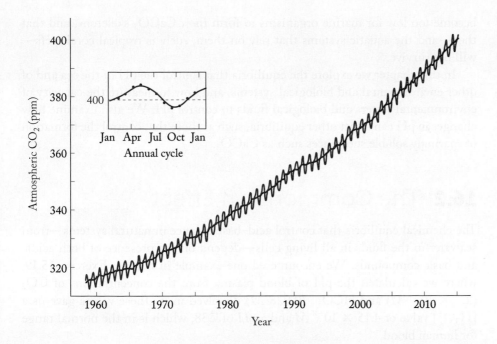

FIGURE 16.2 This graph of CO_2 concentrations in the atmosphere is based on analyses done since 1958 at the Mauna Loa Observatory in Hawaii—work begun by Charles David Keeling (1928–2005). The annual cycle in the data is caused by seasonally changing rates of photosynthesis by trees and other vegetation in the Northern Hemisphere.

Although H_2CO_3 is not stable in aqueous solutions, we will write $[H_2CO_3]$ in this chapter to represent the total amount of $CO_2(g)$ dissolved in water and to remind us that when $CO_2(g)$ dissolves in water, it forms an acidic solution:

$$H_2CO_3(aq) + H_2O(\ell) \rightleftharpoons HCO_3^-(aq) + H_3O^+(aq) \quad K_{a_1} = 4.3 \times 10^{-7} \quad (16.2)$$

We also saw in Chapter 15 that bicarbonate ions are capable of donating a second H^+ ion in aqueous solution:

$$HCO_3^-(aq) + H_2O(\ell) \rightleftharpoons CO_3^{2-}(aq) + H_3O^+(aq) \quad K_{a_2} = 4.7 \times 10^{-11} \quad (16.3)$$

However, the much smaller value of K_{a_2} means that the concentration of CO_3^{2-} ions in slightly basic seawater is much smaller than the concentration of HCO_3^- ions.

The solubility of atmospheric CO_2 increases as its partial pressure increases, so higher concentrations of CO_2 in the atmosphere mean higher concentrations of dissolved CO_2 in the sea. Scientists estimate that about 30% of the "fossil" CO_2 added to the atmosphere dissolves in the sea. Increasing concentrations of dissolved CO_2 have shifted the equilibrium in Equation 16.1 to the right—which, in turn, has shifted the equilibria in both Equations 16.2 and 16.3 to the right as well, resulting in an increase in the acidity of the sea. As a result, the average pH of seawater has dropped by about 0.11 pH units since the 18th century.

Are oceans with lower pH a problem? They are—especially if the pH continues to drop—because many marine organisms such as plankton, shellfish, and corals have exoskeletons made of $CaCO_3$. To make $CaCO_3$, these organisms need a supply of dissolved Ca^{2+} and CO_3^{2-} ions so that the following precipitation reaction can take place inside them:

$$Ca^{2+}(aq) + CO_3^{2-}(aq) \rightleftharpoons CaCO_3(s)$$

Unfortunately, increasing concentrations of H_3O^+ ions in the sea drive down the already small concentrations of CO_3^{2-} ions by shifting the equilibrium in Equation 16.3 to the left. There is a growing concern that the oceanic $[CO_3^{2-}]$ may

CONNECTION Henry's law, discussed in Chapter 11, states that the solubility of a gas is proportional to its partial pressure.

CONNECTION Le Châtelier's principle was introduced in Section 14.7 to discuss how a system at equilibrium responds to stress.

become too low for marine organisms to form their $CaCO_3$ skeletons, and that they—and the aquatic systems that rely on them, such as tropical coral reefs—will not survive.

In this chapter we explore the equilibria that control the pH of the sea and of other environmental and biological systems, and how to measure the capacity of environmental waters and biological fluids to control pH. We also examine how changes in pH can affect other equilibria, such as those that control the formation of sparingly soluble substances such as $CaCO_3$.

16.2 The Common-Ion Effect

The chemical equilibria that control acid–base balance in natural systems—from seawater to the fluids in all living cells—depend on the presence of both acidic and basic compounds. We encountered one example in Sample Exercise 15.19, where we calculated the pH of blood plasma from the concentrations of CO_2 (2.7×10^{-3} M) and HCO_3^- (0.028 M) dissolved in it. These values gave us a $[H_3O^+]$ value of 4.15×10^{-8} M and a pH of 7.38, which is in the normal range for human blood.

Now let's calculate the pH of a solution having the same concentration of dissolved CO_2 as in the blood plasma calculation (2.7×10^{-3} M), but initially containing no bicarbonate ions. We set up the appropriate RICE table with molar concentrations of the reactants and products, using $[H_2CO_3]$ to represent the total amount of dissolved $CO_2(g)$ in solution:

Reaction	$H_2CO_3(aq) + H_2O(\ell)$ $\rightleftharpoons$	$HCO_3^-(aq)$ +	$H_3O^+(aq)$
	$[H_2CO_3]$ (M)	$[HCO_3^-]$ (M)	$[H_3O^+]$ (M)
Initial	2.7×10^{-3}	0	0
Change	$-x$	$+x$	$+x$
Equilibrium	$(2.7 \times 10^{-3}) - x$	x	x

Taking the K_{a_1} value for carbonic acid (4.3×10^{-7}) from Equation 16.2, and solving for x:

$$K_{a_1} = \frac{[HCO_3^-][H_3O^+]}{[H_2CO_3]} = \frac{x \cdot x}{(2.7 \times 10^{-3}) - x} \approx \frac{x^2}{2.7 \times 10^{-3}} = 4.3 \times 10^{-7}$$

$$x = [H_3O^+] = 3.4 \times 10^{-5} \ M$$

Calculating pH:

$$pH = -\log[H_3O^+] = -\log(3.4 \times 10^{-5}) = 4.47$$

Note the difference in pH between this new solution and the blood plasma sample: the new solution has a pH value nearly 3 units lower than the plasma sample, corresponding to a $[H_3O^+]$ value *nearly 10^3 larger*. This difference makes sense because the presence of bicarbonate ions in the plasma sample inhibited the forward reaction in Equation 16.2, as predicted by Le Châtelier's principle. In this new solution, which lacks a second source of bicarbonate ions, the result is a higher $[H_3O^+]$ and lower pH values.

The difference in pH between these two solutions illustrates a principle known as the **common-ion effect**: in any ionic equilibrium, a reaction that produces an ion is suppressed when more of that ion is added to the system. One result of this suppression is that the initial concentrations of the conjugate acid–base pair change very little before equilibrium is achieved. Therefore we can use their initial concentrations to calculate the pH of their solution, as illustrated for this generic acid ionization equilibrium involving a solution of acid HA and its conjugate base, A⁻:

$$K_a = \frac{[H_3O^+][A^-]_{initial}}{[HA]_{initial}}$$

If we take the negative logarithm of both sides of the K_a expression, we transform $[H_3O^+]$ into pH and K_a into pK_a:

$$pK_a = pH - \log\frac{[base]}{[acid]}$$

$$pH = pK_a + \log\frac{[base]}{[acid]} \qquad (16.4)$$

We have generalized this equation by replacing [HA] and [A⁻] with [acid] and [base], respectively. Equation 16.4 is used to calculate the pH of a solution in which there are separate sources of both an acid and its conjugate base (or both a base and its conjugate acid). It is called the **Henderson–Hasselbalch equation**.

Consider what happens to the logarithm term in the Henderson–Hasselbalch equation when the concentrations of acid and base are equal to one another: The numerator and denominator are the same, so the value of the fraction in Equation 16.4 is 1. The log of 1 is 0, so pH = pK_a. This equality serves as a handy reference point in working with solutions of conjugate acid–base pairs. If the concentration of the basic component is greater than that of the acid, the logarithmic term is positive and pH > pK_a. If the concentration of the basic component is less than that of the acid, the logarithmic term is negative, and pH < pK_a.

Suppose the concentration of the base is 10 times the concentration of the acid; that is, [base] = 10[acid]. Substituting this equality into Equation 16.4, we have

$$pH = pK_a + \log\frac{10[acid]}{[acid]}$$

$$= pK_a + \log 10 = pK_a + 1$$

A 10-fold higher concentration of base produces a pH that is one unit above the pK_a value. Similarly, if the concentration of the acid component is 10 times that of the base, then pH = $pK_a - 1$. Sample Exercises 16.1 and 16.2 further illustrate how the Henderson–Hasselbalch equation simplifies pH calculations when we know the concentrations of both components of conjugate pairs.

common-ion effect the shift in the position of an equilibrium caused by the addition of an ion taking part in the reaction.

Henderson–Hasselbalch equation an equation used to calculate the pH of a solution in which the concentrations of acid and conjugate base are known.

C⨀NNECTION In Chapter 15 we introduced pH as the negative logarithm of hydronium ion concentration.

C⨀NNECTION The pK_a notation was introduced in Chapter 15 as an equivalent expression of K_a values but without using scientific notation and negative exponents.

SAMPLE EXERCISE 16.1 Calculating the pH of a Solution   **LO1**
                         of a Weak Acid and Its Conjugate Base

What is the pH of a sample of river water in which $[HCO_3^-] = 1.0 \times 10^{-4}$ M and the concentration of dissolved CO_2 in equilibrium with atmospheric CO_2 is 1.4×10^{-5} M?

Collect and Organize We are given the concentration of $CO_2(aq)$, which we will represent as $[H_2CO_3]$. We also know the concentration of the conjugate base of

H_2CO_3, namely, HCO_3^-. The pH of a solution of a weak acid and its conjugate base can be calculated using the Henderson–Hasselbalch equation:

$$pH = pK_a + \log \frac{[\text{base}]}{[\text{acid}]}$$

Analyze The conjugate acid–base pair relationship in this reaction is described by Equation 16.2:

$$H_2CO_3(aq) + H_2O(\ell) \rightleftharpoons HCO_3^-(aq) + H_3O^+(aq)$$

This reaction has a pK_{a_1} value (Table A5.1) of 6.37. The concentration of the base is nearly 10 times the concentration of the acid, so the log term in the Henderson–Hasselbalch equation should be almost 1, and the calculated pH should be a little higher than 7.

Solve Inserting the concentration and pK_{a_1} values into the Henderson–Hasselbalch equation:

$$pH = pK_{a_1} + \log \frac{[\text{base}]}{[\text{acid}]} = 6.37 + \log \frac{1.0 \times 10^{-4}}{1.4 \times 10^{-5}}$$

$$= 6.37 + 0.85 = 7.22$$

Think About It This result is near the pH value we predicted. Remember that a pH value of 7.22 contains only two significant figures (the 2 and the 2) because there were two significant figures in all three of the values used to calculate it. The 7 in pH 7.22 is not significant because its function is to tell us that the corresponding $[H_3O^+]$ value is between 10^{-7} and 10^{-8} M.

 Practice Exercise What is the pH of a solution in which $[HCOOH] = 2.5 \times 10^{-2}$ M and $[HCOO^-] = 7.8 \times 10^{-3}$ M?

(Answers to Practice Exercises are in the back of the book.)

We can also use the Henderson–Hasselbalch equation to calculate the pH of a solution of a weak base and its conjugate acid. An extra step is usually involved because we may know the pK_b value of the base but not the pK_a value of its conjugate acid, and only pK_a values are used in the Henderson–Hasselbalch equation. However, we learned in Chapter 15 that the equilibrium constants for conjugate acid–base pairs are related by Equation 15.18 ($K_a \times K_b = K_w$). To recast this as an equation with pK terms, we insert the numerical value of K_w at 25°C (1.0×10^{-14}):

$$K_a \times K_b = K_w = 1.0 \times 10^{-14}$$

and take the –log values of all three terms:

$$pK_a + pK_b = 14.00 \tag{16.5}$$

As you can see, converting the pK_b of a base into the pK_a of its conjugate acid is simply a matter of subtracting the pK_b value from 14.00.

SAMPLE EXERCISE 16.2 Calculating the pH of a Solution **LO1**
of a Weak Base and Its Conjugate Acid

What is the pH of a solution that is 0.200 M in NH_3 and 0.300 M in NH_4Cl?

Collect and Organize We are asked to calculate the pH of a solution containing known concentrations of a weak base (NH_3) and a salt of its conjugate acid (NH_4^+).

Analyze We can use the Henderson–Hasselbalch equation to calculate the pH of such a solution from the concentrations of the two components and the pK_a of the acid. Table A5.3 contains K_b and pK_b values of common bases. The pK_a and pK_b values of a conjugate acid–base pair are related by Equation 16.5:

$$pK_a + pK_b = 14.00$$

Our approach involves converting the pK_b value for NH_3 from Table A5.3 (4.75) into a pK_a value. Addition of ammonium ion to a solution of ammonia should produce a solution that is still basic, but not as basic as a solution that contains only ammonia.

Solve Inserting the pK_b value into Equation 16.5, and solving for pK_a:

$$pK_a = 14.00 - pK_b = 14.00 - 4.75 = 9.25$$

We can then use this value and the given concentrations of NH_3 and NH_4^+ in Equation 16.4:

$$pH = pK_a + \log \frac{[\text{base}]}{[\text{acid}]}$$

$$= 9.25 + \log \frac{0.200}{0.300} = 9.07$$

Think About It We predicted the solution would be basic, but not as basic as a solution of ammonia alone. To check whether this prediction was true, we can calculate the pH of $0.200\ M\ NH_3$ by using the approach we followed in Sample Exercise 15.11. The result is a pH of 11.27—more than 2 pH units higher (more basic) than the solution of ammonia and ammonium chloride in this exercise.

 Practice Exercise Calculate the pH of a solution that is $0.150\ M$ in benzoic acid and $0.100\ M$ in sodium benzoate.

16.3 pH Buffers

The common-ion effect plays a key role in controlling the pH of solutions that contain relatively high concentrations of both a weak acid and its conjugate base. These solutions have the capacity to withstand additions of acidic or basic substances with little or no measurable change in their pH, and they are known as **pH buffers**. Such a solution maintains a constant pH because the weak acid component of the buffer gives it the capacity to neutralize additions of basic substances while the conjugate base component gives it the capacity to neutralize acids. Ideally, a buffer has similar concentrations of both components of its conjugate pairs so that it can neutralize additions of acids or bases equally well. When the ratio of the conjugate pair is close to one, the log term in the Henderson–Hasselbalch equation is close to zero, and the pH of the buffer is close to the pK_a of the weak acid in it. Actually, a buffer with different concentrations of its conjugate acid–base pair can still be effective at controlling pH over a range of pH values up to about one unit above or below its acid's pK_a value.

CHEMTOUR
Buffers

SAMPLE EXERCISE 16.3 Building a Buffer for an Acidic pH   **LO2**

Select a weak acid in Table A5.1 of Appendix 5 that, when mixed with the sodium salt of its conjugate base in approximately equimolar proportions, produces a buffer with a pH of 2.80. Will the buffer contain *exactly* the same concentrations of acid and conjugate base, or slightly more acid or base?

pH buffer a solution that resists changes in pH when acids or bases are added to it; typically a solution of a weak acid and its conjugate base.

Collect, Organize, and Analyze The weak acid we seek is one whose pK_a is close to the target pH, 2.80.

Solve Among the acids in Table A5.1 with pK_a values near 2.80 are bromoacetic acid ($pK_a = 2.70$) and chloroacetic acid ($pK_a = 2.85$). Either could be used to prepare a buffer at pH 2.80, though neither would contain exactly the same concentration of the acid and its conjugate base: the bromoacetic acid buffer would require a slightly higher concentration of the conjugate base, and the chloroacetic acid buffer would contain a little more of the acid.

Think About It Another criterion for selecting weak acids for aqueous buffers is that they be soluble in water. Both of the acids selected are carboxylic acids with relatively small molar masses, and they are quite soluble in water. However, organic acids with large hydrocarbon (hydrophobic) regions in their molecular structures, such as octanoic acid and benzoic acid (Figure 16.3), are only slightly soluble in water.

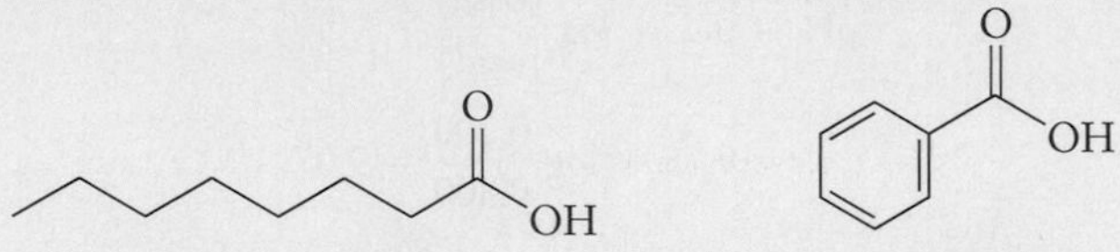

Octanoic acid   Benzoic acid

FIGURE 16.3 Examples of organic acids that are only slightly water soluble.

Practice Exercise Select a weak acid in Table A5.1 that, when mixed with the sodium salt of its conjugate base in approximately equimolar proportions, produces a buffer with a pH of 1.75.

SAMPLE EXERCISE 16.4 Building a Buffer for a Basic pH   **LO2**

Select a weak base in Table A5.3 that, when mixed with the chloride salt of its conjugate acid in approximately equimolar proportions, produces a buffer with a pH of 9.25. Indicate whether the buffer will contain *exactly* the same concentrations of base and conjugate acid, or slightly more base or acid.

Collect, Organize, and Analyze We can follow a strategy similar to that used in Sample Exercise 16.3, but we need to search the table for a weak base *that has a conjugate acid whose* pK_a is close to the target pH of 9.25. This means we are looking for a base with a pK_b value of $14.00 - 9.25 = 4.75$.

Solve A search of the table reveals that NH_3 has a pK_b of 4.75, which exactly matches the value we are looking for. This match means that the buffer should contain equimolar proportions of aqueous ammonia (NH_3) and ammonium chloride (NH_4Cl).

Think About It Both ammonia and ammonium chloride are soluble in water, so they are good candidates for preparing a pH 9.25 buffer, on the assumption that they do not chemically react with other solutes in the solution whose pH we wish to control.

Practice Exercise Select a weak base in Table A5.3 that, when mixed with the chloride salt of its conjugate acid in approximately equimolar proportions, produces a buffer with a pH of 10.75.

Having gained experience in selecting the components that are needed to prepare pH buffers, let's now calculate the quantities of these components that are needed to prepare a particular volume of a buffer with a desired pH.

SAMPLE EXERCISE 16.5 Preparing a Buffer Solution   **LO2**
                    with a Desired pH

A buffer system containing dihydrogen phosphate ($H_2PO_4^-$) and hydrogen phosphate (HPO_4^{2-}) helps regulate the pH of cytoplasm in living cells.

a. What is the mole ratio of HPO_4^{2-} ions to $H_2PO_4^-$ ions in a buffer with a pH of 6.75?
b. If the combined concentration of $H_2PO_4^-$ and HPO_4^{2-} ions in the buffer is to be 0.200 M, how many grams of NaH_2PO_4 (M = 120.0 g/mol) and of Na_2HPO_4 (M = 142.0 g/mol) are needed to prepare 20.0 liters of the buffer?

Collect and Organize We need to determine the mole ratio of HPO_4^{2-} ions to $H_2PO_4^-$ ions in a pH 6.75 buffer, and to calculate the masses of the sodium salts of these ions that are needed to make 20.0 liters of 0.200 M buffer. According to Table A5.1, the pK_a of $H_2PO_4^-$ (actually the pK_{a_2} of H_3PO_4) is 7.19. The Henderson–Hasselbalch equation relates the pH of a buffer to the pK_a of its acid component and the concentrations of that acid and its conjugate base.

Analyze The pH of this buffer is controlled by the acid ionization equilibrium of dihydrogen phosphate ions:

$$H_2PO_4^-(aq) + H_2O(\ell) \rightleftharpoons HPO_4^{2-}(aq) + H_3O^+(aq)$$

The target pH (6.75) is less than the pK_a (7.19), so the buffer must contain a higher concentration of $H_2PO_4^-$ ions than HPO_4^{2-} ions. The sum of $[H_2PO_4^-]$ and $[HPO_4^{2-}]$ is 0.200 M, so if we let $x = [HPO_4^{2-}]$, then $[H_2PO_4^-] = (0.200 - x)$ M.

Solve
a. Let's rearrange the Henderson–Hasselbalch equation to solve for the ratio of base to acid.

$$\log \frac{[\text{base}]}{[\text{acid}]} = pH - pK_a$$

Substituting the values for pH and pK_a gives us:

$$\log \frac{[HPO_4^{2-}]}{[H_2PO_4^-]} = pH - pK_a = 6.75 - 7.19 = -0.44$$

Then the ratio of $[HPO_4^{2-}]$ to $[H_2PO_4^-]$ is

$$\frac{[HPO_4^{2-}]}{[H_2PO_4^-]} = 10^{-0.44} = 0.36$$

b. Calculating the dissolved concentrations of HPO_4^{2-} (x) and $H_2PO_4^-$ (0.200 − x):

$$\frac{x}{0.200 - x} = 0.36$$

$$x = 0.36(0.200 - x)$$

$$1.36x = 0.72$$

$$x = [HPO_4^{2-}] = 0.053 \ M$$

$$(0.200 - x) = [H_2PO_4^-] = 0.147 \ M$$

The masses of their sodium salts in 20.0 liters of buffer are:

$$Na_2HPO_4: \quad 20.0 \ \cancel{L} \times 0.053 \ \frac{\cancel{mol}}{\cancel{L}} \times 142.0 \ \frac{g}{\cancel{mol}} = 150 \ g$$

$$NaH_2PO_4: \quad 20.0 \ \cancel{L} \times 0.147 \ \frac{\cancel{mol}}{\cancel{L}} \times 120.0 \ \frac{g}{\cancel{mol}} = 353 \ g$$

Think About It The buffer contains a higher concentration of $H_2PO_4^-$ ions than HPO_4^{2-} ions, which is consistent with our prediction, and with the ratio of acid to conjugate base of any buffer that has a pH below the pK_a of the acid used to make it.

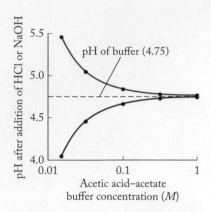

FIGURE 16.4 When strong acid (red line) or strong base (blue line) is added to a buffer solution, the extent to which the pH changes is inversely proportional to buffer concentration: the higher the concentrations of the buffer components, the smaller the change in pH. In this illustration, 100-mL samples of five solutions that are 0.015, 0.030, 0.100, 0.300, and 1.000 M acetic acid and sodium acetate all have an initial pH of 4.75 (dashed line). The graph shows the pH values of these solutions after 1.00 mL of 1.00 M HCl or 1.00 M NaOH has been added.

Buffer Capacity

In addition to selecting the appropriate conjugate acid–base pair to prepare a buffer, chemists also need to decide how concentrated the buffer should be. The greater the concentrations of the conjugate pair components, the greater is its **buffer capacity**—the ability of the buffer to withstand additions of acid or base without a significant change in pH (Figure 16.4).

SAMPLE EXERCISE 16.6 Calculating Buffer Response to **LO3**
Addition of Acid or Base

---

a. What is the change in pH of a 1.00-liter sample of the river water from Sample Exercise 16.1, in which $[H_2CO_3] = 1.4 \times 10^{-5}\,M$ and $[HCO_3^-] = 1.0 \times 10^{-4}\,M$, when 10.0 mL of $1.0 \times 10^{-3}\,M\,HNO_3$ is added to it?
b. Compare the pH change in part (a) to the pH change when the same quantity of strong acid is added to 1.00 liter of pure water (pH = 7.00).

Collect and Organize We know the volume and composition of a pH buffer, and we are asked to determine how much the pH of the buffer changes as a result of adding a strong acid. Quantities (in moles) of buffer components and of the added acid can be calculated by multiplying their volumes by their molar concentrations. We know from Table A5.1 that the pK_a of H_2CO_3 is 6.37. The pH of a solution with known concentrations of a conjugate acid–base pair can be calculated using the Henderson–Hasselbalch equation. Adding a strong acid to pure water is an exercise in dilution, a concept introduced in Section 4.3 and for which we can use Equation 4.3:

$$V_{\text{initial}} \times M_{\text{initial}} = V_{\text{diluted}} \times M_{\text{diluted}}$$

Analyze The river water contains a weak acid (dissolved CO_2) and its conjugate base (HCO_3^- ions); so the water has the capacity to function as a pH buffer. From Table A5.1, we know that the pK_{a_1} of carbonic acid is 6.37. Assuming nitric acid is the limiting reactant, adding x moles of this monoprotic, strong acid to the river water sample means adding x moles of H_3O^+ ions, which reacts with x moles of HCO_3^- ions, forming x moles of carbonic acid:

$$HCO_3^-(aq) + H_3O^+(aq) \rightarrow H_2CO_3(aq) + H_2O(\ell)$$

This sample decomposes to CO_2 gas:

$$H_2CO_3(aq) \rightarrow CO_2(g) + H_2O(\ell)$$

which is released into the atmosphere. Therefore, the concentration of the acid component of this particular buffer (H_2CO_3) is unchanged by the addition of acid.

If only a portion of the initial HCO_3^- concentration is consumed, the value of the [base] term in the Henderson–Hasselbalch equation will decrease somewhat from its original concentration, but the log function should minimize the decrease in pH.

Solve
a. The number of moles of H_3O^+ added to the river water sample is

$$(0.0100\;\text{L}) \times \left(1.0 \times 10^{-3}\,\frac{\text{mol}\,H_3O^+}{\text{L}}\right) = 1.0 \times 10^{-5}\;\text{mol}\,H_3O^+$$

buffer capacity the quantity of acid or base that a pH buffer can neutralize while keeping its pH within a desired range.

This quantity is equal to the number of moles of HCO_3^- in the river water sample consumed by the reverse reaction in Equation 16.2. The number of moles of bicarbonate present initially is

$$\left(1.0 \times 10^{-4}\,\frac{\text{mol}}{\text{L}}\right) \times (1.00\ \text{L}) = 1.0 \times 10^{-4}\ \text{mol}$$

This quantity is dissolved in a final volume of (1.00 L + 10.0 mL) = 1.01 L. Dividing the final quantities by the total sample volume of 1.01 L and inserting them into the Henderson–Hasselbalch equation to calculate pH:

$$\text{pH} = \text{p}K_a + \log\frac{[\text{base}]}{[\text{acid}]} = 6.37 + \log\frac{\left(\dfrac{9.0 \times 10^{-5}\ \text{mol}}{1.01\ \text{L}}\right)}{\left(\dfrac{1.4 \times 10^{-5}\ \text{mol}}{\text{L}}\right)} = 7.18$$

Thus the addition of acid changes the pH of the river water by (7.22 − 7.18) = 0.04 pH units.

b. Solving for M_{diluted}:

$$M_{\text{diluted}} = [\text{H}^+] = \frac{V_{\text{initial}} \times M_{\text{initial}}}{V_{\text{diluted}}}$$

$$= \frac{0.0100\ \text{L} \times (1.0 \times 10^{-3}\ M)}{1.01\ \text{L}} = 9.9 \times 10^{-6}\ M$$

$$\text{pH} = -\log[\text{H}_3\text{O}^+] = -\log(9.9 \times 10^{-6}\ M) = 5.00$$

Therefore the change in pH is (5.00 − 7.00) = −2.00.

Think About It Addition of a strong acid lowered the pH of the river water by only 0.04 pH units because it consumed only 10% of the basic component of the buffer system controlling pH. This decrease in pH is a tiny fraction of the decrease (−2.00 pH units) that happens when the same quantity of acid is added to pure water.

Practice Exercise What is the pH of a buffer that contains 0.225 M acetic acid and 0.375 M sodium acetate? What is the pH of 100.0 mL of the buffer after 1.00 mL of 0.318 M NaOH is added to it?

CONCEPT TEST

Which representation in Figure 16.5 depicts the aqueous solution of formic acid and sodium formate with the largest buffer capacity? The solvent has been omitted for clarity.

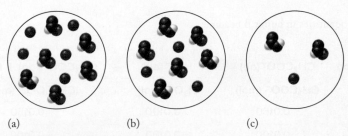

(a)   (b)   (c)

FIGURE 16.5 Possible representations of buffers.

(Answers to Concept Tests are in the back of the book.)

SAMPLE EXERCISE 16.7 Effect of Concentration on Buffer Capacity   **LO3**

Calculate the final pH after 0.0100 mole of H_3O^+ is added to (a) 0.100 L of buffer A, containing 1.00 M acetic acid and 1.00 M sodium acetate, and (b) 0.100 L of buffer B, containing 0.150 M acetic acid and 0.150 M sodium acetate.

Collect and Organize We are asked to calculate the final pH of equal volumes of two buffer solutions after addition of the same quantity of acid (0.0100 mol H_3O^+). We are given the composition of the buffers. From Table A5.1, we know that the pK_a of CH_3COOH is 4.75. The pH of a solution with known concentrations of a conjugate acid–base pair can be calculated using the Henderson–Hasselbalch equation.

Analyze The pH of both buffers is controlled by the acid ionization equilibrium of acetic acid:

$$CH_3COOH(aq) + H_2O(\ell) \rightleftharpoons CH_3COO^-(aq) + H_3O^+(aq)$$

Addition of 0.0100 mol H_3O^+ to the two buffers will shift this equilibrium to the left, consuming 0.0100 mol CH_3COO^- and producing an additional 0.0100 mol CH_3COOH. To evaluate the effect of these changes, we need to calculate the initial numbers of moles of CH_3COOH and CH_3COO^- ions in both buffers. We can predict from Figure 16.4 that the more concentrated buffer will experience a smaller change in pH upon the addition of acid.

Solve The initial quantities of CH_3COO^- and CH_3COOH in buffer A are both

$$0.100 \text{ L} \times \frac{1.00 \text{ mol}}{\text{L}} = 0.100 \text{ mol}$$

The corresponding quantities in buffer B are

$$0.100 \text{ L} \times \frac{0.150 \text{ mol}}{\text{L}} = 0.0150 \text{ mol}$$

We use a modified RICE table to see how the amounts of CH_3COO^- and CH_3COOH in buffer A change as a result of the additional H_3O^+ ions. Note that the quantities in this RICE table are moles, not molarity. After the addition of 0.0100 mol H_3O^+ ions to both buffers, the quantities in buffer A become:

Reaction	$CH_3COO^-(aq)$ +	$H_3O^+(aq)$ →	$CH_3COOH(aq) + H_2O(\ell)$
	CH_3COO^- (mol)	H_3O^+ (mol)	CH_3COOH (mol)
Initial	0.100	0.0100	0.100
Change	−0.0100	−0.0100	+0.0100
Final	0.0900		0.110

And the quantities in buffer B become:

Reaction	$CH_3COO^-(aq)$ +	$H_3O^+(aq)$ →	$CH_3COOH(aq) + H_2O(\ell)$
	CH_3COO^- (mol)	H_3O^+ (mol)	CH_3COOH (mol)
Initial	0.0150	0.0100	0.0150
Change	−0.0100	−0.0100	+0.0100
Final	0.0050		0.0250

The mole ratio of CH_3COO^- to CH_3COOH in each buffer can be substituted for the $[CH_3COO^-]/[CH_3COOH]$ ratio in the Henderson–Hasselbalch equation because both components are dissolved in the same volume of buffer. Using these substitutions and solving for pH:

$$\text{Buffer A: pH} = pK_a + \log \frac{[CH_3COO^-]}{[CH_3COOH]} = 4.75 + \log \frac{0.090}{0.110} = 4.66$$

$$\text{Buffer B: pH} = pK_a + \log \frac{[CH_3COO^-]}{[CH_3COOH]} = 4.75 + \log \frac{0.0050}{0.0250} = 4.05$$

Think About It Adding the same quantity of acid produced pH changes of 0.09 units in buffer A and 0.70 units in an equal volume of buffer B. As predicted, the buffer with a higher concentration was better able to resist pH change because the relative changes in the concentrations of its components were smaller.

Practice Exercise Calculate the change in pH when 0.145 mol OH^- is added to 1.00 L of two buffers: (a) a 1.16 M solution of sodium dihydrogen phosphate containing 1.16 M sodium hydrogen phosphate and (b) a 0.58 M solution of NaH_2PO_4 containing 0.58 M Na_2HPO_4.

Let's address one last question about buffer composition and capacity: does the mole ratio of acid to conjugate base make a difference in how well a buffer controls pH? Suppose, for example, you wish to prepare an acidic buffer and you know that contamination by basic substances is more likely than contamination by acids. You might use a 1:1 mixture if you could find a weak acid with a pK_a that exactly matched the target pH—but what if there were another conjugate acid–base pair that you could use to prepare the buffer, but which required the acid component to be 3 times as concentrated as the base to achieve the desired pH? Would the second buffer, with the greater proportion of acid in it, do a better job of controlling pH against additions of base than the 1:1 buffer? The following exercise provides insights into the answer to this question.

SAMPLE EXERCISE 16.8 Effect of Base: Acid Ratio on Buffer Capacity **LO3**

Two buffers are prepared with pH = 3.75. Buffer A contains equimolar concentrations of formic acid and sodium formate. Buffer B is prepared using the same overall concentration of conjugate pair components, but the pK_a of the weak acid in this buffer is 4.27, so it contains a [base]/[acid] mole ratio of 1:3 in order to achieve the target pH of 3.75. What are the changes in pH of the two buffers if enough NaOH is added to equal volumes of both to neutralize 1/4, 1/2, and 3/4 of the formic acid? Assume the NaOH additions do not significantly increase the volumes of the buffers.

Collect, Organize, and Analyze The composition of the buffer changes due to reaction with the added hydroxide ions:

$$HCOOH(aq) + OH^-(aq) \rightarrow HCOO^-(aq) + H_2O(\ell)$$

We are not given concentrations, but rather mole ratios of the weak acid and its conjugate base in each buffer. If we let x be the initial concentrations of formic acid and sodium formate in buffer A, then the initial concentrations of the acid and base components in buffer B must be 1.5x and 0.5x, respectively, in order to achieve the 1:3 ratio of [base]/[acid]. The three additions of NaOH will reduce the concentrations of the acid components in both buffers by $-0.25x$, $-0.50x$, and $-0.75x$ and increase the concentrations of the base components by $+0.25x$, $+0.50x$, and $+0.75x$.

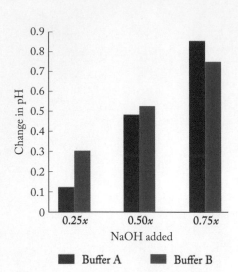

FIGURE 16.6 Changes in pH produced by additions of NaOH to two buffers. Buffer A contains a 1:1 mole ratio of weak acid and its conjugate base; buffer B contains the same total concentration of acid and conjugate base, but the mole ratio of base:acid is 1:3.

Solve Setting up Henderson–Hasselbalch equations with the initial [acid] and [base] values of the two buffers and the three changes in [acid] and [base] due to the three additions of NaOH:

Buffer A, 25% addition: $\quad \text{pH} = \text{p}K_a + \log\dfrac{[\text{base}]}{[\text{acid}]} = 3.75 + \log\dfrac{x(1.00 + 0.25)}{x(1.00 - 0.25)} = 3.97$

50% addition: $\quad \text{pH} = \text{p}K_a + \log\dfrac{[\text{base}]}{[\text{acid}]} = 3.75 + \log\dfrac{x(1.00 + 0.50)}{x(1.00 - 0.50)} = 4.23$

75% addition: $\quad \text{pH} = \text{p}K_a + \log\dfrac{[\text{base}]}{[\text{acid}]} = 3.75 + \log\dfrac{x(1.00 + 0.75)}{x(1.00 - 0.75)} = 4.60$

Buffer B, 25% addition: $\quad \text{pH} = \text{p}K_a + \log\dfrac{[\text{base}]}{[\text{acid}]} = 4.27 + \log\dfrac{x(0.50 + 0.25)}{x(1.50 - 0.25)} = 4.05$

50% addition: $\quad \text{pH} = \text{p}K_a + \log\dfrac{[\text{base}]}{[\text{acid}]} = 4.27 + \log\dfrac{x(0.50 + 0.50)}{x(1.50 - 0.50)} = 4.27$

75% addition: $\quad \text{pH} = \text{p}K_a + \log\dfrac{[\text{base}]}{[\text{acid}]} = 4.27 + \log\dfrac{x(0.50 + 0.75)}{x(1.50 - 0.75)} = 4.49$

The changes in pH compared to the target pH of 3.75 are depicted in Figure 16.6.

Think About It The greater capacity of buffer B to neutralize additions of base is evident when the $0.75x$ addition of NaOH consumes 75% of the acid in buffer A but only 50% of the acid in buffer B. Although buffer B has more capacity to neutralize very large additions of base, Figure 16.6 shows that buffer A does a better job of controlling pH when small to moderate quantities of base are added.

Practice Exercise Small volumes of strong acid, each containing 0.025 mole of $H_3O^+(aq)$ ions, are added to 1.00-liter samples of two basic buffers. Buffer A contains 0.150 mole of diethylamine and 0.150 mole of its conjugate acid. Buffer B contains 0.200 mole of methylamine and 0.100 mole of its conjugate acid. Calculate the changes in pH in the two buffers as a result of adding the strong acid.

16.4 Indicators and Acid–Base Titrations

Swimming pool operators routinely check the pH of pool water to make sure it is close to 7.3, which is the average pH of our eyes. They often add sodium carbonate to raise the water's pH after acidic summertime precipitation sends it below 7.3. To determine how much sodium carbonate to add, they test the pH of the water by using a kit that includes a **pH indicator** (Figure 16.7), which is a substance that changes color as its pH changes.

One such substance is phenol red. It is a weak acid ($\text{p}K_a = 7.6$) that is yellow in its un-ionized form (which, for convenience, we assign the generic formula HIn) and violet in its ionized (In⁻) form. At a pH one unit above the $\text{p}K_a$—at pH 8.6—the ratio [In⁻]/[HIn] is 10:1 and a phenol red solution is violet. At a pH less than 6.6, phenol red is largely un-ionized, and a solution of it is yellow. In the pH range from about 6.8 to 8.6, the color changes from yellow to orange to red to violet with increasing pH (Figure 16.8). These color changes allow pH to be determined to within about ± 0.2 unit.

As with buffers, a particular pH indicator is useful over a particular pH range that spans about 2 pH units. The midpoint of the range is defined by the $\text{p}K_a$ of the

CONNECTION In Section 4.6 we learned that when the number of moles of titrant is stoichiometrically equal to the number of moles of the analyte, the titration is at its equivalence point. When the indicator begins to change color, the titration is at its endpoint.

pH indicator a water-soluble weak organic acid that changes color as it ionizes.

indicator. In addition to their role in determining pH values, indicators are also used to detect the large changes in pH that occur at the equivalence points in acid–base titrations.

Acid–Base Titrations

We introduced acid–base titration methods in Section 4.6. Here we summarize the principal steps involved:

1. Assemble a titration apparatus (such as the one shown in Figure 16.9).
2. Accurately transfer a known volume of sample to the beaker.
3. Either add a few drops of an indicator solution to the sample or insert the probe of a pH meter.
4. Fill the buret with a solution (the *titrant*) of known concentration of a substance that reacts with a solute (the *analyte*) in the sample.
5. Slowly add titrant to the sample, and monitor the change in pH. When the volume of titrant needed to completely consume the analyte has been added, the equivalence point has been reached, as indicated by either a change in indicator color or a large change in pH as sensed by a pH electrode. This volume of titrant is a measure of the concentration of analyte in the sample.

The neutralization titrations in Chapter 4 involved titrating strong acids with strong bases and vice versa. In this section we begin with titrations of aqueous samples containing weak as well as strong monoprotic acids or *monobasic* bases. (A monobasic base accepts one hydrogen ion per molecule.) In the examples that follow, we monitor changes in pH during the titration by using a pH electrode, and we plot the pH of the titration reaction mixture against the volume of titrant added.

(a)

(b)

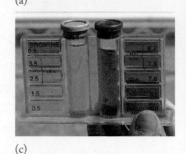
(c)

FIGURE 16.7 Many pool test kits include the pH indicator phenol red. A few drops are added to a sample of pool water collected in the tube with the red cap. (a) After a rainstorm, the pH of the pool water is 6.8 (or less), as indicated by the yellow color of the sample. (b) Sodium carbonate is added to the pool to raise the pH. (c) A follow-up test produces a red-orange color, indicating the pH of the pool has been properly adjusted.

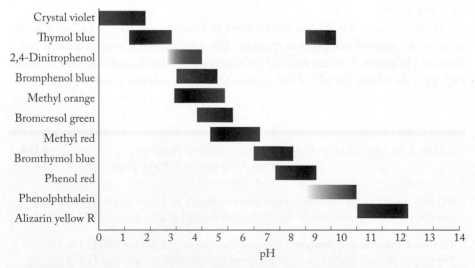

FIGURE 16.8 A pH indicator is useful within a range of 1 pH unit above and below the pK_a value of the indicator. This array of indicators could be used to determine pH values from 0 to 12.

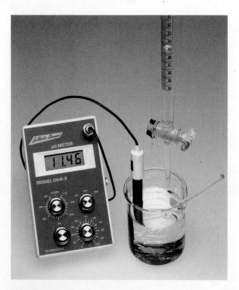

FIGURE 16.9 A digital pH meter is used to measure pH during a titration.

FIGURE 16.10 The titration curves of solutions of HCl and acetic acid with a standard solution of NaOH as the titrant.

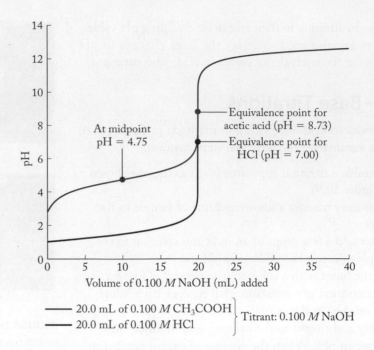

At midpoint pH = 4.75

Equivalence point for acetic acid (pH = 8.73)

Equivalence point for HCl (pH = 7.00)

Volume of 0.100 *M* NaOH (mL) added

—— 20.0 mL of 0.100 *M* CH₃COOH
—— 20.0 mL of 0.100 *M* HCl } Titrant: 0.100 *M* NaOH

CHEMTOUR
Acid–Base Titrations

First, let's compare the titration curves of two 20.0-mL samples: one contains 0.100 *M* HCl and the other contains 0.100 *M* acetic acid (CH₃COOH). Both are titrated with 0.100 *M* NaOH. The two graphs of pH versus titrant volume are shown in Figure 16.10. The initial pH of the HCl solution is lower than that of the acetic acid solution because HCl is a stronger acid, and the HCl titration curve stays below the acetic acid curve until they reach their equivalence points, where both acids have been completely neutralized.

As the base is added, the pH of the HCl sample (the red curve in Figure 16.10) does not change much until nearly all the acid has been consumed. Sample pH is determined only by the [H₃O⁺] remaining in it, which decreases in proportion to the volume of titrant added; however, the increase in pH is not a linear function of the volume of titrant added because of its log dependence on [H₃O⁺]. As the equivalence point is approached, the principal ions present in the sample are Na⁺, Cl⁻, and H₃O⁺.

When enough NaOH has been added to consume all the H₃O⁺ ions in the sample, the equivalence point is reached. The solution now consists only of NaCl dissolved in water. Sodium and chloride ions do not hydrolyze and do not influence pH; therefore, the pH of the sample at the equivalence point is 7.00.

SAMPLE EXERCISE 16.9 Calculating pH during Titration of a Strong Acid with a Strong Base **LO4**

What is the pH of the solution being titrated in Figure 16.10 just before the equivalence point when 19.0 mL of 0.100 *M* NaOH has been added to 20.0 mL of 0.100 *M* HCl?

Collect, Organize, and Analyze We know the volume and molarity of the HCl (a strong acid that completely dissociates), so we can calculate the moles of H₃O⁺ ions in solution. We can also calculate how many moles of OH⁻ ions have been added (NaOH is a strong base and also completely dissociates). Each H₃O⁺ ion reacts with one OH⁻

ion to form H_2O. The pH depends upon the excess number of H_3O^+ ions and can be calculated using pH $= -\log[H_3O^+]$, where $[H_3O^+]$ is calculated using the total volume of acid + added base.

Solve The amount of acid initially in the sample is

$$\text{moles } H_3O^+ = 20.0 \text{ mL} \times \frac{1 \text{ L}}{1000 \text{ mL}} \times \frac{0.100 \text{ mol HCl}}{\text{L solution}} = 2.00 \times 10^{-3} \text{ mol HCl}$$

The amount of base added is

$$\text{moles } OH^- = 19.0 \text{ mL} \times \frac{1 \text{ L}}{1000 \text{ mL}} \times \frac{0.100 \text{ mol NaOH}}{\text{L solution}} = 1.90 \times 10^{-3} \text{ mol NaOH}$$

One mole of HCl produces 1 mole of H_3O^+ ions, and 1 mole of NaOH produces 1 mole of OH^- ions, so there are also 2.00×10^{-3} mol H_3O^+ ions and 1.90×10^{-3} mol OH^- ions.

Each H_3O^+ ion reacts with one OH^- ion, so there is an excess of H_3O^+ ions:

$$\text{excess } H_3O^+ \text{ ions} = 2.00 \times 10^{-3} \text{ mol } H_3O^+ \text{ ions} - 1.90 \times 10^{-3} \text{ mol } H_3O^+ \text{ ions reacted}$$

$$= 1.0 \times 10^{-4} \text{ mol } H_3O^+ \text{ remain in solution}$$

These H_3O^+ ions are dissolved in (20.0 mL + 19.0 mL) = 39.0 mL solution. Therefore the molarity of the unreacted ions is

$$\frac{1.0 \times 10^{-4} \text{ moles } H_3O^+}{0.0390 \text{ L solution}} = 2.6 \times 10^{-3} \ M$$

So the pH $= -\log[2.6 \times 10^{-3}] = 2.59$.

Think About It Note that after the addition of just 1.00 mL of NaOH, the pH jumps to equal 7.00. The pH of the solution before the equivalence point in a strong acid–strong base titration is determined by the number of excess H_3O^+ ions.

Practice Exercise What is the pH of the solution in Figure 16.10 just after the equivalence point when 21.0 mL of 0.100 M NaOH has been added to 20.0 mL of 0.100 M HCl?

CONCEPT TEST

The representations in Figure 16.11 depict the particles present during the titration of LiOH with HI. Which representation depicts the starting solution? The equivalence point? Put the representations in order from the beginning to the end of the titration. The solvent has been omitted for clarity.

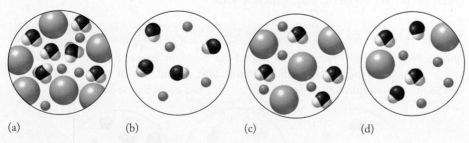

(a)   (b)   (c)   (d)

FIGURE 16.11 Particles present during titration.

The pH of the acetic acid sample (the blue curve in Figure 16.10) changes abruptly with the first few drops of added base, but then the changes become smaller and the titration curve levels out. In the nearly flat region before the equivalence point, the sample acts like a pH buffer as additions of NaOH titrant neutralize acetic acid, forming the sodium salt of its conjugate base:

$$CH_3COOH(aq) + NaOH(aq) \rightarrow CH_3COONa(aq) + H_2O(\ell)$$

As long as there are significant concentrations of both acetic acid and acetate ions in the sample, the changes in pH with added titrant are small.

When enough titrant has been added to consume nearly all the acid in both samples, pH rises sharply. At their equivalence points, the same volume of NaOH titrant has been added to both because the initial concentrations of the two acids and their sample volumes were the same. Therefore the number of moles of acid in each sample was the same, and that is the quantity that defines the number of moles of base needed to reach the equivalence point. It does not matter whether the acids were initially completely ionized because even a weak acid such as acetic acid is completely neutralized at its equivalence point. The reason why is that adding NaOH to a solution of acetic acid (or any weak acid) consumes H_3O^+ ions, which, in accordance with Le Châtelier's principle, shifts the position of the acid ionization equilibrium to the right:

$$CH_3COOH(aq) + H_2O(\ell) \rightleftharpoons CH_3COO^-(aq) + H_3O^+(aq)$$

Even though the volume of titrant needed to reach the equivalence points of the red and blue curves in Figure 16.10 is the same, the pH values at the two equivalence points are not the same: it is 7.00 (neutral) in the HCl titration but 8.73 in the acetic acid titration. The latter solution is slightly basic because the product of the titration reaction is a solution of sodium acetate (CH_3COONa), which is a basic salt. Acetate ions in solution react with water, forming molecules of acetic acid and OH^- ions. As a general rule, the pH at the equivalence point in the titration of any weak acid with a strong base is greater than 7 because the hydrolysis of the anion of the salt formed in the neutralization reaction produces OH^- ions. Beyond their equivalence points, the two curves overlap because the solutions' pH in this region is controlled only by the increasing concentration of NaOH.

Another important point in the acetic acid titration curve lies halfway to the equivalence point. At this *midpoint* in the titration, half of the acetic acid initially in the sample has been converted into its conjugate base, acetate ions. Therefore the concentration of acetate ions produced by the neutralization reaction and the concentration of acetic acid remaining from the original sample *are the same*. If we insert this equality ([base] = [acid]) into the log term of the Henderson–Hasselbalch equation, the value of the term is zero. Therefore the midpoint pH is equal to the pK_a of acetic acid, which is 4.75.

CHEMTOUR
Titrations of Weak Acids

CONCEPT **TEST**

The representations in Figure 16.12 depict the particles present during a titration of HClO with KOH. Which represents the weak acid at the start of the titration? At the midpoint? At the equivalence point? The solvent has been omitted for clarity.

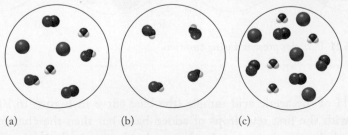

(a)   (b)   (c)

FIGURE 16.12 Particles present during titration.

SAMPLE EXERCISE 16.10 Calculating pH during Titration **LO4**
of a Weak Acid with a Strong Base

What is the pH of the solution being titrated in Figure 16.10 at the midpoint when 10.0 mL of 0.100M NaOH has been added to 20.0 mL of 0.100M CH_3COOH?

Collect, Organize, and Analyze As in Sample Exercise 16.9, we can calculate the moles of weak acid present and the moles of OH^- ions added from their respective volumes and molarities. As the NaOH is added to the acetic acid solution, this reaction occurs:

$$CH_3COOH(aq) + OH^-(aq) \rightarrow CH_3COO^-(aq) + H_2O(\ell)$$

We know how much of the weak acid remains and how much of its conjugate base is produced because 1 mole of acid reacts with 1 mole of OH^- ions to produce 1 mole of acetate ions. We can calculate $[CH_3COO^-]$ and then use the Henderson–Hasselbalch equation to determine the pH of the solution. The pK_a for acetic acid, which can be found in Appendix A5.1, equals 4.75.

Solve The amount of acid in the original sample is

$$\text{moles } CH_3COOH = 20.0 \text{ mL} \times \frac{1 \text{ L}}{1000 \text{ mL}} \times \frac{0.100 \text{ mol } CH_3COOH}{\text{L solution}}$$

$$= 2.00 \times 10^{-3} \text{ mol } CH_3COOH$$

The amount of base added is

$$\text{moles } OH^- = 10.0 \text{ mL} \times \frac{1 \text{ L}}{1000 \text{ mL}} \times \frac{0.100 \text{ mol NaOH}}{\text{L solution}}$$

$$= 1.00 \times 10^{-3} \text{ mol NaOH}$$

We can use a RICE table to determine how many moles of CH_3COOH remain after reacting with all of the NaOH added, and how many moles of CH_3COO^- are produced:

Reaction	$CH_3COOH(aq)$	+	$OH^-(aq)$	$\rightarrow$	$CH_3COO^-(aq) + H_2O(\ell)$
	CH_3COOH (moles)		OH^- (moles)		CH_3COO^- (moles)
Initial	2.00×10^{-3}		1.00×10^{-3}		0
Change	-1.00×10^{-3}		-1.00×10^{-3}		$+1.00 \times 10^{-3}$
Final	1.00×10^{-3}		0		1.00×10^{-3}

The total sample volume is (20.0 mL + 10.0 mL) = 30.0 mL of solution. Therefore

$$[CH_3COOH] = \frac{1.00 \times 10^{-3} \text{ moles } CH_3COOH}{0.030 \text{ L solution}} = 3.33 \times 10^{-2} \, M$$

$$[CH_3COO^-] = \frac{1.00 \times 10^{-3} \text{ moles } CH_3COO^-}{0.030 \text{ L solution}} = 3.33 \times 10^{-2} \, M$$

Using these values in the Henderson–Hasselbalch equation gives us

$$pH = pK_a + \log \frac{[CH_3COO^-]}{[CH_3COOH]}$$

$$= 4.75 + \log\left(\frac{3.33 \times 10^{-2}}{3.33 \times 10^{-2}}\right) = 4.75 + \log(1) = 4.75 + 0 = 4.75$$

Think About It The midpoint of a titration of a weak acid with a strong base is defined as the point when a titration reaction mixture contains equal concentrations of the weak acid and its conjugate base. Therefore the pH at the midpoint, according to the Henderson–Hasselbalch equation, will always be equal to the pK_a of the weak acid.

Practice Exercise What is the pH at the midpoint of the titration with a strong base of an aqueous sample that contains an unknown concentration of benzoic acid?

CONCEPT **TEST**

Why is it better to titrate a weak acid with a strong base instead of titrating a weak acid with a weak base?

Titration curves of weakly or strongly basic analytes with strong acid titrants resemble the curves in Figure 16.10 but are inverted, starting with high initial pH values and ending with low ones. Figure 16.13 illustrates two such titrations. One sample contains a strong base, 0.100 M NaOH; the other contains a weak base, 0.100 M NH$_3$. Both are titrated with 0.100 M HCl. The initial pH of the NaOH titration curve is higher than that of the NH$_3$ sample because NaOH is a stronger base. The pH of the NaOH sample does not change much as acid is added until near the equivalence point. As the equivalence point is approached, the principal ions present in the reaction mixture are Na$^+$, Cl$^-$, and decreasing concentrations of OH$^-$. The sample pH, which is determined only by the [OH$^-$] still present, starts to fall steeply. When enough acid has been added to consume all the OH$^-$ ions in the sample, the equivalence point is reached. The solution now consists of water and NaCl, which is neutral, so pH = 7.00 at the equivalence point.

The pH of the NH$_3$ solution changes abruptly with the first few drops of added acid, but then the changes become smaller and the titration curve levels out. In this nearly flat region, additions of acidic titrant are consumed as NH$_3$ reacts with H$_3$O$^+$

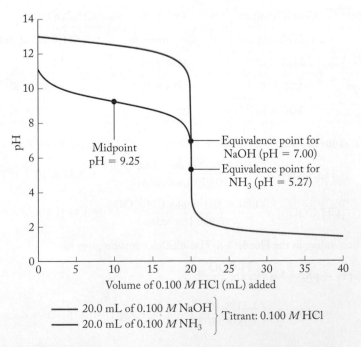

FIGURE 16.13 The titration curves of solutions of NaOH and NH$_3$ with a standard solution of HCl as the titrant.

ions to form NH_4^+ ions. In this region of the titration curve, the sample acts like a pH buffer, just as the mixture of acetic acid and acetate ions did in the titration in Figure 16.10. The pH value at the equivalence point in the ammonia titration is 5.27 (slightly acidic) because the product of the titration reaction is an acidic salt, NH_4Cl, which reacts with water, forming ammonia molecules and H_3O^+ ions. Beyond their equivalence points, the NaOH and NH_3 curves are identical because the pH is determined only by the moles of HCl added after the equivalence point and the total volume of the titration reaction mixtures.

As in the acetic acid titration, the midpoint in the NH_3 titration curve is significant because it is the point at which half of the NH_3 initially in the sample has been converted into NH_4^+ ions. Therefore $[NH_3] = [NH_4^+]$. This concentration equality can be used in the Henderson–Hasselbalch equation based on the hydrolysis of ammonium ions:

$$NH_4^+(aq) + H_2O(\ell) \rightleftharpoons NH_3(aq) + H_3O^+(aq)$$

$$pH = pK_a + \log\frac{[NH_3]}{[NH_4^+]}$$

This pK_a value can be calculated from the pK_b value for ammonia, which is 4.75 (see Table A5.3), using Equation 16.4: $pK_a = 14.00 - 4.75 = 9.25$. Because the value of the log term in the Henderson–Hasselbalch equation is zero, the pH at the midpoint of the titration is equal to this pK_a, or 9.25.

CONCEPT TEST

What is the pH at the midpoint of the titration of an aqueous sample of methylamine with a standard solution of hydrochloric acid?

CONCEPT TEST

The representations in Figure 16.14 depict the particles present during a titration of NH_3 with HCl. Which represents the start of the titration? The region where the solution best functions as a buffer? The equivalence point? The solvent has been omitted for clarity.

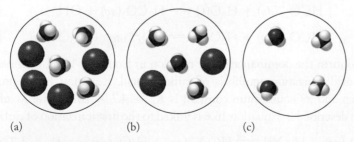

(a)　　　　(b)　　　　(c)

FIGURE 16.14 Particles present during titration.

Titrations with Multiple Equivalence Points

So far, the titration curves in this chapter have each had only one equivalence point. Let's now consider an important one with two equivalence points: an alkalinity titration (Figure 16.15). In environmental science, the alkalinity of a water sample means the capacity of the water to neutralize additions of acid. Titration of the sample with strong acid provides a way to determine that capacity.

FIGURE 16.15 An alkalinity titration curve can have two equivalence points. The first marks the complete conversion of any carbonate in the sample into bicarbonate, and the second marks the conversion of bicarbonate into carbonic acid.

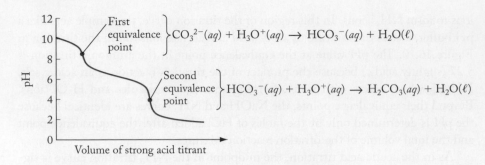

If carbonate is present in the sample, the first additions of titrant convert carbonate into bicarbonate:

$$CO_3{}^{2-}(aq) + H_3O^+(aq) \rightarrow HCO_3{}^-(aq) + H_2O(\ell)$$

This reaction continues until all of the carbonate initially in the sample has been converted to bicarbonate. This represents the first equivalence point in the titration and is marked by a sharp drop in pH. In the second stage of the titration, the bicarbonate ions formed in the first stage plus any bicarbonate present in the original sample react with additional acidic titrant, forming carbonic acid, most of which decomposes into carbon dioxide and water:

$$HCO_3{}^-(aq) + H_3O^+(aq) \rightarrow 2\,CO_2(aq) + H_2O(\ell)$$

The solubility of CO_2 in water is limited, so its production in the second step of an alkalinity titration often results in formation of bubbles of carbon dioxide as it comes out of solution:

$$CO_2(aq) \rightarrow CO_2(g)$$

In the first stage of the alkalinity titration, the titration curve has a region in which added acid has little effect on pH. This is a buffering region where $HCO_3{}^-$ and $CO_3{}^{2-}$ function as a weak acid–conjugate base pair. At the first equivalence point, the dominant carbon-containing species is $HCO_3{}^-$ and the sample is still slightly basic (pH > 8). This basic pH tells us that the bicarbonate is more effective as a base (Equation 16.6) than as an acid (Equation 16.7):

$$HCO_3{}^-(aq) + H_2O(\ell) \rightleftharpoons H_2CO_3(aq) + OH^-(aq) \qquad (16.6)$$

$$HCO_3{}^-(aq) + H_2O(\ell) \rightleftharpoons CO_3{}^{2-}(aq) + H_3O^+(aq) \qquad (16.7)$$

We can confirm the dominance of the reaction in Equation 16.6 by considering the two-step acid ionization equilibria of carbonic acid. Equation 16.7 represents the second step and its equilibrium constant is $K_{a_2} = 4.7 \times 10^{-11}$. The K_b of the base ionization described in Equation 16.6 is linked to the first ionization of carbonic acid:

$$H_2CO_3(aq) + H_2O(\ell) \rightleftharpoons HCO_3{}^-(aq) + H_3O^+(aq) \qquad K_{a_1} = 4.3 \times 10^{-7}$$

Converting K_{a_1} into K_b by using Equation 15.18:

$$K_a \times K_b = K_w$$

$$K_b = \frac{K_w}{K_{a_1}} = \frac{1.0 \times 10^{-14}}{4.3 \times 10^{-7}} = 2.3 \times 10^{-8}$$

This K_b value is much greater than the value of K_{a_2}:

$$\frac{2.3 \times 10^{-8}}{4.7 \times 10^{-11}} = 4.9 \times 10^2$$

Thus bicarbonate is nearly 500 times stronger a base than it is an acid.

During the second stage of the titration, conversion of HCO_3^- ions to H_2CO_3 produces a second pH buffer and another plateau of slowly changing pH. When the HCO_3^- ions are completely consumed, pH drops sharply for a second time at the second equivalence point. Its pH is below 7, reflecting the acidic character and K_{a_1} value of carbonic acid.

Note that the initial pH of the sample in Figure 16.15 is slightly above 10, which is quite basic and above the pH range tolerated by many species of aquatic life. Such highly basic water may be found in arid regions such as the U.S. Southwest, where rocks containing $CaCO_3$ and other basic compounds are in contact with water. The pH of most freshwater is lower than 8, which means that the dominant carbonate species is bicarbonate. The alkalinity titration curves of these waters have only one equivalence point, coinciding with the pH of the second equivalence point in Figure 16.15.

SAMPLE EXERCISE 16.11 Interpreting the Results of a Titration with Multiple Equivalence Points I   **LO4**

If the volume of titrant needed to reach the first equivalence point in the titration shown in Figure 16.15 is 9.00 mL, and *a total of* 27.00 mL is required to reach the second equivalence point, what was the ratio of carbonate to bicarbonate in the original sample?

Collect, Organize, and Analyze Carbonate (CO_3^{2-}) ions in the sample combine with H_3O^+ ions from the titrant and are converted to bicarbonate (HCO_3^-) ions in the first stage of the titration. Bicarbonate ions—including any in the original sample plus all those produced in the first-stage reaction—are converted to carbonic acid, H_2CO_3, in the second stage.

Solve The volume of titrant needed to titrate the CO_3^{2-} in the sample (9.00 mL) is exactly half the additional volume (27.00 − 9.00 = 18.00 mL) needed to titrate the HCO_3^- in the reaction mixture. However, 9.00 mL of the titrant consumed in the second stage was needed just to titrate the HCO_3^- produced in the first stage. This means the volume of titrant needed to titrate the HCO_3^- that was in the original sample was only (18.00 − 9.00) = 9.00 mL. Therefore the original sample contained equal concentrations of CO_3^{2-} and HCO_3^- ions.

Think About It It would be tempting to interpret the volumes of titrant consumed in the two stages to mean that there was twice as much bicarbonate as carbonate in the original sample. It's important to remember that the HCO_3^- ions titrated in the second stage come from the original sample *and* from HCO_3^- ions produced from CO_3^{2-} ions during the first stage of the titration.

 Practice Exercise Is the midpoint pH in the first stage of an alkalinity titration always the same as the pK_{a_2} of carbonic acid? Explain why or why not.

We can detect both equivalence points in an alkalinity titration by using a pH electrode, or we can use appropriate indicators. Phenol red would not be a good choice for the titration in Figure 16.15 because it changes color between pH 6.8 and 8.4. This range is just below the pH of the first equivalence point and well above the pH of the second equivalence point. To detect the first equivalence point, we need an indicator with a pK_a near the pH of the solution at the first equivalence point, which is about 8.5. We can see from Figure 16.8 that one candidate is phenolphthalein (pK_a = 9.7), which is pink in its basic form and colorless at low pH.

To detect the second equivalence point, we could add bromcresol green (pK_a = 4.6) after the first equivalence point has been reached. We would not add it earlier because its blue-green color in basic solutions would obscure the pink-to-colorless transition of phenolphthalein. We do not need to be concerned about the phenolphthalein obscuring the bromcresol green color change because phenolphthalein is colorless in acidic solutions.

SAMPLE EXERCISE 16.12 Interpreting the Results of a Titration with Multiple Equivalence Points II

LO4

A 100.0-mL sample of water from the Sapphire Pool in Yellowstone National Park (Figure 16.16) is titrated with 0.0300 M HCl. A few drops of phenolphthalein are added at the beginning of the titration, and the solution turns pink. It takes 5.91 mL of titrant to reach the pink-to-clear equivalence point. Then a few drops of bromcresol green are added, and it takes an additional 26.02 mL of titrant before the blue-green color changes to yellow. What were the initial concentrations of carbonate and bicarbonate in the sample?

Collect and Organize We are asked to determine the concentrations of CO_3^{2-} and HCO_3^- ions in a water sample from the results of a titration. These determinations are based on the volumes of titrant needed to reach two equivalence points. In the first stage, CO_3^{2-} ions in the sample are converted to HCO_3^- ions:

$$(1) \quad H_3O^+(aq) + CO_3^{2-}(aq) \rightarrow HCO_3^-(aq) + H_2O(\ell)$$

In the second stage, HCO_3^- ions are converted to H_2CO_3:

$$(2) \quad H_3O^+(aq) + HCO_3^-(aq) \rightarrow H_2CO_3(aq) + H_2O(\ell)$$

Analyze The stoichiometries of the reactions tell us it takes 1 mole of HCl to titrate 1 mole of carbonate to bicarbonate in the first stage, and it takes 1 mole of HCl to titrate 1 mole of bicarbonate to carbonic acid in the second stage. The HCO_3^- ions titrated in stage 2 include any in the original sample plus all the HCO_3^- ions produced by reaction 1. The difference between the titrant volumes needed to reach the two equivalence points, (26.02 − 5.91) = 20.11 mL, is the volume of acid required to react with the HCO_3^- that was present in the original sample. This difference is between 3 and 4 times the volume of titrant consumed in stage 1, so we can predict that there was 3 to 4 times as much HCO_3^- in the original sample as CO_3^{2-}.

Solve Calculating the concentrations of CO_3^{2-} and HCO_3^- in the original sample from the volumes and molarity of the HCl titrant consumed in stages 1 and 2:

$$[CO_3^{2-}] = \frac{\text{mol } CO_3^{2-}}{\text{L solution}}$$

$$= \frac{5.91 \text{ mL titrant} \times \dfrac{0.0300 \text{ mmol HCl}}{\text{mL titrant}} \times \dfrac{1 \text{ mol HCl}}{1000 \text{ mmol HCl}} \times \dfrac{1 \text{ mol } CO_3^{2-}}{1 \text{ mol HCl}}}{100.0 \text{ mL solution} \times \dfrac{1 \text{ L}}{1000 \text{ mL}}}$$

$$= 1.77 \times 10^{-3} \ M$$

$$[HCO_3^-] = \frac{\text{mol } HCO_3^-}{\text{L solution}}$$

$$= \frac{20.11 \text{ mL titrant} \times \dfrac{0.0300 \text{ mmol HCl}}{\text{mL titrant}} \times \dfrac{1 \text{ mol HCl}}{1000 \text{ mmol HCl}} \times \dfrac{1 \text{ mol } CO_3^{2-}}{1 \text{ mol HCl}}}{100.0 \text{ mL solution} \times \dfrac{1 \text{ L}}{1000 \text{ mL}}}$$

$$= 6.03 \times 10^{-3} \ M$$

FIGURE 16.16 The water in Sapphire Pool in Yellowstone National Park is slightly alkaline due mostly to carbonate and bicarbonate ions. It is also crystal clear and hot.

Think About It The titration results confirm that the bicarbonate concentration in the original sample was just over 3 times the carbonate concentration.

⊛ **Practice Exercise** Suppose you titrate a 100.0-mL sample of water from another pool in Yellowstone National Park with a ratio of carbonate to bicarbonate of 3:1. If the total volume of titrant used was 24.04 mL, approximately what volume of titrant was used to reach the first equivalence point? What additional volume of titrant was used to reach the second equivalence point?

CONCEPT TEST

In a titration that initially contains both CO_3^{2-} and HCO_3^-, the volume of titrant required to reach the first equivalence point is less than that required to titrate from the first equivalence point to the second. Why?

16.5 Lewis Acids and Bases

Until now we have used the Brønsted–Lowry definitions of acids (H^+ ion donors) and bases (H^+ ion acceptors). However, the time has come to expand our concept of acids and bases to include acid–base interactions that may or may not involve the transfer of H^+ ions. Let's begin by revisiting what happens when ammonia gas dissolves in water:

$$NH_3(g) + H_2O(\ell) \rightleftharpoons NH_4^+(aq) + OH^-(aq)$$

Figure 16.17(a) shows a Brønsted–Lowry interpretation of this reaction: in donating H^+ ions to ammonia, H_2O acts as a Brønsted–Lowry acid. In accepting H^+ ions, NH_3 acts as a Brønsted–Lowry base.

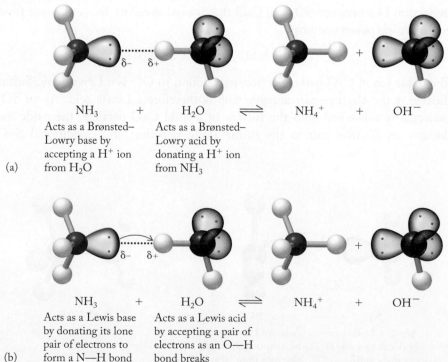

(a)

NH_3 + H_2O $\rightleftharpoons$ NH_4^+ + OH^-

Acts as a Brønsted–Lowry base by accepting a H^+ ion from H_2O

Acts as a Brønsted–Lowry acid by donating a H^+ ion from NH_3

(b)

NH_3 + H_2O $\rightleftharpoons$ NH_4^+ + OH^-

Acts as a Lewis base by donating its lone pair of electrons to form a N—H bond

Acts as a Lewis acid by accepting a pair of electrons as an O—H bond breaks

FIGURE 16.17 (a) Brønsted–Lowry view of the reaction between H_2O (proton donor) and NH_3 (proton acceptor). (b) Lewis view of the reaction: H_2O acts as a Lewis acid (electron-pair acceptor) and NH_3 acts as a Lewis base (electron-pair donor).

Lewis base a substance that *donates* a lone pair of electrons in a chemical reaction.

Lewis acid a substance that *accepts* a lone pair of electrons in a chemical reaction.

CONNECTION Lewis's pioneering theories of the nature of covalent bonding were described in Section 8.2.

Another way to view this reaction is illustrated in Figure 16.17(b). Instead of focusing on the transfer of hydrogen ions, consider the two reactants as a donor and an acceptor of a *pair of electrons*. In this view, the N atom in NH_3 donates its lone pair of electrons to one of the H atoms in H_2O. In the process, one of the H—O bonds in H_2O is broken in such a way that the bonding pair of electrons remains with the O atom. The donated lone pair from the N atom forms a fourth N—H covalent bond. The result is the same as in the Brønsted–Lowry model of acid–base behavior: a molecule of NH_3 bonds to a H^+ ion, forming an NH_4^+ ion, and a molecule of H_2O loses a H^+ ion, becoming a OH^- ion.

Viewing this process as the donation and acceptance of an electron pair provides the following basis for defining acids and bases:

- A **Lewis base** is a substance that *donates* a lone pair of electrons in a chemical reaction.
- A **Lewis acid** is a substance that *accepts* a lone pair of electrons in a chemical reaction.

These definitions are named after their developer, Gilbert N. Lewis, who also pioneered research into the nature of chemical bonds. The Lewis definition of a base is consistent with the Brønsted–Lowry model we have used because a substance must be able to donate a pair of electrons if it is to bond with a H^+ ion. However, the same parallelism is not true for acids. The Brønsted–Lowry model defines an acid as a hydrogen-ion donor, but the Lewis definition includes species that have no hydrogen ions to donate, but that can still accept electrons.

One such compound is boron trifluoride, BF_3. With only six valence electrons, the boron atom in BF_3 can accept another pair to complete its octet. NH_3 is a suitable electron-pair donor, as shown in Figure 16.18. There is no transfer of H^+ ions in this reaction, so it is not an acid–base reaction according to the Brønsted–Lowry model. However, NH_3 donates a lone pair of electrons and BF_3 accepts them, so it is an acid–base reaction according to the broader Lewis model.

Many important Lewis bases are anions, including the halide ions, OH^-, and O^{2-}. To see how O^{2-} functions as a Lewis base, let's revisit the reaction described in Section 14.6 between SO_2 and CaO that is used to reduce SO_2 emissions from coal-burning power stations:

$$CaO(s) + SO_2(g) \rightleftharpoons CaSO_3(s)$$

The oxide ion in CaO is the electron-pair donor, so O^{2-} is a Lewis base. Sulfur dioxide is the electron-pair acceptor and is therefore a Lewis acid. As an SO_2 molecule is adsorbed onto the surface of a solid CaO particle, the oxide ion donates an electron pair to the sulfur atom, resulting in an additional S–O

FIGURE 16.18 In the reaction between NH_3 and BF_3, NH_3 acts as a Lewis base and BF_3 acts as a Lewis acid. This is an acid–base reaction in the Lewis sense because we focus on the acceptance and donation of an electron pair, not the transfer of a proton, as in the Brønsted–Lowry system.

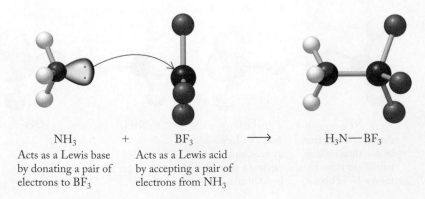

NH_3 + BF_3 $\longrightarrow$ H_3N—BF_3

Acts as a Lewis base by donating a pair of electrons to BF_3

Acts as a Lewis acid by accepting a pair of electrons from NH_3

covalent bond and the formation of a sulfite anion, SO_3^{2-}. The formal charges on the S atom and the double-bonded O atom in SO_3^{2-} are both zero; the formal charges on the two single-bonded O atoms are -1, giving an ion with an overall charge of $2-$.

CONNECTION The concept of formal charge and its calculation were described in Section 8.5.

$$Ca^{2+} \left[:\ddot{O}: \right]^{2-} \quad \overset{:\ddot{O}:}{\underset{:\ddot{O}:}{S:}} \rightarrow Ca^{2+} \left[:\ddot{O}-\overset{:\ddot{O}:}{\underset{:\ddot{O}:}{S:}} \right]^{2-}$$

SAMPLE EXERCISE 16.13 Identifying Lewis Acids and Bases   **LO5**

In the following reaction, which species is a Lewis acid and which is a Lewis base?

$$AlCl_3 + Cl^- \rightarrow AlCl_4^-$$

Collect and Organize We are given a chemical reaction and asked to identify the Lewis acid and base: the reactant that accepts a pair of electrons and the reactant that donates that pair of electrons, respectively.

Analyze A Cl^- ion has four lone pairs of electrons in its valence shell, so it has the capacity to *donate* one of them to form a covalent bond to aluminum. To analyze the capacity of $AlCl_3$ to act as a Lewis acid, we need to draw its Lewis structure, which includes a total of (3×7) electrons from 3 Cl atoms, plus 3 from 1 Al atom, or 24 electrons in all. Using 6 of them to draw three Al—Cl bonds leaves 18 with which to complete the octets around each of the three Cl atoms:

$$\overset{:\ddot{C}l:}{\underset{:\ddot{C}l \quad \ddot{C}l:}{\underset{Al}{|}}}$$

This structure accounts for all the valence electrons, but it leaves Al with only 6 valence electrons and the capacity to accept one more pair, that is, to act as a Lewis acid.

Solve In this reaction, $AlCl_3$ is a Lewis acid and the Cl^- ion is a Lewis base as shown in these Lewis structures:

$$\overset{:\ddot{C}l:}{\underset{:\ddot{C}l \quad \ddot{C}l:}{\underset{Al}{|}}} \leftarrow \left[:\ddot{C}l: \right]^- \rightarrow \left[\overset{:\ddot{C}l:}{\underset{:\ddot{C}l:}{:\ddot{C}l-\underset{|}{Al}-\ddot{C}l:}} \right]^-$$

Think About It Drawing the Lewis structure of $AlCl_3$ and determining that the central Al atom has an incomplete octet is the key to identifying its capacity to act as a Lewis acid. Note that these Lewis structures incorporate the assumption that $AlCl_3$ is a molecular compound and not, like most other metal chlorides, an ionic compound. This assumption is supported by the physical properties of $AlCl_3$ and particularly by the fact that it sublimes at 178°C and 1 atm. Most metal halides do not even melt until heated to temperatures many hundreds of degrees higher than that.

 Practice Exercise In the following reaction, which reactant is the Lewis acid and which is the Lewis base?

$$CO_2(g) + CaO(s) \rightarrow CaCO_3(s)$$

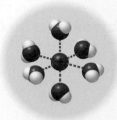

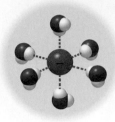

FIGURE 16.19 Ion–dipole interactions in a hydrated cation and a hydrated anion.

CONNECTION Spheres of hydration around cations and anions were introduced in Section 10.3.

16.6 Formation of Complex Ions

In Chapter 10 we described how ions dissolved in water are *hydrated*; that is, they are surrounded by water molecules oriented with the positive ends of their dipoles directed toward anions and their negative ends directed toward cations (Figure 16.19). In some hydrated cations, ion–dipole interactions lead to the sharing of lone-pair electrons on the oxygen atoms of H_2O with empty valence-shell orbitals on the cations. This interaction between a lone pair and an empty orbital is yet another example of Lewis acid–base behavior. These shared electron pairs meet our definition of covalent bonds, but these particular bonds are called *coordinate* covalent bonds, or simply **coordinate bonds**. Such bonds form when either a molecule or an anion donates a lone pair of electrons to an empty valence-shell orbital of an atom, cation, or molecule. Once formed, a coordinate bond is indistinguishable from any other kind of covalent bond. When a cation forms coordinate bonds to one or more molecular or ionic electron-pair donors, the resulting structure is called a **complex ion**. The electron-pair donors in complex ions are called **ligands**. For example, when six molecules of water form coordinate bonds to a Ni^{2+} cation in an aqueous solution, they form a complex ion with the formula $Ni(H_2O)_6^{2+}$ to show that there are six ligands (six water molecules bonded to the metal) in this complex ion.

We can investigate the formation of complex ions by using the mathematical tools we have used in examining acid–base equilibria. These tools are appropriate because complex formation processes are reversible, and many of them reach chemical equilibrium rapidly. We start this investigation with two aqueous solutions, one containing copper(II) sulfate ($CuSO_4$), the other containing NH_3 (Figure 16.20). The $CuSO_4$ solution is robin's-egg blue, the color characteristic of $Cu^{2+}(aq)$ ions dissolved in water, and the ammonia solution is colorless. When the solutions are mixed, the robin's-egg blue turns a dark navy blue, as shown on the right in Figure 16.20. This is the color of $Cu(NH_3)_4^{2+}$ complex ions. The change

FIGURE 16.20 The beaker on the left contains a solution of $Cu^{2+}(aq)$, which is a characteristic robin's-egg blue. As a colorless solution of ammonia is added (from the bottle in the middle), the mixture of the two solutions turns dark blue (beaker on the right), which is the color of the $Cu(NH_3)_4^{2+}$ complex ion.

in color provides visual evidence that the following equilibrium lies far to the right, favoring complex ion formation:

$$Cu^{2+}(aq) + 4\,NH_3(aq) \rightleftharpoons Cu(NH_3)_4{}^{2+}(aq)$$

This conclusion is supported by the large equilibrium constant for the reaction:

$$K_f = \frac{[Cu(NH_3)_4{}^{2+}]}{[Cu^{2+}][NH_3]^4} = 5.0 \times 10^{13}$$

The equilibrium constant K_f is called a **formation constant** because it describes the formation of a complex ion. For the general case in which 1 mole of metal ions (M^{m+}) combines with n moles of ligand (X^{x-}) to form the complex ion $MX_n{}^{(m-nx)+}$, the formation constant expression is

$$K_f = \frac{[MX_n{}^{(m-nx)+}]}{[M^{m+}][X^{x-}]^n}$$

Formation constants can be used to calculate the concentration of free, uncomplexed metal ions, $M^{m+}(aq)$, in equilibrium with a given (usually larger) concentration of a ligand. Because K_f values are usually very large, equilibrium concentrations of uncomplexed metal ions are usually very small. One approach to calculating the concentration of an uncomplexed metal ion is to consider the reverse of the formation reaction and calculate how much of the complex ion, in this case $Cu(NH_3)_4{}^{2+}(aq)$, dissociates. In Chapter 14 we learned that equilibrium constant for the reverse reaction is the reciprocal of the original equilibrium constant.

$$Cu(NH_3)_4{}^{2+}(aq) \rightleftharpoons Cu^{2+}(aq) + 4\,NH_3(aq) \qquad K_c = \frac{1}{K_f} = \frac{[Cu^{2+}][NH_3]^4}{[Cu(NH_3)_4{}^{2+}]}$$

The calculation of concentration of uncomplexed copper ion, $[Cu^{2+}]$, is illustrated in Sample Exercise 16.14.

coordinate bond a covalent bond formed when one anion or molecule donates a pair of electrons to another ion or molecule.

complex ion an ionic species consisting of a metal ion bonded to one or more Lewis bases.

ligand a Lewis base bonded to the central metal ion of a complex ion.

formation constant (K_f) an equilibrium constant describing the formation of a metal complex from a free metal ion and its ligands.

SAMPLE EXERCISE 16.14 Calculating the Concentration of Free   **LO6**
Metal in Equilibrium with a Complex Ion

Ammonia gas is dissolved in a $1.00 \times 10^{-4}\,M$ solution of $CuSO_4$ to give an equilibrium concentration of $[NH_3] = 1.60 \times 10^{-3}\,M$. Calculate the concentration of $Cu^{2+}(aq)$ ions in the solution.

Collect and Organize The concentration of $CuSO_4$ means that $[Cu^{2+}]$ before complex formation is $1.00 \times 10^{-4}\,M$. The equilibrium concentration of the ligand (NH_3) is $1.60 \times 10^{-3}\,M$. The values of $[NH_3]$, $[Cu^{2+}]$, and $[Cu(NH_3)_4{}^{2+}]$ at equilibrium are related by the formation constant expression:

$$K_f = \frac{[Cu(NH_3)_4{}^{2+}]}{[Cu^{2+}][NH_3]^4} = 5.0 \times 10^{13}$$

Analyze Because K_f is large, we can assume that essentially all the Cu^{2+} ions are converted to complex ions and $[Cu(NH_3)_4{}^{2+}] = 1.00 \times 10^{-4}\,M$. Only a tiny concentration of free Cu^{2+} ions, x, remains at equilibrium. We can calculate the $[Cu^{2+}]$ by considering how much $Cu(NH_3)_4{}^{2+}$ dissociates:

$$Cu(NH_3)_4{}^{2+}(aq) \rightleftharpoons Cu^{2+}(aq) + 4\,NH_3(aq)$$

The equilibrium constant, K_c, for this equilibrium is the reciprocal of K_f:

$$K_c = \frac{1}{K_f} = \frac{[Cu^{2+}][NH_3]^4}{[Cu(NH_3)_4{}^{2+}]} = \frac{1}{5.0 \times 10^{13}} = 2.0 \times 10^{-14}$$

We can construct a RICE table incorporating the concentrations and concentration changes of the products and reactants. Given the small value of K_c, we should obtain a $[Cu^{2+}]$ value at equilibrium that is much less than 1.00×10^{-4} M.

Solve First, we complete the row of equilibrium concentrations of Cu^{2+}, NH_3, and $Cu(NH_3)_4{}^{2+}$ in the following RICE table:

Reaction	$Cu(NH_3)_4{}^{2+}(aq)$ $\rightleftharpoons$	$Cu^{2+}(aq)$ +	$4\,NH_3(aq)$
	$[Cu(NH_3)_4{}^{2+}]$ (M)	$[Cu^{2+}]$ (M)	$[NH_3]$ (M)
Initial	1.00×10^{-4}	0	
Change	$-x$	$+x$	
Equilibrium	$1.00 \times 10^{-4} - x$	x	1.60×10^{-3}

Next, we make the simplifying assumption that x is much smaller than 1.00×10^{-4} M. Therefore, we can ignore the x terms in the equilibrium value of $[Cu(NH_3)_4{}^{2+}]$ and use the simplified values in the K_c expression:

$$K_c = \frac{[Cu^{2+}][NH_3]^4}{[Cu(NH_3)_4{}^{2+}]}$$

$$= \frac{[x][1.60 \times 10^{-3}]^4}{[1.0 \times 10^{-4}]} = 2.0 \times 10^{-14}$$

$$x = 3.1 \times 10^{-7} = [Cu^{2+}]$$

Think About It This result validates our simplifying assumption and confirms our prediction that $[Cu^{2+}]$ at equilibrium is much less than $[Cu^{2+}]$ initially. In fact, more than 99% of the copper(II) in the solution is present as $Cu(NH_3)_4{}^{2+}$.

Practice Exercise Calculate the equilibrium concentration of $Ag^+(aq)$ in a solution that is initially 0.100 M $AgNO_3$ and 0.800 M NH_3 after this reaction takes place:

$$Ag^+(aq) + 2\,NH_3(aq) \rightleftharpoons Ag(NH_3)_2{}^+(aq) \qquad K_f = 1.7 \times 10^7$$

16.7 Hydrated Metal Ions as Acids

In Chapter 15 we saw that the strength of an oxoacid depends on the electronegativity of its central atom. For example, the relative strengths of the three hypohalous acids, HOCl > HOBr > HOI, align with the relative electronegativities of their halogen atoms: Cl > Br > I. This order makes sense because the more electronegative the halogen, the more it draws electron density away from the oxygen in the polar –OH bond (see Figure 15.13). These shifts make the negative ion formed by dissociation better able to bear a negative charge because of increased delocalization.

CONNECTION We discussed periodic trends in electronegativity in Chapter 8.

A similar shift in electron density occurs in hydrated metal ions having the generic formula $M(H_2O)_6{}^{n+}$ when $n \geq 2$. The electrons in the O—H bonds of the water molecules surrounding the metal ions are attracted to the positively charged ions. The resulting distortion in electron density increases the likelihood of one of these O—H bonds ionizing and donating a H^+ ion to a neighboring molecule of water:

$$M(H_2O)_6{}^{n+}(aq) + H_2O(\ell) \rightleftharpoons M(H_2O)_5(OH)^{(n-1)+}(aq) + H_3O^+(aq)$$

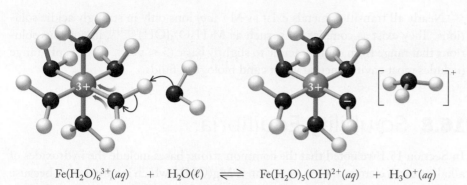

$$Fe(H_2O)_6^{3+}(aq) \quad + \quad H_2O(\ell) \quad \rightleftharpoons \quad Fe(H_2O)_5(OH)^{2+}(aq) \quad + \quad H_3O^+(aq)$$

FIGURE 16.21 A hydrated Fe^{3+} cation draws electron density away from the water molecules of its inner coordination sphere, which makes it possible for one or more of these molecules to donate a H^+ ion to a water molecule outside the sphere. The hydrated Fe^{3+} ion has one fewer H_2O ligands with an electrostatic attraction between the complex ion and the OH^- ion.

Figure 16.21 provides a molecular view of this reaction, using $Fe^{3+}(aq)$ as the central ion. Similar reactions allow other hydrated metal ions, particularly those with charges of 3+, to function as Brønsted–Lowry acids. The K_a values of several metal ions are listed in Table 16.1. Note how much stronger the 3+ ions are than the 2+ ions, as we would expect given the greater electron-withdrawing power of the more highly charged central ions.

Figure 16.21 shows how one of the six water molecules of hydration surrounding a Fe^{3+} ion is converted into a hydroxide ion as a result of the acid ionization reaction. This reduces the charge of the complex ion from 3+ to 2+. If the pH of a solution of $Fe(H_2O)_5(OH)^{2+}$ ions is raised by adding a small quantity of a strong base such as NaOH, the ions undergo additional acid ionization reactions to form $Fe(H_2O)_5(OH)_2^+$:

$$Fe(H_2O)_5(OH)^{2+}(aq) + OH^-(aq) \rightleftharpoons Fe(H_2O)_4(OH)_2^+(aq) + H_2O(\ell)$$

and, at still higher pH, solid iron(III) hydroxide:

$$Fe(H_2O)_4(OH)_2^+(aq) + OH^-(aq) \rightleftharpoons Fe(H_2O)_3(OH)_3(s) + H_2O(\ell)$$

For simplicity, we usually write the formula of iron(III) hydroxide as $Fe(OH)_3(s)$, even though each formula unit also contains three water molecules of hydration.

TABLE 16.1 K_a Values of Hydrated Metal Ions

Ion	K_a
$Fe^{3+}(aq)$	3×10^{-3}
$Cr^{3+}(aq)$	1×10^{-4}
$Al^{3+}(aq)$	1×10^{-5}
$Cu^{2+}(aq)$	3×10^{-5}
$Pb^{2+}(aq)$	3×10^{-5}
$Zn^{2+}(aq)$	1×10^{-9}
$Co^{2+}(aq)$	2×10^{-10}
$Ni^{2+}(aq)$	1×10^{-10}

Acid strength ↑

CONCEPT TEST

Would $Fe^{2+}(aq)$ be a stronger or weaker Lewis acid than $Fe^{3+}(aq)$? Where should the K_a of $Fe^{2+}(aq)$ be listed in Table 16.1 relative to that of $Fe^{3+}(aq)$?

Two other 3+ cations, Cr^{3+} and Al^{3+}, display similar behavior, but they differ from Fe^{3+} in that they are more soluble in strongly basic solutions than in weakly basic solutions. Why? Because solid $Cr(OH)_3$ and $Al(OH)_3$ may accept additional OH^- ions at high pH, forming soluble anionic complex ions:

$$Cr(OH)_3(s) + OH^-(aq) \rightleftharpoons Cr(OH)_4^-(aq) = Cr(H_2O)_2(OH)_4^-(aq)$$

$$Al(OH)_3(s) + OH^-(aq) \rightleftharpoons Al(OH)_4^-(aq) = Al(H_2O)_2(OH)_4^-(aq)$$

Zinc hydroxide, $Zn(OH)_2$, is the only other transition metal hydroxide that is soluble at high pH, because it forms $Zn(OH)_4^{2-}$ ions.

Nearly all transition metals exist as $M^{n+}(aq)$ ions only in strongly acidic solutions. They exist as complex ions, such as $M(H_2O)_5(OH)^{(n-1)+}$, in aqueous solutions that range from slightly acidic to slightly basic ($3 < pH < 9$). This pH range includes most environmental waters and biological fluids.

16.8 Solubility Equilibria

In Section 15.1 we noted that the common strong bases include the hydroxides of alkaline earth elements. One exception is $Mg(OH)_2$, which is a weak base because it has limited solubility in water. Magnesium hydroxide is the active ingredient in a product found in many medicine cabinets: the antacid called *milk of magnesia*. This liquid appears "milky" because it is an aqueous *suspension* (not solution) of solid, white $Mg(OH)_2$. We can express the limited solubility of solid $Mg(OH)_2$ by the following equation:

$$Mg(OH)_2(s) \rightleftharpoons Mg^{2+}(aq) + 2\,OH^-(aq)$$

Because $Mg(OH)_2$ is a solid, its effective concentration does not change as long as some of it is present in the system. Therefore the equilibrium constant for the dissolution of $Mg(OH)_2$ is

$$K_{sp} = [Mg^{2+}][OH^-]^2$$

where K_{sp} represents an equilibrium constant called the **solubility-product constant** or simply the **solubility product**.

The K_{sp} values of $Mg(OH)_2$ and other slightly soluble compounds are listed in Table A5.4. We can use these values to calculate the concentrations of these compounds in aqueous solutions. Two terms are widely used to describe how much of a solid dissolves in a solvent: *solubility*, which is often expressed in grams of solute per 100 mL of solution, and *molar solubility*, which is expressed in moles of solute per liter of solution. In the following Sample Exercise we use the K_{sp} of $Mg(OH)_2$ to calculate its solubility and molar solubility.

SAMPLE EXERCISE 16.15 Calculating the Solubility of an Ionic Compound from K_{sp} **LO8**

What are the solubility (in grams per 100 mL of solution) and the molar solubility of $Mg(OH)_2$ at 25°C?

Collect, Organize, and Analyze The K_{sp} of $Mg(OH)_2$ from Table A5.4 is 5.6×10^{-12}, and the K_{sp} expression is

$$K_{sp} = [Mg^{2+}][OH^-]^2 = 5.6 \times 10^{-12}$$

For every mole of $Mg(OH)_2$ that dissolves, 1 mole of Mg^{2+} ions and 2 moles of OH^- ions go into solution. If we let x be the molar solubility (mol $Mg(OH)_2$/L of solution) at 25°C, then $[Mg^{2+}] = x$ and $[OH^-] = 2x$. Converting molar solubility to g/100 mL is a matter of multiplying molar solubility by the molar mass of $Mg(OH)_2$, 58.32 g/mol, and factoring in the smaller volume.

Solve Inserting the x terms in the K_{sp} expression and solving for x:

$$K_{sp} = (x)(2x)^2 = (x)(4x^2) = 4x^3 = 5.6 \times 10^{-12}$$

$$x = 1.1 \times 10^{-4}\ M$$

solubility-product constant (also called **solubility product, K_{sp}**) an equilibrium constant that describes the formation of a saturated solution of a slightly soluble salt.

Converting this molar solubility to g/100 mL:

$$\frac{1.1 \times 10^{-4} \ \text{mol}}{\text{L}} \times \frac{58.32 \ \text{g}}{1 \ \text{mol}} \times \frac{1 \ \text{L}}{1000 \ \text{mL}} \times 100 \ \text{mL} = 6.4 \times 10^{-4} \ \text{g}$$

Think About It Note that the entire algebraic expression for $[OH^-]$, $2x$, is squared in this calculation: $(2x)^2 = 4x^2$. Forgetting to square the coefficient is a common mistake. Also note that the molar solubility of $Mg(OH)_2$ is a much larger value than its solubility product. This difference is true for all sparingly soluble ionic compounds because K_{sp} values are the products of small concentration values multiplied together, producing even smaller K_{sp} values.

 Practice Exercise What are the molar solubility in water and the solubility (in g/100 mL of solution) of $Ca_3(PO_4)_2$ at 25°C?

CONCEPT TEST

In Sample Exercise 16.15 we calculated molar solubility from K_{sp}, but the reverse calculation is also possible. If the molar solubility of BaF_2 is 7.5×10^{-3} M, what is the K_{sp} for barium fluoride?

SAMPLE EXERCISE 16.16 Evaluating the Common-Ion   **LO8**
Effect on Solubility

The mineral barite is mostly barium sulfate ($BaSO_4$) and is widely used in industry and in medical imaging. Calculate the molar solubility at 25°C of $BaSO_4$ in (a) pure water and (b) seawater in which the concentration of sulfate ions is 2.8 g/L.

Collect and Organize We want to calculate the molar solubility of $BaSO_4$ in both pure water and in seawater that already contains sulfate ions. The K_{sp} value of $BaSO_4$ given in Table A5.4 is 9.1×10^{-11}.

Analyze The dissolution reaction for barium sulfate

$$BaSO_4(s) \rightleftharpoons Ba^{2+}(aq) + SO_4{}^{2-}(aq)$$

tells us that 1 mole of Ba^{2+} ions and 1 mole of $SO_4{}^{2-}$ ions form from each mole of $BaSO_4$ that dissolves. If x mol/L of $BaSO_4$ dissolves in pure water, then $[Ba^{2+}] = [SO_4{}^{2-}] = x$. However, seawater has a background concentration of $SO_4{}^{2-}$. According to Le Châtelier's principle and the common-ion effect, the sulfate ion already in seawater should shift the dissolution equilibrium to the left, which means that less $BaSO_4$ should dissolve in seawater than in pure water.

Solve
a. In pure water,

$$K_{sp} = [Ba^{2+}][SO_4{}^{2-}] = (x)(x) = 1.08 \times 10^{-10}$$
$$x = 1.04 \times 10^{-5} \ M$$

b. In seawater, we first need to calculate the value of $[SO_4{}^{2-}]$ before any $BaSO_4$ dissolves:

$$[SO_4{}^{2-}]_{initial} = \frac{2.8 \ \text{g}}{\text{L}} \times \frac{1 \ \text{mol}}{96.06 \ \text{g}} = \frac{0.029 \ \text{mol}}{\text{L}}$$

The value of $[SO_4{}^{2-}]$ at equilibrium is the sum of the background concentration (0.029 mol/L) and the additional $SO_4{}^{2-}$ ions from the dissolution of $BaSO_4$ (x).

Incorporating this value into the $[SO_4^{2-}]$ term in the K_{sp} expression gives

$$K_{sp} = [Ba^{2+}][SO_4^{2-}] = (x)(0.029 + x) = 1.08 \times 10^{-10}$$

Solving for x is simplified if we assume that the K_{sp} of $BaSO_4$ is so small that we can ignore its contribution to the total SO_4^{2-} concentration. Therefore

$$(x)(0.029 + x) \approx (x)(0.029) = 1.08 \times 10^{-10}$$

$$x = 3.7 \times 10^{-9} \, M$$

Think About It The calculated molar solubility of $BaSO_4$ in seawater is much less than the initial $[SO_4^{2-}]$ value, so our simplifying assumption was justified. The lower solubility of $BaSO_4$ in seawater is another illustration of the common-ion effect: the dissolution of $BaSO_4$ is suppressed by the SO_4^{2-} ions already present in seawater.

 Practice Exercise What is the molar solubility of $MgCO_3$ in alkaline spring water at 25°C in which $[CO_3^{2-}] = 0.0075 \, M$?

In the preceding Sample Exercise we saw how the common-ion effect can suppress the solubility of an ionic compound. Other perturbations to solubility equilibria can actually promote solubility, as happens when the anion of the compound is the conjugate base of a weak acid. The molar solubilities of such compounds increase in acidic solutions. We see why as we solve the following Sample Exercise.

SAMPLE EXERCISE 16.17 Calculating the Effect of pH on Solubility   **LO8**

What is the molar solubility of CaF_2 at 25°C in (a) pure water and (b) an acidic buffer in which $[H_3O^+]$ is a constant 0.050 M?

Collect and Organize We are asked to calculate the solubility of CaF_2 in both pure water and in an acidic buffer. The dissolution process is described by the following equilibrium:

$$(1) \qquad CaF_2(s) \rightleftharpoons Ca^{2+}(aq) + 2 \, F^-(aq) \qquad K_{sp} = 5.3 \times 10^{-9}$$

Analyze To account for the effect of acid on the solubility of a fluoride salt, we need to consider the chemical equilibrium in which the fluoride ion acts as a Brønsted–Lowry base (H^+ ion acceptor):

$$(2) \qquad H_3O^+(aq) + F^-(aq) \rightleftharpoons HF(aq) + H_2O(\ell)$$

This reaction is the reverse of the acid ionization reaction:

$$HF(aq) + H_2O(\ell) \rightleftharpoons H_3O^+(aq) + F^-(aq)$$

Therefore the equilibrium constant for reaction 2 is the reciprocal of the K_a of HF:

$$K_2 = \frac{[HF]}{[H_3O^+][F^-]} = \frac{1}{6.8 \times 10^{-4}} = 1.47 \times 10^3$$

As reaction 2 proceeds, F^- ions are consumed, which shifts the equilibrium in reaction 1 to the right, increasing CaF_2 solubility. Therefore we can anticipate an increase in the solubility of CaF_2 when acid is present.

Solve

a. Let x be the molar solubility of CaF_2 in pure water. According to the stoichiometry of reaction 1, $[Ca^{2+}] = x$ mol/L and $[F^-] = 2x$ mol/L. Inserting these symbols in the K_{sp} expression:

$$K_{sp} = [Ca^{2+}][F^-]^2 = (x)(2x)^2 = 5.3 \times 10^{-9}$$

$$4x^3 = 5.3 \times 10^{-9}$$

$$x = 1.1 \times 10^{-3}\ M$$

b. In the acidic buffer, $[H_3O^+] = 0.050\ M$. The F^- and H_3O^+ ions combine as shown in reaction 2. Assuming $[H_3O^+]$ is a constant 0.050 M, then:

$$K_2 = \frac{[HF]}{[0.050][F^-]} = 1.47 \times 10^3$$

$$\frac{[HF]}{[F^-]} = 73.5$$

$$(3) \quad [HF] = 73.5[F^-]$$

This calculation tells us that most of the F^- produced when calcium fluoride dissolves is converted into HF. The tiny fraction that remains free F^- ions is defined by this ratio:

$$(4) \quad \frac{[F^-]}{[F^-] + [HF]}$$

Here the numerator is the concentration of free F^- ions at equilibrium and the denominator is the concentration of all the F^- ions produced when CaF_2 dissolved. Combining expressions 3 and 4 to calculate the fraction of dissolved fluoride ions that are free F^- ions and not molecules of HF:

$$\frac{[F^-]}{[F^-] + [HF]} = \frac{[F^-]}{[F^-] + 73.5[F^-]} = \frac{[F^-]}{74.5[F^-]} = 0.0134$$

If x is the molar solubility of CaF_2 in the acid, x mol/L Ca^{2+} and $2x$ mol/L F^- are produced. However, most of the fluoride ions are converted into HF, and the free F^- ion concentration is only $(0.0134 \times 2x) = 0.0268\ x$. Inserting this value in the K_{sp} expression:

$$K_{sp} = [Ca^{2+}][F^-]^2 = (x)(0.0268\ x)^2 = 7.18 \times 10^{-4}\ x^3$$

$$7.18 \times 10^{-4}\ x^3 = 3.9 \times 10^{-11}$$

$$x = 3.8 \times 10^{-3}\ M$$

Think About It A comparison of the results from parts (a) and (b) reveals that the molar solubility of CaF_2 is about 4 times higher in the acidic buffer, as we predicted, because most of the F^- ions produced when CaF_2 dissolves in the buffer are converted to molecules of HF. This conversion removes a product from the K_{sp} equilibrium mixture. Le Châtelier's principle tells us that the result will be a shift in the position of the equilibrium to the right, in favor of forming product and increasing solubility.

Practice Exercise What is the molar solubility of $ZnCO_3$ at 25°C in pure water and in a pH 7.00 buffer solution? Assume the pH of the buffer is unaffected by the presence of $ZnCO_3$.

CONCEPT TEST

Consider two aqueous solutions of sodium fluoride and nitric acid. Which would increase the solubility of PbF_2? Which would decrease the solubility of PbF_2? Which would have little effect?

An inspection of the K_{sp} values in Table A5.4 discloses that the hydroxides of many metals, including all transition metals, have very small K_{sp} values. However, we must keep in mind that these tiny K_{sp} values apply to equilibrium concentrations of *free metal ions*. As we have seen in this chapter, many metals form stable complex ions in aqueous solution, reducing their free metal ion concentration. Let's consider, for example, the solubility of Al^{3+} ions in pH 7.00 water. The K_{sp} of $Al(OH)_3$ is

$$K_{sp} = [Al^{3+}][OH^-]^3 = 1.3 \times 10^{-33}$$

Solving the K_{sp} expression for $[Al^{3+}]$ and inserting $[OH^-] = 1.0 \times 10^{-7}$, we get

$$[Al^{3+}] = \frac{K_{sp}}{[OH^-]^3} = \frac{1.3 \times 10^{-33}}{(1.0 \times 10^{-7})^3} = 1.3 \times 10^{-12} \, M \quad (16.8)$$

This very small value seems to imply that no aluminum salt is soluble in water because the $Al^{3+}(aq)$ ions that it releases as it dissolves would immediately precipitate as $Al(OH)_3$. However, that conclusion is incorrect. For example, $Al(NO_3)_3$, like all nitrate salts, is quite soluble in water. This solubility can be explained by the acidic properties of hydrated Al^{3+} ions, $Al(H_2O)_6^{3+}$, which make $Al(NO_3)_3$ an acidic salt. The pH of 0.1 M $Al(NO_3)_3$ is about 3.0. At this pH, $[OH^-] = 1 \times 10^{-11}$. Inserting this value in Equation 16.8 and solving for $[Al^{3+}]$, we get:

$$[Al^{3+}] = \frac{K_{sp}}{[OH^-]^3} = \frac{1.3 \times 10^{-33}}{(1.0 \times 10^{-11})^3} = 1.3 \, M$$

The maximum value is more than 10 times the concentration of Al^{3+} ions in a 0.1 M solution of $Al(OH)_3$.

K_{sp} and Q

We can also use K_{sp} values to predict whether a particular concentration of an ionic compound is possible or whether a precipitate will form when the solutions of two salts are mixed. In making these predictions, it is convenient to use the concept of the reaction quotient Q that we developed in Chapter 14. When applied to the equilibrium governing a slightly soluble salt, Q is sometimes called the *ion product*, because it is the product of the concentrations of the ions in solution after each is raised to a power equal to its subscript in the formula of the compound. If the calculated Q value is greater than the K_{sp} of the compound ($Q > K_{sp}$), the reaction will favor reactant formation, and the compound will precipitate (or never dissolve in the first place). If $Q < K_{sp}$, the reaction will favor product formation, and the compound will be soluble and will not precipitate.

SAMPLE EXERCISE 16.18 Predicting Whether a Precipitate **LO8**
Forms When Two Solutions Are Mixed

Lead(II) chloride is a white pigment used in 15th-century European painting. Will $PbCl_2$ precipitate when 275 mL of a 0.134 M solution of $Pb(NO_3)_2$ is added to 125 mL of a 0.0339 M solution of NaCl?

Collect and Organize We are asked whether $PbCl_2$ will precipitate when two solutions containing Pb^{2+} ions and Cl^- ions are mixed. The solubility product of $PbCl_2$ is

$$K_{sp} = [Pb^{2+}][Cl^-]^2 = 1.70 \times 10^{-5}$$

Analyze To determine whether a precipitate forms, we need to calculate Q and compare its value to K_{sp}. If $Q > K_{sp}$, $PbCl_2$ will precipitate; if $Q < K_{sp}$, it will not precipitate. Q has the same form as the equilibrium constant, $[Pb^{2+}][Cl^-]^2$, but the concentration values used in it are the values for a given system that may or may not be at equilibrium.

Solve First we calculate the concentrations of the lead ions and chloride ions in the two solutions immediately after they are mixed. Mixing the two solutions dilutes both, so the volumes of the solutions (275 mL and 125 mL) must be added to get the final solution volume (400 mL = 0.400 L).

$$Pb^{2+}(aq): \quad 0.134\,\frac{mol}{L} \times 0.275\,L = 0.0369\,mol$$

$$[Pb^{2+}] = \frac{0.0369\,mol}{0.400\,L} = 0.0921\,M$$

$$Cl^-(aq): \quad 0.0339\,\frac{mol}{L} \times 0.125\,L = 0.00424\,mol$$

$$[Cl^-] = \frac{0.00424\,mol}{0.400\,L} = 0.0106\,M$$

The value of Q is

$$Q = [Pb^{2+}][Cl^-]^2 = (0.0921)(0.0106)^2 = 1.03 \times 10^{-5}$$

Because Q is smaller than K_{sp}, no precipitate forms.

Think About It Lead(II) chloride was categorized as an *insoluble* compound in Table 4.5. In this scenario, $PbCl_2$ does not precipitate because the solutions of Pb^{2+} and Cl^- ions are too dilute to provide the concentrations required for the precipitate to form.

Practice Exercise Will calcium fluoride ($K_{sp} = 5.3 \times 10^{-11}$) precipitate when 175 mL of a $4.78 \times 10^{-3}\,M$ solution of $Ca(NO_3)_2$ is added to 135 mL of a $7.35 \times 10^{-3}\,M$ solution of KF?

We can use differences in the solubilities of ionic compounds to selectively separate ions, particularly cations, in solution. For example, suppose an aqueous solution contains $0.10\,M\,Ca^{2+}$ ion and $0.020\,M\,Mg^{2+}$ ion. Is it possible to selectively remove the Mg^{2+} ions from solution by precipitating them as $Mg(OH)_2$ while leaving the Ca^{2+} ions in solution? This approach might work because $Mg(OH)_2$ is much less soluble than $Ca(OH)_2$, as indicated by their K_{sp} values:

$$K_{sp} = [Mg^{2+}][OH^-]^2 = 5.6 \times 10^{-12}$$
$$K_{sp} = [Ca^{2+}][OH^-]^2 = 5.5 \times 10^{-6}$$

One way to answer this ion separation question involves calculating the maximum concentration of OH^- ions that will *not* cause the $0.10\,M\,Ca^{2+}$ ion to precipitate and then determining whether that concentration is high enough to precipitate all of the Mg^{2+} ions. We can calculate the target $[OH^-]$ value from the K_{sp} of calcium hydroxide:

$$K_{sp} = 5.5 \times 10^{-6} = [Ca^{2+}][OH^-]^2 = (0.10)(x)^2$$

$$x = \sqrt{\frac{5.5 \times 10^{-6}}{0.10}} = 7.4 \times 10^{-3}\,M$$

Now we need to determine whether all of the Mg^{2+} ions in solution would precipitate as $Mg(OH)_2$ if $[OH^-]$ were less than 7.4×10^{-3} M. For example, we can calculate the $[Mg^{2+}]$ that would be in equilibrium with 6.9×10^{-3} M OH^- ions:

$$K_{sp} = 5.6 \times 10^{-12} = [Mg^{2+}][OH^-]^2 = (x)(6.9 \times 10^{-3})^2$$

$$x = \frac{5.6 \times 10^{-12}}{(6.9 \times 10^{-3})^2} = 1.2 \times 10^{-7} \, M$$

The concentration of Mg^{2+} ions in the original solution is 0.020 M. The calculated concentration of Mg^{2+} ions at equilibrium is 1.2×10^{-7} M, which represents only 0.00060% of the original amount of Mg^{2+} ions in the sample. This tiny amount remaining means the removal of Mg^{2+} is essentially complete. (In general, reduction of a solute's concentration to 0.1% of its original value or less is considered complete removal.) This separation works as long as we keep $[OH^-]$ below about 10^{-3} M; that is, we don't let the pH go much above 11. This approach of selective precipitation of $Mg(OH)_2$ has been used to separate these ions in seawater, where $[Ca^{2+}] = 0.0106$ M and $[Mg^{2+}] = 0.054$ M.

CONCEPT TEST

In the analysis of using a precipitation reaction to separate magnesium ions from calcium ions in solution, we did not specify the concentration of the hydroxide ion solution used to form the precipitates. Why did we not need this value?

SAMPLE EXERCISE 16.19 Separating Anions in Solution   **LO8, LO9**

Both lead(II) chloride and lead(II) fluoride are slightly soluble salts. A solution of lead(II) nitrate is added to a solution that is 0.275 M in both $Cl^-(aq)$ and $F^-(aq)$. Can we use this method to separate the two halide ions? If "complete precipitation" is defined as there being less than 0.1% of a particular ion left in solution, is the precipitation of the first salt complete before the second salt begins to precipitate?

Collect and Organize We are given a solution that contains two ions that form slightly soluble lead(II) salts, and we are asked whether one ion can be completely removed before the second one starts to precipitate when lead(II) ion is added to the solution. We have the K_{sp} values for both salts and the initial concentrations of both ions.

Analyze The equilibrium constant expressions for both ions are

$$K_{sp} = [Pb^{2+}][Cl^-]^2 = 1.7 \times 10^{-5}$$

$$K_{sp} = [Pb^{2+}][F^-]^2 = 3.3 \times 10^{-8}$$

The K_{sp} of $PbCl_2$ is $(1.7 \times 10^{-5}/3.2 \times 10^{-8}) = 500$ times the K_{sp} of PbF_2, so PbF_2 should precipitate first when Pb^{2+} ions are added to a solution containing the same concentrations of F^- and Cl^- ions. We need to determine the maximum $[Pb^{2+}]$ that could be added to precipitate PbF_2 and not cause $PbCl_2$ to precipitate. When we determine that value, we can calculate $[F^-]$ remaining in solution to determine whether the precipitation of F^- as PbF_2 was complete.

Solve The maximum concentration of Pb^{2+} in the solution that will not cause the chloride ion to precipitate is

$$K_{sp} = 1.7 \times 10^{-5} = [Pb^{2+}][Cl^-]^2 = (x)(0.275)^2$$

$$x = \frac{1.7 \times 10^{-5}}{(0.275)^2} = 2.25 \times 10^{-4} \, M$$

The concentration of $F^-(aq)$ in the solution at this concentration of lead(II) ion is

$$K_{sp} = 3.3 \times 10^{-8} = (2.25 \times 10^{-4})(x)^2$$

$$x = \sqrt{\frac{3.3 \times 10^{-8}}{2.25 \times 10^{-4}}} = 0.0121 \, M$$

The original solution was 0.275 M in F^- ions. A residual concentration of 0.0121 M F^- represents $(0.0123/0.275) \times 100\% = 4.5\%$ of the original $[F^-]$, which is greater than our 0.1% residual value that represents complete removal. Therefore precipitation of $PbF_2(s)$ is not complete and we cannot use this method to separate the two ions.

Think About It This attempt at selective precipitation did not work because the ratio of the K_{sp} values was only 500 *and* because the halide concentration terms in the K_{sp} expressions were both squared. If you look closely at the calculations, you can see that these squared concentration terms had the effect of producing a F^-/Cl^- ion ratio at equilibrium that was only the square root of their K_{sp} ratio: $(1/500)^{0.5} = 0.045$, or 4.5%.

Practice Exercise A water sample contains barium ions (0.0375 M) and calcium ions (0.0667 M). Can they be completely separated by selective precipitation of CaF_2? See Table A5.4 for the appropriate K_{sp} values.

We end this chapter with a Sample Exercise that integrates acid–base and solubility equilibria by revisiting a topic introduced in Section 16.1: ocean acidification. In this exercise we use the ionization equilibria of carbonic acid, as we have done in several other exercises, but this time we use different K_{a_1} and K_{a_2} values. Why? Because the K values we have used until now have all been *theoretical* values, which means they apply to ideal solutions in which each solute ion behaves as freely and independently as if it were the only ion present. As you might guess, theoretical K values work best for very dilute solutions. They don't perform as well for solutions as concentrated (salty) as seawater. They do not, for example, account for ion pair formation. Therefore, we use *apparent* K'_{a_1} and K'_{a_2} values at 25°C in Sample Exercise 16.20 that apply to chemical equilibria in typical seawater, which contains 35 grams of dissolved sea salts per kilogram of seawater. Their symbols contain a prime (′) after the K to indicate that their values apply only to that particular sample.

CONNECTION Ion pair formation in 1.0 M solutions of ionic compounds significantly reduces the number of free ions in these solutions, as described in Chapter 11.

SAMPLE EXERCISE 16.20 Integrating Concepts: Evaluating the Impact of Ocean Acidification

As we discussed at the beginning of this chapter, increasing concentrations of atmospheric CO_2 have increased the concentration of CO_2 dissolved in the sea. As we saw in Equations 16.1–16.3, an increase in $[CO_2]$ will then shift the following equilibria to the right:

(1) $\quad H_2CO_3(aq) + H_2O(\ell) \rightleftharpoons HCO_3^-(aq) + H_3O^+(aq)$
$$pK'_{a_1} = 5.85$$

(2) $\quad HCO_3^-(aq) + H_2O(\ell) \rightleftharpoons CO_3^{2-}(aq) + H_3O^+(aq)$
$$pK'_{a_2} = 9.00$$

a. The average pH of the Pacific Ocean near Hawaii dropped from 8.12 to 8.07 between 1989 and 2014. By how much did the average acidity ($[H_3O^+]$) of the ocean increase? Express you answer as a percentage of the 1989 acidity.

b. Ocean pH is expected to drop by at least another 0.30 units by the end of this century. Assuming the concentrations of most other dissolved ions remain the same as today, that is, $[Ca^{2+}] = 0.0106 \, M$ and $[HCO_3^-] + [CO_3^{2-}] = 0.00211 \, M$, will the skeletal structures of marine organisms that are composed of $CaCO_3$ be soluble or insoluble in seawater in 2100? The K'_{sp} value for aragonite (the crystalline form of $CaCO_3$ in coral and sea shells) is 6.46×10^{-7}.

Collect and Organize We are asked to convert a decrease in pH into an increase in acidity and to evaluate the impact of that increase on the solubility of $CaCO_3$. We know the concentration of Ca^{2+} ions and the total ($[CO_3^{2-}] + [HCO_3^-]$) value, as well as the appropriate K'_{sp} value. The last two ions are a conjugate acid–base pair whose concentrations are linked by reaction (2) and the Henderson–Hasselbalch equation based on it:

$$pH = pK_a + \log \frac{[\text{base}]}{[\text{acid}]} = 9.00 + \log \frac{[CO_3^{2-}]}{[HCO_3^-]}$$

Analyze The ratio of the acidity of the seawater near Hawaii in 2014 compared with that in 1989 can be calculated by applying the definition $[H_3O^+] = 10^{-pH}$. To determine whether $CaCO_3$ dissolves in seawater in 2100, we need to calculate the $[CO_3^{2-}]/[HCO_3^{2-}]$ ratio in equilibrium with pH $(8.07 - 0.30 = 7.77)$ seawater and then use that ratio and the total carbonate and bicarbonate value to calculate $[CO_3^{2-}]$. The product of that value and $[Ca^{2+}]$ will give us a Q value that can be compared with the K'_{sp} of $CaCO_3$ to determine whether seawater will be saturated with $CaCO_3$.

Solve

a. Converting the decrease in pH values to an increase in $[H_3O^+]$, expressed as a ratio of the 2014 value to the 1989 value:

$$\frac{[H^+]_{2014}}{[H^+]_{1989}} = \frac{10^{-8.07}\ M}{10^{-8.12}\ M} = 10^{0.05} = 1.12$$

Therefore the 2014 sample contains 1.12 times as much acidity as the 1989 sample, which means acidity increased 12% during the 25 years before 2014.

b. The $[CO_3^{2-}]/[HCO_3^{2-}]$ ratio in 2100 will be

$$7.77 = 9.00 + \log \frac{[CO_3^{2-}]}{[HCO_3^-]}$$

$$\frac{[CO_3^{2-}]}{[HCO_3^-]} = 0.0589$$

If we let x be $[CO_3^{2-}]$, then

$$\frac{x}{(0.00211 - x)} = 0.0589$$

$$x = 1.17 \times 10^{-4}\ M$$

Calculating Q for $CaCO_3$:

$$Q = [Ca^{2+}][CO_3^{2-}] = 0.0106 \times 1.17 \times 10^{-4} = 1.24 \times 10^{-6}$$

This value of Q is $(1.24 \times 10^{-6})/(6.46 \times 10^{-7}) = 1.9$ times the value of K'_{sp}, which means seawater will still be supersaturated with respect to $CaCO_3$.

Think About It The results of the calculations suggest that corals and seashells should still be able to form their skeletal structures in pH 7.77 seawater because it will be supersaturated with respect to $CaCO_3$. However, the predicted degree of supersaturation (essentially 100%) in 2100 will have decreased from 400% in 1989 [you are invited to repeat the part (b) calculation, using pH 8.12 to confirm this value]. Moreover, some climate change models predict even greater decreases in oceanic pH by 2100. There appears to be reason for concern.

SUMMARY

LO1 As predicted by Le Châtelier's principle, adding conjugate base to a solution of a weak acid inhibits ionization of the acid, causing the pH of the solution to rise. Adding conjugate acid to a solution of a weak base lowers the pH of the solution. These shifts are examples of the **common-ion effect**. (Section 16.2)

LO2 A **pH buffer** is a solution that contains either a weak acid and a salt of its conjugate base or a weak base and a salt of its conjugate acid. The acidic component should have a pK_a close to the desired pH of the buffer. (Section 16.3)

LO3 pH buffers have the capacity to resist pH change because their acid components neutralize additions of bases and their base components neutralize additions of acids. (Section 16.3)

LO4 Color **pH indicators** or pH electrodes are used to detect the equivalence points in pH titrations, which are used to determine the concentrations of acids or bases in aqueous samples. (Section 16.4)

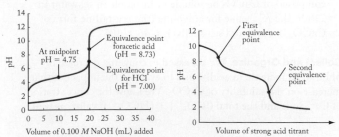

LO5 A **Lewis base** is a substance that donates pairs of electrons to a **Lewis acid**, defined as an electron-pair acceptor. The donated electron pair forms a **coordinate bond**. In some Lewis acid–Lewis base reactions, other bonds must break to accommodate the new one. (Sections 16.5 and 16.6)

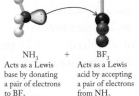

NH₃
Acts as a Lewis base by donating a pair of electrons to BF₃

+

BF₃
Acts as a Lewis acid by accepting a pair of electrons from NH₃

LO6 The stability of any **complex ion** is expressed mathematically by its **formation constant (K_f)**, which can be used to calculate the equilibrium concentration of free metal ions in a solution of complex ions. (Section 16.6)

LO7 Highly charged (e.g., 3+) hydrated metal ions are weak acids because of the ionization of water molecules covalently bonded to them. (Section 16.7)

LO8 The solubility of slightly soluble ionic compounds is described by their K_{sp}, or **solubility product**. Their solubility can be influenced by the common-ion effect, complex ion formation, and pH, especially if the anion is the conjugate base of a weak acid. (Section 16.8)

LO9 The relative solubilities of two slightly soluble ionic compounds can be used to selectively precipitate one from solution while the other remains soluble. (Section 16.8)

PARTICULATE **PREVIEW WRAP-UP**

The proton transfer equilibria that occur are

$$HF(aq) + H_2O(\ell) \rightleftharpoons F^-(aq) + H_3O^+(aq)$$

and

$$F^-(aq) + H_2O(\ell) \rightleftharpoons HF(aq) + OH^-(aq)$$

Only the first equilibrium occurs in the first solution, but both equilibria occur in the second solution. Because the $NaF(aq)$ solution contains equal amounts of F^- ions and HF molecules, it is buffered against large changes in pH no matter whether H^+ ions or OH^- ions are added.

PROBLEM-SOLVING SUMMARY

Type of Problem	Concepts and Equations	Sample Exercises
Calculating pH of a solution of a weak base and its conjugate acid (or a weak acid and its conjugate base)	Insert the concentrations of the base and acid components and the acid's pK_a in the Henderson–Hasselbalch equation: $$pH = pK_a + \log \frac{[\text{base}]}{[\text{acid}]} \qquad (16.4)$$	16.1, 16.2
Preparing a buffer of given pH	Select a weak acid with a pK_a within 1 pH unit of the target pH. Add a Na^+ salt of its conjugate base in a proportion calculated using the Henderson–Hasselbalch equation.	16.3, 16.4, 16.5
Evaluating the effect of concentration and [base]:[acid] ratio on buffer capacity	Assume additions of strong acid or base react completely with the base or acid components of the buffer. Calculate pH from the concentrations of the components that remain by using the Henderson–Hasselbalch equation.	16.6, 16.7, 16.8
Interpreting results of acid–base titrations	Use the volume and molarity of the titrant needed to reach the equivalence point to calculate the number of moles of it consumed by the analyte, and, from the stoichiometry of the titration reaction, the number of moles of analyte in the sample.	16.9, 16.10, 16.11, 16.12
Identifying Lewis acids and bases	Lewis bases donate pairs of electrons, and Lewis acids accept these pairs of electrons.	16.13
Using formation constants to calculate the concentration of free or complexed ion	Set up a RICE table based on the complex formation reaction. Let x be the concentration of free (not complexed) metal ion. Solve for x, which Is usually much smaller than the concentration of the complex.	16.14
Calculating the solubility of an ionic compound from its K_{sp}	Express the concentrations of the cation and anion in the K_{sp} expression in terms of x moles of the compound that dissolve in 1 liter of solution.	16.15, 16.16
Calculating the effect of pH on solubility	If A^- is the conjugate base of a weak acid HA, calculate the fraction of A^- that remains as the free ion. Use this fraction as the coefficient for molar solubility in the K_{sp} expression.	16.17
Determining whether a precipitate forms when solutions are mixed, and which precipitate forms first if more than one is possible	Compare the ion product Q to K_{sp} to determine whether a precipitate will form; use K_{sp} expressions to calculate maximum concentrations of one ion in solution that will not cause another ion to precipitate.	16.18, 16.19

VISUAL PROBLEMS

(Answers to boldface end-of-chapter questions and problems are in the back of the book.)

16.1. The graph in Figure P16.1 shows the titration curves of a 1 *M* solution of a weak acid with a strong base and a 1 *M* solution of a strong acid with the same base. Which curve is which?

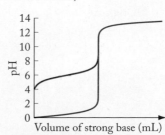

FIGURE P16.1

16.2. Estimate (to two significant figures) the pK_a of the weak acid in Problem 16.1.

16.3. Suppose you have four color indicators to choose from to detect the equivalence point of the titration reaction represented by the red curve in Figure P16.1. The pK_a values of the four indicators are 3.3, 5.0, 7.0, and 9.0. Which indicator would be the best one to choose?

16.4. Explain why the slope of the red titration curve in Figure P16.1 is nearly flat in the region extending about halfway from the start of the titration to its equivalence point.

16.5. One of the titration curves in Figure P16.5 represents the titration of an aqueous sample of Na_2CO_3 with strong acid; the other represents the titration of an aqueous sample of $NaHCO_3$ with the same acid. Which curve is which?

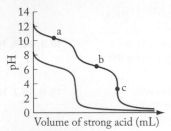

FIGURE P16.5

16.6. Identify the principal carbon-containing species in solution at points a, b, and c on the red titration curve in Figure P16.5.

16.7. Consider the three beakers in Figure P16.7. Each contains a few drops of the color indicator bromthymol blue, which is yellow in acidic solutions and blue in basic solutions. One beaker contains a solution of ammonium chloride, one contains ammonium acetate, and the third contains sodium acetate. Which beaker contains which salt?

FIGURE P16.7

*16.8. The graphs in Figure P16.8 show the conductivity of a solution as a function of the volume of titrant added. Which of the graphs best represents the titration of (a) a strong acid with a strong base and (b) a weak acid with a strong base?

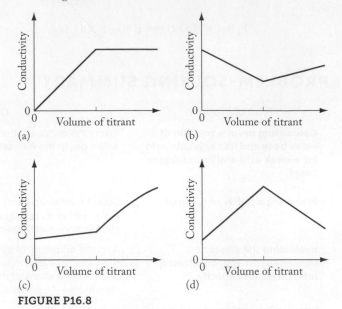

FIGURE P16.8

16.9. (a) What kind of aqueous solution is represented in Figure P16.9: a weak acid, a weak base, or a buffer? The solvent molecules and cations have been omitted for clarity. (b) What particles depicted in Figure P16.9 will change, and how will they change, upon addition of a strong base such as NaOH?

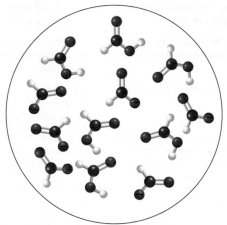

FIGURE P16.9

16.10. Use representations [A] through [I] in Figure P16.10 to answer questions a–f.

a. Images [C] and [G] depict the different molecular structures of the indicator methyl red at high and low pH. Which is which?

b. If methyl red is red at low pH and yellow at higher pH, match the flasks in images [A] and [I] to the species in [C] and [G].

c. Which image represents a buffer with equal concentrations of a weak acid and its conjugate base?

d. Which image represents the buffer from your answer to part (c) after the addition of strong acid, and which represents that buffer after the addition of strong base?

e. Saturated solutions of two sparingly soluble salts, calcium sulfide and calcium fluoride, are shown in [D] and [F]. Which is which?

f. Will adding nitric acid to [D] increase the solubility of the solid? Will adding nitric acid to [F] increase the solubility of the solid?

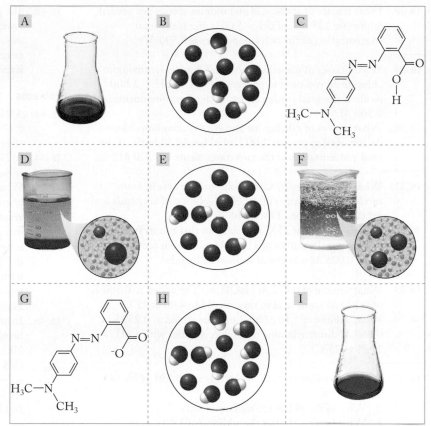

FIGURE P16.10

QUESTIONS AND PROBLEMS

Note: Tables A5.1 and A5.3 in Appendix 5 contain K_a, K_b, pK_a, pK_b, and K_f values that may be useful in answering the questions and solving the problems in this chapter.

The Common-Ion Effect and pH Buffers

Concept Review

16.11. Why does a solution of a weak acid and its conjugate base control pH better than a solution of the weak acid alone?

16.12. Why does a solution of a weak base and its conjugate acid control pH better than a solution of the weak base alone?

16.13. Identify a suitable buffer system to maintain a pH of 3.0 in an aqueous solution.

16.14. Identify a suitable buffer system to maintain a pH of 9.0 in an aqueous solution.

16.15. What does "buffer capacity" mean?

16.16. What effect does adding more NaF have on the pH and buffer capacity of an aqueous solution that is initially 1.0 *M* HF and 0.50 *M* NaF?

16.17. Three buffers are prepared using equal concentrations of formic acid and sodium formate, hydrofluoric acid and sodium fluoride, and acetic acid and sodium acetate. Rank the three buffers from highest to lowest pH.

16.18. Equal volumes of two buffers are prepared with equal concentrations of acid and conjugate base, but they use different weak acids with different pK_a values. Do the two buffers have the same buffer capacity?

16.19. How does diluting a pH 4.00 buffer with an equal volume of pure water affect its pH?

*16.20. Buffer A contains nearly equal concentrations of its conjugate acid–base pair. Buffer B contains the same total concentration of acidic and basic components as buffer A, but B has twice as much of its weak acid as its conjugate base. Which buffer experiences a smaller change in pH when:

a. the same small quantity of strong base is added to both?

b. the same small quantity of strong acid is added to both?

Problems

16.21. What is the pH of a buffer that is 0.200 *M* chloroacetic acid and 0.100 *M* sodium chloroacetate at 25°C?

16.22. What is the pH of a buffer that is 0.100 *M* methylamine and 0.175 *M* methylammonium chloride at 25°C?

16.23. What is the pH of a buffer that is 0.110 *M* HPO_4^{2-} and 0.220 *M* $H_2PO_4^-$ at 25°C?

16.24. What is the pH of a buffer that is 0.200 *M* H_2SO_3 and 0.250 *M* $NaHSO_3$ at 25°C?

16.25. What is the mole ratio of sodium acetate to acetic acid in a buffer with a pH of 5.75?

16.26. What is the mole ratio of ammonia to ammonium chloride in a buffer with a pH of 9.00?

16.27. What masses of bromoacetic acid and sodium bromoacetate are needed to prepare 1.00 L of pH = 3.00 buffer if the total concentration of the two components is 0.200 *M*?

16.28. What masses of acetic acid and sodium acetate are needed to prepare 125 mL of pH = 5.00 buffer if the total concentration of the two components is 0.500 M?

16.29. What masses of dimethylamine and dimethylammonium chloride do you need to prepare 0.500 L of pH = 12.00 buffer if the total concentration of the two components is 0.300 M?

16.30. What masses of ethylamine and ethylammonium chloride do you need to prepare 1.00 L of pH = 10.50 buffer if the total concentration of the two components is 0.250 M?

16.31. What is the pH at 25°C of a solution that results from mixing equal volumes of a 0.05 M solution of ammonia and a 0.025 M solution of hydrochloric acid?

16.32. What is the pH at 25°C of a solution that results from mixing equal volumes of a 0.05 M solution of acetic acid and a 0.025 M solution of sodium hydroxide?

*16.33. What volume of 0.422 M NaOH must be added to 0.500 L of 0.300 M acetic acid to raise its pH to 4.00 at 25°C?

*16.34. What volume of 1.16 M HCl must be added to 0.250 L of 0.350 M dimethylamine to produce a buffer with a pH of 10.75 at 25°C?

*16.35. A buffer consists of 0.120 M HNO_2 and 0.150 M $NaNO_2$ at 25°C.
   a. What is the pH of the buffer?
   b. What is the pH after the addition of 1.00 mL of 11.6 M HCl to 1.00 L of the buffer solution?

*16.36. A buffer is prepared by mixing 50.0 mL of 0.200 M NaOH with 100.0 mL of 0.175 M acetic acid.
   a. What is the pH of the buffer?
   b. What is the pH of the buffer after 1.00 g of NaOH is dissolved in it?

Indicators and Acid–Base Titrations

Concept Review

16.37. Do all titrations of samples of strong monoprotic acids with solutions of strong bases have the same pH at their equivalence points? Explain why or why not.

16.38. Do all titrations of samples of weak monoprotic acids with solutions of strong bases have the same pH at their equivalence points? Explain why or why not.

16.39. Describe two properties of phenolphthalein that make it a good choice of indicator for detecting the first equivalence point in an alkalinity titration.

*16.40. Phenolphthalein can be used as a color indicator to detect the equivalence points of titrations of samples containing either weak or strong acids even though the pH values of the equivalence point vary depending on the identity of the acid. Explain how this is possible.

16.41. In the titration of a solution of a weak monoprotic acid with a standard solution of NaOH, the pH halfway to the equivalence point was 4.44. In the titration of a second solution of the same acid, exactly twice as much of the standard solution of NaOH was needed to reach the equivalence point. What was the pH halfway to the equivalence point in this titration?

16.42. The pH of a solution of a strong monoprotic acid is lower than the pH of an equal concentration of a weak monoprotic acid, yet equal volumes of both require the same volume of basic titrant to reach the equivalence point. Explain why.

Problems

16.43. A 25.0 mL sample of 0.100 M acetic acid is titrated with 0.125 M NaOH at 25°C. What is the pH of the solution after 10.0, 20.0, and 30.0 mL of the base have been added?

16.44. A 25.0 mL sample of a 0.100 M solution of aqueous trimethylamine is titrated with a 0.125 M solution of HCl. What is the pH of the solution after 10.0, 20.0, and 30.0 mL of acid have been added?

*16.45. **Window Cleaner** (a) What is the concentration of ammonia in a popular window cleaner if 25.34 mL of 1.162 M HCl is needed to titrate a 10.00 mL sample of the cleaner? (b) Suppose that the sample was diluted to about 50 mL with deionized water prior to the titration to make it easier to mount a pH electrode in it. What effect did this dilution have on the volume of titrant needed?

*16.46. In an alkalinity titration of a 100.0 mL sample of water from a hot spring, 2.56 mL of a 0.0355 M solution of HCl is needed to reach the first equivalence point (pH = 8.3) and another 10.42 mL is needed to reach the second equivalence point (pH = 4.0). If the alkalinity of the spring water is due only to the presence of carbonate and bicarbonate, what are the concentrations of each?

16.47. What volume of 0.0100 M HCl is required to titrate 250 mL of 0.0100 M Na_2CO_3 to the first equivalence point?

16.48. What volume of 0.0100 M HCl is required to titrate 250 mL of 0.0100 M Na_2CO_3 and 250 mL of 0.0100 M HCO_3^-?

16.49. Sketch a titration curve for the titration of 50.0 mL of 0.250 M HNO_2 with 1.00 M NaOH. What is the pH at the equivalence point?

16.50. Sketch a titration curve for the titration of 40.0 mL of a 0.100 M solution of oxalic acid with a 0.100 M solution of NaOH. What is the pH of the titration reaction mixture at the last equivalence point?

16.51. For each titration, predict whether the equivalence point is less than, equal to, or greater than pH = 7.
   a. Quinine titrated with nitric acid
   b. Pyruvic acid titrated with calcium hydroxide
   c. Hydrobromic acid titrated with strontium hydroxide

16.52. For each titration, predict whether the equivalence point is less than, equal to, or greater than pH = 7.
   a. HCN titrated with $Ca(OH)_2$
   b. LiOH titrated with HI
   c. C_5H_5N titrated with KOH

16.53. When 100 mL of 0.0125 M ascorbic acid is titrated with 0.010 M NaOH, how many equivalence points will the titration curve have, and what pH indicator(s) could be used? Refer to Figure 16.8 for colors of indicators.

16.54. Red cabbage juice is a sensitive acid–base indicator; its colors range from red at acidic pH to yellow in alkaline solutions. What color would red cabbage juice have at the equivalence point when 25 mL of a 0.10 M solution of acetic acid is titrated with 0.10 M NaOH?

Lewis Acids and Bases

Concept Review

16.55. Are all Lewis bases also Brønsted–Lowry bases? Explain why or why not.

16.56. Are all Brønsted–Lowry bases also Lewis bases? Explain why or why not.

16.57. Are all Brønsted–Lowry acids also Lewis acids? Explain why or why not.

16.58. Why is BF_3 a Lewis acid but not a Brønsted–Lowry acid?

Problems

16.59. Draw Lewis structures that show how electron pairs move and bonds form and break during the autoionization of water. Label the appropriate H_2O molecules as the Lewis acid and Lewis base.

16.60. Draw Lewis structures that show how electron pairs move and bonds form and break in this reaction, and identify the Lewis acid and Lewis base.

$$MgO(s) + CO_2(g) \rightarrow MgCO_3(s)$$

16.61. Draw Lewis structures that show how electron pairs move and bonds form and break in this reaction, and identify the Lewis acid and Lewis base.

$$SO_2(g) + H_2O(\ell) \rightarrow H_2SO_3(aq)$$

16.62. Draw Lewis structures that show how electron pairs move and bonds form and break in this reaction, and identify the Lewis acid and Lewis base.

$$SeO_3(g) + H_2O(\ell) \rightarrow H_2SeO_4(aq)$$

16.63. Draw Lewis structures that show how electron pairs move and bonds form and break in this reaction, and identify the Lewis acid and Lewis base.

$$B(OH)_3(aq) + H_2O(\ell) \rightleftharpoons B(OH)_4^-(aq) + H^+(aq)$$

*16.64. Draw Lewis structures that show how electron pairs move and bonds form and break in this reaction, and identify the Lewis acid and Lewis base. (*Note:* $HSbF_6$ is an ionic compound and one of the strongest Brønsted–Lowry acids known.)

$$SbF_5(s) + HF(g) \rightarrow HSbF_6(s)$$

Formation of Complex Ions

Concept Review

16.65. When $CaCl_2$ dissolves in water, which molecules or ions occupy the inner coordination sphere around the Ca^{2+} ions?

16.66. When $AgNO_3$ dissolves in water, which molecules or ions occupy the inner coordination sphere around the Ag^+ ions?

*16.67. A lab technician cleaning glassware that contains residues of AgCl washes the glassware with an aqueous solution of ammonia. The AgCl, which is insoluble in water, rapidly dissolves in the ammonia solution. Why?

*16.68. The procedure used in the previous question dissolves AgCl but not AgI. Why?

Problems

16.69. A solution is prepared in which 0.00100 mol of $Ni(NO_3)_2$ and 0.500 mol of NH_3 are dissolved in a total volume of 1.00 L. What is the concentration of $Ni(H_2O)_6^{2+}$ ions in the solution at equilibrium?

16.70. A 1.00 L solution contains 5.00×10^{-5} M $Cu(NO_3)_2$ and 1.00×10^{-3} M ethylenediamine. What is the concentration of $Cu(H_2O)_6^{2+}$ ions in the solution at equilibrium?

*16.71. Suppose a solution contains 1.00 mmol of $Co(NO_3)_2$, 0.100 mol of NH_3, and 0.100 mol of ethylenediamine in a total volume of 0.250 L. What is the concentration of $Co(H_2O)_6^{2+}$ ions in the solution?

*16.72. If 1.00 mL of 0.0100 M $AgNO_3$, 1.00 mL of 0.100 M NaBr, and 1.00 mL of 0.100 M NaCN are diluted to 250 mL with deionized water in a volumetric flask and shaken vigorously, will the contents of the flask be cloudy or clear? Support your answer with the appropriate calculations. (*Hint:* The K_{sp} of AgBr is 5.4×10^{-13}.)

Hydrated Metal Ions as Acids

Concept Review

16.73. Which, if any, aqueous solutions of the following chloride compounds are acidic? (a) $CaCl_2$; (b) $CrCl_3$; (c) NaCl; (d) $FeCl_3$

16.74. If 0.100 M aqueous solutions of each of these compounds were prepared, which one would have the lowest pH? (a) $BaCl_2$; (b) LiCl; (c) KCl; (d) $TiCl_4$

16.75. When ozone is bubbled through an aqueous solution of Fe^{2+} ions, the ions are oxidized to Fe^{3+} ions. How does the oxidation process affect the pH of the solution?

16.76. As an aqueous solution of KOH is slowly added to a stirred solution of $AlCl_3$, the mixture becomes cloudy but then clears when more KOH is added.
   a. Explain the chemical changes responsible for the changes in the appearance of the mixture.
   b. Would you expect to observe the same changes if KOH were added to a solution of $FeCl_3$? Explain why or why not.

16.77. Chromium(III) hydroxide is amphiprotic. Write chemical equations showing how an aqueous suspension of this compound reacts to the addition of a strong acid and a strong base.

16.78. Zinc hydroxide is amphiprotic. Write chemical equations showing how an aqueous suspension of this compound reacts to the addition of a strong acid and a strong base.

16.79. **Refining Aluminum** To remove impurities such as calcium and magnesium carbonates and iron(III) oxides from aluminum ore (which is mostly Al_2O_3), the ore is treated with a strongly basic solution. In this treatment, Al^{3+} dissolves but the other metal ions do not. Why?

*16.80. Exactly 1.00 g of $FeCl_3$ is dissolved in each of four 0.500 L samples: 1 M HNO_3, 1 M HNO_2, 1 M CH_3COOH, and pure water. Is the concentration of $Fe(H_2O)_6^{3+}$ ions the same in all four solutions? Explain why or why not.

Problems

16.81. What is the pH of 0.25 M Al(NO$_3$)$_3$?

16.82. What is the pH of 0.50 M CrCl$_3$?

16.83. What is the pH of 0.100 M Fe(NO$_3$)$_3$?

16.84. What is the pH of 1.00 M Cu(NO$_3$)$_2$?

16.85. Sketch the titration curve (pH versus volume of 0.50 M NaOH) for a 25 mL sample of 0.25 M FeCl$_3$.

16.86. Sketch the titration curve that results from the addition of 0.50 M NaOH to a sample containing 0.25 M KFe(SO$_4$)$_2$.

Solubility Equilibria

Concept Review

16.87. What is the difference between *molar solubility* and *solubility product*?

16.88. Give an example of how the common-ion effect limits the dissolution of a sparingly soluble ionic compound.

16.89. Which cation will precipitate first as a carbonate mineral from an equimolar solution of Mg^{2+}, Ca^{2+}, and Sr^{2+}?

16.90. If the solubility of a compound increases with increasing temperature, does K_{sp} increase or decrease?

16.91. The K_{sp} of strontium sulfate increases from 2.8×10^{-7} at 37°C to 3.8×10^{-7} at 77°C. Is the dissolution of strontium sulfate endothermic or exothermic?

16.92. Identify any of the following solids that are more soluble in acidic solution than in neutral water: CaCl$_2$, Ba(HCO$_3$)$_2$, PbSO$_4$, Cu(OH)$_2$. Explain your choices.

16.93. **Chemistry of Tooth Decay** Tooth enamel is composed of a mineral known as hydroxyapatite with the formula Ca$_5$(PO$_4$)$_3$(OH). Explain why tooth enamel can be eroded by acidic substances released by bacteria growing in the mouth.

16.94. **Fluoride and Dental Hygiene** Fluoride ions in drinking water and toothpaste convert hydroxyapatite in tooth enamel into fluorapatite:

$$\text{Ca}_5(\text{PO}_4)_3(\text{OH})(s) + \text{F}^-(aq) \rightleftharpoons \text{Ca}_5(\text{PO}_4)_3\text{F}(s) + \text{OH}^-(aq)$$

Why is fluorapatite less susceptible than hydroxyapatite to erosion by acids?

Problems

16.95. At a particular temperature the [Ba^{2+}] in a saturated solution of barium sulfate is 1.04×10^{-5} M. Starting with this information, calculate the K_{sp} value of barium sulfate at this temperature.

16.96. If only 0.160 g of Ca(OH)$_2$ dissolves in 0.100 L of water, what is the K_{sp} value for calcium hydroxide at that temperature?

16.97. What are the equilibrium concentrations of Cu$^+$ and Cl$^-$ in a saturated solution of copper(I) chloride at 25°C?

16.98. What are the equilibrium concentrations of Pb^{2+} and F$^-$ in a saturated solution of lead fluoride at 25°C?

16.99. What is the solubility of calcite (CaCO$_3$) in grams per milliliter at a temperature at which its $K_{sp} = 9.9 \times 10^{-9}$?

16.100. What is the solubility of silver iodide in grams per milliliter at 25°C?

16.101. What is the pH at 25°C of a saturated solution of silver hydroxide?

16.102. **pH of Milk of Magnesia** What is the pH at 25°C of a saturated solution of magnesium hydroxide (the active ingredient in the antacid milk of magnesia)?

16.103. Suppose you have 100 mL of each of the following solutions. In which will the most CaCO$_3$ dissolve? (a) 0.1 M NaCl; (b) 0.1 M Na$_2$CO$_3$; (c) 0.1 M NaOH; (d) 0.1 M HCl

16.104. In which of the following solutions will CaF$_2$ be most soluble? (a) 0.010 M Ca(NO$_3$)$_2$; (b) 0.01 M NaF; (c) 0.001 M NaF; (d) 0.10 M Ca(NO$_3$)$_2$

16.105. **Composition of Seawater** The average concentration of sulfate in surface seawater is about 0.028 M. The average concentration of Sr^{2+} is 9×10^{-5} M. Is the concentration of strontium in the sea significantly controlled by the insolubility of its sulfate salt?

16.106. **Fertilizing the Sea to Combat Climate Change** Some scientists have proposed adding iron(III) compounds to large expanses of the open ocean to promote the growth of phytoplankton that would in turn remove CO$_2$ from the atmosphere through photosynthesis. Assuming the average pH of open ocean water is 8.13, what is the maximum value of [Fe^{3+}] in seawater if the K_{sp} value of Fe(OH)$_3$ is 1.1×10^{-36}?

16.107. Will calcium fluoride precipitate when 125 mL of 0.375 M Ca(NO$_3$)$_2$ is added to 245 mL of 0.255 M NaF at 25°C?

16.108. Will lead(II) chloride precipitate if 185 mL of 0.025 M sodium chloride is added to 235 mL of 0.165 M lead(II) perchlorate at 25°C?

16.109. A solution is 0.010 M in both Br$^-$ and SO$_4^{2-}$. A 0.250 M solution of lead(II) nitrate is slowly added to it with a burette.
  a. Which anion will precipitate first?
  b. What is the concentration in the solution of the first ion when the second one starts to precipitate at 25°C?

16.110. Solution A is 0.0200 M in Ag$^+$ ions and Pb^{2+} ions. You have access to two other solutions: (B) 0.250 M NaCl and (C) 0.250 M NaBr.
  a. Which solution, B or C, would be the better one to add to Solution A to separate Ag$^+$ ions from Pb^{2+} by selective precipitation?
  b. Using the solution you selected in part (a), is the separation of the two ions complete?

Additional Problems

16.111. **Fluoride in Drinking Water** Hydrogen fluoride (HF) behaves as a weak acid in aqueous solution. Two equilibria influence which fluorine-containing species are present in solution.

$$\text{HF}(aq) + \text{H}_2\text{O}(\ell) \rightleftharpoons \text{H}_3\text{O}^+(aq) + \text{F}^-(aq) \qquad K_a = 1.1 \times 10^{-3}$$

$$\text{F}^-(aq) + \text{HF}(aq) \rightleftharpoons \text{HF}_2^-(aq) \qquad K = 2.6 \times 10^{-1}$$

  a. Is fluoride in pH 7.00 drinking water more likely to be present as F$^-$ or HF$_2^-$?

b. What is the equilibrium constant for this equilibrium?

$$2\,HF(aq) + H_2O(\ell) \rightleftharpoons H_3O^+(aq) + HF_2^-(aq)$$

c. What are the pH and equilibrium concentration of HF_2^- in a 0.150 M solution of HF?

16.112. pH of Natural Waters Between 1993 and 1995, sodium phosphate was added to Seathwaite Tarn in the English Lake District to increase its pH. Explain why addition of this compound increased pH.

*__16.113.__ **pH of Baking Soda** A cook dissolves a teaspoon of baking soda ($NaHCO_3$) in a cup of water and then discovers that the recipe calls for a tablespoon, not a teaspoon. So the cook adds two more teaspoons of baking soda to make up the difference. Does the additional baking soda change the pH of the solution? Explain why or why not.

*__16.114.__ **Antacid Tablets** Antacids contain a variety of bases such as $NaHCO_3$, $MgCO_3$, $CaCO_3$, and $Mg(OH)_2$. Only $NaHCO_3$ has appreciable solubility in water.
  a. Write a net ionic equation for the reaction of each base with aqueous HCl.
  b. Explain how substances insoluble in water can act as effective antacids.

16.115. When silver oxide dissolves in water, the following reaction occurs:

$$Ag_2O(s) + H_2O(\ell) \rightarrow 2\,Ag^+(aq) + 2\,OH^-(aq)$$

If a saturated aqueous solution of silver oxide is 1.6×10^{-4} M in hydroxide ion, what is the K_{sp} of silver oxide?

*__16.116.__ Why does adding $CaCl_2$ to a HPO_4^{2-}/PO_4^{3-} buffer increase the ratio of HPO_4^{2-} ions to PO_4^{3-} ions?

*__16.117.__ **Greenhouse Gases and Ocean pH** Some climate models predict the pH of the oceans will decrease by as much as 0.77 pH units as a result of increases in atmospheric carbon dioxide.
  a. Explain, by using the appropriate chemical reactions and equilibria, how an increase in atmospheric CO_2 could produce a decrease in oceanic pH.

  b. How much more acidic (in terms of $[H_3O^+]$) would the oceans be if their pH dropped this much?
  c. Oceanographers are concerned about how a drop in oceanic pH would affect the survival of oysters. Why?

*__16.118.__ A 125.0 mg sample of an unknown monoprotic acid was dissolved in 100.0 mL of distilled water and titrated with a 0.050 M solution of NaOH. The pH of the solution was monitored throughout the titration, and the following data were collected.
  a. What is the K_a value for the acid?
  b. What is the molar mass of the acid?

Volume of OH⁻ Added (mL)	pH	Volume of OH⁻ Added (mL)	pH
0	3.09	22	5.93
5	3.65	22.2	6.24
10	4.10	22.6	9.91
15	4.50	22.8	10.2
17	4.55	23	10.4
18	4.71	24	10.8
19	4.94	25	11.0
20	5.11	30	11.5
21	5.37	40	11.8

17

Thermodynamics

Spontaneous and Nonspontaneous Reactions and Processes

CORROSION Most metal objects in the sea, including the remains of this shipwreck, corrode as a result of spontaneous chemical reactions. Iron metal is converted into iron(III) oxide.

◄ PARTICULATE **REVIEW**

Endothermic or Exothermic?

In Chapter 17 we examine fundamental ideas about why some reactions happen spontaneously but others do not. In doing so, we revisit the concepts of thermochemistry from Chapter 5. Propane (shown here) is a common fuel used for cooking on outdoor grills.

- Is the combustion of propane an endothermic or exothermic process?

- What bonds are broken in the combustion of propane? Is the breaking of bonds an endothermic or exothermic process?

- What bonds are formed in the combustion of propane? Is the forming of bonds an endothermic or exothermic process?

   (Review Sections 5.3 and 8.7 if you need help.)

(Answers to Particulate Review questions are in the back of the book.)

Motion and Dispersion of Energy

We learned in Chapter 5 that the total energy of atoms and molecules is the sum of their kinetic energy (due to random motion of particles) and potential energy (due to their arrangement). Consider these three elements at room temperature: copper, bromine, and helium. As you read Chapter 17, look for ideas that will help you answer these questions:

Copper   Bromine   Helium

- Identify each element as a solid, liquid, or gas at room temperature from the images shown here.

- Compare the possible number of arrangements of the particles in 1 mole of each element. Which has the most possible arrangements? The fewest?

- Which element(s) have vibrational motion of bonds at room temperature?

- Which element can disperse its energy in the most ways?

833

Learning Outcomes

LO1 Predict the signs of entropy changes for spontaneous and nonspontaneous chemical reactions and physical processes
Sample Exercise 17.1

LO2 Predict the relative entropies of substances on the basis of their molecular structures
Sample Exercise 17.2

LO3 Calculate entropy changes in chemical reactions by using standard molar entropies
Sample Exercise 17.3

LO4 Calculate free-energy changes and standard free-energy changes in chemical reactions
Sample Exercises 17.4, 17.5

LO5 Predict the spontaneity of a chemical reaction as a function of temperature
Sample Exercise 17.6

LO6 Relate the value of the equilibrium constant of a reaction to its change in free energy under standard conditions
Sample Exercises 17.7, 17.8

LO7 Use the van 't Hoff equation to calculate the values of the equilibrium constant of a reaction at different temperatures
Sample Exercise 17.8

LO8 Calculate the net change in free energy of coupled spontaneous and nonspontaneous reactions
Sample Exercise 17.9

LO9 Use microstates to explain why a perfect crystalline solid has zero entropy

17.1 Spontaneous Processes

Some processes are so familiar that we rarely consider why they happen. If a car tire is punctured, the air inside rushes out and the tire goes flat; air does not rush back into a punctured tire and reinflate it. Objects made of iron left on the ground or underwater for months or years become clumps of rust; they do not turn back into shiny metal after even more time. A tray of ice cubes accidentally left on a kitchen counter melts into a tray of liquid water. As long as the kitchen remains at the same temperature, there is no way the water in the tray will resolidify into ice cubes.

All of these processes have something in common: they are all **spontaneous**. This means that they all happen without any ongoing intervention and without work being done on the system. In each case, the reverse process is **nonspontaneous**—it can happen only as long as energy is continually added. Why are some processes spontaneous and others not?

As we learned in Chapter 5, the first law of thermodynamics tells us that energy cannot be created or destroyed. This means that when we play the game of energy conversion, we can't win. Furthermore, the second law of thermodynamics—which we explore more closely in the next section—says, in effect, that not all of the energy released by a spontaneous reaction, such as burning gasoline in a car engine, is available to do useful work. In other words, in the game of energy conversion, not only can we not win—we can't even break even.

If energy cannot be destroyed, what happens to the energy that is unavailable to do useful work? The answer is that this energy spreads out, becoming less concentrated. In this chapter we explore the meaning and some of the impacts of the second law of thermodynamics on familiar processes. We also explore the component of the energy that is available to do work and how it is related to reaction spontaneity and equilibrium. We will see how spontaneous reactions can be coupled with nonspontaneous reactions to make multistep chemical and biochemical processes happen. This coupling is important to us because it powers the molecular processes that sustain life.

spontaneous process a process that occurs without outside intervention.

nonspontaneous process a process that occurs only as long as energy is continually added to the system.

The word "spontaneous" can be a bit misleading because many spontaneous reactions do not start all by themselves; they need a little energy boost, such as a spark or external flame. For example, before Thomas Edison invented the light bulb in 1879, city streets were often illuminated at night by gas-fueled lamps. Once lit, they burned through the night until their fuel was cut off as dawn approached. Chemical reactions such as the combustion of gas-lamp fuel are examples of spontaneous reactions: once started, they proceed without outside intervention, as long as reactants are available. It is the self-sustaining nature of these reactions that earns them the label *spontaneous*. The reverse reaction—in this case, converting carbon dioxide and water into fuel and oxygen—is nonspontaneous: it cannot happen on its own without the continual addition of energy from an external source.

In addition, spontaneous does not necessarily mean rapid. Though combustion reactions certainly are fast, other spontaneous reactions, such as the formation of a layer of rust on an object made of iron, can take a very long time, depending on temperature and the rate at which O_2 reaches the iron surface:

$$4 \, Fe(s) + 3 \, O_2(g) \rightarrow 2 \, Fe_2O_3(s) \qquad \Delta H° = -1648 \text{ kJ}$$

Spontaneity has nothing to do with kinetics. Still, rust formation does proceed without intervention and meets the definition of thermodynamic spontaneity.

Spontaneous is also *not* a synonym for exothermic, although many scientists once thought so. In the mid-19th century, many chemists thought that exothermic reactions, in which high-enthalpy reactants formed low-enthalpy (more stable) products, should always be spontaneous. It is true that many exothermic reactions, including combustion reactions and metal corrosion, *are* spontaneous. However, some exothermic processes may not be spontaneous, and some endothermic processes may be spontaneous, depending on reaction conditions.

Familiar examples of spontaneous endothermic processes include the phase changes that occur when ice melts and water boils. Both are endothermic processes, but both can be spontaneous, depending on the temperature and pressure. As another example of a spontaneous endothermic reaction, consider what happens when baking soda (sodium bicarbonate) is added to room-temperature vinegar (dilute acetic acid), as shown in Figure 17.1. The foaming mixture tells us that a reaction is taking place, and a decrease in temperature of the mixture tells us that the reaction is endothermic ($\Delta H > 0$).

Yet another spontaneous endothermic process makes instant cold packs cold (Figure 17.2). An instant cold pack has two compartments. One is filled with

CONNECTION We learned in Chapter 5 that enthalpy change (ΔH) is a thermodynamic quantity that describes energy flow into or out of a system.

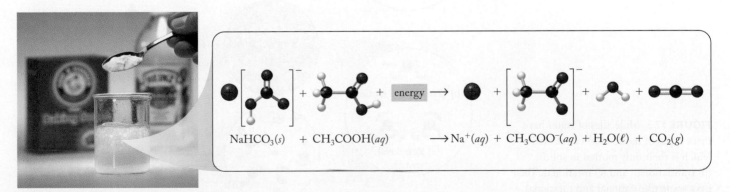

$$NaHCO_3(s) + CH_3COOH(aq) \longrightarrow Na^+(aq) + CH_3COO^-(aq) + H_2O(\ell) + CO_2(g)$$

FIGURE 17.1 Adding baking soda to vinegar produces sodium acetate, water, and carbon dioxide in an endothermic ($\Delta H° = +48.5$ kJ), yet spontaneous, reaction.

FIGURE 17.2 Instant cold packs get cold when the water-filled pouch inside is ruptured and the water-soluble compound (e.g., ammonium nitrate) within the pack dissolves. The NH_4^+ and NO_3^- ions in solid NH_4NO_3 experience increased freedom of motion as they form hydrated NH_4^+ and NO_3^- ions in solution. The temperature of the solution drops because the dissolution process is endothermic.

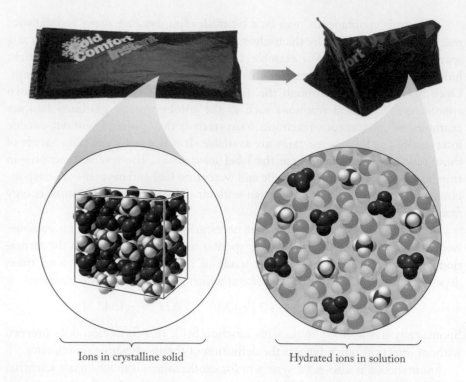

Ions in crystalline solid   Hydrated ions in solution

CONNECTION The ion–dipole interactions that promote the solubility of ionic compounds in water were described in Chapter 10.

water; the other contains a water-soluble compound such as ammonium nitrate that has a positive enthalpy of solution ($\Delta H_{soln} > 0$). When the membrane separating the two compartments is ruptured, the solid compound (NH_4NO_3) mixes with and dissolves in the water. The resulting solution gets very cold, becoming an effective anti-inflammation treatment for bruises and muscle sprains.

Why are endothermic processes such as these spontaneous? The answer to this question lies in something these processes have in common: the particles that make up their products are more spread out and have more freedom of motion than the particles in their starting materials. Consider the molecular changes that

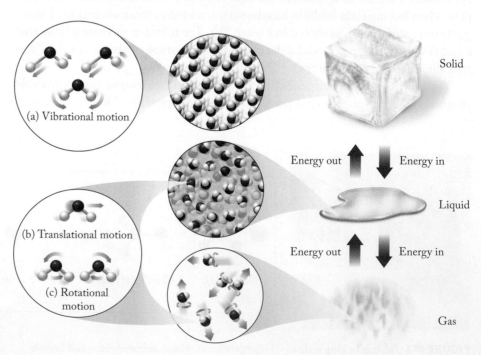

(a) Vibrational motion

(b) Translational motion

(c) Rotational motion

Solid

Energy out   Energy in

Liquid

Energy out   Energy in

Gas

FIGURE 17.3 Molecules of water have three types of motion: (a) vibrational—which is their only motion in solids; (b) translational; and (c) rotational. They have some translational and rotational motion in the liquid phase, and much more of both in the gas phase.

accompany ice melting and water boiling (Figure 17.3). The molecules of H_2O in ice occupy fixed positions. Their motion (Figure 17.3a), like that of particles in all crystalline solids, is limited to vibrating in place, not going anywhere. When ice melts, its H_2O molecules become more mobile, acquiring translational and rotational motion, as shown in Figure 17.3(b) and (c). When liquid water evaporates and the molecules enter the gas phase, their translational and rotational motion increases even more. The water vapor molecules are now also able to expand to fill their container, as are all gas particles at atmospheric pressure, which is why gases are compressible, whereas liquids and solids are not.

The particles that make up the reaction mixtures in the two endothermic chemical reactions described above also experience gains in freedom of motion. In Figure 17.1, a portion of the bicarbonate ions in the solid reactant ($NaHCO_3$) becomes liberated as a gaseous product of the reaction (CO_2). In the instant cold pack in Figure 17.2, the particles of the solid solute acquire more freedom of motion and can move throughout the resulting solution. These particles experience increases in motion similar to those of a melting solid.

CONCEPT **TEST**

In which of the following processes do particles experience an increase in freedom of motion?

a. A glass of water evaporates.

b. Dew forms overnight on grass and other surfaces.

c. Table salt is added to water for cooking spaghetti.

d. A log of wood burns in a fireplace.

(Answers to Concept Tests are in the back of the book.)

17.2 Thermodynamic Entropy

When particles spread out and gain freedom of motion, the kinetic energy associated with their motion also spreads out. This spreading out, or dispersion, of energy turns out to be a key characteristic of all spontaneous processes. It even has a name: **entropy (S)**. Entropy is a thermodynamic property that provides a measure of the dispersal of energy in a system at a specific temperature. The **second law of thermodynamics** states that entropy of an *isolated* thermodynamic system *always increases* during a spontaneous process.

The second law also covers thermodynamic systems that are not isolated (most systems are not isolated; they are either open or closed, as discussed in Section 5.3). Recall from Chapter 5 that in thermodynamics we divide the universe into two parts: the part we are interested in (the system) and everything else (the system's surroundings). Mathematically:

$$\text{Universe} = \text{System} + \text{Surroundings}$$

Logically, then, the overall change in the entropy of the universe is the sum of the entropy changes experienced by the system and by its surroundings:

$$\Delta S_{\text{univ}} = \Delta S_{\text{sys}} + \Delta S_{\text{surr}} \tag{17.1}$$

For a spontaneous process in an isolated system, ΔS_{sys} is greater than zero ($\Delta S_{\text{sys}} > 0$) and the entropy of its surroundings is unchanged ($\Delta S_{\text{surr}} = 0$). Therefore according to Equation 17.1, ΔS_{univ} must also be greater than zero ($\Delta S_{\text{univ}} > 0$).

CHEMTOUR
Entropy

entropy (S) a measure of the dispersion of energy in a system at a specific temperature.

second law of thermodynamics the principle that the total entropy of the universe increases in any spontaneous process.

CONNECTION In Chapter 5 we defined an isolated thermodynamic system as one that exchanges neither energy nor matter with its surroundings; a closed system exchanges energy but not matter; and an open system exchanges both.

The positive value of ΔS_{univ} is the basis for another way of expressing the second law of thermodynamics that applies to all systems (not just isolated ones): *a spontaneous process produces an increase in the entropy of the universe.*

This version of the second law is built on the assumption that a physical or chemical change in a closed or open thermodynamic system can alter the entropies of both the system and its surroundings. The second law says that a process is spontaneous when one or the other of these entropy changes is greater than zero, so that their sum, ΔS_{univ}, is also greater than zero. The second law provides a thermodynamic requirement for reaction spontaneity as well as a criterion for *non*spontaneity: a process that results in a *decrease* in the entropy of the universe does not occur spontaneously. Summarizing these relationships:

- If $\Delta S_{univ} > 0$, then a process is spontaneous.
- If $\Delta S_{univ} < 0$, then a process is nonspontaneous.

To see how a process affects the entropy of its surroundings, let's focus on a familiar exothermic reaction, the combustion of natural gas (methane):

$$CH_4(g) + 2\,O_2(g) \rightarrow CO_2(g) + 2\,H_2O(\ell) \qquad \Delta H° = -890 \text{ kJ}$$

As written, the reaction consumes 3 moles of gases and produces 2 moles of a liquid product and 1 mole of CO_2 gas. Note that there are fewer moles of gas on the product side of the reaction equation than on the reactant side. Given the much greater freedom of motion of particles in the gas phase, we can accurately predict that there will be a decrease in entropy of the reaction mixture, which is our thermodynamic system:

$$\Delta S_{sys} < 0$$

However, we know that the reaction is spontaneous, which means

$$\Delta S_{univ} > 0$$

How do we reconcile these opposing inequalities? Equation 17.1 supplies an explanation. If ΔS_{univ} is greater than zero, then the sum of ΔS_{sys} and ΔS_{surr} must also be greater than zero. The fact that ΔS_{sys} is less than zero simply means that ΔS_{surr} is not only greater than zero, it must have a large enough positive value to more than offset the negative value of ΔS_{sys}. Expressing this relationship in terms of the absolute values of ΔS_{surr} and ΔS_{sys}:

$$|\Delta S_{surr}| > |\Delta S_{sys}|$$

Is the combustion of methane likely to produce a large, positive ΔS_{surr}? Absolutely, because the reaction is highly exothermic: combustion of only 1 mole (16 grams) of methane generates 890 kJ of thermal energy. As energy flows from the system into its surroundings, a dispersion of energy occurs, producing a positive ΔS_{surr} that more than compensates for the unfavorable (negative) value of ΔS_{sys}.

All exothermic reactions have the capacity to increase the entropy of their surroundings. The more energy (q) that flows into the surroundings, or into any collection of particles, the greater the dispersion of kinetic energy among the particles and the greater the increase in their entropy (ΔS). However, adding energy to particles that are already hot produces a smaller change in entropy than adding the same quantity of energy to the same particles at a lower temperature. This inverse relationship between entropy change and temperature is reflected in Equation 17.2:

$$\Delta S = \frac{q_{rev}}{T} \qquad (17.2)$$

The subscript "rev" means that the heating process is reversible. Theoretically, a **reversible process** happens so slowly that equilibrium is constantly maintained. For example, after an incremental change in the system has occurred in the forward direction, the process can be reversed with a tiny change in process conditions so that the original state of the system can be restored with no net flow of energy into or out of the system. In other words, everything about the system goes back to exactly as it was before the process began.

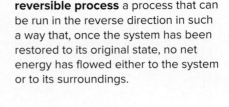

reversible process a process that can be run in the reverse direction in such a way that, once the system has been restored to its original state, no net energy has flowed either to the system or to its surroundings.

A reversible process is an idealization; it describes a theoretical limit. All real chemical reactions are irreversible, but many of them approximate reversibility closely enough that we can adapt Equation 17.2 to calculate ΔS_{sys}:

$$\Delta S_{sys} = \frac{q_{sys}}{T} \tag{17.3}$$

For example, it takes 6.01 kJ (or 6.01×10^3 J) of energy to melt 1.00 mol of ice at 0°C. Assuming the process occurs reversibly, then

$$\Delta S_{sys} = \frac{q_{sys}}{T} = \frac{(1.00 \ \text{mol})(6.01 \times 10^3 \ \text{J/mol})}{273 \ \text{K}} = 22.0 \ \text{J/K}$$

Because energy flows into the ice, q_{sys} is positive, which also makes ΔS_{sys} positive, as we would expect given the greater freedom of motion of particles in the liquid phase. Note that the units of entropy in this calculation are joules per kelvin. We use these units in all entropy calculations in this chapter.

Suppose 1.00 mol of ice melts as 6.01×10^3 J of energy flows into it from room-temperature (22°C or 295 K) surroundings. We can calculate the change in entropy of the surroundings by using another adaptation of Equation 17.2:

$$\Delta S_{surr} = \frac{q_{surr}}{T} = \frac{(1.00 \ \text{mol})(-6.01 \times 10^3 \ \text{J/mol})}{295 \ \text{K}} = -20.4 \ \text{J/K}$$

Note that the sign of q_{surr} is negative because energy flows from the surroundings into the system (ice). Also note that (1) the value of ΔS_{surr} is less than zero, and (2) the magnitude of the decrease in ΔS_{surr} is less than the increase in ΔS_{sys} because the same absolute value of q was divided by a higher temperature to calculate ΔS_{surr}. Therefore when we sum ΔS_{sys} and ΔS_{surr} (Figure 17.4a), we get a positive value of ΔS_{univ}:

$$\Delta S_{univ} = \Delta S_{sys} + \Delta S_{surr} = (22.0 - 20.4) \ \text{J/K} = 1.6 \ \text{J/K}$$

The result of this calculation is one example of a general truth about energy flow that will come as no surprise: energy flows spontaneously into a system that is cooler than its surroundings. Even more generally, energy flows spontaneously from a warm object to an adjacent cooler object. Entropy and the second law of thermodynamics simply provide a mathematical explanation of why.

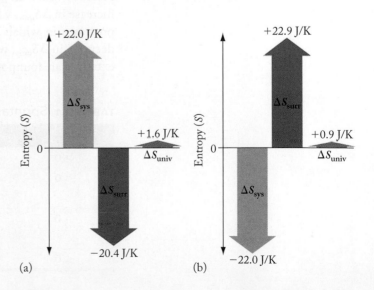

FIGURE 17.4 (a) If the surroundings experience a decrease in entropy ($\Delta S_{surr} = -20.4$ J/K), the system (ice) must experience an increase in entropy ($\Delta S_{sys} = +22.0$ J/K) that more than offsets the decrease if the process (melting) is spontaneous ($\Delta S_{univ} = +1.6$ J/K). (b) If the surroundings are at a lower temperature than the system, energy spontaneously flows into the surroundings ($\Delta S_{univ} = +0.9$ J/K) because of a decrease in entropy for the system ($\Delta S_{sys} = -22.0$ J/K) and an increase in entropy for the surroundings ($\Delta S_{surr} = +22.9$ J/K).

You may wonder why we did not consider the change in temperature of the surroundings as energy flowed from it into the melting ice. The answer lies in the sheer size of the surroundings (the universe minus a small cube of ice). The temperature of such an enormous mass is not likely to change significantly.

Now let's consider how entropy changes when the temperature of the surroundings is *lower* than the temperature of the system. Suppose a tray containing 1.00 mol of liquid water at 0°C (273 K) is placed in a freezer at −10°C (263 K). We know the water will freeze. Because the temperature of the surroundings (the freezer) is lower than the temperature of the liquid water, energy spontaneously flows from the water into its surroundings. The net entropy change for this process is:

$$\Delta S_{univ} = \Delta S_{sys} + \Delta S_{surr}$$

$$= \frac{(1.00 \text{ mol})(-6.01 \times 10^3 \text{ J/mol})}{273 \text{ K}} + \frac{(1.00 \text{ mol})(+6.01 \times 10^3 \text{ J/mol})}{263 \text{ K}}$$

$$= (-22.0 \text{ J/K}) + (+22.9 \text{ J/K})$$

$$= 0.9 \text{ J/K}$$

Once again the entropy of the universe increases (Figure 17.4b) as energy flows spontaneously from the warmer object (liquid water at 0°C) into the colder surroundings (the freezer at −10°C).

FIGURE 17.5 Water vapor exhaled by this Inuit hunter in the Northwest Territories of Canada was spontaneously deposited on his facial hair as frost—evidence of how cold it was when this photo was taken.

CONCEPT **TEST**

Is ΔS_{univ} greater than, less than, or equal to zero when water vapor exhaled by the Inuit hunter in Figure 17.5 is deposited as crystals of ice on his beard?

Entropy-change calculations based on Equation 17.2 assume process reversibility, which, as we have discussed, is an idealized, theoretical concept. In reality, the ΔS values calculated in this way are *minimum* ΔS values. When processes take place in the real world, the accompanying changes in entropy are inevitably greater than those calculated using Equation 17.2.

According to the second law, spontaneous processes *always* produce an increase in the entropy of the universe. Table 17.1 shows how the spontaneity of a process depends on the sign and magnitude of ΔS_{sys} and ΔS_{surr}. Note how processes in which ΔS_{sys} and ΔS_{surr} are both greater than zero inevitably result in an increase in ΔS_{univ}, which means they are always spontaneous. On the other hand, processes in which ΔS_{sys} and ΔS_{surr} are both less than zero always produce a decrease in ΔS_{univ}, which means they are always nonspontaneous. Between these extremes are four possible combinations of ΔS_{sys} and ΔS_{surr} in which these changes

TABLE 17.1 Spontaneity of Process as a Function of ΔS_{sys} and ΔS_{surr}

ΔS_{sys}	ΔS_{surr}	Spontaneity of Process
> 0	> 0	Always spontaneous
< 0	> 0	Spontaneous if $|\Delta S_{sys}| < |\Delta S_{surr}|$
		Nonspontaneous if $|\Delta S_{sys}| > |\Delta S_{surr}|$
> 0	< 0	Spontaneous if $|\Delta S_{sys}| > |\Delta S_{surr}|$
		Nonspontaneous if $|\Delta S_{sys}| < |\Delta S_{surr}|$
< 0	< 0	Always nonspontaneous

have opposite signs. These combinations may or may not produce positive ΔS_{univ} values, depending on the absolute values of ΔS_{sys} and ΔS_{surr}. If the larger of the two is the one with the positive value, then ΔS_{univ} increases, and the process is spontaneous.

SAMPLE EXERCISE 17.1 Predicting the Sign of Entropy Change   **LO1**

Predict whether ΔS_{sys} is greater or less than zero when each of these processes occurs at constant temperature:

a. $H_2O(\ell) \rightarrow H_2O(g)$
b. $NH_3(g) + HCl(g) \rightarrow NH_4Cl(s)$
c. $C_{12}H_{22}O_{11}(s) \rightarrow C_{12}H_{22}O_{11}(aq)$

Collect, Organize, and Analyze To predict the signs of the accompanying entropy changes, we can compare the freedom of motion of the reactant particles to the freedom of motion of the product particles:

a. One mole of liquid water molecules becomes 1 mole of water vapor molecules, which increases the molecules' freedom of motion.
b. Two moles of gaseous substances form 1 mole of a solid compound.
c. One mole of a solid dissolves, forming 1 mole of molecules dispersed in an aqueous solution.

Solve

a. $\Delta S_{sys} > 0$, because the kinetic energies of water molecules are more dispersed in the gas state than they are in the liquid state.
b. $\Delta S_{sys} < 0$, because formation of a solid causes the loss of the translational and rotational motion the gas-phase particles had.
c. $\Delta S_{sys} > 0$, because particles in a solution have translational and rotational motion that particles in crystalline solids do not have.

Think About It Entropy increases when solids melt and liquids vaporize because of the increased freedom of motion of their particles and the increased dispersion of their particles' kinetic energies. When gases combine to form a solid, they lose freedom of motion and entropy decreases. When solids dissolve in liquids, however, they gain freedom of motion and experience an increase in entropy.

Practice Exercise Predict whether these chemical reactions result in an increase or decrease in the entropy of the system. Assume the reactants and products are at the same temperature and pressure.

a. $CaCO_3(s) + 2\, HCl(aq) \rightarrow CaCl_2(aq) + CO_2(g) + H_2O(\ell)$
b. $NH_3(g) + BF_3(g) \rightarrow NH_3BF_3(s)$

(Answers to Practice Exercises are in the back of the book.)

17.3 Absolute Entropy and the Third Law of Thermodynamics

We have seen that the entropy of a system depends on temperature. Higher temperatures mean higher particle kinetic energies, which mean the particles have more vibrational, rotational, and translational motion—and more entropy. Conversely, lower temperatures mean that all these quantities are smaller. If we lower the temperature of a substance to absolute zero, in principle all motion should cease. If we assume that the substance forms a perfect crystalline solid (Figure 17.6), where

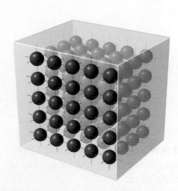

FIGURE 17.6 A perfect crystal at 0 K has zero entropy because all particles have zero freedom of motion and no kinetic energy.

TABLE 17.2 Standard States of Pure Substances and Solutions[a]

Physical State	Standard State	Pressure[b]
Solid	Pure solid, most stable allotrope of an element	1 bar
Liquid	Pure liquid	1 bar
Gas	Pure gas	1 bar
Solution	1 M	1 bar

[a]The thermodynamic data in Appendices 4–6 and used elsewhere in this book are based on a temperature of 298.15 K (25°C). *Note*: This temperature is not the STP temperature we use for gases (see Chapter 6), which is 273 K.

[b]Since 1982, 1 bar has been the standard pressure for tabulating all thermodynamic data. Before 1982, standard pressure was 1 atmosphere (atm) = 1.01325 bar.

each particle is locked in one and only one site within the crystal, then each particle has zero freedom of motion. There is no dispersion of kinetic energy because there is none to disperse. This situation leads to the conclusion that the *absolute* entropy of a perfect crystalline solid is zero at absolute zero (0 K). This conclusion is known as the **third law of thermodynamics**.

Setting a zero point on the entropy scale allows scientists to establish absolute entropy values for pure substances at any temperature. The absolute entropy of a substance is often expressed as its **standard molar entropy ($S°$)**, the entropy of 1 mole of the substance at 298 K and 1 bar (~1 atm) of pressure in its standard state (Table 17.2). For example, the standard molar entropy of solid NaCl is 72.1 J/(mol · K) at 298 K. Absolute entropies are determined from careful determinations of the molar heat capacity (or specific heat) of substances as a function of temperature. Our ability to determine absolute entropy values of substances and systems contrasts with the observation in Chapter 5 that we cannot measure absolute enthalpy (H), and the best we can do is calculate *changes* in enthalpy, ΔH.

The $S°$ values for liquid water and water vapor in Table 17.3 illustrate an important difference between the entropies of liquids and gases that we have discussed before in this chapter: the molecules in a gas under standard conditions are

CONNECTION Molar heat capacity and specific heat were defined in Section 5.5.

third law of thermodynamics the entropy of a perfect crystal is zero at absolute zero.

standard molar entropy ($S°$) the absolute entropy of 1 mole of a substance in its standard state.

TABLE 17.3 Selected Standard Molar Entropy Values[a]

Substance	$S°$, J/(mol · K)	Substance	Name	$S°$, J/(mol · K)
$Br_2(g)$	245.5	$CH_4(g)$	Methane	186.2
$Br_2(\ell)$	152.2	$CH_3CH_3(g)$	Ethane	229.5
$C_{diamond}(s)$	2.4	$CH_3OH(g)$	Methanol	239.9
$C_{graphite}(s)$	5.7	$CH_3OH(\ell)$		126.8
$CO(g)$	197.7	$CH_3CH_2OH(g)$	Ethanol	282.6
$CO_2(g)$	213.8	$CH_3CH_2OH(\ell)$		160.7
$H_2(g)$	130.6	$CH_3CH_2CH_3(g)$	Propane	269.9
$N_2(g)$	191.5	$CH_3(CH_2)_2CH_3(g)$	Butane	310.0
$O_2(g)$	205.0	$CH_3(CH_2)_2CH_3(\ell)$		231.0
$H_2O(g)$	188.8	$C_6H_6(g)$	Benzene	269.2
$H_2O(\ell)$	69.9	$C_6H_6(\ell)$		172.9
$NH_3(g)$	192.5	$C_{12}H_{22}O_{11}(s)$	Sucrose	360.2

[a]Values for additional substances are given in Appendix 4.

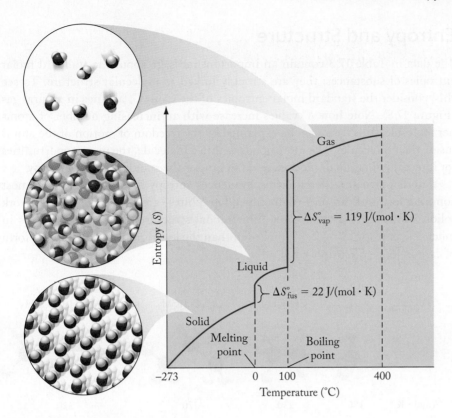

FIGURE 17.7 Abrupt increases in entropy accompany changes of state, with the greater increase occurring during the transition from liquid to gas.

much more widely dispersed than the molecules in a liquid, and the entropies of the different phases of a given substance at a given temperature follow the order $S_{solid} < S_{liquid} < S_{gas}$.

The entropy changes that occur as 1 mole of ice at 0 K is heated are shown in Figure 17.7. Note the jump in entropy as the ice melts and the even bigger jump as the liquid water vaporizes. Also note that the lines between the phase changes are curved. The change in entropy, ΔS, with temperature is not linear because, as described by Equation 17.2, heating a substance at a higher temperature produces a smaller entropy increase than adding the same quantity of heat to the same substance at a lower temperature.

Let's summarize the factors that affect entropy change:

1. Entropy increases when temperature increases.
2. Entropy increases when volume increases.[1]
3. Entropy increases when the number of independent particles increases.

In all three cases, entropy increases because each change increases the dispersion of the kinetic energy of a system's particles. We can often make qualitative predictions about entropy changes that accompany chemical reactions on the basis of these three factors, even if we have no thermodynamic data about the reactants and products. For example, when propane burns in air, 6 moles of gaseous reactants react to form 7 moles of gaseous products:

$$CH_3CH_2CH_3(g) + 5\,O_2(g) \rightarrow 3\,CO_2(g) + 4\,H_2O(g)$$

The number of moles of gaseous particles increases as the reaction proceeds, so entropy also increases ($\Delta S_{sys} > 0$).

CONNECTION In Chapter 6, Avogadro's law told us that the number of moles of a gas is directly proportional to the volume occupied by the gas at constant temperature and pressure.

[1]Water is a notable exception to this observation; the molar volume of ice is larger than for liquid water even though the absolute entropy of ice is less than the absolute entropy of liquid water.

Entropy and Structure

The data in Table 17.3 contain an important message about the standard molar entropies of substances: they are strongly linked to molecular structure. To see this, consider the standard molar entropies of the C_1 to C_4 alkanes in natural gas (Figure 17.8). Note how $S°$ values increase with an increasing number of atoms per molecule. This trend can be explained by the freedom of motion of the atoms inside their molecules. The more bonds within a molecule, the more opportunities for internal (vibrational) motion, and the greater the standard molar entropy.

Another structural feature that influences entropy is rigidity. The two most common forms of carbon—diamond and graphite—are both polymeric network solids. However, the rigid three-dimensional structure of diamonds results in much less entropy [$S° = 2.4$ J/(mol · K)] than the less rigid, layered graphite form of carbon [$S° = 5.7$ J/(mol · K)].

FIGURE 17.8 Among methane, ethane, propane, and butane, standard molar entropies increase as the numbers of atoms and chemical bonds in the molecules increase.

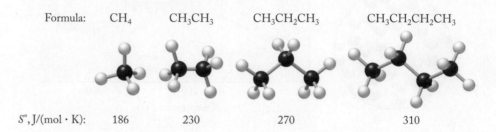

Formula:	CH_4	CH_3CH_3	$CH_3CH_2CH_3$	$CH_3CH_2CH_2CH_3$
$S°$, J/(mol · K):	186	230	270	310

SAMPLE EXERCISE 17.2 Comparing Absolute Entropy Values   **LO2**

Select the component in each of the following pairs of compounds that has the greater absolute entropy per mole at a pressure of 1 bar and 298 K.
 a. HCl(*g*), HCl(*aq*)
 b. $CH_3OH(\ell)$, $CH_3CH_2OH(\ell)$
 c.

Obsidian
(amorphous SiO_2)

Quartz
(crystalline SiO_2)

Collect, Organize, and Analyze Particles in the vapor state have more freedom of motion and entropy than they do in the liquid state, and particles in the liquid state have more freedom of motion and entropy than they do in the solid state. Substances composed of more particles, contributing to more freedom of motion, have more absolute entropy than substances made of fewer particles with less freedom of motion.

Solve
 a. HCl gas loses freedom of motion when it dissolves in water, so HCl gas has a greater $S°$ value at 298 K.

b. Both compounds are liquids with similar formulas, but 1 mole of ethanol (CH_3CH_2OH) has more atoms and more covalent bonds between atoms than 1 mole of methanol (CH_3OH). Therefore ethanol has a greater $S°$ value at 298 K.

c. Both forms of SiO_2 are solids with extended covalent networks of atoms. However, quartz has a crystalline structure and obsidian has an irregular arrangement of bonds and atoms. The randomness of the obsidian structure gives it a greater $S°$ value at 298 K.

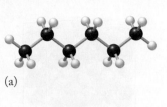

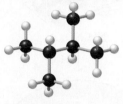
(a)

Think About It The comparison in part (a) is complicated by the fact that HCl(aq) is a strong acid, which means that each molecule produces two ions in solution. However, they are aqueous ions and have less combined entropy than half as many gas-phase HCl molecules. (Compare the $S°$ values of HCl(g), $H^+(aq)$, and $Cl^-(aq)$ in Appendix 4 to see for yourself.)

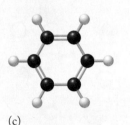
(b)

⊛ **Practice Exercise** Ball-and-stick models of four hydrocarbons that each contain six carbon atoms per molecule are shown in Figure 17.9. Rank these compounds in order of decreasing $S°$ values.

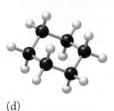
(c)

17.4 Calculating Entropy Changes

The entropy of a system (like its enthalpy and internal energy) is a state function, which means that the change in entropy that accompanies a process depends only on the initial and final states of the system, not on the pathway of the process. Therefore the change in entropy experienced by a system is simply the difference between its initial and final absolute entropy levels:

(d)

FIGURE 17.9 Four six-carbon hydrocarbons.

$$\Delta S_{sys} = S_{final} - S_{initial} \quad (17.4)$$

We can adapt Equation 17.4 to calculate the change in entropy that accompanies a chemical reaction under standard conditions, $\Delta S°_{rxn}$, from the difference in the standard molar entropies of moles of reactants, $n_{reactants}$ (the equivalent of $S_{initial}$ in Equation 17.4), and the molar entropies of moles of products, $n_{products}$, (that is, S_{final}):

CONNECTION State functions were defined in Chapter 5.

$$\Delta S°_{rxn} = \sum n_{products}S°_{products} - \sum n_{reactants}S°_{reactants} \quad (17.5)$$

Each $S°$ value for a product or reactant is multiplied by the appropriate number of moles from the balanced chemical equation. In other words, just as we saw for $\Delta H°$ in Chapter 5, entropy is an extensive thermodynamic property that depends on the quantities of substances consumed or produced in a reaction. Standard molar entropies of selected substances are listed in Appendix 4.

SAMPLE EXERCISE 17.3 Calculating Entropy Changes **LO3**

What is $\Delta S°$ when ammonium nitrate dissolves (Figure 17.10, $\Delta H° > 0$) under standard conditions, given the following standard molar entropy values:

$$NH_4NO_3(s) \rightarrow NH_4^+(aq) + NO_3^-(aq)$$

	$NH_4NO_3(s)$	$NH_4^+(aq)$	$NO_3^-(aq)$
$S°$ [J/(mol·K)]	151.1	113.4	146.4

Collect, Organize, and Analyze We are given the standard molar entropy values of a solid ionic compound and its ions in aqueous solution, and we want to calculate

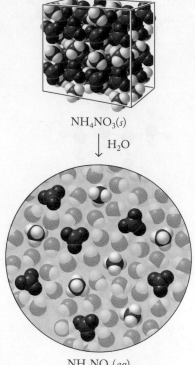

$NH_4NO_3(s)$

$\downarrow H_2O$

$NH_4NO_3(aq)$

FIGURE 17.10 Dissolution of solid ammonium nitrate into NH_4^+ and NO_3^- ions is endothermic but spontaneous under standard conditions because ΔS° is greater than zero, which contributes to $\Delta S_{univ} > 0$.

CONNECTION In Chapter 5 we defined a change in enthalpy (ΔH) as the heat gained or lost in a reaction carried out at constant pressure.

the change in entropy that occurs when the compound dissolves under standard conditions. Entropy changes associated with chemical reactions depend on the entropies of the reactants and products, as described in Equation 17.5. Dissolving an ionic solid in water increases the freedom of motion of the solute ions, so ΔS° should be greater than zero.

Solve

$$\Delta S^\circ = \sum n_{products} S^\circ_{products} - \sum n_{reactants} S^\circ_{reactants}$$

$$= \left[1\text{ mol} \times \left(\frac{113.4\text{ J}}{\text{mol} \cdot \text{K}} \right) + 1\text{ mol} \times \left(\frac{146.4\text{ J}}{\text{mol} \cdot \text{K}} \right) \right] - 1\text{ mol} \times \left(\frac{151.1\text{ J}}{\text{mol} \cdot \text{K}} \right)$$

$$= 108.7\text{ J/K}$$

Think About It As predicted, ΔS° is greater than zero because the freedom of motion and the dispersion of the kinetic energy of solute particles increase when the solid solute dissolves. The dissolution of ammonium nitrate is an example of an endothermic reaction that results in an increase in entropy.

Practice Exercise Calculate the standard molar entropy change for the combustion of methane gas by using S° values from Appendix 4. Before carrying out the calculation, predict whether the entropy of the system increases or decreases. Assume that liquid water is one of the products.

17.5   Free Energy

In Sample Exercise 17.3 we used standard molar entropy values to calculate ΔS°. The solute, solvent, and resulting solution together constituted a closed thermodynamic system, which meant that energy could flow into it as its temperature dropped, but matter was not exchanged with its surroundings. Therefore the ΔS° value that we calculated applied only to the system. There was no evaluation of the change in the entropy of the system's surroundings accompanying the dissolution process, though we might predict that energy flowing from the surroundings into the system would produce a decrease in S_{surr}. (Had the dissolution process been exothermic, energy would have flowed from the system into its surroundings, and ΔS_{surr} would have been greater than zero.)

Thus the entropy change experienced by the surroundings of any chemical thermodynamic system depends on whether the process occurring in the system is exothermic or endothermic. When heat flows from an exothermic process occurring at constant pressure into a system's surroundings, the quantity of heat is equal in magnitude but opposite in sign to the enthalpy change of the system:

$$q_{surr} = -\Delta H_{sys} \tag{17.6}$$

When the system hosts an endothermic process, as in Sample Exercise 17.3, the direction of energy flow is reversed, but Equation 17.6 still applies. Assuming these transfers of energy occur reversibly, we can calculate the value of ΔS_{surr} by using this modification of Equation 17.2:

$$\Delta S_{surr} = \frac{q_{surr}}{T} \tag{17.7}$$

Now let's combine Equations 17.6 and 17.7:

$$\Delta S_{surr} = -\frac{\Delta H_{sys}}{T}$$

We can substitute this expression into Equation 17.1 ($\Delta S_{univ} = \Delta S_{sys} + \Delta S_{surr}$):

$$\Delta S_{univ} = \Delta S_{sys} - \frac{\Delta H_{sys}}{T} \qquad (17.8)$$

The beauty of Equation 17.8 is that it allows us to predict whether a process is spontaneous at a particular temperature once we calculate the enthalpy and entropy changes accompanying the process. The downside of Equation 17.8 is that spontaneity relies on the value of a parameter (ΔS_{univ}) that is impossible to determine directly and that has little physical meaning. It would be great if we could substitute a thermodynamic parameter for ΔS_{univ} that is based only on the system and not the entire universe. Such a parameter exists and is called the change in the system's *free energy*.

In chemistry we focus on a particular kind of free energy called **Gibbs free energy (G)** in honor of American scientist J. Willard Gibbs (1839–1903). Gibbs free energy is the energy released by processes happening at constant temperature and pressure that is available to do useful work.

Like many of the thermodynamic properties we have examined, absolute free energy values of substances are often of less interest than the *changes* in free energy that accompany chemical reactions and other processes. Gibbs proposed that the change in free energy (ΔG_{sys}) of a process occurring at constant temperature and pressure is linked directly to that temperature and ΔS_{univ}:

$$\Delta G_{sys} = -T\Delta S_{univ}$$

Because of the $-T$ multiplier, *negative* values of ΔG_{sys} correspond to *positive* values of ΔS_{univ}. Therefore

- If $\Delta G_{sys} < 0$, then $\Delta S_{univ} > 0$, and the reaction is spontaneous.
- If $\Delta G_{sys} > 0$, then $\Delta S_{univ} < 0$, and the reaction is nonspontaneous. Instead, the reaction running in reverse is spontaneous.
- If $\Delta G_{sys} = 0$, then $\Delta S_{univ} = 0$, and the composition of the reaction mixture does not change. In other words, the reaction has reached chemical equilibrium.

We can combine Gibbs' equation with Equation 17.8 by multiplying all of the terms in Equation 17.8 by $-T$:

$$-T\Delta S_{univ} = -T\Delta S_{sys} + \Delta H_{sys} \qquad (17.9)$$

The left side of Equation 17.9 is equal to ΔG_{sys}. Making that substitution and rearranging the terms on the right side gives us

$$\Delta G_{sys} = \Delta H_{sys} - T\Delta S_{sys}$$

Because all of the parameters in this equation apply to the system, we typically simplify the equation by eliminating them:

$$\Delta G = \Delta H - T\Delta S \qquad (17.10)$$

Gibbs free energy (G) the maximum energy released by a process occurring at constant temperature and pressure that is available to do useful work.

 **CHEMTOUR**
Gibbs Free Energy

CONCEPT TEST

(a) Given the thermodynamic data in Table A4.3 in the Appendix, is the conversion of diamond to graphite spontaneous? Explain your answer. (b) If yes, does knowing that the conversion is spontaneous tell you how rapid the conversion is?

Equation 17.10 highlights the two thermodynamic driving forces that contribute to a decrease in free energy and to making a process spontaneous:

1. The system experiences an increase in entropy ($\Delta S > 0$).
2. The process is exothermic ($\Delta H < 0$).

One or both of these conditions must be true for a reaction to be spontaneous.

Using Equation 17.10, we can calculate the change in Gibbs free energy of a process if we first calculate the values of ΔH and ΔS. For a chemical reaction occurring under standard conditions, we can calculate the *standard* change in Gibbs free energy ΔG_{rxn}° by using the following modified version of Equation 17.10:

$$\Delta G_{rxn}^{\circ} = \Delta H_{rxn}^{\circ} - T\Delta S_{rxn}^{\circ} \tag{17.11}$$

In Chapter 5 we calculated ΔH_{rxn}° values from the differences in standard enthalpies of formation ΔH_{f}° of products and reactants:

$$\Delta H_{rxn}^{\circ} = \sum n_{products}\Delta H_{f,products}^{\circ} - \sum n_{reactants}\Delta H_{f,reactants}^{\circ} \tag{5.19}$$

The value of ΔS_{rxn}° can be calculated using Equation 17.5:

$$\Delta S_{rxn}^{\circ} = \sum n_{products}S_{products}^{\circ} - \sum n_{reactants}S_{reactants}^{\circ} \tag{17.5}$$

We can combine the results of these two calculations in Equation 17.11 to calculate ΔG_{rxn}°, as illustrated in Sample Exercise 17.4.

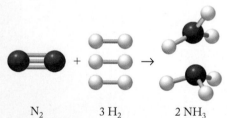

N₂   3 H₂   2 NH₃

FIGURE 17.11 In the synthesis of ammonia, one molecule of nitrogen reacts with three molecules of hydrogen to yield two molecules of ammonia. All substances are gases.

SAMPLE EXERCISE 17.4 Predicting Reaction Spontaneity   **LO4**
under Standard Conditions

Consider the reaction of nitrogen gas and hydrogen gas (Figure 17.11) at 298 K to make ammonia at the same temperature:

$$N_2(g) + 3\,H_2(g) \rightarrow 2\,NH_3(g)$$

a. Before doing any calculations, predict the sign of ΔS_{rxn}°.
b. What is the actual value of ΔS_{rxn}°?
c. What is the value of ΔH_{rxn}°?
d. What is the value of ΔG_{rxn}° at 298 K?
e. Is the reaction spontaneous at 298 K and 1 bar of pressure?

Collect and Organize For a given reaction, we want to predict the sign of the standard entropy change and then calculate the changes in entropy, enthalpy, and Gibbs free energy. We can then determine the spontaneity of the reaction under standard conditions. Standard molar entropies and standard heats of formation of the reactants and product are found in Table 17.3 and Appendix 4, respectively. Figure 17.11 reinforces the point that there are more molecules of gaseous reactants than products in the reaction.

Analyze We can use Equation 17.5 to calculate entropy changes under standard conditions, Equation 5.19 to calculate ΔH_{rxn}° from standard enthalpy of formation values (in Appendix 4), and then use these values in Equation 17.11 to calculate the free energy change in the reaction. The sign of ΔG_{rxn}° will indicate whether the reaction is spontaneous under standard conditions ($T = 298$ K).

Solve
a. The number of gas-phase molecules decreases as the reaction proceeds, so ΔS_{rxn}° is probably less than zero.

b. We use data from Table 17.3 in Equation 17.5 to calculate ΔS°_{rxn}:

$$\Delta S^{\circ}_{rxn} = \sum n_{products}S^{\circ}_{products} - \sum n_{reactants}S^{\circ}_{reactants} = \Delta S^{\circ}_{sys}$$

$$= \left\{\left[2 \text{ mol} \times \left(\frac{192.5 \text{ J}}{\text{mol} \cdot \text{K}}\right)\right] - \left[1 \text{ mol} \times \left(\frac{191.5 \text{ J}}{\text{mol} \cdot \text{K}}\right) + 3 \text{ mol} \times \left(\frac{130.6 \text{ J}}{\text{mol} \cdot \text{K}}\right)\right]\right\}$$

$$= -198.3 \text{ J/K}$$

The entropy change is negative, as predicted in part (a).

c. The change in enthalpy that accompanies the reaction under standard conditions is

$$\Delta H^{\circ}_{rxn} = \sum n_{products}\Delta H^{\circ}_{f,products} - \sum n_{reactants}\Delta H^{\circ}_{f,reactants}$$

$$= \left\{\left[2 \text{ mol} \times \left(\frac{-46.1 \text{ kJ}}{\text{mol}}\right)\right] - \left[1 \text{ mol} \times \left(\frac{0.0 \text{ kJ}}{\text{mol}}\right) + 3 \text{ mol} \times \left(\frac{0.0 \text{ kJ}}{\text{mol}}\right)\right]\right\}$$

$$= -92.2 \text{ kJ}$$

d. We insert the values of ΔS°_{rxn} and ΔH°_{rxn} calculated in parts (a) and (b) into Equation 17.11:

$$\Delta G^{\circ}_{rxn} = \Delta H^{\circ}_{rxn} - T\Delta S^{\circ}_{rxn}$$

$$= -92.2 \text{ kJ} - \left[(298 \text{ K}) \times \left(-198.3 \frac{\text{J}}{\text{K}} \times \frac{1 \text{ kJ}}{1000 \text{ J}}\right)\right]$$

$$= -33.1 \text{ kJ}$$

e. The decrease in Gibbs free energy tells us that the reaction is spontaneous under standard conditions.

Think About It In this reaction, a decrease in entropy ($\Delta S^{\circ}_{rxn} < 0$) is more than offset by a favorable enthalpy change ($\Delta H^{\circ}_{rxn} < 0$) so that overall, $\Delta G^{\circ}_{rxn} < 0$. Therefore, the reaction is spontaneous at 298 K and $P = 1$ bar.

 Practice Exercise
For the reaction $2 H_2(g) + O_2(g) \rightarrow 2 H_2O(\ell)$,

a. Predict the sign of the entropy change for the reaction.
b. What is the value of ΔS°_{rxn}?
c. What is the value of ΔH°_{rxn}?
d. Is the reaction spontaneous at 298 K and 1 bar pressure?

CONCEPT TEST

The preparation of ammonia from nitrogen and hydrogen In Sample Exercise 17.4 is determined to be spontaneous under standard conditions, yet if we mix the two gases at 298 K, no reaction is observed. Suggest a reason why.

Another way to calculate the change in Gibbs free energy of a reaction under standard conditions is based on another thermodynamic property of substances listed in Appendix 4: **standard free energy of formation (ΔG°_f)**. A compound's ΔG°_f value is the change in free energy associated with the formation of 1 mole of it in its standard state from its elements in their standard states.

In Chapter 5 we calculated standard enthalpies of reaction (ΔH°_{rxn}) from the difference in the standard enthalpies of formation (ΔH°_f) of their products and reactants. We can also calculate the change in standard free energy of a reaction under standard conditions from the difference in the standard free energies of formation of its products and reactants. As with standard

standard free energy of formation (ΔG°_f) the change in free energy associated with the formation of 1 mole of a compound in its standard state from its component elements.

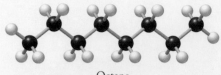

Octane
$\Delta G_f^\circ = 16.3$ kJ/mol

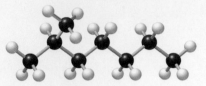

2-Methylheptane
$\Delta G_f^\circ = 11.7$ kJ/mol

3,3-Dimethylhexane
$\Delta G_f^\circ = 12.6$ kJ/mol

FIGURE 17.12 Molecular structures and ΔG_f° values of three C_8H_{18} isomers.

enthalpies of formation, standard free energies of formation of the most stable forms of elements in their standard states are zero. The similarities between the two calculations can be seen from the similar formats of the equations used to calculate ΔG_{rxn}°,

$$\Delta G_{rxn}^\circ = \sum n_{products}\Delta G_{f,products}^\circ - \sum n_{reactants}\Delta G_{f,reactants}^\circ \quad (17.12)$$

and ΔH_{rxn}°,

$$\Delta H_{rxn}^\circ = \sum n_{products}\Delta H_{f,products}^\circ - \sum n_{reactants}\Delta H_{f,reactants}^\circ \quad (5.19)$$

Sample Exercise 17.5 illustrates just how similar the two calculations are.

CONCEPT TEST

Molecular models and standard free energies of formation for three structural isomers with the molecular formula C_8H_{18} are shown in Figure 17.12. All three isomers burn in air, as described by the same chemical equation:

$$2\ C_8H_{18}(\ell) + 25\ O_2(g) \rightarrow 16\ CO_2(g) + 18\ H_2O(g)$$

Are the ΔG_{rxn}° values for the three combustion reactions also the same? Why or why not?

SAMPLE EXERCISE 17.5 Calculating ΔG_{rxn}° by Using   **LO4**
Appropriate ΔG_f° Values

Use the appropriate standard free energy of formation values from Appendix 4 to calculate the change in Gibbs free energy as ethanol burns under standard conditions. Assume the reaction proceeds as described by the following chemical equation:

$$CH_3CH_2OH(\ell) + 3\ O_2(g) \rightarrow 2\ CO_2(g) + 3\ H_2O(\ell)$$

Collect and Organize We can calculate the value of ΔG_{rxn}° for the combustion of ethanol according to Equation 17.12, using the ΔG_f° values of the reactants and products in the combustion reaction:

Substance	$CH_3CH_2OH(\ell)$	$O_2(g)$	$CO_2(g)$	$H_2O(\ell)$
ΔG_f° (kJ/mol)	−174.9	0	−394.4	−237.2

Analyze The reaction consumes 1 mole of liquid ethanol and 3 moles of oxygen, and it produces 2 moles of CO_2 gas and 3 moles of liquid H_2O. Substituting the above ΔG_f° values and the appropriate numbers of moles into Equation 17.12 yields the value of ΔG_{rxn}°. The combustion of ethanol, a common additive in gasoline in the United States and Canada, is spontaneous, so ΔG_{rxn}° should be less than zero.

Solve We insert the appropriate numbers of moles and ΔG_f° values into Equation 17.12:

$$\Delta G_{rxn}^\circ = \sum n_{products}\Delta G_{f,products}^\circ - \sum n_{reactants}\Delta G_{f,reactants}^\circ$$

$$= [2\ \text{mol } CO_2 \times (-394.4\ \text{kJ/mol}) + 3\ \text{mol } H_2O \times (-237.2\ \text{kJ/mol})]$$

$$-1\ \text{mol } CH_3CH_2OH \times (-174.9\ \text{kJ/mol}) + 3\ \text{mol } O_2 \times (0\ \text{kJ/mol})$$

$$= -1325.5\ \text{kJ}$$

Think About It The calculated value represents that part of the total energy released by the combustion of 1 mole of ethanol under standard conditions that is available to do useful work.

Practice Exercise Use the appropriate standard free energy of formation values in Appendix 4 to calculate the value of ΔG°_{rxn} for the steam-reforming reaction used to produce H_2 gas:

$$CH_4(g) + H_2O(g) \rightarrow CO(g) + 3\,H_2(g)$$

What exactly does "energy available to do useful work" mean? Let's attempt to answer this question, using as our model the internal combustion (gasoline) engines used to power most automobiles.

A combustion reaction is a thermodynamic system that experiences a decrease in internal energy (ΔE) as energy in the form of heat, q, flows from it into its surroundings and as it does work, w, on its surroundings. These three variables are related by Equation 5.5:

$$\Delta E = q + w \qquad (5.5)$$

From the perspective of the system, all three quantities are less than zero. Internal combustion engines have cooling systems to manage the dissipation of heat, which is wasted energy that does nothing to power the car. The energy that moves the car is derived from the rapid expansion of the gaseous products of combustion in the cylinders of the engine. As Figure 17.13 shows, this expansion pushes down on the piston of a cylinder, increasing the volume of the reaction mixture. The pressure exerted by the reacting gases and their products on the pistons in a car's engine, multiplied by the volumes they displace, constitutes the P–V work (which we discussed in Section 5.3), which propels the car:

$$w = -P\Delta V$$

Gibbs free energy is a measure of the *maximum* amount of work that can be done by the energy released during combustion. To see how this theoretical quantity of work compares with the total energy released, let's rearrange Equation 17.10 by isolating the ΔH term:

$$\Delta H = \Delta G + T\Delta S \qquad (17.13)$$

In this form, the equation tells us that the enthalpy change that accompanies making and breaking chemical bonds during a chemical reaction may be divided into two parts. One part, ΔG, is the energy that can theoretically be converted into motion and other useful work (such as propelling a car or generating electricity for its electrical system). The other part, $T\Delta S$, is not usable: it is the portion of energy that spreads out when, for example, hot gases flow out of an automobile exhaust pipe. This part of ΔH is wasted.

Consequently, conversion of chemical energy into useful mechanical energy (ΔG) is never 100% efficient. A portion of ΔG is also wasted because combustion and the energy conversion happen quickly and therefore irreversibly. Maximum efficiency comes with very slow and reversible reactions (see Section 17.2), but that is not how automobile engines operate. It turns out that gasoline engines convert only about 30% of the energy produced during combustion into useful work.

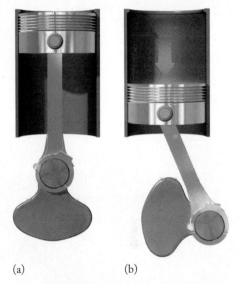

(a)   (b)

FIGURE 17.13 (a) In a car engine, thermal expansion causes the gases in a cylinder to (b) push down on a piston with a pressure, P, represented by the blue arrow. The product of P and the change in volume of the gases, ΔV, is the work done by the expanding gases that propels the car.

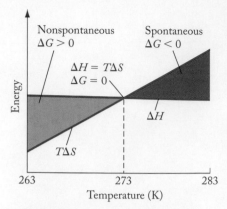

FIGURE 17.14 Changes in the values of ΔH, the quantity $T\Delta S$, and ΔG for ice melting as temperature increases from $-10°C$ (263 K) to $+10°C$ (283 K).

17.6 Temperature and Spontaneity

Let's revisit the process of ice melting, this time focusing on how the values of the three terms ΔH, $T\Delta S$, and ΔG in Equation 17.10 change as the temperature of a mixture of ice and water increases from $-10°C$ to $+10°C$ (Figure 17.14). As temperature rises over this range, there is little impact on the enthalpy of fusion, ΔH, as shown by the nearly flat green line in Figure 17.14. However, it is only reasonable that increasing the value of T increases the value $T\Delta S$ (as shown by the upward slope of the purple line). After all, the ΔS of melting ice (or any melting solid)[2] has a positive value, so $T\Delta S$ must increase as T increases.

The green and purple lines intersect at $0°C$, which means ΔH is equal to $T\Delta S$ at that temperature. Put another way, the difference between ΔH and $T\Delta S$ at $0°C$ is zero, and so is ΔG (remember: $\Delta G = \Delta H - T\Delta S$). The temperature at the point in Figure 17.14 where $\Delta G = 0$ defines the melting (or freezing) point, which by definition is the temperature at which the solid melts at the same rate as the liquid freezes. The two phases are in *equilibrium*. At $0°C$, no net change takes place, and both phases coexist.

We also know that ice melts spontaneously above $0°C$ (273 K), which means that ΔG for the melting process must be less than zero. The graph in Figure 17.14 shows that indeed it is. Above $0°C$, the value of $T\Delta S$ is greater than ΔH. Therefore the difference between them ($\Delta H - T\Delta S$) is less than zero and becomes more negative with increasing temperature, as illustrated by the increasing distance from the purple line to the green line in that region of the graph.

Below $0°C$, ice does not melt spontaneously, which means ΔG is greater than zero. The graph shows why this is true: at $T < 273$ K ($0°C$), $T\Delta S$ values (purple line) are less than ΔH values (green line), which means that $\Delta H - T\Delta S$ (and ΔG) is greater than zero. The positive ΔG values mean the process is nonspontaneous below 273 K, and ice does not melt below its freezing point. However, the opposite process—liquid water freezing—*is* spontaneous because reversing a process keeps the absolute values but switches the signs of ΔH, ΔS, and ΔG. Therefore if the melting process is nonspontaneous, then the exothermic freezing process *is* spontaneous at low temperatures.

CONCEPT TEST

A given process is spontaneous at lower temperature and nonspontaneous at higher temperature. What does a graph like the one in Figure 17.14 look like for such a process?

CONNECTION In Chapter 10 we discussed the equilibrium between phases of a substance and used phase diagrams to illustrate which physical states are stable at various combinations of temperature and pressure.

Equilibrium also exists for water at $100°C$ and 1 atm pressure, which are the temperature and pressure at which $\Delta G_{vaporization}$ and $\Delta G_{condensation}$ both equal zero. At $100°C$, liquid water vaporizes and water vapor condenses at the same rate; the two phases coexist.

The temperature at which $\Delta G = 0$ for a process can be calculated from Equation 17.10 if we know the values of ΔH_{fus} and ΔS_{fus}. For example, the values for the fusion of water are

$$H_2O(s) \rightarrow H_2O(\ell) \qquad \Delta H°_{fus} = 6.01 \times 10^3 \text{ J/mol}; \Delta S°_{fus} = 22.0 \text{ J/(mol} \cdot \text{K)}$$

In doing this calculation, we assume that the values of $\Delta H°$ and $\Delta S°$ do not change significantly with small changes in temperature. For this process, as well as for most other physical and chemical processes, this assumption is acceptable.

---

[2]Helium-3, ^{3}He, is an exception to this observation: $\Delta H_{fus,^3He} < 0$.

TABLE 17.4 Effects of $\Delta H°$ and $\Delta S°$ on $\Delta G°$ and Spontaneity

$\Delta H°$	$\Delta S°$	$\Delta G°$	
−	+	Always < 0	Always spontaneous
−	−	< 0 at lower temperature	Spontaneous at lower temperature
+	+	< 0 at higher temperature	Spontaneous at higher temperature
+	−	Always > 0	Never spontaneous

Therefore we can assume that $\Delta G = \Delta H - T\Delta S \approx \Delta H° - T\Delta S°$. Inserting the values of $\Delta H°$ and $\Delta S°$ and using $\Delta G = 0$ for a process at equilibrium, we get

$$\Delta G = (6.01 \times 10^3 \text{ J/mol}) - T[22.0 \text{ J/(mol} \cdot \text{K)}] = 0$$

$$T = \frac{6.01 \times 10^3 \text{ J/mol}}{22.0 \text{ J/(mol} \cdot \text{K)}} = 273 \text{ K} = 0°\text{C}$$

This is the familiar value for the melting point of ice.

Like physical processes, chemical reactions under standard conditions may be spontaneous below or above a characteristic temperature (T) at which $\Delta G° = 0$, and

$$T = \frac{\Delta H°}{\Delta S°}$$

Such reactions are nonspontaneous at all other temperatures. As summarized in Table 17.4, an exothermic reaction ($\Delta H° < 0$) for which $\Delta S°$ is also less than zero is spontaneous only below a characteristic temperature. An endothermic reaction ($\Delta H° > 0$) for which $\Delta S°$ is also greater than zero is spontaneous only above a characteristic temperature. This pattern makes sense because lower temperatures make the value of the $T\Delta S°$ term in Equation 17.11 less negative if $\Delta S° < 0$. This helps make $\Delta G°$ more negative because of the minus sign in front of the $T\Delta S°$ term. Higher temperatures make the value of $T\Delta S°$ more positive if $\Delta S° > 0$, which again helps make $\Delta G°$ more negative.

SAMPLE EXERCISE 17.6 Relating Reaction Spontaneity to ΔH and ΔS **LO5**

A certain chemical reaction is spontaneous at low temperatures but not at high temperatures. Use Equation 17.10 ($\Delta G = \Delta H - T\Delta S$) to determine the signs of the enthalpy and entropy changes for this reaction.

Collect, Organize, and Analyze We are asked to determine the signs of ΔH (enthalpy change) and ΔS (entropy change) on the basis of the change in spontaneity of a reaction as temperature changes. This means we need to think about not only the signs of ΔH and ΔS but also how the relative magnitude of ΔH and $T\Delta S$ varies with temperature.

Solve The importance of ΔS increases with increasing temperature because the product $T\Delta S$ appears in Equation 17.10. The given reaction is known to be spontaneous ($\Delta G < 0$) at low temperatures, meaning that the magnitude of the enthalpy term is more likely to be larger than the magnitude of $T\Delta S$, in which case ΔH must be negative for the reaction to be spontaneous. The given reaction is known to be nonspontaneous ($\Delta G > 0$) at higher temperatures, where the magnitude of $T\Delta S$ is more likely to be larger than the magnitude of ΔH, in which case ΔS must also be negative in order for $\Delta G > 0$.

Think About It In this reaction, the impact of a negative ΔS value is more than offset by a decrease in enthalpy, a change that favors the reaction. Table 17.4 confirms the observation that a process that is spontaneous only at low temperatures is one in which there is a decrease in both entropy and enthalpy. Recall that the freezing of water below 0°C is an example of an exothermic reaction ($\Delta H < 0$) that is accompanied by a decrease in entropy ($\Delta S < 0$).

 Practice Exercise At high temperatures, ammonia decomposes to nitrogen and hydrogen gases:

$$2\,NH_3(g) \rightarrow N_2(g) + 3\,H_2(g)$$

ΔH for the reaction is positive and ΔS is positive. Predict whether the reaction is spontaneous at all temperatures, or only at high temperatures.

17.7 Free Energy and Chemical Equilibrium

We have just seen that equilibrium exists when the free energy change that accompanies a process is zero. For chemical reactions not at equilibrium, the magnitude of ΔG—how far it is from zero in either a negative or positive direction—indicates how far a system is from its equilibrium position.

In Chapter 14 we discussed how to use the concept of the reaction quotient (Q) to determine whether a reaction is at chemical equilibrium. When the value of Q is much larger or smaller than the value of the reaction's equilibrium constant, K, we know that the reaction is far from chemical equilibrium. For example, we saw in Chapter 14 that when Q is less than K, the ratio of product concentrations to reactant concentrations (raised to the appropriate powers) is less than it would be at equilibrium. To reach equilibrium, the reaction proceeds (spontaneously!) in the forward direction until the ratio of product to reactant concentrations matches the value of K. In other words, when Q is less than K, ΔG_{rxn} must be less than zero.

On the other hand, when Q is greater than K, the ratio of product to reactant concentrations is greater than it would be at equilibrium. To reach equilibrium, the reaction must proceed in reverse, so that products are consumed and reactants form, until the ratio of product to reactant concentrations matches the value of K. Therefore when Q is greater than K, the forward reaction is not spontaneous because the *reverse* reaction is spontaneous, which means ΔG_{rxn} for the forward reaction must be greater than zero.

To summarize these points for any reaction:

- When $Q = K$, $\Delta G_{rxn} = 0$ and the reaction is at equilibrium.
- When $Q < K$, $\Delta G_{rxn} < 0$ and the reaction is spontaneous.
- When $Q > K$, $\Delta G_{rxn} > 0$ and the reaction is nonspontaneous.

The key concept here is that the value of ΔG_{rxn} for any chemical reaction depends not only on the standard free energy of formation values of its reactants and products—that is, on the value of ΔG°_{rxn}—but also on the concentrations of the reactants and products in the reaction mixture. In other words, the value of ΔG_{rxn} depends on the value of Q. This dependency is expressed in Equation 17.14:

$$\Delta G_{rxn} = \Delta G^{\circ}_{rxn} + RT \ln Q \qquad (17.14)$$

To illustrate the dependency of ΔG_{rxn} on Q, let's use as our chemical model the decomposition of N_2O_4:

$$N_2O_4(g) \rightleftharpoons 2\,NO_2(g)$$

CONNECTION In Chapter 14 we defined the reaction quotient (Q) as the ratio of the concentrations (or partial pressures) of products to reactants in a chemical reaction, each term raised to a power equal to the coefficient of that substance in the balanced chemical equation describing the reaction.

 **CHEMTOUR** Equilibrium and Thermodynamics

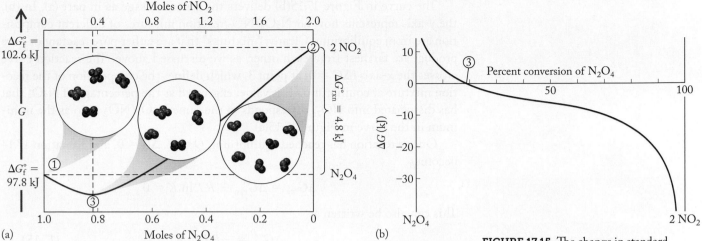

FIGURE 17.15 The change in standard free energy ΔG°_{rxn} is a constant for a given reaction. (a) Point 3, the point of minimum free energy, defines the composition of a system at equilibrium. At point 1, the system consists of reactants only; at point 2, the system consists of products only. (b) ΔG is the "distance" from equilibrium in terms of free energy. As a spontaneous reaction proceeds, the composition of the system changes and ΔG approaches 0 (point 3), at which point no further change in composition occurs.

First we use Equation 17.12 and the ΔG°_f values for the reactant and product from Appendix 4 to calculate the value of ΔG°_{rxn}:

$$\Delta G^\circ_{rxn} = \sum n_{products}\Delta G^\circ_{f,products} - \sum n_{reactants}\Delta G^\circ_{f,reactants}$$

$$= 2 \text{ mol } (51.3 \text{ kJ/mol}) - 1 \text{ mol } (97.8 \text{ kJ/mol}) = +4.8 \text{ kJ}$$

The positive value of ΔG°_{rxn} indicates that the reaction as written is not spontaneous at $T = 298$ K and under standard conditions, which means $P_{NO_2} = P_{N_2O_4} = 1$ bar. Consider a reaction vessel at 298 K that initially contains 1 mole of N_2O_4 at a partial pressure of 1 bar and no (or hardly any) NO_2. Under these conditions, the value of the reaction quotient expressed with partial pressures (Q_p) is zero:

$$Q_p = \frac{(P_{NO_2})^2}{P_{N_2O_4}} = \frac{0}{1} = 0$$

Although we do not know the value of K_p, it has to be greater than zero. Therefore $Q_p < K_p$ and the system should respond by spontaneously forming NO_2 from N_2O_4.

If the reaction in the forward direction is spontaneous, then the change in free energy of the reaction (ΔG) must be less than zero, even though the value of ΔG° is positive (4.8 kJ/mol). The value of ΔG *is* negative because the value of $RT \ln Q$ in Equation 17.14 has an increasingly larger negative value as Q approaches zero. Actually, any reversible reaction, regardless of its ΔG° value, has a negative ΔG value and is spontaneous when the system has only reactant and no (or practically no) product.

The opposite situation occurs if we have 2 moles of NO_2 in the reaction vessel and essentially no N_2O_4. Under these conditions, the denominator of the reaction quotient is nearly zero, making the value of Q_p enormous. Similarly, the $RT \ln Q_p$ term in Equation 17.14 now has a large positive value, guaranteeing that $\Delta G > 0$. This means that the forward reaction is not spontaneous, but the reverse reaction is. We also predict that NO_2 in the reaction vessel should combine to form N_2O_4, on the basis of the fact that the enormous Q_p must be greater than K_p. Therefore the reaction should run in reverse until $Q_p = K_p$.

Figure 17.15(a) shows how free energy changes as the quantities of N_2O_4 and NO_2 in the reaction mixture change. The minimum of the curve (point 3) corresponds to a free energy that is lower than that of either pure N_2O_4 (point 1) or pure NO_2 (point 2). The vertical dashed line, marking the minimum in free energy, crosses the top and bottom axes at values that tell us the reaction mixture at equilibrium contains about 0.83 mol N_2O_4 and about 0.34 mol NO_2.

The curve in Figure 17.15(b) delivers the same message as in part (a). In (b), the y-axis represents how far N_2O_4/NO_2 reaction mixtures of different composition are from equilibrium. Clearly, "mixtures" that are either pure reactant or pure product are farthest from each other, as we discussed above. The reaction curve crosses the x-axis ($\Delta G = 0$) at point 3, which defines the composition of the reaction mixture at equilibrium. This value, expressed as the percentage of N_2O_4 that has dissociated into NO_2, corresponds to the same N_2O_4/NO_2 ratio as the minimum in the curve in Figure 17.15(a).

Once a reaction has reached equilibrium, $Q = K$, $\Delta G = 0$, and Equation 17.14 becomes

$$\Delta G_{rxn} = \Delta G^\circ_{rxn} + RT \ln K = 0$$

This can also be written as

$$\Delta G^\circ_{rxn} = -RT \ln K \qquad (17.15)$$

Rearranging Equation 17.15 allows us to calculate the K value for a reaction from its change in standard free energy and absolute temperature. First we rearrange the terms:

$$\ln K = \frac{-\Delta G^\circ_{rxn}}{RT} \qquad (17.16)$$

Then we take the antilogarithm of both sides:

$$K = e^{-\Delta G^\circ_{rxn}/RT} \qquad (17.17)$$

Equation 17.17 provides the following interpretation of reaction spontaneity under standard conditions: Whenever ΔG° is negative, the exponent $-\Delta G^\circ/RT$ in Equation 17.17 is positive, and $e^{-\Delta G^\circ/RT} > 1$, making $K > 1$. Therefore any reversible reaction with an equilibrium constant greater than 1 is spontaneous under standard conditions, as shown in Figure 17.16(a). This spontaneity has its limits. As reactants are consumed and products are formed, the value of the reaction quotient increases, making the value of ΔG less negative. When it reaches zero, there is no further change in the composition of the reaction mixture because chemical equilibrium has been achieved.

It follows that a reversible reaction with a less negative value of ΔG° (Figure 17.16b) is still spontaneous but has a smaller equilibrium constant, so less reactant is consumed and less product is formed before the value of ΔG reaches 0. Finally, a reaction that has a positive value of ΔG° (Figure 17.16c) has an equilibrium constant that is less than 1, and the reaction is not spontaneous under standard conditions. Instead, the reverse of the reaction is spontaneous.

Let's apply this concept to the equilibrium between N_2O_4 and NO_2:

$$N_2O_4(g) \rightleftharpoons 2\,NO_2(g) \qquad \Delta G^\circ_{rxn} = 4.8\text{ kJ}$$

We focus on the composition of the reaction mixture of these two gases at equilibrium in Figure 17.15. We start by calculating the value of the exponent in Equation 17.17:

$$-\frac{\Delta G^\circ_{rxn}}{RT} = -\frac{\left(\dfrac{4.8\ \cancel{kJ}}{\cancel{mol}}\right)\left(\dfrac{1000\text{ J}}{1\ \cancel{kJ}}\right)}{\left(\dfrac{8.314\text{ J}}{\cancel{mol}\cdot\cancel{K}}\right)(298\ \cancel{K})} = -1.94$$

Inserting this value into Equation 17.17 gives

$$K = e^{-\Delta G^\circ_{rxn}/RT} = e^{-1.94} = 0.14$$

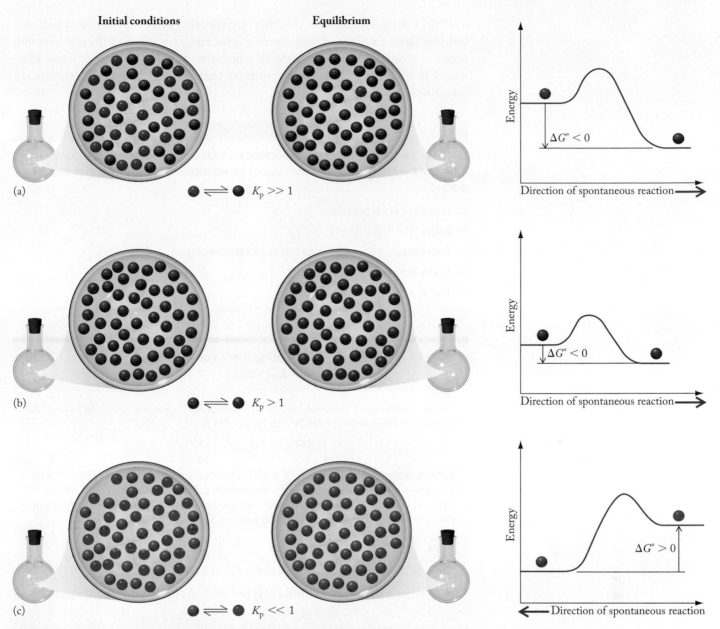

FIGURE 17.16 The equilibrium constant of a chemical reaction is linked to its $\Delta G°$ value. Initially the three flasks on the left each contain equimolar mixtures of different pairs of gases. The initial partial pressure of each gas is 1 atm. The three images on the right represent the mixtures in these same flasks at equilibrium. (a) The value of $\Delta G°$ for the formation of "red" gas from "green" gas has a large negative value, which makes K_p much greater than 1. The reaction proceeds in the forward direction, leaving only a little green gas left over. (b) If the value of $\Delta G°$ for the formation of "orange" gas from "blue" gas has a smaller negative value than in part (a), the value of K_p is smaller, though still greater than 1, and there is more orange gas present at equilibrium than blue gas. (c) If the value of $\Delta G°$ for the formation of "yellow" gas from "purple" gas is positive, K_p is less than 1 and the reaction runs in the reverse direction, forming purple gas from yellow gas.

This result is consistent with the composition of the reaction mixture at equilibrium in Figure 17.15, where $P_{N_2O_4} \approx 0.83$ atm and $P_{NO_2} \approx 0.34$ atm. Inserting these values into the equilibrium constant expression for the reaction gives us an approximate value of K_p that is close to the one we calculated from $\Delta G°_{rxn}$:

$$K_p = \frac{(P_{NO_2})^2}{P_{N_2O_4}} \approx \frac{(0.34)^2}{0.83} = 0.14$$

In this example, a positive $\Delta G°_{rxn}$ value of only few kilojoules corresponds to an equilibrium constant value that is less than one but still greater than 0.1. As a result, the equilibrium reaction mixture contains more reactant than product, but less than an order of magnitude more. Similarly, an equilibrium reaction mixture produced by a reaction with a negative ΔG°_{rxn} value of only a few kilojoules is likely to contain more products than reactants, but with significant quantities of both.

CONCEPT TEST

Suppose the ΔG°_{rxn} value of the hypothetical chemical reaction A ⇌ B is −3.0 kJ/mol. Which of the following statements about an equilibrium mixture of A and B at 298 K is true?

a. There is only A present.

b. There is only B present.

c. There is an equimolar mixture of A and B present.

d. There is more A than B present.

e. There is more B than A present.

SAMPLE EXERCISE 17.7 Calculating the Value of K from ΔG°_f Values   **LO6**

Use ΔG°_f values from Table A4.3 in the Appendix to calculate ΔG°_{rxn} and the value of K_p for the formation of NO_2 from NO and O_2 at 298 K:

$$NO(g) + \tfrac{1}{2} O_2(g) \rightleftharpoons NO_2(g)$$

Collect, Organize, and Analyze The ΔG°_f values we need from Table A4.3 are 51.3 kJ/mol for NO_2 and 86.6 kJ/mol for NO. Because O_2 gas is the most stable form of the element, its ΔG°_f value is 0.0 kJ/mol. We can use Equation 17.12 to calculate ΔG°_{rxn} and then use Equation 17.17 to calculate the value of K.

Solve

$$\Delta G^{\circ}_{rxn} = [\Delta G^{\circ}_f (NO_2)] - [\Delta G^{\circ}_f (NO) + \tfrac{1}{2} \Delta G^{\circ}_f (O_2)]$$

$$= [1 \text{ mol } (51.3 \text{ kJ/mol})] - [1 \text{ mol } (86.6 \text{ kJ/mol}) + \tfrac{1}{2} \text{ mol } (0.0 \text{ kJ/mol})]$$

$$= (51.3 - 86.6) \text{ kJ}$$

$$= -35.3 \text{ kJ, or } -35,300 \text{ J per mol of } NO_2 \text{ produced}$$

The exponent in Equation 17.17 is

$$-\frac{\Delta G^{\circ}_{rxn}}{RT} = -\frac{\left(\dfrac{-35,300 \text{ J}}{\text{mol}}\right)}{\left(\dfrac{8.314 \text{ J}}{\text{mol} \cdot \text{K}}\right)(298 \text{ K})} = 14.2$$

The corresponding value of K_p is

$$K_p = e^{-\Delta G^{\circ}_{rxn}/RT} = e^{14.2} = 1.5 \times 10^6$$

Think About It The exponential relationship between ΔG°_{rxn} and K_p means that a moderately negative free-energy change of −35.3 kJ/mol corresponds to a very large value of K_p: in this case, greater than 10^6.

 Practice Exercise The standard free energy of formation of ammonia at 298 K is −16.5 kJ/mol. What is the value of K_p for the following reaction at 298 K?

$$N_2(g) + 3 H_2(g) \rightleftharpoons 2 NH_3(g)$$

We have yet to explain which kind of equilibrium constant, K_c or K_p, is related to $\Delta G°$ by Equation 17.17. The symbol $\Delta G°$ represents a change in free energy under standard conditions. The standard state of a gaseous reactant or product is one in which its *partial pressure* is 1 bar. Thus the $\Delta G°$ of a reaction *in the gas phase* is linked by Equation 17.17 to its K_p value. However, standard conditions for reactions in solution (the focus of Chapters 15 and 16) mean that all dissolved reactants and products are present at a concentration of 1.00 *M*. Thus the $\Delta G°$ of a reaction *in solution* is related by Equation 17.17 to its K_c value.

17.8 Influence of Temperature on Equilibrium Constants

We noted on many occasions in Chapter 14 that the value of K changes with temperature. In this section we use the thermodynamics of chemical reactions to explain how and why.

Let's begin by considering a simplified version of Equation 17.11 (leaving out all the "rxn" subscripts):

$$\Delta G° = \Delta H° - T\Delta S°$$

We can combine it with a similarly simplified version of Equation 17.16:

$$\ln K = \frac{-\Delta G°}{RT}$$

The result is an equation that relates K to $\Delta H°$ and $\Delta S°$:

$$\ln K = \frac{-\Delta G°}{RT} = -\frac{\Delta H°}{RT} + \frac{T\Delta S°}{RT}$$

$$= -\frac{\Delta H°}{RT} + \frac{\Delta S°}{R} \qquad (17.18)$$

Note how a negative value of $\Delta H°$ or a positive value of $\Delta S°$ contributes to a large value of K. These dependencies make sense because negative values of $\Delta H°$ and positive values of $\Delta S°$ are the two factors that contribute to making reactions spontaneous.

Because we are discussing the influence of temperature on K, let's identify the factors affected by changes in T in Equation 17.18. The $\Delta S°$ term is not affected, but the influence of $\Delta H°$ does depend on temperature: the higher the temperature, the larger the denominator of the $\Delta H°$ term, and the smaller the influence of a favorable $\Delta H°$, that is, a $\Delta H°$ value that is negative. This temperature dependence makes sense according to Le Châtelier's principle and the notion that energy is a product of exothermic reactions. Increasing temperature inhibits exothermic reactions.

If $\Delta H°$ and $\Delta S°$ do not vary much with temperature, then Equation 17.18 predicts that $\ln K$ will be a linear function of $1/T$. Furthermore, we expect a graph of $\ln K$ versus $1/T$ to have a positive slope for an exothermic process and a negative slope for an endothermic process. We can determine $\Delta H°$ from the slope of the line and $\Delta S°$ from its y-intercept. Thus we can calculate fundamental thermodynamic values of a reaction at equilibrium by determining its equilibrium constant at different temperatures.

For example, let's determine the values of $\Delta H°$ and $\Delta S°$ for the reaction

$$2\ CO_2(g) \rightleftharpoons 2\ CO(g) + O_2(g)$$

CONNECTION In Chapter 5 we defined exothermic reactions as those giving off heat; they have a negative enthalpy change ($\Delta H < 0$). Endothermic reactions absorb heat ($\Delta H > 0$).

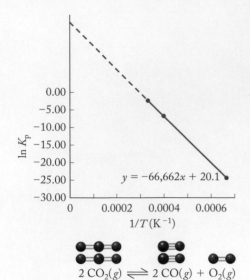

FIGURE 17.17 The graph of $\ln K_p$ versus $1/T$ is a straight line. We calculate $\Delta H°_{rxn}$ from the slope and extrapolate the line to the y-intercept, from which we calculate $\Delta S°_{rxn}$.

We start with the K_p values of 2.75×10^{-11} at 1500 K, 1.42×10^{-3} at 2500 K, and 0.112 at 3000 K. The fact that K_p increases as temperature increases tells us that energy is a reactant and the reaction is endothermic. We can use these data to determine the values of $\Delta H°$ and $\Delta S°$ by plotting $\ln K_p$ values versus $1/T$. To see how this graphical method works, let's rewrite Equation 17.18 so that it fits the form of the equation for a straight line ($y = mx + b$):

$$\ln K = -\frac{\Delta H°_{rxn}}{R}\left(\frac{1}{T}\right) + \frac{\Delta S°_{rxn}}{R} \tag{17.19}$$

The graph of $\ln K_p$ versus $1/T$ (Figure 17.17) is indeed a straight line. The slope of this line is $-66,662$ K, which is equal to $-\Delta H°_{rxn}/R$. The corresponding value of $\Delta H°_{rxn}$ is

$$\Delta H°_{rxn} = -\text{slope} \times R$$
$$= -(-66,662 \text{ K})\left(\frac{8.314 \text{ J}}{\text{mol} \cdot \text{K}}\right)$$
$$= 554,227 \text{ J/mol} = 554 \text{ kJ/mol}$$

The y-intercept ($\Delta S°_{rxn}/R$) of the graph is 20.1. The corresponding value of $\Delta S°_{rxn}$ is

$$\Delta S°_{rxn} = (20.1)\left(\frac{8.314 \text{ J}}{\text{mol} \cdot \text{K}}\right) = 167 \text{ J/(mol} \cdot \text{K)}$$

This positive entropy change is logical because the forward reaction converts 2 moles of gaseous CO_2 into 3 moles of gaseous products.

We can use a modified version of Equation 17.19 to relate the values of K at two different temperatures:

$$\ln\left(\frac{K_2}{K_1}\right) = -\frac{\Delta H°_{rxn}}{R}\left(\frac{1}{T_2} - \frac{1}{T_1}\right) \tag{17.20}$$

Equation 17.20 is called the *van 't Hoff equation* because it was first derived by Jacobus van 't Hoff. It is particularly useful for calculating the value of K at a very high or low temperature if we know what it is at a standard reference temperature (e.g., 298 K). We use such a calculation in the following Sample Exercise.

SAMPLE EXERCISE 17.8 Calculating the Value of K at a Specific Temperature   **LO6, LO7**

Use data from Appendix 4 to calculate the equilibrium constant K_p for the exothermic reaction

$$N_2(g) + 3 H_2(g) \rightleftharpoons 2 NH_3(g)$$

at 298 K and at 773 K, a typical temperature used in the Haber–Bosch process for synthesizing ammonia.

Collect and Organize We are asked to calculate the K_p value for a reaction at two temperatures. One of the temperatures is 298 K, which is the reference temperature for the standard thermodynamic data in Appendix 4. The $\Delta G°_f$ value of NH_3 is -16.5 kJ/mol.

Analyze We can use Equation 17.17 to calculate the value of K_p at 298 K from the value of $\Delta G°_{rxn}$. Then we can use Equation 17.20 to calculate the value of K_p at 773 K from the value of K_p at 298 K and the value of $\Delta H°_{rxn}$. The value of $\Delta H°_{rxn}$ can be calculated from the enthalpy of formation of NH_3 (-46.1 kJ/mol). The negative value of $\Delta H°_f$ means that $\Delta H°_{rxn}$ is also negative, which means that the reaction is exothermic.

Because heat is a product of the reaction, we can predict that the value of K_p is smaller at 773 K than at 298 K.

Solve The value of ΔG°_{rxn} for the reaction is

$$\frac{2 \text{ mol NH}_3}{\text{mol N}_2} \times \frac{-16.5 \text{ kJ}}{\text{mol NH}_3} \times \frac{1000 \text{ J}}{\text{kJ}} = -33,000 \frac{\text{J}}{\text{mol N}_2} = -3.30 \times 10^4 \frac{\text{J}}{\text{mol N}_2}$$

Using this value in Equation 17.17 to calculate the value of K_p gives us

$$K_p = e^{-\Delta G^\circ_{rxn}/RT}$$

$$= e^{\left(\frac{-\left(-3.30 \times 10^4 \frac{\text{J}}{\text{mol}}\right)}{8.314 \frac{\text{J}}{\text{mol} \cdot \text{K}} \times 298 \text{ K}}\right)}$$

$$= 6.09 \times 10^5$$

Once we know the value of K_p at 298 K, we can calculate K_p at 773 K by using Equation 17.20. To do this, we must first calculate the value of ΔH°_{rxn}, which is twice the ΔH°_f of NH_3:

$$\Delta H^\circ_{rxn} = \frac{2 \text{ mol NH}_3}{\text{mol N}_2} \times \frac{-46.1 \text{ kJ}}{\text{mol NH}_3} \times \frac{1000 \text{ J}}{\text{kJ}}$$

$$= -92,200 \frac{\text{J}}{\text{mol N}_2}, \text{ or } -92.2 \text{ kJ}$$

After substituting $K_1 = 6.1 \times 10^5$, $T_1 = 298$ K, and $T_2 = 773$ K into Equation 17.20, we solve for K_2:

$$\ln\left(\frac{K_2}{6.1 \times 10^5}\right) = -\frac{\left(-92,200 \frac{\text{J}}{\text{mol}}\right)}{8.314 \frac{\text{J}}{\text{mol} \cdot \text{K}}}\left(\frac{1}{773 \text{ K}} - \frac{1}{298 \text{ K}}\right)$$

$$\ln\left(\frac{K_2}{6.1 \times 10^5}\right) = -22.87$$

$$\frac{K_2}{6.1 \times 10^5} = e^{-22.87} = 1.17 \times 10^{-10}$$

$$K_2 = 7.1 \times 10^{-5} \text{ at } 773 \text{ K}$$

Think About It The equilibrium constant decreases markedly when the temperature of the reaction is raised from 298 K to 773 K. This decrease fits our prediction for this exothermic reaction. Note that the standard thermodynamic data for the reaction (ΔH°_{rxn} and ΔG°_{rxn}) are expressed per mole of the reactant with a coefficient of 1 in the balanced chemical equation for N_2.

 Practice Exercise Use data from Appendix 4 to calculate the value of K_p for the reaction

$$2 \text{ N}_2(g) + \text{O}_2(g) \rightleftharpoons 2 \text{ N}_2\text{O}(g)$$

at 298 K and 2000 K.

17.9 Driving the Human Engine: Coupled Reactions

The laws of thermodynamics that govern chemical reactions in the laboratory also govern all the chemical reactions that take place in living systems (Figure 17.18). Organisms carry out reactions that release the energy contained in the chemical bonds of food molecules and then use that energy to do work and

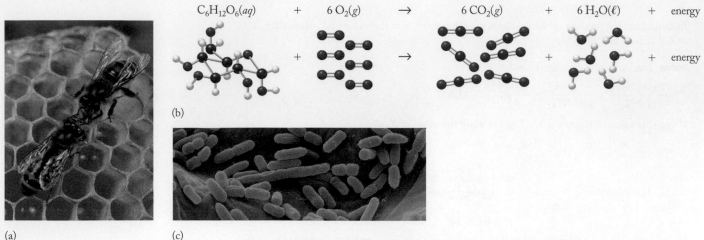

$$C_6H_{12}O_6(aq) \quad + \quad 6\,O_2(g) \quad \rightarrow \quad 6\,CO_2(g) \quad + \quad 6\,H_2O(\ell) \quad + \quad \text{energy}$$

(b)

(a)

(c)

FIGURE 17.18 The rules of thermodynamics apply to all living systems. (a) Honeybees extract energy from nutrients to support life. The bees store this energy as honey (a mixture containing levulose and dextrose), which is then consumed by the bees, by humans, or by other animals to generate energy. (b) When honey is consumed, heat, water, and carbon dioxide are released, increasing the entropy of the universe. (c) Microscopic organisms—such as the *E. coli* that live in our gastrointestinal tracts—consume nutrients to live and generate heat in the process. Their expenditure of energy also increases the entropy of the universe.

CONNECTION In Chapter 5 we defined 1 Calorie, the "calorie" used in discussing food, as equivalent to 1 kcal.

Glucose

↓

↓ Glycolysis

↓

Pyruvate ion   Pyruvate ion

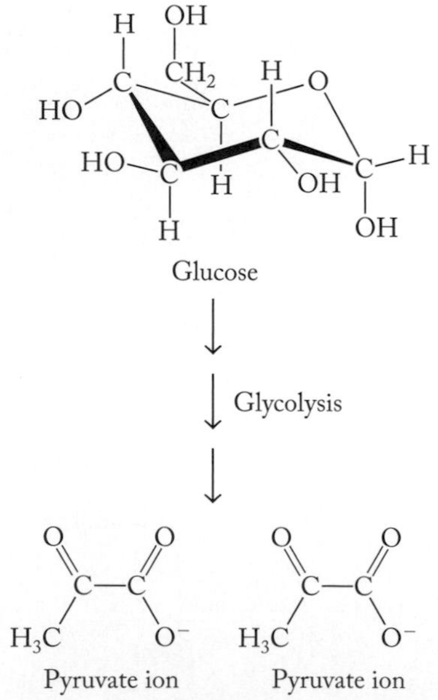

FIGURE 17.19 In glycolysis, molecules of glucose are converted into twice their number of pyruvate ions.

sustain an array of other essential biological functions. Just like mechanical engines, however, humans and other life-forms are far from 100% efficient, which means that life requires a continual input of energy in terms of the caloric content of the food we eat. Thus we must constantly absorb energy in the form of food and release heat and waste products into our surroundings. Young women have an average daily nutritional need of 2100 Cal; for young men the figure is 2900 Cal. This level of caloric intake provides the energy we need in order to function at all levels, from thinking to getting out of bed in the morning.

In living systems, spontaneous reactions ($\Delta G < 0$) typically involve metabolizing food, whereas nonspontaneous reactions ($\Delta G > 0$) involve building molecules needed by the body. Living systems use the energy from spontaneous reactions to run nonspontaneous reactions; we say that the spontaneous reactions are *coupled* to the nonspontaneous reactions. Part of the study of biochemistry involves deciphering the molecular mechanisms that enable reaction coupling. This topic is covered in Chapter 20, but for now it is sufficient to know that elegant molecular processes have evolved to enable living systems to couple chemical reactions so that the energy obtained from spontaneous reactions can be used to drive the nonspontaneous reactions that maintain life. The metabolic chemical reactions we look at here are *not* presented for you to memorize. Rather, the intent is to aid your understanding of how changes in free energy allow spontaneous reactions to drive nonspontaneous reactions and to illustrate operationally what the phrase "coupled reactions" actually means.

As noted at the beginning of Chapter 5, all the energy contained in the food we eat has sunlight as its ultimate source. Green plants store energy from sunlight in their tissues as molecules such as glucose ($C_6H_{12}O_6$), which they produce from CO_2 and H_2O during photosynthesis.

Production of glucose by green plants is a nonspontaneous process, which is why the plants require the energy of sunlight to carry out this reaction. Animals that consume plants use the energy stored in the chemical bonds of glucose and

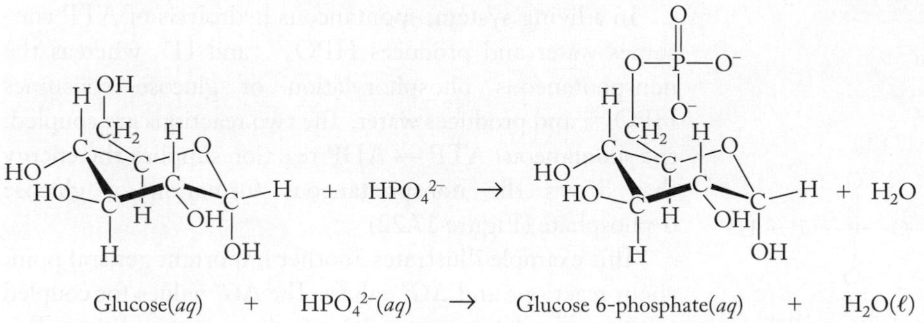

Glucose(aq) + HPO$_4^{2-}$(aq) → Glucose 6-phosphate(aq) + H$_2$O(ℓ)

FIGURE 17.20 The conversion of glucose into glucose 6-phosphate is an early step in glycolysis. This reaction is nonspontaneous, which means energy must be added to make the reaction go: $\Delta G^{\circ}_{rxn} = 13.8$ kJ/mol.

glycolysis a series of reactions that converts glucose into pyruvate; a major anaerobic (no oxygen required) pathway for the metabolism of glucose in the cells of almost all living organisms.

phosphorylation a reaction resulting in the addition of a phosphate group to an organic molecule.

other molecules and release CO$_2$ and H$_2$O back into the environment. This reaction,

$$C_6H_{12}O_6(s) + 6\ O_2(g) \rightarrow 6\ CO_2(g) + 6\ H_2O(\ell)$$

also releases energy, an event that increases the entropy of the universe, and is therefore spontaneous. Thus the processes of life increase the entropy of the universe by converting chemical energy into energy (heat) that flows into the surroundings.

Therefore the free-energy change for the breakdown of glucose must be less than zero, and it is actually $\Delta G^{\circ} = -880$ kJ/mol. This spontaneous process is highly controlled in living systems, however, so that the energy it produces can be directed into the nonspontaneous processes essential to organisms that do not carry out photosynthesis. Figure 17.19 summarizes one portion of glucose metabolism—**glycolysis**—that involves coupled reactions.

In glycolysis, each mole of glucose is converted into 2 moles of pyruvate ion (CH$_3$COCOO$^-$). An early step is the conversion of glucose into glucose 6-phosphate (Figure 17.20), an example of a **phosphorylation** reaction. Glucose reacts with the hydrogen phosphate ion (HPO$_4^{2-}$), producing glucose 6-phosphate and water. This reaction is not spontaneous ($\Delta G^{\circ} = +13.8$ kJ/mol), and the energy needed to make it happen comes from a compound called adenosine triphosphate (ATP). ATP functions in our cells both as a storehouse of energy and as an energy-transfer agent: it hydrolyzes to adenosine diphosphate (ADP) in a reaction (Figure 17.21) that produces a hydrogen phosphate ion and energy: $\Delta G^{\circ}_{rxn} = -30.5$ kJ/mol.

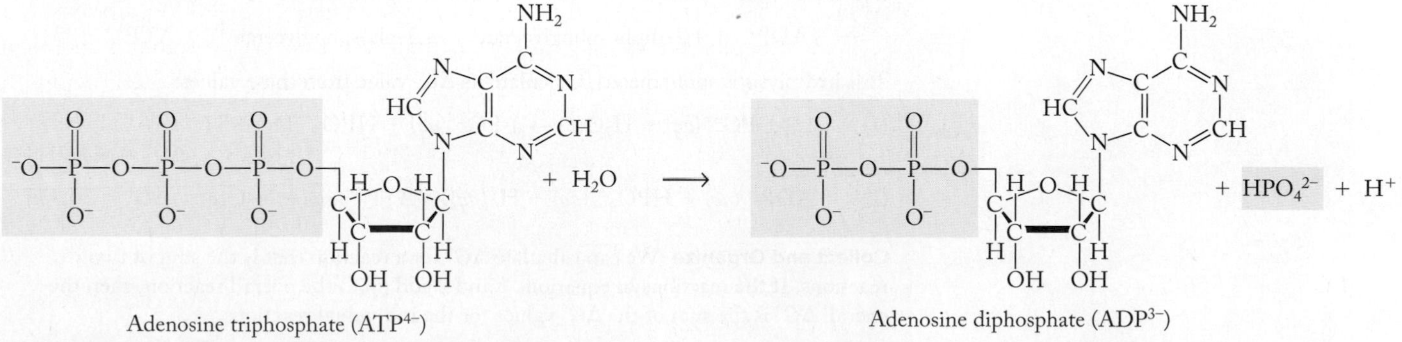

CONNECTION Photosynthesis and the carbon cycle were introduced in Chapter 3.

Adenosine triphosphate (ATP^{4-})

Adenosine diphosphate (ADP^{3-})

FIGURE 17.21 The hydrolysis of ATP to ADP is a spontaneous reaction: $\Delta G^{\circ} = -30.5$ kJ/mol. The body couples this reaction to nonspontaneous reactions so that the energy released can drive them (see Figure 17.22).

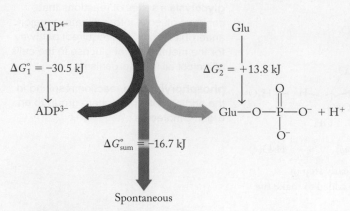

FIGURE 17.22 The spontaneous hydrolysis of ATP is coupled to the nonspontaneous phosphorylation of glucose. The overall reaction—the sum of the two individual reactions—is spontaneous.

C⊕NNECTION In Chapter 5 we used Hess's law (the enthalpy change of a reaction that is the sum of two or more reactions equals the sum of the enthalpy changes of the constituent reactions) to calculate ΔH. We apply a similar principle here when adding $\Delta G°$ values of coupled reactions.

In a living system, spontaneous hydrolysis of ATP consumes water and produces HPO_4^{2-} and H^+, whereas the nonspontaneous phosphorylation of glucose consumes HPO_4^{2-} and produces water. The two reactions are coupled: the spontaneous ATP → ADP reaction supplies the energy that drives the nonspontaneous formation of glucose 6-phosphate (Figure 17.22).

This example illustrates another important general point about reactions and $\Delta G°$ values. The $\Delta G°$ values for coupled reactions (and for sequential reactions) are additive. This is true for any set of reactions, not just those occurring in living systems.

The first two steps in glycolysis are

$$(1) \quad ATP^{4-}(aq) + H_2O(\ell) \rightarrow ADP^{3-}(aq) + HPO_4^{2-}(aq) + H^+(aq)$$
$$\Delta G° = -30.5 \text{ kJ}$$

$$(2) \quad C_6H_{12}O_6(aq) + HPO_4^{2-}(aq) \rightarrow C_6H_{11}O_6PO_3^{2-}(aq) + H_2O(\ell) \quad \Delta G° = 13.8 \text{ kJ}$$

Glucose · · · · · · · Hydrogen · · · · · · · · Glucose
· · · · · · · · · · · phosphate · · · · · · · 6-phosphate

If we add these reactions and their free energies, we get

$$ATP^{4-}(aq) + \cancel{H_2O(\ell)} + C_6H_{12}O_6(aq) + \cancel{HPO_4^{2-}(aq)} \rightarrow$$
$$ADP^{3-}(aq) + \cancel{HPO_4^{2-}(aq)} + C_6H_{11}O_6PO_3^{2-}(aq) + \cancel{H_2O(\ell)} + H^+(aq)$$
$$\Delta G° = (-30.5 + 13.8) \text{ kJ} = -16.7 \text{ kJ}$$

Because equal quantities of H_2O and HPO_4^{2-} appear on both sides of the combined equation, they cancel out, leaving the net reaction

$$C_6H_{12}O_6(aq) + ATP^{4-}(aq) \rightarrow ADP^{3-}(aq) + C_6H_{11}O_6PO_3^{2-}(aq) + H^+(aq)$$
$$\Delta G° = -16.7 \text{ kJ}$$

Because $\Delta G°$ for the net reaction is negative, the reaction is spontaneous.

SAMPLE EXERCISE 17.9 Calculating $\Delta G°$ of Coupled Reactions · · · · · · **LO8**

The body would rapidly run out of ATP if there were not some process for regenerating ATP from ADP. That process is the hydrolysis of 1,3-diphosphoglycerate^{4-} (1,3-DPG^{4-}) to 3-phosphoglycerate^{3-} (3-PG^{3-}; Figure 17.23):

$$ADP^{3-} + 1,3\text{-diphosphoglycerate}^{4-} \rightarrow 3\text{-phosphoglycerate}^{3-} + ATP^{4-}$$

This hydrolysis is spontaneous. Calculate its $\Delta G°$ value from these values:

$$(1) \quad 1,3\text{-DPG}^{4-}(aq) + H_2O(\ell) \rightarrow 3\text{-PG}^{3-}(aq) + HPO_4^{2-}(aq) + H^+(aq)$$
$$\Delta G° = -49.0 \text{ kJ}$$

$$(2) \quad ADP^{3-}(aq) + HPO_4^{2-}(aq) + H^+(aq) \rightarrow ATP^{4-}(aq) + H_2O(\ell) \quad \Delta G° = 30.5 \text{ kJ}$$

Collect and Organize We can calculate $\Delta G°$ for a reaction that is the sum of two reactions. If the reactions in equations 1 and 2 add up to the overall reaction, then the overall $\Delta G°$ is the sum of the $\Delta G°$ values for the individual reactions.

Analyze First we add the reactions described by equations 1 and 2. Assuming that the overall reaction between ADP and 1,3-diphosphoglycerate^{4-} is the sum of the reactions describing the hydrolysis of 1,3-diphosphoglycerate^{4-} and the phosphorylation of

ADP, we know that the sum of ΔG_1° and ΔG_2° will be less than zero because the overall reaction is spontaneous.

Solve We add the reactions in equations 1 and 2 and confirm that they equal the overall reaction:

(1) $\text{1,3-DPG}^{4-}(aq) + H_2O(\ell) \rightarrow \text{3-PG}^{3-}(aq) + HPO_4^{2-}(aq) + H^+(aq)$

(2) $ADP^{3-}(aq) + HPO_4^{2-}(aq) + H^+(aq) \rightarrow ATP^{4-}(aq) + H_2O(\ell)$

$$\text{1,3-DPG}^{4-}(aq) + H_2O(\ell) + ADP^{3-}(aq) + HPO_4^{2-}(aq) + H^+(aq) \rightarrow \text{3-PG}^{3-}(aq) + HPO_4^{2-}(aq) + H^+(aq) + ATP^{4-}(aq) + H_2O(\ell)$$

We sum the ΔG° values for steps 1 and 2 to determine ΔG° for the overall reaction:

$$\Delta G^\circ_{\text{overall}} = \Delta G_1^\circ + \Delta G_2^\circ = [(-49.0) + (30.5)]\ \text{kJ} = -18.5\ \text{kJ}$$

Think About It The hydrolysis of 1,3-diphosphoglycerate^{4-} provides more than sufficient energy for the conversion of ADP into ATP.

Practice Exercise The conversion of glucose into lactic acid drives the phosphorylation of 2 moles of ADP to ATP:

$$\underset{\text{Glucose}}{C_6H_{12}O_6(aq)} + 2\ HPO_4^{2-}(aq) + 2\ ADP^{3-}(aq) + 2\ H^+(aq) \rightarrow \underset{\text{Lactic acid}}{2\ CH_3CH(OH)COOH(aq)} + 2\ ATP^{4-}(aq) + 2\ H_2O(\ell) \qquad \Delta G^\circ = -135\ \text{kJ}$$

What is ΔG° for the conversion of glucose into lactic acid?

$$C_6H_{12}O_6(aq) \rightarrow 2\ CH_3CH(OH)COOH(aq)$$

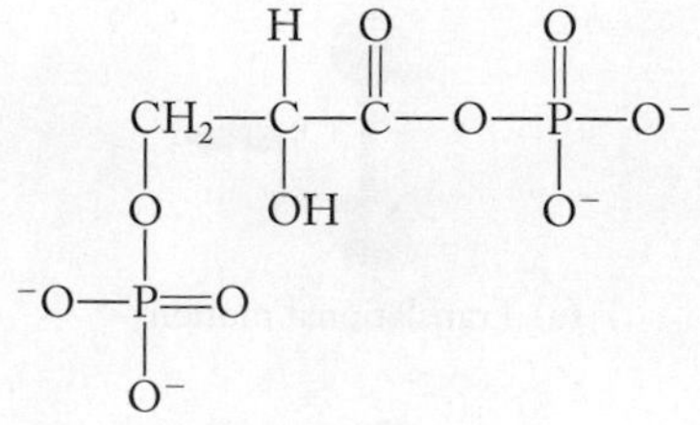

1,3-Diphosphoglycerate^{4-}
(1,3-DPG^{4-})

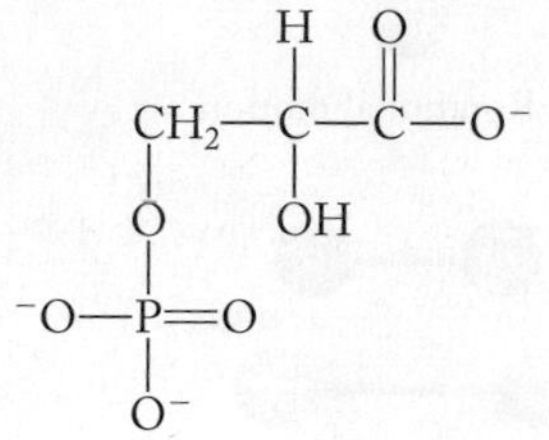

3-Phosphoglycerate^{3-}
(3-PG^{3-})

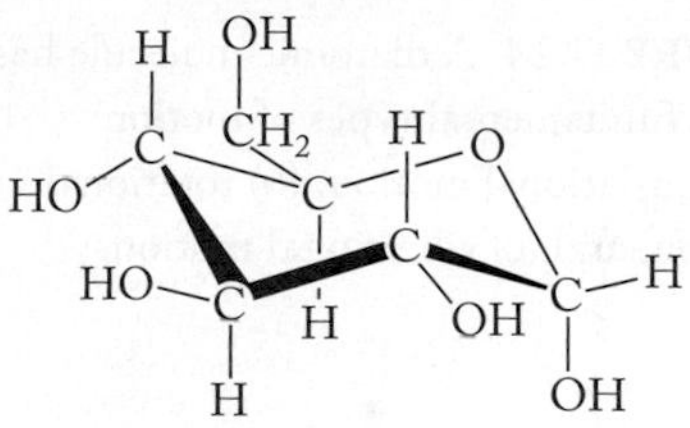

Glucose

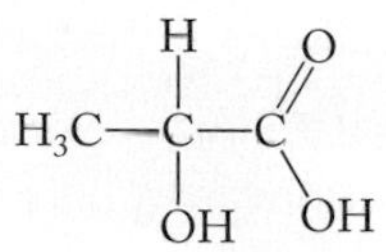

Lactic acid

FIGURE 17.23 Structures of 1,3-diphosphoglycerate, 3-phosphoglycerate, glucose, and lactic acid.

The ATP produced from the breakdown of glucose (as, for example, via the first reaction in the preceding Practice Exercise) is used to drive nonspontaneous reactions in cells. The metabolism of fats and proteins relies on a series of cycles, all of which involve coupled reactions. The ATP–ADP system is a carrier of chemical energy because ADP requires energy to accept a phosphate group and thus is coupled to reactions that yield energy, whereas ATP donates a phosphate group, releases energy, and is coupled to reactions that require energy. All energy changes in living systems are governed by the first and second laws of thermodynamics, as are all the energy changes in the inanimate world.

17.10 Microstates: A Quantized View of Entropy

In this chapter we have used a thermodynamic definition of entropy, describing it as a measure of how energy is distributed throughout a system: the more spread out energy is, the greater the system's entropy. This thermodynamic view evolved in the middle of the 19th century, largely through the work of the German physicist Rudolf Clausius (1822–1888), who was a leader in establishing the field of thermodynamics and who introduced the concept of entropy in 1865.

Clausius's thermodynamic approach views entropy from a macroscopic, systemwide perspective. In much the same way as we use such parameters as pressure, volume, and temperature to describe a quantity of a gaseous substance, the change in a system's entropy is defined in terms of the quantity of energy flowing out from or into the system and its temperature (Equation 17.2). However, just as the kinetic molecular theory of gases provides a particle-based explanation of why gases exhibit the macroscopic properties they do, we can also

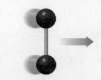

(a) Translational motion

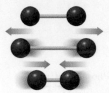

(b) Rotational motion

(c) Vibrational motion

FIGURE 17.24 A diatomic molecule has three fundamental types of motion: (a) translational motion, (b) rotational motion, and (c) vibrational motion.

understand entropy changes from the viewpoint of the dispersion of particles and their energies.

In 1877 an Austrian physicist, Ludwig Boltzmann (1844–1906), developed a microscopic view of Clausius's concept of entropy. His theory would eventually incorporate quantum mechanics to explain how energy was dispersed not only throughout a system but also within its molecules. Let's examine Boltzmann's microscopic definition of entropy by using two models: a single molecule of O_2 and a room full of them.

As we have discussed in this chapter and as shown in Figure 17.24, a molecule of a gas such as O_2 is free to undergo three types of motion: (1) *translational motion* as it zips around the space it occupies; (2) *rotational motion* as it spins about imaginary axes perpendicular to the O$=$O bond; and (3) *vibrational motion* as its bonded atoms move toward and away from each other, like balls on the ends of a spring. All three modes of motion have one thing in common: the greater the thermal energy of the molecule, the greater each type of motion.

In our discussion of quantum mechanics in Chapter 7, we saw that energy is not continuous on the atomic scale. Instead, the energies and motion of atoms and molecules are quantized. At temperatures near room temperature, the differences between the translational energy levels of atoms and molecules in the gas phase are so small that we may consider them a continuum of energy. However, the gaps in vibrational and rotational energy levels are large enough that we must take quantization into account.

To investigate these gaps, let's revisit the change in electrostatic potential energy that occurs when two oxygen atoms approach each other and a covalent bond forms between them (see Figure 8.1). Energy reaches a minimum (Figure 17.25a) when the nuclei of the two atoms are a distance apart that corresponds to the length of the O$=$O bond, or 121 pm. Figure 17.25(a) also contains additional energy levels represented by horizontal red lines. These are vibrational energy levels. Note that the red lines are longer with greater vibrational energy. Greater length corresponds to greater variation in intermolecular distance; that is, more energetic oscillations of the O$=$O bond occur with increasing vibrational energy.

Superimposed on each vibrational energy level is a set of rotational energy levels, shown for the first vibrational energy level in Figure 17.25(b). Similar sets of rotational energy levels are associated with all the other vibrational energy levels, creating a multitude of energy states that are accessible to every O_2 molecule at room temperature.

FIGURE 17.25 Vibrational and rotational energy levels in a molecule of O_2 are quantized. (a) The electrostatic potential energy between two oxygen atoms reaches a minimum when their nuclei are 121 pm apart—the length of an O$=$O bond. (b) Quantized rotational energy levels (represented by the blue horizontal lines) are superimposed on each vibrational energy level, represented by a red horizontal line in part (a). (Not all the accessible rotational and vibrational states are shown, and the gaps between them are not to scale.)

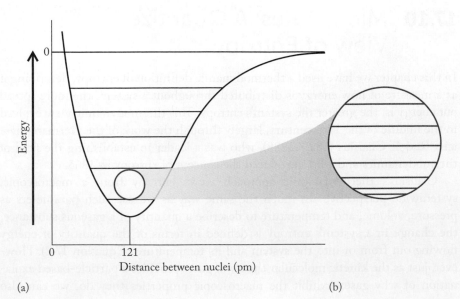

(a)

(b)

Now let's focus on one oxygen molecule as it collides with others, sometimes gaining and other times losing translational kinetic energy. At room temperature and pressure, a single O_2 molecule experiences billions of collisions per second, each collision altering its speed and changing its vibrational and rotational energies. At any instant, the overall quantized energy of the molecule is a particular combination of translational, vibrational, and rotational energy states. The number of possible combinations of energy states accessible to an O_2 molecule is enormous at room temperature, and the number goes up as temperature goes up.

In a room full of O_2 molecules (our second model), the number of different quantized total-molecule energy states is so large (e raised to a power that is many times Avogadro's number) that it is beyond the capacity of your calculator to compute. The entirety of all the quantized energy states occupied by all of the atoms in the room at a particular instant is called a **microstate**. Each microstate represents one discrete way for the system to disperse all the energy in the system. The vast number of microstates to which a system has access defines its entropy: the more microstates, the higher its entropy. This linkage between number of accessible microstates (W) and entropy (S) is defined by a simple mathematical equation developed by Boltzmann:

$$S = k_B \ln W \tag{17.21}$$

Here k_B is the Boltzmann constant, which is equal to the gas constant divided by Avogadro's number:

$$k_B = \frac{R}{N_A} = \frac{8.314 \ \frac{J}{mol \cdot K}}{6.022 \times 10^{23}/mol} = 1.381 \times 10^{-23} \ J/K$$

Boltzmann's definition of entropy, based on access to energy microstates, is conceptually compatible with the thermodynamic, macroscopic view based on energy dispersion. After all, each microstate represents one of the multitude of ways that energy can be dispersed in a thermodynamic system. In fact, some facets of entropy are easier to understand using the microstate model. One of them is the concept of absolute entropy, the notion embedded in the third law of thermodynamics that a perfect crystalline solid has zero entropy at a temperature of absolute zero. If the particles of a crystalline solid are perfectly aligned at 0 K, only one distribution of the particles is possible in the space occupied by the crystal. Therefore the crystal has only one microstate, and, according to Equation 17.21, its entropy is zero:

$$S = k_B (\ln W) = k_B (\ln 1) = 0$$

In this chapter we addressed the reasons why some reactions and processes are spontaneous, whereas others are not. We have seen that the value of free-energy change (ΔG) determines the spontaneity of a chemical reaction or process. Together, enthalpy and entropy changes for chemical reactions allow us to determine the ΔG for a chemical reaction or process. The free-energy change represents the maximum amount of work that can be done by the energy associated with a change. For changes carried out in the real world, the maximum amount of work is always less than ΔG.

microstate a unique distribution of particles among energy levels.

CONCEPT TEST

Predict which of the following values comes closest to the number of microstates accessible to 1 mole of liquid water molecules at 25°C: (a) 1; (b) 10^2; (c) 10^{10}; (d) 10^{23}; (e) $\gg 10^{23}$. Use the appropriate $S°$ value in Appendix 4 and Equation 17.21 to determine whether your selection was the best one.

SAMPLE EXERCISE 17.10 Integrating Concepts: Ötzi's Axe and the Chemistry of Copper Refining

In 1991 two hikers discovered the remains of a prehistoric man frozen in a glacier high in the Ötztal Alps, near the Italy–Austria border. Ötzi the Iceman, as he came to be known, lived about 5300 years ago. Among his possessions was an axe with a head made of nearly pure (99.7%) copper (Figure 17.26). The purity of the copper, as well as other clues, told scientists that the copper had been skillfully extracted and refined. Though copper does occur in its native elemental state in Earth's crust, ancient craftsmen usually had to extract the metal from copper-containing minerals such as chalcocite (Cu_2S). The process required that they find a way to add a lot of energy to a sulfide mineral like Cu_2S because its decomposition reaction has a large, positive $\Delta G°$ value:

$$Cu_2S(s) \rightarrow 2\,Cu(s) + \tfrac{1}{8}\,S_8(s) \qquad \Delta G° = 86.2\ kJ$$

However, when Cu_2S is roasted in a furnace that provides an ample supply of very hot air, any sulfur produced by the decomposition reaction is then oxidized to SO_2 gas in a decidedly spontaneous reaction:

$$\tfrac{1}{8}\,S_8(s) + O_2(g) \rightarrow SO_2(g) \qquad \Delta G° = -300.1\ kJ$$

a. Can these two reactions be coupled so that the spontaneous second reaction drives the nonspontaneous first reaction?
b. Is there likely to be an increase in entropy during the overall reaction?
c. What is the equilibrium constant of the first reaction under standard conditions?
d. How is the equilibrium shifted by the reaction conditions in the furnace?

Collect and Organize We are given two chemical reactions and asked to determine whether the free energy released by the spontaneous one can be coupled to and drive the nonspontaneous reaction. We also need to calculate the equilibrium constant of the nonspontaneous reaction and to decide how reaction conditions

FIGURE 17.26 Copper axe found alongside Ötzi the Iceman, who lived about 5300 years ago.

shift the equilibrium state. Equation 17.17 relates the equilibrium constant of a chemical reaction to its $\Delta G°_{rxn}$ value.

Analyze Sulfur is a product of the first reaction and a reactant in the second one, so it should be possible to couple the two reactions. Given the large positive value of $\Delta G°_{rxn}$ for the first reaction, its K value should be much less than 1. The second reaction consumes a product (sulfur) of the first reaction, which should shift the equilibrium in the first reaction toward forming more products.

Solve
a. One mole of sulfur is produced in the first reaction and consumed in the second, so the two reactions can be coupled. The result is an overall reaction with a $\Delta G°_{rxn}$ value that is the sum of the first two:

$$\begin{array}{ll} Cu_2S(s) \rightarrow 2\,Cu(s) + \tfrac{1}{8}\,S_8(s) & \Delta G°_{rxn} = 86.2\ kJ \\ + \tfrac{1}{8}\,S_8(s) + O_2(g) \rightarrow SO_2(g) & \Delta G°_{rxn} = -300.1\ kJ \\ \hline Cu_2S(s) + O_2(g) \rightarrow 2\,Cu(s) + SO_2(g) & \Delta G°_{rxn} = -213.9\ kJ \end{array}$$

The calculated negative $\Delta G°_{rxn}$ value means that the overall reaction is spontaneous.

b. The reactants and products both contain 1 mole of gas, but SO_2 is triatomic and has a larger standard molar entropy [248.2 J/(mol · K)] than diatomic O_2 [205.0 J/(mol · K)]. Also, the other reactant is 1 mole of a binary ionic solid, but the products include 2 moles of solid metal. Therefore, from the $S°$ values in Appendix 4, we can safely predict that the system entropy increases in the overall reaction.

c. Using Equation 17.17 to calculate the equilibrium constant for the decomposition reaction under standard conditions (and letting $T = 298$ K) gives us:

$$K = e^{-\Delta G°_{rxn}/RT}$$

First calculate the exponent:

$$-\frac{\Delta G°_{rxn}}{RT} = -\frac{86.2\,\dfrac{kJ}{mol} \times 1000\,\dfrac{J}{kJ}}{\left(8.314\,\dfrac{J}{mol \cdot K}\right)(298\ K)} = -34.79$$

Then the value of K is

$$K = e^{-34.79} = 7.8 \times 10^{-16}$$

d. As the reaction mixture in the first reaction comes to equilibrium (after forming an insignificant quantity of product), the removal of sulfur as SO_2 shifts the equilibrium in favor of forming more product, as predicted by Le Châtelier's principle.

Think About It As predicted, the nonspontaneous first reaction has a small K value; however, the reaction becomes spontaneous when coupled to the spontaneous second reaction. This shift toward forming more Cu metal is also predicted by Le Châtelier's principle and the capacity of the second reaction to remove one of the products (sulfur) of the first.

SUMMARY

LO1 **Spontaneous processes** happen on their own without continual outside intervention. **Nonspontaneous processes**, which are spontaneous processes in reverse, do not happen on their own. Spontaneous processes may be exothermic or endothermic and are often accompanied by an increase in the freedom of motion of the particles involved in the process. **Entropy (S)** is a thermodynamic property that provides a measure of how dispersed the energy is in a system at a given temperature. According to the **second law of thermodynamics**, a spontaneous process is accompanied by an increase in the entropy of an isolated system, or an increase in entropy of the universe for a process occurring in any system. A **reversible process** takes place in very small steps and very slowly so that the system can be restored to its initial state with no net flow of energy to or from its surroundings. (Sections 17.1 and 17.2)

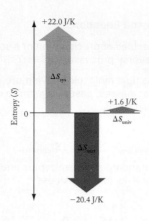

LO2 According to the **third law of thermodynamics**, a perfect crystal of a pure substance has zero entropy at absolute zero. All substances have positive entropies at temperatures above absolute zero. **Standard molar entropies ($S°$)** are entropy values for substances in their standard states. The entropy of a system increases with increasing molecular complexity and with increasing temperature. (Section 17.3)

LO3 The entropy change in a reaction under standard conditions can be calculated from the standard entropies of the products and reactants and their coefficients in the balanced chemical equation. (Section 17.4)

LO4 **Free energy (G)** is defined as the energy available to do useful work. The change in free energy of a process is a state function defining the maximum useful work the system can do on its surroundings. When a process results in a decrease in free energy of a system ($\Delta G < 0$), the process is spontaneous; when $\Delta G > 0$ the process is nonspontaneous. Reversing a process changes the sign of ΔG, and $\Delta G = 0$ for a process at equilibrium. The change in free energy of a reaction under standard conditions can be calculated either from the

standard free energies of formation ($\Delta G_f°$) of the products and reactants or from the enthalpy and entropy changes. (Section 17.5)

LO5 The temperature range over which a process is spontaneous depends on the relative magnitudes of ΔH and ΔS. (Section 17.6)

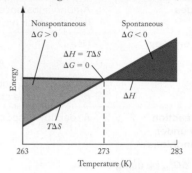

LO6 Negative values of $\Delta G°$ correspond to $K > 1$ and equilibrium reaction mixtures composed mostly of products. Positive values of $\Delta G°$ correspond to $K < 1$. As spontaneous reactions proceed, the free energy of the reaction mixture increases from an initial negative value and reaches zero when the reaction comes to chemical equilibrium. (Section 17.7)

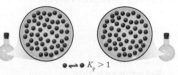

LO7 Higher reaction temperatures increase the equilibrium constant of an endothermic reaction but decrease the equilibrium constant of an exothermic reaction. The slope of a plot of ln K versus $1/T$ for an equilibrium system is used to determine the standard enthalpy for the equilibrium, and the y-intercept of the plot is used to determine the standard entropy change. (Section 17.8)

LO8 Many important biochemical processes, including **glycolysis** and **phosphorylation**, are made possible by coupled spontaneous and nonspontaneous reactions. The free energy released in the spontaneous processes going on in the body is used to drive nonspontaneous processes. (Section 17.9)

LO9 As temperature increases, the translational, rotational, and vibrational motion of molecules increases. A **microstate** is defined as one discrete way for a system to disperse all its energy, and as the number of microstates increases, so does the entropy of a system. (Section 17.10)

PARTICULATE **PREVIEW WRAP-UP**

Cu is a solid, Br_2 is a liquid, and He is a gas. The number of possible arrangements increases from the copper atoms (with the fewest) to the bromine molecules to the helium atoms (with the most). Br_2 molecules in the liquid phase and Cu atoms in the solid phase have vibrational

motion, whereas atoms of He(g) do not. Br_2 molecules also have rotational and translational motion (Cu atoms do not), so Br_2 molecules have more ways to disperse their energy than the atoms of the other two elements.

PROBLEM-SOLVING SUMMARY

Type of Problem	Concepts and Equations	Sample Exercises
Predicting the sign of entropy change	Look for the net removal of gas molecules or precipitation of a solute from solution ($\Delta S < 0$ for both). For the reverse processes, $\Delta S > 0$.	**17.1**
Comparing absolute entropy values	Substances composed of larger, less rigid molecules have more entropy. Among substances with similar molar masses, gases have more entropy than liquids, which have more entropy than solids.	**17.2**
Calculating entropy changes	Use $$\Delta S^\circ_{rxn} = \sum n_{products}S^\circ_{products} - \sum n_{reactants}S^\circ_{reactants} \qquad (17.5)$$ where $S^\circ_{products}$ and $S^\circ_{reactants}$ are the standard molar entropies and $n_{products}$ and $n_{reactants}$ are the stoichiometric coefficients for the process.	**17.3**
Predicting reaction spontaneity under standard conditions	A reaction is spontaneous if $$\Delta S_{univ} = (\Delta S_{sys} + \Delta S_{surr}) > 0$$	**17.4**
Calculating ΔG°_{rxn} by using appropriate ΔG°_f values	Use $$\Delta G^\circ_{rxn} = \sum n_{products}\Delta G^\circ_{f,products} - \sum n_{reactants}\Delta G^\circ_{f,reactants} \qquad (17.12)$$ where $\Delta G^\circ_{f,products}$ and $\Delta G^\circ_{f,reactants}$ are the standard molar free energies of formation, and $n_{products}$ and $n_{reactants}$ are the stoichiometric coefficients for the process.	**17.5**
Relating reaction spontaneity to ΔH and ΔS	Use $$\Delta G^\circ_{rxn} = \Delta H^\circ_{rxn} - T\Delta S^\circ_{rxn} \qquad (17.11)$$	**17.6**
Calculating the value of K from ΔG°_f values	$$K = e^{-\Delta G^\circ_{rxn}/RT} \qquad (17.17)$$	**17.7**
Calculating the value of K at a specific temperature	Use $$\ln\left(\frac{K_2}{K_1}\right) = -\frac{\Delta H^\circ_{rxn}}{R}\left(\frac{1}{T_2} - \frac{1}{T_1}\right) \qquad (17.20)$$ Convert ΔH°_{rxn} from kilojoules per mole to joules per mole to match the units of R.	**17.8**
Calculating ΔG° of coupled reactions	Free-energy changes are additive. If adding two reactions gives the desired overall reaction, add the free-energy changes of the two reactions to obtain the free-energy change of the overall reaction.	**17.9**

VISUAL PROBLEMS

(Answers to boldface end-of-chapter questions and problems are in the back of the book.)

17.1. Two balloons are inflated at the same temperature to the same volume (Figure P17.1), though it takes more gas to inflate the balloon on the right. In which balloon is the gas under greater internal pressure? In which balloon does the gas have greater entropy?

FIGURE P17.1

17.2. Two cubic containers (Figure P17.2) contain the same quantity of gas at the same temperature. Which cube contains gas with more entropy? If the sample in part (b) is left unchanged but the sample in part (a) is cooled so that it condenses, which sample has the higher entropy?

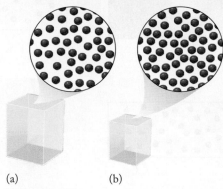

(a)   (b)
FIGURE P17.2

17.3. Figure P17.3 shows a tank that has just been filled with a mixture of two ideal gases: A (red spheres) and B (blue spheres). If the molar mass of A is twice that of B, will the atoms of A eventually fill the bottom of the tank and the atoms of B fill the top of the tank? Why or why not? Does the tank represent an isolated system?

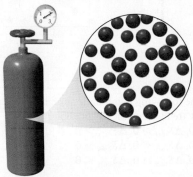

FIGURE P17.3

17.4. The box on the left of Figure P17.4 represents a mixture of two diatomic gases: A_2 (red spheres) and B_2 (blue spheres). As a result of the process depicted by the arrow, how do the entropies of A_2 and B_2 change?

FIGURE P17.4

17.5. Is the process in Figure P17.4 more likely to be spontaneous at high temperature or low temperature, or is it unaffected by changing temperature?

17.6. Figure P17.6 shows the plots of ΔH and $T\Delta S$ for a phase change as a function of temperature.
   a. What is the status of the process at the point where the two lines intersect?
   b. Over what temperature range is the process spontaneous?

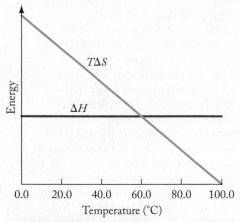

FIGURE P17.6

17.7. Figure P17.7 presents the ΔG_f° values of several elements and compounds selected from Appendix 4. Which of the following conversions are spontaneous? (a) $C_6H_6(\ell)$ to $CO_2(g)$ and $H_2O(\ell)$; (b) $CO_2(g)$ to $C_2H_2(g)$; (c) $H_2(g)$ and $O_2(g)$ to $H_2O(\ell)$. Explain your reasoning. *Hint*: Begin by writing a balanced equation describing the conversion. Add other substances as needed to complete the equation.

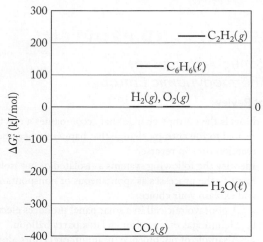

FIGURE P17.7

17.8. Use representations [A] through [I] in Figure P17.8 to answer questions a–f.

a. What is the sign of ΔS for image [F]?

b. The slime in image [H] is a cross-linked polymer of polyvinyl alcohol and borax. What is the sign of ΔS for the formation of slime?

c. What is the sign of ΔS for the sublimation of dry ice in image [B]? Under what conditions would this process *not* be spontaneous?

d. If you mix the particulate substances in [A] and [C] to create image [E], does the entropy increase, decrease, or remain unchanged? When you mix the particulate substances in [G] and [I] to create image [E], how does the change in volume affect the entropy?

e. Image [D] shows a shaken bottle of oil and water. After some time passes, what will happen to the oil and water? Will this process be spontaneous? Will the entropy increase or decrease?

f. Does all mixing lead to an increase in entropy?

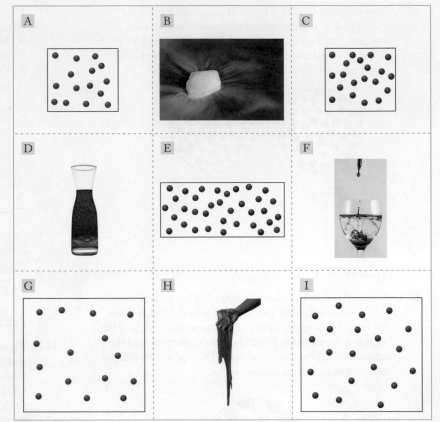

FIGURE P17.8

QUESTIONS AND PROBLEMS

Spontaneous Processes and Thermodynamic Entropy

Concept Review

17.9. How is the entropy change that accompanies a reaction related to the entropy change that happens when the reaction runs in reverse?

17.10. Identify the following systems as isolated or not isolated, identify the processes as spontaneous or nonspontaneous, and explain your choice.

a. A photovoltaic cell in a solar panel produces electricity.

b. Helium gas escapes from a latex party balloon.

c. A sample of pitchblende (uranium ore) emits alpha particles.

17.11. Ice cubes melt in a glass of lemonade, cooling the lemonade from 10.0°C to 0.0°C. If the ice cubes are the system, what are the signs of ΔS_{sys} and ΔS_{surr}?

17.12. Adding sidewalk deicer (calcium chloride) to water causes the temperature of the water to increase. If solid $CaCl_2$ is the system, what are the signs of ΔS_{sys} and ΔS_{surr}?

Problems

17.13. Which of the following combinations of entropy changes for a process are mathematically possible?

a. $\Delta S_{sys} > 0$, $\Delta S_{surr} > 0$, $\Delta S_{univ} > 0$

b. $\Delta S_{sys} > 0$, $\Delta S_{surr} < 0$, $\Delta S_{univ} > 0$

c. $\Delta S_{sys} > 0$, $\Delta S_{surr} > 0$, $\Delta S_{univ} < 0$

17.14. Which of the following combinations of entropy changes for a process are mathematically possible?

a. $\Delta S_{sys} < 0$, $\Delta S_{surr} > 0$, $\Delta S_{univ} > 0$

b. $\Delta S_{sys} < 0$, $\Delta S_{surr} < 0$, $\Delta S_{univ} > 0$

c. $\Delta S_{sys} < 0$, $\Delta S_{surr} > 0$, $\Delta S_{univ} < 0$

17.15. What are the signs of ΔS_{sys} and ΔS_{univ} for the photosynthesis of glucose from carbon dioxide and water?

17.16. What are the signs of ΔS_{sys}, ΔS_{surr}, and ΔS_{univ} for the complete combustion of propane in which the products include water vapor and carbon dioxide?

17.17. The nonspontaneous reaction D + E → F decreases the system entropy by 66.0 J/K. What is the maximum value of the entropy change of the surroundings?

17.18. The spontaneous reaction A → B + C increases the system entropy by 72.0 J/K. What is the minimum value of the entropy change of the surroundings?

17.19. For the following reactions, indicate whether the entropy of the system increases, decreases, or remains nearly the same.

a. $Al^{3+}(aq) + 3\ OH^-(aq) \rightarrow Al(OH)_3(s)$

b. $CaCO_3(s) \rightarrow CaO(s) + CO_2(g)$

c. $Mg(s) + Cu^{2+}(aq) \rightarrow Mg^{2+}(aq) + Cu(s)$

17.20. Which of the following processes would *not* result in an entropy increase for the indicated system?

a. Melting of an ice cube

b. Evaporation of a sample of an alcohol

c. Sublimation of a mothball

d. Cooling of hot water to room temperature

Absolute Entropy and the Third Law of Thermodynamics

Concept Review

17.21. Which component in each of the following pairs has the greater entropy?
   a. 1 mole of $S_2(g)$ or 1 mole of $S_8(g)$
   b. 1 mole of $S_2(g)$ or 1 mole of $S_8(s)$
   c. 1 mole of $O_2(g)$ or 1 mole of $O_3(g)$
   d. 1 gram of $O_2(g)$ or 1 gram of $O_3(g)$

17.22. Digestion During digestion, complex carbohydrates decompose into simple sugars. Do the carbohydrates experience an increase or decrease in entropy?

17.23. Diamond and the fullerenes are two allotropes of carbon. On the basis of their different structures and properties, predict which has the higher standard molar entropy.

17.24. Superfluids The 1996 Nobel Prize in Physics was awarded to Douglas Osheroff, Robert Richardson, and David Lee for discovering *superfluidity* (apparently frictionless flow) in ^{3}He. When ^{3}He is cooled to 2.7 mK, the liquid settles into an *ordered* superfluid state. What is the predicted sign of the entropy change for the conversion of liquid ^{3}He into its superfluid state?

17.25. Rank the compounds in each of the following groups in order of increasing standard molar entropy ($S°$):
   a. $CH_4(g)$, $CF_4(g)$, and $CCl_4(g)$
   b. $CH_2O(g)$, $CH_3CHO(g)$, and $CH_3CH_2CHO(g)$
   c. $HF(g)$, $H_2O(g)$, and $NH_3(g)$

17.26. The hydrocarbon C_5H_{12} has three isomers, shown in Figure P17.26. Which isomer has the largest molar entropy, $S°$?

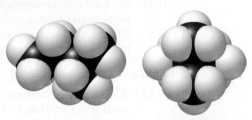

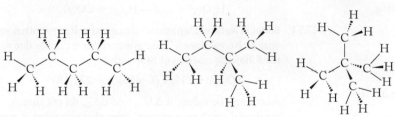

Pentane   2-Methylbutane   2,2-Dimethylpropane

FIGURE P17.26

Calculating Entropy Changes

Concept Review

17.27. Under standard conditions, the products of a reaction have, overall, greater entropy than the reactants. What is the sign of $\Delta S°_{rxn}$?

17.28. Do polymerization reactions tend to have $\Delta S°_{rxn}$ values that are greater than zero or less than zero? Why?

17.29. Do precipitation reactions tend to have $\Delta S°_{rxn}$ values that are greater than zero or less than zero? Why?

*17.30. Suppose compound A(s) decomposes to substances B(ℓ) and C(ℓ) at moderately high temperatures. How would running the decomposition reaction at even higher temperatures (above the melting point of A, but still below the boiling points of B and C) affect the value of $\Delta S°_{rxn}$?

Problems

17.31. Smog Use the standard molar entropies in Appendix 4 to calculate $\Delta S°$ values for each of the following atmospheric reactions that contribute to the formation of photochemical smog.
   a. $N_2(g) + O_2(g) \rightarrow 2\,NO(g)$
   b. $2\,NO(g) + O_2(g) \rightarrow 2\,NO_2(g)$
   c. $NO(g) + \frac{1}{2}\,O_2(g) \rightarrow NO_2(g)$
   d. $2\,NO_2(g) \rightarrow N_2O_4(g)$

17.32. Use the standard molar entropies in Appendix 4 to calculate the $\Delta S°$ value for each of the following reactions.
   a. $CH_4(g) + N_2(g) \rightarrow HCN(g) + NH_3(g)$
   b. $Cu_2S(s) + O_2(g) \rightarrow 2\,Cu(s) + SO_2(g)$
   c. $SO_3(g) + H_2O(\ell) \rightarrow H_2SO_4(aq)$
   d. $S(g) + O_2(g) \rightarrow SO_2(g)$

17.33. What is the entropy change to the surroundings when a small decorative ice sculpture at a temperature of 0°C and weighing 456 g melts on a granite tabletop, if the temperature of the granite is 12°C and the process occurs reversibly? Assume a final temperature for the water of 0°C. The heat of fusion of ice is 6.01 kJ/mol.

17.34. Another decorative "ice" sculpture is carved from dry ice (solid CO_2) and held at its sublimation point of −78.5°C. What is the entropy change to the universe when the CO_2 sculpture, weighing 389 g, sublimes on a granite tabletop if the temperature of the granite is 12°C and the process occurs reversibly? Assume a final temperature for the CO_2 vapor of −78.5°C. The heat of sublimation of CO_2 is 26.1 kJ/mol.

Free Energy

Concept Review

17.35. What single criterion allows us to determine whether a process is spontaneous?
   a. The sign of the equilibrium constant, K
   b. The sign of the enthalpy change, ΔH
   c. The sign of the free-energy change, ΔG
   d. The sign of the entropy change, ΔS

17.36. If ΔG for a reaction is negative, then which of the following is true?
   a. The reaction is spontaneous in the forward direction as written.
   b. The reaction system is at equilibrium.
   c. The reverse of the given reaction is spontaneous.
   d. No reaction would ever have a negative ΔG.

*17.37. Many 19th-century scientists believed that all exothermic reactions were spontaneous. Why did so many scientists share this belief?

17.38. In which direction does a reaction proceed when its ΔG value is (a) less than zero; (b) equal to zero; (c) greater than zero?

17.39. What are the signs of ΔS, ΔH, and ΔG for the sublimation of dry ice (solid CO_2) at 25°C?

17.40. What are the signs of ΔS, ΔH, and ΔG for the formation of dew on a cool night?

17.41. Which of the following processes is/are spontaneous?
a. A hurricane forms.
b. A corpse decomposes.
c. You get an A in this course.
d. Ice cream melts on a hot summer day.

17.42. Which of the following processes is/are spontaneous?
a. Wood burns in air.
b. Water vapor condenses on the sides of a glass of iced tea.
c. Salt dissolves in water.
d. Photosynthesis.

Problems

17.43. Calculate the free-energy change for the dissolution in water of 1 mole of NaBr and 1 mole of NaI at 298 K from the values in the table below.

	ΔH°_{soln} (kJ/mol)	ΔS°_{soln} [J/(mol · K)]
NaBr	−0.86	57
NaI	−7.5	74

17.44. The values of ΔH°_{rxn} and ΔS°_{rxn} for the reaction
$$2\,NO(g) + O_2(g) \rightarrow 2\,NO_2(g)$$
are −12 kJ and −146 J/K, respectively.
a. Use these values to calculate ΔG°_{rxn} at 298 K.
b. Explain why the value of ΔG°_{rxn} is negative.

17.45. A mixture of $CO(g)$ and $H_2(g)$ is produced by passing steam over hot charcoal:
$$H_2O(g) + C(s) \rightarrow H_2(g) + CO(g)$$
Calculate the ΔG°_{rxn} value for the reaction from the appropriate ΔG°_f data in Appendix 4.

17.46. Using appropriate data from Appendix 4, calculate ΔG° for the following reaction:
$$N_2O_3(g) \rightarrow NO(g) + NO_2(g)$$

17.47. Determine the value of ΔG° for the reduction of iron ore with hydrogen gas:
$$Fe_2O_3(s) + 3\,H_2(g) \rightarrow 2\,Fe(s) + 3\,H_2O(g)$$
given the following thermodynamic properties at 25°C:

	$Fe_2O_3(s)$	$H_2(g)$	$Fe(s)$	$H_2O(g)$
ΔH°_f (kJ/mol)	−824.2	0	0	−241.8
S° [J/(mol · K)]	87.4	130.6	27.3	188.8

17.48. Determine the value of ΔG° for the reaction of the fuel gas acetylene with hydrogen gas:
$$C_2H_2(g) + 2\,H_2(g) \rightarrow C_2H_6(g)$$
given the following thermodynamic properties at 25°C:

	$C_2H_2(g)$	$H_2(g)$	$C_2H_6(g)$
ΔH°_f (kJ/mol)	226.7	0	−84.7
S° [J/(mol · K)]	200.8	130.6	229.5

17.49. **Acid Precipitation** Aerosols (fine droplets) of sulfuric acid form in the atmosphere as a result of the reaction below. Use the appropriate ΔG°_f data from Appendix 4 to calculate ΔG°_{rxn} for the combination reaction:
$$SO_3(g) + H_2O(g) \rightarrow H_2SO_4(\ell)$$

17.50. One source of sulfuric acid aerosols in the atmosphere is combustion of high-sulfur fuels, which releases SO_2 gas that then is further oxidized to SO_3:
$$2\,SO_2(g) + O_2(g) \rightarrow 2\,SO_3(g)$$
Use the appropriate ΔG°_f data from Appendix 4 to calculate ΔG°_{rxn} for this combination reaction at 25°C. Is it spontaneous under standard conditions?

Temperature and Spontaneity

Concept Review

17.51. Are exothermic reactions spontaneous only at low temperature? Explain your answer.

17.52. Are endothermic reactions never spontaneous at low temperature? Explain your answer.

Problems

17.53. What is the lowest temperature at which the following reaction (see Problem 17.45) is spontaneous?
$$H_2O(g) + C(s) \rightarrow H_2(g) + CO(g)$$

17.54. **Sulfur in Nature** Deposits of elemental sulfur are often seen near active volcanoes. Their presence there may be due to the following reaction of SO_2 with H_2S:
$$SO_2(g) + 2\,H_2S(g) \rightarrow \tfrac{3}{8}\,S_8(s) + 2\,H_2O(g)$$
Assuming the values of ΔH°_{rxn} and ΔS°_{rxn} do not change appreciably with temperature, over what temperature range is the reaction spontaneous?

17.55. Use the data in Appendix 4 to calculate ΔH° and ΔS° for the vaporization of hydrogen peroxide:
$$H_2O_2(\ell) \rightarrow H_2O_2(g)$$
Assuming that the calculated values are independent of temperature, what is the boiling point of hydrogen peroxide at $P = 1.00$ atm?

17.56. Determine the normal melting point of carbon tetrachloride in degrees Celsius given the following data:
$$\Delta H^{\circ}_{fus} = 2.67 \text{ kJ/mol} \qquad \Delta S^{\circ}_{fus} = 10.86 \text{ J/(mol · K)}$$

17.57. Which of the following reactions is spontaneous (i) only at low temperatures; (ii) only at high temperatures; (iii) at all temperatures?
a. $2\,NO(g) + O_2(g) \rightarrow 2\,NO_2(g)$
b. $2\,NH_3(g) + 2\,O_2(g) \rightarrow N_2O(g) + 3\,H_2O(g)$
c. $NH_4NO_3(s) \rightarrow 2\,H_2O(g) + N_2O(g)$

17.58. Which of the following reactions is spontaneous (i) only at low temperatures; (ii) only at high temperatures; (iii) at all temperatures?
a. $2\,H_2S(g) + 3\,O_2(g) \rightarrow 2\,H_2O(g) + 2\,SO_2(g)$
b. $SO_2(g) + H_2O_2(\ell) \rightarrow H_2SO_4(\ell)$
c. $S(g) + O_2(g) \rightarrow SO_2(g)$

Free Energy and Chemical Equilibrium

Concept Review

17.59. If the value of K for a reaction is less than 1, what is the sign of ΔG°_{rxn}?

*17.60. The equation $\Delta G^\circ = -RT \ln K$ relates the value of K_p, not K_c, to ΔG° for gas-phase reactions. Explain why.

17.61. If a reaction mixture contains only reactants and no products, will the reaction proceed in the forward direction even if $\Delta G^\circ > 0$? Explain why or why not.

*17.62. If a gas-phase reaction mixture contains 1 mole of each reactant and product and has a 1:1 stoichiometry, is the ΔG°_{rxn} of the reaction mixture the same as ΔH°_{rxn}? Explain why or why not.

Problems

17.63. Which of the following reactions has the largest K_p value at 25°C?
a. $Cl_2(g) + F_2(g) \rightleftharpoons 2\,ClF(g)$ $\Delta G^\circ_{rxn} = 115.4$ kJ
b. $Cl_2(g) + Br_2(g) \rightleftharpoons 2\,ClBr(g)$ $\Delta G^\circ_{rxn} = -2.0$ kJ
c. $Cl_2(g) + I_2(g) \rightleftharpoons 2\,ICl(g)$ $\Delta G^\circ_{rxn} = -27.9$ kJ

17.64. Use the appropriate ΔG°_f value from Appendix 4 to calculate the value of K_p at 298 K for the reaction
$$N_2(g) + 2\,O_2(g) \rightleftharpoons 2\,NO_2(g)$$

17.65. Use the appropriate equilibrium constant from Appendix 5 to calculate the value of ΔG°_{rxn} for the reaction
$$NH_3(g) + H_2O(\ell) \rightleftharpoons NH_4^+(aq) + OH^-(aq)$$

17.66. Use the appropriate equilibrium constant from Appendix 5 to calculate the value of ΔG°_{rxn} for the reaction
$$HClO(aq) + H_2O(\ell) \rightleftharpoons ClO^-(aq) + H_3O^+(aq)$$

Influence of Temperature on Equilibrium Constants

Concept Review

17.67. The value of the equilibrium constant of a reaction decreases with increasing temperature. Is the reaction endothermic or exothermic?

17.68. The reaction
$$2\,CO(g) + O_2(g) \rightleftharpoons 2\,CO_2(g)$$
is exothermic. Does the value of K_p increase or decrease with increasing temperature?

17.69. The value of K_p for the water–gas shift reaction
$$CO(g) + H_2O(g) \rightleftharpoons H_2(g) + CO_2(g)$$
increases as the temperature decreases. Is the reaction exothermic or endothermic?

17.70. Does the value of K_p for the reaction
$$CH_4(g) + H_2O(g) \rightleftharpoons 3\,H_2(g) + CO(g) \qquad \Delta H^\circ = 206\ \text{kJ}$$
increase, decrease, or remain unchanged as temperature increases?

Problems

17.71. **Air Pollution** Automobiles and trucks pollute the air with NO. At 2000°C, the value of K_c for the reaction
$$N_2(g) + O_2(g) \rightleftharpoons 2\,NO(g)$$
is 4.10×10^{-4}, and $\Delta H^\circ = 180.6$ kJ. What is the value of K_c at 25°C?

17.72. At 400 K the value of K_p for the reaction
$$N_2(g) + 3\,H_2(g) \rightleftharpoons 2\,NH_3(g)$$
is 41, and $\Delta H^\circ = -92.2$ kJ. What is the value of K_p at 700 K?

17.73. The equilibrium constant for the reaction
$$NO(g) + O_2(g) \rightleftharpoons 2\,NO_2(g)$$
decreases from 1.5×10^5 at 430°C to 23 at 1000°C. From these data, calculate the value of ΔH° for the reaction.

17.74. The value of K_c for the reaction $A \rightleftharpoons B$ is 0.455 at 50°C and 0.655 at 100°C. Calculate ΔH° for the reaction.

Driving the Human Engine: Coupled Reactions

Concept Review

17.75. Describe the ways in which two chemical reactions must complement each other so that the decrease in free energy of the spontaneous reaction can drive the nonspontaneous reaction.

17.76. Why is it important that at least some of the spontaneous steps in glycolysis convert ADP to ATP?

17.77. The second step in glycolysis converts glucose 6-phosphate into fructose 6-phosphate (Figure P17.77). Suggest a reason why ΔG° for this reaction is close to zero.

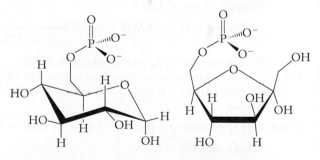

Glucose 6-phosphate   Fructose 6-phosphate
FIGURE P17.77

Problems

17.78. **Uses of Methane** The methane in natural gas is an important starting material, or feedstock, for producing industrial chemicals, including H_2 gas.
a. Use the appropriate ΔG°_f value(s) from Appendix 4 to calculate ΔG°_{rxn} for the reaction known as *steam–methane reforming*:
$$CH_4(g) + H_2O(g) \rightarrow CO(g) + 3\,H_2(g)$$
b. To drive this nonspontaneous reaction, the CO that is produced can be oxidized to CO_2 by using more steam:
$$CO(g) + H_2O(g) \rightarrow CO_2(g) + H_2(g)$$

Use the appropriate ΔG_f° value(s) from Appendix 4 to calculate ΔG_{rxn}° for this reaction, which is known as the *water–gas shift reaction*.

c. Combine these two reactions and write the chemical equation of the overall reaction in which methane and steam combine to produce hydrogen gas and carbon dioxide.

d. Calculate the ΔG_{rxn}° value of the overall reaction. Is it spontaneous under standard conditions?

17.79. In addition to its role in the reactions described in Problem 17.78, methane can, in theory, be used to produce hydrogen gas by a process in which it decomposes into elemental carbon and hydrogen:

$$CH_4(g) \rightarrow C(s) + 2\,H_2(g)$$

The carbon produced in the first step is then oxidized to CO_2:

$$C(s) + O_2(g) \rightarrow CO_2(g)$$

a. Calculate the ΔG_{rxn}° values of the above two reactions.

b. Write a balanced chemical equation describing the overall reaction obtained by coupling the above two reactions, and calculate its ΔG_{rxn}° value. Is the coupled reaction spontaneous under standard conditions?

17.80. Which of the following steps in glycolysis has the largest equilibrium constant?

a. Fructose 1,6-diphosphate $\rightleftharpoons$ 2 glyceraldehyde-3-phosphate $\Delta G_{rxn}^\circ = 24$ kJ

b. 3-Phosphoglycerate $\rightleftharpoons$ 2-phosphoglycerate $\Delta G_{rxn}^\circ = 4.4$ kJ

c. 2-Phosphoglycerate $\rightleftharpoons$ phosphoenolpyruvate $\Delta G_{rxn}^\circ = 1.8$ kJ

17.81. The value of ΔG° for the phosphorylation of glucose in glycolysis is 13.8 kJ/mol. What is the value of the equilibrium constant for the reaction at 298 K?

17.82. In glycolysis, the hydrolysis of ATP to ADP drives the phosphorylation of glucose:

$$Glucose + ATP \rightleftharpoons ADP + glucose\ 6\text{-phosphate}$$
$$\Delta G_{rxn}^\circ = -17.7\ kJ$$

What is the value of K_c for this reaction at 298 K?

17.83. Sucrose enters the series of reactions in glycolysis after the reaction in which it is hydrolyzed, forming glucose and fructose:

$$Sucrose + H_2O \rightleftharpoons glucose + fructose \quad K_c = 5.3 \times 10^{12} \text{ at 298 K}$$

What is the value of ΔG_{rxn}°?

Microstates: A Quantized View of Entropy

Concept Review

17.84. You flip three coins, assigning the values +1 for heads and −1 for tails. Each outcome of the three flips constitutes a microstate. How many different microstates are possible from flipping the three coins? Which value or values for the sums in the microstates are most likely? (*Hint*: The sequence HHT [+1 +1 −1] is one possible outcome, or microstate. Note, however, that this outcome differs from THH [−1 +1 +1], even though the two sequences sum to the same value.)

17.85. Imagine you have four identical chairs to arrange on four steps leading up to a stage, one chair on each step. The chairs have numbers on their backs: 1, 2, 3, and 4. How many different microstates for the chairs are possible? (Notice that when viewed from the front, all the microstates look the same. Viewed from the back, you can identify the different microstates because you can distinguish the chairs by their numbers.)

Problems

17.86. Use the appropriate standard molar entropy value from Appendix 4 to calculate how many microstates are accessible to a single molecule of N_2 at 298 K.

17.87. Use the appropriate standard molar entropy value from Appendix 4 to calculate how many microstates are accessible to a single molecule of liquid H_2O at 298 K.

Additional Problems

17.88. Methanogenic bacteria convert aqueous acetic acid (CH_3COOH) into $CO_2(g)$ and $CH_4(g)$.

a. Is this process endothermic or exothermic under standard conditions?

b. Is the reaction spontaneous under standard conditions?

17.89. Chlorofluorocarbons (CFCs) are no longer used as refrigerants because they catalyze the decomposition of stratospheric ozone. Trichlorofluoromethane (CCl_3F) boils at 23.8°C, and its molar heat of vaporization is 24.8 kJ/mol. What is the molar entropy of vaporization of $CCl_3F(\ell)$?

17.90. Consider the precipitation reactions described by the following net ionic equations:

$$Mg^{2+}(aq) + 2\,OH^-(aq) \rightarrow Mg(OH)_2(s)$$
$$Ag^+(aq) + Cl^-(aq) \rightarrow AgCl(s)$$

a. Predict the sign of ΔS_{rxn}° for the reactions.

b. Using the appropriate values for S° from Appendix 4, calculate ΔS° for these reactions.

c. Do your calculations support your prediction?

17.91. (a) Using the appropriate values for S° from Appendix 4, calculate ΔS° for the reaction of $Na(s)$ and $Cl_2(g)$ to form $NaCl(s)$. (b) Explain why this reaction takes place given your answer to part (a).

*17.92. At what temperature is the free-energy change for the following reaction equal to zero?

$$NH_4Cl(s) \rightarrow NH_3(g) + HCl(g)$$

17.93. Which of these processes result in an entropy decrease of the system?

a. Diluting hydrochloric acid with water

b. Boiling water

c. $2\,NO(g) + O_2(g) \rightarrow 2\,NO_2(g)$

d. Making ice cubes in the freezer

*17.94. Calculate the standard free-energy change of the following reaction. Is it spontaneous?

$$2\,NO(g) + 2\,H_2(g) \rightarrow N_2(g) + 2\,H_2O(g)$$

*17.95. Estimate the free-energy change of the following reaction at 225°C:

$$C_2H_4(g) + 3\,O_2(g) \rightarrow 2\,CO_2(g) + 2\,H_2O(g)$$

*17.96. Show that hydrogen cyanide (HCN) is a gas at 25°C by estimating its normal boiling point from the following data:

	ΔH_f° (kJ/mol)	S° [J/(mol · K)]
HCN(ℓ)	108.9	113
HCN(g)	135.1	202

17.97. **Making Methanol** The element hydrogen (H_2) is not abundant on Earth, but it is a useful reagent in, for example, the potential synthesis of the liquid fuel methanol from gaseous carbon monoxide:

$$2\,H_2(g) + CO(g) \rightarrow CH_3OH(\ell)$$

Under what temperature conditions is this reaction spontaneous?

17.98. **Lightbulb Filaments** Tungsten (W) is the favored metal for lightbulb filaments, partly because of its high melting point of 3422°C. The enthalpy of fusion of tungsten is 35.4 kJ/mol. What is its entropy of fusion?

*17.99. Over what temperature range is the reduction of tungsten(VI) oxide by hydrogen to give metallic tungsten and water spontaneous? The standard heat of formation of $WO_3(s)$ is −843 kJ/mol, and its standard molar entropy is 76 J/(mol · K).

17.100. Two allotropes (A and B) of sulfur interconvert at 369 K and 1 atm of pressure:

$$S_8(s, A) \rightarrow S_8(s, B)$$

The enthalpy change in this transition is 297 J/mol. What is the entropy change?

*17.101. Copper forms two oxides, Cu_2O and CuO.
   a. Name these oxides.
   b. Predict over what temperature range this reaction

$$Cu_2O(s) \rightarrow CuO(s) + Cu(s)$$

is spontaneous by using the following thermodynamic data:

	ΔH_f° (kJ/mol)	S° [J/(mol · K)]
$Cu_2O(s)$	−170.7	92.4
$CuO(s)$	−156.1	42.6

   c. Why is the standard molar entropy of $Cu_2O(s)$ larger than that of $CuO(s)$?

*17.102. **Lime** Enormous amounts of lime (CaO) are used in steel industry blast furnaces to remove impurities from iron. Lime is made by heating limestone and other solid forms of $CaCO_3(s)$. Why is the standard molar entropy of $CaCO_3(s)$ higher than that of $CaO(s)$? At what temperature is the pressure of $CO_2(g)$ over $CaCO_3(s)$ equal to 1.0 atm?

	ΔH_f° (kJ/mol)	S° [J/(mol · K)]
$CaCO_3(s)$	−1207	93
$CaO(s)$	−636	40
$CO_2(g)$	−394	214

*17.103. *Trouton's rule* says that the ratio $\Delta H_{vap}^\circ/T_b$ for a liquid is approximately 88 J/(mol · K). Here, ΔH_{vap}° is the molar enthalpy of vaporization of a liquid, and T_b is its normal boiling point.
   a. What idea suggests that $\Delta H_{vap}^\circ/T_b$ for a range of liquids should be approximately constant?
   b. Check Trouton's rule against the data in Figure P17.103. Which liquids deviate from Trouton's rule, and why?

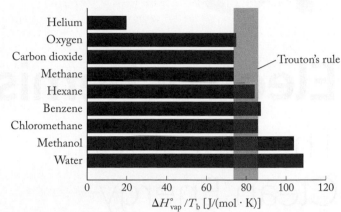

FIGURE P17.103

*17.104. **Melting DNA** When a solution of DNA in water is heated, the DNA double helix separates into two single strands:

$$1\ DNA\ double\ helix \rightleftharpoons 2\ single\ strands$$

   a. What is the sign of ΔS for the forward process as written?
   b. The DNA double helix reforms as the system cools. What is the sign of ΔS for the process by which two single strands reform the double helix?
   c. The melting point of DNA is defined as the temperature at which $\Delta G = 0$. At that temperature, the forward reaction produces two single strands as fast as two single strands recombine to form the double helix. Write an equation that defines the melting temperature (T) of DNA in terms of ΔH and ΔS.

*17.105. **Melting Organic Compounds** When dicarboxylic acids (compounds with two –COOH groups in their structures) melt, they frequently decompose to produce 2 moles of CO_2 gas for every 1 mole of dicarboxylic acid melted (shown in Figure P17.105).
   a. What are the signs of ΔH and ΔS for the process as written?
   b. Problem 17.104 describes the DNA double helix reforming when the system cools after melting. Do you think the dicarboxylic acid will reform when the melted material cools? Why or why not?

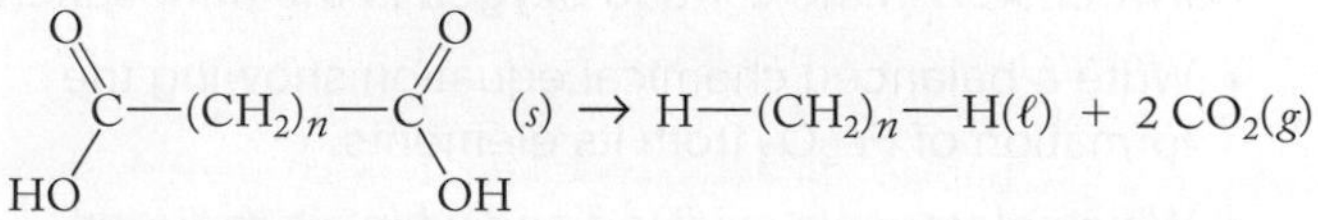

$$\underset{HO}{\overset{O}{\underset{\|}{C}}}-(CH_2)_n-\underset{OH}{\overset{O}{\underset{\|}{C}}}\ (s) \rightarrow H-(CH_2)_n-H(\ell) + 2\,CO_2(g)$$

FIGURE P17.105

18

Electrochemistry

The Quest for Clean Energy

AN ELECTRIFYING SEDAN The 2017 model Chevrolet Volt has an 18.4-kWh lithium-ion battery, which gives the Volt an all-electric driving range of about 85 km (53 mi). A gasoline-powered generator extends the Volt's overall driving range to about 680 km (420 mi).

PARTICULATE **REVIEW**

Redox: Metal versus Nonmetal

In Chapter 18 we investigate the transformation of chemical energy via redox reactions into electrical energy. Rust (mostly Fe_2O_3) forms on old cars such as the one in this photo through a series of reactions between iron in the car and oxygen in the atmosphere.

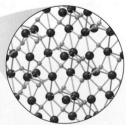

- Write a balanced chemical equation showing the formation of Fe_2O_3 from its elements.

- Which element is oxidized and which is reduced during this reaction?

- Describe the direction of the electron transfer in the formation of Fe_2O_3.

 (Review Section 4.9 if you need help.)

(Answers to Particulate Review questions are in the back of the book.)

Redox: Electricity and Clean Fuel

Not all redox reactions involve the oxidation and reduction of metals. The redox chemistry used to create fuel cells involves hydrogen gas and oxygen gas as pictured here (the electrons are not drawn). As you read Chapter 18, look for ideas that will help you answer these questions:

- Determine which half-reaction represents oxidation and which represents reduction. Write an overall equation for the chemistry of fuel cells that is balanced both in mass and charge.

- How do chemists manipulate the electron transfer in this reaction to generate electricity to power cars?

- Why do chemists call hydrogen a "clean" fuel in this chemical reaction?

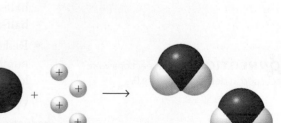

Learning Outcomes

LO1 Combine the appropriate half-reactions to write net ionic equations of spontaneous redox reactions
Sample Exercises 18.1, 18.2, 18.3, 18.4

LO2 Draw cell diagrams and describe the components of electrochemical cells and their roles in interconverting chemical and electrical energy
Sample Exercise 18.2

LO3 Calculate standard cell potentials from standard reduction potentials
Sample Exercise 18.3

LO4 Interconvert a cell's potential and the change in free energy of the cell reaction
Sample Exercise 18.4

LO5 Use the Nernst equation to calculate cell potentials
Sample Exercise 18.5

LO6 Relate standard cell potentials to the equilibrium constants of the cell reactions
Sample Exercise 18.6

LO7 Use the Faraday constant to relate quantity of charge to changes in the quantities of reactants and products in cell reactions
Sample Exercises 18.7, 18.8

18.1 Running on Electrons: Redox Chemistry Revisited

In previous chapters we explored some of the environmental impacts of producing energy by burning fossil fuels. Concerns about these impacts have spurred the development of innovative propulsion systems for cars and trucks that reduce or eliminate the use of fossil fuels. Some vehicles are hybrids, which means that they are propelled by a combination of gasoline engines and electric motors powered by rechargeable batteries. Plug-in hybrids (such as the one in this chapter's opening photograph) are powered exclusively by electric motors, but they have a small gasoline engine to generate electricity and extend their driving range. Completely electric vehicles are powered by banks of high-performance batteries or by fuel cells. If these all-electric cars are to replace most of the vehicles powered by fossil fuels, scientists and engineers will need to develop lighter-weight, higher-capacity, more powerful, more reliable, and less expensive batteries and fuel cells. It will not be easy, but progress is being made.

Batteries and fuel cells are devices based on **electrochemistry**, the branch of chemistry that links chemical reactions to the production or consumption of electrical energy. At the heart of electrochemistry are chemical reactions in which electrons are transferred between substances. In other words, electrochemistry is based on *red*uction and *ox*idation reactions, or *redox* reactions, which were introduced in Section 4.9. There we noted the following:

- Each redox reaction is the sum of two half-reactions: a reduction half-reaction, in which a reactant gains electrons, and an oxidation half-reaction, in which a reactant loses electrons.

- Reduction and oxidation half-reactions happen simultaneously, and the number of electrons gained during reduction must exactly match the number lost during oxidation.

We begin our exploration of electrochemistry by reviewing the oxidation and reduction half-reactions involved in a redox reaction between copper and zinc. When a strip of Zn metal is placed in a solution of $CuSO_4$ as shown in Figure 18.1, electrons spontaneously transfer from Zn atoms to Cu^{2+} ions, forming Zn^{2+} ions

CONNECTION Redox reactions were discussed in detail in Section 4.9.

FIGURE 18.1 A strip of zinc is immersed in an aqueous solution of blue copper(II) sulfate. Over time it becomes encrusted with a dark layer of copper as Cu^{2+} ions are reduced to Cu atoms and Zn atoms are oxidized to colorless Zn^{2+} ions.

and Cu atoms. The shiny zinc surface turns dark brown as a textured layer of copper metal accumulates on it, and the distinctive blue color of $Cu^{2+}(aq)$ ions fades as these ions gain electrons and become atoms of copper metal.

The electron transfer in this spontaneous reaction can be represented through two half-reactions that occur simultaneously—the oxidation half-reaction of zinc atoms and the reduction half-reaction of copper ions:

$$Zn(s) \rightarrow Zn^{2+}(aq) + 2\ e^-$$

$$Cu^{2+}(aq) + 2\ e^- \rightarrow Cu(s)$$

For 1 mole of zinc atoms, 2 moles of electrons are "lost," or transferred to 1 mole of copper ions. Because the number of moles of electrons lost and gained are the same in the two half-reactions, writing a net ionic equation to describe the overall redox reaction is simply a matter of adding the two half-reactions together:

$$Zn(s) \rightarrow Zn^{2+}(aq) + 2\ e^-$$

$$\underline{Cu^{2+}(aq) + 2\ e^- \rightarrow Cu(s)}$$

$$Zn(s) + Cu^{2+}(aq) + \cancel{2\ e^-} \rightarrow Cu(s) + Zn^{2+}(aq) + \cancel{2\ e^-}$$

Canceling out the equal numbers of electrons gained and lost, we get

$$Zn(s) + Cu^{2+}(aq) \rightarrow Cu(s) + Zn^{2+}(aq) \qquad (18.1)$$

Combining half-reactions is a convenient way to write net ionic equations for redox reactions. In this chapter we use a valuable resource in this equation-writing process: a table of common half-reactions in Appendix 6. Note that all the half-reactions are written as reduction half-reactions. There is no need for a separate table of *oxidation* half-reactions because any reduction half-reaction can always be reversed to obtain the corresponding oxidation half-reaction.

Having established that the redox reaction between zinc metal and $Cu^{2+}(aq)$ ions is the result of two distinct half-reactions, let's examine how chemists can *physically separate* the two half-reactions in order to produce electricity. To do this we use a device called an **electrochemical cell** (Figure 18.2). Like nearly all electrochemical cells, it consists of two compartments. One compartment contains a strip of zinc metal immersed in 1.00 *M* ZnSO₄; the other compartment has a strip of copper metal immersed in 1.00 *M* CuSO₄. Sulfate ions are spectator ions in this reaction and are not included in the net ionic equation, although we will see that they do play an important role in cell function. The two metal strips function as the *electrodes* of the cell, providing

CHEMTOUR
Zinc–Copper Cell

electrochemistry the branch of chemistry that examines the transformations between chemical and electrical energy.

electrochemical cell an apparatus that converts chemical energy into electrical work or electrical work into chemical energy.

FIGURE 18.2 This electrochemical cell consists of two compartments: the one on the left contains a zinc metal anode immersed in a 1.00 *M* solution of $ZnSO_4$; the one on the right contains a copper metal cathode immersed in a 1.00 *M* solution of $CuSO_4$. A porous bridge made of either glass or plastic provides an electrical connection through which ions (and their charges) can migrate from one compartment to the other.

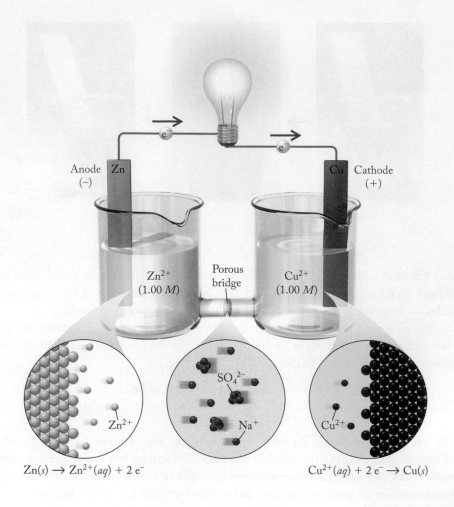

$$Zn(s) \rightarrow Zn^{2+}(aq) + 2\ e^-$$ $$Cu^{2+}(aq) + 2\ e^- \rightarrow Cu(s)$$

anode an electrode at which an oxidation half-reaction (loss of electrons) takes place.

cathode an electrode at which a reduction half-reaction (gain of electrons) takes place.

voltaic cell an electrochemical cell in which chemical energy is transformed into electrical work by a spontaneous cell reaction.

electrolysis a process in which electrical energy is used to drive a nonspontaneous chemical reaction.

electrolytic cell a device in which an external source of electrical energy does work on a chemical system, turning reactant(s) into higher-energy product(s).

pathways through which the electrons produced and consumed in the two half-reactions flow to and from an external circuit, converting chemical energy into electrical work.

As the cell reaction proceeds, oxidation of Zn atoms produces electrons, which travel from the Zn electrode through the external circuit to the surface of the Cu electrode, where they combine with Cu^{2+} ions, forming atoms of Cu metal. In an electrochemical cell, the electrode at which the oxidation half-reaction takes place (the zinc electrode in this case) is called the **anode**, and the electrode at which the reduction half-reaction takes place is called the **cathode**.

You might think that production of Zn^{2+} ions in the left compartment in Figure 18.2 would result in a buildup of positive charge on that side of the cell, and that conversion of Cu^{2+} ions to Cu metal would result in an excess of SO_4^{2-} ions and thus a negative charge in the Cu compartment. However, no such buildup of charge occurs because the two compartments are connected by a porous bridge that allows ions to migrate from one side of the cell to the other and because the solutions surrounding both electrodes contain a *background electrolyte* made of ions that are not involved in either half-reaction. In the Zn/Cu^{2+} cell, Na_2SO_4 is the background electrolyte. Migration of Na^+ ions toward the Cu compartment and SO_4^{2-} ions toward the Zn compartment through the porous bridge shown in Figure 18.2 balances the flow of electrons in the external circuit and eliminates any accumulation of ionic charge in either compartment.

As the cell reaction in Figure 18.2 proceeds, is the increase in mass of the copper strip the same as the decrease in mass of the zinc strip?

(Answers to Concept Tests are in the back of the book.)

18.2 Voltaic and Electrolytic Cells

In this chapter we examine the chemistry of several cells like the Zn/Cu^{2+} cell, in which the chemical reactions inside the cell pump electrons from the anode, through an external circuit, and into the cathode. A cell that operates in this way is called a **voltaic cell** in honor of the Italian physicist Alessandro Volta (1745–1827), who is credited with building the first battery (Figure 18.3). His battery and all of the modern batteries that power familiar electric devices, from cell phones to flashlights to laptop computers, are examples of voltaic cells.

We also investigate cells in which an external electrical power supply drives a nonspontaneous chemical reaction inside the cell. A reaction that is driven by the consumption of electrical energy is called **electrolysis**, and a cell in which electrolysis occurs is called an **electrolytic cell** (Figure 18.4). In electrolytic cells, electrons are pumped into the cathodes (making them the negative electrodes) and flow out the anodes (making them the positive electrodes). As we see in Section 18.9, many of the batteries used to power familiar electronic devices are rechargeable, which means that not only do their cell reactions produce electricity, but also that these same reactions can be forced to run in reverse when connected to an external power supply. When this happens, these voltaic cells become electrolytic cells as the products of the voltaic reaction become the reactants in the electrolytic reaction used to "recharge" the battery.

FIGURE 18.3 Alessandro Volta and his battery, which consisted of a stack of alternating layers of zinc, blotter paper soaked in salt water, and silver.

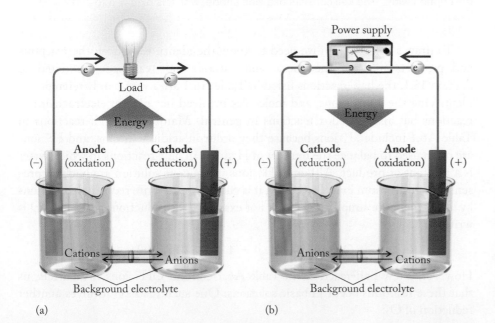

(a) (b)

FIGURE 18.4 Voltaic versus electrolytic cells. (a) In a voltaic cell, a spontaneous reaction produces electrical energy and does electrical work on its surroundings, such as lighting a light bulb. (b) In an electrolytic cell, an external supply of electrical energy does work on the chemical system in the cell, driving a nonspontaneous reaction.

cell diagram symbols that show how the components of an electrochemical cell are connected.

Cell Diagrams

Figure 18.2 depicts a Zn/Cu^{2+} electrochemical cell. Chemists have developed a notation to efficiently convey what components are present in any such cell: A **cell diagram** consists of both chemical formulas and symbols to show how the components of the cell are connected. A cell diagram does not convey stoichiometry, so any coefficients in the balanced equation for the cell reaction do not appear in the cell diagram. A cell diagram has these components:

anode | anode solution || cathode solution | cathode

To draw a cell diagram, we follow the three steps listed below. The symbols used to represent the cell diagram for the Zn/Cu^{2+} electrochemical cell in Figure 18.2 are included as an example with each step.

1. Write the chemical symbol of the anode at the far left of the diagram, the symbol of the cathode at the far right, and double vertical lines for the connecting bridge halfway between them:

$$Zn(s) \dots\dots \| \dots\dots Cu(s)$$

2. Work inward from the electrodes toward the connecting bridge, using vertical lines to indicate phase changes (such as that between a solid metal electrode and an aqueous solution). Represent the electrolytes surrounding the electrode by using the symbols of the ions or compounds that are changed by the cell reaction. Use commas to separate species in the same phase:

$$Zn(s) \,|\, Zn^{2+}(aq) \,\|\, Cu^{2+}(aq) \,|\, Cu(s)$$

3. If known, use the concentrations of the dissolved species in place of (aq) phase symbols, and add the partial pressures of any gases within their (g) phase symbols:

$$Zn(s) \,|\, Zn^{2+}(1.00 \; M) \,\|\, Cu^{2+}(1.00 \; M) \,|\, Cu(s)$$

CONCEPT TEST

Describe in your own words the meaning of the cell diagram for Figure 18.2. Start your description with: "The cell consists of a zinc anode, which is oxidized to . . ."

To draw a cell diagram we need to know the identities of both the reactants and products in both the anode and cathode half-reactions. As noted in Section 18.1, the half-reactions listed in Table A6.1 are a very handy reference in identifying the atoms, ions, and molecules involved not only in electrochemical reactions but also in redox reactions in general. Many of the half-reactions in Table A6.1 include H^+ ions because they occur in acidic solutions, and H^+ ions are often used to balance the number of H atoms in half-reactions in which water is a reactant or product. Although H^+ ions in aqueous solution are better represented as hydronium ions, $H_3O^+(aq)$, it is customary to write redox half-reactions by using the more simplistic $H^+(aq)$. For example, the reduction of O_2 to H_2O is written

$$O_2(g) + 4\,H^+(aq) + 4\,e^- \rightarrow 2\,H_2O(\ell)$$

However, a few half-reactions in Table A6.1 contain OH^- ions, which tells us that these reactions occur in basic solutions. One such reaction involves another reduction of O_2:

$$O_2(g) + 2\,H_2O(\ell) + 4\,e^- \rightarrow 4\,OH^-(aq)$$

In the following Sample and Practice Exercises, we need to select half-reactions that match the pH conditions of the reaction.

SAMPLE EXERCISE 18.1 Writing Net Ionic Equations of Cell   **LO1**
Reactions by Combining Half-Reactions

Identify two half-reactions listed in Table A6.1 that could be combined to produce a net ionic equation describing the electrolysis of water into hydrogen gas and oxygen gas. Assume the pH of the solution is 12.0.

Collect and Organize The products of electrolysis for water are $H_2(g)$ and $O_2(g)$, meaning that our target net ionic equation is

$$2\,H_2O(\ell) \rightarrow 2\,H_2(g) + O_2(g)$$

Table A6.1 contains four half-reactions in which H_2O, H_2, and O_2 are either reactants or products:

$$2\,H^+(aq) + 2\,e^- \rightarrow H_2(g)$$

$$2\,H_2O(\ell) + 2\,e^- \rightarrow H_2(g) + 2\,OH^-(aq)$$

$$O_2(g) + 2\,H_2O(\ell) + 4\,e^- \rightarrow 4\,OH^-(aq)$$

$$O_2(g) + 4\,H^+(aq) + 4\,e^- \rightarrow 2\,H_2O(\ell)$$

Analyze The solution has a basic pH, so the appropriate half-reactions from Table A6.1 are those that contain OH^-, not H^+, ions. This condition eliminates the first and last half-reactions listed above, leaving the middle two to use in writing the net ionic equation:

$$2\,H_2O(\ell) + 2\,e^- \rightarrow H_2(g) + 2\,OH^-(aq)$$

$$O_2(g) + 2\,H_2O(\ell) + 4\,e^- \rightarrow 4\,OH^-(aq)$$

Water is a reactant and H_2 is a product in the first of these half-reactions, as is the case in the target electrolysis reaction. However, O_2 is a reactant in the remaining half-reaction, but a product in the target electrolysis reaction. Therefore, the remaining half-reaction must not happen as a reduction as written, but rather as an oxidation:

$$4\,OH^-(aq) \rightarrow O_2(g) + 2\,H_2O(\ell) + 4\,e^-$$

The number of electrons gained and lost in the two half-reactions must balance, which means we must multiply the reduction half-reaction by 2, so that both half-reactions involve 4 moles of electrons.

Solve Multiplying the reduction half-reaction by 2 and adding it to the oxidation half-reaction gives:

$$4\,H_2O(\ell) + 4\,e^- \rightarrow 2\,H_2(g) + 4\,OH^-(aq)$$

$+$ $4\,OH^-(aq) \rightarrow O_2(g) + 2\,H_2O(\ell) + 4\,e^-$

$$4\,H_2O(\ell) + 4\,OH^-(aq) + 4\,e^- \rightarrow 2\,H_2(g) + 4\,OH^-(aq) + O_2(g) + 2\,H_2O(\ell) + 4\,e^-$$

Simplifying by eliminating the terms common to both sides of the reaction arrow:

$$\overset{2}{\cancel{4}\,H_2O(\ell)} + \cancel{4\,OH^-(aq)} + \cancel{4\,e^-} \rightarrow 2\,H_2(g) + \cancel{4\,OH^-(aq)} + O_2(g) + \cancel{2\,H_2O(\ell)} + \cancel{4\,e^-}$$

or,

$$2\,H_2O(\ell) \rightarrow 2\,H_2(g) + O_2(g)$$

Think About It Even though some half-reactions occur in acidic or basic solutions, the ionic components of the half-reactions cancelled out, leaving a balanced molecular equation describing the decomposition of liquid water into gaseous hydrogen and oxygen.

FIGURE 18.5 A Cu/Ag electrochemical cell.

SAMPLE EXERCISE 18.2 Diagramming an Electrochemical Cell   **LO1, LO2**

Figure 18.5 depicts an electrochemical cell in which a copper electrode immersed in a 1.00 M solution of Cu^{2+} ions is connected to a silver electrode immersed in a 1.00 M solution of Ag^+ ions. Write a balanced chemical equation for this cell reaction and draw a cell diagram for this cell.

Collect and Organize We have a cell reaction in which electrons spontaneously flow from a copper electrode in contact with a solution of Cu^{2+} ions through an external circuit to a silver electrode in contact with a solution of Ag^+ ions. The half-reactions in Table A6.1 involving these metals and ions are

$$Cu^{2+}(aq) + 2\,e^- \rightarrow Cu(s)$$
$$Ag^+(aq) + e^- \rightarrow Ag(s)$$

In a cell diagram, the anode and the species involved in the oxidation half-reaction are written on the left, and the cathode and the species involved in the reduction half-reaction are written on the right. We use single lines to separate phases and a double line to represent the porous bridge separating the two compartments of the cell.

Analyze In an electrochemical cell, electrons flow from the anode through an external circuit to the cathode. Therefore in the cell in Figure 18.5, copper is the anode and silver is the cathode. This means that the Cu reduction half-reaction must run in reverse as an oxidation half-reaction:

$$Cu(s) \rightarrow Cu^{2+}(aq) + 2\,e^-$$

Two moles of electrons are produced in the anode half-reaction, but only 1 mole of electrons is consumed in the cathode half-reaction at the Ag electrode. We therefore need to multiply the silver half-reaction by 2 before combining the two equations.

Solve Multiplying the Ag^+ half-reaction by 2 and adding it to the Cu half-reaction, we get

$$2\,Ag^+(aq) + 2\,e^- \rightarrow 2\,Ag(s)$$
$$\underline{Cu(s) \rightarrow Cu^{2+}(aq) + 2\,e^-}$$
$$\overline{2\,Ag^+(aq) + 2\,\cancel{e^-} + Cu(s) \rightarrow 2\,Ag(s) + Cu^{2+}(aq) + 2\,\cancel{e^-}}$$
$$2\,Ag^+(aq) + Cu(s) \rightarrow 2\,Ag(s) + Cu^{2+}(aq)$$

The equation is balanced, and we are finished with this portion of our task.
Applying the rules for drawing a cell diagram:

1. Anode on the left, cathode on the right, bridge in the middle:

$$Cu(s) \qquad\qquad || \qquad\qquad Ag(s)$$

2. Adding electrode–solution boundaries and the formulas of the ions produced and consumed in the cell reaction:

$$Cu(s)\,|\,Cu^{2+}(aq)\,||\,Ag^+(aq)\,|\,Ag(s)$$

3. Adding concentration terms:

$$Cu(s)\,|\,Cu^{2+}(1.00\,M)\,||\,Ag^+(1.00\,M)\,|\,Ag(s)$$

Think About It To test the validity of the cell diagram, let's translate it into a sentence: *A copper anode is oxidized to aqueous Cu^{2+} ions, which are separated by a porous bridge from an aqueous solution of Ag^+ ions that are reduced to Ag atoms at a silver cathode.* This description matches the net ionic equation for the reaction and is consistent with the cell layout and flow of electrons in Figure 18.5.

Practice Exercise Write a balanced chemical equation, and draw the cell diagram for an electrochemical cell that has a copper cathode immersed in a solution of Cu^{2+} ions and an aluminum anode immersed in a solution of Al^{3+} ions.

18.3 Standard Potentials

Table A6.1 lists half-reactions in order of their **standard reduction potentials ($E°$)**. The superscript (°) has its usual thermodynamic meaning—all reactants and products are in their standard states, that is, the concentrations of all dissolved substances are 1 M and the partial pressures of all gases are 1 bar ($\approx$ 1 atm). The more positive the value of $E°$, the greater the probability that the reduction half-reaction will couple with an oxidation half-reaction to produce a spontaneous redox reaction. The most positive $E°$ value in Table A6.1 is for the reduction of fluorine:

$$F_2(g) + 2\,e^- \rightarrow 2\,F^-(aq) \qquad E° = +2.866\text{ V}$$

meaning that fluorine is the most easily reduced substance in Table A6.1. It also means that F_2 is the strongest oxidizing agent in the table. It can oxidize any of the substances on the product side of the half-reactions lower in the table. Likewise, F^- is the weakest reducing agent.

The half-reaction at the very bottom of the table with the most negative $E°$ value

$$Li^+(aq) + e^- \rightarrow Li(s) \qquad E° = -3.05\text{ V}$$

is least likely to proceed as written because Li^+ is the weakest oxidizing agent in the table. However, that also means that Li is the strongest reducing agent in this table, so the reverse reaction:

$$Li(s) \rightarrow Li^+(aq) + e^-$$

occurs more readily than the reverse of any other half-reaction in Table A6.1. Lithium metal is a very powerful reducing agent that can reduce any of the substances on the reactant side of the half-reactions in Table A6.1.

Substances with very negative $E°$ values at the bottom of Table A6.1 include the major cations in biological systems and environmental waters: Na^+, K^+, Mg^{2+}, and Ca^{2+}. Their negative $E°$ values tell us that these ions are not easily reduced to their free metals in aqueous solutions. They are chemically very stable (which explains the presence of these cations in nature).

CONNECTION Oxidizing and reducing agents were introduced in Chapter 4.

standard reduction potential ($E°$) the potential of a reduction half-reaction in which all reactants and products are in their standard states at 25°C.

CONCEPT TEST

Compare the order of the metals listed in the Activity Series in Table 4.6 to the order of the metals listed in Table A6.1. Describe the relationship between these two tables.

standard cell potential (E°_{cell})
a measure of how forcefully an electrochemical cell in which all reactants and products are in their standard states can pump electrons through an external circuit.

CHEMTOUR
Cell Potential

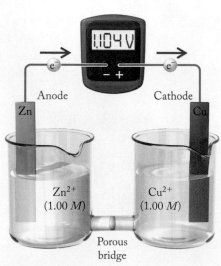

FIGURE 18.6 A voltmeter displays a cell potential of 1.104 V between a Zn electrode immersed in a 1.00 M solution of Zn^{2+} ions and a Cu electrode immersed in a 1.00 M solution of Cu^{2+} ions.

CONCEPT TEST

Use the order of the half-reactions listed in Table A6.1 to predict which of the following reactions is/are spontaneous under standard conditions:

a. $Cu(s) + 2\,Fe^{3+}(aq) \rightarrow Cu^{2+}(aq) + 2\,Fe^{2+}(aq)$

b. $2\,Ag(s) + Zn^{2+}(aq) \rightarrow 2\,Ag^+(aq) + Zn(s)$

c. $Hg(\ell) + 2\,H^+(aq) \rightarrow Hg^{2+}(aq) + H_2(g)$

We can use the standard reduction potentials in Table A6.1 to calculate the **standard cell potentials (E°_{cell})** of electrochemical cells. Standard cell potentials, as their name suggests, are related to the chemical (potential) energy stored in a cell and specifically to how forcefully the cell can pump electrons out from their anodes, through external circuits, and into their cathodes.

Consider the hypothetical case where two half-reactions both have the same standard reduction potential. That is, there would be no difference between the abilities of their reactants to function as oxidizing agents, and no electrons would spontaneously flow. For two half-reactions with different standard reduction potentials, the larger the difference between those potentials, the larger the E°_{cell} of the electrochemical cell that could be built from them. Therefore E°_{cell} can be determined by calculating the difference between the standard reduction potentials of a voltaic cell's cathode and anode:

$$E^\circ_{cell} = E^\circ_{cathode} - E^\circ_{anode} \qquad (18.2)$$

Let's use Equation 18.2 to calculate E°_{cell} for the Zn/Cu^{2+} voltaic cell in Figure 18.2. The standard reduction potential of the cathode half-reaction is

$$Cu^{2+}(aq) + 2\,e^- \rightarrow Cu(s) \qquad E^\circ = 0.342\ V$$

To obtain the standard potential for the oxidation half-reaction at the zinc anode, we find the standard reduction potential of Zn^{2+} ions in Table A6.1:

$$Zn^{2+}(aq) + 2\,e^- \rightarrow Zn(s) \qquad E^\circ = -0.762\ V$$

Now we use Equation 18.2 to calculate E°_{cell}:

$$E^\circ_{cell} = E^\circ_{cathode} - E^\circ_{anode}$$
$$= 0.342 - (-0.762) = 1.104\ V$$

This is the cell potential we would measure if we connected a device called a voltmeter across the two electrodes, as shown in Figure 18.6, at 25°C under standard conditions.

To use Equation 18.2, we need to know which half-reaction occurs at the cathode (reduction) and which occurs at the anode (oxidation). In other words, we need to know which component of the spontaneous cell reaction is more likely to be oxidized and which is more likely to be reduced. As we have seen, this decision can be made based on the data in Table A6.1. In our Zn/Cu^{2+} cell, for instance, the value of E° for the reduction of Cu^{2+} ions to Cu metal is 0.342 V, which is greater than the value of E° for reducing Zn^{2+} ions to Zn metal (-0.762 V). Therefore in the Zn/Cu^{2+} voltaic cell, Cu^{2+} ions are reduced and Zn metal is oxidized. We can generalize this observation to the cell reaction of any voltaic cell: the half-reaction with the more positive value of E° runs as a reduction and the other half-reaction occurs as an oxidation. Note that this means E°_{cell} will always be a positive value for a spontaneous cell reaction.

SAMPLE EXERCISE 18.3 Identifying Anode and Cathode **LO1, LO3**
Half-Reactions and Calculating
the Value of $E°_{cell}$

The standard reduction potentials of the half-reactions in typical AA or AAA
single-use alkaline batteries are

$$ZnO(s) + H_2O(\ell) + 2\,e^- \rightarrow Zn(s) + 2\,OH^-(aq) \qquad E° = -1.25\ V$$

$$2\,MnO_2(s) + H_2O(\ell) + 2\,e^- \rightarrow Mn_2O_3(s) + 2\,OH^-(aq) \qquad E° = 0.15\ V$$

What is the net ionic equation for the cell reaction and the value of $E°_{cell}$?

Collect and Organize We can calculate $E°_{cell}$ by using Equation 18.2:

$$E°_{cell} = E°_{cathode} - E°_{anode}$$

However, first we need to decide which half-reaction occurs at the cathode and which at
the anode. The equation for a cell reaction is written by combining half-reactions once
the loss or gain of electrons in the two half-reactions is balanced.

Analyze The MnO_2 half-reaction has the more positive $E°$ value, making it our
reduction half-reaction. We must reverse the ZnO half-reaction, turning it into an
oxidation half-reaction. The two half-reactions both involve the transfer of 2 moles of
electrons, so we may combine them by simply adding them together.

Solve The oxidation half-reaction at the anode is

$$Zn(s) + 2\,OH^-(aq) \rightarrow ZnO(s) + H_2O(\ell) + 2\,e^-$$

The reduction half-reaction at the cathode is

$$2\,MnO_2(s) + H_2O(\ell) + 2\,e^- \rightarrow Mn_2O_3(s) + 2\,OH^-(aq)$$

Combining these half-reactions to obtain the overall cell reaction, we get

$$2\,MnO_2(s) + H_2O(\ell) + Zn(s) + 2\,OH^-(aq) + 2\,e^- \rightarrow$$
$$Mn_2O_3(s) + 2\,OH^-(aq) + ZnO(s) + H_2O(\ell) + 2\,e^-$$

Simplifying gives us the net ionic equation for the cell reaction:

$$2\,MnO_2(s) + Zn(s) \rightarrow Mn_2O_3(s) + ZnO(s)$$

The overall $E°_{cell}$ for this reaction is obtained by using Equation 18.2:

$$E°_{cell} = E°_{cathode} - E°_{anode}$$
$$= 0.15\ V - (-1.25\ V) = 1.40\ V$$

Think About It The $E°_{cell}$ value is reasonable because the potential of most alkaline
batteries is nominally 1.5 V. In this particular cell reaction, the net ionic equation is also
the complete molecular equation.

 Practice Exercise
The half-reactions in nicad (nickel–cadmium) batteries are

$$Cd(OH)_2(s) + 2\,e^- \rightarrow Cd(s) + 2\,OH^-(aq) \qquad E° = -0.81\ V$$

$$2\,NiO(OH)(s) + 2\,H_2O(\ell) + 2\,e^- \rightarrow 2\,Ni(OH)_2(s) + 2\,OH^-(aq) \qquad E° = 0.52\ V$$

Write the net ionic equation for the cell reaction and calculate the value of $E°_{cell}$.

CHEMTOUR
Alkaline Battery

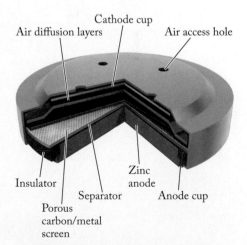

FIGURE 18.7 Most of the internal volume
of a zinc–air battery is occupied by the
anode: a paste of Zn particles in an aqueous
solution of KOH, surrounded by a metal
cup that serves as the negative terminal of
the battery. Oxygen from the air is reduced
at the cathode.

Let's now examine what happens when two half-reactions in which different
numbers of electrons are gained and lost are combined into an electrochemical
cell. This combination occurs in the zinc–air battery (Figure 18.7), a type of battery that has a limitless supply of one of its reactants. This battery powers devices

in which small battery size and low mass are high priorities, such as hearing aids. Most of the internal volume of one of these batteries is occupied by an anode consisting of a paste of zinc particles packed in an aqueous solution of KOH. As in alkaline batteries (Sample Exercise 18.3), the anode half-reaction is

$$Zn(s) + 2\,OH^-(aq) \rightarrow ZnO(s) + H_2O(\ell) + 2\,e^-$$

which is the reverse of the reaction in Table A6.1:

$$ZnO(s) + H_2O(\ell) + 2\,e^- \rightarrow Zn(s) + 2\,OH^-(aq) \qquad E° = -1.25\ V$$

The cathode consists of porous carbon supported by a metal screen. Air diffuses through small holes in the battery and across a layer of Teflon that lets gases pass through but keeps electrolyte from leaking out. As air passes through the cathode, oxygen is reduced to hydroxide ions:

$$O_2(g) + 2\,H_2O(\ell) + 4\,e^- \rightarrow 4\,OH^-(aq) \qquad E°_{cathode} = 0.401\ V$$

To write the overall cell reaction, we need to multiply the oxidation half-reaction by 2 before combining it with the reduction half-reaction:

$$2[Zn(s) + 2\,OH^-(aq) \rightarrow ZnO(s) + H_2O(\ell) + 2\,e^-]$$
$$O_2(g) + 2\,H_2O(\ell) + 4\,e^- \rightarrow 4\,OH^-(aq)$$

$$2\,Zn(s) + 4\,\cancel{OH^-(aq)} + O_2(g) + 2\,\cancel{H_2O(\ell)} + \cancel{4\,e^-} \rightarrow$$
$$2\,ZnO(s) + 2\,\cancel{H_2O(\ell)} + 4\,\cancel{OH^-(aq)} + \cancel{4\,e^-}$$

This simplifies to

$$2\,Zn(s) + O_2(g) \rightarrow 2\,ZnO(s)$$

$$E°_{cell} = E°_{cathode} - E°_{anode} = 0.401\ V - (-1.25\ V) = 1.65\ V$$

Note that when we multiply the anode half-reaction by 2 and add it to the cathode half-reaction, *we do not multiply* the $E°$ of the anode half-reaction by 2. The reason we do not is that $E°$ is an *intensive* property of a half-reaction or a complete cell reaction. It does not change when the quantities of reactants and products change. Thus a zinc–air battery the size of a pea has the same $E°$ as one the size of a book (such as those being developed for electric vehicles). On the other hand, the amount of electrical work a zinc–air battery can do *does* depend on how much zinc is inside it because, as we are about to see, the electrical work that a voltaic cell can do depends on both cell potential *and* the quantity of charge it can deliver at that potential.

18.4 Chemical Energy and Electrical Work

When we connect the Zn and Cu electrodes in Figure 18.6 to a digital voltmeter—the Zn electrode to the negative terminal of the meter and the Cu electrode to the positive terminal—the meter reads 1.104 V. These connections tell us that under standard conditions, the battery can pump electrons from the Zn electrode through an external circuit to the Cu electrode with a potential of 1.104 V.

Where does the energy come from to pump electrons this forcefully through an external circuit? A hint at the answer comes from the fact that these moving electrons can do electrical work (w_{elec}), such as lighting a light bulb or turning an electric motor. We saw in Chapter 17 that the ability of a chemical reaction to do work (different from expansion or compression) is expressed by the change in free energy (ΔG_{cell}) that accompanies the cell reaction. Recall that when a thermodynamic system does work (w) on its surroundings, w is less than zero. Similarly, the change in free energy of the reaction (system) that did this work is also less than zero. Thus the two quantities would be identical if this energy conversion process were 100% efficient:

$$\Delta G_{cell} = w_{elec} \qquad (18.3)$$

The work done by a voltaic cell (in this case the Zn/Cu^{2+} cell reaction) on its surroundings is defined as the product of the quantity of electrical charge (C) the cell pumps through an external circuit times the cell potential:

$$w_{elec} = -CE_{cell} \qquad (18.4)$$

The negative sign reflects the fact that work done *by* a voltaic cell on its surroundings (the external circuit) corresponds to free energy lost by the cell. Keep in mind that the charge of an electron is $1-$, so the passage of 1 mole of electrons through an external circuit in one direction is matched by the passage of 1 mole of positive charge (C) in the opposite direction. The difference is that electrons are real, and charge is derived from a hypothetical quantity of charge carriers that have the same magnitude but opposite sign.

However, quantities of electrical charge are not typically expressed in moles but in coulombs (C). As noted in Chapter 2, the magnitude of the charge on a single electron is 1.602×10^{-19} coulombs (C). Also note the distinction between italic C, a symbol for the variable "charge," and nonitalic C, the abbreviation for the unit "coulomb." The magnitude of electrical charge on 1 mole of electrons is

$$\frac{1.602 \times 10^{-19}\ C}{e^-} \times \frac{6.022 \times 10^{23}\ e^-}{mol\ e^-} = \frac{9.65 \times 10^4\ C}{mol\ e^-}$$

This quantity of charge, 9.65×10^4 C/mol e^-, is called the **Faraday constant (F)** after Michael Faraday (1791–1867), the English chemist and physicist who discovered that redox reactions take place when electrons are transferred from one species to another. The quantity of charge, C, flowing through an electrical circuit is the product of the number of moles (n) of electrons times the Faraday constant:

$$C = nF \qquad (18.5)$$

Combining Equations 18.4 and 18.5 gives us an equation relating w_{elec} and E_{cell}:

$$w_{elec} = -nFE_{cell} \qquad (18.6)$$

If we combine Equations 18.3 and 18.6, we connect the quantity of electrical work a voltaic cell can do on its surroundings with the change in free energy in the cell:

$$\Delta G_{cell} = -nFE_{cell} \qquad (18.7)$$

Perhaps you are wondering how the product on the right side of Equation 18.7 is the equivalent of energy. Note that the units on the right side are

$$\text{mole } e^- \times \frac{\text{coulomb}}{\text{mole } e^-} \times \text{volt} = \text{coulomb-volt}$$

A coulomb-volt is the same quantity of energy as a joule:

$$1 \text{ coulomb-volt} = 1 \text{ joule}$$

$$1\ C \cdot V = 1\ J$$

The negative sign on the right side of Equation 18.7 tells us that the E_{cell} of any voltaic cell must have a positive value because the sign of ΔG_{cell} for the spontaneous chemical reaction inside the cell must be negative.

Let's use Equation 18.7 to calculate the change in standard free energy of the Zn/Cu^{2+} cell reaction. We start with the standard cell potential calculated in Section 18.3:

CONNECTION The sign conventions used for work done *on* a thermodynamic system (+) and the work done *by* the system (−) were explained in Section 5.3.

Faraday constant (F) the magnitude of electrical charge in 1 mole of electrons. Its value to three significant figures is 9.65×10^4 C/mol e^-.

$$E^{\circ}_{cell(Zn/Cu^{2+})} = 1.104 \text{ V}$$

We can convert this standard cell potential into a change in standard free energy (ΔG°_{cell}) by using Equation 18.7 under standard conditions, so that $\Delta G = \Delta G^{\circ}$ and $E = E^{\circ}$:

$$\Delta G^{\circ}_{cell} = -nFE^{\circ}_{cell}$$

$$= -\left(2 \text{ mol e}^- \times \frac{9.65 \times 10^4 \text{ C}}{\text{mol e}^-} \times 1.104 \text{ V}\right) = -2.13 \times 10^5 \text{ C} \cdot \text{V}$$

$$= -2.13 \times 10^5 \text{ J} = -213 \text{ kJ}$$

To put this value in perspective, the Zn/Cu^{2+} reaction produces nearly as much useful energy as the combustion of 1 mole of hydrogen gas:

$$H_2(g) + \tfrac{1}{2} O_2(g) \rightarrow H_2O(g) \qquad \Delta G^{\circ} = -228.6 \text{ kJ}$$

CONCEPT **TEST**

When a rechargeable battery, such as the one used to start a car's engine, is recharged, an external source of electrical power forces the voltaic cell reaction to run in reverse. What are the signs of ΔG_{cell} and E_{cell} during the recharging process?

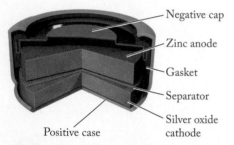

FIGURE 18.8 Many of the button batteries that power small electronic devices incorporate a Zn anode and a Ag_2O cathode separated by a membrane containing KOH electrolyte.

SAMPLE EXERCISE 18.4 Relating ΔG°_{cell} and E°_{cell} **LO1, LO4**

Many of the "button" batteries used in electric watches consist of a Zn anode and a Ag_2O cathode, separated by a membrane soaked in a concentrated solution of KOH (Figure 18.8). At the cathode, Ag_2O is reduced to Ag metal; at the anode, Zn is oxidized to solid $Zn(OH)_2$. Write the net ionic equation for the reaction, and use the appropriate standard reduction potentials from Table A6.1 to calculate the values of E°_{cell} and ΔG°_{cell}.

Collect and Organize We know the reactants and products of the anode and cathode reactions and that the reaction occurs in a basic solution. The following equations should be useful in calculating E°_{cell} and ΔG°_{cell} from the appropriate standard potentials:

$$E^{\circ}_{cell} = E^{\circ}_{cathode} - E^{\circ}_{anode}$$

$$\Delta G^{\circ}_{cell} = -nFE^{\circ}_{cell}$$

Analyze The half-reaction at the cathode is based on the reduction of Ag_2O to Ag. The appropriate half-reaction in Table A6.1 is

$$Ag_2O(s) + H_2O(\ell) + 2 \text{ e}^- \rightarrow 2 \text{ Ag}(s) + 2 \text{ OH}^-(aq) \qquad E^{\circ}_{cathode} = 0.342 \text{ V}$$

We must find an entry in Table A6.1 in which $Zn(OH)_2$ is the reactant and Zn is the product:

$$Zn(OH)_2(s) + 2 \text{ e}^- \rightarrow Zn(s) + 2 \text{ OH}^-(aq) \qquad E^{\circ}_{anode} = -1.249 \text{ V}$$

We must reverse this half-reaction to represent the oxidation occurring at the anode before combining it with the cathode half-reaction. The two half-reactions involve the transfer of the same number of electrons ($n = 2$), so combining them simply means adding them together. The value of E°_{cell} is about $[0.35 - (-1.25)]$, or about 1.60 V. This value is half again as large as the E°_{cell} of the Zn/Cu^{2+} cell. Therefore the magnitude of its ΔG°_{cell} value should be half again as large as -212 kJ/mol, or about -300 kJ/mol.

Solve Reversing the $Zn(OH)_2$ half-reaction and adding it to the Ag_2O half-reaction, we get

$$Zn(s) + 2 OH^-(aq) \rightarrow Zn(OH)_2(s) + 2 e^-$$
$$Ag_2O(s) + H_2O(\ell) + 2 e^- \rightarrow 2 Ag(s) + 2 OH^-(aq)$$
$$\overline{Ag_2O(s) + H_2O(\ell) + Zn(s) + \cancel{2 OH^-(aq)} + \cancel{2e^-} \rightarrow}$$
$$2 Ag(s) + \cancel{2 OH^-(aq)} + Zn(OH)_2(s) + \cancel{2 e^-}$$

This simplifies to

$$Ag_2O(s) + H_2O(\ell) + Zn(s) \rightarrow 2 Ag(s) + Zn(OH)_2(s)$$

Then we calculate E°_{cell}:

$$E^\circ_{cell} = E^\circ_{cathode} - E^\circ_{anode}$$
$$= 0.342 \text{ V} - (-1.249 \text{ V})$$
$$= 1.591 \text{ V}$$

From this value, we can determine ΔG°_{cell}:

$$\Delta G^\circ_{cell} = -nFE^\circ_{cell}$$
$$= -(2 \text{ mol e}^- \times 9.65 \times 10^4 \text{ C/mol e}^- \times 1.591 \text{ V})$$
$$= -3.07 \times 10^5 \text{ C} \cdot \text{V} = -3.07 \times 10^5 \text{ J} = -307 \text{ kJ}$$

Think About It The positive value of E°_{cell} and negative value of ΔG°_{cell} are expected because voltaic cell reactions are spontaneous. The calculated values are close to those we estimated.

 Practice Exercise If a standard alkaline battery produces a cell potential of 1.50 V, what is the value of ΔG°_{cell}?

Some final thoughts about the silver oxide battery reaction in Sample Exercise 18.4: The ΔG°_{cell} value of −307 kJ is based on the reaction of 1 mole of Ag_2O and 1 mole of Zn, which correspond to 232 g of Ag_2O and 65 g of Zn. The energy stored in a button battery (Figure 18.8), which has a mass of only 1 or 2 grams, would be a tiny fraction of this calculated value. Also, note that no ions appear in the net ionic equation. Because all the reactants and products are solids, the net ionic equation and molecular equation are identical.

18.5 A Reference Point: The Standard Hydrogen Electrode

We can measure the value of E_{cell} by using a voltmeter, but can we measure the individual electrode potentials of the cathode and anode? The answer to this question is that we can assign potentials to an individual electrode by arbitrarily assigning a value of zero volts to a half-reaction that serves as a reference point, namely, the standard potential for the reduction of hydrogen ions to hydrogen gas:

$$2 H^+(aq) + 2 e^- \rightarrow H_2(g) \qquad E^\circ = 0.000 \text{ V} \qquad (18.8)$$

An electrode that generates this reference potential, called the **standard hydrogen electrode (SHE)**, consists of a platinum electrode in contact with a solution of a strong acid ([H^+] = 1.00 M) and hydrogen gas at a pressure of 1.00 atm (Figure 18.9). The platinum is not changed by the electrode reaction. Rather, it serves as a chemically inert conveyor of electrons. Electrons are conveyed to the electrode surface when H^+ ions are reduced to hydrogen gas; electrons are

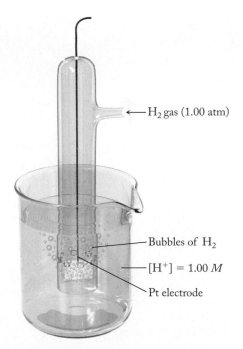

← H₂ gas (1.00 atm)

← Bubbles of H₂

← [H⁺] = 1.00 M

Pt electrode

FIGURE 18.9 The standard hydrogen electrode consists of a platinum electrode immersed in a 1.00 M solution of $H^+(aq)$ and bathed in a stream of pure H_2 gas at a pressure of 1.00 atm. Its potential is the same (0.000 V) whether $H^+(aq)$ ions are reduced or H_2 gas is oxidized.

standard hydrogen electrode (SHE) a reference electrode based on the half-reaction $2 H^+(aq) + 2 e^- \rightarrow H_2(g)$ that produces a standard electrode potential of 0.000 V.

FIGURE 18.10 (a) When a standard hydrogen electrode is coupled to a Zn electrode under standard conditions, the SHE is the cathode (H⁺ is reduced) and the Zn electrode is the anode. When the SHE is connected to the positive terminal of a voltmeter and the Zn electrode to the negative terminal, the meter measures a cell potential of 0.762 V. (b) When coupled to a Cu electrode under standard conditions, the SHE is the anode (H₂ is oxidized), the Cu electrode is the cathode, and the meter measures a cell potential of 0.342 V.

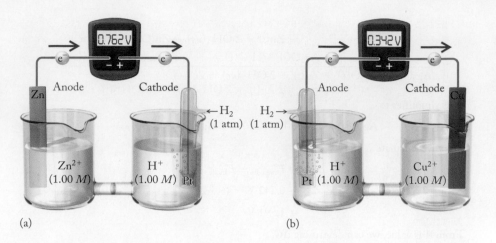

conveyed away from the electrode surface when hydrogen gas is oxidized to H⁺ ions. The potential of the SHE is the same for both half-reactions: 0.000 V.

To draw the cell diagram for a cell in which the SHE serves as the anode, we represent the SHE half of the cell as follows:

$$\text{Pt}(s)\mid \text{H}_2(g,\ 1.00\ \text{atm})\mid \text{H}^+(1.00\ M)\parallel$$

This indicates that the anode half-reaction involves the oxidation of H₂ gas to H⁺ ions. If the SHE is the cathode, then we diagram its half of the cell this way:

$$\parallel \text{H}^+(1.00\ M)\mid \text{H}_2(g,\ 1.00\ \text{atm})\mid \text{Pt}(s)$$

This indicates that the cathode half-reaction involves the reduction of H⁺ ions to H₂ gas.

Because the standard reduction (or oxidation) potential of the SHE is assigned a reference potential of 0.000 V, the measured E_{cell}° of any voltaic cell in which a SHE is one of the two electrodes—either cathode or anode—can be considered the potential produced by the other electrode. This means that if we attach a voltmeter to the cell, the meter reading is the electrode potential of the other electrode. Suppose, for example, that a voltaic cell consists of a strip of zinc metal immersed in a 1.00 M solution of Zn²⁺ ions in one compartment and a SHE in the other (Figure 18.10a). Also suppose that a voltmeter is connected to the cell so that it measures the potential at which the cell pumps electrons from the zinc electrode to the SHE. This direction of electron flow means that the zinc electrode is the cell's anode and the SHE is the cathode of the cell. At 25°C the meter reads 0.762 V. We know that the value of $E_{\text{cathode}}^{\circ}$ is that of the SHE (0.000 V) and that E_{anode}° is E_{Zn}°.

Inserting these values and symbols into Equation 18.2,

$$E_{\text{cell}}^{\circ} = E_{\text{cathode}}^{\circ} - E_{\text{anode}}^{\circ}$$

$$E_{\text{cell}}^{\circ} = E_{\text{SHE}}^{\circ} - E_{\text{Zn}}^{\circ}$$

$$0.762\ \text{V} = 0.000\ \text{V} - E_{\text{Zn}}^{\circ}$$

$$E_{\text{Zn}}^{\circ} = -0.762\ \text{V}$$

This value is equal to the standard reduction potential of Zn²⁺ in Table A6.1:

$$\text{Zn}^{2+}(aq) + 2\ \text{e}^- \rightarrow \text{Zn}(s) \qquad E^{\circ} = -0.762\ \text{V}$$

In Figure 18.10(b), the SHE is coupled to a copper electrode immersed in a 1.00 M solution of Cu²⁺ ions. In this cell, electrons flow from the SHE through an external circuit to the copper electrode at a cell potential of 0.342 V at 25°C. The direction of current flow means that the electrons are consumed at the copper electrode, making it the cathode. The value of $E°$ for the copper half-reaction is calculated as follows:

$$E°_{cell} = E°_{cathode} - E°_{anode}$$

$$= E°_{Cu} - E°_{SHE}$$

$$0.342 \text{ V} = E°_{Cu} - 0.000 \text{ V}$$

$$E°_{Cu} = 0.342 \text{ V}$$

This half-reaction potential matches the value of $E°$ for the reduction of Cu^{2+} to Cu metal in Table A6.1.

FIGURE 18.11 A SHE and a Ni electrode connected by a voltmeter.

CONCEPT TEST

A cell consists of a SHE in one compartment and a Ni electrode immersed in a 1.00 M solution of Ni^{2+} ions in the other. If a voltmeter is connected to the electrodes as shown in Figure 18.11, what will be the value on the voltmeter's display?

18.6 The Effect of Concentration on E_{cell}

Reactions stop when one of the reactants is completely consumed. This concept was the basis for our discussion of limiting reactants in Chapter 3. However, a commercial battery usually stops operating at its rated cell potential—1.5 V for a flashlight battery—before its reactants are completely consumed. This happens because the cell potential of a voltaic cell depends on the concentrations of the reactants and products.

CHEMTOUR
Cell Potential, Equilibrium, and Free Energy

The Nernst Equation

In 1889, the German chemist Walther Nernst (1864–1941) derived an expression, now called the **Nernst equation**, that describes how cell potentials depend on reactant and product concentrations. We can reconstruct his derivation starting with Equation 17.14, which relates the change in free energy ΔG of any reaction to its change in free energy under standard conditions $\Delta G°$:

$$\Delta G = \Delta G° + RT \ln Q \qquad (17.14)$$

As a spontaneous reaction proceeds, concentrations of products increase and concentrations of reactants decrease until the positive value of $RT \ln Q$ offsets the negative value of $\Delta G°$. At that point, $\Delta G = 0$ and the reaction has reached chemical equilibrium.

Now let's write an expression analogous to Equation 17.14 that relates the **cell potential** under nonstandard conditions, that is, E_{cell}, to $E°_{cell}$. We start by substituting $-nFE_{cell}$ for ΔG_{cell} and $-nFE°_{cell}$ for $\Delta G°_{cell}$:

$$-nFE_{cell} = -nFE°_{cell} + RT \ln Q$$

Dividing all terms by $-nF$ gives

$$E_{cell} = E°_{cell} - \frac{RT \ln Q}{nF} \qquad (18.9)$$

CONNECTION In Chapter 17 we discussed the relationship between change in free energy and the reaction quotient Q.

Nernst equation an equation relating the potential of a cell (or half-cell) reaction to its standard potential ($E°$) and to the concentrations of its reactants and products.

cell potential (E_{cell}) the force with which an electrochemical cell can pump electrons through an external circuit.

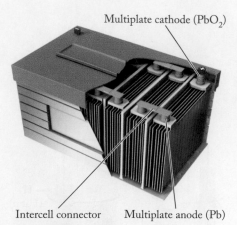

Multiplate cathode (PbO_2)

Intercell connector   Multiplate anode (Pb)

FIGURE 18.12 The lead–acid battery that provides power to start most motor vehicles contains six cells. Each has an anode made of lead and a cathode made of PbO_2 immersed in a background electrolyte of 4.5 M H_2SO_4. The electrodes are formed into plates and held in place by grids made of a lead alloy. The grids connect the cells in series so that the operating potential of the battery (12.0 V) is the sum of six E_{cell} values (each 2.0 V).

This is the equation Walther Nernst developed in 1889. We can obtain a very useful form of Equation 18.9 if we insert values for R [8.314 J/(mol · K)] and F (9.65×10^4 C/mol), assume $T = 298$ K, and convert the natural logarithm to a base-10 logarithm: $\ln Q = 2.303 \log Q$. With these changes, the Nernst equation becomes

$$E_{cell} = E^{\circ}_{cell} - \frac{0.0592 \text{ V}}{n} \log Q \qquad (18.10)$$

Equation 18.10 allows us to predict how the potential (E_{cell} in V) of a voltaic cell at 298 K changes as the concentrations of products inside the cell increase and the concentrations of reactants decrease. As they do, Q increases and so does the log term in Equation 18.10. The negative sign in front of this term means that the value of E_{cell} decreases as reactants are converted into products. Eventually, E_{cell} approaches zero. When it reaches zero, the cell reaction has achieved chemical equilibrium. The cell can no longer pump electrons through an external circuit. In other words, it's dead.

CONCEPT TEST

We can also use Equation 18.10 to calculate the potential of a single electrode. Consider the half-reaction at the Ag/Ag^+ electrode:

$$Ag^+(aq) + e^- \rightarrow Ag(s) \qquad E^{\circ} = 0.800 \text{ V}$$

What is the potential of this half-reaction at 25°C when the concentration of Ag^+ is 0.100 M?

Batteries are voltaic cells, so their cell potential should drop with usage. Let's consider how much a cell potential can drop by focusing on the *lead–acid* battery used to start most car engines. These batteries each contain six electrochemical cells. Their anodes are made of Pb and their cathodes are made of PbO_2. Both electrodes are immersed in 4.5 M H_2SO_4 (Figure 18.12). The value of a fully charged cell is about 2.0 V. The six cells are connected in series so that the operating potential of the battery is the sum of the six cell potentials, or about 12 V.

As the battery discharges, $PbO_2(s)$ is reduced to $PbSO_4(s)$ at the cathodes:

$$PbO_2(s) + 3 \text{ H}^+(aq) + HSO_4^-(aq) + 2 \text{ e}^- \rightarrow PbSO_4(s) + 2 \text{ H}_2O(\ell)$$
$$E^{\circ} = 1.685 \text{ V}$$

Also, Pb(s) is oxidized to $PbSO_4(s)$ at the anodes:

$$Pb(s) + HSO_4^-(aq) \rightarrow PbSO_4(s) + \text{H}^+(aq) + 2 \text{ e}^-$$

The reduction half-reaction consumes 2 moles of electrons, and the oxidation half-reaction involves the loss of 2 moles of electrons for each mole of lead.

The net ionic equation for the overall cell reaction is the sum of the two half-reactions:

$$PbO_2(s) + Pb(s) + 2 \text{ H}^+(aq) + 2 HSO_4^-(aq) \rightarrow 2 PbSO_4(s) + 2 \text{ H}_2O(\ell)$$

To calculate the value of E°_{cell}, we find the reduction potential for the half-reaction that corresponds to the reverse reaction at the anode in Appendix 6, Table A6.1:

$$PbSO_4(s) + \text{H}^+(aq) + 2 \text{ e}^- \rightarrow Pb(s) + HSO_4^-(aq) \qquad E^{\circ} = -0.356 \text{ V}$$

Therefore

$$E^{\circ}_{cell} = E^{\circ}_{cathode} - E^{\circ}_{anode} = 1.685 \text{ V} - (-0.356 \text{ V}) = 2.041 \text{ V}$$

As the battery discharges, the concentration of sulfuric acid decreases, and so does the value of E_{cell} calculated from the Nernst equation:

$$E_{cell} = 2.041 \text{ V} - \frac{0.0592 \text{ V}}{2} \log \frac{1}{[\text{H}^+]^2 [\text{HSO}_4^-]^2}$$

However, the decrease in E_{cell} is very gradual, not falling below 2.0 V until the battery is about 97% discharged, as shown in Figure 18.13. The gradual decrease makes sense because of the logarithmic relationship between Q and E_{cell}. If, for example, the concentration of sulfuric acid decreased by an order of magnitude, say, from 1.00 M to 0.100 M, the value of E_{cell} would decrease by less than 6%—from 2.041 V to

$$E_{cell} = 2.041 \text{ V} - \frac{0.0592 \text{ V}}{2} \log \frac{1}{0.100^2 \times 0.100^2} = 1.923 \text{ V}$$

The logarithmic relationship between Q and E_{cell} means that most batteries can deliver current at a cell potential close to their "design" potential until they are almost completely discharged.

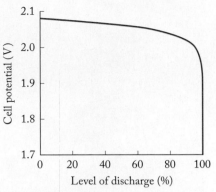

FIGURE 18.13 The potential of a cell in a lead–acid battery decreases as reactants are converted into products, but the change in potential is small until the battery is nearly completely discharged. While the value of E_{cell}° is 2.041 V, a fully charged commercial battery has a slightly higher potential (shown here as 2.08 V) due to the use of more concentrated sulfuric acid.

SAMPLE EXERCISE 18.5 Calculating E_{cell} from E_{cell}° and the   **LO5**
Concentrations of Reactants and Products

The standard potential (E_{cell}°) of a voltaic cell based on the Zn/Cu^{2+} ion reaction:

$$\text{Zn}(s) + \text{Cu}^{2+}(aq) \rightarrow \text{Zn}^{2+}(aq) + \text{Cu}(s)$$

is 1.104 V. What is the value of E_{cell} at 25°C when the concentration of Cu^{2+} is 0.100 M and the concentration of Zn^{2+} is 1.90 M?

Collect and Organize We are given the standard cell potential and the concentrations of Cu^{2+} and Zn^{2+}, and we are asked to determine the value of E_{cell}. The Nernst equation enables us to calculate E_{cell} values for different concentrations of reactants and products. This equation requires us to work with the reaction quotient Q, which we know from Section 14.5 to be the mass action expression for the reaction. Solid copper and zinc are also part of the reaction system, but no terms for pure solids appear in reaction quotients.

Analyze The only term in the numerator of the Q expression for this cell reaction is [Zn^{2+}], and the only one in the denominator is [Cu^{2+}]. Each Cu^{2+} ion gains two electrons, and each Zn atom donates two electrons, so the value of n in the Nernst equation is 2. The value of [Zn^{2+}] is greater than [Cu^{2+}], which makes $Q > 1$. The negative sign in front of the $0.0592/n \times \log Q$ term in Equation 18.10 means that the calculated value of E_{cell} should be less than the value of E_{cell}°.

Solve Substituting the values of [Zn^{2+}] and [Cu^{2+}] in the Nernst equation gives

$$E_{cell} = E_{cell}^\circ - \frac{0.0592 \text{ V}}{n} \log Q = 1.104 \text{ V} - \frac{0.0592 \text{ V}}{2} \log \frac{1.90}{0.100}$$

$$E_{cell} = 1.104 \text{ V} - \frac{0.0592 \text{ V}}{2}(1.279) = 1.066 \text{ V}$$

Think About It The calculated E_{cell} value is only 0.038 V less than E_{cell}° because the logarithmic dependence of cell potential on reactant and product concentrations minimizes the impact of changing concentrations.

Practice Exercise The standard cell potential of the zinc–air battery (Figure 18.7) is 1.65 V. If the partial pressure of oxygen in the air at 25°C diffusing through its cathode is 0.21 atm, what is the cell potential? Assume the cell reaction is

$$2 \text{ Zn}(s) + \text{O}_2(g) \rightarrow 2 \text{ ZnO}(s)$$

$E°$ and K

When the cell reaction of a voltaic cell reaches chemical equilibrium, $\Delta G_{cell} = E_{cell} = 0$ and $Q = K$. Therefore Equation 18.10 becomes

$$0 = E°_{cell} - \frac{0.0592 \text{ V}}{n} \log K$$

We can rearrange this equation to

$$\log K = \frac{nE°_{cell}}{0.0592 \text{ V}} \tag{18.11}$$

We can use Equation 18.11 to calculate the equilibrium constant for any redox reaction at 25°C, not just those in electrochemical cells. For the more general case, we substitute $E°_{rxn}$ for $E°_{cell}$:

$$\log K = \frac{nE°_{rxn}}{0.0592 \text{ V}} \tag{18.12}$$

SAMPLE EXERCISE 18.6 Calculating K for a Redox Reaction from the Standard Potentials of Its Half-Reactions   **LO6**

Many procedures for determining mercury levels in environmental samples involve reducing Hg^{2+} ions to elemental Hg by using Sn^{2+}. Use the appropriate $E°$ values from Table A6.1 to calculate the equilibrium constant at 25°C for the reaction

$$Sn^{2+}(aq) + Hg^{2+}(aq) \rightarrow Sn^{4+}(aq) + Hg(\ell)$$

Collect and Organize Equation 18.12 relates the equilibrium constant for any redox reaction to the standard potential $E°_{rxn}$. To calculate $E°_{rxn}$, we need to find the difference between the appropriate standard reduction potentials. Table A6.1 lists two half-reactions involving our reactants and products:

$$Hg^{2+}(aq) + 2 e^- \rightarrow Hg(\ell) \qquad E° = 0.851 \text{ V}$$
$$Sn^{4+}(aq) + 2 e^- \rightarrow Sn^{2+}(aq) \qquad E° = 0.154 \text{ V}$$

Analyze The problem states that Hg^{2+} is reduced by Sn^{2+}, so Sn^{2+} is the reducing agent in the reaction, which means that it must be oxidized. Therefore the mercury half-reaction occurs at the cathode and the second reaction occurs at the anode, so we must subtract its standard potential from that of the mercury half-reaction. The difference between the two half-reaction potentials is about +0.7 V. Therefore the right side of Equation 18.12 will be about $(2 \times 0.7)/0.06 \approx 23$, and the value of K should be about 10^{23}.

Solve We obtain the standard potential for the reaction from a modified version of Equation 18.2:

$$E°_{rxn} = E°_{cathode} - E°_{anode} = 0.851 \text{ V} - 0.154 \text{ V} = 0.697 \text{ V}$$

Using this value for $E°_{rxn}$ in Equation 18.12 and a value of 2 for n, we have

$$\log K = \frac{nE°_{rxn}}{0.0592 \text{ V}} = \frac{2(0.697 \text{ V})}{0.0592 \text{ V}} = 23.5$$

$$K = 10^{23.5} = 3.16 \times 10^{23}$$

Think About It The calculated value is quite close to what we estimated. Note how a $E°_{rxn}$ value of less than 1 V corresponds to a huge equilibrium constant, indicating that the reaction essentially goes to completion and can therefore be reliably used to determine the concentrations of mercury in samples containing Hg^{2+} ions.

Practice Exercise Use the appropriate standard reduction potentials from Table A6.1 to calculate the value of K at 25°C for the reaction

$$5\,Fe^{2+}(aq) + MnO_4^-(aq) + 8\,H^+(aq) \rightarrow 5\,Fe^{3+}(aq) + Mn^{2+}(aq) + 4\,H_2O(\ell)$$

Before ending our discussion of how to derive equilibrium constant values at 25°C from E°_{cell} values, we should note that measuring the potential of an electrochemical reaction allows us to calculate equilibrium constant values that may be too large or too small to determine from the equilibrium concentrations of reactants and products. A value of K as large as that calculated in Sample Exercise 18.6 could not be obtained by analyzing the composition of an equilibrium reaction mixture because the concentrations of the reactants would be too small to be determined accurately. Similarly, a cell potential of about −1 V would correspond to a tiny K value and concentrations of products that are too small to be determined quantitatively.

Table 18.1 summarizes how the values of K and E°_{cell} are related to each other and to the change in free energy (ΔG°_{cell}) of a cell reaction under standard conditions. If we know any one of the quantities in Table 18.1, we can calculate the other two. Note that spontaneous electrochemical reactions have E°_{cell} values greater than zero and K values greater than 1. The connection between positive cell potential (E_{cell}) and reaction spontaneity applies even under nonstandard conditions. Also keep in mind that small positive values of E°_{cell} (only a fraction of a volt, for example) correspond to very large K values and to cell reactions that go nearly to completion.

18.7 Relating Battery Capacity to Quantities of Reactants

An important performance characteristic of a battery is its capacity to do electrical work, that is, to deliver electrical charge at the designed cell potential. This capacity—the amount of electrical work done—is defined by Equation 18.4,

$$w_{elec} = -CE_{cell}$$

Here C is the quantity of electrical charge delivered in coulombs (C).

Another important unit in electricity is the *ampere* (A), which is the SI base unit of electrical current. An ampere is defined as a current of 1 coulomb per second:

$$1\ ampere = 1\ coulomb/second$$

which we can rearrange to

$$1\ coulomb = 1\ ampere\text{-}second$$

Multiplying both sides of this equation by volts, and recalling that 1 joule of electrical energy is equivalent to 1 coulomb-volt of electrical work, we get

$$1\ (coulomb)(volt) = 1\ joule = 1\ (ampere\text{-}second)(volt) \quad (18.13)$$

However, joules are small energy units, so battery capacities are usually expressed in energy units with intervals longer than seconds. For example, the energy ratings of rechargeable AA batteries (Figure 18.14) are often expressed in ampere-hours at the rated cell potential.

TABLE 18.1 Relationships between K, E°_{cell}, and ΔG°_{cell} Values of Electrochemical Reactions

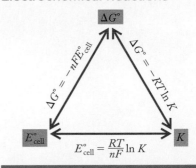

K	E°_{cell}	ΔG°_{cell}	Favors Formation of
<1	<0	>0	Reactants
>1	>0	<0	Products
1	0	0	Neither

FIGURE 18.14 The electrical energy rating of these rechargeable nickel–metal hydride AA batteries is 2500 milliampere-hours at 1.2 V.

We need even bigger units to express the power and energy capacities of the large battery packs used in hybrid vehicles. They are the *watt* (W), the SI unit of power, and the *kilowatt-hour*, a unit of energy equal to more than 3 million joules, as shown in the following unit conversions:

$$1 \text{ watt} = 1 \text{ joule/second}$$

$$1 \text{ kilowatt} = 1000 \text{ W} = 1000 \text{ J/s}$$

$$1 \text{ kilowatt} \cdot \text{hour} = (1000 \text{ W})(1 \text{ h})$$
$$= (1000 \text{ J/s} \times 60 \text{ s/min} \times 60 \text{ min/h})(1 \text{ h})$$
$$= 3.6 \times 10^6 \text{ J}$$

FIGURE 18.15 The 2016 Toyota Prius *v* is powered by a combination of a 73-kW (98-horsepower) gasoline engine and a 60-kW electric motor. Electricity for the motor comes from a 1.3 kWh nickel–metal hydride battery pack.

Nickel–Metal Hydride Batteries

As we noted at the beginning of this chapter, hybrid vehicles such as the Toyota Prius are powered by combinations of small gasoline engines and electric motors. Electricity for the motors comes from battery packs (Figure 18.15) made of dozens of nickel–metal hydride (NiMH) cells (Figure 18.16). At the cathodes in these cells, NiO(OH) is reduced to $Ni(OH)_2$, and at the anodes, made of one or more transition metals, hydrogen atoms are oxidized to H^+ ions. The electrodes are separated by aqueous KOH.

The cathode half-reaction is

$$NiO(OH)(s) + H_2O(\ell) + e^- \rightarrow Ni(OH)_2(s) + OH^-(aq)$$

At the anode, hydrogen is present as a metal hydride. To write the anode half-reaction, we use the generic formula MH, where M stands for a transition metal or metal alloy that forms a hydride. In a basic background electrolyte, the anode oxidation half-reaction is

$$MH(s) + OH^-(aq) \rightarrow M(s) + H_2O(\ell) + e^-$$

FIGURE 18.16 In the cells of a nickel–metal hydride battery pack, NiO(OH) is reduced to $Ni(OH)_2$ at the cathodes (red plates). The OH^- ions produced by the cathode half-reaction migrate across a KOH-soaked porous membrane and are consumed in the anode half-reaction. At the anodes (blue plates), oxidation of the MH hydrogen atoms produces H^+ ions that combine with OH^- ions, forming H_2O.

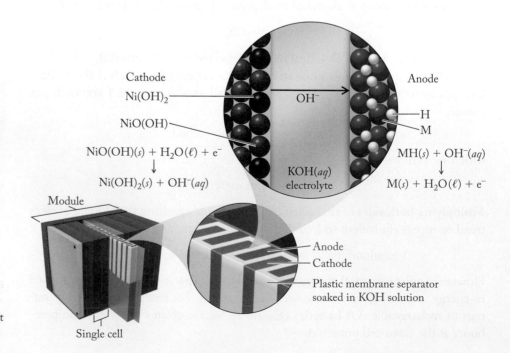

The standard potential of this half-reaction depends on the chemical properties of MH, but generally the value is near that of the SHE, or about 0.0 V.

The overall cell reaction from these two half-reactions is

$$MH(s) + NiO(OH)(s) \rightarrow M(s) + Ni(OH)_2(s)$$

The value of $E°_{cell}$ for the NiMH battery cannot be calculated precisely because we have only an approximate value of $E°_{anode}$. Most NiMH cells are rated at about 1.2 V.

CONCEPT TEST

In a NiMH battery, what are the oxidation states of (a) Ni in NiO(OH), (b) H in MH, (c) M in MH, and (d) H in H_2O?

Now let's relate the electrical energy stored in a battery (in other words, its capacity) to the quantities of reactants needed to produce that energy. Consider a rechargeable AA NiMH battery rated to deliver 2.5 ampere-hours of electrical charge at 1.2 V. How much NiO(OH) has to be converted to $Ni(OH)_2$ to deliver this much charge? To answer this question, we need to relate the quantity of charge to a number of moles of electrons, then convert that to an equivalent number of moles of reactant, and finally to a mass of reactant. Recall that an ampere is defined as a coulomb per second, which means the quantity of electrical charge delivered is

$$2.5 \ A \cdot h \times \frac{1 \ C}{A \cdot s} \times \frac{60 \ min}{1 \ h} \times \frac{60 \ s}{1 \ min} = 9.0 \times 10^3 \ C$$

The Faraday constant tells us that 1 mole of charge is equivalent to 9.65×10^4 C, so the number of moles of charge, which is equal to the number of moles of electrons that flow from the battery, is

$$9.0 \times 10^3 \ C\left(\frac{1 \ mol \ e^-}{9.65 \times 10^4 \ C}\right) = 0.0933 \ mol \ e^-$$

The stoichiometry of the cathode half-reaction tells us that the mole ratio of NiO(OH) to electrons is 1:1. Therefore the mass of NiO(OH) consumed is

$$0.0933 \ mol \ e^-\left(\frac{1 \ mol \ NiO(OH)}{1 \ mol \ e^-}\right)\left(\frac{91.70 \ g \ NiO(OH)}{1 \ mol \ NiO(OH)}\right) = 8.6 \ g \ NiO(OH)$$

The mass of an AA battery is about 30 g, so this mass for the NiO(OH) is reasonable if we allow for the mass of the anode, background electrolyte, and exterior shell.

Lithium-Ion Batteries

The NiMH batteries used in hybrid vehicles do not have the capacity to power them at highway speeds or for extended distances. Nor do these batteries have the energy capacity to power plug-in hybrids such as the Chevrolet Volt on the opening page of this chapter, or all-electric vehicles, such as the Nissan Leaf (Figure 18.17). The electrical power demands of these vehicles require batteries with much greater ratios of energy capacity to battery size. The technology of choice in these applications is the lithium-ion battery (Figure 18.18), the same kind of battery that powers laptop computers, cell phones, and digital cameras.

FIGURE 18.17 The 2016 Nissan Leaf has a 30-kWh lithium-ion battery pack that fits under the seats and gives the car an operating range of about 170 km (105 mi).

FIGURE 18.18 As this lithium-ion battery discharges, Li$^+$ ions stored in graphite layers of the anode travel to the cathode, which contains a transition metal compound such as CoO$_2$. During recharging, the direction of ion migration reverses. The crystal structure of the cathode is a cubic closest-packed array of oxide ions in which the Co^{4+} ions occupy half the octahedral holes. Li$^+$ ions move in and out of the remaining holes.

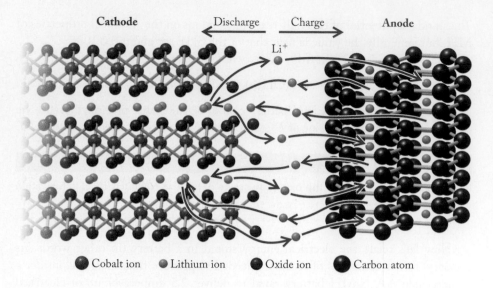

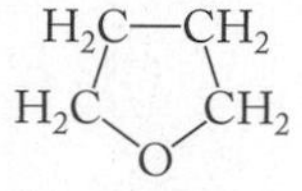

FIGURE 18.19 Polar organic compounds such as these are the solvents for the background electrolytes in a lithium-ion battery.

Tetrahydrofuran

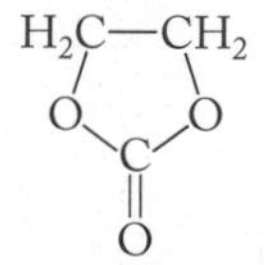

Ethylene carbonate

Propylene carbonate

In a lithium-ion battery, Li$^+$ ions are stored in a graphite or silicon anode. During discharge, these ions migrate through a nonaqueous electrolyte to a porous cathode. These cathodes are made of transition metal oxides or phosphates that can form stable complexes with Li$^+$ ions. One popular cathode material is cobalt(IV) oxide. Lithium-ion batteries with these cathodes have cell potentials of about 3.6 V (three times that of a NiMH battery). The cell reaction for a lithium-ion battery with a cobalt oxide cathode is

$$Li_{1-x}CoO_2(s) + Li_xC_6(s) \rightarrow 6\ C(s) + LiCoO_2(s) \qquad (18.14)$$

In a fully charged cell, $x = 1$, which makes the cathode lithium-free CoO$_2$. As the cell discharges and Li$^+$ ions migrate from the carbon anode to the cobalt oxide cathode, the value of x falls toward zero. To balance this flow of positive charges inside the cell, electrons flow from the anode to the cathode through an external circuit. When fully discharged, the cathode is LiCoO$_2$, and the oxidation number of Co is reduced to +3. The electrodes in a lithium-ion battery may react with oxygen and water, so the background electrolytes (for example, LiPF$_6$) are dissolved in polar organic solvents, such as tetrahydrofuran, ethylene carbonate, or propylene carbonate (Figure 18.19).

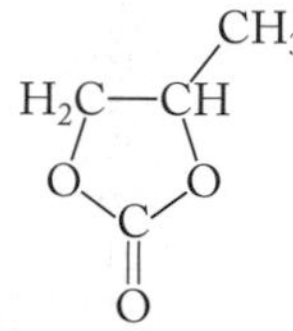

CONCEPT **TEST**

Which element is oxidized and which is reduced in the Li$^+$ ion cell reaction (Equation 18.14)?

SAMPLE EXERCISE 18.7 Relating Mass of Reactant in an Electrochemical Reaction to Quantity of Electrical Charge **LO7**

The capacity of the lithium-ion battery in a digital camera is 3.4 W · h at 3.6 V. How many grams of Li$^+$ ions must migrate from anode to cathode to produce this much electrical energy?

Collect and Organize We are asked to relate the electrical energy generated by an electrochemical cell to the mass of the ions involved in generating that energy. We know the cell potential and its capacity in the energy unit of watt-hours. Given these starting points and the eventual need to calculate moles and then grams of Li^+ ions, the following equivalencies may be useful:

$$1 \text{ watt} = 1 \text{ ampere-volt (A} \cdot \text{V)}$$

$$1 \text{ coulomb (C)} = 1 \text{ ampere-second (A} \cdot \text{s)}$$

We may also need to use the Faraday constant, 9.65×10^4 C/mol e^-.

Analyze We know the energy capacity of the battery, which is the product of the charge (electrons) it can deliver times the cell potential pumping that charge. This exercise focuses on the quantity of charge, so we need to separate the cell potential's contribution to the energy rating from the charge's contribution (Equation 18.4). To do that we need to divide the energy rating in watt-hours by the battery's cell potential in volts, and then follow that division with these unit conversions to get to grams of Li^+ ions.

$$\boxed{\dfrac{\text{watt} \cdot \text{hour}}{\text{volt}}} \xrightarrow{\dfrac{\text{amp} \cdot \text{volt}}{\text{watt}}} \xrightarrow{\dfrac{\text{coulomb}}{\text{amp} \cdot \text{second}}} \xrightarrow{\dfrac{3600 \text{ second}}{\text{hour}}} \xrightarrow{\dfrac{1 \text{ mol } e^-}{9.65 \times 10^4 \text{ C}}} \dfrac{1 \text{ mol Li}^+}{1 \text{ mol } e^-}$$

$$\xrightarrow{\dfrac{\text{molar mass Li}^+}{}} = \boxed{\text{g Li}^+}$$

The ratio of the initial values, 3.4 watt-hours and 3.6 volts, is nearly one, so the result of the calculation can be estimated based on the approximate values of the combined conversion factors, or roughly $(4000 \times 7)/10^5$, or about 0.3.

Solve Using the given energy and cell-potential values in the above unit conversion series, we get

$$\frac{3.4 \text{ W} \cdot \text{h}}{3.6 \text{ V}} \times \frac{1 \text{ A} \cdot \text{V}}{1 \text{ W}} \times \frac{1 \text{ C}}{1 \text{ A} \cdot \text{s}} \times \frac{3600 \text{ s}}{1 \text{ h}} \times \frac{1 \text{ mol } e^-}{9.65 \times 10^4 \text{ C}}$$

$$\times \frac{1 \text{ mol Li}^+}{1 \text{ mol } e^-} \times \frac{6.941 \text{ g Li}^+}{1 \text{ mol Li}^+} = 0.24 \text{ g Li}^+$$

Think About It The battery that is the subject of this exercise has a mass of about 22 grams, so Li^+ ions make up only about 1% of the mass of the battery. This small percentage is not surprising given the masses of the other required components of the cell, including an anode where Li^+ ions are surrounded by hexagons of six carbon atoms and a cathode made of CoO_2, for example, which has 13 times the molar mass of Li.

 Practice Exercise Magnesium metal is produced by passing an electric current through molten $MgCl_2$. The reaction at the cathode is

$$Mg^{2+}(\ell) + 2 e^- \rightarrow Mg(s)$$

How many grams of magnesium metal are produced if an average current of 63.7 A flows for 4.50 hours? Assume all of the current is consumed by the half-reaction shown.

18.8 Corrosion: Unwanted Electrochemical Reactions

We began this chapter with a Particulate Review question based on the corrosion of the metal surface of a car. In this section we examine this process, and some of the half-reactions that contribute to it, in more detail. We have seen how half-reactions are physically separated in electrochemical cells. It turns out they are often separated in corrosion reactions as well; in this respect, the chemistry of corrosion is

FIGURE 18.20 The light green patina of the Statue of Liberty is caused by the accumulation of copper(II) compounds on the surface of the copper sheets that make up its exterior. When the statue was built, these sheets were supported by an iron skeleton that corroded near the points of contact with the sheets and the copper saddles that held the sheets and skeleton together.

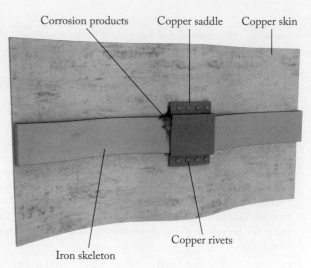

Corrosion products   Copper saddle   Copper skin

Copper rivets

Iron skeleton

much like the electrochemical reactions in voltaic cells. In fact, we can define **corrosion** as the deterioration of metals due to spontaneous electrochemical reactions. This definition is reflected in several of the factors that promote corrosion:

1. *The presence of water.* Metals that are left out in the rain and snow rust or corrode more rapidly than those that are under cover. For example, objects made from iron spontaneously rust in moist air, as iron reacts with O_2 from the air in a series of reactions that includes

$$4 \, Fe(s) + 3 \, O_2(g) + 2 \, H_2O(\ell) \rightarrow 4 \, FeO(OH)(s)$$

2. *The presence of electrolytes.* Just as electrolytes carry electrical current between anodes and cathodes and facilitate cell reactions, corrosion is much more rapid in, for example, seawater than in freshwater.

3. *Contact between dissimilar metals.* Metals corrode more rapidly when in contact with other metals that are less likely to be oxidized; that is, other metals that have higher reduction potentials.

Let's explore the impact of these factors using the Statue of Liberty as a model. As the statue was built, its exterior copper sheets were attached to and supported by an interior network of iron beams (Figure 18.20). The French designers of the statue knew that these two metals in contact with each other might someday pose a corrosion problem because the two have very different electrochemical properties. The difference is reflected in the standard reduction potentials of these elements when they oxidize under neutral to slightly basic conditions, as occur in marine environments:

$$Cu(OH)_2(s) + 2 \, e^- \rightarrow Cu(s) + 2 \, OH^-(aq) \qquad E° = -0.230 \, V$$

$$FeO(OH)(s) + H_2O(\ell) + 3 \, e^- \rightarrow Fe(s) + 3 \, OH^-(aq) \qquad E° = -0.87 \, V$$

As we discussed in Section 18.3, the greater $E°$ of $Cu(OH)_2$ means that it is more easily reduced under standard conditions than $FeO(OH)$, and Fe is more easily oxidized than Cu. We can confirm this by reversing the iron half-reaction and combining it with the copper half-reaction. We also need to multiply the iron half-reaction by 2 and the copper half-reaction by 3 to balance the loss and gain of electrons:

corrosion a process in which a metal is oxidized by substances in its environment.

$$2[Fe(s) + 3\ OH^-(aq) \rightarrow FeO(OH)(s) + H_2O(\ell) + 3\ e^-]$$
$$3[Cu(OH)_2(s) + 2\ e^- \rightarrow Cu(s) + 2\ OH^-(aq)]$$

$$2\ Fe(s) + 3\ Cu(OH)_2(s) + \cancel{6\ OH^-}(aq) + \cancel{6\ e^-} \rightarrow$$
$$2\ FeO(OH)(s) + 3\ Cu(s) + 2\ H_2O(\ell) + \cancel{6\ OH^-}(aq) + \cancel{6\ e^-}$$

or

$$3\ Cu(OH)_2(s) + 2\ Fe(s) \rightarrow 3\ Cu(s) + 2\ FeO(OH)(s) + 2\ H_2O(\ell) \quad (18.15)$$

The resulting equation could be that of an electrochemical cell that has a copper cathode coated in $Cu(OH)_2$ and an iron anode. We obtain the standard potential for the reaction by substituting the reduction potentials for each half-reaction (Appendix A6.1) into Equation 18.2:

$$E^\circ_{rxn} = E^\circ_{cathode} - E^\circ_{anode} = -0.230\ V - (-0.87\ V) = 0.64\ V$$

The reaction has a positive standard cell potential, which means that under standard conditions, iron in contact with copper will spontaneously oxidize to FeO(OH) as $Cu(OH)_2$ is reduced.

To suppress the reaction in Equation 18.15, insulators made of asbestos mats soaked in shellac were used to separate the statue's iron skeleton from its copper exterior (Figure 18.20). Unfortunately, these insulators did not stand up to the humid, marine environment of New York Harbor. They absorbed water vapor and seawater spray, eventually turning into electrolyte-soaked sponges that actually promoted rather than retarded iron oxidation.

What are the sources of the $Cu(OH)_2$ that are the oxidizing agents in Equation 18.15? Recall from Figure 18.20 that the light green patina of the Statue of Liberty is due to the presence of copper(II) compounds such as $Cu(OH)_2$, formed by the oxidation of Cu metal by atmospheric oxygen:

$$O_2(g) + 2\ H_2O(\ell) + 2\ Cu(s) \rightarrow 2\ Cu(OH)_2(s) \quad (18.16)$$

This reaction occurs on an expansive surface of copper metal, which is also an excellent conductor of electricity. Unfortunately, this copper surface was in contact with the statue's iron skeleton through the water-logged, ion-rich asbestos mats. This connection meant the reactions in Equations 18.15 and 18.16 were linked together. Equation 18.17 describes the resulting interior and exterior reactions:

$$2[3\ Cu(OH)_2(s) + 2\ Fe(s) \rightarrow 3\ Cu(s) + 2\ FeO(OH)(s) + 2\ H_2O(\ell)] \quad (18.15)$$
$$3[O_2(g) + 2\ H_2O(\ell) + 2\ Cu(s) \rightarrow 2\ Cu(OH)_2(s)] \quad (18.16)$$

$$\overline{}$$

$$\cancel{6\ Cu(OH)_2(s)} + 4\ Fe(s) + 3\ O_2(g) + \cancel{6}\ H_2O(\ell) + \cancel{6\ Cu(s)} \rightarrow$$
$$\cancel{6\ Cu(s)} + 4\ FeO(OH)(s) + 4\ H_2O(\ell) + \cancel{6\ Cu(OH)_2(s)}$$

or

$$4\ Fe(s) + 3\ O_2(g) + 2\ H_2O(\ell) \rightarrow 4\ FeO(OH)(s) \quad (18.17)$$

Note how the copper half-reaction has disappeared from the overall reaction. Thus, the statue's copper exterior functions as a giant electron delivery system, allowing electrons to flow from iron atoms—as they oxidize to FeO(OH) inside the statue—to molecules of atmospheric O_2 on the exterior surface. The overall effect was a dramatic increase in the oxidation of the original iron skeleton.

The reaction in Equation 18.17 resulted in severe deterioration of the skeletal network that held up the Statue of Liberty, so much so that in the 1980s the iron skeleton had to be replaced with one made of corrosion-resistant stainless steel.

Deterioration of the Statue of Liberty is not an isolated incident. According to one industrial estimate, the direct and indirect cost of corrosion to the U.S. economy exceeds $1 trillion per year. Worldwide, the cost is over $2 trillion. These

FIGURE 18.21 (a) Oceangoing metal structures are likely to have serious corrosion problems unless (b) "sacrificial" anodes are attached to them that corrode (oxidize) before the structure does.

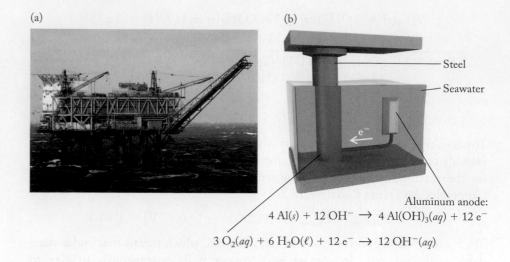

(a)

(b)

Steel

Seawater

Aluminum anode:
$$4 \, Al(s) + 12 \, OH^- \rightarrow 4 \, Al(OH)_3(aq) + 12 \, e^-$$

$$3 \, O_2(aq) + 6 \, H_2O(\ell) + 12 \, e^- \rightarrow 12 \, OH^-(aq)$$

figures include the costs to repair or replace corroded equipment and structures and to protect them against corrosion. The latter category includes money spent on protective coatings, including paint and chemical or electrochemical modification of metal surfaces to make them less reactive.

Another widely used method for inhibiting corrosion involves chemically bonding metal oxide coatings to metal surfaces. For example, a method called *bluing* is widely used to protect steel tools, gun barrels, wood-burning stoves, and other steel materials. The name comes from the distinctive very dark blue color of a surface layer of magnetite (Fe_3O_4), which can be reaction-bonded to steel. Formation of a protective oxide layer is also the mechanism that makes stainless steel resistant to corrosion. Although the main ingredient in all forms of stainless steel is iron, chromium is also present. Oxidation of surface Cr atoms forms a durable protective layer of Cr_2O_3. Similarly, objects made of aluminum are protected by a surface layer of Al_2O_3 that inhibits further oxidation and strongly adheres to the underlying metal.

Another way to protect metal structures, especially those that come in contact with seawater (Figure 18.21), involves attaching to them objects made of even more reactive metals. These objects are called *sacrificial* anodes. As their name implies, their role is to form a voltaic cell with the protected structure in which the object is the anode and the structure is the cathode, so that the sacrificial anodes oxidize and the structure does not. This preservation technique is called *cathodic protection*. Many of the sacrificial anodes used in the marine industry are made of zinc or aluminum/magnesium alloys.

$$Zn(s) + 2 \, OH^-(aq) \rightarrow ZnO(s) + H_2O(\ell) + 2 \, e^-$$

$$Mg(s) + 2 \, OH^-(aq) \rightarrow MgO(s) + H_2O(\ell) + 2 \, e^-$$

$$2 \, Al(s) + 6 \, OH^-(aq) \rightarrow Al_2O_3(s) + 3 \, H_2O(\ell) + 6 \, e^-$$

These materials have the benefit of forming oxide coatings as they oxidize, which partially protect them and slow the rate at which they oxidize further, thereby extending their life span.

18.9 Electrolytic Cells and Rechargeable Batteries

Lead–acid, NiMH, and lithium-ion batteries are rechargeable, which means that their spontaneous ($\Delta G < 0$) cell reactions that convert chemical energy

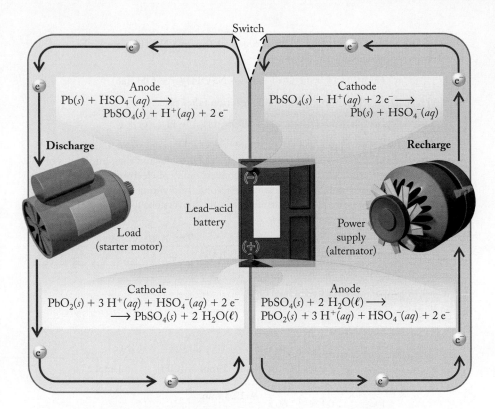

FIGURE 18.22 The lead–acid battery used in many vehicles is based on oxidation of Pb and reduction of PbO_2. As the battery discharges (circuit on left), Pb is oxidized to $PbSO_4$ and PbO_2 is reduced to $PbSO_4$. When the engine is running, a device called an alternator generates electricity that flows into the battery, as shown in the circuit on the right, recharging the battery as both electrode reactions are reversed: $PbSO_4$ is oxidized to PbO_2, and $PbSO_4$ is reduced to Pb.

into electrical work can be forced to run in reverse. Recharging happens when external sources of electrical energy are applied to the batteries. This electrical energy is converted into chemical energy as it drives nonspontaneous ($\Delta G > 0$) reverse cell reactions, reforming reactants from products. To make this possible, the products of the original cell reactions must be substances that either adhere to, or are embedded in, the electrodes and are available to react with the electrons supplied to the cathodes and drawn away from the anodes by the external power supply.

Figure 18.22 shows the discharge/recharge cycle of a lead–acid battery. Note that electrons flow in one direction when the battery discharges—out of the negative terminal and into the positive terminal—but in the opposite direction when the battery is recharging. Thus, the Pb electrodes, which are connected to the negative battery terminal in Figure 18.22, serve as anodes during discharge but as cathodes during recharge. Any $PbSO_4$ that forms on these Pb electrodes during discharge is reduced back to Pb metal during recharge:

$$PbSO_4(s) + H^+(aq) + 2\ e^- \rightarrow Pb(s) + HSO_4^-(aq)$$

Similarly, the PbO_2 electrodes at the positive terminal serve as cathodes during discharge but as anodes during recharge. Any $PbSO_4$ that forms on these PbO_2 electrodes during discharge is oxidized back to PbO_2 during recharge:

$$PbSO_4(s) + 2\ H_2O(\ell) \rightarrow PbO_2(s) + 3\ H^+(aq) + HSO_4^-(aq) + 2\ e^-$$

SAMPLE EXERCISE 18.8 Calculating the Time Required to Oxidize a Quantity of Reactant **LO7**

If a battery charger for AA NiMH batteries supplies a charging current of 1.00 A, how many minutes does it take to oxidize 0.649 g of $Ni(OH)_2$ to $NiO(OH)$?

Collect, Organize, and Analyze During the discharge of a NiMH battery, the spontaneous cathode half-reaction is

$$NiO(OH)(s) + H_2O(\ell) + e^- \rightarrow Ni(OH)_2(s) + OH^-(aq)$$

During recharging, this reverse of this half-reaction takes place:

$$Ni(OH)_2(s) + OH^-(aq) \rightarrow NiO(OH)(s) + H_2O(\ell) + e^-$$

One mole of electrons is produced for each mole of $Ni(OH)_2$ consumed. Our first steps are to convert 0.649 g of $Ni(OH)_2$ into moles of $Ni(OH)_2$ and then into moles of electrons. We can use the Faraday constant to convert moles of electrons to coulombs of charge. A coulomb is the same as an ampere-second, so dividing by the charging current gives us seconds of charging current. The unit conversions to minutes of charging time are summarized as follows:

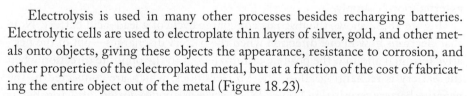

$$= \text{charging time (min)}$$

Solve

$$0.649 \text{ g Ni(OH)}_2 \times \frac{1 \text{ mol Ni(OH)}_2}{92.71 \text{ g Ni(OH)}_2} \times \frac{1 \text{ mol e}^-}{1 \text{ mol Ni(OH)}_2} \times \frac{9.65 \times 10^4 \text{ C}}{1 \text{ mol e}^-}$$

$$\times \frac{1 \text{ A} \cdot \text{s}}{1 \text{ C}} \times \frac{1}{1.00 \text{ A}} \times \frac{1 \text{ min}}{60 \text{ s}} = 11.3 \text{ min}$$

Think About It A charging time of 11.3 min may seem short, but the quantity of the $Ni(OH)_2$ to be oxidized (0.649 g) is much less than the total quantity of $Ni(OH)_2$ in a fully charged AA NiMH battery (8.6 g, as calculated in Section 18.7).

 Practice Exercise Suppose that a car's starter motor draws 230 A of current for 6.0 s to start the car. What mass of Pb is oxidized in the battery to supply this much electricity?

Electrolysis is used in many other processes besides recharging batteries. Electrolytic cells are used to electroplate thin layers of silver, gold, and other metals onto objects, giving these objects the appearance, resistance to corrosion, and other properties of the electroplated metal, but at a fraction of the cost of fabricating the entire object out of the metal (Figure 18.23).

In the chemical industry, electrolysis of molten salts is used to produce highly reactive substances, such as sodium, chlorine, and fluorine; alkali and alkaline earth metals; and aluminum. When NaCl, for instance, is heated to just above its melting point (above 800°C), it becomes an ionic liquid that can conduct electricity. If a sufficiently large potential is applied to carbon electrodes immersed in the molten NaCl, sodium ions are attracted to the negative electrode and are reduced to sodium metal, while chloride ions are attracted to the positive electrode and oxidized to Cl_2 gas:

$$2 \text{ Na}^+(\ell) + 2 \text{ Cl}^-(\ell) \rightarrow 2 \text{ Na}(\ell) + \text{Cl}_2(g)$$

A final note about how the anode and cathode vary in voltaic and electrolytic cells is in order. As we saw in Section 18.2, the reactions in voltaic cells are spontaneous. These cells pump electric current through external circuits and electrical devices with a force equal to their cell potentials. The anode in a voltaic cell is negative because an oxidation half-reaction supplies negatively charged electrons to the device powered by the cell. Electrons flow from the device into the positive battery terminal that is connected to the cathode, where these electrons are consumed in a reduction half-reaction.

FIGURE 18.23 The Oscar statuettes given out at the annual Academy Awards are made of an alloy of tin, antimony, and copper that is electroplated with three different materials: copper, nickel silver (a silvery alloy of copper, nickel, and zinc), and a final layer of 24-karat gold.

The reactions in electrolytic cells are nonspontaneous. They require electrical energy from an external power supply. When the negative terminal of such a power supply is connected to the cathode of the battery, the power supply pumps electrons into the cathode, where they are consumed in reduction half-reactions. Electrons are pumped away from the anode, where they were generated in an oxidation half-reaction, toward the positive terminal of the power supply. Thus the cathode of an electrolytic cell is the negative electrode, but the cathode of a voltaic cell is the positive electrode. Similarly, the anode of an electrolytic cell is the positive electrode, but the anode of a voltaic cell is the negative electrode. These electrode differences between voltaic and electrolytic cells make sense if we keep in mind the fundamental definitions:

- Anodes are electrodes where oxidation takes place.
- Cathodes are electrodes where reduction takes place.

CONCEPT TEST

The electrolysis of molten NaCl produces liquid Na metal at the cathode and Cl_2 gas at the anode. However, the electrolysis of an aqueous solution of NaCl produces gases at both cathode and anode. On the basis of the standard potentials in Table A6.1, predict the gas formed at each electrode.

18.10 Fuel Cells

Fuel cells are promising energy conversion devices for many applications, from powering office buildings to cruise ships to electric vehicles. Fuel cells are voltaic cells, but they differ from batteries in that their supplies of reactants are constantly renewed. Fuel cells are energy conversion devices that continue to supply electricity as long as fuel and oxygen are delivered to the anode and cathode, respectively. Therefore, they do not "discharge": they never run down, and they don't die unless their fuel supply is cut off. From a thermodynamic perspective, most batteries are closed systems and fuel cells are open systems.

In a typical fuel cell, electrons are supplied to an external circuit by the oxidation of H_2 at the anode. Electrons are consumed by the reaction of O_2 with protons migrating through the electrolyte to the cathode. Chemical energy from the reaction of hydrogen with oxygen is converted directly into electrical energy, and the net fuel cell reaction is

$$2 H_2(g) + O_2(g) \rightarrow 2 H_2O(\ell)$$

The fuel cells used to power electric vehicles consist of metallic or graphite electrodes separated by a hydrated polymeric material called a *proton-exchange membrane* (PEM). The PEM serves as both an electrolyte and a barrier that prevents crossover and mixing of the fuel and oxidant (Figure 18.24). The electrodes are carbon-supported transition metal catalysts dispersed in a layer upon the PEM. The catalytic layers speed up the electrode half-reactions. Platinum catalysts promote H—H bond breaking during the oxidation of H_2 gas to H^+ ions at the anode:

$$H_2(g) \rightarrow 2 H^+(aq) + 2 e^- \qquad E° = 0.000 \text{ V}$$

At the cathode, a platinum–nickel alloy with the nominal composition Pt_3Ni is particularly effective in catalyzing the formation of free O atoms from O_2 molecules that are part of the reduction half-reaction:

$$O_2(g) + 4 H^+(aq) + 4 e^- \rightarrow 2 H_2O(\ell) \qquad E° = 1.229 \text{ V}$$

CHEMTOUR
Fuel Cell

fuel cell a voltaic cell based on the oxidation of a continuously supplied fuel. The reaction is the equivalent of combustion, but chemical energy is converted directly into electrical energy.

FIGURE 18.24 Most fuel cells used in vehicles have a proton-exchange membrane between the two halves of the cell. Hydrogen gas diffuses to the anode, and oxygen gas diffuses to the cathode. These electrodes are made of porous material, such as carbon nanofibers, that has a relatively high surface area for a given mass of material. Catalysts on the electrode surfaces also increase the rate of the half-reactions at the anode and the cathode.

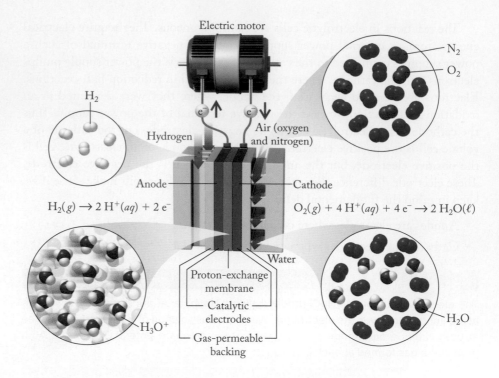

Hydrogen ions that form at the anode migrate through the PEM to the cathode, where they combine with the free O atoms formed on the Pt_3Ni catalyst surface and electrons from the external circuit. This migration of positive charges across the PEM electrolyte is concurrent with the flow of electrons through the device in the external circuit.

A single PEM fuel cell typically delivers useful currents at cell potentials of about 0.8 V. The operating cell voltage is less than the expected E_{cell}° value of 1.23 V because some of the energy of combustion is wasted as heat. When these cells are assembled into fuel cell *stacks*, they can produce 100 kW of electrical power. That is enough to give a midsize car, such as the one in Figure 18.25, a top speed of 160 km/h (100 mi/h).

PEM fuel cells are well suited for use in vehicles because they are compact and lightweight, and they operate at fairly low temperatures of 60°C to 80°C. It turns out that the performance of nearly all fuel cells is better at above-ambient temperatures because the rates of the half-reactions increase with temperature, as predicted by the Arrhenius equation. The increase in rate results in higher efficiency and more power.

FIGURE 18.25 The Honda Clarity is powered by a 100-kW fuel-cell stack and 100-kW electric motor that give the car a top speed of about 160 km/h (100 mph).

CONNECTION The temperature dependence of reaction rates (the Arrhenius equation) was discussed in Chapter 13.

Some fuel cells use basic electrolytes such as concentrated KOH. Pure O_2 is supplied to a cathode made of porous graphite containing a nickel catalyst, and H_2 gas is supplied to a graphite anode containing nickel(II) oxide. Hydroxide ions formed during O_2 reduction at the cathode,

$$O_2(g) + 2\,H_2O(\ell) + 4\,e^- \rightarrow 4\,OH^-(aq) \qquad E^{\circ} = 0.401\text{ V}$$

migrate through the cell to the anode, where they combine with H_2 as it is oxidized to water:

$$H_2(g) + 2\,OH^-(aq) \rightarrow 2\,H_2O(\ell) + 2\,e^-$$

This oxidation half-reaction is the reverse of the following reduction half-reaction in Table A6.1:

$$2 H_2O(\ell) + 2 e^- \rightarrow H_2(g) + 2 OH^-(aq) \qquad E° = -0.828 V$$

Note that this basic pair of standard electrode potentials yields the same $E°_{cell}$ value:

$$E°_{cell} = 0.401 V - (-0.828 V) = 1.229 V$$

as the acidic pair described earlier:

$$E°_{cell} = 1.229 V - 0.000 V = 1.229 V$$

This equality is logical because the energy, $\Delta G°$, released under standard conditions by the oxidation of hydrogen gas to form liquid water should have only one value, which means that $E°_{cell}$ should have only one value, independent of electrolyte pH. What, then, is the advantage of the alkaline fuel cell? Although the thermodynamics of the cell are independent of pH, it turns out that the kinetics of oxygen reduction are much better in an alkaline environment.

CONCEPT TEST

During the operation of molten alkali-metal-carbonate fuel cells, carbonate ions are generated at one electrode, migrate across the cell, and are consumed at the other electrode. Do the carbonate ions migrate toward the cathode or the anode?

The same chemical energy that is released in fuel cells could also be obtained by burning hydrogen gas in an internal combustion engine. However, typically only about 20–25% of the chemical energy in the fuel burned in such an engine is converted into mechanical energy; most is lost to the surroundings as heat. In contrast, fuel-cell technologies can convert up to about 80% of the energy released in a fuel-cell redox reaction into electrical energy. The electric motors they power are also about 80% efficient at converting electrical energy into mechanical energy. Thus, the overall conversion efficiency of a fuel-cell–powered car is theoretically as high as (80% × 80%), or 64%. Actually, the measured efficiency of the propulsion system of the car in Figure 18.25 is about 60%, which is still more than twice that of an internal combustion engine. In addition, H_2-fueled vehicles emit only water vapor; they produce no oxides of nitrogen, no carbon monoxide, and no CO_2.

As fuel cells become even more efficient and less expensive, the principal limit on their use in passenger cars will be the availability and cost of hydrogen fuel. Some people worry about the safety of storing hydrogen in a high-pressure tank in a car. Actually, H_2 has a higher ignition temperature than gasoline and spreads through the air more quickly, reducing the risk of fire. Still, hydrogen in air burns over a much wider range of concentrations than gasoline, and its flame is almost invisible.

Because of the lack of an extensive hydrogen distribution network, many fuel-cell–powered vehicles are buses and fleet vehicles operating from a central location where hydrogen gas is available. With their very large fuel tanks, buses powered by fuel cells have an operating range of 400 km ($\approx$250 mi). Longer ranges would be possible if better methods for storing hydrogen gas were available. Currently under development are mobile chemical-processing plants that can extract hydrogen from gasoline, methane, or methanol. Onboard production of hydrogen from fossil fuels also generates carbon dioxide as a by-product, so these fuels still produce greenhouse gases. However, the efficiencies of fuel cells and the electric motors they power mean that much less of these fuels will be needed and less CO_2 will be emitted.

CONNECTION Processes for producing hydrogen gas by reacting methane and other fuels with steam were discussed in Chapter 15.

SAMPLE EXERCISE 18.9 Integrating Concepts: Corrosion at Sea

In the 18th century, the British Royal Navy began the practice of adding copper sheathing to the bottoms of its sailing vessels to inhibit the growth of barnacles and seaweed that slowed them down. Unfortunately, the first copper sheets they used were fastened with iron nails. (a) Explain why using iron nails was not a good idea. (b) Which metal would corrode?

Modern navies continue to encounter shipboard corrosion problems, sometimes with disastrous results. In 1963, the USS *Thresher*, the lead boat in a new class of nuclear submarines, was conducting deep diving tests about 200 miles off the coast of Cape Cod when it suddenly sank. All 129 sailors and civilians aboard perished. The cause of the sinking was later determined to be corrosion that occurred where silver solder had been used to connect metal pipes that transported seawater to cool the *Thresher*'s power plant. Metal pipes that carry seawater are made of copper/nickel alloys. (c) Why might silver solder have been considered a logical choice for connecting them? (d) How is the oxidation of silver in seawater influenced by the presence of chloride ions? (e) Use the appropriate half-reactions and standard reduction potentials from Table A6.1 in the Appendix to calculate the cell potential for the air-oxidation of Ag in pH 8.00 seawater in which $[Cl^-] = 0.56\ M$. Assume $P_{O_2} = 0.21$ atm. Is the reaction spontaneous?

Collect and Organize First we are asked to explain why nailing one metal to another is a bad idea when they will be immersed in seawater, and to identify which one will corrode. Then we have another corrosion problem involving Ag and Cu/Ni alloys, where we need to address their electrochemical properties and to explain how the oxidation of Ag in seawater depends on the presence of Cl^- ions. Finally, we are asked to determine whether Ag spontaneously oxidizes in aerated seawater, using half-reactions from Table A6.1 in the Appendix to calculate E_{cell}.

Analyze The problem with nailing copper sheets with iron nails is similar to the one encountered in the Statue of Liberty (see Figure 18.20), in which the iron skeleton of the statue rusted so badly it had to be replaced.

The logic behind using silver solder to connect pipes made of another metal (or metal alloy) must have been based on their similar electrochemical properties. To determine similarity we need to use the appropriate half-reaction for silver. Table A6.1 lists two half-reactions: one that is simply $Ag^+(aq) + e^- \rightarrow Ag(s)$ and the other in which the half-reaction occurs in a solution of chloride ions: $AgCl(s) + e^- \rightarrow Ag(s) + Cl^-(aq)$. Given the high $[Cl^-]$ in seawater and the limited solubility of AgCl, we should use the latter to compare electrochemical properties and to determine whether the oxidation of Ag is spontaneous.

Solve

a. Iron and copper should not be allowed to contact each other in seawater because they are dissimilar metals, as shown by the standard reduction potentials of these half-reactions:

$$Fe^{2+}(aq) + 2\,e^- \rightarrow Fe(s) \qquad E° = -0.447\ V$$
$$Cu^{2+}(aq) + 2\,e^- \rightarrow Cu(s) \qquad E° = 0.342\ V$$

When Cu and Fe are in contact with each other in an electrolytic solution, the two become the cathode and anode, respectively, of a voltaic cell with the cell reaction:

$$Cu^{2+}(aq) + Fe(s) \rightarrow Cu(s) + Fe^{2+}(aq)$$

This reaction has an $E_{cell} = E°_{cathode} - E°_{anode} = 0.342\ V - (-0.447\ V) = 0.789\ V$.

b. The spontaneous cell reaction tells us that Fe is the metal that is oxidized (corrodes).

c. Table A6.1 contains these half-reactions involving Ag, Cu, and Ni:

$$AgCl(s) + e^- \rightarrow Ag(s) + Cl^-(aq) \qquad E° = 0.222\ V$$
$$Cu^{2+}(aq) + 2\,e^- \rightarrow Cu(s) \qquad E° = 0.342\ V$$
$$Ni^{2+}(aq) + 2\,e^- \rightarrow Ni(s) \qquad E° = -0.257\ V$$

We see that the $E°$ value of the Ag reaction is in between the other two values. If the actual electrochemical properties of the alloy are closer to those of Cu than Ni (which, it turns out, they are), then Ag would have been a logical choice to use in connecting pipes made from the alloy.

d. The difference in the potentials of the two silver half-reactions in Table A6.1,

$$AgCl(s) + e^- \rightarrow Ag(s) + Cl^-(aq) \qquad E° = 0.222\ V$$
$$Ag^+(aq) + e^- \rightarrow Ag(s) \qquad E° = 0.800\ V$$

indicates that the presence of Cl^- ions has a significant impact on the electrochemical properties of Ag. As Ag atoms are oxidized, they precipitate as AgCl, which keeps the concentration of $Ag^+(aq)$ low. The Nernst equation for the Ag^+/Ag half-reaction is

$$E = E° - 0.0592\ V \log (1/[Ag^+])$$

A small $[Ag^+]$ value in the denominator means a large value for the log term and an E value for the Ag^+/Ag half-reaction that is much less than its $E°$. A smaller potential for the reduction half-reaction means a greater potential for the oxidation of Ag. Therefore, Ag is more likely to be oxidized in the presence of Cl^- ions.

e. The appropriate half-reactions for the corrosion of silver in seawater are

$$O_2(g) + 4\,H^+(aq) + 4\,e^- \rightarrow 2\,H_2O(\ell) \qquad E° = 1.229\ V$$
$$AgCl(s) + e^- \rightarrow Ag(s) + Cl^-(aq) \qquad E° = 0.222\ V$$

To assess the spontaneity of the air-oxidation of Ag, we need to reverse the second half-reaction and multiply it by 4 to balance the loss and gain of electrons:

$$O_2(g) + 4\,H^+(aq) + 4\,e^- \rightarrow 2\,H_2O(\ell)$$
$$4[Ag(s) + Cl^-(aq) \rightarrow AgCl(s) + e^-]$$
$$\overline{O_2(g) + 4\,H^+(aq) + 4\,Ag(s) + 4\,Cl^-(aq) \rightarrow 4\,AgCl(s) + 2\,H_2O(\ell)}$$

The silver reaction takes place at the anode and the standard cell potential is

$$E°_{cell} = E°_{cathode} - E°_{anode} = 1.229\ V - 0.222\ V = 1.007\ V$$

Using this E°_{cell} value and setting up the Nernst equation for the reaction gives us

$$E_{cell} = E^\circ_{cell} - \left(\frac{0.0592 \text{ V}}{4}\right) \log \left[\frac{1}{(P_{O_2})[\text{H}^+]^4[\text{Cl}^-]^4}\right]$$

We take the antilog of pH = 8.00 and use the P_{O_2} and $[\text{Cl}^-]$ values for aerated seawater:

$$E_{cell} = 1.007 \text{ V} - \left(\frac{0.0592 \text{ V}}{4}\right) \log \left[\frac{1}{(0.21)(1.00 \times 10^{-8})^4(0.56)^4}\right]$$
$$= 0.51 \text{ V}$$

The positive cell potential means that the oxidation of silver in seawater should be spontaneous.

Think About It The above analysis made it clear why using iron nails to fasten copper sheathing to the hulls of wooden sailing ships was a bad idea. However, we were unable to answer why the silver solder joints failed onboard the *Thresher*. According to modern naval engineering references, silver solder and Cu/Ni alloys should have similar electrochemical properties when immersed in seawater. Unfortunately, "similar" was apparently not close enough to protect the 129 souls aboard the *Thresher*.

SUMMARY

LO1 **Electrochemistry** is the branch of chemistry that links redox reactions to the production or consumption of electrical energy. Any redox reaction can be broken down into oxidation and reduction half-reactions. (Section 18.1)

LO2 In an **electrochemical cell**, the oxidation half-reaction occurs at the **anode** and the reduction half-reaction occurs at the **cathode**. Migration of the ions in the cell's electrolyte allows electrical charges to flow between the cathode and anode compartments as electrons flow through an external electrical circuit. A **cell diagram** shows how the components of the cathodic and anodic compartments of the cell are connected. (Sections 18.1 and 18.2)

LO3 The difference between the **standard reduction potentials (E°)** of a cell's cathode and anode half-reactions is equal to the **standard cell potential (E°_{cell})**. The value of (E°_{cell}) is a measure of how forcefully a voltaic cell, in which all reactants and products are in their standard states, can pump electrons through an external circuit. All standard cell potentials are referenced to the cell potential of the **standard hydrogen electrode** ($E^\circ_{SHE} = 0.000$ V). (Sections 18.3, 18.4, 18.5)

LO4 A voltaic cell has a positive cell potential ($E_{cell} > 0$) and its cell reaction has a negative change in free energy ($\Delta G_{cell} < 0$). This decrease in free energy in a voltaic cell is available to do work in an external electrical circuit. The **Faraday constant** relates the quantity of electrical charge to the number of moles of electrons and indirectly to the number of moles of reactants. (Section 18.4)

LO5 The potential of a voltaic cell decreases as reactants turn into products. The **Nernst equation** describes how cell potential changes with concentration changes. (Section 18.6)

LO6 The potential of a voltaic cell approaches zero as the cell reaction approaches chemical equilibrium, at which point $E_{cell} = 0$ and $Q = K$. We can calculate the equilibrium constant for any redox reaction at 25°C, not just those in electrochemical cells. (Sections 18.6, 18.7, 18.10)

LO7 The quantities of reactants consumed in a voltaic cell reaction are proportional to the coulombs of electrical charge delivered by the cell. Nickel–metal hydride batteries supply electricity when H atoms are oxidized to H^+ ions at the anodes and NiO(OH) is reduced to Ni(OH)_2 at the cathodes. In lithium-ion batteries, electricity is produced when Li^+ ions stored in graphite anodes migrate toward and are incorporated into transition metal oxide or phosphate cathodes. (Sections 18.7, 18.8, 18.9)

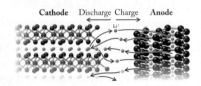

PARTICULATE **PREVIEW WRAP-UP**

The two half-reactions depict the oxidation of hydrogen and the reduction of oxygen:

$$\text{H}_2(g) \rightarrow 2\,\text{H}^+(aq) + 2\,\text{e}^-$$
$$\text{O}_2(g) + 4\,\text{H}^+(aq) + 4\,\text{e}^- \rightarrow 2\,\text{H}_2\text{O}(\ell)$$

which sum to a balanced, overall reaction of

$$2\,\text{H}_2(g) + \text{O}_2(g) \rightarrow 2\,\text{H}_2\text{O}(\ell)$$

When chemists physically separate two half-reactions, electrons can be forced to flow through an external circuit from one half-cell to the other. This external flow of electrons can be harnessed to do work as electricity. Chemists call hydrogen a "clean" fuel because it emits only water vapor, unlike traditional fuels, which when combusted, produce oxides of nitrogen and carbon that pollute the atmosphere.

PROBLEM-SOLVING SUMMARY

Type of Problem	Concepts and Equations	Sample Exercises
Writing net ionic equations of cell reactions by combining half-reactions	Combine the reduction and oxidation half-reactions after balancing the gain and loss of electrons.	**18.1–18.4**
Diagramming an electrochemical cell	Use the format: anode | anode solution || cathode solution | cathode Insert solution concentrations if they are known.	**18.2**
Identifying anode and cathode half-reactions and calculating the value of $E°_{cell}$	The half-reaction with the more positive standard reduction potential is the cathode half-reaction in a voltaic cell. $$E°_{cell} = E°_{cathode} - E°_{anode} \qquad (18.2)$$	**18.3**
Relating $\Delta G°_{cell}$ and $E°_{cell}$	$$\Delta G°_{cell} = -nFE°_{cell} \qquad (18.7)$$ where n is the number of moles of electrons transferred in the cell reaction and F is the Faraday constant, 9.65×10^4 C/mol e$^-$.	**18.4**
Calculating E_{cell} from $E°_{cell}$ and the concentrations of reactants and products	E_{cell} at 25°C is related to $E°_{cell}$ and the cell reaction quotient by the Nernst equation: $$E_{cell} = E°_{cell} - \frac{0.0592\ \text{V}}{n} \log Q \qquad (18.10)$$	**18.5**
Calculating K for a redox reaction from the standard potentials of its half-reactions	$E°_{rxn}$ is related to K at 298 K by $$\log K = \frac{nE°_{rxn}}{0.0592\ \text{V}}$$	**18.6**
Relating mass of reactant in an electrochemical reaction to quantity of electrical charge	Determine the ratio of moles of reactants to moles of electrons transferred; use the Faraday constant to relate coulombs of charge to moles of electrons.	**18.7**
Calculating the time required to oxidize a quantity of reactant	Use the Faraday constant and the relation between moles of electrons and moles of reactants to describe an electrolytic process.	**18.8**

VISUAL PROBLEMS

(Answers to boldface end-of-chapter questions and problems are in the back of the book.)

18.1. In the voltaic cell shown in Figure P18.1, the greater density of a concentrated solution of $CuSO_4$ allows a less concentrated solution of $ZnSO_4$ solution to be (carefully) layered on top of it. Why is a porous bridge not needed in this cell?

FIGURE P18.1

18.2. In the voltaic cell shown in Figure P18.2, the concentrations of Cu^{2+} and Cd^{2+} are 1.00 M. On the basis of the standard potentials in Appendix 6, identify which electrode is the anode and which is the cathode. Indicate the direction of electron flow.

FIGURE P18.2

18.3. In the voltaic cell shown in Figure P18.3, $[Ag^+] = [H^+] = 1.00\ M$ and $P_{H_2} = 1.00$ atm. From the standard potentials in Appendix 6, Table A6.1, identify which electrode is the anode and which is the cathode. Indicate the direction of electron flow.

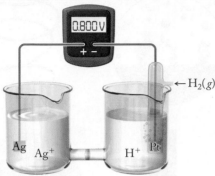

$\leftarrow H_2(g)$

FIGURE P18.3

18.4. In many electrochemical cells the electrodes are metals that carry electrons to and from the cell but are not chemically changed by the cell reaction. Each of the highlighted clusters in the periodic table in Figure P18.4 consists of three metals. Which of the highlighted clusters is best suited to form inert electrodes?

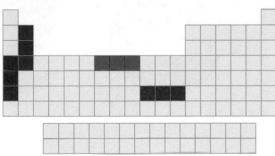

FIGURE P18.4

18.5. Which of the four curves in Figure P18.5 best represents the dependence of the potential of a lead–acid battery on the concentration of sulfuric acid? Note that the scale of the x-axis is logarithmic.

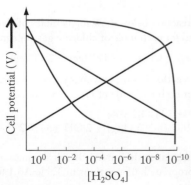

FIGURE P18.5

18.6. Consider the four types of batteries in Figure P18.6. From top to bottom the sizes are AAA, AA, C, and D. The performance of batteries such as these is often expressed in units such as (a) volts, (b) watt-hours, or (c) milliampere-hours. Which of the values differ significantly between the four batteries?

FIGURE P18.6

18.7. The apparatus in Figure P18.7 is used for the electrolysis of water. Hydrogen and oxygen gas are collected in the two inverted burettes. An inert electrode at the bottom of the left burette is connected to the negative terminal of a 6-volt battery; the electrode in the burette on the right is connected to the positive terminal. A small quantity of sulfuric acid is added to speed up the electrolytic reaction.

a. What are the half-reactions at the left and right electrodes and their standard potentials?
b. Why does sulfuric acid make the electrolysis reaction go more rapidly?

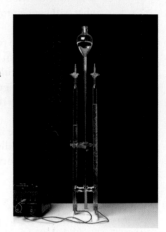

Overall cell reaction
$$H_2O(\ell) \rightarrow H_2(g) + \tfrac{1}{2} O_2(g)$$

FIGURE P18.7

18.8. An electrolytic apparatus identical to the one shown in Problem 18.7 is used to electrolyze water, but the reaction is speeded up by the addition of sodium carbonate instead of sulfuric acid.

a. What are the half-reactions and the standard potentials for the electrodes on the left and right?
b. Why does sodium carbonate make the electrolysis reaction go more rapidly?

18.9. Most classic cars, such as the one shown in Figure P18.9, have chromium-electroplated, or *chrome*, bumpers.

FIGURE P18.9

a. In the electroplating process is the bumper the anode or the cathode?

b. How does the presence of a layer of chromium protect the bumper from corroding?

18.10. Use representations [A] through [I] in Figure P18.10 to answer questions a–f. The photo in image [E] shows silver deposited onto copper. If the materials in [A], [C], [G], and [I] were combined, along with a porous bridge and external circuit, to generate an electrochemical cell that produced [E]:

a. Which metal would be the cathode? Which solution would surround it?

b. Which metal would be the anode? Which solution would surround it?

c. When the reaction in [E] is finished, what will happen to the light blue color of the solution?

d. How could [E] be produced without a porous bridge or an external circuit?

e. Of the four particulate images [B], [D], [F], and [H] in Figure P18.10, which two correspond to the solutions [G] and [I]?

f. Which particulate image represents the copper wire with silver deposited onto it? What does the fourth particulate image depict?

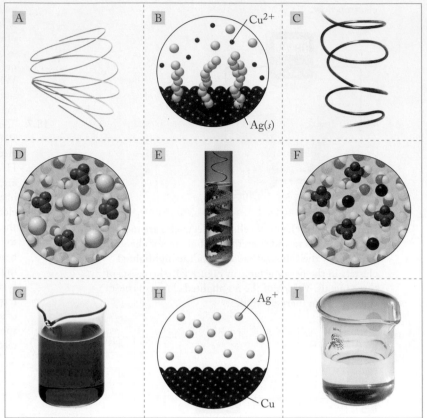

FIGURE P18.10

QUESTIONS AND PROBLEMS

Redox Chemistry Revisited; Electrochemical Cells

Concept Review

18.11. An element with a strong tendency to gain electrons is also _____.
  a. easily oxidized
  b. a good oxidizing agent
  c. a good reducing agent
  d. a reactive metal

18.12. An element that is a good reducing agent is also _____.
  a. easily oxidized
  b. a good oxidizing agent
  c. easily reduced
  d. a noble gas

18.13. Regarding the porous separator between the two halves of an electrochemical cell:
  a. Describe how it allows electrical charge to flow between the two half-cells.
  b. Explain why a piece of wire could not perform the same function.

18.14. The Zn/Cu^{2+} reactions in Figures 18.1 and 18.2 are the same; however, the reaction in the cell in Figure 18.2 generates electricity, whereas the reaction in the beaker in Figure 18.1 does not. Why?

Problems

18.15. In the redox reaction below, how many moles of electrons are transferred for each mole of chlorine gas consumed?

$$2\ Fe^{2+}(aq) + Cl_2(g) \rightarrow 2\ Fe^{3+}(aq) + 2\ Cl^-(aq)$$

18.16. In the redox reaction below, how many electrons are transferred for each molecule of H_2O_2 consumed?

$$2\ MnO_4^-(aq) + 3\ H_2O_2(aq) \rightarrow$$
$$2\ MnO_2(s) + 3\ O_2(g) + 2\ OH^-(aq) + 2\ H_2O(\ell)$$

18.17. Complete and balance the partial chemical equation below, using the appropriate half-reactions in Table A6.1 for acidic solutions.

$$Cr_2O_7^{2-}(aq) + Fe^{2+}(aq) \rightarrow Cr^{3+}(aq) + Fe^{3+}(aq)$$

18.18. Complete and balance the partial net ionic equation below, using the appropriate half-reactions in Table A6.1 for acidic solutions.

$$MnO_4^-(aq) + H_2O_2(aq) \rightarrow MnO_2(s) + O_2(g)$$

18.19. Select the appropriate half-reactions from Appendix 6 to write net ionic equations describing the reaction between:
   a. aluminum metal and Fe^{3+} ions in solution that produces dissolved Al^{3+} and Fe^{2+} ions.
   b. I_2 and NO_2^- ions in an alkaline solution that produces I^- and NO_3^- ions.
   c. MnO_4^- and Cr^{3+} ions in an acidic solution that produces Mn^{2+} and $Cr_2O_7^{2-}$ ions.

18.20. Select the appropriate half-reactions from Appendix 6 to write net ionic equations describing the reaction between:
   a. tin and Ag^+ ions in solution that produces dissolved Sn^{2+} ions and silver metal.
   b. copper and O_2 in an acidic solution that produces Cu^{2+} ions.
   c. solid $Cr(OH)_3$ and O_2 in a basic solution that produces CrO_4^{2-} ions.

18.21. An electrochemical cell with an aqueous electrolyte is based on the reaction between $Ni^{2+}(aq)$ and $Cd(s)$, producing $Ni(s)$ and $Cd^{2+}(aq)$.
   a. Write half-reactions for the anode and cathode.
   b. Write a balanced net ionic equation describing the cell reaction.
   c. Draw the cell diagram.

18.22. A voltaic cell is based on the reaction between $Cu^{2+}(aq)$ and $Ni(s)$, producing $Cu(s)$ and $Ni^{2+}(aq)$.
   a. Write the anode and cathode half-reactions.
   b. Write a balanced cell reaction.
   c. Draw the cell diagram.

18.23. A voltaic cell with a basic aqueous background electrolyte is based on the oxidation of $Cd(s)$ to $Cd(OH)_2(s)$ and the reduction of $MnO_4^-(aq)$ to $MnO_2(s)$.
   a. Write half-reactions for the cell's anode and cathode.
   b. Write a balanced net ionic equation describing the cell reaction.
   c. Draw the cell diagram.

18.24. A voltaic cell is based on the reduction of $Ag^+(aq)$ to $Ag(s)$ and the oxidation of $Sn(s)$ to $Sn^{2+}(aq)$.
   a. Write half-reactions for the cell's anode and cathode.
   b. Write a balanced cell reaction.
   c. Draw the cell diagram.

18.25. Super Iron Batteries In 1999, scientists in Israel developed a battery based on the following cell reaction with iron(VI), nicknamed "super iron":

$$2\ K_2FeO_4(aq) + 3\ Zn(s) \rightarrow Fe_2O_3(s) + ZnO(s) + 2\ K_2ZnO_2(aq)$$

   a. Determine the number of electrons transferred in the cell reaction.
   b. What are the oxidation states of the transition metals in the reaction?
   c. Draw the cell diagram.

18.26. Aluminum–Air Batteries In recent years engineers have been working on an aluminum–air battery as an alternative energy source for electric vehicles. The battery consists of an aluminum anode, which is oxidized to solid aluminum hydroxide, immersed in an electrolyte of aqueous KOH. At the cathode oxygen from the air is reduced to hydroxide ions on an inert metal surface. Write the two half-reactions for the battery and diagram the cell. Use the generic $M(s)$ symbol for the metallic cathode material.

Standard Potentials

Concept Review

18.27. Which is a stronger oxidizing agent under standard conditions: oxygen or chlorine? Use the standard reduction potentials in Table A6.1 to support your answer.

*__18.28.__ Of the group 1 elements Li, K, and Na, which is the strongest reducing agent?

Problems

18.29. Starting with the appropriate standard free energies of formation in Table A4.3 in Appendix 4, calculate the values of $\Delta G°$ and $E°_{cell}$ of the following reactions:
   a. $2\ Cu^+(aq) \rightarrow Cu^{2+}(aq) + Cu(s)$
   b. $Cu(s) + 2\ Fe^{3+}(aq) \rightarrow Cu^{2+}(aq) + 2\ Fe^{2+}(aq)$

18.30. Starting with the appropriate standard free energies of formation in Appendix 4, calculate the values of $\Delta G°$ and $E°_{cell}$ of the following reactions:
   a. $2\ Na(s) + 2\ H_2O(\ell) \rightarrow 2\ NaOH(aq) + H_2(g)$
   b. $2\ Pb(s) + O_2(g) + 2\ H_2SO_4(aq) \rightarrow$ $2\ PbSO_4(s) + 2\ H_2O(\ell)$

18.31. If a piece of silver is placed in a solution in which $[Ag^+] = [Cu^{2+}] = 1.00\ M$, will the following reaction proceed spontaneously?

$$2\ Ag(s) + Cu^{2+}(aq) \rightarrow 2\ Ag^+(aq) + Cu(s)$$

18.32. A piece of cadmium is placed in a solution in which $[Cd^{2+}] = [Sn^{2+}] = 1.00\ M$. Will the following reaction proceed spontaneously?

$$Cd(s) + Sn^{2+}(aq) \rightarrow Cd^{2+}(aq) + Sn(s)$$

18.33. Sometimes the anode half-reaction in the zinc–air battery (Figure 18.7) is written with the zincate ion, $Zn(OH)_4^{2-}$, as the product. Write a balanced equation for the cell reaction based on this product.

*__18.34.__ Sometimes the cell reaction of nickel–cadmium batteries is written with Cd metal as the anode and solid NiO_2 as the cathode. Assuming that the products of the reactions are a solid hydroxide of cadmium(II) at the anode and a solid hydroxide of nickel(II) at the cathode, write balanced equations for the cathode and anode half-reactions and the overall cell reaction.

18.35. In a voltaic cell similar to the Cu–Zn cell in Figure 18.2, the Cu electrode is replaced with one made of Ni immersed in a solution of $NiSO_4$. Will the standard potential of this Ni–Zn cell be greater than, the same as, or less than 1.10 V?

18.36. Suppose the copper half of the Cu–Zn cell in Figure 18.2 were replaced with a silver wire in contact with $1\ M\ Ag^+(aq)$.
   a. What would be the value of $E°_{cell}$?
   b. Which electrode would be the anode?

18.37. Starting with standard potentials listed in Table A6.1, calculate the values of $E°_{cell}$ and $\Delta G°$ of the following reactions.
   a. $Cl_2(g) + 2\ Br^-(aq) \rightarrow Br_2(\ell) + 2\ Cl^-(aq)$
   b. $Zn(s) + Ni^{2+}(aq) \rightarrow Zn^{2+}(aq) + Ni(s)$

18.38. Voltaic cells based on the following pairs of half-reactions are constructed. For each pair, write a balanced equation for the cell reaction, and identify which half-reaction takes place at each anode and cathode.

a. $Cd^{2+}(aq) + 2\,e^- \rightarrow Cd(s)$
   $Ag^+(aq) + e^- \rightarrow Ag(s)$
b. $AgBr(s) + e^- \rightarrow Ag(s) + Br^-(aq)$
   $MnO_2(s) + 4\,H^+(aq) + 2\,e^- \rightarrow Mn^{2+}(aq) + 2\,H_2O(\ell)$
c. $PtCl_4^{2-}(aq) + 2\,e^- \rightarrow Pt(s) + 4\,Cl^-(aq)$
   $AgCl(s) + e^- \rightarrow Ag(s) + Cl^-(aq)$

18.39. Which of the following reductions will occur in the presence of H_2 gas under standard conditions?
   a. Ag^+ to Ag
   b. Mg^{2+} to Mg
   c. Cu^{2+} to Cu
   d. Cd^{2+} to Cd

18.40. Which of the following oxidations will occur in the presence of H_2 gas under standard conditions?
   a. Zn^{2+} to Zn
   b. Fe^{2+} to Fe^{3+}
   c. $Cr(OH)_3$ to CrO_4^{2-}
   d. Ni to Ni^{2+}

18.41. The half-reactions and standard potentials for a nickel–metal hydride battery with a titanium–zirconium anode are as follows:

Cathode: $NiO(OH)(s) + H_2O(\ell) + e^- \rightarrow Ni(OH)_2(s) + OH^-(aq)$
$$E° = 0.52\ \text{V}$$

Anode: $TiZr_2H(s) + OH^-(aq) \rightarrow TiZr_2(s) + H_2O(\ell) + e^-$
$$E° = 0.00\ \text{V}$$

   a. Write the overall cell reaction for this battery.
   b. Calculate the standard cell potential.

*18.42. **Lithium-Ion Batteries** There are lithium-ion batteries that have cathodes composed of $FePO_4$ when fully charged.
   a. What is the formula of the cathode when the battery is fully discharged?
   b. Is Fe oxidized or reduced as the battery discharges?
   c. Is the cell potential of a lithium-ion battery with an iron phosphate cathode likely to differ from one with a cobalt oxide cathode? Explain your answer.

Chemical Energy and Electrical Work

Problems

18.43. The value of $E°_{cell}$ for the reaction below is 0.500 V. What is the value of $\Delta G°_{cell}$?

$$Mn^{3+}(aq) + 2\,H_2O(\ell) \rightarrow Mn^{2+}(aq) + MnO_2(s) + 4\,H^+(aq)$$

18.44. What is the value of $\Delta G°_{cell}$ for an electrochemical cell based on a cell reaction described by the following net ionic equation?

$$Mg(s) + 2\,Cu^+(aq) \rightarrow Mg^{2+}(aq) + 2\,Cu(s)$$

18.45. For many years the 1.50 V batteries used to power flashlights were based on the following cell reaction:

$$Zn(s) + 2\,NH_4Cl(s) + 2\,MnO_2(s) \rightarrow$$
$$Zn(NH_3)_2Cl_2(s) + Mn_2O_3(s) + H_2O(\ell)$$

What is the value of ΔG_{cell}?

18.46. The first generation of laptop computers was powered by nickel–cadmium (nicad) batteries, which generated 1.20 V in accordance with the following cell reaction:

$$Cd(s) + 2\,NiO(OH)(s) + 2\,H_2O(\ell) \rightarrow Cd(OH)_2(s) + 2\,Ni(OH)_2(s)$$

What is the value of ΔG_{cell}?

18.47. The cells in the nickel–metal hydride battery packs used in many hybrid vehicles produce 1.20 V in accordance with the following cell reaction:

$$MH(s) + NiO(OH)(s) \rightarrow M(s) + Ni(OH)_2(s)$$

What is the value of ΔG_{cell}?

18.48. A cell in a lead–acid battery delivers exactly 2.00 V of cell potential in accordance with the following cell reaction:

$$Pb(s) + PbO_2(s) + 2\,H_2SO_4(aq) \rightarrow 2\,PbSO_4(s) + 2\,H_2O(\ell)$$

What is the value of ΔG_{cell}?

A Reference Point: The Standard Hydrogen Electrode

Concept Review

18.49. What is the function of platinum in the standard hydrogen electrode?

18.50. Platinum is very expensive, so why is it used in standard hydrogen standard electrodes where the half–reaction is $2\,H^+(aq) + 2\,e^- \rightarrow H_2(g)$?

18.51. The potential of the standard hydrogen electrode (SHE) is the reference against which other half-reaction potentials are expressed. Why, then, is the SHE not widely used as a reference electrode in electrochemical cells?

*18.52. Suggest a replacement metal for platinum in the standard hydrogen electrode. Explain why you selected the metal you did.

Problems

18.53. An electrochemical cell consists of a standard hydrogen electrode and a second half-cell in which a magnesium electrode is immersed in a 1.00 M solution of Mg^{2+} ions.
   a. What is the value of E_{cell}?
   b. Which electrode is the anode?
   c. Which is a product of the cell reaction: H^+ ions or H_2 gas?

18.54. An electrochemical cell consists of a standard hydrogen electrode and a second half-cell in which a cadmium electrode is immersed in a 1.00 M solution of Cd^{2+} ions.
   a. What is the value of E_{cell}?
   b. Which electrode is the anode?
   c. Which is a product of the cell reaction: Cd^{2+} ions or Cd metal?

The Effect of Concentration on E_cell

Concept Review

18.55. Why does the operating cell potential of most batteries change little until the battery is nearly discharged?

18.56. The standard potential of the Cu–Zn cell reaction,

$$Zn(s) + Cu^{2+}(aq) \rightarrow Zn^{2+}(aq) + Cu(s)$$

is 1.10 V. Would the potential of the Cu–Zn cell differ from 1.10 V if the concentrations of both Cu^{2+} and Zn^{2+} were 0.25 M?

Problems

18.57. If the potential of a hydrogen electrode based on the half-reaction

$$2\,H^+(aq) + 2\,e^- \rightarrow H_2(g)$$

is 0.000 V at pH = 0.00, what is the potential of the same electrode at pH = 7.00?

18.58. Glucose Metabolism The standard potentials for the reduction of nicotinamide adenine dinucleotide (NAD^+) and oxaloacetate (reactants in the multistep metabolism of glucose) are as follows:

$$NAD^+(aq) + 2\,H^+(aq) + 2\,e^- \rightarrow NADH(aq) + H^+(aq)$$
$$E° = -0.320\text{ V}$$

$$\text{Oxaloacetate}^{2-}(aq) + 2\,H^+(aq) + 2\,e^- \rightarrow \text{malate}^{2-}(aq)$$
$$E° = -0.166\text{ V}$$

a. Calculate the standard potential for the following reaction:

$$\text{Oxaloacetate}^{2-}(aq) + NADH(aq) + H^+(aq) \rightarrow$$
$$\text{malate}^-(aq) + NAD^+(aq)$$

b. Calculate the equilibrium constant for the reaction at 25°C.

18.59. Permanganate ion can oxidize sulfite to sulfate in basic solution as follows:

$$2\,MnO_4^-(aq) + 3\,SO_3^{2-}(aq) + H_2O(\ell) \rightarrow$$
$$2\,MnO_2(s) + 3\,SO_4^{2-}(aq) + 2\,OH^-(aq)$$

Determine the potential for the reaction (E_{rxn}) at 25°C when the concentrations of the reactants and products are as follows: $[MnO_4^-] = 0.250\ M$, $[SO_3^{2-}] = 0.425\ M$, $[SO_4^{2-}] = 0.075\ M$, and $[OH^-] = 0.0200\ M$. Will the value of E_{rxn} increase or decrease as the reaction proceeds?

*18.60. A *concentration cell* can be constructed by using the same half-reaction for both the cathode and anode. What is the value of E_{cell} for a concentration cell that combines copper electrodes in contact with 0.35 M copper(II) nitrate and 0.00075 M copper(II) nitrate solutions?

18.61. A copper penny dropped into a solution of nitric acid produces a mixture of nitrogen oxides. The following reaction describes the formation of NO, one of the products:

$$3\,Cu(s) + 8\,H^+(aq) + 2\,NO_3^-(aq) \rightarrow$$
$$2\,NO(g) + 3\,Cu^{2+}(aq) + 4\,H_2O(\ell)$$

a. Starting with the appropriate standard potentials in Table A6.1, calculate $E°_{cell}$ for this reaction.
b. Calculate E_{cell} at 25°C when $[H^+] = 0.500\ M$, $[NO_3^-] = 0.0550\ M$, $[Cu^{2+}] = 0.0500\ M$, and the partial pressure of NO = 0.00250 atm.

18.62. Chlorine dioxide (ClO_2) is produced by the following reaction of chlorate (ClO_3^-) with Cl^- in acid solution:

$$2\,ClO_3^-(aq) + 2\,Cl^-(aq) + 4\,H^+(aq) \rightarrow$$
$$2\,ClO_2(g) + Cl_2(g) + 2\,H_2O(\ell)$$

a. Determine $E°$ for the reaction.
b. The reaction produces a mixture of gases in the reaction vessel in which $P_{ClO_2} = 2.0$ atm; $P_{Cl_2} = 1.00$ atm. Calculate $[ClO_3^-]$ if, at equilibrium ($T = 25°C$), $[H^+] = [Cl^-] = 10.0\ M$.

Relating Battery Capacity to Quantities of Reactants

Concept Review

18.63. One 12-volt lead–acid battery has a higher ampere-hour rating than another. Which of the following parameters are likely to be different for the two batteries?
a. individual cell potentials
b. anode half-reactions
c. total masses of electrode materials
d. number of cells
e. electrolyte composition
f. combined surface areas of their electrodes

18.64. In a voltaic cell based on the Cu–Zn cell reaction

$$Zn(s) + Cu^{2+}(aq) \rightarrow Cu(s) + Zn^{2+}(aq)$$

there is exactly 1 mole of each reactant and product. A second cell based on the Cd–Cu cell reaction

$$Cd(s) + Cu^{2+}(aq) \rightarrow Cu(s) + Cd^{2+}(aq)$$

also has exactly 1 mole of each reactant and product. Which of the following statements about these two cells is true?
a. Their cell potentials are the same.
b. The masses of their electrodes are the same.
c. The quantities of electrical charge that they can produce are the same.
d. The quantities of electrical energy that they can produce are the same.

Problems

18.65. Which of the following voltaic cells will produce the greater quantity of electrical charge per gram of anode material?

$$Cd(s) + 2\,NiO(OH)(s) + 2\,H_2O(\ell) \rightarrow 2\,Ni(OH)_2(s) + Cd(OH)_2(s)$$

or

$$4\,Al(s) + 3\,O_2(g) + 6\,H_2O(\ell) + 4\,OH^-(aq) \rightarrow 4\,Al(OH)_4^-(aq)$$

18.66. Which of the following voltaic cells will produce the greater quantity of electrical charge per gram of anode material?

$$Zn(s) + MnO_2(s) + H_2O(\ell) \rightarrow ZnO(s) + Mn(OH)_2(s)$$

or

$$Li(s) + MnO_2(s) \rightarrow LiMnO_2(s)$$

---

*18.67. Which of the following voltaic cells delivers more electrical energy per gram of anode material at 25°C?

$$Zn(s) + 2\,NiO(OH)(s) + 2\,H_2O(\ell) \rightarrow$$
$$2\,Ni(OH)_2(s) + Zn(OH)_2(s) \qquad E°_{cell} = 1.20\text{ V}$$

or

$$Li(s) + MnO_2(s) \rightarrow LiMnO_2(s) \qquad E°_{cell} = 3.15\text{ V}$$

*18.68. Which of the following voltaic cell reactions delivers more electrical energy per gram of anode material at 25°C?

$$Zn(s) + Ni(OH)_2(s) \rightarrow Zn(OH)_2(s) + Ni(s) \qquad E°_{cell} = 1.50\text{ V}$$

or

$$2\,Zn(s) + O_2(g) \rightarrow 2\,ZnO(s) \qquad E°_{cell} = 2.08\text{ V}$$

Corrosion: Unwanted Electrochemical Reactions

Concept Review

18.69. When the iron skeleton of the Statue of Liberty was replaced with stainless steel, the asbestos mats that had separated the skeleton from the copper exterior were replaced with Teflon spacers. Why was Teflon a good choice?

18.70. What does a sacrificial anode do to protect a metal structure, and why is the process called *cathodic* protection?

18.71. **Aquarium Windows** The windows of the giant ocean tank at the New England Aquarium (Figure P18.71) are held in place with aluminum frames. What would be a good material to use to make sacrificial anodes for the frames?

FIGURE P18.71

18.72. **Corrosion of Copper Pipes** The copper pipes frequently used in household plumbing may corrode and eventually leak. The corrosion reaction is believed to involve the formation of copper(I) chloride:

$$2\,Cu(s) + Cl_2(aq) \rightarrow 2\,CuCl(s)$$

a. Write balanced equations for the half-reactions in this redox reaction.
b. Calculate E°_{rxn} and ΔG°_{rxn} for the reaction.

Electrolytic Cells and Rechargeable Batteries

Concept Review

18.73. The positive terminal of a voltaic cell is the cathode. However, the cathode of an electrolytic cell is connected to the negative terminal of a power supply. Explain this difference in polarity.

18.74. The anode in an electrochemical cell is defined as the electrode where oxidation takes place. Why is the anode in an electrolytic cell connected to the positive (+) terminal of an external supply, whereas the anode in a voltaic cell battery is connected to the negative (−) terminal?

18.75. The salts obtained from the evaporation of seawater can be a source of halogens, principally Cl_2 and Br_2, through the electrolysis of the molten alkali metal halides. As the potential of the anode in an electrolytic cell is increased, which of these two halogens forms first?

18.76. **Quantitative Analysis** Electrolysis can be used to determine the concentration of Cu^{2+} in a given volume of solution by electrolyzing the solution in a cell equipped with a platinum cathode. If all the Cu^{2+} is reduced to Cu metal at the cathode, the increase in mass of the electrode provides a measure of the concentration of Cu^{2+} in the original solution. To ensure the complete (99.99%) removal of the Cu^{2+} from a solution in which $[Cu^{2+}]$ is initially about 1.0 M, will the potential of the cathode (versus SHE)

have to be more or less negative than 0.34 V (the standard potential for $Cu^{2+} + 2\,e^{-} \rightarrow Cu$)?

Problems

18.77. A high school chemistry student wishes to demonstrate how water can be separated into hydrogen and oxygen by electrolysis. She knows that the reaction will proceed more rapidly if an electrolyte is added to the water. She has access to 2.00 M solutions of these compounds: H_2SO_4, HBr, NaI, Na_2SO_4, and Na_2CO_3. Which one(s) should she use? Explain your selection(s).

18.78. A battery charger used to recharge the NiMH batteries used in a digital camera can deliver as much as 0.75 amperes of current to each battery. If it takes 100 min to recharge one battery, how many grams of $Ni(OH)_2$ are oxidized to NiO(OH)?

18.79. A NiMH battery containing 4.10 g of NiO(OH) was 75% discharged when it was connected to a charger with an output of 2.00 A at 1.3 V. How long does it take to recharge the battery?

*18.80. How long does it take to deposit a coating of gold 1.00 μm thick on a disk-shaped medallion 2.0 cm in diameter and 3.0 mm thick at a constant current of 45 A? The density of gold is 18.3 g/cm^3. The gold solution contains gold(III).

*18.81. **Oxygen Supply in Submarines** Nuclear submarines can stay under water nearly indefinitely because they can produce their own oxygen by the electrolysis of water.
   a. How many liters of O_2 at 25°C and 1.00 bar are produced in 1 hour in an electrolytic cell operating at a current of 0.025 A?
   b. Could seawater be used as the source of oxygen in this electrolysis? Explain why or why not.

18.82. In the electrolysis of water, how long will it take to produce 1.00×10^2 L of H_2 at STP (273 K and 1.00 atm) by using an electrolytic cell through which the current is 52 mA?

18.83. Calculate the minimum (least negative) cathode potential (versus SHE) needed to begin electroplating nickel from 0.35 M Ni^{2+} onto a piece of iron.

*18.84. What is the minimum (least negative) cathode potential (versus SHE) needed to electroplate silver onto cutlery in a solution of Ag^{+} and NH_3 in which most of the silver ions are present as the complex, $Ag(NH_3)_2^{+}$, and the concentration of $Ag^{+}(aq)$ is only 2.50×10^{-4} M?

Fuel Cells

Concept Review

18.85. Describe two advantages of hybrid (gasoline engine–electric motor) power systems over all-electric systems based on fuel cells. Describe two disadvantages.

18.86. Describe three factors limiting widespread use of cars powered by fuel cells.

18.87. Methane can serve as the fuel for electric cars powered by fuel cells. Carbon dioxide is a product of the fuel cell reaction. All cars powered by internal combustion engines burning natural gas (mostly methane) produce CO_2. Why are electric vehicles powered by fuel cells likely to produce less CO_2 per mile?

18.88. To make the refueling of fuel cells easier, several manufacturers offer converters that turn readily available fuels—such as natural gas, propane, and methanol—into H_2 for the fuel cells and CO_2. Although vehicles with such power systems are not truly "zero emission," they still offer significant environmental benefits over vehicles powered by internal combustion engines. Describe a few of those benefits.

Problems

18.89. Fuel cells with molten alkali metal carbonates as electrolytes can use methane as a fuel. The methane is first converted into hydrogen in a two-step process:

$$CH_4(g) + H_2O(g) \rightarrow CO(g) + 3\,H_2(g)$$
$$CO(g) + H_2O(g) \rightarrow H_2(g) + CO_2(g)$$

a. Assign oxidation numbers to carbon and hydrogen in the reactants and products.

b. Using the standard free energy of formation values in Table A4.3 in Appendix 4, calculate the standard free-energy changes in the two reactions and the overall $\Delta G°$ for the formation of $H_2 + CO_2$ from methane and steam.

*18.90. A direct methanol fuel cell uses the oxidation of methanol by oxygen to generate electrical energy. The overall reaction, which is given below, has a $\Delta G°$ value of −702.4 kJ/mol of methanol oxidized. What is the standard cell potential for this fuel cell?

$$CH_3OH(\ell) + \tfrac{3}{2}\,O_2(g) \rightarrow CO_2(g) + 2\,H_2O(\ell)$$

Additional Problems

18.91. Calculate the E_{cell} value at 298 K for the cell based on the reaction

$$Fe^{3+}(aq) + Cu^+(aq) \rightarrow Fe^{2+}(aq) + Cu^{2+}(aq)$$

when $[Fe^{3+}] = [Cu^+] = 1.50 \times 10^{-3}\,M$ and $[Fe^{2+}] = [Cu^{2+}] = 2.5 \times 10^{-4}\,M$.

18.92. Calculate the E_{cell} value at 298 K for the cell based on the reaction

$$Cu(s) + 2\,Ag^+(aq) \rightarrow Cu^{2+}(aq) + 2\,Ag(s)$$

when $[Ag^+] = 2.56 \times 10^{-3}\,M$ and $[Cu^{2+}] = 8.25 \times 10^{-4}\,M$.

18.93. Using the appropriate standard potentials in Appendix 6, determine the equilibrium constant for the following reaction at 298 K:

$$Fe^{3+}(aq) + Cr^{2+}(aq) \rightarrow Fe^{2+}(aq) + Cr^{3+}(aq)$$

18.94. Using the appropriate standard potentials in Appendix 6, determine the equilibrium constant at 298 K for the following reaction between MnO_2 and Fe^{2+} in acid solution:

$$4\,H^+(aq) + MnO_2(s) + 2\,Fe^{2+}(aq) \rightarrow$$
$$Mn^{2+}(aq) + 2\,Fe^{3+}(aq) + 2\,H_2O(\ell)$$

*18.95. **Electrolysis of Seawater** Magnesium metal is obtained by the electrolysis of molten Mg^{2+} salts from evaporated seawater.

a. Would elemental Mg form at the cathode or anode?

b. Do you think the principal ingredient in sea salt (NaCl) would need to be separated from the Mg^{2+} salts before electrolysis? Explain your answer.

c. Would electrolysis of an aqueous solution of $MgCl_2$ also produce elemental Mg?

d. If your answer to part (c) was no, what would be the products of electrolysis?

*18.96. **Silverware Tarnish** Low concentrations of hydrogen sulfide in air react with silver to form Ag_2S, more familiar to us as tarnish. Silver polish contains aluminum metal powder in a basic suspension.

a. Write a balanced net ionic equation for the redox reaction between Ag_2S and Al metal that produces Ag metal and $Al(OH)_3$.

b. Calculate $E°$ for the reaction.

18.97. A magnesium battery can be constructed from an anode of magnesium metal and a cathode of molybdenum sulfide, Mo_3S_4. The standard reduction potentials of the electrode half-reactions are

$$Mg^{2+}(aq) + 2\,e^- \rightarrow Mg(s) \qquad E° = -2.37\,V$$
$$Mg^{2+}(aq) + Mo_3S_4(s) + 2\,e^- \rightarrow MgMo_3S_4(s) \qquad E° = ?$$

a. If the standard cell potential for the battery is 1.50 V, what is the value of $E°$ for the reduction of Mo_3S_4?

b. What are the apparent oxidation states of Mo in Mo_3S_4 and in $MgMo_3S_4$?

*c. The electrolyte in the battery contains a complex magnesium salt, $Mg(AlCl_3CH_3)_2$. Why is it necessary to include Mg^{2+} ions in the electrolyte?

18.98. Suppose there were a scale for expressing electrode potentials in which the standard potential for the reduction of water in base

$$2\,H_2O(\ell) + 2\,e^- \rightarrow H_2(g) + 2\,OH^-(aq)$$

is assigned an $E°$ value of 0.000 V. How would the standard potential values on this new scale differ from those in Appendix 6?

*18.99. **Clinical Chemistry** The concentration of Na^+ ions in red blood cells (11 mM) and in the surrounding plasma (140 mM) are quite different. Calculate the potential difference across the cell membrane as a result of this concentration gradient at 37°C.

*18.100. **Waterline Corrosion** The photo in Figure P18.100 shows a phenomenon known as waterline corrosion. Assuming the oxidizing agent in the corrosion process is O_2, propose a reason why metal pilings such as this one tend to corrode the most at the waterline and corrode less at heights above and below the waterline.

FIGURE P18.100

19

Nuclear Chemistry

Applications to Energy and Medicine

POSITRON EMISSION TOMOGRAPHY (PET) In nuclear medicine, PET is a powerful imaging technique based on administering trace concentrations of radioactive nuclides that emit positrons—the antimatter version of electrons. The distribution of these compounds enables physicians to detect abnormal cell activity, for example, in this patient's lymph nodes in the upper chest below the neck. The activity seen in the brain is normal.

PARTICULATE **REVIEW**

Isotopes Revisited

In Chapter 19 we investigate the stability and properties of radioactive nuclei. Three nuclides are depicted here.

- What is the mass of each nuclide in atomic mass units?

- How many protons and neutrons does each nuclide contain?

- Which two nuclides are isotopes of each other?

 (Review Section 2.2 if you need help.)

(Answers to Particulate Review questions are in the back of the book.)

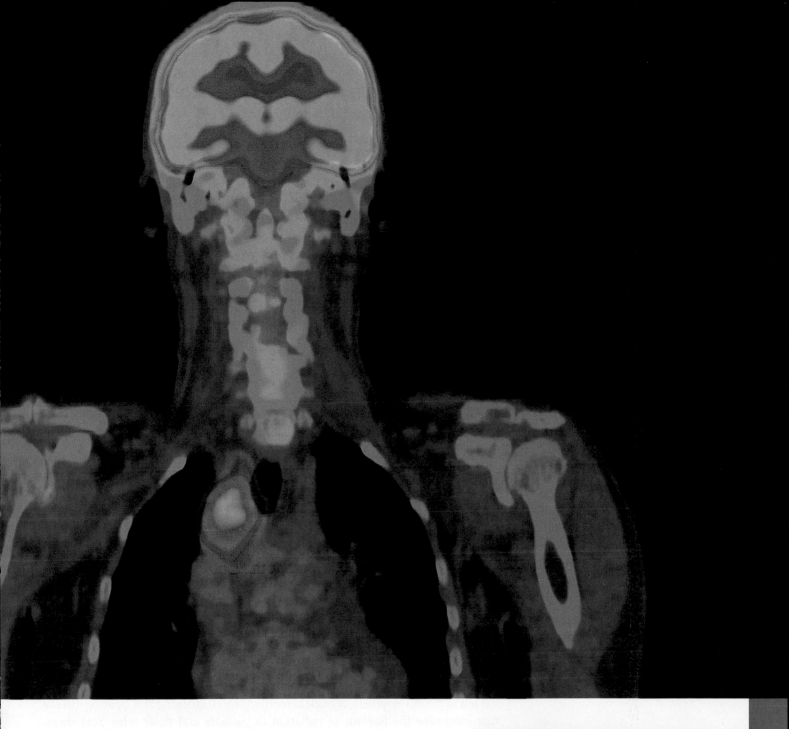

Unstable and Stable Nuclides

Radiocarbon dating measures the amount of carbon-14, the nuclide depicted here, to determine the age of artifacts. Accurate measurements depend upon knowing how this unstable nuclide decays. As you read Chapter 19, look for ideas that will help you answer these questions.

- Does carbon-14 have more neutrons or protons?

- How might the ratio of neutrons to protons affect the decay of an unstable nuclide?

- What nuclide is produced when carbon-14 undergoes decay? What is the neutron-to-proton ratio of this new nuclide?

Learning Outcomes

LO1 Calculate the binding energy of atomic nuclei

LO2 Predict the decay modes of radionuclides
Sample Exercise 19.1

LO3 Calculate the level of radioactivity in a sample of a radionuclide
Sample Exercise 19.2

LO4 Calculate the quantity of a radionuclide remaining after a defined decay time
Sample Exercise 19.3

LO5 Determine the age of a sample by radiometric dating
Sample Exercise 19.4

LO6 Describe the dangers of exposure to nuclear radiation and calculate effective radiation doses
Sample Exercise 19.5

LO7 Calculate the energy released in a nuclear reaction from the masses of the products and reactants
Sample Exercise 19.6

CONNECTION In Section 2.2 we described Henri Becquerel's discovery of radioactive uranium, and we saw how Ernest Rutherford's experiment, using alpha particles from the radioactive decay of uranium, led to the model of a dense atomic nucleus composed of protons and neutrons.

CONNECTION In Section 1.7 we described the rapid transformation of energy into matter that followed the Big Bang. The relationship between the energy and matter involved is described by Einstein's famous equation, $E = mc^2$.

CONNECTION In Section 2.9 we explored the nucleosynthesis responsible for the formation of our early universe, including how early elements such as hydrogen and helium were transformed into more massive nuclei in the cores of the giant stars that populated the first generation of galaxies.

19.1 Energy and Nuclear Stability

In 1898 Marie Curie separated and purified two new radioactive elements from uranium—polonium and radium—helping launch a new branch of chemistry now known as **nuclear chemistry**. Interest in the applications of radium grew rapidly. Radium-containing paint was used to create luminous dials for watches and gauges. Ointments containing radium were prescribed as treatments for skin lesions. People were encouraged to visit health spas such as those in Saratoga Springs, New York, where the waters contained low concentrations of dissolved radium salts.

The proliferation of radium-containing products led to the discovery that nuclear radiation is dangerous. In one tragic example, young women hired to paint the dials of watches during World War I were instructed to "point" the tips of their brushes by licking them, inadvertently introducing radium into their bodies, where it concentrated in their teeth and bones. Within a few years, many of these young women developed bone cancer and died. Marie Curie herself succumbed to aplastic anemia caused by years of research with radioactive materials.

In recent years a better understanding of the biological impacts of radiation has led to the development of nuclear methods for diagnosing and treating disease that minimize the hazards of radiation to patients and those who treat them. Nuclear reactors provide doctors and scientists with radioactive atoms that decay over minutes to hours by predictable pathways. Selective uptake of these nuclides by different organs in the body allows doctors to evaluate organ function and prescribe treatment when function is impaired. Radiation from other nuclides that concentrate in cancerous tissues can be used, often in conjunction with other therapies, to destroy malignant tumors.

Nuclear radiation results from reactions that take place in the unstable nuclei of some atoms. The energy released by these reactions can be used to help meet the world's need for electrical power and to diagnose and treat disease. However, these radioactive substances also pose a threat to human health, so shielding ourselves from them is a concern. We begin by looking at the energy associated with nuclear transformations.

We first explored nuclear transformations in Section 2.9, focusing on the reactions that take place in stars. Beginning with the lightest elements, hydrogen and helium, stars ultimately produce nuclei as heavy as iron atoms, which contain 26 protons. Why does stellar nucleosynthesis end with the formation of iron nuclei? The answer has to do with the energy inside all nuclei that binds nucleons (that is, protons and neutrons) together. This energy was discovered in the 1930s when scientists determined that the masses of stable nuclei are always less than the sum of the free masses of their nucleons. (The masses and symbols of subatomic particles are listed in Table 19.1.) For example, the total mass of the free nucleons in one ^{4}He nucleus—that is, 2 neutrons and 2 protons—is

$$\text{Mass of 2 neutrons} = 2(1.67493 \times 10^{-27}\ \text{kg})$$

$$\underline{\text{Mass of 2 protons} = 2(1.67262 \times 10^{-27}\ \text{kg})}$$

$$\text{Total mass} = 6.69510 \times 10^{-27}\ \text{kg}$$

The difference between this value and the mass of a ^{4}He nucleus (from Table 19.1) is

$$6.69510 \times 10^{-27}\ \text{kg}$$
$$\underline{-\ 6.64465 \times 10^{-27}\ \text{kg}}$$
$$0.05045 \times 10^{-27}\ \text{kg} = 5.045 \times 10^{-29}\ \text{kg}$$

This difference is called the **mass defect (Δm)** of the nucleus. The energy equivalent to the mass defect of a nucleus is called its **binding energy (BE)** and can be calculated using Einstein's equation:

$$\begin{aligned}\text{BE} &= (\Delta m)c^2 \\ &= 5.045 \times 10^{-29}\ \text{kg} \times (2.998 \times 10^8\ \text{m/s})^2 \\ &= 4.534 \times 10^{-12}\ \text{kg} \cdot (\text{m/s})^2 = 4.534 \times 10^{-12}\ \text{J}\end{aligned}$$

This value may seem very small, but it is the binding energy of a single nucleus. Its value per mole of ^{4}He is equivalent to billions of kilojoules:

$$\frac{4.534 \times 10^{-12}\ \text{J}}{\text{atom}} \times \frac{6.022 \times 10^{23}\ \text{atoms}}{\text{mol}} \times \frac{1\ \text{kJ}}{1000\ \text{J}} = 2.730 \times 10^9\ \text{kJ/mol}$$

For comparison, consider that the bond dissociation energy of H_2, 436 kJ/mol, is more than 6,000,000 times *smaller* than the binding energy of a mole of He atoms.

Most of the stable nuclei with more nucleons than helium-4 have even larger binding energies. This makes sense because a nucleus with many protons in proximity must be held together by an enormous energy to overcome the coulombic repulsion that all of these positively charged particles exert on one another. So close together, nucleons come under the influence of a fundamental force of nature known as the **strong nuclear force**. It operates only over very small distances, such as the diameters of atomic nuclei, but it is 100 times stronger than the repulsions the protons experience. The strong nuclear force binds nucleons together and stabilizes atomic nuclei.

To make comparisons of nuclear binding energies meaningful for nuclei of different sizes, the energies are usually divided by the number of nucleons in each nucleus. Expressing binding energy on a per-nucleon basis allows us to compare the stabilities of different nuclides. When we plot binding energy per-nucleon values against mass number, we get the curve shown in Figure 19.1.

TABLE 19.1 Symbols and Masses of Subatomic Particles and Small Nuclei

Particle	Symbol	Mass (kg)
Neutron	^1_0n	1.67493×10^{-27}
Proton	^1_1p or ^1_1H	1.67262×10^{-27}
Electron (β particle)	$^0_{-1}\beta$ or $^0_{-1}\text{e}$	9.10939×10^{-31}
Deuteron	^2_1D or ^2_1H	3.34370×10^{-27}
α Particle	$^4_2\alpha$ or ^4_2He	6.64465×10^{-27}
Positron	$^0_1\beta$	9.10939×10^{-31}

nuclear chemistry a subdiscipline of chemistry that explores the properties and reactivity of radioactive atoms.

mass defect (Δm) the difference between the mass of a stable nucleus and the masses of the individual nucleons that make it up.

binding energy (BE) the energy that would be released when free nucleons combine to form the nucleus of an atom. It is also the energy needed to split the nucleus into free nucleons.

strong nuclear force the fundamental force of nature that keeps nucleons together in atomic nuclei.

FIGURE 19.1 The stability of a nucleus is directly proportional to its binding energy per nucleon. For all nuclides up to ^{56}Fe, fusion reactions lead to products that have greater binding energy per nucleon than do the reactants. This means the fusion reactions release energy. However, fusion of nuclei with atomic numbers greater than 26 produces nuclei that have less binding energy per nucleon than the reactants, and thus consumes energy.

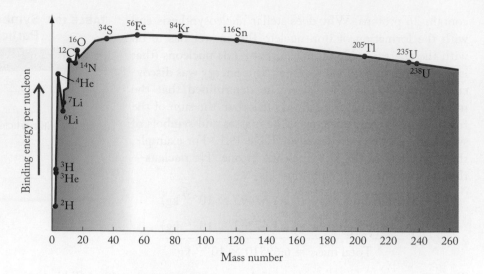

Note that these values reach a maximum with ^{56}Fe. This pattern means that ^{56}Fe is the most stable nuclide in the universe, which fits what we know about stellar nucleosynthesis. The nuclear furnaces of giant stars are fueled by the energy released when lighter nuclei fuse to form heavier nuclei, but only when the heavier ones that form are more stable than the lighter ones. Energy is not released if one of the reacting particles is ^{56}Fe or a heavier nuclide because, according to Figure 19.1, these heavier fusion products are *less stable*, not more stable. Their formation consumes energy instead of producing it. Thus, an aged giant star whose core has turned into iron as a result of multiple nuclear fusion processes has essentially run out of fuel. Without heating from its nuclear furnace, the star's fusion reactions cease and its enormous gravity forces it to collapse into itself. The result is rapid heating as the star's mass is compressed, followed by a gigantic explosion called a supernova.

CONCEPT TEST

Why does a negative change in mass when nucleons combine to form a nucleus produce a positive binding energy?

(Answers to Concept Tests are in the back of the book.)

19.2 Unstable Nuclei and Radioactive Decay

The values of the atomic masses and mass numbers of the elements in the periodic table tell us about the ratios of neutrons to protons in the nuclei of their stable isotopes. The lighter elements have atomic masses that are about twice their atomic numbers and have neutron-to-proton ratios close to unity. For example, ^{12}C has 6 neutrons and 6 protons, and most oxygen atoms have 8 neutrons and 8 protons.

However, with increasing values of Z, the ratio of neutrons to protons increases. This trend is illustrated in Figure 19.2, where the green dots represent combinations of neutrons and protons that form stable nuclides. The band of green dots runs diagonally through the graph, defining the **belt of stability**. Note

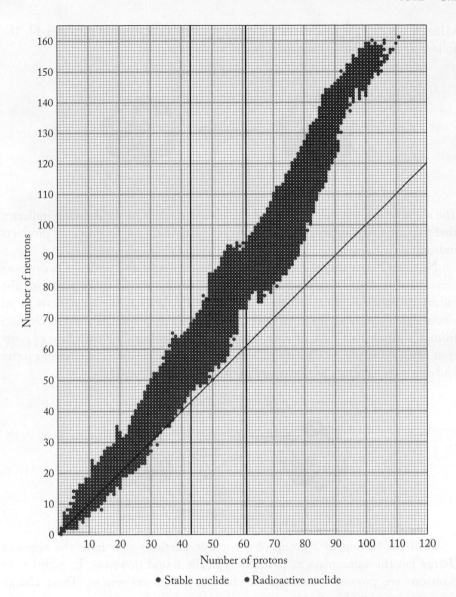

FIGURE 19.2 The belt of stability. Green dots represent stable combinations of protons and neutrons. Orange dots represent known radioactive (unstable) nuclides. Nuclides that fall along the purple line have equal numbers of neutrons and protons. Note that there are no stable nuclides (no green dots) for $Z = 43$ (technetium) and $Z = 61$ (promethium), as indicated by the two vertical red lines. These elements are the only two among the first 83 that are not found in nature.

• Stable nuclide • Radioactive nuclide

how the belt curves upward away from the purple straight line, representing a neutron-to-proton ratio of 1. This curvature shows how the neutron:proton ratio increases from about 1:1 for the lightest stable nuclides to about 1.5:1 for the most massive ones.

The nuclides represented by orange dots in Figure 19.2 are **radionuclides**. They are not stable but instead undergo **radioactive decay**, which is the spontaneous disintegration of radioactive nuclei, accompanied by the release of nuclear radiation. Modes of radioactive decay depend on whether a nuclide is above or below the belt of stability. Those above, such as carbon-14, are *neutron rich* and tend to undergo **β decay**: a **nuclear reaction** in which a neutron disintegrates, producing a proton that remains in the nucleus and a high-speed, high-energy electron, called a β particle, that is emitted from the atom.

For example, a ^{14}C nucleus contains 6 protons and 8 neutrons, which means its neutron-to-proton ratio is 1.333—a large value for a small nucleus. This neutron-rich nucleus undergoes β decay as one of its 8 neutrons disintegrates into a proton, leaving it with $(8 - 1 = 7)$ neutrons and $(6 + 1 = 7)$ protons. These

belt of stability the region on a graph of number of neutrons versus number of protons that includes all stable nuclei.

radionuclide an unstable nuclide that undergoes radioactive decay.

radioactive decay the spontaneous disintegration of unstable particles accompanied by the release of radiation.

beta (β) decay the spontaneous ejection of a β particle by a neutron-rich nucleus.

nuclear reaction a process that changes the number of neutrons or protons in the nucleus of an atom.

CHEMTOUR
Balancing Nuclear Equations

CONNECTION We learned how to write and balance nuclear equations in Section 2.9.

values mean that the product of the decay process is a nucleus of nitrogen-14. The following radiochemical equation describes the reaction:

$$^{14}_{6}C \rightarrow {}^{14}_{7}N + {}^{0}_{-1}\beta$$

Carbon-14 nucleus   β particle ⊖   Nitrogen-14 nucleus

The shimmering red shadow around the carbon nucleus in the drawing indicates that the carbon-14 nucleus is unstable and releases energy in forming the products indicated.

Nuclides below the belt of stability are *neutron poor* and undergo decay processes that *increase* their neutron-to-proton ratio. In one of these processes the radioactive nucleus emits a high-velocity particle that has the same mass as an electron, but that has a positive charge. It is called a **positron ($^{0}_{1}\beta$)**, and its ejection from a neutron-poor nucleus is called **positron emission**. The net effect of positron emission is the production of a nuclide of a nucleus with one fewer proton and one more neutron, as illustrated in the decay of carbon-11:

$$^{11}_{6}C \rightarrow {}^{11}_{5}B + {}^{0}_{1}\beta$$

Carbon-11 nucleus   Positron   Boron-11 nucleus

Positrons belong to a group of subatomic particles that have the opposite charge but the same mass as particles typically found in atoms. In addition to positrons are protons with negative charges, called *antiprotons*. These charge opposites are particles of **antimatter**.

Particles of matter and their antimatter opposites are like mortal enemies. If they collide, they instantly annihilate each other. In their mutual destruction, they cease to exist as matter, and all of their mass is converted to energy in the form of two or more gamma (γ) rays:

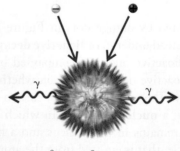

$$^{0}_{1}\beta + {}^{0}_{-1}\beta \rightarrow 2\gamma$$

positron ($^{0}_{1}\beta$) a particle with the mass of an electron but with a positive charge.

positron emission the spontaneous emission of a positron from a neutron-poor nucleus.

antimatter particles that are the charge opposites of normal subatomic particles.

electron capture a nuclear reaction in which a neutron-poor nucleus draws in one of its surrounding electrons, which transforms a proton in the nucleus into a neutron.

The yellow "sunburst" here symbolizes the energy released in the process depicted.

Gamma ray emission accompanies all nuclear reactions, not just positron emission. Gamma rays represent quantities of energy that are equivalent to the loss in mass that occurs when reactants form products in nuclear reactions. They are generated by the nuclear furnaces of stars and permeate outer space. Those that reach Earth are absorbed by the gases in our atmosphere. In the process, molecular gases are broken up into their component atoms, and atomic nuclei may be broken up into subatomic particles.

There is another way to increase the neutron-to-proton ratio of a neutron-poor nucleus: it can capture one of the inner-shell electrons of its atom. When it does, the negatively charged electron combines with a positively charged proton. The product of this combination reaction is a neutron. The effect of this **electron capture** process on the nucleus is the same as positron emission: the number of protons *decreases* by one and the number of neutrons *increases* by one. When a nucleus of carbon-11 undergoes electron capture, the product is the same nuclide formed in positron emission, boron-11:

$$^{11}_{6}\text{C} + ^{0}_{-1}\text{e} \rightarrow ^{11}_{5}\text{B}$$

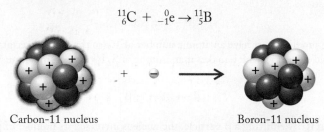

Carbon-11 nucleus   Boron-11 nucleus

Table 19.2 summarizes the effects of being neutron rich, neutron poor, or neither on isotopes of carbon. Note that carbon has two stable isotopes: ^{12}C and ^{13}C. The isotopes with mass numbers greater than 13 are neutron rich and undergo β decay. Those with mass numbers less than 12 are neutron poor and undergo either positron emission or electron capture.

CONNECTION Gamma rays are the highest energy form of electromagnetic radiation (see Figure 7.7).

CHEMTOUR
Radioactive Decay Modes

TABLE 19.2 Isotopes of Carbon and Their Radioactive Decay Products

Name	Symbol	Mass (amu)	Mode(s) of Decay	Half-Life	Natural Abundance (%)
Carbon-10	$^{10}_{6}$C		Positron emission	19.45 s	
Carbon-11	$^{11}_{6}$C		Positron emission, electron capture	20.3 min	
Carbon-12	$^{12}_{6}$C	12.00000	(Stable)		98.89
Carbon-13	$^{13}_{6}$C	13.00335	(Stable)		1.11
Carbon-14	$^{14}_{6}$C		β Decay	5730 yr	
Carbon-15	$^{15}_{6}$C		β Decay	2.4 s	
Carbon-16	$^{16}_{6}$C		β Decay	0.74 s	

SAMPLE EXERCISE 19.1 Predicting the Modes and Products of Radioactive Decay   **LO2**

Predict the mode of radioactive decay of ^{32}P, which is one of the most widely used radionuclides in biomedical research and treatment. Identify the nuclide that is produced in the decay process.

Collect and Organize We are asked to predict the mode of decay of a radionuclide, which depends on whether it is neutron rich or neutron poor. Phosphorus-32 has 17 neutrons and 15 protons per nucleus. Figure 19.2 identifies the stable isotopes of

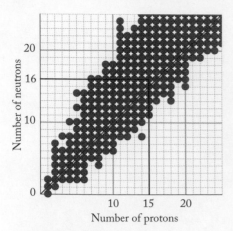

FIGURE 19.3 Nuclides in the belt of stability.

the elements. According to the magnified region of the belt of stability in Figure 19.3, phosphorus ($Z = 15$) has only one stable nuclide, which has 16 neutrons and a mass number of 31.

Analyze Phosphorus-32 is represented by the orange dot directly above the ^{31}P green dot, which means that it is radioactive and neutron rich. Neutron-rich radioisotopes of lighter elements undergo β decay, so the decay of ^{32}P will produce a β particle. We can use a balanced nuclear equation describing the β decay reaction to identify the nuclide that is produced. As we learned in Section 2.9, the sums of the superscripts on the left and right sides must be equal in a balanced nuclear equation, as must the sums of the subscripts.

Solve A β particle must be one product of the decay reaction, giving the incomplete nuclear equation:

$$^{32}_{15}\text{P} \rightarrow ? + {}^{0}_{-1}\beta$$

The missing product must have an atomic number of 16 (so that the subscripts on the right side add up to 15), which makes it an isotope of S. Its mass number must be 32, so the product is sulfur-32:

$$^{32}_{15}\text{P} \rightarrow {}^{32}_{16}\text{S} + {}^{0}_{-1}\beta$$

Think About It By emitting a β particle, the nucleus increased its number of protons by one and decreased its number of neutrons by one, thereby reducing its neutron "richness" and forming a stable isotope of sulfur.

 Practice Exercise What is the mode of radioactive decay of ^{28}Al? Identify the nuclide produced by the decay process.

(Answers to Practice Exercises are in the back of the book.)

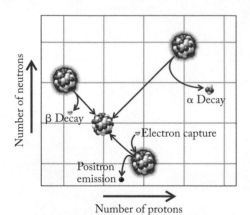

FIGURE 19.4 Radioactive decay results in predictable changes in the number of protons and neutrons in a nucleus. In α decay, the nucleus loses 2 neutrons and 2 protons, resulting in a decrease of 2 in atomic number and 4 in mass number. Beta decay leads to an increase of 1 proton at the expense of 1 neutron, so the atomic number increases by 1, but the mass number is unchanged. In positron emission and electron capture, the number of protons decreases by 1 and the number of neutrons increases by 1, so the atomic number decreases by 1, but the mass number remains the same.

All known nuclides with more than 83 protons are radioactive. Because there is no stable reference point in the pattern of green dots in Figure 19.2, it is hard to say whether any given nuclide with $Z > 83$ is neutron rich or neutron poor. We can make one general statement though: these most massive nuclides tend to undergo either β decay or **alpha (α) decay**. In α decay they produce a nuclide with two fewer protons and two fewer neutrons, as for uranium-238:

$$^{238}_{92}\text{U} \rightarrow {}^{234}_{90}\text{Th} + {}^{4}_{2}\alpha$$

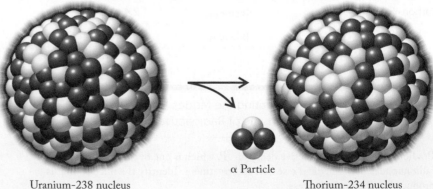

Uranium-238 nucleus α Particle   Thorium-234 nucleus

Figure 19.4 illustrates the changes in atomic number and mass number caused by the various modes of decay.

Among the most massive radionuclides, one radioactive decay process often leads to another in what is referred to as a *radioactive decay series*. Consider, for

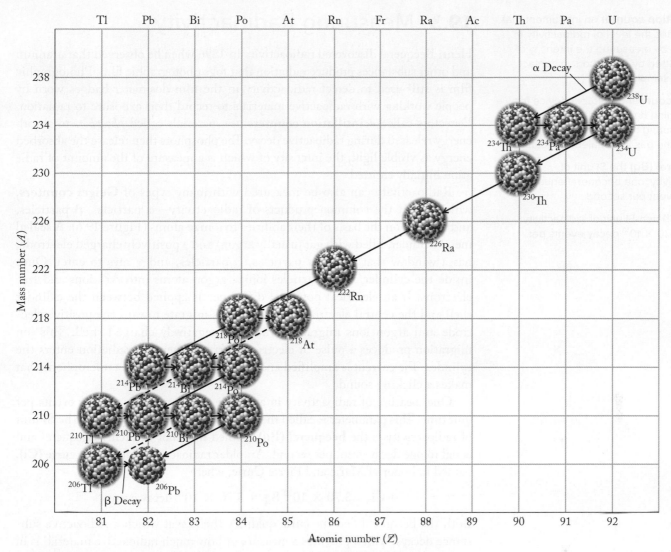

FIGURE 19.5 Uranium-238 radioactive decay series. The long diagonal arrows represent α decay events; the short horizontal ones represent β decay events. The dashed arrows are alternative pathways representing less than 1% of the decay events in this series. Note that whether decay proceeds by the solid-line pathways or the dashed-line pathways, the end product is always stable lead-206.

example, the decay series that begins with the α decay of ^{238}U to ^{234}Th (Figure 19.5). Thorium has no stable isotopes and undergoes two β decay steps to produce ^{234}U. In a series of subsequent α decay steps, ^{234}U turns into thorium-230, radium-226, radon-222, polonium-218, and finally lead-214. Although some isotopes of lead ($Z = 82$) are stable, ^{214}Pb is not one of them. Therefore the radioactive decay series continues, as shown at the bottom left of Figure 19.5, and does not end until the stable nuclide ^{206}Pb is produced.

CONCEPT TEST

In the ^{238}U radioactive decay series, five α decay steps in a row transform ^{234}U into ^{214}Pb. Given the shape of the belt of stability, why does it make sense that the product of these α decay steps would be a neutron-rich nuclide that undergoes β decay?

alpha (α) decay a nuclear reaction in which an unstable nuclide spontaneously emits an alpha particle.

scintillation counter an instrument that determines the level of radioactivity in samples by measuring the intensity of light emitted by phosphors in contact with the samples.

Geiger counter a portable device for determining nuclear radiation levels by measuring how much the radiation ionizes the gas in a sealed detector.

becquerel (Bq) the SI unit of radioactivity; one becquerel equals one decay event per second.

curie (Ci) non-SI unit of radioactivity; 1 Ci = 3.70 × 10¹⁰ decay events per second.

19.3 Measuring Radioactivity

Henri Becquerel discovered radioactivity in 1896 when he observed that uranium and other substances produce radiation that fogs photographic film. Photographic film is still used to detect radioactivity in the film dosimeter badges worn by people working with radioactive materials to record their exposure to radiation. Detectors called **scintillation counters** use materials called *phosphors* to absorb energy released during radioactive decay. The phosphors then release the absorbed energy as visible light, the intensity of which is a measure of the amount of radiation initially emitted.

Radioactivity can also be measured with many types of **Geiger counters**, which detect the common products of radioactivity—α particles, β particles, and γ rays—on the basis of their abilities to ionize atoms (Figure 19.6). A sealed metal cylinder, filled with gas (usually argon) and a positively charged electrode, has a window that allows α particles, β particles, and γ rays to enter. Once inside the cylinder, these particles ionize argon atoms into Ar^+ ions and free electrons. If an electrical potential difference is applied between the cylinder shell and the central electrode, free electrons migrate toward the positive electrode and argon ions migrate toward the negatively charged shell. This ion migration produces a pulse of electrical current whenever radiation enters the cylinder. The current is amplified and read out to a meter and a microphone that makes a clicking sound.

One measure of radioactivity in a sample is the number of decay events per unit time. This parameter is called the radioactivity (*A*) of the sample. The SI unit of radioactivity is the **becquerel (Bq)**, named in honor of Henri Becquerel and equal to one decay event per second. An older radioactivity unit is the **curie (Ci)**, named in honor of Marie and Pierre Curie, where

$$1 \text{ Ci} = 3.70 \times 10^{10} \text{ Bq} = 3.70 \times 10^{10} \text{ decay events/s}$$

Both the becquerel and the curie quantify the *rate* at which a radioactive substance decays, which provides a measure of how much radioactive material is in a sample. All radioactive decay processes follow first-order kinetics, in which the reaction rate is equal to the rate constant, *k*, times the concentration of a reactant, R:

$$\text{Rate} = k[\text{R}]$$

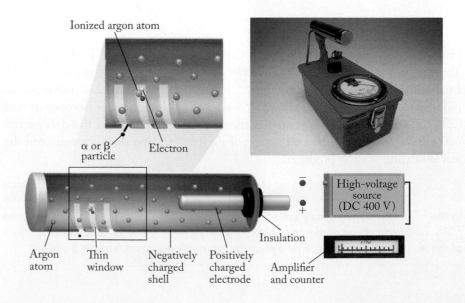

FIGURE 19.6 In a Geiger counter, a particle produced by radioactive decay passes through a thin window, usually made of beryllium or a plastic film. Inside the tube, the particle ionizes atoms of argon gas. The resulting argon cations migrate toward the negatively charged tube housing, and the electrons migrate toward a positive electrode, creating a pulse of current through the tube. The current pulses are amplified and recorded via a meter and a speaker that produces an audible "click" for each pulse.

For radioactive decay processes we refer to radioactivity (A) instead of reaction rate, and to the number of atoms (N) of a radionuclide in a sample instead of its concentration:

$$A = kN \qquad (19.1)$$

Because radioactivity is the number of decay events per second, the units of the *decay rate constant* (k) are decay events per atom per second.

Because radioactive decay is a first-order reaction, the decay rate constant also is related to the half-life of the radionuclide by Equation 13.19:

$$t_{1/2} = \frac{0.693}{k} \qquad (13.19)$$

In the following Sample Exercise, we calculate the radioactivity of a sample on the basis of the quantity and half-life of the radionuclide in the sample.

CONNECTION In Chapter 13 we introduced the concept of *half-life*, $t_{1/2}$, and how its value is inversely proportional to the rate constant, k, of a first-order reaction.

SAMPLE EXERCISE 19.2 Calculating the Radioactivity of a Sample   **LO3**

Radium-223 undergoes decay with a half-life of 11.4 days. What is the radioactivity of a sample that contains 1.00 μg of ^{223}Ra? Express your answer in becquerels and in curies.

Collect and Organize We are given the half-life and quantity of a radioactive substance and are asked to determine its radioactivity, that is, its rate of decay. The decay rate constant, k, is related to half-life by

$$t_{1/2} = \frac{0.693}{k}$$

The radioactivity is the product of the rate constant and the number of atoms of radionuclide in the sample (Equation 19.1, $A = kN$).

Analyze Before using Equation 13.19, we must convert the half-life into seconds because both of the radioactivity units we need to calculate are based on decay events per second. To calculate radioactivity, we determine the number of atoms in 1.00 μg of radium. This will probably be a very large number, which should translate into many decay events per second given the relatively short half-life of ^{223}Ra.

Solve The half-life is

$$11.4 \text{ d} \times \frac{24 \text{ h}}{1 \text{ d}} \times \frac{60 \text{ min}}{1 \text{ h}} \times \frac{60 \text{ s}}{1 \text{ min}} = 9.85 \times 10^5 \text{ s}$$

Using this value in Equation 13.19 and solving for k:

$$k = \frac{0.693}{9.85 \times 10^5 \text{ s}}$$
$$= 7.04 \times 10^{-7} \text{ s}^{-1} = 7.04 \times 10^{-7} \text{ decay events/(atom} \cdot \text{s)}$$

The number of atoms (N) of ^{223}Ra is

$$N = 1.00 \text{ μg} \times \frac{1 \text{ g}}{10^6 \text{ μg}} \times \frac{1 \text{ mol Ra}}{223 \text{ g Ra}} \times \frac{6.022 \times 10^{23} \text{ atoms Ra}}{1 \text{ mol Ra}} = 2.70 \times 10^{15} \text{ atoms Ra}$$

Inserting these values of k and N into Equation 19.1 gives us

$$A = kN = \frac{7.04 \times 10^{-7} \text{ decay events}}{\text{atom} \cdot \text{s}} \times 2.70 \times 10^{15} \text{ atoms Ra} = 1.90 \times 10^9 \text{ decay events/s}$$

Because 1 Bq = 1 decay event/s, the radioactivity of the sample is 1.90×10^9 Bq. Expressing radioactivity in curies:

$$\frac{1.90 \times 10^9 \text{ decay events}}{\text{s}} \times \frac{1 \text{ Ci}}{3.70 \times 10^{10} \text{ decay events/s}} = 0.0514 \text{ Ci}$$

CHEMTOUR
Half-Life

19.4 Rates of Radioactive Decay

In Section 19.2 we examined how radionuclides undergo radioactive decay; in this section we focus on how rapidly they decay. We have defined radioactive decay as the spontaneous disintegration of unstable nuclei, but, as with chemical reactions, "spontaneous" does not necessarily mean "rapid." Because all radioactive decay processes follow first-order kinetics, each one has a characteristic half-life, $t_{1/2}$ (Figure 19.7); the faster the decay process, the shorter the half-life.

To calculate what fraction of radionuclide remains after an amount of time t has elapsed, we first determine how many half-lives, n, have passed during time t:

$$n = \frac{t}{t_{1/2}} \qquad (19.2)$$

Once we know the value of n, we can calculate the fraction of radioactive nuclei remaining at instant t by writing the ratio of the number of nuclei at instant t (N_t) to the number present at t = 0 (N_0):

$$\frac{N_t}{N_0} = 0.5^n \qquad (19.3)$$

In practice, it is often more convenient to relate quantities of radioactive particles directly to time by combining Equations 19.2 and 19.3:

$$\frac{N_t}{N_0} = 0.5^{t/t_{1/2}} \qquad (19.4)$$

In the following exercises and elsewhere in this chapter we apply Equation 19.4 to various radioactive decay processes. We can do so because all of these processes follow first-order reaction kinetics.

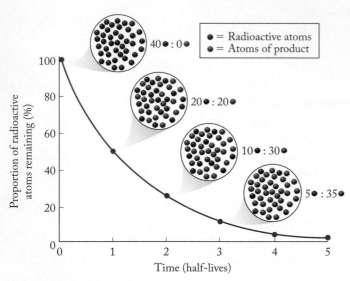

• = Radioactive atoms
• = Atoms of product

FIGURE 19.7 Radioactive decay follows first-order kinetics, which means, for example, that if a sample initially contains 40 radioactive atoms, it will contain only one-half that number after an interval equal to one half-life. Half of the remaining half, or 10 radioactive atoms, remain after two half-lives, and so on.

SAMPLE EXERCISE 19.3 Calculations Involving Half-Lives   **LO4**

Free neutrons are radioactive with a half-life of 12 minutes, undergoing β decay to a proton and an electron. Starting with a population of 6.6×10^5 free neutrons, how many remain after 2.0 min?

Collect, Organize, and Analyze Equation 19.4 relates quantities of radioactive particles to decay times. In this problem the initial number of neutrons (N_0) is 6.6×10^5, t = 2.0 min, $t_{1/2}$ = 12 min, and we need to solve for N_t. The value of t is only a fraction of $t_{1/2}$; far fewer than half of the initial number of neutrons will have decayed after 2.0 min. In other words, most of the neutrons will still be present.

Solve Substituting the data into Equation 19.4 gives

$$\frac{N_t}{6.6 \times 10^5} = 0.5^{2.0 \text{ min}/12 \text{ min}}$$

$$N_t = 5.9 \times 10^5$$

Think About It The value of N_t is reasonable because, as we predicted, only a small fraction of the initial quantity of free neutrons decayed in 2 min.

Practice Exercise Cesium-131 is a short-lived radionuclide ($t_{1/2} = 9.7$ d) used to treat prostate cancer. If the therapeutic strength of the radionuclide is directly proportional to the number of nuclei present, how much therapeutic strength does a cesium-131 source lose over exactly 60 days? Express your answer as a percentage of the strength the source had at the beginning of the first day.

19.5 Radiometric Dating

Radiometric dating is a term used to describe methods for determining the age of objects on the basis of the tiny concentrations of radionuclides that occur naturally in them and the rates of radioactive decay of these nuclides. The concept originated in the early 1900s in the work of Ernest Rutherford, who had already recognized in his pioneering studies on radioactivity that radioactive decay processes have characteristic half-lives. Rutherford proposed to use this concept to determine the age of rocks and even the age of Earth itself. The basis for his initial attempt was the emission of α particles from uranium ore. He correctly suspected that α particles were part of helium atoms, and he proposed to determine the age of uranium ore samples by determining the concentration of helium gas trapped inside them.

Rutherford's helium method did not yield very accurate results, but it did inspire a young American chemist, Bertram Boltwood (1870–1927), who had determined that the decay of radioactive uranium involves a series of decay events ending with the formation of stable lead (see Figure 19.5). In 1907 Boltwood published the results of dating 43 samples of uranium-containing minerals on the basis of the ratio of lead to uranium in them. The ages he reported spanned hundreds of millions to over a billion years and probably represent the first successful attempt at radiometric dating.

The development of the mass spectrometer for accurately determining the abundances of individual isotopes of elements, coupled with more accurate half-life values for decay events such as those in Figure 19.5, has allowed scientists to use the ratio of ^{206}Pb to ^{238}U in geological samples to determine their ages with a precision of about ±1%. Other methods, including one based on the decay of ^{235}U to ^{207}Pb ($t_{1/2} = 7.0 \times 10^6$ yr), may be used to analyze the same samples, providing independent determinations that mutually ensure more accurate results. These analyses have shown that the oldest rocks on Earth are more than 4.0 billion years old and that meteorites that formed as the solar system formed are 4.5 billion years old.

These radiometric methods for dating geological samples yield reliable results only when the sample is a closed system, which means that the only loss of the radionuclide is via radioactive decay, and that all of the nuclides produced by the decay processes remain in the sample. In addition, those decay processes must be the only source of the product nuclides. For these reasons, scientists must exercise

C⚛NNECTION In Section 2.3 we saw that mass spectrometry can be used to determine the abundances of the isotopes of elements in a sample.

radiometric dating a method for determining the age of an object on the basis of the quantity of a radioactive nuclide and/or the products of its decay that the object contains.

radiocarbon dating a method for establishing the age of a carbon-containing object by measuring the amount of radioactive carbon-14 remaining in the object.

care in selecting the types of samples they subject to radiometric dating analysis. For example, the presence of the mineral zircon ($ZrSiO_4$) in a geological sample is good news for scientists interested in using radiometric dating because U^{4+} ions readily substitute for Zr^{4+} ions as crystals of $ZrSiO_4$ solidify from the molten state, but Pb^{2+} ions do not. Therefore the only source of ^{206}Pb and ^{207}Pb in a zircon sample should be the decay of ^{238}U and ^{235}U, respectively.

In 1947 American chemist Willard Libby (1908–1980) developed a radiometric dating technique, called **radiocarbon dating**, for determining the age of artifacts from prehistory and early civilizations. The method is based on determining the carbon-14 content of samples derived from plants or the animals that consumed them. Carbon-14 originates in the upper atmosphere, where cosmic rays break apart the nuclei of atoms, forming free protons and neutrons. When one of these neutrons collides with a nitrogen-14 atom, it forms an atom of radioactive carbon-14 and a proton:

$$^{14}_{7}\text{N} + ^{1}_{0}\text{n} \rightarrow ^{14}_{6}\text{C} + ^{1}_{1}\text{p}$$

Atmospheric carbon-14 combines with oxygen, forming $^{14}CO_2$. The atmospheric concentration of $^{14}CO_2$ amounts to only about 10^{-12} of all the molecules of CO_2 in the air. These traces of radioactive CO_2, along with the stable forms $^{12}CO_2$ and $^{13}CO_2$, are incorporated into the structures of green plants during photosynthesis. The tiny fraction of the plant's mass that is ^{14}C gets even tinier after a plant dies, or after a part of it stops growing and photosynthesizing, because ^{14}C undergoes β decay, as we described in Section 19.2:

$$^{14}_{6}\text{C} \rightarrow ^{14}_{7}\text{N} + ^{0}_{-1}\beta$$

The half-life of this decay process is 5730 years.

If we can determine the ^{14}C content (N_t) of an object of historical interest, such as a piece of wood from an ancient building, charcoal from a prehistoric campfire, or papyrus from an early Egyptian scroll, and if we know (or can predict) its ^{14}C content when the material in it was alive (N_0), then we can apply Equation 19.4 to determine its age:

$$\frac{N_t}{N_0} = 0.5^{t/t_{1/2}}$$

Predicting the value of N_0 is usually done by analyzing samples from growing plants—that is, samples for which the ^{14}C decay time is zero.

To facilitate radiocarbon dating calculations, let's solve Equation 19.4 for t by first taking the natural log of both sides (keeping in mind that $\ln 0.5 = -0.693$):

$$\ln \frac{N_t}{N_0} = -0.693 \frac{t}{t_{1/2}}$$

Rearranging the terms to solve for t gives us a useful equation for determining the radiocarbon age t:

$$t = -\frac{t_{1/2}}{0.693} \ln \frac{N_t}{N_0} \tag{19.5}$$

SAMPLE EXERCISE 19.4 Radiocarbon Dating   **LO5**

The ^{14}C content of a wooden harpoon handle found in the remains of an Inuit encampment in western Alaska is 61.9% of the ^{14}C content of the same type of wood from a recently cut tree. How old is the harpoon?

Collect, Organize, and Analyze The half-life of carbon-14 is 5730 years. Equation 19.5 provides the age t of the artifact if we know the ratio of the ^{14}C in it today to its initial ^{14}C content. The ^{14}C content of the modern sample can be used as a surrogate for the initial ^{14}C content of the artifact. Therefore 61.9% (or 0.619) represents the ratio N_t/N_0. This value is greater than 0.5, which means that the age of the sample is less than one half-life (5730 yr).

Solve Substituting into Equation 19.5:

$$t = -\frac{t_{1/2}}{0.693} \ln \frac{N_t}{N_0}$$

$$= -\frac{5730 \text{ yr}}{0.693} \ln(0.619)$$

$$= 3966 \text{ yr} = 3.97 \times 10^3 \text{ yr}$$

Think About It The resulting age is less than one half-life, which is reasonable because it contained more than half the original carbon-14 content. The result is expressed with three significant figures to match that of the starting composition (61.9%).

Practice Exercise The carbon-14:carbon-12 ratio in papyrus growing along the Nile River today is 1.8 times greater than in a papyrus scroll found near the Great Pyramid at Giza. How old is the scroll?

The accuracy of radiocarbon dating can be checked by determining the ^{14}C content of the annual growth rings of very old trees, such as the bristlecone pines that grow in the American Southwest (Figure 19.8). When scientists plot the radiocarbon ages of these rings against their actual ages obtained by counting rings starting from the outer growth layer of the tree (representing $t = 0$), they find that the two sets of ages do not agree exactly, as shown in Figure 19.9. There are several reasons for this lack of agreement, including variability in the rate of ^{14}C production due to changing intensity of the cosmic rays striking Earth's upper atmosphere. To ensure accurate ^{14}C results, scientists must correct for these and other variations, and they must be careful to avoid contaminating ancient samples with modern carbonaceous

FIGURE 19.8 Radiocarbon dating relies on knowing the atmospheric concentration of carbon-14 over time. Ancient living trees, such as the bristlecone pines in the American Southwest, act as a check of the atmospheric carbon-14 levels over thousands of years. The ages of the rings can be determined by counting them, and their carbon-14 content can be determined by mass spectrometry.

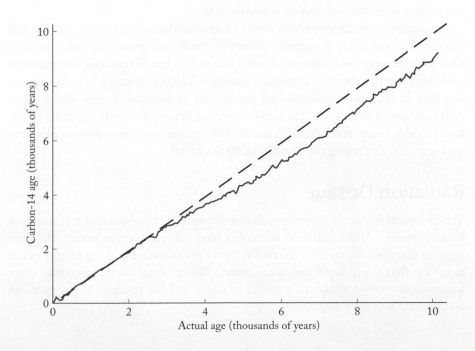

FIGURE 19.9 Calibration curves for radiocarbon dating allow scientists to accurately calculate the ages of archaeological objects. If the rate of ^{14}C production in the upper atmosphere were constant, then the age of objects on the basis of their ^{14}C content would match their actual age—a condition represented by the red dashed line. However, analyses of tree rings, corals, and lake sediments tell us that the rate of ^{14}C production in the upper atmosphere is variable, so a real plot of ^{14}C age versus actual age produces the jagged blue line. This plot allows scientists to convert ^{14}C ages into actual ages.

material. With proper analytical technique, radiocarbon dating results are generally accurate to within 40 years for samples that are 500–50,000 years old.

CONCEPT **TEST**

How might the increased consumption of fossil fuels over the last century affect the ^{14}C content of growing plant tissues?

19.6 Biological Effects of Radioactivity

The γ rays and many of the α and β particles produced by nuclear reactions have more than enough energy to tear chemical bonds apart, producing odd-electron radicals or free electrons and cations. Consequently, these rays and particles are classified as **ionizing radiation**. Other examples include X-rays and short-wavelength ultraviolet rays. The ionization of atoms and molecules in living tissue can lead to radiation sickness, cancer, birth defects, and death. The scientists who first worked with radioactive materials were not aware of these hazards, and some of them suffered for it. Marie Curie died from radiation exposure, and so did her daughter Irène Joliot-Curie, who continued the research program started by her parents.

In medicine, the term *ionizing radiation* is limited to photons and particles that have sufficient energy to remove an electron from water:

$$H_2O(\ell) \xrightarrow{1216 \text{ kJ/mol}} H_2O^+(aq) + e^-$$

The logic behind this definition is that the human body is composed largely of water. Therefore water molecules are the most abundant ionizable targets when we are exposed to nuclear radiation. The cation H_2O^+ reacts with another water molecule in the body to form a hydronium ion and a hydroxyl free radical:

$$H_2O^+(aq) + H_2O(\ell) \rightarrow H_3O^+(aq) + OH(aq)$$

The rapid reactions of free radicals with biomolecules can disrupt cell function, sometimes with life-threatening consequences.

Radiation-induced alterations to the biochemical machinery that controls cell growth are most likely to occur in tissues in which cells grow and divide rapidly. One such tissue is bone marrow, where billions of white blood cells are produced each day to fortify the body's immune system. Molecular damage to bone marrow can lead to leukemia: uncontrolled production of nonfunctioning white blood cells that spread throughout the body, crowding out healthy cells. Ionizing radiation can also cause molecular alterations in the genes and chromosomes of sperm and egg cells, increasing the chances of birth defects.

Radiation Dosage

The biological impact of ionizing radiation depends on how much of it is absorbed by an organism. If the radiation is coming from one radioactive source, then the amount absorbed depends on the radioactivity of the source and the energy of the radiation that is produced per decay event. Tables of radioactive isotopes often include information about their modes of decay and the energies of the particles and gamma rays they emit.

ionizing radiation high-energy products of radioactive decay that can ionize molecules.

gray (Gy) the SI unit of absorbed radiation; 1 Gy = 1 J/kg of tissue.

relative biological effectiveness (RBE) a factor that accounts for the differences in physical damage caused by different types of radiation.

sievert (Sv) SI unit used to express the amount of biological damage caused by ionizing radiation.

Absorbed dose is the quantity of ionizing radiation absorbed by a unit mass of living tissue. The SI unit of absorbed dose is the **gray (Gy)**. One gray is equal to the absorption of 1 J of radiation energy per kilogram of body mass:

$$1 \text{ Gy} = 1 \text{ J/kg}$$

Grays express dosage, but they do not indicate the amount of *tissue damage* caused by that dosage. Different products of nuclear reactions affect living tissue differently. Exposure to 1 Gy of γ rays produces about the same amount of tissue damage as exposure to 1 Gy of β particles. However, 1 Gy of α particles, which move about 10 times slower than β particles but have nearly 10^4 times the mass, causes up to 20 times as much damage as 1 Gy of γ rays. Neutrons cause 3 to 5 times as much damage. To account for these differences, values of **relative biological effectiveness (RBE)** have been established for the various forms of ionizing radiation (Table 19.3). When an absorbed dose in grays is multiplied by an RBE factor, the product is called *effective* dose, a measure of tissue damage. The SI unit of effective dose is the **sievert (Sv)**.

Table 19.4 summarizes the various units used to express quantities of radiation and their biological impact. Two non-SI units are listed that predate their SI counterparts but are still often used. They are *radiation absorbed dose*, or *rad*, which is equivalent to 0.01 Gy, and *rem* for tissue damage, which is an acronym for *roentgen equivalent man*. One rem is the product of one rad of ionization times the appropriate RBE factor. There are 100 rems in 1 Sv.

TABLE 19.3 RBE Values of Nuclear Radiation

Radiation	RBE
γ Rays	1.0
β Particles	1.0–1.5
Neutrons	3–5
Protons	10
α Particles	20

TABLE 19.4 Units for Expressing Quantities of Ionizing Radiation

Parameter	SI Unit	Description	Alternative Common Unit	Description
Radioactivity	Becquerel (Bq)	1 decay event/s	Curie (Ci)	3.70×10^{10} decay events/s
Ionizing energy absorbed	Gray (Gy)	1 J/kg of tlssue	Rad	0.01 J/kg of tissue
Amount of tissue damage	Sievert (Sv)	1 Gy × RBE^a	Rem	1 rad × RBE^a

^{a}RBE, relative biological effectiveness.

The RBE of 20 for α particles may lead you to believe that these particles pose the greatest health threat from radioactivity. Not exactly. Alpha particles are so big that they have little penetrating power; they are stopped by a sheet of paper, clothing, or even a layer of dead skin (Figure 19.10). However, if you ingest or

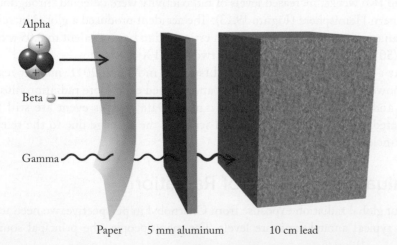

Alpha

Beta

Gamma

Paper   5 mm aluminum   10 cm lead

FIGURE 19.10 The tissue damage caused by α particles, β particles, and γ rays depends on their ability to penetrate materials that shield the tissues. Alpha particles are stopped by paper or clothing but are extremely dangerous if formed inside the body because their low penetrating power traps them inside where they do not have to travel far to cause cell damage. Stopping gamma rays requires a thick layer of lead or several meters of concrete or soil.

FIGURE 19.11 The ruins of the nuclear reactor at Chernobyl, Ukraine, that exploded in 1986.

TABLE 19.5 Acute Effects of Single Whole-Body Effective Doses of Ionizing Radiation

Effective Dose (Sv)	Toxic Effect
0.05–0.25	No acute effect, possible carcinogenic or mutagenic damage to DNA
0.25–1.0	Temporary reduction in white blood cell count
1.0–2.0	Radiation sickness: fatigue, vomiting, diarrhea, impaired immune system
2.0–4.0	Severe radiation sickness: intestinal bleeding, bone marrow destruction
4.0–10.0	Death, usually through infection, within weeks
>10.0	Death within hours

inhale an α emitter, tissue damage can be severe because the relatively massive α particles do not have to travel far to cause cell damage. Gamma rays are considered the most dangerous form of radiation emanating from a source outside the body because they have the greatest penetrating power.

The effects of exposure to different single effective doses of radiation are summarized in Table 19.5. To put these data in perspective, the effective dose from a typical dental X-ray is about 25 μSv, or about 2000 times smaller than the lowest exposure level cited in the table.

Widespread exposure to very high levels of radiation occurred after the 1986 explosion at the Chernobyl nuclear reactor in what is now Ukraine (Figure 19.11). Many plant workers and first responders were exposed to more than 1.0 Sv of radiation. At least 30 of them died in the weeks after the accident. Many of the more than 300,000 workers who cleaned up the area around the reactor exhibited symptoms of radiation sickness, and at least 5 million people in Ukraine, Belarus, and Russia were exposed to fallout in the days following the accident. Studies conducted in the early 1990s uncovered high incidences of thyroid cancer in children in southern Belarus as a result of ^{131}I released in the Chernobyl accident, and children born in the region nearly a decade after the accident had unusually high rates of mutations in their DNA because of their parents' exposure to ionizing radiation. Genetic damage was also widespread among plants and animals living in the region (Figure 19.12).

Radiation exposure was not confined to Ukraine and Belarus. After the accident, a cloud of radioactive material spread rapidly across northern Europe, and within two weeks, increased levels of radioactivity were detected throughout the Northern Hemisphere (Figure 19.13). The accident produced a global increase in human exposure to ionizing radiation estimated to be equivalent to 0.05 mSv per year (50 μSv per year, comparable to two dental X-rays).

As a result of an earthquake and tsunami in March 2011, nuclear reactors at a power station in Fukushima, Japan, released even more radiation into both land and sea than the Chernobyl disaster. Data on this event are still being collected and analyzed, and the full scope of the damage due to the release is still unclear.

Evaluating the Risks of Radiation

To put global radiation exposure from Chernobyl in perspective, we need to consider typical annual exposure levels. For many people, the principal source of

(a)

(b)

FIGURE 19.12 Wildlife surrounding the destroyed nuclear reactor at Chernobyl, Ukraine, was exposed to intense ionizing radiation, which led to deaths and sublethal biological effects such as genetic mutations. One example of the latter is (a) the partially albino barn swallow. (b) A normal swallow has no white feathers directly beneath its beak.

radiation is radon gas in indoor air and in well water (Figure 19.14). Like all noble gases, radon is chemically inert. Unlike the others, all of its isotopes are radioactive. The most common isotope, radon-222, is produced when uranium-238 in rocks and soil decays to lead-206 (Figure 19.5). The radon gas formed in this decay series percolates upward and can enter a building through cracks and pores in its foundation.

If you breathe radon-contaminated air and then exhale before it decays, no harm is done. However, if radon-222 decays inside the lungs, it emits an α particle that can attack lung tissue. The nuclide produced by the α decay of ^{222}Rn is radioactive polonium-218, which may become attached to tissue in the respiratory system and undergo a second α decay, forming lead-214:

$$^{222}_{86}\text{Rn} \rightarrow\ ^{218}_{84}\text{Po} +\ ^{4}_{2}\alpha \qquad t_{1/2} = 3.8\ \text{d}$$

$$^{218}_{84}\text{Po} \rightarrow\ ^{214}_{82}\text{Pb} +\ ^{4}_{2}\alpha \qquad t_{1/2} = 3.1\ \text{min}$$

As we have seen, α particles are the most damaging product of nuclear decay when formed *inside the body*. How big a threat does radon pose to human health? Concentrations of indoor radon depend on local geology (Figure 19.15) and on how gastight building foundations are. The air in many buildings contains concentrations of radon with radioactivity levels in the range of 1 pCi per liter of air. How hazardous are such tiny concentrations? There appears to be no simple answer. The U.S. Environmental Protection Agency has established 4 pCi/L as an "action level," meaning that people occupying houses with higher concentrations should take measures to minimize their exposure.

This action level is based on studies of the incidence of lung cancer in workers in uranium mines. These workers are exposed to radon concentrations (and concentrations of other radionuclides) that are much higher than the concentrations in homes and other buildings. However, many scientists believe that people exposed to very low levels of radon for many years are as much at risk as miners exposed to high levels of radiation for shorter periods. Some researchers use a model that assumes a linear relation between radon exposure and incidence of lung cancer. This model is represented by the red line in Figure 19.16, which

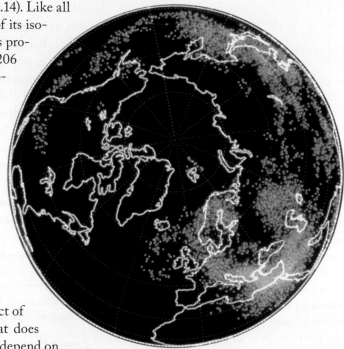

FIGURE 19.13 Radioactive fallout (shown in pink) from the Chernobyl accident in 1986 was detected throughout the Northern Hemisphere.

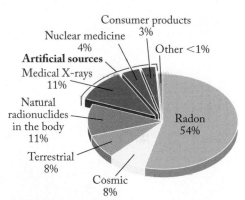

FIGURE 19.14 Sources of radiation exposure of the U.S. population. On average, a person living in the United States is exposed to 0.0036 Sv of radiation each year. More than 80% of this exposure comes from natural sources, mainly radon in the air and water. Artificial sources account for about 18% of the total exposure.

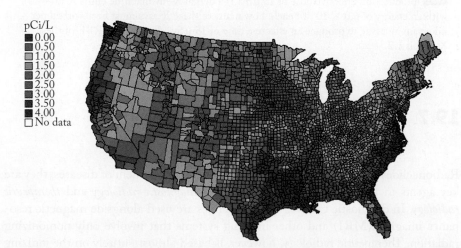

FIGURE 19.15 Levels of radon gas in soils and rocks across the United States.

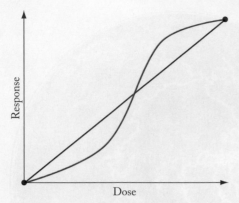

FIGURE 19.16 The risk of death from radiation-induced cancer may follow one of two models. In the linear response model (red line), risk is directly proportional to the radiation exposure. In the S-shaped model (blue line), risk remains low below a critical threshold and then increases rapidly as the exposure increases. In the S-shaped model, the risk is less than for the linear model at low doses but is higher at higher doses.

graphs risk of cancer deaths as a function of radiation absorbed. On the basis of this dose–response model, an estimated 15,000 Americans die of lung cancer each year because of exposure to indoor radon. This number comprises 10% of all lung-cancer fatalities and 30% of those among nonsmokers.

Is this linear model valid? Perhaps—but some scientists believe that there may be a threshold of exposure below which radon poses no significant threat to public health. They advocate an S-shaped dose–response curve, shown by the blue line in Figure 19.16. Notice that the risk of death from cancer in the S-shaped curve is much lower than in the linear response model at low radiation exposure but rises rapidly above a critical value.

SAMPLE EXERCISE 19.5 Calculating Effective Dose   **LO6**

It has been estimated that a person living in a home where the air radon concentration is 4.0 pCi/L receives an annual absorbed dose of ionizing radiation equivalent to 0.40 mGy. What is the person's annual effective dose, in millisieverts, from this radon? Use information from Figure 19.14 to compare this annual effective dose from radon with the average annual effective dose from radon estimated for persons living in the United States.

Collect, Organize, and Analyze We are given an absorbed radiation dose of 0.40 mGy. Radon isotopes emit α particles, which we know from Table 19.3 have a relative biological effectiveness of 20. The effective dose caused by an absorbed dose of ionizing radiation is the absorbed dose multiplied by the RBE of the radiation.

Solve

$$0.40 \text{ mGy} \times 20 = 8.0 \text{ mSv}$$

The caption to Figure 19.14 tells us that the average American is exposed to 3.6 mSv of radiation per year, and the graph tells us that 54% of that amount, or 1.9 mSv, comes from radon. A person living in the home described in this problem has an effective dose from radon that is slightly more than four times the average value.

Think About It The calculated value is more than twice the average annual effective dose of 3.6 mSv from all sources of radiation. The U.S. National Research Council has estimated that a nonsmoker living in air contaminated with 4.0 pCi/L of radon has a 1% chance of dying from lung cancer due to this exposure. The cancer risk for a smoker is close to 5%.

Practice Exercise A dental X-ray for imaging impacted wisdom teeth produces an effective dose of 15 μSv. If a dental X-ray machine emits X-rays with an energy of 6.0×10^{-17} J each, how many of these X-rays must be absorbed per kilogram of tissue to produce an effective dose of 15 μSv? Assume the RBE of these X-rays is 1.2.

19.7 Medical Applications of Radionuclides

Radionuclides are used in both the detection and the treatment of diseases: they are key agents in the respective medical fields of *diagnostic radiology* and *therapeutic radiology*. In diagnostic radiology, radionuclides are used alongside magnetic resonance imaging (MRI) and other imaging systems that involve only nonionizing radiation. Therapeutic radiology, however, is based almost entirely on the ionizing radiation that comes from nuclear processes.

Therapeutic Radiology

Because ionizing radiation causes the most damage to cells that grow and divide rapidly, it is a powerful tool in the fight *against* cancer. Radiation therapy consists of exposing cancerous tissue to γ radiation.

Surgically inaccessible tumors can be treated with beams of γ rays from a radiation source outside the body. Unfortunately, γ radiation destroys both cancer cells and healthy ones. Thus, patients receiving radiation therapy frequently suffer symptoms of radiation sickness, including nausea and vomiting (the tissues that make up intestinal walls are especially susceptible to radiation-induced damage), and hair loss. To reduce the severity of these side effects, radiologists must carefully control the dosage a patient receives.

Often the radiation source is external to the patient, but sometimes it is encased in a platinum capsule and surgically implanted in a cancerous tumor. The platinum provides a chemically inert outer layer and acts as a filter, absorbing α and β particles emitted by the radionuclide but allowing γ rays to pass into the tumor.

A nuclide's chemical properties can sometimes be exploited to direct it to a tumor site. For example, most iodine in the body is concentrated in the thyroid gland, so an effective therapy against thyroid cancer starts with ingestion of potassium iodide containing radioactive iodine-131. Some radionuclides used in cancer therapy are listed in Table 19.6.

TABLE 19.6   Some Radionuclides Used in Radiation Therapy

Nuclide	Radiation	Half-Life	Treatment
^{32}P	β	14.3 d	Leukemia therapy
^{60}Co	β, γ	5.27 yr	Cancer therapy
^{131}I	β	8.1 d	Thyroid therapy
^{131}Cs	γ	9.7 d	Prostate cancer therapy
^{192}Ir	β, γ	74 d	Coronary disease

Diagnostic Radiology

The transport of radionuclides in the body and their accumulation in certain organs also provide ways to assess organ function. In some applications, a tiny quantity of a radioactive isotope is used, together with a much larger amount of a stable isotope of the same element. The radioactive isotope is called a *tracer*, and the stable isotope is the *carrier*. For example, the circulatory system can be imaged by injecting into the blood a solution of sodium chloride containing a trace amount of ^{24}NaCl. Circulation is monitored by measuring the γ rays emitted by ^{24}Na as it decays.

The ideal isotope for medical imaging is one that has a half-life about equal to the length of time required to perform the imaging measurements. It should emit moderate-energy γ rays but no α particles or β particles that might cause tissue damage. Sodium-24 (a γ emitter with a half-life of 15 h) meets both these criteria. Table 19.7 lists several other radionuclides that are used in medical imaging.

Positron emission tomography (PET) is a powerful tool for diagnosing organ and cell function. PET uses short-lived, neutron-poor, positron-emitting radionuclides such as carbon-11, oxygen-15, and fluorine-18. A patient might be administered a solution of glucose in which some of the sugar molecules contain

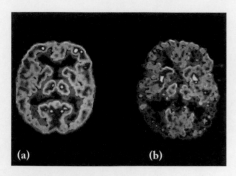

FIGURE 19.17 Positron emission tomography (PET) is used to monitor cell activity in organs such as the brain. (a) Brain function in a healthy person. The red and yellow regions indicate high brain activity; blue and black indicate low activity. (b) Brain function in a patient with Alzheimer's disease.

TABLE 19.7 Selected Radionuclides Used for Medical Imaging

Nuclide	Radiation	Half-Life (h)	Use
^{99m}Tc	γ	6.0	Bones, circulatory system, various organs
^{67}Ga	γ	78	Tumors in the brain and other organs
^{201}Tl	γ	73	Coronary arteries, heart muscle
^{123}I	γ	13.2	Thyroxine production in thyroid gland

atoms of ^{11}C, ^{15}O, or ^{18}F. The rate at which glucose is metabolized in various regions of the brain is monitored by detecting the γ rays produced by positron–electron annihilations. Unusual patterns in PET images of brains (Figure 19.17) can indicate schizophrenia, bipolar disorder, Alzheimer's disease, damage from strokes, and even nicotine addiction in tobacco smokers.

19.8 Nuclear Fission

The energy stored in atomic nuclei can also be put to practical use. When an atom of uranium-235 captures a neutron, the nucleus of the unstable product, uranium-236, splits into two lighter nuclei in a process called **nuclear fission**. Several uranium-235 fission reactions can occur, including these three:

$$^{235}_{92}\text{U} + ^{1}_{0}\text{n} \rightarrow ^{141}_{56}\text{Ba} + ^{92}_{36}\text{Kr} + 3\,^{1}_{0}\text{n}$$

$$^{235}_{92}\text{U} + ^{1}_{0}\text{n} \rightarrow ^{137}_{52}\text{Te} + ^{97}_{40}\text{Zr} + 2\,^{1}_{0}\text{n}$$

$$^{235}_{92}\text{U} + ^{1}_{0}\text{n} \rightarrow ^{138}_{55}\text{Cs} + ^{96}_{37}\text{Rb} + 2\,^{1}_{0}\text{n}$$

In all these reactions, the sums of the masses of the products are slightly less than the sums of the masses of the reactants. As we observed for nucleosynthesis in Section 19.1, this difference in mass is converted to energy in accordance with Einstein's equation ($E = mc^2$).

These reactions also produce additional neutrons, which can smash into other uranium-235 nuclei and initiate more fission events in a **chain reaction** (Figure 19.18). The reaction proceeds as long as there are enough uranium-235 nuclei present to absorb the neutrons being produced. On average, at least one neutron from each fission event must cause another nucleus to split apart for the chain reaction to be self-sustaining. The quantity of fissionable material needed to ensure that every fission event produces another is called the **critical mass**. For uranium-235, the critical mass is about 1 kg of the pure isotope.

Uranium-235 is the most abundant fissionable isotope, but it makes up only 0.72% of the uranium in the principal uranium ore, pitchblende (Figure 19.19a). The uranium in nuclear reactors must be at least 3% to 4% uranium-235, and enrichment to about 85% is needed for nuclear weapons. The most common method for enriching uranium ore involves extracting the uranium in a process that yields a

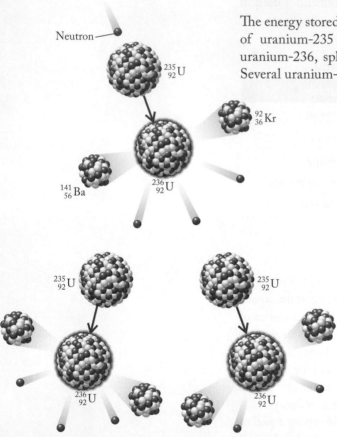

FIGURE 19.18 Each fission event in the chain reaction of a uranium-235 nucleus begins when the nucleus captures a neutron, forming an unstable uranium-236 nucleus that then splits apart (fissions) in one of several ways. In the first process shown here, the uranium-236 nucleus splits into krypton-92, barium-141, and three neutrons. If, on average, at least one of the three neutrons from each fission event causes the fission of another uranium-235 nucleus, then the process is sustained in a chain reaction.

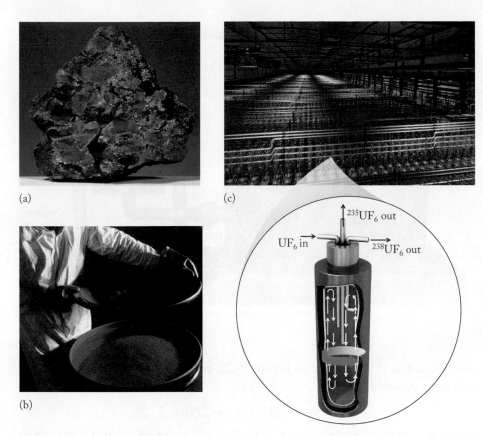

(a)

(b)

(c)

$^{235}UF_6$ out

UF_6 in

$^{238}UF_6$ out

FIGURE 19.19 Preparing uranium fuel. (a) A piece of pitchblende, source of the uranium fuel for nuclear reactors. Pitchblende ore is ground up and extracted with strong acid. (b) The uranium compounds (mostly U_3O_8) obtained from the extract are called yellowcake. (c) Uranium oxides are converted to volatile UF_6, which is centrifuged at very high speed to separate $^{235}UF_6$ from $^{238}UF_6$. The less dense and less abundant $^{235}UF_6$ is enriched near the center of the centrifuge cylinder and separated from the heavier $^{238}UF_6$.

nuclear fission a nuclear reaction in which the nucleus of an element splits into two lighter nuclei. The process is usually accompanied by the release of one or more neutrons and energy.

chain reaction a self-sustaining series of fission reactions in which the neutrons released when nuclei split apart initiate additional fission events and sustain the reaction.

critical mass the minimum quantity of fissionable material needed to sustain a chain reaction.

breeder reactor a nuclear reactor in which fissionable material is produced during normal reactor operation.

material called yellowcake, which is mostly U_3O_8 (Figure 19.19b). This oxide is then converted to UF_6, which, despite a molar mass of more than 300 g per mole, is a volatile solid that sublimes at 56°C. The volatility of this nonpolar molecular compound can be explained by the relatively weak London dispersion forces experienced by its compact, symmetrical molecules (Figure 19.20). Fissionable $^{235}UF_6$ is separated from $^{238}UF_6$ on the basis of their slightly different densities. Elaborate centrifuge systems are used to exploit this difference (Figure 19.19c).

Harnessing the energy released by nuclear fission to generate electricity began in the middle of the 20th century. In a typical nuclear power plant (Figure 19.21), fuel rods containing 3% to 4% uranium-235 are interspersed with rods of boron or cadmium that control the rate of the chain reaction by absorbing some of the neutrons produced during fission. Pressurized water flows around the fuel and control rods, removing the heat created during fission and transferring it to a steam generator. The water also acts as a moderator, slowing down the neutrons and thereby allowing for their more efficient capture by ^{235}U atoms.

In 1952 the first **breeder reactor** was built, so called because in addition to producing energy to make electricity, the reactor makes ("breeds") its own fuel. The reactor starts out with a mixture of plutonium-239 and uranium-238. As the plutonium fissions and the energy from those reactions is collected to produce electricity, some of the neutrons that are produced sustain the fission chain

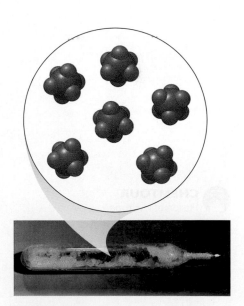

FIGURE 19.20 Uranium hexafluoride is a volatile solid that sublimes at only 56°C because it is composed of compact, symmetrical molecules that experience relatively weak London dispersion forces despite their considerable mass.

FIGURE 19.21 A pressurized, water-cooled nuclear power plant uses fuel rods containing uranium enriched to about 4% uranium-235. The fission chain reaction is regulated with control rods and a moderator that is either water or liquid sodium. The moderator slows down the neutrons released by fission so that they can be more efficiently captured by other uranium-235 nuclei. It also transfers the heat produced by the fission reaction to a steam generator. The steam generated by this heat drives a turbine that generates electricity. Nuclear power plants are usually situated near coasts or large rivers so large amounts of coolant water are readily available.

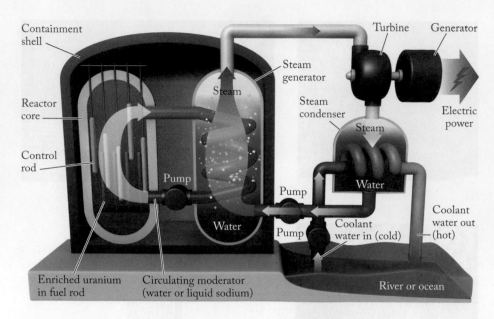

reaction just as in the reactor of Figure 19.21, while others convert the uranium into more plutonium fuel:

$$^{238}_{92}\text{U} + ^{1}_{0}\text{n} \rightarrow ^{239}_{92}\text{U} + \gamma \rightarrow ^{239}_{94}\text{Pu} + 2\ ^{0}_{-1}\beta$$

In less than 10 years of operation, a breeder reactor can make enough plutonium-239 to refuel itself *and* another reactor. Unfortunately, plutonium-239 is a carcinogen and one of the most toxic substances known. Only about half a kilogram is needed to make an atomic bomb, and it has a long half-life: 2.4×10^4 years. Understandably, extreme caution and tight security surround the handling of plutonium fuel and the transportation and storage of nuclear wastes containing even small amounts of plutonium. Health and safety matters related to reactor operation and spent-fuel disposal are the principal reasons there are no breeder power stations in the United States, although they have been built in at least seven other countries.

19.9 Nuclear Fusion and the Quest for Clean Energy

The energy of the Sun is derived from the high-speed collision and combining of hydrogen nuclei to form helium. This **nuclear fusion** process involves more steps than the process that probably took place during primordial nucleosynthesis (see Chapter 2), when protons (hydrogen nuclei) and neutrons fused to form deuterons (nuclei of the hydrogen isotope deuterium, D):

CHEMTOUR
Fusion of Hydrogen

nuclear fusion a nuclear reaction in which subatomic particles or atomic nuclei collide with each other at very high speeds and fuse, forming more massive nuclei and releasing energy.

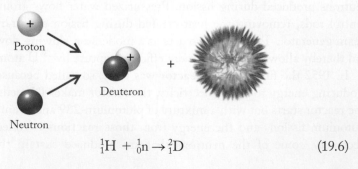

$$^{1}_{1}\text{H} + ^{1}_{0}\text{n} \rightarrow ^{2}_{1}\text{D} \tag{19.6}$$

Once deuterons formed, they also collided with each other and fused, forming α particles, which are the nuclei of helium-4 atoms:

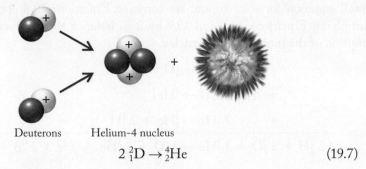

Deuterons   Helium-4 nucleus

$$2\,{}^{2}_{1}\text{D} \rightarrow {}^{4}_{2}\text{He} \tag{19.7}$$

Recall from Chapter 2 that each subscript in a nuclear equation represents the electrical charge of a particle and each superscript represents the particle's mass number. When a particle is the nucleus of an atom, its electrical charge is equal to the number of protons in it, that is, its atomic number.

Hydrogen fusion in our Sun follows a different path because free neutron concentrations are far lower there than they were in the primordial universe. In the Sun, colliding protons may fuse to form a deuteron and a positron:

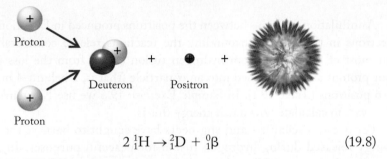

Proton

Deuteron   Positron

Proton

$$2\,{}^{1}_{1}\text{H} \rightarrow {}^{2}_{1}\text{D} + {}^{0}_{1}\beta \tag{19.8}$$

In the second stage of solar fusion, protons fuse with deuterons to form helium-3 nuclei:

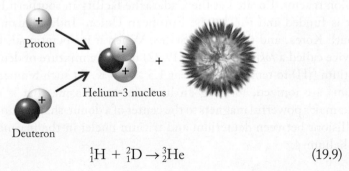

Proton

Helium-3 nucleus

Deuteron

$${}^{1}_{1}\text{H} + {}^{2}_{1}\text{D} \rightarrow {}^{3}_{2}\text{He} \tag{19.9}$$

Finally, fusion of two helium-3 nuclei produces a helium-4 nucleus and 2 protons:

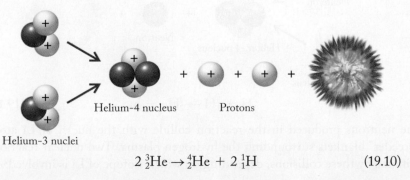

Helium-4 nucleus   Protons

Helium-3 nuclei

$$2\,{}^{3}_{2}\text{He} \rightarrow {}^{4}_{2}\text{He} + 2\,{}^{1}_{1}\text{H} \tag{19.10}$$

Deuterium and ^{3}He nuclei are intermediates in the hydrogen-fusion process because they are made in one step but then consumed in another. To write an overall equation for solar fusion, we combine Equations 19.8, 19.9, and 19.10, multiplying Equations 19.8 and 19.9 by 2 to balance the production and consumption of the intermediate particles:

$$2[2\,^1_1\text{H} \rightarrow\,^2_1\text{D} +\,^0_1\beta]$$

$$+\,2[^1_1\text{H} +\,^2_1\text{D} \rightarrow\,^3_2\text{He}]$$

$$+\qquad 2\,^3_2\text{He} \rightarrow\,^4_2\text{He} + 2\,^1_1\text{H}$$

$$\overline{4\,6\,^1_1\text{H} + 2\,^2_1\text{D} + 2\,^3_2\text{He} \rightarrow 2\,^2_1\text{D} + 2\,^3_2\text{He} +\,^4_2\text{He} + 2\,^0_1\beta + 2\,^1_1\text{H}}$$

This equation reduces to:

$$4\,^1_1\text{H} \rightarrow\,^4_2\text{He} + 2\,^0_1\beta \qquad\qquad (19.11)$$

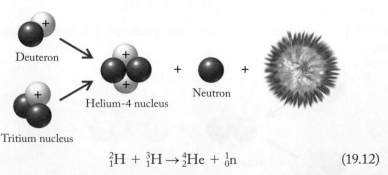

Protons   Helium-4 nucleus   Positrons

Annihilation reactions between the positrons produced in Equation 19.11 and electrons in the matter surrounding the reactants release considerable energy, but most of the energy from hydrogen fusion comes from the loss in mass as four protons are transformed into an α particle (that is, a helium-4 nucleus) and two positrons (Table 19.1). In Sample Exercise 19.6 we use Einstein's equation, $E = mc^2$, to calculate how much energy this is.

For decades scientists and engineers have sought to harness the enormous energy released during hydrogen fusion for peaceful purposes. In 2010 construction began on ITER (originally an acronym for International Thermonuclear Experimental Reactor), a project to build the world's largest nuclear fusion reactor. Located at the Cadarache facility in southern France, the project is funded and run by the European Union, India, Japan, China, Russia, South Korea, and the United States. When it is operational, ITER will use a device called a *tokamak* (Figure 19.22) to heat a mixture of deuterium ($^2_1$H) and tritium ($^3_1$H) to temperatures near 1.5×10^8 K. At such temperatures, all these atoms are ionized, forming an incandescent plasma that is confined by the tokamak's powerful magnets to the center of a donut-shaped tunnel. High-speed collisions between deuterium and tritium nuclei in the plasma produce nuclei of helium-4:

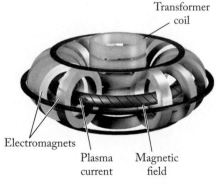

Transformer coil

Electromagnets

Plasma current   Magnetic field

FIGURE 19.22 A tokamak transmits electrical energy into a toroidal (donut-shaped) chamber containing deuterium and tritium, causing these isotopes of hydrogen to ionize and form a plasma of nuclei and free electrons with a temperature above 10^8 K. Combinations of electromagnets confine the plasma to the interior of the torus, where collisions between ^{2}H and ^{3}H nuclei result in fusion, forming ^{4}He nuclei and free neutrons. The neutrons then collide with Li atoms in the walls of the chamber, initiating additional nuclear reactions that produce more tritium fuel.

Deuteron

Helium-4 nucleus   Neutron

Tritium nucleus

$$^2_1\text{H} +\,^3_1\text{H} \rightarrow\,^4_2\text{He} +\,^1_0\text{n} \qquad\qquad (19.12)$$

The neutrons produced in the reaction collide with the nuclei of Li atoms in "breeder" blankets surrounding the hydrogen plasma. Two nuclear reactions are initiated by these collisions, depending on which isotope of Li is involved:

$$_0^1n + {}_3^6Li \rightarrow {}_2^4He + {}_1^3H \qquad (19.13)$$

$$_0^1n + {}_3^7Li \rightarrow {}_2^4He + {}_1^3H + {}_0^1n \qquad (19.14)$$

Note that the reactions in Equations 19.13 and 19.14 produce tritium nuclei. In this way the reactions supply more fuel for the primary fusion reaction. The world's supply of deuterium, the other fuel, is enormous (seawater contains about 15 mg of deuterium per kilogram). On the other hand, tritium is not abundant in nature because it is radioactive, with a half-life of only 12.3 years. One disadvantage of these reactions is their reliance on lithium during a time when expanding production of lithium-ion batteries (see Section 18.7) is increasing our demand for the element.

CONCEPT **TEST**

Nuclear reactors powered by the energy released by the fission of uranium-235 have been operating since the 1950s, but a reactor powered by the energy released by the fusion of hydrogen has yet to be built. Why is it taking so long to build a fusion reactor?

SAMPLE EXERCISE 19.6 Calculating the Energy Released   **LO7**
in a Nuclear Reaction

How much energy in joules is released by the overall fusion process in which four protons undergo nuclear fusion, producing an α particle and two positrons (Equation 19.11)?

Collect and Organize We are asked to calculate the energy released in the nuclear reaction:

$$4\,_1^1H \rightarrow {}_2^4He + 2\,_1^0\beta \qquad (19.11)$$

The energies associated with nuclear reactions are related to differences in the masses of the reactant and product particles and Einstein's equation, $E = mc^2$. The masses of the particles in Table 19.1 are given in kilograms, which is convenient because the relationship between energy and mass is linked to the unit conversion

$$1\,J = 1\,kg \cdot (m/s)^2$$

Analyze Given the value of the masses in Table 19.1, the difference in mass will probably be less than 10^{-27} kg. When multiplied by the square of the speed of light, $(2.998 \times 10^8 \text{ m/s})^2 \approx 10^{17}$, the calculated value of E should be less than 10^{-10} J.

Solve First we calculate the change in mass:

$$\Delta m = (m_{\alpha \text{ particle}} + 2\,m_{\text{positron}}) - 4\,m_{\text{proton}}$$

$$= [6.64465 \times 10^{-27} + (2 \times 9.10939 \times 10^{-31})]\,kg - (4 \times 1.67262 \times 10^{-27})\,kg$$

$$= -4.40081 \times 10^{-29}\,kg$$

The energy corresponding to this loss in mass is calculated using Einstein's equation where $m = -4.40081 \times 10^{-29}$ kg:

$$E = mc^2$$
$$= -4.40081 \times 10^{-29} \text{ kg} \times (2.998 \times 10^8 \text{ m/s})^2$$
$$= -3.955 \times 10^{-12} \text{ kg} \cdot \text{(m/s)}^2 = -3.955 \times 10^{-12} \text{ J}$$

Think About It The decrease in mass translates into energy lost by the reaction system to its surroundings. As we predicted, the absolute value of this energy is less (actually much less) than 10^{-10} J, which seems like an awfully small value compared with the world's energy needs. However, this value applies to the formation of a single α particle. If we multiply it by Avogadro's number and convert it to a value in kilojoules per mole, a unit we typically use in thermochemistry, we get

$$\frac{-3.955 \times 10^{-12} \text{ J}}{\text{α-particle}} \times \frac{6.022 \times 10^{23} \text{ α-particles}}{\text{mol}} \times \frac{1 \text{ kJ}}{1000 \text{ J}} = -2.382 \times 10^9 \text{ kJ/mol}$$

To put this value in perspective, it is about 10^7 times the change in free energy from the combustion of 1 mole of hydrogen gas.

Practice Exercise How much energy is released in the nuclear reaction described by Equation 19.12? Express your answer in kilojoules per mole. (*Note*: The mass of a tritium nucleus is 5.00827×10^{-27} kg.)

SAMPLE EXERCISE 19.7 Integrating Concepts: Radium Girls and Safety in the Workplace

Radium was discovered by Pierre and Marie Curie in 1898. By 1902, the new element had its first practical use: radium compounds were mixed with zinc sulfide (ZnS) to make paint that glowed in the dark as α particles emitted by the decay of ^{226}Ra ($t_{1/2} = 1.6 \times 10^3$ years) caused ZnS crystals to emit a greenish fluorescence (Figure 19.23). The paint was used to make dials for watches, clocks, and instruments used on naval vessels and, a few years later, in military and civilian airplanes.

By 1914, U.S. companies were making radium-painted dials and employing young women in their late teens and early 20s as dial painters. Soon after their employment, many of the women became very sick, suffering from anemia and other symptoms we now associate with exposure to nuclear radiation. Some developed bone cancer and more than one hundred of them, who became known around the world as the Radium Girls, died. Their deaths were linked to the practice of "pointing" the fine paint brushes they used, which meant using their lips to make fine points on the brushes to help them paint the tiny numerals and hands on watch faces (Figure 19.24). In the process, they ingested some of the radioactive paint. Tests later determined that about 20% of the radium ingested was incorporated into their bones, where it attacked bone marrow and caused malignancies known as osteosarcomas.

a. Suggest a reason why radium was concentrated in the victims' bones.

b. Studies of radiation levels and incidence of cancer in more than 1000 female dial painters yielded the results in the table below. Which of the two dose–response curves in Figure 19.16 best fits these results?

Radium Exposure (mg ingested)	Occurrence of Malignancy (% of workers exposed)
1	0
3	0
10	0
30	0
100	5
300	53
1000	85

c. The green luminescence of radium watch dials began to fade after a few years. Was this loss in luminosity due to decreased radioactivity in the paint? Explain why or why not.

d. In one study, the levels of radioactivity in pocket watches with radium-painted dials were found to be between 0.6 and 1.39 μCi per watch. How many micrograms of ^{226}Ra produce 1.39 μCi of radioactivity?

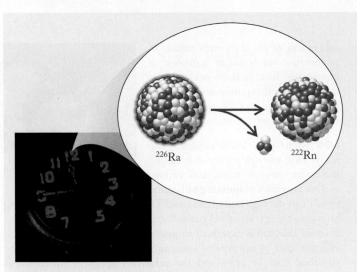

FIGURE 19.23 During the 20th century, many millions of watches and clocks had dials that glowed in the dark as high-energy α particles emitted by ^{226}Ra caused crystals of ZnS to fluoresce.

FIGURE 19.24 This editorial cartoon appeared in Sunday newspapers on February 28, 1926. It portrayed the deadly consequences of young women "pointing" their brushes with their lips as they painted the dials of watches with paint that contained radioactive radium.

Collect and Organize We are asked (a) why ingested ^{226}Ra concentrates in bones; (b) whether malignancy in dial painters was proportional to their exposure to ^{226}Ra radiation or followed an S-shaped dose–response curve; (c) why the luminosity of radium-activated paint fades after a few years; and (d) how many micrograms of ^{226}Ra are needed to produce 1.39 mCi of radiation. One curie (Ci) is equal to 3.70×10^{10} decay events/s. The half-life ($t_{1/2}$) of ^{226}Ra is 1600 years and is related to the first-order rate constant (k) of the decay reaction by the equation $t_{1/2} = 0.693/k$. The level of radioactivity (A) in a sample of radium is equal to the product of the rate constant and the number (N) of ^{226}Ra atoms: $A = kN$.

Analyze Radium is a group 2 element and should have chemical and biochemical properties that are similar to those of the other elements in that group, including its association with biological tissues. High concentrations of another group 2 element, calcium, occur in teeth and bones. Worker exposure levels in the above table cover a wide range, but exposure up to nearly 100 mg ^{226}Ra caused few malignancies, whereas concentrations above 100 mg caused many. The half-life of ^{226}Ra is 1600 years, so the radioactivity of a sample decreases little over a few years or even over many decades. Relating a half-life expressed in years to a level of radioactivity expressed in a multiple of decay events per second requires converting units of time and then quantities of radioactive atoms to moles and then micrograms.

Solve

a. Radium probably accumulates in bones because its chemistry is similar to that of calcium, which means that ^{226}Ra^{2+} ions are likely to take the place of Ca^{2+} ions in bone tissue.

b. Malignancies did not occur among the dial painters who ingested less than 100 mg of ^{226}Ra; however, the percentage of the women who suffered malignancies increased sharply with exposure between 100 and 1000 mg. This pattern is described by the S-shaped (blue) curve in Figure 19.16.

c. Given the 1600-year half-life of ^{226}Ra, the loss of watch dial luminescence was not the result of depleted radioactivity.

Rather, it must have been due to less efficient conversion of the energy of radioactive decay into visible light by ZnS crystals.

d. Let's first convert the half-life of ^{226}Ra into a decay rate constant in units of s^{-1}:

$$k = \frac{0.693}{t_{1/2}} = \frac{0.693}{1.6 \times 10^3 \text{ yr}} \times \frac{1 \text{ yr}}{365.25 \text{ d}} \times \frac{1 \text{ d}}{24 \text{ h}} \times \frac{1 \text{ h}}{3600 \text{ s}}$$
$$= 1.372 \times 10^{-11} \text{ s}^{-1}$$

Next, we solve the equation $A = kN$ for N, and we use the above rate constant and the radioactivity of the watch to calculate the number of ^{226}Ra atoms in the dial:

$$N = \frac{A}{k} = \frac{1.39 \ \mu\text{Ci}}{1.372 \times 10^{-11} \text{ s}^{-1}} \times \frac{1 \text{ Ci}}{10^6 \ \mu\text{Ci}}$$
$$\times \frac{3.70 \times 10^{10} \text{ atoms Ra s}^{-1}}{\text{Ci}} = 3.75 \times 10^{15} \text{ atoms of Ra}$$

The corresponding mass in micrograms is

$$3.75 \times 10^{15} \text{ atoms Ra} \times \frac{1 \text{ mol Ra}}{6.022 \times 10^{23} \text{ atoms Ra}}$$
$$\times \frac{226 \text{ g Ra}}{1 \text{ mol Ra}} \times \frac{10^6 \ \mu\text{g}}{1 \text{ g}} = 1.4 \ \mu\text{g of Ra}$$

Think About It Did you notice the similarity in the level of radioactivity (1.39 μCi) in the watch dial and the mass of radium (1.41 μg) producing it? This is not a coincidence. When the curie was adopted as the standard unit of radioactivity in the early 20th century, it was chosen to honor the pioneering work of Marie and Pierre Curie, and it was based on what was then believed to be the level of radioactivity in 1 g of radium. Newspaper articles published in the 1920s made clear that Marie Curie was deeply troubled by the tragedy of the Radium Girls. Sadly, in 1934 she herself died from aplastic anemia—a disease caused by the inability of bone marrow to produce red blood cells.

SUMMARY

LO1 Nuclear chemistry is the study and application of reactions that involve changes in atomic nuclei. The **mass defect (Δm)** of a nucleus is the difference between its mass and the sum of the masses of its nucleons. **Binding energy (BE)** is the energy released when the nucleons combine to form a nucleus. It is also the energy needed to split the nucleus into its nucleons. Binding energy per nucleon is a measure of the relative stability of a nucleus. When a particle of matter encounters a particle of antimatter, both are converted into energy (they annihilate one another), yielding γ rays. (Section 19.1)

LO2 Stable nuclei have neutron-to-proton ratios that fall within a range of values called the **belt of stability**. Unstable nuclides undergo **radioactive decay**. Neutron-rich nuclides (mass number greater than the average atomic mass) undergo β **decay**; neutron-poor nuclides undergo **positron emission** or **electron capture**. Very large nuclides ($Z > 83$) may undergo β decay or α **decay**. (Section 19.2)

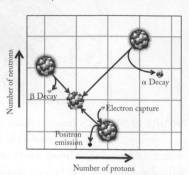

LO3 Scintillation counters and **Geiger counters** are used to measure levels of nuclear radiation. Radioactivity is the number of decay events per unit time. Common units are the **becquerel (Bq**; 1 decay event/s) and the **curie (Ci**; 1 Ci = 3.70×10^{10} Bq). (Section 19.3)

LO4 Radioactive decay follows first-order kinetics, so the half-life ($t_{1/2}$) of a radionuclide is a characteristic value of the decay process. (Section 19.4)

LO5 Radiometric dating is used to determine the age of an object from its content of a radionuclide and/or its decay product. **Radiocarbon dating** involves measuring the amount of radioactive carbon-14 that remains in an object derived from plant or animal tissue to calculate the age of the object. The accuracy of the technique relies on

calibration of the data with results of radiometric analyses of samples of known age, such as the trunks of trees that have lived for thousands of years and whose age can be confirmed by their growth rings. (Section 19.5)

LO6 Alpha particles, β particles, and γ rays have enough energy to break up molecules into electrons and cations and are examples of **ionizing radiation**, which can damage body tissue and DNA. The quantity of ionizing radiation energy absorbed per kilogram of body mass is called the *absorbed dose* and is expressed in **grays (Gy**; 1 Gy = 1.00 J/kg). The effective dose of any type of ionizing radiation is the product of the absorbed dose in grays and the **relative biological effectiveness (RBE)** of the radiation; the unit of effective dose is the **sievert (Sv)**. Alpha particles have a larger RBE than do β particles and γ rays but have the least penetrating power of these three types of ionizing radiation. Selected radioactive isotopes are useful as tracers in the human body to map biological activity and diagnose diseases. Other radioactive isotopes are used to treat cancers. (Sections 19.6, 19.7)

LO7 Neutron absorption by uranium-235 and a few other massive isotopes may lead to **nuclear fission**, creating lighter nuclei, accompanied by the release of energy that can be harnessed to generate electricity. A **chain reaction** happens when the neutrons released during fission collide with other fissionable nuclei. They require a **critical mass** of a fissionable isotope. A **breeder reactor** is used to make plutonium-239 from uranium-238, while also producing energy to make electricity. **Nuclear fusion** occurs when subatomic particles or atomic nuclei collide and fuse. Facilities that harness the enormous energy of hydrogen fusion must operate at temperatures greater than 10^8 K. (Sections 19.8, 19.9)

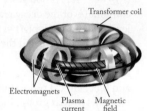

PARTICULATE **PREVIEW WRAP-UP**

Carbon-14 has 6 protons and 8 neutrons; therefore, with a neutron-to-proton ratio greater than 1, it will decay to reduce that ratio. When ^{14}C radioactively decays by emitting a beta particle, it produces a nitrogen nuclide with a 1:1 neutron:proton ratio (7 protons and 7 neutrons).

PROBLEM-SOLVING SUMMARY

Type of Problem	Concepts and Equations		Sample Exercises
Predicting the modes and products of radioactive decay	Neutron-rich nuclides tend to undergo β decay; neutron-poor nuclides undergo positron emission or electron capture.		**19.1**
Calculating the radioactivity of a sample	$A = kN$ where $$k = \frac{0.693}{t_{1/2}}$$	(19.1)	**19.2**

Type of Problem	Concepts and Equations		Sample Exercises
Calculations involving half-lives and radiocarbon dating	$$\dfrac{N_t}{N_0} = 0.5^{t/t_{1/2}}$$	(19.4)	**19.3, 19.4**
	$$t = -\dfrac{t_{1/2}}{0.693} \ln \dfrac{N_t}{N_0}$$	(19.5)	
	where N_t/N_0 is the ratio of the quantity of radionuclide present in a sample at time t (N_t) to the quantity at $t = 0$ (N_0) and is the half-life of the radionuclide.		
Calculating effective dose	Effective dose = absorbed dose × RBE		**19.5**
Calculating the energy released in a nuclear reaction	$E = mc^2$, where m is the loss in mass as reactants form products.		**19.6**

VISUAL PROBLEMS

(Answers to boldface end-of-chapter questions and problems are in the back of the book.)

19.1. Which of the highlighted elements in Figure P19.1 currently plays a key role in the controlled fusion of hydrogen?

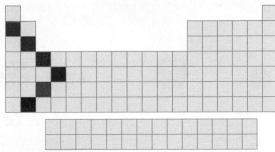

FIGURE P19.1

19.2. Exposure to which of the highlighted elements in Figure P19.1 could cause anemia and bone disease?

19.3. Which of the highlighted elements in Figure P19.1 is produced by the decay of uranium-238?

19.4. What radioactive decay processes are represented by the graphs in Figure P19.4?

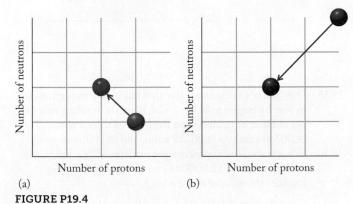

(a)   (b)

FIGURE P19.4

19.5. Which of the graphs in Figure P19.5 illustrates β decay?

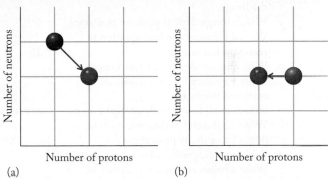

(a)   (b)

FIGURE P19.5

19.6. Which of the curves in Figure P19.6 represents the decay of an isotope that has a half-life of 2.0 days?

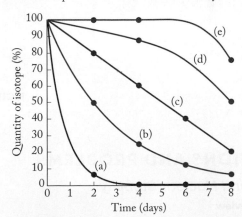

FIGURE P19.6

19.7. Which of the curves in Figure P19.6 do(es) not represent a radioactive decay curve?

19.8. Which of the models in Figure P19.8 represents fission and which represents fusion?

(1)   (2)

FIGURE P19.8

19.9. Isotopes in a nuclear decay series emit particles with a positive charge and particles with a negative charge. The two kinds of particles penetrate a column of water as shown in Figure P19.9. Is the "X" particle the positive or the negative one?

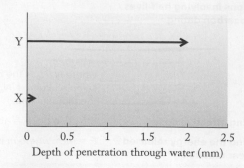

Depth of penetration through water (mm)

FIGURE P19.9

19.10. Use representations [A] through [I] in Figure P19.10 to answer questions a–f.
  a. Which image depicts beta decay? Write a balanced nuclear equation to represent the image.
  b. Which image depicts fusion? Write a balanced nuclear equation to represent the image.
  c. Which image depicts positron emission? Write a balanced nuclear equation to represent the image.
  d. Which image depicts nuclear fission? Write a balanced nuclear equation to represent the image.
  e. Which nuclide will undergo alpha decay? Write a balanced nuclear equation to depict these decay processes.
  f. Which nuclide will undergo beta decay? Write a balanced nuclear equation to depict these decay processes.

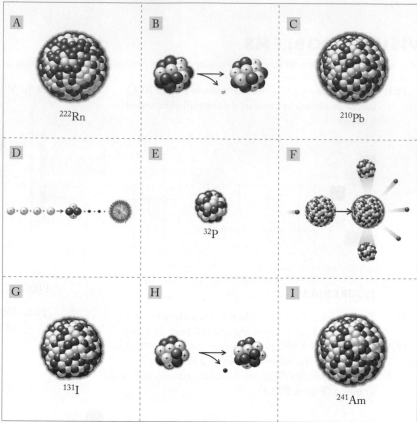

FIGURE P19.10

QUESTIONS AND PROBLEMS

Energy and Nuclear Stability

Concept Review

19.11. What do the terms *mass defect* and *binding energy* mean?
19.12. Why is energy released in a nuclear fusion process when the product is an element preceding iron in the periodic table?

Problems

19.13. What is the binding energy of a deuteron?
19.14. What is the binding energy of ^{6}Li, which has a nuclear mass of 9.98561×10^{-27} kg?

19.15. Our Sun is a fairly small star that has barely enough mass to fuse hydrogen to helium. Calculate the binding energy per nucleon of helium-4 on the basis of these masses: $^4_2$He (4.00260 amu), $^1_1$p (1.00728 amu), and $^1_0$n (1.00866 amu).
19.16. What is the binding energy per nucleon of ^{12}C, the atomic mass of which is 12.00000 amu? (*Note*: Atomic mass includes the mass of 6 electrons.)

Unstable Nuclei and Radioactive Decay

Concept Review

19.17. How can the belt of stability be used to predict the probable decay mode of an unstable nuclide?

19.18. Compare positron-emission and electron-capture processes.

19.19. The ratio of neutrons to protons in stable nuclei increases with increasing atomic number. Use this trend to explain why multiple α decay steps in the ^{238}U decay series are often followed by β decay.

19.20. Copper-64 is an unusual radionuclide in that it may undergo β decay, positron emission, or electron capture. What are the products of these decay processes?

Problems

19.21. Iodine-137 decays to give xenon-137, which decays to give cesium-137. What are the modes of decay in these two reactions?

19.22. Write a balanced nuclear equation describing (a) α decay of curium-242; (b) β decay of magnesium-28; (c) positron emission by xenon-118; (d) electron capture by cadmium-104.

19.23. If the mass number of an isotope is more than twice the atomic number, is the neutron-to-proton ratio less than, greater than, or equal to 1?

19.24. In each of the following pairs of isotopes, select the isotope that has more protons and the isotope that has more neutrons. Also indicate which pairs of isotopes have the same number of neutrons or protons. (a) ^{63}Cu and ^{65}Cu; (b) ^{71}Ga and ^{71}Ge; (c) ^{39}K and ^{40}Ar.

19.25. Identify the type of radiation emitted or consumed in the following processes:
   a. $^{222}_{86}Rn \rightarrow ^{218}_{84}Po + ?$
   b. $^{131}_{53}I \rightarrow ^{131}_{54}Xe + ?$
   c. $^{11}_{6}C \rightarrow ^{11}_{5}B + ?$
   d. $^{7}_{4}Be + ? \rightarrow ^{7}_{3}Li$

19.26. Write a balanced nuclear equation describing the β decay of cesium-137, which is produced in nuclear power plants.

19.27. Predict the mode(s) of decay for the following radioactive isotopes: (a) ^{10}C; (b) ^{19}Ne; (c) ^{50}Ti.

19.28. Predict the mode(s) of decay of the following radionuclides: (a) ^{24}Ne; (b) ^{38}K; (c) ^{45}Ti; (d) ^{237}Np.

19.29. **Elements in a Supernova** The isotopes ^{56}Co and ^{44}Ti were detected in supernova SN 1987A. Predict the decay pathway for these radioactive isotopes.

19.30. There are isotopes of nitrogen that have as few as 5 or as many as 11 neutrons in each of their nuclei. Write a balanced nuclear equation describing the decay of the isotope with 11 neutrons.

Measuring Radioactivity; Rates of Radioactive Decay

Concept Review

*19.31. Chlorine has isotopes with mass numbers from 32 through 39. Two of them, ^{35}Cl and ^{37}Cl, are stable.
   a. Which three of the other isotopes emit positrons?
   b. Which of the other isotopes emit β particles?
   c. Which one of the other isotopes can emit *either* positrons or β particles?

*19.32. Bromine has isotopes with mass numbers from 74 through 90. Two of them, ^{79}Br and ^{81}Br, are stable.
   a. How many of the others emit positrons or undergo electron capture?
   b. How many of the others emit β particles?
   c. Which one of the other isotopes can emit *either* positrons or β particles?

19.33. What percentage of a sample's original radioactivity remains after two half-lives?

19.34. Explain why rates of nuclear decay are independent of temperature.

Problems

19.35. The half-life of radon-222, a radioactive gas found in some basements, is 3.82 days. Calculate the decay rate constant of radon-222.

19.36. The decay rate constant of sodium-24, a tracer used in blood studies, is 4.6×10^{-2} h^{-1}. What is the value of its half-life?

19.37. Explosions at a disabled nuclear power station in Fukushima, Japan, in 2011 may have released more cesium-137 ($t_{1/2} = 30.2$ years) into the ocean than any other single event. How long will it take the radioactivity of this radionuclide to decay to 5.0% of the level released in 2011?

19.38. Spent fuel removed from nuclear power stations contains plutonium-239 ($t_{1/2} = 2.41 \times 10^4$ years). How long will it take a sample of this radionuclide to reach a level of radioactivity that is 2.5% of the level it had when it was removed from a reactor?

Radiometric Dating

Concept Review

19.39. Explain why radiocarbon dating is reliable only for artifacts and fossils younger than about 50,000 years.

19.40. Which of the following statements about ^{14}C dating are true?
   a. The amount of ^{14}C in all objects is the same.
   b. Carbon-14 is unstable and is readily lost from the atmosphere.
   c. The ratio of ^{14}C to ^{12}C in the atmosphere is a constant.
   d. Living tissue will absorb ^{12}C but not ^{14}C.

19.41. Why is ^{40}K dating ($t_{1/2} = 1.28 \times 10^9$ years) useful only for rocks older than 300,000 years?

19.42. Where does the ^{14}C found in plants come from?

Problems

19.43. **First Humans in South America** Archeologists continue to debate the arrival of the first humans in the Western Hemisphere. Radiocarbon dating of charcoal from a cave in Chile was used to establish the earliest date of human habitation in South America as 8700 years ago. What fraction of the ^{14}C initially present remained in the charcoal after 8700 years?

19.44. **Early Financial Records** For thousands of years Native Americans living along the north coast of Peru used knotted cotton strands called *quipu* (Figure P19.44) to record financial transactions and governmental actions. A particular quipu sample is 4800 years old. Compared with the fibers of cotton plants growing today, what is the ratio of carbon-14 to carbon-12 in the sample?

FIGURE P19.44

*19.45. **Rings in Sequoia Trees** Figure P19.45 is a close-up of the center of a giant sequoia tree cut down in 1891 in what is now Kings Canyon National Park. It contained 1342 annual growth rings. If samples of the tree were removed for radiocarbon dating today, what would be the difference in $^{14}C/^{12}C$ ratio in the innermost (oldest) ring compared with that ratio in the youngest ring?

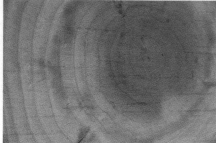

FIGURE P19.45

*19.46. **Dating Volcanic Eruptions** Geologists who study volcanoes can develop historical profiles of previous eruptions by determining the $^{14}C/^{12}C$ ratios of charred plant remains entrapped in old magma and ash flows. If the uncertainty in determining these ratios is 0.1%, could radiocarbon dating distinguish between debris from the eruptions of Mt. Vesuvius that occurred in the years 472 and 512? (*Hint*: Calculate the $^{14}C/^{12}C$ ratios for samples from the two dates.)

19.47. **Age of Mammoth Tusk** Figure P19.47 shows a carved mammoth tusk that was uncovered at an ancient campsite in the Ural Mountains in 2001. The $^{14}C/^{12}C$ ratio in the tusk was only 1.19% of that in modern elephant tusks. How old is the mammoth tusk?

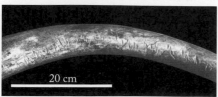

20 cm

FIGURE P19.47

19.48. **Destruction of Jericho** The Bible describes the Exodus as a period of 40 years that began with plagues in Egypt and ended with the destruction of Jericho. Archeologists seeking to establish the exact dates of these events have proposed that the plagues coincided with a huge eruption of the volcano Thera in the Aegean Sea.
   a. Radiocarbon dating suggests that the eruption occurred around 1360 BCE, although other records place the eruption of Thera in the year 1628 BCE. What is the percent difference in the ^{14}C decay rate in biological samples from these two dates?
   b. Radiocarbon dating of blackened grains from the site of ancient Jericho provides a date of 1315 BCE ± 13 years for the fall of the city. What is the $^{14}C/^{12}C$ ratio in the blackened grains compared with that of grain harvested last year?

Biological Effects of Radioactivity

Concept Review

19.49. What is the difference between a *level* of radioactivity and a *dose* of radioactivity?
19.50. What are some of the molecular effects of exposure to radioactivity?
19.51. Describe the dangers of exposure to radon-222.
19.52. **Food Safety** Periodic outbreaks of food poisoning from *E. coli*–contaminated meat have renewed the debate about irradiation as an effective treatment of food. In one newspaper article on the subject, the following statement appeared: "Irradiating food destroys bacteria by breaking apart their molecular structure." How would you improve or expand on this explanation?

Problems

19.53. **Radiation Exposure from Dental X-rays** Dental X-rays expose patients to about 5 μSv of radiation. Given an RBE of 1 for X-rays, how many grays of radiation does 5 μSv represent? For a 50-kg person, how much energy does 5 μSv correspond to?
*19.54. **Radiation Exposure at Chernobyl** Some workers responding to the explosion at the Chernobyl nuclear power plant were exposed to 5 Sv of radiation, killing many of them. If the exposure was primarily in the form of γ rays with an energy of 3.3×10^{-14} J and an RBE of 1, how many γ rays did an 80-kg person absorb?

*19.55. **Strontium-90 in Milk** In the years immediately following the explosion at the Chernobyl nuclear power plant, the concentration of ^{90}Sr in cow's milk in southern Europe was slightly elevated. Some samples contained as much as 1.25 Bq/L of ^{90}Sr radioactivity. The half-life of strontium-90 is 28.8 years.
   a. Write a balanced nuclear equation describing the decay of ^{90}Sr.
   b. How many atoms of ^{90}Sr are in a 200-mL glass of milk with 1.25 Bq/L of ^{90}Sr radioactivity?
   c. Why would strontium-90 be more concentrated in milk than other foods, such as grains, fruits, or vegetables?
*19.56. **Radium Watch Dials** If exactly 1.00 μg of ^{226}Ra was used to paint the glow-in-the-dark dial of a wristwatch made

in 1914, how radioactive is the watch today? Express your answer in microcuries and becquerels. The half-life of ^{226}Ra is 1.60×10^3 years.

19.57. In 1999 the U.S. Environmental Protection Agency set a maximum radon level for drinking water at 4.0 pCi per milliliter.
   a. How many decay events occur per second in a milliliter of water for this level of radon radioactivity?
   b. If the above radioactivity were due to decay of ^{222}Rn ($t_{1/2} = 3.8$ days), how many ^{222}Rn atoms would there be in 1.0 mL of water?

19.58. Death of a Spy A former Russian spy died from radiation sickness in 2006 after dining at a London restaurant where he apparently ingested polonium-210. The other people at his table did not suffer from radiation sickness, even though they were very near the radioactive tea the victim drank. Why were they not affected?

Medical Applications of Radionuclides

Concept Review

19.59. How does the selection of an isotope for radiotherapy relate to (a) its half-life, (b) its mode of decay, and (c) the properties of the products of decay?

19.60. Are the same radioactive isotopes likely to be used for both imaging and cancer treatment? Why or why not?

Problems

19.61. Predict the most likely mode of decay for the following isotopes used as imaging agents in nuclear medicine:
(a) ^{197}Hg (kidney); (b) ^{75}Se (parathyroid gland); (c) ^{18}F (bone).

19.62. Predict the most likely mode of decay for the following isotopes used as imaging agents in nuclear medicine:
(a) ^{133}Xe (cerebral blood flow); (b) ^{57}Co (tumor detection); (c) ^{51}Cr (red blood cell mass); (d) ^{67}Ga (tumor detection).

19.63. A 1.00-mg sample of ^{192}Ir was inserted into the artery of a heart patient. After 30 days, 0.756 mg remained. What is the half-life of ^{192}Ir?

19.64. In a treatment that decreases pain and reduces inflammation of the lining of the knee joint, a sample of dysprosium-165 with a radioactivity of 1100 counts per second was injected into the knee of a patient suffering from rheumatoid arthritis. After 24 h, the radioactivity had dropped to 1.14 counts per second. Calculate the half-life of ^{165}Dy.

19.65. Study of Tourette's Syndrome Tourette's syndrome is a condition whose symptoms include sudden movements and vocalizations. Iodine isotopes are used in brain imaging of people with Tourette's syndrome. To study the uptake and distribution of iodine in cells, researchers treated mammalian brain cells in culture with a solution containing ^{131}I with an initial radioactivity of 108 counts per minute. The cells were removed after 30 days, and the remaining solution was found to have a radioactivity of 4.1 counts per minute. Did the brain cells absorb any ^{131}I ($t_{1/2} = 8.1$ days)?

19.66. Mercury Test of Kidney Function A patient is administered mercury-197 to evaluate kidney function. Mercury-197 has a half-life of 65 h. What fraction of an initial dose of mercury-197 remains after 6 days?

19.67. Carbon-11 is an isotope used in positron emission tomography and has a half-life of 20.4 min. How long will it take for 99% of the ^{11}C injected into a patient to decay?

19.68. Leukemia Treatment Using Sodium Sodium-24 is used to treat leukemia and has a half-life of 15 h. In a patient injected with a salt solution containing sodium-24, what percentage of the ^{24}Na remains after 48 h?

*19.69. Boron Neutron-Capture Therapy In boron neutron-capture therapy (BNCT), a patient is given a compound containing ^{10}B that accumulates inside cancer tumors. Then the tumors are irradiated with neutrons, which are absorbed by ^{10}B nuclei. The product of neutron capture is an unstable form of ^{11}B that undergoes α decay to ^{7}Li.
   a. Write a balanced nuclear equation for the neutron absorption and α decay process.
   b. Calculate the energy released by each nucleus of boron-10 that captures a neutron and undergoes α decay, given the following masses of the particles in the process: ^{10}B (10.0129 amu), ^{7}Li (7.01600 amu), ^{4}He (4.00260 amu), and ^{1}n (1.00866 amu).
   c. Why is the formation of a nuclide that undergoes α decay a particularly effective cancer therapy?

19.70. Balloon Angioplasty and Arteriosclerosis Balloon angioplasty is a common procedure for unclogging arteries in patients suffering from arteriosclerosis. Iridium-192 therapy is being tested as a treatment to prevent reclogging of the arteries. In the procedure, a thin ribbon containing pellets of ^{192}Ir is threaded into the artery. The half-life of ^{192}Ir is 74 days. How long will it take for 99% of the radioactivity from 1.00 mg of ^{192}Ir to disappear?

Nuclear Fission

Concept Review

19.71. How is the rate of energy release controlled in a nuclear reactor?

19.72. How does a breeder reactor create fuel and energy at the same time?

*19.73. Why are neutrons always by-products of the fission of the most massive nuclides? (*Hint:* Look closely at the neutron-to-proton ratios shown in Figure 19.2.)

19.74. Seaborgium (Sg, element 106) is prepared by the bombardment of curium-248 with neon-22, which produces two isotopes, ^{265}Sg and ^{266}Sg. Write balanced nuclear reactions for the formation of both isotopes. Are these reactions better described as fusion or fission processes?

Problems

19.75. The fission of uranium produces dozens of isotopes. For each of the following fission reactions, determine the identity of the unknown nuclide:
   a. $^{235}U + {}_0^1n \rightarrow {}^{96}Zr + ? + 2\,{}_0^1n$
   b. $^{235}U + {}_0^1n \rightarrow {}^{99}Nb + ? + 4\,{}_0^1n$
   c. $^{235}U + {}_0^1n \rightarrow {}^{90}Rb + ? + 3\,{}_0^1n$

19.76. For each of the following fission reactions, determine the identity of the unknown nuclide:

a. $^{235}U + {}^{1}_{0}n \rightarrow {}^{137}I + ? + 2\,{}^{1}_{0}n$

b. $^{235}U + {}^{1}_{0}n \rightarrow {}^{137}Cs + ? + 3\,{}^{1}_{0}n$

c. $^{235}U + {}^{1}_{0}n \rightarrow {}^{141}Ce + ? + 2\,{}^{1}_{0}n$

19.77. For each of the following fission reactions, determine the identity of the unknown nuclide:

a. $^{235}U + {}^{1}_{0}n \rightarrow {}^{131}I + ? + 2\,{}^{1}_{0}n$

b. $^{235}U + {}^{1}_{0}n \rightarrow {}^{103}Ru + ? + 3\,{}^{1}_{0}n$

c. $^{235}U + {}^{1}_{0}n \rightarrow {}^{95}Zr + ? + 3\,{}^{1}_{0}n$

19.78. For each of the following fission reactions, determine the identity of the unknown nuclide:

a. $^{235}U + {}^{1}_{0}n \rightarrow {}^{147}Pm + ? + 2\,{}^{1}_{0}n$

b. $^{235}U + {}^{1}_{0}n \rightarrow {}^{94}Kr + ? + 2\,{}^{1}_{0}n$

c. $^{235}U + {}^{1}_{0}n \rightarrow {}^{95}Sr + ? + 3\,{}^{1}_{0}n$

Nuclear Fusion and the Quest for Clean Energy

Concept Review

19.79. In what ways are the fusion reactions that formed α particles during primordial nucleosynthesis different from those that fuel our Sun today?

19.80. How are the fusion reactions that are the basis for power production in the tokamak described in Section 19.9 different from those that power our Sun?

Problems

19.81. All of the following fusion reactions produce ^{28}Si. Calculate the energy released in each reaction from the masses of the isotopes: ^{2}H (2.0146 amu), ^{4}He (4.00260 amu), ^{10}B (10.0129 amu), ^{12}C (12.00000 amu), ^{14}N (14.00307 amu), ^{16}O (15.99491 amu), ^{24}Mg (23.98504 amu), ^{28}Si (27.97693 amu).

a. $^{14}N + {}^{14}N \rightarrow {}^{28}Si$

b. $^{10}B + {}^{16}O + {}^{2}H \rightarrow {}^{28}Si$

c. $^{16}O + {}^{12}C \rightarrow {}^{28}Si$

d. $^{24}Mg + {}^{4}He \rightarrow {}^{28}Si$

19.82. All of the following fusion reactions produce ^{32}S. Calculate the energy released in each reaction from the masses of the isotopes: ^{4}He (4.00260 amu), ^{6}Li (6.01512 amu), ^{12}C (12.00000 amu), ^{14}N (14.00307 amu), ^{16}O (15.99491 amu), ^{24}Mg (23.98504 amu), ^{28}Si (27.97693 amu), ^{32}S (31.97207 amu).

a. $^{16}O + {}^{16}O \rightarrow {}^{32}S$

b. $^{28}Si + {}^{4}He \rightarrow {}^{32}S$

c. $^{14}N + {}^{12}C + {}^{6}Li \rightarrow {}^{32}S$

d. $^{24}Mg + 2\,{}^{4}He \rightarrow {}^{32}S$

19.83. Tokamak Radiochemistry How much energy is released per nucleus of tritium produced during the following reactions?

a. $^{1}_{0}n + {}^{6}_{3}Li \rightarrow {}^{4}_{2}He + {}^{3}_{1}H$

b. $^{1}_{0}n + {}^{7}_{3}Li \rightarrow {}^{4}_{2}He + {}^{3}_{1}H + {}^{1}_{0}n$

19.84. It has been proposed that electrical power production in the future might be based on the fusion of deuterium to helium-4.

a. Write a radiochemical equation describing the reaction (assume that ^{4}He is the only product).

b. Calculate how much energy is released during the formation of 1 mole of ^{4}He.

Additional Problems

19.85. Powering a Starship Thirty years before the creation of antihydrogen, television producer Gene Roddenberry (1921–1991) proposed to use this form of antimatter to fuel the powerful "warp" engines of the fictional starship *Enterprise*.

a. Why would antihydrogen have been a particularly suitable fuel?

b. Describe the challenges of storing such a fuel on a starship.

19.86. Accuracy in Labeling Tiny concentrations of radioactive tritium ($^{3}_{1}H$) occur naturally in rain and groundwater. The half-life of $^{3}_{1}H$ is 12 years. Assuming that tiny concentrations of tritium can be determined accurately, could the isotope be used to determine whether a bottle of wine with the year 1969 on its label actually contained wine made from grapes that were grown in 1969? Explain your answer.

19.87. The energy released during the fission of ^{235}U is about 3.2×10^{-11} J per atom of the isotope. Compare this quantity of energy with that released by the fusion of four hydrogen atoms to make an atom of helium-4:

$$4\,{}^{1}_{1}H \rightarrow {}^{4}_{2}He + 2\,{}^{0}_{1}\beta$$

Assume that the positrons are annihilated in collisions with electrons so that the masses of the positrons are converted into energy. In your comparison, express the energies released by the fission and fusion processes in joules per nucleon for ^{235}U and ^{4}He, respectively.

19.88. How much energy is required to remove a neutron from the nucleus of an atom of carbon-13 (mass = 13.00335 amu)? (*Hint*: The mass of an atom of carbon-12 is exactly 12.00000 amu.)

19.89. Smoke Detectors Americium-241 ($t_{1/2}$ = 433 yr) is used in smoke detectors. The α particles from this isotope ionize nitrogen and oxygen in the air, creating an electric current. When smoke is present, the current decreases, setting off the alarm.

a. Does a smoke detector bear a closer resemblance to a Geiger counter or to a scintillation counter?

b. How long will it take for the radioactivity of a sample of ^{241}Am to drop to 1% of its original radioactivity?

c. Why are smoke detectors containing ^{241}Am safe to handle without protective equipment?

19.90. Colorectal Cancer Treatment Cancer therapy with radioactive rhenium-188 shows promise in patients with colorectal cancer.

a. Write the symbol for rhenium-188 and determine the number of neutrons, protons, and electrons.

b. Are most rhenium isotopes likely to have fewer neutrons than rhenium-188?

c. The half-life of rhenium-188 is 17 h. If it takes 30 min to bind the isotope to an antibody that delivers the rhenium to the tumor, what percentage of the rhenium remains after binding to the antibody?

d. The effectiveness of rhenium-188 is thought to result from penetration of β particles as deep as 8 mm into the tumor. Why wouldn't an α emitter be more effective?

e. Using an appropriate reference text, such as the *CRC Handbook of Chemistry and Physics*, pick out the two most abundant isotopes of rhenium. List their natural abundances and explain why the one that is radioactive decays by the pathway that it does.

19.91. Synthesis of a New Element In 2006 an international team of scientists confirmed the synthesis of a total of three atoms of $^{294}_{118}$Og in experiments run in 2002 and 2005. They had bombarded a ^{249}Cf target with ^{48}Ca nuclei.

a. Write a balanced nuclear equation describing the synthesis of $^{294}_{118}$Og.

b. The synthesized isotope of Og undergoes α decay ($t_{1/2} = 0.9$ ms). What nuclide is produced by the decay process?

c. The nuclide produced in part (b) also undergoes α decay ($t_{1/2} = 10$ ms). What nuclide is produced by this decay process?

d. The nuclide produced in part (c) also undergoes α decay ($t_{1/2} = 0.16$ s). What nuclide is produced by this decay process?

e. If you had to select an element that occurs in nature and that has physical and chemical properties similar to Og, which element would it be?

*19.92. Consider the following decay series:

$$A\ (t_{1/2} = 4.5\text{ s}) \rightarrow B\ (t_{1/2} = 15.0\text{ days}) \rightarrow C$$

If we start with 10^6 atoms of A, how many atoms of A, B, and C are there after 30 days?

19.93. Which element in the following series will be present in the greatest amount after one year?

$$^{214}_{83}\text{Bi} \xrightarrow{\alpha} {}^{210}_{81}\text{Tl} \xrightarrow{\beta} {}^{210}_{82}\text{Pb} \xrightarrow{\beta} {}^{210}_{83}\text{Bi} \rightarrow$$
$$t_{1/2} = \quad 20\text{ min}\ \ 1.3\text{ min}\ \ 20\text{ yr}\quad 5\text{ d}$$

*19.94. **Dating Cave Paintings** Cave paintings in Gua Saleh Cave in Borneo have been dated by measuring the amount of ^{14}C in calcium carbonate that formed over the pigments used in the paint. The source of the carbonate ion was atmospheric CO_2.

a. What is the ratio of the ^{14}C radioactivity in calcium carbonate that formed 9900 years ago to that in calcium carbonate formed today?

b. The archeologists also used a second method, uranium–thorium dating, to confirm the age of the paintings by measuring trace quantities of these elements present as contaminants in the calcium carbonate. Shown below are two candidates for the U–Th dating method. Which isotope of uranium do you suppose was chosen? Explain your answer.

$$^{235}_{92}\text{U} \quad\rightarrow\quad {}^{231}_{90}\text{Th} \quad\rightarrow\quad {}^{231}_{91}\text{Pa} \quad\rightarrow$$
$$t_{1/2} = \quad 7.04 \times 10^8\text{ yr} \quad 25.6\text{ h} \quad 3.25 \times 10^4\text{ yr}$$

$$^{234}_{92}\text{U} \quad\rightarrow\quad {}^{230}_{90}\text{Th} \quad\rightarrow\quad {}^{226}_{88}\text{Pa} \quad\rightarrow$$
$$t_{1/2} = \quad 2.44 \times 10^5\text{ yr} \quad 7.7 \times 10^4\text{ h} \quad 1600\text{ yr}$$

19.95. The synthesis of new elements and specific isotopes of known elements in linear accelerators involves the fusion of smaller nuclei.

a. An isotope of platinum can be prepared from nickel-64 and tin-124. Write a balanced equation for this nuclear reaction. (You may assume that no neutrons are ejected in the fusion reaction.)

b. Substituting tin-132 for tin-124 increases the rate of the fusion reaction 10 times. Which isotope of Pt is formed in this reaction?

19.96. Radon in Drinking Water A sample of drinking water collected from a suburban Boston municipal water system in 2002 contained 0.5 pCi/L of radon. Assume that this level of radioactivity was due to the decay of ^{222}Rn ($t_{1/2} = 3.8$ days).

a. What was the level of radioactivity (Bq/L) of this nuclide in the sample?

b. How many decay events per hour would occur in 2.5 L of the water?

19.97. Stone Age Skeletons The discovery of six skeletons in an Italian cave at the beginning of the 20th century was considered a significant find in Stone Age archaeology. The age of these bones has been debated. The first attempt at radiocarbon dating indicated an age of 15,000 years. Redetermination of the age in 2004 indicated an older age for two bones, between 23,300 and 26,400 years. What is the ratio of ^{14}C in a sample 15,000 years old to one 25,000 years old?

*19.98. Atmospheric testing of nuclear weapons in the 1950s and '60s produced an increase in the concentration of carbon-14 in the atmosphere. Use one or more balanced nuclear equations to explain how this could have happened.

19.99. Dating Prehistoric Bones In 1997 anthropologists uncovered three partial skulls of prehistoric humans in the Ethiopian village of Herto. From the amount of ^{40}Ar in the volcanic ash in which the remains were buried, their age was estimated at between 154,000 and 160,000 years.

a. ^{40}Ar is produced by the decay of ^{40}K ($t_{1/2} = 1.28 \times 10^9$ yr). Propose a decay mechanism for ^{40}K to ^{40}Ar.

b. Why did the researchers choose ^{40}Ar rather than ^{14}C as the isotope for dating these remains?

*19.100. **Biblical Archeology** The Old Testament describes the construction of the Siloam Tunnel, used to carry water into Jerusalem under the reign of King Hezekiah (727–698 BCE). An inscription on the tunnel has been interpreted as evidence that the tunnel was not built until 200–100 BCE. ^{14}C dating (in 2003) indicated a date close to 700 BCE. What is the ratio of ^{14}C in a wooden object made in 100 BCE to one made from the same kind of wood in 700 BCE?

20

Organic and Biological Molecules

The Compounds of Life

SIMILARITY, DIVERSITY, INTERDEPENDENCE
The abundant diversity of life on Earth arises from different combinations of 40 or 50 small molecules. These basic building blocks of life are responsible for the similarity, diversity, and interdependence of all forms of life.

PARTICULATE **REVIEW**

Finding Functional Groups

In Chapter 20 we explore the relationship between structure and function for carbon-containing compounds.

- Name the functional group present in each molecule shown here.

- Which compound functions as a Brønsted–Lowry acid in water?

- Which compound functions as a Brønsted–Lowry base in water?

- Which functional group is formed when the Brønsted–Lowry acid and base react?

    (Review Sections 2.8, 4.5, and 15.2 if you need help.)

(Answers to Particulate Review questions are in the back of the book.)

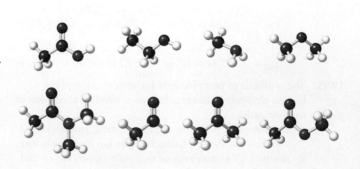

Protonated or Deprotonated?

Glutamic acid is used in the biosynthesis of proteins and contains three functional groups, as shown here. As you read Chapter 20, look for ideas that will help you answer these questions:

- What are the three functional groups in glutamic acid?

- Which functional groups are deprotonated at physiological pH? Which ones are protonated at physiological pH?

- Does glutamic acid exist as a negatively charged, neutral, or positively charged molecule at physiological pH?

Learning Outcomes

LO1 Identify isomers and chiral molecules
Sample Exercises 20.1, 20.2, 20.3

LO2 Interpret titration curves for amino acids and use them to define pK_a and pI values
Sample Exercise 20.4

LO3 Name and draw the structures of small peptides and determine the peptide sequence of a polypeptide from its fragments
Sample Exercises 20.5, 20.6

LO4 Describe the four levels of protein structure and how intermolecular forces and covalent bonds stabilize these structures

LO5 Describe the molecular structures of simple sugars and the types of bonds linking them in polysaccharides, and explain how these compounds are used as energy sources and for energy storage

LO6 Describe the molecular structure and the physical and chemical properties of lipids
Sample Exercise 20.7

LO7 Describe the structures of DNA and RNA and how they function together to translate genetic information
Sample Exercise 20.8

20.1 Molecular Structure and Functional Groups

For all the stunning diversity of the biosphere, from single-cell organisms to elephants, whales, and giant redwoods, all life forms consist of substances made from only about 40 or 50 different small molecules. This small number of starting materials can link together to form large molecules that display astonishing variations of structure and function. In this chapter we explore the composition, structure, and function of molecules associated with life.

The great bulk of these compounds are organic molecules containing C—C and C—H bonds. The designation *organic* for carbon-containing compounds was first applied to substances produced by living organisms, but that definition has been broadened for two reasons. First, scientists have learned how to synthesize many materials previously thought to be the products only of living systems. Second, chemists also synthesize many carbon-based materials that have *never* been produced by living systems. Today the study of *organic chemistry* encompasses the chemistry of all compounds containing carbon–carbon and carbon–hydrogen bonds, regardless of their origin.

In contrast to organic molecules, **biomolecules** retain a link to life: a biomolecule is any molecule produced by a living organism. This definition encompasses the four classes of large molecules that we describe in this chapter—proteins, carbohydrates, lipids, and nucleic acids—as well as small molecules that result from biochemical processes such as metabolism. **Biochemistry**, or biological chemistry, is the study of the chemical processes taking place in living organisms, with a focus on understanding how biomolecules participate in the reactions occurring in living cells. In this chapter we concentrate on the composition and structure of organic molecules and biomolecules that compose our bodies and the other organisms with which we share this planet, as well as the foods we eat and the medicines we take. Many of these molecules are large and complex, but our knowledge of the behavior of small molecules can serve as a framework for understanding the functions and reactions of these compounds in living systems.

CONNECTION We first introduced organic compounds and their functional groups in Section 2.8.

biomolecule any molecule produced by a living organism.

biochemistry the study of chemical processes taking place in living organisms.

An important aspect of studying the biochemical processes associated with life is finding out what happens when something goes wrong. It has been estimated that 70% of inherited diseases in people are caused by the absence of particular proteins or some other defect that causes proteins to function improperly. For example, the malformation or the total absence of the protein dystrophin causes several debilitating diseases collectively referred to as muscular dystrophy (MD). In mild forms of MD, misshapen molecules of dystrophin are unable to build muscle fibers sufficient to function normally. As a result, muscle tissue wastes away and the patient becomes physically disabled. In Duchenne MD, a severe form of the disease, functional dystrophin is absent, and the disability often results in early death. Knowledge of the mechanisms that produce dystrophin has led to promising therapies. Similar approaches based on knowledge of mechanisms have resulted in the development of new drugs and cell therapies for other inherited and contagious diseases. Scientists frequently use information about how biomolecules behave in healthy individuals to develop drugs and treatments for all manner of medical problems.

Families Based on Functional Groups

To understand biomolecules and their reactions, we need to review some fundamental concepts and develop additional ideas about the composition, structure, and reactivity of organic molecules. Much of the variety in the chemistry of carbon arises from the carbon atom's ability to form covalent bonds to other carbon atoms. These carbon atoms, in turn, can bond to more carbon atoms or atoms of other elements, yielding compounds containing a few atoms or many thousands of atoms. Additional variety is introduced because these molecules may contain a multitude of linear and branched structures as well as rings. In addition to carbon and hydrogen, organic compounds can also contain nonmetallic heteroatoms such as nitrogen, oxygen, sulfur, phosphorus, or the halogens.

CONNECTION We first encountered heteroatoms and functional groups in Section 2.8.

Learning to understand the structures and reactivity of millions of organic compounds requires some organizing concepts. One organizing principle is based on the chemical composition and structure of subunits within molecules. Chemists group organic compounds into families on the basis of *functional groups*, subunits in a molecule that confer particular chemical and physical properties.

We have previously encountered many of the most important functional groups in organic chemistry in earlier chapters. These functional groups are reviewed in Table 20.1. Remember that when discussing functional groups, the convention is to use R to represent the entire molecule except the functional group. All the functional groups in Table 20.1 appear in biologically important molecules, and being able to discuss the structure and properties of biomolecules based on the functional groups within them is a fundamental skill.

A second organizing principle for organic molecules and biomolecules is based on molecular size. In Section 12.6 we discussed the structure of polymers and the influence of molecular size on the physical properties of large molecules. All of the principles introduced with respect to polymeric materials in Chapter 12 are pertinent to the discussion of large *biopolymers* in this chapter.

TABLE 20.1 Functional Groups of Organic Compounds

Name	Structural Formula of Group[a]	Example and Name	
Alkane (Section 2.8)	R—H	$CH_3CH_2CH_3$	Propane
Alkene (Section 2.8)	C=C	H₂C=CH₂	Ethylene (ethene)
Alkyne (Section 2.8)	—C≡C—	H—C≡C—H	Acetylene (ethyne)
Alcohol (Section 2.8)	R—OH	H_3C—OH	Methanol
Carboxylic acid (Section 4.5)	R—C(=O)—OH	H_3C—C(=O)—OH	Acetic acid
Amine (Section 4.5)	R—NH₂ R—NHR′ R—NRR′	H_3C—NH₂	Methylamine
Aldehyde (Section 8.2)	R—C(=O)—H	H_3C—C(=O)—H	Acetaldehyde
Ketone (Section 9.5)	R—C(=O)—R′	H_3C—C(=O)—CH₃	Acetone
Aromatic (Section 9.5)	e.g., (benzene ring)	(benzene ring)	Benzene
Ether (Section 10.4)	R—O—R′	H_3C—O—CH₃	Dimethyl ether
Ester (Section 12.6)	R—C(=O)—OR′	H_3C—C(=O)—OCH₃	Methyl acetate
Amide (Section 12.6)	R—C(=O)—NH₂[b]	H_3C—C(=O)—NH₂[b]	Acetamide

[a]In compounds with two R groups, the two may be the same (R = R′) or they may be different (R ≠ R′).
[b]One or both H atoms may be replaced by R– groups: –NHR or NRR′.
Note: Functional groups are highlighted in yellow.

20.2 Organic Molecules, Isomers, and Chirality

At the beginning of Chapter 3 we saw that Friedrich Wöhler's discovery of urea, H_2NCONH_2, occurred during his attempt to synthesize ammonium cyanate, NH_4NCO (Figure 20.1). The chemical formula of both structures can be written CH_4N_2O, but because their structures are not the same, they represent different compounds with different physical and chemical properties. We saw that such compounds, which have the same number and same kinds of atoms but differ in how those atoms are arranged, are called *isomers*. The existence of isomers is partly responsible for the enormous number and wide variety of organic compounds in the world.

To explore isomers further, let's revisit hydrocarbons, the organic compounds composed exclusively of carbon atoms and hydrogen atoms. We have seen that hydrocarbons in which each carbon atom forms four single bonds to four other atoms are called alkanes. In Table 5.5 we compared the properties of two four-carbon alkanes: butane, $CH_3CH_2CH_2CH_3$, and 2-methylpropane, $CH_3CH(CH_3)CH_3$. The molecular formula of both structures is C_4H_{10}, but their atoms are connected to each other in different ways, making them **constitutional isomers** (also called *structural isomers*).

When we look at the condensed structures or carbon-skeleton structures of two alkanes, we often wish to determine whether they represent two compounds with different molecular formulas, a single compound, or two constitutional isomers. To make this comparison, we begin by translating each structure into a molecular formula. If the molecular formulas are different, then the structures represent two compounds. If the molecular formulas are the same, then we must compare the bonds in the two structures; that is, we compare the way the atoms are connected to one another. There are two ways we can do this:

1. Simply inspecting the two structures may reveal that they are identical; however, we may have to reverse or rotate one of the structures. For example, structures (1) and (2) below may seem to be constitutional isomers, but rotating structure (2) by 180° shows that it is the same as structure (1). The two represent the same compound because they have the same composition and the same structure (bonds): a four-carbon *main chain* with a one-carbon *side chain* connected to the carbon atom next to a terminal carbon.

(1)   (2)   180° rotation   After rotation

2. A second method to determine whether two structures are identical involves drawing the structures to be compared with the longest chain horizontal. The longest chains in structures (3) and (4) below are highlighted in yellow. Number the carbons in the longest chain so that the branches have the lowest possible numbers, and then check to see whether the same side chains are attached at the same numbered positions along the longest

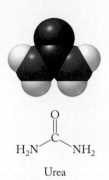

FIGURE 20.1 Urea and ammonium cyanate can be written with the same chemical formula, CH_4N_2O.

Urea

CONNECTION We first mentioned isomers in Section 3.1 in the context of Wöhler's discovery of two compounds with the same composition.

CONNECTION In Section 2.8 we defined hydrocarbons, alkanes, alkenes, and alkynes. In Section 5.9 we explored straight-chain and branched-chain hydrocarbons.

constitutional isomers molecules with the same chemical formula but different connections between the atoms.

chain. In this example, the longest chain in both structures is 5 carbon atoms long, and both chains have methyl (–CH$_3$) groups on carbon 2 and carbon 3, so they are identical. Note that it is acceptable to number some chains right-to-left, as in structure (3), or left-to-right, as in structure (4).

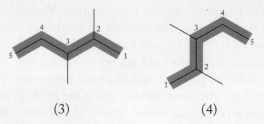

(3)   (4)

SAMPLE EXERCISE 20.1 Recognizing Constitutional Isomers   **LO1**

Do the two structures in each set describe the same compound, constitutional isomers, or compounds with different molecular formulas?

a.   (CH$_3$)$_2$CHCH$_2$CH(CH$_3$)$_2$

b.

c.

Collect and Organize We have structures showing connectivity of the atoms and from which we can determine molecular formulas. Identical compounds have the same molecular formula and the same connectivity of the atoms. Constitutional isomers have the same molecular formula but different connectivity.

Analyze In each pair, we check first to see whether the molecular formulas are the same. That means we count the number of carbon atoms and hydrogen atoms. If the molecular formulas are the same, we may have either one compound drawn two ways or a pair of constitutional isomers. If the carbon skeletons are the same in any pair, the drawings represent the same hydrocarbon.

Solve
a. The molecular formulas are the same: C$_7$H$_{16}$. Converting the condensed structure to a carbon-skeleton structure gives us

(CH$_3$)$_2$CHCH$_2$CH(CH$_3$)$_2$ →

This structure is identical to the carbon-skeleton structure in this set. Therefore the condensed structure and carbon-skeleton structure represent the same molecule.
b. Both molecules in this set have the molecular formula C$_8$H$_{18}$. If we rotate the second structure 180° about a vertical axis, we get the first structure, so the two compounds are identical.

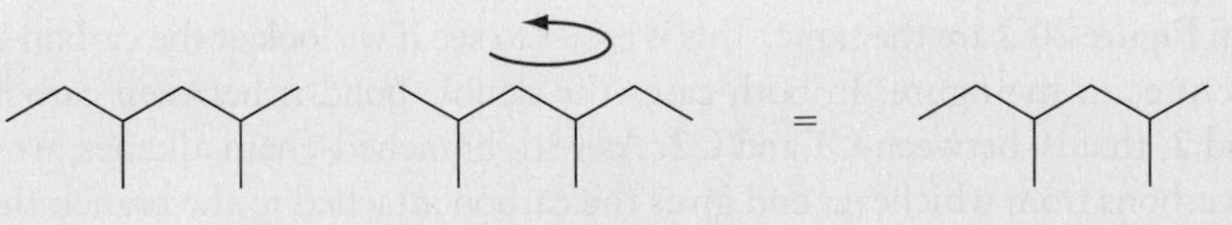

(1)   (2) 180° rotation   After rotation

c. The left structure contains nine carbon atoms, but the right structure contains only eight. Therefore the two structures represent different compounds.

Think About It Just because two compounds have the same molecular formula, they are not necessarily identical. We must determine whether they are constitutional isomers.

Practice Exercise Do the two structures in each set describe the same compound, constitutional isomers, or compounds with different molecular formulas?

a.

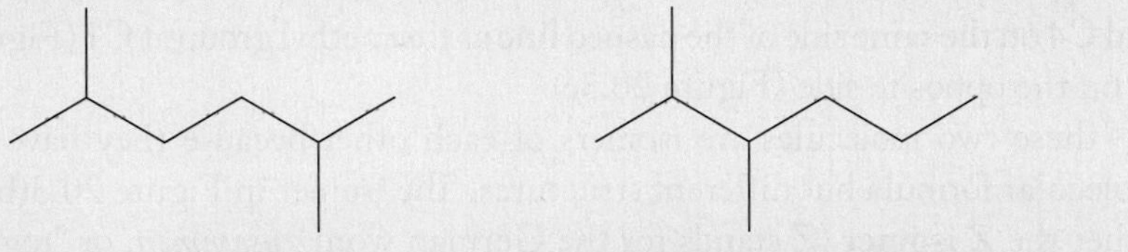

b.

c.

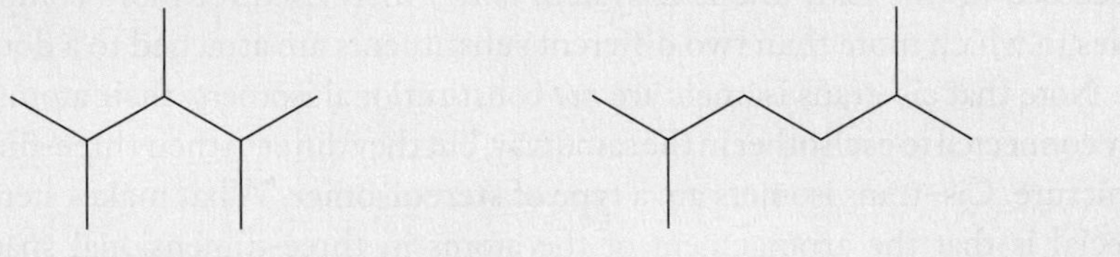

(Answers to Practice Exercises are in the back of the book.)

Hydrocarbons containing one or more carbon–carbon double bonds (alkenes) or triple bonds (alkynes) may also have branches, and their structures are treated just like alkanes to determine their identity. However, another situation arises with alkenes that involves different orientations of groups about the double bond. For example, consider the straight-chain isomers of the alkene that contains five carbon atoms and one double bond (Figure 20.2).

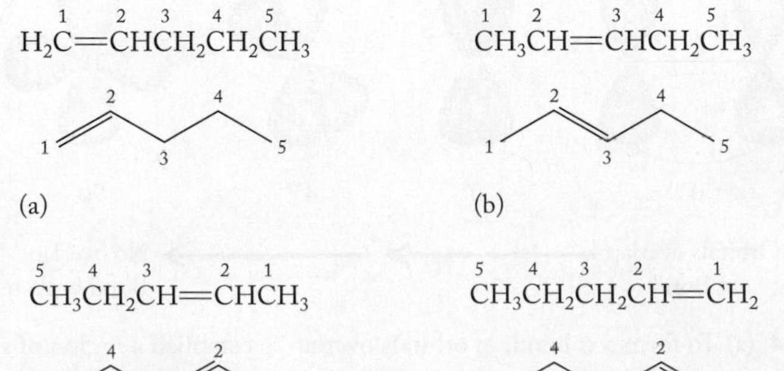

$$\overset{1}{H_2}C\overset{2}{=}\overset{3}{CH}\overset{4}{CH_2}\overset{5}{CH_2}CH_3$$

(a)

$$\overset{1}{CH_3}\overset{2}{CH}\overset{3}{=}\overset{4}{CH}\overset{5}{CH_2}CH_3$$

(b)

$$\overset{5}{CH_3}\overset{4}{CH_2}\overset{3}{CH}\overset{2}{=}\overset{1}{CH}CH_3$$

(c)

$$\overset{5}{CH_3}\overset{4}{CH_2}\overset{3}{CH_2}\overset{2}{CH}\overset{1}{=}CH_2$$

(d)

FIGURE 20.2 Four possible constitutional isomers of pentene, C_5H_{10}. Notice that structures (a) and (d) are identical, as are (b) and (c).

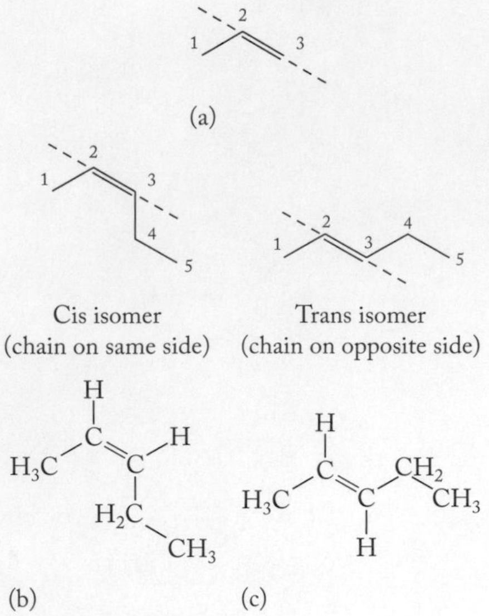

(a)

Cis isomer
(chain on same side)

Trans isomer
(chain on opposite side)

(b)   (c)

FIGURE 20.3 (a) The first three atoms of a carbon chain with a double bond between C2 and C3. (b) The chain continues on the same side of the double bond as C1 in the cis isomer. (c) The chain continues on the opposite side of the double bond from C1 in the trans isomer.

CONNECTION In Section 5.3 we introduced rotations about bonds as one of the types of motion molecules experience as part of their overall kinetic energy.

cis isomer (also called **Z isomer**) a molecule with two like groups (such as two R groups or two hydrogen atoms) on the same side of the molecule.

trans isomer (also called **E isomer**) a molecule with two like groups (such as two R groups or two hydrogen atoms) on opposite sides of the molecule.

stereoisomers isomers created by differences in the orientations of the bonds in molecules.

optical activity a property of a compound that refers to the rotation of a beam of plane-polarized light when it passes through a solution of the compound.

optical isomers molecules that are not superimposable on their mirror images.

Applying the second method we used for alkanes, we see that structures (a) and (d) in Figure 20.2 are the same. This is easier to see if we look at the carbon-skeleton structures in the figure. In both cases the double bond is between carbon atoms 1 and 2, that is, between C1 and C2. As with branched-chain alkanes, we number the carbons from whichever end gives the carbon attached to the branch the lowest number. The same holds for functional groups, as shown here, where the carbons in structure (d) must be numbered from right to left. Structures (b) and (c) are equivalent to each other and are constitutional isomers of (a) and (d) because they have the same chemical formula but their double bond is in a different location.

Drawing the carbon-skeleton structures of (b) and (c) in Figure 20.2, however, presents us with a new situation. After we draw the first three atoms of structure (b), we see that we have two options for how to orient the rest of the molecule with respect to the double bond. If we draw a straight dashed line through the double bond, as in Figure 20.3(a), we can place the bond between C3 and C4 on the same side of the dashed line as the methyl group at C1 (Figure 20.3b) or on the opposite side (Figure 20.3c).

These two molecules are isomers of each other because they have the same molecular formula but different structures. The isomer in Figure 20.3(b) is called either the **Z isomer** (Z stands for the German word *zusammen*, or "together") or the **cis isomer** (*cis* is Latin for "on this side"), which in this case translates to "the methyl group and the chain after the double bond are both *together* or *on this side* of the structure." The isomer in Figure 20.3(c) is called either the **E isomer** (E for *entgegen*, or "opposite") or **trans isomer** (*trans* is Latin for "across"). The system of naming using cis and trans is in wide use and is sufficient for the simple molecules discussed in this text. The *E/Z* system is routinely used for more complex molecules in which more than two different substituents are attached to a double bond.

Note that cis–trans isomers are *not* constitutional isomers: their atoms are actually connected to each other in the same way, but they differ in their three-dimensional structure. Cis–trans isomers are a type of **stereoisomer**. What makes stereoisomers special is that the arrangement of the atoms in three-dimensional space differs. That is why the prefix "stereo-" is used in the name; it means we need to consider the three-dimensional shape of the molecule.

The stereoisomers in Figure 20.3 exist because there is no free rotation about the double bond. In Chapter 9 we learned that a double bond forms from the overlap of two unhybridized *p* orbitals on adjacent carbon atoms. As Figure 20.4 shows, if the carbon atoms joined in a double bond were to rotate freely, they

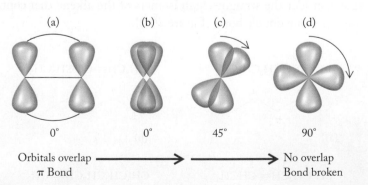

(a)	(b)	(c)	(d)
0°	0°	45°	90°

Orbitals overlap ⟶ No overlap
π Bond   Bond broken

FIGURE 20.4 (a) To form a π bond, p_z orbitals overlap to establish a region of shared electron density above and below the plane of the C—C bond. (b) If you look down the carbon–carbon bond axis, the orbitals line up. (c) If you rotate one carbon atom while keeping the other fixed, the orbitals are no longer parallel and do not overlap. (d) If you rotate one of the two bonded C atoms far enough, the π bond breaks.

would have to twist about the bond axis, eliminating the orbital overlap and breaking the bond. Breaking a π bond in 1 mole of an alkene requires about 290 kJ of energy, and that much energy is not available to the molecules at room temperature. This situation gives rise to restricted rotation about a carbon–carbon double bond and to the existence of stereoisomers.

CONCEPT TEST

Which of the following alkenes has cis and trans isomers?

a. CH_2=$CHCH_2CH_3$

b. CH=CH with CH_2 bridging

c. $(CH_3)_2C$=$C(CH_3)_2$

d. $(CH_3)_2C$=CH_2

e. CH_3CH=$CHCH_3$

(Answers to Concept Tests are in the back of the book.)

Chirality and Optical Activity

Before we launch our discussion of the classes of compounds important in living systems, let's revisit a physical property that characterizes many molecules that are associated with building and maintaining life on Earth. The physical property is called *chirality*, which we introduced in Chapter 9 when we talked about the three-dimensional shapes of molecules, and a tangible measure of chirality is called **optical activity**. As a review, let's look at some familiar items that illustrate the basic principles of chirality and then turn to some important molecules in living systems that have this characteristic. A key point to understand is what is meant by an object being superimposable on its mirror image.

Figure 20.5 illustrates the concept of chirality by using two familiar items. The reflection in a mirror of an object like a plain coffee mug that is *achiral* (not chiral) is an image that can be superimposed on the original object. Your two hands, however, are mirror images that cannot be superimposed. Molecules that have this same property—molecules that cannot be superimposed on their mirror images—are chiral.

Chiral molecules exist as **optical isomers**; a molecule that has one chiral center exists as two optical isomers. The term "optical" refers to the ways these compounds interact with a special kind of light called *plane-polarized* light (Figure 20.6). In plane-polarized light, the electric fields that compose the beam

FIGURE 20.5 A plain coffee mug is superimposable on its mirror image and is achiral. The mirror image of your left hand is your right hand, and the two are not superimposable.

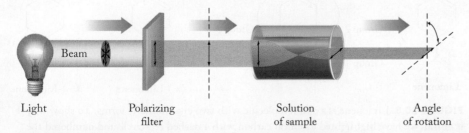

FIGURE 20.6 A beam of plane-polarized light contains an electric field that oscillates in only one direction. The plane of oscillation rotates if the beam passes through a solution of one enantiomer of an optically active compound. The (+) enantiomer causes the beam to rotate clockwise; the (−) enantiomer causes the beam to rotate counterclockwise.

CONNECTION In Section 9.6 we defined chiral molecules as being not superimposable on their mirror images.

CHEMTOUR
Chiral Centers

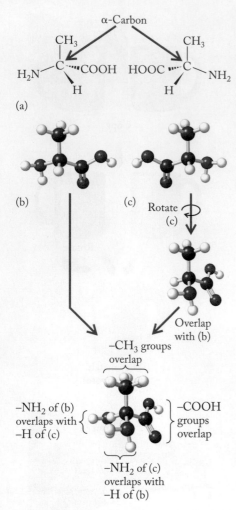

(a)

(b)   (c)   Rotate
              (c)

Overlap
with (b)

–CH₃ groups
overlap

–NH₂ of (b)
overlaps with –COOH
groups
overlap
–H of (c)

–NH₂ of (c)
overlaps with
–H of (b)

FIGURE 20.7 (a) Alanine is an α-amino acid. (b) Like most α-amino acids, alanine is also chiral, which means that the molecular structure of one form is not superimposable on (c) its mirror image. Rotating the mirror image so that the –CH₃ and –COOH groups overlap when the structures are superimposed produces an image in which the –NH₂ and –H do not overlap.

amino acid a molecule that contains at least one amine functional group and one carboxylic acid functional group; in an α-amino acid, the two functional groups are attached to the same (α) carbon atom.

enantiomer one of a pair of optical isomers of a compound.

oscillate in only one plane. When plane-polarized light passes through a solution of one enantiomer of a chiral pair, the light twists in one direction. However, when the light passes through a solution of the other member of the pair, it twists in the opposite direction.

Look carefully at the structure of the molecule alanine in Figure 20.7. Alanine is an **amino acid**, so named because it contains at least one amine (–NH₂) group and at least one carboxylic acid (–COOH) group. Alanine is called an *α-amino acid* because one carbon atom in its structure, called the α-carbon, is bonded to both an –NH₂ group and a –COOH group (Figure 20.7a). Note also that the α-carbon atom in alanine is bonded to four *different* groups, making this α-carbon atom a *chiral center*. It is important to realize that the two molecules shown in Figure 20.7(b) and 20.7(c) are both alanine. They have the same formula and the same bonds; they differ only in the three-dimensional orientation of the groups attached to the chiral center. For the most part they have the same physical properties, such as melting point and boiling point; however, they differ in how they interact with light.

Let's now look at a slightly more complicated molecule, limonene (Figure 20.8), an alkene found in oranges and turpentine. The dashed red circle identifies its chiral center. This chiral carbon atom is part of a six-membered ring, and if we examine the structure of limonene closely, we see that this carbon atom has four different groups bound to it. The three-carbon alkene group and the hydrogen atom are easy to see, but if we follow the carbon atoms around the ring, we find a C=C bond three bonds from the circled carbon on the left side, but C—C single bonds on the right side. The presence of the C=C bond makes the groups attached to the circled carbon different, making limonene chiral. Therefore optical isomers of limonene exist.

The two optical isomers of a single molecule are called **enantiomers**, nonsuperimposable mirror images. In no orientation do all of the groups on the chiral center of superimposed enantiomers coincide, because the molecules have different shapes in three dimensions. Figure 20.7 shows the two enantiomers of alanine; Figure 20.8, the two enantiomers of limonene. If we try to superimpose the two enantiomers of limonene (Figure 20.9), note that the substituents bonded to the chiral carbon do not coincide. Note that because enantiomers differ in their three-dimensional structures, they are a type of stereoisomer.

To distinguish between a pair of enantiomers, (+) and (−) signs are added to their names to refer to the specific effect each isomer has on polarized light. When a beam of plane-polarized light passes through a solution containing one member of an enantiomeric pair, the beam rotates as in Figure 20.6. If the beam rotates to

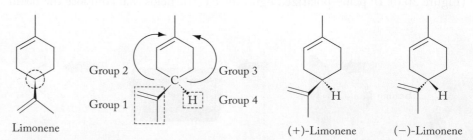

FIGURE 20.8 Limonene is a chiral molecule with two enantiomeric forms. To show why it is chiral, we have highlighted its chiral carbon with a dashed red circle and numbered the four groups bonded to it. Groups 1 and 4 are clearly different. Groups 2 and 3 are two halves of the same ring. They are different because group 2 has a C=C bond and group 3 does not. Therefore the circled carbon atom is bonded to four different groups and is chiral.

the left (counterclockwise), the enantiomer is the *levorotary* form of the molecule and a (−) sign precedes its name. If the beam rotates to the right (clockwise), the enantiomer is the *dextrorotary* form and a (+) sign precedes its name.

The molecules in an enantiomeric pair share the same physical and chemical properties except for those that relate to a few specialized types of behavior. As we saw in Chapter 9, molecular recognition is one of those special behaviors. For example, the (+)-enantiomer of limonene smells like oranges, whereas the (−)-enantiomer smells like turpentine. Part of the process of sensing different aromas involves recognition by receptors in your nasal passages. One receptor is shaped to accommodate (+)-limonene; the other, (−)-limonene, and each enantiomer binds to its own receptor. For similar reasons, certain chiral drugs have different effects according to which enantiomer is administered. The enantiomer of naproxen (Figure 20.10) that rotates light to the right relieves pain and swelling, whereas its mirror image, which rotates light to the left, is toxic to the liver and has no analgesic effects.

CONNECTION In Section 9.5 we saw how molecular recognition of ethylene gas triggers the ripening process for tomatoes.

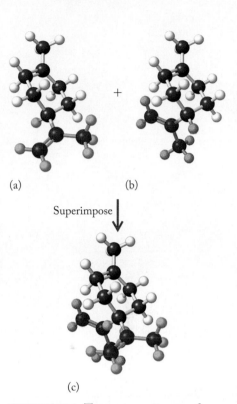

FIGURE 20.9 The two enantiomers of limonene (a) and (b) are mirror images of each other but cannot be superimposed upon each other (c) because the substituents bonded to the chiral carbon do not coincide.

SAMPLE EXERCISE 20.2 Recognizing Chiral Molecules   **LO1**

Identify which of the molecules shown are chiral, and circle the chiral centers. Structures may have more than one chiral center.

a. 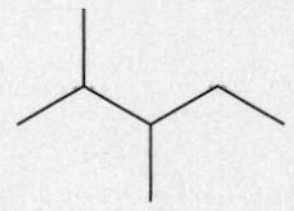   b. 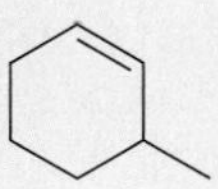   c.

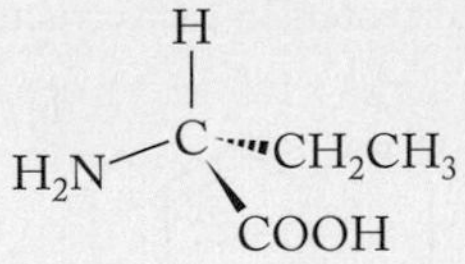

d. 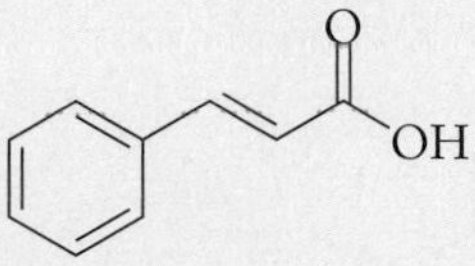   e.

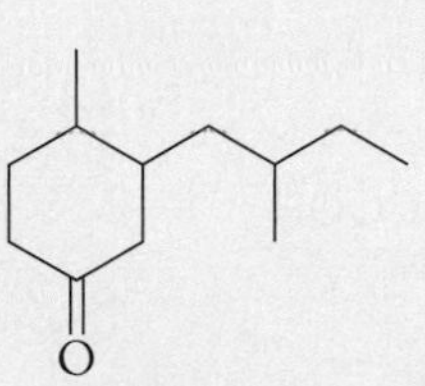

Collect, Organize, and Analzye Chiral molecules are nonsuperimposable on their mirror images. If molecules have carbon atoms with four different groups attached, those carbon atoms are chiral centers and the molecules are chiral.

Solve The chiral carbon atoms in each structure are circled.

a. 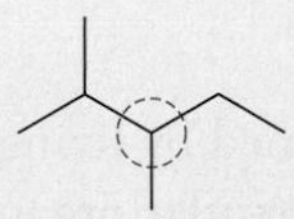   b. 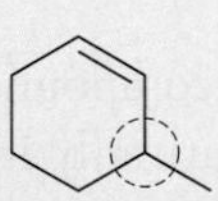   c.

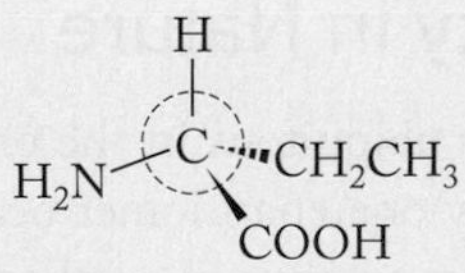

d. 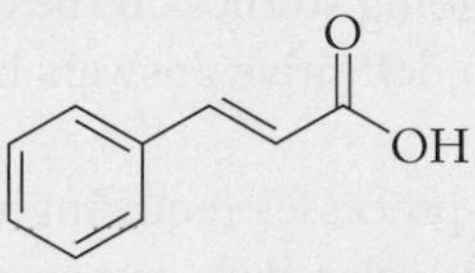   e.

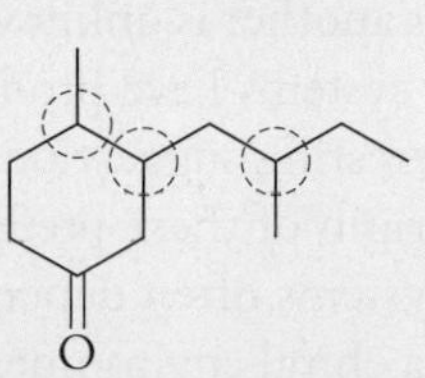

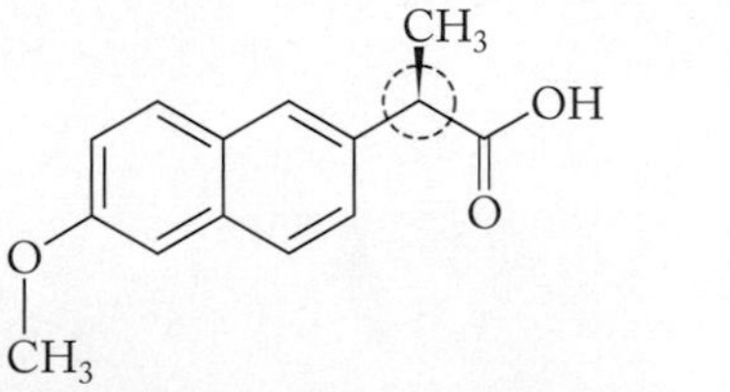

FIGURE 20.10 The chiral carbon in naproxen is indicated by the dashed red circle. Remember all carbon atoms have four bonds and, in this type of structural representation, any bond not shown is understood to be to a hydrogen atom. The chiral carbon in the circle is bonded to four different groups: the ring, the −CH$_3$ group, the −COOH group, and a hydrogen atom that is not explicitly drawn.

The third carbon atom of the pentane chain in compound (a) is bonded to a $-CH_3$ group, a $-CH_2CH_3$ group, a $-CH_2CH_2CH_3$ group, and an H atom, so the compound, 2,3-dimethylpentane, is chiral.

The chiral center in compound (b) is similar to the chiral center in limonene. The chiral carbon is bonded to a methyl group and an H. Tracing your finger around the ring in both directions from the circled carbon reveals the differences in the remaining two groups bonded to the chiral carbon.

The circled carbon atom in (c) is bonded to four different groups: $-H$, $-NH_2$, $-COOH$, and $-CH_2CH_3$, so this compound is chiral.

All of the carbon atoms in compound (d) are sp^2 hybridized, giving (d) a planar molecular geometry. Planar molecules cannot be chiral because they have a superimposable mirror image. Think of a mirror plane that contains all nine carbons in compound (d): the mirror image is exactly the same.

Compound (e) has three chiral centers. Working from right to left, the first chiral center is similar to the chiral center in compound (a). The other two chiral centers are in the cyclohexane ring. In each case, the carbon atom is bonded to four different groups, so compound (e) is chiral.

Think About It The presence of a chiral center in a molecule means that the molecule has two enantiomeric forms. The molecule is not superimposable on its mirror image. We should also note that although structure (d) has no chiral center, it has cis–trans isomers about the double bond in the chain. The isomer shown has the trans configuration.

Practice Exercise Identify which of the molecules below are chiral. Circle the chiral centers in each structure.

a.

b.

c.

d.

e.

Chirality in Nature

Chirality is ubiquitous in the organic compounds formed by living systems, and usually only one enantiomer occurs naturally in a particular organism. A molecule might even have more than one chiral center, which means it could exist as two or more optical isomers. The origin of the fundamental preference of life for one enantiomer over another is unknown. Ongoing studies on the origin of chiral preference in living systems have produced no definitive answers but are providing increasingly interesting suggestions.

Whatever the origin of these preferences, processes requiring molecular recognition in living systems often depend on the selectivity conveyed by chirality. The human body is a chiral environment, so handedness of molecules matters. As many as half of the drugs made by large pharmaceutical companies are chiral and owe their function to recognition by a receptor that favors one enantiomer over the other; in 2014 each of the ten most-prescribed drugs in the United States was chiral. Typically only one of the enantiomers of a chiral drug is active. A classic example of two enantiomers with different activity is the drug ethambutol (Figure 20.11). The (+)-enantiomer is effective against tuberculosis, whereas the (−)-enantiomer causes blindness.

When chiral compounds are produced by living systems, typically only one enantiomer is produced. However, when a molecule with a chiral center is made in a laboratory, both isomers are produced unless special methods are used. When both enantiomers are present in equal amounts in a sample, the material is known as a **racemic mixture**. Because one isomer rotates the plane of polarized light in one direction, and the other to the same extent in the opposite direction, a racemic mixture does not rotate the plane of polarized light at all.

Because any pair of enantiomers may interact differently with receptors, the pharmaceutical industry routinely faces two choices: devise a special synthetic procedure that yields only the isomer of interest, or separate the two isomers at the end of the manufacturing process. Both approaches are widely used. For the chiral drug sotalol (Figure 20.12), one isomer is physiologically active for hypertension and cardiac arrhythmias, whereas the other is inactive; sotalol is routinely prepared as a racemic mixture. For a drug such as ethambutol where the biological consequences of administering the undesired isomer are tragic, it is essential to exclude the incorrect isomer from the preparations administered to a patient.

racemic mixture a sample containing equal amounts of both enantiomers of a compound.

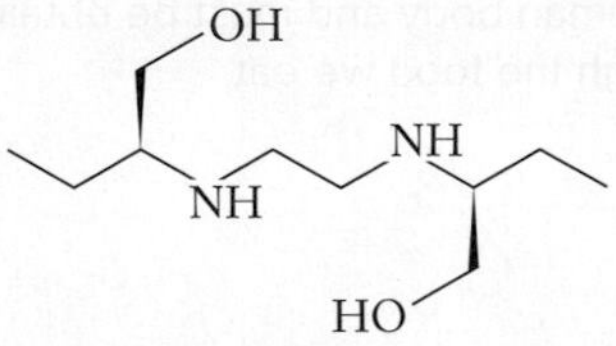

FIGURE 20.11 The (+) enantiomer of ethambutol shown here is active against tuberculosis infections; the (−) enantiomer causes blindness.

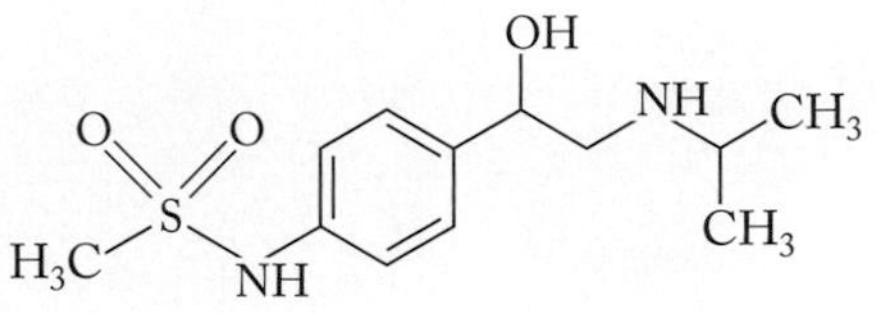

FIGURE 20.12 The chiral drug sotalol is administered as a racemic mixture because only one isomer is active.

SAMPLE EXERCISE 20.3 Recognizing Optical Properties of Chiral Molecules   **LO1**

The structure of the antidepressant drug bupropion is shown in Figure 20.13. When tested, a solution of a sample of the drug that comes directly from the laboratory does not rotate the plane of polarized light, so the analyst rejects the sample as impure, saying that the pure drug is known to cause a beam of polarized light to rotate when it passes through a solution of the compound. The chemist who made the sample tells you that other analytical data (percent composition and mass spectrometry) prove the material is 100% bupropion. Both people are correct in what they say. Explain why.

Collect and Organize We are asked to explain how a sample that is 100% chemically pure is not 100% pure active drug. We are given the structure of the compound and the information that a solution of the material does not rotate the plane of polarized light.

Analyze The sample was rejected by one analyst because a solution did not rotate the plane of polarized light. We must examine the structure to see whether the molecule has a chiral center—that is, a carbon atom bonded to four different groups.

Solve Bupropion is a chiral molecule. The chiral center is circled in the structure shown. The sample must contain both enantiomers of bupropion in equal amounts. They have exactly the same chemical formula, so the sample is chemically pure, but one enantiomer rotates polarized light to the left and the other to the right. No net rotation is observed when polarized light passes through the sample. Only one enantiomer is the active drug. Therefore the sample is chemically pure but not optically pure.

Think About It Enantiomers have the same chemical composition and the same molar mass. Their physical properties are identical except for the direction in which their solutions cause plane-polarized light to rotate when it passes through them.

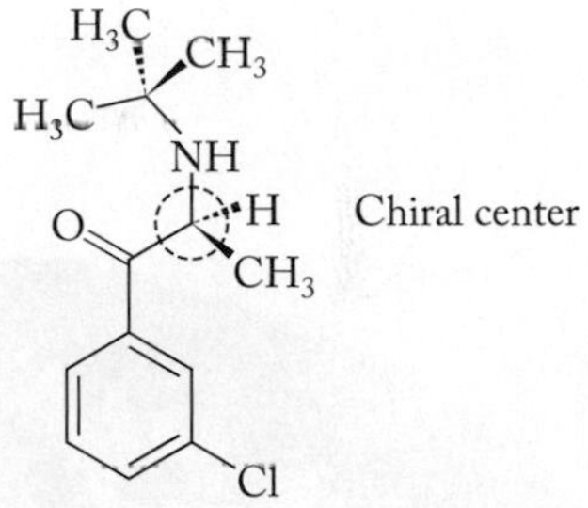

Bupropion

FIGURE 20.13 The antidepressant drug bupropion is a chiral molecule.

protein a biological polymer made of amino acids.

essential amino acid any of the eight amino acids that make up peptides and proteins but are not synthesized in the human body and must be obtained through the food we eat.

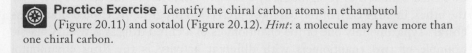

Practice Exercise Identify the chiral carbon atoms in ethambutol (Figure 20.11) and sotalol (Figure 20.12). *Hint*: a molecule may have more than one chiral carbon.

20.3 The Composition of Proteins

We opened our discussion of chirality with the α-amino acid alanine. We chose that molecule because amino acids are the small-molecule building blocks of **proteins**, the most abundant class of biomolecules in all animals, including us. Proteins account for about half the mass of the human body that is not water. They are the major component in skin, muscles, cartilage, hair, and nails. Most of the enzymes that catalyze biochemical reactions are proteins, as are the molecules that transport oxygen to our cells and many of the hormones that regulate cell function and growth. Most proteins are large, with molar masses of 10^5 g/mol or more.

Amino Acids

The molecular structures and biological functions of big biomolecules depend on the identities of their amino acids and the sequence in which those small molecules are connected. About 500 amino acids are known, and about 240 of these occur free in nature, but only 20 common amino acids form proteins in humans (Table 20.2).

In addition to its single bonds to the $-NH_2$ and $-COOH$ groups, the α-carbon atom in each of the amino acids that make up human proteins is bonded to a hydrogen atom and to one of 20 R groups. The structures of these R groups, often called *side-chain groups*, are highlighted in red in the amino acid structures in Table 20.2. The amino acids are arranged in four categories based on their R groups. In the first category, containing 9 amino acids, the R groups contain mostly carbon and hydrogen atoms and are nonpolar. The R groups of the remaining 11 amino acids contain at least one heteroatom (O, N, or S) bonded to an H atom and are polar. Of these amino acids, two (aspartic acid and glutamic acid) have R groups that contain carboxylic acid functional groups, and three (histidine, lysine, and arginine) have R groups with nitrogen atoms that are weakly basic.

> CONCEPT **TEST**

Identify the organic functional groups in the R groups of the following amino acids in Table 20.2: (a) leucine; (b) phenylalanine; (c) serine; (d) glutamine; (e) aspartic acid; (f) lysine.

FIGURE 20.14 A meal of red beans and rice provides all the essential amino acids.

Our bodies can synthesize 12 of the 20 amino acids in Table 20.2, but the other 8 must be present in the food we eat. These 8 are marked with a superscript *b* and are referred to as **essential amino acids**. Most proteins from animal sources, including those in meats, eggs, and dairy products, contain all the essential amino acids needed by the human body, and in close to the correct proportions. These foods are sometimes referred to as *perfect foods* or, more precisely, *complete proteins*. In contrast, most plant proteins from foods such as legumes and vegetables do not contain all the essential amino acids, so vegetarians must be careful to eat a combination of foods that provide all the essential amino acids. A good example is red beans and rice (Figure 20.14), a traditional dish in Latin American cuisine that

TABLE 20.2 Structures and Abbreviations of the 20 Common Amino Acids[a]

Nonpolar R Groups

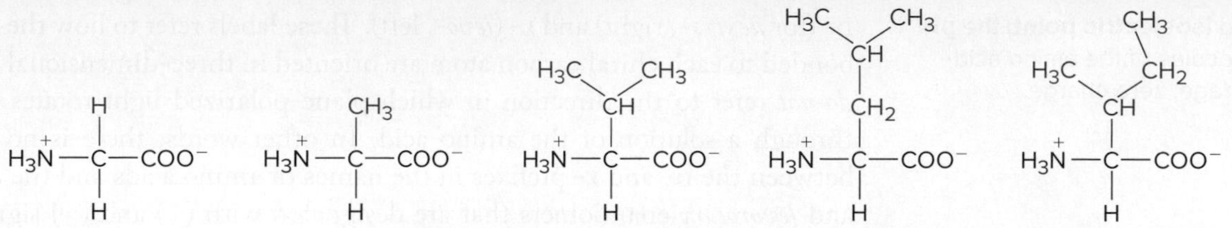

Glycine
(Gly; G)

Alanine
(Ala; A)

Valine[b]
(Val; V)

Leucine[b]
(Leu; L)

Isoleucine[b]
(Ile; I)

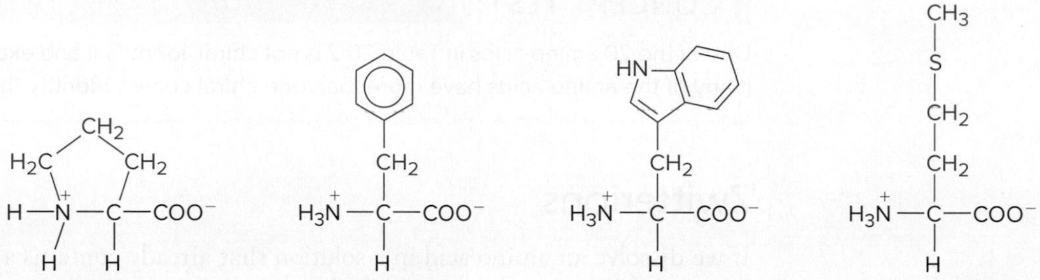

Proline
(Pro; P)

Phenylalanine[b]
(Phe; F)

Tryptophan
(Trp; W)

Methionine[b]
(Met; M)

Polar R Groups

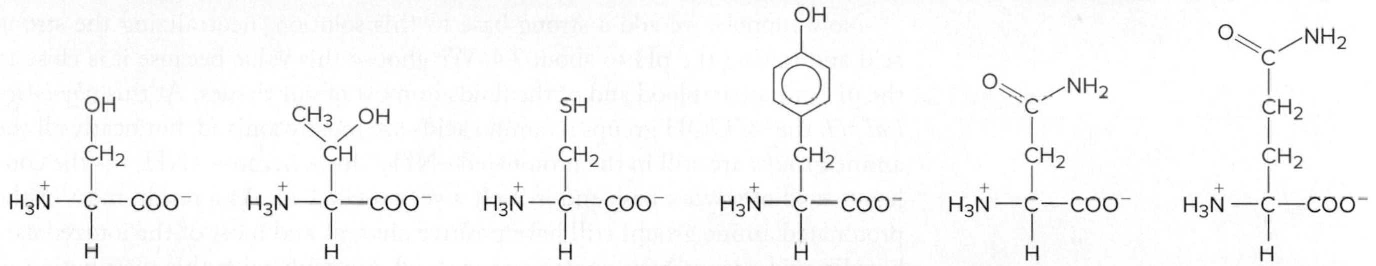

Serine
(Ser; S)

Threonine[b]
(Thr; T)

Cysteine
(Cys; C)

Tyrosine[b]
(Tyr; Y)

Asparagine
(Asn; N)

Glutamine
(Gln; Q)

Acid R Groups

Basic R Groups

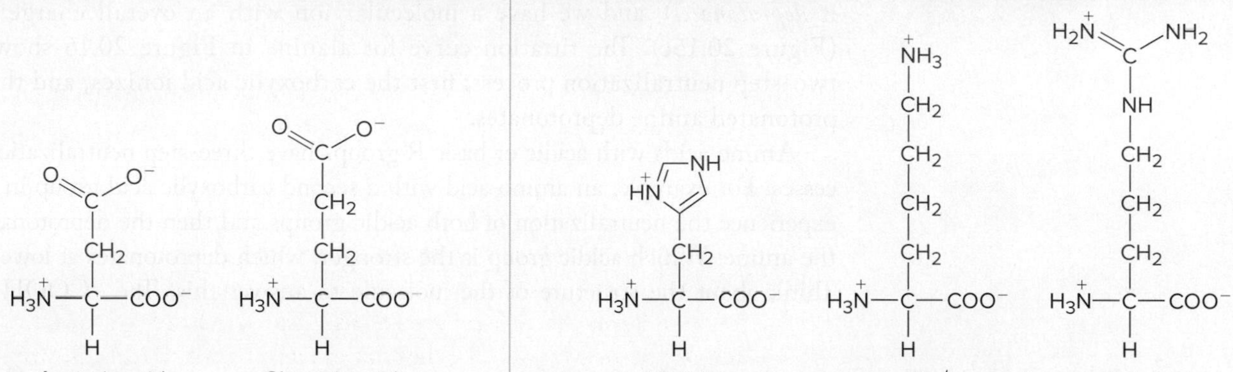

Aspartic acid
(Asp; D)

Glutamic acid
(Glu; E)

Histidine
(His; H)

Lysine[b]
(Lys; K)

Arginine
(Arg; R)

[a]R groups in pink; ionized forms at pH near 7 shown.
[b]The eight essential amino acids for adults (histidine is essential for children).

zwitterion a molecule that has both positively and negatively charged groups in its structure.

p/ (also called **isoelectric point**) the pH at which molecules of the amino acid have, on average, zero charge.

provides a balance of essential amino acids: rice has all of them but lysine, and beans lack only methionine.

For historical reasons, amino acid enantiomers are designated by the prefixes D- (for *dextro-*, right) and L- (*levo-*, left). These labels refer to how the four groups bonded to each chiral carbon atom are oriented in three-dimensional space. They *do not* refer to the direction in which plane-polarized light rotates as it passes through a solution of the amino acid. In other words, there is no connection between the D- and L- prefixes in the names of amino acids and the *dextrorotary* and *levorotary* enantiomers that are designated with (+) and (−) signs based on their optical properties. All the chiral amino acids in the proteins in our bodies are L-enantiomers, even though 9 of the 19 are actually dextrorotary.

CONCEPT TEST

One of the 20 amino acids in Table 20.2 is not chiral. Identify it and explain why it is not. If any of the amino acids have more than one chiral center, identify them.

Zwitterions

If we dissolve an amino acid in a solution that already contains a strong acid and has a low pH, the carboxylic acid group on each molecule does not ionize. In addition, each amine group, being a weak base, accepts a H^+ ion, forming a *protonated* $-NH_3^+$ group. The result is a molecular ion with an overall positive charge, as illustrated for alanine in Figure 20.15(a).

Now suppose we add a strong base to this solution, neutralizing the strong acid and raising the pH to about 7.4. We choose this value because it is close to the pH of human blood and of the fluids in most of our tissues. At this *physiological pH*, the −COOH groups in amino acids are mostly ionized, but nearly all the amine groups are still in the protonated $-NH_3^+$ form because $-NH_3^+$ is the conjugate acid of a weak base and is itself a very weak acid. As a result, most of the protonated amine groups still have positive charges and most of the ionized carboxylic acid groups have negative charges. A molecule with this distribution of charges is called a **zwitterion** (literally, a *hybrid ion*) because it has both a positive and a negative functional group (Figure 20.15b). Its net charge is zero.

If we add more base, alanine loses a H^+ from its NH_3^+ group (we say that it *deprotonates*), and we have a molecular ion with an overall charge of 1− (Figure 20.15c). The titration curve for alanine in Figure 20.16 shows this two-step neutralization process: first the carboxylic acid ionizes, and then the protonated amine deprotonates.

Amino acids with acidic or basic R groups have three-step neutralization processes. For example, an amino acid with a second carboxylic acid group in R will experience the neutralization of both acidic groups and then the deprotonation of the amine. Which acidic group is the stronger; which deprotonates at lowest pH? Think about the structure of the molecule to answer this. The −COOH group

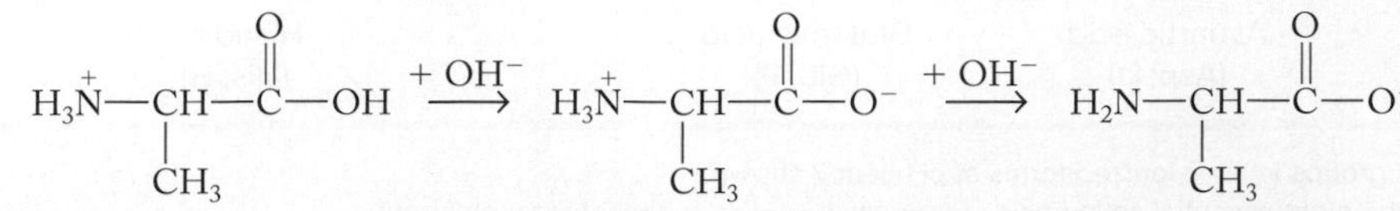

FIGURE 20.15 (a) At low pH, alanine (like many other amino acids) exists as a 1+ ion. (b) At physiological pH (7.4), the −COOH group is ionized and the molecule becomes a zwitterion with an overall charge of zero. (c) In basic solutions, the charge decreases to 1− as the $-NH_3^+$ group deprotonates.

(a) Acidic solution
charge of 1+

(b) pH near 7
charge of 0
zwitterion

(c) Basic solution
charge of 1−

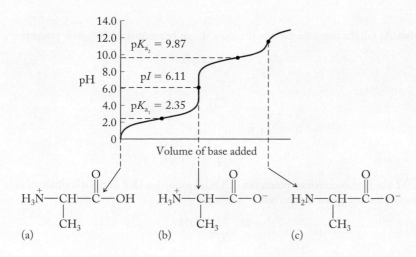

FIGURE 20.16 The titration curve of alanine resembles that of a weak diprotic acid. The carboxylic acid group (pK_{a_1} = 2.35) is neutralized first. The protonated amine group (pK_{a_2} = 9.87) is neutralized in the second step of the titration. The dominant forms of alanine present in solution are shown at (a) the starting point, and at (b) and (c), the two equivalence points.

bonded to the α-carbon atom is only one carbon atom away from a positively charged NH_3^+ group. The side-chain –COOH group is at least two carbon atoms away, and the carboxylate ion that it forms when it ionizes is less stabilized by delocalization of its negative charge toward the cationic group. Therefore the –COOH group bonded to the α-carbon atom is the stronger acid and ionizes first.

For an amino acid with a basic R group, the order of ionization in a titration also depends on the relative strengths of the groups as proton donors. The –COOH always ionizes first as base is added; it is the strongest acid in the molecule. The protonated nitrogen groups deprotonate in order of their relative strengths as proton donors.

Each amino acid has a characteristic **p***I* value, called the **isoelectric point**, representing the pH at which molecules of the amino acid have, on average, zero charge. The pH at the equivalence point between the two pK_a values is the p*I*. Its value is the average of the neighboring pK_a values. At this pH, the amino acid does not migrate in an electric field. If the pH is more acidic than the p*I*, the amino acid is positively charged and migrates to the cathode (negative electrode). At pH values more basic than the p*I*, the amino acid has a net negative charge and migrates toward the anode (positive electrode). This behavior is the basis of several analytical techniques used to characterize and identify amino acids. The p*I* values of neutral amino acids are around pH 7; acidic amino acids have p*I* values much lower than 7, and basic amino acids, much higher.

C⊗NNECTION The inverse relationship between the strengths of acid–base conjugate pairs was discussed in Chapter 15 and illustrated in Figure 15.5.

SAMPLE EXERCISE 20.4 Interpreting Acid–Base Titration Curves of Amino Acids **LO2**

Figure 20.17 shows the titration curve for aspartic acid and its pK_{a_1}, pK_{a_2}, and pK_{a_3} values. Draw the molecular structures of the principal form of aspartic acid that is in solution at the start of the titration and at each equivalence point. Estimate the p*I* of aspartic acid.

Collect, Organize, and Analyze Aspartic acid (Table 20.2) has two carboxylic acid groups and one amine group. At low pH the carboxylic acid groups are not ionized and the amine group is protonated. As pH is raised, the –COOH groups ionize and then, under basic conditions, the protonated amine group releases its proton. The –COOH group bonded to the α-carbon atom should be the stronger acid and ionize first.

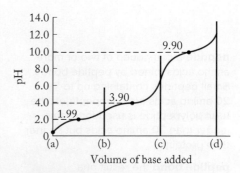

FIGURE 20.17 Titration curve of aspartic acid. The vertical black lines identify the equivalence points.

Solve At (a), the beginning of the titration, the fully protonated 1+ ion is present:

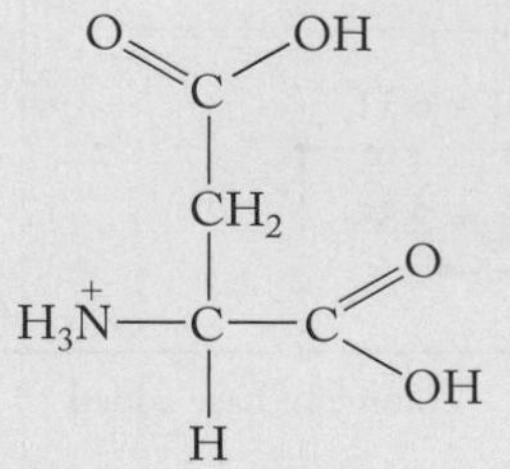

At (b), the first equivalence point, the –COOH group bonded to the α-carbon atom is ionized; the charge on the ion is 0.

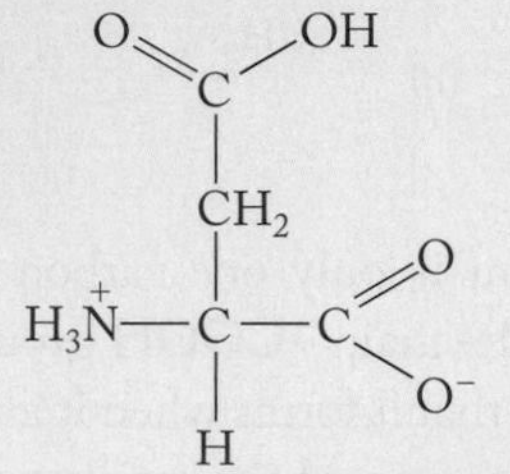

At (c), the second equivalence point, the side-chain –COOH group is ionized; the charge on the ion is 1–:

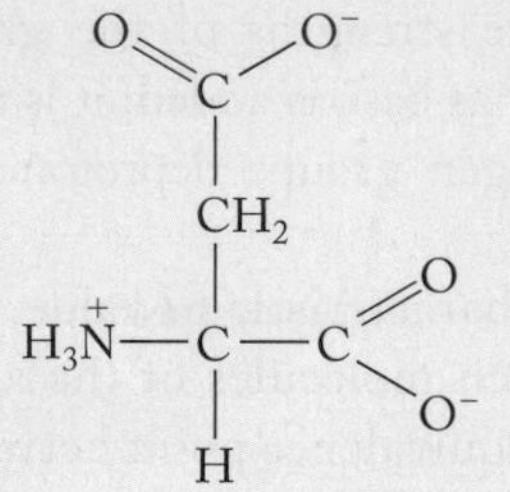

At (d), the third equivalence point in the titration, the amine group is deprotonated, resulting in an ion with a 2– charge:

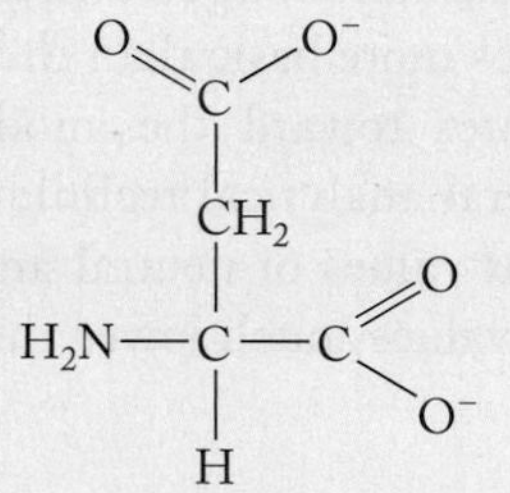

The pI of aspartic acid can be estimated by taking the value midway between pK_{a_1} and pK_{a_2} on Figure 20.17: $(1.99 + 3.90)/2 = 2.94$.

Think About It The values of pK_{a_1} and pK_{a_2} differ by nearly 2 pH units, which means that the α-COOH group is almost 100 times stronger an acid than the side-chain –COOH group, because of the presence of the α-NH$_3^+$ group.

Practice Exercise Sketch the titration curve for lysine starting at pH = 1.0, and draw the molecular structure of the principal form of lysine that is in solution at each equivalence point.

peptide a compound of two or more amino acids joined by peptide bonds. Small peptides containing up to 20 amino acids are oligopeptides; the term polypeptide is used for chains longer than 20 amino acids but shorter than proteins.

peptide bond the result of a condensation reaction between the carboxylic acid group of one amino acid and the amine group of another.

The acid–base characteristics of amino acids are important for their function in enzymes where protonation/deprotonation controls their activity as catalysts.

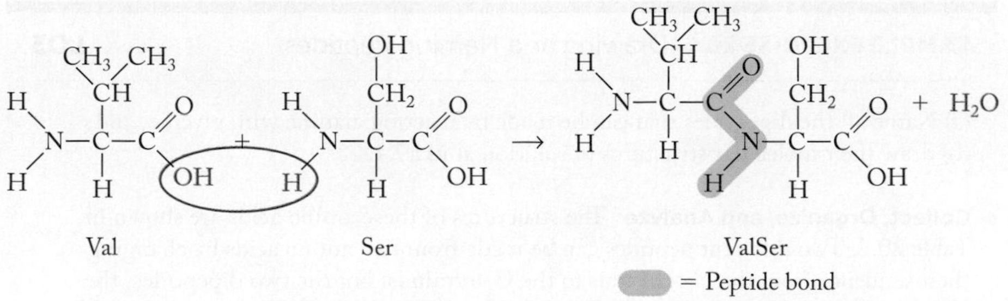

FIGURE 20.18 When the carboxylic acid group of valine reacts with the amine group of serine (blue oval) to form a peptide bond (blue highlight), the products are the dipeptide valylserine and water.

In addition, the structures of proteins depend in part on hydrogen bonding, which can be disrupted by changes in pH. This process is used in food preparation, when acids such as lemon juice are used to "cook" raw fish in dishes such as ceviche.

Peptides

Amino acids bond to form chainlike molecules with a wide range of sizes. Amino acid *residues* (the name we give to amino acids that are part of a larger molecule) make up each link in the chain. The longest chains, some with molar masses as high as 3 million g/mol, are called proteins, but the shortest chains, called **peptides**, are only a few links long. The smallest peptides contain only two or three amino acid residues. These molecules are called *dipeptides* and *tripeptides*, respectively. Peptides up to 20 residues long are called *oligopeptides*. Those made of more than 20 are called *polypeptides*. The size at which a polypeptide becomes a protein, or biopolymer, is arbitrary but is typically set around 50–75 amino acid residues.

The type of bond linking the amino acids in peptides and proteins is called a **peptide bond** (highlighted in blue in Figure 20.18) or *peptide linkage*. It forms when the α-carboxylic acid group of one amino acid condenses with the α-amine group of another, with the loss of a molecule of water. Note that peptide bonds have the same structure as the amide bonds that hold together the monomeric units of synthetic polyamides such as nylon (Chapter 12). This process is not spontaneous and is driven by being coupled to ATP hydrolysis, as we saw in Section 17.9.

The convention for drawing the structures of peptides begins by placing the amino acid that has a free α-amine group at the left end of the peptide chain and the amino acid with a free α-carboxylic acid at the right end. The left end is called the *amine* (or *N-*) *terminus* of the peptide, and the right end is called the *carboxylic acid* (or *C-*) *terminus*. The name of a peptide is based on the names of its amino acids, starting with the one at the N-terminus. The names of all the amino acids except the one at the C-terminus are changed to end in *-yl*. For example, if a peptide bond forms between the α-COOH group of valine (Val) and the α-NH₂ group of serine (Ser), as in Figure 20.18, the dipeptide that is produced is called valylserine, or more typically, using the three-letter or one-letter abbreviations, ValSer or VS.

The artificial sweetener aspartame is the methyl ester of the dipeptide aspartylphenylalanine (Figure 20.19). At pH 7.4, aspartame exists as a zwitterion because the aspartic acid amine group is protonated and the carboxylic acid in its R group is ionized. In contrast, its parent dipeptide has a net charge of 1− because both −COOH groups are ionized at that pH. In an amino acid that has an amine group in its R group, such as lysine or arginine, that amine group is probably protonated at physiological pH, giving the amino acid a net charge of 1+. The overall charge on the peptide at physiological pH is the sum of the positive charges on protonated amine groups and the negative charges of ionized carboxylic acid groups.

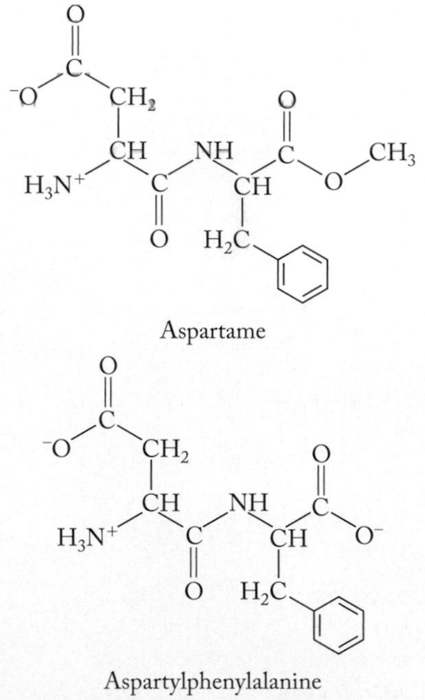

CHEMTOUR
Condensation of Biological Polymers

CONNECTION The formation of amide bonds in reactions between carboxylic acids and amines was described in Section 12.6 and illustrated in Figure 12.42.

Aspartame

Aspartylphenylalanine

FIGURE 20.19 Aspartame is the methyl ester of the dipeptide aspartylphenylalanine and so has one fewer −COOH group than its parent dipeptide. Therefore the overall charge of a molecule of aspartame at pH 7.4 is zero, whereas the charge on aspartylphenylalanine is 1−.

SAMPLE EXERCISE 20.5 Drawing and Naming Peptides LO3

(a) Name all the dipeptides that can be made by reacting alanine with glycine, and (b) draw their molecular structures in solution at pH 7.4.

Collect, Organize, and Analyze The structures of these amino acids are shown in Table 20.2. Two different peptides can be made from two amino acids by changing their sequence from the N-terminus to the C-terminus. For the two dipeptides, the sequences are AlaGly and GlyAla. The R groups in alanine ($-CH_3$) and glycine ($-H$) have no acidic or basic properties. At pH 7.4, the carboxylic acid terminus $-COOH$ groups should be ionized, and the amine terminus $-NH_2$ groups should be protonated.

Solve
a. The names of the two peptides with the amino acid sequence GlyAla and AlaGly are glycylalanine and alanylglycine, respectively.
b. At pH 7.4, the $-NH_2$ groups of amino acids are protonated and the $-COOH$ groups are ionized, so the principal forms of these dipeptides are

Glycylalanine
(GlyAla)

Alanylglycine
(AlaGly)

Think About It Only the N-terminus $-NH_2$ and C-terminus $-COOH$ groups can be protonated or ionized in these dipeptides. Thus AlaGly and GlyAla are both zwitterionic with net charges of $(1+) + (1-) = 0$ at pH 7.4.

Practice Exercise How many different tripeptides can be synthesized from one molecule of each of three different amino acids, Gly, Tyr, and Ser?

Only two sequences are possible for a peptide containing two different amino acids. Amino acid A and amino acid B can make dipeptide AB or BA. For a peptide made from three different amino acids—A, B, and C—six sequences are possible. For longer peptides, the number of possibilities becomes enormous. The sequence matters in determining the properties of the peptides, and figuring out the amino acid sequence of peptides and proteins is a significant challenge in biochemistry. One of the fundamental experimental methods used to determine sequence involves finding the identities of individual component amino acids, and then identifying fragments of the structure containing several amino acids and piecing them together like a puzzle. For example, a tripeptide could be broken down into amino acids A, B, and C. Dipeptide fragments BA and AC could then be identified, which means the overall sequence of the tripeptide must be BAC.

SAMPLE EXERCISE 20.6 Determining the Sequence of a Small Peptide LO3

A small peptide was hydrolyzed completely to its component amino acids: A, B, C, D, E, F, and G. Alternative ways to treat the peptide yielded these larger fragments: BA, CDF, EC, AG, and GEC. What is the sequence of the amino acids in the peptide?

Collect, Organize, and Analyze We know the identity of seven amino acids that make up a peptide, and we have several dipeptide and tripeptide fragments. We need to arrange them to reveal the complete sequence of the peptide.

Solve Arranging the fragments so that like amino acids are in columns:

BA
  AG
    GEC
      CDF
     EC
BAGECDF

The amino acid sequence of the peptide is BAGECDF.

Think About It Longer peptide chains and proteins have many copies of each amino acid, so longer fragments are needed to sequence them.

 Practice Exercise The hydrolysis of a peptide yielded the following amino acids: Gly, Ala, Val, Thr, and 2 Leu; and the following larger fragments: LeuThr, AlaLeuLeu, ThrGly, and ValAla. What is the sequence of the amino acids in the peptide?

20.4 Protein Structure and Function

The structure of a protein is crucial to its function. This fact is easiest to appreciate for structural proteins such as the collagens, which impart flexibility, strength, and elasticity to skin and tendons. The functions of nonstructural proteins, such as the ability of enzymes to catalyze biochemical reactions, are also closely linked to their structures. These large biomolecules must assume particular three-dimensional conformations to interact with other molecules and to function properly. Table 20.3 lists some of the important functional classifications of proteins and gives examples of each.

TABLE 20.3 Functional Classes of Proteins

Class of Protein	Function	Example
Structural	Gives strength and flexibility to tissues	Collagen: a component of connective tissue
Enzymes	Catalyze reactions in cells	Sucrase: promotes conversion of sucrose into simpler sugars
Hormones	Regulate processes in organisms	Insulin: helps cells regulate absorption of glucose
Signaling	Coordinates the activity of cells	GTPase: regulates intracellular dynamics
Immune response	Defense	Antibodies: attack bacteria
Storage	Releases energy during metabolism	Ovalbumin: present in eggs
Transport	Transports molecules from place to place	Hemoglobin: transports oxygen from lungs to cells

primary (1°) structure the sequence in which amino acid monomers occur in a protein chain.

secondary (2°) structure the pattern of arrangement of segments of a protein chain.

α helix a coil in a protein chain's secondary structure.

β-pleated sheet a puckered two-dimensional array of protein strands held together by hydrogen bonds.

Primary Structure

The **primary (1°) structure** of a protein is the sequence of the amino acids in it, starting with the N-terminus (Figure 20.20a). If two proteins are made up of the same number and type of amino acids but have different amino acid sequences, they are different proteins.

Changing only one amino acid can dramatically alter a protein's function. For example, in hemoglobin the sixth amino acid from the N-terminus of a protein strand that is 146 amino acids long is normally glutamic acid. In some people, valine substitutes for glutamic acid at this position. This one substitution alters the solubility of the protein and causes the red blood cell to take on a sickle shape instead of the normal, plump disk shape (Figure 20.21). These sickled cells do not pass through capillaries easily and may impede blood circulation. They also break readily and do not last as long as normal blood cells. These factors lead to a diminished capacity of the blood to carry oxygen, which is one of the characteristics of the disease called sickle-cell anemia.

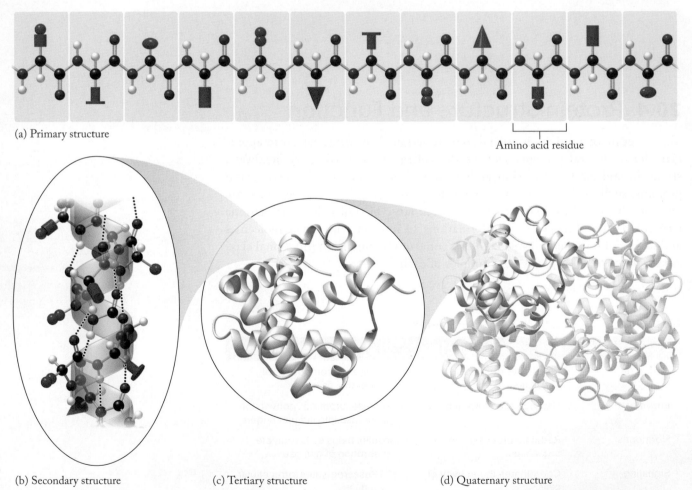

(a) Primary structure

Amino acid residue

(b) Secondary structure   (c) Tertiary structure   (d) Quaternary structure

FIGURE 20.20 The four levels of protein structure. (a) A protein's primary structure is its amino acid sequence. The green shapes represent different R groups. (b) Secondary structure (here, an α helix) describes the three-dimensional pattern adopted by segments of the protein strand. (c) Tertiary structure is the overall shape of the molecule as segments of it bend and fold. (d) Quaternary structure refers to the overall shape adopted by multiple protein strands that assemble into a single unit.

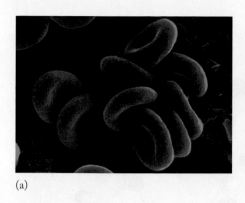

(a)

(b)

FIGURE 20.21 (a) Normal red blood cells are plump disks, whereas (b) those in patients with sickle-cell anemia are distorted and incomplete.

Why does switching valine for glutamic acid affect the solubility of hemoglobin? The answer lies in the R groups of these two amino acids (Figure 20.22). The side-chain –COOH group of glutamic acid is ionized at physiological pH, and strong ion–dipole interactions with water molecules enhance the protein's solubility. However, if valine, with its nonpolar isopropyl R group, replaces glutamic acid, that strong intermolecular interaction with water is lost. The valine creates a hydrophobic patch on the surface of the protein that results in the deoxygenated forms of hemoglobin sticking to each other, producing stiff fibers inside red blood cells. This, in turn, deforms the red blood cell from its usual smooth disk shape (Figure 20.21a) into a sickle shape (Figure 20.21b).

Sickle-cell anemia is a debilitating disease, but it provides a survival advantage in regions where malaria is endemic. Having sickle-cell anemia does not protect people from contracting malaria or make them invulnerable to the parasite that causes it. However, children infected with malaria are more likely to survive the illness if they have sickle-cell anemia. Exactly why sickle-cell anemia has this effect is not completely understood, but recent work points toward sickle-hemoglobin increasing the production of an enzyme that catalyzes the production of carbon monoxide; the carbon monoxide protects the host from succumbing to malaria without actually interfering with the parasite.

CONCEPT **TEST**

Which other amino acids besides valine might result in the sickling of the cell when substituted for glutamic acid?

Primary structure

Normal protein:

    Val - His - Leu - Thr - Pro - Glu - Lys - . . .

Abnormal protein:

    Val - His - Leu - Thr - Pro - Val - Lys - . . .

(a)

$$O=\overset{\displaystyle O^-}{\underset{\displaystyle CH_2}{C}}$$

Glutamic acid   Valine

(b)

FIGURE 20.22 (a) The primary structure of the amine end of a protein in normal hemoglobin and in the abnormal hemoglobin responsible for sickle-cell anemia. (b) The replacement of glutamic acid with its hydrophilic R group by valine with its hydrophobic group is responsible for the disease.

Secondary Structure

The next level of protein structure, the **secondary (2°) structure**, describes the geometric patterns made by segments of amino acid chains. One common pattern is the **α helix**, a coiled arrangement with the R groups pointing outward (Figure 20.20b). The helical structure is maintained by hydrogen bonds between –NH groups on one part of the chain and C=O groups on amino acids four residues away from them in the sequence. The α helix looks very much like a spring. One group of proteins that are mostly α-helical is the keratins, the proteins in hair and fingernails.

Another common pattern of 2° structure is called a **β-pleated sheet**. These sheets are assemblies of multiple amino acid chains aligned side by side. The pleats are caused by the tetrahedral molecular geometries of the atoms along the chains (Figure 20.23). Adjacent chains are linked by hydrogen bonds, and the collection of side-by-side zigzag chains forms a continuous β-pleated sheet. R groups extend

CONNECTION We discussed in Chapter 10 how ion–dipole and dipole–dipole interactions are the key to the solubility of solutes in water.

FIGURE 20.23 Each amino acid chain in a β-pleated sheet is folded in a zigzag pattern. Adjacent chains in a sheet are held together by hydrogen bonds (blue dotted lines). The R groups (green shapes) extend above and below the sheet, linking it to adjacent sheets via noncovalent intermolecular interactions.

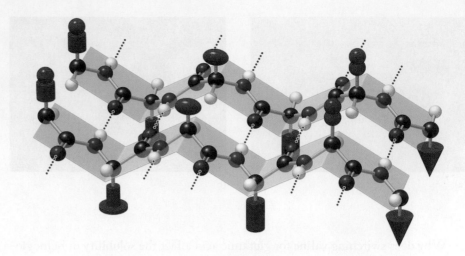

(a)

(b)

(c)

FIGURE 20.24 Intermolecular interactions that influence the secondary and tertiary structures of proteins include (a) ion–ion interactions between acidic and basic R groups, (b) hydrogen bonding, and (c) London forces between nonpolar side chains.

CHEMTOUR
Fiber Strength and Elasticity

above and below the pleats. Sheets may stack on top of one another like two pieces of corrugated roofing. Stacked sheets are held together by the same interactions that hold all proteins together, including ion–ion, hydrogen-bonding, and London forces, depending on the pairs of R groups involved (Figure 20.24).

The proteins that make up strands of silk form thin, planar crystals of β-pleated sheets that are only a few nanometers on a side. Enormous numbers of these crystals form long arrays of sheets stacked together like nanoscale pancakes (Figure 20.25). Hydrogen bonds hold the stacks together and reinforce adjacent sheets. As a result of these interactions, strands of silk are stronger than strands of steel with the same mass; indeed, silk is one of the toughest known materials—natural or synthetic.

Some single-stranded proteins exist as α helices on their own but form β-pleated sheets when they clump together in multistrand aggregates. One consequence of this clumping is the formation of insoluble protein deposits called plaque. Abnormal accumulation of plaque can be a serious health risk: plaque formed by a protein called amyloid β has been linked to the onset of Alzheimer's disease.

If part of a protein is characterized by an irregular or rapidly changing structure, it is said to have a **random coil** 2° structure. The amino acid chain may fold back on itself and around itself, but it has no regular features the way an α helix or a β-pleated sheet does. When proteins *denature*, losing their secondary structure because of heat or change in pH, they may become random coils.

Large protein molecules may contain all three types of 2° structure. In describing a protein, scientists may indicate the percentage of amino acids involved in

FIGURE 20.25 The strength of spider silk comes from the flexible cross-linked chains connecting regions of crystalline stacks of β-sheets, shown here as blue-green boxes, which are only a few nanometers in size. Hydrogen bonds make an extremely strong network, and if a hydrogen bond breaks, many more are left that can maintain the material's overall strength.

each type—for example, 50% α-helical, 30% β-pleated sheet, and 20% random coil. Figure 20.26 shows a model of a protein called carbonic anhydrase, which illustrates all three types.

The aqueous solution of proteins in egg whites turns into a solid mass when eggs are cooked or are dropped into an organic solvent such as acetone.

a. What is the likely secondary structure of the proteins in cooked eggs?

b. Would the solidification process occur as a result of a change in the primary structure of the proteins? Explain your answer.

Tertiary and Quaternary Structure

Large proteins have structure beyond the 1° and 2° levels. Their molecules can fold back on themselves as a result of interactions between R groups on amino acids that are considerable distances apart along the protein chain. These interactions may be ion–ion, ion–dipole, or London forces. They may even involve the formation of covalent bonds. For example, the –SH groups on two cysteine residues may combine to form a disulfide linkage that holds two parts of the protein strand together via an intrastrand –S—S– covalent bond (Figure 20.27). Interactions and reactions such as these determine a protein's **tertiary (3°) structure**, the overall three-dimensional shape of the protein that is key to its biological activity (Figure 20.20c). Because the proteins in living systems exist in an aqueous environment, hydrophobic R groups tend to reside in the interiors of their 3° structures, whereas hydrophilic groups (as we saw in normal hemoglobin) are oriented toward the outside, where they interact with nearby molecules of water. Hydrophobic interactions are the primary force that causes protein folding and compaction, but all the other modes of interaction help a large protein form its unique 3° structure.

Hemoglobin and some other proteins exhibit an even higher order of structure. One hemoglobin unit (Figure 20.28a) contains four protein strands, each of which enfolds a porphyrin ring containing one Fe^{2+} ion. The combination of four protein

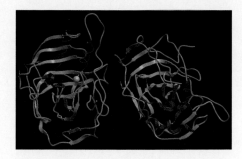

FIGURE 20.26 The structure of the protein carbonic anhydrase has α-helical regions (red), β-pleated sheet regions (green), and random coil (blue). The light blue sphere in the center is a zinc ion.

$$\big|\!-CH_2—SH \quad HS—CH_2-\!\big|$$
$$\downarrow$$
$$\big|\!-CH_2—S—S—CH_2-\!\big|$$
$$+ 2\,H^+(aq) + 2\,e^-$$

FIGURE 20.27 The tertiary structure of some proteins is stabilized by intrastrand –S—S– bonds that form between the –SH groups on the side chains of cysteine residues.

CONNECTION All the types of intermolecular forces described in Chapter 10 are involved in the *intra*molecular interactions that give proteins their unique 2° and 3° structures.

random coil an irregular or rapidly changing part of the secondary structure of a protein.

tertiary (3°) structure the three-dimensional, biologically active structure of a protein that arises because of interactions between the R groups on its amino acids.

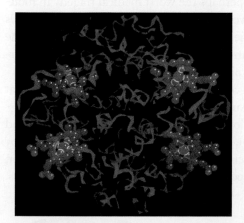

(a) Hemoglobin

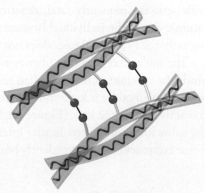

(b) Keratin

FIGURE 20.28 Quaternary structure of proteins. (a) Four protein chains form a single unit in the quaternary structure of hemoglobin. The iron-containing porphyrins are bright green. (b) Pairs of α-helical chains (blue) wound together and linked by interstrand –S—S– bonds (yellow) stabilize the quaternary structure of the keratin in hair and fingernails.

strands to make one hemoglobin assembly is an example of **quaternary (4°) structure** (Figure 20.20d). In the 4° structures of keratins (Figure 20.28b), protein strands with α-helical 2° structures coil around each other to make even larger coils. When the protein strands in keratin structures are held together mostly by London forces, as they are in the keratin in skin tissue, the structures are flexible and elastic. If they are also restrained by many covalent bonds, as in Figure 20.28(b), they produce tissues that are hard and less flexible, like fingernails and the beaks of birds.

Enzymes: Proteins as Catalysts

We introduced *enzymes* as biological catalysts in Chapter 13. As described there, the chemical reactions involved with metabolism—both *catabolism* (breaking down of molecules) and *anabolism* (synthesis of complex materials from simple feedstocks)—are mediated in large part by enzymes. Both catabolism and anabolism are organized in sequences of reactions called *metabolic pathways*, and each step in a metabolic pathway is catalyzed by a specific enzyme.

Enzymes are highly selective: each catalyzes a particular reaction involving a particular reactant. For example, an enzyme called lactase catalyzes only the reaction by which lactose (the sugar in milk) is broken down during digestion. People who lack this enzyme cannot metabolize this sugar; they are said to be *lactose intolerant*. If they consume dairy products, unmetabolized lactose passes into their large intestines where bacteria ferment it, and unpleasant and painful abdominal disturbances result.

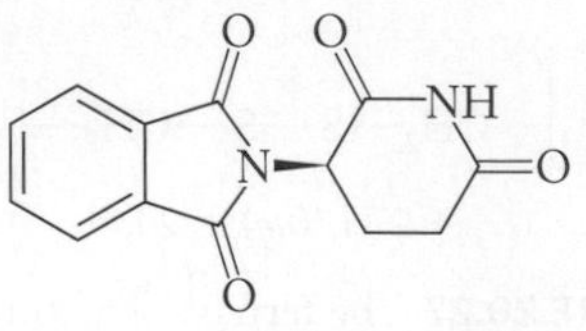

Synthetic reaction pathways catalyzed by enzymes are usually more rapid and involve fewer steps than uncatalyzed pathways to make the same product. The products also tend to be optically pure materials. These advantages can significantly reduce the cost of production of, for example, biological pharmaceuticals. However, biocatalytic reactions often run best in very dilute solutions, which limits production. A key issue driving interest in such processes is that an enantiomerically pure pharmaceutical is likely to be more potent and produce fewer side effects than a racemic mixture of the same product, as we saw with the drug ethambutol in Section 20.2. As another example, the drug thalidomide (Figure 20.29) is a racemic mixture of two enantiomeric forms: the (+)-enantiomer is a sedative and is effective in treating morning sickness, but the (−)-enantiomer is a teratogen (an agent that disturbs the development of a fetus). In perhaps the worst medical tragedy in modern times, more than 10,000 children were born in the late 1950s and early 1960s with serious, frequently fatal, deformities as a result of their mothers taking the racemic drug. The individual isomers can be interconverted *in vivo*, so administering only the (+)-enantiomer does not avoid the serious effects.

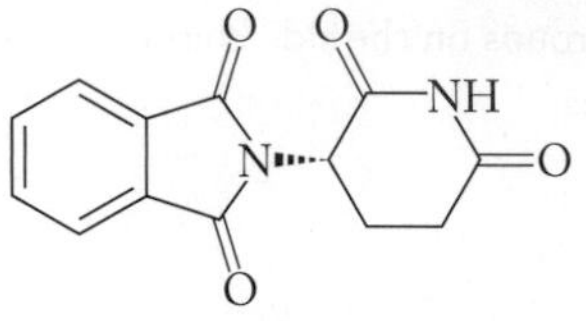

The molecular structure of enzymes contains a region called an **active site** that binds the reactant molecule, called the **substrate**. The action of enzymes was originally explained by a lock-and-key analogy in which the substrate is the key and the active site is the lock (Figure 20.30). The substrate is held in the active site by the same kinds of intermolecular interactions that hold any biomolecules together. Some enzymes become covalently bonded to intermediates in the catalytic process.

CONNECTION In Section 13.6 we described enzymes as homogeneous catalysts that selectively speed up biochemical reactions.

(+)-Thalidomide

(−)-Thalidomide

FIGURE 20.29 Enantiomers of thalidomide.

FIGURE 20.30 In the lock-and-key model of enzyme activity, the substrate (orange) fits exactly into the active site of the enzyme (green) that catalyzes a chemical reaction involving the substrate.

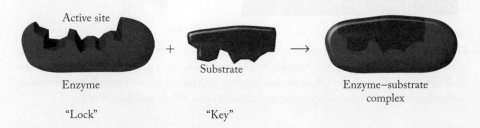

Active site

Enzyme

"Lock"

+

Substrate

"Key"

→

Enzyme–substrate complex

Once in the active site, the substrate is converted into product via a reaction having a lower-energy transition state than it would without the enzyme. The reaction of a substrate S with an enzyme E produces an *enzyme–substrate* (ES) *complex* that decomposes, forming a product P and regenerating the enzyme:

$$E + S \rightleftharpoons ES \rightarrow E + P$$

The lock-and-key analogy, however, does not fully account for enzyme behavior. A more accurate view is provided by the *induced-fit model*, which assumes that the substrate does more than just fit into the existing shape of an active site. This model assumes that as the ES complex forms, the binding site undergoes subtle changes in its shape to more precisely fit the three-dimensional structure of the transition state. The binding energy between the enzyme and the substrate creates a lower-energy transition state, thereby lowering the activation energy barrier. A simplified illustration of such an interaction is shown in Figure 20.31(a).

The induced-fit model helps explain the behavior of compounds called **inhibitors**, which can diminish or destroy the effectiveness of enzymes. An inhibitor may bind to an active site and block it from interacting with the substrate (Figure 20.31b). Alternatively, it may disable the enzyme by preventing it from assuming its active shape. In the latter process the inhibitor may bind to the enzyme at a site other than the active site and, from that position, prevent the enzyme from achieving its active shape (Figure 20.31c).

quaternary (4°) structure the larger structure functioning as a single unit that results when two or more proteins associate.

active site the location on an enzyme where a reactive substance binds.

substrate the reactant that binds to the active site in an enzyme-catalyzed reaction.

inhibitor a compound that diminishes or destroys the ability of an enzyme to catalyze a reaction.

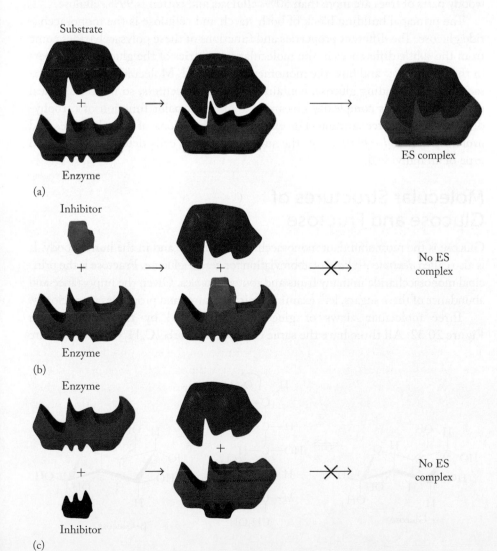

FIGURE 20.31 (a) The induced-fit model assumes that the shape of the enzyme (green) changes to accommodate the substrate (orange) and form the enzyme–substrate (ES) complex. Inhibitors (yellow and red molecules) may (b) block the enzyme's binding site or (c) cause a change elsewhere in the enzyme's structure that prevents the active site from attaining the shape it needs to form the ES complex.

carbohydrate an organic molecule with the generic formula $C_x(H_2O)_y$.

monosaccharide a single-sugar unit and the simplest carbohydrate.

polysaccharide a polymer of monosaccharides.

Natural enzyme inhibitors play important roles in regulating the rates of reactions that are catalyzed by enzymes. For example, in a multistep reaction pathway, the product of a later step may inhibit an enzyme that catalyzes an earlier reaction. This kind of negative feedback keeps the sequence of reactions from running too quickly and perhaps jeopardizing the health of the organism as the result of an accumulation of undesirable products or intermediates. Enzyme inhibitors may also be used to fight disease. Powerful drugs have been developed that inhibit enzymes called proteases involved in virus maturation. Several such drugs have been particularly effective in treating HIV.

20.5 Carbohydrates

Carbohydrates have the generic formula $C_x(H_2O)_y$. This formula gives us a clue about where the name *carbohydrate*, or *hydrate of carbon*, comes from. The smallest carbohydrates are **monosaccharides** (the name means "one sugar"), which bond to form more complex carbohydrates called **polysaccharides**. Many organisms use monosaccharides as their main energy source but convert them to polysaccharides for the purpose of energy storage. Starch is the most abundant energy-storage polysaccharide in plants. Polysaccharides in the form of cellulose also provide structural support in plants. Plants produce over 100 billion tons of cellulose each year—the woody parts of trees are more than 50% cellulose, and cotton is 99% cellulose.

The principal building block of both starch and cellulose is the monosaccharide glucose. The different properties and functions of these polysaccharides come from the subtle differences in the molecular geometries of the glucose monomers in their structures and how the monomers are bonded. Molecules of most monosaccharides, including glucose, contain several chiral centers, so multiple optical isomers exist. This complexity gives rise to another major function of carbohydrates: molecular recognition. For example, combinations of carbohydrates and proteins, called *glycoproteins*, on the surfaces of blood cells determine the blood type of an individual.

Molecular Structures of Glucose and Fructose

Glucose is the most abundant monosaccharide in nature and in the human body. It is also called *dextrose*—a kind of abbreviation for *dextro*-glucose. Fructose is the principal monosaccharide in many fruits and root vegetables. Given the importance and abundance of these sugars, let's examine their structures and properties more closely.

Three molecular views of glucose are provided by the structures in Figure 20.32. All three have the same chemical formula ($C_6H_{12}O_6$), so they are

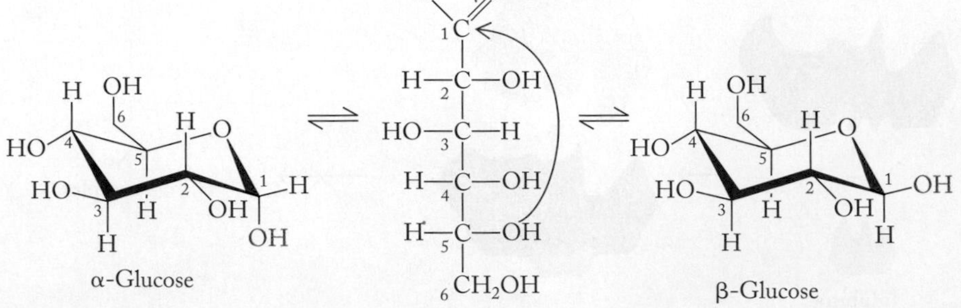

FIGURE 20.32 An equilibrium exists between the linear structure of glucose and the two cyclic forms α-glucose and β-glucose. The difference between the two cyclic structures is the orientation of the −OH group on C1 (highlighted in blue), which is formed by the O atom on C1 as indicated by the blue arrow. The new bond between oxygen and carbon atoms is shown in red in the cyclic structures and is formed by the reaction of the −OH on C5 and C1 indicated by the red arrow.

all isomers of one another. Note that the middle structure contains an aldehyde (–CH=O) group and all three structures contain multiple alcohol (–C—OH) groups. The polarities of these groups and their capacities to form hydrogen bonds give glucose its high molar solubility in water (5.0 *M* at 25°C, or about 9 g in 10 mL of water).

The cyclic structures of glucose form when the carbon backbone of the linear form curls around so that the –OH group bonded to the carbon atom labeled number 5 (C5) comes close to the aldehyde group on C1 (as shown by the red arrow in Figure 20.32). The aldehyde and alcohol groups react to form a six-membered ring made up of five carbon atoms (C1 to C5) and the oxygen atom of the C5 alcohol. Note a small but significant difference in the two cyclic structures. In the molecule on the left, called α-glucose, the –OH group on C1 points down. In the molecule on the right, β-glucose, the C1 –OH group points up. These two orientations are possible because there are two ways that the carbon chain in the middle structure can form a cyclic structure: the –OH group on C5 can approach C1 from either of the two sides of the plane defined by the C1 carbon atom and the O and H atoms bonded to it. When the C5 –OH group approaches C1 as shown in Figure 20.33, the α isomer is produced. Approaching C1 from the other side yields the β isomer. Notice that in the linear form, C1 is an aldehyde and is not a chiral center. The cyclization creates a new chiral center in the cyclic molecule.

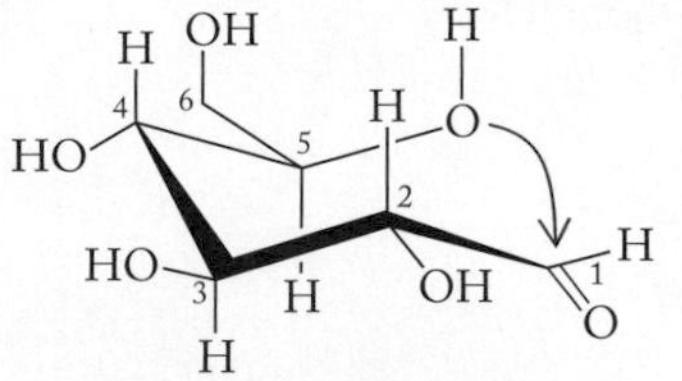

The β form is slightly more stable than the α form and accounts for 64% of glucose molecules in aqueous solution; the α form accounts for the remaining 36%. Both cyclic forms are more stable than the straight-chain form, which exists only as an intermediate between the two cyclic forms. The energy differences are small, however, so glucose molecules in solution are constantly opening and closing in a dynamic structural equilibrium.

Figure 20.34 shows the structures of the linear and cyclic forms of fructose, commonly known as fruit sugar and contained in honey and many fruits. Note that the C=O group in the linear form is not on the terminal carbon atom. In other words, fructose is a ketone rather than an aldehyde. It forms a five-membered ring, not a six-membered ring, when the –OH group at C5 reacts with the carbonyl carbon atom at the C2 position. The product is a ring with two –CH₂OH groups that are either on the same side of the ring (α-fructose) or on opposite sides (β-fructose).

Disaccharides and Polysaccharides

Sucrose, or ordinary table sugar, is a *disaccharide* ("two sugars") that consists of one molecule of α-glucose bonded to one molecule of β-fructose (Figure 20.35). The bond between them forms when the –OH group on the C1 carbon atom of glucose reacts with the C2 carbon atom of fructose, producing a C—O—C

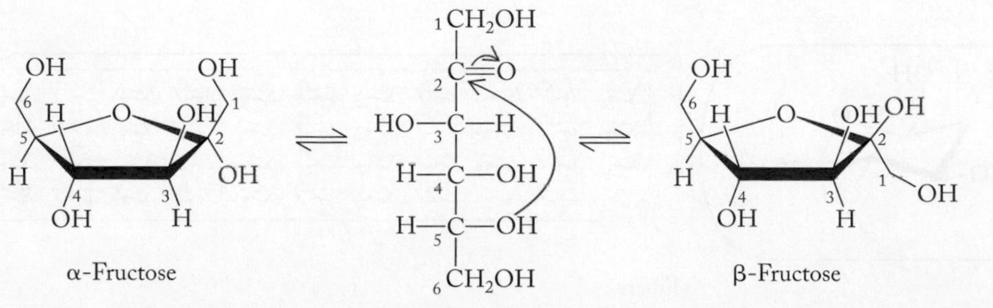

FIGURE 20.33 The –OH group on C5 may approach the aldehyde at C1 from either side of the plane defined by the C, H, and O atoms in the aldehyde group. In the approach shown here the product is the α isomer.

α-Fructose   β-Fructose

FIGURE 20.34 The cyclization of fructose proceeds via the –OH group on C5 and the ketone on C2 indicated by the red arrow. The new bond is shown in red in the cyclic structures. The oxygen atom in the –OH group on C1 comes from the carbonyl as indicated by the blue arrow.

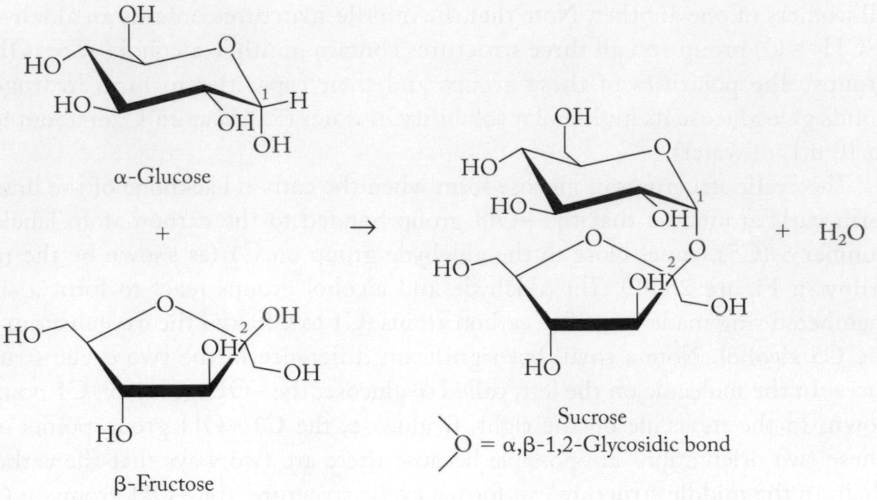

FIGURE 20.35 α-Glucose and β-fructose combine to form a molecule of sucrose, ordinary table sugar. The bond shown in red is the α,β-1,2-glycosidic bond.

α-Glucose

+

β-Fructose

→

Sucrose

○ = α,β-1,2-Glycosidic bond

+ H_2O

CHEMTOUR
Formation of Sucrose

FIGURE 20.36 (a) Starch is a polysaccharide of α-glucose molecules joined by α-1,4-glycosidic bonds, shown in red. Starch molecules form spirals, much like coiled springs, that pack together to form granules. (b) Cellulose is a polysaccharide of β-glucose molecules joined by β-1,4-glycosidic bonds (in red). Cellulose chains pack together much more tightly to form structural fibers.

glycosidic bond or *glycosidic linkage*. Water is also produced, making this reaction another example of a condensation reaction, as is peptide bond formation. The glycosidic bond in sucrose is called an α,β-1,2 linkage because of the orientations (α and β) of the two –OH groups involved and their positions (C1 and C2) in the cyclic structures of the two monosaccharides.

Another important glycosidic bond involves the –OH groups on the C1 and C4 carbon atoms of glucose molecules. When glucose molecules link at these positions they can form long-chain polysaccharides. If the starting monomer is α-glucose, the bonds between them are α-1,4-glycosidic linkages, and the product of bond formation is starch (Figure 20.36a). The conversion of α-glucose into starch is an effective way for plants to store energy because formation of the α-1,4

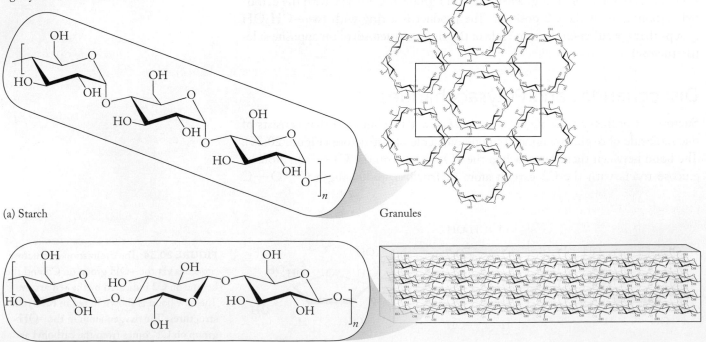

(a) Starch

Granules

(b) Cellulose

Fibers

linkage is reversible. With the aid of digestive enzymes, α-1,4 bonds can be hydrolyzed and starch converted back into glucose by plants that make the starch, or by animals that eat the plants.

The cellulose that plants synthesize to build stems and other organs has a structure (Figure 20.36b) slightly different from that of starch because the building blocks of cellulose are β-glucose instead of α-glucose, so the monomers are linked by β-1,4-glycosidic bonds. This structural difference is important because it enables starch to coil and make granules for efficient energy storage, whereas cellulose forms structural fibers.

Carbohydrates in plants are a major part of the total organic matter in any given ecological system; that is, they make up much of the system's **biomass**. People have used various forms of biomass, including wood and animal dung, as fuel for thousands of years. More recently, we have begun to convert biomass into a liquid *biofuel*, ethanol, for use as a gasoline additive and substitute. Because ethanol contains oxygen, it improves the combustion of a hydrocarbon fuel such as gasoline and reduces carbon monoxide emissions. An important industrial application of starch hydrolysis is the conversion of cornstarch into glucose and then, through fermentation, into ethanol:

$$C_6H_{12}O_6(aq) \rightarrow 2\ CH_3CH_2OH(aq) + 2\ CO_2(g)$$

This exothermic reaction provides energy to the yeast cells whose biological processes drive fermentation.

Unlike grazing animals, humans cannot digest cellulose because we do not have microorganisms in our digestive tracts that have enzymes called cellulases, which catalyze hydrolysis of β-glycosidic bonds. The challenge of reproducing what the cellulose-eating bacteria do in a laboratory or on an industrial scale is the focus of an enormous research effort as scientists try to develop efficient procedures for converting cellulose to glucose and then to ethanol. This research has focused on more efficient, less energy-intensive ways to break apart cellulose fibers. In addition, scientists are genetically engineering microorganisms like those in cattle stomachs to increase the supply of cellulases.

If this research is successful, it will address several major problems associated with ethanol as a gasoline additive or alternative fuel. First of all, it will lower the cost of production. Today it costs more to produce ethanol from cornstarch than to produce gasoline from crude oil. One reason for this is that most of the mass of a corn plant, or any plant, is cellulose, not starch. Ethanol production from cornstarch is also energy intensive: more than 70% of the energy contained in ethanol is expended in producing it; therefore the net energy value of ethanol from corn is less than 30%. If ethanol could be produced from cellulose instead of starch, its net energy value could be as high as 80%. Finally, the use of edible cornstarch in fuel production has driven up the price of foods derived from corn, including livestock feed. The impacts of this inflation have been felt worldwide and have been particularly painful in developing countries. The use of agricultural land and consumption of increasingly scarce water resources for ethanol production raise further concerns.

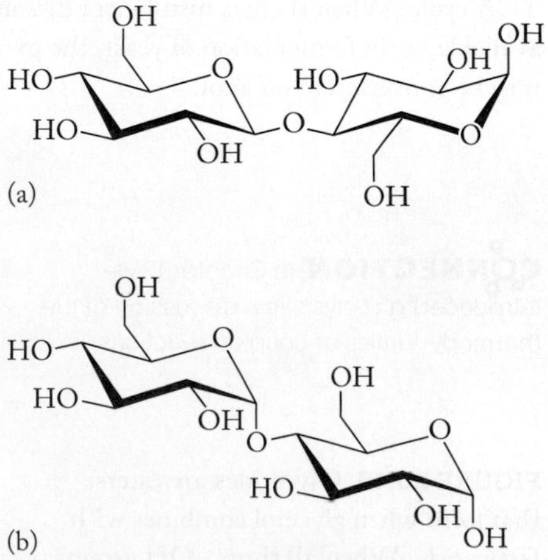

CONCEPT TEST

Cellobiose is a disaccharide made from the degradation of cellulose. We cannot digest cellobiose. Which of the two structures in Figure 20.37 represents a molecule of cellobiose?

glycosidic bond a C—O—C bond between sugar molecules.

biomass the total mass of organic matter in any given ecological system.

FIGURE 20.37 Two disaccharide molecules.

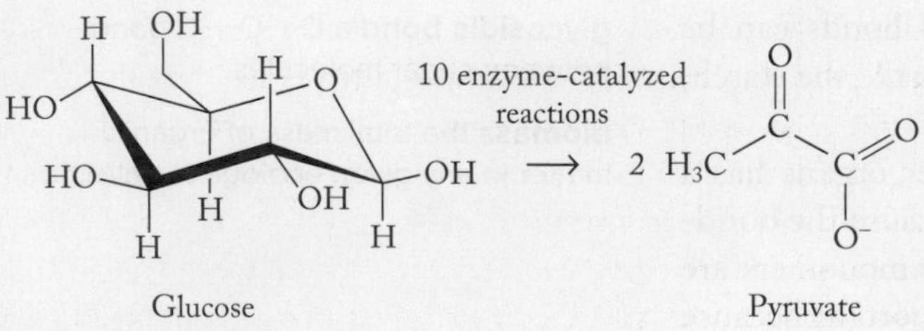

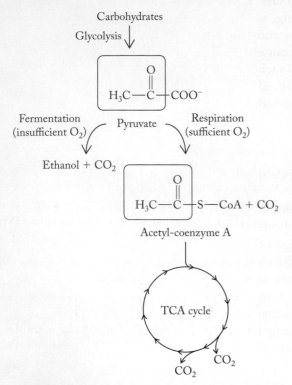

FIGURE 20.38 In glycolysis, molecules of glucose are broken down to pyruvate ions.

Energy from Glucose

Glycolysis is the series of reactions by which most animal cells, including those in our bodies, metabolize glucose. The product of these reactions is pyruvate ions (Figure 20.38), the conjugate base of pyruvic acid. Pyruvate sits at a metabolic crossroads and can be converted into different products, depending on the type of cell in which it is generated, the enzymes present, and the availability of oxygen (Figure 20.39). In yeast cells growing under low-oxygen conditions, pyruvate is converted into ethanol and CO_2.

Another series of reactions, called the **tricarboxylic acid (TCA) cycle** or the *citrate cycle*, occurs in the presence of sufficient dissolved O_2 and is fundamental to the conversion of glucose to energy in humans and other animals. A key step prior to the TCA cycle occurs when pyruvate loses CO_2 and forms an acetyl group, which then combines with coenzyme A. The resulting product, acetyl-coenzyme A, is a reactant in many biosynthetic pathways, including the production of fats.

20.6 Lipids

Lipids differ from carbohydrates and proteins in that they are not biopolymers. Lipids are best described by their physical properties rather than by any common structural subunit: lipids do not dissolve in water but are soluble in nonpolar solvents, and they are oily to the touch. Because they are insoluble in water, they are ideal components of cell membranes, which separate the aqueous solutions within cells from the aqueous environments outside them. An important class of lipids called **glycerides** are esters formed between glycerol and long-chain **fatty acids** (Figure 20.40). The three –OH groups on glycerol allow for mono-, di-, and triglycerides, with the last group being the most abundant. Glycerides account for over 98% of the lipids in the fatty tissues of mammals. When we look at lipids, glycerides, and fatty acids, we will see examples of both kinds of stereoisomers we discussed in Section 20.2 (enantiomers and cis–trans isomers), frequently all in one molecule.

Table 20.4 lists some common fatty acids. The most abundant ones have an even number of carbon atoms because the fatty acids are built *in*

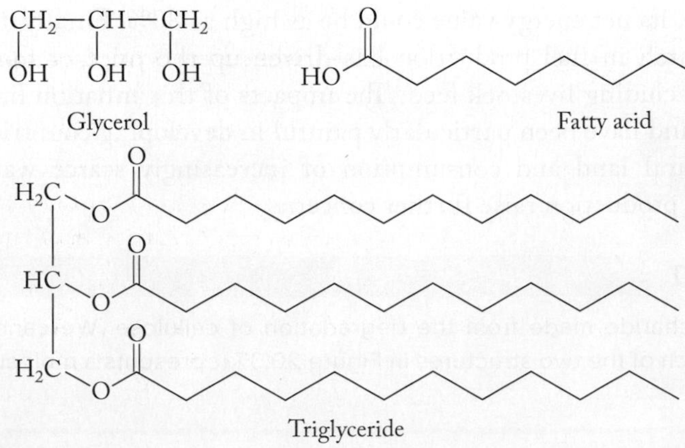

FIGURE 20.39 The fate of the pyruvate ions formed during glycolysis depends on the partial pressure of O_2 in the system. In the presence of sufficient O_2, the oxidation of pyruvate proceeds via the TCA cycle. When there is insufficient dissolved O_2 available, as in fermentation of yeast, the pyruvate may be converted to ethanol.

CONNECTION In Chapter 17 we introduced glycolysis in a discussion of the thermodynamics of coupled reactions.

FIGURE 20.40 Glycerides are esters that form when glycerol combines with fatty acids. When all three –OH groups on glycerol react to form ester bonds, the product is a *tri*glyceride.

TABLE 20.4 Names and Structural Formulas of Common Fatty Acids

Common Name (chemical name) (source)	Formula
SATURATED FATTY ACIDS	
Lauric acid (dodecanoic acid) (coconut oil)	$CH_3(CH_2)_{10}COOH$
Myristic acid (tetradecanoic acid) (nutmeg butter)	$CH_3(CH_2)_{12}COOH$
Palmitic acid (hexadecanoic acid) (animal and vegetable fats)	$CH_3(CH_2)_{14}COOH$
Stearic acid (octadecanoic acid) (animal and vegetable fats)	$CH_3(CH_2)_{16}COOH$
UNSATURATED FATTY ACIDS	
Oleic acid (*cis*-9-octadecenoic acid) (animal and vegetable fats)	$CH_3(CH_2)_7CH{=}CH(CH_2)_7COOH$
Linoleic acid (*cis,cis*-9,12-octadecadienoic acid) (linseed oil, cottonseed oil)	$CH_3(CH_2)_4CH{=}CHCH_2CH{=}CH(CH_2)_7COOH$
α-Linolenic acid (*cis,cis,cis*-9,12,15-octadecatrienoic acid) (linseed oil)	$CH_3CH_2CH{=}CHCH_2CH{=}CHCH_2CH{=}CH(CH_2)_7COOH$

vivo from two-carbon subunits. Their biosynthesis begins with the conversion of pyruvate to acetyl-coenzyme A. Most fatty acids contain between 14 and 22 carbon atoms. Some have no carbon–carbon double or triple bonds and are called *saturated* because they have as much hydrogen in their structures as the carbon atoms can hold. Other fatty acids have carbon–carbon double and/or triple bonds and are called *unsaturated* because they do not have all the hydrogen their carbon atoms could possibly hold; they have the capacity to add hydrogen to those multiple bonds.

A type of fatty acid much in the news because of its alleged health benefits is the family of omega-3 fatty acids. They are *polyunsaturated*, meaning that they have more than one carbon–carbon multiple bond. The name "omega-3" comes from the location of the first double bond counted from the methyl end of the chain, which is known as the omega (ω) end of the molecule. Omega-3 fatty acids are found in fish oils and some plant oils. α-Linolenic acid (see Table 20.4) is an omega-3 fatty acid.

CONNECTION We discussed saturated hydrocarbons in Section 5.9.

tricarboxylic acid (TCA) cycle a series of reactions that continue the oxidation of pyruvate formed in glycolysis.

lipid a class of water-insoluble, oily organic compounds that are common structural materials in cells.

glyceride a lipid consisting of esters formed between fatty acids and the alcohol glycerol.

fatty acid a carboxylic acid with a long hydrocarbon chain, which may be either saturated or unsaturated.

SAMPLE EXERCISE 20.7 Identifying Triglycerides **LO6**

How many different triglycerides (including constitutional isomers and stereoisomers) can be made from glycerol combining with two different fatty acids (symbolized by the letters X and Y to simplify the structures) if each molecule of triglyceride contains at least one molecule of each fatty acid?

Collect and Organize We are asked to determine how many different triglycerides can be made from glycerol and two different fatty acids. A triglyceride contains three fatty acid units.

Analyze Each fatty acid may bond to one of three –OH groups in glycerol. Each triglyceride has at least one X and one Y residue, which makes two formulas possible: X_2Y and Y_2X. Each of these formulas has two constitutional isomers, depending on whether the single fatty acid in the formula is bonded to the middle carbon or to one of the end carbon atoms. Finally, if a structure has an X on one end carbon atom and a Y on the other, then the middle carbon atom is a chiral center, which means that there are two enantiomeric forms of that compound.

Solve We can generate four different molecular structures by attaching X and Y in four different sequences to the glycerol –OH groups:

H_2C—O—X	H_2C—O—X	H_2C—O—Y	H_2C—O—Y
HC—O—X	HC—O—Y	HC—O—Y	HC—O—X
H_2C—O—Y	H_2C—O—X	H_2C—O—X	H_2C—O—Y
(1)	(2)	(3)	(4)

Structures (1) and (3) have chiral centers, so each of these isomers has two enantiomeric forms. Therefore there are a total of six different triglycerides possible.

Think About It The central carbon atoms in structures (1) and (3) are chiral because they are each bonded to a H atom, to an O atom, and to two C atoms, which are themselves bonded to different fatty acids: X and Y.

Practice Exercise How many different triglycerides can be made from glycerol and one molecule each of three different fatty acids? To simplify the drawings, use the letters A, B, and C to symbolize the different fatty acids.

Function and Metabolism of Lipids

Lipids are an important energy source in our diets, providing more energy per gram than carbohydrates or proteins. **Fats** are glycerides composed primarily of saturated fatty acids. They are solids at room temperature because the molecules can pack very tightly together. **Oils** are glycerides composed predominantly of unsaturated fatty acids and are liquids at room temperature because their chains have kinks in them due to the double bonds, which prevent the chains from packing together as closely as the saturated chains can. Oils can be converted into solid, saturated glycerides by hydrogenation. For example, in the hydrogenation of corn oil, which is a mixture of mostly two unsaturated fatty acids (oleic and linoleic acids), hydrogen is added to convert some or all of the –CH=CH– subunits into –CH_2—CH_2– subunits. Hydrogenation converts the oil into a solid at room temperature that is whipped with skim milk, coloring agents, and vitamins to produce the food spread we know as margarine.

The consumption of too much saturated fat is associated with coronary heart disease. One of the purported advantages of the so-called Mediterranean diet is that olive oil, a liquid composed of glycerides containing more than 80% oleic acid, is used in cooking rather than animal fats such as butter and lard. In addition, this diet tends to be richer in fish and vegetables, both of which contain unsaturated fats. Most animal fats, like the marbling in beef that enhances its flavor, are saturated.

Another problem with hydrogenating vegetable oil arises when the oils are only partially hydrogenated. Partial hydrogenation alters the molecular structure

fat a solid triglyceride containing primarily saturated fatty acids.

oil a liquid triglyceride containing primarily unsaturated fatty acids.

around their remaining C=C double bonds, changing them from their natural cis isomers into trans isomers (Figure 20.41). Unsaturated trans fats such as elaidic acid tend to be solids at room temperature because their molecules pack together more uniformly than do molecules of cis unsaturated fatty acids. Consumption of trans fatty acids is associated with increased levels of cholesterol in the blood and other health risks.

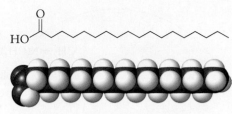

(a) Stearic acid: a saturated fatty acid

CONCEPT TEST

The mixture of lipids in the legs of reindeer living close to the Arctic Circle changes as a function of distance from the hoof: the closer to the hoof, the higher the percentage of unsaturated fatty acids. Why would this be an advantage for these reindeer?

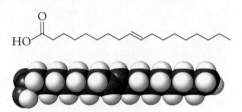

(b) Elaidic acid: a trans unsaturated fatty acid

The lipids in some prepared foods have been modified to reduce their caloric content but still provide the taste, aroma, and "mouth feel" we associate with lipid-rich foods. The active sites of enzymes that break down natural lipids accommodate triglycerides formed from glycerol and fatty acids. However, chemically modified esters made from the same fatty acids, but attached to an alcohol other than glycerol, cannot be metabolized by these enzymes. Such molecules, if they have the appropriate physical properties and are nontoxic, can be incorporated into foods without adding any calories because they are not metabolized.

Olestra is one such product (Figure 20.42). It is an ester made from long-chain fatty acids and the carbohydrate sucrose. (Remember, sugars have –OH groups and technically are alcohols.) Each of the eight –OH groups in a sucrose molecule reacts with a molecule of fatty acid to make the ester in olestra. The resultant material is used to deep-fry potato chips. Any olestra that remains on the chip does not add calories because it cannot be processed by enzymes that recognize only fatty acid esters on a glycerol scaffold.

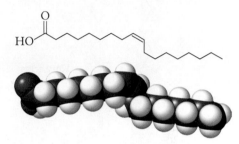

(c) Oleic acid: a cis unsaturated fatty acid

FIGURE 20.41 Types of fatty acids. (a) Stearic acid, a saturated C_{18} fatty acid. (b) Elaidic acid, an unsaturated C_{18} fatty acid (trans isomer). (c) Oleic acid, an unsaturated C_{18} fatty acid (cis isomer).

CONCEPT TEST

Olestra may be "calorie-free" as a food subject to metabolism in the living system, but how would it compare with a common triglyceride in terms of kilojoules of heat released per mole in a calorimeter experiment?

Olestra   Triglyceride

FIGURE 20.42 Olestra has a very different shape from that of the triglycerides typically metabolized by our bodies. Consequently, it cannot be processed by the enzymes that digest triglycerides.

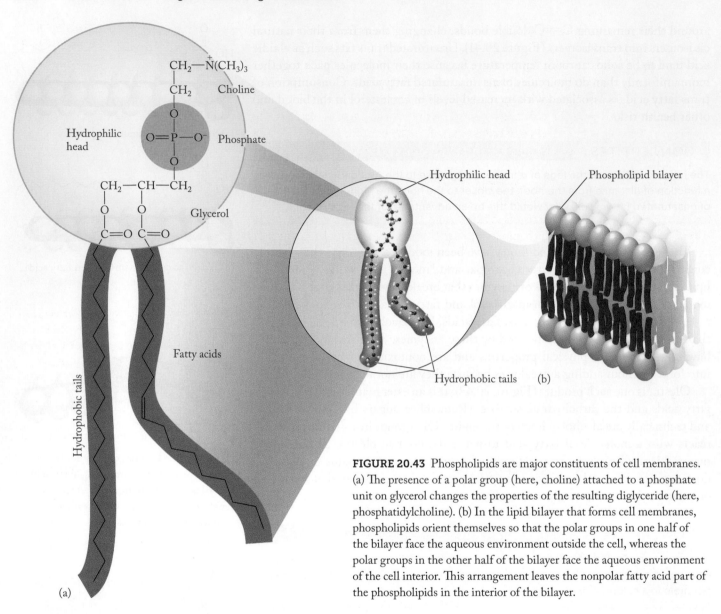

(a)

FIGURE 20.43 Phospholipids are major constituents of cell membranes. (a) The presence of a polar group (here, choline) attached to a phosphate unit on glycerol changes the properties of the resulting diglyceride (here, phosphatidylcholine). (b) In the lipid bilayer that forms cell membranes, phospholipids orient themselves so that the polar groups in one half of the bilayer face the aqueous environment outside the cell, whereas the polar groups in the other half of the bilayer face the aqueous environment of the cell interior. This arrangement leaves the nonpolar fatty acid part of the phospholipids in the interior of the bilayer.

phospholipid a molecule of glycerol with two fatty acid chains and one polar group containing a phosphate; phospholipids are major constituents of cell membranes.

lipid bilayer a double layer of molecules whose polar head groups interact with water molecules and whose nonpolar tails interact with each other.

nucleic acid one of a family of large molecules, which includes deoxyribonucleic acid (DNA) and ribonucleic acid (RNA), that stores the genetic blueprint of an organism and controls the production of proteins.

nucleotide a monomer unit from which nucleic acids are made.

The enzymes that metabolize triglycerides hydrolyze the esters and release glycerol and the fatty acids that were bonded to it. Glycerol enters the metabolic pathway for glucose. The fatty acids are oxidized in a series of reactions known as β-oxidation: a process that removes two carbon atoms at a time. For example, stearic acid (the saturated C_{18} fatty acid) is transformed into a C_2 fragment and the C_{16} acid, palmitic acid. Palmitic acid yields another C_2 fragment and myristic acid, and so forth, until the fatty acid is completely degraded. Electrons released from this oxidative process eventually are donated to O_2. The energy released by this process powers metabolism.

Other Types of Lipids

Cells contain other types of lipids in addition to triglycerides. One type, **phospholipids** (Figure 20.43a), plays a key role in cell structure. A phospholipid molecule consists of a glycerol molecule bonded to two fatty acid chains and to one phosphate group that is also bonded to polar substituents. The presence of nonpolar fatty acid chains and a polar region in the same molecule makes phospholipids ideal for forming cell membranes. In an aqueous medium, phospholipids form a **lipid bilayer**, a double layer enclosing each cell and isolating its interior from the outside environment. The phospholipid molecules of the bilayer align so that the nonpolar groups interact with each other inside the membrane, whereas the polar groups interact with water molecules outside it (Figure 20.43b). Membranes exist both to isolate the contents of cells and to serve as the locus of communication between processes that occur within and outside the cells.

Cholesterol is a lipid and a key component in the structure of cell membranes. It is also a precursor of the bile acids that aid in digestion and of steroid hormones, which regulate the development of the sex organs and secondary sexual traits, stimulate the biosynthesis of proteins, and regulate the balance of electrolytes in the kidneys. Biosynthesis of cholesterol, like the biosynthesis of fatty acids, begins with the conversion of pyruvate to acetyl-coenzyme A (Figure 20.39). In a healthy person, synthesis and use of cholesterol are tightly regulated to prevent overaccumulation and consequent deposition of cholesterol in coronary arteries. We clearly need cholesterol, but deposition in the arteries can lead to serious coronary disease (Figure 20.44).

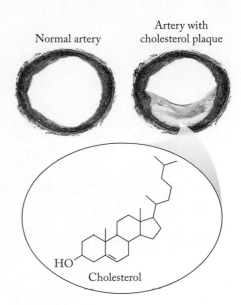

FIGURE 20.44 Cholesterol deposits called plaques are responsible for restricted blood flow, which results in a variety of sometimes catastrophic cardiovascular problems.

20.7 Nucleotides and Nucleic Acids

Nucleic acids are our fourth class of biomolecules and third class of biopolymers. We focus on two types: deoxyribonucleic acid (DNA) and ribonucleic acid (RNA). Even though nucleic acids make up only about 1% of a higher organism's mass, they control the metabolic activity of all its cells. The fraction is much higher in yeast and bacteria because those organisms are packed with ribosomes, which are themselves half RNA. DNA carries the genetic blueprint of an organism, and a variety of RNAs use that DNA blueprint to guide the production of proteins.

A nucleic acid is a polymer composed of monomeric units called **nucleotides**. Each nucleotide unit is in turn composed of three subunits: a five-carbon sugar, a phosphate group, and a nitrogen-containing base (Figure 20.45). The phosphate group in each nucleotide is attached to a carbon in the side chain of the sugar called the $5'$ carbon atom (the prime number refers to the position of the carbon atom in the sugar molecule). The nitrogen-containing base is attached to the $1'$ carbon atom in each sugar molecule. The sugar in Figure 20.45 is called *ribose*, which makes this a nucleotide in a strand of *ribo*nucleic acid, or RNA. If the sugar were *deoxyribose* instead, there would be a H atom in place of the −OH group on the $2'$ carbon atom, and the nucleotide would be a building block of *deoxyribo*nucleic acid, or DNA. Because of the ionized phosphate groups, both DNA and RNA are polyanions. Their anionic character is important because it causes them to interact with proteins while not allowing them to pass through cell membranes.

The nitrogen-containing base in Figure 20.45 is called adenine (A). Structural formulas of adenine and the other four bases in nucleic acids—cytosine (C), guanine (G), thymine (T), and uracil (U)—are shown in Figure 20.46. The point of attachment of the sugar–phosphate groups on each base is indicated by −R. In addition to the difference in their sugars, RNA and DNA also differ in one of the bases in their nucleotides: RNA contains A, C, G, and U, but DNA contains A, C, G, and T.

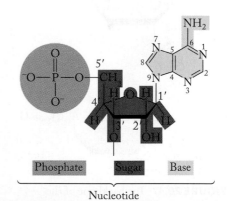

FIGURE 20.45 A nucleotide consists of a phosphate group and a nitrogen-containing base that are both bonded to a five-carbon sugar.

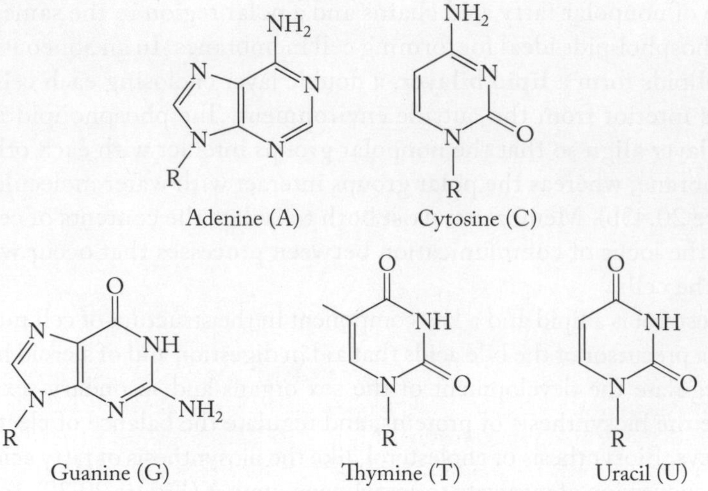

Adenine (A)

Cytosine (C)

Guanine (G)

Thymine (T)

Uracil (U)

FIGURE 20.46 Structural formulas of the five nitrogen-containing bases in nucleotides. The nucleotides in DNA contain A, C, G, and T; those in RNA contain A, C, G, and U. The R group identifies the point of attachment of the sugar residue.

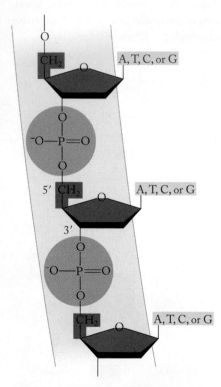

FIGURE 20.47 The backbone of the polymer chain in DNA consists of alternating sugar units (yellow) and phosphate units (pink). The bases (blue) are attached to the backbone through the 1′ carbon atom of the sugar unit.

replication the process by which one double-stranded DNA forms two new DNA molecules, each one containing one strand from the original molecule and one new strand.

In a polymeric strand of nucleic acid, each phosphate is also linked to the 3′ carbon atom in the sugar of the monomer that precedes it in the chain, as shown for a strand of DNA in Figure 20.47. Both DNA and RNA strands are synthesized in the cell from the 5′ to the 3′ direction (downward in Figure 20.47). The structures of DNA and RNA are frequently written using only the single-letter labels of their bases, beginning with the free phosphate group on the 5′ end of the chain and reading toward the free 3′ hydroxyl group at the other terminus.

When scientists first isolated DNA and began to analyze its composition, they made a pivotal observation about the abundance of the nitrogen-containing bases. A typical molecule of DNA consists of thousands of nucleotides, and the percentages of the four bases in different samples of DNA can vary over a wide range. However, the percentage of A in a sample always matches the percentage of T. Likewise, the percentage of C always matches that of G. This result makes sense if the bases are paired because a molecule of A can form two hydrogen bonds to a molecule of T, whereas a molecule of C can form three hydrogen bonds with a molecule of G. Therefore A–T and G–C pairings maximize the number of hydrogen bonds possible (Figure 20.48a). More important, the base pairs assembled this way all have the same width and fit together in a regular structure.

The normal structure of DNA has two strands of nucleotides wrapped around each other in a form that is called a *double helix* (Figure 20.48b). The nucleotide backbone is on the outside of the spiraling strands, with hydrogen bonds between the complementary bases keeping the two strands together. Notice also that the base pairs are parallel to each other and perpendicular to the helical axis. The fidelity of this base-pairing—A always with T, and C always with G—gives DNA the ability to copy itself. If a pair of complementary strands is unzipped into two single strands, each strand provides a template on which a new complementary strand can be synthesized via the process called **replication** (Figure 20.49).

During replication, the two strands are separated, forming a structure called the *replication fork*. The fork advances through the DNA as replication proceeds.

(a)

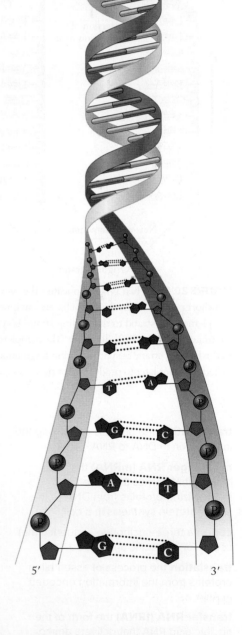

Adenine   Thymine

Guanine   Cytosine

FIGURE 20.48 The nitrogen-containing bases on one strand of DNA pair with the bases on a second strand by hydrogen bonding. (a) Adenine and thymine pair via two hydrogen bonds; guanine and cytosine pair via three hydrogen bonds. (b) DNA as a double helix with the sugar–phosphate backbone on the outside and the base pairs on the inside.

One of the two resulting strands, the leading strand, is unzipped in the 3′–5′ direction, which allows the new complementary strand to be continuously synthesized in the 5′–3′ direction. The replication of the second strand, called the lagging strand, is more complicated. It proceeds in short fragments, which are then assembled into a continuous strand later in the process.

SAMPLE EXERCISE 20.8 Using Base Complementarity in DNA   **LO7**

If 31.6% of the nucleotides in a sample of DNA are adenine, what are the percentages of cytosine, guanine, and thymine?

Collect, Organize, and Analyze We know how much adenine is in a DNA sample and need to calculate the rest of the nucleotide composition. We also know that nucleotides are paired so that A always pairs with T, and C always pairs with G. So the percentage of T must equal the percentage of A in the sample, and the percentage of C must equal the percentage of G.

Solve If A = 31.6%, then T = 31.6%. This leaves (100 − 2 × 31.6) = 36.8% left to be equally distributed between G and C. Therefore G = C = 18.4%.

Think About It The percentages should total 100%, and they do.

 Practice Exercise Indicate the sequence of the complementary strand on the double helix formed by each of these sequences of nucleotides:

a. CGGTATCCGAT
b. TTAAGCCGCTAG

DNA's double-stranded structure is also the key to its ability to preserve genetic information. The two strands carry the same information, much like an old-fashioned photograph and its negative. Genetic information is duplicated every time a DNA molecule is replicated, a process that is essential whenever a cell divides into two new cells.

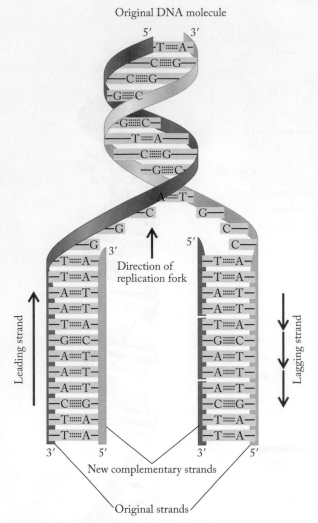

Original DNA molecule

Direction of replication fork

Leading strand

Lagging strand

New complementary strands

Original strands

FIGURE 20.49 When DNA replicates, the two strands of a short portion of the double helix are unzipped. The complementary strand to the leading strand is synthesized continuously in the 5′–3′ direction. The complementary strand to the lagging strand is synthesized in fragments, which are joined together later to produce a continuous strand.

From DNA to New Proteins

Proteins are formed from amino acids in accordance with the *genetic code* contained in the base sequences of DNA strands. The bases A, T, G, and C are the alphabet in this code, and the "words" in the code are three-letter combinations of these four letters, with each word representing a particular amino acid or a signal to begin or end protein synthesis. Using four letters to write three-letter words means there are $4^3 = 64$ combinations possible, more than enough to encode the 20 amino acids found in cells. The function of the genetic code is to specify the protein's primary structure—the sequence of amino acids in proteins. The flow of genetic information goes from DNA to RNA to proteins, a sequence sometimes called the *central dogma of molecular biology*.

Protein synthesis begins with a process called **transcription** (Figure 20.50a), in which double-stranded DNA unwinds and its genetic information guides the synthesis of a single strand of a molecule called **messenger RNA (mRNA)**. This strand of mRNA has the complementary base sequence of the original DNA. It carries the 3-letter words of that DNA, in the form of three-base sequences called **codons** (Table 20.5), from the nucleus of the cell into the cytoplasm, where the mRNA binds with a cellular structure called a ribosome. Keep in mind that a sequence of A, C, G, and T in the original DNA is transcribed into the following sequence in mRNA:

DNA: ...ACGT...

mRNA: ...UGCA...

At the ribosome the genetic information in the messenger RNA directs the synthesis of particular proteins in a process called **translation**. Another type of RNA, called **transfer RNA (tRNA)**, plays a key role in translation. There are 20 different forms of tRNA in the cell, one for each amino acid. To see how tRNA works, let's look at Figure 20.50(b). The first codon in this piece of an mRNA strand is AUG, which codes for the amino acid methionine (see Table 20.5).

transcription the process of copying the information in DNA to RNA.

messenger RNA (mRNA) the form of RNA that carries the code for synthesizing proteins from DNA to the site of protein synthesis in a cell.

codon a three-nucleotide sequence that codes for a specific amino acid.

translation the process of assembling proteins from the information encoded in RNA.

transfer RNA (tRNA) the form of the nucleic acid RNA that delivers amino acids, one at a time, to polypeptide chains being assembled by the ribosome–mRNA complex.

TABLE 20.5 mRNA Codons

Amino Acid	Codons	Amino Acid	Codons
Ala	GCU, GCC, GCA, GCG	Leu	UUA, UUG, CUU, CUC, CUA, CUG
Arg	CGU, CGC, CGA, CGG, AGA, AGG	Lys	AAA, AAG
Asn	AAU, AAC	Met	AUG
Asp	GAU, GAC	Phe	UUU, UUC
Cys	UGU, UGC	Pro	CCU, CCC, CCA, CCG
Gln	CAA, CAG	Ser	UCU, UCC, UCA, UCG, AGU, AGC
Glu	GAA, GAG	Thr	ACU, ACC, ACA, ACG
Gly	GGU, GGC, GGA, GGG	Trp	UGG
His	CAU, CAC	Tyr	UAU, UAC
Ile	AUU, AUC, AUA	Val	GUU, GUC, GUA, GUG
Start	AUG	Stop	UAG, UGA, UAA

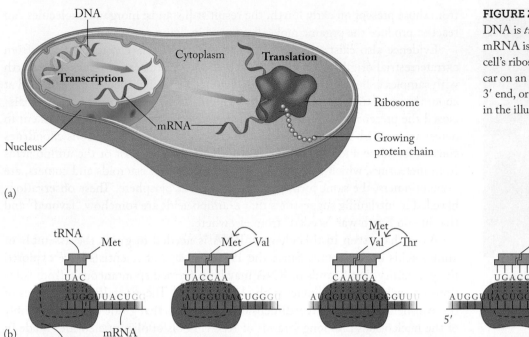

(a)

(b)

FIGURE 20.50 (a) In protein synthesis, DNA is *transcribed* into mRNA, and (b) the mRNA is then *translated* into proteins on the cell's ribosomes. The ribosome moves like a car on an mRNA track from the 5′ end to the 3′ end, or from left to right along the mRNA in the illustration.

In the cytoplasm surrounding the ribosome, molecules of tRNA are reversibly bonded to molecules of every amino acid. The particular tRNA molecules that are bonded to methionine also contain the sequence UAC, the complement of AUG, at a site that allows the tRNA to interact with mRNA. As Figure 20.50(b) shows, the segment of mRNA with the AUG codon links with the complementary strand on the tRNA molecule bonded to methionine. In doing so, the methionine is put into a position to unlink from the tRNA and to be the first amino acid residue in the protein being synthesized.

The sequence of events in the preceding paragraph is repeated many times. In Figure 20.50(b), the second codon, GUU, links up with a molecule of tRNA that has a CAA binding site and a molecule of valine in tow. In this way valine moves into position to become the next amino acid residue in the protein and to form a peptide bond with the N-terminal methionine. Valine is followed by threonine, which is followed by glycine, and so on, until a Stop codon finally signals the end of the translation process.

CONCEPT TEST

If a GUU codon attracts valine to the translation site, does a UUG codon do the same thing? Explain your answer.

20.8 From Biomolecules to Living Cells

We end this chapter by addressing two fundamental questions about the major classes of biomolecules and their roles in sustaining life: (1) how were they first formed on prebiotic Earth, and (2) how did they assemble into living cells? Experiments conducted in the 1950s at the University of Chicago by chemistry professor Harold Urey (1893–1981) and his student Stanley Miller (1930–2007) showed that amino acids could form from H_2O, CH_4, NH_3, and H_2 (Figure 20.51). Although the reactants the two scientists chose are now thought to be different

FIGURE 20.51 The apparatus used by Miller (shown) and Urey to simulate the synthesis of amino acids in the atmosphere of early (prebiotic) Earth. Discharges between the tungsten electrodes were meant to provide the sort of energy that might have come from lightning.

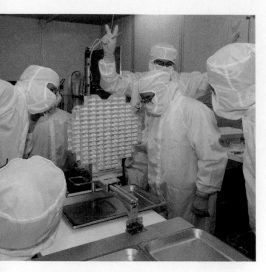

FIGURE 20.52 NASA scientists who analyzed the samples collected by the *Stardust* spacecraft were careful to avoid contaminating it with biological material from Earth. Doing so meant isolating themselves from the sample as they prepared it for analysis.

FIGURE 20.53 Black clouds of transition metal oxides and sulfides flow into the sea through chimneys like this one at a deep-ocean hydrothermal vent. Some scientists believe that these particles may have guided and catalyzed the formation of the first self-replicating molecules on Earth.

from those present on early Earth, the result still stands: inorganic molecules can react to produce the organic molecules found in living systems.

Evidence also exists that some biomolecules may have reached Earth from extraterrestrial origins. In 2006 the NASA spacecraft *Stardust* returned to Earth with samples collected from the tail of a comet that is believed to have formed at about the same time as the solar system (Figure 20.52). Subsequent analyses disclosed the presence of glycine in the comet. This was not the first experiment to detect amino acids in space. A class of meteorites called carbonaceous chondrites contain isovaline and other amino acids. Interestingly, most of the amino acids from meteorites, which are believed to be fragments of asteroids and comets, are L-enantiomers, the same form that dominates our biosphere. These observations have led to intriguing suggestions that L-amino acids are somehow "favored" and that life on Earth was "seeded" from elsewhere.

As we have seen in this chapter, RNA is needed to guide the assembly of amino acids into proteins. Since the 1990s, groups of scientists have explored the possibility that strands of RNA may have formed spontaneously from solutions of nucleotides in contact with clay minerals. The crystalline structures of these minerals provide three-dimensional templates that guide the self-assembly of the nucleotides into long strands of RNA. Pools of oligonucleotides made in this way usually contain many chains with random sequences, but some can actually catalyze their own replication. This ability to self-replicate is crucial to life, and the observation that molecules can speed up their own replication on a clay surface suggests that processes essential to the formation of living cells could have happened spontaneously. Much controversy still exists about these ideas, but the fact that RNA can act as both a source of information and a catalyst is part of the *RNA world hypothesis*. This hypothesis proposes that a world filled with life based on RNA predates the current world of life based on DNA and proteins. The capacity of RNA to both store information like DNA *and* act as a catalyst like an enzyme suggests that RNA alone could have supported cellular or precellular life forms.

Current research also is testing the hypothesis that life on Earth may have evolved near deep-ocean hydrothermal vents. Entire ecosystems have been discovered at these locations since they were first explored in the 1970s. They are sustained by geothermal and chemical energy rather than energy from the Sun. It may be that life actually began in such environments, with hydrothermal energy driving reactions in which inorganic compounds such as carbon dioxide and hydrogen sulfide formed small organic compounds. As with the reactions on the surfaces of clay minerals, the synthesis reactions at hydrothermal vents may have been catalyzed and guided by reactants adsorbed on solid compounds such as FeS and MnO_2, which pour into the sea in dense black clouds near some vents (Figure 20.53). Among the known products of these reactions are acetate ions (CH_3COO^-). Acetate is a key intermediate in many biosynthetic pathways in living organisms. In modern bacteria, the systems that make acetate depend on a catalyst made of iron, nickel, and sulfur that has a structure much like that of particles produced by "black smokers" on the ocean floor.

To take the next step toward forming living cells, large biomolecules must have assembled themselves into even larger structures, such as membranes, that allow cells and structures within them to collect materials and retain them at concentrations different from those in the surrounding medium. Molecules in these assemblies are not necessarily connected by covalent bonds, but rather are held together by the intermolecular interactions we have discussed in this chapter.

SAMPLE EXERCISE 20.9 Integrating Concepts: Liquid Oil to Solid Fat

Imagine we have a research project to turn soybean oil into a solid for use in a butter substitute. We can accomplish this by hydrogenating the oil. Suppose we are working at the level of a small-scale industrial facility to test the feasibility of this process. A particular sample of soybean oil contains 54% linoleic acid, 24% oleic acid, and 18% palmitic acid, by mass.

$$CH_3(CH_2)_4CH\!=\!CHCH_2CH\!=\!CH(CH_2)_7COOH$$

$$CH_3(CH_2)_7CH\!=\!CH(CH_2)_7COOH$$

$$CH_3(CH_2)_{14}COOH$$

We need to hydrogenate 10.0 kg of soybean oil. Our source of hydrogen is the steam-reforming of methane:

$$H_2O(g) + CH_4(g) \rightarrow CO(g) + 3\,H_2(g)$$

a. What volume of methane at 20°C and 1.00 atm of pressure do we need to produce the amount of hydrogen required to completely hydrogenate 10.0 kg of soybean oil?

b. How many different compounds could result from the hydrogenation of a triglyceride that contained one of each of the three fatty acids in this study: palmitic acid, linoleic acid, and oleic acid?

Collect and Organize We need to determine how much methane is needed in the steam-reforming reaction to supply enough hydrogen to react completely with the unsaturated fats present in a 10.0-kg sample of soybean oil. We know the oil contains three fatty acids: linoleic, oleic, and palmitic acid. Table 20.4 contains the molecular structures of all three of these fatty acids.

Analyze According to the structures in Table 20.4, palmitic acid is a saturated fatty acid, which means it does not react with hydrogen. Oleic acid has one $C\!=\!C$ double bond per molecule and linoleic acid has two, which means 1 mole of oleic acid combines with 1 mole of H_2, and 1 mole of linoleic acid combines with 2 moles of H_2. We can use the given weight percentages (24% oleic acid and 54% linoleic acid) to calculate that a 10.0-kg sample contains 2.4 kg of oleic acid and 5.4 kg of linoleic acid. The structures of these two fatty acids can be used to write their molecular formulas, and from those formulas we can calculate their molar masses. Converting the above masses into grams and dividing these masses by the molar masses of the two fatty acids yields the moles of each in the sample. We then use the 1:1 and 1:2 hydrogenation reaction stoichiometries to calculate the number of moles of H_2 we need, and we use the stoichiometry of the steam-reforming reaction (1 mol CH_4:3 mol H_2) to convert moles of H_2 into moles of CH_4. The number of moles of CH_4 needed, along with the given temperature and pressure of the gas, can then be used in the ideal gas law equation to calculate the volume of CH_4 needed.

Solve

a. Using the condensed structure of the two unsaturated fatty acids to determine their molecular formulas and calculate their molar masses:

Fatty Acid	Molecular Formula	Molar Mass (g/mol)
Oleic acid	$C_{18}H_{34}O_2$	282.46
Linoleic acid	$C_{18}H_{32}O_2$	280.45

Converting the masses of these fatty acids in the sample into moles:

$$2.4\ \text{kg oleic acid} \times \frac{1000\ g}{1\ kg} \times \frac{1\ mol}{282.46\ g} = 8.5\ \text{mol oleic acid}$$

$$5.4\ \text{kg linoleic acid} \times \frac{1000\ g}{1\ kg} \times \frac{1\ mol}{280.45\ g} = 19\ \text{mol linoleic acid}$$

Calculating the total moles of CH_4 needed:

$$8.5\ \text{mol oleic acid} \times \frac{1\ mol\ H_2}{1\ mol\ oleic\ acid} \times \frac{1\ mol\ CH_4}{3\ mol\ H_2} = 2.8\ \text{mol } CH_4$$

$$+\,19\ \text{mol oleic acid} \times \frac{2\ mol\ H_2}{1\ mol\ oleic\ acid} \times \frac{1\ mol\ CH_4}{3\ mol\ H_2} = 13\ \text{mol } CH_4$$

Total moles of CH_4 = 2.8 + 13 = 16 mol CH_4

$$PV = nRT$$

$$(1.00\ \text{atm})V = 16\ \text{mol} \times \frac{0.08206\ L \cdot atm}{K \cdot mol} \times 293\ K$$

$$V = \frac{16\ \text{mol} \times \dfrac{0.08206\ L \cdot atm}{K \cdot mol} \times 293\ K}{1.00\ \text{atm}} = 380\ L$$

Reported to the correct number of significant figures, the reaction requires 3.8×10^2 L of methane.

b. If a single triglyceride contained one of each of the saturated fatty acids that result from this process, it would contain one unit of palmitic acid, which was saturated in the original material and unchanged by the hydrogenation, and two units of stearic acid, the C_{18} saturated fatty acid that is the product of hydrogenation of oleic acid and linoleic acid. We can represent this symbolically, letting A = palmitic acid and B = stearic acid. Two constitutional isomers of triglycerides result:

$$\begin{array}{ccc}
CH_2{-}CH{-}CH_2 & \qquad & CH_2{-}CH{-}CH_2 \\
|\quad\ |\quad\ | & & |\quad\ |\quad\ | \\
A\quad B\quad B & & B\quad A\quad B
\end{array}$$

The structure on the left also has two enantiomeric forms because the carbon in the –CH– unit is a chiral center. Therefore the products are two constitutional isomers, one of which also has two enantiomers (stereoisomers), for a total of three different triglycerides.

Think About It The number of moles of methane required seems reasonable according to our estimate. Actual soybean oil probably has several different triglycerides in it that are a combination of saturated and unsaturated fatty acid components, resulting in the observed composition by weight. Natural oils can be quite complex mixtures.

SUMMARY

LO1 Isomers may be **constitutional isomers**, which are distinguished by the connectivity of their atoms, or they may be **stereoisomers**, which are distinguished by their arrangement in three-dimensional space. Many biologically important molecules are chiral, resulting in **optical isomers**, a type of stereoisomer. (Section 20.2)

LO2 The acid–base properties of **amino acids** are important for structural and functional reasons. We can define these properties by using data derived from titrations. (Sections 20.2 and 20.3)

LO3 The **proteins** and **peptides** in the human body are composed of 20 α-amino acids reversibly linked by **peptide bonds**. The sequence of amino acids in peptide and protein chains matters in determining their properties. (Section 20.3)

LO4 The structure of a protein is crucial to its function. It is defined by the sequence of amino acids in the chain, the geometric pattern that segments of a chain adopt, the overall three-dimensional shape of the protein, and any larger structure formed when two or more proteins interact and function as a single unit. (Section 20.4)

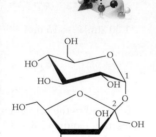

LO5 **Carbohydrates** are produced from CO_2 and H_2O, and organisms derive energy from glycolysis and the **tricarboxylic acid (TCA) cycle**, the reaction pathways by which glucose is oxidized to CO_2 and H_2O. **Monosaccharides** are joined through **glycosidic bonds** into **polysaccharides** such as starch for energy storage and cellulose for structural support in plants. (Section 20.5)

LO6 **Lipids** include important families of molecules, including **glycerides**, **fats**, and **oils**. Some lipids are a major source of energy in our diet, and others play key roles in cell structure. (Section 20.6)

Olestra

LO7 The **nucleic acids** DNA and RNA contain an organism's genetic information and control protein synthesis through **transcription** and **translation**. Living cells may have formed as a result of chemical reactions in which inorganic molecules combined to form small organic molecules such as amino acids and **nucleotides**. (Section 20.7)

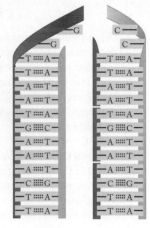

PARTICULATE **PREVIEW WRAP-UP**

Glutamic acid contains an amine ($R-NH_2$) functional group and two carboxylic acid ($R-COOH$) functional groups. At physiological pH (~7.4), the carboxylic acid groups are deprotonated ($R-COO^-$), but the amine is still protonated ($R-NH_3^+$), meaning that the glutamic acid molecule has an overall negative charge at a physiological pH.

PROBLEM-SOLVING SUMMARY

Type of Problem	Concepts and Equations	Sample Exercises
Recognizing constitutional isomers	Establish that the compounds have the same molecular formula, and if they do, look for different arrangements of bonds.	**20.1**
Identifying chiral molecules	Identify carbon atoms with four different groups attached.	**20.2, 20.3**
Interpreting acid–base titration curves of amino acids	At low pH all amino acids have at least two ionizable H atoms, one each from $-COOH$ and $-NH_3^+$. Side-chain carboxylic acid and amine groups may also impart acidic and basic strength to amino acids.	**20.4**

Type of Problem	Concepts and Equations	Sample Exercises
Drawing and naming peptides	Connect the α-amine of one amino acid to the α-carboxylic acid of another with a peptide bond. Starting with the free amine (N-) terminus on the left, name each amino acid residue by changing the ending of the name of the parent amino acid to -yl in all but the last (C-terminal) amino acid.	**20.5**
Determining the sequence of a small peptide	Arrange fragments of the peptide in a way that reveals the sequence of the peptide.	**20.6**
Identifying triglycerides	The –COOH groups of fatty acids react with the –OH groups in glycerol to form triglycerides and water.	**20.7**
Using base complementarity in DNA	Identify the base pairs: A pairs with T; G pairs with C. The percentage of T should equal the percentage of A, and the percentage of C should equal the percentage of G.	**20.8**

VISUAL PROBLEMS

(Answers to boldface end-of-chapter questions and problems are in the back of the book.)

20.1. The nucleotides in DNA contain the bases with the structures shown in Figure P20.1. Identify the basic functional groups in the structures.

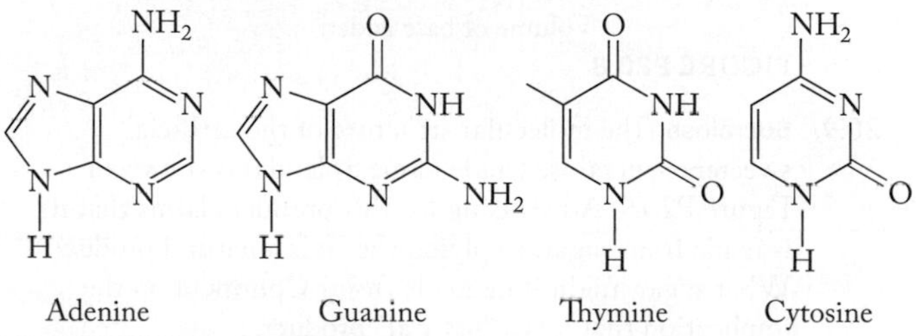

Adenine   Guanine   Thymine   Cytosine

FIGURE P20.1

20.2. **Experimental Agent** Bentiromide (Figure P20.2) is a peptide that was once evaluated as an agent to monitor the function of the pancreas during therapy. Draw the structures of the amino acids that form bentiromide, and indicate whether they are α-amino acids.

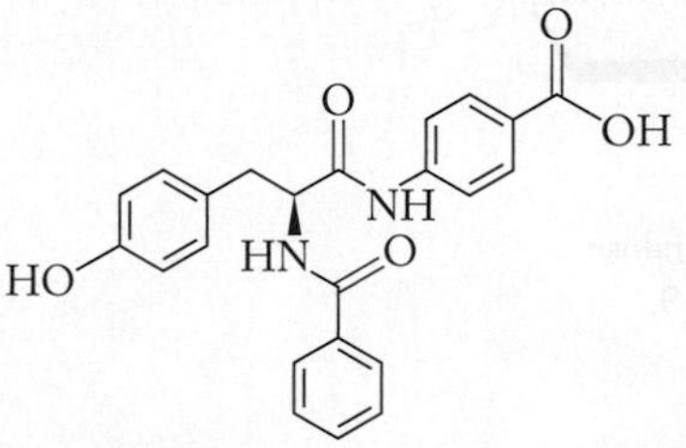

FIGURE P20.2

20.3. **Olive Oil** Olive oil contains triglycerides such as those shown in Figure P20.3. Which of the fatty acids in these triglycerides is/are saturated?

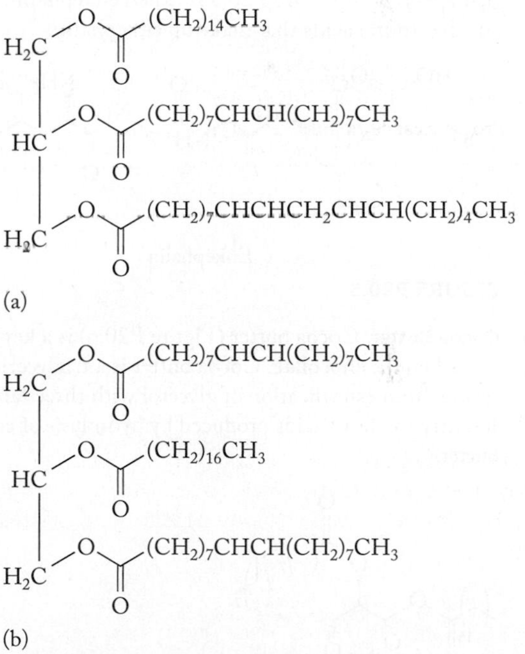

(a)

(b)

FIGURE P20.3

20.4. Treating Infections The major component of the antibiotic ointment bacitracin is a cyclic polypeptide called bacitracin A. It was first isolated in 1943 from a knee scrape from a girl named Margaret Tracy, after whom it is named. Bacitracin is effective topically and is used to treat skin, eye, and wound infections. Figure P20.4 shows the structure of bacitracin A. Identify the amino acids found in human proteins that are also part of the structure of bacitracin A.

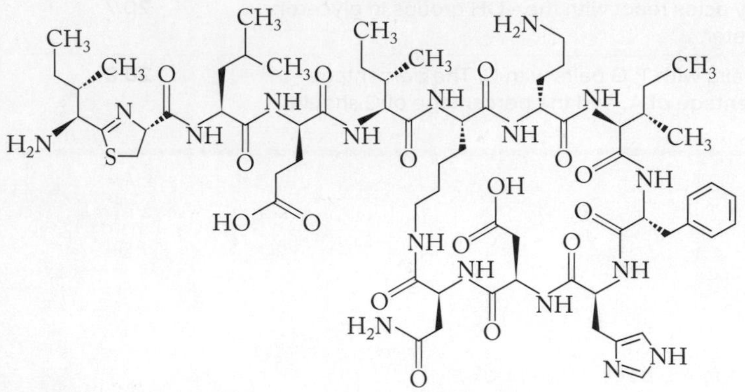

FIGURE P20.4

20.5. Natural Painkillers The human brain produces polypeptides called *endorphins* that help in controlling pain. The pentapeptide in Figure P20.5 is called enkephalin. Identify the five amino acids that make up enkephalin.

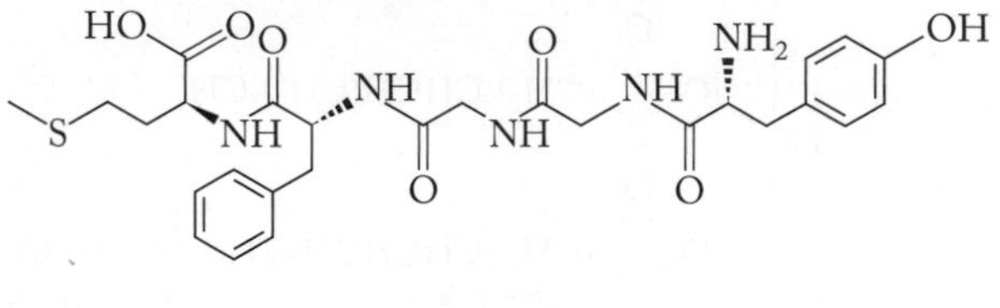

Enkephalin

FIGURE P20.5

20.6. Cocoa Butter Cocoa butter (Figure P20.6) is a key ingredient in chocolate. Cocoa butter is a triglyceride that results from esterification of glycerol with three fatty acids. Identify the fatty acids produced by hydrolysis of cocoa butter.

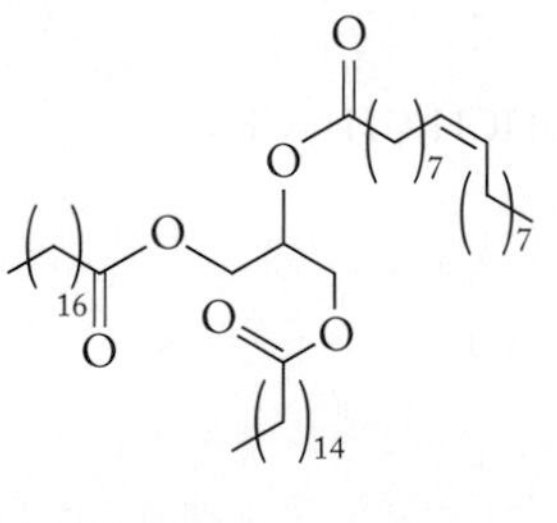

Cocoa butter

FIGURE P20.6

20.7. Trans Fats The role of "trans fats" in human health has been extensively debated both in the scientific community and in the popular press. What type of isomerism does the word "trans fat" refer to? Which of the molecules in Figure P20.7 are considered trans fats?

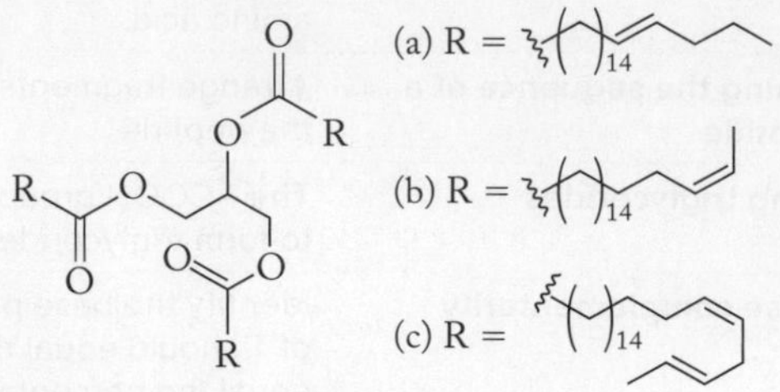

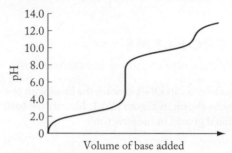

FIGURE P20.7

20.8. Figure P20.8 shows the titration curve of which of these amino acids: leucine, histidine, or lysine?

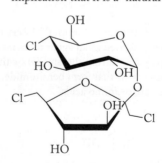

FIGURE P20.8

20.9. Sucralose The molecular structure of the artificial sweetener sucralose (trade name Splenda) is shown in Figure P20.9. Advertising for this product claims that it is made from sugar, implying that it is a natural product. What sugar might it be made from? Comment on the implication that it is a "natural" product.

Sucralose

FIGURE P20.9

20.10. Covalent bonding leads to primary structure and intermolecular forces, both of which ultimately lead to secondary, tertiary, and quaternary structure. Use representations [A] through [I] in Figure P20.10 to answer questions a–f.

a. [A] depicts two strands of DNA in a double helix. Are the two strands held together by covalent bonds, intermolecular forces, or both? [B] depicts two alpha helices linked together. Are the two helices held together by covalent bonds, intermolecular forces, or both?

b. [C] depicts a phospholipid bilayer. Is the bilayer held together by covalent bonds, intermolecular forces, or both?

c. [D] depicts a disaccharide. Are the two sugars each held together by covalent bonds, intermolecular forces, or both?

d. Which of the single molecules depicted are an important structural component of [C]?

e. Which of the single molecules depicted react to form [E]?

f. What intermolecular forces are typically involved for the amino acid depicted in [F] when it is incorporated into a protein?

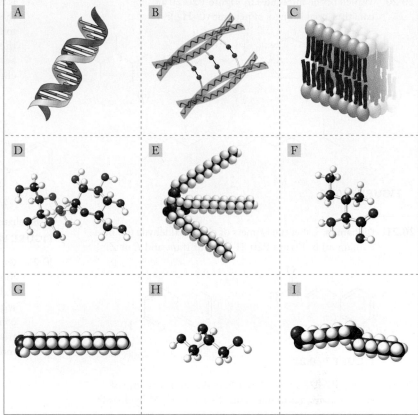

FIGURE P20.10

QUESTIONS AND PROBLEMS

Organic Molecules, Isomers, and Chirality

Concept Review

20.11. Can all of the terms *enantiomer*, *achiral*, and *optically active* be used to describe a single compound? Explain.

20.12. Two compounds have the same structure and the same physical properties but also have the same optical activity. Are they enantiomers or the same molecule?

20.13. How do constitutional isomers differ from stereoisomers?

*20.14. Could a racemic mixture be distinguished from an achiral compound on the basis of optical activity? Explain your answer.

20.15. Why is the amino acid glycine (Figure P20.15) achiral?

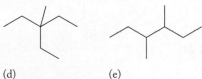

FIGURE P20.15

20.16. Can stereoisomers of molecules such as cis and trans RHC=CHR also have optical isomers? (R may be any of the functional groups we have encountered in this textbook.) Explain your answer.

20.17. Which type of hybrid orbitals on a carbon atom, sp, sp^2, or sp^3, can give rise to enantiomers?

*20.18. Could an oxygen atom in an alcohol, ketone, or ether ever be a chiral center in the molecule?

Problems

20.19. Which of the molecules in Figure P20.19 are constitutional isomers of octane (C_8H_{18})?

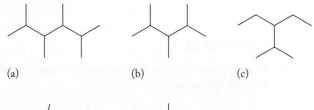

(a)   (b)   (c)

(d)   (e)

FIGURE P20.19

20.20. Which of the molecules in Figure P20.20 are constitutional isomers of heptane (C_7H_{16})?

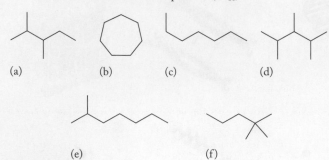

(a)   (b)   (c)   (d)

(e)   (f)

FIGURE P20.20

20.21. **Cinnamon** Label the isomers of cinnamaldehyde (oil of cinnamon) in Figure P20.21 as cis or trans and *E* or *Z*.

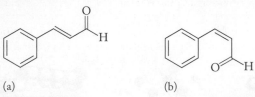

(a)   (b)

FIGURE P20.21

*20.22. Figure P20.22 shows the carbon-skeleton structure of carvone, which is found in oil of spearmint. Why doesn't the molecule carvone have cis and trans isomers?

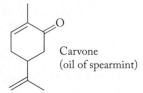

Carvone
(oil of spearmint)

FIGURE P20.22

20.23. Which of the molecules in Figure P20.23 are chiral?

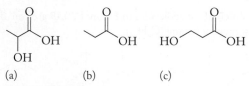

(a)   (b)   (c)

FIGURE P20.23

20.24. Which, if any, of the molecules shown in Figure P20.24 contains a chiral center?

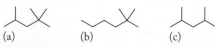

(a)   (b)   (c)

FIGURE P20.24

20.25. **Artificial Sweeteners** Artificial sweeteners are fundamental to the diet food industry. Figure P20.25 shows three artificial sweeteners that have been used in food. Saccharin

is the oldest, dating to 1879. Cyclamates were banned in the United States in 1969 after research suggested they led to tumors. Aspartame may be more familiar to you under the name NutraSweet. Each of these sweeteners contains between zero and two chiral carbon atoms. Circle the chiral center(s) in each compound.

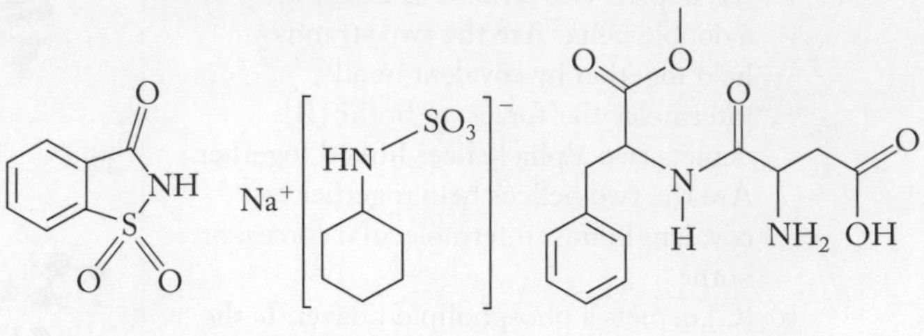

Saccharin   Sodium cyclamate   Aspartame
FIGURE P20.25

20.26. **Smoother Skin** Researchers at Tufts University have found that a fungal toxin called cytochalasin B (Figure P20.26) can restore elasticity and shrink skin cells in mice without causing any ill effects. Mice treated with a cream containing this molecule had smoother skin than mice treated with a plain cream. Circle the chiral centers in the molecule.

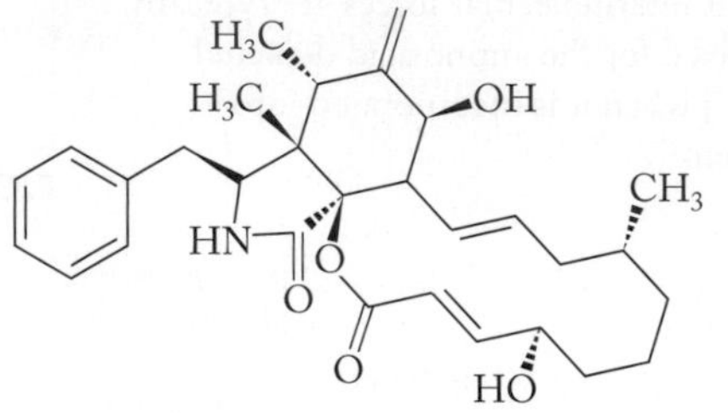

FIGURE P20.26

20.27. **Lowering Cholesterol** Crestor (shown as the carboxylic acid anion in Figure P20.27) is one of the top-ten best-selling pharmaceutical agents in the world.
   a. Is the anion chiral? If so, how many chiral centers does it have?
   b. Are any other kinds of constitutional isomers possible for the anion? If so, draw them.

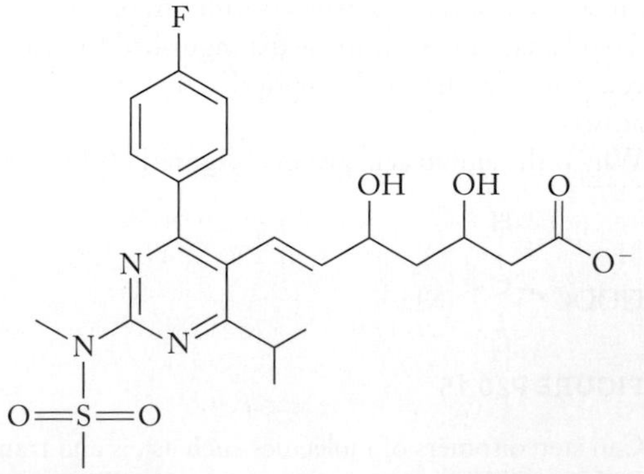

FIGURE P20.27

*20.28. **Preventing Blood Clots** In 2009 Plavix (Figure P20.28) was the second-highest-selling pharmaceutical in the world. It is prescribed to prevent blood clots in heart attack and stroke patients.
   a. Is Plavix optically active? If so, identify the chiral center(s) in the structure.
   b. Plavix contains an ester group. Identify it. Draw the carboxylic acid and the alcohol you could react together to make Plavix.

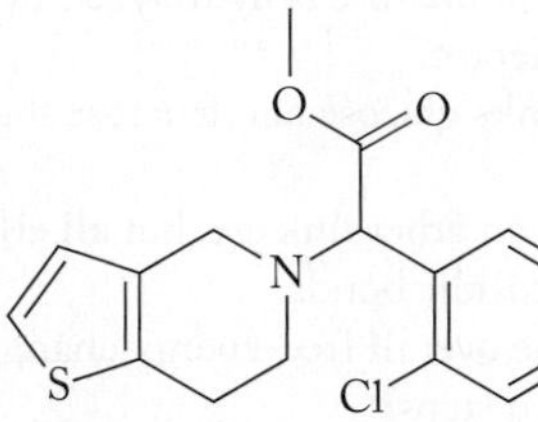

FIGURE P20.28

The Composition of Proteins

Concept Review

20.29. In living cells, amino acids combine to make peptides and proteins. Are these processes accompanied by increases or decreases in entropy of the reaction system?

20.30. What is the difference between a peptide bond and an amide bond?

20.31. **Into the Wild** The author of the book *Into the Wild* postulated that the main character in the book died in the Alaskan wilderness because he accidentally ate seeds containing a toxin called β-ODAP (Figure P20.31). This substance is a neurotoxin; its mechanism of action is not fully understood, but it is referred to as a "glutamic acid mimic." Compare the structure of β-ODAP to that of glutamic acid and describe the similarities between the two molecules.

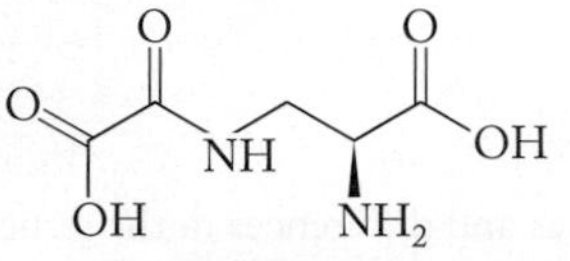

FIGURE P20.31

20.32. **Metabolic Disease in Dogs** Cystinuria is a metabolic disease that occurs in some breeds of dogs. It is characterized by the presence of kidney stones made of cystine, a dimer formed when two cysteine residues covalently link through an –S—S– bond (Figure 20.27).
   a. Draw the structure of the dipeptide Cys-Cys.
   b. Is cystine a dipeptide? Why or why not?

20.33. Meteorites contain more L-amino acids, which are the forms that make up the proteins in our bodies, than D-amino acids. What do the prefixes L- and D- mean?

20.34. Do any of the amino acids in Table 20.2 have more than one chiral carbon atom per molecule?

20.35. Which of the compounds in Figure P20.35 is not an α-amino acid?

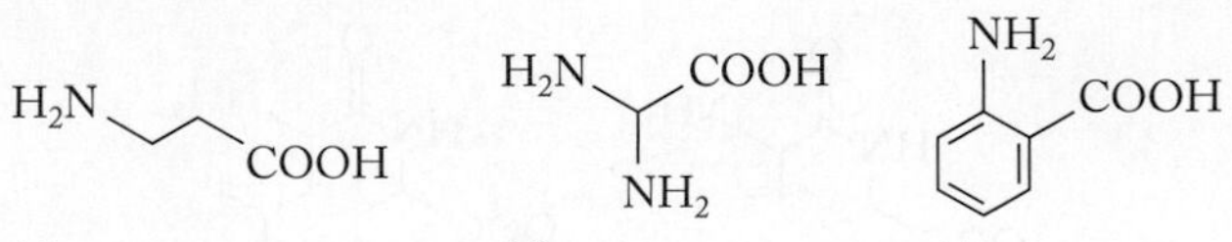

FIGURE P20.35

20.36. Which of the compounds in Figure P20.36 are α-amino acids?

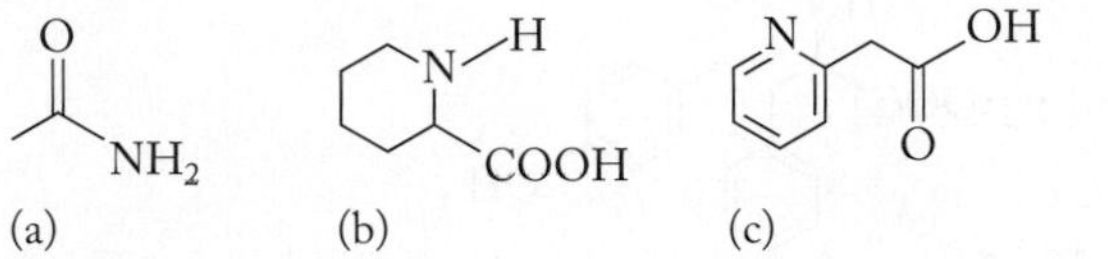

FIGURE P20.36

20.37. Why do most amino acids exist in the zwitterionic form at physiological pH (pH ≈ 7.4)?

20.38. Draw the condensed structural formulas of the amino acid tyrosine that you would expect to predominate in aqueous solution under the following conditions:
   a. in strongly acidic solution
   b. in strongly basic solution
   c. in a solution in which pH = pI

Protein Structure and Function

Concept Review

20.39. When protein strands fold back on themselves in forming stable tertiary structures, lysine residues are often paired up with glutamic acid residues. Why?

20.40. Ion–ion interactions are particularly effective at stabilizing tertiary structures of proteins. Suggest a pair of amino acid residues that would be attracted to each other via ion–ion interactions at pH 7.4.

Problems

20.41. Draw structures and name all possible dipeptides produced from condensation reactions of the following L-amino acids:
   a. alanine + isoleucine
   b. serine + tyrosine
   c. valine + phenylalanine

20.42. Draw structures of the peptides produced from condensation reactions of the following L-amino acids. Assume the amino acids bond in the order given.
   a. glycine + tyrosine + aspartic acid
   b. serine + threonine + leucine
   c. asparagine + lysine + histidine

20.43. Identify the amino acids in the dipeptides shown in Figure P20.43.

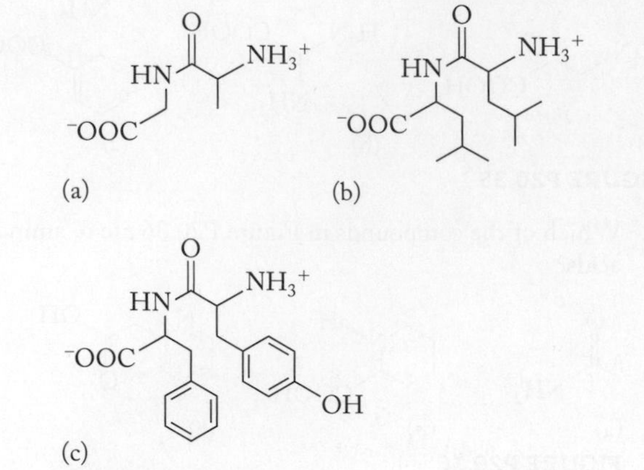

(a)   (b)

(c)
FIGURE P20.43

20.44. Identify the amino acids in the tripeptides in Figure P20.44.

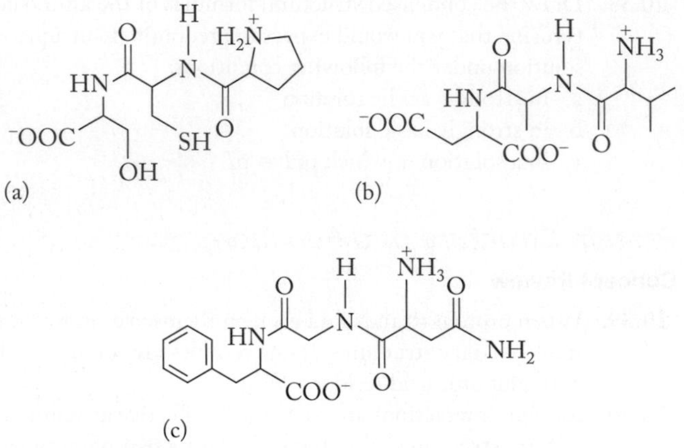

(a)   (b)

(c)
FIGURE P20.44

20.45. Identify the missing product in the metabolic reaction shown in Figure P20.45.

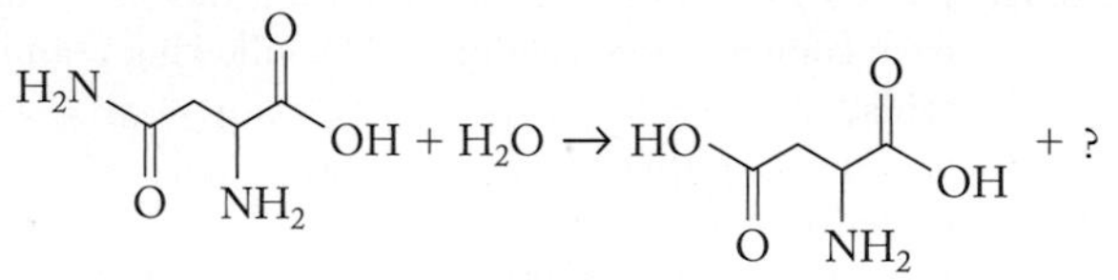

FIGURE P20.45

20.46. The molecular formula for glycine is $C_2H_5NO_2$. What is the molecular formula of the linear peptide formed when ten glycine molecules are linked together in peptide bonds?

Carbohydrates

Concept Review

20.47. What are the structural differences between starch and cellulose?

20.48. Why is the discovery of enzymes that catalyze cellulose hydrolysis a worthwhile objective?

20.49. Is the fuel value (see Chapter 5) of glucose in the linear form the same as that in the cyclic form?

*20.50. Without doing the actual calculation, estimate the fuel values of glucose and starch by considering average bond energies. Do you predict the fuel values of the two substances to be the same or different?

20.51. The second step in glycolysis converts glucose 6-phosphate into fructose 6-phosphate. Can you think of a reason why $\Delta G°$ for this reaction is close to zero?

*20.52. Which of the following statements are correct about glycosidic bonds in carbohydrates?
   a. The glycosidic bond in maltose is hydrolyzed by people who are lactose intolerant.
   b. A glycosidic bond links glucose and fructose together to form sucrose.
   c. A glycosidic bond is an ether linkage, but all ether linkages are not glycosidic bonds.

20.53. How do we calculate the overall free-energy change of a process consisting of two steps?

20.54. During glycolysis a monosaccharide is converted to pyruvate. Do you think this process produces an increase or decrease in the entropy of the system? Explain your answer.

Problems

20.55. Describe the similarities and differences in the structures of the α and β isomers formed when galactose (Figure P20.55) forms a six-membered ring.

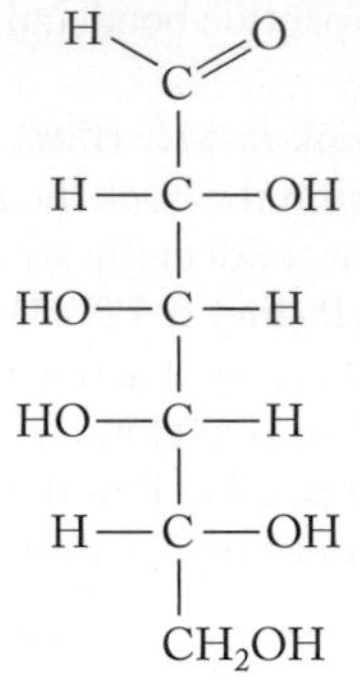

Galactose
FIGURE P20.55

20.56. Describe the similarities and differences in the structures of the α and β isomers formed when ribose (Figure P20.56) forms a five-membered ring.

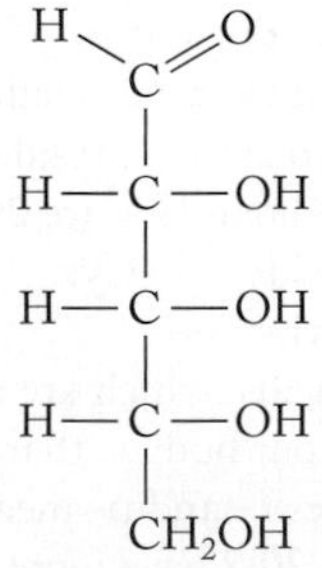

Ribose
FIGURE P20.56

20.57. Which, if any, of the structures in Figure P20.57 are β isomers of a monosaccharide?

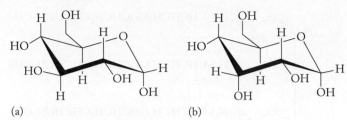

(a)

(b)

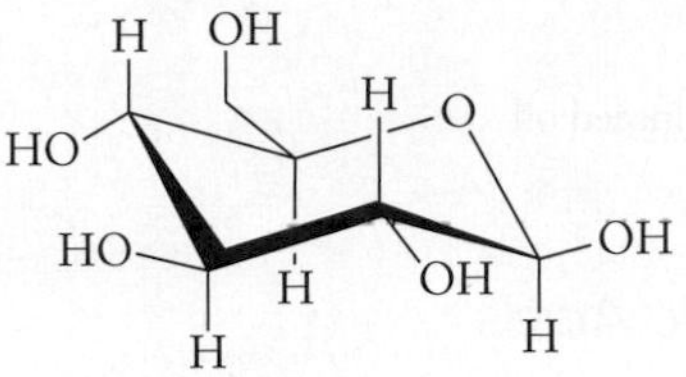

(c)

FIGURE P20.57

20.58. Which, if any, of the structures in Figure P20.58 are α isomers?

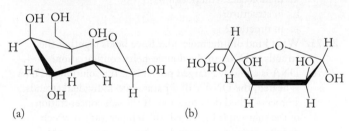

(a)

(b)

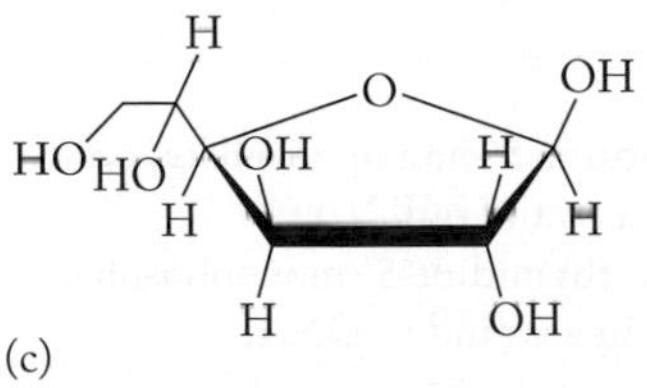

(c)

FIGURE P20.58

20.59. Which of the saccharides in Figure P20.59 is digestible by humans?

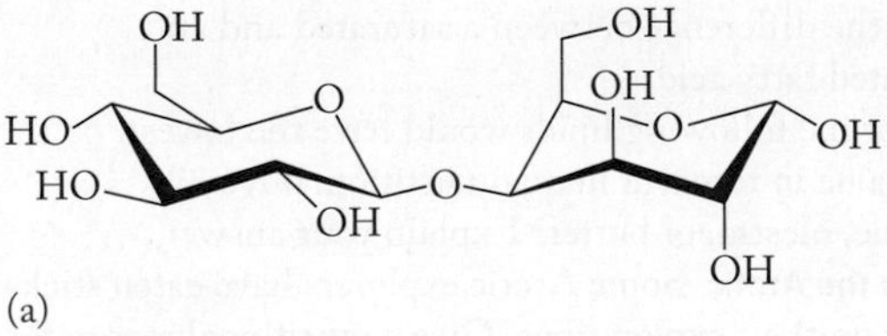

(a)

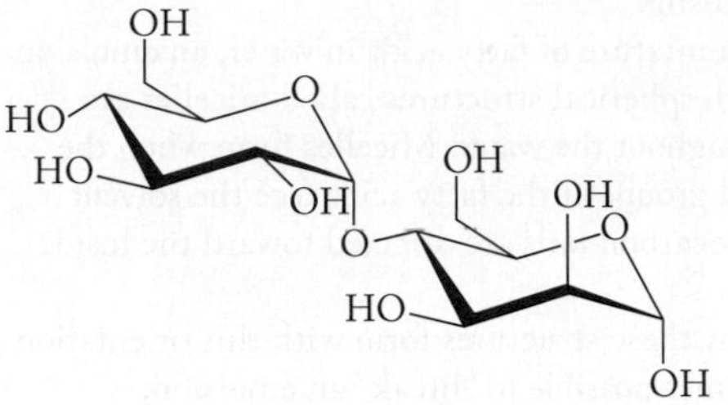

(b)

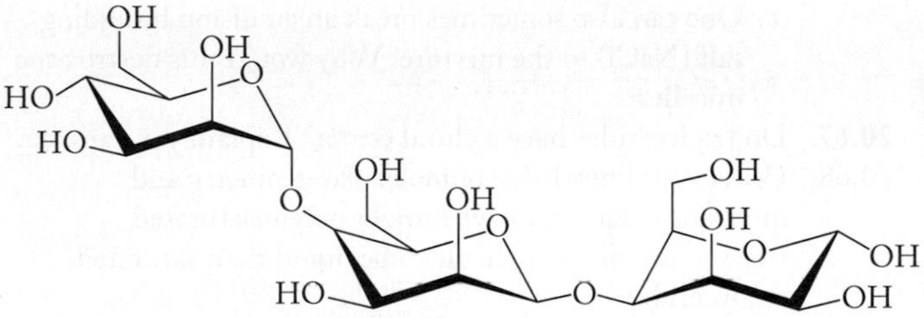

(c)

FIGURE P20.59

*20.60. For any of the disaccharides in Problem 20.59 that are not digestible by humans, draw an isomer that would be digested.

20.61. The structure of the disaccharide maltose appears in Figure P20.61. Hydrolysis of 1 mole of maltose ($\Delta G_f^\circ = -2246.6$ kJ/mol) produces 2 moles of glucose ($\Delta G_f^\circ = -1274.4$ kJ/mol):

$$\text{Maltose} + H_2O \rightarrow 2 \text{ glucose}$$

If the value of ΔG_f° for water is -285.8 kJ/mol, what is the change in free energy of the hydrolysis reaction?

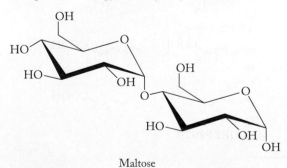

Maltose

FIGURE P20.61

20.62. If the maltose in Problem 20.61 were replaced by another disaccharide, would you expect the free-energy change for the hydrolysis to be exactly the same or just similar in value? Explain your answer.

Lipids

Concept Review

20.63. What is the difference between a saturated and an unsaturated fatty acid?

20.64. Which of the following lipids would have the lowest energy value in terms of human nutrition: olive oil, margarine, olestra, or butter? Explain your answer.

20.65. **Dining in the Arctic** Some Arctic explorers have eaten sticks of butter on their explorations. Give a nutritional reason for this unusual cuisine.

20.66. If you agitate a mixture of fatty acids in water, an emulsion forms, in which spherical structures called micelles are dispersed throughout the water. Micelles form when the carboxylic acid groups of the fatty acids face the solvent and their hydrocarbon tails are directed toward the inside of the sphere.
  a. Explain why these structures form with this orientation.
  b. It is sometimes possible to "break" an emulsion, destroying the micelles by adding a strong acid to the mixture. Why would this destroy the micelles?
  c. One can also sometimes break an emulsion by adding salt (NaCl) to the mixture. Why would this destroy the micelles?

20.67. Do triglycerides have a chiral center? Explain your answer.

20.68. Using your knowledge of molecular geometry and intermolecular forces, why might polyunsaturated triglycerides be more likely to be liquid than saturated triglycerides?

Problems

20.69. Oleic acid and α-linolenic acid (Table 20.4) are both unsaturated fats and are liquids at room temperature. They can be converted into solid saturated fats by hydrogenation.
  a. Which would consume more hydrogen: 1.0 kg of oleic acid or 0.50 kg of α-linolenic acid?
  b. Could you distinguish between the two fatty acids by determining the identity of their hydrogenation products? Explain your answer.

20.70. For each of the pairs of fatty acids in Figure P20.70, indicate whether they are constitutional isomers, stereoisomers, or unrelated compounds.

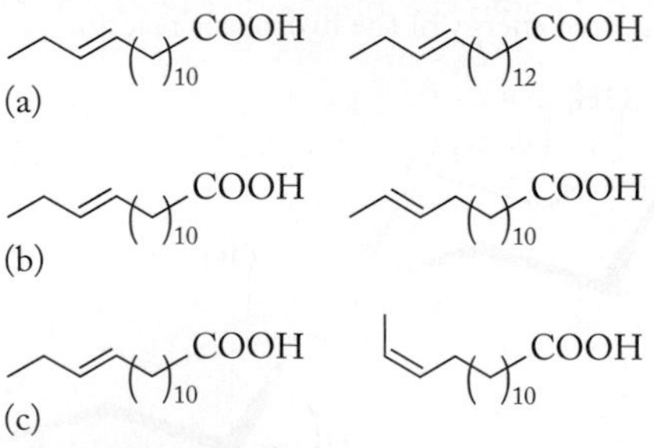

(a) [structures with COOH, 10 and 12]

(b) [structures with COOH, 10 and 10]

(c) [structures with COOH, 10 and 10]

FIGURE P20.70

20.71. Draw the structures of the three fats formed by reaction of glycerol with (a) octanoic acid ($C_7H_{15}COOH$), (b) decanoic acid ($C_9H_{19}COOH$), and (c) dodecanoic acid ($C_{11}H_{23}COOH$).

20.72. **Oil-Based Paints** Oil-based paints contain linseed oil, a triglyceride formed by esterification of glycerol with linolenic acid (Figure P20.72).

a. Draw the line structure of linolenic acid.
*b. Are the double bonds in linolenic acid conjugated?

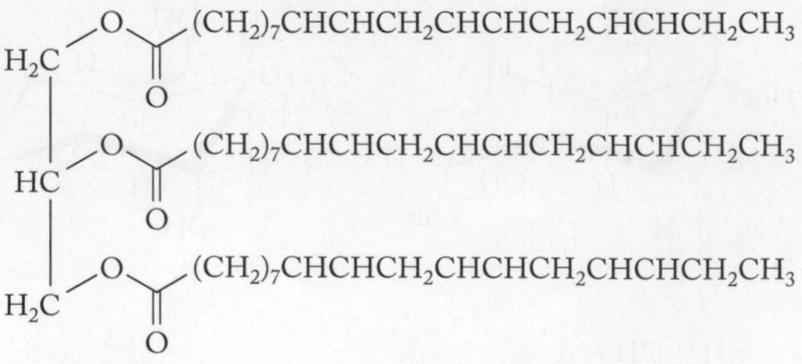

H_2C—O—$(CH_2)_7CHCHCH_2CHCHCH_2CHCHCH_2CH_3$

HC—O—$(CH_2)_7CHCHCH_2CHCHCH_2CHCHCH_2CH_3$

H_2C—O—$(CH_2)_7CHCHCH_2CHCHCH_2CHCHCH_2CH_3$

Linseed oil

FIGURE P20.72

Nucleotides and Nucleic Acids

Concept Review

20.73. What are the three kinds of molecular subunits in DNA? Which two form the "backbone" of DNA strands?

20.74. How do DNA and RNA differ:
  a. in molecular composition?
  b. in structure?
  c. in function?

20.75. What kind of intermolecular force holds together the strands of DNA in the double-helix configuration?

20.76. DNA is a highly charged polyanion. If a solution of DNA is heated, the DNA will separate into individual strands, a process called denaturation. If the salt concentration of the solution is increased, the temperature at which denaturation occurs increases. Suggest a reason why.

Problems

20.77. Draw the structure of adenosine 5′-monophosphate, one of the four ribonucleotides in a strand of RNA.

20.78. Draw the structure of deoxythymidine 5′-monophosphate, one of the four nucleotides in a strand of DNA.

20.79. In the replication of DNA, a segment of an original strand has the sequence T-C-G.
  a. What is the sequence of the opposite strand?
  b. Draw the structure of this section of the double helix, clearly showing all hydrogen bonds.

20.80. If the sequence of one strand of DNA is 5′ ATTGCCA 3′, what is the sequence (in the 5′ to 3′ direction) of the other strand?

Additional Problems

20.81. **Salsa** Salsa has antibacterial properties because it contains dodecenal (Figure P20.81), a compound found in the cilantro used to make salsa.
  a. How many carbon atoms are in dodecenal?
  b. What functional groups are present in dodecenal?
  c. What types of isomerism are possible in dodecenal?

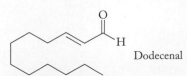

Dodecenal

FIGURE P20.81

20.82. Turmeric Turmeric is commonly used as a spice in Indian and Southeast Asian dishes. Turmeric contains a high concentration of curcumin (Figure P20.82), a potential anticancer drug and a possible treatment for cystic fibrosis.
  a. Are the substituents on the C=C double bonds highlighted in red in cis or trans configurations?
  b. Draw two other stereoisomers of this compound.

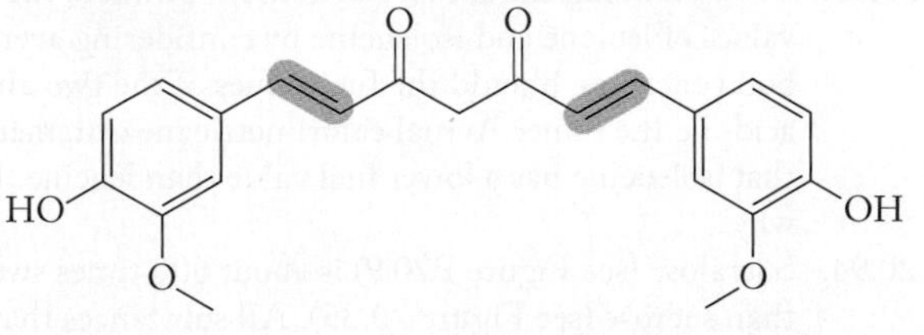

Curcumin
FIGURE P20.82

20.83. Fat Substitutes Olestra is a calorie-free fat substitute. The core of the olestra molecule (Figure P20.83) is a disaccharide that has reacted with a carboxylic acid; this results in the conversion of hydroxyl groups on the disaccharide into the depicted structure.
  a. What is the name of the disaccharide core of the olestra molecule?
  b. What functional group has replaced the hydroxyl groups on the disaccharide?
  c. What is the formula of the carboxylic acid used to make olestra?

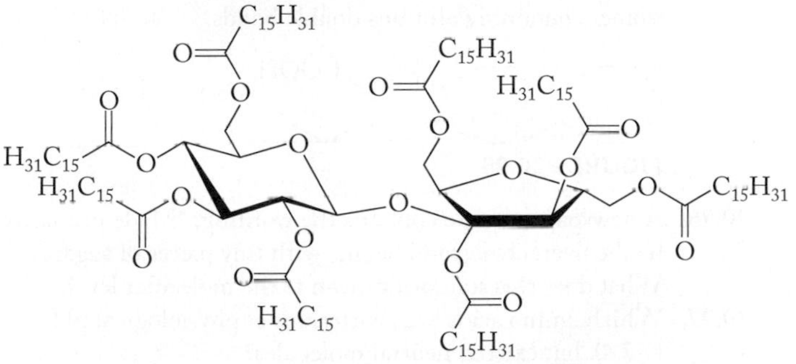

Olestra
FIGURE P20.83

20.84. When scientists at UC Santa Cruz directed UV radiation at an ice crystal containing methanol, ammonia, and hydrogen cyanide, three amino acids (glycine, alanine, and serine) were detected among the products of photochemical reactions. The formation of these amino acids suggests that they may also be synthesized in comets approaching the Sun (and Earth). Determine the standard free-energy change of the hypothetical formation of glycine in comets, using standard free energies of formation of the reactants and products in this reaction [ΔG_f° for HCN(g) is +125 kJ/mol and for solid glycine is −368.4 kJ/mol; other ΔG_f° values are in Appendix 4].

$$CH_3OH(\ell) + HCN(g) + H_2O(\ell) \rightarrow H_2NCH_2COOH(s) + H_2(g)$$

20.85. Homocysteine (Figure P20.85) is formed during the metabolism of amino acids. A mutation in some people's genes leads to high concentrations of homocysteine in the blood and a consequent increase in their risk of heart disease and incidence of bone fractures in old age.
  a. What is the structural difference between homocysteine and cysteine?
  b. Cysteine is a chiral compound. Is homocysteine chiral?

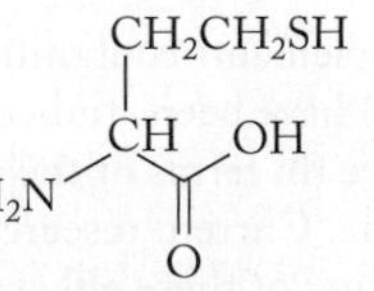

Homocysteine
FIGURE P20.85

20.86. Amino Acids in Comets Some scientists believe life on Earth can be traced to amino acids and other molecules brought to Earth by comets and meteorites. In 2004, a new class of amino acids called diamino acids (Figure P20.86) were found in the Murchison meteorite.
  a. Which of these diamino acids is not an α-amino acid?
  b. Which of these amino acids is chiral?

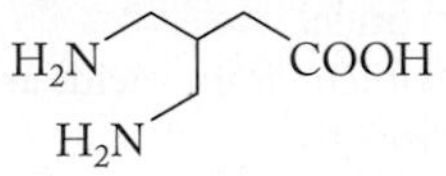

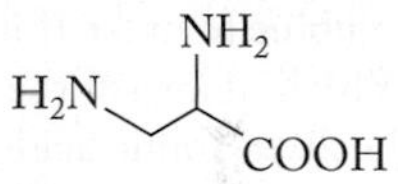

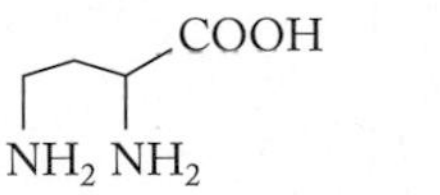

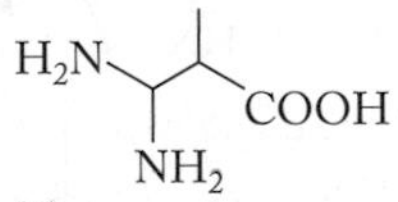

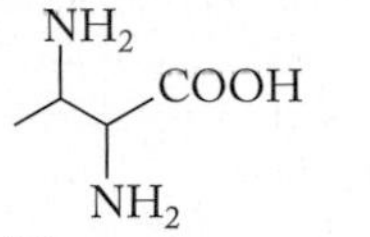

FIGURE P20.86

20.87. Jamaican Fruit Ackee, the national fruit of Jamaica, is a staple in many Jamaican diets. Unfortunately, a potentially fatal sickness known as Jamaican vomiting disease is caused by the consumption of unripe ackee fruit, which contains the amino acid hypoglycin (Figure P20.87). Is hypoglycin an α-amino acid?

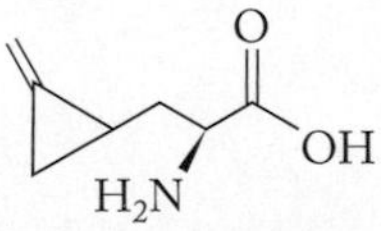

Hypoglycin
FIGURE P20.87

*20.88. In Section 20.2 we discussed the rotation of a beam of polarized light when it passes through a solution containing an optically active molecule. Equimolar solutions of α-glucose and β-glucose rotate plane-polarized light by +112° and +18.7°, respectively. If these two solutions are then mixed and allowed to reach equilibrium, the solution then rotates the polarized light by +53.4°. Calculate the percent glucose in the α and β forms in this solution.

20.89. **Amino Acid Supplements** Supplements containing BCAAs (branched-chain amino acids) have been studied for their effects on athletic performance (in terms of the perception of exertion) and mental fatigue. Current research does not support that they have either of these effects, but they are used medically to slow muscle wasting in bedridden patients. BCAAs are amino acids having hydrocarbon side chains with a branch, a carbon atom bound to more than two other carbon atoms. Consult Table 20.2 and draw the structures of the BCAAs among the common amino acids.

20.90. In response to specific neural messages, the human hypothalamus may secrete several polypeptides, including the tripeptide shown in Figure P20.90.
   a. Sketch the structures of the three amino acids that combine to make this tripeptide.
   b. Which, if any, of the constituent amino acids are among the 20 α-amino acids in proteins?

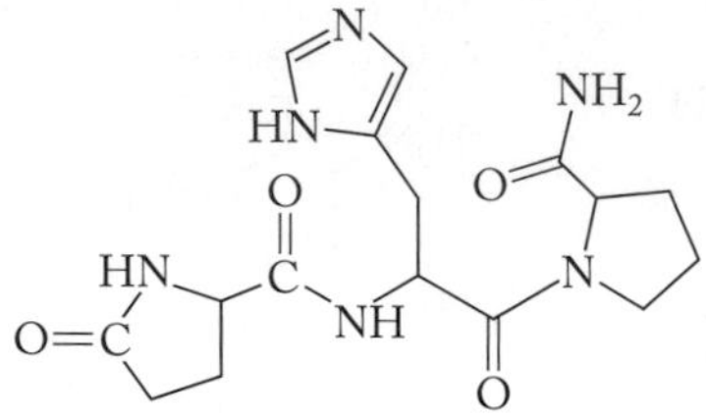

Thyrotropin-releasing factor
FIGURE P20.90

20.91. Glutathione (Figure P20.91) is an essential molecule in the human body. It acts as an activator for enzymes and protects lipids from oxidation. Which three amino acids combine to make glutathione?

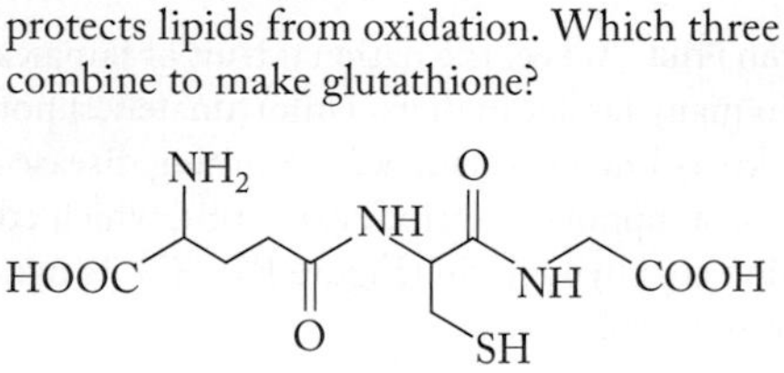

Glutathione
FIGURE P20.91

*20.92. Three amino acids—glutamic acid, arginine, and tryptophan—are dissolved in a gel that is buffered at a pH of 5.9. Two electrodes are placed in the gel and an electric current is applied.
   a. Toward which electrode does each amino acid migrate?
   b. Draw the forms of each amino acid present in the gel at a pH of 5.9.

*20.93. Without doing the actual calculation, estimate the fuel values of leucine and isoleucine by considering average bond energies. Should the fuel values of the two amino acids be the same? Actual calorimetric measurements show that isoleucine has a lower fuel value than leucine. Explain why.

20.94. Sucralose (see Figure P20.9) is about 600 times sweeter than sucrose (see Figure 20.35). All substances that taste sweet have functional groups that form hydrogen bonds with "sweetness" receptor sites on the tongue. What does the difference in sweetness between sucralose and sucrose tell you about additional intermolecular interactions between sweet compounds and receptor sites that contribute to their sweet taste?

*20.95. **Inflammation-Causing Molecules** Prostaglandins, naturally occurring compounds in our bodies that cause inflammation and other physiological responses, are formed from arachidonic acid, an unsaturated hydrocarbon containing four C=C double bonds and a carboxylic acid functional group. The stereoisomer containing all cis double bonds is shown in Figure P20.95. How many stereoisomers other than this one are possible? Draw the isomer containing all trans double bonds.

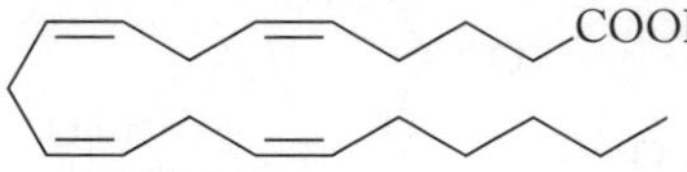

Arachidonic acid
FIGURE P20.95

20.96. A newspaper article contains the wording, "Made primarily by the liver, cholesterol begins with tiny pieces of sugar…" What does this statement mean at the molecular level?

20.97. Which amino acids are zwitterions at physiological pH (~7.4), but exist as neutral molecules?

*20.98. **Opioids** The structures of morphine and demerol, two powerful pain medicines, are shown in Figure P20.98. They interact similarly with receptors because they share similar structural features, including one part of the molecule that is flat or planar and another part that can bind to the receptor site. Identify these two structural features.

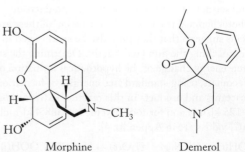

Morphine   Demerol
FIGURE P20.98

*20.99. **Antihistamines** Figure P20.99 shows the structure of the histamine molecule that causes sneezing and itching in allergy sufferers. Antihistamines are a class of molecules that reduce allergy symptoms by preferentially binding to the same receptor sites as histamines. What structural features would permit these molecules to bind to the same receptor?

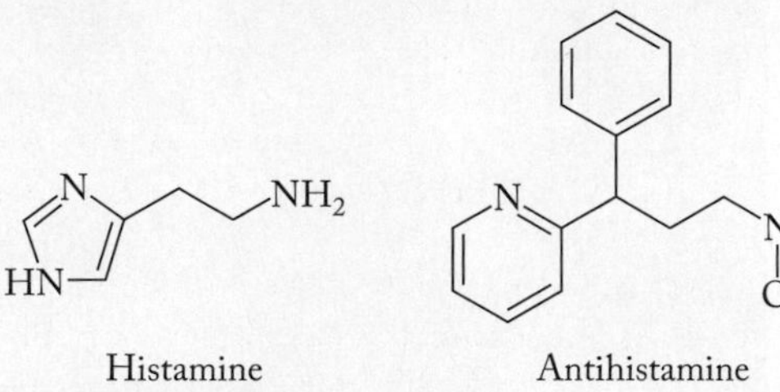

Histamine   Antihistamine
FIGURE P20.99

*20.100. **Pain Relievers** The structures of aspirin, acetaminophen, and ibuprofen are shown in Figure P20.100. What structural features do these molecules share?

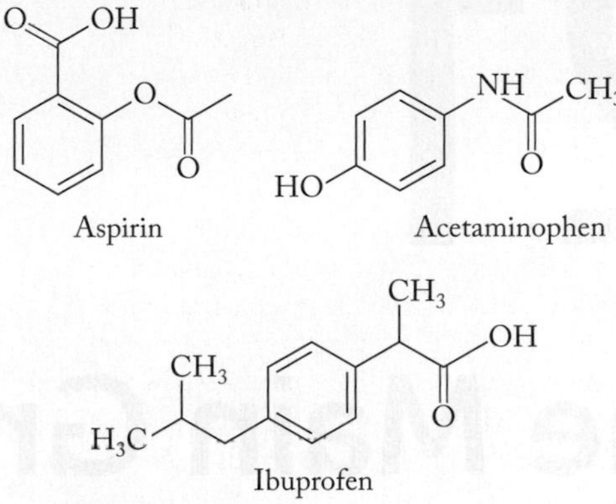

Aspirin   Acetaminophen

Ibuprofen
FIGURE P20.100

21

The Main Group Elements
Life and the Periodic Table

DIETARY SUPPLEMENTS Calcium is included in many vitamins and dietary supplements to promote healthy bones and teeth.

PARTICULATE **REVIEW**

Nitrogen, Sulfur, and Phosphorus: One, Two, or All Three?

In Chapter 21 we survey the roles of main group elements in living organisms. In addition to carbon, hydrogen, and oxygen, atoms of nitrogen, sulfur, and phosphorus are fundamental building blocks for important biological molecules in plants and animals.

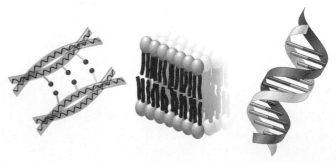

- Which of these structures contains sulfur atoms? Where are the sulfur atoms?

- Which two structures contain phosphorus atoms? Where are the phosphorus atoms?

- All three structures contain nitrogen atoms. List the functional group(s) that contains nitrogen atoms in each structure.

    (Review Chapter 20 if you need help.)

(Answers to Particulate Review questions are in the back of the book.)

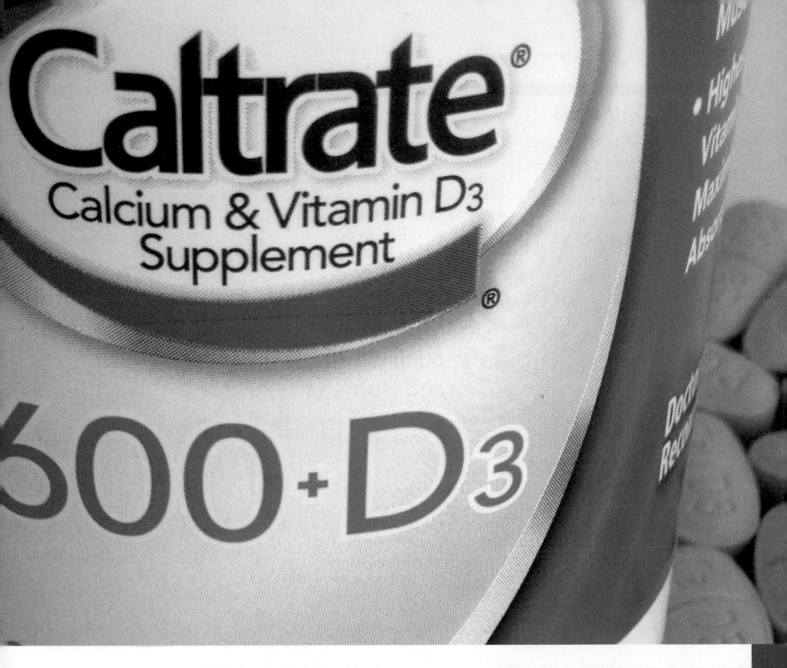

Caltrate®

Calcium & Vitamin D3
Supplement

®

600+D3

Charge versus Size: Selective Transport

Shown here are representations of four essential cations involved in selective ion transport through channels in cell membranes: Ca^{2+}, H_3O^+, K^+, and Na^+. The size of each cation is listed. As you read Chapter 21, look for ideas that will help you answer these questions:

100 pm   102 pm   113 pm   138 pm

- Identify each of the four ions pictured.

- The Na^+ channel is quite selective, allowing only Na^+ cations and one of the other three cations to pass through it. Which other ion—Ca^{2+}, H_3O^+, or K^+—is most likely to undergo selective transport through the sodium channel?

- Which of the other three ions, if any, is likely to pass through the K^+ channel?

Learning Outcomes

LO1 Distinguish between essential and nonessential elements and between major, trace, and ultratrace elements

LO2 Summarize the pathways and calculate the free energy for ion transport across cell membranes
Sample Exercises 21.1

LO3 Balance equations, draw structures, and carry out calculations relevant to the behavior of major, trace, and ultratrace elements and their compounds *in vivo*
Sample Exercises 21.2, 21.3, 21.4

LO4 Describe the function of the essential and nonessential group 14–16 elements in the human body

LO5 Explain how radioactive isotopes of the main group elements are used in the diagnosis of disease
Sample Exercise 21.5

LO6 Describe how compounds containing main group elements are used in therapy and in other applications

21.1 Main Group Elements and Human Health

Have you ever wondered how many of the elements in the periodic table are in the human body? Or wondered which of them are important to the health of humans?

Roughly one-third of the 90 naturally occurring elements have an identifiable role in human health and in organisms in general. **Essential elements** are defined as those that have a beneficial physiological effect, including those whose absence impairs functioning of the organism (Table 21.1). Some—like carbon, hydrogen, oxygen, nitrogen, sulfur, and phosphorus—are the principal constituents of all plants and animals. The alkali metal cations Na^+ and K^+ act as charge carriers, maintain osmotic pressure, and transmit nerve impulses. The alkaline earth cation Mg^{2+} is important in photosynthesis, and its family member Ca^{2+} forms structural materials such as bones and teeth. Chloride ions balance the charge of Na^+ and K^+ ions to maintain electrical neutrality in living cells. Other main group elements, such as iodine and selenium, are required in tiny amounts in our bodies to regulate metabolism and as components of the enzymes that catalyze biochemical reactions.

Many **nonessential elements** are also present in the body but have no known function (Table 21.2). Some are useful in medicine as either diagnostic tools or therapeutic agents. Some radioactive isotopes may be used as imaging agents for organs and tumors, for example, while others are used to treat disease. Drugs containing lithium ions are used to treat depression. Compounds of bismuth act as mild antibacterial agents for treatment of diarrhea, and antimony compounds represent one of the few options for patients suffering from the tropical disease leishmaniasis, caused by protozoan parasites.

In some cases, the presence of a nonessential element has a **stimulatory effect**, which means that the consumption of small amounts of the element causes increased activity or growth in an organism. The effect may be beneficial or not, and often the mechanism of the effect is not understood. For example, small amounts of the nonessential element antimony promote growth in some mammals when added to

TABLE 21.1 Essential Elements Found in the Human Body

Major (>1 mg/g of body mass)	Trace (1–1000 μg/g of body mass)	Ultratrace (<1 μg/g of body mass)
Calcium	Fluorine	Chromium
Carbon	Iodine	Cobalt
Chlorine	Iron	Copper
Hydrogen	Silicon	Manganese
Magnesium	Zinc	Molybdenum
Nitrogen		Nickel
Oxygen		Selenium
Phosphorus		Vanadium
Potassium		
Sodium		
Sulfur		

their diets. Nonessential elements, and even toxic elements, are often incorporated into our bodies because their chemical properties are similar to those of an essential element. For example, Rb^+ ions are retained by the human body because they are similar to K^+ ions in size, charge, and chemistry, making rubidium the most abundant nonessential element in humans.

Oxygen, in the form of O_2 gas, occurs in the body in elemental form; oxygen is also incorporated into many compounds, such as H_2O, and many ions, such as HCO_3^-. When we speak of any element in the body other than oxygen, however, we usually refer to an ion or compound containing that element rather than the pure element. For example, when we describe calcium as an essential element, we are referring to calcium ions, Ca^{2+}, not calcium metal.

The essential elements are further classified as **major**, **trace**, or **ultratrace essential elements**. Major essential elements are present in gram quantities in the human body and are required in large amounts in our diet. Almost all foods are rich in compounds containing carbon, hydrogen, oxygen, nitrogen, sulfur, and phosphorus. Salt is perhaps the most familiar dietary source of sodium and chloride ions, although both are ubiquitous in food. Vegetables such as broccoli and Brussels sprouts and fruits such as bananas are rich in potassium. Calcium is found in dairy products and is often added to orange juice as a dietary supplement.

Table 21.3 compares the elemental compositions of the human body, the universe, Earth's crust, and seawater. Note that the composition of our bodies most

TABLE 21.2 Nonessential Elements Found in the Human Body

Stimulatory	Unknown Role	No Role
Boron	Antimony	Barium
Titanium	Arsenic	Bromine
		Cesium
		Germanium
		Rubidium
		Strontium

TABLE 21.3 Comparative Composition^a of the Universe, Earth's Crust, Seawater, and the Human Body

Element	Universe (%)	Crust (%)	Seawater (%)	Human Body (%)
Hydrogen	91	0.22	66	63
Oxygen	0.57	47	33	25.5
Carbon	0.021	0.019	0.0014	9.5
Nitrogen	0.042			1.4
Calcium		3.5	0.006	0.31
Phosphorus				0.22
Chlorine			0.33	0.03
Potassium		2.5	0.006	0.06
Sulfur	0.001	0.034	0.017	0.05
Sodium		2.5	0.28	0.01
Magnesium	0.002	2.2	0.033	0.01
Helium	9.1			
Silicon	0.003	28		
Aluminum		7.9		
Neon	0.003			
Iron	0.002	6.2		
Bromine			0.0005	
Titanium		0.46		
All other elements	< 0.1	< 0.1	< 0.1	< 0.1

^{a}Compositions are expressed as the percentage of the total number of atoms. Because of rounding, the totals do not equal exactly 100%.

essential element element present in tissue, blood, or other body fluids that has a physiological function.

nonessential element element present in humans that has no known function.

stimulatory effect increased activity, growth, or other biological response to the presence of a nonessential element.

major essential element essential element present in the body in average concentrations greater than 1 mg of element per gram of body mass.

trace essential element essential element present in the body in average concentrations between 1 and 1000 μg of element per gram of body mass.

ultratrace essential element essential element present in the body in average concentrations less than 1 μg of element per gram of body mass.

TABLE 21.4 Dietary Reference Intakes (DRI) and Recommended Dietary Allowances (RDA) for Selected Essential Elements[a]

Element	DRI	RDA
Calcium	1000 mg	1200 mg
Chlorine	2300 mg	2300 mg
Chromium	25–35 µg	35 µg
Copper	900 µg	900 µg
Fluorine	3–4 mg	4 mg
Iodine	150 µg	150 µg
Iron	8–18 mg	18 mg
Magnesium	420 mg	320–400 mg
Manganese	1.8–2.3 mg	2–5 mg
Molybdenum	45 µg	45 µg
Phosphorus	700 mg	700 mg
Potassium	4700 mg	4700 mg
Selenium	55 µg	55 µg
Sodium	1500 mg	1500 mg
Zinc	8–11 mg	11 mg

[a]DRI and RDA values in milligrams or micrograms per day from the U.S. Department of Agriculture (2009) and from the Council on Responsible Nutrition (CRN) for 19- to 30-year-olds.

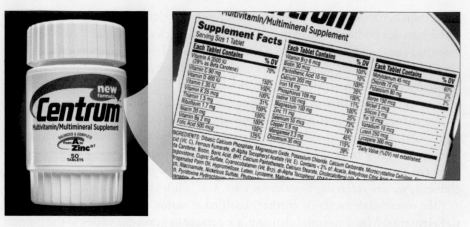

FIGURE 21.1 The labels on multivitamin supplements may not list DRI or RDA values but rather *% daily values* (DVs). Daily values are based on RDA or DRI values, but there can be inconsistencies, particularly among the ultratrace essential elements.

closely resembles the composition of seawater. The match would be even closer if it were not for the biological processes in the sea that remove essential elements such as nitrogen and phosphorus, and that store others in solid structures like the $CaCO_3$ that makes up corals and mollusk shells.

Our diet should supply us with sufficient quantities of all essential elements. In the United States and Canada, these quantities are called *dietary reference intake* (DRI) values. They are based on the recommendations of the Food and Nutrition Board of the National Academy of Sciences and are frequently updated in response to research. For many essential elements, DRI values have replaced the *recommended dietary allowance* (RDA) values you may be familiar with from labels on food and vitamin packages (Figure 21.1). Table 21.4 compares the DRI and RDA values for several major, trace, and ultratrace essential elements. Among the major essential elements, DRI/RDA values range from 0.32 to 0.42 g of magnesium per day to 4.7 grams per day of potassium. DRI/RDA values for trace essential elements, including iron and zinc, are in the range of 10–20 milligrams per day. There are DRI/RDA values for some, but not all, of the ultratrace essential elements, ranging from 55 micrograms per day for selenium and 45 micrograms per day for molybdenum up to 5 milligrams per day for manganese. The roles of transition metal ions such as Fe^{2+}, Fe^{3+}, and Zn^{2+} in biology are addressed in Chapter 22.

Some elements, such as radon, beryllium, and lead, are toxic. As described in Chapter 19, inhaled radon gas poses serious health hazards from α decay taking place inside the body. Beryllium toxicity is most often encountered in industrial settings where beryllium-contaminated dust is inhaled; the Be^{2+} ion replaces Mg^{2+} in the body, where it inhibits Mg^{2+}-catalyzed RNA and DNA synthesis in cells. Lead ions, Pb^{2+}, are incorporated into teeth and bones because they are similar to Ca^{2+} ions in size and charge. Lead also interferes with the functioning of enzymes that require calcium ions and thereby causes chronic neurological problems and blood-based disorders, especially in children. This issue became a major news story early in 2016 when it was discovered that a budget-cutting decision to switch drinking-water sources in Flint, Michigan, had exposed as many as 8000 children under age 6 to unsafe levels of lead. In June 2014, the city's source of water had been changed from Lake Huron to the Flint River. The river water proved more corrosive than the lake water and leached lead from old water pipes, causing lead levels to rise to more than 100 ppb; the EPA's standard for drinking water sets the action level at 15 ppb. Alarmingly, the crisis in Flint is not isolated, as unsafe levels of lead have been identified in other cities with aging water systems.

In this chapter we review the periodic properties of the main group elements and survey the roles of selected elements in the human body as well as their importance to good health. At the same time we call on the knowledge and skills you have acquired in your study of general chemistry to solve problems that link concepts from prior chapters to the central question of this chapter: What are the roles of the main group elements in the chemistry of life?

21.2 Periodic and Chemical Properties of Main Group Elements

The main group or representative elements, as defined in Chapter 2, are found in groups 1, 2, and 13–18 in the periodic table. These eight groups, comprising 44 naturally occurring elements, include five for which no stable isotopes are known: Fr, Ra, Po, At, and Rn. Relatively little is known about the chemistry of francium or astatine, both of which are exceedingly rare. Estimates indicate that the outermost kilometer of Earth's crust contains at most 44 mg of At and only 15 g of Fr. They have no proven uses, so we will largely ignore them in our discussions. Before turning to the role of the main group elements in life, it is useful to review what we have already learned about the physical and chemical properties of these elements and to look for periodic trends in these properties.

CONNECTION The organization of the periodic table was introduced in Chapter 2. Periodic properties were discussed in Chapter 7, and the stability of nuclei was addressed in Chapter 19.

You may recall that the atomic radii of the main group elements increase as we descend a group and decrease across a period. Their first ionization energies and electronegativities decrease down a group and increase across a period. An overall trend toward more negative electron affinities is seen across a period, although several anomalies are also observed. Each of these periodic trends reflects the changes in the effective nuclear charge and shielding of the valence electrons by core electrons as atomic numbers increase and inner shells fill with electrons.

A survey of the main group elements reveals a variety of physical properties. The alkali metal and alkaline earth elements are all solids, whereas all the group 18 elements are gases at standard temperature and pressure. All group 17 elements exist as diatomic molecules. Group 17 is also the only group that contains elements in all three phases of matter at room temperature: bromine is one of two liquid elements in the periodic table, iodine is a volatile solid, and the remaining elements are gases. Groups 13–16 include seven elements classified as semimetals: B, Si, Ge, As, Sb, Te, and At, as well as elements with metallic properties: Al, Ga, In, Tl, Sn, Pb, Bi, and Po. The remaining elements in these groups behave as nonmetals with the properties described in Chapter 2: they are gases (N_2, O_2) or brittle solids that are poor conductors of electricity (C, P, S, and Se). For groups 13–16, metallic properties increase down a group.

The melting and boiling points of the metallic and semimetallic main group elements decrease down groups 1 and 2, but this trend is reversed for the nonmetals in groups 17 and 18 (Table 21.5). These trends can be understood in terms of the different types of forces holding the atoms together. The increase in the size of the metallic elements when descending groups 1 and 2 leads to weaker metallic bonds and lower boiling points. In groups 17 and 18, however, increasing size leads to greater London dispersion forces as the polarizability of the atoms increases. The trends in melting points for groups 13–16 do not fit a clear pattern,

TABLE 21.5 Summary of Periodic Trends for the Main Group Elements

Property	Group 1	Group 2	Group 13	Group 14	Group 15	Group 16	Group 17	Group 18
Melting point	Decreases down the group	Decreases down the group	No single trend	No single trend	No single trend	No single trend	Increases down the group	Increases down the group
Boiling point	Decreases down the group	Decreases down the group	No single trend	Decreases down the group	No single trend	No single trend	Increases down the group	Increases down the group

in part because the properties of the elements change from nonmetallic to metallic down these groups.

Of the main group elements, only hydrogen, carbon, nitrogen, oxygen, sulfur and the six noble gases exist in nature in elemental form. The remaining elements are found exclusively in ionic and covalent compounds. The group 1 and 2 elements readily lose their valence electrons, forming 1+ and 2+ cations, respectively, which can combine with group 17 anions to give familiar ionic compounds such as NaCl and KI. Halide compounds of groups 13–17 generally contain covalent bonds; however, the heavier group 13 and 14 elements form insoluble ionic salts such as $PbCl_2$ and TlCl. The metallic elements in groups 1, 2, and 13 combine with oxygen to yield oxides, including K_2O, CaO (quicklime, used in the manufacture of steel), and Al_2O_3 (alumina, used in orthodontics).

The lighter elements of groups 13–17 tend to share electrons to form covalent bonds rather than transfer electrons to form ionic bonds. The result is a library of more than 50 million organic compounds, covalently bonded substances composed primarily of carbon, nitrogen, hydrogen, and oxygen. More than 95% of the new compounds registered in a given year are classified as organic compounds.

Hydrogen, the smallest and lightest element, is difficult to classify. Most periodic tables include hydrogen in group 1 on the basis of its electron configuration of a half-filled *s* orbital and on the dissociation of acids to protons and anions, analogous to the dissolution of alkali metal salts to 1+ cations and anions. Hydrogen also has a complete valence shell after gaining an electron to form a hydride ion, H^-, which is isoelectronic with He, a noble gas. Compounds called metal hydrides form between hydrogen and group 1, 2, or 13 metals, and they behave as salts containing a metal cation and a hydride anion. However, hydrogen is most commonly covalently bonded in compounds to oxygen (as water, H_2O) or to another group 14–16 element.

CONNECTION Ionic compounds were defined in Section 2.6, and organic compounds were introduced in Section 2.8. Covalent bonding was discussed in detail in Chapters 8 and 9.

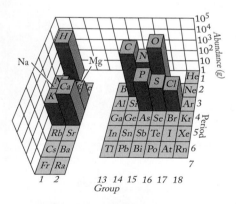

FIGURE 21.2 The 11 elements shown in red are the major essential elements. Their abundances range from 35 g of magnesium to 46 kg of oxygen in a 70-kg adult human.

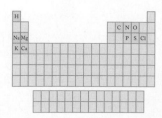

21.3 Major Essential Elements

The 11 elements shown in red in Figure 21.2 and listed in the first column of Table 21.1 are the major essential elements. Together they account for more than 99% of the mass of the human body. Oxygen is the most abundant element by mass, followed by carbon and hydrogen. Although life depends on the presence of elemental oxygen in the form of O_2 gas, much of the oxygen in our bodies is combined with hydrogen in water molecules.

The most abundant elements in the human body include seven nonmetals: C, H, O, Cl, S, P, and N. They are the building blocks for most of the body's molecular compounds and its principal polyatomic ions, HCO_3^-, SO_4^{2-}, and $H_2PO_4^-$, which are dissolved in body fluids. The average concentrations of the four major metals in the human body—Ca^{2+}, K^+, Na^+, and Mg^{2+}—are listed in Table 21.6. In this section we explore some of the roles that sodium, potassium, magnesium, calcium, chlorine, nitrogen, phosphorus, and sulfur play in the biochemistry of the human body. As we do, we revisit several of the chemical principles discussed in earlier chapters.

Sodium and Potassium

Regulated concentrations of sodium and potassium ions are crucial to cell function. For example, too much Na^+ has been linked to hypertension (high blood pressure). To maintain a constant concentration of these two alkali metal ions in body fluids, the ions must be able to move into and out of cells. As noted in Section 20.6, the membrane surrounding a typical cell is a lipid bilayer, with polar groups containing phosphate groups on the two surfaces of the cell membrane and nonpolar fatty acids oriented toward the interior of the membrane. Direct diffusion of Na^+ and K^+ through the lipid bilayer is difficult because these polar cations do not dissolve in the nonpolar interior.

As Figure 21.3 shows, the cell membrane is pierced by **ion channels**, which are groups of protein complexes that allow selective transport of ions. The ion channels control which ions pass through the membrane, on the basis of the size and charge of the ion as well as the shape of the protein. For example, the protein of the ion channel for potassium ions has its amino acids oriented in such a fashion that favorable ion–dipole interactions occur only for ions with the radius of a K^+ ion (138 pm) and not for Na^+ (102 pm) or any other cation. The sodium ion channel is also selective, excluding K^+ and Ca^{2+} even though the radii of Na^+ and Ca^{2+} (100 pm) differ by only 2 pm. Another difference between the Na^+ and K^+ channels is the ability of H_3O^+ (hydronium ion, radius 113 pm) to pass through sodium ion channels but not potassium ion channels.

Living organisms also contain oxygen-rich molecules such as nonactin (Figure 21.4) that bond to Ca^{2+}, K^+, Na^+, and Mg^{2+} ions through strong ion–dipole forces. The resulting complex ions consist of a polar, charged alkali metal ion encapsulated in a nonpolar exterior. Because the complex has both a polar portion and a nonpolar portion, it does not require a channel for passage through a cell membrane. Instead, the complex carries its alkali metal cation through both the polar and nonpolar regions of the bilayer, providing an alternative to ion channels for the transport of metal ions.

In addition to ion channels and diffusion of complex ions, alkali metal cations can be transported by a third mechanism, one involving Na^+–K^+ ion pumps. An

ion channel a group of helical proteins that penetrate cell membranes and allow selective transport of ions.

TABLE 21.6 Average Concentration of Four Metallic Elements in the Human Body

Element	mg/g of Body Mass
Calcium	15.0
Potassium	2.0
Sodium	1.5
Magnesium	0.5

C⌬NNECTION In Chapter 10 we described how alkali metal cations dissolved in water are surrounded by six water molecules. Each water molecule is oriented so that the oxygen atoms point toward the cation. In Chapter 16 we described this interaction as an example of a Lewis acid (cation, electron-pair acceptor) interacting with a Lewis base or ligand (water, electron-pair donor).

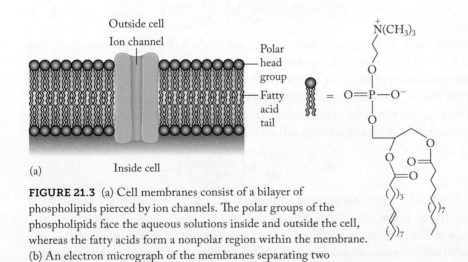

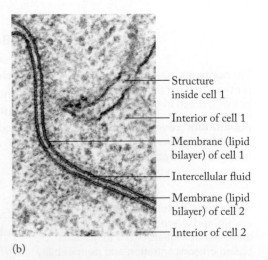

FIGURE 21.3 (a) Cell membranes consist of a bilayer of phospholipids pierced by ion channels. The polar groups of the phospholipids face the aqueous solutions inside and outside the cell, whereas the fatty acids form a nonpolar region within the membrane. (b) An electron micrograph of the membranes separating two adjacent cells.

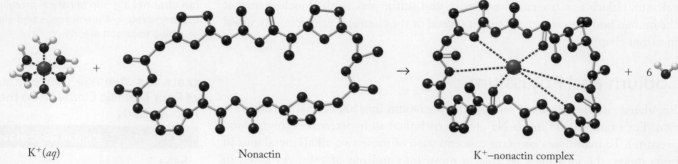

$K^+(aq)$ Nonactin K^+–nonactin complex

FIGURE 21.4 In living organisms, ligands such as nonactin can form a complex with any one of the four alkali metal and alkaline earth major essential ions and carry the ion directly through a cell membrane. An ion channel is not required in this transport pathway.

ion pump a system of membrane proteins that exchange ions inside the cell (for example, Na^+) with those in the intercellular fluid (for example, K^+). Unlike diffusion or transport through ion channels, transport via the Na^+–K^+ pump requires energy, which is provided by the hydrolysis of ATP to ADP. An example of how the Na^+–K^+ ion pump works is the response of a nerve cell to touch. Stimulation of the nerve cell causes Na^+ to flow into the cell and K^+ to flow out via ion channels; this two-way flow of ions produces the nerve impulse. The ion pump then "recharges" the system by pumping Na^+ out of the cell and K^+ into the cell so that another impulse can immediately be transmitted along the nerve.

The unequal concentrations of Na^+, K^+, and other ions on opposite sides of a cell membrane result in an electrochemical **equilibrium potential** or **reversal potential**, E_{ion}, for a particular ion. The equilibrium potential due to a particular ion can be determined from the Nernst equation (introduced in Chapter 18), in which n represents the charge and Q is the reaction quotient:

$$E_{cell} = E^\circ_{cell} - \frac{RT \ln Q}{nF} \tag{18.9}$$

Ion transport across a membrane (from inside the cell to the outside) is described by the following equilibrium:

$$M^{x+}_{inside} \rightleftharpoons M^{x+}_{outside}$$

$$Q = \frac{[M^{x+}_{outside}]}{[M^{x+}_{inside}]} \tag{21.1}$$

Substitution of this expression into Equation 18.9 yields:

$$E_{cell} = E^\circ_{cell} - \frac{RT}{nF} \ln\left(\frac{[M^{x+}_{outside}]}{[M^{x+}_{inside}]}\right) \tag{21.2}$$

Because there is no actual electron transfer (redox) during ion transport, the variable n in Equation 21.2 represents *the moles of charge on 1 mole of the ion* rather than the number of moles of electrons. This equation can be simplified further by noting that $E^\circ_{cell} = 0$ during ion transport because the half-reactions that contribute to E°_{cell} are the same. The result is an expression for E_{ion} that relates it to the concentrations of an ion across a cell membrane:

$$E_{ion} = -\frac{RT}{nF} \ln\left(\frac{[M^{x+}_{outside}]}{[M^{x+}_{inside}]}\right) \tag{21.3}$$

A related overall **membrane potential**, $E_{membrane}$, represents contributions from all the major ions, both cations and anions, inside and outside a cell. The values of the membrane potential reflect the different concentrations of the ions inside the cell versus outside the cell, and their relative ability to pass through the membrane, or *permeability*. $E_{membrane}$ values typically range from -50 to -70 mV.

ion pump a system of membrane proteins that exchange ions inside the cell with those in the intercellular fluid.

equilibrium (reversal) potential, E_{ion} an electrochemical potential that results from a concentration gradient of a particular ion on opposite sides of a cell membrane.

membrane potential, $E_{membrane}$ a weighted average of the equilibrium (reversal) potentials of the major ions based on concentration and permeability of the individual ions.

The driving force $E_{transport}$ that pushes an ion across a cell membrane can be calculated from the following equation:

$$E_{transport} = E_{membrane} - E_{ion} \qquad (21.4)$$

You may recall that free energy (ΔG) represents the work involved in a thermodynamic process, such as transporting an ion across a cell membrane. We can derive an equation connecting the potential of an electrochemical cell to ΔG for the process, allowing us to calculate the work done during the transport of a specific ion across a cell membrane:

$$\Delta G_{transport} = -nFE_{transport} \qquad (21.5)$$

Ion channels, transporters, and pumps, and the ion concentrations they control, are a finely tuned system that regulates the membrane potential of cells. Rapidly multiplying normal cells such as fertilized eggs and differentiating stem cells have membrane potentials much less negative than the typical values and may be as low as 0 to -10 mV. It has recently been suggested that membrane potential plays a role in wound healing, and it is well established that some cancer cells have depolarized cell membranes and that their membrane potentials are also in the 0 to -10 mV range.

CONNECTION The Nernst equation and the relationship between E and ΔG were introduced in Chapter 18.

SAMPLE EXERCISE 21.1 Calculating E_{ion} for Ion Transport   **LO2**

The concentration of Na^+ inside a squid axon is $0.050\ M$ (50 mM), in comparison with $[Na^+] = 0.440\ M$ (440 mM) in the fluid surrounding the cell.

a. Calculate the equilibrium potential E_{ion} for the Na^+ ion at 310 K.
b. If the membrane potential $E_{membrane} = -0.050$ V, is the transport of Na^+ from inside to outside the cell membrane spontaneous?

Collect, Organize, and Analyze We are given the concentrations of Na^+ on opposite sides of a cell membrane, and we are asked to calculate E_{ion} for Na^+ and to determine whether the transport across the membrane is spontaneous. Equation 21.3 allows us to calculate the equilibrium potential that arises from an unequal concentration of an ion on either side of a permeable membrane. If $[Na^+_{outside}] > [Na^+_{inside}]$, then the ln term in the equation will be greater than 1 and E_{ion} will be negative. Using Equations 21.4 and 21.5, we can calculate the work done during transport; if ΔG is positive, the process is nonspontaneous; if negative, it is spontaneous.

Solve
a. Substitution into Equation 21.3 gives:

$$E_{Na^+} = -\frac{RT}{nF} \ln \frac{[Na^+_{outside}]}{[Na^+_{inside}]} = -\frac{[8.314\ \text{J/(mol} \cdot \text{K)}](310\ \text{K})}{(1/\text{mol})(96{,}500\ \text{J/V})} \ln\left(\frac{440\ \text{m}M}{50\ \text{m}M}\right)$$

$$= -0.058\ \text{V} = -58\ \text{mV}$$

b. Using Equation 21.4, we can calculate $E_{transport}$, which is the driving force available to push the ions across the membrane:

$$E_{transport} = E_{membrane} - E_{ion} = -0.050\ \text{V} - (-0.058\ \text{V}) = 0.008\ \text{V} = 8\ \text{mV}$$

Substituting 8 mV into Equation 21.5:

$$\Delta G_{transport} = -nFE_{transport} = -(1/\text{mol})(96{,}500\ \text{J/V})(0.008\ \text{V}) = -772\ \text{J/mol}$$
$$= -8 \times 10^2\ \text{J/mol}$$

Because $\Delta G < 0$, the transport is spontaneous.

Think About It The equilibrium potential for Na^+ under these conditions is negative, consistent with our prediction. The transport of Na^+ from the inside to the outside of the cell membrane is spontaneous and goes against the concentration gradient under these conditions. The membrane potential (-0.050 V) would not have to vary by much to change the transport from spontaneous to nonspontaneous, and research is ongoing to find ways to modify cell potentials as a means to inhibit both tumor growth and spread.

Practice Exercise The concentration of K^+ inside a frog muscle cell is 124 mM, in comparison with $[K^+]$ = 2.3 mM in the fluid surrounding the cell. Calculate the equilibrium potential for the K^+ ion at 310 K. If the membrane potential $E_{membrane}$ is -73 mV, how much work must be done to transport K^+ into the cell at 310 K?

(Answers to Practice Exercises are in the back of the book.)

CONCEPT **TEST**

Do ion pumps represent spontaneous or nonspontaneous processes?

(Answers to Concept Tests are in the back of the book.)

Magnesium and Calcium

The biological roles of Mg^{2+} and Ca^{2+} are more varied than those of Na^+ and K^+. We have mentioned that calcium is a major component of teeth and bones. A prolonged deficiency of calcium can lead to osteoporosis (a disease characterized by low bone density), whereas high concentrations of calcium in muscle cells contribute to cramps. Most kidney stones are made of calcium oxalate or calcium phosphate. Magnesium deficiencies can reduce physical and mental capacity because of the role of Mg^{2+} in the transfer of phosphate groups to and from ATP; slowing this transfer diminishes the amount of energy available to cells. The cellular concentrations of Mg^{2+} and Ca^{2+} are maintained by ion pumps.

Magnesium is a component of chlorophyll, which is one of several molecules used by plants to collect and capture light energy across the visible portion (400 to 700 nm) of the electromagnetic spectrum (Figure 21.5). Chlorophylls from different plants vary slightly in composition, but all of them contain magnesium coordinated to four nitrogen atoms. The presence of magnesium in chlorophyll does not account for the green color of the molecule, nor does it play a direct role in absorption of sunlight. The function of the Mg^{2+} ion is to orient the molecules in positions that allow energy to be transferred to the reaction centers where H_2O is consumed and O_2 is produced during photosynthesis. Carotene and related compounds are responsible for the orange colors of autumn leaves on deciduous trees when chlorophyll production ceases. Mg^{2+} ions play important roles in ATP hydrolysis and ADP phosphorylation. The many Mg^{2+}-mediated ATP → ADP processes include transferring phosphate to glucose in the conversion of glucose to pyruvate and driving Na^+–K^+ ion pumps.

To some extent, calcium ions also can mediate ATP hydrolysis, but these ions play other roles in the cell. They are necessary to trigger muscle contractions, for example—the calcium ions used for this purpose are stored in proteins. Recall that the action of Na^+–K^+ pumps is responsible for the generation of nerve impulses. One effect of nerve impulses is to trigger the release of Ca^{2+} ions from their storage proteins into the intracellular fluid. In a multistep process, muscle cells contract and relax as calcium ions are released. Once the muscle action is complete, the ions are returned to their storage proteins in a process coupled to Mg^{2+}-mediated ATP hydrolysis.

CONNECTION The role of ATP and ADP in metabolism was described in Chapter 20.

CONNECTION The catabolism of glucose was described in Chapter 20.

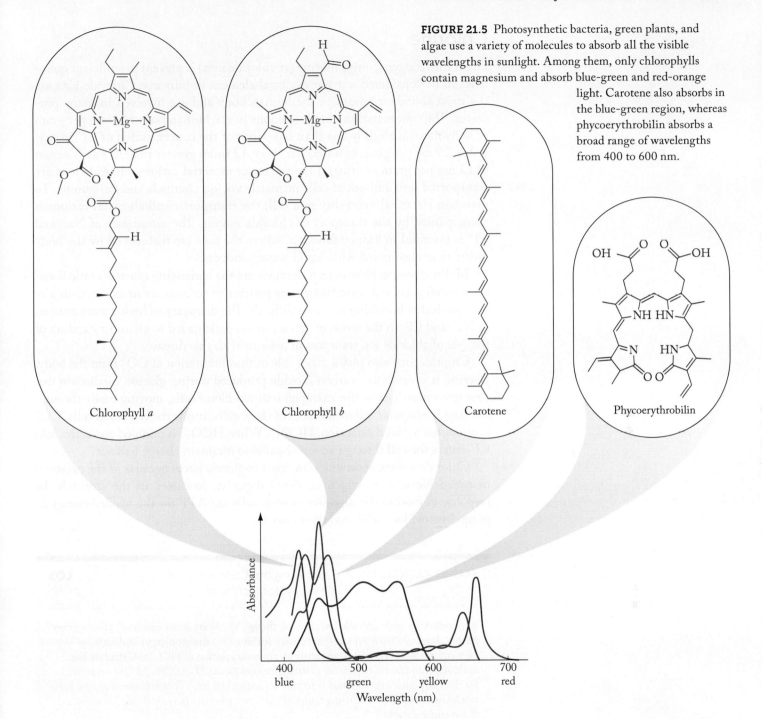

FIGURE 21.5 Photosynthetic bacteria, green plants, and algae use a variety of molecules to absorb all the visible wavelengths in sunlight. Among them, only chlorophylls contain magnesium and absorb blue-green and red-orange light. Carotene also absorbs in the blue-green region, whereas phycoerythrobilin absorbs a broad range of wavelengths from 400 to 600 nm.

Of the four alkali metal and alkaline earth major essential elements, only calcium plays a major role in the formation of teeth and bones. Mammalian bones are a *composite material*, defined as a material containing a mixture of different substances. About 30% of dry bone mass is elastic protein fibers. The rest of the mass consists of calcium compounds, including the mineral hydroxyapatite, $Ca_5(PO_4)_3(OH)$, which is also a principal component of teeth. Hydroxyapatite crystals are bound to the protein fibers in bone through phosphate groups.

The shells of marine organisms are mostly calcium carbonate ($CaCO_3$) in a matrix of proteins and polysaccharides. Some magnesium is incorporated into the calcium carbonate outer shell of marine organisms that are capable of photosynthesis, such as algae and phytoplankton.

Chlorine

Of all the halogens, only chlorine (as chloride ions) is present in sufficient quantities to be considered a major essential element in humans. Chloride ions are the most abundant anions in the human body and are involved in many processes. The concentration of chloride ions in the human body (1.5 mg per gram of body mass) is slightly less than one-tenth of the concentration of Cl^- in seawater (19 mg per gram of water) but about 12 times greater than in Earth's crust (0.13 mg per gram of crust). Like the major essential cations, chloride ions are transported into and out of cells primarily via ion channels and ion pumps. To maintain electrical neutrality in a cell, the transport of alkali metal cations is accompanied by the transport of chloride anions. The *cotransport* of Na^+ and Cl^- is essential in kidney function, where the ions are reabsorbed by the body rather than eliminated with liquid waste products.

Malfunctioning chloride ion channels are the underlying cause of cystic fibrosis, a lethal genetic disease that causes patients to accumulate mucus in their airways such that breathing becomes difficult. The discovery of high concentrations of Na^+ and Cl^- in the sweat of cystic fibrosis patients led to an understanding of the role of chloride ion transport in patients with this disease.

Chloride ions also play a major role in the elimination of CO_2 from the body. Because it is nonpolar, carbon dioxide produced during glucose catabolism can pass from muscle cells (for example) into red blood cells, moving easily through the largely nonpolar cell membranes of these cells. Inside the red blood cells, CO_2 is converted to bicarbonate ion, HCO_3^-. When HCO_3^- is pumped out of the cell, Cl^- enters the cell through an ion channel to maintain charge balance.

Chloride ion concentrations are high in gastric juices because of the presence of hydrochloric acid, which catalyzes digestive processes in the stomach. In response to food in the digestive system, cells tap ATP for the needed energy to pump hydrochloric acid into the stomach.

SAMPLE EXERCISE 21.2 Calculating the Concentration   **LO3**
of HCl in Stomach Acid

Acid reflux (sometimes called heartburn, though the heart is not involved) affects many people. It results from acid in the stomach leaking into the esophagus and causing discomfort. Stomach acid is primarily an aqueous solution of HCl. (a) Calculate the molarity of hydrochloric acid in gastric juice that has a pH of 0.80. (b) One treatment for the symptoms of acid reflux is to take an antacid tablet. What volume of gastric juice can be neutralized by a 750-mg tablet of calcium carbonate (a typical size for an over-the-counter antacid)?

Collect and Organize We are given the pH of a solution and are asked to calculate the concentration of HCl that corresponds to that pH. As we saw in Chapter 15, $pH = -\log[H_3O^+]$. Hydrochloric acid is a strong acid and ionizes completely to H_3O^+ and Cl^- in water:

$$HCl(aq) + H_2O(\ell) \rightarrow H_3O^+(aq) + Cl^-(aq)$$

We are also asked to calculate the volume of HCl solution that can be neutralized by a 750-mg tablet of calcium carbonate, $CaCO_3$. We need to write a balanced chemical equation for the neutralization reaction.

Analyze The equation describing the ionization of hydrochloric acid indicates that 1 mole of H_3O^+ ions is formed for every mole of HCl present. The pH of gastric juice falls between 1 and 0, so $[H_3O^+]$ will be between $10^{-1} (= 0.1)$ M and $10^0 (= 1)$ M.

The neutralization reaction is

$$CaCO_3(s) + 2\,H_3O^+(aq) \rightarrow Ca^{2+}(aq) + CO_2(g) + 3\,H_2O(\ell)$$

This equation indicates that 2 moles of H_3O^+ are consumed for every mole of $CaCO_3$. We are told that the tablet size is typical of an antacid tablet, so common sense leads us to predict that the volume of acid this tablet can neutralize will not be excessively large (greater than 1 L) or small (less than 10 mL): too large a tablet would be a waste of antacid, and too small a tablet would not relieve the symptoms.

Solve

a. Substitution into Equation 15.10 gives

$$pH = -\log[H_3O^+] = 0.80$$

We take the antilog of both sides to solve for $[H_3O^+]$:

$$[H_3O^+] = 10^{-0.80} = 0.16\,M\,H_3O^+$$

Therefore the concentration of HCl is

$$0.16\,M\,H_3O^+ \times \frac{1\,\text{mol HCl}}{1\,\text{mol }H_3O^+} = 0.16\,M\,\text{HCl}$$

b. First we calculate the number of moles of $CaCO_3$ present in 750 mg:

$$0.750\,\text{g CaCO}_3 \times \frac{1\,\text{mol CaCO}_3}{100.09\,\text{g CaCO}_3} = 7.49 \times 10^{-3}\,\text{mol CaCO}_3$$

Next we use the stoichiometry of the neutralization reaction to calculate the volume of 0.16 M HCl this quantity of $CaCO_3$ can neutralize:

$$7.49 \times 10^{-3}\,\text{mol CaCO}_3 \times \frac{2\,\text{mol }H_3O^\pm}{1\,\text{mol CaCO}_3} \times \frac{1\,\text{L}}{0.16\,\text{mol }H_3O^\pm}$$

$$= 9.36 \times 10^{-2}\,\text{L} = 94\,\text{mL of }0.16\,M\,\text{HCl}$$

Think About It A concentration of 0.16 M seems reasonable because it is indeed within the range of values predicted for a solution with pH < 1. The volume of 0.16 M acid that a 750-mg tablet of $CaCO_3$ can neutralize is also reasonable; 94 mL represents about 3 ounces of gastric juice.

Practice Exercise Calculate the pH of a solution prepared by mixing 10.0 mL of 0.160 M HCl with 15.0 mL of water. How much antacid containing 4.00×10^2 mg of $Mg(OH)_2$ in 5.00 mL of water is needed to neutralize this volume of acid?

CONCEPT **TEST**

Taking an antacid tablet is often sufficient to treat an occasional case of mild acid reflux. Another remedy is a drug such as Prilosec, which inhibits a cell's proton pumps by binding to the site of the pump and disabling it for more than 24 hours. Is the equilibrium constant for the binding of a proton pump inhibitor likely to be less than or greater than 1?

Nitrogen

Nitrogen is a major essential element found primarily in proteins but also in DNA and RNA. Nitrogen is available in the atmosphere as N_2, and soil and water contain nitrate ions, but neither of these forms of nitrogen can be directly incorporated into amino acids, the building blocks of proteins. The biosynthesis of amino acids requires ammonia or ammonium ions. For example, glycine (NH_2CH_2COOH),

FIGURE 21.6 The nitrogen cycle: Enzymes interconvert the nitrogen-containing molecules and ions found in nature. Bacteria convert atmospheric nitrogen to ammonium ion, which is oxidized to nitrite (NO_2^-) and nitrate (NO_3^-) ions before being reduced back to N_2.

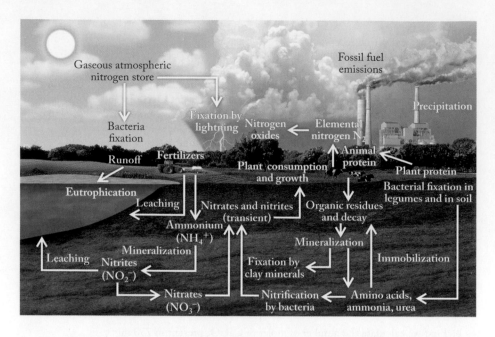

the simplest amino acid, is formed by reaction of CO_2 and ammonia in the presence of the appropriate enzyme. Interconversion of nitrogen-containing compounds in the environment is described by the nitrogen cycle illustrated in Figure 21.6. Certain bacteria use enzymes called *nitrogenases* to convert N_2 to ammonia. Plants convert NO_3^- ions to NO_2^- and then to NH_3 by using enzymes called *reductases*. Ultimately, the chemical reactions in these organisms begin a food chain that supplies the essential amino acids for human diets.

The reactions in Figure 21.7 are all redox reactions that illustrate the wide range of oxidation numbers found among nitrogen compounds in the nitrogen cycle. By definition, each atom in nitrogen gas, N_2, is assigned an oxidation number of zero. When N_2 is converted to NH_3, the oxidation number of N decreases to -3, a reduction. The oxidation numbers of N in nitrite (NO_2^-) and nitrate (NO_3^-) are $+3$ and $+5$, respectively, as a result of oxidation.

In Chapter 13 we encountered a different cycle for nitrogen in the environment: the conversion of N_2 and O_2 to NO and NO_2 in the engines of automobiles. Dinitrogen monoxide, N_2O, is a greenhouse gas. The nitrogen atoms in these volatile nitrogen oxides are assigned oxidation numbers of $+1$, $+2$, and $+4$ for N_2O, NO, and NO_2, respectively.

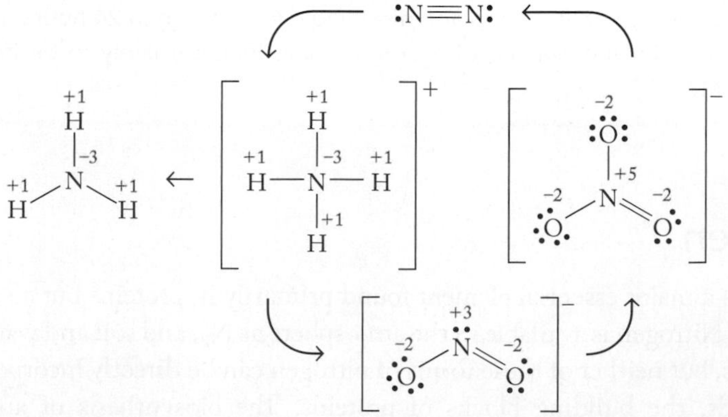

FIGURE 21.7 Nitrogen-containing molecules and ions in the nitrogen cycle exhibit oxidation numbers ranging from -3 to $+5$.

SAMPLE EXERCISE 21.3 Writing a Balanced Chemical   **LO3**
Equation Describing the Reaction
of Nitrate Reductases

Nitrate ion can be reduced to ammonia by enzymes called nitrate reductases. The first step is conversion of nitrate ion to nitrite ion. Assign oxidation numbers to the elements in these ions, and write a balanced equation for the following half-reaction in a basic solution:

$$NO_3^-(aq) \rightarrow NO_2^-(aq)$$

Collect, Organize, and Analyze We need to assign oxidation numbers according to the guidelines in Section 4.9. We know that the oxidation numbers of nitrogen and oxygen in each polyatomic ion must add up to the charge on the ion. Oxygen in compounds usually has an oxidation number of −2.

Solve The oxidation number of nitrogen is unknown, so we call it x. The oxidation number of nitrogen in NO_3^- is

$$x + 3(-2) = -1$$
$$x = +5$$

The oxidation number of nitrogen in NO_2^- is

$$x + 2(-2) = -1$$
$$x = +3$$

The half-reaction is

$$NO_3^-(aq) \rightarrow NO_2^-(aq)$$

The nitrogen is balanced; we balance oxygen by adding water:

$$NO_3^-(aq) \rightarrow NO_2^-(aq) + H_2O(\ell)$$

Then we balance hydrogen by adding hydrogen ions:

$$2\,H^+(aq) + NO_3^-(aq) \rightarrow NO_2^-(aq) + H_2O(\ell)$$

We balance charge by adding electrons:

$$2\,e^- + 2\,H^+(aq) + NO_3^-(aq) \rightarrow NO_2^-(aq) + H_2O(\ell)$$

We switch to a basic solution by adding the same number of OH^- ions to both sides of the equation:

$$2\,OH^-(aq) + 2\,e^- + 2\,H^+(aq) + NO_3^-(aq) \rightarrow NO_2^-(aq) + H_2O(\ell) + 2\,OH^-(aq)$$

The hydrogen ions combine with the hydroxide ions to form water, and we cancel the species that are the same on both sides:

$$2\,e^- + 2\,H_2O(\ell) + NO_3^-(aq) \rightarrow NO_2^-(aq) + \cancel{H_2O(\ell)} + 2\,OH^-(aq)$$

This gives us a final equation for the reduction half-reaction:

$$2\,e^- + H_2O(\ell) + NO_3^-(aq) \rightarrow NO_2^-(aq) + 2\,OH^-(aq)$$

Think About It Assigning oxidation numbers is a convenient way of identifying which element is reduced or oxidized in a half-reaction and of determining how many electrons are gained or lost. The balanced half-reaction confirms that electrons are added to nitrate to reduce it to nitrite ion.

Practice Exercise The reduction half-reaction catalyzed by one type of nitrogenase produces 1 mole of $H_2(g)$ for every 2 moles of $NH_4^+(aq)$ under acidic conditions. Write a balanced equation for this half-reaction.

In humans and other mammals, excess nitrogen is converted to urea in the liver and excreted via the kidneys. Plants use urea as a source of ammonia by the action of *ureases* via the reaction:

$$\text{H}_2\text{N}-\underset{\overset{\|}{\text{O}}}{\text{C}}-\text{NH}_2 + \text{H}_2\text{O} \rightarrow 2\,\text{NH}_3 + \text{CO}_2$$

Unlike reactions catalyzed by nitrogenases and nitrate reductases, the conversion of urea to ammonia and carbon dioxide is not a redox reaction. It is a hydrolysis reaction, similar to the reaction of nonmetal oxides with water described in Chapter 4. Acidic solutions are observed when NH_4^+, NO_2, and N_2O_5 dissolve in water:

$$\text{NH}_4^+(aq) + \text{H}_2\text{O}(\ell) \rightleftharpoons \text{NH}_3(aq) + \text{H}_3\text{O}^+(aq)$$

$$2\,\text{NO}_2(g) + \text{H}_2\text{O}(\ell) \rightarrow \text{HNO}_2(aq) + \text{HNO}_3(aq)$$

$$\text{N}_2\text{O}_5(g) + \text{H}_2\text{O}(\ell) \rightarrow 2\,\text{HNO}_3(aq)$$

Hydrolysis of nitrite ion, however, leads to weakly basic solutions:

$$\text{NO}_2^-(aq) + \text{H}_2\text{O}(\ell) \rightleftharpoons \text{HNO}_2(aq) + \text{OH}^-(aq)$$

CONNECTION Hydrolysis of nitrite ion is described in Chapter 15.

Phosphorus and Sulfur

Phosphorus and sulfur are major essential elements found primarily in proteins and DNA, but they are also present in polyatomic anions prevalent in the environment. In comparison with the nitrogen cycle, the biological phosphorus cycle contains only a single major species: the phosphate ion, PO_4^{3-}, and its conjugate acids HPO_4^{2-} and $H_2PO_4^-$, along with phosphate esters. Reduction of phosphorus(V) in PO_4^{3-} to phosphine (PH_3) does occur in swamps. An oxygen-free environment is needed because PH_3 spontaneously ignites in humid air, yielding phosphoric acid:

$$\text{PH}_3(g) + 2\,\text{O}_2(g) \rightarrow \text{H}_3\text{PO}_4(\ell)$$

Gases analogous to those in the nitrogen cycle, such as NO and NO_2, are absent from the phosphorus cycle. Slow weathering of insoluble phosphate minerals by weak acids in soil introduces phosphate ion into the environment, where it is eventually taken up by plants. Some of this phosphate is incorporated into biominerals such as hydroxyapatite $[Ca_5(PO_4)_3(OH)]$, as seen previously in our discussion of calcium.

CONCEPT TEST

Why are phosphates more likely to precipitate with cations from aqueous solution than nitrates?

You may recall from Chapter 16 that phosphate ion is in equilibrium with its conjugate acid, HPO_4^{2-}, which in turn hydrolyzes to $H_2PO_4^-$ as shown in the following equations:

$$\text{PO}_4^{3-}(aq) + \text{H}_2\text{O}(\ell) \rightleftharpoons \text{HPO}_4^{2-}(aq) + \text{OH}^-(aq)$$

$$\text{HPO}_4^{2-}(aq) + \text{H}_2\text{O}(\ell) \rightleftharpoons \text{H}_2\text{PO}_4^-(aq) + \text{OH}^-(aq)$$

Aqueous phosphate ion can be transported into cells, where it is incorporated into familiar organic molecules such as ATP, glucose-6-phosphate, and nucleic acids

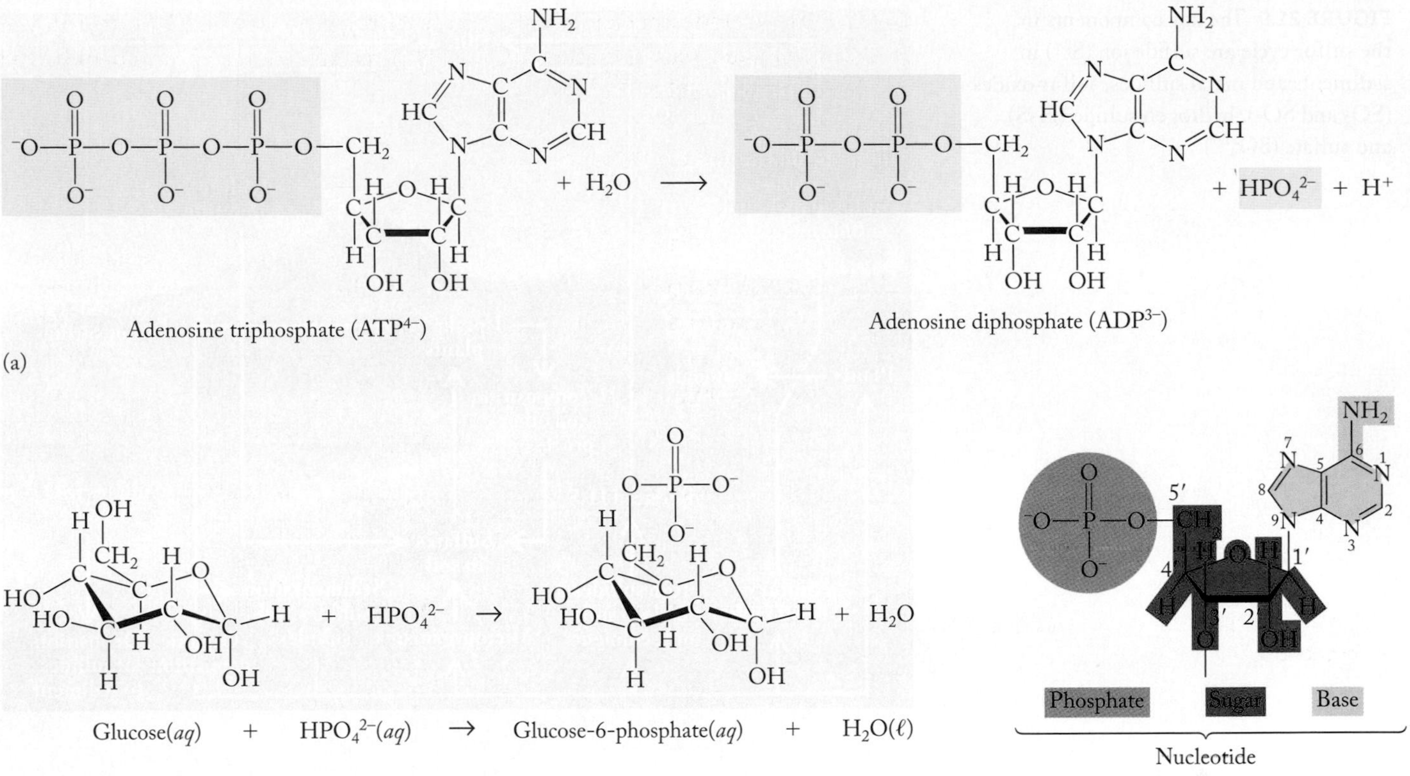

(a)

(b)

Glucose(*aq*) + HPO_4^{2-}(*aq*) → Glucose-6-phosphate(*aq*) + $H_2O(\ell)$

(c)

FIGURE 21.8 Condensation reactions between HPO_4^{2-} (*aq*) and −OH groups yield phosphate esters such as (a) ATP, (b) glucose-6-phosphate, and (c) nucleic acids.

(Figure 21.8). Note the presence of a carbon-oxygen bond between a monosaccharide and the phosphate group in all three of the molecules in Figure 21.8. In glucose-6-phosphate and nucleic acids, this new bond forms through a condensation reaction of an −OH group on the sugar molecule and HPO_4^- that produces water as a product. Adenosine diphosphate (ADP) is converted to ATP through a condensation reaction between HPO_4^{2-} and a phosphate group on ADP.

The processes in Figure 21.8 are all reversible; the P–O bond can be hydrolyzed, releasing HPO_4^{2-}. For ATP, this reaction is exothermic and provides the energy for many cellular processes. Given the importance of phosphorus to life, significant amounts of phosphates are used as fertilizer in agriculture. Agricultural runoff may stimulate rapid growth of algae in freshwater ponds and lakes. The explosive growth of algae can use up all the dissolved oxygen, killing other higher aquatic organisms.

The sulfur cycle in Figure 21.9 illustrates the array of sulfur compounds found in the environment. We have already encountered volatile sulfur oxides, SO_2 and SO_3, in the context of acid rain on early Earth in Chapter 3. Like the nonmetal oxides of groups 14 and 15, SO_2 and SO_3 dissolve in water to produce the weak acid H_2SO_3 and the strong acid H_2SO_4, respectively. The sulfate ion produced by the dissociation of H_2SO_4, the sulfate ions derived from minerals in soil, and the sulfate ions produced by the oxidation of H_2S are all absorbed by plants.

The enzyme ATP sulfurylase promotes the introduction of SO_4^{2-} into ATP as adenosine-5′-phosphosulfate, APS^{2-} (Figure 21.10). The sulfur in APS^{2-} eventually finds its way into sulfur-containing amino acids (cysteine and methionine) and organic sulfur-containing compounds. It is also released to the environment as H_2S, the sulfur analogue of water.

Compounds of hydrogen with oxygen, sulfur, and the other group 16 elements provide interesting contrasts with respect to molecular shape and properties. The

FIGURE 21.9 The key components in the sulfur cycle are sulfide ion (S^{2-}) in sediments and metal sulfides, sulfur oxides (SO_2 and SO_3), hydrogen sulfide (H_2S), and sulfate (SO_4^{2-}).

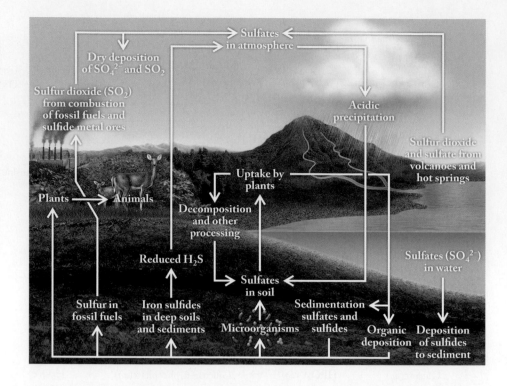

data in Table 21.7 show how different water is from the other compounds. Water is a liquid at ordinary temperatures and pressures, has a large negative heat of formation, and has a much larger bond angle than the other three hydrides. We also know that water is odorless and is absolutely essential for life. The hydrides of sulfur (S), selenium (Se), and tellurium (Te) are all gases under standard conditions; have bond angles close to 90°; and are foul-smelling and poisonous. Hydrogen sulfide is responsible for the smell of rotten eggs. It is especially dangerous because it tends to very quickly fatigue the nasal sensory sites responsible for detecting it. This means that the intensity of the odor is a very poor indicator of the concentration of H_2S in the air. Headache and nausea begin at air concentrations of H_2S as low as 5 ppm, and exposure to 100 ppm leads to paralysis and may result in death.

Similarly, organic compounds containing sulfur have properties that differ from those of their oxygen-containing counterparts. Many of them also have characteristic odors. Methanol is an alcohol with the formula CH_3OH. It is a liquid at room temperature and has an odor usually described as slightly alcoholic. Methanethiol, CH_3—SH, is a gas at room temperature and has the pungent odor of rotten cabbage. It is produced in the intestinal tract of animals by the action of bacteria on proteins and is one of the sulfur compounds responsible for the characteristic aroma of a feedlot or a barnyard.

FIGURE 21.10 The reduction of sulfate ion begins by substitution of a phosphate (PO_4^{3-}) group on ATP by sulfate (SO_4^{2-}) catalyzed by the enzyme ATP-sulfurylase.

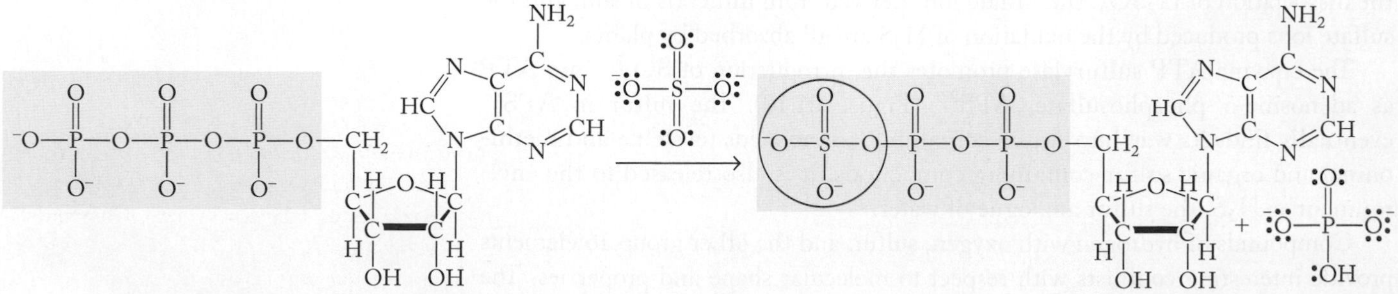

TABLE 21.7 Comparison of the Properties of Group 16 Dihydrides

Hydride[a]	Melting Point (°C)	Boiling Point (°C)	Bond Length (pm)	Bond Angle (degrees)	Heat of Formation (kJ/mol)
H_2O	0	100	96	104.5	−285.8
H_2S	−86	−60	134	92	−20.17
H_2Se	−66	−41	146	91	73.0
H_2Te	−51	−4	169	90	99.6

[a]H_2Po is excluded; too little is known of its chemistry. Polonium has no stable isotopes and is present on Earth only in very small quantities.

If we examine compounds with two carbon atoms, ethanol (beverage-grade alcohol), CH_3CH_2OH, is a liquid at room temperature. The corresponding sulfur compound is a very low-boiling liquid at room temperature called ethanethiol, which has a penetrating and unpleasant odor, like very powerful green onions. The human nose can detect the presence of ethanethiol at levels as low as 1 ppb (part per billion) in the air. This gives rise to its use as an odorant in natural gas. Natural gas has no odor, and natural gas leaks are such enormous fire hazards that ethanethiol is added to natural gas streams to make even small leaks detectable.

If we rearrange the atoms in ethanol and ethanethiol, we produce two new compounds. For ethanol we get dimethyl ether, CH_3—O—CH_3, a colorless gas used in refrigeration systems. Its counterpart, dimethyl sulfide, CH_3—S—CH_3, is one of the compounds responsible for the "low-tide" smell of ocean shorelines.

Three of the sulfur compounds described—hydrogen sulfide, methanethiol, and dimethyl sulfide—are referred to as volatile sulfur compounds (VSCs) by dentists. They are produced by bacteria in the mouth and are the principal compounds responsible for bad breath. One of the reasons the odors of these compounds differ from those of their oxygen counterparts is that their molecular sizes and shapes are slightly different. Also, their polarities differ because of the electronegativity difference between oxygen and sulfur. Recall that we discussed the importance of molecular shape in determining the extent of interaction of a compound with receptors in nasal membranes in Chapter 9. In part, the vast differences in odor and sensory detectability of these compounds are due to their shapes and electron distributions.

Not all sulfur compounds have an odor, but many odiferous compounds do contain sulfur. The characteristic and unpleasant smell of urine produced by some people after eating asparagus results from the inability of their bodies to convert sulfur compounds (Figure 21.11) into odor-free sulfate ions. Not all people can convert the sulfur compounds in asparagus to sulfate, and not all people can smell the odiferous sulfur compounds. Apparently, genetic differences determine how we metabolize these compounds and how well we can sense their odors.

H_3C—S H_3C—S—S—CH_3 H_3C\—S—CH_3
      H

Methanethiol   Dimethyl disulfide   Bis(methylthio)methane

FIGURE 21.11 Structures of some of the volatile sulfur compounds responsible for the smell of "asparagus" urine. Compounds shown here to the left have stronger (and more unpleasant) odors.

FIGURE 21.12 The pungent smell of skunk spray is due to butanethiol, $CH_3(CH_2)_3SH$.

The odor of skunk is due mostly to butanethiol (Figure 21.12), and the odor of well-used athletic shoes is primarily due to the presence of sulfur compounds produced by bacteria. Not all sulfur compounds have aromas as unpleasant as these, however. A compound with the formula $C_{10}H_{18}S$ is responsible for the aroma of grapefruit. If the orientation of two atoms on one of the carbon atoms in the molecule is switched, the resulting molecule has the same Lewis structure but no aroma at all. The carbon atom in question is a chiral site, and the two compounds are an enantiomeric pair.

SAMPLE EXERCISE 21.4 Drawing Lewis Structures for Molecules in the Sulfur Cycle **LO3**

Dimethyl sulfide, CH_3SCH_3, is one of the products of the sulfur cycle in Figure 21.9. In marine environments, dimethyl sulfide can be oxidized by bacteria to dimethyl sulfoxide, $(CH_3)_2SO$.

a. Draw Lewis structures for CH_3SCH_3 and $(CH_3)_2SO$, and determine the molecular geometry about the S atom in each.
b. Methanethiol, CH_3SH, is the simplest of the thiols and has a boiling point of 6°C. Dimethyl sulfide has a boiling point of 38°C. Explain the difference in boiling point between methanethiol and dimethyl sulfide.
c. Which hybrid orbitals does sulfur use in bonding to carbon in CH_3SCH_3 and CH_3SH?

Collect and Organize We need to draw Lewis structures for two sulfur compounds, determine their geometry, and compare their boiling points. Guidelines for drawing Lewis structures were discussed in Chapter 8. We are also asked to describe the hybrid orbitals of S in these compounds. Hybrid orbitals were described in Chapter 9. The effect of structure on boiling points was discussed in Chapter 10.

Analyze The Lewis structure for a molecule depends on the total number of valence electrons available, distributed over the atoms so that each atom has a complete octet (except H, which has a duet). Atoms with $Z \geq 13$ may have an expanded octet if such an arrangement leads to lower formal charges on the atoms. The arrangement of the bonding and lone pairs on the central atom allow us to predict intermolecular forces such as dipole–dipole interactions, hydrogen bonds, and dispersion forces. We can account for observed molecular geometries by combining s, p, and sometimes d orbitals to form hybrid atomic orbitals such as sp, sp^2, sp^3, sp^3d, and sp^3d^2.

Solve (a) Dimethyl sulfide has a total of 20 valence electrons: 6 from the S atom, 6 from the H atoms, and 8 from the C atoms. These electrons can be distributed in six C—H single bonds and two S—C single bonds, with four electrons remaining as two nonbonding pairs on S. Dimethyl sulfoxide has an oxygen atom bonded to the sulfur in dimethyl sulfide. This oxygen brings an additional six valence electrons to the molecule, giving $(CH_3)_2SO$ a total of 26 valence electrons. Sharing one of the S lone pairs with O will complete the octets of both S and O but will leave O with a formal charge of −1 and sulfur with a formal charge of +1. Forming a S=O double bond makes the formal charges on both S and O equal to zero but requires that sulfur have an expanded octet. The two structures for $(CH_3)_2SO$ represent resonance forms:

(b) Methanethiol has 14 valence electrons: 6 from the S atom, 4 from the H atoms, and 4 from the C atom. The electrons are distributed in three C—H single bonds, one C—S single bond, and one S—H single bond. As in dimethyl sulfide, there are two nonbonding pairs left on S. Both dimethyl sulfide and methanethiol contain a sulfur atom surrounded by two bonding pairs and two nonbonding pairs of electrons for a total of four electron pairs. The electron-pair geometry is tetrahedral, and the molecular geometry is bent:

$$H-\overset{\displaystyle H}{\underset{\displaystyle H}{C}}-\ddot{\underset{\displaystyle\cdot\cdot}{S}}-H$$

Both methanethiol and dimethyl sulfide are polar as a result of their molecular geometry and experience dipole–dipole interactions. Both molecules interact through dispersion forces as well. One might expect that the stronger dipole–dipole forces in CH_3SH would lead to a higher boiling point than for CH_3SCH_3, but we observe the opposite. We conclude that the dispersion forces in CH_3SCH_3 have a greater effect than the dipole–dipole interactions in CH_3SH, leading to dimethyl sulfide boiling at a temperature about 32°C higher than methanethiol.

(c) The similar geometry for CH_3SCH_3 and CH_3SH, with each molecule's central sulfur atom having two bonded atoms and two lone pairs, is consistent with sp^3 hybrid orbitals on the sulfur atom in both cases.

Think About It Methanethiol and dimethyl sulfide differ only in that the latter has a CH_3 group in place of a hydrogen atom. (This relationship between thiols and sulfides corresponds to the relationship between alcohols and ethers.) Sulfur does not expand its octet in these compounds because using the lone pairs on S to form multiple bonds is not needed. The boiling points of CH_3SCH_3 and CH_3SH reveal an important observation: many weaker bonds (dispersion forces) in CH_3SCH_3 can outweigh a few stronger forces (dipole–dipole forces) in CH_3SH.

Practice Exercise The nitrogen cycle in Figure 21.6 involves both neutral compounds such as NO_2 and polyatomic ions such as NO_2^-. Draw Lewis structures for both species and determine whether they have the same molecular geometry about nitrogen. Identify which hybrid orbitals contain nitrogen lone pairs in NO_2 and NO_2^-.

21.4 Trace and Ultratrace Essential Elements

Figure 21.13 shows the DRI values of four main group essential elements: silicon, selenium, fluorine, and iodine. Silicon, fluorine, and iodine are present in the body in average concentrations between 1 and 1000 μg of element per gram of body mass and are considered trace elements. Selenium is considered ultratrace, meaning that it is present in an average concentration of less than 1 μg of element per gram of body mass.

Selenium

The volatile selenium analogue to water, H_2Se, is toxic; however, selenium is considered an ultratrace essential element. The average concentration of Se in the human body is 0.3 μg per gram of body mass. Mounting scientific evidence points to a need for a minimum daily dose of approximately 55 μg of selenium. The effects of selenium toxicity, however, are apparent in people who ingest

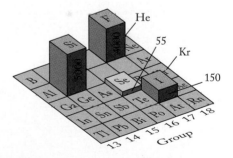

FIGURE 21.13 Silicon, fluorine, and iodine are trace essential elements, and selenium is an ultratrace essential element. The remaining labeled elements are nonessential. The vertical bars show DRI values for these elements in micrograms per day.

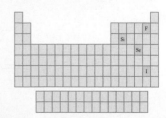

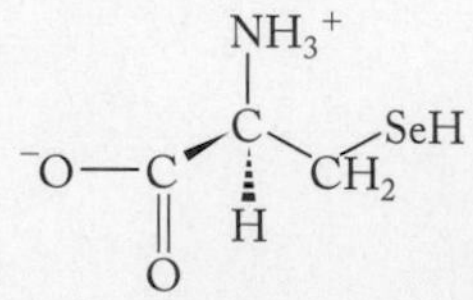

FIGURE 21.14 Selenocysteine is the selenium-containing analogue of the amino acid cysteine. Much of the selenium in the human body is found in proteins containing selenocysteine.

CONNECTION The solubility product, K_{sp}, was introduced in Chapter 16. Le Châtelier's principle was discussed in Chapter 14.

more than 500 μg per day. Most of the selenium we need is obtained from selenium-rich produce such as garlic, mushrooms, and asparagus, or from fish. Selenium occurs in the body as the amino acid selenocysteine (Figure 21.14) and is incorporated into enzymes.

Selenocysteine is an antioxidant. Our bodies need oxygen to survive, yet living in an oxygen-rich atmosphere can lead to the formation of potentially dangerous oxidizing agents in cells. For example, metabolism of fatty acids forms oxidizing agents called alkyl hydroperoxides, which can attack the lipid bilayer of cell membranes. It is believed that aging is related to the inability of the body to inhibit oxidative degradation of tissue. Selenocysteine participates in a series of reactions that result in the decomposition of these alkyl hydroperoxides.

Fluorine and Iodine

Fluorine is a trace essential element, and fluoride ions have significant benefits for dental health. Tooth enamel is composed of the mineral hydroxyapatite, $Ca_5(PO_4)_3(OH)$, which is essentially insoluble in water:

$$Ca_5(PO_4)_3(OH)(s) \rightleftharpoons Ca_5(PO_4)_3^+(aq) + OH^-(aq) \qquad K_{sp} \approx 2.4 \times 10^{-59}$$

When hydroxyapatite comes into contact with weak acids in your mouth, this equilibrium shifts to the right as the acid reacts with the hydroxide ions. This shift effectively increases the solubility of hydroxyapatite, so that your tooth enamel becomes pitted, and dental caries form. This is an example of Le Châtelier's principle. Fluoride ions reduce the likelihood of caries by displacing the OH^- ions in hydroxyapatite to form fluorapatite:

$$Ca_5(PO_4)_3(OH)(s) + F^-(aq) \rightleftharpoons Ca_5(PO_4)_3F(s) + OH^-(aq) \qquad K = 8.48$$

The solubility of fluorapatite is less dependent on pH than is the solubility of hydroxyapatite, so changing tooth enamel to fluorapatite makes your teeth more resistant to decay. This is why toothpaste contains fluoride compounds and why fluoride is added to drinking water in many communities in North America and Europe.

Of all the trace essential elements, iodine may have the best-defined role in human health. The body concentrates iodide ions in the thyroid gland, where they are incorporated into two hormones—thyroxine and 3,5,3'-triiodothyronine (Figure 21.15)—whose role is to regulate energy production and use. The conversion of thyroxine to 3,5,3'-triiodothyronine is catalyzed by selenocysteine-containing proteins. A deficiency of iodine or of either hormone can cause fatigue or feeling cold and can ultimately lead to an enlarged thyroid gland, a condition known as goiter. To help prevent iodine deficiency, table salt sold in the United States and many other countries is "iodized" with a small amount of sodium iodide. An excess of either hormone can cause a person to feel hot and is linked to Graves' disease, an autoimmune

FIGURE 21.15 Thyroxine and 3,5,3'-triiodothyronine, two iodine-containing hormones found in the thyroid gland, regulate metabolism.

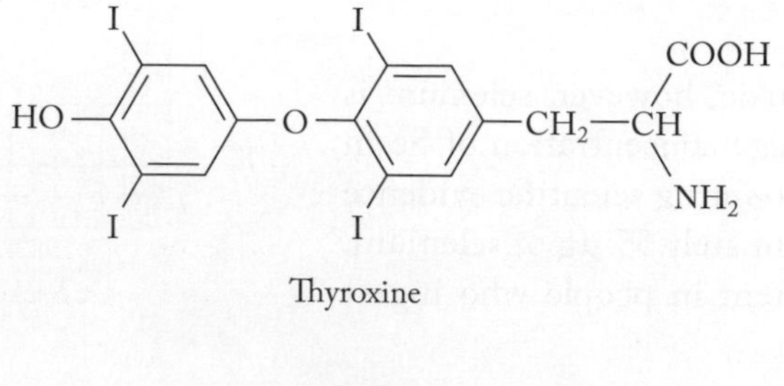

Thyroxine

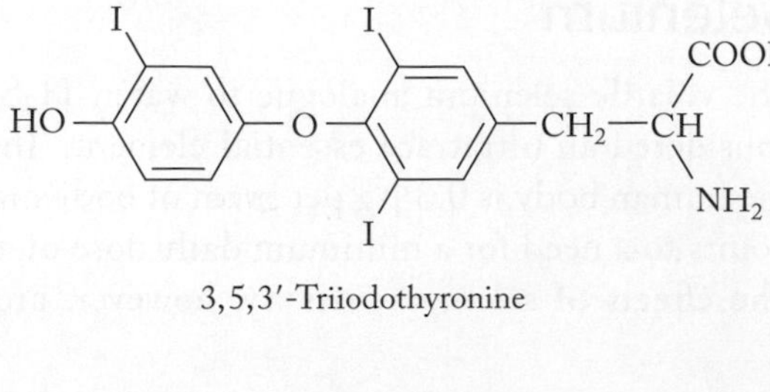

3,5,3'-Triiodothyronine

disease. The immune system in a patient with Graves' disease attacks the thyroid gland and causes it to overproduce the two hormones.

Silicon

The role of silicon in biological systems is less clear than for selenium and the halides. In mammals, a lack of the trace essential element silicon stunts growth. The presence of silicon as silicic acid, $Si(OH)_4$, is believed to reduce the toxicity of Al^{3+} ions in organisms by precipitating the aluminum as aluminosilicate minerals. Amorphous silica, SiO_2, is found in the exoskeletons of diatoms and in the cell membranes of some plants, such as the tips of stinging nettles.

21.5 Nonessential Elements

The ten elements listed in Table 21.2 are found in the human body but are classified as nonessential. In this section we discuss how some of these elements may end up in our bodies, working our way from left to right across the periodic table.

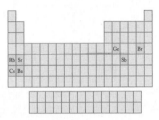

Rubidium and Cesium

Rubidium is generally regarded as nonessential in humans, yet it is the 15th most abundant element in the body. It is believed that Rb^+ is retained by the body because of the similarity of its size and chemistry to that of K^+. Like the other cations of group 1, cesium ions (Cs^+) are also readily absorbed by the body. Cesium cations have no known function, although they can substitute for K^+ and interfere with potassium-dependent functions. Usually, the concentration of cesium in the environment is low, so exposure to Cs^+ is not a health concern. The nuclear accident at Chernobyl in 1986, however, released significant quantities of radioactive ^{137}Cs into the environment. The ability of Cs^+ to substitute for K^+ led to the incorporation of $^{137}Cs^+$ into plants, which rendered crops grown in the immediate area unfit for human consumption because of the radiation hazard posed by this long-lived $(t_{1/2} \approx 30 \text{ yr})$ β emitter.

CONNECTION The biological effects of different types of nuclear radiation were described in Chapter 19.

CONNECTION We discussed the St. Louis Baby Tooth Survey and the effect of nuclear testing on children's teeth in Chapter 2.

Strontium and Barium

Some single-celled organisms build exoskeletons made with $SrSO_4$ and $BaSO_4$, but the human body appears to have no use for Sr^{2+} and Ba^{2+} ions. These ions do find their way into human bones, where they replace Ca^{2+} ions. At the low concentrations of Sr^{2+} and Ba^{2+} that are typically present in the human body, these elements appear to be benign. However, as with radioactive ^{137}Cs, incorporation of ^{90}Sr $(t_{1/2} = 29 \text{ yr})$ in bones can lead to leukemia. Atmospheric testing of nuclear weapons over the Pacific Ocean and in sparsely populated regions of the American West in the 1950s released ^{90}Sr into the environment. The full extent of the toxic effects of the fallout from these tests did not become apparent for several decades.

Germanium

It is generally agreed that germanium is a nonessential element and is barely detectable in the human body. Bis(carboxyethyl)germanium sesquioxide (Figure 21.16) has been touted as a nutritional supplement, but its efficacy remains controversial.

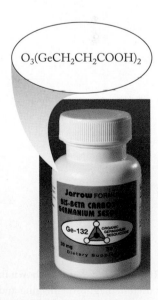

$$O_3(GeCH_2CH_2COOH)_2$$

FIGURE 21.16 Bis(carboxyethyl) germanium sesquioxide has been sold as a nutrition supplement, but its benefits are not well established.

Antimony

The role of antimony is also poorly understood. Most antimony compounds are toxic because they cause liver damage. However, ultratrace amounts of antimony may have a stimulatory effect, and selected antimony compounds have been used medically as antiparasitic agents, as discussed in the next section.

Bromine

Bromine has no known function in the human body but is consumed in foods such as grains, nuts, and fish in amounts ranging from 2 to 8 mg per day, leading to average concentrations of Br^- in blood of about 6 mg/L. Br^- has sedative and anticonvulsive properties but becomes toxic at concentrations around 100 mg/L, limiting its use to veterinary medicine. Bromide ion concentrations in seawater typically range from 65 to 80 mg/L. A select group of aquatic species can metabolize Br^- into bromomethane, CH_3Br, and other brominated organic compounds.

CONCEPT TEST

Looking at groups 1, 2, 14, 15, and 17, what periodic trend do you see in the location of the nonessential elements compared with the essential elements in the same group?

21.6 Elements for Diagnosis and Therapy

So far we have talked about the biological roles of several essential and nonessential main group elements found in our bodies. Some of these elements are also useful in diagnosing or treating diseases, as are some of the other elements in groups 1–2 and 13–18 that we have not mentioned (Figure 21.17). In this section we describe some of the applications of radioactive isotopes in the diagnosis of diseases. We also explore how compounds of essential and nonessential elements have found application in the treatment of a wide variety of illnesses.

Any diagnostic or therapeutic compound that is injected intravenously must be sufficiently soluble in blood to be delivered to the target. While in transit, the

FIGURE 21.17 The elements shown in red are used in diagnostic imaging and those shown in green are used in therapy. Gallium is used in both diagnostic imaging and therapy.

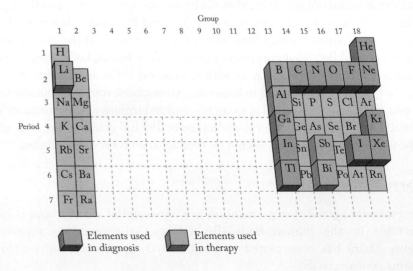

compound must be stable enough not to undergo chemical reactions that result in its precipitation or rapid elimination from the body. A medicinal chemist can also take advantage of substances that occur naturally in the body, such as antibodies, to carry a diagnostic or therapeutic metal ion to its target. Examples of elements and compounds containing elements from all 18 groups of the periodic table have been identified. In this section we focus on the main group elements (groups 1–2 and 13–18). Applications of the transition elements (groups 3–12) are described in Chapter 22.

Diagnostic Applications

Physicians in the 21st century have an array of imaging agents to help in diagnosing disease. Some methods use radionuclides with short half-lives that emit easily detectable gamma rays. Examples include the use of iodine-131 to image the thyroid gland (Figure 21.18) and of neutron-poor isotopes such as carbon-11 and fluorine-18 for positron emission tomography (PET). Not all imaging depends on radionuclides, however. In magnetic resonance imaging (MRI), for instance, which can diagnose soft-tissue injuries, stable isotopes of gadolinium are used as contrast agents to enhance images.

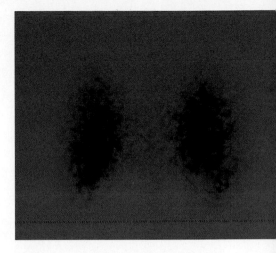

FIGURE 21.18 Gamma radiation that accompanies the decay of iodine-131 can be used to image the two butterfly-shaped lobes of the thyroid gland.

Imaging with Radionuclides The radionuclides used in medicine have short half-lives to limit the patient's exposure to ionizing radiation. If the half-life is too short, however, the nuclide may either decay before it can be administered or not reach the target organ rapidly enough to provide an image. Emission of relatively low-energy γ rays is essential to preventing collateral tissue damage.

Nuclide selection is also governed by the toxicity of both the parent element and the daughter nuclide. The speed at which the imaging agent is eliminated from the body can help mitigate toxic effects. Naturally, the cost and availability of a particular nuclide also factor into its usefulness in a clinical setting.

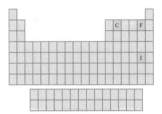

CONNECTION Chapter 19 gave a more detailed discussion of nuclear chemistry and nuclear medicine, including an assessment of the effects of different types of radiation on living tissue.

CONCEPT TEST

Why is it important to consider the nature of the decay products—α, β, or γ particles, or positrons—when choosing a radionuclide for medical imaging?

Gallium, Indium, and Thallium Gallium-66, gallium-67, gallium-68, and indium-111 are used as imaging agents for tumors and leukemia. All four nuclides decay by electron capture, and the γ radiation emitted in this nuclear reaction produces the images. All three gallium isotopes also decay by positron emission, which makes compounds containing these isotopes attractive for positron-emission tomography. Their half-lives range from just over 1 h for gallium-68 to 78 h for gallium-67. The discovery that indium-111–containing compounds can image a variety of cancers has led to the development of the drug Zevalin, currently used to treat some forms of non-Hodgkin's lymphoma.

The use of the gamma emitter thallium-201 ($t_{1/2}$ = 73 h) in diagnosing heart disease presents an interesting case for balancing the risks and benefits of using a particular isotope in medicine. Although thallium compounds are among the most toxic metal-containing compounds known, the nanogram quantities required for diagnosis pose few, if any, health hazards, meaning that the benefits outweigh the risks.

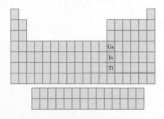

SAMPLE EXERCISE 21.5 Calculating Quantities of   **LO5**
Radioactive Isotopes

Indium-111 ($t_{1/2}$ = 2.805 d) and gallium-67 ($t_{1/2}$ = 3.26 d) are both used in radioimaging to diagnose chronic infections. Which isotope decays faster? If we start with 10.0 mg of each isotope, how much of each remains after 24 h?

Collect and Organize We are given the half-lives of two radionuclides and asked to predict which one will decay faster, and to calculate how much of each isotope remains after 24 h of decay.

Analyze An isotope with a shorter half-life decays faster. Radioactive decay follows first-order kinetics. Quantitatively, the relationship between half-life and the amount of material remaining is described by the following equation from Chapter 19:

$$\ln \frac{N_t}{N_0} = -0.693 \frac{t}{t_{1/2}}$$

Here N_0 and N_t refer to the amount of material present initially and the amount at time t, respectively. If our prediction for the relative decay rates of the two isotopes is correct, then more of the isotope with the longer half-life should remain after 24 h. We need to complete two calculations to determine the amount of each sample present after 24 h.

Solve Indium-111 has the shorter half-life, so it should decay faster.
For the amount of indium-111 remaining after 24 h, we have

$$\ln \frac{N_t}{10.0 \text{ mg}} = \frac{(-0.693)(24 \text{ h})}{(24 \text{ h/d})(2.805 \text{ d})} = -0.247$$

Taking the antilog of both sides, we get

$$\frac{N_t}{10.0 \text{ mg}} = 0.781$$

$$N_t = (0.781)(10.0 \text{ mg}) = 7.81 \text{ mg}$$

For gallium-67:

$$\ln \frac{N_t}{10.0 \text{ mg}} = \frac{(-0.693)(24 \text{ h})}{(24 \text{ h/d})(3.26 \text{ d})} = -0.213$$

$$\frac{N_t}{10.0 \text{ mg}} = 0.808$$

$$N_t = (0.808)(10.0 \text{ mg}) = 8.08 \text{ mg}$$

Think About It We predicted that indium-111 would decay faster, which means that after 24 h the quantity of this isotope should be less than the quantity of gallium-67, and it is.

Practice Exercise Two radioactive isotopes of bismuth are used to treat cancer. The half-lives are 61 min for bismuth-212 and 46 min for bismuth-213. If we start with 25.0 mg of each isotope, how much of each sample remains after 24 h?

Imaging with Noble Gases and MRI So far in this chapter we have had little opportunity to mention the noble gas elements. None of these elements are essential to the human body, although the World Anti-Doping Agency has added both Xe and Ar to the list of banned substances for athletes at Olympic and other sporting events since 2014. Apparently, in addition to behaving as an anesthetic, xenon increases the oxygen-carrying capacity of blood, providing an advantage in aerobically demanding sports. Similar chemistry is believed to occur with argon.

The lack of chemical reactivity of the group 18 elements and their ease of introduction into the body by inhalation, however, make selected isotopes of the noble gases—including helium-3, krypton-83, and xenon-129—attractive as agents for enhancing MRI images, particularly of the lungs. Krypton-83 provides greater sensitivity than xenon-129, allowing for better resolution in the images and, in principle, requiring the use of less gas.

Helium-3 has no known side effects and is preferable to xenon-129 for MRI, but it is present in only trace natural abundance. This isotope is obtained from β decay of tritium (^{3}H):

$$^3_1\text{H} \rightarrow {}^3_2\text{He} + {}^{0}_{-1}\beta$$

Neon-19 has been used in PET despite its short half-life (17.5 s). A patient positioned in a PET scanner breathes air containing a small amount of this isotope. Positron emission from the neon is recorded, and an image is created.

Therapeutic Applications

In this section we examine therapeutic agents that contain metallic and heavier main group elements in addition to carbon, hydrogen, nitrogen, oxygen, and sulfur.

Lithium, Boron, Aluminum, and Gallium The similar size of Li$^+$ (76 pm) and Mg^{2+} (72 pm) means that lithium ions can compete with magnesium ions in biological systems. The substitution of lithium for magnesium may account for its toxicity at high concentrations. Nevertheless, lithium carbonate is used to treat bipolar disorder, and other lithium compounds have been used to treat hyperactivity. In all cases, however, the use of lithium-containing drugs must be carefully monitored.

Of the elements of group 13, only boron and aluminum have been detected in humans. The role of boron in our bodies is not fully understood, but this element appears to play a role in nucleic acid synthesis and carbohydrate metabolism. Selected boron compounds appear to concentrate in human brain tumors. This property has opened the door to a treatment known as boron neutron-capture therapy (BNCT). Once a suitable boron compound has been injected and has made its way to a tumor, irradiation of the tumor with low-energy neutrons leads to the nuclear reaction

$$^{10}_5\text{B} + {}^1_0\text{n} \rightarrow {}^7_3\text{Li} + {}^4_2\text{He}$$

The α particles generated in the reaction have a short penetration depth but high relative biological effectiveness (RBE), so they can kill the tumor cells without harming surrounding tissue. The identification of compounds suitable for BNCT remains an area of active research.

Aluminum is found in some antacids as aluminum hydroxide, Al(OH)$_3$, or aluminum carbonate, Al$_2$(CO$_3$)$_3$. Some brands of baking powder contain sodium aluminum sulfate, NaAl(SO$_4$)$_2$ · 12 H$_2$O. Most of the aluminum in the human body can be traced to these sources. Aluminum is not considered essential to humans, but low-aluminum diets have been observed to harm goats and chickens. High concentrations of aluminum are clearly toxic; the effects are most noticeable in patients with impaired kidney function. The role of aluminum in Alzheimer's disease has been extensively debated but remains unresolved.

Simple gallium compounds such as gallium(III) nitrate and gallium(III) chloride, either alone or in combination with other drugs, have shown activity on bladder and ovarian cancers. The similar ionic radii of Ga^{3+} (62 pm) and Fe^{3+}

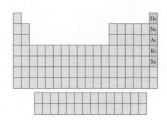

CONNECTION Tritium is produced in nuclear reactors by bombarding ^{6}Li and ^{7}Li with high-energy neutrons. The reactions involved are discussed in Chapter 19.

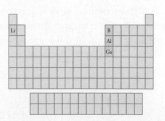

CONNECTION The relative biological effectiveness (RBE) of radioactive particles was introduced in Chapter 19.

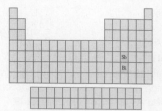

(64.5 pm) allow gallium to block DNA synthesis by replacing iron in a protein called transferrin and in other enzymes. Because gallium compounds accumulate in tumors at a higher rate than in healthy tissue, the disruption of DNA synthesis in the tumor cells inhibits tumor growth.

Antimony and Bismuth Antimony compounds are generally considered toxic. It has been reported, for instance, that exposure of infants to antimony compounds used as fire retardants in mattresses may contribute to sudden infant death syndrome (SIDS). However, this element does appear to have a medical use. Leishmaniasis, an insect-borne disease characterized by the formation of boils or skin lesions, is resistant to most treatments, but some patients have been successfully treated with sodium stibogluconate, one of the few applications of antimony compounds in human health.

Popular over-the-counter remedies for indigestion, diarrhea, and other gastrointestinal disorders contain bismuth subsalicylate (Figure 21.19). The bismuth in these compounds acts as a mild antibacterial agent that reduces the number of diarrhea-causing bacteria.

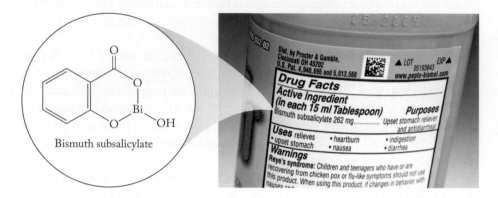

Bismuth subsalicylate

FIGURE 21.19 Bismuth subsalicylate is found in some antacids.

The human body requires about 30 elements to function properly. To manage the problems of disease and injury, scientists, physicians, engineers, and scores of other people have turned to the properties of these and many other elements on the periodic table to develop treatments and to enhance quality of life. This chapter has briefly introduced the roles of the main group elements in establishing and maintaining living systems.

SUMMARY

LO1 **Essential elements** have a physiological function in the body. **Nonessential elements** are present in the body but have no known functions. Some may have **stimulatory effects**. Essential elements are categorized as **major, trace,** or **ultratrace essential elements** depending on their concentrations in the body. (Sections 21.1 and 21.2)

LO2 Transport of Na^+ and K^+ across cell membranes involves **ion pumps** or selective transport through **ion channels**. Chloride ion is the most abundant anion in the human body, facilitating transport of alkali metal cations and elimination of CO_2. Differences in concentrations of ions inside and outside cell membranes give rise to a **membrane potential**, which determines the permeability of ions through the membrane. (Section 21.3)

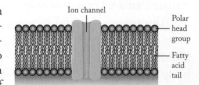

LO3 Acid–base chemistry and redox reactions are of great significance in living systems. The principles of structure and electron distribution in molecules apply to interactions *in vivo* just as they do in the laboratory. (Sections 21.3, 21.4, 21.5, and 21.7)

LO4 The most abundant elements in the human body include seven nonmetals: C, H, O, Cl, S, P, and N; the four major metals in the human body are Ca^{2+}, K^+, Na^+, and Mg^{2+}. Nonessential elements from groups 1 and 2 may be absorbed into cells or substituted into bone or other tissues because they are similar in size and charge to essential elements. Some nonessential elements are used in nutritional supplements, whereas others have therapeutic properties in low concentrations but are toxic at higher levels. (Sections 21.3, 21.4, and 21.5)

LO5 Radionuclides with short half-lives that emit low-energy γ rays are used in assessing function and diagnosing disease. The selection of a radionuclide for medical use is governed by the toxicities of the element and its daughter nuclides, its radioactive half-life, and the speed at which it is eliminated from the body. (Section 21.6)

LO6 Main group elements beyond the essential elements are useful in a wide variety of compounds with therapeutic value. Lithium salts are used in treating depression, whereas aluminum compounds find use as antacids. Fluoride in toothpaste helps prevent cavities. (Section 21.6)

PARTICULATE **PREVIEW WRAP-UP**

From left to right, the ions are Ca^{2+}, Na^+, H_3O^+, and K^+. The H_3O^+ ion is most likely to undergo selective transport through the sodium ion channel as a result of its similar charge. None of the other three ions are likely to pass through the K^+ ion channel because they are quite different in size from the K^+ ion.

PROBLEM-SOLVING SUMMARY

Type of Problem	Concepts and Equations		Sample Exercises
Calculating equilibrium potential, E_{ion}, for ions and ΔG for ion transport	Calculate E_{ion} by using a modified form of the Nernst equation: $$E_{ion} = -\frac{RT}{nF}\ln\left(\frac{[M^{x+}_{outside}]}{[M^{x+}_{inside}]}\right)$$	(21.3)	21.1
	Calculate ΔG by using the relationships between ΔG and E: $$E_{transport} = E_{membrane} - E_{ion}$$	(21.4)	
	$$\Delta G_{transport} = -nFE_{transport}$$	(21.5)	
Calculating an acid concentration from its pH	Relate the pH of a solution to the $[H_3O^+]$ by the equation $$pH = -\log[H_3O^+]$$	(15.10)	21.2
Assigning oxidation numbers and writing half-reactions	Use the guidelines in Section 4.9.		21.3
Drawing Lewis structures for molecules	Use the guidelines in Sections 8.2 and 9.4.		21.4
Calculating quantities of radioactive isotopes	Use the equation $$\ln\frac{N_t}{N_0} = -0.693\frac{t}{t_{1/2}}$$ where N_0 and N_t are the amounts of material present initially and at time t, respectively.		21.5

VISUAL PROBLEMS

(Answers to boldface end-of-chapter questions and problems are in the back of the book.)

21.1. Which part of Figure P21.1 best describes the periodic trend in monatomic cation radii moving up or down a group or across a period in the periodic table? (Arrows point in the direction of increasing radii.)

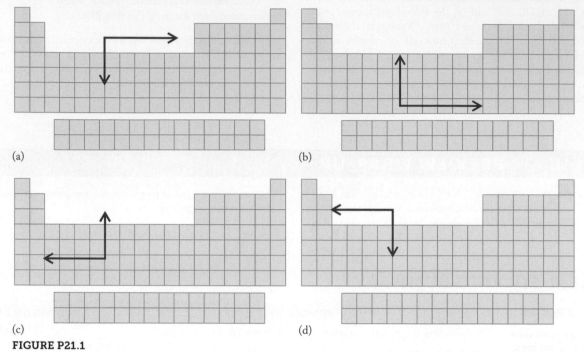

(a)   (b)

(c)   (d)

FIGURE P21.1

21.2. Which part of Figure P21.1 best describes the periodic trend in monatomic anion radii moving up or down a group or across a period in the periodic table? (Arrows point in the direction of increasing radii.)

21.3. Which of the two groups highlighted in the periodic table in Figure P21.3 typically forms ions that have larger radii than those of the corresponding neutral atoms?

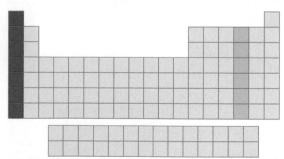

FIGURE P21.3

21.4. Which of the two groups highlighted in the periodic table in Figure P21.4 typically forms ions that have smaller radii than those of the corresponding neutral atoms?

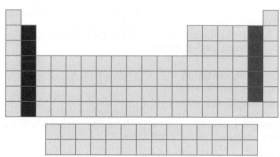

FIGURE P21.4

*21.5. As we saw in Chapter 18, the free energy (ΔG) of a reaction is related to the cell potential by the equation $\Delta G = -nFE$. In Figure P21.5, two solutions of Na^+ of different concentrations are separated by a semipermeable membrane. Calculate ΔG for the transport of Na^+ from the side with higher concentration to the side with lower concentration. (*Hint*: See Problem 21.45.)

Membrane
[Na+] = 150 mM
[Na+] = 10 mM

FIGURE P21.5

21.6. Two solutions of K^+ are separated by a semipermeable membrane in Figure P21.6. Calculate ΔG for the transport of K^+ from the side with lower concentration to the side with higher concentration.

Membrane
[K+] = 100 mM
[K+] = 12 mM

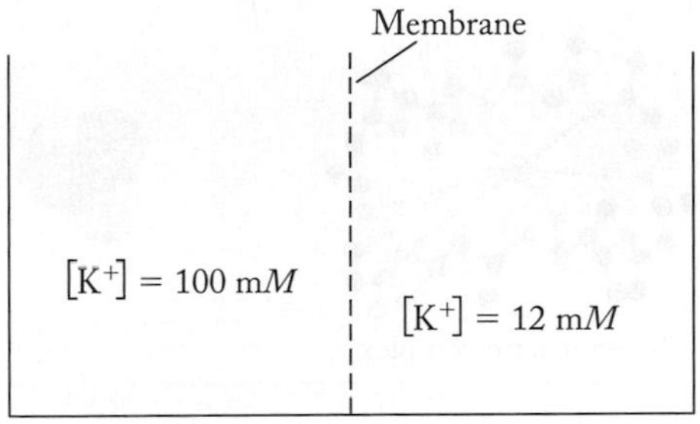

FIGURE P21.6

21.7. Describe the molecular geometry around each germanium atom in the compound shown in Figure P21.7.

O
O
CH2
Ge
Ge
CH2
HOOC
CH2
O
CH2
COOH

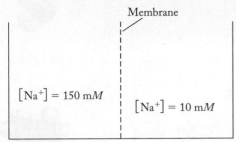

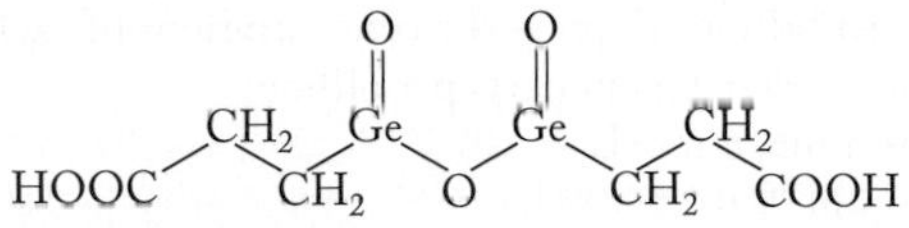

FIGURE P21.7

21.8. Selenocysteine can exist as two enantiomers (stereoisomers). Identify the atom in Figure P21.8 responsible for the two enantiomers.

SeH
H2C
CH
O−
+H3N
C
O

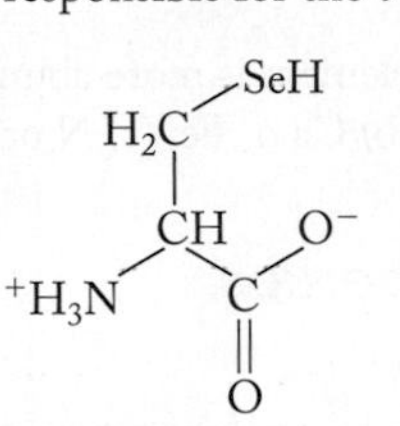

FIGURE P21.8

***21.9.** The relative sizes of the main group atoms and ions are shown in Figure P21.9. Using this figure as a guide, which of the following polyatomic ions is likely to be the largest: sulfate, phosphate, or perchlorate?

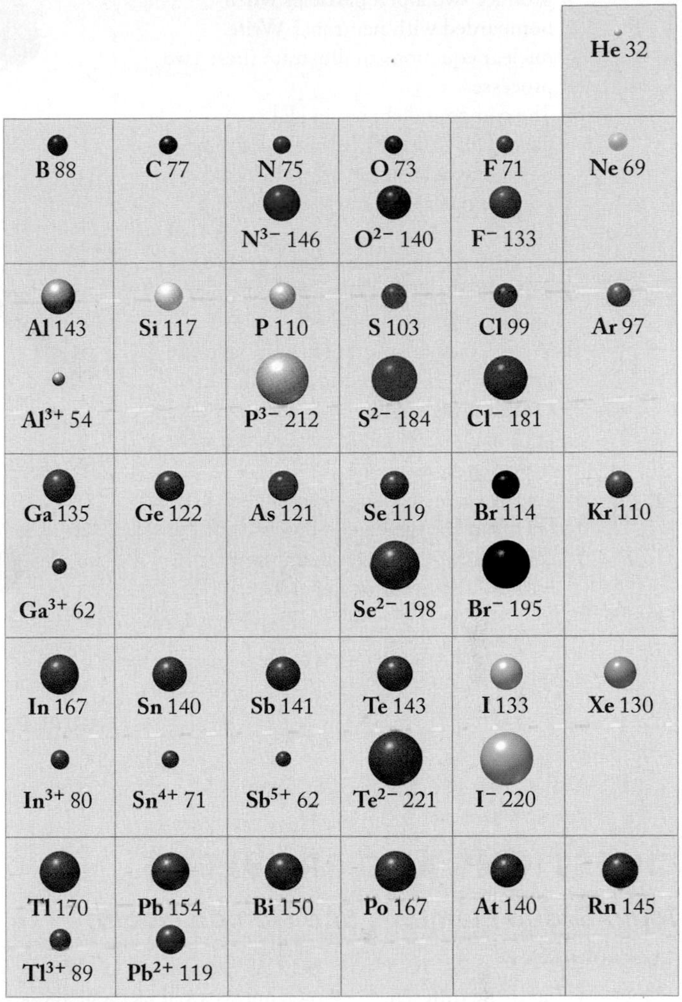

He 32
B 88
C 77
N 75
O 73
F 71
Ne 69
N3− 146
O2− 140
F− 133
Al 143
Si 117
P 110
S 103
Cl 99
Ar 97
Al3+ 54
P3− 212
S2− 184
Cl− 181
Ga 135
Ge 122
As 121
Se 119
Br 114
Kr 110
Ga3+ 62
Se2− 198
Br− 195
In 167
Sn 140
Sb 141
Te 143
I 133
Xe 130
In3+ 80
Sn4+ 71
Sb5+ 62
Te2− 221
I− 220
Tl 170
Pb 154
Bi 150
Po 167
At 140
Rn 145
Tl3+ 89
Pb2+ 119

FIGURE P21.9

21.10. Use representations [A] through [I] in Figure P21.10 to answer questions a–f.

a. Which nuclide is the product of beta decay from tritium? Which will produce two alpha particles when bombarded with neutrons? Write nuclear equations to illustrate these two processes.

b. The compound shown in [E] is also part of [D] and [F]. What metal binds to the nitrogen atoms in [D]? What metal binds to the nitrogen atoms in [F]?

c. Identify the important structural features in [B] and label each as polar or nonpolar.

d. What kinds of substances are likely to pass through the channel in [B]?

e. Of the three substances shown in [G], [H], and [I], which must pass through a channel to enter a cell?

f. Of the three substances shown in [G], [H], and [I], which can be transported directly through a cell membrane, without the need for a channel?

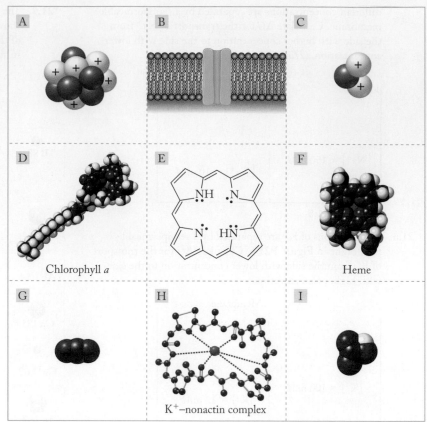

FIGURE P21.10

QUESTIONS AND PROBLEMS

Main Group Elements and Human Health

Concept Review

21.11. What is the difference between an essential element and a nonessential element?

21.12. Are all essential elements major essential elements?

21.13. What is the main criterion that distinguishes major, trace, and ultratrace essential elements from one another?

21.14. Should trace essential elements also be considered to be stimulatory?

Problems

21.15. The concentrations of very dilute solutions are sometimes expressed as parts per million. Express the concentration of each of the following trace and ultratrace essential elements in parts per million:
a. Fluorine, 110 mg in 70 kg
b. Silicon, 525 mg/kg
c. Iodine, 0.043 g in 100 kg

21.16. In the human body, the concentrations of ultratrace essential elements are even lower than those of trace essential elements and therefore are sometimes expressed in parts per billion. Express the concentrations of each of the following elements in parts per billion:
a. Bromine, 6 mg/L
b. Boron, 0.014 g/100 kg
c. Selenium, 5.0 mg/70 kg

21.17. In the following pairs, which element is more abundant in the human body? (a) silicon or oxygen; (b) iron or oxygen; (c) carbon or aluminum

21.18. In the following pairs, which element is more abundant in the human body? (a) H or Si; (b) Ca or Fe; (c) N or Cr

Periodic and Chemical Properties of Main Group Elements

Concept Review

21.19. In Chapter 2 we defined main group elements as those elements found in groups 1, 2, and 13–18 in the periodic table. Why do some chemists refer to these as the "s-block" and "p-block" elements?

21.20. Why do we classify the main group elements by group rather than by period?

21.21. Lithium oxide (Li_2O) and carbon monoxide (CO) have nearly the same molar mass. Why is Li_2O a solid with a high melting point, whereas CO is a gas?

21.22. The nonradioactive group 17 elements are found as diatomic molecules, X_2 (X = F, Cl, Br, I). Why is Br_2 a liquid at room temperature, whereas Cl_2 is a gas?

21.23. Which of the following properties can be used to distinguish a metallic element from a semimetallic element: atomic radius, electrical conductivity, and/or molar mass?

21.24. Which of the following cannot be measured: ionization energy, electron affinity, ionic radius, atomic radius, or electronegativity?

21.25. Why is Be^{2+} more likely than Ca^{2+} to displace Mg^{2+} in biomolecules?

21.26. PbS, $PbCO_3$, and PbCl(OH) have limited solubility in water. Which of them is/are more likely to dissolve in acidic solutions?

Problems

21.27. Which ion channel must accommodate the larger cation, a potassium or a sodium ion channel?

21.28. Which ion is larger: Cl^- or I^-?

_____

21.29. Place the following ions in order of increasing ionic radius: Mg^{2+}, Li^+, Al^{3+}, and Cl^-.

21.30. Place the following ions in order of increasing ionic radius: Br^-, O^{2-}, K^+, and Ca^{2+}.

_____

21.31. Place the following elements in order of increasing electronegativity: K, S, F, and Mg.

21.32. When we compare any two main group elements, is the element with the smaller atomic radius always more electronegative?

_____

21.33. Why can we estimate the electron affinity of Cl atoms by measuring the ionization energy of a Cl^- anion?

21.34. Place the following ions in order of increasing ionization energy: Na^+, S^{2-}, F^+, and Mg^+.

Major Essential Elements

Concept Review

21.35. **Ion Transport in Cells** Describe three ways in which ions of major essential elements (such as Na^+ and K^+) enter and exit cells.

21.36. Which transport mechanism for ions requires ATP: diffusion, ion channels, or ion pumps?

21.37. Why is it difficult for ions to diffuse across cell membranes?

21.38. Why does Sr^{2+} substitute for Ca^{2+} in bones?

21.39. Which alkali metal ion is Rb^+ most likely to substitute for?

21.40. Why don't alkaline earth metal cations substitute for alkali metal cations in cases where the ionic radii are similar?

*21.41. Why might nature have selected calcium carbonate over calcium sulfate as the major exoskeleton material in shells?

21.42. Bromide ion and fluoride ion are nonessential elements in the body. Do you expect their concentrations to be more similar to the concentrations of major essential elements or to the concentrations of ultratrace essential elements?

Problems

21.43. **Osmotic Pressure of Red Blood Cells** One of the functions of the alkali metal cations Na^+ and K^+ in cells is to maintain the cells' osmotic pressure. The concentration of NaCl in red blood cells is approximately 11 mM. Calculate the osmotic pressure of this solution at body temperature (37°C). (*Hint*: See Equation 11.13.)

21.44. Calculate the osmotic pressure exerted by a 92 mM solution of KCl in a red blood cell at body temperature (37°C). (*Hint*: See Equation 11.13.)

_____

*21.45. **Electrochemical Potentials across Cell Membranes** Very different concentrations of Na^+ ions exist in red blood cells (11 mM) and the blood plasma (160 mM) surrounding those cells. Solutions with two different concentrations separated by a membrane constitute a concentration cell. Calculate the electrochemical potential created by the unequal concentrations of Na^+.

21.46. The concentration of K^+ in red blood cells is 92 mM, and the concentration of K^+ in plasma is 10 mM. Calculate the electrochemical potential created by the two concentrations of K^+.

_____

21.47. If the transport of K^+ across a cell membrane requires 5 kJ/mol, how many moles of ATP must be hydrolyzed to provide the necessary energy? The hydrolysis of ATP is described by the equation

$$ATP^{4-} + H_2O \rightarrow ADP^{3-} + HPO_4{}^{2-} + H^+ \qquad \Delta G° = -34.5 \text{ kJ}$$

*21.48. Removing excess Na^+ from a cell by an ion pump requires energy. How many moles of ATP must be hydrolyzed to overcome a cell potential of −0.07 V? The hydrolysis of 1 mole of ATP provides 34.5 kJ of energy.

_____

21.49. **Plankton Exoskeletons** Exoskeletons of planktonic acantharia contain strontium sulfate. Calculate the solubility in moles per liter of $SrSO_4$ in water at 25°C given that $K_{sp} = 3.44 \times 10^{-7}$.

21.50. Algae in the genus *Closterium* contain structures built from barium sulfate (barite). Calculate the solubility in moles per liter of $BaSO_4$ in water at 25°C given that $K_{sp} = 1.08 \times 10^{-10}$.

Trace and Ultratrace Essential Elements; Nonessential Elements

Concept Review

21.51. What danger to human health is posed by ^{137}Cs ($t_{1/2} \approx 30$ yr)?

21.52. Why is ^{137}Cs ($t_{1/2} \approx 30$ yr) considered dangerous to human health, whereas naturally occurring ^{40}K ($t_{1/2} = 1.28 \times 10^6$ yr) is benign?

21.53. What are the likely signs of ΔS and ΔG for the dissolution of tooth enamel?

21.54. Why does fluorapatite resist acid better than hydroxyapatite if both are insoluble in water?

*21.55. Why do superoxide ions ($O_2{}^-$) act as strong oxidizing agents?

*21.56. Why are thyroxine and 3,5,3′-triiodothyronine (Figure 21.15) considered amino acids? Why aren't they *essential* amino acids?

Problems

21.57. What are the products of radioactive decay of ^{137}Cs? Write a balanced equation for the nuclear decay reaction.

21.58. Potassium-40 decays by three pathways: β decay, positron emission, and electron capture. Write balanced equations for each of these processes.

21.59. Calculate the pH of a 1.00×10^{-3} M solution of selenocysteine ($pK_{a_1} = 2.21$, $pK_{a_2} = 5.43$).

21.60. Calculate the pH of a 1.00×10^{-3} M solution of cysteine ($pK_{a_1} = 1.7$, $pK_{a_2} = 8.3$). Is selenocysteine a stronger acid than cysteine?

21.61. **Composition of Tooth Enamel** Tooth enamel contains the mineral hydroxyapatite. Hydroxyapatite reacts with fluoride ion in toothpaste to form fluorapatite. The equilibrium constant for the reaction between hydroxyapatite and fluoride ion is $K = 8.48$. Write the equilibrium constant expression for the following reaction. In which direction does the equilibrium lie?

$$Ca_5(PO_4)_3(OH)(s) + F^-(aq) \rightleftharpoons Ca_5(PO_4)_3(F)(s) + OH^-(aq)$$

*21.62. **Effects of Excess Fluoridation on Teeth** Too much fluoride might lead to the formation of calcium fluoride according to the reaction

$$Ca_5(PO_4)_3(OH)(s) + 10 \, F^-(aq) \rightleftharpoons$$
$$5 \, CaF_2(s) + 3 \, PO_4^{3-}(aq) + OH^-(aq)$$

Write the equilibrium constant expression for the reaction. Given the K_{sp} values for the following two reactions, calculate K for the reaction between $Ca_5(PO_4)_3(OH)$ and fluoride ion that forms CaF_2.

$$Ca_5(PO_4)_3(OH)(s) \rightleftharpoons 5 \, Ca^{2+}(aq) + 3 \, PO_4^{3-}(aq) + OH^-(aq)$$
$$K_{sp} = 2.3 \times 10^{-59}$$

$$CaF_2(s) \rightleftharpoons Ca^{2+}(aq) + 2 \, F^-(aq) \qquad K_{sp} = 3.9 \times 10^{-11}$$

21.63. Tooth enamel is actually a composite material containing both hydroxyapatite and a calcium phosphate, $Ca_8(HPO_4)_2(PO_4)_4 \cdot 6 \, H_2O$ ($K_{sp} = 1.1 \times 10^{-47}$).
   a. Is this calcium mineral more or less soluble than hydroxyapatite ($K_{sp} = 2.3 \times 10^{-59}$)?
   b. Calculate the solubility in moles per liter of hydroxyapatite, $Ca_5(PO_4)_3(OH)$, $K_{sp} = 2.3 \times 10^{-59}$ in water at 25°C and pH = 7.00.
   c. What is the solubility of hydroxyapatite at pH = 5.00?

21.64. The K_{sp} of actual tooth enamel is reported to be 1×10^{-58}.
   a. Does this mean that tooth enamel is more soluble than pure hydroxyapatite ($K_{sp} = 2.3 \times 10^{-59}$)?

   b. Does the measured value of K_{sp} for tooth enamel support the idea that tooth enamel is a mixture of hydroxyapatite, $Ca_5(PO_4)_3(OH)$, and a calcium phosphate, $Ca_8(HPO_4)_2(PO_4)_4 \cdot 6 \, H_2O$ ($K_{sp} = 1.1 \times 10^{-47}$)?
   c. Calculate the solubility in moles per liter of $Ca_8(HPO_4)_2(PO_4)_4 \cdot 6 \, H_2O$ ($K_{sp} = 1.1 \times 10^{-47}$) in water at 25°C and pH = 7.00.

21.65. Some sources give the formula of hydroxyapatite as $Ca_{10}(PO_4)_6(OH)_2$. If the K_{sp} of $Ca_5(PO_4)_3(OH)$ is 2.3×10^{-59}, what is the K_{sp} of $Ca_{10}(PO_4)_6(OH)_2$?

21.66. The same sources mentioned in the previous problem cite the formula of fluorapatite as $Ca_{10}(PO_4)_6F_2$. If the K_{sp} of $Ca_5(PO_4)_3F$ is 3.2×10^{-60}, what is the K_{sp} of $Ca_{10}(PO_4)_6F_2$?

21.67. All the group 16 elements form compounds with the generic formula H_2E (E = O, S, Se, or Te). Which compound is the most polar? Which compound is the least polar?

*21.68. All the group 15 elements form compounds with the generic formula H_3E (E = N, P, As, Sb, and Bi). Which compound is the most polar? Which compound do you predict to have the smallest H–E–H bond angle?

Elements for Diagnosis and Therapy
Concept Review

21.69. When choosing an isotope for imaging, why is it important to consider the decay mode of the isotope as well as the half-life?

21.70. Why might an α emitter be a good choice for radiation therapy?

*21.71. What advantage does a β emitter have over an α emitter for imaging?

21.72. Why do we sometimes use radioisotopes of toxic elements, such as thallium, for imaging?

21.73. Why does ^{213}Bi undergo β-decay but ^{111}In decays by electron capture?

21.74. Several isotopes of arsenic are used in medical imaging. Which isotope, ^{72}As or ^{77}As, is more likely to be useful for PET imaging?

21.75. The World Anti-Doping Agency (WADA) added xenon and argon to the list of banned substances in 2014. Which intermolecular forces account for the solubility of Xe and Ar in blood?

21.76. Helium is used in SCUBA gear to prevent nitrogen narcosis. Do you expect the solubility of He in blood to be greater than or less than the solubility of Xe and Ar in blood?

Problems

21.77. PET Imaging with Gallium A patient is injected with a 5 μM solution of gallium citrate containing ^{68}Ga ($t_{1/2}$ = 9.4 h) for a PET study. How long is it before the activity of the ^{68}Ga drops to 5% of its initial value?

21.78. Iodine-123 ($t_{1/2}$ = 13.3 h) has replaced iodine-131 ($t_{1/2}$ = 8.1 d) for diagnosis of thyroid conditions. How long is it before the activity of ^{123}I drops to 5% of its initial value?

21.79. The bismuth in over-the-counter antacids is found as BiO^+. Draw the Lewis structure for the BiO^+ cation.

21.80. Some medicines used in treating depression contain lithium carbonate. Draw the Lewis structure for Li_2CO_3.

21.81. Aluminum hydroxide is used in some antacids. Write a balanced net ionic equation for the reaction of aluminum hydroxide with HCl.

21.82. Aluminum carbonate is used in some antacids. Write a balanced net ionic equation for the reaction of aluminum carbonate with hydrochloric acid.

21.83. The antacid known as Maalox contains a mixture of magnesium and aluminum hydroxides. Which substance will neutralize more acid on a per-mole basis? Does the same substance also neutralize more acid on a per-gram basis?

21.84. Sodium bicarbonate and calcium carbonate both act as antacids and are found in common stomach remedies. Which substance will neutralize more acid on a per-mole basis? Does the same substance also neutralize more acid on a per-gram basis?

21.85. How many grams of magnesium hydroxide are needed to neutralize 115 mL of 0.75 M stomach acid?

21.86. How many grams of aluminum hydroxide are needed to neutralize 115 mL of 0.75 M stomach acid?

22

Transition Metals

Biological and Medical Applications

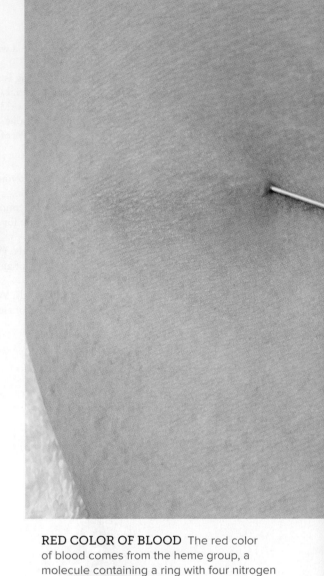

RED COLOR OF BLOOD The red color of blood comes from the heme group, a molecule containing a ring with four nitrogen atoms that bind to a central Fe^{2+} ion.

PARTICULATE **REVIEW**

Lewis Acid or Lewis Base?

In Chapter 22 we discuss the bonding and structure of compounds and complex ions formed by transition metals and learn about their presence in biological systems and their applications to medicine. The compounds shown here are ammonia, borane, and water.

- Draw the Lewis structure for each compound.
- Which compound(s) has/have a lone pair of electrons on the central atom?
- Which compound(s) can function as a Lewis acid? Which can function as a Lewis base?

    (Review Chapters 4 and 16 if you need help.)

(Answers to Particulate Review questions are in the back of the book.)

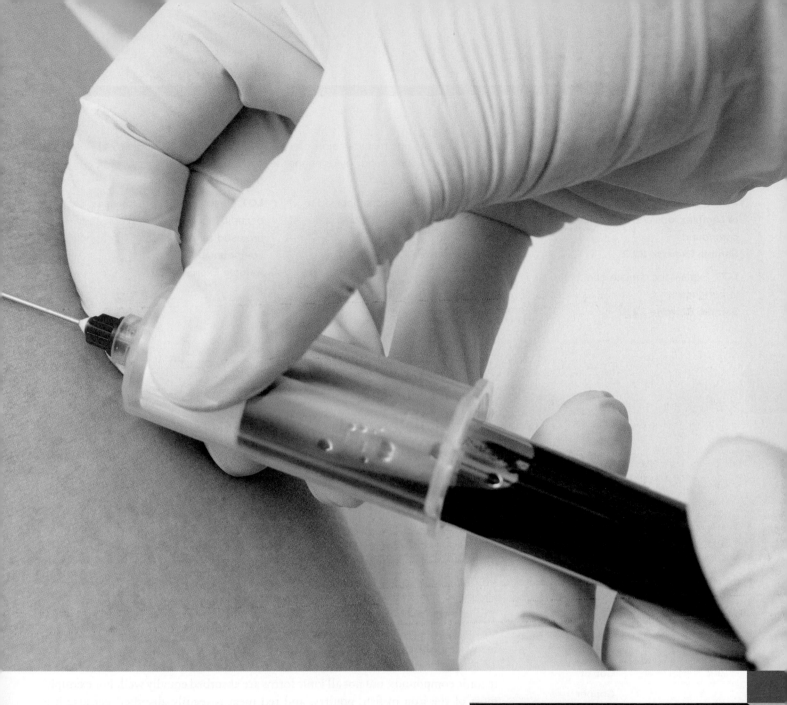

One Molecule, One Bond versus One Molecule, Two Bonds

Here are two complex ions that contain metal cations bonded to different molecules. As you read Chapter 22, look for ideas that will help you answer these questions:

- Atoms of the same element form coordinate bonds to both of the central metals in the complex ions. Which element is this?

- What molecule is bonded to the copper cation? How many of these molecules form coordinate bonds with copper?

- What molecule is bonded to the nickel cation? How many of these molecules form coordinate bonds with the nickel cation?

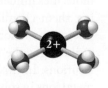

Cu^{2+} complex

Ni^{2+} complex

Learning Outcomes

LO1 Recognize complex ions and their counterions in chemical formulas
Sample Exercise 22.1

LO2 Interconvert the names and formulas of complex ions and coordination compounds
Sample Exercise 22.2

LO3 Explain the chelate effect and its importance
Sample Exercise 22.3

LO4 Explain the origin of the colors of transition metal compounds by using the spectrochemical series

LO5 Describe the factors that lead to high-spin or low-spin electronic states of complex ions
Sample Exercise 22.4

LO6 Identify stereoisomers of coordination compounds
Sample Exercise 22.5

LO7 Describe where metal complexes occur in biochemistry and how they are used as diagnostic or therapeutic compounds
Sample Exercise 22.6

TABLE 22.1 Essential Transition Elements Found in the Human Body

Trace (1–1000 μg/g Body Mass)	Ultratrace (<1 μg/g Body Mass)
Iron	Chromium
Zinc	Cobalt
	Copper
	Manganese
	Molybdenum
	Nickel
	Vanadium

inner coordination sphere the ligands that are bound directly to a metal via coordinate bonds.

22.1 Transition Metals in Biology: Complex Ions

Many of the metallic elements in the periodic table are essential to good health. For example, copper, zinc, and cobalt play key roles in protein function. Iron is needed to transport oxygen from our lungs to all the cells of our body. These and other essential metallic elements, several of which we discussed in Chapter 21, should be present either in our diets or in the supplements many of us rely on for balanced nutrition. Table 22.1 lists the transition metals essential to our bodies in trace and ultratrace amounts. However, the mere presence of these elements is not sufficient—they must be in a form that our cells can use. Swallowing an 18-mg steel pellet as if it were an aspirin tablet would not be a good way for you to get your recommended daily allowance of iron. If we are to benefit from consuming essential metals in food and nutritional supplements, the metals need to be in compounds, not free elements, and the compounds must be bioavailable to the body.

All the metallic transition elements essential to human health occur in nature in ionic compounds, but not all ionic forms are absorbed equally well. For example, most of the iron in fish, poultry, and red meat is readily absorbed because it is present in a form called *heme iron*. However, the iron in plants is mostly nonheme and is not as readily absorbed. Eating a meal that includes both meat and vegetables improves the absorption of the nonheme iron in the vegetables, as does consuming foods high in vitamin C. All these dietary factors work together at the molecular level to provide us with the nutrients we need to survive.

Interactions between transition metals and accompanying nonmetal ions and molecules influence the solubility of the metals, which is a key factor toward making them chemically reactive and biologically available to plants and animals. These interactions also influence other properties, including the wavelengths of visible light the metals absorb and therefore the colors of their compounds and solutions. In this chapter we explore how the chemical environment of transition metal ions in solids and in solutions affects their physical, chemical, and biological properties. We answer questions such as why many, but not all, metal compounds have distinctive colors, and how, through the formation of complex ions with biomolecules, transition metals play key roles in many biological processes.

We start this chapter by examining the interactions between transition metal ions and the other ions and molecules that surround them in solids and solutions.

To understand these interactions, we need to review the definitions of Lewis acids and Lewis bases we used in Chapter 16:

- A *Lewis base* is a substance that *donates* a lone pair of electrons in a chemical reaction.

- A *Lewis acid* is a substance that *accepts* a lone pair of electrons in a chemical reaction.

In Chapter 10, we described how ions dissolved in water are *hydrated*, that is, surrounded by water molecules oriented with their positive dipoles directed toward anions and their negative dipoles directed toward cations. When these ion–dipole interactions lead to the sharing of lone-pair electrons with empty valence-shell orbitals on the cations, they meet our definition of covalent bonds, and in this case are called *coordinate covalent bonds*, or simply *coordinate bonds*. Much of the chemistry of the transition metals is associated with their ability to form coordinate bonds with molecules or anions. Review Chapter 10 if you need help with these concepts.

As we saw in Chapter 16, molecules or anions that function as Lewis bases and form coordinate bonds with metal cations are called *ligands*. The resulting species, which are composed of central metal ions and the surrounding ligands, are called *complex ions* or simply *complexes*. Direct bonding to a central cation means that the

CONNECTION Lewis's pioneering theories of the nature of covalent bonding were described in Chapter 8, and the formation of complex ions between metal ions (Lewis acids) and ligands (Lewis bases) was discussed in Chapter 16.

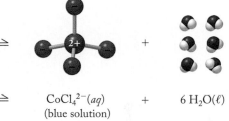

$$Co(H_2O)_6{}^{2+}(aq) \quad + \quad 4\,Cl^-(aq) \quad \rightleftharpoons \quad CoCl_4{}^{2-}(aq) \quad + \quad 6\,H_2O(\ell)$$

(pink solution)   (blue solution)

FIGURE 22.1 Equilibrium between two forms of cobalt in aqueous HCl solution.

ligands in a complex occupy the **inner coordination sphere** of the cation. Take another look at Sample Exercise 14.12, in which we discussed the equilibrium between two forms of cobalt(II), one pink and one blue, in an aqueous solution of HCl (Figure 22.1). Both the pink and blue forms are complexes; the pink cobalt(II) species has six water ligands in its inner coordination sphere, and the blue species has four chloride ions. The charge on each complex ion is the sum of the charges of the metal ion and the ligands: 2+ for $Co(H_2O)_6{}^{2+}$ because the charge of the cobalt ion is 2+ and water molecules are neutral, and 2− for $CoCl_4{}^{2-}$ because the sum of the 2+ charge on the cobalt ion and four 1− charges on the chloride ions is 2−.

The foundation of our understanding of the bonding and structure of complex ions comes from the pioneering research of Swiss chemist Alfred Werner (1866–1919), for which he was awarded the Nobel Prize in Chemistry in 1913. Some of Werner's research addressed the unusual behavior of different compounds formed by dissolving cobalt(II) chloride in aqueous ammonia and oxidizing it to cobalt(III) by bubbling air through the solution. One of the redox reactions produces an orange compound (Figure 22.2a) that contains 3 moles of Cl^- ions for every 1 mole of Co^{3+} ions and 6 moles of ammonia. Another reaction produces a purple compound (Figure 22.2b) that has the same proportions of Cl^- and Co^{3+} ions, but with only 5 moles of ammonia per mole of cobalt(III). Both compounds are water-soluble solids that react with aqueous solutions of $AgNO_3$, forming solid AgCl. However, 1 mole of the orange compound produces 3 moles of solid AgCl, whereas 1 mole of the purple one produces only 2 moles of AgCl. Results like these inspired Werner to study the electrical conductivity of aqueous solutions of the two compounds. He found that the orange compound was the better conductor, indicating that it produced more ions in solution than the purple one.

Werner concluded that the differences in the two compounds' composition, in their capacities to react with Ag^+ ions, and in their electrolytic properties are all caused by the presence of two types of Co–Cl bonds inside them. He proposed

(a)

(b)

FIGURE 22.2 Two compounds of cobalt(III) chloride and ammonia. (a) $[Co(NH_3)_6]Cl_3$, (b) $[Co(NH_3)_5Cl]Cl_2$.

coordination number the number of sites occupied by ligands around a metal ion in a complex.

coordination compound a compound made up of at least one complex ion.

counterion an ion whose charge balances the charge of a complex ion in a coordination compound.

CONNECTION We discussed in Chapter 4 how the electrical conductivity of aqueous solutions depends on the concentrations of dissolved ions in the solutions.

that these bonds are all ionic in the orange compound, but that only two-thirds of them are ionic in the purple compound. The remaining one-third are coordinate covalent bonds.

This capacity to form two kinds of bonds means that Co^{3+} and other transition metal ions have two kinds of bonding capacity, or *valence*. The first kind involves ionic bonds and is based on the number of electrons a metal atom loses when it forms an ion. This valence is equivalent to its oxidation number, which is +3 for cobalt in both the orange and purple compounds. The second kind of valence is based on the capacity of metal ions to form coordinate bonds. This property corresponds to an ion's **coordination number**, the number of sites around a central metal ion where bonds to ligands form.

The formulas of the reactants in the two chemical equations below fit the chemical composition and explain the properties of the orange and purple cobalt(III) compounds. Note how brackets in the formulas of these **coordination compounds** set off the complex ions from the ionically bonded chloride **counterions**, which balance the charges on the complex ions: 3+ for the orange one (Figure 22.3a) but only 2+ for the purple one because it contains a negatively charged Cl^- ion within its inner coordination sphere (Figure 22.3b).

The release of 3 moles of chloride ions from the orange compound, but only 2 from the purple compound, explains the compounds' chemical and electrolytic properties.

Werner proposed, correctly, that the six ligands in the inner coordination sphere of cobalt(III) ions are arranged in an octahedral geometry, as shown in the structures in Figure 22.3. This geometry is consistent with the formation of six bonds to a central atom (or central ion in this case), as we learned in Chapter 9. Many other transition metal ions form six coordinate bonds and octahedral complex ions. Some are listed in Table 22.2, as are complex ions that have only two or four ligands. Those with two have linear structures, whereas those with four may be tetrahedral or square planar. Both geometries occur in cobalt(II) ions: $CoCl_4^{2-}$ is tetrahedral, whereas $Co(CN)_4^{2-}$ is square planar.

> ### CONCEPT **TEST**
>
> In the coordination compound $Na_3[Fe(CN)_6]$, which ions occupy the inner coordination sphere of the Fe^{3+} ion, and which ions are counterions? Of the four cobalt complexes described in this section, which would have conductivity in aqueous solution similar to that of $Na_3[Fe(CN)_6]$?
>
> *(Answers to Concept Tests are in the back of the book.)*

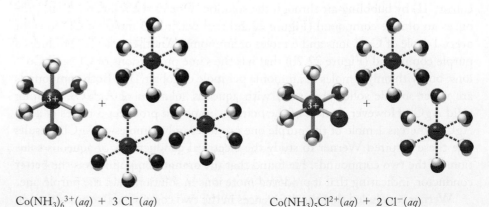

FIGURE 22.3 Structures of cobalt(III) complex ions and counterions in aqueous solution.

$Co(NH_3)_6{}^{3+}(aq) + 3\ Cl^-(aq)$
(a) Orange solution

$Co(NH_3)_5Cl^{2+}(aq) + 2\ Cl^-(aq)$
(b) Purple solution

TABLE 22.2 Common Coordination Numbers and Shapes of Complex Ions

Coordination Number	Steric Number	Shape	Structure	Examples
6	6	Octahedral		$Fe(H_2O)_6^{3+}$ $Ni(H_2O)_6^{2+}$ $Co(H_2O)_6^{3+}$
4	6	Square planar		$Pt(NH_3)_4^{2+}$
4	4	Tetrahedral		$Zn(H_2O)_4^{2+}$
2	2	Linear		$Ag(NH_3)_2^+$

SAMPLE EXERCISE 22.1 Writing Formulas of Coordination Compounds   **LO1**

The first modern synthetic color, Prussian blue, was made in 1704 in Berlin, Germany. It was heavily used in the 19th-century Japanese woodblock print *The Great Wave off Kanagawa* (Figure 22.4), and it continues to be popular with artists worldwide. The formula for the insoluble pigment is $Fe_4[Fe(CN)_6]_3$.

a. What is the formula of the complex ion in this compound and what is its charge if the counterion is Fe^{3+}?
b. What is the oxidation state of iron in the complex ion?

Collect, Organize, and Analyze We have the formula of a coordination compound and are asked to identify the complex ion and the counterion. The complex ion is contained in brackets, the four Fe^{3+} counterions are outside the brackets, and the compound must be electrically neutral.

Solve
a. The combined charge of four Fe^{3+} is 12+. To achieve electrical neutrality, the combined charge of the three complex anions must be 12−. Therefore the charge of each anion is (12−)/3 = 4−, and the formula for the anion is $[Fe(CN)_6]^{4-}$.

b. Six cyanide ligands, CN^-, occupy the inner coordination sphere for a total charge of 6−. Therefore the charge on Fe must be 2+ for the charge on the complex ion to equal 4−.

Think About It Complex ions may have positive or negative charges. Correspondingly, counterions will have charges opposite to those of the complex ions to ensure electrical neutrality.

Practice Exercise A coordination compound of ruthenium, $[Ru(NH_3)_4Cl_2]Cl$, has shown some activity against leukemia in animal studies. Identify the complex ion, determine the oxidation number of the metal, and identify the counterion.

(Answers to Practice Exercises are in the back of the book.)

FIGURE 22.4 The source of the blue color in *The Great Wave off Kanagawa*, a famous woodblock print by the Japanese artist Hokusai, is the pigment Prussian blue.

22.2 Naming Complex Ions and Coordination Compounds

The names of complex ions and coordination compounds tell us the identity and oxidation state of the central ion, the names and numbers of ligands, the charge in the case of complex ions, and the identity of counterions. To convey all this information, we need to follow some naming rules.

Complex Ions with a Positive Charge

1. Start with the identities of the ligand(s). Names of common ligands appear in Table 22.3. If there is more than one kind of ligand, list the names alphabetically.

2. Use the usual prefix(es) in front of the name(s) written in step 1 to indicate the number of each type of ligand (Table 22.4).

3. Write the name of the metal ion with a Roman numeral indicating its oxidation state.

Examples:

Formula	Name	Structure
$Ni(H_2O)_6{}^{2+}$	Hexaaquanickel(II)	
$Co(NH_3)_6{}^{3+}$	Hexaamminecobalt(III)	
$Cu(NH_3)_4(H_2O)_2{}^{2+}$	Tetraamminediaquacopper(II)	

It may seem strange having two *a*'s together in these names, but it is consistent with current naming rules. Prefixes are ignored in determining alphabetical order, which is why *ammine* comes before *aqua* rather than *di* before *tetra* in tetraamminediaquacopper(II).

TABLE 22.3 Names and Structures of Common Ligands

Ligand	Name within Complex Ion	Structure	Charge	Number of Donor Groups
Iodide	Iodo	$:\!\ddot{I}\!:^{-}$	1–	1
Bromide	Bromo	$:\!\ddot{Br}\!:^{-}$	1–	1
Chloride	Chloro	$:\!\ddot{Cl}\!:^{-}$	1–	1
Fluoride	Fluoro	$:\!\ddot{F}\!:^{-}$	1–	1
Nitrite	Nitro	$\left[O\!\!\cdots\!\!N\!\!\cdots\!\!O \right]^{-}$	1–	1
Hydroxide	Hydroxo	$[O\!-\!H]^{-}$	1–	1
Water	Aqua	$H\!-\!O\!-\!H$	0	1
Pyridine (py)	Pyridyl	(pyridine ring structure)	0	1
Ammonia	Ammine	NH_3	0	1
Ethylenediamine (en)	(same)[a]	$H_2N\!-\!CH_2\!-\!CH_2\!-\!NH_2$	0	2
2,2'-Bipyridine (bipy)	Bipyridyl	(bipyridine ring structure)	0	2
1,10-Phenanthroline (phen)	(same)[a]	(phenanthroline ring structure)	0	2
Cyanide[b]	Cyano	$[C\!\equiv\!N]^{-}$	1–	1
Carbon monoxide[b]	Carbonyl	$C\!\equiv\!O$	0	1

[a]The names of some electrically neutral ligands in complexes are the same as the names of the molecules.
[b]Carbon atoms are the lone pair donors in these ligands.

TABLE 22.4 Common Prefixes Used in the Names of Complex Ions

Number of Ligands	Prefix
2	di-
3	tri-
4	tetra-
5	penta-
6	hexa-

In these three examples the ligands are all electrically neutral. This makes determining the oxidation state of the central metal ion a simple task because the charge on the complex ion is the same as the charge on the metal ion, which is the oxidation state of the metal. When the ligands are anions, determining the oxidation state of the central metal ion requires us to account for these charges.

Complex Ions with a Negative Charge

1. Follow the steps for naming positively charged complexes.

2. Add *-ate* to the name of the central metal ion to indicate that the complex ion carries a negative charge (just as we use *-ate* to end the names of oxoanions). For some metals, the base name changes, too. The two most common examples are iron, which becomes *ferrate*, and copper, which becomes *cuprate*.

Examples:

Formula	Name	Structure
$Fe(CN)_6^{3-}$	Hexacyanoferrate(III)	
$[Fe(H_2O)(CN)_5]^{3-}$	Aquapentacyanoferrate(II)	
$[Al(H_2O)_2(OH)_4]^-$	Diaquatetrahydroxoaluminate	

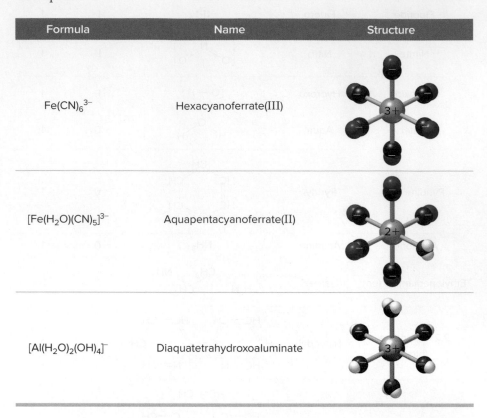

In the first two examples we must determine the oxidation state of Fe. We start with the charge on the complex ion and then take into account the charges on the ligand anions to calculate the charge on the metal ion. For example, the overall charge of the aquapentacyanoferrate(II) ion is 3−. It contains five CN^- ions. To reduce the combined charge of 5− from these cyanide ions to an overall charge of 3−, the charge on Fe must be 2+.

CONCEPT **TEST**

What is the name of the complex anion with the formula $PtCl_4^{2-}$?

Coordination Compounds

1. If the counterion of the complex ion is a cation, the cation's name goes first, followed by the name of the anionic complex ion.

2. If the counterion of the complex ion is an anion, the name of the cationic complex ion goes first, followed by the name of the anion.

Examples:

Formula	Name	Structure
$[Ni(NH_3)_6]Cl_2$	Hexaamminenickel(II) chloride	
$K_3[Fe(CN)_6]$	Potassium hexacyanoferrate(III)	
$[Co(NH_3)_5(H_2O)]Br_2$	Pentaammineaquacobalt(II) bromide	

A key to naming coordination compounds is to recognize from their formulas that they are coordination compounds. For help with this, look for formulas that have the atomic symbols of a metallic element and one or more ligands, all in brackets, either followed by the atomic symbol of an anion, as in $[Co(NH_3)_5(H_2O)]$ Br_2, or preceded by the symbol of a cation, as in $K_3[Fe(CN)_6]$.

SAMPLE EXERCISE 22.2 Naming Coordination Compounds **LO2**

Name the coordination compounds (a) $Na_4[Co(CN)_6]$ and (b) $[Co(NH_3)_5Cl](NO_3)_2$.

Collect and Organize We are asked to write a name for each compound that unambiguously identifies its composition. The formulas of the complex ions appear in brackets in both compounds. Because cobalt, the central metal ion in both, is a transition metal, we express its oxidation state by using Roman numerals. The names of common ligands are given in Table 22.3.

Analyze It is useful to take an inventory of the ligands and counterions:

Compound	Counterion	LIGAND			
		Formula	Name	Number	Prefix
$Na_4[Co(CN)_6]$	Na^+	CN^-	Cyano	6	Hexa-
$[Co(NH_3)_5Cl](NO_3)_2$	NO_3^-	NH_3	Ammine	5	Penta-
		Cl^-	Chloro	1	—

The oxidation state of each cobalt ion can be calculated by setting the sum of the charges on all the ions in both compounds equal to zero:

a. Ions: $(4\ Na^+\ ions) + (1\ Co^x\ ion) + (6\ CN^-\ ions)$

Charges: $4+\ \ +\ \ x\ \ +\ \ 6-\ \ \ \ \ = 0$

$$x = 2+$$

b. Ions: $(1\ Co^x\ ion) + (1\ Cl^-\ ion) + (2\ NO_3^-\ ions)$

Charges: $x\ \ +\ \ 1-\ \ +\ \ 2-\ \ \ \ \ = 0$

$$x = 3+$$

Solve

a. Because the counterion, sodium, is a cation, its name comes first. The complex ion is an anion. To name it, we begin with the ligand cyano, to which we add the prefix *hexa-* and write *hexacyano*. This is followed by the name of the transition metal ion: hexacyano*cobalt*. We add *-ate* to the ending of the name of the complex ion because it is an anion: hexacyanocobalt*ate*. We then add a Roman numeral to indicate the oxidation state of the cobalt: hexacyanocobaltate(*II*). Putting it all together, we get sodium hexacyanocobaltate(II).

b. The complex ion is the cation in this compound, and we begin by naming the ligands directly attached to the metal ion in alphabetical order: ammine and chloro. We indicate the number (5) of NH_3 ligands with the appropriate prefix: *penta*amminechloro. We name the metal next and indicate its oxidation state with a Roman numeral: pentaamminechloro*cobalt(III)*. Finally we name the anionic counterion: *nitrate*. Putting it all together, we obtain the name: pentaamminechlorocobalt(III) nitrate.

Think About It Naming coordination compounds requires us to (1) distinguish between ligands and counterions and (2) recall which ligands are electrically neutral and which are anions. The structures of the complex ions in the named coordination compounds are shown in Figure 22.5.

 Practice Exercise Identify the ligands and counterions in (a) $[Zn(NH_3)_4]Cl_2$ and (b) $[Co(NH_3)_4(H_2O)_2](NO_2)_2$, and name each compound.

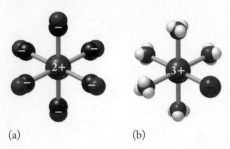

(a)   (b)

FIGURE 22.5 The structures of the complex ions in (a) sodium hexacyanocobaltate(II) and (b) pentaamminechlorocobalt(III) nitrate.

22.3 Polydentate Ligands and Chelation

We have seen that ligands are electron-pair donors—that is, Lewis bases. Let's explore the strengths of several ligands as Lewis bases by considering their affinity for $Ni^{2+}(aq)$ ions. Suppose we dissolve crystals of nickel(II) chloride hexahydrate, $NiCl_2 \cdot 6\ H_2O$, in water. The dot connecting the two halves of the formula and the prefix *hexa* indicate that each Ni^{2+} ion in crystals of nickel(II) chloride is surrounded by six water molecules. Lime green crystals of $NiCl_2 \cdot 6\ H_2O$ form green aqueous solutions (Figure 22.6). The fact that the solid and its solution have the same color tells us that the same arrangement of water molecules around Ni^{2+} ions occurs in both solid $NiCl_2 \cdot 6\ H_2O$ and aqueous solutions of Ni^{2+} ions. Thus the Ni^{2+} ion in a solution of nickel(II) chloride is most likely $Ni(H_2O)_6^{2+}$.

Now let's bubble colorless ammonia gas through a green solution of $Ni(H_2O)_6^{2+}$ ions. As shown in the middle test tube in Figure 22.6(b), the green solution turns blue. The color change means that different ligands are bonded to the Ni^{2+} ions. We may conclude that NH_3 molecules have displaced at least some H_2O molecules

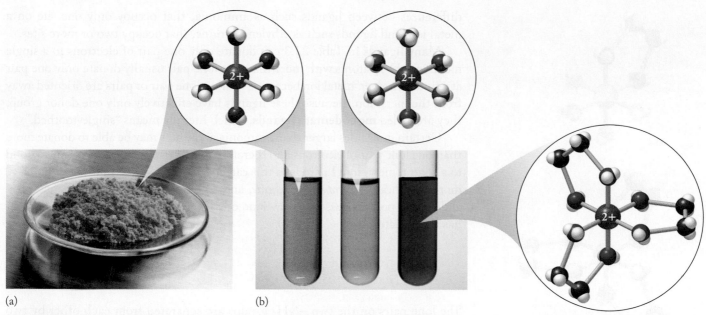

FIGURE 22.6 Structures of the complex ions in solid and dissolved nickel(II) chloride. (a) Solid nickel(II) chloride hexahydrate. (b) When it dissolves in water, the resulting solution has the same green color, telling us that each Ni^{2+} ion (gold sphere) must be surrounded by H_2O molecules both in the solid and in the solution. When ammonia gas is bubbled through a solution of $Ni(H_2O)_6^{2+}$, the color changes to blue as NH_3 replaces H_2O in the Ni^{2+} ion's inner coordination sphere. When ethylenediamine is added to a solution of $Ni(NH_3)_6^{2+}$, the color turns from blue to purple as the ethylenediamine displaces the ammonia ligands and the $Ni(en)_3^{2+}$ complex forms, where "en" represents ethylenediamine.

around the Ni^{2+} ions. If all the molecules of H_2O are displaced, the complex $Ni(NH_3)_6^{2+}$ is formed. The following chemical equation describes this change:

$$Ni(H_2O)_6^{2+}(aq) + 6\,NH_3(g) \rightleftharpoons Ni(NH_3)_6^{2+}(aq) + 6\,H_2O(\ell)$$
$$K_f = 5 \times 10^8$$

Keep in mind that the hydrated ion $Ni(H_2O)_6^{2+}$ is often expressed as $Ni^{2+}(aq)$.

This *ligand displacement* reaction illustrates that Ni^{2+} ions have a greater affinity for molecules of NH_3 than for molecules of H_2O. Many other transition metal ions also have a greater affinity for ammonia than for water. We may conclude that ammonia is inherently a better electron-pair donor and hence a stronger Lewis base than water. This conclusion is reasonable because we saw in Chapter 15 that ammonia was also a stronger Brønsted–Lowry base than H_2O.

Next we add the compound ethylenediamine (see Table 22.3) to the blue solution of $Ni(NH_3)_6^{2+}$ ions. The solution changes color again, from blue to purple (Figure 22.6b), indicating yet another change in the ligands surrounding the Ni^{2+} ions. Molecules of ethylenediamine displace ammonia molecules from the inner coordination sphere of Ni^{2+} ions. This affinity of Ni^{2+} ions for ethylenediamine molecules is reflected in the large formation constant for $Ni(en)_3^{2+}$ (where "en" represents ethylenediamine):

$$Ni(H_2O)_6^{2+}(aq) + 3\,en(aq) \rightleftharpoons Ni(en)_3^{2+}(aq) + 6\,H_2O(\ell) \qquad K_f = 1.1 \times 10^{18}$$

This value is more than 10^9 times the K_f value for $Ni(NH_3)_6^{2+}$. Why should the affinity of Ni^{2+} ions for ethylenediamine be so much greater than their affinity for ammonia? After all, in both ligands the coordinate bonds are formed by lone pairs of electrons on N atoms. To answer this question, we need to look at the

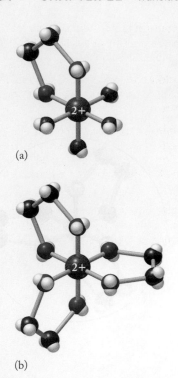

(a)

(b)

FIGURE 22.7 (a) The bidentate ligand ethylenediamine has two N atoms that can each donate a pair of electrons to empty orbitals of adjacent octahedral bonding sites on the same $Ni^{2+}(aq)$ ion (gold sphere), displacing two molecules of water. (b) Three ethylenediamine molecules occupy all six octahedral coordination sites of a Ni^{2+} ion.

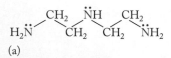

(a)

(b)

FIGURE 22.8 Tridentate chelation. (a) The three amine groups in the tridentate ligand diethylenetriamine are all potential electron-pair donor groups. (b) When these groups donate their lone pairs of electrons to a $Ni^{2+}(aq)$ ion (gold sphere), they occupy three of the six coordination sites on the ion.

differences between ligands such as ammonia, that occupy only one site on a metal ion, and ligands such as ethylenediamine, that occupy two or more sites.

Many ligands in Table 22.3 can donate only one pair of electrons to a single metal ion. Even atoms with more than one lone pair usually donate only one pair at a time to a given metal ion because the other lone pair or pairs are oriented away from the metal ion. Because these ligands have effectively only one donor group, they are called **monodentate ligands**, which literally means "single-toothed."

Certain molecules larger than ammonia and water may be able to donate more than one lone pair of electrons and therefore form more than one coordinate bond to a central metal ion. Ligands in this category are called **polydentate ligands**, or more specifically *bidentate*, *tridentate*, and so on. One group of polydentate ligands is the polyamines, which include ethylenediamine, a bidentate ligand that has the structure:

$$H_2\ddot{N} \qquad \ddot{N}H_2$$
$$H_2C - CH_2$$

The lone pairs on the two $-NH_2$ groups are separated from each other by two $-CH_2-$ groups. This combination means that a molecule of ethylenediamine can partially encircle a metal ion so that both lone pairs can bond to the same metal ion.

The structure of an ethylenediamine complex of $Ni^{2+}(aq)$ is shown in Figure 22.7(a). The two orbitals that share the lone pairs of electrons from a molecule of ethylenediamine must be on the same side of the Ni^{2+} ion. However, two more ethylenediamine molecules can bond to other pairs of bonding sites, displacing additional pairs of water molecules and forming a complex in which the Ni^{2+} ion is surrounded by three bidentate ethylenediamine molecules, as shown in Figure 22.7(b).

Note that each ethylenediamine molecule forms a five-atom ring with the metal ion. If the ring were a regular pentagon (meaning all bond lengths and bond angles were exactly the same), each of its bond angles would be 108°. These pentagons are not perfect, but each ring's preferred octahedral bond angles of 90° for the N–Ni–N bond, and of 107° to 109° for all the other bonds, are accommodated with only a little strain on the ideal bond angles.

An even larger ligand, diethylenetriamine ($H_2NCH_2CH_2NHCH_2CH_2NH_2$), is shown in Figure 22.8(a). The lone pairs of electrons on its three nitrogen atoms give diethylenetriamine the capacity to form three coordinate bonds to a metal ion, meaning this is a tridentate ligand (Figure 22.8b).

As you may imagine, larger molecules may have even more atoms per molecule that can bond to a single metal ion. The interaction of a metal ion with a ligand having multiple donor atoms is called **chelation** (pronounced *key-LAY-shun*). The word comes from the Greek *chele*, meaning "claw." The polydentate ligands that take part in these interactions are called *chelating agents*.

When we added ethylenediamine to the blue solution of $Ni(NH_3)_6^{2+}$ ions in Figure 22.6(b), molecules of ethylenediamine (en) displaced ammonia molecules from the inner coordination sphere of Ni^{2+} ions as described in the following chemical equation:

$$Ni(NH_3)_6^{2+}(aq) + 3\ en(aq) \rightleftharpoons Ni(en)_3^{2+}(aq) + 6\ NH_3(aq) \qquad (22.1)$$

The color change tells us that this reaction as written is spontaneous. As we discussed in Chapter 17, spontaneous reactions are those in which free energy decreases ($\Delta G < 0$). Furthermore, under standard conditions the change in free energy ($\Delta G°$) is related to the changes in enthalpy and entropy that accompany the reaction:

$$\Delta G° = \Delta H° - T\Delta S°$$

The displacement of NH_3 by ethylenediamine is exothermic, but only slightly ($\Delta H° = -12$ kJ/mol). More important, $\Delta S° = +185$ J/(mol · K). This means that at 25°C,

$$T\Delta S° = 298\ \mathrm{K} \times \frac{185\ \mathrm{J}}{\mathrm{mol} \cdot \mathrm{K}} \times \frac{1\ \mathrm{kJ}}{1000\ \mathrm{J}} = 55.1\ \mathrm{kJ/mol}$$

To understand why there is such a large increase in entropy, consider that there are 4 moles of reactants but 7 moles of products in Equation 22.1. Nearly doubling the number of moles of aqueous products over reactants translates into a large gain in entropy. It is this positive $\Delta S°$, more than the negative $\Delta H°$ value, that drives the reaction and makes it spontaneous. Entropy gains drive many complexation reactions that involve polydentate ligands. The entropy-driven affinity of metal ions for polydentate ligands is called the **chelate effect**.

Many chelating agents have more than one kind of electron-pair–donating group. *Aminocarboxylic acids* represent one family of such compounds. The most important of them is ethylenediaminetetraacetic acid, EDTA, the molecular structure of which is shown in Figure 22.9(a). Note that one molecule of EDTA contains two amine (nitrogen-containing) groups and four carboxylic acid (–COOH) groups. When the acid groups release their H^+ ions, they form four carboxylate anions, $-COO^-$, in which either of the O atoms can donate a pair of electrons to a central metal ion. When O atoms on all four groups do so and the two amine groups do as well, six octahedral bonding sites around the metal ion can be occupied, as shown in Figure 22.9(b).

EDTA forms very stable complex ions and is used as a metal ion *sequestering agent*, that is, a chelating agent that binds metal ions so tightly that they are "sequestered" and prevented from reacting with other substances. For example, EDTA is used as a preservative in many beverages and prepared foods because it sequesters iron, copper, zinc, manganese, and other transition metal ions often present in these foods that can catalyze the degradation of ingredients in the foods. Many foods are fortified with ascorbic acid (vitamin C), which is particularly vulnerable to metal-catalyzed degradation because it is also a polydentate ligand and is more likely to be oxidized when chelated to one of the above metal ions. EDTA effectively shields vitamin C from these ions.

monodentate ligand a species that forms only a single coordinate bond to a metal ion in a complex.

polydentate ligand a species that can form more than one coordinate bond per molecule.

chelation the interaction of a metal with a polydentate ligand (chelating agent); pairs of electrons on one molecule of the ligand occupy two or more coordination sites on the central metal.

chelate effect the greater affinity of metal ions for polydentate ligands than for monodentate ligands.

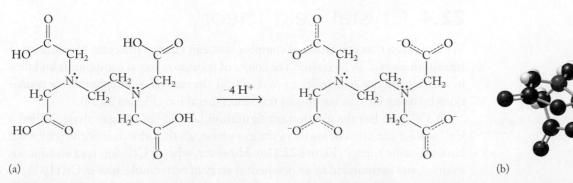

(a)   (b)

FIGURE 22.9 (a) In the hexadentate ligand EDTA, the six donor groups are the two amine groups and the four carboxylic acid groups. The acid groups ionize to form carboxylate anions. (b) All six Lewis base groups in ionized EDTA can form a coordinate bond with the same metal ion, such as Co^{3+} (the gold sphere) shown here. In the process they form four 5-membered rings.

SAMPLE EXERCISE 22.3 Identifying the Potential
Electron-Pair–Donor Groups in a Molecule

How many donor groups does this polydentate ligand, nitrilotriacetic acid (NTA), have?

Collect, Organize, and Analyze We need to examine this molecular structure to find electron pairs that can be donated. Because there are three single bonds around the N atom, the atom's fourth sp^3 orbital must contain a lone pair of electrons. When all three carboxylic acid groups are ionized, there are three carboxylate groups in the molecule, and each carboxylate group can donate one nonbonding pair of electrons from one of its oxygen atoms to a metal atom.

Solve The central N atom and an O atom from each of the three carboxylate groups form a total of four coordinate bonds. Therefore NTA is potentially a tetradentate ligand with four donor groups.

Think About It The tetradentate capacity of NTA is reasonable because, like EDTA, it is an aminocarboxylic acid. It has one fewer amino group and one fewer carboxylic acid group than the hexadentate EDTA.

 Practice Exercise How many potential donor groups are there in citric acid, a component of citrus fruits and a widely used preservative in the food industry?

22.4 Crystal Field Theory

We have seen that formation of complex ions can change the color of solutions of transition metals. Why is this? The colors of transition metal compounds and ions in solution are due to transitions of d-orbital electrons. Let's explore these transitions by using Cr^{3+} as our model transition metal ion (Figure 22.10).

A Cr^{3+} ion has the electron configuration $[Ar]3d^3$ in the gas phase. When a Cr^{3+} ion (or any atom or ion) is in the gas phase, all the orbitals in a given subshell have the same energy (Figure 22.11a). However, when a Cr^{3+} ion is in an aqueous solution and surrounded by an octahedral array of water molecules in $Cr(H_2O)_6^{3+}$, the energies of its $3d$ orbitals are no longer all the same. The $3d_{xy}$, $3d_{yz}$, and $3d_{xz}$ orbitals experience some increase in energy, but the energies of the $3d_{x^2-y^2}$ and $3d_{z^2}$ orbitals increase even more (Figure 22.11b) because the lobes of the $3d_{x^2-y^2}$ and

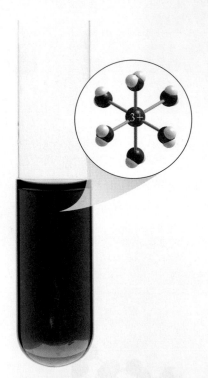

FIGURE 22.10 When chromium(III) nitrate dissolves in water, the resulting solution has a distinctive violet color due to the presence of $Cr(H_2O)_6^{3+}$ ions.

CONNECTION Crystal field theory is an example of molecular orbital theory, which was introduced in Chapter 9.

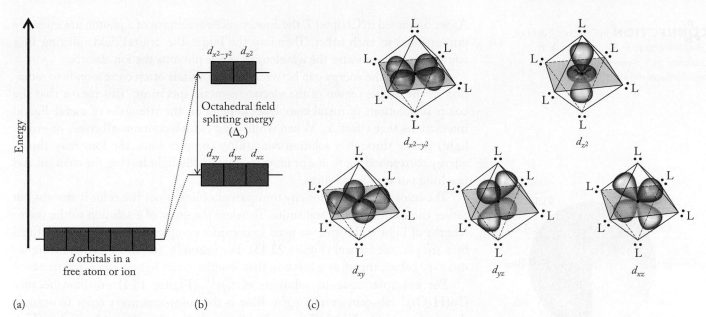

(a)   (b)   (c)

FIGURE 22.11 Octahedral crystal field splitting. (a) In an atom or ion in the gas phase, all orbitals in a subshell are degenerate, as shown here for the five $3d$ orbitals. (b) When an ion is part of a complex ion in a compound or solution, repulsions between electrons in the ion's d orbitals and ligand electrons raise the energy of the orbitals, as shown here for an octahedral field. (c) The greatest repulsion is experienced by electrons in the $d_{x^2-y^2}$ and d_{z^2} orbitals because the lobes of these orbitals are directed toward the corners of the octahedron and so are closest to the lone pairs on the ligands (L). The lobes of the lower-energy d_{xy}, d_{yz}, and d_{xz} orbitals are directed toward points that lie between the corners of the octahedron, so electrons in them experience less repulsion.

$3d_{z^2}$ orbitals point directly toward the H_2O molecules' oxygen atoms at the corners of the octahedron formed by the ligands and are repelled by the electrons on those O atoms (Figure 22.11c). The energies of the $3d_{xy}$, $3d_{yz}$, and $3d_{xz}$ orbitals are not raised as much because the lobes of these three orbitals do not point directly toward the corners of the octahedron, so the electron repulsion they experience is weaker.

This process of changing degenerate (equal-energy) d-orbitals into orbitals with different energies is known as **crystal field splitting**, and the difference in energy created by crystal field splitting is called **crystal field splitting energy (Δ)**. The name was originally used to describe splitting of d-orbital energies in minerals, but the theory is also routinely applied to species in aqueous solutions.

In a $Cr(H_2O)_6^{3+}$ ion, three electrons are distributed among five $3d$ orbitals. According to Hund's rule, each of the three electrons should occupy one of the three lower-energy orbitals, leaving the two higher-energy orbitals unoccupied, as shown in Figure 22.12(a). The energy difference between the two subsets of orbitals is symbolized by Δ_o, where the subscript "o" indicates that the energy split was caused by an *o*ctahedral array of electron repulsions.

What if an aqueous Cr^{3+} ion absorbs a photon whose energy is exactly equal to Δ_o? As the photon is absorbed, a $3d$ electron moves from a lower-energy orbital to a higher-energy orbital (Figure 22.12b). The wavelength λ of the absorbed photon is related to the energy difference between the two groups of orbitals—in other words, to the crystal field splitting energy—as follows:

$$E = \frac{hc}{\lambda} = \Delta_o \qquad (22.2)$$

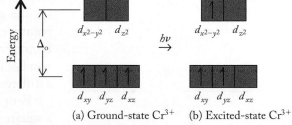

(a) Ground-state Cr^{3+} (b) Excited-state Cr^{3+}

FIGURE 22.12 A Cr^{3+} ion, $[Ar]3d^3$, in an octahedral field can absorb a photon of light that has energy ($h\nu$) equal to Δ_o. This energy raises a $3d$ electron from (a) one of the lower-energy d orbitals to (b) one of the higher-energy d orbitals.

CHEMTOUR
Crystal Field Splitting

crystal field splitting the separation of a set of d orbitals into subsets with different energies as a result of interactions between electrons in those orbitals and lone pairs of electrons in ligands.

crystal field splitting energy (Δ) the difference in energy between subsets of d orbitals split by interactions in a crystal field.

CONNECTION In Chapter 7 we first used the equation $E = h\nu = hc/\lambda$ in discussing the energy of light in the electromagnetic spectrum.

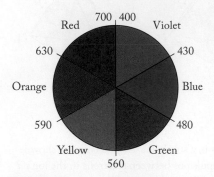

FIGURE 22.13 A color wheel. Colors on opposite sides of the wheel are complementary to each other. When we look at a solution or object that absorbs light corresponding to a given color, we see the complementary color. Wavelengths are in nanometers.

As we discussed in Chapter 7, the energy and wavelength of a photon are inversely proportional to each other. Therefore the larger the crystal field splitting in a complex ion, the shorter the wavelength of the photons the ion absorbs.

The size of the energy gap between split d orbitals often corresponds to radiation in the visible region of the electromagnetic spectrum. This means that the colors of solutions of metal complexes depend on the strengths of metal–ligand interactions that affect Δ_o. When white light (which contains all colors of visible light) passes through a solution containing complex ions, the ions may absorb energy corresponding to one or more colors. So, the light leaving the solution and reaching our eyes is missing.

The color we perceive for any transparent object is not the color it absorbs but rather the color(s) that it transmits. To relate the color of a solution to the wavelengths of light it absorbs, we need to consider complementary colors as defined by a simple color wheel (Figure 22.13). For example, red and green are complementary colors; therefore a solution that absorbs green light appears red to us.

For example, aqueous solutions of Cu^{2+} (Figure 18.1) are blue because $Cu(H_2O)_6^{2+}$ absorbs orange light. Blue is the complementary color to orange. A solution of $Cu(NH_3)_4^{2+}$ ions also has a distinctive deep blue color (Figure 22.14). The spectrum of a solution containing $Cu(NH_3)_4^{2+}$ features a rather broad absorption band between 580 and 590 nm that spans yellow, orange, and red wavelengths. Once again, the complementary color is in the blue-violet region. If more than one color is absorbed, then our brain processes the bands of transmitted colors and signals to us as the average of these colors. The violet color of the solution containing $Cr^{3+}(aq)$ in Figure 22.10 is the result of color averaging, as are the colors of aqueous Ni^{2+}, $Ni(NH_3)_6^{2+}$, and $Ni(en)_3^{2+}$ solutions in Figure 22.6.

The nickel(II) and chromium(III) complexes we have examined up to this point were octahedral, and their central ions had a coordination number of 6. However, in the solution of $Cu(NH_3)_4^{2+}$, the copper(II) ion has a coordination number of 4. This means that the deep blue color of this complex ion is caused by a different crystal field.

Four ligands around a central metal have either a tetrahedral arrangement or a square planar arrangement (Table 22.2). Square planar geometries tend to be limited to the transition metal ions with nearly filled valence-shell d orbitals, particularly those with d^8 or d^9 electron configurations. Cu^{2+} has the electron configuration $[Ar]3d^9$, and the $Cu(NH_3)_4^{2+}$ complex is square planar—which means that the strongest interactions occur between the $3d$ orbitals on the

FIGURE 22.14 The visible light transmitted by a solution of $Cu(NH_3)_4^{2+}$ ions is missing much of the yellow, orange, and red portions of the visible spectrum because of a broad absorption band centered at 590 nm. Our eyes and brain perceive the transmitted colors as navy blue.

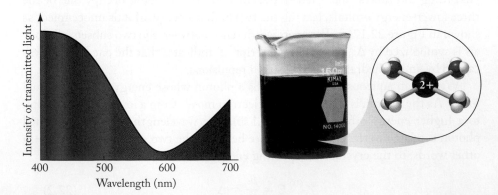

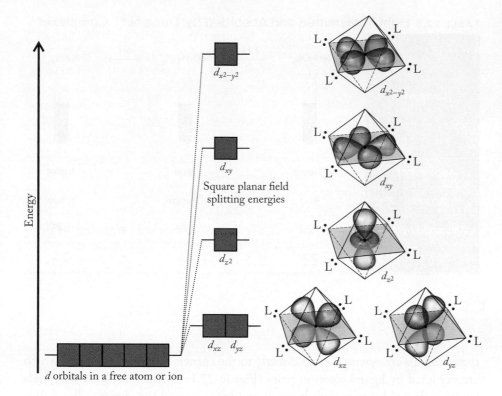

FIGURE 22.15 Square planar crystal field splitting. The *d* orbitals of a transition metal ion in a square planar field are split into several energy levels depending on the relative orientations of the metal orbitals and ligand electrons at the four corners of the square. The $d_{x^2-y^2}$ orbital has the highest energy because its lobes are directed right at the four corners of the square plane.

central ion and the nitrogen atom lone pairs at the four corners of the equatorial plane of the octahedron, as shown in Figure 22.15. The $3d_{x^2-y^2}$ orbital has the strongest interactions and the highest energy because its lobes are oriented directly at the four corners of the plane. The d_{xy} orbital has slightly less energy because its lobes, although in the *xy* plane, are directed 45° away from the corners. Electrons in the three *d* orbitals with most of their electron density out of the *xy* plane interact even less with the lone pairs of the ligand and thus have even lower energies.

Finally, let's consider the *d* orbital crystal field splitting that occurs in a tetra-hedral complex (Figure 22.16a). In this geometry, the greatest electron–electron repulsions are experienced in the d_{xy}, d_{yz}, and d_{xz} orbitals because the lobes of

FIGURE 22.16 (a) In a tetrahedral complex ion, such as $Zn(NH_3)_4^{2+}$, the *d* orbitals of the metal ion are split by a tetrahedral crystal field. (b) The lobes of the higher-energy orbitals—d_{xy}, d_{yz}, and d_{xz}—are closer to the ligands at the four corners of the tetrahedron than the lobes of the lower-energy orbitals are. (One of the four corners of the tetrahedron is hidden in these drawings.)

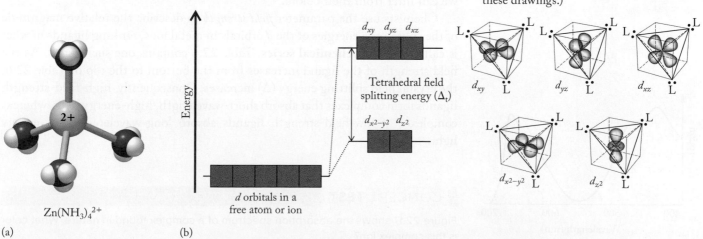

$Zn(NH_3)_4^{2+}$

(a)

(b)

spectrochemical series a list of ligands rank-ordered by their ability to split the energies of the d orbitals of transition metal ions.

TABLE 22.5 Light Transmitted and Absorbed by Three Ni^{2+} Complexes

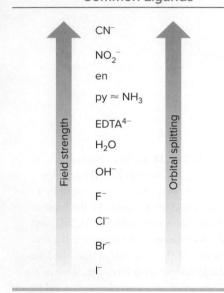

Complex	$Ni(H_2O)_6^{2+}$ $\xrightarrow{NH_3}$		$Ni(NH_3)_6^{2+}$ $\xrightarrow{en}$		$Ni(en)_3^{2+}$
Appearance	Green		Blue		Violet
Absorbs	Red		Orange		Yellow
Absorbed λ (nm)	725	>	570	>	545
$E = hc/\lambda$ (J × 10^{19})	2.7	<	3.5	<	3.6

TABLE 22.6 Spectrochemical Series of Some Common Ligands

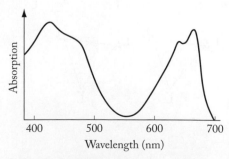

these orbitals are oriented most directly to the corners of the tetrahedron, which are occupied by ligand electron pairs (Figure 22.16b). The two other d orbitals are less affected because their lobes do not point toward the corners. The difference in energy between the two subsets of d orbitals in a tetrahedral geometry is labeled Δ_t.

Before ending this discussion on the colors of transition metal ions, let's revisit the color changes we saw in Figure 22.6, when first ammonia and then ethylenediamine were added to a solution of Ni^{2+} ions. Let's think in terms of the colors these solutions *absorb*. A solution of $Ni^{2+}(aq)$ ions absorbs colors at the red end of the spectrum. Similarly, a solution of $Ni(NH_3)_6^{2+}$ absorbs colors opposite blue, which are centered on orange, and a solution of $Ni(en)_3^{2+}$ absorbs green light. Note how the colors these three solutions absorb are in the sequence red, yellow, and green. This sequence runs from longest wavelength to shortest, from about 725 nm for red to about 545 nm for green. Radiant energy is inversely proportional to wavelength (Equation 22.2); therefore the crystal field splitting of the d orbitals of Ni^{2+} ions is en > NH_3 > H_2O. Table 22.5 summarizes the observed colors of these Ni^{2+} ion complexes and the meaning we can infer from their colors.

Chemists use the parameter *field strength* to describe the relative magnitude of the split in the energies of the d orbitals in metal ions, ranking ligands in what is called a **spectrochemical series**. Table 22.6 contains one such series. As the field strength of the ligand increases from the bottom to the top of Table 22.6, the crystal field splitting energy (Δ) increases. Consequently, high-field-strength ligands form complexes that absorb short-wavelength, high-energy light, whereas complexes of low-field-strength ligands absorb long-wavelength, low-energy light.

CONCEPT TEST

Figure 22.17 shows the absorption spectrum of a complex found in nature. What color is this complex ion?

FIGURE 22.17 Absorption spectrum.

22.5 Magnetism and Spin States

In addition to contributing to the color of transition metal ions, crystal field splitting influences their magnetic properties because these properties depend on the number of unpaired electrons in the valence shell d orbitals. The more unpaired electrons, the more paramagnetic the ion. For example, an Fe^{3+} ion has five $3d$ electrons (Figure 22.18a). In an octahedral field there are two ways to distribute the five $3d$ electrons among these orbitals. One arrangement conforms to Hund's rule and has a single electron in each orbital, leaving them all unpaired (Figure 22.18b). However, when Δ_o is large, as shown in Figure 22.18(c), all five electrons occupy the three lower-energy orbitals. This pattern of electron distribution occurs when the energy of repulsion between two electrons in the same orbital is less than the energy needed to promote an electron to a higher-energy orbital. In this configuration, only one electron is unpaired.

CONNECTION We introduced the magnetic behavior of matter in Chapter 9 in our discussion of molecular orbital theory.

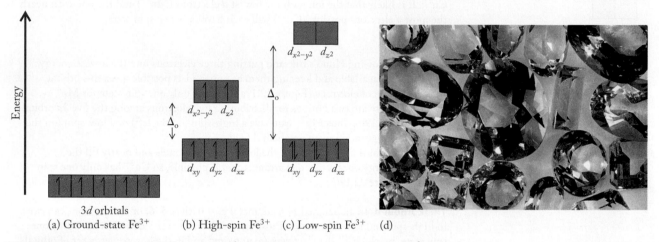

(a) Ground-state Fe^{3+} (b) High-spin Fe^{3+} (c) Low-spin Fe^{3+} (d)

FIGURE 22.18 Low-spin and high-spin complexes. (a) The ground state of a free Fe^{3+} ion has a degenerate, half-filled set of $3d$ orbitals. (b) A weak octahedral field ($\Delta_o <$ electron-pairing energy) produces the high-spin state: five unpaired electrons, each in its own orbital. (c) In a strong octahedral field ($\Delta_o >$ electron-pairing energy), the energies of the $3d$ orbitals are split enough to produce the low-spin state: two sets of paired electrons; one unpaired electron; and two empty, higher energy orbitals. (d) The Fe^{3+} ions in crystals of aquamarine are high spin.

The configuration with all five electrons unpaired is called the *high-spin state* because the spin on all five electrons is in the same direction, resulting in the maximum magnetic field produced by the spins. The configuration with only one electron unpaired is called the *low-spin state*. Both configurations are paramagnetic because both have at least one unpaired electron, but material containing high-spin iron(III) ions would be more strongly attracted to an external magnet than a material containing low-spin iron(III) ions.

Not all transition metal ions can have both high-spin and low-spin states. Consider, for example, Cr^{3+} ions in an octahedral field. Because each Cr^{3+} ion has only three $3d$ electrons (Figure 22.12), each electron is unpaired whether the orbital energies are split a lot or only a little. Therefore Cr^{3+} ions and any ion with less than four d electrons will have only one spin state. Metal ions with eight or more electrons in d orbitals also have only one spin state because there cannot be more than one or two unpaired electrons in these orbitals no matter how the electrons are distributed.

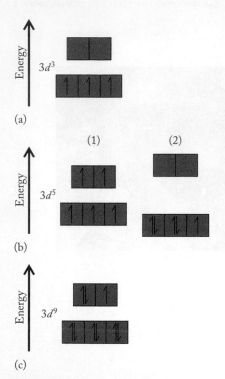

FIGURE 22.19 Options for electron distribution.

SAMPLE EXERCISE 22.4 Predicting Spin States **LO5**

Determine which of these ions can have high-spin and low-spin configurations when part of an octahedral complex: (a) Mn^{4+}; (b) Mn^{2+}; (c) Cu^{2+}.

Collect and Organize Mn and Cu are in groups 7 and 11 of the periodic table, so their atoms have 7 and 11 valence electrons, respectively. In an octahedral field, a set of five d orbitals splits into a low-energy subset of three orbitals and a high-energy subset of two orbitals.

Analyze To determine whether high-spin and low-spin states are possible, we need to determine the number of d electrons in each ion. Then we need to distribute them among sets of d orbitals split by an octahedral field to see whether it is possible for the ions to have different spin states. The electron configurations of the atoms of the three elements are $[Ar]3d^5 4s^2$ for Mn and $[Ar]3d^{10} 4s^1$ for Cu. When they form cations, the atoms of these transition metals lose their $4s$ electrons first and then their $3d$ electrons. Therefore the numbers of d electrons in the ions are 3 in Mn^{4+}, 5 in Mn^{2+}, and 9 in Cu^{2+}. It is likely that the ion with the fewest d electrons (Mn^{4+}) and the one with nearly the most d electrons possible (Cu^{2+}) will each have only one spin state.

Solve
a. Mn^{4+}: Following Hund's rule and putting three electrons into the lowest-energy $3d$ orbitals available and keeping them as unpaired as possible gives this orbital distribution of electrons (Figure 22.19a). There is only one spin state for Mn^{4+}.
b. Mn^{2+}: There are two options for distributing five electrons among the five $3d$ orbitals (Figure 22.19b). Thus Mn^{2+} can have a high-spin (on the left) or a low-spin (on the right) configuration in an octahedral field.
c. Cu^{2+}: The nine $3d$ electrons fill the lower-energy orbitals and nearly fill the higher-energy ones. Only one arrangement is possible, so Cu^{2+} has only one spin state (Figure 22.19c).

Think About It In an octahedral field, metal ions with 4, 5, 6, or 7 d electrons can exist in high-spin and low-spin states. Those ions with 3 or fewer d electrons have only one spin state, in which all the electrons are unpaired and in the lower-energy set of orbitals. Ions with 8 or more d electrons have only one spin state because their lower-energy set of orbitals is filled. The magnitude of the crystal field splitting energy, Δ_o, determines which spin state an ion with 4, 5, 6, or 7 d electrons occupies.

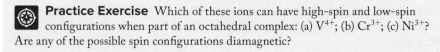 **Practice Exercise** Which of these ions can have high-spin and low-spin configurations when part of an octahedral complex: (a) V^{4+}; (b) Cr^{3+}; (c) Ni^{3+}? Are any of the possible spin configurations diamagnetic?

As noted earlier, whether a transition metal ion is in a high-spin state or a low-spin state depends on whether less energy is needed to promote an electron to a higher-energy orbital or to overcome the repulsion experienced by two electrons sharing the same lower-energy orbital. Several factors affect the size of Δ_o. We have already discussed a major one in the context of the spectrochemical series shown in Table 22.6: the different field strengths of different ligands. Because molecules containing nitrogen atoms with lone pairs of electrons are stronger field-splitting ligands than H_2O molecules, hydrated metal ions (small Δ_o) are more likely to be in high-spin states, and metals surrounded by nitrogen-containing ligands (large Δ_o) are more likely to be in low-spin states. Another factor affecting spin state is the oxidation state of the metal ion. The higher the oxidation number (and ionic charge), the stronger the attraction of the electron pairs on the ligands for the ion. Greater attraction leads to more ligand–d orbital interaction and therefore to a larger Δ_o.

Complexes of transition metals in the fifth and sixth rows tend to be low spin because their $4d$ and $5d$ orbitals extend farther from the nucleus than do $3d$ orbitals. These larger d orbitals overlap more and interact more strongly with the lone pairs of electrons on the ligands, leading to greater crystal field splitting.

Our discussion of high-spin and low-spin states has focused entirely on d orbitals split by octahedral fields. What about spin states in tetrahedral fields? Almost all tetrahedral complexes are high spin because tetrahedral fields are weaker than octahedral fields. Weaker field strength means less d-orbital splitting—not enough to offset the energies associated with pairing two electrons in the same orbitals. Therefore Hund's rule is obeyed.

CONCEPT **TEST**

Explain the following:

a. $Mn(py)_6^{2+}$ is a high-spin complex ion, but $Mn(CN)_6^{4-}$ is low spin.

b. $Fe(NH_3)_6^{2+}$ is high spin, but $Ru(NH_3)_6^{2+}$ is low spin.

22.6 Isomerism in Coordination Compounds

The coordination compound $Pt(NH_3)_2Cl_2$ was first described in 1849 and is now widely used as a chemotherapeutic agent to treat several types of cancer. Both molecules of ammonia and the two chloride ions are coordinately bonded to the Pt^{2+} ion, so the name of the compound is diamminedichloroplatinum(II). No counterions are present because the sum of the charges on the ligands and the platinum is zero and the complex is neutral. The compound is square planar, and when we draw it, there are two ways to arrange the ligands about the central metal ion: the two chloride ions and the two ammonia ligands can be at adjacent corners of the square (Figure 22.20a) or at opposite corners (Figure 22.20b). Coordination compounds like these two that have the same composition and the same connections between parts but differ in the three-dimensional arrangement of those parts are stereoisomers.

The two molecules in Figure 22.20 have different physical and chemical properties, and we must find a way to distinguish between them when we name them. Isomer (a), with two members of each pair of ligands at adjacent corners, is called *cis*-diamminedichloroplatinum(II). Isomer (b), with pairs of ligands at opposite corners, is *trans*-diamminedichloroplatinum(II). By analogy to the stereoisomers of alkenes we saw in Chapter 20, these types of stereoisomers are called cis–trans isomers.

To illustrate the importance of stereoisomerism in coordination compounds, consider this: *cis*-diamminedichloroplatinum(II) is a widely used anticancer drug with the common name *cisplatin*, but the trans isomer is much less effective in fighting cancer. The therapeutic power of cisplatin comes from its structurally specific reactions with DNA. During these reactions the two Cl atoms of $Pt(NH_3)_2Cl_2$ are replaced by two nitrogen-containing bases on a strand of DNA in the nucleus of a cell, as shown in Figure 22.21. The ability of cisplatin to

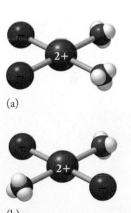

(a)

(b)

FIGURE 22.20 Two ways to orient the Cl^- ions and NH_3 molecules around a Pt^{2+} ion (gold sphere) in the square planar coordination compound $Pt(NH_3)_2Cl_2$. (a) The two members of each pair of ligands are on the same side of the square in *cis*-diamminedichloroplatinum(II). (b) The two members of each pair are at opposite corners in *trans*-diamminedichloroplatinum(II).

FIGURE 22.21 The attack of *cis*-diamminedichloroplatinum(II) on DNA. (a) After the drug is administered, one of its chloride ions is displaced by a molecule of water, turning each molecule into a 1+ ion. (b) The ion is attracted to a nitrogen-containing base on a strand of DNA, which displaces the water molecule and forms a coordinate bond to Pt. (c) A nearby base forms a bond to Pt by displacing the chloride ion. Forming bonds to the two DNA bases distorts the DNA molecule so much that it cannot function properly. (d) A magnified view of the bonds that form between platinum(II) and nitrogen atoms in DNA. We explored the structure and function of DNA in Chapter 20.

(a)   (b)   (c)

(d)

CONNECTION We introduced the concept of isomers in Chapter 3 and again in Chapter 9 when we discussed chirality. Enantiomers, also called optical isomers, are one type of stereoisomer; they are molecules that are not superimposable on their mirror image and that rotate the plane of polarized light, as we saw in Chapter 20.

cross-link these bases distorts the molecular shape of the DNA (Figure 22.21d) and disrupts its normal function. Most important, this kind of cell DNA damage inhibits the ability of cancerous cells to grow and replicate. The trans isomer forms complexes with other intracellular compounds more readily than with DNA so it is much less effective as an anticancer agent.

Cis–trans isomers are also possible in octahedral complexes containing more than one type of ligand. For example, there are two possible stereoisomers of $[Co(NH_3)_4Cl_2]Cl$ (Figure 22.22). The two chloro ligands in $[Co(NH_3)_4Cl_2]^+$ are either on the same side of the complex with a 90° Cl–Co–Cl bond angle (the cis isomer) or across from each other so that the Cl–Co–Cl angle is 180° (the trans isomer) *cis*-Tetraamminedichlorocobalt(III) chloride is violet, and *trans*-tetraamminedichlorocobalt(III) chloride is green.

FIGURE 22.22 Structures of the complex ions in the two stereoisomers of the coordination compound with the formula $[Co(NH_3)_4Cl_2]Cl$: (a) *cis*-tetraamminedichlorocobalt(III) chloride and (b) *trans*-tetraamminedichlorocobalt(III) chloride.

(a) *cis*-Tetraamminedichlorocobalt(III) chloride

(b) *trans*-Tetraamminedichlorocobalt(III) chloride

SAMPLE EXERCISE 22.5 Identifying Stereoisomers of Coordination Compounds

LO6

Sketch the structures and name the stereoisomers of $Ni(NH_3)_4Cl_2$.

Collect and Organize We are given the formula and asked to name and draw the structures of the stereoisomers of a coordination compound. The Ni^{2+} complexes we have seen so far in the chapter have all been octahedral. Ammonia molecules and Cl^- ions are both monodentate ligands.

Analyze The formula contains no brackets, so the Cl^- ions are not counterions; they must be covalently bonded to the Ni ion. There are two Cl^- ions, so the charge on Ni must be 2+. There are a total of six ligands, which confirms that the compound is octahedral.

Solve There are two ways to orient the chloride ions: opposite each other with a Cl–Ni–Cl bond angle of 180° or on the same side of the octahedron with a Cl–Ni–Cl bond angle of 90°:

$$
\begin{array}{ccc}
& \text{Cl} & \\
& | \quad \text{NH}_3 & \\
\text{H}_3\text{N}\!-\!&\text{Ni}\!-\!\text{NH}_3 & \\
& | & \\
\text{H}_3\text{N} & \text{Cl} &
\end{array}
\qquad
\begin{array}{ccc}
\text{H}_3\text{N} & & \text{Cl} \\
& \diagdown & \\
\text{H}_3\text{N}\!-\!&\text{Ni}\!-\!\text{Cl} & \\
& | & \\
\text{H}_3\text{N} & \text{NH}_3 &
\end{array}
$$

The first isomer is *trans*-tetraamminedichloronickel(II); the second is *cis*-tetraamminedichloronickel(II).

Think About It We can draw other tetraamminedichloronickel(II) structures that do not look exactly like these two structures. If we flip or rotate them, however, they will match one of the two structures shown.

 Practice Exercise Sketch the stereoisomers of $[CoBr_2(en)(NH_3)_2]^+$ and name them.

Enantiomers and Linkage Isomers

Consider the octahedral cobalt(III) complex ion containing two ethylenediamine molecules and two chloride ions. There are two ways to arrange the chloride ions: on adjacent bonding sites in a cis isomer or on opposite sides of the octahedron in a trans isomer. These two molecules are cis–trans isomers, but the cis isomer (Figure 22.23) illustrates another kind of stereoisomerism that is possible in complex ions and coordination compounds.

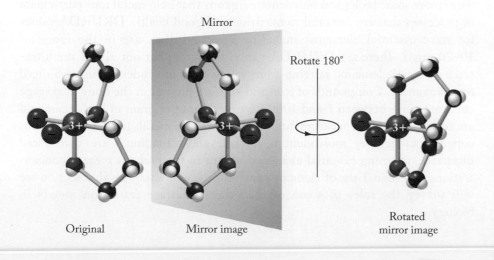

Mirror

Rotate 180°

Original   Mirror image   Rotated mirror image

FIGURE 22.23 The complex ion *cis*-dichlorobis(ethylenediamine)cobalt(III) is chiral, which means that its mirror image is not superimposable on the original complex. To illustrate this point, we rotate the mirror image 180° about its vertical axis so that it looks as much like the original as possible. However, note that the top ethylenediamine ligand is located behind the plane of the page in the original but in front of the plane of the page in the rotated mirror image. Thus the mirror images are not superimposable.

Figure 22.23 shows that *cis*-$Co(en)_2Cl_2^+$ is chiral: it has a mirror image that is not identical to the original. The difference is demonstrated by the fact that there is no way to rotate the mirror image so that its atoms align exactly with those in the original. In other words, the two structures are not superimposable. We encountered this phenomenon in Chapters 9 and 20 and noted that such nonsuperimposable stereoisomers are called *enantiomers*.

Naming this cis isomer also requires an additional rule. The complex ion is named *cis*-dichlorobis(ethylenediamine)cobalt(III). Note the prefix *bis*- just before (ethylenediamine). In naming complex ions containing a polydentate ligand, we use *bis*- instead of *di*- to indicate that two molecules of the ligand are present and to avoid the use of two *di*- prefixes in the same ligand name. Other prefixes used in this fashion are *tris*- for three polydentate ligands and *tetrakis*- for four.

(a) $[Co(CN)_5SCN]^{3-}$

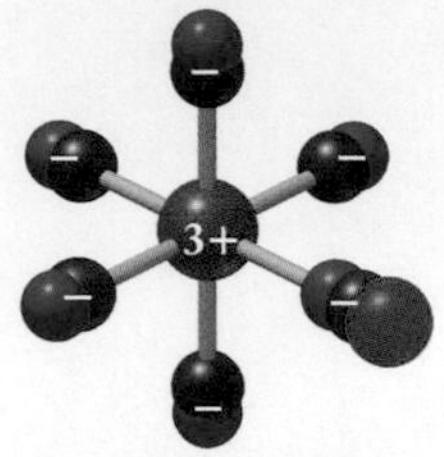

(b) $[Co(CN)_5NCS]^{3-}$

FIGURE 22.24 The SCN^- ligand is bound to the Co^{3+} ion through the S atom in (a) the pentacyanothiocyanatocobaltate(III) ion and through the N atom in (b) the pentacyanoisothiocyanatocobaltate(III) ion.

CONCEPT TEST

Would four different ligands arranged in a square planar geometry produce a chiral complex ion? What about four different ligands in a tetrahedral geometry?

A third kind of isomerism called *linkage isomerism* occurs in coordination compounds when a ligand can bind to a central metal ion by using either of two possible electron-pair–donating atoms. The two complexes of Co^{3+} shown in Figure 22.24 are a pair of linkage isomers. One complex contains the thiocyanate (SCN^-) ion as a ligand where the S atom covalently bonds to the metal (Figure 22.24a). The other contains the isothiocyanate (NCS^-) ion as a ligand where the N atom forms the bond (Figure 22.24b). Another ligand that forms linkage isomers with metals is NO_2^-, called nitro when the N atom donates an electron pair and nitrito when an O atom is the donor. Note that because these pairs of molecules have different connectivities, linkage isomers are *not* stereoisomers. Instead, linkage isomers are a type of constitutional isomer.

CONNECTION We discussed constitutional isomers, which have the same molecular formula but different connections between their atoms, in Chapters 3 and 20.

22.7 Coordination Compounds in Biochemistry

At the beginning of the chapter, we noted that metals essential to human health must be present in foods in forms the body can absorb. In this section we explore some biological polydentate ligands that help metal ions participate in processes that are essential to nutrition and good health. DRI/RDA values for trace essential elements, including iron and zinc, are in the range of 10–20 mg/d. There are DRI/RDA values for some, but not all, of the ultratrace essential elements, ranging from 45 μg/d for molybdenum up to 5 mg/d for manganese. Compounds of iron and zinc are present in the body in average concentrations between 1 and 1000 μg of element per gram of body mass and are considered to be trace elements. Other transition metals (chromium, cobalt, copper, manganese, molybdenum, nickel, and vanadium) are considered ultratrace, meaning essential elements present in the body in average concentrations less than 1 μg of element per gram of body mass. In this section we will survey the roles of some of the other ultratrace transition metals in biology.

CONNECTION Dietary reference intake (DRI) and recommended dietary allowance (RDA) were introduced in Chapter 21 in the discussion of essential elements in the human diet.

Manganese and Photosynthesis

Let's begin with photosynthesis, a chemical process at the foundation of our food chain. Green plants can harness solar energy because they contain large biomolecules we collectively call *chlorophyll*. All molecules of chlorophyll contain ring-shaped tetradentate ligands called *chlorins* (Figure 22.25a). The structures of chlorins are similar to those of **porphyrins**, another class of tetradentate ligands found in biological systems (Figure 22.25b). We saw porphyrins in Chapter 20 as the Fe^{2+}-binding molecules in hemoglobin. Chlorins and porphyrins are members of a larger category of polydentate compounds known as **macrocyclic ligands**. (*Macrocycle* means, literally, "big ring.")

Two of the four nitrogen atoms in porphyrins and chlorins are sp^3 hybridized and bound to hydrogen atoms, whereas the other two are sp^2 hybridized with no hydrogen atoms. When either compound forms coordinate bonds with a metal ion M^{n+} (Figure 22.25c), the two hydrogen atoms ionize, giving the ring a charge of $2-$ and the complex ion an overall charge of $(n - 2)$. The lone pairs of electrons on the N atoms in the ionized structure are oriented toward the ring center. These lone pairs can occupy either the four equatorial coordination sites in an octahedral complex ion or all four coordination sites in a square planar complex ion. In octahedral complex ions, each central metal ion still has two axial sites available for bonding to other ligands. Depending on the charge of the central ion, the coordination compound may be either ionic (a complex ion) or electrically neutral.

Porphyrin and chlorin rings are widespread in nature and play many biochemical roles. Their chemical and physical properties depend on

1. the identity of the central metal ion
2. the species that occupy the axial coordination sites of octahedral complexes
3. the number and identity of organic groups attached to the outside of the ring

The role of the major essential element Mg^{2+} in photosynthesis was described in Chapter 21. Another metal, the ultratrace essential element manganese, also plays a role in the production of oxygen during photosynthesis. To examine that role, let's write an equation for photosynthesis that is slightly different from the equation we are used to seeing. Instead of writing the formula $C_6H_{12}O_6$ for glucose, we use the generic carbohydrate formula $(CH_2O)_n$ so that the coefficient is 1 for all other species in the reaction (rather than 6):

$$H_2O(\ell) + CO_2(g) \rightarrow \tfrac{1}{n}(CH_2O)_n(aq) + O_2(g)$$

Writing the equation in this form makes it easier to see how the overall reaction between water and carbon dioxide involves electron transfer: photosynthesis is a redox reaction where oxygen is oxidized and carbon is reduced.

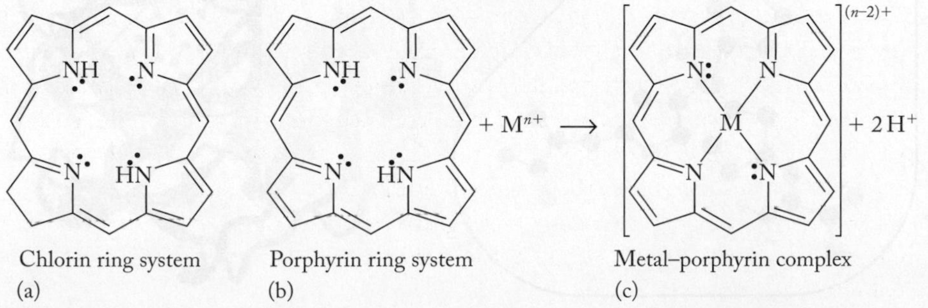

porphyrin a type of tetradentate macrocyclic ligand.

macrocyclic ligand a ring containing multiple electron-pair donors that bind to a metal ion.

Chlorin ring system
(a)

Porphyrin ring system
(b)

Metal–porphyrin complex
(c)

FIGURE 22.25 Skeletal structures of (a) chlorin and (b) porphyrin. (In these structures the bonds formed by carbon atoms are shown, but not the atoms themselves nor their bonds to hydrogen atoms.) The principal difference between the structures is a $C=C$ in the porphyrin structure (shown in red) that is a single bond in chlorin rings. The innermost atoms in each ring are four nitrogen atoms with lone pairs of electrons. All four N atoms form coordinate bonds with a metal ion, as shown with the porphyrin ring in (c).

This redox process requires metalloenzymes. Manganese-containing biomolecules in which Mn is in the +3 and +4 oxidation states mediate the transfer of electrons from water in photosynthesis. Although the exact structure of these manganese compounds remains undetermined, it is believed that two manganese(III) ions and two manganese(IV) ions are present at the site of O_2 production. Recall that copper-containing enzymes are also involved in the series of reactions that make up photosynthesis.

CONCEPT TEST

Manganese ions involved in photosynthesis are often surrounded by six ligands in an octahedral geometry. Is reduction of a manganese(IV) ion to manganese(III) in an octahedral complex (see Section 22.5) accompanied by a change in spin state of the Mn ion?

Transition Metals in Enzymes

Transition metal cations can form complexes by bonding to the nitrogen atoms of the amino acids in proteins and other Lewis bases in biological systems. Many of the enzymes in our bodies contain zinc or iron ions; these enzymes, along with others containing transition metal ions, are called *metalloenzymes*.

Zinc The enzyme carbonic anhydrase catalyzes the reaction between water and carbon dioxide to form bicarbonate ions:

$$H_2O(\ell) + CO_2(aq) \rightleftharpoons HCO_3^-(aq) + H^+(aq)$$

The α-form of carbonic anhydrase contains 260 amino acid residues and a zinc ion at the active site. Note in Figure 22.26 that the zinc ion is coordinately bonded to three nitrogen atoms on histidine side chains and to one molecule of water. The presence of these ligands and a fourth histidine nearby facilitates ionization of the water molecule. Ionization leaves a OH^- ion attached to the Zn^{2+} ion and a H^+ ion bonded to the side-chain nitrogen atom of the fourth histidine. In addition, a space just the right size and shape to accommodate a CO_2 molecule is next to the active site. When a CO_2 molecule in this space bonds to a hydroxide ion, an HCO_3^- ion forms. As the bicarbonate ion pulls away, another water molecule occupies the fourth coordination site on the Zn^{2+} ion, another CO_2 molecule

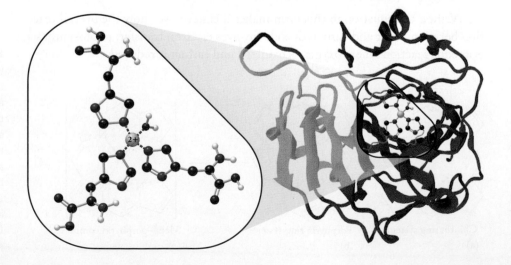

FIGURE 22.26 The active site of one form of carbonic anhydrase consists of a zinc ion (gray sphere) bonded to three histidine molecules and one H_2O molecule. The OH^- ion produced when this H_2O ionizes combines with a molecule of CO_2, forming an HCO_3^- ion.

enters, the histidine is protonated, and the catalytic cycle repeats. This reaction is important because it helps eliminate CO_2 from cells during respiration and mediates the uptake of CO_2 during photosynthesis in some plants.

Iron Iron-containing peroxidases and catalases are integral to the transfer of oxygen to biomolecules. For example, plants use a fatty acid peroxidase to catalyze the stepwise degradation of fatty acids. One $-CH_2-$ group at a time is removed from the fatty acid by using hydrogen peroxide:

$$R-CH_2-COOH(aq) + 2\,H_2O_2(aq) \xrightarrow{\text{fatty acid peroxidase}} 3\,H_2O(\ell) + R-CHO(aq) + CO_2(aq)$$

Fatty acid   Hydrogen peroxide   Aldehyde

Subsequent oxidation of the aldehyde back to a carboxylic acid yields a new fatty acid with one fewer $-CH_2-$ group:

$$2\,R-CHO(aq) + O_2(aq) \rightarrow 2\,RCHOOH(aq)$$

Aldehyde   Fatty acid

Iron-containing enzymes also catalyze the reduction of nitrite (NO_2^-) and sulfite (SO_3^{2-}) ions:

$$NO_2^-(aq) + 6\,e^- + 8\,H^+(aq) \xrightarrow{\text{nitrite reductase}} NH_4^+(aq) + 2\,H_2O(\ell)$$

$$SO_3^{2-}(aq) + 6\,e^- + 7\,H^+(aq) \xrightarrow{\text{sulfite reductase}} HS^-(aq) + 3\,H_2O(\ell)$$

In each case, the iron in the enzyme is oxidized, providing the electrons needed for reduction.

Proteins called *cytochromes* also contain one or more heme groups (Figure 22.27a). Cytochromes mediate oxidation and reduction processes connected with energy production in cells. The heme group conveys electrons as the half-reaction

$$Fe^{3+} + e^- \rightleftharpoons Fe^{2+}$$

rapidly and reversibly consumes or releases electrons needed in the biochemical reactions that sustain life. Cytochromes catalyze electron transport in photosynthesis and the metabolism of glucose to CO_2 and water. Their role in electron transport makes them key participants in *in vivo* reactions that produce ATP, which is required for intracellular energy production and transfer in living things. As catalysts for oxidation and reduction, cytochromes are essential for the removal of toxic substances and for the process of programmed cell death, both of which are required for the health of multicellular organisms. Cytochrome *c* (Figure 22.27b) is a component of the electron transport chain in mitochondria that is ubiquitous in life forms ranging in complexity from microbes to human beings. Its amino acid sequence is very similar or the same across a spectrum of plants, animals, and many unicellular organisms. Consequently, studies of the cytochrome *c* molecule are seminal in evolutionary biology.

Many other kinds of cytochrome proteins have different substituents on the porphyrin rings and different axial ligands, each of which influences the function of the

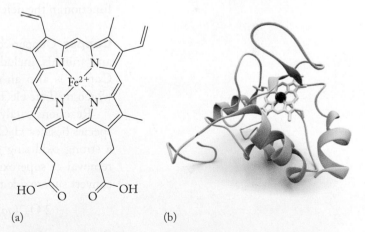

(a)   (b)

FIGURE 22.27 (a) Heme is the specific porphyrin that binds iron in the oxygen-transport protein hemoglobin and also in many enzymes of great significance in all life on Earth. (b) The structure of cytochrome proteins, such as cytochrome *c* shown here, includes one or more heme complexes (shown in gray) that mediate energy production and redox reactions in living cells. Different cytochromes have different axial ligands occupying the fifth and sixth octahedral coordination sites, and different groups in the protein may be attached to the porphyrin ring.

complex. This last point has been repeated several times in this chapter: the chemical properties and biological functions of transition metals that are essential to living organisms are linked to their molecular environments and to the formation of stable complex ions with ligands that are strong electron donors—that is, strong Lewis bases.

> **CONCEPT TEST**
>
> Which of the following small molecules could not function as a Lewis base and hence would not be expected to bond to iron in a heme protein? CO; H_2; NO_2; H_2S

Molybdenum and Vanadium Many metalloenzymes are involved in transformations of nitrogen. Molybdenum-containing reductases are responsible for converting NO_3^- ions to NO_2^- ions and then to NH_3 in the nitrogen cycle. The active site of sulfite oxidase, which converts SO_3^{2-} ions to SO_4^{2-} ions in the sulfur cycle, also contains molybdenum. Sometimes more than one type of transition metal is found in a metalloenzyme. For example, both molybdenum and iron are required by xanthine oxidase, an important enzyme along the pathway for degradation of excess nucleic acids (adenine and guanine) to xanthine and then to uric acid for elimination through the kidneys.

Xanthine   Uric acid

The combination of iron and vanadium is essential to the function of haloperoxidases, a class of enzymes found in some algae, lichens, and fungi that replace C—H bonds with carbon–halogen bonds. The reaction products are thought to function in the defense systems of these organisms.

Copper Copper-containing proteins perform several functions in both plants and animals, including oxygen transport in mollusks such as clams and oysters. Copper is also an essential element in the enzymes azurin and plastocyanin, which mediate electron transfer during photosynthesis.

Reactions catalyzed by xanthine oxidase may produce other reactive oxygen species besides H_2O_2. One of them is the superoxide ion, O_2^-. Superoxide ion is a strong oxidizing agent and must be eliminated to prevent cell damage. The removal of superoxide begins with the action of superoxide dismutases, which convert superoxide to hydrogen peroxide:

$$2\,O_2^-(aq) + 2\,H^+(aq) \xrightarrow{\text{superoxide dismutase}} H_2O_2(aq) + O_2(aq)$$

Researchers have isolated superoxide dismutases containing a variety of transition metals, including a copper–zinc enzyme. The hydrogen peroxide produced in this reaction is decomposed to water and oxygen by iron-containing catalase.

Nickel Ureases, the enzymes responsible for the conversion of urea to ammonia in plants, contain nickel(II). Nickel is also found with iron in enzymes called *hydrogenases*. Hydrogenases oxidize hydrogen gas to protons:

$$H_2(g) \xrightarrow{\text{hydrogenase}} 2\,H^+(aq) + 2\,e^-$$

Nickel and iron also combine in CO dehydrogenase, an enzyme that catalyzes the formation of acetyl-CoA, which is a key component in the tricarboxylic acid cycle discussed in Chapter 20. Finally, nickel-containing enzymes are found among the enzymes responsible for methane generation by bacteria.

Cobalt and Coenzymes Many enzymes require the presence of a **coenzyme**, which is an organic compound that cocatalyzes a biochemical reaction. The coenzyme B_{12} (Figure 22.28) contains the ultratrace essential element cobalt(III) and is a derivative of vitamin B_{12}. The cobalt(III) in coenzyme B_{12} is easily reduced to cobalt(II) and even cobalt(I) in the course of enzyme-catalyzed redox reactions. The change in oxidation state of Co allows for facile transfer of methyl groups, as in the conversion of methionine to homocysteine:

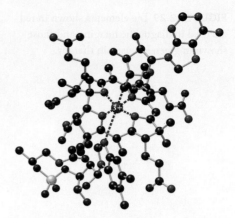

FIGURE 22.28 Coenzyme B_{12} contains cobalt(III) (gold sphere), an ultratrace essential metal.

$$\overset{+}{N}H_3$$

Methionine $\longrightarrow$ Homocysteine

Coenzyme B_{12} is also critical to the function of *mutases*, which are enzymes that catalyze the rearrangement of the skeleton of a molecule, as in the interconversion of glutamate and methylaspartate:

Glutamate $\underset{\text{coenzyme } B_{12}}{\overset{\text{glutamate mutase}}{\rightleftharpoons}}$ Methylaspartate

Chromium Chromium in the +3 oxidation state is an ultratrace essential element in our diets. It is involved in regulating glucose levels in the blood through a molecule called chromodulin. Chromodulin is a polypeptide incorporating only 4 of the 20 naturally occurring amino acids: glycine, cysteine, glutamic acid, and aspartic acid. Four Cr^{3+} ions are bound to the peptide chain. Cereals and grains contain enough chromium for our daily needs, but certain plants (such as shepherd's purse) concentrate chromium and have been used as herbal remedies in diabetes treatment.

Chromium in the +6 oxidation state, as found in chromate ions (CrO_4^{2-}), is acutely toxic and also carcinogenic. Chromate ion enters cells through ion channels that transport SO_4^{2-} ions. Once inside, CrO_4^{2-} is reduced to Cr^{3+}, which binds to the phosphate backbone of DNA.

22.8 Coordination Compounds in Medicine

Transition metal coordination compounds are becoming important in both the diagnosis and treatment of diseases (Figure 22.29). Any diagnostic or therapeutic compound that is injected intravenously must be sufficiently soluble in blood to be delivered to the target. While in transit, the compound must be stable enough not to undergo chemical reactions that result in its precipitation or rapid elimination from the body. Occasionally, the compound can be in the form of a simple salt, but more often a metal ion is introduced as a coordination complex or coordination

coenzyme organic molecule that, like an enzyme, accelerates the rate of biochemical reactions.

FIGURE 22.29 The elements shown in red are used in diagnostic imaging and those shown in green are used in therapy.

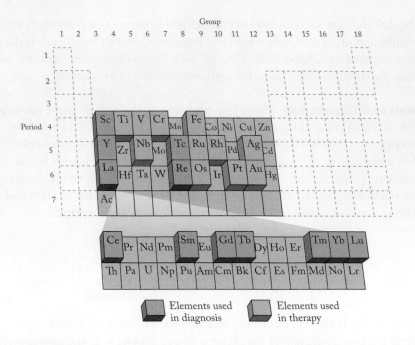

CONNECTION Magnetic resonance imaging (MRI) was mentioned in Chapter 3 in our discussion of the use of cryogenic helium for superconducting magnets.

CONNECTION Chapter 19 gave a more detailed discussion of nuclear chemistry and nuclear medicine, including an assessment of how different types of radiation affect living tissue.

compound. Ligands used in forming biologically active coordination complexes include amino acids and simple anions such as the citrate ion. Chelating ligands like diethylenetriaminepentaacetate ($DTPA^{5-}$; Figure 22.30) are often used in biological applications. A medicinal chemist can also take advantage of substances that occur naturally in the body, such as antibodies, to transport a diagnostic or therapeutic metal ion to its target.

Transition Metals in Diagnosis

As discussed in Chapter 19, radionuclides of main group elements with short half-lives that emit easily detectable gamma rays are useful in positron emission tomography (PET). An equally wide array of transition metal isotopes are available for imaging. Magnetic resonance imaging (MRI), for instance, can be used to diagnose soft-tissue injuries and uses stable isotopes of gadolinium to enhance images.

Technetium and Rhenium Technetium, just below manganese in group 7 of the periodic table, is the most widely used radioactive isotope in imaging. Technetium is unusual in that it does not exist naturally on Earth in easily measurable amounts because it has no stable isotopes; in other words, all technetium isotopes

FIGURE 22.30
Diethylenetriaminepentaacetate ($DTPA^{5-}$), citrate^{3-}, and ethylenediaminetetraacetate ($EDTA^{4-}$) are often used as chelating ligands for diagnostic and therapeutic agents based on transition metals. These ions form stable complex ions with 2+ and 3+ metal cations. The solubilities of the complex ions are typically much greater than those of the hydrated ions at physiological pH (7.4).

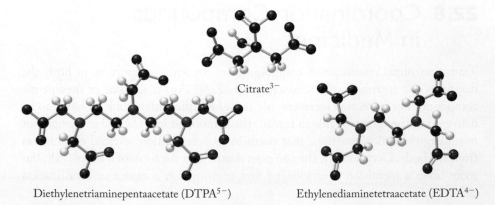

Diethylenetriaminepentaacetate ($DTPA^{5-}$)   Citrate^{3-} Ethylenediaminetetraacetate ($EDTA^{4-}$)

are radioactive. These isotopes can be produced in nuclear reactors for use in medicine. The ^{99m}Tc used in hospitals for imaging is prepared in technetium generators in which a stable isotope of molybdenum, ^{98}Mo (23.78% natural abundance), is bombarded with neutrons. Technetium-99 has a half-life greater than 20,000 years; however, when it is produced in a nuclear reactor, its nucleus is in an excited state, called a *metastable* nucleus. The metastable state, designated by adding the letter "m" to the mass number, as in technetium-99m or ^{99m}Tc, has a half-life of six hours ($t_{1/2}$ of ^{99m}Tc = 6.0 h) and decays to the more stable technetium-99 nucleus.

Technetium-99m has been widely used as an imaging agent because it has a short half-life and emits low-energy γ rays. Patients can be injected with a variety of technetium compounds, depending on the target organ. For imaging the heart, the coordination compounds shown in Figure 22.31 are used.

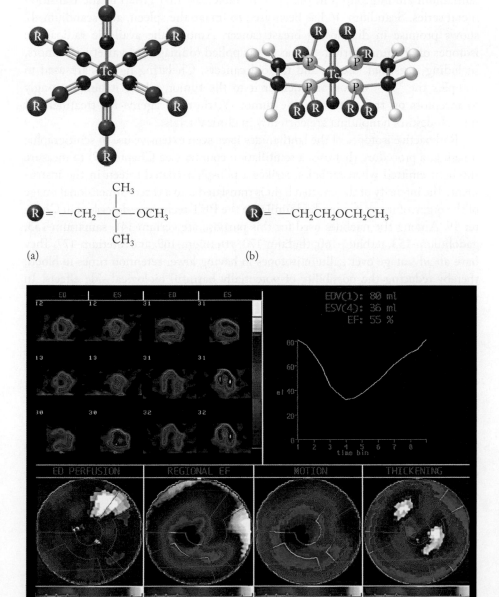

(a)

$$R = -CH_2-\underset{\underset{CH_3}{|}}{\overset{\overset{CH_3}{|}}{C}}-OCH_3$$

(b)

$$R = -CH_2CH_2OCH_2CH_3$$

(c)

FIGURE 22.31 (a) Cardiolite [Tc(CNR)$_6$] and (b) Myoview [TcO$_2$(RPCH$_2$CH$_2$PR)$_2$] are used for imaging the heart. (c) Images of a patient's heart after intravenous injection with a technetium-containing drug. The images along the bottom show four measurements of heart function. Radioactive technetium compounds emit gamma rays that allow the blood to be tracked as it is pumped through the heart.

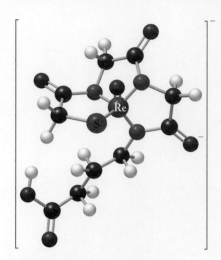

FIGURE 22.32 This rhenium–mercaptoacetylglycylglycyl-γ-amino acid complex is used to attach ^{186}Re or ^{188}Re to monoclonal antibodies or peptides.

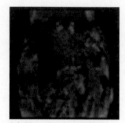

(a) Unenhanced

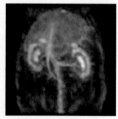

(b) Enhanced

FIGURE 22.33 MRI scans made (a) without a contrast agent and (b) enhanced by a gadolinium contrast agent.

The ability to target the delivery of radionuclides to a particular organ opens the possibility for selective irradiation of a tumor located in that organ. Therefore some radionuclides can be used to not only image but also treat cancers. Certain tumors have highly selective receptor sites for particular molecules on their surfaces. Rhenium, for example, is being studied as both an imaging and a therapeutic agent for some tumors, including breast, liver, and skin cancer. Rhenium-186 and rhenium-188 undergo β decay with half-lives of 3.72 d and 17.0 h, respectively. By including a radioactive rhenium ion in a molecule that binds strongly and specifically to these receptors, physicians can deliver both an imaging agent and a therapeutic agent to the tumor. In principle, the β particles destroy the tumor. Several patents have been issued for the use of compounds containing rhenium isotopes, but therapies based on these compounds remain in the experimental stage. One example of a rhenium compound used in these applications is shown in Figure 22.32.

Scandium, Yttrium, and Lanthanide Elements Scandium, yttrium, and lanthanum are in group 3 in the periodic table, the first group in the transition metal series. Scandium-46 has been used to image the spleen, and scandium-47 shows promise in diagnosing breast cancer. Among the available radioactive isotopes of yttrium, yttrium-90 has been applied to imaging a variety of tumors, including intestinal, breast, and thyroid cancers. Chelating ligands are used to complex the ^{90}Y^{3+} cation and deliver it to the tumor, where it binds strongly to receptors on the surface of the tumor. Yttrium-90 agents for treatment of non-Hodgkin's lymphoma are currently in clinical trials.

Radioactive isotopes of the lanthanides have seen extensive use in scintigraphic imaging, a procedure that uses a scintillation counter (see Chapter 19) to measure the light emitted when radiation strikes a phosphor-coated screen in the instrument. The intensity of the emitted light is translated into a three-dimensional image of the organ of interest, a method similar to the PET technique described in Chapter 19. Among the nuclides used for this purpose are cerium-141, samarium-153, gadolinium-153, terbium-160, thulium-170, ytterbium-169, and lutetium-177. They have an advantage over gallium isotopes in having lower retention times in blood, thereby reducing the possibility of potentially harmful biological side effects. In some cases, lutetium-177 derivatives have replaced yttrium-90 compounds under investigation as radioactive antitumor agents.

As noted earlier, the quality of an MRI scan (Figure 22.33) can be improved when gadolinium is used as a contrast agent. Before the procedure, the patient is injected with a gadolinium compound. Many ligands have been investigated for Gd^{3+} MRI contrast agents. The gadolinium used is a mixture of naturally occurring isotopes of gadolinium. Coordination compounds of other lanthanide ions (Dy^{3+} and Ho^{3+}) have been evaluated but do not work as well as those of gadolinium.

CONCEPT TEST

What is wrong with the statement, "A radioactive isotope with twice the half-life of another will allow an image to be collected in half the time"?

Transition Metals in Therapy

Like main group metal compounds described in Chapter 21, coordination compounds of the transition metals are finding their way into our expanding arsenal of useful therapeutic agents. We have already described the use of platinum compounds to treat cancer and the potential for chromium compounds to help mitigate diabetes. In this section we examine additional therapeutic agents that contain transition metal elements.

Iron The body must regulate the amount of iron in cells in order to produce enough hemoglobin to maintain good health. Deficiencies in iron lead to several diseases broadly classified as anemia. Mild forms of anemia are common among women of child-bearing age and are treated with oral iron supplements. However, a more serious genetic form of anemia, known as thalassemia, must be treated with blood transfusions that leave patients with too much iron in their blood. To remove the excess iron, these patients are treated with chelating ligands that complex some of the iron and transport it out of the red blood cells and eventually out of the body in what is known as chelation therapy. Chelation therapy relies on a large equilibrium constant for the reaction:

$$FeL(aq) + L'(aq) \rightleftharpoons FeL'(aq) + L(aq)$$

where iron coordinated to ligand L (such as hemoglobin) in the blood is treated with ligand L', which also binds iron. If the formation constant for the complex between iron and L' is greater than for iron and L, then the equilibrium constant for the equation is greater than one, and formation of the iron–L' complex is favored. The ligand L' is chosen so that the complex formed between iron and L' is eliminated from the body, thereby reducing the concentration of iron and relieving the symptoms caused by thalassemia treatments. Often the chelating ligand leads to chiral metal complexes of the kind described in Section 22.6. One enantiomer of an iron complex often works better than the other in the treatment of diseases such as thalassemia. Chelation therapy is also the treatment of choice for heavy metal poisoning from toxic metals, including lead and mercury.

SAMPLE EXERCISE 22.6 Calculating the Equilibrium Constant **LO7**
of a Ligand Exchange Reaction

The protein transferrin is involved in the transport of iron into cells. Iron accumulation in the human body has been implicated in diseases such as Parkinson's, Alzheimer's, and thalassemia. The chelating ligand deferoxamine (DFO) is used to treat thalassemia; the complex between iron and DFO is eliminated from the body. Given the formation constants for the complexation of iron(III) by transferrin (Equation 1) and the reaction of DFO with iron (Equation 2), calculate the equilibrium constant for the ligand exchange reaction between DFO and transferrin (Equation 3).

$$Fe^{3+}(aq) + transferrin(aq) \rightleftharpoons Fe(transferrin)^{3+}(aq)$$
$$K_{f,1} = 4.7 \times 10^{20} \quad (1)$$

$$Fe^{3+}(aq) + DFO(aq) \rightleftharpoons Fe(DFO)^{3+}(aq)$$
$$K_{f,2} = 4.0 \times 10^{30} \quad (2)$$

$$Fe(transferrin)^{3+}(aq) + DFO(aq) \rightleftharpoons Fe(DFO)^{3+}(aq) + transferrin(aq)$$
$$K_3 = ? \quad (3)$$

Collect and Organize We are given the formation constants of two reactions involving iron and different ligands and are asked to calculate the equilibrium constant for the ligand exchange reaction. You may wish to refer to Chapter 16 for calculations involving formation constants of complexes, as well as to Chapters 14 and 15, where we manipulated equilibrium constants.

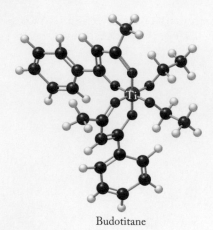

Budotitane

FIGURE 22.34 Budotitane is a titanium(IV) complex that shows activity against colon cancer.

(a)

(b)

FIGURE 22.35 (a) Vanadium(IV) compounds such as bis(allixinato) oxovanadium(IV) can act as insulin mimics. (b) The chromium(III) complex in a glucose tolerance factor contributes to our bodies' ability to regulate insulin levels. The light brown spheres are the side-chain R groups of the amino acids in the structure. (To simplify the structures, the hydrogen atoms bonded to carbon atoms are not shown.)

Analyze Transferrin is a reactant in Equation 1 and a product in the exchange reaction (3). Therefore we need to reverse Equation 1 before adding it to Equation 2 to obtain Equation 3. Reversing a reaction means taking the reciprocal of its equilibrium constant. Combining the reverse of Equation 1 with Equation 2 to obtain Equation 3 means multiplying $(1/K_{f,1})$ and $K_{f,2}$ together to obtain K_3. The reciprocal of $K_{f,1}$ has a value of about 10^{-20}. Therefore the value of K_3 should be about $10^{-20} \times 10^{30}$, or about 10^{10}.

Solve Multiplying $(1/K_{f,1})$ by $K_{f,2}$ to obtain K_3:

$$K_3 = \frac{1}{4.7 \times 10^{20}} \times (4.0 \times 10^{30}) = 8.5 \times 10^9$$

Think About It The equilibrium constant for the ligand exchange between the Fe(transferrin)$^{3+}$ complex and DFO is indeed $\sim 10^{10}$ and illustrates why DFO is an effective treatment for thalassemia—the large value for the equilibrium constant means that the equilibrium in Equation 3 lies toward the products (to the right), removing iron from the blood.

 Practice Exercise The equilibrium constants for the reactions of penicillamine and methionine with methyl mercury are

$$CH_3Hg^+ + \text{penicillamine} \rightleftharpoons CH_3Hg(\text{penicillamine})^+ \qquad K_f = 6.3 \times 10^{13}$$

$$CH_3Hg^+ + \text{methionine} \rightleftharpoons CH_3Hg(\text{methionine})^+ \qquad K_f = 2.5 \times 10^7$$

Calculate the equilibrium constant for the exchange reaction:

$$CH_3Hg(\text{methionine})^+ + \text{penicillamine} \rightleftharpoons CH_3Hg(\text{penicillamine})^+ + \text{methionine}$$

Titanium, Vanadium, and Niobium Cancer Drugs Compounds of a surprisingly large number of transition metals demonstrate antitumor activity. Titanium(IV) complexes such as budotitane (Figure 22.34) show promise against colon cancer. Encouraging results against breast, lung, and colon cancers were observed with a series of compounds with formulas $(C_5H_5)_2TiCl_2$, $(C_5H_5)_2VCl$, and $(C_5H_5)_2NbCl_2$. Unfortunately, at therapeutically useful doses the potential for liver damage outweighs the benefits of these compounds. These compounds contain carbon–metal bonds and belong to a class of substances called **organometallic compounds**.

Chromium, Vanadium, and Controlling Blood Sugar Insulin is a hormone needed for proper glucose metabolism and protein synthesis. People with diabetes either cannot produce insulin or produce the hormone but cannot use it effectively. The suggestion that vanadium plays a role in insulin production has prompted investigation of vanadium compounds as oral diabetes drugs. Encouraging results from animal studies using bis(allixinato)oxovanadium(IV) (Figure 22.35a) have been reported, but the compound is not currently approved for use in humans. Chromium(III) compounds have been shown to lower fasting blood sugar levels and reduce the amount of insulin required by some diabetics (Figure 22.35b).

Platinum Group and Coinage Metals The period 5 and period 6 elements in groups 8, 9, and 10 (Ru, Rh, Pd, Os, Ir, and Pt) are often referred to as the *platinum group metals*. The group 11 elements of these two periods (Ag and Au) along with Cu are called the *coinage metals*. In this section we explore examples of soluble compounds of these metals used in medications for arthritis, cancer, and other diseases.

The serendipitous discovery in the 1960s that *cis*-diamminedichloro-platinum(II) (cisplatin in Figure 22.36) is effective in treating testicular, ovarian, and other cancers spurred the development of a host of cancer drugs based on both platinum group metals and coinage metals. The results of this research include a compound known as carboplatin that shows the same activity as cisplatin, whose mechanism of activity is shown in Figure 22.21, but has fewer side effects. In addition to platinum compounds, the antitumor activities of complexes of gold, rhodium, ruthenium, and silver have been explored. The effectiveness of all of these drugs lies in their ability to bind to the nitrogen-containing bases in DNA and inhibit cell replication. If their cells cannot divide, tumors cannot grow. The greater toxicity of rhodium compounds than those of platinum and ruthenium has limited their clinical application.

organometallic compound a molecule containing direct carbon–metal covalent bonds.

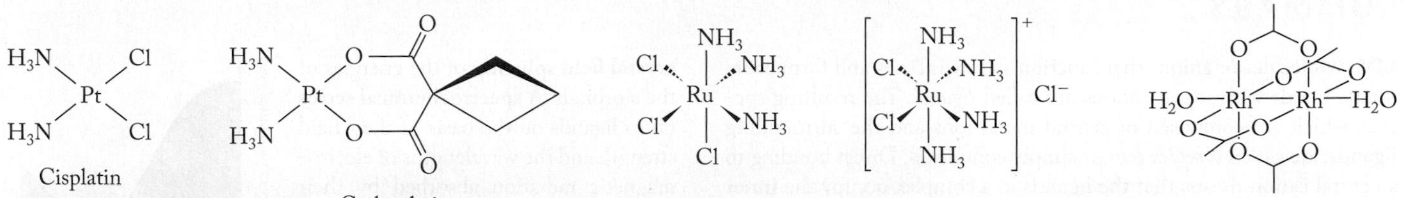

Cisplatin
(Platinol)

Carboplatin

FIGURE 22.36 Cisplatin and carboplatin are effective antitumor agents. The ruthenium and rhodium compounds show activity against leukemia but have not yet seen widespread use.

Selected osmium compounds reduce inflammation in joints resulting from arthritis, although the use of such compounds has diminished with the development of other anti-inflammatory agents. The therapeutic effects of aqueous solutions of osmium tetroxide, OsO_4, were first investigated in the 1950s. The use of osmium tetroxide was superseded by the use of glucose polymers containing osmium, known as osmarins. The use of osmarins reduces the toxic effects of osmium and illustrates how even toxic metals can be adapted to therapy.

Although the historic use of gold for medicinal purposes dates back millennia, the effective use of gold-containing pharmaceuticals originated with the discovery in the 1920s and 1930s that a gold thiosulfate compound, sanochrysin, alleviates the symptoms of rheumatoid arthritis. The most commonly used gold drugs for the treatment of arthritis today are sold under the names myochrysine and auranofin (Figure 22.37). Myochrysine is injected; auranofin can be taken orally.

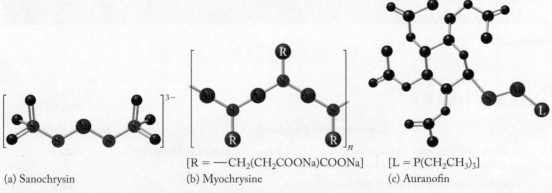

[R = —CH₂(CH₂COONa)COONa]   [L = P(CH₂CH₃)₃]

(a) Sanochrysin   (b) Myochrysine   (c) Auranofin

FIGURE 22.37 (a) Sanochrysin, (b) myochrysine, and (c) auranofin are three gold-containing drugs used to treat arthritis.

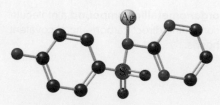

Silver sulfadiazine

FIGURE 22.38 Silver sulfadiazine is an effective antibiotic when applied to burns. (H atoms are not shown.)

Eye drops containing silver salts are used to treat eye infections. Silver sulfadiazine is a broad-spectrum "sulfa" drug used to prevent and treat bacterial and fungal infections (Figure 22.38). It is also an active ingredient in creams used to treat thermal and chemical burns.

Although present in smaller amounts in the human body than the main group elements, the transition metals play crucial roles in our physiological system and in those of many organisms. Their rich and diverse chemistry has led to many applications of both essential and nonessential transition metal compounds to maintaining health and diagnosing and treating disease. Without a doubt, the smorgasbord of the periodic table will continue to provide us with useful substances and materials.

SUMMARY

LO1 Molecules or anions that function as Lewis bases and form coordinate bonds with metal cations are called *ligands*. The resulting species, which are composed of central metal ions and the surrounding ligands, are called *complex ions* or simply complexes. Direct bonding to a central cation means that the ligands in a complex occupy the **inner coordination sphere** of the cation. (Section 22.1)

LO2 The names of complex ions and coordination compounds provide information about the identities and numbers of ligands, the identity and oxidation state of the central metal ion, and the identity and number of counterions. (Section 22.2)

LO3 A **monodentate ligand** donates one pair of electrons in a complex ion; a **polydentate ligand** donates more than one pair in a process called **chelation**. Polydentate ligands are particularly effective at forming complex ions. This phenomenon is called the **chelate effect**. EDTA is a particularly effective sequestering agent, which is a chelating agent that prevents metal ions in solution from reacting with other substances. (Section 22.3)

LO4 The colors of transition metal compounds can be explained by the interactions between electrons in different *d* orbitals and the lone pairs of electrons on surrounding ligands. These interactions create

crystal field splitting of the energies of the *d* orbitals. A **spectrochemical series** ranks ligands on the basis of their field strength and the wavelengths of electromagnetic radiation absorbed by their complex ions. (Section 22.4)

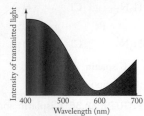

LO5 Strong repulsions and large values of crystal field splitting energy can lead to electron pairing in lower-energy orbitals and an electron configuration called a low-spin state. Metals and their ions with their *d* electrons evenly distributed across all the *d* orbitals in the valence shell represent a high-spin state. (Section 22.5)

LO6 Complex metal ions containing more than one type of ligand may form stereoisomers. When one type occupies two adjacent corners of a square planar complex, the complex is a *cis* isomer; when the same ligand occupies opposite corners, it is a *trans* isomer. (Sections 22.6 and 22.7)

LO7 Many enzymes contain transition metal ions. Soluble coordination compounds and **organometallic compounds** of the transition metals appear in enzymes. Such compounds are used for imaging and in medications for arthritis, cancer, and other diseases. (Section 22.8)

PARTICULATE **PREVIEW WRAP-UP**

Nitrogen; ammonia, 4; ethylenediamine, 3.

PROBLEM-SOLVING SUMMARY

Type of Problem	Concepts and Equations	Sample Exercises
Writing formulas of coordination compounds	Write formulas that distinguish ligands in the inner and outer coordination spheres. Use square brackets to contain a complex ion.	**22.1**
Naming coordination compounds	Follow the naming rules in Section 22.2.	**22.2**

Type of Problem	Concepts and Equations	Sample Exercises
Identifying the potential electron-pair–donor groups in a molecule	Examine the molecular structure of a compound and find lone pairs that can be donated to a metal.	**22.3**
Predicting spin states	Sketch a *d*-orbital diagram based on crystal field splitting. Fill the lowest energy orbitals with one valence electron each. If there are two ways of adding the remaining valence electrons to the diagram, then multiple spin states are possible.	**22.4**
Identifying stereoisomers of coordination compounds	If ligands of one type are all on the same side of the complex ion, it is a *cis* isomer. If ligands of one type are on opposite sides, it is a *trans* isomer.	**22.5**
Calculating the equilibrium constant for a ligand exchange reaction	Use the rules for manipulating equilibrium constant expressions to calculate the equilibrium constant for a ligand exchange reaction.	**22.6**

VISUAL PROBLEMS

(Answers to boldface end-of-chapter questions and problems are in the back of the book.)

22.1. Two of the four highlighted elements in Figure P22.1 have cations that form colored compounds with Cl^-. Which ones?

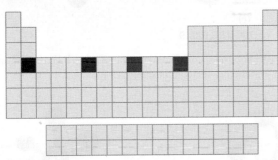

FIGURE P22.1

22.2. Which of the highlighted transition metals in Figure P22.2 form M^{2+} cations that cannot have high-spin and low-spin states?

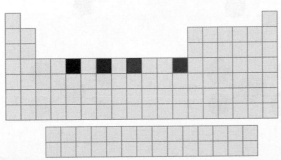

FIGURE P22.2

22.3. Which of the highlighted transition metals in Figure P22.2 have M^{2+} cations that form colorless tetrahedral complex ions?

22.4. Smoky quartz has distinctive lavender and purple colors due to the presence of manganese impurities in crystals of silicon dioxide. Which of the orbital diagrams in Figure P22.4 best describes the Mn^{2+} ion in a tetrahedral field?

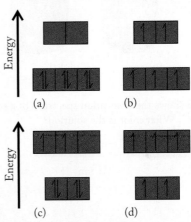

FIGURE P22.4

22.5. Chelation Therapy I The compound with the structure shown in Figure P22.5 is widely used in chelation therapy to remove excessive lead or mercury in patients exposed to these metals. How many electron-pair–donor groups ("teeth") does the sequestering agent have when the carboxylic acid groups are ionized?

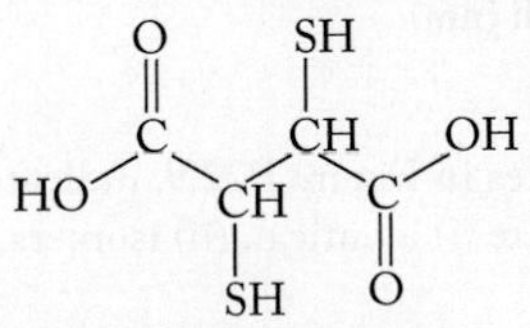

FIGURE P22.5

22.6. Chelation Therapy II The compound with the structure shown in Figure P22.6 has been used to treat people exposed to plutonium, americium, and other actinide metal ions. How many electron pair donor groups does the sequestering agent have when the carboxylic acid groups are ionized?

FIGURE P22.6

22.7. The three beakers shown in Figure P22.7 contain solutions of $[CoF_6]^{3-}$, $[Co(NH_3)_6]^{3+}$, and $[Co(CN)_6]^{3-}$. According to the colors of the three solutions, which compound is present in each of the beakers?

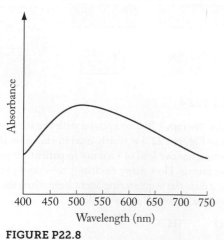

(a)   (b)   (c)

FIGURE P22.7

22.8. Figure P22.8 shows the absorption spectrum of a solution of $Ti(H_2O)_6^{3+}$. What color is the solution?

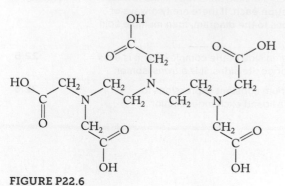

FIGURE P22.8

22.9. For each pair of complexes in Figure P22.9, indicate whether the complexes are (i) identical, (ii) isomers, or (iii) neither.

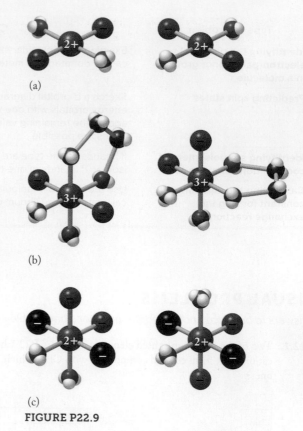

(a)

(b)

(c)

FIGURE P22.9

22.10. For each pair of complexes in Figure P22.10, indicate whether the complexes are (i) identical, (ii) isomers, or (iii) neither.

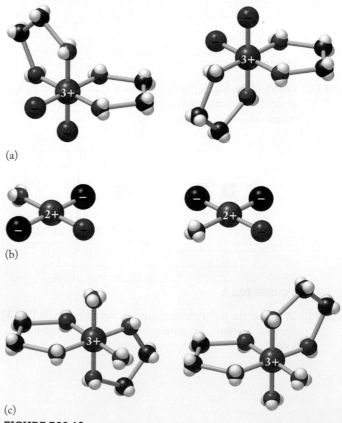

(a)

(b)

(c)

FIGURE P22.10

*22.11. The periodic trend for iodide, bromide, and chloride in the spectrochemical series is $Cl^- > Br^- > I^-$. Which curve in Figure P22.11 represents the spectrum for $CoCl_4{}^{2-}$? Which curve represents $CoI_4{}^{2-}$?

22.12. Use representations [A] through [I] in Figure P22.12 to answer questions a–f.
  a. What are the donor atoms in each ligand depicted?
  b. Which ligands are monodentate?
  c. Which ligands are bidentate?
  d. What is the coordination number for nickel in $Ni(CN)_4{}^{2-}$? In $Ni(dmg)_2$?
  e. If the field strength of cyanide is greater than that of dmg, which complex forms the yellow solution in [E]: $Ni(CN)_4{}^{2+}$ or $Ni(dmg)_2$? Which forms the red solution?
  f. Which orbital diagram depicts the d-orbital energies for $Ni(CN)_4{}^{2-}$? For $Ni(dmg)_2$?

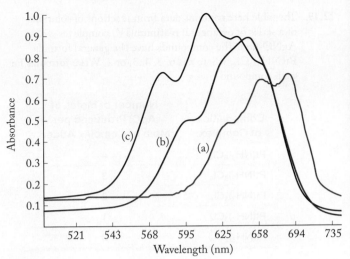

FIGURE P22.11

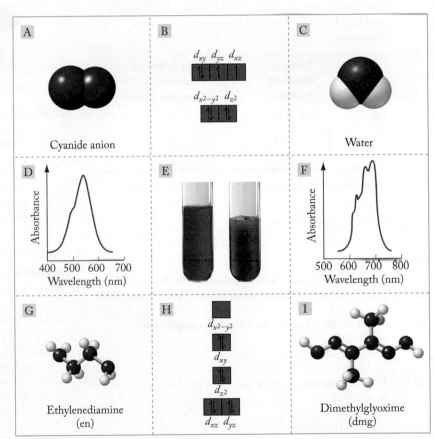

FIGURE P22.12

QUESTIONS AND PROBLEMS

Complex Ions

Concept Review

22.13. When NaCl dissolves in water, which molecules or ions occupy the inner coordination sphere around the Na^+ ions?

22.14. When $Cr(NO_3)_3$ dissolves in water, which of the following species are nearest the Cr^{3+} ions?
  a. Other Cr^{3+} ions
  b. $NO_3{}^-$ ions

  c. H_2O molecules with the O atoms closest to the Cr^{3+}
  d. H_2O molecules with the H atoms closest to the Cr^{3+}

22.15. When $Ni(NO_3)_2$ dissolves in water, what molecules or ions occupy the inner coordination sphere around the Ni^{2+} ions?

22.16. When $[Ni(NH_3)_6]Cl_2$ dissolves in water, which molecules or ions occupy the inner coordination sphere around the Ni^{2+} ions?

22.17. Which ion is the counterion in the coordination compound $Na_2[Zn(CN)_4]$?

22.18. Which ion is the counterion in the coordination compound $[Co(NH_3)_4Cl_2]NO_3$?

22.19. The table here contains data from reactions of solutions of a series of octahedral platinum(IV) complexes with $AgNO_3(aq)$. The compounds have the general formula $Pt(NH_3)_xCl_4$, where x is 6, 5, 4, 3, or 2. Write formulas for each compound.

Composition of Complex	Number of Moles of AgCl Produced per Mole of Complex Added
$Pt(NH_3)_6Cl_4$	4
$Pt(NH_3)_5Cl_4$	3
$Pt(NH_3)_4Cl_4$	2
$Pt(NH_3)_3Cl_4$	1
$Pt(NH_3)_2Cl_4$	0

22.20. The compositions of three compounds of chromium(III) and chloride ion are known: $Cr(H_2O)_6Cl_3$, $Cr(H_2O)_5Cl_3$, and $Cr(H_2O)_4Cl_3$. The table here summarizes their properties.

Compound	Color of Aqueous Solution	Number of Moles of AgCl Produced per Mole of Complex Added
A	Dark green	1
B	Gray-blue	3
C	Light green	2

Write the correct formulas for compounds A, B, and C.

Naming Complex Ions and Coordination Compounds

Problems

22.21. Name the complex ions of platinum(IV) in Problem 22.19.

22.22. Name the complex ions of chromium(III) in Problem 22.20.

22.23. What are the names of the following complex ions?
a. $Cr(NH_3)_6^{3+}$
b. $Co(H_2O)_6^{3+}$
c. $[Fe(NH_3)_5Cl]^{2+}$

22.24. What are the names of the following complex ions?
a. $Cu(NH_3)_2^{+}$
b. $Ti(H_2O)_4(OH)_2^{2+}$
c. $Ni(NH_3)_4(H_2O)_2^{2+}$

22.25. What are the names of the following complex ions?
a. $CoBr_4^{2-}$
b. $Zn(H_2O)(OH)_3^{-}$
c. $Ni(CN)_5^{3-}$

22.26. What are the names of the following complex ions?
a. CoI_4^{2-}
b. $CuCl_4^{2-}$
c. $[Cr(en)(OH)_4]^{-}$

22.27. What are the names of the following coordination compounds?
a. $[Zn(en)_2]SO_4$
b. $[Ni(NH_3)_5(H_2O)]Cl_2$
c. $K_4[Fe(CN)_6]$

22.28. What are the names of the following coordination compounds?
a. $(NH_4)_3[Co(CN)_6]$
b. $[Co(en)_2Cl](NO_3)_2$
c. $[Fe(H_2O)_4(OH)_2]Cl$

Polydentate Ligands and Chelation

Concept Review

22.29. What is meant by the term *sequestering agent*? What properties make a substance an effective sequestering agent?

22.30. The structures of two compounds that each contain two $-NH_2$ groups are shown in Figure P22.30. The one on the left is ethylenediamine, a bidentate ligand. Does the molecule on the right have the same ability to donate two pairs of electrons to a metal ion? Explain why you think it does or does not.

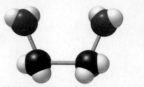

FIGURE P22.30

22.31. How does the chelating ability of an aminocarboxylic acid vary with changing pH?

22.32. **Food Preservative** The EDTA that is widely used as a food preservative is added to food, not as the undissociated acid, but rather as the calcium disodium salt: $Na_2[CaEDTA]$. This salt is actually a coordination compound with a Ca^{2+} ion at the center of a complex ion. Draw the structure of this compound.

Crystal Field Theory

Concept Review

22.33. Explain why the compounds of most of the first-row transition metals are colored.

22.34. Unlike the compounds of most transition metal ions, those of Ti^{4+} are colorless. Why?

22.35. Why is the d_{xy} orbital higher in energy than the d_{xz} and d_{yz} orbitals in a square planar crystal field?

22.36. On average, the d orbitals of a transition metal ion in an octahedral field are higher in energy than they are when the ion is in the gas phase. Why?

Problems

22.37. Aqueous solutions of one of the following complex ions of chromium(III) are violet; solutions of the other are yellow. Which is which? (a) $Cr(H_2O)_6^{3+}$; (b) $Cr(NH_3)_6^{3+}$

22.38. Which of the following complex ions should absorb the shortest wavelengths of electromagnetic radiation? (a) $CuCl_4^{2-}$; (b) CuF_4^{2-}; (c) CuI_4^{2-}; (d) $CuBr_4^{2-}$

22.39. The octahedral crystal field splitting energy Δ_o of $Co(phen)_3^{3+}$ is 5.21×10^{-19} J/ion. What is the color of a solution of this complex ion?

22.40. The octahedral crystal field splitting energy Δ_o of $Co(CN)_6^{3-}$ is 6.74×10^{-19} J/ion. What is the color of a solution of this complex ion?

22.41. Solutions of $NiCl_4^{2-}$ and $NiBr_4^{2-}$ absorb light at 702 nm and 756 nm, respectively. In which ion is the split of *d*-orbital energies greater?

22.42. Chromium(III) chloride forms six-coordinate complexes with bipyridine, including *cis*-$[Cr(bipy)_2Cl_2]^+$, which reacts slowly with water to produce two products, *cis*-$[Cr(bipy)_2(H_2O)Cl]^{2+}$ and *cis*-$[Cr(bipy)_2(H_2O)_2]^{3+}$. In which of these complexes should Δ_o be the largest?

Magnetism and Spin States

Concept Review

22.43. What determines whether a transition metal ion is in a *high-spin* configuration or a *low-spin* configuration?

22.44. Would you expect a solution of a high-spin complex of a transition metal ion to be the same color as a solution of a low-spin complex? Why?

Problems

22.45. How many unpaired electrons are there in the following transition metal ions in an octahedral field? High-spin Fe^{2+}, Cu^{2+}, Co^{2+}, and Mn^{3+}

22.46. Which of the following cations can have either a high-spin or a low-spin electron configuration in an octahedral field? Fe^{2+}, Co^{3+}, Mn^{2+}, and Cr^{3+}

22.47. Which of the following cations can, in principle, have either a high-spin or a low-spin electron configuration in a tetrahedral field? Co^{2+}, Cr^{3+}, Ni^{2+}, and Zn^{2+}

22.48. How many unpaired electrons are in the following transition metal ions in an octahedral crystal field? High-spin Fe^{3+}, Rh^+, V^{3+}, and low-spin Mn^{3+}

22.49. The manganese minerals pyrolusite, MnO_2, and hausmannite, Mn_3O_4, contain manganese ions surrounded by oxide ions.
   a. What are the charges of the Mn ions in each mineral?
   b. In which of these compounds could there be high-spin and low-spin Mn ions?

22.50. **Dietary Supplement** Chromium picolinate is an over-the-counter diet aid sold in many pharmacies. The Cr^{3+} ions in this coordination compound are in an octahedral field. Is the compound paramagnetic or diamagnetic?

*22.51. **Refining Cobalt** One method for refining cobalt involves the formation of the complex ion $CoCl_4^{2-}$. This anion is tetrahedral. Is this complex paramagnetic or diamagnetic?

*22.52. Why is it that $Ni(CN)_4^{2-}$ is diamagnetic, but $NiCl_4^{2-}$ is paramagnetic?

Isomerism in Coordination Compounds

Concept Review

22.53. What do the prefixes *cis-* and *trans-* mean in the context of an octahedral complex ion?

22.54. What do the prefixes *cis-* and *trans-* mean in the context of a square planar complex?

22.55. What minimum number of donor groups are required in order to have stereoisomers of a square planar complex?

22.56. With respect to your answer to the previous question, do all square planar complexes with this many types of donor groups have stereoisomers?

Problems

22.57. Does the complex $Co(en)(H_2O)_2Cl_2$ have stereoisomers?

22.58. Does the complex ion $Fe(en)_3^{3+}$ have stereoisomers?

*22.59. Sketch the stereoisomers of the square planar complex ion $CuCl_2Br_2^{2-}$. Are any of these isomers chiral?

*22.60. Sketch the stereoisomers of the complex ion $Ni(en)Cl_2(CN)_2^{2-}$. Are any of these isomers chiral?

Coordination Compounds in Biochemistry

Concept Review

22.61. Enzymes are large proteins.
   a. What is the function of enzymes?
   b. Are all proteins enzymes?

*22.62. Why is Cd^{2+} more likely than Cr^{2+} to replace Zn^{2+} in an enzyme such as carbonic anhydrase?

*22.63. What effect does an enzyme have on the activation energy of a biochemical reaction?

*22.64. Why might reductases also be described as reducing agents?

*22.65. When a transition metal ion such as Cu^{2+} is incorporated into a metalloenzyme, is the formation constant likely to be much greater than one ($K \gg 1$) or much less than one ($K \ll 1$)?

$$Cu^{2+} + \text{protein} \rightleftharpoons \text{metalloenzyme} \qquad K = \frac{[\text{metalloenzyme}]}{[Cu^{2+}][\text{protein}]}$$

*22.66. Carbon monoxide poisoning derives from competitive binding of CO versus O_2 to the Fe^{2+} in the coordination sphere of hemoglobin. If CO can be displaced by breathing pure O_2, which of the following is likely to be correct under these conditions: $K_{O_2}/K_{CO} < 1$ or $K_{O_2}/K_{CO} > 1$?

22.67. What is the likely sign of ΔS for the reaction in Figure P22.67 catalyzed by carboxypeptidase?

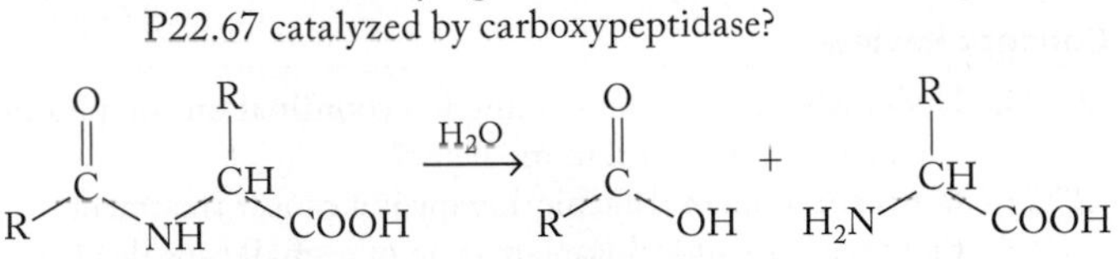

FIGURE P22.67

22.68. What is the likely sign of ΔG for the reaction in question 22.67?

Problems

*22.69. The activation energy for the uncatalyzed decomposition of hydrogen peroxide at 20°C is 75.3 kJ/mol. In the presence of the enzyme catalase, the activation energy is reduced to 29.3 kJ/mol. By using the following form of the Arrhenius equation, $RT\ln(k_1/k_2) = E_{a_2} - E_{a_1}$, how much faster is the catalyzed reaction?

*22.70. **Enzymatic Activity of Urease** Urease catalyzes the decomposition of urea to ammonia and carbon dioxide (Figure P22.70). The rate constant for the uncatalyzed reaction at 20°C and pH = 8 is $k = 3 \times 10^{-10}$ s^{-1}. A urease isolated from the jack bean increases the rate constant to $k = 3 \times 10^4$ s^{-1}. By using the $RT\ln(k_1/k_2) = E_{a_2} - E_{a_1}$ form of the Arrhenius equation, calculate the difference between the activation energies.

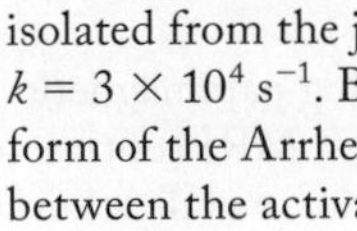

FIGURE P22.70

*22.71. The initial rate of an enzyme-catalyzed reaction depends on the concentration of substrate as shown in Figure P22.71. What is the apparent reaction order at the far right side of the graph?

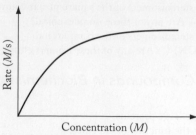

FIGURE P22.71

22.72. The rate of an enzyme-catalyzed reaction depends on the concentration of substrate. Which line in Figure P22.72 has the highest reaction rate?

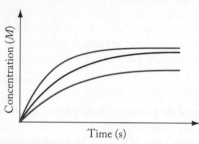

FIGURE P22.72

Coordination Compounds in Medicine

Concept Review

22.73. Under what circumstances might a coordination complex of a toxic metal be useful in medicine?

22.74. **Cancer Treatment** Brachiotherapy is a cancer treatment that involves surgical implantation of a small capsule of ^{192}Ir into a tumor. Iridium-192 lies between two stable isotopes of iridium, ^{193}Ir and ^{191}Ir. How does this account for the fact that ^{192}Ir decays by β decay, electron capture, and positron emission?

22.75. Gadolinium-153 decays by electron capture. What type of radiation does ^{153}Gd produce that makes it useful for imaging?

22.76. Gadolinium-153 and samarium-153 both have the same mass number. Why might ^{153}Gd decay by electron capture, whereas ^{153}Sm decays by emitting β particles?

22.77. How do platinum- and ruthenium-containing drugs fight cancer?

*22.78. Many transition metal complexes are brightly colored. Why might the titanium(IV) compound budotitane be colorless?

*22.79. Is the glucose tolerance factor that contains chromium(III) paramagnetic or diamagnetic?

22.80. Coordination complexes containing paramagnetic transition metal ions make the best MRI contrast agents.
   a. Which of the first-row transition metal cations with a 2+ charge have the most unpaired valence electrons in the gas phase?
   b. Which of the first-row transition metal cations with a 2+ charge have the most unpaired valence electrons in octahedral coordination complexes?

Problems

22.81. The complexation of mercury(II) ion with methionine

$$Hg^{2+} + \text{methionine} \rightleftharpoons Hg(\text{methionine})^{2+}$$

has a formation constant of log K_f = 14.2, whereas the formation constant for the Hg^{2+} complex with penicillamine

$$Hg^{2+} + \text{penicillamine} \rightleftharpoons Hg(\text{penicillamine})^{2+}$$

is log K_f = 16.3. Calculate the equilibrium constant for the reaction

$$Hg(\text{methionine})^{2+} + \text{penicillamine} \rightleftharpoons$$
$$Hg(\text{penicillamine})^{2+} + \text{methionine}$$

22.82. The complexation of mercury(II) ion with cysteine in aqueous solution

$$Hg^{2+} + \text{cysteine} \rightleftharpoons Hg(\text{cysteine})^{2+}$$

has a formation constant of log K_f = 14.2, whereas the formation constant for the Hg^{2+} complex with glycine

$$Hg^{2+} + \text{glycine} \rightleftharpoons Hg(\text{glycine})^{2+}$$

is log K_f = 10.3. Calculate the equilibrium constant for the reaction

$$Hg(\text{cysteine})^{2+} + \text{glycine} \rightleftharpoons Hg(\text{glycine})^{2+} + \text{cysteine}$$

22.83. The equilibrium constant of the reaction

$$CH_3Hg(\text{penicillamine})^+(aq) + \text{cysteine}(aq) \rightleftharpoons$$
$$CH_3Hg(\text{cysteine})^+(aq) + \text{penicillamine}(aq)$$

is K = 0.633. Calculate the equilibrium concentrations of cysteine and penicillamine if we start with a 1.00 M solution of cysteine and a 1.00 mM solution of $CH_3Hg(\text{penicillamine})^+$.

22.84. The equilibrium constant of the reaction

$$CH_3Hg(\text{glutathione})^+(aq) + \text{cysteine}(aq) \rightleftharpoons$$
$$CH_3Hg(\text{cysteine})^+(aq) + \text{glutathione}(aq)$$

is K = 5.0. Calculate the equilibrium concentrations of cysteine and glutathione if we start with a 1.20 mM solution of cysteine and a 1.20 mM solution of $CH_3Hg(\text{glutathione})^+$.

22.85. The two platinum compounds shown in Figure P22.85 have been studied in cancer therapy.
   a. What is the difference between the two compounds?
   b. Name the two complex ions.
   c. Do both compounds have the same orbital diagram?
   d. Sketch the orbital diagram for the platinum compound(s), including the appropriate number of electrons.

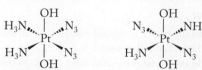

FIGURE P22.85

22.86. Anticancer treatments based on coordination complexes of ruthenium may have less severe side effects than cisplatin and other platinum-based medications. The complex shown in Figure P22.86 was tested against a colon cancer cell line with promising results.

a. Draw the other stereoisomer of $[Ru(py)_2Cl_4]^-$.

b. Name both isomers.

c. Sketch the orbital diagram for the ruthenium compound, including the appropriate number of electrons.

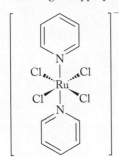

FIGURE P22.86

*22.87. Two square planar gold and platinum complexes with medicinal properties are shown in Figure P22.87.

a. Do these two complexes have the same number of d electrons in their orbital diagram?

b. Are these compounds diamagnetic or paramagnetic?

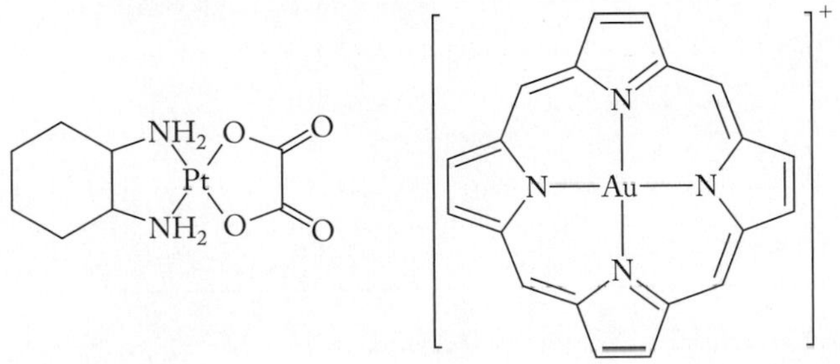

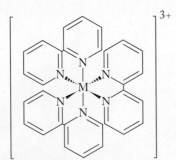

FIGURE P22.87

*22.88. In Chapter 9 we saw that polycyclic aromatic compounds can intercalate into DNA. Compounds such as the rhodium and ruthenium complexes shown in Figure P22.88 behave in a similar fashion toward DNA.

a. What is the hydridization at the nitrogen and the carbon atoms in these ligands?

b. Why are these ligands planar?

c. Are these compounds diamagnetic or paramagnetic?

$M = Ru, Rh$

FIGURE P22.88

Additional Problems

22.89. Dissolving cobalt(II) nitrate in water gives a beautiful purple solution. There are three unpaired electrons in this cobalt(II) complex. When cobalt(II) nitrate is dissolved in aqueous ammonia and oxidized with air, the resulting yellow complex has no unpaired electrons. Which cobalt complex has the larger crystal field splitting energy Δ_o?

22.90. **Lead Poisoning** Children used to be treated for lead poisoning with intravenous injections of EDTA. If the concentration of EDTA in the blood of a patient is 2.5×10^{-8} M and the formation constant for the complex Pb(EDTA)$^{2-}$ is 2.0×10^{18}, what is the concentration ratio of the free (and potentially toxic) Pb^{2+}(aq) in the blood to the much less toxic Pb(EDTA)$^{2-}$ complex?

*22.91. A solid compound containing iron(II) in an octahedral crystal field has four unpaired electrons at 298 K. When the compound is cooled to 80 K, the same sample appears to have no unpaired electrons. How do you explain this change in the compound's properties?

22.92. When Ag$_2$O reacts with peroxodisulfate (S$_2$O$_8$$^{2-}$) ion (a powerful oxidizing agent), AgO is produced. Crystallographic and magnetic analyses of AgO suggest that it is not simply silver(II) oxide, but rather a blend of silver(I) and silver(III) in a square planar environment. The Ag^{2+} ion is paramagnetic but, like AgO, Ag$^+$ and Ag^{3+} are diamagnetic. Explain why.

22.93. The iron(II) compound Fe(bipy)$_2$(SCN)$_2$ is paramagnetic, but the corresponding cyanide compound Fe(bipy)$_2$(CN)$_2$ is diamagnetic. Why do these two compounds have different magnetic properties?

22.94. Aqueous solutions of copper(II)–ammonia complexes are dark blue. Will the color of the series of complexes Cu(H$_2$O)$_{(6-x)}$(NH$_3$)$_x$$^+$ shift toward shorter or longer wavelengths as the value of x increases from 0 to 6?

22.95. Manganese(II) chloride was one of the first compounds to be investigated as MRI contrast agents. How many unpaired electrons does the complex have when MnCl$_2$ dissolves in water to form the coordination compound MnCl$_2$(H$_2$O)$_4$?

22.96. Gadolinium(III) ions are used in contrast agents for MRI because they have unpaired electrons. What is the electron configuration for Gd^{3+} and how many unpaired electrons does it have?

Appendix 1

Mathematical Procedures

Working with Scientific Notation

Quantities that scientists work with are sometimes very large, such as Earth's mass, and other times very small, such as the mass of an electron. It is easier to work with these values when they are expressed in scientific notation.

The general form of standard scientific notation is a value between 1 and 10 multiplied by 10 raised to an integral power. According to this definition, 598×10^{22} kg (Earth's mass) is not in standard scientific notation, but 5.98×10^{24} kg is. It is good practice to use and report values in standard scientific notation.

1. **To convert an "ordinary" number to standard scientific notation,** move the decimal point to the left for a large number, or to the right for a small one, so that the decimal point is located after the first nonzero digit.

   A. For example, to express Earth's average density (5517 kg/m^3) in scientific notation requires moving the decimal point three places to the left. Doing so is the same as dividing the number by 1000, or 10^3. To keep the value the same we multiply it by 10^3. So, Earth's density in standard scientific notation is $5.517 \times 10^3 \text{ kg/m}^3$.

   B. If we move the decimal point of a value less than 1 to the right to express it in scientific notation, then the exponent is a negative integer equal to the number of places we moved the decimal point to the right. For example, the value of R used in solving ideal gas law problems is $0.08206 \text{ L} \cdot \text{atm/(mol} \cdot \text{K)}$. Moving the decimal point two places to the right converts the value of R to scientific notation: $8.206 \times 10^{-2} \text{ L} \cdot \text{atm/(mol} \cdot \text{K)}$.

   C. Another value of R, $8.314 \text{ J/(mol} \cdot \text{K)}$, does not need an exponent, though it could be written $8.314 \times 10^0 \text{ J/(mol} \cdot \text{K)}$ because any value raised to the zero power is equal to 1.

2. **For calculations with numbers in scientific notation,** most calculators have a function key for entering the exponents of values expressed in scientific notation. In many calculators it is labeled "E" or "EE" or "Exp." To enter, for example, the speed of light in meters per second, 2.998×10^8, we enter the value before the exponent, 2.998, followed by the exponent key and then the value of the exponent (8). To enter a value with a negative exponent, use the sign-change key. Sometimes it is labeled "(−)" or "+/−" or "±."

Working with Logarithms

A logarithm to the base 10 has the following form:

$$\log_{10} x = \log x = a, \text{ where } x = 10^a$$

We usually abbreviate the logarithm function "log" if the logarithm is to the base 10, which means the scale in which $\log 10 = 1$.

A logarithm to the base e, called a *natural logarithm*, has the following form:

$$\log_e x = \ln x = b, \text{ where } x = e^b$$

Scientific calculators have (LOG) and (LN) keys, so it is easy to convert a number into its log or ln form. The directions below apply to most calculators.

Sample Exercise 1 Find the logarithm to the base 10 of 2.247 (log 2.247).

Solution In some calculators you enter 2.247 first and then press the (LOG) key. In others, such as the TI 84/89 series, you press the (LOG) key first followed by 2.247 and then the (ENTER) key. Either way, the answer should be 0.3516 (to four significant figures).

Sample Exercise 2 Find the natural logarithm of 2.247 (ln 2.247).

Solution Follow the same procedure as in Sample Exercise 1 except use the (LN) key. The answer should be 0.8096. This answer is $(0.8096/0.3516) = 2.303$ times larger than log value. That is,

$$\ln x = 2.303 \log x$$

Sample Exercise 3 Find the log of 6.0221×10^{23}.

Solution Following the procedures described above for entering a value with an exponent into your calculator and then taking its log to the base 10, you should obtain the value 23.77974796. We know the original value to five significant figures; the log value should have the same precision, but how do we express it? You might think the log value should be 23.780; however, the 23 to the left of the decimal point reflects the value of the exponent in the original value, and exponents don't count in determining the number of significant figures in a value. Therefore, the value of the logarithm to five significant figures is 23.77975. To understand better why this is so, calculate the log of 6.0221. You should get 0.77975 to five significant figures. Note how the difference between the two log values (23.77975 and 0.77975) is simply the value of the exponent in the first value.

Combining Logs The following equations summarize how logarithms of the products or quotients of two or more values are related to the individual logs of those values:

$$\text{logarithm } (ab) = \text{logarithm } a + \text{logarithm } b$$

and

$$\text{logarithm } (a/b) = \text{logarithm } a - \text{logarithm } b$$

Converting Logarithms into Numbers

If we know the value of $\log x$, what is the value of x? This question is frequently asked when working with pH, which is the negative log of the concentration of hydrogen ions, $[H_3O^+]$, in solution:

$$pH = -\log[H_3O^+]$$

Suppose the pH of a solution of a weak acid is 2.50. The concentration of H_3O^+ is related to this pH value as follows:

$$2.50 = -\log[H_3O^+]$$

or

$$-2.50 = \log[H_3O^+]$$

To find the value of $[H_3O^+]$, we enter 2.5 into the calculator and press the appropriate key to change its sign to -2.5. The next step depends on the type of calculator. If yours has a (10^x) key (often accessible using a second function key), use it to find the value of $10^{-2.5}$, which is the value we are looking for. The corresponding keystrokes with many graphing calculators are (10^x), $((-))$, 2.5, (ENTER). Some calculators, including the virtual one in many Windows operating systems, have an (x^y) key. We can use it for this problem by entering 10 and pushing the (x^y) key, then entering 2.5 and pushing the $(+/-)$ key followed by the $(=)$ key. All of these approaches do the same calculation, taking 10 to the -2.50 power, and give the same answer, $[H_3O^+] = 3.2 \times 10^{-3}$ to two significant figures. (Remember, pH is a log value, so the digit before the decimal point is not significant.)

Solving Quadratic Equations

If the terms in an equation can be rearranged so that they take the form

$$ax^2 + bx + c = 0$$

they have the form of a quadratic equation. The value(s) of x can be determined from the values of the coefficients a, b, and c by using the equation

$$x = \frac{-b \pm \sqrt{b^2 - 4ac}}{2a}$$

For example, if the solution to a problem yields the following expression where x is the concentration of a solute:

$$x^2 + 0.112x - 1.2 \times 10^{-3} = 0$$

Then the value of x can be determined as follows:

$$x = \frac{-b \pm \sqrt{b^2 - 4ac}}{2a}$$

$$= \frac{-0.112 \pm \sqrt{(0.112)^2 - 4(1)(-1.2 \times 10^{-3})}}{2(1)}$$

$$= \frac{-0.112 \pm \sqrt{0.01254 + 0.0048}}{2}$$

$$= \frac{-0.112 \pm 0.132}{2} = +0.010 \text{ or } -0.122$$

In this example, the negative value for x satisfies the equation, but it has no meaning because we cannot have negative concentration values; therefore we use only the $+0.010$ value.

Expressing Data in Graphical Form

Fitting curves to plots of experimental data is a powerful tool in determining the relationships between variables. Many natural phenomena obey exponential functions. For example, the rate constant (k) of a chemical reaction increases exponentially with increasing absolute temperature (T). This relationship is described by the Arrhenius equation:

$$k = A\, e^{-E_a/RT}$$

where A is a constant for a particular reaction (called the frequency factor), E_a is the activation energy of the reaction, and R is the ideal gas constant. Taking the natural logarithms of both sides of the Arrhenius equation gives

$$\ln k = \ln A - \left(\frac{E_a}{RT}\right)$$

This equation fits the general equation of a straight line ($y = mx + b$) if ($\ln k$) is the y-variable and ($1/T$) is the x-variable. Plotting ($\ln k$) versus ($1/T$) should give a straight line with a slope equal to $-E_a/R$. The slopes of these plots are negative because the activation energies, E_a, of chemical reactions are positive. The data for a reaction given in columns 2 and 4 of Table A1.1 are plotted in Figure A1.1. The program that generated the graph also gives us the equation of the straight line that best fits the data. The slope (-1281 K) of this line is used to calculate the value of E_a:

$$-1281 \text{ K} = -\frac{E_a}{R}$$

$$E_a = -(-1281 \text{ K})[8.314 \text{ J/(mol} \cdot \text{K)}]$$

$$= 10{,}650 \text{ J/mol} = 10.65 \text{ kJ/mol}$$

Expressing Precision and Accuracy

The precision in the results of a set of replicate measurements or analyses is determined by calculating the arithmetic mean ($\bar{x}$) of the data set and its standard deviation (s), which is a measure of the variation between the mean and the results of each measurement or analysis (x_i). The equation for calculating s is

$$s = \sqrt{\frac{\sum_i (x_i - \bar{x})^2}{n - 1}}$$

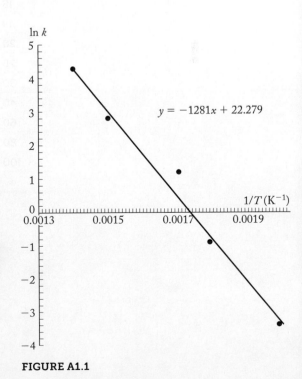

TABLE A1.1 Rate Constant k as a Function of Temperature T

Temperature T (K)	$1/T$ (K^{-1})	Rate Constant k	$\ln k$
500	0.0020	0.030	-3.51
550	0.0018	0.38	-0.97
600	0.0017	2.9	1.06
650	0.0015	17	2.83
700	0.0014	75	4.32

FIGURE A1.1

where n is the number of data points. To calculate the confidence interval, which is the range that is predicted to contain the true mean value (μ) of the data set, we use this equation:

$$\mu = \bar{x} \pm \frac{t\,s}{\sqrt{n}}$$

where t is a statistic that depends on the value of n and the degree of certainty, or *confidence level*, in the prediction. Table A1.2 contains values of t for confidence levels of 90.0, 95.0, 99.0, and 99.9%.

TABLE A1.2 Values of t

	CONFIDENCE LEVEL			
$n - 1$	90.0%	95.0%	99.0%	99.9%
1	6.314	12.71	63.66	636.62
2	2.920	4.303	9.925	31.599
3	2.353	3.182	5.841	12.924
4	2.132	2.776	4.604	8.610
5	2.015	2.571	4.032	6.869
6	1.943	2.447	3.707	5.959
7	1.895	2.365	3.499	5.408
8	1.860	2.306	3.355	5.041
9	1.833	2.262	3.250	4.781
10	1.812	2.228	3.169	4.587
11	1.796	2.201	3.106	4.437
12	1.782	2.179	3.055	4.318
13	1.771	2.160	3.012	4.221
14	1.761	2.145	2.977	4.140
15	1.753	2.131	2.947	4.073
16	1.746	2.120	2.921	4.015
17	1.740	2.110	2.898	3.965
18	1.734	2.101	2.878	3.922
19	1.729	2.093	2.861	3.883
20	1.725	2.086	2.845	3.850
25	1.708	2.060	2.787	3.725
30	1.697	2.042	2.750	3.646
40	1.684	2.021	2.704	3.551
60	1.671	2.000	2.660	3.460
80	1.664	1.990	2.639	3.416
100	1.660	1.984	2.626	3.390
∞	1.645	1.960	2.576	3.291

Appendix 2

SI Units and Conversion Factors

TABLE A2.1 Six SI Base Units

SI Base Quantity	Unit	Symbol
length	meter	m
mass	kilogram	kg
time	second	s
amount of substance	mole	mol
temperature	kelvin	K
electric current	ampere	A

TABLE A2.2 Some SI-Derived Units

SI-Derived Quantity	Unit	Symbol	Dimensions
electric charge	coulomb	C	$A \cdot s$
electric potential	volt	V	J/C
force	newton	N	$kg \cdot m/s^2$
frequency	hertz	Hz	s^{-1}
momentum	newton-second	$N \cdot s$	$kg \cdot m/s$
power	watt	W	J/s
pressure	pascal	Pa	N/m^2
radioactivity	becquerel	Bq	s^{-1}
speed or velocity	meter per second	m/s	m/s
energy	joule (newton-meter)	$J (N \cdot m)$	$kg \cdot m^2/s^2$

TABLE A2.3 SI Prefixes

Prefix	Symbol	Multiplier	Prefix	Symbol	Multiplier
deci	d	10^{-1}	deka	da	10^1
centi	c	10^{-2}	hecto	h	10^2
milli	m	10^{-3}	kilo	k	10^3
micro	μ	10^{-6}	mega	M	10^6
nano	n	10^{-9}	giga	G	10^9
pico	p	10^{-12}	tera	T	10^{12}
femto	f	10^{-15}	peta	P	10^{15}
atto	a	10^{-18}	exa	E	10^{18}
zepto	z	10^{-21}	zetta	Z	10^{21}

TABLE A2.4 Special Units and Conversion Factors

Quantity	Unit	Symbol	Conversion[a]
energy[a]	electron-volt	eV	$1\ \text{eV} = 1.6022 \times 10^{-19}\ \text{J}$
energy[a]	kilowatt-hour	kWh	$1\ \text{kWh} = 3600\ \text{kJ}$
energy	calorie	cal	$1\ \text{cal} = 4.184\ \text{J}$
mass	pound	lb	$1\ \text{lb} = 453.59\ \text{g}$
mass[a]	atomic mass unit	amu	$1\ \text{amu} = 1.66054 \times 10^{-27}\ \text{kg}$
length	angstrom	Å	$1\ \text{Å} = 10^{-8}\ \text{cm} = 10^{-10}\ \text{m}$
length	inch	in	$1\ \text{in} = 2.54\ \text{cm}$
length	mile	mi	$1\ \text{mi} = 5280\ \text{ft} = 1.6093\ \text{km}$
pressure	atmosphere	atm	$1\ \text{atm} = 1.01325 \times 10^5\ \text{Pa}$
pressure	torr	torr	$1\ \text{torr} = 1/760\ \text{atm}$
temperature	Celsius scale	°C	$T(\text{°C}) = T(\text{K}) - 273.15$
temperature	Fahrenheit scale	°F	$T(\text{°F}) = \frac{9}{5}T(\text{°C}) + 32$
volume	liter	L	$1\ \text{L} = 1\ \text{dm}^3 = 10^{-3}\ \text{m}^3$
volume	cubic centimeter	cm³, cc	$1\ \text{cm}^3 = 1\ \text{mL} = 10^{-3}\ \text{L}$
volume	cubic foot	ft³	$1\ \text{ft}^3 = 7.4805\ \text{gal}$
volume	gallon (U.S.)	gal	$1\ \text{gal} = 3.785\ \text{L}$

[a]From http://physics.nist.gov/cuu/constants/.

TABLE A2.5 Physical Constants[a]

Quantity	Symbol	Value
acceleration due to gravity (Earth)	g	$9.807\ \text{m/s}^2$
Avogadro's number	N_A	$6.0221 \times 10^{23}\ \text{mol}^{-1}$
Bohr radius	a_0	$5.29 \times 10^{-11}\ \text{m}$
Boltzmann constant	k_B	$1.3806 \times 10^{-23}\ \text{J/K}$
electron charge-to-mass ratio	$-e/m_e$	$1.7588 \times 10^{11}\ \text{C/kg}$
elementary charge	e	$1.602 \times 10^{-19}\ \text{C}$
Faraday constant	F	$9.65 \times 10^4\ \text{C/mol}$
mass of an electron	m_e	$9.10938 \times 10^{-31}\ \text{kg}$
mass of a neutron	m_n	$1.67493 \times 10^{-27}\ \text{kg}$
mass of a proton	m_p	$1.67262 \times 10^{-27}\ \text{kg}$
molar volume of ideal gas at 0°C and 1 atm	V_m	$22.4\ \text{L/mol}$
Planck constant	h	$6.626 \times 10^{-34}\ \text{J} \cdot \text{s}$
speed of light in vacuum	c	$2.998 \times 10^8\ \text{m/s}$
universal gas constant	R	$8.314\ \text{J/(mol} \cdot \text{K)}$ $0.08206\ \text{L} \cdot \text{atm/(mol} \cdot \text{K)}$

[a]From http://physics.nist.gov/cuu/constants/.

Appendix 3

The Elements and Their Properties

TABLE A3.1 Ground-State Electron Configurations, Atomic Radii, and First Ionization Energies of the Elements

Element	Symbol	Atomic Number Z	Ground-State Configuration	Atomic Radius (pm)	First Ionization Energy (kJ/mol)
hydrogen	H	1	$1s^1$	37	1312.0
helium	He	2	$1s^2$	32	2372.3
lithium	Li	3	$[He]2s^1$	152	520.2
beryllium	Be	4	$[He]2s^2$	112	899.5
boron	B	5	$[He]2s^22p^1$	88	800.6
carbon	C	6	$[He]2s^22p^2$	77	1086.5
nitrogen	N	7	$[He]2s^22p^3$	75	1402.3
oxygen	O	8	$[He]2s^22p^4$	73	1313.9
fluorine	F	9	$[He]2s^22p^5$	71	1681.0
neon	Ne	10	$[He]2s^22p^6$	69	2080.7
sodium	Na	11	$[Ne]3s^1$	186	495.3
magnesium	Mg	12	$[Ne]3s^2$	160	737.7
aluminum	Al	13	$[Ne]3s^23p^1$	143	577.5
silicon	Si	14	$[Ne]3s^23p^2$	117	786.5
phosphorus	P	15	$[Ne]3s^23p^3$	110	1011.8
sulfur	S	16	$[Ne]3s^23p^4$	103	999.6
chlorine	Cl	17	$[Ne]3s^23p^5$	99	1251.2
argon	Ar	18	$[Ne]3s^23p^6$	97	1520.6
potassium	K	19	$[Ar]4s^1$	227	418.8
calcium	Ca	20	$[Ar]4s^2$	197	589.8
scandium	Sc	21	$[Ar]4s^23d^1$	162	633.1
titanium	Ti	22	$[Ar]4s^23d^2$	147	658.8
vanadium	V	23	$[Ar]4s^23d^3$	135	650.9
chromium	Cr	24	$[Ar]4s^13d^5$	128	652.9
manganese	Mn	25	$[Ar]4s^23d^5$	127	717.3
iron	Fe	26	$[Ar]4s^23d^6$	126	762.5
cobalt	Co	27	$[Ar]4s^23d^7$	125	760.4
nickel	Ni	28	$[Ar]4s^23d^8$	124	737.1
copper	Cu	29	$[Ar]4s^13d^{10}$	128	745.5
zinc	Zn	30	$[Ar]4s^23d^{10}$	134	906.4
gallium	Ga	31	$[Ar]4s^23d^{10}4p^1$	135	578.8
germanium	Ge	32	$[Ar]4s^23d^{10}4p^2$	122	762.2
arsenic	As	33	$[Ar]4s^23d^{10}4p^3$	121	947.0
selenium	Se	34	$[Ar]4s^23d^{10}4p^4$	119	941.0

Continued on next page

TABLE A3.1 Ground-State Electron Configurations, Atomic Radii, and First Ionization Energies of the Elements *(Continued)*

Element	Symbol	Atomic Number Z	Ground-State Configuration	Atomic Radius (pm)	First Ionization Energy (kJ/mol)
bromine	Br	35	$[Ar]4s^23d^{10}4p^5$	114	1139.9
krypton	Kr	36	$[Ar]4s^23d^{10}4p^6$	110	1350.8
rubidium	Rb	37	$[Kr]5s^1$	247	403.0
strontium	Sr	38	$[Kr]5s^2$	215	549.5
yttrium	Y	39	$[Kr]5s^24d^1$	180	599.8
zirconium	Zr	40	$[Kr]5s^24d^2$	160	640.1
niobium	Nb	41	$[Kr]5s^14d^4$	146	652.1
molybdenum	Mo	42	$[Kr]5s^14d^5$	139	684.3
technetium	Tc	43	$[Kr]5s^24d^5$	136	702.4
ruthenium	Ru	44	$[Kr]5s^14d^7$	134	710.2
rhodium	Rh	45	$[Kr]5s^14d^8$	134	719.7
palladium	Pd	46	$[Kr]4d^{10}$	137	804.4
silver	Ag	47	$[Kr]5s^14d^{10}$	144	731.0
cadmium	Cd	48	$[Kr]5s^24d^{10}$	151	867.8
indium	In	49	$[Kr]5s^24d^{10}5p^1$	167	558.3
tin	Sn	50	$[Kr]5s^24d^{10}5p^2$	140	708.6
antimony	Sb	51	$[Kr]5s^24d^{10}5p^3$	141	833.6
tellurium	Te	52	$[Kr]5s^24d^{10}5p^4$	143	869.3
iodine	I	53	$[Kr]5s^24d^{10}5p^5$	133	1008.4
xenon	Xe	54	$[Kr]5s^24d^{10}5p^6$	130	1170.4
cesium	Cs	55	$[Xe]6s^1$	265	375.7
barium	Ba	56	$[Xe]6s^2$	222	502.9
lanthanum	La	57	$[Xe]6s^25d^1$	187	538.1
cerium	Ce	58	$[Xe]6s^24f^15d^1$	182	534.4
praseodymium	Pr	59	$[Xe]6s^24f^3$	182	527.2
neodymium	Nd	60	$[Xe]6s^24f^4$	181	533.1
promethium	Pm	61	$[Xe]6s^24f^5$	183	535.5
samarium	Sm	62	$[Xe]6s^24f^6$	180	544.5
europium	Eu	63	$[Xe]6s^24f^7$	208	547.1
gadolinium	Gd	64	$[Xe]6s^24f^75d^1$	180	593.4
terbium	Tb	65	$[Xe]6s^24f^9$	177	565.8
dysprosium	Dy	66	$[Xe]6s^24f^{10}$	178	573.0
holmium	Ho	67	$[Xe]6s^24f^{11}$	176	581.0
erbium	Er	68	$[Xe]6s^24f^{12}$	176	589.3
thulium	Tm	69	$[Xe]6s^24f^{13}$	176	596.7
ytterbium	Yb	70	$[Xe]6s^24f^{14}$	193	603.4
lutetium	Lu	71	$[Xe]6s^24f^{14}5d^1$	174	523.5
hafnium	Hf	72	$[Xe]6s^24f^{14}5d^2$	159	658.5
tantalum	Ta	73	$[Xe]6s^24f^{14}5d^3$	146	761.3
tungsten	W	74	$[Xe]6s^24f^{14}5d^4$	139	770.0
rhenium	Re	75	$[Xe]6s^24f^{14}5d^5$	137	760.3
osmium	Os	76	$[Xe]6s^24f^{14}5d^6$	135	839.4
iridium	Ir	77	$[Xe]6s^24f^{14}5d^7$	136	878.0

TABLE A3.1 Ground-State Electron Configurations, Atomic Radii, and First Ionization Energies of the Elements *(Continued)*

Element	Symbol	Atomic Number Z	Ground-State Configuration	Atomic Radius (pm)	First Ionization Energy (kJ/mol)
platinum	Pt	78	$[Xe]6s^14f^{14}5d^9$	139	868.4
gold	Au	79	$[Xe]6s^14f^{14}5d^{10}$	144	890.1
mercury	Hg	80	$[Xe]6s^24f^{14}5d^{10}$	151	1007.1
thallium	Tl	81	$[Xe]6s^24f^{14}5d^{10}6p^1$	170	589.4
lead	Pb	82	$[Xe]6s^24f^{14}5d^{10}6p^2$	154	715.6
bismuth	Bi	83	$[Xe]6s^24f^{14}5d^{10}6p^3$	150	703.3
polonium	Po	84	$[Xe]6s^24f^{14}5d^{10}6p^4$	167	812.1
astatine	At	85	$[Xe]6s^24f^{14}5d^{10}6p^5$	140	924.6
radon	Rn	86	$[Xe]6s^24f^{14}5d^{10}6p^6$	145	1037.1
francium	Fr	87	$[Rn]7s^1$	242	380
radium	Ra	88	$[Rn]7s^2$	211	509.3
actinium	Ac	89	$[Rn]7s^26d^1$	188	499
thorium	Th	90	$[Rn]7s^26d^2$	179	587
protactinium	Pa	91	$[Rn]7s^25f^26d^1$	163	568
uranium	U	92	$[Rn]7s^25f^36d^1$	156	587
neptunium	Np	93	$[Rn]7s^25f^46d^1$	155	597
plutonium	Pu	94	$[Rn]7s^25f^6$	159	585
americium	Am	95	$[Rn]7s^25f^7$	173	578
curium	Cm	96	$[Rn]7s^25f^76d^1$	174	581
berkelium	Bk	97	$[Rn]7s^25f^9$	170	601
californium	Cf	98	$[Rn]7s^25f^{10}$	186	608
einsteinium	Es	99	$[Rn]7s^25f^{11}$	186	619
fermium	Fm	100	$[Rn]7s^25f^{12}$	167	627
mendelevium	Md	101	$[Rn]7s^25f^{13}$	173	635
nobelium	No	102	$[Rn]7s^25f^{14}$	176	642
lawrencium	Lr	103	$[Rn]7s^25f^{14}6d^1$	161	—
rutherfordium	Rf	104	$[Rn]7s^25f^{14}6d^2$	157	—
dubnium	Db	105	$[Rn]7s^25f^{14}6d^3$	149	—
seaborgium	Sg	106	$[Rn]7s^25f^{14}6d^4$	143	—
bohrium	Bh	107	$[Rn]7s^25f^{14}6d^5$	141	—
hassium	Hs	108	$[Rn]7s^25f^{14}6d^6$	134	—
meitnerium	Mt	109	$[Rn]7s^25f^{14}6d^7$	129	—
darmstadtium	Ds	110	$[Rn]7s^25f^{14}6d^8$	128	—
roentgenium	Rg	111	$[Rn]7s^25f^{14}6d^9$	121	—
copernicium	Cn	112	$[Rn]7s^25f^{14}6d^{10}$	122	—
nihonium[a]	Nh	113	$[Rn]7s^25f^{14}6d^{10}7p^1$	136	—
flerovium	Fl	114	$[Rn]7s^25f^{14}6d^{10}7p^2$	143	—
moscovium[a]	Mc	115	$[Rn]7s^25f^{14}6d^{10}7p^3$	162	—
livermorium	Lv	116	$[Rn]7s^25f^{14}6d^{10}7p^4$	175	—
tennessine[a]	Ts	117	$[Rn]7s^25f^{14}6d^{10}7p^5$	165	—
oganesson[a]	Og	118	$[Rn]7s^25f^{14}6d^{10}7p^6$	157	—

[a]Names and symbols for these were recommended by the IUPAC in June 2016.

TABLE A3.2 Miscellaneous Physical Properties of the Naturally Occurring Elements[a]

Element	Symbol	Atomic Number	Physical State[b,c]	Density[d] (g/cm^3)	Melting Point (°C)	Boiling Point (°C)
hydrogen	H	1	gas	0.000090	−259.14	−252.87
helium	He	2	gas	0.000179	<−272.2	−268.93
lithium	Li	3	solid	0.534	180.5	1347
beryllium	Be	4	solid	1.848	1283	2484
boron	B	5	solid	2.34	2300	3650
carbon	C	6	solid (gr)	1.9–2.3	~3350	sublimes
nitrogen	N	7	gas	0.00125	−210.00	−195.8
oxygen	O	8	gas	0.00143	−218.8	−182.95
fluorine	F	9	gas	0.00170	−219.62	−188.12
neon	Ne	10	gas	0.00090	−248.59	−246.08
sodium	Na	11	solid	0.971	97.72	883
magnesium	Mg	12	solid	1.738	650	1090
aluminum	Al	13	solid	2.6989	660.32	2467
silicon	Si	14	solid	2.33	1414	2355
phosphorus	P	15	solid (wh)	1.82	44.15	280
sulfur	S	16	solid	2.07	115.21	444.60
chlorine	Cl	17	gas	0.00321	−101.5	−34.04
argon	Ar	18	gas	0.00178	−189.3	−185.9
potassium	K	19	solid	0.862	63.28	759
calcium	Ca	20	solid	1.55	842	1484
scandium	Sc	21	solid	2.989	1541	2380
titanium	Ti	22	solid	4.54	1668	3287
vanadium	V	23	solid	6.11	1910	3407
chromium	Cr	24	solid	7.19	1857	2671
manganese	Mn	25	solid	7.3	1246	1962
iron	Fe	26	solid	7.874	1538	2750
cobalt	Co	27	solid	8.9	1495	2870
nickel	Ni	28	solid	8.902	1455	2730
copper	Cu	29	solid	8.96	1084.6	2562
zinc	Zn	30	solid	7.133	419.53	907
gallium	Ga	31	solid	5.904	29.76	2403
germanium	Ge	32	solid	5.323	938.25	2833
arsenic	As	33	solid (gy)	5.727	614	sublimes
selenium	Se	34	solid (gy)	4.79	221	685
bromine	Br	35	liquid	3.12	−7.2	58.78
krypton	Kr	36	gas	0.00373	−157.36	−153.22
rubidium	Rb	37	solid	1.532	39.31	688
strontium	Sr	38	solid	2.64	777	1382
yttrium	Y	39	solid	4.469	1526	3336
zirconium	Zr	40	solid	6.506	1855	4409
niobium	Nb	41	solid	8.57	2477	4744
molybdenum	Mo	42	solid	10.22	2623	4639

TABLE A3.2   Miscellaneous Physical Properties of
the Naturally Occurring Elements[a] *(Continued)*

Element	Symbol	Atomic Number	Physical State[b,c]	Density[d] (g/cm³)	Melting Point (°C)	Boiling Point (°C)
technetium	Tc	43	solid	11.50	2157	4538
ruthenium	Ru	44	solid	12.41	2334	3900
rhodium	Rh	45	solid	12.41	1964	3695
palladium	Pd	46	solid	12.02	1555	2963
silver	Ag	47	solid	10.50	961.78	2212
cadmium	Cd	48	solid	8.65	321.07	767
indium	In	49	solid	7.31	156.60	2072
tin	Sn	50	solid (wh)	7.31	231.9	2270
antimony	Sb	51	solid	6.691	630.63	1750
tellurium	Te	52	solid	6.24	449.5	998
iodine	I	53	solid	4.93	113.7	184.4
xenon	Xe	54	gas	0.00589	−111.75	−108.0
cesium	Cs	55	solid	1.873	28.44	671
barium	Ba	56	solid	3.5	727	1640
lanthanum	La	57	solid	6.145	920	3455
cerium	Ce	58	solid	6.770	799	3424
praseodymium	Pr	59	solid	6.773	931	3510
neodymium	Nd	60	solid	7.008	1016	3066
promethium	Pm	61	solid	7.264	1042	~3000
samarium	Sm	62	solid	7.520	1072	1790
europium	Eu	63	solid	5.244	822	1596
gadolinium	Gd	64	solid	7.901	1314	3264
terbium	Tb	65	solid	8.230	1359	3221
dysprosium	Dy	66	solid	8.551	1411	2561
holmium	Ho	67	solid	8.795	1472	2694
erbium	Er	68	solid	9.066	1529	2862
thulium	Tm	69	solid	9.321	1545	1946
ytterbium	Yb	70	solid	6.966	824	1194
lutetium	Lu	71	solid	9.841	1663	3393
hafnium	Hf	72	solid	13.31	2233	4603
tantalum	Ta	73	solid	16.654	3017	5458
tungsten	W	74	solid	19.3	3422	5660
rhenium	Re	75	solid	21.02	3186	5596
osmium	Os	76	solid	22.57	3033	5012
iridium	Ir	77	solid	22.42	2446	4130
platinum	Pt	78	solid	21.45	1768.4	3825
gold	Au	79	solid	19.3	1064.18	2856
mercury	Hg	80	liquid	13.546	−38.83	356.73
thallium	Tl	81	solid	11.85	304	1473
lead	Pb	82	solid	11.35	327.46	1749
bismuth	Bi	83	solid	9.747	271.4	1564

Continued on next page

TABLE A3.2   Miscellaneous Physical Properties of
the Naturally Occurring Elements[a] *(Continued)*

Element	Symbol	Atomic Number	Physical State[b,c]	Density[d] (g/cm^3)	Melting Point (°C)	Boiling Point (°C)
polonium	Po	84	solid	9.32	254	962
astatine	At	85	solid	unknown	302	337
radon	Rn	86	gas	0.00973	−71	−61.7
francium	Fr	87	solid	unknown	27	677
radium	Ra	88	solid	5	700	1737
actinium	Ac	89	solid	10.07	1051	~3200
thorium	Th	90	solid	11.72	1750	4788
protactinium	Pa	91	solid	15.37	1572	unknown
uranium	U	92	solid	19.05	1132	3818

[a]For relative atomic masses and alphabetical listing of the elements, see the flyleaf at the front of this volume.
[b]Normal state at 25°C and 1 atm.
[c]Allotropes: gr = graphite, gy = gray, wh = white.
[d]Liquids and solids at 25°C and 1 atm; gases at 0°C and 1 atm (STP).

TABLE A3.3   A Selection of Stable Isotopes[a]

Isotope ^AX	Natural Abundance (%)	Atomic Number Z	Neutron Number N	Mass Number A	Atomic Mass (amu)	Binding Energy per Nucleon (MeV)[b]
^1H	99.985	1	0	1	1.007825	—
^2H	0.015	1	1	2	2.014000	1.160
^3He	0.000137	2	1	3	3.016030	2.572
^4He	99.999863	2	2	4	4.002603	7.075
^6Li	7.5	3	3	6	6.015121	5.333
^7Li	92.5	3	4	7	7.016003	5.606
^9Be	100.0	4	5	9	9.012182	6.463
^{10}B	19.9	5	5	10	10.012937	6.475
^{11}B	80.1	5	6	11	11.009305	6.928
^{12}C	98.90	6	6	12	12.000000	7.680
^{13}C	1.10	6	7	13	13.003355	7.470
^{14}N	99.634	7	7	14	14.003074	7.476
^{15}N	0.366	7	8	15	15.000108	7.699
^{16}O	99.762	8	8	16	15.994915	7.976
^{17}O	0.038	8	9	17	16.999131	7.751
^{18}O	0.200	8	10	18	17.999160	7.767
^{19}F	100.0	9	10	19	18.998403	7.779
^{20}Ne	90.48	10	10	20	19.992435	8.032
^{21}Ne	0.27	10	11	21	20.993843	7.972
^{22}Ne	9.25	10	12	22	21.991383	8.081
^{23}Na	100.0	11	12	23	22.989770	8.112
^{24}Mg	78.99	12	12	24	23.985042	8.261
^{25}Mg	10.00	12	13	25	24.985837	8.223
^{26}Mg	11.01	12	14	26	25.982593	8.334

TABLE A3.3   A Selection of Stable Isotopes[a] *(Continued)*

Isotope [A]X	Natural Abundance (%)	Atomic Number Z	Neutron Number N	Mass Number A	Atomic Mass (amu)	Binding Energy per Nucleon (MeV)[b]
[27]Al	100.0	13	14	27	26.981538	8.331
[28]Si	92.23	14	14	28	27.976927	8.448
[29]Si	4.67	14	15	29	28.976495	8.449
[30]Si	3.10	14	16	30	29.973770	8.521
[31]P	100.0	15	16	31	30.973761	8.481
[32]S	95.02	16	16	32	31.972070	8.493
[33]S	0.75	16	17	33	32.971456	8.498
[34]S	4.21	16	18	34	33.967866	8.584
[36]S	0.02	16	20	36	35.967080	8.575
[35]Cl	75.77	17	18	35	34.968852	8.520
[37]Cl	24.23	17	20	37	36.965903	8.570
[36]Ar	0.337	18	18	36	35.967545	8.520
[38]Ar	0.063	18	20	38	37.962732	8.614
[40]Ar	99.600	18	22	40	39.962384	8.595
[39]K	93.258	19	20	39	38.963707	8.557
[41]K	6.730	19	22	41	40.961825	8.576
[40]Ca	96.941	20	20	40	39.962591	8.551
[42]Ca	0.647	20	22	42	41.958618	8.617
[43]Ca	0.135	20	23	43	42.958766	8.601
[44]Ca	2.086	20	24	44	43.955480	8.658
[46]Ca	0.004	20	26	46	45.953689	8.669
[48]Ca	0.187	20	28	48	47.952533	8.666
[45]Sc	100.0	21	24	45	44.955910	8.619
[46]Ti	8.0	22	24	46	45.952629	8.656
[47]Ti	7.3	22	25	47	46.951764	8.661
[48]Ti	73.8	22	26	48	47.947947	8.723
[49]Ti	5.5	22	27	49	48.947871	8.711
[50]Ti	5.4	22	28	50	49.944792	8.756
[51]V	99.750	23	28	51	50.943962	8.742
[50]Cr	4.345	24	26	50	49.946046	8.701
[52]Cr	83.789	24	28	52	51.940509	8.776
[53]Cr	9.501	24	29	53	52.940651	8.760
[54]Cr	2.365	24	30	54	53.938882	8.778
[55]Mn	100.0	25	30	55	54.938049	8.765
[54]Fe	5.9	26	28	54	53.939612	8.736
[56]Fe	91.72	26	30	56	55.934939	8.790
[57]Fe	2.1	26	31	57	56.935396	8.770
[58]Fe	0.28	26	32	58	57.933277	8.792
[59]Co	100.0	27	32	59	58.933200	8.768
[204]Pb	1.4	82	122	204	203.973020	7.880

Continued on next page

TABLE A3.3   A Selection of Stable Isotopes^a (Continued)

Isotope ^{A}X	Natural Abundance (%)	Atomic Number Z	Neutron Number N	Mass Number A	Atomic Mass (amu)	Binding Energy per Nucleon (MeV)^b
^{206}Pb	24.1	82	124	206	205.974440	7.875
^{207}Pb	22.1	82	125	207	206.975872	7.870
^{208}Pb	52.4	82	126	208	207.976627	7.868
^{209}Bi	100.0	83	126	209	208.980380	7.848

^aSelection is complete through cobalt-59. Where natural abundances do not add to 100%, the differences are made up by radioactive isotopes with exceedingly long half-lives: potassium-40 (0.0117%, $t_{1/2} = 1.3 \times 10^9$ yr); vanadium-50 (0.250%, $t_{1/2} > 1.4 \times 10^{17}$ yr).
^b1 MeV (mega electron-volt) = 1.6022×10^{-13} J.

TABLE A3.4   A Selection of Radioactive Isotopes

Isotope ^{A}X	Decay Mode^a	Half-Life $t_{1/2}$	Atomic Number Z	Neutron Number N	Mass Number A	Atomic Mass (amu)	Binding Energy per Nucleon (MeV)^b
^{3}H	β^-	12.3 yr	1	2	3	3.01605	2.827
^{8}Be	α	$\sim 7 \times 10^{-17}$ s	4	4	8	8.005305	7.062
^{14}C	β^-	5.7×10^3 yr	6	8	14	14.003241	7.520
^{22}Na	β^+	2.6 yr	11	11	22	21.994434	7.916
^{24}Na	β^-	15.0 hr	11	13	24	23.990961	8.064
^{32}P	β^-	14.3 d	15	17	32	31.973907	8.464
^{35}S	β^-	87.2 d	16	19	35	34.969031	8.538
^{59}Fe	β^-	44.5 d	26	33	59	58.934877	8.755
^{60}Co	β^-	5.3 yr	27	33	60	59.933819	8.747
^{90}Sr	β^-	29.1 yr	38	52	90	89.907738	8.696
^{99}Tc	β^-	2.1×10^5 yr	43	56	99	98.906524	8.611
^{109}Cd	EC	462 d	48	61	109	108.904953	8.539
^{125}I	EC	59.4 d	53	72	125	124.904620	8.450
^{131}I	β^-	8.04 d	53	78	131	130.906114	8.422
^{137}Cs	β^-	30.3 yr	55	82	137	136.907073	8.389
^{222}Rn	α	3.82 d	86	136	222	222.017570	7.695
^{226}Ra	α	1600 yr	88	138	226	226.025402	7.662
^{232}Th	α	1.4×10^{10} yr	90	142	232	232.038054	7.615
^{235}U	α	7.0×10^8 yr	92	143	235	235.043924	7.591
^{238}U	α	4.5×10^9 yr	92	146	238	238.050784	7.570
^{239}Pu	α	2.4×10^4 yr	94	145	239	239.052157	7.560

^aModes of decay include alpha emission (α), beta emission (β^-), positron emission (β^+), and electron capture (EC).
^b1 MeV (mega electron-volt) = 1.6022×10^{-13} J.

Appendix 4

Chemical Bonds and Thermodynamic Data

TABLE A4.1 Average Lengths and Energies of Covalent Bonds

Atom	Bond	Bond Length (pm)	Bond Energy (kJ/mol)
H	H—H	75	436
	H—F	92	567
	H—Cl	127	431
	H—Br	141	366
	H—I	161	299
C	C—C	154	348
	C=C	134	614
	C≡C	120	839
	C—H	110	413
	C—N	147	293
	C=N	127	615
	C≡N	116	891
	C—O	143	358
	C=O	123	743^a
	C≡O	113	1072
	C—F	133	485
	C—Cl	177	328
	C—Br	179	276
	C—I	215	238
N	N—N	147	163
	N=N	124	418
	N≡N	110	945
	N—H	104	391
	N—O	136	201
	N=O	122	607
	N≡O	106	678
O	O—O	148	146
	O=O	121	498
	O—H	96	463
S	S—O	151	265
	S=O	143	523
	S—S	204	266
	S—H	134	347
F	F—F	143	155
Cl	Cl—Cl	200	243
Br	Br—Br	228	193
I	I—I	266	151

^{a}The bond energy of C=O in CO_2 is 799 kJ/mol.

TABLE A4.2 Critical Temperatures (T_c) and van der Waals Parameters (a, b) of Real Gases

Gas[a]	Molar Mass (g/mol)	T_c (K)	a ($L^2 \cdot atm/mol^2$)	b (L/mol)
H_2O	18.015	647.14	5.46	0.0305
Br_2	159.808	588	9.75	0.0591
CCl_3F	137.367	471.2	14.68	0.1111
Cl_2	70.906	416.9	6.343	0.0542
CO_2	44.010	304.14	3.59	0.0427
Kr	83.798	209.41	2.325	0.0396
CH_4	16.043	190.53	2.25	0.0428
O_2	31.999	154.59	1.36	0.0318
Ar	39.948	150.87	1.34	0.0322
F_2	37.997	144.13	1.171	0.0290
CO	28.010	132.91	1.45	0.0395
N_2	28.013	126.21	1.39	0.0391
H_2	2.016	32.97	0.244	0.0266
He	4.003	5.19	0.0341	0.0237

[a]Listed in descending order of critical temperature.

TABLE A4.3 Thermodynamic Properties at 25°C

Substance[a,b]	Molar Mass (g/mol)	ΔH_f° (kJ/mol)	S° [J/(mol · K)]	ΔG_f° (kJ/mol)
ELEMENTS AND MONATOMIC IONS				
$Ag^+(aq)$	107.87	105.6	72.7	77.1
$Ag(g)$	107.87	284.9	173.0	246.0
$Ag(s)$	107.87	0.0	42.6	0.0
$Al^{3+}(aq)$	26.982	−531	−321.7	−485
$Al(g)$	26.982	330.0	164.6	289.4
$Al(s)$	26.982	0.0	28.3	0.0
$Al(\ell)$	26.982	10.6	39.6	−1.2
$Ar(g)$	39.948	0.0	154.8	0.0
$Au(g)$	196.97	366.1	180.5	326.3
$Au(s)$	196.97	0.0	47.4	0.0
$B(g)$	10.811	565.0	153.4	521.0
$B(s)$	10.811	0.0	5.9	0.0
$Ba^{2+}(aq)$	137.33	−537.6	9.6	−560.8
$Ba(g)$	137.33	180.0	170.2	146.0
$Ba(s)$	137.33	0.0	62.8	0.0
$Be(g)$	9.0122	324.0	136.3	286.6
$Be(s)$	9.0122	0.0	9.5	0.0
$Br^-(aq)$	79.904	−121.6	82.4	−104.0
$Br(g)$	79.904	111.9	175.0	82.4
$Br_2(g)$	159.808	30.9	245.5	3.1
$Br_2(\ell)$	159.808	0.0	152.2	0.0
$C(g)$	12.011	716.7	158.1	671.3
$C(s, diamond)$	12.011	1.9	2.4	2.9

TABLE A4.3 Thermodynamic Properties at 25°C *(Continued)*

Substance^{a,b}	Molar Mass (g/mol)	ΔH_f° (kJ/mol)	S° [J/(mol · K)]	ΔG_f° (kJ/mol)
C(s, graphite)	12.011	0.0	5.7	0.0
Ca^{2+}(aq)	40.078	−542.8	−55.3	−553.6
Ca(g)	40.078	177.8	154.9	144.0
Ca(s)	40.078	0.0	41.6	0.0
Cl^-(aq)	35.453	−167.2	56.5	−131.2
Cl(g)	35.453	121.3	165.2	105.3
Cl_2(g)	70.906	0.0	223.0	0.0
Co^{2+}(aq)	58.933	−58.2	−113	−54.4
Co^{3+}(aq)	58.933	92	−305	134
Co(g)	58.933	424.7	179.5	380.3
Co(s)	58.933	0.0	30.0	0.0
Cr(g)	51.996	396.6	174.5	351.8
Cr(s)	51.996	0.0	23.8	0.0
Cs^+(aq)	132.91	−258.3	133.1	−292.0
Cs(g)	132.91	76.5	175.6	49.6
Cs(s)	132.91	0.0	85.2	0.0
Cu^+(aq)	63.546	71.7	40.6	50.0
Cu^{2+}(aq)	63.546	64.8	−99.6	65.5
Cu(g)	63.546	337.4	166.4	297.7
Cu(s)	63.546	0.0	33.2	0.0
F^-(aq)	18.998	−332.6	−13.8	−278.8
F(g)	18.998	79.4	158.8	62.3
F_2(g)	37.997	0.0	202.8	0.0
Fe^{2+}(aq)	55.845	−89.1	−137.7	−78.9
Fe^{3+}(aq)	55.845	−48.5	−315.9	−4.7
Fe(g)	55.845	416.3	180.5	370.7
Fe(s)	55.845	0.0	27.3	0.0
H^+(aq)	1.0079	0.0	0.0	0.0
H(g)	1.0079	218.0	114.7	203.3
H_2(g)	2.0158	0.0	130.6	0.0
He(g)	4.0026	0.0	126.2	0.0
Hg_2^{2+}(aq)	401.18	172.4	84.5	153.5
Hg^{2+}(aq)	200.59	171.1	−32.2	164.4
Hg(g)	200.59	61.4	175.0	31.8
Hg(ℓ)	200.59	0.0	75.9	0.0
I^-(aq)	126.90	−55.2	111.3	−51.6
I(g)	126.90	106.8	180.8	70.2
I_2(g)	253.81	62.4	260.7	19.3
I_2(s)	253.81	0.0	116.1	0.0
K^+(aq)	39.098	−252.4	102.5	−283.3
K(g)	39.098	89.0	160.3	60.5
K(s)	39.098	0.0	64.7	0.0
Li^+(aq)	6.941	−278.5	13.4	−293.3

Continued on next page

TABLE A4.3   Thermodynamic Properties at 25°C *(Continued)*

Substance[a,b]	Molar Mass (g/mol)	ΔH_f° (kJ/mol)	S° [J/(mol · K)]	ΔG_f° (kJ/mol)
Li(g)	6.941	159.3	138.8	126.6
Li$^+$(g)	6.941	685.7	133.0	648.5
Li(s)	6.941	0.0	29.1	0.0
Mg^{2+}(aq)	24.305	−466.9	−138.1	−454.8
Mg(g)	24.305	147.1	148.6	112.5
Mg(s)	24.305	0.0	32.7	0.0
Mn^{2+}(aq)	54.938	−220.8	−73.6	−228.1
Mn(g)	54.938	280.7	173.7	238.5
Mn(s)	54.938	0.0	32.0	0.0
N(g)	14.007	472.7	153.3	455.5
N$_2$(g)	28.013	0.0	191.5	0.0
Na$^+$(aq)	22.990	−240.1	59.0	−261.9
Na(g)	22.990	107.5	153.7	77.0
Na$^+$(g)	22.990	609.3	148.0	574.3
Na(s)	22.990	0.0	51.3	0.0
Ne(g)	20.180	0.0	146.3	0.0
Ni^{2+}(aq)	58.693	−54.0	−128.9	−45.6
Ni(g)	58.693	429.7	182.2	384.5
Ni(s)	58.693	0.0	29.9	0.0
O(g)	15.999	249.2	161.1	231.7
O$_2$(g)	31.999	0.0	205.0	0.0
O$_3$(g)	47.998	142.7	238.8	163.2
P(g)	30.974	314.6	163.1	278.3
P$_4$(s, red)	123.895	−17.6	22.8	−12.1
P$_4$(s, white)	123.895	0.0	41.1	0.0
Pb^{2+}(aq)	207.2	−1.7	10.5	−24.4
Pb(g)	207.2	195.2	162.2	175.4
Pb(s)	207.2	0.0	64.8	0.0
Rb$^+$(aq)	85.468	−251.2	121.5	−284.0
Rb(g)	85.468	80.9	170.1	53.1
Rb(s)	85.468	0.0	76.8	0.0
S(g)	32.065	277.2	167.8	236.7
S$_8$(g)	256.520	102.3	430.2	49.1
S$_8$(s)	256.520	0.0	32.1	0.0
Sc(g)	44.956	377.8	174.8	336.0
Sc(s)	44.956	0.0	34.6	0.0
Si(g)	28.086	450.0	168.0	405.5
Si(s)	28.086	0.0	18.8	0.0
Sn(g)	118.71	301.2	168.5	266.2
Sn(s, gray)	118.71	−2.1	44.1	0.1
Sn(s, white)	118.71	0.0	51.2	0.0
Sr^{2+}(aq)	87.62	−545.8	−32.6	−559.5
Sr(g)	87.62	164.4	164.6	130.9
Sr(s)	87.62	0.0	52.3	0.0

TABLE A4.3   Thermodynamic Properties at 25°C *(Continued)*

Substance[a,b]	Molar Mass (g/mol)	ΔH_f° (kJ/mol)	S° [J/(mol · K)]	ΔG_f° (kJ/mol)
Ti(g)	47.867	473.0	180.3	428.4
Ti(s)	47.867	0.0	30.7	0.0
V(g)	50.942	514.2	182.2	468.5
V(s)	50.942	0.0	28.9	0.0
W(s)	183.84	0.0	32.6	0.0
Zn^{2+}(aq)	65.38	−153.9	−112.1	−147.1
Zn(g)	65.38	130.4	161.0	94.8
Zn(s)	65.38	0.0	41.6	0.0
POLYATOMIC IONS				
CH_3COO^-(aq)	59.045	−486.0	86.6	−369.3
CO_3^{2-}(aq)	60.009	−677.1	−56.9	−527.8
$C_2O_4^{2-}$(aq)	88.020	−825.1	45.6	−673.9
CrO_4^{2-}(aq)	115.994	−881.2	50.2	−727.8
$Cr_2O_7^{2-}$(aq)	215.988	−1490.3	261.9	−1301.1
$HCOO^-$(aq)	45.018	−425.6	92	−351.0
HCO_3^-(aq)	61.017	−692.0	91.2	−586.8
HSO_4^-(aq)	97.072	−887.3	131.8	−755.9
MnO_4^-(aq)	118.936	−541.4	191.2	−447.2
NH_4^+(aq)	18.038	−132.5	113.4	−79.3
NO_3^-(aq)	62.005	−205.0	146.4	−108.7
OH^-(aq)	17.007	−230.0	−10.8	−157.2
PO_4^{3-}(aq)	94.971	−1277.4	−222	−1018.7
SO_4^{2-}(aq)	96.064	−909.3	20.1	−744.5
INORGANIC COMPOUNDS				
AgCl(s)	143.32	−127.1	96.2	−109.8
AgI(s)	234.77	−61.8	115.5	−66.2
$AgNO_3$(s)	169.87	−124.4	140.9	−33.4
Al_2O_3(s)	101.961	−1675.7	50.9	−1582.3
B_2H_6(g)	27.669	35.0	232.0	86.6
B_2O_3(s)	69.622	−1263.6	54.0	−1184.1
$BaCO_3$(s)	197.34	−1216.3	112.1	−1137.6
$BaSO_4$(s)	233.39	−1473.2	132.2	−1362.2
$CaCO_3$(s)	100.087	−1206.9	92.9	−1128.8
$CaCl_2$(s)	110.984	−795.4	108.4	−748.8
CaF_2(s)	78.075	−1228.0	68.5	−1175.6
CaO(s)	56.077	−634.9	38.1	−603.3
$Ca(OH)_2$(s)	74.093	−985.2	83.4	−897.5
$CaSO_4$(s)	136.142	−1434.5	106.5	−1322.0
CO(g)	28.010	−110.5	197.7	−137.2
CO_2(g)	44.010	−393.5	213.8	−394.4
CO_2(aq)	44.010	−412.9	121.3	−386.2
CS_2(g)	76.143	115.3	237.8	65.1
CS_2(ℓ)	76.143	87.9	151.0	63.6

Continued on next page

TABLE A4.3 Thermodynamic Properties at 25°C *(Continued)*

Substance[a,b]	Molar Mass (g/mol)	ΔH_f° (kJ/mol)	S° [J/(mol · K)]	ΔG_f° (kJ/mol)
$CsCl(s)$	168.358	−443.0	101.2	−414.6
$CuSO_4(s)$	159.610	−771.4	109.2	−662.2
$Cu_2S(s)$	159.16	−79.5	120.9	−86.2
$FeCl_2(s)$	126.750	−341.8	118.0	−302.3
$FeCl_3(s)$	162.203	−399.5	142.3	−334.0
$FeO(s)$	71.844	−271.9	60.8	−255.2
$Fe_2O_3(s)$	159.688	−824.2	87.4	−742.2
$HBr(g)$	80.912	−36.3	198.7	−53.4
$HCl(g)$	36.461	−92.3	186.9	−95.3
$HCN(g)$	27.02	135.1	201.81	124.7
$HF(g)$	20.006	−273.3	173.8	−275.4
$HI(g)$	127.912	26.5	206.6	1.7
$HNO_2(g)$	47.014	−79.5	254.1	−46.0
$HNO_3(g)$	63.013	−135.1	266.4	−74.7
$HNO_3(\ell)$	63.013	−174.1	155.6	−80.7
$HNO_3(aq)$	63.013	−206.6	146.0	−110.5
$HgCl_2(s)$	271.50	−224.3	146.0	−178.6
$Hg_2Cl_2(s)$	472.09	−265.4	191.6	−210.7
$H_2O(g)$	18.015	−241.8	188.8	−228.6
$H_2O(\ell)$	18.015	−285.8	69.9	−237.2
$H_2S(g)$	34.082	−20.17	205.6	−33.01
$H_2O_2(g)$	34.015	−136.3	232.7	−105.6
$H_2O_2(\ell)$	34.015	−187.8	109.6	−120.4
$H_2SO_4(\ell)$	98.079	−814.0	156.9	−690.0
$H_2SO_4(aq)$	98.079	−909.2	20.1	−744.5
$KBr(s)$	119.002	−393.8	95.9	−380.7
$KCl(s)$	74.551	−436.5	82.6	−408.5
$KHCO_3(s)$	100.115	−963.2	115.5	−863.6
$K_2CO_3(s)$	138.205	−1151.0	155.5	−1063.5
$LiBr(s)$	86.845	−351.2	74.3	−342.0
$LiCl(s)$	42.394	−408.6	59.3	−384.4
$Li_2CO_3(s)$	73.891	−1215.9	90.4	−1132.1
$MgCl_2(s)$	95.211	−641.3	89.6	−591.8
$Mg(OH)_2(s)$	58.320	−924.5	63.2	−833.5
$MgSO_4(s)$	120.369	−1284.9	91.6	−1170.6
$MnO_2(s)$	86.937	−520.0	53.1	−465.1
$CH_3COONa(s)$	82.034	−708.8	123.0	−607.2
$NaBr(s)$	102.894	−361.1	86.82	−349.0
$NaCl(s)$	58.443	−411.2	72.1	−384.2
$NaCl(g)$	58.443	−181.4	229.8	−201.3
$Na_2CO_3(s)$	105.989	−1130.7	135.0	−1044.4
$NaHCO_3(s)$	84.007	−950.8	101.7	−851.0
$NaNO_3(s)$	84.995	−467.9	116.5	−367.0
$NaOH(s)$	39.997	−425.6	64.5	−379.5

TABLE A4.3 Thermodynamic Properties at 25°C *(Continued)*

Substance^{a,b}	Molar Mass (g/mol)	ΔH_f° (kJ/mol)	S° [J/(mol · K)]	ΔG_f° (kJ/mol)
$Na_2SO_4(s)$	142.043	−1387.1	149.6	−1270.2
$NF_3(g)$	71.002	−132.1	260.8	−90.6
$NH_3(aq)$	17.031	−80.3	111.3	−26.50
$NH_3(g)$	17.031	−46.1	192.5	−16.5
$NH_4Cl(s)$	53.491	−314.4	94.6	−203.0
$NH_4NO_3(s)$	80.043	−365.6	151.1	−183.9
$N_2H_4(g)$	32.045	95.35	238.5	159.4
$N_2H_4(\ell)$	32.045	50.63	121.52	149.3
$NiCl_2(s)$	129.60	−305.3	97.7	−259.0
$NiO(s)$	74.60	−239.7	38.0	−211.7
$NO(g)$	30.006	90.3	210.7	86.6
$NO_2(g)$	46.006	33.2	240.0	51.3
$N_2O(g)$	44.013	82.1	219.9	104.2
$N_2O_3(g)$	76.01	86.6	314.7	142.4
$N_2O_4(g)$	92.011	9.2	304.2	97.8
$NOCl(g)$	65.459	51.7	261.7	66.1
$PCl_3(g)$	137.33	−288.07	311.7	−269.6
$PCl_3(\ell)$	137.33	−319.6	217	−272.4
$PF_5(g)$	125.96	−1594.4	300.8	−1520.7
$PH_3(g)$	33.998	5.4	210.2	13.4
$PbCl_2(s)$	278.1	−359.4	136.0	−314.1
$PbSO_4(s)$	303.3	−920.0	148.5	−813.0
$SO_2(g)$	64.065	−296.8	248.2	−300,1
$SO_3(g)$	80.064	−395.7	256.8	−371.1
$ZnCl_2(s)$	136.30	−415.1	111.5	−369.4
$ZnO(s)$	81.37	−348.0	43.9	−318.2
$ZnSO_4(s)$	161.45	−982.8	110.5	−871.5
ORGANIC COMPOUNDS				
$CCl_4(g)$	153.823	−102.9	309.7	−60.6
$CCl_4(\ell)$	153.823	−135.4	216.4	−65.3
$CH_4(g)$	16.043	−74.8	186.2	−50.8
$CH_3COOH(g)$	60.053	−432.8	282.5	−374.5
$CH_3COOH(\ell)$	60.053	−484.5	159.8	−389.9
$CH_3OH(g)$	32.042	−200.7	239.9	−162.0
$CH_3OH(\ell)$	32.042	−238.7	126.8	−166.4
$C_2H_2(g)$	26.038	226.7	200.8	209.2
$C_2H_4(g)$	28.054	52.4	219.5	68.1
$C_2H_6(g)$	30.070	−84.67	229.5	−32.9
$CH_3CH_2OH(g)$	46.069	−235.1	282.6	−168.6
$CH_3CH_2OH(\ell)$	46.069	−277.7	160.7	−174.9
$CH_3CHO(g)$	44.053	−166	266	−133.7
$C_3H_8(g)$	44.097	−103.8	269.9	−23.5
$CH_3(CH_2)_2CH_3(g)$	58.123	−125.6	310.0	−15.7

Continued on next page

TABLE A4.3 Thermodynamic Properties at 25°C *(Continued)*

Substance^{a,b}	Molar Mass (g/mol)	ΔH_f° (kJ/mol)	S° [J/(mol · K)]	ΔG_f° (kJ/mol)
$CH_3(CH_2)_2CH_3(\ell)$	58.123	−147.6	231.0	−15.0
$CH_3COCH_3(\ell)$	58.079	−248.4	199.8	−155.6
$CH_3COCH_3(g)$	58.079	−217.1	295.3	−152.7
$CH_3(CH_2)_2CH_2OH(\ell)$	74.122	−327.3	225.8	
$(CH_3CH_2)_2O(\ell)$	74.122	−279.6	172.4	
$(CH_3CH_2)_2O(g)$	74.122	−252.1	342.7	
$(CH_3)_2C{=}C(CH_3)_2(\ell)$	84.161	66.6	362.6	−69.2
$(CH_3)_2NH(\ell)$	45.084	−43.9	182.3	
$(CH_3)_2NH(g)$	45.084	−18.5	273.1	
$(C_2H_5)_2NH(\ell)$	73.138	−103.3		
$(C_2H_5)_2NH(g)$	73.138	−71.4		
$(CH_3)_3N(\ell)$	59.111	−46.0	208.5	
$(CH_3)_3N(g)$	59.111	−23.6	287.1	
$(CH_3CH_2)_3N(\ell)$	101.191	−134.3		
$(CH_3CH_2)_3N(g)$	101.191	−95.8		
$C_6H_6(g)$	78.114	82.9	269.2	129.7
$C_6H_6(\ell)$	78.114	49.0	172.9	124.5
$C_6H_{12}O_6(s)$	180.158	−1274.4	212.1	−910.1
$CH_3(CH_2)_6\,CH_3(\ell)$	114.231	−249.9	361.1	6.4
$CH_3(CH_2)_6\,CH_3(g)$	114.231	−208.6	466.7	16.4
$C_{12}H_2O_{11}(s)$	342.300	−2221.7	360.2	−1543.8
$HCOOH(\ell)$	46.026	−424.7	129.0	−361.4

^{a}Substances are arranged alphabetically by chemical formula within each class: (1) elements and monatomic ions; (2) polyatomic ions; (3) inorganic compounds (including CO and CO_2); (4) organic compounds (hydrocarbon-based).
^{b}Symbols denote standard enthalpy of formation (ΔH_f°), standard third-law entropy (S°), and standard Gibbs free energy of formation (ΔG_f°). Entropies in aqueous solution are referred to $S^\circ[H^+(aq)] = 0$, not to absolute zero.

TABLE A4.4 Vapor Pressure of Water as a Function of Temperature

T (°C)	P (torr)
0.0	4.579
10.0	9.209
20.0	17.535
25.0	23.756
30.0	31.824
40.0	55.324
60.0	149.4
70.0	233.7
90.0	525.8
100	760.0
105	906.0

Appendix 5

Equilibrium Constants

TABLE A5.1 Ionization Constants of Selected Acids at 25°C

Acid	Step	Aqueous Equilibrium^a	K_a	pK_a
acetic	1	$CH_3COOH(aq) + H_2O(\ell) \rightleftharpoons H_3O^+(aq) + CH_3COO^-(aq)$	1.76×10^{-5}	4.75
arsenic	1	$H_3AsO_4(aq) + H_2O(\ell) \rightleftharpoons H_3O^+(aq) + H_2AsO_4^-(aq)$	5.5×10^{-3}	2.26
	2	$H_2AsO_4^-(aq) + H_2O(\ell) \rightleftharpoons H_3O^+(aq) + HAsO_4^{2-}(aq)$	1.7×10^{-7}	6.77
	3	$HAsO_4^{2-}(aq) + H_2O(\ell) \rightleftharpoons H_3O^+(aq) + AsO_4^{3-}(aq)$	5.1×10^{-12}	11.29
ascorbic	1	$H_2C_6H_6O_6(aq) + H_2O(\ell) \rightleftharpoons H_3O^+(aq) + HC_6H_6O_6^-(aq)$	9.1×10^{-5}	4.04
	2	$HC_6H_6O_6^-(aq) + H_2O(\ell) \rightleftharpoons H_3O^+(aq) + C_6H_6O_6^{2-}(aq)$	5×10^{-12}	11.3
benzoic	1	$C_6H_5COOH(aq) + H_2O(\ell) \rightleftharpoons H_3O^+(aq) + C_6H_5COO^-(aq)$	6.25×10^{-5}	4.20
boric	1	$H_3BO_3(aq) + H_2O(\ell) \rightleftharpoons H_3O^+(aq) + H_2BO_3^-(aq)$	5.4×10^{-10}	9.27
	2	$H_2BO_3^-(aq) + H_2O(\ell) \rightleftharpoons H_3O^+(aq) + HBO_3^{2-}(aq)$	$<10^{-14}$	>14
bromoacetic	1	$CH_2BrCOOH(aq) + H_2O(\ell) \rightleftharpoons H_3O^+(aq) + CH_2BrCOO^-(aq)$	2.0×10^{-3}	2.70
butanoic	1	$CH_3CH_2CH_2COOH(aq) + H_2O(\ell) \rightleftharpoons H_3O^+(aq) + CH_3CH_2CH_2COO^-(aq)$	1.5×10^{-5}	4.82
carbonic	1	$H_2CO_3(aq) + H_2O(\ell) \rightleftharpoons H_3O^+(aq) + HCO_3^-(aq)$	4.3×10^{-7}	6.37
	2	$HCO_3^-(aq) + H_2O(\ell) \rightleftharpoons H_3O^+(aq) + CO_3^{2-}(aq)$	4.7×10^{-11}	10.33
chloric	1	$HClO_3(aq) + H_2O(\ell) \rightleftharpoons H_3O^+(aq) + ClO_3^-(aq)$	~1	~0
chloroacetic	1	$CH_2ClCOOH(aq) + H_2O(\ell) \rightleftharpoons H_3O^+(aq) + CH_2ClCOO^-(aq)$	1.4×10^{-3}	2.85
chlorous	1	$HClO_2(aq) + H_2O(\ell) \rightleftharpoons H_3O^+(aq) + ClO_2^-(aq)$	1.1×10^{-2}	1.96
citric	1	$CH_2(COOH)C(OH)(COOH)CH_2COOH(aq) + H_2O(\ell) \rightleftharpoons$ $H_3O^+(aq) + CH_2(COOH)C(OH)(COO^-)CH_2COOH(aq)$	7.4×10^{-4}	3.13
	2	$CH_2(COOH)C(OH)(COO^-)CH_2COOH(aq) + H_2O(\ell) \rightleftharpoons$ $H_3O^+(aq) + CH_2(COO^-)C(OH)(COO^-)CH_2COOH(aq)$	1.7×10^{-5}	4.77
	3	$CH_2(COO^-)C(OH)(COO^-)CH_2COOH(aq) + H_2O(\ell) \rightleftharpoons$ $H_3O^+(aq) + CH_2(COO^-)C(OH)(COO^-)CH_2COO^-(aq)$	4.0×10^{-7}	6.40
dichloroacetic	1	$CHCl_2COOH(aq) + H_2O(\ell) \rightleftharpoons H_3O^+(aq) + CHCl_2COO^-(aq)$	5.5×10^{-2}	1.26
ethanol	1	$CH_3CH_2OH(aq) + H_2O(\ell) \rightleftharpoons H_3O^+(aq) + CH_3CH_2O^-(aq)$	1.3×10^{-16}	15.9
fluoroacetic	1	$CH_2FCOOH(aq) + H_2O(\ell) \rightleftharpoons H_3O^+(aq) + CH_2FCOO^-(aq)$	2.6×10^{-3}	2.59
formic	1	$HCOOH(aq) + H_2O(\ell) \rightleftharpoons H_3O^+(aq) + HCOO^-(aq)$	1.77×10^{-4}	3.75
germanic	1	$H_2GeO_3(aq) + H_2O(\ell) \rightleftharpoons H_3O^+(aq) + HGeO_3^-(aq)$	9.8×10^{-10}	9.01
	2	$HGeO_3^-(aq) + H_2O(\ell) \rightleftharpoons H_3O^+(aq) + GeO_3^{2-}(aq)$	5×10^{-13}	12.3
hydr(o)azoic	1	$HN_3(aq) + H_2O(\ell) \rightleftharpoons H_3O^+(aq) + N_3^-(aq)$	1.9×10^{-5}	4.72
hydrobromic	1	$HBr(aq) + H_2O(\ell) \rightleftharpoons H_3O^+(aq) + Br^-(aq)$	$\gg1$ (strong)	<0
hydrochloric	1	$HCl(aq) + H_2O(\ell) \rightleftharpoons H_3O^+(aq) + Cl^-(aq)$	$\gg1$ (strong)	<0
hydrocyanic	1	$HCN(aq) + H_2O(\ell) \rightleftharpoons H_3O^+(aq) + CN^-(aq)$	6.2×10^{-10}	9.21
hydrofluoric	1	$HF(aq) + H_2O(\ell) \rightleftharpoons H_3O^+(aq) + F^-(aq)$	6.8×10^{-4}	3.17
hydr(o)iodic	1	$HI(aq) + H_2O(\ell) \rightleftharpoons H_3O^+(aq) + I^-(aq)$	$\gg1$ (strong)	<0

Continued on next page

TABLE A5.1   Ionization Constants of Selected Acids at 25°C (Continued)

Acid	Step	Aqueous Equilibrium[a]	K_a	pK_a
hydrosulfuric	1	$H_2S(aq) + H_2O(\ell) \rightleftharpoons H_3O^+(aq) + HS^-(aq)$	8.9×10^{-8}	7.05
	2	$HS^-(aq) + H_2O(\ell) \rightleftharpoons H_3O^+(aq) + S^{2-}(aq)$	$\sim10^{-19}$	~19
hypobromous	1	$HBrO(aq) + H_2O(\ell) \rightleftharpoons H_3O^+(aq) + BrO^-(aq)$	2.3×10^{-9}	8.64
hypochlorous	1	$HClO(aq) + H_2O(\ell) \rightleftharpoons H_3O^+(aq) + ClO^-(aq)$	2.9×10^{-8}	7.54
hypoiodous	1	$HIO(aq) + H_2O(\ell) \rightleftharpoons H_3O^+(aq) + IO^-(aq)$	2.3×10^{-11}	10.64
iodic	1	$HIO_3(aq) + H_2O(\ell) \rightleftharpoons H_3O^+(aq) + IO_3^-(aq)$	1.7×10^{-1}	0.77
iodoacetic	1	$CH_2ICOOH(aq) + H_2O(\ell) \rightleftharpoons H_3O^+(aq) + CH_2ICOO^-(aq)$	7.6×10^{-4}	3.12
lactic	1	$CH_3CHOHCOOH(aq) + H_2O(\ell) \rightleftharpoons H_3O^+(aq) + CH_3CHOHCOO^-(aq)$	1.4×10^{-4}	3.85
maleic (cis-butenedioic)	1	$HOOCCH{=}CHCOOH(aq) + H_2O(\ell) \rightleftharpoons H_3O^+(aq) + HOOCCH{=}CHCOO^-(aq)$	1.2×10^{-2}	1.92
	2	$HOOCCH{=}CHCOO^-(aq) + H_2O(\ell) \rightleftharpoons H_3O^+(aq) + {}^-OOCCH{=}CHCOO^-(aq)$	4.7×10^{-7}	6.33
malonic	1	$HOOCCH_2COOH(aq) + H_2O(\ell) \rightleftharpoons H_3O^+(aq) + HOOCCH_2COO^-(aq)$	1.5×10^{-3}	2.82
	2	$HOOCCH_2COO^-(aq) + H_2O(\ell) \rightleftharpoons H_3O^+(aq) + {}^-OOCCH_2COO^-(aq)$	2.0×10^{-6}	5.70
nitric	1	$HNO_3(aq) + H_2O(\ell) \rightleftharpoons H_3O^+(aq) + NO_3^-(aq)$	$\gg1$ (strong)	<0
nitrous	1	$HNO_2(aq) + H_2O(\ell) \rightleftharpoons H_3O^+(aq) + NO_2^-(aq)$	4.0×10^{-4}	3.40
oxalic	1	$HOOCCOOH(aq) + H_2O(\ell) \rightleftharpoons H_3O^+(aq) + HOOCCOO^-(aq)$	5.9×10^{-2}	1.23
	2	$HOOCCOO^-(aq) + H_2O(\ell) \rightleftharpoons H_3O^+(aq) + {}^-OOCCOO^-(aq)$	6.4×10^{-5}	4.19
perchloric	1	$HClO_4(aq) + H_2O(\ell) \rightleftharpoons H_3O^+(aq) + ClO_4^-(aq)$	$\gg1$ (strong)	<0
periodic	1	$HIO_4(aq) + H_2O(\ell) \rightleftharpoons H_3O^+(aq) + IO_4^-(aq)$	2.3×10^{-2}	1.64
phenol	1	$C_6H_5OH(aq) + H_2O(\ell) \rightleftharpoons H_3O^+(aq) + C_6H_5O^-(aq)$	1.3×10^{-10}	9.89
phosphoric	1	$H_3PO_4(aq) + H_2O(\ell) \rightleftharpoons H_3O^+(aq) + H_2PO_4^-(aq)$	6.9×10^{-3}	2.16
	2	$H_2PO_4^-(aq) + H_2O(\ell) \rightleftharpoons H_3O^+(aq) + HPO_4^{2-}(aq)$	6.4×10^{-8}	7.19
	3	$HPO_4^{2-}(aq) + H_2O(\ell) \rightleftharpoons H_3O^+(aq) + PO_4^{3-}(aq)$	4.8×10^{-13}	12.32
propanoic	1	$CH_3CH_2COOH(aq) + H_2O(\ell) \rightleftharpoons H_3O^+(aq) + CH_3CH_2COO^-(aq)$	1.4×10^{-5}	4.85
pyruvic	1	$CH_3C(O)COOH(aq) + H_2O(\ell) \rightleftharpoons H_3O^+(aq) + CH_3C(O)COO^-(aq)$	2.8×10^{-3}	2.55
sulfuric	1	$H_2SO_4(aq) + H_2O(\ell) \rightleftharpoons H_3O^+(aq) + HSO_4^-(aq)$	$\gg1$ (strong)	<0
	2	$HSO_4^-(aq) + H_2O(\ell) \rightleftharpoons H_3O^+(aq) + SO_4^{2-}(aq)$	1.2×10^{-2}	1.92
sulfurous	1	$H_2SO_3(aq) + H_2O(\ell) \rightleftharpoons H_3O^+(aq) + HSO_3^-(aq)$	1.7×10^{-2}	1.77
	2	$HSO_3^-(aq) + H_2O(\ell) \rightleftharpoons H_3O^+(aq) + SO_3^{2-}(aq)$	6.2×10^{-8}	7.21
thiocyanic	1	$HSCN(aq) + H_2O(\ell) \rightleftharpoons H_3O^+(aq) + SCN^-(aq)$	$\gg1$ (strong)	<0
trichloroacetic	1	$CCl_3COOH(aq) + H_2O(\ell) \rightleftharpoons H_3O^+(aq) + CCl_3COO^-(aq)$	2.3×10^{-1}	0.64
trifluoroacetic	1	$CF_3COOH(aq) + H_2O(\ell) \rightleftharpoons H_3O^+(aq) + CF_3COO^-(aq)$	5.9×10^{-1}	0.23
water	1	$H_2O(aq) + H_2O(\ell) \rightleftharpoons H_3O^+(aq) + OH^-(aq)$	1.0×10^{-14}	14.00

[a]The formulas of most carboxylic acids are written in an RCOOH format to highlight their molecular structures.

TABLE A5.2 Acid Ionization Constants of Hydrated Metal Ions at 25°C

Free Ion	Hydrated Ion	K_a
Fe^{3+}	$Fe(H_2O)_6^{3+}$	3×10^{-3}
Sn^{2+}	$Sn(H_2O)_6^{2+}$	4×10^{-4}
Cr^{3+}	$Cr(H_2O)_6^{3+}$	1×10^{-4}
Al^{3+}	$Al(H_2O)_6^{3+}$	1×10^{-5}
Cu^{2+}	$Cu(H_2O)_6^{2+}$	3×10^{-8}
Pb^{2+}	$Pb(H_2O)_6^{2+}$	3×10^{-8}
Zn^{2+}	$Zn(H_2O)_6^{2+}$	1×10^{-9}
Co^{2+}	$Co(H_2O)_6^{2+}$	2×10^{-10}
Ni^{2+}	$Ni(H_2O)_6^{2+}$	1×10^{-10}

TABLE A5.3 Ionization Constants of Selected Bases at 25°C

Base	Aqueous Equilibrium	K_b	pK_b
ammonia	$NH_3(aq) + H_2O(\ell) \rightleftharpoons NH_4^+(aq) + OH^-(aq)$	1.76×10^{-5}	4.75
aniline	$C_6H_5NH_2(aq) + H_2O(\ell) \rightleftharpoons C_6H_5NH_3^+(aq) + OH^-(aq)$	4.0×10^{-10}	9.40
diethylamine	$(CH_3CH_2)_2NH(aq) + H_2O(\ell) \rightleftharpoons (CH_3CH_2)_2NH_2^+(aq) + OH^-(aq)$	8.6×10^{-4}	3.07
dimethylamine	$(CH_3)_2NH(aq) + H_2O(\ell) \rightleftharpoons (CH_3)_2NH_2^+(aq) + OH^-(aq)$	5.9×10^{-4}	3.23
ethylamine	$CH_3CH_2NH_2(aq) + H_2O(\ell) \rightleftharpoons CH_3CH_2NH_3^+(aq) + OH^-(aq)$	5.6×10^{-4}	3.25
methylamine	$CH_3NH_2(aq) + H_2O(\ell) \rightleftharpoons CH_3NH_3^+(aq) + OH^-(aq)$	4.4×10^{-4}	3.36
nicotine (1)		1.0×10^{-6}	6.0
(2)		1.3×10^{-11}	10.9
pyridine	$C_5H_5N(aq) + H_2O(\ell) \rightleftharpoons C_5H_5NH^+(aq) + OH^-(aq)$	1.7×10^{-9}	8.77
quinine (1)		3.3×10^{-6}	5.48
(2)		1.4×10^{-10}	9.9
trimethylamine	$(CH_3)_3N(aq) + H_2O(\ell) \rightleftharpoons (CH_3)_3NH^+(aq) + OH^-(aq)$	6.46×10^{-5}	4.19
urea	$H_2NCONH_2(aq) + H_2O(\ell) \rightleftharpoons H_2NCONH_3^+(aq) + OH^-(aq)$	1.3×10^{-14}	13.9

TABLE A5.4 Solubility-Product Constants at 25°C

Cation	Anion	Heterogeneous Equilibrium[a]	K_{sp}[b]
aluminum	hydroxide	$Al(OH)_3(s) \rightleftharpoons Al^{3+}(aq) + 3\,OH^-(aq)$	1.3×10^{-33}
	phosphate	$AlPO_4(s) \rightleftharpoons Al^{3+}(aq) + PO_4^{3-}(aq)$	9.84×10^{-21}
barium	carbonate	$BaCO_3(s) \rightleftharpoons Ba^{2+}(aq) + CO_3^{2-}(aq)$	2.58×10^{-9}
	fluoride	$BaF_2(s) \rightleftharpoons Ba^{2+}(aq) + 2\,F^-(aq)$	1.84×10^{-7}
	sulfate	$BaSO_4(s) \rightleftharpoons Ba^{2+}(aq) + SO_4^{2-}(aq)$	1.08×10^{-10}
calcium	carbonate	$CaCO_3(s) \rightleftharpoons Ca^{2+}(aq) + CO_3^{2-}(aq)$	2.8×10^{-9}
	fluoride	$CaF_2(s) \rightleftharpoons Ca^{2+}(aq) + 2\,F^-(aq)$	5.3×10^{-9}
	hydroxide	$Ca(OH)_2(s) \rightleftharpoons Ca^{2+}(aq) + 2\,OH^-(aq)$	5.5×10^{-6}
	phosphate	$Ca_3(PO_4)_2(s) \rightleftharpoons 3\,Ca^{2+}(aq) + 2\,PO_4^{3-}(aq)$	2.07×10^{-29}
	sulfate	$CaSO_4(s) \rightleftharpoons Ca^{2+}(aq) + SO_4^{2-}(aq)$	4.93×10^{-5}
cobalt(II)	carbonate	$CoCO_3(s) \rightleftharpoons Co^{2+}(aq) + CO_3^{2-}(aq)$	1.4×10^{-3}
	phosphate	$Co_3(PO_4)_2(s) \rightleftharpoons 3\,Co^{2+}(aq) + 2\,PO_4^{3-}(aq)$	2.05×10^{-7}
	sulfide	$CoS(s) \rightleftharpoons Co^{2+}(aq) + S^{2-}(aq)$	2.0×10^{-25}
copper(I)	bromide	$CuBr(s) \rightleftharpoons Cu^+(aq) + Br^-(aq)$	6.27×10^{-9}
	chloride	$CuCl(s) \rightleftharpoons Cu^+(aq) + Cl^-(aq)$	1.72×10^{-7}
	iodide	$CuI(s) \rightleftharpoons Cu^+(aq) + I^-(aq)$	1.27×10^{-12}
copper(II)	phosphate	$Cu_3(PO_4)_2(s) \rightleftharpoons 3\,Cu^{2+}(aq) + 2\,PO_4^{3-}(aq)$	1.4×10^{-37}
	hydroxide	$Cu(OH)_2(s) \rightleftharpoons Cu^{2+}(aq) + 2\,OH^-(aq)$	2.2×10^{-20}
iron(II)	carbonate	$FeCO_3(s) \rightleftharpoons Fe^{2+}(aq) + CO_3^{2-}(aq)$	3.13×10^{-11}
	fluoride	$FeF_2(s) \rightleftharpoons Fe^{2+}(aq) + 2\,F^-(aq)$	2.36×10^{-6}
	hydroxide	$Fe(OH)_2(s) \rightleftharpoons Fe^{2+}(aq) + 2\,OH^-(aq)$	4.87×10^{-17}
	sulfide	$FeS(s) \rightleftharpoons Fe^{2+}(aq) + S^{2-}(aq)$	6.3×10^{-18}
lead	bromide	$PbBr_2(s) \rightleftharpoons Pb^{2+}(aq) + 2\,Br^-(aq)$	6.60×10^{-6}
	carbonate	$PbCO_3(s) \rightleftharpoons Pb^{2+}(aq) + CO_3^{2-}(aq)$	7.4×10^{-14}
	chloride	$PbCl_2(s) \rightleftharpoons Pb^{2+}(aq) + 2\,Cl^-(aq)$	1.7×10^{-5}
	fluoride	$PbF_2(s) \rightleftharpoons Pb^{2+}(aq) + 2\,F^-(aq)$	3.3×10^{-8}
	iodide	$PbI_2(s) \rightleftharpoons Pb^{2+}(aq) + 2\,I^-(aq)$	9.8×10^{-9}
	sulfate	$PbSO_4(s) \rightleftharpoons Pb^{2+}(aq) + SO_4^{2-}(aq)$	2.53×10^{-8}
lithium	carbonate	$Li_2CO_3(s) \rightleftharpoons 2\,Li^+(aq) + CO_3^{2-}(aq)$	2.5×10^{-2}
magnesium	carbonate	$MgCO_3(s) \rightleftharpoons Mg^{2+}(aq) + CO_3^{2-}(aq)$	6.82×10^{-6}
	fluoride	$MgF_2(s) \rightleftharpoons Mg^{2+}(aq) + 2\,F^-(aq)$	5.16×10^{-11}
	hydroxide	$Mg(OH)_2(s) \rightleftharpoons Mg^{2+}(aq) + 2\,OH^-(aq)$	5.61×10^{-12}
manganese(II)	carbonate	$MnCO_3(s) \rightleftharpoons Mn^{2+}(aq) + CO_3^{2-}(aq)$	2.34×10^{-11}
	hydroxide	$Mn(OH)_2(s) \rightleftharpoons Mn^{2+}(aq) + 2\,OH^-(aq)$	1.9×10^{-13}
mercury(I)	bromide	$Hg_2Br_2(s) \rightleftharpoons Hg_2^{2+}(aq) + 2\,Br^-(aq)$	6.40×10^{-23}
	carbonate	$Hg_2CO_3(s) \rightleftharpoons Hg_2^{2+}(aq) + CO_3^{2-}(aq)$	3.6×10^{-17}
	chloride	$Hg_2Cl_2(s) \rightleftharpoons Hg_2^{2+}(aq) + 2\,Cl^-(aq)$	1.43×10^{-18}
	iodide	$Hg_2I_2(s) \rightleftharpoons Hg_2^{2+}(aq) + 2\,I^-(aq)$	5.2×10^{-29}
	sulfate	$Hg_2SO_4(s) \rightleftharpoons Hg_2^{2+}(aq) + SO_4^{2-}(aq)$	6.5×10^{-7}
mercury(II)	hydroxide	$Hg(OH)_2(s) \rightleftharpoons Hg^{2+}(aq) + 2\,OH^-(aq)$	3.2×10^{-26}
	iodide	$HgI_2(s) \rightleftharpoons Hg^{2+}(aq) + 2\,I^-(aq)$	2.9×10^{-29}
nickel(II)	carbonate	$NiCO_3(s) \rightleftharpoons Ni^{2+}(aq) + CO_3^{2-}(aq)$	1.42×10^{-7}
	phosphate	$Ni_3(PO_4)_2(s) \rightleftharpoons 3\,Ni^{2+}(aq) + 2\,PO_4^{3-}(aq)$	4.74×10^{-32}
	sulfide	$NiS(s) \rightleftharpoons Ni^{2+}(aq) + S^{2-}(aq)$	1×10^{-24}
silver	bromide	$AgBr(s) \rightleftharpoons Ag^+(aq) + Br^-(aq)$	5.35×10^{-13}
	carbonate	$Ag_2CO_3(s) \rightleftharpoons 2\,Ag^+(aq) + CO_3^{2-}(aq)$	8.46×10^{-12}
	chloride	$AgCl(s) \rightleftharpoons Ag^+(aq) + Cl^-(aq)$	1.77×10^{-10}
	chromate	$Ag_2CrO_4(s) \rightleftharpoons 2\,Ag^+(aq) + CrO_4^{2-}(aq)$	1.12×10^{-12}
	hydroxide	$AgOH(s) \rightleftharpoons Ag^+(aq) + OH^-(aq)$	2.0×10^{-8}
	iodide	$AgI(s) \rightleftharpoons Ag^+(aq) + I^-(aq)$	8.52×10^{-17}
	phosphate	$Ag_3PO_4(s) \rightleftharpoons 3\,Ag^+(aq) + PO_4^{3-}(aq)$	8.89×10^{-17}
	sulfate	$Ag_2SO_4(s) \rightleftharpoons 2\,Ag^+(aq) + SO_4^{2-}(aq)$	1.20×10^{-5}
	sulfide	$Ag_2S(s) \rightleftharpoons 2\,Ag^+(aq) + S^{2-}(aq)$	6.3×10^{-50}
strontium	carbonate	$SrCO_3(s) \rightleftharpoons Sr^{2+}(aq) + CO_3^{2-}(aq)$	5.60×10^{-10}
	fluoride	$SrF_2(s) \rightleftharpoons Sr^{2+}(aq) + 2\,F^-(aq)$	4.33×10^{-9}
	sulfate	$SrSO_4(s) \rightleftharpoons Sr^{2+}(aq) + SO_4^{2-}(aq)$	3.44×10^{-7}
zinc	carbonate	$ZnCO_3(s) \rightleftharpoons Zn^{2+}(aq) + CO_3^{2-}(aq)$	1.46×10^{-10}
	hydroxide	$Zn(OH)_2(s) \rightleftharpoons Zn^{2+}(aq) + 2\,OH^-(aq)$	3.0×10^{-17}

[a]Equilibrium is between solid phase and aqueous solution.
[b]From Dean, J. *Lange's Handbook of Chemistry* (The McGraw-Hill Companies, 1998).

TABLE A5.5 Formation Constants of Complex Ions at 25°C

Complex Ion	Aqueous Equilibrium	K_f
$[Ag(NH_3)_2]^+$	$Ag^+(aq) + 2\ NH_3(aq) \rightleftharpoons Ag(NH_3)_2^+(aq)$	1.7×10^7
$[AgCl_2]^-$	$Ag^+(aq) + 2\ Cl^-(aq) \rightleftharpoons AgCl_2^-(aq)$	2.5×10^5
$[Ag(CN)_2]^-$	$Ag^+(aq) + 2\ CN^-(aq) \rightleftharpoons Ag(CN)_2^-(aq)$	1.0×10^{21}
$[Ag(S_2O_3)_2]^{3-}$	$Ag^+(aq) + 2\ S_2O_3^{2-}(aq) \rightleftharpoons Ag(S_2O_3)_2^{3-}(aq)$	4.7×10^{13}
$[AlF_6]^{3-}$	$Al^{3+}(aq) + 6\ F^-(aq) \rightleftharpoons AlF_6^{3-}(aq)$	4.0×10^{19}
$[Al(OH)_4]^-$	$Al^{3+}(aq) + 4\ OH^-(aq) \rightleftharpoons Al(OH)_4^-(aq)$	7.7×10^{33}
$[Au(CN)_2]^-$	$Au^+(aq) + 2\ CN^-(aq) \rightleftharpoons Au(CN)_2^-(aq)$	2.0×10^{38}
$[Co(NH_3)_6]^{2+}$	$Co^{2+}(aq) + 6\ NH_3(aq) \rightleftharpoons Co(NH_3)_6^{2+}(aq)$	7.7×10^4
$[Co(NH_3)_6]^{3+}$	$Co^{3+}(aq) + 6\ NH_3(aq) \rightleftharpoons Co(NH_3)_6^{3+}(aq)$	5.0×10^{31}
$[Co(en)_3]^{2+}$	$Co^{2+}(aq) + 3\ en(aq) \rightleftharpoons Co(en)_3^{2+}(aq)$	8.7×10^{13}
$[Co(C_2O_4)_3]^{4-}$	$Co^{2+}(aq) + 3\ C_2O_4^{2-}(aq) \rightleftharpoons Co(C_2O_4)_3^{4-}(aq)$	4.5×10^6
$[Cu(NH_3)_4]^{2+}$	$Cu^{2+}(aq) + 4\ NH_3(aq) \rightleftharpoons Cu(NH_3)_4^{2+}(aq)$	5.0×10^{13}
$[Cu(en)_2]^{2+}$	$Cu^{2+}(aq) + 2\ en(aq) \rightleftharpoons Cu(en)_2^{2+}(aq)$	3.2×10^{19}
$[Cu(CN)_4]^{2-}$	$Cu^{2+}(aq) + 4\ CN^-(aq) \rightleftharpoons Cu(CN)_4^{2-}(aq)$	1.0×10^{25}
$[Cu(C_2O_4)_2]^{2-}$	$Cu^{2+}(aq) + 2\ C_2O_4^{2-}(aq) \rightleftharpoons Cu(C_2O_4)_2^{2-}(aq)$	1.7×10^{10}
$[Fe(C_2O_4)_3]^{4-}$	$Fe^{2+}(aq) + 3\ C_2O_4^{2-}(aq) \rightleftharpoons Fe(C_2O_4)_3^{4-}(aq)$	6×10^6
$[Fe(C_2O_4)_3]^{3-}$	$Fe^{3+}(aq) + 3\ C_2O_4^{2-}(aq) \rightleftharpoons Fe(C_2O_4)_3^{3-}(aq)$	3.3×10^{20}
$[HgCl_4]^{2-}$	$Hg^{2+}(aq) + 4\ Cl^-(aq) \rightleftharpoons HgCl_4^{2-}(aq)$	1.2×10^{15}
$[Ni(NH_3)_6]^{2+}$	$Ni^{2+}(aq) + 6\ NH_3(aq) \rightleftharpoons Ni(NH_3)_6^{2+}(aq)$	5.5×10^8
$[PbCl_4]^{2-}$	$Pb^{2+}(aq) + 4\ Cl^-(aq) \rightleftharpoons PbCl_4^{2-}(aq)$	2.5×10^1
$[Zn(NH_3)_4]^{2+}$	$Zn^{2+}(aq) + 4\ NH_3(aq) \rightleftharpoons Zn(NH_3)_4^{2+}(aq)$	2.9×10^9
$[Zn(OH)_4]^{2-}$	$Zn^{2+}(aq) + 4\ OH^-(aq) \rightleftharpoons Zn(OH)_4^{2-}(aq)$	2.8×10^{15}

Appendix 6

Standard Reduction Potentials

TABLE A6.1 Standard Reduction Potentials at 25°C

Half-Reaction	n	$E°$ (V)
$F_2(g) + 2\,e^- \rightarrow 2\,F^-(aq)$	2	2.866
$H_2N_2O_2(s) + 2\,H^+(aq) + 2\,e^- \rightarrow N_2(g) + 2\,H_2O(\ell)$	2	2.65
$O(g) + 2\,H^+(aq) + 2\,e^- \rightarrow H_2O(\ell)$	2	2.421
$Cu^{3+}(aq) + e^- \rightarrow Cu^{2+}(aq)$	1	2.4
$XeO_3(s) + 6\,H^+(aq) + 6\,e^- \rightarrow Xe(g) + 3\,H_2O(\ell)$	6	2.10
$O_3(g) + 2\,H^+(aq) + 2\,e^- \rightarrow O_2(g) + H_2O(\ell)$	2	2.076
$OH(g) + e^- \rightarrow OH^-(aq)$	1	2.02
$Co^{3+}(aq) + e^- \rightarrow Co^{2+}(aq)$	1	1.92
$H_2O_2(\ell) + 2\,H^+(aq) + 2\,e^- \rightarrow 2\,H_2O(\ell)$	2	1.776
$N_2O(g) + 2\,H^+(aq) + 2\,e^- \rightarrow N_2(g) + H_2O(\ell)$	2	1.766
$Ce(OH)^{3+}(aq) + H^+(aq) + e^- \rightarrow Ce^{3+}(aq) + H_2O(\ell)$	1	1.70
$Au^+(aq) + e^- \rightarrow Au(s)$	1	1.692
$PbO_2(s) + SO_4^{2-}(aq) + 4\,H^+(aq) + 2\,e^- \rightarrow PbSO_4(s) + 2\,H_2O(\ell)$	2	1.691
$PbO_2(s) + HSO_4^-(aq) + 3\,H^+(aq) + 2\,e^- \rightarrow PbSO_4(s) + 2\,H_2O(\ell)$	2	1.685
$MnO_4^-(aq) + 4\,H^+(aq) + 3\,e^- \rightarrow MnO_2(s) + 2\,H_2O(\ell)$	3	1.673
$NiO_2(s) + 4\,H^+(aq) + 2\,e^- \rightarrow Ni^{2+}(aq) + 2\,H_2O(\ell)$	2	1.678
$HClO(\ell) + H^+(aq) + e^- \rightarrow \frac{1}{2}\,Cl_2(g) + H_2O(aq)$	1	1.63
$Ce^{4+}(aq) + e^- \rightarrow Ce^{3+}(aq)$	1	1.61
$Mn^{3+}(aq) + e^- \rightarrow Mn^{2+}(aq)$	1	1.542
$MnO_4^-(aq) + 8\,H^+(aq) + 5\,e^- \rightarrow Mn^{2+}(aq) + 4\,H_2O(\ell)$	5	1.507
$BrO_3^-(aq) + 6\,H^+(aq) + 5\,e^- \rightarrow \frac{1}{2}\,Br_2(\ell) + 3\,H_2O(\ell)$	5	1.52
$ClO_3^-(aq) + 6\,H^+(aq) + 5\,e^- \rightarrow \frac{1}{2}\,Cl_2(g) + 3\,H_2O(\ell)$	5	1.47
$PbO_2(s) + 4\,H^+(aq) + 2\,e^- \rightarrow Pb^{2+}(aq) + 2\,H_2O(\ell)$	2	1.455
$Au^{3+}(aq) + 3\,e^- \rightarrow Au(s)$	3	1.40
$Cl_2(g) + 2\,e^- \rightarrow 2\,Cl^-(aq)$	2	1.358
$Cr_2O_7^{2-}(aq) + 14\,H^+(aq) + 6\,e^- \rightarrow 2\,Cr^{3+}(aq) + 7\,H_2O(\ell)$	6	1.33
$MnO_2(s) + 4\,H^+(aq) + 2\,e^- \rightarrow Mn^{2+}(aq) + 2\,H_2O(\ell)$	2	1.23
$O_2(g) + 4\,H^+(aq) + 4\,e^- \rightarrow 2\,H_2O(\ell)$	4	1.229
$IO_3^-(aq) + 6\,H^+(aq) + 5\,e^- \rightarrow \frac{1}{2}\,I_2(s) + 3\,H_2O(\ell)$	5	1.195
$IO_3^-(aq) + 6\,H^+(aq) + 6\,e^- \rightarrow I^-(aq) + 3\,H_2O(\ell)$	6	1.085
$Br_2(\ell) + 2\,e^- \rightarrow 2\,Br^-(aq)$	2	1.066
$HNO_2(\ell) + H^+(aq) + e^- \rightarrow NO(g) + H_2O(\ell)$	1	1.00
$VO_2^+(aq) + 2\,H^+(aq) + e^- \rightarrow VO^{2+}(aq) + H_2O(\ell)$	1	1.00
$NO_3^-(aq) + 4\,H^+(aq) + 3\,e^- \rightarrow NO(g) + 2\,H_2O(\ell)$	3	0.96
$2\,Hg^{2+}(aq) + 2\,e^- \rightarrow Hg_2^{2+}(aq)$	2	0.92

TABLE A6.1   Standard Reduction Potentials at 25°C *(Continued)*

Half-Reaction	n	$E°$ (V)
$ClO^-(aq) + H_2O(\ell) + 2\,e^- \rightarrow Cl^-(aq) + 2\,OH^-(aq)$	2	0.89
$HO_2^-(aq) + H_2O(\ell) + 2\,e^- \rightarrow 3\,OH^-(aq)$	2	0.88
$Hg^{2+}(aq) + 2\,e^- \rightarrow Hg(\ell)$	2	0.851
$Ag^+(aq) + e^- \rightarrow Ag(s)$	1	0.800
$Hg_2^{2+}(aq) + 2\,e^- \rightarrow 2\,Hg(\ell)$	2	0.797
$Fe^{3+}(aq) + e^- \rightarrow Fe^{2+}(aq)$	1	0.770
$PtCl_4^{2-}(aq) + 2\,e^- \rightarrow Pt(s) + 4\,Cl^-(aq)$	2	0.73
$O_2(g) + 2\,H^+(aq) + 2\,e^- \rightarrow H_2O_2(\ell)$	2	0.68
$MnO_4^-(aq) + 2\,H_2O(\ell) + 3\,e^- \rightarrow MnO_2(s) + 4\,OH^-(aq)$	3	0.59
$H_3AsO_4(s) + 2\,H^+(aq) + 2\,e^- \rightarrow H_3AsO_3(aq) + H_2O(\ell)$	2	0.559
$I_2(s) + 2\,e^- \rightarrow 2\,I^-(aq)$	2	0.536
$Cu^+(aq) + e^- \rightarrow Cu(s)$	1	0.521
$2\,NiO(OH)(s) + 2\,H_2O(\ell) + 2\,e^- \rightarrow 2\,Ni(OH)_2(s) + 2\,OH^-(aq)$	2	0.52
$H_2SO_3(\ell) + 4\,H^+(aq) + 4\,e^- \rightarrow S(s) + 3\,H_2O(\ell)$	4	0.449
$Ag_2CrO_4(s) + 2\,e^- \rightarrow 2\,Ag(s) + CrO_4^{2-}(aq)$	2	0.447
$O_2(g) + 2\,H_2O(\ell) + 4\,e^- \rightarrow 4\,OH^-(aq)$	4	0.401
$Fe(CN)_6^{3-}(aq) + e^- \rightarrow Fe(CN)_6^{4-}(aq)$	1	0.36
$Ag_2O(s) + H_2O(\ell) + 2\,e^- \rightarrow 2\,Ag(s) + 2\,OH^-(aq)$	2	0.342
$Cu^{2+}(aq) + 2\,e^- \rightarrow Cu(s)$	2	0.342
$BiO^+(aq) + 2\,H^+(aq) + 3\,e^- \rightarrow Bi(s) + H_2O(\ell)$	3	0.32
$AgCl(s) + e^- \rightarrow Ag(s) + Cl^-(aq)$	1	0.222
$HSO_4^-(aq) + 3\,H^+(aq) + 2\,e^- \rightarrow H_2SO_3(\ell) + H_2O(\ell)$	2	0.17
$Sn^{4+}(aq) + 2\,e^- \rightarrow Sn^{2+}(aq)$	2	0.154
$Cu^{2+}(aq) + e^- \rightarrow Cu^+(aq)$	1	0.153
$2\,MnO_2(s) + H_2O(\ell) + 2\,e^- \rightarrow Mn_2O_3(s) + 2\,OH^-(aq)$	2	0.15
$S(s) + 2\,H^+(aq) + 2\,e^- \rightarrow H_2S(g)$	2	0.141
$HgO(s) + H_2O(\ell) + 2\,e^- \rightarrow Hg(\ell) + 2\,OH^-(aq)$	2	0.0977
$AgBr(s) + e^- \rightarrow Ag(s) + Br^-(aq)$	1	0.095
$Ag(S_2O_3)_2^{3-}(aq) + e^- \rightarrow Ag(s) + 2\,S_2O_3^{2-}(aq)$	1	0.01
$NO_3^-(aq) + H_2O(\ell) + 2\,e^- \rightarrow NO_2^-(aq) + 2\,OH^-(aq)$	2	0.01
$2\,H^+(aq) + 2\,e^- \rightarrow H_2(g)$	2	0.000
$Pb^{2+}(aq) + 2\,e^- \rightarrow Pb(s)$	2	−0.126
$CrO_4^{2-}(aq) + 4\,H_2O(\ell) + 3\,e^- \rightarrow Cr(OH)_3(s) + 5\,OH^-(aq)$	3	−0.13
$Sn^{2+}(aq) + 2\,e^- \rightarrow Sn(s)$	2	−0.136
$AgI(s) + e^- \rightarrow Ag(s) + I^-(aq)$	1	−0.152
$CuI(s) + e^- \rightarrow Cu(s) + I^-(aq)$	1	−0.185
$N_2(g) + 5\,H^+(aq) + 4\,e^- \rightarrow N_2H_5^+(aq)$	4	−0.23
$Ni^{2+}(aq) + 2\,e^- \rightarrow Ni(s)$	2	−0.257
$PbSO_4(s) + H^+(aq) + 2\,e^- \rightarrow Pb(s) + HSO_4^-(aq)$	2	−0.356
$Co^{2+}(aq) + 2\,e^- \rightarrow Co(s)$	2	−0.277
$Ag(CN)_2^-(aq) + e^- \rightarrow Ag(s) + 2\,CN^-(aq)$	1	−0.31
$Cd^{2+}(aq) + 2\,e^- \rightarrow Cd(s)$	2	−0.403
$Cr^{3+}(aq) + e^- \rightarrow Cr^{2+}(aq)$	1	−0.41

Continued on next page

TABLE A6.1 Standard Reduction Potentials at 25°C *(Continued)*

Half-Reaction	n	$E°$ (V)
$Fe^{2+}(aq) + 2\,e^- \rightarrow Fe(s)$	2	−0.447
$2\,CO_2(g) + 2\,H^+(aq) + 2\,e^- \rightarrow H_2C_2O_4(s)$	2	−0.49
$Ni(OH)_2(s) + 2\,e^- \rightarrow Ni(s) + 2\,OH^-(aq)$	2	−0.72
$Cr^{3+}(aq) + 3\,e^- \rightarrow Cr(s)$	3	−0.74
$Zn^{2+}(aq) + 2\,e^- \rightarrow Zn(s)$	2	−0.762
$Cd(OH)_2(s) + 2\,e^- \rightarrow Cd(s) + 2\,OH^-(aq)$	2	−0.81
$2\,H_2O(\ell) + 2\,e^- \rightarrow H_2(g) + 2\,OH^-(aq)$	2	−0.828
$SO_4^{2-}(aq) + H_2O(\ell) + 2\,e^- \rightarrow SO_3^{2-}(aq) + 2\,OH^-(aq)$	2	−0.92
$N_2(g) + 4\,H_2O(\ell) + 4\,e^- \rightarrow 4\,OH^-(aq) + N_2H_4(\ell)$	4	−1.16
$Mn^{2+}(aq) + 2\,e^- \rightarrow Mn(s)$	2	−1.185
$Zn(OH)_2(s) + 2\,e^- \rightarrow Zn(s) + 2\,OH^-(aq)$	2	−1.249
$ZnO(s) + H_2O(\ell) + 2\,e^- \rightarrow Zn(s) + 2\,OH^-(aq)$	2	−1.25
$Al^{3+}(aq) + 3\,e^- \rightarrow Al(s)$	3	−1.662
$Mg^{2+}(aq) + 2\,e^- \rightarrow Mg(s)$	2	−2.37
$Na^+(aq) + e^- \rightarrow Na(s)$	1	−2.71
$Ca^{2+}(aq) + 2\,e^- \rightarrow Ca(s)$	2	−2.868
$Ba^{2+}(aq) + 2\,e^- \rightarrow Ba(s)$	2	−2.912
$K^+(aq) + e^- \rightarrow K(s)$	1	−2.95
$Li^+(aq) + e^- \rightarrow Li(s)$	1	−3.05

Appendix 7

Naming Organic Compounds

While organic chemistry was becoming established as a discipline within chemistry, many compounds were given trivial names that are still commonly used and recognized. We refer to many of these compounds by their nonsystematic names throughout this book, and their names and structures are listed in Table A7.1.

TABLE A7.1 Organic Compounds and Their Commonly Used Nonsystematic Names

Name	Formula	Structure
ethylene	C_2H_4	
acetylene	C_2H_2	$HC\equiv CH$
benzene	C_6H_6	
toluene	$C_6H_5CH_3$	
ethyl alcohol	CH_3CH_2OH	
acetone	CH_3COCH_3	
acetic acid	CH_3COOH	
formaldehyde	CH_2O	

The International Union of Pure and Applied Chemistry (IUPAC) has proposed a set of rules for the systematic naming of organic compounds. When naming compounds or drawing structures based on names, we need to keep in mind that the IUPAC system of nomenclature is based on two fundamental ideas: (1) the name of a compound must indicate how the carbon atoms in the skeleton are bonded together, and (2) the name must identify the location of any functional groups in the molecule.

Alkanes

Table A7.2 contains the prefixes used for carbon chains ranging in size from C_1 to C_{20} and gives the names for compounds consisting of unbranched chains. The name of a compound consists of a prefix identifying the number of carbons in the chain and a suffix defining the type of hydrocarbon. The suffix *-ane* indicates that the compounds are alkanes and that all carbon–carbon bonds are single bonds.

TABLE A7.2   Prefixes for Naming Carbon Chains

Prefix	Example	Name	Prefix	Example	Name
meth	CH_4	methane	undec	$C_{11}H_{24}$	undecane
eth	C_2H_6	ethane	dodec	$C_{12}H_{26}$	dodecane
pro	C_3H_8	propane	tridec	$C_{13}H_{28}$	tridecane
but	C_4H_{10}	butane	tetradec	$C_{14}H_{30}$	tetradecane
pent	C_5H_{12}	pentane	pentadec	$C_{15}H_{32}$	pentadecane
hex	C_6H_{14}	hexane	hexadec	$C_{16}H_{34}$	hexadecane
hept	C_7H_{16}	heptane	heptadec	$C_{17}H_{36}$	heptadecane
oct	C_8H_{18}	octane	octadec	$C_{18}H_{38}$	octadecane
non	C_9H_{20}	nonane	nonadec	$C_{19}H_{40}$	nonadecane
dec	$C_{10}H_{22}$	decane	eicos	$C_{20}H_{42}$	eicosane

Branched-Chain Alkanes

The alkane drawn here is used to illustrate each step in the naming rules:

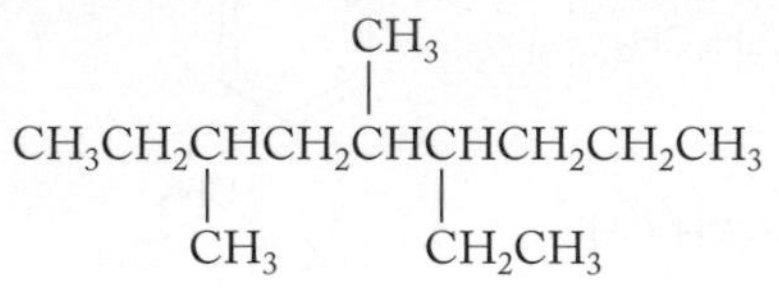

$$CH_3$$
$$|$$
$$CH_3CH_2CHCH_2CHCHCH_2CH_2CH_3$$
$$|\qquad\qquad|$$
$$CH_3\qquad CH_2CH_3$$

1. **Identify and name the longest continuous carbon chain.**

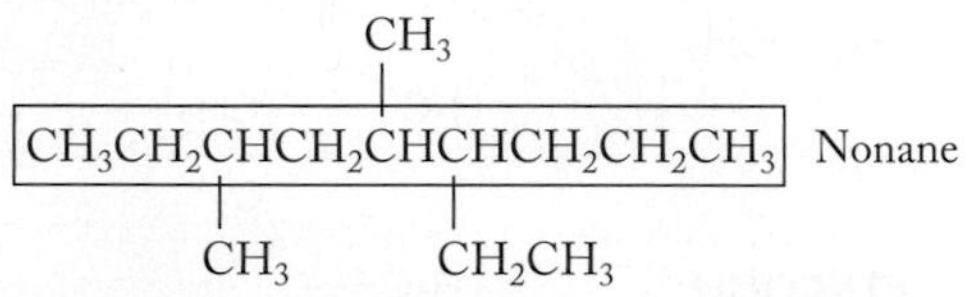

$$CH_3$$
$$|$$
$$\boxed{CH_3CH_2CHCH_2CHCHCH_2CH_2CH_3}\quad \text{Nonane}$$
$$|\qquad\qquad|$$
$$CH_3\qquad CH_2CH_3$$

2. **Identify the groups attached to this chain and name them.** Names of substituent groups consist of the prefix from Table A7.2 that identifies the length of the group and the suffix *-yl* that identifies it as an alkyl group.

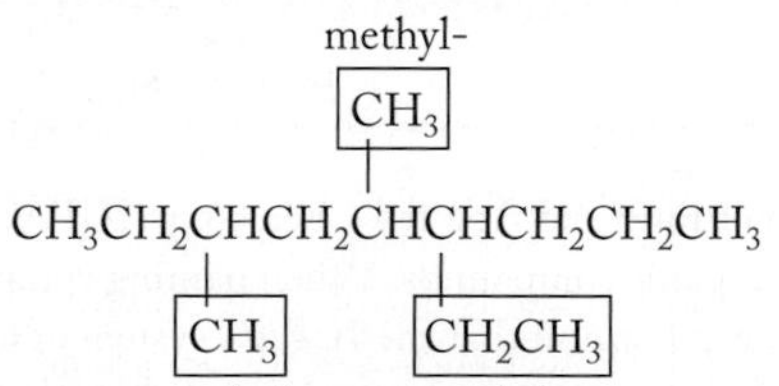

$$\text{methyl-}$$
$$\boxed{CH_3}$$
$$|$$
$$CH_3CH_2CHCH_2CHCHCH_2CH_2CH_3$$
$$|\qquad\qquad|$$
$$\boxed{CH_3}\qquad \boxed{CH_2CH_3}$$
$$\text{methyl-}\qquad \text{ethyl-}$$

3. **Number the carbon atoms in the longest chain,** starting at the end nearest a substituent group. Doing this identifies the points of attachment of the alkyl groups with the lowest possible numbers.

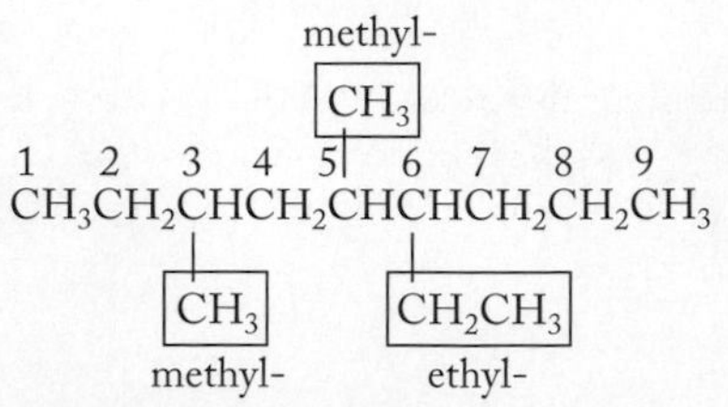

$$\text{methyl-}$$
$$\boxed{CH_3}$$
$$1\ \ 2\ \ 3\ \ 4\ \ 5|\ \ 6\ \ 7\ \ 8\ \ 9$$
$$CH_3CH_2CHCH_2CHCHCH_2CH_2CH_3$$
$$|\qquad\qquad|$$
$$\boxed{CH_3}\qquad \boxed{CH_2CH_3}$$
$$\text{methyl-}\qquad \text{ethyl-}$$

4. **Designate the location and identity of each substituent group with a number,** followed by a hyphen, and its name.

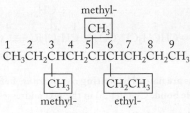

methyl-

$$
\begin{array}{ccccccccc}
1 & 2 & 3 & 4 & 5 & 6 & 7 & 8 & 9 \\
\end{array}
$$
CH$_3$CH$_2$CHCH$_2$CHCHCH$_2$CH$_2$CH$_3$

CH$_3$ CH$_2$CH$_3$
methyl-   ethyl-

3-methyl-, 5-methyl-, 6-ethyl-

5. **Put together the complete name by listing the substituent groups in alphabetical order.** If more than one of a given type of substituent group is present, prefixes *di-*, *tri-*, *tetra-*, and so forth, are appended to the names, but these numerical prefixes are not considered when determining the alphabetical order. The name of the last substituent group is written together with the name identifying the longest carbon chain.

CH$_3$

CH$_3$CH$_2$CHCH$_2$CHCHCH$_2$CH$_2$CH$_3$

CH$_3$ CH$_2$CH$_3$

6-Ethyl-3,5-dimethylnonane

Cycloalkanes

The simplest examples of this class of compounds consist of one unsubstituted ring of carbon atoms. The IUPAC names of these compounds consist of the prefix *cyclo-* followed by the parent name from Table A7.2 to indicate the number of carbon atoms in the ring. As an illustration, the names, formulas, and line structures of the first three cycloalkanes in the homologous series are

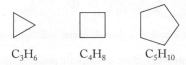

C$_3$H$_6$ C$_4$H$_8$ C$_5$H$_{10}$

Cyclopropane Cyclobutane Cyclopentane

Alkenes and Alkynes

Alkenes have carbon–carbon double bonds and alkynes have carbon–carbon triple bonds as functional groups. The names of these types of compounds consist of (1) a parent name that identifies the longest carbon chain that includes the double or triple bond, (2) a suffix that identifies the class of compound, and (3) names of any substituent groups attached to the longest carbon chain. The suffix *-ene* identifies an alkene; *-yne* identifies an alkyne.

The alkene and alkyne drawn here are used to illustrate each step in the naming rules:

CH$_3$ CH$_3$

CH$_3$CHCH$=$CHCH$_2$CH$_2$CH$_3$ CH$_3$CHC$\equiv$CCH$_2$CH$_2$CH$_3$

1. **To determine the parent name,** identify the longest chain that contains the unsaturation. Name the parent compound with the prefix that defines the number of carbons in that chain and the suffix that identifies the class of compound.

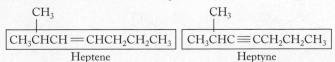

CH$_3$ CH$_3$

CH$_3$CHCH$=$CHCH$_2$CH$_2$CH$_3$ CH$_3$CHC$\equiv$CCH$_2$CH$_2$CH$_3$

Heptene   Heptyne

2. **Number the parent chain from the end nearest the unsaturation so that the first carbon in the double or triple bond has the lowest number possible.** (If the unsaturation is in the middle of a chain, the location of any substituent group is used to determine where the

numbering starts.) The smaller of the two numbers identifying the carbon atoms involved in the unsaturation is used as the locator of the multiple bond.

$$
\underset{\text{3-Heptene}}{\overset{\displaystyle\overset{\text{CH}_3}{|}}{\underset{1\ \ 2\ \ 3\ \ \ \ \ 4\ \ 5\ \ 6\ \ 7}{\text{CH}_3\text{CHCH}=\text{CHCH}_2\text{CH}_2\text{CH}_3}}}
\qquad
\underset{\text{3-Heptyne}}{\overset{\displaystyle\overset{\text{CH}_3}{|}}{\underset{1\ \ 2\ \ 3\ \ \ \ 4\ 5\ \ 6\ \ 7}{\text{CH}_3\text{CHC}\equiv\text{CCH}_2\text{CH}_2\text{CH}_3}}}
$$

3. **Stereoisomers of alkenes are named by writing** *cis-* **or** *trans-* **before the number identifying the location of the double bond.** Chapter 20 in the text addresses naming stereoisomers.

4. **The rules for naming substituted alkanes are followed to name and locate any other groups on the chain.**

$$
\underset{\text{2-Methyl-3-heptene}}{\overset{\displaystyle\overset{\text{CH}_3}{|}}{\text{CH}_3\text{CHCH}=\text{CHCH}_2\text{CH}_2\text{CH}_3}}
\qquad
\underset{\text{2-Methyl-3-heptyne}}{\overset{\displaystyle\overset{\text{CH}_3}{|}}{\text{CH}_3\text{CHC}\equiv\text{CCH}_2\text{CH}_2\text{CH}_3}}
$$

Halogens attached to an alkane, alkene, or alkyne are named as fluoro- (F–), chloro- (Cl–), bromo- (Br–), or iodo- (I–) and are located by using the same numbering system described for alkyl groups.

Benzene Derivatives

Naming compounds containing substituted benzene rings is less systematic than naming hydrocarbons. Many compounds have common names that are incorporated into accepted names, but for simple substituted benzene rings, the following rules may be applied.

1. **For monosubstituted benzene rings,** a prefix identifying the group is appended to the parent name benzene:

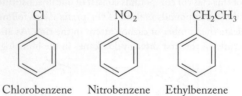

Chlorobenzene   Nitrobenzene   Ethylbenzene

2. **For disubstituted benzene rings,** three isomers are possible. The relative position of the substituent groups is indicated by numbers in IUPAC nomenclature, but the set of prefixes shown are very commonly used as well:

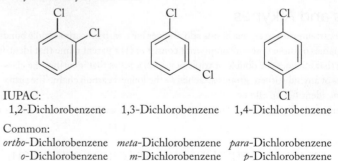

IUPAC:
   1,2-Dichlorobenzene   1,3-Dichlorobenzene   1,4-Dichlorobenzene

Common:
ortho-Dichlorobenzene   *meta*-Dichlorobenzene   *para*-Dichlorobenzene
   o-Dichlorobenzene   *m*-Dichlorobenzene   *p*-Dichlorobenzene

3. **When three or more groups are attached to a benzene ring,** the lowest possible numbers are assigned to locate the groups with respect to each other.

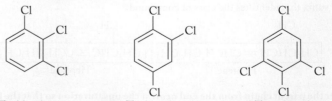

1,2,3-Trichlorobenzene   1,2,4-Trichlorobenzene   1,2,3,5-Tetrachlorobenzene
(*Note*: Not 1,3,4-trichlorobenzene; and not 1,3,4,5-tetrachlorobenzene.)

Hydrocarbons Containing Other Functional Groups

The same basic principles developed for naming alkanes apply to naming hydrocarbons with functional groups other than alkyl groups The name must identify the carbon skeleton, locate the functional group, and contain a suffix that defines the class of compound. The following examples give the suffixes for some common functional groups; when suffixes are used, they replace the final -e in the name of the parent alkane. Other functional groups may be identified by including the name of the class of compounds in the name of the molecule.

Alcohols: Suffix -ol

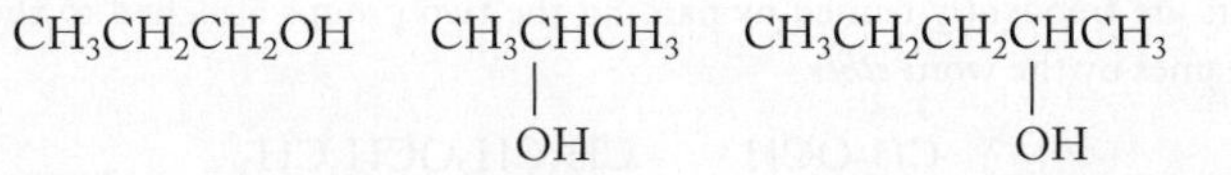

$CH_3CH_2CH_2OH$ CH_3CHCH_3 $CH_3CH_2CH_2CHCH_3$
                           | |
                          OH   OH

1-Propanol   2-Propanol   2-Pentanol

Aldehydes: Suffix -al

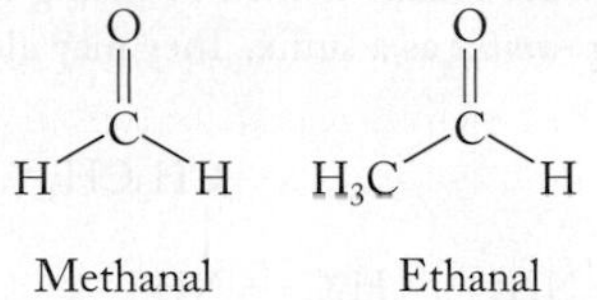

IUPAC: Methanal   Ethanal

Common: Formaldehyde   Acetaldehyde

Because the aldehyde group can only be on a terminal carbon, no number is necessary to locate it on the carbon chain.

Ketones: Suffix -one   The location of the carbonyl is given by a number, and the chain is numbered so that the carbonyl carbon has the lowest possible value. Many ketones also have common names generated by identifying the hydrocarbon groups on both sides of the carbonyl group.

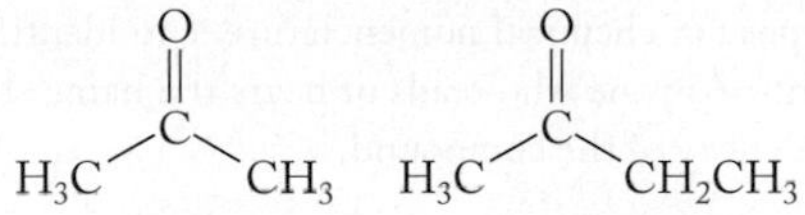

IUPAC: Propan-2-one   Butan-2-one

Common: Acetone   Methyl ethyl ketone

Carboxylic Acids: Suffix -oic acid   The carboxylic acid group is by definition carbon 1, so no number identifying its location is included in the name.

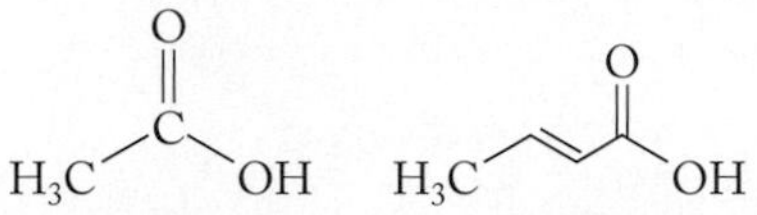

IUPAC: Ethanoic acid   trans-2-Butenoic acid

Common: Acetic acid

Salts of Carboxylic Acids   Salts are named with the cation first, followed by the anion name of the acid from which -ic acid is dropped and the suffix -ate is added. The sodium salt of acetic acid is sodium acetate.

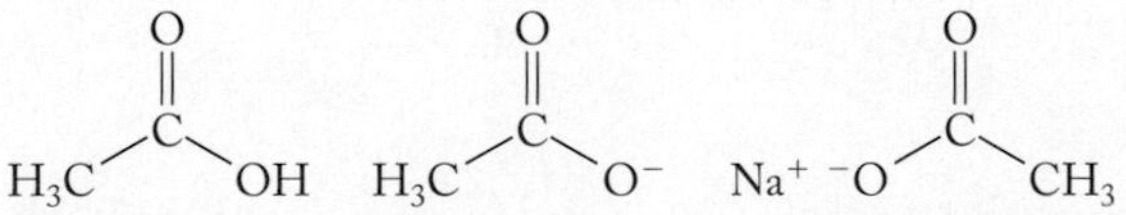

Acetic acid   Acetate ion   Sodium acetate

Esters   Esters are viewed as derivatives of carboxylic acids. They are named in a manner analogous to that of salts. The alkyl group comes first, followed by the name of the carboxylate anion.

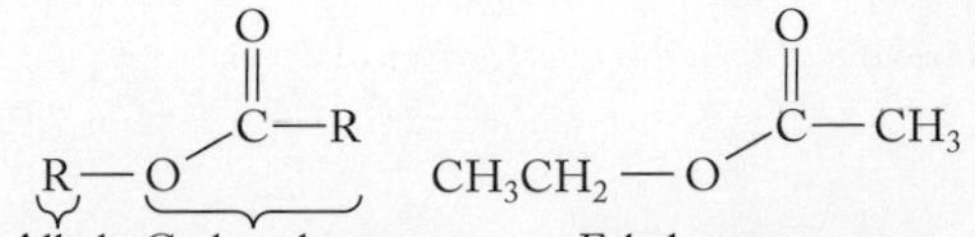

Alkyl Carboxylate   Ethyl acetate

Amides   Amides are also derivatives of carboxylic acids. They are named by replacing *-ic acid* (of the common names) or *-oic acid* of the IUPAC names with *-amide*.

$$
\underbrace{R - \overset{\displaystyle O}{\overset{\|}{C}} \diagdown_{NH_2}}_{\text{Parent acid -amide}}
\qquad
\underset{\substack{\text{Acetamide}\\ \text{(ethanamide)}}}{H_3C - \overset{\displaystyle O}{\overset{\|}{C}} \diagdown_{NH_2}}
$$

Ethers   Ethers are frequently named by naming the two groups attached to the oxygen and following those names by the word *ether*.

$$
\underset{\text{Dimethyl ether}}{CH_3OCH_3}
\qquad
\underset{\text{Diethyl ether}}{CH_3CH_2OCH_2CH_3}
$$

Amines   Aliphatic amines are usually named by listing the group or groups attached to the nitrogen and then appending *-amine* as a suffix. They may also be named by prefixing *amino-* to the name of the parent chain.

$$
\underset{\text{Methylamine}}{H_3C - NH_2}
\qquad
\underset{\text{Ethylmethylamine}}{H_3C - \overset{\overset{\displaystyle CH_2CH_3}{\big|}}{N}H}
\qquad
\underset{\text{2-Aminoethanol}}{H_2NCH_2CH_2OH}
$$

This brief summary will enable you to understand the names of organic compounds used in this book. IUPAC rules are much more extensive than this and can be applied to all varieties of carbon compounds, including those with multiple functional groups. It is important to recognize that the rules of systematic nomenclature do not necessarily lead to a unique name for each compound, but they do always lead to an unambiguous one. Furthermore, common names are still used frequently in organic chemistry because the systematic alternatives do not improve communication. Remember that the main purpose of chemical nomenclature is to identify a chemical species by means of written or spoken words. Anyone who reads or hears the name should be able to deduce the structure and thereby the identity of the compound.

Glossary

A

absolute temperature   Temperature expressed in kelvins on the absolute (Kelvin) temperature scale, on which 0 K is the lowest possible temperature. (Ch. 6)

absolute zero (0 K)   The zero point on the Kelvin temperature scale; theoretically the lowest temperature possible. (Ch. 1)

absorbance (A)   A measure of the quantity of light absorbed by a sample. (Ch. 4)

accuracy   Agreement between one or more experimental values and the true value. (Ch. 1)

acid (Brønsted–Lowry acid)   A proton donor. (Ch. 4)

activated complex   A species formed in a chemical reaction when molecules have both the proper orientation and enough energy to react with each other. (Ch. 13)

activation energy (E_a)   The minimum energy molecules need to react when they collide. (Ch. 13)

active site   The location on an enzyme where a reactive substance binds. (Ch. 20)

activity series   A qualitative ordering of the oxidizing ability of metals and their cations. (Ch. 4)

actual yield   The amount of product obtained from a chemical reaction, which is often less than the theoretical yield. (Ch. 3)

addition polymer   A polymer formed by an addition reaction. (Ch. 12)

addition reaction   A reaction in which two reactants couple to form one product without the loss of any atoms or molecules. (Ch. 12)

alcohol   An organic compound containing the OH functional group. (Ch. 2)

aldehyde   An organic compound having a carbonyl group with a single bond to a hydrogen atom and a single bond to another atom or group of atoms designated as R– in the general formula RCHO. (Ch. 8)

alkali metals   The elements in group 1 of the periodic table. (Ch. 2)

alkaline earth metals   The elements in group 2 of the periodic table. (Ch. 2)

alkane   A hydrocarbon in which all the bonds are single bonds. (Ch. 2)

alkene   A hydrocarbon containing one or more carbon–carbon double bonds. (Ch. 2)

alkyne   A hydrocarbon containing one or more carbon–carbon triple bonds. (Ch. 2)

allotropes   Different molecular forms of the same element, such as oxygen (O_2) and ozone (O_3). (Ch. 8)

alloy   A blend of a host metal and one or more other elements, which may or may not be metals, that are added to change the properties of the host metal. (Ch. 12)

alpha (α) decay   A nuclear reaction in which an unstable nuclide spontaneously emits an alpha particle. (Ch. 19)

α helix   A coil in a protein chain's secondary structure. (Ch. 20)

alpha (α) particle   A radioactive emission with a charge of 2+ and a mass equivalent to that of a helium nucleus. (Ch. 2)

amide   An organic compound in which the same carbon atom is single bonded to a nitrogen atom and double bonded to an oxygen atom. (Ch. 12)

amine   An organic compound that functions as a base and has the general formula RNH_2, R_2NH, or R_3N, where R is any organic subgroup. (Ch. 4)

amino acid   A molecule that contains at least one amine functional group and one carboxylic acid functional group; in an α-amino acid, the two functional groups are attached to the same (α) carbon atom. (Ch. 20)

Amontons's law   The principle that the pressure of a quantity of gas at constant volume is directly proportional to its absolute temperature. (Ch. 6)

amorphous solid   A solid that lacks long-range order for the atoms, ions, or molecules in its structure. (Ch. 12)

amphiprotic   Describes a substance that can behave as either a proton acceptor or a proton donor. (Ch. 4)

amplitude   The height of the crest or depth of the trough of a wave with respect to the center line of the wave. (Ch. 7)

angular   (also called *bent*) Molecular geometry about a central atom with a steric number of 3 and one lone pair or a steric number of 4 and two lone pairs. (Ch. 9)

angular momentum quantum number (ℓ)   An integer having any value from 0 to $n-1$ that defines the shape of an orbital. (Ch. 7)

anion   An ion with a negative charge. (Ch. 1)

anode   An electrode at which an oxidation half-reaction (loss of electrons) takes place. (Ch. 18)

antibonding orbital   Term in MO theory describing regions of electron density in a molecule that destabilize the molecule because they decrease the electron density between nuclear centers. (Ch. 9)

antimatter   Particles that are the charge opposites of normal subatomic particles. (Ch. 19)

aromatic compound   A cyclic, planar compound with delocalized π (pi) electrons above and below the plane of the molecule. (Ch. 9)

Arrhenius acid   A compound that produces H_3O^+ ions in aqueous solution. (Ch. 15)

Arrhenius base   A compound that produces OH^- ions in aqueous solution. (Ch. 15)

Arrhenius equation   An equation relating the rate constant of a reaction to absolute temperature (T), the activation energy of the reaction (E_a), and the frequency factor (A). (Ch. 13)

atmosphere (1 atm)   A unit of pressure based on Earth's average atmospheric pressure at sea level. (Ch. 6)

atmospheric pressure (P_{atm})   The force exerted by the gases surrounding Earth on Earth's surface and on all surfaces of all objects. (Ch. 6)

atom   The smallest particle of an element that cannot be chemically or mechanically divided into smaller particles. (Ch. 1)

atomic absorption spectrum   A characteristic series of dark lines produced when free, gaseous atoms are illuminated by a continuous source of radiation. (Ch. 7)

atomic emission spectrum   A characteristic series of bright lines produced by high-temperature atoms. (Ch. 7)

atomic mass unit (amu)   Unit used to express the relative masses of atoms and subatomic particles; it is exactly 1/12 the mass of one atom of carbon with six protons and six neutrons in its nucleus. (Ch. 2)

atomic number (Z)   The number of protons in the nucleus of an atom. (Ch. 2)

atomic solid A solid formed by weak attractions between noble gas atoms. (Ch. 12)

aufbau principle The method of building electron configurations of atoms by adding one electron at a time as atomic number increases across the rows of the periodic table; each electron goes into the lowest-energy orbital available. (Ch. 7)

autoionization The process that produces equal and very small concentrations of H_3O^+ and OH^- ions in pure water. (Ch. 15)

average atomic mass A weighted average of the masses of all the isotopes of an element, calculated by multiplying the natural abundance of each isotope by its mass in atomic mass units and then summing these products. (Ch. 2)

Avogadro's law The principle that the volume of a gas at a given temperature and pressure is proportional to the quantity of the gas. (Ch. 6)

Avogadro's number (N_A) The number of carbon atoms in exactly 12 grams of the carbon-12 isotope; $N_A = 6.022 \times 10^{23}$. It is the number of particles in one mole. (Ch. 3)

B

band gap (E_g) The energy gap between the valence and conduction bands. (Ch. 9)

band theory An extension of molecular orbital theory that describes bonding in solids. (Ch. 9)

barometer An instrument that measures atmospheric pressure. (Ch. 6)

base (Brønsted–Lowry base) A proton acceptor. (Ch. 4)

becquerel (Bq) The SI unit of radioactivity; one becquerel equals one decay event per second. (Ch. 19)

Beer's Law The relation of the absorbance of a solution (A) to concentration (c), the light's path length (b), and the solute's molar absorptivity (ε) by the equation $A = \varepsilon bc$. (Ch. 4)

belt of stability The region on a graph of number of neutrons versus number of protons that includes all stable nuclei. (Ch. 19)

bent See *angular*.

beta (β) decay The spontaneous ejection of a β particle by a neutron-rich nucleus. (Ch. 2, Ch. 19)

beta (β) particle A radioactive emission equivalent to a high-energy electron. (Ch. 2)

β-pleated sheet A puckered two-dimensional array of protein strands held together by hydrogen bonds. (Ch. 20)

bimolecular step A step in a reaction mechanism involving a collision between two molecules. (Ch. 13)

binding energy (BE) The energy that would be released when free nucleons combine to form the nucleus of an atom. It is also the energy needed to split the nucleus into free nucleons. (Ch. 19)

biocatalysis The use of enzymes to catalyze reactions on a large scale; it is becoming especially important in processes that involve chiral materials. (Ch. 13)

biochemistry The study of chemical processes taking place in living organisms. (Ch. 20)

biomass The total mass of organic matter in any given ecological system. (Ch. 20)

biomolecule Any molecule produced by a living organism. (Ch. 20)

body-centered cubic (bcc) unit cell An array of atoms or molecules with one particle at each of the eight corners of a cube and one at the center of the cell. (Ch. 12)

bomb calorimeter A constant-volume device used to measure the energy released during a combustion reaction. (Ch. 5)

bond angle The angle (in degrees) defined by lines joining the centers of two atoms to the center of a third atom to which they are chemically bonded. (Ch. 9)

bond dipole Separation of electrical charge created when atoms with different electronegativities form a covalent bond. (Ch. 9)

bond energy The energy needed to break 1 mole of a particular covalent bond in a molecule or in a polyatomic ion in the gas phase. (Ch. 8)

bond length The distance between the nuclear centers of two atoms joined in a bond. (Ch. 8)

bond order The number of bonds between atoms: 1 for a single bond, 2 for a double bond, and 3 for a triple bond. (Ch. 8)

bond polarity A measure of the extent to which bonding electrons are unequally shared due to differences in electronegativity of the bonded atoms. (Ch. 8)

bonding capacity The number of covalent bonds an atom forms to have an octet of electrons in its valence shell. (Ch. 8)

bonding orbital Term in MO theory describing regions of increased electron density between nuclear centers that serve to hold atoms together in molecules. (Ch. 9)

bonding pair A pair of electrons shared between two atoms. (Ch. 8)

Born–Haber cycle A series of steps with corresponding enthalpy changes that describes the formation of an ionic solid from its constituent elements. (Ch. 11)

Boyle's law The principle that the volume of a given amount of gas at constant temperature is inversely proportional to its pressure. (Ch. 6)

branched-chain hydrocarbon A hydrocarbon in which the chain of carbon atoms is not linear. (Ch. 5)

breeder reactor A nuclear reactor in which fissionable material is produced during normal reactor operation. (Ch. 19)

buffer capacity The quantity of acid or base that a pH buffer can neutralize while keeping its pH within a desired range. (Ch. 16)

C

calibration curve A graph showing how a measurable property, such as absorbance, varies for a set of standard samples of known concentration that can later be used to identify an unknown concentration from a measured absorbance. (Ch. 4)

calorie (cal) The amount of energy necessary to raise the temperature of 1 gram of water by 1°C, from 14.5°C to 15.5°C. (Ch. 5)

calorimeter A device used to measure the absorption or release of energy by a physical change or chemical process. (Ch. 5)

calorimeter constant ($C_{calorimeter}$) The heat capacity of a calorimeter. (Ch. 5)

calorimetry The measurement of the quantity of energy transferred during a physical change or chemical process. (Ch. 5)

capillary action The rise of a liquid in a narrow tube as a result of adhesive forces between the liquid and the tube and cohesive forces within the liquid. (Ch. 10)

carbohydrate An organic molecule with the generic formula $C_x(H_2O)_y$. (Ch. 20)

carbonyl group A carbon atom with a double bond to an oxygen atom. (Ch. 8)

carboxylic acid An organic compound containing the –COOH group. (Ch. 4)

catalyst A substance added to a reaction that increases the rate of the reaction but is not consumed in the process. (Ch. 13)

cathode An electrode at which a reduction half-reaction (gain of electrons) takes place. (Ch. 18)

cathode rays Streams of electrons emitted by the cathode in a partially evacuated tube. (Ch. 2)

cation An ion with a positive charge. (Ch. 1)

cell diagram Symbols that show how the components of an electrochemical cell are connected. (Ch. 18)

cell potential (E_{cell})   The force with which an electrochemical cell can pump electrons through an external circuit. (Ch. 18)

chain reaction   A self-sustaining series of fission reactions in which the neutrons released when nuclei split apart initiate additional fission events and sustain the reaction. (Ch. 19)

Charles's law   The principle that the volume of a fixed quantity of gas at constant pressure is directly proportional to its absolute temperature. (Ch. 6)

chelate effect   The greater affinity of metal ions for polydentate ligands than for monodentate ligands. (Ch. 22)

chelation   The interaction of a metal with a polydentate ligand (chelating agent); pairs of electrons on one molecule of the ligand occupy two or more coordination sites on the central metal. (Ch. 22)

chemical bond   A force that holds two atoms or ions in a compound together. (Ch. 1)

chemical equation   Notation in which chemical formulas express the identities and their coefficients express the quantities of substances involved in a chemical reaction. (Ch. 1)

chemical equilibrium   A dynamic process in which the concentrations of reactants and products remain constant over time and the rate of the reaction in the forward direction matches its rate in the reverse direction. (Ch. 14)

chemical formula   Notation for representing elements and compounds; consists of the symbols of the constituent elements and subscripts identifying the number of atoms of each element present. (Ch. 1)

chemical kinetics   The study of the rates of change of concentrations of substances involved in chemical reactions. (Ch. 13)

chemical property   A property of a substance that can be observed only by reacting it to form another substance. (Ch. 1)

chemical reaction   The transformation of one or more substances into different substances. (Ch. 1)

chemistry   The study of the composition, structure, and properties of matter, and of the energy consumed or given off when matter undergoes a change. (Ch. 1)

chirality   Property of a molecule that is not superimposable on its mirror image. (Ch. 9)

cis isomer   (also called *Z isomer*) A molecule with two like groups (such as two R groups or two hydrogen atoms) on the same side of the molecule. (Ch. 20)

Clausius–Clapeyron equation   A relationship between the vapor pressure of a substance at two temperatures and its heat of vaporization. (Ch. 10)

closed system   A system that exchanges energy but not matter with the surroundings. (Ch. 5)

codon   A three-nucleotide sequence that codes for a specific amino acid. (Ch. 20)

coenzyme   Organic molecule that, like an enzyme, accelerates the rate of biochemical reactions. (Ch. 22)

colligative properties   Characteristics of solutions that depend on the concentration and not the identity of particles dissolved in the solvent. (Ch. 11)

combination reaction   A reaction in which two (or more) substances combine to form one product. (Ch. 3)

combined gas law   (also called *general gas equation*) The principle relating the pressure, volume, and temperature of a quantity of an ideal gas:

$$\frac{P_1 V_1}{T_1} = \frac{P_2 V_2}{T_2}$$

(Ch. 6)

combustion analysis   A laboratory procedure for determining the composition of a substance by burning it completely in oxygen to produce known compounds whose masses are used to determine the composition of the original material. (Ch. 3)

combustion reaction   A heat-producing reaction between oxygen and another element or compound. (Ch. 3)

common-ion effect   The shift in the position of an equilibrium caused by the addition of an ion taking part in the reaction. (Ch. 16)

complex ion   An ionic species consisting of a metal ion bonded to one or more Lewis bases. (Ch. 16)

compound   A pure substance that is composed of two or more elements bonded together in fixed proportions and that can be broken down into those elements by a chemical reaction. (Ch. 1)

concentration   The amount of a solute in a particular amount of solvent or solution. (Ch. 4)

condensation polymer   A polymer formed by a condensation reaction. (Ch. 12)

condensation reaction   A reaction in which two molecules combine to form a larger molecule and a small molecule, typically water. (Ch. 12)

conduction band   In metals, an unoccupied band higher in energy than a valence band, in which electrons are free to migrate. (Ch. 9)

confidence interval   A range of values that has a specified probability of containing the true value of a measurement. (Ch. 1)

conjugate acid   The acid formed when a Brønsted–Lowry base accepts a H^+ ion. (Ch. 15)

conjugate acid–base pair   A Brønsted–Lowry acid and base that differ from each other only by a H^+ ion: acid ⇌ conjugate base + H^+. (Ch. 15)

conjugate base   The base formed when a Brønsted–Lowry acid donates a H^+ ion. (Ch. 15)

constitutional isomers   Molecules with the same chemical formula but different connections between the atoms. (Ch. 20)

conversion factor   A fraction in which the numerator is equivalent to the denominator, even though they are expressed in different units, making the value of the fraction one. (Ch. 1)

coordinate bond   A covalent bond formed when one anion or molecule donates a pair of electrons to another ion or molecule. (Ch. 16)

coordination compound   A compound made up of at least one complex ion. (Ch. 22)

coordination number   The number of sites occupied by ligands around a metal ion in a complex. (Ch. 22)

copolymer   A polymer formed from the chemical reaction of two different monomers. (Ch. 12)

core electrons   Electrons in the filled, inner shells of an atom or ion that are not involved in chemical reactions. (Ch. 7)

corrosion   A process in which a metal is oxidized by substances in its environment. (Ch. 18)

counterion   An ion whose charge balances the charge of a complex ion in a coordination compound. (Ch. 22)

covalent bond   A bond between two atoms created by sharing one or more pairs of electrons. (Ch. 2)

covalent network solid   A solid formed by covalent bonds among nonmetal atoms in an extended array. (Ch. 12)

critical mass   The minimum quantity of fissionable material needed to sustain a chain reaction. (Ch. 19)

critical point   A specific temperature and pressure at which the liquid and gas phases of a substance have the same density and are indistinguishable from each other. (Ch. 10)

crystal field splitting   The separation of a set of *d* orbitals into subsets with different energies as a result of interactions between electrons in those orbitals and lone pairs of electrons in ligands. (Ch. 22)

crystal field splitting energy (Δ)   The difference in energy between subsets of *d* orbitals split by interactions in a crystal field. (Ch. 22)

crystal lattice   A three-dimensional repeating array of particles (atoms, ions, or molecules) in a crystalline solid. (Ch. 12)

crystal structure   A particular arrangement in three-dimensional space that specifies the positions of the particles (atoms, ions, or molecules) in relation to one another in a crystalline solid. (Ch. 12)

crystalline solid   A solid made of an ordered array of atoms, ions, or molecules. (Ch. 12)

cubic closest-packed (ccp)   A crystal lattice in which the layers of atoms or ions in face-centered cubic unit cells have an *abcabc . . .* stacking pattern. (Ch. 12)

curie (Ci)   Non-SI unit of radioactivity; 1 Ci = 3.70×10^{10} decay events per second. (Ch. 19)

D

dalton (Da)   A unit of mass identical to 1 atomic mass unit. (Ch. 2)

Dalton's law of partial pressures   The principle that the total pressure of any mixture of gases equals the sum of the partial pressure of each gas in the mixture. (Ch. 6)

degenerate (orbitals)   Describes orbitals of the same energy. (Ch. 7)

degree of ionization   The ratio of the quantity of a substance that is ionized to the concentration of the substance before ionization; when expressed as a percentage, called *percent ionization*. (Ch. 15)

delocalized electrons   Electrons that are shared among more than two atoms. (Ch. 8)

density (*d*)   The ratio of the mass (*m*) of an object to its volume (*V*). (Ch. 1)

deposition   Transformation of a vapor (gas) directly into a solid. (Ch. 1)

diamagnetic   Describes a substance with no unpaired electrons that is weakly repelled by a magnetic field. (Ch. 9)

diffusion   The spread of one substance (usually a gas or liquid) through another. (Ch. 6)

dilution   The process of lowering the concentration of a solution by adding more solvent. (Ch. 4)

dipole moment (*μ*)   A measure of the degree to which a molecule aligns itself in a strong electric field; a quantitative expression of the polarity of a molecule. (Ch. 9)

dipole–dipole interaction   An attractive force between polar molecules. (Ch. 10)

dipole–induced dipole interaction   An attraction between a polar molecule and the oppositely charged pole it temporarily induces in another molecule. (Ch. 10)

dispersion force   (also called *London force*) An intermolecular force between nonpolar molecules caused by the presence of temporary dipoles in the molecules. (Ch. 10)

double bond   A chemical bond in which two atoms share two pairs of electrons. (Ch. 8)

E

***E* isomer**   See *trans isomer*.

effective nuclear charge (*Z*_{eff})   The attraction toward the nucleus experienced by an electron in an atom, equal to the positive charge on the nucleus reduced by the extent to which other electrons in the atom shield the electron from the nucleus. (Ch. 7)

effusion   The process by which a gas escapes from its container through a tiny hole into a region of lower pressure. (Ch. 6)

electrochemical cell   An apparatus that converts chemical energy into electrical work or electrical work into chemical energy. (Ch. 18)

electrochemistry   The branch of chemistry that examines the transformations between chemical and electrical energy. (Ch. 18)

electrolysis   A process in which electrical energy is used to drive a nonspontaneous chemical reaction. (Ch. 18)

electrolyte   A substance that dissociates into ions when it dissolves in water, enhancing the conductivity of the solvent. (Ch. 4)

electrolytic cell   A device in which an external source of electrical energy does work on a chemical system, turning reactant(s) into higher-energy product(s). (Ch. 18)

electromagnetic radiation   Any form of radiant energy in the electromagnetic spectrum. (Ch. 7)

electromagnetic spectrum   A continuous range of radiant energy that includes radio waves, microwaves, infrared radiation, visible light, ultraviolet radiation, X-rays, and gamma rays. (Ch. 7)

electron   A subatomic particle that has a negative charge and little mass. (Ch. 2)

electron affinity (EA)   The energy change that occurs when 1 mole of electrons combines with 1 mole of atoms or monatomic cations in the gas phase. (Ch. 7)

electron capture   A nuclear reaction in which a neutron-poor nucleus draws in one of its surrounding electrons, which transforms a proton in the nucleus into a neutron. (Ch. 19)

electron configuration   The distribution of electrons among the orbitals of an atom or ion. (Ch. 7)

electron transition   Movement of an electron between energy levels. (Ch. 7)

electronegativity   A relative measure of the ability of an atom to attract electrons in a bond to itself. (Ch. 8)

electron-pair geometry   The three-dimensional arrangement of bonding pairs and lone pairs of electrons about a central atom. (Ch. 9)

electrostatic potential energy (*E*_{el})   The energy a particle has because of its electrostatic charge and its position with respect to another particle; it is directly proportional to the product of the charges of the particles and inversely proportional to the distance between them. (Ch. 5)

element   A pure substance that cannot be separated into simpler substances. (Ch. 1)

elementary step   A molecular view of a single process taking place in a chemical reaction. (Ch. 13)

empirical formula   A formula showing the smallest whole-number ratio of the elements in a compound. (Ch. 2)

enantiomer   One of a pair of optical isomers of a compound. (Ch. 20)

end point   The point in a titration that is reached when just enough standard solution has been added to cause the indicator to change color. (Ch. 4)

endothermic process   A thermochemical process in which energy flows from the surroundings into the system. (Ch. 5)

energy   The capacity to do work. (Ch. 1)

energy profile   A graph showing the changes in energy for a reaction as a function of the progress of the reaction from reactants to products. (Ch. 13)

enthalpy (*H*)   The sum of the internal energy and the pressure–volume product of a system; $H = E + PV$. (Ch. 5)

enthalpy change (Δ*H*)   The heat absorbed by an endothermic process or given off by an exothermic process occurring at constant pressure. (Ch. 5)

enthalpy of hydration (Δ*H*_{hydration})   The energy change when gas-phase ions dissolve in a solvent. (Ch. 11)

enthalpy of reaction (Δ*H*_{rxn})   The energy absorbed or given off by a chemical reaction under conditions of constant pressure; also called *heat of reaction*. (Ch. 5)

enthalpy of solution (Δ*H*_{solution})   The overall energy change when a solute is dissolved in a solvent. (Ch. 11)

entropy (*S*)   A measure of the dispersion of energy in a system at a specific temperature. (Ch. 17)

enzyme   A protein that catalyzes a reaction. (Ch. 13)

equilibrium constant (*K*)   The numerical value of the equilibrium constant expression of a reversible chemical reaction at a particular temperature. (Ch. 14)

equilibrium constant expression   The ratio of the concentrations or partial pressures of products to reactants at equilibrium, with each term raised to a power equal to the coefficient of that substance in the balanced chemical equation for the reaction. (Ch. 14)

equilibrium (reversal) potential (*E*$_{ion}$)   An electrochemical potential that results from a concentration gradient of a particular ion on opposite sides of a cell membrane. (Ch. 21)

equivalence point   The point in a titration at which just enough titrant has been added to completely react the substance being analyzed. (Ch. 4)

essential amino acid   Any of the eight amino acids that make up peptides and proteins but are not synthesized in the human body and must be obtained through the food we eat. (Ch. 20)

essential element   Element present in tissue, blood, or other body fluids that has a physiological function. (Ch. 21)

ester   An organic compound in which the –OH of a carboxylic acid group is replaced by –OR, where R can be any organic group. (Ch. 12)

ether   Organic compound with the general formula R—O—R', where R is any alkyl group or aromatic ring; the two R groups may be different. (Ch. 10)

excited state   Any energy state above the ground state in an atom or ion. (Ch. 7)

exothermic process   A thermochemical process in which energy flows from a system into its surroundings. (Ch. 5)

extensive property   A property that varies with the quantity of the substance present. (Ch. 1)

F

face-centered cubic (fcc) unit cell   An array of closest-packed particles that has one particle at each of the eight corners of a cube and one more at the center of each face of the cube. (Ch. 12)

Faraday constant (*F*)   The magnitude of electrical charge in 1 mole of electrons. Its value to three significant figures is 9.65×10^4 C/mol e$^-$. (Ch. 18)

fat   A solid triglyceride containing primarily saturated fatty acids. (Ch. 20)

fatty acid   A carboxylic acid with a long hydrocarbon chain, which may be either saturated or unsaturated. (Ch. 20)

first law of thermodynamics   The principle that the energy gained or lost by a system must equal the energy lost or gained by the surroundings. (Ch. 5)

food value   The quantity of energy produced when a material consumed by an organism for sustenance is burned completely; it is typically reported in Calories (kilocalories) per gram of food. (Ch. 5)

formal charge (FC)   Value calculated for an atom in a molecule or polyatomic ion by determining the difference between the number of valence electrons in the free atom and the sum of lone-pair electrons plus half of the electrons in the atom's bonding pairs. (Ch. 8)

formation constant (*K*$_f$)   An equilibrium constant describing the formation of a metal complex from a free metal ion and its ligands. (Ch. 16)

formation reaction   A reaction in which 1 mole of a substance is formed from its component elements in their standard states. (Ch. 5)

formula mass   The mass of one formula unit of an ionic compound. (Ch. 3)

formula unit   The smallest electrically neutral unit of an ionic compound. (Ch. 2)

fractional distillation   A method of separating a mixture of compounds on the basis of their different boiling points. (Ch. 11)

Fraunhofer lines   A set of dark lines in the otherwise continuous solar spectrum. (Ch. 7)

free radical   An odd-electron molecule with an unpaired electron in its Lewis structure. (Ch. 8)

frequency (*ν*)   The number of crests of a wave that pass a stationary point of reference per second. (Ch. 7)

frequency factor (*A*)   The product of the frequency of molecular collisions and a factor that expresses the probability that the orientation of the molecules is appropriate for a reaction to occur. (Ch. 13)

fuel cell   A voltaic cell based on the oxidation of a continuously supplied fuel. The reaction is the equivalent of combustion, but chemical energy is converted directly into electrical energy. (Ch. 18)

fuel density   The quantity of energy released during the complete combustion of 1 liter of a liquid fuel. (Ch. 5)

fuel value   The quantity of energy released during the complete combustion of 1 gram of a substance. (Ch. 5)

functional group   A group of atoms in the molecular structure of an organic compound that imparts characteristic chemical and physical properties. (Ch. 2)

G

gas   A form of matter that has neither definite volume nor shape and that expands to fill its container; also called *vapor*. (Ch. 1)

Geiger counter   A portable device for determining nuclear radiation levels by measuring how much the radiation ionizes the gas in a sealed detector. (Ch. 19)

general gas equation   See *combined gas law*.

Gibbs free energy (*G*)   The maximum energy released by a process occurring at constant temperature and pressure that is available to do useful work. (Ch. 17)

glyceride   A lipid consisting of esters formed between fatty acids and the alcohol glycerol. (Ch. 20)

glycolysis   A series of reactions that converts glucose into pyruvate; a major anaerobic (no oxygen required) pathway for the metabolism of glucose in the cells of almost all living organisms. (Ch. 17)

glycosidic bond   A C—O—C bond between sugar molecules. (Ch. 20)

Graham's law of effusion   The principle that the rate of effusion of a gas is inversely proportional to the square root of its molar mass. (Ch. 6)

gray (Gy)   The SI unit of absorbed radiation; 1 Gy = 1 J/kg of tissue. (Ch. 19)

ground state   The most stable, lowest-energy state available to an atom or ion. (Ch. 7)

group   All the elements in the same column of the periodic table; also called *family*. (Ch. 2)

Grubbs' test   A statistical test used to detect an outlier in a set of data. (Ch. 1)

H

half-life (*t*$_{1/2}$)   The time in the course of a chemical reaction during which the concentration of a reactant decreases by half. (Ch. 13)

half-reaction   One of the two halves of an oxidation–reduction reaction; one half-reaction is the oxidation component, and the other is the reduction component. (Ch. 4)

halogens   The elements in group 17 of the periodic table. (Ch. 2)

heat   The energy transferred between objects because of a difference in their temperatures. (Ch. 5)

heat capacity (C_P)   The quantity of energy needed to raise the temperature of an object by 1°C at constant pressure. (Ch. 5)

Heisenberg uncertainty principle   The principle that we cannot determine both the position and the momentum of an electron in an atom at the same time. (Ch. 7)

Henderson–Hasselbalch equation   An equation used to calculate the pH of a solution in which the concentrations of acid and conjugate base are known. (Ch. 16)

Henry's law   The principle that the concentration of a sparingly soluble, chemically unreactive gas in a liquid is proportional to the partial pressure of the gas. (Ch. 10)

hertz (Hz)   The SI unit of frequency, equivalent to 1 cycle per second, or simply 1/s. (Ch. 7)

Hess's law   The principle that the enthalpy of reaction ΔH_{rxn} for a reaction that is the sum of two or more reactions is equal to the sum of the ΔH_{rxn} values of the constituent reactions; also called *Hess's law of constant heat of summation*. (Ch. 5)

heteroatom   Atom of an element other than carbon and hydrogen within a molecule of an organic compound. (Ch. 2)

heterogeneous catalyst   A catalyst in a different phase from the reactants. (Ch. 13)

heterogeneous equilibria   Equilibria that involve reactants and products in more than one phase. (Ch. 14)

heterogeneous mixture   A mixture in which the components are not distributed uniformly, so that the mixture contains distinct regions of different compositions. (Ch. 1)

heteropolymer   A polymer made of three or more different monomer units. (Ch. 12)

hexagonal closest-packed (hcp)   A crystal lattice in which the layers of atoms or ions in hexagonal unit cells have an *ababab . . .* stacking pattern. (Ch. 12)

hexagonal unit cell   An array of closest-packed particles that includes parts of four particles on the top and four on the bottom faces of a hexagonal prism and one particle in a middle layer. (Ch. 12)

homogeneous catalyst   A catalyst in the same phase as the reactants. (Ch. 13)

homogeneous equilibria   Equilibria that involve reactants and products in the same phase. (Ch. 14)

homogeneous mixture   A mixture in which the components are distributed uniformly throughout and have no visible boundaries or regions. (Ch. 1)

homopolymer   A polymer composed of only one kind of monomer. (Ch. 12)

Hund's rule   The lowest-energy electron configuration of an atom has the maximum number of unpaired electrons, all of which have the same spin, in degenerate orbitals. (Ch. 7)

hybrid atomic orbital   In valence bond theory, one of a set of equivalent orbitals about an atom created when specific atomic orbitals are mixed. (Ch. 9)

hybridization   In valence bond theory, the mixing of atomic orbitals to generate new sets of orbitals that are then available to form covalent bonds with other atoms. (Ch. 9)

hydrocarbon   An organic compound whose molecules are composed only of carbon and hydrogen atoms. (Ch. 2)

hydrogen bond   The strongest dipole–dipole interaction. It occurs between a hydrogen atom bonded to a small, highly electronegative element (O, N, F) and an atom of oxygen or nitrogen in another molecule. Molecules of HF also form hydrogen bonds. (Ch. 10)

hydrolysis   The reaction of water with another material. The hydrolysis of nonmetal oxides produces acids. (Ch. 4)

hydronium ion (H_3O^+)   A H^+ ion bonded to a molecule of water, H_2O; the form in which the hydrogen ion is found in an aqueous solution. (Ch. 4)

hydrophilic   A "water-loving," or attractive, interaction between a solute and water that promotes water solubility. (Ch. 10)

hydrophobic   A "water-fearing," or repulsive, interaction between a solute and water that diminishes water solubility. (Ch. 10)

hypothesis   A tentative and testable explanation for an observation or a series of observations. (Ch. 1)

I

i **factor**   See *van't Hoff factor*.

ideal gas   A gas whose behavior is predicted by the linear relationships defined by Boyle's, Charles's, Avogadro's, and Amontons's laws. (Ch. 6)

ideal gas equation   (also called *ideal gas law*) The principle relating the pressure, volume, number of moles, and temperature of an ideal gas; expressed as $PV = nRT$, where R is the universal gas constant. (Ch. 6)

ideal gas law   See *ideal gas equation*.

ideal solution   A solution that obeys Raoult's law. (Ch. 11)

induced dipole   The separation of charge produced in an atom or molecule by a momentary uneven distribution of electrons. (Ch. 10)

inhibitor   A compound that diminishes or destroys the ability of an enzyme to catalyze a reaction. (Ch. 20)

initial rate   The rate of a reaction at $t = 0$, immediately after the reactants are mixed. (Ch. 13)

inner coordination sphere   The ligands that are bound directly to a metal via coordinate bonds. (Ch. 22)

integrated rate law   A mathematical expression that describes the change in concentration of a reactant in a chemical reaction with time. (Ch. 13)

intensive property   A property that is independent of the amount of substance present. (Ch. 1)

intermediate   A species produced in one step of a reaction and consumed in a subsequent step. (Ch. 13)

internal energy (E)   The sum of all the kinetic and potential energies of all the components of a system. (Ch. 5)

interstitial alloy   An alloy in which atoms of the added element occupy the spaces between atoms of the host. (Ch. 12)

ion   A particle consisting of one or more atoms that has a net positive or negative charge. (Ch. 1)

ion channel   A group of helical proteins that penetrate cell membranes and allow selective transport of ions. (Ch. 21)

ion exchange   A process by which one ion is displaced by another. (Ch. 4)

ion pair   A cluster formed when a cation and an anion associate with each other in solution. (Ch. 11)

ion pump   A system of membrane proteins that exchange ions inside the cell with those in the intercellular fluid. (Ch. 21)

ion–dipole interaction   An attractive force between an ion and a molecule that has a permanent dipole moment. (Ch. 10)

ionic bond   A chemical bond that results from the electrostatic attraction between a cation and an anion. (Ch. 8)

ionic character   An estimate of the magnitude of charge separation in a covalent bond. (Ch. 8)

ionic compound   A compound composed of positively and negatively charged ions held together by electrostatic attraction. (Ch. 2)

ionic solid   A solid formed by ionic bonds among monatomic and/or polyatomic ions. (Ch. 12)

ionization energy (IE)   The quantity of energy needed to remove 1 mole of electrons from 1 mole of ground-state atoms or ions in the gas phase. (Ch. 7)

ionizing radiation   High-energy products of radioactive decay that can ionize molecules. (Ch. 19)

isoelectric point   See *pI*.

isoelectronic   Describes atoms or ions that have identical electron configurations. (Ch. 7)

isolated system   A system that exchanges neither energy nor matter with the surroundings. (Ch. 5)

isomers   Compounds with the same molecular formula but different arrangements of the atoms in their molecules. (Ch. 3)

isotopes   Atoms of an element containing different numbers of neutrons. (Ch. 2)

J

joule (J)   The SI unit of energy; 4.184 J = 1 cal. (Ch. 5)

K

kelvin (K)   The SI unit of temperature. (Ch. 1)

ketone   Organic molecule containing a carbonyl group bonded to a carbon atom on each side of the carbonyl carbon. (Ch. 9)

kinetic energy (KE)   The energy of an object in motion due to its mass (m) and its speed (u): $KE = \frac{1}{2}mu^2$. (Ch. 5)

kinetic molecular theory   A model that describes the behavior of ideal gases; all equations defining relationships between pressure, volume, temperature, and number of moles of gases can be derived from the theory. (Ch. 6)

L

lattice energy (U)   The enthalpy change that occurs when 1 mole of an ionic compound forms from its free ions in the gas phase. (Ch. 11)

law of conservation of energy   The principle that energy cannot be created or destroyed but can be converted from one form into another. (Ch. 5)

law of conservation of mass   The principle that the sum of the masses of the reactants in a chemical reaction is equal to the sum of the masses of the products. (Ch. 3)

law of constant composition   The principle that all samples of a particular compound contain the same elements combined in the same proportions. (Ch. 1)

law of mass action   The principle relating the balanced chemical equation of a reversible reaction to its mass action expression (or equilibrium constant expression). (Ch. 14)

law of multiple proportions   The principle that, when two masses of one element react with a given mass of another element to form two different compounds, the two masses of the first element have a ratio of two small whole numbers. (Ch. 2)

Le Châtelier's principle   The principle that a system at equilibrium responds to a stress in such a way that it relieves that stress. (Ch. 14)

leveling effect   The observation that all strong acids have the same strength in water and are completely converted into solutions of H_3O^+ ions; strong bases are likewise leveled in water and are completely converted into solutions of OH^- ions. (Ch. 15)

Lewis acid   A substance that *accepts* a lone pair of electrons in a chemical reaction. (Ch. 16)

Lewis base   A substance that *donates* a lone pair of electrons in a chemical reaction. (Ch. 16)

Lewis dot symbol   See *Lewis symbol*.

Lewis structure   A two-dimensional representation of the bonds and lone pairs of valence electrons in a molecule or polyatomic ion. (Ch. 8)

Lewis symbol   (also called *Lewis dot symbol*) The chemical symbol for an atom surrounded by one or more dots representing the valence electrons. (Ch. 8)

ligand   A Lewis base bonded to the central metal ion of a complex ion. (Ch. 16)

limiting reactant   A reactant that is consumed completely in a chemical reaction. The amount of product formed depends on the amount of the limiting reactant available. (Ch. 3)

linear   Molecular geometry about a central atom with a steric number of 2 and no lone pairs of electrons; the bond angle is 180°. (Ch. 9)

lipid   A class of water-insoluble, oily organic compounds that are common structural materials in cells. (Ch. 20)

lipid bilayer   A double layer of molecules whose polar head groups interact with water molecules and whose nonpolar tails interact with each other. (Ch. 20)

liquid   A form of matter that occupies a definite volume but flows to assume the shape of its container. (Ch. 1)

London force   See *dispersion force*.

lone pair   A pair of electrons that is not shared. (Ch. 8)

M

macrocyclic ligand   A ring containing multiple electron-pair donors that bind to a metal ion. (Ch. 22)

macromolecule   See *polymer*.

magnetic quantum number (m_ℓ)   Defines the orientation of an orbital in space; an integer that may have any value from $-\ell$ to $+\ell$, where ℓ is the angular momentum quantum number. (Ch. 7)

main group elements   (also called *representative elements*) The elements in groups 1, 2, and 13 through 18 of the periodic table. (Ch. 2)

major essential element   Essential element present in the body in average concentrations greater than 1 mg of element per gram of body mass. (Ch. 21)

manometer   An instrument for measuring the pressure exerted by a gas. (Ch. 6)

mass   The property that defines the quantity of matter in an object. (Ch. 1)

mass action expression   An expression equivalent in form to the equilibrium constant expression, but applied to reaction mixtures that may or may not be at equilibrium. (Ch. 14)

mass defect (Δm)   The difference between the mass of a stable nucleus and the masses of the individual nucleons that make it up. (Ch. 19)

mass number (A)   The number of nucleons in an atom. (Ch. 2)

mass spectrometer   An instrument that separates and counts ions according to their mass-to-charge ratio. (Ch. 3)

mass spectrum   A graph of the data from a mass spectrometer, where m/z ratios of the deflected particles are plotted against the number of particles with a particular mass. (Ch. 3)

matter   Anything that has mass and occupies space. (Ch. 1)

matter wave   The wave associated with any particle. (Ch. 7)

mean (arithmetic mean, $\bar{x}$)   An average calculated by summing a set of related values and dividing the sum by the number of values in the set. (Ch. 1)

membrane potential ($E_{membrane}$)   A weighted average of the equilibrium (reversal) potentials of the major ions based on concentration and permeability of the individual ions. (Ch. 21)

meniscus   The concave or convex surface of a liquid in a small-diameter tube. (Ch. 10)

messenger RNA (mRNA)   The form of RNA that carries the code for synthesizing proteins from DNA to the site of protein synthesis in a cell. (Ch. 20)

metallic bond   A chemical bond consisting of metal nuclei surrounded by a "sea" of shared electrons. (Ch. 8)

metallic solid   A solid formed by metallic bonds among atoms of metallic elements. (Ch. 12)

metalloids   (also called *semimetals*) Elements along the border between metals and nonmetals in the periodic table; they have some metallic and some nonmetallic properties. (Ch. 2)

metals   The elements on the left side of the periodic table that are typically shiny solids that conduct heat and electricity well and are malleable and ductile. (Ch. 2)

meter   The standard unit of length, equivalent to 39.37 inches. (Ch. 1)

methyl group   ($-CH_3$) A structural unit that can make only one bond. (Ch. 5)

methylene group   ($-CH_2-$) A structural unit that can make two bonds. (Ch. 5)

microstate   A unique distribution of particles among energy levels. (Ch. 17)

millimeters of mercury (mmHg)   (also called *torr*) A unit of pressure where 1 atm = 760 mmHg = 760 torr. (Ch. 6)

miscible   Capable of being mixed in any proportion (without reacting chemically). (Ch. 6)

mixture   A combination of pure substances in variable proportions in which the individual substances retain their chemical identities and can be separated from one another by a physical process. (Ch. 1)

molality (m)   Concentration expressed as the number of moles of solute per kilogram of solvent. (Ch. 11)

molar absorptivity (ε)   A measure of how well a compound or ion absorbs light. (Ch. 4)

molar enthalpy of fusion (ΔH_{fus})   The energy required to convert 1 mole of a solid substance at its melting point into the liquid state. (Ch. 5)

molar enthalpy of vaporization (ΔH_{vap})   The energy required to convert 1 mole of a liquid substance at its boiling point to the vapor state. (Ch. 5)

molar heat capacity (c_P)   The quantity of energy required to raise the temperature of 1 mole of a substance by 1°C at constant pressure. (Ch. 5)

molar mass ($\mathcal{M}$)   The mass of 1 mole of a substance. The molar mass of an element in grams per mole is numerically equal to that element's average atomic mass in atomic mass units. (Ch. 3)

molar volume   The volume occupied by 1 mole of an ideal gas at STP; 22.4 L. (Ch. 6)

molarity (M)   The concentration of a solution expressed in moles of solute per liter of solution ($M = n/V$). (Ch. 4)

mole (mol)   An amount of material (atoms, ions, or molecules) that contains Avogadro's number ($N_A = 6.022 \times 10^{23}$) of particles. (Ch. 3)

mole fraction (X_x)   The ratio of the number of moles of a component in a mixture to the total number of moles in the mixture. (Ch. 6)

molecular compound   A compound composed of molecules that contain the atoms of two or more elements. (Ch. 2)

molecular equation   A balanced equation describing a reaction in solution in which the reactants and products are written as undissociated compounds. (Ch. 4)

molecular formula   A notation showing the number and type of atoms present in one molecule of a molecular compound. (Ch. 2)

molecular geometry   The three-dimensional arrangement of the atoms in a molecule. (Ch. 9)

molecular ion (M$^+$)   The peak of highest mass in a mass spectrum; it has the same mass as the molecule from which it came. (Ch. 3)

molecular mass   The mass of one molecule of a molecular compound. (Ch. 3)

molecular orbital   A region of characteristic shape and energy where electrons in a molecule are delocalized over two or more atoms in a molecule. (Ch. 9)

molecular orbital diagram   In MO theory, an energy-level diagram showing the relative energies and electron occupancy of the molecular orbitals for a molecule. (Ch. 9)

molecular orbital (MO) theory   A bonding theory based on the mixing of atomic orbitals of similar shapes and energies to form molecular orbitals that extend across two or more atoms. (Ch. 9)

molecular recognition   The process by which molecules interact with other molecules in living tissues to produce a biological effect. (Ch. 9)

molecular solid   A solid formed by intermolecular attractive forces among neutral, covalently bonded molecules. (Ch. 12)

molecularity   The number of ions, atoms, or molecules involved in an elementary step in a reaction. (Ch. 13)

molecule   A collection of atoms chemically bonded together in characteristic proportions. (Ch. 1)

monodentate ligand   A species that forms only a single coordinate bond to a metal ion in a complex. (Ch. 22)

monomer   A small molecule that bonds with others like it to form polymers. (Ch. 12)

monoprotic acid   An acid that has one ionizable hydrogen atom per molecule. (Ch. 15)

monosaccharide   A single-sugar unit and the simplest carbohydrate. (Ch. 20)

N

nanoparticle   An approximately spherical sample of matter with dimensions smaller than 100 nanometers (1×10^{-7} m). (Ch. 12)

natural abundance   The proportion of a particular isotope, usually expressed as a percentage, relative to all the isotopes of that element in a natural sample. (Ch. 2)

Nernst equation   An equation relating the potential of a cell (or half-cell) reaction to its standard potential ($E°$) and to the concentrations of its reactants and products. (Ch. 18)

net ionic equation   A balanced equation that describes the actual reaction taking place in aqueous solution; it is obtained by eliminating the spectator ions from the overall ionic equation. (Ch. 4)

neutralization reaction   A reaction that takes place when an acid reacts with a base and produces a solution of a salt in water. (Ch. 4)

neutron   An electrically neutral (uncharged) subatomic particle found in the nucleus of an atom. (Ch. 2)

neutron capture   The absorption of a neutron by a nucleus. (Ch. 2)

noble gases   The elements in group 18 of the periodic table. (Ch. 2)

node   A location in a standing wave that experiences no displacement. In the context of orbitals, nodes are locations at which electron density goes to zero. (Ch. 7)

nonelectrolyte   A substance that does not dissociate into ions and therefore does not result in conductivity when dissolved in water. (Ch. 4)

nonessential element   Element present in humans that has no known function. (Ch. 21)

nonmetals   Elements with properties opposite those of metals, including poor conductivity of heat and electricity. (Ch. 2)

nonpolar covalent bond   A bond characterized by an even distribution of charge; electrons in the bonds are shared equally by the two atoms. (Ch. 8)

nonspontaneous process A process that occurs only as long as energy is continually added to the system. (Ch. 17)

normal boiling point The temperature at which the vapor pressure of a liquid equals 1 atm (760 torr). (Ch. 10)

n-type semiconductor Semiconductor containing electron-rich dopant atoms that contribute excess electrons. (Ch. 9)

nuclear chemistry A subdiscipline of chemistry that explores the properties and reactivity of radioactive atoms. (Ch. 19)

nuclear fission A nuclear reaction in which the nucleus of an element splits into two lighter nuclei. The process is usually accompanied by the release of one or more neutrons and energy. (Ch. 19)

nuclear fusion A nuclear reaction in which subatomic particles or atomic nuclei collide with each other at very high speeds and fuse, forming more massive nuclei and releasing energy. (Ch. 19)

nuclear reaction A process that changes the number of neutrons or protons in the nucleus of an atom. (Ch. 19)

nucleic acid One of a family of large molecules, which includes deoxyribonucleic acid (DNA) and ribonucleic acid (RNA), that stores the genetic blueprint of an organism and controls the production of proteins. (Ch. 20)

nucleon Either a proton or a neutron in a nucleus. (Ch. 2)

nucleosynthesis The natural formation of nuclei as a result of fusion and other nuclear processes. (Ch. 2)

nucleotide A monomer unit from which nucleic acids are made. (Ch. 20)

nucleus The positively charged center of an atom that contains nearly all the atom's mass. (Ch. 2)

nuclide An atom with particular numbers of neutrons and protons in its nucleus. (Ch. 2)

O

octahedral Molecular geometry about a central atom with a steric number of 6 and no lone pairs of electrons, in which all six sites are equivalent; four equatorial sites are 90° apart; two axial sites are 180° apart. (Ch. 9)

octet rule The tendency of atoms of main group elements to make bonds by gaining, losing, or sharing electrons to achieve a valence shell containing eight electrons, or four electron pairs. (Ch. 8)

oil A liquid triglyceride containing primarily unsaturated fatty acids. (Ch. 20)

open system A system that exchanges both energy and matter with the surroundings. (Ch. 5)

optical activity A property of a compound that refers to the rotation of a beam of plane-polarized light when it passes through a solution of the compound. (Ch. 20)

optical isomers Molecules that are not superimposable on their mirror images. (Ch. 20)

orbital A region around the nucleus of an atom where the probability of finding an electron is high; each orbital is defined by the square of the wave function (ψ^2) and identified by a unique combination of three quantum numbers. (Ch. 7)

orbital diagram Depiction of the arrangement of electrons in an atom or ion, using boxes to represent orbitals. (Ch. 7)

organic chemistry The study of organic compounds. (Ch. 2)

organic compound A molecule containing carbon atoms whose structure typically consists of carbon–carbon bonds and carbon–hydrogen bonds, and may include one or more heteroatoms such as oxygen, nitrogen, sulfur, phosphorus, or the halogens. (Ch. 2)

organometallic compound A molecule containing direct carbon–metal covalent bonds. (Ch. 22)

osmosis The flow of a fluid through a semipermeable membrane to balance the concentration of solutes in solutions on the two sides of the membrane. The solvent particles' flow proceeds from the more dilute solution into the more concentrated one. (Ch. 11)

osmotic pressure (Π) The pressure applied across a semipermeable membrane to stop the flow of solvent from the compartment containing pure solvent or a less concentrated solution to the compartment containing a more concentrated solution. The osmotic pressure of a solution increases with solute concentration, M, and with solution temperature, T. (Ch. 11)

outlier A data point that is distant from the other observations. (Ch. 1)

overall ionic equation A balanced equation that shows all the species, both ionic and molecular, present in a reaction occurring in aqueous solution. (Ch. 4)

overall reaction order The sum of the exponents of the concentration terms in the rate law. (Ch. 13)

overlap A term in valence bond theory describing bonds arising from two orbitals on different atoms that occupy the same region of space. (Ch. 9)

oxidation A chemical change in which a species loses electrons; the oxidation number of the species increases. (Ch. 4)

oxidation number (O.N.) (also called *oxidation state*) A positive or negative number based on the number of electrons an atom gains or loses when it forms an ion, or that it shares when it forms a covalent bond with another element; pure elements have an oxidation number of zero. (Ch. 4)

oxidation state See *oxidation number (O.N.)*.

oxidizing agent A substance in a redox reaction that contains the element being reduced. (Ch. 4)

oxoanion A polyatomic ion that contains oxygen in combination with one or more other elements. (Ch. 2)

P

packing efficiency The percentage of the total volume of a unit cell occupied by the atoms, ions, or molecules. (Ch. 12)

paramagnetic Describes a substance with unpaired electrons that is attracted to a magnetic field. (Ch. 9)

partial pressure The contribution to the total pressure made by one gas in a mixture of gases. (Ch. 6)

Pauli exclusion principle The principle that no two electrons in an atom can have the same set of four quantum numbers. (Ch. 7)

peptide A compound of two or more amino acids joined by peptide bonds. Small peptides containing up to 20 amino acids are oligopeptides; the term polypeptide is used for chains longer than 20 amino acids but shorter than proteins. (Ch. 20)

peptide bond The result of a condensation reaction between the carboxylic acid group of one amino acid and the amine group of another. (Ch. 20)

percent composition The composition of a compound expressed in terms of the percentage by mass of each element in the compound. (Ch. 3)

percent ionization See *degree of ionization*.

percent yield The ratio, expressed as a percentage, of the actual yield of a chemical reaction to the theoretical yield. (Ch. 3)

period A horizontal row in the periodic table. (Ch. 2)

periodic table of the elements A chart of the elements arranged in order of their atomic numbers and in a pattern based on their physical and chemical properties. (Ch. 2)

pH The negative logarithm of the hydronium ion concentration in an aqueous solution. (Ch. 15)

pH buffer   A solution that resists changes in pH when acids or bases are added to it; typically a solution of a weak acid and its conjugate base. (Ch. 16)

pH indicator   A water-soluble weak organic acid that changes color as it ionizes. (Ch. 16)

phase diagram   A graphical representation of how the stabilities of the physical states of a substance depend on temperature and pressure. (Ch. 10)

phospholipid   A molecule of glycerol with two fatty acid chains and one polar group containing a phosphate; phospholipids are major constituents of cell membranes. (Ch. 20)

phosphorylation   A reaction resulting in the addition of a phosphate group to an organic molecule. (Ch. 17)

photochemical smog   A mixture of gases formed in the lower atmosphere when sunlight interacts with compounds produced in internal combustion engines and other pollutants. (Ch. 13)

photoelectric effect   The phenomenon of light striking a metal surface and producing an electric current (a flow of electrons). (Ch. 7)

photon   A quantum of electromagnetic radiation. (Ch. 7)

physical process   A transformation of a sample of matter, such as a change in its physical state, that does not alter the chemical identity of any substance in the sample. (Ch. 1)

physical property   A property of a substance that can be observed without changing it into another substance. (Ch. 1)

pI   (also called *isoelectric point*) The pH at which molecules of the amino acid have, on average, zero charge. (Ch. 20)

pi (π) bond   A covalent bond in which electron density is greatest around—not along—the bonding axis. (Ch. 9)

pi (π) molecular orbitals   In MO theory, molecular orbitals formed by the mixing of atomic orbitals oriented above and below, or in front of and behind, the bonding axis. (Ch. 9)

Planck constant (h)   The proportionality constant between the energy and frequency of electromagnetic radiation expressed in $E = h\nu$; $h = 6.626 \times 10^{-34}$ J · s. (Ch. 7)

pOH   The negative logarithm of the hydroxide ion concentration in an aqueous solution. (Ch. 15)

polar covalent bond   A bond characterized by unequal sharing of bonding pairs of electrons between atoms. (Ch. 8)

polarizability   The relative ease with which the electron cloud in a molecule, ion, or atom can be distorted, inducing a temporary dipole. (Ch. 10)

polyatomic ion   A charged group of two or more atoms joined by covalent bonds. (Ch. 2)

polydentate ligand   A species that can form more than one coordinate bond per molecule. (Ch. 22)

polymer   (also called *macromolecule*) A very large molecule with high molar mass formed by bonding many small molecules of low molecular mass. (Ch. 12)

polyprotic acid   An acid that has two or more ionizable hydrogen atoms per molecule. (Ch. 15)

polysaccharide   A polymer of monosaccharides. (Ch. 20)

porphyrin   A type of tetradentate macrocyclic ligand. (Ch. 22)

positron ($_1^0\beta$)   A particle with the mass of an electron but with a positive charge. (Ch. 19)

positron emission   The spontaneous emission of a positron from a neutron-poor nucleus. (Ch. 19)

potential energy (PE)   The energy stored in an object because of its position. (Ch. 5)

precipitate   A solid product formed from a reaction in solution. (Ch. 4)

precipitation reaction   A reaction that produces an insoluble product upon mixing two solutions. (Ch. 4)

precision   Agreement between the results of multiple measurements that were carried out in the same way. (Ch. 1)

pressure (P)   The ratio of force to the surface area over which the force is applied. (Ch. 6)

pressure–volume (P–V) work   The work associated with the expansion or compression of a gas. (Ch. 5)

primary (1°) structure   The sequence in which amino acid monomers occur in a protein chain. (Ch. 20)

principal quantum number (n)   A positive integer describing the relative size and energy of an atomic orbital or group of orbitals in an atom. (Ch. 7)

product   A substance formed during a chemical reaction. (Ch. 3)

protein   A biological polymer made of amino acids. (Ch. 20)

proton   A positively charged subatomic particle present in the nucleus of an atom. (Ch. 2)

p-type semiconductor   Semiconductor containing electron-poor dopant atoms that cause a reduction in the number of electrons, which is equivalent to the presence of positively charged holes. (Ch. 9)

Q

quantized   Having values restricted to whole-number multiples of a specific base value. (Ch. 7)

quantum   (plural *quanta*) The smallest discrete quantity of a particular form of energy. (Ch. 7)

quantum mechanics   See *wave mechanics*.

quantum number   A number that specifies the energy, the probable location or orientation of an orbital, or the spin of an electron within an orbital. (Ch. 7)

quantum theory   A model based on the idea that energy is absorbed and emitted in discrete quantities of energy called quanta. (Ch. 7)

quarks   Elementary particles that combine to form neutrons and protons. (Ch. 2)

quaternary (4°) structure   The larger structure functioning as a single unit that results when two or more proteins associate. (Ch. 20)

R

racemic mixture   A sample containing equal amounts of both enantiomers of a compound. (Ch. 20)

radioactive decay   The spontaneous disintegration of unstable particles accompanied by the release of radiation. (Ch. 19)

radioactivity   The spontaneous emission of high-energy radiation and particles by materials. (Ch. 2)

radiocarbon dating   A method for establishing the age of a carbon-containing object by measuring the amount of radioactive carbon-14 remaining in the object. (Ch. 19)

radiometric dating   A method for determining the age of an object on the basis of the quantity of a radioactive nuclide and/or the products of its decay that the object contains. (Ch. 19)

radionuclide   An unstable nuclide that undergoes radioactive decay. (Ch. 19)

random coil   An irregular or rapidly changing part of the secondary structure of a protein. (Ch. 20)

Raoult's law   The principle that the vapor pressure of the solvent in a solution is equal to the vapor pressure of the pure solvent multiplied by the mole fraction of the solvent in the solution. (Ch. 11)

rate constant   The proportionality constant that relates the rate of a reaction to the concentrations of reactants. (Ch. 13)

rate law   An equation that defines the experimentally determined relation between the concentrations of reactants in a chemical reaction and the rate of that reaction. (Ch. 13)

rate-determining step   The slowest step in a multistep chemical reaction. (Ch. 13)

reactant   A substance consumed during a chemical reaction. (Ch. 3)

reaction mechanism   A set of steps that describe how a reaction occurs at the molecular level; the mechanism must be consistent with the experimentally determined rate law for the reaction. (Ch. 13)

reaction order   An experimentally determined number defining the dependence of the reaction rate on the concentration of a reactant. (Ch. 13)

reaction quotient (Q)   The numerical value of the mass action expression based on the concentrations or partial pressures of the reactants and products present at any time during a reaction. At equilibrium, $Q = K$. (Ch. 14)

reducing agent   A substance in a redox reaction that contains the element being oxidized. (Ch. 4)

reduction   A chemical change in which a species gains electrons; the oxidation number of the species decreases. (Ch. 4)

relative biological effectiveness (RBE)   A factor that accounts for the differences in physical damage caused by different types of radiation. (Ch. 19)

replication   The process by which one double-stranded DNA forms two new DNA molecules, each one containing one strand from the original molecule and one new strand. (Ch. 20)

representative elements   See *main group elements*.

resonance   A characteristic of electron distributions when two or more equivalent Lewis structures can be drawn for one compound. (Ch. 8)

resonance structure   One of two or more Lewis structures with the same arrangement of atoms but different arrangements of bonding pairs of electrons. (Ch. 8)

reverse osmosis   A process in which solvent is forced through semipermeable membranes, leaving a more concentrated solution behind. (Ch. 11)

reversible process   A process that can be run in the reverse direction in such a way that, once the system has been restored to its original state, no net energy has flowed either to the system or to its surroundings. (Ch. 17)

root-mean-square speed (u_{rms})   The square root of the average of the squared speeds of all the molecules in a population of gas molecules; a molecule possessing the average kinetic energy moves at this speed. (Ch. 6)

S

salt   The product of a neutralization reaction; it is made up of the cation of the base in the reaction plus the anion of the acid. (Ch. 4)

saturated hydrocarbon   An alkane; compounds containing the maximum ratio of hydrogen atoms to carbon atoms. (Ch. 5)

saturated solution   A solution that contains the maximum concentration of a solute possible at a given temperature. (Ch. 4)

Schrödinger wave equation   A description of how the electron matter wave varies with location and time around the nucleus of a hydrogen atom. (Ch. 7)

scientific method   An approach to acquiring knowledge based on observation of phenomena, development of a testable hypothesis, and additional experiments that test the validity of the hypothesis. (Ch. 1)

scientific theory (model)   A general explanation of a widely observed phenomenon that has been extensively tested and validated. (Ch. 1)

scintillation counter   An instrument that determines the level of radioactivity in samples by measuring the intensity of light emitted by phosphors in contact with the samples. (Ch. 19)

second law of thermodynamics   The principle that the total entropy of the universe increases in any spontaneous process. (Ch. 17)

secondary (2°) structure   The pattern of arrangement of segments of a protein chain. (Ch. 20)

seesaw   Molecular geometry about a central atom with a steric number of 5 and one lone pair of electrons in an equatorial position. (Ch. 9)

semiconductor   A semimetal (metalloid) with electrical conductivity between that of metals and insulators that can be chemically altered to increase its electrical conductivity. (Ch. 9)

semimetals   See *metalloids*.

sievert (Sv)   SI unit used to express the amount of biological damage caused by ionizing radiation. (Ch. 19)

sigma (σ) bond   A covalent bond in which the highest electron density lies between the two atoms along the bond axis. (Ch. 9)

sigma (σ) molecular orbital   In MO theory, a molecular orbital in which the greatest electron density is concentrated along an imaginary line drawn through the bonded atom centers. (Ch. 9)

significant figures   All the certain digits in a measured value plus one estimated digit. The greater the number of significant figures, the greater the certainty with which the value is known. (Ch. 1)

simple cubic (sc) unit cell   An array of atoms or molecules with one particle at the eight corners of a cube. (Ch. 12)

single bond   A chemical bond that results when two atoms share one pair of electrons. (Ch. 8)

solid   A form of matter that has a definite shape and volume. (Ch. 1)

solubility   The maximum amount of a substance that dissolves in a given quantity of solvent at a given temperature. (Ch. 4)

solubility product, K_{sp}   See *solubility-product constant*.

solubility-product constant   (also called *solubility product, K_{sp}*) An equilibrium constant that describes the formation of a saturated solution of a slightly soluble salt. (Ch. 16)

solute   Any component in a solution other than the solvent. A solution may contain one or more solutes. (Ch. 4)

solution   Another name for homogeneous mixture. Solutions are often liquids, but they may also be solids or gases. (Ch. 1)

solvent   The component of a solution that is present in the largest number of moles. (Ch. 4)

sp hybrid orbitals   Two hybrid orbitals oriented 180° from one another, formed by mixing one s and one p orbital. (Ch. 9)

sp^2 hybrid orbitals   Three hybrid orbitals in a trigonal planar orientation, formed by mixing one s and two p atomic orbitals. (Ch. 9)

sp^3 hybrid orbitals   A set of four hybrid orbitals with a tetrahedral orientation, formed by mixing one s and three p atomic orbitals. (Ch. 9)

sp^3d hybrid orbitals   Five equivalent hybrid orbitals with lobes pointing toward the vertices of a trigonal bipyramid, formed by mixing one s orbital, three p orbitals, and one d orbital from the same shell. (Ch. 9)

sp^3d^2 hybrid orbitals   Six equivalent hybrid orbitals with lobes pointing toward the vertices of an octahedron, formed by mixing one s orbital, three p orbitals, and two d orbitals from the same shell. (Ch. 9)

specific heat (c_s)   The quantity of energy required to raise the temperature of 1 gram of a substance by 1°C at constant pressure. (Ch. 5)

spectator ion   An ion that is present in a reaction vessel when a chemical reaction takes place but is unchanged by the reaction; spectator ions appear in an overall ionic equation but not in a net ionic equation. (Ch. 4)

spectrochemical series   A list of ligands rank-ordered by their ability to split the energies of the d orbitals of transition metal ions. (Ch. 22)

sphere of hydration   The cluster of water molecules surrounding an ion in aqueous solution; the general term applied to such a cluster forming in any solvent is *sphere of solvation*. (Ch. 10)

spin magnetic quantum number (m_s) Either $+\frac{1}{2}$ or $-\frac{1}{2}$, indicating that the spin orientation of an electron is either up or down. (Ch. 7)

spontaneous process A process that occurs without outside intervention. (Ch. 17)

square planar Molecular geometry about a central atom with a steric number of 6 and two lone pairs of electrons that occupy axial sites; the atoms occupy four equatorial positions. (Ch. 9)

square pyramidal Molecular geometry about a central atom with a steric number of 6 and one lone pair of electrons; as typically drawn, the atoms occupy four equatorial and one axial site. (Ch. 9)

standard cell potential ($E°_{cell}$) A measure of how forcefully an electrochemical cell in which all reactants and products are in their standard states can pump electrons through an external circuit. (Ch. 18)

standard conditions In thermodynamics: a pressure of 1 bar (~1 atm) and some specified temperature, assumed to be 25°C unless otherwise stated; for solutions, a concentration of 1 M is specified. (Ch. 5)

standard deviation (s) A measure of the amount of variation, or dispersion, in a set of related values. (Ch. 1)

standard enthalpy of formation ($\Delta H°_f$) The enthalpy change of a formation reaction; also called *standard heat of formation* or *heat of formation*. (Ch. 5)

standard enthalpy of reaction ($\Delta H°_{rxn}$) The energy associated with a reaction that takes place under standard conditions; also called *standard heat of reaction*. (Ch. 5)

standard free energy of formation ($\Delta G°_f$) The change in free energy associated with the formation of 1 mole of a compound in its standard state from its component elements. (Ch. 17)

standard hydrogen electrode (SHE) A reference electrode based on the half-reaction $2\,H^+(aq) + 2\,e^- \rightarrow H_2(g)$ that produces a standard electrode potential of 0.000 V. (Ch. 18)

standard molar entropy ($S°$) The absolute entropy of 1 mole of a substance in its standard state. (Ch. 17)

standard reduction potential ($E°$) The potential of a reduction half-reaction in which all reactants and products are in their standard states at 25°C. (Ch. 18)

standard solution A solution of known concentration used in titrations. (Ch. 4)

standard state The most stable form of a substance under 1 bar pressure and some specified temperature (25°C unless otherwise stated). (Ch. 5)

standard temperature and pressure (STP) 0°C and 1 bar as defined by IUPAC; 0°C and 1 atm are commonly used in the United States. (Ch. 6)

standing wave A wave confined to a given space, with a wavelength (λ) related to the length L of the space by $L = n(\lambda/2)$, where n is a whole number. (Ch. 7)

state function A property of an entity based solely on its chemical or physical state or both, but not on how it achieved that state. (Ch. 5)

stereoisomers Isomers created by differences in the orientations of the bonds in molecules. (Ch. 20)

steric number (SN) The sum of both the number of atoms bonded to a central atom and the number of lone pairs of electrons on the central atom. (Ch. 9)

stimulatory effect Increased activity, growth, or other biological response to the presence of a nonessential element. (Ch. 21)

stock solution A concentrated solution of a substance used to prepare solutions of lower concentration. (Ch. 4)

stoichiometry The quantitative relation between reactants and products in a chemical reaction. (Ch. 3)

straight-chain hydrocarbon A hydrocarbon in which the carbon atoms are bonded together in one continuous carbon chain. (Ch. 5)

strong acid An acid that completely dissociates into ions in aqueous solution. (Ch. 4)

strong base A base that completely dissociates into ions in aqueous solution. (Ch. 4)

strong electrolyte A substance that dissociates completely into ions when it dissolves in water. (Ch. 4)

strong nuclear force The fundamental force of nature that keeps nucleons together in atomic nuclei. (Ch. 19)

sublimation Transformation of a solid directly into a vapor (gas). (Ch. 1)

substance Matter that has a constant composition and cannot be broken down to simpler matter by any physical process; also called *pure substance*. (Ch. 1)

substitutional alloy An alloy in which atoms of the nonhost metal replace host atoms in the crystal lattice. (Ch. 12)

substrate The reactant that binds to the active site in an enzyme-catalyzed reaction. (Ch. 20)

supercritical fluid A substance at conditions above its critical temperature and pressure, where the liquid and vapor phases are indistinguishable and have some characteristics of both a liquid and a gas. (Ch. 10)

supersaturated solution A solution that contains more than the maximum quantity of solute predicted to be soluble in a given volume of solution at a given temperature. (Ch. 4)

surface tension The energy needed to separate the molecules at the surface of a liquid. (Ch. 10)

surroundings Everything that is not part of the system. (Ch. 5)

system The part of the universe that is the focus of a thermochemical study. (Ch. 5)

T

termolecular step A step in a reaction mechanism involving a collision among three molecules. (Ch. 13)

tertiary (3°) structure The three-dimensional, biologically active structure of a protein that arises because of interactions between the R groups on its amino acids. (Ch. 20)

tetrahedral Molecular geometry about a central atom with a steric number of 4 and no lone pairs of electrons; the bond angles are all 109.5°. (Ch. 9)

theoretical yield The maximum amount of product possible in a chemical reaction for given quantities of reactants; also called *stoichiometric yield*. (Ch. 3)

thermal energy The kinetic energy of atoms, ions, and molecules. (Ch. 5)

thermal equilibrium A condition in which temperature is uniform throughout a material and no energy flows from one point to another. (Ch. 5)

thermochemical equation The chemical equation of a reaction that includes the change in enthalpy that accompanies that reaction. (Ch. 5)

thermochemistry The study of the relation between chemical reactions and changes in energy. (Ch. 5)

thermodynamics The study of energy and its transformations. (Ch. 5)

third law of thermodynamics The entropy of a perfect crystal is zero at absolute zero. (Ch. 17)

threshold frequency (ν_0) The minimum frequency of light required to produce the photoelectric effect. (Ch. 7)

titrant The standard solution added to the sample in a titration. (Ch. 4)

titration An analytical method for determining the concentration of a solute in a sample by reacting the solute with a standard solution of known concentration. (Ch. 4)

torr   See *millimeters of mercury (mmHg)*.

trace essential element   Essential element present in the body in average concentrations between 1 and 1000 μg of element per gram of body mass. (Ch. 21)

trans isomer   (also called *E isomer*) A molecule with two like groups (such as two R groups or two hydrogen atoms) on opposite sides of the molecule. (Ch. 20)

transcription   The process of copying the information in DNA to RNA. (Ch. 20)

transfer RNA (tRNA)   The form of the nucleic acid RNA that delivers amino acids, one at a time, to polypeptide chains being assembled by the ribosome–mRNA complex. (Ch. 20)

transition metals   The elements in groups 3 through 12 of the periodic table. (Ch. 2)

transition state   A high-energy state between reactants and products in a chemical reaction. (Ch. 13)

translation   The process of assembling proteins from the information encoded in RNA. (Ch. 20)

tricarboxylic acid (TCA) cycle   A series of reactions that continue the oxidation of pyruvate formed in glycolysis. (Ch. 20)

trigonal bipyramidal   Molecular geometry about a central atom with a steric number of 5 and no lone pairs of electrons; three atoms occupy equatorial sites (bond angle 120°) and two other atoms occupy axial sites (bond angle 180°) above and below the equatorial plane; the bond angle between the axial and equatorial bonds is 90°. (Ch. 9)

trigonal planar   Molecular geometry about a central atom with a steric number of 3 and no lone pairs of electrons; the bond angles are all 120°. (Ch. 9)

trigonal pyramidal   Molecular geometry about a central atom with a steric number of 4 and one lone pair of electrons. (Ch. 9)

triple bond   A chemical bond in which two atoms share three pairs of electrons. (Ch. 8)

triple point   The temperature and pressure at which all three phases of a substance coexist. Under these conditions, freezing and melting, boiling and condensation, and sublimation and deposition all proceed at the same rate. (Ch. 10)

T-shaped   Molecular geometry about a central atom with a steric number of 5 and two lone pairs of electrons that occupy equatorial positions; the outer atoms occupy two axial sites and one equatorial site. (Ch. 9)

U

ultratrace essential element   Essential element present in the body in average concentrations less than 1 μg of element per gram of body mass. (Ch. 21)

unimolecular step   A step in a reaction mechanism involving only one molecule on the reactant side. (Ch. 13)

unit cell   The repeating unit of the arrangement of atoms, ions, or molecules in a crystal lattice. (Ch. 12)

universal gas constant   The constant R in the ideal gas equation; its value and units depend on the units used for the variables in the equation. (Ch. 6)

V

valence band   A band of orbitals that are filled or partially filled by valence electrons. (Ch. 9)

valence bond theory   A quantum mechanics–based theory of bonding incorporating the assumption that covalent bonds form when half-filled orbitals on different atoms overlap or occupy the same region in space. (Ch. 9)

valence electrons   Electrons in the outermost occupied shell of an atom. These are the electrons that are transferred or shared in chemical reactions. (Ch. 7)

valence shell   The outermost occupied shell of an atom. (Ch. 7)

valence-shell electron-pair repulsion (VSEPR) theory   A model predicting that the arrangement of valence electron pairs around a central atom minimizes their mutual repulsion to produce the lowest-energy orientations. (Ch. 9)

van der Waals equation   An equation that includes experimentally determined factors a and b that quantify the contributions of non-negligible molecular volume and non-negligible intermolecular interactions to the behavior of real gases with respect to changes in P, V, and T. (Ch. 6)

van't Hoff factor   (also called *i factor*) The ratio of the experimentally measured value of a colligative property to the theoretical value expected for that property if the solute were a nonelectrolyte. (Ch. 11)

vapor pressure   The pressure exerted by a gas at a given temperature in equilibrium with its liquid phase. (Ch. 10)

vinyl group   The subgroup $CH_2{=}CH-$. (Ch. 12)

viscosity   The resistance to flow of a liquid. (Ch. 10)

voltaic cell   An electrochemical cell in which chemical energy is transformed into electrical work by a spontaneous cell reaction. (Ch. 18)

W

wave function (ψ)   A solution to the Schrödinger wave equation. (Ch. 7)

wave mechanics   (also called *quantum mechanics*) A mathematical description of the wavelike behavior of particles on the atomic level. (Ch. 7)

wavelength (λ)   The distance from crest to crest or trough to trough on a wave. (Ch. 7)

wave–particle duality   The behavior of an object that exhibits the properties of both a wave and a particle. (Ch. 7)

weak acid   An acid that is a weak electrolyte and so has a limited capacity to donate protons to the medium. (Ch. 4)

weak base   A base that is a weak electrolyte and so has a limited capacity to accept protons. (Ch. 4)

weak electrolyte   A substance that dissolves in water with most molecules staying intact, but with a small fraction dissociating into ions. (Ch. 4)

work   A form of energy: the energy required to move an object through a given distance. (Ch. 5)

work function (ϕ)   The amount of energy needed to remove an electron from the surface of a metal. (Ch. 7)

Z

Z isomer   See *cis isomer*.

zeolites   Natural crystalline minerals or synthetic materials consisting of three-dimensional networks of channels that contain sodium or other 1+ cations. (Ch. 4)

zwitterion   A molecule that has both positively and negatively charged groups in its structure. (Ch. 20)

Answers
to Particulate Review, Concept Tests, and Practice Exercises

Chapter 1
Particulate Review
8 hydrogen atoms
4 hydrogen molecules
5 helium atoms
Molecules are composed of atoms.

Concept Tests
p. 6 (b) and (c) show a physical process (the melting of a solid), whereas (a) and (d) show a chemical reaction (the burning of fuel).

p. 8 1:1

p. 13 (b)

p. 15 (b) Less energy, because fewer intermolecular interactions are broken when melting ice than when boiling water.

p. 18 (b) Decreasing

p. 23 (1) 3 significant figures; (2) 3 significant figures; (4) 4 significant figures

p. 26 Speed = distance/time. Distance, because actual distance traveled will vary depending on whether a car stays closer to the inside or the outside of the oval.

p. 29 The average deviation is 0.0076, which is smaller than the standard deviation. One reason: taking the squares in the standard deviation calculation weights the bigger deviations more than the smaller ones.

p. 31 2.80 g. No penny would have a mass equal to the average mass.

p. 36 Hypothesis, because his explanation had not been thoroughly tested yet.

Practice Exercises
1.1. Properties (a) and (c) are physical properties; property (d) is a chemical property; property (b) is both physical and chemical.

1.2. a. The diatomic particles in the box on the left represent a gas because the particles are widely spaced and fill the box. The particles in the box on the right represent a solid because the diatomic particles are ordered and do not take the shape of the box. The change of state (physical process) represented is deposition.

b. Sublimation

1.3. 0.324 km; 3.24×10^4 cm

1.4. 9.45×10^{12} km (using 365 days/yr)

1.5. 1.14

1.6. Statistics (a) and (e) are exact numbers; statistics (b)–(d) have inherent uncertainty.

1.7. Mean = 36.38
STDEV = 0.38
95% CL = 36.38 ± 0.47 or 36.4 ± 0.5

1.8. The mean and standard deviation are 194 ± 18.6 mg/dL. Testing 215 mg/dL as an outlier gives us a calculated Z value of 1.1 mg/dL, which is less than the reference Z values for $n = 3$ from Table 1.8 at both the 95% and 99% confidence levels. The high value of 215 should *not* be considered an outlier.

1.9. K_{low} = 40 K and K_{high} = 396 K; $°F_{low}$ = −387°F and $°F_{high}$ = 253°F

Chapter 2
Particulate Review
Helium, He, element
Nitrogen, N_2, element
Gold, Au, element
Ethanol, CH_3CH_2OH, compound
Sodium chloride, NaCl, compound

Concept Tests
p. 49 Matter is neutral, so some positive particle must be present to provide electrical neutrality.

p. 53 $^{87}_{38}Q$ is an isotope of strontium-90 because they both contain 38 protons (i.e., they are both atoms of the same element, strontium). $^{90}_{40}X$ is an atom of zirconium with a mass number of 90 (40 protons, 50 neutrons), and $^{234}_{90}Z$ is an atom of thorium with an atomic number of 90 (90 protons).

p. 56 Because the next heaviest elements known in the 1870s did not have chemical properties similar to the other elements in the same column as the skipped cells.

p. 61 CsN

p. 62 H_2O_2 and HO

p. 70 (a) aldehyde; (b) ether; (c) amine

Practice Exercises
2.1. a. ^{56}Fe
b. ^{15}N
c. ^{37}Cl
d. ^{39}K

2.2. (a) 27 p, 33 n; (b) 53 p, 78 n; (c) 77 p, 115 n

2.3. ^{107}Ag = 51.85%; ^{109}Ag = 48.15%

2.4. a. As, arsenic
b. K, potassium
c. Cd, cadmium
d. Se, selenium

2.5. 40.0 g

2.6. (a)–(d) are molecular; (e) is ionic.

2.7. a. Tetraphosphorus decoxide
b. Carbon monoxide
c. Nitrogen trichloride

    d. SF_6

    e. ICl

    f. Br_2O

2.8. a. $SrCl_2$

    b. MgO

    c. NaF

    d. $CaBr_2$

2.9. $MnCl_2$ and MnO_2

2.10. a. $Sr(NO_3)_2$

    b. K_2SO_4

    c. Sodium hypochlorite

    d. Potassium permanganate

2.11. a. Hypobromous acid

    b. Bromic acid

    c. Carbonic acid

Chapter 3

Particulate Review

CO_2, carbon dioxide

H_2O, water

N_2, nitrogen

CO, carbon monoxide

H_2S, hydrogen sulfide

Concept Tests

p. 89 Both the mole and a gross represent a specific number of particles independent of the size, mass, or identity of an individual particle.

p. 91 One gram of Ag has more atoms than one gram of Au because Ag has a smaller molar mass.

p. 100 When balancing equations we cannot change the subscripts because doing so changes the identity of the substance.

p. 111 (a) C_2H_4 and (b) $C_{20}H_{40}$ have the same empirical formula (CH_2) and the same percent composition (85.6% C, 14.4% H); (c) C_2H_2 and (d) C_6H_6 have the same empirical formula (CH) and the same percent composition (92.3% C, 7.7% H).

p. 114 The empirical formula of (b) C_3H_8O is identical to its molecular formula; (a) $C_2H_6O_2$ and (c) $C_6H_{12}O_6$ have empirical formulas that differ from the given molecular formulas. The empirical formula of $C_2H_6O_2$ is CH_3O and the empirical formula of $C_6H_{12}O_6$ is CH_2O.

p. 116 No, because they have different molecular formulas.

p. 117 Molecular formula

Practice Exercises

3.1. 1.5×10^{10} atoms

3.2. 6.80×10^{24} electrons

3.3. 5.41×10^{-2} moles C

3.4. 49.2 g Au

3.5. CO_2 = 44.01 g/mol; O_2 = 32.00 g/mol; $C_6H_{12}O_6$ = 180.16 g/mol

3.6. 5.00×10^{-3} moles; 3.01×10^{21} formula units

3.7. 129 g Cr

3.8.

3.9. (a) $P_4(s) + 5\ O_2(g) \rightarrow P_4O_{10}(s)$; (b) $P_4O_{10}(s) + 6\ H_2O(\ell) \rightarrow 4\ H_3PO_4(aq)$

3.10. $2\ CO(g) + O_2(g) \rightarrow 2\ CO_2(g)$

3.11. $C_3H_8(g) + 5\ O_2(g) \rightarrow 3\ CO_2(g) + 4\ H_2O(g)$

3.12. $2\ C_4H_{10}(g) + 13\ O_2(g) \rightarrow 8\ CO_2(g) + 10\ H_2O(g)$; 3.03 g CO_2 produced

3.13. 321 g Cu_2S; 129 g SO_2

3.14. 25.99% C; 74.01% F

3.15. NO

3.16. Li_2CO_3

3.17. $C_{20}H_{40}$

3.18. C_5H_8

3.19. $C_8H_8O_3$

3.20. This fuel–oxygen mixture is lean.

3.21. 3.74 g $C_2H_4N_2$

3.22. 89.9%

3.23. 1.10×10^3 g Al_2O_3; 194 g C

Chapter 4

Particulate Review

Ethylene glycol is a covalent compound with covalent bonds, whereas sodium sulfate is an ionic compound with ionic bonds. The space-filling representations of the molecules of ethylene glycol remain "intact" and do not separate into smaller particles, whereas the particles shown for sodium sulfate are separate sodium cations and sulfate anions. (Note that the space-filling sulfate anions contain covalent bonds between the sulfur and oxygen atoms.)

Concept Tests

p. 145 (c) Clear cough syrup; (d) Filtered dry air

p. 148 (d) 56,000 nM NaCl is the least concentrated solution.

p. 150 0.25 ppm is equivalent to 250 ppb. The water from neither home would be safe to drink.

p. 156 Solvent is added to dilute the stock solution, which decreases the concentration.

p. 159 (top) Equivalent molar concentrations mean that the same number of solute moles are dissolved in a given solution volume. Differences in conductivity are due to the different numbers of ions the solutes make when they dissolve.

p. 159 (bottom) Drawing (a) represents a weak electrolyte and (b) represents a strong electrolyte.

p. 164 (a) H_2E^{2-}; (b) H_4E

p. 165 (a) weak acid; (b) neither; (c) weak base; (d) strong acid

p. 174 React aqueous lead(II) nitrate with the stoichiometric amount of aqueous potassium dichromate. Stir the mixture for a minute or so; let it stand for 10 min; filter off the yellow $PbCr_2O_7$ precipitate; wash it with water in the filter; and allow the washed solid to air-dry overnight.

p. 181 In the reaction of Fe_3O_4 with oxygen, the iron is oxidized, because the product contains more oxygen than the reactant: 3 Fe ions are present for every 4 oxygen ions (1.33 oxygen ions per iron ion) versus 2 Fe ions for every 3 oxygen ions (1.5 oxygen ions per iron ion). If iron is oxidized, the other substance involved in the reaction must be reduced; oxygen is reduced.

p. 183 Reactions (a) and (b) are redox reactions. In reaction (a), the O.N. of Br goes from 0 to −1 while the O.N. of Sn goes from +2 to +4. In reaction (b), the O.N. of F goes from 0 to −1 while the O.N. of O goes from −2 to 0.

Practice Exercises

4.1. The well water is 120 times more concentrated in arsenic than the WHO standard.

4.2. 1.88 M $MgCl_2$

4.3. 0.109 M KCl

4.4. 4.48 g $NaC_3H_5O_3$

4.5. $V_{initial} = 12.5$ mL

4.6. 1.16×10^{-4} M

4.7. a. $CH_3COOH(aq) + NaOH(aq) \rightarrow$
$CH_3COONa(aq) + H_2O(\ell)$
b. $CH_3COOH(aq) + Na^+(aq) + OH^-(aq) \rightarrow$
$Na^+(aq) + CH_3COO^-(aq) + H_2O(\ell)$
c. $CH_3COOH(aq) + OH^-(aq) \rightarrow CH_3COO^-(aq) + H_2O(\ell)$

4.8. (a) strong electrolyte, neither; (b) strong acid, strong electrolyte; (c) weak base, weak electrolyte; (d) nonelectrolyte, neither

4.9. The lemon juice is 0.4162 M $C_6H_8O_7$; 100 mL of juice contains 8.00 g $C_6H_8O_7$.

4.10. 0.0987 M

4.11. $3\,Ba^{2+}(aq) + 6\,OH^-(aq) + 2\,H_3PO_4(aq) \rightarrow$
$Ba_3(PO_4)_2(s) + 6\,H_2O(\ell)$

4.12. a. No precipitate forms.
b. Hg_2Cl_2 precipitates from the mixture.
c. $Hg_2^{2+}(aq) + 2\,Cl^-(aq) \rightarrow Hg_2Cl_2(s)$

4.13. 0.174 g HgS

4.14. 1.31×10^{-4} M SO_4^{2-}

4.15. Yes

4.16. a. +4
b. +1
c. +5

4.17. Oxygen is reduced and is the oxidizing agent; S is oxidized and SO_2 is the reducing agent.

4.18. a. Yes. Because the oxidation numbers for iron and palladium change from reactants to products, this reaction is a redox reaction.
b. $2\,Fe(s) + 3\,Pd^{2+}(aq) \rightarrow 2\,Fe^{3+}(aq) + 3\,Pd(s)$

4.19. Iron

4.20. Aluminum nitrate and metallic silver
Oxidation reaction: $Al(s) \rightarrow Al^{3+}(aq) + 3\,e^-$
Reduction reaction: $1\,e^- + Ag^+(aq) \rightarrow Ag(s)$
Molecular equation: $Al(s) + 3\,AgNO_3(aq) \rightarrow$
$Al(NO_3)_3(aq) + 3\,Ag(s)$
Net ionic equation: $Al(s) + 3\,Ag^+(aq) \rightarrow Al^{3+}(aq) + 3\,Ag(s)$

4.21. $3\,HO_2^-(aq) + 2\,MnO_4^-(aq) + H_2O(\ell) \rightarrow$
$3\,O_2(g) + 2\,MnO_2(s) + 5\,OH^-(aq)$

Chapter 5

Particulate Review

The molecule is H_2O (water). The ions are H_3O^+ (hydronium), Cl^- (chloride), Na^+ (sodium), and OH^- (hydroxide). The buret contains H_2O molecules, OH^- ions, and Na^+ ions. The colorless solution in the flask on the left contains H_2O molecules, Cl^- ions, and H_3O^+ ions. The pink solution in the flask on the right contains H_2O molecules, Cl^- ions, and Na^+ ions.

Concept Tests

p. 213 No; skier 1 has more potential energy: $m_1gh =$
$(PE)_{skier\ 1} > (PE)_{skier\ 2} = m_2gh$

p. 214 The potential energy of skier 2 is less than the potential energy of skier 1; skier 1 has greater kinetic energy:
$$\tfrac{1}{2}\,m_1u^2 = (KE)_{skier\ 1} > (KE)_{skier\ 2} = \tfrac{1}{2}\,m_2u^2$$

p. 215 There is far more water in the pool than in the cup. The thermal energy in the cup is "less than," even if the pool temperature is, say, 20°C.

p. 219 (a) open; (b, c) closed, because the bottle and the sandwich wrap can conduct heat; (d) open

p. 233 The mass of the aluminum is much less than the mass of the water, and the molar heat capacity of aluminum is much less than that of water.

p. 251 +2219.9 kJ; we are now asked about the reverse of a reaction for which we calculated a negative ΔH, so we just need to change the sign on the value.

p. 253 Each alkane has two terminal $-CH_3$ groups, which have a mass of 15 g/mol each. Therefore, the mass of the $-CH_2-$ groups is 114 g/mol $-$ 2(15 g/mol) = 84 g/mol. Each $-CH_2-$ group has a mass of 14 g/mol. So, 84 g/mol would be the mass of the CH_2- units, so there are 84 g/14 g/mol = 6 $-CH_2-$ units. Therefore, $n = 2 + 6 = 8$.

p. 255 (a) 1 mole CH_4; (b) 1 g H_2

Practice Exercises

5.1. About 13% slower

5.2. a. The match is the system, $q < 0$, and the process is exothermic.
b. The wax is the system, $q < 0$, and the process is exothermic.
c. The liquid is the system, $q > 0$, and the process is endothermic.

5.3. $\Delta E = 32$ J

5.4. $w = 1.58 \times 10^6$ kJ

5.5. 1.13×10^3 g or 1.13 kg

5.6. ΔH_{sys} is negative; q_{surr} is positive.

5.7. -321 kJ

5.8. 0.0°C

5.9. 51°C

5.10. -4.6 kJ/mol

5.11. When 0.500 g of the hydrocarbon mixture is burned, the energy released is 24.6 kJ. When 1.000 g of the hydrocarbon mixture is burned, the energy released is 49.2 kJ.

5.12. Reverse B and then add the three equations.

5.13.
$$2\,CH_4(g) + 3\,O_2(g) \rightarrow 2\,\cancel{CO(g)} + 4\,H_2O(g)$$
$$\Delta H_{comb} = -802\ kJ$$
$$\underline{2\,\cancel{CO(g)} + O_2(g) \rightarrow 2\,CO_2(g) \qquad \Delta H_{comb} = -566\ kJ}$$
$$2\,CH_4(g) + 4\,O_2(g) \rightarrow 2\,CO_2(g) + 4\,H_2O(g)$$
$$\Delta H_{comb} = -1604\ kJ$$

For 1 mole CH_4, $\Delta H_{comb} = -802$ kJ

5.14. a. $Ca(s) + C(s, graphite) + \tfrac{3}{2}O_2(g) \rightarrow CaCO_3(s)$
b. $2\,C(s, graphite) + 2\,H_2(g) + O_2(g) \rightarrow CH_3COOH(\ell)$
c. $K(s) + Mn(s) + 2\,O_2(g) \rightarrow KMnO_4(s)$

5.15. $\Delta H^\circ_{rxn} = -41.2$ kJ

5.16. Fuel value of kerosene = 41.40 kJ/g; fuel density of kerosene = 3.10×10^4 kJ/L

5.17. $C_{calorimeter} = 11.2$ kJ/°C

Chapter 6

Particulate Review

The balanced chemical equation is $2 H_2O(g) \rightarrow 2 H_2(g) + O_2(g)$. The product is a mixture of two elements.

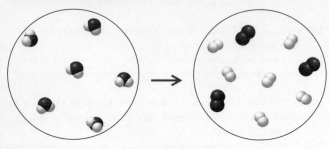

Concept Tests

p. 279   the gas in the flask

p. 282   (c)

p. 284   The two graphs have different slopes because the slope depends on P. The slope of graph 1 is twice the slope of graph 2.

p. 286   (c)

p. 290   a and c

p. 296   Kr

p. 297   (top) (a) i; (b) ii

p. 297   (bottom) (c) Low T and high P give greatest d.

p. 300   Because $P = F/A$, the force results from all components combined, so one cannot measure the partial pressure because one cannot measure the partial force.

p. 301   Yes

p. 305   Lower

p. 307 $UF_6 < SF_6 < Kr < CO_2 < Ar < H_2$

p. 309 T cancels out.

p. 312 CO_2, CH_4

Practice Exercises

6.1. $P = 0.41$ kPa

6.2.   At the end of the experiment the mercury levels will be separated by $\Delta h = 144$ mmHg.

6.3. $V_2 = 10.5$ L

6.4. $V_1/V_2 = 0.622$

6.5. $P_2 = 34$ psi

6.6. $V_2 = 2.10 \times 10^3$ L

6.7. $V = 1.8 \times 10^3$ L

6.8. 1.10×10^2 g

6.9.   259 mL of N_2O

6.10.   The balloon will sink to the floor.

6.11. $M = 44.0$ g/mol; CO_2
$HCl(aq) + NaHCO_3(aq) \rightarrow NaCl(aq) + CO_2(g) + H_2O(\ell)$

6.12. $O_2 = 0.1201$, $He = 0.8799$

6.13. $X_{O_2} = 0.042$

6.14.   2.2 mg of H_2

6.15. $u_{rms,He} = 1.367 \times 10^3$ m/s, or 2.65 times faster than N_2

6.16.   Ar

6.17. N_2 behaves more ideally.

Chapter 7

Particulate Review

The ions from left to right are K^+, S^{2-}, Al^{3+}, and F^-. Potassium has 19 electrons in a neutral atom and 18 electrons in its cation, K^+. Sulfur has 16 electrons in a neutral atom and 18 electrons in its anion, S^{2-}. Aluminum has 13 electrons in a neutral atom and 10 electrons in its cation, Al^{3+}. Fluorine has 9 electrons in a neutral atom and 10 electrons in its anion, F^-.

Concept Tests

p. 334   (c)

p. 336   Higher frequency

p. 338   (a) continuous; (b) quantized; (c) continuous; (d) quantized

p. 341   Yes

p. 344   (c) > (a) > (b) > (d)

p. 345   No, because He atoms have two electrons.

p. 357   Five

p. 367   Ground-state potassium atom: $[Ar]4s^1$, excited state potassium atom: $[Ar]3p^1$ (others would be possible as well), potassium ion, K^+: $[Ar]$. The potassium atom has one more electron than the potassium ion.

p. 369   A half-filled set of f orbitals is more stable because adding another f electron requires pairing energy.

p. 372 $1s > 2s > 2p > 3s > 3p > 4s > 4p$

p. 375   Because the valence electrons are farther from the nucleus and shielded from it by more inner-shell electrons

p. 376   The magnitude of IE and EA both increase with increasing Z across a row (except for group 18). EA values do not display clear trends within groups, whereas IE values decrease with increasing Z.

Practice Exercises

7.1. $\lambda = 3.30$ m

7.2. $E_{450} = 4.39 \times 10^{-19}$ J; $E_{470} = 4.12 \times 10^{-19}$ J

7.3. $\lambda = 2.62 \times 10^{-7}$ m, or 262 nm

7.4.   486 nm

7.5.   Prediction: Less energy is required to remove an electron from the hydrogen atom in the $n = 3$ state than for a hydrogen atom in the $n = 1$ state. Calculated value: 2.420×10^{-19} J

7.6. $\lambda = 3.3 \times 10^{-10}$ m

7.7. $\Delta x \geq 7 \times 10^{-11}$ m

7.8.   25

7.9.

n	ℓ	m_ℓ	m_s
3	1	-1	$+\frac{1}{2}$
3	1	0	$+\frac{1}{2}$
3	1	1	$+\frac{1}{2}$

7.10.   Ga: $[Ar]3d^{10}4s^24p^1$ As: $[Ar]3d^{10}4s^24p^3$

7.11.   Co $= [Ar]3d^74s^2$

7.12. $K^+ = [Ar]$; $I^- = [Kr]4d^{10}5s^25p^6 = [Xe]$; $Ba^{2+} = [Xe]$; $Rb^+ = [Kr]$; $O^{2-} = [He]2s^22p^6 = [Ne]$; $Al^{3+} = [Ne]$; $Cl^- = [Ne]3s^23p^6 = [Ar]$. K^+ and Cl^- are isoelectronic with Ar.

7.13. $Mn = [Ar]3d^54s^2$; $Mn^{3+} = [Ar]3d^4$; $Mn^{4+} = [Ar]3d^3$

7.14.   a. $Li^+ < F^- < Cl^-$
          b. $Al^{3+} < Mg^{2+} < P^{3-}$

7.15.   Ne > Ca > Cs

Chapter 8

Particulate Review

C—H bonds and O=O bonds are broken; C=O bonds and H—O bonds are formed. For each mole of CH_4 that is combusted, 4 moles of C—H single bonds and 2 moles of O=O double bonds are broken.

$$CH_4(g) + 2\,O_2(g) \rightarrow CO_2(g) + 2\,H_2O(g)$$

Two moles of C=O double bonds and 4 moles of O—H bonds are formed.

Concept Tests

p. 400 $\longleftrightarrow$ $\longleftrightarrow$

$\ddot{O}=C=\ddot{O}$

p. 402   No, stretching in N≡N or O=O does not result in infrared absorptions because the bonds are nonpolar, and no change in polarity occurs when they stretch.

p. 410   −1

p. 420 $N_2O > NO_2 > NO$

p. 422   The O=O bond is not as strong as the N≡N bond and is more easily broken.

Practice Exercises

8.1.
H
|
H—C—H
|
H

8.2. $:\ddot{C}l—\overset{\displaystyle P}{\underset{\displaystyle :\ddot{C}l:}{|}}—\ddot{C}l:$

8.3. $\ddot{O}=C=\ddot{O}$

8.4. $\left[:\ddot{F}:\right]^{-} Mg^{2+} \left[:\ddot{F}:\right]^{-}$

8.5.   Be—Cl; the bond is not ionic; it is a polar covalent bond.

8.6. $\ddot{O}=\ddot{S}—\ddot{O}: \longleftrightarrow :\ddot{O}—\ddot{S}=\ddot{O}$

8.7.   Resonance forms for N_3^-:

$\left[:\ddot{N}=N=\ddot{N}:\right]^{-} \longleftrightarrow \left[:N≡N—\ddot{N}:\right]^{-} \longleftrightarrow \left[:\ddot{N}—N≡N:\right]^{-}$

Resonance forms for NO_2^+:

$\left[:\ddot{O}=N=\ddot{O}:\right]^{+} \longleftrightarrow \left[:O≡N—\ddot{O}:\right]^{+} \longleftrightarrow \left[:\ddot{O}—N≡O:\right]^{+}$

8.8. $\left[:\ddot{O}=N=\ddot{O}:\right]^{+}$

8.9.

8.10.

8.11.   Sodium atoms do not have an octet in either structure.

Na—$\ddot{\underset{..}{C}l}$:
|
:$\ddot{C}l$—Na

8.12. $\Delta H_{rxn} = -79$ kJ

Chapter 9

Particulate Review

Malic acid is $C_4H_6O_5$ and contains two carboxylic acid functional groups (one at each end of the molecule) and one alcohol functional group in the middle. Putrescine is $C_4H_{12}N_2$ and contains two amine functional groups.

Concept Tests

p. 440   Molecular geometry is derived from electron-pair geometry. They are the same if there are no lone pairs around the central atom but different if there are lone pairs around the central atom.

p. 444   (b) > (a) > (c)

p. 446   One lone pair on the central P atom in PH_3 and two lone pairs on the central S atom in H_2S repel bonding pairs and reduce bond angles. There are no lone pairs on the central Si atom in SiH_4.

p. 448   Slightly smaller

p. 451 SCl_4 has a seesaw molecular geometry, so the bond dipoles do not cancel. $XeCl_4$ has a square planar molecular geometry, so the bond dipoles do cancel.

p. 453   Because the electronegativity difference between H and S is less than the difference between H and O

p. 461   It needs unhybridized p orbitals to form π bonds.

p. 468   (a) and (c), because there are alternating double and single bonds at sp^2 hybridized carbons with empty p orbitals

p. 470   (b) and (c) are chiral

p. 472   Similar: they are the products of mixing atomic orbitals. Different: MOs are delocalized over the entire molecule.

p. 477   Yes, because O_2 is paramagnetic and N_2 is not.

p. 480 No, the bond orders in the ground states are higher because each has one more electron in a bonding orbital and one fewer in an antibonding orbital.

p. 482 Overlapping conduction and valence bands, because bands from filled Mg 3s orbitals overlap with empty p orbitals.

Practice Exercises

9.1. Tetrahedral:

$$
\begin{array}{c}
Cl \\
| \\
H - C \cdots Cl \\
| \\
Cl
\end{array}
$$

9.2. The O—S—O angle in SO_3 is greater than the O—S—O angle in SO_2.

9.3. Tetrahedral; bond angles ~109.5°

9.4. No, because its molecular geometry is linear.

9.5. $5p$ on I and $4p$ on Br

9.6. (a) CCl_4 and (d) PH_3

9.7. sp^3d^2

9.8. Each N in diazene is trigonal planar. With one lone pair on each N atom, the molecular geometry around the N atoms is bent with H—N—N bond angles of less than 120°. The N atoms are sp^2 hybridized and the molecule is flat. Each N in hydrazine is tetrahedral. With one lone pair on each N atom, the molecular geometry around the N atoms is trigonal pyramidal with H—N—N bond angles of less than 109.5°. The N atoms are sp^3 hybridized and the molecule is nonplanar.

$$H - \ddot{N} = \ddot{N} - H$$
Diazene

$$
H - \underset{\underset{H}{|}}{\ddot{N}} - \underset{\underset{H}{|}}{\ddot{N}} - H
$$
Hydrazine

9.9. Hexane: ⌇⌇⌇⌇⌇ ; heptane: $CH_3(CH_2)_5CH_3$

9.10. H_2^+ can exist; its bond order is 0.5.

9.11. The bond order increases on the addition of an electron for B_2 to B_2^- and C_2 to C_2^-.

9.12.

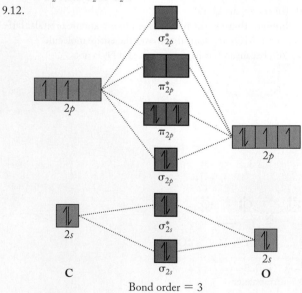

Bond order = 3

9.13. Selenium

Chapter 10

Particulate Review

Carbon dioxide is nonpolar despite containing polar bonds. Water is polar and contains polar bonds. Ozone is polar even though it contains nonpolar bonds.

Concept Tests

p. 500 H < Ne < Ar < Kr

p. 501 Molecules of CCl_4 are larger and experience larger dispersion forces.

p. 503 Acetone has the higher boiling point. The larger dipole results in stronger dipole–dipole interactions, meaning that a higher temperature (more energy) is required to overcome those interactions.

p. 506 (top) (a) CH_3Br; (b) CH_3CH_2OH; (c) $(CH_3)_3N$

p. 506 (bottom) Yes. Ion–induced dipole forces are slightly stronger than dipole–induced dipole interactions and weaker than dipole–dipole interactions.

p. 518 Gasoline

p. 522 Yes, it's possible that increased pressure will cause the ice to melt, according to the slope of the blue line in Figure 10.28 and the footprint of the tires.

p. 523 No

p. 525 No, the adhesive forces are weak.

Practice Exercises

10.1. To enter the vapor phase from the liquid phase (to boil), ethylene glycol would need to break two hydrogen bonds compared with isopropanol's one hydrogen bond; therefore ethylene glycol has a higher boiling point.

10.2. Largest dipole–dipole forces: (a) H_2NNH_2; largest dispersion forces: (d) $CH_3CH_2CH_2CH_2CH_3$; lowest boiling point: (c) Ne

10.3. Helium, being smaller and with fewer electrons than nitrogen, is less soluble in blood because it is less polarizable in its interaction with the polar molecules in blood (water).

10.4. Methanol. Both methanol and hexanol have an —OH group that can form hydrogen bonds with the oxygen on acetone. Hexanol has a much longer hydrocarbon chain than methanol, and its molecules would interact via dispersion forces, which might cause hexanol to interact more strongly with itself than with acetone and hence cause it to be less soluble in acetone.

10.5. 9.5×10^{-5} mol/L

10.6. 28.4 kJ/mol

10.7. At 25 atm of pressure and −100°C, the sample of CO_2 is a solid. As the temperature is increased to about −50°C the solid melts into a liquid, and as the temperature is raised further (to about −20°C) the liquid CO_2 boils to form a gas.

Chapter 11

Particulate Review

Potassium and sodium are both metals, whereas sulfur and fluorine are nonmetals. We learned in Chapter 7 that atomic radius increases down a group (so potassium is larger than sodium) and across a period (so metals are larger than nonmetals). We also know that cations are smaller than their neutral parent atoms, and anions are larger than their neutral parent atoms. Given that the image is drawn with neutral atoms above their ions, the atoms and ions are

Concept Tests

p. 539 (a) LiF < NaF < KF; (b) CaF_2 < $CaCl_2$ < $CaBr_2$

p. 552 The vapor pressure of a pure solvent is an intensive property. The vapor pressure of a solution is an extensive property.

p. 555 Dimethyl ether will distill first because its smaller dipole moment means its intermolecular attractions are weaker than those of acetone.

p. 558 (c)

p. 560 The solution is mostly water, which has a density of 1 kg/L, so $volume_{H_2O} \approx mass_{H_2O}$.

p. 566 (b)

p. 567 (b) and (d) are not possible.

p. 570 Toward the KCl

p. 571 Doubling the concentration

p. 575 Although more opposing pressure is needed at 50°C than at 20°C, the greater molecular motion would cause osmosis to occur faster.

Practice Exercises

11.1. $BaO > CaCl_2 > KCl$

11.2. TiO_2

11.3. $U = -3792$ kJ/mol

11.4. $U = -670$ kJ/mol

11.5. $P_{solution} = 0.848$ atm or 644 torr

11.6.

[Graph: Temperature (°C) on y-axis from 0 to 160 versus Volume of distillate (mL) on x-axis from 0 to 100. The curve shows a step pattern: constant at about 70°C from 0 to ~30 mL, then rising to a plateau at about 98°C from ~30 to ~80 mL, then rising to a plateau at about 150°C from ~80 to 100 mL.]

11.7. 1.4

11.8. 0.840 m

11.9. 4.4 m

11.10. −15.1°C

11.11. $i = 3$; 102.7°C

11.12. −0.12°C

11.13. 28.1 atm

11.14. 3.7 atm

11.15. 27.5 atm

11.16. 180 g/mol

11.17. 6.40×10^4 g/mol

Chapter 12

Particulate Review

Potassium iodide exists because of the ion–ion electrostatic attractions between K^+ cations and I^- anions that form as a result of the transfer of electrons from potassium atoms to iodine atoms. Platinum atoms are attracted to one another through London forces due to electrostatic attractions between nuclei in neighboring platinum atoms and the valence electrons in those same atoms.

Concept Tests

p. 595 A unit cell is the smallest repeating pattern within a crystal lattice, which extends in three dimensions.

p. 601 (top) Yes, because it is homogeneous; the tin atoms are uniformly distributed among the copper atoms.

p. 601 (bottom) Both alloys could be either.

p. 607 A group 15 element such as N or P

p. 608 Dispersion forces

p. 610 Different molecular structures and properties mean LDPE and HDPE must be recycled separately.

p. 614 The longer the hydrocarbon portion of the chain, the lower the water solubility.

Practice Exercises

12.1. The atomic radius of both silver and gold is 144 pm as listed in Appendix 3. For silver, $d = 10.57$ g/mL, close to the 10.50 g/mL value for the density of silver from Appendix 3; for gold, $d = 19.41$ g/mL, close to the 19.3 g/mL value for the density of gold from Appendix 3.

12.2. The sizes of the titanium and molybdenum atoms relative to niobium both fall within the 15% radii guideline. However, niobium and molybdenum both form body-centered cubic structures whereas titanium is hexagonal closest-packed. Therefore molybdenum will form an alloy, but titanium will not.

12.3. 101 pm; 2.16 g/cm^3

12.4. The carbon skeleton of the monomer is

[Structural formula of a monomer with a carbon double bond and an OCH$_3$ ester group]

The condensed structure of the monomer is
$H_2C\!=\!C(CH_3)C(O)OCH_3$

12.5. London dispersion and dipole–induced dipole interactions

12.6.

12.7. The polar fibers of cotton and polyester repel very nonpolar greases and oils but attract water molecules, so perspiration wicks out of the gloves to cool the skin.

12.8. The carbon skeleton structures of the monomers are

The repeating unit in the polymer is

Chapter 13

Particulate Review

Cylinder 1 contains the more concentrated solution (0.01 M). The concentration of cylinder 2 is 0.005 M. The two cylinders contain equivalent amounts of dissolved solid.

Concept Tests

p. 644 (d)

p. 647 (a) 1; (b) 0

p. 649 Four: rate $= k[A][B]^2$; rate $= k[A]^2[B]$; rate $= k[A]^3$; rate $= k[B]^3$

p. 655 Fast

p. 660 Molecules are moving faster, so the likelihood of collisions is greater.

p. 665 (c)

p. 667 n

p. 669 No

p. 672 Similar rate laws indicate similar reaction mechanisms.

p. 674 (a)

p. 675 Intermediate

Practice Exercises

13.1. The rate of formation of CO_2 (a product) is twice the rate of consumption of O_2.

13.2. $\dfrac{\Delta[N_2]}{\Delta t} = 10.8$ M/s

$\dfrac{\Delta[H_2O]}{\Delta t} = 21.5$ M/s

13.3. 1.2×10^{-6} M/s

13.4. Rate $= k[NO][NO_3]$; $k = 1.57 \times 10^{10}/(M \cdot s)$

13.5. The decomposition of H_2O_2 is first order; $k = 8.30 \times 10^{-4}$ s^{-1}

13.6. 0.430 M

13.7. $k = 2.5 \times 10^{-2}$/day

13.8. This reaction is second order in $[NO_2]$; $k = 0.751/(M \cdot s)$

13.9. 6.9 kJ

13.10. 306 K

13.11. Rate $= k_{overall}[A]^2[B]^2$

13.12. Because none of the rate laws possible with this mechanism match the experimental rate law, this proposed mechanism cannot be valid.

13.13. Yes, NO_2 acts as a catalyst in this reaction.

Chapter 14

Particulate Review

NO and O_2 combine to produce NO_2; NO and O_2 are produced when NO_2 molecules collide. The chemical equation for the two simultaneous reactions is

$$2\,NO(g) + O_2(g) \rightleftharpoons 2\,NO_2(g)$$

The rate would be increased by any change that increases the number of collisions between NO molecules and O_2 molecules, such as increasing temperature or decreasing the volume of the container in which the reaction takes place. These changes would also increase collisions between NO_2 molecules, that is, increase the rate of the reverse reaction.

Concept Tests

p. 696 Increased temperature or pressure for a gas-phase reaction increases the number of collisions.

p. 700 (b) $[CO_2] = [H_2] > [CO] = [H_2O]$

p. 712 Zone (a)

p. 718 The number of moles of gaseous reactants and products is the same.

Practice Exercises

14.1. $K_c = \dfrac{[CO][H_2]^3}{[CH_4][H_2O]}$ $K_p = \dfrac{(P_{CO})(P_{H_2})^3}{(P_{CH_4})(P_{H_2O})}$

14.2. $K_c = \dfrac{[CH_3OH]}{[CO][H_2]^2} = 2.9 \times 10^2$

14.3. $K_p = 2.7 \times 10^4$

14.4. $K_c = 3.6 \times 10^8$

14.5. $K_{p,reverse} = 2.3 \times 10^2$

14.6. $K_c = 0.13$

14.7. $K_{c,overall} = 1.7 \times 10^2$

14.8. This reaction is not at equilibrium and proceeds to the right.

14.9. a. $K_p = \dfrac{(P_{CO})^2}{(P_{CO_2})}$

b. $K_p = \dfrac{(P_{CO})}{(P_{CO_2})(P_{H_2})}$

14.10. a. When the reaction is cooled and water vapor condenses, one product is removed from the reaction mixture and the equilibrium shifts to the right, forming more SO_2.

b. When SO_2 gas dissolves in liquid water as it condenses, products are removed and the equilibrium shifts to the right, forming more products.

c. When O_2 is added, the concentration of one reactant increases and the equilibrium shifts to the right, forming more products.

14.11. Increasing the pressure shifts the equilibrium in the reaction to the products, the side of the reaction that has the fewest moles of gas.

14.12. The value of K for the endothermic reaction increases with increasing reaction temperature.

14.13. $P_{PCl_5} = 1.13$ atm; $P_{PCl_3} = 0.216$ atm; $P_{Cl_2} = 0.216$ atm

14.14. $P_{Cl_2} = 0.0902$ atm; $P_{Br_2} = 0.0902$ atm; $P_{BrCl} = 0.0196$ atm

14.15. $P_{CO} = 3.48 \times 10^{-5}$ atm; $P_{Cl_2} = 3.48 \times 10^{-5}$ atm

Chapter 15

Particulate Review

Only HCl and CH_3COOH produce H_3O^+ ions when dissolved in water. The HCl molecule dissociates completely, whereas only a small fraction of the CH_3COOH molecules dissociate. The acidic hydrogen in CH_3COOH is the hydrogen atom bonded to oxygen; the hydrogen atoms in the $-CH_3$ group are not acidic.

Concept Tests

p. 741   As CO_2 is removed more rapidly than it can be produced, both equilibria shift to the left.

p. 750   pH 0.22 = strongly acidic; 4.37 = weakly acidic; 10.03 = weakly basic; 13.77 = strongly basic; 7.00 = neutral

p. 752   Higher

p. 754   The larger its K_a, the greater the percent ionization for a given concentration of HA.

p. 756   Most: C; least: A

p. 763   C and D

p. 765 H_3PO_4, $H_2PO_4^-$, HPO_4^{2-}, PO_4^{3-}, H_3O^+, OH^-, H_2O

p. 768   No

Practice Exercises

15.1. $HClO_2 > CH_2BrCOOH > Lactic > C_6H_5COOH > HN_3$

15.2. $CH_3COOH(aq) + H_2O(\ell) \rightleftharpoons CH_3COO^-(aq) + H_3O^+(aq)$
            acid   base   conjugate base   conjugate
                                                          acid

15.3. $S^{2-} > ClO^- > CH_3COO^- > HSO_3^- > Br^-$

15.4.   Greatest increase: b; greatest decrease: d

15.5.   The pH at the start of the industrial revolution was 8.18, so the value has dropped by 0.11 pH units. The ocean has a higher hydrogen ion concentration now and a lower pH.

15.6. $[H_3O^+] = 2.0 \times 10^{-12}\ M$; $[OH^-] = 5.0 \times 10^{-3}\ M$

15.7.   pH = 2.23; percent ionization = 12%; $K_a = 7.9 \times 10^{-4}$

15.8.   7.2% ionized, $K_b = 5.7 \times 10^{-4}$

15.9.   12.7

15.10.   pH = 3.11

15.11.   pH = 11.97

15.12.   pH = 7.52

15.13.   pH = 0.68. Note that doubling the concentration decreased the pH by only 0.28 units.

15.14.   pH = 5.69

15.15.   Strongest: H_2SO_4; weakest: H_2SeO_3

15.16. $SO_4^{2-}(aq) + H_2O(\ell) \rightleftharpoons HSO_4^-(aq) + OH^-(aq)$

15.17.   pH = 11.66

15.18.   pH = 5.62

Chapter 16

Particulate Review

The compounds are (a) PbS, (b) Li_2S, (c) CaF_2, and (d) KF. PbS and CaF_2 are much less soluble than Li_2S and KF.

Concept Tests

p. 795   (b)

p. 801   (c) is the equivalence point. The titration starts at (b) and proceeds through (d), (c), and then (a).

p. 802   (b) depicts $HClO(aq)$ before the titration, (a) depicts the midpoint, and (c) depicts the equivalence point.

p. 804   There would be too small a change in pH at the equivalence point to detect it precisely.

p. 805   (top) 10.64

p. 805   (bottom) (c) depicts $NH_3(aq)$ before the titration, (b) the buffer region, and (a) the equivalence point.

p. 809   Because neutralization of the CO_3^{2-} produces more HCO_3^-, which adds to the HCO_3^- present initially in the sample to make the second plateau wider than the first

p. 815 $Fe^{2+}(aq)$ should be a weaker Lewis acid than $Fe^{3+}(aq)$ because of its smaller charge and larger size. Therefore the K_a for $Fe^{2+}(aq)$ would be below that (smaller) of $Fe^{3+}(aq)$.

p. 817 $K_{sp} = 1.7 \times 10^{-6}$

p. 819   NaF: decrease, HNO_3: increase

p. 822   The value $[OH^-]$ is defined by the initial $[Ca^{2+}]$ and the K_{sp} of $Ca(OH)_2$.

Practice Exercises

16.1.   pH = 3.25

16.2.   pH = 4.02

16.3. H_2SO_3 ($pK_a = 1.77$)

16.4. $(CH_3)_2NH$ ($pK_b = 3.23$)

16.5.   930 g of sodium ascorbate + 51 g of ascorbic acid

16.6.   pH = 4.97; pH = 4.98

16.7.   (a) change in pH of +0.11; (b) change in pH of +0.22

16.8.   Both Buffer A and Buffer B decrease in pH by 0.15.

16.9.   pH = 11.39

16.10.   pH = 4.20

16.11.   No, not always, because the hydrogencarbonate ion contributes to the pH and causes it to be lower than the pK_a.

16.12.   10.3 mL followed by an additional 13.7 mL

16.13.   CaO acts as a Lewis base and CO_2 acts as a Lewis acid.

16.14. $[Ag^+(aq)] = 1.6 \times 10^{-8}\ M$

16.15. 3.4×10^{-6} g/100 mL; $1.1 \times 10^{-7}\ M$

16.16. $9.1 \times 10^{-4}\ M$

16.17. $1.1 \times 10^{-5}\ M$

16.18.   Yes

16.19.   Yes, Ba^{2+} and Ca^{2+} ions in solution can be completely separated by selective precipitation with F^-.

Chapter 17

Particulate Review

Combustion is an exothermic process resulting from the endothermic breaking of bonds in propane (C—H single bonds and C—C single bonds) and oxygen (O=O double bonds) along with the exothermic formation of bonds in carbon dioxide (C=O double bonds) and water (H—O single bonds).

Concept Tests

p. 837   (a), (c), and (d)

p. 840 $\Delta S_{univ} > 0$

p. 847   (a) Yes, it is spontaneous because $\Delta G < 0$; (b) thermodynamics says nothing about the rate of reaction.

p. 849   The reaction is very slow.

p. 850   No. The sum of $\Delta G^\circ_{f,prod}$ are all the same, but the reactants have different ΔG°_f values, so each reaction will have a different ΔG° value.

p. 852

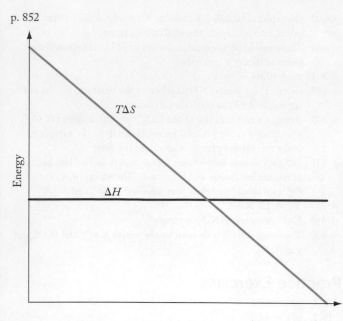

p. 858 (e)
p. 867 (e)

Practice Exercises

17.1. (a) increase in entropy; (b) decrease in entropy
17.2. (a) > (b) > (d) > (c)
17.3. Prediction: S_{sys} decreases; $\Delta S^{\circ}_{rxn} = -243$ J/K
17.4. a. ΔS_{rxn} is expected to be negative.
       b. $\Delta S^{\circ}_{rxn} = -326$ J/K
       c. $\Delta H^{\circ}_{rxn} = -571.6$ kJ/mol
       d. $\Delta G^{\circ}_{rxn} = -474.3$ kJ, so the reaction is spontaneous under standard conditions.
17.5. 142.2 kJ/mol
17.6. The reaction is spontaneous only at high temperatures.
17.7. 6.09×10^5
17.8. 2.95×10^{-37} at 298 K; 9.20×10^{-13} at 2000 K
17.9. $\Delta G^{\circ} = -196$ kJ

Chapter 18

Particulate Review

In the reaction $4\,Fe(s) + 3\,O_2(g) \rightarrow 2\,Fe_2O_3(s)$, iron is oxidized and oxygen is reduced as electrons transfer from the iron to the oxygen.

Concept Tests

p. 883 No, because the atomic masses of Cu and Zn are different.
p. 884 The cell consists of a zinc anode, which is oxidized to Zn^{2+} ions as Cu^{2+} ions are reduced to Cu metal at the cathode.
p. 887 The most "active" metals listed at the top of the activity series are the metals most easily oxidized, as shown by their position at the bottom of Table A6.1.
p. 888 (a)
p. 892 $\Delta G_{cell} > 0$, positive; $E_{cell} < 0$, negative
p. 895 0.257 V
p. 896 $E = 0.741$ V
p. 901 (a) +3; (b) −1; (c) +1; (d) +1

p. 902 C is oxidized; Co is reduced.
p. 909 Cathode: H_2; anode: Cl_2
p. 911 Anode

Practice Exercises

18.1. $O_2(g) + 2\,NO_2^-(aq) \rightarrow 2\,NO_3^-(aq)$
18.2. The balanced redox reaction is

$$3\,Cu^{2+}(aq) + 2\,Al(s) \rightarrow 3\,Cu(s) + 2\,Al^{3+}(aq)$$

The cell diagram is

$$Al(s)\,|\,Al^{3+}(aq)\,\|\,Cu^{2+}(aq)\,|\,Cu(s)$$

18.3. The net ionic equation is $Cd(s) + 2\,NiO(OH)(s) + 2\,H_2O(\ell) \rightarrow Cd(OH)_2(s) + 2\,Ni(OH)_2(s)$. $E^{\circ}_{cell} = 1.33$ V.
18.4. $\Delta G_{cell} = -2.90 \times 10^2$ kJ
18.5. $E_{cell} = 1.62$ V
18.6. $K = 1.8 \times 10^{62}$
18.7. 1.30×10^2 g
18.8. 1.5 g

Chapter 19

Particulate Review

From left to right: Nuclide 1 has a mass of 11 amu (5 protons + 6 neutrons), as does nuclide 2 (6 protons + 5 neutrons). Nuclide 3 has a mass of 13 amu (6 protons + 7 neutrons). Nuclides 2 and 3 are isotopes.

Concept Tests

p. 926 The loss in mass becomes an energy that must be added to separate the nucleons that are bound together.
p. 931 The neutron-to-proton ratio for stable isotopes increases with atomic number. For heavy isotopes it is about 1.5 to 1. Losing α particles increases this ratio, creating neutron-rich nuclides that undergo β decay.
p. 938 Increased CO_2 (^{14}C depleted) in the air from burned fossil fuels will reduce the $^{14}C/^{12}C$ ratio in the air and in plant tissues.
p. 949 To fuse nuclei, their coulombic repulsion must be overcome, which requires they collide at very high velocities, requiring very high temperatures. Man-made fusion has not been carried out except in a hydrogen bomb. We cannot make a fusion reactor until the process itself can be safely carried out.

Practice Exercises

19.1. Beta decay; $^{28}_{14}Si$
19.2. $A = 7.02 \times 10^{12}$ Bq; $A = 190$ mCi
19.3. 98.6%
19.4. 4900 years old
19.5. 2.1×10^{11} X-rays
19.6. 1.753×10^9 kJ/mol

Chapter 20

Particulate Review

The functional groups from left to right in the top row are carboxylic acid, alcohol, amine, and ether. In the bottom row from left to right they are amide, aldehyde, ketone, and ester. The carboxylic acid and the amine function as a Brønsted–Lowry acid and base, respectively. When they react, they form the amide.

Concept Tests

p. 969 (e)

p. 974 (a) alkane; (b) aromatic; (c) alcohol; (d) amide; (e) carboxylic acid; (f) amine

p. 976 Glycine is not chiral because the alpha carbon is not bonded to four different groups (there are two hydrogen atoms). Isoleucine and threonine each have two chiral centers.

p. 983 Those with nonpolar R groups, e.g., alanine, leucine, and isoleucine.

p. 985 (a) Random coil; (b) no, a change in the 2° structure is involved.

p. 991 Figure 20.37a

p. 994 (top) At low temperatures the unsaturated fatty acids remain liquid.

p. 994 (bottom) Olestra is a much larger molecule with many more C—C and C—H bonds, so on a per-mole basis it would give off much more energy.

p. 1001 No, the direction in which the code is read matters; UUG codes for leucine.

Practice Exercises

20.1. (a) Constitutional isomers; (b) two different compounds; (c) constitutional isomers

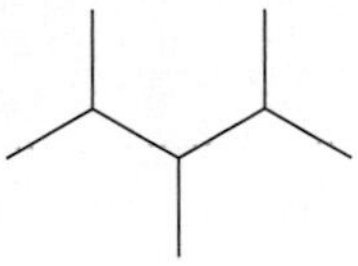

20.2.

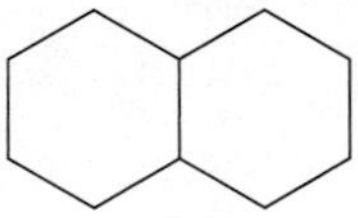

20.3.

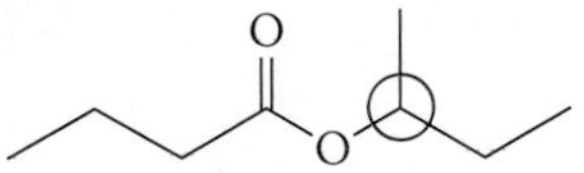

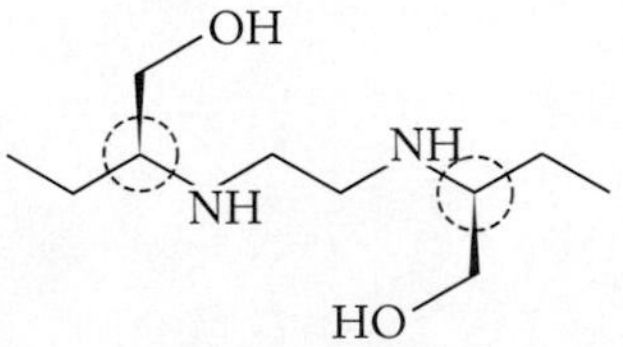

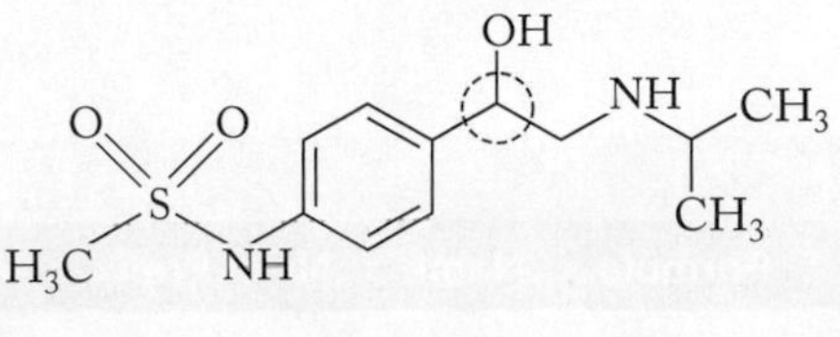

20.4.

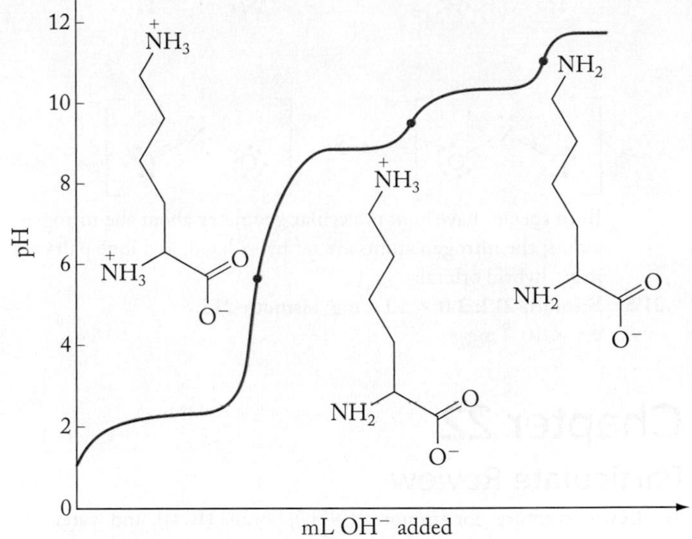

20.5. Six

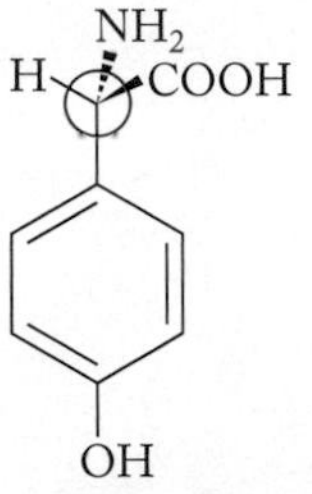

20.6. Val-Ala-Leu-Leu-Thr-Gly

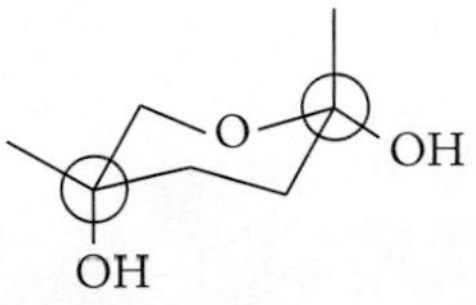

20.7. Six

20.8. (a) GCCATAGGCTA; (b) AATTCGGCGATC

Chapter 21

Particulate Review

Sulfur atoms are in keratin: disulfide bonds are located between the α-helices.

Phosphorus atoms are in the phospholipid bilayer in the phosphate units in the hydrophilic head-groups.

Phosphorus atoms are also in DNA in the phosphate units connecting the ribose units in the backbone.

Keratin has nitrogen atoms in the amide groups of amino acids; phospholipids contain nitrogen atoms in the quaternary amine groups in the hydrophilic head-group; DNA contains nitrogen atoms in the nucleotide bases.

Concept Tests

p. 1026 Nonspontaneous—they require energy to pump ions

p. 1029 $K > 1$ for the drug to be effective

p. 1032 Solubility rules and K_{sp} values indicate that all nitrates are soluble but many phosphates are not. This is borne out by the presence of phosphate in biominerals such as bone and teeth.

p. 1040 Essential elements tend to be the first or second elements in a group. Nonessential elements are generally located toward the bottom of the periodic table. They have larger atomic numbers and are less abundant than the essential elements in the same group.

p. 1041 To avoid tissue damage, the nuclide should not emit high-energy α or β particles.

Practice Exercises

21.1. $E_{K^+} = -110$ mV; $\Delta G_{transport} = +17$ kJ/mol; because $\Delta G < 0$, the transport is non-spontaneous.

21.2. pH = 1.194; the volume of Mg(OH)$_2$ solution required to neutralize the acid solution is 0.584 mL.

21.3. $N_2(g) + 10\,H^+(aq) + 8\,e^- \rightarrow 2\,NH_4^+(aq) + H_2(g)$

21.4.

Both species have bent molecular geometry about the nitrogen atoms; the nitrogen atoms are sp^2 hybridized, and lone pairs are in sp^2 hybrid orbitals.

21.5. Bismuth-212: 2.0×10^{-6} mg; bismuth-213: 9.5×10^{-9} mg

Chapter 22

Particulate Review

The Lewis structures for ammonia (NH_3), borane (BH_3), and water (H_2O) are

Ammonia has one lone pair on its central atom, and water has two lone pairs on its central atom; therefore both can function as Lewis bases. Borane has no lone pairs on the central boron atom and an incomplete octet; it can function as a Lewis acid.

Concept Tests

p. 1056 CN^- ions occupy the inner coordination of Fe^{3+}; Na^+ is the counterion. $Na_3[Fe(CN)_6]$ would have the same conductivity as $[Co(NH_3)_6]Cl_3$.

p. 1060 Tetrachloroplatinate(II)

p. 1070 Green

p. 1073 (a) CN^- is a stronger field ligand than pyridine.
(b) Ru^{2+} ions are larger than Fe^{2+} and have a larger Δ_o; their $4d$ electrons interact more with ligand lone pairs than do the $3d$ electrons of Fe^{2+}.

p. 1076 No for square planar; yes for tetrahedral

p. 1078 Yes. Manganese(IV) has three unpaired d electrons with only one possible distribution; there is no low-spin option. Manganese(III) has four d electrons; depending on the strength of the ligands, it can be either high spin (4 unpaired electrons) or low spin (2 unpaired electrons).

p. 1080 H_2

p. 1084 A longer half-life means slower decay; collecting a sufficient signal to get an image will take longer.

Practice Exercises

22.1. $[Ru(NH_3)_4Cl_2]^+$ is the complex ion; Ru is present as Ru^{3+}; chloride is the counterion.

22.2.

a. $[Zn(NH_3)_4]Cl_2$ = tetraamminezinc(II) chloride
b. $[Co(NH_3)_4(H_2O)_2](NO_2)_2$ = tetraamminediaquacobalt(II) nitrite

22.3. Three

22.4. Ni^{3+} can have either a high-spin or a low-spin configuration; none of the ions is diamagnetic.

22.5.

cis-Diammine-trans-dibromo(ethylenediamine)cobalt(III)
cis-Diammine-cis-dibromo(ethylenediamine)cobalt(III)
trans-Diammine-cis-dibromo(ethylenediamine)cobalt(III)

22.6. 2.5×10^6

		LIGAND								
Compound	Counterion	Formula	Name	Number	Prefix	Formula	Name	Nr	Prefix	M^{n+}
$[Zn(NH_3)_4]Cl_2$	Cl^- (chloride)	NH_3	ammine	4	tetra-					2+
$[Co(NH_3)_4(H_2O)_2](NO_2)_2$	NO_2^- (nitrite)	NH_3	ammine	4	tetra-	H_2O	aqua	2	di-	2+

Answers
to Selected End-of-Chapter Questions and Problems

Chapter 1

1.1. (a) A pure compound in the gas phase; (b) a heterogeneous mixture of blue element atoms and red element atoms: blue atoms are in the gas phase, red atoms are in the liquid phase.

1.3. (b)

1.5. CH_2O_2

1.7. Sample A is both accurate and precise. Sample B is precise but not accurate.

1.9. Compounds are different from elements in that they are made up of two or more elements that can be separated from each other (but elements cannot be separated further), and compounds have different chemical and physical properties than the elements that compose them. Compounds are more commonly found in nature than pure elements. Compounds are similar to elements in that they are composed of atoms, have definite physical and chemical properties, and can be isolated in pure form.

1.11. Particles are most free to move about in the gas phase and least free in the solid phase.

1.13. The snow sublimed to form water vapor.

1.15. Density, melting point, thermal and electrical conductivity, and softness (a–d) are all physical properties, whereas both tarnishing and reaction with water (e and f) are chemical properties.

1.17. (a)

1.19. (a)

1.21. (a)

1.23. (b)

1.25. (c)

1.27. Extensive properties will change with the size of the sample and therefore cannot be used to identify a substance.

1.29. To form a hypothesis we need at least one observation, experiment, or idea (from examining nature).

1.31. Yes

1.33. *Theory* in normal conversation refers to someone's idea or opinion or speculation that can be changed.

1.35. SI units can be easily converted into a larger or smaller unit by multiplying or dividing by multiples of 10. English units are based on other number multiples and thus are more complicated to manipulate.

1.37. 7.48×10^2 g

1.39. 9.24×10^3 mL

1.41. Peter is taller than Paul.

1.43. 23 g

1.45. (a) $\dfrac{1\ ps}{1 \times 10^3\ fs}$; (b) $\dfrac{1\ kg}{1 \times 10^6\ mg}$; (c) $\dfrac{1\ kg\ Ti}{2.20 \times 10^{-4}\ m^3}$

1.47. 279 cm

1.49. (a) 7.34 mi/h; (b) 3.28 m/s

1.51. 1330 Cal

1.53. 58.0 cm^3

1.55. 73.8 mL

1.57. (a) 5.8×10^3 g; (b) 2.1×10^2 oz

1.59. 0.28 cm^3

1.61. Only one data point may be eliminated.

1.63. (a) The standard deviation (*s*) is larger than the range described by the 95% confidence interval.

1.65. (b) and (c), depending on the significance of the zeros in (c).

1.67. (b) and (c), depending on the significance of the zeros in (c).

1.69. (b), (c), and (d), depending on the significance of the zeros in (d).

1.71. (a) 17.4; (b) 1×10^{-13}; (c) 5.70×10^{-23}; (d) 3.58×10^{-3}

1.73. (a) Manufacturer 1 has $\bar{x} = 0.511$ and s = 0.005; manufacturer 2 has $\bar{x} = 0.513$ and s = 0.001; manufacturer 3 has $\bar{x} = 0.501$ and s = 0.001; (b) manufacturer 3; (c) manufacturer 3 is precise and accurate; manufacturer 2 is precise but not accurate.

1.75. Yes, it is an outlier.

1.77. Yes, −40°C is equal to −40°F.

1.79. −38°F; 230°F

1.81. The T_c for $YBa_2Cu_3O_7$ is already expressed in kelvin, $T_c = 93.0$ K. The T_c of Nb_3Ge converted to K is 23.2 K. The T_c of $HgBa_2CaCu_2O_6$ converted to K is 127.0 K. The superconductor with the highest T_c is $HgBa_2CaCu_2O_6$.

1.83. −269.0°C

1.85. 39.2°C

1.87. −89.2°C; 183.9 K

1.89. Both mixtures (a) and (b) react so that there is neither sodium nor chlorine left over.

1.91. (a) Peanuts $\bar{x} = 63\%$, $s = 3.9$; raisins $\bar{x} = 63\%$, $s = 3.9$; (b) Peanuts $\mu = 63 \pm 4.6\%$; raisins $\mu = 63 \pm 4.6\%$

1.93. Spring B is stronger.

1.95. 5.2×10^2 mg

1.97. 10 tablets

Chapter 2

2.1. The positively charged alpha particle will be attracted to the negative (left) plate, whereas the negatively charged beta particle will be attracted to the positive (right) plate.

2.3. (c)

2.5. (a) Chlorine (Cl_2) (yellow); (b) neon (Ne) (red); (c) sodium (Na) (dark blue)

2.7. (a) Mg (green) will form MgO; (b) K (red) will form K_2O; (c) Ti (yellow) will form TiO_2; (d) Al (dark blue) will form Al_2O_3.

2.9. The element shaded in red (Zn) is too heavy to have been produced by stellar fusion.

2.11. Rutherford concluded that the positive charge in the atom could not be spread out (as in a plum pudding) in the atom, but must result from a concentration of charge in the center of the atom (the nucleus). Most of the particles were deflected only slightly or passed directly through the gold foil, so he reasoned that the nucleus must be small compared to the size of the entire atom. The negatively charged electrons do not deflect the particles, and Rutherford reasoned that the electrons took up the remainder of the space of the atom outside the nucleus.

2.13. The fact that cathode rays were deflected by a magnetic field indicated that the rays were streams of charged particles.

2.15. A *weighted average* takes into account the proportion of each value in the group of values to be averaged.

2.17. The element symbol (X) and the atomic number (Z) provide the same information. Any change to the number of protons will change the identity of the element.

2.19.

Atom	Mass Number	Atomic Number = Number of Protons	Number of Neutrons = Mass Number − Atomic Number	Number of Electrons = Number of Protons
(a) ^{14}C	14	6	8	6
(b) ^{59}Fe	59	26	33	26
(c) ^{90}Sr	90	38	52	38
(d) ^{210}Pb	210	82	128	82

2.21. Greater than 1

2.23.

Symbol	^{16}O	^{56}Fe	^{118}Sn	^{197}Au
Number of protons	8	26	50	79
Number of neutrons	8	30	68	118
Number of electrons	8	26	50	79
Mass number	16	56	118	197

2.25. (a) and (b)

2.27. Yes

2.29. 47.95 amu

2.31. Mendeleev knew only the masses of the elements at the time he arranged the elements into his periodic table.

2.33. Group 2, RO; Group 3, R_2O_3; Group 4, RO_2

2.35. (c)

2.37. (a), (b), and (d)

2.39. (a) metal; (b) metal; (c) metalloid; (d) nonmetal; (e) nonmetal

2.41. (a) Bromine (Br) is a halogen; (b) calcium (Ca) is an alkaline earth metal; (c) potassium (K) is an alkali metal; (d) krypton (Kr) is a noble gas; (e) vanadium (V) is a transition metal

2.43. C, N, and O, respectively

2.45. (a) Palladium (Pd); (b) rhodium (Rh); (c) platinum (Pt)

2.47. Dalton's atomic theory states that, because atoms are indivisible, the ratio of the atoms (elements) in a compound is a ratio of whole numbers. Thus in water, the ratio of the volume of hydrogen to oxygen is 2:1, a whole-number ratio, because the atoms in water are present in the ratio of 2:1.

2.49. Molecular compounds are composed of nonmetals, whereas ionic compounds are composed of ions derived from a metal and a nonmetal.

2.51. 2 Co:3s

2.53. 7.5 g

2.55.

Symbol	$^{37}Cl^-$	$^{23}Na^+$	$^{81}Br^-$	$^{210}Pb^{2+}$
Number of protons	17	11	35	82
Number of neutrons	18	12	46	128
Number of electrons	18	10	36	80
Mass number	35	23	81	210

2.57. Compounds (a) and (d) consist of molecules; compounds (b) and (c) consist of ions.

2.59. (a) ionic bonds; (b) covalent bonds; (c) covalent bonds; (d) ionic bonds.

2.61. (a) 4 atoms; (b) 5 atoms; (c) 4 atoms; (d) 5 atoms

2.63. XO_2^{2-}

2.65. Roman numerals indicate the charge on the transition metal cation.

2.67. (a) nitrogen trioxide; (b) dinitrogen pentoxide; (c) dinitrogen tetroxide; (d) nitrogen dioxide

2.69. (a) Na_2S, sodium sulfide; (b) $SrCl_2$, strontium chloride; (c) Al_2O_3, aluminum oxide; (d) LiH, lithium hydride

2.71. Magnesium hydroxide

2.73. (a) sodium oxide; (b) sodium sulfide; (c) sodium sulfate; (d) sodium nitrate; (e) sodium nitrite

2.75. (a) K_2S; (b) K_2Se; (c) Rb_2SO_4; (d) $RbNO_2$; (e) $MgSO_4$

2.77. (a) $NaBrO$; (b) K_2SO_4; (c) $LiIO_3$; (d) $Mg(NO_2)_2$

2.79. (b)

2.81. $LiSO_4$ is incorrect (should be Li_2SO_4).

2.83.

	Na^+ cation	K^+ cation	Mg^{2+} cation	Ca^{2+} cation
Cl^- anion	NaCl	KCl	$MgCl_2$	$CaCl_2$
$H_2PO_4^-$ anion	NaH_2PO_4	KH_2PO_4	$Mg(H_2PO_4)_2$	$Ca(H_2PO_4)_2$

2.85. (a) Chromium(III) telluride; (b) vanadium(III) sulfate; (c) iron(I) chromate; (d) manganese(II) oxide

2.87. (a) $ZrCr_2O_7$; (b) $Fe(CH_3COO)_3$; (c) Hg_2O_2; (d) $Sc(SCN)_3$

2.89. (a) Cu_2O; (b) Cu_2S; (c) CuS

2.91. (a) hydrofluoric acid; (b) bromic acid; (c) hydrobromic acid; (d) periodic acid

2.93. Alkanes, alkenes, and alkynes

2.95. Amines and amides contain nitrogen atoms. Amides contain both nitrogen and oxygen atoms.

2.97. (a) alkane; (b) alkyne

2.99. R–C(O)–R, the ester functional group

2.101. Chemistry is the study of the composition, structure, properties, and reactivity of matter. Cosmology is the study of the history, structure, and dynamics of the universe. A few of the ways that these two sciences are related include the following: (1) because the universe is composed of matter and the study of matter is chemistry, the study of the universe is chemistry; (2) the changing universe is driven by chemical and atomic or nuclear reactions, which are also studied in chemistry; and (3) cosmology often asks what the universe (including stars, black holes, etc.) are made of at the atomic level.

2.103. Because quarks combine to make up the three particles that are important to the properties and reactivity of atoms: protons, neutrons, and electrons

2.105. The density of the universe is decreasing.

2.107. $HClO_3$, chloric acid

2.109. $^{21}_{10}Ne + ^4_2\alpha \rightarrow ^1_0n + ^{24}_{12}Mg$

2.111. (a) Two; (b) the spots would all be in different locations, one at the center of the screen, one closer to the negatively charged plate, and one closer to the positively charged plate. The spot closest to the negatively charged plate would be the α particle, the spot closest to the positively charged plate would be the β particle.

2.113. (a) 12 H atoms for every 1 He atom; (b) the value is smaller so there is more helium present now; (c) stars consume hydrogen and produce helium; (d) look at the elemental composition of older galaxies, and compare it to our own galaxy.

2.115. 17 Cu:1 Sn

2.117. (a) Scandium (Sc) with an average mass of 44.956 amu, gallium (Ga) with an average mass of 69.723 amu, and germanium (Ge) with an average mass 72.61 amu; (b) ekaaluminum is gallium, ekaboron is scandium, and ekasilicon is germanium; (c) scandium was discovered in 1879 by Lars Fredrik Nilson in Sweden, gallium was discovered in 1875 by Paul-Emile Lecoq de Boisbaudran in France, and germanium was discovered in 1886 by Clemens Winkler in Germany.

2.119. 60.11%

2.121. (a) $^{79}Br-^{79}Br = 157.8366$ amu, $^{79}Br-^{81}Br = 159.8346$ amu, $^{81}Br-^{81}Br = 161.8326$ amu; (b) $^{79}Br-^{79}Br$, 25.41% abundant, $^{79}Br-^{81}Br$, 50.00% abundant, $^{81}Br-^{81}Br$, 24.59% abundant

2.123. (a) CO_2, carbon dioxide; (b) Li_3N, lithium nitride

2.125. Radium will adopt a 2+ charge to form Ra^{2+} ions. It will likely be malleable, be relatively dense, conduct heat and electric current, and melt at a fairly high temperature in its metallic state.

	Melting Point (°C)		Melting Point (°C)
$CaCl_2$	772	CaO	2572
$SrCl_2$	874	SrO	2531
$BaCl_2$	962	BaO	1923
$RaCl_2$	(950 to 1050)	RaO	(1700 to 2000)

2.127. Despite being heavier (on average), argon contains 18 protons, whereas potassium contains 19 protons. Because the modern periodic table is organized by increasing atomic number, argon is placed before potassium.

Chapter 3

3.1. a. $4 X(g) + 4 Y(g) \rightarrow 4 XY(g)$
b. $4 X(g) + 4 Y(g) \rightarrow 4 XY(s)$
c. $4 X(g) + 4 Y(g) \rightarrow 2 XY_2(g) + 2 X(g)$
d. $4 X_2(g) + 4 Y_2(g) \rightarrow 8 XY(g)$

3.3. (b)

3.5. (a) and (c) are the same (NO_2), and (b) and (d) are the same (N_2O).

3.7. (a) Water; (b) carbon dioxide

3.9. 75%

3.11. Less than

3.13. $Fe(\ell)$

3.15. It is too small a unit to conveniently express the very large number of atoms, ions, or molecules present in laboratory quantities such as a mole.

3.17. No, the molar mass of a substance does not directly correlate to the number of atoms in a molecular compound. The statement would be true only if the two compounds were composed of the same element.

3.19. (a) 7.3×10^{-10} mol Ne; (b) 7.0×10^{-11} mol CH_4; (c) 4.2×10^{-12} mol O_3; (d) 8.1×10^{-15} mol NO_2

3.21. (a) 7.53×10^{22} Ti atoms; (b) 7.53×10^{22} Ti atoms; (c) 1.51×10^{23} Ti atoms; (d) 2.26×10^{23} Ti atoms

3.23. (a) Fe_2S_3; (b) CaS; (c) both molecules contain the same number of S atoms.

3.25. (a) 3.00 mol; (b) 3.00 mol; (c) 4.50 mol

3.27. 41.63 mol

3.29. 7.0×10^{19} Ir atoms

3.31. (a) 1 mol; (b) 2 mol; (c) 1 mol; (d) 3 mol

3.33. (a) 64.06 g/mol; (b) 48.00 g/mol; (c) 44.01 g/mol; (d) 108.02 g/mol

3.35. (a) 152.16 g/mol; (b) 164.22 g/mol; (c) 148.22 g/mol; (d) 132.17 g/mol

3.37. (a) NO; (b) CO_2; (c) O_2

3.39. 0.752 mol SiO_2

3.41. 10.3 g

3.43. Diamond

3.45. No

3.47. No

3.49.
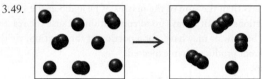

3.51. 4

3.53. a. $3 FeSiO_3(s) + 4 H_2O(\ell) \rightarrow Fe_3Si_2O_5(OH)_4(s) + H_4SiO_4(aq)$
b. $Fe_2SiO_4(s) + 2 CO_2(g) + 2 H_2O(\ell) \rightarrow 2 FeCO_3(s) + H_4SiO_4(aq)$
c. $Fe_3Si_2O_5(OH)_4(s) + 3 CO_2(g) + 2 H_2O(\ell) \rightarrow 3 FeCO_3(s) + 2 H_4SiO_4(aq)$

3.55. a. $N_2(g) + O_2(g) \rightarrow 2 NO(g)$
b. $2 NO(g) + O_2(g) \rightarrow 2 NO_2(g)$
c. $NO(g) + NO_3(g) \rightarrow 2 NO_2(g)$
d. $2 N_2(g) + O_2(g) \rightarrow 2 N_2O(g)$

3.57. a. $N_2O_5(g) + Na(s) \rightarrow NaNO_3(s) + NO_2(g)$
b. $N_2O_4(g) + H_2O(\ell) \rightarrow HNO_3(aq) + HNO_2(aq)$
c. $3 NO(g) \rightarrow N_2O(g) + NO_2(g)$
d. $2 C_2H_2(g) + 5 O_2(g) \rightarrow 4 CO_2(g) + 2 H_2O(g)$

3.59. Yes, if the masses were unequal, then the equation would be missing either some reactants or products.

3.61. (a) $2 NaHCO_3(s) \rightarrow CO_2(g) + H_2O(g) + Na_2CO_3(s)$; (b) 6.55 g CO_2

3.63. 1.17 kg

3.65. 1.5 t

3.67. $2 C_8H_{18} + 25 O_2 \rightarrow 16 CO_2 + 18 H_2O$; $C_2H_6O + 3 O_2 \rightarrow 2 CO_2 + 3 H_2O$; octane produces more CO_2.

3.69. 346 g

3.71. An empirical formula shows the lowest whole-number ratio of atoms in a substance. A molecular formula shows the actual numbers of each kind of atom that compose one molecule of the substance.

3.73. No, lighter elements may be present in sufficient quantities to be of a greater percent mass than a heavier element.

3.75. Yes. All of these have the same empirical and molecular formulas.

3.77. The empirical formulas for the given compounds are

Molecular Formula	C_6H_{14}	C_7H_{16}	C_8H_{18}	C_9H_{20}
Empirical Formula	C_3H_7	C_7H_{16}	C_4H_9	C_9H_{20}

3.79. (a) 74.19% Na, 25.81% O; (b) 57.48% Na, 40.00% O, 2.52% H; (c) 27.37% Na, 1.20% H, 14.30% C, 57.13% O; (d) 43.38% Na, 11.33% C, 45.28% O

3.81. Pyrene, $C_{16}H_{10}$, has the greatest percent carbon by mass. Empirical formulas differ.

3.83. CH_4

3.85. Ti_6Al_4V

3.87. Citric acid

3.89. $Mg_3Si_2H_4O_9$

3.91. The empirical formula is CHN; the molecular formula is $C_5H_5N_5$.

3.93. The excess of oxygen is required in combustion analysis to ensure the complete reaction of the hydrogen and carbon to form water and carbon dioxide.

3.95. Yes

3.97. NO_2

3.99. The empirical formula is C_2H_3; the molecular formula is $C_{20}H_{30}$.

3.101. $C_{10}H_{18}O$

3.103. $C_9H_8O_4$

3.105. (c) Less than the sum of the masses of Fe and S at the start

3.107. Reactions do not always go to completion because the reaction may be slow or may have, for a portion of the reaction, yielded different products than expected.

3.109. 3 cups

3.111. (a) 5.0 g Li_3N; (b) 34.5 g H_3PO_4; (c) 8.0 g SO_3

3.113. $NH_3(g) + HCl(g) \rightarrow NH_4Cl(s)$; 0.7 g NH_3

3.115. (a) 2 $PbO(s)$ + 2 $NaCl(s)$ + $H_2O(\ell)$ + $CO_2(g) \rightarrow$ $Pb_2Cl_2CO_3(s)$ + 2 $NaOH(aq)$; (b) 12.2 g; (c) 22.3%

3.117. 56%

3.119. (a) $C_6H_{12}O_6(aq) \rightarrow 2\ C_2H_5OH(\ell)$ + 2 $CO_2(g)$; (b) 77.1%

3.121. (a) 36.09%; (b) 36.07%

3.123. (a) 870 g; (b) 1020 g; (c) 1400 g

3.125. (a) a = 1, b = 3, charge on U is 6+; (b) c = 3, d = 8, charge on U is 5.33+; (c) x = 2, y = 2, z = 6

3.127. (a) 5.838×10^{20} molecules of $C_{13}H_{18}O_2$; (b) 3.008×10^{21} molecules of $CaCO_3$; (c) 9×10^{18} molecules of $C_{16}H_{19}N_2Cl$

3.129. (a) Mn_2O_3 is manganese(III) oxide and MnO_2 is manganese(IV) oxide; (b) % Mn in Mn_2O_3 = 69.60%, % Mn in MnO_2 = 63.19%; (c) these compounds contain the same elements, but in different atom ratios.

3.131. (a) 0.966 g; (b) $C_{10}H_{20}O_{10}$

3.133. 1×10^{-8} mol

3.135. 55 mol ethanol

3.137. Re

3.139. 82.4%

3.141. (a) 6.0 metric tons; (b) 2 $SO_2(g)$ + 2 $H_2O(g)$ + $O_2(g) \rightarrow$ 2 $H_2SO_4(\ell)$; (c) 9.2 metric tons

3.143. 3.06 g H_2SO_4

3.145. Mg_2SiO_4

3.147. (a) 3.05×10^3 g KO_2; (b) more Na_2O_2; (c) Na_2O_2

Chapter 4

4.1. Yellow

4.3. (a) Cl (purple); (b) S (orange); (c) N (green); (d) P (blue)

4.5. Strong electrolyte: (a) and (c); weak acid: (b); weak electrolyte: (b); nonelectrolyte: (d)

4.7. Hydronium and nitrate ions will remain in solution.

4.9. The solvent is usually the liquid component of the solution. If both the solvent and solute are liquids or solids, the solvent is that component present in the greatest amount.

4.11. 1.00 M

4.13. (a) 5.6 M $BaCl_2$; (b) 1.00 M Na_2CO_3; (c) 1.30 M $C_6H_{12}O_6$; (d) 5.92 M KNO_3

4.15. (a) 0.14 M Na^+; (b) 0.11 M Cl^-; (c) 0.096 M SO_4^{2-}; (d) 0.20 M Ca^{2+}

4.17. (a) 11.7 g NaCl; (b) 4.99 g $CuSO_4$; (c) 6.41 g CH_3OH

4.19. 2.72 g

4.21. (a) 9.6×10^{-3} mol; (b) 7.80×10^{-4} mol; (c) 8.8×10^{-2} mol; (d) 4.22 mol

4.23. Orchard sample: 1.2×10^{-7} ppm, 3.4×10^{-4} mmol/L; residential area sample: 2.0×10^{-8} ppm, 5.6×10^{-5} mmol/L; after-storm sample: 1.1×10^{-5} ppm, 3.2×10^{-2} mmol/L

4.25. 4.54×10^{-5} μg NF_3 per kg air

4.27. 4.4×10^{-5} M

4.29. (a) 0.0210 M Na^+; (b) 1.28×10^{-2} mM LiCl; (c) 1.28×10^{-2} mM Zn^{2+}

4.31. 58.6 mL

4.33. 0.40 M

4.35. 12.3 mL

4.37. A = 0.75

4.39. Table salt produces Na^+ and Cl^- ions in solution when it dissolves. Sugar does not dissociate into ions because it is not a salt. Ions are required to conduct electricity.

4.41. The lack of ions in methanol means that the liquid is nonconductive. Molten NaOH, however, has freely moving Na^+ and OH^- ions, which can conduct electricity.

4.43. In order of decreasing conductivity, 1.0 M Na_2SO_4 (c) > 1.2 M KCl (b) > 1.0 M NaCl (a) > 0.75 M LiCl (d).

4.45. (a) 0.025 M; (b) 0.050 M; (c) 0.075 M

4.47. (b)

4.49. Acid

4.51. (a) Strong acid; (b) weak acid; (c) weak acid; (d) strong acid

4.53. Base

4.55. (a) Strong base; (b) weak base; (c) weak base; (d) strong base

4.57. (a) Ionic and net ionic equation: $H^+(aq)$ + $HSO_4^-(aq)$ + $Ca^{2+}(aq)$ + 2 $OH^-(aq) \rightarrow CaSO_4(s)$ + 2 $H_2O(\ell)$, the acid is H_2SO_4 and the base is $Ca(OH)_2$; (b) ionic and net ionic equation: $PbCO_3(s)$ + 2 $H^+(aq)$ + $SO_4^{2-}(aq) \rightarrow PbSO_4(s)$ + $CO_2(g)$ + $H_2O(\ell)$, $PbCO_3$ is the base and sulfuric acid is the acid; (c) ionic equation: $Ca^{2+}(aq)$ + 2 $OH^-(aq)$ + 2 $CH_3COOH(aq) \rightarrow Ca^{2+}(aq)$ + 2 $CH_3COO^-(aq)$ + 2 $H_2O(\ell)$, calcium is a spectator ion, $Ca(OH)_2$ is the base, and CH_3COOH is the acid; net ionic equation: $OH^-(aq)$ + $CH_3COOH(aq) \rightarrow CH_3COO^-(aq)$ + $H_2O(\ell)$

4.59. (a) Molecular equation: $Mg(OH)_2(s)$ + $H_2SO_4(aq) \rightarrow$ $MgSO_4(aq)$ + 2 $H_2O(\ell)$, net ionic equation: $Mg(OH)_2(s)$ + $HSO_4^-(aq)$ + $H^+(aq) \rightarrow Mg^{2+}(aq)$ + $SO_4^{2-}(aq)$ + 2 $H_2O(\ell)$; (b) molecular equation: $MgCO_3(s)$ + 2 $HCl(aq) \rightarrow$ $MgCl_2(aq)$ + $H_2CO_3(aq)$, net ionic equation: $MgCO_3(s)$ + 2 $H^+(aq) \rightarrow Mg^{2+}(aq)$ + $H_2O(\ell)$ + $CO_2(g)$; (c) molecular equation as well as net ionic equation: $NH_3(g)$ + $HCl(g) \rightarrow$ $NH_4Cl(s)$; (d) molecular equation: $SO_3(g)$ + 2 $NaOH(aq) \rightarrow$ $Na_2SO_4(aq)$ + $H_2O(\ell)$, net ionic equation: $SO_3(g)$ + 2 $OH^-(aq) \rightarrow SO_4^{2-}(aq)$ + $H_2O(\ell)$

4.61. $PbCO_3(s)$ + 2 $H^+(aq) \rightarrow Pb^{2+}(aq)$ + $CO_2(g)$ + $H_2O(\ell)$; $Pb(OH)_2(s)$ + 2 $H^+(aq) \rightarrow Pb^{2+}(aq)$ + 2 $H_2O(\ell)$

4.63. (a) 18.0 mL; (b) 31.6 mL; (c) 114 mL

4.65. 5.00×10^2 mL

4.67. 290 mL

4.69. A saturated solution contains the maximum concentration of a solute. A supersaturated solution *temporarily* contains *more* than the maximum concentration of a solute at a given temperature.

4.71. $CaCO_3$

4.73. A saturated solution may not be a concentrated solution if the solute is only sparingly or slightly soluble in the solution. In that case, the solution is a saturated dilute solution.

4.75. (a) Barium sulfate is insoluble; (e) lead hydroxide is insoluble; (f) calcium phosphate is insoluble.

4.77. (a) Balanced reaction: $Pb(NO_3)_2(aq) + Na_2SO_4(aq) \rightarrow PbSO_4(s) + 2\,NaNO_3(aq)$, net ionic equation: $Pb^{2+}(aq) + SO_4^{2-}(aq) \rightarrow PbSO_4(s)$; (b) no precipitation reaction occurs; (c) balanced reaction: $FeCl_2(aq) + Na_2S(aq) \rightarrow FeS(s) + 2\,NaCl\,(aq)$, net ionic equation: $Fe^{2+}(aq) + S^{2-}(aq) \rightarrow FeS(s)$; (d) balanced reaction: $MgSO_4(aq) + BaCl_2(aq) \rightarrow MgCl_2(aq) + BaSO_4(s)$, net ionic equation: $Ba^{2+}(aq) + SO_4^{2-}(aq) \rightarrow BaSO_4(s)$

4.79. 2.11×10^{-2} g

4.81. 5.4×10^{-2} g

4.83. 1.3×10^2 kg

4.85. (a) $[Na^+]$ and $[NO_3^-]$ remain the same, and $[Ag^+]$ and $[Cl^-]$ decrease; (b) $[Na^+]$ and $[Cl^-]$ remain the same, and $[H^+]$ and $[OH^-]$ decrease; (c) all ionic concentrations remain the same.

4.87. To deionize water, cations such as Na^+ and Ca^{2+} are exchanged for H^+ at cation-exchange sites. Anions such as Cl^- and SO_4^{2-} are exchanged for OH^- at anion-exchange sites. The released ions (H^+ and OH^-) at these sites combine to form H_2O.

4.89. (a) 1: $Cl^-(aq) + Ag^+(aq) \rightarrow AgCl(s)$, 2: NR, 3: $Mg^{2+}(aq) + 2\,Cl^-(aq) \rightarrow MgCl_2(s)$; (b) reactions (1) and (3) will give visual evidence (a precipitate).

4.91. The number of electrons gained or lost is directly related to the change in oxidation number of a species.

4.93. (a) -1; (b) $+1$; (c) -2; (d) -3

4.95. Silver

4.97. $Na^+ + e^- \rightarrow Na(s)$, $2\,Cl^- \rightarrow Cl_2(g) + 2\,e^-$

4.99. (a) $+3$; (b) $+3$; (c) $+3$

4.101. a. $2\,e^- + Br_2(\ell) \rightarrow 2\,Br^-(aq)$: reduction

b. $Pb(s) + 2\,Cl^-(aq) \rightarrow PbCl_2(s) + 2\,e^-$: oxidation

c. $2\,e^- + O_3(g) + 2\,H^+(aq) \rightarrow O_2(g) + H_2O(\ell)$: reduction

d. $2\,H_2SO_3(aq) + H^+(aq) + 2\,e^- \rightarrow HS_2O_4^-(aq) + 2\,H_2O\,(\ell)$: reduction

4.103. a. $2\,H^+(aq) + MnO_2(s) + 2\,HCl(aq) \rightarrow Mn^{2+}(aq) + Cl_2(g) + 2\,H_2O(\ell)$, manganese is reduced and chlorine is oxidized.

b. $I_2(s) + 2\,S_2O_3^{2-}(aq) \rightarrow S_4O_6^{2-}(aq) + 2\,I^-(aq)$, sulfur is oxidized and iodine is reduced.

c. $8\,H^+(aq) + MnO_4^-(aq) + Fe^{2+}(aq) \rightarrow Mn^{2+}(aq) + Fe^{3+}(aq) + 4\,H_2O(\ell)$, iron is oxidized and manganese is reduced.

4.105. a.

Reactants	Products
SiO_2: Si = +4, O = −2	Fe_2SiO_4: Fe = +2, Si = +4, O = −2
Fe_3O_4: Fe = +8/3, O = −2	O_2: O = 0

Oxygen is oxidized (O^{2-} to O_2) and iron is reduced (Fe^{3+} to Fe^{2+}).

b.

Reactants	Products
SiO_2: Si = +4, O = −2	Fe_2SiO_4: Fe = +2, Si = +4, O = −2
Fe: Fe = 0	
O_2: O = 0	

Iron is oxidized (Fe^0 to Fe^{2+}) and oxygen is reduced (O_2 to O^{2-}).

c.

Reactants	Products
FeO: Fe = +2, O = −2	$Fe(OH)_3$: Fe = +3, O = −2, H = +1
O_2: O = 0	
H_2O: H = +1, O = −2	

Iron is oxidized (Fe^{2+} to Fe^{3+}) and oxygen is reduced (O_2 to O^{2-}).

4.107. a. $O_2(aq) + 4\,FeCO_3(s) \rightarrow 2\,Fe_2O_3(s) + 4\,CO_2(g)$

b. $O_2(aq) + 6\,FeCO_3\,(s) \rightarrow 2\,Fe_3O_4(s) + 6\,CO_2(g)$

c. $O_2(aq) + 4\,Fe_3O_4(s) \rightarrow 2\,Fe_2O_3(s)$

4.109. $NH_4^+(aq) + 2\,O_2(g) \rightarrow NO_3^-(aq) + 2\,H^+(aq) + H_2O\,(\ell)$

4.111. (a)

Reactants	Products
$HCrO_4^-$: H = +1, Cr = +6, O = −2	Cr_2O_3: Cr = +3, O = −2
H_2S: H = +1, S = −2	SO_4^{2-}: S = +6, O = −2

(b) $3\,H_2S(aq) + 8\,HCrO_4(aq) + 2\,H^+ \rightarrow 3\,SO_4^{2-}(aq) + 4\,Cr_2O_3(s) + 8\,H_2O(\ell)$; (c) 3

4.113. $2\,Fe(OH)_2^+(aq) + Mn^{2+}(aq) \rightarrow 2\,Fe^{2+}(aq) + 2\,H_2O(\ell) + MnO_2(s)$

4.115. Zinc and aluminum

4.117. Vanadium is placed below aluminum, and scandium is placed above aluminum on the activity series. We could use magnesium to test vanadium's position. If magnesium is oxidized by Sc^{3+} ions, scandium must lie between aluminum and magnesium.

4.119. (a) $Cr_2O_7^{2-}(aq) + 14\,H^+(aq) + 6\,Fe^{2+}(aq) \rightarrow 2\,Cr^{3+}(aq) + 7\,H_2O(\ell) + 6\,Fe^{3+}(aq)$; (b) $0.123\,M$

4.121. $1.95\,M$

4.123. (a) $11.7\,M$; (b) 42.7 mL; (c) 1.72 kg

4.125. (a) $2\,OH^-(aq) + 2\,H_2O(\ell) + 3\,S_2O_4^{2-}(aq) + 2\,CrO_4^{2-}(aq) \rightarrow 6\,SO_3^{2-}(aq) + 2\,Cr(OH)_3(s)$; (b) sulfur is oxidized and chromium is reduced; (c) the oxidizing agent is CrO_4^{2-} and the reducing agent is $S_2O_4^{2-}$; (d) 38.7 g

4.127. a. Balanced equation: $2\,Ag(s) + H_2S(g) \rightarrow Ag_2S(s) + H_2(g)$
Oxidation numbers:
Ag = 0 Ag_2S: Ag = +1, S = −2
H_2S: H = +1, S = −2 H_2: H = 0
One mole of electrons is transferred per mole of Ag.

b. $3\,Ag_2S(s) + 12\,H_2O(\ell) + 4\,Al(s) \rightarrow 6\,Ag(s) + 3\,H_2S(g) + 3\,H_2(g) + 4\,Al(OH)_3(s)$

4.129. (a) Balanced equation: $HC_2H_3O_2(aq) + KOH(aq) \rightarrow H_2O(\ell) + KC_2H_3O_2(aq)$, net ionic equation: $HC_2H_3O_2(aq) + OH^-(aq) \rightarrow H_2O(\ell) + C_2H_3O_2^-(aq)$; (b) balanced equation: $Na_2CO_3(aq) + CaCl_2(aq) \rightarrow CaCO_3(s) + 2\,NaCl(aq)$, net ionic equation: $CO_3^{2-}(aq) + Ca^{2+}(aq) \rightarrow CaCO_3(s)$; (c) balanced equation: $CaO(s) + H_2O(\ell) \rightarrow Ca(OH)_2(aq)$, net ionic equation: $CaO(s) + H_2O(\ell) \rightarrow Ca^{2+}(aq) + 2\,OH^-(aq)$

4.131. (a) $NaClO_4$, NH_4ClO_4; (b) 427 kg; (c) 2.80×10^{10} gal; (d) the MA lab

4.133. (a) $3\,CH_2O \rightarrow CO_2 + C_2H_5OH$; (b) $C_2H_5OH + O_2 \rightarrow HC_2H_3O_2 + H_2O$; (c) $CH_2O: C = 0$, $CO_2: C = +4$, $C_2H_5OH: C = -4$ over two carbon atoms so oxidation number on each carbon $= -2$, $HC_2H_3O_2: C = 0$; (d) 66.7 g acetic acid

4.135. (a) $H^+(aq) + HSO_4^-(aq) + Ba^{2+}(aq) + 2\,OH^-(aq) \rightarrow BaSO_4(s) + 2\,H_2O(\ell)$; (b) graph (c)

4.137. $HC_2H_3O_2(aq) + OH^-(aq) \rightarrow H_2O(\ell) + C_2H_3O_2^-(aq)$; the bulb will become brighter.

4.139. (a) Oxidation: $O_2^-(aq) \rightarrow O_2(aq) + 1\,e^-$, reduction: $2\,H^+(aq) + O_2^-(aq) + 1\,e^- \rightarrow H_2O_2(aq)$; (b) $2\,H^+(aq) + 2\,O_2^-(aq) \rightarrow H_2O_2(aq) + O_2(aq)$

4.141. (a) The first reaction is a redox reaction in which eight electrons are transferred; (b) $H^+(aq) + HSO_4^-(aq) + CaCO_3(s) \rightarrow CaSO_4(s) + H_2O(\ell) + CO_2(g)$; (c) $H^+(aq) + HSO_4^-(aq) + CaCO_3(s) \rightarrow CaSO_4(s) + H_2CO_3(aq)$

4.143. (c) and (d)

4.145. No, this is not an example of ion exchange because the ions are not simply being swapped. Rather, they are changing oxidation states, so this is a redox reaction.

Chapter 5

5.1. At 35 ft above street level KE = 150 J; just before hitting the street KE = 500 J

5.3. (a)

(b) The piston is higher in the cylinder; (c) yes; (d) the system did work on the surroundings.

5.5. (a) No; (b) the internal energy of the system increases.

5.7. (a) $2\,SO_2(g) + O_2(g) \rightarrow 2\,SO_3(g)$; (b) positive; (c) -98.9 kJ/mol

5.9. Energy makes work possible.

5.11. The value of a state function is independent of the path—only the initial and final values are important.

5.13. (a) The potential energy in a battery consists of the chemicals that can react via a redox reaction; (b) the potential energy in a gallon of gasoline consists of the chemical bonds in the fuel that release heat as the fuel is combusted; (c) the potential energy of the crest of a wave is due to its position above the ground.

5.15. The system is that part of the universe that we are interested in. The surroundings are everything else, extending to the entire universe.

5.17. The sign of ΔE depends on the magnitude of q and ω.

5.19. (a) Exothermic; (b) endothermic; (c) exothermic

5.21. Energy is absorbed from the surroundings. Thus q increases and E increases.

5.23. $w = -0.500\,L \cdot atm = -50.7$ J

5.25. (a) 80.0 J; (b) 9.2 kJ; (c) -940 J

5.27. (b)

5.29. (a) 127 J; (b) -2.47 kJ

5.31. A change in enthalpy is the sum of the change of internal energy and the product of the system's pressure and change in volume.

5.33. The symbol that refers to a physical change is ΔH_{fus}.

5.35. Negative

5.37. Positive

5.39. Negative

5.41. Specific heat is specified for a gram of the substance, while molar heat capacity is specified for one mole of the substance.

5.43. No

5.45. Water's high heat capacity compared to air means that water carries away more energy from the engine for every Celsius degree rise in temperature, so water is a good choice to cool automobile engines.

5.47. 29.3 kJ

5.49.

5.51. 886 g

5.53. Gold

5.55. $-47.5°C$

5.57. To know how much energy (generated or absorbed by the system) is required to change the temperature of the surroundings (the calorimeter) in order to calculate the heat capacity or final temperature of the system in an experiment

5.59. The heat capacity of the new liquid is different from that of water. The liquid is part of the calorimeter and therefore part of the surroundings. The calorimeter constant depends on the heat capacity of the surroundings, so it must be redetermined.

5.61. NH_4NO_3 provides the greatest temperature drop.

5.63. 6.12 kJ/°C

5.65. -5129 kJ/mol

5.67. 23.29°C

5.69. (a) $Mg(s) + 2\,H^+(aq) \rightarrow H_2(g) + Mg^{2+}(aq)$; (b) $\Delta H_{rxn} = -44.4$ kJ/mol

5.71. 0.145 g

5.73. When we apply Hess's law, all the heat is accounted for in the reaction—energy is neither created nor destroyed.

5.75. Reverse the direction of the second reaction, and add the first reaction to obtain the third reaction (after canceling species common to both sides).

5.77. 54.08 kJ

5.79. If we write out the chemical equations for the ΔH_f° for the reactants and products, these formation reactions will add up to the overall reaction.

5.81. Because ozone and elemental oxygen are different forms of oxygen, their standard enthalpies of formation are different.

5.83. (a) and (d)

5.85. 52 kJ

5.87. -35.9 kJ; $NH_4NO_3(s) \rightarrow N_2O(g) + 2\,H_2O(g)$

5.89.   −7198 kJ

5.91.   Reverse the direction of the combustion reaction (change the sign of ΔH_{comb}), and add the reaction describing the formation of $CO_2(g)$ from pure $C(s)$ and $O_2(g)$ to obtain the reaction for the formation of $CO(g)$:
$\Delta H_f^{\circ}(CO) = \Delta H_f^{\circ}(CO_2) - \Delta H_{comb}(CO)$

5.93.   12.6 kJ

5.95.   The energy per gram a fuel releases on burning

5.97.   The fuel value (kJ/g) is obtained by dividing the molar heat of combustion (kJ/mol) by the molar mass (mol/g).

5.99.   Gasoline

5.101. (a) 48.99 kJ/g; (b) 4.90×10^4 kJ; (c) 5.97 g

5.103. Diethyl ether has a fuel value of 36.781 kJ/g and a fuel density of 2.624×10^4 kJ/L. Both of these values are lower than those of diesel.

5.105. 12.9 kJ

5.107. 3.54×10^4 mol

5.109. (a) −92.2 kJ; (b) 46.1 kJ

5.111. 58.1 g

5.113. 33.3°C

5.115. (a) $6\,FeO(s) + O_2(g) \rightarrow 2\,Fe_3O_4(s)$; (b) $\Delta H_{rxn} = -636$ kJ

5.117. (a) The reaction is exothermic; (b) bonds must be broken, which requires more energy than breaking intermolecular forces; (c) 793 kJ

5.119. The balanced chemical equation for the reaction of iron(II) oxide with oxygen to form iron(III) oxide is
$4\,FeO(s) + O_2(g) \rightarrow 2\,Fe_2O_3(s)$ $\Delta H_{rxn}^{\circ} = -560.8$ kJ

5.121. -3.27×10^3 kJ

5.123. (a) Exothermic; (b) −471.8 kJ; (c) −882.3 kJ; (d) the reaction would be less exothermic.

5.125. −74.9 kJ

5.127. -1.39×10^3 kJ

5.129. −1234.7 kJ

5.131. 38.5°C

5.133.

Element	$\mathcal{M}$ (g/mol)	c_s[J/(g · °C)]	$\mathcal{M} \cdot c_s$
Bismuth	210.8	0.120	25.3
Lead	207.2	0.123	25.5
Gold	197.0	0.125	24.6
Platinum	195.1	0.130	25.3
Tin	118.7	0.215	25.5
Silver	108.6	0.233	25.3
Zinc	65.38	0.388	25.4
Copper	63.5	0.397	25.2
Cobalt/nickel	58.4	0.433	25.3
Iron	55.8	0.460	25.7
Sulfur	32.1	0.788	25.3
Average value in column 4			25.3

(a) The units are J/(mol · °C); (c) although silver is the likely element with a mass of 108.6 g/mol, there is more uncertainty about the element with a mass of 58.4 g/mol. This value lies between the accepted molar masses for cobalt and nickel.

5.135. Heat the piece of metal by immersing a dry test tube containing the metal into a boiling water bath long enough for the metal to come to thermal equilibrium. The temperature of the boiling water will be the T_i of the metal. After recording T_i of the water in the Styrofoam cup, quickly transfer the metal from the test tube to the water in the cup. Record the highest temperature of the water as it heats up from the transfer of energy from the hot metal to the water. This temperature will be T_f for both the metal and the water. Because all the energy lost by the metal is gained by the water, $-q_{lost,metal} = q_{gained,water}$ and $-mc_{s,metal}(T_f - T_{i,metal}) = nc_{P,water}(T_f - T_{i,water})$, where the mass m of the metal, T_f, $T_{i,water}$, $T_{i,metal}$, moles n of the water, and the specific heat of water (75.3 J/mol · °C) are all known. This leaves $c_{s,metal}$ as the only unknown in the equation.

5.137. The reaction is endothermic, $\Delta H_{comb} = 68$ kJ

5.139. a. For $H_2(g)$, the fuel density is 10.78 kJ/L, and for $H_2(\ell)$, the fuel density is 8492 kJ/L.
b. $H_3NBH_3(g) \rightarrow NH_3(g) + BH_3(g)$ $\Delta H_{rxn}^{\circ} = 102.2$ kJ
$H_3NBH_3(g) \rightarrow H_2(g) + H_2NBH_2(g)$ $\Delta H_{rxn}^{\circ} = -28.40$ kJ
$H_2NBH_2(g) \rightarrow H_2(g) + HNBH(g)$ $\Delta H_{rxn}^{\circ} = 123.4$ kJ
c. 76.6 kg

5.141. Yes

5.143. Exothermic

Chapter 6

6.1.   The height of the mercury in a barometer depends on the atmospheric pressure: The higher the pressure, the higher the column of mercury. Because Denver has lower atmospheric pressure due to its altitude, the barometer on the left (a) reflects the pressure in Denver, CO.

6.3.   (a)

6.5.   (c)

6.7.   Line 1; no

6.9.   Line 2

6.11.   The molar mass of helium is 4 g/mol. The line on the graph in Figure P6.10 for this gas goes below line 1. The molar mass of NO is 30 g/mol, so the line on the graph for this gas goes above line 2. All lines will converge at $P = 0$ and $d = 0$.

6.13.   (b)

6.15.   Curve 1 represents SO_2. Curve 2 could represent CO_2 or C_3H_8.

6.17. Br_2 (orange)

6.19.   (c)

6.21.   (a)

6.23.   Force is the product of the mass of an object and the acceleration due to gravity. Pressure uses force in its definition: It is the force an object exerts over a given area.

6.25.   The weight of the column of mercury lowers the overall height of the mercury in the column, creating a space at the closed end of the column that is under a vacuum (no gas molecules). As the pressure of the atmosphere increases over the open pool of mercury, the column of mercury rises in the tube to indicate rising pressure.

6.27.   1 millibar = 100 pascals

6.29.   A sharpened blade has a smaller area over which the force is distributed than a dull blade.

6.31.   As we go up in altitude, the overlying mass of the atmosphere above us decreases, so the pressure also decreases.

6.33. 3.9×10^3 Pa

6.35.   (a) 0.020 atm; (b) 0.739 atm

6.37.   10 atm

6.39.   (a) 814.6 mmHg; (b) 1.072 atm; (c) 1086 mbar

6.41.   (a) 91 atm; (b) CO_2 is heavier than N_2, so the atmosphere has a greater mass than that on Earth.

6.43. The higher the temperature, the faster the gas molecules move. The faster they move, the more often they collide with the walls of the container and the greater the force with which gas molecules hit the walls. Both of these result in increased pressure as temperature is raised.

6.45. Decrease

6.47. Increase

6.49. (a) 2.00 atm; (b) 0.664 atm; (c) 1.67 atm

6.51. 2.30 atm; 13.0 m

6.53.

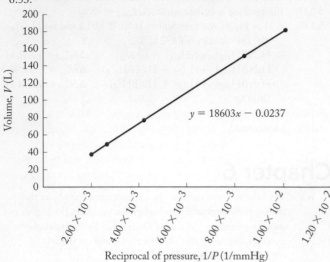

$y = 18603x - 0.0237$

Yes, the graph is exactly the same for the same number of moles of argon gas.

6.55.

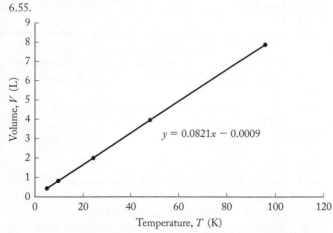

$y = 0.0821x - 0.0009$

If the amount of gas were halved, the graph would still be linear, but the slope of the line would be halved.

6.57. 596 K or 323°C

6.59. (a) 4.27 L; (b) 2.02 L; (c) 2.41 L

6.61. −4.22 kJ

6.63. (b)

6.65. (a) No change; (b) decrease to 1/4 the original volume; (c) increase of 17%

6.67. 615.4 mL

6.69. 6.6 atm

6.71. STP is defined as 1 atm and 0°C (273 K); $V = 22.4$ L

6.73. The product of the number of moles of gas in the sample, the temperature, and the gas constant

6.75. 0.67 mol

6.77. 1.50 atm

6.79. (a) 1.8×10^3 L; (b) 1.8×10^5 L; (c) the balloon will break under both sets of conditions.

6.81. (a) 0.0419 mol/h; (b) 10.7 g

6.83. 1.5×10^3 g $(NH_4)_2Cr_2O_7$ and 1.2×10^2 g $KClO_3$

6.85. 715 g

6.87. The densities of different gases are not necessarily the same for a particular temperature and pressure. Gases with different molar masses will exhibit different densities.

6.89. Density (a) increases with increasing pressure and (b) increases with decreasing temperature.

6.91. (a) 9.08 g/L; (b) in the basement

6.93. SO_2

6.95. 78.10 g/mol

6.97. The pressure that a particular gas individually contributes to the total pressure

6.99. (c)

6.101. Oxygen

6.103. $P_{total} = 5.56$ atm and $P_{N_2} = 1.12$ atm

6.105. (a) $4\ NH_3(g) + 7\ O_2(g) \rightarrow 4\ NO_2(g) + 6\ H_2O(g)$; (b) $4\ NH_3{:}7\ O_2$

6.107. (a) 0.0190 mol; (b) no, there will be fewer moles of O_2 if collected over ethanol.

6.109. (a) Greater than; (b) lower than; (c) greater than; (d) greater than

6.111. 1.7 times more

6.113. 680 mmHg

6.115. 25%

6.117. The speed of a molecule in a gas that has the average kinetic energy of all the molecules of the sample

6.119. (a) As the molar mass increases, the u_{rms} decreases; (b) as temperature increases, the u_{rms} increases.

6.121. To determine the molar mass of an unknown gas, measure the rate of its effusion (r_x) relative to the rate of effusion of a known gas (r_y). Because we know the molar mass of the known gas, $\mathcal{M}_y$, we can use the equation to solve for the unknown $\mathcal{M}_x$.

6.123. Diffusion is the spread of one substance into another. Effusion is the escape of a gas from its container through a tiny hole. The gas is escaping from a region of higher pressure to one of lower pressure.

6.125. The rank order in terms of increasing root-mean-square speed is $SO_2 < NO_2 < CO_2$.

6.127. Gas C

6.129.

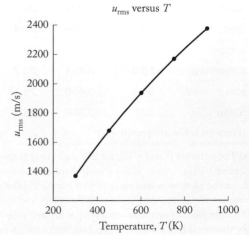

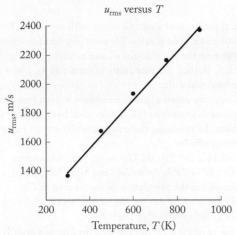

u_{rms} versus T

No, the graph is not linear.

6.131. 0.711

6.133. No, the molar masses are nearly identical.

6.135. 128.0 g/mol

6.137. (a) $r(^{12}CO_2)/r(^{13}CO_2) = 1.01$; (b) $^{12}CO_2$

6.139. The smaller balloon

6.141. $P = \dfrac{nRT}{(V - nb)} - \dfrac{n^2 a}{V^2}$

Real gases exhibit lower pressures than we would expect for an ideal gas because of the slight attractive forces between particles. Other particles exert a force that cancels out a small amount of the force exerted by the gas particle on the container wall, resulting in a lower pressure.

6.143. Krypton (Kr)

6.145. H_2

6.147. (a) $P = 910$ atm; (b) $P = 476$ atm

6.149. 126 L

6.151. 2.8×10^3 psi

6.153. 3.31 L

6.155. 120 mmHg = 120 torr = 0.158 atm = 0.160 bar = 16.0 kPa, 80 mmHg = 80 torr = 0.11 atm = 0.11 bar = 11 kPa

6.157. 21 L

6.159. $V = 2.59$ L; $d = 7.33$ g/L

6.161. (a) CH_3CO_2H and $(CH_3)_3N$; (b) HCl and $(CH_3)_3N$; (c) no

6.163. Yes

6.165. 1.23×10^3 torr

6.167. 79.3 g/mol

6.169. 137 g/mol

6.171. $u_{rms} = \sqrt{\dfrac{3P}{d}}$

6.173. 2.7 atm

6.175. 38.4 L N_2, 192 L CO_2, 1.74 g/L

6.177. 382 L

6.179. 54.9 L

6.181. 0.00781 mol

Chapter 7

7.1. (a) True purple (Na), red (Cr), and orange (Au); (b) blue (Ne); (c) orange (Au); (d) red (Cr); (e) blue (Ne) and green (Cl)

7.3. (a) One electron would be lost from the element shaded blue (Rb) to become the [Kr] core electron configuration; (b) two electrons would be lost from the element shaded green (Sr) to become the [Kr] core electron configuration; (c) three

electrons would be lost from the element shaded orange (Y) to become the [Kr] core electron configuration; (d) the element shaded gray (I) would gain one electron to become the [Xe] core configuration; (e) the element shaded red (Te) would gain two electrons to become the [Xe] core configuration.

7.5. (a) Gray (I) < red (Te) < orange (Y) < green (Sr) < blue (Rb); (b) orange (Y^{3+}) < green (Sr^{2+}) < blue (Rb^+) < gray (I^-) < red (Te^{2-})

7.7. (a) Na; (b) K; (c) Na^+

7.9. (a) Wave (b); (b) wave (a); (c) wave (b)

7.11. The hydrogen absorption spectrum consists of dark lines at wavelengths specific to hydrogen. The emission spectrum has bright lines on a dark background with the lines appearing at the exact same wavelengths as the dark lines in the absorption spectrum.

7.13. Because each element shows distinctive and unique absorption and emission lines, the bright emission lines observed for the pure elements could be matched to the many dark absorption lines in the spectrum of sunlight. This approach can be used to deduce the Sun's elemental composition.

7.15. All these forms of light have perpendicular, oscillating electric and magnetic fields that travel together through space.

7.17. The lead shield must protect the parts of the body that might be exposed to the highly energetic X-rays but are not being imaged.

7.19. X-rays and ultraviolet radiation

7.21. (a) Ultraviolet radiation; (b) gamma rays; (c) microwave radiation

7.23. The energy of UV light is twice that of the red light.

7.25. 4.87×10^{14} s^{-1}

7.27. (a) 2.88 m; (b) 2.95 m; (c) 2.98 m

7.29. (a)

7.31. 1.28 s

7.33. A quantum is the smallest indivisible amount of radiant energy that an atom can absorb or emit. A photon is the smallest indivisible packet or particle of light energy.

7.35. As the intensity (amplitude) of the light increases, more photons strike the surface as a function of time, meaning more electrons are emitted.

7.37. (b) is quantized, as the elevator can stop only at specified floors. Both the height of the escalator stairs (a) and the speed of the automobile (c) are continuously variable and so are not considered quantized.

7.39. Potassium; 8.04×10^5 m/s

7.41. No

7.43. 3.17×10^{18} photons/s

7.45. The single electron interacts only with the proton in the nucleus; there are no other electrons to repel it.

7.47. The difference between n levels determines emission energy.

7.49. (a)

7.51. No

7.53. At $n = 7$, the wavelength of the electron's transition ($n = 7$ to $n = 2$) has moved out of the visible region.

7.55. 1875 nm; infrared

7.57. (a) Decreases; (b) no

7.59. 72.9 nm

7.61. In the de Broglie equation, λ is the wavelength the particle of mass m exhibits as it travels at speed u, where h is Planck's constant. This equation states that (1) any moving particle has wavelike properties because a wavelength can be calculated through the equation, and (2) the wavelength of the particle is inversely related to its momentum (mass multiplied by velocity).

7.63.   No

7.65.   (a) 10.8 nm; (b) 0.180 nm; (c) 1.24×10^{-27} nm;
        (d) 3.68×10^{-54} nm

7.67.   (c)

7.69.   (a) 1.77×10^3 m/s; (b) 0.0565 nm

7.71. $\Delta x \geq 1.3 \times 10^{-13}$ m

7.73.   The Bohr model orbit showed the quantized nature of the
        electron in the atom as a particle moving around the nucleus
        in concentric orbits. In quantum theory, an orbital is a region
        of space where the probability of finding the electron is high.
        The electron is not viewed as a particle, but as a wave, and it is
        not confined to a clearly defined orbit; rather, we refer to the
        probability of the electron being at various locations around
        the nucleus.

7.75.   Three: n, ℓ, and m_ℓ

7.77.   (a) 1; (b) 4; (c) 9; (d) 16; (e) 25

7.79.   3, 2, 1, 0

7.81.   (a) $2s$, two electrons; (b) $3p$, six electrons; (c) $4d$, ten electrons;
        (d) $1s$, two electrons

7.83.   (b)

7.85.   Degenerate orbitals have the same energy and are indistin-
        guishable from each other.

7.87.   As we start from an argon core of electrons, we move to
        potassium and calcium, which are located in the s block on
        the periodic table. It is not until Sc, Ti, V, and so on, that we
        begin to fill electrons into the $3d$ shell.

7.89.   Electrons may be promoted to many different orbitals with en-
        ergy higher than the ground state; each of these corresponds
        to an excited state with a different energy.

7.91.   (c) $3s$ < (a) $3d$ < (d) $4p$ < (b) $5p$

7.93.   Li: $[\text{He}]2s^1$, Li$^+$: $1s^2$ or [He], Ca: $[\text{Ar}]4s^2$, F$^-$: $[\text{He}]2s^22p^6$ or
        [Ne], Na$^+$: $[\text{He}]2s^22p^6$ or [Ne], Mg^{2+}: $[\text{He}]2s^22p^6$ or [Ne],
        Al^{3+}: $[\text{He}]2s^22p^6$ or [Ne]

7.95.   K: $[\text{Ar}]4s^1$, K$^+$: [Ar], S^{2-}: [Ar] or $[\text{Ne}]3s^23p^6$, N: $[\text{He}]2s^22p^3$,
        Ba: $[\text{Xe}]6s^2$, Ti^{4+}: $[\text{Ne}]3s^23p^6$ or [Ar], Al: $[\text{Ne}]3s^23p^1$

7.97.   (a) 3; (b) 2; (c) 0; (d) 0

7.99.   Ti, two unpaired electrons

7.101. Cl$^-$, no unpaired electrons

7.103. No

7.105. Al^{3+}, N^{3-}, Mg^{2+}, and Cs$^+$

7.107. (a) and (d)

7.109. (a) B: $1s^22s^22p^1$ or $[\text{He}]2s^22p^1$, O: $1s^22s^22p^4$ or $[\text{He}]2s^22p^4$;
        (b) H (+1), B (+3), O(−2), H$^+$: $1s^0$, B^{3+}: $1s^22s^0$ or [He], O^{2-}:
        $1s^22s^22p^6$ or [Ne]

7.111. $5p$, yes

7.113. If electrons do not repel each other as much in Na$^+$ as they do
        in Na, they will have lower energy and be, on average, closer
        to the nucleus, resulting in a smaller size. When electrons are
        added to an atom (Cl), the e$^-$–e$^-$ repulsion increases, so the
        electrons have higher energy and they will be, on average, far-
        ther from the nucleus, thereby creating a larger species (Cl$^-$).

7.115. Rb; the size of atoms increases down a group because electrons
        have been added to higher n levels.

7.117. (a) As the atomic number increases down a group, electrons
        are added to higher n levels, leading to a decrease in ionization
        energy; (b) as the atomic number increases across a period, the
        effective nuclear charge increases, leading to an increase in
        ionization energy across a period of elements.

7.119. Fluorine, with a higher nuclear charge, exerts a higher Z_{eff}
        on the $2p$ electrons than boron, resulting in higher ionization
        energy.

7.121. Sr

7.123. No, the statement about all atoms with negative EA values
        existing as anions is false. For example, fluorine atoms are less
        stable than the fluoride anion, and exhibits a negative value
        for EA. Rather, fluorine exists in nature as F$_2$, a homonuclear
        diatomic molecule.

7.125. As we move down a group, electron–electron repulsions
        decrease as a result of the valence shell being farther from the
        nucleus. Minimizing the energy penalty results in a greater
        electron affinity.

7.127. (a) -1.11×10^{-26} J; (b) 17.9 m; (c) a radio telescope

7.129. (a) Cu: $1s^22s^22p^63s^23p^64s^13d^{10}$ or $[\text{Ar}]4s^13d^{10}$; (b) the
        difference in the population of the $3d$ orbital: Cu$^+$:
        $1s^22s^22p^63s^23p^64s^03d^{10}$ or $[\text{Ar}]4s^03d^{10}$ and Cu^{2+}:
        $1s^22s^22p^63s^23p^64s^03d^9$ or $[\text{Ar}]4s^03d^9$

7.131. (a) Silver forms a 1+ ion through the loss of a high-lying $5s$
        electron and cadmium loses two $5s$ electrons to form a 2+
        cation; (b) the heavier group 13 elements may form 1+ cations
        through the loss of the np electron and form 3+ cations
        through the loss of np electrons and the two ns electrons;
        (c) the heavier group 14 elements may form 2+ cations
        through the loss of two np electrons and form 4+ cations
        through the loss of both np electrons and the two ns electrons,
        whereas the group 4 elements may lose the two ns electrons
        to form 2+ cations and may lose both the ns electrons and the
        two $(n − 1)d$ electrons to form 4+ cations.

7.133. (a) Cl$^-$: $[\text{Ne}]3s^23p^6$ or [Ar], Cl: $[\text{Ne}]3s^23p^5$, Ag: $[\text{Kr}]4d^{10}5s^1$,
        Ag$^+$: $[\text{Kr}]4d^{10}$; (b) an excited state occurs when an electron
        occupies a higher energy orbital. The electron is not in its
        lowest energy state; (c) more energy is needed to remove
        an electron from Cl$^-$ than from Br$^-$ because the electron
        removed from Cl$^-$ is at a lower n (principal quantum number)
        level and is held more tightly by the nucleus; (d) if AgBr were
        used in place of AgCl, longer wavelength (lower energy)
        light would remove the electron. The AgBr sunglasses would
        darken perhaps in the infrared region.

7.135. (a) Sn^{2+}: $[\text{Kr}]4d^{10}5s^2$, Sn^{4+}: $[\text{Kr}]4d^{10}$, Mg^{2+}: $[\text{He}]2s^22p^6$ or
        [Ne]; (b) cadmium has the same electron configuration as
        Sn^{2+} and neon has the same electron configuration as Mg^{2+};
        (c) Cd^{2+}

7.137. (a) Ne, 5.76 and Ar, 6.76; (b) the outermost electron in argon
        is a $3p$ electron, which is mostly shielded by the electrons in
        the $n = 2$ level (10 electrons) and the $n = 1$ level (2 electrons),
        whereas the outermost electron in neon is a $2p$ electron,
        which is shielded only by the electrons in the $n = 1$ level
        (2 electrons).

7.139. When we think of the electron as a wave, we can envision the
        node between the two lobes as a wave of zero amplitude and
        the p orbital as a standing wave.

7.141. No, it will be in the infrared portion.

Chapter 8

8.1.   (a) Group 1 (red); (b) group 14 (blue); (c) group 16 (purple)

8.3.   Mg^{2+}

8.5.   Group 14 (blue, carbon)

8.7.   Lithium (red) and fluorine (lilac)

8.9.   (b)

8.11.   The arrangement of the atoms in two of the structures is
        S—O—S, and in the other two structures it is S—S—O.

Because the arrangement of atoms differs, they are not resonance structures. Also, for each arrangement, the structures do not show a different arrangement of electrons on the atoms; only the bonds are drawn bent, not straight. The "bent form" and "linear form" are not resonance forms of each other if the numbers of lone pairs and bonding pairs of electrons on each atom are the same.

8.13. (a) The red shading should be located on the more electronegative element: oxygen.

8.15. Fluorine (purple) and oxygen (light blue)

8.17. The structure with fewer formal charges is favored in each case. These structures are:

:F:
|
Al
/   \
:F: :F:

:S=N:

:S=N:
| \
:I: N—I:

8.19. Yes, for hydrogen and helium

8.21. Yes

8.23. In the diatomic molecule XY shown here:

:X:Y:

Lewis counts 6 e^- in three lone pairs on both X and Y. He also counts the 2 e^- shared between X and Y separately (2 e^- for X and 2 e^- for Y). However, 4 e^- are being not shared, only 2 e^-. It seems that the Lewis counting scheme counts the shared electrons twice.

8.25. To fill the 1s orbital, hydrogen forms a duet (two valence electrons). In the arrangement X—H—X, hydrogen would have four valence electron in bonds, violating the Lewis formalism.

8.27. Li· ·Mg· ·Al·

8.29. The corrected Lewis symbols are $[K]^+$; $[·Pb·]^{2+}$; $[:\ddot{S}·]^{2-}$; $[Al]^{3+}$

8.31. $[·In·]^+$ $[:\ddot{I}:]^-$ $[Ca]^{2+}$ $[·Sn·]^{2+}$

I^- and Ca^{2+} have a complete valence-shell octet.

8.33. (a) $[X·]^+$; (b) $[X]^{3+}$

8.35. (a) 8; (b) 8; (c) 8; (d) 10

8.37. (a) :C≡O:; (b) :Ö=Ö:; (c) $[:\ddot{Cl}—\ddot{O}:]^-$; (d) $[:C≡N:]^-$

8.39. Groups 1, 13, 15, and 17

8.41.

(a)
:Cl:
|
:F—C—Cl:
|
:F:

(b)
:Cl: :F:
| |
:F—C—C—F:
| |
:Cl: :Cl:

(c)
:Cl   :F:
\   /
C=C
/   \
:F: :F:

8.43.

H   H   H   H
| | | |
H—C—C—C—C—S—H   H—S—H
| | | |
H   H   H   H

8.45.

:Ö:
|
:Cl—Cl—Ö: [:Ö—Cl—Ö:]^-

8.47. If there is an electronegativity difference of 2.0 or greater, the bond between the atoms is ionic; below 2.0, the bond is covalent.

8.49. The size of an atom is the result of the nucleus pulling on the electrons. The higher the nuclear charge, the stronger the pull on the electrons within a given valence shell. This is why the size of atoms generally decreases across a period. A small atom

will form a shorter bond with another atom, and the electrons in the bond will feel a strong pull from the nucleus of a smaller atom because the bonding electrons will be "closer" to the nucleus. This stronger pull results in a higher electronegativity for smaller atoms.

8.51. A polar covalent bond is one in which the electrons are shared, but not equally, by the atoms.

8.53. Like the panes of glass in a greenhouse, the greenhouse gases in the atmosphere are transparent to visible light. Once the visible light warms the surface of Earth and is reemitted as infrared (lower energy) light, the greenhouse gases absorb the infrared light, preventing the warmed air from mixing with the cool air outside in the same way that the panes of glass do not allow the heat from inside the greenhouse to escape.

8.55. An infrared-active stretch must change the electric field of the molecule. Stretching the nonpolar N–N bond will not alter the dipole moment, whereas stretching the polar N–O bond will. Only the N–O bond stretch contributes to the absorption of infrared radiation by N_2O.

8.57. The polar bonds and the atoms with the greater electronegativity (underlined) are <u>C</u>—Se, C—<u>O</u>, <u>N</u>—H, and <u>C</u>—H.

8.59. CsI

8.61. The F–N bond

8.63. The O–H bond

8.65. PF_3

8.67. Resonance occurs when two or more valid Lewis structures may be drawn for a molecular species. The true structure of the species is a hybrid of the structures drawn.

8.69. A molecule or ion shows resonance when there is more than one correct Lewis structure: that is, when the electrons in the correct Lewis structure may be distributed in more than one way. Often, when the central atom has both a single and a double bond, resonance is possible.

8.71. Either N–O bond in the NO_2 structure could be double-bonded, and the formal charges for each structure are identical, so there is more than one correct Lewis structure, and NO_2 will exhibit resonance.

$$:\overset{-1}{\ddot{O}}—\overset{+1}{\ddot{N}}=\overset{0}{\ddot{O}} \longleftrightarrow \overset{0}{\ddot{O}}=\overset{+1}{\ddot{N}}—\overset{-1}{\ddot{O}}:$$

The resonance forms of CO_2 show that one is dominant (the one in which all formal charges are zero) and so the other forms contribute little to the true structure of CO_2.

$$\overset{0}{\ddot{O}}=\overset{0}{C}=\overset{0}{\ddot{O}} \longleftrightarrow :\overset{+1}{O}≡\overset{0}{C}—\overset{-1}{\ddot{O}}: \longleftrightarrow :\overset{-1}{\ddot{O}}—\overset{0}{C}≡\overset{+1}{O}:$$

8.73.

$$H—C≡N—\ddot{O}: \longleftrightarrow H—\ddot{C}—N≡O: \longleftrightarrow H—\ddot{C}=\ddot{N}=O:$$

8.75. N_2O_2:

$$\ddot{O}=\ddot{N}—\ddot{N}=\ddot{O} \longleftrightarrow :O≡N—\ddot{N}—\ddot{O}: \longleftrightarrow \ddot{O}=N=\ddot{N}—\ddot{O}: \longleftrightarrow$$

$$\ddot{O}—N≡N—\ddot{O}: \longleftrightarrow :\ddot{O}—\ddot{N}—N≡O: \longleftrightarrow :\ddot{O}—\ddot{N}=N=\ddot{O}$$

N_2O_3:

$$:\ddot{O}=\ddot{N}—N \overset{\ddot{O}:}{\underset{\ddot{O}:}{}} \longleftrightarrow :\ddot{O}=\ddot{N}—N \overset{\ddot{O}:}{\underset{\ddot{O}:}{}} \longleftrightarrow$$

$$:\ddot{O}=N=N \overset{\ddot{O}:}{\underset{\ddot{O}:}{}} \longleftrightarrow :O≡N—\ddot{N} \overset{\ddot{O}:}{\underset{\ddot{O}:}{}}$$

8.77. (a) :F—Ö—Ö—F:; H—Ö—Ö—H

(b) The presence of an O=O double bond shortens the bond length.

$$:\ddot{F}-\ddot{O}-\ddot{O}-\ddot{F}: \;\longrightarrow\; :\ddot{F}-\ddot{O}-\ddot{O}^+ \;\longrightarrow\; :\ddot{F}-\ddot{O}=\overset{+}{\ddot{O}} \quad :\ddot{F}:^-$$

Incomplete octet
on oxygen

8.79.
$$:\ddot{Cl}-\ddot{Se}-\ddot{N}=\ddot{S}-\ddot{O}: \longleftrightarrow :\ddot{Cl}-\ddot{Se}=\ddot{N}-\ddot{S}-\ddot{O}:$$

$$:\ddot{Cl}-\ddot{Se}-\ddot{N}-\ddot{S}=\ddot{O} \longleftrightarrow :\ddot{Cl}=\ddot{Se}-\ddot{N}-\ddot{S}-\ddot{O}:$$

$$:\ddot{Cl}-\ddot{Se}-\ddot{N}=\ddot{S}=\ddot{O}$$

8.81. The best possible structure for a molecule, judging by formal charges, is the structure in which the formal charges are minimized and the negative formal charges are on the most electronegative atoms in the structure.

8.83. No. The electronegativity of oxygen (3.5) is higher than that of sulfur (2.5), so the negative formal charge must be on the O atom in the structure that contributes most to the bonding.

8.85.
$$\overset{0}{H}-\overset{+1}{N}\equiv\overset{-1}{C}: \qquad \overset{0}{H}-\overset{0}{C}\equiv\overset{0}{N}:$$

The formal charges are zero for all the atoms in HCN, whereas in HNC the carbon atom, with a lower electronegativity than N, has a −1 formal charge.

8.87.
$$\overset{0}{\underset{0}{H}}\!\!>\!\overset{+1}{N}=\overset{0}{C}=\overset{-1}{\ddot{N}}: \longleftrightarrow \overset{0}{\underset{0}{H}}\!\!>\!\overset{0}{N}-\overset{0}{C}\equiv\overset{0}{N}:$$

The preferred structure is the one with the C triple bonded to N.

8.89. (a)

$$\underset{H}{\overset{:\ddot{O}:}{\underset{}{\underset{H}{\overset{\|}{C}}}}} \quad \text{and} \quad \underset{H}{\overset{:\ddot{C}\cdot}{\underset{}{\underset{H}{\overset{\|}{O}}}}}$$

The Lewis structure with carbon as the central atom is more likely, as it minimizes formal charges. (b) No

8.91.
$$:\overset{-2}{\ddot{N}}-\overset{+2}{\ddot{O}}\equiv\overset{0}{N}: \longleftrightarrow :\overset{-1}{\ddot{N}}=\overset{+2}{\ddot{O}}=\overset{-1}{\ddot{N}}: \longleftrightarrow :\overset{0}{N}\equiv\overset{+2}{\ddot{O}}-\overset{-2}{\ddot{N}}:$$

Because oxygen is more electronegative than nitrogen, none of these structures is likely to be stable because the formal charge on O is positive.

8.93. Yes

8.95. In order for the atom to accommodate more than 8 e⁻ in covalently bonded molecules, it would require the use of orbitals beyond s and p. The d orbitals are not available to the small elements in the second period.

8.97. (a), (b), and (c)

8.99. (a) 12; (b) 8; (c) 12; (d) 10

8.101.

$$\underset{:\ddot{F}:}{\overset{\overset{-1}{:\ddot{O}:}}{\overset{|}{\underset{0}{\overset{+1}{N}}}}}\!\!\!-\overset{0}{\ddot{F}}: \qquad \underset{:\ddot{F}:}{\overset{\overset{0}{\cdot\ddot{O}\cdot}}{\overset{\|}{\underset{0}{\overset{0}{P}}}}}\!\!\!-\overset{0}{\ddot{F}}:$$

In POF_3 there is a double bond and no formal charges; in NOF_3 there are only single bonds, and formal charges are present on N and O.

8.103.

$$\underset{:\ddot{F}\cdot \quad \cdot\ddot{F}:}{\overset{:\ddot{F}:}{\underset{}{\overset{|}{\underset{Se}{\overset{}{:\ddot{F}}}}}}} \qquad \left[\underset{:\ddot{F}\cdot \quad \cdot\ddot{F}:}{\overset{:\ddot{F}:}{\underset{}{\overset{|}{\underset{Se}{\overset{}{:\ddot{F}}}}}}}\!\!\!\overset{}{\ddot{F}:}\right]^-$$

In both structures Se has more than eight valence electrons.

8.105.
$$\underset{\overset{\|}{O}}{\overset{:\ddot{O}:^0}{\underset{}{\overset{}{Cl}}}}\!\!\!-\overset{0}{Cl}\overset{0}{\underset{\|}{}}$$

The central chlorine atom has an expanded octet.

8.107. (c), (d), and (e)

8.109. (d)

8.111. The formation of a dimer completes the octet on Al.

$$2\;\; \underset{H}{\overset{H}{\underset{|}{\overset{|}{H-C-Al-\ddot{Cl}:}}}} \;\longrightarrow\; \text{[dimer structure]}$$

8.113.

$$\underset{:\ddot{F}:}{\overset{:\ddot{F}:}{\underset{}{\overset{|}{:\ddot{F}-\overset{2+}{S}-\ddot{F}:}}}} \qquad \left[\text{ionic form with }:\ddot{F}:^-\right]$$

The ionic form is less favorable because of the presence of formal charges on S and F (no formal charges are required with an expanded octet on S).

8.115. Each bond in the nitrate ion is 1.33 bonds because of resonance.

$$\left[\underset{:\ddot{O}: \quad :\ddot{O}:}{\overset{:\ddot{O}:}{\underset{}{\overset{\|}{N}}}}\right]^- \longleftrightarrow \left[\underset{:\ddot{O}: \quad \ddot{O}:}{\overset{:\ddot{O}:}{\underset{}{\overset{}{N}}}}\right]^- \longleftrightarrow \left[\underset{:\ddot{O}: \quad :\ddot{O}:}{\overset{:\ddot{O}:}{\underset{}{\overset{}{N}}}}\right]^-$$

Each bond in the nitrite ion is 1.5 bonds because of resonance.

$$\left[:\ddot{O}=N-\ddot{O}:\right]^- \longleftrightarrow \left[:\ddot{O}-\ddot{N}=\ddot{O}:\right]^-$$

No, the nitrogen–oxygen bond lengths in NO_3^- and NO_2^- are not the same; they are different.

8.117. The nitrogen–oxygen bond in N_2O_4 has a bond order of 1.5 due to four equivalent resonance forms:

$$[\text{four resonance forms of } N_2O_4]$$

The nitrogen–oxygen bond in N_2O has a bond order of 1.5 due to resonance between three resonance forms (where the last resonance structure shown does not significantly contribute to the structure of the molecule because of the buildup of too much formal charge):

$$:\overset{-1}{\ddot{N}}=\overset{+1}{N}=\overset{0}{\ddot{O}}: \longleftrightarrow :\overset{0}{N}\equiv\overset{+1}{N}-\overset{-1}{\ddot{O}}: \longleftrightarrow :\overset{-2}{\ddot{N}}-\overset{+1}{N}\equiv\overset{+1}{O}:$$

Therefore, owing to resonance, N_2O_4 and N_2O are expected to have nearly equal bond lengths.

8.119. (a) $NO^+ < NO_2^- < NO_3^-$; (b) $NO_3^- < NO_2^- < NO^+$

8.121. No

8.123. We must account for all the bonds that break and all the bonds that form in the reaction. In order to do so, we must have a balanced chemical reaction.

8.125. If the compounds are in the solid or liquid phase, interactions between molecules may slightly change the bond energy for a given bond.

8.127. (a) 862 kJ; (b) 98 kJ; (c) 93 kJ

8.129. 1068 kJ/mol

8.131. 278 kJ less

8.133. −667 kJ

8.135. 552 kJ/mol

8.137. $\ddot{\text{O}}=\text{C}=\text{C}=\ddot{\text{O}}$

Both carbon–oxygen bonds are equal.

8.139. (a) ·Be·; (b) ·Al·; (c) ·Ċ·; (d) He:

8.141. $\overset{0}{:\!\ddot{\text{S}}}=\text{C}=\overset{0}{\ddot{\text{S}}:}$ $\overset{0}{:\!\ddot{\text{S}}}=\overset{+2}{\text{S}}=\overset{-2}{\ddot{\text{C}}:}$

The preferred structure for carbon disulfide is when C is the central atom.

8.143. (a)

[Lewis structure: O double bonded to central C, with two Cl atoms bonded to C]

(b)

H—Ö—H + [Lewis structure of COCl₂] → Ö=C=Ö + 2 H—Ċl:

8.145.
(a)

$:\!\text{O}\!\equiv\!\text{C}\!-\!\ddot{\text{N}}\!-\!\ddot{\text{N}}\!-\!\text{C}\!\equiv\!\text{O}:\leftrightarrow\;\ddot{\text{O}}\!=\!\text{C}\!=\!\ddot{\text{N}}\!-\!\ddot{\text{N}}\!=\!\text{C}\!=\!\ddot{\text{O}}\leftrightarrow$

$\ddot{\text{O}}\!=\!\text{C}\!-\!\text{N}\!\equiv\!\text{N}\!-\!\text{C}\!=\!\ddot{\text{O}}\leftrightarrow:\!\ddot{\text{O}}\!-\!\text{C}\!\equiv\!\text{N}\!-\!\text{N}\!\equiv\!\text{C}\!-\!\ddot{\text{O}}:\leftrightarrow$

$\ddot{\text{O}}\!=\!\text{C}\!=\!\text{N}\!-\!\text{N}\!\equiv\!\text{C}\!-\!\ddot{\text{O}}:\leftrightarrow:\!\ddot{\text{O}}\!-\!\text{C}\!\equiv\!\text{N}\!-\!\text{N}\!=\!\text{C}\!=\!\ddot{\text{O}}$

(b)

$:\!\ddot{\text{Br}}\!-\!\ddot{\text{N}}\!=\!\ddot{\text{O}}:$

$:\!\text{O}\!\equiv\!\text{C}\!-\!\ddot{\text{N}}\!=\!\text{N}\!=\!\ddot{\text{O}}:\leftrightarrow\;\ddot{\text{O}}\!=\!\text{C}\!-\!\ddot{\text{N}}\!=\!\text{N}\!=\!\ddot{\text{O}}\leftrightarrow$

$\ddot{\text{O}}\!-\!\text{C}\!=\!\text{N}\!-\!\text{N}\!\equiv\!\text{O}:$

(c)

[Resonance structures with O, N, C chain]

$:\!\text{O}\!\equiv\!\text{C}\!-\!\ddot{\text{N}}\!-\!\text{C}\!-\!\ddot{\text{N}}\!-\!\text{C}\!\equiv\!\text{O}:\leftrightarrow\;\ddot{\text{O}}\!=\!\text{C}\!=\!\text{N}\!-\!\text{C}\!-\!\text{N}\!=\!\text{C}\!=\!\ddot{\text{O}}\leftrightarrow$

$\ddot{\text{O}}\!=\!\text{C}\!=\!\text{N}\!-\!\text{C}\!=\!\text{N}\!=\!\text{C}\!=\!\ddot{\text{O}}\leftrightarrow:\!\ddot{\text{O}}\!-\!\text{C}\!\equiv\!\text{N}\!-\!\text{C}\!-\!\text{N}\!\equiv\!\text{C}\!-\!\ddot{\text{O}}:\leftrightarrow$

$:\!\ddot{\text{O}}\!-\!\text{C}\!\equiv\!\text{N}\!-\!\text{C}\!-\!\text{N}\!=\!\text{C}\!=\!\ddot{\text{O}}\leftrightarrow\;\ddot{\text{O}}\!=\!\text{C}\!=\!\text{N}\!-\!\text{C}\!-\!\text{N}\!\equiv\!\text{C}\!-\!\ddot{\text{O}}:$

8.147. For Cl₂O₆ with a Cl—Cl bond:

[Lewis structure: O=Cl(O)(O)—Cl(O)(O)=O]

For Cl₂O₆ with a Cl—O—Cl bond:

[Lewis structure with Cl—O—Cl bridge]

For ClO₂:

$\overset{0}{:\!\ddot{\text{O}}}=\overset{0}{\ddot{\text{Cl}}}=\overset{0}{\ddot{\text{O}}}:$

8.149. (a) ·C≡N: The more likely structure for cyanogen is the one that contains the C—C bond. (b) It would be expected that oxalic acid would retain the C—C bond from the cyanogen from which it is formed in the reaction of cyanogen with water. This is consistent with the structure for cyanogen predicted by formal charge analysis.

8.151.

[Lewis structure with N, C, S, F atoms]

8.153.

[Lewis structure of TeF₆O²⁻ type ion with 2− charge]

8.155. (a) 2; (b) 4; (c) 1; (d) 5; (e) 8

8.157. (a) Assuming A and X have different electronegativities, we would be able to distinguish X—A—A from A—X—A. Bond lengths would only be useful if we were able to look for a characteristic A—A bond, which would be found in X—A—A but not A—X—A. (b) Electrostatic potential mapping cannot distinguish resonance forms. If the resonance forms for A—X—A shown are all equally weighted (none is more preferred than another), we would expect the average X—A bond to be a double bond, as in A=X=A. If one resonance form contributes more to the bonding in a molecule, we would see that resonance form reflected in the bond distances being shortened or lengthened.

8.159.

[Lewis structure of NH₄⁺ and SH⁻]

There cannot be a nitrogen–sulfur covalent bond because the nitrogen atom in NH₄⁺ has a complete octet through its bonding with hydrogen and because it cannot expand its octet because it is a second period element.

8.161. (a) and (b)

$:\!\text{N}\!\equiv\!\text{N}\!-\!\ddot{\text{N}}\!=\!\ddot{\text{N}}:\leftrightarrow\;:\!\ddot{\text{N}}\!=\!\text{N}\!=\!\text{N}\!=\!\ddot{\text{N}}:\leftrightarrow$

$:\!\ddot{\text{N}}\!=\!\text{N}\!-\!\text{N}\!\equiv\!\text{N}:$

The middle structure has the most nonzero formal charges separated over three bond lengths, so this one is least preferred. The first and last resonance structures are preferred and are indistinguishable from each other.

(c)

[Resonance structures of N₄ ring]

$\overset{0}{\cdot\ddot{\text{N}}}\!-\!\overset{0}{\ddot{\text{N}}\cdot}\;\leftrightarrow\;\overset{0}{\cdot\ddot{\text{N}}}\!=\!\overset{0}{\ddot{\text{N}}\cdot}$

8.163.

$$\left[\ :\overset{\cdot\cdot}{\underset{\cdot\cdot}{Cl}}:^{0}\ \right]$$
:Ḟ:—Al—Ċl:
with formal charges 0, 0, 0

8.165. (b) and (c)

8.167. (a) and (b)

$$\left[:\overset{-2}{N}-\overset{0}{N}=\overset{0}{N}-\overset{+1}{N}\equiv\overset{0}{N}:\right]^{-}\longleftrightarrow\left[:\overset{-2}{N}-\overset{0}{N}=\overset{+1}{N}=\overset{+1}{N}=\overset{-1}{N}:\right]^{-}\longleftrightarrow$$

$$\left[:\overset{-2}{N}-\overset{+1}{N}\equiv\overset{0}{N}-\overset{0}{N}=\overset{-1}{N}:\right]^{-}\longleftrightarrow\left[:\overset{-1}{N}=\overset{0}{N}-\overset{+1}{N}=\overset{+1}{N}=\overset{-2}{N}:\right]^{-}\longleftrightarrow$$

$$\left[:N\equiv\overset{+1}{N}-\overset{0}{N}=\overset{0}{N}=\overset{-2}{N}:\right]^{-}\longleftrightarrow\left[:\overset{-1}{N}=\overset{0}{N}-\overset{0}{N}=\overset{+1}{N}=\overset{-1}{N}:\right]^{-}\longleftrightarrow$$

$$\left[:\overset{-1}{N}=\overset{+1}{N}=\overset{0}{N}-\overset{0}{N}=\overset{-1}{N}:\right]^{-}\longleftrightarrow\left[:\overset{0}{N}\equiv\overset{+1}{N}-\overset{-1}{N}=\overset{0}{N}=\overset{-1}{N}:\right]^{-}\longleftrightarrow$$

$$\left[:\overset{-1}{N}=\overset{0}{N}-\overset{-1}{N}-\overset{+1}{N}\equiv\overset{0}{N}:\right]^{-}$$

The structures that contribute most have the lowest formal charges (last four structures shown). (c) N_3^- has the Lewis structures

$$\left[:\overset{-2}{N}-\overset{+1}{N}\equiv\overset{0}{N}:\right]^{+}\longleftrightarrow\left[:\overset{-1}{N}=\overset{+1}{N}=\overset{-1}{N}:\right]^{-}\longleftrightarrow\left[:\overset{0}{N}\equiv\overset{+1}{N}-\overset{-2}{N}:\right]^{-}$$

From these resonance structures, we see that each bond is predicted to be of double bond character in N_3^-. Therefore in N_5^- there are two longer N–N bonds than in N_3^-. N_3^- has the higher average bond order.

8.169.

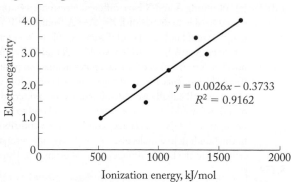

Using the equation for the best-fit line where x = the ionization energy of neon gives a value of y (electronegativity) of neon: $y = 5.0$

8.171. (a) Isoelectronic means that the two species have the same number of electrons.

(b) $\left[:\overset{0}{N}=\overset{+1}{N}-\overset{0}{\underset{\cdot\cdot}{F}}:\right]^{+}\longleftrightarrow\left[:\overset{-1}{N}=\overset{+1}{N}=\overset{+1}{\underset{\cdot\cdot}{F}}:\right]^{+}\longleftrightarrow\left[:\overset{-2}{N}-\overset{+1}{N}\equiv\overset{+2}{F}:\right]^{+}$

(c) The central nitrogen atom in all the resonance structures always carries a +1 formal charge. (d) The second and third resonance forms shown are unacceptable because they have greater than the minimal formal charges on the atoms. (e) Yes, the fluorine could be the central atom in the molecule, but this would place significant positive formal charge on the fluorine atom (the most electronegative element). These structures are unlikely:

$\left[:\overset{0}{N}\equiv\overset{+3}{F}-\overset{-2}{N}:\right]^{+}\longleftrightarrow\left[:\overset{-1}{N}=\overset{+3}{F}=\overset{-1}{N}:\right]^{+}\longleftrightarrow\left[:\overset{-2}{N}-\overset{+3}{F}\equiv\overset{0}{N}:\right]^{+}$

8.173. This molecule is nonpolar overall because the individual bond dipoles are equal in magnitude and, as vectors, they cancel each other out.

F   ↖↗   F
    C=C
F   ↙↘   F

8.175. (a)

:Ċl—Ö: ↔ :Ċl—Ö· ↔ :Ċl·

:Ḟ:
|
:Ḟ—C—Ċl: ↔ etc.

In order of increasing reactivity, $CF_2Cl\cdot < ClO\cdot < Cl\cdot$

(b) The products are

:Ċl—Ċl: :Ċl—Ċl=Ö: :Ċl—Ö—Ċl: :Ḟ—C—Ċl:
                                          |
                                        :Ċl:

Chapter 9

9.1. Yes

9.3. N_2F_2 and NCCN are planar; there are no delocalized π electrons in any of these molecules.

9.5. More

9.7. The axial F–Re–axial F bond angle is 180°. The axial F–Re–equatorial F angle is 90°. The equatorial F–Re–equatorial F bonds are all 72°.

9.9. Because the electrons take up most of the space in the atom and because the nucleus is located in the center of the electron cloud, the electron clouds repel each other before the nuclei get close to each other.

9.11. Both have three atoms bonded with no lone pairs on the central atom.

9.13. Because the lone pair feels attraction from only one nucleus, it is less confined than bonding pairs and therefore occupies more space around the central N atom.

9.15. The seesaw geometry has only two lone pair–bond pair interactions at 90° (compared with trigonal pyramidal's three), so it has lower energy.

9.17. (b) < (c) < (a)

9.19. Trigonal bipyramidal and seesaw

9.21. (a)

9.23. Pentagonal pyramidal and distorted octahedral

9.25. (a) Tetrahedral; (b) trigonal pyramidal; (c) bent; (d) tetrahedral

9.27. (a) 109.5°; (b) <109.5°; (c) ~120°; (d) 90°, 180°

9.29. (a) Tetrahedral; (b) tetrahedral; (c) trigonal planar; (d) linear

9.31. O_3, SO_2, and S_2O are bent (120°), whereas N_2O and CO_2 are linear.

9.33.

$$\left[:N\equiv C-\overset{\cdot\cdot}{C}-C\equiv N:\right]^{-}$$
with a C (double bond to N) substituent

$$\longleftrightarrow$$

$$\left[:\overset{\cdot\cdot}{N}=O=C\begin{smallmatrix}C\equiv N:\\C\equiv N:\end{smallmatrix}\right]^{-}$$

The structure on the right contributes most.

9.35. Silicon may exceed its octet, changing the geometry around the nitrogen atom to trigonal planar.

9.37.

Square planar   Pentagonal bipyramidal

9.39.

ClO_2^+ ClO_2 ClO_2^-

9.41. A polar bond is between only two atoms in a molecule. Molecular polarity takes into account all the individual bond polarities and the geometry of the molecule. A polar molecule has a permanent, measurable dipole moment.

9.43. Yes

9.45. (a) All five molecules contain polar bonds; (b) $CHCl_3$, H_2S, and SO_2; (c) CCl_4 and CO_2.

9.47. All of the molecules (a)–(c) are polar.

9.49. (a) i; (b) i; (c) ii

9.51. (a) Tetrahedral; (b) nonpolar

9.53. A sigma bond is formed from the overlap of two atomic or hybridized atomic orbitals along a bonding axis. A pi bond is formed from unhybridized p or d orbitals above and below a bonding axis.

9.55. (a) If we consider the phosphorus atom to contain sp^3 hybrid orbitals, we would expect an angle of <109.5°, but the additional steric bulk from lone pairs on the fluorine atoms interacting with the lone pair on phosphorus might lead to a smaller bond angle. (b) If we consider the phosphorus atom to be unhybridized, we would expect an angle of 90°. The large fluorine atoms would interact with one another, leading to a repulsion and therefore a greater bond angle.

9.57. (a) sp hybridized; (b) sp^2 hybridized; (c) sp hybridized; (d) sp^2 hybridized; (e) sp^2 hybridized

9.59. Both Lewis structures have three resonance forms: one with two N=N bonds and two with an N≡N and an N—N bond. For both molecules, the central nitrogen atoms are sp hybridized.

9.61. The hybridization is SF_2 (sp^3), SF_4 (sp^3d), and SF_6 (sp^3d^2).

9.63. The N—O bond is a single bond. There is also an N—C single bond and an N=C double bond.

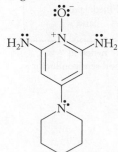

9.65.

The ion has tetrahedral molecular geometry. At first glance this would mean that the hybridization would be assigned as sp^3. However, notice that the Cl forms three π bonds to three of the oxygen atoms. This requires that three of the p orbitals on Cl not be involved in the hybridization so that it can form π bonds. Therefore Cl must use low-lying d orbitals in place of the p orbitals for sd^3 hybridization to form the 4 σ bonds to oxygen.

9.67. H—Ar—F sp^3d hybridized

9.69. (a)

Tetrahedral C and N

SN = 4
sp^3 hybridized
Electron-pair geometry = tetrahedral
Molecular geometry = tetrahedral
All C—N bonds are an overlap of C–sp^3 and N–sp^3 hybrid orbitals.

(b)

(c) Trigonal bipyramidal

9.71. All types of structures convey the correct atom stoichiometry within the molecule, though the condensed structure does not show connectivity. Lewis, Kekulé, and carbon-skeleton structures are all capable of displaying resonance structures, and functional groups may be identified in each. A certain degree of chemical knowledge is required to correctly interpret a carbon-skeleton structure because the identity or presence of C and H atoms is to be assumed.

9.73. Yes, molecules with more than one central atom can indeed have resonance forms. Resonance forms are Lewis structures that show alternative (yet still valid) electron distributions in a molecule.

9.75. Yes; in resonance structures the electron distribution is blurred across all the resonance forms, which, in essence, defines the delocalization of electrons.

9.77. (b) and (c)

9.79. $CH_3(CH_2)_2CH=CH—CH=CH(CH_2)_9OH$

9.81. Both hormones contain alkane, alkene, and ketone functional groups.

9.83. $CH_3(CH_2)_7CH=CH(CH_2)_7C(O)OH$

9.85.

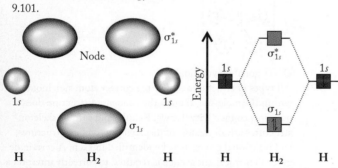

9.87. All three carbon atoms are trigonal planar. The nitrogen and oxygen atoms both adopt a bent (104.5°) geometry, and the sulfur atom is tetrahedral. The ring C–O bond is an overlap of C_{sp^2}/O_{sp^3}, while the C=O bond is an overlap of C_{sp^2}/O_{sp^2} and C_{2p}/O_{2p}. The C–N bond is formed by an overlap of C_{sp^2}/N_{sp^3}. The extra electron on N is located in an sp^3 hybrid orbital.

9.89. (a) and (c) are chiral; (b) and (d) are not chiral.

9.91. No. Although it is true that s–s overlap always gives σ molecular orbitals, there are other orbitals that may overlap to also give σ bonds such as s–p, p–p, and d–p.

9.93. No; the overlap of $1s$ and $2s$ orbitals is not as efficient as $1s$–$1s$ or $2s$–$2s$ overlaps, so the match in size and energy is poor.

9.95. If a molecule or ion has an even number of valence electrons distributed over an odd number of orbitals, or if the highest occupied molecular orbital is a π orbital, paramagnetic species could result.

9.97. Application of an electrical potential across a metal causes its mobile valence electrons to move toward the positive potential.

9.99. Doping a metalloid generates a semiconductor with a smaller bandgap than the starting material. N-type semiconductors have "extra" electrons higher in energy than the valence band for the undoped material. P-type semiconductors have "extra" holes lower in energy than the conduction band for the undoped material. In both types of doped semiconductor, the effect is similar; the band gap is lowered, and electrons flow with a smaller energy barrier.

9.101.

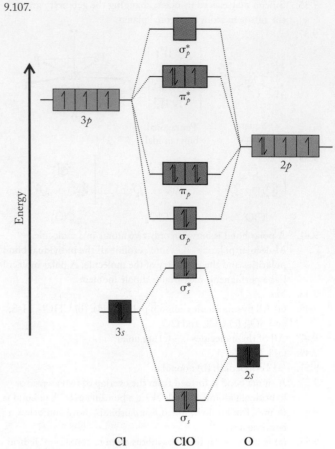

9.103. (a) N_2^+: $(\sigma_{2s})^2 (\sigma^*_{2s})^2 (\pi_{2p})^4 (\sigma_{2p})^1$, O_2^+: $(\sigma_{2s})^2 (\sigma^*_{2s})^2 (\sigma_{2p})^2 (\pi_{2p})^4$ $(\pi^*_{2p})^1$, C_2^+: $(\sigma_{2s})^2 (\sigma^*_{2s})^2 (\pi_{2p})^3$, Br_2^{2-}: $(\sigma_{4s})^2 (\sigma^*_{4s})^2 (\sigma_{4p})^2 (\pi_{4p})^4$ $(\pi^*_{4p})^4 (\sigma^*_{4p})^2$; (b) N_2^+: BO = 2.5, O_2^+: BO = 2.5, C_2^+: BO = 1.5, Br_2^{2-}: BO = 0; (c) all species with nonzero bond order (N_2^+, O_2^+, and C_2^+) are expected to exist.

9.105. (a), (b), (c), (e), (g), and (h).

9.107.

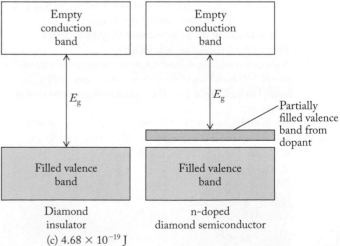

The odd electron is located in an antibonding orbital.

9.109. (a) and (b)

9.111. No

9.113. (a) n–type

(b)

Empty conduction band	Empty conduction band
E_g	E_g
	Partially filled valence band from dopant
Filled valence band	Filled valence band
Diamond insulator	n-doped diamond semiconductor

(c) 4.68×10^{-19} J

9.115.

$$\left[\begin{array}{c} H \\ | \\ H-N-H \\ | \\ H \end{array} \right]^{+} \quad \text{Tetrahedral}$$

$$\left[\begin{array}{c} :\overset{..}{O}:^{0} \\ \| \\ :\overset{..}{O}=\overset{0}{Cl}-\overset{..}{O}:^{-1} \\ \| \\ :\overset{..}{O}:^{0} \end{array} \right]^{-} \quad \text{Tetrahedral}$$

9.117.

→SN = 3
Electron-pair geometry = trigonal planar
O—C—O bond angle = 120°

→SN = 4
Electron-pair geometry = tetrahedral
C—O—H bond angle = 109.5°

SN = 4
Electron-pair geometry = tetrahedral
N—C—C bond angle = 109.5°

9.119. (a) All of the central atoms (O—O and O—Cl) have SN = 4, so their electron-pair geometries are tetrahedral. Each of the central atoms also has two lone pairs and two bonding pairs, which gives them a bent molecular geometry. Therefore, neither of the two molecules is linear. (b) Only one structure (ClOOCl) is symmetrical, and would be expected to have bond dipoles that cancel one another, generating a nonpolar molecule. The other molecule (ClOClO) would instead be polar, as the O—Cl bond dipoles would not cancel. The two molecules do not have the same dipole moment.

9.121. (a) $\left[:\overset{..}{Cl}=\overset{..}{O}: \right]^{+}$; (b) BO = 2

9.123.

Molecular geometry around P = tetrahedral

9.125. :B—B≡C=Ö:

All formal charges = 0
Both B atoms have incomplete octets
On both central B and C atoms SN = 2
Molecular geometry = linear

:Ö=C=B—B=C=Ö:

All formal charges = 0
Both B atoms have incomplete octets
On all central B and C atoms SN = 2
Molecular geometry = linear

9.127.

$$H^{0}-\overset{\overset{H^0}{|}}{\underset{\underset{H^0}{|}}{C^0}}-\overset{..}{N}=C^0=\overset{..}{\underset{..}{S}}:^{0} \longleftrightarrow H^{0}-\overset{\overset{H^0}{|}}{\underset{\underset{H^0}{|}}{C^0}}-\overset{..}{N}^{-1}-C^0\equiv S:^{+1} \longleftrightarrow$$

(1)   (2)

$$H^{0}-\overset{\overset{H^0}{|}}{\underset{\underset{H^0}{|}}{C^0}}-N^{+1}\equiv C^0-\overset{..}{\underset{..}{S}}:^{-1}$$

(3)

Structure 1 is likely to contribute the most to bonding. The methyl (CH₃) carbon is tetrahedral. The geometry around the isothiocyanate (NCS) carbon is linear.

9.129. ⌁⌁⌁SH

9.131. Yes

9.133. BO = 0

σ_p^*

π_p^*

π_p

σ_p

σ_s^*

σ_s

9.135. N₂O₅ and N₂O₃ as well as N₂O₂, depending on its actual structure

9.137. Diamagnetic

σ_p^*

π_p^*

π_p

σ_p

σ_s^*

σ_s

9.139. sp^3

9.141. This molecule is polar because, although the oxygen–oxygen bonds themselves are nonpolar, the lone pair has its own "pull" on the electrons in the molecule. Also, the π bonds between the oxygen atoms place slightly more electron density on the terminal O atoms and makes the "nonpolar O—O bond" actually polar.

9.143. (a)

Alliin

→ SN = 4, trigonal pyramidal molecular geometry

Allicin

→ SN = 4, bent molecular geometry

→ SN = 4, trigonal pyramidal molecular geometry

(b) The C–S–S bond angle is predicted to be the same as in H_2S, CH_3SH, and $(CH_3)_2S$.

Chapter 10

10.1. Coulomb's law states that as the distance between ions increases, the energy of the interaction decreases. Because I^- is larger than F^-, the KI interaction is weaker. Therefore the stronger ion–ion interaction results in a higher melting point for KF than KI.

10.3. Because the electronegativity of P is equal to that of H, the phosphine, PH_3, is predicted to be less polar. Ammonia, NH_3, on the other hand, is more polar; moreover, because H is bonded to the very electronegative N atom, it forms strong hydrogen bonds with other NH_3 molecules. Ammonia therefore has the higher boiling point ($-33°C$ YH_3 in Figure P10.3). Phosphine, with much weaker forces between its atoms, has a low ($-88°C$) boiling point and is represented by XH_3 in Figure P10.3.

10.5. The slope of the trend line in each graph is proportional to the size of ΔH_{vap}, so the graph with the larger slope has a larger enthalpy of vaporization. A large value of ΔH_{vap} corresponds to strong intermolecular forces, as seen in the graph for the liquid on the left.

10.7. (a) Increases; (b) no; as pressure is increased, the liquid phase is converted into the solid at a given temperature. This means that the solid is more dense than the liquid and so the solid will not float in its liquid phase.

10.9. Dispersion forces

10.11. Because real gases have nonzero volume, and when cooled they have less kinetic energy to overcome their intermolecular forces

10.13. (a) CCl_4; (b) C_3H_8

10.15. PCl_5 is a solid at room temperature, whereas PCl_3 is a liquid. Despite being nonpolar, PCl_5 has a higher molar mass and a larger surface area, so the dispersion forces for PCl_5 outweigh the weaker dispersion forces and dipole–dipole interactions for PCl_3.

10.17. Water molecules are oriented around Cl^- so as to point the partially positive hydrogen atoms toward the Cl^- ion.

10.19. Because of the full positive or negative charge on the ion, the ion–dipole interaction is stronger than the dipole–dipole interaction.

10.21. The charge buildup on H (partially positive) and the electronegative element (partially negative) means that the X–H bond is polar. It is still a dipole–dipole interaction except that its strength is noticeably higher than in other dipole–dipole interactions.

10.23. The greater dispersion forces of CH_2Cl_2 add to the dipole–dipole interactions to give stronger intermolecular forces between the CH_2Cl_2 molecules than those of CH_2F_2 molecules.

10.25. H_2O_2 has a greater enthalpy of vaporization, and as a result, greater intermolecular forces.

10.27. CH_3F is a polar molecule and therefore has stronger intermolecular forces than the nonpolar molecules of CH_4, which have only the weak dispersion forces. Because it takes more energy to overcome strong intermolecular forces, CH_3F has a higher melting point than does CH_4.

10.29. The carbon–hydrogen bond in CH_3F is not polar enough to exhibit hydrogen bonding. In HF, however, the H atom is bonded to fluorine, which is the most electronegative element. It is this H that shows hydrogen bonding in HF.

10.31. (b)

10.33. (a) and (d)

10.35. Miscible solutes and solvents dissolve completely in each other; an insoluble solute does not dissolve at all.

10.37. 1,1-dichloroethane is polar, and soluble in water. 1,2-dichloroethane is nonpolar.

10.39. Hydrophilic substances dissolve in water. Hydrophobic substances do not dissolve, or are immiscible, in water.

10.41. (a) $CHCl_3$; (b) CH_3OH; (c) NaF; (d) BaF_2

10.43. (a), (b), (c), and (d)

10.45. (b)

10.47. (d)

10.49. (b) and (c) are ethers, whereas (a) and (d) are alcohols. The expected ordering of boiling points is (b) < (c) < (a) < (d).

10.51. At low temperatures, the solubility of a gas increases because fewer gas molecules dissolved in a solvent have sufficient (kinetic) energy to overcome the intermolecular forces between the solvent molecules.

10.53. No, the Henry's Law constant for air is not simply the sum of k_H for N_2 and O_2.

$$C_{air} = k_{H,air}P_{air} = C_{O_2} + C_{N_2}$$
$$k_{H,air}P_{air} = k_{H,O_2}P_{O_2} + k_{H,N_2}P_{N_2}$$
$$= k_{H,O_2}(0.22P_{air}) + k_{H,N_2}(0.78P_{air})$$
$$= 0.22k_{H,O_2}P_{air} + 0.78k_{H,N_2}P_{air}$$
$$= (0.22k_{H,O_2} + 0.78k_{H,N_2})P_{air}$$
$$k_{H,air} = 0.22k_{H,O_2} + 0.78k_{H,N_2}$$

10.55. SO_3

10.57. 3.7×10^{-2} mol/(L · atm)

10.59. (a) 2.74×10^{-3} M; (b) 2.34×10^{-2} M

10.61. 2.18 g

10.63. When the average kinetic energy of the liquid molecules increases, more of the molecules can escape the liquid phase and enter the gas phase. More molecules in the gas phase increase the vapor pressure.

10.65. Intensive

10.67. (a) < (b) < (c)

10.69. (a) 41.0 kJ/mol; (b) 5.6 torr.

10.71. In sublimation the solid does not first liquefy before evaporating. In evaporation a liquid becomes a gas.

10.73. (a) Solid, liquid, and gas; (b) liquid and gas

10.75. (a) Solid phase; (b) gas phase

10.77. Yes. From the phase diagram for water we see that above the triple point, the solid phase must change to the liquid phase to enter the gaseous phase. Below the triple point, changing the temperature at a given pressure will sublime solid water into the gas phase.

10.79. Approximately 55°C to 60°C

10.81. Ethylene glycol has stronger intermolecular forces. We can tell this from the higher normal boiling point.

10.83. Reduce the temperature from 25°C to 0.01°C, then reduce the pressure from 1 atm to 0.006 atm.

10.85. The water vaporizes from liquid to gas.

10.87. −57°C

10.89. (a) Liquid; (b) gas; (c) gas

10.91. A needle floats on water but not on methanol because of the high surface tension of water. This is because water can hydrogen-bond through two O–H bonds with other water molecules, whereas methanol has only one O–H bond through which to form strong hydrogen bonds.

10.93. The expansion of water in the pipes on freezing may create sufficient pressure on the wall of the pipes to cause them to burst.

10.95. The cohesive forces in mercury are stronger than the adhesive forces of the mercury to the glass.

10.97. Molecules in the bulk liquid are "pulled" by all the other liquid molecules surrounding them and they are, therefore, "suspended" in the bulk liquid. Molecules on the surface of a liquid, however, are pulled only by molecules under and beside them, creating a tight film of molecules on the surface.

10.99. As temperature increases, the surface tension decreases because the intermolecular forces acting on the molecular "film" on the surface of the liquid are weaker. Likewise, the viscosity decreases as the temperature increases because molecules have more energy to readily break the intermolecular forces to enable them to slide past each other more freely.

10.101. Water is able to form more hydrogen bonds with the glass surface (composed of Si—O—H bonds) than ethanol, so water rises higher in the capillary tube.

10.103. Liquid B is expected to have the higher surface tension and higher viscosity because its higher boiling point is indicative of stronger intermolecular forces.

10.105. Alcohols exhibit stronger intermolecular forces (hydrogen bonds) than ethers (dipole–dipole).

10.107. Increase; the equilibrium line between the solid and vapor phases of water increases upward and to the right of the phase diagram. As the pressure increases, the temperature at which sublimation occurs will also increase until the ice reaches the triple point (0.0060 atm and 0.010°C).

10.109. Yes

10.111.

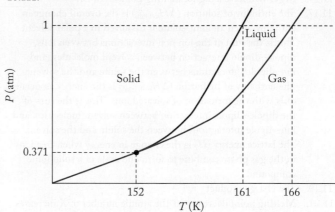

10.113. Inside a narrow capillary tube, the mass of water that rises to a particular height in the column equals the mass of water that can rise in a wider diameter tube, but the level in the test tube is lower because the volume (mass) is spread out across a wider area. It is the same mass that feels the force of gravity being balanced by the adhesive and cohesive forces.

10.115. As the temperature of the pond water increases, the oxygen becomes less soluble, and the fish do not have enough oxygen to survive.

10.117. (a)

10.119. Evaporation is endothermic; energy is transferred to the ethanol as it evaporates.

Chapter 11

11.1. (a)

11.3. (a) (I) = NaCl, (II) = $MgCl_2$, (III) = $C_6H_{12}O_6$, and (IV) = K_3PO_4; (b) (i) (III), (ii) (IV), (iii) (III), (iv) (IV)

11.5. (a) Both substances are nonelectrolytes: if they were electrolytes, we would expect to observe deviations from linearity at the higher molalities due to ion-pair formation. (b) Given that the K_f value is the slope of the line in Figure P11.5, and because both substances give the same $\Delta T_f/m$ value, we conclude that the K_f value is the same for aqueous solutions of both substances.

11.7. (a) In diagram (i) water will pass through the membrane from the pure water side to the protein solution side, and in diagram (ii) water will pass from the 0.2 M KCl solution side to the protein solution side. (b) In diagram (i) Na^+ ions will flow from the 0.5 M NaCl (protein solution) side to the pure water side of the membrane, and in diagram (ii) Na^+ ions will flow from the 0.5 M NaCl (protein solution) side to the aqueous 0.2 M KCl side of the membrane. (c) In diagram (i) K^+ ions will flow from the 0.1 M KCl (protein solution) side to the pure water side of the membrane, and in diagram (ii) K^+ ions will flow from the 0.2 M KCl (aqueous solution) side to the 0.1 M KCl (protein solution) side of the membrane.

11.9. (c)

11.11. The ion–ion bond in $CaSO_4$ is stronger than in NaCl because of the higher charges on the cation and anion. For $CaSO_4$, this is greater than the ion–dipole interactions that would occur when Ca^{2+} and SO_4^{2-} dissolve, so $CaSO_4$ is not very soluble in water. NaCl has a lower ion–ion bond strength and its ion–dipole interactions with water are strong, so it dissolves in water.

11.13. $CsBr < KBr < SrBr_2$

11.15. MgO will have a higher melting point than LiF.

11.17. The enthalpy of solution ($\Delta H_{solution}$) is the overall change in enthalpy when an ionic solute is dissolved in a polar solvent. This is the sum of the ion–ion interactions between ions, dipole–dipole interaction between solvent molecules, and ion–dipole interactions between the solute and the solvent. The enthalpy of hydration ($\Delta H_{hydration}$) is the energy associated only with the formation of solvated ions. This is the sum of the dipole–dipole interaction between solvent molecules, and ion–dipole interactions between the solute and the solvent. The lattice energy (U) is the change in energy when free ions in the gas phase combine to form one mole of a solid ionic compound.

11.19. The charge product

11.21. Melting point decreases as the atomic number of X increases.

11.23. (a)

11.25. -723 kJ/mol

11.27. -690 kJ/mol

11.29. A nonvolatile solute is a compound that dissolves into a solvent and does not, under conditions to maintain the solution, enter appreciably into the gas phase.

11.31. When the average kinetic energy of the liquid molecules increases, more of the molecules can escape the liquid phase and enter the gas phase. More molecules in the gas phase increase the vapor pressure.

11.33. The vapor pressure of pure ethanol is greater than the vapor pressure of the sugar solution of ethanol. The rate of evaporation will be higher for the pure ethanol. On the other hand, the rate of condensation in the beaker containing the sugar solution is higher than the rate of evaporation, so eventually the beaker containing the sugar solution will overflow.

11.35. Mole fraction of water in solution: $X = 0.70$; vapor pressure of solution at 25°C: $P_{solution} = 17$ torr

11.37.
$$P_{soln} = X_{solvent} \cdot P°_{solvent}$$
$$X_{solvent} + X_{solute} = 1$$
$$X_{solvent} = 1 - X_{solute}$$
$$P_{soln} = (1 - X_{solute}) \cdot P°_{solvent}$$
$$P_{soln} = P°_{solvent} - P°_{solvent} \cdot (X_{solute})$$
$$P°_{solvent} \cdot (X_{solute}) = P°_{solvent} - P_{soln}$$
$$X_{solute} = \frac{(P°_{solvent} - P_{soln})}{P°_{solvent}}$$

11.39. C_5H_{12}

11.41. Both components of the solution are of similar size, and both are nonpolar. It is reasonable that the magnitude of the intermolecular forces would be similar, and that the solution would behave ideally.

11.43. 60 torr

11.45. (a) 23 styrene to 27 ethylbenzene; (b) (ii)

11.47. The van't Hoff factor is an experimentally determined value and cannot be used to determine the molality (and thus the formula mass) for an ionic solute.

11.49. The mass of the solvent, and thus the molality, does not change with T. Density, and thus molarity of the solution, decreases as T increases. The difference between molarity and molality will increase.

11.51. Whereas a molecular compound will not dissociate, an ionic compound may dissociate in the solvent. This dissociation

yields two or more particles in solution from one dissolved solute particle. This results in greater changes in the melting and boiling points than for those of a molecular solute that does not dissociate.

11.53. The theoretical value of i for CH_3OH is 1 because methanol is molecular and does not dissociate in a solvent such as water. NaBr has a theoretical value of $i = 2$ because it dissociates into two particles on dissolution (Na^+ and Br^-). K_2SO_4 has a theoretical value of $i = 3$ because it dissociates into three particles on dissolution ($2 K^+$ and SO_4^{2-}).

11.55. A semipermeable membrane is a boundary between two solutions through which some molecules may pass but others cannot. Usually, small molecules may pass through but large molecules are excluded.

11.57. Solvent flows across a semipermeable membrane from the more dilute solution side to the more concentrated solution side to balance the concentration of solutes on both sides of the membrane.

11.59. Reverse osmosis transfers solvent across a semipermeable membrane from a region of higher solute concentration to a region of lower solute concentration. Because reverse osmosis goes against the natural flow of solvent across the membrane, the key component needed is a pump to apply pressure to the more concentrated side of the membrane. Other components needed include a containment system, piping to introduce and remove the solutions, and a tough semipermeable membrane that can withstand the high pressures needed.

11.61. Because $NaCl(aq)$ dissociates to form $Na^+(aq)$ and $Cl^-(aq)$, we need twice as many moles of dextrose to achieve the same molar concentration.

11.63. (a) 0.21 m; (b) 0.572 m; (c) 0.440 m

11.65. (a) 749 g; (b) 31.7 g; (c) 104.56 g

11.67. 6.5×10^{-5} m NH_3, 8.7×10^{-6} m NO_2^-, 2.195×10^{-2} m NO_3^-

11.69. 3.81°C

11.71. 2.52×10^{-2} m

11.73. -1.89°C

11.75. 0.5 m $CaCl_2$

11.77. 0.0100 m $Mg(NO_3)_2$

11.79. (a) < (b) < (c)

11.81. (a) From side A to side B; (b) from side B to side A; (c) from side A to side B

11.83. (a) 57.5 atm; (b) 0.682 atm; (c) 52.9 atm; (d) 46.5 atm

11.85. (a) 2.75×10^{-2} M; (b) 1.11×10^{-3} M; (c) 1.00×10^{-2} M

11.87. False; the molarity of the NaCl solution would be greater by 1.5 times than the molarity of $CaCl_2$.

11.89. (a) Osmotic pressure increases; (b) freezing point decreases; (c) boiling point increases

11.91. 94.1 g/mol

11.93. Molar mass = 164 g/mol; the molecular formula of eugenol is $C_{10}H_{12}O_2$.

11.95. (a)

11.97. Yes

11.99. For 0.0935 m NH_4Cl, $i = 1.85$; for 0.0378 m $(NH_4)_2SO_4$, $i = 2.46$

11.101. 2.3 atm

11.103. 4270 g/mol

11.105.
$$[Na^+]_{saline} = [Na^+]_{RL} + [K^+]_{RL} + [Ca^{2+}]_{RL}$$
$$[Cl^-]_{saline} = [Cl^-]_{RL} + [C_3H_5O_3^-]_{RL}$$

Chapter 12

12.1. (b) and (d) are crystalline; (a) and (c) are amorphous.

12.3.

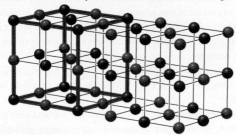

The chemical formula is A_4B_4 or AB.

12.5. A_4B_2C

12.7. Li_2S

12.9. Cs (blue) and Sr (purple)

12.11. MgB_2

12.13. (c)

12.15. (a) Hydrogen bonds. (b) Amine and carboxylic acid functional groups form an amide bond.

12.17. Cubic closest-packed structures have an *abcabc* . . . pattern and hexagonal closest-packed structures have an *abab* . . . pattern.

12.19. Body-centered cubic

12.21. These structural forms are not allotropes because iron is not molecular.

12.23. 104.2 pm

12.25. 513 pm

12.27. (c)

12.29. 0.963 g/cm³

12.31. 4.50 g/cm³

12.33. In a substitutional alloy, some of the atoms of one lattice are replaced with atoms of another element. One example of a substitutional alloy is bronze, in which up to 30% of the Cu atoms in a copper lattice have been replaced with tin atoms. The tin atoms in this alloy are randomly distributed throughout the lattice, occupying any lattice position normally occupied by a copper atom. An interstitial alloy contains solutes in the spaces (or "holes") between atoms when the solvent lattice forms. One example of an interstitial alloy is austenite, an alloy containing carbon in the octahedral holes of the iron fcc lattice.

12.35. The edge length will be smaller.

12.37. The significant increase in mass when hafnium is substituted for magnesium will result in an alloy of greater density than that of pure magnesium.

12.39. Yes; both unit cells produce the same ratio of Ni to Ti, so both are valid.

12.41. (a) 90 pm; (b) substitutional alloy

12.43. substitutional alloy

12.45. V_2C

12.47. 5.78 g/cm³

12.49. (a) A substitutional alloy with a bcc lattice; (b) $CuZn_3$

12.51. Yes, the alloy is denser. Because tin has a greater molar mass than copper, the mass of the unit cell in the alloy would be greater.

12.53. K^+ is large and so does not fit well into the octahedral holes of the fcc lattice.

12.55. The radius of Cl^- is 181 pm and the radius of Cs^+ is 170 pm, so their radii are very similar. The Cs^+ ion at the center of

Figure P12.55 occupies the center of the cubic cell, so CsCl could be viewed as a body-centered cubic structure when taking into account the ions' slight difference in size. The CsCl structure can also be viewed as a simple cubic lattice of Cl^- ions with Cs^+ ions in the cubic holes.

12.57. The rock salt structure would not provide the correct ratio of ions for $CaCl_2$.

12.59. Less than

12.61. $MgFe_2O_4$

12.63. (a) Octahedral; (b) half

12.65. (a) The rock salt structure is an fcc lattice of anions with cations in the octahedral holes. The cesium chloride structure is a simple cubic lattice of anions with cations in the cubic holes. (b) Both structures have the same ratio of Mg^{2+} to Se^{2-} ions (1:1).

12.67. 5.25 g/cm³

12.69. 421 pm

12.71. The hybridization of the phosphorus atom in black phosphorus is sp^3 with bond angles of 102° to give a puckered ring. In graphite, the carbon atoms are sp^2 hybridized with bond angles of 120°, which gives graphite a flat geometry.

12.73. (a) 1.708 g/cm³; (b) 498.6 pm

12.75. 3.81 g/cm³

12.77. Monomers must have appropriate functional groups to react with one another in a polymerization. Provided they have appropriate functional groups, different monomers can react to form a polymer.

12.79. 3564 monomer units

12.81.

12.83.

12.85. (a) Polymer I monomers

$H_2N(CH_2)_8NH_2$ and $HOC(CH_2)_8COH$

Polymer II monomers

$H_2N(CH_2)_6NH_2$ and $HOC(CH_2)_{10}COH$

(b) Because of the longer chain length between C=O bonds in Polymer II, we might expect that this polymer would be more flexible along that portion of the polymer chain.

12.87. (a) HCl; (b) the CO_3 or $-OCOO-$ linkage gives the term *polycarbonate* to Lexan.

12.89. XYZ_3

12.91. (a) bcc; (b) two phase changes, hcp → bcc → hex (3)

12.93. (a) 2.3×10^5 unit cells/particle; (b) 9.1×10^5 Ag atoms; (c) 8.4×10^{12} Ag particles

12.95. (a) 7.53 g/cm³; (b) 3.42 g/cm³; (c) 3.36 g/cm³

12.97. AuZn

12.99. Substitutional alloys

12.101. A bcc Cu lattice with four corner atoms replaced by Al is consistent with the formula.

12.103.

(a)

[structure: 3-aminobenzoic acid — HO–C(=O)–benzene ring with NH₂]

[structure: repeating polymer unit with amide linkages, HO–C(=O)–benzene–N(H)–C(=O)–benzene–N(H)–C(=O)–benzene–NH₂]

Repeating unit depicted in P12.103

(b)

[structure: HO–C(=O)–C(CH₃)₂–OH and polymer: HO–C(=O)–C(CH₃)₂–O–C(=O)–C(CH₃)₂–O–C(=O)–C(CH₃)₂–OH with dashed box around repeating unit]

12.105. This polymer has one monomeric unit with seven carbon atoms; nylon-6 has a single monomeric unit as well, but it consists of six carbon atoms.

[structure: nylon polymer repeating unit with N–H and C=O groups, bracketed with subscript n]

12.107. Cross-linking increases the strength, hardness, melting (softening) point, and chemical resistance of a polymer.

12.109.

[structure: polymer with CN and C(=O)O groups, bracketed with subscript n]

Chapter 13

13.1. [N₂O]: green line; [O₂]: red line

13.3. (b)

13.5. (d)

13.7. (c)

13.9. (a) 3*; (b) a, red; (c) b, black; (d) c, blue

13.11. In the absence of sufficient sunlight, NO_2 will be formed from NO (generated by vehicles), but will not be broken down, so $[NO_2]$ will increase over time, while [NO] will slowly decrease. The absence of sunlight will also ensure that $[O_3]$ stays low over most of the day. VOCs will be formed throughout the day, but not broken down by sunlight, meaning [VOC] will continue to rise over the day.

13.13. As the concentration of NO increases in the atmosphere, more NO_2 is generated. This process is not immediate, and it results in a decrease in the amount of free NO in the atmosphere as it proceeds. The maximum concentration of NO_2 will occur several hours after the maximum concentration of NO is reached as a result of this delay.

13.15. O atoms are more reactive as a result of the incomplete octet and unpaired electrons on each atom.

13.17. The rate of a reaction is the overall change in the concentration of reactants and products, whereas the rate constant for a reaction is the proportionality between the rate of the reaction and the concentration of each reactant, raised to the order of each reactant.

13.19. Tracking the partial pressure of CH_3CHO, CH_4, or CO, or following the total pressure. Yes, the rate data would be the same.

13.21. In solution, ionic compounds can dissociate and mix with one another. As solids, ions from the two reactants are still bound to one another, so it is very unlikely that a reaction will occur on any significant scale.

13.23. The rate of formation of B and consumption of A are equal, but opposite in sign.

13.25. Decrease

13.27. a. $\dfrac{\Delta[N_2]}{\Delta t} = \dfrac{\Delta[O_2]}{\Delta t}$

b. $\dfrac{\Delta[N_2]}{\Delta t} = -\dfrac{1}{2}\dfrac{\Delta[NO]}{\Delta t}$

13.29. a. $\text{Rate} = -\dfrac{\Delta[F_2]}{\Delta t} = +\dfrac{\Delta[HOF]}{\Delta t} = +\dfrac{\Delta[HF]}{\Delta t}$

b. $\text{Rate} = -\dfrac{1}{3}\dfrac{\Delta[HCl]}{\Delta t} = +\dfrac{\Delta[H_2]}{\Delta t}$

c.

$\text{Rate} = -\dfrac{1}{4}\dfrac{\Delta[NH_3]}{\Delta t} = -\dfrac{1}{4}\dfrac{\Delta[O_2]}{\Delta t} = +\dfrac{1}{2}\dfrac{\Delta[N_2]}{\Delta t} = +\dfrac{1}{6}\dfrac{\Delta[H_2O]}{\Delta t}$

13.31. 5.81×10^{-6} M/s

13.33. a. $\text{Rate} = \dfrac{\Delta[CO_2]}{\Delta t} = -\dfrac{2}{3}\dfrac{\Delta[CO]}{\Delta t}$

b. $\text{Rate} = \dfrac{\Delta[COS]}{\Delta t} = -\dfrac{\Delta[SO_2]}{\Delta t}$

c. $\text{Rate} = \dfrac{\Delta[CO]}{\Delta t} = 3\dfrac{\Delta[SO_2]}{\Delta t}$

13.35. (a) 1.2×10^7 M/s; (b) 2.9×10^4 M/s

13.37. Between 0 and 100 μs: 1.4×10^{-5} $M/μs$; between 200 and 300 μs: 5.5×10^{-6} $M/μs$

13.39. For the change in concentration of ClO versus time we obtain the following plot:

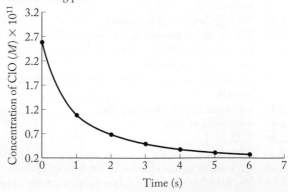

The instantaneous rate at 1 s is 8.28×10^{10} molecules $cm^{-3} s^{-1}$. For the change in concentration of Cl_2O_2 versus time we obtain the following plot:

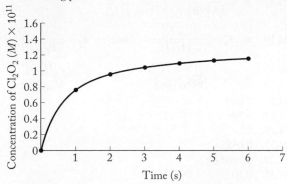

The instantaneous rate at 1 s is 4.13×10^{10} molecules $cm^{-3} s^{-1}$.

13.41. The concentration of reactants decreases over time, so the likelihood of collisions decreases, and the rate of reaction will decrease.

13.43. Zero order

13.45. The half-life will be halved.

13.47. (a) First order in both A and B, and second order overall; (b) second order in A, first order in B, and third order overall; (c) first order in A, third order in B, and fourth order overall

13.49. (a) Rate = $k[O][NO_2]$; k units = $M^{-1}s^{-1}$; (b) rate = $k[NO]^2[Cl_2]$; k units = $M^{-2}s^{-1}$; (c) rate = $k[CHCl_3][Cl_2]^{1/2}$; k units = $M^{-1/2}s^{-1}$; (d) rate = $k[O_3]^2[O]^{-1}$; k units = s^{-1}

13.51. (a) Rate = $k[BrO]$; (b) rate = $k[BrO]^2$; (c) rate = $k[BrO]$; (d) rate = $k[BrO]^0 = k$

13.53. No; the same behavior would be observed if the reaction were second order in one reactant and zero order in the other.

13.55. (a) Rate = $k[NO_2][O_3]$; (b) $4.9 \times 10^{-11} M/s$; (c) $4.9 \times 10^{-11} M/s$; (d) the rate doubles.

13.57. (c)

13.59. Rate = $3.4 \times 10^{-5} s^{-1}[N_2O_5]$

13.61. Rate = $k[ClO_2][OH^-]$; $k = 14 M^{-1}s^{-1}$

13.63. Rate = $k[NO]^2[H_2]$; $k = 6.32 M^{-2}s^{-1}$

13.65. (a) $-3.56 \times 10^{-4} M \cdot min^{-1}$; (b) $-3.56 \times 10^{-4} M \cdot min^{-1}$

13.67. 0.293 M

13.69. (a) $k = 8.31 \times 10^{-5} s^{-1}$; (b) 0.616 M

13.71. (a) Rate = $k[N_2O]$; (b) 4

13.73. (a) Rate = $k[^{32}P]$; (b) 0.0485 day^{-1}; (c) 14.3 days

13.75. $k = 5.40 \times 10^{-12} cm^3$ molecules$^{-1} s^{-1}$; $t_{1/2} = 0.712$ s

13.77. Reverse

13.79. An increase in temperature increases the frequency and the kinetic energy at which the reactants collide. This speeds up the reaction. The order of the reaction is unaffected.

13.81. The reaction with the larger activation energy (150 kJ/mol)

13.83. Reaction 2

13.85. $E_a = 17.1$ kJ/mol; $A = 1.002$

13.87. (a) $E_a = 314$ kJ/mol; (b) $A = 5.03 \times 10^{10}$; (c) $k = 1.06 \times 10^{-44} M^{-1/2}s^{-1}$

13.89. $E_a = 39.1$ kJ/mol; $A = 1.27 \times 10^{12}$

13.91. Yes, though it doesn't have to. If a catalyst or intermediate is present, the reaction could proceed in two or more steps.

13.93. The statement might be true but does not necessarily have to be true.

13.95.

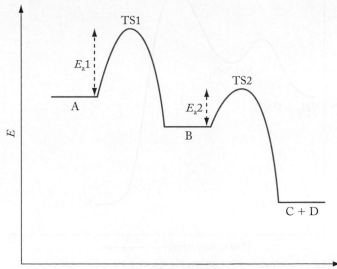

13.97. (a) Rate = $k[SO_2Cl_2]$; unimolecular; (b) rate = $k[NO_2][CO]$; bimolecular; (c) rate = $k[NO_2]^2$; bimolecular

13.99. $N_2O_5(g) + O(g) \rightarrow 2 NO_2(g) + O_2(g)$

13.101. The second step

13.103. The first step

13.105. (b) or (c)

13.107. Yes. Because the rate of the reaction is faster (affecting the rate) and the activation energy is lowered (affecting the value of k), a catalyst affects both the rate of the reaction and the value of the rate constant.

13.109. Yes

13.111. No

13.113. NO is the catalyst.

13.115. The reaction of O_3 with Cl

13.117. When the concentration of a reactant (O_2 for the combustion reaction) increases, the rate of combustion increases.

13.119. The bodily reactions that use O_2 are slower at colder temperatures.

13.121. Yes, we could use other times, just not $t = 0$, as long as the rate of the reverse reaction is still much slower than the forward reaction. At $t = 0$, we use reactant concentrations.

13.123. An elementary step with a molecularity of zero would have no molecules colliding.

13.125. 5.0 M/s

13.127. (a) Rate $= k[NH_3][HNO_2]^2$; the units of k are $M^{-2}s^{-1}$; (b) yes; (c) -482.4 kJ; (d)

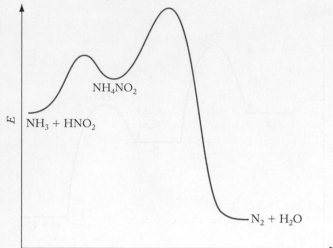

13.129. a. Rate $= k[NO][ONOO^-]$; $k = 1.30 \times 10^{-3}\ M^{-1}s^{-1}$

b.

$$\left[\overset{0}{\ddot{O}}=\overset{0}{\ddot{N}}-\overset{0}{\ddot{O}}-\overset{-1}{\ddot{O}\!:}\right] \longleftrightarrow \left[:\overset{-1}{\ddot{O}}-\overset{0}{\ddot{N}}=\overset{+1}{\ddot{O}}-\overset{-1}{\ddot{O}\!:}\right]^- \longleftrightarrow$$
(preferred)

$$\left[:\overset{-1}{\ddot{O}}-\overset{-1}{\ddot{N}}-\overset{+1}{\ddot{O}}=\overset{0}{\ddot{O}}\right]^-$$

c. -55 kJ

Chapter 14

14.1. a. $K_c = \dfrac{[A_3B]^2}{[A_2]^3[B_2]}$

b. $K_c = 0.87$

14.3. This reaction is not at equilibrium because Q (0.023) $\neq K$. Because $Q < K$, the reaction shifts to the right.

14.5. No—at 20 μs the concentrations of A and B are still changing.

14.7. A system is at equilibrium when the rate of the forward reaction equals the rate of the reverse reaction.

14.9. No. The equilibrium may favor either products or reactants, depending on the thermodynamic energy difference. The rate at which the reversible reaction occurs does not change the energy of the products or reactants.

14.11. As the reaction approaches equilibrium, the forward reaction occurs more rapidly than the reverse reaction, so more B is formed than A.

14.13.

Molar Mass	Compound	How Present
28	$^{14}N_2$	Originally present
29	$^{15}N^{14}N$	From decomposition of $^{15}N^{14}NO$
30	$^{15}N_2$	From decomposition of $^{15}N_2O$
32	O_2	Originally present
44	$^{14}N_2O$	From combination of $^{14}N_2$ and O_2
45	$^{15}N^{14}NO$	From combination of $^{15}N^{14}N$ and O_2
46	$^{15}N_2O$	Originally present

14.15. When $\Delta n = 0$; when the number of moles of gaseous products equals the number of moles of gaseous reactants

14.17. $K_c = \dfrac{[C]^2}{[A]^2[B]}$

14.19. a. $K_c = \dfrac{[C_2H_6]}{[C_2H_4][H_2]}$ and $K_p = \dfrac{(P_{C_2H_6})}{(P_{C_2H_4})(P_{H_2})}$

b. $K_c = \dfrac{[SO_3]^2}{[SO_2]^2[O_2]}$ and $K_p = \dfrac{(P_{SO_3})^2}{(P_{SO_2})^2(P_{O_2})}$

14.21. 0.068

14.23. 0.50

14.25. 27

14.27. 0.0583

14.29. (a) and (c)

14.31. 0.10

14.33. When scaling the coefficients of a reaction up or down, the new value of the equilibrium constant is the first K raised to the power of the scaling constant.

14.35. 11.0

14.37. $K_{c,forward} = \dfrac{[NO_2]^2}{[NO][NO_3]}$; $K_{c,reverse} = \dfrac{[NO][NO_3]}{[NO_2]^2}$;

$K_{c,reverse} = \dfrac{1}{K_{c,forward}}$

14.39. 1.7×10^{-2}

14.41. (a) 0.049; (b) 420; (c) 20

14.43. 1.1×10^{-18}

14.45. The reaction quotient, Q, is for a reaction mixture not necessarily at equilibrium. Only if the system is at equilibrium is Q equal to K.

14.47. No; $Q < K$, so the reaction proceeds to the right to reach equilibrium.

14.49. $Q > K$, so the reaction will proceed to the left.

14.51. The reaction is not at equilibrium. It will proceed to the left.

14.53. $K_p = \dfrac{(P_{SO_2})}{(P_{O_2})}$

14.55. $K_c = \dfrac{[NaCl]^2[CO_2]}{[NaHCO_3]^2[CaCl_2]}$

14.57. $K_p = \dfrac{1}{[Fe(OH)_2]^4[O_2]}$

14.59. No. The relative concentrations of the reactants and products will adjust until they achieve the value of K. This value is affected only by temperature.

14.61. As the concentration of O_2 increases, the reaction shifts to the right and the CO on the hemoglobin is displaced.

14.63. According to Le Châtelier's principle, an increase in the partial pressure (or concentration) of O_2 above the water shifts the equilibrium to the right so that more oxygen becomes dissolved in the water. This is consistent with Henry's law.

14.65. (b) and (d)

14.67. (a) Increasing the concentration of the reactant O_3 shifts the equilibrium to the right, increasing the concentration of the product O_2; (b) increasing the concentration of the product O_2 shifts the equilibrium to the left, increasing the concentration of the reactant O_3; (c) decreasing the volume of the reaction to 1/10 its original volume shifts the equilibrium to the left, increasing the concentration of the reactant O_3.

14.69. The equilibrium shifts to the left.

14.71. (b)

14.73. When K is small, the amount of reactants that are transformed into products may be so small that at equilibrium the concentrations of the reactants are approximately equal to the initial concentrations. This means that we can make an approximation in the K expression to make our calculations easier.

14.75. (a) $P_{PCl_5} = 0.024$ atm, $P_{PCL_3} = 1.036$ atm, $P_{Cl_2} = 0.536$ atm; (b) the partial pressure of PCl_3 decreases and the partial pressure of PCl_5 increases.

14.77. $[H_2O] = [Cl_2O] = 3.76 \times 10^{-3}$ M, $[HOCl] = 1.13 \times 10^{-3}$ M

14.79. 6.9×10^5:1

14.81. $P_{CO} = 2.4$ atm, $P_{CO_2} = 3.8$ atm

14.83. (a) $P_{NO} = 0.272$ atm, $P_{NO_2} = 7.98 \times 10^{-3}$ atm; (b) $P_T = 0.416$ atm

14.85. $P_{O_2} = 0.17$ atm, $P_{N_2} = 0.75$ atm, $P_{NO} = 0.080$ atm

14.87. 5.75 M

14.89. $P_{CO} = P_{Cl_2} = 0.258$ atm, $P_{COCl_2} = 0.00680$ atm

14.91. $[CO] = [H_2O] = 0.031$ M, $[CO_2] = [H_2] = 0.069$ M

14.93. The reaction is endothermic: at 1500 K, $K_p = 5.5 \times 10^{-11}$; at 2500 K, $K_p = 4.0 \times 10^{-3}$; and at 3000 K, $K_p = 0.40$. This reaction does not favor products even at very high temperature, so it is not a viable source of CO and is not a remedy to decrease CO_2 as a contributor to global warming. Also, the process produces poisonous CO gas.

14.95. 9×10^{-22} M

14.97. (a) $P_{SO_2} = 9.2 \times 10^{-74}$ atm; (b) 2.63×10^{-47} molecules

14.99. 19.5

Chapter 15

15.1. Red line

15.3. The right (smallest) bar

15.5. (a) Basic; (b) the amine (R_2NH)

15.7. Piperidine is the stronger base. The electronegative O atom in morpholine pulls electron density away from the N atom, rendering this the less basic molecule. By comparison piperidine, with CH_2 instead of an O atom, is a stronger base.

15.9. The electron-withdrawing chlorine atoms pull electron density away from the OH bond, leading to greater ionization.

15.11. HF is the acid and H_2O is the base.

15.13. NH_3 is the base and H_2O is the acid.

15.15. (a) HCl is the acid and NaOH is the base; (b) HCl is the acid and $MgCO_3$ is the base; (c) H_2SO_4 is the acid and NH_3 is the base.

15.17. NO_2^-, OCl^-, $H_2PO_4^-$, NH_2^-

15.19. Conjugate acid is H_2SO_4 and conjugate base is SO_4^{2-}

15.21. 0.65 M

15.23. 0.0410 M

15.25. Because the pH function is a $-\log$ function, as $[H_3O^+]$ increases, the value of $-\log[H_3O^+]$ decreases.

15.27. $[H_3O^+]$ is greater than 1 M. One example is 2.5 M HCl, which has a pH of -0.40.

15.29.

15.31. (a) pH = 2.28, pOH = 11.72, acidic; (b) pH = 8.42, pOH = 8.42, basic; (c) pH = 5.14, pOH = 8.86, acidic; (d) pH = 0.00, pOH = 14.00, acidic

15.33. (a) 1.2×10^{-11} M; (b) 7.7×10^{-11} M; (c) 2.22×10^{-12} M; (d) 3.4×10^{-10} M

15.35. (a) pH = 0.810, pOH = 13.190; (b) pH = 2.301, pOH = 11.699; (c) pH = 1.903, pOH = 12.097; (d) pH = 1.200, pOH = 12.800

15.37. 7.006

15.39. (a) $HCl > HNO_2 > CH_3COOH > HClO$; (b) $HClO < CH_3COOH < HNO_2 < HCl$

15.41. HNO_3 is a strong acid, and as such, is completely ionized in aqueous solution. HNO_2 is a weak acid, so only a small fraction of the molecules are ionized in solution. HNO_3, therefore, with more dissolved ions in solution, is a better conductor of electricity.

15.43. $K_a = \dfrac{[H_3O^+][F^-]}{[HF]}$

15.45. (a) Water; (b) water

15.47. H_2O is the acid, and CH_3NH_2 is the base.

15.49. 8.91×10^{-4}

15.51. 3.26%, $K_a = 1.37 \times 10^{-4}$

15.53. 2.51

15.55. 2.3 times

15.57. (a) Weaker; (b) 10.857; (c) 1.00×10^{-4}

15.59. (a) 10.635; (b) 9.356

15.61. With each successive ionization, it becomes more difficult to remove H^+ from a species that is more negatively charged.

15.63. Ge is less electronegative than C, so even less electron density is pulled away from the acidic proton in H_2GeO_3 than in H_2CO_3, resulting in a lower percent ionization and K_a for H_2GeO_3.

15.65. 0.12

15.67. 2.80

15.69. 9.50

15.71. 10.27

15.73. Sulfur is more electronegative than selenium. The higher electronegativity on the sulfur atom stabilizes the anion HSO_4^- more than the anion $HSeO_4^-$.

15.75. (a) H_2SO_3; (b) H_2SeO_4

15.77. Increase

15.79. Ammonium nitrate

15.81. The citric acid in the lemon juice neutralizes the volatile trimethylamine to make a nonvolatile dissolved salt.

15.83. 3.32

15.85. 7.35

15.87. (a) $HClO_4$; (b) $Ca(OH)_2$; (c) CH_3COOH; (d) CH_3NH_2

15.89. Yes, all Arrhenius bases are Brønsted–Lowry bases. No, not all Brønsted-Lowry bases are Arrhenius bases. For example, NH_3 is a Brønsted–Lowry base but does not contain OH^-, and so is not an Arrhenius base.

15.91. In burning S-containing fuels:
$$S_8(s) + 8\,O_2(g) \xrightarrow{+4\,O_2(g)} 8\,SO_2(g) \rightarrow 8\,SO_3(g)$$
In the presence of water, these produce a weak acid, H_2SO_3, and a strong acid, H_2SO_4:
$$SO_2(g) + H_2O(\ell) \rightarrow H_2SO_3(aq)$$
$$SO_3(g) + H_2O(\ell) \rightarrow H_2SO_4(aq)$$
These acids dissolve $CaCO_3$:
$$2\,H^+(aq) + CaCO_3(s) \rightarrow H_2CO_3(aq) + Ca^{2+}$$

15.93. 1.4×10^9 L

15.95. (a) In water, amines show basic character. They pick up a proton from water to form ammonium cations and OH^-. Therefore dissolving Prozac in water gives a slightly basic solution. (b) The secondary amine (N atom) on Prozac is more likely to react with HCl than the O atom. (c) The HCl salt of Prozac is more soluble because it is charged and water molecules form stronger ion–dipole forces around the molecule than the dipole–induced dipole forces between the neutral molecule and water.

15.97. (a)

$4n + 2 = 6$ electrons, so this ring is aromatic. (b) C_5F_5H is very acidic because the presence of five very electronegative F atoms on the carbon ring stabilizes the anion formed when the proton is lost.

15.99. (a)
$$\overset{+1+5-2}{HNO_3} + 2\ \overset{+1+6-2}{H_2SO_4} \rightarrow \overset{+5-2}{NO_2^+} + \overset{+1-2}{H_3O^+} + 2\ \overset{+1+6-2}{HSO_4^-}$$
No oxidation number change in this reaction, so this is not a redox reaction.
(b)

Acid = H_2SO_4, conjugate base = HSO_4^-; base = HNO_3 (the OH group), conjugate acid = H_3O^+

Chapter 16

16.1. The blue titration curve represents the titration of a 1 M solution of strong acid; the red titration curve represents the titration of a 1 M solution of weak acid.

16.3. The indicator with a pK_a of 9.0

16.5. The red titration curve represents the titration of Na_2CO_3; the blue titration curve represents the titration of $NaHCO_3$.

16.7. NH_4Cl is dissolved in the yellow solution, $NaC_2H_3O_2$ is dissolved in the blue solution, and $NH_4C_2H_3O_2$ is dissolved in the light green solution.

16.9. (a) Buffer; (b) some of the HCOOH will be converted to $HCOO^-$

16.11. A buffer is formed, which can resist changes in pH as a result of adding acid or base to the solution.

16.13. Iodoacetic acid/sodium iodoacetate

16.15. The quantity of an external acid or base that can be added to a buffer while maintaining a pH that is ±1 unit from the pK_a of the acid component

16.17. The acetic acid/acetate buffer will have the highest pH, the formic acid/formate buffer will lie in the middle, and the hydrofluoric acid/fluoride buffer will have the lowest pH.

16.19. The pH will not change.

16.21. 2.55

16.23. 6.89

16.25. 10:1

16.27. 9.25 g bromoacetic acid and 21.4 g sodium bromoacetate

16.29. 6.39 g dimethylamine and 0.680 g dimethylammonium chloride

16.31. 9.25

16.33. 53.7 mL

16.35. (a) 3.50; (b) 3.42

16.37. Yes; strong acid–strong base neutralizations generate water and a neutral salt, which will have a pH of 7.

16.39. The endpoint is a highly visible colored to colorless transition that occurs at pK_a of 8.8, near the first equivalence point of pH = 8.5.

16.41. 4.44

16.43. After 10.0 mL of OH^- has been added, the pH = 4.754; after 20.0 mL of OH^- has been added, the pH = 8.750; after 30.0 mL of OH^- has been added, the pH = 12.356.

16.45. (a) 2.945 M; (b) no effect

16.47. 250 mL

16.49.

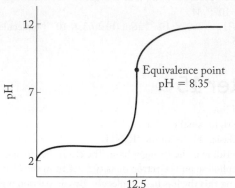

16.51. (a) Less than 7; (b) greater than 7; (c) equal to 7

16.53. Two equivalence points; phenolphthalein and alizarin yellow R

16.55. No; some Lewis bases donate an electron pair but do not accept a proton while doing so.

16.57. Yes; Brønsted–Lowry acids donate a proton, meaning that they can also accept an electron pair.

16.59.

16.61. The sulfur atom in SO_2 and the hydrogen atom in H_2O are acting as Lewis acids, while the oxygen atoms in SO_2 and H_2O are acting as Lewis bases.

16.63. $B(OH)_3$ is the Lewis acid and H_2O is the Lewis base.

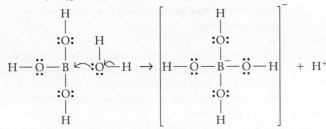

16.65. Water

16.67. Ag^+ forms a soluble complex with NH_3, removing Ag^+ from solution and shifting the equilibrium for the dissolution of AgCl to the right.

16.69. $1.25 \times 10^{-10}\ M$

16.71. $2.48 \times 10^{-13}\ M$

16.73. (b) and (d)

16.75. The solution will become more acidic.

16.77. In basic solution: $Cr(OH)_3(s) + OH^-(aq) \rightleftharpoons Cr(OH)_4^-(aq)$
        In acidic solution: $Cr(OH)_3(s) + 3\ H^+(aq) \rightleftharpoons$
        $Cr^{3+}(aq) + 3\ H_2O(\ell)$

16.79. $Al(OH)_3$ reacts with OH^- in solution to form soluble $Al(OH)_4^-$. The other ions do not form this type of soluble complex ion.

16.81. 2.80

16.83. 1.80

16.85.

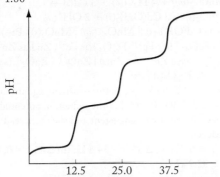

Volume of $0.50\ M$ NaOH (mL)

16.87. Molar solubility is the quantity (moles) of substance that dissolves in a liter of solution. The solubility product is the equilibrium constant for the dissolution of a substance.

16.89. Mg^{2+}

16.91. Endothermic

16.93. Acidic substances react with the OH^- released on dissolution of hydroxyapatite. The equilibrium is shifted to the right, dissolving more hydroxyapatite.

16.95. 1.08×10^{-10}

16.97. $[Cu^+] = [Cl^-] = 1.01 \times 10^{-3}\ M$

16.99. 9.96×10^{-6} g/mL

16.101. 10.091

16.103. (d)

16.105. No

16.107. Yes

16.109. (a) SO_4^{2-}; (b) $1.34 \times 10^{-4}\ M$

16.111. (a) Because HF is weak, the $[F^-]$ is lower than that of water and so HF reacts with H_2O to form F^- as the major anionic species; (b) $K_{overall} = 2.9 \times 10^{-4}$; (c) pH = 1.91; $[HF_2^-]_{eq} = 4.40 \times 10^{-4}\ M$

16.113. Subsequent additions of HCO_3^- react with water to form bicarbonate's conjugate acid (H_2CO_3) and its conjugate base (CO_3^{2-}) in the same proportions as the first addition, so pH does not change.

16.115. 6.6×10^{-16}

16.117. a. The acidity of the oceans increases (lower pH) because of increased ionization of the weak acid carbonic acid. An increase in the partial pressure of CO_2 in the atmosphere would cause the following equilibria to shift to the right (Le Châtelier's principle):
$$CO_2(g) \rightleftharpoons CO_2(aq)$$
$$CO_2(aq) + H_2O(\ell) \rightleftharpoons H_2CO_3(aq)$$
$$H_2CO_3(aq) \rightleftharpoons H^+(aq) + HCO_3^-(aq)$$
$$HCO_3^-(aq) \rightleftharpoons H^+(aq) + CO_3^{2-}(aq)$$

b. Six times greater

c. Because oyster shells are composed of $CaCO_3$ they are likely to dissolve to a greater extent as the pH decreases.

Chapter 17

17.1. The gas in the balloon on the right is under greater pressure and has greater entropy.

17.3. No. This is a low probability because each gas would then be confined to a smaller volume and would have more order, so this change would involve a decrease in entropy. The tank is an isolated system as long as no heat is transferred through the cylinder wall.

17.5. At low temperature

17.7. (a) and (c) are spontaneous because both have a negative value for ΔG_{rxn}. The balanced equations are
        (a) $2\ C_6H_6(\ell) + 15\ O_2(g) \rightarrow 12\ CO_2(g) + 6\ H_2O(\ell)$
        (c) $2\ H_2(g) + O_2(g) \rightarrow 2\ H_2O(\ell)$

17.9. They are equal in magnitude, but the sign is reversed.

17.11. ΔS_{sys} is positive and ΔS_{surr} is negative.

17.13. (a) and (b)

17.15. ΔS_{sys} is negative and ΔS_{univ} is negative.

17.17. ΔS_{surr} must be less than $+66.0$ J/K.

17.19. (a) decreases; (b) increases; (c) approximately the same

17.21. (a) $S_8(s)$; (b) $S_2(g)$; (c) $O_3(g)$; (d) $O_2(g)$

17.23. Fullerenes

17.25. (a) $CH_4(g) < CF_4(g) < CCl_4(g)$;
        (b) $CH_2O(g) < CH_3CHO(g) < CH_3CH_2CHO(g)$;
        (c) $HF(g) < H_2O(g) < NH_3(g)$

17.27. Positive

17.29. Solids have a lower entropy (are a more ordered phase) than particles dissolved in aqueous solution. As a result of the precipitation reaction, fewer particles are present in any phase of the reaction. Both of these factors lead to ΔS_{rxn}'s being less than zero.

17.31. (a) 24.9 J/K; (b) -146.4 J/K; (c) -73.2 J/K; (d) -175.8 J/K

17.33. 218.9 J/(mol · K)

17.35. (c)

17.37. In general, exothermic reactions have a corresponding increase in entropy. Both an increase in entropy and decrease in enthalpy lead to a spontaneous reaction, though now we realize that it is possible for an exothermic reaction to be spontaneous at low temperatures if disfavored by entropy.

17.39. ΔS is positive, ΔH is positive, and ΔG is negative.

17.41. (b) and (d)

17.43. For NaBr, -18 kJ/mol; for NaI, -30 kJ/mol

17.45. $+91.4$ kJ

17.47. $+56.5$ kJ

17.49. -81.7 kJ

17.51. No; if ΔS_{rxn} were positive, then the exothermic reaction would be spontaneous at all temperatures.

17.53. 981.3 K

17.55. $\Delta H_{rxn} = 44.0$ kJ; $\Delta S_{rxn} = 118.9$ J/K or 0.1189 kJ/K; 370.1 K or 96.9°C

17.57. (a) Low temperatures; (b) low temperatures; (c) all temperatures

17.59. Positive

17.61. No. When $\Delta G° > 0$, $K < 1$. The reaction favors the formation of reactants, so the reaction will not shift to the right.

17.63. (c)

17.65. +27.124 kJ/mol

17.67. Exothermic

17.69. Exothermic

17.71. 1.3×10^{-31}

17.73. −115 kJ/mol

17.75. Some of the products of one reaction must be reactants of the second reaction, and the overall value of ΔG when the reactions are summed must be negative.

17.77. The bond arrangements are only slightly different between the two structures.

17.79. (a) +50.8 kJ and −394.4 kJ; (b) $CH_4(g) + O_2(g) \rightarrow 2\,H_2(g) + CO_2(g)$, $\Delta G_{rxn} = -343.6$ kJ, spontaneous

17.81. 3.81×10^{-3}

17.83. −73 kJ/mol

17.85. 24

17.87. 4.47×10^3

17.89. 83.5 J/(mol · K)

17.91. (a) −90.7 J/(mol · K); (b) a large negative value of ΔH will cause ΔG to be negative and thus spontaneous.

17.93. (c) and (d)

17.95. −1308.1 kJ

17.97. Below 386.0 K

17.99. Above 896 K

17.101. (a) Cu_2O: copper(I) oxide and CuO: copper(II) oxide; (b) never spontaneous; (c) Cu_2O is more complex than CuO because it has more bonds.

17.103. (a) Trouton's rule predicts that $\Delta S°_{vap}$ for liquids is relatively constant because the change in entropy from liquid to gas should be similar for similar molecules; (b) helium exhibits a lower $\Delta S°_{vap}$ because its intermolecular forces are weak, causing more randomness than usual in the liquid phase. Methanol and water deviate because these liquids form hydrogen bonds, causing more order in the liquid state.

17.105. (a) ΔH is negative and ΔS is positive; (b) no: the reverse reaction with a positive ΔH and a negative ΔS would never be spontaneous.

Chapter 18

18.1. Because of the careful layering, each half-cell has its metal in contact with its cation solution. The solutions are not mixing, but nevertheless the layers allow the ions needed to balance the charge in each half-cell to pass.

18.3. Ag is the cathode; Pt in the SHE is the anode; electrons flow from the SHE to Ag.

18.5. Blue line

18.7. a. $2\,H_2O(\ell) + 2\,e^- \rightarrow H_2(g) + 2\,OH^-(aq)$
$$E°_{cathode} = -0.8277\ V$$
$2\,H_2O(\ell) \rightarrow O_2(g) + 4\,H^+(aq) + 4\,e^-$
$$E°_{anode} = 1.229\ V$$
   b.   It increases the conductivity of the solution.

18.9. (a) Cathode; (b) chromium forms the passivating oxide Cr_2O_3, which prevents the metal beneath it from being oxidized in the presence of air.

18.11. (b)

18.13. (a) To allow nonreactive ions to pass through the separator to maintain electrical neutrality; (b) the wire would allow electrons to pass, but not other ions, and because electrons cannot travel through the solution, the wire would not complete the circuit.

18.15. 2

18.17. $Cr_2O_7^{2-}(aq) + 14\,H^+(aq) + 6\,Fe^{2+}(aq) \rightarrow 2\,Cr^{3+}(aq) + 7\,H_2O(\ell) + 6\,Fe^{3+}(aq)$

18.19. a. $Fe^{3+}(aq) + 1\,e^- \rightarrow Fe^{2+}(aq)$
$Al^{3+}(aq) + 3\,e^- \rightarrow Al(s)$
   b. $I_2(aq) + 2\,e^- \rightarrow 2\,I^-(aq)$
$NO_3^-(aq) + H_2O(\ell) + 2\,e^- \rightarrow NO_2^-(aq) + 2\,OH^-(aq)$
   c. $MnO_4^-(aq) + 8\,H^+(aq) + 5\,e^- \rightarrow Mn^{2+}(aq) + 4\,H_2O(\ell)$
$Cr_2O_7^{2-}(aq) + 14\,H^+(aq) + 6\,e^- \rightarrow 2\,Cr^{3+}(aq) + 7\,H_2O(\ell)$

18.21. a.   Cathode: $Ni^{2+}(aq) + 2\,e^- \rightarrow Ni(s)$
Anode: $Cd(s) \rightarrow Cd^{2+}(aq) + 2\,e^-$
   b. $Ni^{2+}(aq) + Cd(s) \rightarrow Cd^{2+}(aq) + Ni(s)$
   c. $Cd(s)\,|\,Cd^{2+}(aq)\,||\,Ni^{2+}(aq)\,|\,Ni(s)$

18.23. a.   Cathode: $MnO_4^-(aq) + 2\,H_2O(\ell) + 3\,e^- \rightarrow MnO_2(s) + 4\,OH^-(aq)$
Anode: $Cd(s) + 2\,OH^-(aq) \rightarrow Cd(OH)_2(s) + 2\,e^-$
   b. $2\,MnO_4^-(aq) + 4\,H_2O(\ell) + 3\,Cd(s) \rightarrow 2\,MnO_2(s) + 3\,Cd(OH)_2(s) + 2\,OH^-(aq)$
   c. $Cd(s)\,|\,Cd(OH)_2(s)\,||\,MnO_4^-(aq)\,|\,MnO_2(s)\,|\,Pt(s)$

18.25. (a) 6; (b) FeO_4^{2-} has Fe^{6+}, Fe_2O_3 has Fe^{3+}, Zn has Zn^0, ZnO and ZnO_2^{2-} have Zn^{2+}; (c) $Zn(s)\,|\,ZnO(s)\,|\,ZnO_2^{2-}(aq)\,||\,FeO_4^{2-}(aq)\,|\,Fe_2O_3(s)\,|\,Pt(s)$

18.27. Cl_2 ($\Delta E°_{red} = 1.3583$ V) is a better oxidizing agent than O_2 ($\Delta E°_{red} = 1.229$ V). The higher the reduction potential, the more powerful the oxidizing agent and the more readily it is itself reduced.

18.29. (a) $\Delta E°_{cell} = 0.358$ V; $\Delta G° = -34.5$ kJ; (b) $\Delta E°_{cell} = 0.430$ V; $\Delta G° = -82.9$ kJ

18.31. No

18.33. $O_2(g) + 2\,H_2O(\ell) + 2\,Zn(s) + 4\,OH^-(aq) \rightarrow 2\,Zn(OH)_4^{2-}(aq)$

18.35. Less than 1.10 V

18.37. (a) $\Delta E°_{cell} = 0.292$ V; $\Delta G° = -56.4$ kJ; (b) $\Delta E°_{cell} = 0.505$ V; $\Delta G° = -97.4$ kJ

18.39. (a) and (c)

18.41. (a) $NiO(OH)(s) + TiZr_2H(s) \rightarrow TiZr_2(s) + Ni(OH)_2(s)$; (b) 1.32 V

18.43. −48.3 kJ

18.45. −290 kJ

18.47. −116 kJ

18.49. The platinum electrode transfers electrons to the half-cell; it is inert and not involved in the reaction.

18.51. It is difficult to maintain the correct concentration and pressure in the standard hydrogen electrode, and any H_2 gas that is evolved could combust.

18.53. (a) 2.37 V; (b) Mg; (c) $H_2(g)$

18.55. Voltage of a battery (a voltaic cell) is governed by the Nernst equation:
$$E_{cell} = E°_{cell} - \frac{RT}{nF}\ln Q$$
As a battery discharges, the value of Q, the reaction quotient, changes:
$$Q = \frac{[products]^x}{[reactants]^y}$$

At the start of the reaction, Q is very small because [reactants] $\gg$ [products]. As the reaction proceeds, [products] grows and Q increases but does not increase significantly until significant amounts of products form, that is, when the battery is nearly discharged.

18.57. -0.414 V

18.59. $E_{rxn} = 1.55$ V; decrease

18.61.   (a) 0.62 V; (b) 0.66 V

18.63.   (c) and (f)

18.65.   Al–O$_2$

18.67.   Li–MnO$_2$

18.69.   Teflon is a polymeric material composed of poly(tetrafluoro-ethylene). Because no metals or ionizable ions are present in teflon (the C–F bond is extremely stable), no redox activity is expected. Teflon is also water repellent, which keeps the metal structure drier than it would be with the asbestos pad.

18.71.   The material used must be more readily oxidized than aluminum. Magnesium is a stable, nonreactive metal that would serve as a sacrificial anode for aluminum.

18.73.   In a voltaic cell, the electrons are produced at the anode so a negative ($-$) charge builds up there; in an electrolytic cell, electrons are being forced onto the cathode so that it builds up negative ($-$) charge. The flow of electrons in the outside circuit is reversed in an electrolytic cell compared to the flow in a voltaic cell.

18.75.   Br$_2$

18.77.   The student's best choice is Na$_2$CO$_3$, as it dissociates in solution to form three ions, with an effective concentration of 6.00 M. The other solutions would have smaller concentrations of ions.

18.79.   27.0 minutes

18.81.   (a) 5.78×10^{-3} L; (b) no, because some Cl$_2$ and Br$_2$ would be produced

18.83. -0.270 V

18.85.   A hybrid vehicle uses a relatively inexpensive fuel (gasoline) in the internal combustion engine and has good fuel economy, but still gives off emissions. A fuel-cell vehicle does not give off emissions (the reaction produces H$_2$O) but requires a more expensive and explosive fuel (hydrogen); moreover, current battery technologies incorporate materials that are still very expensive and bulky.

18.87.   Electric engines are more efficient, converting more of the energy into motion instead of losing it as heat.

18.89.   a. $\overset{-4+1}{CH_4}(g) + \overset{+1}{H_2O}(g) \rightarrow \overset{+2}{CO}(g) + 3\ \overset{0}{H_2}(g)$

   $\overset{+2}{CO}(g) + \overset{+1}{H_2O}(g) \rightarrow H_2(g) + \overset{+4}{CO_2}(g)$

   b.   For the reaction of CH$_4$ with H$_2$O, $\Delta G^{\circ}_{rxn} = 142.2$ kJ, for the reaction of CO with H$_2$O, $\Delta G^{\circ}_{rxn} = -28.6$ kJ, and for the overall reaction, $\Delta G^{\circ}_{overall} = \Delta G^{\circ}_{rxn_1} + \Delta G^{\circ}_{rxn_2} = 113.6$ kJ.

18.91.   0.617 V

18.93. 8.56×10^{19}

18.95.   (a) Cathode; (b) no, because Mg^{2+}, with a higher positive charge, has a lower (less negative) reduction potential than Na$^+$; (c) no; (d) H$_2$ and O$_2$

18.97.   (a) -0.87 V; (b) Mo$_3$S$_4$: Mo $= +2.67$; MgMoS$_4$: Mo $= +2$; (c) Mg^{2+} is added to the electrolyte to better carry the charge in the cell. This cation is produced at the anode and consumed at the cathode.

18.99.   0.030 V

Chapter 19

19.1.   Purple (lithium)

19.3.   Red (radium)

19.5.   (a)

19.7.   (c), (d), and (e)

19.9.   Positive

19.11.   The *mass defect* is the difference between the mass of the nucleus of an isotope and the sum of the masses of the individual nuclear particles that make up that isotope. The *binding energy* is the energy released when individual nucleons combine to form the nucleus of an isotope.

19.13. 3.46×10^{-13} J

19.15. 1.09×10^{-12} J/nucleon

19.17.   If the nuclide lies in the belt of stability (green dots on the plot in Figure 19.2), it is not radioactive and is stable. If it lies above the belt of stability, then it is neutron-rich and tends to undergo β decay to increase the number of protons and reduce the number of neutrons in its nucleus. If it lies below the belt of stability, it is neutron-poor and tends to undergo positron emission or electron capture to increase the number of neutrons and reduce the number of protons in its nucleus.

19.19.   Alpha decay increases the neutron-to-proton ratio to produce less stable isotopes, which can then be made more stable through β emission to decrease the neutron-to-proton ratio.

19.21.   Both of these processes are β decays.

19.23.   Greater than 1

19.25.   (a) $^4_2\alpha$; (b) $^0_{-1}\beta$; (c) $^0_{+1}\beta$; (d) $^0_{-1}\beta$

19.27.   (a) Electron capture or positron emission; (b) electron capture or positron emission; (c) this isotope is stable.

19.29. ^{56}Co has 27 protons and 29 neutrons and is neutron-poor; it may undergo electron capture or positron emission. ^{44}Ti has 22 protons and 22 neutrons and is neutron-poor; it may undergo electron capture or positron emission.

19.31.   (a) ^{32}Cl, ^{33}Cl, and ^{34}Cl will emit positrons; (b) ^{38}Cl and ^{39}Cl will emit β particles; (c) ^{36}Cl will emit either positrons or β particles.

19.33.   25%

19.35.   0.181 d^{-1}

19.37.   131 years

19.39.   After 8.726 half-lives the ratio of ^{14}C present to that originally in an artifact is $N_t/N_0 = 0.50^{8.726} = 0.00236$, or 0.236%. This is too little to detect.

19.41.   After 0.00023 half-lives the ratio of ^{40}K present to that originally in a sample is $N_t/N_0 = 0.50^{0.00023} = 0.9998$, or 99.98%. This level is the point at which we can detect the difference in amounts of ^{40}K.

19.43.   35%

19.45.   The oldest ring would have a 15% lower ^{14}C/^{12}C ratio.

19.47.   36,640 y

19.49.   The level of radioactivity is the amount of radioactive particles present in a given instant of time. The dose is the accumulation of exposure over a length of time.

19.51.   When radon-222 decays to polonium-218 while in the lungs, the ^{218}Po, a reactive solid that is chemically similar to oxygen, lodges in the lung tissue, where it continues to emit α radiation. Alpha radiation is one of the most damaging kinds of radiation when in contact with biological tissues. The result of exposure to high levels of radon is an increased risk for lung cancer.

19.53.   5 μSv $= 5$ μGy; 250 μJ

19.55. (a) $^{90}_{38}Sr \rightarrow {^{0}_{-1}}\beta + {^{90}_{39}}Y$; (b) 3.28×10^8 ^{90}Sr atoms; (c) strontium-90 is more concentrated in milk than in many other foods because it is chemically similar to calcium, and milk is rich in calcium.

19.57. (a) 0.15 decays/s; (b) 7.0×10^4

19.59. (a) The half-life should be long enough to effect treatment of the cancerous cells but not so long as to cause damage to healthy tissues; (b) because α radiation does not penetrate far beyond a tumor, the α decay mode is best; (c) products should be nonradioactive, if possible, or have short half-lives and be able to be flushed from the body by normal cellular and biological processes.

19.61. (a) Positron emission or electron capture; (b) positron emission or electron capture; (c) positron emission or electron capture

19.63. 74.3 days

19.65. Yes

19.67. 136 min

19.69. (a) $^{10}_{5}B + {^{1}_{0}}n \rightarrow {^{7}_{3}}Li + {^{4}_{2}}\alpha$; (b) 4.43×10^{-13} J; (c) alpha particles have a high RBE and they do not penetrate into healthy tissue if the radionuclide is placed inside a tumor.

19.71. Control rods made of boron or cadmium are used to absorb the excess neutrons to control the rate of energy release.

19.73. The neutron-to-proton ratio for heavy nuclei is high and when the nuclide undergoes fission to form smaller nuclides, it must emit neutrons because the fission products require a lower neutron-to-proton ratio for stability.

19.75. (a) $^{138}_{52}Te$; (b) $^{133}_{51}Sb$; (c) $^{143}_{55}Cs$

19.77. (a) $^{103}_{39}Y$; (b) $^{130}_{48}Cd$; (c) $^{138}_{52}Se$

19.79. Today's Sun has fewer neutrons than the primordial Sun. The primordial Sun formed α particles via the fusion of a proton and a neutron:

$$2\,{^{1}_{1}}p + 2\,{^{1}_{0}}n \rightarrow 2\,{^{2}_{1}}H \rightarrow {^{4}_{2}}He$$

Today's Sun forms α particles via the fusion of two protons:

$$2\,{^{1}_{1}}p \rightarrow {^{2}_{1}}H + {^{0}_{+1}}\beta$$
$$2\,{^{2}_{1}}H + 2\,{^{1}_{1}}p \rightarrow 2\,{^{3}_{2}}He \rightarrow {^{4}_{2}}He + 2\,{^{1}_{1}}p$$

19.81. (a) 4.37×10^{-12} J; (b) 6.80×10^{-12} J; (c) 2.69×10^{-12} J; (d) 1.60×10^{-12} J

19.83. (a) 7.67×10^{-13} J; (b) -3.96×10^{-13} J

19.85. (a) Besides releasing a large amount of energy to power the starship *Enterprise*, hydrogen is an abundant fuel in the universe and therefore could easily react with any antihydrogen produced; (b) if any of the antimatter fuel came into contact with conventional matter (such as the ship, the crew, or the warp engine), the two would annihilate each other, releasing a large quantity of energy in the form of an explosion.

19.87. The energy released in the fusion reaction is $\Delta E = 9.91 \times 10^{-13}$ J/nucleon. The energy released in the fission reaction is $\Delta E = 1.4 \times 10^{-13}$ J/nucleon. On a per-nucleon basis, the fusion reaction generates more energy.

19.89. (a) Geiger counter; (b) 2877 years; (c) ^{241}Am is an α emitter, and α particles do not travel more than a few inches in air and cannot penetrate the first layer of skin.

19.91. (a) $^{249}_{98}Cf + {^{48}_{20}}Ca \rightarrow {^{294}_{118}}Og + 3{^{1}_{0}}n$; (b) $^{290}_{116}Lv$; (c) $^{286}_{114}Fl$; (d) $^{282}_{112}Cn$; (e) because ^{294}Og is a member of the noble gas family, it has chemical and physical properties similar to those of naturally occurring radon.

19.93. ^{210}Pb

19.95. (a) $^{64}_{28}Ni + {^{124}_{50}}Sn \rightarrow {^{188}_{78}}Pt$; (b) $^{196}_{78}Pt$

19.97. 3.35

19.99. (a) $^{40}_{19}K \rightarrow {^{40}_{18}}Ar + {^{0}_{1}}\beta$; (b) because the half-life of ^{40}K is so much longer than that of ^{14}C

Chapter 20

20.1.

Adenine   Guanine   Thymine   Cytosine

20.3. (a) Palmitic acid; (b) stearic acid

20.5. Tyrosine, glycine, glycine, phenylalanine, and methionine

20.7. Trans fats exhibit geometric isomerism around the C=C bond where similar groups on the two carbon atoms are situated on opposite sides of the double bond. Structures (a) and (c) contain trans fats.

20.9. Sucrose; the difference in the structures is that in sucralose, three –OH groups on sucrose have been replaced by Cl atoms. Being derived from sucrose implies that the sugar is natural, but the presence of Cl atoms on sugars is not natural.

20.11. No; the terms *enantiomer* and *optically active* can be used to describe the same chiral molecule, but *achiral* cannot.

20.13. Constitutional isomers have the same molecular formula, but different bonds to the atoms, or even different functional groups. Stereoisomers have the same bonds to each atom, but different arrangements of those bonds.

20.15. Glycine has no chiral carbon centers.

20.17. sp^3

20.19. (b) 2,3,4-trimethylpentane; (c) 3-ethyl-2-methylpentane; (d) 3-ethyl-3-methylpentane; (e) 3,4-dimethylhexane

20.21. (a) trans, *E*; (b) cis, *Z*

20.23. (a)

20.25.

Saccharin   Sodium cyclamate   Aspartame

20.27. (a) Yes, 2 chiral centers; (b) yes, the *cis* isomer is shown here:

20.29. Decreases

20.31. Both amino acids have acidic, polar side chains, with a C=O bond approximately the same distance from the amine and acid functional groups.

20.33. D- and L- refer to how the four groups on a chiral carbon are oriented.

20.35. (a) and (c)

20.37. Most amino acids are zwitterions at pH ≈ 7.4 because the amino group will be protonated and the carboxylic acid group will be deprotonated, giving $H_3\overset{+}{N}-CH(R)-COO^-$.

20.39. Lysine contains two amino groups, one of which is on a long carbon tail. This can react with the carboxylic acid on the carbon tail of glutamic acid to form a salt bridge.

20.41. (a) Alanine + serine = Alaser (or AS); (b) alanine + phenylalanine = Alaphe (or AF); (c) alanine + valine = Alaval (or AV).

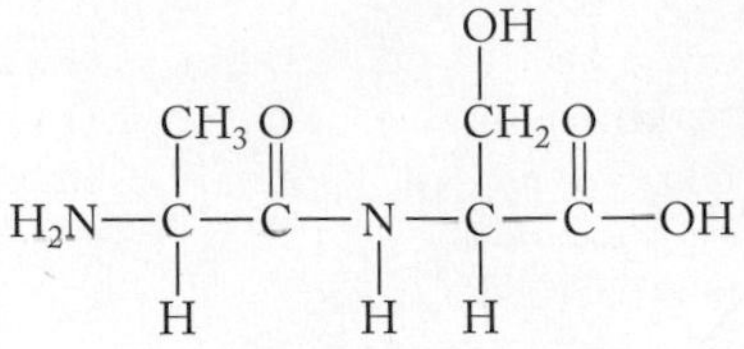

(a)

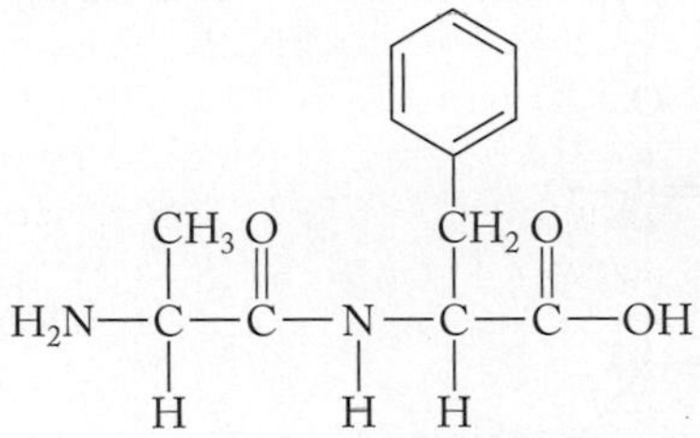

(b)

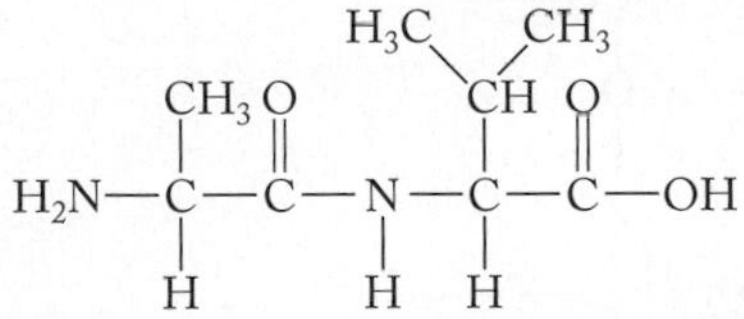

(c)

20.43. (a) Alanine + glycine; (b) leucine + leucine; (c) tyrosine + phenylalanine

20.45. NH_3

20.47. Starch has α-glycosidic bonds, but cellulose has β-glycosidic bonds. Starch coils into granules, but cellulose forms linear molecules.

20.49. No

20.51. The bonding in fructose and glucose is nearly the same.

20.53. To calculate the free-energy change for a two-step process we need only to sum the individual ΔG values for each reaction.

20.55. The position of the hydroxy group on carbon 1 differs between the α and β forms of galactose. The relative positions of the hydroxy groups on carbons 2, 3, and 4 are the same on both isomers.

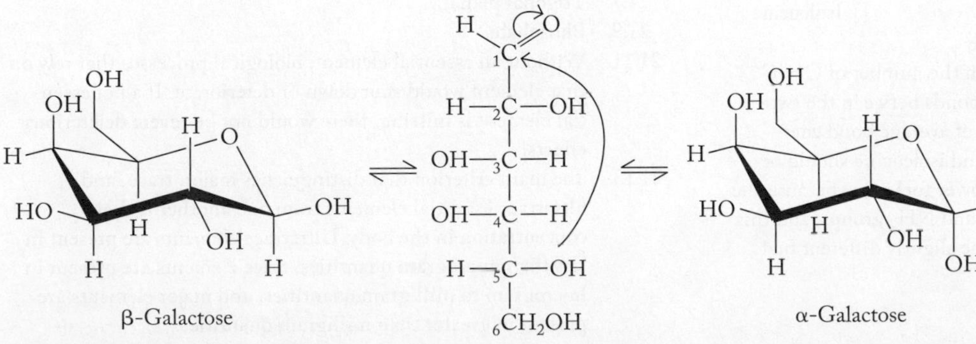

β-Galactose   α-Galactose

20.57. (c)

20.59. (b)

20.61. −16.4 kJ

20.63. Saturated fatty acids have all C—C single bonds in their structure, but unsaturated fatty acids have C=C double bonds.

20.65. Fatty acids have a high fuel value, and eating sticks of butter affords Arctic explorers with more energy per gram of food compared to carbohydrates or proteins.

20.67. If the two fatty acids linked to the glycerol at C-1 and C-3 are different then, yes, the triglyceride has a chiral center.

20.69. (a) 0.50 kg of α-linolenic acid would consume more hydrogen than 1.0 kg of oleic acid; (b) no; the hydrogenation product of both species is stearic acid.

20.71.

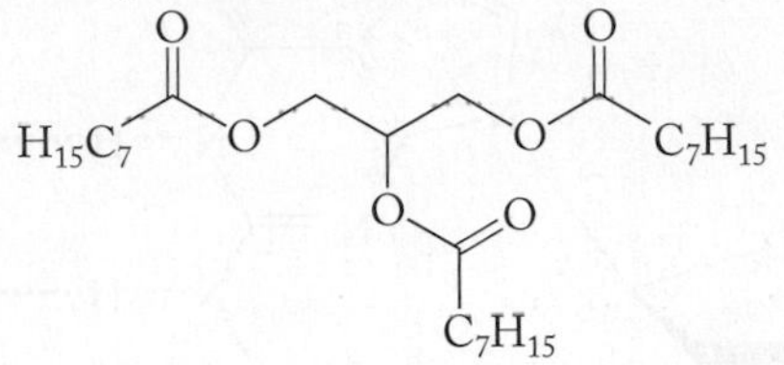

(a) Glycerol with octanoic acid

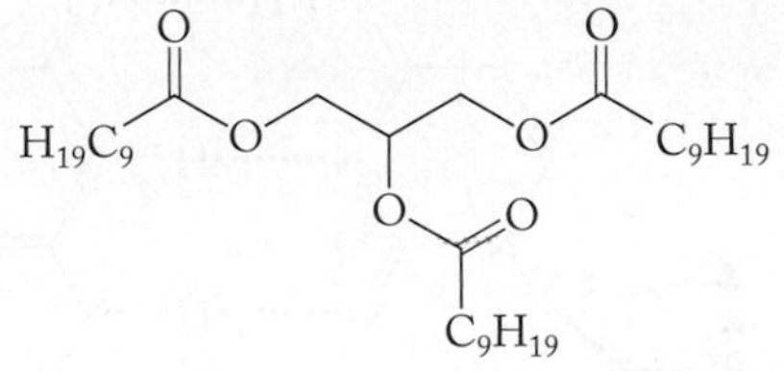

(b) Glycerol with decanoic acid

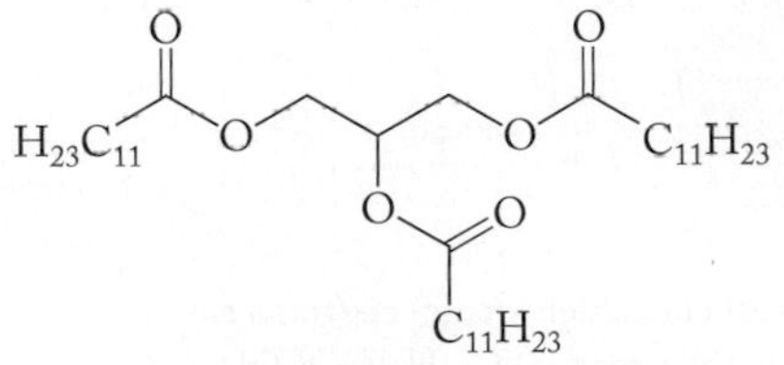

(c) Glycerol with dodecanoic acid

20.73. A phosphate group, five-carbon sugar, and nitrogen base; alternating sugar residues and phosphate groups

20.75. Hydrogen bonds

20.77.

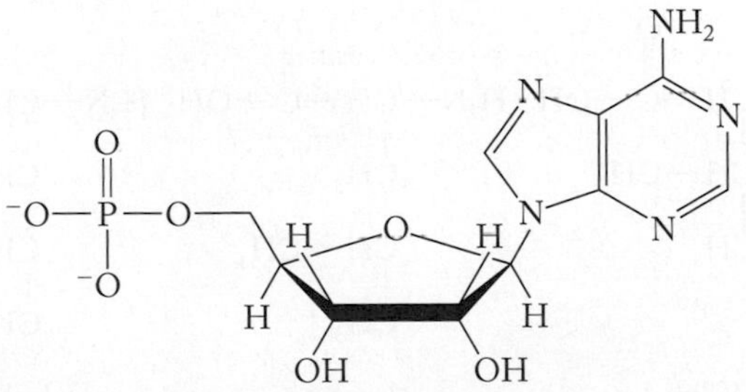

20.79. a. A-G-C
        b.

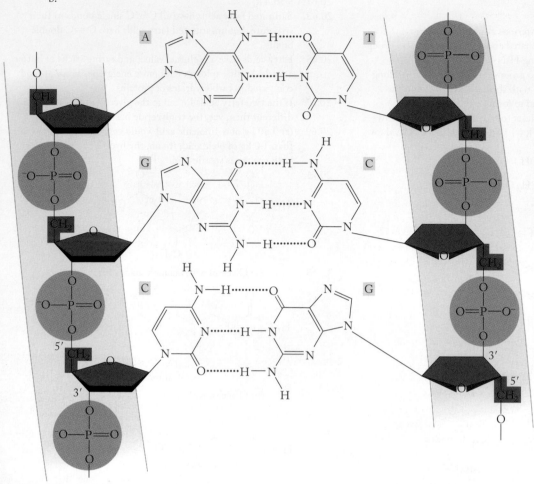

20.81. (a) 12; (b) alkene, aldehyde; (c) cis/trans isomers
20.83. (a) Sucrose; (b) esters; (c) $C_{15}H_{31}COOH$
20.85. (a) There is an extra –CH_2– group in homocysteine's
        sulfur-containing side chain; (b) yes
20.87. Yes
20.89.

$$H_2N-CH-C-OH \quad\quad H_2N-CH-C-OH \quad\quad H_2N-CH-C-OH$$

Valine   Leucine   Isoleucine

20.91. Glutamic acid, cysteine, and glycine
20.93. Yes; because there is no difference in the number of C—C,
        C—H, C=O, C—O, or N—H bonds between the two
        compounds, we expect on the basis of average bond ener-
        gies that the fuel values of leucine and isoleucine should be
        identical. Isoleucine might have a lower fuel value because the
        CH_3 group is closer to the COOH and NH_2 groups, and this
        difference in shape contributes to the slightly different fuel
        values.

20.95. 16

20.97. All nine amino acids with nonpolar R groups (G, A, V, L, I,
        P, F, W, and M), and the six polar amino acids without acidic
        or basic R groups (S, T, C, Y, N, and Q)
20.99. Both molecules contain a planar ring with a nitrogen atom,
        and an alkyl-substituted nitrogen donor.

Chapter 21

21.1.   (d)
21.3.   Group 16 (lavender)
21.5.   −6.72 kJ
21.7.   Trigonal planar
21.9.   Phosphate
21.11. Without an essential element, biological processes that rely on
        that element would shut down or deteriorate. If a nonessen-
        tial element is missing, there would not be severe deleterious
        effects.
21.13. The main criterion that distinguishes major, trace, and
        ultratrace essential elements from one another is their
        concentration in the body. Ultratrace elements are present in
        less than microgram quantities, trace elements are present in
        microgram to milligram quantities, and major elements are
        present in greater than milligram quantities.

21.15. (a) 1.6 ppm; (b) 525 ppm; (c) 0.43 ppm
21.17. (a) Oxygen; (b) oxygen; (c) carbon
21.19. The valence electrons for the neutral atoms from the "s-block" are ns, and those for the "p-block" are np.
21.21. Li_2O is an ionic compound, whereas the bonding in CO is covalent.
21.23. Electrical conductivity
21.25. Ca^{2+} is larger than Mg^{2+} and may not fit into the active site of a biomolecule.
21.27. Potassium
21.29. $Al^{3+} < Mg^{2+} < Li^+ < Cl^-$
21.31. $K < Mg < S < F$
21.33. The two processes are the reverse of one another.
21.35. Osmosis, ion channels, and ion pumps
21.37. The hydrophobic interior of the cell membrane makes it difficult to transport charged ions through the cell membrane.
21.39. Potassium
21.41. The greater insolubility of $CaCO_3$ than of $CaSO_4$ makes calcium carbonate a better structural material. Also, the partial pressure of CO_2 in the atmosphere is higher than SO_3, so the carbonate solubility equilibrium is shifted more to the left by Le Châtelier's principle than is the sulfate equilibrium.
21.43. 0.56 atm
21.45. 0.0688 V
21.47. 0.15 mol
21.49. $5.8 \times 10^{-4}\ M$
21.51. $^{137}Cs^+$ may substitute for K^+ in cells; as a β emitter with a relatively long half-life, it may cause cancer.
21.53. ΔS is probably positive, and ΔG is probably negative.
21.55. The superoxide anion contains an unpaired electron, and an incomplete octet on oxygen. This radical will react to complete the octet and pair the electron, being itself reduced in the process.
21.57. The products are a β particle, and barium-137.
$$^{137}_{55}Cs \rightarrow\ ^{0}_{-1}\beta +\ ^{137}_{56}Ba$$
21.59. 3.06
21.61. $K = [OH^-]/[F^-] = 8.48$; the equilibrium lies to the right.
21.63. (a) More soluble; (b) $8.5 \times 10^{-8}\ M$; (c) $1.5 \times 10^{-7}\ M$
21.65. 5.3×10^{-118}
21.67. H_2O is the most polar, and H_2Te is the least polar.
21.69. We must consider the decay mode to ensure that α-emitters are not used inside the body. γ-Emitters are preferred because γ radiation has relatively low energy and can easily escape the body, minimizing damage to tissue and organs.
21.71. Because β particles have low charge and relatively small mass, they penetrate tissues better than α particles do. When the β particles exit the body, they are imaged on a detector screen to create the image of the organ.
21.73. ^{111}In is proton-rich (or neutron-poor), so electron capture, which changes a proton into a neutron, helps the nucleus become more stable. ^{213}Bi is proton-poor (or neutron-rich), so β-decay, which changes a neutron into a proton, helps the nucleus become more stable.
21.75. Dispersion forces
21.77. 41 h
21.79. $[:Bi\equiv O:]^+$
21.81. $Al(OH)_3(s) + 3\ H^+(aq) \rightarrow 3\ H_2O(\ell) + Al^{3+}(aq)$
21.83. $Al(OH)_3$ will neutralize more acid on a per-mole and a per-gram basis than will $Mg(OH)_2$.
21.85. 2.52 g

Chapter 22

22.1. Chromium (green) and cobalt (yellow)
22.3. Zinc (blue)
22.5. Four
22.7. (a) Yellow: $[Co(CN)_6]^{3-}$; (b) blue: $[CoF_6]^{3-}$; orange: $[Co(NH_3)_6]^{3+}$
22.9. (a) Identical; (b) isomers; (c) identical
22.11. (a) Blue: CoI_4^{2-}; (c) black: $CoCl_4^{2-}$
22.13. Water
22.15. Water
22.17. Na^+
22.19. $[Pt(NH_3)_6]Cl_4$; $[Pt(NH_3)_5Cl]Cl_3$; $[Pt(NH_3)_4Cl_2]Cl_2$; $[Pt(NH_3)_3Cl_3]Cl$; $[Pt(NH_3)_2Cl_4]$
22.21. $[Pt(NH_3)_6]Cl_4$ - hexammineplatinum(IV) chloride; $[Pt(NH_3)_5Cl]Cl_3$ - pentamminechloroplatinum(IV) chloride; $[Pt(NH_3)_4Cl_2]Cl_2$ - tetramminedichloroplatinum(IV) chloride; $[Pt(NH_3)_3Cl_3]Cl$ - triamminetrichloroplatinum(IV) chloride; $[Pt(NH_3)_2Cl_4]$ - diamminetetrachloroplatinum(IV) chloride
22.23. (a) Hexamminechromium(III); (b) hexaaquacobalt(III); (c) pentamminechloroiron(III)
22.25. (a) Tetrabromocolbaltate(II); (b) aquatrihydroxozincate(II); (c) pentacyanonickelate(II)
22.27. (a) Ethylenediaminezinc(II) sulfate; (b) pentaammineaquanickel(II) chloride; (c) potassium hexacyanoferrate(II)
22.29. A sequestering agent is a multidentate ligand that separates metal ions from other substances so that they can no longer react. Properties that make a sequestering agent effective include strong bonds formed between the metal and the ligand and large formation constants.
22.31. As pH increases, the chelating ability increases because OH^- removes the H on the carboxylic acid groups, providing an additional site for binding to the metal cation.
22.33. The energy difference between d-orbitals lies within the visible range of the spectrum. For most transition metals, transitions occur between d-orbitals, and thus these complexes appear colored.
22.35. The repulsions due to the ligands in a square-planar crystal field are highest for the d_{xy} orbital, and its energy is higher because this orbital lies in the plane of the ligands.
22.37. The yellow solution contains (b) $Cr(NH_3)_6^{3+}$, and the violet solution contains (a) $Cr(H_2O)_6^{3+}$.
22.39. Colorless
22.41. $NiCl_4^{2-}$
22.43. Ligand field strength determines whether a complex is high spin or low spin. Stronger ligand fields induce low-spin arrangements.
22.45. Fe^{2+} has four unpaired electrons; Cu^{2+} has one unpaired electron; Co^{2+} has three unpaired electrons; Mn^{3+} has four unpaired electrons.
22.47. Co^{2+} and Cr^{3+}
22.49. (a) Mn^{4+} in MnO_2, two Mn^{3+} and one Mn^{2+} in Mn_3O_4; (b) both low-spin and high-spin configurations are possible in Mn_3O_4 (d^4 and d^5) but not in MnO_2 (d^3).
22.51. Paramagnetic
22.53. For an octahedral geometry, *cis*- means that two ligands are side by side and have a 90° bond angle between them. Ligands that are *trans*- to each other have a 180° bond angle between them.
22.55. At least two different ligands

22.57. Yes

22.59.

$$\begin{bmatrix} Br & Cl \\ & Cu \\ Br & Cl \end{bmatrix}^{2-} \quad \begin{bmatrix} Cl & Br \\ & Cu \\ Br & Cl \end{bmatrix}^{2-}$$

       Cis   Trans

Neither isomer is chiral.

22.61. (a) Enzymes catalyze biochemical reactions; (b) no

22.63. Lowers the activation energy

22.65. Much greater

22.67. Positive

22.69. 1.59×10^8 times

22.71. Zero order

22.73. Toxic metals that are sensitive imaging agents may be used at extremely low concentrations, such that we could essentially ignore any toxicity. The complex ion could also exhibit different toxicity from that of the metal ion alone. The metal ion might also be taken up preferentially by, for example, cancer cells, thus targeting any damage to unhealthy cells.

22.75. γ radiation

22.77. By binding to the nitrogen atoms in DNA to stop the division of cells

22.79. Paramagnetic

22.81. 130

22.83. $[\text{penicillamine}]_{eq} = 4.43 \times 10^{-4}\ M$ and $[\text{cysteine}]_{eq} = 5.6 \times 10^{-4}\ M$

22.85. (a) The ammonia and azide ligands are each *cis* to one another in one molecule and *trans* to one another in the other molecule; (b) *cis*-diamminediazidodihydroxyplatinum(IV) and *trans*-diamminediazidodihydroxyplatinum(IV); (c) yes; (d)

22.87. (a) Yes; (b) diamagnetic

22.89. The yellow complex containing Co^{3+}

22.91. As the sample is cooled, the Δ_o must increase, and there is a transition from a high-spin to low-spin state.

22.93. CN^- is a strong-field ligand with a very large Δ_o, which leaves d^6 Fe^{2+} diamagnetic and with no unpaired electrons. SCN^-, however, must be a weak-field ligand, which leaves d^6 Fe^{2+} paramagnetic and with four unpaired electrons.

22.95. Five

Credits

Chapter 1

pp. 2–3: NASA; p. 5: (left) (a) Photographer's Choice/Punchstock; (middle, left) Shutterstock; (middle, right) Getty Images; (right) Bloomberg/Contributor/Getty; p. 6: (a) Sami Sarkis/Getty Images; (b) Charles D. Winter/Getty Images; p. 7: (top) (a) Anton Vakhlachev/Shutterstock; (bottom) (a) Science Photo Library—Zephyr; (b) IBM; p. 9: Lester V. Bergman/Getty Images; p. 10: A. Benoist/Science Source; p. 11: (a) Nicholas Rjabow/Thinkstock; (b) Courtesy of IQ Air; (c) Huntstock/Getty Images; (bottom) Andrea Pistolesi/ZUMA Press/Newscom; p. 14: Jim Brandenburg/Getty Images; p. 17: W. Langdon Kihin/National Geographic Creative; p. 18: Edwin Hubble, © 2008 United States Postal Service. All Rights Reserved. Used with Permission; p. 21: World History Archive/Ann Ronan Collection/Age footstock; p. 22: Cappi Thompson/Getty Images; p. 23: (both) Courtesy of A&D Weighing, San Jose, CA, www.andweighing.com; p. 24: © 2009 Richard Megna/Fundamental Photographs; p. 26: aeropix/Alamy; p. 34: (top) Bettmann/Contributor; (bottom) Science Photo Library/Alamy; p. 35: (top) AP Images; (a and b) Courtesy of NASA/WMAP Science Team; p. 36 NASA/JPL-Caltech/MSSS; p. 37: A. Benoist/Science Source; Jim Brandenburg/Getty Images; p. 39: Shutterstock; Charles D Winters/Getty Images; Charles D Winters/Getty Images.

Chapter 2

pp. 44–45: Roberto Machado Noa/LightRocket via Getty Images; p. 46: (left) Greater St. Louis Committee for Nuclear Information; (right) Greater St. Louis Committee for Nuclear Information; p. 47: The Royal Institution, London/Bridgeman Art Library; p. 49: Astrid & Hanns-Frieder Michler/Science Source; p. 50: Photo by Universal History Archive/Getty Images; p. 51: (a) Grasseto/iStockphoto/Getty Images; p. 57: (left) (a) Igor Stevanovic/Shutterstock; (middle) (a) GIPhotoStock/Science Source; (right) (a) Shutterstock; (left) (b) Foto-Ruhrgebiet/Shutterstock; (middle) (b) Ian Miles-Flashpoint Pictures/Alamy Stock Photo; (right) (b) SPL/Science Source; (left) (c) GI PhotoStock/Science Source; (middle) (c) Phil Degginger/Science Source; (right) (c) Clive Streeter/Dorling Kindersley/Science Museum, London/Science Source; p. 61: (b) Dirk Wiersma/Science Photo Library/Science Source; p. 72: Courtesy of NASA/HST/J. Morse/K. Davidson; p. 73: (top) Courtesy of NASA/CXC/SAO; (bottom) ISM/Phototake; p. 74: Courtesy of NASA/CXC/SAO.

Chapter 3

pp. 82–83: Mark Thiessen/National Geographic Stock; p. 84: Getty Images; p. 87: AFP/Getty Images; p. 88: Richard Megna/Fundamental Photographs; p. 89: icefront/iStockphoto; (a) Zach Holmes/Alamy; p. 90: (top) Rona Tuccillo; (bottom) Vladimir Mucibabic/Dreamstime; p. 91: Francois Lenoir/Reuters via ZUMA Press; p. 92: Kei Shooting/Shutterstock; p. 96: Roger Ressmeyer/Corbis/VCG/Getty; p. 102: vovan13/iStockphoto; p. 107: Davids1993 | Dreamstime.com; p. 108: Gilbert S Grant/Photo Researchers RM/Getty Images; p. 109: (top) vupulepe/Shutterstock; (bottom) michaeljung/Shutterstock; p. 113: (a) Biophoto Associates/Science Source; (b) Krzysztof Stanek/

Dreamstime.com; (c) FocusTechnology/Alamy; p. 115: (a) RoJo Images/Shutterstock; (b) Thinkstock; p. 119: Edyta Pawlowska/Shutterstock; p. 120: (top) Shutterstock; (a) Shutterstock; (b) Bochkarev Photography/Shutterstock; p. 121: The Picture Pantry/Alamy; p. 128: Robin Bush/Oxford Scientific/Getty Images; p. 129: Richard Megna/Fundamental Photographs; p. 133: (Figure P3.8) From *Interpretation of Mass Spectra, 3rd Edition* by Fred W. McLafferty and Frantisek Turecek, © 1980. Reproduced with permission from University Science Books, all rights reserved; (c) Shutterstock; (d) Cherkas/Shutterstock; (e) iStockphoto; (g) Shane Maritch/Shutterstock; p. 134: Courtesy of Purest Colloids, Inc.; p. 137: aneese/Thinkstock; p. 140: REX/Shutterstock.

Chapter 4

pp. 142–143: Randy Wells/Corbis; p. 143: Richard Megna/Fundamental Photographs; p. 144: (a) NASA; (b) Time Life Pictures/Getty Images; (c) Steve Schmeissner/Science Source; p. 145: (a) Courtesy of NASA/JPL-Caltech/MSSS and PSI; (b) Courtesy of NASA/JPL-Caltech/MSSS and PSI; p. 146: (left) Richard Megna/Fundamental Photographs; (right) E. R. Degginger/www.color-pic.com; p. 147: Colin Anderson/Getty Images/Brand X; p. 151: Leigh Smith Images/Alamy; p. 153: (a–d) Richard Megna/Fundamental Photographs, NYC; (bottom) Ton Koene/Agefotostock/Design Pics; p. 157: Courtesy of The University of North Carolina at Pembroke; p. 158: (a–e) 2009 Richard Megna, Fundamental Photographs; p. 162: (top) Richard Thom/Visuals Unlimited; (bottom, left) TopFoto/The Image Works; (bottom, right) TopFoto/The Image Works; p. 167: (a) Richard Megna/Fundamental Photographs; (b) Richard Megna/Fundamental Photographs; p. 168: Dirk Ercken/Shutterstock; p. 170: 1994 Richard Megna, Fundamental Photographs; p. 171: (a) 2009 Richard Megna, Fundamental Photographs; (b) Richard Megna/Fundamental Photographs; p. 177: Richard Megna, Fundamental Photographs, NYC; p. 178: (all) Fundamental Photographs; p. 180: (a) Joel Arem/Science Source; (b) W. W. Norton; p. 181: (a) Getty Images; (b) Courtesy NASA; p. 184: (a) Peticolas/Megna/Fundamental Photographs; (b) Peticolas/Megna/Fundamental Photographs; p. 187: (all) Richard Megna/Fundamental Photographs; p. 190: Bill Ross/Getty Images; (a and b) From Wetlands Field Manual/Courtesy USDA; p. 191: (all) Richard Megna, Fundamental Photographs; p. 195: (bottom, left) Richard Megna/Fundamental Photographs; (top, right) Richard Megna/Fundamental Photographs; (bottom, right) Peticolas/Megna/Fundamental Photographs; p. 198: (a) Charles D Winters/Getty Images/Photo Researchers; (e) Photo Researchers; (i) Andrew Lambert Photography/Science Source; p. 201: (left) David R. Frazier/Photo Library, Inc./Alamy; p. 205: Courtesy Richard Sugarek/Environmental Protection Agency.

Chapter 5

pp. 208–209: franckreporter/iStockphoto; p. 208: (bottom) Richard Megna/Fundamental Photographs; p. 213: (left) Yutaka/AFLO/Newscom; (right) AP Photo/Matthias Schrader; p. 214: Kristen Brochmann/Fundamental Photographs; p. 217: (top) Marmaduke St. John/Alamy; (bottom) Ben Cooper/Getty Images; p. 224: Newscom; p. 227: Richard Megna/Fundamental Photographs; p. 228: (top) Sprokop/Dreamstime.

com; (bottom) Johner Images/Getty Images; p. 232: Rtimages/Dreamstime.com; p. 255: Dr. Keith Wheeler/Science Source; AP Photo; p. 257: (a) SKrow/Istockphoto; (b) nito/Shutterstock; (c) jfmdesign/iStockPhoto; p. 263: (i) Thinkstock/Getty Images; (g) Charles D. Winters/Science Source; (f) Samo Trebizan/Shutterstock; (b) Shutterstock; p. 265: (all) Courtesy of Grabber Inc.

Chapter 6

pp. 272–273: Khoroshunova Olga/Shutterstock; p. 273: (bottom) Doug Martin/Science Source; p. 276: (a) Bill Ross/Getty Images; (b) Christian Kohler/Shutterstock; Ignacio Salaverria/iStockphoto; p. 277: Courtesy of NOAA; p. 279: Sam Ogden/Science Source; p. 280: David R. Frazier Photography, Inc./Alamy; p. 289: Richard Megna, Fundamental Photographs; p. 293: Thinkstock/Getty Images; p. 294: (bottom) Romilly Lockyer/Getty Images; p. 295: (top) Jim Zuckerman/Alamy Stock Photo; (bottom) David Samuel Robbins/Getty Images; p. 296: Joel Gordon; p. 298: (all) Richard Megna, Fundamental Photographs; p. 302: (top) Kevin Arnold/Getty Images; (bottom) Richard Megna/Fundamental Photographs; p. 315: Courtesy of NOAA; p. 316: (middle) Richard Megna/Fundamental Photographs; (top) Richard Megna/Fundamental Photographs; (bottom) Doug Martin/Science Source; p. 321: Kevin Arnold/Getty Images; p. 322: Jonathan Blair/Getty Images; p. 325: F. Jack Jackson/Alamy; p. 326: Barry Bishop/National Geographic Creative; p. 329: Stocktrek/Getty Images.

Chapter 7

pp. 330–331: Thegoodly/Getty Images; p. 332: North Wind Picture Archives/Alamy Stock Photo; p. 333: (a, b, c) Richard Megna/Fundamental Photographs; p. 334: (a) James Cavallini/Science Source; (b) Courtesy of Blacklight.com; (c) Construction Photography/Alamy; (d) vaeenma/iStockphoto; (e) Alex Genovese/Alamy Stock Photo; p. 337: (all) Richard Megna/Fundamental Photographs; (bottom) Science Source; p. 338: James Leynse/Contributor/Getty Images; p. 354: Margrethe Bohr Collection/American Institute of Physics/Science Photo Library/Science Source; p. 355: Jeff J. Daly/Alamy; p. 362: (a) David Taylor/Science Source; (b) Dorling Kindersley/Getty Images; p. 365: Paul Silverman/Fundamental Photographs; p. 376: Shutterstock; p. 377: Richard Megna/Fundamental Photographs; p. 383: (top) Richard Megna, Fundamental Photographs; (b) Richard Megna/Fundamental Photographs; p. 384: Richard Megna/Fundamental Photographs.

Chapter 8

pp. 386–387: Albert Lleal/Minden Pictures; p. 390: (a, b, c) Charles D. Winters/Science Source; p. 402: W. Perry Conway/Getty Images; p. 411: Steve Nichols/Alamy; p. 413: Bob Rowan/Getty Images; p. 430: Education Images/UIG via Getty Images; p. 432: (left) Charles D Winters/Getty Images; (middle) Scott Eells/Bloomberg via Getty Images; p. 434: (left) sbretz/Shutterstock; (right) NASA/Damian Peach, Amateur Astonomer; p. 438: (right) GIPhotoStock/Getty Images.

Chapter 9

pp. 436–437: Shutterstock; p. 441: (all) Richard Megna/Fundamental Photographs; p. 466: (right) Portland Press Herald/Getty Images; p. 470: Shutterstock; p. 477: Yoav Levy/Phototake; p. 485: Richard Megna/Fundamental Photographs; p. 488: (a) John T. Fowler/Alamy Stock Photo; (b) Shutterstock; p. 493: Smithsonian Institution, Washington, D.C./Bridgeman Art Library; p. 494: Johnny Habell/Shutterstock.

Chapter 10

pp. 496–497: Sean Davey/Aurora Photos; p. 512: CC Holdings, Inc.; p. 525: (a) Tom Pantages; (b) Adam Hart-Davis/Fundamental Photographs; (bottom) Larry Stepanowicz/Fundamental Photographs; p. 526: iStockphoto; p. 528: (a) Jeff Daly/Visuals Unlimited; (b) Martin Shields/Science Source; (bottom) Martin Shields/Science Source; p. 533: (left) Shutterstock; (right) Mark Bolton/Getty.

Chapter 11

pp. 536–537: Susumu Nishinaga/Getty Images/Science Photo Library; p. 544: (a) Andrew Lambert Photography/Science Source; (b) George Resch/Fundamental Photographs; (c) Paul Whitehill/Science Photo Library/Science Source; p. 569: (all) Dr. David M. Phillips/Visuals Unlimited; p. 572: Richard T. Nowitz/Getty Images; p. 575: (a) AP Photo; (b) Courtesy of RODI Systems Corp.; (c) John Kasawa/Dreamstime.com; p. 580: Dr. David M. Phillips/Visuals Unlimited.

Chapter 12

pp. 588–589: Courtesy of Takao Someya, University of Tokyo; p. 590: (a) Goruppa/Dreamstime.com; (b) Raja Rc/Dreamstime.com; (c) Purest Colloids, Inc.; p. 593: (a) Dorling Kindersley/Getty Images; (b) imagebroker/Alamy; p. 594: Paul Silverman/Fundamental Photographs; p. 595; Zimmer Biomet; p. 599: (a) Andrew Taylor/Alamy Stock Photo; (b) AlonsoAguilar/Getty Images/iStockphoto; Kris Mercer/Alamy; p. 600: (a, b) Philippe Plailly/Science Source; p. 603: (a) Richard Megna/Fundamental Photographs; p. 604: Andrew Silver/U.S. Geological Survey; p. 606: (a) Jon Stokes/Science Source; (b) Shutterstock; (c) Andre Geim & Kostya Novoselov/Science Source; (d) David McCarthy/Science Source; p. 608: Dieter Klein/Getty Images; p. 609: Millard H. Sharp/Sicence Source; (top) Bsip/Photoshot/ZUMA Press/Newscom; (bottom) Jeffrey Hamilton/Getty Images; p. 611: (left) Emily Spence/Lexington Herald-Leader/MCT/Newscom; (middle, left) Fotosearch; (middle, right) studiomode/Alamy; (right) Joel Koyama/Minneapolis Star Tribune/ZUMAPRESS.com; p. 612: Photo by Luis Carlos Rubino; (right) Stephen Stickler/Getty Images; p. 613: Richard Megna/Fundamental Photographs; p. 615: (bottom, right) Joel Koyama/Minneapolis Star Tribune/ZUMAPRESS.com/Alamy; p. 616: Barry Slaven/PMODE Photography; p. 617: Creatas/Jupiterimages; p. 620: Courtesy DuPont; p. 623: Photo12/The Image Works; p. 625: Getty Images/All Canada Photos; p. 628: Corbis RF; p. 631: (Figure P12.92) From *Phase Diagrams of the Elements* by David A. Young, p. 94, Fig. 7.7. Copyright © 1991 by the Regents of the University of California. Reprinted by permission of the University of California Press; Greenshoots Communications/Alamy.

Chapter 13

pp. 634–635: Robert Landau/Getty Images; p. 634: Richard Megna/Fundamental Photographs; p. 636: Mirrorpix/Newscom; p. 638: (a) © Kateleigh/Dreamstime.com; (b) Richard Megna/Fundamental Photographs; (c) Charles D Winters/Getty Images; p. 676: Clive Streeter/Getty Images; p. 688: Lawrence Migdale/Science Source; p. 692: Andrew Lambert Photography/Science Photo Library.

Chapter 14

pp. 694–695: Federico Rostagno/iStockphoto/Getty Images; p. 703: Universal History Archive/UIG via Getty Images; p. 713: (a) Stan Pritchard/Alamy; (b) Robert Jones/Alamy; (bottom) (a) Thermo Scientific Model

48000/Courtesy of Thermo Fisher Scientific; p. 715: Chip Clark/Fundamental Photographs; p. 717: (a, b) Richard Megna/Fundamental Photographs; p. 720: (a, b) Richard Megna/Fundamental Photographs; p. 731: (b, d, f, h) Richard Megna/Fundamental Photographs.

Chapter 15

pp. 738–739: Shutterstock; p. 759: Shutterstock; p. 760: Shutterstock; p. 772: Shutterstock.

Chapter 16

pp. 784–785: Shutterstock; p. 799: (all) Larry Stepanowicz/Visuals Unlimited; (bottom) Richard Megna/Fundamental Photographs; p. 808: Shutterstock; p. 812: Richard Megna/Fundamental Photographs; p. 826: Richard Megna/Fundamental Photographs; p. 827: (a, d, i) Martyn F. Chillmaid/Science Photo Library; (d) Charles D. Winters/SciSource.

Chapter 17

pp. 832–833: iStockphoto; p. 835: (top) Phil Degginger/Alamy; p. 836: (bottom) Richard Megna/Fundamental Photographs; p. 840: (b, c) Alexander/Science Source; p. 862: (a) Reinhard, H./picture alliance/Arco Images G/Newscom; (c) SCIMAT/Science Photo Library/Science Source; p. 868: Kenneth Garrett/National Geographic/Getty Images; p. 872: (b) Shutterstock; (d) Bon Appetit/Alamy; (f) novzzvon/iStockphoto/Getty Images; (h) Trish Gant/Getty Images.

Chapter 18

pp. 878–879: Raymond Boyd/Getty Images; p. 878: Jeff Greenberg/UIG via Getty Images; p. 881: E. R. Degginger/www.color-pic.com; p. 883: Mary Evans Picture Library/Alamy; p. 899: Tom Gilbert; p. 900: Tomohiro Ohsumi/Getty Images; p. 901: philipus/Alamy; p. 904: (a) iStockphoto; p. 906: (a) Richard Goldberg | Dreamstime.com; p. 908: Reuters/Alamy; p. 910: Yoshio Tsunoda/AFLO/Alamy; p. 913: E. R. Degginger/www.color-pic.com; p. 915: (top, right) Alix/Science Source; (middle, right) Science Photos/Alamy; (bottom, right) Shutterstock; p. 916: (c) Getty Images/Thinkstock; (e) Peticolas/Megna/Fundamental Photographs; (g, i) Richard Megna/Fundamental Photographs; p. 920: Franz Marc Frei; p. 921: Glenda Powers/Dreamstime.com.

Chapter 19

pp. 922–923: Centre Jean Perrin/ISM/Phototake; p. 932: (top, right) Photodisc/Alamy; p. 937: Tom Bean; p. 940: Igor Kostin/Sygma via Getty Images; (bottom, both) © 2006 T. A. Mousseau and A. P. Moller;

p. 941: (top) DOE Photo; p. 944: (top) Dr. Robert Friedland/Science Source; p. 945: (c) Shutterstock; (a) Astrid & Hanns-Frieder Michler/Science Photo Library/Science Source; (b) Tom Tracey Photography/Alamy; (right) Argonne National Laboratory, US DOE Office of Environmental Management; p. 951: (right) The American Weekly, 1926; (left) Richard Megna/Fundamental Photographs; p. 952: (top, right) Tom Bean; p. 956: (top) Mireille Vautier/Alamy; (middle) blickwinkel/Alamy; (bottom) Reprinted by permission from Macmillan Publishers Ltd.: Human Presence in the European Arctic Nearly 40,000 Years Ago, *Nature* 413 (September 6, 2001): 64‰ 67, Figure 4‰ mammoth tusk showing human markings, © 2001.

Chapter 20

pp. 960–961: Shutterstock; p. 974: Valueline/Punchstock; p. 983: (a) Dennis Kunkel Microscopy, Inc./Visuals Unlimited, Inc.; (b) Omikron/Science Source; p. 984: © Kokodrill/Dreamstime.com; p. 985: (top) Courtesy of N.I.S.T.; (bottom) (a) Chemical Design/Science Photo Library/Science Source; p. 1001: Roger Ressmeyer/Getty Images; p. 1002: (top) NASA; (bottom) © W.R. Normak, courtesy USGS.

Chapter 21

pp. 1016–1017: Getty Images; p. 1020: (both) E. R. Degginger/www.color-pic.com; p. 1023: (b) Don W. Fawcett/Science Source; p. 1035: Dreamstime; p. 1036: Tom Brakefield/Stockbyte/Getty Images; p. 1039: E. R. Degginger/www.color-pic.com; p. 1044: E. R. Degginger/www.color-pic.com; p. 1045: SIU/Visuals Unlimited, Inc.

Chapter 22

pp. 1052–1053: Getty Images/iStockphoto; p. 1055: (a) Photo by Albris, May 2011, http://creativecommons.org/licenses/by-sa/3.0/deed.en; p. 1057: Image copyright © The Metropolitan Museum of Art. Image source: Art Resource, NY; p. 1063: (a, b) Richard Megna, Fundamental Photographs; p. 1066: Richard Megna/Fundamental Photographs, NYC; p. 1069: (middle) Richard Megna/Fundamental Photographs, NYC; p. 1070: Richard Megna/Fundamental Photographs; p. 1071: (d) Vaughan Fleming/Science Source; p. 1074: (a, b) E. R. Degginger/www.color-pic.com; p. 1079: Phantatomix/Science Source; p. 1083: (c) Zephry/Science Photo Library/Science Source; p. 1084: (a, b) From Small molecular gadolinium (III) complexes as MRI contrast agents for diagnostic imaging, Chan Kannie Wai-Yan and Wong wing-Tak, Coordination Chemistry Reviews, Sept., 2007, Elsevier B.V., Copyright Clearance Center; p. 1091: (right) (e) Richard Megna/Fundamental Photographs; (left) (e) Richard Megna/Fundamental Photographs.

Index

Note: Material in figures or tables is indicated by *italic* page numbers. Footnotes are indicated by n after the page number.

A

absolute entropy, 841–43, 844–46
  microstates and, 867
absolute temperature
  average kinetic energy of gas molecules and, 304, 307
  definition of, 283
  Kelvin scale of, 32–34
  volume of gas and, 283–85
absolute zero (0 K), 33
  determined from Charles's law, 285
  zero entropy of crystal at, 841–42, 867
absorbance (*A*), 156–57
absorbed dose, 939
acceptor band, 483
accuracy, 27, 28–30
acetaldehyde, in photochemical smog, 637
acetate, at hydrothermal vents, 1002
acetic acid
  converted to methane by bacteria, 243
  dimer of, *505*
  endothermic reaction with sodium bicarbonate, 835, 837
  reaction with ammonia, *618*
  vinegar as solution of, 167–68, 578
  as weak acid, 163
  as weak electrolyte, 159
acetone
  boiling point of, 504
  dipole moment of, 503
  hydrogen bonding with water, 507
acetyl-coenzyme A, 992, 997, 1081
acetylene (ethyne), 68
  hybrid orbitals in, 458–59
  Lewis structure of, 396
  mass spectrum of, 117
  torch using, 226
acid rain
  chemical weathering caused by, 162, 176–77
  on early Earth, 87
  nonmetal oxides producing acid in, 162–63
  sulfur trioxide and, 87, *162*, 414, 1033
  sulfur-containing fuels and, 99, 764
acid–base equilibria
  autoionization of water and, 748–49, 762–63, 767
  in blood, 740–41, 774–75

conjugate acid–base pairs and, 746–48, 772, 789–91, *790*
  See also K_a; K_b; pH
acid–base reactions, 146, 159–65
  according to Lewis model, 810
  antacid tablets and, 147, 168–69
  precipitate formed in, 170
  See also acid–base titrations
acid–base titrations, 166–69, 799–809
  of amino acids, 976–78
  with multiple equivalence points, 805–9
  of strong acid with strong base, 166–67, 800–801
  of weak acid with strong base, 167–68, 800, 801–4
  of weak or strong base with strong acid, 168–69, 804–5
acidic salts, 770–72, 773–74
acids
  Arrhenius concept of, 740, 741
  Brønsted–Lowry concept of, 160, 746, 809
  carboxylic, *69*, 163–64
  formulas of, *65*, 66–67
  hydrated metal ions as, 814–16, 820
  Lewis concept of, 809–11, *812*, 1055
  naming of, *65*, 66–67
  oxoacids, *65*, 66–67, 768–69, 814
  polyprotic, 163, 764–68
  See also amino acids; strong acids; weak acids
acrolein, 467
actinides, *58*, 364, 369
activated complex, 661
  elementary steps and, 666, 667
activation energy (E_a), 659–65
  calculation of, 662–64
  of catalyzed reaction, 721
  of enzyme-catalyzed reaction, 677
active site of enzyme, 986–87
activity series, 188–90
actual yield, 126–29
addition polymers, 609
  common examples of, *620*
addition reactions, 609
adenine, 997, *999*, 1000, 1080
adenosine triphosphate (ATP), 863–65
  calcium ions and, 1026
  conversion between ADP and, 1032–33
  cytochromes and, 1079
  magnesium ions and, 1026
  Na^+–K^+ pumps and, 1024, 1026

adenosine-5′-phosphosulfate, 1033
adhesive forces, 524–25
adipic acid, *618*
adsorption, by heterogeneous catalyst, 674–75
air, 274–75
  composition of, *274*
  density of, 297
air bags, automobile, 293, 294
airplanes
  oxygen generators on, 293–94
  oxygen partial pressure outside of, 301–2
alanine
  as chiral compound, 970
  titration curve for, 976
  as zwitterion at physiological pH, 976
alcohols, 69
  in ester formation, 615
  hydrogen bonding in, 509, 513, *514*
  in polyester formation, 615–17
  polymers of, 612–13
  water solubility of, 513, *514*
aldehydes, 69
  Lewis structures for, 395
  See also acetaldehyde; formaldehyde
alkali metals (group 1), 58
  bcc unit cells of, 595
alkaline batteries, 889
alkaline earth metals (group 2), 58
alkalinity titration, 805–9
alkanes, 67–68, 252–55
  isomers of, 965–67
  nomenclature of, 252–53
  prefixes for naming, *253*
  straight-chain, boiling points of, *253*, 500–501
  straight-chain, melting points of, *253*
  uses of, *253*, 254–55
  valence bond theory of, 459
alkenes, 68
  isomers of, 967–69
  polymers of, 609–11
  valence bond theory of, 459
alkynes, 68
  valence bond theory of, 459
allotropes, 403
  of carbon, 606–7
alloys, 592, 599–603
  definition of, 600
  interstitial, 601–2
  substitutional, 600–601, 602–3

alpha (α) decay, 930, 931
alpha (α) particles
    biological effects of, 939–40, 941
    early experiments with, 49–51
    in Geiger counter, 932
    in nucleosynthesis, 947, 948
    from radium, 950
α helix, 983
alternating copolymer, 613
altitude
    atmospheric pressure and, 277, 515, 516
    partial pressure of oxygen and, 301–2, 515, 516
aluminum
    in AlGaAs₂ semiconductors, 483
    alloys with magnesium, 906
    in human body, 1043
    production from ore, 259–60
    recycling of, 259–60
    surface layer of oxide on, 906
    toxicity of, 1039, 1043
aluminum chloride
    bonding in, 417, 419
    as Lewis acid, 811
    as molecular compound, 811
aluminum hydroxide, 815, 820
aluminum nitrate, 820
Alzheimer's disease, 944, 984, 1043
amides, 69, 618
    polyamides, 618–20
amine (N-) terminus, 979
amines, 69
    amides from carboxylic acids and, 618
    polyamides from acids and, 618–20
    as weak bases, 164
amino acids, 974–76
    acid–base properties of, 976–79
    biosynthesis of, 1029–30
    chiral center of, 970
    D- and L- enantiomers of, 976, 1002
    essential, 974–76
    genetic code for, 1000–1001
    incorporation of sulfur into, 1033
    in meteorites, 84, 1002
    origin of life and, 84, 96, 1001
    selenocysteine, 1038
    in space, 1002
    structures and abbreviations of, 975
aminocarboxylic acids, 1065–66
ammonia
    in amino acid biosynthesis, 1029–30
    from animal waste, 180
    in fertilizer production, 694, 696
    geometry of, 446
    hydrogen bonds in, 504–5, 512
    K_b value of, 756
    as Lewis base, 809, 810, 1063
    Lewis structure of, 394
    as ligand in complex ions, 812–14, 1057,
        1058, 1059, 1061–62, 1073–75
    in nitric acid synthesis, 728
    in nitrogen cycle, 1030, 1080
    pH of solution of, 761–62

    reaction with carboxylic acid, 618
    solubility in water, 512
    sp^3 hybrid orbitals in, 455–56
    from urea in plants, 1032, 1080
    as weak base, 164, 745, 756, 761–62
ammonia synthesis
    catalyst in, 721
    equilibrium constant for, 860–61
    free energy change in, 848–49
    relative rates in, 640–41
    temperature change and, 720
ammonium acetate, as neutral salt, 771
ammonium chloride, weakly acidic solutions of,
        770, 773–74
ammonium cyanate, 85, 965
ammonium nitrate, dissolution in water
    as endothermic process, 836
    entropy change in, 845–46
Amontons, Guillaume, 287
Amontons's law, 287–88
    kinetic molecular theory and, 304, 305
    summary of, 291
amorphous solids, 590, 591
ampere (A), 899
amphiprotic substance, 165, 749
amplitude of wave, 335
amu (atomic mass units), 51
amyloid β, 984
anabolism, 986
analyte, 799
angioplasty, 600
angular (bent) molecular geometry
    hybrid orbitals and, 462
    with one lone pair, 444–45, 447
    of sulfur dioxide, 444–45
    with two lone pairs, 446, 447
    of water, 446, 450
angular momentum quantum number (ℓ), 351
anions
    definition of, 9
    enthalpy of hydration for, 549
    oxoanions, 65–66, 768
    of p block elements, 366–67
anode
    of cell diagram, 884
    of electrochemical cell, 882
    sacrificial, 906
    of voltaic cell vs. electrolytic cell, 908–9
antacid tablets, 147, 168–69, 1028
    aluminum in, 1043
antibonding orbitals, 471
anti-fluorite structure, 604
antifreeze solutions, 538, 553, 558
antimatter, 928
antimony
    in human body, 1040
    in leishmaniasis treatment, 1044
    toxicity of, 1044
aqueous solutions, 145
    acid–base reactions in, 146, 147, 159–65,
        168–69, 170
    boiling point elevation of, 561–62

    on Earth's surface, 145
    electrolytes and nonelectrolytes, 158–59,
        564–68
    equilibrium constant expressions for reactions
        in, 714
    intermolecular interactions and, 498, 502–3
    in living things, 145–46, 498
    phase diagram for, 558, 559
    phase symbol for, 10
    precipitation reactions in, 169–78
    of salts, acid–base properties of, 770–74
    See also concentration; hydrated ions; solutions
aragonite, 823
argon, banned for athletes, 1042
arithmetic mean, 28
aromatic compounds, 407, 468
    polymers of, 612
Arrhenius, Svante, 660, 740
Arrhenius acids, 740, 741
Arrhenius bases, 740, 741
Arrhenius equation, 660–65
arrow
    bond polarity and, 398
    in chemical equation, 8
    of equilibrium, 159
    of reversible reaction, 697
aspartame, 979
aspartic acid, titration curve of, 977–78
astatine, 1021
Aston, Francis W., 52
atmosphere (1 atm), 275–76
    different pressure units and, 277
atmosphere of Earth
    composition of, 274
    thickness of, 275
    See also greenhouse gases
atmospheric pollutants. See acid rain; nitrogen
        dioxide; nitrogen monoxide; smog; sulfur
        dioxide; sulfur trioxide
atmospheric pressure (P_{atm}), 222
    altitude and, 277, 515, 516
    calculation of, 278
    definition of, 275
    normal boiling point and, 518
    oxygen in blood and, 516
    as sum of partial pressures, 299
atom, definition of, 7
atomic absorption spectra, 333
atomic emission spectra, 333
atomic mass, average, 54–55
atomic mass units (amu), 51
atomic number (Z), 52–54
atomic orbitals. See orbitals, atomic
atomic radii
    alloy formation and, 602–3
    periodic trends in, 369–72, 1021
    unit cells and, 596–99
atomic solids, 590, 591
atomic theory
    Dalton's theory, 51–52, 59
    Rutherford's model, 50–51, 343
    Thomson's plum-pudding model, 49–50

ATP sulfurylase, 1033
aufbau principle, 358, 362
   molecular orbitals and, 475
auroras, 470, 471, 480
austenite, 601–2
autoionization of water, 748–49
   in rainwater, 767
   in very dilute strong acid, 762–63
automobiles
   air bags of, 293, 294
   all-electric, 112–13, 901
   catalytic converters of, 637, 638, 676–77
   coolant of, 552, 553, 558
   fuel cells for, 909–11
   hybrid, 112–13, 878, 880, 900–901
   lead–acid batteries of, 896–97, 906–7
   nitrogen monoxide formed in, 636, 637,
      639–40, 642, 676–77, 1030
   useful work done by engines of, 851
average
   calculation of, 27–28
   weighted, 54
average atomic mass, 54–55
average speed of molecules, 306
Avogadro, Amedeo, 87, 286
Avogadro's law, 285–86, 289
   kinetic molecular theory and, 304–5
   summary of, *291*
Avogadro's number (N_A), 87

B

Baby Tooth Survey, 46
background electrolyte
   in lithium-ion battery, 902
   in zinc/copper voltaic cell, 882
Bacon, Francis, 17
baking soda. *See* sodium bicarbonate
balanced chemical equations, 87, 96–101
   for combustion reactions, 102–4
   for electrochemical cell, 886
   equilibrium constant and, 699
   reaction mechanism and, 667
   relative reaction rates and, 640–42
   in RICE table, 721
ball-and-stick molecular models, 9
balloon angioplasty, 600
balloons
   Avogadro's law and, 285–86
   Boyle's law and, *281*, 282–83
   Charles's law and, 284
   density of gas and, 296
   helium-filled, 224, 282–83, 286, *296*
   hot-air, 222–23, 284–85, 296
   mimicking electron-pair geometry, 441
   weather balloons, 280, 291–92
Balmer, Johann, 341–42, 343
band gap (E_g), 482–83
band theory, 481–84
barium, in human body, 1039
barium sulfate, solubility of, 817–18
barograph, *279*

barometer, 275, 277, 279
bases
   Arrhenius concept of, 740, 741
   Brønsted–Lowry concept of, 160, 746, 809
   Lewis concept of, 809–11, 812, 1055, 1062
   monobasic, 799
   See also acid–base equilibria; acid–base
      reactions; acid–base titrations; strong
      bases; weak bases
basic salts, 770–73
batteries
   alkaline, 889
   button batteries, 892–93
   capacity of, 899–903
   lead–acid, 896–97, 906–7
   lithium-ion, 112–13, 901–3, 906–7
   nickel–metal hydride, *899*, 900–901, 907–8
   work done by, 891
   zinc/copper voltaic cell, 881–83, 888, 890–92,
      897
BCNU (1,3-bis(2-chloroethyl)-1-nitrosourea), 527
becquerel (Bq), 932
Becquerel, Henri, 49, 932
Beer's law, 156–57
bell curve, 30
belt of stability, 926–27
bent molecule. *See* angular (bent) molecular
   geometry
benzene
   boiling-point-elevation constant for, *577*
   freezing point depression of, 575–76, *577*
   mass spectrum of, 117
   pi (π) bonds in, 467–68
   resonance in, 407, 467
   solubility properties of, 514
beryllium, electron configuration of, 360
beryllium chloride, 417
beryllium toxicity, 1020
Berzelius, Jöns Jacob, 85
beta (β) decay, 927–28
   atomic number and mass number in, *930*
   of carbon isotopes, *929*
   of free neutrons, 934–35
   of massive nuclides, 930, 931
   of phosphorus-32, 929–30
   in stellar nucleosynthesis, 73
beta (β) particles, 927–28
   biological effects of, 939
   early experiments with, 49
   in Geiger counter, 932
   in stellar nucleosynthesis, 73
β-pleated sheet, 983–84
bicarbonate
   in alkalinity titration, 805–9
   in blood, 740–41, 774–75
   carbonic anhydrase and, 1078
   ocean acidification and, 823–24
   in seawater, 787
   See also sodium bicarbonate
Big Bang, 17–18, 32, 34–36
   cosmic microwave background and, 34–35
   primordial nucleosynthesis and, 70–72

bimolecular elementary step, 666
binary ionic compounds, 60–62
   crystal structures of, 603–6
   formulas of, 63–64
   Lewis structures of, 397
   naming of, 63–64
binding energy (BE), 925–26
biocatalysis, 678
biochemistry, 962
   See also carbohydrates; fatty acids; human
     body; life; nucleic acids; proteins
biomass, 991
biomolecules, 962
biopolymers, 963
bismuth subsalicylate, 1044
block copolymer, 613
blood
   acid–base equilibria in, 740–41, 774–75
   as heterogeneous mixture, 6–7
   osmosis and, *569*, 570–72
   oxygen in, 516
   pH of, 774–75
   separation of, 9–10
   solutes in, 145–46
   solutes in serum component of, 148–49
blood type, 988
blue cheese, 466
bluing, 906
body-centered cubic (bcc) unit cell, 595, 597–98
   of ferrite, 602
   molten iron crystallizing in, 601
Bohr, Niels, 343, 350, *354*
Bohr model, 343–45
   de Broglie wavelength and, 347–48
boiling
   entropy change in, 843
   spontaneous, 835, 837
boiling point elevation, 558, 561–62, 563, 564,
     566–67
   constants for selected solvents, *577*
   molar mass determined from, 576–77
boiling points
   of binary hydrides, 504–5
   of ethers, 511
   of halogens, 500
   hydrogen bonding and, 507–10
   intermolecular forces and, 499–502
   molecular shape and, 501–2
   of noble gases, 499–500
   normal, 518
   periodic trends in, 1021–22
   on phase diagram, 520–21
   of straight-chain hydrocarbons, *253*,
     500–501
   temperature scales and, 32–33
Boltwood, Bertram, 935
Boltzmann, Ludwig, 866
Boltzmann constant, 867
Boltzmann's definition of entropy, 867
bomb calorimeter, 241–42
   food value and, 257
bond. *See* chemical bonds

bond angles
    in carbon dioxide, 439
    definition of, 439
    double bond and, 445
    hybrid orbitals and, *462*
    lone pairs and, 445, *447*
    in methane, 439
    VSEPR theory and, 444–49
bond dipole, 398, 450
    See also polar covalent bonds
bond energies, 420–22
    in hydrogen molecule, 390
    of selected covalent bonds, *419*, 420–21
bond length, 390
bond order, 419–20
    bond energy and, 421
    in molecular orbital theory, 473–74, 477–78,
        479–80
bond polarity, 398–402
    See also polar covalent bonds
bond strength. *See* bond energies
bonding capacity, 391–92
    formal charge and, 411
    resonance and, 404, 411
bonding orbitals, 471
bonding pairs, in Lewis structures, 392
bones, mammalian, 1027
    strontium and barium in, 1039
Born, Max, 350
Born–Haber cycle
    enthalpy of hydration and, 548
    lattice energy and, 545–48
boron
    electron configuration of, 360
    in human body, 1043
    paramagnetic diatomic molecule of, 477
boron neutron-capture therapy (BNCT), 1043
boron trifluoride
    bonding in, 416–17, 419
    as Lewis acid, 810
    trigonal planar geometry of, 442
Bosch, Carl, 721
boundary–surface representation, of *s* orbital,
    356
Boyle, Robert, 280–81
Boyle's law, 280–83, 288–89
    breathing and, *290*
    kinetic molecular theory and, 304
    summary of, *291*
branched-chain hydrocarbons, 254
brass, *599*, 600
breathing, and Boyle's law, *290*
breeder reactor, 945–46
bristlecone pines, 937
bromcresol green, 808
bromine
    in human body, 1040
    normal boiling point of, 518
bromochlorofluoromethane, 469–70
Brønsted–Lowry acids, 160
    of conjugate pair, 746

hydrated metal ions as, 814–16, 820
    Lewis acids and, 809
Brønsted–Lowry bases, 160
    of conjugate pair, 746, 748
    Lewis bases and, 809
    weak, 757
bronze, *599*, 600, 601
buckyballs, 607
buffer capacity, 794–98
    base:acid mole ratio and, 797–98
    change in pH and, 794–95
    effect of concentration on, 796–97
buffers. *See* pH buffers
Bunsen, Robert Wilhelm, 333
bupropion, 973
buret, 166–67, 799
butane, *254*
    standard molar entropy of, *844*
butanethiol, 1036
butanone, 465
button batteries, 892–93
butyric acid, 615

C

calcium
    electron configuration of, 362–63
    in living organisms, 1026–27
    strontium-90 substituting for, 46–47, 1039
calcium carbonate
    in antacid tablet, 1028–29
    in aqueous acetic acid, 579
    decomposition of, 713
    dissolved by carbonic acid, 162
    in skeletons of marine organisms, 787–88,
        823–24, 1027
calcium fluoride
    crystal structure of, 604
    lattice energy of, 546–47, 548
    pH effect on solubility of, 818–19
calcium ions, separating from magnesium ions,
    821–22
calcium oxide
    from decomposition of calcium carbonate, 713
    reaction with sulfur dioxide, 810–11
calcium sulfate, lattice energy of, 549–50
calibration curve, in spectrophotometry, 157
calorie (cal), 223
Calorie (Cal), 223
    food value and, 257–58
calorimeter, 235
    bomb, 241–42, 257
    coffee-cup, 239–40
calorimeter constant, 241–42
calorimetry, 235–42
    coffee-cup, 239–40
    constant-volume, 241–42
    enthalpy of reaction measured with, 239–40,
        242
camping lanterns, 621
cancer
    cell membrane potentials in, 1025

chromium contributing to, 1081
    cisplatin treatment for, 1073–74, 1087
    gallium compounds with activity against,
        1043–44
    imaging of, 1041
    ionizing radiation contributing to, 938, 940,
        941–42, 950–51, 1039
    partition coefficients of drugs for, 527
    polycyclic aromatic hydrocarbons and, 468
    radiation therapy for, 943, 1043
    in Radium Girls, 950–51
    transition metals in treatment of, 1073–74,
        1084, 1086, 1087
    See also medical imaging
capillary action, 525
caraway, 438, 468, 470
carbohydrates, 988–92
    See also glucose
carbon
    allotropes of, 606–7
    as charcoal, 293
    chiral compounds of, 469–70, 485
    diatomic molecule of, 477
    electron configuration of, 360–61
    isotopes of, *929*
    pi (π) bonds formed by, 457, 459
    radioactive isotopes of, *929*
    with *sp* hybrid orbitals, 458–60
    with *sp*^{2} hybrid orbitals, 456–57
    with *sp*^{3} hybrid orbitals, 455
    in steel, 601–2
    stellar nucleosynthesis of, 72
    See also organic compounds
carbon-11
    electron capture in, 929
    positron emission from, 928, 1041
carbon-14
    beta (β) decay of, 927–28, 936
    radiocarbon dating with, 936–38
carbon cycle, 104
carbon dioxide
    atmospheric increase in, 295, 388, 786–88
    in blood, 740–41, 774–75
    bond angle in, 439
    bond length in, 420
    chloride ions in human body and, 1028
    from combustion of fossil fuels, 104–7, 388,
        786–87
    from combustion of hydrocarbons, 102
    from decomposition of calcium carbonate,
        713
    density of, 295–96
    dissolved in rainwater, 162, 765–67
    dissolved in river water, 789–90
    of early atmosphere, 86
    enzyme-catalyzed hydrolysis of, 677
    for fighting fires, 296
    formal charge calculations for, 410
    as greenhouse gas, 388, 402
    linear geometry of, 389, 441
    ocean acidification and, 823–24
    phase diagram of, 522

carbon dioxide *(cont.)*
  polar bonds in, 450
  radiocarbon dating and, 936
  from reaction of baking soda and vinegar, 835, 837
  stability of, at 25°C, 703
  supercritical, 522
  valence bond representation of, 460
  from volcanoes, 295
  See also water–gas shift reaction
carbon monoxide
  atmospheric, 101
  bond length in, 419–20
  of early atmosphere, 86
  reaction with oxygen, 102
  See also water–gas shift reaction
carbon monoxide dehydrogenase, 1081
carbon nanotubes, 607
carbon tetrachloride
  boiling-point-elevation constant for, *577*
  freezing-point-depression constant for, *577*
  limited water solubility of, 512
  tetrahedral geometry of, 442
carbonate
  alkalinity titration and, 805–9
  ocean acidification and, 823–24
carbonic acid
  in alkalinity titration, 805–9
  in blood, 740–41, 774–75
  dissolving calcium carbonate, 162
  ocean acidification and, 786–87, 823–24
  pH of solution of, 765–67
  in rainwater, 765–67
carbonic anhydrase, 677, 774
  structure of, *985*
  zinc ion in, 1078–79
carbon-skeleton structure, 465–67
carbonyl group, 395
  polarization of, 401
  sp^2 hybrid orbitals for, 456–57
carboplatin, 1087
carboxylic acid (C-) terminus, 979
carboxylic acids, *69*, 163–64
  esterification of, 615
  polyamides from amines and, 618–20
  polyesters from alcohols and, 615–17
carotene, *1027*
cars. *See* automobiles
carvone, *469*
catabolism, 986
catalases, *638*
  iron-containing, 1079, 1080
catalysts, 638–39, 672–79
  bacterial, 1002
  biocatalysis in pharmaceutical industry, 678
  definition of, 672
  enzymes, *638*, 677–78, 774, *985*, 986–88, 1078–81
  equilibrium and, 721
  in fuel cells, 909, 910
  heterogeneous, 675, 678
  homogeneous, 674

in hydrogen gas production, 696–97
in Ostwald process, 728
ozone layer and, 672–75
RNA acting as, 1002
catalytic converters, 637, 638, 676–77
cathode
  of cell diagram, 884
  of electrochemical cell, 882
  of voltaic cell vs. electrolytic cell, 908–9
cathode rays, 47–48, 52
cathode-ray tubes (CRTs), 47–48
cathodic protection, 906
cations
  definition of, 9
  enthalpy of hydration for, *549*
  of *s* block elements, 366, 367
  of transition metals, 368–69
  See also complex ions
cell, electrochemical. *See* electrochemical cells
cell diagrams, 884–87
cell membrane
  membrane potential of, 1024–26
  osmosis and, 568, *569*, 570–71
  structure of, 997, *1023*
cell potential (E_{cell})
  effect of concentration on, 895–97
  See also standard cell potentials ($E°_{cell}$)
cellobiose, 991
cellulose, 988, 991
Celsius temperature scale, 32–34
cembrene A, 120
centrifugation, 10
cesium, in human body, 1039
Chadwick, James, 51
chain reaction, nuclear, 944, 945–46
chalcocite, 868
changes of state. *See* phase changes
Charles, Jacques, 283
Charles's law, 283–85, 288–89
  kinetic molecular theory and, 304, 305–6
  summary of, *291*
chelate effect, 1065
chelation, 1064–65
  medical applications of, 1082, 1084, 1085–86
chelation therapy, 1085–86
chemical bonds
  coordinate bonds, 812, 1055
  definition of, 7
  energies of, *419*, 420–22
  energy of reactions and, 217
  limitations of models of, 418–19
  in molecular models, 9
  stability of compounds and, 389
  in structural formulas, 9
  types of, 389–90
  vibration of, *221*, 401–2, 844
  See also bond angles; bond order; covalent bonds; ionic bonds; Lewis structures
chemical equations, 8
  phase symbols in, 8, 10
  See also balanced chemical equations

chemical equilibrium
  basic concept of, 696–98
  catalysts and, 721
  in complex ion formation, 812–14
  far to right or left, 703, 726–27
  free energy and, 847, 854–59
  heterogeneous, 713–14
  homogeneous, 713
  Le Châtelier's principle and, 714–21
  response to stresses, *721*
  in voltaic cell, 896, 898
  See also equilibrium constant (K); water–gas shift reaction
chemical formulas
  of common acids, *65*, 66–67
  of complex ions, 1056, 1057–62
  definition of, 8
  of ionic compounds, 63–65
  of molecular compounds, 8, 60, 62–63
  of polyatomic ions, *64*, 65–66
  of transition metal compounds, 64–65, 1056, 1057–62
chemical kinetics
  definition of, 638
  half-life in, 653–55, 658, 659
  See also reaction mechanisms; reaction order; reaction rates
chemical properties, 13–14
chemical reactions
  bond energies and, 420–22
  chemical equations for, 8
  definition of, 6
  energy profile for, 661–62, 667, 721
  gases in, 293–95, 302–3
  in living systems, 861–65
  See also acid–base reactions; combustion reactions; condensation reactions; formation reactions; oxidation–reduction reactions; precipitation reactions; stoichiometry
chemical weathering, 162, 176–77
chemistry, 5
Chernobyl nuclear disaster, 940, 1039
Chevrolet Volt, 878–79, 901
chiral centers, 970
  recognizing, 971–72
chirality
  of carbon in japonilure, 485
  in living systems, 972–73
  molecular recognition and, 468–70, 971, 988
  of monosaccharides, 988, 989
  in octahedral complex ions, 1075–76
  optical activity and, 969–71
  recognizing, 971–72
chloric acid, *769*
chloride ion
  in human body, 1028–29
  size of, 371–72
chlorine
  as catalyst for ozone destruction, 674
  oxoacids of, *769*

chlorine monoxide, decomposition of, 656–57, 663–64
chlorins, 1077
chlorofluorocarbons (CFCs), 673–74
chloroform
    dipole moment of, 451–52
    Lewis structure of, 393
chlorophylls, 1026, 1077
chlorous acid, *769*
cholesterol, 997
chromium
    complex ions of, 1066–68, 1071
    diabetes and, 1081, 1086
    electron configuration of, 364
    as essential dietary element, 1081
    in stainless steel, 601, 906
    toxic and carcinogenic oxidation state of, 1081
chromium hydroxide, 815
chromodulin, 1081
cis isomer, 968, 1073–75
cisplatin, 1073–74, 1087
citrate cycle, 992
citrate ion, medical applications of, *1082*
citric acid, 168, 767–68, 1066
Clausius, Rudolf, 865
Clausius–Clapeyron equation, 519–20
climate change
    carbon dioxide and, 295
    See also global warming
close packing, 593–94, *595*
    holes in, 601–2
closed system, 218
clusters, 607
coal, sulfur impurities in, 636, 713, 764, 810
COAST problem-solving framework, 11–12
cobalt(II) chloride, 720, 1056
cobalt(II) complex ions, 1056
cobalt(III)
    in coenzyme B_{12}, 1081
    in complex ions, 1055–56, 1074, 1075–76
cobalt(IV) oxide, in lithium-ion battery, 902
codons, 1000–1001
coenzyme B_{12}, 1081
coenzymes, 1081
coffee-cup calorimetry, 239–40
cohesive forces, 524–25
coinage metals, 1086, 1087–88
cold pack, chemical, 226–27, 835–36
colligative properties, 558–75
    boiling point elevation, 558, 561–62, 563, 564, 566–67, 576–77
    definition of, 558
    freezing point depression, 558, 562–68, 575–77
    molality units used for, 558–60
    molar mass of solute and, 575–78
    osmotic pressure, 568–73
    reverse osmotic pressure, 573–75
    van't Hoff factor and, 564–68, 570, 573
    vapor pressure, 550–51, 558, 563
color wheel, *1068*
colors
    of auroras, 470, 471, 480

complementary, 1068
    spectrophotometry and, 156–57
    of transition metal solutions, 1066–70
combination reaction, 87
combined gas law, 290–92
combustion analysis, 117–21
combustion reactions, 101–4
    of alkane fuels, 254
    carbon monoxide generated in, 243–44, 245–46
    of charcoal, 293
    of ethanol, 850–51
    as exothermic reactions, 219
    free energy change in, 850–51
    as oxidation–reduction reactions, 181
    as spontaneous reactions, 835
    See also hydrocarbon fuels; methane combustion
common-ion effect, 788–91
    solubility and, 817–18
complementary colors, 1068
complete proteins, 974
complex ions
    basic concepts of, 1054–57
    crystal field theory and, 1066–70
    formation of, 812–14
    formulas of, 1056, 1057–62
    hydrated, 812, 814–16, 820, 1055, 1062–63, 1066–67, 1072
    names of, 1058–60
    See also ligands
complexes, 1055
    See also complex ions
compounds
    chemical formulas of, 8, 60, 62–67
    definition of, 6
    naming, 62–70
    organic, 67–70, 1022
    See also coordination compounds; ionic compounds; molecular compounds
concentrated solution, 147
concentration
    definition of, 147
    determining by titration, 166–69
    determining with spectrophotometry, 156–57
    in equilibrium constant expressions, 699–702
    Le Châtelier's principle and, 715–17
    of major constituents of seawater and serum, 148–49
    of pure liquid, 714
    of pure solid, 713
    reaction rates and, 638, 645–59
    units of, 147–53, 558–60
condensation, 15, *219*
    in equilibrium with vaporization, 517–18
    free energy and, 852
    phase diagram and, 520
condensation polymers, 615–20
    common examples of, *620*
condensation reactions, definition of, 615
condensed electron configuration, 360
condensed structures, 465, 466

conduction band, 482, 483
confidence intervals, 28–30
conjugate acid–base pair, 746–48
    pH of solution containing, 789–91
    product of K_a and K_b for, 772, 790
    See also pH buffers
conservation of energy, 213, 222
conservation of mass, 95
constant composition, law of, 8
constant pressure, 225
constant temperature, 229
constant-volume calorimetry, 241–42
constitutional isomers, 965–68
    linkage isomers, 1076
control sample, 27–29
conversion factors, 20–22
coordinate bonds, 812, 1055–56
coordination compounds, 1056
    in biochemistry, 1076–81
    isomerism in, 1073–76
    in medicine, 1081–88
    names of, 1060–62
coordination number, 1056
    shapes corresponding to, *1057*
copolymer, 613
copper
    alloys of, *599*, 600, 601
    antibacterial properties of, 599, 600
    bonding in, 481
    corrosion of iron in contact with, 904–5, 912–13
    electron configuration of, 364
    in photosynthesis, 1078
    proteins containing, 1080
    redox reaction between zinc and, 880–83
    refining of, 868
    See also zinc/copper voltaic cell
copper ions (Cu^{2+})
    spectrophotometric measurement of, 157
    spin state of, 1072
copper(II) sulfate, complex formation with ammonia, 812–14
core electrons, 360
corrosion, 903–6
    as rust, 190, 832, 835, 904
    in seawater, 904, 905, 906, 912–13
    See also rust
cosmic microwave background radiation, 34–35
cotton, water absorption by, 617
coulomb (C), 48–49
Coulomb, Charles Augustin de, 215
coulombic interaction. *See* electrostatic potential energy (E_{el})
Coulomb's law, 215
counterions, 1056
coupled reactions
    in copper refining, 868
    in living systems, 861–65
covalent bonds, 60
    coordinate bonds as, 812, 1055–56
    electrostatic potential energy and, 389–90

covalent bonds *(cont.)*
    lengths of, 419–20
    polar, 398–402, 450–53
    in solids, *591*
    strengths of, *419*, 420–22
    See also Lewis structures
covalent network solids, 590, *591*
    allotropes of carbon, 606–7
covalent radius, *369*
critical mass, 944
critical point, 522
crude oil, 254
    distillation of, 553
    formation water accompanying, 562
    Raoult's law and, 556, 557
crystal field splitting energy (Δ), 1067
    factors affecting size of, 1072–73
    magnetic properties and, 1071–73
    spectrochemical series and, 1070
crystal field theory, 1066–70
crystal lattice, 592–93
    lattice energy, 543–50
    See also unit cells
crystal structure, definition of, 593
crystalline solids, 590, *591*
    zero entropy of, at 0 K, 841–42, 867
C-terminus, 979
cubic closest-packed (ccp) structure, 594
    of anions in anti-fluorite structure, 604
    of gold and platinum, 602
    malleability of metals and, 599
    packing efficiency of, 594, 599
    summary of, *595*
cubic packing, 595
cuproine, 157
curie (Ci), 932
Curie, Marie, 49, 924, 932, 938, 950, 951
Curie, Pierre, 49, 932, 950, 951
cycles per second, 335
cymene, 115–16
cysteine residues, disulfide linkage of, 985
cystic fibrosis, 1028
cytochromes, 1079–80
cytosine, 997, *999*, 1000

D

d block elements, 364, 365
d orbitals, 351, 358
    colors of transition metal complexes and,
        1068, 1070
    crystal field theory and, 1066–73
    filling of, 362–66
    magnetic properties and, 1071–73
D5W, 573
da Vinci, Leonardo, 180
Dacron, 617
dalton (Da), 51
Dalton, John, 51–52, 59
Dalton's law of partial pressures, 299
    kinetic molecular theory and, 304–5
    in laboratory, 302–3

de Broglie, Louis, 345–46
de Broglie wavelength, 346–48
debye (D), 451
deferoxamine (DFO), 1085–86
degenerate orbitals, 360–61
degree of ionization
    of weak acids, 753–56
    of weak bases, 756–58
delocalized electrons, 390
    with alternating single and double bonds, 467
    in anions of oxoacids, 768–69
    in aromatic compounds, 468
    in benzene, 467–68
    in molecular orbitals, 471
delta (Δ)
    as change, 221
    as crystal field splitting energy, 1067
delta (δ), as partial electrical charge, 398
denatured proteins, 984
density
    calculation of, 23–26
    of common materials, *20*
    definition of, 13
    of gases, 295–99
    of metals, 596, 598
    unit cell dimensions and, 596, 598, 605–6,
        621
    units of, 20
deoxyribose, 997
deposition, 15, *219*
    on phase diagram, 520
desalination of water, 573–75
detritus, 104
deuterium
    abundance of, 949
    in experimental fusion reactions, 948–49
deuterons, 70–71, 946–48
dextrorotatory enantiomer, 971, 976
dextrose, 573, 988
    See also glucose
diabetes, 1081, 1086
diagnostic radiology. *See* medical imaging
diamagnetic substances, 477
diamond, 606–7
    standard molar entropy of, 844
diamond crystal lattice, *594*, 606
diatomic molecules
    elements existing as, 7–8, 98
    heteronuclear, 478–80
    homonuclear, 474–78
    magnetic properties of, 476–77
    molecular orbitals of, 474–80
diazene, hybrid orbitals in, *458*
Dicke, Robert, 34, 35
Dieng Plateau disaster, 295–96
diesel engine, P–V work in, 222
diesel fuel, 256
dietary reference intake (DRI), 1020
diethyl ether, 510–11
    normal boiling point of, 518
diethylenetriamine ligand, 1064
diethylenetriaminepentaacetate ($DTPA^{5-}$), 1082

diffusion, 310–11
digested food, pH of, 751, 752–53
dihydrogen phosphate/hydrogen phosphate
    buffer, 793
dilute solution, 147
dilutions, 154–56
dimensional analysis, 20–22
dimethyl ether, 1035
    boiling point of, 507–8
    dipole moment of, 503
dimethyl sulfide, 1035, 1036–37
dimethyl sulfoxide, 1036
dinitrogen monoxide
    as greenhouse gas, 1030
    half-life for decomposition of, 653
    production of, 295, 407–8
    structure of, 407–11
dinitrogen pentoxide
    decomposition of, 652–53, 671
    hydrolysis of, 163
1,3-diphosphoglycerate^{4-} (1,3-DPG^{4-}),
    864–65
dipole moment (μ), 451–53
    See also polar molecules
dipole–dipole interactions, 503–4, *506*
    of polar solutes in polar solvents, 510
    in solids, *591*
    viscosity and, 525
    See also hydrogen bonds
dipole–induced dipole interactions, 504, *506*
    between water and gases, 514, 515
dipoles
    bond dipole, 398, 450
    induced (temporary), 500–502
    permanent, 451–53
    See also ion–dipole interactions
diprotic acids, 163, 764–67
disaccharides, 989–90, 991
dispersion forces (London forces), 499–502, *506*
    boiling points and, 509, 1021
    of long hydrocarbon chains, 513
    protein structures and, *984*, 985, 986
    in solids, *591*
    in uranium hexafluoride, 945
    viscosity and, 525
dissociation
    compared to dissolving, 159
    of strong electrolyte, 158
    of weak electrolyte, 159
distillation, 10
    energy flow in, 220
disulfide linkage, 985
DNA (deoxyribonucleic acid), 997–1001
    antitumor transition metal compounds and,
        1073–74, 1087
    carcinogenic chromium ions and, 1081
    hydrogen bonds in, 506–7, 998, *999*
    polycyclic aromatic hydrocarbons and, 468
    radiation damage to, 940
dopant, 482
doping, 482

double bonds
    definition of, 394
    Lewis structures with, 394–96
    restricted rotation around, 968–69
double helix, 998
doublets, 353
drawing
    of larger molecules, 465–67
    of Lewis structures, 392–94
    of molecular geometry, 442
drinking water
    contaminated by rock salt, 174
    produced by desalination, 573–74
    safe levels for contaminants in, 149–51
    treated with zeolites, 180
drugs
    chiral, 972–73, 986
    fullerenes for transport of, 607
    partition coefficients of, 527
    PEGs attached to, 614
    protease inhibitors, 988
    shelf-stability of, 193–94
    transition metals in, 1084, 1085–88
dry gas, 302
dry ice, 522
dynamic equilibrium, in weak electrolyte
        solution, 159
dystrophin, 963

E

E isomer, 968
    See also cis isomer
Earth
    age of oldest rocks on, 935
    atmosphere of, 274, 275
    early atmosphere of, 84, 85–87, 144
    elemental composition of, 85–86, 1019
    insoluble compounds in crust of, 169
Edison, Thomas, 835
effective dose of radiation
    biological effects and, 940
    from radon gas, 942
    units of, 939
effective nuclear charge (Z_{eff}), 359
    ionization energies and, 373–75
    size of atoms and, 370–71
effusion, 309–10
egg, osmotic pressure experiment with, 578–79
eicosene, 575–76
Einstein, Albert
    de Broglie theory and, 348
    mass-energy equivalence, 18, 346, 925, 944,
        949–50
    photoelectric effect and, 339, 340
    quantum indeterminacy and, 354–55
electric field, dipoles aligned in, 451
electrical charge
    flowing through a circuit, 891
    See also partial electrical charge
electrical conductivity
    of diamond, 606

of graphite, 607
of metals, 481–82
of semiconductors, 482–83
electrical work, 890–91, 899
electrochemical cells, 881–83
    electrolytic, 883, 908–9
    free-energy change in, 890–93
    See also standard cell potentials (E°_{cell}); voltaic
        cells
electrochemistry
    basic concepts of, 880–82
    See also corrosion; electrochemical cells
electrodes
    of electrochemical cell, 881–82, 884, 908–9
    in electrolyte, 158
electrolysis, 883, 906–9
    of water, 885
electrolytes, 158–59
    background electrolyte, 882, 902
    colligative properties and, 564–68
    corrosion promoted by, 904
electrolytic cell, 883, 908–9
electromagnetic radiation, 335–36
    emitted and absorbed by elements, 333–34
    emitted by hot objects, 337
    emitted by semiconductors, 483
    Planck's quantum theory of, 337–39
electromagnetic spectrum, 334
electron affinity (EA), 375–76
    Born–Haber cycle for calculation of, 546
    periodic trends in, 1021
electron capture, 929
    atomic number and mass number in, 930
    in gallium and indium nuclide decay, 1041
electron configurations of atoms, 358–66
    condensed, 360
    definition of, 359
electron configurations of ions, 366
    condensed, 368–69
    of isoelectronic species, 367
    of main group elements, 366–67
    of transition metals, 368–69
electron spin, 353–54
electron transitions, 344–45
    between molecular orbitals, 471, 480
    in sodium atom, 362
electron waves, 345–50
electron-deficient compounds, 391
electronegativity, 398–400
    formal charge and, 409
    Lewis structures and, 400, 404
    periodic trends in, 1021
electronegativity difference (ΔEN), 400–401
    limitations of bonding models and, 419
electron-pair geometry, 439
    summary of, 447
electrons
    Big Bang and, 70
    as cathode rays, 48
    charge on one mole of, 891
    mass-to-charge ratio of, 47–49
    Millikan's determination of mass, 49

properties of, 51
    See also beta (β) particles
electrophoresis, 10
electroplating, 908
    of Academy Awards statuettes, 908
electrostatic attractions in solids, 591
electrostatic potential energy (E_{el}), 215–16
    bond formation and, 389–90
    energy levels of diatomic molecule and, 866
    of ion–ion interactions, 215–16, 539
    of sodium chloride structure, 603
elementary steps, 666–68
    in enzyme kinetics, 678
elements
    allotropes of, 403
    atoms of, 7, 51–52
    definition of, 6
    emission and absorption spectra of, 333–34
    isotopes of, 52–55
    origin of, 70–73
    transmutation of, 50
    See also periodic table
emission spectrum
    atomic, 333
    of heated metal filament, 333
empirical formulas, 60
    compared to molecular formulas, 113–16
    from percent composition data, 110–13
    See also ionic compounds
enantiomers, 970–72
    of amino acids, 976, 1002
    of complex ions, 1075–76
    in living systems, 972–73
    racemic mixture of, 973, 986
end point of titration, 167
endothermic processes, 219–22
    enthalpy changes in, 225–26
    spontaneous, 835–37, 846, 853
energy
    of Big Bang, 18
    in chemical equation, 8
    of chemical reaction, 217
    conservation of, 213, 222
    definition of, 8
    from dietary glucose, 992
    from dietary lipids, 994, 996
    dispersion of, 837, 838, 842, 843, 865–66,
        867
    Einstein's equivalence to mass, 18, 346, 925,
        944, 949–50
    flowing from warm to cooler object, 211, 839,
        840
    of photon, 338–39, 344
    thermodynamics as study of, 210–11
    See also activation energy (E_a); free energy
        (G); internal energy (E); kinetic energy
        (KE); potential energy (PE); work
energy microstates, 867
energy profile for reaction, 661–62
    with catalyst, 721
    with two maxima, 667

energy-level diagram
  electron transitions shown on, 344
  of subshells in multielectron atoms, *363*
enthalpy (H)
  definition of, 225
  as a state function, 225, 244
enthalpy change (ΔH), 225–27
  bond energy as, 420–22
  free energy change and, 846–54
enthalpy of combustion (ΔH_{comb}), 238–39, 241–42
  Hess's law and, 243–44, 245–46
  methane bond energy and, 421
  standard, 255–56
enthalpy of formation. *See* standard enthalpy of
    formation (ΔH_f°)
enthalpy of fusion, molar (ΔH_{fus}), 229, 231
enthalpy of hydration ($\Delta H_{hydration}$), 543, 548–50
  for selected cations and anions, *549*
enthalpy of reaction (ΔH_{rxn}), 238–40
  from bomb calorimetry, 242
  from bond energies, 421–22
  from Hess's law, 243–46, 250, 251–52
  path independence of, 250
  See also standard enthalpy of reaction (ΔH_{rxn}°)
enthalpy of solution ($\Delta H_{solution}$), 542–43, 548, 550
enthalpy of vaporization (ΔH_{vap}), 519–20
  molar, 229–30
entropy (S)
  absolute, 841–43, 844–46, 867
  calculating changes in, 845–46
  chelate effect and, 1065
  factors affecting changes in, 843
  living systems and, 863
  macroscopic thermodynamic view of, 865–66
  microstate view of, 866–67
  of reaction, 845–46, 848–49
  spontaneous processes and, 837–38, 840–41,
    847–49, 1065
  standard molar, 842–43, 844–46
  as state function, 845
  temperature and, 838–40, 841–42, 843
  units of, 839
enzymes, 677–78, 986–88
  carbonic anhydrase, 677, 774, *985*
  catalase, *638*
  in photosynthesis, 1078
  transition metals in, 1078–81
enzyme–substrate complex, 678, 987
equations. *See* chemical equations; net ionic
    equations; nuclear equations
equilibrium. *See* chemical equilibrium; phase
    equilibria; solubility equilibria
equilibrium arrow, 159
equilibrium constant (K)
  catalyst and, 721
  definition of, 699
  equilibrium concentrations or pressures and,
    721–27
  for ionization reactions of acids, 743–44
  for ionization reactions of bases, 745
  for redox reactions, 898–99
  at specific temperature, 860–61

standard free energy of reaction and, 856–59,
  860–61
  temperature dependence of, 720, 859–61
  very large, 703, 899
  very small, 703, 726–27, 899
  whether reaction is at equilibrium and, 712,
    721
  See also chemical equilibrium; K_a; K_b; K_c; K_f;
    K_p; K_{sp}; K_w
equilibrium constant expressions, 699–702
  compared to reaction quotient, 711
  definition of, 699
  for heterogeneous equilibria, 713–14
  multiplying or dividing an equation and, 707–8
  pure liquids in, 714
  pure solids in, 713
  relationships between K_c and K_p, 703–5
  for reverse reactions, 706–7
  for sum of two or more reactions, 708–10
  summary for manipulation of, 710
  units and, 700
equilibrium lines, 520
equilibrium potential (E_{ion}), 1024–25
equimolar amounts, 122
equivalence point of titration, 167, 799
  multiple, 805–9
essential elements, 1018
  major, 1019, 1020, 1022–37
  trace, 1019, 1020, 1037, 1038–39, *1054*, 1076
  transition metals among, *1054*, 1076,
    1080–81
  ultratrace, 1019, 1020, 1037–38, *1054*, 1076,
    1080–81
esters, *69*, 615
  polyesters, 615–17
ethambutol, 973
ethane, 68, 254
  standard molar entropy of, *844*
  valence bond theory and, 463–64
ethanethiol, 1035
ethanol, 69
  absolute entropy of, 845
  as biofuel, 991
  boiling point of, 507–8, 518
  boiling-point-elevation constant for, *577*
  in esterification of acid, 615
  from fermentation of glucose, 991, 992
  freezing-point-depression constant for, *577*
  miscibility in water, 513
  as nonelectrolyte, 158
  standard free energy of combustion for,
    850–51
  vapor pressure of water mixed with, 556–57
ethers, *69*, 511
  polymers of, 613–14
ethylene (ethene), 68
  copolymer with vinyl acetate, 613
  polymers of, 608, 609–10
  ripening of tomatoes and, 463, 464
  valence bond theory and, 463–64
ethylene glycol
  in antifreeze, 538, 558, 562–63

  boiling point of, 508, 518
  freezing point of solution of, 562–63
  miscibility in water, 514
  polymer of, 614
  vapor pressure of solution of, 552
ethylene oxide, polymer of, 614
ethylenediamine ligand, 1063–64, 1070
  in chiral complex, 1075–76
ethylenediaminetetraacetic acid (EDTA), 1065,
  1082
eugenol, 120–21
evaporation
  in equilibrium with condensation, 517–18
  rate of, 517
  volatility and, 519
  See also vaporization
exact values, 26–27
excited states
  of hydrogen atom, 344
  indeterminate lifetimes of, 354–55
  of sodium atom, 362
exclusion principle, 353–54
exothermic processes, 219–22
  enthalpy changes in, 225–26
  entropy of surroundings and, 838
  free energy changes in, 848–49, 853
  not necessarily spontaneous, 835
extensive properties, 13

F

f block elements, 364, 365
f orbitals, 351, 358
  filling of, 364, 369
face-centered cubic (fcc) unit cell, 594, **596–97**,
  599
  of austenite, 601, 602
  of gold and platinum, 602
  of lithium chloride, 605–6
  of sodium chloride, 603–4
  of sphalerite, 604
Fahrenheit temperature scale, 32–34
families of elements, 56
Faraday, Michael, 891
Faraday constant (F), 891–93
fats, 994
fatty acids, 992–94
  β-oxidation of, 996
  names and formulas of, *993*
  of phospholipids, 996–97
  plants' degradation of, 1079
femtochemistry, 666, 666n
ferrite, 602
fertilizers
  nitrogen-based, 694, 696
  phosphates in, 1033
filled shell, 359
filtration of mixtures, 10
fireworks
  magnesium oxide in, 548
  red, 376

first ionization energy (IE₁), 372–75
  electronegativity and, 398–400
  periodic trends in, 1021
first law of thermodynamics, 222, 223
first-order reactions
  distinguishing from second-order, 656–57
  half-life of, 653–55
  integrated rate law for, 650–53
  radioactive decay as, 932–35
  summary of how to distinguish, *659*
fluorapatite, 1038
fluorine
  electron configuration of, 361
  electron configuration of anion of, 366–67
  as strong oxidizing agent, 887
  as trace essential element, 1038
fluorite, 604
fluorite structure, 604
  of thorite, 621
5-fluorouracil (5-FU), 527
food value, 257–58
formal charge (FC), 407–11
  in exceptions to octet rule, 412–17
  as model with limitations, 419
formaldehyde
  bond lengths in, 420
  dipole moment of, 452–53
  formula of, 114
  Lewis structure for, 395
  trigonal planar geometry of, 443–44
  valence bond theory for, 456–57
formation constant (K_f), 813
  chelation therapy and, 1085–86
formation reactions, 246, 247–48
formation water, 562
formic acid, pH of, 760–61
formula mass, 92
formula unit, 60
formulas. *See* chemical formulas; empirical
    formulas; molecular formulas; structural
    formulas
fossil fuels
  carbon dioxide from combustion of, 104–7,
    388, 786–87
  coal with sulfur content, 636, 713, 764, 810
  definition of, 104
  production of hydrogen from, 249–50, 911
  replacement by electric vehicle, 880
  See also gasoline; hydrocarbon fuels
fractional distillation, 553–56
francium, 1021
Fraunhofer, Joseph von, 333
Fraunhofer lines, 333, 334
free energy (G), 846–51
  chemical equilibrium and, 847, 854–59
  in coupled reactions, 864–65
  reactions in living systems and, 862–65
  temperature effect on spontaneity and,
    852–54
free energy change (ΔG)
  additive, 864–65
  as electrical work, 890–93

electrochemical equilibrium constant and, 899
  for ion transport across cell membrane, 1025
  as maximum available work, 847, 851, 867
  standard cell potential and, 899
free energy of formation. *See* standard free
    energy of formation (ΔG_f°)
free energy of reaction. *See* standard free energy
    of reaction (ΔG_{rxn}°)
free radicals, 412
  photochemical smog and, 637
  produced by ionizing radiation, 938
freezing, 15, 16, *219*
  enthalpy change in, 225–26
  entropy change in, 840
freezing point. *See* melting point
freezing point depression, 558, 562–64
  constants for selected solvents, *577*
  molar mass determined from, 575–77
  van't Hoff factor and, 564–68
frequency (ν), 335–36
  photon energy and, 338
frequency factor, in rate constant, 660
freshwater
  pH of, 770, 807
  See also alkalinity titration; drinking water
fructose, 988, 989
fuel cells, 249, 880, 909–11
fuel density, 255–56
fuel value, 255–56
fuels. *See* fossil fuels; hydrocarbon fuels
Fukushima nuclear disaster, 940
Fuller, R. Buckminster, 607
fullerenes, 606, 607
functional groups, 68–70, 963, *964*
  multiple, 467–68

G

gadolinium contrast agents, 1041, 1084
Galileo Galilei, 4
gallium
  medical imaging with radioisotopes of, 1041,
    1042
  silicon doped with, 483
gallium arsenide (GaAs), 483
gamma (γ) rays, 928–29
  biological effects of, 939, 940
  in electromagnetic spectrum, *334*
  in Geiger counter, 932
  in medical imaging, 943–44, 1041
  in radiation therapy, 943
gas constant (R), 289
gas laws, 280–92
  Amontons's law, 287–88, 304, 305
  Avogadro's law, 285–86, 289, 304–5
  Boyle's law, 280–83, 288–89, *290*, 304
  Charles's law, 283–85, 288–89, 304, 305–6
  ideal gas law, 288–92, 293–95, 296
  summary of, *291*
gases
  in chemical reactions, 293–95, 302–3
  collected over water, 302–3

compared to liquids and solids, 14–16, 274–75
  definition of, 14
  density of, 295–99
  kinetic molecular theory of, 304–11
  mixtures of, 299–303
  molecular speeds in, 306–9
  phase symbol for, 8
  real, 311–15
  standard states of, *842*
  water solubility of, 514–17
  wet vs. dry, 302–3
  See also noble gases (group 18); pressure
gasoline
  alkanes in, 252, *253*, 254
  from distillation of crude oil, 553
  ethanol added to, 991
  fuel value and fuel density of, 256
  hydrogen fuel compared to, 911
  photochemical smog and, 636
Geiger, Hans, 49–50
Geiger counters, 932
general gas equation, 290–92
genetic code, 1000–1001
Gerlach, Walther, 353
germanium
  diamond crystal lattice of, 606
  in human body, 1039
Gibbs, J. Willard, 847
Gibbs free energy (G). *See* free energy (G)
global warming, 388, 422–23
  See also climate change; greenhouse gases
glucose
  α and β isomers of, 989, 990–91
  chromium in regulation of, 1081
  cytochromes in metabolism of, 1079
  in D5W solution, 573
  empirical formula of, 114
  ethanol from fermentation of, 991, 992
  food value of, 258
  glycosidic bonds involving, 989–91
  magnesium ions in metabolism of, 1026
  metabolism of, 862–65
  polysaccharides formed from, 990–91
  in positron emission tomography, 943–44
  redox reaction in breakdown of, 146
  structure of, 988–89
  See also photosynthesis
glucose 6-phosphate, 863–64, 1033
glutamate, 1081
glycerides, 992, 993–96
glycerol, 992, 996
glycoaldehyde, 113–14
glycolic acid, copolymer with lactic acid, 615–16
glycolysis, 863–64, 992
glycoproteins, 988
glycosidic bond, 990
gold, 590, 594
  alloyed with platinum, 602–3
  antitumor complexes of, 1087
  arthritis drugs containing, 1087–88
gold-foil experiments, 49–50
Goudsmit, Samuel, 352–53

Graham, Thomas, 309
Graham's law of diffusion, 310–11
Graham's law of effusion, 309–10
graphene, 607
graphite, 606, 607
    standard molar entropy of, 844
gray (Gy), 939
greenhouse gases, 388
    carbon dioxide as, 388, 402
    climate stability and, 422
    dinitrogen monoxide as, 1030
    infrared absorption by, 388, 401–2, 413
    nitrous oxide as, 408, 653
    sulfur hexafluoride as, 413
ground state, of hydrogen atom, 344
groups of elements, 56, 57–58
Grubbs' test, 31–32
guanine, 997, *999*, 1000, 1080
Guldberg, Cato, 699

H

Haber, Fritz, 721
Haber–Bosch process, 696, 721
    equilibrium constant for, 860–61
    hydrogen production for, 725
half-life ($t_{1/2}$) of radionuclide, 933–37
half-life ($t_{1/2}$) of reaction
    first-order, 653–55
    second-order, 658
    summary of, for single reactant, *659*
    zero-order, 659
half-reactions, 185–88
    combined to give net ionic equation, 881,
        885–86
    in electrochemistry, 880
    pH conditions of, 884–85, 910–11
    See also electrochemical cells
halite structure, 111
Hall–Héroult process, 259
halogens (group 17)
    boiling points of, 500
    common acids containing, 66
    in periodic table, 58
haloperoxidases, 1080
hard water, softening of, 178–79
heat (q)
    from car engine, 851
    at constant pressure, 225
    definition of, 211
    sign of, 219, 223
heat capacity (C_P)
    of calorimeter, 241
    See also molar heat capacity (c_P)
heat of reaction. *See* enthalpy of reaction (ΔH_{rxn})
heat sinks, 232
heat transfer, 211
heating curves
    of solution of volatile liquids, 554
    of water, 227–28
Heisenberg, Werner, 348, 350
Heisenberg uncertainty principle, 348–50

Heliox, 301
helium
    balloons inflated with, 224, 282–83, 286, *296*
    binding energy of nucleus, 925
    electron configuration of, 359
    in experimental fusion reactions, 948–49
    molecular orbitals of, 473
    primordial nucleosynthesis of, 71–72, 947
    in scuba diving mixture, 300–301
    stellar nucleosynthesis of, 72, 947–48
helium-3
    enthalpy of fusion of, 852n
    in magnetic resonance imaging, 1043
    in nuclear fusion, 947–48
hematite, 87, 190
heme groups, in cytochromes, 1079
heme iron, dietary, 1054
hemoglobin, 516
    quaternary structure of, 985–86
    sickle-cell, 982–83
Henderson–Hasselbalch equation, 789–91
    buffers and, 791, 793, 795, 796–97, 798
    titration of weak acid with strong base and,
        802
    titration of weak base with strong acid and,
        805
Henry, William, 516
Henry's law, 516–17
hertz (Hz), 335
Hess's law, 243–46, 250, 251–52
heteroatoms, 67, 68–70
heterogeneous alloys, 592, 600
heterogeneous catalysts, 675
    in biocatalysis, 678
heterogeneous equilibria, 713–14
heterogeneous mixtures, 7
heteronuclear diatomic molecules, molecular
        orbitals of, 478–80
heteropolymer, 613
hexagonal closest-packed (hcp) structure, 593–
        94, 599
hexagonal unit cell, 594
hexamethylenediamine, *618*
high-density polyethylene (HDPE), 609–10
high-spin states, 1071–73
Hindenburg, 703
HIV, 988
homocysteine, 1081
homogeneous alloys, 600, 601
homogeneous catalyst, 674
homogeneous equilibria, 713
homogeneous mixture, 7
homonuclear diatomic molecules
    elements existing as, 7–8, 98
    magnetic properties of, 476–77
    molecular orbitals of, 474–78
homopolymer, 609
Honda Clarity, *910*
hot-air balloon
    buoyancy of, 296
    Charles's law and, 284–85
    P–V work done by, 222–23

Hoyle, Fred, 18
Hubble, Edwin, 18
human body
    classes of chemical reactions in, 146
    coupled reactions in, 861–65
    elemental composition of, 1019–20
    nitric oxide in, 411
    See also blood; cancer; essential elements;
        medical imaging; medical therapy
Hund, Friedrich, 360
Hund's rule, 360–61
    molecular orbitals and, 475, 476
hybrid atomic orbitals, 454
    sp, 458–60
    sp^2, 456–58
    sp^3, 455–56
    sp^3*d*, 461–62
    sp^3*d*^{2}, 461–62
    summary of, *462*
hybrid vehicles, 112–13, 878, 880, 900–901
hybridization, 454
hydrangea, colors of, 740
hydrated complex ions, 812, 814–16, 820, 1055,
        1062–63, 1066–67
    spin states of, 1072
hydrated ions, 502–3
    enthalpy of hydration and, 543, 548–50
    enthalpy of solution and, 542–43, 548, 550
hydration, sphere of, 502–3, *815*
hydride ion, 1022
hydrocarbon fuels
    alkanes as, *253*, 254
    balanced equations for combustion of, 101–4
    disadvantages of, 251
    enthalpies of combustion of, 250–51
    enthalpies of formation of, 248
    ethanol added to, 991
    polycyclic aromatic hydrocarbons from
        burning of, 468
    See also crude oil; fossil fuels; methane
hydrocarbons, 67–68
    atmospheric, 108
    branched-chain, 254
    combustion analysis of, 117–20
    as nonpolar solvents, 512
    straight-chain, 253, *254*, 500–501
    viscosities of, 525
    See also alkanes; hydrocarbon fuels
hydrochloric acid, 159–61
    activity series of metals and, 189–90
    gastric, 751, 1028–29
    pH of 7.4 M solution, 759
    pH of very dilute solution, 762–63
    See also hydrogen chloride
hydrofluoric acid, burns caused by, 174–75
hydrogen
    1s orbital of, 355–56
    in activity series, 189
    bacterial production of, 243
    Bohr model of, 343–45
    combustion of, 217, 703, 911

hydrogen (cont.)
   covalent H–H bond formed by, 389–90,
      453–54
   deuterium isotope of, 948–49
   electron configuration of, 359
   elemental properties of, 1022
   in fertilizer production, 696
   as fuel, 217, 249–50, 909–11
   Hindenburg disaster and, *703*
   ionization energy of, 344, 345
   molecular orbitals of, 472
   primordial nucleosynthesis of, 70–72
   valence bond theory of, 453–54
   See also steam–methane reforming reaction;
      water–gas shift reaction
hydrogen bonds, 504–10
   of acetone in water, 507
   in alcohols, 509, 513, 514
   in amides, 618
   in ammonia, 504–5, 512
   of glucose in water, 989
   in Kevlar, 619
   in nucleic acids, 506–7, 998, *999*
   in protein structure, 983–84
   in solids, *591*
   in water, 504–5, 522, 523–26, 548
hydrogen chloride
   absolute entropy of, 844, 845
   in aqueous solution, 743
   overlapping orbitals in, 454
   See also hydrochloric acid
hydrogen emission spectrum, 341–42
   Bohr model and, 343–45
   pair of red lines in, 352
hydrogen fluoride
   hydrogen bonds in, 504–5, 512
   solubility in water, 512
hydrogen halides, polarity of bonds in, 400
hydrogen iodide, equilibrium in formation of,
      722–23
hydrogen ions, meaning hydronium ions, 160
hydrogen peroxide
   Lewis structure of, 396
   rate of decomposition of, *638*
   superoxide dismutase and, 1080
   xanthine oxidase and, 1080
hydrogen sulfide, 1033, 1034, 1035
hydrogenases, 1080
hydrolysis
   of nitrogen compounds, 1032
   of nonmetal oxides, 163
hydronium ions, 146, 160
   Arrhenius acids and, 740
   pathways for conversions of, *758*
   sodium ion channels and, 1023
hydrophilic interactions, 513
hydrophobic interactions, 513
hydrothermal vents, deep-ocean, 1002
hydroxide ions
   Arrhenius bases and, 740, 745
   Brønsted–Lowry bases and, 164, 745

   as Lewis bases, 810
   redox reactions involving, 192–93
hydroxyapatite, 146, 1027, 1032
   fluoride and, 1038
hydroxyl radicals
   photochemical smog and, 637
   produced by ionizing radiation, 938
hypertonic solution, *569*, 572
hypobromous acid, 768, *769*
hypochlorous acid, 768, *769*
hypohalous acids, 768, *769*
   relative strengths of, 814
hypoiodous acid, 768, *769*
hypothesis, 17
hypotonic solution, *569*, 573
hypoxia, 515

I

i factor, 564–68
   osmotic pressure and, 570, 573
ice
   aquatic life and, 526
   chilling beverages with, 232–35
   energy in melting of, 228, 229, 230, 232–35,
      238–39
   entropy change in freezing and, 840
   entropy change in melting of, 839–40, 843
   hydrogen bonds in, 522, 525–26
   less dense than water, 522, 523, 525–26
   melting/freezing behavior of, 521–22, 525–26
   temperature effect on melting/freezing of,
      852–53
ice-resurfacing machine, 226
ideal gas law, 288–92
   chemical reactions and, 293–95
   density of gas and, 296
   deviations from, 311–14
ideal gases
   definition of, 288
   kinetic molecular theory of, 304–11
ideal solutions
   Raoult's law and, 551–52, 557
   theoretical values of K_a in, 823
   van't Hoff factor and, 567
indicator solution, 167
indium-111, 1041, 1042
induced dipoles, 500–502
   interacting with permanent dipoles, 504, *506*,
      514, 515
induced-fit model of enzyme action, 987
inertness, chemical, 361
infrared active vibrations, 402
infrared inactive vibrations, 402
infrared radiation
   in electromagnetic spectrum, *334*
   greenhouse effect and, 388, 401–2, 413
   from hot objects, 337
   in hydrogen emission spectrum, 342, *344*
   photoelectric applications of, 340
   vibration of bonds and, 401–2
inhibitors of enzymes, 987–88

initial rate of a reaction, 645
inner coordination sphere, *815*, 1055
insect pheromones, 484–85
insignificant digits, 24
insoluble compound, definition of, 170
integrated rate law
   for first-order reaction, 650–53
   for second-order reaction, 655–58
   summary of, for single reactant, *659*
   for zero-order reaction, 658
intensive properties, 12
intercalation in DNA, 468
intermediate, 666
intermetallic compound, 600
intermolecular forces, 498
   boiling points and, 499–502
   changes of state and, *498*, 499
   combinations of, 513–14
   of neutral atoms and molecules, 499–502
   of polar molecules, 502–10
   in protein structures, *984*
   in real gases, 312, 313
   relative strengths of, *506*
   vaporization and, 517
   in water, 498–99
   See also hydrogen bonds
internal energy (*E*)
   changes in, 221–24
   definition of, 221
   enthalpy and, 225
International Union of Pure and Applied
      Chemistry (IUPAC), 68
interstitial alloys, 601–2
intramolecular forces, 498
intravenous medications, 572–73
iodine, in human health, 1038–39
iodine-131
   for thyroid cancer, 943
   in thyroid imaging, 1041
ion channels, 1023
   for chloride, 1028
ion exchange, 10, 178–80
ion pairs, 566–67, 568
   apparent values of K_a and, 823
   in concentrated solutions, 823
ion product, 820
ion pump, 1023–24
ion–dipole interactions, 502–3, *506*
   coordinate bonds and, 1055
   membrane transport and, 1023
   protein structures and, 985
ionic bonds, 389
   electronegativity difference and, 400, *401*
   in solids, *591*
   strength of, 539
ionic character, 398
ionic compounds, 9
   binary, 60–62, 63–64, 397, 603–6
   dissolving in water, 502–3, 542–43
   enthalpy of solution, 542–43, 548, 550
   entropy change on dissolving in water,
      845–46

ionic compounds (cont.)
  formulas of, 63–64
  inter-ion distance in, 539
  lattice energy of, 543–50
  naming of, 63–64
  as solids, 539
  of transition metals, 64–65
ionic equations, 160, 161–62
  See also net ionic equations
ionic radii, 369–72
  in crystal structures, 603, 605–6, 621
  lattice energies and, 544
  of main group elements, 540
  melting points and, 544
  of polyatomic ions, 539
  strength of ion–ion interactions and, 539,
    541–42
ionic solids, 590, 591, 592, 603–6
ion–ion interactions
  coulombic, 215–16, 539
  protein structures and, 984, 985
  strength of, 539–42, 543
  See also lattice energy (U)
ionization energy (IE), 372–75
  electronegativity and, 398–400
  of hydrogen atom, 344, 345
  periodic trends in, 1021
ionizing radiation
  biological effects of, 938–42
  dosage of, 938–40
  radon gas and, 940–42
  sources of U.S. exposure to, 941
  units of, 939
ions, 9
  monatomic, 60
  See also anions; cations; polyatomic ions
iron
  available to the body, 1054
  chelation therapy for excess of, 1085–86
  complexed with heme, 1054, 1079
  enzymes containing, 1079–81
  redox reactions of compounds of, 190–93
  steel as alloy of, 601–2, 906
  stellar nucleosynthesis and, 72–73, 924–25, 926
  See also rust
iron carbide, 602
iron(III) hydroxide, 815
iron ions
  electron configuration of Fe^{3+}, 368
  spectrophotometric measurement of Fe^{2+}, 157
  spin states of iron(III), 1071
iron(III) oxide
  of red-orange rock formations, 190
  as rust, 190, 832, 835
iron sulfate, 87
isobars, 277
isoelectric point (pI), 977
isoelectronic species, 367
isolated system, 218
isomers
  constitutional isomers, 965–68, 1076
  of coordination compounds, 1073–76

  definition of, 85
  linkage isomers, 1076
  stereoisomers, 968–74, 1073–76
  See also chirality
isopropanol, boiling point of, 508
isotonic solution, 569, 572
isotopes, 52–54
  average atomic mass and, 54–55
  See also radionuclides

J

Japanese beetle, 484–85
japonilure, 484–85
joints, artificial
  polyethylene in knee replacement, 610
  tantalum metal in, 595
Joliot-Curie, Irène, 938
joule (J), 223

K

K. See equilibrium constant (K)
K_a, 743–44, 752
  for diprotic acid, 764, 766
  of hydrated metal ions, 815
  theoretical values of, 823
  for triprotic acid, 767–68
  for weak acid, 754–56, 760
K_b, 745, 752
  for weak base, 756–58, 761
K_c, 700–702, 703
  compared to Q_c, 711
  free energy change and, 859
  K_p and, 703–5
K_f (formation constant), 813
K_p, 700–701, 702–3
  free energy change and, 859, 860–61
  K_c and, 703–5
K_{sp} (solubility product), 816–24
K_w, 749, 752, 758
  for conjugate acid–base pairs, 772, 790
  very dilute strong acid and, 762–63
Kalkwasser
  concentration of, 152–53
  pH of, 759
Kekulé, August, 465
Kekulé structures, 465
kelvin (K), 33
Kelvin temperature scale, 32–34
  Charles's law and, 283
keratins, 986
kerosene, 254, 256
ketones, 69, 465
Kevlar, 619–20
kilowatt-hour, 900
kinetic energy (KE)
  average, 304, 307, 308
  definition of, 213
  dispersion of, 837, 838, 842, 843
  entropy and, 837, 838, 841
  of gas molecules, 304, 305, 306–9

  at molecular level, 214–15, 216–17
  as part of internal energy, 221
  of photoelectron, 339
  reaction rates and, 638, 660
  solubility of gases in liquids and, 515
  states of matter and, 499
  vaporization and, 517
kinetic molecular theory of gases, 304–11
  assumptions of, 304
  reaction rates and, 645, 660
  solubility of gases and, 515
kinetics, chemical
  definition of, 638
  half-life in, 653–55, 658, 659
  See also reaction mechanisms; reaction order;
    reaction rates
Kirchhoff, Gustav Robert, 333
krypton-83, in MRI, 1043
Kwolek, Stephanie, 619

L

lactase, 986
lactic acid
  conversion of glucose into, 865
  copolymer with glycolic acid, 615–16
  structure of, 865
lactose intolerance, 986
landfill, gases emitted by, 298
lanterns, for camping, 621
lanthanides, 58, 364, 369
  in scintigraphic imaging, 1084
lasers, 355
lattice energy (U), 543–44
  calculating, 545–48
  of common binary ionic compounds, 543
  enthalpy of hydration and, 548–50
laughing gas. See dinitrogen monoxide
law of conservation of energy, 213, 222
law of conservation of mass, 95
law of constant composition, 8
law of mass action, 699
law of multiple proportions, 59
Le Châtelier, Henri Louis, 715
Le Châtelier's principle, 714–21
  adding or removing reactants or products,
    715–17
  percent ionization and, 755
  pressure and volume changes, 717–19
  summary of, 721
  temperature changes and, 719–20
lead
  electron configuration of, 365
  toxicity of, 1020–21, 1085
  in uranium-238 decay series, 931, 935, 941
lead–acid battery, 896–97
  recharging of, 906–7
lead(II) chloride
  attempt at precipitation of, 820–21
  attempt at selective precipitation of, 822–23
lean mixture, 124
Lemaître, Georges-Henri, 18

length, units of, 19, *20*
Leonardo da Vinci, 180
leveling effect, 747–48
levorotatory enantiomer, 971, 976
Lewis, Gilbert N., 391, 398, 419, 810
Lewis acids, 809–11
Lewis bases, 809–11
    as ligands, 812, 1055, 1062
Lewis dot symbols, 391–92
Lewis structures, 391–97
    of binary ionic compounds, 397
    definition of, 392
    with double and triple bonds, 394–96
    electronegativity and, 400, 404
    exceptions to octet rule and, 411–19
    formal charge and, 407–11, 412–17
    with less than an octet, 416–18
    with more than an octet, 413–16
    octet rule and, 391, 392–93
    with odd number of electrons, 411–13
    overview of, 392
    resonance and, 403–7
    steps to follow when drawing, 392–94
    steric number determined from, 440, 441
Lewis symbols, 391–92
Libby, Willard, 936
life
    coordination compounds in, 1076–81
    origin of, 84–85, 96, 144, 1001–2
    water on Mars and, 144–45
    See also human body; photosynthesis
ligand displacement reaction, 1063
ligand exchange reaction, 1085–86
ligands, 812, 1055
    as Lewis bases, 812, 1055, 1062
    macrocyclic, 1077
    names and structures of, *1059*
    polydentate, 1064–66, 1076–81
    prefixes for number of, *1059*, 1076
    spectrochemical series of, 1070
    spin states and, 1072
light
    emitted by hot objects, 337
    spectrum of, 332–33, *334*
    speed of, 335
    wave–particle duality of, 340
    See also colors; sunlight
lightning, ozone production by, 403
like dissolves like, 510–11
limestone. *See* calcium carbonate
limiting reactants, 122–25
limonene, 119–20, 970–71
linear complex ions, *1057*
linear geometry
    hybrid orbitals and, 458–60, *462*
    with no lone pairs, 440
    with three lone pairs, *447*, 448
linkage isomerism, 1076
lipid bilayer, 997, *1023*
lipids, 992–97
liquids
    compared to solids and gases, 14–16

definition of, 14
    phase symbol for, 8
    standard states of, *842*
    vapor pressure of, 517–20
    volatilities of, 10, 519
liter (L), 20
lithium
    electron configuration of, 359–60
    in experimental fusion reactions, 948–49
    ionization energy of, 374
    as strong reducing agent, 887
    therapeutic applications of, 1043
lithium aluminum hydride, 417–18
lithium carbonate, of red fireworks, 376
lithium chloride, crystal structure of, 605–6
lithium perchlorate, freezing point depression
        by, 565–66
lithium-ion batteries, 112–13, 901–3
    recharging of, 906–7
London, Fritz, 499, 500
London dispersion forces. *See* dispersion forces
lone pairs
    Lewis acids and bases and, 809–11, 812,
        1055
    in Lewis structures, 392–93
    VSEPR theory and, 439, 444–49
low-density polyethylene (LDPE), 609–10
low-spin states, 1071–73
lycopene, 115–16
Lyman, Theodore, 342

M

macrocyclic ligands, 1077
macromolecules, 608
    See also polymers
macroscopic phenomena, 45
magnesium
    alloys with aluminum, 906
    electron configuration of, 362
    in living organisms, 1026
magnesium hydroxide, 816–17
magnesium ions, separating from calcium ions,
        821–22
magnesium oxide, lattice energy of, 548
magnetic properties
    molecular orbital theory and, 477
    of transition metal ions, 1071–73
magnetic quantum number (m_ℓ), 351
magnetic resonance imaging (MRI), 1041,
        1042–43, 1084
magnetite, bonded to steel, 906
main group elements, 57
    for diagnosis and therapy, 1040–44
    electron configurations of ions of, 366–67
    first ionization energies of, 373–75
    ionic radii of, *540*
    Lewis symbols of, 391
    periodic trends in, 1021–22
    properties of, 1021–22
    See also major essential elements
major essential elements, 1019, 1020, 1022–37

malleability
    of bronze vs. copper, 601
    of metals with ccp structure, 599
    of molecular solids, 592
maltose, as human energy source, 244–45
manganese, and photosynthesis, 1077–78
manganese dioxide
    in alkaline batteries, 889
    hydrogen peroxide decomposition and, *638*
manganese ions, spin states of, 1072
manometers, 278–80
marble, chemical weathering of, 162, 176–77
Mars, water on, 144–45
Marsden, Ernest, 49–50
marsh gas, 254
mass
    conservation of, 95
    definition of, 5
    density and, 13
    Einstein's equivalence to energy, 18, 346, 925,
        944, 949–50
    units of, *20*
mass action expression, 699
    as reaction quotient, 710–12
    See also reaction quotient (Q)
mass defect (Δm), 925
mass number (A), 52
mass spectra, 116–17
    in radiometric dating, 935
mass spectrometers, 116
mass-to-mass ratios, 149–50
mass-to-volume ratios, 147–48
matter
    classes of, 5–7
    definition of, 5
    properties of, 12–14
    states of, 14–16
matter waves, 346–48
Maxwell, James Clerk, 335
mean, 28
measurements
    accuracy of, 27, 28–30
    precision of, 27–32
    significant figures in, 23–27
    units of, 18–22
medical imaging
    of heart, 1083
    main group elements used in, 1041–43
    with MRI, 1041, 1042–43, 1084
    with PET, 922–23, 943–44, 1041, 1043
    radionuclides used in, 942, 943–44
    with technetium, 73–74, *927*, 1082–83
    transition metals in, 1081–84
medical therapy
    artificial joints, 595, 610
    for burn patients, 615–16
    diamond films on prosthetic implants, 607
    dissolving sutures, 616
    PETE in, 615
    Teflon tubing in vascular surgery, 610
    transition metals in, 1081–82, 1084–88
    See also drugs

melting, 15, 16, *219*
  enthalpy change in, 225–26
  entropy change in, 839–40, 843
  spontaneous, 835, 837
melting point
  intermolecular forces and, 499
  lattice energy and, 543–44
  periodic trends in, 1021–22
  on phase diagram for water, 520, 521–22
  temperature scales and, 32–33
  types of solids and, *591*, 592
  See also freezing point depression
membrane. *See* cell membrane; osmotic pressure
  (Π)
membrane potential ($E_{membrane}$), 1024–26
Mendeleev, Dmitri, 56, 58, 59, 73
meniscus, 524–25
mercury
  determining ionic concentrations of, 898
  meniscus of, 524–25
  therapy for poisoning from, 1085, 1086
messenger RNA (mRNA), 1000–1001
metabolic pathways, enzymes of, 677, 986, 988
metal hydrides, 1022
metallic bonds, 390, 480–82, *591*
metallic radius, 369
metallic solids, 590, *591*
metalloenzymes, 1078–81
metalloids
  in periodic table, 56
  as semiconductors, 482–84
  unit cells in, *594*
metals
  activity series for, 188–90
  conductivity of, 481–82
  densities of, 596, 598
  major metals in human body, 1022–27
  in periodic table, 56, 57, 58
  photoelectric effect of, 339
  properties of, 590
  structures of, 592–99
  unit cells of, 593–94
  See also transition metals
metastable nucleus, 1083
meteorites
  age of, 935
  amino acids in, 84, 1002
  heating of early atmosphere by, 86
  Martian, 144–45
  organic compounds in, 113–14
meter (m), 19, 20
methane
  as alkane, 68, 252–55
  atmospheric, sources of, 101, 108
  bacterial generation of, 243, 1081
  bond angles in, 439
  bond length in, 420
  from degradation of organic matter, 254, 298
  diamond films grown from, 607
  in hydrogen gas production, 249, 696–97,
    705, 1003
  in natural gas, 101–2, 254

sp^3 hybrid orbitals in, 455
  standard molar entropy of, *844*
methane combustion, 102
  bond energies in, 420–21
  entropy change in, 838
  oxidation numbers in, 183
methanethiol, 1034
  in "asparagus" urine, *1035*
  boiling point of, 1036, 1037
methanol, 69
  absolute entropy of, 845
methionine, 1081
methyl group, 253
  enzymatic transfer of, 1081
methylaspartate, 1081
methylene group, 253
methylpropane, boiling point of, 504
micromolarity, 148
microstates, 867
microwave background radiation, 34–35
milk of magnesia, 816
Miller, Stanley, 84, 96, 98, 1001
millibars, 277
Millikan, Robert, 48–49, 52
millimeters of mercury (mmHg), 276–77
millimolarity, 148
miscibility
  definition of, 275
  of ethanol and water, 513
  of gases, 275
  predicting, 514
mixtures, 6–7
  definition of, 6
  of gases, 299–303
  heterogeneous, 7
  homogeneous, 7
  separation of, 9–10
  of volatile solutes, 553–58
models
  molecular, 9
  scientific, 17
molality (*m*), 558–60
molar absorptivity (*ε*), 156–57
molar enthalpy of fusion (ΔH_{fus}), 229, 231
molar enthalpy of vaporization (ΔH_{vap}), 229–30
molar heat capacity (c_P), 228
  measurement of, 235–37
  water's high value of, 232
  of water's physical states, 228, 229, 230, 231
molar mass, 89–94
  calculated from gas density, 298–99
  from colligative properties, 575–78
  definition of, 90
  Graham's law of effusion and, 309–10
  of ionic compound, 91
  molecular formula and, 115
  root-mean-square speed and, 307–10
  van der Waals constants and, 313
molar solubility, 816–19
molar volume, 289
molarity (*M*), 148, 148n
  appropriate uses of, 558–59

mole (mol), 87–89
mole fraction (X_x)
  definition of, 299
  of gases, 299–302
mole ratio, 105–8
molecular compounds, 8, 60, 61–63
molecular equations, 160, 161
molecular formulas, 8, 60
  compared to empirical formulas, 113–16
  names of compounds and, 62–63
molecular geometry, 439
  conventions for drawing of, 442
  hybrid orbitals and, *462*
  summary of, *447*
molecular ion (H_2^-), bond order of, 473–74
molecular ion (M^+), in mass spectrometry,
  116–17
molecular mass, 91
  from mass spectrum, 117
molecular models, 9
molecular motion, types of, *221*
molecular orbital, definition of, 471
molecular orbital diagram, 472
molecular orbital (MO) theory, 470–84
  colors of auroras and, 470, 471, 480
  electron transitions in, 471, 480
  guidelines for, 474–75
  of heteronuclear diatomic molecules, 478–80
  of homonuclear diatomic molecules, 474–78
  metallic bonding and, 480–82
  semiconductors and, 482–84
molecular recognition
  carbohydrates in, 988
  chirality and, 468–70, 971, 988
  definition of, 463
  by sense of smell, 466, 468–69, 971, 1035
molecular solids, 590, *591*, 592
molecular spectra, 471
molecularity of elementary step, 666
molecules, 7–9
Molina, Mario, 673
molybdenum
  in enzymes, 1080
  in technetium generator, 1083
monatomic ions, 60
monobasic bases, 799
monodentate ligands, 1064
monomers, 608
monoprotic acids, 163, 764
monosaccharides, 988–89
  phosphate groups bonded to, 1033
  See also glucose
most probable speed of molecules, 306, *307*
moth balls, 423
motion detectors, 340
motor vehicles
  fuel cells for, 909–11
  See also automobiles
multiple proportions, law of, 59
muscle contraction, 1026
muscular dystrophy (MD), 963
mutases, 1081

N

Na⁺–K⁺ pump, 1023–24
names of complex ions, 1058–60
names of compounds
    common acids, 65, 66–67
    coordination compounds, 1060–62
    ionic compounds, 63–65
    molecular compounds, 62–63
    organic compounds, 67–70
    Roman numerals in, 64–65, 1058
    transition metal compounds, 64–65, 1060–62
names of polyatomic ions, 64, 65–66
nanomolarity, 148
nanoparticles, 590
    of gold–platinum alloy, 602–3
nanotubes, carbon, 607
naproxen, 971
natural gas
    combustion of, 101–3
    compounds in, 254
    electric power plants and, 106–7
    entropies of alkanes in, 844
    odorant in, 1035
    See also methane
neon
    electron configuration of, 361
    neon-19 in PET imaging, 1043
Nernst, Walther, 895
Nernst equation, 895–97
    cell membrane potential and, 1024
nerve impulses, 1024, 1026
net ionic equations, 161–62
    by combining half-reactions, 881, 885–86
    for precipitation reactions, 172–73
    for redox reactions, 187, 189–93
neutral salts, 770, 771
neutralization reactions, 160–62
    precipitate formed in, 170
    See also acid–base reactions
neutron capture, 72
neutrons
    Big Bang and, 70
    discovery of, 51
    in nucleosynthesis, 70–73
    properties of, 51
    in Rutherford's atomic model, 51
    symbol for, 71
Newton, Isaac, 332
nickel, enzymes containing, 1080–81
nickel(II) complex ions, 1062–63, 1070, 1075
nickel–metal hydride batteries, 900–901
    rechargeable AA, 899, 901, 907–8
    recharging of, 907–8
niobium, in complex with antitumor activity, 1086
Nissan Leaf, 901
NiTi (nitinol), 600, 601
nitrate ion
    in nitrogen cycle, 1030, 1080
    reduction to nitrite, 1030, 1031, 1080
    resonance structures of, 406–7

nitrate reductases, 1030, 1031, 1080
nitric acid
    from dinitrogen pentoxide, 163
    industrial production of, 728
    as strong acid, 163, 742
    structural reasons for strength of, 768
nitric oxide, 411–12
    molecular orbital theory of, 478–80
    See also nitrogen monoxide
nitrilotriacetic acid (NTA), 1066
nitrite ions, enzyme-catalyzed reduction of, 1079, 1080
nitrogen
    in automobile air bags, 293, 294
    bonding in diatomic molecule of, 460
    deviation from ideality, 313–14
    electron configuration of, 361
    in living systems, 1029–32
    as main component of air, 275, 299–300
    molecular orbitals for, 475–76
    molecular speed of, 307–8
    in scuba diving mixture, 300–301, 302
    with sp hybrid orbitals, 460
    with sp² hybrid orbitals, 458
    with sp³ hybrid orbitals, 455–56
nitrogen cycle, 1030, 1080
nitrogen dioxide
    dimerization of, 701–2, 704, 705, 714–15, 717–18
    equilibrium constant for formation of, 858
    as odd-electron molecule, 411, 412–13
    in Ostwald process, 728
    in photochemical smog, 636
    rate of formation of, 641, 642–45, 668–69
    thermal decomposition of, 655–56, 658, 665–68
    zero-order reaction with carbon monoxide, 658–59, 672
nitrogen monoxide
    as catalyst in ozone decomposition, 675
    equilibrium in production of, 726
    equilibrium in reaction with oxygen, 706–7
    formed in engines, 636, 637, 639–40, 642, 676–77, 1030
    in Ostwald process, 728
    photochemical smog and, 636–37
    rate of production of, 649–50
    rate of reaction with oxygen, 641, 642–45, 645–46, 648
    rate of reaction with ozone, 646–47, 661–62
    See also nitric oxide
nitrogen narcosis, 300, 302, 314–15
nitrogenases, 1030
nitrous acid
    conjugate base of, 747
    percent ionization of, 755
    structural reasons for weakness of, 768
    as weak acid, 742–43, 753–54
nitrous oxide
    production of, 295, 407–8
    structure of, 407–11
    See also dinitrogen monoxide

Nobel Prize
    Arrhenius, Svante, 740
    de Broglie, Louis, 348
    Einstein, Albert, 340
    Planck, Max, 337
    Rutherford, Ernest, 50
    Werner, Alfred, 1055
noble gases (group 18), 58
    as atomic solids, 590
    banned for athletes, 1042
    boiling points of, 499–500
    electron configurations of, 361
    in medical imaging, 1043
    radon exposure, 940–42, 1020
nodes
    in p orbital, 357
    in s orbital, 356
    of standing wave, 347
nomenclature. See names of compounds
nonactin, 1023
nonelectrolytes, 158
nonessential elements in human body, 1018–19, 1039–40
nonmetal oxides, 60
    forming acids in water, 162–63
nonmetals, in periodic table, 56
nonpolar covalent bond, 398
    electronegativity difference (ΔEN) and, 400, 401
nonpolar solvents, 512–13
nonspontaneous processes
    definition of, 834
    in electrolytic cells, 909
    entropy changes in, 838, 840–41
    free energy changes in, 847
    in living systems, 862, 863, 864, 865
    recharging of batteries, 907
normal boiling point, 518
normal distribution, 30
N-terminus, 979
n-type semiconductor, 482–84
nuclear charge. See effective nuclear charge (Z_eff)
nuclear chemistry, 924
nuclear equations, 71
nuclear fission, 944–46
nuclear fusion
    efforts to produce energy with, 948–49
    in primordial nucleosynthesis, 70–72, 946–47
    solar, 947–48
    stellar, 72–73, 924–25, 926
nuclear power plants, 945–46
nuclear reactions, 927–31
nuclear submarine, sinking of, 912–13
nuclear weapons testing, 46–47, 1039
nucleic acids, 997–1001
    bonding of phosphate to, 1033
    See also DNA (deoxyribonucleic acid); RNA (ribonucleic acid)
nucleons, 52
    See also neutrons; protons

nucleosynthesis
   primordial, 70–72, 946–47
   stellar, 72–73, 924–25, 926
nucleotides, 997–99
   enzyme-catalyzed degradation of, 1080
   self-assembly of, 1002
nucleus
   metastable, 1083
   in Rutherford atomic model, 50–51
nuclides, 53–54
   See also radionuclides
nylon, 618–19

O

obsidian, absolute entropy of, 844
ocean acidification, 786–88, 823–24
oceans
   of early Earth, 84, 144
   as heat sinks, 232
   See also seawater
octahedral complex ions, 1056, *1057*
   chirality in, 1075–76
   cis–trans isomers in, 1074–75
   crystal field splitting and, 1066–67, 1071–73
   with macrocyclic ligands, 1077
octahedral electron-pair geometry, 448–49
octahedral holes, 601–2
   in lithium chloride, 605
   in sodium chloride, 603–4
octahedral molecular geometry, 441, *447*
   sp^3d^2 hybrid orbitals and, 461, *462*
octane
   distillation of heptane and, 554–56
   heating curve of, 554
   normal boiling point of, 518
   vapor pressure of, 519–20
octanol, and partition coefficient, **527**
octet rule, 391, 392–93
   exceptions to, 411–19
   limitations of models using, 418–19
odd-electron molecules, 411–13
oils, 994
   hydrogenation of, 994–95, 1003
olestra, 995
oligomers, 608
oligopeptides, 979
omega-3 fatty acids, 993
open system, 218–19
optical activity, 969–71
optical isomers, 469–70, 969–71
orbital diagrams, 359
orbitals, atomic
   degenerate, 360
   filling of, 358–66
   overlap of, 453
   periodic table and, 358–66
   quantum numbers and, 350–54
   sizes and shapes of, 355–58
orbitals, molecular. *See* molecular orbital (MO)
     theory
organic chemistry, 67, 962

organic compounds, 67–70
   functional groups of, 68–70, 467–68, 963, *964*
   large number of, 1022
   Wöhler's early synthesis of, 85
   See also hydrocarbons
organometallic compounds, 1086
osmium compounds, anti-inflammatory, 1087
osmosis, 568–73
osmotic pressure (Π), 568–73
   in egg experiment, 578–79
   molar mass determined from, 577–78
Ostwald process, 728
Ötzi the Iceman, 868
outlier, 30–32
overall ionic equation, 160, 161
overall reaction order, 648–49
overlap of atomic orbitals, 453
oxidation
   definition of, 184
   original meaning of, 181
oxidation half-reaction, 880
oxidation number (O.N.), 181–83
   of metal in complex ion, 1057, 1058, 1059, 1060, 1072–73
oxidation state. *See* oxidation number (O.N.)
oxidation–reduction reactions, 180–94
   in acidic solutions, 190–92
   activity series and, 188–90
   annotation of equations for, 183, 186
   balancing with half-reactions, 185–88, 886
   in basic solutions, 190, 192–93, 1031
   changes in oxidation number in, 183
   coenzyme B_{12} in, 1081
   cytochromes as catalysts for, 1079
   electrochemistry and, 880–82
   electron transfer in, 184–85
   equilibrium constants for, 898–99
   glucose metabolism and, 146, 1079
   in nature, 190–94
   net ionic equations for, 187, 189–93, 881, 885–86
   in nitrogen cycle, 1030
   original definition of, 181
   photosynthesis as, 1077–78
   spontaneous, 887, 888, 899, 904, 908
   See also electrochemical cells; half-reactions
oxide ion, as Lewis base, 810–11
oxidizing agent, 184–85
oxoacids, 65, 66–67
   acid strength of, 768–69, 814
   of chlorine, *769*
oxoanions, 65–66, 768
oxygen
   as component of air, 275, 299–300
   electron configuration of, 361
   hemoglobin binding of, 516
   at high altitude, 301–2, 515, 516
   molecular orbitals for, 475–77
   as paramagnetic substance, 477
   in scuba diving mixture, 300–301, 302
   solubility in liquids, 512, 514, 515, 517

   with sp^2 hybrid orbitals, 457–58
   toxicity of, 315
oxygen generators on airplanes, 293–94
ozone, 403–4
   bond lengths in, 403, 419
   nitric oxide and decomposition of, 675
   photochemical decomposition of, 650–52, 672–73
   in photochemical smog, 637, 679
ozone depletion over Antarctica
   catalysts and, 672–75
   Rowland and Molina's prediction of, 673

P

p block elements, 364, 365
   monatomic anions of, 366–67
p orbitals, 351, 357
packing efficiency, 594–95, 599
paclitaxel (Taxol), 128–29
paraffins, 252
paramagnetic substances
   molecular orbital theory and, 477
   transition metal ions, 1071–73
partial electrical charge
   on induced dipole, 500
   of polar covalent bond, 398
partial ionic character. *See* polar covalent bonds
partial pressure
   calculating equilibrium value of, 725–26
   calculation of, 300
   definition of, 299
   in equilibrium constant expressions, 700–701, 702–3, 704–5
   of gas collected over water, 302–3
   Le Châtelier's principle and, 717–19
   solubility of gas in liquid and, 515–17
   of water vapor at selected temperatures, *303*
particulate representations, 4–5, 7–9
partition coefficient, 527
parts per billion (ppb), 149–50
parts per million (ppm), 149
pascal (Pa), 277
Pascal, Blaise, 277
Paschen, Friedrich, 342
Pauli, Wolfgang, 353, *354*
Pauli exclusion principle, 353–54
Pauling, Linus, 398, 400, 453
PDB (*para*-dichlorobenzene), 423
pentane, 465
   boiling point of, *501*, 502
Penzias, Arno A., 35
peptide bond, 979
peptides, 979–81
percent composition, 108–13
percent ionization
   of weak acids, 754–56
   of weak bases, 757–58
percent yield, 126–29
perchloric acid, *769*
periodic table, 52, 55–58
   electron configurations and, 358–66

periodic table (cont.)
electronegativity and, 398–99
orbital filling sequence in, 363
symbols in, 54
unit cells of metals and metalloids in, 594
periodic trends
in atomic radii, 369–72, 1021
in boiling points, 1021–22
in electron affinity (EA), 1021
in electronegativity, 1021
in ionization energy (IE), 1021
in melting points, 1021–22
periods, 56
permanent dipoles, 451–53
See also polar molecules
permanganate ion, as oxidizing agent, 191
peroxidases, 1079
peroxyacetyl nitrate (PAN), 637, 679
pH, 749–53
of aqueous salt solutions, 771–74
calculation of, for acidic and basic solutions, 759–63
of common aqueous solutions, 750
of human blood, 774–75
ionization of weak acid and, 755–56
ionization of weak base and, 757–58
negative value of, 759
pathways for conversions involving, 758
significant figures in, 751
solubility and, 818–20
of solution with conjugate acid–base pair, 789–91
of strong diprotic acid solution, 764–65
of very dilute strong acid, 762–63
of weak acid or base, 760–62
of weak diprotic acid solution, 766–67
pH buffers, 791–98
buffer capacity of, 794–98
preparing with desired pH, 793–94
selecting components for, 791–92
water solubility required for, 792
pH indicators, 798–99
for alkalinity titration, 807–8
array of, for pH 0 to 12, 799
pH meter, 799
phase changes, 219, 221
at constant temperature and pressure, 228
enthalpy changes in, 225–26
entropy changes in, 843
on heating curve, 228
phase diagrams, 520–23
for aqueous solution of nonvolatile solute, 558
for water, 520–22, 558
phase equilibria, 852–53
phase symbols, 8
for aqueous solution, 10
phenanthroline, 157
phenol red, 798, 807
phenolphthalein, 167, 807
pheromones, of insects, 484–85
phosgene, equilibrium in decomposition of, 727
phosphate esters, 1033

phosphate ion, 1032
Lewis structure for, 415–16
phosphine, 1032
3-phosphoglycerate^{3-} (3-PG^{3-}), 864–65
phospholipids, 996–97, 1023
phosphoric acid, 767–68, 1032
phosphors, 932
phosphorus, 1032–33
phosphorus cycle, 1032
phosphorus pentafluoride
sp^3d hybrid orbitals in, 461
trigonal bipyramidal geometry of, 442, 461
phosphorus-32, radioactive decay of, 929–30
phosphorus-doped silicon, 482
phosphorylation of glucose, 863–64
photochemical smog, 636–37, 679
photoelectric effect, 339–41
photon
definition of, 338
energy of, 338–39, 344
photosynthesis, 12, 104, 340
carbon dioxide consumed in, 388
copper-containing enzymes in, 1080
cytochromes in, 1079
light-absorbing molecules in, 1026
manganese and, 1077–78
as nonspontaneous process, 862
radiocarbon dating and, 936
phycoerythrobilin, 1027
physical process, definition of, 5
physical properties, 13–14
physiological saline, 156, 572, 573
pI (isoelectric point), 977
pi (π) bonds
in benzene, 467–68
between carbon atoms, 459
definition of, 457
in formaldehyde, 457–58
formed by nitrogen, 458
in stereoisomers, 968–69
pi (π) molecular orbitals, 475
picomolarity, 148
pinene, 108–9
piston and cylinder, 222, 223–24
pK_a, 752
Henderson-Hasselbalch equation and, 789–91
pK_b, 752
pK_w, 752, 758
Planck, Max, 337
Planck constant, 338
Bohr model and, 344
de Broglie wavelength and, 346
uncertainty principle and, 349
plane-polarized light, 469, 969–71
plaque
of cholesterol deposits, 997
of protein deposits, 984
plaster of Paris, 549
platinum
gold alloyed with, 602–3
microscope image of atoms of, 7

platinum group metals, 1086–87
β-pleated sheet, 983–84
plum-pudding model, 49–50
plutonium-239, 945–46
pOH, 752–53
ionization of weak base and, 757–58
pathways for conversions involving, 758
polar covalent bonds, 398–402, 450–53
electronegativity difference (ΔEN) and, 400–401, 419
greenhouse effect and, 401–2
vibrational frequencies of, 401–2
polar molecules, 450–53
dipole–dipole interactions between, 503–4
solubility and, 510–14
polar solvents
dissolution of ionic compounds in, 542–43
organic, 902
polarizability, of neutral atoms and molecules, 500–501, 504
polarized light, 469, 969–71
polonium, 924, 931, 941
polyamides, 618–20
polyatomic ions, 64, 65–66
formal charge of, 409
in human body, 1022
Lewis structures of, 392, 396
oxoanions, 65–66, 768
radii of, 539, 541–42
resonance structures of, 405–7, 409
polycyclic aromatic hydrocarbons (PAHs), 468
polydentate ligands, 1064–66
biological, 1076–81
macrocyclic, 1077
prefixes for number of, 1076
polyesters, 615–17
polyethylene (PE), 608, 609–10
poly(ethylene glycol) (PEG), 613–14
poly(ethylene oxide) (PEO), 613–14
poly(ethylene terephthalate) (PETE), 613, 615
poly(ethylene-co-vinyl alcohol) (EVAL), 613
polymers, 592, 607–20
of alcohols and ethers, 612–14
of alkenes, 609–11
with aromatic rings, 612, 619–20
compared to small molecules, 608
copolymers, 613
heteropolymers, 613
medical applications of, 610, 615–16
molar masses of, 608
physical properties of, 608
polyamides, 618–20
polyesters, 615–17
polyvinyl chloride, 151
summary of common polymers, 620
polypeptides, 979
polypropylene, 611
polyprotic acids, 163, 764–68
polysaccharides, 988, 990–91
polystyrene (PS), 612
polytetrafluoroethylene (Teflon), 610
polyunsaturated fatty acids, 993

poly(vinyl alcohol) (PVAL), 612–13
polyvinyl chloride, pipes made of, 151
pool test kits, 798
porphyrins, 1077
  of cytochromes, 1079–80
positron, 928
positron emission, 928
  atomic number and mass number in, *930*
  by carbon isotopes, 928, *929*
  in gallium nuclide decay, 1041
positron emission tomography (PET), 922–23,
    943–44, 1041, 1043
potassium
  electron configuration of, 362–63
  in living organisms, 1023–25, 1026
potential energy (PE), 211–13
  at molecular level, 214–17
  as part of internal energy, 221
  See also electrostatic potential energy (E_{el})
precipitate, 170
precipitation, and solubility, 177–78
precipitation reactions, 169–77
  in analytical chemistry, 174–77
  definition of, 170
  net ionic equations for, 172–73
  solubility and, 169–71, 820–23
  for synthesizing insoluble salts, 173–74
  tooth enamel formed by, 146
precision, 27–32
pressure
  boiling point and, 521
  definition of, 275
  measurement of, 278–80
  phase diagrams and, 520–23
  units of, 275–77
  volume of gas and, 280–83
  water solubility of gases and, 514–17
  See also atmospheric pressure (P_{atm}); partial
    pressure; vapor pressure
pressure–volume (*P–V*) work, 222–24
  by car engine, 851
  units of, 223
primary (1°) structure of proteins, 982–83
principal quantum number (*n*), 351
problem-solving framework, 11–12
product, 86
  removing or adding, to shift reaction, 715–17
promethium, *927*
propane
  as alkane, 252–55
  enthalpy of combustion, 250–51, 252
  entropy of combustion, 843
  in natural gas, 254
  standard molar entropy of, *844*
propylene (propene), 68
  polymer of, 611
  simulated smog and, 679
propyne, 68
protease inhibitors, 988
proteins
  abundance of, 974
  acid–base characteristics of, 978–79

denatured, 984
functional classes of, *981*
hydrogen bonds in, 506–7
inherited diseases and, 963
primary (1°) structure of, 982–83
quaternary (4°) structure of, 985–86
secondary (2°) structure of, 983–85
separation by electrophoresis, 10
synthesis of, 1000–1001
tertiary (3°) structure of, 985
proton acceptor (base), 160
proton donor (acid), 160
proton-exchange membrane (PEM), 909–10
protons
  Big Bang and, 70
  meaning hydronium ions, 160
  properties of, *51*
  in Rutherford's atomic model, 51
  symbol for, 70–71
Prussian blue, 1057
p-type semiconductor, 483–84
pure substances, 5–6
P–V (pressure–volume) work, 222–24
  of car engine, 851
  units of, 223
pyruvate ions, *862*, 863, 992, 1026

Q

quantized property
  definition of, 338
  energy of hydrogen atom, 344
quantum, definition of, 337
quantum numbers, 350–54
  periodic table and, 364–65
quantum theory
  electron waves and, 345–50
  entropy and, 866–67
  Heisenberg uncertainty principle, 348–50
  hydrogen spectrum and, 341–45
  indeterminacy in, 354–55
  lasers and, 355
  photoelectric effect and, 339–41
  of Planck, 337–38, 339
  quantum numbers, 350–54, 364–65
  summary of development of, *355*
quarks, 70
quartz, absolute entropy of, 844
quaternary (4°) structure of proteins, 985–86

R

racemic mixture, 973, 986
rad (radiation absorbed dose), 939
radial distribution profiles
  of 1*s* orbitals, 356, *357*
  of 2*s* and 2*p* orbitals, 359–60
radioactive decay, 927–31
  See also alpha (α) decay; beta (β) decay;
    gamma (γ) rays
radioactive decay series, 930–31

radioactivity
  biological effects of, 924, 938–42
  definition of, 46
  early experiments on, 49–50
  measurement of, 932–34
  from nuclear weapons testing, 46–47, 1039
  sources of U.S. exposure to, *941*
  units of, 932
radiocarbon dating, 936–38
radiology
  diagnostic, 942, 943–44
  therapeutic, 942–43
  See also medical imaging
radiometric dating, 935–38
radionuclides, 927–31
  medical applications of, 73–74, 942–44,
    1041–43, 1082–84
radium, 924, 931, 933–34
Radium Girls, 950–51
radon gas, 940–42, 1020
rainbows, 332, 336
rainwater
  carbonic acid in, 765–67
  pH changes in, by soils, 770
random coil protein structure, 984
random copolymer, 613
randomly distributed data, 30
Raoult, François Marie, 551
Raoult's law, 551–53
  deviations from, 557–58
  for mixtures of volatile compounds, 556–58
rate constant, 647, 648
  Arrhenius equation for, 660–65
  for first-order reaction, 651–52
  half-life and, 653–55
  from initial reaction data, 648–50
  for radioactive decay, 932–33
  for second-order reaction, 656–57
  temperature and, 663, 664–65
  units of, 648
  for zero-order reaction, 658
rate law
  definition of, 647
  for first-order reaction, 650–53
  from initial reaction rate data, 649–50
  reaction mechanisms and, 667–72
  reaction order and, 647–50
  for second-order reaction, 655–58
  summary of, for single reactant, *659*
  for zero-order reaction, 658
  See also reaction order; reaction rates
rate-determining step, 668–69
  zero-order reactions and, 672
reactants, 86
  limiting reactant, 122–25
  removing or adding, to shift reaction, 715–17
reaction mechanisms, 665–72
  difficulty of proving, 669
  elementary steps in, 666–68
  rate laws and, 667–72
  rate-determining step in, 668–69, 672
  zero-order reactions and, 672

reaction order, 645–50
  definition of, 645
  determination of, 647
  distinguishing between first and second
    order, 656–57
  first-order, 650–55, 656–57
  overall, 648–49
  second-order, 655–58
  summary of, for single reactant, *659*
  zero-order, 658–59, 672
reaction quotient (*Q*), 710–12
  cell potential and, 895–97
  free energy change and, 854–56
  Le Châtelier's principle and, 716, 718, 719
  solubility equilibria and, 820–21
reaction rates, 638–40
  average, 642
  concentration and, 638, 645–59
  at equilibrium, 698
  experimentally determined, 640–42
  factors affecting, 638–39
  instantaneous, 642–45
  rate-determining step and, 668–69, 672
  relative, 640–42
  sign of, 639
  temperature and, 638, 648, 660, 662–65, 677
  units of, 640
  See also rate constant; rate law
reactions. *See* chemical reactions; nuclear
    reactions
real gases, 311–15
receptors, biological, 463
recommended dietary allowance (RDA), 1020
red beans and rice, 974, 976
red blood cells, and osmosis, *569*, 570–72
red fireworks, 376
redox reactions, 146, 181
  See also oxidation–reduction reactions
reducing agent, 184–85
reductases, 1030
reduction
  definition of, 184
  original meaning of, 181
  See also oxidation–reduction reactions
reduction half-reaction, 880
reduction potentials (*E°*), 887–90
  corrosion and, 904–5
  as intensive property, 890
relative biological effectiveness (RBE), 939
rem (roentgen equivalent man), 939
replication of DNA, 998–99
representative elements. *See* main group
    elements
residues, amino acid, 979
resonance, 403–7
  bond lengths and, 419
  exceptions to octet rule and, 411–19
  formal charge and, 407–11, 412–17, 419
respiration, 104, 124
respiratory acidosis, 741
reversal potential (*E*$_{\text{ion}}$), 1024–25
reverse osmosis, 573–75

reverse reactions, *K* values for, 706–7
reversible process, 839, 840
reversible reactions, 697–98
  See also chemical equilibrium
rhenium, 1084
rhodium, antitumor complexes of, 1087
ribose, 997
ribosome, 1000–1001
RICE table
  buffer capacity and, 794–95, 796–97
  equilibrium concentrations or pressures,
    721–27
  pH of salt solutions, 772–74
  pH of strong diprotic acid solution, 764–65
  pH of very dilute strong acid, 762–63
  pH of weak acid or base, 760–62
  pH of weak diprotic acid solution, 766–67,
    775
  titration of weak acid with strong base, 803–4
  uncomplexed metal ion concentration, 813–14
rich mixture, 124
Ringer's lactate solution, 153
RNA (ribonucleic acid), 997–98, 1000–1001
RNA world hypothesis, 1002
rock candy, 177–78
rock salt structure, 604, 621
Roman numerals, 64–65, 1058
Röntgen, Wilhelm Conrad, 49, 49n
root-mean-square speed (*u*$_{\text{rms}}$), 306–9
rotational energy levels, 866–67
rotational motion, *221*, 837, 866
rounding off, 24
  See also significant figures
Rowland, Sherwood, 673
rubidium, in human body, 1019, 1039
rust, 190, 832, 835, 904
  See also corrosion
ruthenium, antitumor complexes of, 1087
Rutherford, Ernest, 49–51, 52, 343, 935
Rydberg, Johannes Robert, 342, 343

S

s block elements, 364, 365
  monatomic cations of, 366, 367
s orbitals, 351, 355–56
sacrificial anodes, 906
saline solution
  physiological, 156, *572*, 573
  preparation of, 156
  as strong electrolyte, 158
salts
  acid–base properties of, 770–74
  definition of, 160
  insoluble, 170–74
Sapphire Pool in Yellowstone National Park,
    808–9
saturated fatty acids, 993, 994, *995*
saturated hydrocarbons, 252
saturated solution, 177–78
scandium, 1084
scanning tunneling microscope (STM), 7

Schrödinger, Erwin, 350
Schrödinger wave equation, 350, 351, 352
scientific method, 17
scientific theory (or model), 17
scintillation counters, 932, 1084
scuba diving, 314–15
  gas mixtures for, 300–301, 302, 315
seawater
  acidification of, 786–88, 823–24
  apparent *K*$_a$ values in, 823–24
  bromide ions in, 1040
  common-ion effect on solubility in, 817–18
  concentrations of major constituents, 148–49
  corrosion in, 904, 905, 906, 912–13
  desalination of, 573–74
  deuterium in, 949
  diluted in estuaries, 155
  elemental composition of, 1019–20
  freezing point depression of, 563–64
  ion pairs in, 823
  osmosis and, 568, 569, 571–72
  pH of, 770, 787–88
  vapor pressure of, 550–51
second ionization energy (IE$_2$), 373–74
second law of thermodynamics, 837–38, 839,
    840
secondary (2°) structure of proteins, 983–85
second-order reactions
  distinguishing from first-order, 656–57
  half-life of, 658
  integrated rate law for, 655–58
  summary of how to distinguish, *659*
seed crystal, 178
seesaw molecular geometry, *447*, 448, 449
  sp^3*d* hybrid orbitals and, *462*
selective precipitation, 821–23
selenium, 1037–38
selenocysteine, 1038
semiconductors, 482–84
  carbon allotropes with properties of, 607
semimetals
  in periodic table, 56
  as semiconductors, 482–84
semipermeable membrane, 568
  of chicken egg, 578–79
  See also osmosis; reverse osmosis
sequestering agent, 1065
serum, human, major constituents of, 148–49
shape memory alloys, 600
shapes
  of larger molecules, 463–64
  London dispersion forces and, 501–2
  See also molecular recognition
shells, 351–52, 364–65
SI units, 18–20
  base units, *19*
  conversions involving, 20–22
  derived units, 19–20
  prefixes for, *19*
sickle-cell anemia, 982–83
side-chain groups, of amino acids, 974
sievert (Sv), 939

sigma (σ) bonds
   between carbon atoms, 459
   definition of, 454
   between hydrogen atoms, 454
sigma (σ) molecular orbital, 472
significant figures, 23–27
   of pH values, 751
silica, amorphous, 1039
silicic acid, 1039
silicon
   in biological systems, 1039
   diamond crystal lattice of, 606
   as semiconductor, 482–83
silicon dioxide, absolute entropy of, 844, 845
silk, 984
silver
   antitumor complexes of, 1087
   drugs containing, 1088
   electron configuration of, 365–66
   electron configuration of cation of, 368
   oxidation in seawater, 912–13
silver sulfadiazine, 1088
simple cubic (sc) unit cell, 595, 596, *597*
single bond, 392
skeletal structure, 392
skunk odor, 1036
smell, molecular recognition by, 466, 468–69,
      971, 1035
smog
   pea-soup, 636
   photochemical, 636–37, 679
   simulations of, 679
sodium
   electron configuration of, 361–62
   emission spectrum of, 333, 362
   in living organisms, 1023–26
sodium-24, in medical imaging, 943
sodium bicarbonate
   in antacid tablets, 147, 168–69
   endothermic reaction with acetic acid, 835,
      837
   See also bicarbonate
sodium chloride
   boiling point elevation due to, 561–62, 564,
      567
   Born–Haber cycle for formation of, 545–46
   crystal structure, 603–4
   dissolving in water, 502–3
   electrolysis of, 908
   enthalpy of hydration, 548
   freezing point depression due to, 564–65,
      567–68
   lattice energy, 545–46
   molality vs. molarity of solution, 559–60
   neutral solution of, 770
   osmotic pressure and, *569*, 571
   as strong electrolyte, 158, 567
   van't Hoff factor for, 564–65, 567
   See also saline solution; seawater
sodium fluoride
   lattice energy of, 546–47, 548
   weakly basic solution of, 770

sodium hydroxide, calculating pH of, 759
sodium hypochlorite, acid–base properties of,
      771, 772–73
sodium ion
   electron configuration of, 366, 367
   size of, 371
solid solutions, 592, 600
solids
   classification of, 590, *591*
   compared to liquids and gases, 14–16
   definition of, 14
   equilibrium constant expressions and, 713
   phase symbol for, 8
   standard states of, *842*
solubility
   of alcohols in water, 513, *514*
   of common ionic compounds, 169–71
   common-ion effect and, 817–18
   definition of, 177, 816
   of gases in water, 514–17
   lattice energy and, 548
   molar, 816–19
   pH and, 818–20
   polarity and, 510–14
   of polymer PEG, 614
   saturated solutions and, 177–78
   separating ions in solution and, 821–23
solubility equilibria, 816–23
   reaction quotient and, 820–21
solubility product (K_{sp}), 816–24
   reaction quotient and, 820–21
solubility rules, *170, 171*
solute, definition of, 145
solutions
   components of, 145
   definition of, 7
   dilution of, 154–56
   ideal, 551–52, 557, 567, 823
   preparing with known concentration, *153*
   solid, 592, 600
   standard states of, *842*
   vapor pressure of, 550–53
   See also aqueous solutions; colligative
      properties; concentration
solvated ions, 502–3
   See also hydrated ions
solvent, definition of, 145
Sørensen, Søren, 749
sotalol, 973
sp hybrid orbitals, 458–60
sp^2 hybrid orbitals, 456–58
sp^3 hybrid orbitals, 455–56
$sp^3 d$ hybrid orbitals, 461–62
$sp^3 d^2$ hybrid orbitals, 461–62
space-filling molecular models, 9
spearmint, 438, 468, 470
specific heat (c_s), 229
   measurement of, 235–37
spectator ions, 161, 162
   in electrochemical cell, 881
   in redox reactions, 187
spectrochemical series, 1070

spectrophotometry, 156–57
spectrum
   electromagnetic, *334*
   of emission and absorption by elements,
      333–34
   of heated metal filament, 337
   molecular, 471
speed of light, 335
sphalerite, 604
sphalerite structure, 604
sphere of hydration, 502–3
   of complex ion, *815*
   ion pairing and, 568
sphere of solvation, 502–3
spin magnetic quantum number (m_s), 353–54,
      476
spin–paired electrons, 359
spodumene, 113
spontaneous processes, 834–37
   chelate effect and, 1065
   definition of, 834
   endothermic, 835–37, 846, 853
   entropy changes in, 837–38, 840–41, 847–49,
      1065
   free energy changes in, 847–49, 852–56
   in living systems, 862, 863–64, 1025–26
   temperature and, 852–54
spontaneous redox reactions
   corrosion as, 904
   standard cell potential and, 888, 899
   standard reduction potential and, 887
   in voltaic cells, 908
square packing, 595
square planar complex ions, *1057*, 1068–69
   cis–trans isomers in, 1073–74
   with macrocyclic ligands, 1077
square planar molecular geometry, *447*, 449
   $sp^3 d^2$ hybrid orbitals and, *462*
square pyramidal molecular geometry, *447*, 449
   $sp^3 d^2$ hybrid orbitals and, *462*
stacking patterns, 592–95
   summary of, *595*
stainless steel, 601, 906
stalactites and stalagmites, 162, 178
standard cell potentials (E°_{cell}), 888–90
   equilibrium constant and, 898–99
   of fuel cells, 911
   hydrogen electrode and, 893–95
   nonstandard conditions and, 895–97
   standard free energy change and, 892–93
standard conditions, 247
standard deviation (s), 28
standard enthalpy of combustion (ΔH°_{comb})
   for common fuels, *255*
   fuel value and, 255–56
   See also enthalpy of combustion (ΔH_{comb})
standard enthalpy of formation (ΔH°_f), 246–52
   lattice energy calculated from, 545
   for selected substances, *247*
standard enthalpy of reaction (ΔH°_{rxn}), 247
   equilibrium constant at specific temperature
      and, 860–61

standard enthalpy of reaction (ΔH_{rxn}°) *(cont.)*
  free energy change and, 848–50
  for reverse reaction, 250
  from standard enthalpies of formation, 248–49, 250–52
  See also enthalpy of reaction (ΔH_{rxn})
standard entropy of reaction (ΔS_{rxn}°), 845–46
  free energy change and, 848–49
standard free energy of formation (ΔG_f°), 849–51
  equilibrium constant and, 858
standard free energy of reaction (ΔG_{rxn}°), 848–51
  equilibrium constant and, 856–59, 860–61
  reaction quotient and, 854–56
  temperature effect on spontaneity and, 853–54
standard heat of formation. *See* standard enthalpy of formation (ΔH_f°)
standard heat of reaction. *See* standard enthalpy of reaction (ΔH_{rxn}°)
standard hydrogen electrode (SHE), 893–95
standard molar entropy (S°), 842–43
  chemical reaction and, 845–46
  molecular structure and, 844–45
  selected values of, *842*
standard reduction potentials (E°), 887–90
  corrosion and, 904–5
  as intensive property, 890
standard solution, definition of, 166
standard states, 247
  free energy changes and, 859
  of pure substances and solutions, *842*
standard temperature and pressure (STP), 289
standing wave, 347–48
starch, 988, 990–91
state function
  definition of, 212
  potential energy as, 212
states of matter, 14–16
  kinetic energy and, 499
  reaction rates and, 638
  standard entropy values of, 842–43
  See also phase changes
Statue of Liberty, 904–5
steam–methane reforming reaction, 249, 696–97, 705, 1003
steel, 601–2
  inhibiting corrosion of, 906
  stainless, 601, 906
stellar nucleosynthesis, 72–73, 924–25, 926
stents, arterial, 600
stereoisomers, 968–74
  coordination compounds with, 1073–76
  See also chirality
steric number (SN), 440, 441
  hybridization and, 455, *462*
Stern, Otto, 353
stimulatory effect, 1018–19
stock solutions
  definition of, 154
  dilution of, 154, 156

Stock system, 64–65
stoichiometric yield, 124
stoichiometry
  balancing chemical equations in, 96–104
  definition of, 95
  gases in, 293–95
  limiting reactants in, 122–25
  masses of products and reactants in, 104–8
  mole ratio in, 105–8
  moles in, 87–89
  percent yield in, 126–29
  See also balanced chemical equations
straight-chain hydrocarbons, 253
  boiling points of, 500–501
  compared to branched, *254*
strong acids, 163
  activity series of metals and, 189–90
  common examples of, *742*, 743
  conjugate bases of, 747
  diprotic, 764–65
  leveling effect of, 747
  oxoacid strength and, 768–69
  pH of 7.4 *M* solution, 759
  pH of very dilute solution, 762–63
strong bases, 164, 745
  calculating pH of, 759
  conjugate acids of, 747
  leveling effect of, 747
strong electrolytes, 158
  strong acids as, 163
  van't Hoff factor for, 564–68
strong nuclear force, 925
strontium
  electron configuration of, 365
  in human body, 1039
strontium-90, 46–47, 1039
structural formulas, 9
structural isomers, 965–68
styrene, 612
Styrofoam, 612
subatomic particles
  Big Bang and, 70
  early experiments on, 47–52
  masses of, *925*
  properties of, *51*
  symbols of, 71, *925*
  See also electrons; neutrons; protons
sublimation, 15, *219*
  on phase diagram, 520, 521
sublimation points, 520
subshells, 351–52, 364–65
substitutional alloys, 600–601, 602–3
substrate, 677–78, 986–87
sucrose, 989–90
sulfate ion
  Lewis structure of, 414–15
  in sulfur cycle, 1033, 1080
sulfite ions, enzyme-catalyzed reduction of, 1079
sulfite oxidase, 1080
sulfur, 1032, 1033–37
sulfur cycle, 1033, 1080

sulfur dioxide
  from coal-burning power plants, 713, 764, 810–11
  in copper refining, 868
  geometry of, 444–45
  as Lewis acid, 810
  in sulfur cycle, 1033
sulfur hexafluoride, 413–14
  octahedral geometry of, 443, 461
  sp^3d^2 hybrid orbitals in, 461
sulfur tetrafluoride, 449
sulfur trioxide
  acid rain and, 87, *162*, 414, 1033
  reaction with water vapor, 86–87, 95, 122–23, 163, 414, 764
  resonance structures of, 405
  from sulfur dioxide and oxygen, 99–100, 414, 708, 764
sulfuric acid
  from abandoned coal mines, 166
  as diprotic acid, 163, 764
  dissociating in dilute solution, 414
  in early atmosphere, 86–87
  Lewis structure for, 415
  pH of solution of, 764–65
  as strong acid, 163
  structural reasons for strength of, 768
  from sulfur trioxide and water, 86–87, 95, 122–23, 163, 414, 764
  See also acid rain
sulfurous acid, 768
sunlight
  chemical energy from, 210
  energy in food derived from, 862
  photochemical decomposition of ozone and, 650–52
  photochemical smog and, 636, 679
  spectrum of, 332–33, 334
sunscreens, eicosene in, 575–76
supercritical fluid, 522
supercritical region, 520
supernovas, 73, 926
superoxide dismutases, 1080
superoxide ion, 1080
supersaturated solution, 178
surface tension, 524–25
surgical steel, 601
surroundings, 217
  entropy and, 837–41, 846–47
swamp gas, 254
swimming pool water, pH of, 798
system
  definition of, 217
  entropy and, 837–41, 846–47
  types of, 218–19

T

tantalum, 595, 598
t-distribution, 29
technetium, 73–74, *927*, 1082–83

teeth
    enamel of, 146
    fluoride and, 1038
    hydroxyapatite in, 146, 1027, 1038
    strontium-90 in, 46–47
Teflon (polytetrafluoroethylene), 610
temperature
    density of a gas and, 296–98
    of Earth, 388
    energy of particles and, 214–15, 221
    entropy change and, 838–40, 841–42, 843
    equilibria shifting due to changes in, 719–20
    equilibrium constant and, 720, 859–61
    heat and, 211, 219
    molar heat capacity and, 228
    phase diagrams and, 520–23
    radiation from hot objects and, 337
    rate constant and, 663, 664–65
    reaction rates and, 638, 648, 660, 662–65,
        677
    solubility and, 177–78
    spontaneous processes and, 852–54
    vapor pressure and, 517, 518, 519–20
    volume of a gas and, 283–85
    water solubility of gases and, 514–15
temperature scales, 32–34
temporary dipoles, 500–502
    interacting with permanent dipoles, 504, 506,
        514, 515
termolecular elementary step, 666
tertiary (3°) structure of proteins, 985
tetrahedral complex ions, 1057, 1069–70, 1073
tetrahedral geometry, 440, 447
    nonpolar substances with, 450
    sp^3 hybrid orbitals and, 455–56, 462
tetrahedral holes, 602
    in binary ionic solids, 604, 621
thalassemia, 1085–86
thalidomide, 986
thallium-201, 1041
theoretical yield, 124
    vs. actual yield, 126–29
theory, scientific, 17
therapeutic radiology, 942–43
thermal energy, 215
thermal equilibrium, 211
thermochemical equation, 238
thermochemistry, 211
thermocline, 526
thermodynamics
    definition of, 210–11
    first law of, 222, 223
    second law of, 837–38, 839, 840
    third law of, 842, 867
thermos bottle, 218
third law of thermodynamics, 842, 867
Thomson, Joseph John (J. J.), 47–48, 49, 52
thorite, 621
thorium, oxide of, 621
Thresher, USS, 912–13
threshold frequency (ν_0), 339

thymine, 997, 999, 1000
tin
    in bronze, 599, 600, 601
    diamond crystal lattice of, 606
titanium
    cation of, 368
    in complexes with antitumor activity, 1086
    hcp crystal structure of, 593, 594
titrant, 166, 799
titrations. See acid–base titrations
tokamak, 948–49
torr, 276–77
Torricelli, Evangelista, 276–77, 279
Toyota Prius, 900
trace essential elements, 1019, 1020, 1037,
        1038–39
    transition metals among, 1054, 1076
trans fats, 995
trans isomer, 968, 1073–75
transcription, 1000
transfer RNA (tRNA), 1000–1001
transferrin, 1085–86
transition metals
    bcc unit cells in, 595
    electron configurations of, 363–64, 365–66
    electron configurations of cations of, 368–69
    in enzymes, 1078–81
    essential for human body, 1054, 1076,
        1080–81
    in medicine, 1081–88
    naming compounds of, 64–65, 1060–62
    in periodic table, 57
    writing formulas containing, 64–65, 1056,
        1057–62
    See also complex ions; coordination
        compounds
transition state, 661
    of enzyme-catalyzed reaction, 678, 987
    two-step reaction and, 667
translation, 1001
translational energy levels, 866, 867
translational motion, 221, 837, 866
transport rate constant, 655
tricarboxylic acid (TCA) cycle, 992, 1081
triglycerides, 992, 993–94, 995–96
trigonal bipyramidal geometry
    with lone pairs, 446–48, 449
    with no lone pairs, 440–41
    sp^3d hybrid orbitals and, 461–62
    summary of, 447
trigonal planar geometry, 440, 447
    of ethylene, 463
    sp^2 hybrid orbitals and, 456–58
trigonal pyramidal molecular geometry, 446,
        447
    sp^3 hybrid orbitals and, 462
trimethylamine, 757–58
Trimix, 300, 315
triple bonds
    definition of, 394
    Lewis structures with, 394–96

triple point, 522
triprotic acids, 163, 767–68
tritium
    in experimental fusion reactions, 948–49
    helium-3 from decay of, 1043
T-shaped molecular geometry, 447, 448
    sp^3d hybrid orbitals and, 462
turnover number for enzyme, 677
TV remote control, 340

U

Uhlenbeck, George, 352–53
ultrahigh molecular weight PE (UHMWPE),
        610
ultratrace essential elements, 1019, 1020,
        1037–38
    transition metals among, 1054, 1076,
        1080–81
ultraviolet (UV) radiation, 334
    in hydrogen emission spectrum, 342, 344
uncertainty principle, 348–50
unimolecular elementary step, 666
unit cells, 593–95
    body-centered cubic (bcc), 595, 597–98, 601,
        602
    dimensions of, 596–99
    face-centered cubic (fcc), 594, 596–97, 599,
        601, 602–6
    periodic table and, 594
    summary of, 595
unit factor method, 20–22
units
    conversions of, 20–22
    SI units, 18–20
    U.S. Customary units, 20
universal gas constant (R), 289
universe
    elemental composition of, 1019–20
    entropy of, 837–38, 847
    origin of elements in, 70–73
    timeline of energy and matter in, 71
    See also Big Bang
unsaturated fatty acids, 993, 994, 995
uracil, 997
uranium
    dating geological samples containing, 935–36
    early research with, 924, 932, 935
uranium-235, fission reactions of, 944–45
uranium-238
    in breeder reactor, 945–46
    decay series of, 930–31
    in geological samples, 935, 936
    radon from decay of, 941
urea
    in living systems, 1032, 1080
    Wöhler's synthesis of, 85, 965
ureases, 1032, 1080
Urey, Harold, 84, 96, 98, 1001
uric acid, 1080
U.S. Customary units, 20

V

valence, two kinds of, 1056
valence band, 481–83
valence bond theory, 453–63
valence electrons, 360
   in Lewis structures, 392
   repulsion between, 371
   See also Lewis structures
valence shell, 360, 361, 362
   expanded, 413–16
valence-shell electron-pair repulsion (VSEPR)
     theory, 439–49
   central atoms with lone pairs, 444–49
   central atoms with no lone pairs, 440–44
van der Waals constants, *313*
van der Waals equation, 313–14
van't Hoff, Jacobus, 564, 660, 860
van't Hoff equation, 860–61
van't Hoff factor (*i* factor), 564–68
   osmotic pressure and, 570, 573
vanadium
   in complex with antitumor activity, 1086
   diabetes and, 1086
   in haloperoxidases, 1080
vanillin, 121
vapor, 14
vapor pressure
   as colligative property, 550–51, 558, 563
   definition of, 518
   of mixtures of volatile solutes, 553–58
   of pure liquids, 517–20
   of solutions of nonvolatile solutes, 550–53
   temperature and, 517, 518–20
vaporization, 15, *219*
   free energy and, 852
   See also evaporation
vermilion (Chinese red), 175
vibration of bonds, *221*
   in crystalline solids, 837
   infrared frequencies and, 401–2
   standard molar entropy and, 844
vibrational energy levels, 866–67
vinegar, 167–68
   chicken egg submerged in, 578
   See also acetic acid
vinyl acetate, copolymer with ethylene, 613
vinyl alcohol, polymer of, 612–13
vinyl chloride, in drinking water, 151
vinyl group, 612
viscosity, 525
vitamin C
   iron absorption and, 1054
   sequestered by EDTA, 1065
volatile organic compounds (VOCs), in smog,
   637
volatile sulfur compounds (VSCs), 1035
volatility
   definition of, 519
   distillation and, 10
   vapor pressure and, 519

volcanic activity
   carbon dioxide from, 295
   early atmosphere and, 86
   origin of life and, 84, 96
Volta, Alessandro, 883
voltaic cells, 883
   chemical equilibrium in, 896, 898
   work done by, 891
   See also batteries; fuel cells; zinc/copper
     voltaic cell
voltmeter, 888
volume
   Boyle's law and, 280–83
   density and, 13, 23–26
   entropy changes and, 843
   gas-phase equilibria and, 717–19
   units of, *20*

W

Waage, Peter, 699
water
   from abandoned coal mines, 166
   alkalinity of sample of, 805–9
   as amphiprotic substance, 165, 746, 749
   angular (bent) geometry of, 446, 450
   aquatic life and, 514, 526
   autoionization of, 748–49, 762–63, 767
   boiling-point-elevation constant for, 576–77
   corrosion promoted by, 904–5
   density changes with temperature, 525–26
   electrolysis of, 885
   entropy and change of state in, 843, 843n
   freezing-point-depression constant for,
     576–77
   group 16 dihydrides compared to, 1033–34,
     1035
   hard, 178–79
   heat capacity of, 228, 229, 230, 231, 232
   heating curve of, 227–28
   heating of, 227–35
   hot object dropped into, 237–38
   hydrated complex ions and, 812, 814–16, 820,
     1055, 1062–63, 1066–67, 1072
   hydrated ions and, 502–3, 542–43, 548–50
   hydrogen bonding in, 504–5, 522, 523–26,
     548
   hydrogen bonds with acetone, 507
   in hydrolysis reactions, 163, 1032
   intermolecular interactions of, 498–99
   ionizing radiation and, 938
   as Lewis base, 1063
   on Mars, 144–45
   melting point of, 523
   melting/freezing behavior of, 521–22, 525–26
   normal boiling point of, 518, 523
   origin of life and, 144–45
   partition coefficient between octanol and, 527
   phase diagrams for, 520–22
   phase equilibria of, 852–53
   phases of, *498*, 499

   as polar molecule, 450–51, 498–99
   purification by distillation, 220
   remarkable properties of, 523–26
   sp^3 hybrid orbitals in, 456
   spontaneous phase changes in, 837
   temporary dipole induced by, *504*
   viscosity of, 525
   See also aqueous solutions; drinking water;
     freshwater; ice; seawater
water softening, 178–79
water vapor
   of early atmosphere, 86, 144
   entropy of, 843
   partial pressure of, *303*
water–gas shift reaction
   achievement of equilibrium in, 697–98
   equilibrium constant for, 698–700
   equilibrium partial pressure calculation, 725
   K_p equals K_c for, 704–5
   reaction quotient for, 711
   removing carbon dioxide to shift equilibrium,
     715–16
watt (W), 900
wave functions (ψ), 350
wave mechanics, 350
wave number (1/λ), 342
wavelength (λ), 335–36
   de Broglie, 346–48
   photon energy and, 338–39
wave–particle duality, 340–41
waves
   characteristics of, 335
   electromagnetic, 335–36
weak acids, 163
   calculating pH of, 760–61
   common examples of, 743, *744*
   conjugate bases of, 747
   degree of ionization of, 753–56
   diprotic, 765–67
   K_a values of, *744*
   oxoacid strength and, 768–69
   pH of solution with conjugate base and,
     789–90
   salt solutions with property of, 770–72,
     773–74
weak bases, 164, 745–46
   ammonia as, 164, 745, 756, 761–62
   calculating pH of, 760, 761
   common examples of, *745*
   conjugate acids of, 747
   degree of ionization of, 756–58
   pH of solution of conjugate acid and, 790–91
   salt solutions with property of, 770–73
weak electrolytes, 159
weak-link principle, 24
weather balloons, 280, 291–92
weighted average, 54
welding, nitric oxide produced in, 411
Werner, Alfred, 1055–56
wet gas, 302
wetlands, soil color in, 190

Wilson, Robert W., 35
Wöhler, Friedrich, 67, 85, 965
Wollaston, William Hyde, 332
work
    definition of, 211
    electrical, 890–91, 899
    energy and, 8
    internal energy and, 222–24
    maximum energy available for, 847, 851, 867
    pressure–volume (P–V) work, 222–24, 851
work function (φ), 339

X

xanthine oxidase, 1080
xenon, banned for athletes, 1042
xenon-129, in magnetic resonance imaging, 1043
X-rays, discovery of, 49, 49n

Y

yield
    actual, 126–29
    percent, 126–29
    theoretical, 124
yttrium, 1084

Z

Z isomer, 968
    See also trans isomer
zeolites, 179–80
zero-order reactions, 658–59
    half-life of, 659
    rate law for, 658
    reaction mechanisms and, 672
    summary of how to distinguish, *659*
Zewail, Ahmed H., 666n

zinc
    band theory and, 482
    in carbonic anhydrase, 1078–79
    conductivity of, 482
    electron configuration of cation of, 368
    redox reaction between copper and, 880–83
    in sacrificial anodes, 906
    in superoxide dismutase, 1080
zinc hydroxide, 815
zinc oxide, in alkaline batteries, 889
zinc sulfide, crystal structure of, 604
zinc–air battery, 889–90
zinc/copper voltaic cell, 881–83
    ion concentrations and, 897
    standard cell potential for, 888, 890
    standard free-energy change of, 891–92
    work done by, 890–91
zircon, in dating geological samples, 936
zwitterions, 976–79

ATOMIC COLOR PALETTE

Orbitals
- Molecular
- Hybrid
  - s
  - p
  - d

Proton
Neutron
Positron
Electron
Miscellaneous
Muted water molecule in solution

	1	2	8	11	12	13	14	15	16	17	18
1	1 **H** Hydrogen 1.0079										2 **He** Helium 4.0026
2	3 **Li** Lithium 6.941					5 **B** Boron 10.811	6 **C** Carbon 12.011	7 **N** Nitrogen 14.007	8 **O** Oxygen 15.999	9 **F** Fluorine 18.998	10 **Ne** Neon 20.180
3	11 **Na** Sodium 22.990	12 **Mg** Magnesium 24.305				13 **Al** Aluminum 26.982	14 **Si** Silicon 28.086	15 **P** Phosphorus 30.974	16 **S** Sulfur 32.065	17 **Cl** Chlorine 35.453	18 **Ar** Argon 39.948
4	19 **K** Potassium 39.098	20 **Ca** Calcium 40.078	26 **Fe** Iron 55.845	29 **Cu** Copper 63.546	30 **Zn** Zinc 65.38			33 **As** Arsenic 74.922		35 **Br** Bromine 79.904	36 **Kr** Krypton 83.798
5				47 **Ag** Silver 107.87						53 **I** Iodine 126.90	54 **Xe** Xenon 131.29
6		56 **Ba** Barium 137.33		79 **Au** Gold 196.97			82 **Pb** Lead 207.2				

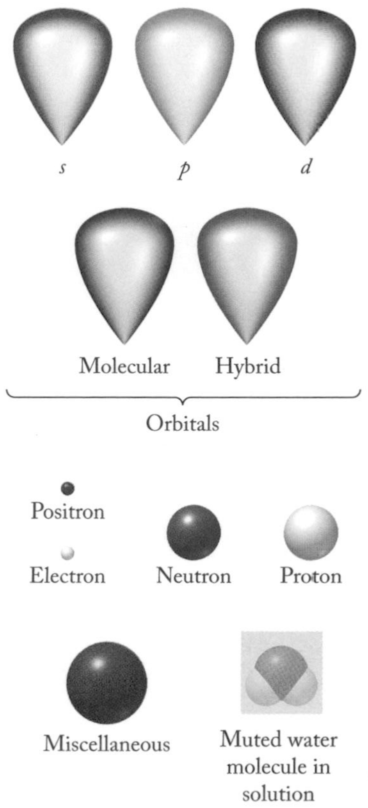

Special Units and Conversion Factors

Quantity	Unit	Symbol	Conversion
energy[a]	electron-volt	eV	$1\ eV = 1.6022 \times 10^{-19}\ J$
energy[a]	kilowatt-hour	kWh	$1\ kWh = 3600\ kJ$
energy	calorie	cal	$1\ cal = 4.184\ J$
mass	pound	lb	$1\ lb = 453.59\ g$
mass[a]	atomic mass unit	amu	$1\ amu = 1.66054 \times 10^{-27}\ kg$
length	angstrom	Å	$1\ Å = 10^{-8}\ cm = 10^{-10}\ m$
length	inch	in	$1\ in = 2.54\ cm$
length	mile	mi	$1\ mi = 5280\ ft = 1.6093\ km$
pressure	atmosphere	atm	$1\ atm = 1.01325 \times 10^5\ Pa$
pressure	torr	torr	$1\ torr = 1/760\ atm$
temperature	Celsius scale	°C	$T(°C) = T(K) - 273.15$
temperature	Fahrenheit scale	°F	$T(°F) = \frac{9}{5} T(°C) + 32$
volume	liter	L	$1\ L = 1\ dm^3 = 10^{-3}\ m^3$
volume	cubic centimeter	cm^3, cc	$1\ cm^3 = 1\ mL = 10^{-3}\ L$
volume	cubic foot	ft^3	$1\ ft^3 = 7.4805\ gal$
volume	gallon (U.S.)	gal	$1\ gal = 3.785\ L$

[a]From http://physics.nist.gov/cuu/constants/.

Physical Constants[a]

Quantity	Symbol	Value
Avogadro's number	N_A	$6.0221 \times 10^{23}\ mol^{-1}$
Bohr radius	a_0	$5.29 \times 10^{-11}\ m$
Boltzmann constant	k_B	$1.3806 \times 10^{-23}\ J/K$
electron charge-to-mass ratio	$-e/m_e$	$1.7588 \times 10^{11}\ C/kg$
elementary charge	e	$1.602 \times 10^{-19}\ C$
Faraday constant	F	$9.65 \times 10^4\ C/mol$
mass of an electron	m_e	$9.10938 \times 10^{-31}\ kg$
mass of a neutron	m_n	$1.67493 \times 10^{-27}\ kg$
mass of a proton	m_p	$1.67262 \times 10^{-27}\ kg$
molar volume of ideal gas at 0°C and 1 atm	V_m	$22.4\ L/mol$
Planck constant	h	$6.626 \times 10^{-34}\ J \cdot s$
speed of light in vacuum	c	$2.998 \times 10^8\ m/s$
universal gas constant	R	$8.314\ J/(mol \cdot K)$ $0.08206\ L \cdot atm/(mol \cdot K)$

[a]From http://physics.nist.gov/cuu/constants/.